M000266685

# INTE RATED MATH 2


John A. Carter

Gilbert Cuevas

Roger Day

Carol Malloy

Berchie Holliday

Beatrice Moore Luchin

Jerry Cummins

Ruth Casey

Viken Hovsepian

Dinah Zike

mheonline.com

The McGraw·Hill Companies

Education



Send all inquiries to:
McGraw-Hill Education
8787 Orion Place
Columbus, OH 43240

ISBN: 978-0-07-663861-1
MHID: 0-07-663861-8

Printed in the United States of America.

4 5 6 7 8 9 QVS 17 16 15

# Contents in Brief


**Chapter 0** Preparing for Integrated Math II

**Chapter 1** Quadratic Expressions and Equations

**Chapter 2** Quadratic Functions and Equations

**Chapter 3** Quadratic Functions and Relations

**Chapter 4** Exponential and Logarithmic Functions and Relations

**Chapter 5** Reasoning and Proof

**Chapter 6** Congruent Triangles

**Chapter 7** Relationships in Triangles

**Chapter 8** Quadrilaterals

**Chapter 9** Proportions and Similarity

**Chapter 10** Right Triangles and Trigonometry

**Chapter 11** Circles

**Chapter 12** Extending Surface Area and Volume

**Chapter 13** Probability and Measurement

**Student Handbook**

# Authors

Our lead authors ensure that the Macmillan/McGraw-Hill and Glencoe/McGraw-Hill mathematics programs are truly vertically aligned by beginning with the end in mind—success in Integrated Math I and beyond. By "backmapping" the content from the high school programs, all of our mathematics programs are well articulated in their scope and sequence.

## Lead Authors


**John A. Carter, Ph.D.**
Principal
Adlai E. Stevenson High School
Lincolnshire, Illinois


Areas of Expertise: Using technology and manipulatives to visualize concepts; mathematics achievement of English-language learners


**Gilbert J. Cuevas, Ph.D.**
Professor of Mathematics Education
Texas State University—San Marcos
San Marcos, Texas


Areas of Expertise: Applying concepts and skills in mathematically rich contexts; mathematical representations


**Roger Day, Ph.D., NBCT**
Mathematics Department Chairperson
Pontiac Township High School
Pontiac, Illinois


Areas of Expertise: Understanding and applying probability and statistics; mathematics teacher education


**Carol Malloy, Ph.D.**
Associate Professor
University of North Carolina at Chapel Hill
Chapel Hill, North Carolina


Areas of Expertise: Representations and critical thinking; student success in Algebra 1

## Program Authors


**Ruth Casey**
Mathematics Consultant
Regional Teacher Partner
University of Kentucky
Lexington, Kentucky


Areas of Expertise: Graphing technology and mathematics


**Jerry Cummins**
Mathematics Consultant
Former President, National Council of Supervisors of Mathematics
Western Springs, Illinois


Areas of Expertise: Graphing technology and mathematics


**Dr. Berchie Holliday, Ed.D.**
National Mathematics Consultant
Silver Spring, Maryland


Areas of Expertise: Using mathematics to model and understand real-world data; the effect of graphics on mathematical understanding


**Beatrice Moore Luchin**
Mathematics Consultant
Houston, Texas


Areas of Expertise: Mathematical literacy; working with English language learners

## Contributing Author


**Dinah Zike** FOLDABLES
Educational Consultant
Dinah-Might Activities, Inc.
San Antonio, Texas

These professionals were instrumental in providing valuable input and suggestions for improving the effectiveness of the mathematics instruction.

## Lead Consultants

**Viken Hovsepian**
Professor of Mathematics
Rio Hondo College
Whittier, California

**Jay McTiche**
Educational Author and Consultant
Columbia, Maryland

## Consultants

### Mathematical Content

**Grant A. Fraser, Ph.D.**
Professor of Mathematics
California State University, Los Angeles
Los Angeles, California

**Arthur K. Wayman, Ph.D.**
Professor of Mathematics Emeritus
California State University, Long Beach
Long Beach, California

### Gifted and Talented

**Shelbi K. Cole**
Research Assistant
University of Connecticut
Storrs, Connecticut

### College Readiness

**Robert Lee Kimball, Jr.**
Department Head, Math and Physics
Wake Technical Community College
Raleigh, North Carolina

### Differentiation for English-Language Learners

**Susana Davidenko**
State University of New York
Cortland, New York

**Alfredo Gómez**
Mathematics/ESL Teacher
George W. Fowler High School
Syracuse, New York

### Graphing Calculator

**Ruth M. Casey**
$T^3$ National Instructor
Frankfort, Kentucky

**Jerry Cummins**
Former President
National Council of Supervisors of Mathematics
Western Springs, Illinois

### Mathematical Fluency

**Robert M. Capraro**
Associate Professor
Texas A&M University
College Station, Texas

### Pre-AP

**Dixie Ross**
Lead Teacher for Advanced Placement Mathematics
Pflugerville High School
Pflugerville, Texas

### Reading and Writing

**ReLeah Cossett Lent**
Author and Educational Consultant
Morgantown, GA

**Lynn T. Havens**
Director of Project CRISS
Kalispell, Montana

CHAPTER 0

# Preparing for Integrated Math II

**Get Started** on Chapter 0 ..... P2

■ **Pretest** ..... P3

**0-1** **Changing Units of Measure Within Systems** ..... P4

**0-2** **Changing Units of Measure Between Systems** ..... P6

**0-3** **Simple Probability** ..... P8

**0-4** **Algebraic Expressions** ..... P10

**0-5** **Linear Equations** ..... P11

**0-6** **Linear Inequalities** ..... P13

**0-7** **Inverse Linear Functions** ..... P15

**Explore: Algebra Lab** Drawing Inverses ..... P22

**0-8** **Ordered Pairs** ..... P23

**0-9** **Systems of Linear Equations** ..... P25

**0-10** **Square Roots and Simplifying Radicals** ..... P27

■ **Posttest** ..... P29


connectED.mcgraw-hill.com **Your Digital Math Portal**

**Vocabulary** p. P2

**Multilingual eGlossary** p. P2

**Personal Tutor** p. P18

**Foldables** p. P2

CHAPTER 1

# Quadratic Expressions and Equations

**Get Ready** for Chapter 1 . . . . . 3

Explore: Algebra Lab Adding and Subtracting Polynomials . . . . . 5

1-1 **Adding and Subtracting Polynomials** . . . . . 7

1-2 **Multiplying a Polynomial by a Monomial** . . . . . 14

Explore: Algebra Lab Multiplying Polynomials . . . . . 20

1-3 **Multiplying Polynomials** . . . . . 22

1-4 **Special Products** . . . . . 28

■ **Mid-Chapter Quiz** . . . . . 34

Explore: Algebra Lab Factoring Using the Distributive Property . . . . . 35

1-5 **Using the Distributive Property** . . . . . 36

Explore: Algebra Lab Factoring Trinomials . . . . . 43

1-6 **Solving $x^2 + bx + c = 0$** . . . . . 45

1-7 **Solving $ax^2 + bx + c = 0$** . . . . . 52

1-8 **Differences of Squares** . . . . . 58

1-9 **Perfect Squares** . . . . . 64

1-10 **Roots and Zeros** . . . . . 72

**ASSESSMENT**

■ **Study Guide and Review** . . . . . 80

■ **Practice Test** . . . . . 86

■ **Preparing for Standardized Tests** . . . . . 87

■ **Standardized Test Practice, Chapter 1** . . . . . 89


connectED.mcgraw-hill.com **Your Digital Math Portal**

**Animation** pp. 2, 90

**Vocabulary** pp. 4, 92

**Multilingual eGlossary** pp. 4, 92

**Personal Tutor** pp. 7, 98

CHAPTER 2

# Quadratic Functions and Equations

**Get Ready** for Chapter 2 ........ 91

**2-1 Graphing Quadratic Functions** ........ 93

**Extend: Algebra Lab** Rate of Change of a Quadratic Function ........ 104

**2-2 Solving Quadratic Equations by Graphing** ........ 105

**Extend: Graphing Technology Lab** Quadratic Inequalities ........ 111

**Explore: Graphing Technology Lab** Family of Quadratic Functions ........ 112

**2-3 Transformations of Quadratic Functions** ........ 114

**Extend: Graphing Technology Lab** Systems of Linear and Quadratic Equations ........ 122

**2-4 Solving Quadratic Equations by Completing the Square** ........ 124

**Extend: Algebra Lab** Finding the Maximum or Minimum Value ........ 130

■ **Mid-Chapter Quiz** ........ 132

**2-5 Solving Quadratic Equations by Using the Quadratic Formula** ........ 133

**2-6 Analyzing Functions with Successive Differences** ........ 140

**Extend: Graphing Technology Lab** Curve Fitting ........ 146

**2-7 Special Functions** ........ 148

**Extend: Graphing Technology Lab** Piecewise-Linear Functions ........ 156

**ASSESSMENT**

■ **Study Guide and Review** ........ 157

■ **Practice Test** ........ 161

■ **Preparing for Standardized Tests** ........ 162

■ **Standardized Test Practice, Chapters 1–2** ........ 164

**Virtual Manipulatives** pp. 29, 98

**Graphing Calculator** pp. 111, 156

**Foldables** pp. 4, 92

**Self-Check Practice** pp. 3, 91

CHAPTER 3

# Quadratic Functions and Relations

**Get Ready** for Chapter 3 . . . . . . 167

3-1 Extend: Graphing Technology Lab Modeling Real-World Data . . . . . . 169

3-2 Solving Quadratic Equations by Factoring . . . . . . 170

3-3 Complex Numbers . . . . . . 178
Extend: Algebra Lab The Complex Plane . . . . . . 185
Extend: Graphing Technology Lab Solving Quadratic Equations . . . . . . 187

■ Mid-Chapter Quiz . . . . . . 188

3-4 The Quadratic Formula and the Discriminant . . . . . . 189

Explore: Graphing Technology Lab Families of Parabolas . . . . . . 198

3-5 Transformations of Quadratic Graphs . . . . . . 200
Extend: Algebra Lab Quadratics and Rate of Change . . . . . . 206

3-6 Quadratic Inequalities . . . . . . 207
Extend: Graphing Technology Lab Quadratic Inequalities . . . . . . 214
Extend: Graphing Technology Lab Modeling Motion . . . . . . 215

ASSESSMENT

■ Study Guide and Review . . . . . . 216
■ Preparing for Standardized Tests . . . . . . 220
■ Standardized Test Practice, Chapters 1–3 . . . . . . 222




connectED.mcgraw-hill.com Your Digital Math Portal

Animation pp. 166, 224

Vocabulary pp. 168, 226

Multilingual eGlossary pp. 168, 226

Personal Tutor pp. 179, 253

CHAPTER 4

# Exponential and Logarithmic Functions and Relations

**Get Ready** for Chapter 4 ........ 225

**4-1 Graphing Exponential Functions** ........ 227

**Explore: Graphing Technology Lab** Solving Exponential Equations and Inequalities ........ 235

**4-2 Solving Exponential Equations and Inequalities** ........ 237

**Extend: Algebra Lab** Transforming Exponential Expressions ........ 244

**4-3 Simplifying Radical Expressions** ........ 245

**Extend: Algebra Lab** Rational and Irrational Numbers ........ 251

**4-4 Operations with Radical Expressions** ........ 252

**Extend: Algebra Lab** Simplifying *n* th Root Expressions ........ 257

**4-5 Radical Equations** ........ 259

**ASSESSMENT**

- **Study Guide and Review** ........ 264
- **Practice Test** ........ 267
- **Preparing for Standardized Tests** ........ 268
- **Standardized Test Practice, Chapters 1–4** ........ 270

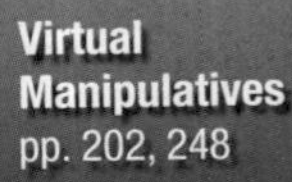

**Virtual Manipulatives** pp. 202, 248

**Graphing Calculator** pp. 169, 235

**Foldables** pp. 168, 226

**Self-Check Practice** pp. 167, 225

Purestock/Getty Images

CHAPTER 5

# Reasoning and Proof

**Get Ready** for Chapter 5 ..... 273

**Explore: Geometry Lab** Necessary and Sufficient Conditions ..... 275

**5-1 Postulates and Paragraph Proofs** ..... 276

**5-2 Algebraic Proof** ..... 284

**5-3 Proving Segment Relationships** ..... 292

**5-4 Proving Angle Relationships** ..... 299

**5-5 Angles and Parallel lines** ..... 308

**5-6 Proving Lines Parallel** ..... 315

**ASSESSMENT**

- **Study Guide and Review** ..... 323
- **Practice Test** ..... 327
- **Preparing for Standardized Tests** ..... 328
- **Standardized Test Practice, Chapters 1–5** ..... 330


connectED.mcgraw-hill.com **Your Digital Math Portal**

**Animation** pp. 272, 332

**Vocabulary** pp. 274, 334

**Multilingual eGlossary** pp. 274, 334

**Personal Tutor** pp. 277, 342

# CHAPTER 6 Congruent Triangles

**Get Ready** for Chapter 6 .......... 333

**6-1 Angles of Triangles** .......... 335

**6-2 Congruent Triangles** .......... 344

**6-3 Proving Triangles Congruent—SSS, SAS** .......... 353

**Extend: Geometry Lab** Proving Constructions .......... 362

**Mid-Chapter Quiz** .......... 363

**6-4 Proving Triangles Congruent—ASA, AAS** .......... 364

**Extend: Geometry Lab** Congruence in Right Triangles .......... 372

**6-5 Isosceles and Equilateral Triangles** .......... 374

**6-6 Triangles and Coordinate Proof** .......... 383

**Extend: Geometry Lab** Two-Dimensional Representations of Three-Dimensional Objects 390

**ASSESSMENT**

**Study Guide and Review** .......... 393

**Practice Test** .......... 397

**Preparing for Standardized Tests** .......... 398

**Standardized Test Practice, Chapters 1–6** .......... 400




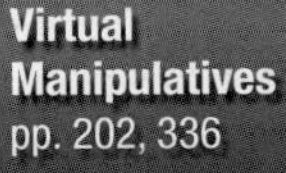
**Virtual Manipulatives** pp. 202, 336

**Graphing Calculator** pp. 235, 342

**Foldables** pp. 272, 334

**Self-Check Practice** pp. 273, 333

CHAPTER 7

# Relationships in Triangles

**Get Ready** for Chapter 7 ..... 403

Explore: Geometry Lab Constructing Bisectors ..... 405

**7-1 Bisectors of Triangles** ..... 406

Explore: Geometry Lab Constructing Medians and Altitudes ..... 416

**7-2 Medians and Altitudes of Triangles** ..... 417

**7-3 Inequalities in One Triangle** ..... 426

■ **Mid-Chapter Quiz** ..... 434

Explore: Geometry Lab Matrix Logic ..... 435

**7-4 Indirect Proof** ..... 437

Explore: Graphing Technology Lab The Triangle Inequality ..... 445

**7-5 The Triangle Inequality** ..... 446

**7-6 Inequalities in Two Triangles** ..... 453

ASSESSMENT

■ **Study Guide and Review** ..... 463

■ **Practice Test** ..... 467

■ **Preparing for Standardized Tests** ..... 468

■ **Standardized Test Practice, Chapters 1–7** ..... 470


connectED.mcgraw-hill.com Your Digital Math Portal

Animation pp. 402, 472

Vocabulary pp. 404, 474

Multilingual eGlossary pp. 404, 474

Personal Tutor pp. 407, 473

CHAPTER 8

# Quadrilaterals

**Get Ready** for Chapter 8 ... 473

8-1 **Angles of Polygons** ... 475

**Extend: Spreadsheet Lab** Angles of Polygons ... 484

8-2 **Parallelograms** ... 485

**Explore: Graphing Technology Lab** Parallelograms ... 494

8-3 **Tests for Parallelograms** ... 495

**Mid-Chapter Quiz** ... 504

8-4 **Rectangles** ... 505

8-5 **Rhombi and Squares** ... 512

8-6 **Trapezoids and Kites** ... 521

**ASSESSMENT**

- **Study Guide and Review** ... 531
- **Practice Test** ... 535
- **Preparing for Standardized Tests** ... 536
- **Standardized Test Practice, Chapters 1–8** ... 538

**Virtual Manipulatives** pp. 418, 488

**Foldables** pp. 404, 474

**Self-Check Practice** pp. 403, 473

CHAPTER 9

# Proportions and Similarity

**Get Ready** for Chapter 9 ..... 541

9-1 **Ratios and Proportions** ..... 543

**Extend: Graphing Technology Lab** Fibonacci Sequence and Ratios ..... 550

9-2 **Similar Polygons** ..... 551

9-3 **Similar Triangles** ..... 560

**Extend: Geometry Lab** Proofs of Perpendicular and Parallel Lines ..... 570

9-4 **Parallel Lines and Proportional Parts** ..... 572

■ **Mid-Chapter Quiz** ..... 582

9-5 **Parts of Similar Triangles** ..... 583

**Extend: Geometry Lab** Fractals ..... 591

9-6 **Similarity Transformations** ..... 593

9-7 **Scale Drawings and Models** ..... 600

**ASSESSMENT**

■ **Study Guide and Review** ..... 606

■ **Practice Test** ..... 611

■ **Preparing for Standardized Tests** ..... 612

■ **Standardized Test Practice, Chapters 1–9** ..... 614




connectED.mcgraw-hill.com **Your Digital Math Portal**

**Animation** pp. 540, 616

**Vocabulary** pp. 542, 618

**Multilingual eGlossary** pp. 542, 618

**Personal Tutor** pp. 543, 619

# CHAPTER 10 Right Triangles and Trigonometry

**Get Ready** for Chapter 10 . . . . . 617

10-1 **Geometric Mean** . . . . . 619

**Explore: Geometry Lab** Proofs Without Words . . . . . 628

10-2 **The Pythagorean Theorem and Its Converse** . . . . . 629

**Extend: Geometry Lab** Coordinates in Space . . . . . 638

10-3 **Special Right Triangles** . . . . . 640

**Explore: Graphing Technology Lab** Trigonometry . . . . . 649

10-4 **Trigonometry** . . . . . 650

**Extend: Graphing Technology Lab** Secant, Cosecant, and Cotangent . . . . . 660

■ **Mid-Chapter Quiz** . . . . . 661

10-5 **Angles of Elevation and Depression** . . . . . 662

10-6 **The Law of Sines and Law of Cosines** . . . . . 670

**Extend: Geometry Lab** The Ambiguous Case . . . . . 680

10-7 **Vectors** . . . . . 682

**Extend: Geometry Lab** Adding Vectors . . . . . 691

**Explore: Graphing Technology Lab** Dilations . . . . . 692

10-8 **Dilations** . . . . . 694

**ASSESSMENT**

■ **Study Guide and Review** . . . . . 702

■ **Practice Test** . . . . . 707

■ **Preparing for Standardized Tests** . . . . . 708

■ **Standardized Test Practice, Chapters 1–10** . . . . . 710

**Virtual Manipulatives**
pp. 591

**Graphing Calculator**
pp. 550, 649

**Foldables**
pp. 542, 618

**Self-Check Practice**
pp. 541, 617

CHAPTER 11

# Circles

**Get Ready** for Chapter 11 ... 713

**11-1** **Circles and Circumference** ... 715

**11-2** **Measuring Angles and Arcs** ... 724

**11-3** **Arcs and Chords** ... 733

**11-4** **Inscribed Angles** ... 741

■ **Mid-Chapter Quiz** ... 749

**11-5** **Tangents** ... 750

**Extend: Geometry Lab** Inscribed and Circumscribed Circles ... 758

**11-6** **Secants, Tangents, and Angle Measures** ... 759

**11-7** **Special Segments in a Circle** ... 768

**11-8** **Equations of Circles** ... 775

**Extend: Geometry Lab** Parabolas ... 782

**11-9** **Areas of Circles and Sectors** ... 784

**ASSESSMENT**

■ **Study Guide and Review** ... 791

■ **Practice Test** ... 797

■ **Preparing for Standardized Tests** ... 798

■ **Standardized Test Practice, Chapters 1–11** ... 800


James Randklev/Photographer's Choice RF/Getty Images

connectED.mcgraw-hill.com **Your Digital Math Portal**

**Animation** pp. 712, 802

**Vocabulary** pp. 714, 804

**Multilingual eGlossary** pp. 714, 804

**Personal Tutor** pp. 715, 748

CHAPTER 12

# Extending Surface Area and Volume

**Get Ready** for Chapter 12 .......... 803

**Explore: Geometry Lab** Solids Formed by Translation .......... 805

**12-1 Representations of Three-Dimensional Figures** .......... 807

**Extend: Geometry Lab** Topographic Maps .......... 813

**12-2 Surface Areas of Prisms and Cylinders** .......... 814

**12-3 Surface Areas of Pyramids and Cones** .......... 822

**12-4 Volumes of Prisms and Cylinders** .......... 831

**Extend: Graphing Technology Lab** Changing Dimensions .......... 839

■ **Mid-Chapter Quiz** .......... 840

**12-5 Volumes of Pyramids and Cones** .......... 841

**12-6 Surface Areas and Volumes of Spheres** .......... 848

**Extend: Geometry Lab** Locus and Spheres .......... 856

**12-7 Spherical Geometry** .......... 857

**Extend: Geometry Lab** Navigational Coordinates .......... 863

**12-8 Congruent and Similar Solids** .......... 864

**ASSESSMENT**

■ **Study Guide and Review** .......... 871

■ **Practice Test** .......... 875

■ **Preparing for Standardized Tests** .......... 876

■ **Standardized Test Practice, Chapters 1–12** .......... 878

**Virtual Manipulatives** pp. 775

**Graphing Calculator** pp. 839

**Foldables** pp. 714, 804

**Self-Check Practice** pp. 713, 803

CHAPTER 13

# Probability and Measurement

**Get Ready** for Chapter 13 . . . 881

**13-1** **Representing Sample Spaces** . . . 883

**13-2** **Probability with Permutations and Combinations** . . . 890

**13-3** **Geometric Probability** . . . 899

■ **Mid-Chapter Quiz** . . . 906

**13-4** **Simulations** . . . 907

**13-5** **Probabilities of Independent and Dependent Events** . . . 915

**Extend: Geometry Lab** Two-Way Frequency Tables . . . 922

**13-6** **Probabilities of Mutually Exclusive Events** . . . 924

**Extend: Geometry Lab** Graph Theory . . . 932

**ASSESSMENT**

■ **Study Guide and Review** . . . 934

■ **Practice Test** . . . 937

■ **Preparing for Standardized Tests** . . . 938

■ **Standardized Test Practice, Chapters 1–13** . . . 940


connectED.mcgraw-hill.com **Your Digital Math Portal**

**Animation** pp. 672, 793

**Vocabulary** pp. 674, 757

**Multilingual eGlossary** pp. 674, 750

**Personal Tutor** pp. 697, 748

Image Source/SuperStock

# Student Handbook

**Reference**


Glossary/Glosario

Index

Formulas and Measures Symbols and Properties

# **Get Started** on the Chapter

**You will review several concepts, skills, and vocabulary terms as you study Chapter 0. To get ready, identify important terms and organize your resources.**

## FOLDABLES® StudyOrganizer

Throughout this text, you will be invited to use Foldables to organize your notes.

### Why should you use them?

- They help you organize, display, and arrange information.
- They make great study guides, specifically designed for you.
- You can use them as your math journal for recording main ideas, problem-solving strategies, examples, or questions you may have.
- They give you a chance to improve your math vocabulary.

### How should you use them?

- Write general information — titles, vocabulary terms, concepts, questions, and main ideas — on the front tabs of your Foldable.
- Write specific information — ideas, your thoughts, answers to questions, steps, notes, and definitions — under the tabs.
- Use the tabs for:
  - math concepts in parts, like types of triangles,
  - steps to follow, or
  - parts of a problem, like *compare* and *contrast* (2 parts) or *what, where, when, why,* and *how* (5 parts).
- You may want to store your Foldables in a plastic zipper bag that you have three-hole punched to fit in your notebook.

### When should you use them?

- Set up your Foldable as you begin a chapter, or when you start learning a new concept.
- Write in your Foldable every day.
- Use your Foldable to review for homework, quizzes, and tests.

## ReviewVocabulary

| English | | Español |
|---|---|---|
| experiment | p. P8 | experimento |
| trial | p. P8 | prueba |
| outcome | p. P8 | resultado |
| event | p. P8 | evento |
| probability | p. P8 | probabilidad |
| theoretical probability | p. P9 | probabilidad teórica |
| experimental probability | p. P9 | probabilidad experimental |
| ordered pair | p. P23 | par ordenado |
| *x*-coordinate | p. P23 | coordenada *x* |
| *y*-coordinate | p. P23 | coordenada *y* |
| quadrant | p. P23 | cuadrante |
| origin | p. P23 | origen |
| system of equations | p. P25 | sistema de ecuaciones |
| substitution | p. P25 | sustitución |
| elimination | p. P26 | eliminación |
| Product Property | p. P27 | Propriedad de Producto |
| Quotient Property | p. P27 | Propriedad de Cociente |

# CHAPTER 0 Pretest

**State which metric unit you would probably use to measure each item.**

1. length of a computer keyboard
2. mass of a large dog

**Complete each sentence.**

3. 4 ft = ___?___ in.
4. 21 ft = ___?___ yd
5. 180 g = ___?___ kg
6. 3 T = ___?___ lb
7. 32 g ≈ ___?___ oz
8. 3 mi ≈ ___?___ km
9. 35 yd ≈ ___?___ m
10. 5.1 L ≈ ___?___ qt

11. **TUNA** A can of tuna is 6 ounces. About how many grams is it?

12. **CRACKERS** A box of crackers is 453 grams. About how many pounds is it? Round to the nearest pound.

13. **DISTANCE** A road sign in Canada gives the distance to Toronto as 140 kilometers. What is this distance to the nearest mile?

**PROBABILITY** **A bag contains 3 blue chips, 7 red chips, 4 yellow chips, and 5 green chips. A chip is randomly drawn from the bag. Find each probability.**

14. $P(\text{yellow})$
15. $P(\text{green})$
16. $P(\text{red or blue})$
17. $P(\text{not red})$

**Evaluate each expression if $r = 3$, $q = 1$, and $w = -2$.**

18. $4r + q$
19. $rw - 6$
20. $\dfrac{r + 3q}{4r}$
21. $\dfrac{5w}{3r + q}$
22. $|2 - r| + 17$
23. $8 + |q - 5|$

**Solve each equation.**

24. $k + 3 = 14$
25. $a - 7 = 9$
26. $5c = 20$
27. $n + 2 = -11$
28. $6t - 18 = 30$
29. $4x + 7 = -1$
30. $\frac{r}{4} = -8$
31. $\frac{3}{5}b = -2$
32. $-\frac{w}{2} = -9$
33. $3y - 15 = y + 1$
34. $27 - 6d = 7 + 4d$
35. $2(m - 16) = 44$

**Solve each inequality.**

36. $y - 13 < 2$
37. $t + 8 \geq 19$
38. $\frac{n}{4} > -6$
39. $9a \leq 45$
40. $x + 12 > -14$
41. $-2w < 24$
42. $-\frac{n}{7} \geq 3$
43. $-\frac{b}{5} \leq -6$

**Write the ordered pair for each point shown.**

44. $F$
45. $H$
46. $A$
47. $D$

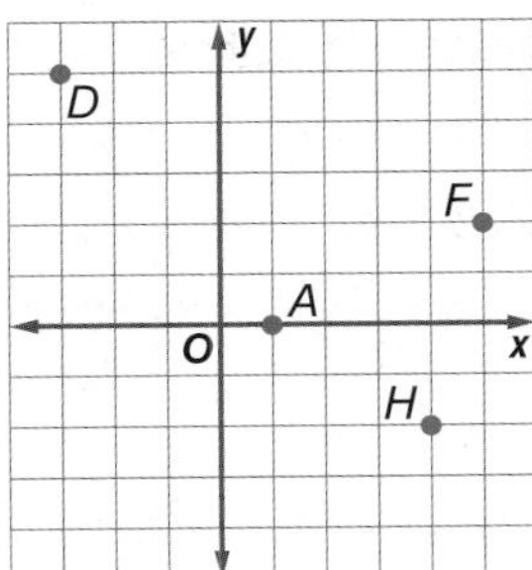


**Graph and label each point on the coordinate plane above.**

48. $B(4, 1)$
49. $G(0, -3)$
50. $R(-2, -4)$
51. $P(-3, 3)$

52. Graph the triangle with vertices $J(1, -4)$, $K(2, 3)$, and $L(-1, 2)$.

53. Graph four points that satisfy the equation $y = 2x - 1$.

**Solve each system of equations.**

54. $y = 2x$
$y = -x + 6$

55. $-3x - y = 4$
$4x + 2y = -8$

56. $y = 2x + 1$
$y = 3x$

57. $\frac{1}{2}x - y = -1$
$x - 2y = 5$

58. $x + y = -6$
$2x - y = 3$

59. $\frac{1}{3}x - 3y = -4$
$x - 9y = -12$

**Simplify.**

60. $\sqrt{18}$
61. $\sqrt{\frac{25}{49}}$
62. $\sqrt{24x^2y^3}$
63. $\dfrac{3}{4 - \sqrt{5}}$

# LESSON 0-1 Changing Units of Measure Within Systems

## Objective

- Convert units of measure within the customary and metric systems.

### Example 1 Choose Best Unit of Measure

**State which metric unit you would use to measure the length of your pen.**

A pen has a small length, but not very small. The *centimeter* is the appropriate unit of measure.

| Metric Units of Length |
|---|
| 1 kilometer (km) = 1000 meters (m) |
| 1 m = 100 centimeters (cm) |
| 1 cm = 10 millimeters (mm) |

| Customary Units of Length |
|---|
| 1 foot (ft) = 12 inches (in.) |
| 1 yard (yd) = 3 ft |
| 1 mile (mi) = 5280 ft |

- To convert from larger units to smaller units, multiply.
- To convert from smaller units to larger units, divide.
- To use dimensional analysis, multiply by the ratio of the units.

### Example 2 Convert from Larger Units to Smaller Units of Length

**Complete each sentence.**

**a. 4.2 km = __?__ m**

There are 1000 meters in a kilometer.
4.2 km × 1000 = 4200 m

**b. 13 yd = __?__ ft**

There are 3 feet in a yard.
13 yd × 3 = 39 ft

### Example 3 Convert from Smaller Units to Larger Units of Length

**Complete each sentence.**

**a. 17 mm = __?__ m**

There are 100 centimeters in a meter. First change *millimeters* to *centimeters.*

17 mm = __?__ cm — smaller unit → larger unit

17 mm ÷ 10 = 1.7 cm — Since 10 mm = 1 cm, divide by 10.

Then change *centimeters* to *meters.*

1.7 cm = __?__ m — smaller unit → larger unit

1.7 cm ÷ 100 = 0.017 m — Since 100 cm = 1 m, divide by 100.

**b. 6600 yd = __?__ mi**

Use dimensional analysis.

$$6600\ \cancel{\text{yd}} \times \frac{3\ \cancel{\text{ft}}}{1\ \cancel{\text{yd}}} \times \frac{1\ \text{mi}}{5280\ \cancel{\text{ft}}} = 3.75\ \text{mi}$$

| Metric Units of Capacity |
|---|
| 1 liter (L) = 1000 milliliters (mL) |

| Customary Units of Capacity | |
|---|---|
| 1 cup (c) = 8 fluid ounces (fl oz) | 1 quart (qt) = 2 pt |
| 1 pint (pt) = 2 c | 1 gallon (gal) = 4 qt |

**StudyTip**

**Dimensional Analysis** You can use dimensional analysis for any conversion in this lesson.

PT

### Example 4 Convert Units of Capacity

**Complete each sentence.**

**a.** **3.7 L = __?__ mL**

There are 1000 milliliters in a liter.

3.7 L × 1000 = 3700 mL

**b.** **16 qt = __?__ gal**

There are 4 quarts in a gallon.

16 qt ÷ 4 = 4 gal

**c.** **7 pt = __?__ fl oz**

There are 8 fluid ounces in a cup.

First change *pints* to *cups*.

7 pt = __?__ c

7 pt × 2 = 14 c

Then change *cups* to *fluid ounces*.

14 c = __?__ fl oz

14 c × 8 = 112 fl oz

**d.** **4 gal = __?__ pt**

There are 4 quarts in a gallon.

First change *gallons* to *quarts*.

4 gal = __?__ qt

4 gal × 4 = 16 qt

Then change *quarts* to *pints*.

16 qt = __?__ pt

16 qt × 2 = 32 pt

The mass of an object is the amount of matter that it contains.

| Metric Units of Mass |
|---|
| 1 kilogram (kg) = 1000 grams (g) |
| 1 g = 1000 milligrams (mg) |

| Customary Units of Weight |
|---|
| 1 pound (lb) = 16 ounces (oz) |
| 1 ton (T) = 2000 lb |

### Example 5 Convert Units of Mass

**Complete each sentence.**

**a.** **5.47 kg = __?__ mg**

There are 1000 milligrams in a gram.

Change *kilograms* to *grams*.

5.47 kg = __?__ g

5.47 kg × 1000 = 5470 g

Then change *grams* to *milligrams*.

5470 g = __?__ mg

5470 g × 1000 = 5,470,000 mg

**b.** **5 T = __?__ oz**

There are 16 ounces in a pound.

Change *tons* to *pounds*.

5 T = __?__ lb

5 T × 2000 = 10,000 lb

Then change *pounds* to *ounces*.

10,000 lb = __?__ oz

10,000 lb × 16 = 160,000 oz

## Exercises

**State which metric unit you would probably use to measure each item.**

**1.** radius of a tennis ball
**2.** length of a notebook
**3.** mass of a textbook
**4.** mass of a beach ball
**5.** liquid in a cup
**6.** water in a bathtub

**Complete each sentence.**

**7.** 120 in. = __?__ ft
**8.** 18 ft = __?__ yd
**9.** 10 km = __?__ m
**10.** 210 mm = __?__ cm
**11.** 180 mm = __?__ m
**12.** 3100 m = __?__ km
**13.** 90 in. = __?__ yd
**14.** 5280 yd = __?__ mi
**15.** 8 yd = __?__ ft
**16.** 0.62 km = __?__ m
**17.** 370 mL = __?__ L
**18.** 12 L = __?__ mL
**19.** 32 fl oz = __?__ c
**20.** 5 qt = __?__ c
**21.** 10 pt = __?__ qt
**22.** 48 c = __?__ gal
**23.** 4 gal = __?__ qt
**24.** 36 mg = __?__ g
**25.** 13 lb = __?__ oz
**26.** 130 g = __?__ kg
**27.** 9.05 kg = __?__ g

LESSON 0-2

# Changing Units of Measure Between Systems

## Objective

- Convert units of measure between the customary and metric systems.

The table below shows approximate equivalents between customary units of length and metric units of length.

| Units of Length | |
|---|---|
| Customary → Metric | Metric → Customary |
| 1 in. ≈ 2.5 cm | 1 cm ≈ 0.4 in. |
| 1 yd ≈ 0.9 m | 1 m ≈ 1.1 yd |
| 1 mi ≈ 1.6 km | 1 km ≈ 0.6 mi |

### Example 1 Convert Units of Length Between Systems

**Complete each sentence.**

**a. 30 in. ≈ __?__ cm**

There are approximately 2.5 centimeters in an inch.

30 in. × 2.5 = 75 cm

**b. 5 km ≈ __?__ mi**

There is approximately 0.6 mile in a kilometer.

5 km × 0.6 = 3 mi

### Example 2 Convert Units of Length Between Systems

**Complete: 2000 yd ≈ __?__ km.**

There is approximately 0.9 meter in a yard. First find the number of meters in 2000 yards.

2000 yd × 0.9 = 1800 m

Then change *meters* to *kilometers*. There are 1000 meters in a kilometer.

1800 m ÷ 1000 = 1.8 km

The table below shows approximate equivalents between customary units of capacity and metric units of capacity.

| Units of Capacity | |
|---|---|
| Customary → Metric | Metric → Customary |
| 1 qt ≈ 0.9 L | 1 L ≈ 1.1 qt |
| 1 pt ≈ 0.5 L | 1 L ≈ 2.1 pt |

### Example 3 Convert Units of Capacity Between Systems

**Complete each sentence.**

**a. 7 qt ≈ __?__ L**

There is approximately 0.9 liter in a quart.

7 qt × 0.9 = 6.3 L

**b. 2 L ≈ __?__ pt**

There are approximately 2.1 pints in a liter.

2 L × 2.1 = 4.2 pt

### Example 4 Convert Units of Capacity Between Systems

**Complete: 10 L ≈ __?__ gal.**

There are approximately 1.1 quarts in a liter. First find the number of quarts in 10 liters.

$10 \text{ L} \times 1.1 = 11 \text{ qt}$

Then change *quarts* to *gallons*. There are 4 quarts in a gallon.

$11 \text{ qt} \div 4 = 2.75 \text{ gal}$

You can also use dimensional analysis.

$$10 \cancel{\text{L}} \times \frac{1.1 \cancel{\text{qt}}}{1 \cancel{\text{L}}} \times \frac{1 \text{ gal}}{4 \cancel{\text{qt}}} = 2.75 \text{ gal}$$

**StudyTip**

**Dimensional Analysis** If the unit that you want to eliminate is in the numerator, make sure it is in the denominator of the ratio when you multiply. If it is in the denominator, make sure that it is in the numerator of the ratio.

The table below shows approximate equivalents between customary units of weight and metric units of mass.

| Units of Weight/Mass | |
|---|---|
| Customary → Metric | Metric → Customary |
| 1 oz ≈ 28.3 g | 1 g ≈ 0.04 oz |
| 1 lb ≈ 0.5 kg | 1 kg ≈ 2.2 lb |

### Example 5 Convert Units of Mass Between Systems

**Complete each sentence.**

**a. 58.5 kg ≈ __?__ lb**

There are approximately 2.2 pounds in a kilogram.

$58.5 \text{ kg} \times 2.2 = 128.7 \text{ lb}$

**b. 14 oz ≈ __?__ g**

There are approximately 28.3 grams in an ounce.

$14 \text{ oz} \times 28.3 = 396.2 \text{ g}$

## Exercises

**Complete each sentence.**

**1.** 8 in. ≈ __?__ cm

**2.** 15 m ≈ __?__ yd

**3.** 11 qt ≈ __?__ L

**4.** 25 oz ≈ __?__ g

**5.** 10 mi ≈ __?__ km

**6.** 32 cm ≈ __?__ in.

**7.** 20 km ≈ __?__ mi

**8.** 9.5 L ≈ __?__ qt

**9.** 6 yd ≈ __?__ m

**10.** 4.3 kg ≈ __?__ lb

**11.** 10.7 L ≈ __?__ pt

**12.** 82.5 g ≈ __?__ oz

**13.** $2\frac{1}{4}$ lb ≈ __?__ kg

**14.** 10 ft ≈ __?__ m

**15.** $1\frac{1}{2}$ gal ≈ __?__ L

**16.** 350 g ≈ __?__ lb

**17.** 600 in. ≈ __?__ m

**18.** 2.1 km ≈ __?__ yd

**19.** **CEREAL** A box of cereal is 13 ounces. About how many grams is it?

**20.** **FLOUR** A bag of flour is 2.26 kilograms. How much does it weigh? Round to the nearest pound.

**21.** **SAUCE** A jar of tomato sauce is 1 pound 10 ounces. About how many grams is it?

LESSON 0-3

# Simple Probability

## Objective

- Find the probability of simple events.

### NewVocabulary

experiment
trial
outcome
event
probability
theoretical probability
experimental probability

### Common Core State Standards

**Content Standards**

S.MD.6 (+) Use probabilities to make fair decisions (e.g., drawing by lots, using a random number generator).

S.MD.7 (+) Analyze decisions and strategies using probability concepts (e.g., product testing, medical testing, pulling a hockey goalie at the end of a game).

**Mathematical Practices**

4 Model with mathematics.

A situation involving chance such as flipping a coin or rolling a die is an **experiment**. A single performance of an experiment such as rolling a die one time is a **trial**. The result of a trial is called an **outcome**. An **event** is one or more outcomes of an experiment.

When each outcome is equally likely to happen, the **probability** of an event is the ratio of the number of favorable outcomes to the number of possible outcomes. The probability of an event is always between 0 and 1, inclusive.

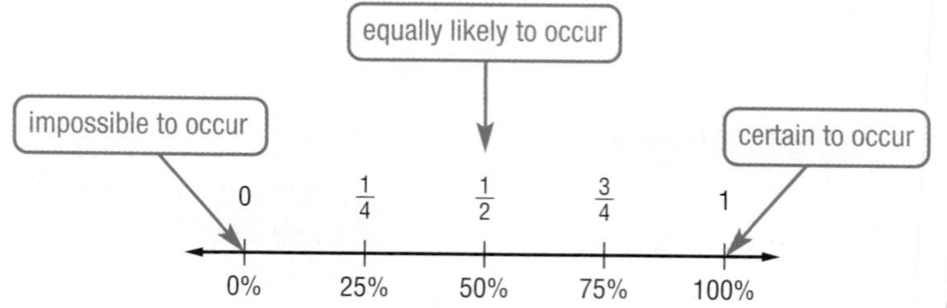

### Example 1 Find Probability

**Suppose a die is rolled. What is the probability of rolling an odd number?**

There are 3 odd numbers on a die: 1, 3, and 5.

There are 6 possible outcomes: 1, 2, 3, 4, 5, and 6.

$$P(\text{odd}) = \frac{\text{number of favorable outcomes}}{\text{number of possible outcomes}}$$

$$= \frac{3}{6} \text{ or } \frac{1}{2}$$

The probability of rolling an odd number is $\frac{1}{2}$ or 50%.

For a given experiment, the sum of the probabilities of all possible outcomes must sum to 1.

### Example 2 Find Probability

**Suppose a bag contains 4 red, 3 green, 6 blue, and 2 yellow marbles. What is the probability a randomly chosen marble will not be yellow?**

Since the sum of the probabilities of all of the colors must sum to 1, subtract the probability that the marble will be yellow from 1.

The probability that the marble will be yellow is $\frac{2}{15}$ because there are 2 yellow marbles and 15 total marbles.

$$P(\text{not yellow}) = 1 - P(\text{yellow})$$

$$= 1 - \frac{2}{15}$$

$$= \frac{13}{15}$$

The probability that the marble will not be yellow is $\frac{13}{15}$ or about 87%.

The probabilities in Examples 1 and 2 are called theoretical probabilities. The **theoretical probability** is what *should* occur. The **experimental probability** is what *actually* occurs when a probability experiment is repeated many times.

**StudyTip**

Experimental Probability The experimental probability of an experiment is not necessarily the same as the theoretical probability, but when an experiment is repeated many times, the experimental probability should be close to the theoretical probability.

PT

### Example 3 Find Experimental Probability

**The table shows the results of an experiment in which a number cube was rolled. Find the experimental probability of rolling a 3.**

| Outcome | Tally | Frequency |
|---|---|---|
| 1 | 𝍸 I | 6 |
| 2 | IIII | 4 |
| 3 | 𝍸 II | 7 |
| 4 | III | 3 |
| 5 | IIII | 4 |
| 6 | I | 1 |

$P(3) = \frac{\text{number of times 3 occurs}}{\text{total number of outcomes}}$ or $\frac{7}{25}$

The experimental probability for getting a 3 in this case is $\frac{7}{25}$ or 28%.

## Exercises

**A die is rolled. Find the probability of each outcome.**

1. $P$(less than 3)
2. $P$(even)
3. $P$(greater than 2)
4. $P$(prime)
5. $P$(4 or 2)
6. $P$(integer)

**A jar contains 65 pennies, 27 nickels, 30 dimes, and 18 quarters. A coin is randomly selected from the jar. Find each probability.**

7. $P$(penny)
8. $P$(quarter)
9. $P$(not dime)
10. $P$(penny or dime)
11. $P$(value greater than \$0.15)
12. $P$(not nickel)
13. $P$(nickel or quarter)
14. $P$(value less than \$0.20)

PRESENTATIONS **The students in a class are randomly drawing cards numbered 1 through 28 from a hat to determine the order in which they will give their presentations. Find each probability.**

15. $P$(13)
16. $P$(1 or 28)
17. $P$(less than 14)
18. $P$(not 1)
19. $P$(not 2 or 17)
20. $P$(greater than 16)

**The table shows the results of an experiment in which three coins were tossed.**

| Outcome | HHH | HHT | HTH | THH | TTH | THT | HTT | TTT |
|---|---|---|---|---|---|---|---|---|
| Tally | 𝍸 | 𝍸 | 𝍸 I | 𝍸 I | 𝍸 II | 𝍸 | 𝍸 III | 𝍸 III |
| Frequency | 5 | 5 | 6 | 6 | 7 | 5 | 8 | 8 |

21. What is the experimental probability that all three of the coins will be heads? The theoretical probability?
22. What is the experimental probability that at least two of the coins will be heads? The theoretical probability?
23. DECISION MAKING You and two of your friends have pooled your money to buy a new video game. Describe a method that could be used to make a fair decision as to who gets to play the game first.
24. DECISION MAKING A new study finds that the incidence of heart attack while taking a certain diabetes drug is less than 5%. Should a person with diabetes take this drug? Should they take the drug if the risk is less than 1%? Explain your reasoning.

# LESSON 0-4 Algebraic Expressions

## Objective

- Use the order of operations to evaluate algebraic expressions.

An expression is an algebraic expression if it contains sums and/or products of variables and numbers. To evaluate an algebraic expression, replace the variable or variables with known values, and then use the order of operations.

| Order of Operations | |
|---|---|
| **Step 1** | Evaluate expressions inside grouping symbols. |
| **Step 2** | Evaluate all powers. |
| **Step 3** | Do all multiplications and/or divisions from left to right. |
| **Step 4** | Do all additions and/or subtractions from left to right. |

### Example 1 Addition/Subtraction Algebraic Expressions

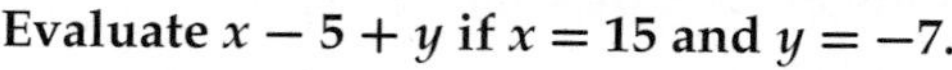

**Evaluate $x - 5 + y$ if $x = 15$ and $y = -7$.**

$x - 5 + y = 15 - 5 + (-7)$ Substitute.

$= 10 + (-7)$ or $3$ Subtract.

### Example 2 Multiplication/Division Algebraic Expressions

**Evaluate each expression if $k = -2$, $n = -4$, and $p = 5$.**

**a.** $\frac{2k + n}{p - 3}$

$\frac{2k + n}{p - 3} = \frac{2(-2) + (-4)}{5 - 3}$ Substitute.

$= \frac{-4 - 4}{5 - 3}$ Multiply.

$= \frac{-8}{2}$ or $-4$ Subtract.

**b.** $-3(k^2 + 2n)$

$-3(k^2 + 2n) = -3[(-2)^2 + 2(-4)]$

$= -3[4 + (-8)]$

$= -3(-4)$ or $12$

### Example 3 Absolute Value Algebraic Expressions

**Evaluate $3|a - b| + 2|c - 5|$ if $a = -2$, $b = -4$, and $c = 3$.**

$3|a - b| + 2|c - 5| = 3|-2 - (-4)| + 2|3 - 5|$ Substitute for $a$, $b$, and $c$.

$= 3|2| + 2|-2|$ Simplify.

$= 3(2) + 2(2)$ or $10$ Find absolute values.

## Exercises

**Evaluate each expression if $a = 2$, $b = -3$, $c = -1$, and $d = 4$.**

**1.** $2a + c$
**2.** $\frac{bd}{2c}$
**3.** $\frac{2d - a}{b}$
**4.** $3d - c$
**5.** $\frac{3b}{5a + c}$
**6.** $5bc$
**7.** $2cd + 3ab$
**8.** $\frac{c - 2d}{a}$

**Evaluate each expression if $x = 2$, $y = -3$, and $z = 1$.**

**9.** $24 + |x - 4|$
**10.** $13 + |8 + y|$
**11.** $|5 - z| + 11$
**12.** $|2y - 15| + 7$

# LESSON 0-5 Linear Equations

**Objective**

- Use algebra to solve linear equations.

If the same number is added to or subtracted from each side of an equation, the resulting equation is true.

### Example 1 Addition/Subtraction Linear Equations

**Solve each equation.**

**a.** $\mathbf{x - 7 = 16}$

| | |
|---|---|
| $x - 7 = 16$ | Original equation |
| $x - 7 + 7 = 16 + 7$ | Add 7 to each side. |
| $x = 23$ | Simplify. |

**b.** $\mathbf{m + 12 = -5}$

| | |
|---|---|
| $m + 12 = -5$ | Original equation |
| $m + 12 + (-12) = -5 + (-12)$ | Add −12 to each side. |
| $m = -17$ | Simplify. |

**c.** $\mathbf{k + 31 = 10}$

| | |
|---|---|
| $k + 31 = 10$ | Original equation |
| $k + 31 - 31 = 10 - 31$ | Subtract 31 from each side. |
| $k = -21$ | Simplify. |

If each side of an equation is multiplied or divided by the same number, the resulting equation is true.

### Example 2 Multiplication/Division Linear Equations

**Solve each equation.**

**a.** $\mathbf{4d = 36}$

| | |
|---|---|
| $4d = 36$ | Original equation |
| $\frac{4d}{4} = \frac{36}{4}$ | Divide each side by 4. |
| $x = 9$ | Simplify. |

**b.** $\mathbf{-\frac{t}{8} = -7}$

| | |
|---|---|
| $-\frac{t}{8} = -7$ | Original equation |
| $-8\left(-\frac{t}{8}\right) = -8(-7)$ | Multiply each side by −8. |
| $t = 56$ | Simplify. |

**c.** $\mathbf{\frac{3}{5}x = -8}$

| | |
|---|---|
| $\frac{3}{5}x = -8$ | Original equation |
| $\frac{5}{3}\left(\frac{3}{5}\right)x = \frac{5}{3}(-8)$ | Multiply each side by $\frac{5}{3}$. |
| $x = -\frac{40}{3}$ | Simplify. |

To solve equations with more than one operation, often called *multi-step equations*, undo operations by working backward.

### Example 3 Multi-step Linear Equations

**Solve each equation.**

**a.** $8q - 15 = 49$

$8q - 15 = 49$ — Original equation

$8q = 64$ — Add 15 to each side.

$q = 8$ — Divide each side by 8.

**b.** $12y + 8 = 6y - 5$

$12y + 8 = 6y - 5$ — Original equation

$12y = 6y - 13$ — Subtract 8 from each side.

$6y = -13$ — Subtract $6y$ from each side.

$y = -\frac{13}{6}$ — Divide each side by 6.

**WatchOut!**

**Order of Operations** Remember that the order of operations applies when you are solving linear equations.

When solving equations that contain grouping symbols, first use the Distributive Property to remove the grouping symbols.

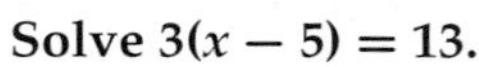

### Example 4 Multi-step Linear Equations

**Solve $3(x - 5) = 13$.**

$3(x - 5) = 13$ — Original equation

$3x - 15 = 13$ — Distributive Property

$3x = 28$ — Add 15 to each side.

$x = \frac{28}{3}$ — Divide each side by 3.

## Exercises

**Solve each equation.**

**1.** $r + 11 = 3$
**2.** $n + 7 = 13$
**3.** $d - 7 = 8$
**4.** $\frac{8}{5}a = -6$
**5.** $-\frac{p}{12} = 6$
**6.** $\frac{x}{4} = 8$
**7.** $\frac{12}{5}f = -18$
**8.** $\frac{y}{7} = -11$
**9.** $\frac{6}{7}y = 3$
**10.** $c - 14 = -11$
**11.** $t - 14 = -29$
**12.** $p - 21 = 52$
**13.** $b + 2 = -5$
**14.** $q + 10 = 22$
**15.** $-12q = 84$
**16.** $5t = 30$
**17.** $5c - 7 = 8c - 4$
**18.** $2\ell + 6 = 6\ell - 10$
**19.** $\frac{m}{10} + 15 = 21$
**20.** $-\frac{m}{8} + 7 = 5$
**21.** $8t + 1 = 3t - 19$
**22.** $9n + 4 = 5n + 18$
**23.** $5c - 24 = -4$
**24.** $3n + 7 = 28$
**25.** $-2y + 17 = -13$
**26.** $-\frac{t}{13} - 2 = 3$
**27.** $\frac{2}{9}x - 4 = \frac{2}{3}$
**28.** $9 - 4g = -15$
**29.** $-4 - p = -2$
**30.** $21 - b = 11$
**31.** $-2(n + 7) = 15$
**32.** $5(m - 1) = -25$
**33.** $-8a - 11 = 37$
**34.** $\frac{7}{4}q - 2 = -5$
**35.** $2(5 - n) = 8$
**36.** $-3(d - 7) = 6$

# LESSON 0-6 Linear Inequalities

**Objective**

- Use algebra to solve linear inequalities.

Statements with greater than (>), less than (<), greater than or equal to (≥), or less than or equal to (≤) are inequalities. If any number is added or subtracted to each side of an inequality, the resulting inequality is true.

**Example 1 Addition/Subtraction Linear Inequalities**

**Solve each inequality.**

**a.** $x - 17 > 12$

| | |
|---|---|
| $x - 17 > 12$ | Original inequality |
| $x - 17 + 17 > 12 + 17$ | Add 17 to each side. |
| $x > 29$ | Simplify. |

The solution set is $\{x \mid x > 29\}$.

**b.** $y + 11 \leq 5$

| | |
|---|---|
| $y + 11 \leq 5$ | Original inequality |
| $y + 11 - 11 \leq 5 - 11$ | Subtract 11 from each side. |
| $y \leq -6$ | Simplify. |

The solution set is $\{y \mid y \leq -6\}$.

If each side of an inequality is multiplied or divided by a positive number, the resulting inequality is true.

**Example 2 Multiplication/Division Linear Inequalities**

**Solve each inequality.**

**a.** $\frac{t}{6} \geq 11$

| | |
|---|---|
| $\frac{t}{6} \geq 11$ | Original inequality |
| $(6)\frac{t}{6} \geq (6)11$ | Multiply each side by 6. |
| $t \geq 66$ | Simplify. |

The solution set is $\{t \mid t \geq 66\}$.

**b.** $8p < 72$

| | |
|---|---|
| $8p < 72$ | Original inequality |
| $\frac{8p}{8} < \frac{72}{8}$ | Divide each side by 8. |
| $p < 9$ | Simplify. |

The solution set is $\{p \mid p < 9\}$.

If each side of an inequality is multiplied or divided by the same negative number, the direction of the inequality symbol must be *reversed* so that the resulting inequality is true.

**Example 3 Multiplication/Division Linear Inequalities**

**Solve each inequality.**

**a.** $-5c > 30$

| | |
|---|---|
| $-5c > 30$ | Original inequality |
| $\frac{-5c}{-5} < \frac{30}{-5}$ | Divide each side by −5. Change > to <. |
| $c < -6$ | Simplify. |

The solution set is $\{c \mid c < -6\}$.

*(continued on the next page)*

**b.** $-\frac{d}{13} \leq -4$

| | |
|---|---|
| $-\frac{d}{13} \leq -4$ | Original inequality |
| $(-13)\left(\frac{-d}{13}\right) \geq (-13)(-4)$ | Multiply each side by −13. Change ≤ to ≥. |
| $d \geq 52$ | Simplify. |

The solution set is $\{d \mid d \geq 52\}$.

Inequalities involving more than one operation can be solved by undoing the operations in the same way you would solve an equation with more than one operation.

PT

## Example 4 Multi-Step Linear Inequalities

**Solve each inequality.**

**a.** $-6a + 13 < -7$

| | |
|---|---|
| $-6a + 13 < -7$ | Original inequality |
| $-6a + 13 - 13 < -7 - 13$ | Subtract 13 from each side. |
| $-6a < -20$ | Simplify. |
| $\frac{-6a}{-6} > \frac{-20}{-6}$ | Divide each side by −6. Change < to >. |
| $a > \frac{10}{3}$ | Simplify. |

The solution set is $\left\{a \middle| a > \frac{10}{3}\right\}$.

**WatchOut!**

**Dividing by a Negative** Remember that any time you divide an inequality by a negative number you reverse the direction of the sign.

**b.** $4z + 7 \geq 8z - 1$

| | |
|---|---|
| $4z + 7 \geq 8z - 1$ | Original inequality |
| $4z + 7 - 7 \geq 8z - 1 - 7$ | Subtract 7 from each side. |
| $4z \geq 8z - 8$ | Simplify. |
| $4z - 8z \geq 8z - 8 - 8z$ | Subtract 8*z* from each side. |
| $-4z \geq -8$ | Simplify. |
| $\frac{-4z}{-4} \leq \frac{-8}{-4}$ | Divide each side by −4. Change ≥ to ≤. |
| $z \leq 2$ | Simplify. |

The solution set is $\{z \mid z \leq 2\}$.

## Exercises

1. $x - 7 < 6$
2. $a + 7 \geq -5$
3. $4y < 20$
4. $-\frac{a}{8} < 5$
5. $\frac{t}{6} > -7$
6. $\frac{a}{11} \leq 8$
7. $d + 8 \leq 12$
8. $m + 14 > 10$
9. $12k \geq -36$
10. $6t - 10 \geq 4t$
11. $3z + 8 < 2$
12. $4c + 23 \leq -13$
13. $m - 21 < 8$
14. $x - 6 \geq 3$
15. $-3b \leq 48$
16. $-\frac{p}{5} \geq 14$
17. $2z - 9 < 7z + 1$
18. $-4h > 36$
19. $\frac{2}{5}b - 6 \leq -2$
20. $\frac{8}{3}t + 1 > -5$
21. $7q + 3 \geq -4q + 25$
22. $-3n - 8 > 2n + 7$
23. $-3w + 1 \leq 8$
24. $-\frac{4}{5}k - 17 > 11$

# LESSON 0-7 Inverse Linear Functions

**Then**

- You represented relations as tables, graphs, and mappings.

**Now**

1. Find the inverse of a relation.
2. Find the inverse of a linear function.

**Why?**

- Randall is writing a report on Santiago, Chile, and he wants to include a brief climate analysis. He found a table of temperatures recorded in degrees Celsius. He knows that a formula for converting degrees Fahrenheit to degrees Celsius is $C(x) = \frac{5}{9}(x - 32)$. He will need to find the *inverse* function to convert from degrees Celsius to degrees Fahrenheit.

| Average Temp (°C) | | |
|---|---|---|
| Month | Min | Max |
| Jan | 12 | 29 |
| March | 9 | 27 |
| May | 5 | 18 |
| July | 3 | 15 |
| Sept | 6 | 29 |
| Nov | 9 | 26 |

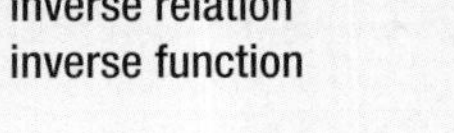

**NewVocabulary**
inverse relation
inverse function

**Common Core State Standards**

**Content Standards**
A.CED.2 Create equations in two or more variables to represent relationships between quantities; graph equations on coordinate axes with labels and scales.
F.BF.4a Solve an equation of the form $f(x) = c$ for a simple function $f$ that has an inverse and write an expression for the inverse.

**Mathematical Practices**
6 Attend to precision.

## 1 Inverse Relations

An **inverse relation** is the set of ordered pairs obtained by exchanging the $x$-coordinates with the $y$-coordinates of each ordered pair in a relation. If (5, 3) is an ordered pair of a relation, then (3, 5) is an ordered pair of the inverse relation.

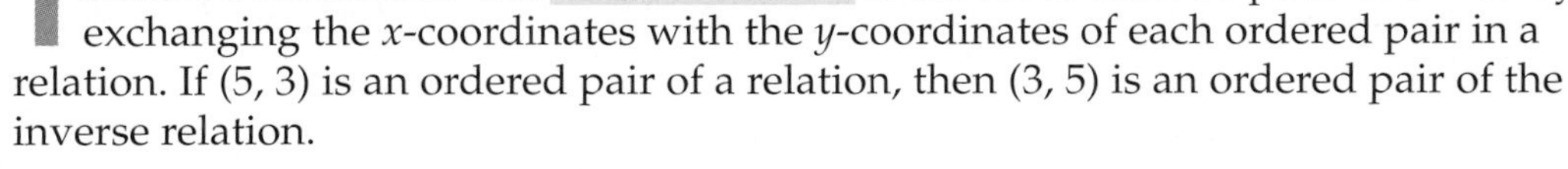

**KeyConcept** Inverse Relations

**Words** If one relation contains the element $(a, b)$, then the inverse relation will contain the element $(b, a)$.

**Example** $A$ and $B$ are inverse relations.

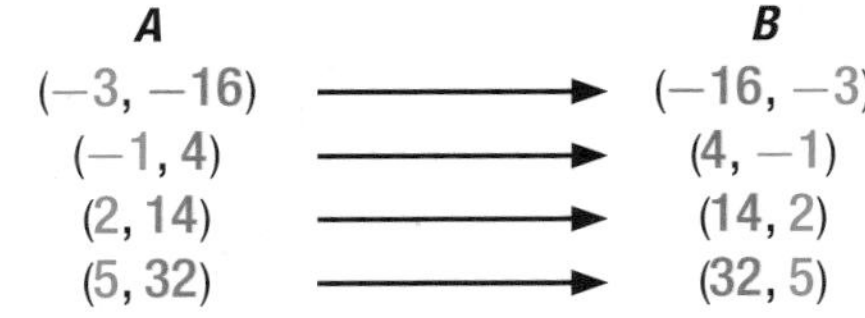

| A | | B |
|---|---|---|
| (−3, −16) | → | (−16, −3) |
| (−1, 4) | → | (4, −1) |
| (2, 14) | → | (14, 2) |
| (5, 32) | → | (32, 5) |

Notice that the domain of a relation becomes the range of its inverse, and the range of the relation becomes the domain of its inverse.

### Example 1 Inverse Relations

**Find the inverse of each relation.**

**a.** **{(4, −10), (7, −19), (−5, 17), (−3, 11)}**

To find the inverse, exchange the coordinates of the ordered pairs.

(4, −10) → (−10, 4)    (−5, 17) → (17, −5)
(7, −19) → (−19, 7)    (−3, 11) → (11, −3)

The inverse is {(−10, 4), (−19, 7), (17, −5), (11, −3)}.

**b.**

| x | −4 | −1 | 5 | 9 |
|---|---|---|---|---|
| y | −13 | −8.5 | 0.5 | 6.5 |

Write the coordinates as ordered pairs. Then exchange the coordinates of each pair.

(−4, −13) → (−13, −4)    (5, 0.5) → (0.5, 5)
(−1, −8.5) → (−8.5, −1)    (9, 6.5) → (6.5, 9)

The inverse is {(−13, −4), (−8.5, −1), (0.5, 5), (6.5, 9)}.

TongRo Image Stock/Alamy

### GuidedPractice

**1A.** $\{(-6, 8), (-15, 11), (9, 3), (0, 6)\}$

**1B.**

| x | −10 | −4 | −3 | 0 |
|---|---|---|---|---|
| y | 5 | 11 | 12 | 15 |

The graphs of relations can be used to find and graph inverse relations.

## Example 2 Graph Inverse Relations

**Graph the inverse of the relation.**

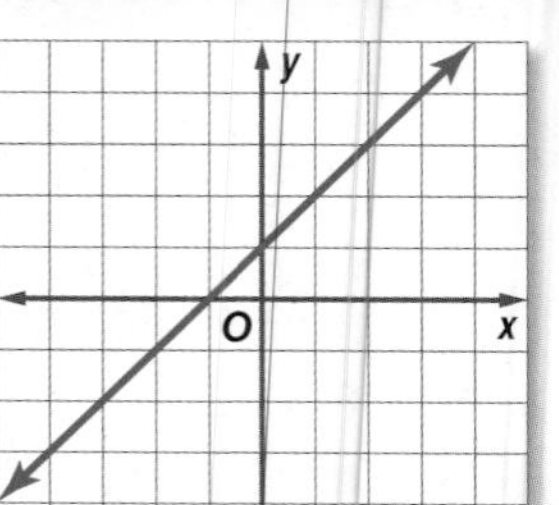


> **StudyTip**
>
> **CCSS Precision** Only two points are necessary to graph the inverse of a line, but several should be used to avoid possible error.

The graph of the relation passes through the points at $(-4, -3)$, $(-2, -1)$, $(0, 1)$, $(2, 3)$, and $(3, 4)$. To find points through which the graph of the inverse passes, exchange the coordinates of the ordered pairs. The graph of the inverse passes through the points at $(-3, -4)$, $(-1, -2)$, $(1, 0)$, $(3, 2)$, and $(4, 3)$. Graph these points and then draw the line that passes through them.

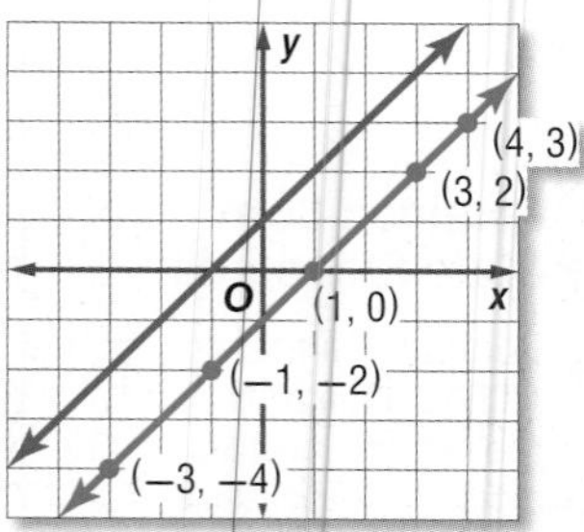


### GuidedPractice

**Graph the inverse of each relation.**

**2A.**

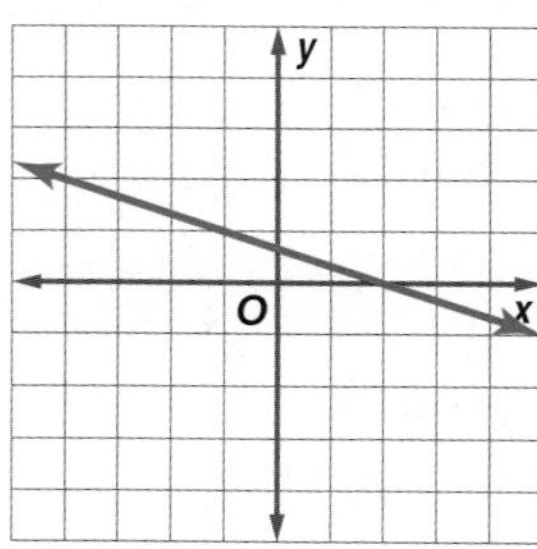


**2B.**

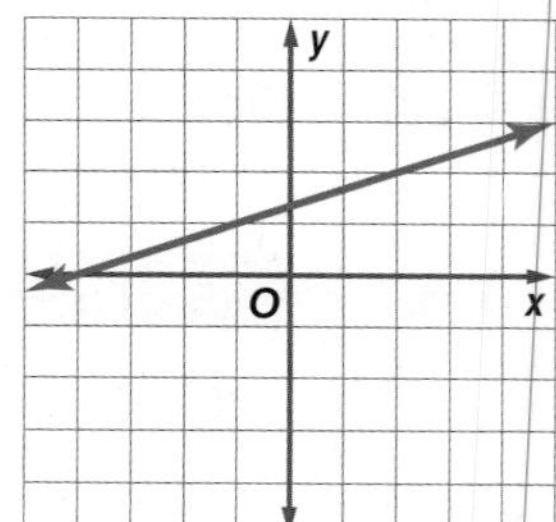


The graphs from Example 2 are graphed on the right with the line $y = x$. Notice that the graph of an inverse is the graph of the original relation reflected in the line $y = x$. For every point $(x, y)$ on the graph of the original relation, the graph of the inverse will include the point $(y, x)$.

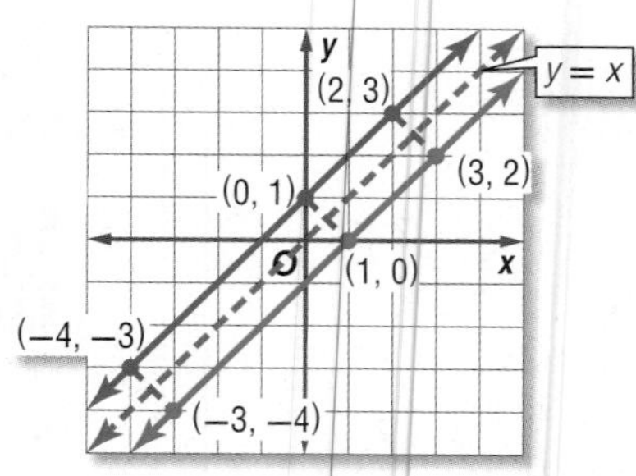


## 2 Inverse Functions

A linear relation that is described by a function has an **inverse function** that can generate ordered pairs of the inverse relation The inverse of the linear function $f(x)$ can be written as $f^{-1}(x)$ and is read *f of x inverse* or *the inverse of f of x*.

## KeyConcept Finding Inverse Functions

To find the inverse function $f^{-1}(x)$ of the linear function $f(x)$, complete the following steps.

**Step 1** Replace $f(x)$ with $y$ in the equation for $f(x)$.

**Step 2** Interchange $y$ and $x$ in the equation.

**Step 3** Solve the equation for $y$.

**Step 4** Replace $y$ with $f^{-1}(x)$ in the new equation.

## Example 3 Find Inverse Linear Functions

**Find the inverse of each function.**

**a.** $f(x) = 4x - 8$

| | | |
|---|---|---|
| **Step 1** | $f(x) = 4x - 8$ | Original equation |
| | $y = 4x - 8$ | Replace $f(x)$ with $y$. |
| **Step 2** | $x = 4y - 8$ | Interchange $y$ and $x$. |
| **Step 3** | $x + 8 = 4y$ | Add 8 to each side. |
| | $\frac{x+8}{4} = y$ | Divide each side by 4. |
| **Step 4** | $\frac{x+8}{4} = f^{-1}(x)$ | Replace $y$ with $f^{-1}(x)$. |

The inverse of $f(x) = 4x - 8$ is $f^{-1}(x) = \frac{x+8}{4}$ or $f^{-1}(x) = \frac{1}{4}x + 2$.

**WatchOut!**
**Notation** The $-1$ in $f^{-1}(x)$ is *not* an exponent.

**CHECK** Graph both functions and the line $y = x$ on the same coordinate plane. $f^{-1}(x)$ appears to be the reflection of $f(x)$ in the line $y = x$. ✓

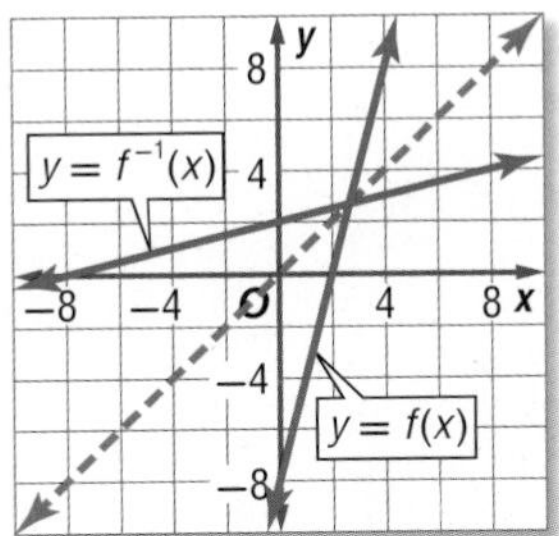


**b.** $f(x) = -\frac{1}{2}x + 11$

| | | |
|---|---|---|
| **Step 1** | $f(x) = -\frac{1}{2}x + 11$ | Original equation |
| | $y = -\frac{1}{2}x + 11$ | Replace $f(x)$ with $y$. |
| **Step 2** | $x = -\frac{1}{2}y + 11$ | Interchange $y$ and $x$. |
| **Step 3** | $x - 11 = -\frac{1}{2}y$ | Subtract 11 from each side. |
| | $-2(x - 11) = y$ | Multiply each side by $-2$. |
| | $-2x + 22 = y$ | Distributive Property |
| **Step 4** | $-2x + 22 = f^{-1}(x)$ | Replace $y$ with $f^{-1}(x)$. |

The inverse of $f(x) = -\frac{1}{2}x + 11$ is $f^{-1}(x) = -2x + 22$.

### GuidedPractice

**3A.** $f(x) = 4x - 12$

**3B.** $f(x) = \frac{1}{3}x + 7$

## Real-World Example 4 Use an Inverse Function

**TEMPERATURE Refer to the beginning of the lesson. Randall wants to convert the temperatures from degrees Celsius to degrees Fahrenheit.**

**a. Find the inverse function $C^{-1}(x)$.**

**Step 1** $C(x) = \frac{5}{9}(x - 32)$ Original equation

$y = \frac{5}{9}(x - 32)$ Replace $C(x)$ with $y$.

**Step 2** $x = \frac{5}{9}(y - 32)$ Interchange $y$ and $x$.

**Step 3** $\frac{9}{5}x = y - 32$ Multiply each side by $\frac{9}{5}$.

$\frac{9}{5}x + 32 = y$ Add 32 to each side.

**Step 4** $\frac{9}{5}x + 32 = C^{-1}(x)$ Replace $y$ with $C^{-1}(x)$.

The inverse function of $C(x)$ is $C^{-1}(x) = \frac{9}{5}x + 32$.

**b. What do $x$ and $C^{-1}(x)$ represent in the context of the inverse function?**

$x$ represents the temperature in degrees Celsius. $C^{-1}(x)$ represents the temperature in degrees Fahrenheit.

**c. Find the average temperatures for July in degrees Fahrenheit.**

The average minimum and maximum temperatures for July are 3° C and 15° C, respectively. To find the average minimum temperature, find $C^{-1}(3)$.

$C^{-1}(x) = \frac{9}{5}x + 32$ Original equation

$C^{-1}(3) = \frac{9}{5}(3) + 32$ Substitute 3 for $x$.

$= 37.4$ Simplify.

To find the average maximum temperature, find $C^{-1}(15)$.

$C^{-1}(x) = \frac{9}{5}x + 32$ Original equation

$C^{-1}(15) = \frac{9}{5}(15) + 32$ Substitute 15 for $x$.

$= 59$ Simplify.

The average minimum and maximum temperatures for July are 37.4° F and 59° F, respectively.

### Real-WorldLink

The winter months in Chile occur during the summer months in the U.S. due to Chile's location in the southern hemisphere. The average daily high temperature of Santiago during its winter months is about 60° F.

**Source:** World Weather Information Service

### GuidedPractice

**4. RENTAL CAR** Peggy rents a car for the day. The total cost $C(x)$ in dollars is given by $C(x) = 19.99 + 0.3x$, where $x$ is the number of miles she drives.

**A.** Find the inverse function $C^{-1}(x)$.

**B.** What do $x$ and $C^{-1}(x)$ represent in the context of the inverse function?

**C.** How many miles did Peggy drive if her total cost was $34.99?

## Check Your Understanding

**Example 1** **Find the inverse of each relation.**

1. $\{(4, -15), (-8, -18), (-2, -16.5), (3, -15.25)\}$

2.

| $x$ | −3 | 0 | 1 | 6 |
|---|---|---|---|---|
| $y$ | 11.8 | 3.7 | 1 | −12.5 |

**Example 2** **Graph the inverse of each relation.**

3. 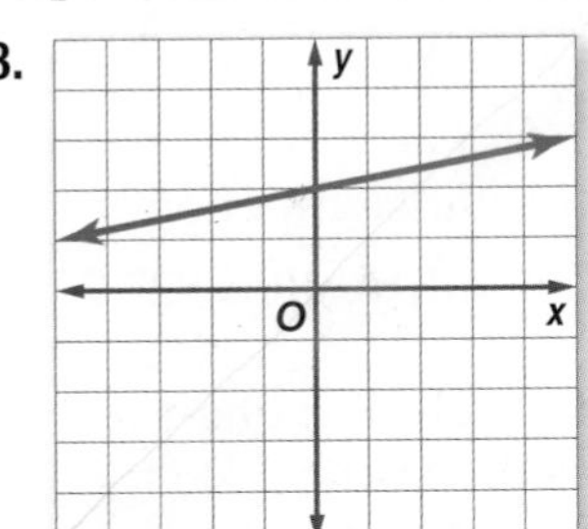


4. 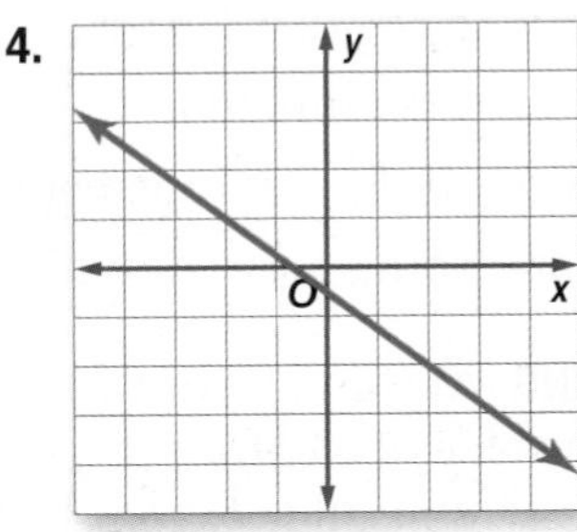


**Example 3** **Find the inverse of each function.**

5. $f(x) = -2x + 7$

6. $f(x) = \frac{2}{3}x + 6$

**Example 4**

7. **CCSS REASONING** Dwayne and his brother purchase season tickets to the Cleveland Crusaders games. The ticket package requires a one-time purchase of a personal seat license costing $1200 for two seats. A ticket to each game costs $70. The cost $C(x)$ in dollars for Dwayne for the first season is $C(x) = 600 + 70x$, where $x$ is the number of games Dwayne attends.
   a. Find the inverse function.
   b. What do $x$ and $C^{-1}(x)$ represent in the context of the inverse function?
   c. How many games did Dwayne attend if his total cost for the season was $950?

## Practice and Problem Solving

**Example 1** **Find the inverse of each relation.**

8. $\{(-5, 13), (6, 10.8), (3, 11.4), (-10, 14)\}$

9. $\{(-4, -49), (8, 35), (-1, -28), (4, 7)\}$

10.

| $x$ | $y$ |
|---|---|
| −8 | −36.4 |
| −2 | −15.4 |
| 1 | −4.9 |
| 5 | 9.1 |
| 11 | 30.1 |

11.

| $x$ | $y$ |
|---|---|
| −3 | 7.4 |
| −1 | 4 |
| 1 | 0.6 |
| 3 | −2.8 |
| 5 | −6.2 |

**Example 2** **Graph the inverse of each relation.**

12. 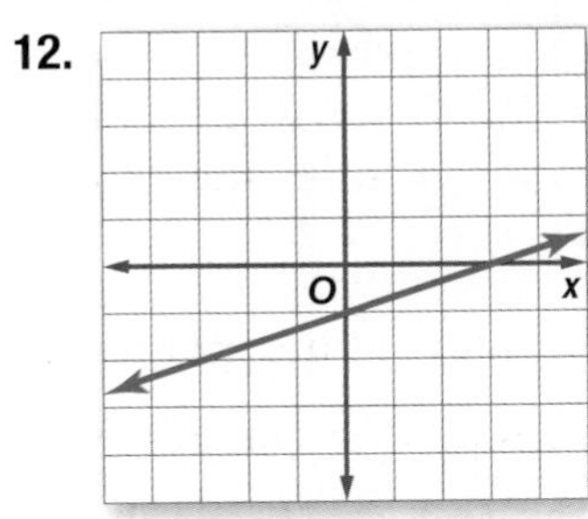


13. 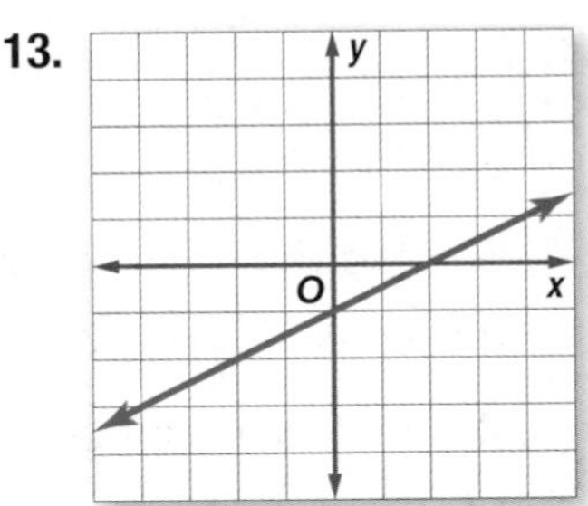

**Example 3** **Find the inverse of each function.**

14. $f(x) = 25 + 4x$
15. $f(x) = 17 - \frac{1}{3}x$
16. $f(x) = 4(x + 17)$
17. $f(x) = 12 - 6x$
18. $f(x) = \frac{2}{5}x + 10$
19. $f(x) = -16 - \frac{4}{3}x$

**Example 4**

20. **DOWNLOADS** An online music subscription service allows members to download songs for \$0.99 each after paying a monthly service charge of \$3.99. The total monthly cost $C(x)$ of the service in dollars is $C(x) = 3.99 + 0.99x$, where $x$ is the number of songs downloaded.
    a. Find the inverse function.
    b. What do $x$ and $C^{-1}(x)$ represent in the context of the inverse function?
    c. How many songs were downloaded if a member's monthly bill is \$27.75?

21. **LANDSCAPING** At the start of the mowing season, Chuck collects a one-time maintenance fee of \$10 from his customers. He charges the Fosters \$35 for each cut. The total amount collected from the Fosters in dollars for the season is $C(x) = 10 + 35x$, where $x$ is the number of times Chuck mows the Fosters' lawn.
    a. Find the inverse function.
    b. What do $x$ and $C^{-1}(x)$ represent in the context of the inverse function?
    c. How many times did Chuck mow the Fosters' lawn if he collected a total of \$780 from them?

**Write the inverse of each equation in $f^{-1}(x)$ notation.**

22. $3y - 12x = -72$
23. $x + 5y = 15$
24. $-42 + 6y = x$
25. $3y + 24 = 2x$
26. $-7y + 2x = -28$
27. $3y - x = 3$

**CCSS TOOLS** **Match each function with the graph of its inverse.**

A.
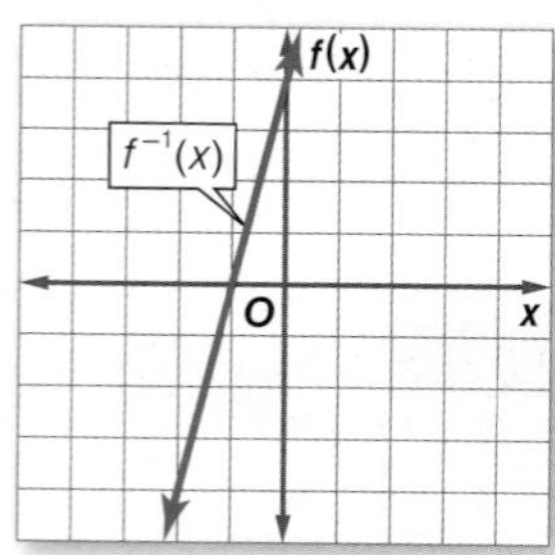


B.
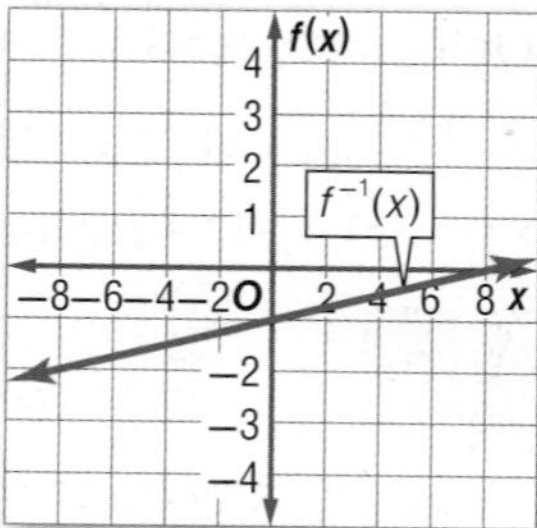


C.
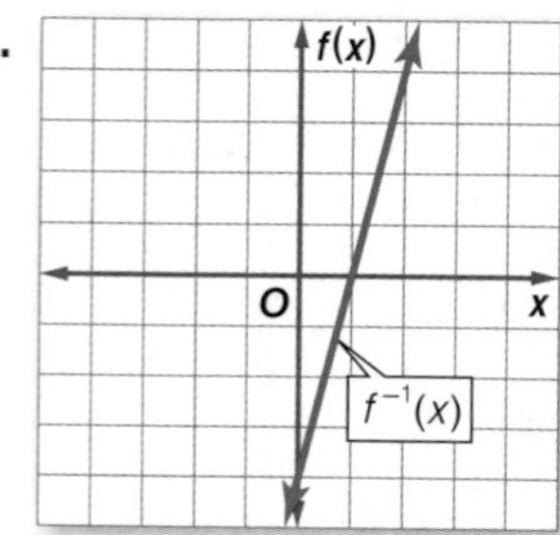


D.
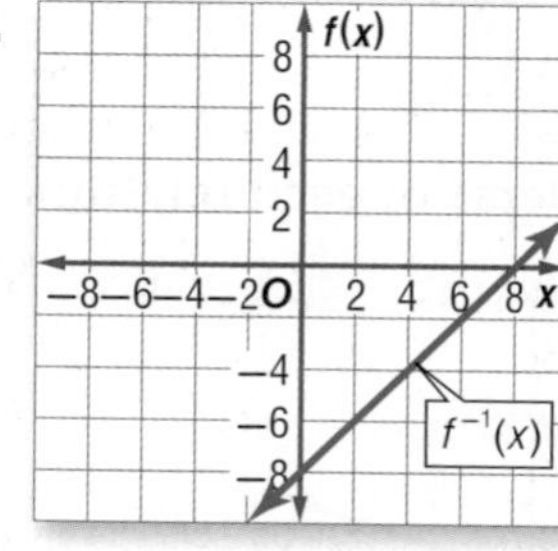


28. $f(x) = x + 4$
29. $f(x) = 4x + 4$
30. $f(x) = \frac{1}{4}x + 1$
31. $f(x) = \frac{1}{4}x - 1$

**Write an equation for the inverse function $f^{-1}(x)$ that satisfies the given conditions.**

**32.** slope of $f(x)$ is 7; graph of $f^{-1}(x)$ contains the point (13, 1)

**33.** graph of $f(x)$ contains the points (−3, 6) and (6, 12)

**34.** graph of $f(x)$ contains the point (10, 16); graph of $f^{-1}(x)$ contains the point (3, −16)

**35.** slope of $f(x)$ is 4; $f^{-1}(5) = 2$

**36. CELL PHONES** Mary Ann pays a monthly fee for her cell phone package which includes 700 minutes. She gets billed an additional charge for every minute she uses the phone past the 700 minutes. During her first month, Mary Ann used 26 additional minutes and her bill was $37.79. During her second month, Mary Ann used 38 additional minutes and her bill was $41.39.

**a.** Write a function that represents the total monthly cost $C(x)$ of Mary Ann's cell phone package, where $x$ is the number of additional minutes used.

**b.** Find the inverse function.

**c.** What do $x$ and $C^{-1}(x)$ represent in the context of the inverse function?

**d.** How many additional minutes did Mary Ann use if her bill for her third month was $48.89?

**37. MULTIPLE REPRESENTATIONS** In this problem, you will explore the domain and range of inverse functions.

**a. Algebraic** Write a function for the area $A(x)$ of the rectangle shown.

8 | Area = $A(x)$ | $(x - 3)$

**b. Graphical** Graph $A(x)$. Describe the domain and range of $A(x)$ in the context of the situation.

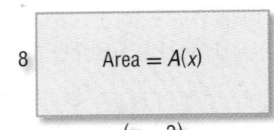

**c. Algebraic** Write the inverse of $A(x)$. What do $x$ and $A^{-1}(x)$ represent in the context of the situation?

**d. Graphical** Graph $A^{-1}(x)$. Describe the domain and range of $A^{-1}(x)$ in the context of the situation.

**e. Logical** Determine the relationship between the domains and ranges of $A(x)$ and $A^{-1}(x)$.

## H.O.T. Problems Use Higher-Order Thinking Skills

**38. CHALLENGE** If $f(x) = 5x + a$ and $f^{-1}(10) = -1$, find $a$.

**39. CHALLENGE** If $f(x) = \frac{1}{a}x + 7$ and $f^{-1}(x) = 2x - b$, find $a$ and $b$.

**CCSS ARGUMENTS Determine whether the following statements are *sometimes*, *always*, or *never* true. Explain your reasoning.**

**40.** If $f(x)$ and $g(x)$ are inverse functions, then $f(a) = b$ and $g(b) = a$.

**41.** If $f(a) = b$ and $g(b) = a$, then $f(x)$ and $g(x)$ are inverse functions.

**42. OPEN ENDED** Give an example of a function and its inverse. Verify that the two functions are inverses by graphing the functions and the line $y = x$ on the same coordinate plane.

**43. WRITING IN MATH** Explain why it may be helpful to find the inverse of a function.

EXPLORE 0-7

# Algebra Lab
# Drawing Inverses

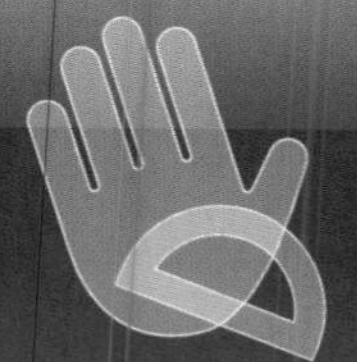

You can use patty paper to draw the graph of an inverse relation by reflecting the original graph in the line $y = x$.

**CCSS Common Core State Standards**
**Content Standards**
**F.BF.4a** Solve an equation of the form $f(x) = c$ for a simple function $f$ that has an inverse and write an expression for the inverse.

## Activity Draw an Inverse

**Consider the graphs shown.**

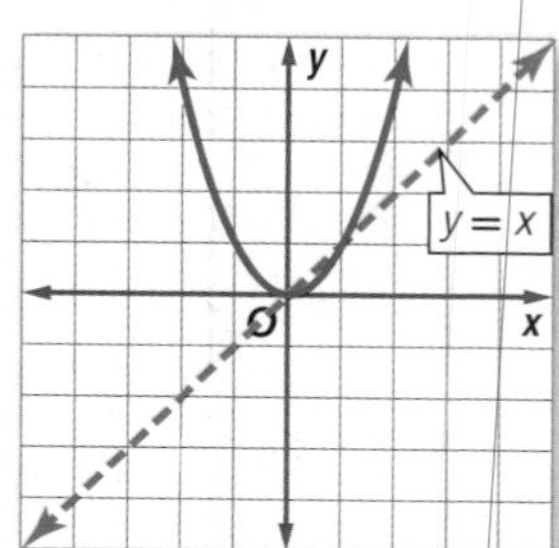


**Step 1** Trace the graphs onto a square of patty paper, waxed paper, or tracing paper.

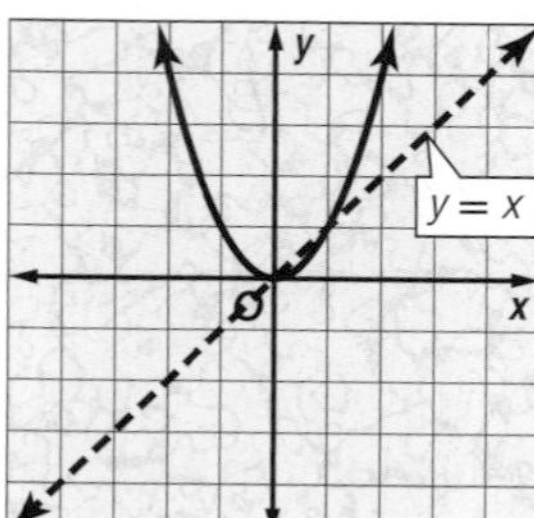


**Step 2** Flip the patty paper over and lay it on the original graph so that the traced $y = x$ is on the original $y = x$.

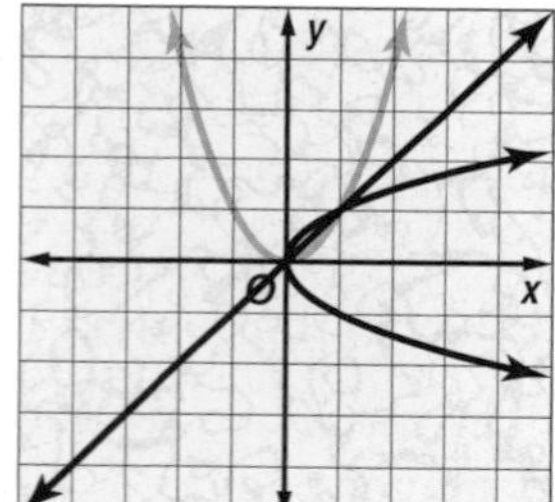


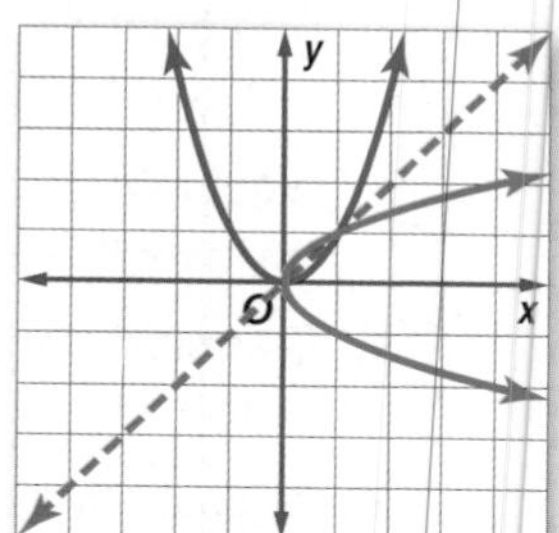


Notice that the result is the reflection of the graph in the line $y = x$ or the inverse of the graph.

## Analyze The Results

1. Is the graph of the original relation a function? Explain.
2. Is the graph of the inverse relation a function? Explain.
3. What are the domain and range of the original relation? of the inverse relation?
4. If the domain of the original relation is restricted to $D = \{x \mid x \geq 0\}$, is the inverse relation a function? Explain.
5. If the graph of a relation is a function, what can you conclude about the graph of its inverse?
6. **CHALLENGE** The vertical line test can be used to determine whether a relation is a function. Write a rule that can be used to determine whether a function has an inverse that is also a function.

# LESSON 0-8 Ordered Pairs

## Objective

- Name and graph points in the coordinate plane.

### NewVocabulary

ordered pair
*x*-coordinate
*y*-coordinate
quadrant
origin

Points in the coordinate plane are named by **ordered pairs** of the form $(x, y)$. The first number, or **$x$-coordinate**, corresponds to a number on the $x$-axis. The second number, or **$y$-coordinate**, corresponds to a number on the $y$-axis.

### Example 1 Writing Ordered Pairs

**Write the ordered pair for each point.**

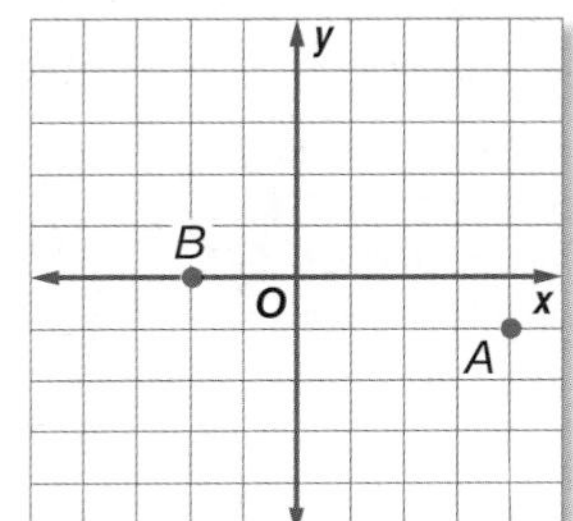


**a.** $A$

The $x$-coordinate is 4.

The $y$-coordinate is $-1$.

The ordered pair is $(4, -1)$.

**b.** $B$

The $x$-coordinate is $-2$.

The point lies on the $x$-axis, so its $y$-coordinate is 0.

The ordered pair is $(-2, 0)$.

The $x$-axis and $y$-axis separate the coordinate plane into four regions, called **quadrants**. The point at which the axes intersect is called the **origin**. The axes and points on the axes are not located in any of the quadrants.

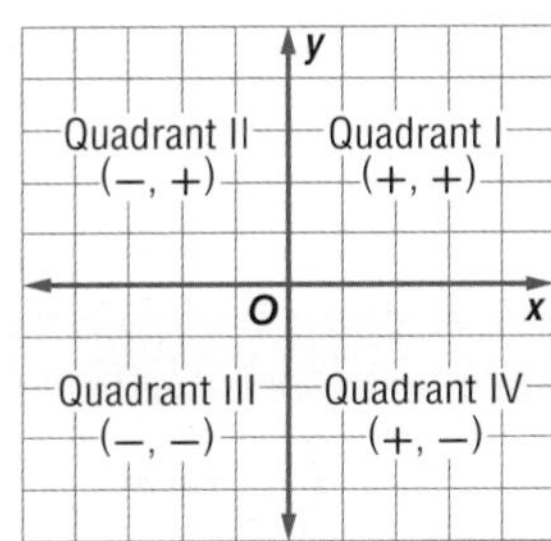

### Example 2 Graphing Ordered Pairs

**Graph and label each point on a coordinate plane. Name the quadrant in which each point is located.**

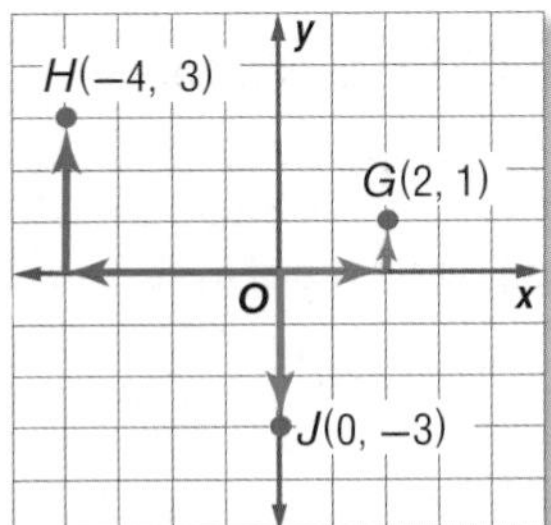


**a.** $G(2, 1)$

Start at the origin. Move 2 units right, since the $x$-coordinate is 2. Then move 1 unit up, since the $y$-coordinate is 1. Draw a dot, and label it $G$. Point $G(2, 1)$ is in Quadrant I.

**b.** $H(-4, 3)$

Start at the origin. Move 4 units left, since the $x$-coordinate is $-4$. Then move 3 units up, since the $y$-coordinate is 3. Draw a dot, and label it $H$. Point $H(-4, 3)$ is in Quadrant II.

**c.** $J(0, -3)$

Start at the origin. Since the $x$-coordinate is 0, the point lies on the $y$-axis. Move 3 units down, since the $y$-coordinate is $-3$. Draw a dot, and label it $J$. Because it is on one of the axes, point $J(0, -3)$ is not in any quadrant.

## Example 3 Graphing Multiple Ordered Pairs

**Graph a polygon with vertices $A(-3, 3)$, $B(1, 3)$, $C(0, 1)$, and $D(-4, 1)$.**

Graph the ordered pairs on a coordinate plane. Connect each pair of consecutive points. The polygon is a parallelogram.

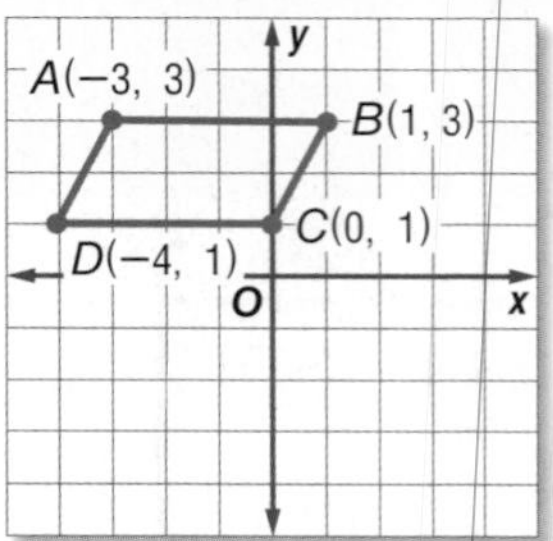


**StudyTip**

Lines There are infinitely many points on a line, so when you are asked to find points on a line, there are many answers.

## Example 4 Graphing and Solving for Ordered Pairs

**Graph four points that satisfy the equation $y = 4 - x$.**

Make a table.

Choose four values for $x$.

Evaluate each value of $x$ for $4 - x$.

| $x$ | $4 - x$ | $y$ | $(x, y)$ |
|---|---|---|---|
| 0 | 4 − 0 | 4 | (0, 4) |
| 1 | 4 − 1 | 3 | (1, 3) |
| 2 | 4 − 2 | 2 | (2, 2) |
| 3 | 4 − 3 | 1 | (3, 1) |

Plot the points.

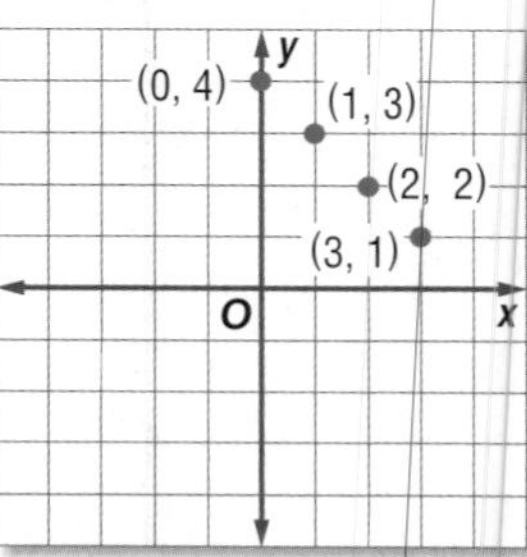


## Exercises

**Write the ordered pair for each point shown at the right.**

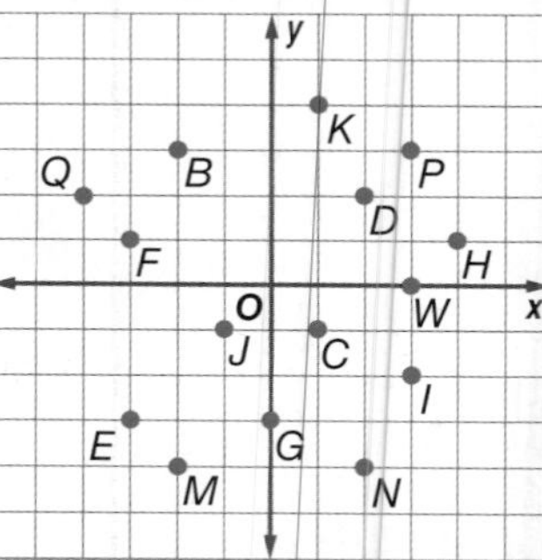


1. $B$
2. $C$
3. $D$
4. $E$
5. $F$
6. $G$
7. $H$
8. $I$
9. $J$
10. $K$
11. $W$
12. $M$
13. $N$
14. $P$
15. $Q$

**Graph and label each point on a coordinate plane. Name the quadrant in which each point is located.**

16. $M(-1, 3)$
17. $S(2, 0)$
18. $R(-3, -2)$
19. $P(1, -4)$
20. $B(5, -1)$
21. $D(3, 4)$
22. $T(2, 5)$
23. $L(-4, -3)$

**Graph the following geometric figures.**

24. a square with vertices $W(-3, 3)$, $X(-3, -1)$, $Z(1, 3)$, and $Y(1, -1)$
25. a polygon with vertices $J(4, 2)$, $K(1, -1)$, $L(-2, 2)$, and $M(1, 5)$
26. a triangle with vertices $F(2, 4)$, $G(-3, 2)$, and $H(-1, -3)$

**Graph four points that satisfy each equation.**

27. $y = 2x$
28. $y = 1 + x$
29. $y = 3x - 1$
30. $y = 2 - x$

# LESSON 0-9 Systems of Linear Equations

## Objective

- Use graphing, substitution, and elimination to solve systems of linear equations.

**NewVocabulary**
system of equations
substitution
elimination

Two or more equations that have common variables are called a **system of equations**. The solution of a system of equations in two variables is an ordered pair of numbers that satisfies both equations. A system of two linear equations can have zero, one, or an infinite number of solutions. There are three methods by which systems of equations can be solved: graphing, elimination, and substitution.

### Example 1 Graphing Linear Equations

**Solve each system of equations by graphing. Then determine whether each system has *no* solution, *one* solution, or *infinitely many* solutions.**

**a.** $y = -x + 3$
$y = 2x - 3$

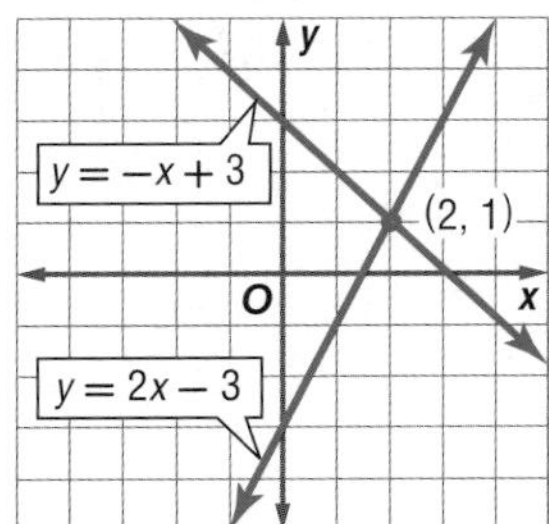


The graphs appear to intersect at (2, 1). Check this estimate by replacing $x$ with 2 and $y$ with 1 in each equation.

**CHECK** $y = -x + 3$ | $y = 2x - 3$

$1 \stackrel{?}{=} -2 + 3$ | $1 \stackrel{?}{=} 2(2) - 3$

$1 = 1$ ✓ | $1 = 1$ ✓

The system has one solution at (2, 1).

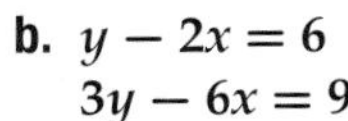

**b.** $y - 2x = 6$
$3y - 6x = 9$

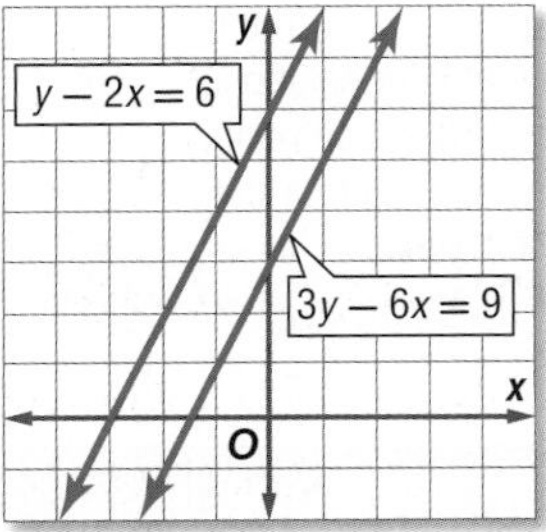


The graphs of the equations are parallel lines. Since they do not intersect, there are no solutions of this system of equations. Notice that the lines have the same slope but different $y$-intercepts. Equations with the same slope *and* the same $y$-intercepts have an infinite number of solutions.

It is difficult to determine the solution of a system when the two graphs intersect at noninteger values. There are algebraic methods by which an exact solution can be found. One such method is **substitution**.

### Example 2 Substitution

**Use substitution to solve the system of equations.**

$y = -4x$
$2y + 3x = 8$

Since $y = -4x$, substitute $-4x$ for $y$ in the second equation.

| | |
|---|---|
| $2y + 3x = 8$ | Second equation |
| $2(-4x) + 3x = 3$ | $y = -4x$ |
| $-8x + 3x = 8$ | Simplify. |
| $-5x = 8$ | Combine like terms. |
| $\frac{-5x}{-5} = \frac{8}{-5}$ | Divide each side by −5. |
| $x = -\frac{8}{5}$ | Simplify. |

Use $y = -4x$ to find the value of $y$.

| | |
|---|---|
| $y = -4x$ | First equation |
| $= -4\left(-\frac{8}{5}\right)$ | $x = -\frac{8}{5}$ |
| $= \frac{32}{5}$ | Simplify. |

The solution is $\left(-\frac{8}{5}, \frac{32}{5}\right)$.

Sometimes adding or subtracting two equations together will eliminate one variable. Using this step to solve a system of equations is called **elimination**.

PT

### Example 3 Elimination

**Use elimination to solve the system of equations.**

$\mathbf{3x + 5y = 7}$

$\mathbf{4x + 2y = 0}$

Either $x$ or $y$ can be eliminated. In this example, we will eliminate $x$.

$3x + 5y = 7$ → Multiply by 4. → $12x + 20y = 28$

$4x + 2y = 0$ → Multiply by −3. → $+ (-12x) - 6y = 0$

$14y = 28$ Add the equations.

$\frac{14y}{14} = \frac{28}{14}$ Divide each side by 14.

$y = 2$ Simplify.

Now substitute 2 for $y$ in either equation to find the value of $x$.

$4x + 2y = 0$ Second equation

$4x + 2(2) = 0$ $y = 2$

$4x + 4 = 0$ Simplify.

$4x + 4 - 4 = 0 - 4$ Subtract 4 from each side.

$4x = -4$ Simplify.

$\frac{4x}{4} = \frac{-4}{4}$ Divide each side by 4.

$x = -1$ Simplify.

The solution is $(-1, 2)$.

**StudyTip**

Checking Solutions You can confirm that your solutions are correct by substituting the values into both of the original equations.

## Exercises

**Solve by graphing.**

1. $y = -x + 2$
   $y = -\frac{1}{2}x + 1$

2. $y = 3x - 3$
   $y = x + 1$

3. $y - 2x = 1$
   $2y - 4x = 1$

**Solve by substitution.**

4. $-5x + 3y = 12$
   $x + 2y = 8$

5. $x - 4y = 22$
   $2x + 5y = -21$

6. $y + 5x = -3$
   $3y - 2x = 8$

**Solve by elimination.**

7. $-3x + y = 7$
   $3x + 2y = 2$

8. $3x + 4y = -1$
   $-9x - 4y = 13$

9. $-4x + 5y = -11$
   $2x + 3y = 11$

**Name an appropriate method to solve each system of equations. Then solve the system.**

10. $4x - y = 11$
    $2x - 3y = 3$

11. $4x + 6y = 3$
    $-10x - 15y = -4$

12. $3x - 2y = 6$
    $5x - 5y = 5$

13. $3y + x = 3$
    $-2y + 5x = 15$

14. $4x - 7y = 8$
    $-2x + 5y = -1$

15. $x + 3y = 6$
    $4x - 2y = -32$

# LESSON 0-10 Square Roots and Simplifying Radicals

## Objective

- Evaluate square roots and simplify radical expressions.

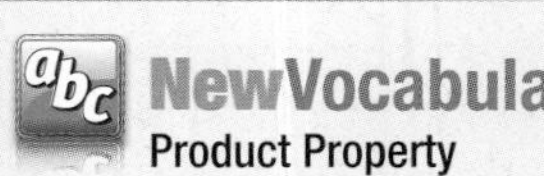

**NewVocabulary**
Product Property
Quotient Property

A radical expression is an expression that contains a square root. The expression is in simplest form when the following three conditions have been met.

- No radicands have perfect square factors other than 1.
- No radicands contain fractions.
- No radicals appear in the denominator of a fraction.

The **Product Property** states that for two numbers $a$ and $b \geq 0$, $\sqrt{ab} = \sqrt{a} \cdot \sqrt{b}$.

### Example 1 Product Property

**Simplify.**

**a.** $\sqrt{45}$

$\sqrt{45} = \sqrt{3 \cdot 3 \cdot 5}$ — Prime factorization of 45

$= \sqrt{3^2} \cdot \sqrt{5}$ — Product Property of Square Roots

$= 3\sqrt{5}$ — Simplify.

**b.** $\sqrt{6} \cdot \sqrt{15}$

$\sqrt{6} \cdot \sqrt{15} = \sqrt{6 \cdot 15}$ — Product Property

$= \sqrt{3 \cdot 2 \cdot 3 \cdot 5}$ — Prime factorization

$= \sqrt{3^2} \cdot \sqrt{10}$ — Product Property

$= 3\sqrt{10}$ — Simplify.

For radical expressions in which the exponent of the variable inside the radical is *even* and the resulting simplified exponent is *odd*, you must use absolute value to ensure nonnegative results.

### Example 2 Product Property

**Simplify** $\sqrt{20x^3y^5z^6}$.

$\sqrt{20x^3y^5z^6} = \sqrt{2^2 \cdot 5 \cdot x^3 \cdot y^5 \cdot z^6}$ — Prime factorization

$= \sqrt{2^2} \cdot \sqrt{5} \cdot \sqrt{x^3} \cdot \sqrt{y^5} \cdot \sqrt{z^6}$ — Product Property

$= 2 \cdot \sqrt{5} \cdot x \cdot \sqrt{x} \cdot y^2 \cdot \sqrt{y} \cdot |z^3|$ — Simplify.

$= 2xy^2|z^3|\sqrt{5xy}$ — Simplify.

The **Quotient Property** states that for any numbers $a$ and $b$, where $a \geq 0$ and $b \geq 0$, $\sqrt{\frac{a}{b}} = \frac{\sqrt{a}}{\sqrt{b}}$.

### Example 3 Quotient Property

**Simplify** $\sqrt{\frac{25}{16}}$.

$\sqrt{\frac{25}{16}} = \frac{\sqrt{25}}{\sqrt{16}}$ — Quotient Property

$= \frac{5}{4}$ — Simplify.

Rationalizing the denominator of a radical expression is a method used to eliminate radicals from the denominator of a fraction. To rationalize the denominator, multiply the expression by a fraction equivalent to 1 such that the resulting denominator is a perfect square.

PT

### Example 4 Rationalize the Denominator

**Simplify.**

**a.** $\frac{2}{\sqrt{3}}$

$\frac{2}{\sqrt{3}} = \frac{2}{\sqrt{3}} \cdot \frac{\sqrt{3}}{\sqrt{3}}$ Multiply by $\frac{\sqrt{3}}{\sqrt{3}}$.

$= \frac{2\sqrt{3}}{3}$ Simplify.

**b.** $\frac{\sqrt{13y}}{\sqrt{18}}$

$\frac{\sqrt{13y}}{\sqrt{18}} = \frac{\sqrt{13y}}{\sqrt{2 \cdot 3 \cdot 3}}$ Prime factorization

$= \frac{\sqrt{13y}}{3\sqrt{2}}$ Product Property

$= \frac{\sqrt{13y}}{3\sqrt{2}} \cdot \frac{\sqrt{2}}{\sqrt{2}}$ Multiply by $\frac{\sqrt{2}}{\sqrt{2}}$.

$= \frac{\sqrt{26y}}{6}$ Product Property

**WatchOut!**

**Rationalizing the Denominator** Don't forget to multiply both the numerator and denominator by the radical when you rationalize the denominator.

Sometimes, conjugates are used to simplify radical expressions. Conjugates are binomials of the form $p\sqrt{q} + r\sqrt{t}$ and $p\sqrt{q} - r\sqrt{t}$.

### Example 5 Conjugates

**Simplify** $\frac{3}{5 - \sqrt{2}}$.

$\frac{3}{5 - \sqrt{2}} = \frac{3}{5 - \sqrt{2}} \cdot \frac{5 + \sqrt{2}}{5 + \sqrt{2}}$ $\frac{5 + \sqrt{2}}{5 + \sqrt{2}} = 1$

$= \frac{3(5 + \sqrt{2})}{5^2 - (\sqrt{2})^2}$ $(a - b)(a + b) = a^2 - b^2$

$= \frac{15 + 3\sqrt{2}}{25 - 2}$ Multiply. $(\sqrt{2})^2 = 2$

$= \frac{15 + 3\sqrt{2}}{23}$ Simplify.

## Exercises

**Simplify.**

1. $\sqrt{32}$
2. $\sqrt{75}$
3. $\sqrt{50} \cdot \sqrt{10}$
4. $\sqrt{12} \cdot \sqrt{20}$
5. $\sqrt{6} \cdot \sqrt{6}$
6. $\sqrt{16} \cdot \sqrt{25}$
7. $\sqrt{98x^3y^6}$
8. $\sqrt{56a^2b^4c^5}$
9. $\sqrt{\frac{81}{49}}$
10. $\sqrt{\frac{121}{16}}$
11. $\sqrt{\frac{63}{8}}$
12. $\sqrt{\frac{288}{147}}$
13. $\frac{\sqrt{10p^3}}{\sqrt{27}}$
14. $\frac{\sqrt{108}}{\sqrt{2q^6}}$
15. $\frac{4}{5 - 2\sqrt{3}}$
16. $\frac{7\sqrt{3}}{5 - 2\sqrt{6}}$
17. $\frac{3}{\sqrt{48}}$
18. $\frac{\sqrt{24}}{\sqrt{125}}$
19. $\frac{3\sqrt{5}}{2 - \sqrt{2}}$
20. $\frac{3}{-2 + \sqrt{13}}$

# CHAPTER 0 Posttest

State which metric unit you would probably use to measure each item.

1. mass of a book
2. length of a highway

Complete each sentence.

3. 8 in. = __?__ ft
4. 6 yd = __?__ ft
5. 24 fl oz = __?__ pt
6. 3.7 kg = __?__ lb
7. 4.2 km = __?__ m
8. 285 g = __?__ kg
9. 0.75 kg = __?__ mg
10. 1.9 L = __?__ qt

11. **PROBABILITY** The table shows the results of an experiment in which a number cube was rolled. Find the experimental probability of rolling a 4.

| Outcome | Tally | Frequency |
|---|---|---|
| 1 | \|\|\|\| | 4 |
| 2 | ~~\|\|\|\|~~ \| | 6 |
| 3 | ~~\|\|\|\|~~ | 5 |
| 4 | \|\|\| | 3 |
| 5 | ~~\|\|\|\|~~ \|\| | 7 |

**CANDY** A bag of candy contains 3 lollipops, 8 peanut butter cups, and 4 chocolate bars. A piece of candy is randomly drawn from the bag. Find each probability.

12. $P$(peanut butter cup)
13. $P$(lollipop or peanut butter cup)
14. $P$(not chocolate bar)
15. $P$(chocolate bar or lollipop)

Evaluate each expression if $x = 2$, $y = -3$, and $z = 4$.

16. $6x - z$
17. $6y + xz$
18. $3yz$
19. $\frac{6z}{xy}$
20. $\frac{y + 2x}{10z}$
21. $7 + |y - 11|$

Solve each equation.

22. $9 + s = 21$
23. $h - 8 = 12$
24. $\frac{4m}{14} = 18$
25. $\frac{2}{9}d = 10$
26. $3(20 - b) = 36$
27. $37 + w = 5w - 27$
28. $\frac{x}{6} = 7$
29. $\frac{1}{4}(n + 5) = 16$

Solve each inequality.

30. $4y - 9 > 1$
31. $-2z + 15 \geq 4$
32. $3r + 7 < r - 8$
33. $-\frac{2}{5}k - 20 \leq 10$
34. $-3(b - 4) > 33$
35. $2 - m \leq 6m - 12$
36. $8 \leq r - 14$
37. $\frac{2}{3}n < \frac{3}{9}n - 5$

Write the ordered pair for each point shown.

38. $M$
39. $N$
40. $P$
41. $Q$

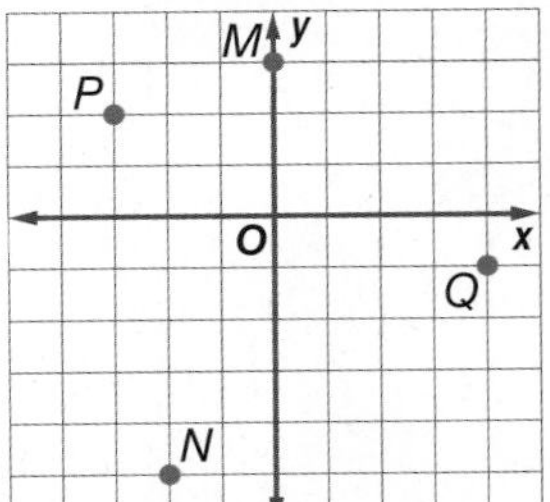


Name and label each point on the coordinate plane above.

42. $A(-2, 0)$
43. $C(1, 3)$
44. $D(-4, -4)$
45. $F(3, -5)$

46. Graph the quadrilateral with vertices $R(2, 0)$, $S(4, -2)$, $T(4, 3)$, and $W(2, 5)$.

47. Graph three points that satisfy the equation $y = \frac{1}{2}x - 5$.

Solve each system of equations.

48. $2r + m = 11$
    $6r - 2m = -2$
49. $2x + 4y = 6$
    $7x = 4 + 3y$
50. $2c + 6d = 14$
    $-\frac{7}{3} + \frac{1}{3}c = -d$
51. $5a - b = 17$
    $3a + 2b = 5$
52. $6d + 3f = 12$
    $2d = 8 - f$
53. $4x - 5y = 17$
    $3x + 4y = 5$

Simplify.

54. $\sqrt{80}$
55. $\sqrt{\frac{128}{5}}$
56. $\sqrt{36} \cdot \sqrt{81}$
57. $\sqrt{\frac{7x^3}{3}}$
58. $\sqrt{\frac{5}{81}}$
59. $\sqrt{12x^5y^2}$

CHAPTER 1

# Quadratic Expressions and Equations

## Then

You applied the laws of exponents and explored exponential functions.

## Now

In this chapter, you will:

- Add, subtract, and multiply polynomials.
- Factor trinomials.
- Factor differences of squares.
- Graph quadratic functions.
- Solve quadratic equations.

## Why?

**ARCHITECTURE** Quadratic equations can be used to model the shape of architectural structures such as the tallest memorial in the United States, the Gateway Arch in St. Louis, Missouri.

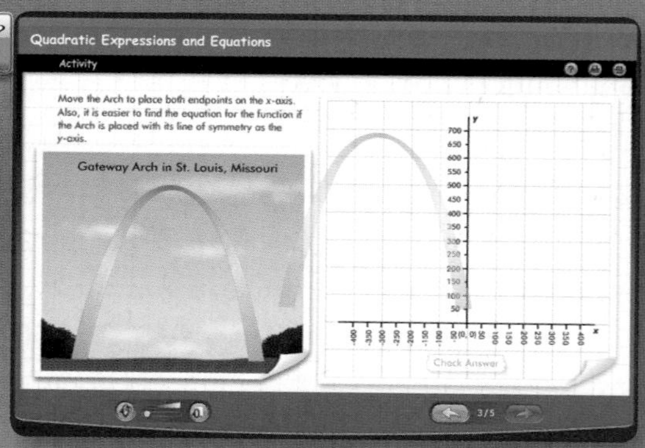


connectED.mcgraw-hill.com Your Digital Math Portal

Animation 
Vocabulary 
eGlossary 
Personal Tutor 
Virtual Manipulatives 
Graphing Calculator 
Audio 
Foldables 
Self-Check Practice 
Worksheets 

Panoramic Images/Getty Images

# Get Ready for the Chapter

**Diagnose** Readiness | You have two options for checking prerequisite skills.

**1 Textbook Option** Take the Quick Check below. Refer to the Quick Review for help.

## QuickCheck

Rewrite each expression using the Distributive Property. Then simplify.

1. $a(a + 5)$
2. $2(3 + x)$
3. $n(n - 3n^2 + 2)$
4. $-6(x^2 - 5x + 6)$
5. **FINANCIAL LITERACY** Five friends will pay \$9 per ticket, \$3 per drink, and \$6 per popcorn at the movies. Write an expression that could be used to determine the cost for them to go to the movies.

Simplify each expression. If not possible, write *simplified.*

6. $3u + 10u$
7. $5a - 2 + 6a$
8. $6m^2 - 8m$
9. $4w^2 + w + 15w^2$
10. $2x^2 + 5 - 11x^2$
11. $8v^3 - 27$
12. $4k^2 + 2k - 2k + 1$
13. $a^2 - 4a - 4a + 16$
14. $6y^2 + 2y - 3y - 1$
15. $9g^2 - 3g - 6g + 2$

Simplify.

16. $b(b^6)$
17. $4n^3(n^2)$
18. $8m(4m^2)$
19. $-5z^4(3z^5)$
20. $5xy(4x^3y)$
21. $(-2a^4c^5)(7ac^4)$
22. **GEOMETRY** A square is $6x^3$ inches on each side. What is the area of the square?

## QuickReview

**Example 1**

Rewrite $6x(-3x - 5x - 5x^2 + x^3)$ using the Distributive Property. Then simplify.

$$\begin{aligned}&6x(-3x - 5x - 5x^2 + x^3)\\&= 6x(-3x) + 6x(-5x) + 6x(-5x^2) + 6x(x^3)\\&= -18x^2 - 30x^2 - 30x^3 + 6x^4\\&= -48x^2 - 30x^3 + 6x^4\end{aligned}$$

**Example 2**

Simplify $8c + 6 - 4c + 2c^2$.

$$\begin{aligned}8c + 6 - 4c + 2c^2 &= 2c^2 + 8c - 4c + 6\\&= 2c^2 + (8 - 4)c + 6\\&= 2c^2 + 4c + 6\end{aligned}$$

**Example 3**

Simplify $(-2y^3)(9y^4)$.

$$\begin{aligned}(9y^3)(-2y^4) &= (-2 \cdot 9)(y^3 \cdot y^4)\\&= (-2 \cdot 9)(y^{3+4})\\&= -18y^7\end{aligned}$$

**2 Online Option** Take an online self-check Chapter Readiness Quiz at connectED.mcgraw-hill.com.

# Get Started on the Chapter

You will learn several new concepts, skills, and vocabulary terms as you study Chapter 1. To get ready, identify important terms and organize your resources. You may wish to refer to Chapter 0 to review prerequisite skills.

## FOLDABLES StudyOrganizer

**Quadratic Expressions and Equations** Make this Foldable to help you organize your Chapter 1 notes about quadratic expressions and equations. Begin with five sheets of grid paper.

**1** **Fold** in half along the width. On the first three sheets, cut 5 centimeters along the fold at the ends. On the second two sheets cut in the center, stopping 5 centimeters from the ends.

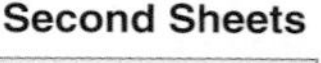

First Sheets

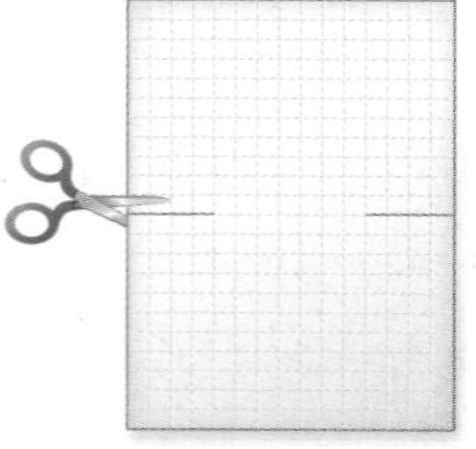

Second Sheets

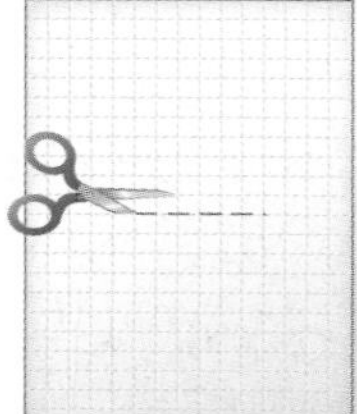

**2** **Insert** the first sheets through the second sheets and align the folds. Label the front Chapter 8, Quadratic Expressions and Equations. Label the pages with lesson numbers and the last page with vocabulary.

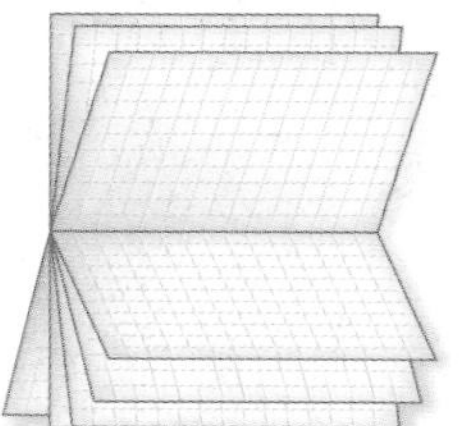

## NewVocabulary

| English | | Español |
|---|---|---|
| polynomial | p. 7 | polinomio |
| binomial | p. 7 | binomio |
| trinomial | p. 7 | trinomio |
| degree of a monomial | p. 7 | grado de un monomio |
| degree of a polynomial | p. 7 | grado de un polinomio |
| standard form of a polynomial | p. 8 | forma estándar de polinomio |
| leading coefficient | p. 8 | coeficiente líder |
| FOIL method | p. 23 | método foil |
| quadratic expression | p. 23 | expression cuadrática |
| factoring | p. 36 | factorización |
| factoring by grouping | p. 37 | factorización por agrupamiento |
| Zero Product Property | p. 38 | propiedad del producto de cero |
| quadratic equation | p. 48 | ecuación cuadrática |
| prime polynomial | p. 54 | polinomio primo |
| difference of two squares | p. 58 | diferencia de cuadrados |
| perfect square trinomial | p. 64 | trinomio cuadrado perfecto |
| Square Root Property | p. 67 | Propiedad de la raíz cuadrada |

## ReviewVocabulary

absolute value valor absoluto the absolute value of any number $n$ is the distance the number is from zero on a number line and is written $|n|$

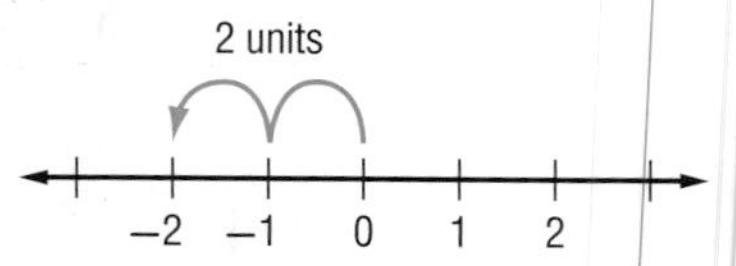


The absolute value of −2 is 2 because it is 2 units from 0.

perfect square cuadrado perfecto a number with a square root that is a rational number

EXPLORE 1-1

# Algebra Lab
# Adding and Subtracting Polynomials

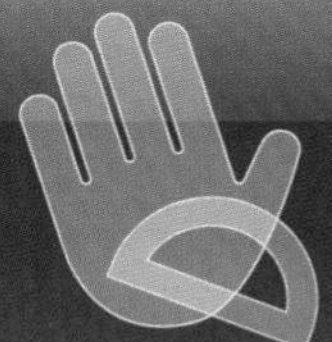

Algebra tiles can be used to model polynomials. A polynomial is a monomial or the sum of monomials. The diagram below shows the models.

**CCSS Common Core State Standards**
**Content Standards**
**A.APR.1** Understand that polynomials form a system analogous to the integers, namely, they are closed under the operations of addition, subtraction, and multiplication; add, subtract, and multiply polynomials.

**Polynomial Models**

- Polynomials are modeled using three types of tiles.
- Each tile has an opposite.

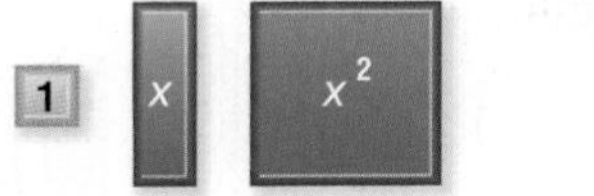

## Activity 1 Model Polynomials

**Use algebra tiles to model each polynomial.**

- $5x$

  To model this polynomial, you will need 5 green $x$-tiles.

- $-2x^2 + x + 3$

  To model this polynomial, you will need 2 red $-x^2$-tiles, 1 green $x$-tile, and 3 yellow 1-tiles.

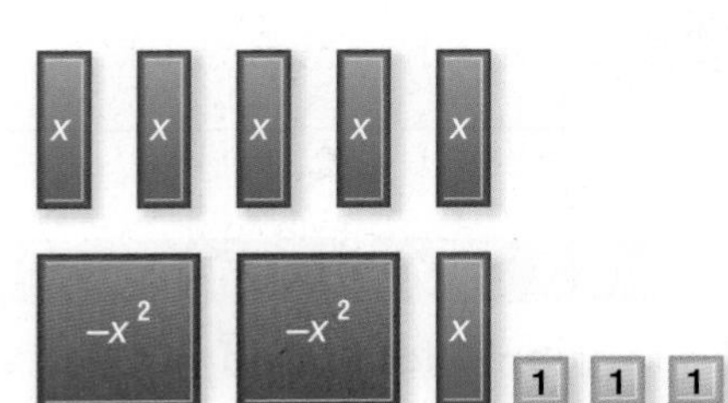


Monomials such as $3x$ and $-2x$ are called *like terms* because they have the same variable to the same power.

**Polynomial Models**

- Like terms are represented by tiles that have the same shape and size.
- A *zero pair* may be formed by pairing one tile with its opposite. You can remove or add zero pairs without changing the polynomial.

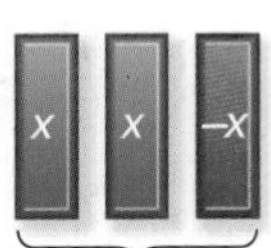


like terms

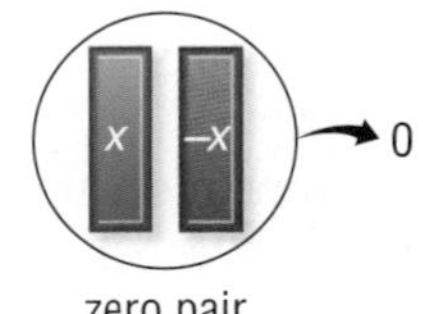


zero pair

## Activity 2 Add Polynomials

**Use algebra tiles to find $(2x^2 - 3x + 5) + (x^2 + 6x - 4)$.**

**Step 1**

Model each polynomial.

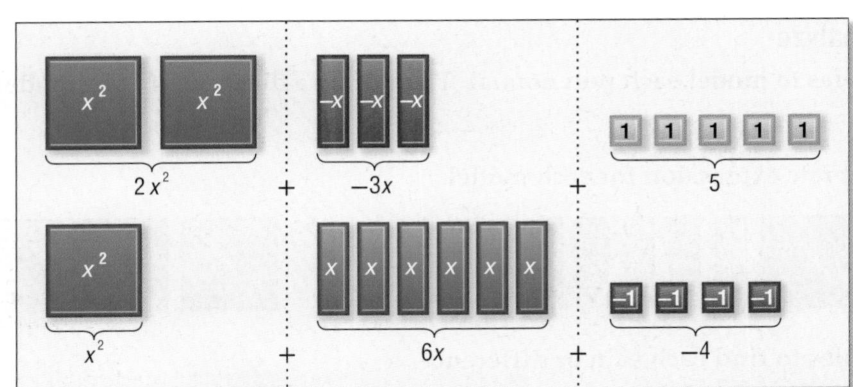


(continued on the next page)

# Algebra Lab
# Adding and Subtracting Polynomials *Continued*

**Step 2** Combine like terms and remove zero pairs.

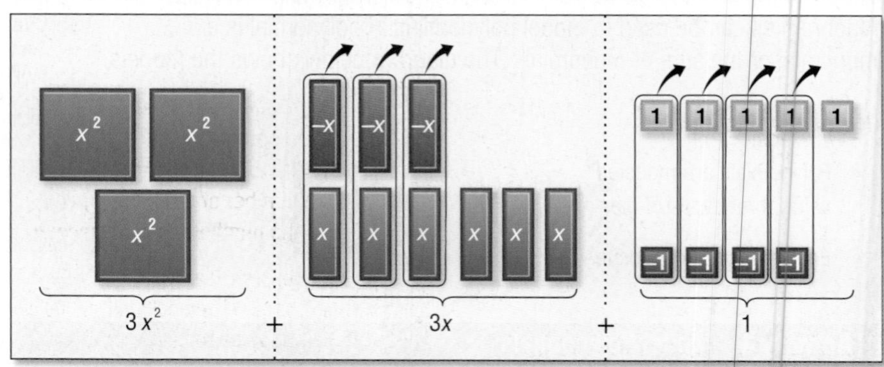


**Step 3** Write the polynomial. $(2x^2 - 3x + 5) + (x^2 + 6x - 4) = 3x^2 + 3x + 1$

**Activity 3** Subtract Polynomials

**Use algebra tiles to find $(4x + 5) - (-3x + 1)$.**

**Step 1** Model the polynomial $4x + 5$.

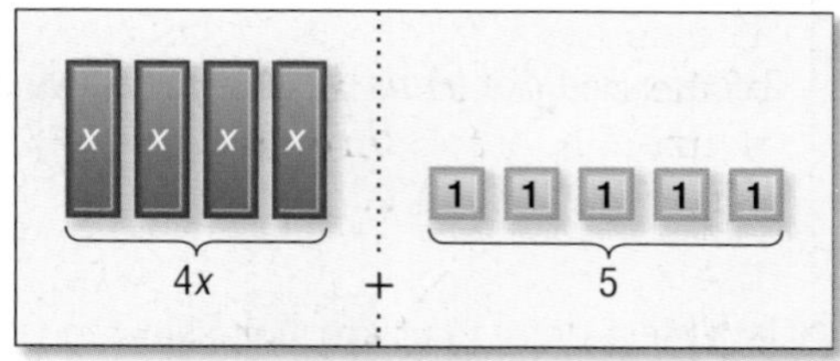


**Step 2** To subtract $-3x + 1$, remove 3 red $-x$-tiles and 1 yellow 1-tile. You can remove the 1-tile, but there are no $-x$-tiles. Add 3 zero pairs of $x$-tiles. Then remove the 3 red $-x$-tiles.

**Step 3** Write the polynomial.
$(4x + 5) - (-3x + 1) = 7x + 4$

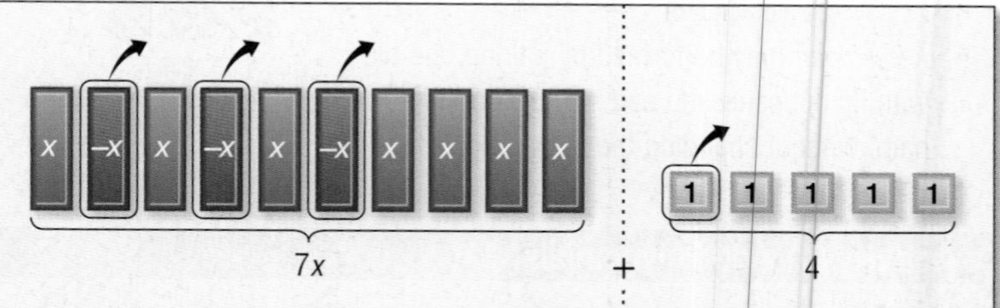


## Model and Analyze

**Use algebra tiles to model each polynomial. Then draw a diagram of your model.**

**1.** $-2x^2$ **2.** $5x - 4$ **3.** $x^2 - 4x$

**Write an algebraic expression for each model.**

**4.** 

**5.** 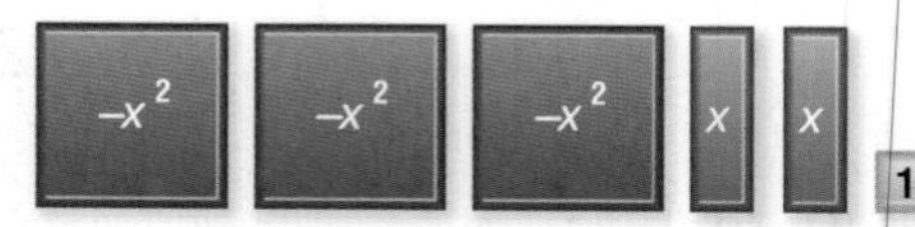


**Use algebra tiles to find each sum or difference.**

**6.** $(x^2 + 5x - 2) + (3x^2 - 2x + 6)$ **7.** $(2x^2 + 8x + 1) - (x^2 - 4x - 2)$ **8.** $(-4x^2 + x) - (x^2 + 5x)$

# LESSON 1-1 Adding and Subtracting Polynomials

**Then**

- You identified monomials and their characteristics.

**Now**

1. Write polynomials in standard form.
2. Add and subtract polynomials.

**Why?**

- In 2017, sales of digital audio players are expected to reach record numbers. The sales data can be modeled by the equation $U = -2.7t^2 + 49.4t + 128.7$, where $U$ is the number of units shipped in millions and $t$ is the number of years since 2005.

  The expression $-2.7t^2 + 49.4t + 128.7$ is an example of a polynomial. Polynomials can be used to model situations.

**New Vocabulary**

polynomial
binomial
trinomial
degree of a monomial
degree of a polynomial
standard form of a polynomial
leading coefficient

**Common Core State Standards**

**Content Standards**

**A.SSE.1a** Interpret parts of an expression, such as terms, factors, and coefficients.

**A.APR.1** Understand that polynomials form a system analogous to the integers, namely, they are closed under the operations of addition, subtraction, and multiplication; add, subtract, and multiply polynomials.

**Mathematical Practices**

3 Construct viable arguments and critique the reasoning of others.

## 1 Polynomials in Standard Form

A **polynomial** is a monomial or the sum of monomials, each called a *term* of the polynomial. Some polynomials have special names. A **binomial** is the sum of *two* monomials, and a **trinomial** is the sum of *three* monomials.

| Monomial | Binomial | Trinomial |
|---|---|---|
| $5x$ | $2x^2 + 7$ | $x^3 - 10x + 1$ |

The **degree of a monomial** is the sum of the exponents of all its variables. A nonzero constant term has degree 0, and zero has no degree.

The **degree of a polynomial** is the greatest degree of any term in the polynomial. You can find the degree of a polynomial by finding the degree of each term. Polynomials are named based on their degree.

| Degree | Name |
|---|---|
| 0 | constant |
| 1 | linear |
| 2 | quadratic |
| 3 | cubic |
| 4 | quartic |
| 5 | quintic |
| 6 or more | 6th degree, 7th degree, and so on |

### Example 1 Identify Polynomials

**Determine whether each expression is a polynomial. If it is a polynomial, find the degree and determine whether it is a *monomial, binomial,* or *trinomial*.**

| Expression | Is it a polynomial? | Degree | Monomial, binomial, or trinomial? |
|---|---|---|---|
| **a.** $4y - 5xz$ | Yes; $4y - 5xz$ is the sum of $4y$ and $-5xz$. | 2 | binomial |
| **b.** $-6.5$ | Yes; $-6.5$ is a real number. | 0 | monomial |
| **c.** $7a^{-3} + 9b$ | No; $7a^{-3} = \frac{7}{a^3}$, which is not a monomial. | — | —— |
| **d.** $6x^3 + 4x + x + 3$ | Yes; $6x^3 + 4x + x + 3 = 6x^3 + 5x + 3$, the sum of three monomials. | 3 | trinomial |

**Guided Practice**

**1A.** $x$

**1B.** $-3y^2 - 2y + 4y - 1$

**1C.** $5rx + 7tuv$

**1D.** $10x^{-4} - 8x^a$

4 Eyes Photography/Taxi/Getty Images

The terms of a polynomial can be written in any order. However, polynomials in one variable are usually written in standard form. The **standard form of a polynomial** has the terms in order from greatest to least degree. In this form, the coefficient of the first term is called the **leading coefficient**.

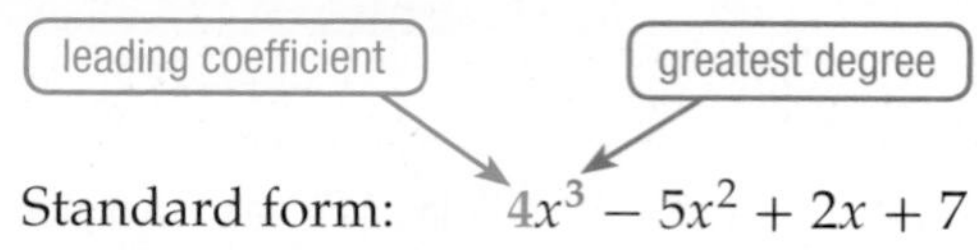


Standard form: $4x^3 - 5x^2 + 2x + 7$

**Example 2** Standard Form of a Polynomial

**Write each polynomial in standard form. Identify the leading coefficient.**

**a.** $3x^2 + 4x^5 - 7x$

Find the degree of each term.

Degree: 2, 5, 1

Polynomial: $3x^2 + 4x^5 - 7x$

The greatest degree is 5. Therefore, the polynomial can be rewritten as $4x^5 + 3x^2 - 7x$, with a leading coefficient of 4.

**b.** $5y - 9 - 2y^4 - 6y^3$

Find the degree of each term.

Degree: 1, 0, 4, 3

Polynomial: $5y - 9 - 2y^4 - 6y^3$

The greatest degree is 4. Therefore, the polynomial can be rewritten as $-2y^4 - 6y^3 + 5y - 9$, with a leading coefficient of $-2$.

**Guided**Practice

**2A.** $8 - 2x^2 + 4x^4 - 3x$

**2B.** $y + 5y^3 - 2y^2 - 7y^6 + 10$

## 2 Add and Subtract Polynomials

Adding polynomials involves adding like terms. You can group like terms by using a horizontal or vertical format.

**Example 3** Add Polynomials

**Find each sum.**

**a.** $(2x^2 + 5x - 7) + (3 - 4x^2 + 6x)$

**Horizontal Method**

Group and combine like terms.

$$(2x^2 + 5x - 7) + (3 - 4x^2 + 6x)$$

$$= [2x^2 + (-4x^2)] + [5x + 6x] + [-7 + 3] \quad \text{Group like terms.}$$

$$= -2x^2 + 11x - 4 \quad \text{Combine like terms.}$$

**b.** $(3y + y^3 - 5) + (4y^2 - 4y + 2y^3 + 8)$

**Vertical Method**

Align like terms in columns and combine.

$$\begin{array}{rl} y^3 + 0y^2 + 3y - 5 & \text{Insert a placeholder to help align the terms.} \\ (+)\ 2y^3 + 4y^2 - 4y + 8 & \text{Align and combine like terms.} \\ \hline 3y^3 + 4y^2 - \ \ y + 3 & \end{array}$$

**Study**Tip

Vertical Method Notice that the polynomials are written in standard form with like terms aligned. Since there is no $y^2$-term in the first polynomial, $0y^2$ is used as a placeholder.

**Guided**Practice

**3A.** $(5x^2 - 3x + 4) + (6x - 3x^2 - 3)$

**3B.** $(y^4 - 3y + 7) + (2y^3 + 2y - 2y^4 - 11)$

**StudyTip**

**Additive Inverse** When finding the additive inverse of a polynomial, you are multiplying every term by −1.

You can subtract a polynomial by adding its additive inverse. To find the additive inverse of a polynomial, write the opposite of each term, as shown.

$$-(3x^2 + 2x - 6) = \underbrace{-3x^2 - 2x + 6}_{\text{Additive Inverse}}$$

## Example 4 Subtract Polynomials

**Find each difference.**

**a.** $(3 - 2x + 2x^2) - (4x - 5 + 3x^2)$

**Horizontal Method**

Subtract $4x - 5 + 3x^2$ by adding its additive inverse.

$(3 - 2x + 2x^2) - (4x - 5 + 3x^2)$

$= (3 - 2x + 2x^2) + (-4x + 5 - 3x^2)$ — The additive inverse of $4x - 5 + 3x^2$ is $-4x + 5 - 3x^2$.

$= [2x^2 + (-3x^2)] + [(-2x) + (-4x)] + [3 + 5]$ — Group like terms.

$= -x^2 - 6x + 8$ — Combine like terms.

**b.** $(7p + 4p^3 - 8) - (3p^2 + 2 - 9p)$

**Vertical Method**

Align like terms in columns and subtract by adding the additive inverse.

$$\begin{array}{r} 4p^3 + 0p^2 + 7p - 8 \\ (-) \qquad 3p^2 - 9p + 2 \\ \hline \end{array} \quad \xrightarrow{\text{Add the opposite.}} \quad \begin{array}{r} 4p^3 + 0p^2 + 7p - 8 \\ (+) \quad -3p^2 + 9p - 2 \\ \hline 4p^3 - 3p^2 + 16p - 10 \end{array}$$

**GuidedPractice**

**4A.** $(4x^3 - 3x^2 + 6x - 4) - (-2x^3 + x^2 - 2)$

**4B.** $(8y - 10 + 5y^2) - (7 - y^3 + 12y)$

Adding or subtracting integers results in an integer, so the set of integers is closed under addition and subtraction. Similarly, adding or subtracting polynomials results in a polynomial, so the set of polynomials is closed under addition and subtraction.

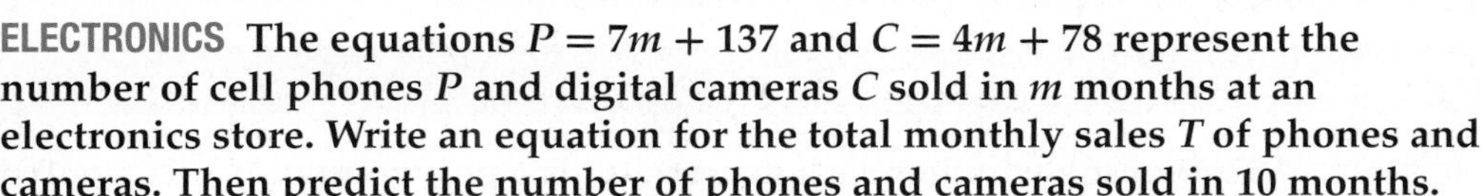

## Real-World Example 5 Add and Subtract Polynomials

**ELECTRONICS The equations $P = 7m + 137$ and $C = 4m + 78$ represent the number of cell phones $P$ and digital cameras $C$ sold in $m$ months at an electronics store. Write an equation for the total monthly sales $T$ of phones and cameras. Then predict the number of phones and cameras sold in 10 months.**

To write an equation that represents the total sales $T$, add the equations that represent the number of cell phones $P$ and digital cameras $C$.

$T = 7m + 137 + 4m + 78$

$= 11m + 215$

Substitute 10 for $m$ to predict the number of phones and cameras sold in 10 months.

$T = 11(10) + 215$

$= 110 + 215$ or $325$

Therefore, a total of 325 cell phones and digital cameras will be sold in 10 months.

**GuidedPractice**

**5.** Use the information above to write an equation that represents the difference in the monthly sales of cell phones and the monthly sales of digital cameras. Use the equation to predict the difference in monthly sales in 24 months.

**Real-WorldLink**

Sales of digital cameras recently increased by 42% in one year. Sales are expected to increase by at least 15% each year as consumers upgrade their cameras.

**Source:** Big Planet Marketing Company

Digital Vision/Alamy

## Check Your Understanding

**Example 1** **Determine whether each expression is a polynomial. If it is a polynomial, find the degree and determine whether it is a *monomial, binomial,* or *trinomial.***

**1.** $7ab + 6b^2 - 2a^3$

**2.** $2y - 5 + 3y^2$

**3.** $3x^2$

**4.** $\frac{4m}{3p}$

**5.** $5m^2p^3 + 6$

**6.** $5q^{-4} + 6q$

**Example 2** **Write each polynomial in standard form. Identify the leading coefficient.**

**7.** $2x^5 - 12 + 3x$

**8.** $-4d^4 + 1 - d^2$

**9.** $4z - 2z^2 - 5z^4$

**10.** $2a + 4a^3 - 5a^2 - 1$

**Examples 3–4** **Find each sum or difference.**

**11.** $(6x^3 - 4) + (-2x^3 + 9)$

**12.** $(g^3 - 2g^2 + 5g + 6) - (g^2 + 2g)$

**13.** $(4 + 2a^2 - 2a) - (3a^2 - 8a + 7)$

**14.** $(8y - 4y^2) + (3y - 9y^2)$

**15.** $(-4z^3 - 2z + 8) - (4z^3 + 3z^2 - 5)$

**16.** $(-3d^2 - 8 + 2d) + (4d - 12 + d^2)$

**17.** $(y + 5) + (2y + 4y^2 - 2)$

**18.** $(3n^3 - 5n + n^2) - (-8n^2 + 3n^3)$

**Example 5** **19.** CCSS **SENSE-MAKING** The total number of students $T$ who traveled for spring break consists of two groups: students who flew to their destinations $F$ and students who drove to their destination $D$. The number (in thousands) of students who flew and the total number of students who flew or drove can be modeled by the following equations, where $n$ is the number of years since 1995.

$$T = 14n + 21 \qquad F = 8n + 7$$

**a.** Write an equation that models the number of students who drove to their destination for this time period.

**b.** Predict the number of students who will drive to their destination in 2012.

**c.** How many students will drive or fly to their destination in 2015?

## Practice and Problem Solving

**Example 1** **Determine whether each expression is a polynomial. If it is a polynomial, find the degree and determine whether it is a *monomial, binomial,* or *trinomial.***

**20.** $\frac{5y^3}{x^2} + 4x$

**21.** $21$

**22.** $c^4 - 2c^2 + 1$

**23.** $d + 3d^c$

**24.** $a - a^2$

**25.** $5n^3 + nq^3$

**Example 2** **Write each polynomial in standard form. Identify the leading coefficient.**

**26.** $5x^2 - 2 + 3x$

**27.** $8y + 7y^3$

**28.** $4 - 3c - 5c^2$

**29.** $-y^3 + 3y - 3y^2 + 2$

**30.** $11t + 2t^2 - 3 + t^5$

**31.** $2 + r - r^3$

**32.** $\frac{1}{2}x - 3x^4 + 7$

**33.** $-9b^2 + 10b - b^6$

**Examples 3–4** **Find each sum or difference.**

**34.** $(2c^2 + 6c + 4) + (5c^2 - 7)$

**35.** $(2x + 3x^2) - (7 - 8x^2)$

**36.** $(3c^3 - c + 11) - (c^2 + 2c + 8)$

**37.** $(z^2 + z) + (z^2 - 11)$

**38.** $(2x - 2y + 1) - (3y + 4x)$

**39.** $(4a - 5b^2 + 3) + (6 - 2a + 3b^2)$

**40.** $(x^2y - 3x^2 + y) + (3y - 2x^2y)$

**41.** $(-8xy + 3x^2 - 5y) + (4x^2 - 2y + 6xy)$

**42.** $(5n - 2p^2 + 2np) - (4p^2 + 4n)$

**43.** $(4rxt - 8r^2x + x^2) - (6rx^2 + 5rxt - 2x^2)$

**Example 5**

**44.** **PETS** From 1999 through 2009, the number of dogs $D$ and the number of cats $C$ (in hundreds) adopted from animal shelters in the United States are modeled by the equations $D = 2n + 3$ and $C = n + 4$, where $n$ is the number of years since 1999.

**a.** Write a function that models the total number $T$ of dogs and cats adopted in hundreds for this time period.

**b.** If this trend continues, how many dogs and cats will be adopted in 2013?

**Classify each polynomial according to its degree and number of terms.**

**45.** $4x - 3x^2 + 5$

**46.** $11z^3$

**47.** $9 + y^4$

**48.** $3x^3 - 7$

**49.** $-2x^5 - x^2 + 5x - 8$

**50.** $10t - 4t^2 + 6t^3$

**51.** **ENROLLMENT** In a rapidly growing school system, the numbers (in hundreds) of total students is represented by $N$ and the number of students in Kindergarten through 5th grade is represented by $P$. The equations $N = 1.25t^2 - t + 7.5$ and $P = 0.7t^2 - 0.95t + 3.8$, model the number of students enrolled from 2000 to 2009, where $t$ is the number of years since 2000.

**a.** Write an equation modeling the number of students $S$ in grades 6 through 12 enrolled for this time period.

**b.** How many students were enrolled in grades 6 through 12 in the school system in 2007?

**52.** **CCSS REASONING** The perimeter of the triangle can be represented by the expression $3x^2 - 7x + 2$. Write a polynomial that represents the measure of the third side.

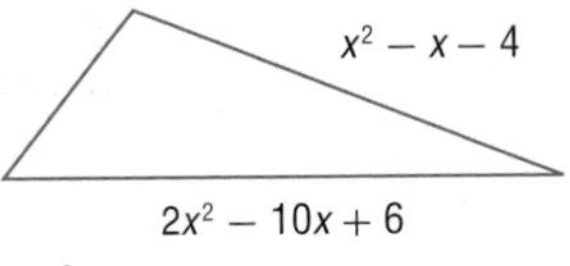


**53.** **GEOMETRY** Consider the rectangle.

**a.** What does $(4x^2 + 2x - 1)(2x^2 - x + 3)$ represent?

**b.** What does $2(4x^2 + 2x - 1) + 2(2x^2 - x + 3)$ represent?

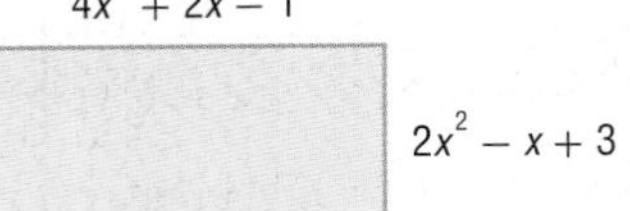


**Find each sum or difference.**

**54.** $(4x + 2y - 6z) + (5y - 2z + 7x) + (-9z - 2x - 3y)$

**55.** $(5a^2 - 4) + (a^2 - 2a + 12) + (4a^2 - 6a + 8)$

**56.** $(3c^2 - 7) + (4c + 7) - (c^2 + 5c - 8)$

**57.** $(3n^3 + 3n - 10) - (4n^2 - 5n) + (4n^3 - 3n^2 - 9n + 4)$

**58.** **FOOTBALL** The National Football League is divided into two conferences, the American $A$ and the National $N$. From 2002 through 2009, the total attendance $T$ (in thousands) for both conferences and for the American Conference games can be modeled by the following equations, where $x$ is the number of years since 2002.

$T = -0.69x^3 + 55.83x^2 + 643.31x + 10{,}538$ $\qquad A = -3.78x^3 + 58.96x^2 + 265.96x + 5257$

Estimate how many people attended National Conference football games in 2009.

**59** **CAR RENTAL** The cost to rent a car for a day is $15 plus $0.15 for each mile driven.

a. Write a polynomial that represents the cost of renting a car for $m$ miles.

b. If a car is driven 145 miles, how much would it cost to rent?

c. If a car is driven 105 miles each day for four days, how much would it cost to rent a car?

d. If a car is driven 220 miles each day for seven days, how much would it cost to rent a car?

**60.** **MULTIPLE REPRESENTATIONS** In this problem, you will explore perimeter and area.

a. **Geometric** Draw three rectangles that each have a perimeter of 400 feet.

b. **Tabular** Record the width and length of each rectangle in a table like the one shown below. Find the area of each rectangle.

| Rectangle | Length | Width | Area |
|---|---|---|---|
| 1 | 100 ft | | |
| 2 | 50 ft | | |
| 3 | 75 ft | | |
| 4 | $x$ ft | | |

c. **Graphical** On a coordinate system, graph the area of rectangle 4 in terms of the length, $x$. Use the graph to determine the largest area possible.

d. **Analytical** Determine the length and width that produce the largest area.

## H.O.T. Problems Use Higher-Order Thinking Skills

**61.** **CCSS CRITIQUE** Cheyenne and Sebastian are finding $(2x^2 - x) - (3x + 3x^2 - 2)$. Is either of them correct? Explain your reasoning.

| Cheyenne | Sebastian |
|---|---|
| $(2x^2 - x) - (3x + 3x^2 - 2)$ | $(2x^2 - x) - (3x + 3x^2 - 2)$ |
| $= (2x^2 - x) + (-3x + 3x^2 - 2)$ | $= (2x^2 - x) + (-3x - 3x^2 - 2)$ |
| $= 5x^2 - 4x - 2$ | $= -x^2 - 4x - 2$ |

**62.** **REASONING** Determine whether each of the following statements is *true* or *false*. Explain your reasoning.

a. A binomial can have a degree of zero.

b. The order in which polynomials are subtracted does not matter.

**63.** **CHALLENGE** Write a polynomial that represents the sum of an odd integer $2n + 1$ and the next two consecutive odd integers.

**64.** **WRITING IN MATH** Why would you add or subtract equations that represent real-world situations? Explain.

**65.** **WRITING IN MATH** Describe how to add and subtract polynomials using both the vertical and horizontal formats.

## Standardized Test Practice

**66.** Three consecutive integers can be represented by $x$, $x + 1$, and $x + 2$. What is the sum of these three integers?

**A** $x(x + 1)(x + 2)$ **C** $3x + 3$

**B** $x^3 + 3$ **D** $x + 3$

**67. SHORT RESPONSE** What is the perimeter of a square with sides that measure $2x + 3$ units?

**68.** Jim cuts a board in the shape of a regular hexagon and pounds in a nail at each vertex, as shown. How many rubber bands will he need to stretch a rubber band across every possible pair of nails?

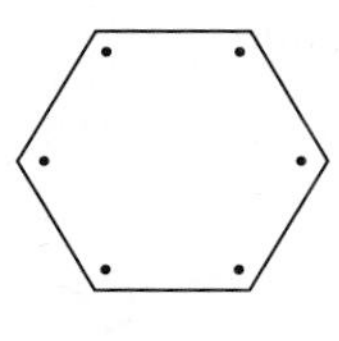

**F** 15 **G** 14 **H** 12 **J** 9

**69.** Which ordered pair is in the solution set of the system of inequalities shown in the graph?

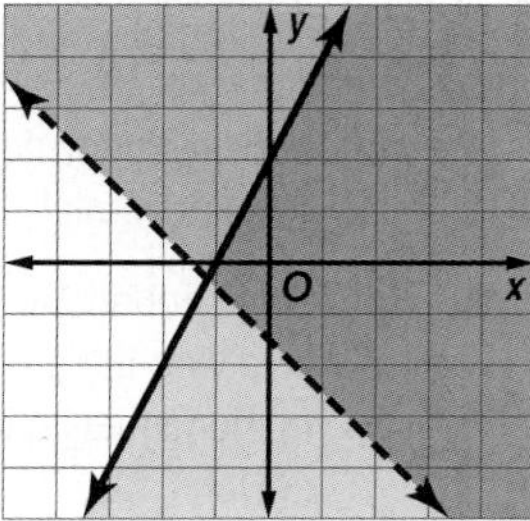


**A** $(-3, 0)$ **C** $(5, 0)$

**B** $(0, -3)$ **D** $(0, 5)$

## Spiral Review

**Simplify.** (Lesson 0-10)

**70.** $\sqrt{72}$ **71.** $\sqrt{18} \cdot \sqrt{14}$ **72.** $\sqrt{44x^4y^3}$ **73.** $\frac{3}{\sqrt{18}}$

**74.** $\sqrt{\frac{28}{75}}$ **75.** $\frac{\sqrt{8a^6}}{\sqrt{108}}$ **76.** $\frac{5}{4 - \sqrt{2}}$ **77.** $\frac{4\sqrt{3}}{2 + \sqrt{5}}$

**78. FINANCIAL LITERACY** Suppose you buy 3 shirts and 2 pairs of slacks on sale at a clothing store for \$72. The next day, a friend buys 2 shirts and 4 pairs of slacks for \$96. If the shirts you each bought were all the same price and the slacks were also all the same price, then what was the cost of each shirt and each pair of slacks? (Lesson 0-9)

**Graph the following points, and connect them in order to form a figure.** (Lesson 0-8)

**79.** $A(-5, 3)$, $B(3, -4)$, and $C(-2, -3)$

**80.** $P(-2, 1)$, $Q(3, 4)$, $R(5, 1)$, and $S(0, -2)$

**GROCERIES** **Find an approximate metric weight for each item.** (Lesson 0-2)

**81.**

Net Wt: 15 oz

**82.**

Net Wt: 8.2 oz

**83.**

Net Wt: 2.5 lb

## Skills Review

**Simplify.**

**84.** $t(t^5)(t^7)$ **85.** $n^3(n^2)(-2n^3)$ **86.** $(5t^5v^2)(10t^3v^4)$ **87.** $(-8u^4z^5)(5uz^4)$

**88.** $[(3)^2]^3$ **89.** $[(2)^3]^2$ **90.** $(2m^4k^3)^2(-3mk^2)^3$ **91.** $(6xy^2)^2(2x^2y^2z^2)^3$

# LESSON 1-2 Multiplying a Polynomial by a Monomial

**Then**

- You multiplied monomials.

**Now**

1. Multiply a polynomial by a monomial.
2. Solve equations involving the products of monomials and polynomials.

**Why?**

- Charmaine Brooks is opening a fitness club. She tells the contractor that the length of the fitness room should be three times the width plus 8 feet.

  To cover the floor with mats for exercise classes, Ms. Brooks needs to know the area of the floor. So she multiplies the width times the length, $w(3w + 8)$.

**Common Core State Standards**

**Content Standards**

A.APR.1 Understand that polynomials form a system analogous to the integers, namely, they are closed under the operations of addition, subtraction, and multiplication; add, subtract, and multiply polynomials.

**Mathematical Practices**

5 Use appropriate tools strategically.

## 1 Polynomial Multiplied by Monomial

To find the product of a polynomial and a monomial, you can use the Distributive Property.

### Example 1 Multiply a Polynomial by a Monomial

**Find $-3x^2(7x^2 - x + 4)$.**

**Horizontal Method**

| | |
|---|---|
| $-3x^2(7x^2 - x + 4)$ | Original expression |
| $= -3x^2(7x^2) - (-3x^2)(x) + (-3x^2)(4)$ | Distributive Property |
| $= -21x^4 - (-3x^3) + (-12x^2)$ | Multiply. |
| $= -21x^4 + 3x^3 - 12x^2$ | Simplify. |

**Vertical Method**

$$\begin{array}{r} 7x^2 - x + 4 \\ (\times) \qquad\qquad -3x^2 \\ \hline -21x^4 + 3x^3 - 12x^2 \end{array}$$

Distributive Property

Multiply.

**Guided Practice**

**Find each product.**

**1A.** $5a^2(-4a^2 + 2a - 7)$

**1B.** $-6d^3(3d^4 - 2d^3 - d + 9)$

We can use this same method more than once to simplify large expressions.

### Example 2 Simplify Expressions

**Simplify $2p(-4p^2 + 5p) - 5(2p^2 + 20)$.**

| | |
|---|---|
| $2p(-4p^2 + 5p) - 5(2p^2 + 20)$ | Original expression |
| $= (2p)(-4p^2) + (2p)(5p) + (-5)(2p^2) + (-5)(20)$ | Distributive Property |
| $= -8p^3 + 10p^2 - 10p^2 - 100$ | Multiply. |
| $= -8p^3 + (10p^2 - 10p^2) - 100$ | Commutative and Associative Properties |
| $= -8p^3 - 100$ | Combine like terms. |

David Schmidt/Masterfile

**GuidedPractice**

**Simplify each expression.**

**2A.** $3(5x^2 + 2x - 4) - x(7x^2 + 2x - 3)$ **2B.** $15t(10y^3t^5 + 5y^2t) - 2y(yt^2 + 4y^2)$

We can use the Distributive Property to multiply monomials by polynomials and solve real world problems.

### Standardized Test Example 3 Write and Evaluate a Polynomial Expression

PT

**GRIDDED RESPONSE The theme for a school dance is "Solid Gold." For one decoration, Kana is covering a trapezoid-shaped piece of poster board with metallic gold paper to look like a bar of gold. If the height of the poster board is 18 inches, how much metallic paper will Kana need in square inches?**

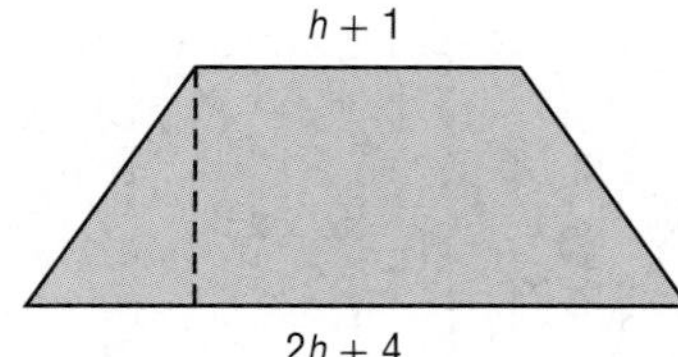


**Read the Test Item**

The question is asking you to find the area of the trapezoid with a height of $h$ and bases of $h + 1$ and $2h + 4$.

**Solve the Test Item**

Write an equation to represent the area of the trapezoid.
Let $b_1 = h + 1$, let $b_2 = 2h + 4$ and let $h$ = height of the trapezoid.

| | |
|---|---|
| $A = \frac{1}{2}h(b_1 + b_2)$ | Area of a trapezoid |
| $= \frac{1}{2}h[(h + 1) + (2h + 4)]$ | $b_1 = h + 1$ and $b_2 = 2h + 4$ |
| $= \frac{1}{2}h(3h + 5)$ | Add and simplify. |
| $= \frac{3}{2}h^2 + \frac{5}{2}h$ | Distributive Property |
| $= \frac{3}{2}(18)^2 + \frac{5}{2}(18)$ | $h = 18$ |
| $= 531$ | Simplify. |

| 5 | 3 | 1 | | |
|---|---|---|---|---|

Kana will need 531 square inches of metallic paper.
Grid in your response of 531.

**Test-TakingTip**

**CCSS Tools** Many standardized tests provide formula sheets with commonly used formulas. If you are unsure of the correct formula, check the sheet before beginning to solve the problem.

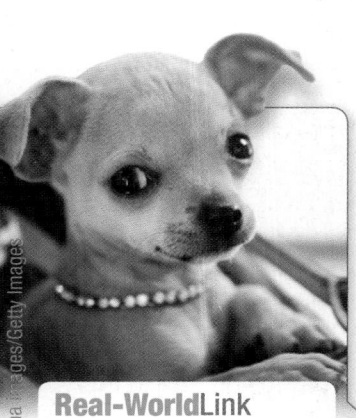

Noriyuki Yamagashira/ANYONE/amana images/Getty Images

**Real-WorldLink**

In a recent year, the pet supply business hit an estimated $7.05 billion in sales. This business ranges from gourmet food to rhinestone tiaras, pearl collars, and cashmere coats.

**Source:** *Entrepreneur Magazine*

**GuidedPractice**

**3. SHORT RESPONSE** Kachima is making triangular bandanas for the dogs and cats in her pet club. The base of the bandana is the length of the collar with 4 inches added to each end to tie it on. The height is $\frac{1}{2}$ of the collar length.

**A.** If Kachima's dog has a collar length of 12 inches, how much fabric does she need in square inches?

**B.** If Kachima makes a bandana for her friend's cat with a 6-inch collar, how much fabric does Kachima need in square inches?

## 2 Solve Equations with Polynomial Expressions

We can use the Distributive Property to solve equations that involve the products of monomials and polynomials.

**StudyTip**

**Combining Like Terms** When simplifying a long expression, it may be helpful to put a circle around one set of like terms, a rectangle around another set, a triangle around another set, and so on.

### Example 4 Equations with Polynomials on Both Sides

**Solve $2a(5a - 2) + 3a(2a + 6) + 8 = a(4a + 1) + 2a(6a - 4) + 50$.**

$2a(5a - 2) + 3a(2a + 6) + 8 = a(4a + 1) + 2a(6a - 4) + 50$ Original equation

$10a^2 - 4a + 6a^2 + 18a + 8 = 4a^2 + a + 12a^2 - 8a + 50$ Distributive Property

$16a^2 + 14a + 8 = 16a^2 - 7a + 50$ Combine like terms.

$14a + 8 = -7a + 50$ Subtract $16a^2$ from each side.

$21a + 8 = 50$ Add $7a$ to each side.

$21a = 42$ Subtract 8 from each side.

$a = 2$ Divide each side by 21.

**CHECK**

$2a(5a - 2) + 3a(2a + 6) + 8 = a(4a + 1) + 2a(6a - 4) + 50$

$2(2)[5(2) - 2] + 3(2)[2(2) + 6] + 8 \stackrel{?}{=} 2[4(2) + 1] + 2(2)[6(2) - 4] + 50$

$4(8) + 6(10) + 8 \stackrel{?}{=} 2(9) + 4(8) + 50$ Simplify.

$32 + 60 + 8 \stackrel{?}{=} 18 + 32 + 50$ Multiply.

$100 = 100$ ✓ Add and subtract.

**GuidedPractice**

**Solve each equation.**

**4A.** $2x(x + 4) + 7 = (x + 8) + 2x(x + 1) + 12$

**4B.** $d(d + 3) - d(d - 4) = 9d - 16$

## Check Your Understanding

**Example 1** **Find each product.**

1. $5w(-3w^2 + 2w - 4)$
2. $6g^2(3g^3 + 4g^2 + 10g - 1)$
3. $4km^2(8km^2 + 2k^2m + 5k)$
4. $-3p^4r^3(2p^2r^4 - 6p^6r^3 - 5)$
5. $2ab(7a^4b^2 + a^5b - 2a)$
6. $c^2d^3(5cd^7 - 3c^3d^2 - 4d^3)$

**Example 2** **Simplify each expression.**

7. $t(4t^2 + 15t + 4) - 4(3t - 1)$
8. $x(3x^2 + 4) + 2(7x - 3)$
9. $-2d(d^3c^2 - 4dc^2 + 2d^2c) + c^2(dc^2 - 3d^4)$
10. $-5w^2(8w^2x - 11wx^2) + 6x(9wx^4 - 4w - 3x^2)$

**Example 3**

11. **GRIDDED RESPONSE** Marlene is buying a new plasma television. The height of the screen of the television is one half the width plus 5 inches. The width is 30 inches. Find the height of the screen in inches.

**Example 4** **Solve each equation.**

12. $-6(11 - 2c) = 7(-2 - 2c)$
13. $t(2t + 3) + 20 = 2t(t - 3)$
14. $-2(w + 1) + w = 7 - 4w$
15. $3(y - 2) + 2y = 4y + 14$
16. $a(a + 3) + a(a - 6) + 35 = a(a - 5) + a(a + 7)$
17. $n(n - 4) + n(n + 8) = n(n - 13) + n(n + 1) + 16$

## Practice and Problem Solving

**Example 1** **Find each product.**

**18.** $b(b^2 - 12b + 1)$

**19.** $f(f^2 + 2f + 25)$

**20.** $-3m^3(2m^3 - 12m^2 + 2m + 25)$

**21.** $2j^2(5j^3 - 15j^2 + 2j + 2)$

**22.** $2pr^2(2pr + 5p^2r - 15p)$

**23.** $4t^3u(2t^2u^2 - 10tu^4 + 2)$

**Example 2** **Simplify each expression.**

**24.** $-3(5x^2 + 2x + 9) + x(2x - 3)$

**25.** $a(-8a^2 + 2a + 4) + 3(6a^2 - 4)$

**26.** $-4d(5d^2 - 12) + 7(d + 5)$

**27.** $-9g(-2g + g^2) + 3(g^2 + 4)$

**28.** $2j(7j^2k^2 + jk^2 + 5k) - 9k(-2j^2k^2 + 2k^2 + 3j)$

**29.** $4n(2n^3p^2 - 3np^2 + 5n) + 4p(6n^2p - 2np^2 + 3p)$

**Example 3**

**30.** **DAMS** A new dam being built has the shape of a trapezoid. The length of the base at the bottom of the dam is 2 times the height. The length of the base at the top of the dam is $\frac{1}{5}$ times the height minus 30 feet.

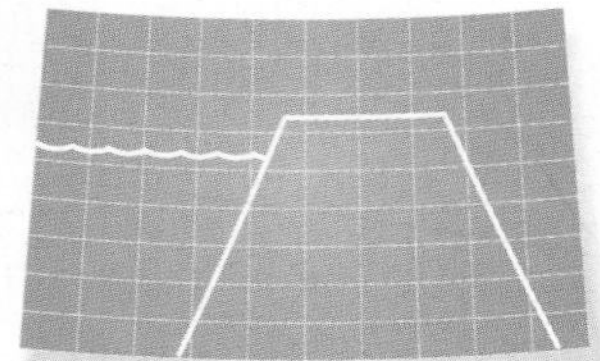

**a.** Write an expression to find the area of the trapezoidal cross section of the dam.

**b.** If the height of the dam is 180 feet, find the area of this cross section.

**Example 4** **Solve each equation.**

**31.** $7(t^2 + 5t - 9) + t = t(7t - 2) + 13$

**32.** $w(4w + 6) + 2w = 2(2w^2 + 7w - 3)$

**33.** $5(4z + 6) - 2(z - 4) = 7z(z + 4) - z(7z - 2) - 48$

**34.** $9c(c - 11) + 10(5c - 3) = 3c(c + 5) + c(6c - 3) - 30$

**35.** $2f(5f - 2) - 10(f^2 - 3f + 6) = -8f(f + 4) + 4(2f^2 - 7f)$

**36.** $2k(-3k + 4) + 6(k^2 + 10) = k(4k + 8) - 2k(2k + 5)$

**Simplify each expression.**

**37.** $\frac{2}{3}np^2(30p^2 + 9n^2p - 12)$

**38.** $\frac{3}{5}r^2t(10r^3 + 5rt^3 + 15t^2)$

**39.** $-5q^2w^3(4q + 7w) + 4qw^2(7q^2w + 2q) - 3qw(3q^2w^2 + 9)$

**40.** $-x^2z(2z^2 + 4xz^3) + xz^2(xz + 5x^3z) + x^2z^3(3x^2z + 4xz)$

**41.** **PARKING** A parking garage charges \$30 per month plus \$0.50 per daytime hour and \$0.25 per hour during nights and weekends. Suppose Trent parks in the garage for 47 hours in January and $h$ of those are night and weekend hours.

**a.** Find an expression for Trent's January bill.

**b.** Find the cost if Trent had 12 hours of night and weekend hours.

$h + 4$

$h$

$3h + 1$

**42.** **CCSS MODELING** Che is building a dog house for his new puppy. The upper face of the dog house is a trapezoid. If the height of the trapezoid is 12 inches, find the area of the face of this piece of the dog house.

43 **TENNIS** The tennis club is building a new tennis court with a path around it.

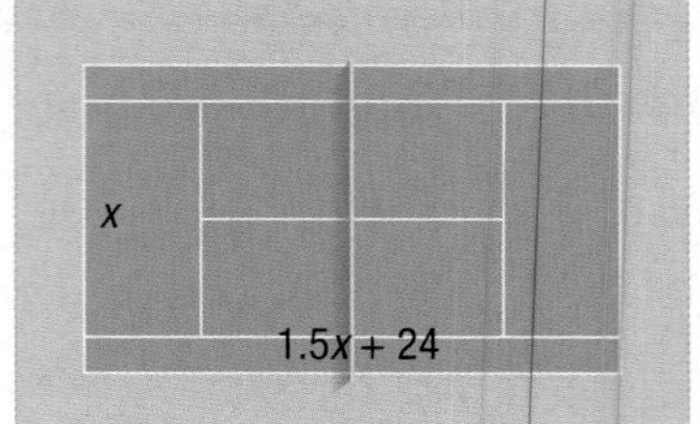


**a.** Write an expression for the area of the tennis court.

**b.** Write an expression for the area of the path.

**c.** If $x = 36$ feet, what is the perimeter of the outside of the path?

**44.** **MULTIPLE REPRESENTATIONS** In this problem, you will investigate the degree of the product of a monomial and a polynomial.

**a. Tabular** Write three monomials of different degrees and three polynomials of different degrees. Determine the degree of each monomial and polynomial. Multiply the monomials by the polynomials. Determine the degree of each product. Record your results in a table like the one shown below.

| Monomial | Degree | Polynomial | Degree | Product of Monomial and Polynomial | Degree |
|---|---|---|---|---|---|
| | | | | | |
| | | | | | |
| | | | | | |

**b. Verbal** Make a conjecture about the degree of the product of a monomial and a polynomial. What is the degree of the product of a monomial of degree $a$ and a polynomial of degree $b$?

## H.O.T. Problems Use Higher-Order Thinking Skills

**45.** **ERROR ANALYSIS** Pearl and Ted both worked on this problem. Is either of them correct? Explain your reasoning.

| Pearl | Ted |
|---|---|
| $2x^2(3x^2 + 4x + 2)$ | $2x^2(3x^2 + 4x + 2)$ |
| $6x^4 + 8x^2 + 4x^2$ | $6x^4 + 8x^3 + 4x^2$ |
| $6x^4 + 12x^2$ | |

**46.** **CCSS PERSEVERANCE** Find $p$ such that $3x^p(4x^{2p+3} + 2x^{3p-2}) = 12x^{12} + 6x^{10}$.

**47.** **CHALLENGE** Simplify $4x^{-3}y^2(2x^5y^{-4} + 6x^{-7}y^6 - 4x^0y^{-2})$.

**48.** **REASONING** Is there a value for $x$ that makes the statement $(x + 2)^2 = x^2 + 2^2$ true? If so, find a value for $x$. Explain your reasoning.

**49.** **OPEN ENDED** Write a monomial and a polynomial using $n$ as the variable. Find their product.

**50.** **WRITING IN MATH** Describe the steps to multiply a polynomial by a monomial.

## Standardized Test Practice

**51.** Every week a store sells $j$ jeans and $t$ T-shirts. The store makes \$8 for each T-shirt and \$12 for each pair of jeans. Which of the following expressions represents the total amount of money, in dollars, the store makes every week?

A $8j + 12t$
B $12j + 8t$
C $20(j + t)$
D $96jt$

**52.** If $a = 5x + 7y$ and $b = 2y - 3x$, what is $a + b$?

F $2x - 9y$
G $3y + 4x$
H $2x + 9y$
J $2x - 5y$

**53.** **GEOMETRY** A triangle has sides of length 5 inches and 8.5 inches. Which of the following cannot be the length of the third side?

A 3.5 inches
B 4 inches
C 5.5 inches
D 12 inches

**54.** **SHORT RESPONSE** Write an equation in which $x$ varies directly as the cube of $y$ and inversely as the square of $z$.

## Spiral Review

**Find each sum or difference.** (Lesson 1-1)

**55.** $(2x^2 - 7) + (8 - 5x^2)$

**56.** $(3z^2 + 2z - 1) + (z^2 - 6)$

**57.** $(2a - 4a^2 + 1) - (5a^2 - 2a - 6)$

**58.** $(a^3 - 3a^2 + 4) - (4a^2 + 7)$

**59.** $(2ab - 3a + 4b) + (5a + 4ab)$

**60.** $(8c^3 - 3c^2 + c - 2) - (3c^3 + 9)$

**61.** **CLOCKS** The period of a pendulum is the time required for it to make one complete swing back and forth. The formula of the period $P$ in seconds of a pendulum is $P = 2\pi\sqrt{\frac{\ell}{32}}$, where $\ell$ is the length of the pendulum in feet. (Lesson 0-10)

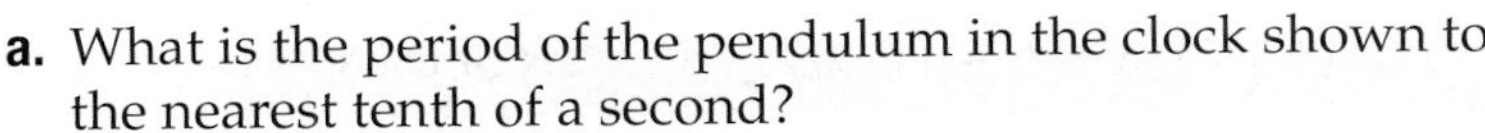

**a.** What is the period of the pendulum in the clock shown to the nearest tenth of a second?

**b.** About how many inches long should the pendulum be in order for it to have a period of 1 second?

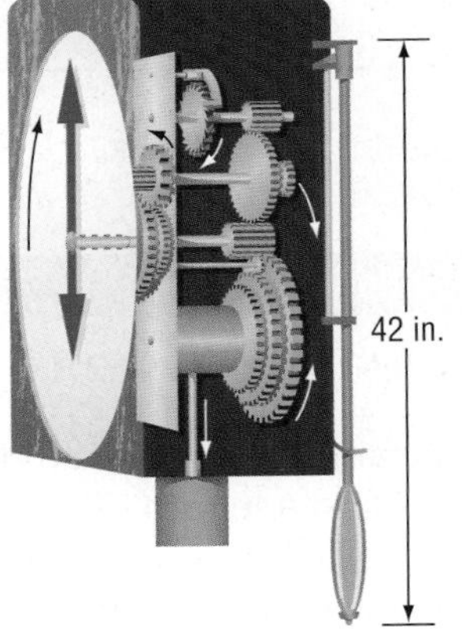


**Solve each inequality.** (Lesson 0-6)

**62.** $-14n \geq 42$

**63.** $p + 6 > 15$

**64.** $-2a - 5 < 20$

**65.** $5x \leq 3x - 26$

## Skills Review

**Simplify.**

**66.** $b(b^2)(b^3)$

**67.** $2y(3y^2)$

**68.** $-y^4(-2y^3)$

**69.** $-3z^3(-5z^4 + 2z)$

**70.** $2m(-4m^4) - 3(-5m^3)$

**71.** $4p^2(-2p^3) + 2p^4(5p^6)$

EXPLORE 1-3

# Algebra Lab
# Multiplying Polynomials

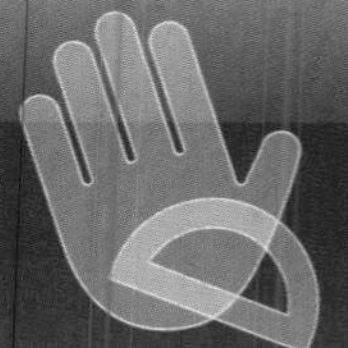

You can use algebra tiles to find the product of two binomials.

**CCSS Common Core State Standards**
**Content Standards**
**A.APR.1** Understand that polynomials form a system analogous to the integers, namely, they are closed under the operations of addition, subtraction, and multiplication; add, subtract, and multiply polynomials.

## Activity 1 Multiply Binomials

**Use algebra tiles to find $(x + 3)(x + 4)$.**

The rectangle will have a width of $x + 3$ and a length of $x + 4$. Use algebra tiles to mark off the dimensions on a product mat. Then complete the rectangle with algebra tiles.

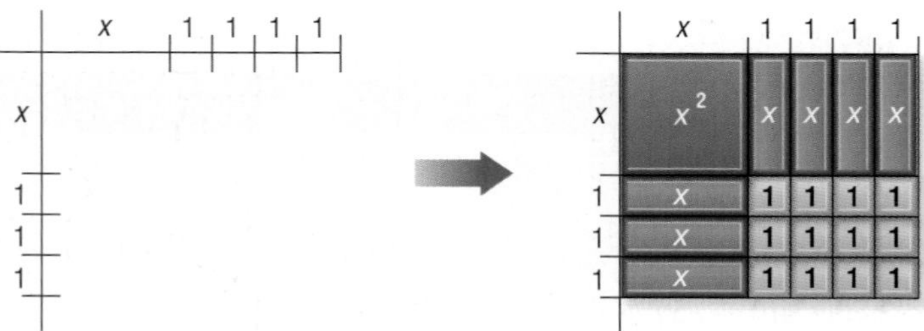

The rectangle consists of 1 blue $x^2$-tile, 7 green $x$-tiles, and 12 yellow 1-tiles. The area of the rectangle is $x^2 + 7x + 12$. So, $(x + 3)(x + 4) = x^2 + 7x + 12$.

## Activity 2 Multiply Binomials

**Use algebra tiles to find $(x - 2)(x - 5)$.**

**Step 1** The rectangle will have a width of $x - 2$ and a length of $x - 5$. Use algebra tiles to mark off the dimensions on a product mat. Then begin to make the rectangle with algebra tiles.

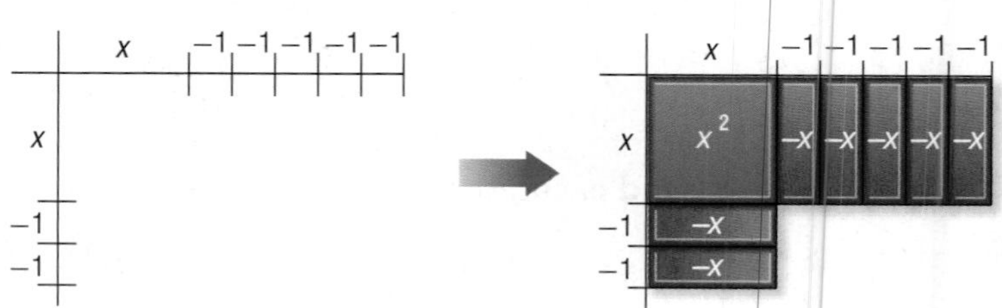

**Step 2** Determine whether to use 10 yellow 1-tiles or 10 red $-1$-tiles to complete the rectangle. The area of each yellow tile is the product of $-1$ and $-1$. Fill in the space with 10 yellow 1-tiles to complete the rectangle.

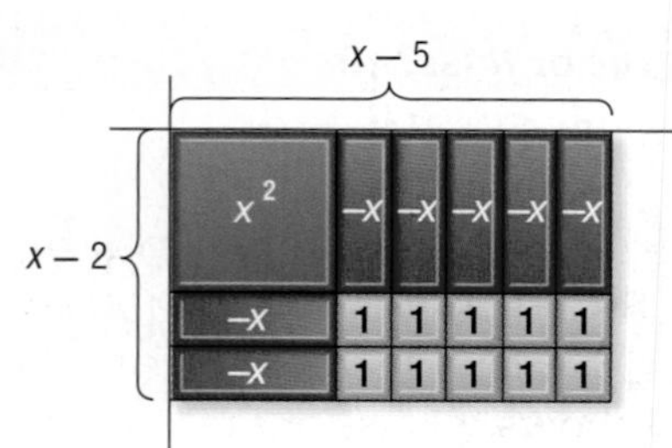

The rectangle consists of 1 blue $x^2$-tile, 7 red $-x$-tiles, and 10 yellow 1-tiles. The area of the rectangle is $x^2 - 7x + 10$. So, $(x - 2)(x - 5) = x^2 - 7x + 10$.

## Activity 3 Multiply Binomials

**Use algebra tiles to find $(x - 4)(2x + 3)$.**

**Step 1** The rectangle will have a width of $x - 4$ and a length of $2x + 3$. Use algebra tiles to mark off the dimensions on a product mat. Then begin to make the rectangle with algebra tiles.

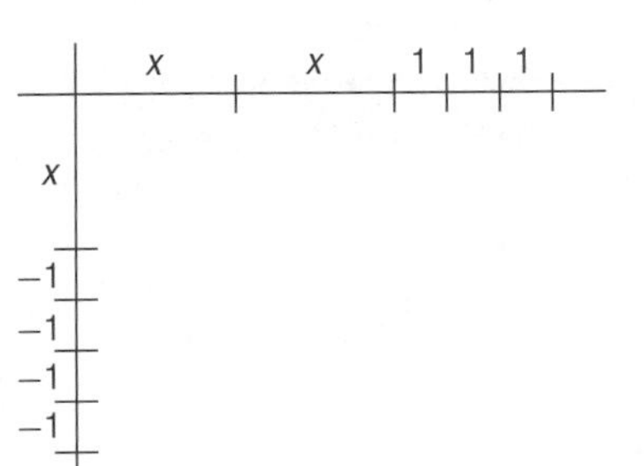

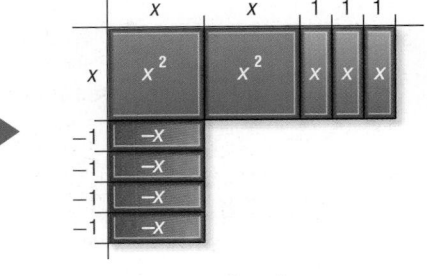

**Step 2** Determine what color $x$-tiles and what color 1-tiles to use to complete the rectangle. The area of each red $x$-tile is the product of $x$ and $-1$. The area of each red $-1$-tile is represented by $1(-1)$ or $-1$.

Complete the rectangle with 4 red $x$-tiles and 12 red $-1$-tiles.

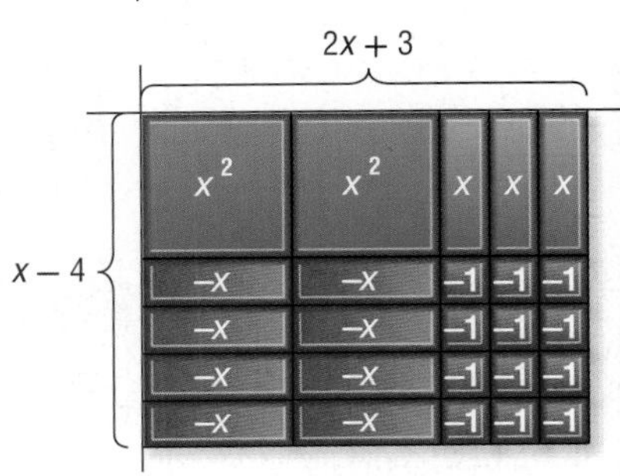


**Step 3** Rearrange the tiles to simplify the polynomial you have formed. Notice that a 3 zero pair are formed by three positive and three negative $x$-tiles.

There are 2 blue $x^2$-tiles, 5 red $-x$-tiles, and 12 red $-1$-tiles left. In simplest form, $(x - 4)(2x + 3) = 2x^2 - 5x - 12$.

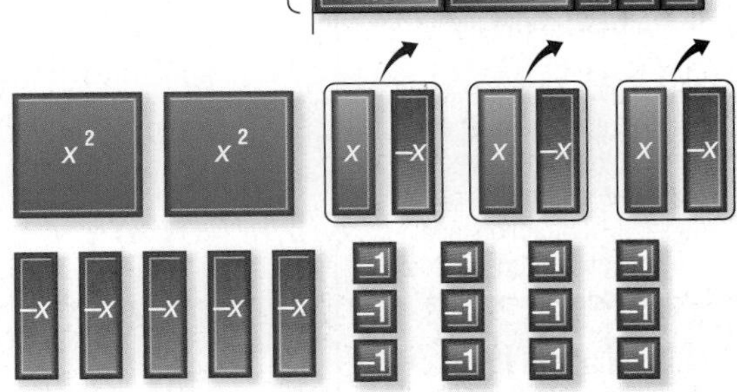


## Model and Analyze

**Use algebra tiles to find each product.**

1. $(x + 1)(x + 4)$
2. $(x - 3)(x - 2)$
3. $(x + 5)(x - 1)$
4. $(x + 2)(2x + 3)$
5. $(x - 1)(2x - 1)$
6. $(x + 4)(2x - 5)$

**Is each statement *true* or *false*? Justify your answer with a drawing of algebra tiles.**

7. $(x - 4)(x - 2) = x^2 - 6x + 8$
8. $(x + 3)(x + 5) = x^2 + 15$

9. **WRITING IN MATH** You can also use the Distributive Property to find the product of two binomials. The figure at the right shows the model for $(x + 4)(x + 5)$ separated into four parts. Write a sentence or two explaining how this model shows the use of the Distributive Property.

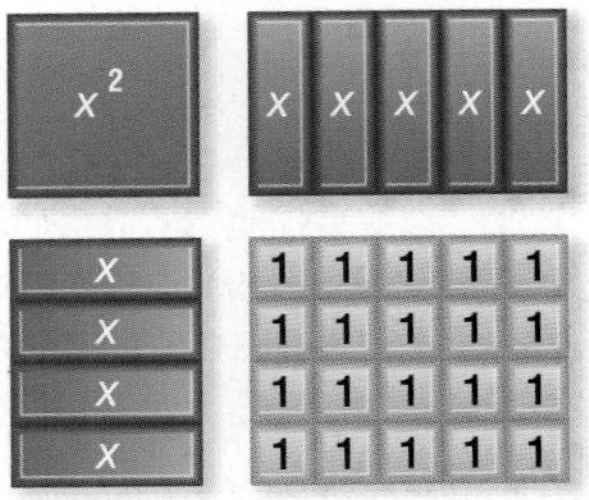

LESSON 1-3

# Multiplying Polynomials

**Then**

- You multiplied polynomials by monomials.

**Now**

1. Multiply binomials by using the FOIL method.
2. Multiply polynomials by using the Distributive Property.

**Why?**

- Bodyboards, which are used to ride waves, are made of foam and are more rectangular than surfboards. A bodyboard's dimensions are determined by the height and skill level of the user.

The length of Ann's bodyboard should be Ann's height $h$ minus 32 inches or $h - 32$. The board's width should be half of Ann's height plus 11 inches or $\frac{1}{2}h + 11$. To approximate the area of the bodyboard, you need to find $(h - 32)\left(\frac{1}{2}h + 11\right)$.

**NewVocabulary**
FOIL method
quadratic expression

**CCSS Common Core State Standards**

**Content Standards**
A.APR.1 Understand that polynomials form a system analogous to the integers, namely, they are closed under the operations of addition, subtraction, and multiplication; add, subtract, and multiply polynomials.

**Mathematical Practices**
7 Look for and make use of structure.

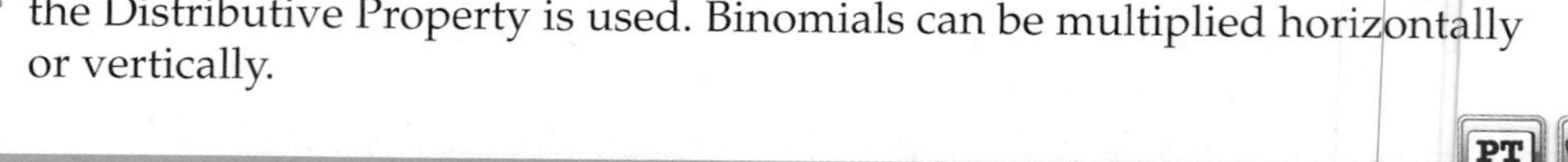

## 1 Multiply Binomials

To multiply two binomials such as $h - 32$ and $\frac{1}{2}h + 11$, the Distributive Property is used. Binomials can be multiplied horizontally or vertically.

### Example 1 The Distributive Property

**Find each product.**

**a.** $(2x + 3)(x + 5)$

**Vertical Method**

Multiply by 5.

$$\begin{array}{r} 2x + 3 \\ (\times)\ x + 5 \\ \hline 10x + 15 \end{array}$$

$5(2x + 3) = 10x + 15$

Multiply by $x$.

$$\begin{array}{r} 2x + 3 \\ (\times)\ x + 5 \\ \hline 10x + 15 \\ 2x^2 + 3x \quad\ \ \end{array}$$

$x(2x + 3) = 2x^2 + 3x$

Combine like terms.

$$\begin{array}{r} 2x + 3 \\ (\times)\ x + 5 \\ \hline 10x + 15 \\ 2x^2 + 3x \quad\ \ \\ \hline 2x^2 + 13x + 15 \end{array}$$

**Horizontal Method**

$$\begin{aligned} (2x + 3)(x + 5) &= 2x(x + 5) + 3(x + 5) && \text{Rewrite as the sum of two products.} \\ &= 2x^2 + 10x + 3x + 15 && \text{Distributive Property} \\ &= 2x^2 + 13x + 15 && \text{Combine like terms.} \end{aligned}$$

**b.** $(x - 2)(3x + 4)$

**Vertical Method**

Multiply by 4.

$$\begin{array}{r} x - 2 \\ (\times)\ 3x + 4 \\ \hline 4x - 8 \end{array}$$

$4(x - 2) = 4x - 8$

Multiply by $3x$.

$$\begin{array}{r} x - 2 \\ (\times)\ 3x + 4 \\ \hline 4x - 8 \\ 3x^2 - 6x \quad\ \ \end{array}$$

$3x(x - 2) = 3x^2 - 6x$

Combine like terms.

$$\begin{array}{r} x - 2 \\ (\times)\ 3x + 4 \\ \hline 4x - 8 \\ 3x^2 - 6x \quad\ \ \\ \hline 3x^2 - 2x - 8 \end{array}$$

**Horizontal Method**

$$\begin{aligned} (x - 2)(3x + 4) &= x(3x + 4) - 2(3x + 4) && \text{Rewrite as the difference of two products.} \\ &= 3x^2 + 4x - 6x - 8 && \text{Distributive Property} \\ &= 3x^2 - 2x - 8 && \text{Combine like terms.} \end{aligned}$$

Peter Cade/Iconica/Getty Images

**GuidedPractice**

**1A.** $(3m + 4)(m + 5)$

**1B.** $(5y - 2)(y + 8)$

A shortcut version of the Distributive Property for multiplying binomials is called the **FOIL method**.

## KeyConcept FOIL Method

**Words** To multiply two binomials, find the sum of the products of *F* the *First* terms, *O* the *Outer* terms, *I* the *Inner* terms, *L* and the *Last* terms.

**Example**

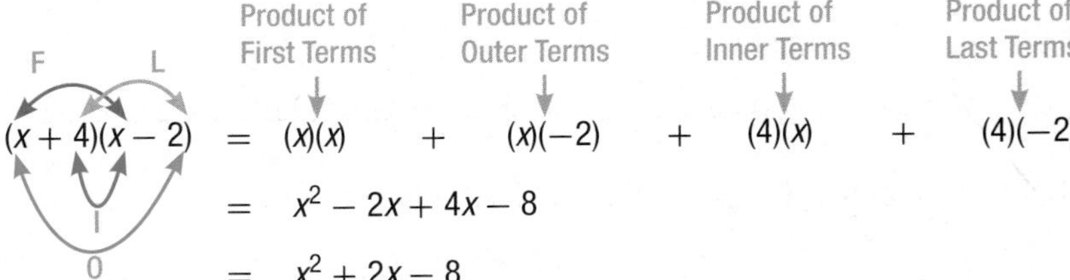


$$(x + 4)(x - 2) = (x)(x) + (x)(-2) + (4)(x) + (4)(-2)$$
$$= x^2 - 2x + 4x - 8$$
$$= x^2 + 2x - 8$$

**ReadingMath**

**Polynomials as Factors** The expression $(x + 4)(x - 2)$ is read *the quantity x plus 4 times the quantity x minus 2.*

## Example 2 FOIL Method

**Find each product.**

**a.** $(2y - 7)(3y + 5)$

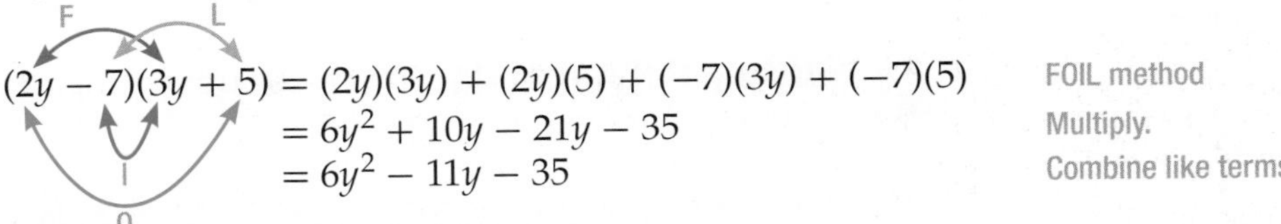


$(2y - 7)(3y + 5) = (2y)(3y) + (2y)(5) + (-7)(3y) + (-7)(5)$ FOIL method

$= 6y^2 + 10y - 21y - 35$ Multiply.

$= 6y^2 - 11y - 35$ Combine like terms.

**b.** $(4a - 5)(2a - 9)$

$(4a - 5)(2a - 9)$

$= (4a)(2a) + (4a)(-9) + (-5)(2a) + (-5)(-9)$ FOIL method

$= 8a^2 - 36a - 10a + 45$ Multiply.

$= 8a^2 - 46a + 45$ Combine like terms.

**GuidedPractice**

**2A.** $(x + 3)(x - 4)$

**2B.** $(4b - 5)(3b + 2)$

**2C.** $(2y - 5)(y - 6)$

**2D.** $(5a + 2)(3a - 4)$

Notice that when two linear expressions are multiplied, the result is a quadratic expression. A **quadratic expression** is an expression in one variable with a degree of 2. When three linear expressions are multiplied, the result has a degree of 3.

The FOIL method can be used to find an expression that represents the area of a rectangular object when the lengths of the sides are given as binomials.

**Real-WorldLink**

The cost of a swimming pool depends on many factors, including the size of the pool, whether the pool is an above-ground or an in-ground pool, and the material used.

**Source:** American Dream Homes

## Real-World Example 3 FOIL Method

**SWIMMING POOL A contractor is building a deck around a rectangular swimming pool. The deck is $x$ feet from every side of the pool. Write an expression for the total area of the pool and deck.**

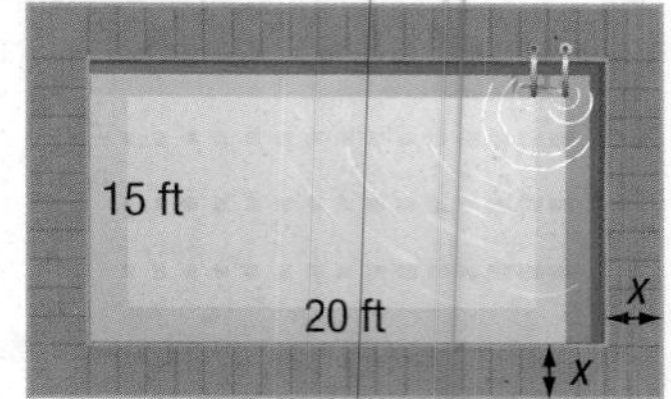


**Understand** We need to find an expression for the total area of the pool and deck.

**Plan** Find the product of the length and width of the pool with the deck.

**Solve** Since the deck is the same distance from every side of the pool, the length and width of the pool are $2x$ longer. So, the length can be represented by $2x + 20$ and the width can be represented by $2x + 15$.

| | |
|---|---|
| Area = length · width | Area of a rectangle |
| $= (2x + 20)(2x + 15)$ | Substitution |
| $= (2x)(2x) + (2x)(15) + (20)(2x) + (20)(15)$ | FOIL Method |
| $= 4x^2 + 30x + 40x + 300$ | Multiply. |
| $= 4x^2 + 70x + 300$ | Combine like terms. |

So, the total area of the deck and pool is $4x^2 + 70x + 300$.

**Check** Choose a value for $x$. Substitute this value into $(2x + 20)(2x + 15)$ and $4x^2 + 70x + 300$. The result should be the same for both expressions.

**GuidedPractice**

**3.** If the pool is 25 feet long and 20 feet wide, find the area of the pool and deck.

## 2 Multiply Polynomials

The Distributive Property can also be used to multiply any two polynomials.

## Example 4 The Distributive Property

**Find each product.**

**a.** $(6x + 5)(2x^2 - 3x - 5)$

| | |
|---|---|
| $(6x + 5)(2x^2 - 3x - 5)$ | |
| $= 6x(2x^2 - 3x - 5) + 5(2x^2 - 3x - 5)$ | Distributive Property |
| $= 12x^3 - 18x^2 - 30x + 10x^2 - 15x - 25$ | Multiply. |
| $= 12x^3 - 8x^2 - 45x - 25$ | Combine like terms. |

**b.** $(2y^2 + 3y - 1)(3y^2 - 5y + 2)$

| | |
|---|---|
| $(2y^2 + 3y - 1)(3y^2 - 5y + 2)$ | |
| $= 2y^2(3y^2 - 5y + 2) + 3y(3y^2 - 5y + 2) - 1(3y^2 - 5y + 2)$ | Distributive Property |
| $= 6y^4 - 10y^3 + 4y^2 + 9y^3 - 15y^2 + 6y - 3y^2 + 5y - 2$ | Multiply. |
| $= 6y^4 - y^3 - 14y^2 + 11y - 2$ | Combine like terms. |

**StudyTip**

Multiplying Polynomials If a polynomial with $c$ terms and a polynomial with $d$ terms are multiplied together, there will be $c \cdot d$ terms before simplifying. In Example 4a, there are $2 \cdot 3$ or 6 terms before simplifying.

**GuidedPractice**

**4A.** $(3x - 5)(2x^2 + 7x - 8)$

**4B.** $(m^2 + 2m - 3)(4m^2 - 7m + 5)$

## Check Your Understanding

**Examples 1–2** **Find each product.**

1. $(x + 5)(x + 2)$
2. $(y - 2)(y + 4)$
3. $(b - 7)(b + 3)$
4. $(4n + 3)(n + 9)$
5. $(8h - 1)(2h - 3)$
6. $(2a + 9)(5a - 6)$

**Example 3**

7. **FRAME** Hugo is designing a frame as shown at the right. The frame has a width of $x$ inches all the way around. Write an expression that represents the total area of the picture and frame.

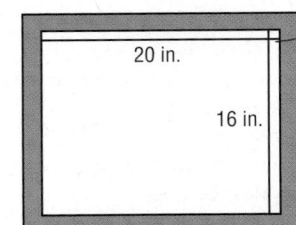

**Example 4** **Find each product.**

8. $(2a - 9)(3a^2 + 4a - 4)$
9. $(4y^2 - 3)(4y^2 + 7y + 2)$
10. $(x^2 - 4x + 5)(5x^2 + 3x - 4)$
11. $(2n^2 + 3n - 6)(5n^2 - 2n - 8)$

## Practice and Problem Solving

**Examples 1–2** **Find each product.**

12. $(3c - 5)(c + 3)$
13. $(g + 10)(2g - 5)$
14. $(6a + 5)(5a + 3)$
15. $(4x + 1)(6x + 3)$
16. $(5y - 4)(3y - 1)$
17. $(6d - 5)(4d - 7)$
18. $(3m + 5)(2m + 3)$
19. $(7n - 6)(7n - 6)$
20. $(12t - 5)(12t + 5)$
21. $(5r + 7)(5r - 7)$
22. $(8w + 4x)(5w - 6x)$
23. $(11z - 5y)(3z + 2y)$

**Example 3**

24. **GARDEN** A walkway surrounds a rectangular garden. The width of the garden is 8 feet, and the length is 6 feet. The width $x$ of the walkway around the garden is the same on every side. Write an expression that represents the total area of the garden and walkway.

**Example 4** **Find each product.**

25. $(2y - 11)(y^2 - 3y + 2)$
26. $(4a + 7)(9a^2 + 2a - 7)$
27. $(m^2 - 5m + 4)(m^2 + 7m - 3)$
28. $(x^2 + 5x - 1)(5x^2 - 6x + 1)$
29. $(3b^3 - 4b - 7)(2b^2 - b - 9)$
30. $(6z^2 - 5z - 2)(3z^3 - 2z - 4)$

**Simplify.**

31. $(m + 2)[(m^2 + 3m - 6) + (m^2 - 2m + 4)]$
32. $[(t^2 + 3t - 8) - (t^2 - 2t + 6)](t - 4)$

**CCSS STRUCTURE** **Find an expression to represent the area of each shaded region.**

33.

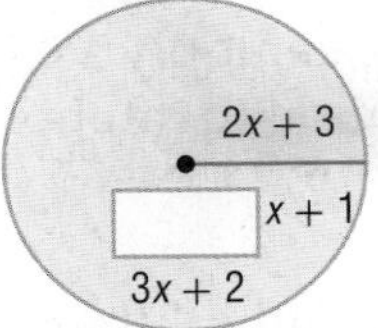


34.

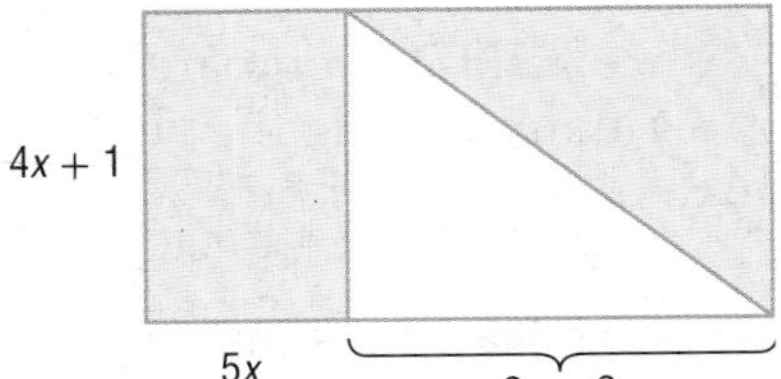

**35** **VOLLEYBALL** The dimensions of a sand volleyball court are represented by a width of $6y - 5$ feet and a length of $3y + 4$ feet.

**a.** Write an expression that represents the area of the court.

**b.** The length of a sand volleyball court is 31 feet. Find the area of the court.

**36.** **GEOMETRY** Write an expression for the area of a triangle with a base of $2x + 3$ and a height of $3x - 1$.

**Find each product.**

**37.** $(a - 2b)^2$ **38.** $(3c + 4d)^2$ **39.** $(x - 5y)^2$

**40.** $(2r - 3t)^3$ **41.** $(5g + 2h)^3$ **42.** $(4y + 3z)(4y - 3z)^2$

**43.** **CONSTRUCTION** A sandbox kit allows you to build a square sandbox or a rectangular sandbox as shown.

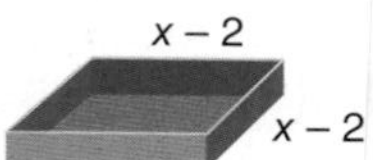


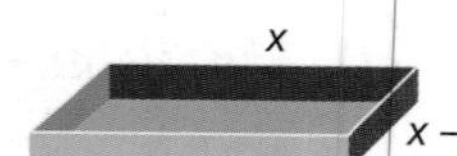


**a.** What are the possible values of $x$? Explain.

**b.** Which shape has the greater area?

**c.** What is the difference in areas between the two?

**44.** **MULTIPLE REPRESENTATIONS** In this problem, you will investigate the square of a sum.

**a. Tabular** Copy and complete the table for each sum.

| Expression | (Expression)$^2$ |
|---|---|
| $x + 5$ | |
| $3y + 1$ | |
| $z + q$ | |

**b. Verbal** Make a conjecture about the terms of the square of a sum.

**c. Symbolic** For a sum of the form $a + b$, write an expression for the square of the sum.

## H.O.T. Problems Use Higher-Order Thinking Skills

**45.** **REASONING** Determine if the following statement is *sometimes*, *always*, or *never* true. Explain your reasoning.

*The FOIL method can be used to multiply a binomial and a trinomial.*

**46.** **CHALLENGE** Find $(x^m + x^p)(x^{m-1} - x^{1-p} + x^p)$.

**47.** **OPEN ENDED** Write a binomial and a trinomial involving a single variable. Then find their product.

**48.** **CCSS REGULARITY** Compare and contrast the procedure used to multiply a trinomial by a binomial using the vertical method with the procedure used to multiply a three-digit number by a two-digit number.

**49.** **WRITING IN MATH** Summarize the methods that can be used to multiply polynomials.

## Standardized Test Practice

**50.** What is the product of $2x - 5$ and $3x + 4$?

A $5x - 1$

B $6x^2 - 7x - 20$

C $6x^2 - 20$

D $6x^2 + 7x - 20$

**51.** Which statement is correct about the symmetry of this design?

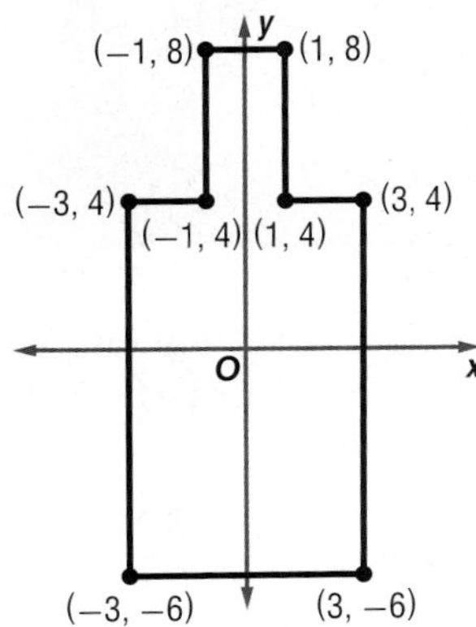


F The design is symmetrical only about the $y$-axis.

G The design is symmetrical only about the $x$-axis.

H The design is symmetrical about both the $y$- and the $x$-axes.

J The design has no symmetry.

**52.** Which point on the number line represents a number that, when cubed, will result in a number greater than itself?

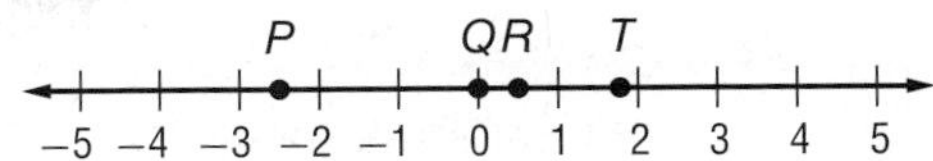


A $P$

B $Q$

C $R$

D $T$

**53. SHORT RESPONSE** For a science project, Jodi selected three bean plants of equal height. Then, for five days, she measured their heights in centimeters and plotted the values on the graph below.

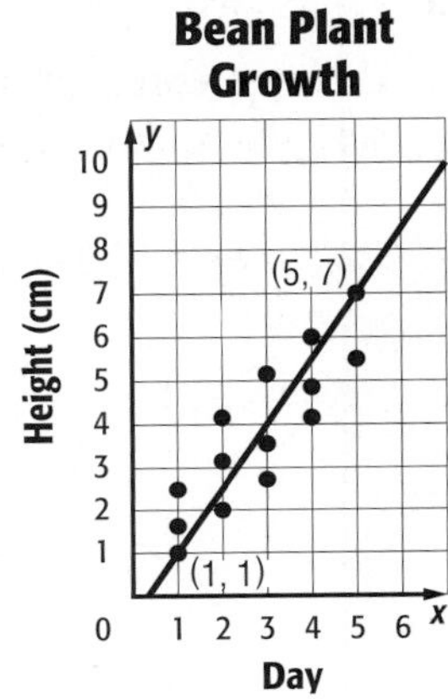


She drew a line of best fit on the graph. What is the slope of the line that she drew?

## Spiral Review

**54. SAVINGS** Carrie has \$6000 to invest. She puts $x$ dollars of this money into a savings account that earns 2% interest per year. She uses the rest of the money to purchase a certificate of deposit that earns 4% interest. Write an equation for the amount of money that Carrie will have in one year. (Lesson 1-2)

**Find each sum or difference.** (Lesson 1-1)

**55.** $(7a^2 - 5) + (-3a^2 + 10)$

**56.** $(8n - 2n^2) + (4n - 6n^2)$

**57.** $(4 + n^3 + 3n^2) + (2n^3 - 9n^2 + 6)$

**58.** $(-4u^2 - 9 + 2u) + (6u + 14 + 2u^2)$

**59.** $(b + 4) + (c + 3b - 2)$

**60.** $(3a^3 - 6a) - (3a^3 + 5a)$

**61.** $(-4m^3 - m + 10) - (3m^3 + 3m^2 - 7)$

**62.** $(3a + 4ab + 3b) - (2b + 5a + 8ab)$

## Skills Review

**Simplify.**

**63.** $(-2t^4)^3 - 3(-2t^3)^4$

**64.** $(-3h^2)^3 - 2(-h^3)^2$

**65.** $2(-5y^3)^2 + (-3y^3)^3$

**66.** $3(-6n^4)^2 + (-2n^2)^2$

LESSON 1-4

# Special Products

**Then**

- You multiplied binomials by using the FOIL method.

**Now**

1. Find squares of sums and differences.
2. Find the product of a sum and a difference.

**Why?**

- Colby wants to attach a dartboard to a square piece of corkboard. If the radius of the dartboard is $r + 12$, how large does the square corkboard need to be?

Colby knows that the diameter of the dartboard is $2(r + 12)$ or $2r + 24$. Each side of the square also measures $2r + 24$. To find how much corkboard is needed, Colby must find the area of the square: $A = (2r + 24)^2$.

**Common Core State Standards**

**Content Standards**

**A.APR.1** Understand that polynomials form a system analogous to the integers, namely, they are closed under the operations of addition, subtraction, and multiplication; add, subtract, and multiply polynomials.

**Mathematical Practices**

8 Look for and express regularity in repeated reasoning.

## 1 Squares of Sums and Differences

Some pairs of binomials, such as squares like $(2r + 24)^2$, have products that follow a specific pattern. Using the pattern can make multiplying easier. The square of a sum, $(a + b)^2$ or $(a + b)(a + b)$, is one of those products.

$$(a + b)^2 = a^2 + ab + ab + b^2$$

### KeyConcept Square of a Sum

**Words** The square of $a + b$ is the square of $a$ plus twice the product of $a$ and $b$ plus the square of $b$.

**Symbols** $(a + b)^2 = (a + b)(a + b)$
$= a^2 + 2ab + b^2$

**Example** $(x + 4)^2 = (x + 4)(x + 4)$
$= x^2 + 8x + 16$

### Example 1 Square of a Sum

**Find $(3x + 5)^2$.**

| | |
|---|---|
| $(a + b)^2 = a^2 + 2ab + b^2$ | Square of a sum |
| $(3x + 5)^2 = (3x)^2 + 2(3x)(5) + 5^2$ | $a = 3x$, $b = 5$ |
| $= 9x^2 + 30x + 25$ | Simplify. Use FOIL to check your solution. |

**GuidedPractice**

**Find each product.**

**1A.** $(8c + 3d)^2$

**1B.** $(3x + 4y)^2$

There is also a pattern for the *square of a difference*. Write $a - b$ as $a + (-b)$ and square it using the square of a sum pattern.

$$(a - b)^2 = [a + (-b)]^2$$
$$= a^2 + 2(a)(-b) + (-b)^2 \quad \text{Square of a sum}$$
$$= a^2 - 2ab + b^2 \quad \text{Simplify.}$$

## KeyConcept Square of a Difference

**Words** The square of $a - b$ is the square of $a$ minus twice the product of $a$ and $b$ plus the square of $b$.

**Symbols** $(a - b)^2 = (a - b)(a - b) = a^2 - 2ab + b^2$

**Example** $(x - 3)^2 = (x - 3)(x - 3) = x^2 - 6x + 9$

**WatchOut!**

**CCSS Regularity** Remember that $(x - 7)^2$ does not equal $x^2 - 7^2$, or $x^2 - 49$.

$(x - 7)^2$
$= (x - 7)(x - 7)$
$= x^2 - 14x + 49$

### Example 2 Square of a Difference

**Find $(2x - 5y)^2$.**

$$(a - b)^2 = a^2 - 2ab + b^2 \quad \text{Square of a difference}$$
$$(2x - 5y)^2 = (2x)^2 - 2(2x)(5y) + (5y)^2 \quad a = 2x \text{ and } b = 5y$$
$$= 4x^2 - 20xy + 25y^2 \quad \text{Simplify.}$$

**GuidedPractice**

**Find each product.**

**2A.** $(6p - 1)^2$

**2B.** $(a - 2b)^2$

The product of the square of a sum or the square of a difference is called a *perfect square trinomial*. We can use these to find patterns to solve real-world problems.

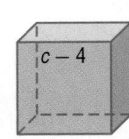

### Real-World Example 3 Square of a Difference

**PHYSICAL SCIENCE Each edge of a cube of aluminum is 4 centimeters less than each edge of a cube of copper. Write an equation to model the surface area of the aluminum cube.**

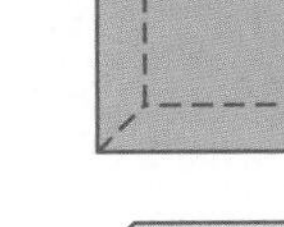


Let $c$ = the length of each edge of the cube of copper. So, each edge of the cube of aluminum is $c - 4$.

$SA = 6s^2$ Formula for surface area of a cube

$SA = 6(c - 4)^2$ Replace $s$ with $c - 4$.

$SA = 6[c^2 - 2(4)(c) + 4^2]$ Square of a difference

$SA = 6(c^2 - 8c + 16)$ Simplify.

**GuidedPractice**

**3. GARDENING** Alano has a garden that is $g$ feet long and $g$ feet wide. He wants to add 3 feet to the length and the width.

**A.** Show how the new area of the garden can be modeled by the square of a binomial.

**B.** Find the square of this binomial.

## 2 Product of a Sum and a Difference

Now we will see what the result is when we multiply a sum and a difference, or $(a + b)(a - b)$. Recall that $a - b$ can be written as $a + (-b)$.

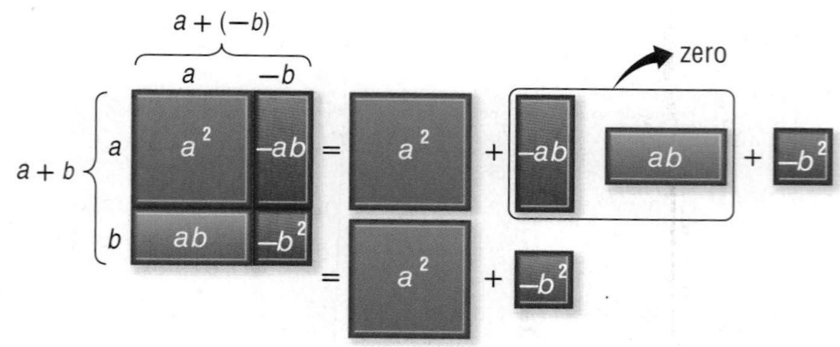


Notice that the middle terms are opposites and add to a zero pair. So $(a + b)(a - b) = a^2 - ab + ab - b^2 = a^2 - b^2$.

**StudyTip**

**Patterns** When using any of these patterns, $a$ and $b$ can be numbers, variables, or expressions with numbers and variables.

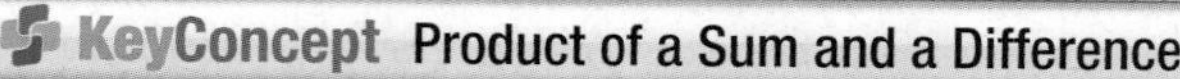

### KeyConcept Product of a Sum and a Difference

**Words** The product of $a + b$ and $a - b$ is the square of $a$ minus the square of $b$.

**Symbols**

$$(a + b)(a - b) = (a - b)(a + b) = a^2 - b^2$$

### Example 4 Product of a Sum and a Difference

**Find $(2x^2 + 3)(2x^2 - 3)$.**

$(a + b)(a - b) = a^2 - b^2$ — Product of a sum and a difference

$(2x^2 + 3)(2x^2 - 3) = (2x^2)^2 - (3)^2$ — $a = 2x^2$ and $b = 3$

$= 4x^4 - 9$ — Simplify.

**GuidedPractice**

**Find each product.**

**4A.** $(3n + 2)(3n - 2)$

**4B.** $(4c - 7d)(4c + 7d)$

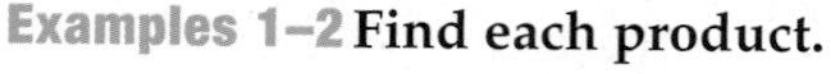

## Check Your Understanding

**Examples 1–2** **Find each product.**

**1.** $(x + 5)^2$

**2.** $(11 - a)^2$

**3.** $(2x + 7y)^2$

**4.** $(3m - 4)(3m - 4)$

**5.** $(g - 4h)(g - 4h)$

**6.** $(3c + 6d)^2$

**Example 3**

**7. GENETICS** The color of a Labrador retriever's fur is genetic. Dark genes $D$ are dominant over yellow genes $y$. A dog with genes $DD$ or $Dy$ will have dark fur. A dog with genes $yy$ will have yellow fur. Pepper's genes for fur color are $Dy$, and Ramiro's are $yy$.

| | D | y |
|---|---|---|
| D | DD | Dy |
| y | Dy | yy |

**a.** Write an expression for the possible fur colors of Pepper's and Ramiro's puppies.

**b.** What is the probability that a puppy will have yellow fur?

**Example 4** **Find each product.**

8. $(a - 3)(a + 3)$
9. $(x + 5)(x - 5)$
10. $(6y - 7)(6y + 7)$
11. $(9t + 6)(9t - 6)$

## Practice and Problem Solving

**Examples 1–2** **Find each product.**

12. $(a + 10)(a + 10)$
13. $(b - 6)(b - 6)$
14. $(h + 7)^2$
15. $(x + 6)^2$
16. $(8 - m)^2$
17. $(9 - 2y)^2$
18. $(2b + 3)^2$
19. $(5t - 2)^2$
20. $(8h - 4n)^2$

**Example 3**

21. **GENETICS** The ability to roll your tongue is inherited genetically from parents if either parent has the dominant trait $T$. Children of two parents without the trait will not be able to roll their tongues.

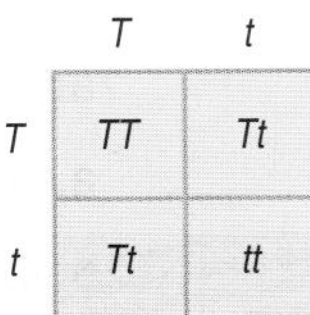


   a. Show how the combinations can be modeled by the square of a sum.
   b. Predict the percent of children that will have both dominant genes, one dominant gene, and both recessive genes.

**Example 4** **Find each product.**

22. $(u + 3)(u - 3)$
23. $(b + 7)(b - 7)$
24. $(2 + x)(2 - x)$
25. $(4 - x)(4 + x)$
26. $(2q + 5r)(2q - 5r)$
27. $(3a^2 + 7b)(3a^2 - 7b)$
28. $(5y + 7)^2$
29. $(8 - 10a)^2$
30. $(10x - 2)(10x + 2)$
31. $(3t + 12)(3t - 12)$
32. $(a + 4b)^2$
33. $(3q - 5r)^2$
34. $(2c - 9d)^2$
35. $(g + 5h)^2$
36. $(6y - 13)(6y + 13)$
37. $(3a^4 - b)(3a^4 + b)$
38. $(5x^2 - y^2)^2$
39. $(8a^2 - 9b^3)(8a^2 + 9b^3)$
40. $\left(\frac{3}{4}k + 8\right)^2$
41. $\left(\frac{2}{5}y - 4\right)^2$
42. $(7z^2 + 5y^2)(7z^2 - 5y^2)$
43. $(2m + 3)(2m - 3)(m + 4)$
44. $(r + 2)(r - 5)(r - 2)(r + 5)$

45. **CCSS SENSE-MAKING** Write a polynomial that represents the area of the figure at the right.

$x - 1$ $x + 2$

$x - 1$ $x + 2$

46. **FLYING DISKS** A flying disk shaped like a circle has a radius of $x + 3$ inches.
   a. Write an expression representing the area of the flying disk.
   b. A hole with a radius of $x - 1$ inches is cut in the center of the disk. Write an expression for the remaining area.

**GEOMETRY** **Find the area of each shaded region.**

47.

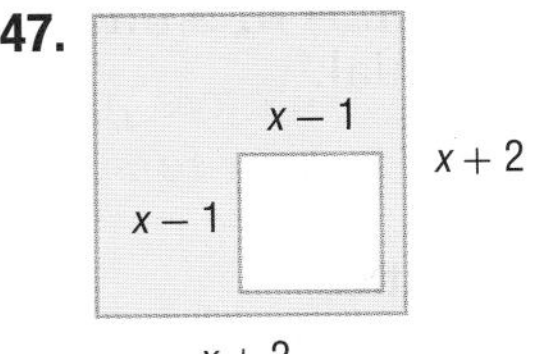


48.

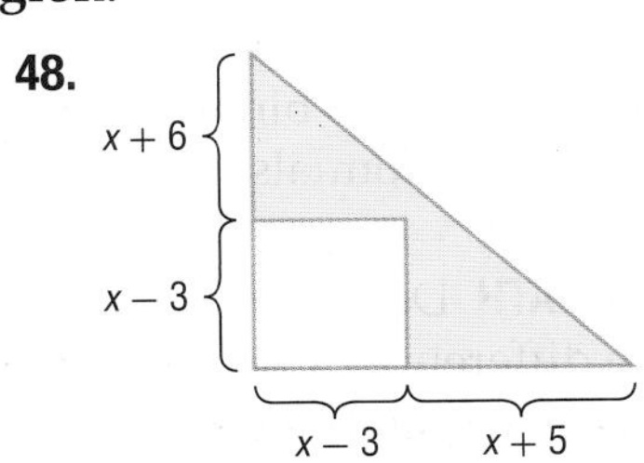

CHAPTER 1

# Mid-Chapter Quiz

## Lessons 1-1 through 1-4

**Determine whether each expression is a polynomial. If it is a polynomial, find the degree and determine whether it is a *monomial*, *binomial*, or *trinomial*.** (Lesson 1-1)

1. $3y^2 - 2$
2. $4t^5 + 3t^2 + t$
3. $\frac{3x}{5y}$
4. $ax^{-3}$
5. $3b^2$
6. $2x^{-3} - 4x + 1$
7. **POPULATION** The table shows the population density for Nevada for various years. (Lesson 1-1)

| Year | Years Since 1930 | People/ Square Mile |
|---|---|---|
| 1930 | 0 | 0.8 |
| 1960 | 30 | 2.6 |
| 1980 | 50 | 7.3 |
| 1990 | 60 | 10.9 |
| 2000 | 70 | 18.2 |

a. The population density $d$ of Nevada from 1930 to 2000 can be modeled by $d = 0.005n^2 - 0.127n + 1$, where $n$ represents the number of years since 1930. Identify the type of polynomial for $0.005n^2 - 0.127n + 1$.

b. What is the degree of the polynomial?

c. Predict the population density of Nevada for 2020 and for 2030. Explain your method.

**Find each sum or difference.** (Lesson 1-1)

8. $(y^2 + 2y + 3) + (y^2 + 3y - 1)$
9. $(3n^3 - 2n + 7) - (n^2 - 2n + 8)$
10. $(5d + d^2) - (4 - 4d^2)$
11. $(x + 4) + (3x + 2x^2 - 7)$
12. $(3a - 3b + 2) - (4a + 5b)$
13. $(8x - y^2 + 3) + (9 - 3x + 2y^2)$

**Find each product.** (Lesson 1-2)

14. $6y(y^2 + 3y + 1)$
15. $3n(n^2 - 5n + 2)$
16. $d^2(-4 - 3d + 2d^2)$
17. $-2xy(3x^2 + 2xy - 4y^2)$
18. $ab^2(12a + 5b - ab)$
19. $x^2y^4(3xy^2 - x + 2y^2)$
20. **MULTIPLE CHOICE** Simplify $x(4x + 5) + 3(2x^2 - 4x + 1)$. (Lesson 1-2)

A $10x^2 + 17x + 3$ C $2x^2 - 7x + 3$
B $10x^2 - 7x + 3$ D $2x^2 + 17x + 3$

**Find each product.** (Lesson 1-3)

21. $(x + 2)(x + 5)$
22. $(3b - 2)(b - 4)$
23. $(n - 5)(n + 3)$
24. $(4c - 2)(c + 2)$
25. $(k - 1)(k - 3k^2)$
26. $(8d - 3)(2d^2 + d + 1)$
27. **MANUFACTURING** A company is designing a box for dry pasta in the shape of a rectangular prism. The length is 2 inches more than twice the width, and the height is 3 inches more than the length. Write an expression in terms of the width for the volume of the box. (Lesson 1-3)

**Find each product.** (Lesson 1-4)

28. $(x + 2)^2$
29. $(n - 11)^2$
30. $(4b - 2)^2$
31. $(6c + 3)^2$
32. $(5d - 3)(5d + 3)$
33. $(9k + 1)(9k - 1)$
34. **DISC GOLF** The discs approved for use in disc golf vary in size. (Lesson 1-4)

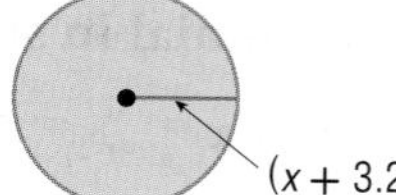


a. Write two different expressions for the area of the largest disc.

b. If $x$ is 10.5, what are the areas of the smallest and largest discs?

EXPLORE 1-5

# Algebra Lab: Factoring Using the Distributive Property

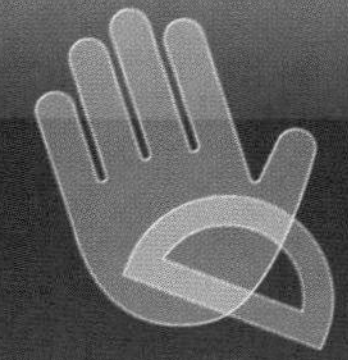

When two or more numbers are multiplied, these numbers are *factors* of the product. Sometimes you know the product of binomials and are asked to find the factors. This is called factoring. You can use algebra tiles and a product mat to factor binomials.

**CCSS Common Core State Standards**
**Content Standards**
**A.SSE.2** Use the structure of an expression to identify ways to rewrite it.

## Activity 1 Use Algebra Tiles to Factor $2x - 8$

**Step 1** Model $2x - 8$.

**Step 2** Arrange the tiles into a rectangle. The total area of the rectangle represents the product, and its length and width represent the factors.

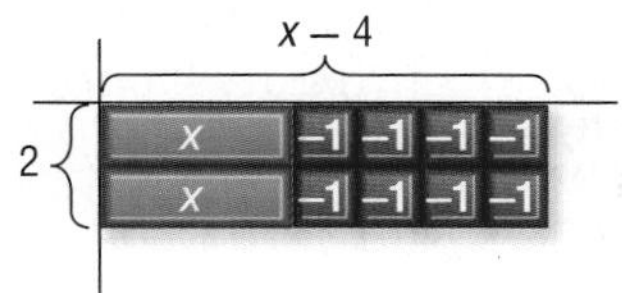


The rectangle has a width of 2 and a length of $x - 4$. Therefore, $2x - 8 = 2(x - 4)$.

## Activity 2 Use Algebra Tiles to Factor $x^2 + 3x$

**Step 1** Model $x^2 + 3x$.

**Step 2** Arrange the tiles into a rectangle.

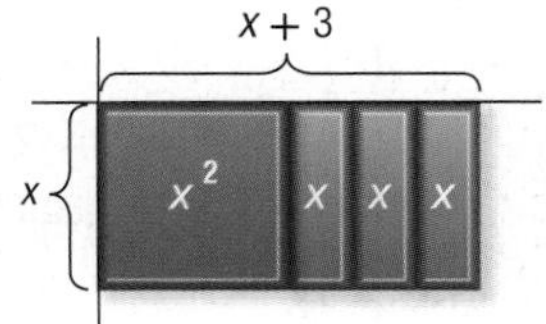


The rectangle has a width of $x$ and a length of $x + 3$. Therefore, $x^2 + 3x = x(x + 3)$.

## Model and Analyze

**Use algebra tiles to factor each binomial.**

**1.** $4x + 12$ **2.** $4x - 6$ **3.** $3x^2 + 4x$ **4.** $10 - 2x$

**Determine whether each binomial can be factored. Justify your answer with a drawing.**

**5.** $6x - 9$ **6.** $5x - 4$ **7.** $4x^2 + 7$ **8.** $x^2 + 3x$

**9.** **WRITING IN MATH** Write a paragraph that explains how you can use algebra tiles to determine whether a binomial can be factored. Include an example of one binomial that can be factored and one that cannot.

LESSON

# 1-5 Using the Distributive Property

**Then**

- Used the Distributive Property to evaluate expressions.

**Now**

1. Use the Distributive Property to factor polynomials.
2. Solve equations of the form $ax^2 + bx = 0$.

**Why?**

- The cost of rent for Mr. Cole's store is determined by the square footage of the space. The area of the store can be modeled by the equation $A = 1.6w^2 + 6w$, where $w$ is the width of the store in feet. We can use factoring and the Zero Product Property to find possible dimensions of the store.

**NewVocabulary**

factoring
factoring by grouping
Zero Product Property

**Common Core State Standards**

**Content Standards**

A.SSE.2 Use the structure of an expression to identify ways to rewrite it.

A.SSE.3a Factor a quadratic expression to reveal the zeros of the function it defines.

**Mathematical Practices**

2 Reason abstractly and quantitatively.

## 1 Use the Distributive Property to Factor

You have used the Distributive Property to multiply a monomial by a polynomial. You can work backward to express a polynomial as the product of a monomial factor and a polynomial factor.

$$1.6w^2 + 6w = 1.6w(w) + 6(w)$$
$$= w(1.6w + 6)$$

So, $w(1.6w + 6)$ is the *factored form* of $1.6w^2 + 6w$. **Factoring** a polynomial involves finding the *completely* factored form.

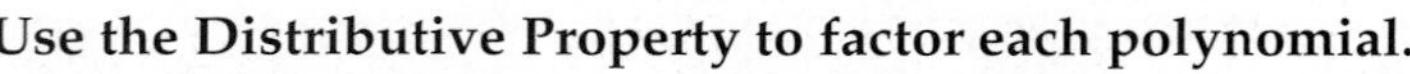

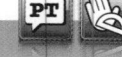

**Example 1** Use the Distributive Property

**Use the Distributive Property to factor each polynomial.**

**a.** $27y^2 + 18y$

Find the GCF of each term.

$27y^2 = 3 \cdot 3 \cdot 3 \cdot y \cdot y$ — Factor each term.

$18y = 2 \cdot 3 \cdot 3 \cdot y$ — Circle common factors.

$\text{GCF} = 3 \cdot 3 \cdot y$ or $9y$

Write each term as the product of the GCF and the remaining factors. Use the Distributive Property to *factor out* the GCF.

$27y^2 + 18y = 9y(3y) + 9y(2)$ — Rewrite each term using the GCF.

$= 9y(3y + 2)$ — Distributive Property

**b.** $-4a^2b - 8ab^2 + 2ab$

$-4a^2b = -1 \cdot 2 \cdot 2 \cdot a \cdot a \cdot b$ — Factor each term.

$-8ab^2 = -1 \cdot 2 \cdot 2 \cdot 2 \cdot a \cdot b \cdot b$ — Circle common factors.

$2ab = 2 \cdot a \cdot b$

$\text{GCF} = 2 \cdot a \cdot b$ or $2ab$

$-4a^2b - 8ab^2 + 2ab = 2ab(-2a) - 2ab(4b) + 2ab(1)$ — Rewrite each term using the GCF.

$= 2ab(-2a - 4b + 1)$ — Distributive Property

**GuidedPractice**

**1A.** $15w - 3v$

**1B.** $7u^2t^2 + 21ut^2 - ut$

Inti St Clair/Blend Images/Getty Images

Using the Distributive Property to factor polynomials with four or more terms is called **factoring by grouping** because terms are put into groups and then factored. The Distributive Property is then applied to a common binomial factor.

### KeyConcept Factoring by Grouping

**Words** A polynomial can be factored by grouping only if all of the following conditions exist.

- There are four or more terms.
- Terms have common factors that can be grouped together.
- There are two common factors that are identical or additive inverses of each other.

**Symbols**

$$\begin{aligned} ax + bx + ay + by &= (ax + bx) + (ay + by) \\ &= x(a + b) + y(a + b) \\ &= (x + y)(a + b) \end{aligned}$$

PT

### Example 2 Factor by Grouping

**Factor $4qr + 8r + 3q + 6$.**

| | |
|---|---|
| $4qr + 8r + 3q + 6$ | Original expression |
| $= (4qr + 8r) + (3q + 6)$ | Group terms with common factors. |
| $= 4r(q + 2) + 3(q + 2)$ | Factor the GCF from each group. |

Notice that $(q + 2)$ is common in both groups, so it becomes the GCF.

| | |
|---|---|
| $= (4r + 3)(q + 2)$ | Distributive Property |

**GuidedPractice**

**Factor each polynomial.**

**2A.** $rn + 5n - r - 5$

**2B.** $3np + 15p - 4n - 20$

It can be helpful to recognize when binomials are additive inverses of each other. For example $6 - a = -1(a - 6)$.

**StudyTip**

Check To check your factored answers, multiply your factors out. You should get your original expression as a result.

PT

### Example 3 Factor by Grouping with Additive Inverses

**Factor $2mk - 12m + 42 - 7k$.**

| | |
|---|---|
| $2mk - 12m + 42 - 7k$ | |
| $= (2mk - 12m) + (42 - 7k)$ | Group terms with common factors. |
| $= 2m(k - 6) + 7(6 - k)$ | Factor the GCF from each group. |
| $= 2m(k - 6) + 7[(-1)(k - 6)]$ | $6 - k = -1(k - 6)$ |
| $= 2m(k - 6) - 7(k - 6)$ | Associative Property |
| $= (2m - 7)(k - 6)$ | Distributive Property |

**GuidedPractice**

**Factor each polynomial.**

**3A.** $c - 2cd + 8d - 4$

**3B.** $3p - 2p^2 - 18p + 27$

## 2 Solve Equations by Factoring

Some equations can be solved by factoring. Consider the following.

$3(0) = 0$ $\qquad$ $0(2 - 2) = 0$ $\qquad$ $-312(0) = 0$ $\qquad$ $0(0.25) = 0$

Notice that in each case, at least one of the factors is 0. These examples are demonstrations of the **Zero Product Property**.

### KeyConcept Zero Product Property

| | |
|---|---|
| Words | If the product of two factors is 0, then at least one of the factors must be 0. |
| Symbols | For any real numbers $a$ and $b$, if $ab = 0$, then $a = 0$, $b = 0$, or both $a$ and $b$ equal zero. |

Recall that a solution or root of an equation is any value that makes the equation true.

**WatchOut!**

Unknown Value It may be tempting to solve an equation by dividing each side by the variable. However, the variable has an unknown value, so you may be dividing by 0, which is undefined.

PT

### Example 4 Solve Equations

**Solve each equation. Check your solutions.**

**a.** $(2d + 6)(3d - 15) = 0$

| | | |
|---|---|---|
| $(2d + 6)(3d - 15) = 0$ | | Original equation |
| $2d + 6 = 0$ or | $3d - 15 = 0$ | Zero Product Property |
| $2d = -6$ | $3d = 15$ | Solve each equation. |
| $d = -3$ | $d = 5$ | Divide. |

The roots are $-3$ and 5.

**CHECK** Substitute $-3$ and 5 for $d$ in the original equation.

| | |
|---|---|
| $(2d + 6)(3d - 15) = 0$ | $(2d + 6)(3d - 15) = 0$ |
| $[2(-3) + 6][3(-3) - 15] \stackrel{?}{=} 0$ | $[2(5) + 6][3(5) - 15] \stackrel{?}{=} 0$ |
| $(-6 + 6)(-9 - 15) \stackrel{?}{=} 0$ | $(10 + 6)(15 - 15) \stackrel{?}{=} 0$ |
| $(0)(-24) \stackrel{?}{=} 0$ | $16(0) \stackrel{?}{=} 0$ |
| $0 = 0$ ✓ | $0 = 0$ ✓ |

**b.** $c^2 = 3c$

| | |
|---|---|
| $c^2 = 3c$ | Original equation |
| $c^2 - 3c = 0$ | Subtract $3c$ from each side to get 0 on one side of the equation. |
| $c(c - 3) = 0$ | Factor by using the GCF to get the form $ab = 0$. |
| $c = 0$ or $c - 3 = 0$ | Zero Product Property |
| $c = 3$ | Solve each equation. |
| The roots are 0 and 3. | Check by substituting 0 and 3 for $c$. |

**GuidedPractice**

**4A.** $3n(n + 2) = 0$ $\qquad$ **4B.** $8b^2 - 40b = 0$ $\qquad$ **4C.** $x^2 = -10x$

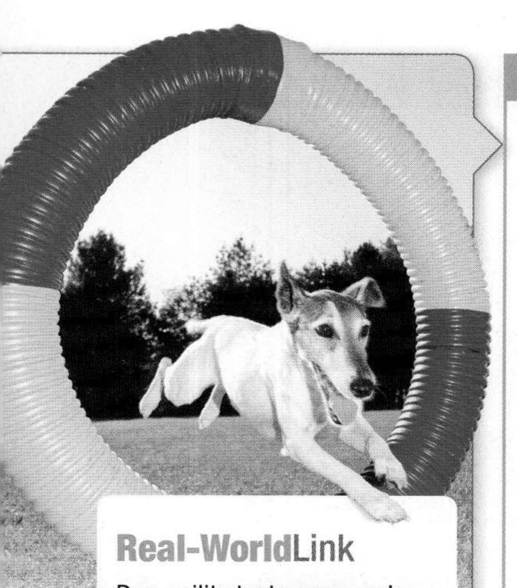

### Real-WorldLink

Dog agility tests a person's skills as a trainer and handler. Competitors race through an obstacle course that includes hurdles, tunnels, a see-saw, and line poles.

**Source:** United States Dog Agility Association

## Real-World Example 5 Use Factoring

**AGILITY Penny is a Fox Terrier who competes with her trainer in the agility course. Within the course, Penny must leap over a hurdle. Penny's jump can be modeled by the equation $h = -16t^2 + 20t$, where $h$ is the height of the leap in inches at $t$ seconds. Find the values of $t$ when $h = 0$.**

| | |
|---|---|
| $h = -16t^2 + 20t$ | Original equation |
| $0 = -16t^2 + 20t$ | Substitution, $h = 0$ |
| $0 = 4t(-4t + 5)$ | Factor by using the GCF. |
| $4t = 0$ or $-4t + 5 = 0$ | Zero Product Property |
| $t = 0$ $\quad -4t = -5$ | Solve each equation. |
| $t = \frac{5}{4}$ or 1.25 | Divide each side by $-4$. |

Penny's height is 0 inches at 0 seconds and 1.25 seconds into the jump.

### GuidedPractice

**5.** **KANGAROOS** The hop of a kangaroo can be modeled by $h = 24t - 16t^2$ where $h$ represents the height of the hop in meters and $t$ is the time in seconds. Find the values of $t$ when $h = 0$.

## Check Your Understanding

**Example 1** **Use the Distributive Property to factor each polynomial.**

**1.** $21b - 15a$

**2.** $14c^2 + 2c$

**3.** $10g^2h^2 + 9gh^2 - g^2h$

**4.** $12jk^2 + 6j^2k + 2j^2k^2$

**Examples 2–3** **Factor each polynomial.**

**5.** $np + 2n + 8p + 16$

**6.** $xy - 7x + 7y - 49$

**7.** $3bc - 2b - 10 + 15c$

**8.** $9fg - 45f - 7g + 35$

**Example 4** **Solve each equation. Check your solutions.**

**9.** $3k(k + 10) = 0$

**10.** $(4m + 2)(3m - 9) = 0$

**11.** $20p^2 - 15p = 0$

**12.** $r^2 = 14r$

**Example 5**

**13.** **SPIDERS** Jumping spiders can commonly be found in homes and barns throughout the United States. A jumping spider's jump can be modeled by the equation $h = 33.3t - 16t^2$, where $t$ represents the time in seconds and $h$ is the height in feet.

**a.** When is the spider's height at 0 feet?

**b.** What is the spider's height after 1 second? after 2 seconds?

**14.** **CCSS REASONING** At a Fourth of July celebration, a rocket is launched straight up with an initial velocity of 125 feet per second. The height $h$ of the rocket in feet above sea level is modeled by the formula $h = 125t - 16t^2$, where $t$ is the time in seconds after the rocket is launched.

**a.** What is the height of the rocket when it returns to the ground?

**b.** Let $h = 0$ in the equation and solve for $t$.

**c.** How many seconds will it take for the rocket to return to the ground?

## Practice and Problem Solving

**Example 1** **Use the Distributive Property to factor each polynomial.**

15. $16t - 40y$
16. $30v + 50x$
17. $2k^2 + 4k$
18. $5z^2 + 10z$
19. $4a^2b^2 + 2a^2b - 10ab^2$
20. $5c^2v - 15c^2v^2 + 5c^2v^3$

**Examples 2–3** **Factor each polynomial.**

21. $fg - 5g + 4f - 20$
22. $a^2 - 4a - 24 + 6a$
23. $hj - 2h + 5j - 10$
24. $xy - 2x - 2 + y$
25. $45pq - 27q - 50p + 30$
26. $24ty - 18t + 4y - 3$
27. $3dt - 21d + 35 - 5t$
28. $8r^2 + 12r$
29. $21th - 3t - 35h + 5$
30. $vp + 12v + 8p + 96$
31. $5br - 25b + 2r - 10$
32. $2nu - 8u + 3n - 12$
33. $5gf^2 + g^2f + 15gf$
34. $rp - 9r + 9p - 81$
35. $27cd^2 - 18c^2d^2 + 3cd$
36. $18r^3t^2 + 12r^2t^2 - 6r^2t$
37. $48tu - 90t + 32u - 60$
38. $16gh + 24g - 2h - 3$

**Example 4** **Solve each equation. Check your solutions.**

39. $3b(9b - 27) = 0$
40. $2n(3n + 3) = 0$
41. $(8z + 4)(5z + 10) = 0$
42. $(7x + 3)(2x - 6) = 0$
43. $b^2 = -3b$
44. $a^2 = 4a$

**Example 5**

45. **CCSS SENSE-MAKING** Use the drawing at the right.
    a. Write an expression in factored form to represent the area of the blue section.
    b. Write an expression in factored form to represent the area of the region formed by the outer edge.
    c. Write an expression in factored form to represent the yellow region.

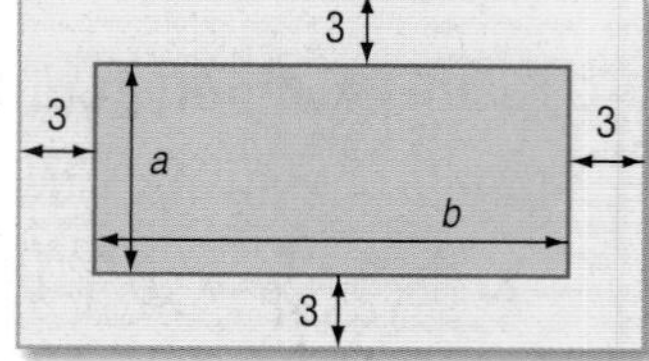


46. **FIREWORKS** A ten-inch fireworks shell is fired from ground level. The height of the shell in feet is given by the formula $h = 263t - 16t^2$, where $t$ is the time in seconds after launch.
    a. Write the expression that represents the height in factored form.
    b. At what time will the height be 0? Is this answer practical? Explain.
    c. What is the height of the shell 8 seconds and 10 seconds after being fired?
    d. At 10 seconds, is the shell rising or falling?

47. **ARCHITECTURE** The frame of a doorway is an arch that can be modeled by the graph of the equation $y = -3x^2 + 12x$, where $x$ and $y$ are measured in feet. On a coordinate plane, the floor is represented by the $x$-axis.
    a. Make a table of values for the height of the arch if $x = 0, 1, 2, 3$, and 4 feet.
    b. Plot the points from the table on a coordinate plane and connect the points to form a smooth curve to represent the arch.
    c. How high is the doorway?

48. **RIDES** Suppose the height of a rider after being dropped can be modeled by $h = -16t^2 - 96t + 160$, where $h$ is the height in feet and $t$ is time in seconds.

a. Write an expression to represent the height in factored form.

b. From what height is the rider initially dropped?

c. At what height will the rider be after 3 seconds of falling? Is this possible? Explain.

49. **ARCHERY** The height $h$ in feet of an arrow can be modeled by the equation $h = 64t - 16t^2$, where $t$ is time in seconds. Ignoring the height of the archer, how long after the arrow is released does it hit the ground?

50. **TENNIS** A tennis player hits a tennis ball upward with an initial velocity of 80 feet per second. The height $h$ in feet of the tennis ball can be modeled by the equation $h = 80t - 16t^2$, where $t$ is time in seconds. Ignoring the height of the tennis player, how long does it take the ball to hit the ground?

51. **MULTIPLE REPRESENTATIONS** In this problem, you will explore the *box method* of factoring. To factor $x^2 + x - 6$, write the first term in the top left-hand corner of the box, and then write the last term in the lower right-hand corner.

| | ? | ? |
|---|---|---|
| ? | $x^2$ | ? |
| ? | ? | $-6$ |

a. **Analytical** Determine which two factors have a product of $-6$ and a sum of 1.

b. **Symbolic** Write each factor in an empty square in the box. Include the positive or negative sign and variable.

c. **Analytical** Find the factor for each row and column of the box. What are the factors of $x^2 + x - 6$?

d. **Verbal** Describe how you would use the box method to factor $x^2 - 3x - 40$.

## H.O.T. Problems Use Higher-Order Thinking Skills

52. **CCSS CRITIQUE** Hernando and Rachel are solving $2m^2 = 4m$. Is either of them correct? Explain your reasoning.

**Hernando**

$2m^2 = 4m$

$\frac{2m^2}{m} = \frac{4m^2}{2m}$

$2m = 2$

$m = 1$

**Rachel**

$2m^2 = 4m$

$2m^2 - 4m = 0$

$2m(m - 2) = 0$

$2m = 0$ or $m - 2 = 0$

$m = 0$ or $2$

53. **CHALLENGE** Given the equation $(ax + b)(ax - b) = 0$, solve for $x$. What do we know about the values of $a$ and $b$?

54. **OPEN ENDED** Write a four-term polynomial that can be factored by grouping. Then factor the polynomial.

55. **REASONING** Given the equation $c = a^2 - ab$, for what values of $a$ and $b$ does $c = 0$?

56. **WRITING IN MATH** Explain how to solve a quadratic equation by using the Zero Product Property.

## Standardized Test Practice

**57.** Which is a factor of $6z^2 - 3z - 2 + 4z$?

A $2z + 1$
B $3z - 2$
C $z + 2$
D $2z - 1$

**58. PROBABILITY** Hailey has 10 blocks: 2 red, 4 blue, 3 yellow, and 1 green. What is the probability that a randomly chosen block will be either red or yellow?

F $\frac{3}{10}$
G $\frac{1}{5}$
H $\frac{1}{2}$
J $\frac{7}{10}$

**59. GRIDDED RESPONSE** Cho is making a 140-inch by 160-inch quilt with quilt squares that measure 8 inches on each side. How many will be needed to make the quilt?

**60. GEOMETRY** The area of the right triangle shown below is $5h$ square centimeters. What is the height of the triangle?

A 2 cm
B 5 cm
C 8 cm
D 10 cm

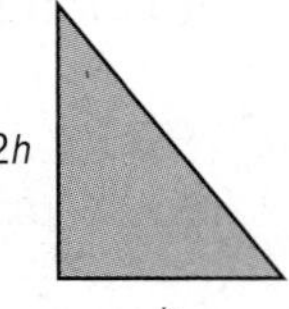


## Spiral Review

**61. GENETICS** Brown genes $B$ are dominant over blue genes $b$. A person with genes $BB$ or $Bb$ has brown eyes. Someone with genes $bb$ has blue eyes. Elisa has brown eyes with $Bb$ genes, and Bob has blue eyes. Write an expression for the possible eye coloring of Elisa and Bob's children. Determine the probability that their child would have blue eyes. (Lesson 1-4)

**Find each product.** (Lesson 1-2)

**62.** $n(n^2 - 4n + 3)$

**63.** $2b(b^2 + b - 5)$

**64.** $-c(4c^2 + 2c - 2)$

**65.** $-4x(x^3 + x^2 + 2x - 1)$

**66.** $2ab(4a^2b + 2ab - 2b^2)$

**67.** $-3xy(x^2 + xy + 2y^2)$

**68. CLASS TRIP** Mr. Wong's American History class will take taxis from their hotel in Washington, D.C., to the Lincoln Memorial. The fare is \$2.75 for the first mile and \$1.25 for each additional mile. If the distance is $m$ miles and $t$ taxis are needed, write an expression for the cost to transport the group. (Lesson 1-2)

**Find the degree of each polynomial.** (Lesson 1-1)

**69.** 2

**70.** $-3a$

**71.** $5x^2 + 3x$

**72.** $d^4 - 6c^2$

**73.** $2x^3 - 4z + 8xz$

**74.** $3d^4 + 5d^3 - 4c^2 + 1$

## Skills Review

**Find each product.**

**75.** $(a + 2)(a + 5)$

**76.** $(d + 4)(d + 10)$

**77.** $(z - 1)(z - 8)$

**78.** $(c + 9)(c - 3)$

**79.** $(x - 7)(x - 6)$

**80.** $(g - 2)(g + 11)$

# Algebra Lab
# Factoring Trinomials

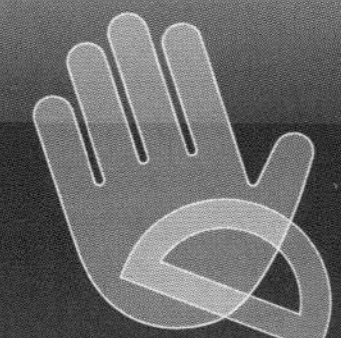

You can use algebra tiles to factor trinomials. If a polynomial represents the area of a rectangle formed by algebra tiles, then the rectangle's length and width are *factors* of the area. If a rectangle cannot be formed to represent the trinomial, then the trinomial is not factorable.

**CCSS Common Core State Standards**
**Content Standards**
**A.SSE.2** Use the structure of an expression to identify ways to rewrite it.

## Activity 1 Factor $x^2 + bx + c$

**Use algebra tiles to factor $x^2 + 4x + 3$.**

**Step 1** Model $x^2 + 4x + 3$.

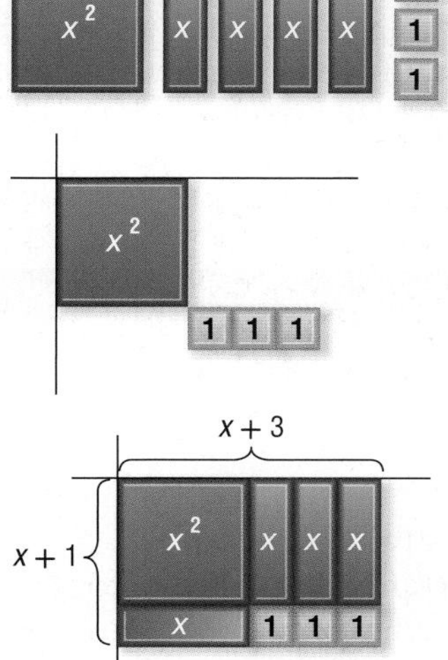


**Step 2** Place the $x^2$-tile at the corner of the product mat. Arrange the 1-tiles into a rectangular array. Because 3 is prime, the 3 tiles can be arranged in a rectangle in one way, a 1-by-3 rectangle.

**Step 3** Complete the rectangle with the $x$-tiles.

The rectangle has a width of $x + 1$ and a length of $x + 3$.

Therefore, $x^2 + 4x + 3 = (x + 1)(x + 3)$.

## Activity 2 Factor $x^2 + bx + c$

**Use algebra tiles to factor $x^2 + 8x + 12$.**

**Step 1** Model $x^2 + 8x + 12$.

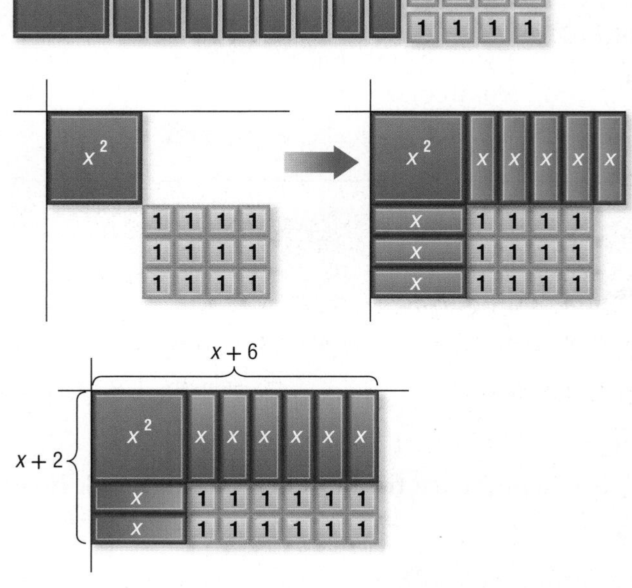


**Step 2** Place the $x^2$-tile at the corner of the product mat. Arrange the 1-tiles into a rectangular array. Since $12 = 3 \times 4$, try a 3-by-4 rectangle. Try to complete the rectangle. Notice that there is an extra $x$-tile.

**Step 3** Arrange the 1-tiles into a 2-by-6 rectangular array. This time you can complete the rectangle with the $x$-tiles.

The rectangle has a width of $x + 2$ and a length of $x + 6$.

Therefore, $x^2 + 8x + 12 = (x + 2)(x + 6)$.

Algebra Lab

# Factoring Trinomials *Continued*

## Activity 3 Factor $x^2 - bx + c$

**Use algebra tiles to factor $x^2 - 5x + 6$.**

**Step 1** Model $x^2 - 5x + 6$.

**Step 2** Place the $x^2$-tile at the corner of the product mat. Arrange the 1-tiles into a 2-by-3 rectangular array as shown.

**Step 3** Complete the rectangle with the $x$-tiles. The rectangle has a width of $x - 2$ and a length of $x - 3$.

Therefore, $x^2 - 5x + 6 = (x - 2)(x - 3)$.

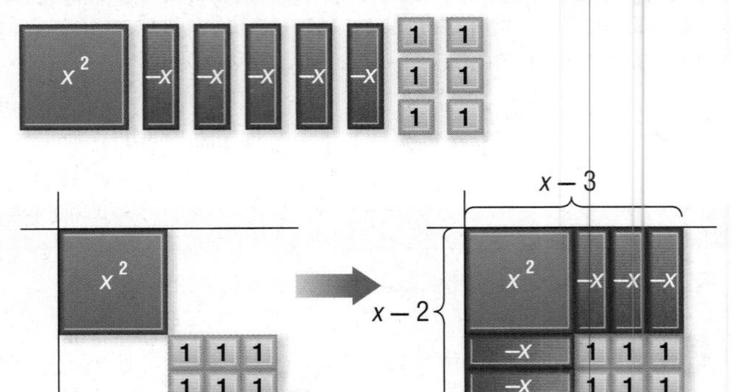


## Activity 4 Factor $x^2 - bx - c$

**Use algebra tiles to factor $x^2 - 4x - 5$.**

**Step 1** Model $x^2 - 4x - 5$.

**Step 2** Place the $x^2$-tile at the corner of the product mat. Arrange the 1-tiles into a 1-by-5 rectangular array as shown.

**Step 3** Place the $x$-tile as shown. Recall that you can add zero pairs without changing the value of the polynomial. In this case, add a zero pair of $x$-tiles.

The rectangle has a width of $x + 1$ and a length of $x - 5$.

Therefore, $x^2 - 4x - 5 = (x + 1)(x - 5)$.

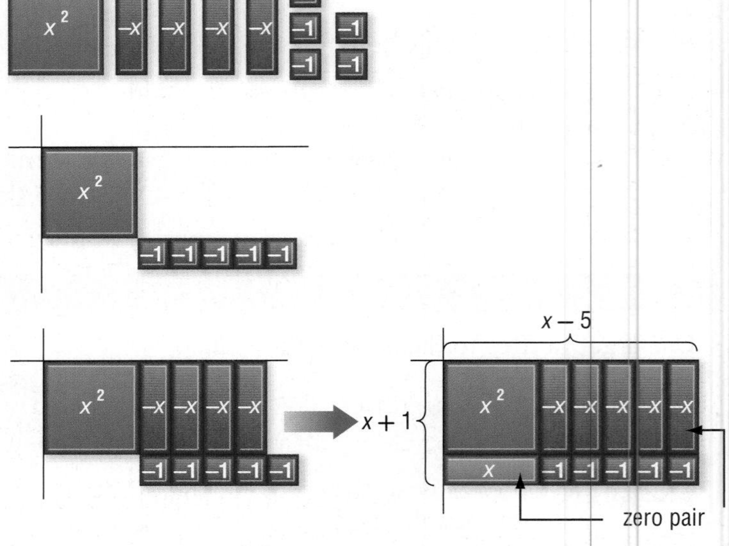


### Model and Analyze

**Use algebra tiles to factor each trinomial.**

**1.** $x^2 + 3x + 2$ **2.** $x^2 + 6x + 8$ **3.** $x^2 + 3x - 4$ **4.** $x^2 - 7x + 12$

**5.** $x^2 + 7x + 10$ **6.** $x^2 - 2x + 1$ **7.** $x^2 + x - 12$ **8.** $x^2 - 8x + 15$

**Tell whether each trinomial can be factored. Justify your answer with a drawing.**

**9.** $x^2 + 3x + 6$ **10.** $x^2 - 5x - 6$ **11.** $x^2 - x - 4$ **12.** $x^2 - 4$

**13.** **WRITING IN MATH** How can you use algebra tiles to determine whether a trinomial can be factored?

# LESSON 1-6 Solving $x^2 + bx + c = 0$

**Then**

- You multiplied binomials by using the FOIL method.

**Now**

1. Factor trinomials of the form $x^2 + bx + c$.
2. Solve equations of the form $x^2 + bx + c = 0$.

**Why?**

- Diana is having a rectangular in-ground swimming pool installed and she wants to include a 24-foot fence around the pool. The pool requires a space of 36 square feet. What dimensions should the pool have?

To solve this problem, the landscape architect needs to find two numbers that have a product of 36 and a sum of 12, half the perimeter of the pool.

**NewVocabulary**
quadratic equation

**Common Core State Standards**

**Content Standards**

**A.SSE.3a** Factor a quadratic expression to reveal the zeros of the function it defines.

**A.REI.4b** Solve quadratic equations by inspection (e.g., for $x^2 = 49$), taking square roots, completing the square, the quadratic formula and factoring, as appropriate to the initial form of the equation. Recognize when the quadratic formula gives complex solutions and write them as $a \pm bi$ for real numbers $a$ and $b$.

**Mathematical Practices**

7 Look for and make use of structure.

8 Look for and express regularity in repeated reasoning.

Corbis InsideOutPix/Photolibrary/Index Stock Imagery

## 1 Factor $x^2 + bx + c$

You have learned how to multiply two binomials using the FOIL method. Each of the binomials was a factor of the product. The pattern for multiplying two binomials can be used to factor certain types of trinomials.

$$\begin{aligned}(x+3)(x+4) &= x^2 + 4x + 3x + 3 \cdot 4 && \text{Use the FOIL method.}\\ &= x^2 + (4+3)x + 3 \cdot 4 && \text{Distributive Property}\\ &= x^2 + 7x + 12 && \text{Simplify.}\end{aligned}$$

Notice that the coefficient of the middle term, $7x$, is the sum of 3 and 4, and the last term, 12, is the product of 3 and 4.

Observe the following pattern in this multiplication.

$$\begin{aligned}(x+3)(x+4) &= x^2 + (4+3)x + (3 \cdot 4) &&\\ (x+m)(x+p) &= x^2 + (p+m)x + mp && \text{Let } 3 = m \text{ and } 4 = p.\\ &= x^2 + \underbrace{(m+p)x}_{bx} + \underbrace{mp}_{c} && \text{Commutative } (+)\\ & \quad x^2 + bx + c && b = m + p \text{ and } c = mp\end{aligned}$$

Notice that the coefficient of the middle term is the sum of $m$ and $p$, and the last term is the product of $m$ and $p$. This pattern can be used to factor trinomials of the form $x^2 + bx + c$.

**KeyConcept Factoring $x^2 + bx + c$**

| | |
|---|---|
| **Words** | To factor trinomials in the form $x^2 + bx + c$, find two integers, $m$ and $p$, with a sum of $b$ and a product of $c$. Then write $x^2 + bx + c$ as $(x + m)(x + p)$. |
| **Symbols** | $x^2 + bx + c = (x + m)(x + p)$ when $m + p = b$ and $mp = c$. |
| **Example** | $x^2 + 6x + 8 = (x + 2)(x + 4)$, because $2 + 4 = 6$ and $2 \cdot 4 = 8$. |

When $c$ is positive, its factors have the same signs. Both of the factors are positive or negative based upon the sign of $b$. If $b$ is positive, the factors are positive. If $b$ is negative, the factors are negative.

## Example 1 $b$ and $c$ are Positive

**Problem-SolvingTip**

Guess and Check When factoring a trinomial, make an educated guess, check for reasonableness, and then adjust the guess until the correct answer is found.

**Factor $x^2 + 9x + 20$.**

In this trinomial, $b = 9$ and $c = 20$. Since $c$ is positive and $b$ is positive, you need to find two positive factors with a sum of 9 and a product of 20. Make an organized list of the factors of 20, and look for the pair of factors with a sum of 9.

| Factors of 20 | Sum of Factors |
|---|---|
| 1, 20 | 21 |
| 2, 10 | 12 |
| 4, 5 | 9 |

The correct factors are 4 and 5.

$x^2 + 9x + 20 = (x + m)(x + p)$ Write the pattern.

$= (x + 4)(x + 5)$ $m = 4$ and $p = 5$

**CHECK** You can check this result by multiplying the two factors. The product should be equal to the original expression.

$(x + 4)(x + 5) = x^2 + 5x + 4x + 20$ FOIL Method

$= x^2 + 9x + 20$ ✓ Simplify.

**GuidedPractice**

**Factor each polynomial.**

**1A.** $d^2 + 11x + 24$

**1B.** $9 + 10t + t^2$

When factoring a trinomial in which $b$ is negative and $c$ is positive, use what you know about the product of binomials to narrow the list of possible factors.

## Example 2 $b$ is Negative and $c$ is Positive

**Factor $x^2 - 8x + 12$. Confirm your answer using a graphing calculator.**

In this trinomial, $b = -8$ and $c = 12$. Since $c$ is positive and $b$ is negative, you need to find two negative factors with a sum of $-8$ and a product of 12.

| Factors of 12 | Sum of Factors |
|---|---|
| −1, −12 | −13 |
| −2, −6 | −8 |
| −3, −4 | −7 |

The correct factors are −2 and −6.

**StudyTip**

CCSS Regularity Once the correct factors are found, it is not necessary to test any other factors. In Example 2, −2 and −6 are the correct factors, so −3 and −4 do not need to be tested.

$x^2 - 8x + 12 = (x + m)(x + p)$ Write the pattern.

$= (x - 2)(x - 6)$ $m = -2$ and $p = -6$

**CHECK** Graph $y = x^2 - 8x + 12$ and $y = (x - 2)(x - 6)$ on the same screen. Since only one graph appears, the two graphs must coincide. Therefore, the trinomial has been factored correctly. ✓

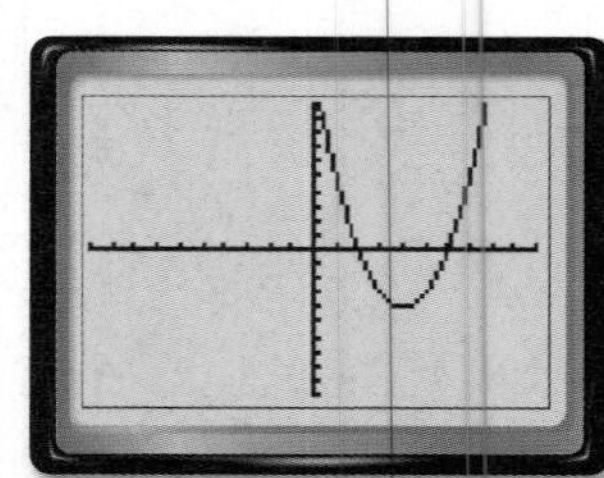

[−10, 10] scl: 1 by [−10, 10] scl: 1

**GuidedPractice**

**Factor each polynomial.**

**2A.** $21 - 22m + m^2$

**2B.** $w^2 - 11w + 28$

**ReviewVocabulary**

**absolute value** the distance a number is from zero on a number line, written $|n|$

When $c$ is negative, its factors have opposite signs. To determine which factor is positive and which is negative, look at the sign of $b$. The factor with the greater absolute value has the same sign as $b$.

## Example 3 $c$ is Negative

**Factor each polynomial. Confirm your answers using a graphing calculator.**

**a.** $x^2 + 2x - 15$

In this trinomial, $b = 2$ and $c = -15$. Since $c$ is negative, the factors $m$ and $p$ have opposite signs. So either $m$ or $p$ is negative, but not both. Since $b$ is positive, the factor with the greater absolute value is also positive.

List the factors of $-15$, where one factor of each pair is negative. Look for the pair of factors with a sum of 2.

| Factors of −15 | Sum of Factors |
|---|---|
| −1, 15 | 14 |
| −3, 5 | 2 |

The correct factors are −3 and 5.

$x^2 + 2x - 15 = (x + m)(x + p)$ Write the pattern.

$= (x - 3)(x + 5)$ $m = -3$ and $p = 5$

**CHECK** $(x - 3)(x + 5) = x^2 + 5x - 3x - 15$ FOIL Method

$= x^2 + 2x - 15$ ✓ Simplify.

**b.** $x^2 - 7x - 18$

In this trinomial, $b = -7$ and $c = -18$. Either $m$ or $p$ is negative, but not both. Since $b$ is negative, the factor with the greater absolute value is also negative.

List the factors of $-18$, where one factor of each pair is negative. Look for the pair of factors with a sum of $-7$.

| Factors of −18 | Sum of Factors |
|---|---|
| 1, −18 | −17 |
| 2, −9 | −7 |
| 3, −6 | −3 |

The correct factors are 2 and −9.

$x^2 - 7x - 18 = (x + m)(x + p)$ Write the pattern.

$= (x + 2)(x - 9)$ $m = 2$ and $p = -9$

**CHECK** Graph $y = x^2 - 7x - 18$ and $y = (x + 2)(x - 9)$ on the same screen.

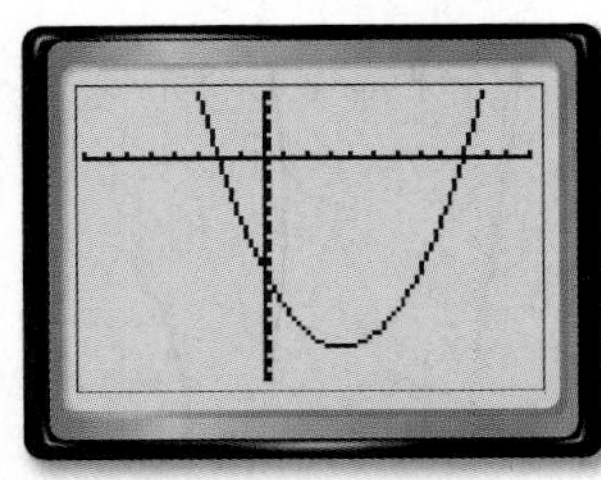

[−10, 15] scl: 1 by [−40, 20] scl: 1

The graphs coincide. Therefore, the trinomial has been factored correctly. ✓

**GuidedPractice**

**3A.** $y^2 + 13y - 48$

**3B.** $r^2 - 2r - 24$

## 2 Solve Equations by Factoring

A **quadratic equation** can be written in the standard form $ax^2 + bx + c = 0$, where $a \neq 0$. Some equations of the form $x^2 + bx + c = 0$ can be solved by factoring and then using the Zero Product Property.

PT

### Example 4 Solve an Equation by Factoring

**Solve $x^2 + 6x = 27$. Check your solutions.**

| | |
|---|---|
| $x^2 + 6x = 27$ | Original equation |
| $x^2 + 6x - 27 = 0$ | Subtract 27 from each side. |
| $(x - 3)(x + 9) = 0$ | Factor. |
| $x - 3 = 0$ or $x + 9 = 0$ | Zero Product Property |
| $x = 3$ $\quad x = -9$ | Solve each equation. |

The roots are 3 and −9.

**CHECK** Substitute 3 and −9 for $x$ in the original equation.

$$x^2 + 6x = 27$$
$$(3)^2 + 6(3) \stackrel{?}{=} 27$$
$$9 + 18 \stackrel{?}{=} 27$$
$$27 = 27 \checkmark$$

$$x^2 + 6x = 27$$
$$(-9)^2 + 6(-9) \stackrel{?}{=} 27$$
$$81 - 54 \stackrel{?}{=} 27$$
$$27 = 27 \checkmark$$

**StudyTip**

**Solving an Equation By Factoring** Remember to get 0 on one side of the equation before factoring.

**GuidedPractice**

**Solve each equation. Check your solutions.**

**4A.** $z^2 - 3z = 70$

**4B.** $x^2 + 3x - 18 = 0$

Factoring can be useful when solving real-world problems.

PT

### Real-World Example 5 Solve a Problem by Factoring

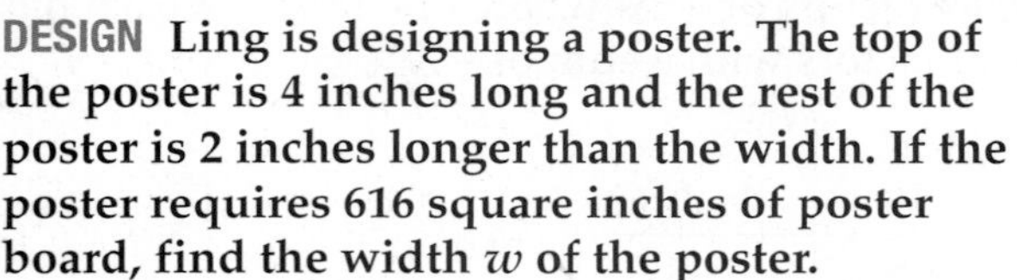

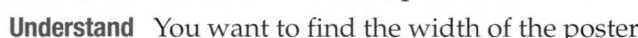

**DESIGN Ling is designing a poster. The top of the poster is 4 inches long and the rest of the poster is 2 inches longer than the width. If the poster requires 616 square inches of poster board, find the width $w$ of the poster.**

**Understand** You want to find the width of the poster.

**Plan** Since the poster is a rectangle, width • length = area.

**Solve** Let $w$ = the width of the poster. The length is $w + 2 + 4$ or $w + 6$.

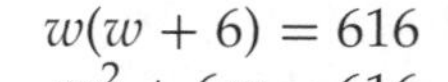

| | |
|---|---|
| $w(w + 6) = 616$ | Write the equation. |
| $w^2 + 6w = 616$ | Multiply. |
| $w^2 + 6w - 616 = 0$ | Subtract 616 from each side. |
| $(w + 28)(w - 22) = 0$ | Factor. |
| $w + 28 = 0$ or $w - 22 = 0$ | Zero Product Property |
| $w = -28$ $\quad w = 22$ | Solve each equation. |

Since dimensions cannot be negative, the width is 22 inches.

**Check** If the width is 22 inches, then the area of the poster is $22 \cdot (22 + 6)$ or 616 square inches, which is the amount the poster requires. ✓

GO TIGERS

4

$w + 2$

$w$

**Real-WorldLink**

A company that produces event signs recommends foamcore boards for event signs that will be used only once. For signs used more than once, use a stronger type of foamcore board.

**Source:** MegaPrint Inc.

GLG3/Photodisc/Getty Images

**GuidedPractice**

**5. GEOMETRY** The height of a parallelogram is 18 centimeters less than its base. If the area is 175 square centimeters, what is its height?

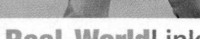

## Check Your Understanding

**Examples 1–3** **Factor each polynomial. Confirm your answers using a graphing calculator.**

**1.** $x^2 + 14x + 24$

**2.** $y^2 - 7y - 30$

**3.** $n^2 + 4n - 21$

**4.** $m^2 - 15m + 50$

**Example 4** **Solve each equation. Check your solutions.**

**5.** $x^2 - 4x - 21 = 0$

**6.** $n^2 - 3n + 2 = 0$

**7.** $x^2 - 15x + 54 = 0$

**8.** $x^2 + 12x = -32$

**9.** $x^2 - x - 72 = 0$

**10.** $x^2 - 10x = -24$

**Example 5** **11. FRAMING** Tina bought a frame for a photo, but the photo is too big for the frame. Tina needs to reduce the width and length of the photo by the same amount. The area of the photo should be reduced to half the original area. If the original photo is 12 inches by 16 inches, what will be the dimensions of the smaller photo?

## Practice and Problem Solving

**Examples 1–3** **Factor each polynomial. Confirm your answers using a graphing calculator.**

**12.** $x^2 + 17x + 42$

**13.** $y^2 - 17y + 72$

**14.** $a^2 + 8a - 48$

**15.** $n^2 - 2n - 35$

**16.** $44 + 15h + h^2$

**17.** $40 - 22x + x^2$

**18.** $-24 - 10x + x^2$

**19.** $-42 - m + m^2$

**Example 4** **Solve each equation. Check your solutions.**

**20.** $x^2 - 7x + 12 = 0$

**21.** $y^2 + y = 20$

**22.** $x^2 - 6x = 27$

**23.** $a^2 + 11a = -18$

**24.** $c^2 + 10c + 9 = 0$

**25.** $x^2 - 18x = -32$

**26.** $n^2 - 120 = 7n$

**27.** $d^2 + 56 = -18d$

**28.** $y^2 - 90 = 13y$

**29.** $h^2 + 48 = 16h$

**Example 5** **30. GEOMETRY** A triangle has an area of 36 square feet. If the height of the triangle is 6 feet more than its base, what are its height and base?

**31. GEOMETRY** A rectangle has an area represented by $x^2 - 4x - 12$ square feet. If the length is $x + 2$ feet, what is the width of the rectangle?

**32. SOCCER** The width of a high school soccer field is 45 yards shorter than its length.

**a.** Define a variable, and write an expression for the area of the field.

**b.** The area of the field is 9000 square yards. Find the dimensions.

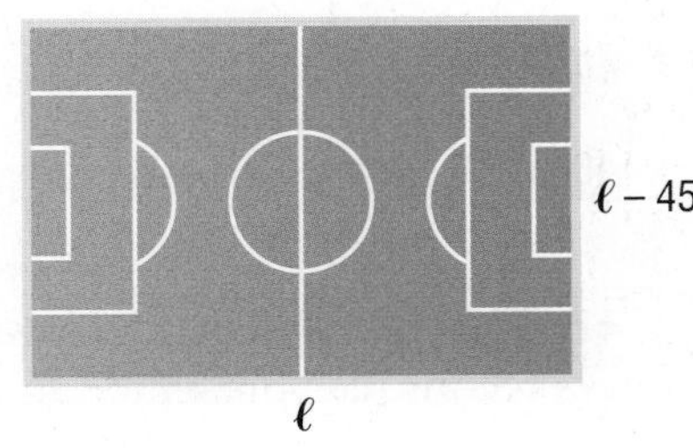


**CCSS STRUCTURE** **Factor each polynomial.**

**33.** $q^2 + 11qr + 18r^2$

**34.** $x^2 - 14xy - 51y^2$

**35.** $x^2 - 6xy + 5y^2$

**36.** $a^2 + 10ab - 39b^2$

**37** **SWIMMING** The length of a rectangular swimming pool is 20 feet greater than its width. The area of the pool is 525 square feet.

a. Define a variable and write an equation for the area of the pool.

b. Solve the equation.

c. Interpret the solutions. Do both solutions make sense? Explain.

**GEOMETRY** **Find an expression for the perimeter of a rectangle with the given area.**

**38.** $A = x^2 + 24x - 81$

**39.** $A = x^2 + 13x - 90$

**40.** **MULTIPLE REPRESENTATIONS** In this problem, you will explore factoring when the leading coefficient is not 1.

a. **Tabular** Copy and complete the table below.

| Product of Two Binomials | $ax^2 + mx + px + c$ | $ax^2 + bx + c$ | $m \times p$ | $a \times c$ |
|---|---|---|---|---|
| $(2x + 3)(x + 4)$ | $2x^2 + 8x + 3x + 12$ | $2x^2 + 11x + 12$ | 24 | 24 |
| $(x + 1)(3x + 5)$ | | | | |
| $(2x - 1)(4x + 1)$ | | | | |
| $(3x + 5)(4x - 2)$ | | | | |

b. **Analytical** How are $m$ and $p$ related to $a$ and $c$?

c. **Analytical** How are $m$ and $p$ related to $b$?

d. **Verbal** Describe a process you can use for factoring a polynomial of the form $ax^2 + bx + c$.

## H.O.T. Problems Use Higher-Order Thinking Skills

**41.** **ERROR ANALYSIS** Jerome and Charles have factored $x^2 + 6x - 16$. Is either of them correct? Explain your reasoning.

| Jerome | Charles |
|---|---|
| $x^2 + 6x - 16 = (x + 2)(x - 8)$ | $x^2 + 6x - 16 = (x - 2)(x + 8)$ |

**CCSS ARGUMENTS** **Find all values of $k$ so that each polynomial can be factored using integers.**

**42.** $x^2 + kx - 19$

**43.** $x^2 + kx + 14$

**44.** $x^2 - 8x + k, k > 0$

**45.** $x^2 - 5x + k, k > 0$

**46.** **REASONING** For any factorable trinomial, $x^2 + bx + c$, will the absolute value of $b$ *sometimes*, *always*, or *never* be less than the absolute value of $c$? Explain.

**47.** **OPEN ENDED** Give an example of a trinomial that can be factored using the factoring techniques presented in this lesson. Then factor the trinomial.

**48.** **CHALLENGE** Factor $(4y - 5)^2 + 3(4y - 5) - 70$.

**49.** **WRITING IN MATH** Explain how to factor trinomials of the form $x^2 + bx + c$ and how to determine the signs of the factors of $c$.

## Standardized Test Practice

**50.** Which inequality is shown in the graph below?

A $y \leq -\frac{3}{4}x + 3$

B $y < -\frac{3}{4}x + 3$

C $y > -\frac{3}{4}x + 3$

D $y \geq -\frac{3}{4}x + 3$

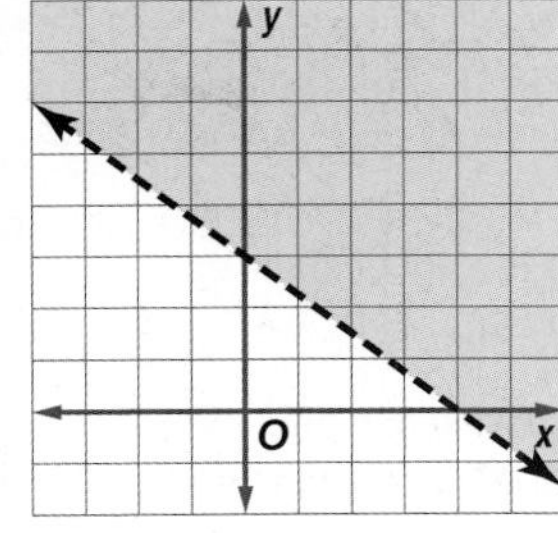


**51. SHORT RESPONSE** Olivia must earn more than \$254 from selling candy bars in order to go on a trip with the National Honor Society. If each candy bar is sold for \$1.25, what is the fewest candy bars she must sell?

**52. GEOMETRY** Which expression represents the length of the rectangle?

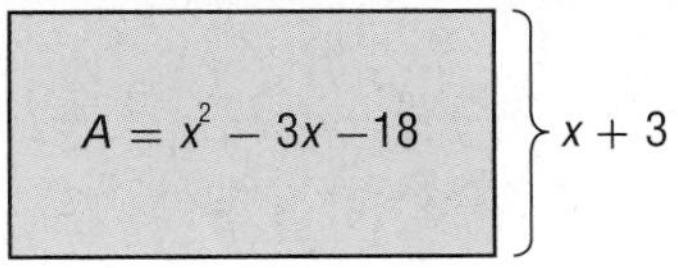

$A = x^2 - 3x - 18$ ; width $x + 3$

F $x + 5$

G $x + 6$

H $x - 6$

J $x - 5$

**53.** The difference of 21 and a number $n$ is 6. Which equation shows the relationship?

A $21 - n = 6$

B $21 + n = 6$

C $21n = 6$

D $6n = -21$

## Spiral Review

**Factor each polynomial.** (Lesson 1-5)

**54.** $10a^2 + 40a$

**55.** $11x + 44x^2y$

**56.** $2m^3p^2 - 16mp^2 + 8mp$

**57.** $2ax + 6xc + ba + 3bc$

**58.** $8ac - 2ad + 4bc - bd$

**59.** $x^2 - xy - xy + y^2$

**60.** Write a polynomial that represents the area of the shaded region in the figure at the right. (Lesson 1-4)

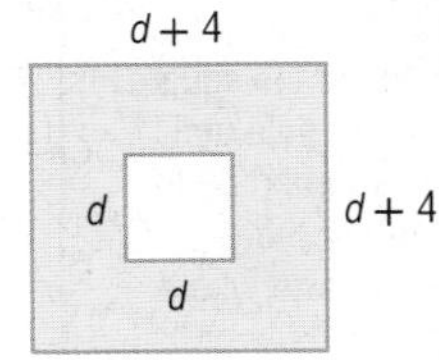


**Refer to the conversion charts inside the back cover of your textbook and in Lesson 0-2.** (Lesson 0-2)

**61. RUNNING** Ling is participating in a 5-kilometer charity run next weekend. About how many miles is the race?

**62. NATURE** An African elephant weighs about 9 tons. About how many kilograms is this?

**63. SPORTS** A football field is 100 yards long from one end zone to the other. How many feet long is a football field?

## Skills Review

**Factor each polynomial.**

**64.** $6mx - 4m + 3rx - 2r$

**65.** $3ax - 6bx + 8b - 4a$

**66.** $2d^2g + 2fg + 4d^2h + 4fh$

LESSON 1-7

# Solving $ax^2 + bx + c = 0$

**Then**

- You factored trinomials of the form $x^2 + bx + c$.

**Now**

1. Factor trinomials of the form $ax^2 + bx + c$.
2. Solve equations of the form $ax^2 + bx + c = 0$.

**Why?**

- The path of a rider on the amusement park ride shown at the right can be modeled by $16t^2 - 5t + 120$.

  Factoring this expression can help the ride operators determine how long a rider rides on the initial swing.

**NewVocabulary**
prime polynomial

**Common Core State Standards**

**Content Standards**

A.SSE.3a Factor a quadratic expression to reveal the zeros of the function it defines.

A.REI.4b Solve quadratic equations by inspection (e.g., for $x^2 = 49$), taking square roots, completing the square, the quadratic formula and factoring, as appropriate to the initial form of the equation. Recognize when the quadratic formula gives complex solutions and write them as $a \pm bi$ for real numbers $a$ and $b$.

**Mathematical Practices**

4 Model with mathematics.

## 1 Factor $ax^2 + bx + c$

In the last lesson, you factored quadratic expressions of the form $ax^2 + bx + c$, where $a = 1$. In this lesson, you will apply the factoring methods to quadratic expressions in which $a$ is not 1.

The dimensions of the rectangle formed by the algebra tiles are the factors of $2x^2 + 5x + 3$. The factors of $2x^2 + 5x + 3$ are $x + 1$ and $2x + 3$.

You can also use the method of factoring by grouping to solve this expression.

**Step 1** Apply the pattern: $2x^2 + 5x + 3 = 2x^2 + mx + px + 3$.

**Step 2** Find two numbers that have a product of $2 \cdot 3$ or 6 and a sum of 5.

| Factors of 6 | Sum of Factors |
|---|---|
| 1, 6 | 7 |
| 2, 3 | 5 |

**Step 3** Use grouping to find the factors.

$$2x^2 + 5x + 3 = 2x^2 + mx + px + 3 \quad \text{Write the pattern.}$$
$$= 2x^2 + 2x + 3x + 3 \quad m = 2 \text{ and } p = 3$$
$$= (2x^2 + 2x) + (3x + 3) \quad \text{Group terms with common factors.}$$
$$= 2x(x + 1) + 3(x + 1) \quad \text{Factor the GCF.}$$
$$= (2x + 3)(x + 1) \quad x + 1 \text{ is the common factor.}$$

Therefore, $2x^2 + 5x + 3 = (2x + 3)(x + 1)$.

### KeyConcept Factoring $ax^2 + bx + c$

**Words** To factor trinomials of the form $ax^2 + bx + c$, find two integers, $m$ and $p$, with a sum of $b$ and a product of $ac$. Then write $ax^2 + bx + c$ as $ax^2 + mx + px + c$, and factor by grouping.

**Example**

$$5x^2 - 13x + 6 = 5x^2 - 10x - 3x + 6 \quad m = -10 \text{ and } p = -3$$
$$= 5x(x - 2) + (-3)(x - 2)$$
$$= (5x - 3)(x - 2)$$

ARCO/W Dieterich/age fotostock

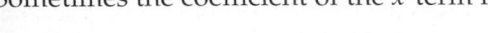

## Example 1 Factor $ax^2 + bx + c$

**Factor each trinomial.**

**a.** $7x^2 + 29x + 4$

In this trinomial, $a = 7$, $b = 29$, and $c = 4$. You need to find two numbers with a sum of 29 and a product of $7 \cdot 4$ or 28. Make a list of the factors of 28 and look for the pair of factors with the sum of 29.

| Factors of 28 | Sum of Factors |
|---|---|
| 1, 28 | 29 |

The correct factors are 1 and 28.

$7x^2 + 29x + 4 = 7x^2 + mx + px + 4$ — Write the pattern.

$= 7x^2 + 1x + 28x + 4$ — $m = 1$ and $p = 28$

$= (7x^2 + 1x) + (28x + 4)$ — Group terms with common factors.

$= x(7x + 1) + 4(7x + 1)$ — Factor the GCF.

$= (x + 4)(7x + 1)$ — $7x + 1$ is the common factor.

**b.** $3x^2 + 15x + 18$

The GCF of the terms $3x^2$, $15x$, and 18 is 3. Factor this first.

$3x^2 + 15x + 18 = 3(x^2 + 5x + 6)$ — Distributive Property

$= 3(x + 3)(x + 2)$ — Find two factors of 6 with a sum of 5.

**StudyTip**

**Greatest Common Factor** Always look for a GCF of the terms of a polynomial before you factor.

**GuidedPractice**

**1A.** $5x^2 + 13x + 6$

**1B.** $6x^2 + 22x - 8$

Sometimes the coefficient of the $x$-term is negative.

**Real-WorldCareer**

**Urban Planner** Urban planners design the layout of an area. They take into consideration the available land and geographical and environmental factors to design an area that benefits the community the most. City planners have a bachelor's degree in planning and almost half have a master's degree.

## Example 2 Factor $ax^2 - bx + c$

**Factor $3x^2 - 17x + 20$.**

In this trinomial, $a = 3$, $b = -17$, and $c = 20$. Since $b$ is negative, $m + p$ will be negative. Since $c$ is positive, $mp$ will be positive.

To determine $m$ and $p$, list the negative factors of $ac$ or 60. The sum of $m$ and $p$ should be $-17$.

| Factors of 60 | Sum of Factors |
|---|---|
| −2, −30 | −32 |
| −3, −20 | −23 |
| −4, −15 | −19 |
| −5, −12 | −17 |

The correct factors are −5 and −12.

$3x^2 - 17x + 20 = 3x^2 - 12x - 5x + 20$ — $m = -12$ and $p = -5$

$= (3x^2 - 12x) + (-5x + 20)$ — Group terms with common factors.

$= 3x(x - 4) + (-5)(x - 4)$ — Factor the GCF.

$= (3x - 5)(x - 4)$ — Distributive Property

**GuidedPractice**

**2A.** $2n^2 - n - 1$

**2B.** $10y^2 - 35y + 30$

LWA/Dann Tardif/Blend Images/Getty Images

A polynomial that cannot be written as a product of two polynomials with integral coefficients is called a **prime polynomial**.

PT

### Example 3 Determine Whether a Polynomial is Prime

**Factor $4x^2 - 3x + 5$, if possible. If the polynomial cannot be factored using integers, write *prime*.**

In this trinomial, $a = 4$, $b = -3$, and $c = 5$. Since $b$ is negative, $m + p$ is negative. Since $c$ is positive, $mp$ is positive. So, $m$ and $p$ are both negative. Next, list the factors of 20. Look for the pair with a sum of $-3$.

| Factors of 20 | Sum of Factors |
|---|---|
| $-20, -1$ | $-21$ |
| $-4, -5$ | $-9$ |
| $-2, -10$ | $-12$ |

There are no factors with a sum of $-3$. So the quadratic expression cannot be factored using integers. Therefore, $4x^2 - 3x + 5$ is prime.

**GuidedPractice**

**Factor each polynomial, if possible. If the polynomial cannot be factored using integers, write *prime*.**

**3A.** $4r^2 - r + 7$

**3B.** $2x^2 + 3x - 5$

**Real-WorldLink**

Cheetahs are the fastest land animals in the world, reaching speeds of up to 70 mph. It can accelerate from 0 to 40 mph in 3 strides. It takes just seconds for the cheetah to reach the full speed of 70 mph.

**Source:** Cheetah Conservation Fund

## 2 Solve Equations by Factoring

A model for the height of a projectile is given by $h = -16t^2 + vt + h_0$, where $h$ is the height in feet, $t$ is the time in seconds, $v$ is the initial upward velocity in feet per second, and $h_0$ is the initial height in feet. Equations of the form $ax^2 + bx + c = 0$ can be solved by factoring and by using the Zero Product Property.

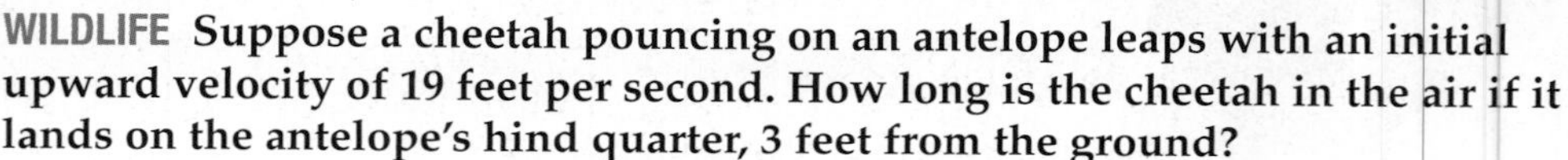

### Real-World Example 4 Solve Equations by Factoring

PT

**WILDLIFE Suppose a cheetah pouncing on an antelope leaps with an initial upward velocity of 19 feet per second. How long is the cheetah in the air if it lands on the antelope's hind quarter, 3 feet from the ground?**

| | |
|---|---|
| $h = -16t^2 + vt + h_0$ | Equation for height |
| $3 = -16t^2 + 19t + 0$ | $h = 3$, $v = 19$, and $h_0 = 0$ |
| $0 = -16t^2 + 19t - 3$ | Subtract 3 from each side. |
| $0 = 16t^2 - 19t + 3$ | Multiply each side by $-1$. |
| $0 = (16t - 3)(t - 1)$ | Factor $16t^2 - 19t + 3$. |
| $16t - 3 = 0$ or $t - 1 = 0$ | Zero Product Property |
| $16t = 3$ $\quad$ $t = 1$ | Solve each equation. |
| $t = \frac{3}{16}$ | |

The solutions are $\frac{3}{16}$ and 1 seconds. It takes the cheetah $\frac{3}{16}$ second to reach a height of 3 feet on his way up. It takes the cheetah 1 second to reach a height of 3 feet on his way down. So, the cheetah is in the air 1 second before he catches the antelope.

**WatchOut!**

**Keep the −1** Do not forget to carry the −1 that was factored out through the rest of the steps or multiply both sides by −1.

**GuidedPractice**

**4. PHYSICAL SCIENCE** A person throws a ball upward from a 506-foot tall building. The ball's height $h$ in feet after $t$ seconds is given by the equation $h = -16t^2 + 48t + 506$. The ball lands on a balcony that is 218 feet above the ground. How many seconds was it in the air?

Anup Shah/Riser/Getty Images

## Check Your Understanding

Examples 1–3 **Factor each polynomial, if possible. If the polynomial cannot be factored using integers, write *prime*.**

1. $3x^2 + 17x + 10$
2. $2x^2 + 22x + 56$
3. $5x^2 - 3x + 4$
4. $3x^2 - 11x - 20$

Example 4 **Solve each equation. Confirm your answers using a graphing calculator.**

5. $2x^2 + 9x + 9 = 0$
6. $3x^2 + 17x + 20 = 0$
7. $3x^2 - 10x + 8 = 0$
8. $2x^2 - 17x + 30 = 0$

9. **CCSS MODELING** Ken throws the discus at a school meet.
   a. What is the initial height of the discus?
   b. After how many seconds does the discus hit the ground?

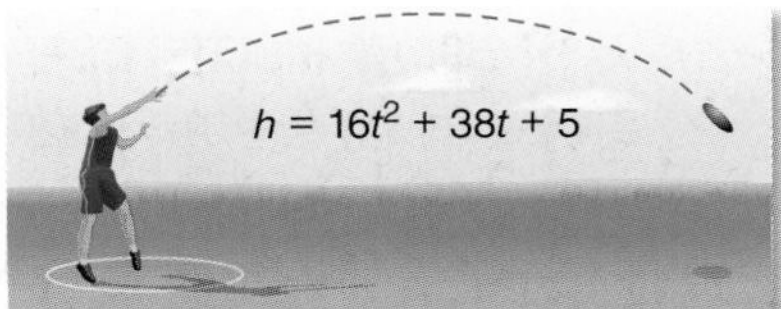


## Practice and Problem Solving

Examples 1–3 **Factor each polynomial, if possible. If the polynomial cannot be factored using integers, write *prime*.**

10. $5x^2 + 34x + 24$
11. $2x^2 + 19x + 24$
12. $4x^2 + 22x + 10$
13. $4x^2 + 38x + 70$
14. $2x^2 - 3x - 9$
15. $4x^2 - 13x + 10$
16. $2x^2 + 3x + 6$
17. $5x^2 + 3x + 4$
18. $12x^2 + 69x + 45$
19. $4x^2 - 5x + 7$
20. $5x^2 + 23x + 24$
21. $3x^2 - 8x + 15$

Example 4

22. **SHOT PUT** An athlete throws a shot put with an initial upward velocity of 29 feet per second and from an initial height of 6 feet.
   a. Write an equation that models the height of the shot put in feet with respect to time in seconds.
   b. After how many seconds will the shot put hit the ground?

**Solve each equation. Confirm your answers using a graphing calculator.**

23. $2x^2 + 9x - 18 = 0$
24. $4x^2 + 17x + 15 = 0$
25. $-3x^2 + 26x = 16$
26. $-2x^2 + 13x = 15$
27. $-3x^2 + 5x = -2$
28. $-4x^2 + 19x = -30$

29. **BASKETBALL** When Jerald shoots a free throw, the ball is 6 feet from the floor and has an initial upward velocity of 20 feet per second. The hoop is 10 feet from the floor.
   a. Use the vertical motion model to determine an equation that models Jerald's free throw.
   b. How long is the basketball in the air before it reaches the hoop?
   c. Raymond shoots a free throw that is 5 foot 9 inches from the floor with the same initial upward velocity. Will the ball be in the air more or less time? Explain.

30. **DIVING** Ben dives from a 36-foot platform. The equation $h = -16t^2 + 14t + 36$ models the dive. How long will it take Ben to reach the water?

31 **NUMBER THEORY** Six times the square of a number $x$ plus 11 times the number equals 2. What are possible values of $x$?

**Factor each polynomial, if possible. If the polynomial cannot be factored using integers, write *prime*.**

32. $-6x^2 - 23x - 20$
33. $-4x^2 - 15x - 14$
34. $-5x^2 + 18x + 8$
35. $-6x^2 + 31x - 35$
36. $-4x^2 + 5x - 12$
37. $-12x^2 + x + 20$

38. **URBAN PLANNING** The city has commissioned the building of a rectangular park. The area of the park can be expressed as $660x^2 + 524x + 85$. Factor this expression to find binomials with integer coefficients that represent possible dimensions of the park. If $x = 8$, what is a possible perimeter of the park?

39. **MULTIPLE REPRESENTATIONS** In this problem, you will explore factoring a special type of polynomial.

  a. **Geometric** Draw a square and label the sides $a$. Within this square, draw a smaller square that shares a vertex with the first square. Label the sides $b$. What are the areas of the two squares?

  b. **Geometric** Cut and remove the small square. What is the area of the remaining region?

  c. **Analytical** Draw a diagonal line between the inside corner and outside corner of the figure, and cut along this line to make two congruent pieces. Then rearrange the two pieces to form a rectangle. What are the dimensions?

  d. **Analytical** Write the area of the rectangle as the product of two binomials.

  e. **Verbal** Complete this statement: $a^2 - b^2 = \ldots$ Why is this statement true?

## H.O.T. Problems Use Higher-Order Thinking Skills

40. **CCSS CRITIQUE** Zachary and Samantha are solving $6x^2 - x = 12$. Is either of them correct? Explain your reasoning.

| Zachary | Samantha |
|---|---|
| $6x^2 - x = 12$ | $6x^2 - x = 12$ |
| $x(6x - 1) = 12$ | $6x^2 - x - 12 = 0$ |
| $x = 12$ or $6x - 1 = 12$ | $(2x - 3)(3x + 4) = 0$ |
| $6x = 13$ | $2x - 3 = 0$ or $3x + 4 = 0$ |
| $x = \frac{13}{6}$ | $x = \frac{3}{2}$  $x = -\frac{4}{3}$ |

41. **REASONING** A square has an area of $9x^2 + 30xy + 25y^2$ square inches. The dimensions are binomials with positive integer coefficients. What is the perimeter of the square? Explain.

42. **CHALLENGE** Find all values of $k$ so that $2x^2 + kx + 12$ can be factored as two binomials using integers.

43. **WRITING IN MATH** What should you consider when solving a quadratic equation that models a real-world situation?

44. **WRITING IN MATH** Explain how to determine which values should be chosen for $m$ and $p$ when factoring a polynomial of the form $ax^2 + bx + c$.

## Standardized Test Practice

**45. Gridded Response** Savannah has two sisters. One sister is 8 years older than her and the other sister is 2 years younger than her. The product of Savannah's sisters' ages is 56. How old is Savannah?

**46.** What is the product of $\frac{2}{3}a^3b^5$ and $\frac{3}{5}a^5b^2$?

A $\frac{2}{5}a^8b^7$

B $\frac{2}{5}a^2b^3$

C $\frac{2}{5}a^8b^3$

D $\frac{2}{5}a^2b^7$

**47.** What is the solution set of $x^2 + 2x - 24 = 0$?

F $\{-4, 6\}$ H $\{-3, 8\}$

G $\{3, -8\}$ J $\{4, -6\}$

**48.** Which is the solution set of $x \geq -2$?

A 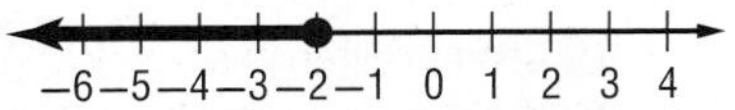


B 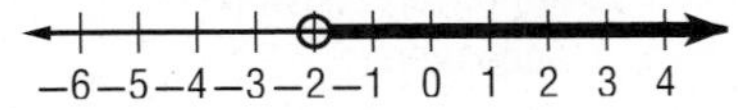


C 

D 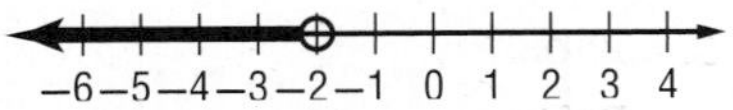


## Spiral Review

**Factor each polynomial.** (Lesson 1-6)

**49.** $x^2 - 9x + 14$

**50.** $n^2 - 8n + 15$

**51.** $x^2 - 5x - 24$

**52.** $z^2 + 15z + 36$

**53.** $r^2 + 3r - 40$

**54.** $v^2 + 16v + 63$

**Solve each equation. Check your solutions.** (Lesson 1-5)

**55.** $a(a - 9) = 0$

**56.** $(2y + 6)(y - 1) = 0$

**57.** $10x^2 - 20x = 0$

**58.** $8b^2 - 12b = 0$

**59.** $15a^2 = 60a$

**60.** $33x^2 = -22x$

**Name an appropriate method to solve each system of equations. Then solve the system.** (Lesson 0-9)

**61.** $-5x + 2y = 13$
$2x + 3y = -9$

**62.** $y = -5x + 7$
$y = 3x - 17$

**63.** $x - 8y = 16$
$7x - 4y = -18$

**Complete each sentence.** (Lesson 0-1)

**64.** 54 in. = __?__ ft

**65.** 275 mm = __?__ m

**66.** 7 gal = __?__ pt

**67. TRUCKS** A sport-utility vehicle has a maximum load limit of 75 pounds for its roof. You want to place a 38-pound cargo carrier and 4 pieces of luggage on top of the roof. Write and solve an inequality to find the average allowable weight for each piece of luggage. (Lesson 0-6)

## Skills Review

**Find the principal square root of each number.**

**68.** 16

**69.** 36

**70.** 64

**71.** 81

**72.** 121

**73.** 100

# LESSON 1-8 Differences of Squares

| Then | Now | Why? |
|---|---|---|
| You factored trinomials into two binomials. | **1** Factor binomials that are the difference of squares.<br>**2** Use the difference of squares to solve equations. | Computer graphics designers use a combination of art and mathematics skills to design images and videos. They use equations to form shapes and lines on computers. Factoring can help to determine the dimensions and shapes of the figures. |

**NewVocabulary**
difference of two squares

**Common Core State Standards**

**Content Standards**

**A.SSE.3a** Factor a quadratic expression to reveal the zeros of the function it defines.

**A.REI.4b** Solve quadratic equations by inspection (e.g., for $x^2 = 49$), taking square roots, completing the square, the quadratic formula and factoring, as appropriate to the initial form of the equation. Recognize when the quadratic formula gives complex solutions and write them as $a \pm bi$ for real numbers $a$ and $b$.

**Mathematical Practices**

1 Make sense of problems and persevere in solving them.

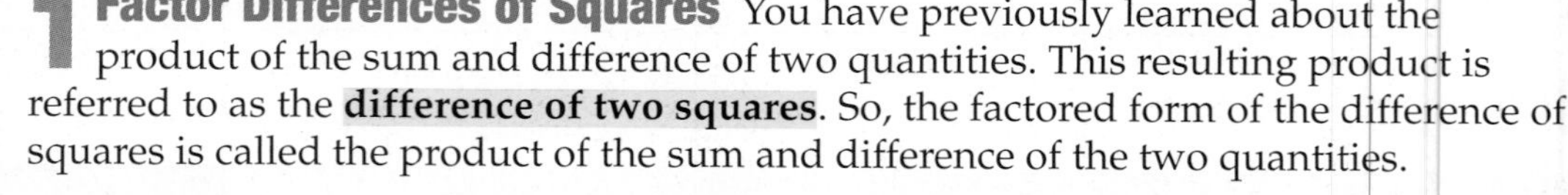

## 1 Factor Differences of Squares

You have previously learned about the product of the sum and difference of two quantities. This resulting product is referred to as the **difference of two squares**. So, the factored form of the difference of squares is called the product of the sum and difference of the two quantities.

**KeyConcept** Difference of Squares

**Symbols** $a^2 - b^2 = (a + b)(a - b)$ or $(a - b)(a + b)$

**Examples** $x^2 - 25 = (x + 5)(x - 5)$ or $(x - 5)(x + 5)$

$t^2 - 64 = (t + 8)(t - 8)$ or $(t - 8)(t + 8)$

PT

### Example 1 Factor Differences of Squares

**Factor each polynomial.**

**a.** $16h^2 - 9a^2$

$16h^2 - 9a^2 = (4h)^2 - (3a)^2$ Write in the form of $a^2 - b^2$.

$= (4h + 3a)(4h - 3a)$ Factor the difference of squares.

**b.** $121 - 4b^2$

$121 - 4b^2 = (11)^2 - (2b)^2$ Write in the form of $a^2 - b^2$.

$= (11 - 2b)(11 + 2b)$ Factor the difference of squares.

**c.** $27g^3 - 3g$

Because the terms have a common factor, factor out the GCF first. Then proceed with other factoring techniques.

$27g^3 - 3g = 3g(9g^2 - 1)$ Factor out the GCF of $3g$.

$= 3g[(3g)^2 - (1)^2]$ Write in the form $a^2 - b^2$.

$= 3g(3g - 1)(3g + 1)$ Factor the difference of squares.

**GuidedPractice**

**1A.** $81 - c^2$

**1B.** $64g^2 - h^2$

**1C.** $9x^3 - 4x$

**1D.** $-4y^3 + 9y$

Ciaran Griffin/Lifesize/Getty Images

To factor a polynomial completely, a technique may need to be applied more than once. This also applies to the difference of squares pattern.

PT

## Example 2 Apply a Technique More than Once

**Factor each polynomial.**

**a.** $b^4 - 16$

$b^4 - 16 = (b^2)^2 - (4)^2$ Write $b^4 - 16$ in $a^2 - b^2$ form.

$= (b^2 + 4)(b^2 - 4)$ Factor the difference of squares.

$= (b^2 + 4)(b^2 - 2^2)$ $b^2 - 4$ is also a difference of squares.

$= (b^2 + 4)(b + 2)(b - 2)$ Factor the difference of squares.

**b.** $625 - x^4$

$625 - x^4 = (25)^2 - (x^2)^2$ Write $625 - x^4$ in $a^2 - b^2$ form.

$= (25 + x^2)(25 - x^2)$ Factor the difference of squares.

$= (25 + x^2)(5^2 - x^2)$ Write $25 - x^2$ in $a^2 - b^2$ form.

$= (25 + x^2)(5 - x)(5 + x)$ Factor the difference of squares.

**WatchOut!**

**Sum of Squares** The sum of squares, $a^2 + b^2$, does not factor into $(a + b)(a + b)$. The sum of squares is a prime polynomial and cannot be factored.

### GuidedPractice

**2A.** $y^4 - 1$

**2B.** $4a^4 - b^4$

**2C.** $81 - x^4$

**2D.** $16y^4 - 1$

Sometimes more than one factoring technique needs to be applied to ensure that a polynomial is factored completely.

PT

## Example 3 Apply Different Techniques

**Factor each polynomial.**

**a.** $5x^5 - 45x$

$5x^5 - 45x = 5x(x^4 - 9)$ Factor out GCF.

$= 5x[(x^2)^2 - (3)^2]$ Write $x^4 - 9$ in the form $a^2 - b^2$.

$= 5x(x^2 - 3)(x^2 + 3)$ Factor the difference of squares.

$x^2 - 3$ is not a difference of squares because 3 is not a perfect square.

**b.** $7x^3 + 21x^2 - 7x - 21$

$7x^3 + 21x^2 - 7x - 21$ Original expression

$= 7(x^3 + 3x^2 - x - 3)$ Factor out GCF.

$= 7[(x^3 + 3x^2) - (x + 3)]$ Group terms with common factors.

$= 7[x^2(x + 3) - 1(x + 3)]$ Factor each grouping.

$= 7(x + 3)(x^2 - 1)$ $x + 3$ is the common factor.

$= 7(x + 3)(x + 1)(x - 1)$ Factor the difference of squares.

### GuidedPractice

**3A.** $2y^4 - 50$

**3B.** $6x^4 - 96$

**3C.** $2m^3 + m^2 - 50m - 25$

**3D.** $r^3 + 6r^2 + 11r + 66$

## 2 Solve Equations by Factoring

After factoring, you can apply the Zero Product Property to an equation that is written as the product of factors set equal to 0.

### Standardized Test Example 4 Solve an Equation by Factoring

**In the equation $y = x^2 - \frac{9}{16}$, which is a value of $x$ when $y = 0$?**

A $-\frac{9}{4}$  B 0  C $\frac{3}{4}$  D $\frac{9}{4}$

**Test-Taking Tip**

CCSS Sense-Making Another method that can be used to solve this equation is to substitute each answer choice into the equation.

**Read the Test Item**

Replace $y$ with 0 and then solve.

**Solve the Test Item**

| | |
|---|---|
| $y = x^2 - \frac{9}{16}$ | Original equation |
| $0 = x^2 - \frac{9}{16}$ | Replace $y$ with 0. |
| $0 = x^2 - \left(\frac{3}{4}\right)^2$ | Write in the form $a^2 - b^2$. |
| $0 = \left(x + \frac{3}{4}\right)\left(x - \frac{3}{4}\right)$ | Factor the difference of squares. |
| $0 = x + \frac{3}{4}$ or $0 = x - \frac{3}{4}$ | Zero Product Property |
| $x = -\frac{3}{4}$  $x = \frac{3}{4}$ | The correct answer is C. |

**Guided Practice**

4. Which are the solutions of $18x^3 = 50x$?

F $0, \frac{5}{3}$  G $-\frac{5}{3}, \frac{5}{3}$  H $-\frac{5}{3}, \frac{5}{3}, 0$  J $-\frac{5}{3}, \frac{5}{3}, 1$

## Check Your Understanding

**Examples 1–3** **Factor each polynomial.**

1. $x^2 - 9$
2. $4a^2 - 25$
3. $9m^2 - 144$
4. $2p^3 - 162p$
5. $u^4 - 81$
6. $2d^4 - 32f^4$
7. $20r^4 - 45n^4$
8. $256n^4 - c^4$
9. $2c^3 + 3c^2 - 2c - 3$
10. $f^3 - 4f^2 - 9f + 36$
11. $3t^3 + 2t^2 - 48t - 32$
12. $w^3 - 3w^2 - 9w + 27$

**Example 4** **EXTENDED RESPONSE** **During an accident, skid marks may result from sudden breaking. The formula $\frac{1}{24}s^2 = d$ approximates a vehicle's speed $s$ in miles per hour given the length $d$ in feet of the skid marks on dry concrete.**

13. If skid marks on dry concrete are 54 feet long, how fast was the car traveling when the brakes were applied?
14. If the skid marks on dry concrete are 150 feet long, how fast was the car traveling when the brakes were applied?

## Practice and Problem Solving

Examples 1–3 **Factor each polynomial.**

**15.** $q^2 - 121$

**16.** $r^4 - k^4$

**17.** $6n^4 - 6$

**18.** $w^4 - 625$

**19.** $r^2 - 9t^2$

**20.** $2c^2 - 32d^2$

**21.** $h^3 - 100h$

**22.** $h^4 - 256$

**23.** $2x^3 - x^2 - 162x + 81$

**24.** $x^2 - 4y^2$

**25.** $7h^4 - 7p^4$

**26.** $3c^3 + 2c^2 - 147c - 98$

**27.** $6k^2h^4 - 54k^4$

**28.** $5a^3 - 20a$

**29.** $f^3 + 2f^2 - 64f - 128$

**30.** $3r^3 - 192r$

**31.** $10q^3 - 1210q$

**32.** $3xn^4 - 27x^3$

**33.** $p^3r^5 - p^3r$

**34.** $8c^3 - 8c$

**35.** $r^3 - 5r^2 - 100r + 500$

**36.** $3t^3 - 7t^2 - 3t + 7$

**37.** $a^2 - 49$

**38.** $4m^3 + 9m^2 - 36m - 81$

**39.** $3m^4 + 243$

**40.** $3x^3 + x^2 - 75x - 25$

**41.** $12a^3 + 2a^2 - 192a - 32$

**42.** $x^4 + 6x^3 - 36x^2 - 216x$

**43.** $15m^3 + 12m^2 - 375m - 300$

Example 4

**44.** **GEOMETRY** The drawing at the right is a square with a square cut out of it.

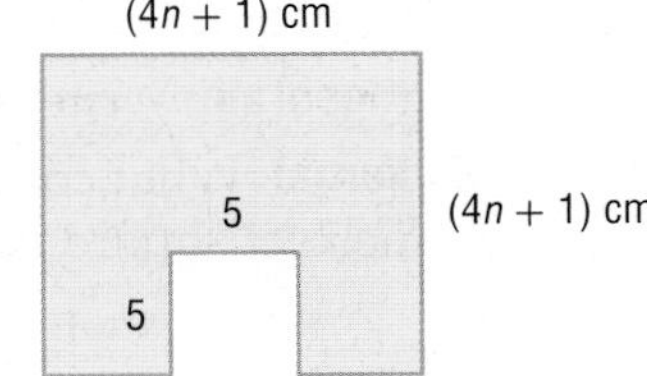


**a.** Write an expression that represents the area of the shaded region.

**b.** Find the dimensions of a rectangle with the same area as the shaded region in the drawing. Assume that the dimensions of the rectangle must be represented by binomials with integral coefficients.

**45.** **DECORATIONS** An arch decorated with balloons was used to decorate the gym for the spring dance. The shape of the arch can be modeled by the equation $y = -0.5x^2 + 4.5x$, where $x$ and $y$ are measured in feet and the $x$-axis represents the floor.

**a.** Write the expression that represents the height of the arch in factored form.

**b.** How far apart are the two points where the arch touches the floor?

**c.** Graph this equation on your calculator. What is the highest point of the arch?

**46.** CCSS **SENSE-MAKING** Zelda is building a deck in her backyard. The plans for the deck show that it is to be 24 feet by 24 feet. Zelda wants to reduce one dimension by a number of feet and increase the other dimension by the same number of feet. If the area of the reduced deck is 512 square feet, what are the dimensions of the deck?

**47** **SALES** The sales of a particular CD can be modeled by the equation $S = -25m^2 + 125m$, where $S$ is the number of CDs sold in thousands, and $m$ is the number of months that it is on the market.

**a.** In what month should the music store expect the CD to stop selling?

**b.** In what month will CD sales peak?

**c.** How many copies will the CD sell at its peak?

**Solve each equation by factoring. Confirm your answers using a graphing calculator.**

**48.** $36w^2 = 121$

**49** $100 = 25x^2$

**50.** $64x^2 - 1 = 0$

**51.** $4y^2 - \frac{9}{16} = 0$

**52.** $\frac{1}{4}b^2 = 16$

**53.** $81 - \frac{1}{25}x^2 = 0$

**54.** $9d^2 - 81 = 0$

**55.** $4a^2 = \frac{9}{64}$

**56.** **MULTIPLE REPRESENTATIONS** In this problem, you will investigate perfect square trinomials.

**a.** **Tabular** Copy and complete the table below by factoring each polynomial. Then write the first and last terms of the given polynomials as perfect squares.

**b.** **Analytical** Write the middle term of each polynomial using the square roots of the perfect squares of the first and last terms.

| Polynomial | Factored Polynomial | First Term | Last Term | Middle Term |
|---|---|---|---|---|
| $4x^2 + 12x + 9$ | $(2x + 3)(2x + 3)$ | $4x^2 = (2x)^2$ | $9 = 3^2$ | |
| $9x^2 - 24x + 16$ | | | | |
| $4x^2 - 20x + 25$ | | | | |
| $16x^2 + 24x + 9$ | | | | |
| $25x^2 + 20x + 4$ | | | | |

**c.** **Algebraic** Write the pattern for a perfect square trinomial.

**d.** **Verbal** What conditions must be met for a trinomial to be classified as a perfect square trinomial?

## H.O.T. Problems Use Higher-Order Thinking Skills

**57.** **ERROR ANALYSIS** Elizabeth and Lorenzo are factoring an expression. Is either of them correct? Explain your reasoning.

| Elizabeth | Lorenzo |
|---|---|
| $16x^4 - 25y^2 =$ | $16x^4 - 25y^2 =$ |
| $(4x - 5y)(4x + 5y)$ | $(4x^2 - 5y)(4x^2 + 5y)$ |

**58.** **CHALLENGE** Factor and simplify $9 - (k + 3)^2$, a difference of squares.

**59.** CCSS **PERSEVERANCE** Factor $x^{16} - 81$.

**60.** **REASONING** Write and factor a binomial that is the difference of two perfect squares and that has a greatest common factor of $5mk$.

**61.** **REASONING** Determine whether the following statement is *true* or *false*. Give an example or counterexample to justify your answer.

*All binomials that have a perfect square in each of the two terms can be factored.*

**62.** **OPEN ENDED** Write a binomial in which the difference of squares pattern must be repeated to factor it completely. Then factor the binomial.

**63.** **WRITING IN MATH** Describe why the difference of squares pattern has no middle term with a variable.

## Standardized Test Practice

**64.** One of the roots of $2x^2 + 13x = 24$ is $-8$. What is the other root?

A $-\frac{3}{2}$ C $\frac{2}{3}$

B $-\frac{2}{3}$ D $\frac{3}{2}$

**65.** Which of the following is the sum of both solutions of the equation $x^2 + 3x = 54$?

F $-21$ H 3

G $-3$ J 21

**66.** What are the $x$-intercepts of the graph of $y = -3x^2 + 7x + 20$?

A $\frac{5}{3}, -4$ C $-\frac{5}{3}, 4$

B $-\frac{5}{3}, -4$ D $\frac{5}{3}, 4$

**67. EXTENDED RESPONSE** Two cars leave Cleveland at the same time from different parts of the city and both drive to Cincinnati. The distance in miles of the cars from the center of Cleveland can be represented by the two equations below, where $t$ represents the time in hours.

Car A: $65t + 15$ Car B: $60t + 25$

**a.** Which car is faster? Explain.

**b.** Find an expression that models the distance between the two cars.

**c.** How far apart are the cars after $2\frac{1}{2}$ hours?

## Spiral Review

**Factor each trinomial, if possible. If the trinomial cannot be factored using integers, write *prime*.** (Lesson 1-7)

**68.** $5x^2 - 17x + 14$ **69.** $5a^2 - 3a + 15$ **70.** $10x^2 - 20xy + 10y^2$

**Solve each equation. Check your solutions.** (Lesson 1-6)

**71.** $n^2 - 9n = -18$ **72.** $10 + a^2 = -7a$ **73.** $22x - x^2 = 96$

**Solve each equation. Check the solutions.** (Lesson 1-4)

**74.** $2x^2 = 32$ **75.** $(x - 4)^2 = 25$ **76.** $4x^2 - 4x + 1 = 16$

**77.** $2x^2 + 16x = -32$ **78.** $(x + 3)^2 = 5$ **79.** $4x^2 - 12x = -9$

**Find each sum or difference.** (Lesson 1-1)

**80.** $(3n^2 - 3) + (4 + 4n^2)$ **81.** $(2d^2 - 7d - 3) - (4d^2 + 7)$

**82.** $(2b^3 - 4b^2 + 4) - (3b^4 + 5b^2 - 9)$ **83.** $(8 - 4h^2 + 6h^4) + (5h^2 - 3 + 2h^3)$

## Skills Review

**Find each product.**

**84.** $(x - 6)^2$ **85.** $(x - 2)(x - 2)$ **86.** $(x + 3)(x + 3)$

**87.** $(2x - 5)^2$ **88.** $(6x - 1)^2$ **89.** $(4x + 5)(4x + 5)$

# LESSON 1-9 Perfect Squares

**Then**

- You found the product of a sum and difference.

**Now**

1. Factor perfect square trinomials.
2. Solve equations involving perfect squares.

**Why?**

- In a vacuum, a feather and a piano would fall at the same speed, or velocity. To find about how long it takes an object to hit the ground if it is dropped from an initial height of $h_0$ feet above ground, you would need to solve the equation $0 = -16t^2 + h_0$, where $t$ is time in seconds after the object is dropped.

**NewVocabulary**
perfect square trinomial

**Common Core State Standards**

**Content Standards**

A.SSE.3a Factor a quadratic expression to reveal the zeros of the function it defines.

A.REI.1 Explain each step in solving a simple equation as following from the equality of numbers asserted at the previous step, starting from the assumption that the original equation has a solution. Construct a viable argument to justify a solution method.

**Mathematical Practices**

6 Attend to precision.

## 1 Factor Perfect Square Trinomials

You have learned the patterns for the products of the binomials $(a + b)^2$ and $(a - b)^2$. Recall that these are special products that follow specific patterns.

$$\begin{aligned}(a + b)^2 &= (a + b)(a + b)\\ &= a^2 + ab + ab + b^2\\ &= a^2 + 2ab + b^2\end{aligned}$$

$$\begin{aligned}(a - b)^2 &= (a - b)(a - b)\\ &= a^2 - ab - ab + b^2\\ &= a^2 - 2ab + b^2\end{aligned}$$

These products are called **perfect square trinomials**, because they are the squares of binomials. The above patterns can help you factor perfect square trinomials.

For a trinomial to be factorable as a perfect square, the first and last terms must be perfect squares and the middle term must be two times the square roots of the first and last terms.

The trinomial $16x^2 + 24x + 9$ is a perfect square trinomial, as illustrated below.

$$16x^2 + 24x + 9$$

- Is the first term a perfect square? Yes, because $16x^2 = (4x)^2$.
- Is the middle term twice the product of the square roots of the first and last terms? Yes, because $24x = 2(4x)(3)$.
- Is the last term a perfect square? Yes, because $9 = 3^2$.

**KeyConcept** Factoring Perfect Square Trinomials

| | |
|---|---|
| Symbols | $a^2 + 2ab + b^2 = (a + b)(a + b) = (a + b)^2$ |
| | $a^2 - 2ab + b^2 = (a - b)(a - b) = (a - b)^2$ |
| Examples | $x^2 + 8x + 16 = (x + 4)(x + 4)$ or $(x + 4)^2$ |
| | $x^2 - 6x + 9 = (x - 3)(x - 3)$ or $(x - 3)^2$ |

**StudyTip**

**Recognizing Perfect Square Trinomials** If the constant term of the trinomial is negative, the trinomial is not a perfect square trinomial, so it is not necessary to check the other conditions.

## Example 1 Recognize and Factor Perfect Square Trinomials

**Determine whether each trinomial is a perfect square trinomial. Write *yes* or *no*. If so, factor it.**

**a.** $4y^2 + 12y + 9$

1. Is the first term a perfect square? Yes, $4y^2 = (2y)^2$.
2. Is the last term a perfect square? Yes, $9 = 3^2$.
3. Is the middle term equal to $2(2y)(3)$? Yes, $12y = 2(2y)(3)$

Since all three conditions are satisfied, $4y^2 + 12y + 9$ is a perfect square trinomial.

$4y^2 + 12y + 9 = (2y)^2 + 2(2y)(3) + 3^2$ Write as $a^2 + 2ab + b^2$.

$= (2y + 3)^2$ Factor using the pattern.

**b.** $9x^2 - 6x + 4$

1. Is the first term a perfect square? Yes, $9x^2 = (3x)^2$.
2. Is the last term a perfect square? Yes, $4 = 2^2$.
3. Is the middle term equal to $-2(3x)(2)$? No, $-6x \neq -2(3x)(2)$.

Since the middle term does not satisfy the required condition, $9x^2 - 6x + 4$ is not a perfect square trinomial.

**GuidedPractice**

**1A.** $9y^2 + 24y + 16$ **1B.** $2a^2 + 10a + 25$

A polynomial is completely factored when it is written as a product of prime polynomials. More than one method might be needed to factor a polynomial completely. When completely factoring a polynomial, the Concept Summary can help you decide where to start.

Remember, if the polynomial does not fit any pattern or cannot be factored, the polynomial is prime.

**ConceptSummary Factoring Methods**

| Steps | Number of Terms | Examples |
|---|---|---|
| **Step 1** Factor out the GCF. | any | $4x^3 + 2x^2 - 6x = 2x(2x^2 + x - 3)$ |
| **Step 2** Check for a difference of squares or a perfect square trinomial. | 2 or 3 | $9x^2 - 16 = (3x + 4)(3x - 4)$<br>$16x^2 + 24x + 9 = (4x + 3)^2$ |
| **Step 3** Apply the factoring patterns for $x^2 + bx + c$ or $ax^2 + bx + c$ (general trinomials), or factor by grouping. | 3 or 4 | $x^2 - 8x + 12 = (x - 2)(x - 6)$<br>$2x^2 + 13x + 6 = (2x + 1)(x + 6)$<br>$12y^2 + 9y + 8y + 6$<br>$= (12y^2 + 9y) + (8y + 6)$<br>$= 3y(4y + 3) + 2(4y + 3)$<br>$= (4y + 3)(3y + 2)$ |

## Example 2 Factor Completely

**Factor each polynomial, if possible. If the polynomial cannot be factored, write *prime*.**

**a.** $5x^2 - 80$

**Step 1** The GCF of $5x^2$ and $-80$ is 5, so factor it out.

**Step 2** Since there are two terms, check for a difference of squares.

| | |
|---|---|
| $5x^2 - 80 = 5(x^2 - 16)$ | 5 is the GCF of the terms. |
| $= 5(x^2 - 4^2)$ | $x^2 = x \cdot x$ and $16 = 4 \cdot 4$ |
| $= 5(x - 4)(x + 4)$ | Factor the difference of squares. |

**b.** $9x^2 - 6x - 35$

**Step 1** The GCF of $9x^2$, $-6x$, and $-35$ is 1.

**Step 2** Since 35 is not a perfect square, this is not a perfect square trinomial.

**Step 3** Factor using the pattern $ax^2 + bx + c$. Are there two numbers with a product of $9(-35)$ or $-315$ and a sum of $-6$? Yes, the product of 15 and $-21$ is $-315$, and the sum is $-6$.

| | |
|---|---|
| $9x^2 - 6x - 35 = 9x^2 + mx + px - 35$ | Write the pattern. |
| $= 9x^2 + 15x - 21x - 35$ | $m = 15$ and $n = -21$ |
| $= (9x^2 + 15x) + (-21x - 35)$ | Group terms with common factors. |
| $= 3x(3x + 5) - 7(3x + 5)$ | Factor out the GCF from each grouping. |
| $= (3x + 5)(3x - 7)$ | $3x + 5$ is the common factor. |

**StudyTip**

Check Your Answer You can check your answer by:

- Using the FOIL method.
- Using the Distributive Property.
- Graphing the original expression and factored expression and comparing the graphs.

If the product of the factors does not match the original expression exactly, the answer is incorrect.

### GuidedPractice

**2A.** $2x^2 - 32$

**2B.** $12x^2 + 5x - 25$

## 2 Solve Equations with Perfect Squares

When solving equations involving repeated factors, it is only necessary to set one of the repeated factors equal to zero.

## Example 3 Solve Equations with Repeated Factors

**Solve $9x^2 - 48x = -64$.**

| | |
|---|---|
| $9x^2 - 48x = -64$ | Original equation |
| $9x^2 - 48x + 64 = 0$ | Add 64 to each side. |
| $(3x)^2 - 2(3x)(8) + (8)^2 = 0$ | Recognize $9x^2 - 48x + 64$ as a perfect square trinomial. |
| $(3x - 8)^2 = 0$ | Factor the perfect square trinomial. |
| $(3x - 8)(3x - 8) = 0$ | Write $(3x - 8)^2$ as two factors. |
| $3x - 8 = 0$ | Set the repeated factor equal to zero. |
| $3x = 8$ | Add 8 to each side. |
| $x = \frac{8}{3}$ | Divide each side by 3. |

### GuidedPractice

**Solve each equation. Check your solutions.**

**3A.** $a^2 + 12a + 36 = 0$

**3B.** $y^2 - \frac{4}{3}y + \frac{4}{9} = 0$

You have solved equations like $x^2 - 16 = 0$ by factoring. You can also use the definition of a square root to solve the equation.

| | |
|---|---|
| $x^2 - 16 = 0$ | Original equation |
| $x^2 = 16$ | Add 16 to each side. |
| $x = \pm\sqrt{16}$ | Take the square root of each side. |

**ReadingMath**

**Square Root Solutions** $\pm\sqrt{16}$ is read as plus or minus the square root of 16.

Remember that there are two square roots of 16, namely 4 and −4. Therefore, the solution set is $\{-4, 4\}$. You can express this as $\{\pm 4\}$.

## KeyConcept Square Root Property

| | |
|---|---|
| **Words** | To solve a quadratic equation in the form $x^2 = n$, take the square root of each side. |
| **Symbols** | For any number $n \geq 0$, if $x^2 = n$, then $x = \pm\sqrt{n}$. |
| **Example** | $x^2 = 25$<br>$x = \pm\sqrt{25}$ or $\pm 5$ |

In the equation $x^2 = n$, if $n$ is not a perfect square, you need to approximate the square root. Use a calculator to find an approximation. If $n$ is a perfect square, you will have an exact answer.

## Example 4 Use the Square Root Property

**StudyTip**

**Solving by Inspection** Equations involving square roots can often be solved mentally. For $x^2 = n$, think: *The square of what number is n?* When $n$ is a perfect square, $x$ is rational. Otherwise, $x$ is irrational.

**Solve each equation. Check your solutions.**

**a.** $(y - 6)^2 = 81$

| | |
|---|---|
| $(y - 6)^2 = 81$ | Original equation |
| $y - 6 = \pm\sqrt{81}$ | Square Root Property |
| $y - 6 = \pm 9$ | $81 = 9 \cdot 9$ |
| $y = 6 \pm 9$ | Add 6 to each side. |
| $y = 6 + 9$ or $y = 6 - 9$ | Separate into two equations. |
| $= 15$ $\quad = -3$ | Simplify. |
| The roots are 15 and −3. | Check in the original equation. |

**b.** $(x + 6)^2 = 12$

| | |
|---|---|
| $(x + 6)^2 = 12$ | Original equation |
| $x + 6 = \pm\sqrt{12}$ | Square Root Property |
| $x = -6 \pm \sqrt{12}$ | Subtract 6 from each side. |

The roots are $-6 \pm \sqrt{12}$ or $-6 + \sqrt{12}$ and $-6 - \sqrt{12}$.

Using a calculator, $-6 + \sqrt{12} \approx -2.54$ and $-6 - \sqrt{12} \approx -9.46$.

**GuidedPractice**

**4A.** $(a - 10)^2 = 121$ **4B.** $(z + 3)^2 = 26$

## Real-World Example 5 Solve an Equation

**PHYSICAL SCIENCE During an experiment, a ball is dropped from a height of 205 feet. The formula $h = -16t^2 + h_0$ can be used to approximate the number of seconds $t$ it takes for the ball to reach height $h$ from an initial height of $h_0$ in feet. Find the time it takes the ball to reach the ground.**

At ground level, $h = 0$ and the initial height is 205, so $h_0 = 205$.

| | |
|---|---|
| $h = -16t^2 + h_0$ | Original Formula |
| $0 = -16t^2 + 205$ | Replace $h$ with 0 and $h_0$ with 205. |
| $-205 = -16t^2$ | Subtract 205 from each side. |
| $12.8125 = t^2$ | Divide each side by $-16$. |
| $\pm 3.6 \approx t$ | Use the Square Root Property. |

Since a negative number does not make sense in this situation, the solution is 3.6. It takes about 3.6 seconds for the ball to reach the ground.

**Math HistoryLink**

Galileo Galilei (1564–1642) Galileo was the first person to prove that objects of different weights fall at the same velocity by dropping two objects of different weights from the top of the Leaning Tower of Pisa in 1589.

### GuidedPractice

**5.** Find the time it takes a ball to reach the ground if it is dropped from a bridge that is half as high as the one described above.

## Check Your Understanding

**Example 1** **Determine whether each trinomial is a perfect square trinomial. Write *yes* or *no*. If so, factor it.**

**1.** $25x^2 + 60x + 36$

**2.** $6x^2 + 30x + 36$

**Example 2** **Factor each polynomial, if possible. If the polynomial cannot be factored, write *prime*.**

**3.** $2x^2 - x - 28$

**4.** $6x^2 - 34x + 48$

**5.** $4x^2 + 64$

**6.** $4x^2 + 9x - 16$

**Examples 3–4** **Solve each equation. Confirm your answers using a graphing calculator.**

**7.** $4x^2 = 36$

**8.** $25a^2 - 40a = -16$

**9.** $64y^2 - 48y + 18 = 9$

**10.** $(z + 5)^2 = 47$

**Example 5** **11.** CCSS **REASONING** While painting his bedroom, Nick drops his paintbrush off his ladder from a height of 6 feet. Use the formula $h = -16t^2 + h_0$ to approximate the number of seconds it takes for the paintbrush to hit the floor.

## Practice and Problem Solving

**Example 1** **Determine whether each trinomial is a perfect square trinomial. Write *yes* or *no*. If so, factor it.**

**12.** $4x^2 - 42x + 110$

**13.** $16x^2 - 56x + 49$

**14.** $81x^2 - 90x + 25$

**15.** $x^2 + 26x + 168$

**Example 2** **Factor each polynomial, if possible. If the polynomial cannot be factored, write *prime*.**

16. $24d^2 + 39d - 18$
17. $8x^2 + 10x - 21$
18. $2b^2 + 12b - 24$
19. $8y^2 - 200z^2$
20. $16a^2 - 121b^2$
21. $12m^3 - 22m^2 - 70m$
22. $8c^2 - 88c + 242$
23. $12x^2 - 84x + 147$
24. $w^4 - w^2$
25. $12p^3 - 3p$
26. $16q^3 - 48q^2 + 36q$
27. $4t^3 + 10t^2 - 84t$
28. $x^3 + 2x^2y - 4x - 8y$
29. $2a^2b^2 - 2a^2 - 2ab^3 + 2ab$
30. $2r^3 - r^2 - 72r + 36$
31. $3k^3 - 24k^2 + 48k$
32. $4c^4d - 10c^3d + 4c^2d^3 - 10cd^3$
33. $g^2 + 2g - 3h^2 + 4h$

**Examples 3–4** **Solve each equation. Confirm your answers using a graphing calculator.**

34. $4m^2 - 24m + 36 = 0$
35. $(y - 4)^2 = 7$
36. $a^2 + \frac{10}{7}a + \frac{25}{49} = 0$
37. $x^2 - \frac{3}{2}x + \frac{9}{16} = 0$
38. $x^2 + 8x + 16 = 25$
39. $5x^2 - 60x = -180$
40. $4x^2 = 80x - 400$
41. $9 - 54x = -81x^2$
42. $4c^2 + 4c + 1 = 15$
43. $x^2 - 16x + 64 = 6$

44. **PHYSICAL SCIENCE** For an experiment in physics class, a water balloon is dropped from the window of the school building. The window is 40 feet high. How long does it take until the balloon hits the ground? Round to the nearest hundredth.

45. **SCREENS** The area $A$ in square feet of a projected picture on a movie screen can be modeled by the equation $A = 0.25d^2$, where $d$ represents the distance from a projector to a movie screen. At what distance will the projected picture have an area of 100 square feet?

**Example 5**

46. **GEOMETRY** The area of a square is represented by $9x^2 - 42x + 49$. Find the length of each side.

47. **GEOMETRY** The area of a square is represented by $16x^2 + 40x + 25$. Find the length of each side.

48. **GEOMETRY** The volume of a rectangular prism is represented by the expression $8y^3 + 40y^2 + 50y$. Find the possible dimensions of the prism if the dimensions are represented by polynomials with integer coefficients.

49. **POOLS** Ichiro wants to buy an above-ground swimming pool for his yard. Model A is 42 inches deep and holds 1750 cubic feet of water. The length of the rectangular pool is 5 feet more than the width.

    a. What is the surface area of the water?

    b. What are the dimensions of the pool?

    c. Model B pool holds twice as much water as Model A. What are some possible dimensions for this pool?

    d. Model C has length and width that are both twice as long as Model A, but the height is the same. What is the ratio of the volume of Model A to Model C?

**50. GEOMETRY** Use the rectangular prism at the right.

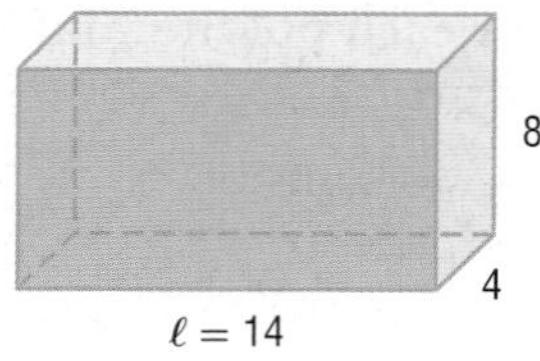


**a.** Write an expression for the height and width of the prism in terms of the length, $\ell$.

**b.** Write a polynomial for the volume of the prism in terms of the length.

**51. CCSS PRECISION** A zoo has an aquarium shaped like a rectangular prism. It has a volume of 180 cubic feet. The height of the aquarium is 9 feet taller than the width, and the length is 4 feet shorter than the width. What are the dimensions of the aquarium?

**52. ELECTION** For the student council elections, Franco is building the voting box shown with a volume of 96 cubic inches. What are the dimensions of the voting box?

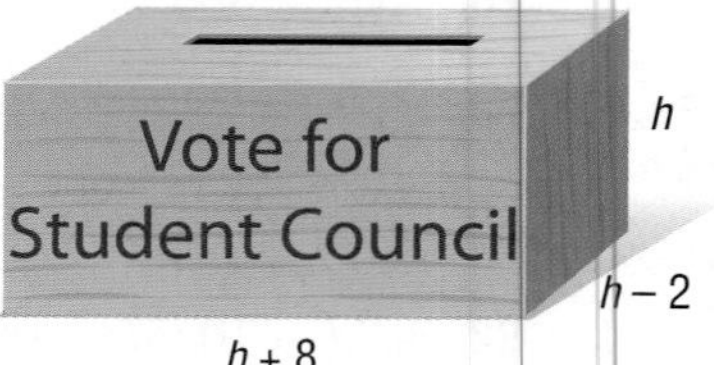


## H.O.T. Problems Use Higher-Order Thinking Skills

**53. ERROR ANALYSIS** Debbie and Adriano are factoring the expression $x^8 - x^4$ completely. Is either of them correct? Explain your reasoning.

| Debbie | Adriano |
|---|---|
| $x^8 - x^4 = x^4(x^2 + 1)(x^2 - 1)$ | $x^8 - x^4 = x^4(x^2 + 1)(x - 1)(x + 1)$ |

**54. CHALLENGE** Factor $x^{n+6} + x^{n+2} + x^n$ completely.

**55. OPEN ENDED** Write a perfect square trinomial equation in which the coefficient of the middle term is negative and the last term is a fraction. Solve the equation.

**56. REASONING** A counterexample is a specific case in which a statement is false. Find a counterexample to the following statement.

*A polynomial equation of degree three always has three real solutions.*

**57. CCSS REGULARITY** Explain how to factor a polynomial completely.

**58. WHICH ONE DOESN'T BELONG?** Identify the trinomial that does not belong. Explain.

| $4x^2 - 36x + 81$ | $25x^2 + 10x + 1$ | $4x^2 + 10x + 4$ | $9x^2 - 24x + 16$ |
|---|---|---|---|

**59. OPEN ENDED** Write a binomial that can be factored using the difference of two squares twice. Set your binomial equal to zero and solve the equation.

**60. WRITING IN MATH** Explain how to determine whether a trinomial is a perfect square trinomial.

## Standardized Test Practice

**61.** What is the solution set for the equation $(x - 3)^2 = 25$?

A $\{-8, 2\}$ C $\{4, 14\}$
B $\{-2, 8\}$ D $\{-4, 14\}$

**62. SHORT RESPONSE** Write an equation in slope-intercept form for the graph shown below.

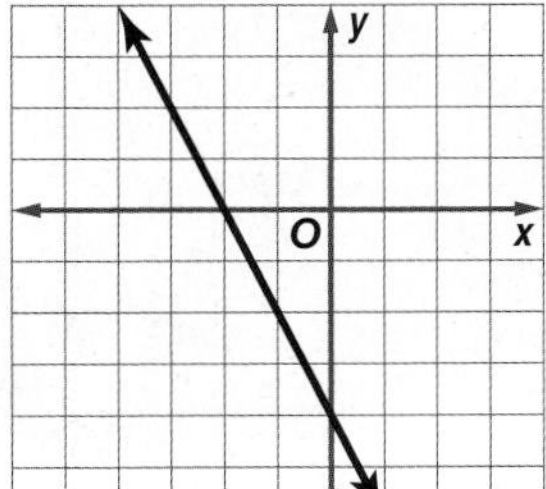


**63.** At an amphitheater, the price of 2 lawn seats and 2 pavilion seats is \$120. The price of 3 lawn seats and 4 pavilion seats is \$225. How much do lawn and pavilion seats cost?

F \$20 and \$41.25
G \$10 and \$50
H \$15 and \$45
J \$30 and \$30

**64. GEOMETRY** The circumference of a circle is $\frac{6\pi}{5}$ units. What is the area of the circle?

A $\frac{9\pi}{25}$ units$^2$ C $\frac{6\pi}{5}$ units$^2$
B $\frac{3\pi}{5}$ units$^2$ D $\frac{12\pi}{5}$ units$^2$

## Spiral Review

**Factor each polynomial, if possible. If the polynomial cannot be factored, write *prime*.** (Lesson 1-8)

**65.** $x^2 - 16$ **66.** $4x^2 - 81y^2$ **67.** $1 - 100p^2$

**68.** $3a^2 - 20$ **69.** $25n^2 - 1$ **70.** $36 - 9c^2$

**Solve each equation. Confirm your answers using a graphing calculator.** (Lesson 1-7)

**71.** $4x^2 - 8x - 32 = 0$ **72.** $6x^2 - 48x + 90 = 0$ **73.** $14x^2 + 14x = 28$

**74.** $2x^2 - 10x = 48$ **75.** $5x^2 - 25x = -30$ **76.** $8x^2 - 16x = 192$

**77. AMUSEMENT RIDE** The height $h$ in feet of a car above the exit ramp of a free-fall ride can be modeled by $h(t) = -16t^2 + s$. $t$ is the time in seconds after the car drops, and $s$ is the starting height of the car in feet. If the designer wants the ride to last 3 seconds, what should be the starting height in feet? (Lesson 1-9)

**Factor each polynomial. If the polynomial cannot be factored, write *prime*.** (Lesson 1-8)

**78.** $x^2 - 81$ **79.** $a^2 - 121$ **80.** $n^2 + 100$

**81.** $-25 + 4y^2$ **82.** $p^4 - 16$ **83.** $4t^4 - 4$

## Skills Review

**Find the slope of the line that passes through each pair of points.**

**84.** $(5, 7), (-2, -3)$ **85.** $(2, -1), (5, -3)$ **86.** $(-4, -1), (-3, -3)$

**87.** $(-3, -4), (5, -1)$ **88.** $(-2, 3), (8, 3)$ **89.** $(-5, 4), (-5, -1)$

# LESSON 1-10 Roots and Zeros

**Then**

- You used complex numbers to describe solutions of quadratic equations.

**Now**

1. Determine the number and type of roots for a polynomial equation.
2. Find the zeros of a polynomial function.

**Why?**

- The function $g(x) = 1.384x^4 - 0.003x^3 + 0.28x^2 - 0.078x + 1.365$ can be used to model the average price of a gallon of gasoline in a given year if $x$ is the number of years since 1990. To find the average price of gasoline in a specific year, you can use the roots of the related polynomial equation.

**Common Core State Standards**

**Content Standards**

**N.CN.9** Know the Fundamental Theorem of Algebra; show that it is true for quadratic polynomials.

**A.APR.3** Identify zeros of polynomials when suitable factorizations are available, and use the zeros to construct a rough graph of the function defined by the polynomial.

**Mathematical Practices**

6 Attend to precision.

## 1 Synthetic Types of Roots

Previously, you learned that a zero of a function $f(x)$ is any value $c$ such that $f(c) = 0$. When the function is graphed, the real zeros of the function are the $x$-intercepts of the graph.

### ConceptSummary Zeros, Factors, Roots, and Intercepts

**Words** Let $P(x) = a_nx^n + \cdots + a_1x + a_0$ be a polynomial function. Then the following statements are equivalent.

- $c$ is a zero of $P(x)$.
- $c$ is a root or solution of $P(x) = 0$.
- $x - c$ is a factor of $a_nx^n + \cdots + a_1x + a_0$.
- If $c$ is a real number, then $(c, 0)$ is an $x$-intercept of the graph of $P(x)$.

**Example** Consider the polynomial function $P(x) = x^4 + 2x^3 - 7x^2 - 8x + 12$.

The zeros of $P(x) = x^4 + 2x^3 - 7x^2 - 8x + 12$ are $-3, -2, 1$, and $2$.

The roots of $x^4 + 2x^3 - 7x^2 - 8x + 12 = 0$ are $-3, -2, 1$, and $2$.

The factors of $x^4 + 2x^3 - 7x^2 - 8x + 12$ are $(x + 3), (x + 2), (x - 1)$, and $(x - 2)$.

The $x$-intercepts of the graph of $P(x) = x^4 + 2x^3 - 7x^2 - 8x + 12$ are $(-3, 0), (-2, 0), (1, 0)$, and $(2, 0)$.

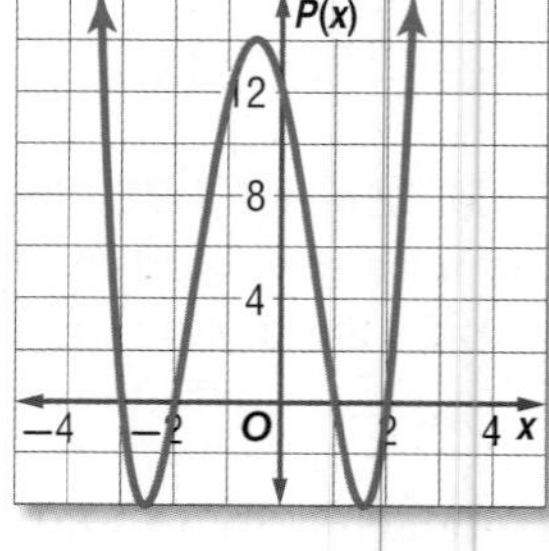


When solving a polynomial equation with degree greater than zero, there may be one or more real roots or no real roots (the roots are imaginary numbers). Since real numbers and imaginary numbers both belong to the set of complex numbers, all polynomial equations with degree greater than zero will have at least one root in the set of complex numbers. This is the **Fundamental Theorem of Algebra**.

### KeyConcept Fundamental Theorem of Algebra

Every polynomial equation with degree greater than zero has at least one root in the set of complex numbers.

## Example 1 Determine Number and Type of Roots

**Solve each equation. State the number and type of roots.**

**a.** $x^2 + 6x + 9 = 0$

| | |
|---|---|
| $x^2 + 6x + 9 = 0$ | Original equation |
| $(x + 3)^2 = 0$ | Factor. |
| $x + 3 = 0$ | Take the root of each side. |
| $x = -3$ | Solve for $x$. |

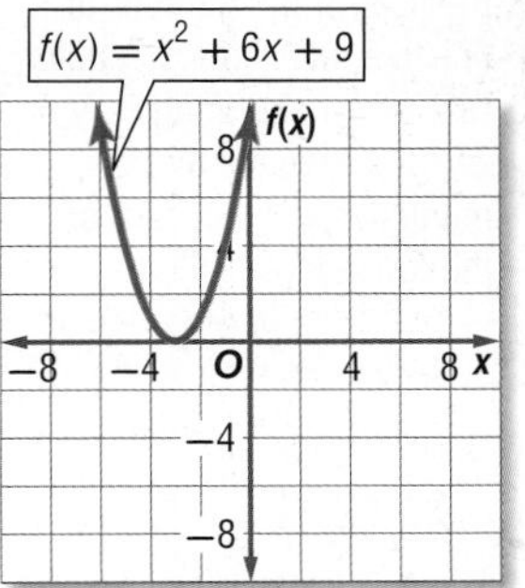


Because $(x + 3)$ is twice a factor of $x^2 + 6x + 9$, $-3$ is a double root. Thus, the equation has one real repeated root, $-3$.

**CHECK** The graph of the equation touches the $x$-axis at $x = -3$. Since $-3$ is a double root, the graph does not cross the axis. ✓

> **ReadingMath**
>
> **Repeated Roots** Polynomial equations can have double roots, triple roots, quadruple roots, and so on. In general, these are referred to as *multiple roots*.

**b.** $x^3 + 25x = 0$

| | |
|---|---|
| $x^3 + 25x = 0$ | Original equation |
| $x(x^2 + 25) = 0$ | Factor. |
| $x = 0$ or $x^2 + 25 = 0$ | |
| $x^2 = -25$ | |
| $x = \pm\sqrt{-25}$ or $\pm 5i$ | |

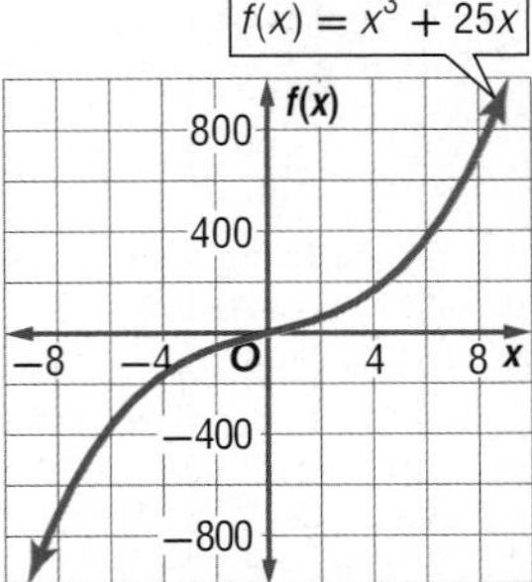


This equation has one real root, 0, and two imaginary roots, $5i$ and $-5i$.

**CHECK** The graph of this equation crosses the $x$-axis at only one place, $x = 0$. ✓

### GuidedPractice

**1A.** $x^3 + 2x = 0$

**1B.** $x^4 - 16 = 0$

**1C.** $3x^3 - x^2 + 9x - 3 = 0$

Examine the solutions for each equation in Example 1. Notice that the number of solutions for each equation is the same as the degree of each polynomial. The following corollary to the Fundamental Theorem of Algebra describes this relationship between the degree and the number of roots of a polynomial equation.

### KeyConcept Corollary to the Fundamental Theorem of Algebra

**Words** A polynomial equation of degree $n$ has exactly $n$ roots in the set of complex numbers, including repeated roots.

**Example**

| $x^3 + 2x^2 + 6$ | $4x^4 - 3x^3 + 5x - 6$ | $-2x^5 - 3x^2 + 8$ |
|---|---|---|
| 3 roots | 4 roots | 5 roots |

Similarly, an $n$th degree polynomial function has exactly $n$ zeros.

Additionally, French mathematician René Descartes discovered a relationship between the signs of the coefficients of a polynomial function and the number of positive and negative real zeros.

**StudyTip**

**Zero at the Origin** If a zero of a function is at the origin, the sum of the number of positive real zeros, negative real zeros, and imaginary zeros is reduced by how many times 0 is a zero of the function.

### KeyConcept Descartes' Rule of Signs

Let $P(x) = a_n x^n + \cdots + a_1 x + a_0$ be a polynomial function with real coefficients. Then

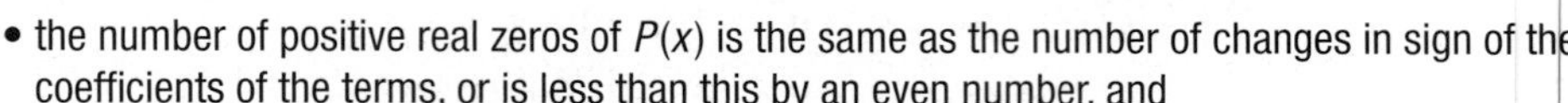

- the number of positive real zeros of $P(x)$ is the same as the number of changes in sign of the coefficients of the terms, or is less than this by an even number, and

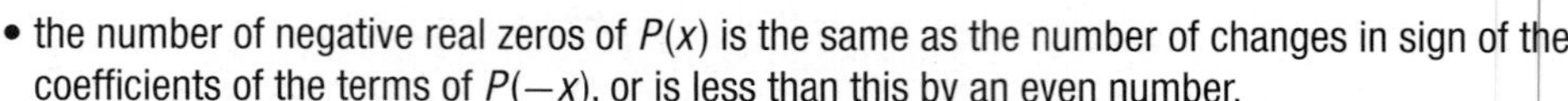

- the number of negative real zeros of $P(x)$ is the same as the number of changes in sign of the coefficients of the terms of $P(-x)$, or is less than this by an even number.

### Example 2 Find Numbers of Positive and Negative Zeros

**State the possible number of positive real zeros, negative real zeros, and imaginary zeros of $f(x) = x^6 + 3x^5 - 4x^4 - 6x^3 + x^2 - 8x + 5$.**

Because $f(x)$ has degree 6, it has six zeros, either real or imaginary. Use Descartes' Rule of Signs to determine the possible number and type of *real* zeros.

Count the number of changes in sign for the coefficients of $f(x)$.

$$f(x) = x^6 + 3x^5 - 4x^4 - 6x^3 + x^2 - 8x + 5$$

no (+ to +), yes (+ to −), no (− to −), yes (− to +), yes (+ to −), yes (− to +)

There are 4 sign changes, so there are 4, 2, or 0 positive real zeros.

Count the number of changes in sign for the coefficients of $f(-x)$.

$$f(-x) = (-x)^6 + 3(-x)^5 - 4(-x)^4 - 6(-x)^3 + (-x)^2 - 8(-x) + 5$$
$$= x^6 - 3x^5 - 4x^4 + 6x^3 + x^2 + 8x + 5$$

yes (+ to −), no (− to −), yes (− to +), no (+ to +), no (+ to +), no (+ to +)

There are 2 sign changes, so there are 2, or 0 negative real zeros.
Make a chart of the possible combinations of real and imaginary zeros.

| Number of Positive Real Zeros | Number of Negative Real Zeros | Number of Imaginary Zeros | Total Number of Zeros |
|---|---|---|---|
| 4 | 2 | 0 | $4 + 2 + 0 = 6$ |
| 4 | 0 | 2 | $4 + 0 + 2 = 6$ |
| 2 | 2 | 2 | $2 + 2 + 2 = 6$ |
| 2 | 0 | 4 | $2 + 0 + 4 = 6$ |
| 0 | 2 | 4 | $0 + 2 + 4 = 6$ |
| 0 | 0 | 6 | $0 + 0 + 6 = 6$ |

**GuidedPractice**

**2.** State the possible number of positive real zeros, negative real zeros, and imaginary zeros of $h(x) = 2x^5 + x^4 + 3x^3 - 4x^2 - x + 9$.

## 2 Find Zeros

You can use the various strategies and theorems you have learned to find all of the zeros of a function.

## Example 3 Use Synthetic Substitution to Find Zeros

**Find all of the zeros of $f(x) = x^4 - 18x^2 + 12x + 80$.**

**Step 1** Determine the total number of zeros.

Since $f(x)$ has degree 4, the function has 4 zeros.

**Step 2** Determine the type of zeros.

Examine the number of sign changes for $f(x)$ and $f(-x)$.

$f(x) = x^4 - 18x^2 + 12x + 80$ — yes, yes, no

$f(-x) = x^4 - 18x^2 - 12x + 80$ — yes, no, yes

Because there are 2 sign changes for the coefficients of $f(x)$, the function has 2 or 0 positive real zeros. Because there are 2 sign changes for the coefficients of $f(-x)$, $f(x)$ has 2 or 0 negative real zeros. Thus, $f(x)$ has 4 real zeros, 2 real zeros and 2 imaginary zeros, or 4 imaginary zeros.

**StudyTip**

Testing for Zeros If a value is not a zero for a polynomial, then it will not be a zero for the depressed polynomial either, so it does not need to be checked again.

**Step 3** Determine the real zeros.

List some possible values, and then use synthetic substitution to evaluate $f(x)$ for real values of $x$.

| x | 1 | 0 | −18 | 12 | 80 |
|---|---|---|---|---|---|
| −3 | 1 | −3 | −9 | 39 | −37 |
| −2 | 1 | −2 | −14 | 40 | 0 |
| −1 | 1 | −1 | −17 | 29 | 51 |
| 0 | 1 | 0 | −18 | 12 | 80 |
| 1 | 1 | 1 | −17 | −5 | 75 |
| 2 | 1 | 2 | −14 | −2 | 76 |

Each row shows the coefficients of the depressed polynomial and the remainder.

From the table, we can see that one zero occurs at $x = -2$. Since there are 2 negative real zeros, use synthetic substitution with the depressed polynomial function $f(x) = x^3 - 2x^2 - 14x + 40$ to find a second negative zero.

A second negative zero is at $x = -4$. Since the depressed polynomial $x^2 - 6x + 10$ is quadratic, use the Quadratic Formula to find the remaining zeros of $f(x) = x^2 - 6x + 10$.

| x | 1 | −2 | −14 | 40 |
|---|---|---|---|---|
| −4 | 1 | −6 | 10 | 0 |
| −5 | 1 | −7 | 21 | −65 |
| −6 | 1 | −8 | 34 | −164 |

$$x = \frac{-b \pm \sqrt{b^2 - 4ac}}{2a}$$ Quadratic Formula

$$= \frac{-(-6) \pm \sqrt{(-6)^2 - 4(1)(10)}}{2(1)}$$ Replace $a$ with 1, $b$ with −6, and $c$ with 10.

$$= 3 \pm i$$ Simplify.

The function has zeros at $-4$, $-2$, $3 + i$, and $3 - i$.

**StudyTip**

Locating Zeros Refer to Lesson 4-2 on how to use the CALC menu to locate a zero on your calculator.

**CHECK** Graph the function on a graphing calculator. The graph crosses the $x$-axis two times, so there are two real zeros. Use the zero function under the **CALC** menu to locate each zero. The two real zeros are −4 and −2.

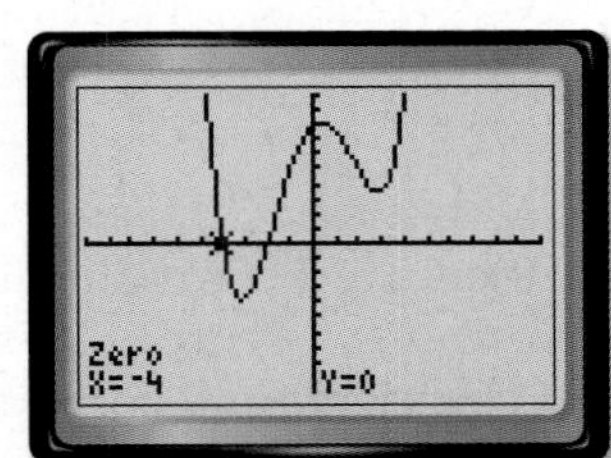


[−10, 10] scl: 1 by [−100, 100] scl: 10

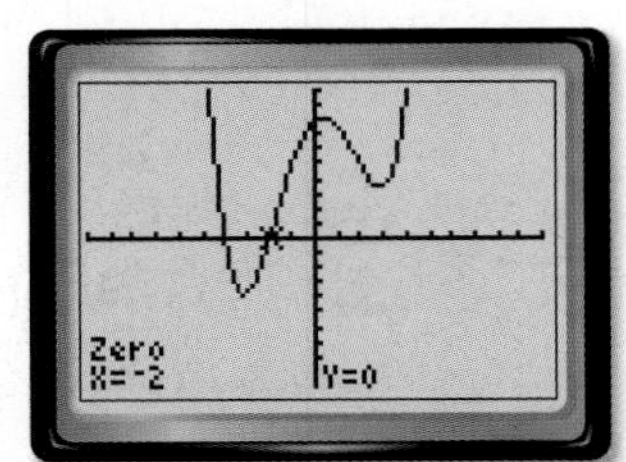


[−10, 10] scl: 1 by [−100, 100] scl: 10

### GuidedPractice

**3.** Find all of the zeros of $h(x) = x^3 + 2x^2 + 9x + 18$.

**ReviewVocabulary**

complex conjugates two complex numbers of the form $a + bi$ and $a - bi$

Previously you learned that the product of complex conjugates is always a real number and that complex roots always come in conjugate pairs. For example, if one root of $x^2 - 8x + 52 = 0$ is $4 + 6i$, then the other root is $4 - 6i$.

This applies to the zeros of polynomial functions as well. For any polynomial function with real coefficients, if an imaginary number is a zero of that function, its conjugate is also a zero. This is called the **Complex Conjugates Theorem**.

### KeyConcept Complex Conjugates Theorem

**Words** Let $a$ and $b$ be real numbers, and $b \neq 0$. If $a + bi$ is a zero of a polynomial function with real coefficients, then $a - bi$ is also a zero of the function.

**Example** If $3 + 4i$ is a zero of $f(x) = x^3 - 4x^2 + 13x + 50$, then $3 - 4i$ is also a zero of the function.

When you are given all of the zeros of a polynomial function and are asked to determine the function, convert the zeros to factors and then multiply all of the factors together. The result is the polynomial function.

PT

### Example 4 Use Zeros to Write a Polynomial Function

**Write a polynomial function of least degree with integral coefficients, the zeros of which include −1 and 5 − *i*.**

**Understand** If $5 - i$ is a zero, then $5 + i$ is also a zero according to the Complex Conjugates Theorem. So, $x + 1$, $x - (5 - i)$, and $x - (5 + i)$ are factors of the polynomial.

**Plan** Write the polynomial function as a product of its factors.

$P(x) = (x + 1)[x - (5 - i)][x - (5 + i)]$

**Solve** Multiply the factors to find the polynomial function.

| | |
|---|---|
| $P(x) = (x + 1)\,[x - (5 - i)][x - (5 + i)]$ | Write the equation. |
| $= (x + 1)\,[(x - 5) + i][(x - 5) - i]$ | Regroup terms. |
| $= (x + 1)\,[(x - 5)^2 - i^2]$ | Difference of squares |
| $= (x + 1)\,[(x^2 - 10x + 25 - (-1)]$ | Square terms. |
| $= (x + 1)\,(x^2 - 10x + 26)$ | Simplify. |
| $= x^3 - 10x^2 + 26x + x^2 - 10x + 26$ | Multiply. |
| $= x^3 - 9x^2 + 16x + 26$ | Combine like terms. |

**Check** Because there are 3 zeros, the degree of the polynomial function must be 3, so $P(x) = x^3 - 9x^2 + 16x + 26$ is a polynomial function of least degree with integral coefficients and zeros of $-1$, $5 - i$, and $5 + i$.

**GuidedPractice**

4. Write a polynomial function of least degree with integral coefficients having zeros that include $-1$ and $1 + 2i$.

## Check Your Understanding

**Example 1** **Solve each equation. State the number and type of roots.**

**1.** $x^2 - 3x - 10 = 0$

**2.** $x^3 + 12x^2 + 32x = 0$

**3.** $16x^4 - 81 = 0$

**4.** $0 = x^3 - 8$

**Example 2** **State the possible number of positive real zeros, negative real zeros, and imaginary zeros of each function.**

**5.** $f(x) = x^3 - 2x^2 + 2x - 6$

**6.** $f(x) = 6x^4 + 4x^3 - x^2 - 5x - 7$

**7.** $f(x) = 3x^5 - 8x^3 + 2x - 4$

**8.** $f(x) = -2x^4 - 3x^3 - 2x - 5$

**Example 3** **Find all of the zeros of each function.**

**9.** $f(x) = x^3 + 9x^2 + 6x - 16$

**10.** $f(x) = x^3 + 7x^2 + 4x + 28$

**11.** $f(x) = x^4 - 2x^3 - 8x^2 - 32x - 384$

**12.** $f(x) = x^4 - 6x^3 + 9x^2 + 6x - 10$

**Example 4** **Write a polynomial function of least degree with integral coefficients that have the given zeros.**

**13.** $4, -1, 6$

**14.** $3, -1, 1, 2$

**15.** $-2, 5, -3i$

**16.** $-4, 4 + i$

## Practice and Problem Solving

**Example 1** **Solve each equation. State the number and type of roots.**

**17.** $2x^2 + x - 6 = 0$

**18.** $4x^2 + 1 = 0$

**19.** $x^3 + 1 = 0$

**20.** $2x^2 - 5x + 14 = 0$

**21.** $-3x^2 - 5x + 8 = 0$

**22.** $8x^3 - 27 = 0$

**23.** $16x^4 - 625 = 0$

**24.** $x^3 - 6x^2 + 7x = 0$

**25.** $x^5 - 8x^3 + 16x = 0$

**26.** $x^5 + 2x^3 + x = 0$

**Example 2** **State the possible number of positive real zeros, negative real zeros, and imaginary zeros of each function.**

**27.** $f(x) = x^4 - 5x^3 + 2x^2 + 5x + 7$

**28.** $f(x) = 2x^3 - 7x^2 - 2x + 12$

**29.** $f(x) = -3x^5 + 5x^4 + 4x^2 - 8$

**30.** $f(x) = x^4 - 2x^2 - 5x + 19$

**31.** $f(x) = 4x^6 - 5x^4 - x^2 + 24$

**32.** $f(x) = -x^5 + 14x^3 + 18x - 36$

**Example 3** **Find all of the zeros of each function.**

**33.** $f(x) = x^3 + 7x^2 + 4x - 12$

**34.** $f(x) = x^3 + x^2 - 17x + 15$

**35.** $f(x) = x^4 - 3x^3 - 3x^2 - 75x - 700$

**36.** $f(x) = x^4 + 6x^3 + 73x^2 + 384x + 576$

**37.** $f(x) = x^4 - 8x^3 + 20x^2 - 32x + 64$

**38.** $f(x) = x^5 - 8x^3 - 9x$

**Example 4** **Write a polynomial function of least degree with integral coefficients that have the given zeros.**

**39.** $5, -2, -1$

**40.** $-4, -3, 5$

**41.** $-1, -1, 2i$

**42.** $-3, 1, -3i$

**43.** $0, -5, 3 + i$

**44.** $-2, -3, 4 - 3i$

**45.** **CCSS REASONING** A computer manufacturer determines that the profit for producing $x$ computers per day is $P(x) = -0.006x^4 + 0.15x^3 - 0.05x^2 - 1.8x$.

**a.** How many positive real zeros, negative real zeros, and imaginary zeros exist?

**b.** What is the meaning of the zeros in this situation?

**Sketch the graph of each function using its zeros.**

46. $f(x) = x^3 - 5x^2 - 2x + 24$

47. $f(x) = 4x^3 + 2x^2 - 4x - 2$

48. $f(x) = x^4 - 6x^3 + 7x^2 + 6x - 8$

49. $f(x) = x^4 - 6x^3 + 9x^2 + 4x - 12$

**Match each graph to the given zeros.**

a. $-3, 4, i, -i$

b. $-4, 3$

c. $-4, 3, i, -i$

50.

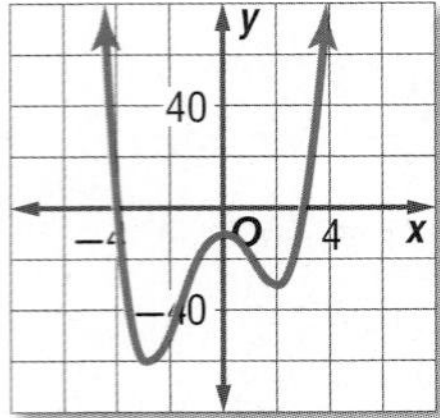


51.

52.

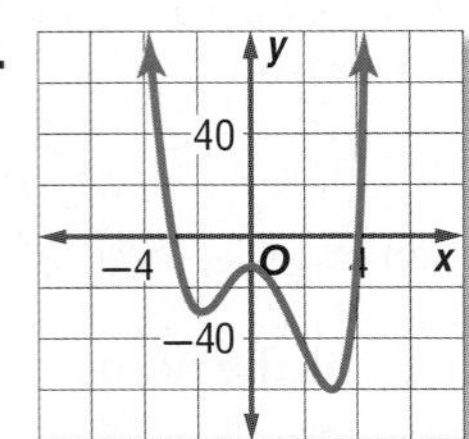


53. **CONCERTS** The amount of money Hoshi's Music Hall took in from 2003 to 2010 can be modeled by $M(x) = -2.03x^3 + 50.1x^2 - 214x + 4020$, where $x$ is the years since 2003.

a. How many positive real zeros, negative real zeros, and imaginary zeros exist?

b. Graph the function using your calculator.

c. Approximate all real zeros to the nearest tenth. What is the significance of each zero in the context of the situation?

**Determine the number of positive real zeros, negative real zeros, and imaginary zeros for each function. Explain your reasoning.**

54.

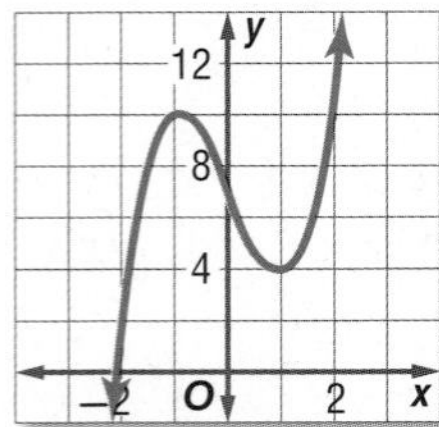


degree: 3

55.

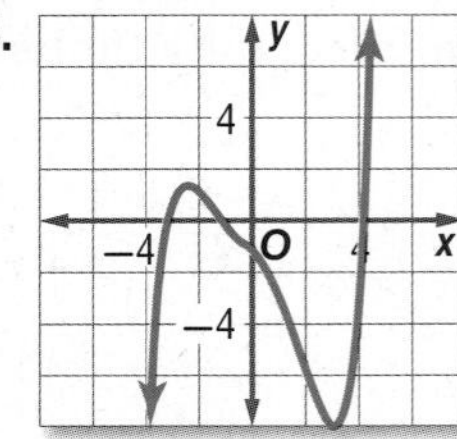


degree: 5

## H.O.T. Problems Use Higher-Order Thinking Skills

56. **OPEN ENDED** Sketch the graph of a polynomial function with:

a. 3 real, 2 imaginary zeros  b. 4 real zeros  c. 2 imaginary zeros

57. **CHALLENGE** Write an equation in factored form of a polynomial function of degree 5 with 2 imaginary zeros, 1 nonintegral zero, and 2 irrational zeros. Explain.

58. **CCSS ARGUMENTS** Determine which equation is not like the others. Explain.

$r^4 + 1 = 0$  $r^3 + 1 = 0$  $r^2 - 1 = 0$  $r^3 - 8 = 0$

59. **REASONING** Provide a counterexample for each statement.

a. All polynomial functions of degree greater than 2 have at least 1 negative real root.

b. All polynomial functions of degree greater than 2 have at least 1 positive real root.

60. **WRITING IN MATH** Explain to a friend how you would use Descartes' Rule of Signs to determine the number of possible positive real roots and the number of possible negative roots of the polynomial function $f(x) = x^4 - 2x^3 + 6x^2 + 5x - 12$.

## Standardized Test Practice

**61.** Use the graph of the polynomial function below. Which is not a factor of the polynomial $x^5 + x^4 - 3x^3 - 3x^2 - 4x - 4$?

**A** $x - 2$
**B** $x + 2$
**C** $x - 1$
**D** $x + 1$

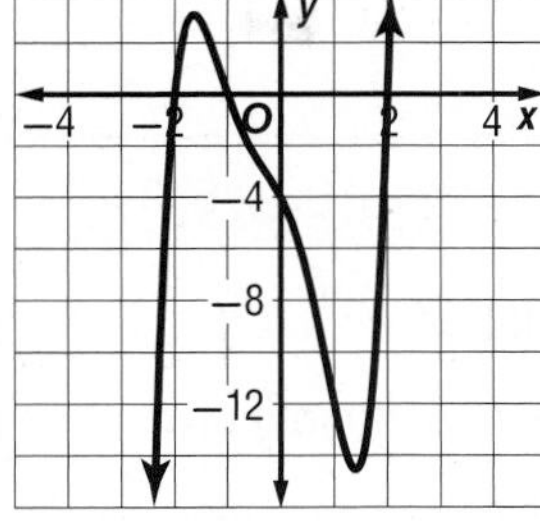


**62. SHORT RESPONSE** A window is in the shape of an equilateral triangle. Each side of the triangle is 8 feet long. The window is divided in half by a support from one vertex to the midpoint of the side of the triangle opposite the vertex. Approximately how long is the support?

**63. GEOMETRY** In rectangle $ABCD$, $\overline{AD}$ is 8 units long. What is the length of $\overline{AB}$?

**F** 4 units
**G** 8 units
**H** $8\sqrt{3}$ units
**J** 16 units

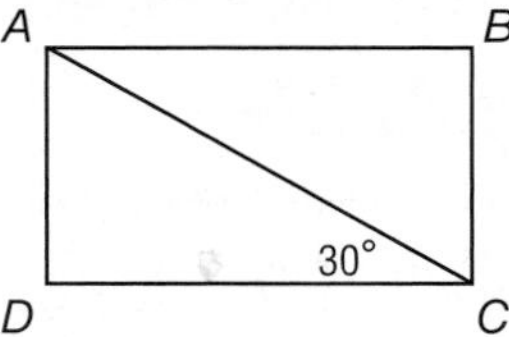


**64. SAT/ACT** The total area of a rectangle is $25a^4 - 16b^2$ square units. Which factors could represent the length and width?

**A** $(5a^2 + 4b)$ units and $(5a^2 + 4b)$ units
**B** $(5a^2 + 4b)$ units and $(5a^2 - 4b)$ units
**C** $(5a^2 - 4b)$ units and $(5a^2 - 4b)$ units
**D** $(5a - 4b)$ units and $(5a - 4b)$ units
**E** $(5a + 4b)$ units and $(5a - 4b)$ units

## Spiral Review

**65. GENETICS** Brown genes $B$ are dominant over blue genes $b$. A person with genes $BB$ or $Bb$ has brown eyes. Someone with genes $bb$ has blue eyes. Mrs. Dunn has brown eyes with genes $Bb$, and Mr. Dunn has blue eyes. Write an expression for the possible eye coloring of their children. Then find the probability that a child would have blue eyes. (Lesson 1-4)

**Factor each polynomial.** (Lesson 1-6)

**66.** $x^2 - 4x - 21$

**67.** $11x + x^2 + 30$

**68.** $32 + x^2 - 12x$

**69.** $-36 - 9x + x^2$

**70.** $x^2 + 12x + 20$

**71.** $-x + x^2 - 42$

**72. MANUFACTURING** A company is designing a box for dry pasta in the shape of a rectangular prism. The length is 2 inches more than twice the width, and the height is 3 inches more than the length. Write an expression for the volume of the box. (Lesson 1-3)

## Skills Review

**Find all of the possible values of $\pm\frac{b}{a}$ for each replacement set.**

**73.** $a = \{1, 2, 4\}$; $b = \{1, 2, 3, 6\}$

**74.** $a = \{1, 5\}$; $b = \{1, 2, 4, 8\}$

**75.** $a = \{1, 2, 3, 6\}$; $b = \{1, 7\}$

CHAPTER 1

# Study Guide and Review

## Study Guide

### KeyConcepts

**Operations with Polynomials** (Lessons 1-1 through 1-4)

- To add or subtract polynomials, add or subtract like terms.
- To multiply polynomials, use the Distributive Property.
- Special products: $(a + b)^2 = a^2 + 2ab + b^2$
  $(a - b)^2 = a^2 - 2ab + b^2$
  $(a + b)(a - b) = a^2 - b^2$

**Factoring Using the Distributive Property** (Lesson 1-5)

- Using the Distributive Property to factor polynomials with four or more terms is called factoring by grouping.
  $ax + bx + ay + by = x(a + b) + y(a + b)$
  $= (a + b)(x + y)$

**Solving Quadratic Equations by Factoring** (Lessons 1-6 through 1-8)

- To factor $x^2 + bx + c$, find $m$ and $p$ with a sum of $b$ and a product of $c$. Then write $x^2 + bx + c$ as $(x + m)(x + p)$.
- To factor $ax^2 + bx + c$, find $m$ and $p$ with a sum of $b$ and a product of $ac$. Then write as $ax^2 + mx + px + c$ and factor by grouping.
- $a^2 - b^2 = (a - b)(a + b)$

**Perfect Squares and Factoring** (Lesson 1-9)

- For a trinomial to be a perfect square, the first and last terms must be perfect squares, and the middle term must be twice the product of the square roots of the first and last terms.
- For any number $n \geq 0$, if $x^2 = n$, then $x = \pm\sqrt{n}$.

### FOLDABLES StudyOrganizer

Be sure the Key Concepts are noted in your Foldable.

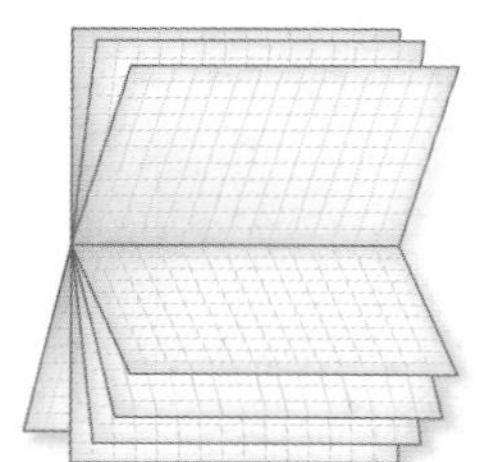

### KeyVocabulary

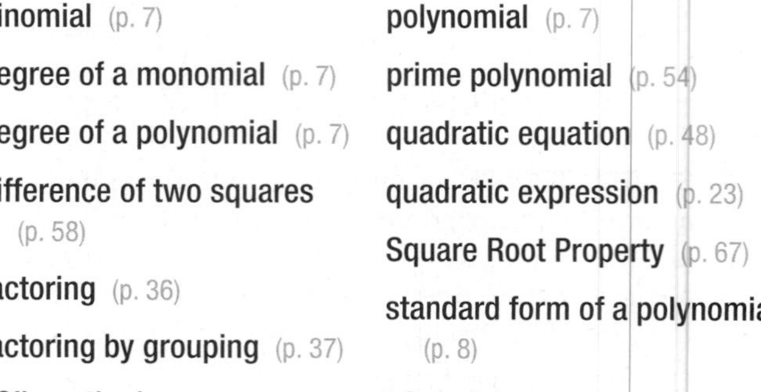

binomial (p. 7)
degree of a monomial (p. 7)
degree of a polynomial (p. 7)
difference of two squares (p. 58)
factoring (p. 36)
factoring by grouping (p. 37)
FOIL method (p. 23)
leading coefficient (p. 8)
perfect square trinomial (p. 64)
polynomial (p. 7)
prime polynomial (p. 54)
quadratic equation (p. 48)
quadratic expression (p. 23)
Square Root Property (p. 67)
standard form of a polynomial (p. 8)
trinomial (p. 7)
Zero Product Property (p. 38)

### VocabularyCheck

**State whether each sentence is *true* or *false*. If *false*, replace the underlined phrase or expression to make a true sentence.**

1. $x^2 + 5x + 6$ is an example of a prime polynomial.
2. $(x + 5)(x - 5)$ is the factorization of a difference of squares.
3. $4x^2 - 2x + 7$ is a polynomial of degree 2.
4. $(x + 5)(x - 2)$ is the factored form of $x^2 - 3x - 10$.
5. Expressions with four or more unlike terms can sometimes be factored by grouping.
6. The Zero Product Property states that if $ab = 1$, then $a$ or $b$ is 1.
7. $x^2 - 12x + 36$ is an example of a perfect square trinomial.
8. The leading coefficient of $1 + 6a + 9a^2$ is 1.
9. $x^2 - 16$ is an example of a perfect square trinomial.
10. The FOIL method is used to multiply two trinomials.

## 1-1 Adding and Subtracting Polynomials

Write each polynomial in standard form.

**11.** $x + 2 + 3x^2$

**12.** $1 - x^4$

**13.** $2 + 3x + x^2$

**14.** $3x^5 - 2 + 6x - 2x^2 + x^3$

Find each sum or difference.

**15.** $(x^3 + 2) + (-3x^3 - 5)$

**16.** $a^2 + 5a - 3 - (2a^2 - 4a + 3)$

**17.** $(4x - 3x^2 + 5) + (2x^2 - 5x + 1)$

**18.** **PICTURE FRAMES** Jean is framing a painting that is a rectangle. What is the perimeter of the frame?

### Example 1

**Write $3 - x^2 + 4x$ in standard form.**

**Step 1** Find the degree of each term.

3: degree 0

$-x^2$: degree 2

$4x$: degree 1

**Step 2** Write the terms in descending order of degree.

$3 - x^2 + 4x = -x^2 + 4x + 3$

### Example 2

**Find $(8r^2 + 3r) - (10r^2 - 5)$.**

$(8r^2 + 3r) - (10r^2 - 5)$

$= (8r^2 + 3r) + (-10r^2 + 5)$ Use the additive inverse.

$= (8r^2 - 10r^2) + 3r + 5$ Group like terms.

$= -2r^2 + 3r + 5$ Add like terms.

## 1-2 Multiplying a Polynomial by a Monomial

Solve each equation.

**19.** $x^2(x + 2) = x(x^2 + 2x + 1)$

**20.** $2x(x + 3) = 2(x^2 + 3)$

**21.** $2(4w + w^2) - 6 = 2w(w - 4) + 10$

**22.** **GEOMETRY** Find the area of the rectangle.

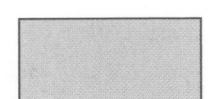


### Example 3

**Solve $m(2m - 5) + m = 2m(m - 6) + 16$.**

$m(2m - 5) + m = 2m(m - 6) + 16$

$2m^2 - 5m + m = 2m^2 - 12m + 16$

$2m^2 - 4m = 2m^2 - 12m + 16$

$-4m = -12m + 16$

$8m = 16$

$m = 2$

## 1-3 Multiplying Polynomials

Find each product.

**23.** $(x - 3)(x + 7)$

**24.** $(3a - 2)(6a + 5)$

**25.** $(3r - 7t)(2r + 5t)$

**26.** $(2x + 5)(5x + 2)$

**27.** **PARKING LOT** The parking lot shown is to be paved. What is the area to be paved?

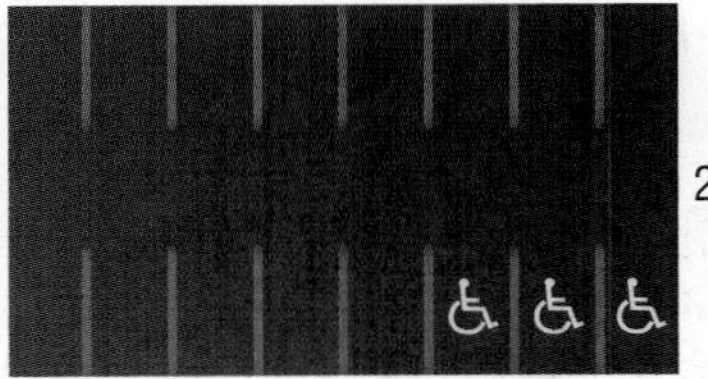


### Example 4

**Find $(6x - 5)(x + 4)$.**

$(6x - 5)(x + 4)$

F O I L

$= (6x)(x) + (6x)(4) + (-5)(x) + (-5)(4)$

$= 6x^2 + 24x - 5x - 20$ Multiply.

$= 6x^2 + 19x - 20$ Combine like terms.

# Study Guide and Review *Continued*

## 1-4 Special Products

**Find each product.**

**28.** $(x + 5)(x - 5)$

**29.** $(3x - 2)^2$

**30.** $(5x + 4)^2$

**31.** $(2x - 3)(2x + 3)$

**32.** $(2r + 5t)^2$

**33.** $(3m - 2)(3m + 2)$

**34.** **GEOMETRY** Write an expression to represent the area of the shaded region.

$2x + 5$

$x + 2$

$x - 2$

$2x - 5$

### Example 5

**Find $(x - 7)^2$.**

$(a - b)^2 = a^2 - 2ab + b^2$ — Square of a Difference

$(x - 7)^2 = x^2 - 2(x)(7) + (-7)^2$ — $a = x$ and $b = 7$

$= x^2 - 14x + 49$ — Simplify.

### Example 6

**Find $(5a - 4)(5a + 4)$.**

$(a + b)(a - b) = a^2 - b^2$ — Product of a Sum and Difference

$(5a - 4)(5a + 4) = (5a)^2 - (4)^2$ — $a = 5a$ and $b = 4$

$= 25a^2 - 16$ — Simplify.

## 1-5 Using the Distributive Property

**Use the Distributive Property to factor each polynomial.**

**35.** $12x + 24y$

**36.** $14x^2y - 21xy + 35xy^2$

**37.** $8xy - 16x^3y + 10y$

**38.** $a^2 - 4ac + ab - 4bc$

**39.** $2x^2 - 3xz - 2xy + 3yz$

**40.** $24am - 9an + 40bm - 15bn$

**Solve each equation. Check your solutions.**

**41.** $x(3x - 6) = 0$

**42.** $6x^2 = 12x$

**43.** $x^2 = 3x$

**44.** $3x^2 = 5x$

**45.** **GEOMETRY** The area of the rectangle shown is $x^3 - 2x^2 + 5x$ square units. What is the length?

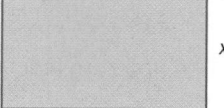

### Example 7

**Factor $12y^2 + 9y + 8y + 6$.**

$12y^2 + 9y + 8y + 6$

$= (12y^2 + 9y) + (8y + 6)$ — Group terms with common factors.

$= 3y(4y + 3) + 2(4y + 3)$ — Factor the GCF from each group.

$= (4y + 3)(3y + 2)$ — Distributive Property

### Example 8

**Solve $x^2 - 6x = 0$. Check your solutions.**

Write the equation so that it is of the form $ab = 0$.

$x^2 - 6x = 0$ — Original equation

$x(x - 6) = 0$ — Factor by using the GCF.

$x = 0$ or $x - 6 = 0$ — Zero Product Property

$x = 6$ — Solve.

The roots are 0 and 6. Check by substituting 0 and 6 for $x$ in the original equation.

## 1-6 Solving $x^2 + bx + c = 0$

**Factor each trinomial. Confirm your answers using a graphing calculator.**

**46.** $x^2 - 8x + 15$

**47.** $x^2 + 9x + 20$

**48.** $x^2 - 5x - 6$

**49.** $x^2 + 3x - 18$

**Solve each equation. Check your solutions.**

**50.** $x^2 + 5x - 50 = 0$

**51.** $x^2 - 6x + 8 = 0$

**52.** $x^2 + 12x + 32 = 0$

**53.** $x^2 - 2x - 48 = 0$

**54.** $x^2 + 11x + 10 = 0$

**55.** **ART** An artist is working on a painting that is 3 inches longer than it is wide. The area of the painting is 154 square inches. What is the length of the painting?

### Example 9

**Factor $x^2 + 10x + 21$**

$b = 10$ and $c = 21$, so $m + p$ is positive and $mp$ is positive. Therefore, $m$ and $p$ must both be positive. List the positive factors of 21, and look for the pair of factors with a sum of 10.

| Factors of 21 | Sum of 10 |
|---|---|
| 1, 21 | 22 |
| 3, 7 | 10 |

The correct factors are 3 and 7.

$$x^2 + 10x + 21 = (x + m)(x + p) \quad \text{Write the pattern.}$$
$$= (x + 3)(x + 7) \quad m = 3 \text{ and } p = 7$$

## 1-7 Solving $ax^2 + bx + c = 0$

**Factor each trinomial, if possible. If the trinomial cannot be factored, write *prime*.**

**56.** $12x^2 + 22x - 14$

**57.** $2y^2 - 9y + 3$

**58.** $3x^2 - 6x - 45$

**59.** $2a^2 + 13a - 24$

**Solve each equation. Confirm your answers using a graphing calculator.**

**60.** $40x^2 + 2x = 24$

**61.** $2x^2 - 3x - 20 = 0$

**62.** $-16t^2 + 36t - 8 = 0$

**63.** $6x^2 - 7x - 5 = 0$

**64.** **GEOMETRY** The area of the rectangle shown is $6x^2 + 11x - 7$ square units. What is the width of the rectangle?

### Example 10

**Factor $12a^2 + 17a + 6$**

$a = 12$, $b = 17$, and $c = 6$. Since $b$ is positive, $m + p$ is positive. Since $c$ is positive, $mp$ is positive. So, $m$ and $p$ are both positive. List the factors of 12(6) or 72, where both factors are positive.

| Factors of 72 | Sum of 17 |
|---|---|
| 1, 72 | 73 |
| 2, 36 | 38 |
| 3, 24 | 27 |
| 4, 18 | 22 |
| 6, 12 | 18 |
| 8, 9 | 17 |

The correct factors are 8 and 9.

$$12a^2 + 17a + 6 = 12a^2 + ma + pa + 6$$
$$= 12a^2 + 8a + 9a + 6$$
$$= (12a^2 + 8a) + (9a + 6)$$
$$= 4a(3a + 2) + 3(3a + 2)$$
$$= (3a + 2)(4a + 3)$$

So, $12a^2 + 17a + 6 = (3a + 2)(4a + 3)$.

# Study Guide and Review *Continued*

## 1-8 Differences of Squares

Factor each polynomial.

**65.** $y^2 - 81$

**66.** $64 - 25x^2$

**67.** $16a^2 - 21b^2$

**68.** $3x^2 - 3$

Solve each equation by factoring. Confirm your answers using a graphing calculator.

**69.** $a^2 - 25 = 0$

**70.** $9x^2 - 25 = 0$

**71.** $81 - y^2 = 0$

**72.** $x^2 - 5 = 20$

**73.** **EROSION** A boulder falls down a mountain into water 64 feet below. The distance $d$ that the boulder falls in $t$ seconds is given by the equation $d = 16t^2$. How long does it take the boulder to hit the water?

**Example 11**

Solve $x^2 - 4 = 12$ by factoring.

| | |
|---|---|
| $x^2 - 4 = 12$ | Original equation |
| $x^2 - 16 = 0$ | Subtract 12 from each side. |
| $x^2 - (4)^2 = 0$ | $16 = 4^2$ |
| $(x + 4)(x - 4) = 0$ | Factor the difference of squares. |
| $x + 4 = 0$ or $x - 4 = 0$ | Zero Product Property |
| $x = -4$ $\quad$ $x = 4$ | Solve each equation. |

The solutions are $-4$ and 4.

## 1-9 Perfect Squares

Factor each polynomial, if possible. If the polynomial cannot be factored write *prime*.

**74.** $x^2 + 12x + 36$

**75.** $x^2 + 5x + 25$

**76.** $9y^2 - 12y + 4$

**77.** $4 - 28a + 49a^2$

**78.** $x^4 - 1$

**79.** $x^4 - 16x^2$

Solve each equation. Confirm your answers using a graphing calculator.

**80.** $(x - 5)^2 = 121$

**81.** $4c^2 + 4c + 1 = 9$

**82.** $4y^2 = 64$

**83.** $16d^2 + 40d + 25 = 9$

**84.** **LANDSCAPING** A sidewalk of equal width is being built around a square yard. What is the width of the sidewalk?

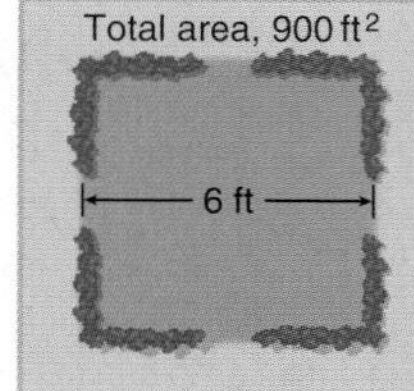


**Example 12**

Solve $(x - 9)^2 = 144$.

| | |
|---|---|
| $(x - 9)^2 = 144$ | Original equation |
| $x - 9 = \pm\sqrt{144}$ | Square Root Property |
| $x - 9 = \pm 12$ | $144 = 12 \cdot 12$ |
| $x = 9 \pm 12$ | Add 9 to each side. |
| $x = 9 + 12$ or $x = 9 - 12$ | Zero Product Property |
| $x = 21$ $\quad$ $x = -3$ | Solve. |

**CHECK**

| | |
|---|---|
| $(x - 9)^2 = 144$ | $(x - 9)^2 = 144$ |
| $(21 - 9)^2 \stackrel{?}{=} 144$ | $(-3 - 9)^2 \stackrel{?}{=} 144$ |
| $(12)^2 \stackrel{?}{=} 144$ | $(-12)^2 \stackrel{?}{=} 144$ |
| $144 = 144$ ✓ | $144 = 144$ ✓ |

## 1-10 Roots and Zeros

State the possible number of positive real zeros, negative real zeros, and imaginary zeros of each function.

**85.** $f(x) = -2x^3 + 11x^2 - 3x + 2$

**86.** $f(x) = -4x^4 - 2x^3 - 12x^2 - x - 23$

**87.** $f(x) = x^6 - 5x^3 + x^2 + x - 6$

**88.** $f(x) = -2x^5 + 4x^4 + x^2 - 3$

**89.** $f(x) = -2x^6 + 4x^4 + x^2 - 3x - 3$

### Example 13

**State the possible number of positive real zeros, negative real zeros, and imaginary zeros of $f(x) = 3x^4 + 2x^3 - 2x^2 - 26x - 48$.**

$f(x)$ has one sign change, so there is 1 positive real zero.

$f(-x)$ has 3 sign changes, so there are 3 or 1 negative real zeros.

There are 0 or 2 imaginary zeros.

CHAPTER 1

# Practice Test

Find each sum or difference.

1. $(x + 5) + (x^2 - 3x + 7)$
2. $(7m - 8n^2 + 3n) - (-2n^2 + 4m - 3n)$

3. **MULTIPLE CHOICE** Antonia is carpeting two of the rooms in her house. The dimensions are shown. Which expression represents the total area to be carpeted?

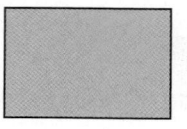

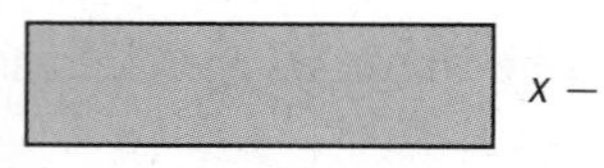


A $x^2 + 3x$
B $2x^2 + 6x - 10$
C $x^2 + 3x - 5$
D $8x + 12$

Find each product.

4. $a(a^2 + 2a - 10)$
5. $(2a - 5)(3a + 5)$
6. $(x - 3)(x^2 + 5x - 6)$
7. $(x + 3)^2$
8. $(2b - 5)(2b + 5)$

9. **FINANCIAL LITERACY** Suppose you invest \$4000 in a 2-year certificate of deposit (CD).
   a. If the interest rate is 5% per year, the expression $4000(1 + 0.05)^2$ can be evaluated to find the total amount of money after two years. Explain the numbers in this expression.
   b. Find the amount at the end of two years.
   c. Suppose you invest \$10,000 in a CD for 4 years at an annual rate of 6.25%. What is the total amount of money you will have after 4 years?

10. **MULTIPLE CHOICE** The area of the rectangle shown below is $2x^2 - x - 15$ square units. What is the width of the rectangle?

F $x - 5$
G $x + 3$
H $x - 3$
J $2x - 3$

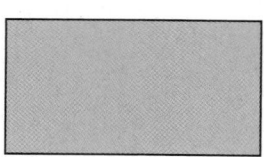

Solve each equation.

11. $5(t^2 - 3t + 2) = t(5t - 2)$
12. $3x(x + 2) = 3(x^2 - 2)$

Factor each polynomial.

13. $5xy - 10x$
14. $7ab + 14ab^2 + 21a^2b$
15. $4x^2 + 8x + x + 2$
16. $10a^2 - 50a - a + 5$

Solve each equation. Confirm your answers using a graphing calculator.

17. $y(y - 14) = 0$
18. $3x(x + 6) = 0$
19. $a^2 = 12a$

20. **MULTIPLE CHOICE** Chantel is carpeting a room that has an area of $x^2 - 100$ square feet. If the width of the room is $x - 10$ feet, what is the length of the room?

A $x - 10$ ft
B $x + 10$ ft
C $x - 100$ ft
D 10 ft

Factor each trinomial.

21. $x^2 + 7x + 6$
22. $x^2 - 3x - 28$
23. $10x^2 - x - 3$
24. $15x^2 + 7x - 2$
25. $x^2 - 25$
26. $4x^2 - 81$
27. $9x^2 - 12x + 4$
28. $16x^2 + 40x + 25$

Solve each equation. Confirm your answers using a graphing calculator.

29. $x^2 - 4x = 21$
30. $x^2 - 2x - 24 = 0$
31. $6x^2 - 5x - 6 = 0$
32. $2x^2 - 13x + 20 = 0$

33. **MULTIPLE CHOICE** Which choice is a factor of $x^4 - 1$ when it is factored completely?

F $x^2 - 1$
G $x - 1$
H $x$
J 1

# Preparing for Standardized Tests

## Solve Multi-Step Problems

Some problems that you will encounter on standardized tests require you to solve multiple parts in order to come up with the final solution. Use this lesson to practice these types of problems.

### Strategies for Solving Multi-Step Problems

**Step 1**

Read the problem statement carefully.

**Ask yourself:**

- What am I being asked to solve? What information is given?
- Are there any intermediate steps that need to be completed before I can solve the problem?

**Step 2**

Organize your approach.

- List the steps you will need to complete in order to solve the problem.
- Remember that there may be more than one possible way to solve the problem.

**Step 3**

Solve and check.

- Work as efficiently as possible to complete each step and solve.
- If time permits, check your answer.

### Standardized Test Example

**Read the problem. Identify what you need to know. Then use the information in the problem to solve.**

A florist has 80 roses, 50 tulips, and 20 lilies that he wants to use to create bouquets. He wants to create the maximum number of bouquets possible and use all of the flowers. Each bouquet should have the same number of each type of flower. How many roses will be in each bouquet?

**A** 4 roses

**B** 8 roses

**C** 10 roses

**D** 15 roses

Read the problem carefully. You are given the number of roses, tulips, and lilies and told that bouquets will be made using the same number of flowers in each. You need to find the number of roses that will be in each bouquet.

**Step 1** Find the GCF of the number of roses, tulips, and lilies.

**Step 2** Use the GCF to determine how many bouquets will be made.

**Step 3** Divide the total number of roses by the number of bouquets.

**Step 1** Write the prime factorization of each number of flowers to find the GCF.

$80 = \mathbf{2} \cdot 2 \cdot 2 \cdot 2 \cdot \mathbf{5}$

$50 = \mathbf{2} \cdot \mathbf{5} \cdot 5$

$20 = \mathbf{2} \cdot 2 \cdot \mathbf{5}$

$\text{GCF} = 2 \cdot 5 = 10$

**Step 2** The GCF of the number of roses, tulips, and lilies tells you how many bouquets can be made because each bouquet will contain the same number of flowers. So, the florist can make a total of 10 bouquets.

**Step 3** Divide the number of roses by the number of bouquets to find the number of roses in each bouquet.

$\frac{80}{10} = 8$

So, there will be 8 roses in each bouquet. The answer is B.

## Exercises

**Read each problem. Identify what you need to know. Then use the information in the problem to solve.**

1. Which of the following values is not a solution to $x^3 - 3x^2 - 25x + 75 = 0$?

   A $x = 5$
   B $x = 3$
   C $x = -3$
   D $x = -5$

2. There are 12 teachers, 90 students, and 36 parent volunteers going on a field trip. Mrs. Bartholomew wants to divide everyone into equal groups with the same number of teachers, students, and parents in each group. If she makes as many groups as possible, how many students will be in each group?

   F 6
   G 9
   H 12
   J 15

3. What is the area of the square?

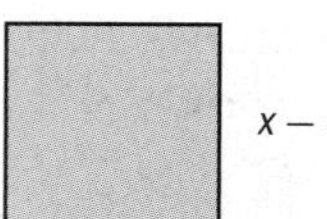

$x - 4$

   A $x^2 + 16$
   B $4x - 16$
   C $x^2 - 8x - 16$
   D $x^2 - 8x + 16$

4. Students are selling magazines to raise money for a field trip. They make $2.75 for each magazine they sell. If they want to raise $600, what is the least amount of magazines they need to sell?

   F 121
   G 177
   H 202
   J 219

CHAPTER 1

# Standardized Test Practice

## Cumulative, Chapter 1

### Multiple Choice

**1.** Which of the following is a solution to $x^2 + 6x - 112 = 0$?

F $-14$

G $-8$

H 6

J 12

**2.** Which of the following polynomials is prime?

A $5x^2 + 34x + 24$

B $4x^2 + 22x + 10$

C $4x^2 + 38x + 70$

D $5x^2 + 3x + 4$

**3.** Which of the following is not a factor of the polynomial $45a^2 - 80b^2$?

F 5

G $3a - 4b$

H $2a - 5b$

J $3a + 4b$

**4.** A rectangular gift box has dimensions that can be represented as shown in the figure. The volume of the box is $56w$ cubic inches. Which of the following is *not* a dimension of the box?

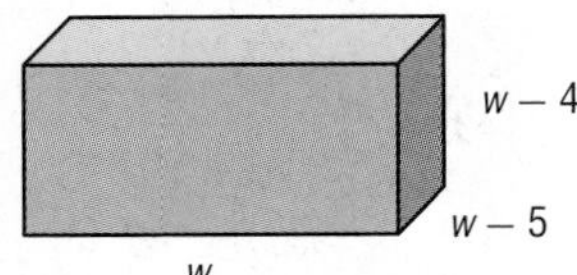


A 6 in.

B 7 in.

C 8 in.

D 12 in.

### Short Response/Gridded Response

**Record your answers on the answer sheet provided by your teacher or on a sheet of paper.**

**5.** The equation $h = -16t^2 + 40t + 3$ models the height $h$ in feet of a soccer ball after $t$ seconds. What is the height of the ball after 2 seconds?

**6.** Factor $2x^4 - 32$ completely.

**7.** **GRIDDED RESPONSE** Peggy is having a cement walkway installed around the perimeter of her swimming pool with the dimensions shown below. If $x = 3$ find the area, in square feet, of the pool and walkway.

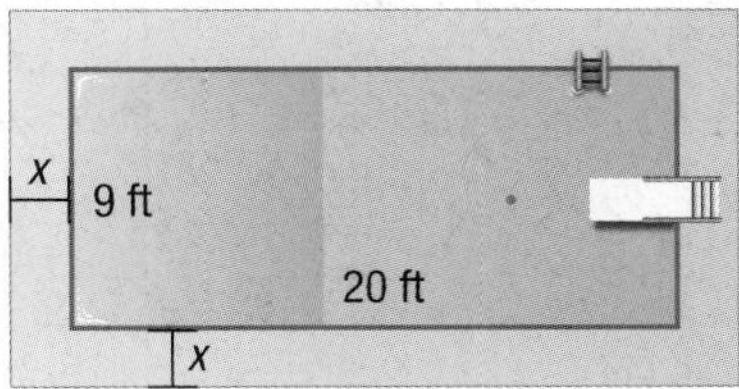


### Extended Response

**Record your answers on a sheet of paper. Show your work.**

**8.** The height in feet of a model rocket $t$ seconds after being launched into the air is given by the function $h(t) = -16t^2 + 200t$.

**a.** Write the expression that shows the height of the rocket in factored form.

**b.** At what time(s) is the height of the rocket equal to zero feet above the ground? Describe the real world meaning of your answer.

**c.** What is the greatest height reached by the model rocket? When does this occur?

**Need ExtraHelp?**

| If you missed Question... | 1 | 2 | 3 | 4 | 5 | 6 | 7 | 8 |
|---|---|---|---|---|---|---|---|---|
| Go to Lesson... | 1-6 | 1-3 | 1-7 | 1-9 | 1-7 | 1-8 | 1-3 | 1-5 |

CHAPTER 2

# Quadratic Functions and Equations

## Then

You solved quadratic equations by factoring and by using the Square Root Property.

## Now

In this chapter, you will:

- Solve quadratic equations by graphing, completing the square, and using the Quadratic Formula.
- Analyze functions with successive differences and ratios.
- Identify and graph special functions.

## Why?

**FINANCE** The value of a certain company's stock can be modeled by the function $f(x) = x^2 - 12x + 75$. By graphing this quadratic function, we can make an educated guess as to how the stock will perform in the near future.

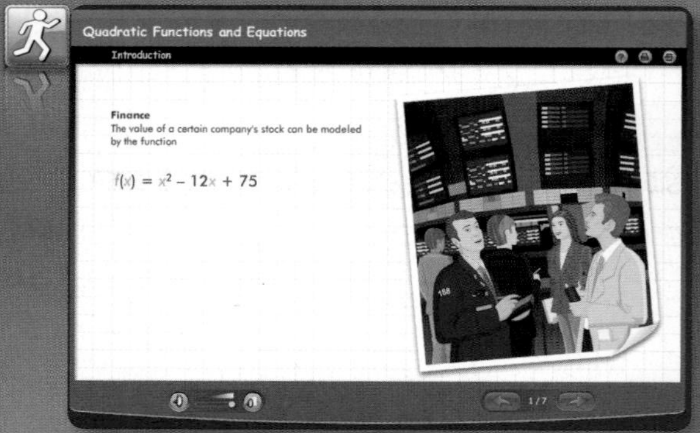


connectED.mcgraw-hill.com Your Digital Math Portal

Animation  Vocabulary  eGlossary  Personal Tutor  Virtual Manipulatives  Graphing Calculator  Audio  Foldables  Self-Check Practice  Worksheets   

Stephen Chernin/Stringer/Getty Images Entertainment/Getty Images

# Get Ready for the Chapter

**Diagnose** Readiness | You have two options for checking prerequisite skills.

**1 Textbook Option** Take the Quick Check below. Refer to the Quick Review for help.

## QuickCheck

Use a table of values to graph each equation.

1. $y = x + 3$
2. $y = 2x + 2$
3. $y = -2x - 3$
4. $y = 0.5x - 1$
5. $4x - 3y = 12$
6. $3y = 6 + 9x$
7. **SAVINGS** Jack has \$100 to buy a game system. He plans to save \$10 each week. Graph an equation to show the total amount $T$ Jack will have in $w$ weeks.

Determine whether each trinomial is a perfect square trinomial. Write *yes* or *no*. If so, factor it.

8. $a^2 + 12a + 36$
9. $x^2 + 5x + 25$
10. $x^2 - 12x + 32$
11. $x^2 + 20x + 100$
12. $4x^2 + 28x + 49$
13. $k^2 - 16k + 64$
14. $a^2 - 22a + 121$
15. $5t^2 - 12t + 25$

Evaluate each expression if $a = -2$, $b = -1$, $c = 0$, and $d = 2.5$.

16. $|a - 3|$
17. $|2a + 1|$
18. $|4 - b|$
19. $\left|\frac{1}{2}b - 2\right|$
20. $|12 - 4c|$
21. $|2c - 3| + 1$
22. $|4d - 6|$
23. $|3d - 2| - 8$

## QuickReview

**Example 1**

Use a table of values to graph $y = 3x + 1$.

| $x$ | $y = 3x + 1$ | $y$ |
|---|---|---|
| −1 | 3(−1) + 1 | −2 |
| 0 | 3(0) + 1 | 1 |
| 1 | 3(1) + 1 | 4 |
| 2 | 3(2) + 1 | 7 |

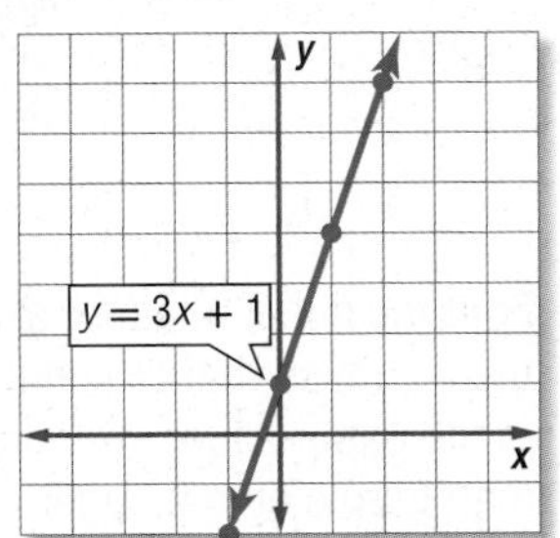


**Example 2**

Determine whether $x^2 - 10x + 25$ is a perfect square trinomial. Write *yes* or *no*. If so, factor it.

1. Is the first term a perfect square? yes
2. Is the last term a perfect square? yes
3. Is the middle term equal to $-2(1x)(5)$? yes

$x^2 - 10x + 25 = (x - 5)^2$

**Example 3**

Evaluate $|2x + 1| - 7$ if $x = -1$.

| | |
|---|---|
| $|2x + 1| - 7 = |2(-1) + 1| - 7$ | $x = -1$ |
| $= |-2 + 1| - 7$ | Multiply. |
| $= |-1| - 7$ | Add. |
| $= 1 - 7$ | $|-1| = 1$ |
| $= -6$ | Subtract. |

**2 Online Option** Take an online self-check Chapter Readiness Quiz at connectED.mcgraw-hill.com.

# Get Started on the Chapter

You will learn several new concepts, skills, and vocabulary terms as you study Chapter 2. To get ready, identify important terms and organize your resources. You may wish to refer to Chapter 0 to review prerequisite skills.

## FOLDABLES® StudyOrganizer

**Quadratic Functions and Equations** Make this Foldable to help you organize your Chapter 2 notes about quadratic functions. Begin with a sheet of notebook paper.

1 **Fold** the sheet of paper along the length so that the edge of the paper aligns with the margin rule on the paper.

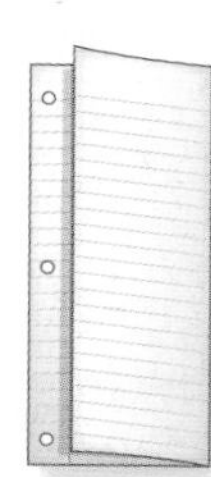

2 **Fold** the sheet twice widthwise to form four sections.

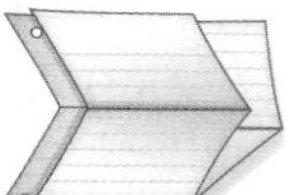

3 **Unfold** the sheet, and cut along the folds on the front flap only.

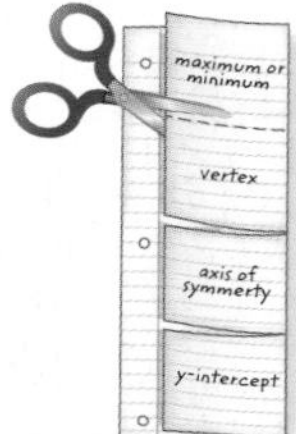


4 **Label** each section as shown.

## NewVocabulary

| English | | Español |
|---|---|---|
| quadratic function | p. 93 | función cuadrática |
| parabola | p. 93 | parábola |
| axis of symmetry | p. 93 | eje de simetría |
| vertex | p. 93 | vértice |
| minimum | p. 93 | mínimo |
| maximum | p. 93 | máximo |
| double root | p. 106 | doble raíz |
| transformation | p. 114 | transformación |
| completing the square | p. 124 | completar el cuadrado |
| Quadratic Formula | p. 133 | Formula cuadrática |
| discriminant | p. 136 | discriminante |
| step function | p. 148 | función etapa |
| greatest integer function | p. 148 | función del máximo entero |
| absolute value function | p. 149 | función del valor absoluto |

## ReviewVocabulary

domain dominio all the possible values of the independent variable, $x$

leading coefficient coeficiente delantero the coefficient of the first term of a polynomial written in standard form

range rango all the possible values of the dependent variable, $y$

In the function represented by the table, the domain is {0, 2, 4, 6}, and the range is {3, 5, 7, 9}.

| $x$ | $y$ |
|---|---|
| 0 | 3 |
| 2 | 5 |
| 4 | 7 |
| 6 | 9 |

LESSON 2-1

# Graphing Quadratic Functions

| Then | Now | Why? |
|---|---|---|
| You graphed linear and exponential functions. | 1 Analyze the characteristics of graphs of quadratic functions.<br>2 Graph quadratic functions. | The Innovention Fountain in Epcot's Futureworld in Orlando, Florida, is an elaborate display of water, light, and music. The sprayers shoot water in shapes that can be modeled by quadratic equations. You can use graphs of these equations to show the path of the water. |

**NewVocabulary**
- quadratic function
- standard form
- parabola
- axis of symmetry
- vertex
- minimum
- maximum

**Common Core State Standards**

**Content Standards**

**F.IF.4** For a function that models a relationship between two quantities, interpret key features of graphs and tables in terms of the quantities, and sketch graphs showing key features given a verbal description of the relationship.

**F.IF.7a** Graph linear and quadratic functions and show intercepts, maxima, and minima.

**Mathematical Practices**

2 Reason abstractly and quantitatively.

## 1 Characteristics of Quadratic Functions

**Quadratic functions** are nonlinear and can be written in the form $f(x) = ax^2 + bx + c$, where $a \neq 0$. This form is called the **standard form** of a quadratic function.

The shape of the graph of a quadratic function is called a **parabola**. Parabolas are symmetric about a central line called the **axis of symmetry**. The axis of symmetry intersects a parabola at only one point, called the **vertex**.

### KeyConcept Quadratic Functions

| | |
|---|---|
| Parent Function: | $f(x) = x^2$ |
| Standard Form: | $f(x) = ax^2 + bx + c$ |
| Type of Graph: | parabola |
| Axis of Symmetry: | $x = -\frac{b}{2a}$ |
| $y$-intercept: | $c$ |

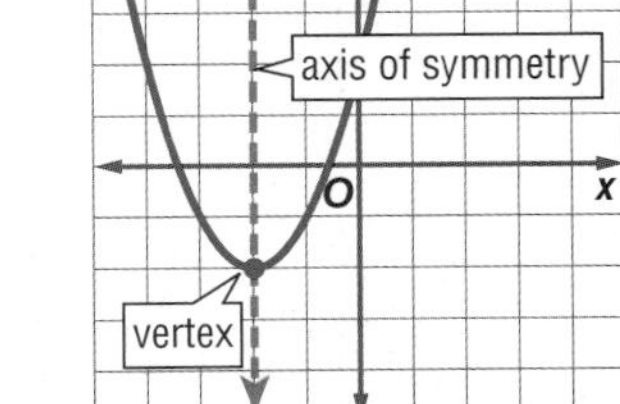


When $a > 0$, the graph of $y = ax^2 + bx + c$ opens upward. The lowest point on the graph is the **minimum**. When $a < 0$, the graph opens downward. The highest point is the **maximum**. The maximum or minimum is the vertex.

### Example 1 Graph a Parabola

**Use a table of values to graph $y = 3x^2 + 6x - 4$. State the domain and range.**

| $x$ | $y$ |
|---|---|
| 1 | 5 |
| 0 | −4 |
| −1 | −7 |
| −2 | −4 |
| −3 | 5 |

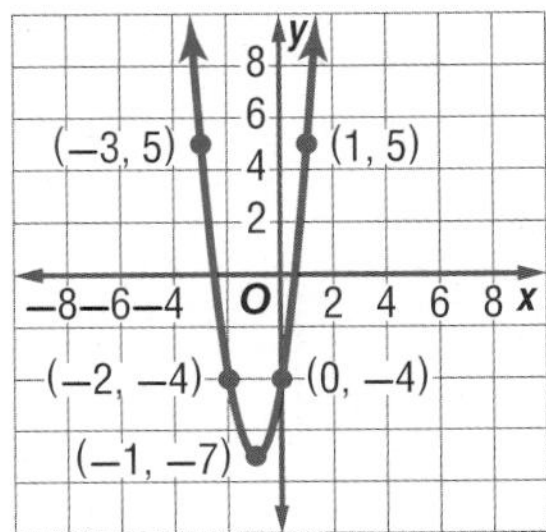


Graph the ordered pairs, and connect them to create a smooth curve. The parabola extends to infinity. The domain is all real numbers. The range is $\{y \mid y \geq -7\}$, because −7 is the minimum.

**GuidedPractice**

1. Use a table of values to graph $y = x^2 + 3$. State the domain and range.

Recall that figures with symmetry are those in which each half of the figure matches exactly.

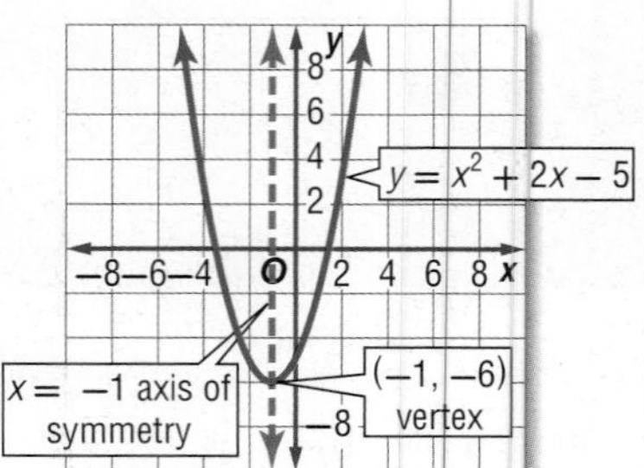


A parabola is symmetric about the axis of symmetry. Every point on the parabola to the left of the axis of symmetry has a corresponding point on the other half. The function is increasing on one side of the axis of symmetry and decreasing on the other side.

When identifying characteristics from a graph, it is often easiest to locate the vertex first. It is either the maximum or minimum point of the graph.

## Example 2 Identify Characteristics from Graphs

**Find the vertex, the equation of the axis of symmetry, and the $y$-intercept of each graph.**

**a.**

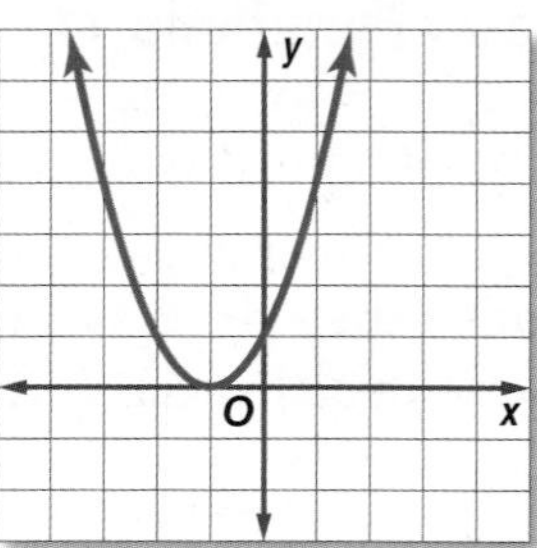

**Step 1** Find the vertex.
Because the parabola opens upward, the vertex is located at the minimum point of the parabola. It is located at $(-1, 0)$.

**Step 2** Find the axis of symmetry.
The axis of symmetry is the line that goes through the vertex and divides the parabola into congruent halves. It is located at $x = -1$.

**Step 3** Find the $y$-intercept.
The $y$-intercept is the point where the graph intersects the $y$-axis. It is located at $(0, 1)$, so the $y$-intercept is 1.

**b.**

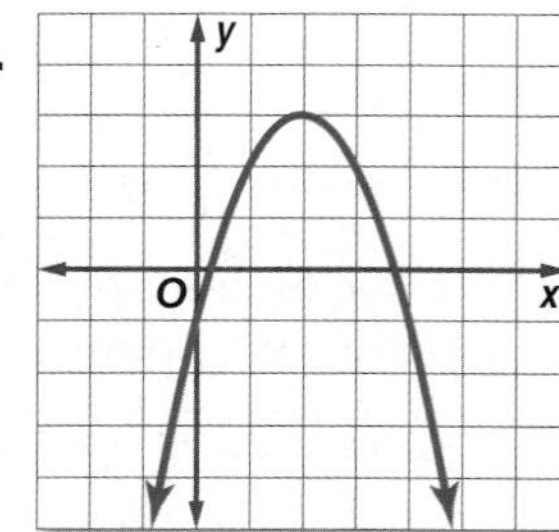

**Step 1** Find the vertex.
The parabola opens downward, so the vertex is located at its maximum point, $(2, 3)$.

**Step 2** Find the axis of symmetry.
The axis of symmetry is located at $x = 2$.

**Step 3** Find the $y$-intercept.
The $y$-intercept is where the parabola crosses the $y$-axis. It is located at $(0, -1)$, so the $y$-intercept is $-1$.

### Guided Practice

**2A.**

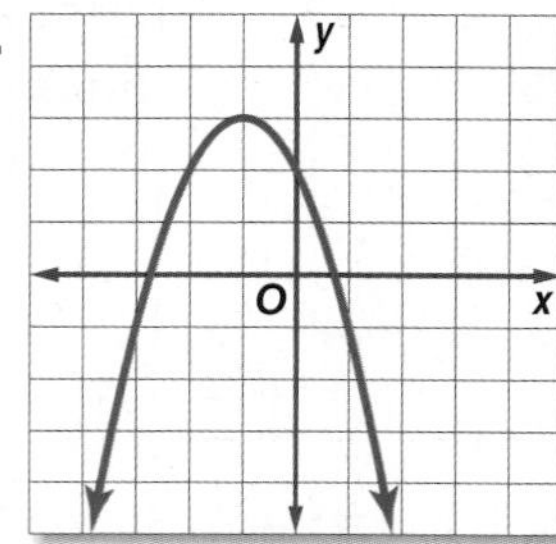

**2B.**

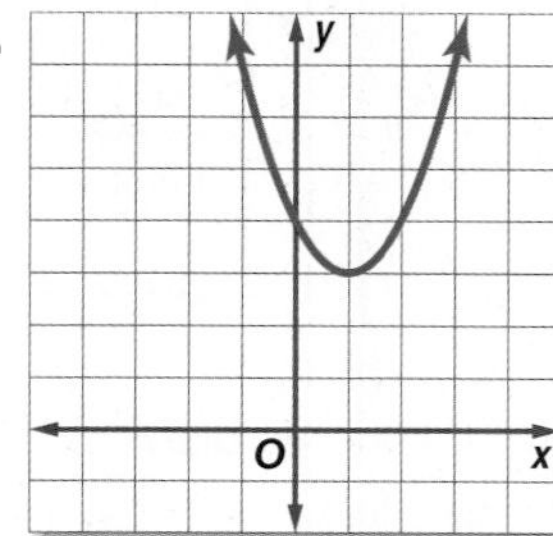

> **StudyTip**
> Function Characteristics When identifying characteristics of a function, it is often easiest to locate the axis of symmetry first.

## Example 3 Identify Characteristics from Functions

**Find the vertex, the equation of the axis of symmetry, and the $y$-intercept of each function.**

**a.** $y = 2x^2 + 4x - 3$

$x = -\frac{b}{2a}$ — Formula for the equation of the axis of symmetry

$x = -\frac{4}{2 \cdot 2}$ — $a = 2$ and $b = 4$

$x = -1$ — Simplify.

The equation for the axis of symmetry is $x = -1$.

To find the vertex, use the value you found for the axis of symmetry as the $x$-coordinate of the vertex. Find the $y$-coordinate using the original equation.

$y = 2x^2 + 4x - 3$ — Original equation

$= 2(-1)^2 + 4(-1) - 3$ — $x = -1$

$= -5$ — Simplify.

The vertex is at $(-1, -5)$.

The $y$-intercept always occurs at $(0, c)$. So, the $y$-intercept is $-3$.

**b.** $y = -x^2 + 6x + 4$

$x = -\frac{b}{2a}$ — Formula for the equation of the axis of symmetry

$x = -\frac{6}{2(-1)}$ — $a = -1$ and $b = 6$

$x = 3$ — Simplify.

The equation of the axis of symmetry is $x = 3$.

$y = -x^2 + 6x + 4$ — Original equation

$= -(3)^2 + 6(3) + 4$ — $x = 3$

$= 13$ — Simplify.

The vertex is at $(3, 13)$.

The $y$-intercept is 4.

> **StudyTip**
> *y*-intercept The $y$-coordinate of the $y$-intercept is also the constant term ($c$) of the quadratic function in standard form.

### GuidedPractice

**3A.** $y = -3x^2 + 6x - 5$

**3B.** $y = 2x^2 + 2x + 2$

Next you will learn how to identify whether the vertex is a maximum or a minimum.

## KeyConcept Maximum and Minimum Values

**Words**

The graph of $f(x) = ax^2 + bx + c$, where $a \neq 0$:

- opens upward and has a minimum value when $a > 0$, and
- opens downward and has a maximum value when $a < 0$.
- The range of a quadratic function is all real numbers greater than or equal to the minimum, or all real numbers less than or equal to the maximum.

**Examples**

$a$ is positive.

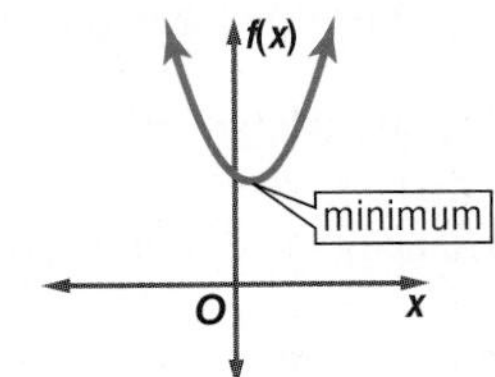


$a$ is negative.

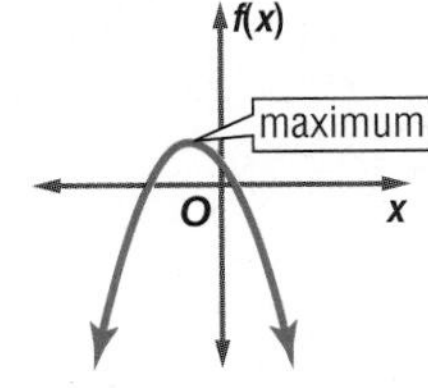

## Example 4 Maximum and Minimum Values

**Consider $f(x) = -2x^2 - 4x + 6$.**

**a. Determine whether the function has a *maximum* or *minimum* value.**

For $f(x) = -2x^2 - 4x + 6$, $a = -2$, $b = -4$, and $c = 6$.

Because $a$ is negative the graph opens down, so the function has a maximum value.

**WatchOut!**

**Minimum and Maximum Values** Don't forget to find both coordinates of the vertex $(x, y)$. The minimum or maximum value is the $y$-coordinate.

**b. State the maximum or minimum value of the function.**

The maximum value is the $y$-coordinate of the vertex.

The $x$-coordinate of the vertex is $\frac{-b}{2a}$ or $\frac{4}{2(-2)}$ or $-1$.

| | |
|---|---|
| $f(x) = -2x^2 - 4x + 6$ | Original function |
| $f(-1) = -2(-1)^2 - 4(-1) + 6$ | $x = -1$ |
| $f(-1) = 8$ | Simplify. |

The maximum value is 8.

**ReviewVocabulary**

**Domain and Range** The domain is the set of all of the possible values of the independent variable $x$. The range is the set of all of the possible values of the dependent variable $y$.

**c. State the domain and range of the function.**

The domain is all real numbers. The range is all real numbers less than or equal to the maximum value, or $\{y \mid y \leq 8\}$.

**GuidedPractice**

**Consider $g(x) = 2x^2 - 4x - 1$.**

**4A.** Determine whether the function has a *maximum* or *minimum* value.

**4B.** State the maximum or minimum value.

**4C.** State the domain and range of the function.

## 2 Graph Quadratic Functions

You have learned how to find several important characteristics of quadratic functions.

### KeyConcept Graph Quadratic Functions

**Step 1** Find the equation of the axis of symmetry.

**Step 2** Find the vertex, and determine whether it is a maximum or minimum.

**Step 3** Find the $y$-intercept.

**Step 4** Use symmetry to find additional points on the graph, if necessary.

**Step 5** Connect the points with a smooth curve.

**StudyTip**

**Symmetry and Points** When locating points that are on opposite sides of the axis of symmetry, not only are the points equidistant from the axis of symmetry, they are also equidistant from the vertex.

## Example 5 Graph Quadratic Functions

**Graph $f(x) = x^2 + 4x + 3$.**

**Step 1** Find the equation of the axis of symmetry.

$x = \frac{-b}{2a}$ Formula for the equation of the axis of symmetry

$x = \frac{-4}{2 \cdot 1}$ or $-2$ $a = 1$ and $b = 4$

**Step 2** Find the vertex, and determine whether it is a maximum or minimum.

$f(x) = x^2 + 4x + 3$ Original equation

$= (-2)^2 + 4(-2) + 3$ $x = -2$

$= -1$ Simplify.

The vertex lies at $(-2, -1)$. Because $a$ is positive the graph opens up, and the vertex is a minimum.

**Step 3** Find the $y$-intercept.

$f(x) = x^2 + 4x + 3$ Original equation

$= (0)^2 + 4(0) + 3$ $x = 0$

$= 3$ The $y$-intercept is 3.

**Step 4** The axis of symmetry divides the parabola into two equal parts. So if there is a point on one side, there is a corresponding point on the other side that is the same distance from the axis of symmetry and has the same $y$-value.

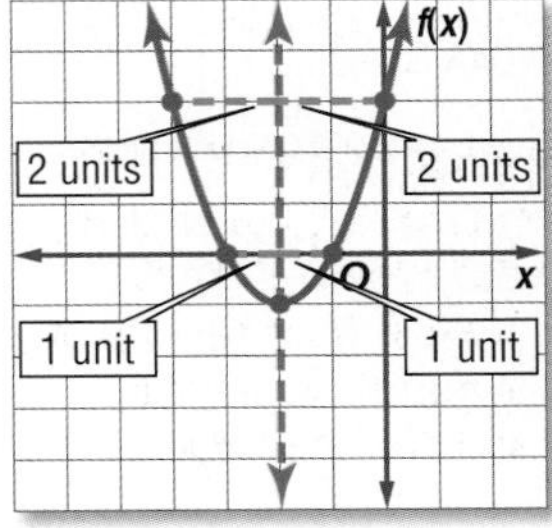


**Step 5** Connect the points with a smooth curve.

**GuidedPractice** **Graph each function.**

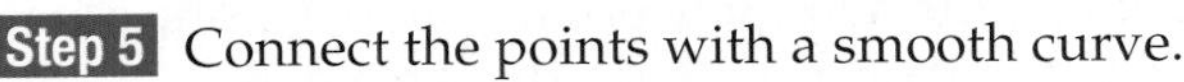

**5A.** $f(x) = -2x^2 + 2x - 1$

**5B.** $f(x) = 3x^2 - 6x + 2$

There are general differences between linear, exponential, and quadratic functions.

| | Linear Functions | Exponential Functions | Quadratic Functions |
|---|---|---|---|
| **Equation** | $y = mx + b$ | $y = ab^x$, $a \neq 0$, $b > 0$, $b \neq 1$ | $y = ax^2 + bx + c$, $a \neq 0$ |
| **Degree** | 1 | $x$ | 2 |
| **Graph** | line | curve | parabola |
| **Increasing / Decreasing** | $m > 0$: $y$ is increasing on the entire domain.<br>$m < 0$: $y$ is decreasing on the entire domain. | $a > 0$, $b > 1$ or $a < 0$, $0 < b < 1$: $y$ is increasing on the entire domain.<br>$a > 0$, $0 < b < 1$ or $a < 0$, $b > 1$: $y$ is decreasing on the entire domain. | $a > 0$: $y$ is decreasing to the left of the axis of symmetry and increasing on the right.<br>$a < 0$: $y$ is increasing to the left of the axis of symmetry and decreasing on the right. |
| **End Behavior** | $m > 0$: as $x$ increases, $y$ increases; as $x$ decreases, $y$ decreases<br>$m < 0$: as $x$ increases, $y$ decreases; as $x$ decreases, $y$ increases | $b > 1$: as $x$ decreases, $y$ approaches 0; $a > 0$, as $x$ increases, $y$ increases; $a < 0$, as $x$ increases, $y$ decreases.<br>$0 < b < 1$: as $x$ increases, $y$ approaches 0; $a > 0$, as $x$ decreases, $y$ increases; $a < 0$, as $x$ decreases, $y$ decreases. | $a > 0$: as $x$ increases, $y$ increases; as $x$ decreases, $y$ increases.<br>$a < 0$: as $x$ increases, $y$ decreases; as $x$ decreases, $y$ decreases |

You have used what you know about quadratic functions, parabolas, and symmetry to create graphs. You can analyze these graphs to solve real-world problems.

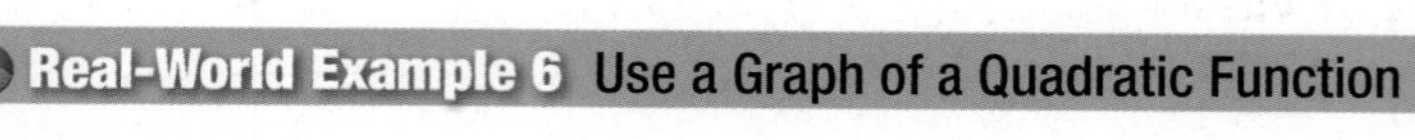

## Real-World Example 6 Use a Graph of a Quadratic Function

**SCHOOL SPIRIT The cheerleaders at Lake High School launch T-shirts into the crowd every time the Lakers score a touchdown. The height of the T-shirt can be modeled by the function $h(x) = -16x^2 + 48x + 6$, where $h(x)$ represents the height in feet of the T-shirt after $x$ seconds.**

**a. Graph the function.**

$x = -\frac{b}{2a}$ Equation of the axis of symmetry

$x = -\frac{48}{2(-16)}$ or $\frac{3}{2}$ $a = -16$ and $b = 48$

The equation of the axis of symmetry is $x = \frac{3}{2}$. Thus, the $x$-coordinate for the vertex is $\frac{3}{2}$.

$y = -16x^2 + 48x + 6$ Original equation

$= -16\left(\frac{3}{2}\right)^2 + 48\left(\frac{3}{2}\right) + 6$ $x = \frac{3}{2}$

$= -16\left(\frac{9}{4}\right) + 48\left(\frac{3}{2}\right) + 6$ $\left(\frac{3}{2}\right)^2 = \frac{9}{4}$

$= -36 + 72 + 6$ or $42$ Simplify.

The vertex is at $\left(\frac{3}{2}, 42\right)$.

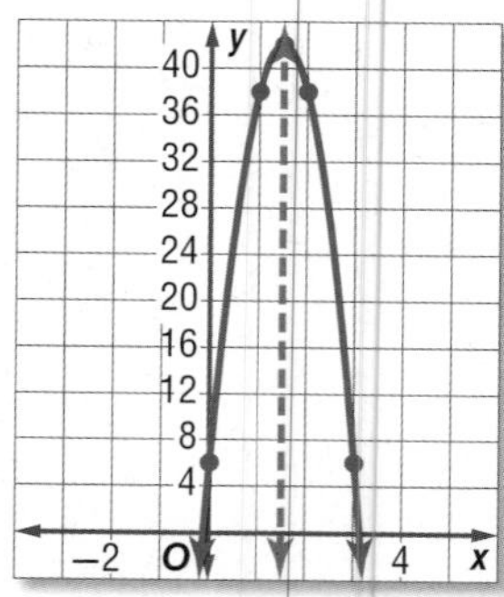


Let's find another point. Choose an $x$-value of 0 and substitute. Our new point is at (0, 6). The point paired with it on the other side of the axis of symmetry is (3, 6).

Repeat this and choose an $x$-value of 1 to get (1, 38) and its corresponding point (2, 38). Connect these points and create a smooth curve.

**b. At what height was the T-shirt launched?**

The T-shirt is launched when time equals 0, or at the $y$-intercept.

So, the T-shirt was launched 6 feet from the ground.

**c. What is the maximum height of the T-shirt? When was the maximum height reached?**

The maximum height of the T-shirt occurs at the vertex.

So the T-shirt reaches a maximum height of 42 feet. The time was $\frac{3}{2}$ or 1.5 seconds after launch.

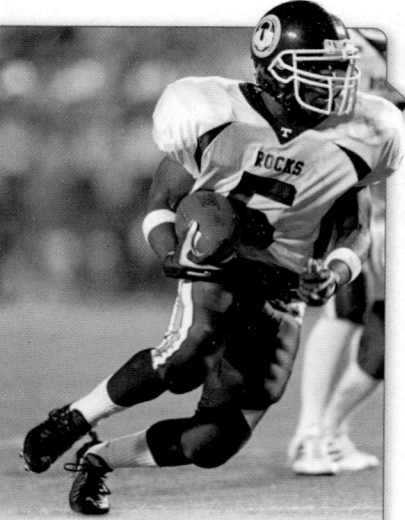

### Real-WorldLink

About 1 in 17 high school seniors playing football will go on to play football at an NCAA school.

**Source:** National Collegiate Athletic Association

### GuidedPractice

**6. TRACK** Emilio is competing in the javelin throw. The height of the javelin can be modeled by the equation $y = -16x^2 + 64x + 6$, where $y$ represents the height in feet of the javelin after $x$ seconds.

**A.** Graph the path of the javelin.

**B.** At what height is the javelin thrown?

**C.** What is the maximum height of the javelin?

Andy Lyons/Getty Images Sport/Getty Images

## Check Your Understanding

**Example 1** Use a table of values to graph each equation. State the domain and range.

1. $y = 2x^2 + 4x - 6$
2. $y = x^2 + 2x - 1$
3. $y = x^2 - 6x - 3$
4. $y = 3x^2 - 6x - 5$

**Example 2** Find the vertex, the equation of the axis of symmetry, and the $y$-intercept of each graph.

5.
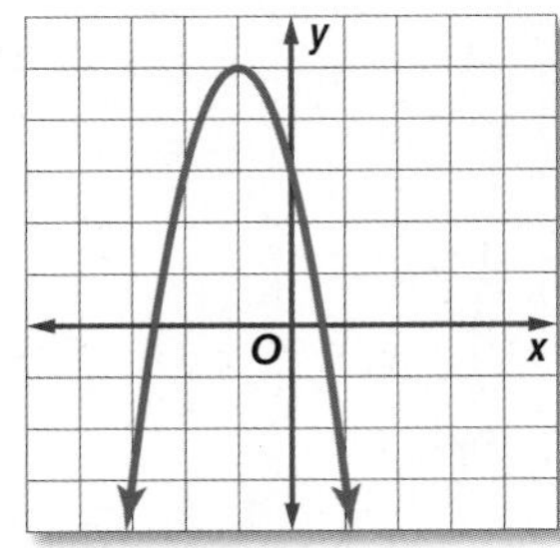


6.
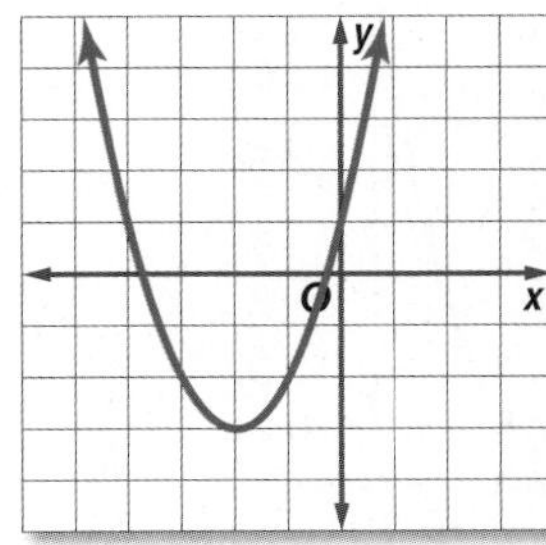


7.
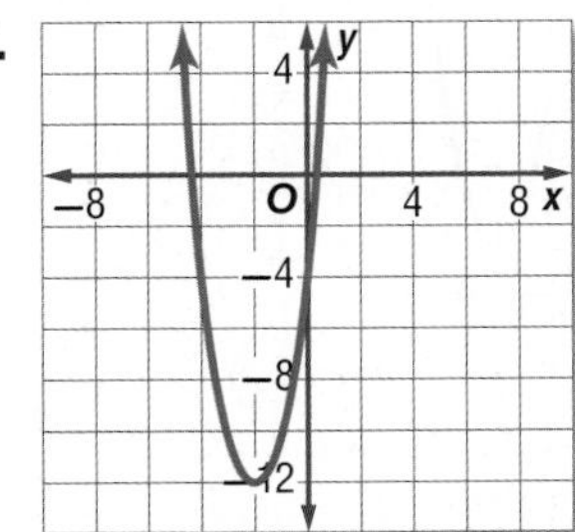


8.
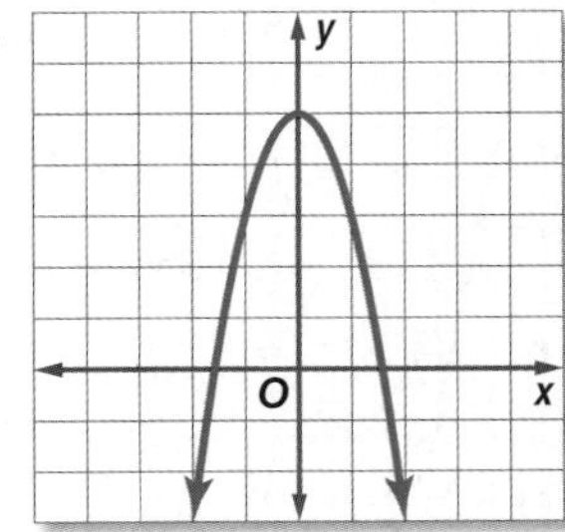


**Example 3** Find the vertex, the equation of the axis of symmetry, and the $y$-intercept of the graph of each function.

9. $y = -3x^2 + 6x - 1$
10. $y = -x^2 + 2x + 1$
11. $y = x^2 - 4x + 5$
12. $y = 4x^2 - 8x + 9$

**Example 4** Consider each function.

a. Determine whether the function has *maximum* or *minimum* value.

b. State the maximum or minimum value.

c. What are the domain and range of the function?

13. $y = -x^2 + 4x - 3$
14. $y = -x^2 - 2x + 2$
15. $y = -3x^2 + 6x + 3$
16. $y = -2x^2 + 8x - 6$

**Example 5** Graph each function.

17. $f(x) = -3x^2 + 6x + 3$
18. $f(x) = -2x^2 + 4x + 1$
19. $f(x) = 2x^2 - 8x - 4$
20. $f(x) = 3x^2 - 6x - 1$

**Example 6** 21. CCSS **REASONING** A juggler is tossing a ball into the air. The height of the ball in feet can be modeled by the equation $y = -16x^2 + 16x + 5$, where $y$ represents the height of the ball at $x$ seconds.

a. Graph this equation.

b. At what height is the ball thrown?

c. What is the maximum height of the ball?

## Practice and Problem Solving

**Example 1** Use a table of values to graph each equation. State the domain and range.

**22.** $y = x^2 + 4x + 6$ **23.** $y = 2x^2 + 4x + 7$ **24.** $y = 2x^2 - 8x - 5$

**25.** $y = 3x^2 + 12x + 5$ **26.** $y = 3x^2 - 6x - 2$ **27.** $y = x^2 - 2x - 1$

**Example 2** Find the vertex, the equation of the axis of symmetry, and the $y$-intercept of each graph.

**28.**

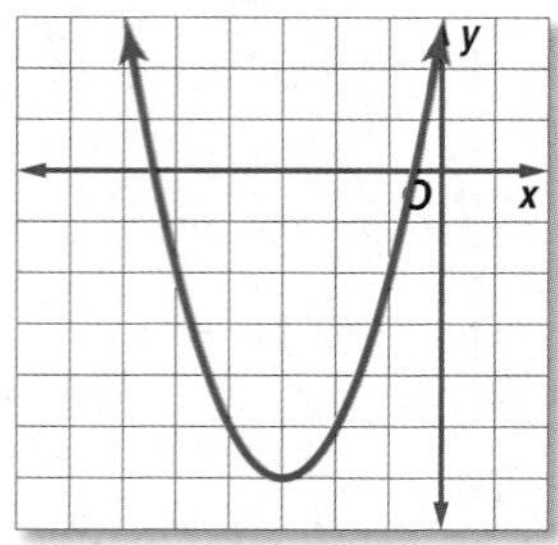


**29.**

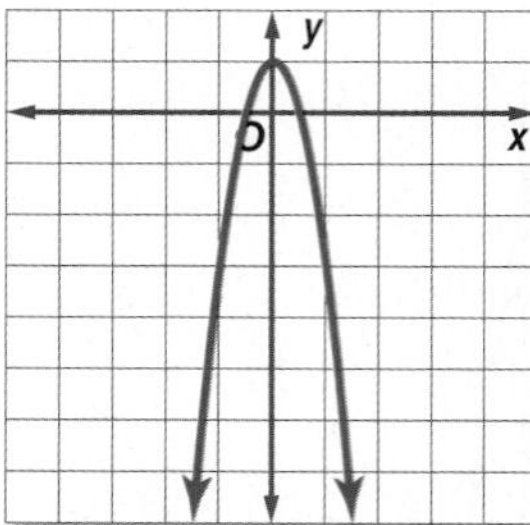


**30.**

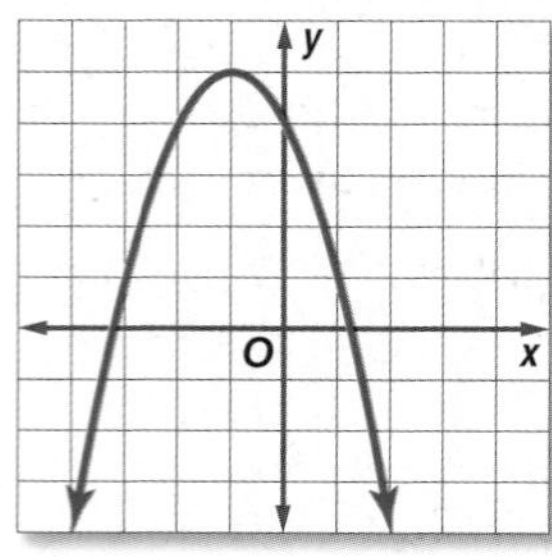


**31.**

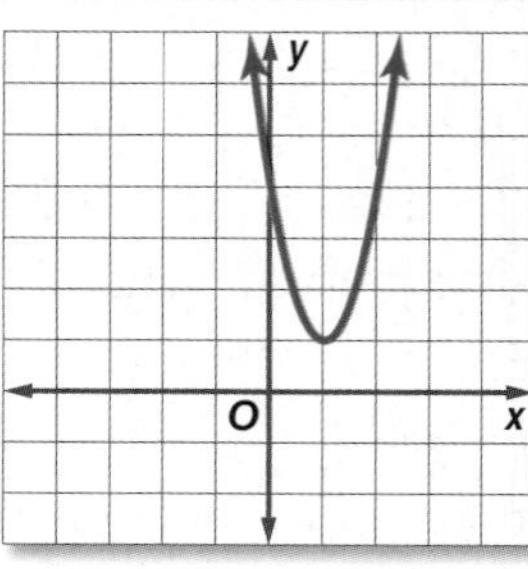


**32.**

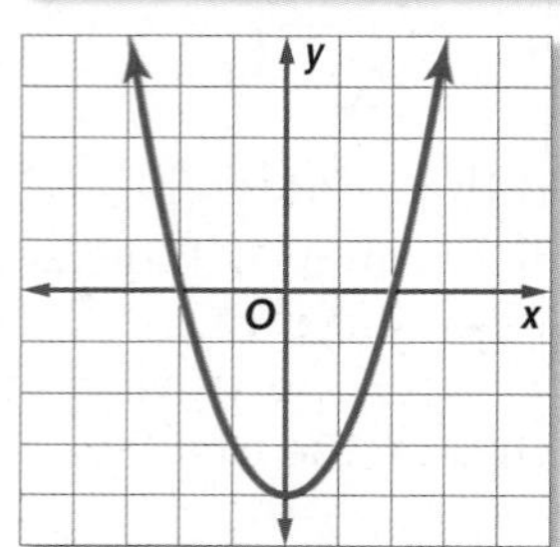


**33.**

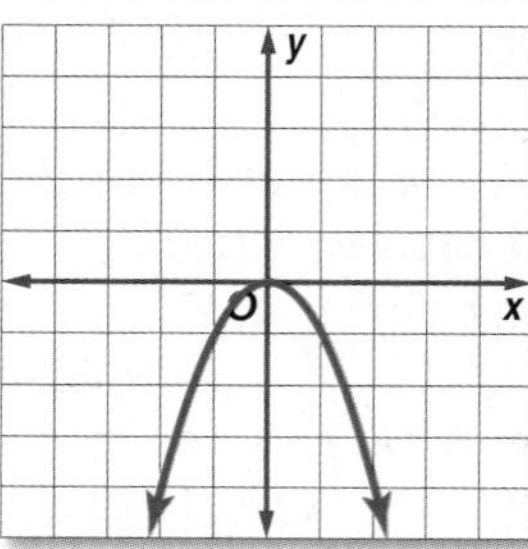


**Example 3** Find the vertex, the equation of the axis of symmetry, and the $y$-intercept of each function.

**34.** $y = x^2 + 8x + 10$ **35.** $y = 2x^2 + 12x + 10$ **36.** $y = -3x^2 - 6x + 7$

**37.** $y = -x^2 - 6x - 5$ **38.** $y = 5x^2 + 20x + 10$ **39.** $y = 7x^2 - 28x + 14$

**40.** $y = 2x^2 - 12x + 6$ **41.** $y = -3x^2 + 6x - 18$ **42.** $y = -x^2 + 10x - 13$

**Example 4** Consider each function.

**a.** Determine whether the function has a *maximum* or *minimum* value.

**b.** State the maximum or minimum value.

**c.** What are the domain and range of the function?

**43.** $y = -2x^2 - 8x + 1$ **44.** $y = x^2 + 4x - 5$ **45.** $y = 3x^2 + 18x - 21$

**46.** $y = -2x^2 - 16x + 18$ **47.** $y = -x^2 - 14x - 16$ **48.** $y = 4x^2 + 40x + 44$

**49.** $y = -x^2 - 6x - 5$ **50.** $y = 2x^2 + 4x + 6$ **51.** $y = -3x^2 - 12x - 9$

**Example 5** Graph each function.

**52.** $y = -3x^2 + 6x - 4$ **53.** $y = -2x^2 - 4x - 3$ **54.** $y = -2x^2 - 8x + 2$

**55.** $y = x^2 + 6x - 6$ **56.** $y = x^2 - 2x + 2$ **57.** $y = 3x^2 - 12x + 5$

**Example 6**

58. **BOATING** Miranda has her boat docked on the west side of Casper Point. She is boating over to the Casper Marina. The distance traveled by Miranda over time can be modeled by the equation $d = -16t^2 + 66t$, where $d$ is the number of feet she travels in $t$ minutes.

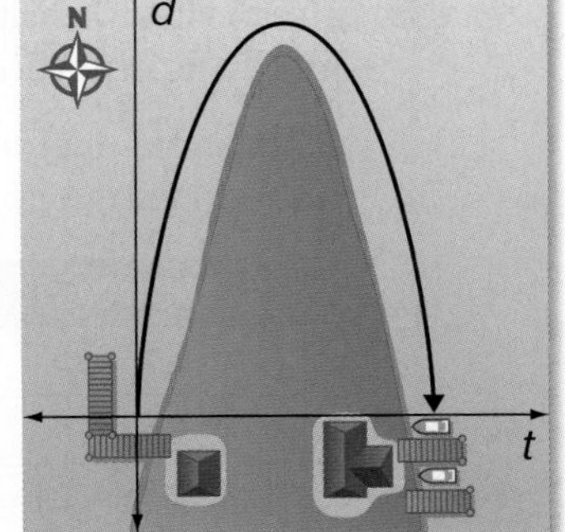


   a. Graph this equation.

   b. What is the maximum number of feet north that she traveled?

   c. How long did it take her to reach Casper Marina?

**GRAPHING CALCULATOR Graph each equation. Use the TRACE feature to find the vertex on the graph. Round to the nearest thousandth if necessary.**

59. $y = 4x^2 + 10x + 6$

60. $y = 8x^2 - 8x + 8$

61. $y = -5x^2 - 3x - 8$

62. $y = -7x^2 + 12x - 10$

63. **GOLF** The average amateur golfer can hit a ball with an initial upward velocity of 31.3 meters per second. The height can be modeled by the equation $h = -4.9t^2 + 31.3t$, where $h$ is the height of the ball, in meters, after $t$ seconds.

   a. Graph this equation. What do the portions of the graph where $h > 0$ represent in the context of the situation? What does the end behavior of the graph represent?

   b. At what height is the ball hit?

   c. What is the maximum height of the ball?

   d. How long did it take for the ball to hit the ground?

   e. State a reasonable range and domain for this situation.

64. **FUNDRAISING** The marching band is selling poinsettias to buy new uniforms. Last year the band charged \$5 each, and they sold 150. They want to increase the price this year, and they expect to lose 10 sales for each \$1 increase. The sales revenue $R$, in dollars, generated by selling the poinsettias is predicted by the function $R = (5 + p)(150 - 10p)$, where $p$ is the number of \$1 price increases.

   a. Write the function in standard form.

   b. Find the maximum value of the function.

   c. At what price should the poinsettias be sold to generate the most sales revenue? Explain your reasoning.

65. **FOOTBALL** A football is kicked up from ground level at an initial upward velocity of 90 feet per second. The equation $h = -16t^2 + 90t$ gives the height $h$ of the football after $t$ seconds.

   a. What is the height of the ball after one second?

   b. When is the ball 126 feet high?

   c. When is the height of the ball 0 feet? What do these points represent in the context of the situation?

66. **CCSS STRUCTURE** Let $f(x) = x^2 - 9$.

   a. What is the domain of $f(x)$?

   b. What is the range of $f(x)$?

   c. For what values of $x$ is $f(x)$ negative?

   d. When $x$ is a real number, what are the domain and range of $f(x) = \sqrt{x^2 - 9}$?

**67** **MULTIPLE REPRESENTATIONS** In this problem, you will investigate solving quadratic equations using tables.

a. **Algebraic** Determine the related function for each equation. Copy and complete the first two columns of the table below.

| Equation | Related Function | Zeros | $y$-Values |
|---|---|---|---|
| $x^2 - x = 12$ | | | |
| $x^2 + 8x = 9$ | | | |
| $x^2 = 14x - 24$ | | | |
| $x^2 + 16x = -28$ | | | |

b. **Graphical** Graph each related function with a graphing calculator.

c. **Analytical** The number of zeros is equal to the degree of the related function. Use the table feature on your calculator to determine the zeros of each related function. Record the zeros in the table above. Also record the values of the function one unit less than and one unit more than each zero.

d. **Verbal** Examine the function values for $x$-values just before and just after a zero. What happens to the sign of the function value before and after a zero?

## H.O.T. Problems Use Higher-Order Thinking Skills

**68. OPEN ENDED** Write a quadratic function for which the graph has an axis of symmetry of $x = -\frac{3}{8}$. Summarize your steps.

**69. ERROR ANALYSIS** Jade thinks that the parabolas represented by the graph and the description have the same axis of symmetry. Chase disagrees. Who is correct? Explain your reasoning.

a parabola that opens downward, passing through (0, 6) and having a vertex at (2, 2)

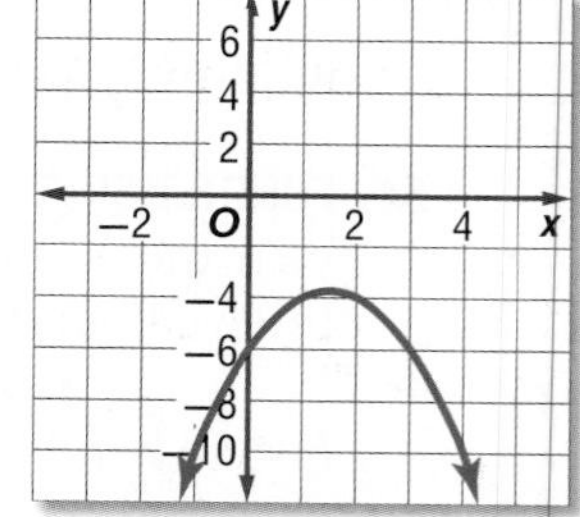


**70. CHALLENGE** Using the axis of symmetry, the $y$-intercept, and one $x$-intercept, write an equation for the graph shown.

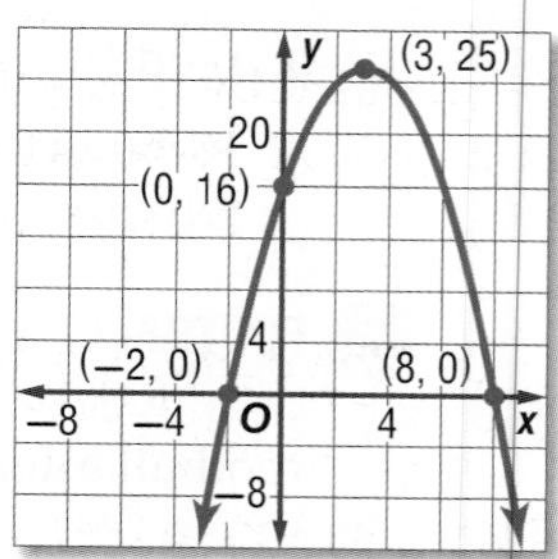


**71. CCSS STRUCTURE** The graph of a quadratic function has a vertex at (2, 0). One point on the graph is (5, 9). Find another point on the graph. Explain how you found it.

**72. OPEN ENDED** Describe a real-world situation that involves a quadratic equation. Explain what the vertex represents.

**73. REASONING** Provide a counterexample that is a specific case to show that the following statement is false. *The vertex of a parabola is always the minimum of the graph.*

**74. WRITING IN MATH** Use tables and graphs to compare and contrast an exponential function $f(x) = ab^x + c$, where $a \neq 0$, $b > 0$, and $b \neq 1$, a quadratic function $g(x) = ax^2 + c$, and a linear function $h(x) = ax + c$. Include intercepts, portions of the graph where the functions are increasing, decreasing, positive, or negative, relative maxima and minima, symmetries, and end behavior. Which function eventually exceeds the others?

## Standardized Test Practice

**75.** Which of the following is an equation for the line that passes through (2, −5) and is perpendicular to $2x + 4y = 8$?

**A** $y = 2x + 10$ **C** $y = 2x - 9$

**B** $y = -\frac{1}{2}x - 4$ **D** $y = -2x - 1$

**76. GEOMETRY** The area of the circle is $36\pi$ square units. If the radius is doubled, what is the area of the new circle?

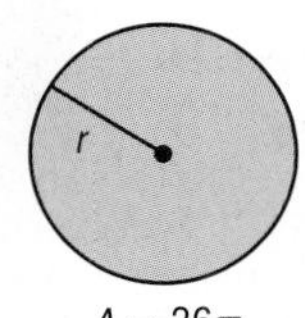


$A = 36\pi$

**F** $1296\pi$ units$^2$ **H** $72\pi$ units$^2$

**G** $144\pi$ units$^2$ **J** $9\pi$ units$^2$

**77.** What is the range of the function $f(x) = -4x^2 - \frac{1}{2}$?

**A** $\left\{\text{all integers less than or equal to } \frac{1}{2}\right\}$

**B** {all nonnegative integers}

**C** {all real numbers}

**D** $\left\{\text{all real numbers less than or equal to } -\frac{1}{2}\right\}$

**78. SHORT RESPONSE** Dylan delivers newspapers for extra money. He starts delivering the newspapers at 3:15 P.M. and finishes at 5:05 P.M. How long does it take Dylan to complete his route?

## Spiral Review

**Determine whether each trinomial is a perfect square trinomial. Write *yes* or *no*. If so, factor it.** (Lesson 1-9)

**79.** $4x^2 + 4x + 1$ **80.** $4x^2 - 20x + 25$ **81.** $9x^2 + 8x + 16$

**Factor each polynomial if possible. If the polynomial cannot be factored, write *prime*.** (Lesson 1-8)

**82.** $n^2 - 16$ **83.** $x^2 + 25$ **84.** $9 - 4a^2$

**Find each product.** (Lesson 1-3)

**85.** $(b - 7)(b + 3)$ **86.** $(c - 6)(c - 5)$ **87.** $(2x - 1)(x + 9)$

**88. MULTIPLE BIRTHS** The number of quadruplet births $Q$ in the United States in recent years can be modeled by $Q = -0.5t^3 + 11.7t^2 - 21.5t + 218.6$, where $t$ represents the number of years since 2002. What is the expected number of quadruplet births in the United States in 2017? (Lesson 1-1)

**89. FORESTRY** The number of board feet $B$ that a log will yield can be estimated by using the formula $B = \frac{L}{16}\left(D^2 - 8D + 16\right)$, where $D$ is the diameter in inches and $L$ is the log length in feet. For logs that are 16 feet long, what diameter will yield approximately 256 board feet? (Lesson 1-9)

## Skills Review

**Find the $x$-intercept of the graph of each equation.**

**90.** $x + 2y = 10$ **91.** $2x - 3y = 12$ **92.** $3x - y = -18$

EXTEND 2-1

# Algebra Lab
# Rate of Change of a Quadratic Function

**Common Core State Standards**
**Content Standards**
**F.IF.6** Calculate and interpret the average rate of change of a function (presented symbolically or as a table) over a specified interval. Estimate the rate of change from a graph.

A model rocket is launched from the ground with an upward velocity of 144 feet per second. The function $y = -16x^2 + 144x$ models the height $y$ of the rocket in feet after $x$ seconds. Using this function, we can investigate the rate of change of a quadratic function.

## Activity

**Step 1** Copy the table below.

| $x$ | 0 | 0.5 | 1.0 | 1.5 | ... | 9.0 |
|---|---|---|---|---|---|---|
| $y$ | 0 | | | | | |
| Rate of Change | – | | | | | |

**Step 2** Find the value of $y$ for each value of $x$ from 0 through 9.

**Step 3** Graph the ordered pairs $(x, y)$ on grid paper. Connect the points with a smooth curve. Notice that the function *increases* when $0 < x < 4.5$ and *decreases* when $4.5 < x < 9$.

**Step 4** Recall that the *rate of change* is the change in $y$ divided by the change in $x$. Find the rate of change for each half second interval of $x$ and $y$.

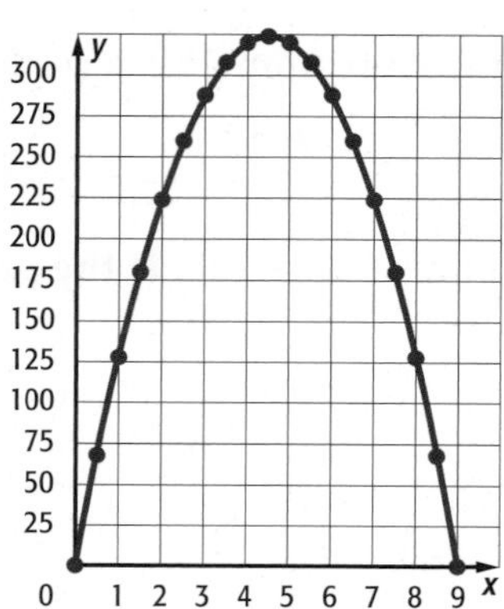


## Exercises

**Use the quadratic function $y = x^2$.**

1. Make a table, similar to the one in the Activity, for the function using $x = -4, -3, -2, -1, 0, 1, 2, 3$, and 4. Find the values of $y$ for each $x$-value.
2. Graph the ordered pairs on grid paper. Connect the points with a smooth curve. Describe where the function is increasing and where it is decreasing.
3. Find the rate of change for each column starting with $x = -3$. Compare the rates of change when the function is increasing and when it is decreasing.
4. **CHALLENGE** If an object is dropped from 100 feet in the air and air resistance is ignored, the object will fall at a rate that can be modeled by the function $f(x) = -16x^2 + 100$, where $f(x)$ represents the object's height in feet after $x$ seconds. Make a table like that in Exercise 1, selecting appropriate values for $x$. Fill in the $x$-values, the $y$-values, and rates of change. Compare the rates of change. Describe any patterns that you see.

Ron Brown/age fotostock

LESSON 2-2

# Solving Quadratic Equations by Graphing

**Then**

- You solved quadratic equations by factoring.

**Now**

1. Solve quadratic equations by graphing.
2. Estimate solutions of quadratic equations by graphing.

**Why?**

- Dorton Arena at the state fairgrounds in Raleigh, North Carolina, has a shape created by two intersecting parabolas. The shape of one of the parabolas can be modeled by $y = -x^2 + 127x$, where $x$ is the width of the parabola in feet, and $y$ is the length of the parabola in feet. The $x$-intercepts of the graph of this function can be used to find the distance between the points where the parabola meets the ground.

**NewVocabulary**

double root

**Common Core State Standards**

**Content Standards**

**A.REI.4b** Solve quadratic equations by inspection (e.g., for $x^2 = 49$), taking square roots, completing the square, the quadratic formula and factoring, as appropriate to the initial form of the equation. Recognize when the quadratic formula gives complex solutions and write them as $a \pm bi$ for real numbers $a$ and $b$.

**F.IF.7a** Graph linear and quadratic functions and show intercepts, maxima, and minima.

**Mathematical Practices**

3 Construct viable arguments and critique the reasoning of others.

6 Attend to precision.

## 1 Solve by Graphing

A quadratic equation can be written in the standard form $ax^2 + bx + c = 0$, where $a \neq 0$. To write a quadratic function as an equation, replace $y$ or $f(x)$ with 0. Recall that the solutions or roots of an equation can be identified by finding the $x$-intercepts of the related graph.

### KeyConcept Solutions of Quadratic Equations

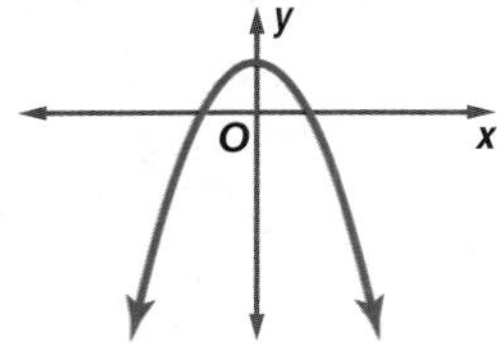


*two* unique real solutions

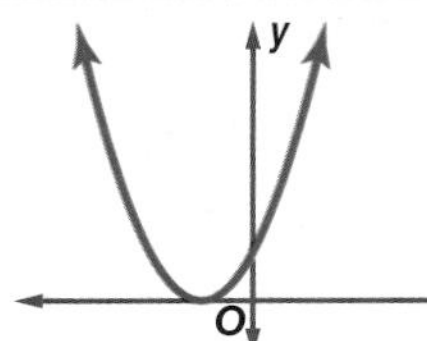


*one* unique real solution

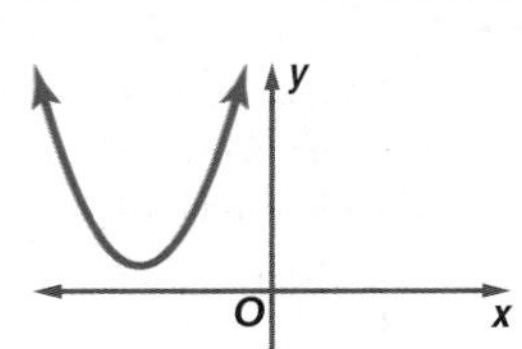


*no* real solutions

### Example 1 Two Roots

**Solve $x^2 - 2x - 8 = 0$ by graphing.**

Graph the related function $f(x) = x^2 - 2x - 8$.

The $x$-intercepts of the graph appear to be at $-2$ and 4, so the solutions are $-2$ and 4.

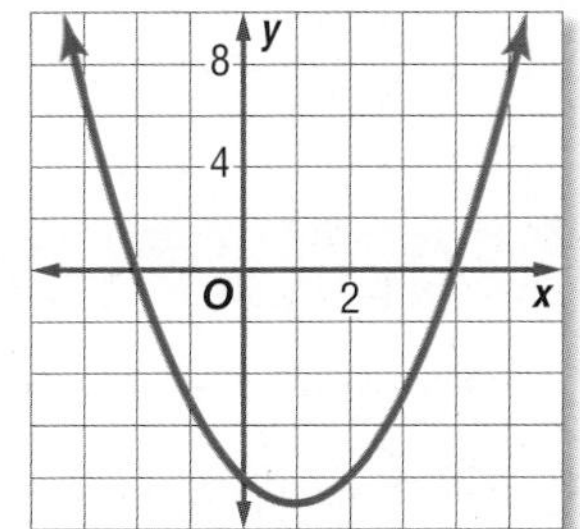


**CHECK** Check each solution in the original equation.

| | | |
|---|---|---|
| $x^2 - 2x - 8 = 0$ | Original equation | $x^2 - 2x - 8 = 0$ |
| $(-2)^2 - 2(-2) - 8 \stackrel{?}{=} 0$ | $x = -2$ or $x = 4$ | $(4)^2 - 2(4) - 8 \stackrel{?}{=} 0$ |
| $0 = 0$ ✓ | Simplify. | $0 = 0$ ✓ |

**GuidedPractice** **Solve each equation by graphing.**

**1A.** $-x^2 - 3x + 18 = 0$

**1B.** $x^2 - 4x + 3 = 0$

N.C. Department of Agriculture and Consumer Services

The solutions in Example 1 were two distinct numbers. Sometimes the two roots are the same number, called a **double root**.

### Example 2 Double Root

**Solve $x^2 - 6x = -9$ by graphing.**

**Step 1** Rewrite the equation in standard form.

$x^2 - 6x = -9$ Original equation

$x^2 - 6x + 9 = 0$ Add 9 to each side.

**Step 2** Graph the related function $f(x) = x^2 - 6x + 9$.

**Step 3** Locate the $x$-intercepts of the graph. Notice that the vertex of the parabola is the only $x$-intercept. Therefore, there is only one solution, 3.

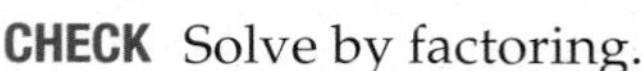

**CHECK** Solve by factoring.

$x^2 - 6x + 9 = 0$ Original equation

$(x - 3)(x - 3) = 0$ Factor.

$x - 3 = 0$ or $x - 3 = 0$ Zero Product Property

$x = 3$ $x = 3$ Add 3 to each side.

The only solution is 3.

**WatchOut!**

**CCSS Precision** Solutions found from the graph of an equation may appear to be exact. Check them in the original equation to be sure.

**GuidedPractice**

**Solve each equation by graphing.**

**2A.** $x^2 + 25 = 10x$

**2B.** $x^2 = -8x - 16$

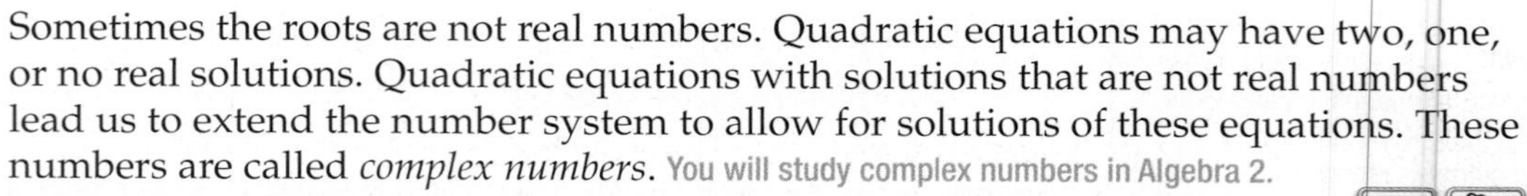

Sometimes the roots are not real numbers. Quadratic equations may have two, one, or no real solutions. Quadratic equations with solutions that are not real numbers lead us to extend the number system to allow for solutions of these equations. These numbers are called *complex numbers*. You will study complex numbers in Algebra 2.

### Example 3 No Real Roots

**Solve $2x^2 - 3x + 5 = 0$ by graphing.**

**Step 1** Rewrite the equation in standard form.
This equation is written in standard form.

**Step 2** Graph the related function $f(x) = 2x^2 - 3x + 5$.

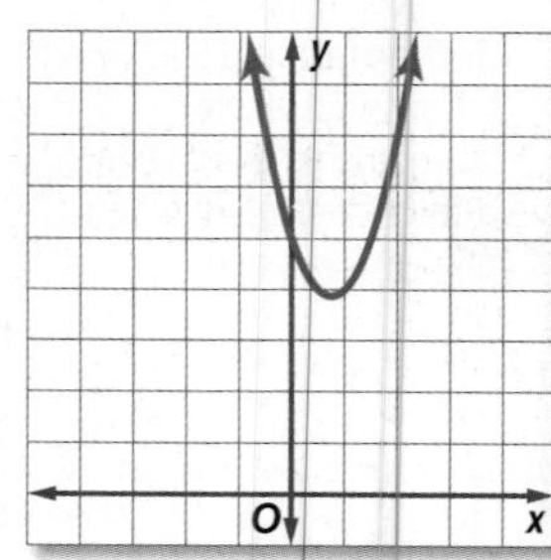


**Step 3** Locate the $x$-intercepts of the graph. This graph has no $x$-intercepts. Therefore, this equation has no real number solutions. The solution set is $\varnothing$.

**GuidedPractice**

**Solve each equation by graphing.**

**3A.** $-x^2 - 3x = 5$

**3B.** $-2x^2 - 8 = 6x$

**2 Estimate Solutions** The real roots found thus far have been integers. However, the roots of quadratic equations are usually not integers. In these cases, use estimation to approximate the roots of the equation.

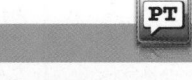

### Example 4 Approximate Roots with a Table

**StudyTip**

Location of Zeros
Since quadratic functions are continuous, there must be a zero between two $x$-values for which the corresponding $y$-values have opposite signs.

**Solve $x^2 + 6x + 6 = 0$ by graphing. If integral roots cannot be found, estimate the roots to the nearest tenth.**

Graph the related function $f(x) = x^2 + 6x + 6$.

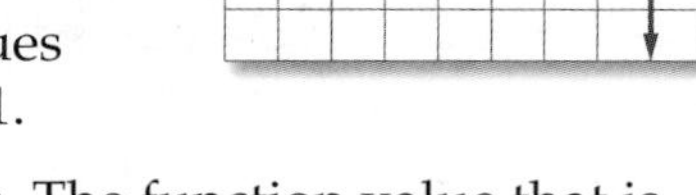

The $x$-intercepts are located between $-5$ and $-4$ and between $-2$ and $-1$.

Make a table using an increment of 0.1 for the $x$-values located between $-5$ and $-4$ and between $-2$ and $-1$.

Look for a change in the signs of the function values. The function value that is closest to zero is the best approximation for a zero of the function.

| $x$ | −4.9 | −4.8 | −4.7 | −4.6 | −4.5 | −4.4 | −4.3 | −4.2 | −4.1 |
|---|---|---|---|---|---|---|---|---|---|
| $y$ | 0.61 | 0.24 | −0.11 | −0.44 | −0.75 | −1.04 | −1.31 | −1.56 | −1.79 |

| $x$ | −1.9 | −1.8 | −1.7 | −1.6 | −1.5 | −1.4 | −1.3 | −1.2 | −1.1 |
|---|---|---|---|---|---|---|---|---|---|
| $y$ | −1.79 | −1.56 | −1.31 | −1.04 | −0.75 | −0.44 | −0.11 | 0.24 | 0.61 |

For each table, the function value that is closest to zero when the sign changes is $-0.11$. Thus, the roots are approximately $-4.7$ and $-1.3$.

**GuidedPractice**

**4.** Solve $2x^2 + 6x - 3 = 0$ by graphing. If integral roots cannot be found, estimate the roots to the nearest tenth.

**Real-WorldLink**

The game of soccer, called "football" outside of North America, began in 1863 in Britain when the Football Association was founded. Soccer is played on every continent of the world.

**Source:** Sports Know How

moodboard/Alamy

Approximating the $x$-intercepts of graphs is helpful for real-world applications.

### Real-World Example 5 Approximate Roots with a Calculator

**SOCCER A goalie kicks a soccer ball with an upward velocity of 65 feet per second, and her foot meets the ball 1 foot off the ground. The quadratic function $h = -16t^2 + 65t + 1$ represents the height of the ball $h$ in feet after $t$ seconds. Approximately how long is the ball in the air?**

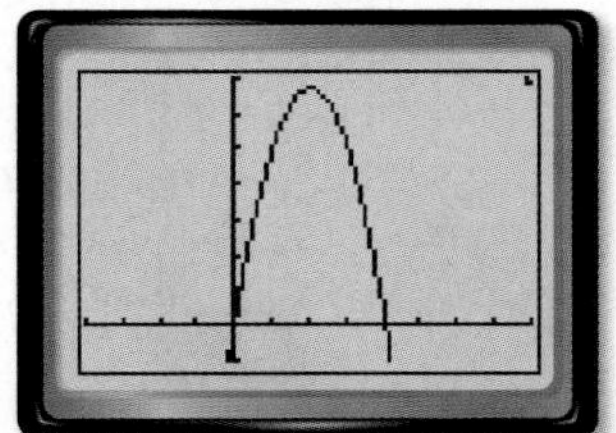

[−4, 7] scl: 1 by [−10, 70] scl: 10

You need to find the roots of the equation $-16t^2 + 65t + 1 = 0$. Use a graphing calculator to graph the related function $f(x) = -16t^2 + 65t + 1$.

The positive $x$-intercept of the graph is approximately 4. Therefore, the ball is in the air for approximately 4 seconds.

**GuidedPractice**

**5.** If the goalie kicks the soccer ball with an upward velocity of 55 feet per second and his foot meets the ball 2 feet off the ground, approximately how long is the ball in the air?

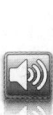

## Check Your Understanding

**Examples 1–3** **Solve each equation by graphing.**

1. $x^2 + 3x - 10 = 0$
2. $2x^2 - 8x = 0$
3. $x^2 + 4x = -4$
4. $x^2 + 12 = -8x$

**Example 4** **Solve each equation by graphing. If integral roots cannot be found, estimate the roots to the nearest tenth.**

5. $-x^2 - 5x + 1 = 0$
6. $-9 = x^2$
7. $x^2 = 25$
8. $x^2 - 8x = -9$

**Example 5**

9. **SCIENCE FAIR** Ricky built a model rocket. Its flight can be modeled by the equation shown, where $h$ is the height of the rocket in feet after $t$ seconds. About how long was Ricky's rocket in the air?

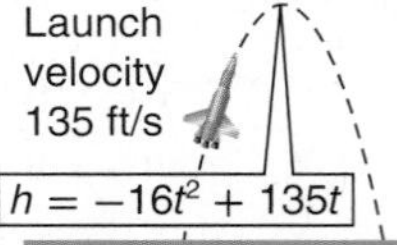


## Practice and Problem Solving

**Examples 1–3** **Solve each equation by graphing.**

10. $x^2 + 7x + 14 = 0$
11. $x^2 + 2x - 24 = 0$
12. $x^2 - 16x + 64 = 0$
13. $x^2 - 5x + 12 = 0$
14. $x^2 + 14x = -49$
15. $x^2 = 2x - 1$
16. $x^2 - 10x = -16$
17. $-2x^2 - 8x = 13$
18. $2x^2 - 16x = -30$
19. $2x^2 = -24x - 72$
20. $-3x^2 + 2x = 15$
21. $x^2 = -2x + 80$

**Example 4** **Solve each equation by graphing. If integral roots cannot be found, estimate the roots to the nearest tenth.**

22. $x^2 + 2x - 9 = 0$
23. $x^2 - 4x = 20$
24. $x^2 + 3x = 18$
25. $2x^2 - 9x = -8$
26. $3x^2 = -2x + 7$
27. $5x = 25 - x^2$

**Example 5**

28. **SOFTBALL** The equation $h = -16t^2 + 47t + 3$ models the height $h$, in feet, of a ball that Sofia hits after $t$ seconds. How long is the ball in the air?

29. **RIDES** The Terror Tower launches riders straight up and returns straight down. The equation $h = -16t^2 + 122t$ models the height $h$, in feet, of the riders from their starting position after $t$ seconds. How long is it until the riders return to the bottom?

**Use factoring to determine how many times the graph of each function intersects the $x$-axis. Identify each zero.**

30. $y = x^2 - 8x + 16$
31. $y = x^2 + 4x + 4$
32. $y = x^2 + 2x - 24$
33. $y = x^2 + 12x + 32$

34. **NUMBER THEORY** Use a quadratic equation to find two numbers that have a sum of 9 and a product of 20.

35. **NUMBER THEORY** Use a quadratic equation to find two numbers that have a sum of 1 and a product of −12.

36. **CCSS MODELING** The height of a golf ball in the air can be modeled by the equation $h = -16t^2 + 76t$, where $h$ is the height in feet of the ball after $t$ seconds.
    - **a.** How long was the ball in the air?
    - **b.** What is the ball's maximum height?
    - **c.** When will the ball reach its maximum height?

**37** **SKIING** Stefanie is in a freestyle aerial competition. The equation $h = -16t^2 + 30t + 10$ models Stefanie's height $h$, in feet, $t$ seconds after leaving the ramp.

**a.** How long is Stefanie in the air?

**b.** When will Stefanie reach a height of 15 feet?

**c.** To earn bonus points in the competition, you must reach a height of 20 feet. Will Stefanie earn bonus points?

**38.** **MULTIPLE REPRESENTATIONS** In this problem, you will explore how to further interpret the relationship between quadratic functions and graphs.

**a. Graphical** Graph $y = x^2$.

**b. Analytical** Name the vertex and two other points on the graph.

**c. Graphical** Graph $y = x^2 + 2$, $y = x^2 + 4$, and $y = x^2 + 6$ on the same coordinate plane as the previous graph.

**d. Analytical** Name the vertex and two points from each of these graphs that have the same $x$-coordinates as the first graph.

**e. Analytical** What conclusion can you draw from this?

**GRAPHING CALCULATOR** **Solve each equation by graphing.**

**39.** $x^3 - 3x^2 - 6x + 8 = 0$

**40.** $x^3 - 8x^2 + 15x = 0$

## H.O.T. Problems Use Higher-Order Thinking Skills

**41.** **CCSS CRITIQUE** Iku and Zachary are finding the number of real zeros of the function graphed at the right. Iku says that the function has no real zeros because there are no $x$-intercepts. Zachary says that the function has one real zero because the graph has a $y$-intercept. Is either of them correct? Explain your reasoning.

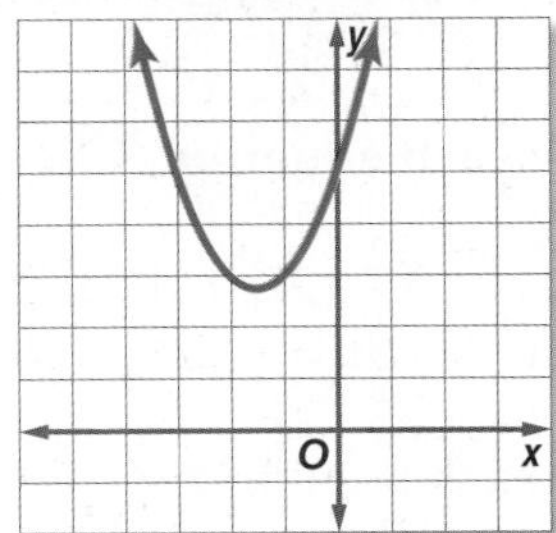


**42.** **OPEN ENDED** Describe a real-world situation in which a thrown object travels in the air. Write an equation that models the height of the object with respect to time, and determine how long the object travels in the air.

**43.** **REASONING** The graph shown is that of a *quadratic inequality*. Analyze the graph, and determine whether the $y$-value of a solution of the inequality is *sometimes*, *always*, or *never* greater than 2. Explain.

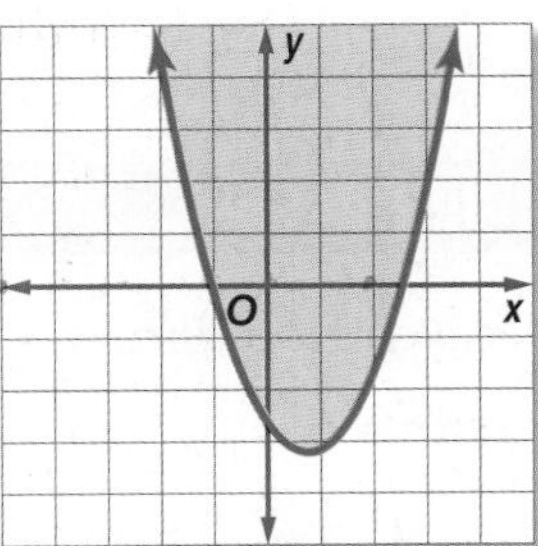


**44.** **CHALLENGE** Write a quadratic equation that has the roots described.

**a.** one double root

**b.** one rational (nonintegral) root and one integral root

**c.** two distinct integral roots that are additive opposites.

**45.** **CHALLENGE** Find the roots of $x^2 = 2.25$ without using a calculator. Explain your strategy.

**46.** **WRITING IN MATH** Explain how to approximate the roots of a quadratic equation when the roots are not integers.

## Standardized Test Practice

**47.** Adrahan earned 50 out of 80 points on a test. What percentage did Adrahan score on the test?

A 62.5%
B 16%
C 6.25%
D 1.6%

**48.** Ernesto needs to loosen a bolt. He needs a wrench that is smaller than a $\frac{7}{8}$ inch wrench, but larger than a $\frac{3}{4}$ inch wrench. Which of the following sizes should Ernesto use?

F $\frac{3}{8}$ inch
G $\frac{5}{8}$ inch
H $\frac{13}{16}$ inch
J $\frac{15}{16}$ inch

**49. EXTENDED RESPONSE** Two boats leave a dock. One boat travels 4 miles east and then 5 miles north. The second boat travels 12 miles south and 9 miles west. Draw a diagram that represents the paths traveled by the boats. How far apart are the boats in miles?

**50.** The formula $s = \frac{1}{2}at^2$ represents the distance $s$ in meters that a free-falling object will fall on a planet or moon in a given time $t$ in seconds. Solve the formula for $a$, the acceleration due to gravity.

A $a = \frac{1}{2}t^2 - s$
B $a = 2s - t^2$
C $a = s - \frac{1}{2}t^2$
D $a = \frac{2s}{t^2}$

## Spiral Review

**Write the equation of the axis of symmetry, and find the coordinates of the vertex of the graph of each function. Identify the vertex as a maximum or minimum. Then graph the function.** (Lesson 2-1)

**51.** $y = 3x^2$
**52.** $y = -4x^2 - 5$
**53.** $y = -x^2 + 4x - 7$
**54.** $y = x^2 - 6x - 8$
**55.** $y = 3x^2 + 2x + 1$
**56.** $y = -4x^2 - 8x + 5$

**Solve each equation. Check the solutions.** (Lesson 1-9)

**57.** $2x^2 = 32$
**58.** $(x - 4)^2 = 25$
**59.** $4x^2 - 4x + 1 = 16$
**60.** $2x^2 + 16x = -32$
**61.** $(x + 3)^2 = 5$
**62.** $4x^2 - 12x = -9$

**Find each sum or difference.** (Lesson 1-1)

**63.** $(3n^2 - 3) + (4 + 4n^2)$
**64.** $(2d^2 - 7d - 3) - (4d^2 + 7)$
**65.** $(2b^3 - 4b^2 + 4) - (3b^4 + 5b^2 - 9)$
**66.** $(8 - 4h^2 + 6h^4) + (5h^2 - 3 + 2h^3)$

## Skills Review

**Graph each function.**

**67.** $y = x^2 + 5$
**68.** $y = x^2 - 8$
**69.** $y = 2x^2 - 7$
**70.** $y = -x^2 + 2$
**71.** $y = -0.5x^2 - 3$
**72.** $y = (-x)^2 + 1$

EXTEND 2-2

# Graphing Technology Lab
# Quadratic Inequalities

Recall that the graph of a linear inequality consists of the boundary and the shaded half plane. The solution set of the inequality lies in the shaded region of the graph. Graphing quadratic inequalities is similar to graphing linear inequalities.

## Activity 1 Shade Inside a Parabola

**Graph $y \geq x^2 - 5x + 4$ in a standard viewing window.**

First, clear all functions from the Y= list.

To graph $y \geq x^2 - 5x + 4$, enter the equation in the Y= list. Then use the left arrow to select =. Press [ENTER] until shading above the line is selected.

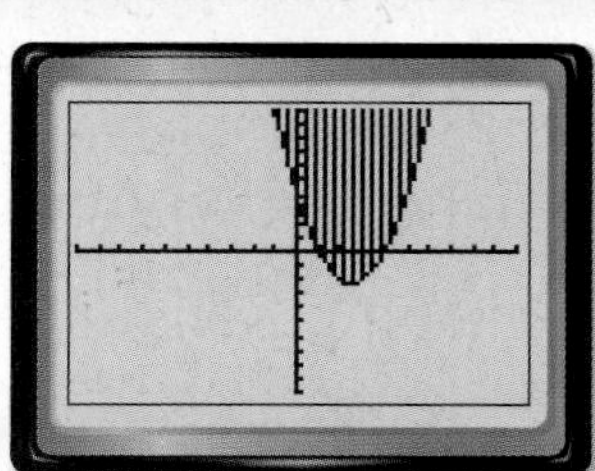

[−10, 10] scl: 1 by [−10, 10] scl: 1

**KEYSTROKES:** [◀] [◀] [ENTER] [ENTER] [▶] [▶] [X,T,θ,n] [$x^2$] [−] 5 [X,T,θ,n] [+] 4 [ZOOM] 6

All ordered pairs for which $y$ is *greater than or equal* to $x^2 - 5x + 4$ lie *above or on* the line and are solutions.

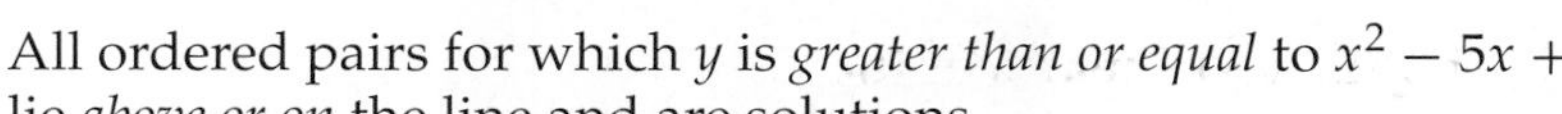

A similar procedure will be used to graph an inequality in which the shading is outside of the parabola.

## Activity 2 Shade Outside a Parabola

**Graph $y - 4 \leq x^2 - 5x$ in a standard viewing window.**

First, clear the graph that is displayed.

**KEYSTROKES:** [Y=] [CLEAR]

Then rewrite $y - 4 \leq x^2 - 5x$ as $y \leq x^2 - 5x + 4$, and graph it.

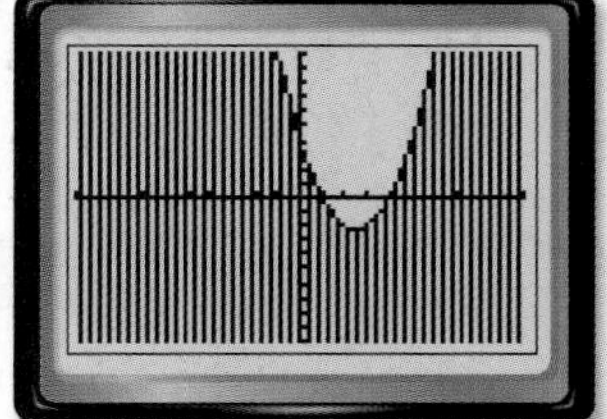

[−10, 10] scl: 1 by [−10, 10] scl: 1

**KEYSTROKES:** [◀] [◀] [ENTER] [ENTER] [ENTER] [▶] [▶] [X,T,θ,n] [$x^2$] [−] 5 [X,T,θ,n] [+] 4 [GRAPH]

All ordered pairs for which $y$ is *less than or equal* to $x^2 - 5x + 4$ lie *below or on* the line and are solutions.

## Exercises

1. Compare and contrast the two graphs shown above.
2. Graph $y - 2x + 6 \geq 5x^2$ in the standard viewing window. Name three solutions of the inequality.
3. Graph $y - 6x \leq -x^2 - 3$ in the standard viewing window. Name three solutions of the inequality.

EXPLORE

# 2-3 Graphing Technology Lab
# Family of Quadratic Functions

You have studied the effects of changing parameters in the equations of linear and exponential functions. You can use a graphing calculator to analyze how changing the parameters of the equation of a quadratic function affects the graphs in the family of quadratic functions.

**CCSS Common Core State Standards**
**Content Standards**
**F.IF.7a** Graph linear and quadratic functions and show intercepts, maxima, and minima.
**F.BF.3** Identify the effect on the graph of replacing $f(x)$ by $f(x) + k$, $kf(x)$, $f(kx)$, and $f(x + k)$ for specific values of $k$ (both positive and negative); find the value of $k$ given the graphs. Experiment with cases and illustrate an explanation of the effects on the graph using technology.

PT

## Activity 1 Change $k$ in $y = a(x - h)^2 + k$

**Graph the set of equations on the same screen in the standard viewing window. Describe any similarities and differences among the graphs.**

$y = x^2$, $y = x^2 + 2$, $y = x^2 - 4$

Enter the equations in the **Y =** list and graph in the standard viewing window. Use the **ZOOM** feature to investigate the key features of the graphs.

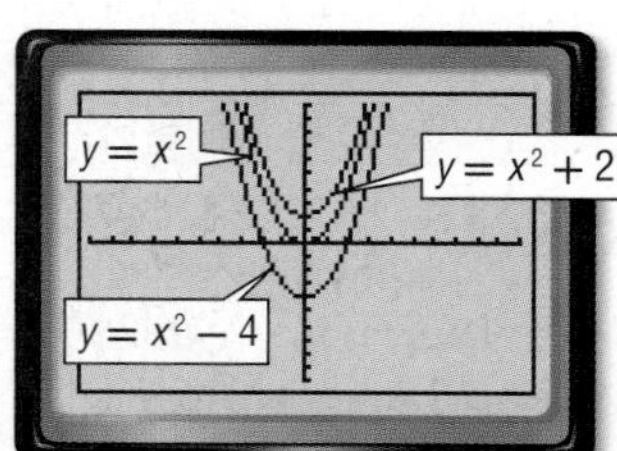

[−10, 10] scl: 1 by [−10, 10] scl: 1

The graphs have the same shape, and all open up. The vertex of each graph is on the $y$-axis, which is the axis of symmetry.

However, the graphs have different vertical positions. The graph of $y = x^2 + 2$ is shifted up 2 units. The graph of $y = x^2 - 4$ is shifted down 4 units.

Changing the value of $h$ in $y = a(x - h)^2 + k$ affects the graphs in a different way than changing $k$.

## Activity 2 Change $h$ in $y = a(x - h)^2 + k$

**Graph the set of equations on the same screen in the standard viewing window. Describe any similarities and differences among the graphs.**

$y = x^2$, $y = (x + 2)^2$, $y = (x - 4)^2$

The graphs have the same shape, and all open up. The vertex of each graph is on the $x$-axis.

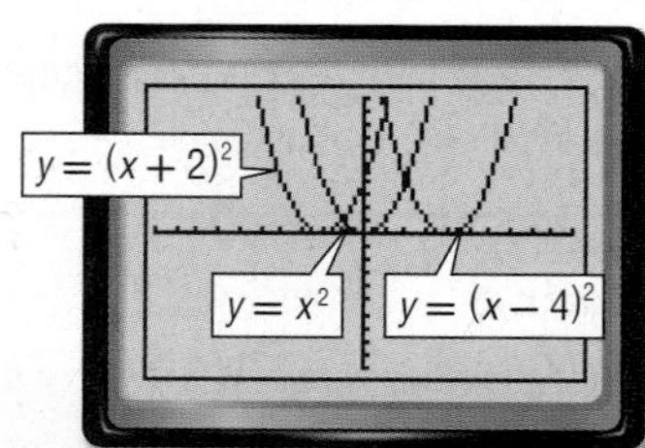

[−10, 10] scl: 1 by [−10, 10] scl: 1

However, the graphs have different horizontal positions. Each has a different axis of symmetry. The graph of $y = (x + 2)^2$ is shifted to the left 2 units. The graph of $y = (x - 4)^2$ is shifted to the right 4 units.

It appears that changing the values of $h$ and $k$ in $y = a(x - h)^2 + k$ moves the graph vertically or horizontally. How does changing the value of $a$ affect the graphs?

**Activity 3** Change $a$ in $y = a(x - h)^2 + k$

**Graph each set of equations on the same screen in the standard viewing window. Describe any similarities and differences among the graphs.**

**a.** $y = x^2, y = 2x^2, y = \frac{1}{3}x^2$

The graphs have the same vertex, they have the same axis of symmetry, and all open up.

However, the graphs have different widths. The graph of $y = 2x^2$ is narrower than the graph of $y = x^2$. The graph of $y = \frac{1}{3}x^2$ is wider than the graph of $y = x^2$.

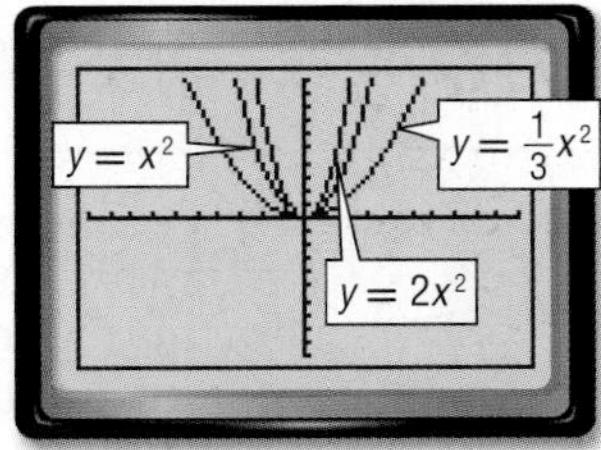

[−10, 10] scl: 1 by [−10, 10] scl: 1

**b.** $y = x^2, y = -\frac{1}{3}x^2, y = -2x^2$

The graphs have the same vertex and the same axis of symmetry.

However, the graphs of $y = -\frac{1}{3}x^2$ and $y = -2x^2$ open down. Also the graph of $y = -2x^2$ is narrower than the graph of $y = x^2$. The graph of $y = -\frac{1}{3}x^2$ is wider than the graph of $y = x^2$.

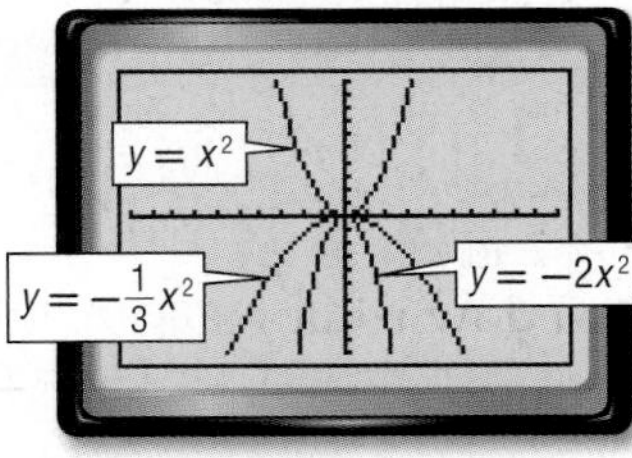

[−10, 10] scl: 1 by [−10, 10] scl: 1

## Model and Analyze

**How does each parameter affect the graph of $y = a(x - h)^2 + k$? Give examples.**

**1.** $k$

**2.** $h$

**3.** $a$

**Examine each pair of equations and predict the similarities and differences in their graphs. Use a graphing calculator to confirm your predictions. Write a sentence or two comparing the two graphs.**

**4.** $y = x^2, y = x^2 + 3$

**5.** $y = \frac{1}{2}x^2, y = 3x^2$

**6.** $y = x^2, y = (x - 5)^2$

**7.** $y = 3x^2, y = -3x^2$

**8.** $y = x^2, y = -4x^2$

**9.** $y = x^2 - 1, y = x^2 + 2$

**10.** $y = \frac{1}{2}x^2 + 3, y = -2x^2$

**11.** $y = x^2 - 4, y = (x - 4)^2$

LESSON 2-3

# Transformations of Quadratic Functions

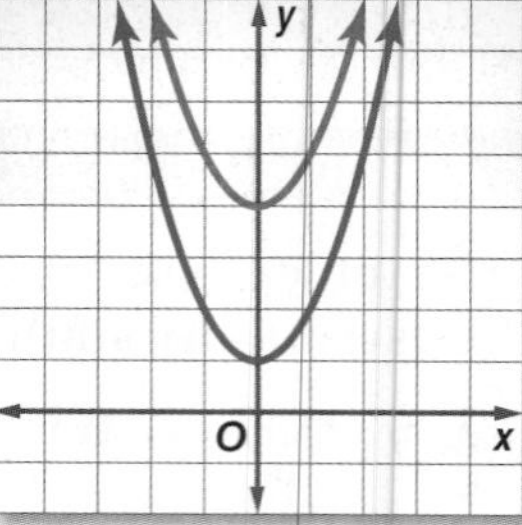


**Then**

- You graphed quadratic functions by using the vertex and axis of symmetry.

**Now**

1. Apply translations to quadratic functions.
2. Apply dilations and reflections to quadratic functions.

**Why?**

- The graphs of the parabolas shown at the right are the same size and shape, but notice that the vertex of the red parabola is higher on the *y*-axis than the vertex of the blue parabola. Shifting a parabola up and down is an example of a transformation.

**NewVocabulary**

transformation
translation
dilation
reflection
vertex form

**Common Core State Standards**

**Content Standards**

A.SSE.3b Complete the square in a quadratic expression to reveal the maximum or minimum value of the function it defines.

F.IF.7a Graph linear and quadratic functions and show intercepts, maxima, and minima.

**Mathematical Practices**

1 Make sense of problems and persevere in solving them.

8 Look for and express regularity in repeated reasoning.

## 1 Translations

A **transformation** changes the position or size of a figure. One transformation, a **translation**, moves a figure up, down, left, or right. When a constant $k$ is added to or subtracted from the parent function, the graph of the resulting function $f(x) \pm k$ is the graph of the parent function translated up or down.

The parent function of the family of quadratics is $f(x) = x^2$. All other quadratic functions have graphs that are transformations of the graph of $f(x) = x^2$.

### KeyConcept Vertical Translations

The graph of $f(x) = x^2 + k$ is the graph of $f(x) = x^2$ translated vertically.

If $k > 0$, the graph of $f(x) = x^2$ is translated $|k|$ units **up**.

If $k < 0$, the graph of $f(x) = x^2$ is translated $|k|$ units **down**.

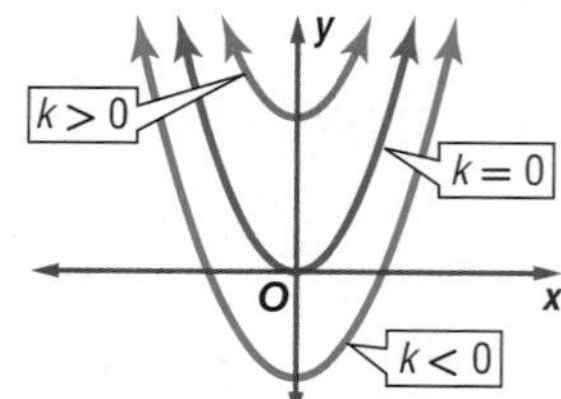

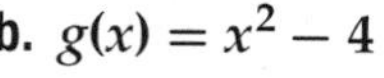

### Example 1 Describe and Graph Translations

**Describe how the graph of each function is related to the graph of $f(x) = x^2$.**

**a.** $h(x) = x^2 + 3$

$k = 3$ and $3 > 0$
$h(x)$ is a translation of the graph of $f(x) = x^2$ up 3 units.

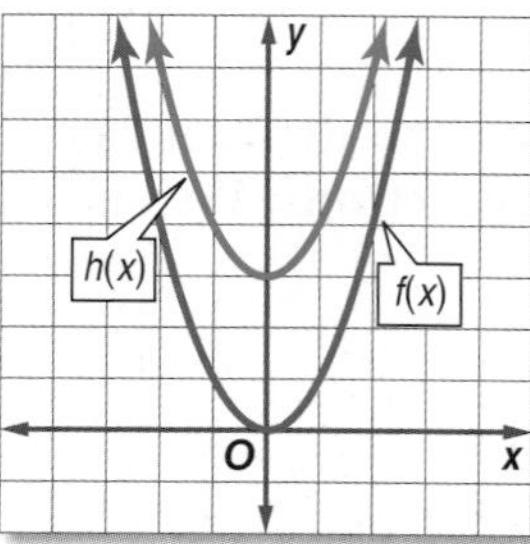


**b.** $g(x) = x^2 - 4$

$k = -4$ and $-4 < 0$
$g(x)$ is a translation of the graph of $f(x) = x^2$ down 4 units.

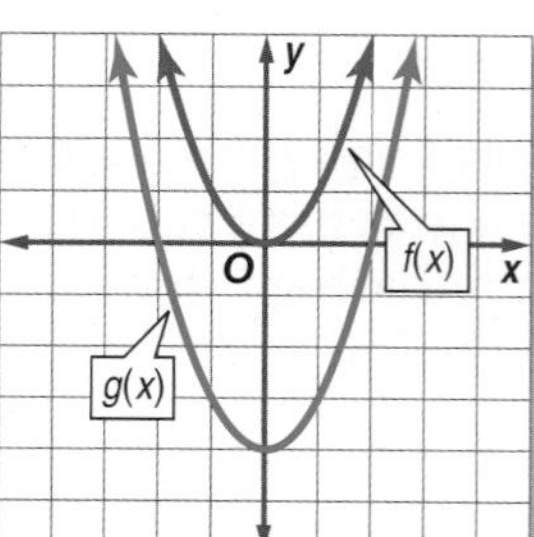


**GuidedPractice**

**1A.** $f(x) = x^2 - 7$ **1B.** $g(x) = 5 + x^2$ **1C.** $h(x) = -5 + x^2$ **1D.** $f(x) = x^2 + 1$

A quadratic graph can be translated horizontally by subtracting an $h$ term from $x$.

## KeyConcept Horizontal Translations

The graph of $g(x) = (x - h)^2$ is the graph of $f(x) = x^2$ translated horizontally.

If $h > 0$, the graph of $f(x) = x^2$ is translated $h$ units to the **right**.

If $h < 0$, the graph of $f(x) = x^2$ is translated $|h|$ units to the **left**.

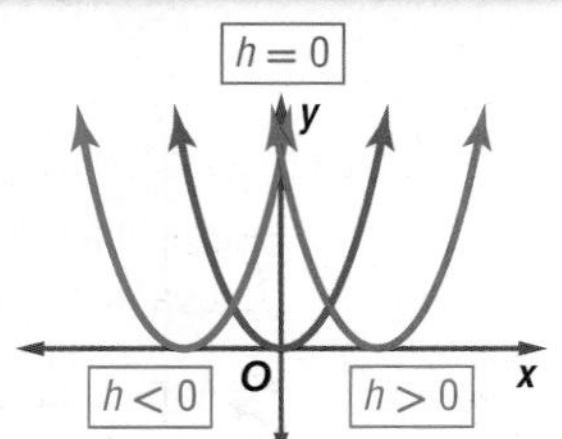


## Example 2 Horizontal Translations

**Describe how the graph of each function is related to the graph of $f(x) = x^2$.**

**a.** $g(x) = (x - 2)^2$

$k = 0, h = 2$ and $2 > 0$

$g(x)$ is a translation of the graph of $f(x) = x^2$ to the right 2 units.

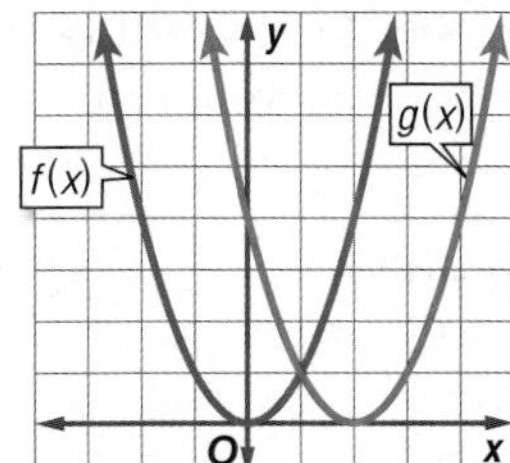


**b.** $g(x) = (x + 1)^2$

$k = 0, h = -1$ and $-1 < 0$

$g(x)$ is a translation of the graph of $f(x) = x^2$ to the left 1 unit.

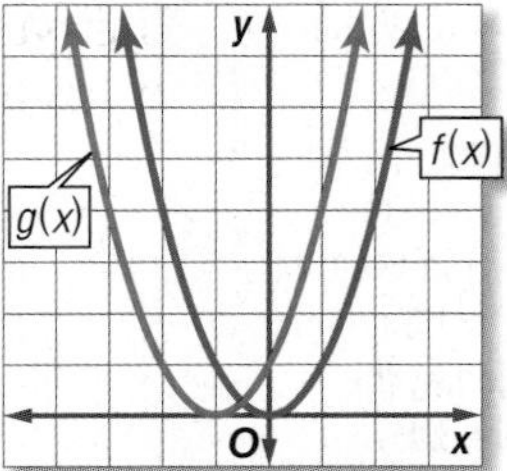


### GuidedPractice

**2A.** $g(x) = (x - 3)^2$

**2B.** $g(x) = (x + 2)^2$

A quadratic graph can be translated both horizontally and vertically.

## Example 3 Horizontal and Vertical Translations

**Describe how the graph of each function is related to the graph of $f(x) = x^2$.**

**a.** $g(x) = (x - 3)^2 + 2$

$k = 2, h = 3$ and $3 > 0$

$g(x)$ is a translation of the graph of $f(x) = x^2$ to the right 3 units and up 2 units.

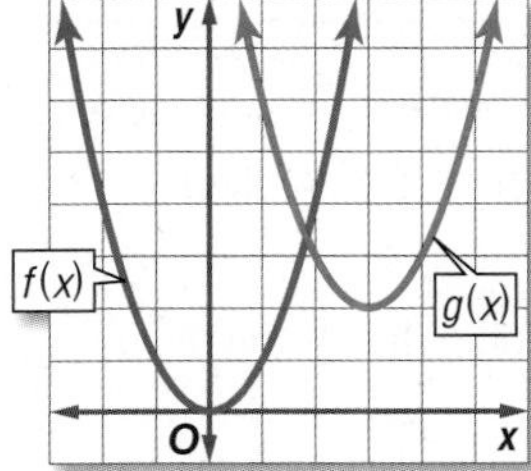


**b.** $g(x) = (x + 3)^2 - 1$

$k = -1, h = -3$ and $-3 < 0$

$g(x)$ is a translation of the graph of $f(x) = x^2$ to the left 3 units and down 1 unit.

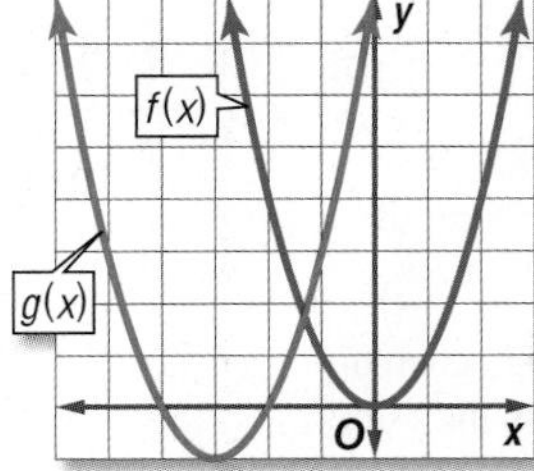


### GuidedPractice

**3A.** $g(x) = (x + 2)^2 + 3$

**3B.** $g(x) = (x - 4)^2 - 4$

## 2 Dilations and Reflections

Another type of transformation is a dilation. A **dilation** makes the graph narrower than the parent graph or wider than the parent graph. When the parent function $f(x) = x^2$ is multiplied by a constant $a$, the graph of the resulting function $f(x) = ax^2$ is either stretched or compressed vertically.

### KeyConcept Dilations

The graph of $g(x) = ax^2$ is the graph of $f(x) = x^2$ stretched or compressed vertically.

If $|a| > 1$, the graph of $f(x) = x^2$ is stretched vertically.

If $0 < |a| < 1$, the graph of $f(x) = x^2$ is compressed vertically.

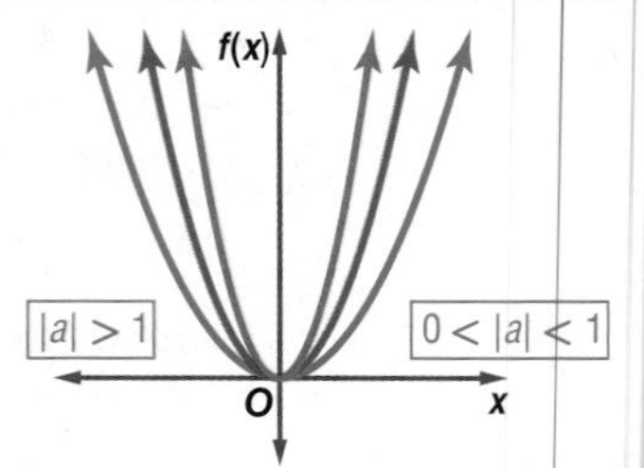


**StudyTip**

**CCSS Sense-Making** When the graph of a quadratic function is stretched vertically, the shape of the graph is narrower than that of the parent function. When it is compressed vertically, the graph is wider than the parent function.

### Example 4 Describe and Graph Dilations

**Describe how the graph of each function is related to the graph of $f(x) = x^2$.**

**a.** $h(x) = \frac{1}{2}x^2$

$a = \frac{1}{2}$ and $0 < \frac{1}{2} < 1$

$h(x)$ is a dilation of the graph of $f(x) = x^2$ that is compressed vertically.

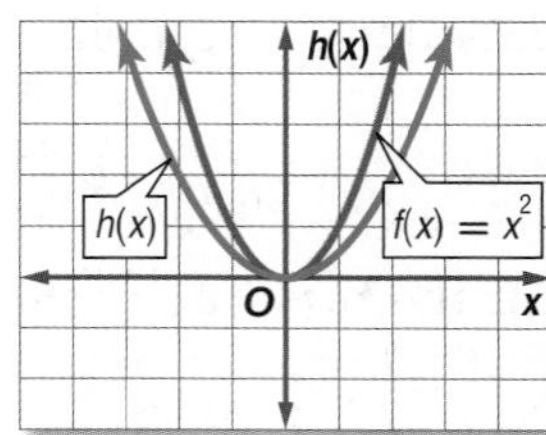


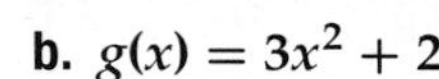

**b.** $g(x) = 3x^2 + 2$

$a = 3$ and $3 > 1$, $k = 2$ and $2 > 0$

$g(x)$ is a dilation of the graph of $f(x) = x^2$ that is stretched vertically and translated up 2 units.

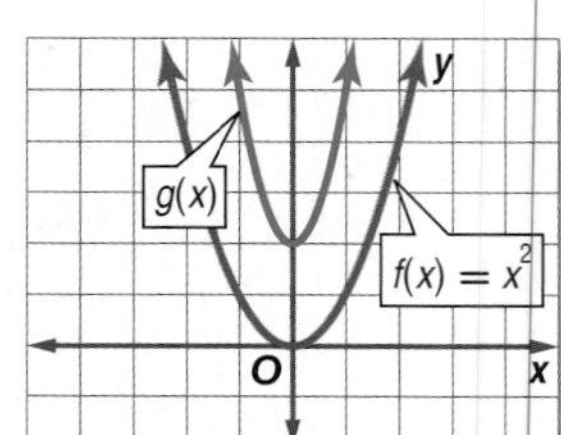


**GuidedPractice**

**4A.** $j(x) = 2x^2$

**4B.** $h(x) = 5x^2 - 2$

**4C.** $g(x) = \frac{1}{3}x^2 + 2$

A **reflection** flips a figure across a line.

### KeyConcept Reflections

The graph of $-f(x)$ is the reflection of the graph of $f(x) = x^2$ across the $x$-axis.

The graph of $f(-x)$ is the reflection of the graph of $f(x) = x^2$ across the $y$-axis.

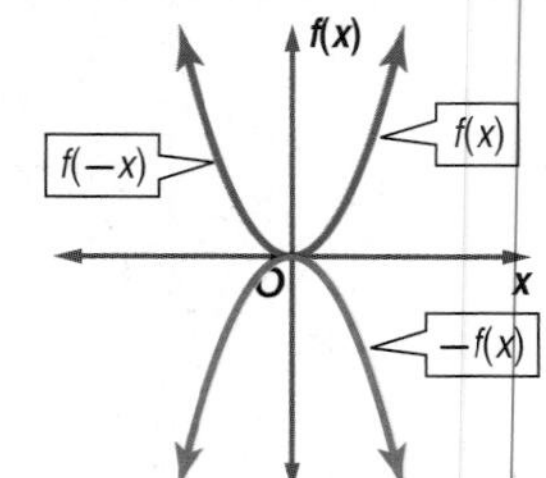


**StudyTip**

**Reflection** A reflection of $f(x) = x^2$ across the $y$-axis results in the same function, because $f(-x) = (-x)^2 = x^2$.

**WatchOut!**

**Transformations** The graph of $f(x) = -ax^2$ can result in two transformations of the graph of $f(x) = x^2$: a reflection across the $x$-axis if $a > 0$ and either a compression or expansion depending on the absolute value of $a$.

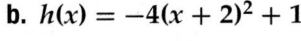

## Example 5 Describe and Graph Transformations

**Describe how the graph of each function is related to the graph of $f(x) = x^2$.**

**a.** $g(x) = -2x^2 - 3$

- $a = -2$, $-2 < 0$, and $|-2| > 1$, so there is a reflection across the $x$-axis and the graph is vertically stretched.
- $k = -3$ and $-3 < 0$, so there is a translation down 3 units.

**b.** $h(x) = -4(x + 2)^2 + 1$

- $a = -4$, $-4 < 0$, and $|-4| > 1$, so there is a reflection across the $x$-axis and the graph is vertically stretched.
- $h = -2$ and $-2 < 0$, so there is a translation 2 units to the left.
- $k = 1$ and $1 > 0$, so there is a translation up 1 unit.

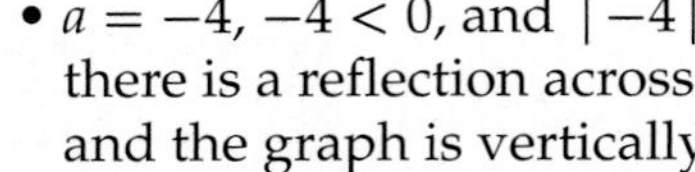

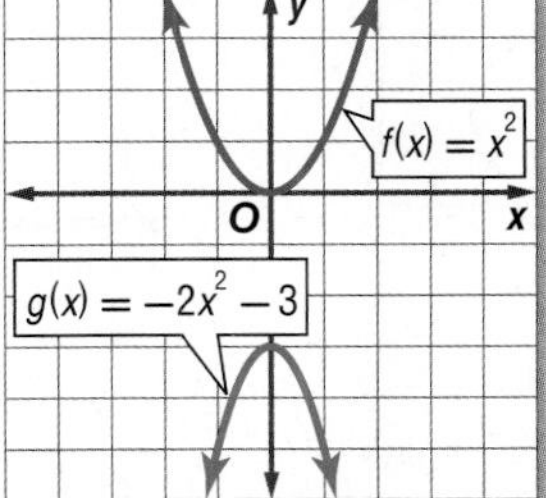


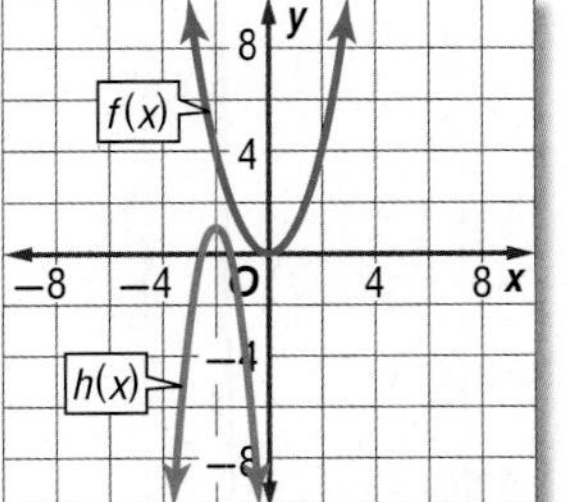


**GuidedPractice**

**5A.** $h(x) = 2(-x)^2 - 9$ **5B.** $g(x) = \frac{1}{5}x^2 + 3$ **5C.** $j(x) = -2(x - 1)^2 - 2$

You can use what you know about the characteristics of graphs of quadratic equations to match an equation with a graph.

## Standardized Test Example 6 Identify an Equation for a Graph

**Which is an equation for the function shown in the graph?**

**A** $y = \frac{1}{2}x^2 - 5$ **C** $y = -\frac{1}{2}x^2 + 5$

**B** $y = -2x^2 - 5$ **D** $y = 2x^2 + 5$

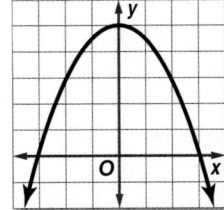

### Read the Test Item

You are given a graph. You need to find its equation.

### Solve the Test Item

The graph opens downward, so the graph of $y = x^2$ has been reflected across the $x$-axis. The leading coefficient should be negative, so eliminate choices A and D.

The parabola is translated up 5 units, so $k = 5$. Look at the equations. Only choices C and D have $k = 5$. The answer is C.

**GuidedPractice**

**6.** Which is the graph of $y = -3x^2 + 1$?

**F**

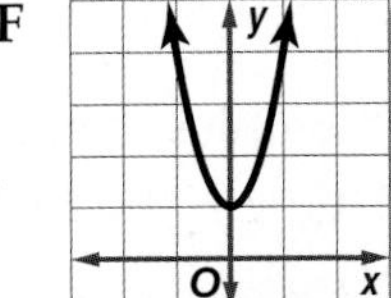


**G**

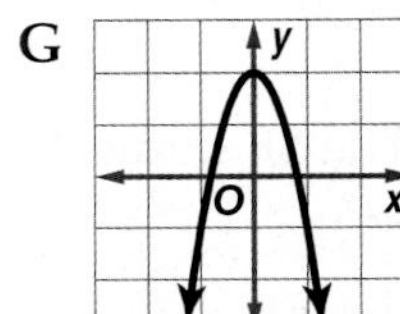


**H**

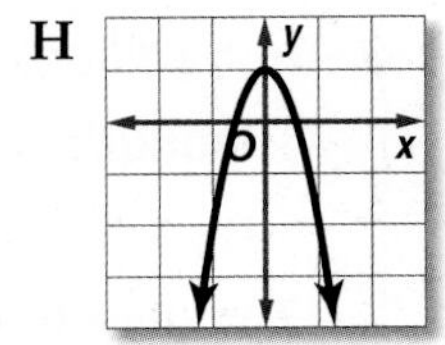


**J**

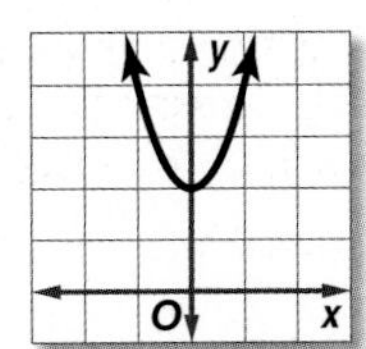

A quadratic function written in the form $f(x) = a(x - h)^2 + k$ is said to be in **vertex form**. Transformations of the parent graph are easily found from an equation in vertex form.

## ConceptSummary Transformations of Quadratic Functions

$f(x) = a(x - h)^2 + k$

**$h$, Horizontal Translation**
$h$ units to the right if $h$ is positive
$|h|$ units to the left if $h$ is negative

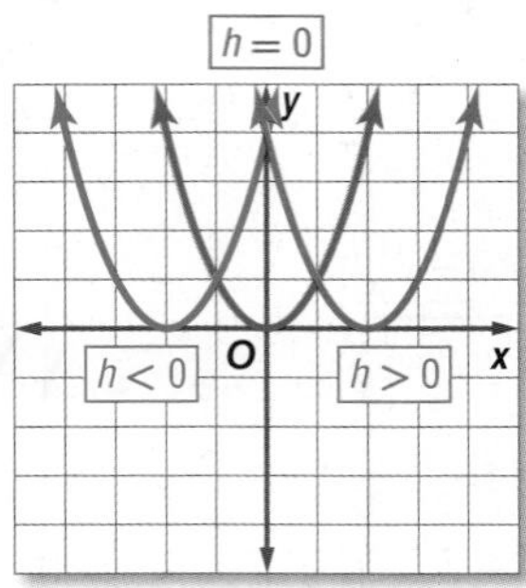


**$k$, Vertical Translation**
$k$ units up if $k$ is positive
$|k|$ units down if $k$ is negative

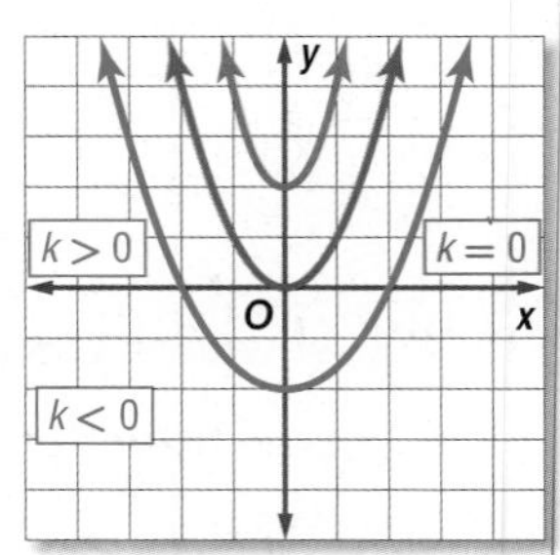


**$a$, Reflection**
If $a > 0$, the graph opens up.
If $a < 0$, the graph opens down.

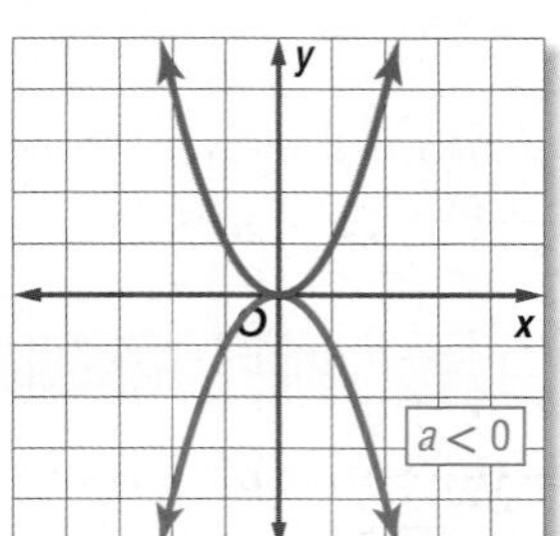


**$a$, Dilation**
If $|a| > 1$, the graph is stretched vertically. If $0 < |a| < 1$, the graph is compressed vertically.

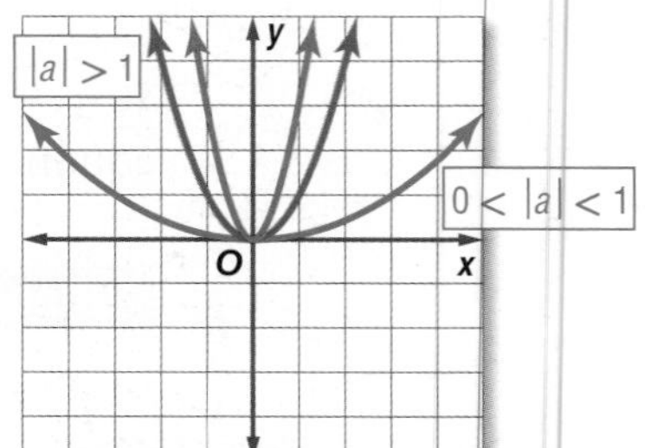

## Real-World Example 7 Transformations with a Calculator

**FIREWORKS During a firework show, the height $h$ in meters of a specific rocket after $t$ seconds can be modeled by $h(t) = -4.6(t - 3)^2 + 75$. Graph the function. How is it related to the graph of $f(x) = x^2$?**

Four separate transformations are occurring.

The negative sign of the coefficient of $x^2$ causes a reflection across the $x$-axis. A dilation occurs, which stretches the graph vertically. There are also translations up 75 units and to the right 3 units.

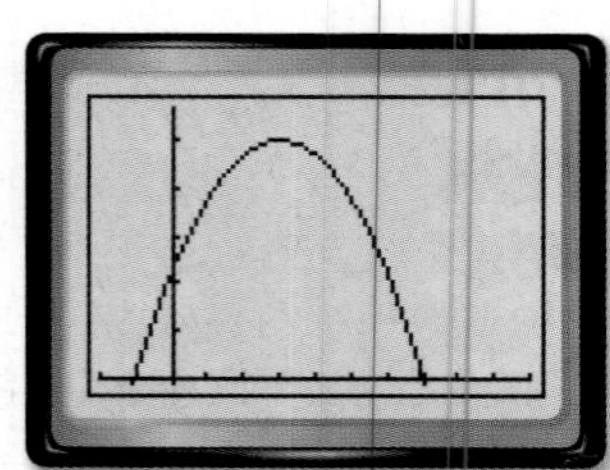

[−2, 10] scl: 1 by [−2, 85] scl: 15

### GuidedPractice

**7. MONUMENTS** The St. Louis Arch resembles a quadratic and can be modeled by $h(x) = -\frac{2}{315}x^2 + 630$. Graph the function. How is it related to the graph of $f(x) = x^2$?

## Check Your Understanding

**Examples 1–5, 7** Describe how the graph of each function is related to the graph of $f(x) = x^2$.

1. $g(x) = x^2 - 11$
2. $h(x) = \frac{1}{2}(x - 2)^2$
3. $h(x) = -x^2 + 8$
4. $g(x) = x^2 + 6$
5. $g(x) = -4(x + 3)^2$
6. $h(x) = -x^2 - 2$

**Example 6**

7. **MULTIPLE CHOICE** Which is an equation for the function shown in the graph?

A $g(x) = \frac{1}{5}x^2 + 2$
C $g(x) = \frac{1}{5}x^2 - 2$
B $g(x) = -5x^2 - 2$
D $g(x) = -\frac{1}{5}x^2 - 2$

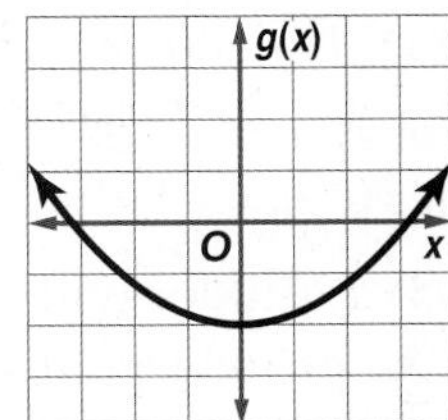


## Practice and Problem Solving

**Examples 1–5, 7** Describe how the graph of each function is related to the graph of $f(x) = x^2$.

8. $g(x) = -10 + x^2$
9. $h(x) = -7 - x^2$
10. $g(x) = 2(x - 3)^2 + 8$
11. $h(x) = 6 + \frac{2}{3}x^2$
12. $g(x) = -5 - \frac{4}{3}x^2$
13. $h(x) = 3 + \frac{5}{2}x^2$
14. $g(x) = 0.25x^2 - 1.1$
15. $h(x) = 1.35(x + 1)^2 + 2.6$
16. $g(x) = \frac{3}{4}x^2 + \frac{5}{6}$
17. $h(x) = 1.01x^2 - 6.5$

**Example 6** Match each equation to its graph.

A
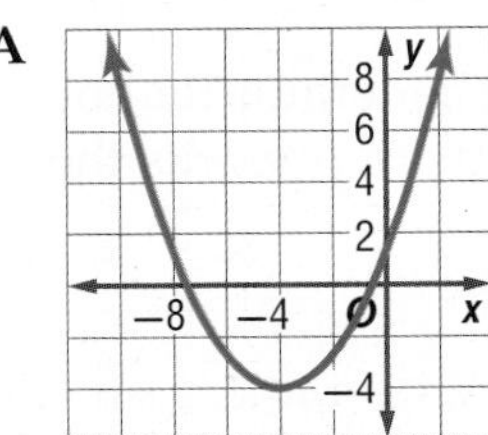


B
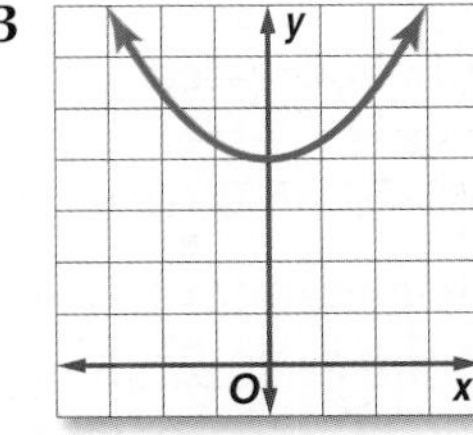


C
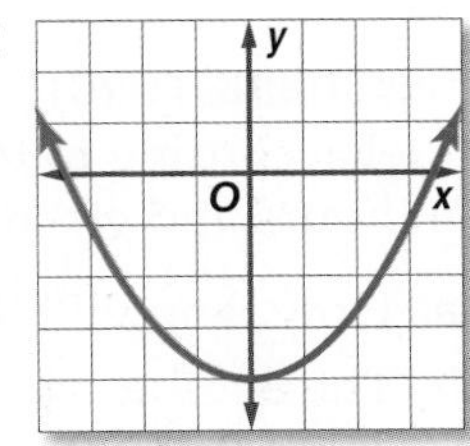


D
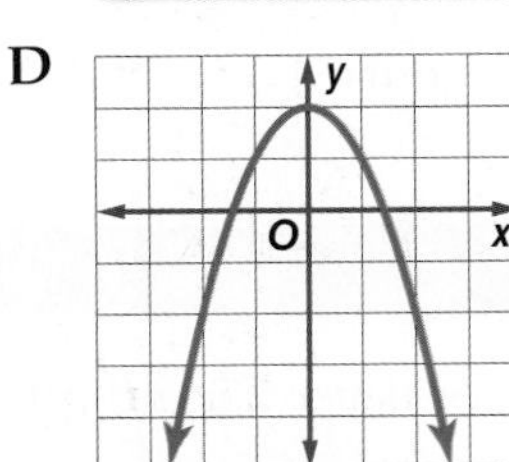


E
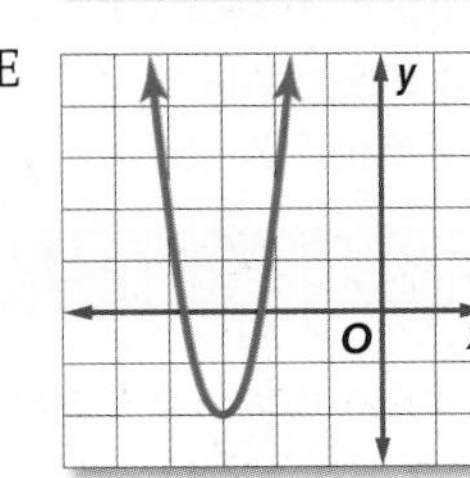


F
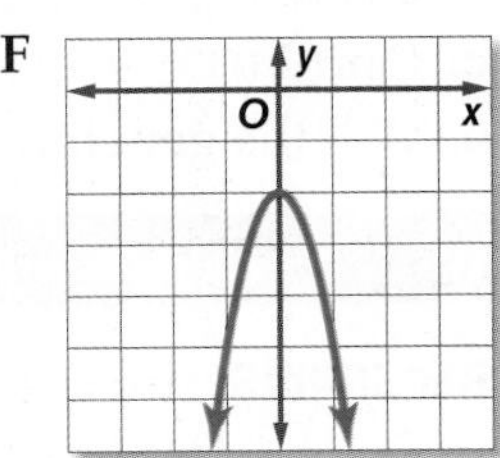


18. $y = \frac{1}{3}x^2 - 4$
19. $y = \frac{1}{3}(x + 4)^2 - 4$
20. $y = \frac{1}{3}x^2 + 4$
21. $y = -3x^2 - 2$
22. $y = -x^2 + 2$
23. $y = (2x + 6)^2 + 2$

24. **SQUIRRELS** A squirrel 12 feet above the ground drops an acorn from a tree. The function $h = -16t^2 + 12$ models the height of the acorn above the ground in feet after $t$ seconds. Graph the function, and compare this graph to the graph of its parent function.

**CCSS REGULARITY** List the functions in order from the most stretched vertically to the least stretched vertically graph.

25. $g(x) = 2x^2, h(x) = \frac{1}{2}x^2$
26. $g(x) = -3x^2, h(x) = \frac{2}{3}x^2$
27. $g(x) = -4x^2, h(x) = 6x^2, f(x) = 0.3x^2$
28. $g(x) = -x^2, h(x) = \frac{5}{3}x^2, f(x) = -4.5x^2$

**29. ROCKS** A rock falls from a cliff 300 feet above the ground. At the same time, another rock falls from a cliff 700 feet above the ground.

**a.** Write functions that model the height $h$ of each rock after $t$ seconds.

**b.** If the rocks fall at the same time, how much sooner will the first rock reach the ground?

**30. SPRINKLERS** The path of water from a sprinkler can be modeled by quadratic functions. The following functions model paths for three different sprinklers.

Sprinkler A: $y = -0.35x^2 + 3.5$ Sprinkler B: $y = -0.21x^2 + 1.7$
Sprinkler C: $y = -0.08x^2 + 2.4$

**a.** Which sprinkler will send water the farthest? Explain.

**b.** Which sprinkler will send water the highest? Explain.

**c.** Which sprinkler will produce the narrowest path? Explain.

**31. GOLF** The path of a drive can be modeled by a quadratic function where $g(x)$ is the vertical distance in yards of the ball from the ground and $x$ is the horizontal distance in yards.

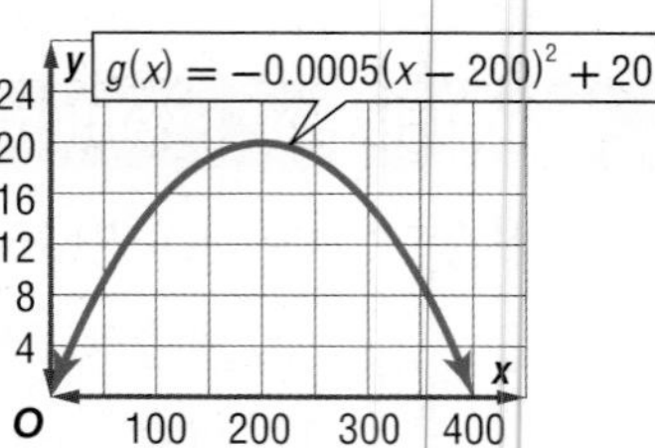


**a.** How can you obtain $g(x)$ from the graph of $f(x) = x^2$.

**b.** A second golfer hits a ball from the red tee, which is 30 yards closer to the hole. What function $h(x)$ can be used to describe the second golfer's shot?

**Describe the transformations to obtain the graph of $g(x)$ from the graph of $f(x)$.**

**32.** $f(x) = x^2 + 3$
$g(x) = x^2 - 2$

**33.** $f(x) = x^2 - 4$
$g(x) = (x - 2)^2 + 7$

**34.** $f(x) = -6x^2$
$g(x) = -3x^2$

**35. COMBINING FUNCTIONS** An engineer created a self-refueling generator that burns fuel according to the function $g(t) = -t^2 + 10t + 200$, where $t$ represents the time in hours and $g(t)$ represents the number of gallons remaining.

**a.** How long will it take for the generator to run out of fuel?

**b.** The engine self-refuels at a rate of 40 gallons per hour. Write a linear function $h(t)$ to represent the refueling of the generator.

**c.** Find $T(t) = g(t) + h(t)$. What does this new function represent?

**d.** Will the generator run out of fuel? If so, when?

## H.O.T. Problems Use Higher-Order Thinking Skills

**36. REASONING** Are the following statements *sometimes*, *always*, or *never* true? Explain.

**a.** The graph of $y = x^2 + k$ has its vertex at the origin.

**b.** The graphs of $y = ax^2$ and its reflection over the $x$-axis are the same width.

**c.** The graph of $y = x^2 + k$, where $k \geq 0$, and the graph of a quadratic with vertex at $(0, -3)$ have the same maximum or minimum point.

**37. CHALLENGE** Write a function of the form $y = ax^2 + k$ with a graph that passes through the points $(-2, 3)$ and $(4, 15)$.

**38. CCSS ARGUMENTS** Determine whether all quadratic functions that are reflected across the $y$-axis produce the same graph. Explain your answer.

**39. OPEN ENDED** Write a quadratic function that opens downward and is wider than the parent graph.

**40. WRITING IN MATH** Describe how the values of $a$ and $k$ affect the graphical and tabular representations for the functions $y = ax^2$, $y = x^2 + k$, and $y = ax^2 + k$.

## Standardized Test Practice

**41. SHORT RESPONSE** A tutor charges a flat fee of \$55 and \$30 for each hour of work. Write a function that represents the total charge $C$, in terms of the number of hours $h$ worked.

**42.** Which *best* describes the graph of $y = 2x^2$?

**A** a line with a $y$-intercept of 2 and an $x$-intercept at the origin

**B** a parabola with a minimum point at (0, 0) and that is wider than the graph of $y = x^2$

**C** a parabola with a maximum point at (0, 0) and that is narrower than the graph of $y = x^2$

**D** a parabola with a minimum point at (0, 0) and that is narrower than the graph of $y = x^2$

**43.** Candace is 5 feet tall. If 1 inch is about 2.54 centimeters, how tall is Candace to the nearest centimeter?

**F** 13 cm　　**H** 123 cm

**G** 26 cm　　**J** 152 cm

**44.** While in England, Imani spent 49.60 British pounds on a pair of jeans. If this is equivalent to \$100 in U.S. currency, how many British pounds would Imani have spent on a sweater that cost \$60?

**A** 2976 pounds

**B** 29.76 pounds

**C** 19.84 pounds

**D** 8.26 pounds

## Spiral Review

**Solve each equation by graphing.** (Lesson 2-2)

**45.** $x^2 + 6 = 0$　　**46.** $x^2 - 10x = -24$　　**47.** $x^2 + 5x + 4 = 0$

**48.** $2x^2 - x = 3$　　**49.** $2x^2 - x = 15$　　**50.** $12x^2 = -11x + 15$

**Find the vertex, the equation of the axis of symmetry, and the $y$-intercept of each graph.** (Lesson 2-1)

**51.**

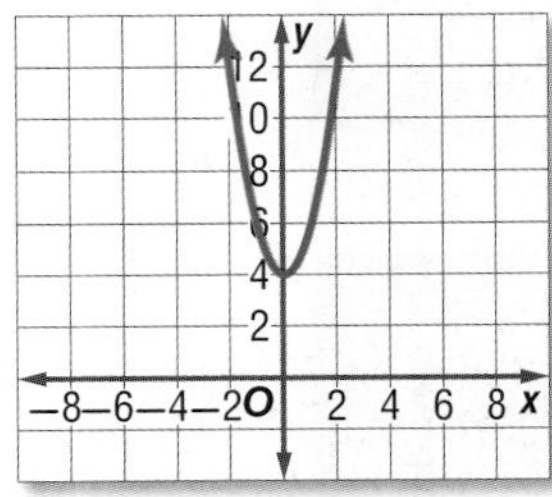


**52.**

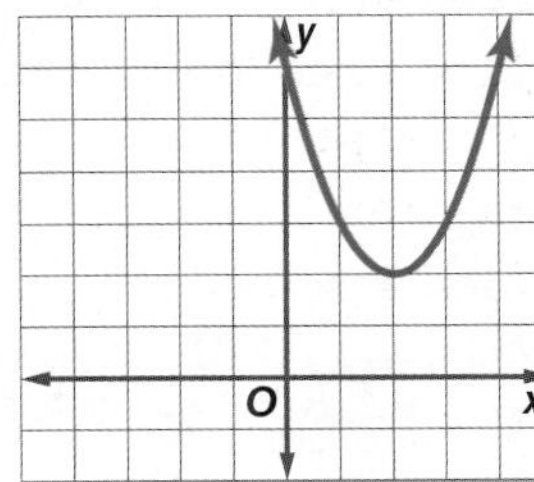


**53.**

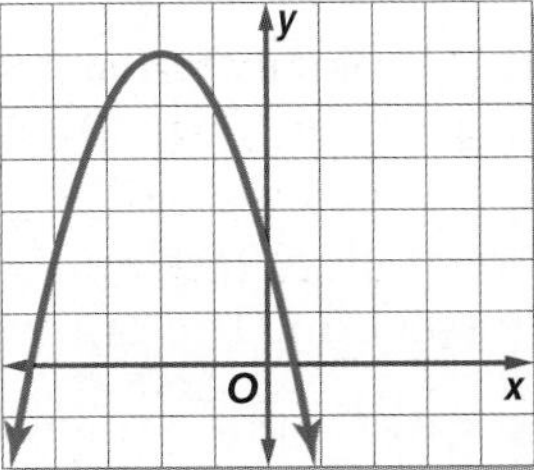


**54. CLASS TRIP** Mr. Wong's American History class will take taxis from their hotel in Washington, D.C., to the Lincoln Memorial. The fare is \$2.75 for the first mile and \$1.25 for each additional mile. If the distance is $m$ miles and $t$ taxis are needed, write an expression for the cost to transport the group. (Lesson 1-2)

## Skills Review

**Determine whether each trinomial is a perfect square trinomial. If so, factor it.**

**55.** $16x^2 - 24x + 9$　　**56.** $9x^2 + 6x + 1$　　**57.** $25x^2 - 60x + 36$

**58.** $x^2 - 8x + 81$　　**59.** $36x^2 - 84x + 49$　　**60.** $4x^2 - 3x + 9$

EXTEND

# 2-3 Graphing Technology Lab
# Systems of Linear and Quadratic Equations

You can use a graphing calculator to solve systems involving linear and quadratic equations.

**CCSS Common Core State Standards**
**Content Standards**
**A.REI.7** Solve a simple system consisting of a linear equation and a quadratic equation in two variables algebraically and graphically.
**F.IF.7a** Graph linear and quadratic functions and show intercepts, maxima, and minima.

## Activity 1 Solve a System of Equations Graphically

**Use a graphing calculator to solve the system of equations.**

$y = x^2 - x - 6$
$y = x - 3$

**Step 1** Enter each equation in the **Y=** list.

KEYSTROKES: [X,T,θ,n] [$x^2$] [−] [X,T,θ,n] [−] 6 [ENTER] [X,T,θ,n] [−] 3

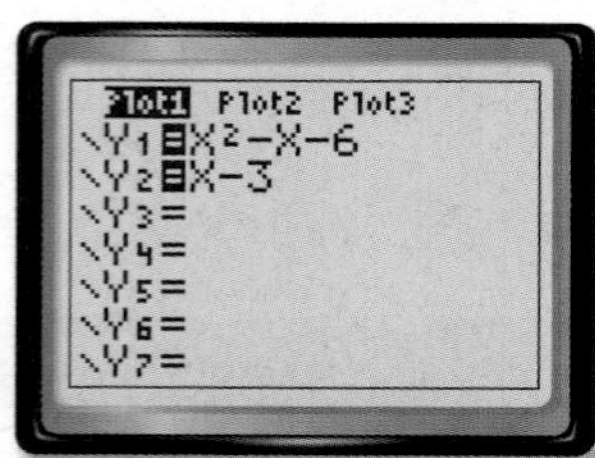


**Step 2** Graph the system. KEYSTROKES: [GRAPH]

The graphs intersect at two points. So, there are two solutions.

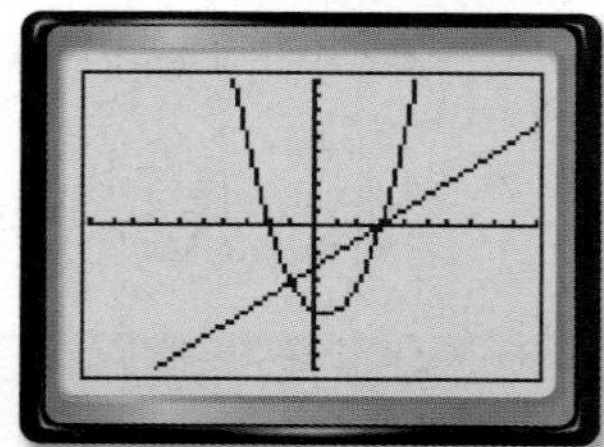

[−10, 10] scl: 1 by [−10, 10] scl: 1

**Step 3** Find the intersection on the left by using the **CALC** menu.

KEYSTROKES: [2nd] [CALC] 5 [ENTER] [ENTER]

Use the arrow keys to move the cursor close to the intersection on the left. Press [ENTER] again.

The graphs intersect at $(-1, -4)$.

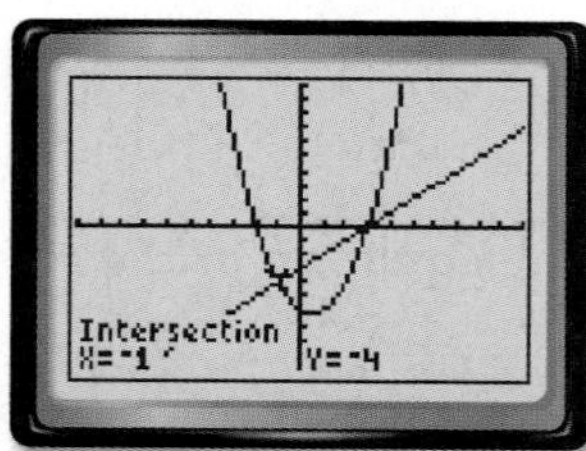


[−10, 10] scl: 1 by [−10, 10] scl: 1

**Step 4** Repeat Step 3 but move the cursor to the other intersection. The graphs intersect at $(3, 0)$.

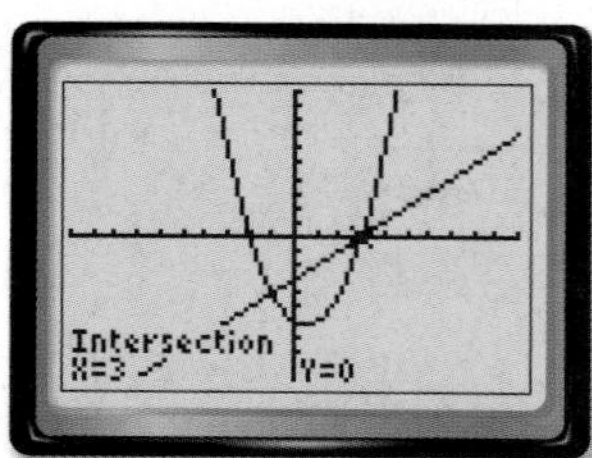


[−10, 10] scl: 1 by [−10, 10] scl: 1

You can use a graphing calculator to verify solutions of systems found algebraically.

## Activity 2 Check Solutions Graphically

**Solve the system of equations algebraically. Use a graphing calculator to check your solutions.**

$y = 2x - 6$
$y = x^2 - 8x + 19$

**Step 1** Set the expressions equal to each other, and solve for $x$.

$x^2 - 8x + 19 = 2x - 6$ Substitute $x^2 - 8x + 19$ for $y$.
$x^2 - 10x + 25 = 0$ Simplify.
$(x - 5)^2 = 0$ Factor.
$x = 5$ Solve for $x$.

**Step 2** Substitute the $x$-value into either equation to find the $y$-value: $y = 2(5) - 6$ or 4.

**Step 3** Graph the system and find the point(s) of intersection as in Activity 1.

The graphs intersect at (5, 4). Thus, the solution of the system of equations is (5, 4).

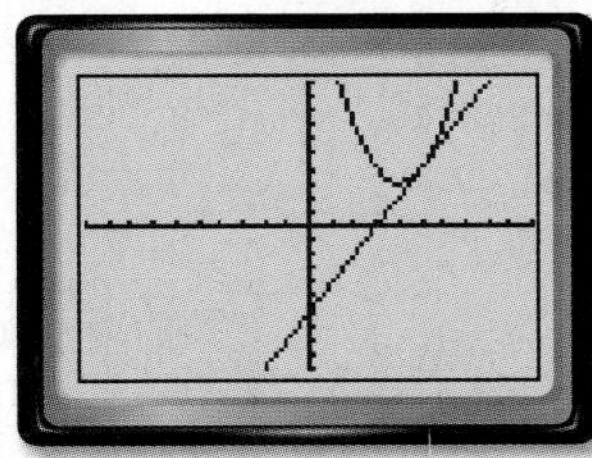

[−10, 10] scl: 1 by [−10, 10] scl: 1

You can solve a quadratic equation graphically by writing each side of the equation as a separate function. The $x$-coordinate of the point(s) of intersection will be the solution of the equation, since at that point(s) the original equations are true.

## Activity 3 Use a System to Solve an Equation

**Use a system of equations to solve $x^2 - 3x + 1 = \frac{11}{4}x - 6$.**

**Step 1** Write as a system of equations. $y = x^2 - 3x + 1$

$y = \frac{11}{4}x - 6$

**Step 2** Enter the equations into the graphing calculator, and graph them.

**Step 3** Use the **CALC** menu to find the two points of intersection.

The graphs intersect at (1.75, −1.1875) and (4, 5). Thus, the solutions of $x^2 - 3x + 1 = \frac{11}{4}x - 6$ are 1.75 and 4.

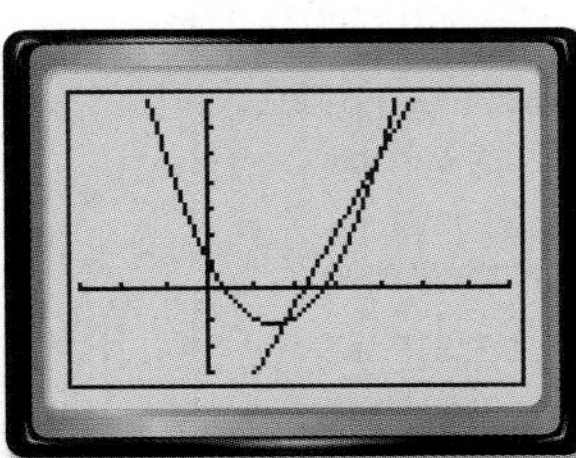

[−3, 7] scl: 1 by [−3, 7] scl: 1

## Exercises

**Use a graphing calculator to solve each system of equations.**

1. $y = x^2$
   $y = 2x$

2. $y = -2x^2 + 7x - 2$
   $y = 3 - 4x$

3. $y = -x^2 + 4$
   $y = \frac{1}{2}x + 5$

**Solve each system of equations algebraically. Use a graphing calculator to check your solutions.**

4. $y = x^2 + 7x + 12$
   $y = 2x + 8$

5. $y = x^2 - x - 20$
   $y = 3x + 12$

6. $y = 3x^2 - x - 2$
   $y = -2x + 2$

**Use a system of equations to solve each equation.**

7. $x^2 = -2x - 1$

8. $\frac{1}{2}x^2 - 4 = 3x + 4$

9. $x^2 + 5x + 5 = -x - 8$

**CHALLENGE Use a graphing calculator to solve other types of systems.**

10. $y = x^2 + 3x - 5$
    $y = -x^2$

11. $y = \frac{3}{4}x$
    $x^2 + y^2 = 1$ (*Hint:* Enter as two functions, $y = \sqrt{1 - x^2}$ and $y = -\sqrt{1 - x^2}$.)

LESSON 2-4

# Solving Quadratic Equations by Completing the Square

**Then**

- You solved quadratic equations by using the square root property.

**Now**

1. Complete the square to write perfect square trinomials.
2. Solve quadratic equations by completing the square.

**Why?**

- In competitions, skateboarders may launch themselves from a half pipe into the air to perform tricks. The equation $h = -16t^2 + 20t + 12$ can be used to model their height, in feet, after $t$ seconds.

To find how long a skateboarder is in the air if he is 25 feet above the half pipe, you can solve $25 = -16t^2 + 20t + 12$ by using a method called completing the square.

**NewVocabulary**
completing the square

**Common Core State Standards**

**Content Standards**

A.REI.4 Solve quadratic equations in one variable.

a. Use the method of completing the square to transform any quadratic equation in $x$ into an equation of the form $(x - p)^2 = q$ that has the same solutions. Derive the quadratic formula from this form.

b. Solve quadratic equations by inspection (e.g., for $x^2 = 49$), taking square roots, completing the square, the quadratic formula and factoring, as appropriate to the initial form of the equation. Recognize when the quadratic formula gives complex solutions and write them as $a \pm bi$ for real numbers $a$ and $b$.

F.IF.8a Use the process of factoring and completing the square in a quadratic function to show zeros, extreme values, and symmetry of the graph, and interpret these in terms of a context.

**Mathematical Practices**

4 Model with mathematics.

## 1 Complete the Square

You have previously solved equations by taking the square root of each side. This method worked only because the expression on the left-hand side was a perfect square. In perfect square trinomials in which the leading coefficient is 1, there is a relationship between the **coefficient of the $x$-term** and the **constant term**.

$$(x + 5)^2 = x^2 + 2(5)(x) + 5^2$$
$$= x^2 + 10x + 25$$

Notice that $\left(\frac{10}{2}\right)^2 = 25$. To get the constant term, divide the coefficient of the $x$-term by 2 and square the result. Any quadratic expression in the form $x^2 + bx$ can be made into a perfect square by using a method called **completing the square**.

### KeyConcept Completing the Square

**Words** To complete the square for any quadratic expression of the form $x^2 + bx$, follow the steps below.

**Step 1** Find one half of $b$, the coefficient of $x$.

**Step 2** Square the result in Step 1.

**Step 3** Add the result of Step 2 to $x^2 + bx$.

**Symbols** $x^2 + bx + \left(\frac{b}{2}\right)^2 = \left(x + \frac{b}{2}\right)^2$

### Example 1 Complete the Square

**Find the value of $c$ that makes $x^2 + 4x + c$ a perfect square trinomial.**

**Method 1** Use algebra tiles.

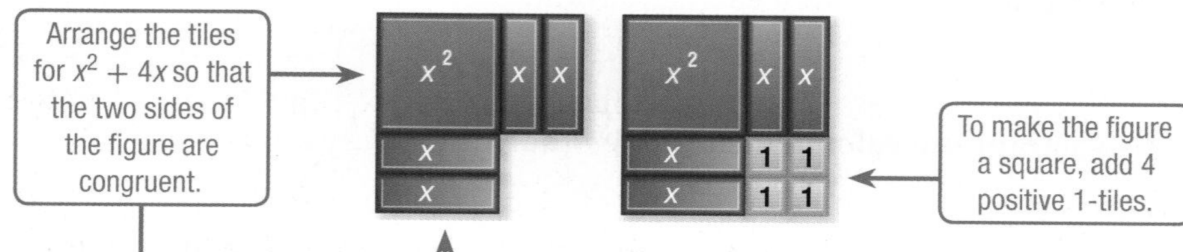

Arrange the tiles for $x^2 + 4x$ so that the two sides of the figure are congruent.
x 2
x
x
x
x
1 1
1 1
To make the figure a square, add 4 positive 1-tiles.


Erik Isakson/Blend Images/age footstock

**StudyTip**

**Algorithms** An algorithm is a series of steps for carrying out a procedure or solving a problem.

**Method 2** Use complete the square algorithm.

**Step 1** Find $\frac{1}{2}$ of 4. $\quad \frac{4}{2} = 2$

**Step 2** Square the result in Step 1. $\quad 2^2 = 4$

**Step 3** Add the result of Step 2 to $x^2 + 4x$. $\quad x^2 + 4x + 4$

Thus, $c = 4$. Notice that $x^2 + 4x + 4 = (x + 2)^2$.

**GuidedPractice**

**1.** Find the value of $c$ that makes $r^2 - 8r + c$ a perfect square trinomial.

## 2 Solve Equations by Completing the Square

You can complete the square to solve quadratic equations. First, you must isolate the $x^2$- and $bx$-terms.

### Example 2 Solve an Equation by Completing the Square

**Solve $x^2 - 6x + 12 = 19$ by completing the square.**

| | |
|---|---|
| $x^2 - 6x + 12 = 19$ | Original equation |
| $x^2 - 6x = 7$ | Subtract 12 from each side. |
| $x^2 - 6x + 9 = 7 + 9$ | Since $\left(\frac{-6}{2}\right)^2 = 9$, add 9 to each side. |
| $(x - 3)^2 = 16$ | Factor $x^2 - 6x + 9$. |
| $x - 3 = \pm 4$ | Take the square root of each side. |
| $x = 3 \pm 4$ | Add 3 to each side. |
| $x = 3 + 4$ or $x = 3 - 4$ | Separate the solutions. |
| $= 7 \qquad\quad = -1$ | The solutions are 7 and $-1$. |

**GuidedPractice**

**2.** Solve $x^2 - 12x + 3 = 8$ by completing the square.

To solve a quadratic equation in which the leading coefficient is not 1, divide each term by the coefficient. Then isolate the $x^2$- and $x$-terms and complete the square.

PT

### Example 3 Equation with $a \neq 1$

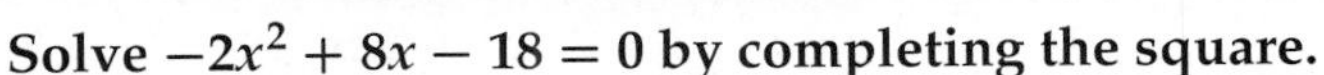

**Solve $-2x^2 + 8x - 18 = 0$ by completing the square.**

**WatchOut!**

**Leading Coefficient** Remember that the leading coefficient has to be 1 before you can complete the square.

| | |
|---|---|
| $-2x^2 + 8x - 18 = 0$ | Original equation |
| $\frac{-2x^2 + 8x - 18}{-2} = \frac{0}{-2}$ | Divide each side by $-2$. |
| $x^2 - 4x + 9 = 0$ | Simplify. |
| $x^2 - 4x = -9$ | Subtract 9 from each side. |
| $x^2 - 4x + 4 = -9 + 4$ | Since $\left(\frac{-4}{2}\right)^2 = 4$, add 4 to each side. |
| $(x - 2)^2 = -5$ | Factor $x^2 - 4x + 4$. |

No real number has a negative square. So, this equation has no real solutions.

**GuidedPractice**

**3.** Solve $3x^2 - 9x - 3 = 21$ by completing the square.

## Real-World Example 4 Use a Graph of a Quadratic Function

**JERSEYS The senior class at Bay High School buys jerseys to wear to the football games. The cost of the jerseys can be modeled by the equation $C = 0.1x^2 + 2.4x + 25$, where $C$ is the amount it costs to buy $x$ jerseys. How many jerseys can they purchase for \$430?**

The seniors have \$430, so set the equation equal to 430 and complete the square.

$0.1x^2 + 2.4x + 25 = 430$ — Original equation

$\frac{0.1x^2 + 2.4x + 25}{0.1} = \frac{430}{0.1}$ — Divide each side by 0.1.

$x^2 + 24x + 250 = 4300$ — Simplify.

$x^2 + 24x + 250 - \mathbf{250} = 4300 - \mathbf{250}$ — Subtract 250 from each side.

$x^2 + 24x = 4050$ — Simplify.

$x^2 + 24x + \mathbf{144} = 4050 + \mathbf{144}$ — Since $\left(\frac{24}{2}\right)^2 = 144$, add 144 to each side.

$x^2 + 24x + 144 = 4194$ — Simplify.

$(x + 12)^2 = 4194$ — Factor $x^2 + 24x + 144$.

$x + 12 = \pm\sqrt{4194}$ — Take the square root of each side.

$x = -12 \pm\sqrt{4194}$ — Subtract 12 from each side.

Use a calculator to approximate each value of $x$.

$x = -12 + \sqrt{4194}$ or $x = -12 - \sqrt{4194}$ — Separate the solutions.

$\approx 52.8$ $\approx -76.8$ — Evaluate.

Since you cannot buy a negative number of jerseys, the negative solution is not reasonable. The seniors can afford to buy 52 jerseys.

**Real-WorldLink**

The oldest public high school rivalry takes place between Wellesley High School and Needham Heights High School in Massachusetts. The first football game between them took place on Thanksgiving morning in 1882 in Needham.

**Source:** USA Football

### GuidedPractice

**4.** If the senior class were able to raise \$620, how many jerseys could they buy?

## Check Your Understanding

**Example 1** **Find the value of $c$ that makes each trinomial a perfect square.**

**1.** $x^2 - 18x + c$

**2.** $x^2 + 22x + c$

**3.** $x^2 + 9x + c$

**4.** $x^2 - 7x + c$

**Examples 2–3** **Solve each equation by completing the square. Round to the nearest tenth if necessary.**

**5.** $x^2 + 4x = 6$

**6.** $x^2 - 8x = -9$

**7.** $4x^2 + 9x - 1 = 0$

**8.** $-2x^2 + 10x + 22 = 4$

**Example 4** **9.** **CCSS MODELING** Collin is building a deck on the back of his family's house. He has enough lumber for the deck to be 144 square feet. The length should be 10 feet more than its width. What should the dimensions of the deck be?

DreamPictures/Photodisc/Getty Images

## Practice and Problem Solving

**Example 1** **Find the value of $c$ that makes each trinomial a perfect square.**

**10.** $x^2 + 26x + c$ **11.** $x^2 - 24x + c$ **12.** $x^2 - 19x + c$

**13.** $x^2 + 17x + c$ **14.** $x^2 + 5x + c$ **15.** $x^2 - 13x + c$

**16.** $x^2 - 22x + c$ **17.** $x^2 - 15x + c$ **18.** $x^2 + 24x + c$

**Examples 2–3** **Solve each equation by completing the square. Round to the nearest tenth if necessary.**

**19.** $x^2 + 6x - 16 = 0$ **20.** $x^2 - 2x - 14 = 0$

**21.** $x^2 - 8x - 1 = 8$ **22.** $x^2 + 3x + 21 = 22$

**23.** $x^2 - 11x + 3 = 5$ **24.** $5x^2 - 10x = 23$

**25.** $2x^2 - 2x + 7 = 5$ **26.** $3x^2 + 12x + 81 = 15$

**27.** $4x^2 + 6x = 12$ **28.** $4x^2 + 5 = 10x$

**29.** $-2x^2 + 10x = -14$ **30.** $-3x^2 - 12 = 14x$

**Example 4** **31. FINANCIAL LITERACY** The price $p$ in dollars for a particular stock can be modeled by the quadratic equation $p = 3.5t - 0.05t^2$, where $t$ represents the number of days after the stock is purchased. When is the stock worth $60?

**GEOMETRY** **Find the value of $x$ for each figure. Round to the nearest tenth if necessary.**

**32.** $A = 45 \text{ in}^2$

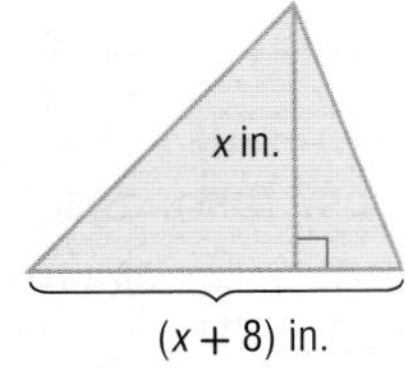


**33.** $A = 110 \text{ ft}^2$

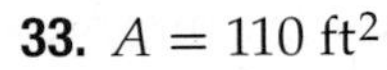

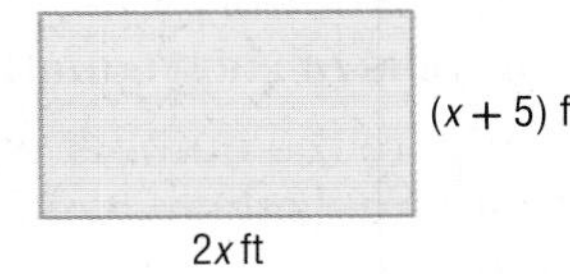


**34. NUMBER THEORY** The product of two consecutive even integers is 224. Find the integers.

**35. CCSS PRECISION** The product of two consecutive negative odd integers is 483. Find the integers.

**36. GEOMETRY** Find the area of the triangle below.

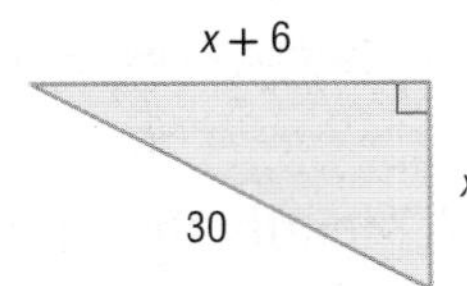


**Solve each equation by completing the square. Round to the nearest tenth if necessary.**

**37.** $0.2x^2 - 0.2x - 0.4 = 0$ **38.** $0.5x^2 = 2x - 0.3$

**39.** $2x^2 - \frac{11}{5}x = -\frac{3}{10}$ **40.** $\frac{2}{3}x^2 - \frac{4}{3}x = \frac{5}{6}$

**41.** $\frac{1}{4}x^2 + 2x = \frac{3}{8}$ **42.** $\frac{2}{5}x^2 + 2x = \frac{1}{5}$

**43** **ASTRONOMY** The height of an object $t$ seconds after it is dropped is given by the equation $h = -\frac{1}{2}gt^2 + h_0$, where $h_0$ is the initial height and $g$ is the acceleration due to gravity. The acceleration due to gravity near the surface of Mars is $3.73 \text{ m/s}^2$, while on Earth it is $9.8 \text{ m/s}^2$. Suppose an object is dropped from an initial height of 120 meters above the surface of each planet.

**a.** On which planet would the object reach the ground first?

**b.** How long would it take the object to reach the ground on each planet? Round each answer to the nearest tenth.

**c.** Do the times that it takes the object to reach the ground seem reasonable? Explain your reasoning.

**44.** Find all values of $c$ that make $x^2 + cx + 100$ a perfect square trinomial.

**45.** Find all values of $c$ that make $x^2 + cx + 225$ a perfect square trinomial.

**46.** **PAINTING** Before she begins painting a picture, Donna stretches her canvas over a wood frame. The frame has a length of 60 inches and a width of 4 inches. She has enough canvas to cover 480 square inches. Donna decides to increase the dimensions of the frame. If the increase in the length is 10 times the increase in the width, what will the dimensions of the frame be?

**47.** **MULTIPLE REPRESENTATIONS** In this problem, you will investigate a property of quadratic equations.

| Trinomial | $b^2 - 4ac$ | Number of Roots |
|---|---|---|
| $x^2 - 8x + 16$ | 0 | 1 |
| $2x^2 - 11x + 3$ | | |
| $3x^2 + 6x + 9$ | | |
| $x^2 - 2x + 7$ | | |
| $x^2 + 10x + 25$ | | |
| $x^2 + 3x - 12$ | | |

**a.** **Tabular** Copy the table shown and complete the second column.

**b.** **Algebraic** Set each trinomial equal to zero, and solve the equation by completing the square. Complete the last column of the table with the number of roots of each equation.

**c.** **Verbal** Compare the number of roots of each equation to the result in the $b^2 - 4ac$ column. Is there a relationship between these values? If so, describe it.

**d.** **Analytical** Predict how many solutions $2x^2 - 9x + 15 = 0$ will have. Verify your prediction by solving the equation.

## H.O.T. Problems Use Higher-Order Thinking Skills

**48.** CCSS **PERSEVERANCE** Given $y = ax^2 + bx + c$ with $a \neq 0$, derive the equation for the axis of symmetry by completing the square and rewriting the equation in the form $y = a(x - h)^2 + k$.

**49.** **REASONING** Determine the number of solutions $x^2 + bx = c$ has if $c < -\left(\frac{b}{2}\right)^2$. Explain.

**50.** **WHICH ONE DOESN'T BELONG?** Identify the expression that does not belong with the other three. Explain your reasoning.

| $n^2 - n + \frac{1}{4}$ | $n^2 + n + \frac{1}{4}$ | $n^2 - \frac{2}{3}n + \frac{1}{9}$ | $n^2 + \frac{1}{3}n + \frac{1}{9}$ |
|---|---|---|---|

**51.** **OPEN ENDED** Write a quadratic equation for which the only solution is 4.

**52.** **WRITING IN MATH** Compare and contrast the following strategies for solving $x^2 - 5x - 7 = 0$: completing the square, graphing, and factoring.

## Standardized Test Practice

**53.** The length of a rectangle is 3 times its width. The area of the rectangle is 75 square feet. Find the length of the rectangle in feet.

A 25 B 15 C 10 D 5

**54. PROBABILITY** At a festival, winners of a game draw a token for a prize. There is one token for each prize. The prizes include 9 movie passes, 8 stuffed animals, 5 hats, 10 jump ropes, and 4 glow necklaces. What is the probability that the first person to draw a token will win a movie pass?

F $\frac{1}{36}$ G $\frac{1}{9}$ H $\frac{9}{61}$ J $\frac{1}{4}$

**55. GRIDDED RESPONSE** The population of a town can be modeled by $P = 22{,}000 + 125t$, where $P$ represents the population and $t$ represents the number of years from 2000. How many years after 2000 will the population be 26,000?

**56.** Percy delivers pizzas for Pizza King. He is paid \$6 an hour plus \$2.50 for each pizza he delivers. Percy earned \$280 last week. If he worked a total of 30 hours, how many pizzas did he deliver?

A 250 pizzas
B 184 pizzas
C 40 pizzas
D 34 pizzas

## Spiral Review

**Describe how the graph of each function is related to the graph of $f(x) = x^2$.** (Lesson 2-3)

**57.** $g(x) = -12 + x^2$ **58.** $h(x) = (x + 2)^2$ **59.** $g(x) = 2x^2 + 5$

**60.** $h(x) = \frac{2}{3}(x - 6)^2$ **61.** $g(x) = 6 + \frac{4}{3}x^2$ **62.** $h(x) = -1 - \frac{3}{2}x^2$

**63. RIDES** A popular amusement park ride whisks riders to the top of a 250-foot tower and drops them. A function for the height of a rider is $h = -16t^2 + 250$, where $h$ is the height and $t$ is the time in seconds. The ride stops the descent of the rider 40 feet above the ground. Write an equation that models the drop of the rider. How long does it take to fall from 250 feet to 40 feet? (Lesson 2-2)

**Describe how the graph of each function is related to the graph of $f(x) = x^2$.** (Lesson 2-3)

**64.** $g(x) = x^2 - 8$ **65.** $h(x) = \frac{1}{4}x^2$ **66.** $h(x) = -x^2 + 5$

**67.** $g(x) = (x + 10)^2$ **68.** $g(x) = -2x^2$ **69.** $h(x) = -x^2 - \frac{4}{3}$

## Skills Review

**Evaluate $\sqrt{b^2 - 4ac}$ for each set of values. Round to the nearest tenth if necessary.**

**70.** $a = 2, b = -5, c = 2$ **71.** $a = 1, b = 12, c = 11$ **72.** $a = -9, b = 10, c = -1$

**73.** $a = 1, b = 7, c = -3$ **74.** $a = 2, b = -4, c = -6$ **75.** $a = 3, b = 1, c = 2$

# EXTEND 2-4 Algebra Lab: Finding the Maximum or Minimum Value

In Lesson 9-3, we learned about the vertex form of the equation of a quadratic function. You will now learn how to write equations in vertex form and use them to identify key characteristics of the graphs of quadratic functions.

**CCSS Common Core State Standards**
**Content Standards**
**A.SSE.3b** Complete the square in a quadratic expression to reveal the maximum or minimum value of the function it defines.
**F.IF.8a** Use the process of factoring and completing the square in a quadratic function to show zeros, extreme values, and symmetry of the graph, and interpret these in terms of a context.

## Activity 1 Find a Minimum

**Write $y = x^2 + 4x - 10$ in vertex form. Identify the axis of symmetry, extrema, and zeros. Then graph the function.**

**Step 1** Complete the square to write the function in vertex form.

| | |
|---|---|
| $y = x^2 + 4x - 10$ | Original function |
| $y + 10 = x^2 + 4x$ | Add 10 to each side. |
| $y + 10 + 4 = x^2 + 4x + 4$ | Since $\left(\frac{4}{2}\right)^2 = 4$, add 4 to each side. |
| $y + 14 = (x + 2)^2$ | Factor $x^2 + 4x + 4$. |
| $y = (x + 2)^2 - 14$ | Subtract 14 from each side to write in vertex form. |

**Step 2** Identify the axis of symmetry and extrema based on the equation in vertex form. The vertex is at $(h, k)$ or $(-2, -14)$. Since there is no negative sign before the $x^2$-term, the parabola opens up and has a minimum at $(-2, -14)$. The equation of the axis of symmetry is $x = -2$.

**Step 3** Solve for $x$ to find the zeros.

| | |
|---|---|
| $(x + 2)^2 - 14 = 0$ | Vertex form, $y = 0$ |
| $(x + 2)^2 = 14$ | Add 14 to each side. |
| $x + 2 = \pm\sqrt{14}$ | Take square root of each side. |
| $x \approx -5.74$ or $1.74$ | Subtract 2 from each side. |

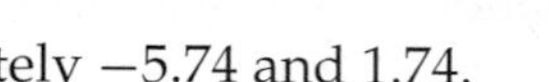

The zeros are approximately $-5.74$ and $1.74$.

**Step 4** Use the key features to graph the function.

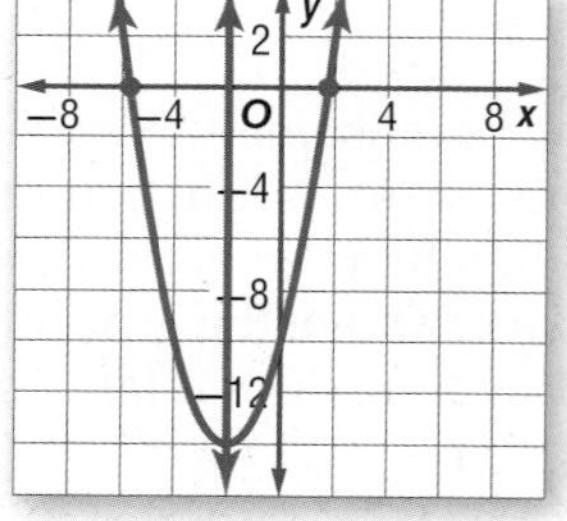


There may be a negative coefficient before the quadratic term. When this is the case, the parabola will open down and have a maximum.

## Activity 2 Find a Maximum

**Write $y = -x^2 + 6x - 5$ in vertex form. Identify the axis of symmetry, extrema, and zeros. Then graph the function.**

**Step 1** Complete the square to write the equation of the function in vertex form.

| | |
|---|---|
| $y = -x^2 + 6x - 5$ | Original function |
| $y + 5 = -x^2 + 6x$ | Add 5 to each side. |
| $y + 5 = -(x^2 - 6x)$ | Factor out $-1$. |
| $y + 5 - 9 = -(x^2 - 6x + 9)$ | Since $\left(\frac{6}{2}\right)^2 = 9$, add $-9$ to each side. |
| $y - 4 = -(x - 3)^2$ | Factor $x^2 - 6x + 9$. |
| $y = -(x - 3)^2 + 4$ | Add 4 to each side to write in vertex form. |

**Step 2** Identify the axis of symmetry and extrema based on the equation in vertex form. The vertex is at $(h, k)$ or $(3, 4)$. Since there is a negative sign before the $x^2$-term, the parabola opens down and has a maximum at $(3, 4)$. The equation of the axis of symmetry is $x = 3$.

**Step 3** Solve for $x$ to find the zeros.

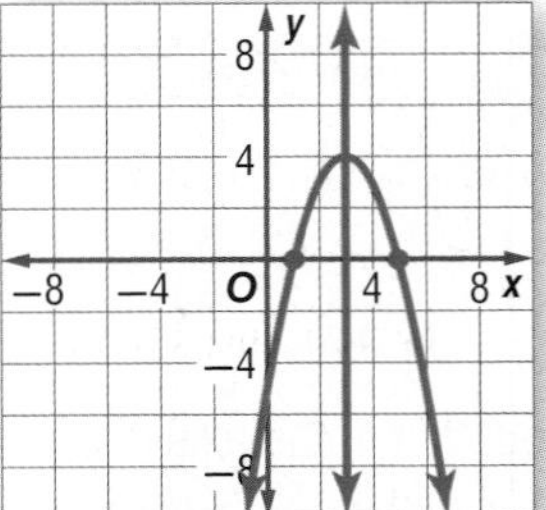


| | |
|---|---|
| $0 = -(x-3)^2 + 4$ | Vertex form, $y = 0$ |
| $(x-3)^2 = 4$ | Add $(x-3)^2$ to each side. |
| $x - 3 = \pm 2$ | Take the square root of each side. |
| $x = 5$ or $1$ | Add 3 to each. |

**Step 4** Use the key features to graph the function.

## Analyze the Results

1. Why do you need to complete the square to write the equation of a quadratic function in vertex form?

**Write each function in vertex form. Identify the axis of symmetry, extrema, and zeros. Then graph the function.**

2. $y = x^2 + 6x$
3. $y = x^2 - 8x + 6$
4. $y = x^2 + 2x - 12$
5. $y = x^2 + 6x + 8$
6. $y = x^2 - 4x + 3$
7. $y = x^2 - 2.4x - 2.2$
8. $y = -4x^2 + 16x - 11$
9. $y = 3x^2 - 12x + 5$
10. $y = -x^2 + 6x - 5$

## Activity 3 Use Extrema in the Real World

**DIVING Alexis jumps from a diving platform upward and outward before diving into the pool. The function $h = -9.8t^2 + 4.9t + 10$, where $h$ is the height of the diver in meters above the pool after $t$ seconds approximates Alexis's dive. Graph the function, then find the maximum height that she reaches and the equation of the axis of symmetry.**

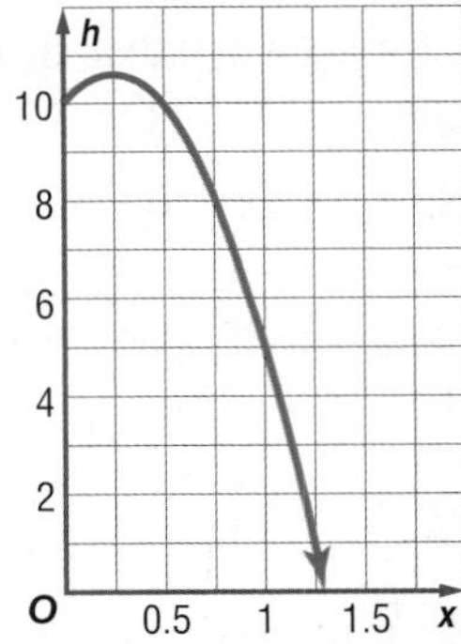


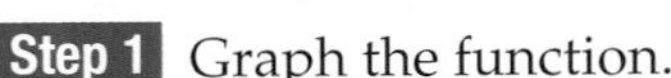

**Step 1** Graph the function.

**Step 2** Complete the square to write the eqution of the function in vertex form.
$h = -9.8t^2 + 4.9t + 10$
$h = -9.8(t - 0.25)^2 + 10.6125$

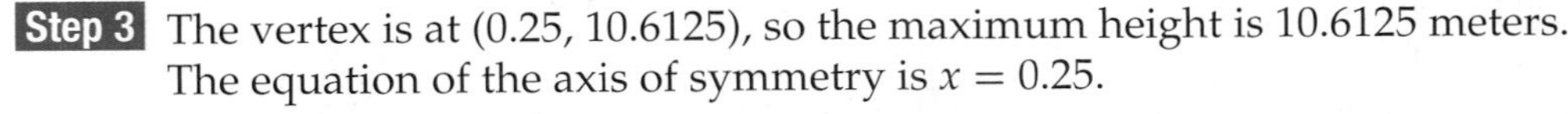

**Step 3** The vertex is at $(0.25, 10.6125)$, so the maximum height is 10.6125 meters. The equation of the axis of symmetry is $x = 0.25$.

## Exercise

11. **SOFTBALL** Jenna throws a ball in the air. The function $h = -16t^2 + 40t + 5$, where $h$ is the height in feet and $t$ represents the time in seconds, approximates Jenna's throw. Graph the function, then find the maximum height of the ball and the equation of the axis of symmetry. When does the ball hit the ground?

CHAPTER 2

# Mid-Chapter Quiz

## Lessons 2-1 through 2-4

**Use a table of values to graph each equation. State the domain and range.** (Lesson 2-1)

1. $y = x^2 + 3x + 1$
2. $y = 2x^2 - 4x + 3$
3. $y = -x^2 - 3x - 3$
4. $y = -3x^2 - x + 1$

Consider $y = x^2 - 5x + 4$. (Lesson 2-1)

5. Write the equation of the axis of symmetry.
6. Find the coordinates of the vertex. Is it a maximum or minimum point?
7. Graph the function.
8. **SOCCER** A soccer ball is kicked from ground level with an initial upward velocity of 90 feet per second. The equation $h = -16t^2 + 90t$ gives the height $h$ of the ball after $t$ seconds. (Lesson 2-1)
   a. What is the height of the ball after one second?
   b. How many seconds will it take for the ball to reach its maximum height?
   c. When is the height of the ball 0 feet? What do these points represent in this situation?

**Solve each equation by graphing. If integral roots cannot be found, estimate the roots to the nearest tenth.** (Lesson 2-2)

9. $x^2 + 5x + 6 = 0$
10. $x^2 + 8 = -6x$
11. $-x^2 + 3x - 1 = 0$
12. $x^2 = 12$
13. **BASEBALL** Juan hits a baseball. The equation $h = -16t^2 + 120t$ models the height $h$, in feet, of the ball after $t$ seconds. How long is the ball in the air? (Lesson 2-2)
14. **CONSTRUCTION** Christopher is repairing the roof on a shed. He accidentally dropped a box of nails from a height of 14 feet. This is represented by the equation $h = -16t^2 + 14$, where $h$ is the height in feet and $t$ is the time in seconds. Describe how the graph is related to $h = t^2$. (Lesson 2-3)
15. **PARTIES** Della's parents are throwing a Sweet 16 party for her. At 10:00, a ball will slide 25 feet down a pole and light up. A function that models the drop is $h = -t^2 + 5t + 25$, where $h$ is height in feet of the ball after $t$ seconds. How many seconds will it take for the ball to reach the bottom of the pole? (Lesson 2-2)

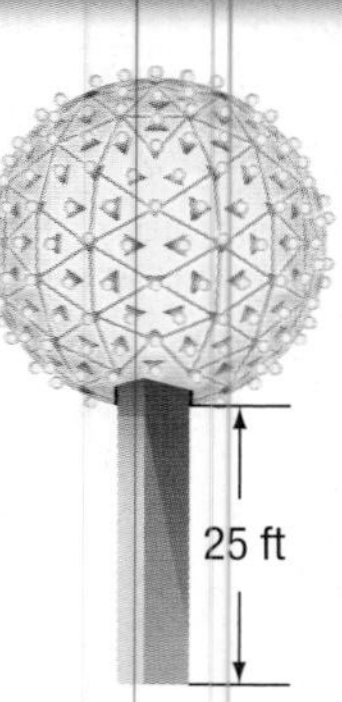


**Describe how the graph of each function is related to the graph of $f(x) = x^2$.** (Lesson 2-3)

16. $g(x) = x^2 + 3$
17. $h(x) = 2x^2$
18. $g(x) = x^2 - 6$
19. $h(x) = \frac{1}{5}x^2$
20. $g(x) = -x^2 + 1$
21. $h(x) = -\frac{5}{8}x^2$
22. **MULTIPLE CHOICE** Which is an equation for the function shown in the graph? (Lesson 2-3)

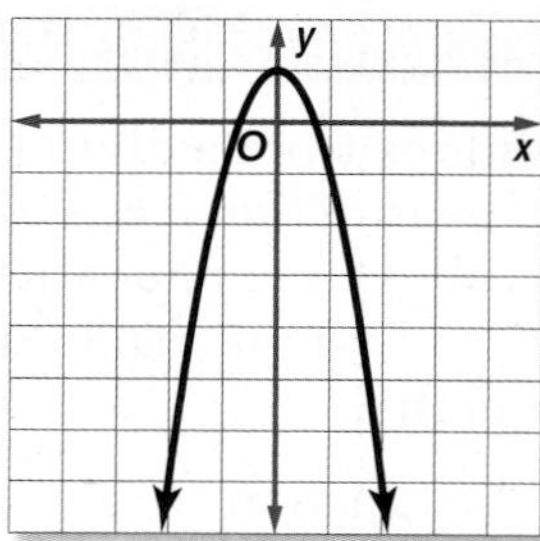


A $y = -2x^2$

B $y = 2x^2 + 1$

C $y = x^2 - 1$

D $y = -2x^2 + 1$

**Solve each equation by completing the square. Round to the nearest tenth.** (Lesson 2-4)

23. $x^2 + 4x + 2 = 0$
24. $x^2 - 2x - 10 = 0$
25. $2x^2 + 4x - 5 = 7$

# LESSON 2-5 Solving Quadratic Equations by Using the Quadratic Formula

**Then**

- You solved quadratic equations by completing the square.

**Now**

1. Solve quadratic equations by using the Quadratic Formula.
2. Use the discriminant to determine the number of solutions of a quadratic equation.

**Why?**

- For adult women, the normal systolic blood pressure $P$ in millimeters of mercury (mm Hg) can be modeled by $P = 0.01a^2 + 0.05a + 107$, where $a$ is age in years. This equation can be used to approximate the age of a woman with a certain systolic blood pressure. However, it would be difficult to solve by factoring, graphing, or completing the square.

**NewVocabulary**
Quadratic Formula
discriminant

**Common Core State Standards**

**Content Standards**
A.REI.4 Solve quadratic equations in one variable.
a. Use the method of completing the square to transform any quadratic equation in $x$ into an equation of the form $(x - p)^2 = q$ that has the same solutions. Derive the quadratic formula from this form.
b. Solve quadratic equations by inspection (e.g., for $x^2 = 49$), taking square roots, completing the square, the quadratic formula and factoring, as appropriate to the initial form of the equation. Recognize when the quadratic formula gives complex solutions and write them as $a \pm bi$ for real numbers $a$ and $b$.

**Mathematical Practices**
6 Attend to precision.

## 1 Quadratic Formula

Completing the square of the quadratic equation $ax^2 + bx + c = 0$ produces a formula that allows you to find the solutions of *any* quadratic equation. This formula is called the **Quadratic Formula**.

### KeyConcept The Quadratic Formula

The solutions of a quadratic equation $ax^2 + bx + c = 0$, where $a \neq 0$, are given by the Quadratic Formula.

$$x = \frac{-b \pm \sqrt{b^2 - 4ac}}{2a}$$

You will derive this formula in Lesson 10-2.

### Example 1 Use the Quadratic Formula

**Solve $x^2 - 12x = -20$ by using the Quadratic Formula.**

**Step 1** Rewrite the equation in standard form.

$x^2 - 12x = -20$ Original equation

$x^2 - 12x + 20 = 0$ Add 20 to each side.

**Step 2** Apply the Quadratic Formula.

$x = \frac{-b \pm \sqrt{b^2 - 4ac}}{2a}$ Quadratic Formula

$= \frac{-(-12) \pm \sqrt{(-12)^2 - 4(1)(20)}}{2(1)}$ $a = 1$, $b = -12$, and $c = 20$

$= \frac{12 \pm \sqrt{144 - 80}}{2}$ Multiply.

$= \frac{12 \pm \sqrt{64}}{2}$ or $\frac{12 \pm 8}{2}$ Subtract and take the square root.

$x = \frac{12 - 8}{2}$ or $x = \frac{12 + 8}{2}$ Separate the solutions.

$= 2$ $= 10$ The solutions are 2 and 10.

**GuidedPractice**

1. Solve $2x^2 + 9x = 18$ by using the Quadratic Formula.

Creatas/PunchStock

The solutions of quadratic equations are not always integers.

PT

## Example 2 Use the Quadratic Formula

**Solve each equation by using the Quadratic Formula. Round to the nearest tenth if necessary.**

**a.** $3x^2 + 5x - 12 = 0$

For this equation, $a = 3$, $b = 5$, and $c = -12$.

$$x = \frac{-b \pm \sqrt{b^2 - 4ac}}{2a}$$ Quadratic Formula

$$= \frac{-(5) \pm \sqrt{(5)^2 - 4(3)(-12)}}{2(3)}$$ $a = 3$, $b = 5$, and $c = -12$

$$= \frac{-5 \pm \sqrt{25 + 144}}{6}$$ Multiply.

$$= \frac{-5 \pm \sqrt{169}}{6} \text{ or } \frac{-5 \pm 13}{6}$$ Add and simplify.

$$x = \frac{-5 - 13}{6} \text{ or } x = \frac{-5 + 13}{6}$$ Separate the solutions.

$$= -3 \qquad = \frac{4}{3}$$ Simplify.

The solutions are $-3$ and $\frac{4}{3}$.

**b.** $10x^2 - 5x = 25$

**Step 1** Rewrite the equation in standard form.

$$10x^2 - 5x = 25$$ Original equation

$$10x^2 - 5x - 25 = 0$$ Subtract 25 from each side.

**Step 2** Apply the Quadratic Formula.

$$x = \frac{-b \pm \sqrt{b^2 - 4ac}}{2a}$$ Quadratic Formula

$$= \frac{-(-5) \pm \sqrt{(-5)^2 - 4(10)(-25)}}{2(10)}$$ $a = 10$, $b = -5$, and $c = -25$

$$= \frac{5 \pm \sqrt{25 + 1000}}{20}$$ Multiply.

$$= \frac{5 \pm \sqrt{1025}}{20}$$ Add.

$$= \frac{5 - \sqrt{1025}}{20} \text{ or } \frac{5 + \sqrt{1025}}{20}$$ Separate the solutions.

$$\approx -1.4 \qquad \approx 1.9$$ Simplify.

The solutions are about $-1.4$ and $1.9$.

### GuidedPractice

**2A.** $4x^2 - 24x + 35 = 0$

**2B.** $3x^2 - 2x - 9 = 0$

**StudyTip**

**CCSS Precision** In Example 2, the number $\sqrt{1025}$ is irrational, so the calculator can only give you an approximation of its value. So, the exact answer in Example 2 is $\frac{5 \pm \sqrt{1025}}{20}$. The numbers $-1.4$ and $1.9$ are approximations.

You can solve quadratic equations by using many different methods. No one way is always best.

**WatchOut!**

**Solutions** No matter what method is used to solve a quadratic equation, all of the methods should produce the same solution(s).

## Example 3 Solve Quadratic Equations Using Different Methods

**Solve $x^2 - 4x = 12$.**

**Method 1** Graphing

Rewrite the equation in standard form.

$x^2 - 4x = 12$ Original equation

$x^2 - 4x - 12 = 0$ Subtract 12 from each side.

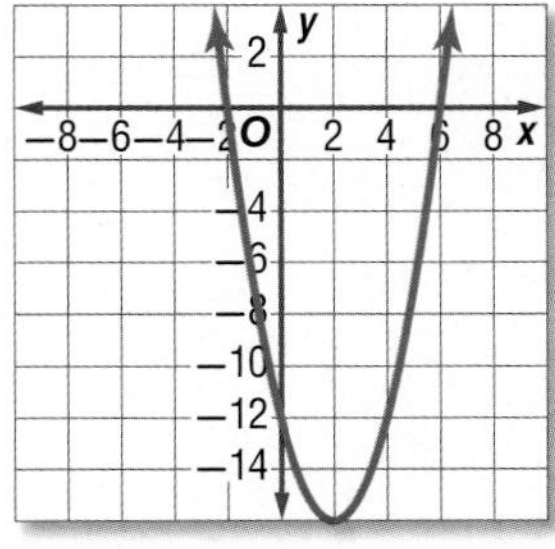


Graph the related function $f(x) = x^2 - 4x - 12$. Locate the $x$-intercepts of the graph.

The solutions are $-2$ and $6$.

**Method 2** Factoring

$x^2 - 4x = 12$ Original equation

$x^2 - 4x - 12 = 0$ Subtract 12 from each side.

$(x - 6)(x + 2) = 0$ Factor.

$x - 6 = 0$ or $x + 2 = 0$ Zero Product Property

$x = 6$ $\quad$ $x = -2$ Solve for $x$.

**Method 3** Completing the Square

The equation is in the correct form to complete the square, since the leading coefficient is 1 and the $x^2$ and $x$ terms are isolated.

$x^2 - 4x = 12$ Original equation

$x^2 - 4x + 4 = 12 + 4$ Since $\left(\frac{-4}{2}\right)^2 = 4$, add 4 to each side.

$(x - 2)^2 = 16$ Factor $x^2 - 4x + 4$.

$x - 2 = \pm 4$ Take the square root of each side.

$x = 2 \pm 4$ Add 2 to each side.

$x = 2 + 4$ or $x = 2 - 4$ Separate the solutions.

$= 6$ $\quad$ $= -2$ Simplify.

**Method 4** Quadratic Formula

From Method 1, the standard form of the equation is $x^2 - 4x - 12 = 0$.

$x = \frac{-b \pm \sqrt{b^2 - 4ac}}{2a}$ Quadratic Formula

$= \frac{-(-4) \pm \sqrt{(-4)^2 - 4(1)(-12)}}{2(1)}$ $a = 1$, $b = -4$, and $c = -12$

$= \frac{4 \pm \sqrt{16 + 48}}{2}$ Multiply.

$= \frac{4 \pm \sqrt{64}}{2}$ or $\frac{4 \pm 8}{2}$ Add and simplify.

$x = \frac{4 - 8}{2}$ or $x = \frac{4 + 8}{2}$ Separate the solutions.

$= -2$ $\quad$ $= 6$ Simplify.

**GuidedPractice**

**Solve each equation.**

**3A.** $2x^2 - 17x + 8 = 0$

**3B.** $4x^2 - 4x - 11 = 0$

| **ConceptSummary** Solving Quadratic Equations | |
|---|---|
| **Method** | **When to Use** |
| Factoring | Use when the constant term is 0 or if the factors are easily determined. Not all equations are factorable. |
| Graphing | Use when an approximate solution is sufficient. |
| Using Square Roots | Use when an equation can be written in the form $x^2 = n$. Can only be used if the equation has no $x$-term. |
| Completing the Square | Can be used for any equation $ax^2 + bx + c = 0$, but is simplest to apply when $b$ is even and $a = 1$. |
| Quadratic Formula | Can be used for any equation $ax^2 + bx + c = 0$. |

## 2 The Discriminant

In the Quadratic Formula, the expression under the radical sign, $b^2 - 4ac$, is called the **discriminant**. The discriminant can be used to determine the number of real solutions of a quadratic equation.

**StudyTip**

**Discriminant** Recall that when the left side of the standard form of an equation is a perfect square trinomial, there is only one solution. Therefore, the discriminant of a perfect square trinomial will always be zero.

| **KeyConcept** Using the Discriminant | | | |
|---|---|---|---|
| Equation | $x^2 + 2x + 5 = 0$ | $x^2 + 10x + 25 = 0$ | $2x^2 - 7x + 2 = 0$ |
| Discriminant | $b^2 - 4ac = -16$ negative | $b^2 - 4ac = 0$ zero | $b^2 - 4ac = 33$ positive |
| Graph of Related Function | 0 $x$-intercepts | 1 $x$-intercept | 2 $x$-intercepts |
| Real Solutions | 0 | 1 | 2 |

### Example 4 Use the Discriminant

PT

**State the value of the discriminant of $4x^2 + 5x = -3$. Then determine the number of real solutions of the equation.**

**Step 1** Rewrite in standard form. $4x^2 - 5x = -3 \longrightarrow 4x^2 - 5x + 3 = 0$

**Step 2** Find the discriminant.

$b^2 - 4ac = (-5)^2 - 4(4)(3)$ — $a = 4$, $b = -5$, and $c = 3$

$= -23$ — Simplify.

Since the discriminant is negative, the equation has no real solutions.

**GuidedPractice**

**4A.** $2x^2 + 11x + 15 = 0$

**4B.** $9x^2 - 30x + 25 = 0$

## Check Your Understanding

**Examples 1–2** **Solve each equation by using the Quadratic Formula. Round to the nearest tenth if necessary.**

**1.** $x^2 - 2x - 15 = 0$ **2.** $x^2 - 10x + 16 = 0$ **3.** $x^2 - 8x = -10$

**4.** $x^2 + 3x = 12$ **5.** $10x^2 - 31x + 15 = 0$ **6.** $5x^2 + 5 = -13x$

**Example 3** **Solve each equation. State which method you used.**

**7.** $2x^2 + 11x - 6 = 0$ **8.** $2x^2 - 3x - 6 = 0$

**9.** $9x^2 = 25$ **10.** $x^2 - 9x = -19$

**Example 4** **State the value of the discriminant for each equation. Then determine the number of real solutions of the equation.**

**11.** $x^2 - 9x + 21 = 0$ **12.** $2x^2 - 11x + 10 = 0$

**13.** $9x^2 + 24x = -16$ **14.** $3x^2 - x = 8$

**15.** **TRAMPOLINE** Eva springs from a trampoline to dunk a basketball. Her height $h$ in feet can be modeled by the equation $h = -16t^2 + 22.3t + 2$, where t is time in seconds. Use the discriminant to determine if Eva will reach a height of 10 feet. Explain.

## Practice and Problem Solving

**Examples 1–2** **Solve each equation by using the Quadratic Formula. Round to the nearest tenth if necessary.**

**16.** $4x^2 + 5x - 6 = 0$ **17.** $x^2 + 16 = 0$ **18.** $6x^2 - 12x + 1 = 0$

**19.** $5x^2 - 8x = 6$ **20.** $2x^2 - 5x = -7$ **21.** $5x^2 + 21x = -18$

**22.** $81x^2 = 9$ **23.** $8x^2 + 12x = 8$ **24.** $4x^2 = -16x - 16$

**25.** $10x^2 = -7x + 6$ **26.** $-3x^2 = 8x - 12$ **27.** $2x^2 = 12x - 18$

**28.** **AMUSEMENT PARKS** The Demon Drop at Cedar Point in Ohio takes riders to the top of a tower and drops them 60 feet. A function that approximates this ride is $h = -16t^2 + 64t - 60$, where $h$ is the height in feet and $t$ is the time in seconds. About how many seconds does it take for riders to drop 60 feet?

**Example 3** **Solve each equation. State which method you used.**

**29.** $2x^2 - 8x = 12$ **30.** $3x^2 - 24x = -36$ **31.** $x^2 - 3x = 10$

**32.** $4x^2 + 100 = 0$ **33.** $x^2 = -7x - 5$ **34.** $12 - 12x = -3x^2$

**Example 4** **State the value of the discriminant for each equation. Then determine the number of real solutions of the equation.**

**35.** $0.2x^2 - 1.5x + 2.9 = 0$ **36.** $2x^2 - 5x + 20 = 0$ **37.** $x^2 - \frac{4}{5}x = 3$

**38.** $0.5x^2 - 2x = -2$ **39.** $2.25x^2 - 3x = -1$ **40.** $2x^2 = \frac{5}{2}x + \frac{3}{2}$

**41.** **CCSS MODELING** The percent of U.S. households with high-speed Internet $h$ can be estimated by $h = -0.2n^2 + 7.2n + 1.5$, where $n$ is the number of years since 1990.

**a.** Use the Quadratic Formula to determine when 20% of the population will have high-speed Internet.

**b.** Is a quadratic equation a good model for this information? Explain.

42. **TRAFFIC** The equation $d = 0.05v^2 + 1.1v$ models the distance $d$ in feet it takes a car traveling at a speed of $v$ miles per hour to come to a complete stop. If Hannah's car stopped after 250 feet on a highway with a speed limit of 65 miles per hour, was she speeding? Explain your reasoning.

**Without graphing, determine the number of $x$-intercepts of the graph of the related function for each equation.**

43. $4.25x + 3 = -3x^2$

44. $x^2 + \frac{2}{25} = \frac{3}{5}x$

45. $0.25x^2 + x = -1$

**Solve each equation by using the Quadratic Formula. Round to the nearest tenth if necessary.**

46. $-2x^2 - 7x = -1.5$

47. $2.3x^2 - 1.4x = 6.8$

48. $x^2 - 2x = 5$

49. **POSTER** Bartolo is making a poster for the dance. He wants to cover three fourths of the area with text.

a. Write an equation for the area of the section with text.

b. Solve the equation by using the Quadratic Formula.

c. What should be the margins of the poster?



50. **MULTIPLE REPRESENTATIONS** In this problem, you will investigate writing a quadratic equation with given roots. If $p$ is a root of $0 = ax^2 + bx + c$, then $(x - p)$ is a factor of $ax^2 + bx + c$.

a. **Tabular** Copy and complete the first two columns of the table.

b. **Algebraic** Multiply the factors to write each equation with integral coefficients. Use the equations to complete the last column of the table. Write each equation.

c. **Analytical** How could you write an equation with three roots? Test your conjecture by writing an equation with roots 1, 2, and 3. Is the equation quadratic? Explain.

| Roots | Factors | Equation |
|---|---|---|
| 2, 5 | $(x - 2), (x - 5)$ | $(x - 2)(x - 5) = 0$<br>$x^2 - 7x + 10 = 0$ |
| 1, 9 | | |
| $-1, 3$ | | |
| 0, 6 | | |
| $\frac{1}{2}, 7$ | | |
| $-\frac{2}{3}, 4$ | | |

## H.O.T. Problems Use Higher-Order Thinking Skills

51. **CHALLENGE** Find all values of $k$ such that $2x^2 - 3x + 5k = 0$ has two solutions.

52. **REASONING** Use factoring techniques to determine the number of real zeros of $f(x) = x^2 - 8x + 16$. Compare this method to using the discriminant.

**CCSS STRUCTURE Determine whether there are *two*, *one*, or *no* real solutions of each equation.**

53. The graph of the related quadratic function does not have an $x$-intercept.

54. The graph of the related quadratic function touches but does not cross the $x$-axis.

55. The graph of the related quadratic function intersects the $x$-axis twice.

56. Both $a$ and $b$ are greater than 0 and $c$ is less than 0 in a quadratic equation.

57. **WRITING IN MATH** Why can the discriminant be used to confirm the number of real solutions of a quadratic equation?

58. **WRITING IN MATH** Describe the advantages and disadvantages of each method of solving quadratic equations. Which method do you prefer, and why?

## Standardized Test Practice

**59.** If $n$ is an even integer, which expression represents the product of three consecutive even integers?

**A** $n(n + 1)(n + 2)$
**B** $(n + 1)(n + 2)(n + 3)$
**C** $3n + 2$
**D** $n(n + 2)(n + 4)$

**60. SHORT RESPONSE** The triangle shown is an isosceles triangle. What is the value of $x$?

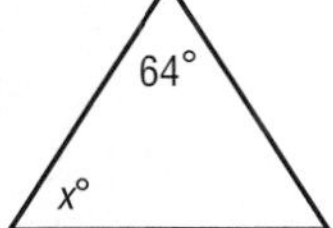


**61.** Which statement best describes the graph of $x = 5$?

**F** It is parallel to the $x$-axis.
**G** It is parallel to the $y$-axis.
**H** It passes through the point (2, 5).
**J** It has a $y$-intercept of 5.

**62.** What are the solutions of the quadratic equation $6h^2 + 6h = 72$?

**A** 3 or −4
**B** −3 or 4
**C** no solution
**D** 12 or −48

## Spiral Review

**Solve each equation by completing the square. Round to the nearest tenth if necessary.** (Lesson 2-4)

**63.** $6x^2 - 17x + 12 = 0$
**64.** $x^2 - 9x = -12$
**65.** $4x^2 = 20x - 25$

**Describe the transformations needed to obtain the graph of $g(x)$ from the graph of $f(x)$.** (Lesson 2-3)

**66.** $f(x) = 4x^2$
$g(x) = 2x^2$

**67.** $f(x) = x^2 + 5$
$g(x) = x^2 - 1$

**68.** $f(x) = x^2 - 6$
$g(x) = x^2 + 3$

**Solve each equation by using the Quadratic Formula. Round to the nearest tenth if necessary.** (Lesson 2-5)

**69.** $v^2 + 12v + 20 = 0$
**70.** $3t^2 - 7t - 20 = 0$
**71.** $5y^2 - y - 4 = 0$
**72.** $2x^2 + 98 = 28x$
**73.** $2n^2 - 7n - 3 = 0$
**74.** $2w^2 = -(7w + 3)$

**75. THEATER** The drama club is building a backdrop using arches with a shape that can be represented by the function $f(x) = -x^2 + 2x + 8$, where $x$ is the length of the arch in feet. The region under each arch is to be covered with fabric. (Lesson 2-2)

**a.** Graph the quadratic function and determine its $x$-intercepts.
**b.** What is the height of the arch?

## Skills Review

**Determine whether each sequence is *arithmetic*, *geometric*, or neither. Explain.**

**76.** 20, 25, 30, …
**77.** 1000, 950, 900, …
**78.** 200, 350, 650, …
**79.** 6, 18, 54, …
**80.** 2, 4, 16, …
**81.** 8, −4, 2 …

# LESSON 2-6 Analyzing Functions with Successive Differences

| Then | Now | Why? |
|---|---|---|
| You graphed linear, quadratic, and exponential functions. | 1 Identify linear, quadratic, and exponential functions from given data.<br>2 Write equations that model data. | Every year the golf team sells candy to raise money for charity. By knowing what type of function models the sales of the candy, they can determine the best price of the candy. |

**Common Core State Standards**

**Content Standards**

F.IF.6 Calculate and interpret the average rate of change of a function (presented symbolically or as a table) over a specified interval. Estimate the rate of change from a graph.

F.LE.1 Distinguish between situations that can be modeled with linear functions and with exponential functions.

a. Prove that linear functions grow by equal differences over equal intervals, and that exponential functions grow by equal factors over equal intervals.

b. Recognize situations in which one quantity changes at a constant rate per unit interval relative to another.

c. Recognize situations in which a quantity grows or decays by a constant percent rate per unit interval relative to another.

**Mathematical Practices**

7 Look for and make use of structure.

## 1 Identify Functions

You can use linear functions, quadratic functions, and exponential functions to model data. The general forms of the equations and a graph of each function type are listed below.

**ConceptSummary Linear and Nonlinear Functions**

| Linear Function | Quadratic Function | Exponential Function |
|---|---|---|
| $y = mx + b$ | $y = ax^2 + bx + c$ | $y = ab^x$, when $b > 0$ |

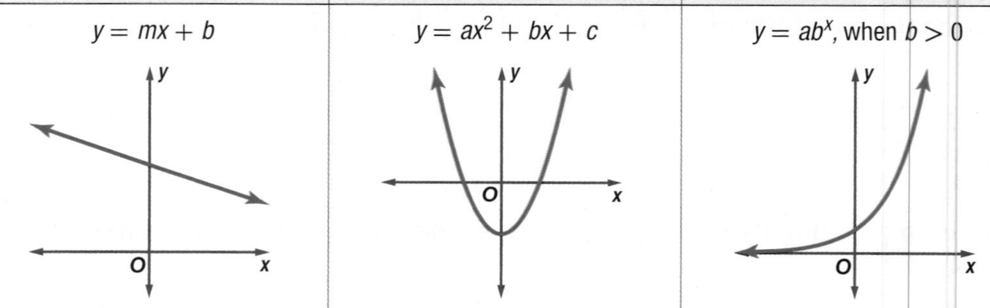

**Example 1 Choose a Model Using Graphs**

**Graph each set of ordered pairs. Determine whether the ordered pairs represent a *linear* function, a *quadratic* function, or an *exponential* function.**

**a.** $\{(-2, 5), (-1, 2), (0, 1), (1, 2), (2, 5)\}$

The ordered pairs appear to represent a quadratic function.

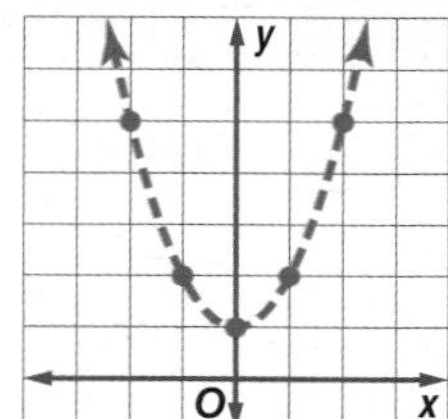


**b.** $\left\{\left(-2, \frac{1}{4}\right), \left(-1, \frac{1}{2}\right), (0, 1), (1, 2), (2, 4)\right\}$

The ordered pairs appear to represent an exponential function.

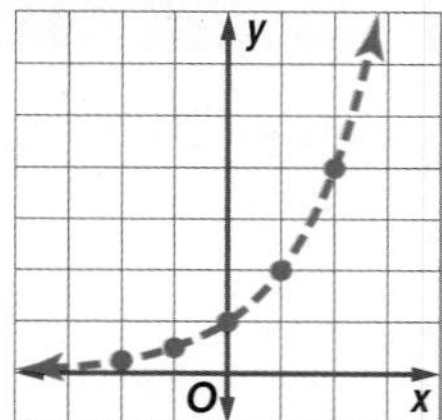


**GuidedPractice**

**1A.** $(-2, -3), (-1, -1), (0, 1), (1, 3)$

**1B.** $(-1, 0.25), (0, 1), (1, 4), (2, 16)$

Elan Fleisher/Look/age fotostock

Another way to determine which model best describes data is to use patterns. The differences of successive $y$-values are called *first differences*. The differences of successive first differences are called *second differences*.

- If the differences of successive $y$-values are all equal, the data represent a linear function.
- If the second differences are all equal, but the first differences are not equal, the data represent a quadratic function.
- If the ratios of successive $y$-values are all equal and $r \neq 1$, the data represent an exponential function.

**WatchOut!**

**$x$-Values** Before you check for successive differences or ratios, make sure the $x$-values are increasing by the same amount.

## Example 2 Choose a Model Using Differences or Ratios

**Look for a pattern in each table of values to determine which kind of model best describes the data.**

**a.**

| $x$ | −2 | −1 | 0 | 1 | 2 |
|---|---|---|---|---|---|
| $y$ | −8 | −3 | 2 | 7 | 12 |

First differences: −8 −3 2 7 12; differences: 5 5 5 5

Since the first differences are all equal, the table of values represents a linear function.

**b.**

| $x$ | −1 | 0 | 1 | 2 | 3 |
|---|---|---|---|---|---|
| $y$ | 8 | 4 | 2 | 1 | 0.5 |

First differences: 8 4 2 1 0.5; differences: −4 −2 −1 −0.5

The first differences are not all equal. So, the table of values does not represent a linear function. Find the second differences and compare.

First differences: −4 −2 −1 −0.5

Second differences: 2 1 0.5

The second differences are not all equal. So, the table of values does not represent a quadratic function. Find the ratios of the $y$-values and compare.

8 4 2 1 0.5

Ratios: $\frac{4}{8} = \frac{1}{2}$ $\frac{2}{4} = \frac{1}{2}$ $\frac{1}{2}$ $\frac{0.5}{1} = \frac{1}{2}$

The ratios of successive $y$-values are equal. Therefore, the table of values can be modeled by an exponential function.

**Guided Practice**

**2A.**

| $x$ | −3 | −2 | −1 | 0 | 1 |
|---|---|---|---|---|---|
| $y$ | −3 | −7 | −9 | −9 | −7 |

**2B.**

| $x$ | −2 | −1 | 0 | 1 | 2 |
|---|---|---|---|---|---|
| $y$ | −18 | −13 | −8 | −3 | 2 |

## 2 Write Equations

Once you find the model that best describes the data, you can write an equation for the function. For a quadratic function in this lesson, the equation will have the form $y = ax^2$.

## Example 3 Write an Equation

**Determine which kind of model best describes the data. Then write an equation for the function that models the data.**

**Step 1** Determine which model fits the data.

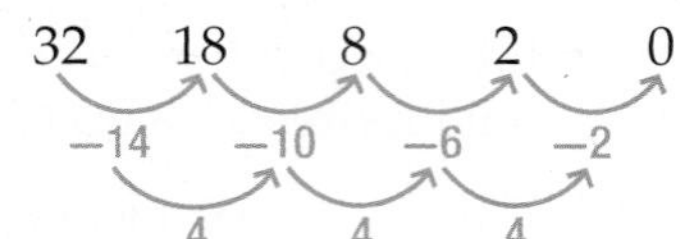


First differences:

Second differences:

Since the second differences are equal, a quadratic function models the data.

**Step 2** Write an equation for the function that models the data.

The equation has the form $y = ax^2$. Find the value of $a$ by choosing one of the ordered pairs from the table of values. Let's use $(-1, 2)$.

$y = ax^2$ Equation for quadratic function

$2 = a(-1)^2$ $x = -1$ and $y = 2$

$2 = a$ An equation that models the data is $y = 2x^2$.

**WatchOut!**

**Finding *a*** In Example 3, the point (0, 0) cannot be used to find the value of $a$. You will have to divide each side by 0, giving you an undefined value for $a$.

### GuidedPractice

**3A.**

| $x$ | −2 | −1 | 0 | 1 | 2 |
|---|---|---|---|---|---|
| $y$ | 11 | 7 | 3 | −1 | −5 |

**3B.**

| $x$ | −3 | −2 | −1 | 0 | 1 |
|---|---|---|---|---|---|
| $y$ | 0.375 | 0.75 | 1.5 | 3 | 6 |

## Real-World Example 4 Write an Equation for a Real-World Situation

**BOOK CLUB The table shows the number of book club members for four consecutive years. Determine which model best represents the data. Then write a function that models the data.**

**Understand** We need to find a model for the data, and then write a function.

| Time (years) | 0 | 1 | 2 | 3 | 4 |
|---|---|---|---|---|---|
| Members | 5 | 10 | 20 | 40 | 80 |

**Plan** Find a pattern using successive differences or ratios. Then use the general form of the equation to write a function.

**Solve** The constant ratio is 2. This is the value of the base. An exponential function of the form $y = ab^x$ models the data.

$y = ab^x$ Equation for exponential function

$5 = a(2)^0$ $x = 0$, $y = 5$, and $b = 2$

$5 = a$ The equation that models the data is $y = 5 \cdot 2^x$.

**Check** You used (0, 5) to write the function. Verify that every other ordered pair satisfies the equation.

**Real-WorldLink**

A poll by the National Education Association found that 87% of all teens polled found reading relaxing, 85% viewed reading as rewarding, and 79% found reading exciting.

**Source:** *American Demographics*

### GuidedPractice

**4. ADVERTISING** The table shows the cost of placing an ad in a newspaper. Determine a model that best represents the data and write a function that models the data.

| No. of Lines | 5 | 6 | 7 | 8 |
|---|---|---|---|---|
| Total Cost ($) | 14.50 | 16.60 | 18.70 | 20.80 |

DK Stock/David Deas/Collection Mix: Subjects/Getty Images

## Check Your Understanding

**Example 1** **Graph each set of ordered pairs. Determine whether the ordered pairs represent a *linear* function, a *quadratic* function, or an *exponential* function.**

1. (−2, 8), (−1, 5), (0, 2), (1, −1)
2. (−3, 7), (−2, 3), (−1, 1), (0, 1), (1, 3)
3. (−3, 8), (−2, 4), (−1, 2), (0, 1), (1, 0.5)
4. (0, 2), (1, 2.5), (2, 3), (3, 3.5)

**Example 2** **Look for a pattern in each table of values to determine which kind of model best describes the data.**

5.

| x | 0 | 1 | 2 | 3 | 4 |
|---|---|---|---|---|---|
| y | 5 | 8 | 17 | 32 | 53 |

6.

| x | −3 | −2 | −1 | 0 |
|---|---|---|---|---|
| y | −6.75 | −7.5 | −8.25 | −9 |

7.

| x | −1 | 0 | 1 | 2 | 3 |
|---|---|---|---|---|---|
| y | 3 | 6 | 12 | 24 | 48 |

8.

| x | 3 | 4 | 5 | 6 | 7 |
|---|---|---|---|---|---|
| y | −1.5 | 0 | 2.5 | 6 | 10.5 |

**Example 3** **Determine which kind of model best describes the data. Then write an equation for the function that models the data.**

9.

| x | −1 | 0 | 1 | 2 | 3 |
|---|---|---|---|---|---|
| y | 1 | 3 | 9 | 27 | 81 |

10.

| x | −5 | −4 | −3 | −2 | −1 |
|---|---|---|---|---|---|
| y | 125 | 80 | 45 | 20 | 5 |

11.

| x | −3 | −2 | −1 | 0 | 1 |
|---|---|---|---|---|---|
| y | 1 | 1.5 | 2 | 2.5 | 3 |

12.

| x | −1 | 0 | 1 | 2 |
|---|---|---|---|---|
| y | −1.25 | −1 | −0.75 | −0.5 |

**Example 4** 13. **PLANTS** The table shows the height of a plant for four consecutive weeks. Determine which kind of function best models the height. Then write a function that models the data.

| Week | 0 | 1 | 2 | 3 | 4 |
|---|---|---|---|---|---|
| Height (in.) | 3 | 3.5 | 4 | 4.5 | 5 |

## Practice and Problem Solving

**Example 1** **Graph each set of ordered pairs. Determine whether the ordered pairs represent a *linear* function, a *quadratic* function, or an *exponential* function.**

14. (−1, 1), (0, −2), (1, −3), (2, −2), (3, 1)
15. (1, 2.75), (2, 2.5), (3, 2.25), (4, 2)
16. (−3, 0.25), (−2, 0.5), (−1, 1), (0, 2)
17. (−3, −11), (−2, −5), (−1, −3), (0, −5)
18. (−2, 6), (−1, 1), (0, −4), (1, −9)
19. (−1, 8), (0, 2), (1, 0.5), (2, 0.125)

**Examples 2–3** **Look for a pattern in each table of values to determine which kind of model best describes the data. Then write an equation for the function that models the data.**

20.

| x | −3 | −2 | −1 | 0 |
|---|---|---|---|---|
| y | −8.8 | −8.6 | −8.4 | −8.2 |

21.

| x | −2 | −1 | 0 | 1 | 2 |
|---|---|---|---|---|---|
| y | 10 | 2.5 | 0 | 2.5 | 10 |

22.

| x | −1 | 0 | 1 | 2 | 3 |
|---|---|---|---|---|---|
| y | 0.75 | 3 | 12 | 48 | 192 |

23.

| x | −2 | −1 | 0 | 1 | 2 |
|---|---|---|---|---|---|
| y | 0.008 | 0.04 | 0.2 | 1 | 5 |

24.

| x | 0 | 1 | 2 | 3 | 4 |
|---|---|---|---|---|---|
| y | 0 | 4.2 | 16.8 | 37.8 | 67.2 |

25.

| x | −3 | −2 | −1 | 0 | 1 |
|---|---|---|---|---|---|
| y | 14.75 | 9.75 | 4.75 | −0.25 | −5.25 |

Example 4

26. **WEB SITES** A company tracked the number of visitors to its Web site over 4 days. Determine which kind of model best represents the number of visitors to the Web site with respect to time. Then write a function that models the data.

| Day | 0 | 1 | 2 | 3 | 4 |
|---|---|---|---|---|---|
| Visitors (in thousands) | 0 | 0.9 | 3.6 | 8.1 | 14.4 |

27. **CALLING** The cost of an international call depends on the length of the call. The table shows the cost for up to 6 minutes.

| Length of call (min) | 1 | 2 | 3 | 4 | 5 | 6 |
|---|---|---|---|---|---|---|
| Cost ($) | 0.12 | 0.24 | 0.36 | 0.48 | 0.60 | 0.72 |

**a.** Graph the data and determine which kind of function best models the data.

**b.** Write an equation for the function that models the data.

**c.** Use your equation to determine how much a 10-minute call would cost.

28. **DEPRECIATION** The value of a car depreciates over time. The table shows the value of a car over a period of time.

| Year | 0 | 1 | 2 | 3 | 4 |
|---|---|---|---|---|---|
| Value ($) | 18,500 | 15,910 | 13,682.60 | 11,767.04 | 10,119.65 |

**a.** Determine which kind of function best models the data.

**b.** Write an equation for the function that models the data.

**c.** Use your equation to determine how much the car is worth after 7 years.

29. **BACTERIA** A scientist estimates that a bacteria culture with an initial population of 12 will triple every hour.

**a.** Make a table to show the bacteria population for the first 4 hours.

**b.** Which kind of model best represents the data?

**c.** Write a function that models the data.

**d.** How many bacteria will there be after 8 hours?

30. **PRINTING** A printing company charges the fees shown to print flyers. Write a function that models the total cost of the flyers, and determine how much 30 flyers would cost.

## H.O.T. Problems Use Higher-Order Thinking Skills

31. **CHALLENGE** Write a function that has constant second differences, first differences that are not constant, a $y$-intercept of $-5$, and contains the point (2, 3).

32. **CCSS ARGUMENTS** What type of function will have constant third differences but not constant second differences? Explain.

33. **OPEN ENDED** Write a linear function that has a constant first difference of 4.

34. **PROOF** Write a paragraph proof to show that linear functions grow by equal differences over equal intervals, and exponential functions grow by equal factors over equal intervals. (*Hint:* Let $y = ax$ represent a linear function and let $y = a^x$ represent an exponential function.)

35. **WRITING IN MATH** How can you determine whether a given set of data should be modeled by a *linear* function, a *quadratic* function, or an *exponential* function?

## Standardized Test Practice

**36. SHORT RESPONSE** Write an equation that models the data in the table.

| x | 0 | 1 | 2 | 3 | 4 |
|---|---|---|---|---|---|
| y | 3 | 6 | 12 | 24 | 48 |

**37.** What is the equation of the line below?

**A** $y = \frac{2}{5}x + 2$

**B** $y = \frac{2}{5}x - 2$

**C** $y = \frac{5}{2}x + 2$

**D** $y = \frac{5}{2}x - 2$

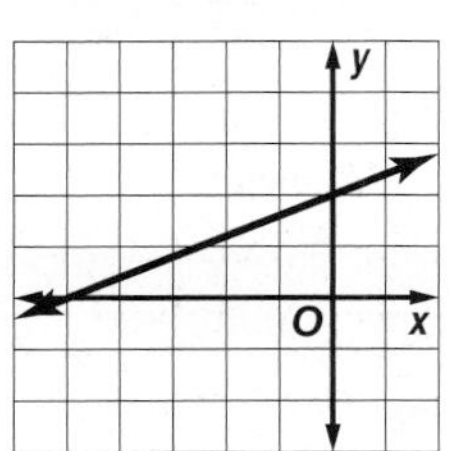


**38.** The point $(r, -4)$ lies on a line with an equation of $2x + 3y = -8$. Find the value of $r$.

**F** $-10$ **H** 2

**G** 0 **J** 8

**39. GEOMETRY** The rectangle has an area of 220 square feet. Find the length $\ell$.

**A** 8 feet

**B** 10 feet

**C** 22 feet

**D** 34 feet

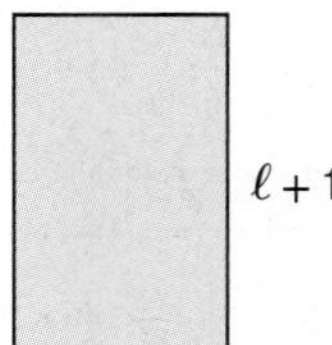


## Spiral Review

**Solve each equation by using the Quadratic Formula. Round to the nearest tenth if necessary.** (Lesson 2-5)

**40.** $6x^2 - 3x - 30 = 0$ **41.** $4x^2 + 18x = 10$ **42.** $2x^2 + 6x = 7$

**Solve each equation by taking the square root of each side. Round to the nearest tenth if necessary.** (Lesson 2-4)

**43.** $x^2 = 25$ **44.** $x^2 + 6x + 9 = 16$ **45.** $x^2 - 14x + 49 = 15$

**Look for a pattern in each table of values to determine which kind of model best describes the data.** (Lesson 2-6)

**46**

| x | 0 | 1 | 2 | 3 | 4 |
|---|---|---|---|---|---|
| y | 4 | 5 | 6 | 7 | 8 |

**47.**

| x | 1 | 2 | 3 | 4 | 5 |
|---|---|---|---|---|---|
| y | 2 | 4 | 8 | 16 | 32 |

**48.**

| x | −3 | −2 | −1 | 0 | 1 |
|---|---|---|---|---|---|
| y | 14 | 9 | 6 | 5 | 6 |

**49.**

| x | 3 | 4 | 5 | 6 | 7 |
|---|---|---|---|---|---|
| y | 3 | 5 | 7 | 9 | 11 |

**50. PHYSICAL SCIENCE** A projectile is shot straight up from ground level. Its height $h$, in feet, after $t$ seconds is given by $h = 96t - 16t^2$. Find the value(s) of $t$ when $h$ is 96 feet. (Lesson 2-5)

## Skills Review

**Evaluate each expression if $x = -3$, $y = -1$, and $z = 4$.**

**51.** $|x - 4|$ **52.** $|2y + 1|$ **53.** $|4 - z|$

**54.** $\left|\frac{1}{2}x + 2\right|$ **55.** $|12 - 4z|$ **56.** $|2y - 3| - 6$

# EXTEND 2-6 Graphing Technology Lab: Curve Fitting

If there is a constant increase or decrease in data values, there is a linear trend. If the values are increasing or decreasing more and more rapidly, there may be a quadratic or exponential trend.

**CCSS Common Core State Standards**
**Content Standards**
**F.LE.2** Construct linear and exponential functions, including arithmetic and geometric sequences, given a graph, a description of a relationship, or two input-output pairs (include reading these from a table).
**S.ID.6a** Fit a function to the data; use functions fitted to data to solve problems in the context of the data. Use given functions or choose a function suggested by the context. Emphasize linear, quadratic, and exponential models.

Linear Trend

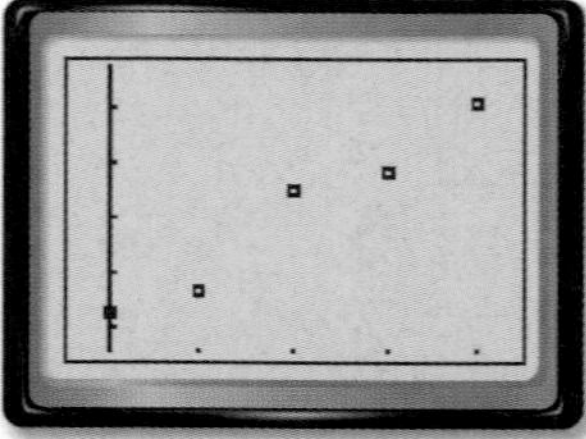

[0, 5] scl: 1 by [0, 6] scl: 1

Quadratic Trend

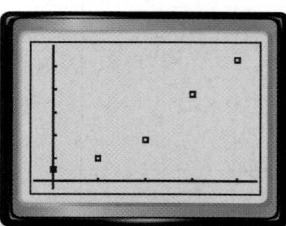

[0, 5] scl: 1 by [0, 6] scl: 1

Exponential Trend

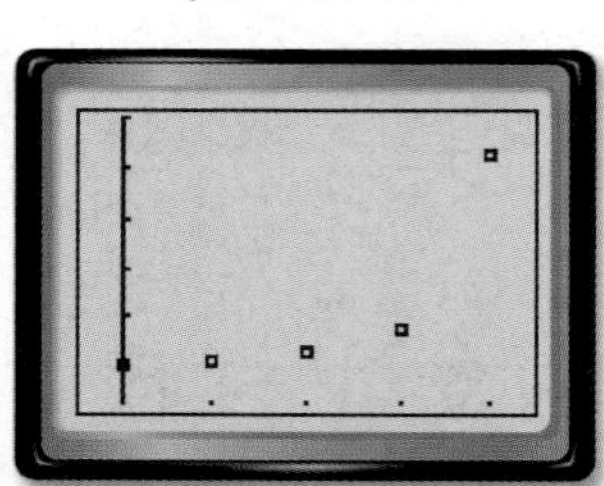

[0, 5] scl: 1 by [0, 6] scl: 1

With a graphing calculator, you can find the appropriate regression equation.

## Activity

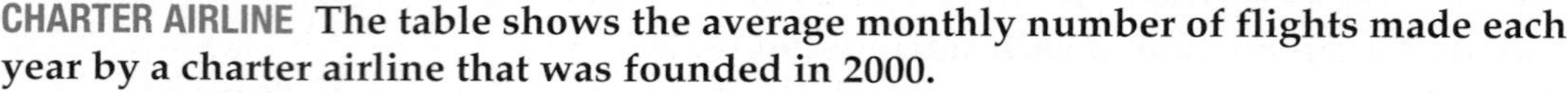

**CHARTER AIRLINE The table shows the average monthly number of flights made each year by a charter airline that was founded in 2000.**

| Year | 2000 | 2001 | 2002 | 2003 | 2004 | 2005 | 2006 | 2007 |
|---|---|---|---|---|---|---|---|---|
| Flights | 17 | 20 | 24 | 28 | 33 | 38 | 44 | 50 |

**Step 1** Make a scatter plot.

- Enter the number of years since 2000 in **L1** and the number of flights in **L2**.

**KEYSTROKES:** *Review entering a list on page 255.*

- Use **STAT PLOT** to graph the scatter plot.

**KEYSTROKES:** *Review statistical plots on page 256.*

Use ZOOM 9 to graph.

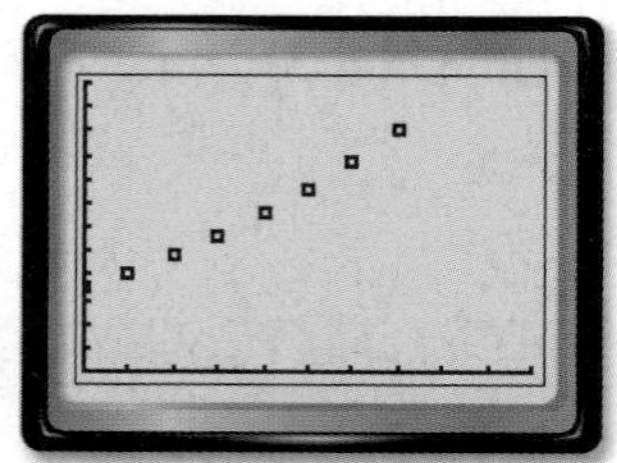

[0, 10] scl: 1 by [0, 60] scl: 5

From the scatter plot we can see that the data may have either a quadratic trend or an exponential trend.

**Step 2** Find the regression equation.

We will check both trends by examining their regression equations.

- Select **DiagnosticOn** from the **CATALOG**.
- Select **QuadReg** on the **STAT** menu.

**KEYSTROKES:** STAT ▶ 5 ENTER ENTER

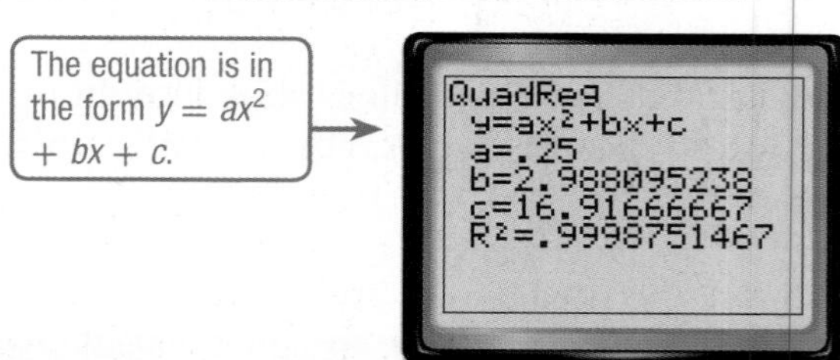


The equation is about $y = 0.25x^2 + 3x + 17$.

$R^2$ is the **coefficient of determination**. The closer $R^2$ is to 1, the better the model. To acquire the exponential equation select **ExpReg** on the **STAT** menu. To choose a quadratic or exponential model, fit both and use the one with the $R^2$ value closer to 1.

**Step 3** Graph the quadratic regression equation.

- Copy the equation to the **Y=** list and graph.

**KEYSTROKES:** [Y=] [VARS] 5 [▶] [▶] 1 [ZOOM] 9

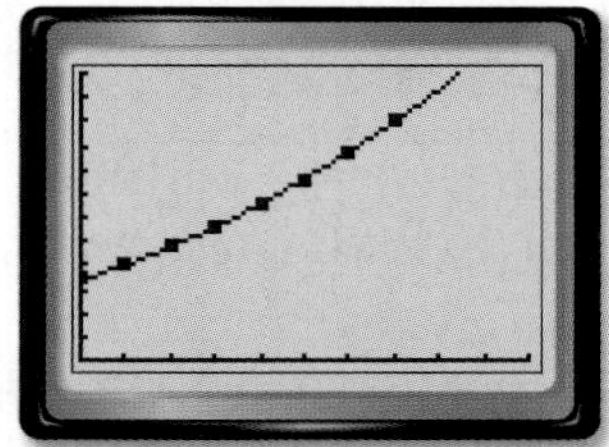

[0, 10] scl: 1 by [0, 60] scl: 5

**Step 4** Predict using the equation.

If this trend continues, we can use the graph of our equation to predict the monthly number of flights the airline will make in a specific year. Let's check the year 2020. First adjust the window.

**KEYSTROKES:** [2nd] [CALC] 1 At $x$ = enter 20 [ENTER].

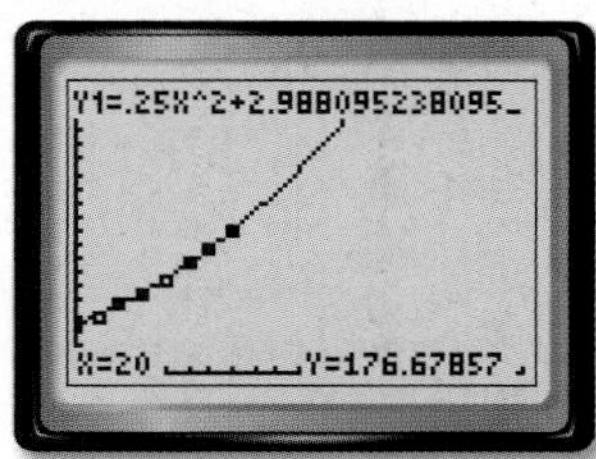


[0, 25] scl: 1 by [0, 200] scl: 5

There will be approximately 177 flights per month if this trend continues.

## Exercises

**Plot each set of data points. Determine whether to use a *linear, quadratic* or *exponential* regression equation. State the coefficient of determination.**

**1.**

| x | y |
|---|---|
| 1 | 30 |
| 2 | 40 |
| 3 | 50 |
| 4 | 55 |
| 5 | 50 |
| 6 | 40 |

**2.**

| x | y |
|---|---|
| 0.0 | 12.1 |
| 0.1 | 9.6 |
| 0.2 | 6.3 |
| 0.3 | 5.5 |
| 0.4 | 4.8 |
| 0.5 | 1.9 |

**3.**

| x | y |
|---|---|
| 0 | 1.1 |
| 2 | 3.3 |
| 4 | 2.9 |
| 6 | 5.6 |
| 8 | 11.9 |
| 10 | 19.8 |

**4.**

| x | y |
|---|---|
| 1 | 1.67 |
| 5 | 2.59 |
| 9 | 4.37 |
| 13 | 6.12 |
| 17 | 5.48 |
| 21 | 3.12 |

**5. BAKING Alyssa baked a cake and is waiting for it to cool so she can ice it. The table shows the temperature of the cake every 5 minutes after Alyssa took it out of the oven.**

| Time (min) | Temperature (°F) |
|---|---|
| 0 | 350 |
| 5 | 244 |
| 10 | 178 |
| 15 | 137 |
| 20 | 112 |
| 25 | 96 |
| 30 | 89 |

**a.** Make a scatter plot of the data.

**b.** Which regression equation has an $R^2$ value closest to 1? Is this the equation that best fits the context of the problem? Explain your reasoning.

**c.** Find an appropriate regression equation, and state the coefficient of determination. What is the domain and range?

**d.** Alyssa will ice the cake when it reaches room temperature (70°F). Use the regression equation to predict when she can ice her cake.

# LESSON 2-7 Special Functions

**Then**

- You identified and graphed linear, exponential, and quadratic functions.

**Now**

1. Identify and graph step functions.
2. Identify and graph absolute value and piecewise-defined functions.

**Why?**

- Kim is ordering books online. The site charges for shipping based on the amount of the order. If the order is less than \$10, shipping costs \$3. If the order is at least \$10 but less than \$20, it will cost \$5 to ship it.

**NewVocabulary**

step function
piecewise-linear function
greatest integer function
absolute value function
piecewise-defined function

**Common Core State Standards**

**Content Standards**

**F.IF.4** For a function that models a relationship between two quantities, interpret key features of graphs and tables in terms of the quantities, and sketch graphs showing key features given a verbal description of the relationship.

**F.IF.7b** Graph square root, cube root, and piecewise-defined functions, including step functions and absolute value functions.

**Mathematical Practices**

4 Model with mathematics.

## 1 Step Functions

The graph of a **step function** is a series of line segments. Because each part of a step function is linear, this type of function is called a **piecewise-linear function**. One example of a step function is the **greatest integer function**, written as $f(x) = [\![x]\!]$, where $f(x)$ is the greatest integer not greater than $x$. For example, $[\![6.8]\!] = 6$ because 6 is the greatest integer that is not greater than 6.8.

### KeyConcept Greatest Integer Function

Parent function: $f(x) = [\![x]\!]$

Type of graph: disjointed line segments

Domain: all real numbers

Range: all integers

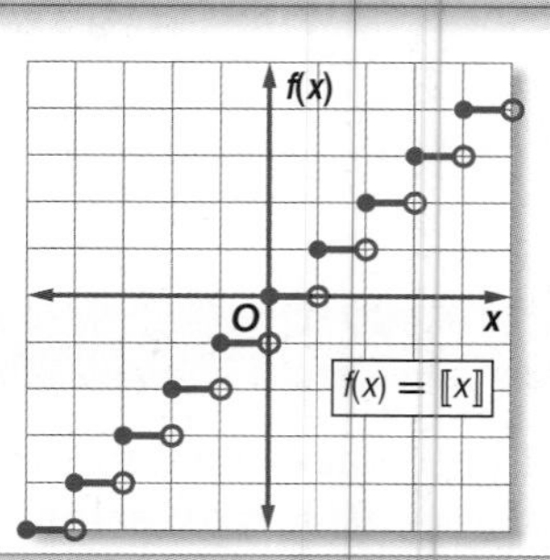


### Example 1 Greatest Integer Function

**Graph $f(x) = [\![x + 2]\!]$. State the domain and range.**

First, make a table. Select a few values between integers. On the graph, dots represent included points. Circles represent points not included.

| $x$ | $x + 2$ | $[\![x + 2]\!]$ |
|---|---|---|
| 0 | 2 | 2 |
| 0.25 | 2.25 | 2 |
| 0.5 | 2.5 | 2 |
| 1 | 3 | 3 |
| 1.25 | 3.25 | 3 |
| 1.5 | 3.5 | 3 |
| 2 | 4 | 4 |
| 2.25 | 4.25 | 4 |

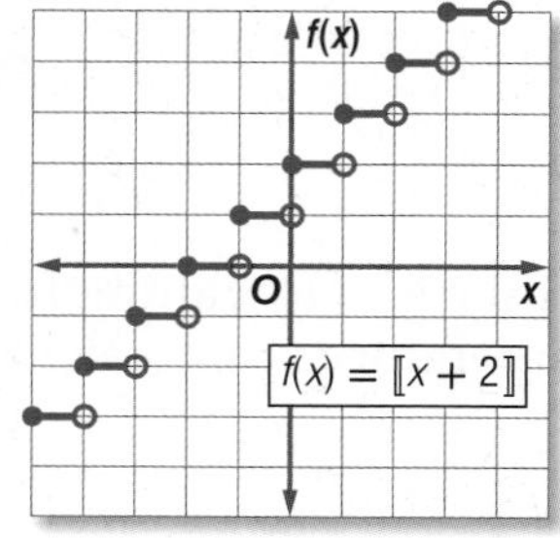


Note that this is the graph of $f(x) = [\![x]\!]$ shifted 2 units to the left.

Because the dots and circles overlap, the domain is all real numbers. The range is all integers. Notice that the graph has no symmetry and no maximum or minimum values. As $x$ increases, $f(x)$ increases, and as $x$ decreases, $f(x)$ decreases.

**GuidedPractice**

**1.** Graph $g(x) = 2[\![x]\!]$. State the domain and range.

P. Broze/Getty Images

Step functions can be used to represent many real-world situations involving money.

### Real-World Example 2 Step Function

**CELL PHONE PLANS Cell phone companies charge by the minute, not by the second. A cell phone company charges $0.45 per minute or any fraction thereof for exceeding the number of minutes allotted on each plan. Draw a graph that represents this situation.**

The total cost for the extra minutes will be a multiple of $0.45, and the graph will be a step function. If the time is greater than 0 but less than or equal to 1 minute, the charge will be $0.45. If the time is greater than 2 but is less than or equal to 3 minutes, you will be charged for 3 minutes or $1.35.

| $x$ | $f(x)$ |
|---|---|
| $0 < x \leq 1$ | 0.45 |
| $1 < x \leq 2$ | 0.90 |
| $2 < x \leq 3$ | 1.35 |
| $3 < x \leq 4$ | 1.80 |
| $4 < x \leq 5$ | 2.25 |
| $5 < x \leq 6$ | 2.70 |
| $6 < x \leq 7$ | 3.15 |

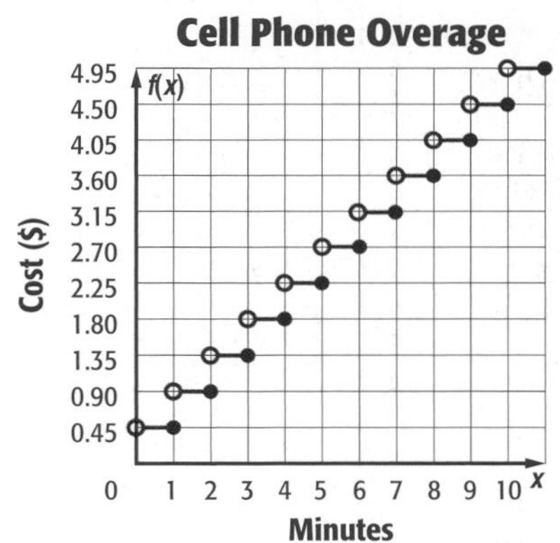


**Real-WorldLink**

North Americans are the most likely to have cell phones with 93.2% of the population owning phones.

**Source:** IT Facts

#### GuidedPractice

**2. PARKING** A garage charges $4 for the first hour and $1 for each additional hour. Draw a graph that represents this situation.

**ReviewVocabulary**

absolute value the distance a number is from zero on a number line; written $|n|$

## 2 Absolute Value Functions

Another type of piecewise-linear function is the **absolute value function**. Recall that the absolute value of a number is always nonnegative. So in the absolute value parent function, written as $f(x) = |x|$, all of the values of the range are nonnegative.

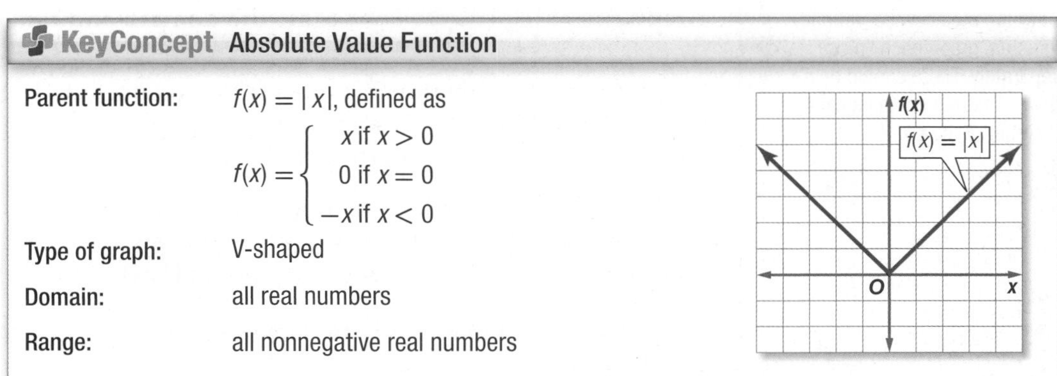

### KeyConcept Absolute Value Function

**Parent function:** $f(x) = |x|$, defined as

$$f(x) = \begin{cases} x \text{ if } x > 0 \\ 0 \text{ if } x = 0 \\ -x \text{ if } x < 0 \end{cases}$$

**Type of graph:** V-shaped

**Domain:** all real numbers

**Range:** all nonnegative real numbers



The absolute value function is called a **piecewise-defined function** because it is defined using two or more expressions.

**Math HistoryLink**

**Florence Nightingale David** (1909–1993) A renowned statistician born in Ivington, England, Florence Nightingale David received the first Elizabeth L. Scott Award for her "efforts in opening the door to women in statistics; ... for research contributions to ... statistical methods ...; and her spirit as a lecturer and as a role model."

## Example 3 Absolute Value Function

**Graph $f(x) = |x - 4|$. State the domain and range.**

Since $f(x)$ cannot be negative, the minimum point of the graph is where $f(x) = 0$.

| | |
|---|---|
| $f(x) = \|x - 4\|$ | Original function |
| $0 = x - 4$ | Replace $f(x)$ with 0 and $\|x - 4\|$ with $x - 4$. |
| $4 = x$ | Add 4 to each side. |

Next make a table of values. Include values for $x > 4$ and $x < 4$.

| $f(x) = \|x - 4\|$ | |
|---|---|
| $x$ | $f(x)$ |
| −2 | 6 |
| 0 | 4 |
| 2 | 2 |
| 4 | 0 |
| 5 | 1 |
| 6 | 2 |
| 7 | 3 |
| 8 | 4 |

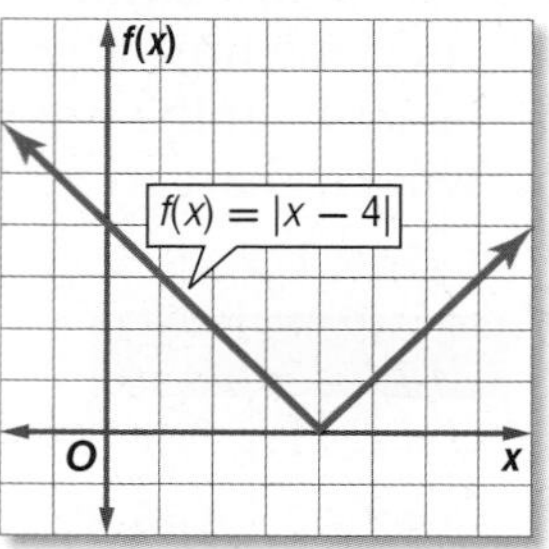


The domain is all real numbers. The range is all real numbers greater than or equal to 0. Note that this is the graph of $f(x) = |x|$ shifted 4 units to the right. Notice that the graph is symmetric about the line $x = 4$, and the minimum value of the function is 0 at $x = 4$. As $x$ increases, $f(x)$ increases, and as $x$ decreases, $f(x)$ increases.

**GuidedPractice**

**3.** Graph $f(x) = |2x + 1|$. State the domain and range.

Not all piecewise-defined functions are absolute value functions. Step functions are also piecewise-defined functions. In fact, all piecewise-linear functions are piecewise-defined.

**StudyTip**

**Piecewise Functions** To graph a piecewise-defined function, graph each "piece" separately. There should be a dot or line that contains each member of the domain.

## Example 4 Piecewise-Defined Function

**Graph $f(x) = \begin{cases} -2x \text{ if } x > 1 \\ x + 3 \text{ if } x \leq 1 \end{cases}$. State the domain and range.**

Graph the first expression. Create a table of values for when $x > 1$, $f(x) = -2x$ and draw the graph. Since $x$ is not equal to 1, place a circle at $(1, -2)$.

Next, graph the second expression. Create a table of values for when $x \leq 1$, $f(x) = x + 3$ and draw the graph. If $x = 1$, then $f(x) = 4$; place a dot at $(1, 4)$.

The domain is all real numbers. The range is $y \leq 4$.

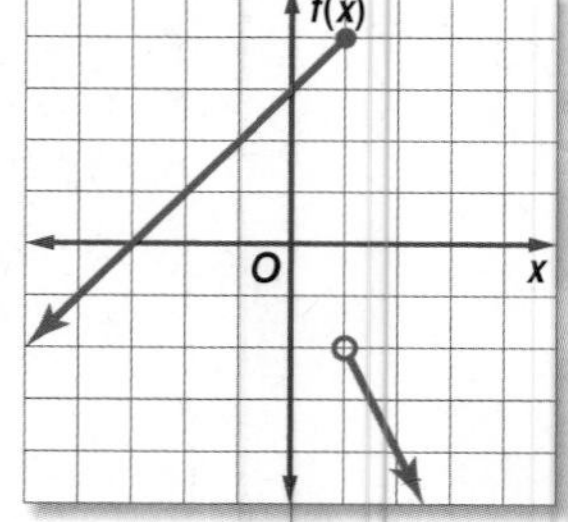


**GuidedPractice**

**4.** Graph $f(x) = \begin{cases} 2x + 1 \text{ if } x > 0 \\ 3 \text{ if } x \leq 0 \end{cases}$. State the domain and range.

**StudyTip**

**Nonlinear Functions** Like exponential and quadratic functions, the greatest integer function, absolute value function, and piecewise defined functions are nonlinear functions.

## ConceptSummary Special Functions

| Step Function | Absolute Value Function | Piecewise-Defined Function |
|---|---|---|
| | | 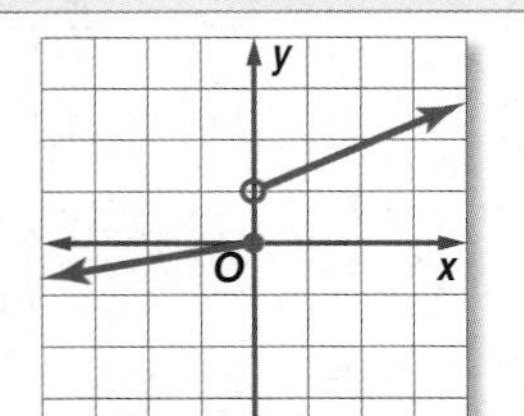 |

## Check Your Understanding

**Example 1** **Graph each function. State the domain and range.**

1. $f(x) = \frac{1}{2}[\![x]\!]$
2. $g(x) = -[\![x]\!]$
3. $h(x) = [\![2x]\!]$

**Example 2** 4. **SHIPPING** Elan is ordering a gift for his dad online. The table shows the shipping rates. Graph the step function.

| Order Total ($) | Shipping Cost ($) |
|---|---|
| 0–15 | 3.99 |
| 15.01–30 | 5.99 |
| 30.01–50 | 6.99 |
| 50.01–75 | 7.99 |
| 75.01–100 | 8.99 |
| Over $100 | 9.99 |

**Examples 3–4** **Graph each function. State the domain and range.**

5. $f(x) = |x - 3|$
6. $g(x) = |2x + 4|$
7. $f(x) = \begin{cases} 2x - 1 \text{ if } x > -1 \\ -x \text{ if } x \leq -1 \end{cases}$
8. $g(x) = \begin{cases} -3x - 2 \text{ if } x > -2 \\ -x + 1 \text{ if } x \leq -2 \end{cases}$

## Practice and Problem Solving

**Example 1** **Graph each function. State the domain and range.**

9. $f(x) = 3[\![x]\!]$
10. $f(x) = [\![-x]\!]$
11. $g(x) = -2[\![x]\!]$
12. $g(x) = [\![x]\!] + 3$
13. $h(x) = [\![x]\!] - 1$
14. $h(x) = \frac{1}{2}[\![x]\!] + 1$

**Example 2** 15. **CAB FARES** Lauren wants to take a taxi from a hotel to a friend's house. The rate is $3 plus $1.50 per mile after the first mile. Every fraction of a mile is rounded up to the next mile.

a. Draw a graph to represent the cost of using a taxi cab.

b. What is the cost if the trip is 8.5 miles long?

16. **CCSS MODELING** The United States Postal Service increases the rate of postage periodically. The table shows the cost to mail a letter weighing 1 ounce or less from 1995 through 2009. Draw a step graph to represent the data.

| Year | 1995 | 1999 | 2001 | 2002 | 2006 | 2007 | 2008 | 2009 |
|---|---|---|---|---|---|---|---|---|
| Cost ($) | 0.32 | 0.33 | 0.34 | 0.37 | 0.39 | 0.41 | 0.42 | 0.44 |

**Examples 3–4** **Graph each function. State the domain and range.**

**17.** $f(x) = |2x - 1|$

**18.** $f(x) = |x + 5|$

**19.** $g(x) = |-3x - 5|$

**20.** $g(x) = |-x - 3|$

**21.** $f(x) = \left|\frac{1}{2}x - 2\right|$

**22.** $f(x) = \left|\frac{1}{3}x + 2\right|$

**23.** $g(x) = |x + 2| + 3$

**24.** $g(x) = |2x - 3| + 1$

**25.** $f(x) = \begin{cases} \frac{1}{2}x - 1 \text{ if } x > 3 \\ -2x + 3 \text{ if } x \leq 3 \end{cases}$

**26.** $f(x) = \begin{cases} 2x - 5 \text{ if } x > 1 \\ 4x - 3 \text{ if } x \leq 1 \end{cases}$

**27.** $f(x) = \begin{cases} 2x + 3 \text{ if } x \geq -3 \\ -\frac{1}{3}x + 1 \text{ if } x < -3 \end{cases}$

**28.** $f(x) = \begin{cases} 3x + 4 \text{ if } x \geq 1 \\ x + 3 \text{ if } x < 1 \end{cases}$

**29.** $f(x) = \begin{cases} 3x + 2 \text{ if } x > -1 \\ -\frac{1}{2}x - 3 \text{ if } x \leq -1 \end{cases}$

**30.** $f(x) = \begin{cases} 2x + 1 \text{ if } x < -2 \\ -3x - 1 \text{ if } x \geq -2 \end{cases}$

**Determine the domain and range of each function.**

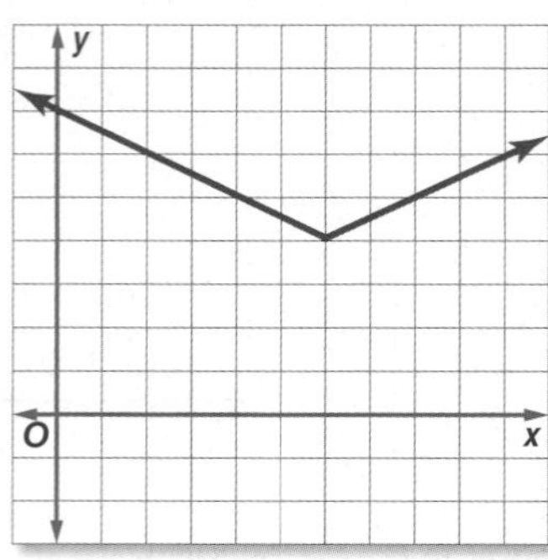


**32.**

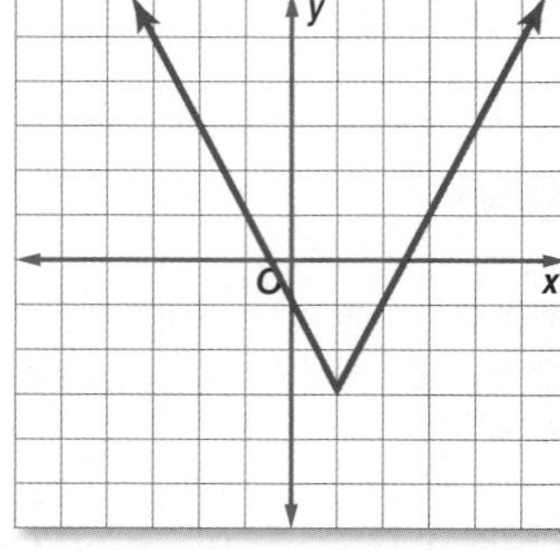


**33.**

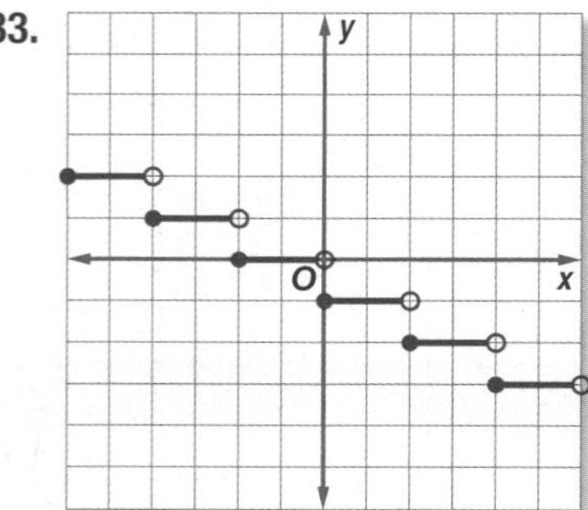


**34.**

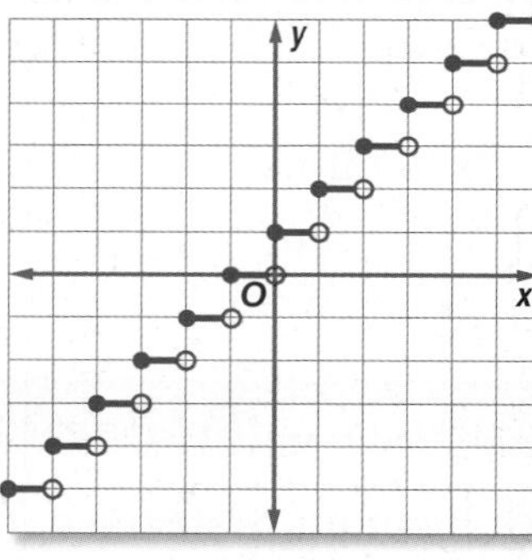


**35.**

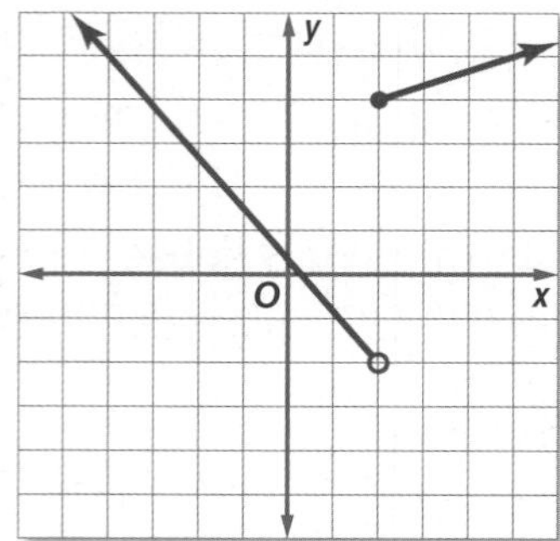


**36.**

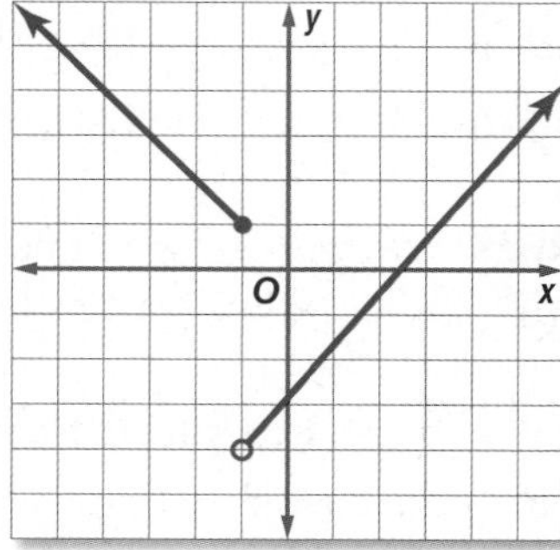


**37. BOATING** According to Boat Minnesota, the maximum number of people that can safely ride in a boat is determined by the boat's length and width. The table shows some guidelines for the length of a boat that is 6 feet wide. Graph this relation.

| Length of Boat (ft) | 18–19 | 20–22 | 23–24 |
|---|---|---|---|
| Number of People | 7 | 8 | 9 |

**For Exercises 38–41, match each graph to one of the following equations.**

| A | B | C | D |
|---|---|---|---|
| $y = 2x - 1$ | $y = [\![2x]\!] - 1$ | $y = \lvert 2x \rvert - 1$ | $y = \begin{cases} 2x + 1 \text{ if } x > 0 \\ -2x + 1 \text{ if } x \leq 0 \end{cases}$ |

**38.**

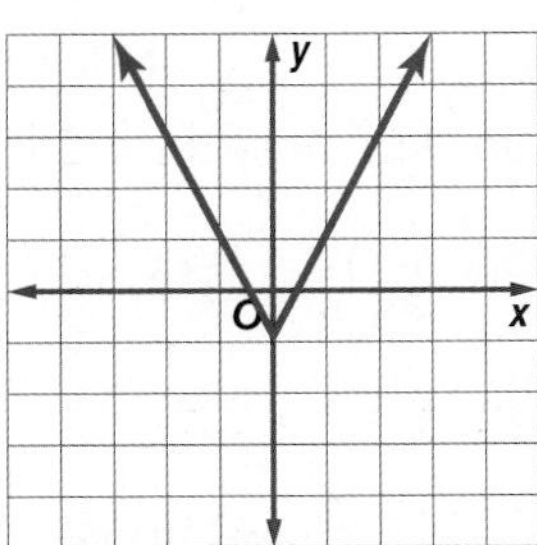


**39.**

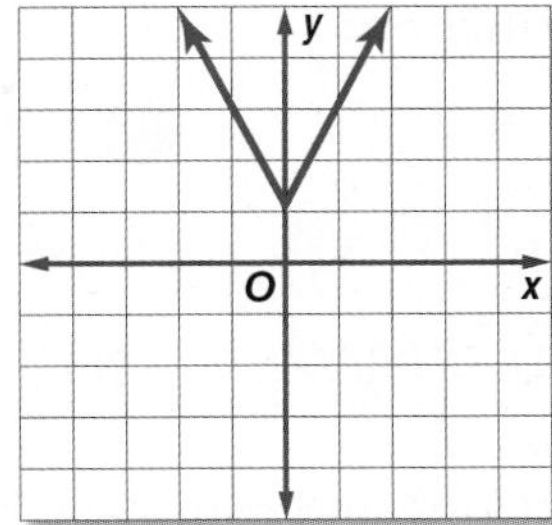


**40.**

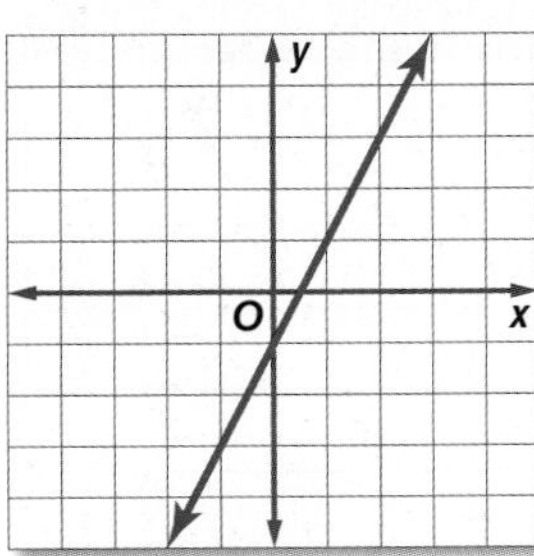


**41.**

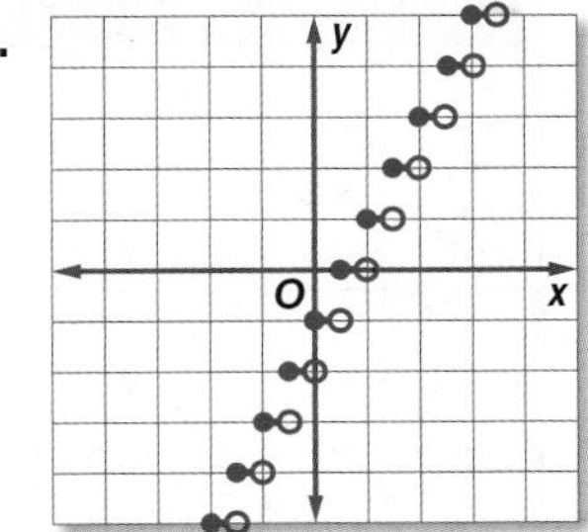


**42. CAR LEASE** As part of Marcus' leasing agreement, he will be charged \$0.20 per mile for each mile over 12,000. Any fraction of a mile is rounded up to the next mile. Make a step graph to represent the cost of going over the mileage.

**43 BASEBALL** A baseball team is ordering T-shirts with the team logo on the front and the players' names on the back. A graphic design store charges \$10 to set up the artwork plus \$10 per shirt, \$4 each for the team logo, and \$2 to print the last name for an order of 10 shirts or less. For orders of 11–20 shirts, a 5% discount is given. For orders of more than 20 shirts, a 10% discount is given.

**a.** Organize the information into a table. Include a column showing the total order price for each size order.

**b.** Write an equation representing the total price for an order of $x$ shirts.

**c.** Graph the piecewise relation.

**44.** Consider the function $f(x) = \lvert 2x + 3 \rvert$.

**a.** Make a table of values where $x$ is all integers from $-5$ to 5, inclusive.

**b.** Plot the points on a coordinate grid.

**c.** Graph the function.

**45.** Consider the function $f(x) = \lvert 2x \rvert + 3$.

**a.** Make a table of values where $x$ is all integers from $-5$ to 5, inclusive.

**b.** Plot the points on a coordinate grid.

**c.** Graph the function.

**d.** Describe how this graph is different from the graph in Exercise 44.

46. **DANCE** A local studio owner will teach up to 4 students by herself. Her instructors can teach up to 5 students each. Draw a step function graph that best describes the number of instructors needed for the different numbers of students.

47. **THEATERS** A community theater will only perform a show if there are at least 250 pre-sale ticket requests. Additional performances will be added for each 250 requests after that. Draw a step function graph that best describes this situation.

**Graph each function.**

48. $f(x) = \frac{1}{2}|x| + 2$

49. $g(x) = \frac{1}{3}|x| + 4$

50. $h(x) = -2|x - 3| + 2$

51. $f(x) = -4|x + 2| - 3$

52. $g(x) = -\frac{2}{3}|x + 6| - 1$

53. $h(x) = -\frac{3}{4}|x - 8| + 1$

54. **MULTIPLE REPRESENTATIONS** In this problem, you will explore piecewise linear functions.

a. **Tabular** Copy and complete the table of values for $f(x) = |[\![x]\!]|$ and $g(x) = [\![|x|]\!]$.

| $x$ | $[\![x]\!]$ | $f(x) = \lvert[\![x]\!]\rvert$ | $\lvert x\rvert$ | $g(x) = [\![\lvert x\rvert]\!]$ |
|---|---|---|---|---|
| −3 | −3 | 3 | 3 | 3 |
| −2.5 | | | | |
| −2 | | | | |
| 0 | | | | |
| 0.5 | | | | |
| 1 | | | | |
| 1.5 | | | | |

b. **Graphical** Graph each function on a coordinate plane.

c. **Analytical** Compare and contrast the graphs of $f(x)$ and $g(x)$.

## H.O.T. Problems Use Higher-Order Thinking Skills

55. **REASONING** Does the piecewise relation $y = \begin{cases} -2x + 4 \text{ if } x \geq 2 \\ -\frac{1}{2}x - 1 \text{ if } x \leq 4 \end{cases}$ represent a function? Why or why not?

**CCSS SENSE-MAKING Refer to the graph.**

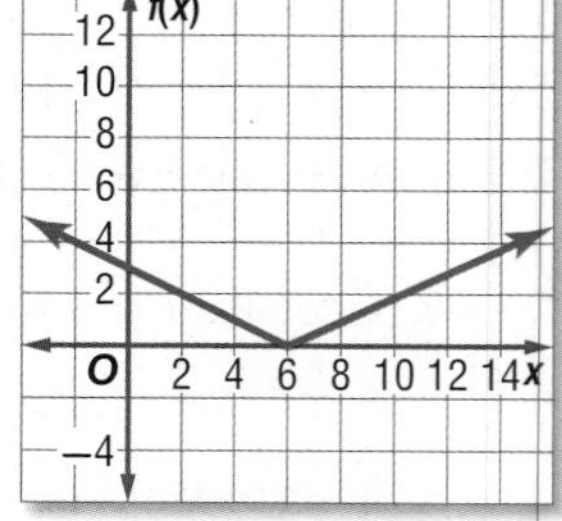


56. Write an absolute value function that represents the graph.

57. Write a piecewise function to represent the graph.

58. What are the domain and range?

59. **WRITING IN MATH** Compare and contrast the graphs of absolute value, step, and piecewise-defined functions with the graphs of quadratic and exponential functions. Discuss the domains, ranges, maxima, minima, and symmetry.

60. **CHALLENGE** A bicyclist travels up and down a hill with a vertical cross section that can be modeled by $y = -\frac{1}{4}|x - 400| + 100$, where $x$ and $y$ are measured in feet.

a. If $0 \leq x \leq 800$, find the slope for the uphill portion of the trip and downhill portion of the trip.

b. Graph this function. What are the domain and range?

## Standardized Test Practice

**61.** Which equation represents a line that is perpendicular to the graph and passes through the point at (2, 0)?

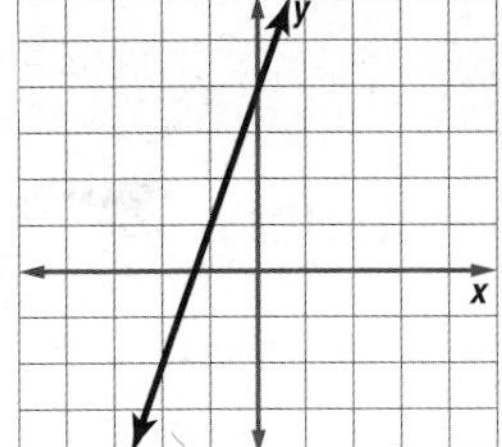


**A** $y = 3x - 6$

**B** $y = -3x + 6$

**C** $y = -\frac{1}{3}x + \frac{2}{3}$

**D** $y = \frac{1}{3}x - \frac{2}{3}$

**62.** A giant tortoise travels at a rate of 0.17 mile per hour. Which equation models the time $t$ it would take the giant tortoise to travel 0.8 mile?

**F** $t = \frac{0.8}{0.17}$ **H** $t = \frac{0.17}{0.8}$

**G** $t = (0.17)(0.8)$ **J** $0.8 = \frac{0.17}{t}$

**63. GEOMETRY** If $\triangle JKL$ is similar to $\triangle JNM$ what is the value $a$?

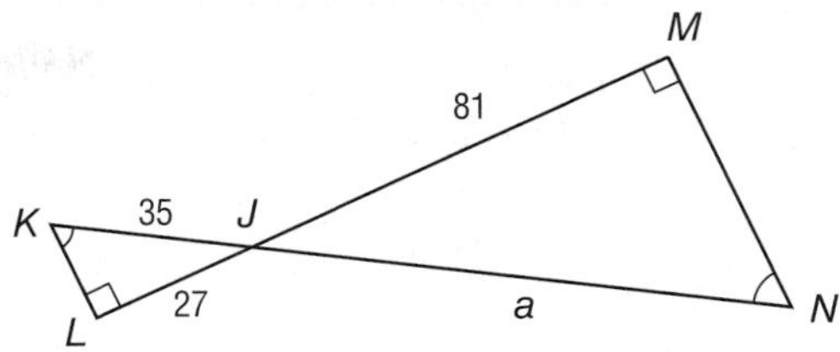


**A** 62.5

**B** 105

**C** 125

**D** 155.5

**64. GRIDDED RESPONSE** What is the difference in the value of $2.1(x + 3.2)$, when $x = 5$ and when $x = 3$?

## Spiral Review

**Look for a pattern in each table of values to determine which model best describes the data.** (Lesson 2-6)

**65.**

| x | 0 | 1 | 2 | 3 | 4 |
|---|---|---|---|---|---|
| y | 1 | 3 | 5 | 7 | 9 |

**66.**

| x | −2 | −1 | 0 | 1 | 2 |
|---|---|---|---|---|---|
| y | 5 | 2 | 1 | 2 | 5 |

**67.**

| x | −1 | 0 | 1 | 2 | 3 |
|---|---|---|---|---|---|
| y | 1 | 2 | 4 | 8 | 16 |

**68.**

| x | 5 | 6 | 7 | 8 | 9 |
|---|---|---|---|---|---|
| y | −2.5 | −1.5 | 1.5 | 6.5 | 13.5 |

**Determine the domain and range for each function.** (Lesson 2-7)

**69.** $f(x) = |2x - 5|$

**70.** $h(x) = [\![x - 1]\!]$

**71.** $g(x) = \begin{cases} -3x + 4 \text{ if } x > 2 \\ x - 1 \text{ if } x \leq 2 \end{cases}$

**Solve each equation by using the Quadratic Formula. Round to the nearest tenth if necessary.** (Lesson 2-5)

**72.** $x^2 - 25 = 0$

**73.** $r^2 + 25 = 0$

**74.** $4w^2 + 100 = 40w$

**75.** $2r^2 + r - 14 = 0$

**76.** $5v^2 - 7v = 1$

**77.** $11z^2 - z = 3$

## Skills Review

**Evaluate each expression. If necessary, round to the nearest hundredth.**

**78.** $\sqrt{9}$

**79.** $\sqrt{12}$

**80.** $\sqrt{4.5}$

**81.** $3\sqrt{16}$

**82.** $2\sqrt{10}$

**83.** $\sqrt{5} - 2$

# 2-7 Graphing Technology Lab
# Piecewise-Linear Functions

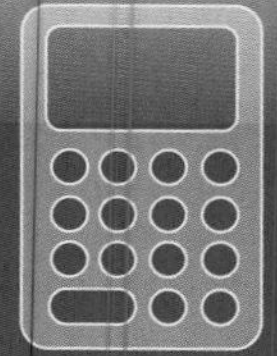

You can use a graphing calculator to graph and analyze various piecewise functions, including greatest integer functions and absolute value functions.

**CCSS Common Core State Standards**
**Content Standards**
**F.IF.7b** Graph square root, cube root, and piecewise-defined functions, including step functions and absolute value functions.

## Activity 1 Greatest Integer Functions

**Graph $f(x) = [\![x]\!]$ in the standard viewing window.**

The calculator may need to be changed to dot mode for the function to graph correctly. Press MODE then use the arrow and ENTER keys to select **DOT**.

Enter the equation in the **Y=** list. Then graph the equation.

**KEYSTROKES:** Y= MATH ▶ 5 X,T,θ,*n* ) ZOOM 6

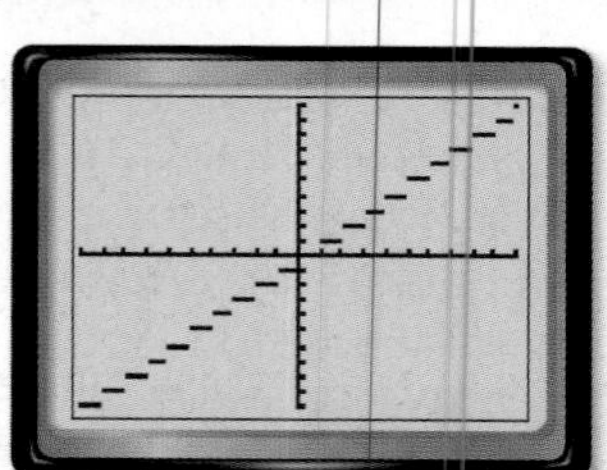

[−10, 10] scl: 1 by [−10, 10] scl: 1

**1A.** How does the graph of $f(x) = [\![x]\!]$ compare to the graph of $f(x) = x$?

**1B.** What are the domain and range of the function $f(x) = [\![x]\!]$? Explain.

The graphs of piecewise functions are affected by changes in parameters.

## Activity 2 Absolute Value Functions

**Graph $y = |x| - 3$ and $y = |x| + 1$ in the standard viewing window.**

Enter the equations in the **Y=** list. Then graph.

**KEYSTROKES:** Y= MATH ▶ 1 X,T,θ,*n* ) − 3 ENTER MATH ▶ 1 X,T,θ,*n* ) + 1 ZOOM 6

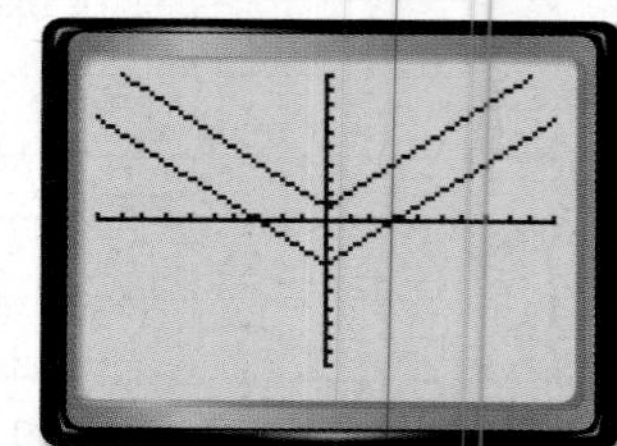

[−10, 10] scl: 1 by [−10, 10] scl: 1

**2A.** Compare and contrast the graphs to the graph of $y = |x|$.

**2B.** How does the value of $k$ affect the graph of $y = |x| + k$?

## Analyze the Results

**1.** A parking garage charges $4 for every hour or fraction of an hour. Is this situation modeled by a *linear* function or a *step* function? Explain your reasoning.

**2.** A maintenance technician is testing an elevator system. The technician starts the elevator at the fifth floor. It is sent to the ground floor, then back to the fifth floor. Assume the elevator travels at a constant rate. Should the height of the elevator be modeled by a step function or an absolute value function? Explain.

**Because the points on a graph are solutions of its equation, the $x$-coordinates of points where $y = f(x)$ and $y = g(x)$ intersect are solutions of $f(x) = g(x)$. For example, the solution of $5x - 2 = |x|$ is the intersection of the graphs of $y = 5x - 2$ and $y = |x|$. Write each equation as a system of equations, and then use a graph or a table to solve it.**

**3.** $5x - 2 = |x|$

**4.** $2|x - 2| = x - 1$

**5.** $|4x + 2| = -|x| + 3$

# Study Guide and Review

## Study Guide

### KeyConcepts

**Graphing Quadratic Functions** (Lesson 2-1)

- A quadratic function can be described by an equation of the form $y = ax^2 + bx + c$, where $a \neq 0$.
- The axis of symmetry for the graph of $y = ax^2 + bx + c$, where $a \neq 0$, is $x = -\frac{b}{2a}$.

**Solving Quadratic Equations** (Lessons 2-2, 2-4, and 2-5)

- Quadratic equations can be solved by graphing. The solutions are the $x$-intercepts or zeros of the related quadratic function.
- Quadratic equations can be solved by completing the square. To complete the square for $x^2 + bx$, find $\frac{1}{2}$ of $b$, square this result, and then add the result to $x^2 + bx$.
- Quadratic equations can be solved by using the Quadratic Formula, $x = \frac{-b \pm \sqrt{b^2 - 4ac}}{2a}$.

**Transformations of Quadratic Functions** (Lesson 2-3)

- $f(x) = x^2 + c$ translates the graph up or down.
- $f(x) = ax^2$ compresses or expands the graph vertically.

**Special Functions** (Lesson 2-7)

- The greatest integer function is written as $f(x) = [\![x]\!]$, where $f(x)$ is the greatest integer not greater than $x$.
- The absolute value function is written as $f(x) = |x|$, where $f(x)$ is the distance from $x$ to 0 on a number line.

### FOLDABLES StudyOrganizer

Be sure the Key Concepts are noted in your Foldable.

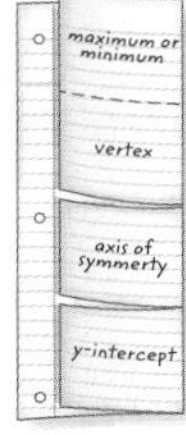


### KeyVocabulary

absolute value function (p. 149)
axis of symmetry (p. 93)
completing the square (p. 124)
dilation (p. 116)
discriminant (p. 136)
double root (p. 106)
greatest integer function (p. 149)
maximum (p. 93)
minimum (p. 93)
parabola (p. 93)
piecewise-defined function (p. 149)
piecewise-linear function (p. 148)
Quadratic Formula (p. 131)
quadratic function (p. 93)
reflection (p. 114)
standard form (p. 93)
step function (p. 148)
transformation (p. 114)
translation (p. 114)
vertex (p. 93)

### VocabularyCheck

**State whether each sentence is *true* or *false*. If *false*, replace the underlined term to make a true sentence.**

1. The <u>axis of symmetry</u> of a quadratic function can be found by using the equation $x = -\frac{b}{2a}$.
2. The <u>vertex</u> is the maximum or minimum point of a parabola.
3. The graph of a quadratic function is a <u>straight line</u>.
4. The graph of a quadratic function has a <u>maximum</u> if the coefficient of the $x^2$-term is positive.
5. A quadratic equation with a graph that has two $x$-intercepts has <u>one</u> real root.
6. The expression $b^2 - 4ac$ is called the <u>discriminant</u>.
7. A function that is defined differently for different parts of its domain is called a <u>piecewise-defined function</u>.
8. The <u>range</u> of the greatest integer function is the set of all real numbers.
9. The solutions of a quadratic equation are called <u>roots</u>.
10. The graph of the parent function is <u>translated down</u> to form the graph of $f(x) = x^2 + 5$.

# Study Guide and Review *Continued*

## Lesson-by-Lesson Review

### 2-1 Graphing Quadratic Functions

Consider each equation.

a. Determine whether the function has a *maximum* or *minimum* value.

b. State the maximum or minimum value.

c. What are the domain and range of the function?

11. $y = x^2 - 4x + 4$
12. $y = -x^2 + 3x$
13. $y = x^2 - 2x - 3$
14. $y = -x^2 + 2.$

15. **ROCKET** A toy rocket is launched with an upward velocity of 32 feet per second. The equation $h = -16t^2 + 32t$ gives the height of the ball $t$ seconds after it is launched.

    a. Determine whether the function has a *maximum* or *minimum* value.

    b. State the maximum or minimum value.

    c. State a reasonable domain and range for this situation.

**Example 1**

Consider $f(x) = x^2 + 6x + 5$.

**a. Determine whether the function has a *maximum* or *minimum* value.**

For $f(x) = x^2 + 6x + 5$, $a = 1$, $b = 6$, and $c = 5$.

Because $a$ is positive, the graph opens up, so the function has a minimum value.

**b. State the *maximum* or *minimum* value of the function.**

The minimum value is the $y$-coordinate of the vertex.

The $x$-coordinate of the vertex is $\frac{-b}{2a}$ or $\frac{-6}{2(1)}$ or $-3$.

$f(x) = x^2 + 6x + 5$ Original function

$f(-3) = (-3)^2 + 6(-3) + 5$ $x = -3$

$f(-3) = -4$ Simplify.

The minimum value is $-4$.

**c. State the domain and range of the function.**

The domain is all real numbers. The range is all real numbers greater than or equal to the minimum value, or $\{y \mid y \geq -4\}$.

### 2-2 Solving Quadratic Equations by Graphing

Solve each equation by graphing. If integral roots cannot be found, estimate the roots to the nearest tenth.

16. $x^2 - 3x - 4 = 0$
17. $-x^2 + 6x - 9 = 0$
18. $x^2 - x - 12 = 0$
19. $x^2 + 4x - 3 = 0$
20. $x^2 - 10x = -21$
21. $6x^2 - 13x = 15$

22. **NUMBER THEORY** Find two numbers that have a sum of 2 and a product of $-15$.

**Example 2**

**Solve $x^2 - x - 6 = 0$ by graphing.**

Graph the related function $f(x) = x^2 - x - 6$.

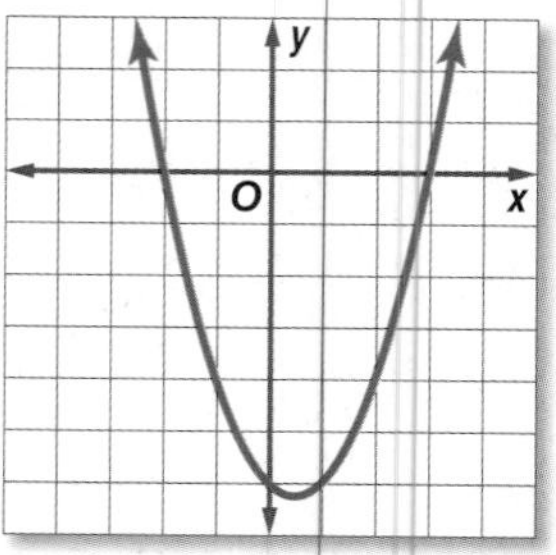


The $x$-intercepts of the graph appear to be at $-2$ and 3, so the solutions are $-2$ and 3.

## 2-3 Transformations of Quadratic Functions

Describe how the graph of each function is related to the graph of $f(x) = x^2$.

**23.** $f(x) = x^2 + 8$

**24.** $f(x) = x^2 - 3$

**25.** $f(x) = 2x^2$

**26.** $f(x) = 4x^2 - 18$

**27.** $f(x) = \frac{1}{3}x^2$

**28.** $f(x) = \frac{1}{4}x^2$

**29.** Write an equation for the function shown in the graph.

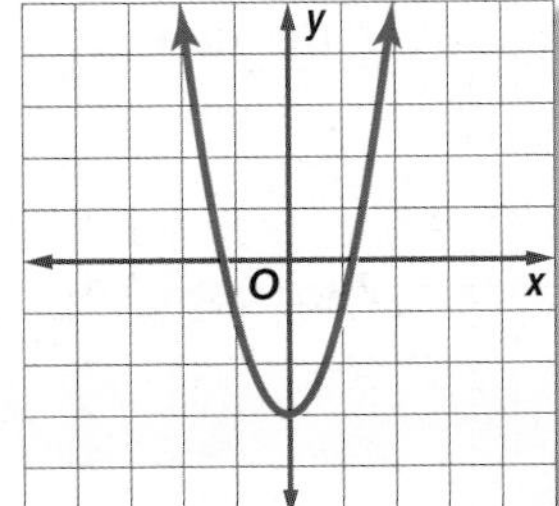


**30. PHYSICS** A ball is dropped off a cliff that is 100 feet high. The function $h = -16t^2 + 100$ models the height $h$ of the ball after $t$ seconds. Compare the graph of this function to the graph of $h = t^2$.

**Example 3**

**Describe how the graph of $f(x) = x^2 - 2$ is related to the graph of $f(x) = x^2$.**

The graph of $f(x) = x^2 + c$ represents a translation up or down of the parent graph.

Since $c = -2$, the translation is down.

So, the graph is translated down 2 units from the parent function.

**Example 4**

**Write an equation for the function shown in the graph.**

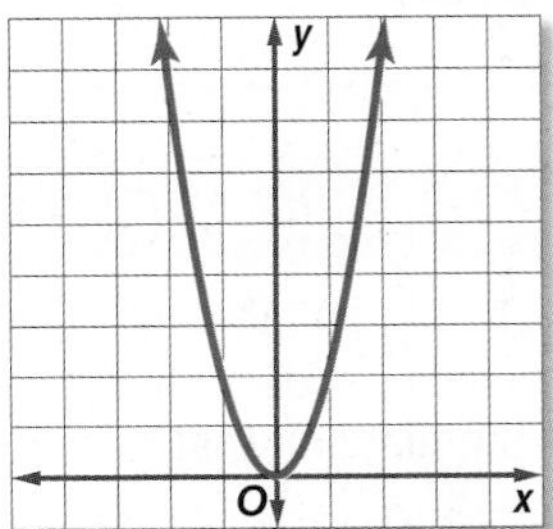


Since the graph opens upward, the leading coefficient must be positive. The parabola has not been translated up or down, so $c = 0$. Since the graph is stretched vertically, it must be of the form of $f(x) = ax^2$ where $a > 1$. The equation for the function is $y = 2x^2$.

## 2-4 Solving Quadratic Equations by Completing the Square

Solve each equation by completing the square. Round to the nearest tenth if necessary.

**31.** $x^2 + 6x + 9 = 16$

**32.** $-a^2 - 10a + 25 = 25$

**33.** $y^2 - 8y + 16 = 36$

**34.** $y^2 - 6y + 2 = 0$

**35.** $n^2 - 7n = 5$

**36.** $-3x^2 + 4 = 0$

**37. NUMBER THEORY** Find two numbers that have a sum of $-2$ and a product of $-48$.

**Example 5**

**Solve $x^2 - 16x + 32 = 0$ by completing the square. Round to the nearest tenth if necessary.**

Isolate the $x^2$- and $x$-terms. Then complete the square and solve.

| | |
|---|---|
| $x^2 - 16x + 32 = 0$ | Original equation |
| $x^2 - 16x = -32$ | Isolate the $x^2$- and $x$-terms. |
| $x^2 - 16x + 64 = -32 + 64$ | Complete the square. |
| $(x - 8)^2 = 32$ | Factor. |
| $x - 8 = \pm\sqrt{32}$ | Take the square root. |
| $x = 8 \pm\sqrt{32}$ | Add 8 to each side. |
| $x = 8 \pm 4\sqrt{2}$ | Simplify. |

The solutions are about 2.3 and 13.7.

# Study Guide and Review *Continued*

## 2-5 Solving Quadratic Equations by Using the Quadratic Formula

Solve each equation by using the Quadratic Formula. Round to the nearest tenth if necessary.

**38.** $x^2 - 8x = 20$

**39.** $21x^2 + 5x - 7 = 0$

**40.** $d^2 - 5d + 6 = 0$

**41.** $2f^2 + 7f - 15 = 0$

**42.** $2h^2 + 8h + 3 = 3$

**43.** $4x^2 + 4x = 15$

**44.** **GEOMETRY** The area of a square can be quadrupled by increasing the side length and width by 4 inches. What is the side length?

### Example 6

Solve $x^2 + 10x + 9 = 0$ by using the Quadratic Formula.

$x = \frac{-b \pm \sqrt{b^2 - 4ac}}{2a}$ — Quadratic Formula

$= \frac{-10 \pm \sqrt{10^2 - 4(1)(9)}}{2(1)}$ — $a = 1, b = 10, c = 9$

$= \frac{-10 \pm \sqrt{64}}{2}$ — Simplify.

$x = \frac{-10 + 8}{2}$ or $x = \frac{-10 - 8}{2}$ — Separate the solutions.

$= -1 \qquad = -9$ — Simplify.

## 2-6 Analyzing Functions with Successive Differences

Look for a pattern in each table of values to determine which kind of model best describes the data. Then write an equation for the function that models the data.

**45.**

| x | 0 | 1 | 2 | 3 | 4 |
|---|---|---|---|---|---|
| y | 0 | 3 | 12 | 27 | 48 |

**46.**

| x | 0 | 1 | 2 | 3 | 4 |
|---|---|---|---|---|---|
| y | 1 | 2 | 4 | 8 | 16 |

**47.**

| x | 0 | 1 | 2 | 3 | 4 |
|---|---|---|---|---|---|
| y | 0 | −1 | −4 | −9 | −16 |

### Example 7

Determine the model that best describes the data. Then write an equation for the function that models the data.

| x | 0 | 1 | 2 | 3 | 4 |
|---|---|---|---|---|---|
| y | 3 | 4 | 5 | 6 | 7 |

**Step 1** First differences:

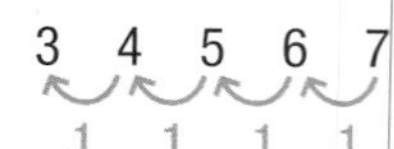


A linear function models the data.

**Step 2** The slope is 1 and the $y$-intercept is 3, so the equation is $y = x + 3$.

## 2-7 Special Functions

Graph each function. State the domain and range.

**48.** $f(x) = [\![x]\!]$

**49.** $f(x) = [\![2x]\!]$

**50.** $f(x) = |x|$

**51.** $f(x) = |2x - 2|$

**52.** $f(x) = \begin{cases} x - 2 \text{ if } x < 1 \\ 3x \text{ if } x \geq 1 \end{cases}$

**53.** $f(x) = \begin{cases} 2x - 3 \text{ if } x \leq 2 \\ x + 1 \text{ if } x > 2 \end{cases}$

### Example 8

Graph $f(x) = |x + 3|$. State the domain and range.

| x | f(x) |
|---|---|
| −5 | 2 |
| −4 | 1 |
| −3 | 0 |
| −2 | 1 |
| −1 | 2 |

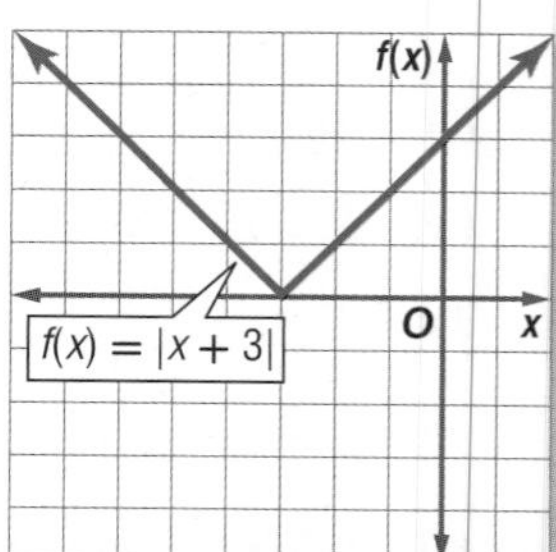


The domain is all real numbers, and the range is $f(x) \geq 0$.

# CHAPTER 2 Practice Test

**Use a table of values to graph the following functions. State the domain and range.**

1. $y = x^2 + 2x + 5$
2. $y = 2x^2 - 3x + 1$

**Consider $y = x^2 - 7x + 6$.**

3. Determine whether the function has a *maximum* or *minimum* value.
4. State the maximum or minimum value.
5. What are the domain and range?

**Solve each equation by graphing. If integral roots cannot be found, estimate the roots to the nearest tenth.**

6. $x^2 + 7x + 10 = 0$
7. $x^2 - 5 = -3x$

**Describe how the graph of each function is related to the graph of $f(x) = x^2$.**

8. $g(x) = x^2 - 5$
9. $g(x) = -3x^2$
10. $h(x) = \frac{1}{2}x^2 + 4$
11. **MULTIPLE CHOICE** Which is an equation for the function shown in the graph?

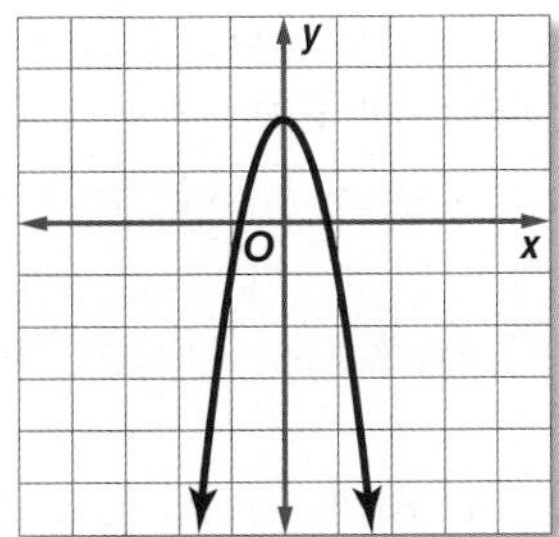


A $y = -3x^2$

B $y = 3x^2 + 1$

C $y = x^2 + 2$

D $y = -3x^2 + 2$

**Solve each equation by completing the square.**

12. $x^2 + 2x + 5 = 0$
13. $x^2 - x - 6 = 0$
14. $2x^2 - 36 = -6x$

**Solve each equation by using the Quadratic Formula. Round to the nearest tenth if necessary.**

15. $x^2 - x - 30 = 0$
16. $x^2 - 10x = -15$
17. $2x^2 + x - 15 = 0$
18. **BASEBALL** Elias hits a baseball into the air. The equation $h = -16t^2 + 60t + 3$ models the height $h$ in feet of the ball after $t$ seconds. How long is the ball in the air?
19. Graph $\{(-2, 4), (-1, 1), (0, 0), (1, 1), (2, 4)\}$. Determine whether the ordered pairs represent a *linear function*, a *quadratic function*, or an *exponential function*.
20. Look for a pattern in the table to determine which kind of model best describes the data.

| x | 0 | 1 | 2 | 3 | 4 |
|---|---|---|---|---|---|
| y | 1 | 3 | 5 | 7 | 9 |

21. **CAR CLUB** The table shows the number of car club members for four consecutive years after it began.

| Time (years) | 0 | 1 | 2 | 3 | 4 |
|---|---|---|---|---|---|
| Members | 10 | 20 | 40 | 80 | 160 |

a. Determine which model best represents the data.

b. Write a function that models the data.

c. Predict the number of car club members after 6 years.

**Graph each function.**

22. $f(x) = |x - 1|$
23. $f(x) = -|2x|$
24. $f(x) = [\![x]\!]$
25. $f(x) = \begin{cases} 2x - 1 \text{ if } x < 2 \\ x - 3 \text{ if } x \geq 2 \end{cases}$
26. Determine the domain and range of the function graphed below.

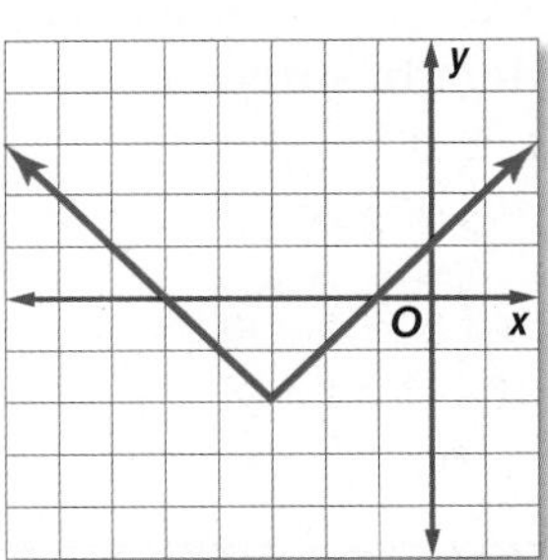

# CHAPTER 2 Preparing for Standardized Tests

## Use a Formula

A *formula* is an equation that shows a relationship among certain quantities. Many standardized test problems will require using a formula to solve them.

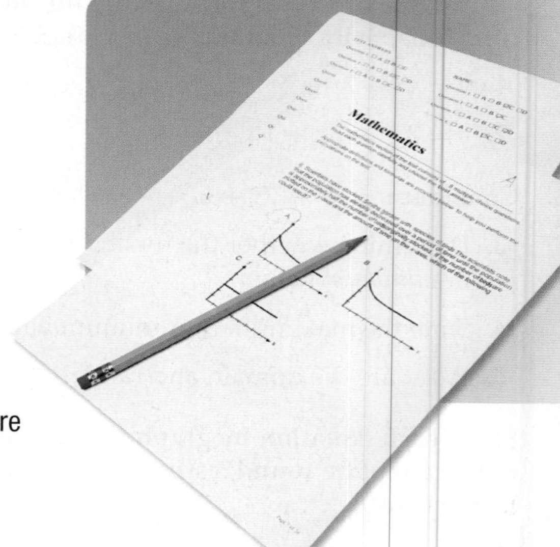

Stockbyte/PunchStock

### Strategies for Using a Formula

**Step 1**

Become familiar with common formulas and their uses. You may or may not be given access to a formula sheet to use during the test.

- **If given a formula sheet,** be sure to practice with the formulas on it before taking the test so you know how to apply them.
- **If not given a formula sheet,** study and practice with common formulas such as perimeter, area, and volume formulas, the Distance Formula, the Pythagorean Theorem, the Midpoint Formula, the Quadratic Formula, and others.

**Step 2**

Choose a formula and solve.

- **Ask Yourself:** What quantities are given in the problem statement?
- **Ask Yourself:** What quantities am I looking for?
- **Ask Yourself:** Is there a formula I know that relates these quantities?
- **Write:** Write the formula out that you have chosen each time.
- **Solve:** Substitute known quantities into the formula and solve for the unknown quantity.
- **Check:** Check your answer if time permits.

### Standardized Test Example

**Read the problem. Identify what you need to know. Then use the information in the problem to solve.**

Find the exact roots of the quadratic equation $-2x^2 + 6x + 5 = 0$.

**A** $\frac{3 \pm \sqrt{17}}{4}$

**B** $\frac{4 \pm \sqrt{17}}{3}$

**C** $\frac{3 \pm \sqrt{19}}{2}$

**D** $\frac{3 \pm \sqrt{19}}{4}$

Read the problem carefully. You are given a quadratic equation and asked to find the exact roots of the equation. Use the **Quadratic Formula** to find the roots.

| | |
|---|---|
| $-2x^2 + 6x + 5 = 0$ | Original equation |
| $a = -2, b = 6, c = 5$ | Identify the coefficients of the equation. |
| $x = \dfrac{-b \pm \sqrt{b^2 - 4ac}}{2a}$ | Quadratic Formula |
| $= \dfrac{-(6) \pm \sqrt{(6)^2 - 4(-2)(5)}}{2(-2)}$ | $a = -2$, $b = 6$, and $c = 5$ |
| $= \dfrac{-6 \pm \sqrt{36 - (-40)}}{-4}$ | Simplify. |
| $= \dfrac{-6 \pm \sqrt{76}}{-4}$ | Subtract. |
| $= \dfrac{-6 \pm 2\sqrt{19}}{-4}$ | $\sqrt{76} = \sqrt{4 \cdot 19}$ or $2\sqrt{19}$. |
| $= \dfrac{-2(3 \pm \sqrt{19})}{-2(2)}$ | Factor out $-2$ from the numerator and denominator. |
| $= \dfrac{3 \pm \sqrt{19}}{2}$ | Simplify. |

The roots of the equation are $\dfrac{3 + \sqrt{19}}{2}$ and $\dfrac{3 - \sqrt{19}}{2}$. The correct answer is C.

## Exercises

**Read each problem. Identify what you need to know. Then use the information in the problem to solve.**

1. Find the exact roots of the quadratic equation $x^2 + 5x - 12 = 0$.

   A $\dfrac{-5 \pm \sqrt{73}}{2}$  C $\dfrac{-3 \pm \sqrt{73}}{4}$

   B $\dfrac{4 \pm \sqrt{61}}{3}$  D $\dfrac{-1 \pm \sqrt{61}}{2}$

2. The area of a triangle in which the length of the base is 4 centimeters greater than twice the height is 80 square centimeters. What is the length of the base of the triangle?

   F $-10$

   G 8

   H 16

   J 20

3. Find the volume of the figure below.

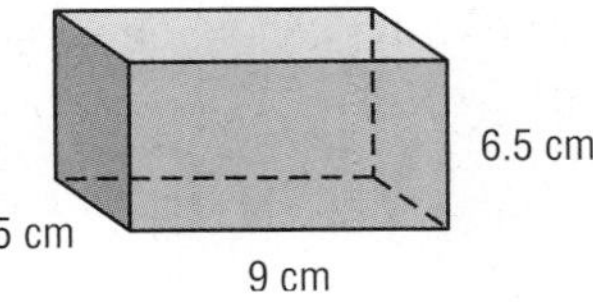


   A 18.5 cm$^3$  C 272 cm$^3$

   B 91 cm$^3$  D 292.5 cm$^3$

4. Myron is traveling 263.5 miles at an average rate of 62 miles per hour. How long will it take Myron to complete his trip?

   F 4 h 10 min

   G 4 h 15 min

   H 5 h 10 min

   J 5 h 25 min

CHAPTER 2

# Standardized Test Practice

## Cumulative, Chapters 1 and 2

### Multiple Choice

**Read each question. Then fill in the correct answer on the answer document provided by your teacher or on a sheet of paper.**

**1.** What is the vertex of the parabola graphed below?

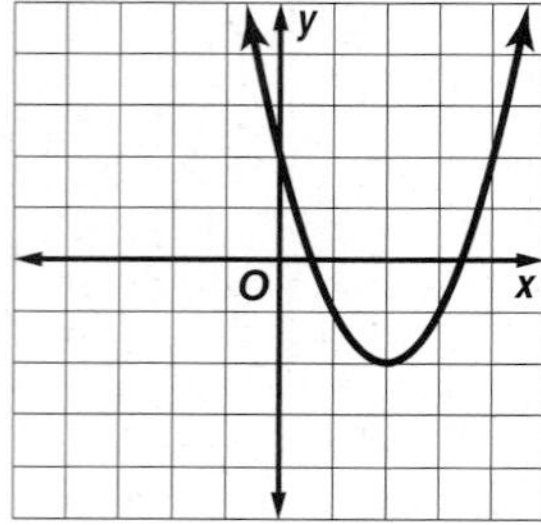


A $(2, 0)$

B $(0, 2)$

C $(-2, 2)$

D $(2, -2)$

**2.** Which of the following is *not* a factor of $x^4 - 6x^2 - 27$?

A $x^2 + 3$

B $x - 3$

C $x + 3$

D $x^2 - 3$

**3.** Use the Quadratic Formula to find the exact solutions of the equation $2x^2 - 6x + 3 = 0$.

A $\frac{3 \pm \sqrt{3}}{2}$

B $\frac{3 \pm \sqrt{2}}{4}$

C $\frac{2 \pm \sqrt{5}}{3}$

D $\frac{5 \pm \sqrt{2}}{2}$

**4.** Write an expression for the area of the rectangle below.

$2b^4c^3 - 3bc$

$5bc^2$

F $10b^5c^5 - 3bc$

G $10b^5c^5 - 15b^2c^3$

H $2b^5c^5 - 3b^2c^3$

J $10b^4c^6 - 15bc^2$

**5.** Solve the quadratic equation below by graphing.

$$x^2 - 2x - 15 = 0$$

A $-1, 4$

B $-3, 5$

C $3, -5$

D $\varnothing$

**6. GRIDDED RESPONSE** How many times does the graph of $y = x^2 - 4x + 10$ cross the $x$-axis?

**Test-Taking Tip**

**Question 5** If permitted, you can use a graphing calculator to quickly graph an equation and find its roots.

## Short Response/Gridded Response

**Record your answers on the answer sheet provided by your teacher or on a sheet of paper.**

**7.** Use the graph of the quadratic equation shown below to answer each question.

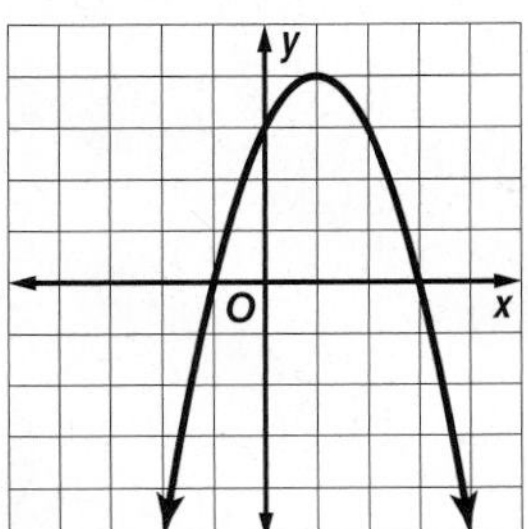


**a.** What is the vertex?

**b.** What is the $y$-intercept?

**c.** What is the axis of symmetry?

**d.** What are the roots of the corresponding quadratic equation?

**Record your answers on a sheet of paper. Show your work.**

**8.** Carl's father is building a tool chest that is shaped like a rectangular prism. He wants the tool chest to have a surface area of 62 square feet. The height of the chest will be 1 foot shorter than the width. The length will be 3 feet longer than the height.

**a.** Sketch a model to represent the problem.

**b.** Write a polynomial that represents the surface area of the tool chest.

**c.** What are the dimensions of the tool chest?

## Extended Response

**Record your answers on a sheet of paper. Show your work.**

**9.** Use the equation and its graph to answer each question.

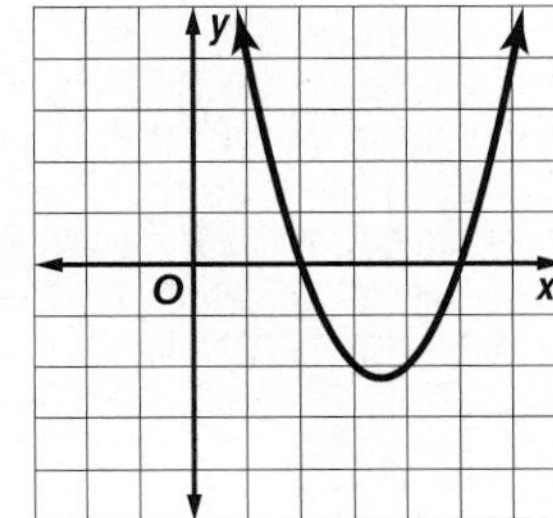


**a.** Factor $x^2 - 7x + 10$.

**b.** What are the solutions of $x^2 - 7x + 10 = 0$?

**c.** What do you notice about the graph of the quadratic equation and where it crosses the $x$-axis? How do these values compare to the solutions of $x^2 - 7x + 10 = 0$? Explain.

**Need ExtraHelp?**

| If you missed Question... | 1 | 2 | 3 | 4 | 5 | 6 | 7 | 8 |
|---|---|---|---|---|---|---|---|---|
| Go to Lesson... | 2-1 | 1-9 | 2-5 | 1-2 | 2-2 | 2-5 | 2-1 | 1-9 |

CHAPTER 3

# Quadratic Functions and Relations

## Then

You graphed linear equations and inequalities.

## Now

You will:

- Graph quadratic functions.
- Solve quadratic equations.
- Perform operations with complex numbers.
- Graph and solve quadratic inequalities.

## Why? ▲

**MOTION** The path that a soccer ball or a firework takes can be modeled by a quadratic function. Quadratic functions can map an object in motion. In this chapter you will look at a pumpkin catapult, an amusement park ride, and a diver in motion.

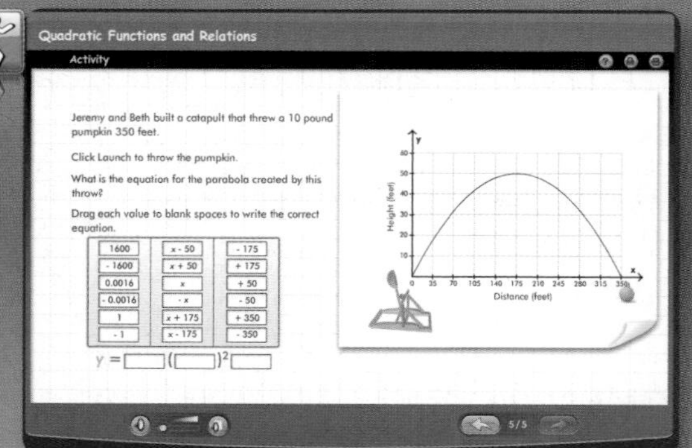


connectED.mcgraw-hill.com Your Digital Math Portal

Animation  Vocabulary  eGlossary  Personal Tutor  Virtual Manipulatives  Graphing Calculator  Audio  Foldables  Self-Check Practice  Worksheets 

# Get Ready for the Chapter

**Diagnose** Readiness | You have two options for checking prerequisite skills.

**Textbook Option** Take the Quick Check below. Refer to the Quick Review for help.

## QuickCheck

Given $f(x) = 2x^2 + 4$ and $g(x) = -x^2 - 2x + 3$, find each value.

**1.** $f(-1)$ **2.** $f(3)$

**3.** $f(0)$ **4.** $g(4)$

**5.** $g(0)$ **6.** $g(-3)$

**7.** **FISH** Tuna swim at a steady rate of 9 miles per hour until they die, and they never stop moving.

**a.** Write a function that is a model for the situation.

**b.** Evaluate the function to estimate how far a 2-year old tuna has traveled.

**8.** **BUDGET** Marla has budgeted \$65 per day on food during a business trip. Write a function that is a model for the situation and evaluate what she would spend on a 2 week business trip.

Factor completely. If the polynomial is not factorable, write *prime*.

**9.** $x^2 + 13x + 40$ **10.** $x^2 - 10x + 21$

**11.** $2x^2 + 7x - 4$ **12.** $2x^2 - 7x - 15$

**13.** $x^2 - 11x + 15$ **14.** $x^2 + 12x + 36$

**15.** **FLOOR PLAN** The rectangular room pictured below has an area of $x^2 + 14x + 48$ square feet. If the width of the room is $(x + 6)$ feet, what is the length?

$A = (x^2 + 14x + 48)$ ft$^2$ $(x + 6)$ ft

## QuickReview

**Example 1**

Given $f(x) = -2x^2 + 3x - 1$ and $g(x) = 3x^2 - 5$, find each value.

**a.** $f(2)$

$f(x) = -2x^2 + 3x - 1$ Original function

$f(2) = -2(2)^2 + 3(2) - 1$ Substitute 2 for $x$.

$= -8 + 6 - 1$ or $-3$ Simplify.

**b.** $g(-2)$

$g(x) = 3x^2 - 5$ Original function

$g(-2) = 3(-2)^2 - 5$ Substitute $-2$ for $x$.

$= 12 - 5$ or 7 Simplify.

**Example 2**

Factor $2x^2 - x - 3$ completely. If the polynomial is not factorable, write *prime*.

To find the coefficients of the $x$-terms, you must find two numbers whose product is $2(-3)$ or $-6$, and whose sum is $-1$. The two coefficients must be 2 and $-3$ since $2(-3) = -6$ and $2 + (-3) = -1$. Rewrite the expression and factor by grouping.

$2x^2 - x - 3$

$= 2x^2 + 2x - 3x - 3$ Substitute $2x - 3x$ for $-x$.

$= (2x^2 + 2x) + (-3x - 3)$ Associative Property

$= 2x(x + 1) + -3(x + 1)$ Factor out the GCF.

$= (2x - 3)(x + 1)$ Distributive Property

**2 Online Option** Take an online self-check Chapter Readiness Quiz at connectED.mcgraw-hill.com.

# Get Started on the Chapter

You will learn several new concepts, skills, and vocabulary terms as you study Chapter 3. To get ready, identify important terms and organize your resources. You may wish to refer to Chapter 0 to review prerequisite skills.

## FOLDABLES StudyOrganizer

**Quadratic Functions** Make this Foldable to help you organize your Chapter 3 notes about quadratic functions and relations. Begin with one sheet of 11″ by 17″ paper.

1 **Fold** in half lengthwise.

2 **Fold** in fourths crosswise.

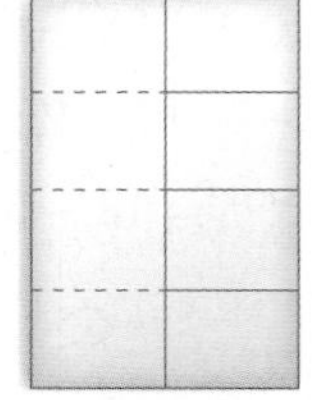

3 **Cut** along the middle fold from the edge to the last crease as shown.

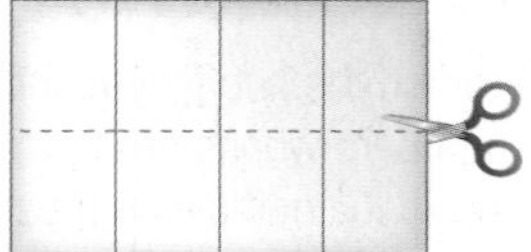

4 **Refold** along the lengthwise fold and tape the uncut section at the top. Label each section with a lesson number and close to form a booklet.

## NewVocabulary

| English | | Español |
|---|---|---|
| imaginary unit | p. 178 | unidad imaginaria |
| pure imaginary number | p. 178 | número imaginario puro |
| complex number | p. 179 | número complejo |
| complex conjugates | p. 181 | conjugados complejos |
| Quadratic Formula | p. 189 | fórmula cuadrática |
| discriminant | p. 192 | discriminante |
| vertex form | p. 200 | forma de vértice |
| quadratic inequality | p. 207 | desigualdad cuadrática |

## ReviewVocabulary

domain dominio the set of all $x$-coordinates of the ordered pairs of a relation

function función a relation in which each $x$-coordinate is paired with exactly one $y$-coordinate

range rango the set of all $y$-coordinates of the ordered pairs of a relation

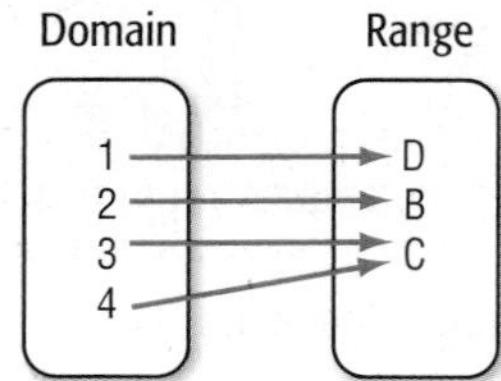

EXTEND 3-1

# Graphing Technology Lab
# Modeling Real-World Data

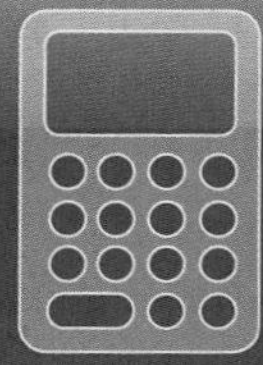

You can use a TI-83/84 Plus graphing calculator to model data points for which a curve of best fit is a quadratic function.

**CCSS Common Core State Standards**

**Content Standards**

F.IF.4 For a function that models a relationship between two quantities, interpret key features of graphs and tables in terms of the quantities, and sketch graphs showing key features given a verbal description of the relationship.

**Mathematical Practices**

5 Use appropriate tools strategically.

**WATER** A bottle is filled with water. The water is allowed to drain from a hole made near the bottom of the bottle. The table shows the level of the water $y$ measured in centimeters from the bottom of the bottle after $x$ seconds.

| Time (s) | 0 | 20 | 40 | 60 | 80 | 100 | 120 | 140 | 160 | 180 | 200 | 220 |
|---|---|---|---|---|---|---|---|---|---|---|---|---|
| Water level (cm) | 42.6 | 40.7 | 38.9 | 37.2 | 35.8 | 34.3 | 33.3 | 32.3 | 31.5 | 30.8 | 30.4 | 30.1 |

Find and graph a linear regression equation and a quadratic regression equation. Determine which equation is a better fit for the data.

## Activity

**Step 1** **Find and graph a linear regression equation.**

- Enter the times in **L1** and the water levels in **L2**. Then find a linear regression equation.

  KEYSTROKES: *Refer to Lesson 2-5.*

- Use **STAT PLOT** to graph a scatter plot. Copy the equation to the **Y=** list and graph.

  KEYSTROKES: *Review statistical plots and graphing a regression equation in Lesson 2-5.*

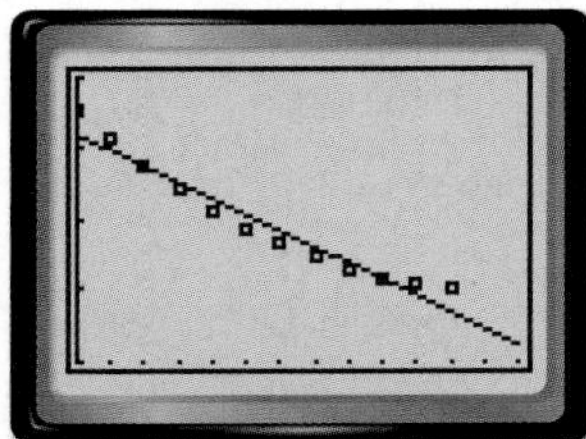

[0, 260] scl: 20 by [25, 45] scl: 5

**Step 2** **Find and graph a quadratic regression equation.**

- Find the quadratic regression equation. Then copy the equation to the **Y=** list and graph.

  KEYSTROKES: STAT ▶ 5 ENTER Y= VARS 5 ▶ ▶ ENTER GRAPH

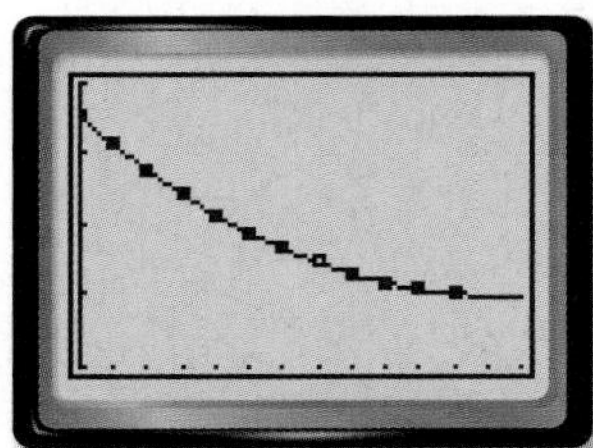

[0, 260] scl: 20 by [25, 45] scl: 5

Notice that the graph of the linear regression equation appears to pass through just two data points. However, the graph of the quadratic regression equation fits the data very well.

## Exercises

**Refer to the table.**

**Height of Player's Feet above Floor**

| Time (s) | Height (in.) |
|---|---|
| 0.1 | 3.04 |
| 0.2 | 5.76 |
| 0.3 | 8.16 |
| 0.4 | 10.24 |
| 0.5 | 12 |
| 0.6 | 13.44 |
| 0.7 | 14.56 |

1. Find and graph a linear regression equation and a quadratic regression equation for the data. Determine which equation is a better fit for the data.
2. Estimate the height of the player's feet after 1 second and 1.5 seconds. Use mental math to check the reasonableness of your estimates.
3. Compare and contrast the estimates you found in Exercise 2.
4. How might choosing a regression equation that does not fit the data well affect predictions made by using the equation?

LESSON

# 3-1 Solving Quadratic Equations by Factoring

**Then**

- You found the greatest common factors of sets of numbers.

**Now**

1. Write quadratic equations in standard form.
2. Solve quadratic equations by factoring.

**Why?**

- The **factored form** of a quadratic equation is $0 = a(x - p)(x - q)$. In the equation, $p$ and $q$ represent the $x$-intercepts of the graph of the equation.

The $x$-intercepts of the graph at the right are 2 and 6. In this lesson, you will learn how to change a quadratic equation in factored form into standard form and vice versa.

| Standard Form | Factored Form |
|---|---|
| $0 = x^2 - 8x + 12$ | $0 = (x - 6)(x - 2)$ |
| | Factors |

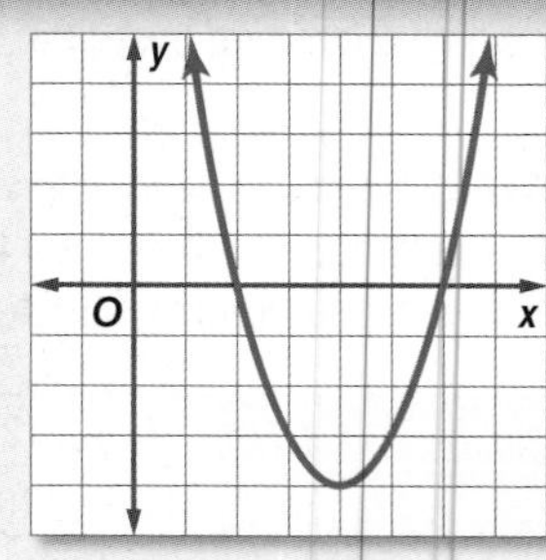

Related Graph
2 and 6 are $x$-intercepts.

**NewVocabulary**

factored form
FOIL method

**Common Core State Standards**

**Content Standards**

A.SSE.2 Use the structure of an expression to identify ways to rewrite it.

F.IF.8.a Use the process of factoring and completing the square in a quadratic function to show zeros, extreme values, and symmetry of the graph, and interpret these in terms of a context.

**Mathematical Practices**

2 Reason abstractly and quantitatively.

## 1 Standard Form

You can use the FOIL method to write a quadratic equation that is in factored form in standard form. The **FOIL method** uses the Distributive Property to multiply binomials.

### KeyConcept FOIL Method for Multiplying Binomials

**Words** To multiply two binomials, find the sum of the products of ***F*** the *First terms*, ***O*** the *Outer terms*, ***I*** the *Inner terms*, and ***L*** the *Last terms*.

**Examples**

| | | Product of **First** Terms | | Product of **Outer** Terms | | Product of **Inner** Terms | | Product of **Last** Terms |
|---|---|---|---|---|---|---|---|---|
| $(x - 6)(x - 2)$ | $=$ | $(x)(x)$ | $+$ | $(x)(-2)$ | $+$ | $(-6)(x)$ | $+$ | $(-6)(-2)$ |

$= x^2 - 2x - 6x + 12$ or $x^2 - 8x + 12$

### Example 1 Translate Sentences into Equations

**Write a quadratic equation in standard form with $-\frac{1}{3}$ and 6 as its roots.**

$(x - p)(x - q) = 0$ Write the pattern.

$\left[x - \left(-\frac{1}{3}\right)\right](x - 6) = 0$ Replace $p$ with $-\frac{1}{3}$ and $q$ with 6.

$\left(x + \frac{1}{3}\right)(x - 6) = 0$ Simplify.

$x^2 - \frac{17}{3}x - 2 = 0$ Multiply.

$3x^2 - 17x - 6 = 0$ Multiply each side by 3 so that $b$ and $c$ are integers.

**GuidedPractice**

**1.** Write a quadratic equation in standard form with $\frac{3}{4}$ and $-5$ as its roots.

## 2 Solve Equations by Factoring

Solving quadratic equations by factoring is an application of the Zero Product Property.

### KeyConcept Zero Product Property

**Words** For any real numbers $a$ and $b$, if $ab = 0$, then either $a = 0$, $b = 0$, or both $a$ and $b$ equal zero.

**Example** If $(x + 3)(x - 5) = 0$, then $x + 3 = 0$ or $x - 5 = 0$.

PT

### Example 2 Factor the GCF

**Solve $16x^2 + 8x = 0$.**

| | |
|---|---|
| $16x^2 + 8x = 0$ | Original equation. |
| $8x(2x) + 8x^2(1) = 0$ | Factor the GCF. |
| $8x(2x + 1) = 0$ | Distributive Property |
| $8x = 0$ or $2x + 1 = 0$ | Zero Product Property |
| $x = 0 \qquad 2x = -1$ | Solve both equations. |
| $x = -\frac{1}{2}$ | |

**GuidedPractice** **Solve each equation.**

**2A.** $20x^2 + 15x = 0$ **2B.** $4y^2 + 16y = 0$ **2C.** $6a^5 + 18a^4 = 0$

**ReviewVocabulary**

perfect square a number with a positive square root that is a whole number

Trinomials and binomials that are perfect squares have special factoring rules. In order to use these rules, the first and last terms need to be perfect squares and the middle term needs to be twice the product of the square roots of the first and last terms.

### Example 3 Perfect Squares and Differences of Squares

**Solve each equation.**

**a. $x^2 + 16x + 64 = 0$**

| | |
|---|---|
| $x^2 = (x)^2$; $64 = (8)^2$ | First and last terms are perfect squares. |
| $16x = 2(x)(8)$ | Middle term equals $2ab$. |

$x^2 + 16x + 64$ is a perfect square trinomial.

| | |
|---|---|
| $x^2 + 16x + 64 = 0$ | Original equation |
| $(x + 8)^2 = 0$ | Factor using the pattern. |
| $x + 8 = 0$ | Take the square root of each side. |
| $x = -8$ | Solve. |

**b. $x^2 = 64$**

| | |
|---|---|
| $x^2 = 64$ | Original equation |
| $x^2 - 64 = 0$ | Subtract 64 from each side. |
| $x^2 - (8)^2 = 0$ | Write in the form $a^2 - b^2$. |
| $(x + 8)(x - 8) = 0$ | Factor the difference of squares. |
| $x + 8 = 0$ or $x - 8 = 0$ | Zero Product Property |
| $x = -8 \qquad x = 8$ | Solve. |

**StudyTip**

Square Roots By inspection, notice that the square root of 64 is $-8$ and 8. Also, for $x^2 = 4$, the solutions would be $-2$ and 2.

**GuidedPractice**

**3A.** $4x^2 - 12x + 9 = 0$ **3B.** $81x^2 - 9x = 0$ **3C.** $6a^2 - 3a = 0$

**StudyTip**

**CCSS Structure** If values for $m$ and $p$ exist, then the trinomial can always be factored.

A special pattern is used when factoring trinomials of the form $ax^2 + bx + c$. First, multiply the values of $a$ and $c$. Then, find two values, $m$ and $p$, such that their product equals $ac$ and their sum equals $b$.

Consider $6x^2 + 13x - 5$: $ac = 6(-5) = -30$.

| Factors of −30 | Sum | Factors of −30 | Sum |
|---|---|---|---|
| 1, −30 | −29 | −1, 30 | 29 |
| 2, −15 | −13 | **−2, 15** | **13** |
| 3, −10 | −7 | −3, 10 | 7 |
| 5, −6 | −1 | −5, 6 | 1 |

Now the middle term, $13x$, can be rewritten as $-2x + 15x$.

This polynomial can now be factored by grouping.

$6x^2 + 13x - 5 = 6x^2 + mx + px - 5$ — Write the pattern.
$= 6x^2 - 2x + 15x - 5$ — $m = -2$ and $p = 15$
$= (6x^2 - 2x) + (15x - 5)$ — Group terms.
$= 2x(3x - 1) + 5(3x - 1)$ — Factor the GCF.
$= (2x + 5)(3x - 1)$ — Distributive Property

## Example 4 Factor Trinomials

PT

**Solve each equation.**

**a.** $\mathbf{x^2 + 9x + 20 = 0}$

$ac = 20$ — $a = 1, c = 20$

| Factors of 20 | Sum | Factors of 20 | Sum |
|---|---|---|---|
| 1, 20 | 21 | −1, −20 | −21 |
| 2, 10 | 12 | −2, −10 | −12 |
| **4, 5** | **9** | −4, −5 | −9 |

**StudyTip**

**Trinomials** It does not matter if the values of $m$ and $p$ are switched when grouping.

$x^2 + 9x + 20 = 0$ — Original expression
$x^2 + mx + px + 20 = 0$ — Write the pattern.
$x^2 + 4x + 5x + 20 = 0$ — $m = 4$ and $p = 5$
$(x^2 + 4x) + (5x + 20) = 0$ — Group terms with common factors.
$x(x + 4) + 5(x + 4) = 0$ — Factor the GCF from each grouping.
$(x + 5)(x + 4) = 0$ — Distributive Property
$x + 5 = 0$ or $x + 4 = 0$ — Zero Product Property
$x = -5$ $\quad x = -4$ — Solve each equation.

**b.** $\mathbf{6y^2 - 23y + 20 = 0}$

$ac = 120$ — $a = 6, c = 20$
$m = -8, p = -15$ — $-8(-15) = 120; -8 + (-15) = -23$
$6y^2 - 23y + 20 = 0$ — Original equation
$6y^2 + my + py + 20 = 0$ — Write the pattern.
$6y^2 - 8y - 15y + 20 = 0$ — $m = -8$ and $p = -15$
$(6y^2 - 8y) + (-15y + 20) = 0$ — Group terms with common factors.
$2y(3y - 4) - 5(3y - 4) = 0$ — Factor the GCF from each grouping.
$(2y - 5)(3y - 4) = 0$ — Distributive Property
$2y - 5 = 0$ or $3y - 4 = 0$ — Zero Product Property
$2y = 5$ $\quad 3y = 4$ — Solve both equations.
$y = \frac{5}{2}$ $\quad y = \frac{4}{3}$

## Guided Practice

**4A.** $x^2 - 11x + 30 = 0$

**4B.** $x^2 - 4x - 21 = 0$

**4C.** $15x^2 - 8x + 1 = 0$

**4D.** $-12x^2 + 8x + 15 = 0$

**Real-World Link**

Cuba's Osleidys Menendez broke the javelin world record in 2002 with a distance of 234 feet 8 inches.

**Source:** *New York Times*

### Real-World Example 5 Solve Equations by Factoring

**TRACK AND FIELD** **The height of a javelin in feet is modeled by $h(t) = -16t^2 + 79t + 5$, where $t$ is the time in seconds after the javelin is thrown. How long is it in the air?**

To determine how long the javelin is in the air, we need to find when the height equals 0. We can do this by solving $-16t^2 + 79t + 5 = 0$.

| | |
|---|---|
| $-16t^2 + 79t + 5 = 0$ | Original equation |
| $m = 80;\ p = -1$ | $-16(5) = -80,\ 80 \cdot (-1) = -80,\ 80 + (-1) = 79$ |
| $-16t^2 + 80t - t + 5 = 0$ | Write the pattern. |
| $(-16t^2 + 80t) + (-t + 5) = 0$ | Group terms with common factors. |
| $16t(-t + 5) + 1(-t + 5) = 0$ | Factor GCF from each group. |
| $(16t + 1)(-t + 5) = 0$ | Distributive Property |
| $16t + 1 = 0$ or $-t + 5 = 0$ | Zero Product Property |
| $16t = -1$    $-t = -5$ | Solve both equations. |
| $t = -\frac{1}{16}$    $t = 5$ | Solve. |

**CHECK** We have two solutions.

- The first solution is negative and since time cannot be negative, this solution can be eliminated.
- The second solution of 5 seconds seems reasonable for the time a javelin spends in the air.
- The answer can be confirmed by substituting back into the original equation.

$$-16t^2 + 79t + 5 = 0$$
$$-16(5)^2 + 79(5) + 5 \stackrel{?}{=} 0$$
$$-400 + 395 + 5 \stackrel{?}{=} 0$$
$$0 = 0 \checkmark$$

The javelin is in the air for 5 seconds.

## Guided Practice

**5.** **BUNGEE JUMPING** Juan recorded his brother bungee jumping from a height of 1100 feet. At the time the cord lifted his brother back up, he was 76 feet above the ground. If Juan started recording as soon as his brother fell, how much time elapsed when the cord snapped back? Use $f(t) = -16t^2 + c$, where $c$ is the height in feet.

## Check Your Understanding

**Example 1** **Write a quadratic equation in standard form with the given root(s).**

1. $-8, 5$
2. $\frac{3}{2}, \frac{1}{4}$
3. $-\frac{2}{3}, \frac{5}{2}$

**Examples 2–4** **Factor each polynomial.**

4. $35x^2 - 15x$
5. $18x^2 - 3x + 24x - 4$
6. $x^2 - 12x + 32$
7. $x^2 - 4x - 21$
8. $2x^2 + 7x - 30$
9. $16x^2 - 16x + 3$

**Example 5** **Solve each equation.**

10. $x^2 - 36 = 0$
11. $12x^2 - 18x = 0$
12. $12x^2 - 2x - 2 = 0$
13. $x^2 - 9x = 0$
14. $x^2 - 3x - 28 = 0$
15. $2x^2 - 24x = -72$

16. **CCSS SENSE-MAKING** Tamika wants to double the area of her garden by increasing the length and width by the same amount. What will be the dimensions of her garden then?

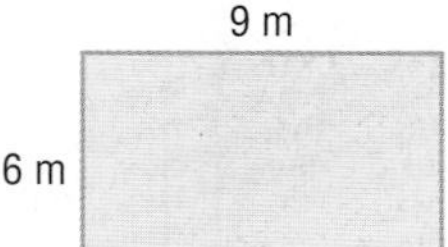


## Practice and Problem Solving

**Example 1** **Write a quadratic equation in standard form with the given root(s).**

17. $7$
18. $-5, \frac{1}{2}$
19. $\frac{1}{5}, 6$

**Examples 2–4** **Factor each polynomial.**

20. $40a^2 - 32a$
21. $51c^3 - 34c$
22. $32xy + 40bx - 12ay - 15ab$
23. $3x^2 - 12$
24. $15y^2 - 240$
25. $48cg + 36cf - 4dg - 3df$
26. $x^2 + 13x + 40$
27. $x^2 - 9x - 22$
28. $3x^2 + 12x - 36$
29. $15x^2 + 7x - 2$
30. $4x^2 + 29x + 30$
31. $18x^2 + 15x - 12$
32. $8x^2z^2 - 4xz^2 - 12z^2$
33. $9x^2 - 25$
34. $18x^2y^2 - 24xy^2 + 36y^2$

**Example 3** **Solve each equation.**

35. $15x^2 - 84x - 36 = 0$
36. $12x^2 + 13x - 14 = 0$
37. $12x^2 - 108x = 0$
38. $x^2 + 4x - 45 = 0$
39. $x^2 - 5x - 24 = 0$
40. $x^2 = 121$
41. $x^2 + 13 = 17$
42. $-3x^2 - 10x + 8 = 0$
43. $-8x^2 + 46x - 30 = 0$
44. **GEOMETRY** The hypotenuse of a right triangle is 1 centimeter longer than one side and 4 centimeters longer than three times the other side. Find the dimensions of the triangle.
45. **NUMBER THEORY** Find two consecutive even integers with a product of 624.

**GEOMETRY** **Find $x$ and the dimensions of each rectangle.**

46. $A = 96$ ft$^2$; sides $x - 2$ ft and $x + 2$ ft

47. $A = 432$ in$^2$; sides $x - 2$ in. and $x + 4$ in.

48. $A = 448$ ft$^2$; sides $3x - 4$ ft and $x + 2$ ft

**Solve each equation by factoring.**

**49.** $12x^2 - 4x = 5$

**50.** $5x^2 = 15x$

**51.** $16x^2 + 36 = -48x$

**52.** $75x^2 - 60x = -12$

**53.** $4x^2 - 144 = 0$

**54.** $-7x + 6 = 20x^2$

**55** **MOVIE THEATER** A company plans to build a large multiplex theater. The financial analyst told her manager that the profit function for their theater was $P(x) = -x^2 + 48x - 512$, where $x$ is the number of movie screens, and $P(x)$ is the profit earned in thousands of dollars. Determine the range of production of movie screens that will guarantee that the company will not lose money.

**Write a quadratic equation in standard form with the given root(s).**

**56.** $-\frac{4}{7}, \frac{3}{8}$

**57.** $3.4, 0.6$

**58.** $\frac{2}{11}, \frac{5}{9}$

**Solve each equation by factoring.**

**59.** $10x^2 + 25x = 15$

**60.** $27x^2 + 5 = 48x$

**61.** $x^2 + 0.25x = 1.25$

**62.** $48x^2 - 15 = -22x$

**63.** $3x^2 + 2x = 3.75$

**64.** $-32x^2 + 56x = 12$

**65.** **DESIGN** A square is cut out of the figure at the right. Write an expression for the area of the figure that remains, and then factor the expression.

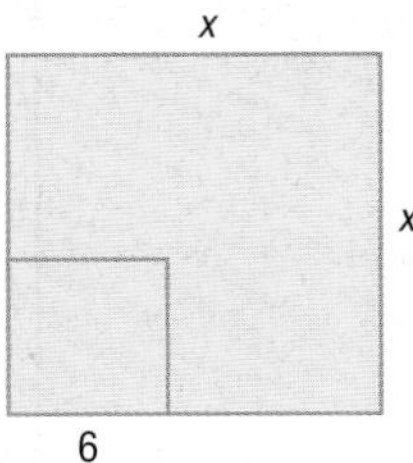


**66.** CCSS **PERSEVERANCE** After analyzing the market, a company that sells Web sites determined the profitability of their product was modeled by $P(x) = -16x^2 + 368x - 2035$, where $x$ is the price of each Web site and $P(x)$ is the company's profit. Determine the price range of the Web sites that will be profitable for the company.

**67.** **PAINTINGS** Enola wants to add a border to her painting, distributed evenly, that has the same area as the painting itself. What are the dimensions of the painting with the border included?

15 in.

10 in.

**68.** **MULTIPLE REPRESENTATIONS** In this problem, you will consider $a(x - p)(x - q) = 0$.

**a. Graphical** Graph the related function for $a = 1$, $p = 2$, and $q = -3$.

**b. Analytical** What are the solutions of the equation?

**c. Graphical** Graph the related functions for $a = 4, -3$, and $\frac{1}{2}$ on the same graph.

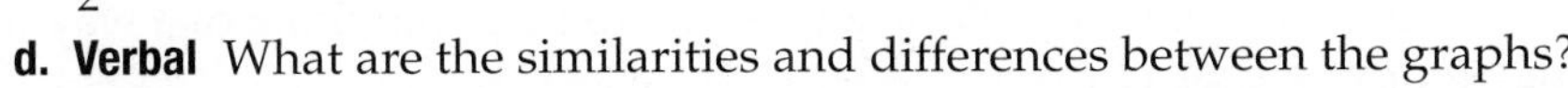

**d. Verbal** What are the similarities and differences between the graphs?

**e. Verbal** What conclusion can you make about the relationship between the factored form of a quadratic equation and its solutions?

**69.** **GEOMETRY** The area of the triangle is 26 square centimeters. Find the length of the base.

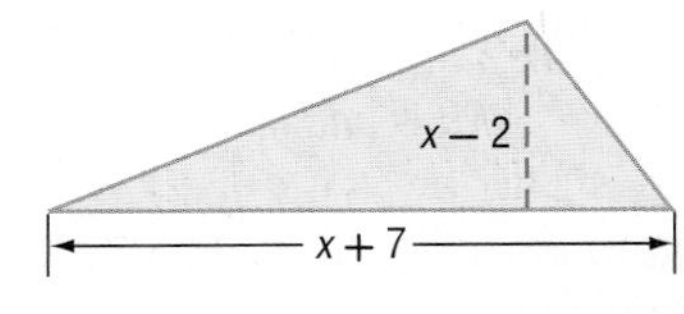

**70. SOCCER** When a ball is kicked in the air, its height in meters above the ground can be modeled by $h(t) = -4.9t^2 + 14.7t$ and the distance it travels can be modeled by $d(t) = 16t$, where $t$ is the time in seconds.

**a.** How long is the ball in the air?

**b.** How far does it travel before it hits the ground? (*Hint*: Ignore air resistance.)

**c.** What is the maximum height of the ball?

**Factor each polynomial.**

**71.** $18a - 24ay + 48b - 64by$

**72.** $3x^2 + 2xy + 10y + 15x$

**73** $6a^2b^2 - 12ab^2 - 18b^3$

**74.** $12a^2 - 18ab + 30ab^3$

**75.** $32ax + 12bx - 48ay - 18by$

**76.** $30ac + 80bd + 40ad + 60bc$

**77.** $5ax^2 - 2by^2 - 5ay^2 + 2bx^2$

**78.** $12c^2x + 4d^2y - 3d^2x - 16c^2y$

## H.O.T. Problems Use Higher-Order Thinking Skills

**79. ERROR ANALYSIS** Gwen and Morgan are solving $-12x^2 + 5x + 2 = 0$. Is either of them correct? Explain your reasoning.

| Gwen | Morgan |
|---|---|
| $-12x^2 + 5x + 2 = 0$ | $-12x^2 + 5x + 2 = 0$ |
| $-12x^2 + 8x - 3x + 2 = 0$ | $-12x^2 + 8x - 3x + 2 = 0$ |
| $4x(-3x + 2) - (3x + 2) = 0$ | $4x(-3x + 2) + (-3x + 2) = 0$ |
| $(4x - 1)(3x + 2) = 0$ | $(4x + 1)(-3x + 2) = 0$ |
| $x = \frac{1}{4}$ or $-\frac{2}{3}$ | $x = -\frac{1}{4}$ or $\frac{2}{3}$ |

**80. CHALLENGE** Solve $3x^6 - 39x^4 + 108x^2 = 0$ by factoring.

**81. CHALLENGE** The rule for factoring a difference of cubes is shown below. Use this rule to factor $40x^5 - 135x^2y^3$.

$$a^3 - b^3 = (a - b)(a^2 + ab + b^2)$$

**82. OPEN ENDED** Choose two integers. Then write an equation in standard form with those roots. How would the equation change if the signs of the two roots were switched?

**83. CHALLENGE** For a quadratic equation of the form $(x - p)(x - q) = 0$, show that the axis of symmetry of the related quadratic function is located halfway between the $x$-intercepts $p$ and $q$.

**84. WRITE A QUESTION** A classmate is using the guess-and-check strategy to factor trinomials of the form $x^2 + bx + c$. Write a question to help him think of a way to use that strategy for $ax^2 + bx + c$.

**85. CCSS ARGUMENTS** Determine whether the following statement is *sometimes*, *always*, or *never* true. Explain your reasoning.

*In a quadratic equation in standard form where a, b, and c are integers, if b is odd, then the quadratic cannot be a perfect square trinomial.*

**86. WRITING IN MATH** Explain how to factor a trinomial in standard form with $a > 1$.

## Standardized Test Practice

**87. SHORT RESPONSE** If $ABCD$ is transformed by $(x, y) \rightarrow (3x, 4y)$, determine the area of $A'B'C'D'$.

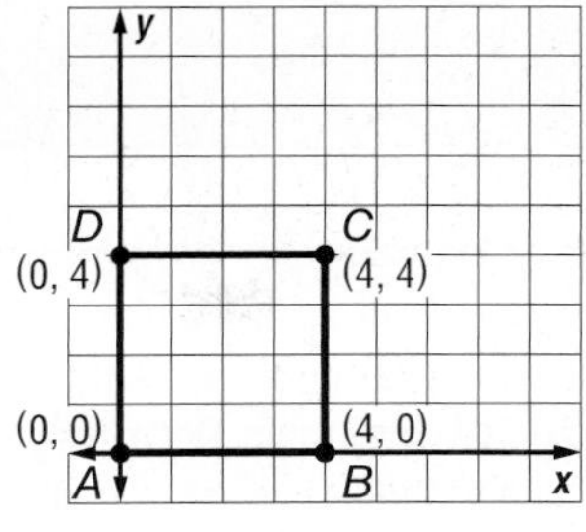


**88.** For $y = 2|6 - 3x| + 4$, which set describes $x$ when $y < 6$?

**A** $\left\{x \middle| \frac{5}{3} < x < \frac{7}{3}\right\}$

**B** $\left\{x \middle| x < \frac{5}{3} \text{ or } x > \frac{7}{3}\right\}$

**C** $\left\{x \middle| x < \frac{5}{3}\right\}$

**D** $\left\{x \middle| x > \frac{7}{3}\right\}$

**89. PROBABILITY** A 5-character password can contain the numbers 0 through 9 and 26 letters of the alphabet. None of the characters can be repeated. What is the probability that the password begins with a consonant?

**F** $\frac{21}{26}$

**G** $\frac{21}{35}$

**H** $\frac{21}{36}$

**J** $\frac{5}{36}$

**90. SAT/ACT** If $c = \frac{8a^3}{b}$, what happens to the value of $c$ when both $a$ and $b$ are doubled?

**A** $c$ is unchanged.

**B** $c$ is halved.

**C** $c$ is doubled.

**D** $c$ is multiplied by 4.

**E** $c$ is multiplied by 8.

## Spiral Review

**Graph each function.** (Lesson 2-7)

**91.** $f(x) = |3x + 2|$

**92.** $f(x) = \begin{cases} x - 2 \text{ if } x > -1 \\ x + 3 \text{ if } x \leq -1 \end{cases}$

**93.** $f(x) = [\![x + 1]\!]$

**94.** $f(x) = \left|\frac{1}{4}x - 1\right|$

**Graph each set of ordered pairs. Determine whether the ordered pairs represent a *linear* function, a *quadratic* function, or an *exponential* function.** (Lesson 2-6)

**95.** $\{(-2, 5), (-1, 3), (0, 1), (1, -1), (2, -3)\}$

**96.** $\{(0, 0), (1, 3), (2, 4), (3, 3), (4, 0)\}$

**97.** $\left\{\left(-2, \frac{1}{4}\right), (0, 1), (1, 2), (2, 4), (3, 8)\right\}$

**98.** $\{(-3, 1), (-2, -5), (-1, -7), (0, -5), (1, 1)\}$

**99. FINANCIAL LITERACY** Determine the amount of an investment if \$250 is invested at an interest rate of 7.3% compounded quarterly for 40 years. (Lesson 2-3)

**100. DIVING** To avoid hitting any rocks below, a cliff diver jumps up and out. The equation $h = -16t^2 + 4t + 26$ describes her height $h$ in feet $t$ seconds after jumping. Find the time at which she returns to a height of 26 feet. (3-1)

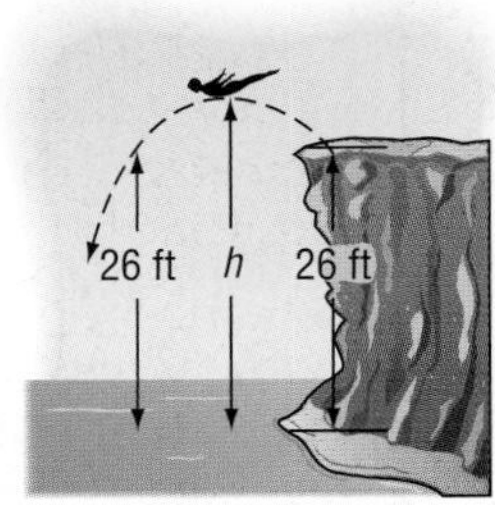


## Skills Review

**Simplify.**

**101.** $\sqrt{5} \cdot \sqrt{15}$

**102.** $\sqrt{8} \cdot \sqrt{32}$

**103.** $2\sqrt{3} \cdot \sqrt{27}$

# LESSON 3-2 Complex Numbers

**Then**

- You simplified square roots.

**Now**

1. Perform operations with pure imaginary numbers.
2. Perform operations with complex numbers.

**Why?**

- Consider the graph of $y = x^2 + 2x + 4$ at the right. Notice how this graph has no $x$-intercepts and therefore does not have any roots. Does this mean there are no solutions to $0 = x^2 + 2x + 4$?

Use the **Solver** function located in the MATH menu of a graphing calculator. Enter the equation and select $x = 2$ as your guess to a solution.

Press [ALPHA] [ENTER] and the calculator will attempt to solve the equation. The calculator indicates there is no solution with the error message. So there are no real solutions. However, there are imaginary solutions.

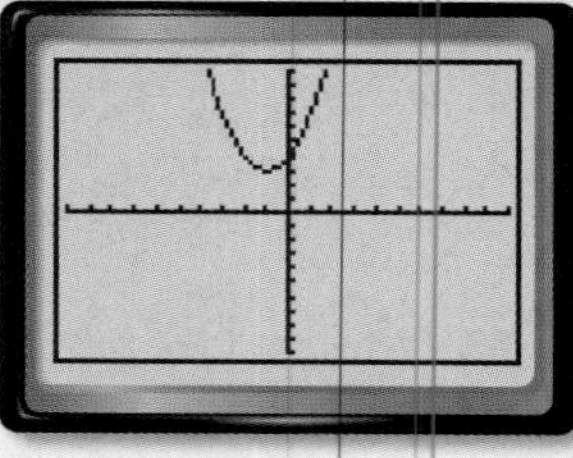

[−10, 10] scl: 1 by [−10, 10] scl: 1

**NewVocabulary**
imaginary unit
pure imaginary number
complex number
complex conjugates

**Common Core State Standards**

**Content Standards**
N.CN.1 Know there is a complex number $i$ such that $i^2 = -1$, and every complex number has the form $a + bi$ with $a$ and $b$ real.
N.CN.2 Use the relation $i^2 = -1$ and the commutative, associative, and distributive properties to add, subtract, and multiply complex numbers.

**Mathematical Practices**
6 Attend to precision.

## 1 Pure Imaginary Numbers

In your math studies so far, you have worked with real numbers. Equations like the one above led mathematicians to define imaginary numbers. The **imaginary unit $i$** is defined to be $i^2 = -1$. The number $i$ is the principal square root of $-1$; that is, $i = \sqrt{-1}$.

Numbers of the form $6i$, $-2i$, and $i\sqrt{3}$ are called **pure imaginary numbers**. Pure imaginary numbers are square roots of negative real numbers. For any positive real number $b$, $\sqrt{-b^2} = \sqrt{b^2} \cdot \sqrt{-1}$ or $bi$.

### Example 1 Square Roots of Negative Numbers

**Simplify.**

**a.** $\sqrt{-27}$

$$\sqrt{-27} = \sqrt{-1 \cdot 3^2 \cdot 3}$$
$$= \sqrt{-1} \cdot \sqrt{3^2} \cdot \sqrt{3}$$
$$= i \cdot 3 \cdot \sqrt{3} \text{ or } 3i\sqrt{3}$$

**b.** $\sqrt{-216}$

$$\sqrt{-216} = \sqrt{-1 \cdot 6^2 \cdot 6}$$
$$= \sqrt{-1} \cdot \sqrt{6^2 \cdot 6}$$
$$= i \cdot 6 \cdot \sqrt{6} \text{ or } 6i\sqrt{6}$$

**GuidedPractice**

**1A.** $\sqrt{-18}$

**1B.** $\sqrt{-125}$

The Commutative and Associative Properties of Multiplication hold true for pure imaginary numbers. The first few powers of $i$ are shown below.

| $i^1 = i$ | $i^2 = -1$ | $i^3 = i^2 \cdot i$ or $-i$ | $i^4 = (i^2)^2$ or 1 |
|---|---|---|---|
| $i^5 = i^4 \cdot i$ or $i$ | $i^6 = i^4 \cdot i^2$ or $-1$ | $i^7 = i^4 \cdot i^3$ or $-i$ | $i^8 = (i^2)^4$ or 1 |

## Example 2 Products of Pure Imaginary Numbers

**Simplify.**

**a.** $-5i \cdot 3i$

$$\begin{aligned} -5i \cdot 3i &= -15i^2 && \text{Multiply.} \\ &= -15(-1) && i^2 = -1 \\ &= 15 && \text{Simplify.} \end{aligned}$$

**b.** $\sqrt{-6} \cdot \sqrt{-15}$

$$\begin{aligned} \sqrt{-6} \cdot \sqrt{-15} &= i\sqrt{6} \cdot i\sqrt{15} && i = \sqrt{-1} \\ &= i^2\sqrt{90} && \text{Multiply.} \\ &= -1 \cdot \sqrt{9} \cdot \sqrt{10} && \text{Simplify.} \\ &= -3\sqrt{10} && \text{Multiply.} \end{aligned}$$

**GuidedPractice**

**2A.** $3i \cdot 4i$ **2B.** $\sqrt{-20} \cdot \sqrt{-12}$ **2C.** $i^{31}$

You can solve some quadratic equations by using the **Square Root Property**. Similar to a difference of squares, the sum of squares can be factored over the complex numbers.

## Example 3 Equation with Pure Imaginary Solutions

**Solve $x^2 + 64 = 0$.**

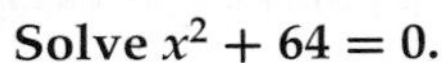

**Method 1** Square Root Property

$$\begin{aligned} x^2 + 64 &= 0 \\ x^2 &= -64 \\ x &= \pm\sqrt{-64} \\ x &= \pm 8i \end{aligned}$$

**Method 2** Factoring

$$\begin{aligned} x^2 + 64 &= 0 \\ x^2 + 8^2 &= 0 \\ x^2 - (-8^2) &= 0 \\ (x + 8i)(x - 8i) &= 0 \end{aligned}$$

$(x + 8i) = 0$ or $(x - 8i) = 0$

$x = -8i$ $\quad$ $x = 8i$

**GuidedPractice**

**Solve each equation.**

**3A.** $4x^2 + 100 = 0$ **3B.** $x^2 + 4 = 0$

**Real-WorldCareer**

**Electrical Engineer** Electrical engineers design, develop, test, and supervise the making of electrical equipment such as digital music players, electric motors, lighting, and radar and navigation systems. A bachelor's degree in engineering is required for almost all entry-level engineering jobs.



## 2 Operations with Complex Numbers

Consider $2 + 3i$. Since 2 is a real number and $3i$ is a pure imaginary number, the terms are not like terms and cannot be combined. This type of expression is called a **complex number**.

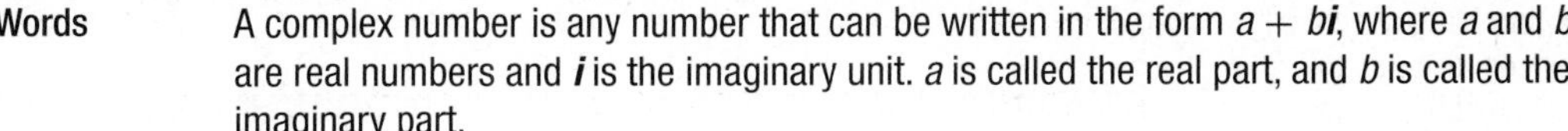

**KeyConcept Complex Numbers**

| | |
|---|---|
| **Words** | A complex number is any number that can be written in the form $a + bi$, where $a$ and $b$ are real numbers and $i$ is the imaginary unit. $a$ is called the real part, and $b$ is called the imaginary part. |
| **Examples** | $5 + 2i$ $\qquad$ $1 - 3i = 1 + (-3)i$ |

The Venn diagram shows the set of complex numbers.

- If $b = 0$, the complex number is a real number.
- If $b \neq 0$, the complex number is imaginary.
- If $a = 0$, the complex number is a pure imaginary number.

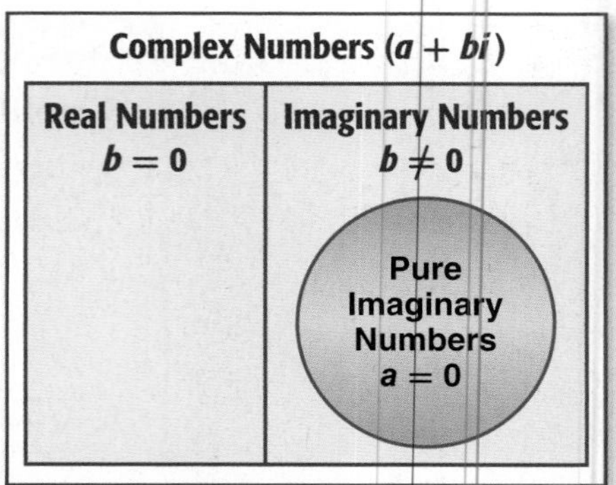


Two complex numbers are equal if and only if their real parts are equal and their imaginary parts are equal. That is, $a + bi = c + di$ if and only if $a = c$ and $b = d$.

**StudyTip**

**Complex Numbers** Whereas all real numbers are also complex, the term *complex number* usually refers to a number that is not real.

### Example 4 Equate Complex Numbers

**Find the values of $x$ and $y$ that make $3x - 5 + (y - 3)i = 7 + 6i$ true.**

Set the real parts equal to each other and the imaginary parts equal to each other.

| | | | |
|---|---|---|---|
| $3x - 5 = 7$ | Real parts | $y - 3 = 6$ | Imaginary parts |
| $3x = 12$ | Add 5 to each side. | $y = 9$ | Add 3 to each side. |
| $x = 4$ | Divide each side by 3. | | |

**GuidedPractice**

**4.** Find the values of $x$ and $y$ that make $5x + 1 + (3 + 2y)i = 2x - 2 + (y - 6)i$ true.

The Commutative, Associative, and Distributive Properties of Multiplication and Addition hold true for complex numbers. To add or subtract complex numbers, combine like terms. That is, combine the real parts, and combine the imaginary parts.

### Example 5 Add and Subtract Complex Numbers

**Simplify.**

**a.** $(5 - 7i) + (2 + 4i)$

$(5 - 7i) + (2 + 4i) = (5 + 2) + (-7 + 4)i$ — Commutative and Associative Properties

$= 7 - 3i$ — Simplify.

**b.** $(4 - 8i) - (3 - 6i)$

$(4 - 8i) - (3 - 6i) = (4 - 3) + [-8 - (-6)]i$ — Commutative and Associative Properties

$= 1 - 2i$ — Simplify.

**GuidedPractice**

**5A.** $(-2 + 5i) + (1 - 7i)$ **5B.** $(4 + 6i) - (-1 + 2i)$

**StudyTip**

**Reading Math** Electrical engineers use $j$ as the imaginary unit to avoid confusion with the $i$ for current.

Complex numbers are used with electricity. In these problems, $j$ usually represents the imaginary unit. In a circuit with alternating current, the voltage, current, and impedance, or hindrance to current, can be represented by complex numbers. To multiply these numbers, use the FOIL method.

## Real-World Example 6 Multiply Complex Numbers

**ELECTRICITY In an AC circuit, the voltage $V$, current $C$, and impedance $I$ are related by the formula $V = C \cdot I$. Find the voltage in a circuit with current $2 + 4j$ amps and impedance $9 - 3j$ ohms.**

| | |
|---|---|
| $V = C \cdot I$ | Electricity formula |
| $= (2 + 4j) \cdot (9 - 3j)$ | $C = 2 + 4j$ and $I = 9 - 3j$ |
| $= 2(9) + 2(-3j) + 4j(9) + 4j(-3j)$ | FOIL Method |
| $= 18 - 6j + 36j - 12j^2$ | Multiply. |
| $= 18 + 30j - 12(-1)$ | $j^2 = -1$ |
| $= 30 + 30j$ | Add. |

The voltage is $30 + 30j$ volts.

### GuidedPractice

**6.** Find the voltage in a circuit with current $2 - 4j$ amps and impedance $3 - 2j$ ohms.

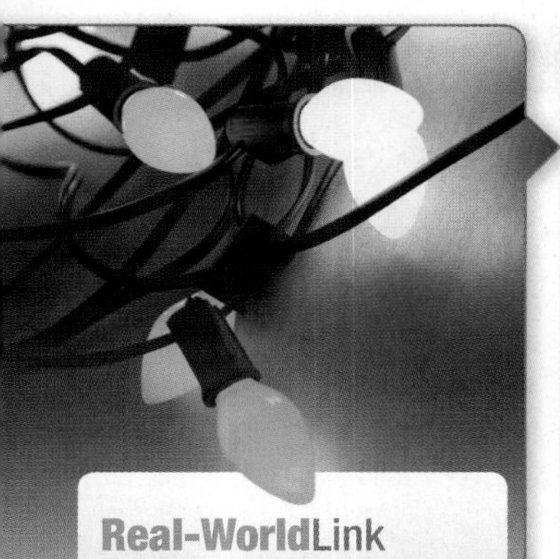

**Real-WorldLink**

An example of a series circuit is a string of holiday lights. The number of bulbs on a circuit affects the strength of the current, which in turn affects the brightness of the lights.

**Source:** *Popular Science*

Two complex numbers of the form $a + bi$ and $a - bi$ are called **complex conjugates**. The product of complex conjugates is always a real number. You can use this fact to simplify the quotient of two complex numbers.

## Example 7 Divide Complex Numbers

**Simplify.**

**a.** $\dfrac{2i}{3 + 6i}$

| | |
|---|---|
| $\dfrac{2i}{3+6i} = \dfrac{2i}{3+6i} \cdot \dfrac{3-6i}{3-6i}$ | $3 + 6i$ and $3 - 6i$ are complex conjugates. |
| $= \dfrac{6i - 12i^2}{9 - 36i^2}$ | Multiply. |
| $= \dfrac{6i - 12(-1)}{9 - 36(-1)}$ | $i^2 = -1$ |
| $= \dfrac{6i + 12}{45}$ | Simplify. |
| $= \dfrac{4}{15} + \dfrac{2}{15}i$ | $a + bi$ form |

**b.** $\dfrac{4 + i}{5i}$

| | |
|---|---|
| $\dfrac{4+i}{5i} = \dfrac{4+i}{5i} \cdot \dfrac{i}{i}$ | Multiply by $\frac{i}{i}$. |
| $= \dfrac{4i + i^2}{5i^2}$ | Multiply. |
| $= \dfrac{4i - 1}{-5}$ | $i^2 = -1$ |
| $= \dfrac{1}{5} - \dfrac{4}{5}i$ | $a + bi$ form |

### GuidedPractice

**7A.** $\dfrac{-2i}{3 + 5i}$

**7B.** $\dfrac{2 + i}{1 - i}$

**StudyTip**

**Technology** Operations with complex numbers can be preformed with a TI-83/84 Plus graphing calculator. Use the [2nd] [$i$] function to enter the expression. Then press [MATH] [ENTER] [ENTER] to view the answer.

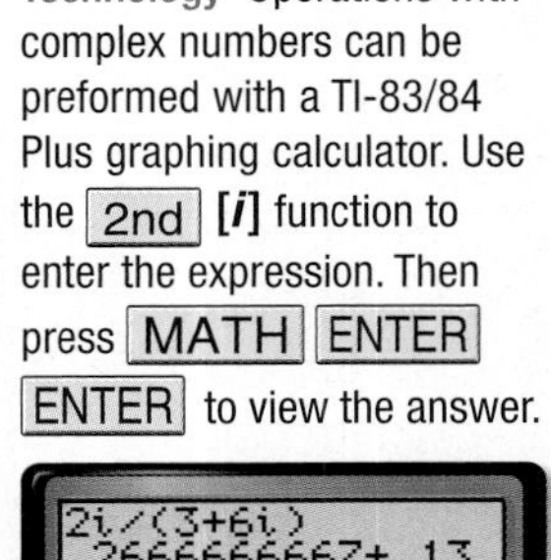


Kelly Redinger/Design Pics Inc./Alamy

## Check Your Understanding

**Examples 1–2** **Simplify.**

1. $\sqrt{-81}$
2. $\sqrt{-32}$
3. $(4i)(-3i)$
4. $3\sqrt{-24} \cdot 2\sqrt{-18}$
5. $i^{40}$
6. $i^{63}$

**Example 3** **Solve each equation.**

7. $4x^2 + 32 = 0$
8. $x^2 + 1 = 0$

**Example 4** **Find the values of $a$ and $b$ that make each equation true.**

9. $3a + (4b + 2)i = 9 - 6i$
10. $4b - 5 + (-a - 3)i = 7 - 8i$

**Examples 5 and 7** **Simplify.**

11. $(-1 + 5i) + (-2 - 3i)$
12. $(7 + 4i) - (1 + 2i)$
13. $(6 - 8i)(9 + 2i)$
14. $(3 + 2i)(-2 + 4i)$
15. $\frac{3 - i}{4 + 2i}$
16. $\frac{2 + i}{5 + 6i}$

**Example 6**

17. **ELECTRICITY** The current in one part of a series circuit is $5 - 3j$ amps. The current in another part of the circuit is $7 + 9j$ amps. Add these complex numbers to find the total current in the circuit.

## Practice and Problem Solving

**Examples 1–2** **CCSS STRUCTURE** **Simplify.**

18. $\sqrt{-121}$
19. $\sqrt{-169}$
20. $\sqrt{-100}$
21. $\sqrt{-81}$
22. $(-3i)(-7i)(2i)$
23. $4i(-6i)^2$
24. $i^{11}$
25. $i^{25}$
26. $(10 - 7i) + (6 + 9i)$
27. $(-3 + i) + (-4 - i)$
28. $(12 + 5i) - (9 - 2i)$
29. $(11 - 8i) - (2 - 8i)$
30. $(1 + 2i)(1 - 2i)$
31. $(3 + 5i)(5 - 3i)$
32. $(4 - i)(6 - 6i)$
33. $\frac{2i}{1 + i}$
34. $\frac{5}{2 + 4i}$
35. $\frac{5 + i}{3i}$

**Example 3** **Solve each equation.**

36. $4x^2 + 4 = 0$
37. $3x^2 + 48 = 0$
38. $2x^2 + 50 = 0$
39. $2x^2 + 10 = 0$
40. $6x^2 + 108 = 0$
41. $8x^2 + 128 = 0$

**Example 4** **Find the values of $x$ and $y$ that make each equation true.**

42. $9 + 12i = 3x + 4yi$
43. $x + 1 + 2yi = 3 - 6i$
44. $2x + 7 + (3 - y)i = -4 + 6i$
45. $5 + y + (3x - 7)i = 9 - 3i$
46. $a + 3b + (3a - b)i = b + bi$
47. $(2a - 4b)i + a + 5b = 15 + 58i$

**Examples 5 and 7** **Simplify.**

**48.** $\sqrt{-10} \cdot \sqrt{-24}$

**49.** $4i\left(\frac{1}{2}i\right)^2(-2i)^2$

**50.** $i^{41}$

**51.** $(4 - 6i) + (4 + 6i)$

**52.** $(8 - 5i) - (7 + i)$

**53.** $(-6 - i)(3 - 3i)$

**54.** $\frac{(5 + i)^2}{3 - i}$

**55.** $\frac{6 - i}{2 - 3i}$

**56.** $(-4 + 6i)(2 - i)(3 + 7i)$

**57.** $(1 + i)(2 + 3i)(4 - 3i)$

**58.** $\frac{4 - i\sqrt{2}}{4 + i\sqrt{2}}$

**59.** $\frac{2 - i\sqrt{3}}{2 + i\sqrt{3}}$

**Example 6**

**60.** **ELECTRICITY** The impedance in one part of a series circuit is $7 + 8j$ ohms, and the impedance in another part of the circuit is $13 - 4j$ ohms. Add these complex numbers to find the total impedance in the circuit.

**ELECTRICITY Use the formula $V = C \cdot I$.**

**61.** The current in a circuit is $3 + 6j$ amps, and the impedance is $5 - j$ ohms. What is the voltage?

**62.** The voltage in a circuit is $20 - 12j$ volts, and the impedance is $6 - 4j$ ohms. What is the current?

**63.** Find the sum of $ix^2 - (4 + 5i)x + 7$ and $3x^2 + (2 + 6i)x - 8i$.

**64.** Simplify $[(2 + i)x^2 - ix + 5 + i] - [(-3 + 4i)x^2 + (5 - 5i)x - 6]$.

**65.** **MULTIPLE REPRESENTATIONS** In this problem, you will explore quadratic equations that have complex roots.

**a. Algebraic** Write a quadratic equation in standard form with $3i$ and $-3i$ as its roots.

**b. Graphical** Graph the quadratic equation found in part **a** by graphing its related function.

**c. Algebraic** Write a quadratic equation in standard form with $2 + i$ and $2 - i$ as its roots.

**d. Graphical** Graph the quadratic equation found in part **c** by graphing its related function.

**e. Analytical** How do you know when a quadratic equation will have only complex solutions?

## H.O.T. Problems Use Higher-Order Thinking Skills

**66.** **CCSS CRITIQUE** Joe and Sue are simplifying $(2i)(3i)(4i)$. Is either of them correct? Explain your reasoning.

| Joe | Sue |
|---|---|
| $24i^3 = -24$ | $24i^3 = -24i$ |

**67.** **CHALLENGE** Simplify $(1 + 2i)^3$.

**68.** **REASONING** Determine whether the following statement is *always*, *sometimes*, or *never* true. Explain your reasoning.

*Every complex number has both a real part and an imaginary part.*

**69.** **OPEN ENDED** Write two complex numbers with a product of 20.

**70.** **WRITING IN MATH** Explain how complex numbers are related to quadratic equations.

## Standardized Test Practice

**71. EXTENDED RESPONSE** Refer to the figure to answer the following.

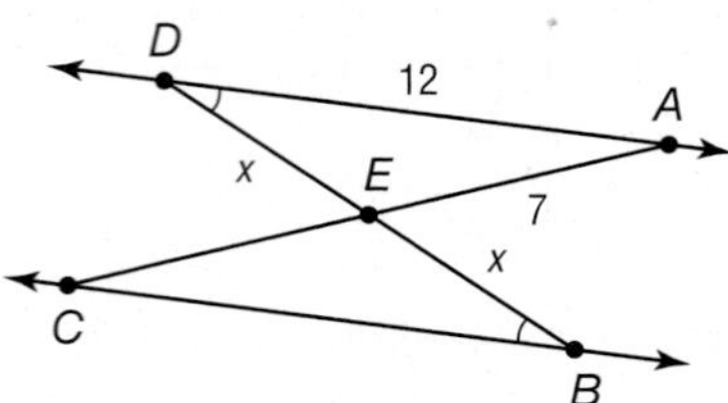


**a.** Name two congruent triangles with vertices in correct order.

**b.** Explain why the triangles are congruent.

**c.** What is the length of $\overline{EC}$? Explain your procedure.

**72.** $(3 + 6)^2 =$

A $2 \times 3 + 2 \times 6$

B $9^2$

C $3^2 + 6^2$

D $3^2 \times 6^2$

**73. SAT/ACT** A store charges \$49 for a pair of pants. This price is 40% more than the amount it costs the store to buy the pants. After a sale, any employee is allowed to purchase any remaining pairs of pants at 30% off the store's cost. How much would it cost an employee to purchase the pants after the sale?

F \$10.50

G \$12.50

H \$13.72

J \$24.50

K \$35.00

**74.** What are the values of $x$ and $y$ when $(5 + 4i) - (x + yi) = (-1 - 3i)$?

A $x = 6, y = 7$

B $x = 4, y = i$

C $x = 6, y = i$

D $x = 4, y = 7$

## Spiral Review

**Solve each equation by factoring.** (Lesson 3-2)

**75.** $2x^2 + 7x = 15$

**76.** $4x^2 - 12 = 22x$

**77.** $6x^2 = 5x + 4$

**78. BASEBALL** A baseball player hits a high pop-up with an initial upward velocity of 30 meters per second, 1.4 meters above the ground. The height $h(t)$ of the ball in meters $t$ seconds after being hit is modeled by $h(t) = -4.9t^2 + 30t + 1.4$. How long does an opposing player have to get under the ball if he catches it 1.7 meters above the ground? Does your answer seem reasonable? Explain. (Lesson 3-2)

**79. ELECTRICITY** The impedance in one part of a series circuit is $3 + 4j$ ohms, and the impedance in another part of the circuit is $2 - 6j$ ohms. Add these complex numbers to find the total impedance of the circuit. (Lesson 3-3)

**Simplify.** (Lesson 3-3)

**80.** $(8 + 5i)^2$

**81.** $4(3 - i) + 6(2 - 5i)$

**82.** $\frac{5 - 2i}{6 + 9i}$

**Write a quadratic equation in standard form with the given root(s).** (Lesson 3-2)

**83.** $\frac{4}{5}, \frac{3}{4}$

**84.** $-\frac{2}{5}, 6$

**85.** $-\frac{1}{4}, -\frac{6}{7}$

## Skills Review

**Determine whether each trinomial is a perfect square trinomial. Write *yes* or *no*.**

**86.** $x^2 + 16x + 64$

**87.** $x^2 - 12x + 36$

**88.** $x^2 + 8x - 16$

**89.** $x^2 - 14x - 49$

**90.** $x^2 + x + 0.25$

**91.** $x^2 + 5x + 6.25$

EXTEND 3-2

# Algebra Lab
# The Complex Plane

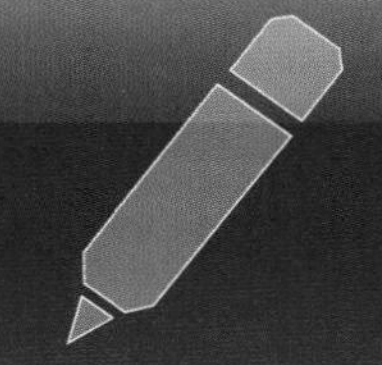

A complex number $a + bi$ can be graphed in the **complex plane** by representing it with the point $(a, b)$. Similar to a coordinate plane, the complex plane is comprised of two axes. The real component is plotted on the **real axis**, which is horizontal. The imaginary component is plotted on the **imaginary axis**, which is vertical. The complex plane may also be referred to as the **Argand (ar GON) plane**.

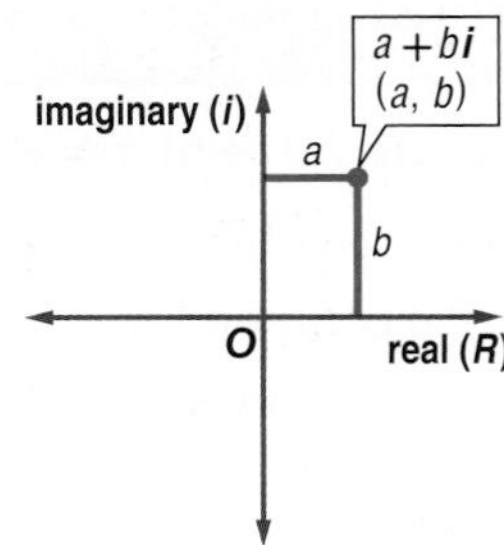


## Example 1 Graph in the Complex Plane

**Graph $z = 3 + 4i$ in the complex plane.**

**Step 1** Represent $z$ with the point $(a, b)$.

The real component $a$ of $z$ is 3.

The imaginary component $bi$ of $z$ is $4i$.

$z$ can be represented by the point $(a, b)$ or $(3, 4)$.

**Step 2** Graph $z$ in the complex plane.

Construct the complex plane and plot the point $(3, 4)$.

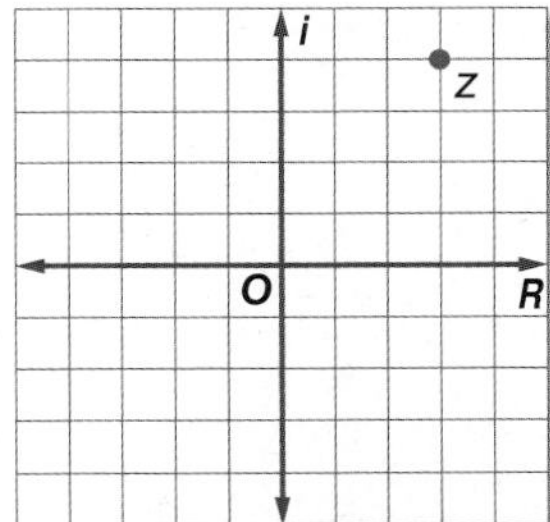


Recall that for a real number, the absolute value is its distance from zero on the number line. Similarly, the **absolute value of a complex number** is its distance from the origin in the complex plane. When $a + bi$ is graphed in the complex plane, the absolute value of $a + bi$ is the distance from $(a, b)$ to the origin. This can be found by using the Distance Formula.

$$\sqrt{(a - 0)^2 + (b - 0)^2} \text{ or } \sqrt{a^2 + b^2}$$

### KeyConcept Absolute Value of a Complex Number

The absolute value of the complex number $z = a + bi$ is

$$|z| = |a + bi| = \sqrt{a^2 + b^2}.$$

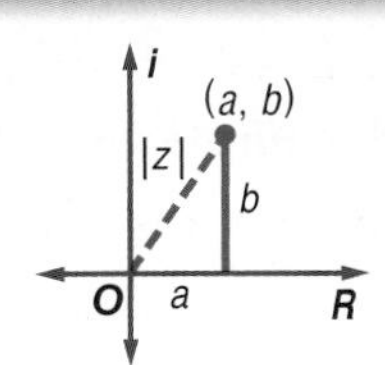

# The Complex Plane *Continued*

## Example 2 Absolute Value of a Complex Number

**Find the absolute value of $z = -5 + 12i$.**

**Step 1** Determine values for $a$ and $b$.

The real component $a$ of $z$ is $-5$. The imaginary component $bi$ of $z$ is $12i$.

Thus, $a = -5$ and $b = 12$.

**Step 2** Find the absolute value of $z$.

$|z| = \sqrt{a^2 + b^2}$ Absolute value of a complex number

$= \sqrt{(-5)^2 + 12^2}$ $a = -5$ and $b = 12$

$= \sqrt{169}$ or 13 Simplify.

The absolute value of $z = -5 + 12i$ is 13.

Addition and subtraction of complex numbers can be performed graphically.

## Example 3 Simplify by Graphing

**Simplify $(1 - 2i) - (-2 - 5i)$ by graphing.**

**Step 1** Write $(1 - 2i) - (-2 - 5i)$ as $(1 - 2i) + (2 + 5i)$.

**Step 2** Graph $1 - 2i$ and $2 + 5i$ on the same complex plane. Connect each point with the origin using a dashed segment.

**Step 3** Complete the parallelogram that has the two segments as two of its sides. Plot a point where the two additional sides meet. The solution of $(1 - 2i) - (-2 - 5i)$ is $3 + 3i$.

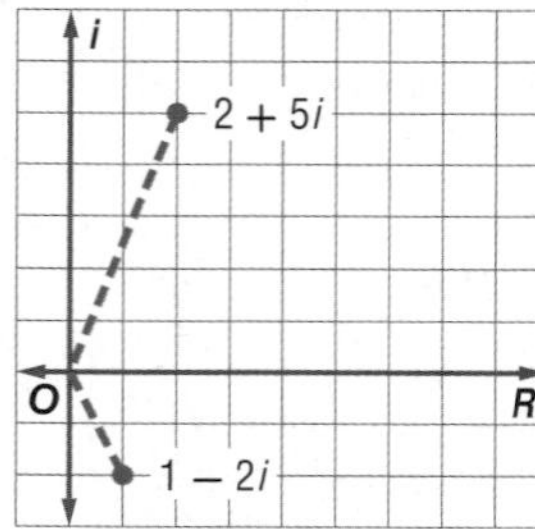

Step 2

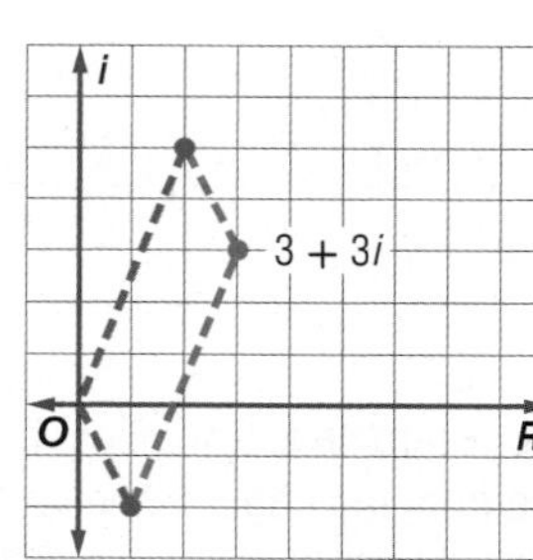

Step 3

## Exercises

**Graph each number in the complex plane.**

1. $z = 3 + i$
2. $z = -4 - 2i$
3. $z = 2 - 2i$

**Find the absolute value of each complex number.**

4. $z = -4 - 3i$
5. $z = 7 - 2i$
6. $z = -6 - i$

**Simplify by graphing.**

7. $(6 + 5i) + (-2 - 3i)$
8. $(8 - 2i) - (4 + 7i)$
9. $(5 + 6i) + (-4 + 3i)$

EXTEND 3-2

# Graphing Technology Lab
# Solving Quadratic Equations

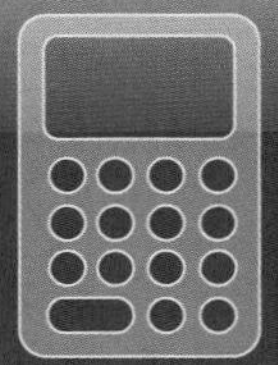

You can use a TI-Nspire™ CAS Technology to solve quadratic equations.

**CCSS Common Core State Standards**
**Content Standards**
**N.CN.7** Solve quadratic equations with real coefficients that have complex solutions.

## Activity Finding Roots

**Solve each equation.**

**a.** $3x^2 - 4x + 1 = 0$

**Step 1** Add a new **Calculator** page.

**Step 2** Select the **Solve** tool from the **Algebra** menu.

**Step 3** Type $3x^2 - 4x + 1 = 0$ followed by a comma, $x$, and then **enter**.

The solutions are $x = \frac{1}{3}$ or $x = 1$.

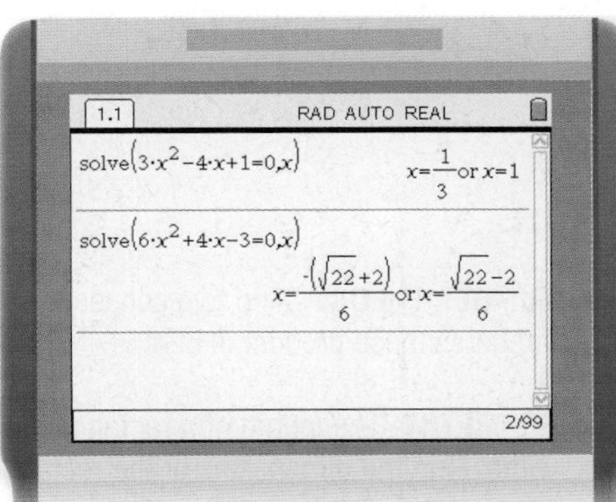


**b.** $6x^2 + 4x - 3 = 0$

**Step 1** Select the **Solve** tool from the **Algebra** menu.

**Step 2** Type $6x^2 - 4x - 3 = 0$ followed by a comma, $x$, and then **enter**.

The solutions are $x = \frac{-2 \pm \sqrt{22}}{6}$.

**c.** $x^2 - 6x + 10 = 0$.

**Step 1** Select the **Solve** tool from the **Algebra** menu.

**Step 2** Type $x^2 - 6x + 10 = 0$ followed by a comma, $x$, and then **enter**.

The calculator returns a value of *false*, meaning that there are no real solutions.

**Step 3** Under menu, select **Algebra**, **Complex**, then **Solve**. Reenter the equation.

The solutions are $x = 3 \pm i$.

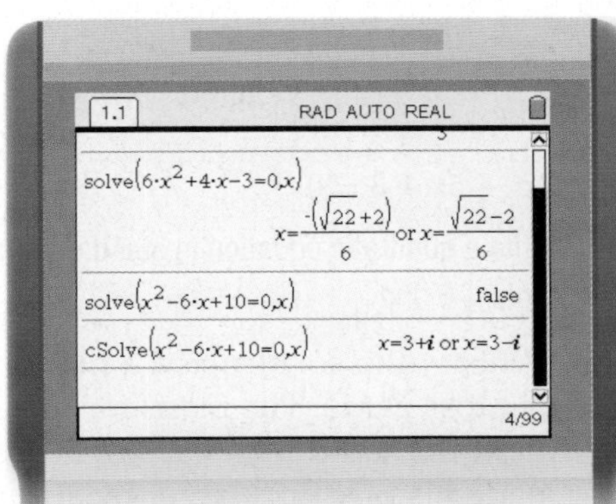


## Exercises

**Solve each equation.**

**1.** $x^2 - 2x - 24 = 0$

**2.** $-x^2 + 4x - 1 = 0$

**3.** $0 = -3x^2 - 6x + 9$

**4.** $x^2 - 2x + 5 = 0$

**5.** $0 = 4x^2 - 8$

**6.** $0 = 2x^2 - 4x + 1$

**7.** $x^2 + 3x + 8 = 5$

**8.** $25 + 4x^2 = -20x$

**9.** $x^2 - x = -6$

# CHAPTER 3 Mid-Chapter Quiz

## Lessons 3-1 through 3-3

**Write a quadratic equation in standard form with the given root(s).** (Lesson 3-2)

1. 7, 2
2. 0, 3
3. $-5, 8$
4. $-7, -8$
5. $-6, -3$
6. $3, -4$
7. $1, \frac{1}{2}$
8. **NUMBER THEORY** Find two consecutive even positive integers whose product is 624. (Lesson 3-2)
9. **GEOMETRY** The length of a rectangle is 2 feet more than its width. Find the dimensions of the rectangle if its area is 63 square feet. (Lesson 3-2)

**Solve each equation by factoring.** (Lesson 3-2)

10. $x^2 - x - 12 = 0$
11. $3x^2 + 7x + 2 = 0$
12. $x^2 - 2x - 15 = 0$
13. $2x^2 + 5x - 3 = 0$
14. Write a quadratic equation in standard form with roots $-6$ and $\frac{1}{4}$. (Lesson 3-2)
15. **GAMES** Julio constructed a platform for a bean bag toss game. The plans for the original platform had dimensions of 3 feet by 5 feet. He made his platform larger by adding $x$ feet to each side. The area of the new platform is 35 square feet. (Lesson 3-2)

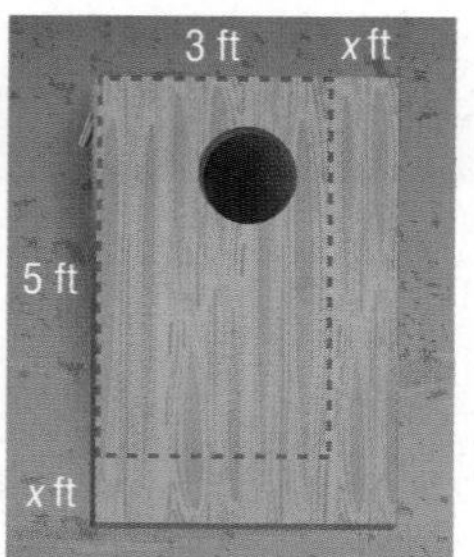


a. Write a quadratic equation that represents the area of his platform.

b. Find the dimensions of the platform Julio made.

16. **TRIANGLES** Find the dimensions of a triangle if the base is $\frac{2}{3}$ the measure of the height and the area is 12 square centimeters. (Lesson 3-2)
17. **PATIO** Eli is putting a cement slab in his backyard. The original slab was going to have dimensions of 8 feet by 6 feet. He decided to make the slab larger by adding $x$ feet to each side. The area of the new slab is 120 square feet. (Lesson 3-2)

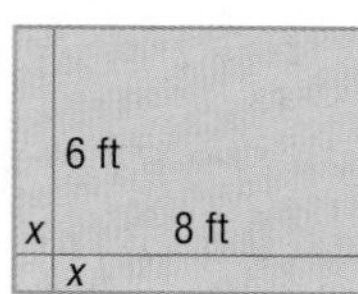


a. Write a quadratic equation that represents the area of the new slab.

b. Find the new dimensions of the slab.

**Simplify.** (Lesson 3-3)

18. $\sqrt{-81}$
19. $\sqrt{-25x^4y^5}$
20. $(15 - 3i) - (4 - 12i)$
21. $i^{37}$
22. $(5 - 3i)(5 + 3i)$
23. $\frac{3 - i}{2 + 5i}$
24. The impedance in one part of a series circuit is $3 + 4j$ ohms and the impedance in another part of the circuit is $6 - 7j$ ohms. Add these complex numbers to find the total impedance in the circuit. (Lesson 3-3)

**Simplify.** (Lesson 3-3)

25. $(3 - 4i) - (9 - 5i)$
26. $\frac{4i}{4 - i}$

LESSON 3-3

# The Quadratic Formula and the Discriminant

**Then**

- You solved equations by completing the square.

**Now**

1. Solve quadratic equations by using the Quadratic Formula.
2. Use the discriminant to determine the number and type of roots of a quadratic equation.

**Why?**

- Pumpkin catapult is an event in which a contestant builds a catapult and launches a pumpkin at a target.

The path of the pumpkin can be modeled by the quadratic function $h = -4.9t^2 + 117t + 42$, where $h$ is the height of the pumpkin and $t$ is the number of seconds.

To predict when the pumpkin will hit the target, you can solve the equation $0 = -4.9t^2 + 117t + 42$. This equation would be difficult to solve using factoring, graphing, or completing the square.

**NewVocabulary**

Quadratic Formula
discriminant

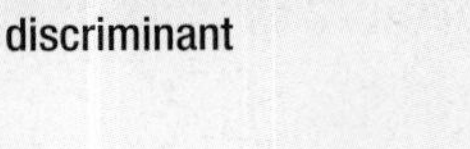

**Common Core State Standards**

**Content Standards**

N.CN.7 Solve quadratic equations with real coefficients that have complex solutions.

A.SSE.1.b Interpret complicated expressions by viewing one or more of their parts as a single entity.

**Mathematical Practices**

8 Look for and express regularity in repeated reasoning.

Izzy Schwartz/Photodisc/Getty Images

## 1 Quadratic Formula

You have found solutions of some quadratic equations by graphing, by factoring, and by using the Square Root Property. There is also a formula that can be used to solve any quadratic equation. This formula can be derived by solving the standard form of a quadratic equation.

| General Case | | Specific Case |
|---|---|---|
| $ax^2 + bx + c = 0$ | Standard quadratic equation | $2x^2 + 8x + 1 = 0$ |
| $x^2 + \frac{b}{a}x + \frac{c}{a} = 0$ | Divide each side by $a$. | $x^2 + 4x + \frac{1}{2} = 0$ |
| $x^2 + \frac{b}{a}x = -\frac{c}{a}$ | Subtract $\frac{c}{a}$ from each side. | $x^2 + 4x = -\frac{1}{2}$ |
| $x^2 + \frac{b}{a}x + \frac{b^2}{4a^2} = -\frac{c}{a} + \frac{b^2}{4a^2}$ | Complete the square. | $x^2 + 4x + \left(\frac{4}{2}\right)^2 = -\frac{1}{2} + \left(\frac{4}{2}\right)^2$ |
| $\left(x + \frac{b}{2a}\right)^2 = -\frac{c}{a} + \frac{b^2}{4a^2}$ | Factor the left side. | $(x + 2)^2 = -\frac{1}{2} + \left(\frac{4}{2}\right)^2$ |
| $\left(x + \frac{b}{2a}\right)^2 = \frac{b^2 - 4ac}{4a^2}$ | Simplify the right side. | $(x + 2)^2 = \frac{7}{2}$ |
| $x + \frac{b}{2a} = \pm\frac{\sqrt{b^2 - 4ac}}{2a}$ | Square Root Property | $x + 2 = \pm\sqrt{\frac{7}{2}}$ |
| $x = -\frac{b}{2a} \pm \frac{\sqrt{b^2 - 4ac}}{2a}$ | Subtract $\frac{b}{2a}$ from each side. | $x = -2 \pm \sqrt{\frac{7}{2}}$ |
| $x = \frac{-b \pm \sqrt{b^2 - 4ac}}{2a}$ | Simplify. | $x = \frac{-4 \pm \sqrt{14}}{2}$ |

The equation $x = \frac{-b \pm \sqrt{b^2 - 4ac}}{2a}$ is known as the **Quadratic Formula**.

**StudyTip**

**Quadratic Formula** Although factoring may be an easier method to solve some of the equations, the Quadratic Formula can be used to solve any quadratic equation.

## KeyConcept Quadratic Formula

**Words** The solutions of a quadratic equation of the form $ax^2 + bx + c = 0$, where $a \neq 0$, are given by the following formula.

$$x = \frac{-b \pm \sqrt{b^2 - 4ac}}{2a}$$

**Example** $x^2 + 5x + 6 = 0 \rightarrow x = \frac{-5 \pm \sqrt{5^2 - 4(1)(6)}}{2(1)}$

PT

## Example 1 Two Rational Roots

**Solve $x^2 - 10x = 11$ by using the Quadratic Formula.**

First, write the equation in the form $ax^2 + bx + c = 0$ and identify $a$, $b$, and $c$.

$$\begin{array}{ccc} & & ax^2 + bx + c = 0 \\ & & \downarrow \quad \downarrow \quad \downarrow \\ x^2 - 10x = 11 & \rightarrow & 1x^2 - 10x - 11 = 0 \end{array}$$

Then, substitute these values into the Quadratic Formula.

$x = \frac{-b \pm \sqrt{b^2 - 4ac}}{2a}$ — Quadratic Formula

$= \frac{-(-10) \pm \sqrt{(-10)^2 - 4(1)(-11)}}{2(1)}$ — Replace $a$ with 1, $b$ with −10, and $c$ with −11.

$= \frac{10 \pm \sqrt{100 + 44}}{2}$ — Multiply.

$= \frac{10 \pm \sqrt{144}}{2}$ — Simplify.

$= \frac{10 \pm 12}{2}$ — $\sqrt{144} = 12$

$x = \frac{10 + 12}{2}$ or $x = \frac{10 - 12}{2}$ — Write as two equations.

$= 11$ $\quad = -1$ — Simplify.

The solutions are −1 and 11.

**CHECK** Substitute both values into the original equation.

$$\begin{aligned} x^2 - 10x &= 11 \\ (-1)^2 - 10(-1) &\stackrel{?}{=} 11 \\ 1 + 10 &\stackrel{?}{=} 11 \\ 11 &= 11 \checkmark \end{aligned} \qquad \begin{aligned} x^2 - 10x &= 11 \\ (11)^2 - 10(11) &\stackrel{?}{=} 11 \\ 121 - 110 &\stackrel{?}{=} 11 \\ 11 &= 11 \checkmark \end{aligned}$$

### GuidedPractice

**Solve each equation by using the Quadratic Formula.**

**1A.** $x^2 + 6x = 16$

**1B.** $2x^2 + 25x + 33 = 0$

**ReviewVocabulary**

**radicand** the value underneath the radical symbol

When the value of the radicand in the Quadratic Formula is 0, the quadratic equation has exactly one rational root.

## Example 2 One Rational Root

**Solve $x^2 + 8x + 16 = 0$ by using the Quadratic Formula.**

Identify $a$, $b$, and $c$. Then, substitute these values into the Quadratic Formula.

$$x = \frac{-b \pm \sqrt{b^2 - 4ac}}{2a}$$ Quadratic Formula

$$= \frac{-(8) \pm \sqrt{(8)^2 - 4(1)(16)}}{2(1)}$$ Replace $a$ with 1, $b$ with 8, and $c$ with 16.

$$= \frac{-8 \pm \sqrt{0}}{2}$$ Simplify.

$$= \frac{-8}{2} \text{ or } -4$$ $\sqrt{0} = 0$

The solution is $-4$.

**CHECK** A graph of the related function shows that there is one solution at $x = -4$.

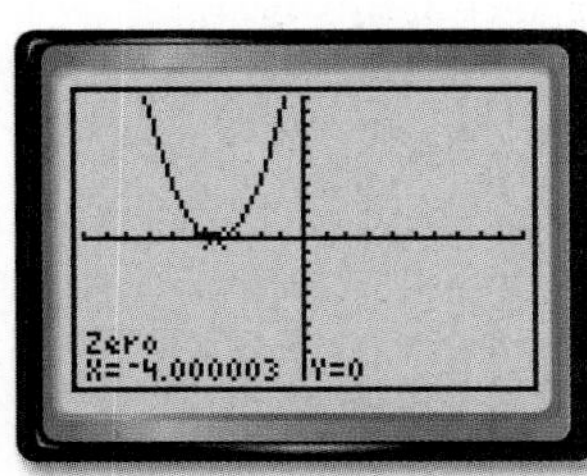


$[-10, 10]$ scl: 1 by $[-10, 10]$ scl: 1

**Math HistoryLink**

**Brahmagupta (598–668)** Indian mathematician Brahmagupta offered the first general solution of the quadratic equation $ax^2 + bx = c$, now known as the Quadratic Formula.

### GuidedPractice

**Solve each equation by using the Quadratic Formula.**

**2A.** $x^2 - 16x + 64 = 0$

**2B.** $x^2 + 34x + 289 = 0$

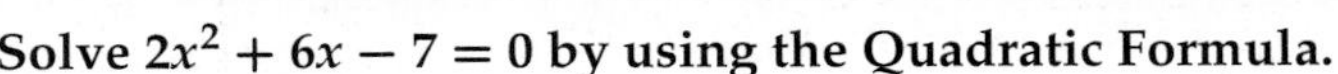

You can express irrational roots exactly by writing them in radical form.

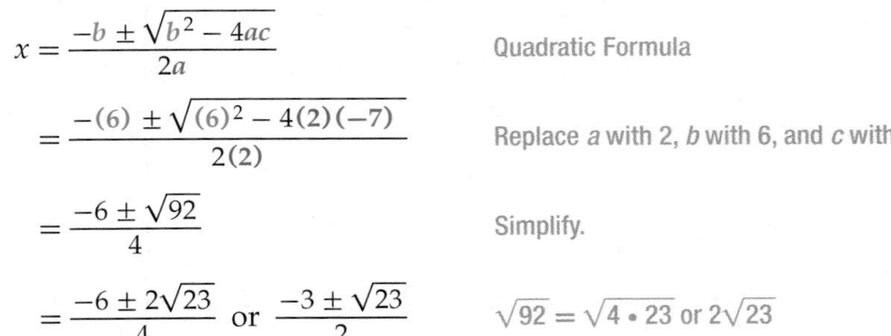

## Example 3 Irrational Roots

**Solve $2x^2 + 6x - 7 = 0$ by using the Quadratic Formula.**

$$x = \frac{-b \pm \sqrt{b^2 - 4ac}}{2a}$$ Quadratic Formula

$$= \frac{-(6) \pm \sqrt{(6)^2 - 4(2)(-7)}}{2(2)}$$ Replace $a$ with 2, $b$ with 6, and $c$ with $-7$.

$$= \frac{-6 \pm \sqrt{92}}{4}$$ Simplify.

$$= \frac{-6 \pm 2\sqrt{23}}{4} \text{ or } \frac{-3 \pm \sqrt{23}}{2}$$ $\sqrt{92} = \sqrt{4 \cdot 23}$ or $2\sqrt{23}$

The approximate solutions are $-3.9$ and $0.9$.

**CHECK** Check these results by graphing the related quadratic function, $y = 2x^2 + 6x - 7$. Using the **ZERO** function of a graphing calculator, the approximate zeros of the related function are $-3.9$ and $0.9$.

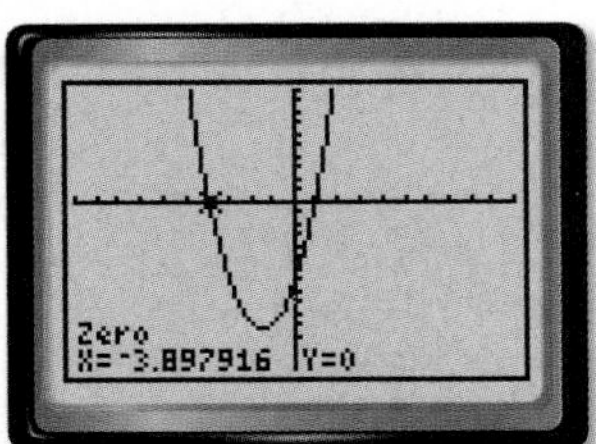


$[-10, 10]$ scl: 1 by $[-10, 10]$ scl: 1

### GuidedPractice

**Solve each equation by using the Quadratic Formula.**

**3A.** $3x^2 + 5x + 1 = 0$

**3B.** $x^2 - 8x + 9 = 0$

British Library/akg-images

**StudyTip**

**Complex Numbers** Remember to write your solutions in the form $a + bi$, sometimes called the *standard form* of a complex number.

When using the Quadratic Formula, if the value of the radicand is negative, the solutions will be complex. Complex solutions always appear in conjugate pairs.

## Example 4 Complex Roots

PT

**Solve $x^2 - 6x = -10$ by using the Quadratic Formula.**

$$x = \frac{-b \pm \sqrt{b^2 - 4ac}}{2a}$$ Quadratic Formula

$$= \frac{-(-6) \pm \sqrt{(-6)^2 - 4(1)(10)}}{2(1)}$$ Replace $a$ with 1, $b$ with −6, and $c$ with 10.

$$= \frac{6 \pm \sqrt{-4}}{2}$$ Simplify.

$$= \frac{6 \pm 2i}{2}$$ $\sqrt{-4} = \sqrt{4 \cdot (-1)}$ or $2i$

$$= 3 \pm i$$ Simplify.

The solutions are the complex numbers $3 + i$ and $3 - i$.

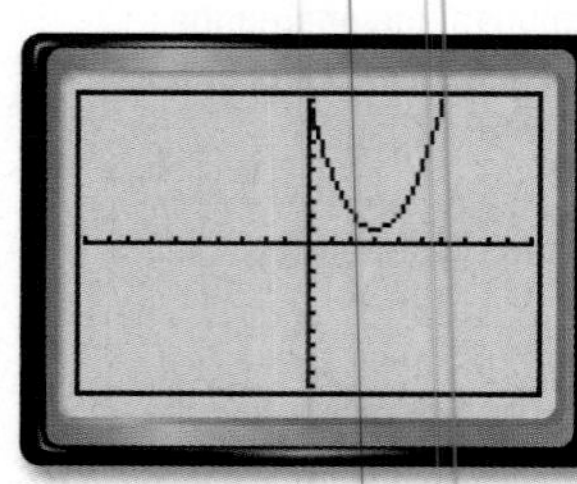

[−10, 10] scl: 1 by [−10, 10] scl: 1

**CHECK** A graph of the related function shows that the solutions are complex, but it cannot help you find them. To check complex solutions, substitute them into the original equation.

$$x^2 - 6x = -10$$ Original equation

$$(3 + i)^2 - 6(3 + i) \stackrel{?}{=} -10$$ $x = 3 + i$

$$9 + 6i + i^2 - 18 - 6i \stackrel{?}{=} -10$$ Square of a sum; Distributive Property

$$-9 + i^2 \stackrel{?}{=} -10$$ Simplify.

$$-9 - 1 = -10 \checkmark$$ $i^2 = -1$

$$x^2 - 6x = -10$$ Original equation

$$(3 - i)^2 - 6(3 - i) \stackrel{?}{=} -10$$ $x = 3 - i$

$$9 - 6i + i^2 - 18 + 6i \stackrel{?}{=} -10$$ Square of a sum; Distributive Property

$$-9 + i^2 \stackrel{?}{=} -10$$ Simplify.

$$-9 - 1 = -10 \checkmark$$ $i^2 = -1$

### GuidedPractice

**Solve each equation by using the Quadratic Formula.**

**4A.** $3x^2 + 5x + 4 = 0$

**4B.** $x^2 - 4x = -13$

## 2 Roots and the Discriminant

In the previous examples, observe the relationship between the value of the expression under the radical and the roots of the quadratic equation. The expression $b^2 - 4ac$ is called the **discriminant**.

$$x = \frac{-b \pm \sqrt{b^2 - 4ac}}{2a}$$ ← **discriminant**

The value of the discriminant can be used to determine the number and type of roots of a quadratic equation. The table on the following page summarizes the possible types of roots.

The discriminant can also be used to confirm the number and type of solutions after you solve the quadratic equation.

**StudyTip**

**Roots** Remember that the solutions of an equation are called *roots* or *zeros* and are the value(s) where the graph crosses the $x$-axis.

## KeyConcept Discriminant

Consider $ax^2 + bx + c = 0$, where $a$, $b$, and $c$ are rational numbers and $a \neq 0$.

| Value of Discriminant | Type and Number of Roots | Example of Graph of Related Function |
|---|---|---|
| $b^2 - 4ac > 0$; $b^2 - 4ac$ is a perfect square. | 2 real, rational roots | |
| $b^2 - 4ac > 0$; $b^2 - 4ac$ is *not* a perfect square. | 2 real, irrational roots | |
| $b^2 - 4ac = 0$ | 1 real rational root | |
| $b^2 - 4ac < 0$ | 2 complex roots | |

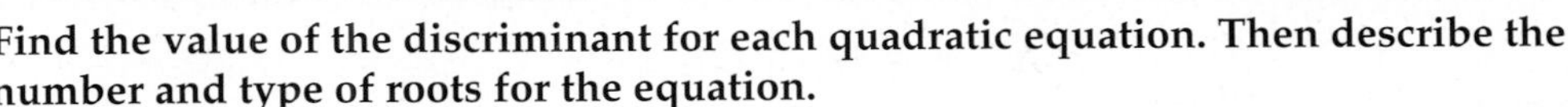

## Example 5 Describe Roots

**Find the value of the discriminant for each quadratic equation. Then describe the number and type of roots for the equation.**

**a.** $7x^2 - 11x + 5 = 0$

$a = 7, b = -11, c = 5$

$$b^2 - 4ac = (-11)^2 - 4(7)(5)$$
$$= 121 - 140$$
$$= -19$$

The discriminant is negative, so there are two complex roots.

**b.** $x^2 + 22x + 121 = 0$

$a = 1, b = 22, c = 121$

$$b^2 - 4ac = (22)^2 - 4(1)(121)$$
$$= 484 - 484$$
$$= 0$$

The discriminant is 0, so there is one rational root.

**GuidedPractice**

**5A.** $-5x^2 + 8x - 1 = 0$

**5B.** $-7x + 15x^2 - 4 = 0$

You have studied a variety of methods for solving quadratic equations. The table below summarizes these methods.

**StudyTip**

**Study Notebook** You may wish to copy this list of methods to your math notebook or Foldable to keep as a reference as you study.

**ConceptSummary** Solving Quadratic Equations

| Method | Can be Used | When to Use |
|---|---|---|
| graphing | sometimes | Use only if an exact answer is not required. Best used to check the reasonableness of solutions found algebraically. |
| factoring | sometimes | Use if the constant term is 0 or if the factors are easily determined. **Example** $x^2 - 7x = 0$ |
| Square Root Property | sometimes | Use for equations in which a perfect square is equal to a constant. **Example** $(x - 5)^2 = 18$ |
| completing the square | always | Useful for equations of the form $x^2 + bx + c = 0$, where $b$ is even. **Example** $x^2 + 6x - 14 = 0$ |
| Quadratic Formula | always | Useful when other methods fail or are too tedious. **Example** $2.3x^2 - 1.8x + 9.7 = 0$ |

## Check Your Understanding

**Examples 1–4** **Solve each equation by using the Quadratic Formula.**

1. $x^2 + 12x - 9 = 0$
2. $x^2 + 8x + 5 = 0$
3. $4x^2 - 5x - 2 = 0$
4. $9x^2 + 6x - 4 = 0$
5. $10x^2 - 3 = 13x$
6. $22x = 12x^2 + 6$
7. $-3x^2 + 4x = -8$
8. $x^2 + 3 = -6x + 8$

**Examples 3–4** 9. **CCSS MODELING** An amusement park ride takes riders to the top of a tower and drops them at speeds reaching 80 feet per second. A function that models this ride is $h = -16t^2 - 64t + 60$, where $h$ is the height in feet and $t$ is the time in seconds. About how many seconds does it take for riders to drop from 60 feet to 0 feet?

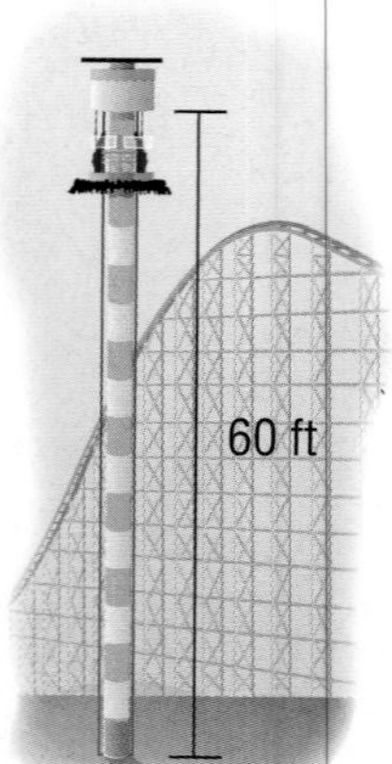


**Example 5** **Complete parts a and b for each quadratic equation.**

**a.** **Find the value of the discriminant.**

**b.** **Describe the number and type of roots.**

10. $3x^2 + 8x + 2 = 0$
11. $2x^2 - 6x + 9 = 0$
12. $-16x^2 + 8x - 1 = 0$
13. $5x^2 + 2x + 4 = 0$

## Practice and Problem Solving

Examples 1–4 **Solve each equation by using the Quadratic Formula.**

**14.** $x^2 + 45x = -200$

**15.** $4x^2 - 6 = -12x$

**16.** $3x^2 - 4x - 8 = -6$

**17.** $4x^2 - 9 = -7x - 4$

**18.** $5x^2 - 9 = 11x$

**19.** $12x^2 + 9x - 2 = -17$

**20.** **DIVING** Competitors in the 10-meter platform diving competition jump upward and outward before diving into the pool below. The height $h$ of a diver in meters above the pool after $t$ seconds can be approximated by the equation $h = -4.9t^2 + 3t + 10$.

**a.** Determine a domain and range for which this function makes sense.

**b.** When will the diver hit the water?

Example 5 **Complete parts a–c for each quadratic equation.**

**a. Find the value of the discriminant.**

**b. Describe the number and type of roots.**

**c. Find the exact solutions by using the Quadratic Formula.**

**21.** $2x^2 + 3x - 3 = 0$

**22.** $4x^2 - 6x + 2 = 0$

**23.** $6x^2 + 5x - 1 = 0$

**24.** $6x^2 - x - 5 = 0$

**25.** $3x^2 - 3x + 8 = 0$

**26.** $2x^2 + 4x + 7 = 0$

**27.** $-5x^2 + 4x + 1 = 0$

**28.** $x^2 - 6x = -9$

**29.** $-3x^2 - 7x + 2 = 6$

**30.** $-8x^2 + 5 = -4x$

**31.** $x^2 + 2x - 4 = -9$

**32.** $-6x^2 + 5 = -4x + 8$

**33.** **VIDEO GAMES** While Darnell is grounded his friend Jack brings him a video game. Darnell stands at his bedroom window, and Jack stands directly below the window. If Jack tosses a game cartridge to Darnell with an initial velocity of 35 feet per second, an equation for the height $h$ feet of the cartridge after $t$ seconds is $h = -16t^2 + 35t + 5$.

**a.** If the window is 25 feet above the ground, will Darnell have 0, 1, or 2 chances to catch the video game cartridge?

**b.** If Darnell is unable to catch the video game cartridge, when will it hit the ground?

**34.** CCSS **SENSE-MAKING** Civil engineers are designing a section of road that is going to dip below sea level. The road's curve can be modeled by the equation $y = 0.00005x^2 - 0.06x$, where $x$ is the horizontal distance in feet between the points where the road is at sea level and $y$ is the elevation. The engineers want to put stop signs at the locations where the elevation of the road is equal to sea level. At what horizontal distances will they place the stop signs?

**Complete parts a–c for each quadratic equation.**

**a. Find the value of the discriminant.**

**b. Describe the number and type of roots.**

**c. Find the exact solutions by using the Quadratic Formula.**

**35.** $5x^2 + 8x = 0$

**36.** $8x^2 = -2x + 1$

**37.** $4x - 3 = -12x^2$

**38.** $0.8x^2 + 2.6x = -3.2$

**39.** $0.6x^2 + 1.4x = 4.8$

**40.** $-4x^2 + 12 = -6x - 8$

**41** **SMOKING** A decrease in smoking in the United States has resulted in lower death rates caused by lung cancer. The number of deaths per 100,000 people $y$ can be approximated by $y = -0.26x^2 - 0.55x + 91.81$, where $x$ represents the number of years after 2000.

| Year | Deaths per 100,000 |
|---|---|
| 2000 | 91.8 |
| 2002 | 89.7 |
| 2004 | 85.5 |
| 2010 | 60.3 |
| 2015 | ? |
| 2017 | ? |

**a.** Calculate the number of deaths per 100,000 people for 2015 and 2017.

**b.** Use the Quadratic Formula to solve for $x$ when $y = 50$.

**c.** According to the quadratic function, when will the death rate be 0 per 100,000? Do you think that this prediction is reasonable? Why or why not?

**42.** **NUMBER THEORY** The sum $S$ of consecutive integers 1, 2, 3, …, $n$ is given by the formula $S = \frac{1}{2}n(n + 1)$. How many consecutive integers, starting with 1, must be added to get a sum of 666?

## H.O.T. Problems Use Higher-Order Thinking Skills

**43.** **CCSS CRITIQUE** Tama and Jonathan are determining the number of solutions of $3x^2 - 5x = 7$. Is either of them correct? Explain your reasoning.

**Tama**

$3x^2 - 5x = 7$

$b^2 - 4ac = (-5)^2 - 4(3)(7)$

$= -59$

Since the discriminant is negative, there are no real solutions.

**Jonathan**

$3x^2 - 5x = 7$

$3x^2 - 5x - 7 = 0$

$b^2 - 4ac = (-5)^2 - 4(3)(-7)$

$= 109$

Since the discriminant is positive, there are two real roots.

**44.** **CHALLENGE** Find the solutions of $4ix^2 - 4ix + 5i = 0$ by using the Quadratic Formula.

**45.** **REASONING** Determine whether each statement is *sometimes, always,* or *never* true. Explain your reasoning.

**a.** In a quadratic equation in standard form, if $a$ and $c$ are different signs, then the solutions will be real.

**b.** If the discriminant of a quadratic equation is greater than 1, the two roots are real irrational numbers.

**46.** **OPEN ENDED** Sketch the corresponding graph and state the number and type of roots for each of the following.

**a.** $b^2 - 4ac = 0$

**b.** A quadratic function in which $f(x)$ never equals zero.

**c.** A quadratic function in which $f(a) = 0$ and $f(b) = 0$; $a \neq b$.

**d.** The discriminant is less than zero.

**e.** $a$ and $b$ are both solutions and can be represented as fractions.

**47.** **CHALLENGE** Find the value(s) of $m$ in the quadratic equation $x^2 + x + m + 1 = 0$ such that it has one solution.

**48.** **WRITING IN MATH** Describe three different ways to solve $x^2 - 2x - 15 = 0$. Which method do you prefer, and why?

## Standardized Test Practice

**49.** A company determined that its monthly profit $P$ is given by $P = -8x^2 + 165x - 100$, where $x$ is the selling price for each unit of product. Which of the following is the best estimate of the maximum price per unit that the company can charge without losing money?

**A** \$10  **B** \$20  **C** \$30  **D** \$40

**50. SAT/ACT** For which of the following sets of numbers is the mean greater than the median?

**F** {4, 5, 6, 7, 8}  **J** {3, 5, 6, 7, 8}
**G** {4, 6, 6, 6, 8}  **K** {2, 6, 6, 6, 6}
**H** {4, 5, 6, 7, 9}

**51. SHORT RESPONSE** In the figure below, $P$ is the center of the circle with radius 15 inches. What is the area of $\triangle APB$?

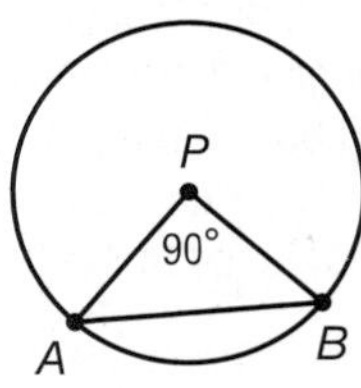


**52.** 75% of 88 is the same as 60% of what number?

**A** 100  **B** 101  **C** 108  **D** 110

## Spiral Review

**Simplify.** (Lesson 3-3)

**53.** $i^{26}$

**54.** $\sqrt{-16}$

**55.** $4\sqrt{-9} \cdot 2\sqrt{-25}$

**56. HIGHWAY SAFETY** Engineers can use the formula $d = 0.05v^2 + 1.1v$ to estimate the minimum stopping distance $d$ in feet for a vehicle traveling $v$ miles per hour. If a car is able to stop after 125 feet, what is the fastest it could have been traveling when the driver first applied the brakes? (Lesson 3-4)

**57. BRIDGES** The supporting cables of the Golden Gate Bridge approximate the shape of a parabola. The parabola can be modeled by the quadratic function $y = 0.00012x^2 + 6$, where $x$ represents the distance from the axis of symmetry and $y$ represents the height of the cables. The related quadratic equation is $0.00012x^2 + 6 = 0$. (Lesson 3-4)

**a.** Calculate the value of the discriminant.

**b.** What does the discriminant tell you about the supporting cables of the Golden Gate Bridge?

## Skills Review

**Write an equation for each graph.**

**58.**

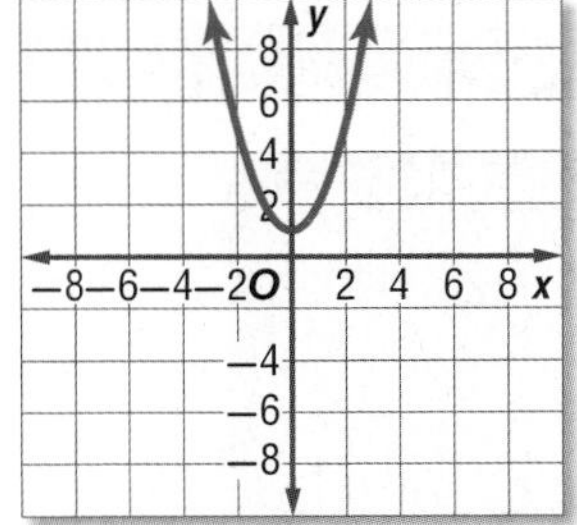

**59.**

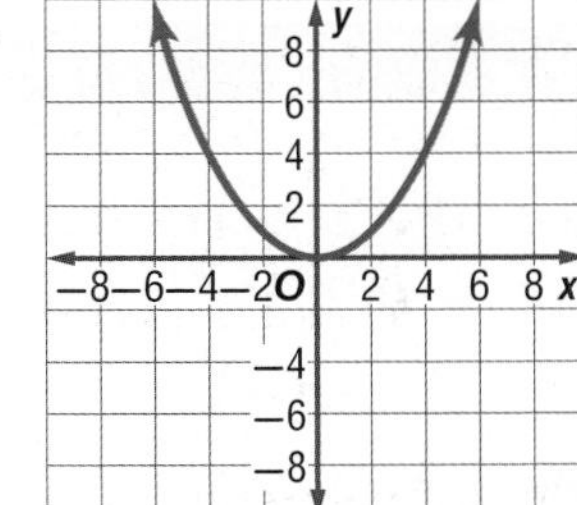

**60.**

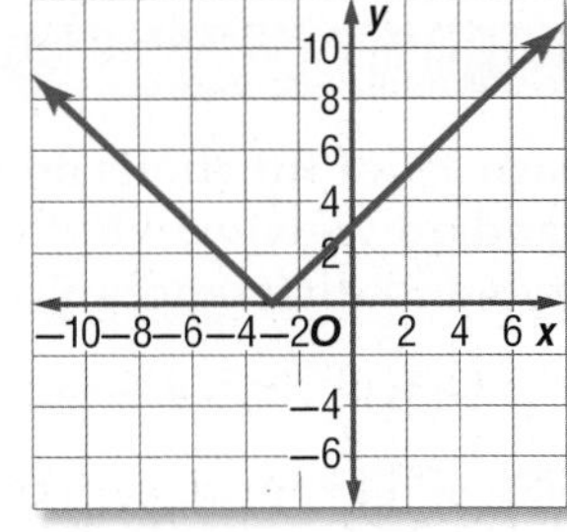

# EXPLORE 3-4 Graphing Technology Lab
# Families of Parabolas

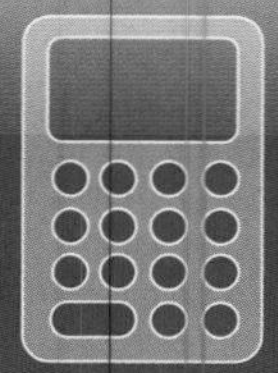

The general form of a quadratic function is $y = a(x - h)^2 + k$. Changing the values of $a$, $h$, and $k$ results in a different parabola in the family of quadratic functions. You can use a TI-83/84 Plus graphing calculator to analyze the effects that result from changing each of these parameters.

**CCSS Common Core State Standards**
**Content Standards**
**F.IF.4** For a function that models a relationship between two quantities, interpret key features of graphs and tables in terms of the quantities, and sketch graphs showing key features given a verbal description of the relationship.
**F.BF.3** Identify the effect on the graph of replacing $f(x)$ by $f(x) + k$, $k\,f(x)$, $f(kx)$, and $f(x + k)$ for specific values of $k$ (both positive and negative); find the value of $k$ given the graphs. Experiment with cases and illustrate an explanation of the effects on the graph using technology.

## Activity 1 Change in $k$

**Graph each set of functions on the same screen in the standard viewing window. Describe any similarities and differences among the graphs.**

$y = x^2, y = x^2 + 4, y = x^2 - 3$

The graphs have the same shape, and all open up. The vertex of each graph is on the $y$-axis. However, the graphs have different vertical positions.

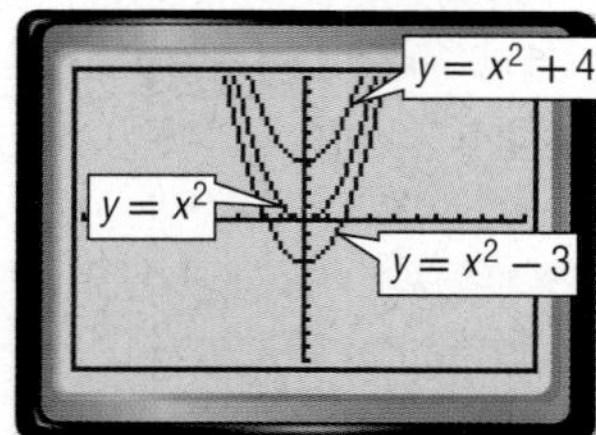


[−10, 10] scl: 1 by [−10, 10] scl: 1

Activity 1 shows how changing the value of $k$ in the function $y = a(x - h)^2 + k$ *translates* the parabola along the $y$-axis. If $k > 0$, the parabola is translated $k$ units up, and if $k < 0$, it is translated $|k|$ units down.

How do you think changing the value of $h$ will affect the graph of $y = a(x - h)^2 + k$?

## Activity 2 Change in $h$

**Graph each set of functions on the same screen in the standard viewing window. Describe any similarities and differences among the graphs.**

$y = x^2, y = (x + 4)^2, y = (x - 3)^2$

These three graphs all open up and have the same shape. The vertex of each graph is on the $x$-axis. However, the graphs have different horizontal positions.

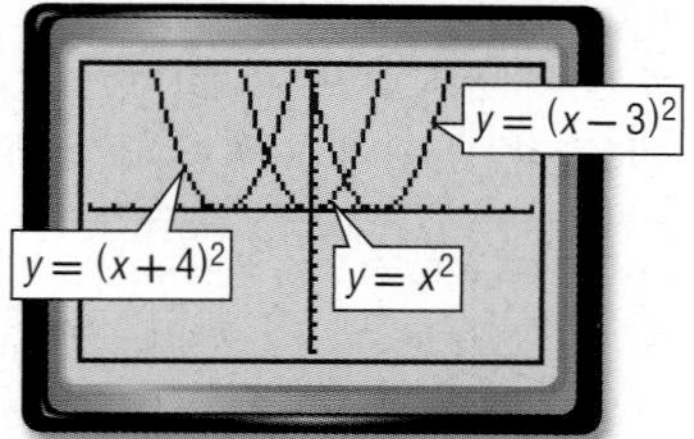


[−10, 10] scl: 1 by [−10, 10] scl: 1

Activity 2 shows how changing the value of $h$ in the equation $y = a(x - h)^2 + k$ *translates* the graph horizontally. If $h > 0$, the graph translates to the right $h$ units. If $h < 0$, the graph translates to the left $|h|$ units.

## Activity 3 Change in $h$ and $k$

**Graph each set of functions on the same screen in the standard viewing window. Describe any similarities and differences among the graphs.**

$y = x^2, y = (x + 6)^2 - 5, y = (x - 4)^2 + 6$

These three graphs all open up and have the same shape. However, the graphs have different horizontal and vertical positions.

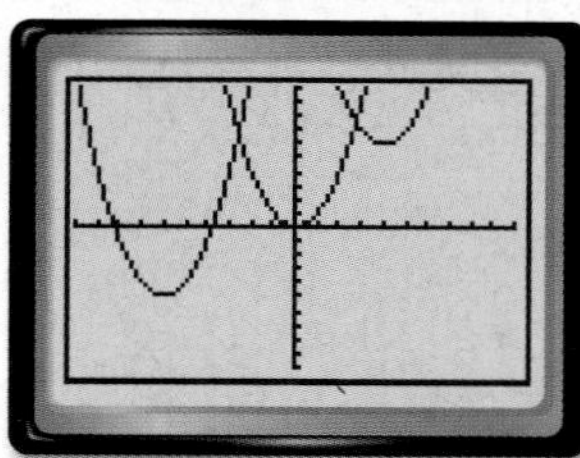

[−10, 10] scl: 1 by [−10, 10] scl: 1

(continued on the next page)

# Graphing Technology Lab
# Families of Parabolas *Continued*

How does the value of $a$ affect the graph of $y = a(x - h)^2 + k$?

## Activity 4 Change in $a$

**Graph each set of functions on the same screen in the standard viewing window. Describe any similarities and differences among the graphs.**

**a.** $y = x^2, y = -x^2$

The graphs have the same vertex and the same shape. However, the graph of $y = x^2$ opens up and the graph of $y = -x^2$ opens down.

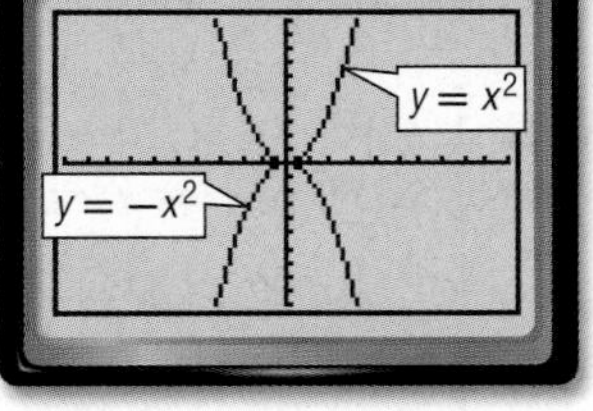


[−10, 10] scl: 1 by [−10, 10] scl: 1

**b.** $y = x^2, y = 5x^2, y = \frac{1}{5}x^2$

The graphs have the same vertex, (0, 0), but each has a different shape. The graph of $y = 5x^2$ is narrower than the graph of $y = x^2$. The graph of $y = \frac{1}{5}x^2$ is wider than the graph of $y = x^2$.

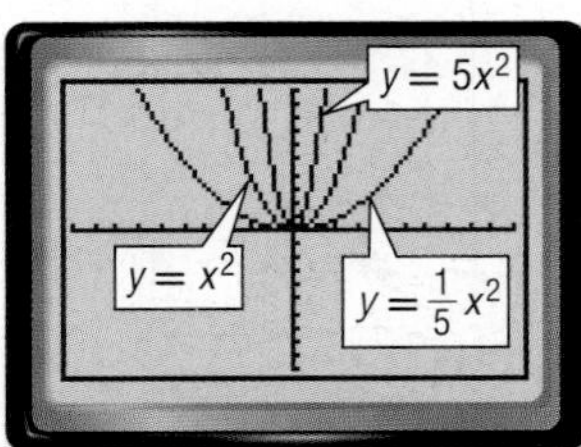


[−10, 10] scl: 1 by [−10, 10] scl: 1

Changing the value of $a$ in the function $y = a(x - h)^2 + k$ can affect the direction of the opening and the shape of the graph. If $a > 0$, the graph opens up, and if $a < 0$, the graph opens down or is *reflected* over the $x$-axis. If $|a| > 1$, the graph is expanded vertically and is narrower than the graph of $y = x^2$. If $|a| < 1$, the graph is compressed vertically and is wider than the graph of $y = x^2$. Thus, a change in the absolute value of $a$ results in a *dilation* of the graph of $y = x^2$.

### Analyze the Results

1. How does changing the value of $h$ in $y = a(x - h)^2 + k$ affect the graph? Give an example.
2. How does changing the value of $k$ in $y = a(x - h)^2 + k$ affect the graph? Give an example.
3. How does using $-a$ instead of $a$ in $y = a(x - h)^2 + k$ affect the graph? Give an example.

**Examine each pair of functions and predict the similarities and differences in their graphs. Use a graphing calculator to confirm your predictions. Write a sentence or two comparing the two graphs.**

4. $y = x^2, y = x^2 + 3.5$
5. $y = -x^2, y = x^2 - 7$
6. $y = x^2, y = 4x^2$
7. $y = x^2, y = -8x^2$
8. $y = x^2, y = (x + 2)^2$
9. $y = -\frac{1}{6}x^2, y = -\frac{1}{6}x^2 + 2$
10. $y = x^2, y = (x - 5)^2$
11. $y = x^2, y = 2(x + 3)^2 - 6$
12. $y = x^2, y = -\frac{1}{8}x^2 + 1$
13. $y = (x + 5)^2 - 4, y = (x + 5)^2 + 7$
14. $y = 2(x + 1)^2 - 4, y = 5(x + 3)^2 - 1$
15. $y = 5(x - 2)^2 - 3, y = \frac{1}{4}(x - 5)^2 - 6$

LESSON 3-4

# Transformations of Quadratic Graphs

**Then**

- You transformed graphs of functions.

**Now**

1. Write a quadratic function in the form $y = a(x - h)^2 + k$.
2. Transform graphs of quadratic functions of the form $y = a(x - h)^2 + k$.

**Why?**

- Recall that a family of graphs is a group of graphs that display one or more similar characteristics. The parent graph is the simplest graph in the family. For the family of quadratic functions, $y = x^2$ is the parent graph.

  Other graphs in the family of quadratic functions, such as $y = (x - 2)^2$ and $y = x^2 - 4$, can be drawn by transforming the graph of $y = x^2$.

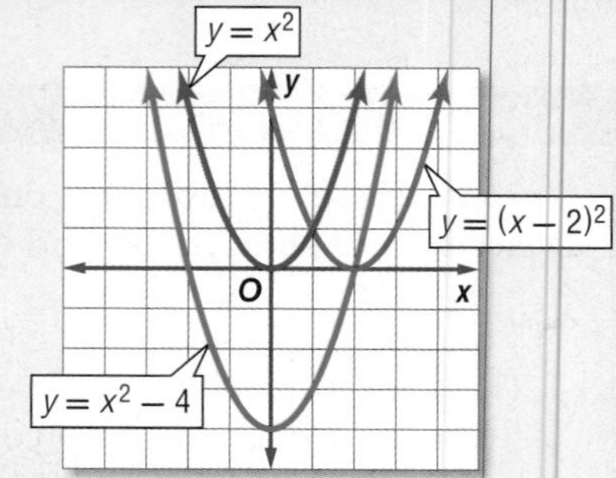

**NewVocabulary**
vertex form

**Common Core State Standards**

**Content Standards**

**F.IF.8.a** Use the process of factoring and completing the square in a quadratic function to show zeros, extreme values, and symmetry of the graph, and interpret these in terms of a context.

**F.BF.3** Identify the effect on the graph of replacing $f(x)$ by $f(x) + k$, $k\,f(x)$, $f(kx)$, and $f(x + k)$ for specific values of $k$ (both positive and negative); find the value of $k$ given the graphs. Experiment with cases and illustrate an explanation of the effects on the graph using technology.

**Mathematical Practices**

7 Look for and make use of structure.

## 1 Write Quadratic Functions in Vertex Form

Each function above is written in **vertex form**, $y = a(x - h)^2 + k$, where $(h, k)$ is the vertex of the parabola, $x = h$ is the axis of symmetry, and $a$ determines the shape of the parabola and the direction in which it opens.

When a quadratic function is in the form $y = ax^2 + bx + c$, you can complete the square to write the function in vertex form. If the coefficient of the quadratic term is not 1, then factor the coefficient from the quadratic and linear terms *before* completing the square. After completing the square and writing the function in vertex form, the value of $k$ indicated a minimum value if $a < 0$ or a maximum value if $a > 0$.

### Example 1 Write Functions in Vertex Form

**Write each function in vertex form.**

**a.** $y = x^2 + 6x - 5$

| | |
|---|---|
| $y = x^2 + 6x - 5$ | Original function |
| $y = (x^2 + 6x + 9) - 5 - 9$ | Complete the square. |
| $y = (x + 3)^2 - 14$ | Simplify. |

**b.** $y = -2x^2 + 8x - 3$

| | |
|---|---|
| $y = -2x^2 + 8x - 3$ | Original function |
| $y = -2(x^2 - 4x) - 3$ | Group $ax^2 + bx$ and factor, dividing by $a$. |
| $y = -2(x^2 - 4x + 4) - 3 - (-2)(4)$ | Complete the square. |
| $y = -2(x - 2)^2 + 5$ | Simplify. |

**GuidedPractice**

**1A.** $y = x^2 + 4x + 6$

**1B.** $y = 2x^2 - 12x + 17$

If the vertex and one additional point on the graph of a parabola are known, you can write the equation of the parabola in vertex form.

## Standardized Test Example 2 Write an Equation Given a Graph

**Which is an equation of the function shown in the graph?**

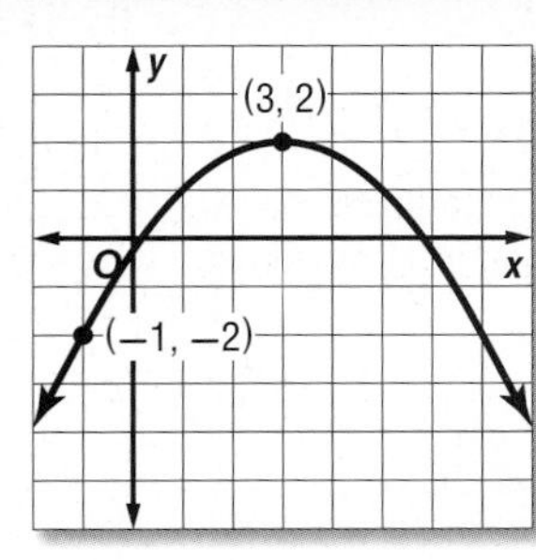


**A** $y = -4(x - 3)^2 + 2$

**B** $y = -\frac{1}{4}(x - 3)^2 + 2$

**C** $y = \frac{1}{4}(x + 3)^2 - 2$

**D** $y = 4(x + 3)^2 - 2$

### Read the Test Item

You are given a graph of a parabola with the vertex and a point on the graph labeled. You need to find an equation of the parabola.

### Solve the Test Item

**Test-TakingTip**

**The Meaning of *a*** The sign of *a* in vertex form does not determine the width of the parabola. The sign indicates whether the parabola opens up or down. The width of a parabola is determined by the absolute value of *a*.

The vertex of the parabola is at (3, 2), so $h = 3$ and $k = 2$. Since $(-1, -2)$ is a point on the graph, let $x = -1$ and $y = -2$. Substitute these values into the vertex form of the equation and solve for $a$.

$y = a(x - h)^2 + k$ — Vertex form

$-2 = a(-1 - 3)^2 + 2$ — Substitute −2 for $y$, −1 for $x$, 3 for $h$ and 2 for $k$.

$-2 = a(16) + 2$ — Simplify.

$-4 = 16a$ — Subtract 2 from each side.

$-\frac{1}{4} = a$ — Divide each side by 16.

The equation of the parabola in vertex form is $y = -\frac{1}{4}(x - 3)^2 + 2$.

The answer is B.

### GuidedPractice

2. Which is an equation of the function shown in the graph?

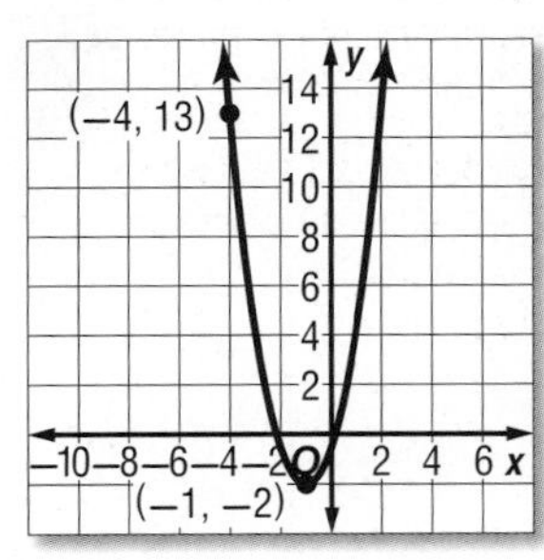


**F** $y = \frac{9}{25}(x - 1)^2 + 2$

**G** $y = \frac{3}{5}(x + 1)^2 - 2$

**H** $y = \frac{5}{3}(x + 1)^2 - 2$

**J** $y = \frac{25}{9}(x - 1)^2 + 2$

## 2 Transformations of Quadratic Graphs

In Lesson 2-7, you learned how different transformations affect the graphs of parent functions. The following summarizes these transformations for quadratic functions.

## ConceptSummary Transformations of Quadratic Functions

$f(x) = a(x - h)^2 + k$

**$h$, Horizontal Translation**
$h$ units to the right if $h$ is positive
$|h|$ units to the left if $h$ is negative

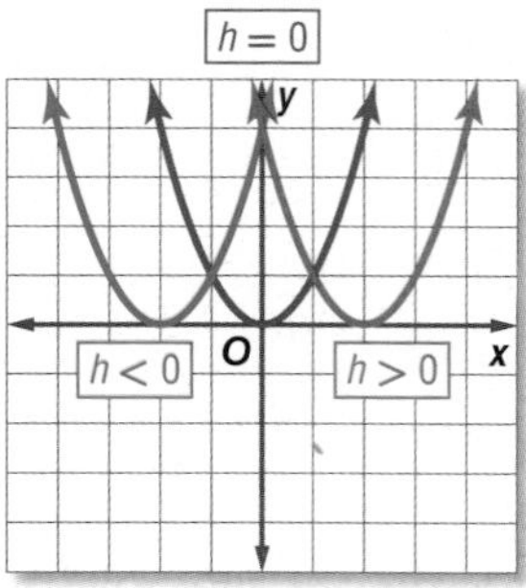


**$k$, Vertical Translation**
$k$ units up if $k$ is positive
$|k|$ units down if $k$ is negative

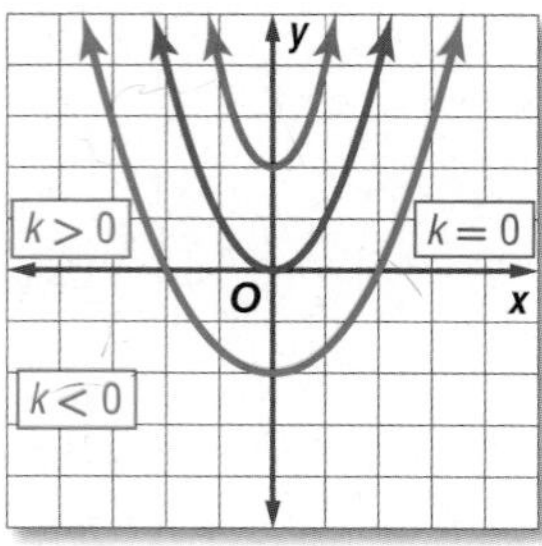


**StudyTip**
Absolute Value
$0 < |a| < 1$ means that $a$ is a rational number between 0 and 1, such as $\frac{3}{4}$, or a rational number between −1 and 0, such as −0.3.

**$a$, Reflection**
If $a > 0$, the graph opens up.
If $a < 0$, the graph opens down.

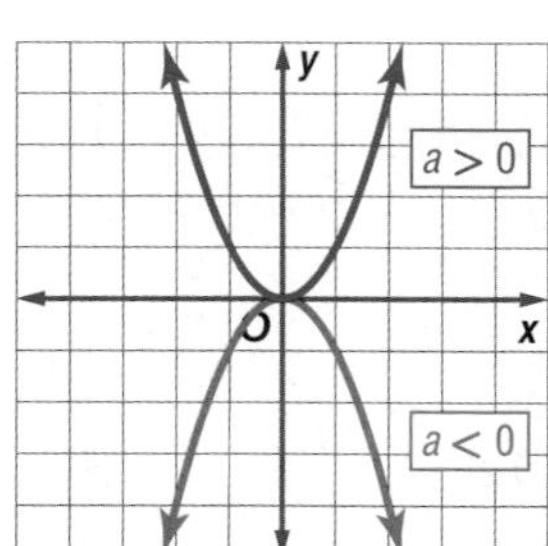


**$a$, Dilation**
**If $|a| > 1$, the graph is stretched vertically. If $0 < |a| < 1$, the graph is compressed vertically.**

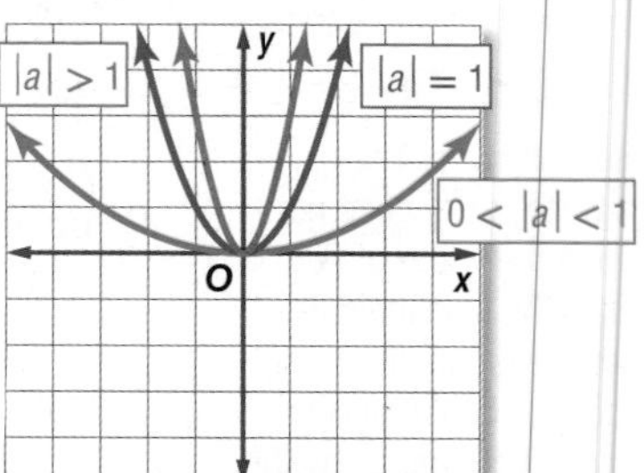

### Example 3 Graph Equations in Vertex Form

**Graph $y = 4x^2 - 16x - 40$.**

**Step 1** Rewrite the equation in vertex form.

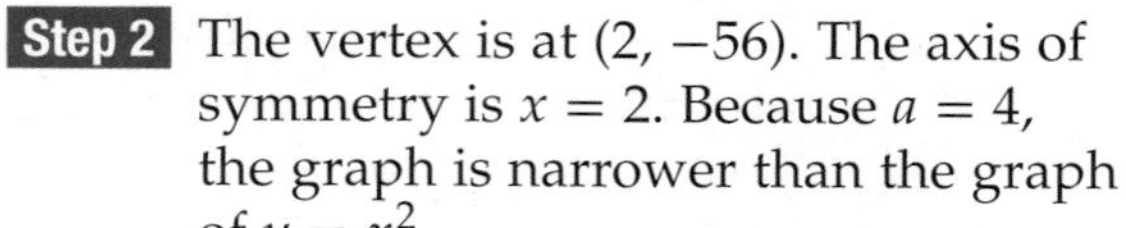

$y = 4x^2 - 16x - 40$ Original equation

$y = 4(x^2 - 4x) - 40$ Distributive Property

$y = 4(x - 4x + 4) - 40 - 4(4)$ Complete the square.

$y = 4(x - 2)^2 - 56$ Simplify.

**Step 2** The vertex is at (2, −56). The axis of symmetry is $x = 2$. Because $a = 4$, the graph is narrower than the graph of $y = x^2$.

**Step 3** Plot additional points to help you complete the graph.

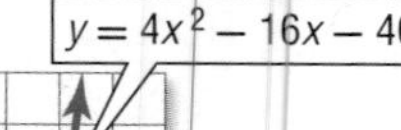


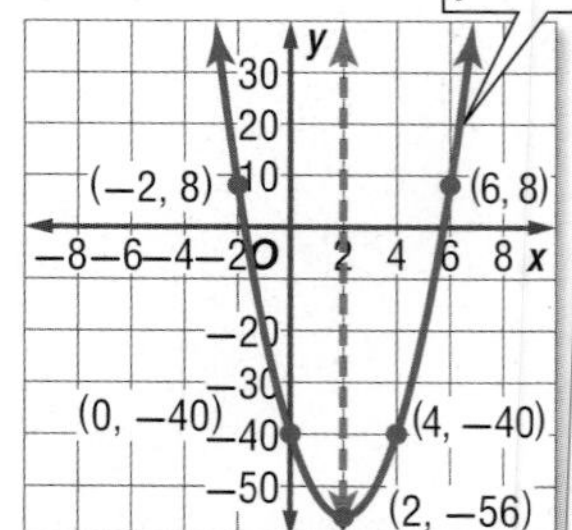


**GuidedPractice**

**3A.** $y = (x - 3)^2 - 2$

**3B.** $y = 0.25(x + 1)^2$

## Check Your Understanding

**Example 1** **Write each function in vertex form.**

1. $y = x^2 + 6x + 2$
2. $y = -2x^2 + 8x - 5$
3. $y = 4x^2 + 24x + 24$

**Example 2**

4. **MULTIPLE CHOICE** Which function is shown in the graph?

   A $y = -(x + 3)^2 + 6$

   B $y = -(x - 3)^2 - 6$

   C $y = -2(x + 3)^2 + 6$

   D $y = -2(x - 3)^2 - 6$

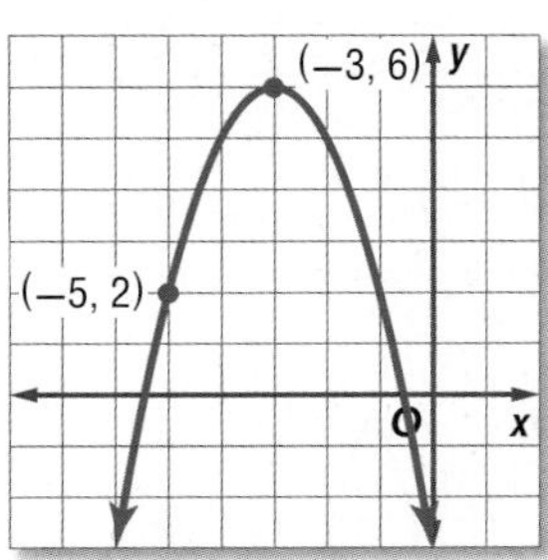


**Example 3** **Graph each function.**

5. $y = (x - 3)^2 - 4$
6. $y = -2x^2 + 5$
7. $y = \frac{1}{2}(x + 6)^2 - 8$

## Practice and Problem Solving

**Example 1** **Write each function in vertex form.**

8. $y = x^2 + 9x + 8$
9. $y = x^2 - 6x + 3$
10. $y = -2x^2 + 5x$
11. $y = x^2 + 2x + 7$
12. $y = -3x^2 + 12x - 10$
13. $y = x^2 + 8x + 16$
14. $y = 2x^2 - 4x - 3$
15. $y = 3x^2 + 10x$
16. $y = x^2 - 4x + 9$
17. $y = -4x^2 - 24x - 15$
18. $y = x^2 - 12x + 36$
19. $y = -x^2 - 4x - 1$

**Example 2**

20. **FIREWORKS** During an Independence Day fireworks show, the height $h$ in meters of a specific rocket after $t$ seconds can be modeled by $h = -4.9(t - 4)^2 + 80$. Graph the function.

21. **FINANCIAL LITERACY** A bicycle rental shop rents an average of 120 bicycles per week and charges $25 per day. The manager estimates that there will be 15 additional bicycles rented for each $1 reduction in the rental price. The maximum income the manager can expect can be modeled by $y = -15x^2 + 255x + 3000$, where $y$ is the weekly income and $x$ is the number of bicycles rented. Write this function in vertex form. Then graph.

**Example 3** **Graph each function.**

22. $y = (x - 5)^2 + 3$
23. $y = 9x^2 - 8$
24. $y = -2(x - 5)^2$
25. $y = \frac{1}{10}(x + 6)^2 + 6$
26. $y = -3(x - 5)^2 - 2$
27. $y = -\frac{1}{4}x^2 - 5$
28. $y = 2x^2 + 10$
29. $y = -(x + 3)^2$
30. $y = \frac{1}{6}(x - 3)^2 - 10$
31. $y = (x - 9)^2 - 7$
32. $y = -\frac{5}{8}x^2 - 8$
33. $y = -4(x - 10)^2 - 10$

34. **CCSS MODELING** A sailboard manufacturer uses an automated process to manufacture the masts for its sailboards. The function $f(x) = \frac{1}{250}x^2 + \frac{3}{5}x$ is programmed into a computer to make one such mast.

    a. Write the quadratic function in vertex form. Then graph the function.

    b. Describe how the manufacturer can adjust the function to make its masts with a greater or smaller curve.

**Write an equation in vertex form for each parabola.**

35. 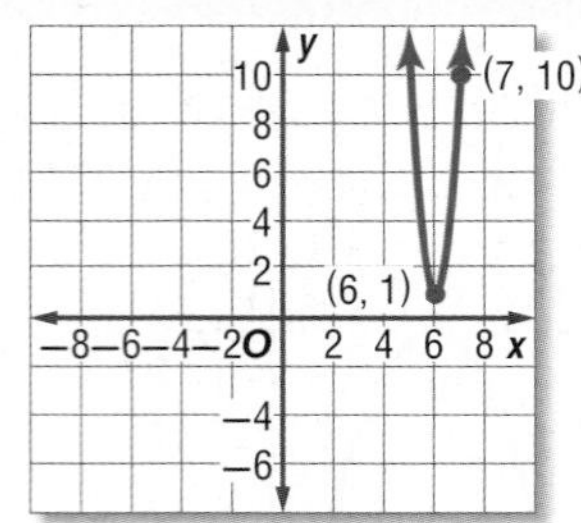


36. 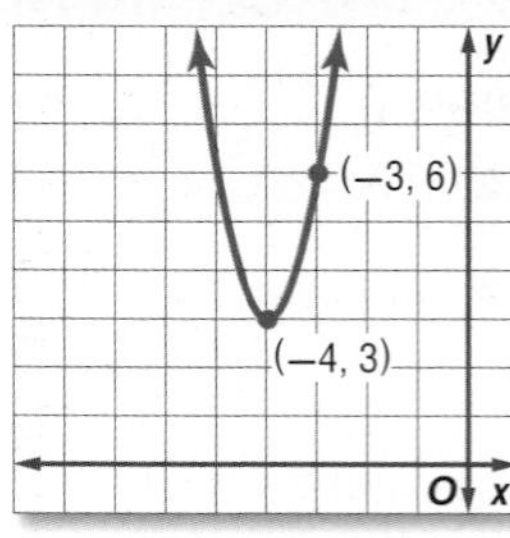


37. 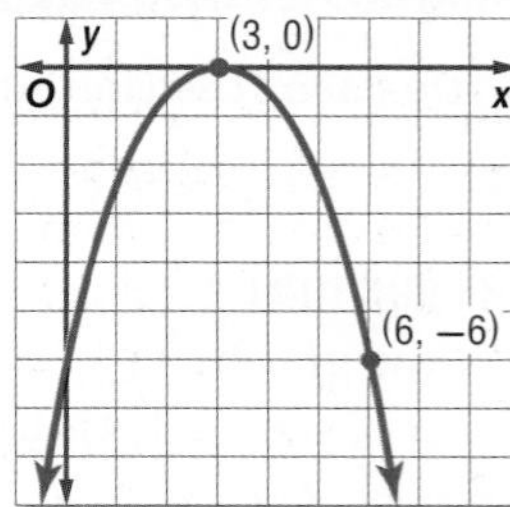


38. 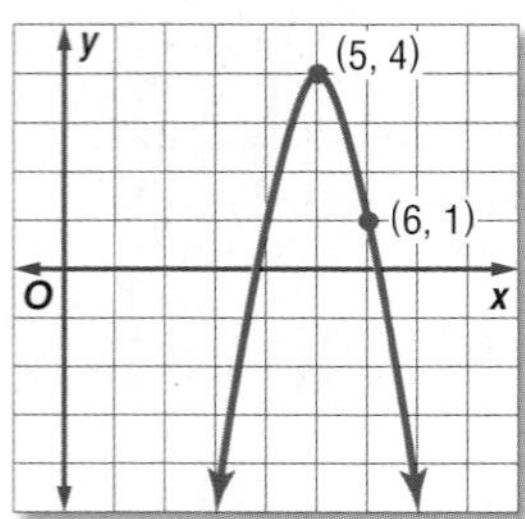


39. 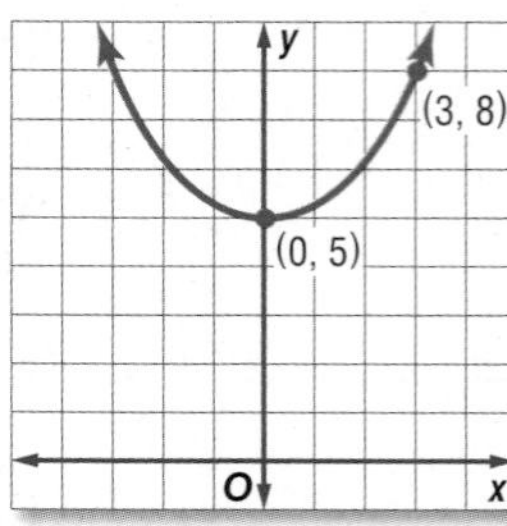


40. 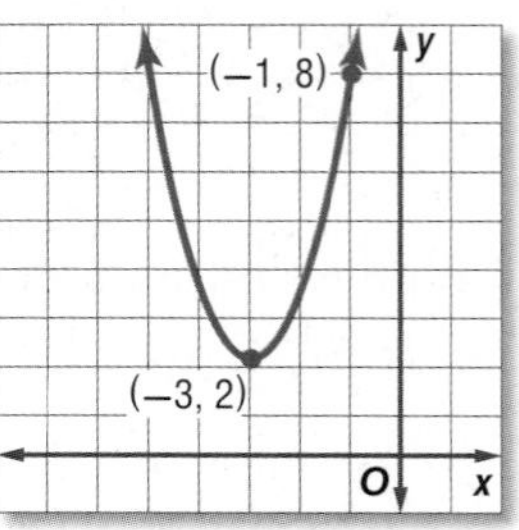


**Write each function in vertex form. Then identify the vertex, axis of symmetry, and direction of opening.**

41. $3x^2 - 4x = 2 + y$

42. $-2x^2 + 7x = y - 12$

43. $-x^2 - 4.7x = y - 2.8$

44. $x^2 + 1.4x - 1.2 = y$

45. $x^2 - \frac{2}{3}x - \frac{26}{9} = y$

46. $x^2 + 7x + \frac{49}{4} = y$

47. **CARS** The formula $S(t) = \frac{1}{2}at^2 + v_0t$ can be used to determine the position $S(t)$ of an object after $t$ seconds at a rate of acceleration $a$ with initial velocity $v_0$. Valerie's car can accelerate 0.002 miles per second squared.

   a. Express $S(t)$ in vertex form as she accelerates from 35 miles per hour to enter highway traffic.

   b. How long will it take Valerie to match the average speed of highway traffic of 68 miles per hour? (*Hint*: Use *acceleration* • *time* = *velocity*.)

   c. If the entrance ramp is $\frac{1}{8}$-mile long, will Valerie have sufficient time to match the average highway speed? Explain.

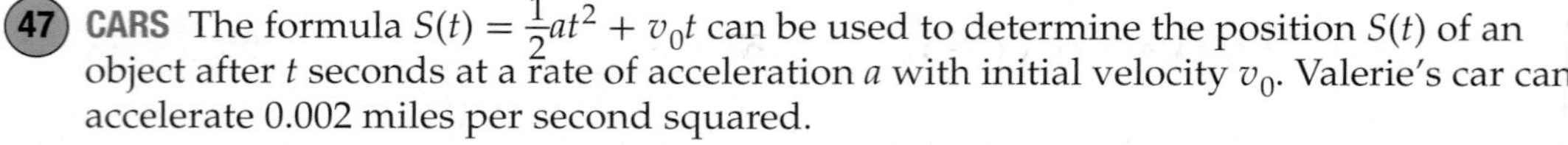

## H.O.T. Problems Use Higher-Order Thinking Skills

48. **OPEN ENDED** Write an equation for a parabola that has been translated, compressed, and reflected in the $x$-axis.

49. **CHALLENGE** Explain how you can find an equation of a parabola using the coordinates of three points on the graph.

50. **CHALLENGE** Write the standard form of a quadratic function $ax^2 + bx + c = y$ in vertex form. Identify the vertex and the axis of symmetry.

51. **REASONING** Describe the graph of $f(x) = a(x - h)^2 + k$ when $a = 0$. Is the graph the same as that of $g(x) = ax^2 + bx + c$ when $a = 0$? Explain.

52. CCSS **ARGUMENTS** Explain how the graph of $y = x^2$ can be used to graph any quadratic function. Include a description of the effects produced by changing $a$, $h$, and $k$ in the equation $y = a(x - h)^2 + k$, and a comparison of the graph of $y = x^2$ and the graph of $y = a(x - h)^2 + k$ using values you choose for $a$, $h$, and $k$.

## Standardized Test Practice

**53.** Flowering bushes need a mixture of 70% soil and 30% vermiculite. About how many buckets of vermiculite should you add to 20 buckets of soil?

A 6.0
B 8.0
C 14.0
D 24.0

**54.** **SAT/ACT** The sum of the integers $x$ and $y$ is 495. The units digit of $x$ is 0. If $x$ is divided by 10, the result is equal to $y$. What is the value of $x$?

F 40
G 45
H 245
J 250
K 450

**55.** What is the solution set of the inequality $|4x - 1| < 9$?

A $\{x \mid 2.5 < x \text{ or } x < -2\}$
B $\{x \mid x < 2.5\}$
C $\{x \mid x > -2\}$
D $\{x \mid -2 < x < 2.5\}$

**56.** **SHORT RESPONSE** At your store, you buy wrenches for \$30.00 a dozen and sell them for \$3.50 each. What is the percent markup for the wrenches?

## Spiral Review

**Solve each equation by using the method of your choice. Find exact solutions.** (Lesson 3-4)

**57.** $4x^2 + 15x = 21$

**58.** $-3x^2 + 19 = 5x$

**59.** $6x - 5x^2 + 9 = 3$

**Simplify.** (Lesson 3-3)

**60.** $(3 + 4i)(5 - 2i)$

**61.** $(\sqrt{6} + i)(\sqrt{6} - i)$

**62.** $\frac{1 + i}{1 - i}$

**63.** $\frac{4 - 3i}{1 + 2i}$

**64.** **FOUNTAINS** The height of a fountain's water stream can be modeled by a quadratic function. Suppose the water from a jet reaches a maximum height of 8 feet at a distance 1 foot away from the jet. (Lesson 3-5)

**a.** If the water lands 3 feet away from the jet, find a quadratic function that models the height $h(d)$ of the water at any given distance $d$ feet from the jet. Then compare the graph of the function to the parent function.

**b.** Suppose a worker increases the water pressure so that the stream reaches a maximum height of 12.5 feet at a distance of 15 inches from the jet. The water now lands 3.75 feet from the jet. Write a new quadratic function for $h(d)$. How do the changes in $h$ and $k$ affect the shape of the graph?

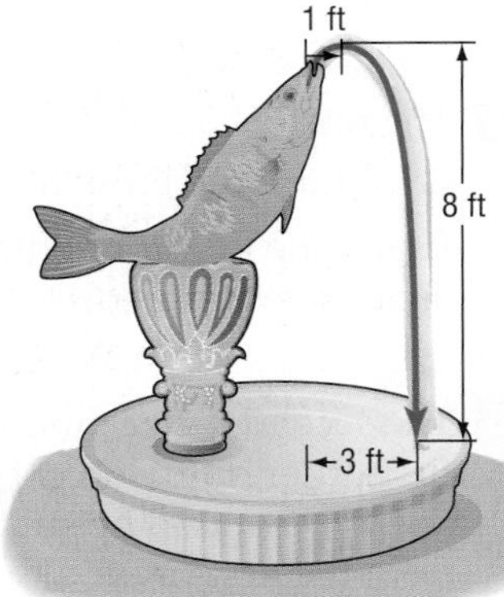


## Skills Review

**Determine whether the given value satisfies the inequality.**

**65.** $3x^2 - 5 > 6; x = 2$

**66.** $-2x^2 + x - 1 < 4; x = -2$

**67.** $4x^2 + x - 3 \leq 36; x = 3$

EXTEND 3-4

# Algebra Lab
# Quadratics and Rate of Change

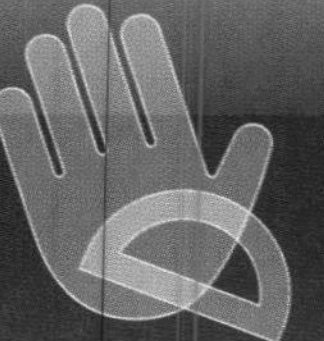

You have learned that a linear function has a constant rate of change. In this lab, you will investigate the rate of change for quadratic functions.

**CCSS Common Core State Standards**
**Content Standards**
**F.IF.4** For a function that models a relationship between two quantities, interpret key features of graphs and tables in terms of the quantities, and sketch graphs showing key features given a verbal description of the relationship.
**F.IF.6** Calculate and interpret the average rate of change of a function (presented symbolically or as a table) over a specified interval. Estimate the rate of change from a graph.

## Activity Determine Rate of Change

**Consider $f(x) = 0.1875x^2 - 3x + 12$.**

**Step 1** Make a table like the one below. Use values from 0 through 16 for $x$.

| $x$ | 0 | 1 | 2 | 3 | ... | 16 |
|---|---|---|---|---|---|---|
| $y$ | 12 | 9.1875 | 6.75 | | | |
| First-Order Differences | | | | | | |
| Second-Order Differences | | | | | | |

**Step 2** Find each $y$-value. For example, when $x = 1$, $y = 0.1875(1)^2 - 3(1) + 12$ or 9.1875.

**Step 3** Graph the ordered pairs $(x, y)$. Then connect the points with a smooth curve. Notice that the function *decreases* when $0 < x < 8$ and *increases* when $8 < x < 16$.

**Step 4** The rate of change from one point to the next can be found by using the slope formula. From (0, 12) to (1, 9.1875), the slope is $\frac{9.1875 - 12}{1 - 0}$ or $-2.8125$. This is the first-order difference at $x = 1$. Complete the table for all the first-order differences. Describe any patterns in the differences.

**Step 5** The second-order differences can be found by subtracting consecutive first-order differences. For example, the second-order difference at $x = 2$ is found by subtracting the first order difference at $x = 1$ from the first-order difference at $x = 2$. Describe any patterns in the differences.

## Exercises

**For each function make a table of values for the given $x$-values. Graph the function. Then determine the first-order and second-order differences.**

1. $y = -x^2 + 2x - 1$ for $x = -3, -2, -1, 0, 1, 2, 3$

2. $y = 0.5x^2 + 2x - 2$ for $x = -5, -4, -3, -2, -1, 0, 1$

3. $y = -3x^2 - 18x - 26$ for $x = -6, -5, -4, -3, -2, -1, 0$

4. **MAKE A CONJECTURE** Repeat the activity for a cubic function. At what order difference would you expect $g(x) = x^4$ to be constant? $h(x) = x^n$?

LESSON 3-5

# Quadratic Inequalities

| :: Then | :: Now | :: Why? |
|---|---|---|
| You solved linear inequalities. | **1** Graph quadratic inequalities in two variables. **2** Solve quadratic inequalities in one variable. | A water balloon launched from a slingshot can be represented by several different quadratic equations and inequalities. Suppose the height of a water balloon $h(t)$ in meters above the ground $t$ seconds after being launched is modeled by the quadratic function $h(t) = -4.9t^2 + 32t + 1.2$. You can solve a quadratic inequality to determine how long the balloon will be a certain distance above the ground. |

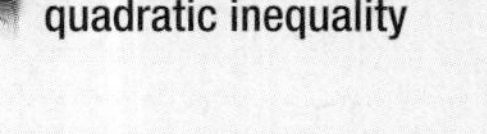

**NewVocabulary**
quadratic inequality

**Common Core State Standards**

**Content Standards**

**A.CED.1** Create equations and inequalities in one variable and use them to solve problems.

**A.CED.3** Represent constraints by equations or inequalities, and by systems of equations and/or inequalities, and interpret solutions as viable or nonviable options in a modeling context.

**Mathematical Practices**

1 Make sense of problems and persevere in solving them.

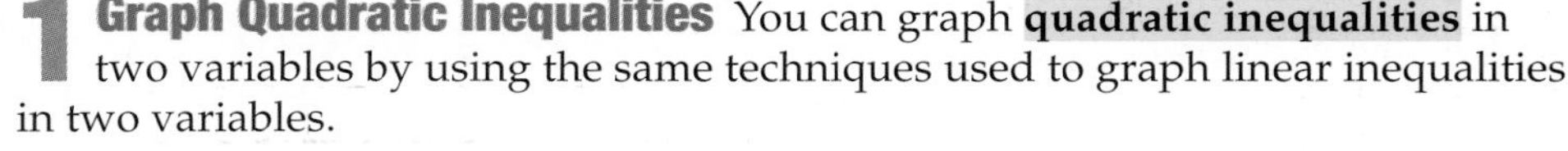

## 1 Graph Quadratic Inequalities

You can graph **quadratic inequalities** in two variables by using the same techniques used to graph linear inequalities in two variables.

**Step 1** Graph the related function.

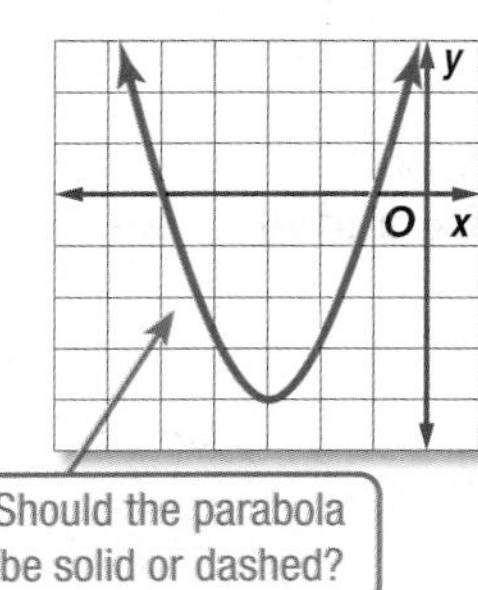


**Step 2** Test a point not on the parabola.

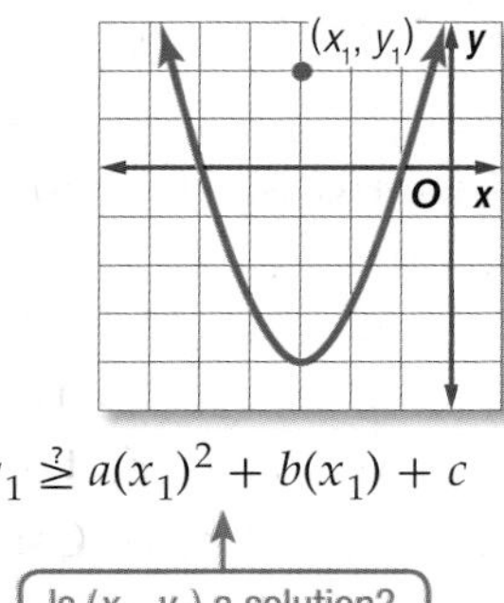


$y_1 \overset{?}{\geq} a(x_1)^2 + b(x_1) + c$

Is $(x_1, y_1)$ a solution?

**Step 3** Shade accordingly.

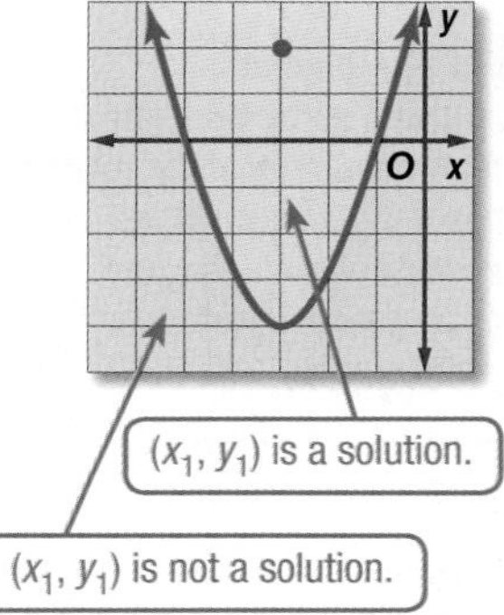

### Example 1 Graph a Quadratic Inequality

**Graph $y > x^2 + 2x + 1$.**

**Step 1** Graph the related function, $y = x^2 + 2x + 1$. The parabola should be dashed.

**Step 2** Test a point not on the graph of the parabola.

$y > x^2 + 2x + 1$

$-1 \overset{?}{>} 0^2 + 2(0) + 1$

$-1 \not> 1$ So, $(0, -1)$ is *not* a solution of the inequality.

**Step 3** Shade the region that does not contain the point $(0, -1)$.

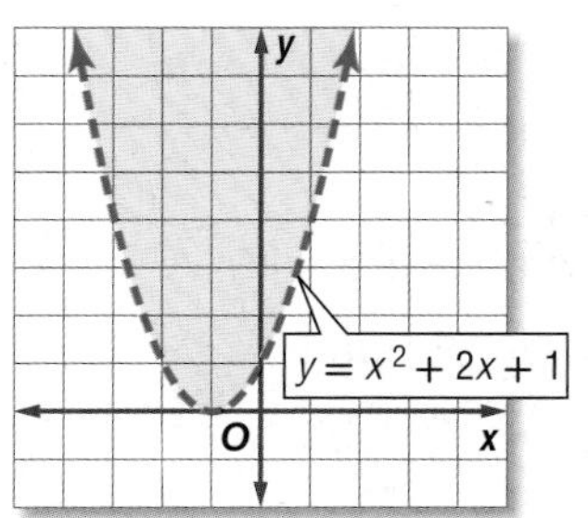


**GuidedPractice**

**Graph each inequality.**

**1A.** $y \leq x^2 + 2x + 4$

**1B.** $y < -2x^2 + 3x + 5$

madelyn mulvaney/Flickr/Getty Images

## 2 Solve Quadratic Inequalities

Quadratic inequalities in one variable can be solved using the graphs of the related quadratic functions.

**$ax^2 + bx + c < 0$**

Graph $y = ax^2 + bx + c$ and identify the $x$-values for which the graph lies *below* the $x$-axis.

For $\leq$, include the $x$-intercepts in the solution.

$a > 0$

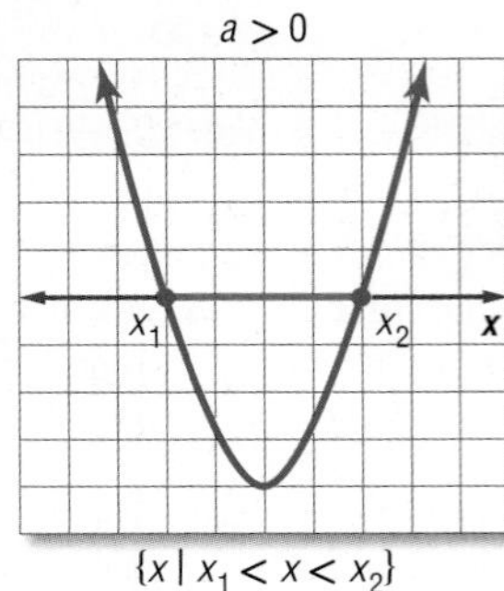


$\{x \mid x_1 < x < x_2\}$

$a < 0$

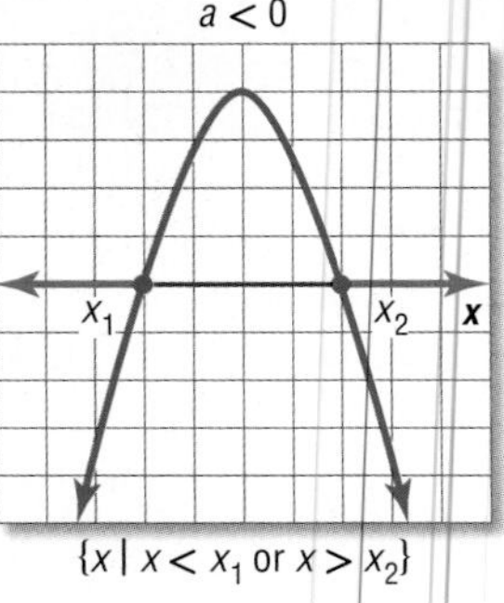


$\{x \mid x < x_1 \text{ or } x > x_2\}$

**$ax^2 + bx + c > 0$**

Graph $y = ax^2 + bx + c$ and identify the $x$-values for which the graph lies *above* the $x$-axis.

For $\geq$, include the $x$-intercepts in the solution.

$a > 0$

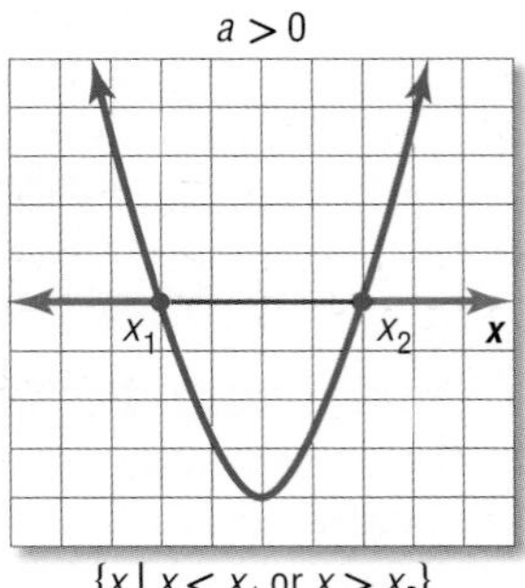


$\{x \mid x < x_1 \text{ or } x > x_2\}$

$a < 0$

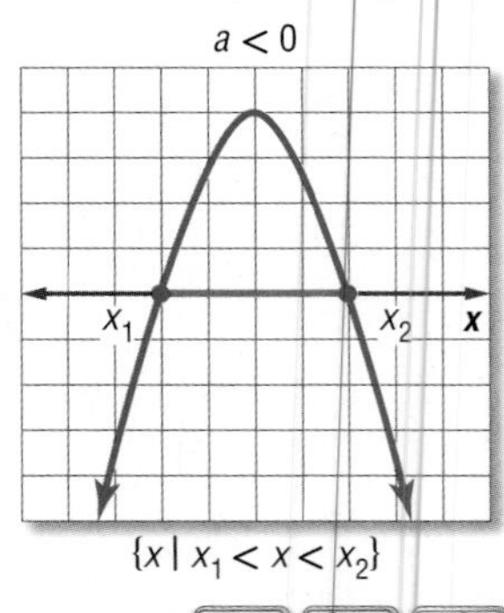


$\{x \mid x_1 < x < x_2\}$

**StudyTip**

**Solving Quadratic Inequalities by Graphing** A precise graph of the related quadratic function is not necessary since the zeros of the function were found algebraically.

### Example 2 Solve $ax^2 + bx + c < 0$ by Graphing

**Solve $x^2 + 2x - 8 < 0$ by graphing.**

The solution consists of $x$-values for which the graph of the related function lies *below* the $x$-axis. Begin by finding the roots of the related function.

| | |
|---|---|
| $x^2 + 2x - 8 = 0$ | Related equation |
| $(x - 2)(x + 4) = 0$ | Factor. |
| $x - 2 = 0$ or $x + 4 = 0$ | Zero Product Property |
| $x = 2$ $\quad$ $x = -4$ | Solve each equation. |

Sketch the graph of a parabola that has $x$-intercepts at $-4$ and 2. The graph should open up because $a > 0$.

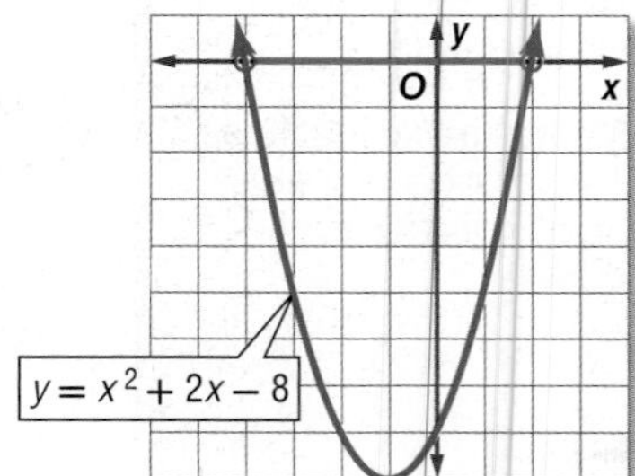


The graph lies below the $x$-axis between $x = -4$ and $x = 2$. Thus, the solution set is $\{x \mid -4 < x < 2\}$ or $(-4, 2)$.

**CHECK** Test one value of $x$ less than $-4$, one between $-4$ and 2, and one greater than 2 in the original inequality.

| Test $x = -6$. | Test $x = 0$. | Test $x = 5$. |
|---|---|---|
| $x^2 + 2x - 8 < 0$ | $x^2 + 2x - 8 < 0$ | $x^2 + 2x - 8 < 0$ |
| $(-6)^2 + 2(-6) - 8 \stackrel{?}{<} 0$ | $0^2 + 2(0) - 8 \stackrel{?}{<} 0$ | $5^2 + 2(5) - 8 \stackrel{?}{<} 0$ |
| $16 < 0$ ✗ | $-8 < 0$ ✓ | $27 < 0$ ✗ |

**GuidedPractice**

**Solve each inequality by graphing.**

**2A.** $0 > x^2 + 5x - 6$

**2B.** $-x^2 + 3x + 10 \leq 0$

## Example 3 Solve $ax^2 + bx + c \geq 0$ by Graphing

**Solve $2x^2 + 4x - 5 \geq 0$ by graphing.**

The solution consists of $x$-values for which the graph of the related function lies *on and above* the $x$-axis. Begin by finding the roots of the related function.

| | |
|---|---|
| $2x^2 + 4x - 5 = 0$ | Related equation |
| $x = \dfrac{-b \pm \sqrt{b^2 - 4ac}}{2a}$ | Use the Quadratic Formula |
| $x = \dfrac{-4 \pm \sqrt{4^2 - 4(2)(-5)}}{2(2)}$ | Replace $a$ with 4, $b$ with 2, and $c$ with −5. |
| $x = \dfrac{-4 + \sqrt{56}}{4}$ or $x = \dfrac{-4 - \sqrt{56}}{4}$ | Simplify and write as two equations. |
| $\approx 0.87$ $\qquad\approx -2.87$ | Simplify. |

Sketch the graph of a parabola with $x$-intercepts at $-2.87$ and $0.87$. The graph opens up since $a > 0$. The graph lies on and above the $x$-axis at about $x \leq -2.87$ and $x \geq 0.87$. Therefore, the solution is approximately $\{x \mid x \leq -2.87 \text{ or } x \geq 0.87\}$ or $(-\infty, -2.87] \cup [0.87, \infty)$.

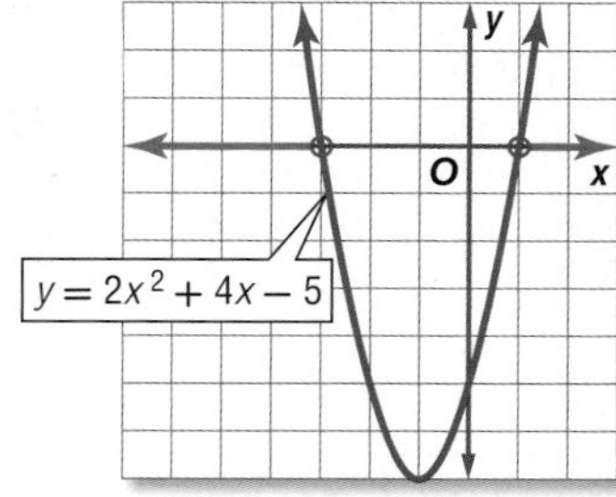


### GuidedPractice

**Solve each inequality by graphing.**

**3A.** $x^2 - 6x + 2 > 0$

**3B.** $-4x^2 + 5x + 7 \geq 0$

Real-world problems can be solved by graphing quadratic inequalities.

### Real-WorldLink

It takes just milliseconds for a water balloon to break. A high-speed camera can capture the impact on the fluid before gravity makes it fall.

**Source:** NASA

## Real-World Example 4 Solve a Quadratic Inequality

**WATER BALLOONS Refer to the beginning of the lesson. At what time will a water balloon be within 3 meters of the ground after it has been launched?**

The function $h(t) = -4.9t^2 + 32t + 1.2$ describes the height of the water balloon. Therefore, you want to find the values of $t$ for which $h(t) \leq 3$.

| | |
|---|---|
| $h(t) \leq 3$ | Original inequality |
| $-4.9t^2 + 32t + 1.2 \leq 3$ | $h(t) = -4.9t^2 + 32t + 1.2$ |
| $-4.9t^2 + 32t - 1.8 \leq 0$ | Subtract 3 from each side. |

Graph the related function $y = -4.9x^2 + 32x - 1.8$ using a graphing calculator. The zeros of the function are about 0.06 and 6.47, and the graph lies below the $x$-axis when $x < 0.06$ and $x > 6.47$.

So, the water balloon is within 3 meters of the ground during the first 0.06 second after being launched and again after about 6.47 seconds until it hits the ground.

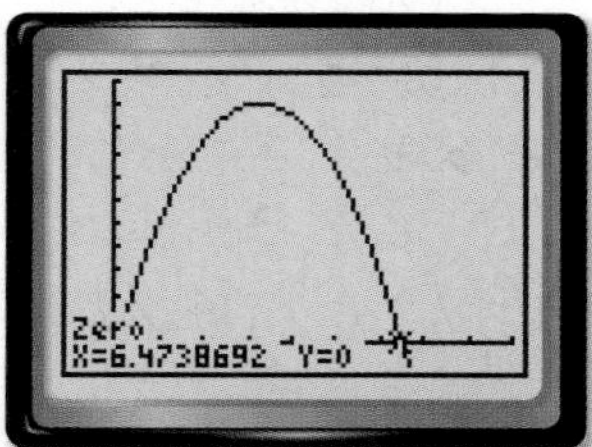


[−1, 9] scl: 1 by [−5, 55] scl: 5

### GuidedPractice

**4. ROCKETS** The height $h(t)$ of a model rocket in feet $t$ seconds after its launch can be represented by the function $h(t) = -16t^2 + 82t + 0.25$. During what interval is the rocket at least 100 feet above the ground?

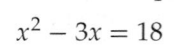

**StudyTip**

**Solving Quadratic Inequalities Algebraically** The solution set of a quadratic inequality is all real numbers when all three test points satisfy the inequality. It is the empty set when none of the test points satisfy the inequality.

## Example 5 Solve a Quadratic Inequality Algebraically

**Solve $x^2 - 3x \leq 18$ algebraically.**

**Step 1** Solve the related quadratic equation $x^2 - 3x = 18$.

| | |
|---|---|
| $x^2 - 3x = 18$ | Related quadratic equation |
| $x^2 - 3x - 18 = 0$ | Subtract 18 from each side. |
| $(x + 3)(x - 6) = 0$ | Factor. |
| $x + 3 = 0$ or $x - 6 = 0$ | Zero Product Property |
| $x = -3$ $\quad$ $x = 6$ | Solve each equation. |

**Step 2** Plot −3 and 6 on a number line. Use dots since these values are solutions of the original inequality. Notice that the number line is divided into three intervals.

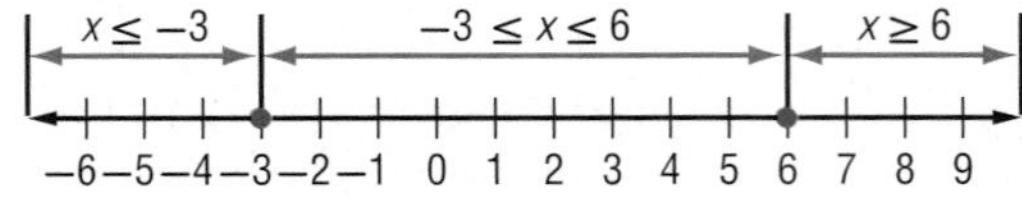


**Step 3** Test a value from each interval to see if it satisfies the original inequality.

| $x \leq -3$ | $-3 \leq x \leq 6$ | $x \geq 6$ |
|---|---|---|
| Test $x = -5$. | Test $x = 0$. | Test $x = 8$. |
| $x^2 - 3x \leq 18$ | $x^2 - 3x \leq 18$ | $x^2 - 3x \leq 18$ |
| $(-5)^2 - 3(-5) \overset{?}{\leq} 18$ | $(0)^2 - 3(0) \overset{?}{\leq} 18$ | $(8)^2 - 3(8) \overset{?}{\leq} 18$ |
| $40 \nleq 18$ | $0 \leq 18$ | $40 \nleq 18$ |

The solution set is $\{x \mid -3 \leq x \leq 6\}$ or $[-3, 6]$.

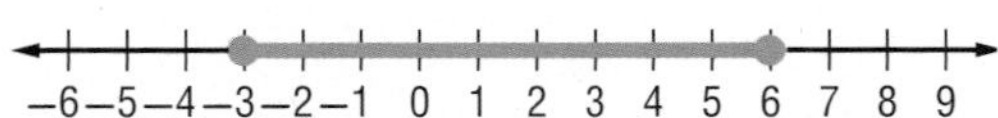


**GuidedPractice**

**Solve each inequality algebraically.**

**5A.** $x^2 + 5x < -6$

**5B.** $x^2 + 11x + 30 \geq 0$

## Check Your Understanding

**Example 1** **Graph each inequality.**

**1.** $y \leq x^2 - 8x + 2$

**2.** $y > x^2 + 6x - 2$

**3.** $y \geq -x^2 + 4x + 1$

**Examples 2–3** 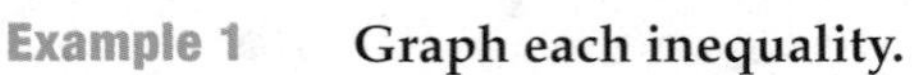 **SENSE-MAKING** **Solve each inequality by graphing.**

**4.** $0 < x^2 - 5x + 4$

**5.** $x^2 + 8x + 15 < 0$

**6.** $-2x^2 - 2x + 12 \geq 0$

**7.** $0 \geq 2x^2 - 4x + 1$

**Example 4**

**8.** **SOCCER** A midfielder kicks a ball toward the goal during a match. The height of the ball in feet above the ground $h(t)$ at time $t$ can be represented by $h(t) = -0.1t^2 + 2.4t + 1.5$. If the height of the goal is 8 feet, at what time during the kick will the ball be able to enter the goal?

**Example 5** **Solve each inequality algebraically.**

**9.** $x^2 + 6x - 16 < 0$

**10.** $x^2 - 14x > -49$

**11** $-x^2 + 12x \geq 28$

**12.** $x^2 - 4x \leq 21$

## Practice and Problem Solving

**Example 1** **Graph each inequality.**

**13.** $y \geq x^2 + 5x + 6$
**14.** $x^2 - 2x - 8 < y$
**15.** $y \leq -x^2 - 7x + 8$
**16.** $-x^2 + 12x - 36 > y$
**17.** $y > 2x^2 - 2x - 3$
**18.** $y \geq -4x^2 + 12x - 7$

**Examples 2–3** **Solve each inequality by graphing.**

**19.** $x^2 - 9x + 9 < 0$
**20.** $x^2 - 2x - 24 \leq 0$
**21.** $x^2 + 8x + 16 \geq 0$
**22.** $x^2 + 6x + 3 > 0$
**23.** $0 > -x^2 + 7x + 12$
**24.** $-x^2 + 2x - 15 < 0$
**25.** $4x^2 + 12x + 10 \leq 0$
**26.** $-3x^2 - 3x + 9 > 0$
**27.** $0 > -2x^2 + 4x + 4$
**28.** $3x^2 + 12x + 36 \leq 0$
**29.** $0 \leq -4x^2 + 8x + 5$
**30.** $-2x^2 + 3x + 3 \leq 0$

**Example 4** **31.** **ARCHITECTURE** An arched entry of a room is shaped like a parabola that can be represented by the equation $f(x) = -x^2 + 6x + 1$. How far from the sides of the arch is its height at least 7 feet?

**32.** **MANUFACTURING** A box is formed by cutting 4-inch squares from each corner of a square piece of cardboard and then folding the sides. If $V(x) = 4x^2 - 64x + 256$ represents the volume of the box, what should the dimensions of the original piece of cardboard be if the volume of the box cannot exceed 750 cubic inches?

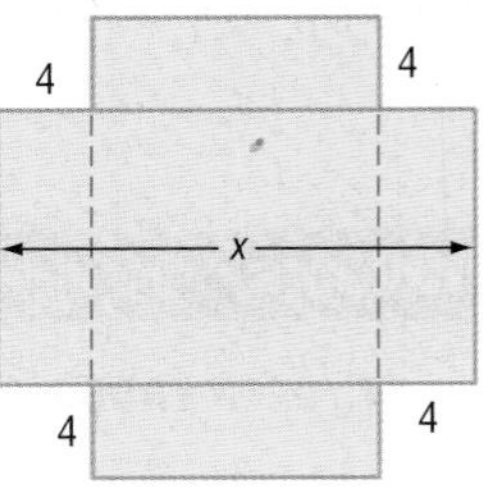


**Example 5** **Solve each inequality algebraically.**

**33.** $x^2 - 9x < -20$
**34.** $x^2 + 7x \geq -10$
**35.** $2 > x^2 - x$
**36.** $-3 \leq -x^2 - 4x$
**37.** $-x^2 + 2x \leq -10$
**38.** $-6 > x^2 + 4x$
**39.** $2x^2 + 4 \geq 9$
**40.** $3x^2 + x \geq -3$
**41.** $-4x^2 + 2x < 3$
**42.** $-11 \geq -2x^2 - 5x$
**43.** $-12 < -5x^2 - 10x$
**44.** $-3x^2 - 10x > -1$

**45.** **CCSS PERSEVERANCE** The Sanchez family is adding a deck along two sides of their swimming pool. The deck width will be the same on both sides and the total area of the pool and deck cannot exceed 750 square feet.

**a.** Graph the quadratic inequality.

**b.** Determine the possible widths of the deck.

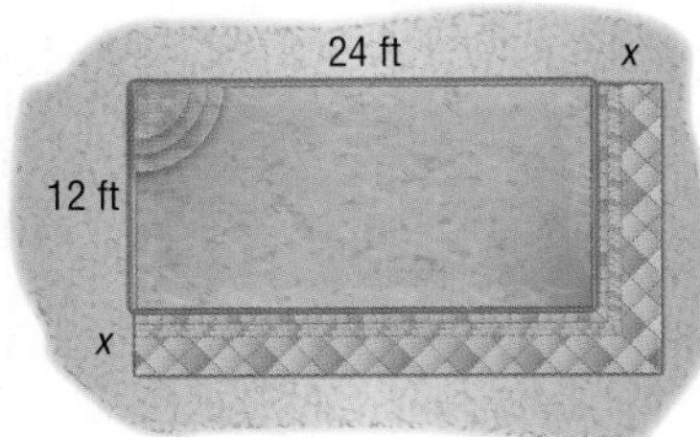


**Write a quadratic inequality for each graph.**

**46.**

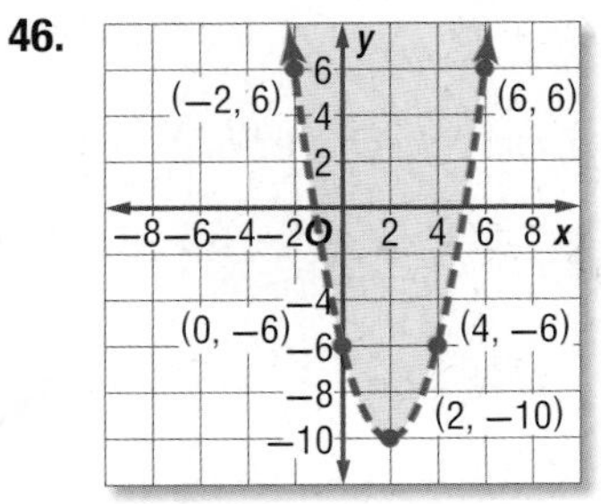


**47.**

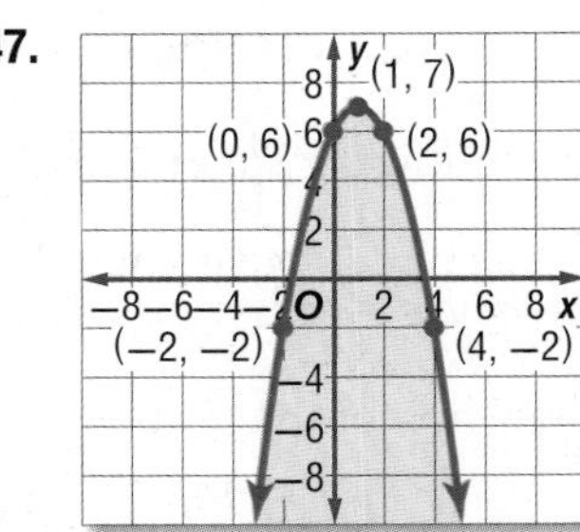


**48.**

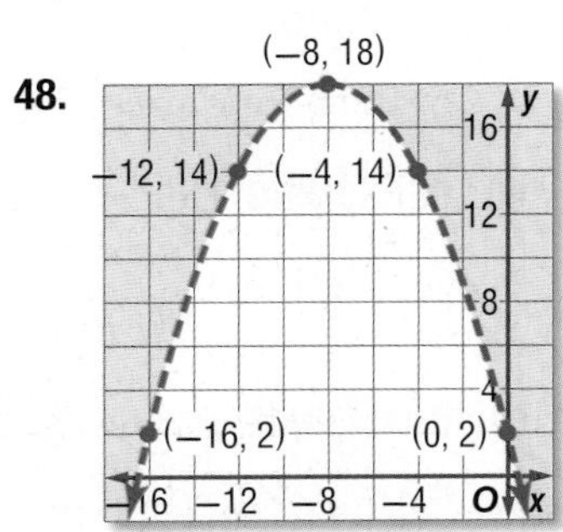

**Solve each quadratic inequality by using a graph, a table, or algebraically.**

**49.** $-2x^2 + 12x < -15$

**50.** $5x^2 + x + 3 \geq 0$

**51** $11 \leq 4x^2 + 7x$

**52.** $x^2 - 4x \leq -7$

**53.** $-3x^2 + 10x < 5$

**54.** $-1 \geq -x^2 - 5x$

**55. BUSINESS** An electronics manufacturer uses the function $P(x) = x(-27.5x + 3520) + 20{,}000$ to model their monthly profits when selling $x$ thousand digital audio players.

**a.** Graph the quadratic inequality for a monthly profit of at least $100,000.

**b.** How many digital audio players must the manufacturer sell to earn a profit of at least $100,000 in a month?

**c.** Suppose the manufacturer has an additional monthly expense of $25,000. Explain how this affects the graph of the profit function. Then determine how many digital audio players the manufacturer needs to sell to have at least $100,000 in profits.

**56. UTILITIES** A contractor is installing drain pipes for a shopping center's parking lot. The outer diameter of the pipe is to be 10 inches. The cross sectional area of the pipe must be at least 35 square inches and should not be more than 42 square inches.

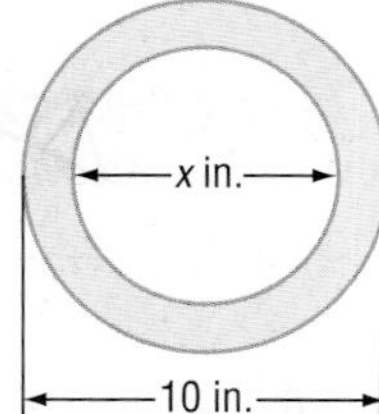


**a.** Graph the quadratic inequalities.

**b.** What thickness of drain pipe can the contractor use?

## H.O.T. Problems Use Higher-Order Thinking Skills

**57. OPEN ENDED** Write a quadratic inequality for each condition.

**a.** The solution set is all real numbers.

**b.** The solution set is the empty set.

**58. CCSS CRITIQUE** Don and Diego used a graph to solve the quadratic inequality $x^2 - 2x - 8 > 0$. Is either of them correct? Explain.

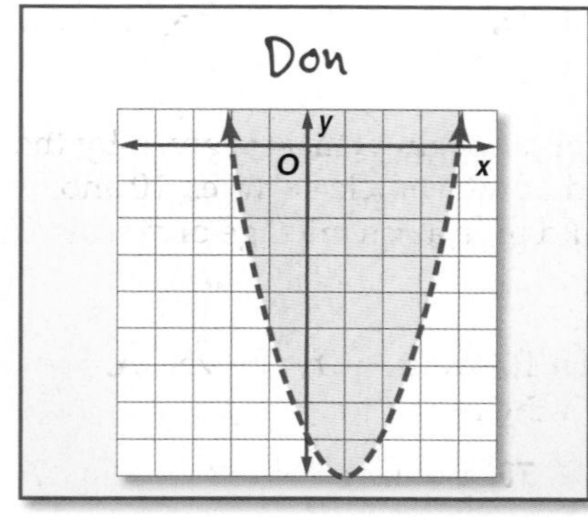


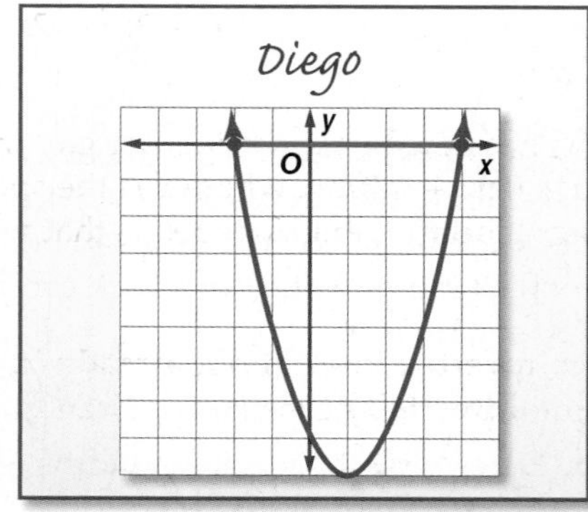


**59. REASONING** Are the boundaries of the solution set of $x^2 + 4x - 12 \leq 0$ twice the value of the boundaries of $\frac{1}{2}x^2 + 2x - 6 \leq 0$? Explain.

**60. REASONING** Determine if the following statement is *sometimes*, *always*, or *never* true. Explain your reasoning.

*The intersection of $y \leq -ax^2 + c$ and $y \geq ax^2 - c$ is the empty set.*

**61. CHALLENGE** Graph the intersection of the graphs of $y \leq -x^2 + 4$ and $y \geq x^2 - 4$.

**62. WRITING IN MATH** How are the techniques used when solving quadratic inequalities and quadratic equations similar? different?

## Standardized Test Practice

**63. GRIDDED RESPONSE** You need to seed an area that is 80 feet by 40 feet. Each bag of seed can cover 25 square yards of land. How many bags of seed will you need?

**64. SAT/ACT** The product of two integers is between 107 and 116. Which of the following *cannot* be one of the integers?

A 5
B 10
C 12
D 15
E 23

**65 PROBABILITY** Five students are to be arranged side by side with the tallest student in the center and the two shortest students on the ends. If no two students are the same height, how many different arrangements are possible?

F 2
G 4
H 5
J 6

**66. SHORT RESPONSE** Simplify $\frac{5+i}{6-3i}$.

## Spiral Review

**Write an equation in vertex form for each parabola.** (Lesson 3-5)

**67.**

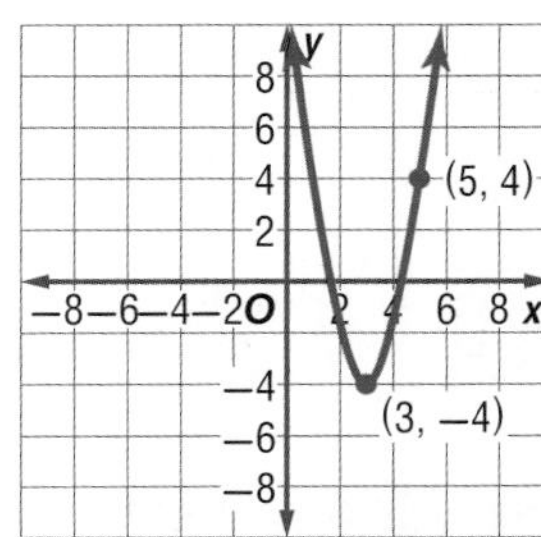


**68.**

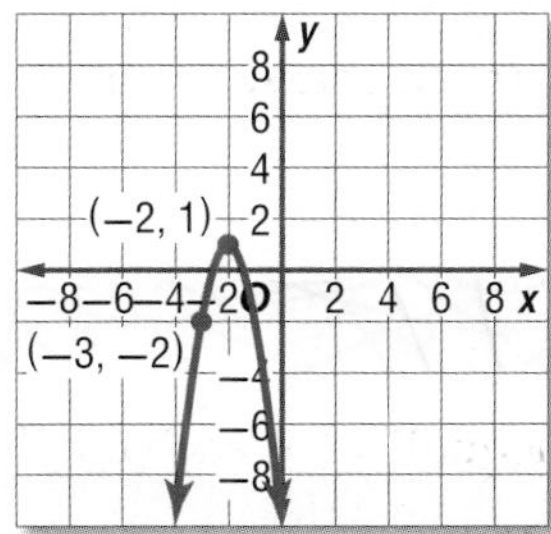


**69.**

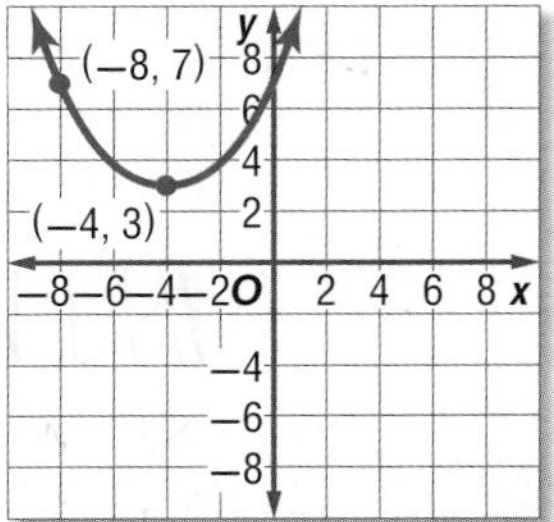


**Complete parts a and b for each quadratic equation.**

**a. Find the value of the discriminant.**

**b. Describe the number and type of roots.** (Lesson 3-4)

**70.** $4x^2 + 7x - 3 = 0$

**71.** $-3x^2 + 2x - 4 = 9$

**72.** $6x^2 + x - 4 = 12$

**73. GAS MILEAGE** The gas mileage $y$ in miles per gallon for a certain vehicle is given by the equation $y = 10 + 0.9x - 0.01x^2$, where $x$ is the speed of the vehicle between 10 and 75 miles per hour. Find the range of speeds that would give a gas mileage of at least 25 miles per gallon. (Lesson 3-6)

**Write each equation in vertex form, if not already in that form. Identify the vertex, axis of symmetry, and direction of opening. Then graph the function.** (Lesson 3-5)

**74.** $y = -6(x + 2)^2 + 3$

**75.** $y = -\frac{1}{3}x^2 + 8x$

**76.** $y = (x - 2)^2 - 2$

**77.** $y = 2x^2 + 8x + 10$

## Skills Review

**Use the Distributive Property to find each product.**

**78.** $-6(x - 4)$

**79.** $8(w + 3x)$

**80.** $-4(-2y + 3z)$

**81.** $-1(c - d)$

**82.** $0.5(5x + 6y)$

**83.** $-3(-6y - 4z)$

EXTEND 3-5

# Graphing Technology Lab
# More Quadratic Inequalities

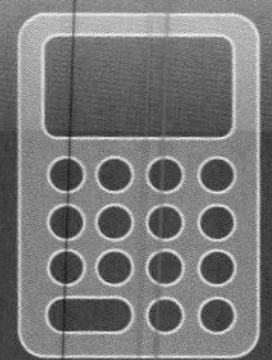

Complete the following for more practice with graphing inequalities.

## Activity 1 Shade Outside a Parabola

**Graph $y \geq -4x^2 + 12x - 7$ in a standard viewing window.**

First, clear all functions from the Y= list.

To graph $y \geq -4x^2 + 12x - 7$, enter the equation in the Y= list. Then use the left arrow to select =. Press [ENTER] until shading above the line is selected.

**KEYSTROKES:** [◀] [◀] [ENTER] [ENTER] [▶] [▶] [–] 4 [X,T,θ,n] [$x^2$] [+] 12 [X,T,θ,n] [–] 7 [ZOOM] 6

All ordered pairs for which $y$ is *greater than or equal* to $-4x^2 + 12x - 7$ lie *above or on* the line and are solutions.

## Activity 2 Shade Inside a Parabola

**Graph $y \leq -x^2 - 7x + 8$ in a standard viewing window.**

First, clear the graph that is displayed.

**KEYSTROKES:** [Y=] [CLEAR]

**KEYSTROKES:** [◀] [◀] [ENTER] [ENTER] [ENTER] [▶] [▶] [–] [X,T,θ,n] [$x^2$] [–] 7 [X,T,θ,n] [+] 8 [GRAPH]

All ordered pairs for which $y$ is *less than or equal* to $-x^2 - 7x + 8$ lie *below or on* the line and are solutions.

## Exercises

1. Compare and contrast the two graphs shown above.
2. For each inequality in the Activities above, write and graph a new inequality that will have the reverse portions of the parabola shaded.

EXTEND 3-5

# Graphing Technology Lab
# Modeling Motion

## Set Up the Lab

- Place a board on a stack of books to create a ramp.
- Connect the data collection device to the graphing calculator. Place at the top of the ramp so that the data collection device can read the motion of the car on the ramp.
- Hold the car still about 6 inches up from the bottom of the ramp and zero the collection device.

**Common Core State Standards**
**Mathematical Practices**
4 Model with mathematics.

### Activity

**Step 1** One group member should press the button to start collecting data.

**Step 2** Another group member places the car at the bottom of the ramp. After data collection begins, gently but quickly push the car so it travels up the ramp toward the motion detector.

**Step 3** Stop collecting data when the car returns to the bottom of the ramp. Save the data as Trial 1.

**Step 4** Remove one book from the stack. Then repeat the experiment. Save the data as Trial 2. For Trial 3, create a steeper ramp and repeat the experiment.

## Analyze the Results

1. What type of function could be used to represent the data? Justify your answer.
2. Use the **CALC** menu to find the vertex of the graph. Record the coordinates in a table like the one at the right.
3. Use the **TRACE** feature of the calculator to find the coordinates of another point on the graph. Then use the coordinates of the vertex and the point to find an equation of the graph.

| Trial | Vertex $(h, k)$ | Point $(x, y)$ | Equation |
|---|---|---|---|
| 1 | | | |
| 2 | | | |
| 3 | | | |

4. Find an equation for each of the graphs of Trials 2 and 3.
5. How do the equations for Trials 1, 2, and 3 compare? Which graph is widest and which is most narrow? Explain what this represents in the context of the situation. How is this represented in the equations?
6. What do the $x$-intercepts and vertex of each graph represent?
7. Why were the values of $h$ and $k$ different in each trial?

# CHAPTER 3 Study Guide and Review

## Study Guide

### KeyConcepts

**Solving Quadratic Equations** (Lessons 3-2)

- Roots of a quadratic equation are the zeros of the related quadratic function. You can find the zeros of a quadratic function by finding the $x$-intercepts of the graph.

**Complex Numbers** (Lesson 3-3)

- $i$ is the imaginary unit; $i^2 = -1$ and $i = \sqrt{-1}$.

**Solving Quadratic Equations** (Lessons 3-4)

- Completing the square: **Step 1** Find one half of $b$, the coefficient of $x$. **Step 2** Square the result in Step 1. **Step 3** Add the result of Step 2 to $x^2 + bx$.
- Quadratic Formula: $x = \dfrac{-b \pm \sqrt{b^2 - 4ac}}{2a}$

**Transformations of Quadratic Graphs** (Lesson 3-5)

- The graph of $y = (x - h)^2 + k$ is the graph of $y = x^2$ translated $|h|$ units left if $h$ is negative or $h$ units right if $h$ is positive and $k$ units up if $k$ is positive or $|k|$ units down if $k$ is negative.
- Consider $y = a(x - h)^2 + k$, $a \neq 0$. If $a > 0$, the graph opens up; if $a < 0$ the graph opens down. If $|a| > 1$, the graph is narrower than the graph of $y = x^2$. If $|a| < 1$, the graph is wider than the graph of $y = x^2$.

**Quadratic Inequalities** (Lesson 3-6)

- Graph the related function, test a point *not* on the parabola and determine if it is a solution, and shade accordingly.

### FOLDABLES StudyOrganizer

Be sure the Key Concepts are noted in your Foldable.

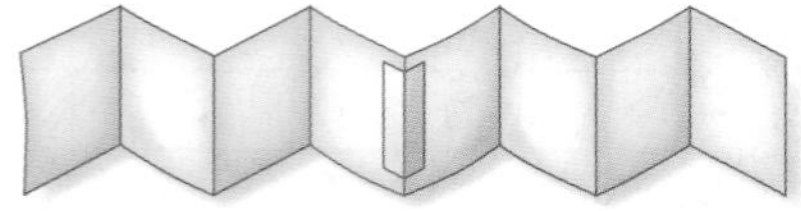

### KeyVocabulary

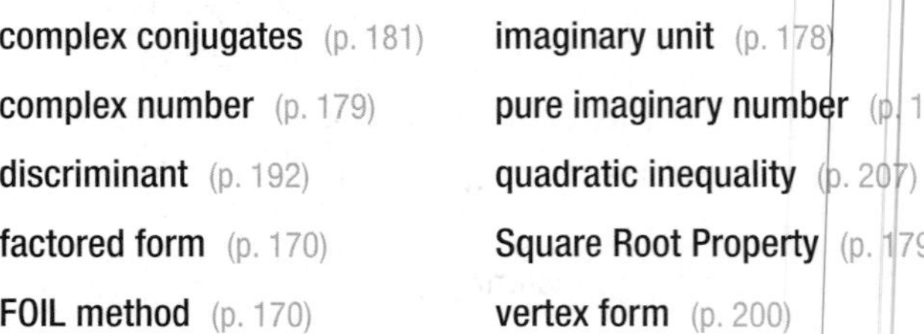

complex conjugates (p. 181)
complex number (p. 179)
discriminant (p. 192)
factored form (p. 170)
FOIL method (p. 170)
imaginary unit (p. 178)
pure imaginary number (p. 178)
quadratic inequality (p. 207)
Square Root Property (p. 179)
vertex form (p. 200)

### VocabularyCheck

**State whether each sentence is *true* or *false*. If *false*, replace the underlined term to make a true sentence.**

1. The axis of symmetry will intersect a parabola in one point called the vertex.
2. A method called FOIL method is used to make a quadratic expression a perfect square in order to solve the related equation.
3. The number $6i$ is called a pure imaginary number.
4. The two numbers $2 + 3i$ and $2 - 3i$ are called complex conjugates.

## 3-1 Solving Quadratic Equations by Factoring

Write a quadratic equation in standard form with the given roots.

**5.** 5, 6

**6.** −3, −7

**7.** −4, 2

**8.** $-\frac{2}{3}, 1$

**9.** $\frac{1}{6}, 5$

**10.** $-\frac{1}{4}, -1$

Solve each equation by factoring.

**11.** $2x^2 - 2x - 24 = 0$

**12.** $2x^2 - 5x - 3 = 0$

**13.** $3x^2 - 16x + 5 = 0$

**14.** Find $x$ and the dimensions of the rectangle below.

$A = 126$ ft$^2$ | $x - 3$ | $x + 2$

### Example 1

**Write a quadratic equation in standard form with $-\frac{1}{2}$ and 4 as its roots.**

$(x - p)(x - q) = 0$ — Write the pattern.

$\left[x - \left(-\frac{1}{2}\right)\right](x - 4) = 0$ — Replace $p$ with $-\frac{1}{2}$ and $q$ with 4.

$\left(x + \frac{1}{2}\right)(x - 4) = 0$ — Simplify.

$x^2 - \frac{7}{2}x - 2 = 0$ — Multiply.

$2x^2 - 7x - 4 = 0$ — Multiply each side by 2 so that $b$ and $c$ are integers.

### Example 2

**Solve $2x^2 - 3x - 5 = 0$ by factoring.**

$2x^2 - 3x - 5 = 0$ — Original equation

$(2x - 5)(x + 1) = 0$ — Factor the trinomial.

$2x - 5 = 0$ or $x + 1 = 0$ — Zero Product Property

$x = \frac{5}{2}$ $\quad$ $x = -1$

The solution set is $\left\{-1, \frac{5}{2}\right\}$ or $\left\{x \mid x = -1, \frac{5}{2}\right\}$.

## 3-2 Complex Numbers

Simplify.

**15.** $\sqrt{-8}$

**16.** $(2 - i) + (13 + 4i)$

**17.** $(6 + 2i) - (4 - 3i)$

**18.** $(6 + 5i)(3 - 2i)$

**19.** **ELECTRICITY** The impedance in one part of a series circuit is $3 + 2j$ ohms, and the impedance in the other part of the circuit is $4 - 3j$ ohms. Add these complex numbers to find the total impedance in the circuit.

Solve each equation.

**20.** $2x^2 + 50 = 0$

**21.** $4x^2 + 16 = 0$

**22.** $3x^2 + 15 = 0$

**23.** $8x^2 + 16 = 0$

**24.** $4x^2 + 1 = 0$

### Example 3

**Simplify $(12 + 3i) - (-5 + 2i)$.**

$(12 + 3i) - (-5 + 2i)$

$= [12 - (-5)] + (3 - 2)i$ — Group the real and imaginary parts.

$= 17 + i$ — Simplify.

### Example 4

**Solve $3x^2 + 12 = 0$.**

$3x^2 + 12 = 0$ — Original equation

$3x^2 = -12$ — Subtract 12 from each side.

$x^2 = -4$ — Divide each side by 3.

$x = \pm\sqrt{-4}$ — Square Root Property

$x = \pm 2i$ — $\sqrt{-4} = \sqrt{4} \cdot \sqrt{-1}$

# Study Guide and Review *Continued*

## 3-3 The Quadratic Formula and the Discriminant

Complete parts **a–c** for each quadratic equation.

**a.** Find the value of the discriminant.

**b.** Describe the number and type of roots.

**c.** Find the exact solutions by using the Quadratic Formula.

**25.** $x^2 - 10x + 25 = 0$

**26.** $x^2 + 4x - 32 = 0$

**27.** $2x^2 + 3x - 18 = 0$

**28.** $2x^2 + 19x - 33 = 0$

**29.** $x^2 - 2x + 9 = 0$

**30.** $4x^2 - 4x + 1 = 0$

**31.** $2x^2 + 5x + 9 = 0$

**32.** **PHYSICAL SCIENCE** Lauren throws a ball with an initial velocity of 40 feet per second. The equation for the height of the ball is $h = -16t^2 + 40t + 5$, where $h$ represents the height in feet and $t$ represents the time in seconds. When will the ball hit the ground?

### Example 5

**Solve $x^2 - 4x - 45 = 0$ by using the Quadratic Formula.**

In $x^2 - 4x - 45 = 0$, $a = 1$, $b = -4$, and $c = -45$.

$x = \frac{-b \pm \sqrt{b^2 - 4ac}}{2a}$ Quadratic Formula

$= \frac{-(-4) \pm \sqrt{(-4)^2 - 4(1)(-45)}}{2(1)}$

$= \frac{4 \pm 14}{2}$

Write as two equations.

$x = \frac{4 + 14}{2}$ or $x = \frac{4 - 14}{2}$

$= 9$ $\quad = -5$

The solution set is $\{-5, 9\}$ or $\{x \mid x = -5, 9\}$.

## 3-4 Transformations of Quadratic Graphs

**Write each quadratic function in vertex form, if not already in that form. Then identify the vertex, axis of symmetry, and direction of opening. Then graph the function.**

**33.** $y = -3(x - 1)^2 + 5$

**34.** $y = 2x^2 + 12x - 8$

**35.** $y = -\frac{1}{2}x^2 - 2x + 12$

**36.** $y = 3x^2 + 36x + 25$

**37.** The graph at the right shows a product of 2 numbers with a sum of 10. Find a function that models this product and use it to determine the two numbers that would give a maximum product.

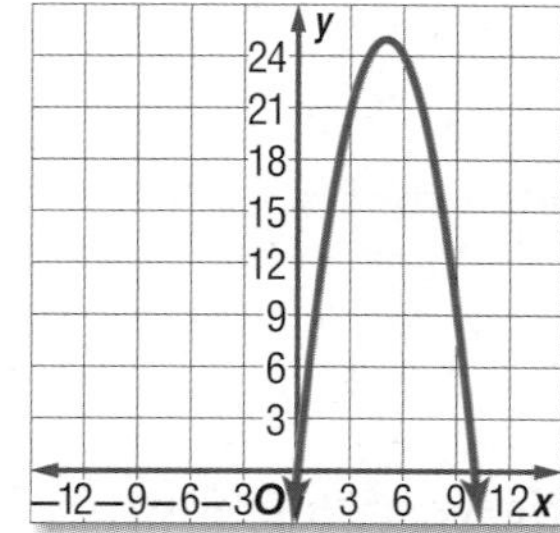


### Example 6

**Write the quadratic function $y = 3x^2 + 24x + 15$ in vertex form. Then identify the vertex, axis of symmetry, and direction of opening.**

| | |
|---|---|
| $y = 3x^2 + 24x + 15$ | Original equation |
| $y = 3(x^2 + 8x) + 15$ | Group and factor. |
| $y = 3(x^2 + 8x + 16) + 15 - 3(16)$ | Complete the square. |
| $y = 3(x + 4)^2 - 33$ | Rewrite $x^2 + 8x + 16$ as a perfect square. |

So, $a = 3$, $h = -4$, and $k = -33$. The vertex is at $(-4, -33)$ and the axis of symmetry is $x = -4$. Since $a$ is positive, the graph opens up.

## 3-5 Quadratic Inequalities

Graph each quadratic inequality.

38. $y \geq x^2 + 5x + 4$

39. $y < -x^2 + 5x - 6$

40. $y > x^2 - 6x + 8$

41. $y \leq x^2 + 10x - 4$

42. Solomon wants to put a deck along two sides of his garden. The deck width will be the same on both sides and the total area of the garden and deck cannot exceed 500 square feet. How wide can the deck be?

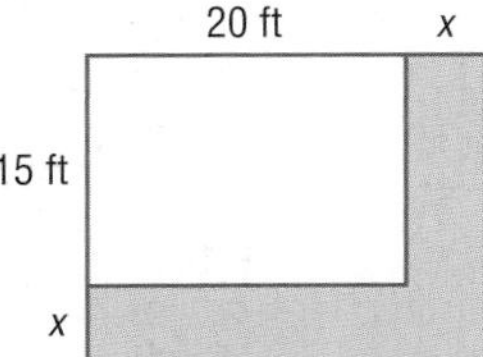


Solve each inequality using a graph or algebraically.

43. $x^2 + 8x + 12 > 0$

44. $6x + x^2 \geq -9$

45. $2x^2 + 3x - 20 > 0$

46. $4x^2 - 3 < -5x$

47. $3x^2 + 4 > 8x$

### Example 7

Graph $y > x^2 + 3x + 2$.

**Step 1** Graph the related function, $y > x^2 + 3x + 2$. Because the inequality symbol $>$ is used, the parabola should be dashed.

**Step 2** Test a point not on the graph of the parabola such as (0, 0).

$$y > x^2 + 3x + 2$$

$$(0) \stackrel{?}{>} (0)^2 + 3(0) + 2$$

$$0 \ngtr 2$$

So, (0, 0) is not a solution of the inequality.

**Step 3** Shade the region that does not contain the point (0, 0).

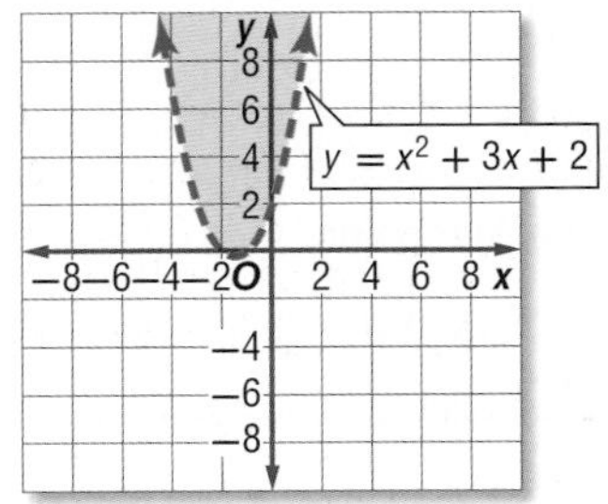


# Practice Test

1. **MULTIPLE CHOICE** Which equation below has roots at $-6$ and $\frac{1}{5}$?

   A $0 = 5x^2 - 29x - 6$

   B $0 = 5x^2 + 31x + 6$

   C $0 = 5x^2 + 29x - 6$

   D $0 = 5x^2 - 31x + 6$

2. **PHYSICS** A ball is thrown into the air vertically with a velocity of 112 feet per second. The ball was released 6 feet above the ground. The height above the ground $t$ seconds after release is modeled by $h(t) = -16t^2 + 112t + 6$.

   a. When will the ball reach 130 feet?

   b. Will the ball ever reach 250 feet? Explain.

   c. In how many seconds after its release will the ball hit the ground?

3. The rectangle below has an area of 104 square inches. Find the value of $x$ and the dimensions of the rectangle.

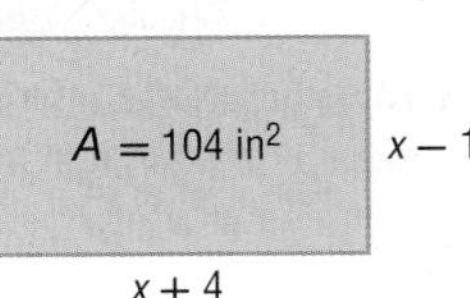


4. **MULTIPLE CHOICE** Which value of $c$ makes the trinomial $x^2 - 12x + c$ a perfect square trinomial?

   F 6

   G 12

   H 36

   J 144

**Solve each inequality by using a graph or algebraically.**

5. $x^2 + 6x > -5$

6. $4x^2 - 19x \leq -12$

# Preparing for Standardized Tests

## Use a Graph

Using a graph can help you solve many different kinds of problems on standardized tests. Graphs can help you solve equations, evaluate functions, and interpret solutions to real-world problems.

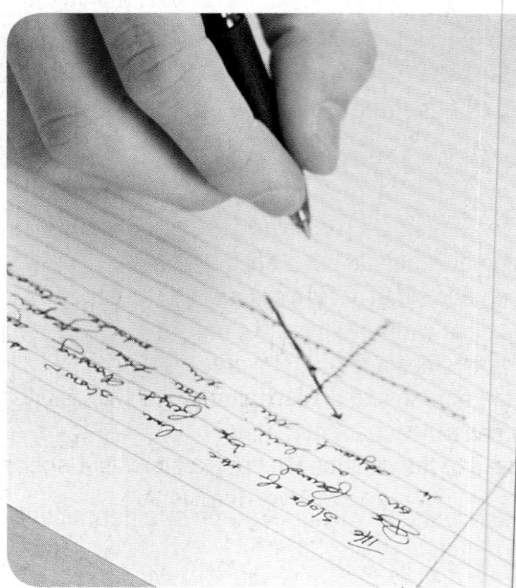

### Strategies for Using a Graph

**Step 1**

Read the problem statement carefully.

**Ask yourself:**

- What am I being asked to solve?
- What information is given in the problem?
- How could a graph help me solve the problem?

**Step 2**

Create your graph.

- Sketch your graph on scrap paper if appropriate.
- If allowed, you can also use a graphing calculator to create the graph.

**Step 3**

Solve the problem.

- Use your graph to help you model and solve the problem.
- Check to be sure your answer makes sense.

### Standardized Test Example

**Read the problem. Identify what you need to know. Then use the information in the problem to solve.**

The students in Mr. Himebaugh's physics class built a model rocket. The rocket is launched in a large field with an initial upward velocity of 128 feet per second. The function $h(t) = -16t^2 + 128t$ models the height of the rocket above the ground (in feet) $t$ seconds after it is launched. How long will it take for the rocket to reach its maximum height?

**A** 4 seconds

**B** 5 seconds

**C** 6 seconds

**D** 8 seconds

Graphing the quadratic function will allow you to determine the peak height of the rocket and when it occurs. A graphing calculator can help you quickly graph the function and analyze it.

**KEYSTROKES:** [Y=] [(−)] 16 [X,T,θ,n] [$x^2$] [+] 128 [X,T,θ,n] [GRAPH]

After graphing the equation, use **maximum** under the **CALC** menu.

Press [2nd] **[CALC]** 4. Then use [◀] to place the cursor to the left of the maximum point and press [ENTER]. Use [▶] to place the cursor to the right of the maximum point and press [ENTER] [ENTER].

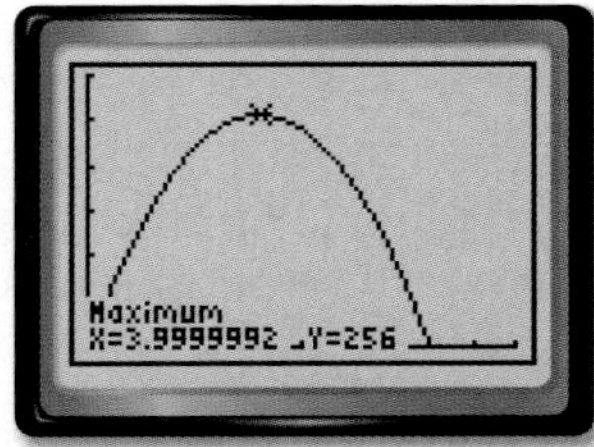

[0, 10] scl: 1 by [0, 300] scl: 50

The graph shows that the rocket takes 4 seconds to reach its maximum height of 256 feet. The correct answer is A.

## Exercises

**Read each problem. Identify what you need to know. Then use the information in the problem to solve.**

1. What are the roots of $y = 2x^2 + 10x - 48$?

   A $-5, 4$

   B $-6, 1$

   C $-8, 3$

   D $2, 3$

2. How many times does the graph of $f(x) = 2x^2 - 3x + 2$ cross the $x$-axis?

   F 0     H 2

   G 1     J 3

3. Which statement best describes the graphs of the two equations?

$$16x - 2y = 24$$
$$12x = 3y - 36$$

   A The lines are parallel.

   B The lines are the same.

   C The lines intersect in only one point.

   D The lines intersect in more than one point, but are not the same.

4. Adrian is using 120 feet of fencing to enclose a rectangular area for her puppy. One side of the enclosure will be her house.

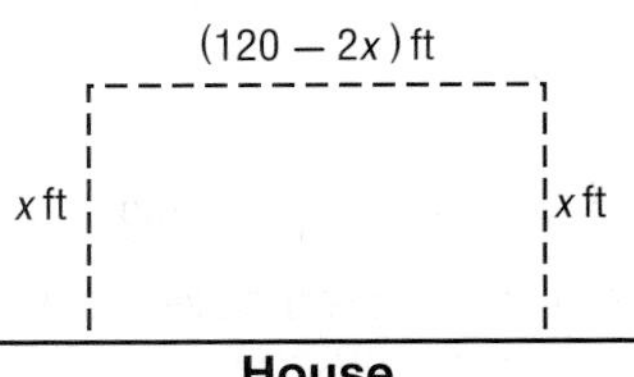


   The function $f(x) = x(120 - 2x)$ represents the area of the enclosure. What is the greatest area Adrian can enclose with the fencing?

   F 1650 ft$^2$     H 1980 ft$^2$

   G 1800 ft$^2$     J 2140 ft$^2$

5. For which equation is the $x$-coordinate of the vertex at 4?

   A $f(x) = x^2 - 8x + 15$     C $f(x) = x^2 + 6x + 8$

   B $f(x) = -x^2 - 4x + 12$     D $f(x) = -x^2 - 2x + 2$

6. For what value of $x$ does $f(x) = x^2 + 5x + 6$ reach its minimum value?

   F $-5$     H $-\frac{5}{2}$

   G $-3$     J $-2$

CHAPTER 3

# Standardized Test Practice

## Cumulative, Chapters 1 through 3

### Multiple Choice

**Read each question. Then fill in the correct answer on the answer document provided by your teacher or on a sheet of paper.**

1. The graph of $g(x) = \frac{2}{5}x^2 - 4x + 2$ is translated down 5 units to produce the graph of the function $h(x)$. Which of the following could be $h(x)$?

   F $h(x) = \frac{2}{5}x^2 - 4x + 7$

   G $h(x) = \frac{2}{5}x^2 - 4x - 3$

   H $h(x) = \frac{2}{5}x^2 - 9x + 2$

   J $h(x) = \frac{2}{5}x^2 + x + 2$

2. The function $P(t) = -0.068t^2 + 7.85t + 56$ can be used to approximate the population, in thousands, of Clarksville between 1960 and 2000. The domain $t$ of the function is the number of years since 1960. According to the model, in what year did the population of Clarksville reach 200,000 people?

   F 1974     H 1981

   G 1977     J 1983

3. What is the effect on the graph of the equation $y = x^2 + 4$ when it is changed to $y = x^2 - 3$?

   F The slope of the graph changes.

   G The graph widens.

   H The graph is the same shape, and the vertex of the graph is moved down.

   J The graph is the same shape, and the vertex of the graph is shifted to the left.

4. Which equation will produce the narrowest parabola when graphed?

   A $y = 3x^2$     C $y = -6x^2$

   B $y = \frac{3}{4}x^2$     D $y = -\frac{3}{4}x^2$

5. Which of the following does *not* accurately describe the graph $y = -2x^2 + 4$?

   A The parabola is symmetric about the $y$-axis.

   B The parabola opens downward.

   C The parabola has the origin as its vertex.

   D The parabola crosses the $x$-axis in two different places.

6. Which of the following is *not* a factor of $x^4 - 6x^2 - 27$?

   A $x^2 + 3$     C $x + 3$

   B $x - 3$     D $x^2 - 3$

7. Graph $f(x) \geq |x - 2|$ on a coordinate grid.

8. **GRIDDED RESPONSE** How many times does the graph of $y = x^2 - 4x + 10$ cross the $x$-axis?

## Short Response/Gridded Response

**Record your answers on the answer sheet provided by your teacher or on a sheet of paper.**

9. **GRIDDED RESPONSE** Simplify $-2i \cdot 5i$.

10. Describe the translation of the graph of $y = (x + 5)^2 - 1$ to the graph of $y = (x - 1)^2 + 3$.

## Extended Response

**Record your answers on a sheet of paper. Show your work.**

11. For a given quadratic equation $y = ax^2 + bx + c$, describe what the discriminant $b^2 - 4ac$ tells you about the roots of the equation.

12. Carl's father is building a tool chest that is shaped like a rectangular prism. He wants the tool chest to have a surface area of 62 square feet. The height of the chest will be 1 foot shorter than the width. The length will be 3 feet longer than the height.
    a. Sketch a model to represent the problem.
    b. Write a polynomial that represents the surface area of the tool chest.
    c. What are the dimensions of the tool chest?

**Need ExtraHelp?**

| If you missed Question... | 1 | 2 | 3 | 4 | 5 | 6 | 7 | 8 | 9 | 10 | 11 | 12 |
|---|---|---|---|---|---|---|---|---|---|---|---|---|
| Go to Lesson... | 3-5 | 3-4 | 3-5 | 3-5 | 2-3 | 1-9 | 2-7 | 2-5 | 3-3 | 3-5 | 3-4 | 1-9 |

CHAPTER 4

# Exponential and Logarithmic Functions and Relations

## Then

You graphed functions and transformations of functions.

## Now

You will:

- Graph exponential and logarithmic functions.
- Solve exponential and logarithmic equations and inequalities.
- Solve problems involving exponential growth and decay.

## Why? ▲

**SCIENCE** Mathematics and science go hand in hand. Whether it is chemistry, biology, paleontology, zoology, or anthropology, you will need strong math skills. In this chapter, you will learn mathematical aspects of science such as computer viruses, populations of insects, bacteria growth, cell division, astronomy, tornados, and earthquakes.

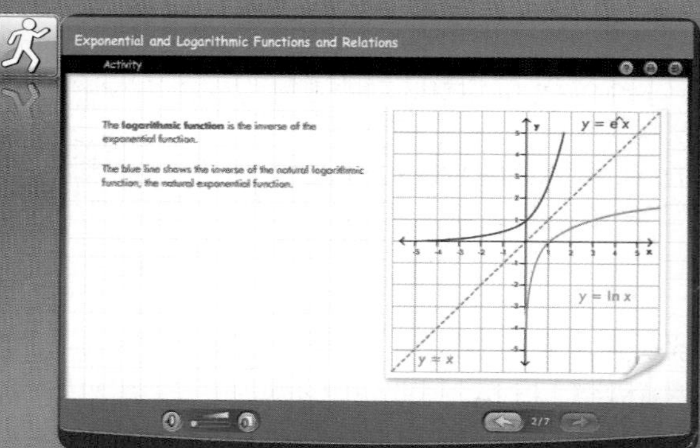


connectED.mcgraw-hill.com Your Digital Math Portal

Animation

Vocabulary

eGlossary

Personal Tutor

Virtual Manipulatives

Graphing Calculator

Audio

Foldables

Self-Check Practice

Worksheets

# Get Ready for the Chapter

**Diagnose** Readiness | You have two options for checking prerequisite skills.

**Textbook Option** Take the Quick Check below. Refer to the Quick Review for help.

## QuickCheck

Simplify. Assume that no variable equals zero.

1. $a^4a^3a^5$
2. $(2xy^3z^2)^3$
3. $\dfrac{-24x^8y^5z}{16x^2y^8z^6}$
4. $\left(\dfrac{-8r^2n}{36n^3t}\right)^2$
5. **DENSITY** The density of an object is equal to the mass divided by the volume. An object has a mass of $7.5 \times 10^3$ grams and a volume of $1.5 \times 10^3$ cubic centimeters. What is the density of the object?

Find the inverse of each function. Then graph the function and its inverse.

6. $f(x) = 2x + 5$
7. $f(x) = x - 3$
8. $f(x) = -4x$
9. $f(x) = \frac{1}{4}x - 3$
10. $f(x) = \frac{x - 1}{2}$
11. $y = \frac{1}{3}x + 4$

Determine whether each pair of functions are inverse functions.

12. $f(x) = x - 6$, $g(x) = x + 6$
13. $f(x) = 2x + 5$, $g(x) = 2x - 5$
14. **FOOD** A pizzeria charges $12 for a medium cheese pizza and $2 for each additional topping. If $f(x) = 2x + 12$ represents the cost of a medium pizza with $x$ toppings, find $f^{-1}(x)$ and explain its meaning.

## QuickReview

**Example 1**

Simplify $\dfrac{(a^3bc^2)^2}{a^4a^2b^2bc^5c^3}$. Assume that no variable equals zero.

$$\frac{(a^3bc^2)^2}{a^4a^2b^2bc^5c^3}$$

$$= \frac{a^6b^2c^4}{a^6b^3c^8}$$ Simplify the numerator by using the Power of a Power Rule and the denominator by using the Product of Powers Rule.

$$= \frac{1}{bc^4} \text{ or } b^{-1}c^{-4}$$ Simplify by using the Quotient of Powers Rule.

**Example 2**

Find the inverse of $f(x) = 3x - 1$.

**Step 1** Replace $f(x)$ with $y$ in the original equation: $f(x) = 3x - 1 \rightarrow y = 3x - 1$.

**Step 2** Interchange $x$ and $y$: $x = 3y - 1$.

**Step 3** Solve for $y$.

| | |
|---|---|
| $x = 3y - 1$ | Inverse |
| $x + 1 = 3y$ | Add 1 to each side. |
| $\frac{x + 1}{3} = y$ | Divide each side by 3. |
| $\frac{1}{3}x + \frac{1}{3} = y$ | Simplify. |

**Step 4** Replace $y$ with $f^{-1}(x)$.

$$y = \frac{1}{3}x + \frac{1}{3} \rightarrow f^{-1}(x) = \frac{1}{3}x + \frac{1}{3}$$

**2 Online Option** Take an online self-check Chapter Readiness Quiz at connectED.mcgraw-hill.com.

# Get Started on the Chapter

You will learn several new concepts, skills, and vocabulary terms as you study Chapter 4. To get ready, identify important terms and organize your resources.

## FOLDABLES StudyOrganizer

**Exponential and Logarithmic Functions and Relations**
Make this Foldable to help you organize your Chapter 4 notes about exponential and logarithmic functions. Begin with two sheets of grid paper.

1. **Fold** in half along the width.

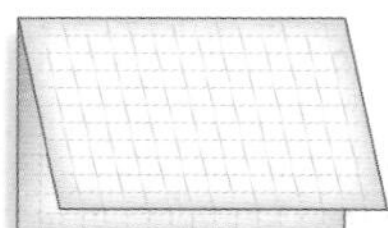

2. **On** the first sheet, cut 5 cm along the fold at the ends.

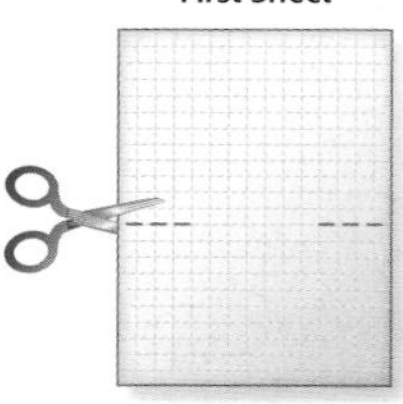


3. **On** the second sheet, cut in the center, stopping 5 cm from the ends.

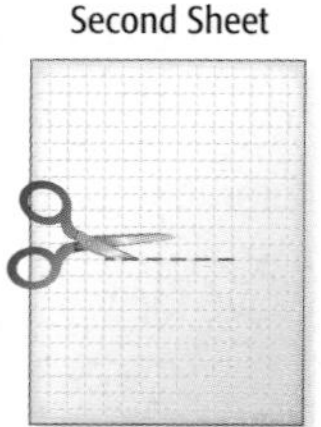


4. **Insert** the first sheet through the second sheet and align the folds. Label the pages with lesson numbers.

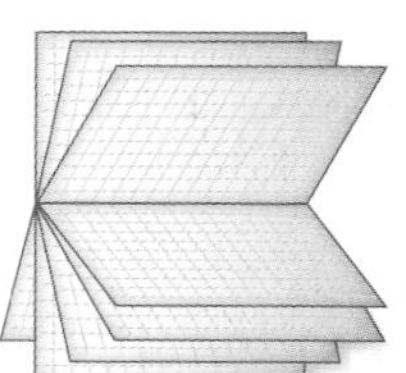

## NewVocabulary

| English | | Español |
|---|---|---|
| exponential function | p. 227 | función exponencial |
| exponential growth | p. 227 | crecimiento exponencial |
| asymptote | p. 227 | asíntota |
| growth factor | p. 229 | factor de crecimiento |
| exponential decay | p. 229 | desintegración exponencial |
| decay factor | p. 230 | factor de desintegración |
| exponential equation | p. 237 | ecuación exponencial |
| compound interest | p. 238 | interés compuesto |
| exponential inequality | p. 239 | desigualdad exponencial |
| conjugate | p. 247 | conjugado |
| radical equations | P. 259 | ecuaciones radicales |

## ReviewVocabulary

domain dominio the set of all $x$-coordinates of the ordered pairs of a relation

function función a relation in which each element of the domain is paired with exactly one element in the range

range rango the set of all $y$-coordinates of the ordered pairs of a relation

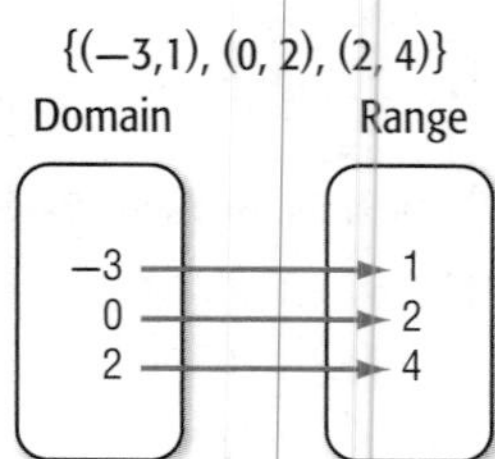

LESSON 4-1

# Graphing Exponential Functions

**Then**

- You graphed polynomial functions.

**Now**

1. Graph exponential growth functions.
2. Graph exponential decay functions.

**Why?**

- Have you ever received an e-mail that tells you to forward it to 5 friends? If each of those 5 friends then forwards it to 5 of their friends, who each forward it to 5 of their friends, the number of people receiving the e-mail is growing exponentially.

The equation $y = 5^x$ can be used to represent this situation, where $x$ is the number of rounds that the e-mail has been forwarded.

**NewVocabulary**

exponential function
exponential growth
asymptote
growth factor
exponential decay
decay factor

**Common Core State Standards**

**Content Standards**

F.IF.7.e Graph exponential and logarithmic functions, showing intercepts and end behavior, and trigonometric functions, showing period, midline, and amplitude.

F.IF.8.b Use the properties of exponents to interpret expressions for exponential functions.

**Mathematical Practices**

3 Construct viable arguments and critique the reasoning of others.

## 1 Exponential Growth

A function like $y = 5^x$, where the base is a constant and the exponent is the independent variable, is an **exponential function**. One type of exponential function is exponential growth. An **exponential growth** function is a function of the form $f(x) = b^x$, where $b > 1$. The graph of an exponential function has an **asymptote**, which is a line that the graph of the function approaches.

### KeyConcept Parent Function of Exponential Growth Functions

Parent Functions: $f(x) = b^x, b > 1$

Type of graph: continuous, one-to-one, and increasing

Domain: all real numbers

Range: all positive real numbers

Asymptote: $x$-axis

Intercept: (0, 1)

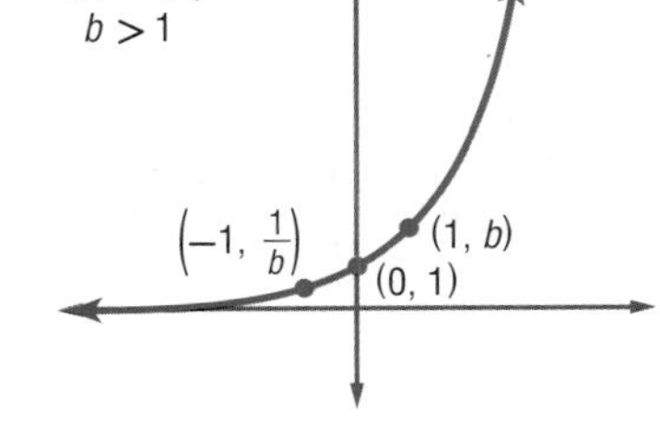


### Example 1 Graph Exponential Growth Functions

**Graph $y = 3^x$. State the domain and range.**

Make a table of values. Then plot the points and sketch the graph.

| $x$ | $-3$ | $-2$ | $-\frac{1}{2}$ | $0$ |
|---|---|---|---|---|
| $y = 3^x$ | $3^{-3} = \frac{1}{27}$ | $3^{-2} = \frac{1}{9}$ | $3^{-\frac{1}{2}} = \frac{\sqrt{3}}{3}$ | $3^0 = 1$ |

| $x$ | $1$ | $\frac{3}{2}$ | $2$ |
|---|---|---|---|
| $y = 3^x$ | $3^1 = 3$ | $3^{\frac{3}{2}} = \sqrt{27}$ | $3^2 = 9$ |

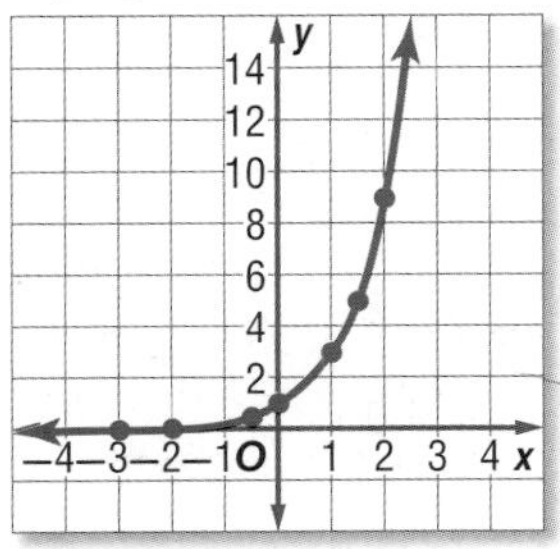


The domain is all real numbers, and the range is all positive real numbers.

**GuidedPractice**

**1.** Graph $y = 4^x$. State the domain and range.

Stockbyte/Alamy

The graph of $f(x) = b^x$ represents a parent graph of the exponential functions. The same techniques used to transform the graphs of other functions you have studied can be applied to the graphs of exponential functions.

**KeyConcept** Transformations of Exponential Functions

$$f(x) = ab^{x-h} + k$$

| $h$ – Horizontal Translation | $k$ – Vertical Translation |
|---|---|
| $h$ units right if $h$ is positive<br>$\lvert h \rvert$ units left if $h$ is negative | $k$ units up if $k$ is positive<br>$\lvert k \rvert$ units down if $k$ is negative |

$a$ – Orientation and Shape

If $a < 0$, the graph is reflected in the $x$-axis.

If $|a| > 1$, the graph is stretched vertically. If $0 < |a| < 1$, the graph is compressed vertically.

**StudyTip**

**CCSS Precision** Remember that end behavior is the action of the graph as $x$ approaches positive infinity or negative infinity. In Example 2a, as $x$ approaches infinity, $y$ approaches infinity. In Example 2b, as $x$ approaches infinity, $y$ approaches negative infinity.

## Example 2 Graph Transformations

**Graph each function. State the domain and range.**

**a.** $y = 2^x + 1$

The equation represents a translation of the graph of $y = 2^x$ one unit up.

| $x$ | $y = 2^x + 1$ |
|---|---|
| −3 | $2^{-3} + 1 = 1.125$ |
| −2 | $2^{-2} + 1 = 1.25$ |
| −1 | $2^{-1} + 1 = 1.5$ |
| 0 | $2^0 + 1 = 2$ |
| 1 | $2^1 + 1 = 3$ |
| 2 | $2^2 + 1 = 5$ |
| 3 | $2^3 + 1 = 9$ |

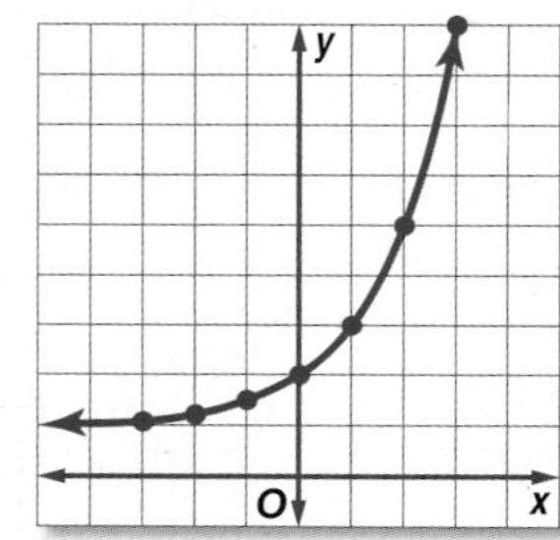


Domain = {all real numbers}; Range = $\{y \mid y > 1\}$

**b.** $y = -\frac{1}{2} \cdot 5^{x-2}$

The equation represents a transformation of the graph of $y = 5^x$.

Graph $y = 5^x$ and transform the graph.

- $a = -\frac{1}{2}$: The graph is reflected in the $x$-axis and compressed vertically.
- $h = 2$: The graph is translated 2 units right.
- $k = 0$: The graph is not translated vertically.

Domain = {all real numbers}

Range = $\{y \mid y < 0\}$

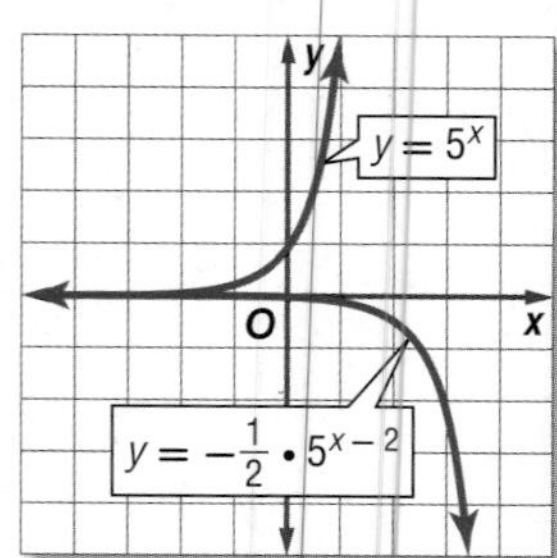


**GuidedPractice**

**2A.** $y = 2^{x+3} - 5$

**2B.** $y = 0.1(6)^x - 3$

You can model exponential growth with a constant percent increase over specific time periods using the following function.

$$A(t) = a(1 + r)^t$$

The function can be used to find the amount $A(t)$ after $t$ time periods, where $a$ is the initial amount and $r$ is the percent of increase per time period. Note that the base of the exponential expression, $1 + r$, is called the **growth factor**.

The exponential growth function is often used to model population growth.

**Real-WorldLink**

The U.S. Census Bureau's American Community Survey is mailed to approximately 1 out of every 480 households.

**Source:** Census Bureau

### Real-World Example 3 Graph Exponential Growth Functions

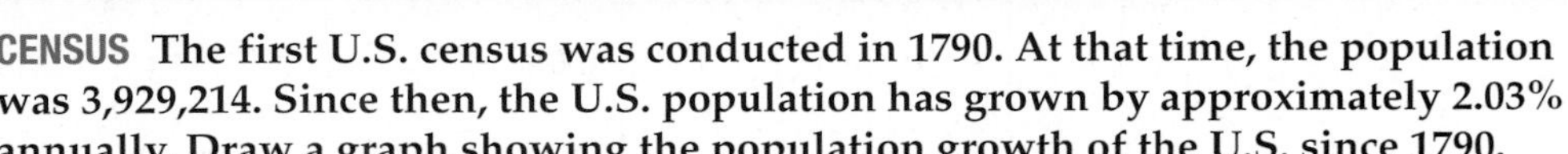

**CENSUS The first U.S. census was conducted in 1790. At that time, the population was 3,929,214. Since then, the U.S. population has grown by approximately 2.03% annually. Draw a graph showing the population growth of the U.S. since 1790.**

First, write an equation using $a = 3{,}929{,}214$, and $r = 0.0203$.

$y = 3{,}929{,}214(1.0203)^t$

Then graph the equation.

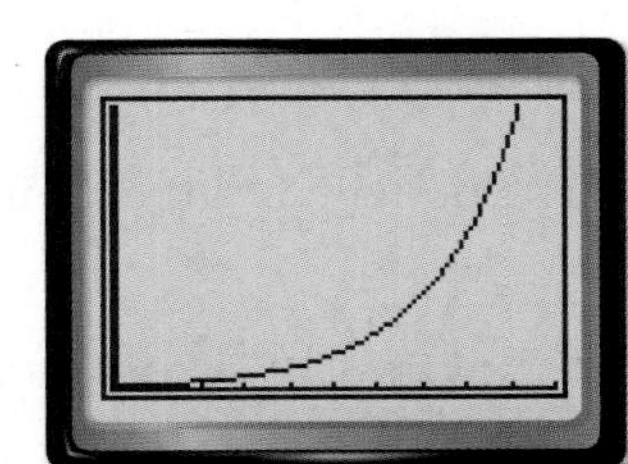

[0, 250] scl: 25 by [0, 400,000,000] scl: 40,000,000

**GuidedPractice**

**3. FINANCIAL LITERACY** Teen spending is expected to grow 3.5% annually from \$79.7 billion in 2006. Draw a graph to show the spending growth.

**StudyTip**

**Interest** The formula for simple interest, $i = prt$, illustrates linear growth over time. However, the formula for compound interest, $A(t) = a(1 + r)^t$, illustrates exponential growth over time. This is why investments with compound interest make more money.

## 2 Exponential Decay

The second type of exponential function is **exponential decay**.

### KeyConcept Parent Function of Exponential Decay Functions

**Parent Functions:** $f(x) = b^x$, $0 < b < 1$

**Type of graph:** continuous, one-to-one, and decreasing

**Domain:** all real numbers

**Range:** positive real numbers

**Asymptote:** $x$-axis

**Intercept:** (0, 1)

**Model**

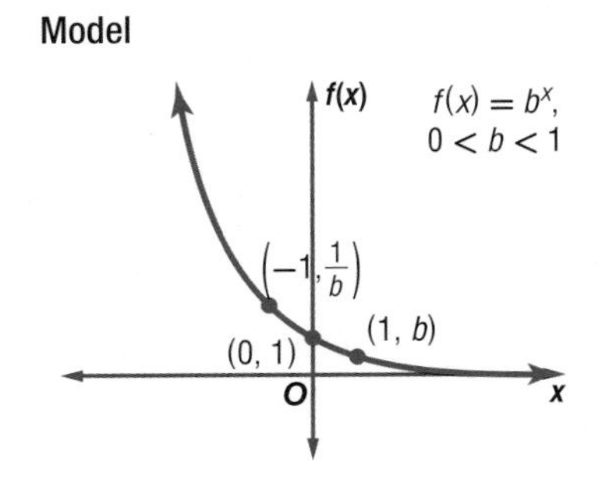


The graphs of exponential decay functions can be transformed in the same manner as those of exponential growth.

Creatas/SuperStock

**StudyTip**

**Exponential Decay** Be sure not to confuse a dilation in which $|a| < 1$ with exponential decay in which $0 < b < 1$.

## Example 4 Graph Exponential Decay Functions

**Graph each function. State the domain and range.**

**a.** $y = \left(\frac{1}{3}\right)^x$

| $x$ | $y = \left(\frac{1}{3}\right)^x$ |
|---|---|
| $-3$ | $\left(\frac{1}{3}\right)^{-3} = 27$ |
| $-2$ | $\left(\frac{1}{3}\right)^{-2} = 9$ |
| $-\frac{1}{2}$ | $\left(\frac{1}{3}\right)^{-\frac{1}{2}} = \sqrt{3}$ |
| $0$ | $\left(\frac{1}{3}\right)^{0} = 1$ |
| $1$ | $\left(\frac{1}{3}\right)^{1} = \frac{1}{3}$ |
| $\frac{3}{2}$ | $\left(\frac{1}{3}\right)^{\frac{3}{2}} = \sqrt{\frac{1}{27}}$ |
| $2$ | $\left(\frac{1}{3}\right)^{2} = \frac{1}{9}$ |

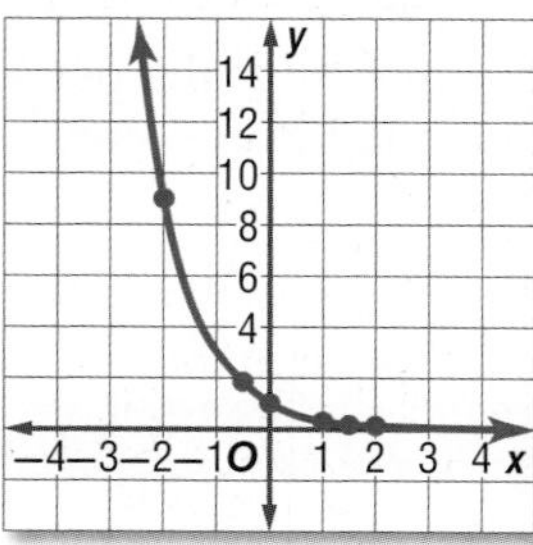


The domain is all real numbers, and the range is all positive real numbers.

**b.** $y = 2\left(\frac{1}{4}\right)^{x+2} - 3$

The equation represents a transformation of the graph of $y = \left(\frac{1}{4}\right)^x$.

Examine each parameter.

- $a = 2$: The graph is stretched vertically.
- $h = -2$: The graph is translated 2 units left.
- $k = -3$: The graph is translated 3 units down.

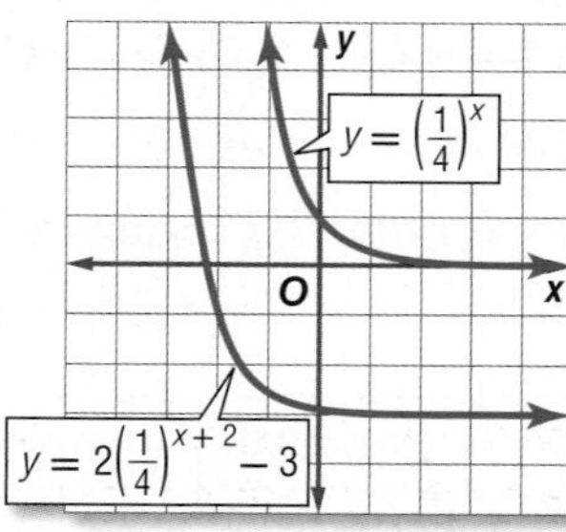


The domain is all real numbers, and the range is all real numbers greater than $-3$.

**GuidedPractice**

**4A.** $y = -3\left(\frac{2}{5}\right)^{x-4} + 2$

**4B.** $y = \frac{3}{8}\left(\frac{5}{6}\right)^{x-1} + 1$

Similar to exponential growth, you can model exponential decay with a constant percent of decrease over specific time periods using the following function.

$$A(t) = a(1 - r)^t$$

The base of the exponential expression, $1 - r$, is called the **decay factor**.

## Real-WorldLink

After water, tea is the most consumed beverage in the U.S. It can be found in over 80% of American households. Just over half the American population drinks tea daily.

**Source:** The Tea Association of the USA

### Real-World Example 5 Graph Exponential Decay Functions

**TEA A cup of green tea contains 35 milligrams of caffeine. The average teen can eliminate approximately 12.5% of the caffeine from their system per hour.**

**a. Draw a graph to represent the amount of caffeine remaining after drinking a cup of green tea.**

$y = a(1 - r)^t$

$= 35(1 - 0.125)^t$

$= 35(0.875)^t$

Graph the equation.

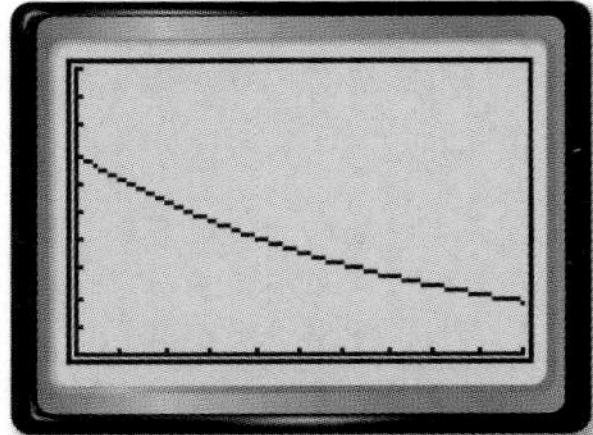

[0, 10] scl: 1 by [0, 50] scl: 5

**b. Estimate the amount of caffeine in a teenager's body 3 hours after drinking a cup of green tea.**

$y = 35(0.875)^t$ Equation from part a

$= 35(0.875)^3$ Replace $t$ with 3.

$\approx 23.45$ Use a calculator.

The caffeine in a teenager will be about 23.45 milligrams after 3 hours.

#### GuidedPractice

**5.** A cup of black tea contains about 68 milligrams of caffeine. Draw a graph to represent the amount of caffeine remaining in the body of an average teen after drinking a cup of black tea. Estimate the amount of caffeine in the body 2 hours after drinking a cup of black tea.

## Check Your Understanding

**Examples 1–2** **Graph each function. State the domain and range.**

**1.** $f(x) = 2^x$

**2.** $f(x) = 5^x$

**3.** $f(x) = 3^{x-2} + 4$

**4.** $f(x) = 2^{x+1} + 3$

**5.** $f(x) = 0.25(4)^x - 6$

**6.** $f(x) = 3(2)^x + 8$

**Example 3**

**7.** CCSS **SENSE-MAKING** A virus spreads through a network of computers such that each minute, 25% more computers are infected. If the virus began at only one computer, graph the function for the first hour of the spread of the virus.

**Example 4** **Graph each function. State the domain and range.**

**8.** $f(x) = 2\left(\frac{2}{3}\right)^{x-3} - 4$

**9.** $f(x) = -\frac{1}{2}\left(\frac{3}{4}\right)^{x+1} + 5$

**10.** $f(x) = -\frac{1}{3}\left(\frac{4}{5}\right)^{x-4} + 3$

**11.** $f(x) = \frac{1}{8}\left(\frac{1}{4}\right)^{x+6} + 7$

**Example 5**

**12. FINANCIAL LITERACY** A new SUV depreciates in value each year by a factor of 15%. Draw a graph of the SUV's value for the first 20 years after the initial purchase.

Core Imagery/Alamy

## Practice and Problem Solving

**Examples 1–2** **Graph each function. State the domain and range.**

**13.** $f(x) = 2(3)^x$ **14.** $f(x) = -2(4)^x$ **15.** $f(x) = 4^{x+1} - 5$

**16.** $f(x) = 3^{2x} + 1$ **17.** $f(x) = -0.4(3)^{x+2} + 4$ **18.** $f(x) = 1.5(2)^x + 6$

**Example 3**

**19** **SCIENCE** The population of a colony of beetles grows 30% each week for 10 weeks. If the initial population is 65 beetles, graph the function that represents the situation.

**Example 4** **Graph each function. State the domain and range.**

**20.** $f(x) = -4\left(\frac{3}{5}\right)^{x+4} + 3$ **21.** $f(x) = 3\left(\frac{2}{5}\right)^{x-3} - 6$ **22.** $f(x) = \frac{1}{2}\left(\frac{1}{5}\right)^{x+5} + 8$

**23.** $f(x) = \frac{3}{4}\left(\frac{2}{3}\right)^{x+4} - 2$ **24.** $f(x) = -\frac{1}{2}\left(\frac{3}{8}\right)^{x+2} + 9$ **25.** $f(x) = -\frac{5}{4}\left(\frac{4}{5}\right)^{x+4} + 2$

**Example 5**

**26.** **ATTENDANCE** The attendance for a basketball team declined at a rate of 5% per game throughout a losing season. Graph the function modeling the attendance if 15 home games were played and 23,500 people were at the first game.

**27.** **PHONES** The function $P(x) = 2.28(0.9^x)$ can be used to model the number of pay phones in millions $x$ years since 1999.

**a.** Classify the function as either exponential *growth* or *decay*, and identify the growth or decay factor. Then graph the function.

**b.** Explain what the $P(x)$-intercept and the asymptote represent in this situation.

**28.** **HEALTH** Each day, 10% of a certain drug dissipates from the system.

**a.** Classify the function representing this situation as either exponential *growth* or *decay*, and identify the growth or decay factor. Then graph the function.

**b.** How much of the original amount remains in the system after 9 days?

**c.** If a second dose should not be taken if more than 50% of the original amount is in the system, when should the label say it is safe to redose? Design the label and explain your reasoning.

**29.** **CCSS REASONING** A sequence of numbers follows a pattern in which the next number is 125% of the previous number. The first number in the pattern is 18.

**a.** Write the function that represents the situation.

**b.** Classify the function as either exponential *growth* or *decay*, and identify the growth or decay factor. Then graph the function for the first 10 numbers.

**c.** What is the value of the tenth number? Round to the nearest whole number.

**For each graph, $f(x)$ is the parent function and $g(x)$ is a transformation of $f(x)$. Use the graph to determine the equation of $g(x)$.**

**30.** $f(x) = 3^x$

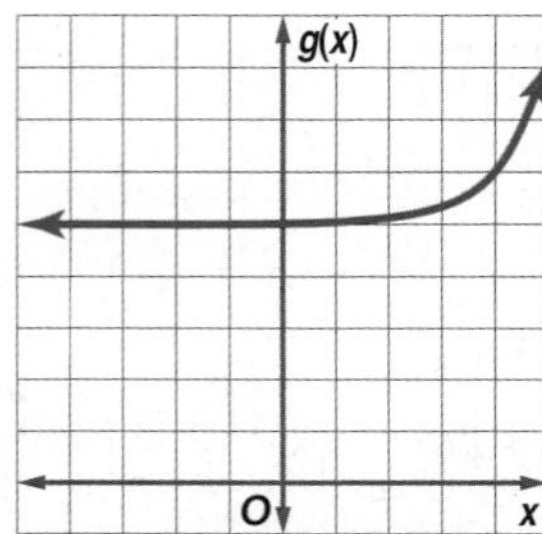


**31.** $f(x) = 2^x$

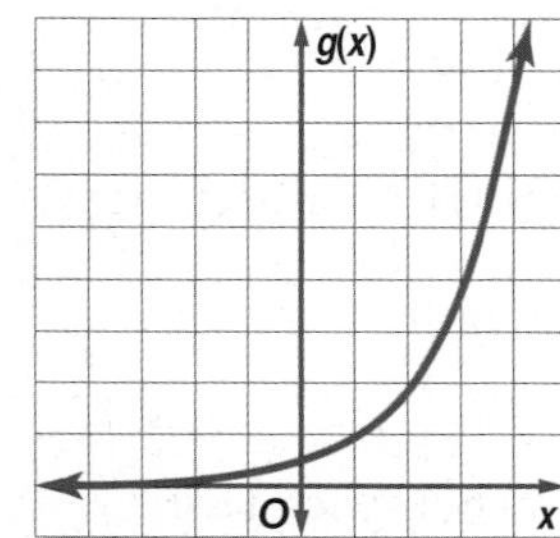


**32.** $f(x) = 4^x$

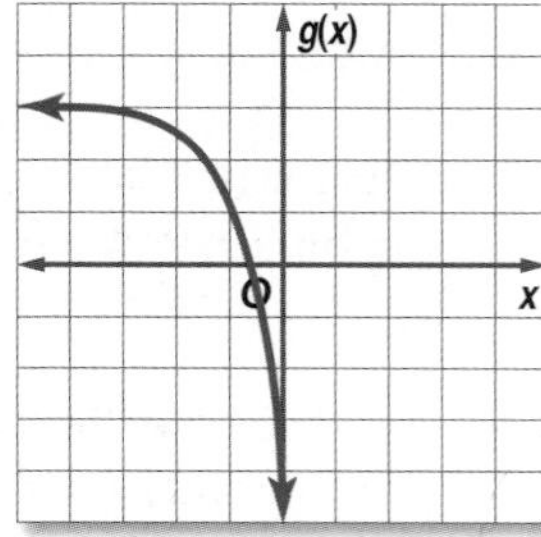

**33** **MULTIPLE REPRESENTATIONS** In this problem, you will use the tables below for exponential functions $f(x)$, $g(x)$, and $h(x)$.

| $x$ | −1 | 0 | 1 | 2 | 3 | 4 | 5 |
|---|---|---|---|---|---|---|---|
| $f(x)$ | 2.5 | 2 | 1 | −1 | −5 | −13 | −29 |

| $x$ | −1 | 0 | 1 | 2 | 3 | 4 | 5 |
|---|---|---|---|---|---|---|---|
| $g(x)$ | 5 | 11 | 23 | 47 | 95 | 191 | 383 |

| $x$ | −1 | 0 | 1 | 2 | 3 | 4 | 5 |
|---|---|---|---|---|---|---|---|
| $h(x)$ | 3 | 2.5 | 2.25 | 2.125 | 2.0625 | 2.0313 | 2.0156 |

**a. Graphical** Graph the functions for $-1 \leq x \leq 5$ on separate graphs.

**b. Verbal** List any function with a negative coefficient. Explain your reasoning.

**c. Analytical** List any function with a graph that is translated to the left.

**d. Analytical** Determine which functions are growth models and which are decay models.

## H.O.T. Problems Use Higher-Order Thinking Skills

**34. REASONING** Determine whether each statement is *sometimes*, *always*, or *never* true. Explain your reasoning.

**a.** An exponential function of the form $y = ab^{x-h} + k$ has a $y$-intercept.

**b.** An exponential function of the form $y = ab^{x-h} + k$ has an $x$-intercept.

**c.** The function $f(x) = |b|^x$ is an exponential growth function if $b$ is an integer.

**35. CCSS CRITIQUE** Vince and Grady were asked to graph the following functions. Vince thinks they are the same, but Grady disagrees. Who is correct? Explain your reasoning.

| x | y |
|---|---|
| 0 | 2 |
| 1 | 1 |
| 2 | 0.5 |
| 3 | 0.25 |
| 4 | 0.125 |
| 5 | 0.0625 |
| 6 | 0.03125 |

an exponential function with rate of decay of $\frac{1}{2}$ and an initial amount of 2

**36. CHALLENGE** A substance decays 35% each day. After 8 days, there are 8 milligrams of the substance remaining. How many milligrams were there initially?

**37. OPEN ENDED** Give an example of a value of $b$ for which $f(x) = \left(\frac{8}{b}\right)^x$ represents exponential decay.

**38. WRITING IN MATH** Write the procedure for transforming the graph of $g(x) = b^x$ to the graph of $f(x) = ab^{x-h} + k$. Justify each step.

## Standardized Test Practice

**39. GRIDDED RESPONSE** In the figure, $\overline{PO} \parallel \overline{RN}$, $ON = 12$, $MN = 6$, and $RN = 4$. What is the length of $\overline{PO}$?

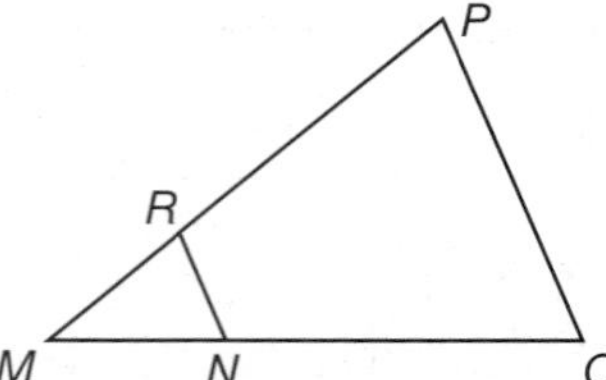


**40.** Ivan has enough money to buy 12 used CDs. If the cost of each CD was $0.20 less, Ivan could buy 2 more CDs. How much money does Ivan have to spend on CDs?

**A** $16.80
**B** $16.40
**C** $15.80
**D** $15.40

**41.** One hundred students will attend the fall dance if tickets cost $30 each. For each $5 increase in price, 10 fewer students will attend. What price will deliver the maximum dollar sales?

**F** $30
**G** $35
**H** $40
**J** $45

**42. SAT/ACT** Javier mows a lawn in 2 hours. Tonya mows the same lawn in 1.5 hours. About how many minutes will it take to mow the lawn if Javier and Tonya work together?

**A** 28 minutes
**B** 42 minutes
**C** 51 minutes
**D** 1.2 hours
**E** 1.4 hours

## Spiral Review

**Match each equation to its graph.** (Lesson 2-3)

**A**

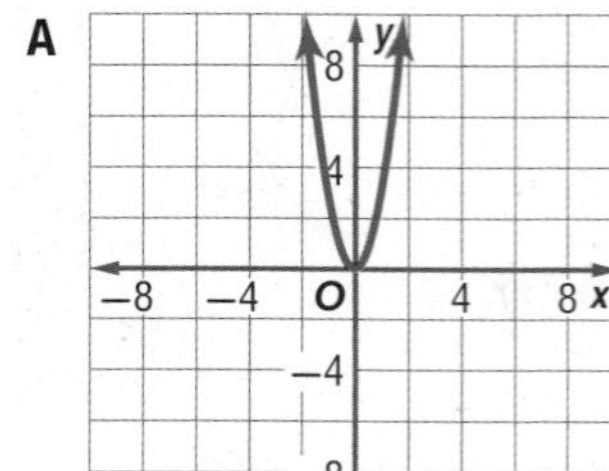

**B**

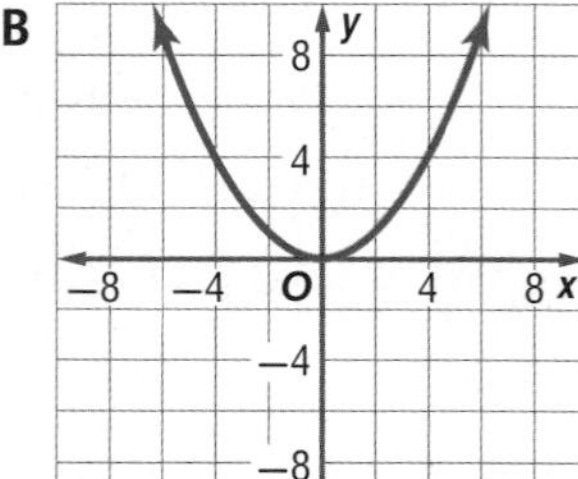

**C**

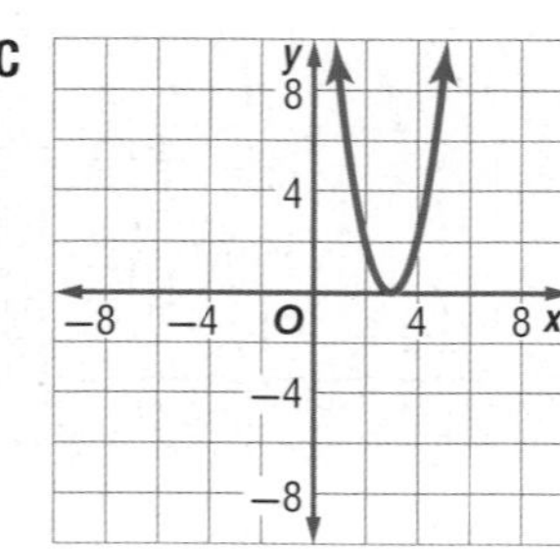

**D**

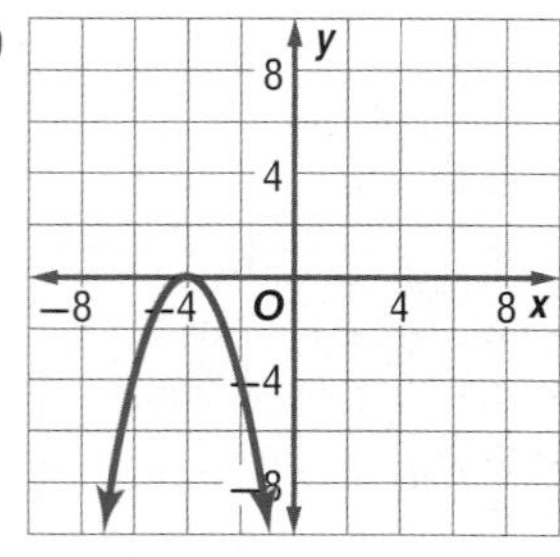

**43.** $y = 3x^2$

**44.** $y = \frac{1}{4}x^2$

**45.** $y = -(x + 4)^2$

**46.** $y = 2(x - 3)^2$

**47. FOOTBALL** The path of a football thrown across a field is given by the equation $y = -0.005x^2 + x + 5$, where $x$ represents the distance, in feet, the ball has traveled horizontally and $y$ represents the height, in feet, of the ball above ground level. About how far has the ball traveled horizontally when it returns to ground level? (Lesson 4-6)

## Skills Review

**Simplify. Assume that no variable equals 0.**

**48.** $f^{-7} \cdot f^4$

**49.** $(3x^2)^3$

**50.** $(2y)(4xy^3)$

**51.** $\left(\frac{3}{5}c^2f\right)\left(\frac{4}{3}cd\right)^2$

EXPLORE 4-2

# Graphing Technology Lab
# Solving Exponential Equations and Inequalities

You can use a TI-83/84 Plus graphing calculator to solve exponential equations by graphing or by using the table feature. To do this, you will write the equations as systems of equations.

**CCSS Common Core State Standards**
**Content Standards**
**A.REI.11** Explain why the $x$-coordinates of the points where the graphs of the equations $y = f(x)$ and $y = g(x)$ intersect are the solutions of the equation $f(x) = g(x)$; find the solutions approximately, e.g., using technology to graph the functions, make tables of values, or find successive approximations. Include cases where $f(x)$ and/or $g(x)$ are linear, polynomial, rational, absolute value, exponential, and logarithmic functions.

## Activity 1

**Solve $3^{x-4} = \frac{1}{9}$.**

**Step 1** Graph each side of the equation as a separate function. Enter $3^{x-4}$ as **Y1**. Be sure to include parentheses around the exponent. Enter $\frac{1}{9}$ as **Y2**. Then graph the two equations.

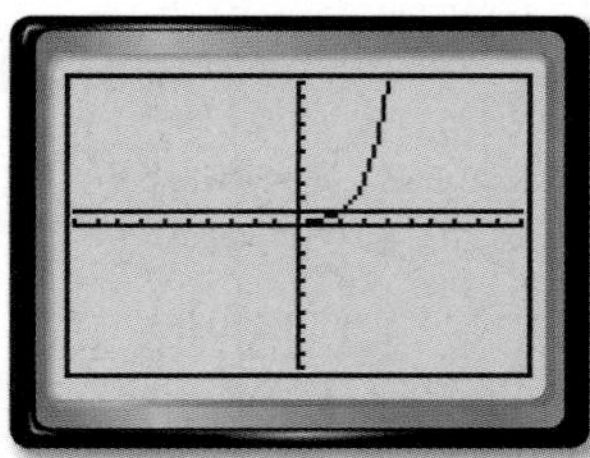

[−10, 10] scl: 1 by [−1, 1] scl: 0.1

**Step 2** Use the **intersect** feature.

You can use the **intersect** feature on the **CALC** menu to approximate the ordered pair of the point at which the graphs cross.

The calculator screen shows that the $x$-coordinate of the point at which the curves cross is 2. Therefore, the solution of the equation is 2.

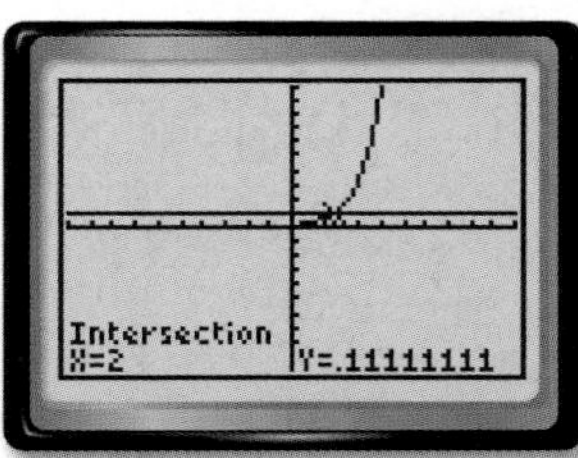


[−10, 10] scl: 1 by [−1, 1] scl: 0.1

**Step 3** Use the **TABLE** feature.

You can also use the **TABLE** feature to locate the point at which the curves intersect.

The table displays $x$-values and corresponding $y$-values for each graph. Examine the table to find the $x$-value for which the $y$-values of the graphs are equal.

At $x = 2$, both functions have a $y$-value of $0.\overline{1}$ or $\frac{1}{9}$. Thus, the solution of the equation is 2.

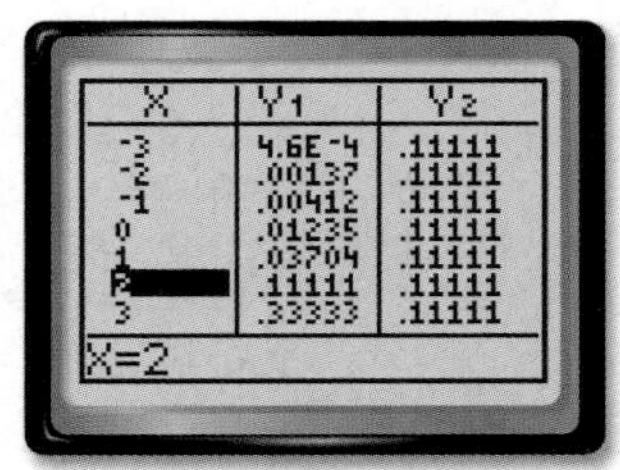

| X | Y1 | Y2 |
|---|---|---|
| -3 | 4.6E-4 | .11111 |
| -2 | .00137 | .11111 |
| -1 | .00412 | .11111 |
| 0 | .01235 | .11111 |
| 1 | .03704 | .11111 |
| 2 | .11111 | .11111 |
| 3 | .33333 | .11111 |

X=2

**CHECK** Substitute 2 for $x$ in the original equation.

$3^{x-4} \stackrel{?}{=} \frac{1}{9}$ Original equation

$3^{2-4} \stackrel{?}{=} \frac{1}{9}$ Substitute 2 for $x$.

$3^{-2} \stackrel{?}{=} \frac{1}{9}$ Simplify.

$\frac{1}{9} = \frac{1}{9}$ ✓ The solution checks.

A similar procedure can be used to solve exponential inequalities.

*(continued on the next page)*

# Graphing Technology Lab
# Solving Exponential Equations and Inequalities *Continued*

## Activity 2 Description

**Solve $2^{x-2} \geq 0.5^{x-3}$.**

**Step 1** Enter the related inequalities.

Rewrite the problem as a system of inequalities.

The first inequality is $2^{x-2} \geq y$ or $y \leq 2^{x-2}$. Since this inequality includes the *less than or equal to* symbol, shade below the curve.

First enter the boundary, and then use the arrow and [ENTER] keys to choose the shade below icon, ◣.

The second inequality is $y \geq 0.5^{x-3}$. Shade above the curve since this inequality contains *greater than or equal to.*

**KEYSTROKES:** [Y=] [◀] [◀] [ENTER] [ENTER] [ENTER] [▶] [▶] 2 [^] [(] [X,T,θ,n] [−] 2 [)] [ENTER] [◀] [◀] [ENTER] [ENTER] [▶] [▶] .5 [^] [(] [X,T,θ,n] [−] 3 [)]

Plot1 Plot2 Plot3
\Y1=2^(X−2)
\Y2=.5^(X−3)
\Y3=
\Y4=
\Y5=
\Y6=
\Y7=

**Step 2** Graph the system.

**KEYSTROKES:** [GRAPH]

The $x$-values of the points in the region where the shadings overlap is the solution set of the original inequality. Using the **intersect** feature, you can conclude that the solution set is $\{x \mid x \geq 2.5\}$.

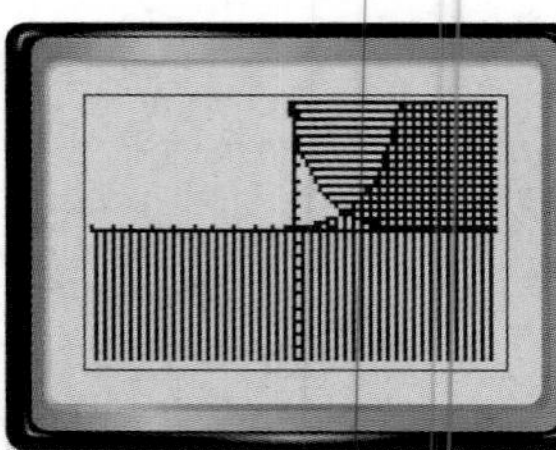

[−10, 10] scl: 1 by [−10, 10] scl: 1

**Step 3** Use the **TABLE** feature.

Verify using the **TABLE** feature. Set up the table to show $x$-values in increments of 0.5.

**KEYSTROKES:** [2nd] [TBLSET] 0 [ENTER] .5 [ENTER] [2nd] [TABLE]

Notice that for $x$-values greater than $x = 2.5$, **Y1** > **Y2**. This confirms that the solution of the inequality is $\{x \mid x \geq 2.5\}$.

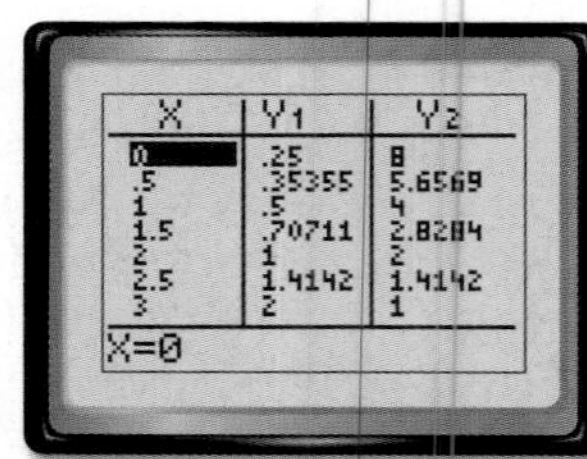

| X | Y1 | Y2 |
|---|---|---|
| 0 | .25 | 8 |
| .5 | .35355 | 5.6569 |
| 1 | .5 | 4 |
| 1.5 | .70711 | 2.8284 |
| 2 | 1 | 2 |
| 2.5 | 1.4142 | 1.4142 |
| 3 | 2 | 1 |

X=0

## Exercises

**Solve each equation or inequality.**

**1.** $9^{x-1} = \frac{1}{81}$

**2.** $4^{x+3} = 2^{5x}$

**3.** $5^{x-1} = 2^x$

**4.** $3.5^{x+2} = 1.75^{x+3}$

**5.** $-3^{x+4} = -0.5^{2x+3}$

**6.** $6^{2-x} - 4 < -0.25^{x-2.5}$

**7.** $16^{x-1} > 2^{2x+2}$

**8.** $3^x - 4 \leq 5^{\frac{x}{2}}$

**9.** $5^{x+3} \leq 2^{x+4}$

**10.** **WRITING IN MATH** Explain why this technique of graphing a system of equations or inequalities works to solve exponential equations and inequalities.

LESSON 4-2

# Solving Exponential Equations and Inequalities

| Then | Now | Why? |
|---|---|---|
| You graphed exponential functions. | **1** Solve exponential equations. **2** Solve exponential inequalities. | Membership on Internet social networking sites tends to increase exponentially. The membership growth of one Web site can be modeled by the equation $y = 2.2(1.37)^x$, where $x$ is the number of years since 2004 and $y$ is the number of members in millions. You can use $y = 2.2(1.37)^x$ to determine how many members there will be in a given year, or to determine the year in which membership was at a certain level. |

**NewVocabulary**
exponential equation
compound interest
exponential inequality

**Common Core State Standards**

**Content Standards**

**A.CED.1** Create equations and inequalities in one variable and use them to solve problems.

**F.LE.4** For exponential models, express as a logarithm the solution to $ab^{ct} = d$ where $a$, $c$, and $d$ are numbers and the base $b$ is 2, 10, or $e$; evaluate the logarithm using technology.

**Mathematical Practices**

2 Reason abstractly and quantitatively.

## 1 Solve Exponential Equations

In an **exponential equation**, variables occur as exponents.

**KeyConcept** Property of Equality for Exponential Functions

| | |
|---|---|
| Words | Let $b > 0$ and $b \neq 1$. Then $b^x = b^y$ if and only if $x = y$. |
| Example | If $3^x = 3^5$, then $x = 5$. If $x = 5$, then $3^x = 3^5$. |

The Property of Equality can be used to solve exponential equations.

### Example 1 Solve Exponential Equations

**Solve each equation.**

**a.** $2^x = 8^3$

| | |
|---|---|
| $2^x = 8^3$ | Original equation |
| $2^x = (2^3)^3$ | Rewrite 8 as $2^3$. |
| $2^x = 2^9$ | Power of a Power |
| $x = 9$ | Property of Equality for Exponential Functions |

**b.** $9^{2x-1} = 3^{6x}$

| | |
|---|---|
| $9^{2x-1} = 3^{6x}$ | Original equation |
| $(3^2)^{2x-1} = 3^{6x}$ | Rewrite 9 as $3^2$. |
| $3^{4x-2} = 3^{6x}$ | Power of a Power |
| $4x - 2 = 6x$ | Property of Equality for Exponential Functions |
| $-2 = 2x$ | Subtract $4x$ from each side. |
| $-1 = x$ | Divide each side by 2. |

**GuidedPractice**

**1A.** $4^{2n-1} = 64$

**1B.** $5^{5x} = 125^{x+2}$

Susan Wides/UpperCut Images/Getty Images

You can use information about growth or decay to write the equation of an exponential function.

PT

## Real-World Example 2 Write an Exponential Function

**SCIENCE Kristin starts an experiment with 7500 bacteria cells. After 4 hours, there are 23,000 cells.**

**a. Write an exponential function that could be used to model the number of bacteria after $x$ hours if the number of bacteria changes at the same rate.**

At the beginning of the experiment, the time is 0 hours and there are 7500 bacteria cells. Thus, the $y$-intercept, and the value of $a$, is 7500.

When $x = 4$, the number of bacteria cells is 23,000. Substitute these values into an exponential function to determine the value of $b$.

| | |
|---|---|
| $y = ab^x$ | Exponential function |
| $23{,}000 = 7500 \cdot b^4$ | Replace $x$ with 4, $y$ with 23,000, and $a$ with 7500. |
| $3.067 \approx b^4$ | Divide each side by 7500. |
| $\sqrt[4]{3.067} \approx b$ | Take the 4th root of each side. |
| $1.323 \approx b$ | Use a calculator. |

An equation that models the number of bacteria is $y \approx 7500(1.323)^x$.

**b. How many bacteria cells can be expected in the sample after 12 hours?**

| | |
|---|---|
| $y \approx 7500(1.323)^x$ | Modeling equation |
| $\approx 7500(1.323)^{12}$ | Replace $x$ with 12. |
| $\approx 215{,}665$ | Use a calculator. |

There will be approximately 215,665 bacteria cells after 12 hours.

### GuidedPractice

2. **RECYCLING** A manufacturer distributed 3.2 million aluminum cans in 2005.
   - **A.** In 2010, the manufacturer distributed 420,000 cans made from the recycled cans it had previously distributed. Assuming that the recycling rate continues, write an equation to model the distribution each year of cans that are made from recycled aluminum.
   - **B.** How many cans made from recycled aluminum can be expected in the year 2050?

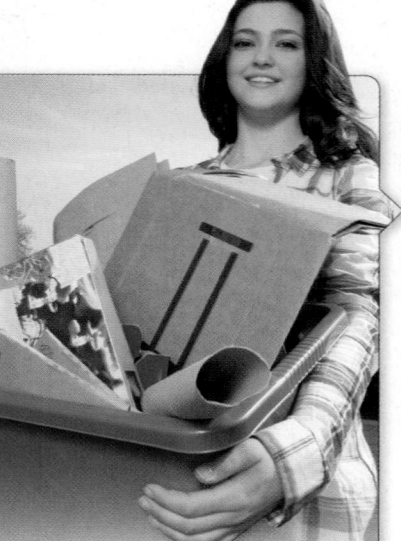

### Real-WorldLink

In 2008, the U.S. recycling rate for metals of 35% prevented the release of approximately 25 million metric tons of carbon into the air—roughly the amount emitted annually by 4.5 million cars.

**Source:** Environmental Protection Agency

Exponential functions are used in situations involving compound interest. **Compound interest** is interest paid on the principal of an investment and any previously earned interest.

### KeyConcept Compound Interest

You can calculate compound interest using the following formula.

$$A = P\left(1 + \frac{r}{n}\right)^{nt},$$

where $A$ is the amount in the account after $t$ years, $P$ is the principal amount invested, $r$ is the annual interest rate, and $n$ is the number of compounding periods each year.

## Example 3 Compound Interest

**An investment account pays 4.2% annual interest compounded monthly. If $2500 is invested in this account, what will be the balance after 15 years?**

**Understand** Find the total amount in the account after 15 years.

**Plan** Use the compound interest formula.
$P = 2500$, $r = 0.042$, $n = 12$, and $t = 15$

**Solve**

$A = P\left(1 + \frac{r}{n}\right)^{nt}$ Compound Interest Formula

$= 2500\left(1 + \frac{0.042}{12}\right)^{12 \cdot 15}$ $P = 2500$, $r = 0.042$, $n = 12$, $t = 15$

$\approx 4688.87$ Use a calculator.

**Check** Graph the corresponding equation $y = 2500(1.0035)^{12t}$. Use **CALC: value** to find $y$ when $x = 15$.

The $y$-value 4688.8662 is very close to 4688.87, so the answer is reasonable.

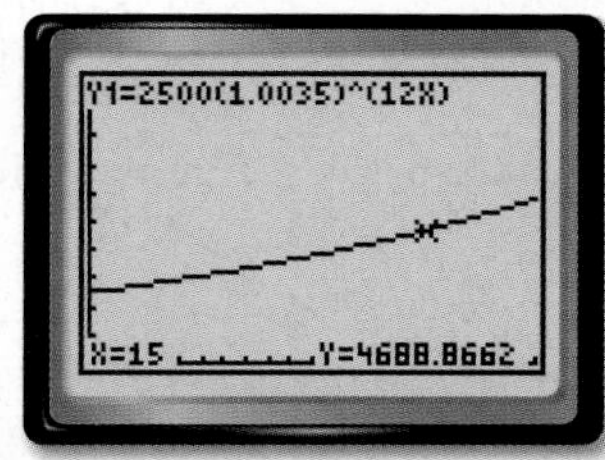


[0, 20] scl: 1 by [0, 10,000] scl: 1000

**WatchOut!**
**Percents** Remember to convert all percents to decimal form; 4.2% is 0.042.

### GuidedPractice

**3.** Find the account balance after 20 years if $100 is placed in an account that pays 1.2% interest compounded twice a month.

## 2 Solve Exponential Inequalities

An **exponential inequality** is an inequality involving exponential functions.

### KeyConcept Property of Inequality for Exponential Functions

| | |
|---|---|
| Words | Let $b > 1$. Then $b^x > b^y$ if and only if $x > y$, and $b^x < b^y$ if and only if $x < y$. |
| Example | If $2^x > 2^6$, then $x > 6$. If $x > 6$, then $2^x > 2^6$. |

This property also holds true for $\leq$ and $\geq$.

## Example 4 Solve Exponential Inequalities

**Solve $16^{2x-3} < 8$.**

| | |
|---|---|
| $16^{2x-3} < 8$ | Original inequality |
| $(2^4)^{2x-3} < 2^3$ | Rewrite 16 as $2^4$ and 8 as $2^3$. |
| $2^{8x-12} < 2^3$ | Power of a Power |
| $8x - 12 < 3$ | Property of Inequality for Exponential Functions |
| $8x < 15$ | Add 12 to each side. |
| $x < \frac{15}{8}$ | Divide each side by 8. |

### GuidedPractice

**Solve each inequality.**

**4A.** $3^{2x-1} \geq \frac{1}{243}$

**4B.** $2^{x+2} > \frac{1}{32}$

## Check Your Understanding

**Example 1** **Solve each equation.**

1. $3^{5x} = 27^{2x-4}$
2. $16^{2y-3} = 4^{y+1}$
3. $2^{6x} = 32^{x-2}$
4. $49^{x+5} = 7^{8x-6}$

**Example 2**

5. **SCIENCE** Mitosis is a process in which one cell divides into two. The *Escherichia coli* is one of the fastest growing bacteria. It can reproduce itself in 15 minutes.
   a. Write an exponential function to represent the number of cells $c$ after $t$ minutes.
   b. If you begin with one *Escherichia coli* cell, how many cells will there be in one hour?

**Example 3**

6. A certificate of deposit (CD) pays 2.25% annual interest compounded biweekly. If you deposit $500 into this CD, what will the balance be after 6 years?

**Example 4** **Solve each inequality.**

7. $4^{2x+6} \leq 64^{2x-4}$
8. $25^{y-3} \leq \left(\frac{1}{125}\right)^{y+2}$

## Practice and Problem Solving

**Example 1** **Solve each equation.**

9. $8^{4x+2} = 64$
10. $5^{x-6} = 125$
11. $81^{a+2} = 3^{3a+1}$
12. $256^{b+2} = 4^{2-2b}$
13. $9^{3c+1} = 27^{3c-1}$
14. $8^{2y+4} = 16^{y+1}$

**Example 2**

15. **CCSS MODELING** In 2009, My-Lien received $10,000 from her grandmother. Her parents invested all of the money, and by 2021, the amount will have grown to $16,960.
   a. Write an exponential function that could be used to model the money $y$. Write the function in terms of $x$, the number of years since 2009.
   b. Assume that the amount of money continues to grow at the same rate. What would be the balance in the account in 2031?

**Write an exponential function for the graph that passes through the given points.**

16. (0, 6.4) and (3, 100)
17. (0, 256) and (4, 81)
18. (0, 128) and (5, 371,293)
19. (0, 144), and (4, 21,609)

**Example 3**

20. Find the balance of an account after 7 years if $700 is deposited into an account paying 4.3% interest compounded monthly.
21. Determine how much is in a retirement account after 20 years if $5000 was invested at 6.05% interest compounded weekly.
22. A savings account offers 0.7% interest compounded bimonthly. If $110 is deposited in this account, what will the balance be after 15 years?
23. A college savings account pays 13.2% annual interest compounded semiannually. What is the balance of an account after 12 years if $21,000 was initially deposited?

**Example 4** **Solve each inequality.**

24. $625 \geq 5^{a+8}$
25. $10^{5b+2} > 1000$
26. $\left(\frac{1}{64}\right)^{c-2} < 32^{2c}$
27. $\left(\frac{1}{27}\right)^{2d-2} \leq 81^{d+4}$
28. $\left(\frac{1}{9}\right)^{3t+5} \geq \left(\frac{1}{243}\right)^{t-6}$
29. $\left(\frac{1}{36}\right)^{w+2} < \left(\frac{1}{216}\right)^{4w}$

**30.** **SCIENCE** A mug of hot chocolate is 90°C at time $t = 0$. It is surrounded by air at a constant temperature of 20°C. If stirred steadily, its temperature in Celsius after $t$ minutes will be $y(t) = 20 + 70(1.071)^{-t}$.

**a.** Find the temperature of the hot chocolate after 15 minutes.

**b.** Find the temperature of the hot chocolate after 30 minutes.

**c.** The optimum drinking temperature is 60°C. Will the mug of hot chocolate be at or below this temperature after 10 minutes?

**31** **ANIMALS** Studies show that an animal will defend a territory, with area in square yards, that is directly proportional to the 1.31 power of the animal's weight in pounds.

**a.** If a 45-pound beaver will defend 170 square yards, write an equation for the area $a$ defended by a beaver weighing $w$ pounds.

**b.** Scientists believe that thousands of years ago, the beaver's ancestors were 11 feet long and weighed 430 pounds. Use your equation to determine the area defended by these animals.

**Solve each equation.**

**32.** $\left(\frac{1}{2}\right)^{4x+1} = 8^{2x+1}$

**33.** $\left(\frac{1}{5}\right)^{x-5} = 25^{3x+2}$

**34.** $216 = \left(\frac{1}{6}\right)^{x+3}$

**35.** $\left(\frac{1}{8}\right)^{3x+4} = \left(\frac{1}{4}\right)^{-2x+4}$

**36.** $\left(\frac{2}{3}\right)^{5x+1} = \left(\frac{27}{8}\right)^{x-4}$

**37.** $\left(\frac{25}{81}\right)^{2x+1} = \left(\frac{729}{125}\right)^{-3x+1}$

**38.** CCSS **MODELING** In 1950, the world population was about 2.556 billion. By 1980, it had increased to about 4.458 billion.

**a.** Write an exponential function of the form $y = abx$ that could be used to model the world population $y$ in billions for 1950 to 1980. Write the equation in terms of $x$, the number of years since 1950. (Round the value of $b$ to the nearest ten-thousandth.)

**b.** Suppose the population continued to grow at that rate. Estimate the population in 2000.

**c.** In 2000, the population of the world was about 6.08 billion. Compare your estimate to the actual population.

**d.** Use the equation you wrote in Part **a** to estimate the world population in the year 2020. How accurate do you think the estimate is? Explain your reasoning.

**39.** **TREES** The diameter of the base of a tree trunk in centimeters varies directly with the $\frac{3}{2}$ power of its height in meters.

**a.** A young sequoia tree is 6 meters tall, and the diameter of its base is 19.1 centimeters. Use this information to write an equation for the diameter $d$ of the base of a sequoia tree if its height is $h$ meters high.

**b.** The General Sherman Tree in Sequoia National Park, California, is approximately 84 meters tall. Find the diameter of the General Sherman Tree at its base.

**40.** **FINANCIAL LITERACY** Mrs. Jackson has two different retirement investment plans from which to choose.

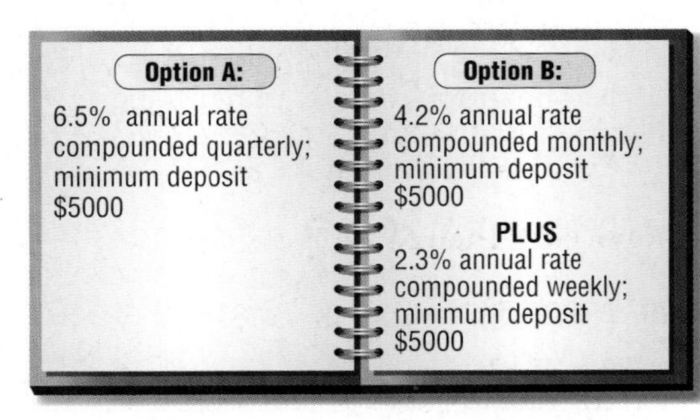
Option A:
6.5% annual rate compounded quarterly; minimum deposit $5000
Option B:
4.2% annual rate compounded monthly; minimum deposit $5000
PLUS
2.3% annual rate compounded weekly; minimum deposit $5000

**a.** Write equations for Option A and Option B given the minimum deposits.

**b.** Draw a graph to show the balances for each investment option after $t$ years.

**c.** Explain whether Option A or Option B is the better investment choice.

41. **MULTIPLE REPRESENTATIONS** In this problem, you will explore the rapid increase of an exponential function. A large sheet of paper is cut in half, and one of the resulting pieces is placed on top of the other. Then the pieces in the stack are cut in half and placed on top of each other. Suppose this procedure is repeated several times.

   a. **Concrete** Perform this activity and count the number of sheets in the stack after the first cut. How many pieces will there be after the second cut? How many pieces after the third cut? How many pieces after the fourth cut?

   b. **Tabular** Record your results in a table.

   c. **Symbolic** Use the pattern in the table to write an equation for the number of pieces in the stack after $x$ cuts.

   d. **Analytical** The thickness of ordinary paper is about 0.003 inch. Write an equation for the thickness of the stack of paper after $x$ cuts.

   e. **Analytical** How thick will the stack of paper be after 30 cuts?

## H.O.T. Problems Use Higher-Order Thinking Skills

42. **WRITING IN MATH** In a problem about compound interest, describe what happens as the compounding period becomes more frequent while the principal and overall time remain the same.

43. **ERROR ANALYSIS** Beth and Liz are solving $6^{x-3} > 36^{-x-1}$. Is either of them correct? Explain your reasoning.

| Beth | Liz |
|---|---|
| $6^{x-3} > 36^{-x-1}$ | $6^{x-3} > 36^{-x-1}$ |
| $6^{x-3} > (6^2)^{-x-1}$ | $6^{x-3} > (6^2)^{-x-1}$ |
| $6^{x-3} > 6^{-2x-2}$ | $6^{x-3} > 6^{-x+1}$ |
| $x - 3 > -2x - 2$ | $x - 3 > -x + 1$ |
| $3x > 1$ | $2x > 4$ |
| $x > \frac{1}{3}$ | $x > 2$ |

44. **CHALLENGE** Solve for $x$: $16^{18} + 16^{18} + 16^{18} + 16^{18} + 16^{18} = 4^x$.

45. **OPEN ENDED** What would be a more beneficial change to a 5-year loan at 8% interest compounded monthly: reducing the term to 4 years or reducing the interest rate to 6.5%?

46. **CCSS ARGUMENTS** Determine whether the following statements are *sometimes*, *always*, or *never* true. Explain your reasoning.

   a. $2^x > -8^{20x}$ for all values of $x$.

   b. The graph of an exponential growth equation is increasing.

   c. The graph of an exponential decay equation is increasing.

47. **OPEN ENDED** Write an exponential inequality with a solution of $x \leq 2$.

48. **PROOF** Show that $27^{2x} \cdot 81^{x+1} = 3^{2x+2} \cdot 9^{4x+1}$.

49. **WRITING IN MATH** If you were given the initial and final amounts of a radioactive substance and the amount of time that passes, how would you determine the rate at which the amount was increasing or decreasing in order to write an equation?

## Standardized Test Practice

**50.** $3 \times 10^{-4} =$

A 0.003
B 0.0003
C 0.00003
D 0.000003

**51.** Which of the following could *not* be a solution to $5 - 3x < -3$?

F 2.5
G 3
H 3.5
J 4

**52. GRIDDED RESPONSE** The three angles of a triangle are $3x$, $x + 10$, and $2x - 40$. Find the measure of the smallest angle in the triangle.

**53. SAT/ACT** Which of the following is equivalent to $(x)(x)(x)(x)$ for all $x$?

A $x + 4$
B $4x$
C $2x^2$
D $4x^2$
E $x^4$

## Spiral Review

**Graph each function.** (Lesson 4-1)

**54.** $y = 2(3)^x$

**55.** $y = 5(2)^x$

**56.** $y = 4\left(\frac{1}{3}\right)^x$

**Use the Distributive Property to factor each polynomial.** (Lesson 1-5)

**57.** $4m^3n^2 + 16m^2n^3 - 8m^3n^4$

**58.** $12j^4k^4 + 36j^3k^2 - 3j^2k^5$

**Factor each polynomial.** (Lesson 0-3)

**59.** $x^2 - 4x + 3xy - 12y$

**60.** $4a - 10ab + 6b - 15b^2$

**Write an equation in vertex form for each parabola.** (Lesson 4-7)

**61.**

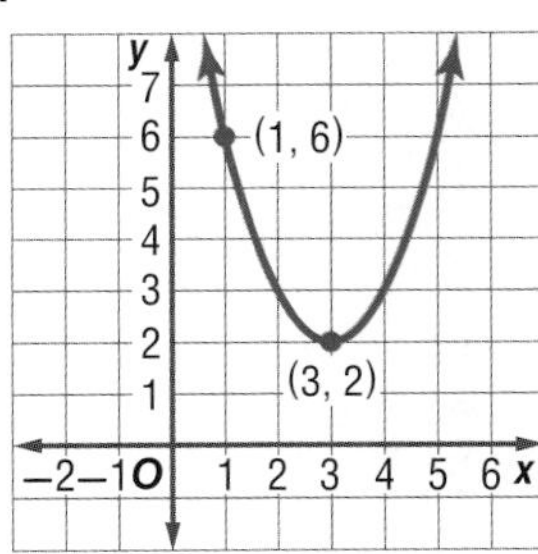


**62.**

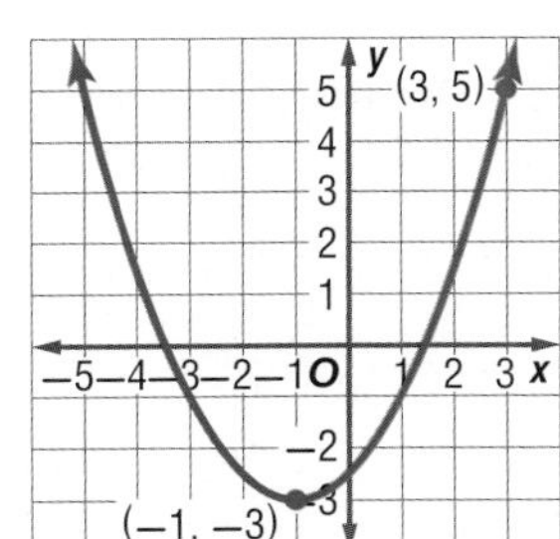


**Graph each function. State the domain and range.** (Lesson 4-1)

**63.** $f(x) = \frac{2}{3}(2^x)$

**64.** $f(x) = 4^x + 3$

**65.** $f(x) = 2\left(\frac{1}{3}\right)^x - 1$

**66. PRIZES** A machine is used to throw T-shirts into the crowd at the Hornets basketball games. (Lesson 1-7)

**a.** What is the initial height of the T-shirt?

**b.** If the T-shirt is caught after 2 seconds, what is the height?

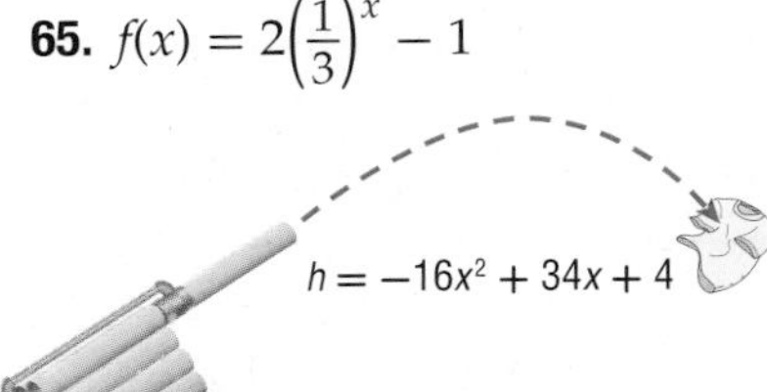


## Skills Review

**Find $[g \circ h](x)$ and $[h \circ g](x)$.**

**66.** $h(x) = 2x - 1$
$g(x) = 3x + 4$

**67.** $h(x) = x^2 + 2$
$g(x) = x - 3$

**68.** $h(x) = x^2 + 1$
$g(x) = -2x + 1$

**69.** $h(x) = -5x$
$g(x) = 3x - 5$

**70.** $h(x) = x^3$
$g(x) = x - 2$

**71.** $h(x) = x + 4$
$g(x) = |x|$

EXTEND 4-2

# Algebra Lab Transforming Exponential Expressions

You can use the properties of rational exponents to transform exponential functions into other forms in order to solve real-world problems.

**CCSS Common Core State Standards**
**Content Standards**
**A.SSE.3c** Use the properties of exponents to transform expressions for exponential functions.
**F.IF.8b** Use the properties of exponents to interpret expressions for exponential functions.

## Activity Write Equivalent Exponential Expressions

**Monique is trying to decide between two savings account plans. Plan A offers a monthly compounding interest rate of 0.25%, while Plan B offers 2.5% interest compounded annually. Which is the better plan? Explain.**

In order to compare the plans, we must compare rates with the same compounding frequency. One way to do this is to compare the approximate monthly interest rates of each plan, also called the *effective* monthly interest rate. While you can use the compound interest formula to find this rate, you can also use the properties of exponents.

Write a function to represent the amount $A$ Monique would earn after $t$ years with Plan B. For convenience, let the initial amount of Monique's investment be \$1.

| | |
|---|---|
| $y = a(1 + r)^t$ | Equation for exponential growth |
| $A(t) = 1(1 + 0.025)^t$ | $y = A(t)$, $a = 1$, $r = 2.5\%$ or $0.025$ |
| $= 1.025^t$ | Simplify. |

Now write a function equivalent to $A(t)$ that represents 12 compoundings per year, with a power of $12t$, instead of 1 per year, with a power of $1t$.

| | |
|---|---|
| $A(t) = 1.025^{1t}$ | Original function |
| $= 1.025^{\left(\frac{1}{12} \cdot 12\right)t}$ | $1 = \frac{1}{12} \cdot 12$ |
| $= \left(1.025^{\frac{1}{12}}\right)^{12t}$ | Power of a Power |
| $\approx 1.0021^{12t}$ | $(1.025)^{\frac{1}{12}} = \sqrt[12]{1.025}$ or about 1.0021 |

From this equivalent function, we can determine that the effective monthly interest by Plan B is about 0.0021 or about 0.21% per month. This rate is less than the monthly interest rate of 0.25% per month offered by Plan A, so Plan A is the better plan.

## Model and Analyze

1. Use the compound interest formula $A = P\left(1 + \frac{r}{n}\right)^{nt}$ to determine the effective monthly interest rate for Plan B. How does this rate compare to the rate calculated using the method in the Activity above?
2. Write a function to represent the amount $A$ Monique would earn after $t$ months by Plan A. Then use the properties of exponents to write a function equivalent to $A(t)$ that represents the amount earned after $t$ years.
3. From the expression you wrote in Exercise 2, identify the effective annual interest rate by Plan A. Use this rate to explain why Plan A is the better plan.
4. Suppose Plan A offered a quarterly compounded interest rate of 1.5%. Use the properties of exponents to explain which is the better plan.

LESSON 4-3

# Simplifying Radical Expressions

**Then**

- You simplified radicals.

**Now**

1. Simplify radical expressions by using the Product Property of Square Roots.
2. Simplify radical expressions by using the Quotient Property of Square Roots.

**Why?**

- The Sunshine Skyway Bridge across Florida's Tampa Bay is supported by 21 steel cables, each 9 inches in diameter.

To find the diameter a steel cable should have to support a given weight, you can use the equation $d = \sqrt{\frac{w}{8}}$, where $d$ is the diameter of the cable in inches and $w$ is the weight in tons.

**NewVocabulary**
radical expression
rationalizing the denominator
conjugate

**Common Core State Standards**

**Content Standards**
A.REI.4a Use the method of completing the square to transform any quadratic equation in $x$ into an equation of the form $(x - p)^2 = q$ that has the same solutions. Derive the quadratic formula from this form.

**Mathematical Practices**
7 Look for and make use of structure.
8 Look for and express regularity in repeated reasoning.

## 1 Product Property of Square Roots

A **radical expression** contains a radical, such as a square root. Recall the expression under the radical sign is called the radicand. A radicand is in simplest form if the following three conditions are true.

- No radicands have perfect square factors other than 1.
- No radicands contain fractions.
- No radicals appear in the denominator of a fraction.

The following property can be used to simplify square roots.

**KeyConcept** Product Property of Square Roots

| | |
|---|---|
| Words | For any nonnegative real numbers $a$ and $b$, the square root of $ab$ is equal to the square root of $a$ times the square root of $b$. |
| Symbols | $\sqrt{ab} = \sqrt{a} \cdot \sqrt{b}$, if $a \geq 0$ and $b \geq 0$ |
| Examples | $\sqrt{4 \cdot 9} = \sqrt{36}$ or 6 $\quad$ $\sqrt{4 \cdot 9} = \sqrt{4} \cdot \sqrt{9} = 2 \cdot 3$ or 6 |

PT

**Example 1** Simplify Square Roots

Simplify $\sqrt{80}$.

$\sqrt{80} = \sqrt{2 \cdot 2 \cdot 2 \cdot 2 \cdot 5}$ — Prime factorization of 80

$= \sqrt{2^2} \cdot \sqrt{2^2} \cdot \sqrt{5}$ — Product Property of Square Roots

$= 2 \cdot 2 \cdot \sqrt{5}$ or $4\sqrt{5}$ — Simplify.

**GuidedPractice**

**1A.** $\sqrt{54}$ $\qquad$ **1B.** $\sqrt{180}$

SuperStock/age fotostock

## Example 2 Multiply Square Roots

**Simplify $\sqrt{2} \cdot \sqrt{14}$.**

$\sqrt{2} \cdot \sqrt{14} = \sqrt{2} \cdot \sqrt{2} \cdot \sqrt{7}$ — Product Property of Square Roots

$= \sqrt{2^2} \cdot \sqrt{7}$ or $2\sqrt{7}$ — Product Property of Square Roots

### GuidedPractice

**2A.** $\sqrt{5} \cdot \sqrt{10}$

**2B.** $\sqrt{6} \cdot \sqrt{8}$

Consider the expression $\sqrt{x^2}$. It may seem that $x = \sqrt{x^2}$, but when finding the principal square root of an expression containing variables, you have to be sure that the result is not negative. Consider $x = -3$.

$\sqrt{x^2} \stackrel{?}{=} x$

$\sqrt{(-3)^2} \stackrel{?}{=} -3$ — Replace $x$ with $-3$.

$\sqrt{9} \stackrel{?}{=} -3$ — $(-3)^2 = 9$

$3 \neq -3$ — $\sqrt{9} = 3$

Notice in this case, if the right hand side of the equation were $|x|$, the equation would be true. For expressions where the exponent of the variable inside a radical is even and the simplified exponent is odd, you must use absolute value.

$\sqrt{x^2} = |x| \qquad \sqrt{x^3} = x\sqrt{x} \qquad \sqrt{x^4} = x^2 \qquad \sqrt{x^6} = |x^3|$

## Example 3 Simplify a Square Root with Variables

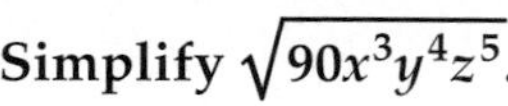

**Simplify $\sqrt{90x^3y^4z^5}$.**

$\sqrt{90x^3y^4z^5} = \sqrt{2 \cdot 3^2 \cdot 5 \cdot x^3 \cdot y^4 \cdot z^5}$ — Prime factorization

$= \sqrt{2} \cdot \sqrt{3^2} \cdot \sqrt{5} \cdot \sqrt{x^2} \cdot \sqrt{x} \cdot \sqrt{y^4} \cdot \sqrt{z^4} \cdot \sqrt{z}$ — Product Property

$= \sqrt{2} \cdot 3 \cdot \sqrt{5} \cdot x \cdot \sqrt{x} \cdot y^2 \cdot z^2 \cdot \sqrt{z}$ — Simplify.

$= 3y^2z^2x\sqrt{10xz}$ — Simplify.

### GuidedPractice

**3A.** $\sqrt{32r^2k^4t^5}$

**3B.** $\sqrt{56xy^{10}z^5}$

## 2 Quotient Property of Square Roots

**Quotient Property of Square Roots** To divide square roots and simplify radical expressions, you can use the Quotient Property of Square Roots.

### KeyConcept Quotient Property of Square Roots

| | |
|---|---|
| Words | For any real numbers $a$ and $b$, where $a \geq 0$ and $b > 0$, the square root of $\frac{a}{b}$ is equal to the square root of $a$ divided by the square root of $b$. |
| Symbols | $\sqrt{\frac{a}{b}} = \frac{\sqrt{a}}{\sqrt{b}}$ |

**ReadingMath**

**Fractions in the Radicand** The expression $\sqrt{\frac{a}{b}}$ is read *the square root of a over b*, or *the square root of the quantity of a over b.*

You can use the properties of square roots to **rationalize the denominator** of a fraction with a radical. This involves multiplying the numerator and denominator by a factor that eliminates radicals in the denominator.

## Standardized Test Example 4 Rationalize a Denominator

**Which expression is equivalent to $\sqrt{\frac{35}{15}}$?**

A $\frac{5\sqrt{21}}{15}$    B $\frac{\sqrt{21}}{3}$    C $\frac{\sqrt{525}}{15}$    D $\frac{\sqrt{35}}{15}$

> **Test-Taking Tip**
>
> **CCSS Structure** Look at the radicand to see if it can be simplified first. This may make your computations simpler.

**Read the Test Item** The radical expression needs to be simplified.

**Solve the Test Item**

$\sqrt{\frac{35}{15}} = \sqrt{\frac{7}{3}}$    Reduce $\frac{35}{15}$ to $\frac{7}{3}$.

$= \frac{\sqrt{7}}{\sqrt{3}}$    Quotient Property of Square Roots

$= \frac{\sqrt{7}}{\sqrt{3}} \cdot \frac{\sqrt{3}}{\sqrt{3}}$    Multiply by $\frac{\sqrt{3}}{\sqrt{3}}$.

$= \frac{\sqrt{21}}{3}$    Product Property of Square Roots

The correct choice is B.

**Guided Practice**

**4.** Simplify $\frac{\sqrt{6y}}{\sqrt{12}}$.

F $\frac{\sqrt{y}}{2}$    G $\frac{\sqrt{y}}{4}$    H $\frac{\sqrt{2y}}{2}$    J $\frac{\sqrt{2y}}{4}$

Binomials of the form $a\sqrt{b} + c\sqrt{d}$ and $a\sqrt{b} - c\sqrt{d}$, where $a$, $b$, $c$, and $d$ are rational numbers, are called **conjugates**. For example, $2 + \sqrt{7}$ and $2 - \sqrt{7}$ are conjugates. The product of two conjugates is a rational number and can be found using the pattern for the difference of squares.

## Example 5 Use Conjugates to Rationalize a Denominator

**Simplify $\frac{3}{5 + \sqrt{2}}$.**

$\frac{3}{5 + \sqrt{2}} = \frac{3}{5 + \sqrt{2}} \cdot \frac{5 - \sqrt{2}}{5 - \sqrt{2}}$    The conjugate of $5 + \sqrt{2}$ is $5 - \sqrt{2}$.

$= \frac{3(5 - \sqrt{2})}{5^2 - (\sqrt{2})^2}$    $(a - b)(a + b) = a^2 - b^2$

$= \frac{15 - 3\sqrt{2}}{25 - 2}$ or $\frac{15 - 3\sqrt{2}}{23}$    $(\sqrt{2})^2 = 2$

**Guided Practice** **Simplify each expression.**

**5A.** $\frac{3}{2 + \sqrt{2}}$    **5B.** $\frac{7}{3 - \sqrt{7}}$

## Check Your Understanding

**Examples 1–3** Simplify each expression.

1. $\sqrt{24}$
2. $3\sqrt{16}$
3. $2\sqrt{25}$
4. $\sqrt{10} \cdot \sqrt{14}$
5. $\sqrt{3} \cdot \sqrt{18}$
6. $3\sqrt{10} \cdot 4\sqrt{10}$
7. $\sqrt{60x^4y^7}$
8. $\sqrt{88m^3p^2r^5}$
9. $\sqrt{99ab^5c^2}$

**Example 4**

10. **MULTIPLE CHOICE** Which expression is equivalent to $\sqrt{\frac{45}{10}}$?

A $\frac{5\sqrt{2}}{10}$   B $\frac{\sqrt{45}}{10}$   C $\frac{\sqrt{50}}{10}$   D $\frac{3\sqrt{2}}{2}$

**Example 5** Simplify each expression.

11. $\frac{3}{3 + \sqrt{5}}$
12. $\frac{5}{2 - \sqrt{6}}$
13. $\frac{2}{1 - \sqrt{10}}$
14. $\frac{1}{4 + \sqrt{12}}$
15. $\frac{4}{6 - \sqrt{7}}$
16. $\frac{6}{5 + \sqrt{11}}$

## Practice and Problem Solving

**Examples 1–3** Simplify each expression.

17. $\sqrt{52}$
18. $\sqrt{56}$
19. $\sqrt{72}$
20. $3\sqrt{18}$
21. $\sqrt{243}$
22. $\sqrt{245}$
23. $\sqrt{5} \cdot \sqrt{10}$
24. $\sqrt{10} \cdot \sqrt{20}$
25. $3\sqrt{8} \cdot 2\sqrt{7}$
26. $4\sqrt{2} \cdot 5\sqrt{8}$
27. $3\sqrt{25t^2}$
28. $5\sqrt{81q^5}$
29. $\sqrt{28a^2b^3}$
30. $\sqrt{75qr^3}$
31. $7\sqrt{63m^3p}$
32. $4\sqrt{66g^2h^4}$
33. $\sqrt{2ab^2} \cdot \sqrt{10a^5b}$
34. $\sqrt{4c^3d^3} \cdot \sqrt{8c^3d}$

35. **ROLLER COASTER** Starting from a stationary position, the velocity $v$ of a roller coaster in feet per second at the bottom of a hill can be approximated by $v = \sqrt{64h}$, where $h$ is the height of the hill in feet.

   a. Simplify the equation.

   b. Determine the velocity of a roller coaster at the bottom of a 134-foot hill.

36. **CCSS PRECISION** When fighting a fire, the velocity $v$ of water being pumped into the air is modeled by the function $v = \sqrt{2hg}$, where $h$ represents the maximum height of the water and $g$ represents the acceleration due to gravity (32 ft/s$^2$).

   a. Solve the function for $h$.

   b. The Hollowville Fire Department needs a pump that will propel water 80 feet into the air. Will a pump advertised to project water with a velocity of 70 feet per second meet their needs? Explain.

   c. The Jackson Fire Department must purchase a pump that will propel water 90 feet into the air. Will a pump that is advertised to project water with a velocity of 77 feet per second meet the fire department's need? Explain.

**Examples 4–5** **Simplify each expression.**

**37.** $\sqrt{\frac{32}{t^4}}$

**38.** $\sqrt{\frac{27}{m^5}}$

**39.** $\frac{\sqrt{68ac^3}}{\sqrt{27a^2}}$

**40.** $\frac{\sqrt{h^3}}{\sqrt{8}}$

**41.** $\sqrt{\frac{3}{16}} \cdot \sqrt{\frac{9}{5}}$

**42.** $\sqrt{\frac{7}{2}} \cdot \sqrt{\frac{5}{3}}$

**43.** $\frac{7}{5 + \sqrt{3}}$

**44.** $\frac{9}{6 - \sqrt{8}}$

**45.** $\frac{3\sqrt{3}}{-2 + \sqrt{6}}$

**46.** $\frac{3}{\sqrt{7} - \sqrt{2}}$

**47.** $\frac{5}{\sqrt{6} + \sqrt{3}}$

**48.** $\frac{2\sqrt{5}}{2\sqrt{7} + 3\sqrt{3}}$

**49. ELECTRICITY** The amount of current in amperes $I$ that an appliance uses can be calculated using the formula $I = \sqrt{\frac{P}{R}}$, where $P$ is the power in watts and $R$ is the resistance in ohms.

**a.** Simplify the formula.

**b.** How much current does an appliance use if the power used is 75 watts and the resistance is 5 ohms?

**50. KINETIC ENERGY** The speed $v$ of a ball can be determined by the equation $v = \sqrt{\frac{2k}{m}}$, where $k$ is the kinetic energy and $m$ is the mass of the ball.

**a.** Simplify the formula if the mass of the ball is 3 kilograms.

**b.** If the ball is traveling 7 meters per second, what is the kinetic energy of the ball in Joules?

**51. SUBMARINES** The greatest distance $d$ in miles that a lookout can see on a clear day is modeled by the formula shown. Determine how high the submarine would have to raise its periscope to see a ship, if the submarine is the given distances away from the ship.

h ft

$d = \sqrt{\frac{3h}{2}}$

| Distance | 3 | 6 | 9 | 12 | 15 |
|---|---|---|---|---|---|
| Height | | | | | |

## H.O.T. Problems Use Higher-Order Thinking Skills

**52. CCSS STRUCTURE** Explain how to solve $\frac{\sqrt{3} + 2}{x} = \frac{\sqrt{3} - 1}{\sqrt{3}}$.

**53. CHALLENGE** Simplify each expression.

**a.** $\sqrt[3]{27}$ **b.** $\sqrt[3]{40}$ **c.** $\sqrt[3]{750}$

**54. REASONING** Marge takes a number, subtracts 4, multiplies by 4, takes the square root, and takes the reciprocal to get $\frac{1}{2}$. What number did she start with? Write a formula to describe the process.

**55. OPEN ENDED** Write two binomials of the form $a\sqrt{b} + c\sqrt{f}$ and $a\sqrt{b} - c\sqrt{f}$. Then find their product.

**56. CHALLENGE** Use the Quotient Property of Square Roots to derive the Quadratic Formula by solving the quadratic equation $ax^2 + bx + c = 0$. (*Hint*: Begin by completing the square.)

**57. WRITING IN MATH** Summarize how to write a radical expression in simplest form.

## Standardized Test Practice

**58.** Jerry's electric bill is $23 less than his natural gas bill. The two bills are a total of $109. Which of the following equations can be used to find the amount of his natural gas bill?

A $g + g = 109$ C $g - 23 = 109$
B $23 + 2g = 109$ D $2g - 23 = 109$

**59.** Solve $a^2 - 2a + 1 = 25$.

F $-4, -6$ H $-4, 6$
G $4, -6$ J $4, 6$

**60.** The expression $\sqrt{160x^2y^5}$ is equivalent to which of the following?

A $16|x|y^2\sqrt{10y}$ C $4|x|y^2\sqrt{10y}$
B $|x|y^2\sqrt{160y}$ D $10|x|y^2\sqrt{4y}$

**61. GRIDDED RESPONSE** Miki earns $10 an hour and 10% commission on sales. If Miki worked 38 hours and had a total sales of $1275 last week, how much did she make?

## Spiral Review

**Find each sum or difference.** (Lesson 1-1)

**62.** $(7g^3 + 2g^2 - 12) - (-2g^3 - 4g)$

**63.** $(-3h^2 + 3h - 6) + (5h^2 - 3h - 10)$

**Solve each equation by using the Quadratic Formula.** (Lesson 3-4)

**64.** $2x^2 - 3x - 9 = 0$

**65.** $4x^2 + 2x - 1 = 0$

**66.** $3x^2 + 4x - 2 = 0$

**67.** $3x^2 - 5x - 10 = 0$

**Factor each polynomial, if possible. If the polynomial cannot be factored using integers, write *prime*.** (Lesson 1-7)

**68.** $6x^2 + 21x - 90$

**69.** $3x^2 - 11x - 42$

**70.** $6x^2 - 13x - 5$

**71.** $5y^2 - 3y + 11$

**Look for a pattern in each table of values to determine which kind of model best describes the data.** (Lesson 2-6)

**72.**

| x | 2 | 3 | 4 | 5 | 6 |
|---|---|---|---|---|---|
| y | $\frac{9}{4}$ | $\frac{27}{8}$ | $\frac{81}{16}$ | $\frac{243}{32}$ | $\frac{729}{64}$ |

**73.**

| x | −2 | −1 | 0 | 1 | 2 |
|---|---|---|---|---|---|
| y | −13 | −6.25 | 0 | 5.75 | 11 |

**Graph each function. State the domain and range.** (Lesson 2-7)

**74.** $f(x) = |x + 5|$

**75.** $f(x) = 2[\![x]\!]$

**Solve each equation.** (Lesson 4-2)

**76.** $\left(\frac{1}{7}\right)^{y-3} = 343$

**77.** $10^{x-1} = 100^{2x-3}$

**78.** $36^{2p} = 216^{p-1}$

## Skills Review

**Write the prime factorization of each number.**

**79.** 24

**80.** 88

**81.** 180

**82.** 31

**83.** 60

**84.** 90

# EXTEND 4-3 Algebra Lab
# Rational and Irrational Numbers

A set is **closed** under an operation if for any numbers in the set, the result of the operation is also in the set. A set may be closed under one operation and not closed under another.

**CCSS Common Core State Standards**
**Content Standards**
**N.RN.3** Explain why the sum or product of two rational numbers is rational; that the sum of a rational number and an irrational number is irrational; and that the product of a nonzero rational number and an irrational number is irrational.
**Mathematical Practices**
7 Look for and make use of structure.

## Activity 1 Closure of Rational Numbers and Irrational Numbers

**Are the sets of rational and irrational numbers closed under multiplication? under addition?**

**Step 1** To determine if each set is closed under multiplication, examine several products of two rational factors and then two irrational factors.

Rational: $5 \times 2 = 10; -3 \times 4 = -12; 3.7 \times 0.5 = 1.85; \frac{3}{4} \times \frac{2}{3} = \frac{1}{2}$

Irrational: $\pi \times \sqrt{2} = \sqrt{2}\pi; \sqrt{3} \times \sqrt{7} = \sqrt{21}; \sqrt{5} \times \sqrt{5} = 5$

The product of each pair of rational numbers is rational. However, the products of pairs of irrational numbers are both irrational and rational. Thus, it appears that the set of rational numbers is closed under multiplication, but the set of irrational numbers is not.

**Step 2** Repeat this process for addition.

Rational: $3 + 8 = 11; -4 + 7 = 3; 3.7 + 5.82 = 9.52; \frac{2}{5} + \frac{1}{4} = \frac{13}{20}$

Irrational: $\sqrt{3} + \pi = \sqrt{3} + \pi; 3\sqrt{5} + 6\sqrt{5} = 9\sqrt{5}; \sqrt{12} + \sqrt{50} = 2\sqrt{3} + 5\sqrt{2}$

The sum of each pair of rational numbers is rational, and the sum of each pair of irrational numbers is irrational. Both sets are closed under addition.

## Activity 2 Rational and Irrational Numbers

**What kind of numbers are the product and sum of a rational and irrational number?**

**Step 1** Examine the products of several pairs of rational and irrational numbers.

$3 \times \sqrt{8} = 6\sqrt{2}; \frac{3}{4} \times \sqrt{2} = \frac{3\sqrt{2}}{4}; 1 \times \sqrt{7} = \sqrt{7}; 0 \times \sqrt{5} = 0$

The product is rational only when the rational factor is 0. The product of each nonzero rational number and irrational number is irrational.

**Step 2** Find the sums of several pairs of a rational and irrational number.

$5 + \sqrt{3} = 5 + \sqrt{3}; \frac{2}{3} + \sqrt{5} = \frac{2 + 3\sqrt{5}}{3}; -4 + \sqrt{6} = -1(4 - \sqrt{6})$

The sum of each rational and irrational number is irrational.

## Analyze the Results

1. What kinds of numbers are the difference of two unique rational numbers, two unique irrational numbers, and a rational and an irrational number?
2. Is the quotient of every rational and irrational number always another rational or irrational number? If not, provide a counterexample.
3. **CHALLENGE** Recall that rational numbers are numbers that can be written in the form $\frac{a}{b}$, where $a$ and $b$ are integers and $b \neq 0$. Using $\frac{a}{b}$ and $\frac{c}{d}$ show that the sum and product of two rational numbers must always be a rational number.

LESSON 4-4

# Operations with Radical Expressions

| :: Then | :: Now | :: Why? |
|---|---|---|
| You simplified radical expressions. | **1** Add and subtract radical expressions. **2** Multiply radical expressions. | Conchita is going to run in her neighborhood to get ready for the soccer season. She plans to run the course that she has laid out three times each day. How far does Conchita have to run to complete the course that she laid out? How far does she run every day? |

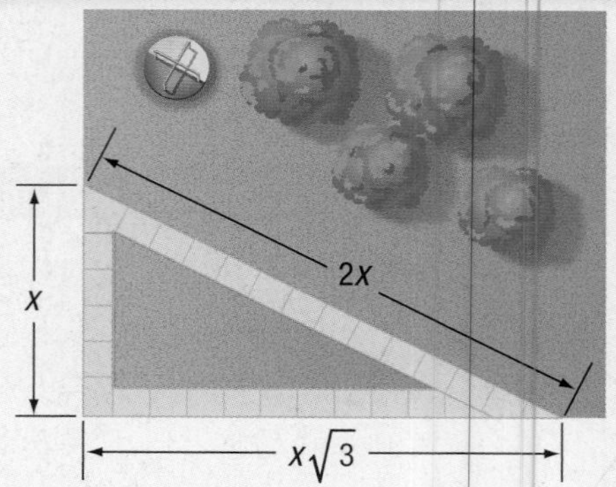

**Common Core State Standards**

**Content Standards**

N.RN.2 Rewrite expressions involving radicals and rational exponents using the properties of exponents.

**Mathematical Practices**

2 Reason abstractly and quantitatively.

## 1 Add or Subtract Radical Expressions

To add or subtract radical expressions, the radicands must be alike in the same way that monomial terms must be alike to add or subtract.

| Monomials | Radical Expressions |
|---|---|
| $4a + 2a = (4 + 2)a$ | $4\sqrt{5} + 2\sqrt{5} = (4 + 2)\sqrt{5}$ |
| $= 6a$ | $= 6\sqrt{5}$ |
| $9b - 2b = (9 - 2)b$ | $9\sqrt{3} - 2\sqrt{3} = (9 - 2)\sqrt{3}$ |
| $= 7b$ | $= 7\sqrt{3}$ |

Notice that when adding and subtracting radical expressions, the radicand does not change. This is the same as when adding or subtracting monomials.

### Example 1 Add and Subtract Expressions with Like Radicands

**Simplify each expression.**

**a.** $\mathbf{5\sqrt{2} + 7\sqrt{2} - 6\sqrt{2}}$

$5\sqrt{2} + 7\sqrt{2} - 6\sqrt{2} = (5 + 7 - 6)\sqrt{2}$ Distributive Property

$= 6\sqrt{2}$ Simplify.

**b.** $\mathbf{10\sqrt{7} + 5\sqrt{11} + 4\sqrt{7} - 6\sqrt{11}}$

$10\sqrt{7} + 5\sqrt{11} + 4\sqrt{7} - 6\sqrt{11} = (10 + 4)\sqrt{7} + (5 - 6)\sqrt{11}$ Distributive Property

$= 14\sqrt{7} - \sqrt{11}$ Simplify.

**GuidedPractice**

**1A.** $3\sqrt{2} - 5\sqrt{2} + 4\sqrt{2}$

**1B.** $6\sqrt{11} + 2\sqrt{11} - 9\sqrt{11}$

**1C.** $15\sqrt{3} - 14\sqrt{5} + 6\sqrt{5} - 11\sqrt{3}$

**1D.** $4\sqrt{3} + 3\sqrt{7} - 6\sqrt{3} + 3\sqrt{7}$

Not all radical expressions have like radicands. Simplifying the expressions may make it possible to have like radicands so that they can be added or subtracted.

**StudyTip**

**Simplify First** Simplify each radical term first. Then perform the operations needed.

## Example 2 Add and Subtract Expressions with Unlike Radicands

**Simplify $2\sqrt{18} + 2\sqrt{32} + \sqrt{72}$.**

$$\begin{aligned} 2\sqrt{18} + 2\sqrt{32} + \sqrt{72} &= 2(\sqrt{3^2} \cdot \sqrt{2}) + 2(\sqrt{4^2} \cdot \sqrt{2}) + (\sqrt{6^2} \cdot \sqrt{2}) && \text{Product Property} \\ &= 2(3\sqrt{2}) + 2(4\sqrt{2}) + (6\sqrt{2}) && \text{Simplify.} \\ &= 6\sqrt{2} + 8\sqrt{2} + 6\sqrt{2} && \text{Multiply.} \\ &= 20\sqrt{2} && \text{Simplify.} \end{aligned}$$

**GuidedPractice**

**2A.** $4\sqrt{54} + 2\sqrt{24}$

**2B.** $4\sqrt{12} - 6\sqrt{48}$

**2C.** $3\sqrt{45} + \sqrt{20} - \sqrt{245}$

**2D.** $\sqrt{24} - \sqrt{54} + \sqrt{96}$

## 2 Multiply Radical Expressions

Multiplying radical expressions is similar to multiplying monomial algebraic expressions. Let $x \geq 0$.

| Monomials | Radical Expressions |
|---|---|
| $(2x)(3x) = 2 \cdot 3 \cdot x \cdot x$ | $(2\sqrt{x})(3\sqrt{x}) = 2 \cdot 3 \cdot \sqrt{x} \cdot \sqrt{x}$ |
| $= 6x^2$ | $= 6x$ |

You can also apply the Distributive Property to radical expressions.

## Example 3 Multiply Radical Expressions

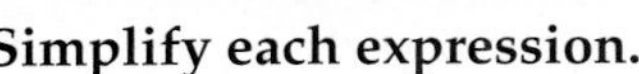
**Simplify each expression.**

**WatchOut!**

**Multiplying Radicands** Make sure that you multiply the radicands when multiplying radical expressions. A common mistake is to add the radicands rather than multiply.

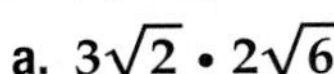
**a.** $3\sqrt{2} \cdot 2\sqrt{6}$

$$\begin{aligned} 3\sqrt{2} \cdot 2\sqrt{6} &= (3 \cdot 2)(\sqrt{2} \cdot \sqrt{6}) && \text{Associative Property} \\ &= 6(\sqrt{12}) && \text{Multiply.} \\ &= 6(2\sqrt{3}) && \text{Simplify.} \\ &= 12\sqrt{3} && \text{Multiply.} \end{aligned}$$

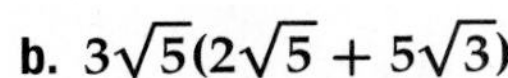
**b.** $3\sqrt{5}(2\sqrt{5} + 5\sqrt{3})$

$$\begin{aligned} 3\sqrt{5}(2\sqrt{5} + 5\sqrt{3}) &= (3\sqrt{5} \cdot 2\sqrt{5}) + (3\sqrt{5} \cdot 5\sqrt{3}) && \text{Distributive Property} \\ &= [(3 \cdot 2)(\sqrt{5} \cdot \sqrt{5})] + [(3 \cdot 5)(\sqrt{5} \cdot \sqrt{3})] && \text{Associative Property} \\ &= [6(\sqrt{25})] + [15(\sqrt{15})] && \text{Multiply.} \\ &= [6(5)] + [15(\sqrt{15})] && \text{Simplify.} \\ &= 30 + 15\sqrt{15} && \text{Multiply.} \end{aligned}$$

**GuidedPractice**

**3A.** $2\sqrt{6} \cdot 7\sqrt{3}$

**3B.** $9\sqrt{5} \cdot 11\sqrt{15}$

**3C.** $3\sqrt{2}(4\sqrt{3} + 6\sqrt{2})$

**3D.** $5\sqrt{3}(3\sqrt{2} - \sqrt{3})$

You can also multiply radical expressions with more than one term in each factor. This is similar to multiplying two algebraic binomials with variables.

## Real-World Example 4 Multiply Radical Expressions

**GEOMETRY Find the area of the rectangle in simplest form.**

$\sqrt{5} + 4\sqrt{3}$

$5\sqrt{2} - \sqrt{3}$

$A = (5\sqrt{2} - \sqrt{3})(\sqrt{5} + 4\sqrt{3})$ $\quad A = \ell \cdot w$

First Terms, Outer Terms, Inner Terms, Last Terms

$= (5\sqrt{2})(\sqrt{5}) + (5\sqrt{2})(4\sqrt{3}) + (-\sqrt{3})(\sqrt{5}) + (\sqrt{3})(4\sqrt{3})$

$= 5\sqrt{10} + 20\sqrt{6} - \sqrt{15} - 4\sqrt{9}$ Multiply.

$= 5\sqrt{10} + 20\sqrt{6} - \sqrt{15} - 12$ Simplify.

**ReviewVocabulary**

**FOIL Method** Multiply two binomials by finding the sum of the products of the First terms, the Outer terms, the Inner terms, and the Last terms.

### GuidedPractice

4. **GEOMETRY** The area $A$ of a rhombus can be found using the equation $A = \frac{1}{2}d_1d_2$, where $d_1$ and $d_2$ are the lengths of the diagonals. What is the area of the rhombus at the right?

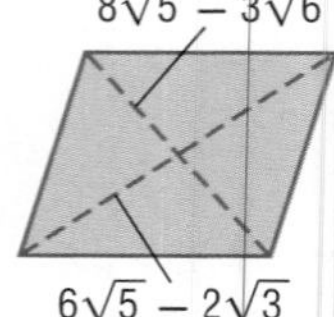


### ConceptSummary Operations with Radical Expressions

| Operation | Symbols | Example |
|---|---|---|
| addition, $b \geq 0$ | $a\sqrt{b} + c\sqrt{b} = (a + c)\sqrt{b}$ like radicands | $4\sqrt{3} + 6\sqrt{3} = (4 + 6)\sqrt{3}$ $= 10\sqrt{3}$ |
| subtraction, $b \geq 0$ | $a\sqrt{b} + c\sqrt{b} = (a - c)\sqrt{b}$ like radicands | $12\sqrt{5} - 8\sqrt{5} = (12 - 8)\sqrt{5}$ $= 4\sqrt{5}$ |
| multiplication, $b \geq 0, g \geq 0$ | $a\sqrt{b}(f\sqrt{g}) = af\sqrt{bg}$ Radicands do not have to be like radicands. | $3\sqrt{2}(5\sqrt{7}) = (3 \cdot 5)(\sqrt{2 \cdot 7})$ $= 15\sqrt{14}$ |

## Check Your Understanding

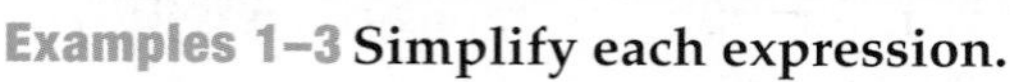

**Examples 1–3** **Simplify each expression.**

1. $3\sqrt{5} + 6\sqrt{5}$
2. $8\sqrt{3} + 5\sqrt{3}$
3. $\sqrt{7} - 6\sqrt{7}$
4. $10\sqrt{2} - 6\sqrt{2}$
5. $4\sqrt{5} + 2\sqrt{20}$
6. $\sqrt{12} - \sqrt{3}$
7. $\sqrt{8} + \sqrt{12} + \sqrt{18}$
8. $\sqrt{27} + 2\sqrt{3} - \sqrt{12}$
9. $9\sqrt{2}(4\sqrt{6})$
10. $4\sqrt{3}(8\sqrt{3})$
11. $\sqrt{3}(\sqrt{7} + 3\sqrt{2})$
12. $\sqrt{5}(\sqrt{2} + 4\sqrt{2})$

**Example 4**

13. **GEOMETRY** The area $A$ of a triangle can be found by using the formula $A = \frac{1}{2}bh$, where $b$ represents the base and $h$ is the height. What is the area of the triangle at the right?

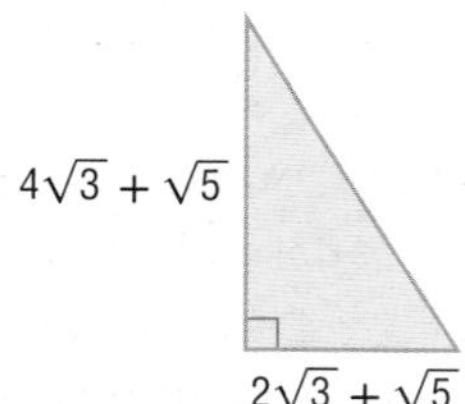

## Practice and Problem Solving

**Examples 1–3** **Simplify each expression.**

14. $7\sqrt{5} + 4\sqrt{5}$
15. $2\sqrt{6} + 9\sqrt{6}$
16. $3\sqrt{5} - 2\sqrt{20}$
17. $3\sqrt{50} - 3\sqrt{32}$
18. $7\sqrt{3} - 2\sqrt{2} + 3\sqrt{2} + 5\sqrt{3}$
19. $\sqrt{5}(\sqrt{2} + 4\sqrt{2})$
20. $\sqrt{6}(2\sqrt{10} + 3\sqrt{2})$
21. $4\sqrt{5}(3\sqrt{5} + 8\sqrt{2})$
22. $5\sqrt{3}(6\sqrt{10} - 6\sqrt{3})$
23. $(\sqrt{3} - \sqrt{2})(\sqrt{15} + \sqrt{12})$
24. $(3\sqrt{11} + 3\sqrt{15})(3\sqrt{3} - 2\sqrt{2})$
25. $(5\sqrt{2} + 3\sqrt{5})(2\sqrt{10} - 5)$

**Example 4**

26. **GEOMETRY** Find the perimeter and area of a rectangle with a width of $2\sqrt{7} - 2\sqrt{5}$ and a length of $3\sqrt{7} + 3\sqrt{5}$.

**Simplify each expression.**

27. $\sqrt{\frac{1}{5}} - \sqrt{5}$
28. $\sqrt{\frac{2}{3}} + \sqrt{6}$
29. $2\sqrt{\frac{1}{2}} + 2\sqrt{2} - \sqrt{8}$
30. $8\sqrt{\frac{5}{4}} + 3\sqrt{20} - 10\sqrt{\frac{1}{5}}$
31. $(3 - \sqrt{5})^2$
32. $(\sqrt{2} + \sqrt{3})^2$

33. **ROLLER COASTERS** The velocity $v$ in feet per second of a roller coaster at the bottom of a hill is related to the vertical drop $h$ in feet and the velocity $v_0$ of the coaster at the top of the hill by the formula $v_0 = \sqrt{v^2 - 64h}$.
    a. What velocity must a coaster have at the top of a 225-foot hill to achieve a velocity of 120 feet per second at the bottom?
    b. Explain why $v_0 = v - 8\sqrt{h}$ is not equivalent to the formula given.

34. **FINANCIAL LITERACY** Tadi invests $225 in a savings account. In two years, Tadi has $232 in his account. You can use the formula $r = \sqrt{\frac{v_2}{v_0}} - 1$ to find the average annual interest rate $r$ that the account has earned. The initial investment is $v_0$, and $v_2$ is the amount in two years. What was the average annual interest rate that Tadi's account earned?

35. **ELECTRICITY** Electricians can calculate the electrical current in amps $A$ by using the formula $A = \frac{\sqrt{w}}{\sqrt{r}}$, where $w$ is the power in watts and $r$ the resistance in ohms. How much electrical current is running through a microwave oven that has 850 watts of power and 5 ohms of resistance? Write the number of amps in simplest radical form, and then estimate the amount of current to the nearest tenth.

## H.O.T. Problems Use Higher-Order Thinking Skills

36. **CHALLENGE** Determine whether the following statement is *true* or *false*. Provide a proof or counterexample to support your answer.

$$x + y > \sqrt{x^2 + y^2} \text{ when } x > 0 \text{ and } y > 0$$

37. CCSS **ARGUMENTS** Make a conjecture about the sum of a rational number and an irrational number. Is the sum *rational* or *irrational*? Is the product of a nonzero rational number and an irrational number *rational* or *irrational*? Explain your reasoning.

38. **OPEN ENDED** Write an equation that shows a sum of two radicals with different radicands. Explain how you could combine these terms.

39. **WRITING IN MATH** Describe step by step how to multiply two radical expressions, each with two terms. Write an example to demonstrate your description.

## Standardized Test Practice

**40. SHORT RESPONSE** The population of a town is 13,000 and is increasing by about 250 people per year. This can be represented by the equation $p = 13{,}000 + 250y$, where $y$ is the number of years from now and $p$ represents the population. In how many years will the population of the town be 14,500?

**41. GEOMETRY** Which expression represents the sum of the lengths of the 12 edges on this rectangular solid?

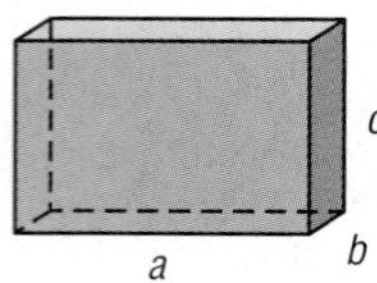


A $2(a + b + c)$
B $3(a + b + c)$
C $4(a + b + c)$
D $12(a + b + c)$

**42.** Evaluate $\sqrt{n - 9}$ and $\sqrt{n} - \sqrt{9}$ for $n = 25$.

F 4; 4 G 4; 2
H 2; 4 J 2; 2

**43.** The current $I$ in a simple electrical circuit is given by the formula $I = \frac{V}{R}$, where $V$ is the voltage and $R$ is the resistance of the circuit. If the voltage remains unchanged, what effect will doubling the resistance of the circuit have on the current?

A The current will remain the same.
B The current will double its previous value.
C The current will be half its previous value.
D The current will be two units more than its previous value.

## Spiral Review

**Simplify.** (Lesson 4-3)

**44.** $\sqrt{18}$ **45.** $\sqrt{24}$ **46.** $\sqrt{60}$

**47.** $\sqrt{50a^3b^5}$ **48.** $\sqrt{169x^4y^7}$ **49.** $\sqrt{63c^3d^4f^5}$

**50. KICKBALL** A kickball is kicked in the air. The equation $h = -16t^2 + 60t$ gives the height $h$ of the ball in feet after $t$ seconds. (Lesson 2-1)

a. What is the height of the ball after one second?
b. When will the ball reach its maximum height?
c. When will the ball hit the ground?

**51. SIDEWALKS** Reynoldsville is repairing sidewalks. If the sidewalk is the same width around a city block, write an expression for the area of the block and the sidewalk. (Lesson 1-3)

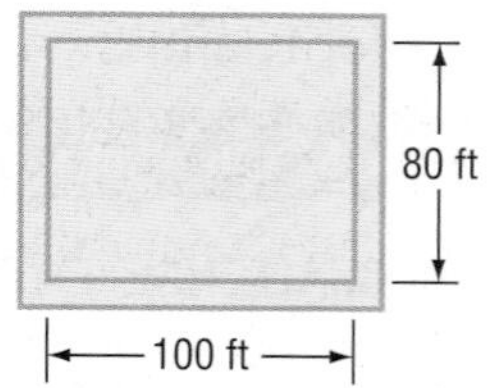


## Skills Review

**Solve each equation. Round each solution to the nearest tenth, if necessary.**

**52.** $-4c - 1.2 = 0.8$ **53.** $-2.6q - 33.7 = 84.1$ **54.** $0.3m + 4 = 9.6$

**55.** $-10 - \frac{n}{5} = 6$ **56.** $\frac{-4h - (-5)}{-7} = 13$ **57.** $3.6t + 6 - 2.5t = 8$

EXTEND 4-4

# Algebra Lab
# Simplifying *n*th Root Expressions

The inverse of raising a number to the *n*th power is finding the **nth root** of a number. The **index** of a radical expression indicates to what root the value under the radicand is being taken. The fourth root of a number is indicated with an index of 4. When simplifying a radical expression in which there is a variable with an exponent in the radicand, divide the exponent by the index.

**CCSS Common Core State Standards**
**Content Standards**
**N.RN.2** Rewrite expressions involving radicals and rational exponents using the properties of exponents.

$13 \div 5 = 2 \text{ R } 3 \longrightarrow \sqrt[5]{x^{13}} = x^2 \cdot \sqrt[5]{x^3}$ (index → 5; quotient → 2; remainder → 3)

### Example 1 Simplify Expressions

**Simplify each expression.**

**a.** $\sqrt[3]{x^7}$

$\sqrt[3]{x^7} = x^2\sqrt[3]{x}$    $7 \div 3 = 2 \text{ R } 1$

**b.** $\sqrt[5]{32x^9}$

$\sqrt[5]{32x^9} = \sqrt[5]{32} \cdot \sqrt[5]{x^9}$    Multiplication Property

$= 2x\sqrt[5]{x^4}$    $9 \div 5 = 1 \text{ R } 4$

The properties of square roots (and *n*th roots) also apply when the radicand contains fractions. Separate the numerator and denominator and then simplify them individually.

### Example 2 Simplify Expressions with Fractions

**Simplify** $\sqrt[3]{\frac{27x^5}{8y^3}}$.

$\sqrt[3]{\frac{27x^5}{8y^3}} = \frac{\sqrt[3]{27}}{\sqrt[3]{8}} \cdot \frac{\sqrt[3]{x^5}}{\sqrt[3]{y^3}}$    Multiplication Property of Radicals

$= \frac{3}{2} \cdot \frac{x\sqrt[3]{x^2}}{y}$    Simplify.

$= \frac{3x\sqrt[3]{x^2}}{2y}$    Multiplication Property of Radicals

The indices *and* the radicands must be alike in order to add or subtract radical expressions.

### Example 3 Combine Like Terms

**Simplify** $8\sqrt[4]{\frac{4}{3}} + \sqrt[4]{\frac{5}{4}} - 3\sqrt[4]{\frac{4}{3}} + \sqrt[3]{\frac{4}{3}}$.

Combine the expressions with identical indices and radicands. Then simplify.

$8\sqrt[4]{\frac{4}{3}} + \sqrt[4]{\frac{5}{4}} - 3\sqrt[4]{\frac{4}{3}} + \sqrt[3]{\frac{4}{3}} = (8 - 3)\sqrt[4]{\frac{4}{3}} + \sqrt[4]{\frac{5}{4}} + \sqrt[3]{\frac{4}{3}}$    Associative Property

$= 5\sqrt[4]{\frac{4}{3}} + \sqrt[4]{\frac{5}{4}} + \sqrt[3]{\frac{4}{3}}$    Simplify.

When multiplying radical expressions, ensure that the indices are the same. Then multiply the radicands and simplify if possible. Once none of the remaining terms can be combined or simplified, the expression is considered simplified.

### Example 4 Simplify Expressions with Products

**Simplify $5\sqrt[4]{6} \cdot 2\sqrt[4]{12} \cdot \sqrt[3]{10}$.**

Multiply the radicands with identical indexes.

$$5\sqrt[4]{6} \cdot 2\sqrt[4]{12} \cdot \sqrt[3]{10} = (5 \cdot 2)(\sqrt[4]{6} \cdot \sqrt[4]{12}) \cdot \sqrt[3]{10} \quad \text{Associative Property}$$
$$= 10 \cdot (\sqrt[4]{6} \cdot \sqrt[4]{12}) \cdot \sqrt[3]{10} \quad \text{Multiply.}$$
$$= 10\sqrt[4]{72}\sqrt[3]{10} \quad \text{Multiply.}$$

The properties of radical expressions still hold when variables are in the radicand.

### Example 5 Simplify Expressions with Several Operations

**Simplify $6\sqrt[4]{x} \cdot \sqrt[4]{x^3} + 3(\sqrt[3]{x} + 2\sqrt[3]{x})$.**

Follow the order of operations and the properties of radical expressions.

$$6\sqrt[4]{x} \cdot \sqrt[4]{x^3} + 3(\sqrt[3]{x} + 2\sqrt[3]{x}) = 6\sqrt[4]{x} \cdot \sqrt[4]{x^3} + 3(3\sqrt[3]{x}) \quad \text{Add like terms.}$$
$$= 6\sqrt[4]{x \cdot x^3} + 3(3\sqrt[3]{x}) \quad \text{Associative Property}$$
$$= 6\sqrt[4]{x^4} + 9\sqrt[3]{x} \quad \text{Multiply.}$$
$$= 6x + 9\sqrt[3]{x} \quad \text{Simplify.}$$

## Exercises

**Simplify each expression.**

1. $\sqrt[3]{c^6}$
2. $\sqrt[4]{16d^9}$
3. $\sqrt[3]{9} \cdot \sqrt[3]{6} \cdot \sqrt[3]{3}$
4. $\sqrt[3]{\dfrac{8a^4}{125b^7}}$
5. $\sqrt[5]{\dfrac{32x^4}{5y^6z^5}}$
6. $\sqrt[4]{\dfrac{3}{2}} + 5\sqrt[4]{\dfrac{3}{2}} - 2\sqrt[4]{\dfrac{2}{3}}$
7. $3\sqrt[4]{6} \cdot 4\sqrt[3]{6} \cdot 5\sqrt[4]{8}$
8. $3\sqrt[4]{x^2} + 2\sqrt[4]{x} \cdot 4\sqrt[4]{x}$
9. $\sqrt[5]{a} \cdot 2\sqrt[5]{a^3} - 2(\sqrt[5]{a} + 4\sqrt[5]{a})$
10. $\sqrt[4]{\dfrac{x}{4}} + 5\sqrt[4]{\dfrac{x}{4}} - 2\sqrt[4]{\dfrac{2x}{3}}$
11. $\sqrt[4]{\dfrac{8a^2}{15b^3}} \cdot 3\sqrt[4]{\dfrac{2a^3}{27b}}$
12. $\sqrt[4]{\dfrac{16x^3}{81y^5}} + 3\sqrt[4]{\dfrac{x^3}{y}} + \sqrt[3]{\dfrac{16x}{y^8}}$

### Think About It

13. Provide an example in which two radical expressions with *unlike* radicands can be combined by addition.
14. Provide an example in which two radical expressions with identical indices and with like variables in the radicand *cannot* be combined by addition.

LESSON 4-5

# Radical Equations

**Then**

- You added, subtracted, and multiplied radical expressions.

**Now**

1. Solve radical equations.
2. Solve radical equations with extraneous solutions.

**Why?**

- The waterline length of a sailboat is the length of the line made by the water's edge when the boat is full. A sailboat's hull speed is the fastest speed that it can travel.

You can estimate hull speed $h$ by using the formula $h = 1.34\sqrt{\ell}$, where $\ell$ is the length of the sailboat's waterline.

**NewVocabulary**
radical equations
extraneous solutions

**Common Core State Standards**

**Content Standards**

N.RN.2 Rewrite expressions involving radicals and rational exponents using the properties of exponents.

A.CED.2 Create equations in two or more variables to represent relationships between quantities; graph equations on coordinate axes with labels and scales.

**Mathematical Practices**

3 Construct viable arguments and critique the reasoning of others.

4 Model with mathematics.

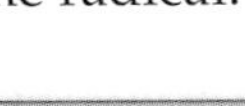

## 1 Radical Equations

Equations that contain variables in the radicand, like $h = 1.34\sqrt{\ell}$, are called **radical equations**. To solve, isolate the desired variable on one side of the equation first. Then square each side of the equation to eliminate the radical.

### KeyConcept Power Property of Equality

| | |
|---|---|
| Words | If you square both sides of a true equation, the resulting equation is still true. |
| Symbols | If $a = b$, then $a^2 = b^2$. |
| Examples | If $\sqrt{x} = 4$, then $(\sqrt{x})^2 = 4^2$. |

### Real-World Example 1 Variable as a Radicand

**SAILING Idris and Sebastian are sailing in a friend's sailboat. They measure the hull speed at 9 nautical miles per hour. Find the length of the sailboat's waterline. Round to the nearest foot.**

**Understand** You know how fast the boat will travel and that it relates to the length.

**Plan** The boat travels at 9 nautical miles per hour. The formula for hull speed is $h = 1.34\sqrt{\ell}$.

**Solve**

$h = 1.34\sqrt{\ell}$ Formula for hull speed

$9 = 1.34\sqrt{\ell}$ Substitute 9 for $h$.

$\frac{9}{1.34} = \frac{1.34\sqrt{\ell}}{1.34}$ Divide each side by 1.34.

$6.72 \approx \sqrt{\ell}$ Simplify.

$(6.72)^2 \approx (\sqrt{\ell})^2$ Square each side of the equation.

$45.16 \approx \ell$ Simplify.

The sailboat's waterline length is about 45 feet.

**Check** Check by substituting the estimate into the original formula.

$h = 1.34\sqrt{\ell}$ Formula for hull speed

$9 \stackrel{?}{=} 1.34\sqrt{45}$ $h = 9$ and $\ell = 45$

$9 \approx 8.98899327$ ✓ Multiply.

David Sanger/The Image Bank/Getty Images

## GuidedPractice

1. **DRIVING** The equation $v = \sqrt{2.5r}$ represents the maximum velocity that a car can travel safely on an unbanked curve when $v$ is the maximum velocity in miles per hour and $r$ is the radius of the turn in feet. If a road is designed for a maximum speed of 65 miles per hour, what is the radius of the turn?

To solve a radical equation, isolate the radical first. Then square both sides of the equation.

PT

### Example 2 Expression as a Radicand

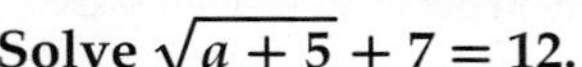

**Solve $\sqrt{a+5} + 7 = 12$.**

| | |
|---|---|
| $\sqrt{a+5} + 7 = 12$ | Original equation |
| $\sqrt{a+5} = 5$ | Subtract 7 from each side. |
| $(\sqrt{a+5})^2 = 5^2$ | Square each side. |
| $a + 5 = 25$ | Simplify. |
| $a = 20$ | Subtract 5 from each side. |

> **WatchOut!**
> **Squaring Each Side** Remember that when you square each side of the equation, you must square the entire side of the equation, even if there is more than one term on the side.

## GuidedPractice

**Solve each equation.**

**2A.** $\sqrt{c-3} - 2 = 4$

**2B.** $4 + \sqrt{h+1} = 14$

## 2 Extraneous Solutions

Squaring each side of an equation sometimes produces a solution that is not a solution of the original equation. These are called **extraneous solutions**. Therefore, you must check all solutions in the original equation.

### Example 3 Variable on Each Side

**Solve $\sqrt{k+1} = k - 1$. Check your solution.**

| | |
|---|---|
| $\sqrt{k+1} = k - 1$ | Original equation |
| $(\sqrt{k+1})^2 = (k-1)^2$ | Square each side. |
| $k + 1 = k^2 - 2k + 1$ | Simplify. |
| $0 = k^2 - 3k$ | Subtract $k$ and 1 from each side. |
| $0 = k(k-3)$ | Factor. |
| $k = 0$ or $k - 3 = 0$ | Zero Product Property |
| $k = 3$ | Solve. |

**CHECK**

| | | | |
|---|---|---|---|
| $\sqrt{k+1} = k - 1$ | Original equation | $\sqrt{k+1} = k - 1$ | Original equation |
| $\sqrt{0+1} \stackrel{?}{=} 0 - 1$ | $k = 0$ | $\sqrt{3+1} \stackrel{?}{=} 3 - 1$ | $k = 3$ |
| $\sqrt{1} \stackrel{?}{=} -1$ | Simplify. | $\sqrt{4} \stackrel{?}{=} 2$ | Simplify. |
| $1 \neq -1$ ✗ | False | $2 = 2$ ✓ | True |

Since 0 does not satisfy the original equation, 3 is the only solution.

> **StudyTip**
> **Extraneous Solutions** When checking solutions for extraneous solutions, we are only interested in principal roots.

## GuidedPractice

**Solve each equation. Check your solution.**

**3A.** $\sqrt{t+5} = t + 3$

**3B.** $x - 3 = \sqrt{x-1}$

## Check Your Understanding

**Example 1**

1. **GEOMETRY** The surface area of a basketball is $x$ square inches. What is the radius of the basketball if the formula for the surface area of a sphere is $SA = 4\pi r^2$?

**Examples 2–3** **Solve each equation. Check your solution.**

2. $\sqrt{10h} + 1 = 21$
3. $\sqrt{7r + 2} + 3 = 7$
4. $5 + \sqrt{g - 3} = 6$
5. $\sqrt{3x - 5} = x - 5$
6. $\sqrt{2n + 3} = n$
7. $\sqrt{a - 2} + 4 = a$

## Practice and Problem Solving

**Example 1**

8. **EXERCISE** Suppose the function $S = \pi\sqrt{\frac{9.8\ell}{1.6}}$, where $S$ represents speed in meters per second and $\ell$ is the leg length of a person in meters, can approximate the maximum speed that a person can run.
   a. What is the maximum running speed of a person with a leg length of 1.1 meters to the nearest tenth of a meter?
   b. What is the leg length of a person with a running speed of 6.7 meters per second to the nearest tenth of a meter?
   c. As leg length increases, does maximum speed increase or decrease? Explain.

**Examples 2–3** **Solve each equation. Check your solution.**

9. $\sqrt{a} + 11 = 21$
10. $\sqrt{t} - 4 = 7$
11. $\sqrt{n - 3} = 6$
12. $\sqrt{c + 10} = 4$
13. $\sqrt{h - 5} = 2\sqrt{3}$
14. $\sqrt{k + 7} = 3\sqrt{2}$
15. $y = \sqrt{12 - y}$
16. $\sqrt{u + 6} = u$
17. $\sqrt{r + 3} = r - 3$
18. $\sqrt{1 - 2t} = 1 + t$
19. $5\sqrt{a - 3} + 4 = 14$
20. $2\sqrt{x - 11} - 8 = 4$

21. **RIDES** The amount of time $t$, in seconds, that it takes a simple pendulum to complete a full swing is called the *period*. It is given by $t = 2\pi\sqrt{\frac{\ell}{32}}$, where $\ell$ is the length of the pendulum, in feet.
   a. The Giant Swing completes a period in about 8 seconds. About how long is the pendulum's arm? Round to the nearest foot.
   b. Does increasing the length of the pendulum increase or decrease the period? Explain.

**Solve each equation. Check your solution.**

22. $\sqrt{6a - 6} = a + 1$
23. $\sqrt{x^2 + 9x + 15} = x + 5$
24. $6\sqrt{\frac{5k}{4}} - 3 = 0$
25. $\sqrt{\frac{5y}{6}} - 10 = 4$
26. $\sqrt{2a^2 - 121} = a$
27. $\sqrt{5x^2 - 9} = 2x$

28. **CCSS REASONING** The formula for the slant height $c$ of a cone is $c = \sqrt{h^2 + r^2}$, where $h$ is the height of the cone and $r$ is the radius of its base. Find the height of the cone if the slant height is 4 units and the radius is 2 units. Round to the nearest tenth.

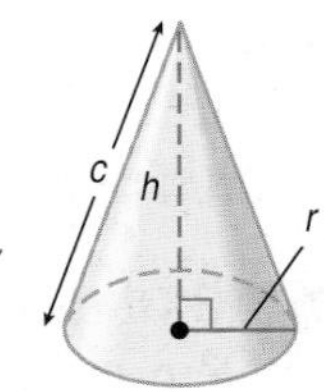

**29.** **MULTIPLE REPRESENTATIONS** Consider $\sqrt{2x - 7} = x - 7$.

a. **Graphical** Clear the **Y=** list. Enter the left side of the equation as **Y1** $= \sqrt{2x - 7}$. Enter the right side of the equation as **Y2** $= x - 7$. Press GRAPH.

b. **Graphical** Sketch what is shown on the screen.

c. **Analytical** Use the **intersect** feature on the **CALC** menu to find the point of intersection.

d. **Analytical** Solve the radical equation algebraically. How does your solution compare to the solution from the graph?

**30.** **PACKAGING** A cylindrical container of chocolate drink mix has a volume of 162 cubic inches. The radius $r$ of the container can be found by using the formula $r = \sqrt{\frac{V}{\pi h}}$, where $V$ is the volume of the container and $h$ is the height.

a. If the radius is 2.5 inches, find the height of the container. Round to the nearest hundredth.

b. If the height of the container is 10 inches, find the radius. Round to the nearest hundredth.

## H.O.T. Problems Use Higher-Order Thinking Skills

**31.** **CCSS CRITIQUE** Jada and Fina solved $\sqrt{6 - b} = \sqrt{b + 10}$. Is either of them correct? Explain.

Jada

$\sqrt{6 - b} = \sqrt{b + 10}$

$(\sqrt{6 - b})^2 = (\sqrt{b + 10})^2$

$6 - b = b + 10$

$-2b = 4$

$b = -2$

Check $\sqrt{6 - (-2)} \stackrel{?}{=} \sqrt{(-2) + 10}$

$\sqrt{8} = \sqrt{8}$ ✓

Fina

$\sqrt{6 - b} = \sqrt{b + 10}$

$(\sqrt{6 - b})^2 = (\sqrt{b + 10})^2$

$6 - b = b + 10$

$2b = 4$

$b = 2$

Check $\sqrt{6 - (2)} \stackrel{?}{=} \sqrt{(2) + 10}$

$\sqrt{4} \neq \sqrt{12}$ X

no solution

**32.** **REASONING** Which equation has the same solution set as $\sqrt{4} = \sqrt{x + 2}$? Explain.

**A.** $\sqrt{4} = \sqrt{x} + \sqrt{2}$ **B.** $4 = x + 2$ **C.** $2 - \sqrt{2} = \sqrt{x}$

**33.** **REASONING** Explain how solving $5 = \sqrt{x} + 1$ is different from solving $5 = \sqrt{x + 1}$.

**34.** **OPEN ENDED** Write a radical equation with a variable on each side. Then solve the equation.

**35.** **REASONING** Is the following equation *sometimes*, *always* or *never* true? Explain.

$$\sqrt{(x - 2)^2} = x - 2$$

**36.** **CHALLENGE** Solve $\sqrt{x + 9} = \sqrt{3} + \sqrt{x}$.

**37.** **WRITING IN MATH** Write some general rules about how to solve radical equations. Demonstrate your rules by solving a radical equation.

## Standardized Test Practice

**38. SHORT RESPONSE** Zack needs to drill a hole at $A$, $B$, $C$, $D$, and $E$ on circle $P$.

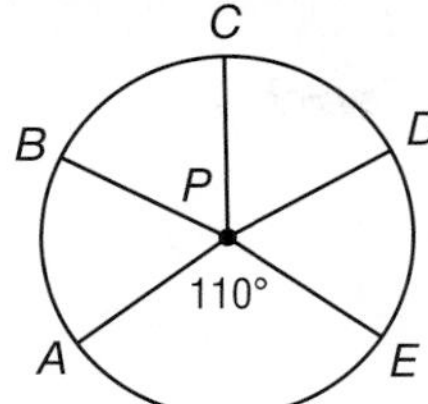


If Zack drills holes so that $m\angle APE = 110°$ and the other four angles are congruent, what is $m\angle CPD$?

**39.** Which expression is undefined when $w = 3$?

A $\frac{w-3}{w+1}$

B $\frac{w^2-3w}{3w}$

C $\frac{w+1}{w^2-3w}$

D $\frac{3w}{3w^2}$

**40.** What is the slope of a line that is parallel to the line?

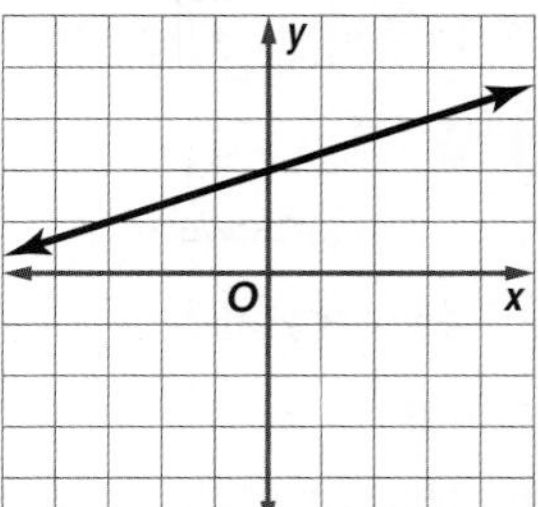


F $-3$

G $-\frac{1}{3}$

H $\frac{1}{3}$

J $3$

**41.** What are the solutions of $\sqrt{x+3} - 1 = x - 4$?

A $1, 6$

B $-1, -6$

C $1$

D $6$

## Spiral Review

**42. ELECTRICITY** The voltage $V$ required for a circuit is given by $V = \sqrt{PR}$, where $P$ is the power in watts and $R$ is the resistance in ohms. How many more volts are needed to light a 100-watt light bulb than a 75-watt light bulb if the resistance of both is 110 ohms? (Lesson 4-4)

**Simplify each expression.** (Lesson 4-3)

**43.** $\sqrt{6} \cdot \sqrt{8}$

**44.** $\sqrt{3} \cdot \sqrt{6}$

**45.** $7\sqrt{3} \cdot 2\sqrt{6}$

**46.** $\sqrt{\frac{27}{a^2}}$

**47.** $\sqrt{\frac{5c^5}{4d^5}}$

**48.** $\frac{\sqrt{9x^3y}}{\sqrt{16x^2y^2}}$

**49. PHYSICAL SCIENCE** A projectile is shot straight up from ground level. Its height $h$, in feet, after $t$ seconds is given by $h = 96t - 16t^2$. Find the value(s) of $t$ when $h$ is 96 feet. (Lesson 2-5)

**Factor each trinomial, if possible. If the trinomial cannot be factored using integers, write *prime*.** (Lesson 1-7)

**50.** $2x^2 + 7x + 5$

**51.** $6p^2 + 5p - 6$

**52.** $5d^2 + 6d - 8$

**53.** $8k^2 - 19k + 9$

**54.** $9g^2 - 12g + 4$

**55.** $2a^2 - 9a - 18$

## Skills Review

**Simplify.**

**56.** $9^2$

**57.** $10^6$

**58.** $4^5$

**59.** $(8v)^2$

**60.** $\left(\frac{w^3}{9}\right)^2$

**61.** $(10y^2)^3$

# CHAPTER 4 Study Guide and Review

## Study Guide

### KeyConcepts

**Exponential Functions** (Lessons 4-1 and 4-2)

- An exponential function is in the form $y = ab^x$, where $a \neq 0$, $b > 0$ and $b \neq 1$.
- Property of Equality for Exponential Functions: If $b$ is a positive number other than 1, then $b^x = b^y$ if and only if $x = y$.
- Property of Inequality for Exponential Functions: If $b > 1$, then $b^x > b^y$ if and only if $x > y$, and $b^x < b^y$ if and only if $x < y$.

**Simplifying Radical Expressions** (Lesson 4-3)

- A radical expression is in simplest form when
  - no radicands have perfect square factors other than 1,
  - no radicals contain fractions,
  - and no radicals appear in the denominator of a fraction.

**Operations with Radical Expressions and Equations** (Lessons 4-4 and 4-5)

- Radical expressions with like radicals can be added or subtracted.
- Use the FOIL method to multiply radical expressions.

### FOLDABLES StudyOrganizer

Be sure the Key Concepts are noted in your Foldable.

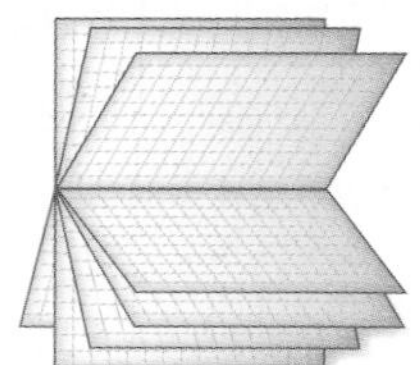

### KeyVocabulary

asymptote (p. 227)
compound interest (p. 238)
conjugate (p. 247)
decay factor (p. 230)
exponential decay (p. 229)
exponential equation (p. 237)
exponential function (p. 227)
exponential growth (p. 227)
exponential inequality (p. 239)
growth factor (p. 229)
radical equation (p. 259)
radical expression (p. 245)
rationalizing the denominator (p. 247)

### VocabularyCheck

**Choose a word or term from the list above that best completes each statement or phrase.**

1. A function of the form $f(x) = b^x$ where $b > 1$ is a(n) ____________ function.
2. A(n) ____________ is an equation in which variables occur as exponents.
3. The base of the exponential function, $A(t) = a(1 - r)^t$, $1 - r$ is called the ____________.
4. The expressions $2 + \sqrt{5}$ and $2 - \sqrt{5}$ are ____________.

## Lesson-by-Lesson Review

### 4-1 Graphing Exponential Functions

Graph each function. State the domain and range.

5. $f(x) = 3^x$
6. $f(x) = -5(2)^x$
7. $f(x) = 3(4)^x - 6$
8. $f(x) = 3^{2x} + 5$
9. $f(x) = 3\left(\frac{1}{4}\right)^{x+3} - 1$
10. $f(x) = \frac{3}{5}\left(\frac{2}{3}\right)^{x-2} + 3$
11. **POPULATION** A city with a population of 120,000 decreases at a rate of 3% annually.
    a. Write the function that represents this situation.
    b. What will the population be in 10 years?

**Example 1**

Graph $f(x) = -2(3)^x + 1$. State the domain and range.

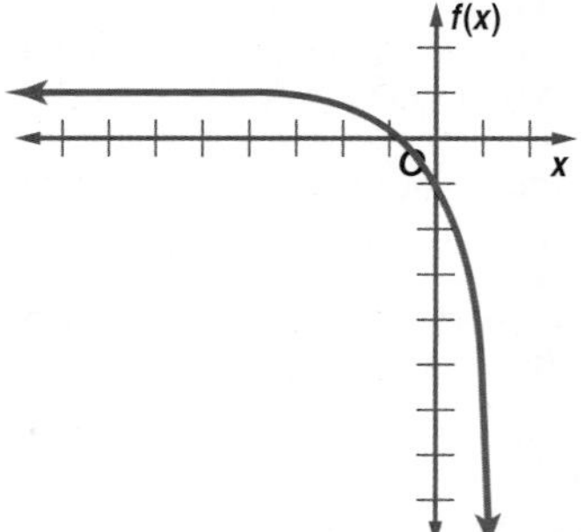


The domain is all real numbers, and the range is all real numbers less than 1.

### 4-2 Solving Exponential Equations and Inequalities

Solve each equation or inequality.

12. $16^x = \frac{1}{64}$
13. $3^{4x} = 9^{3x+7}$
14. $64^{3n} = 8^{2n-3}$
15. $8^{3-3y} = 256^{4y}$
16. $9^{x-2} > \left(\frac{1}{81}\right)^{x+2}$
17. $27^{3x} \leq 9^{2x-1}$
18. **BACTERIA** A bacteria population started with 5000 bacteria. After 8 hours there were 28,000 in the sample.
    a. Write an exponential function that could be used to model the number of bacteria after $x$ hours if the number of bacteria changes at the same rate.

**Example 2**

Solve $4^{3x} = 32^{x-1}$ for $x$.

| | |
|---|---|
| $4^{3x} = 32^{x-1}$ | Original equation |
| $(2^2)^{3x} = (2^5)^{x-1}$ | Rewrite so each side has the same base. |
| $2^{6x} = 2^{5x-5}$ | Power of a Power |
| $6x = 5x - 5$ | Property of Equality for Exponential Functions |
| $x = -5$ | Subtract $5x$ from each side. |

The solution is $-5$.

### 4-3 Simplifying Radical Expressions

Simplify.

19. $\sqrt{36x^2y^7}$
20. $\sqrt{20ab^3}$
21. $\sqrt{3} \cdot \sqrt{6}$
22. $2\sqrt{3} \cdot 3\sqrt{12}$
23. $(4 - \sqrt{5})^2$
24. $(1 + \sqrt{2})^2$
25. $\sqrt{\frac{50}{a^2}}$
26. $\sqrt{\frac{2}{5}} \cdot \sqrt{\frac{3}{4}}$
27. $\frac{3}{2 - \sqrt{5}}$
28. $\frac{5}{\sqrt{7} + 6}$
29. **WEATHER** To estimate how long a thunderstorm will last, use $t = \sqrt{\frac{d^3}{216}}$, where $t$ is the time in hours and $d$ is the diameter of the storm in miles. A storm is 10 miles in diameter. How long will it last?

**Example 2**

Simplify $\frac{2}{4 + \sqrt{3}}$.

| | |
|---|---|
| $\frac{2}{4 + \sqrt{3}}$ | Original expression |
| $= \frac{2}{4 + \sqrt{3}} \cdot \frac{4 - \sqrt{3}}{4 - \sqrt{3}}$ | Rationalize the denominator. |
| $= \frac{2(4) - 2\sqrt{3}}{4^2 - (\sqrt{3})^2}$ | $(a - b)(a + b) = a^2 - b^2$ |
| $= \frac{8 - 2\sqrt{3}}{16 - 3}$ | $(\sqrt{3})^2 = 3$ |
| $= \frac{8 - 2\sqrt{3}}{13}$ | Simplify. |

## 4-4 Operations with Radical Expressions

Simplify each expression.

**29.** $\sqrt{6} - \sqrt{54} + 3\sqrt{12} + 5\sqrt{3}$

**30.** $2\sqrt{6} - \sqrt{48}$

**31.** $4\sqrt{3x} - 3\sqrt{3x} + 3\sqrt{3x}$

**32.** $\sqrt{50} + \sqrt{75}$

**33.** $\sqrt{2}(5 + 3\sqrt{3})$

**34.** $(2\sqrt{3} - \sqrt{5})(\sqrt{10} + 4\sqrt{6})$

**35.** $(6\sqrt{5} + 2)(4\sqrt{2} + \sqrt{3})$

**36.** **MOTION** The velocity of a dropped object when it hits the ground can be found using $v = \sqrt{2gd}$, where $v$ is the velocity in feet per second, $g$ is the acceleration due to gravity, and $d$ is the distance in feet the object drops. Find the speed of a penny when it hits the ground, after being dropped from 984 feet. Use 32 feet per second squared for $g$.

### Example 3

**Simplify $2\sqrt{6} - \sqrt{24}$.**

$2\sqrt{6} - \sqrt{24} = 2\sqrt{6} - \sqrt{4 \cdot 6}$ Product Property

$= 2\sqrt{6} - 2\sqrt{6}$ Simplify.

$= 0$ Simplify.

### Example 4

**Simplify $(\sqrt{3} - \sqrt{2})(\sqrt{3} + 2\sqrt{2})$.**

$(\sqrt{3} - \sqrt{2})(\sqrt{3} + 2\sqrt{2})$

$= (\sqrt{3})(\sqrt{3}) + (\sqrt{3})(2\sqrt{2}) + (-\sqrt{2})(\sqrt{3}) + (\sqrt{2})(2\sqrt{2})$

$= 3 + 2\sqrt{6} - \sqrt{6} + 4$

$= 7 + \sqrt{6}$

## 4-5 Radical Equations

Solve each equation. Check your solution.

**37.** $10 + 2\sqrt{x} = 0$

**38.** $\sqrt{5 - 4x} - 6 = 7$

**39.** $\sqrt{a + 4} = 6$

**40.** $\sqrt{3x} = 2$

**41.** $\sqrt{x + 4} = x - 8$

**42.** $\sqrt{3x - 14} + x = 6$

**43.** **FREE FALL** Assuming no air resistance, the time $t$ in seconds that it takes an object to fall $h$ feet can be determined by $t = \frac{\sqrt{h}}{4}$. If a skydiver jumps from an airplane and free falls for 10 seconds before opening the parachute, how many feet does she free fall?

### Example 5

**Solve $\sqrt{7x + 4} - 18 = 5$.**

$\sqrt{7x + 4} - 18 = 5$ Original equation

$\sqrt{7x + 4} = 23$ Add 18 to each side.

$(\sqrt{7x + 4})^2 = 23^2$ Square each side.

$7x + 4 = 529$ Simplify.

$7x = 525$ Subtract 4 from each side.

$x = 75$ Divide each side by 7.

**CHECK** $\sqrt{7x + 4} - 18 = 5$ Original equation

$\sqrt{7(75) + 4} - 18 \stackrel{?}{=} 5$ $x = 75$

$\sqrt{525 + 4} - 18 \stackrel{?}{=} 5$ Multiply.

$\sqrt{529} - 18 \stackrel{?}{=} 5$ Add.

$23 - 18 \stackrel{?}{=} 5$ Simplify.

$5 = 5$ ✓ True.

CHAPTER 4

# Practice Test

**Graph each function. State the domain and range.**

1. $f(x) = 3^{x-3} + 2$
2. $f(x) = 2\left(\frac{3}{4}\right)^{x+1} - 3$
3. $f(x) = 3(4)^x$
4. $f(x) = -(2)^x + 5$
5. $f(x) = -0.5(3)^{x+2} + 4$
6. $f(x) = -3\left(\frac{2}{3}\right)^{x-1} + 8$

7. **SCIENCE** You are studying a bacteria population. The population originally started with 6000 bacteria cells. After 2 hours, there were 28,000 bacteria cells.

   a. Write an exponential function that could be used to model the number of bacteria after $x$ hours if the number of bacteria changes at the same rate.

   b. How many bacteria cells can be expected after 4 hours?

8. **MULTIPLE CHOICE** Which exponential function has a graph that passes through the points at (0, 125) and (3, 1000)?

   A $f(x) = 125(3)^x$

   B $f(x) = 1000(3)^x$

   C $f(x) = 125(1000)^x$

   D $f(x) = 125(2)^x$

**Solve each equation or inequality. Round to the nearest ten-thousandth if necessary.**

9. $8^{c+1} = 16^{2c+3}$
10. $9^{x-2} > \left(\frac{1}{27}\right)^x$
11. $2^{a+3} = 3^{2a-1}$
12. $\log_2 (x^2 - 7) = \log_2 6x$
13. $\log_5 x > 2$
14. $\log_3 x + \log_3 (x - 3) = \log_3 4$
15. $6^{n-1} \leq 11^n$
16. $4e^{2x} - 1 = 5$
17. $\ln (x + 2)^2 > 2$

**Simplify each expression.**

18. $5\sqrt{36}$
19. $\frac{3}{1 - \sqrt{2}}$
20. $2\sqrt{3} + 7\sqrt{3}$
21. $3\sqrt{6}(5\sqrt{2})$
22. $2\sqrt{25}$
23. $\sqrt{12} \cdot \sqrt{8}$
24. $\sqrt{72xy^5z^6}$
25. $\frac{3}{1 + \sqrt{5}}$
26. $\frac{1}{5 - \sqrt{7}}$

27. **MULTIPLE CHOICE** Find the area of the rectangle.

   F $7\sqrt{2}$

   G 14

   H $14\sqrt{2}$

   J $98\sqrt{2}$

**Solve each equation. Check your solution.**

28. $\sqrt{10x} = 20$
29. $\sqrt{4x - 3} = 6 - x$

30. **PACKAGING** A cylindrical container of chocolate drink mix has a volume of about 162 $in^3$. The radius of the container can be found by using the formula $r = \sqrt{\frac{V}{\pi h}}$, where $r$ is the radius and $h$ is the height. If the height is 8.25 inches, find the radius of the container.

31. **MULTIPLE CHOICE** Which expression is equivalent to $\sqrt{\frac{16}{32}}$?

   F $\frac{1}{2}$

   G $\frac{\sqrt{2}}{2}$

   H 2

   J 4

CHAPTER 4

# Preparing for Standardized Tests

## Draw a Picture

Sometimes it is easier to visualize how to solve a problem if you draw a picture first. You can sketch your picture on scrap paper or in your test booklet (if allowed). Be careful not make any marks on your answer sheet other than your answers.

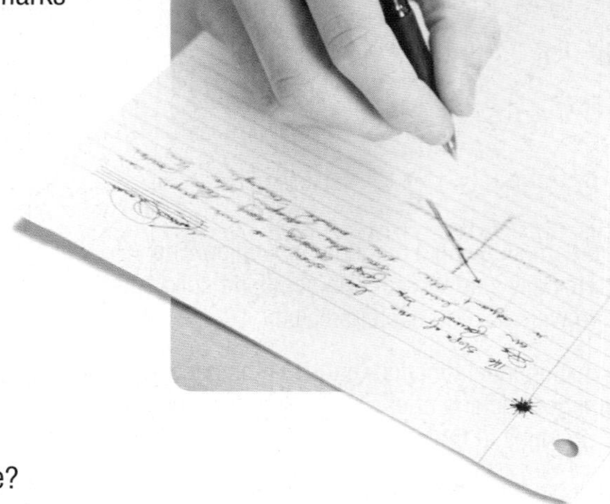

### Strategies for Drawing a Picture

**Step 1**

Read the problem statement carefully.

**Ask yourself:**

- What am I being asked to solve?
- What information is given in the problem?
- What is the unknown quantity for which I need to solve?

**Step 2**

Sketch and label your picture.

- Draw your picture as clearly and accurately as possible.
- Label the picture carefully. Be sure to include all of the information given in the problem statement.

**Step 3**

Solve the problem.

- Use your picture to help you model the problem situation with an equation. Then solve the equation.
- Check your answer to make sure it is reasonable.

### Standardized Test Example

**Read the problem. Identify what you need to know. Then use the information in the problem to solve. Show your work.**

An 18-foot ladder is leaning against a building. For stability, the base of the ladder must be 36 inches away from the wall. How far up the wall does the ladder reach?

Jupiterimages/Polka Dot/Alamy

Read the problem statement carefully. You know the height of the ladder leaning against the building and you know that the base of the ladder must be 36 inches away from the wall. You need to find how far up the wall the ladder reaches.

| Scoring Rubric | |
|---|---|
| **Criteria** | **Score** |
| **Full Credit:** The answer is correct and a full explanation is provided that shows each step. | 2 |
| **Partial Credit:**<br>• The answer is correct, but the explanation is incomplete.<br>• The answer is incorrect, but the explanation is correct. | 1 |
| **No Credit:** Either an answer is not provided or the answer does not make sense. | 0 |

Example of a 2-point response:

First convert all measurements to feet.

36 inches = 3 feet

Use a right triangle to find how high the ladder reaches. Draw and label a triangle to represent the situation.

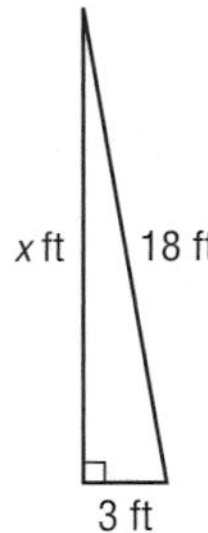


You know the measures of a leg and the hypotenuse, and need to know the length of the other leg. So you can use the Pythagorean Theorem.

$$c^2 = a^2 + b^2$$
$$18^2 = 3^2 + b^2$$
$$324 = 9 + b^2$$
$$315 = b^2$$
$$\pm 315 = b$$
$$17.7 \approx b$$

The ladder reaches about 17.7 feet or about 17 feet 9 inches.

## Exercises

**Read each problem. Identify what you need to know. Then use the information in the problem to solve. Show your work.**

1. A building casts a 15-foot shadow, while a billboard casts a 4.5-foot shadow. If the billboard is 26 feet high, what is the height of the building? Round to the nearest tenth if necessary.

2. A space shuttle is directed toward the Moon, but drifts 1.2° from its intended course. The distance from Earth to the Moon is about 240,000 miles. If the pilot doesn't get the shuttle back on course, how far will the shuttle have drifted from its intended landing position?

CHAPTER 4

# Standardized Test Practice

## Cumulative, Chapters 1 through 4

### Multiple Choice

**Read each question. Then fill in the correct answer on the answer document provided by your teacher or a sheet of paper.**

**1.** Simplify $\frac{1}{4 + \sqrt{2}}$.

F $\frac{4 + \sqrt{2}}{14}$

G $\frac{2 - \sqrt{2}}{7}$

H $\frac{4 - \sqrt{2}}{14}$

J $\frac{2 + \sqrt{2}}{7}$

**2.** What is the area of the triangle below?

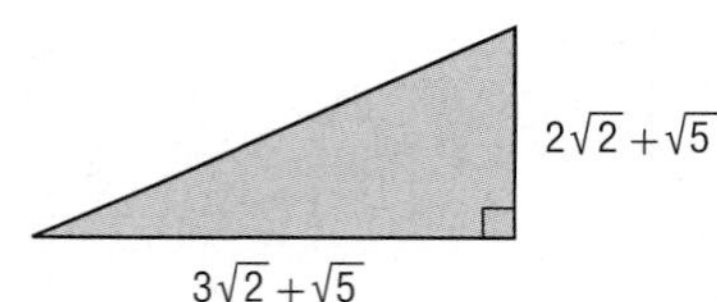


A $3\sqrt{2} + 10\sqrt{5}$

B $17 + 5\sqrt{10}$

C $12\sqrt{2} + 8\sqrt{5}$

D $8.5 + 2.5\sqrt{10}$

**3.** The formula for the slant height $c$ of a cone is $c = \sqrt{h^2 + r^2}$, where $h$ is the height of the cone and $r$ is the radius of its base. What is the radius of the cone below? Round to the nearest tenth.

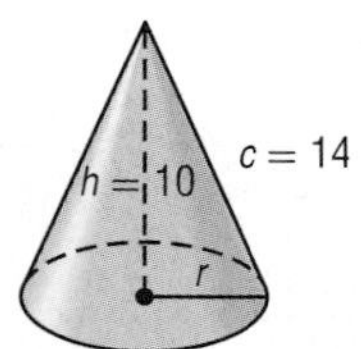


F 4.9 H 9.8

G 6.3 J 10.2

**4.** What is the $y$-intercept of the exponential function below?

$$y = 4^x - 1$$

A 0 B 1 C 2 D 3

**5.** Suppose a certain bacteria duplicates to reproduce itself every 20 minutes. If you begin with one cell of the bacteria, how many will there be after 2 hours?

A 2 B 6 C 32 D 64

### Short Response/Gridded Response

**Record your answers on the answer sheet provided by your teacher or on a sheet of paper.**

**6.** **GRIDDED RESPONSE** How many times does the graph of $y = x^2 - 4x + 10$ cross the $x$-axis?

**7.** Factor $2x^4 - 32$ completely.

**8.** The function $y = \left(\frac{1}{2}\right)^x$ is graphed below. What is the domain of the function?

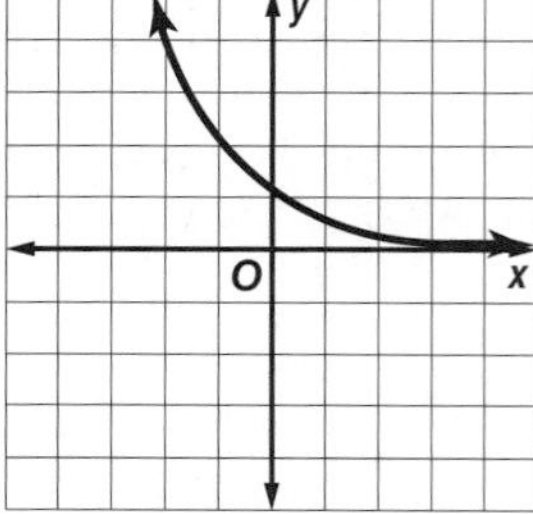


**Test-TakingTip**

**Question 3** Substitute for $c$ and $h$ in the formula. Then solve for $r$.

9. Julio constructed a platform for a bean bag toss game. The plans for the original platform had dimensions of 3 feet by 5 feet. He made his platform larger by adding $x$ feet to each side. The area of the new platform is 35 square feet.

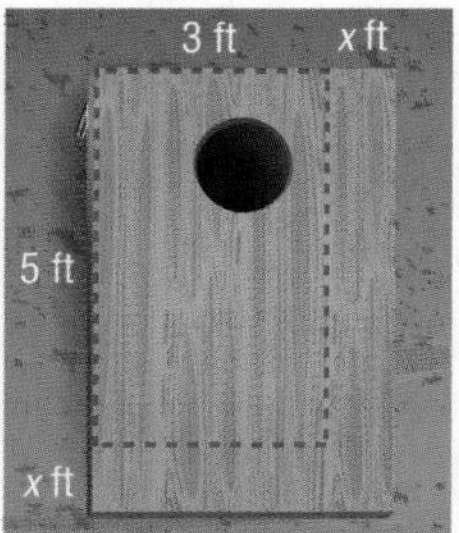


a. Write a quadratic equation that represents the area of his platform.

b. Find the dimensions of the platform Julio made.

10. **GRIDDED RESPONSE** For what value of $x$ would the rectangle below have an area of 48 square units?

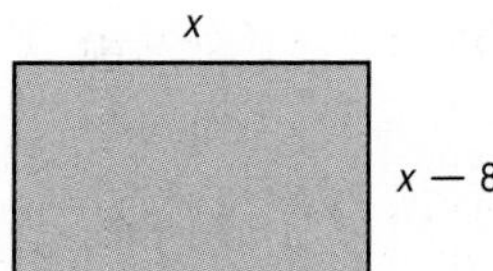


11. Jason shot a marble straight up using a slingshot. The equation $h = -16t^2 + 42t + 5.5$ models the height $h$, in feet, of the marble after $t$ seconds. After how long will the marble hit the

12. The number of cells in a Petri dish can be modeled by the quadratic equation $n = 6t^2 - 4.5t + 74$, where $t$ is the number of hours the cells have been in the dish. When will there be 200 cells in the Petri dish?

13. Scott launches a model rocket from ground level. The rocket's height $h$ in meters is given by the equation $h = -4.9t^2 + 56t$, where $t$ is the time in seconds after the launch.

a. What is the maximum height the rocket will reach? Round to the nearest tenth of a meter. Show each step and explain your method.

b. How long after it is launched will the rocket reach its maximum height? Round to the nearest tenth of a second.

**Need ExtraHelp?**

| If you missed Question... | 1 | 2 | 3 | 4 | 5 | 6 | 7 | 8 | 9 | 10 | 11 | 12 | 13 |
|---|---|---|---|---|---|---|---|---|---|---|---|---|---|
| Go to Lesson... | 4-3 | 4-4 | 4-5 | 4-1 | 4-2 | 2-5 | 1-8 | 4-1 | 3-2 | 3-2 | 2-2 | 2-4 | 3-5 |

CHAPTER 5

# Reasoning and Proof

## Then

You used segment and angle relationships.

## Now

In this chapter, you will:

- Make conjectures and find counterexamples for statements.
- Use deductive reasoning to reach valid conclusions.
- Write proofs involving segment and angle theorems.

## Why?

**SCIENCE AND NATURE** Biologists and other scientists use inductive and deductive reasoning to make decisions and draw logical conclusions about animal populations.

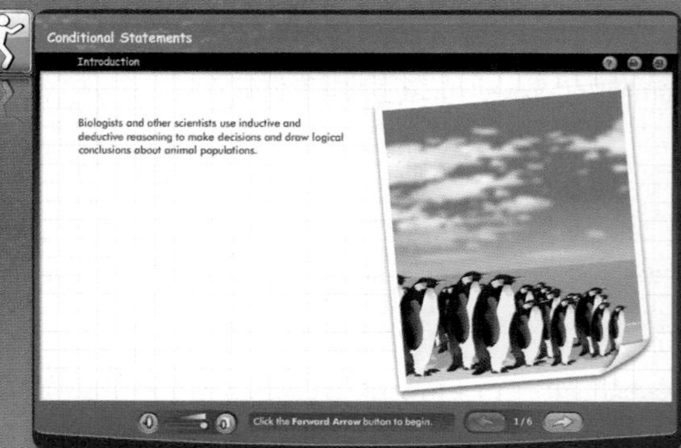
Conditional Statements
Introduction
Biologists and other scientists use inductive and deductive reasoning to make decisions and draw logical conclusions about animal populations.
Click the Forward Arrow button to begin.
1/6

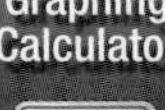

connectED.mcgraw-hill.com Your Digital Math Portal

Animation | Vocabulary | eGlossary | Personal Tutor | Virtual Manipulatives | Graphing Calculator | Audio | Foldables | Self-Check Practice | Worksheets

# Get Ready for the Chapter

**Diagnose** Readiness | You have two options for checking prerequisite skills.

**Textbook Option** Take the Quick Check below. Refer to the Quick Review for help.

## QuickCheck

Evaluate each expression for the given value of $x$.

1. $4x + 7$; $x = 6$
2. $(x - 2)180$; $x = 8$
3. $5x^2 - 3x$; $x = 2$
4. $\frac{x(x-3)}{2}$; $x = 5$
5. $x + (x + 1) + (x + 2)$; $x = 3$

Write each verbal expression as an algebraic expression.

6. eight less than five times a number
7. three more than the square of a number

## QuickReview

**Example 1**

Evaluate $x^2 - 2x + 11$ for $x = 6$.

| | |
|---|---|
| $x^2 - 2x + 11$ | Original expression |
| $= (6)^2 - 2(6) + 11$ | Substitute 6 for $x$. |
| $= 36 - 2(6) + 11$ | Evaluate the exponent. |
| $= 36 - 12 + 11$ | Multiply. |
| $= 35$ | Simplify. |

Solve each equation.

8. $8x - 10 = 6x$
9. $18 + 7x = 10x + 39$
10. $3(11x - 7) = 13x + 25$
11. $3x + 8 = \frac{1}{2}x + 35$
12. $\frac{2}{3}x + 1 = 5 - 2x$
13. **CLOTHING** Nancy bought 4 shirts at the mall for \$52. Write and solve an equation to find the average cost of one shirt.

**Example 2**

Solve $36x - 14 = 16x + 58$.

| | |
|---|---|
| $36x - 14 = 16x + 58$ | Original equation |
| $36x - 14 - 16x = 16x + 58 - 16x$ | Subtract $16x$ from each side. |
| $20x - 14 = 58$ | Simplify. |
| $20x - 14 + 14 = 58 + 14$ | Add 14 to each side. |
| $20x = 72$ | Simplify. |
| $\frac{20x}{20} = \frac{72}{20}$ | Divide each side by 20. |
| $x = 3.6$ | Simplify. |

Refer to the figure in Example 3.

14. Identify a pair of vertical angles that appear to be obtuse.
15. Identify a pair of adjacent angles that appear to be complementary.
16. Identify a linear pair.
17. If $m\angle DXB = 116$ and $m\angle EXA = 3x + 2$, find $x$.
18. If $m\angle BXC = 90$, $m\angle CXD = 6x - 13$, and $m\angle DXE = 10x + 7$, find $x$.

**Example 3**

If $m\angle BXA = 3x + 5$ and $m\angle DXE = 56$, find $x$.

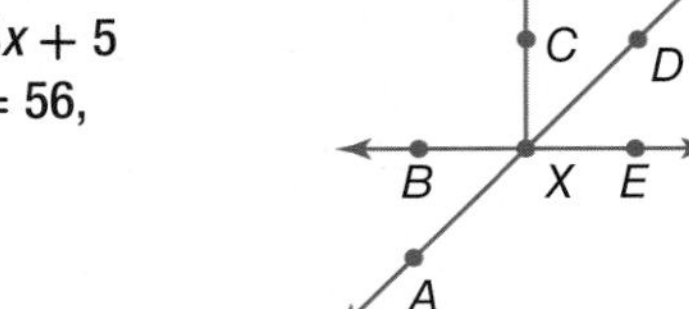


| | |
|---|---|
| $m\angle BXA = m\angle DXE$ | Vertical $\angle$s are $\cong$. |
| $3x + 5 = 56$ | Substitution |
| $3x = 51$ | Subtract 5 from each side. |
| $x = 17$ | Divide each side by 3. |

**2 Online Option** Take an online self-check Chapter Readiness Quiz at connectED.mcgraw-hill.com.

# **Get Started** on the Chapter

You will learn several new concepts, skills, and vocabulary terms as you study Chapter 5. To get ready, identify important terms and organize your resources. You may refer to Chapter 0 to review prerequisite skills.

## FOLDABLES StudyOrganizer

**Reasoning and Proof** Make this Foldable to help you organize your Chapter 5 notes about logic, reasoning, and proof. Begin with one sheet of notebook paper.

**1** **Fold** lengthwise to the holes.

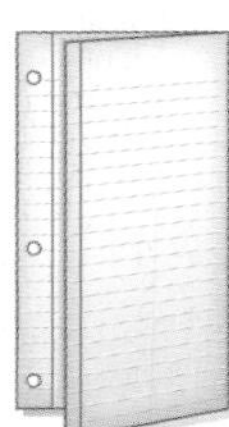

**2** **Cut** five tabs in the top sheet.

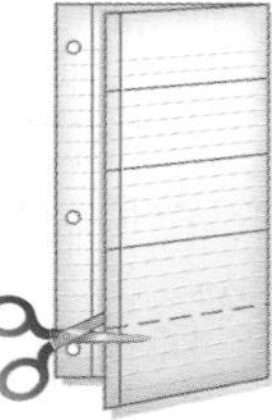

**3** **Label** the tabs as shown.

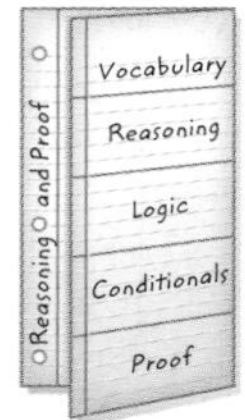


## NewVocabulary

| English | | Español |
|---|---|---|
| postulate | p. 276 | postulado |
| proof | p. 277 | demostración |
| theorem | p. 278 | teorema |

## ReviewVocabulary

complementary angles ángulos complementarios two angles with measures that have a sum of 90

supplementary angles ángulos suplementarios two angles with measures that have a sum of 180

vertical angles ángulos opuestos por el vértice two nonadjacent angles formed by intersecting lines

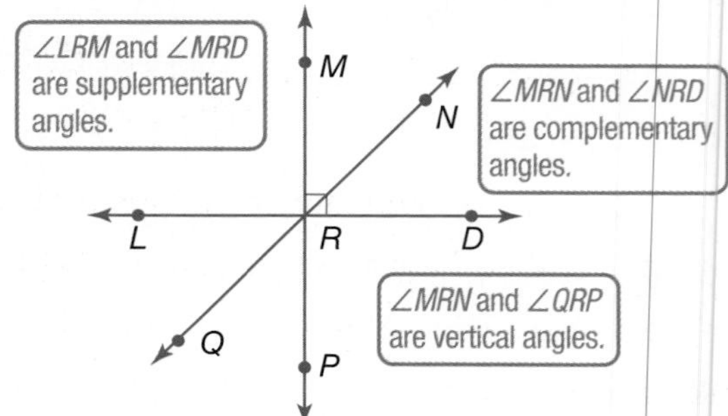

EXPLORE

# 5-1 Geometry Lab
# Necessary and Sufficient Conditions

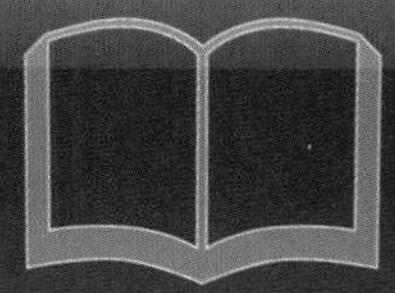

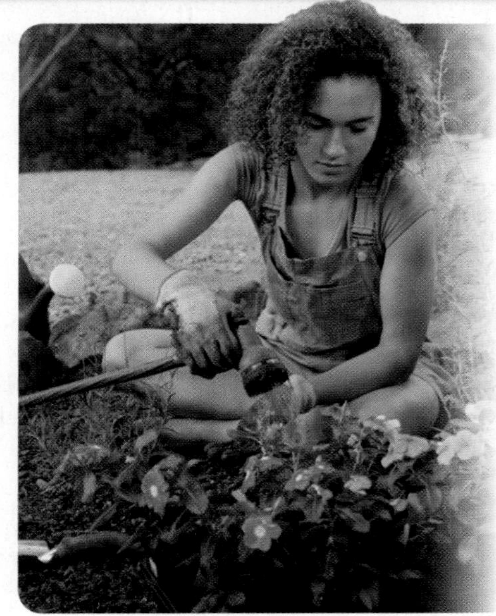

We all know that water is a *necessary* condition for plants to survive. However, it is not a *sufficient* condition. For example, plants also need sunlight to survive.

Necessary and sufficient conditions are important in mathematics. Consider the property of having four sides. While *having four sides* is a necessary condition for something being a square, that single condition is not, by itself, a sufficient condition to guarantee that it is a square. Trapezoids are four-sided figures that are not squares.

| Condition | Definition | Examples |
|---|---|---|
| necessary | A condition *A* is said to be *necessary* for a condition *B*, if and only if the falsity or nonexistence of *A* guarantees the falsity or nonexistence of *B*. | Having opposite sides parallel is a necessary condition for something being a square. |
| sufficient | A condition *A* is said to be *sufficient* for a condition *B*, if and only if the truth or existence of *A* guarantees the truth or existence of *B*. | Being a square is a sufficient condition for something being a rectangle. |

## Exercises

**Determine whether each statement is *true* or *false*. If false, give a counterexample.**

1. Being a square is a necessary condition for being a rectangle.
2. Being a rectangle is a necessary condition for being a square.
3. Being greater than 5 is a necessary condition for being less than 10.
4. Being less than 18 is a sufficient condition for being less than 25.
5. Walking on four legs is a sufficient condition for being a dog.
6. Breathing air is a necessary condition for being a human being.
7. Being an equilateral rectangle is both a necessary and sufficient condition for being a square.

**Determine whether I is a *necessary* condition for II, a *sufficient* condition for II, or *both*. Explain.**

**8.** I. Two points are given.
II. An equation of a line can be written.

**9.** I. Two planes are parallel.
II. Two planes do not intersect.

**10.** I. Two angles are acute.
II. Two angles are complementary.

# LESSON 5-1 Postulates and Paragraph Proofs

**Then**

- You used deductive reasoning by applying the Law of Detachment and the Law of Syllogism.

**Now**

1. Identify and use basic postulates about points, lines, and planes.
2. Write paragraph proofs.

**Why?**

- If a feather and an apple are dropped from the same height in a vacuum chamber, the two objects will fall at the same rate. This demonstrates one of Sir Isaac Newton's laws of gravity and inertia. These laws are accepted as fundamental truths of physics. Some laws in geometry also must be assumed or accepted as true.

**NewVocabulary**
postulate
axiom
proof
theorem
deductive argument
paragraph proof
informal proof

**Common Core State Standards**

**Content Standards**

G.MG.3 Apply geometric methods to solve problems (e.g., designing an object or structure to satisfy physical constraints or minimize cost; working with typographic grid systems based on ratios). ★

**Mathematical Practices**

2 Reason abstractly and quantitatively.
3 Construct viable arguments and critique the reasoning of others.

## 1 Points, Lines, and Planes

A **postulate** or **axiom** is a statement that is accepted as true without proof. Basic ideas about points, lines, and planes can be stated as postulates.

**Postulates** Points, Lines, and Planes

| Words | Example | |
|---|---|---|
| **5.1** Through any two points, there is exactly one line. | P, R, $n$ | Line $n$ is the only line through points $P$ and $R$. |
| **5.2** Through any three noncollinear points, there is exactly one plane. | A, B, C, $\mathcal{K}$ | Plane $\mathcal{K}$ is the only plane through noncollinear points $A$, $B$, and $C$. |
| **5.3** A line contains at least two points. | P, Q, R, $n$ | Line $n$ contains points $P$, $Q$, and $R$. |
| **5.4** A plane contains at least three noncollinear points. | L, B, E, C, $\mathcal{K}$ | Plane $\mathcal{K}$ contains noncollinear points $L$, $B$, $C$, and $E$. |
| **5.5** If two points lie in a plane, then the entire line containing those points lies in that plane. | A, B, $m$, $\mathcal{K}$ | Points $A$ and $B$ lie in plane $\mathcal{K}$, and line $m$ contains points $A$ and $B$, so line $m$ is in plane $\mathcal{K}$. |

**KeyConcept** Intersections of Lines and Planes

| Words | Example | |
|---|---|---|
| **5.6** If two lines intersect, then their intersection is exactly one point. | P, $s$, $t$ | Lines $s$ and $t$ intersect at point $P$. |
| **5.7** If two planes intersect, then their intersection is a line. | $w$, $\mathcal{F}$, $\mathcal{G}$ | Planes $\mathcal{F}$ and $\mathcal{G}$ intersect in line $w$. |

Jim Sugar/CORBIS

These additional postulates form a foundation for proofs and reasoning about points, lines, and planes.

PT

## Real-World Example 1 Identifying Postulates

**ARCHITECTURE Explain how the picture illustrates that each statement is true. Then state the postulate that can be used to show each statement is true.**

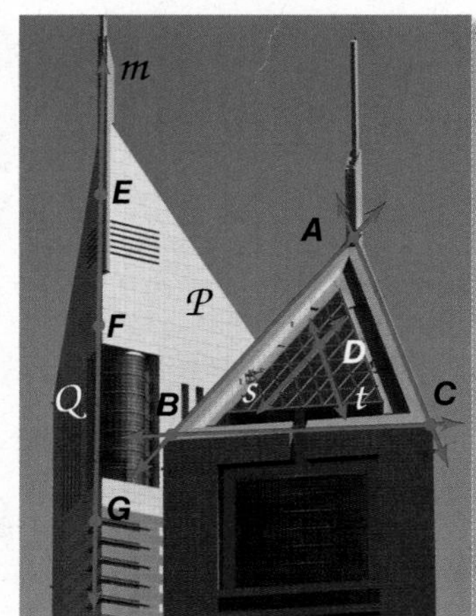


**a. Line *m* contains points *F* and *G*. Point *E* can also be on line *m*.**

The edge of the building is a straight line *m*. Points *E*, *F*, and *G* lie along this edge, so they lie along a line *m*. Postulate 5.3, which states that a line contains at least two points, shows that this is true.

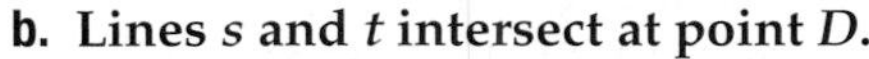

**b. Lines *s* and *t* intersect at point *D*.**

The lattice on the window of the building forms intersecting lines. Lines *s* and *t* of this lattice intersect at only one location, point *D*. Postulate 5.6, which states that if two lines intersect, then their intersection is exactly one point, shows that this is true.

**GuidedPractice**

**1A.** Points *A*, *B*, and *C* determine a plane. **1B.** Planes $\mathcal{P}$ and *Q* intersect in line *m*.

You can use postulates to explain your reasoning when analyzing statements.

## Example 2 Analyze Statements Using Postulates

**Determine whether each statement is *always, sometimes,* or *never* true. Explain your reasoning.**

**a. If two coplanar lines intersect, then the point of intersection lies in the same plane as the two lines.**

Always; Postulate 5.5 states that if two points lie in a plane, then the entire line containing those points lies in that plane. So, since both points lie in the plane, any point on those lines, including their point of intersection, also lies in the plane.

**b. Four points are noncollinear.**

Sometimes; Postulate 5.3 states that a line contains at least two points. This means that a line can contain two *or more* points. So four points can be noncollinear, like *A*, *E*, *C*, and *D*, or collinear, like points *A*, *B*, *C*, and *D*.

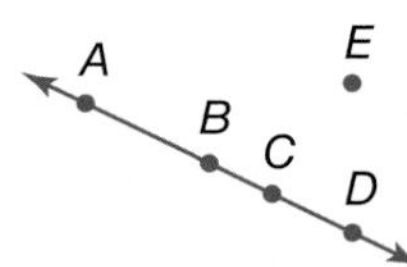


**GuidedPractice**

**2A.** Two intersecting lines determine a plane. **2B.** Three lines intersect in two points.

**StudyTip**

**Axiomatic System** An axiomatic system is a set of axioms, from which some or all axioms can be used to logically derive theorems.

## 2 Paragraph Proofs

To prove a conjecture, you use deductive reasoning to move from a hypothesis to the conclusion of the conjecture you are trying to prove. This is done by writing a **proof**, which is a logical argument in which each statement you make is supported by a statement that is accepted as true.

Once a statement or conjecture has been proven, it is called a **theorem**, and it can be used as a reason to justify statements in other proofs.

**StudyTip**

Proposition A *proposition* is a statement that makes an assertion that is either false or true. In mathematics, a proposition is usually used to mean a true assertion and can be synonymous with theorem.

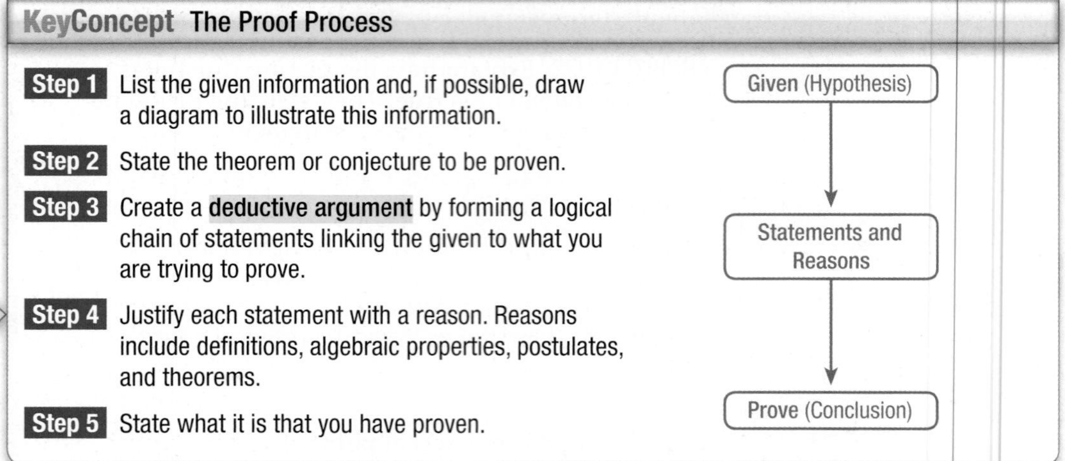

**KeyConcept The Proof Process**

**Step 1** List the given information and, if possible, draw a diagram to illustrate this information.

**Step 2** State the theorem or conjecture to be proven.

**Step 3** Create a **deductive argument** by forming a logical chain of statements linking the given to what you are trying to prove.

**Step 4** Justify each statement with a reason. Reasons include definitions, algebraic properties, postulates, and theorems.

**Step 5** State what it is that you have proven.



One method of proving statements and conjectures, a **paragraph proof**, involves writing a paragraph to explain why a conjecture for a given situation is true. Paragraph proofs are also called **informal proofs**, although the term *informal* is not meant to imply that this form of proof is any less valid than any other type of proof.

## Example 3 Write a Paragraph Proof

**Given that $M$ is the midpoint of $\overline{XY}$ write a paragraph proof to show that $\overline{XM} \cong \overline{MY}$.**

Steps 1 and 2 →

**Given:** $M$ is the midpoint of $\overline{XY}$.

**Prove:** $\overline{XM} \cong \overline{MY}$

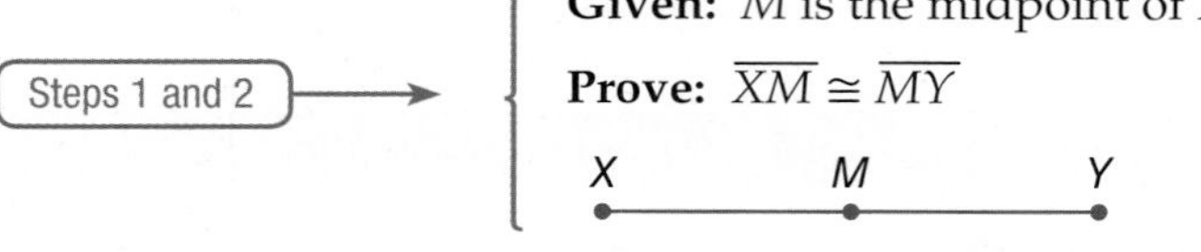


Steps 3 and 4 → If $M$ is the midpoint of $\overline{XY}$, then from the definition of midpoint of a segment, we know that $XM = MY$. This means that $\overline{XM}$ and $\overline{MY}$ have the same measure. By the definition of congruence, if two segments have the same measure, then they are congruent.

Step 5 → Thus, $\overline{XM} \cong \overline{MY}$.

**Problem-SolvingTip**

Work Backward One strategy for writing a proof is to *work backward.* Start with what you are trying to prove, and work backward step by step until you reach the given information.

**GuidedPractice**

**3.** Given that $C$ is between $A$ and $B$ and $\overline{AC} \cong \overline{CB}$, write a paragraph proof to show that $C$ is the midpoint of $\overline{AB}$.

Once a conjecture has been proven true, it can be stated as a theorem and used in other proofs. The conjecture in Example 3 is known as the Midpoint Theorem.

**Theorem 5.1 Midpoint Theorem**

If $M$ is the midpoint of $\overline{AB}$, then $\overline{AM} \cong \overline{MB}$.

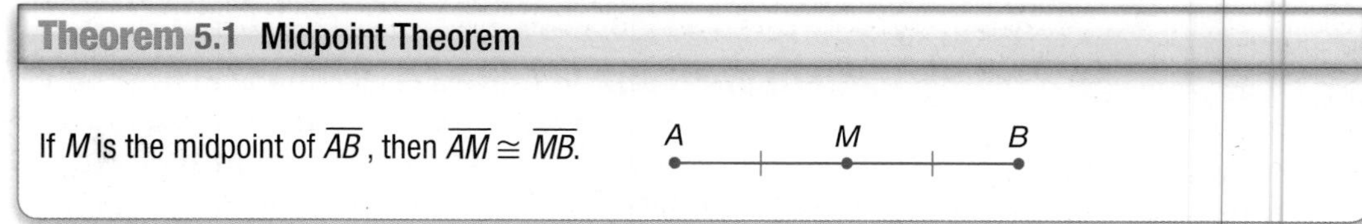

## Check Your Understanding

**Example 1** **Explain how the figure illustrates that each statement is true. Then state the postulate that can be used to show each statement is true.**

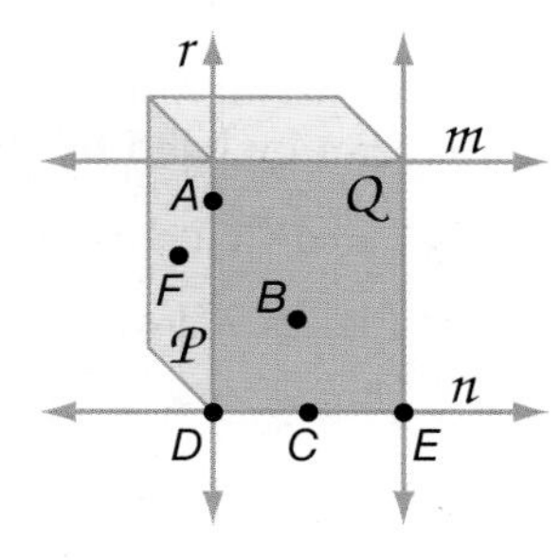


1. Planes $\mathcal{P}$ and $Q$ intersect in line $r$.
2. Lines $r$ and $n$ intersect at point $D$.
3. Line $n$ contains points $C$, $D$, and $E$.
4. Plane $\mathcal{P}$ contains the points $A$, $F$, and $D$.
5. Line $n$ lies in plane $Q$.
6. Line $r$ is the only line through points $A$ and $D$.

**Example 2** **Determine whether each statement is *always*, *sometimes*, or *never* true. Explain your reasoning.**

7. The intersection of three planes is a line.
8. Line $r$ contains only point $\mathcal{P}$.
9. Through two points, there is exactly one line.

**In the figure, $\overrightarrow{AK}$ is in plane $\mathcal{P}$ and $M$ is on $\overleftrightarrow{NE}$. State the postulate that can be used to show each statement is true.**

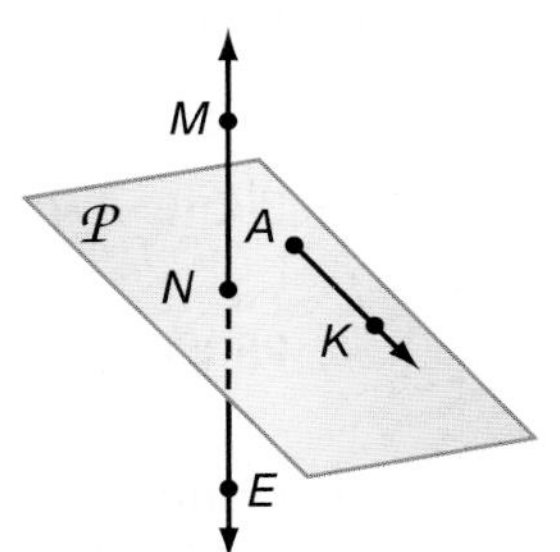


10. $M$, $K$, and $N$ are coplanar.
11. $\overleftrightarrow{NE}$ contains points $N$ and $M$.
12. $N$ and $K$ are collinear.
13. Points $N$, $K$, and $A$ are coplanar.

14. **SPORTS** Each year, Jennifer's school hosts a student vs. teacher basketball tournament to raise money for charity. This year, there are eight teams participating in the tournament. During the first round, each team plays all of the other teams.
    a. How many games will be played in the first round?
    b. Draw a diagram to model the number of first round games. Which postulate can be used to justify your diagram?
    c. Find a numerical method that you could use regardless of the number of the teams in the tournament to calculate the number of games in the first round.

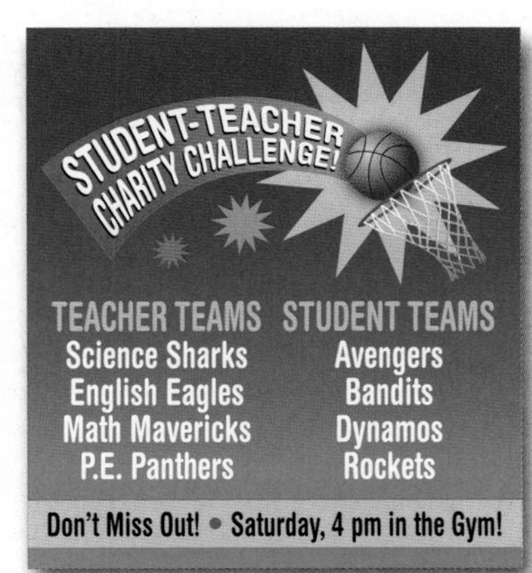


**Example 3** 15. **CCSS ARGUMENTS** In the figure at the right, $\overline{AE} \cong \overline{DB}$ and $C$ is the midpoint of $\overline{AE}$ and $\overline{DB}$. Write a paragraph proof to show that $AC = CB$.

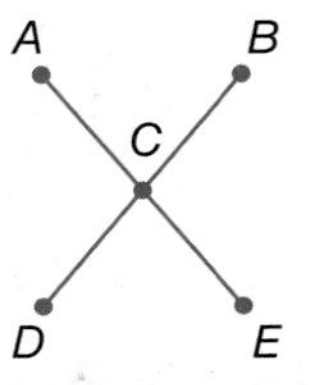

## Practice and Problem Solving

**Example 1** **CAKES Explain how the picture illustrates that each statement is true. Then state the postulate that can be used to show each statement is true.**

16. Lines $n$ and $\ell$ intersect at point $K$.
17. Planes $\mathcal{P}$ and $\mathcal{Q}$ intersect in line $m$.
18. Points $D$, $K$, and $H$ determine a plane.
19. Point $D$ is also on the line $n$ through points $C$ and $K$.
20. Points $D$ and $H$ are collinear.
21. Points $E$, $F$, and $G$ are coplanar.
22. $\overleftrightarrow{EF}$ lies in plane $\mathcal{Q}$.
23. Lines $h$ and $g$ intersect at point $J$.

**Example 2** **Determine whether each statement is *always*, *sometimes*, or *never* true. Explain.**

24. There is exactly one plane that contains noncollinear points $A$, $B$, and $C$.
25. There are at least three lines through points $J$ and $K$.
26. If points $M$, $N$, and $P$ lie in plane $X$, then they are collinear.
27. Points $X$ and $Y$ are in plane $Z$. Any point collinear with $X$ and $Y$ is in plane $Z$.
28. The intersection of two planes can be a point.
29. Points $A$, $B$, and $C$ determine a plane.

**Example 3**

30. **PROOF** Point $Y$ is the midpoint of $\overline{XZ}$. $Z$ is the midpoint of $\overline{YW}$. Prove that $\overline{XY} \cong \overline{ZW}$.

31. **PROOF** Point $L$ is the midpoint of $\overline{JK}$. $\overline{JK}$ intersects $\overline{MK}$ at $K$. If $\overline{MK} \cong \overline{JL}$, prove that $\overline{LK} \cong \overline{MK}$.

32. CCSS **ARGUMENTS** Last weekend, Emilio and his friends spent Saturday afternoon at the park. There were several people there with bikes and skateboards. There were a total of 11 bikes and skateboards that had a total of 36 wheels. Use a paragraph proof to show how many bikes and how many skateboards there were.

33. **DRIVING** Keisha is traveling from point A to point B. Two possible routes are shown on the map. Assume that the speed limit on Southside Boulevard is 55 miles per hour and the speed limit on I–295 is 70 miles per hour.
    a. Which of the two routes covers the shortest distance? Explain your reasoning.
    b. If the distance from point A to point B along Southside Boulevard is 10.5 miles and the distance along I-295 is 11.6 miles, which route is faster, assuming that Keisha drives the speed limit?

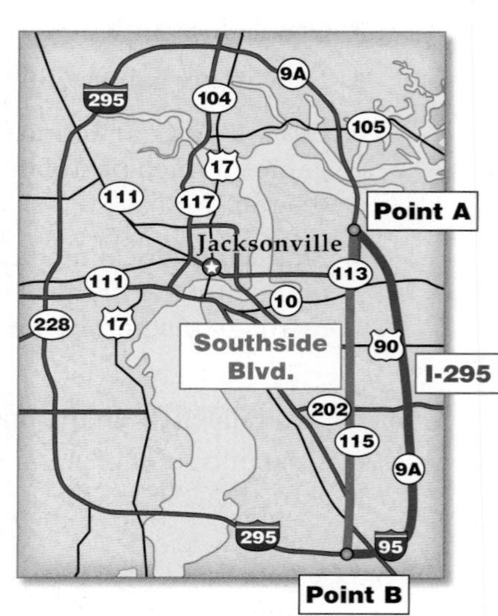

### GuidedPractice

**State the property that justifies each statement.**

**1A.** If $4 + (-5) = -1$, then $x + 4 + (-5) = x - 1$.

**1B.** If $5 = y$, then $y = 5$.

**1C.** Prove that if $2x - 13 = -5$, then $x = 4$. Write a justification for each step.

Example 1 is a proof of the conditional statement *If $-5(x + 4) = 70$, then $x = -18$.* Notice that the column on the left is a step-by-step process that leads to a solution. The column on the right contains the reason for each statement.

**StudyTip**

**CCSS Arguments** An *algorithm* is a series of steps for carrying out a procedure or solving a problem. Proofs can be considered a type of algorithm because they go step by step.

In geometry, a similar format is used to prove conjectures and theorems. A **two-column proof** or **formal proof** contains *statements* and *reasons* organized in two columns.

PT

## Real-World Example 2 Write an Algebraic Proof

**SCIENCE If the formula to convert a Fahrenheit temperature to a Celsius temperature is $C = \frac{5}{9}(F - 32)$, then the formula to convert a Celsius temperature to a Fahrenheit temperature is $F = \frac{9}{5}C + 32$. Write a two-column proof to verify this conjecture.**

Begin by stating what is given and what you are to prove.

**Given:** $C = \frac{5}{9}(F - 32)$

**Prove:** $F = \frac{9}{5}C + 32$

**Proof:**

| Statements | Reasons |
|---|---|
| **1.** $C = \frac{5}{9}(F - 32)$ | **1.** Given |
| **2.** $\frac{9}{5}C = \frac{9}{5} \cdot \frac{5}{9}(F - 32)$ | **2.** Multiplication Property of Equality |
| **3.** $\frac{9}{5}C = F - 32$ | **3.** Substitution Property of Equality |
| **4.** $\frac{9}{5}C + 32 = F - 32 + 32$ | **4.** Addition Property of Equality |
| **5.** $\frac{9}{5}C + 32 = F$ | **5.** Substitution Property of Equality |
| **6.** $F = \frac{9}{5}C + 32$ | **6.** Symmetric Property of Equality |

**StudyTip**

**Mental Math** If your teacher permits you to do so, some steps may be eliminated by performing mental calculations. For example, steps 2 and 4 in Example 2 could be omitted. Then the reason for statement 3 would be Multiplication Property of Equality and the reason for statement 5 would be Addition Property of Equality.

### GuidedPractice

**Write a two-column proof to verify that each conjecture is true.**

**2A.** If $\frac{5x + 1}{2} - 8 = 0$, then $x = 3$.

**2B.** **PHYSICS** If the distance $d$ moved by an object with initial velocity $u$ and final velocity $v$ in time $t$ is given by $d = t \cdot \frac{u + v}{2}$, then $u = \frac{2d}{t} - v$.

## 2 Geometric Proof

Since geometry also uses variables, numbers, and operations, many of the properties of equality used in algebra are also true in geometry. For example, segment measures and angle measures are real numbers, so properties from algebra can be used to discuss their relationships as shown in the table below.

| Property | Segments | Angles |
|---|---|---|
| Reflexive | $AB = AB$ | $m\angle 1 = m\angle 1$ |
| Symmetric | If $AB = CD$, then $CD = AB$. | If $m\angle 1 = m\angle 2$, then $m\angle 2 = m\angle 1$. |
| Transitive | If $AB = CD$ and $CD = EF$, then $AB = EF$. | If $m\angle 1 = m\angle 2$ and $m\angle 2 = m\angle 3$, then $m\angle 1 = m\angle 3$. |

**StudyTip**

**Commutative and Associative Properties** Throughout this text we shall assume that if $a$, $b$, and $c$ are real numbers, then the following properties are true.

**Commutative Property of Addition**
$a + b = b + a$

**Commutative Property of Multiplication**
$a \cdot b = b \cdot a$

**Associative Property of Addition**
$(a + b) + c = a + (b + c)$

**Associative Property of Multiplication**
$(a \cdot b) \cdot c = a \cdot (b \cdot c)$

These properties can be used to write geometric proofs.

### Example 3 Write a Geometric Proof

**If $\angle FGJ \cong \angle JGK$ and $\angle JGK \cong \angle KGH$, then $x = 6$. Write a two-column proof to verify this conjecture.**

**Given:** $\angle FGJ \cong \angle JGK$, $\angle JGK \cong \angle KGH$, $m\angle FGJ = 6x + 7$, $m\angle KGH = 8x - 5$

**Prove:** $x = 6$

**Proof:**

| Statements | Reasons |
|---|---|
| **1.** $m\angle FGH = 6x + 7$, $m\angle KGH = 8x - 5$; $\angle FGJ \cong \angle JGK$; $\angle JGK \cong \angle KGH$ | **1.** Given |
| **2.** $m\angle FGJ = m\angle JGK$; $m\angle JGK = m\angle KGH$ | **2.** Definition of congruent angles |
| **3.** $m\angle FGJ = m\angle KGH$ | **3.** Transitive Property of Equality |
| **4.** $6x + 7 = 8x - 5$ | **4.** Substitution Property of Equality |
| **5.** $6x + 7 + 5 = 8x - 5 + 5$ | **5.** Addition Property of Equality |
| **6.** $6x + 12 = 8x$ | **6.** Substitution Property of Equality |
| **7.** $6x + 12 - 6x = 8x - 6x$ | **7.** Subtraction Property of Equality |
| **8.** $12 = 2x$ | **8.** Substitution Property of Equality |
| **9.** $\frac{12}{2} = \frac{2x}{2}$ | **9.** Division Property of Equality |
| **10.** $6 = x$ | **10.** Substitution Property of Equality |
| **11.** $x = 6$ | **11.** Symmetric Property of Equality |

### GuidedPractice

**Write a two-column proof to verify each conjecture.**

**3A.** If $\angle A \cong \angle B$ and $m\angle A = 37$, then $m\angle B = 37$.

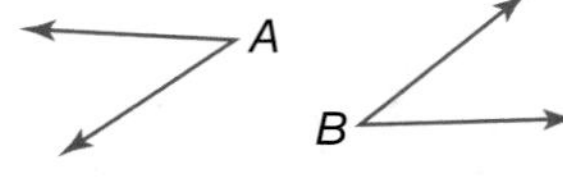


**3B.** If $\overline{CD} \cong \overline{EF}$, then $y = 8$.

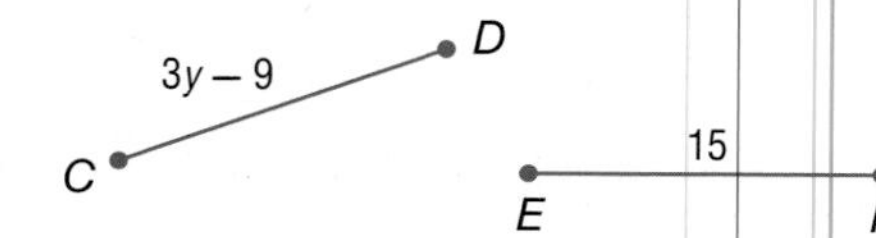

## Check Your Understanding

Example 1

**1.** CCSS **ARGUMENTS** Copy and complete the proof.

**Given:** $\overline{LK} \cong \overline{NM}$, $\overline{KJ} \cong \overline{MJ}$

**Prove:** $\overline{LJ} \cong \overline{NJ}$

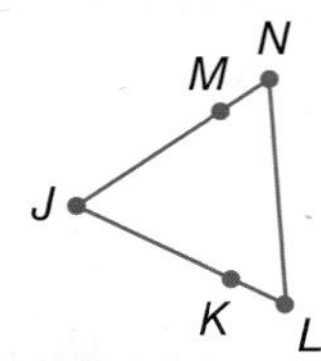


**Proof:**

| Statements | Reasons |
|---|---|
| a. $\overline{LK} \cong \overline{NM}$, $\overline{KJ} \cong \overline{MJ}$ | a. ___?___ |
| b. ___?___ | b. Def. of congruent segments |
| c. $LK + KJ = NM + MJ$ | c. ___?___ |
| d. ___?___ | d. Segment Addition Postulate |
| e. $LJ = NJ$ | e. ___?___ |
| f. $\overline{LJ} \cong \overline{NJ}$ | f. ___?___ |

Example 2

**2.** **PROOF** Prove the following.

**Given:** $\overline{WX} \cong \overline{YZ}$

**Prove:** $\overline{WY} \cong \overline{XZ}$

W X Y Z

**3** **SCISSORS** Refer to the diagram shown. $\overline{AR}$ is congruent to $\overline{CR}$. $\overline{DR}$ is congruent to $\overline{BR}$. Prove that $AR + DR = CR + BR$.

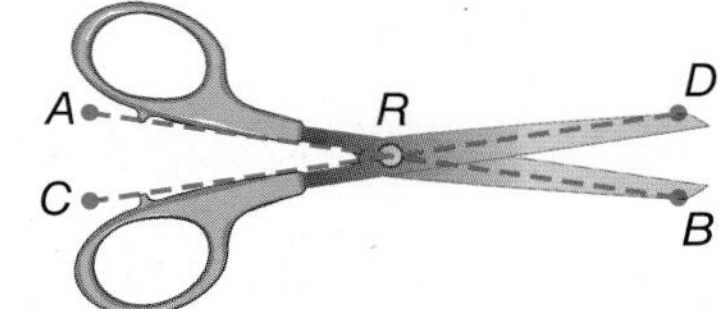


## Practice and Problem Solving

Example 1

**4.** CCSS **ARGUMENTS** Copy and complete the proof.

**Given:** $C$ is the midpoint of $\overline{AE}$.

$C$ is the midpoint of $\overline{BD}$.

$\overline{AE} \cong \overline{BD}$

**Prove:** $\overline{AC} \cong \overline{CD}$

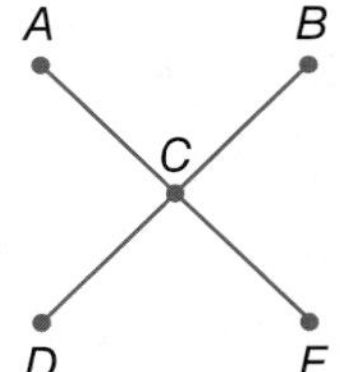


**Proof:**

| Statements | Reasons |
|---|---|
| a. ___?___ | a. Given |
| b. $AC = CE$, $BC = CD$ | b. ___?___ |
| c. $AE = BD$ | c. ___?___ |
| d. ___?___ | d. Segment Addition Postulate |
| e. $AC + CE = BC + CD$ | e. ___?___ |
| f. $AC + AC = CD + CD$ | f. ___?___ |
| g. ___?___ | g. Simplify. |
| h. ___?___ | h. Division Property |
| i. $\overline{AC} \cong \overline{CD}$ | i. ___?___ |

Example 2

5. **TILING** A tile setter cuts a piece of tile to a desired length. He then uses this tile as a pattern to cut a second tile congruent to the first. He uses the first two tiles to cut a third tile whose length is the sum of the measures of the first two tiles. Prove that the measure of the third tile is twice the measure of the first tile.

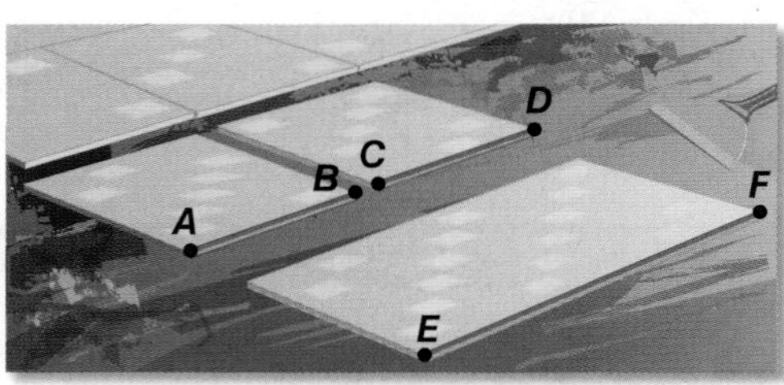


**CCSS ARGUMENTS** **Prove each theorem.**

6. Symmetric Property of Congruence (Theorem 5.2)

7. Reflexive Property of Congruence (Theorem 5.2)

8. **TRAVEL** Four cities in New York are connected by Interstate 90: Buffalo, Utica, Albany, and Syracuse, Buffalo is the farthest west.
   - Albany is 126 miles from Syracuse and 263 miles from Buffalo.
   - Buffalo is 137 miles from Syracuse and 184 miles from Utica.

   a. Draw a diagram to represent the locations of the cities in relation to each other and the distances between each city. Assume that Interstate 90 is straight.

   b. Write a paragraph proof to support your conclusion.

**PROOF** **Prove the following.**

9. If $\overline{SC} \cong \overline{HR}$ and $\overline{HR} \cong \overline{AB}$, then $\overline{SC} \cong \overline{AB}$.

10. If $\overline{VZ} \cong \overline{VY}$ and $\overline{WY} \cong \overline{XZ}$, then $\overline{VW} \cong \overline{VX}$.

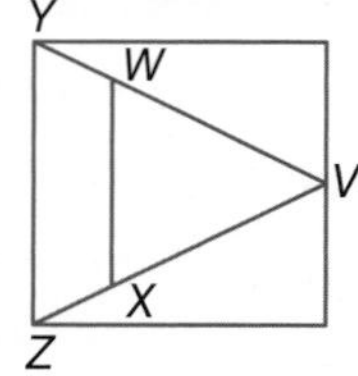


11. If $E$ is the midpoint of $\overline{DF}$ and $\overline{CD} \cong \overline{FG}$, then $\overline{CE} \cong \overline{EG}$.

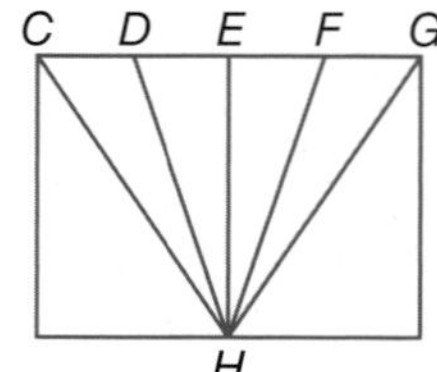


12. If $B$ is the midpoint of $\overline{AC}$, $D$ is the midpoint of $\overline{CE}$, and $\overline{AB} \cong \overline{DE}$, then $AE = 4AB$.

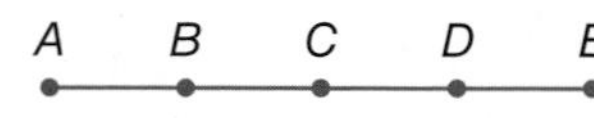


13. **OPTICAL ILLUSION** $\overline{AC} \cong \overline{GI}$, $\overline{FE} \cong \overline{LK}$, and $AC + CF + FE = GI + IL + LK$.

    a. Prove that $\overline{CF} \cong \overline{IL}$.

    b. Justify your proof using measurement. Explain your method.

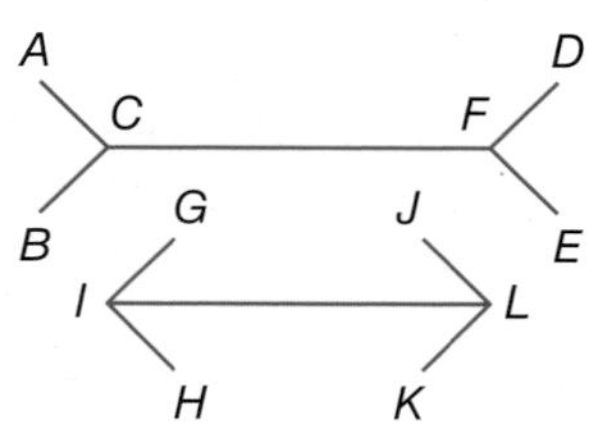

14. **CONSTRUCTION** Construct a segment that is twice as long as $\overline{PQ}$. Explain how the Segment Addition Postulate can be used to justify your construction.

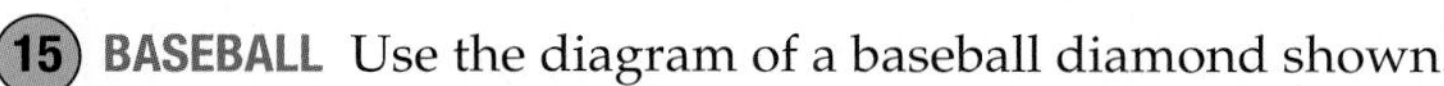

15. **BASEBALL** Use the diagram of a baseball diamond shown.

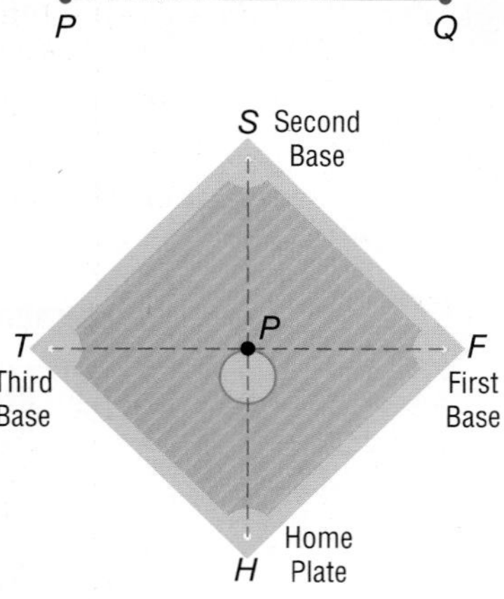


 a. On the diagram, $\overline{SH} \cong \overline{TF}$. $P$ is the midpoint of $\overline{SH}$ and $\overline{TF}$. Using a two-column proof, prove that $\overline{SP} \cong \overline{TP}$.

 b. The distance from home plate to second base is 127.3 feet. What is the distance from first base to second base?

16. **MULTIPLE REPRESENTATIONS** $A$ is the midpoint of $\overline{PQ}$, $B$ is the midpoint of $\overline{PA}$, and $C$ is the midpoint of $\overline{PB}$.

 a. **Geometric** Make a sketch to represent this situation.

 b. **Algebraic** Make a conjecture as to the algebraic relationship between $PC$ and $PQ$.

 c. **Geometric** Copy segment $\overline{PQ}$ from your sketch. Then construct points $B$ and $C$ on $\overline{PQ}$. Explain how you can use your construction to support your conjecture.

 d. **Concrete** Use a ruler to draw a segment congruent to $\overline{PQ}$ from your sketch and to draw points $B$ and $C$ on $\overline{PQ}$. Use your drawing to support your conjecture.

 e. **Logical** Prove your conjecture.

## H.O.T. Problems Use Higher-Order Thinking Skills

17. **CCSS CRITIQUE** In the diagram, $\overline{AB} \cong \overline{CD}$ and $\overline{CD} \cong \overline{BF}$. Examine the conclusions made by Leslie and Shantice. Is either of them correct?

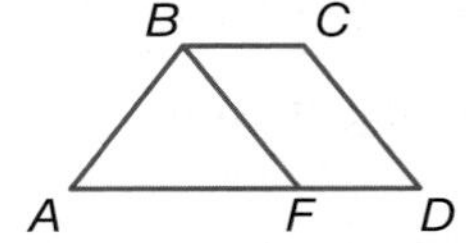


**Leslie**

Since $\overline{AB} \cong \overline{CD}$ and $\overline{CD} \cong \overline{BF}$, then $\overline{AB} \cong \overline{AF}$ by the Transitive Property of Congruence

**Shantice**

*Since $\overline{AB} \cong \overline{CD}$ and $\overline{CD} \cong \overline{BF}$, then $\overline{AB} \cong \overline{BF}$ by the Reflexive Property of Congruence.*

18. **CHALLENGE** $ABCD$ is a square. Prove that $\overline{AC} \cong \overline{BD}$.

19. **WRITING IN MATH** Does there exist an Addition Property of Congruence? Explain.

20. **REASONING** Classify the following statement as *true* or *false*. If false, provide a counterexample.

 *If A, B, C, D, and E are collinear with B between A and C, C between B and D, and D between C and E, and AC = BD = CE, then AB = BC = DE.*

21. **OPEN ENDED** Draw a representation of the Segment Addition Postulate in which the segment is two inches long, contains four collinear points, and contains no congruent segments.

22. **WRITING IN MATH** Compare and contrast paragraph proofs and two-column proofs.

## Standardized Test Practice

**23. ALGEBRA** The chart below shows annual recycling by material in the United States. About how many pounds of aluminum are recycled each year?

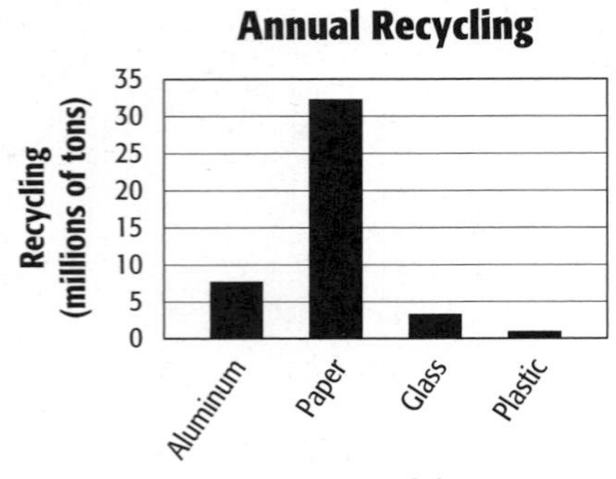


**A** 7.5 **C** 7,500,000

**B** 15,000 **D** 15,000,000,000

**24. ALGEBRA** Which expression is equivalent to $\frac{12x^{-4}}{4x^{-8}}$?

**F** $\frac{1}{3x^4}$ **H** $8x^2$

**G** $3x^4$ **J** $\frac{x^4}{3}$

**25. SHORT RESPONSE** The measures of two complementary angles are in the ratio 4 : 1. What is the measure of the smaller angle?

**26. SAT/ACT** Julie can word process 40 words per minute. How many minutes will it take Julie to word process 200 words?

**A** 0.5 **D** 10

**B** 2 **E** 12

**C** 5

## Spiral Review

**27. PROOF** **Write a two-column proof.** (Lesson 5-2)

**Given:** $AC = DF$

$AB = DE$

**Prove:** $BC = EF$

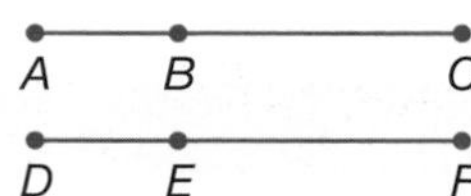


**28. MODELS** Brian is using six squares of cardboard to form a rectangular prism. What geometric figure do the pieces of cardboard represent, and how many lines will be formed by their intersections? (Lesson 2-5)

**29. FRUIT** An apple fell 25 feet from a tree. The formula $h = -16t^2 + 25$ can be used to approximate the number of seconds it will take the apple to hit the ground. (Lesson 1-8)

**a.** How long will it take the apple to hit the ground?

**b.** If you catch it at 4 feet, how long did the apple drop?

**Simplify.** (Lesson 0-10)

**30.** $\sqrt{48}$ **31.** $\sqrt{162}$ **32.** $\sqrt{25a^6b^4}$ **33.** $\sqrt{45xy^8}$

## Skills Review

**ALGEBRA** **Find** $x$.

**34.**

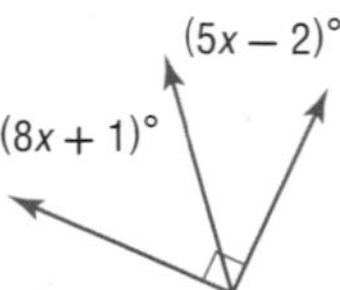


**35.**

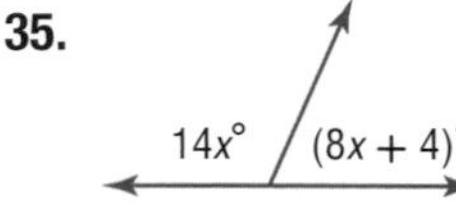


**36.**

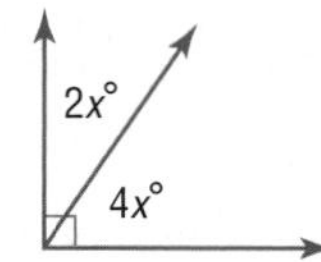

# LESSON 5-4 Proving Angle Relationships

**:: Then**

- You identified and used special pairs of angles.

**:: Now**

1. Write proofs involving supplementary and complementary angles.
2. Write proofs involving congruent and right angles.

**:: Why?**

- Jamal's school is building a walkway that will include bricks with the names of graduates from each class. All of the bricks are rectangular, so when the bricks are laid, all of the angles form linear pairs.

**Common Core State Standards**

**Content Standards**
G.CO.9 Prove theorems about lines and angles.

**Mathematical Practices**
3 Construct viable arguments and critique the reasoning of others.
6 Attend to precision.

## 1 Supplementary and Complementary Angles

The Protractor Postulate illustrates the relationship between angle measures and real numbers.

### Postulate 5.10 Protractor Postulate

**Words** Given any angle, the measure can be put into one-to-one correspondence with real numbers between 0 and 180.

**Example** If $\overrightarrow{BA}$ is placed along the protractor at 0°, then the measure of $\angle ABC$ corresponds to a positive real number.

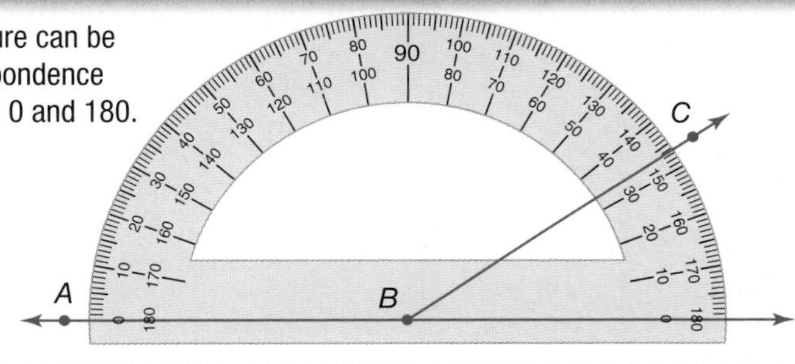


In Lesson 5-3, you learned about the Segment Addition Postulate. A similar relationship exists between the measures of angles.

### Postulate 5.11 Angle Addition Postulate

$D$ is in the interior of $\angle ABC$ if and only if $m\angle ABD + m\angle DBC = m\angle ABC$.

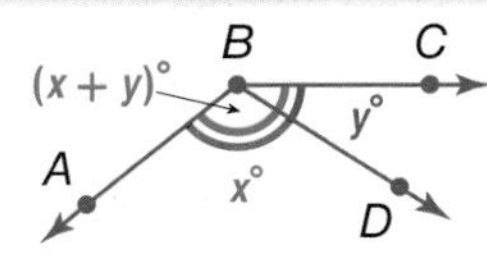


### Example 1 Use the Angle Addition Postulate

**Find $m\angle 1$ if $m\angle 2 = 56$ and $m\angle JKL = 145$.**

| | |
|---|---|
| $m\angle 1 + m\angle 2 = m\angle JKL$ | Angle Addition Postulate |
| $m\angle 1 + 56 = 145$ | $m\angle 2 = 56$ $m\angle JKL = 145$ |
| $m\angle 1 + 56 - 56 = 145 - 56$ | Subtraction Property of Equality |
| $m\angle 1 = 89$ | Substitution |

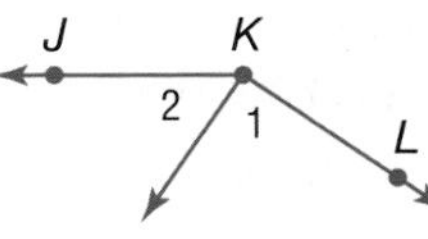


**Guided Practice**

1. If $m\angle 1 = 23$ and $m\angle ABC = 131$, find the measure of $\angle 3$. Justify each step.

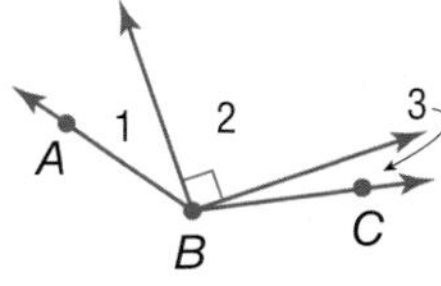

The Angle Addition Postulate can be used with other angle relationships to provide additional theorems relating to angles.

**StudyTip**

**Linear Pair Theorem** The Supplement Theorem may also be known as the *Linear Pair Theorem.*

**Theorems**

**5.3 Supplement Theorem** If two angles form a linear pair, then they are supplementary angles.

**Example** $m\angle 1 + m\angle 2 = 180$

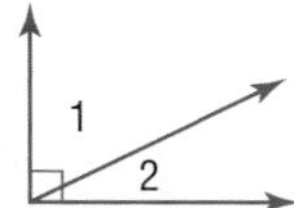

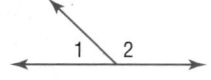


**5.4 Complement Theorem** If the noncommon sides of two adjacent angles form a right angle, then the angles are complementary angles.

**Example** $m\angle 1 + m\angle 2 = 90$

You will prove Theorems 5.3 and 5.4 in Exercises 16 and 17, respectively.

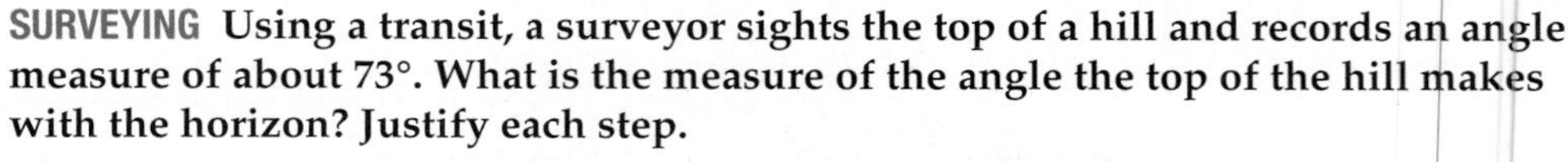

**Real-World Example 2** Use Supplement or Complement

**SURVEYING Using a transit, a surveyor sights the top of a hill and records an angle measure of about 73°. What is the measure of the angle the top of the hill makes with the horizon? Justify each step.**

**Understand** Make a sketch of the situation. The surveyor is measuring the angle of his line of sight below the vertical. Draw a vertical ray and a horizontal ray from the point where the surveyor is sighting the hill, and label the angles formed. We know that the vertical and horizontal rays form a right angle.

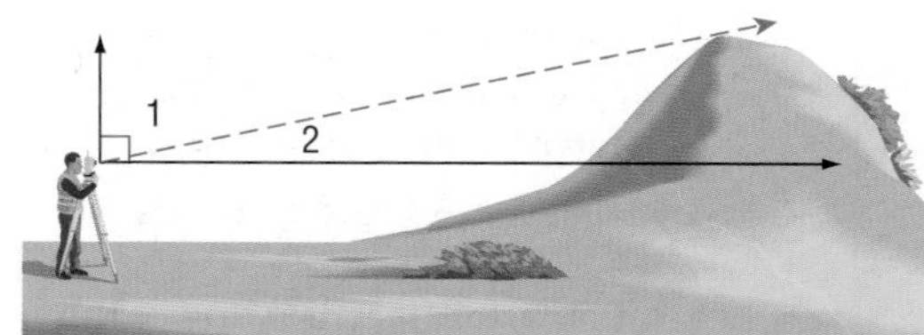


**Plan** Since $\angle 1$, and $\angle 2$ form a right angle, you can use the Complement Theorem.

**Solve**

| | |
|---|---|
| $m\angle 1 + m\angle 2 = 90$ | Complement Theorem |
| $73 + m\angle 2 = 90$ | $m\angle 1 = 73$ |
| $73 + m\angle 2 - 73 = 90 - 73$ | Subtraction Property of Equality |
| $m\angle 2 = 17$ | Substitution |

The top of the hill makes a 17° angle with the horizon.

**Check** Since we know that the sum of the angles should be 90, check your math. The sum of 17 and 73 is 90. ✓

**ReviewVocabulary**

**supplementary angles** two angles with measures that add to 180

**complementary angles** two angles with measures that add to 90

**linear pair** a pair of adjacent angles with noncommon sides that are opposite rays

**GuidedPractice**

2. $\angle 6$ and $\angle 7$ form linear pair. If $m\angle 6 = 3x + 32$ and $m\angle 7 = 5x + 12$, find $x$, $m\angle 6$, and $m\angle 7$. Justify each step.

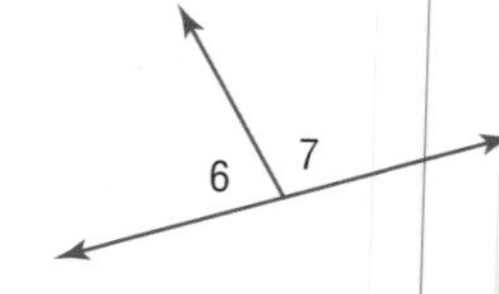

## 2 Congruent Angles

**Congruent Angles** The properties of algebra that applied to the congruence of segments and the equality of their measures also hold true for the congruence of angles and the equality of their measures.

### Theorem 5.5 Properties of Angle Congruence

**Reflexive Property of Congruence**

$\angle 1 \cong \angle 1$

**Symmetric Property of Congruence**

If $\angle 1 \cong \angle 2$, then $\angle 2 \cong \angle 1$.

**Transitive Property of Congruence**

If $\angle 1 \cong \angle 2$ and $\angle 2 \cong \angle 3$, then $\angle 1 \cong \angle 3$.

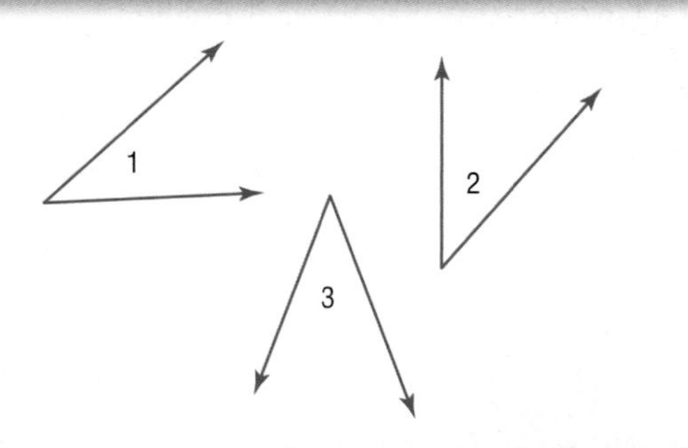


You will prove the Reflexive and Transitive Properties of Congruence in Exercises 18 and 19, respectively.

### Proof Symmetric Property of Congruence

**Given:** $\angle A \cong \angle B$

**Prove:** $\angle B \cong \angle A$

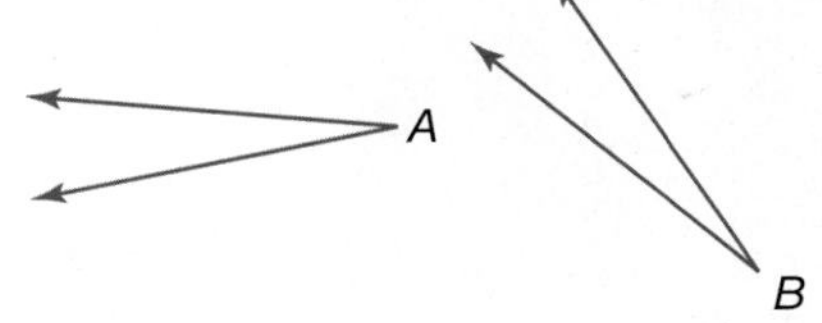


**Paragraph Proof:**

We are given $\angle A \cong \angle B$. By the definition of congruent angles, $m\angle A = m\angle B$. Using the Symmetric Property of Equality, $m\angle B = m\angle A$. Thus, $\angle B \cong \angle A$ by the definition of congruent angles.

Algebraic properties can be applied to prove theorems for congruence relationships involving supplementary and complementary angles.

### Theorems

**5.6 Congruent Supplements Theorem**
Angles supplementary to the same angle or to congruent angles are congruent.

**Abbreviation** *∠s suppl. to same ∠ or ≅ ∠s are ≅.*

**Example** If $m\angle 1 + m\angle 2 = 180$ and $m\angle 2 + m\angle 3 = 180$, then $\angle 1 \cong \angle 3$.

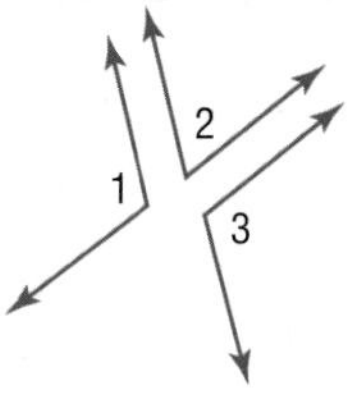

**5.7 Congruent Complements Theorem**
Angles complementary to the same angle or to congruent angles are congruent.

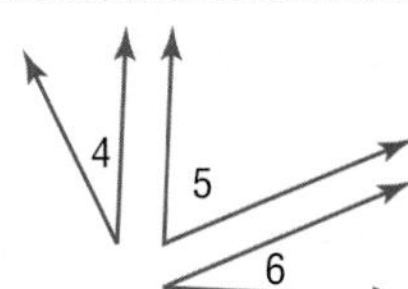


**Abbreviation** *∠s compl. to same ∠ or ≅ ∠s are ≅.*

**Example** If $m\angle 4 + m\angle 5 = 90$ and $m\angle 5 + m\angle 6 = 90$, then $\angle 4 \cong \angle 6$.

You will prove one case of Theorem 5.6 in Exercise 6.

**ReadingMath**

Abbreviations and Symbols
The notation ∠s means *angles.*

## Proof One Case of the Congruent Supplements Theorem

**Given:** $\angle 1$ and $\angle 2$ are supplementary.
$\angle 2$ and $\angle 3$ are supplementary.

**Prove:** $\angle 1 \cong \angle 3$

**Proof:**

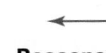

| Statements | Reasons |
|---|---|
| **1.** $\angle 1$ and $\angle 2$ are supplementary. $\angle 2$ and $\angle 3$ are supplementary. | **1.** Given |
| **2.** $m\angle 1 + m\angle 2 = 180$; $m\angle 2 + m\angle 3 = 180$ | **2.** Definition of supplementary angles |
| **3.** $m\angle 1 + m\angle 2 = m\angle 2 + m\angle 3$ | **3.** Substitution |
| **4.** $m\angle 2 = m\angle 2$ | **4.** Reflexive Property |
| **5.** $m\angle 1 = m\angle 3$ | **5.** Subtraction Property |
| **6.** $\angle 1 \cong \angle 3$ | **6.** Definition of congruent angles |

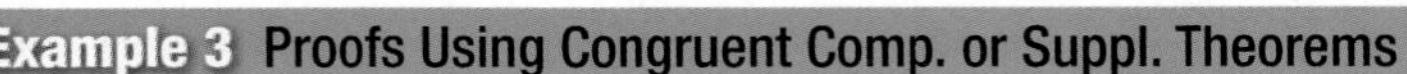

## Example 3 Proofs Using Congruent Comp. or Suppl. Theorems

**Prove that vertical angles 2 and 4 in the photo at the left are congruent.**

**Given:** $\angle 2$ and $\angle 4$ are vertical angles.

**Prove:** $\angle 2 \cong \angle 4$

**Proof:**

| Statements | Reasons |
|---|---|
| **1.** $\angle 2$ and $\angle 4$ are vertical angles. | **1.** Given |
| **2.** $\angle 2$ and $\angle 4$ are nonadjacent angles formed by intersecting lines. | **2.** Definition of vertical angles |
| **3.** $\angle 2$ and $\angle 3$ from a linear pair. $\angle 3$ and $\angle 4$ form a linear pair. | **3.** Definition of a linear pair |
| **4.** $\angle 2$ and $\angle 3$ are supplementary. $\angle 3$ and $\angle 4$ are supplementary. | **4.** Supplement Theorem |
| **5.** $\angle 2 \cong \angle 4$ | **5.** $\measuredangle$ suppl. to same $\angle$ or $\cong$ $\measuredangle$ are $\cong$. |

### GuidedPractice

**3.** In the figure, $\angle ABE$ and $\angle DBC$ are right angles. Prove that $\angle ABD \cong \angle EBC$.

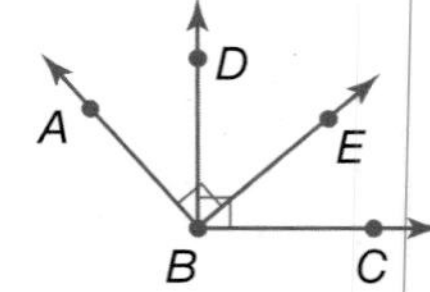


**Real-WorldLink**

The 100-story John Hancock Building uses huge X-braces in its design. These diagonals are connected to the exterior columns, making it possible for strong wind forces to be carried from the braces to the exterior columns and back.

**Source:** PBS

**ReviewVocabulary**

**Vertical Angles** two nonadjacent angles formed by intersecting lines

Note that in Example 3, $\angle 1$ and $\angle 3$ are vertical angles. The conclusion in the example supports the following Vertical Angles Theorem.

## Theorem 5.8 Vertical Angles Theorem

If two angles are vertical angles, then they are congruent.

**Abbreviation** *Vert. $\measuredangle$ are $\cong$.*

**Example** $\angle 1 \cong \angle 3$ and $\angle 2 \cong \angle 4$

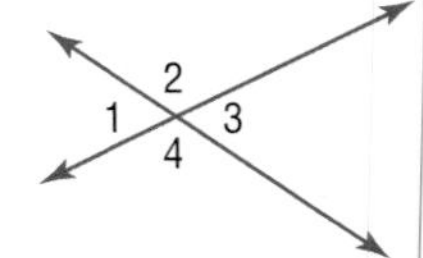


You will prove Theorem 5.8 in Exercise 28.

Yannis Emmanuel Mavromatakis/Alamy

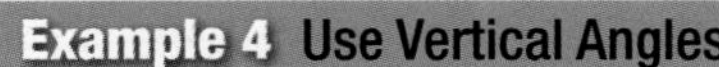

## Example 4 Use Vertical Angles

**Prove that if $\overrightarrow{DB}$ bisects $\angle ADC$, then $\angle 2 \cong \angle 3$.**

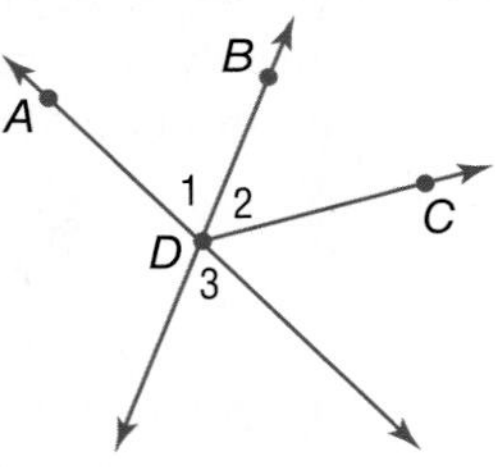


**Given:** $\overrightarrow{DB}$ bisects $\angle ADC$.

**Prove:** $\angle 2 \cong \angle 3$

**Proof:**

| Statements | Reasons |
|---|---|
| **1.** $\overrightarrow{DB}$ bisects $\angle ADC$. | **1.** Given |
| **2.** $\angle 1 \cong \angle 2$ | **2.** Definition of angle bisector |
| **3.** $\angle 1$ and $\angle 3$ are vertical angles. | **3.** Definition of vertical angles |
| **4.** $\angle 3 \cong \angle 1$ | **4.** Vert. ∠s are ≅. |
| **5.** $\angle 3 \cong \angle 2$ | **5.** Transitive Property of Congruence |
| **6.** $\angle 2 \cong \angle 3$ | **6.** Symmetric Property of Congruence |

### GuidedPractice

**4.** If $\angle 3$ and $\angle 4$ are vertical angles, $m\angle 3 = 6x + 2$, and $m\angle 4 = 8x - 14$, find $m\angle 3$ and $m\angle 4$. Justify each step.

The theorems in this lesson can be used to prove the following right angle theorems.

## Theorems Right Angle Theorems

| Theorem | Example |
|---|---|
| **5.9** Perpendicular lines intersect to form four right angles. **Example** If $\overrightarrow{AC} \perp \overrightarrow{DB}$, then $\angle 1$, $\angle 2$, $\angle 3$, and $\angle 4$ are rt. ∠s. | |
| **5.10** All right angles are congruent. **Example** If $\angle 1$, $\angle 2$, $\angle 3$, and $\angle 4$ are rt. ∠s, then $\angle 1 \cong \angle 2 \cong \angle 3 \cong \angle 4$. | |
| **5.11** Perpendicular lines form congruent adjacent angles. **Example** If $\overrightarrow{AC} \perp \overrightarrow{DB}$, then $\angle 1 \cong \angle 2$, $\angle 2 \cong \angle 4$, $\angle 3 \cong \angle 4$, and $\angle 1 \cong \angle 3$. | |
| **5.12** If two angles are congruent and supplementary, then each angle is a right angle. **Example** If $\angle 5 \cong \angle 6$ and $\angle 5$ is suppl. to $\angle 6$, then $\angle 5$ and $\angle 6$ are rt. ∠s. | |
| **5.13** If two congruent angles form a linear pair, then they are right angles. **Example** If $\angle 7$ and $\angle 8$ form a linear pair, then $\angle 7$ and $\angle 8$ are rt. ∠s. | |

You will prove Theorems 5.9–5.13 in Exercises 22–26.

## Check Your Understanding

**Example 1** **Find the measure of each numbered angle, and name the theorems that justify your work.**

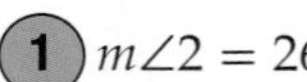

1. $m\angle 2 = 26$
2. $m\angle 2 = x$, $m\angle 3 = x - 16$
3. $m\angle 4 = 2x$, $m\angle 5 = x + 9$
4. $m\angle 4 = 3(x - 1)$, $m\angle 5 = x + 7$

**Example 2**

5. **PARKING** Refer to the diagram of the parking lot at the right. Given that $\angle 2 \cong \angle 6$, prove that $\angle 4 \cong \angle 8$.

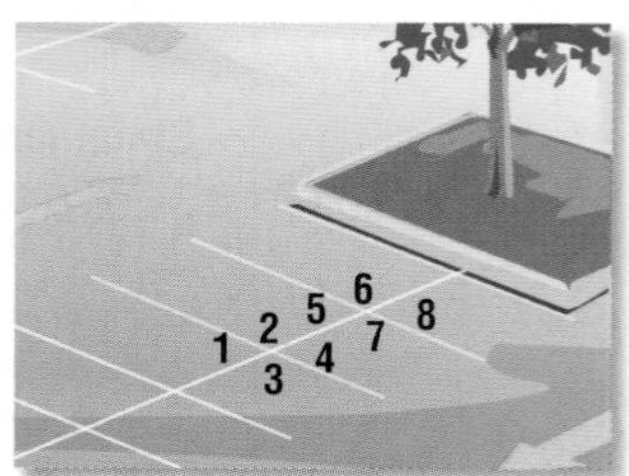


**Example 3**

6. **PROOF** **Copy and complete the proof of one case of Theorem 5.6.**

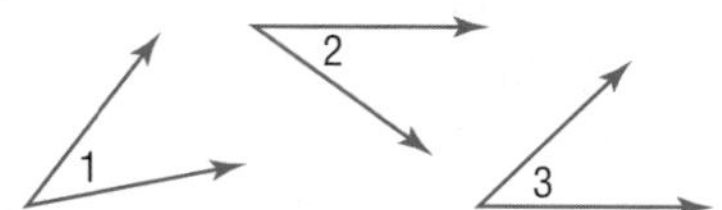


**Given:** $\angle 1$ and $\angle 3$ are complementary.
$\angle 2$ and $\angle 3$ are complementary.

**Prove:** $\angle 1 \cong \angle 2$

**Proof:**

| Statements | Reasons |
|---|---|
| **a.** $\angle 1$ and $\angle 3$ are complementary. $\angle 2$ and $\angle 3$ are complementary. | **a.** ? |
| **b.** $m\angle 1 + m\angle 3 = 90$; $m\angle 2 + m\angle 3 = 90$ | **b.** ? |
| **c.** $m\angle 1 + m\angle 3 = m\angle 2 + m\angle 3$ | **c.** ? |
| **d.** ? | **d.** Reflexive Property |
| **e.** $m\angle 1 = m\angle 2$ | **e.** ? |
| **f.** $\angle 1 \cong \angle 2$ | **f.** ? |

**Example 4**

7. CCSS **ARGUMENTS** Write a two-column proof.

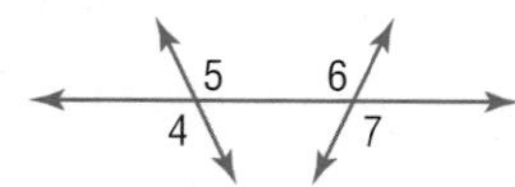


**Given:** $\angle 4 \cong \angle 7$

**Prove:** $\angle 5 \cong \angle 6$

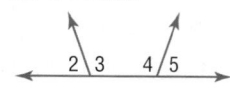

## Practice and Problem Solving

**Examples 1–3** **Find the measure of each numbered angle, and name the theorems used that justify your work.**

8. $m\angle 5 = m\angle 6$

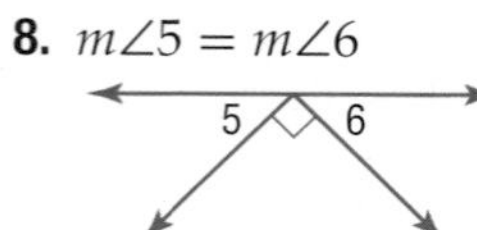


9. $\angle 2$ and $\angle 3$ are complementary. $\angle 1 \cong \angle 4$ and $m\angle 2 = 28$

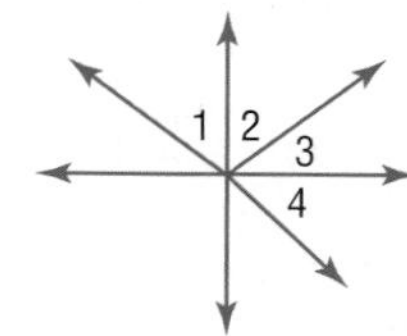


10. $\angle 2$ and $\angle 4$ and $\angle 4$ and $\angle 5$ are supplementary. $m\angle 4 = 105$

**Find the measure of each numbered angle and name the theorems used that justify your work.**

**11.** $m\angle 9 = 3x + 12$
$m\angle 10 = x - 24$

**12.** $m\angle 3 = 2x + 23$
$m\angle 4 = 5x - 112$

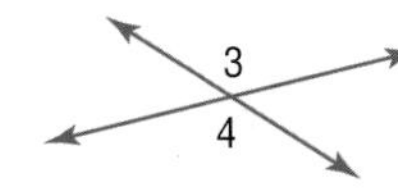


**13.** $m\angle 6 = 2x - 21$
$m\angle 7 = 3x - 34$

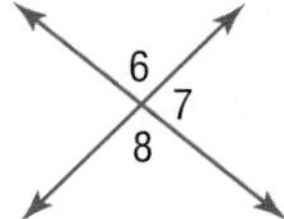


Example 4

**PROOF Write a two-column proof.**

**14. Given:** $\angle ABC$ is a right angle.

**Prove:** $\angle ABD$ and $\angle CBD$ are complementary.

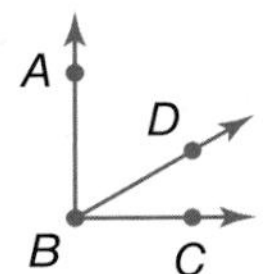


**15. Given:** $\angle 5 \cong \angle 6$

**Prove:** $\angle 4$ and $\angle 6$ are supplementary.

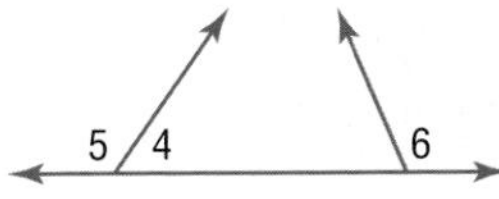


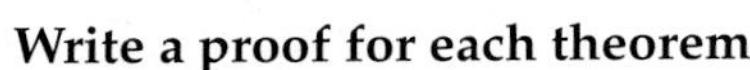

**Write a proof for each theorem.**

**16.** Supplement Theorem

**17.** Complement Theorem

**18.** Reflexive Property of Angle Congruence

**19.** Transitive Property of Angle Congruence

**20.** FLAGS Refer to the Florida state flag at the right. Prove that the sum of the four angle measures is 360.

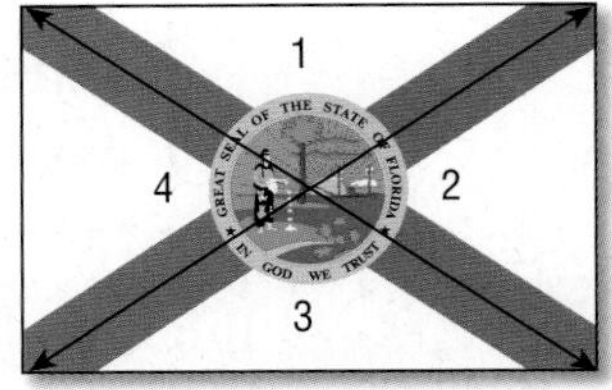


**21.** CCSS ARGUMENTS The diamondback rattlesnake is a pit viper with a diamond pattern on its back. An enlargement of a skin is shown below. If $\angle 1 \cong \angle 4$, prove that $\angle 2 \cong \angle 3$.

**PROOF Use the figure to write a proof of each theorem.**

**22.** Theorem 5.9

**23.** Theorem 5.10

**24.** Theorem 5.11

**25.** Theorem 5.12

**26.** Theorem 5.13

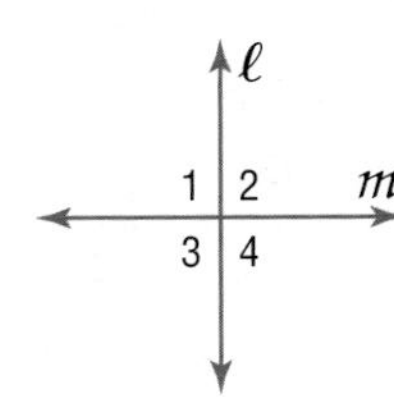

27. CCSS **ARGUMENTS** To mark a specific tempo, the weight on the pendulum of a metronome is adjusted so that it swings at a specific rate. Suppose $\angle ABC$ in the photo is a right angle. If $m\angle 1 = 45$, write a paragraph proof to show that $\overrightarrow{BR}$ bisects $\angle ABC$.

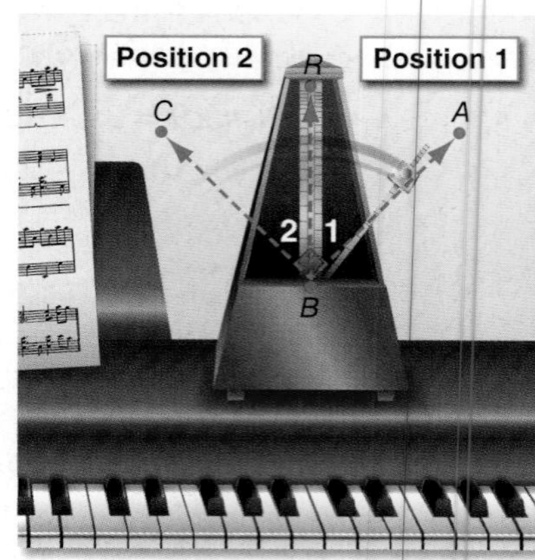


28. **PROOF** Write a proof of Theorem 5.8.

29. **GEOGRAPHY** Utah, Colorado, Arizona, and New Mexico all share a common point on their borders called Four Corners. This is the only place where four states meet in a single point. If $\angle 2$ is a right angle, prove that lines $\ell$ and $m$ are perpendicular.

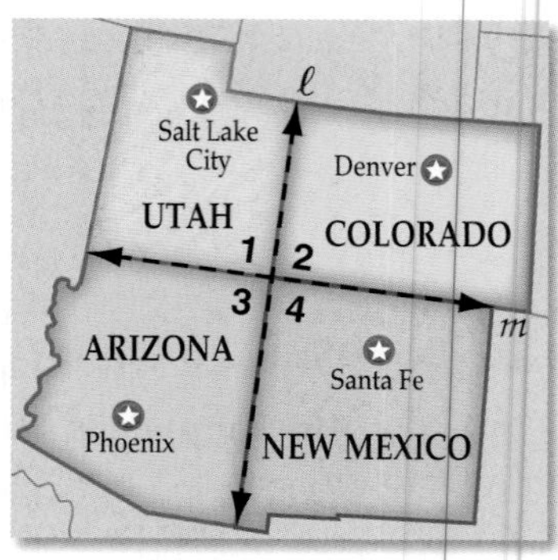


30. **MULTIPLE REPRESENTATIONS** In this problem, you will explore angle relationships.

a. **Geometric** Draw a right angle $ABC$. Place point $D$ in the interior of this angle and draw $\overrightarrow{BD}$. Draw $\overrightarrow{KL}$ and construct $\angle JKL$ congruent to $\angle ABD$.

b. **Verbal** Make a conjecture as to the relationship between $\angle JKL$ and $\angle DBC$.

c. **Logical** Prove your conjecture.

## H.O.T. Problems Use Higher-Order Thinking Skills

31. **OPEN ENDED** Draw an angle $WXZ$ such that $m\angle WXZ = 45$. Construct $\angle YXZ$ congruent to $\angle WXZ$. Make a conjecture as to the measure of $\angle WXY$, and then prove your conjecture.

32. **WRITING IN MATH** Write the steps that you would use to complete the proof below.

**Given:** $\overline{BC} \cong \overline{CD}$, $AB = \frac{1}{2}BD$

**Prove:** $\overline{AB} \cong \overline{CD}$

A B C D

33. **CHALLENGE** In this lesson, one case of the Congruent Supplements Theorem was proven. In Exercise 6, you proved the same case for the Congruent Complements Theorem. Explain why there is another case for each of these theorems. Then write a proof of this second case for each theorem.

34. **REASONING** Determine whether the following statement is *sometimes*, *always*, or *never* true. Explain your reasoning.

*If one of the angles formed by two intersecting lines is acute, then the other three angles formed are also acute.*

35. **WRITING IN MATH** Explain how you can use your protractor to quickly find the measure of the supplement of an angle.

## Standardized Test Practice

**36. GRIDDED RESPONSE** What is the mode of this set of data?

$$4, 3, -2, 1, 4, 0, 1, 4$$

**37.** Find the measure of $\angle CFD$.

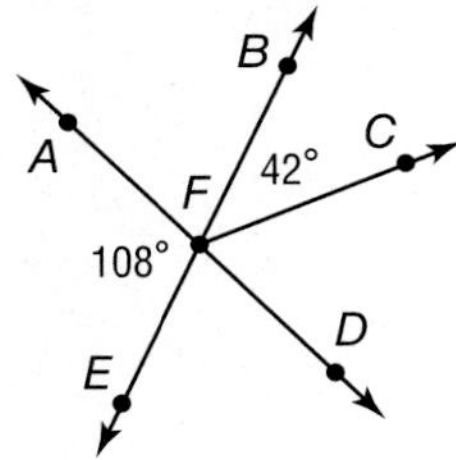


**A** 66° **C** 108°

**B** 72° **D** 138°

**38. ALGEBRA** Simplify.

$$4(3x - 2)(2x + 4) + 3x^2 + 5x - 6$$

**F** $9x^2 + 3x - 14$

**G** $9x^2 + 13x - 14$

**H** $27x^2 + 37x - 38$

**J** $27x^2 + 27x - 26$

**39. SAT/ACT** On a coordinate grid where each unit represents 1 mile, Isabel's house is located at (3, 0) and a mall is located at (0, 4). What is the distance between Isabel's house and the mall?

**A** 3 miles **D** 13 miles

**B** 5 miles **E** 25 miles

**C** 12 miles

## Spiral Review

**40. MAPS** On a U.S. map, there is a scale that lists kilometers on the top and miles on the bottom.

0 km 20 40 50 60 80 100

0 mi 31 62

Suppose $\overline{AB}$ and $\overline{CD}$ are segments on this map. If $AB = 100$ kilometers and $CD = 62$ miles, is $\overline{AB} \cong \overline{CD}$? Explain. (Lesson 5-3)

**State the property that justifies each statement.** (Lesson 5-4)

**41.** If $y + 7 = 5$, then $y = -2$.

**42.** If $MN = PQ$, then $PQ = MN$.

**43.** If $a - b = x$ and $b = 3$, then $a - 3 = x$.

**44.** If $x(y + z) = 4$, then $xy + xz = 4$.

**Graph each function. State the domain and range.** (Lesson 4-1)

**45.** $f(x) = 3(4)^x$

**46.** $f(x) = 2^{3x} - 3$

**Solve each equation.** (Lesson 4-2)

**47.** $2^{x-1} = 8^{x+3}$

**48.** $5^{2x+12} = 25^{10x-12}$

## Skills Review

**Refer to the figure.**

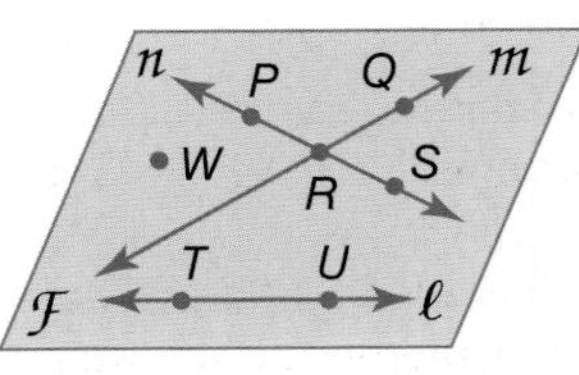


**49.** Name a line that contains point $P$.

**50.** Name the intersection of lines $n$ and $m$.

**51.** Name a point not contained in lines $\ell$, $m$, or $n$.

**52.** What is another name for line $n$?

**53.** Does line $\ell$ intersect line $m$ or line $n$? Explain.

LESSON

# 5-5 Angles and Parallel Lines

**Then**

- You named angle pairs formed by parallel lines and transversals.

**Now**

1. Use theorems to determine the relationships between specific pairs of angles.
2. Use algebra to find angle measurements.

**Why?**

- Construction and maintenance workers often use an access scaffold. This structure provides support and access to elevated areas. The transversal $t$ shown provides structural support to the two parallel working areas.

**Common Core State Standards**

**Content Standards**

**G.CO.1** Know precise definitions of angle, circle, perpendicular line, parallel line, and line segment, based on the undefined notions of point, line, distance along a line, and distance around a circular arc.

**G.CO.9** Prove theorems about lines and angles.

**Mathematical Practices**

1 Make sense of problems and persevere in solving them.

3 Construct viable arguments and critique the reasoning of others.

## 1 Parallel Lines and Angle Pairs

In the photo, line $t$ is a transversal of lines $a$ and $b$, and $\angle 1$ and $\angle 2$ are corresponding angles. Since lines $a$ and $b$ are parallel, there is a special relationship between corresponding angle pairs.

**Postulate 5.12 Corresponding Angles Postulate**

If two parallel lines are cut by a transversal, then each pair of corresponding angles is congruent.

**Examples** $\angle 1 \cong \angle 3$, $\angle 2 \cong \angle 4$, $\angle 5 \cong \angle 7$, $\angle 6 \cong \angle 8$

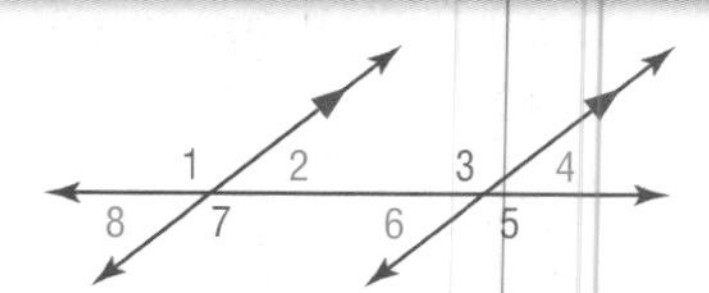

**Example 1 Use Corresponding Angles Postulate**

**In the figure, $m\angle 5 = 72$. Find the measure of each angle. Tell which postulate(s) or theorem(s) you used.**

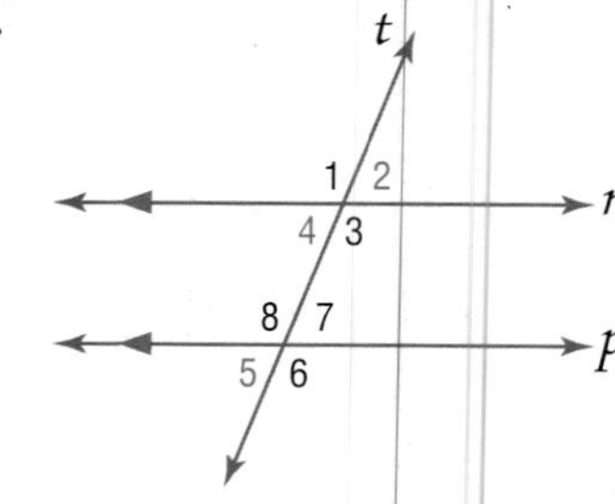


**a.** $\angle 4$

| | |
|---|---|
| $\angle 4 \cong \angle 5$ | Corresponding Angles Postulate |
| $m\angle 4 = m\angle 5$ | Definition of congruent angles |
| $m\angle 4 = 72$ | Substitution |

**b.** $\angle 2$

| | |
|---|---|
| $\angle 2 \cong \angle 4$ | Vertical Angles Theorem |
| $\angle 4 \cong \angle 5$ | Corresponding Angles Postulate |
| $\angle 2 \cong \angle 5$ | Transitive Property of Congruence |
| $m\angle 2 = m\angle 5$ | Definition of congruent angles |
| $m\angle 2 = 72$ | Substitution |

**Guided Practice**

**In the figure, suppose that $m\angle 8 = 105$. Find the measure of each angle. Tell which postulate(s) or theorem(s) you used.**

**1A.** $\angle 1$ **1B.** $\angle 2$ **1C.** $\angle 3$

In Example 1, $\angle 2$ and $\angle 5$ are congruent alternate exterior angles. This and other examples suggest the following theorems about the other angle pairs formed by two parallel lines cut by a transversal.

age fotostock/SuperStock

**StudyTip**

**Angle Relationships** These theorems generalize the relationships between specific pairs of angles. If you get confused about the relationships, you can verify them with the methods you used in Example 1, using only corresponding, vertical, and supplementary angles.

## Theorems Parallel Lines and Angle Pairs

**5.14 Alternate Interior Angles Theorem** If two parallel lines are cut by a transversal, then each pair of alternate interior angles is congruent.

**Examples** $\angle 1 \cong \angle 3$ and $\angle 2 \cong \angle 4$

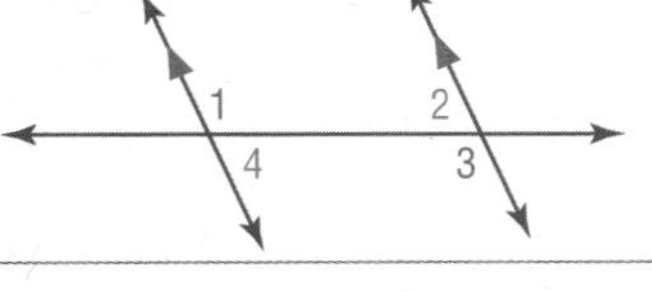


**5.15 Consecutive Interior Angles Theorem** If two parallel lines are cut by a transversal, then each pair of consecutive interior angles is supplementary.

**Examples** $\angle 1$ and $\angle 2$ are supplementary.
$\angle 3$ and $\angle 4$ are supplementary.

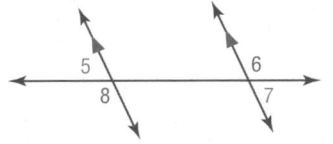

**5.16 Alternate Exterior Angles Theorem** If two parallel lines are cut by a transversal, then each pair of alternate exterior angles is congruent.

**Examples** $\angle 5 \cong \angle 7$ and $\angle 6 \cong \angle 8$

You will prove Theorems 5.15 and 5.16 in Exercises 30 and 35, respectively.

Since postulates are accepted without proof, you can use the Corresponding Angles Postulate to prove each of the theorems above.

## Proof Alternate Interior Angles Theorem

**Given:** $a \parallel b$
$t$ is a transversal of $a$ and $b$.

**Prove:** $\angle 4 \cong \angle 5$, $\angle 3 \cong \angle 6$

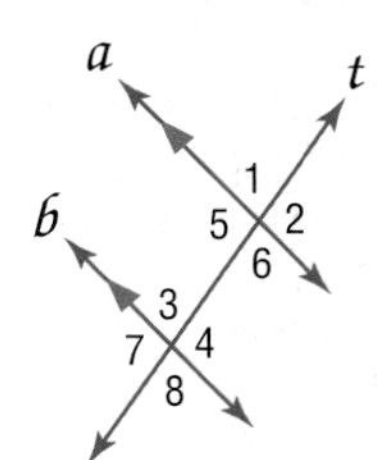


**Paragraph Proof:** We are given that $a \parallel b$ with a transversal $t$. By the Corresponding Angles Postulate, corresponding angles are congruent. So, $\angle 2 \cong \angle 4$ and $\angle 6 \cong \angle 8$. Also, $\angle 5 \cong \angle 2$ and $\angle 8 \cong \angle 3$ because vertical angles are congruent. Therefore, $\angle 5 \cong \angle 4$ and $\angle 3 \cong \angle 6$ since congruence of angles is transitive.

## Real-World Example 2 Use Theorems about Parallel Lines

**COMMUNITY PLANNING Redding Lane and Creek Road are parallel streets that intersect Park Road along the west side of Wendell Park. If $m\angle 1 = 118$, find $m\angle 2$.**

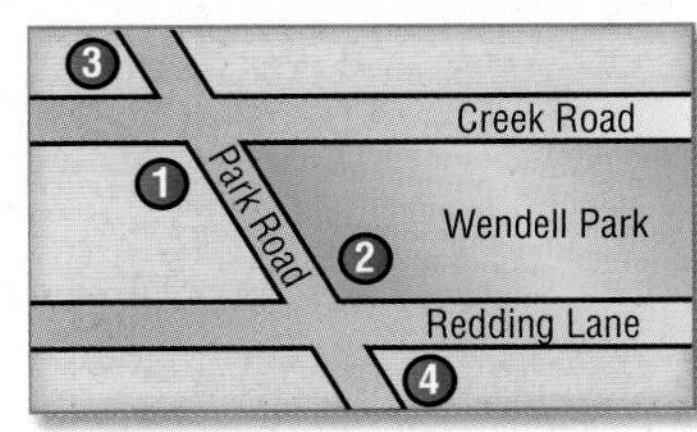


| | |
|---|---|
| $\angle 2 \cong \angle 1$ | Alternate Interior Angles Postulate |
| $m\angle 2 = m\angle 1$ | Definition of congruent angles |
| $m\angle 2 = 118$ | Substitution |

**Real-WorldLink**

Some cities require that streets in newly planned subdivisions intersect at no less than a 60° angle.

Glowimages/Getty Images

### GuidedPractice

**COMMUNITY PLANNING Refer to the diagram above to find each angle measure. Tell which postulate(s) or theorem(s) you used.**

**2A.** If $m\angle 1 = 100$, find $m\angle 4$.

**2B.** If $m\angle 3 = 70$, find $m\angle 4$.

## 2 Algebra and Angle Measures

The special relationships between the angles formed by two parallel lines and a transversal can be used to find unknown values.

### Example 3 Find Values of Variables

**ALGEBRA Use the figure at the right to find the indicated variable. Explain your reasoning.**

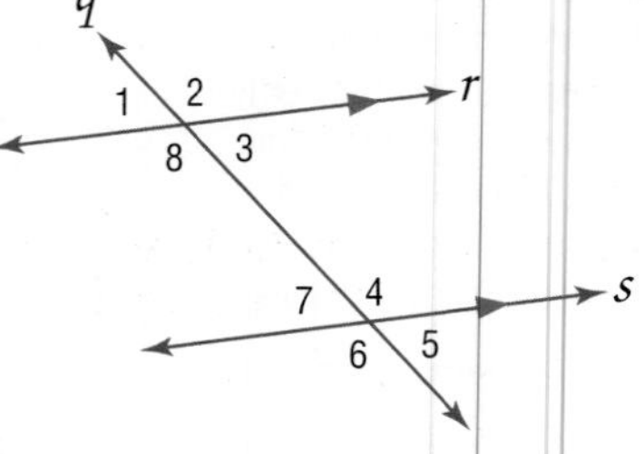


**a. If $m\angle 4 = 2x - 17$ and $m\angle 1 = 85$, find $x$.**

| | |
|---|---|
| $\angle 3 \cong \angle 1$ | Vertical Angles Theorem |
| $m\angle 3 = m\angle 1$ | Definition of congruent angles |
| $m\angle 3 = 85$ | Substitution |

Since lines $r$ and $s$ are parallel, $\angle 4$ and $\angle 3$ are supplementary by the Consecutive Interior Angles Theorem.

| | |
|---|---|
| $m\angle 3 + m\angle 4 = 180$ | Definition of supplementary angles |
| $85 + 2x - 17 = 180$ | Substitution |
| $2x + 68 = 180$ | Simplify. |
| $2x = 112$ | Subtract 68 from each side. |
| $x = 56$ | Divide each side by 2. |

**b. Find $y$ if $m\angle 3 = 4y + 30$ and $m\angle 7 = 7y + 6$.**

| | |
|---|---|
| $\angle 3 \cong \angle 7$ | Alternate Interior Angles Theorem |
| $m\angle 3 = m\angle 7$ | Definition of congruent angles |
| $4y + 30 = 7y + 6$ | Substitution |
| $30 = 3y + 6$ | Subtract 4y from each side. |
| $24 = 3y$ | Subtract 6 from each side. |
| $8 = y$ | Divide each side by 3. |

**StudyTip**

**CCSS Precision** The postulates and theorems you will be studying in this lesson only apply to *parallel* lines cut by a transversal. You should assume that lines are parallel only if the information is given or the lines are marked with parallel arrows.

**GuidedPractice**

**3A.** If $m\angle 2 = 4x + 7$ and $m\angle 7 = 5x - 13$, find $x$.

**3B.** Find $y$ if $m\angle 5 = 68$ and $m\angle 3 = 3y - 2$.

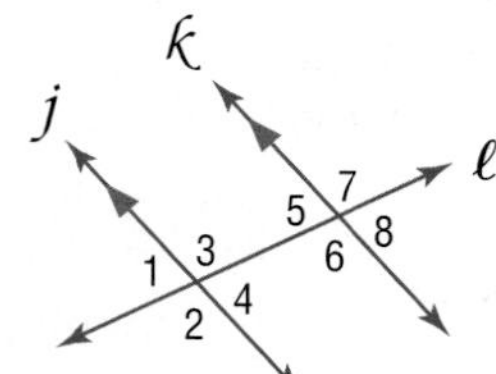


A special relationship exists when the transversal of two parallel lines is a perpendicular line.

### Theorem 5.17 Perpendicular Transversal Theorem

In a plane, if a line is perpendicular to one of two parallel lines, then it is perpendicular to the other.

**Examples** If line $a \parallel$ line $b$ and line $a \perp$ line $t$, then line $b \perp$ line $t$.

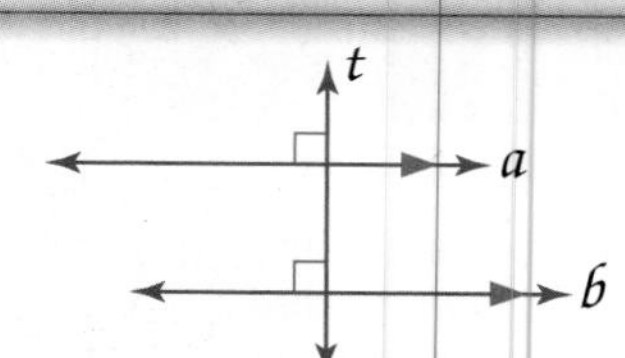


You will prove Theorem 5.17 in Exercise 37.

## Check Your Understanding

**Example 1** **In the figure, $m\angle 1 = 94$. Find the measure of each angle. Tell which postulate(s) or theorem(s) you used.**

**1.** $\angle 3$ **2.** $\angle 5$ **3.** $\angle 4$

**Example 2** **In the figure, $m\angle 4 = 101$. Find the measure of each angle. Tell which postulate(s) or theorem(s) you used.**

**4.** $\angle 6$ **5.** $\angle 7$ **6.** $\angle 5$

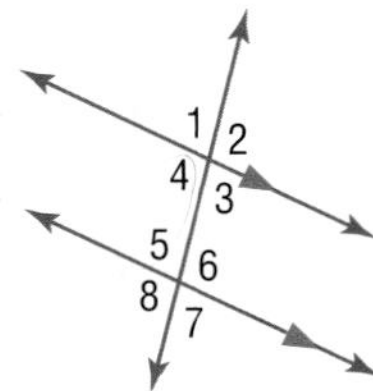


**7. ROADS** In the diagram, the guard rail is parallel to the surface of the roadway and the vertical supports are parallel to each other. Find the measures of angles 2, 3, and 4.

**Example 3** **Find the value of the variable(s) in each figure. Explain your reasoning.**

**8.**

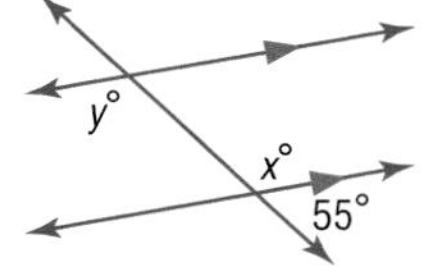


**9.**

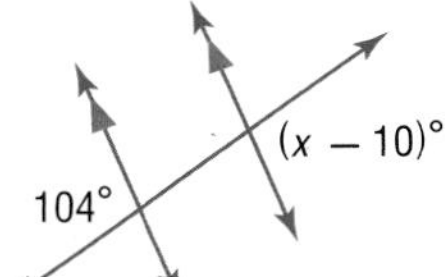


**10.**

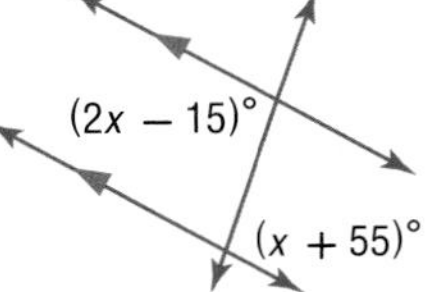


## Practice and Problem Solving

**Examples 1–2** **In the figure, $m\angle 11 = 62$ and $m\angle 14 = 38$. Find the measure of each angle. Tell which postulate(s) or theorem(s) you used.**

**11.** $\angle 4$ **12.** $\angle 3$ **13.** $\angle 12$

**14.** $\angle 8$ **15.** $\angle 6$ **16.** $\angle 2$

**17.** $\angle 10$ **18.** $\angle 5$ **19.** $\angle 1$

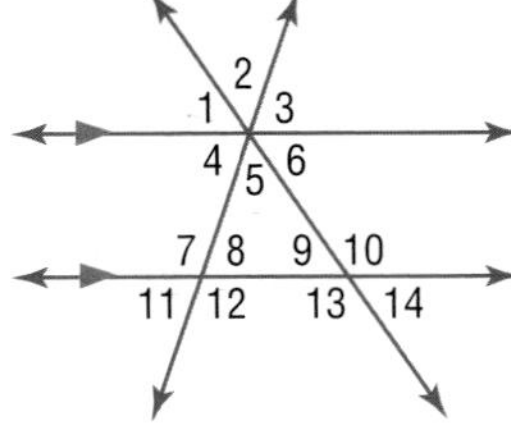


**Example 3** **CCSS MODELING** A solar dish collects energy by directing radiation from the Sun to a receiver located at the focal point of the dish. Assume that the radiation rays are parallel. Determine the relationship between each pair of angles, and explain your reasoning.

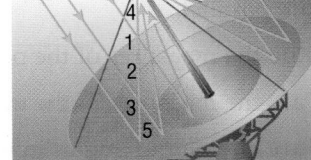

**20.** $\angle 1$ and $\angle 2$  **21** 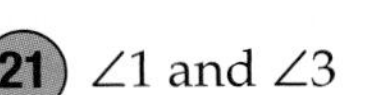 $\angle 1$ and $\angle 3$ **22.** $\angle 4$ and $\angle 5$ **23.** $\angle 3$ and $\angle 4$

## Standardized Test Practice

**47.** Suppose $\angle 4$ and $\angle 5$ form a linear pair. If $m\angle 1 = 2x$, $m\angle 2 = 3x - 20$, and $m\angle 3 = x - 4$, what is $m\angle 3$?

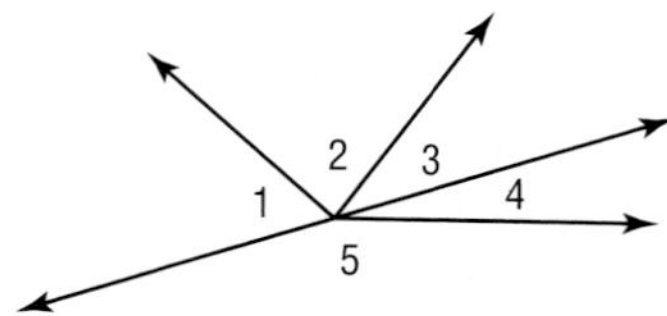


**A** 26°
**B** 28°
**C** 30°
**D** 32°

**48. SAT/ACT** A farmer raises chickens and pigs. If his animals have a total of 120 heads and a total of 300 feet, how many chickens does the farmer have?

**F** 60
**G** 70
**H** 80
**J** 90

**49. SHORT RESPONSE** If $m \parallel n$, then which of the following statements must be true?

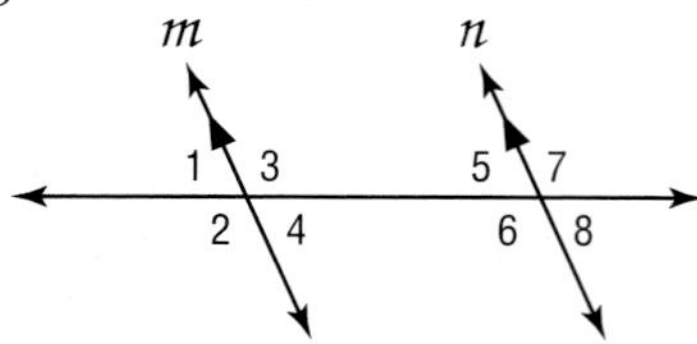


**I.** $\angle 3$ and $\angle 6$ are Alternate Interior Angles.
**II.** $\angle 4$ and $\angle 6$ are Consecutive Interior Angles.
**III.** $\angle 1$ and $\angle 7$ are Alternate Exterior Angles.

**50. ALGEBRA** If $-2 + x = -6$, then $-17 - x = ?$

**A** $-13$
**B** $-4$
**C** 9
**D** 13
**E** 21

## Spiral Review

**Graph each function. State the domain and range.** (Lesson 4-1)

**51.** $y = 2(3)^x$ **52.** $y = 5(2)^x$ **53.** $y = 0.5(4)^x$ **54.** $y = 4\left(\frac{1}{3}\right)^x$

**Use the given statement to find the measure of each numbered angle.** (Lesson 5-4)

**55.** $\angle 1$ and $\angle 2$ form a linear pair and $m\angle 2 = 67$.

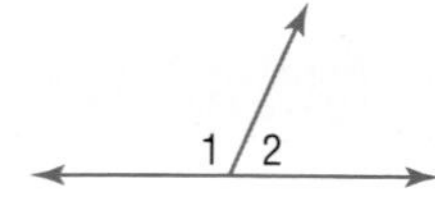


**56.** $\angle 6$ and $\angle 8$ are; complementary $m\angle 8 = 47$.

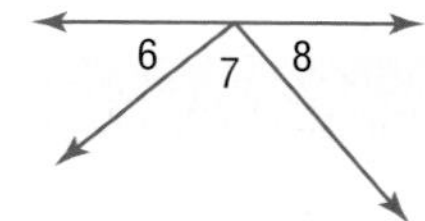


**57.** $m\angle 4 = 32$

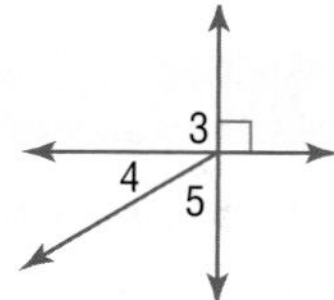


**58. TRAINS** A train company wants to provide routes to New York City, Dallas, Chicago, Los Angeles, San Francisco, and Washington, D.C. An engineer draws lines between each pair of cities on a map. No three of the cities are collinear. How many lines did the engineer draw? (Lesson 5-1)

## Skills Review

**Simplify each expression.**

**59.** $\frac{6 - 5}{4 - 2}$ **60.** $\frac{-5 - 2}{4 - 7}$ **61.** $\frac{-11 - 4}{12 - (-9)}$

**62.** $\frac{16 - 12}{15 - 11}$ **63.** $\frac{10 - 22}{8 - 17}$ **64.** $\frac{8 - 17}{12 - (-3)}$

# LESSON 5-6 Proving Lines Parallel

| Then | Now | Why? |
|---|---|---|
| You used slopes to identify parallel and perpendicular lines. | **1** Recognize angle pairs that occur with parallel lines.<br>**2** Prove that two lines are parallel. | When you see a roller coaster track, the two sides of the track are always the same distance apart, even though the track curves and turns. The tracks are carefully constructed to be parallel at all points so that the car is secure on the track. |

**Common Core State Standards**

**Content Standards**

G.CO.9 Prove theorems about lines and angles.

G.CO.12 Make formal geometric constructions with a variety of tools and methods (compass and straightedge, string, reflective devices, paper folding, dynamic geometric software, etc.).

**Mathematical Practices**

1 Make sense of problems and persevere in solving them.

3 Construct viable arguments and critique the reasoning of others.

## 1 Identify Parallel Lines

The two sides of the track of a roller coaster are parallel, and all of the supports along the track are also parallel. Each of the angles formed between the track and the supports are corresponding angles. We have learned that corresponding angles are congruent when lines are parallel. The converse of this relationship is also true.

### Postulate 5.13 Converse of Corresponding Angles Postulate

If two lines are cut by a transversal so that corresponding angles are congruent, then the lines are parallel.

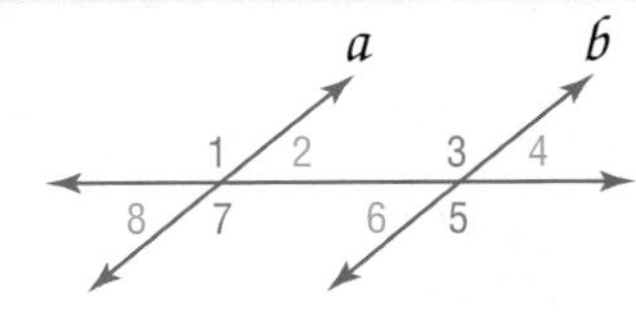


**Examples** If $\angle 1 \cong \angle 3$, $\angle 2 \cong \angle 4$, $\angle 5 \cong \angle 7$, $\angle 6 \cong \angle 8$, then $a \parallel b$.

The Converse of the Corresponding Angles Postulate can be used to construct parallel lines.

### Construction Parallel Line Through a Point Not on the Line

**Step 1** Use a straightedge to draw $\overleftrightarrow{AB}$. Draw a point $C$ that is not on $\overleftrightarrow{AB}$. Draw $\overleftrightarrow{CA}$.

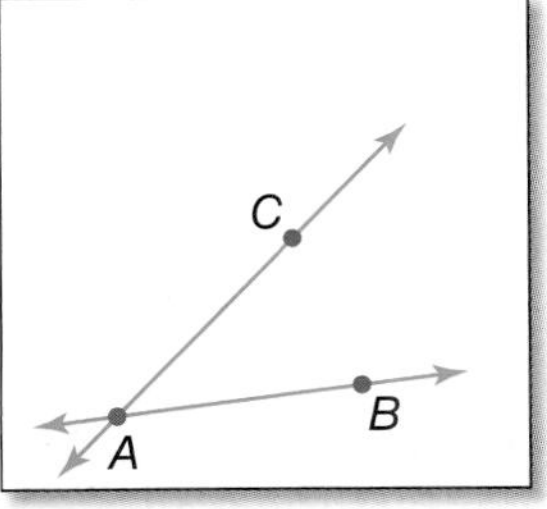


**Step 2** Copy $\angle CAB$ so that $C$ is the vertex of the new angle. Label the intersection points $D$ and $E$.

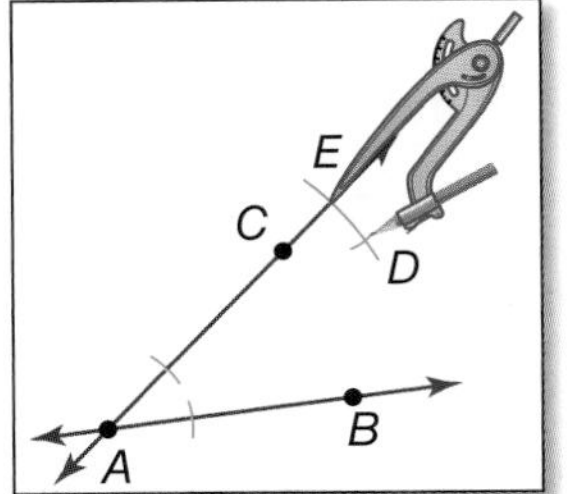


**Step 3** Draw $CD$. Because $\angle ECD \cong \angle CAB$ by construction and they are corresponding angles, $\overleftrightarrow{AB} \parallel \overleftrightarrow{CD}$.

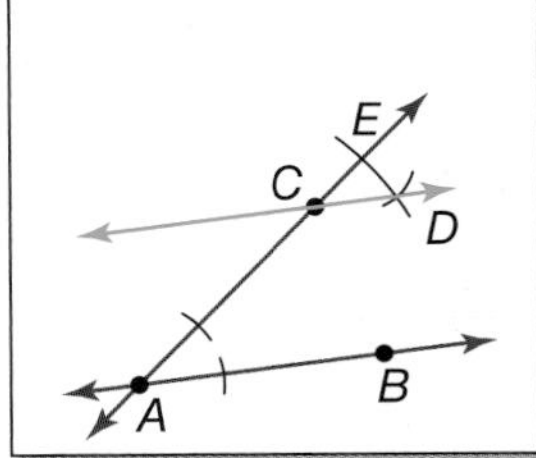

The construction establishes that there is *at least* one line through $C$ that is parallel to $\overleftrightarrow{AB}$. The following postulate guarantees that this line is the *only* one.

**Postulate 5.14 Parallel Postulate**

If given a line and a point not on the line, then there exists exactly one line through the point that is parallel to the given line.

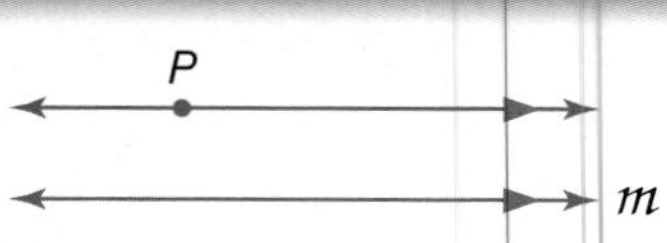


Parallel lines that are cut by a transversal create several pairs of congruent angles. These special angle pairs can also be used to prove that a pair of lines are parallel.

**Theorems Proving Lines Parallel**

| Theorem | Example |
| --- | --- |
| **5.18 Alternate Exterior Angles Converse** If two lines in a plane are cut by a transversal so that a pair of alternate exterior angles is congruent, then the two lines are parallel. | 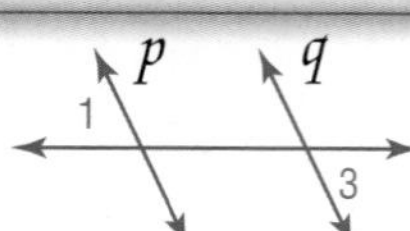  If $\angle 1 \cong \angle 3$, then $p \parallel q$. |
| **5.19 Consecutive Interior Angles Converse** If two lines in a plane are cut by a transversal so that a pair of consecutive interior angles is supplementary, then the lines are parallel. | 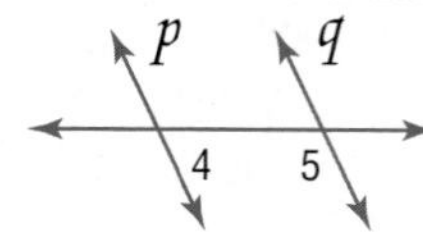  If $m\angle 4 + m\angle 5 = 180$, then $p \parallel q$. |
| **5.20 Alternate Interior Angles Converse** If two lines in a plane are cut by a transversal so that a pair of alternate interior angles is congruent, then the lines are parallel. | 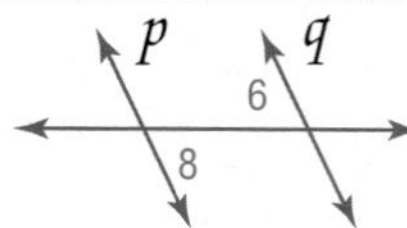  If $\angle 6 \cong \angle 8$, then $p \parallel q$. |
| **5.21 Perpendicular Transversal Converse** In a plane, if two lines are perpendicular to the same line, then they are parallel. | 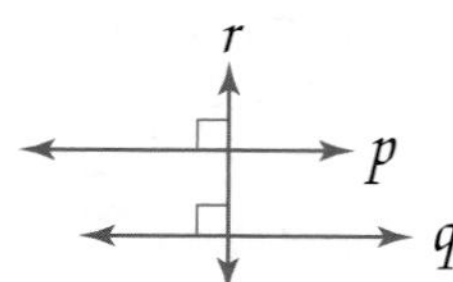  If $p \perp r$ and $q \perp r$, then $p \parallel q$. |

You will prove Theorems 5.18, 5.19, 5.20, and 5.21 in Exercises 6, 23, 31, and 30, respectively.

**Example 1 Identify Parallel Lines**

**Given the following information, determine which lines, if any, are parallel. State the postulate or theorem that justifies your answer.**

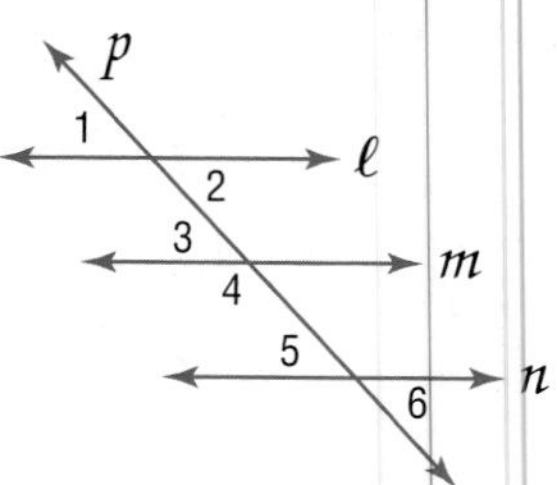


**a.** $\angle 1 \cong \angle 6$

$\angle 1$ and $\angle 6$ are alternate exterior angles of lines $\ell$ and $n$.

Since $\angle 1 \cong \angle 6$, $\ell \parallel n$ by the Converse of the Alternate Exterior Angles Theorem.

**b.** $\angle 2 \cong \angle 3$

$\angle 2$ and $\angle 3$ are alternate interior angles of lines $\ell$ and $m$.

Since $\angle 2 \cong \angle 3$, $\ell \parallel m$ by the Converse of the Alternate Interior Angles Theorem.

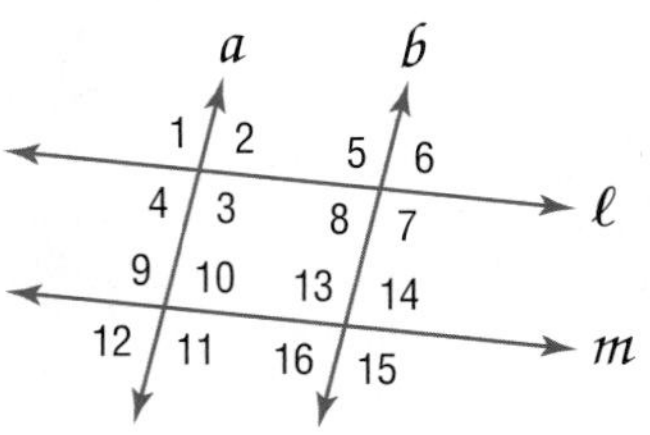


## Guided Practice

**1A.** $\angle 2 \cong \angle 8$

**1B.** $\angle 3 \cong \angle 11$

**1C.** $\angle 12 \cong \angle 14$

**1D.** $\angle 1 \cong \angle 15$

**1E.** $m\angle 8 + m\angle 13 = 180$

**1F.** $\angle 8 \cong \angle 6$

Angle relationships can be used to solve problems involving unknown values.

## Standardized Test Example 2 Use Angle Relationships

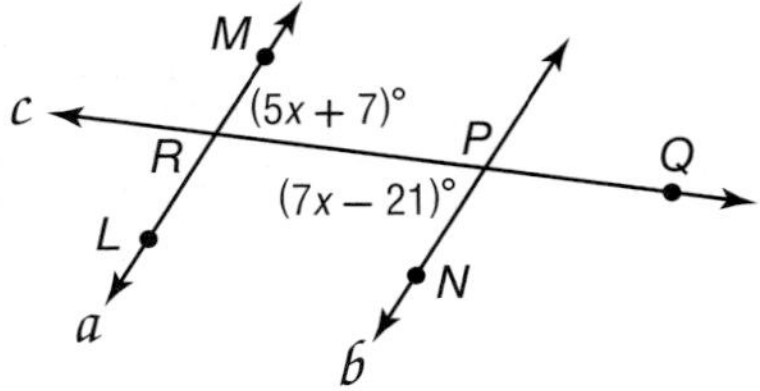


**OPEN ENDED Find $m\angle MRQ$ so that $a \parallel b$. Show your work.**

### Read the Test Item

From the figure, you know that $m\angle MRQ = 5x + 7$ and $m\angle RPN = 7x - 21$. You are asked to find the measure of $\angle MRQ$.

### Solve the Test Item

$\angle MRQ$ and $\angle RPN$ are alternate interior angles. For lines $a$ and $b$ to be parallel, alternate interior angles must be congruent, so $\angle MRQ \cong \angle RPN$. By the definition of congruence, $m\angle MRQ = m\angle RPN$. Substitute the given angle measures into this equation and solve for $x$.

$m\angle MRQ = m\angle RPN$ — Alternate interior angles

$5x + 7 = 7x - 21$ — Substitution

$7 = 2x - 21$ — Subtract $5x$ from each side.

$28 = 2x$ — Add 21 to each side.

$14 = x$ — Divide each side by 2.

Now, use the value of $x$ to find $\angle MRQ$.

$m\angle MRQ = 5x + 7$ — Substitution

$= 5(14) + 7$ — $x = 14$

$= 77$ — Simplify.

**CHECK** Check your answer by using the value of $x$ to find $m\angle RPN$.

$m\angle RP = 7x - 21$

$= 7(14) - 21$ or $77$ ✓

Since $m\angle MRQ = m\angle RPN$, $\angle MRQ \cong \angle RPN$ and $a \parallel b$. ✓

**Study Tip**

**Finding What Is Asked For** Be sure to reread test questions carefully to be sure you are answering the question that was asked. In Example 2, a common error would be to stop after you have found the value of $x$ and say that the solution of the problem is 14.

## Guided Practice

**2.** Find $y$ so that $e \parallel f$. Show your work.

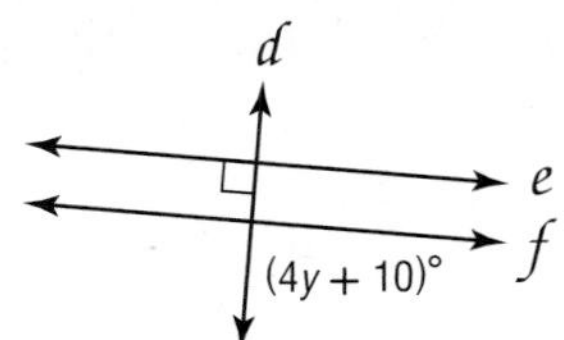

**StudyTip**

**Proving Lines Parallel** When two parallel lines are cut by a transversal, the angle pairs formed are either congruent or supplementary. When a pair of lines forms angles that do not meet this criterion, the lines cannot possibly be parallel.

## 2 Prove Lines Parallel

The angle pair relationships formed by a transversal can be used to prove that two lines are parallel.

### Real-World Example 3 Prove Lines Parallel

**HOME FURNISHINGS In the ladder shown, each rung is perpendicular to the two rails. Is it possible to prove that the two rails are parallel and that all of the rungs are parallel? If so, explain how. If not, explain why not.**

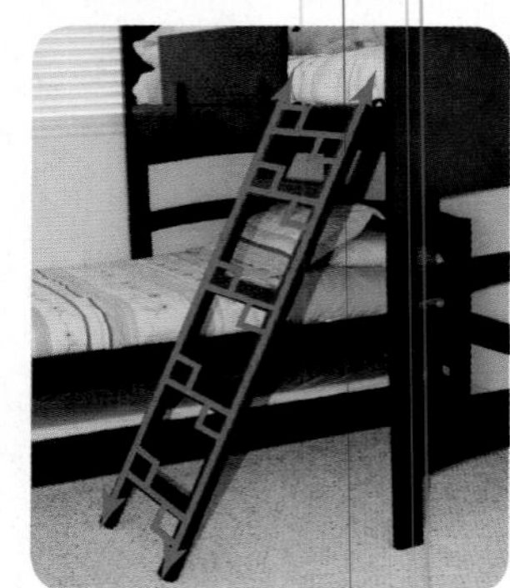

Since both rails are perpendicular to each rung, the rails are parallel by the Perpendicular Transversal Converse. Since any pair of rungs is perpendicular to the rails, they are also parallel.

**GuidedPractice**

3. **ROWING** In order to move in a straight line with maximum efficiency, rower's oars should be parallel. Refer to the photo at the right. Is it possible to prove that any of the oars are parallel? If so, explain how. If not, explain why not.

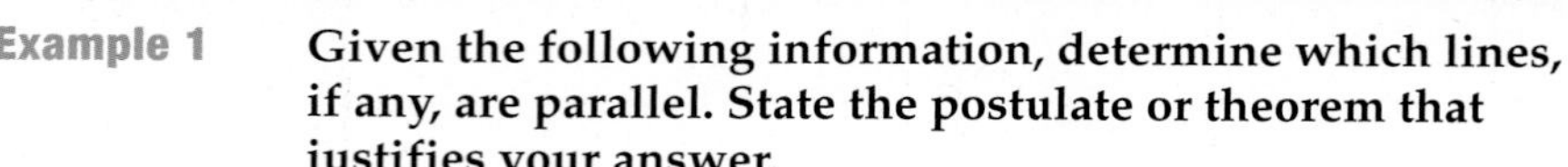


**Example 1** Given the following information, determine which lines, if any, are parallel. State the postulate or theorem that justifies your answer.

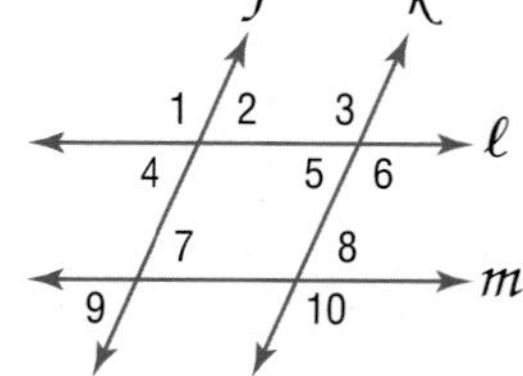


1. $\angle 1 \cong \angle 3$
2. $\angle 2 \cong \angle 5$
3. $\angle 3 \cong \angle 10$
4. $m\angle 6 + m\angle 8 = 180$

**Example 2**

5. **SHORT RESPONSE** Find $x$ so that $m \parallel n$. Show your work.

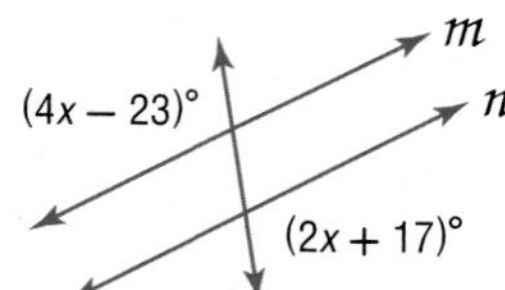


**Example 3**

6. **PROOF** Copy and complete the proof of Theorem 5.18.

**Given:** $\angle 1 \cong \angle 2$

**Prove:** $\ell \parallel m$

**Proof:**

| Statements | Reasons |
|---|---|
| a. $\angle 1 \cong \angle 2$ | a. Given |
| b. $\angle 2 \cong \angle 3$ | b. ___?___ |
| c. $\angle 1 \cong \angle 3$ | c. Transitive Property |
| d. ___?___ | d. ___?___ |

**7. RECREATION** Is it possible to prove that the backrest and footrest of the lounging beach chair are parallel? If so, explain how. If not, explain why not.

## Practice and Problem Solving

**Example 1** **Given the following information, determine which lines, if any, are parallel. State the postulate or theorem that justifies your answer.**

**8.** $\angle 1 \cong \angle 2$

**9.** $\angle 2 \cong \angle 9$

**10.** $\angle 5 \cong \angle 7$

**11.** $m\angle 7 + m\angle 8 = 180$

**12.** $m\angle 3 + m\angle 6 = 180$

**13.** $\angle 3 \cong \angle 5$

**14.** $\angle 3 \cong \angle 7$

**15.** $\angle 4 \cong \angle 5$

**Example 2** **Find $x$ so that $m \parallel n$. Identify the postulate or theorem you used.**

**16.**

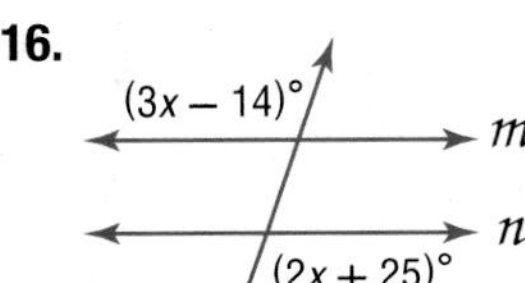


**17.**

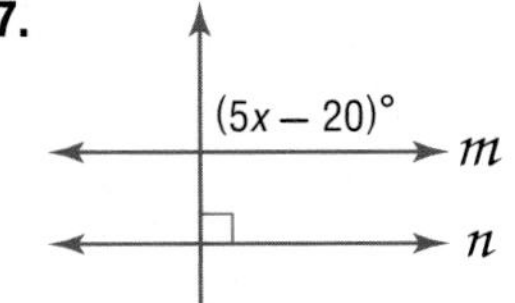


**18.**

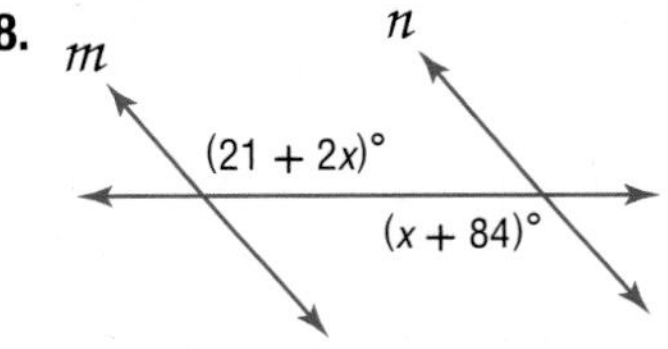


**19.**

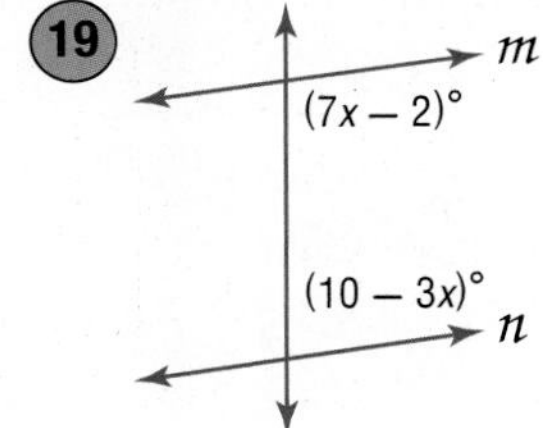


**20.**

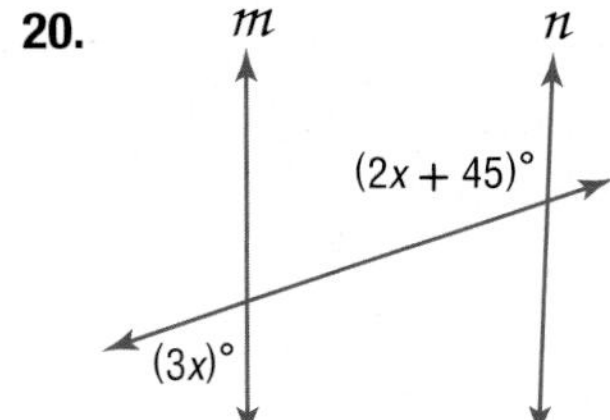


**21.**

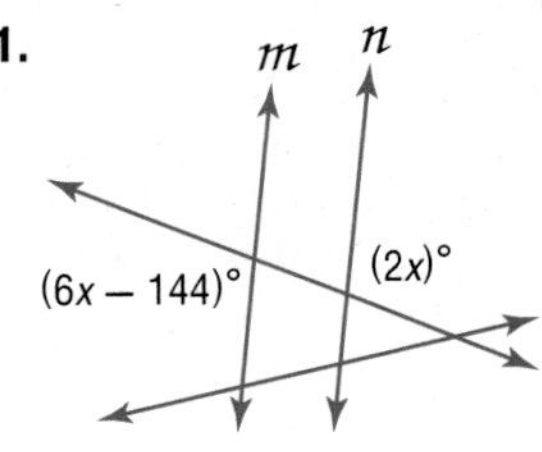


**22. CCSS SENSE-MAKING** Wooden picture frames are often constructed using a miter box or miter saw. These tools allow you to cut at an angle of a given size. If each of the four pieces of framing material is cut at a 45° angle, will the sides of the frame be parallel? Explain your reasoning.

**Example 3** **23. PROOF** Copy and complete the proof of Theorem 5.19.

**Given:** $\angle 1$ and $\angle 2$ are supplementary.

**Prove:** $\ell \parallel m$

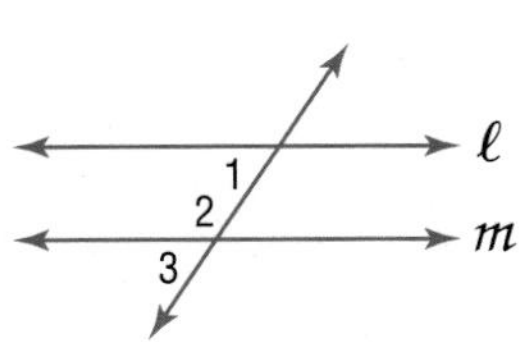


**Proof:**

| Statements | Reasons |
|---|---|
| **a.** ? | **a.** Given |
| **b.** $\angle 2$ and $\angle 3$ form a linear pair. | **b.** ? |
| **c.** ? | **c.** ? |
| **d.** $\angle 1 \cong \angle 3$ | **d.** ? |
| **e.** $\ell \parallel m$ | **e.** ? |

**24. CRAFTS** Jacqui is making a stained glass piece. She cuts the top and bottom pieces at a 30° angle. If the corners are right angles, explain how Jacqui knows that each pair of opposite sides are parallel.

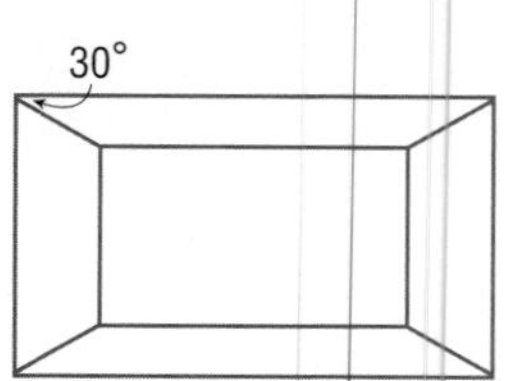


**PROOF** **Write a two-column proof for each of the following.**

**25. Given:** $\angle 1 \cong \angle 3$

$\overline{AC} \parallel \overline{BD}$

**Prove:** $\overline{AB} \parallel \overline{CD}$

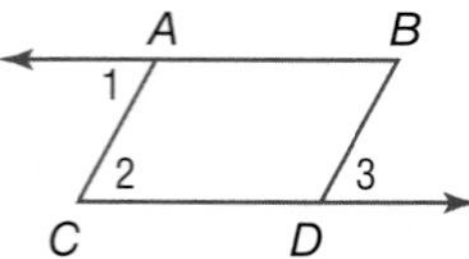


**26. Given:** $\overline{WX} \parallel \overline{YZ}$

$\angle 2 \cong \angle 3$

**Prove:** $\overline{WY} \parallel \overline{XZ}$

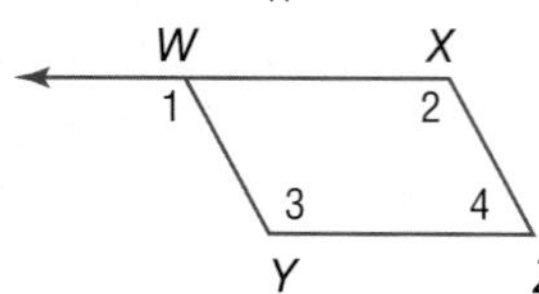


**27. Given:** $\angle ABC \cong \angle ADC$

$m\angle A + m\angle ABC = 180$

**Prove:** $\overline{AB} \parallel \overline{CD}$

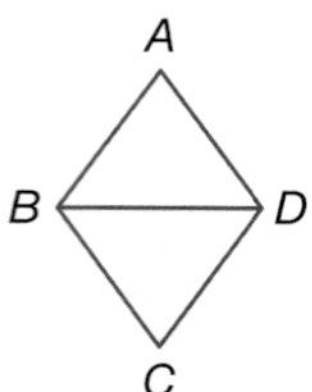


**28. Given:** $\angle 1 \cong \angle 2$

$\overline{LJ} \perp \overline{ML}$

**Prove:** $\overline{KM} \perp \overline{ML}$

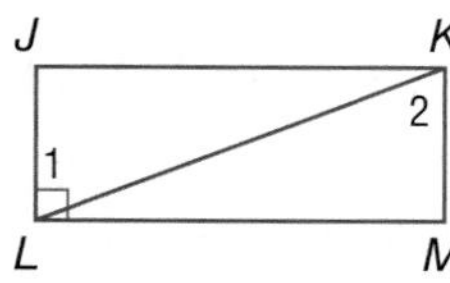


**29** **MAILBOXES** Mail slots are used to make the organization and distribution of mail easier. In the mail slots shown, each slot is perpendicular to each of the sides. Explain why you can conclude that the slots are parallel.

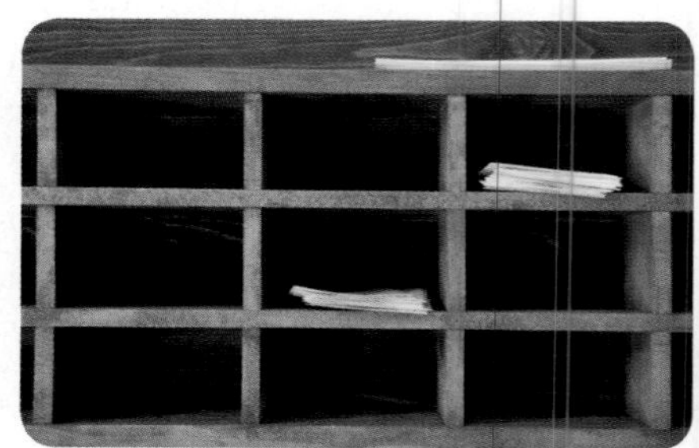

**30. PROOF** Write a paragraph proof of Theorem 5.21.

**31. PROOF** Write a two-column proof of Theorem 5.20.

**32. CCSS REASONING** Based upon the information given in the photo of the staircase at the right, what is the relationship between each step? Explain your answer.

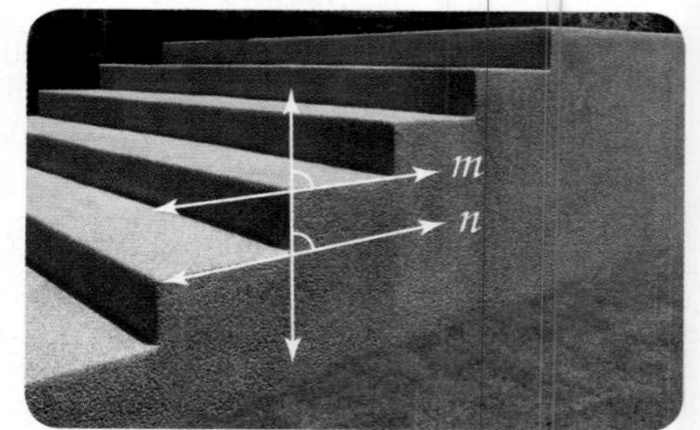


(t)Design Pics Inc./Alamy, (b)Sheltered Images/SuperStock

**Determine whether lines** $r$ **and** $s$ **are parallel. Justify your answer.**

**33.**

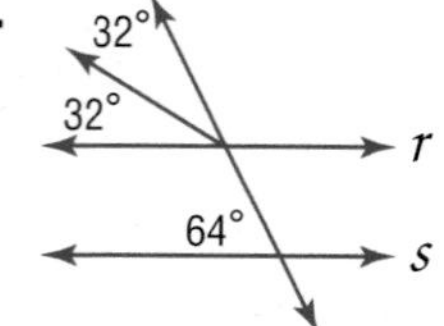


**34.**

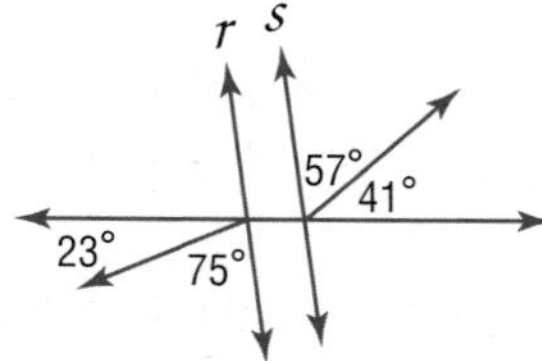


**35.**

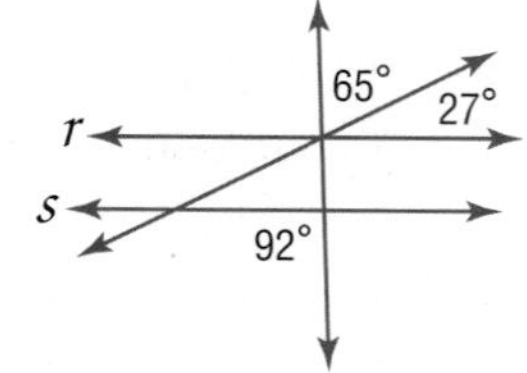

36. MULTIPLE REPRESENTATIONS In this problem, you will explore the shortest distance between two parallel lines.

a. **Geometric** Draw three sets of parallel lines $k$ and $\ell$, $s$ and $t$, and $x$ and $y$. For each set, draw the shortest segment $\overline{BC}$ and label points $A$ and $D$ as shown below.

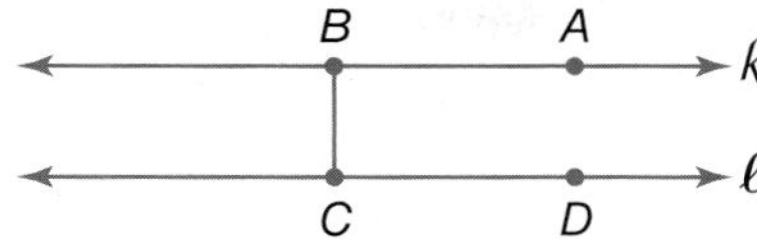


b. **Tabular** Copy the table below, measure $\angle ABC$ and $\angle BCD$, and complete the table.

| Set of Parallel Lines | $m\angle ABC$ | $m\angle BCD$ |
|---|---|---|
| $k$ and $\ell$ | | |
| $s$ and $t$ | | |
| $x$ and $y$ | | |

c. **Verbal** Make a conjecture about the angle the shortest segment forms with both parallel lines.

## H.O.T. Problems Use Higher-Order Thinking Skills

37. ERROR ANALYSIS Sumi and Daniela are determining which lines are parallel in the figure at the right. Sumi says that since $\angle 1 \cong \angle 2$, $\overline{WY} \parallel \overline{XZ}$. Daniela disagrees and says that since $\angle 1 \cong \angle 2$, $\overline{WX} \parallel \overline{YZ}$. Is either of them correct? Explain.

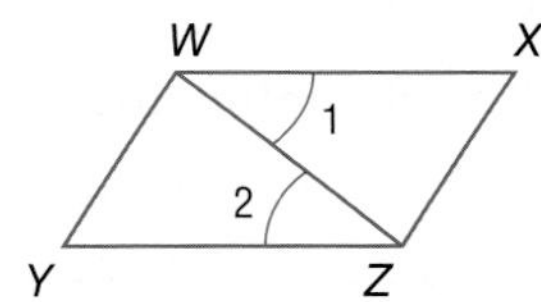


38. CCSS REASONING Is Theorem 5.21 still true if the two lines are not coplanar? Draw a figure to justify your answer.

39. CHALLENGE Use the figure at the right to prove that two lines parallel to a third line are parallel to each other.

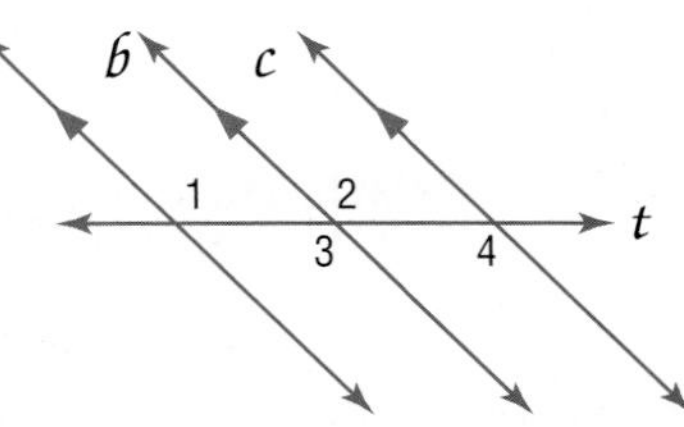


40. OPEN ENDED Draw a triangle $ABC$.

a. Construct the line parallel to $\overline{BC}$ through point $A$.

b. Use measurement to justify that the line you constructed is parallel to $\overline{BC}$.

c. Use mathematics to justify this construction.

41. CHALLENGE Refer to the figure at the right.

a. If $m\angle 1 + m\angle 2 = 180$, prove that $a \parallel c$.

b. Given that $a \parallel c$, if $m\angle 1 + m\angle 3 = 180$, prove that $t \perp c$.

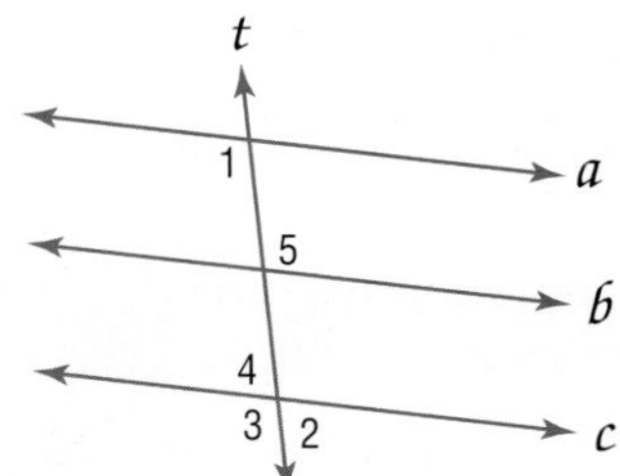


42. WRITING IN MATH Summarize the five methods used in this lesson to prove that two lines are parallel.

43. WRITING IN MATH Can a pair of angles be supplementary and congruent? Explain your reasoning.

## Standardized Test Practice

**44.** Which of the following facts would be sufficient to prove that line $d$ is parallel to $\overline{XZ}$?

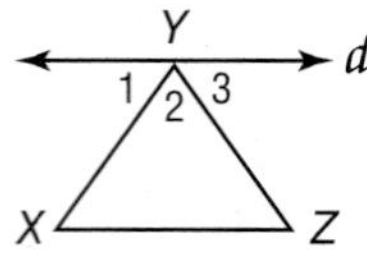


**A** $\angle 1 \cong \angle 3$  **C** $\angle 1 \cong \angle Z$
**B** $\angle 3 \cong \angle Z$  **D** $\angle 2 \cong \angle X$

**45. ALGEBRA** The expression $\sqrt{52} + \sqrt{117}$ is equivalent to

**F** 13  **H** $6\sqrt{13}$
**G** $5\sqrt{13}$  **J** $13\sqrt{13}$

**46.** What is the approximate surface area of the figure?

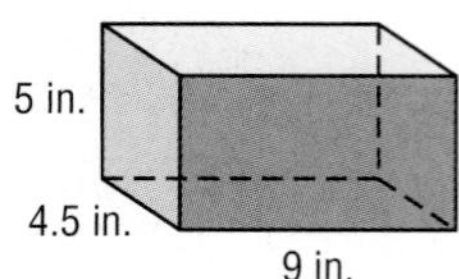


**A** 101.3 in$^2$  **C** 202.5 in$^2$
**B** 108 in$^2$  **D** 216 in$^2$

**47. SAT/ACT** If $x^2 = 25$ and $y^2 = 9$, what is the greatest possible value of $(x - y)^2$?

**F** 4  **J** 64
**G** 16  **K** 70
**H** 58

## Spiral Review

**48.** Refer to the figure at the right. Determine whether $a \parallel b$. Justify your answer. (Lesson 5-6)

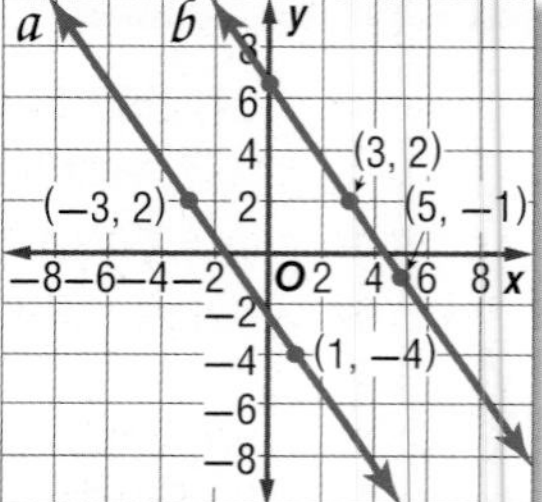


**Prove the following.** (Lesson 5-3)

**49.** If $AB = BC$, then $AC = 2BC$.

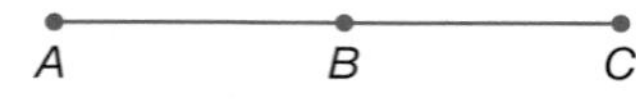


**50. Given:** $\overline{JK} \cong \overline{KL}$, $\overline{HJ} \cong \overline{GH}$, $\overline{KL} \cong \overline{HJ}$
**Prove:** $\overline{GH} \cong \overline{JK}$

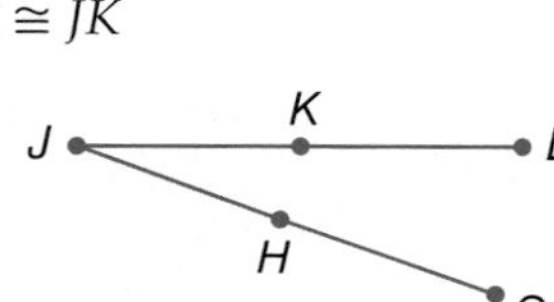


**51. VIDEO GAMES** The daily sales of a video game company can be modeled by the function $S(p) = -2p^2 + 80p + 1250$, where $S(p)$ is the amount of sales and $p$ is the price per game. (Lesson 3-5)

**a.** Write the function in vertex form.

**b.** What is the maximum amount of daily sales the company can expect to make?

**Solve each inequality algebraically.** (Lesson 3-6)

**52.** $x^2 + 3x < 18$

**53.** $x^2 - 3x \leq 28$

**54.** $2x^2 - 13x \geq -20$

## Skills Review

**55.** Find $x$ and $y$ so that $\overline{BE}$ and $\overline{AD}$ are perpendicular.

B
C
A
$6x°$
$3x°$
$(12y - 10)°$
F
D
E

# Study Guide and Review

## Study Guide

### KeyConcepts

**Proof** (Lessons 5-1 through 5-4)

**Step 1** List the given information and draw a diagram, if possible.

**Step 2** State what is to be proved.

**Step 3** Create a deductive argument.

**Step 4** Justify each statement with a reason.

**Step 5** State what you have proved.

**Transversals** (Lesson 5-5)

- When a transversal intersects two lines, the following types of angles are formed: exterior, interior, consecutive interior, alternate interior, alternate exterior, and corresponding.
- If two parallel lines are cut by a transversal, then:
  - each pair of corresponding angles is congruent,
  - each pair of alternate interior angles is congruent,
  - each pair of consecutive interior angles is supplementary, and
  - each pair of alternate exterior angles is congruent.

**Proving Lines Parallel** (Lesson 5-6)

- If two lines in a plane are cut by a transversal so that any one of the following is true, then the two lines are parallel:
  - a pair of corresponding angles is congruent,
  - a pair of alternate exterior angles is congruent,
  - a pair of alternate interior angles is congruent, or
  - a pair of consecutive interior angles is supplementary.
- In a plane, if two lines are perpendicular to the same line, then they are parallel.

### FOLDABLES® StudyOrganizer

Be sure the Key Concepts are noted in your Foldable.

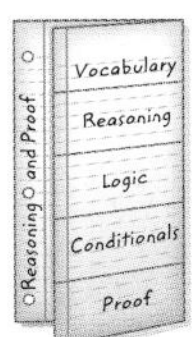


### KeyVocabulary

algebraic proof (p. 284)
axiom (p. 278)
deductive argument (p. 278)
formal proof (p. 285)
informal proof (p. 278)
paragraph proof (p. 278)
postulate (p. 276)
proof (p. 276)
theorem (p. 278)
two-column proof (p. 285)

### VocabularyCheck

**State whether each sentence is *true* or *false*. If *false*, replace the underlined term to make a true sentence.**

1. A <u>postulate</u> is a statement that requires proof.
2. A <u>theorem</u> is a statement that is accepted as true without proof.
3. In a two-column proof, the properties that justify each step are called <u>reasons</u>.
4. An <u>informal proof</u> involves writing a paragraph to explain why a conjecture is true.

# CHAPTER 5 Study Guide and Review

## 5-1 Postulates and Paragraph Proofs

**Determine whether each statement is *always, sometimes,* or *never* true. Explain.**

5. Two planes intersect at a point.
6. Three points are contained in more than one plane.
7. If line $m$ lies in plane $\mathcal{X}$ and line $m$ contains a point $Q$, then point $Q$ lies in plane $\mathcal{X}$.
8. If two angles are complementary, then they form a right angle.
9. **NETWORKING** Six people are introduced at a business convention. If each person shakes hands with each of the others, how many handshakes will be exchanged? Include a model to support your reasoning.

### Example 1

**Determine whether each statement is *always, sometimes,* or *never* true. Explain.**

**a.** ***If points X, Y, and Z lie in plane $\mathcal{R}$, then they are not collinear.***

Sometimes; the fact that $X$, $Y$, and $Z$ are contained in plane $\mathcal{R}$ has no bearing on whether those points are collinear or not.

**b. For any two points $A$ and $B$, there is exactly one line that contains them.**

Always; according to Postulate 5-1, there is exactly one line through any two points.

## 5-2 Algebraic Proof

**State the property that justifies each statement.**

10. If $7(x - 3) = 35$, then $35 = 7(x - 3)$.
11. If $2x + 19 = 27$, then $2x = 8$.
12. $5(3x + 1) = 15x + 5$
13. $7x - 2 = 7x - 2$
14. If $12 = 2x + 8$ and $2x + 8 = 3y$, then $12 = 3y$.
15. Copy and complete the following proof.
    Given: $6(x - 4) = 42$
    Prove: $x = 11$

| Statements | Reasons |
|---|---|
| a. $6(x - 4) = 42$ | a. ? |
| b. $6x - 24 = 42$ | b. ? |
| c. $6x = 66$ | c. ? |
| d. $x = 11$ | d. ? |

16. Write a two-column proof to show that if $PQ = RS$, $PQ = 5x + 9$, and $RS = x - 31$, then $x = -10$.
17. **GRADES** Jerome received the same quarter grade as Paula. Paula received the same quarter grade as Heath. Which property would show that Jerome and Heath received the same grade?

### Example 2

**Write a two-column proof.**

**Given:** $\frac{5x - 3}{6} = 2x + 1$

**Prove:** $x = -\frac{9}{7}$

**Proof:**

| Statements | Reasons |
|---|---|
| 1. $\frac{5x - 3}{6} = 2x + 1$ | 1. Given |
| 2. $5x - 3 = 6(2x + 1)$ | 2. Multiplication Property of Equality |
| 3. $5x - 3 = 12x + 6$ | 3. Distributive Property of Equality |
| 4. $-3 = 7x + 6$ | 4. Subtraction Property of Equality |
| 5. $-9 = 7x$ | 5. Subtraction Property of Equality |
| 6. $-\frac{9}{7} = x$ | 6. Division Property of Equality |
| 7. $x = -\frac{9}{7}$ | 7. Symmetric Property of Equality |

# Study Guide and Review *Continued*

## 5-3 Proving Segment Relationships

Write a two-column proof.

**18.** Given: $X$ is the midpoint of $\overline{WY}$ and $\overline{VZ}$.

Prove: $VW = ZY$

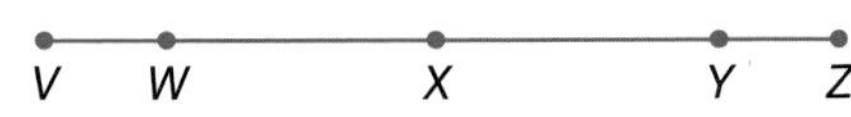


**19.** Given: $AB = DC$

Prove: $AC = DB$

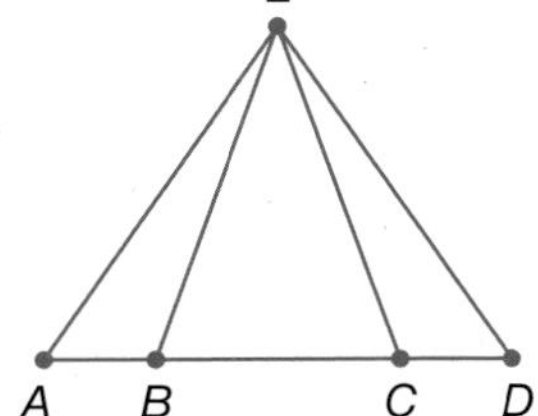


**20.** **GEOGRAPHY** Leandro is planning to drive from Kansas City to Minneapolis along Interstate 35. The map he is using gives the distance from Kansas City to Des Moines as 194 miles and from Des Moines to Minneapolis as 243 miles. What allows him to conclude that the distance he will be driving is 437 miles from Kansas City to Minneapolis? Assume that Interstate 35 forms a straight line.

### Example 3

Write a two-column proof.

Given: $B$ is the midpoint of $\overline{AC}$.

$C$ is the midpoint of $\overline{BD}$.

Prove: $\overline{AB} \cong \overline{CD}$

Proof:

| Statements | Reasons |
|---|---|
| **1.** $B$ is the midpoint of $\overline{AC}$. | **1.** Given |
| **2.** $\overline{AB} \cong \overline{BC}$ | **2.** Definition of midpoint |
| **3.** $C$ is the midpoint of $\overline{BD}$. | **3.** Given |
| **4.** $\overline{BC} \cong \overline{CD}$ | **4.** Definition of midpoint |
| **5.** $\overline{AB} \cong \overline{CD}$ | **5.** Transitive Property of Equality |

## 5-4 Proving Angle Relationships

Find the measure of each angle.

**21.** $\angle 5$

**22.** $\angle 6$

**23.** $\angle 7$

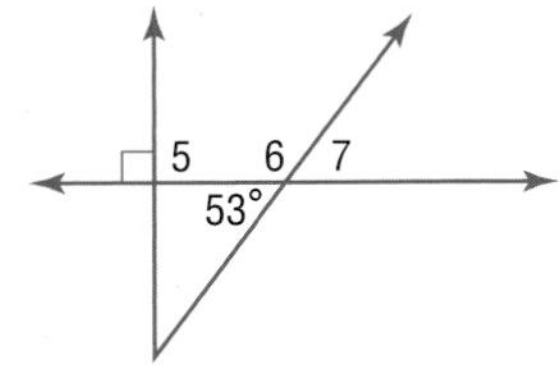


**24.** **PROOF** Write a two-column proof.

Given: $\angle 1 \cong \angle 4$, $\angle 2 \cong \angle 3$

Prove: $\angle AFC \cong \angle EFC$

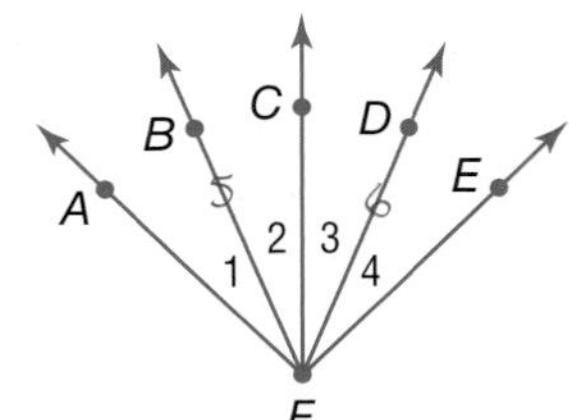


### Example 4

Find the measure of each numbered angle if $m\angle 1 = 72$ and $m\angle 3 = 26$.

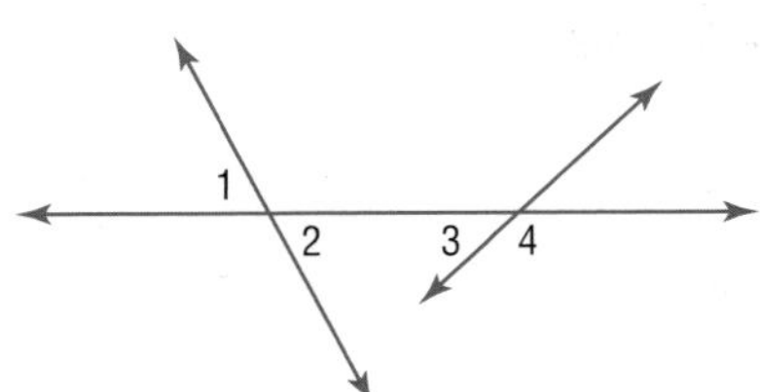


$m\angle 2 = 72$, since $\angle 1$ and $\angle 2$ are vertical angles.

$\angle 3$ and $\angle 4$ form a linear pair and must be supplementary angles.

$26 + m\angle 4 = 180$ Definition of supplementary angles

$m\angle 4 = 154$ Subtract 26 from each side.

# Study Guide and Review

## 5-5 Angles and Parallel Lines

In the figure, $m\angle 1 = 123$. Find the measure of each angle. Tell which postulate(s) or theorem(s) you used.

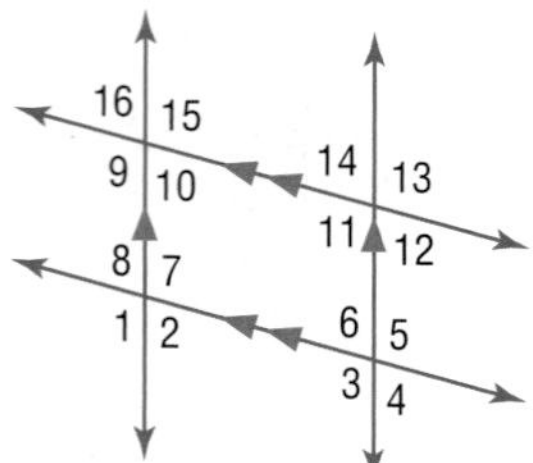


**25.** $\angle 5$ **26.** $\angle 14$ **27.** $\angle 16$

**28.** $\angle 11$ **29.** $\angle 4$ **30.** $\angle 6$

**31.** **MAPS** The diagram shows the layout of Elm, Plum, and Oak streets. Find the value of *x*.

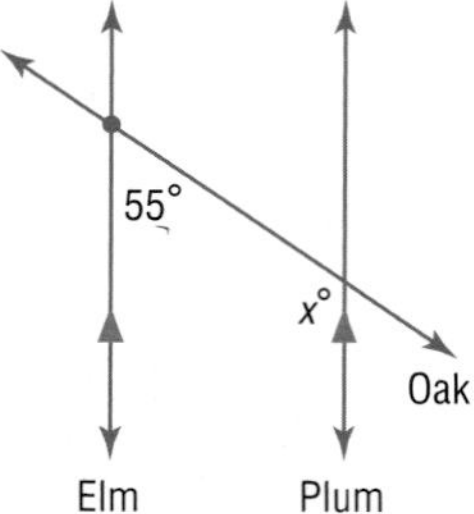


### Example 5

**ALGEBRA** If $m\angle 5 = 7x - 5$ and $m\angle 4 = 2x + 23$, find $x$. Explain your reasoning.

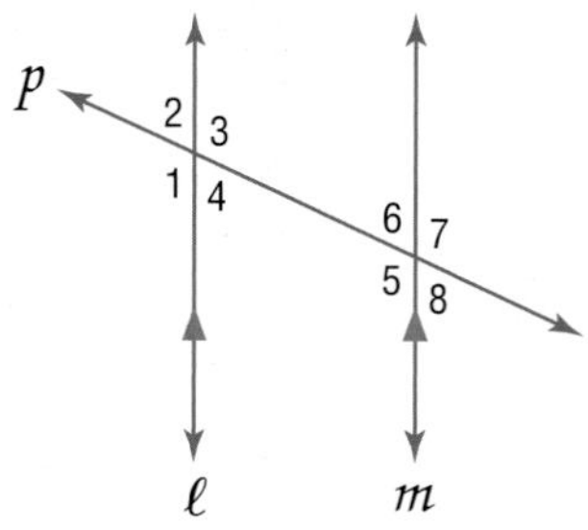


| | |
|---|---|
| $m\angle 4 + m\angle 5 = 180$ | Def. of Supp. ∠s |
| $(2x + 23) + (7x - 5) = 180$ | Substitution |
| $9x + 18 = 180$ | Simplify. |
| $9x = 162$ | Subtract. |
| $x = 18$ | Divide. |

Since lines $\ell$ and $m$ are parallel, $\angle 4$ and $\angle 5$ are supplementary by the Consecutive Interior Angles Theorem.

## 5-6 Proving Lines Parallel

Given the following information, determine which lines, if any, are parallel. State the postulate or theorem that justifies your answer.

**32.** $\angle 7 \cong \angle 10$

**33.** $\angle 2 \cong \angle 10$

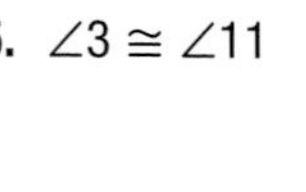

**34.** $\angle 1 \cong \angle 3$

**35.** $\angle 3 \cong \angle 11$

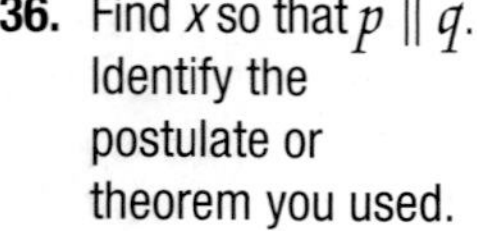

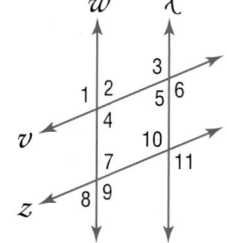

**36.** Find *x* so that $p \parallel q$. Identify the postulate or theorem you used.

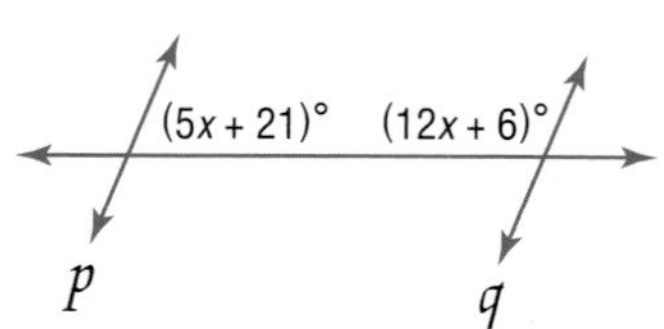


### Example 6

Given the following information, determine which lines, if any, are parallel. State the postulate or theorem that justifies your answer.

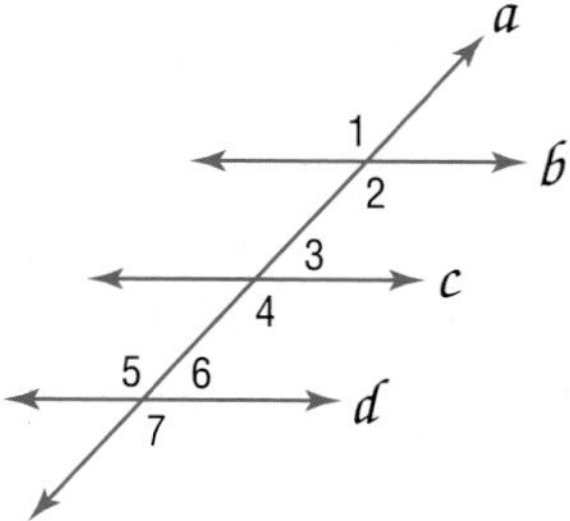


$\angle 1 \cong \angle 7$

$\angle 1$ and $\angle 7$ are alternate exterior angles of lines $b$ and $d$.

Since $\angle 1 \cong \angle 7$, $b \parallel d$ by the Converse of the Alternate Exterior Angles Theorem.

# CHAPTER 5 Practice Test

1. **PROOF** Copy and complete the following proof.
   **Given:** $3(x - 4) = 2x + 7$
   **Prove:** $x = 19$
   **Proof:**

| Statements | Reasons |
|---|---|
| a. $3(x - 4) = 2x + 7$ | a. Given |
| b. $3x - 12 = 2x + 7$ | b. ___?___ |
| c. ___?___ | c. Subtraction Property |
| d. $x = 19$ | d. ___?___ |

**Determine whether each statement is *always*, *sometimes*, or *never* true.**

2. Two angles that are supplementary form a linear pair.
3. If $B$ is between $A$ and $C$, then $AC + AB = BC$.
4. If two lines intersect to form congruent adjacent angles, then the lines are perpendicular.

**Find the measure of each numbered angle, and name the theorems that justify your work.**

5. $m\angle 1 = x$, $m\angle 2 = x - 6$

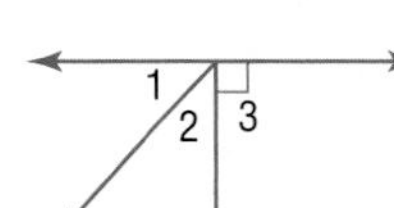


6. $m\angle 7 = 2x + 15$, $m\angle 8 = 3x$

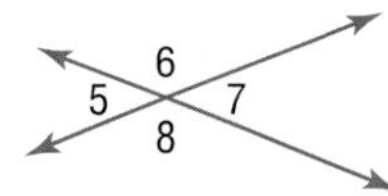


**Write each statement in if-then form.**

7. An acute angle measures less than 90.
8. Two perpendicular lines intersect to form right angles.

9. **MULTIPLE CHOICE** If a triangle has one obtuse angle, then it is an obtuse triangle.

   Which of the following statements is the contrapositive of the conditional above?

   **A** If a triangle is not obtuse, then it has one obtuse angle.

   **B** If a triangle does not have one obtuse angle, then it is not an obtuse triangle.

   **C** If a triangle is not obtuse, then it does not have one obtuse angle.

   **D** If a triangle is obtuse, then it has one obtuse angle.

**Determine the slope of the line that contains the given points.**

10. $G(8, 1)$, $H(8, -6)$
11. $A(0, 6)$, $B(4, 0)$
12. $E(6, 3)$, $F(-6, 3)$
13. $E(5, 4)$, $F(8, 1)$

**In the figure, $m\angle 8 = 96$ and $m\angle 12 = 42$. Find the measure of each angle. Tell which postulate(s) or theorem(s) you used.**

14. $\angle 9$
15. $\angle 11$
16. $\angle 6$

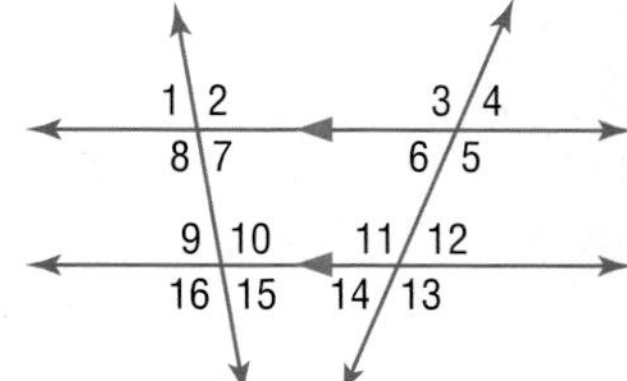


17. Find the value of $x$ in the figure below.

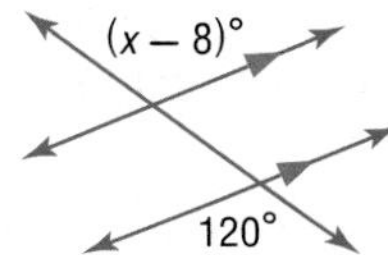


**Given the following information, determine which lines, if any, are parallel. State the postulate or theorem that justifies your answer.**

18. $\angle 4 \cong \angle 10$
19. $\angle 9 \cong \angle 6$
20. $\angle 7 \cong \angle 11$

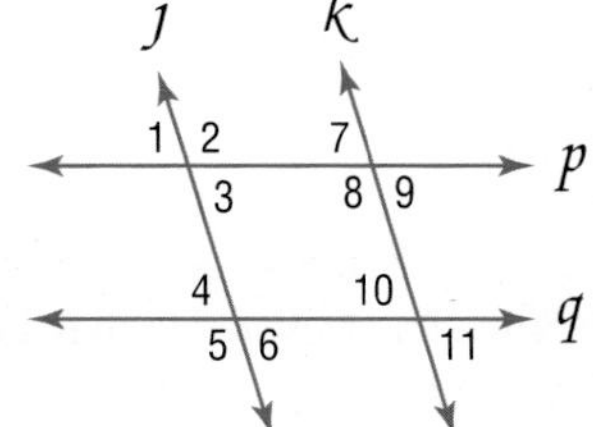

CHAPTER 5

# Preparing for Standardized Tests

## Logical Reasoning

Solving geometry problems frequently requires the use of logical reasoning. You can use the fundamentals of logical reasoning to help you solve problems on standardized tests.

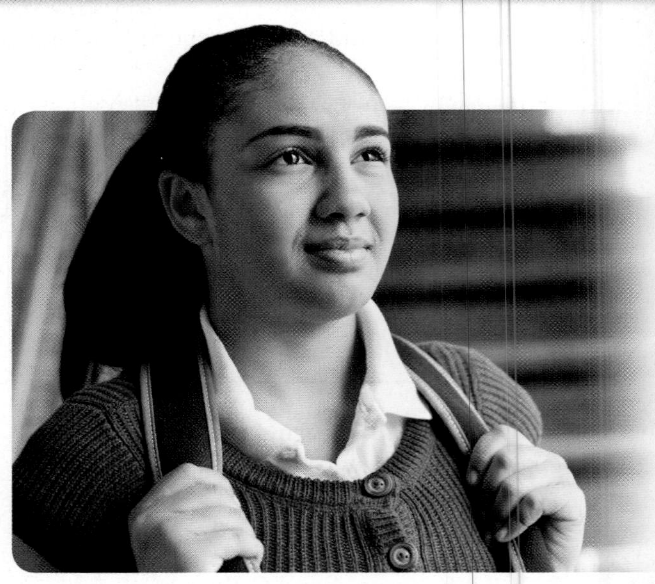

Lauren Nicole/Photographer's Choice RF/Getty Images

### Strategies for Using Logical Reasoning

**Step 1**

Read the problem to determine what information you are given and what you need to find out in order to answer the question.

**Step 2**

Determine if you can apply one of the principles of logical reasoning to the problem.

- **Counterexample**: A counterexample contradicts a statement that is known to be true.

  Identify any answer choices that contradict the problem statement and eliminate them.

- **Postulates:** A postulate is a statement that describes a fundamental relationship in geometry.

  Determine if you can apply a postulate to draw a logical conclusion.

**Step 3**

If you cannot reach a conclusion using only the principles in Step 2, determine if one of the tools below would be helpful.

- **Patterns:** Look for a pattern to make a conjecture.
- **Truth Tables:** Use a truth table to organize the truth values of the statement provided in the problem.
- **Venn Diagrams:** Use a Venn Diagram to clearly represent the relationships between members of groups.
- **Proofs:** Use deductive and inductive reasoning to reach a conclusion in the form of a proof.

**Step 4**

If you still cannot reach a conclusion using the tools in Step 3, make a **conjecture**, or educated guess, about which answer choice is most reasonable. Then mark the problem so that you can return to it if you have extra time at the end of the exam.

## Standardized Test Example

**Read the problem. Identify what you need to know. Then use the information in the problem to solve.**

In a school of 292 students, 94 participate in sports, 122 participate in academic clubs, and 31 participate in both. How many students at the school do not participate in sports or academic clubs?

**A** 95 **C** 122

**B** 107 **D** 138

Read the problem carefully. There are no clear counterexamples, and a postulate cannot be used to draw a logical conclusion. Therefore, consider the tools that you can use to organize the information.

A Venn diagram can be used to show the intersection of two sets. Make a Venn diagram with the information provided in the problem statement.

Determine how many students participate in only sports or academic clubs.

Only sports: $94 - 31 = 63$

Only academic clubs: $122 - 31 = 91$

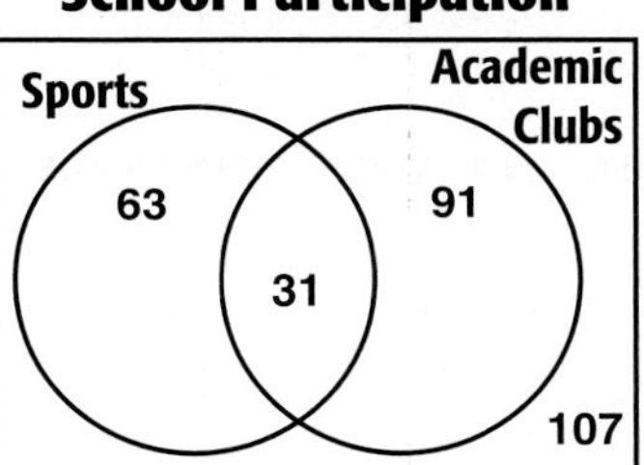


Use the information to calculate the number of students who do not participate in either sports or academic clubs.

$292 - 63 - 91 - 31 = 107$

There are 107 students who do not participate in either sports or academic clubs. The correct answer is B.

## Exercises

**Read each question. Then fill in the correct answer on the answer document provided by your teacher or on a sheet of paper.**

**1.** Determine the truth of the following statement. If the statement is false, give a counterexample.

*The product of two even numbers is even.*

**A** false; $8 \times 4 = 32$

**B** false; $7 \times 6 = 42$

**C** false; $3 \times 10 = 30$

**D** true

**2.** Find the next item in the pattern.

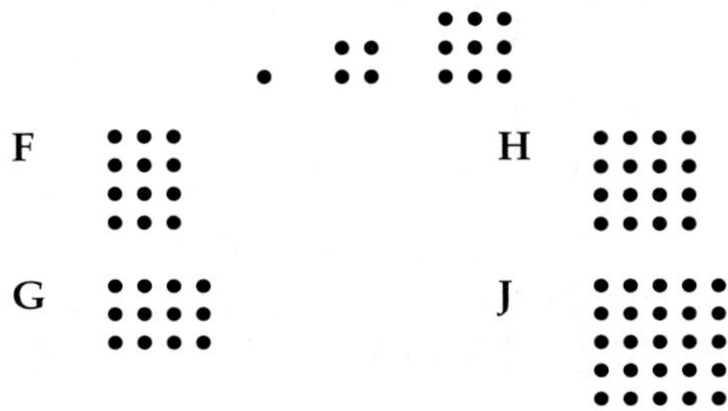

CHAPTER 5

# Standardized Test Practice

## Cumulative, Chapters 1 through 5

### Multiple Choice

**Read each question. Then fill in the correct answer on the answer document provided by your teacher or on a sheet of paper.**

**1.** In the diagram below, $\angle 1 \cong \angle 3$.

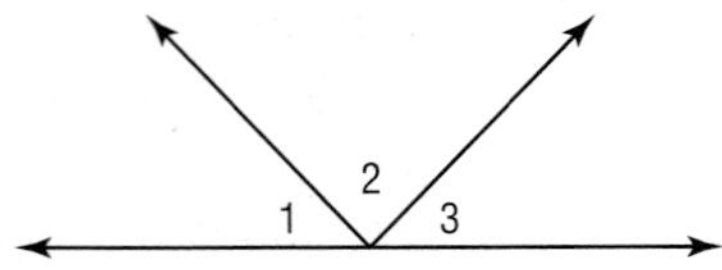


Which of the following conclusions does not have to be true?

**F** $m\angle 1 - m\angle 2 + m\angle 3 = 90$

**G** $m\angle 1 + m\angle 2 + m\angle 3 = 180$

**H** $m\angle 1 + m\angle 2 = m\angle 2 + m\angle 3$

**J** $m\angle 2 - m\angle 1 = m\angle 2 - m\angle 3$

**2.** If $a \parallel b$ in the diagram below, which of the following may *not* true?

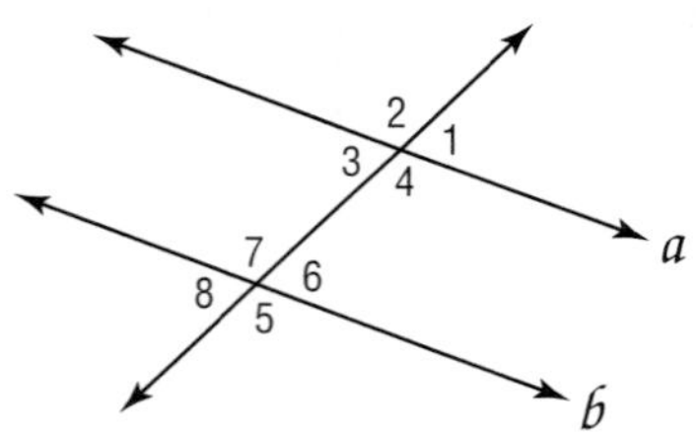


**A** $\angle 1 \cong \angle 3$ **C** $\angle 2 \cong \angle 5$

**B** $\angle 4 \cong \angle 7$ **D** $\angle 8 \cong \angle 2$

**Test-TakingTip**

Question 3 A *counterexample* is an example used to show that a given statement is not always true.

**3.** In the diagram, $\overline{BD}$ intersects $\overline{AE}$ at $C$. Which of the following conclusions does *not* have to be true?

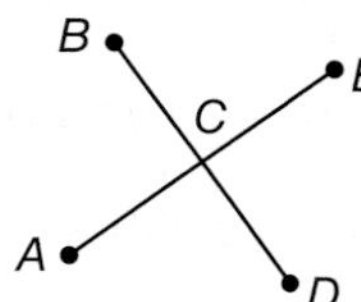


**A** $\angle ACB \cong \angle ECD$

**B** $\angle ACB$ and $\angle ACD$ form a linear pair.

**C** $\angle BCE$ and $\angle ACD$ are vertical angles.

**D** $\angle BCE$ and $\angle ECD$ are complementary angles.

**4.** What is the effect on the graph of the equation $y = x^2 + 4$ when it is changed to $y = x^2 - 3$?

**F** The slope of the graph changes.

**G** The graph widens.

**H** The graph is the same shape, and the vertex of the graph is moved down.

**J** The graph is the same shape, and the vertex of the graph is shifted to the left.

**5.** Which equation will produce the narrowest parabola when graphed?

**A** $y = 3x^2$ **C** $y = -6x^2$

**B** $y = \frac{3}{4}x^2$ **D** $y = -\frac{3}{4}x^2$

**6.** What is the effect on the graph of the equation $y = 3x^2$ when the equation is changed to $y = 2x^2$?

**F** The graph of $y = 2x^2$ is a reflection of the graph of $y = 3x^2$ across the $y$-axis.

**G** The graph is rotated 90 degrees about the origin.

**H** The graph is narrower.

**J** The graph is wider.

## Short Response/Gridded Response

**Record your answers on the answer sheet provided by your teacher or on a sheet of paper.**

**7.** Use the proof to answer the question.

**Given:** $\angle A$ is the complement of $\angle B$.
$m\angle B = 46$

**Prove:** $m\angle A = 44$

**Proof:**

| Statements | Reasons |
|---|---|
| **1.** $A$ is the complement of $\angle B$; $m\angle B = 46$. | **1.** Given |
| **2.** $m\angle A + m\angle B = 90$ | **2.** Def. of comp. angles |
| **3.** $m\angle A + 46 = 90$ | **3.** Substitution Prop. |
| **4.** $m\angle A + 46 - 46 = 90 - 46$ | **4.** ___?___ |
| **5.** $m\angle A = 44$ | **5.** Substitution Prop. |

What reason can be given to justify Statement 4?

**8. HEIGHT** The height $h$ of a bouncing ball at time $t$ seconds can be modeled by the equation $h = -16t^2 + 28.3t$.

**a.** Write the equation that models the height in factored form.

**b.** What is the height of the ball at 1.5 seconds?

**c.** How high will the ball bounce?

## Extended Response

**9.** Scott launches a model rocket from ground level. The rocket's height $h$ in meters is given by the equation $h = -4.9t^2 + 56t$, where $t$ is the time in seconds after the launch.

**a.** What is the maximum height the rocket will reach? Round to the nearest tenth of a meter. Show each step and explain your method.

**b.** How long after it is launched will the rocket reach its maximum height? Round to the nearest tenth of a second.

### Need ExtraHelp?

| If you missed Question... | 1 | 2 | 3 | 4 | 5 | 6 | 7 | 8 | 9 | 10 | 11 | 12 |
|---|---|---|---|---|---|---|---|---|---|---|---|---|
| Go to Lesson... | 5-4 | 5-5 | 5-3 | 3-5 | 3-5 | 3-5 | 5-4 | 1-5 | 2-8 | 2-3 | 1-3 | 2-1 |

CHAPTER 6

# Congruent Triangles

## Then

You learned about segments, angles, and discovered relationships between their measures.

## Now

In this chapter, you will:

- Apply special relationships about the interior and exterior angles of triangles.
- Identify corresponding parts of congruent triangles and prove triangles congruent.
- Learn about the special properties of isosceles and equilateral triangles

## Why? ▲

**FITNESS** Triangles are used to add strength to many structures, including fitness equipment such as bike frames.

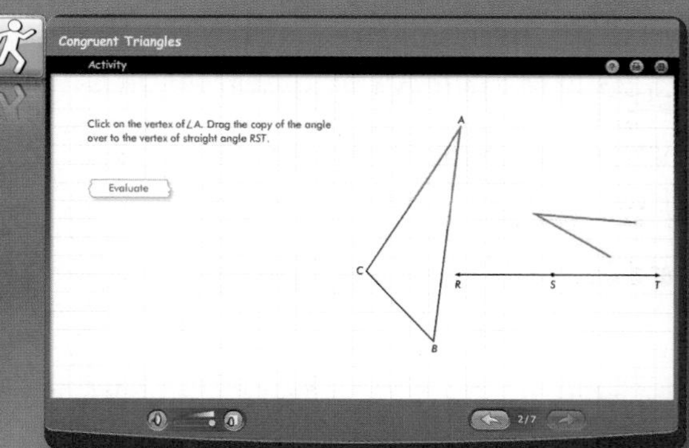


connectED.mcgraw-hill.com Your Digital Math Portal

Animation

Vocabulary

eGlossary

Personal Tutor

Virtual Manipulatives

Graphing Calculator

Audio

Foldables

Self-Check Practice

Worksheets

Scott Markewitz/Getty Ima

# Get Ready for the Chapter

**Diagnose** Readiness | You have two options for checking prerequisite skills.

## 1 Textbook Option

Take the Quick Check below. Refer to the Quick Review for help.

### QuickCheck

Classify each angle as *right*, *acute*, or *obtuse*.

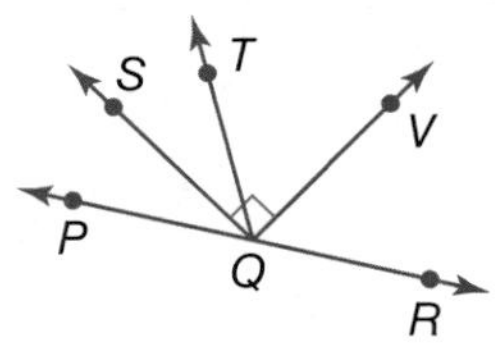


1. $m\angle VQS$
2. $m\angle TQV$
3. $m\angle PQV$
4. **ORIGAMI** The origami fold involves folding a strip of paper so that the lower edge of the strip forms a right angle with itself. Identify each angle as *right*, *acute*, or *obtuse*.

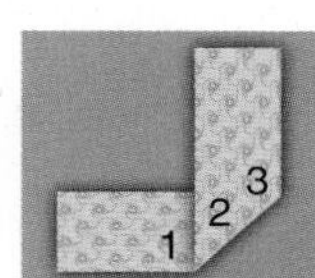


**ALGEBRA** Use the figure to find the indicated variable(s). Explain your reasoning.

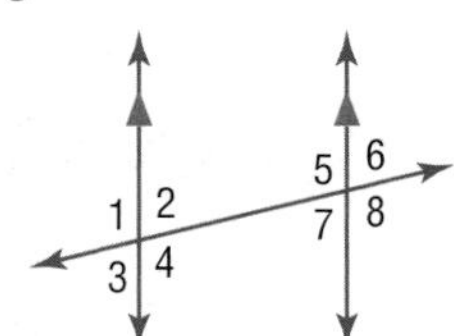


5. Find $x$ if $m\angle 3 = x - 12$ and $m\angle 6 = 72$.
6. If $m\angle 4 = 2y + 32$ and $m\angle 5 = 3y - 3$, find $y$.

Find the distance between each pair of points.

7. $F(3, 6)$, $G(7, -4)$
8. $X(-2, 5)$, $Y(1, 11)$
9. $R(8, 0)$, $S(-9, 6)$
10. $A(14, -3)$, $B(9, -9)$
11. **MAPS** Miranda laid a coordinate grid on a map of a state where each 1 unit is equal to 10 miles. If her city is located at $(-8, -12)$ and the state capital is at $(0, 0)$, find the distance from her city to the capital to the nearest tenth of a mile.

### QuickReview

**Example 1**

Classify each angle as *right*, *acute*, or *obtuse*.

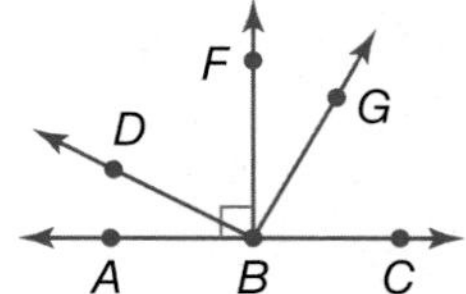


**a.** $m\angle ABG$

Point $G$ on angle $\angle ABG$ lies on the exterior of right angle $\angle ABF$, so $\angle ABG$ is an obtuse angle.

**b.** $m\angle DBA$

Point $D$ on angle $\angle DBA$ lies on the interior of right angle $\angle FBA$, so $\angle DBA$ is an acute angle.

**Example 2**

In the figure, $m\angle 4 = 42$. Find $m\angle 7$.

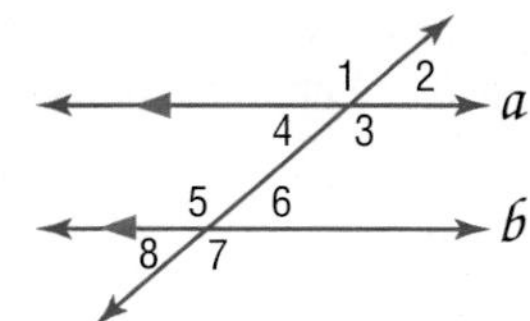


$\angle 7$ and $\angle 1$ are alternate interior angles, so they are congruent. $\angle 1$ and $\angle 4$ are a linear pair, so they are supplementary. Therefore, $\angle 7$ is supplementary to $\angle 1$. The measure of $\angle 7$ is $180 - 42$ or 138.

**Example 3**

Find the distance between $J(5, 2)$ and $K(11, -7)$.

| | |
|---|---|
| $JK = \sqrt{(x_2 - x_1)^2 + (y_2 - y_1)^2}$ | Distance Formula |
| $= \sqrt{(11 - 5)^2 + [(-7) - 2]^2}$ | Substitute. |
| $= \sqrt{6^2 + (-9)^2}$ | Subtract. |
| $= \sqrt{36 + 81}$ or $\sqrt{117}$ | Simplify. |

## 2 Online Option

Take an online self-check Chapter Readiness Quiz at connectED.mcgraw-hill.com.

## Get Started on the Chapter

**You will learn several new concepts, skills, and vocabulary terms as you study Chapter 6. To get ready, identify important terms and organize your resources. You may wish to refer to Chapter 0 to review prerequisite skills.**

### FOLDABLES® StudyOrganizer

**Congruent Triangles** Make this Foldable to help you organize your Chapter 6 notes about congruent triangles. Begin with a sheet of $8\frac{1}{2}$" × 11" paper.

1. **Fold** into a taco forming a square. Cut off the excess paper strip formed by the square.

2. **Open** the fold and refold it the opposite way forming another taco and an X fold pattern.

3. **Open** and fold the corners toward the center point of the X forming a small square.

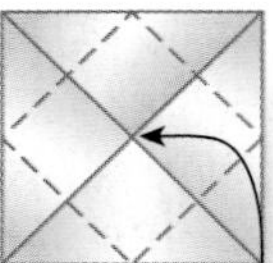

4. **Label** the flaps as shown.

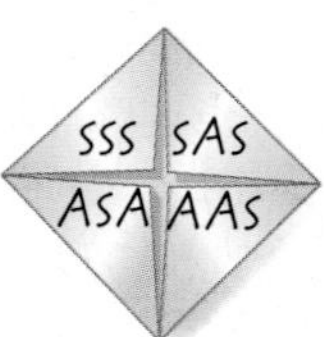
SSS
SAS
ASA
AAS

### NewVocabulary

| English | | Español |
|---|---|---|
| auxiliary line | p. 335 | linea auxiliar |
| congruent | p. 344 | congruente |
| congruent polygons | p. 344 | polígonos congruentes |
| corresponding parts | p. 344 | partes correspondientes |
| included angle | p. 355 | ángulo incluido |
| included side | p. 364 | lado incluido |
| base angle | p. 374 | ángulo de la base |

### ReviewVocabulary

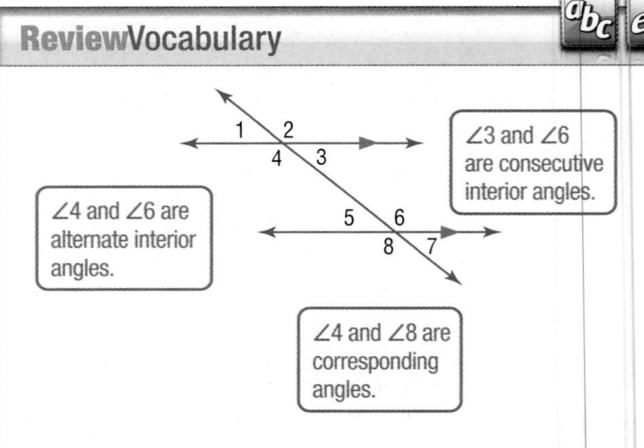
1
2
4
3
5
6
8
7
∠3 and ∠6 are consecutive interior angles.
∠4 and ∠6 are alternate interior angles.
∠4 and ∠8 are corresponding angles.

# LESSON 6-1 Angles of Triangles

**Then**

- You classified triangles by their side or angle measures.

**Now**

1. Apply the Triangle Angle-Sum Theorem.
2. Apply Exterior Angle Theorem.

**Why?**

- Massachusetts Institute of Technology (MIT) sponsors the annual *Design 2.007* contest in which students design and build a robot.

One test of a robot's movements is to program it to move in a triangular path. The sum of the measures of the pivot angles through which the robot must turn will always be the same.

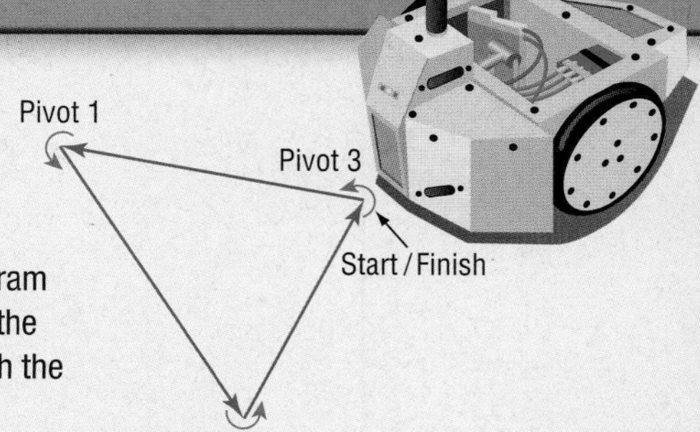

**NewVocabulary**
auxiliary line
exterior angle
remote interior angles
flow proof
corollary

**Common Core State Standards**

**Content Standards**
G.CO.10 Prove theorems about triangles.

**Mathematical Practices**
1 Make sense of problems and persevere in solving them.
3 Construct viable arguments and critique the reasoning of others.

## 1 Triangle Angle-Sum Theorem

The Triangle Angle-Sum Theorem gives the relationship among the interior angle measures of any triangle.

### Theorem 6.1 Triangle Angle-Sum Theorem

**Words** The sum of the measures of the angles of a triangle is 180.

**Example** $m\angle A + m\angle B + m\angle C = 180$

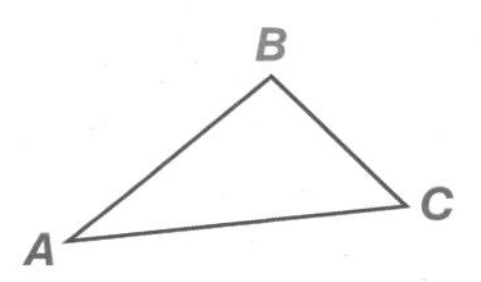


The proof of the Triangle Angle-Sum Theorem requires the use of an auxiliary line. An **auxiliary line** is an extra line or segment drawn in a figure to help analyze geometric relationships. As with any statement in a proof, you must justify any properties of an auxiliary line that you have drawn.

### Proof Triangle Angle-Sum Theorem

**Given:** $\triangle ABC$

**Prove:** $m\angle 1 + m\angle 2 + m\angle 3 = 180$

**Proof:**

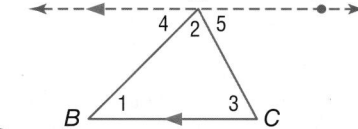

| Statements | Reasons |
|---|---|
| 1. $\triangle ABC$ | 1. Given |
| 2. Draw $\overleftrightarrow{AD}$ through $A$ parallel to $\overline{BC}$. | 2. Parallel Postulate |
| 3. $\angle 4$ and $\angle BAD$ form a linear pair. | 3. Def. of a linear pair |
| 4. $\angle 4$ and $\angle BAD$ are supplementary. | 4. If 2 ∡ form a linear pair, they are supplementary. |
| 5. $m\angle 4 + m\angle BAD = 180$ | 5. Def. of suppl. ∡ |
| 6. $m\angle BAD = m\angle 2 + m\angle 5$ | 6. Angle Addition Postulate |
| 7. $m\angle 4 + m\angle 2 + m\angle 5 = 180$ | 7. Substitution |
| 8. $\angle 4 \cong \angle 1$, $\angle 5 \cong \angle 3$ | 8. Alt. Int. ∡ Theorem |
| 9. $m\angle 4 = m\angle 1$, $m\angle 5 = m\angle 3$ | 9. Def. of ≅ ∡ |
| 10. $m\angle 1 + m\angle 2 + m\angle 3 = 180$ | 10. Substitution |

The Triangle Angle-Sum Theorem can be used to determine the measure of the third angle of a triangle when the other two angle measures are known.

## Real-World Example 1 Use the Triangle Angle-Sum Theorem

**SOCCER The diagram shows the path of the ball in a passing drill created by four friends. Find the measure of each numbered angle.**

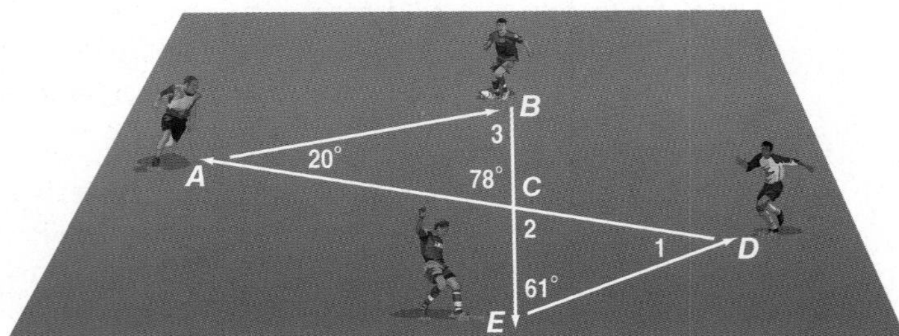


**Real-WorldLink**

The pass-and-move soccer drill incorporates several fundamental aspects of passing. All passes in this drill are made in a triangle, which is the basis of all ball movement. Additionally, the players are forced to move immediately after passing the ball.

**Understand** Examine the information given in the diagram. You know the measures of two angles of one triangle and only one measure of another. You also know that $\angle ACB$ and $\angle 2$ are vertical angles.

**Plan** Find $m\angle 3$ using the Triangle Angle-Sum Theorem, because the measures of two angles of $\angle ABC$ are known. Use the Vertical Angles Theorem to find $m\angle 2$. Then you will have enough information to find the measure of $\angle 1$ in $\triangle CDE$.

**Solve**

| | |
|---|---|
| $m\angle 3 + m\angle BAC + m\angle ACB = 180$ | Triangle Angle-Sum Theorem |
| $m\angle 3 + 20 + 78 = 180$ | Substitution |
| $m\angle 3 + 98 = 180$ | Simplify. |
| $m\angle 3 = 82$ | Subtract 98 from each side. |

$\angle ACB$ and $\angle 2$ are congruent vertical angles. So, $m\angle 2 = 78$.

Use $m\angle 2$ and $\angle CED$ of $\triangle CDE$ to find $m\angle 1$.

| | |
|---|---|
| $m\angle 1 + m\angle 2 + m\angle CED = 180$ | Triangle Angle-Sum Theorem |
| $m\angle 1 + 78 + 61 = 180$ | Substitution |
| $m\angle 1 + 139 = 180$ | Simplify. |
| $m\angle 1 = 41$ | Subtract 139 from each side. |

**Check** The sums of the measures of the angles of $\triangle ABC$ and $\triangle CDE$ should be 180.

$\triangle ABC$: $m\angle 3 + m\angle BAC + m\angle ACB = 82 + 20 + 78$ or 180 ✓

$\triangle CDE$: $m\angle 1 + m\angle 2 + m\angle CED = 41 + 78 + 61$ or 180 ✓

**Problem-SolvingTip**

 **Sense-Making** Often a complex problem can be more easily solved if you first break it into more manageable parts. In Example 1, before you can find $m\angle 1$, you must first find $m\angle 2$.

### GuidedPractice

**Find the measures of each numbered angle.**

**1A.**

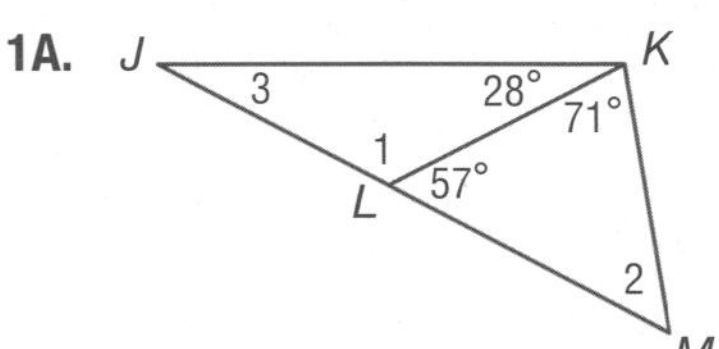


**1B.**

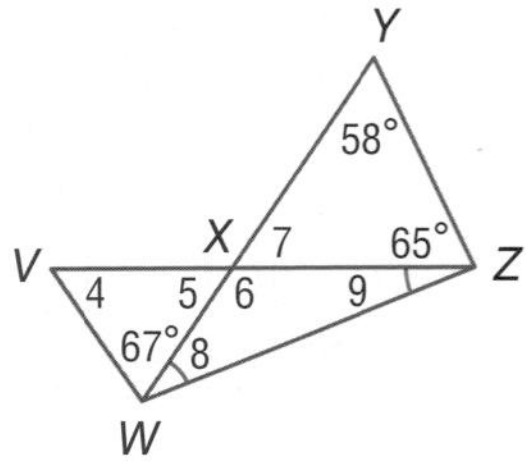


moodboard/CORBIS

## 2 Exterior Angle Theorem

In addition to its three interior angles, a triangle can have **exterior angles** formed by one side of the triangle and the extension of an adjacent side. Each exterior angle of a triangle has two **remote interior angles** that are not adjacent to the exterior angle.

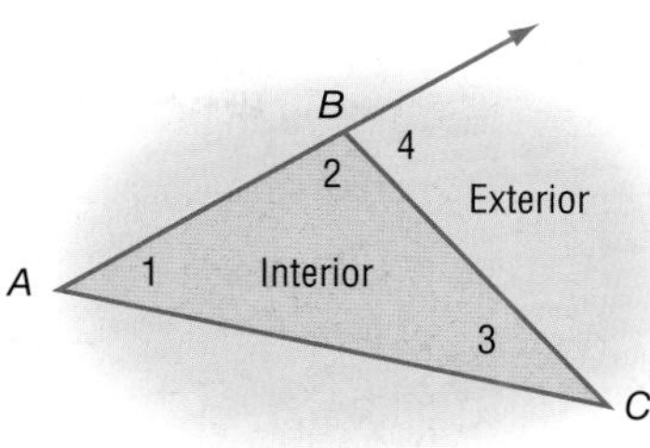


$\angle 4$ is an exterior angle of $\triangle ABC$. Its two remote interior angles are $\angle 1$ and $\angle 3$.

### Theorem 6.2 Exterior Angle Theorem

The measure of an exterior angle of a triangle is equal to the sum of the measures of the two remote interior angles.

**Example** $m\angle A + m\angle B = m\angle 1$

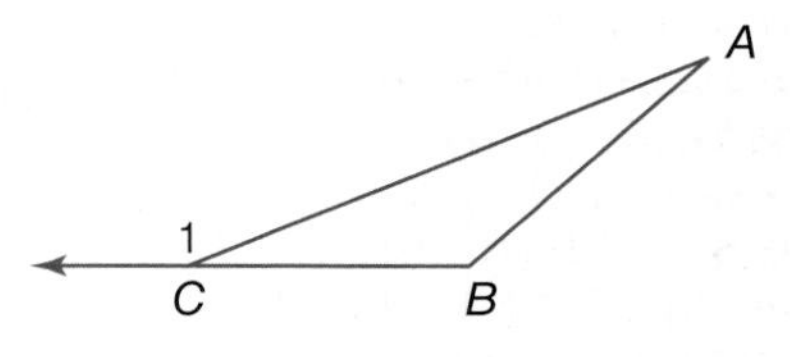


**ReadingMath**

**Flowchart Proof** A flow proof is sometimes called a *flowchart* proof.

A **flow proof** uses statements written in boxes and arrows to show the logical progression of an argument. The reason justifying each statement is written below the box. You can use a flow proof to prove the Exterior Angle Theorem.

### Proof Exterior Angle Theorem

**Given:** $\triangle ABC$

**Prove:** $m\angle A + m\angle B = m\angle 1$

**Flow Proof:**

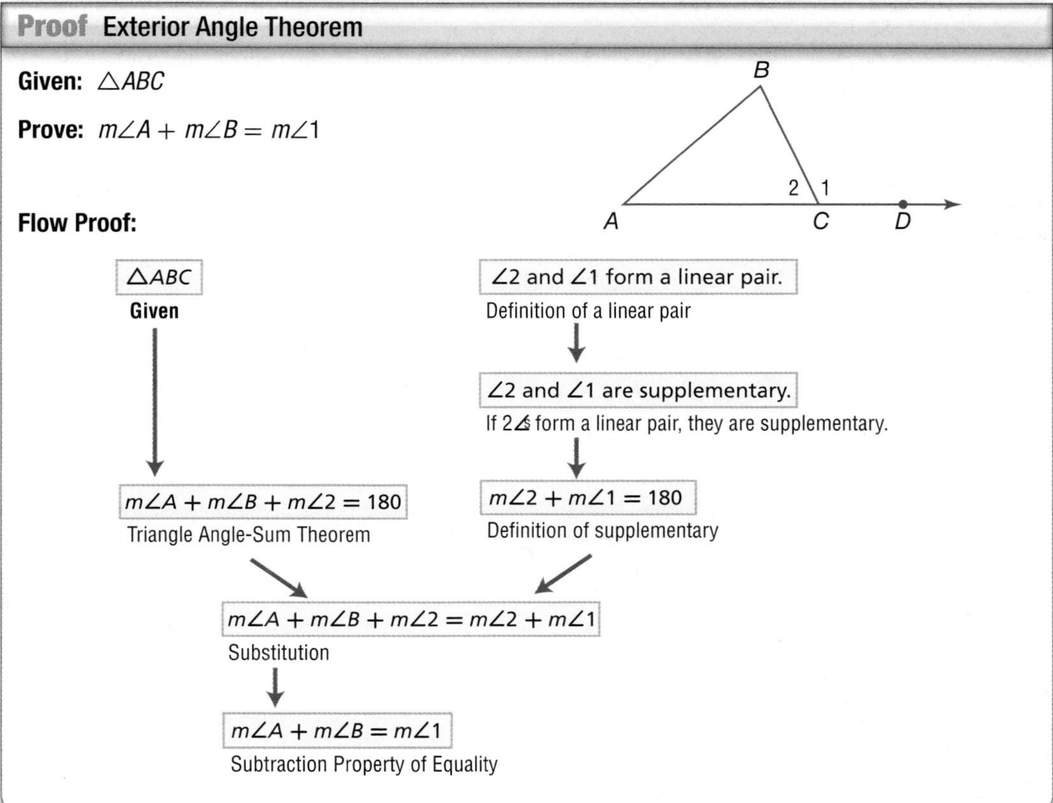


**StudyTip**

**Flow Proofs** Flow proofs can be written vertically or horizontally.

The Exterior Angle Theorem can also be used to find missing measures.

## Real-World Example 2 Use the Exterior Angle Theorem

**FITNESS Find the measure of $\angle JKL$ in the Triangle Pose shown.**

| | |
|---|---|
| $m\angle KLM + m\angle LMK = m\angle JKL$ | Exterior Angle Theorem |
| $x + 50 = 2x - 15$ | Substitution |
| $50 = x - 15$ | Subtract $x$ from each side. |
| $65 = x$ | Add 15 to each side. |

So, $m\angle JKL = 2(65) - 15$ or 115.

### GuidedPractice

**2. CLOSET ORGANIZING** Tanya mounts the shelving bracket shown to the wall of her closet. What is the measure of $\angle 1$, the angle that the bracket makes with the wall?

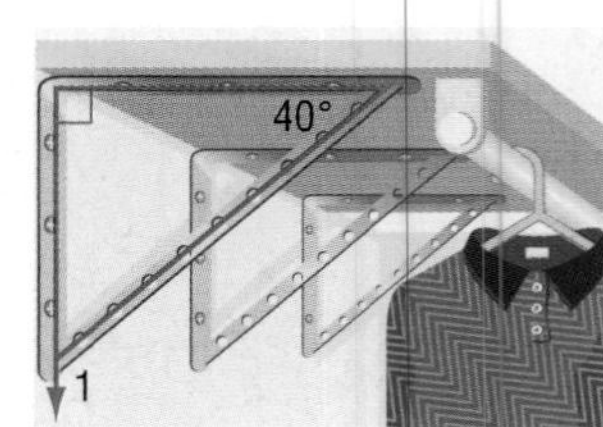


### Real-WorldCareer

**Personal Trainer** Personal trainers instruct and motivate individuals in exercise activities. They demonstrate various exercises and help clients improve their exercise techniques. Personal trainers must obtain certification in the fitness field.

A **corollary** is a theorem with a proof that follows as a direct result of another theorem. As with a theorem, a corollary can be used as a reason in a proof. The corollaries below follow directly from the Triangle Angle-Sum Theorem.

## Corollaries Triangle Angle-Sum Corollaries

**6.1** The acute angles of a right triangle are complementary.

**Abbreviation:** *Acute ∠s of a rt. △ are comp.*

**Example:** If $\angle C$ is a right angle, then $\angle A$ and $\angle B$ are complementary.

**6.2** There can be at most one right or obtuse angle in a triangle.

**Example:** If $\angle L$ is a right or an obtuse angle, then $\angle J$ and $\angle K$ must be acute angles.

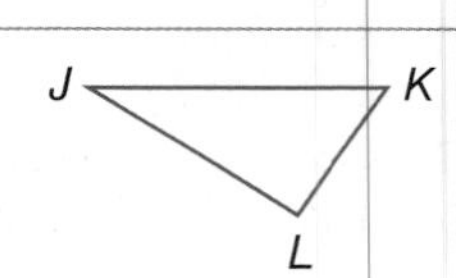

You will prove Corollaries 6.1 and 6.2 in Exercises 34 and 35.

## Example 3 Find Angle Measures in Right Triangles

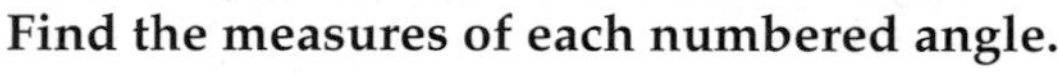

**Find the measures of each numbered angle.**

| | |
|---|---|
| $m\angle 1 + m\angle TYZ = 90$ | Acute ∠s of a rt. △ are comp. |
| $m\angle 1 + 52 = 90$ | Substitution |
| $m\angle 1 = 38$ | Subtract 52 from each side. |

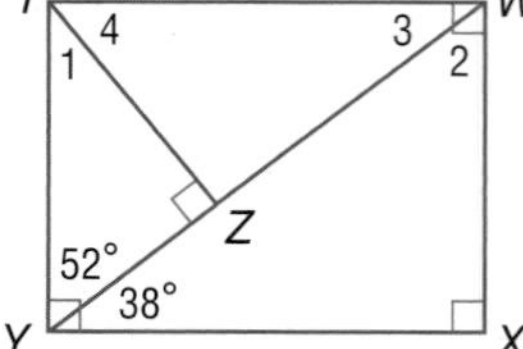


### GuidedPractice

**3A.** $\angle 2$ **3B.** $\angle 3$ **3C.** $\angle 4$

### StudyTip

**Check for Reasonableness** When you are solving for the measure of one or more angles of a triangle, always check to make sure that the sum of the angle measures is 180.

Digital Vision/Getty Images

## Check Your Understanding

**Example 1** **Find the measures of each numbered angle.**

**1.** 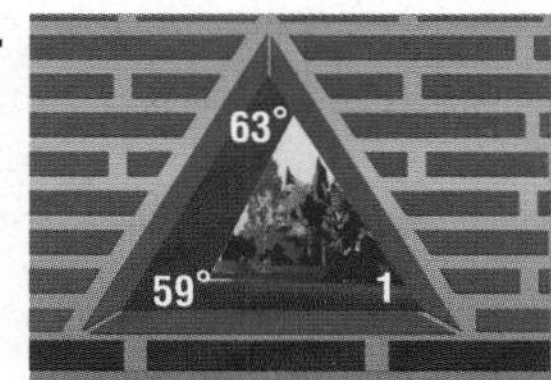


**2.** 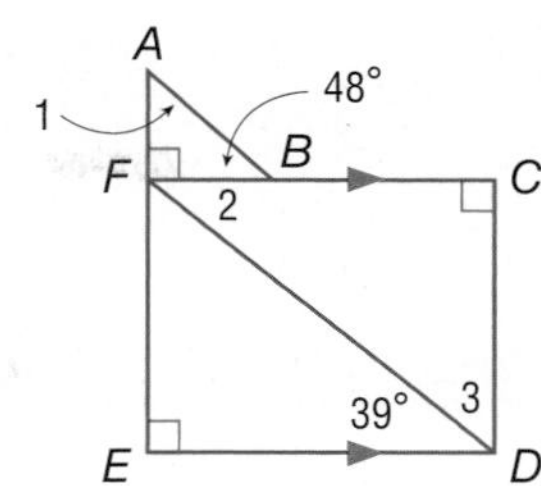


**Example 2** **Find each measure.**

**3.** $m\angle 2$

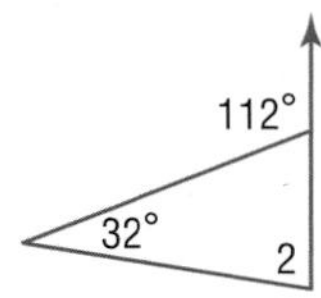


**4.** $m\angle MPQ$

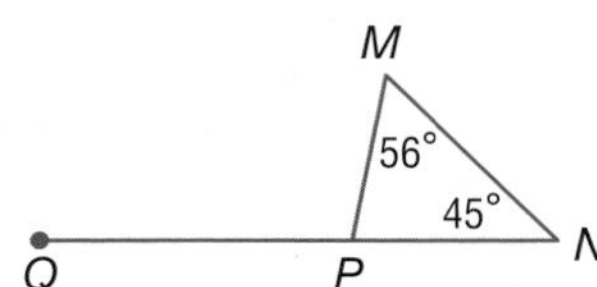


**DECK CHAIRS** **The brace of this deck chair forms a triangle with the rest of the chair's frame as shown. If $m\angle 1 = 102$ and $m\angle 3 = 53$, find each measure.**

**5.** $m\angle 4$

**6.** $m\angle 6$

**7.** $m\angle 2$

**8.** $m\angle 5$

**Example 3** **CCSS REGULARITY** **Find each measure.**

**9.** $m\angle 1$

**10.** $m\angle 3$

**11.** $m\angle 2$

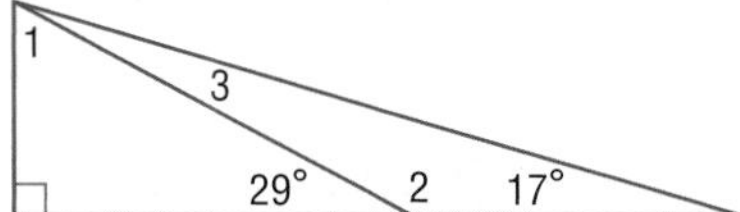


## Practice and Problem Solving

**Example 1** **Find the measure of each numbered angle.**

**12.** 

**13.** 

**14.** 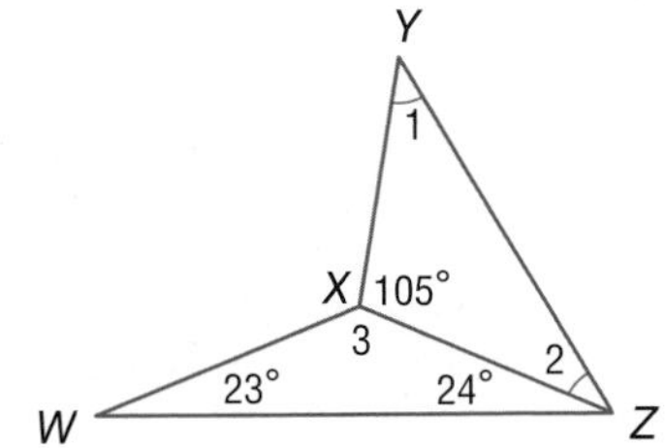


**15** 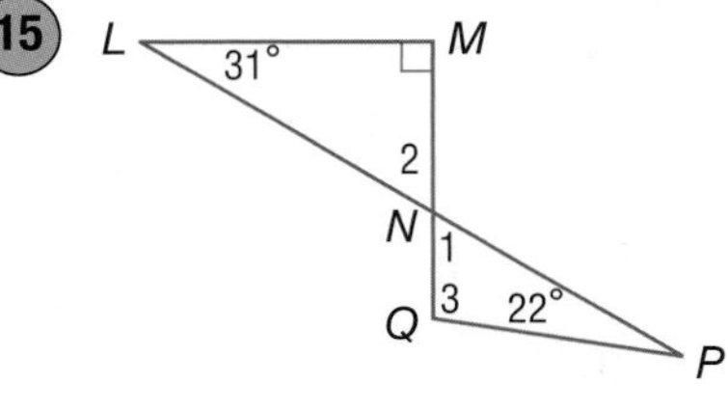

16. **AIRPLANES** The path of an airplane can be modeled using two sides of a triangle as shown. The distance covered during the plane's ascent is equal to the distance covered during its descent.

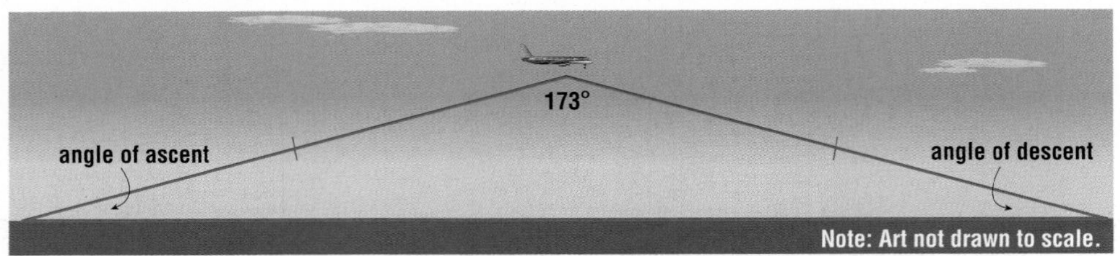


a. Classify the model using its sides and angles.

b. The angles of ascent and descent are congruent. Find their measures.

**Example 2** **Find each measure.**

17. $m\angle 1$

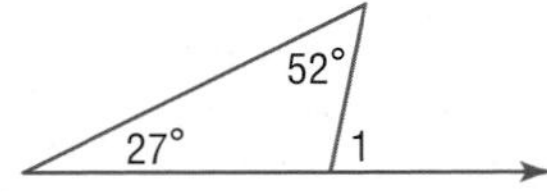


18. $m\angle 3$

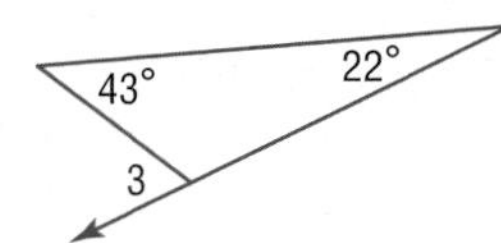


19. $m\angle 2$

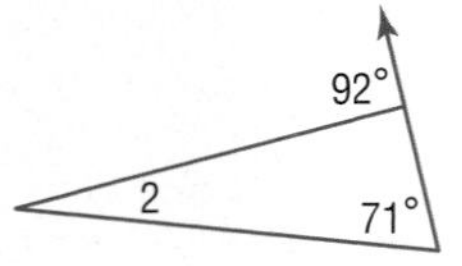


20. $m\angle 4$

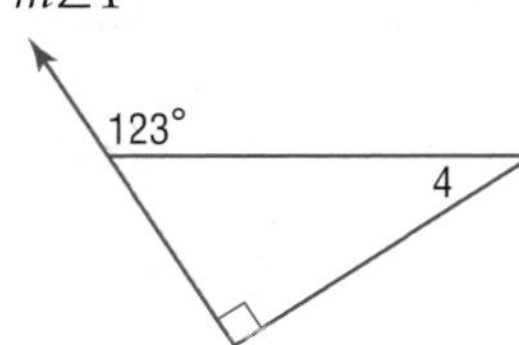


21. $m\angle ABC$

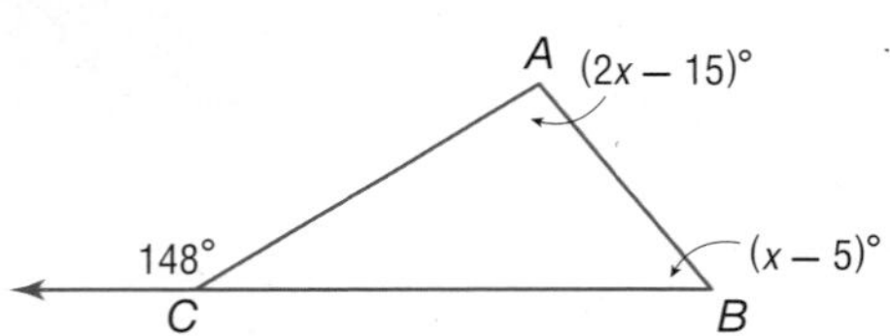


22. $m\angle JKL$

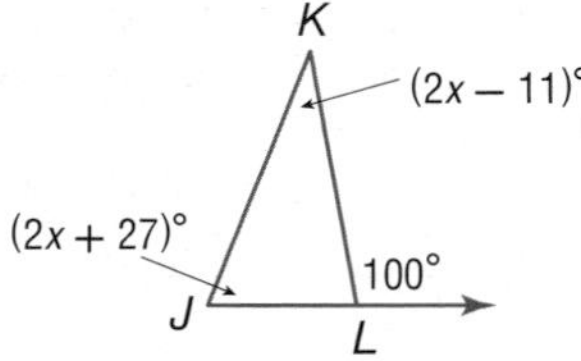


**Example 3** 23. **WHEELCHAIR RAMP** Suppose the wheelchair ramp shown makes a 12° angle with the ground. What is the measure of the angle the ramp makes with the van door?

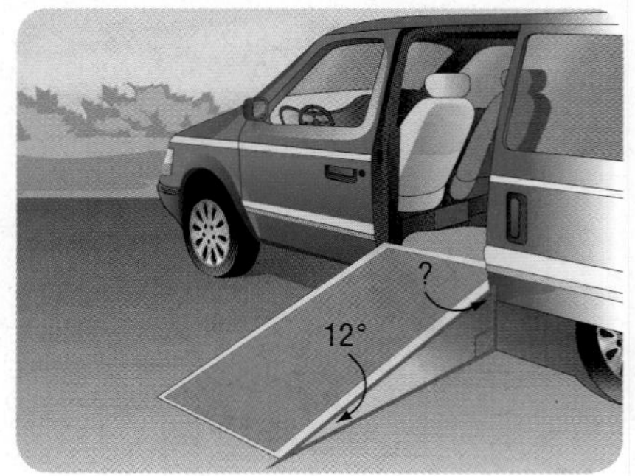


**CCSS REGULARITY** **Find each measure.**

24. $m\angle 1$
25. $m\angle 2$
26. $m\angle 3$
27. $m\angle 4$
28. $m\angle 5$
29. $m\angle 6$

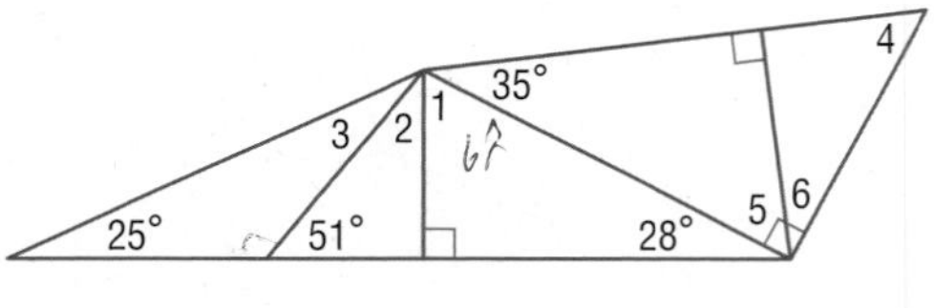

**ALGEBRA Find the value of $x$. Then find the measure of each angle.**

**30.** 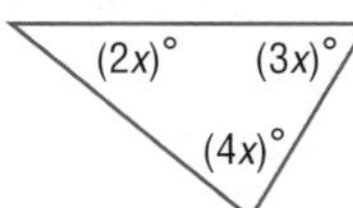


**31.** 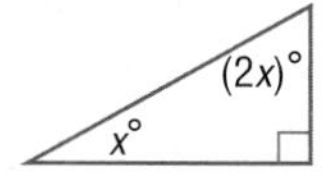


**32.** 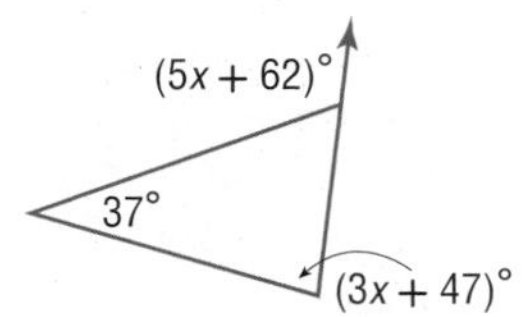


**33) GARDENING** A landscaper is forming an isosceles triangle in a flowerbed using chrysanthemums. She wants $m\angle A$ to be three times the measure of $\angle B$ and $\angle C$. What should the measure of each angle be?

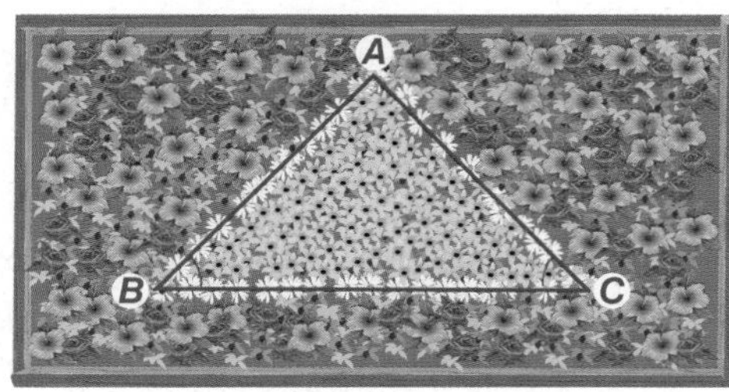


**PROOF Write the specified type of proof.**

**34.** flow proof of Corollary 6.1

**35.** paragraph proof of Corollary 6.2

**CCSS REGULARITY Find the measure of each numbered angle.**

**36.** 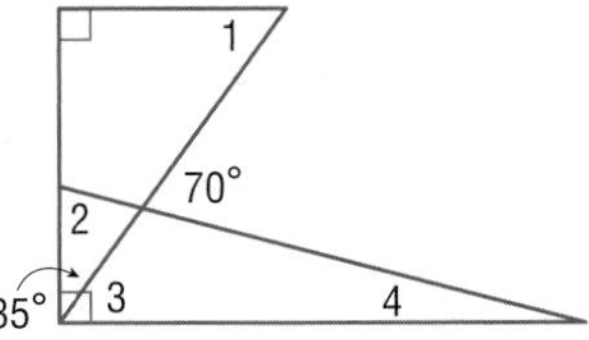


**37.** 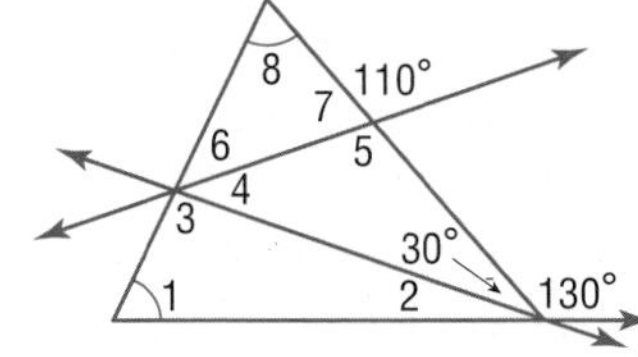


**38. ALGEBRA** Classify the triangle shown by its angles. Explain your reasoning.

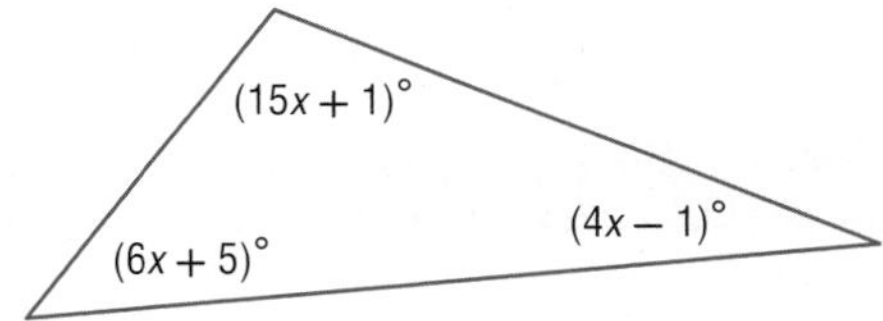


**39. ALGEBRA** The measure of the larger acute angle in a right triangle is two degrees less than three times the measure of the smaller acute angle. Find the measure of each angle.

**40.** Determine whether the following statement is *true* or *false*. If false, give a counterexample. If true, give an argument to support your conclusion.

*If the sum of two acute angles of a triangle is greater than 90, then the triangle is acute.*

**41. ALGEBRA** In $\triangle XYZ$, $m\angle X = 157$, $m\angle Y = y$, and $m\angle Z = z$. Write an inequality to describe the possible measures of $\angle Z$. Explain your reasoning.

**42. CARS** Refer to the photo at the right.

**a.** Find $m\angle 1$ and $m\angle 2$.

**b.** If the support for the hood were shorter than the one shown, how would $m\angle 1$ change? Explain.

**c.** If the support for the hood were shorter than the one shown, how would $m\angle 2$ change? Explain.

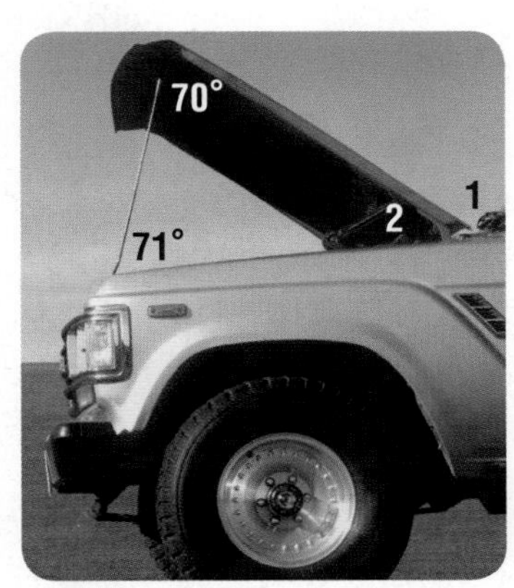


Jan Stromme/Photonica/Getty Images

**PROOF** **Write the specified type of proof.**

**43.** two-column proof
**Given:** *RSTUV* is a pentagon.
**Prove:** $m\angle S + m\angle STU + m\angle TUV + m\angle V + m\angle VRS = 540$

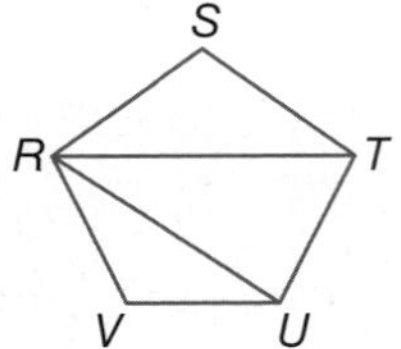


**44.** flow proof
**Given:** $\angle 3 \cong \angle 5$
**Prove:** $m\angle 1 + m\angle 2 = m\angle 6 + m\angle 7$

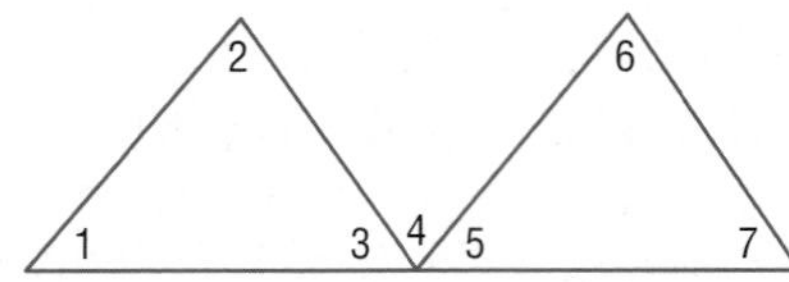


**45.** **MULTIPLE REPRESENTATIONS** In this problem, you will explore the sum of the measures of the exterior angles of a triangle.

**a. Geometric** Draw five different triangles, extending the sides and labeling the angles as shown. Be sure to include at least one obtuse, one right, and one acute triangle.

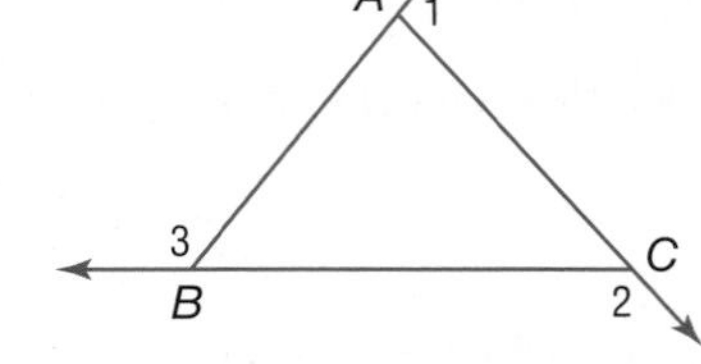


**b. Tabular** Measure the exterior angles of each triangle. Record the measures for each triangle and the sum of these measures in a table.

**c. Verbal** Make a conjecture about the sum of the exterior angles of a triangle. State your conjecture using words.

**d. Algebraic** State the conjecture you wrote in part **c** algebraically.

**e. Analytical** Write a paragraph proof of your conjecture.

## H.O.T. Problems Use Higher-Order Thinking Skills

**46.** **CCSS CRITIQUE** Curtis measured and labeled the angles of the triangle as shown. Arnoldo says that at least one of his measures is incorrect. Explain in at least two different ways how Arnoldo knows that this is true.

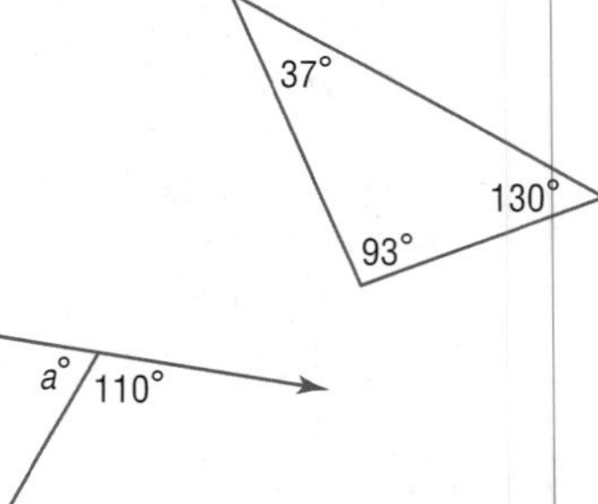


**47.** **WRITING IN MATH** Explain how you would find the missing measures in the figure shown.

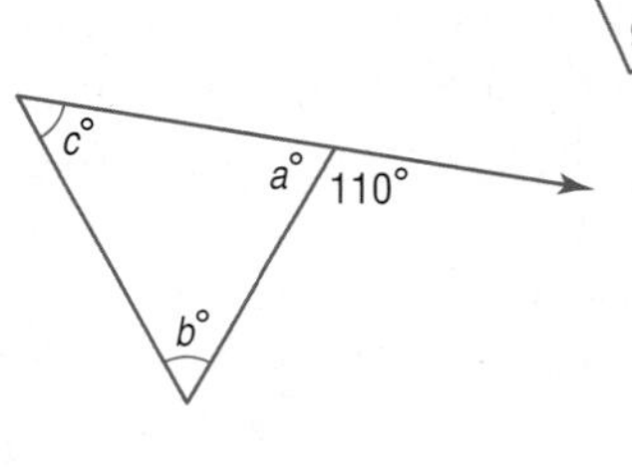


**48.** **OPEN ENDED** Construct a right triangle and measure one of the acute angles. Find the measure of the second acute angle using calculation and explain your method. Confirm your result using a protractor.

**49.** **CHALLENGE** Find the values of $y$ and $z$ in the figure at the right.

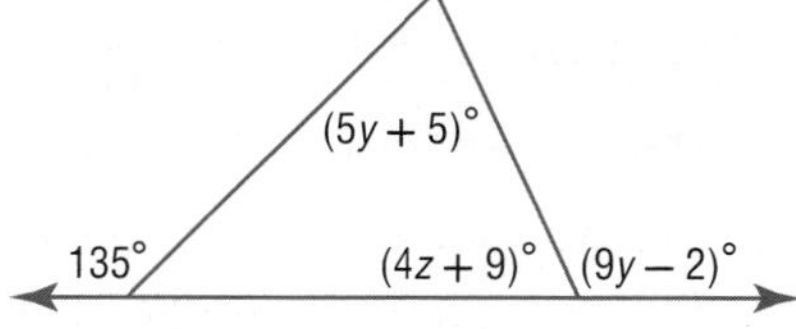


**50.** **REASONING** If an exterior angle adjacent to $\angle A$ is acute, is $\triangle ABC$ acute, right, obtuse, or can its classification not be determined? Explain your reasoning.

**51.** **WRITING IN MATH** Explain why a triangle cannot have an obtuse, acute, and a right exterior angle.

## Standardized Test Practice

**52. PROBABILITY** Mr. Glover owns a video store and wants to survey his customers to find what type of movies he should buy. Which of the following options would be the best way for Mr. Glover to get accurate survey results?

A surveying customers who come in from 9 P.M. until 10 P.M.

B surveying customers who come in on the weekend

C surveying the male customers

D surveying at different times of the week and day

**53. SHORT RESPONSE** Two angles of a triangle have measures of 35° and 80°. Find the values of the exterior angle measures of the triangle.

**54. ALGEBRA** Which equation is equivalent to $7x - 3(2 - 5x) = 8x$?

F $2x - 6 = 8$

G $22x - 6 = 8x$

H $-8x - 6 = 8x$

J $22x + 6 = 8x$

**55. SAT/ACT** Joey has 4 more video games than Solana and half as many as Melissa. If together they have 24 video games, how many does Melissa have?

A 7

B 9

C 12

D 13

E 14

## Spiral Review

**Simplify each expression.** (Lesson 1-2)

**56.** $-\frac{1}{2}n^3p^2(5np^3 - 3n^2p^2 + 8n)$

**57.** $6j^2(-3j + 3k^2) - 2k^2(2j + 10j^2)$

**Solve each equation. Check your solution.** (Lesson 1-6)

**58.** $h^2 + 3h - 4 = 0$

**59.** $a^2 + 9a + 18 = 0$

**60.** $x^2 - x - 6 = 0$

**61.** $y^2 + 2y - 15 = 0$

**Find the value of the variables in each figure. Explain your reasoning.** (Lesson 5-5)

**62.**

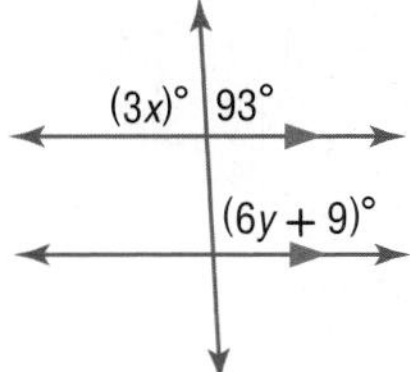


**63.**

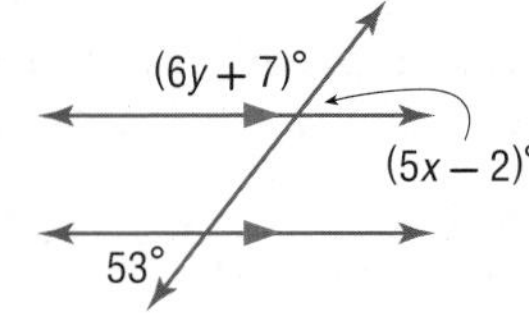


## Skills Review

**State the property that justifies each statement.**

**64.** If $\frac{x}{2} = 7$, then $x = 14$.

**65.** If $x = 5$ and $b = 5$, then $x = b$.

**66.** If $XY - AB = WZ - AB$, then $XY = WZ$.

**67.** If $m\angle A = m\angle B$ and $m\angle B = m\angle C$, $m\angle A = m\angle C$.

**68.** If $m\angle 1 + m\angle 2 = 90$ and $m\angle 2 = m\angle 3$, then $m\angle 1 + m\angle 3 = 90$.

LESSON

# 6-2 Congruent Triangles

| Then | Now | Why? |
|---|---|---|
| • You identified and used congruent angles. | • **1** Name and use corresponding parts of congruent polygons.<br>**2** Prove triangles congruent using the definition of congruence. | • As an antitheft device, many manufacturers make car stereos with removable faceplates. The shape and size of the faceplate and of the space where it fits must be exactly the same for the faceplate to properly attach to the car's dashboard. |

**NewVocabulary**
congruent
congruent polygons
corresponding parts

**Common Core State Standards**

**Content Standards**

**G.CO.7** Use the definition of congruence in terms of rigid motions to show that two triangles are congruent if and only if corresponding pairs of sides and corresponding pairs of angles are congruent.

**G.SRT.5** Use congruence and similarity criteria for triangles to solve problems and to prove relationships in geometric figures.

**Mathematical Practices**

6 Attend to precision.

3 Construct viable arguments and critique the reasoning of others.

## 1 Congruence and Corresponding Parts

If two geometric figures have exactly the same shape and size, they are **congruent**.

| Congruent | Not Congruent |
|---|---|
| 1   | 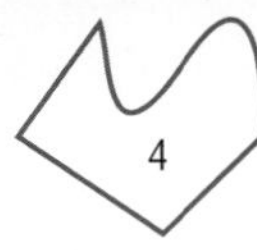  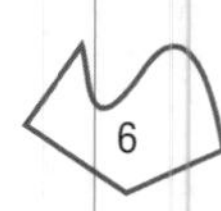 |
| While positioned differently, Figures 1, 2, and 3 are exactly the same shape and size. | Figures 4 and 5 are exactly the same shape but not the same size. Figures 5 and 6 are the same size but not exactly the same shape. |

In two **congruent polygons**, all of the parts of one polygon are congruent to the **corresponding parts** or matching parts of the other polygon. These corresponding parts include *corresponding angles* and *corresponding sides*.

### KeyConcept Definition of Congruent Polygons

**Words** Two polygons are congruent if and only if their corresponding parts are congruent.

**Example**

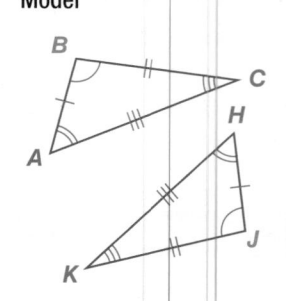

Corresponding Angles

$\angle A \cong \angle H$ $\quad \angle B \cong \angle J$ $\quad \angle C \cong \angle K$

Corresponding Sides

$\overline{AB} \cong \overline{HJ}$ $\quad \overline{BC} \cong \overline{JK}$ $\quad \overline{AC} \cong \overline{HK}$

Congruence Statement

$\triangle ABC \cong \triangle HJK$

**Model**

Other congruence statements for the triangles above exist. Valid congruence statements for congruent polygons list corresponding vertices in the same order.

| Valid Statement | Not a Valid Statement |
|---|---|
| $\triangle BCA \cong \triangle JKH$ | $\triangle ABC \cong \triangle HKJ$ |

Car Culture/Collection Mix: Subjects/Getty Images

## Math HistoryLink

**Johann Carl Friedrich Gauss (1777–1855)** Gauss developed the congruence symbol to show that two sides of an equation were the same even if they weren't equal. He made many advances in math and physics, including a proof of the fundamental theorem of algebra.

**Source:** The Granger Collection, New York

PT

## Example 1 Identify Corresponding Congruent Parts

**Show that the polygons are congruent by identifying all the congruent corresponding parts. Then write a congruence statement.**

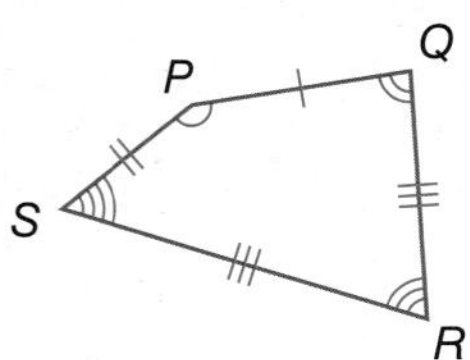


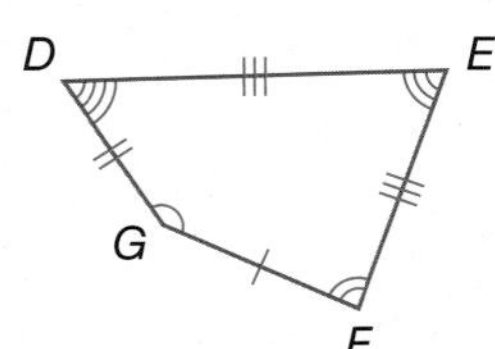


**Angles:** $\angle P \cong \angle G$, $\angle Q \cong \angle F$,
$\angle R \cong \angle E$, $\angle S \cong \angle D$

**Sides:** $\overline{PQ} \cong \overline{GF}$, $\overline{QR} \cong \overline{FE}$,
$\overline{RS} \cong \overline{ED}$, $\overline{SP} \cong \overline{DG}$

All corresponding parts of the two polygons are congruent. Therefore, polygon $PQRS \cong$ polygon $GFED$.

### GuidedPractice

**1A.**

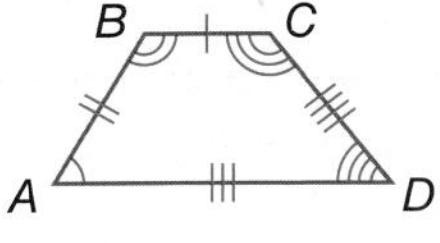


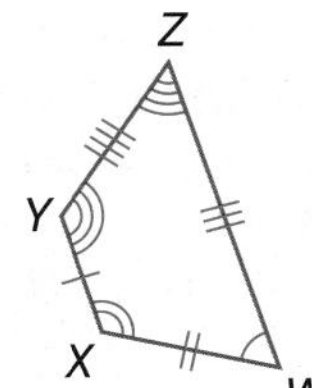


**1B.**

J, K, L; 50, 52°, 39, 50°, 78°, 40

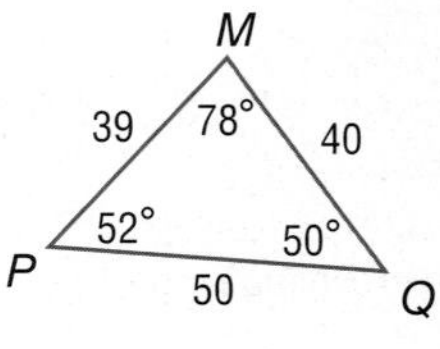


The phrase "if and only if" in the congruent polygon definition means that both the conditional and its converse are true. So, if two polygons are congruent, then their corresponding parts are congruent. For triangles, we say *Corresponding parts of congruent triangles are congruent*, or CPCTC.

## Example 2 Use Corresponding Parts of Congruent Triangles

**In the diagram, $\triangle ABC \cong \triangle DFE$. Find the values of $x$ and $y$.**

B: 99°, A: 45°, C: 36°; AB = 32.1, BC = 38.4, AC = 54

F: $(8y - 5)°$, FE = $2y + x$; D, E

| | |
|---|---|
| $\angle F \cong \angle B$ | CPCTC |
| $m\angle F = m\angle B$ | Definition of congruence |
| $8y - 5 = 99$ | Substitution |
| $8y = 104$ | Add 5 to each side. |
| $y = 13$ | Divide each side by 8. |
| $\overline{FE} \cong \overline{BC}$ | CPCTC |
| $FE = BC$ | Definition of congruence |
| $2y + x = 38.4$ | Substitution |
| $2(13) + x = 38.4$ | Substitution |
| $26 + x = 38.4$ | Simplify. |
| $x = 12.4$ | Subtract 26 from each side. |

### StudyTip

**Using a Congruence Statement** You can use a congruence statement to help you correctly identify corresponding sides.

$\triangle ABC \cong \triangle DFE$

$\overline{BC} \cong \overline{FE}$

### GuidedPractice

**2.** In the diagram, $\triangle RSV \cong \triangle TVS$. Find the values of $x$ and $y$.

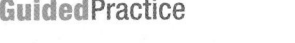

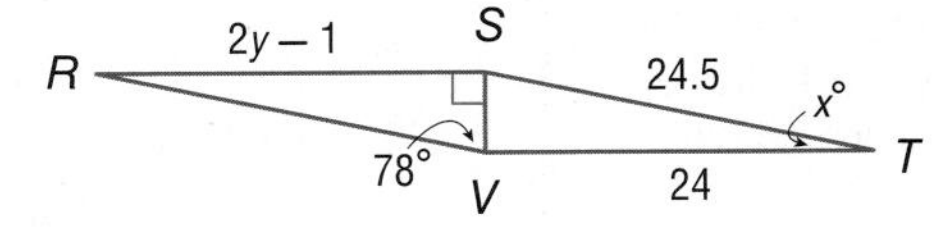


akg-images

## 2 Prove Triangles Congruent

The Triangle Angle-Sum Theorem you learned in Lesson 6-2 leads to another theorem about the angles in two triangles.

### Theorem 6.3 Third Angles Theorem

**Words:** If two angles of one triangle are congruent to two angles of a second triangle, then the third angles of the triangles are congruent.

**Example:** If $\angle C \cong \angle K$ and $\angle B \cong \angle J$, then $\angle A \cong \angle L$.

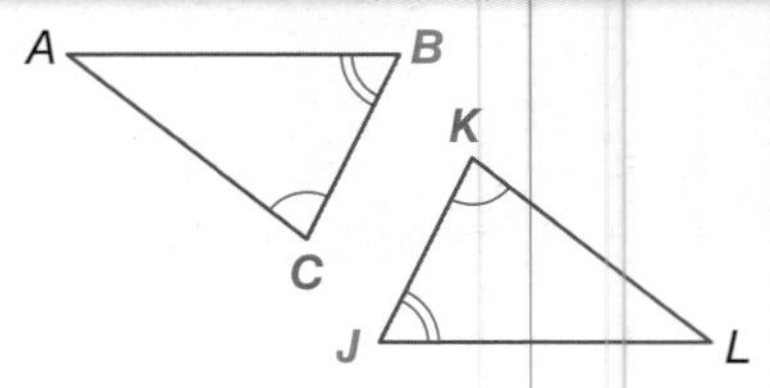


You will prove this theorem in Exercise 21.

### Real-World Example 3 Use the Third Angles Theorem

PT

**PARTY PLANNING The planners of the Senior Banquet decide to fold the dinner napkins using the Triangle Pocket Fold so that they can place a small gift in the pocket. If $\angle NPQ \cong \angle RST$, and $m\angle NPQ = 40$, find $m\angle SRT$.**

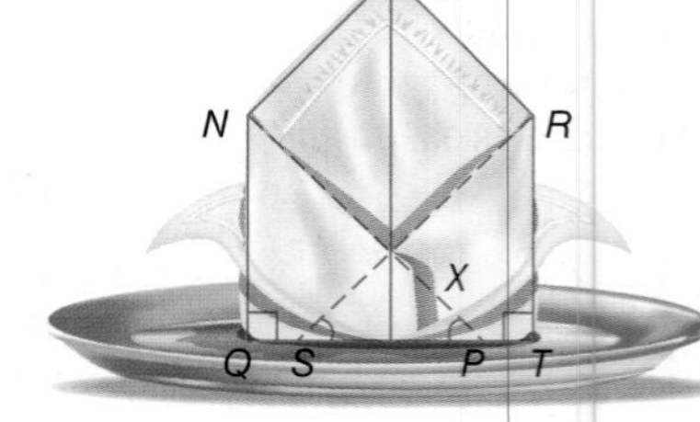


$\angle NPQ \cong \angle RST$, and since all right angles are congruent, $\angle NQP \cong \angle RTS$. So by the Third Angles Theorem, $\angle QNP \cong \angle SRT$. By the definition of congruence, $m\angle QNP = m\angle TRS$.

| | |
|---|---|
| $m\angle QNP + m\angle NPQ = 90$ | The acute angles of a right triangle are complementary. |
| $m\angle QNP + 40 = 90$ | Substitution |
| $m\angle QNP = 50$ | Subtract 40 from each side. |

By substitution, $m\angle SRT = m\angle QNP$ or 50.

#### Real-WorldLink

Using some basic skills with napkin folding can add an elegant touch to any party. Many of the folds use triangles.

#### GuidedPractice

**3.** In the diagram above, if $\angle WNX \cong \angle WRX$, $\overline{WX}$ bisects $\angle NXR$, $m\angle WNX = 88$, and $m\angle NXW = 49$, find $m\angle NWR$. Explain your reasoning.

### Example 4 Prove That Two Triangles are Congruent

**Write a two-column proof.**

**Given:** $\overline{DE} \cong \overline{GE}$, $\overline{DF} \cong \overline{GF}$, $\angle D \cong \angle G$, $\angle DFE \cong \angle GFE$

**Prove:** $\triangle DEF \cong \triangle GEF$

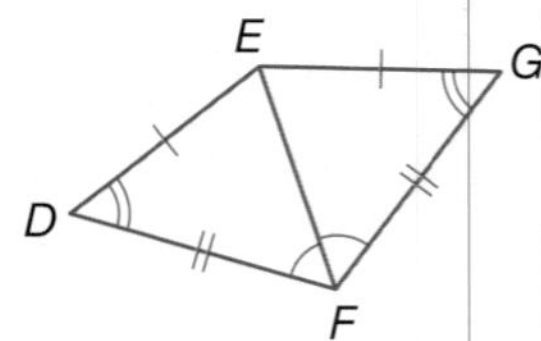


**Proof:**

| Statements | Reasons |
|---|---|
| **1.** $\overline{DE} \cong \overline{GE}$, $\overline{DF} \cong \overline{GF}$ | **1.** Given |
| **2.** $\overline{EF} \cong \overline{EF}$ | **2.** Reflexive Property of Congruence |
| **3.** $\angle D \cong \angle G$, $\angle DFE \cong \angle GFE$ | **3.** Given |
| **4.** $\angle DEF \cong \angle GEF$ | **4.** Third Angles Theorem |
| **5.** $\triangle DEF \cong \triangle GEF$ | **5.** Definition of Congruent Polygons |

#### StudyTip

**Reflexive Property** When two triangles share a common side, use the Reflexive Property of Congruence to establish that the common side is congruent to itself.

## GuidedPractice

**4.** Write a two column proof.

**Given:** $\angle J \cong \angle P$, $\overline{JK} \cong \overline{PM}$, $\overline{JL} \cong \overline{PL}$, and $L$ bisects $\overline{KM}$.

**Prove:** $\triangle JLK \cong \triangle PLM$

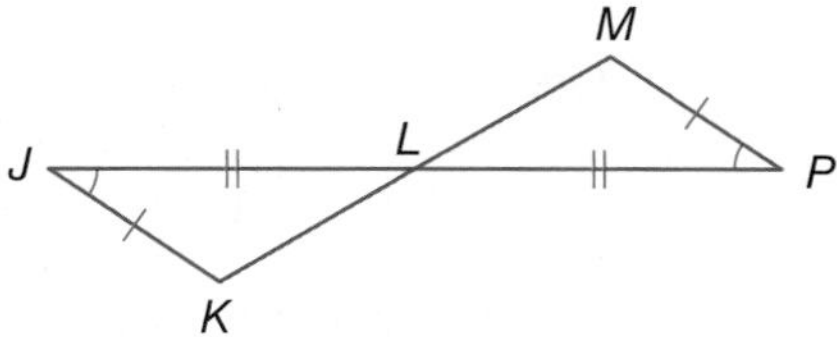


Like congruence of segments and angles, congruence of triangles is reflexive, symmetric, and transitive.

### Theorem 6.4 Properties of Triangle Conguruence

**Reflexive Property of Triangle Congruence**

$\triangle ABC \cong \triangle ABC$

**Symmetric Property of Triangle Congruence**

If $\triangle ABC \cong \triangle EFG$, then $\triangle EFG \cong \triangle ABC$.

**Transitive Property of Triangle Congruence**

If $\triangle ABC \cong \triangle EFG$ and $\triangle EFG \cong \triangle JKL$, then $\triangle ABC \cong \triangle JKL$.

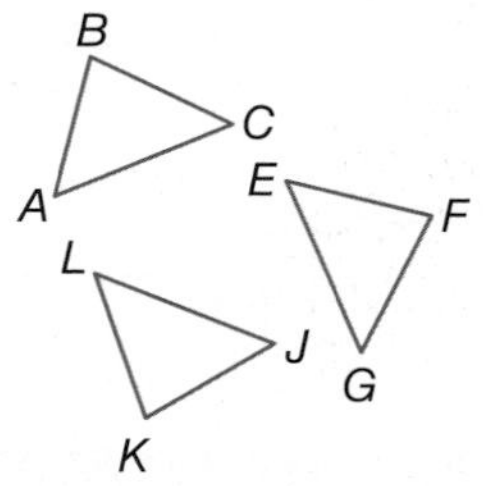


You will prove the reflexive, symmetric, and transitive parts of Theorem 6.4 in Exercises 27, 22, and 26, respectively.

## Check Your Understanding

**Example 1** **Show that polygons are congruent by identifying all congruent corresponding parts. Then write a congruence statement.**

**1.**

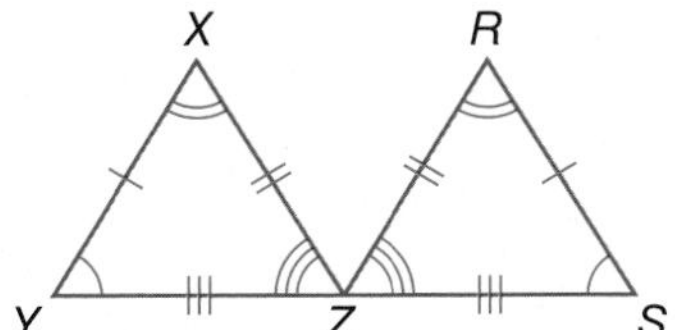


**2.**

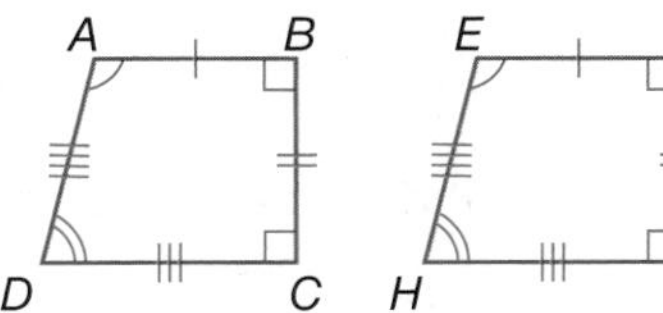


**3. TOOLS** Sareeta is changing the tire on her bike and the nut securing the tire looks like the one shown. Which of the sockets below should she use with her wrench to remove the tire? Explain your reasoning.

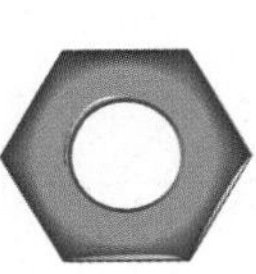

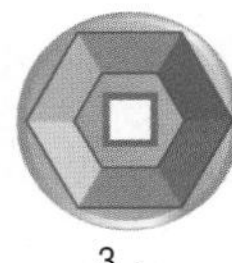

$\frac{3}{8}$ in.

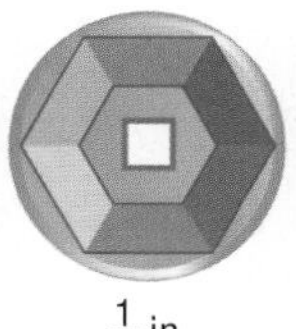

$\frac{1}{2}$ in.

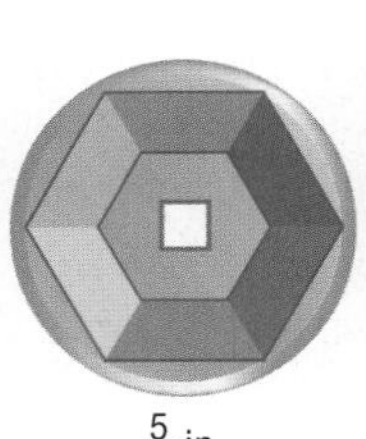

$\frac{5}{8}$ in.

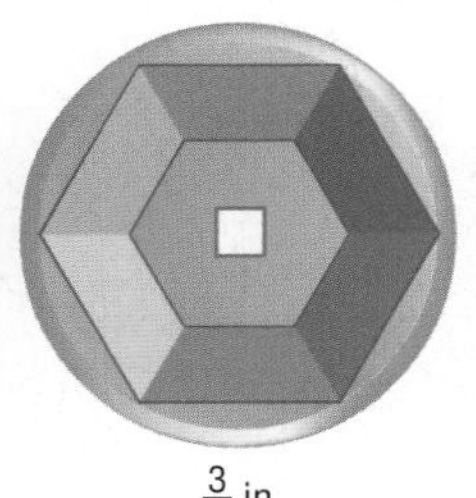

$\frac{3}{4}$ in.

**Example 2** **In the figure, $\triangle LMN \cong \triangle QRS$.**

4. Find $x$.

5. Find $y$.

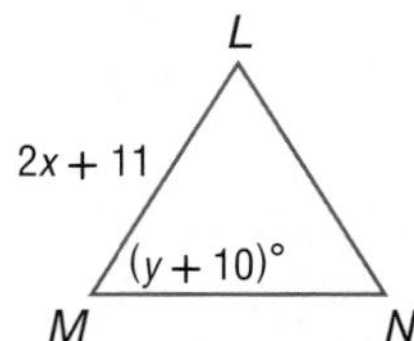


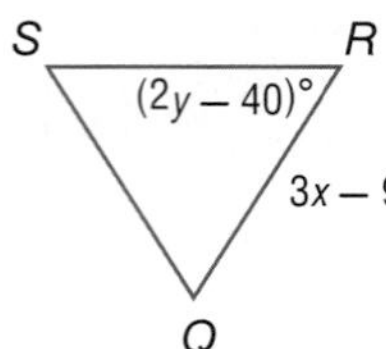


**Example 3** **CCSS REGULARITY** **Find $x$. Explain your reasoning.**

6.

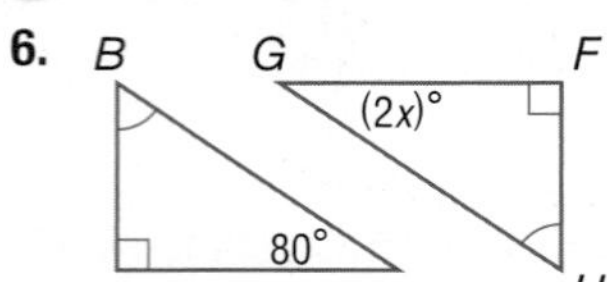


7.

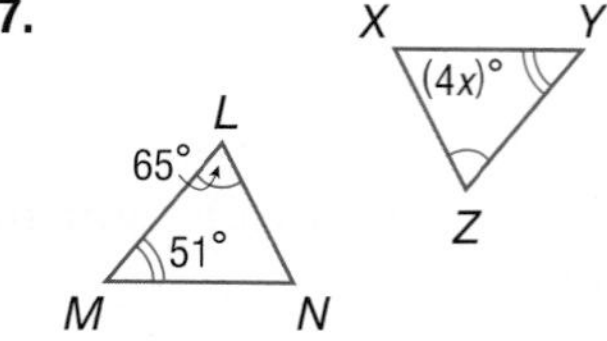


**Example 4** 8. **PROOF** Write a paragraph proof.

**Given:** $\angle WXZ \cong \angle YXZ$, $\angle XZW \cong \angle XZY$, $\overline{WX} \cong \overline{YX}$, $\overline{WZ} \cong \overline{YZ}$

**Prove:** $\triangle WXZ \cong \triangle YXZ$

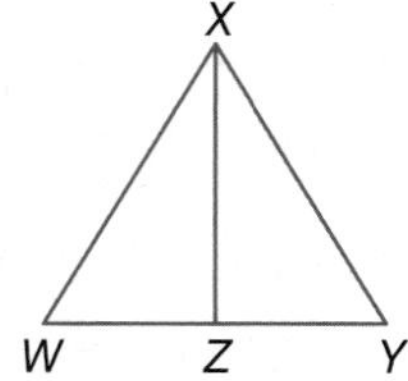


## Practice and Problem Solving

**Example 1** **Show that polygons are congruent by identifying all congruent corresponding parts. Then write a congruence statement.**

9.

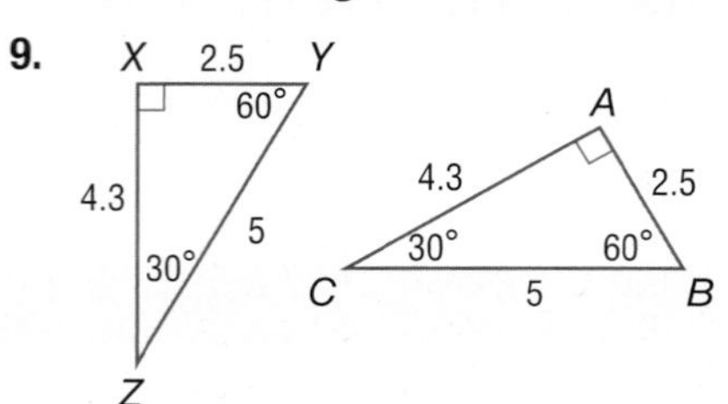


10.

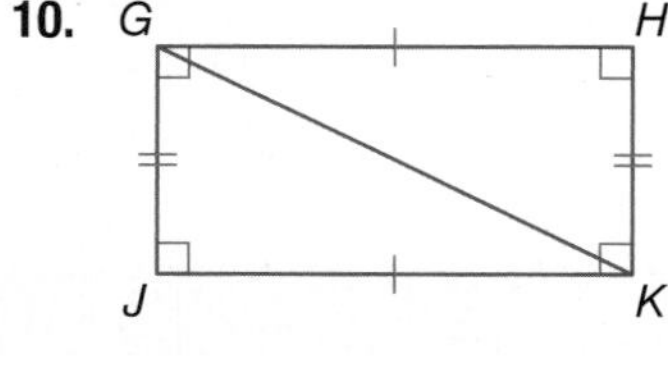


11.

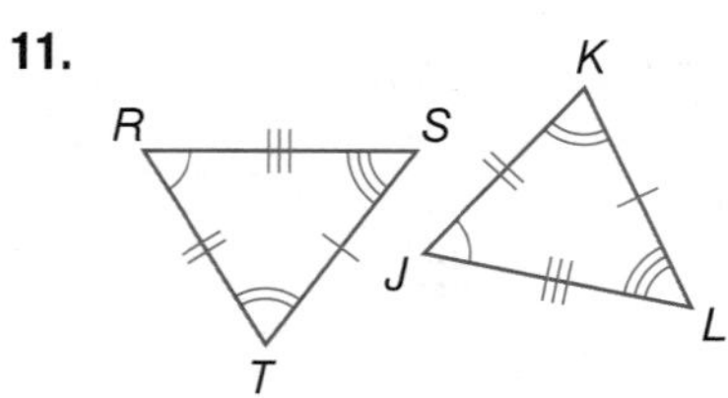


12.

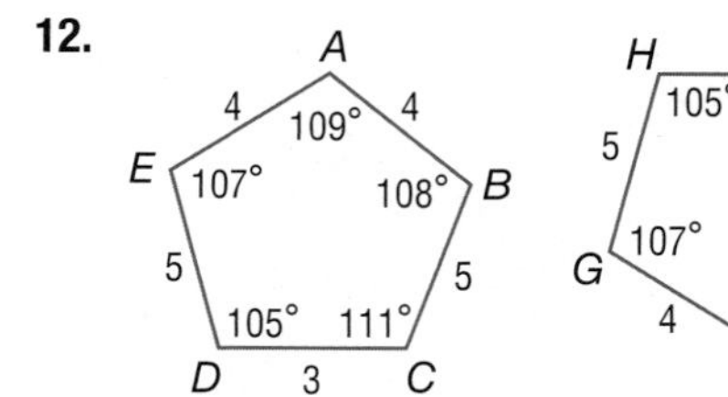

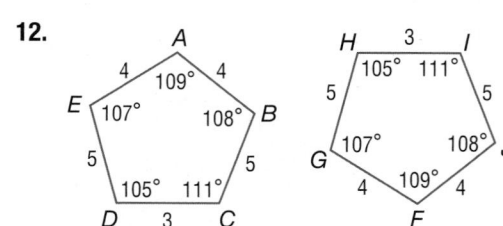


**Example 2** **Polygon $BCDE \cong$ polygon $RSTU$. Find each value.**

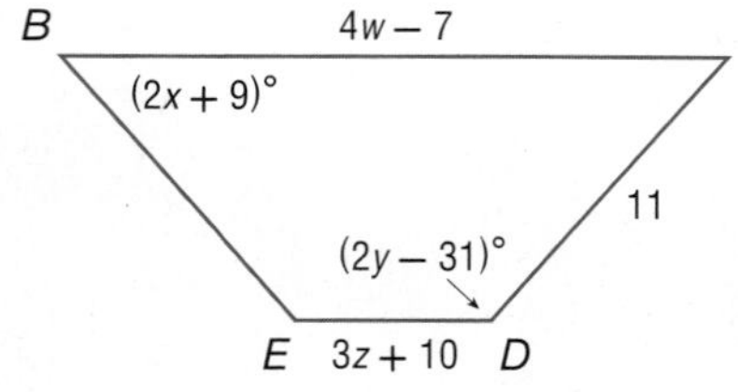


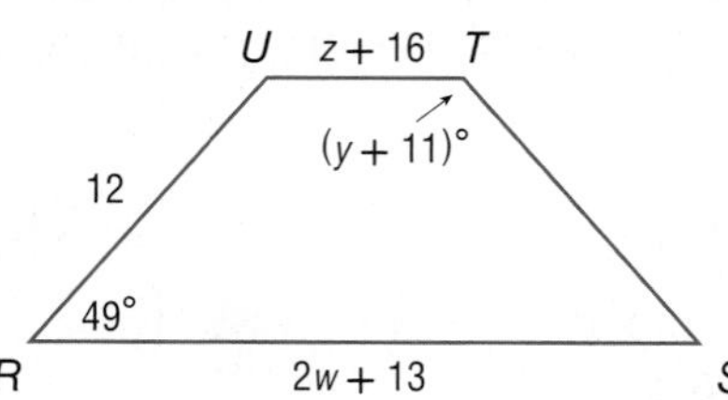


13. $x$

14. $y$

15. $z$

16. $w$

**17. SAILING** To ensure that sailboat races are fair, the boats and their sails are required to be the same size and shape.

**a.** Write a congruence statement relating the triangles in the photo.

**b.** Name six pairs of congruent segments.

**c.** Name six pairs of congruent angles.

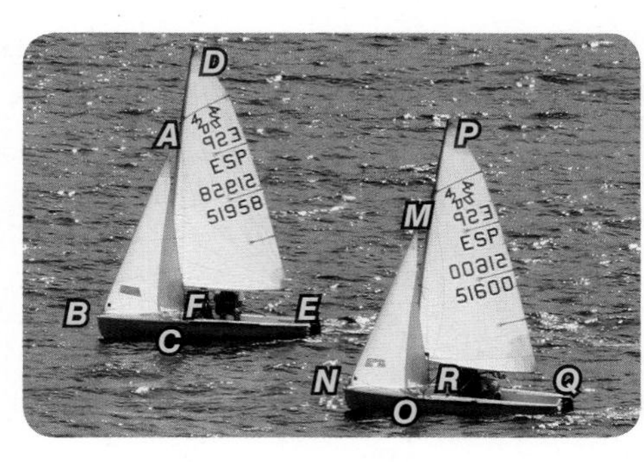


**Example 3** **Find $x$ and $y$.**

**18.**

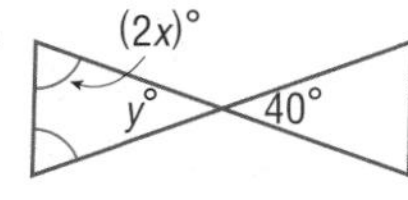


**19.**

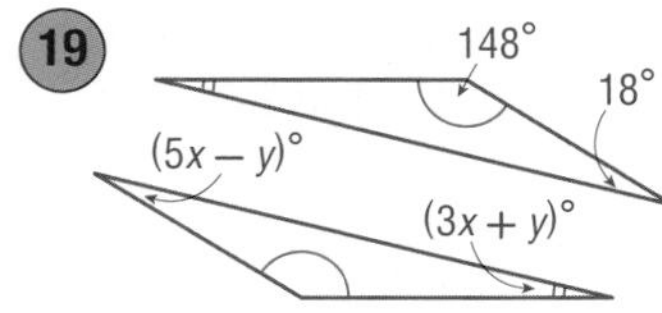


**20.**

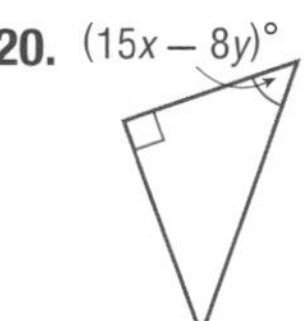


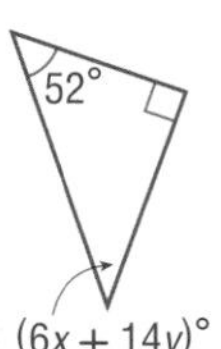


**Example 4** **21. PROOF** Write a two-column proof of Theorem 6.3.

**22. PROOF** Put the statements used to prove the statement below in the correct order. Provide the reasons for each statement.

*Congruence of triangles is symmetric. (Theorem 6.4)*

**Given:** $\triangle RST \cong \triangle XYZ$

**Prove:** $\triangle XYZ \cong \triangle RST$

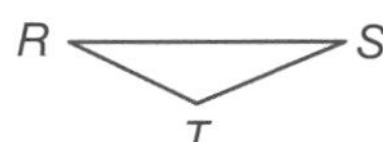

**Proof:**

| $\angle X \cong \angle R$, $\angle Y \cong \angle S$, $\angle Z \cong \angle T$, $\overline{XY} \cong \overline{RS}$, $\overline{YZ} \cong \overline{ST}$, $\overline{XZ} \cong \overline{RT}$ | $\angle R \cong \angle X$, $\angle S \cong \angle Y$, $\angle T \cong \angle Z$, $\overline{RS} \cong \overline{XY}$, $\overline{ST} \cong \overline{YZ}$, $\overline{RT} \cong \overline{XZ}$ | $\triangle RST \cong \triangle XYZ$ | $\triangle XYZ \cong \triangle RST$ |
|---|---|---|---|
| ? | ? | ? | ? |

**CCSS ARGUMENTS** **Write a two-column proof.**

**23. Given:** $\overline{BD}$ bisects $\angle B$.
$\overline{BD} \perp \overline{AC}$

**Prove:** $\angle A \cong \angle C$

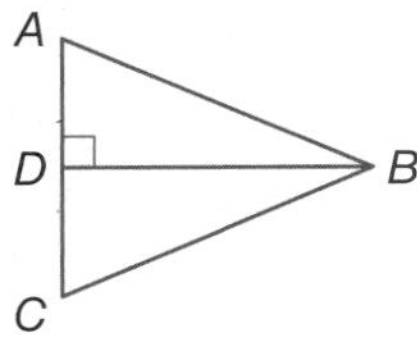


**24. Given:** $\angle P \cong \angle T$, $\angle S \cong \angle Q$
$\overline{TR} \cong \overline{PR}$, $\overline{RP} \cong \overline{RQ}$,
$\overline{RT} \cong \overline{RS}$
$\overline{PQ} \cong \overline{TS}$

**Prove:** $\triangle PRQ \cong \triangle TRS$

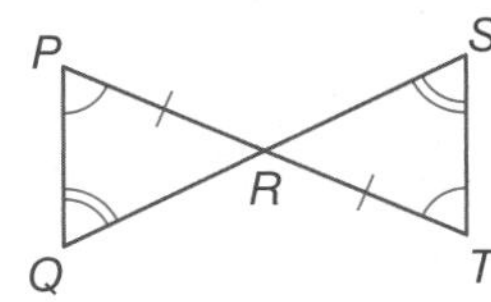


**25. SCRAPBOOKING** Lanie is using a flower-shaped corner decoration punch for a scrapbook she is working on. If she punches the corners of two pages as shown, what property guarantees that the punched designs are congruent? Explain.

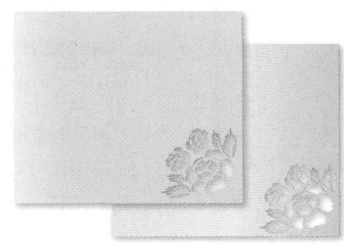

**PROOF** **Write the specified type of proof of the indicated part of Theorem 6.4.**

**26.** Congruence of triangles is transitive. (paragraph proof)

**27.** Congruence of triangles is reflexive. (flow proof)

**ALGEBRA Draw and label a figure to represent the congruent triangles. Then find $x$ and $y$.**

**28.** $\triangle ABC \cong \triangle DEF$, $AB = 7$, $BC = 9$, $AC = 11 + x$, $DF = 3x - 13$, and $DE = 2y - 5$

**29.** $\triangle LMN \cong \triangle RST$, $m\angle L = 49$, $m\angle M = 10y$, $m\angle S = 70$, and $m\angle T = 4x + 9$

**30.** $\triangle JKL \cong \triangle MNP$, $JK = 12$, $LJ = 5$, $PM = 2x - 3$, $m\angle L = 67$, $m\angle K = y + 4$ and $m\angle N = 2y - 15$

**31** **PENNANTS** Scott is in charge of roping off an area of 100 square feet for the band to use during a pep rally. He is using a string of pennants that are congruent isosceles triangles.

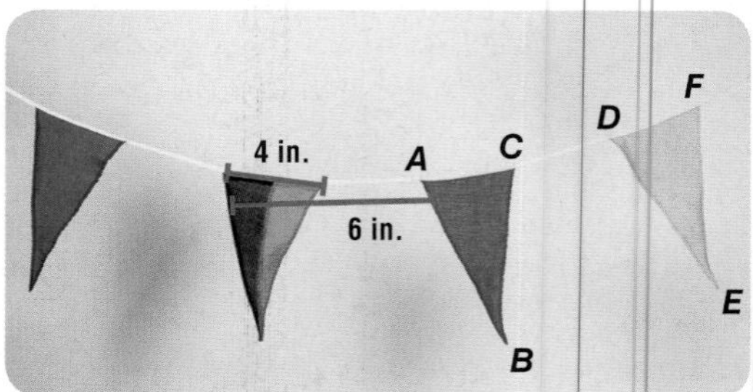


**a.** List seven pairs of congruent segments in the photo.

**b.** If the area he ropes off is a square, how long will the pennant string need to be?

**c.** How many pennants will be on the string?

**32.** **CCSS SENSE-MAKING** In the photo of New York City's Chrysler Building at the right, $\overline{TS} \cong \overline{ZY}$, $\overline{XY} \cong \overline{RS}$, $\overline{TR} \cong \overline{ZX}$, $\angle X \cong \angle R$, $\angle T \cong \angle Z$, $\angle Y \cong \angle S$, and $\triangle HGF \cong \triangle LKJ$.

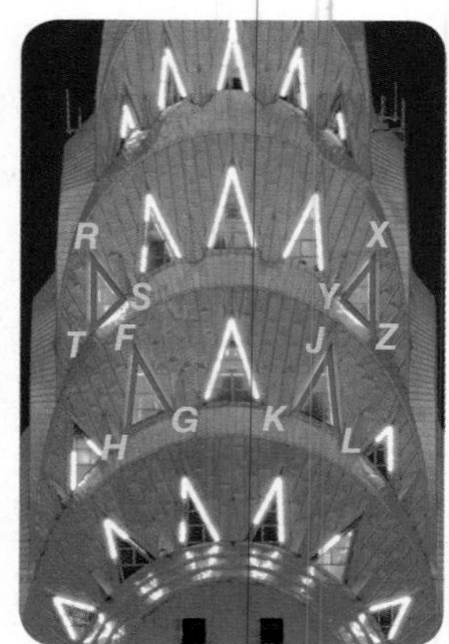


**a.** Which triangle, if any, is congruent to $\triangle YXZ$? Explain your reasoning.

**b.** Which side(s) are congruent to $\overline{JL}$? Explain your reasoning.

**c.** Which angle(s) are congruent to $\angle G$? Explain your reasoning.

**33.** **MULTIPLE REPRESENTATIONS** In this problem, you will explore the statement *The areas of congruent triangles are equal.*

**a. Verbal** Write a conditional statement to represent the relationship between the areas of a pair of congruent triangles.

**b. Verbal** Write the converse of your conditional statement. Is the converse *true* or *false*? Explain your reasoning.

**c. Geometric** If possible, draw two equilateral triangles that have the same area but are not congruent. If not possible, explain why not.

**d. Geometric** If possible, draw two rectangles that have the same area but are not congruent. If not possible, explain why not.

**e. Geometric** If possible, draw two squares that have the same area but are not congruent. If not possible, explain why not.

**f. Verbal** For which polygons will the following conditional and its converse both be true? Explain your reasoning.

*If a pair of ________ are congruent, then they have the same area.*

(t)Karl Shone/Dorling Kindersley/Getty Images, (b)Joseph Pobereskin/Stone/Getty Images

34. **PATTERNS** The pattern shown is created using regular polygons.

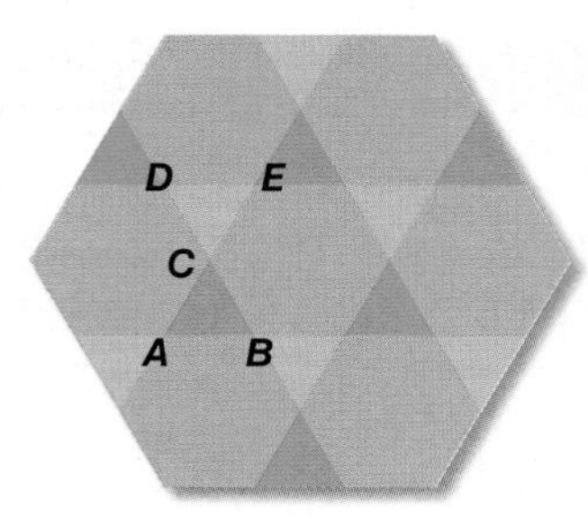


a. What two polygons are used to create the pattern?

b. Name a pair of congruent triangles.

c. Name a pair of corresponding angles.

d. If $CB = 2$ inches, what is $AE$? Explain.

e. What is the measure of $\angle D$? Explain.

35. **FITNESS** A fitness instructor is starting a new aerobics class using fitness hoops. She wants to confirm that all of the hoops are the same size. What measure(s) can she use to prove that all of the hoops are congruent? Explain your reasoning.

## H.O.T. Problems Use Higher-Order Thinking Skills

36. **WRITING IN MATH** Explain why the order of the vertices is important when naming congruent triangles. Give an example to support your answer.

37. **ERROR ANALYSIS** Jasmine and Will are evaluating the congruent figures below. Jasmine says that $\triangle CAB \cong \triangle ZYX$ and Will says that $\triangle ABC \cong \triangle YXZ$. Is either of them correct? Explain.

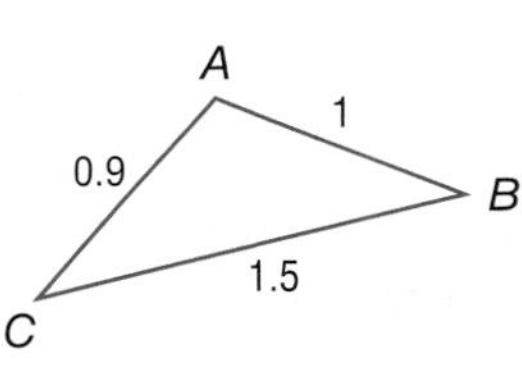


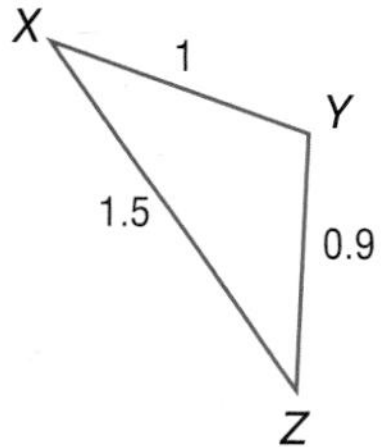


38. **WRITE A QUESTION** A classmate is using the Third Angles Theorem to show that if 2 corresponding pairs of the angles of two triangles are congruent, then the third pair is also congruent. Write a question to help him decide if he can use the same strategy for quadrilaterals.

39. **CHALLENGE** Find $x$ and $y$ if $\triangle PQS \cong \triangle RQS$.

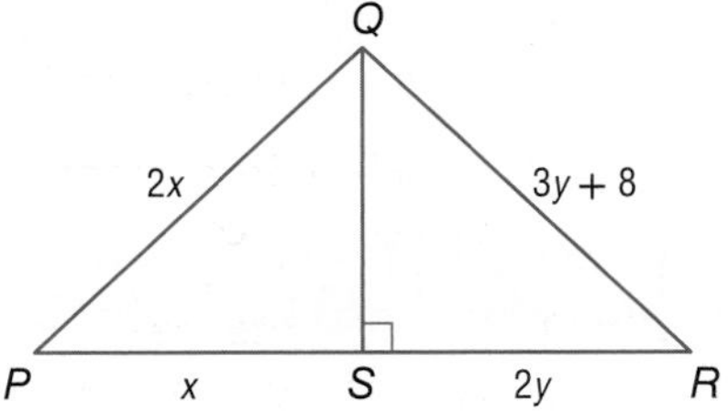


**CCSS ARGUMENTS Determine whether each statement is *true* or *false*. If false, give a counterexample. If true, explain your reasoning.**

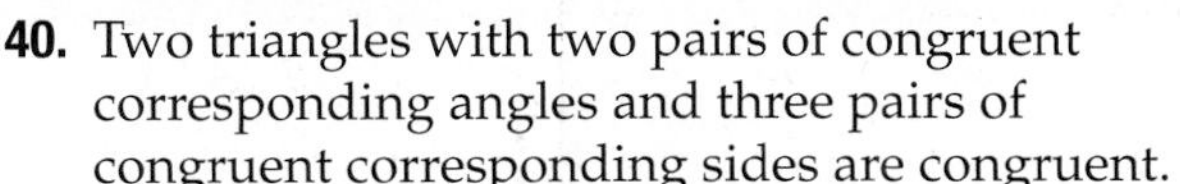

40. Two triangles with two pairs of congruent corresponding angles and three pairs of congruent corresponding sides are congruent.

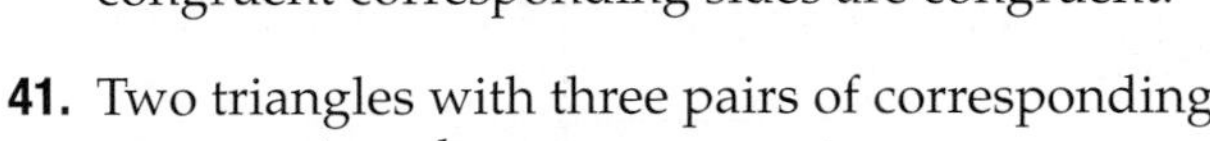

41. Two triangles with three pairs of corresponding congruent angles are congruent.

42. **CHALLENGE** Write a paragraph proof to prove polygon $ABED \cong$ polygon $FEBC$.

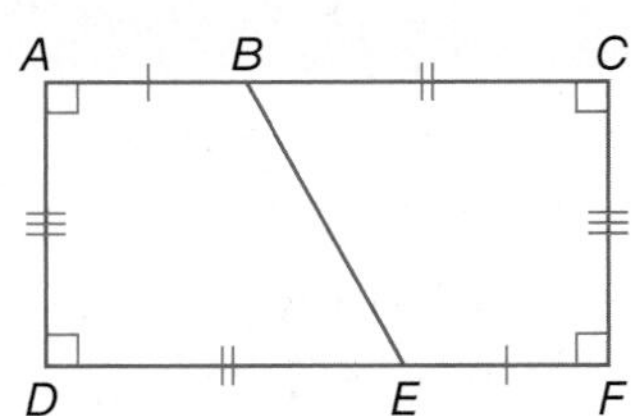


43. **WRITING IN MATH** Determine whether the following statement is *always*, *sometimes*, or *never* true. Explain your reasoning.

*Equilateral triangles are congruent.*

## Standardized Test Practice

**44.** Barrington cut four congruent triangles off the corners of a rectangle to make an octagon as shown below. What is the area of the octagon?

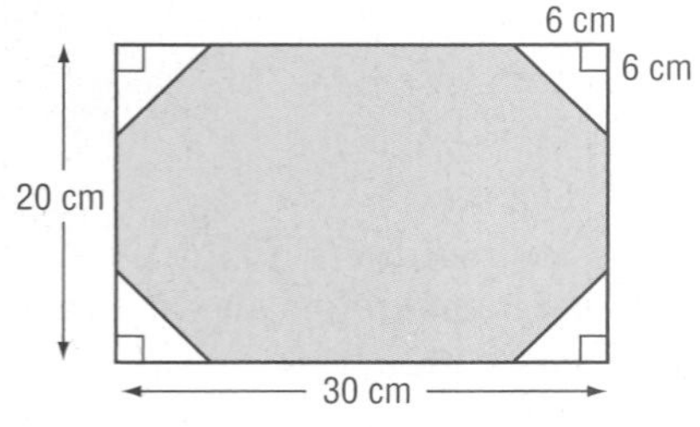


A 456 cm$^2$    C 552 cm$^2$
B 528 cm$^2$    D 564 cm$^2$

**45. GRIDDED RESPONSE** Triangle $ABC$ is congruent to $\triangle HIJ$. The vertices of $\triangle ABC$ are $A(-1, 2)$, $B(0, 3)$ and $C(2, -2)$. What is the measure of side $\overline{HJ}$?

**46. ALGEBRA** Which is a factor of $x^2 + 19x - 42$?

F $x + 14$    H $x - 2$
G $x + 2$    J $x - 14$

**47. SAT/ACT** Mitsu travels a certain distance at 30 miles per hour and returns the same route at 65 miles per hour. What is his average speed in miles per hour for the round trip?

A 32.5    D 47.5
B 35.0    E 55.3
C 41.0

## Spiral Review

**Find each measure in the triangle at the right.** (Lesson 6-1)

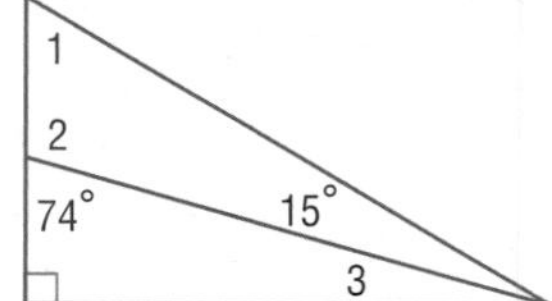


**48.** $m\angle 2$    **49.** $m\angle 1$    **50.** $m\angle 3$

**51. FENCES** Find the measures of $\angle 2$, $\angle 3$, and $\angle 4$ of the fence shown. (Lesson 5-5)

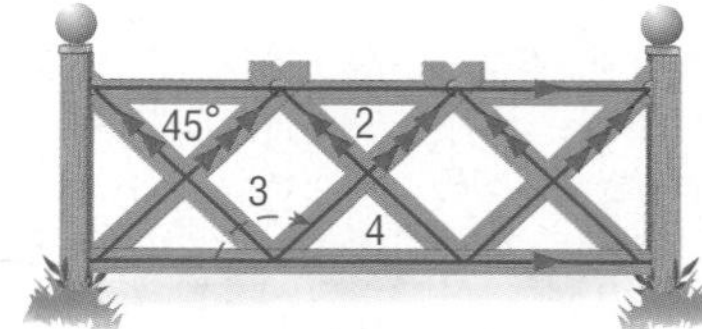


**Determine whether each statement is *always*, *sometimes*, or *never* true. Explain your reasoning.** (Lesson 5-1)

**52.** If planes $\mathcal{A}$ and $\mathcal{B}$ intersect, then their intersection is a line.

**53.** If point $A$ lies in plane $\mathcal{P}$, then $\overleftrightarrow{AB}$ lies in plane $\mathcal{P}$.

**54.** Lines $p$ and $q$ intersect in points $M$ and $N$.

## Skills Review

**55.** Copy and complete the proof.

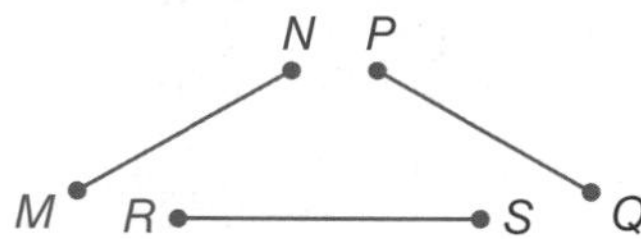


**Given:** $\overline{MN} \cong \overline{PQ}$, $\overline{PQ} \cong \overline{RS}$
**Prove:** $\overline{MN} \cong \overline{RS}$
**Proof:**

| Statements | Reasons |
|---|---|
| **a.** ___?___ | **a.** Given |
| **b.** $MN = PQ$, $PQ = RS$ | **b.** ___?___ |
| **c.** ___?___ | **c.** ___?___ |
| **d.** $\overline{MN} \cong \overline{RS}$ | **d.** Definition of congruent segments |

LESSON 6-3

# Proving Triangles Congruent–SSS, SAS

| Then | Now | Why? |
|---|---|---|
| You proved triangles congruent using the definition of congruence. | 1 Use the SSS Postulate to test for triangle congruence.<br>2 Use the SAS Postulate to test for triangle congruence. | An A-frame sandwich board is a convenient way to display information. Not only does it fold flat for easy storage, but with each sidearm locked into place, the frame is extremely sturdy. With the sidearms the same length and positioned the same distance from the top on either side, the open frame forms two congruent triangles. |

**NewVocabulary**
included angle

**Common Core State Standards**

**Content Standards**

G.CO.10 Prove theorems about triangles.

G.SRT.5 Use congruence and similarity criteria for triangles to solve problems and to prove relationships in geometric figures.

**Mathematical Practices**

3 Construct viable arguments and critique the reasoning of others.

1 Make sense of problems and persevere in solving them.

## 1 SSS Postulate

In Lesson 6-2, you proved that two triangles were congruent by showing that all six pairs of corresponding parts were congruent. It is possible to prove two triangles congruent using fewer pairs.

The sandwich board demonstrates that if two triangles have the same three side lengths, then they are congruent. This is expressed in the postulate below.

### Postulate 6.1 Side-Side-Side (SSS) Congruence

If three sides of one triangle are congruent to three sides of a second triangle, then the triangles are congruent.

**Example** If Side $\overline{AB} \cong \overline{DE}$,
Side $\overline{BC} \cong \overline{EF}$, and
Side $\overline{AC} \cong \overline{DF}$,
then $\triangle ABC \cong \triangle DEF$.

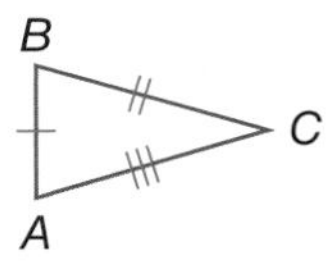


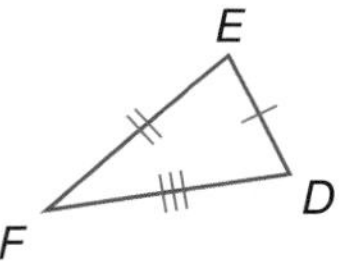

### Example 1 Use SSS to Prove Triangles Congruent

**Write a flow proof.**

**Given:** $\overline{GH} \cong \overline{KJ}$, $\overline{HL} \cong \overline{JL}$, and $L$ is the midpoint of $\overline{GK}$.

**Prove:** $\triangle GHL \cong \triangle KJL$

H J

G L K

**Flow Proof:**

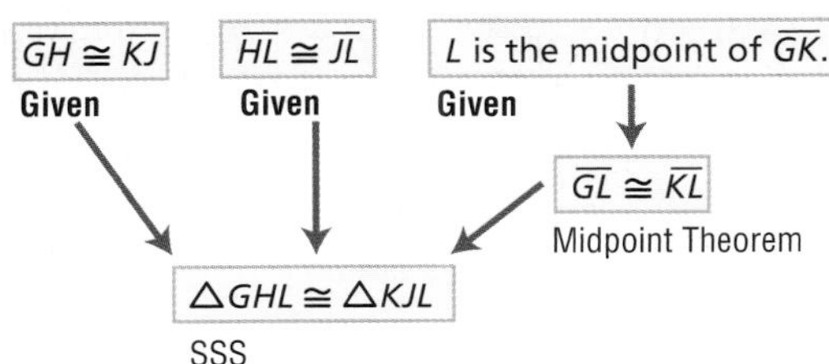


**GuidedPractice**

**1. Write a flow proof.**

**Given:** $\triangle QRS$ is isosceles with $\overline{QR} \cong \overline{SR}$. $\overline{RT}$ bisects $\overline{QS}$ at point $T$.

**Prove:** $\triangle QRT \cong \triangle SRT$

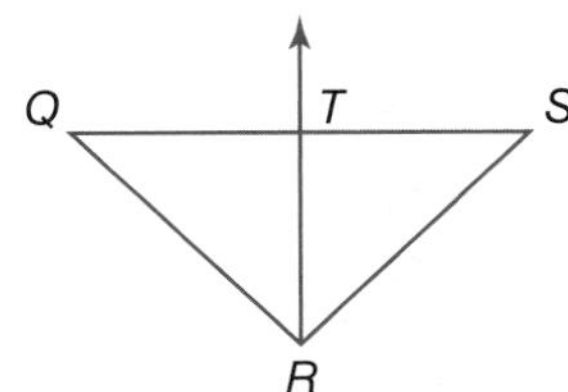

## Standardized Test Example 2 SSS on the Coordinate Plane

**EXTENDED RESPONSE Triangle $ABC$ has vertices $A(1, 1)$, $B(0, 3)$, and $C(2, 5)$. Triangle $EFG$ has vertices $E(1, -1)$, $F(2, -5)$, and $G(4, -4)$.**

**a. Graph both triangles on the same coordinate plane.**

**b. Use your graph to make a conjecture as to whether the triangles are congruent. Explain your reasoning.**

**c. Write a logical argument using coordinate geometry to support the conjecture you made in part b.**

### Read the Test Item

You are asked to do three things in this problem. In part **a**, you are to graph $\triangle ABC$ and $\triangle EFG$ on the same coordinate plane. In part **b**, you should make a conjecture that $\triangle ABC \cong \triangle EFG$ or $\triangle ABC \ncong \triangle EFG$ based on your graph. Finally, in part **c**, you are asked to prove your conjecture.

**Test-TakingTip**

**CCSS Tools** When you are solving problems using the coordinate plane, remember to use tools like the Distance, Midpoint, and Slope Formulas to solve problems and to check your solutions.

### Solve the Test Item

**a.**

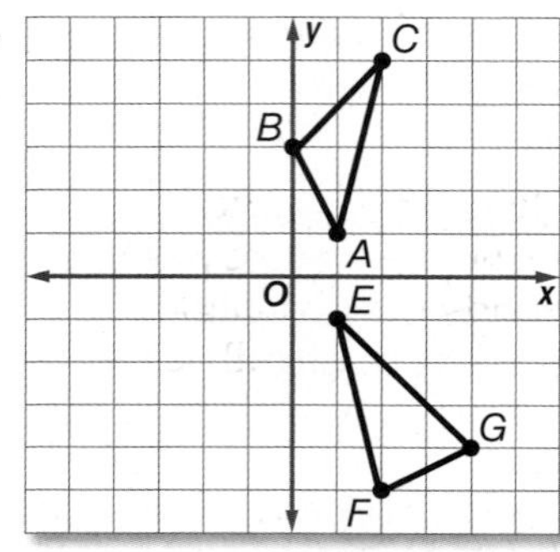


**b.** From the graph, it appears that the triangles do not have the same shape, so we can conjecture that they are not congruent.

**c.** Use the Distance Formula to show that not all corresponding sides have the same measure.

$AB = \sqrt{(0-1)^2 + (3-1)^2}$
$= \sqrt{1+4}$ or $\sqrt{5}$

$EF = \sqrt{(2-1)^2 + [-5-(-1)]^2}$
$= \sqrt{1+16}$ or $\sqrt{17}$

$BC = \sqrt{(2-0)^2 + (5-3)^2}$
$= \sqrt{4+4}$ or $\sqrt{8}$

$FG = \sqrt{(4-2)^2 + [-4-(-5)]^2}$
$= \sqrt{4+1}$ or $\sqrt{5}$

$AC = \sqrt{(2-1)^2 + (5-1)^2}$
$= \sqrt{1+16}$ or $\sqrt{17}$

$EG = \sqrt{(4-1)^2 + [-4-(-1)]^2}$
$= \sqrt{9+9}$ or $\sqrt{18}$

While $AB = FG$ and $AC = EF$, $BC \neq EG$. Since SSS congruence is not met, $\triangle ABC \ncong \triangle EFG$.

**ReadingMath**

Symbols $\triangle ABC \ncong \triangle EFG$ is read as *triangle ABC is not congruent to triangle EFG.*

### GuidedPractice

**2.** Triangle $JKL$ has vertices $J(2, 5)$, $K(1, 1)$, and $L(5, 2)$. Triangle $NPQ$ has vertices $N(-3, 0)$, $P(-7, 1)$, and $Q(-4, 4)$.

**a.** Graph both triangles on the same coordinate plane.

**b.** Use your graph to make a conjecture as to whether the triangles are congruent. Explain your reasoning.

**c.** Write a logical argument using coordinate geometry to support the conjecture you made in part **b**.

## Construction Congruent Triangles Using Sides

Draw a triangle and label it $\triangle ABC$. Then use the SSS Postulate to construct $\triangle XYZ \cong \triangle ABC$.

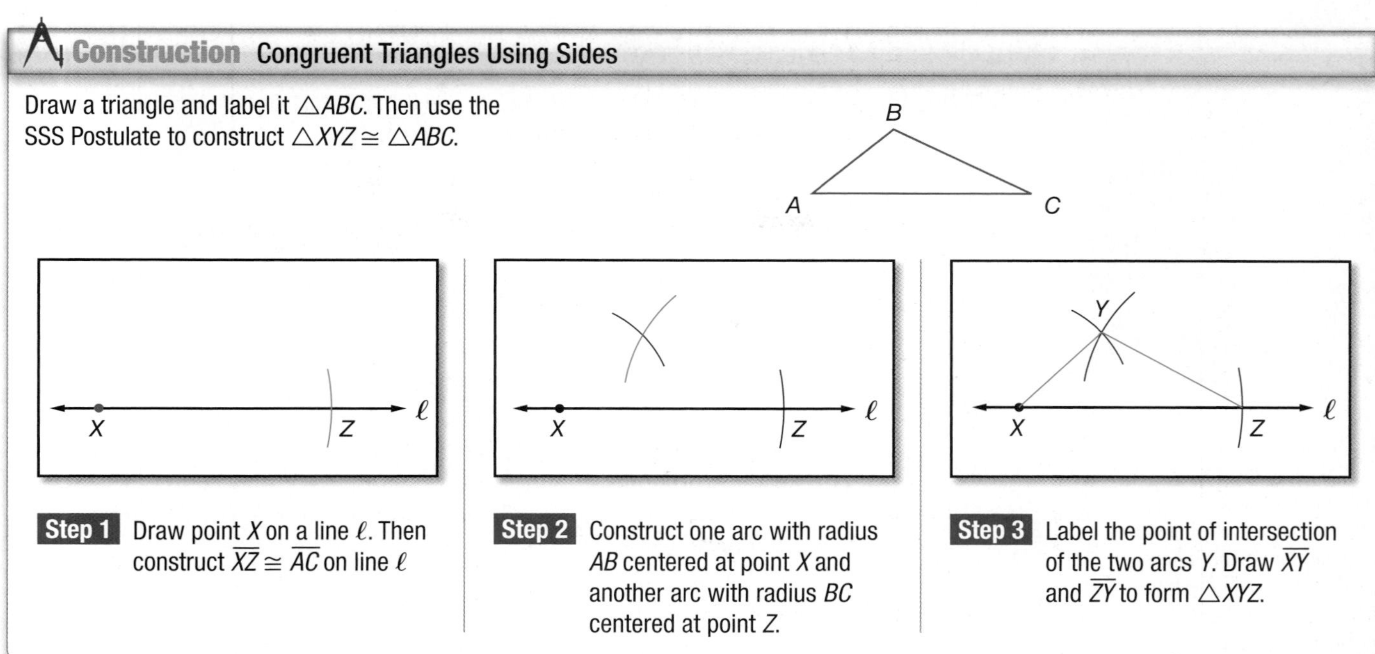


**Step 1** Draw point $X$ on a line $\ell$. Then construct $\overline{XZ} \cong \overline{AC}$ on line $\ell$

**Step 2** Construct one arc with radius $AB$ centered at point $X$ and another arc with radius $BC$ centered at point $Z$.

**Step 3** Label the point of intersection of the two arcs $Y$. Draw $\overline{XY}$ and $\overline{ZY}$ to form $\triangle XYZ$.

## 2 SAS Postulate

The angle formed by two adjacent sides of a polygon is called an **included angle**. Consider included angle $JKL$ formed by the hands on the first clock shown below. Any time the hands form an angle with the same measure, the distance between the ends of the hands $\overline{JL}$ and $\overline{PR}$ will be the same.

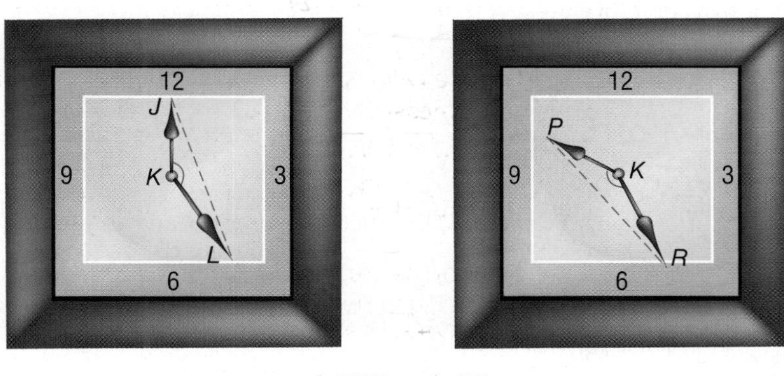


$\triangle PKR \cong \triangle JKL$

Any two triangles formed using the same side lengths and included angle measure will be congruent. This illustrates the following postulate.

**StudyTip**

**Side-Side-Angle** The measures of two sides and a nonincluded angle are not sufficient to prove two triangles congruent.

### Postulate 6.2 Side-Angle-Side (SAS) Congruence

**Words** If two sides and the included angle of one triangle are congruent to two sides and the included angle of a second triangle, then the triangles are congruent.

**Example** If Side $\overline{AB} \cong \overline{DE}$,
Angle $\angle B \cong \angle E$, and
Side $\overline{BC} \cong \overline{EF}$,
then $\triangle ABC \cong \triangle DEF$.

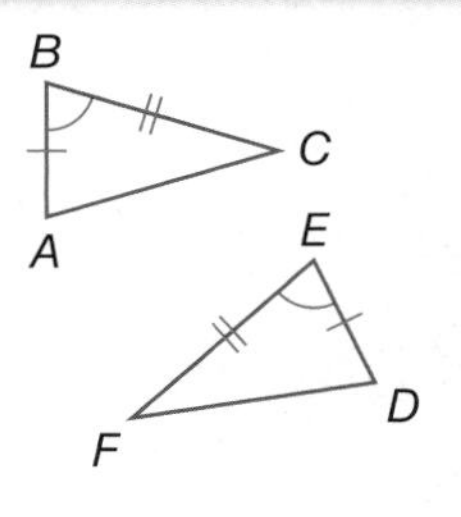

## Real-World Career

**Lighting Technicians** In the motion picture industry, gaffers, or lighting technicians, place the lighting required for a film. Gaffers make sure the angles the lights form are in the correct positions. They may have college or technical school degrees, or they may have completed a formal training program.

## Real-World Example 3 Use SAS to Prove Triangles are Congruent

**LIGHTING The scaffolding for stage lighting shown appears to be made up of congruent triangles. If $\overline{WX} \cong \overline{YZ}$ and $\overline{WX} \parallel \overline{ZY}$, write a two-column proof to prove that $\triangle WXZ \cong \triangle YZX$.**

**Proof:**

| Statements | Reasons |
|---|---|
| **1.** $\overline{WX} \cong \overline{YZ}$ | **1.** Given |
| **2.** $\overline{WX} \parallel \overline{ZY}$ | **2.** Given |
| **3.** $\angle WXZ \cong \angle XZY$ | **3.** Alternate Interior Angle Theorem |
| **4.** $\overline{XZ} \cong \overline{ZX}$ | **4.** Reflexive Property of Congruence |
| **5.** $\triangle WXZ \cong \triangle YZX$ | **5.** SAS |

### Guided Practice

**3. EXTREME SPORTS** The wings of the hang glider shown appear to be congruent triangles. If $\overline{FG} \cong \overline{GH}$ and $\overline{JG}$ bisects $\angle FGH$, prove that $\triangle FGJ \cong \triangle HGJ$.

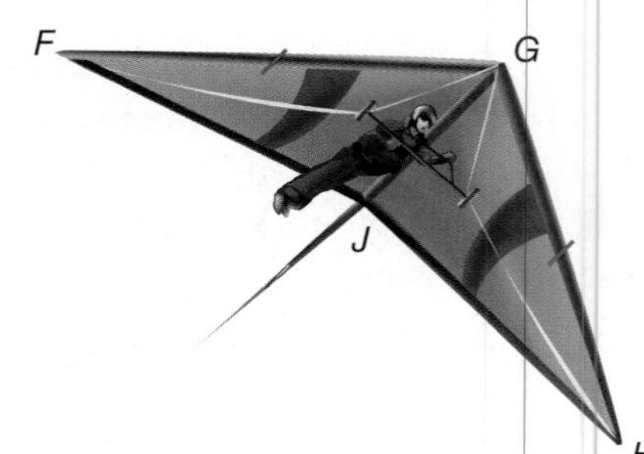


You can also construct congruent triangles given two sides and the included angle.

## Construction Congruent Triangles Using Two Sides and the Included Angle

Draw a triangle and label it $\triangle ABC$. Then use the SAS Postulate to construct $\triangle RST \cong \triangle ABC$.

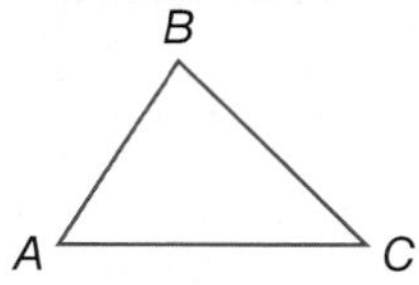


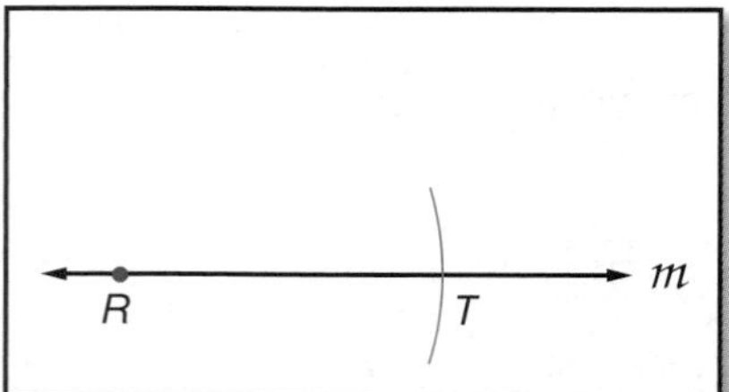


**Step 1** Draw point $R$ on a line $m$. Then construct $\overline{RT} \cong \overline{AC}$ on line $m$.

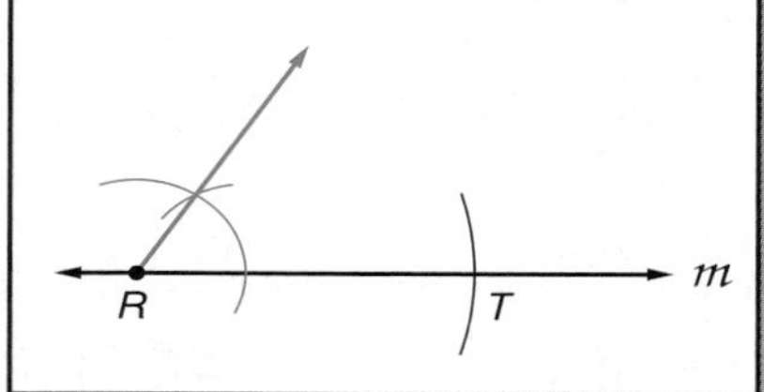


**Step 2** Construct $\angle R \cong \angle A$ using $\overline{RT}$ as a side of the angle and point $R$.

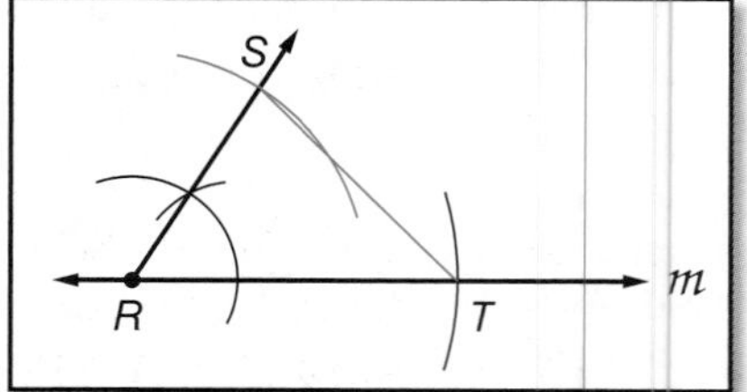


**Step 3** Construct $\overline{RS} \cong \overline{AB}$. Then draw $\overline{ST}$ to form $\triangle RST$.

## Example 4 Use SAS or SSS in Proofs

**StudyTip**

**Overlapping Figures** When triangles overlap, it can be helpful to draw each triangle separately and label the congruent parts. In Example 4, the figure could have been separated as shown.

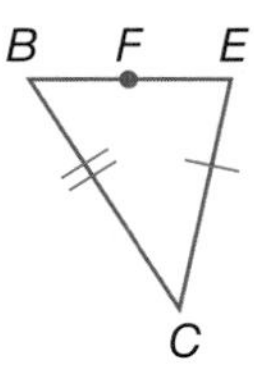


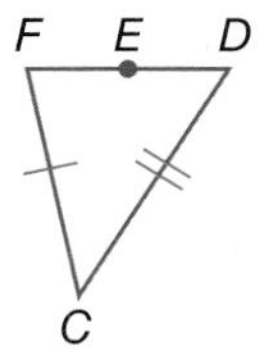


**Write a paragraph proof.**

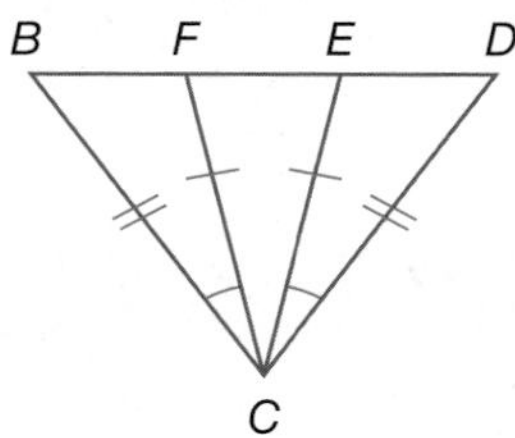


**Given:** $\overline{BC} \cong \overline{DC}$, $\angle BCF \cong \angle DCE$, $\overline{FC} \cong \overline{EC}$

**Prove:** $\angle CFD \cong \angle CEB$

**Proof:**

Since $\overline{BC} \cong \overline{DC}$, $\angle BCF \cong \angle DCE$, and $\overline{FC} \cong \overline{EC}$, then $\triangle BCF \cong \triangle DCE$ by SAS. By CPCTC, $\angle CFB \cong \angle CED$. $\angle CFD$ forms a linear pair with $\angle CFB$, and $\angle CEB$ forms a linear pair with $\angle CED$. By the Congruent Supplements Theorem, $\angle CFD$ is supplementary to $\angle CFB$ and $\angle CEB$ is supplementary to $\angle CED$. Since angles supplementary to the same angle or congruent angles are congruent, $\angle CFD \cong \angle CEB$.

**GuidedPractice**

**4. Write a two-column proof.**

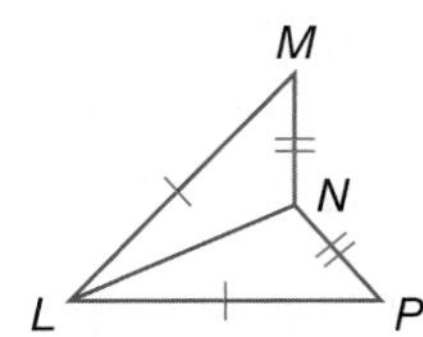


**Given:** $\overline{MN} \cong \overline{PN}$, $\overline{LM} \cong \overline{LP}$

**Prove:** $\angle LNM \cong \angle LNP$

## Check Your Understanding

**Example 1**

1. **OPTICAL ILLUSION** The figure shown is a pattern formed using four large congruent squares and four small congruent squares.

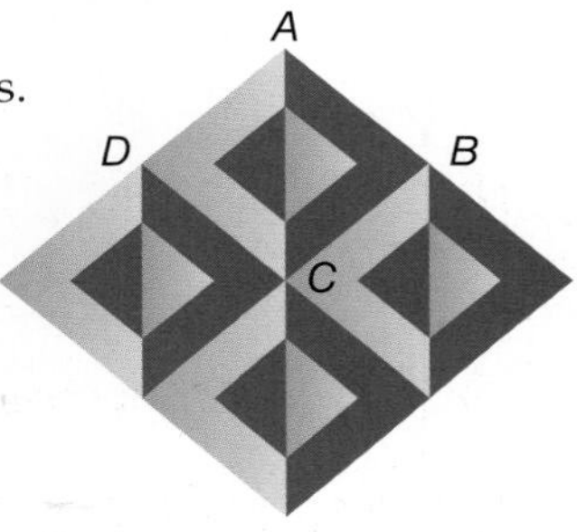


   a. How many different-sized triangles are used to create the illusion?

   b. Use the Side-Side-Side Congruence Postulate to prove that $\triangle ABC \cong \triangle CDA$.

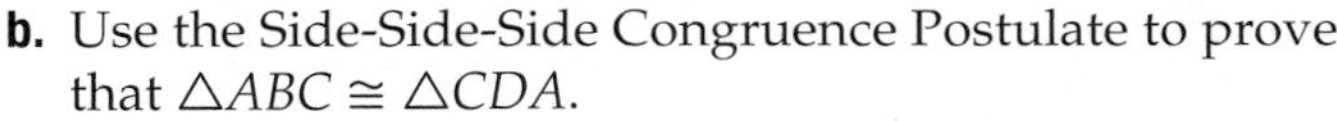

   c. What is the relationship between $\overleftrightarrow{AB}$ and $\overleftrightarrow{CD}$? Explain your reasoning.

**Example 2**

2. **EXTENDED RESPONSE** Triangle $ABC$ has vertices $A(-3, -5)$, $B(-1, -1)$, and $C(-1, -5)$. Triangle $XYZ$ has vertices $X(5, -5)$, $Y(3, -1)$, and $Z(3, -5)$.

   a. Graph both triangles on the same coordinate plane.

   b. Use your graph to make a conjecture as to whether the triangles are congruent. Explain your reasoning.

   c. Write a logical argument using coordinate geometry to support your conjecture.

**Example 3**

3. **EXERCISE** In the exercise diagram, if $\overline{LP} \cong \overline{NO}$, $\angle LPM \cong \angle NOM$, and $\triangle MOP$ is equilateral, write a paragraph proof to show that $\triangle LMP \cong \triangle NMO$.

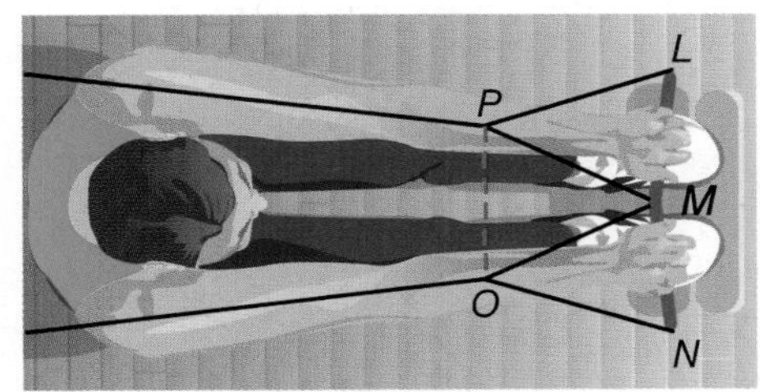

Example 4

**4. Write a two-column proof.**

**Given:** $\overline{BA} \cong \overline{DC}$, $\angle BAC \cong \angle DCA$

**Prove:** $\overline{BC} \cong \overline{DA}$

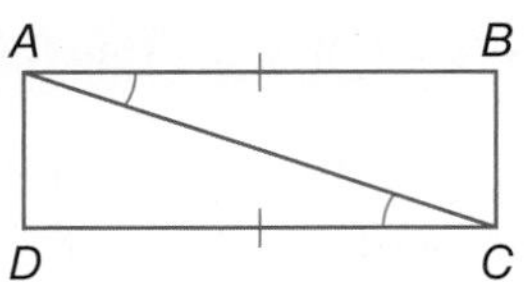


## Practice and Problem Solving

Example 1

**PROOF Write the specified type of proof.**

5. paragraph proof

**Given:** $\overline{QR} \cong \overline{SR}$, $\overline{ST} \cong \overline{QT}$

**Prove:** $\triangle QRT \cong \triangle SRT$

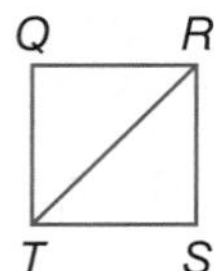


6. two-column proof

**Given:** $\overline{AB} \cong \overline{ED}$, $\overline{CA} \cong \overline{CE}$; $\overline{AC}$ bisects $\overline{BD}$.

**Prove:** $\triangle ABC \cong \triangle EDC$

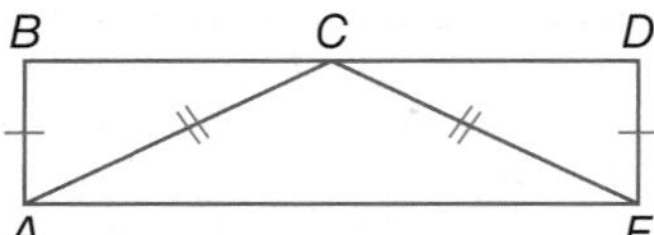


7. **BRIDGES** The Sunshine Skyway Bridge in Florida is the world's longest cable-stayed bridge, spanning 4.1 miles of Tampa Bay. It is supported using steel cables suspended from two concrete supports. If the supports are the same height above the roadway and perpendicular to the roadway, and the topmost cables meet at a point midway between the supports, prove that the two triangles shown in the photo are congruent.

Example 2

**CCSS SENSE-MAKING Determine whether $\triangle MNO \cong \triangle QRS$. Explain.**

8. $M(2, 5)$, $N(5, 2)$, $O(1, 1)$, $Q(-4, 4)$, $R(-7, 1)$, $S(-3, 0)$

9. $M(0, -1)$, $N(-1, -4)$, $O(-4, -3)$, $Q(3, -3)$, $R(4, -4)$, $S(3, 3)$

10. $M(0, -3)$, $N(1, 4)$, $O(3, 1)$, $Q(4, -1)$, $R(6, 1)$, $S(9, -1)$

11. $M(4, 7)$, $N(5, 4)$, $O(2, 3)$, $Q(2, 5)$, $R(3, 2)$, $S(0, 1)$

Example 3

**PROOF Write the specified type of proof.**

12. two-column proof

**Given:** $\overline{BD} \perp \overline{AC}$, $\overline{BD}$ bisects $\overline{AC}$.

**Prove:** $\triangle ABD \cong \triangle CBD$

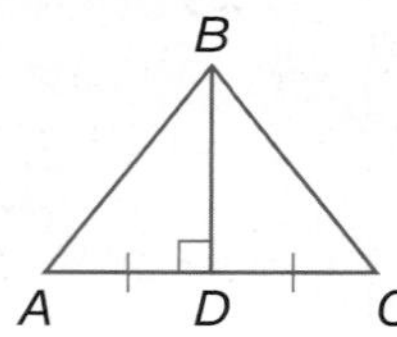


13. paragraph proof

**Given:** $R$ is the midpoint of $\overline{QS}$ and $\overline{PT}$.

**Prove:** $\triangle PRQ \cong \triangle TRS$

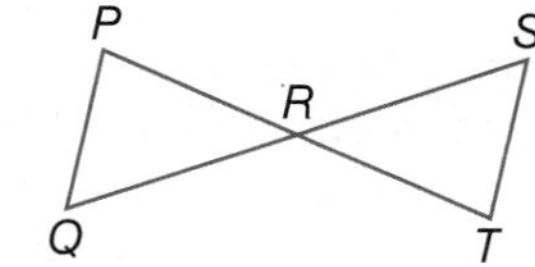

**Example 4** **PROOF** **Write the specified type of proof.**

**14.** flow proof

**Given:** $\overline{JM} \cong \overline{NK}$; $L$ is the midpoint of $\overline{JN}$ and $\overline{KM}$.

**Prove:** $\angle MJL \cong \angle KNL$

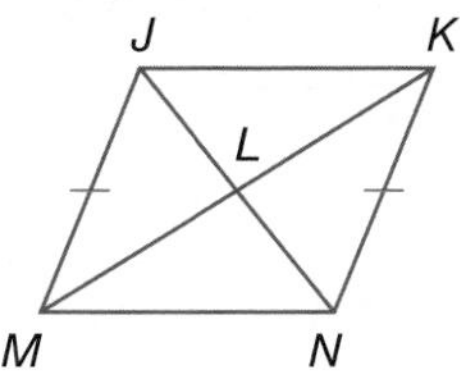

**15.** paragraph proof

**Given:** $\triangle XYZ$ is equilateral. $\overline{WY}$ bisects $\angle XYZ$.

**Prove:** $\overline{XW} \cong \overline{ZW}$

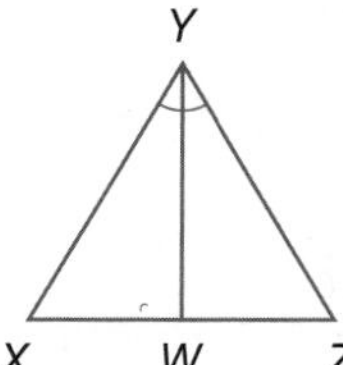

**CCSS ARGUMENTS** **Determine which postulate can be used to prove that the triangles are congruent. If it is not possible to prove congruence, write *not possible*.**

**16.** 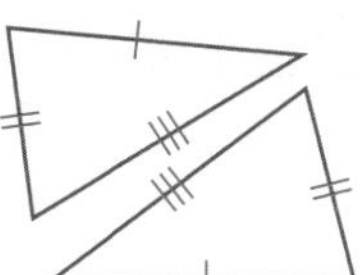

**17.** 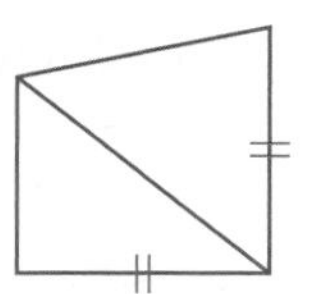

**18.** 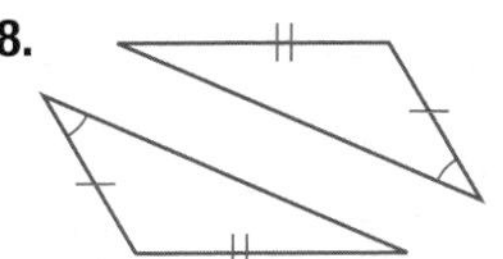

**19.** 

**20.** **SIGNS** Refer to the diagram at the right.

**a.** Identify the three-dimensional figure represented by the wet floor sign.

**b.** If $\overline{AB} \cong \overline{AD}$ and $\overline{CB} \cong \overline{DC}$, prove that $\triangle ACB \cong \triangle ACD$.

**c.** Why do the triangles not look congruent in the diagram?

**PROOF** **Write a flow proof.**

**21. Given:** $\overline{MJ} \cong \overline{ML}$; $K$ is the midpoint of $\overline{JL}$.

**Prove:** $\triangle MJK \cong \triangle MLK$

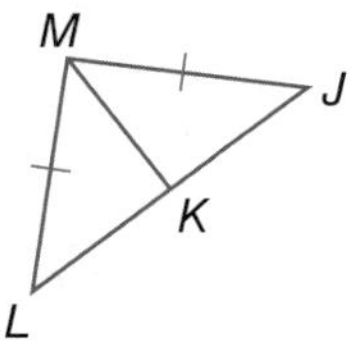

**22. Given:** $\triangle TPQ \cong \triangle SPR$

$\angle TQR \cong \angle SRQ$

**Prove:** $\triangle TQR \cong \triangle SRQ$

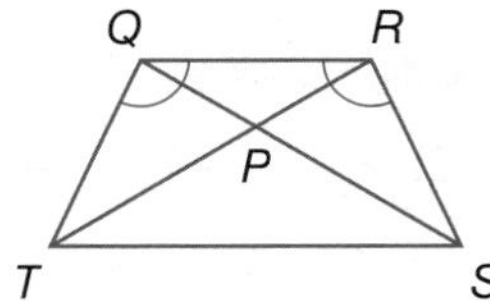

**23.** **SOFTBALL** Use the diagram of a fast-pitch softball diamond shown. Let $F$ = first base, $S$ = second base, $T$ = third base, $P$ = pitching point, and $R$ = home plate.

**a.** Write a two-column proof to prove that the distance from first base to third base is the same as the distance from home plate to second base.

**b.** Write a two-column proof to prove that the angle formed between second base, home plate, and third base is the same as the angle formed between second base, home plate, and first base.

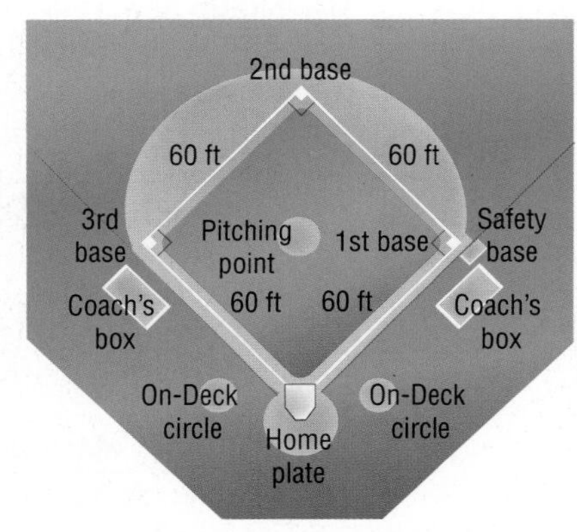

**PROOF** **Write a two-column proof.**

**24.** **Given:** $\overline{YX} \cong \overline{WZ}$, $\overline{YX} \parallel \overline{ZW}$

**Prove:** $\triangle YXZ \cong \triangle WZX$

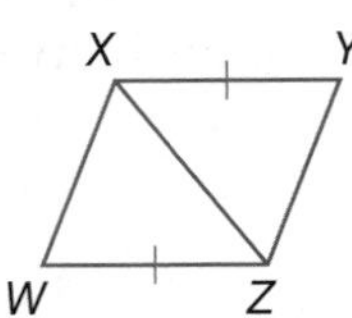


**25.** **Given:** $\triangle EAB \cong \triangle DCB$

**Prove:** $\triangle EAD \cong \triangle DCE$

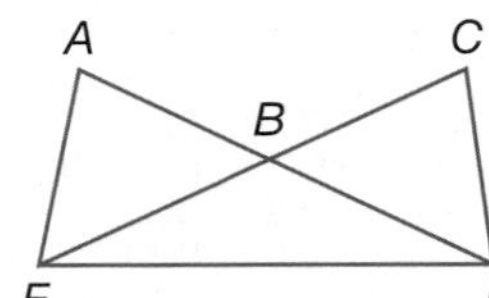


**26.** **CCSS ARGUMENTS** Write a paragraph proof.

**Given:** $\overline{HL} \cong \overline{HM}$, $\overline{PM} \cong \overline{KL}$, $\overline{PG} \cong \overline{KJ}$, $\overline{GH} \cong \overline{JH}$

**Prove:** $\angle G \cong \angle J$

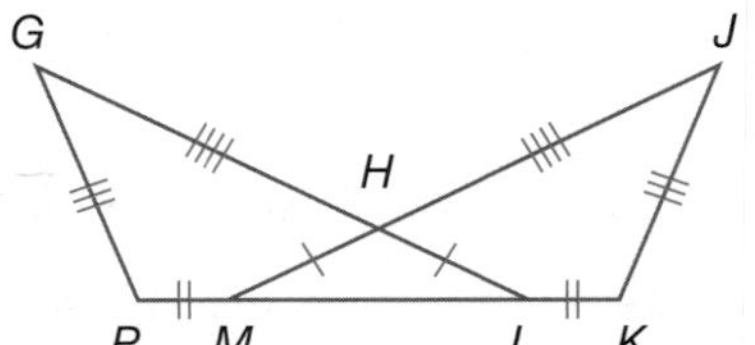


**ALGEBRA** **Find the value of the variable that yields congruent triangles. Explain.**

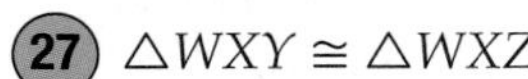

**27.** $\triangle WXY \cong \triangle WXZ$

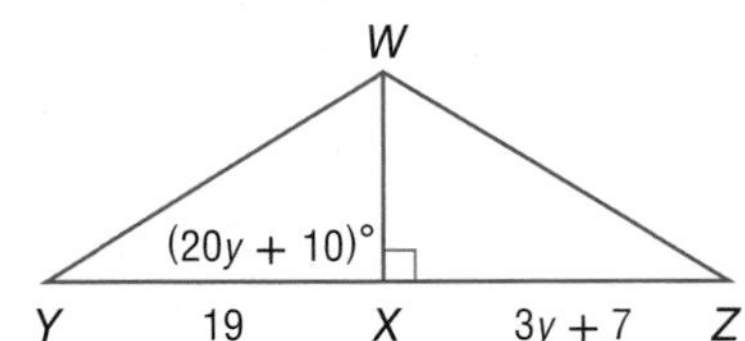


**28.** $\triangle ABC \cong \triangle FGH$

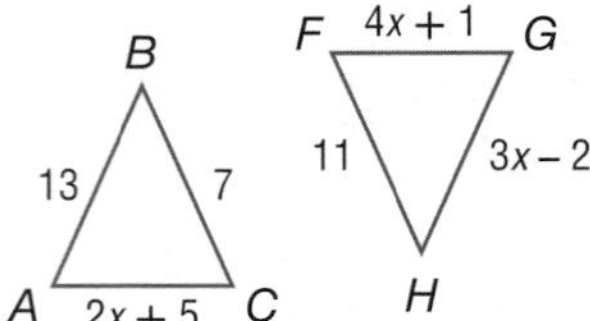


## H.O.T. Problems Use Higher-Order Thinking Skills

**29.** **CHALLENGE** **Refer to the graph shown.**

**a.** Describe two methods you could use to prove that $\triangle WYZ$ is congruent to $\triangle WYX$. You may not use a ruler or a protractor. Which method do you think is more efficient? Explain.

**b.** Are $\triangle WYZ$ and $\triangle WYX$ congruent? Explain your reasoning.

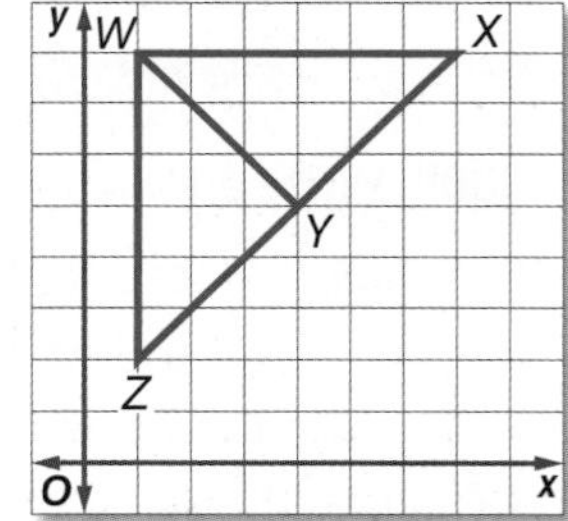


**30.** **REASONING** Determine whether the following statement is *true* or *false*. If true, explain your reasoning. If *false*, provide a counterexample.

*If the congruent sides in one isosceles triangle have the same measure as the congruent sides in another isosceles triangle, then the triangles are congruent.*

**31.** **ERROR ANALYSIS** Bonnie says that $\triangle PQR \cong \triangle XYZ$ by SAS. Shada disagrees. She says that there is not enough information to prove that the two triangles are congruent. Is either of them correct? Explain.

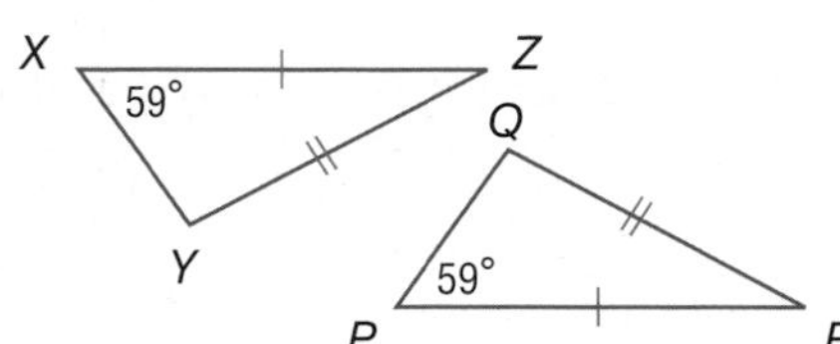


**32.** **OPEN ENDED** Use a straightedge to draw obtuse triangle *ABC*. Then construct $\triangle XYZ$ so that it is congruent to $\triangle ABC$ using either SSS or SAS. Justify your construction mathematically and verify it using measurement.

**33.** **WRITING IN MATH** Two pairs of corresponding sides of two right triangles are congruent. Are the triangles congruent? Explain your reasoning.

## Standardized Test Practice

**34. ALGEBRA** The Ross Family drove 300 miles to visit their grandparents. Mrs. Ross drove 70 miles per hour for 65% of the trip and 35 miles per hour or less for 20% of the trip that was left. Assuming that Mrs. Ross never went over 70 miles per hour, how many miles did she travel at a speed between 35 and 70 miles per hour?

**A** 195 **C** 21

**B** 84 **D** 18

**35.** In the figure, $\angle C \cong \angle Z$ and $\overline{AC} \cong \overline{XZ}$.

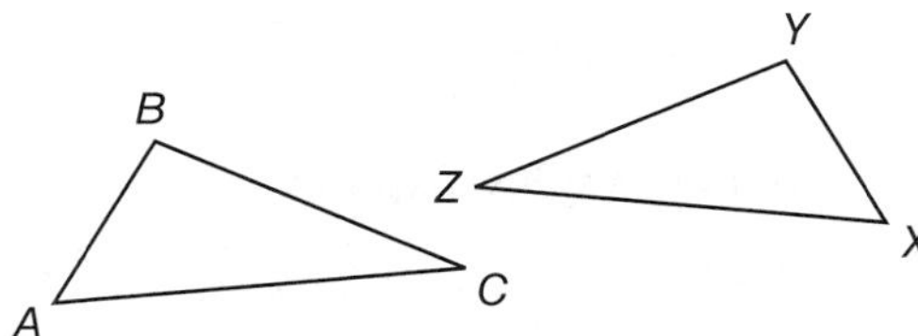


What additional information could be used to prove that $\triangle ABC \cong \triangle XYZ$?

**F** $\overline{BC} \cong \overline{YZ}$

**G** $\overline{AB} \cong \overline{XY}$

**H** $\overline{BC} \cong \overline{XZ}$

**J** $\overline{XZ} \cong \overline{XY}$

**36. EXTENDED RESPONSE** The graph below shows the eye colors of all of the students in a class. What is the probability that a student chosen at random from this class will have blue eyes? Explain your reasoning.

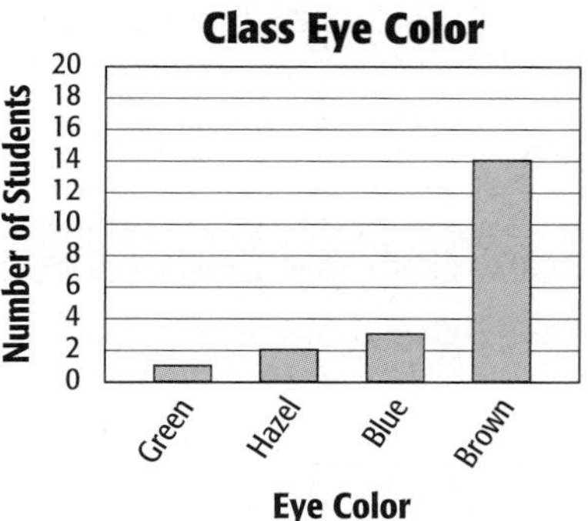


**37. SAT/ACT** If $4a + 6b = 6$ and $-2a + b = -7$, what is the value of $a$?

**A** −2

**B** −1

**C** 2

**D** 3

**E** 4

## Spiral Review

**In the diagram, $\triangle LMN \cong \triangle QRS$.** (Lesson 6-2)

**38.** Find $x$.

**39.** Find $y$.

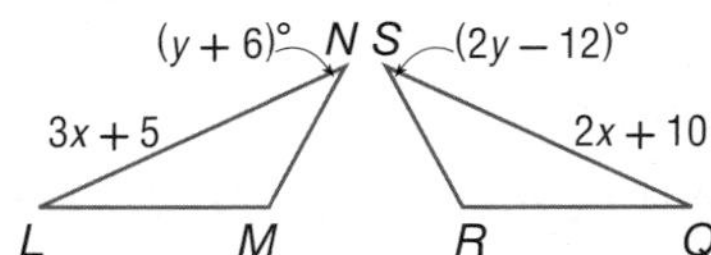


**40. ASTRONOMY** The Big Dipper is a part of the larger constellation Ursa Major. Three of the brighter stars in the constellation form $\triangle RSA$. If $m\angle R = 41$ and $m\angle S = 109$, find $m\angle A$. (Lesson 6-1)

**41. JOBS** Katie mows lawns in the summer to earn extra money. She started with 3 lawns, and she now mows 12 lawns in her fourth summer. (Lesson 3-3)

**a.** Graph the line that models the growth of Katie's business over time.

**b.** What is the slope of the graph? What does it represent?

**c.** Assuming that the business continues to grow at the same rate, how many lawns should Katie plan to mow during her sixth summer?

## Skills Review

**State the property that justifies each statement.**

**42.** $AB = AB$

**43.** If $EF = GH$ and $GH = JK$, then $EF = JK$.

**44.** If $a^2 = b^2 - c^2$, then $b^2 - c^2 = a^2$.

**45.** If $XY + 20 = YW$ and $XY + 20 = DT$, then $YW = DT$.

EXTEND 6-3

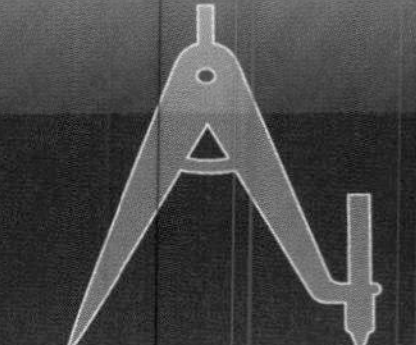

# Geometry Lab
# Proving Constructions

When you perform a construction using a straightedge and compass, you assume that segments constructed using the same compass setting are congruent. You can use this information, along with definitions, postulates, and theorems to prove constructions.

**CCSS Common Core State Standards**
**Content Standards**
**G.CO.12** Make formal geometric constructions with a variety of tools and methods (compass and straightedge, string, reflective devices, paper folding, dynamic geometric software, etc.).
**G.SRT.5** Use congruence and similarity criteria for triangles to solve problems and to prove relationships in geometric figures.
**Mathematical Practices** 3, 5

## Activity

**Follow the steps below to bisect an angle. Then prove the construction.**

**Step 1**

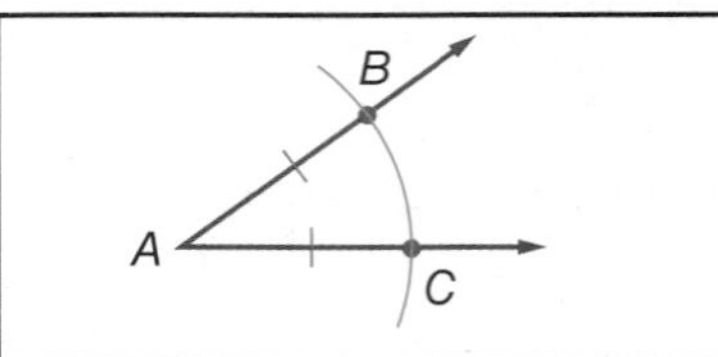


Draw any angle with vertex $A$. Place the compass point at $A$ and draw an arc that intersects both sides of $\angle A$. Label the points $B$ and $C$. Mark the congruent segments.

**Step 2**

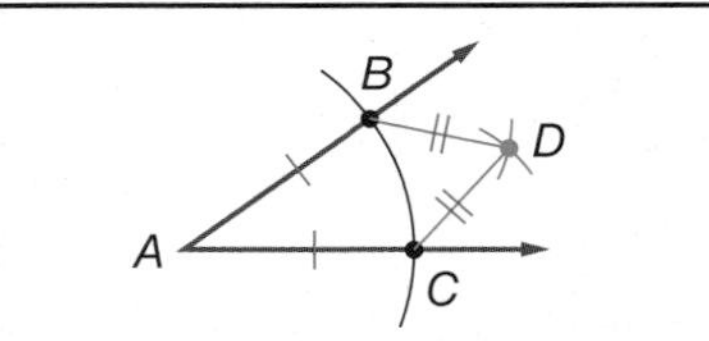


With the compass point at $B$, draw an arc in the interior of $\angle A$. With the same radius, draw an arc from $C$ intersecting the first arc at $D$. Draw the segments $\overline{BD}$ and $\overline{CD}$. Mark the congruent segments.

**Step 3**

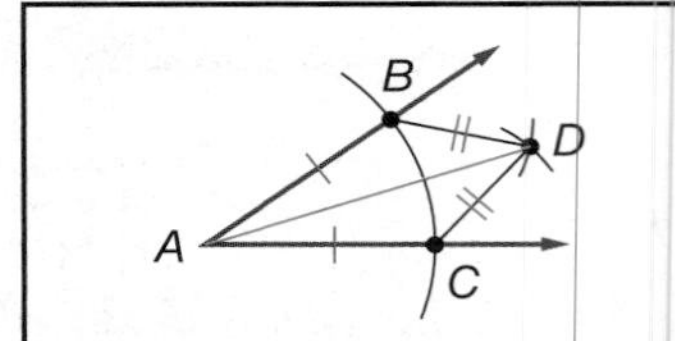


Draw $\overline{AD}$.

**Given:** Description of steps and diagram of construction

**Prove:** $\overline{AD}$ bisects $\angle BAC$.

**Proof:**

| Statements | Reasons |
|---|---|
| 1. $\overline{AB} \cong \overline{AC}$ | 1. The same compass setting was used from point $A$ to construct points $B$ and $C$. |
| 2. $\overline{BD} \cong \overline{CD}$ | 2. The same compass setting was used from points $B$ and $C$ to construct point $D$. |
| 3. $\overline{AD} \cong \overline{AD}$ | 3. Reflexive Property |
| 4. $\triangle ABD \cong \triangle ACD$ | 4. SSS Postulate |
| 5. $\angle BAD \cong \angle CAD$ | 5. CPCTC |
| 6. $\overline{AD}$ bisects $\angle BAC$. | 6. Definition of angle bisector |

## Exercises

1. Construct a line parallel to a given line through a given point. Write a two-column proof of your construction.
2. Construct an equilateral triangle. Write a paragraph proof of your construction.
3. **CHALLENGE** Construct the bisector of a segment that is also perpendicular to the segment and write a two-column proof of your construction. (*Hint:* You will need to use more than one pair of congruent triangles.).

CHAPTER 6

# Mid-Chapter Quiz

## Lessons 6-1 through 6-3

**Find the measure of each angle indicated.** (Lesson 6-1)

**1.** $m\angle 1$

**2.** $m\angle 2$

**3.** $m\angle 3$

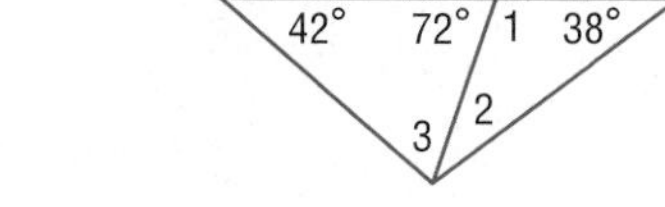


**4. ASTRONOMY** Leo is a constellation that represents a lion. Three of the brighter stars in the constellation form $\triangle LEO$. If the angles have measures as shown in the figure, find $m\angle OLE$. (Lesson 6-1)

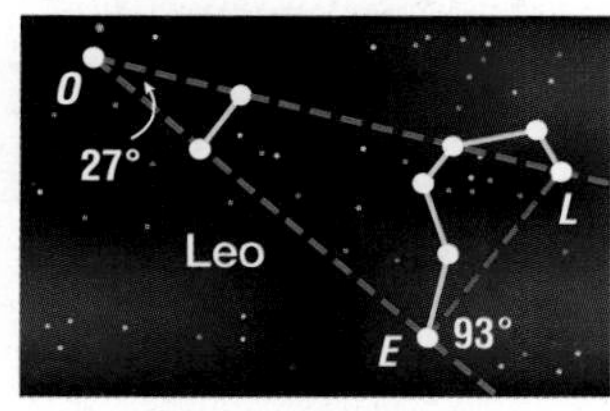


**Find the measure of each numbered angle.** (Lesson 6-1)

**5.** $m\angle 4$

**6.** $m\angle 5$

**7.** $m\angle 6$

**8.** $m\angle 7$

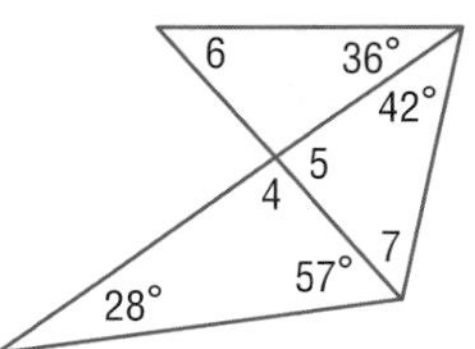


In the diagram, $\triangle RST \cong \triangle ABC$. (Lesson 6-2)

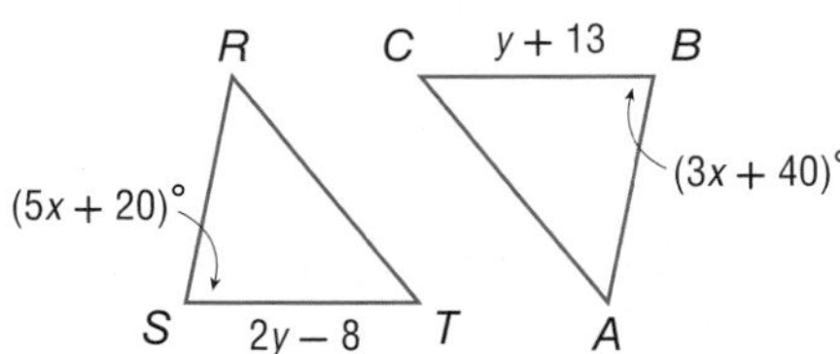


**9.** Find $x$.

**10.** Find $y$.

**11. ARCHITECTURE** The diagram shows an A-frame house with various points labeled. Assume that segments and angles that appear to be congruent in the diagram are congruent. Indicate which triangles are congruent. (Lesson 6-2)

**12. MULTIPLE CHOICE** Determine which statement is true given that $\triangle CBX \cong \triangle SML$. (Lesson 6-2)

**F** $\overline{MO} \cong \overline{SL}$

**G** $\overline{XC} \cong \overline{ML}$

**H** $\angle X \cong \angle S$

**J** $\angle XCB \cong \angle LSM$

**13. BRIDGES** A bridge truss is shown in the diagram below, where $\overline{AC} \perp \overline{BD}$ and $B$ is the midpoint of $\overline{AC}$. What method can be used to prove that $\triangle ABD \cong \triangle CBD$? (Lesson 6-3)

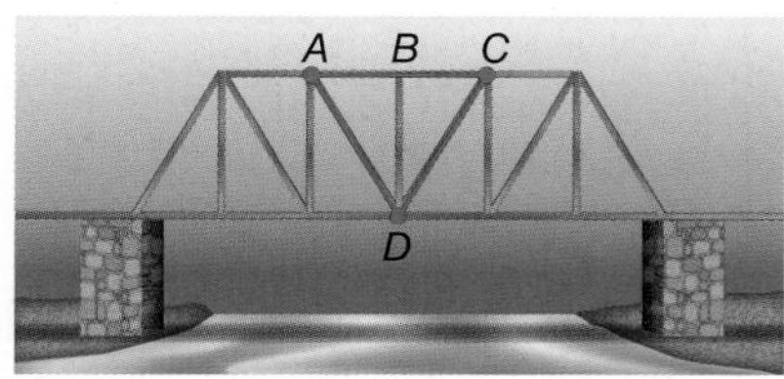


**Determine whether $\triangle PQR \cong \triangle XYZ$.** (Lesson 6-3)

**14.** $P(3, -5)$, $Q(11, 0)$, $R(1, 6)$, $X(5, 1)$, $Y(13, 6)$, $Z(3, 12)$

**15.** $P(-3, -3)$, $Q(-5, 1)$, $R(-2, 6)$, $X(2, -6)$, $Y(3, 3)$, $Z(5, -1)$

**16.** $P(8, 1)$, $Q(-7, -15)$, $R(9, -6)$, $X(5, 11)$, $Y(-10, -5)$, $Z(6, 4)$

**17. Write a two-column proof.** (Lesson 6-3)

**Given:** $\triangle LMN$ is isos. with $\overline{LM} \cong \overline{NM}$, and $\overline{MO}$ bisects $\angle LMN$.

**Prove:** $\triangle MLO \cong \triangle MNO$

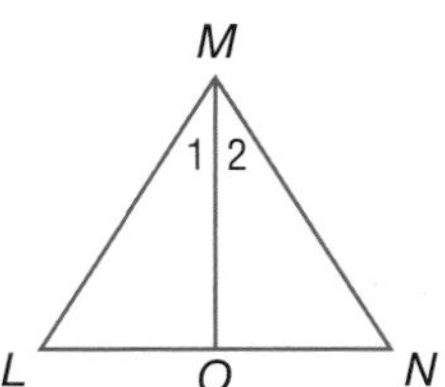

LESSON 6-4

# Proving Triangles Congruent—ASA, AAS

## Then

- You proved triangles congruent using SSS and SAS.

## Now

1. Use the ASA Postulate to test for congruence.
2. Use the AAS Theorem to test for congruence.

## Why?

- Competitive sweep rowing, also called *crew*, involves two or more people who sit facing the stern of the boat, with each rower pulling one oar. In high school competitions, a race, called a *regatta*, usually requires a body of water that is more than 1500 meters long. Congruent triangles can be used to measure distances that are not easily measured directly, like the length of a regatta course.

**NewVocabulary**
included side

**Common Core State Standards**

**Content Standards**

G.CO.10 Prove theorems about triangles.

G.SRT.5 Use congruence and similarity criteria for triangles to solve problems and to prove relationships in geometric figures.

**Mathematical Practices**

3 Construct viable arguments and critique the reasoning of others.

5 Use appropriate tools strategically.

## 1 ASA Postulate

An **included side** is the side located between two consecutive angles of a polygon. In $\triangle ABC$ at the right, $\overline{AC}$ is the included side between $\angle A$ and $\angle C$.

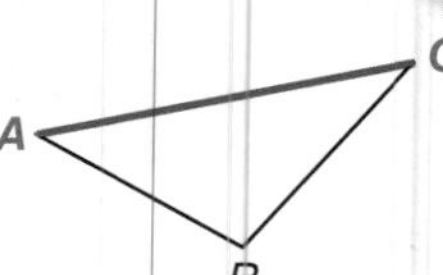


### Postulate 6.3 Angle-Side-Angle (ASA) Congruence

If two angles and the included side of one triangle are congruent to two angles and the included side of another triangle, then the triangles are congruent.

**Example** If **A**ngle $\angle A \cong \angle D$,
**S**ide $\overline{AB} \cong \overline{DE}$, and
**A**ngle $\angle B \cong \angle E$,
then $\triangle ABC \cong \triangle DEF$.

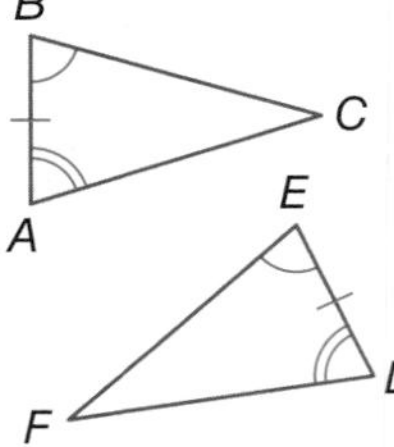


### Construction Congruent Triangles Using Two Angles and Included Side

Draw a triangle and label it $\triangle ABC$. Then use the ASA Postulate to construct $\triangle XYZ \cong \triangle ABC$.

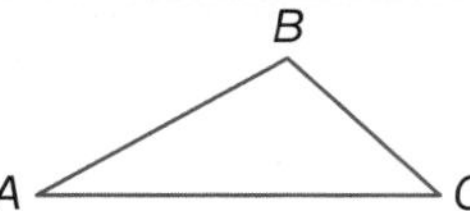


**Step 1**

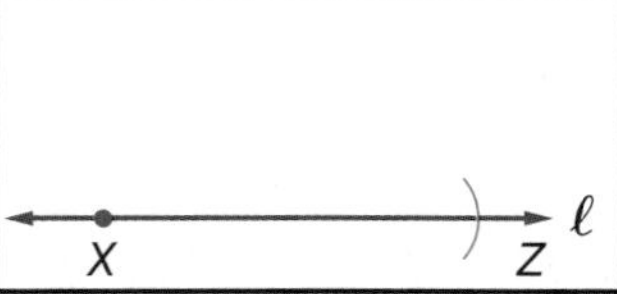


Draw a line $\ell$ and select a point $X$. Construct $\overline{XZ}$ such that $\overline{XZ} \cong \overline{AC}$.

**Step 2**

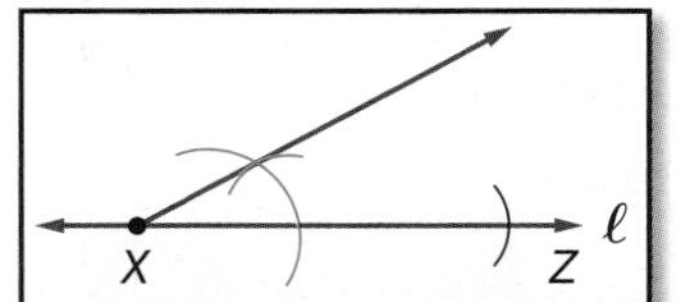


Construct an angle congruent to $\angle A$ at $X$ using $\overleftrightarrow{XZ}$ as a side of the angle.

**Step 3**

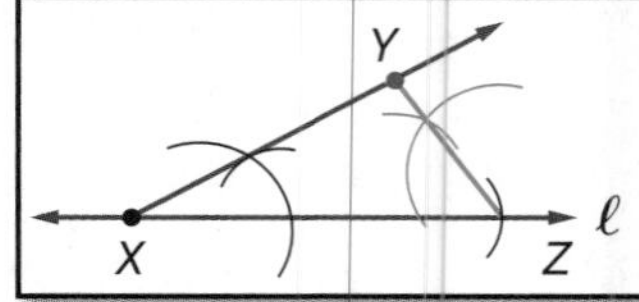


Construct an angle congruent to $\angle C$ at $Z$ using $\overleftrightarrow{XZ}$ as a side of the angle. Label the point where the new sides of the angles meet as $Y$.

## Example 1 Use ASA to Prove Triangles Congruent

**Write a two-column proof.**

**Given:** $\overline{QS}$ bisects $\angle PQR$;
$\angle PSQ \cong \angle RSQ$.

**Prove:** $\triangle PQS \cong \triangle RQS$

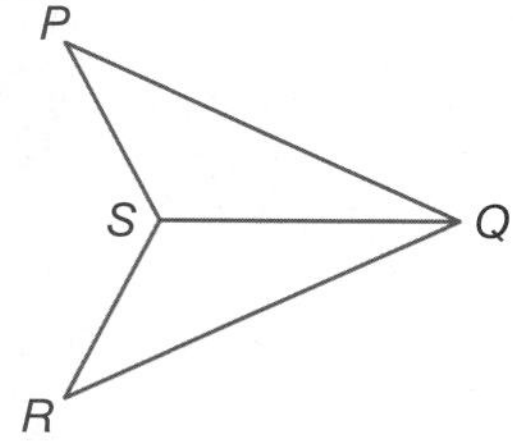


**Proof:**

| Statements | Reasons |
|---|---|
| 1. $\overline{QS}$ bisects $\angle PQR$; $\angle PSQ \cong \angle RSQ$. | 1. Given |
| 2. $\angle PQS \cong \angle RQS$ | 2. Definition of Angle Bisector |
| 3. $\overline{QS} \cong \overline{QS}$ | 3. Reflexive Property of Congruence |
| 4. $\triangle PQS \cong \triangle RQS$ | 4. ASA |

### GuidedPractice

1. Write a flow proof.
   **Given:** $\overline{ZX}$ bisects $\angle WZY$; $\overline{XZ}$ bisects $\angle YXW$.
   **Prove:** $\triangle WXZ \cong \triangle XZY$

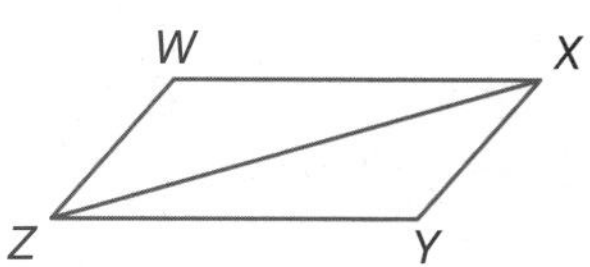


**2 AAS Theorem** The congruence of two angles and a nonincluded side are also sufficient to prove two triangles congruent. This congruence relationship is a theorem because it can be proved using the Third Angles Theorem.

### Theorem 6.5 Angle-Angle-Side (AAS) Congruence

If two angles and the nonincluded side of one triangle are congruent to the corresponding two angles and side of a second triangle, then the two triangles are congruent.

**Example** If Angle $\angle A \cong \angle D$,
Angle $\angle B \cong \angle E$, and
Side $\overline{BC} \cong \overline{EF}$,
then $\triangle ABC \cong \triangle DEF$.

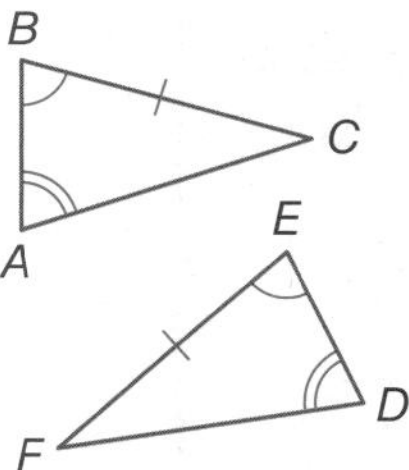


### Proof Angle-Angle-Side Theorem

**Given:** $\angle L \cong \angle Q$, $\angle M \cong \angle R$, $\overline{MN} \cong \overline{RS}$

**Prove:** $\triangle LMN \cong \triangle QRS$

**Proof:**

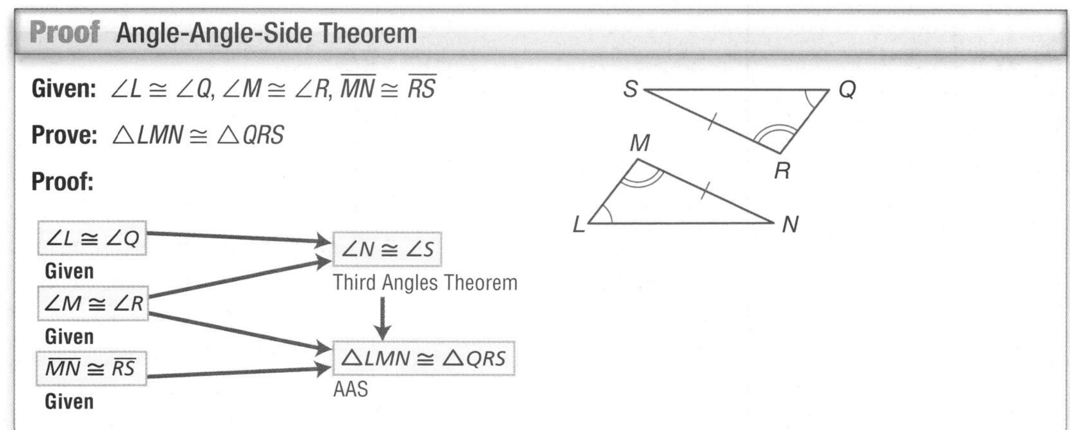

## Example 2 Use AAS to Prove Triangles Congruent

**Write a two-column proof.**

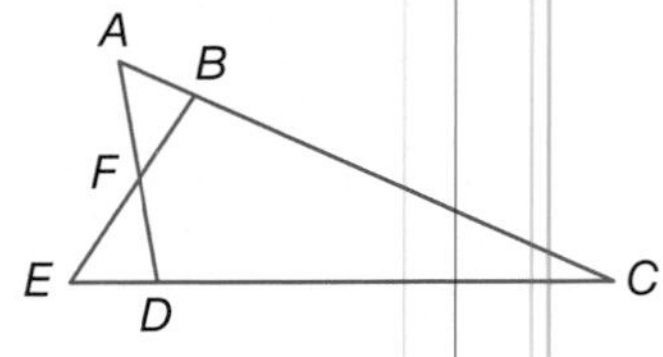


**Given:** $\angle DAC \cong \angle BEC$
$\overline{DC} \cong \overline{BC}$

**Prove:** $\triangle ACD \cong \triangle ECB$

**Proof:** We are given that $\angle DAC \cong \angle BEC$ and $\overline{DC} \cong \overline{BC}$. $\angle C \cong \angle C$ by the Reflexive Property. By AAS, $\triangle ACD \cong \triangle ECB$.

### GuidedPractice

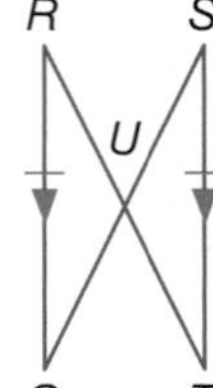


**2.** Write a flow proof.
**Given:** $\overline{RQ} \cong \overline{ST}$ and $\overline{RQ} \parallel \overline{ST}$
**Prove:** $\triangle RUQ \cong \triangle TUS$

You can use congruent triangles to measure distances that are difficult to measure directly.

## Real-World Example 3 Apply Triangle Congruence

**COMMUNITY SERVICE Jeremias is working with a community service group to build a bridge across a creek at a local park. The bridge will span the creek between points *C* and *B*. Jeremias located a fixed point *D* to use as a reference point so that the segments have the relationships shown. *A* is the midpoint of $\overline{CD}$ and *DE* is 15 feet. How long does the bridge need to be?**

In order to determine the length of $\overline{CB}$, we must first prove that the two triangles Jeremias has created are congruent.

- Since $\overline{CD}$ is perpendicular to both $\overline{CB}$ and $\overline{DE}$, the segments form right angles as shown on the diagram.
- All right angles are congruent, so $\angle BCA \cong \angle EDA$.
- Point *A* is the midpoint of $\overline{CD}$, so $\overline{CA} \cong \overline{AD}$.
- $\angle BAC$ and $\angle EAD$ are vertical angles, so they are congruent.

Therefore, by ASA, $\triangle BAC \cong \triangle EAD$.

Since $\triangle BAC \cong \triangle EAD$, $\overline{DE} \cong \overline{CB}$ by CPCTC. Since the measure of $\overline{DE}$ is 15 feet, the measure of $\overline{CB}$ is also 15 feet. Therefore, the bridge needs to be 15 feet long.

**StudyTip**

**Angle-Angle-Angle** In Example 3, $\angle B$ and $\angle E$ are congruent by the Third Angles Theorem. Congruence of all three corresponding angles is not sufficient, however, to prove two triangles congruent.

## GuidedPractice

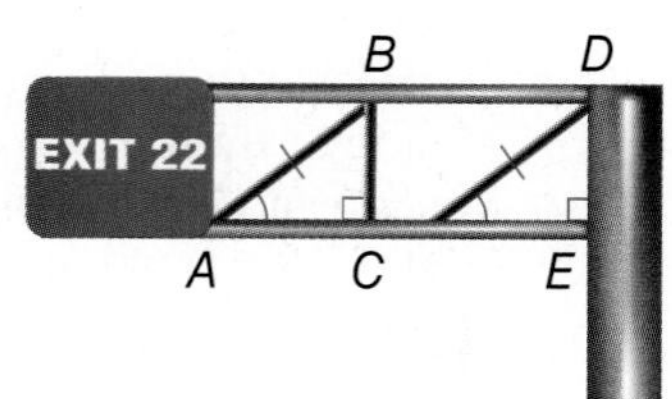


3. In the sign scaffold shown at the right, $\overline{BC} \perp \overline{AC}$ and $\overline{DE} \perp \overline{CE}$. $\angle BAC \cong \angle DCE$, and $\overline{AB} \cong \overline{CD}$. Write a paragraph proof to show that $\overline{BC} \cong \overline{DE}$.

You have learned several methods for proving triangle congruence.

## ConceptSummary Proving Triangles Congruent

| SSS | SAS | ASA | AAS |
|---|---|---|---|
| Three pairs of corresponding sides are congruent. | Two pairs of corresponding sides and their included angles are congruent. | Two pairs of corresponding angles and their included sides are congruent. | Two pairs of corresponding angles and the corresponding nonincluded sides are congruent. |

## Check Your Understanding

**Example 1** **PROOF Write the specified type of proof.**

1. two-column proof

   **Given:** $\overline{CB}$ bisects $\angle ABD$ and $\angle ACD$.

   **Prove:** $\triangle ABC \cong \triangle DBC$

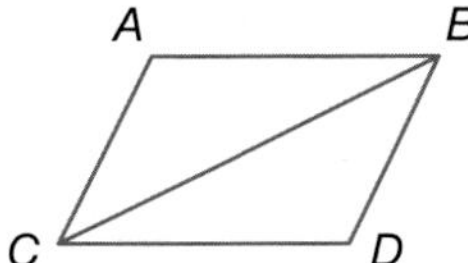


2. flow proof

   **Given:** $\overline{JK} \parallel \overline{LM}$, $\overline{JL} \parallel \overline{KM}$

   **Prove:** $\triangle JML \cong \triangle MJK$

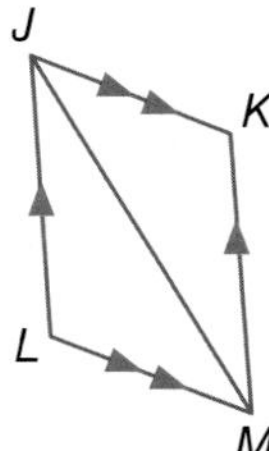


**Example 2**

3. paragraph proof

   **Given:** $\angle K \cong \angle M$, $\overline{JK} \cong \overline{JM}$, $\overline{JL}$ bisects $\angle KLM$.

   **Prove:** $\triangle JKL \cong \triangle JML$

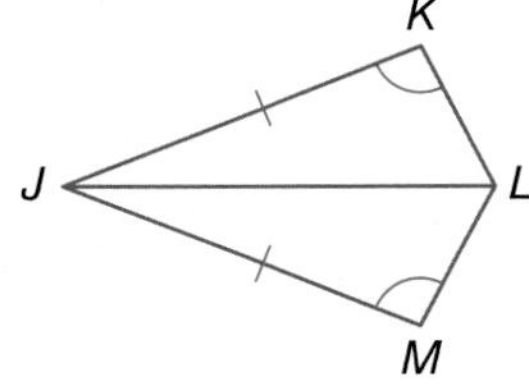


4. two-column proof

   **Given:** $\overline{GH} \parallel \overline{FJ}$

   $m\angle G = m\angle J = 90$

   **Prove:** $\triangle HJF \cong \triangle FGH$

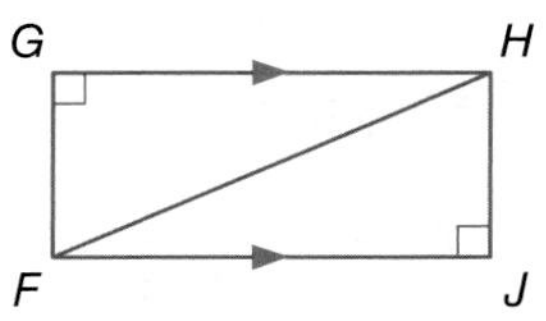

Example 3

5. **BRIDGE BUILDING** A surveyor needs to find the distance from point $A$ to point $B$ across a canyon. She places a stake at $A$, and a coworker places a stake at $B$ on the other side of the canyon. The surveyor then locates $C$ on the same side of the canyon as $A$ such that $\overline{CA} \perp \overline{AB}$. A fourth stake is placed at $E$, the midpoint of $\overline{CA}$. Finally, a stake is placed at $D$ such that $\overline{CD} \perp \overline{CA}$ and $D$, $E$, and $B$ are sited as lying along the same line.

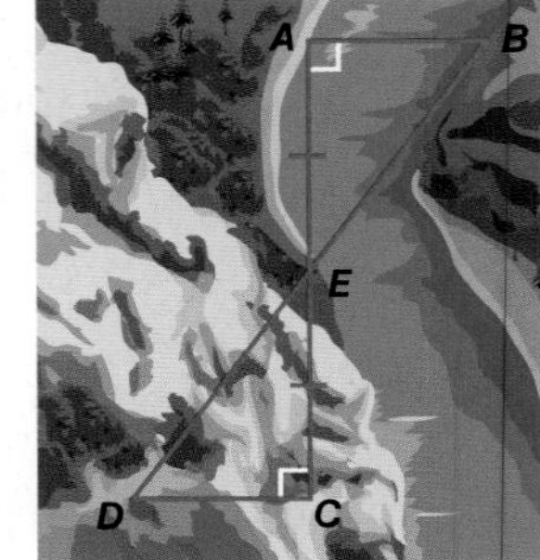


**a.** Explain how the surveyor can use the triangles formed to find $AB$.

**b.** If $AC = 1300$ meters, $DC = 550$ meters, and $DE = 851.5$ meters, what is $AB$? Explain your reasoning.

## Practice and Problem Solving

Example 1

**PROOF** **Write a paragraph proof.**

6. **Given:** $\overline{CE}$ bisects $\angle BED$; $\angle BCE$ and $\angle ECD$ are right angles.

**Prove:** $\triangle ECB \cong \triangle ECD$

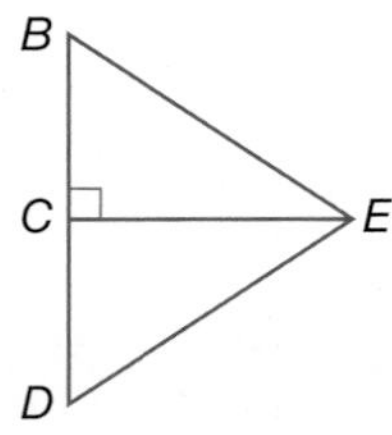


7. **Given:** $\angle W \cong \angle Y$, $\overline{WZ} \cong \overline{YZ}$, $\overline{XZ}$ bisects $\angle WZY$.

**Prove:** $\triangle XWZ \cong \triangle XYZ$

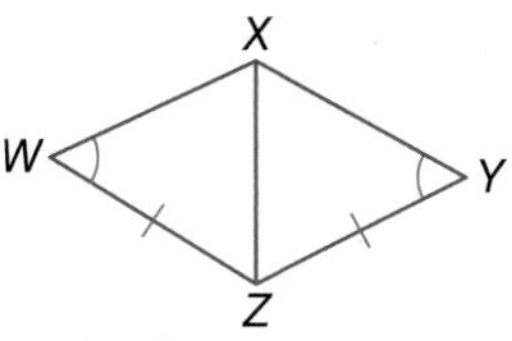


8. **TOYS** The object of the toy shown is to make the two spheres meet and strike each other repeatedly on one side of the wand and then again on the other side. If $\angle JKL \cong \angle MLK$ and $\angle JLK \cong \angle MKL$, prove that $\overline{JK} \cong \overline{ML}$.

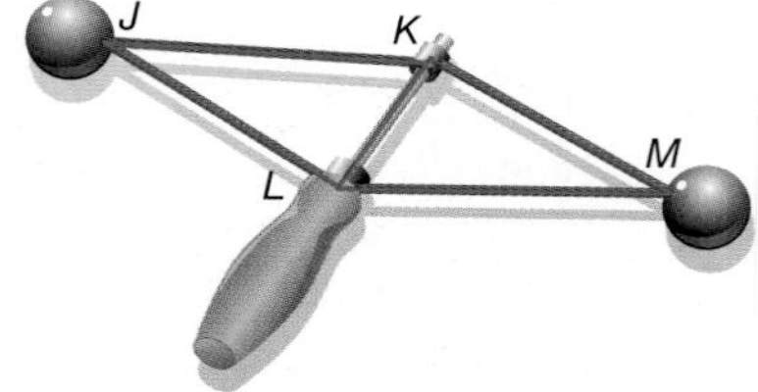


Example 2

**PROOF** **Write a two-column proof.**

9. **Given:** $V$ is the midpoint of $\overline{YW}$; $\overline{UY} \parallel \overline{XW}$.

**Prove:** $\triangle UVY \cong \triangle XVW$

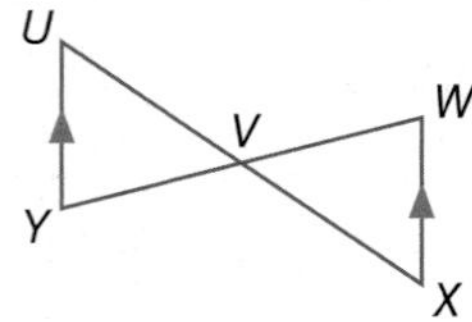


10. **Given:** $\overline{MS} \cong \overline{RQ}$, $\overline{MS} \parallel \overline{RQ}$

**Prove:** $\triangle MSP \cong \triangle RQP$

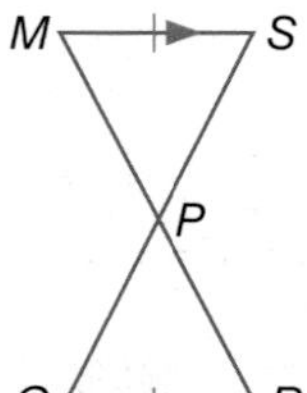


11. **CCSS ARGUMENTS** Write a flow proof.

**Given:** $\angle A$ and $\angle C$ are right angles. $\angle ABE \cong \angle CBD$, $\overline{AE} \cong \overline{CD}$

**Prove:** $\overline{BE} \cong \overline{BD}$

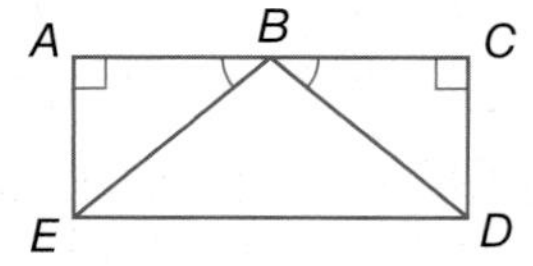

**12. PROOF** Write a flow proof.

**Given:** $\overline{KM}$ bisects $\angle JML$; $\angle J \cong \angle L$.

**Prove:** $\overline{JM} \cong \overline{LM}$

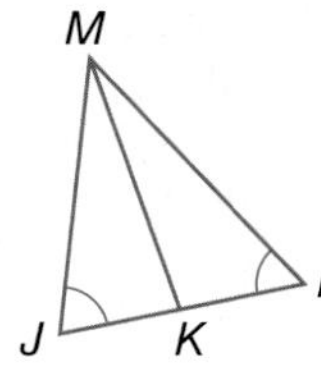


Example 3

**13. CCSS MODELING** A high school wants to hold a 1500-meter regatta on Lake Powell but is unsure if the lake is long enough. To measure the distance across the lake, the crew members locate the vertices of the triangles below and find the measures of the lengths of $\triangle HJK$ as shown below.

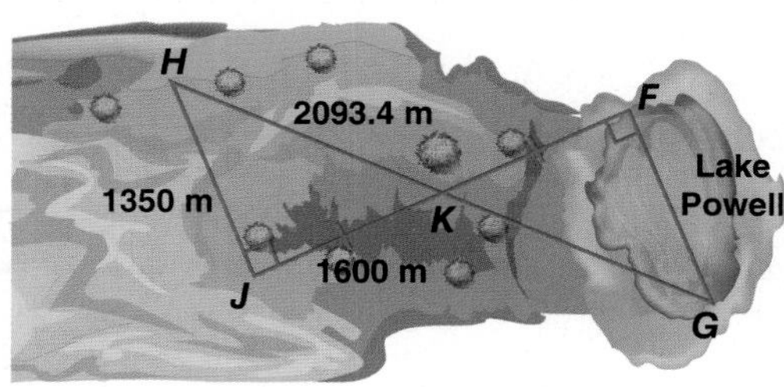


a. Explain how the crew team can use the triangles formed to estimate the distance $FG$ across the lake.

b. Using the measures given, is the lake long enough for the team to use as the location for their regatta? Explain your reasoning.

**ALGEBRA Find the value of the variable that yields congruent triangles.**

**14.** $\triangle BCD \cong \triangle WXY$

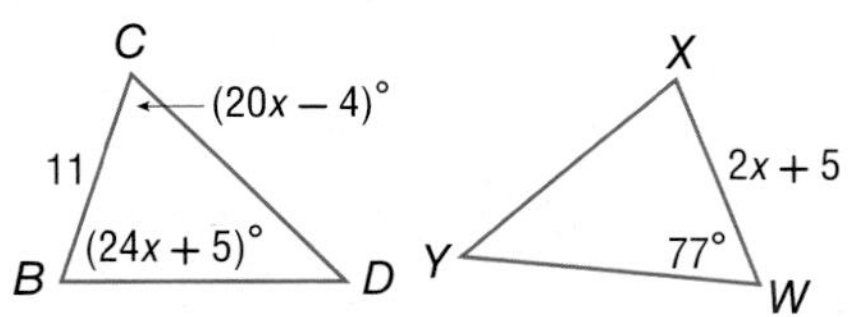


**15** $\triangle MHJ \cong \triangle PQJ$

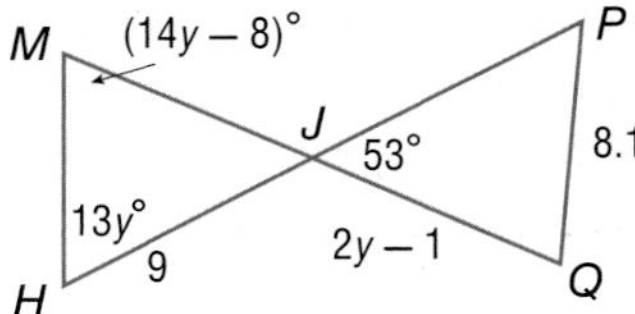


**16. THEATER DESIGN** The trusses of the roof of the outdoor theater shown below appear to be several different pairs of congruent triangles. Assume that trusses that appear to lie on the same line actually lie on the same line.

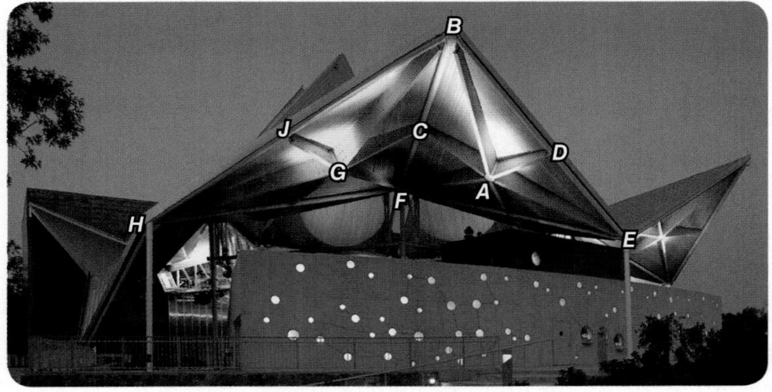


a. If $\overline{AB}$ bisects $\angle CBD$ and $\angle CAD$, prove that $\triangle ABC \cong \triangle ABD$.

b. If $\triangle ABC \cong \triangle ABD$ and $\angle FCA \cong \angle EDA$, prove that $\triangle CAF \cong \triangle DAE$.

c. If $\overline{HB} \cong \overline{EB}$, $\angle BHG \cong \angle BEA$, $\angle HGJ \cong \angle EAD$, and $\angle JGB \cong \angle DAB$, prove that $\triangle BHG \cong \triangle BEA$.

Greg Murphey

**PROOF** **Write a paragraph proof.**

**17.** **Given:** $\overline{AE} \perp \overline{DE}$, $\overline{EA} \perp \overline{AB}$, $C$ is the midpoint of $\overline{AE}$.

**Prove:** $\overline{CD} \cong \overline{CB}$

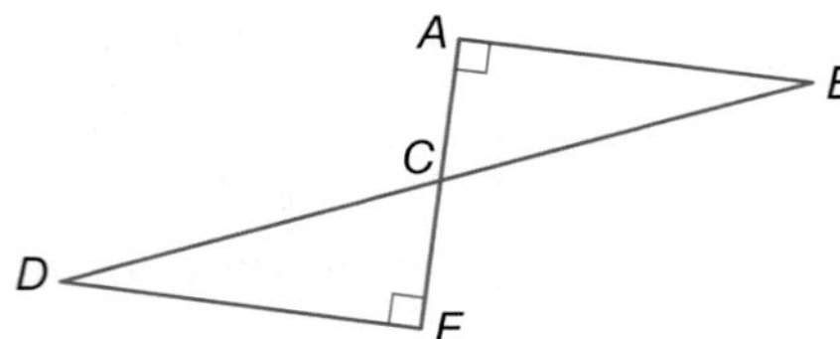


**18.** **Given:** $\angle F \cong \angle J$, $\overline{FH} \parallel \overline{GJ}$

**Prove:** $\overline{FH} \cong \overline{JG}$

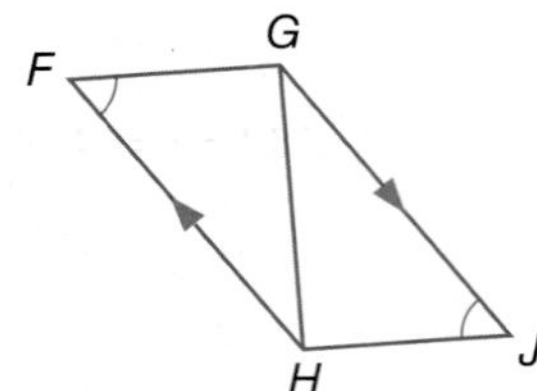


**PROOF** **Write a two-column proof.**

**19.** **Given:** $\angle K \cong \angle M$, $\overline{KP} \perp \overline{PR}$, $\overline{MR} \perp \overline{PR}$

**Prove:** $\angle KPL \cong \angle MRL$

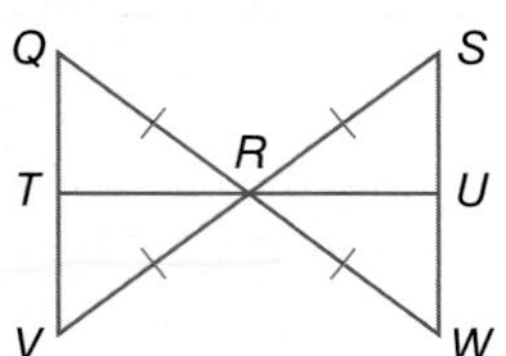


**20.** **Given:** $\overline{QR} \cong \overline{SR} \cong \overline{WR} \cong \overline{VR}$

**Prove:** $\overline{QT} \cong \overline{WU}$

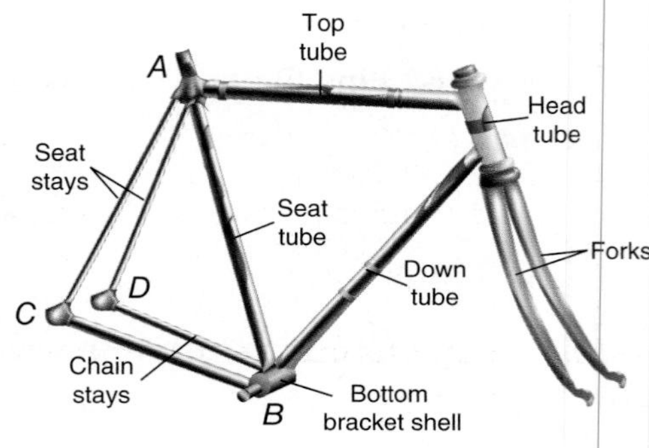


**21** **FITNESS** The seat tube of a bicycle forms a triangle with each seat and chain stay as shown. If each seat stay makes a 44° angle with its corresponding chain stay and each chain stay makes a 68° angle with the seat tube, show that the two seat stays are the same length.



## H.O.T. Problems Use Higher-Order Thinking Skills

**22.** **OPEN ENDED** Draw and label two triangles that could be proved congruent by ASA.

**23.** **CCSS CRITIQUE** Tyrone says it is not possible to show that $\triangle ADE \cong \triangle ACB$. Lorenzo disagrees, explaining that since $\angle ADE \cong \angle ACB$, and $\angle A \cong \angle A$ by the Reflexive Property, $\triangle ADE \cong \triangle ACB$. Is either of them correct? Explain.

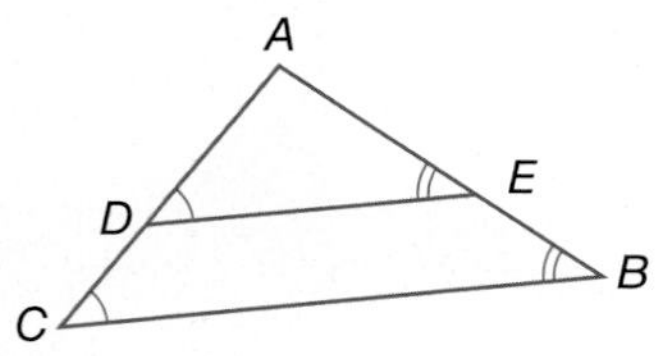


**24.** **REASONING** Find a counterexample to show why SSA (Side-Side-Angle) cannot be used to prove the congruence of two triangles.

**25.** **CHALLENGE** Using the information given in the diagram, write a flow proof to show that $\triangle PVQ \cong \triangle SVT$.

**26.** **WRITING IN MATH** How do you know what method (SSS, SAS, etc.) to use when proving triangle congruence? Use a chart to explain your reasoning.

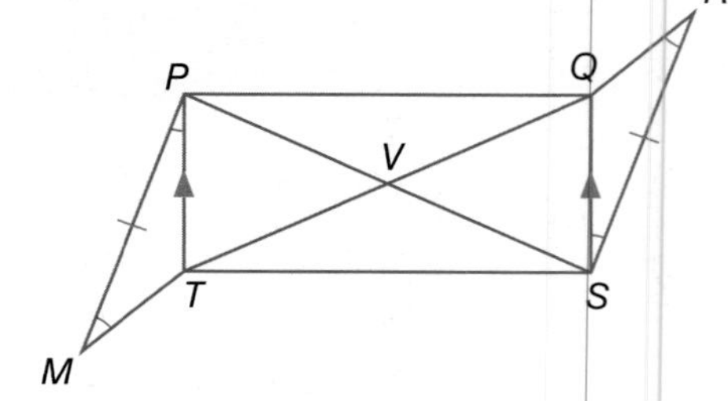

## Standardized Test Practice

**27.** Given: $\overline{BC}$ is perpendicular to $\overline{AD}$; $\angle 1 \cong \angle 2$.

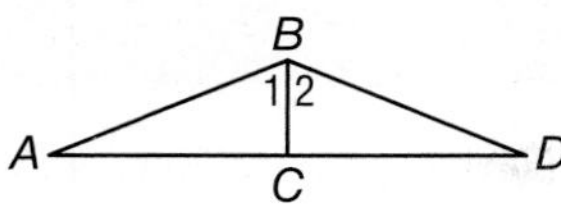


Which theorem or postulate could be used to prove $\triangle ABC \cong \triangle DBC$?

**A** AAS **C** SAS

**B** ASA **D** SSS

**28. SHORT RESPONSE** Write an expression that can be used to find the values of $s(n)$ in the table.

| $n$ | −8 | −4 | −1 | 0 | 1 |
|---|---|---|---|---|---|
| $s(n)$ | 1.00 | 2.00 | 2.75 | 3.00 | 3.25 |

**29. ALGEBRA** If −7 is multiplied by a number greater than 1, which of the following describes the result?

**F** a number greater than 7

**G** a number between −7 and 7

**H** a number greater than −7

**J** a number less than −7

**30. SAT/ACT** $\sqrt{121 + 104} = ?$

**A** 15

**B** 21

**C** 25

**D** 125

**E** 225

## Spiral Review

**Determine whether $\triangle ABC \cong \triangle XYZ$. Explain.** (Lesson 6-3)

**31.** $A(6, 4)$, $B(1, -6)$, $C(-9, 5)$, $X(0, 7)$, $Y(5, -3)$, $Z(15, 8)$

**32.** $A(0, 5)$, $B(0, 0)$, $C(-2, 0)$, $X(4, 8)$, $Y(4, 3)$, $Z(6, 3)$

**33. ALGEBRA** If $\triangle RST \cong \triangle JKL$, $RS = 7$, $ST = 5$, $RT = 9 + x$, $JL = 2x - 10$, and $JK = 4y - 5$, draw and label a figure to represent the congruent triangles. Then find $x$ and $y$. (Lesson 6-2)

**Given the following information, determine which lines, if any, are parallel. State the postulate or theorem that justifies your answer.** (Lesson 5-6)

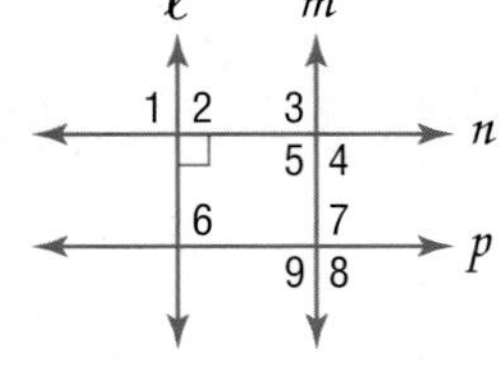


**34.** $m\angle 4 + m\angle 7 = 180$

**35.** $\angle 1 \cong \angle 4$

**36.** $m\angle 6 = 90$

## Skills Review

**PROOF Write a two-column proof for each of the following.**

**37. Given:** $\angle 2 \cong \angle 1$

$\angle 1 \cong \angle 3$

**Prove:** $\overline{AB} \parallel \overline{DE}$

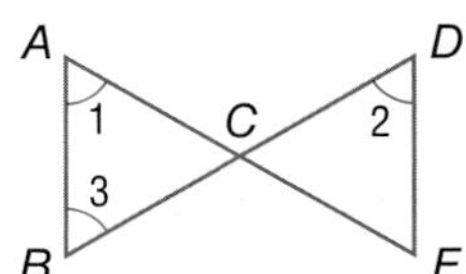


**38. Given:** $\angle MJK \cong \angle KLM$

$\angle LMJ$ and $\angle KLM$ are supplementary.

**Prove:** $\overline{KJ} \parallel \overline{LM}$

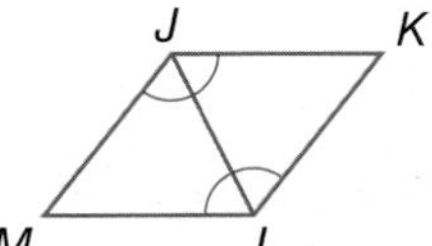

EXTEND 6-4

# Geometry Lab
# Congruence in Right Triangles

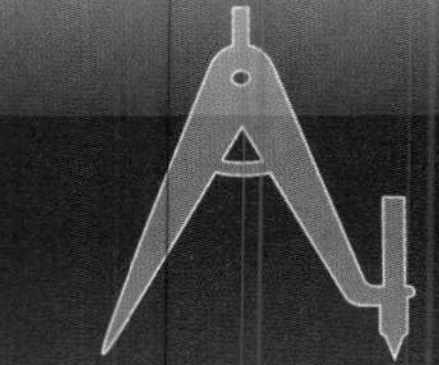

In Lessons 6-3 and 6-4, you learned theorems and postulates to prove triangles congruent. How do these theorems and postulates apply to right triangles?

**CCSS Common Core State Standards**
**Content Standards**
**G.SRT.5** Use congruence and similarity criteria for triangles to solve problems and to prove relationships in geometric figures.
**Mathematical Practices** 5

Study each pair of right triangles.

**a.**
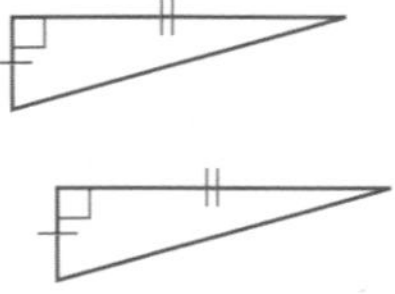

**b.**
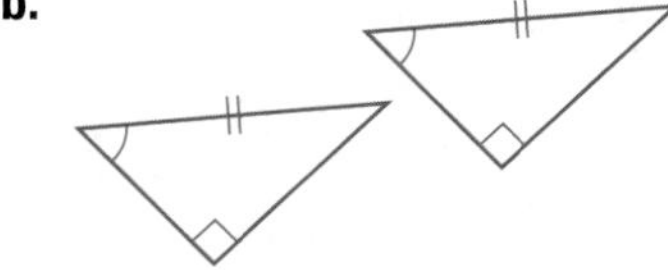

**c.**

## Analyze

1. Is each pair of triangles congruent? If so, which congruence theorem or postulate applies?
2. Rewrite the congruence rules from Exercise 1 using *leg*, (L), or *hypotenuse*, (H), to replace *side*. Omit the *A* for any right angle since we know that all right triangles contain a right angle and all right angles are congruent.
3. **MAKE A CONJECTURE** If you know that the corresponding legs of two right triangles are congruent, what other information do you need to declare the triangles congruent? Explain.

In Lesson 6-4, you learned that SSA is not a valid test for determining triangle congruence. Can SSA be used to prove right triangles congruent?

### Activity SSA and Right Triangles

**Step 1**

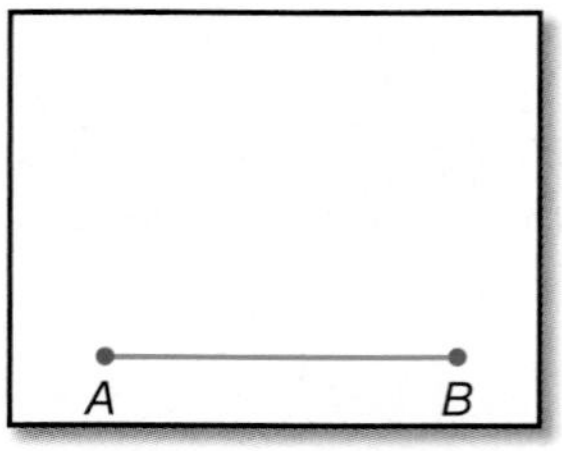

Draw $\overline{AB}$ so that $AB = 6$ centimeters.

**Step 2**

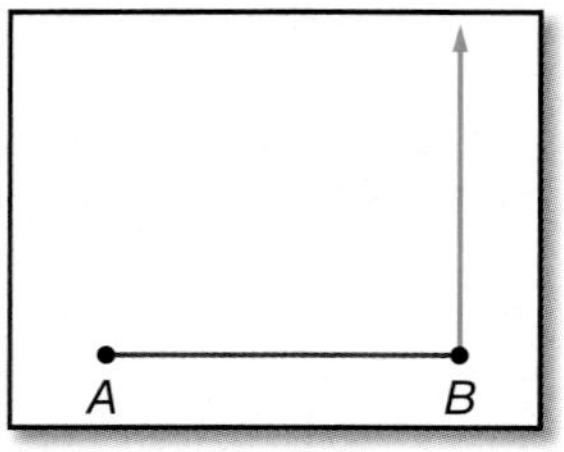

Use a protractor to draw a ray from *B* that is perpendicular to $\overline{AB}$.

**Step 3**

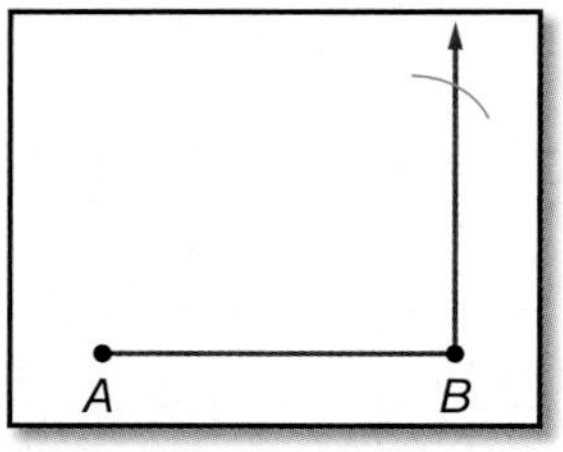

Open your compass to a width of 8 centimeters. Place the point at *A* and draw an arc to intersect the ray.

**Step 4**

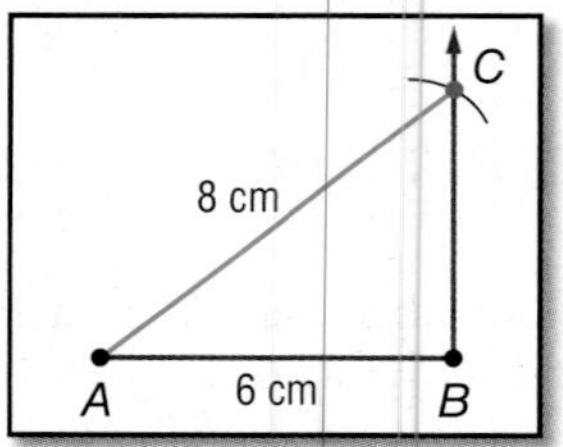

Label the intersection *C* and draw $\overline{AC}$ to complete $\triangle ABC$.

## Analyze

4. Does the model yield a unique triangle?
5. Can you use the lengths of the hypotenuse and a leg to show right triangles are congruent?
6. **Make a conjecture** about the case of SSA that exists for right triangles.

(continued on the next page)

Geometry Lab

# Congruence in Right Triangles *Continued*

Your work on the previous page provides evidence for four ways to prove right triangles congruent.

## Theorem Right Triangle Congruence

**Theorem 6.6 Leg-Leg Congruence**
If the legs of one right triangle are congruent to the corresponding legs of another right triangle, then the triangles are congruent.

**Abbreviation** *LL*

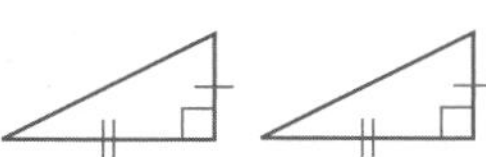

**Theorem 6.7 Hypotenuse-Angle Congruence**
If the hypotenuse and acute angle of one right triangle are congruent to the hypotenuse and corresponding acute angle of another right triangle, then the two triangles are congruent.

**Abbreviation** *HA*

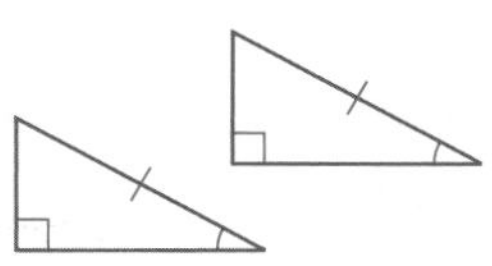

**Theorem 6.8 Leg-Angle Congruence**
If one leg and an acute angle of one right triangle are congruent to the corresponding leg and acute angle of another right triangle, then the triangles are congruent.

**Abbreviation** *LA*

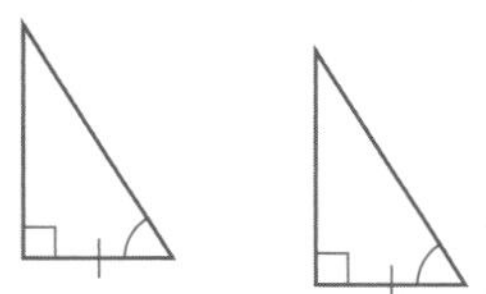

**Theorem 6.9 Hypotenuse-Leg Congruence**
If the hypotenuse and a leg of one right triangle are congruent to the hypotenuse and corresponding leg of another right triangle, then the triangles are congruent.

**Abbreviation** *HL*

## Exercises

**Determine whether each pair of triangles is congruent. If yes, tell which postulate or theorem applies.**

**7.** 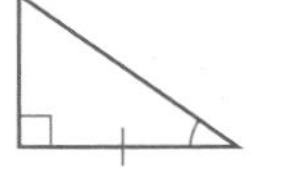

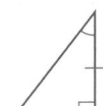

**8.** 

**9.** 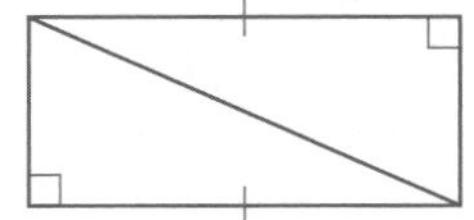

**PROOF Write a proof for each of the following.**

**10.** Theorem 6.6

**11.** Theorem 6.7

**12.** Theorem 6.8 (*Hint*: There are two possible cases.)

**13.** Theorem 6.9 (*Hint*: Use the Pythagorean Theorem.)

**Use the figure at the right.**

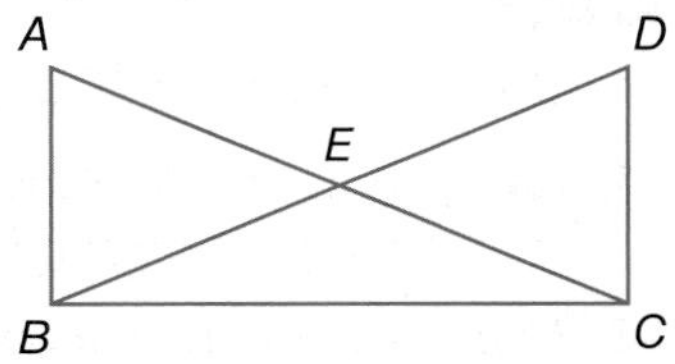


**14. Given:** $\overline{AB} \perp \overline{BC}$, $\overline{DC} \perp \overline{BC}$
$\overline{AC} \cong \overline{BD}$

**Prove:** $\overline{AB} \cong \overline{DC}$

**15. Given:** $\overline{AB} \parallel \overline{DC}$, $\overline{AB} \perp \overline{BC}$
$E$ is the midpoint of $\overline{AC}$ and $\overline{BD}$.

**Prove:** $\overline{AC} \cong \overline{DB}$

LESSON

# 6-5 Isosceles and Equilateral Triangles

| Then | Now | Why? |
|---|---|---|
| You identified isosceles and equilateral triangles. | 1 Use properties of isosceles triangles. 2 Use properties of equilateral triangles. | The tracks on the roller coaster have triangular reinforcements between the tracks for support and stability. The triangle supports in the photo are isosceles triangles. |

**NewVocabulary**

legs of an isosceles triangle
vertex angle
base angles

**Common Core State Standards**

**Content Standards**

G.CO.10 Prove theorems about triangles.

G.CO.12 Make formal geometric constructions with a variety of tools and methods (compass and straightedge, string, reflective devices, paper folding, dynamic geometric software, etc.).

**Mathematical Practices**

2 Reason abstractly and quantitatively.

3 Construct viable arguments and critique the reasoning of others.

## 1 Properties of Isosceles Triangles

Recall that isosceles triangles have at least two congruent sides. The parts of an isosceles triangle have special names.

The two congruent sides are called the **legs of an isosceles triangle**, and the angle with sides that are the legs is called the **vertex angle**. The side of the triangle opposite the vertex angle is called the *base*. The two angles formed by the base and the congruent sides are called the **base angles**.

$\angle 1$ is the vertex angle.

$\angle 2$ and $\angle 3$ are the base angles.

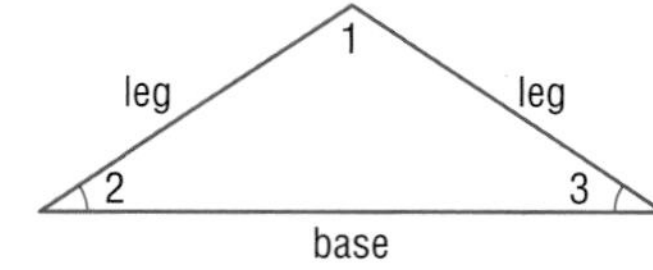


### Theorems Isosceles Triangle

**6.10 Isosceles Triangle Theorem**
If two sides of a triangle are congruent, then the angles opposite those sides are congruent.

**Example** If $\overline{AC} \cong \overline{BC}$, then $\angle 2 \cong \angle 1$.

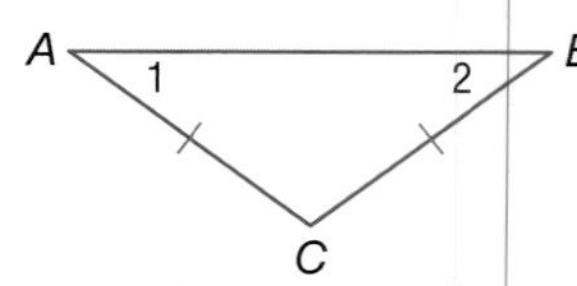


**6.11 Converse of Isosceles Triangle Theorem**
If two angles of a triangle are congruent, then the sides opposite those angles are congruent.

**Example** If $\angle 1 \cong \angle 2$, then $\overline{FE} \cong \overline{DE}$.

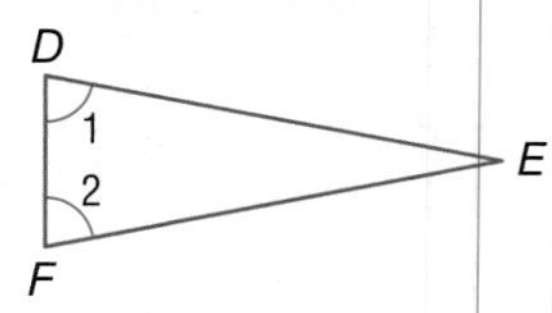


You will prove Theorem 6.11 in Exercise 37.

### Example 1 Congruent Segments and Angles

PT

**a. Name two unmarked congruent angles.**

$\angle ACB$ is opposite $\overline{AB}$ and $\angle B$ is opposite $\overline{AC}$, so $\angle ACB \cong \angle B$.

**b. Name two unmarked congruent segments.**

$\overline{AD}$ is opposite $\angle ACD$ and $\overline{AC}$ is opposite $\angle D$, so $\overline{AD} \cong \overline{AC}$.

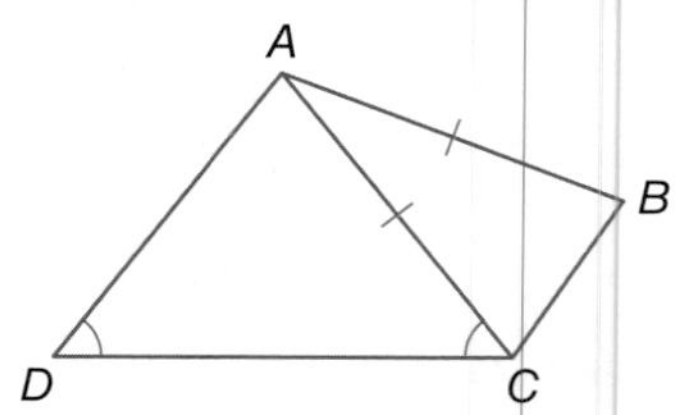

## GuidedPractice

**1A.** Name two unmarked congruent angles.

**1B.** Name two unmarked congruent segments.

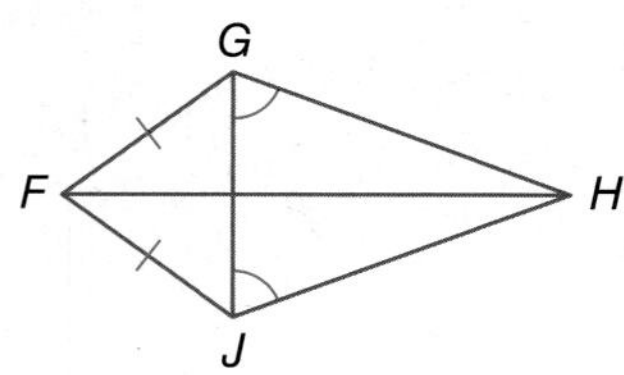


To prove the Isosceles Triangle Theorem, draw an auxiliary line and use the two triangles formed.

### Proof Isosceles Triangle Theorem

**Given:** $\triangle LMP$; $\overline{LM} \cong \overline{LP}$

**Prove:** $\angle M \cong \angle P$

**Proof:**

| Statements | Reasons |
|---|---|
| **1.** Let $N$ be the midpoint of $\overline{MP}$. | **1.** Every segment has exactly one midpoint. |
| **2.** Draw an auxiliary segment $\overline{LN}$. | **2.** Two points determine a line. |
| **3.** $\overline{MN} \cong \overline{PN}$ | **3.** Midpoint Theorem |
| **4.** $\overline{LN} \cong \overline{LN}$ | **4.** Reflexive Property of Congruence |
| **5.** $\overline{LM} \cong \overline{LP}$ | **5.** Given |
| **6.** $\triangle LMN \cong \triangle LPN$ | **6.** SSS |
| **7.** $\angle M \cong \angle P$ | **7.** CPCTC |

## 2 Properties of Equilateral Triangles

The Isosceles Triangle Theorem leads to two corollaries about the angles of an equilateral triangle.

**ReviewVocabulary**

**equilateral triangle** a triangle with three congruent sides

### Corollaries Equilateral Triangle

**6.3** A triangle is equilateral if and only if it is equiangular.

**Example** If $\angle A \cong \angle B \cong \angle C$, then $\overline{AB} \cong \overline{BC} \cong \overline{CA}$.

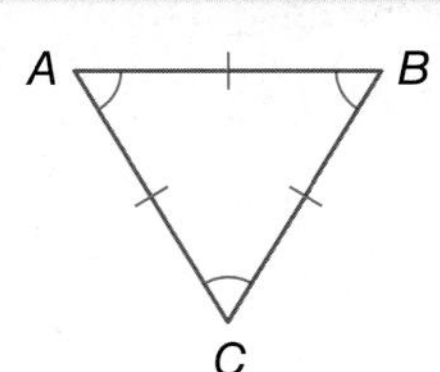


**6.4** Each angle of an equilateral triangle measures 60.

**Example** If $\overline{DE} \cong \overline{EF} \cong \overline{FE}$, then $m\angle A = m\angle B = m\angle C = 60$.

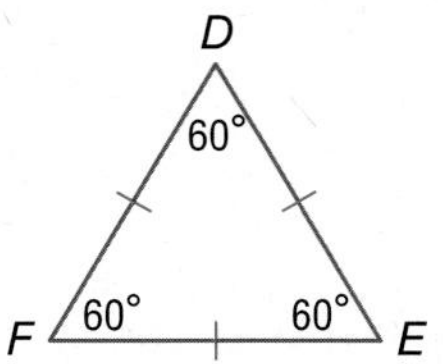


You will prove Corollaries 6.3 and 6.4 in Exercises 35 and 36.

## Example 2 Find Missing Measures

**Find each measure.**

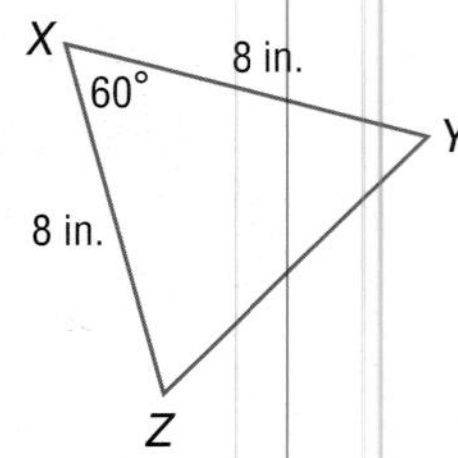


**a.** $m\angle Y$

Since $XY = XZ$, $\overline{XY} \cong \overline{XZ}$. By the Isosceles Triangle Theorem, base angles $Z$ and $Y$ are congruent, so $m\angle Z = m\angle Y$. Use the Triangle Sum Theorem to write and solve an equation to find $m\angle Y$.

| | |
|---|---|
| $m\angle X + m\angle Y + m\angle Z = 180$ | Triangle Sum Theorem |
| $60 + m\angle Y + m\angle Y = 180$ | $m\angle X = 60$, $m\angle Z = m\angle Y$ |
| $60 + 2(m\angle Y) = 180$ | Simplify. |
| $2(m\angle Y) = 120$ | Subtract 60 from each side. |
| $m\angle Y = 60$ | Divide each side by 2. |

**StudyTip**

Isosceles Triangles As you discovered in Example 2, any isosceles triangle that has one 60° angle must be an equilateral triangle.

**b.** $YZ$

$m\angle Z = m\angle Y$, so $m\angle Z = 60$ by substitution. Since $m\angle X = 60$, all three angles measure 60, so the triangle is equiangular. Because an equiangular triangle is also equilateral, $XY = XZ = ZY$. Since $XY = 8$ inches, $YZ = 8$ inches by substitution.

### GuidedPractice

**2A.** $m\angle M$

**2B.** $PN$

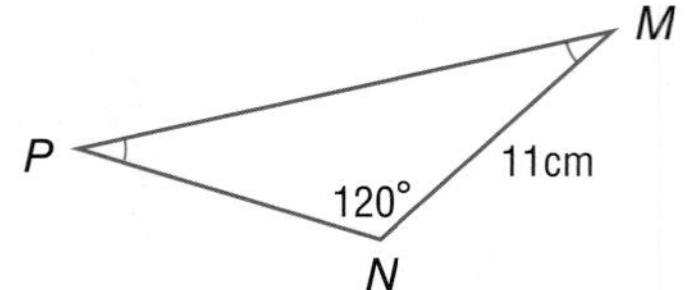

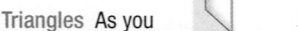

You can use the properties of equilateral triangles and algebra to find missing values.

## Example 3 Find Missing Values

**ALGEBRA Find the value of each variable.**

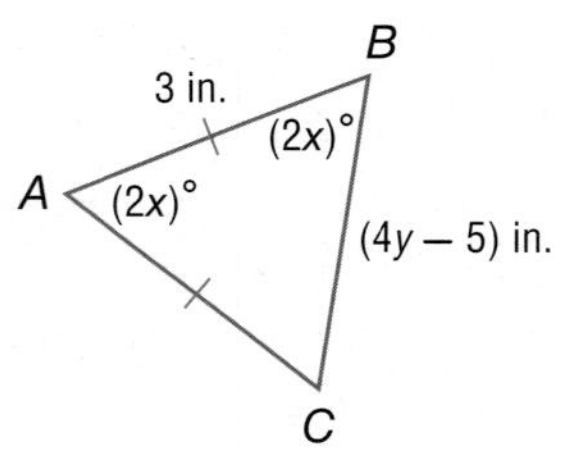


Since $\angle B = \angle A$, $\overline{AC} \cong \overline{BC}$ by the Converse of the Isosceles Triangle Theorem. All of the sides of the triangle are congruent, so the triangle is equilateral. Each angle of an equilateral triangle measures 60°, so $2x = 60$ and $x = 30$.

The triangle is equilateral, so all of the sides are congruent, and the lengths of all of the sides are equal.

| | |
|---|---|
| $AB = BC$ | Definition of equilateral triangle |
| $3 = 4y - 5$ | Substitution |
| $8 = 4y$ | Add 5 to each side. |
| $2 = y$ | Divide each side by 4. |

### GuidedPractice

**3.** Find the value of each variable.

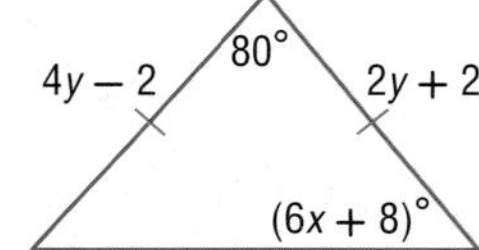

### Real-WorldLink

Biosphere II is the largest totally enclosed ecosystem ever built, covering 3.14 acres in Oracle, Arizona. The controlled-environment facility is 91 feet at its highest point, and it has 6500 windows that enclose a volume of 7.2 million cubic feet.

**Source:** University of Arizona

## Real-World Example 4 Apply Triangle Congruence

**ENVIRONMENT Refer to the photo of Biosphere II at the right. $\triangle ACE$ is an equilateral triangle. $F$ is the midpoint of $\overline{AE}$, $D$ is the midpoint of $\overline{EC}$, and $B$ is the midpoint of $\overline{CA}$. Prove that $\triangle FBD$ is also equilateral.**

**Given:** $\triangle ACE$ is equilateral. $F$ is the midpoint of $\overline{AE}$, $D$ is the midpoint of $\overline{EC}$, and $B$ is the midpoint of $\overline{CA}$.

**Prove:** $\triangle FBD$ is equilateral.

**Proof:**

| Statements | Reasons |
|---|---|
| **1.** $\triangle ACE$ is equilateral. | **1.** Given |
| **2.** $F$ is the midpoint of $AE$, $D$ is the midpoint of $EC$, and $B$ is the midpoint of $CA$. | **2.** Given |
| **3.** $m\angle A = 60$, $m\angle C = 60$, $m\angle E = 60$ | **3.** Each angle of an equilateral triangle measures 60. |
| **4.** $\angle A \cong \angle C \cong \angle E$ | **4.** Definition of congruence and substitution |
| **5.** $\overline{AE} \cong \overline{EC} \cong \overline{CA}$ | **5.** Definition of equilateral triangle |
| **6.** $AE = EC = CA$ | **6.** Definition of congruence |
| **7.** $\overline{AF} \cong \overline{FE}$, $\overline{ED} \cong \overline{DC}$, $\overline{CB} \cong \overline{BA}$ | **7.** Midpoint Theorem |
| **8.** $AF = FE$, $ED = DC$, $CB = BA$ | **8.** Definition of congruence |
| **9.** $AF + FE = AE$, $ED + DC = EC$, $CB + BA = CA$ | **9.** Segment Addition Postulate |
| **10.** $AF + AF = AE$, $FE + FE = AE$, $ED + ED = EC$, $DC + DC = EC$, $CB + CB = CA$, $BA + BA = CA$ | **10.** Substitution |
| **11.** $2AF = AE$, $2FE = AE$, $2ED = EC$, $2DC = EC$, $2CB = CA$, $2BA = CA$ | **11.** Addition Property |
| **12.** $2AF = AE$, $2FE = AE$, $2ED = AE$, $2DC = AE$, $2CB = AE$, $2BA = AE$ | **12.** Substitution Property |
| **13.** $2AF = 2ED = 2CB$, $2FE = 2DC = 2BA$ | **13.** Transitive Property |
| **14.** $AF = ED = CB$, $FE = DC = BA$ | **14.** Division Property |
| **15.** $\overline{AF} \cong \overline{ED} \cong \overline{CB}$, $\overline{FE} \cong \overline{DC} \cong \overline{BA}$ | **15.** Definition of congruence |
| **16.** $\triangle AFB \cong \triangle EDF \cong \triangle CBD$ | **16.** SAS |
| **17.** $\overline{DF} \cong \overline{FB} \cong \overline{BD}$ | **17.** CPCTC |
| **18.** $\triangle FBD$ is equilateral. | **18.** Definition of equilateral triangle |

### GuidedPractice

**4.** Given that $\triangle ACE$ is equilateral, $\overline{FB} \parallel \overline{EC}$, $\overline{FD} \parallel \overline{BC}$, $\overline{BD} \parallel \overline{EF}$, and $D$ is the midpoint of $\overline{EC}$, prove that $\triangle FED \cong \triangle BDC$.

(l)Emily Riddell/age footstock, (r)Horst Mahr/age fotostock

## Check Your Understanding

**Example 1** **Refer to the figure at the right.**

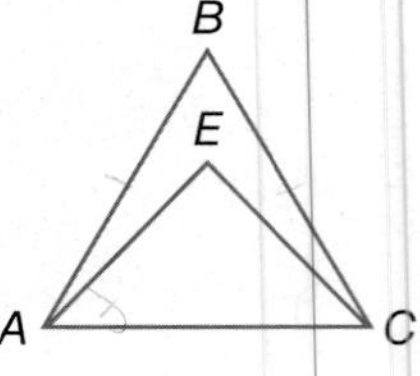


1. If $\overline{AB} \cong \overline{CB}$, name two congruent angles.

2. If $\angle EAC \cong \angle ECA$, name two congruent segments.

**Example 2** **Find each measure.**

3. $FH$

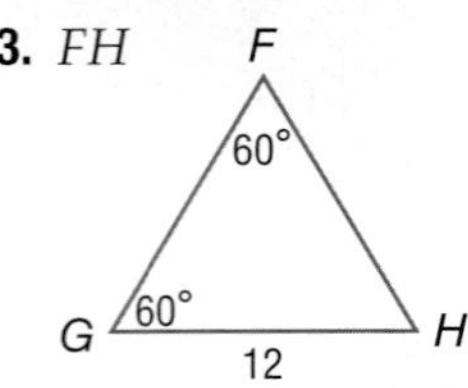


4. $m\angle MRP$

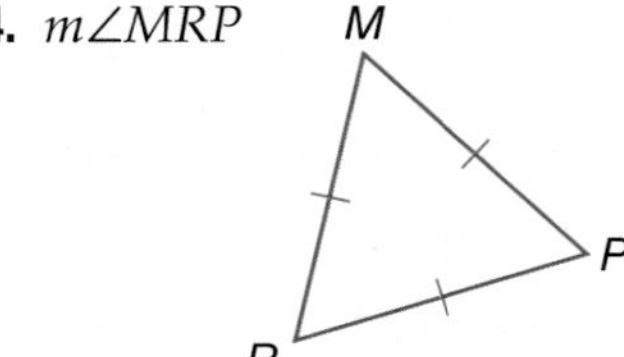


**Example 3** **CCSS SENSE-MAKING** **Find the value of each variable.**

5.

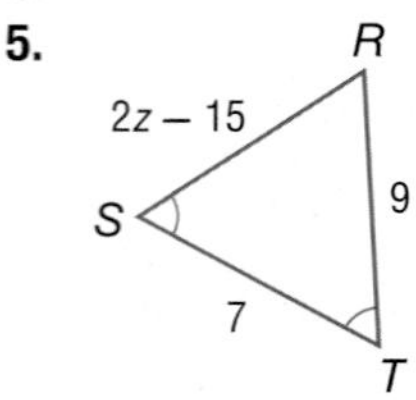


6.

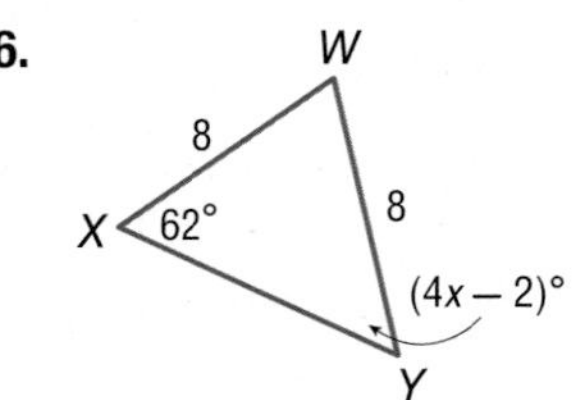


**Example 4**

7. **PROOF** **Write a two-column proof.**

**Given:** $\triangle ABC$ is isosceles; $\overline{EB}$ bisects $\angle ABC$.

**Prove:** $\triangle ABE \cong \triangle CBE$

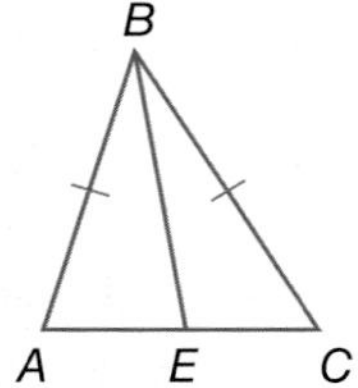


8. **ROLLER COASTERS** The roller coaster track shown in the photo on page 285 appears to be composed of congruent triangles. A portion of the track is shown.

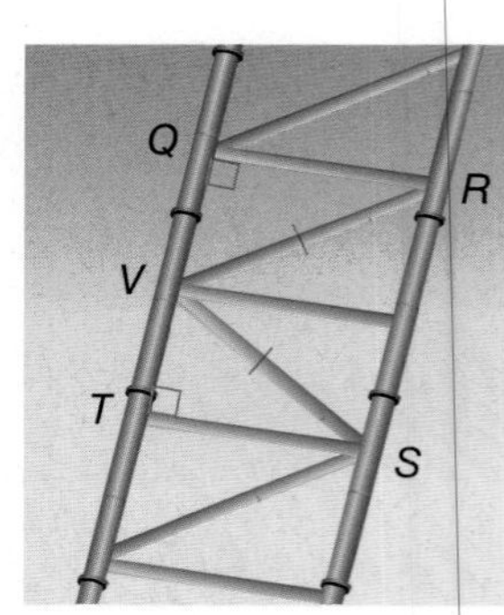


a. If $\overline{QR}$ and $\overline{ST}$ are perpendicular to $\overline{QT}$, $\triangle VSR$ is isosceles with base $\overline{SR}$, and $\overline{QT} \parallel \overline{SR}$, prove that $\triangle RQV \cong \triangle STV$.

b. If $VR = 2.5$ meters and $QR = 2$ meters, find the distance between $\overline{QR}$ and $\overline{ST}$. Explain your reasoning.

## Practice and Problem Solving

**Example 1** **Refer to the figure at the right.**

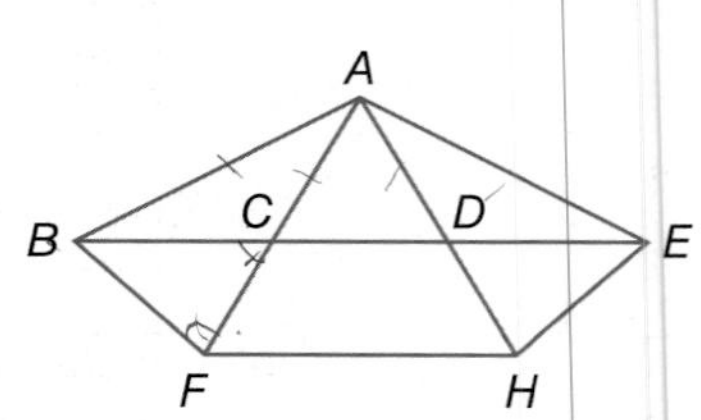


9. If $\overline{AB} \cong \overline{AE}$, name two congruent angles.

10. If $\angle ABF \cong \angle AFB$, name two congruent segments.

11. If $\overline{CA} \cong \overline{DA}$, name two congruent angles.

12. If $\angle DAE \cong \angle DEA$, name two congruent segments.

13. If $\angle BCF \cong \angle BFC$, name two congruent segments.

14. If $\overline{FA} \cong \overline{AH}$, name two congruent angles.

**Example 2** **Find each measure.**

15. $m\angle BAC$

16. $m\angle SRT$

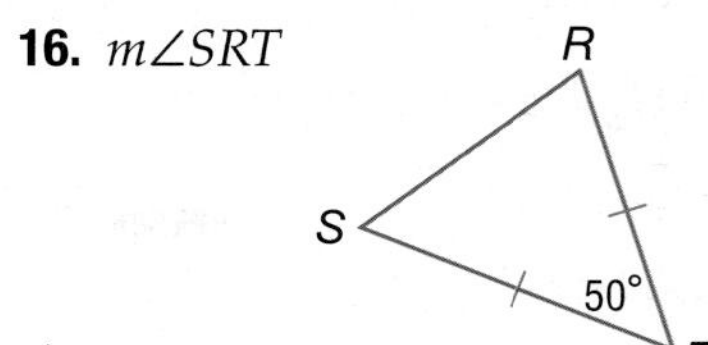


17. $TR$

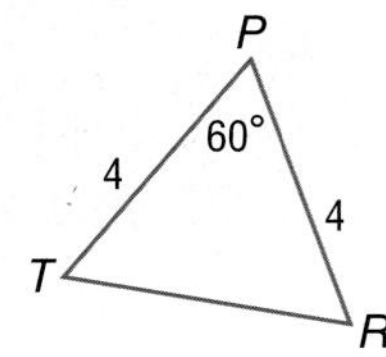


18. $CB$

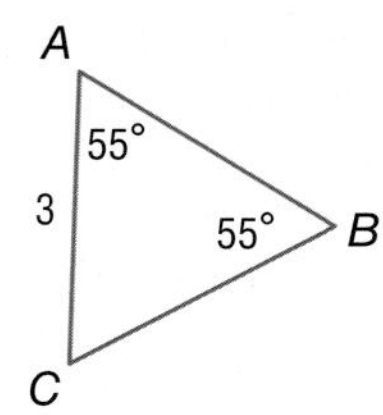


**Example 3** **CCSS REGULARITY** **Find the value of each variable.**

19.

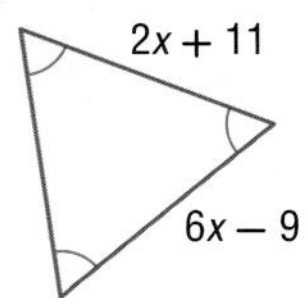


20.

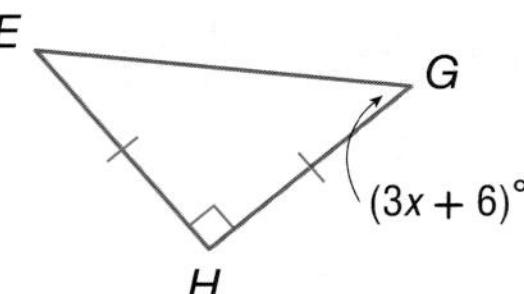


21.

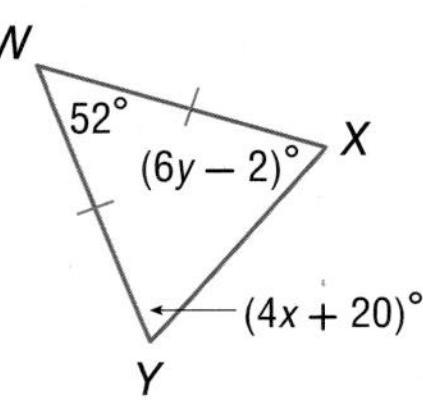


22.

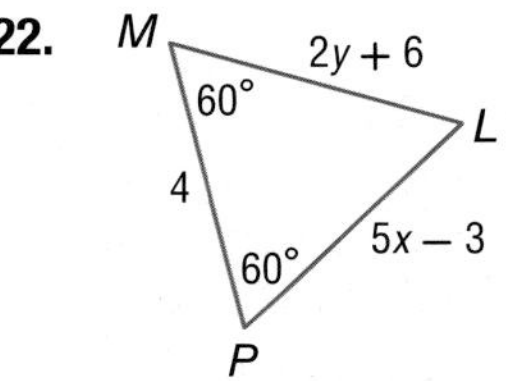


**Example 4** **PROOF** **Write a paragraph proof.**

23. **Given:** $\triangle HJM$ is isosceles, and $\triangle HKL$ is equilateral. $\angle JKH$ and $\angle HKL$ are supplementary and $\angle HLK$ and $\angle MLH$ are supplementary.
**Prove:** $\angle JHK \cong \angle MHL$

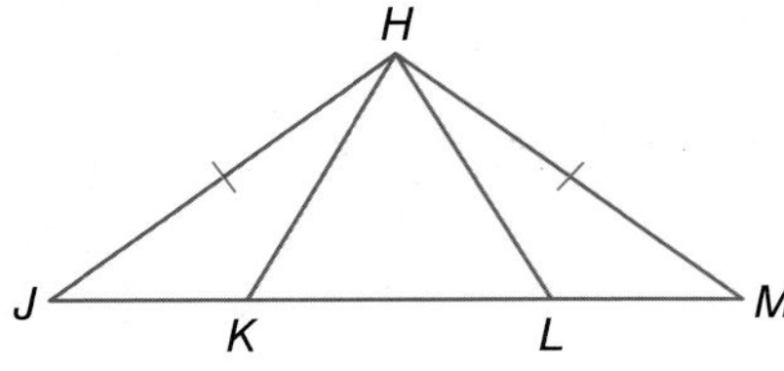


24. **Given:** $\overline{XY} \cong \overline{XZ}$
$W$ is the midpoint of $\overline{XY}$.
$Q$ is the midpoint of $\overline{XZ}$.
**Prove:** $\overline{WZ} \cong \overline{QY}$

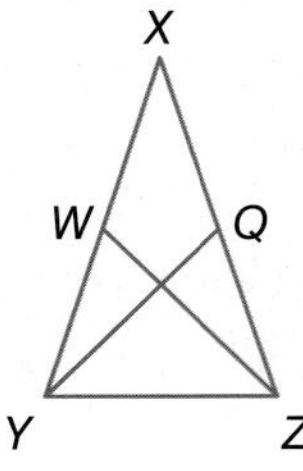


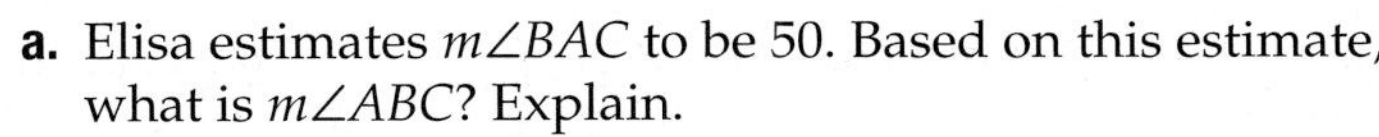

25. **BABYSITTING** While babysitting her neighbor's children, Elisa observes that the supports on either side of a park swing set form two sets of triangles. Using a jump rope to measure, Elisa is able to determine that $\overline{AB} \cong \overline{AC}$, but $\overline{BC} \not\cong \overline{AB}$.

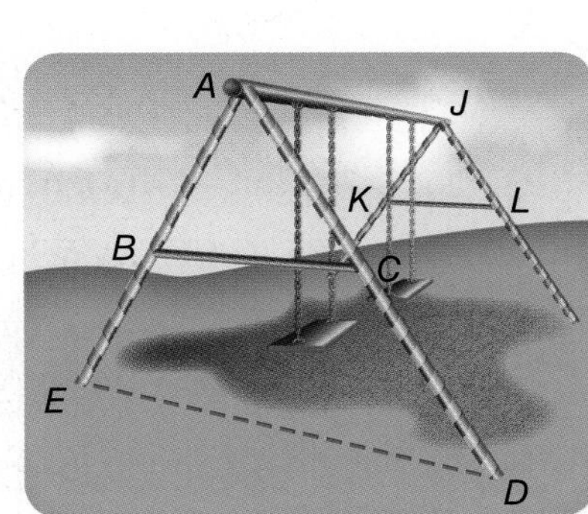


a. Elisa estimates $m\angle BAC$ to be 50. Based on this estimate, what is $m\angle ABC$? Explain.

b. If $\overline{BE} \cong \overline{CD}$, show that $\triangle AED$ is isosceles.

c. If $\overline{BC} \parallel \overline{ED}$ and $\overline{ED} \cong \overline{AD}$, show that $\triangle AED$ is equilateral.

d. If $\triangle JKL$ is isosceles, what is the minimum information needed to prove that $\triangle ABC \cong \triangle JLK$? Explain your reasoning.

**26. CHIMNEYS** In the picture, $\overline{BD} \perp \overline{AC}$ and $\triangle ABC$ is an isosceles triangle with base $\overline{AC}$. Show that the chimney of the house, represented by $\overline{BD}$, bisects the angle formed by the sloped sides of the roof, $\angle ABC$.

**27. CONSTRUCTION** Construct three different isosceles right triangles. Explain your method. Then verify your constructions using measurement and mathematics.

**28. PROOF** Based on your construction in Exercise 27, make and prove a conjecture about the relationship between the base angles of an isosceles right triangle.

**CCSS REGULARITY** **Find each measure.**

**29** $m\angle CAD$

**30.** $m\angle ACD$

**31.** $m\angle ACB$

**32.** $m\angle ABC$

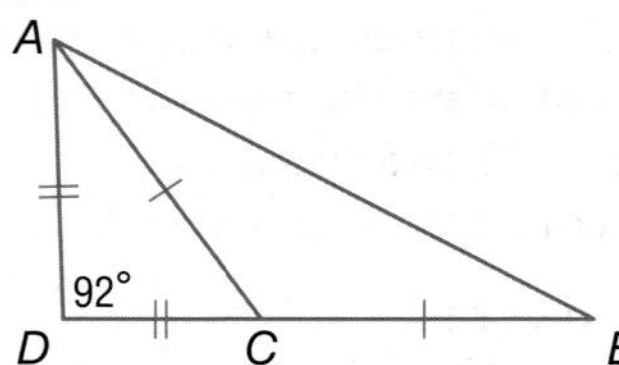


**33. FITNESS** In the diagram, the rider will use his bike to hop across the tops of each of the concrete solids shown. If each triangle is isosceles with vertex angles $G$, $H$, and $J$, and $\overline{BG} \cong \overline{HC}$, $\overline{HD} \cong \overline{JF}$, $\angle G \cong \angle H$, and $\angle H \cong \angle J$, show that the distance from $B$ to $F$ is three times the distance from $D$ to $F$.

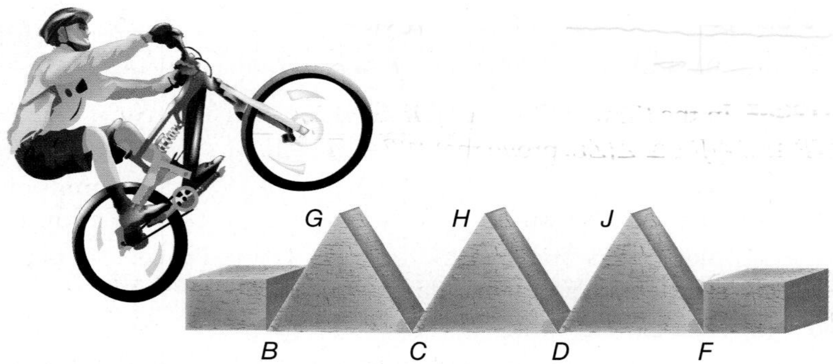


**34. Given:** $\triangle XWV$ is isosceles; $\overline{ZY} \perp \overline{YV}$.

**Prove:** $\angle X$ and $\angle YZV$ are complementary.

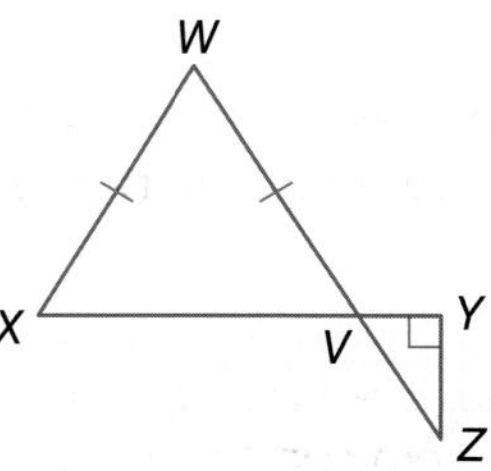


**PROOF** **Write a two-column proof of each corollary or theorem.**

**35.** Corollary 6.3

**36.** Corollary 6.4

**37.** Theorem 6.11

**Find the value of each variable.**

**38.**

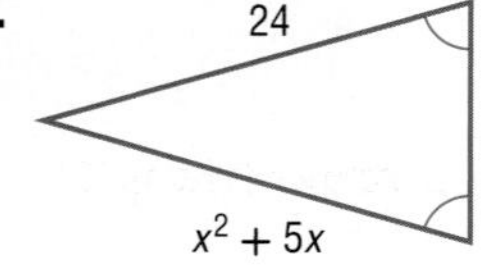


**39.**

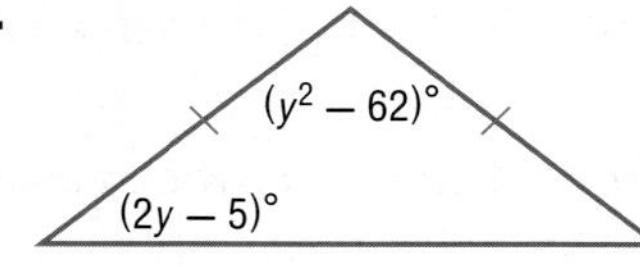

**GAMES Use the diagram of a game timer shown to find each measure.**

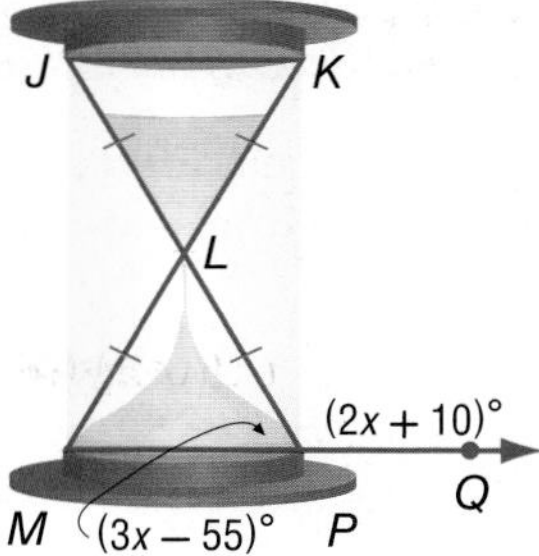


40. $m\angle LPM$

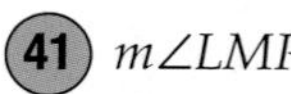

41. $m\angle LMP$

42. $m\angle JLK$

43. $m\angle JKL$

44. **MULTIPLE REPRESENTATIONS** In this problem, you will explore possible measures of the interior angles of an isosceles triangle given the measure of one exterior angle.

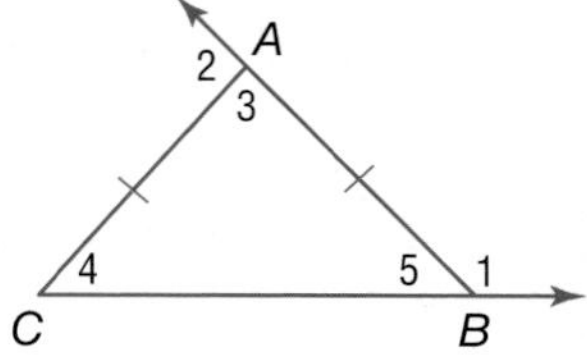


   a. **Geometric** Use a ruler and a protractor to draw three different isosceles triangles, extending one of the sides adjacent to the vertex angle and to one of the base angles, and labeling as shown.

   b. **Tabular** Use a protractor to measure and record $m\angle 1$ for each triangle. Use $m\angle 1$ to calculate the measures of $\angle 3$, $\angle 4$, and $\angle 5$. Then find and record $m\angle 2$ and use it to calculate these same measures. Organize your results in two tables.

   c. **Verbal** Explain how you used $m\angle 1$ to find the measures of $\angle 3$, $\angle 4$, and $\angle 5$. Then explain how you used $m\angle 2$ to find these same measures.

   d. **Algebraic** If $m\angle 1 = x$, write an expression for the measures of $\angle 3$, $\angle 4$, and $\angle 5$. Likewise, if $m\angle 2 = x$, write an expression for these same angle measures.

## H.O.T. Problems Use Higher-Order Thinking Skills

45. **CHALLENGE** In the figure at the right, if $\triangle WJZ$ is equilateral and $\angle ZWP \cong \angle WJM \cong \angle JZL$, prove that $\overline{WP} \cong \overline{ZL} \cong \overline{JM}$.

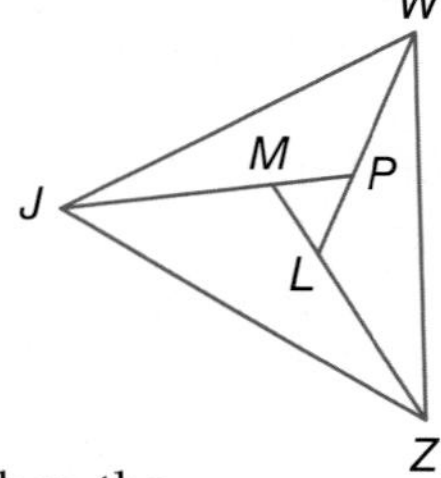


**CCSS PRECISION Determine whether the following statements are *sometimes*, *always*, or *never* true. Explain.**

46. If the measure of the vertex angle of an isosceles triangle is an integer, then the measure of each base angle is an integer.

47. If the measures of the base angles of an isosceles triangle are integers, then the measure of its vertex angle is odd.

48. **ERROR ANALYSIS** Alexis and Miguela are finding $m\angle G$ in the figure shown. Alexis says that $m\angle G = 35$, while Miguela says that $m\angle G = 60$. Is either of them correct? Explain your reasoning.

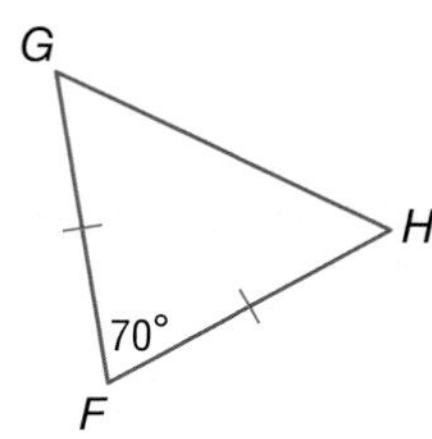


49. **OPEN ENDED** If possible, draw an isosceles triangle with base angles that are obtuse. If it is not possible, explain why not.

50. **REASONING** In isosceles $\triangle ABC$, $m\angle B = 90$. Draw the triangle. Indicate the congruent sides and label each angle with its measure.

51. **WRITING IN MATH** How can triangle classifications help you prove triangle congruence?

## Standardized Test Practice

**52. ALGEBRA** What quantity should be added to both sides of this equation to complete the square?

$$x^2 - 10x = 3$$

**A** −25 **C** 5
**B** −5 **D** 25

**53. SHORT RESPONSE** In a school of 375 students, 150 students play sports and 70 students are involved in the community service club. 30 students play sports and are involved in the community service club. How many students are *not* involved in either sports or the community service club?

**54.** In the figure $\overline{AE}$ and $\overline{BD}$ bisect each other at point *C*.

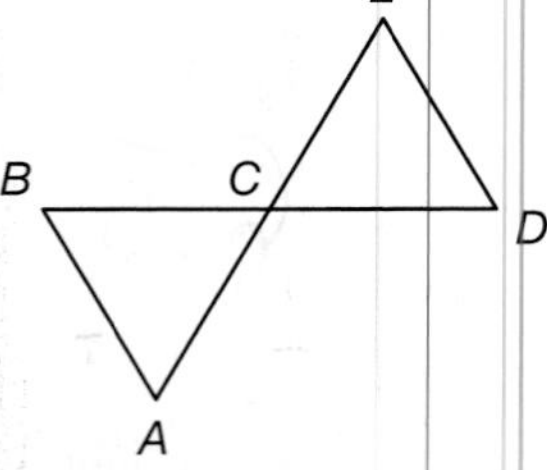


Which additional piece of information would be enough to prove that $\overline{DE} \cong \overline{DC}$?

**F** $\angle A \cong \angle BCA$ **H** $\angle ACB \cong \angle EDC$
**G** $\angle B \cong \angle D$ **J** $\angle A \cong \angle B$

**55. SAT/ACT** If $x = -3$, then $4x^2 - 7x + 5 =$

**A** 2 **C** 20 **E** 62
**B** 14 **D** 42

## Spiral Review

**56.** If $m\angle ADC = 35$, $m\angle ABC = 35$, $m\angle DAC = 26$, and $m\angle BAC = 26$, determine whether $\triangle ADC \cong \triangle ABC$. (Lesson 6-4)

**Determine whether $\triangle STU \cong \triangle XYZ$. Explain.** (Lesson 6-3)

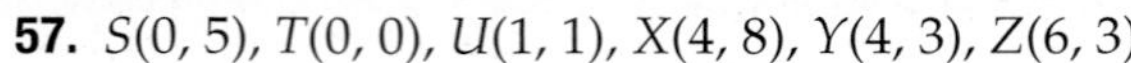

**57.** $S(0, 5)$, $T(0, 0)$, $U(1, 1)$, $X(4, 8)$, $Y(4, 3)$, $Z(6, 3)$

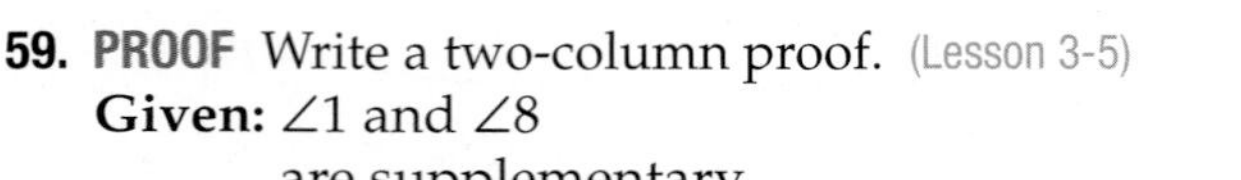

**58.** $S(2, 2)$, $T(4, 6)$, $U(3, 1)$, $X(-2, -2)$, $Y(-4, 6)$, $Z(-3, 1)$

A
C
D
B

**59. PROOF** Write a two-column proof. (Lesson 3-5)
**Given:** $\angle 1$ and $\angle 8$ are supplementary.
**Prove:** $\ell \parallel m$

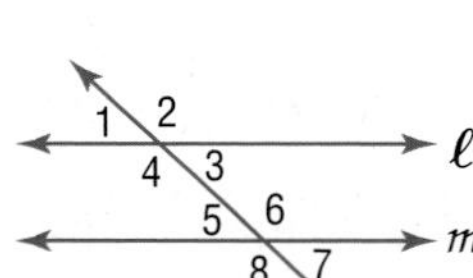


**Find the measure of each numbered angle, and name the theorems that justify your work.** (Lesson 5-4)

**60.** $\angle 2$ and $\angle 3$ are supplementary; $m\angle 2 = 149$

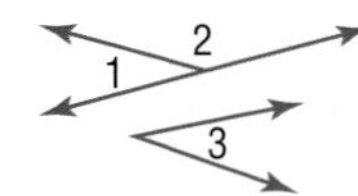


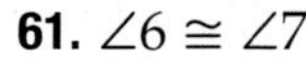

**61.** $\angle 6 \cong \angle 7$

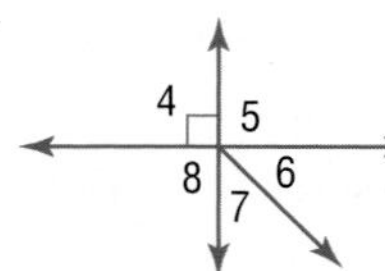


**62. PROOF** Given that $\angle BEC$ is a right angle, prove that $\angle AEB$ and $\angle CED$ are complementary. (Lesson 2-8)

## Skills Review

**63. PROOF** If $\angle ACB \cong \angle ABC$, then $\angle XCA \cong \angle YBA$.

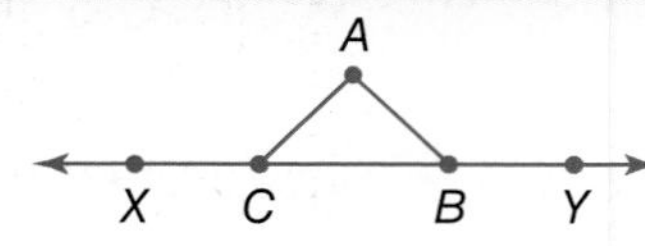

LESSON 6-6

# Triangles and Coordinate Proof

| Then | Now | Why? |
|---|---|---|
| You used coordinate geometry to prove triangle congruence. | 1 Position and label triangles for use in coordinate proofs.<br>2 Write coordinate proofs. | A global positioning system (GPS) receives transmissions from satellites that allow the exact location of a car to be determined. The information can be used with navigation software to provide driving directions. |

**NewVocabulary**
coordinate proof

**Common Core State Standards**

**Content Standards**
G.CO.10 Prove theorems about triangles.
G.GPE.4 Use coordinates to prove simple geometric theorems algebraically.

**Mathematical Practices**
3 Construct viable arguments and critique the reasoning of others.
2 Reason abstractly and quantitatively.

## 1 Position and Label Triangles

As with global positioning systems, knowing the coordinates of a figure in a coordinate plane allows you to explore its properties and draw conclusions about it. **Coordinate proofs** use figures in the coordinate plane and algebra to prove geometric concepts. The first step in a coordinate proof is placing the figure on the coordinate plane.

PT

### Example 1 Position and Label a Triangle

**Position and label right triangle $MNP$ on the coordinate plane so that leg $\overline{MN}$ is $a$ units long and leg $\overline{NP}$ is $b$ units long.**

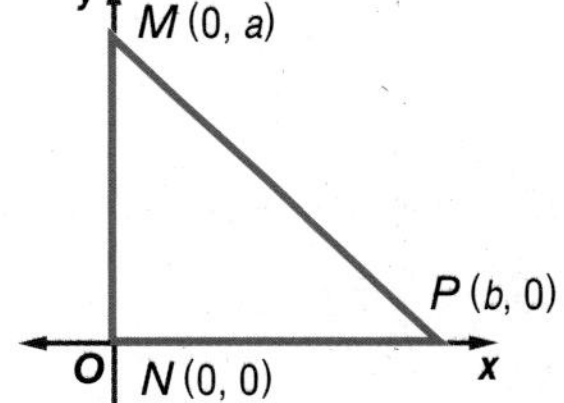


- The length(s) of the side(s) that are along the axes will be easier to determine than the length(s) of side(s) that are not along an axis. Since this is a right triangle, two sides can be located on an axis.
- Placing the right angle of the triangle, $\angle N$, at the origin will allow the two legs to be along the $x$- and $y$-axes.
- Position the triangle in the first quadrant.
- Since $M$ is on the $y$-axis, its $x$-coordinate is 0. Its $y$-coordinate is $a$ because the leg is $a$ units long.
- Since $P$ is on the $x$-axis, its $y$-coordinate is 0. Its $x$-coordinate is $b$ because the leg is $b$ units long.

**GuidedPractice**

1. Position and label isosceles triangle $JKL$ on the coordinate plane so that its base $\overline{JL}$ is $a$ units long, vertex $K$ is on the $y$-axis, and the height of the triangle is $b$ units.

### KeyConcept Placing Triangles on Coordinate Plane

**Step 1** Use the origin as a vertex or center of the triangle.

**Step 2** Place at least one side of a triangle on an axis.

**Step 3** Keep the triangle within the first quadrant if possible.

**Step 4** Use coordinates that make computations as simple as possible.

## Example 2 Identify Missing Coordinates

**Name the missing coordinates of isosceles triangle $XYZ$.**

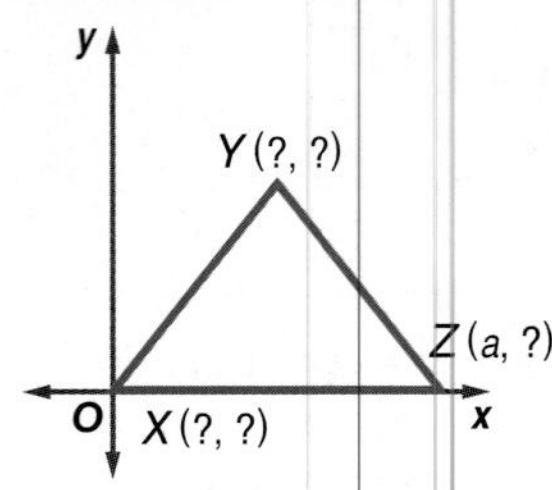


Vertex $X$ is positioned at the origin; its coordinates are $(0, 0)$.

Vertex $Z$ is on the $x$-axis, so its $y$-coordinate is 0. The coordinates of vertex $Z$ are $(a, 0)$.

$\triangle XYZ$ is isosceles, so using a vertical segment from $Y$ to the $x$-axis and the Hypotenuse-Leg Theorem shows that the $x$-coordinate of $Y$ is halfway between 0 and $a$ or $\frac{a}{2}$. We cannot write the $y$-coordinate in terms of $a$, so call it $b$. The coordinates of point $Y$ are $\left(\frac{a}{2}, b\right)$.

**StudyTip**

**Right Angle** The intersection of the $x$- and $y$-axis forms a right angle, so it is a convenient place to locate the right angle of a figure such as a right triangle.

### GuidedPractice

**2.** Name the missing coordinates of isosceles right triangle $ABC$.

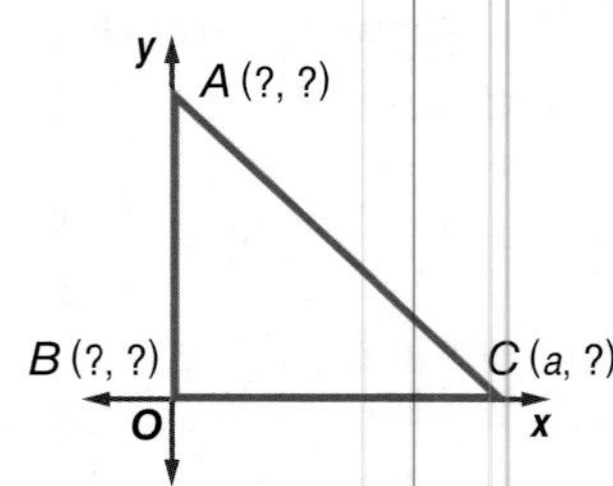


## 2 Write Coordinate Proofs

After a triangle is placed on the coordinate plane and labeled, we can use coordinate proofs to verify properties and to prove theorems.

**StudyTip**

**Coordinate Proof** The guidelines and methods used in this lesson apply to all polygons, not just triangles.

## Example 3 Write a Coordinate Proof

**Write a coordinate proof to show that a line segment joining the midpoints of two sides of a triangle is parallel to the third side.**

Place a vertex at the origin and label it $A$. Use coordinates that are multiples of 2 because the Midpoint Formula involves dividing the sum of the coordinates by 2.

y
C (2b, 2c)
S
T
x
O
A (0, 0)
B (2a, 0)

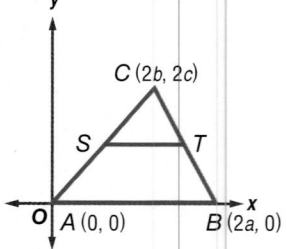

**Given:** $\triangle ABC$
$S$ is the midpoint of $\overline{AC}$.
$T$ is the midpoint of $\overline{BC}$.

**Prove:** $\overline{ST} \parallel \overline{AB}$

**Proof:**

By the Midpoint Formula, the coordinates of $S$ are $\left(\frac{2b + 0}{2}, \frac{2c + 0}{2}\right)$ or $(b, c)$ and the coordinates of $T$ are $\left(\frac{2a + 2b}{2}, \frac{0 + 2c}{2}\right)$ or $(a + b, c)$.

By the Slope Formula, the slope of $\overline{ST}$ is $\frac{c - c}{a + b - b}$ or 0 and the slope of $\overline{AB}$ is $\frac{0 - 0}{2a - 0}$ or 0.

Since $\overline{ST}$ and $\overline{AB}$ have the same slope, $\overline{ST} \parallel \overline{AB}$.

## Guided Practice

3. Write a coordinate proof to show that $\triangle ABX \cong \triangle CDX$.

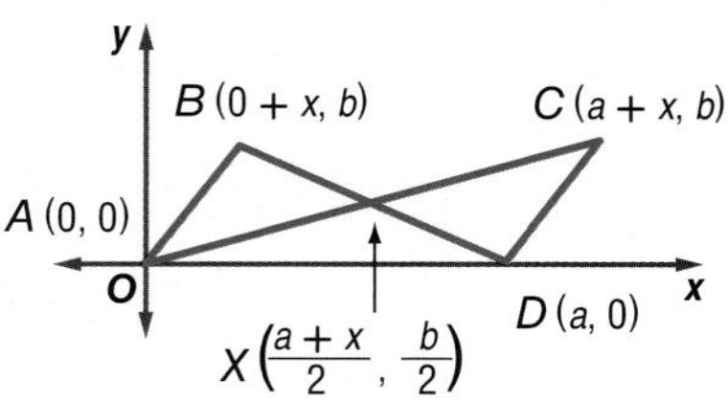


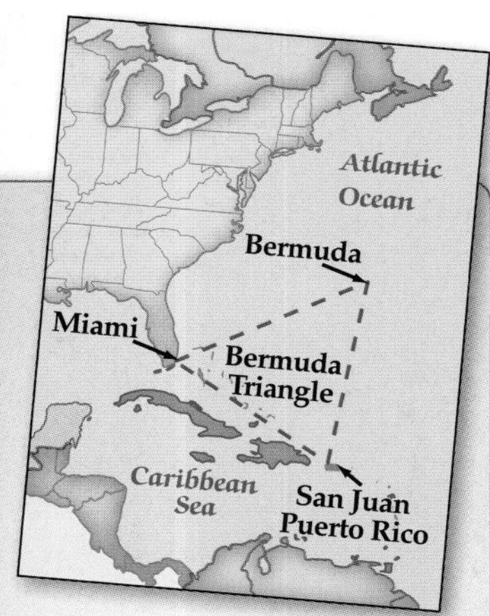


### Real-World Link

More than 50 ships and 20 airplanes have mysteriously disappeared from a section of the North Atlantic Ocean off of North America commonly referred to as the Bermuda Triangle.

**Source:** *Encyclopaedia Britannica*

The techniques used for coordinate proofs can be used to solve real-world problems.

### Real-World Example 4 Classify Triangles

**GEOGRAPHY The Bermuda Triangle is a region formed by Miami, Florida, San Jose, Puerto Rico, and Bermuda. The approximate coordinates of each location, respectively, are 25.8°N 80.27°W, 18.48°N 66.12°W, and 33.37°N 64.68°W. Write a coordinate proof to prove that the Bermuda Triangle is scalene.**

The first step is to label the coordinates of each location. Let $M$ represent Miami, $B$ represent Bermuda, and $P$ represent Puerto Rico.

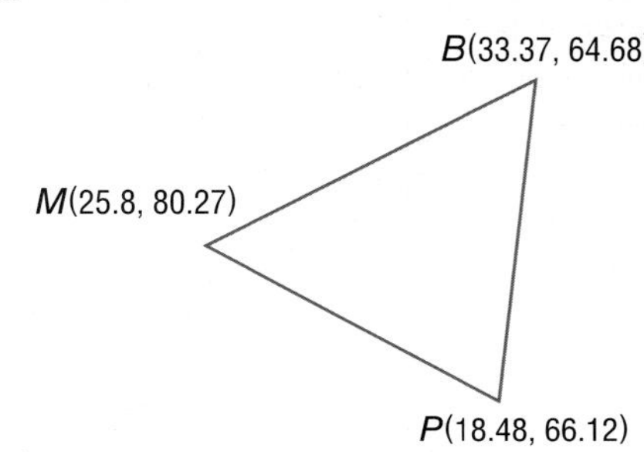


If no two sides of $\triangle MPB$ are congruent, then the Bermuda Triangle is scalene. Use the Distance Formula and a calculator to find the distance between each location.

$$MB = \sqrt{(33.37 - 25.8)^2 + (64.68 - 80.27)^2}$$

$$\approx 17.33$$

$$MP = \sqrt{(25.8 - 18.48)^2 + (80.27 - 66.12)^2}$$

$$\approx 15.93$$

$$PB = \sqrt{(33.37 - 18.48)^2 + (64.68 - 66.12)^2}$$

$$\approx 14.96$$

Since each side is a different length, $\triangle MPB$ is scalene. Therefore, the Bermuda Triangle is scalene.

## Guided Practice

4. **GEOGRAPHY** In 2006, a group of art museums collaborated to form the West Texas Triangle to promote their collections. This region is formed by the cities of Odessa, Albany, and San Angelo. The approximate coordinates of each location, respectively, are 31.9°N 102.3°W, 32.7°N 99.3°W, and 31.4°N 100.5°W. Write a coordinate proof to prove that the West Texas Triangle is approximately isosceles.

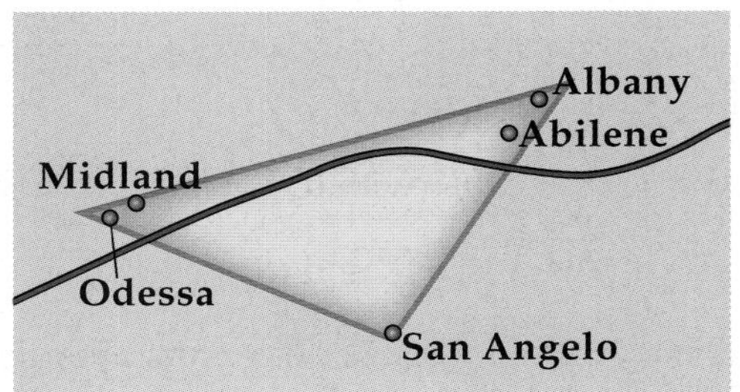

## Check Your Understanding

**Example 1** **Position and label each triangle on the coordinate plane.**

1. right $\triangle ABC$ with legs $\overline{AC}$ and $\overline{AB}$ so that $\overline{AC}$ is $2a$ units long and leg $\overline{AB}$ is $2b$ units long

2. isosceles $\triangle FGH$ with base $\overline{FG}$ that is $2a$ units long

**Example 2** **Name the missing coordinate(s) of each triangle.**

3.

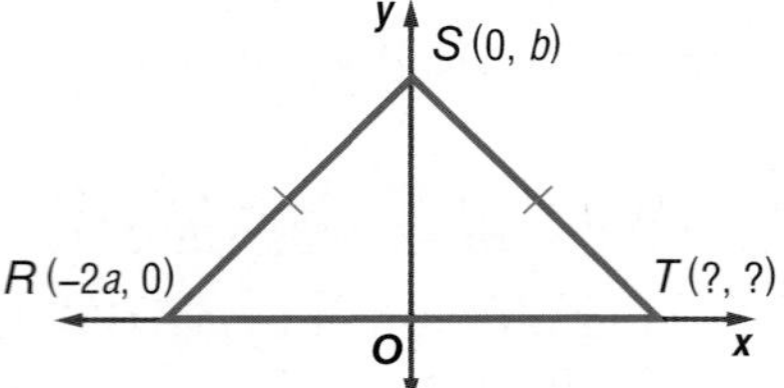


4.

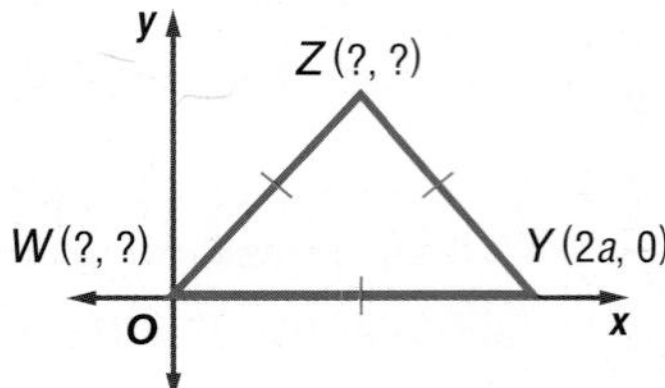


**Example 3** 5. **CCSS ARGUMENTS** Write a coordinate proof to show that $\triangle FGH \cong \triangle FDC$.

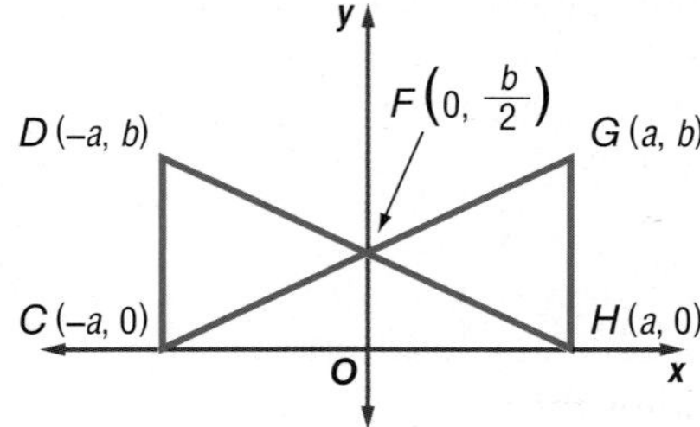


**Example 4** 6. **FLAGS** Write a coordinate proof to prove that the large triangle in the center of the flag is isosceles. The dimensions of the flag are 4 feet by 6 feet and point $B$ of the triangle bisects the bottom of the flag.

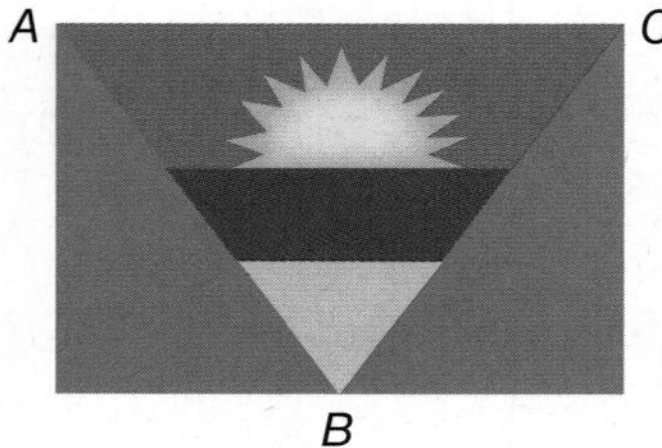


## Practice and Problem Solving

**Example 1** **Position and label each triangle on the coordinate plane.**

7. isosceles $\triangle ABC$ with base $\overline{AB}$ that is $a$ units long

8. right $\triangle XYZ$ with hypotenuse $\overline{YZ}$, the length of $\overline{XY}$ is $b$ units long, and the length of $\overline{XZ}$ is three times the length of $\overline{XY}$

9. isosceles right $\triangle RST$ with hypotenuse $\overline{RS}$ and legs $3a$ units long

10. right $\triangle JKL$ with legs $\overline{JK}$ and $\overline{KL}$ so that $\overline{JK}$ is $a$ units long and leg $\overline{KL}$ is $4b$ units long

11. equilateral $\triangle GHJ$ with sides $\frac{1}{2}a$ units long

12. equilateral $\triangle DEF$ with sides $4b$ units long

**Example 2** **Name the missing coordinate(s) of each triangle.**

13. 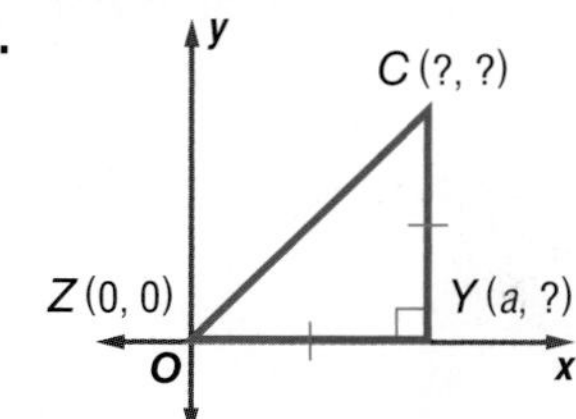


14. 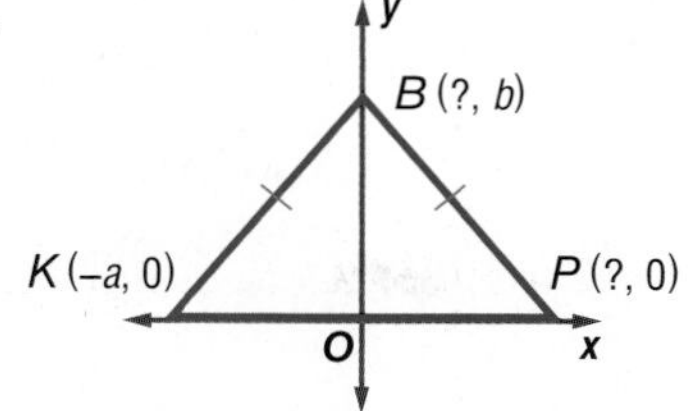


15. 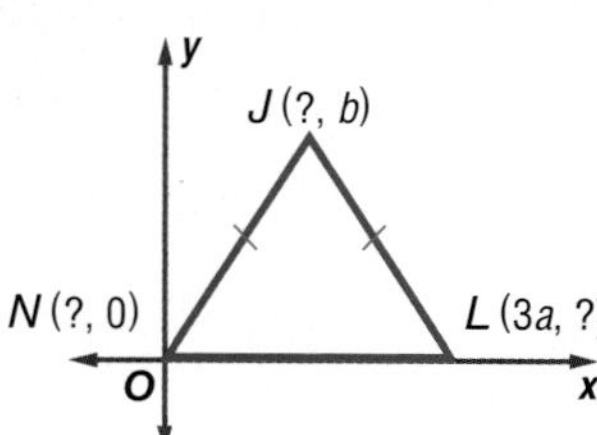


16. 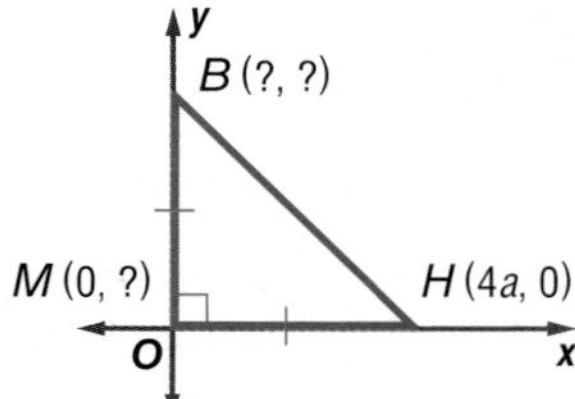


17. 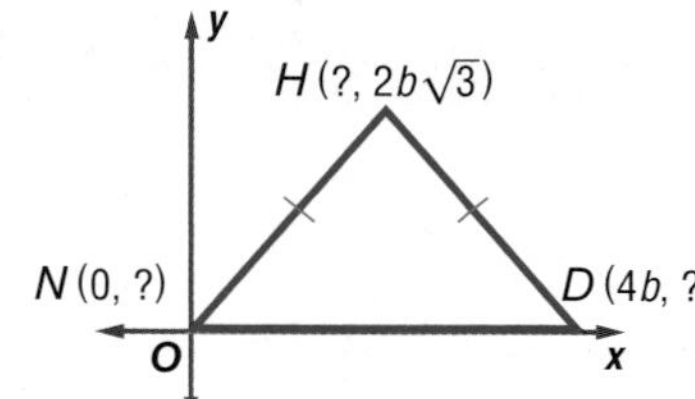


18. 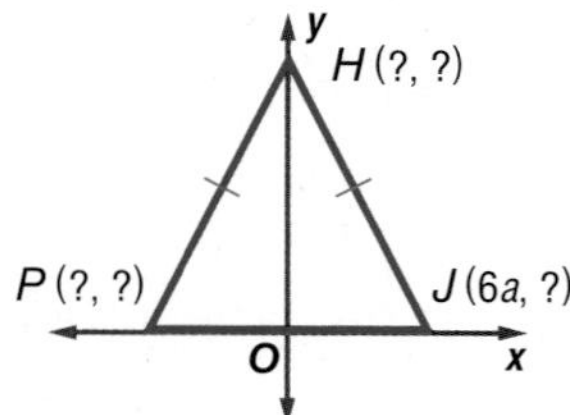


**Example 3** CCSS **ARGUMENTS** **Write a coordinate proof for each statement.**

19. The segments joining the base vertices to the midpoints of the legs of an isosceles triangle are congruent.

20. The three segments joining the midpoints of the sides of an isosceles triangle form another isosceles triangle.

**Example 4** **PROOF** **Write a coordinate proof for each statement.**

21. The measure of the segment that joins the vertex of the right angle in a right triangle to the midpoint of the hypotenuse is one-half the measure of the hypotenuse.

22. If a line segment joins the midpoints of two sides of a triangle, then its length is equal to one half the length of the third side.

23. **RESEARCH TRIANGLE** The cities of Raleigh, Durham, and Chapel Hill, North Carolina, form what is known as the Research Triangle. The approximate latitude and longitude of Raleigh are 35.82°N 78.64°W, of Durham are 35.99°N 78.91 °W, and of Chapel Hill are 35.92°N 79.04°W. Show that the triangle formed by these three cities is scalene.

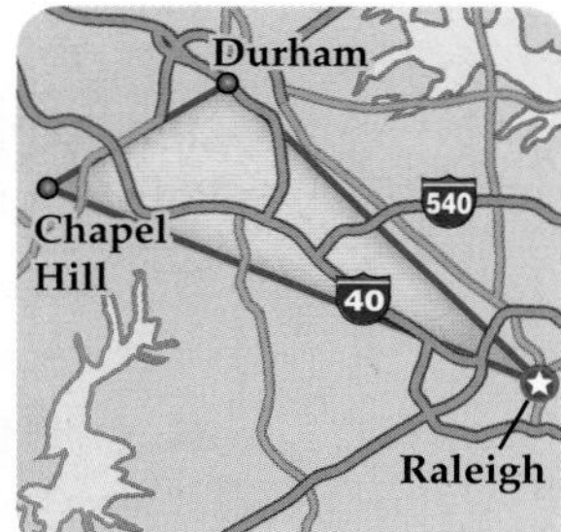


24. **PARTY PLANNING** Three friends live in houses with backyards adjacent to a neighborhood bike path. They decide to have a round-robin party using their three homes, inviting their friends to start at one house and then move to each of the other two. If one friend's house is centered at the origin, then the location of the other homes are (5, 12) and (13, 0). Write a coordinate proof to prove that the triangle formed by these three homes is isosceles.

**Draw $\triangle XYZ$ and find the slope of each side of the triangle. Determine whether the triangle is a right triangle. Explain.**

25. $X(0, 0)$, $Y(2h, 2h)$, $Z(4h, 0)$

26. $X(0, 0)$, $Y(1, h)$, $Z(2h, 0)$

27. **CAMPING** Two families set up tents at a state park. If the ranger's station is located at (0, 0), and the locations of the tents are (0, 25) and (12, 9), write a coordinate proof to prove that the figure formed by the locations of the ranger's station and the two tents is a right triangle.

28. **PROOF** Write a coordinate proof to prove that $\triangle ABC$ is an isosceles triangle if the vertices are $A(0, 0)$, $B(a, b)$, and $C(2a, 0)$.

**29** **WATER SPORTS** Three personal watercraft vehicles launch from the same dock. The first vehicle leaves the dock traveling due northeast, while the second vehicle travels due northwest. Meanwhile, the third vehicle leaves the dock traveling due north.

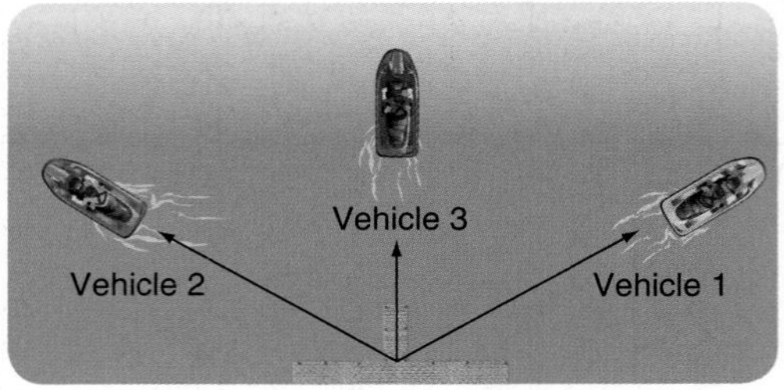


The first and second vehicles stop about 300 yards from the dock, while the third stops about 212 yards from the dock.

**a.** If the dock is located at (0, 0), sketch a graph to represent this situation. What is the equation of the line along which the first vehicle lies? What is the equation of the line along which the second vehicle lies? Explain your reasoning.

**b.** Write a coordinate proof to prove that the dock, the first vehicle, and the second vehicle form an isosceles right triangle.

**c.** Find the coordinates of the locations of all three watercrafts. Explain your reasoning.

**d.** Write a coordinate proof to prove that the positions of all three watercrafts are approximately collinear and that the third watercraft is at the midpoint between the other two.

## H.O.T. Problems Use Higher-Order Thinking Skills

**30. REASONING** The midpoints of the sides of a triangle are located at $(a, 0)$, $(2a, b)$ and $(a, b)$. If one vertex is located at the origin, what are the coordinates of the other vertices? Explain your reasoning.

**CHALLENGE Find the coordinates of point $L$ so $\triangle JKL$ is the indicated type of triangle. Point $J$ has coordinates (0, 0) and point $K$ has coordinates $(2a, 2b)$.**

**31.** scalene triangle **32.** right triangle **33.** isosceles triangle

**34. OPEN ENDED** Draw an isosceles right triangle on the coordinate plane so that the midpoint of its hypotenuse is the origin. Label the coordinates of each vertex.

**35. CHALLENGE** Use a coordinate proof to show that if you add $n$ units to each $x$-coordinate of the vertices of a triangle and $m$ to each $y$-coordinate, the resulting figure is congruent to the original triangle.

**36.** CCSS **REASONING** A triangle has vertex coordinates (0, 0) and $(a, 0)$. If the coordinates of the third vertex are in terms of $a$, and the triangle is isosceles, identify the coordinates and position the triangle on the coordinate plane.

**37. WRITING IN MATH** Explain why following each guideline below for placing a triangle on the coordinate plane is helpful in proving coordinate proofs.

**a.** Use the origin as a vertex of the triangle.

**b.** Place at least one side of the triangle on the $x$- or $y$-axis.

**c.** Keep the triangle within the first quadrant if possible.

## Standardized Test Practice

**38. GRIDDED RESPONSE** In the figure below, $m\angle B = 76$. The measure of $\angle A$ is half the measure of $\angle B$. What is $m\angle C$?

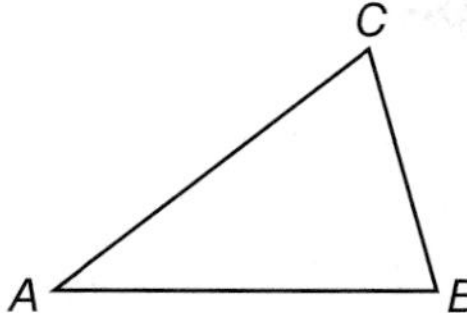


**39. ALGEBRA** What is the $x$-coordinate of the solution to the system of equations shown below?

$$\begin{cases} 2x - 3y = 3 \\ -4x + 2y = -18 \end{cases}$$

A $-6$  
B $-3$  
C 3  
D 6

**40.** What are the coordinates of point $R$ in the triangle?

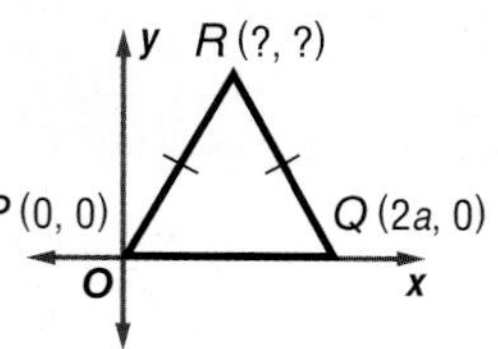


F $\left(\frac{a}{2}, a\right)$  
G $(a, b)$  
H $\left(\frac{b}{2}, a\right)$  
J $\left(\frac{b}{2}, \frac{a}{2}\right)$

**41. SAT/ACT** For all $x$,
$17x^5 + 3x^2 + 2 - (-4x^5 + 3x^3 - 2) =$

A $13x^5 + 3x^3 + 3x^2$  
B $13x^5 + 6x^2 + 4$  
C $21x^5 - 3x^3 + 3x^2 + 4$  
D $21x^5 + 3x^2 + 3x^3$  
E $21x^5 + 3x^3 + 3x^2 + 4$

## Spiral Review

**42. PHYSICS** The formula for pressure under water is $P = \rho gh$, where $\rho$ is the density of water, $g$ is gravity, and $h$ is the height of the water. Given the water pressure, prove that you can calculate the height of the water using $h = \frac{P}{\rho g}$. (Lesson 5-2)

**43. PROOF** Write a two-column proof to verify that if $\overline{MN} \cong \overline{QP}$, then $x = 7$. (Lesson 5-2)

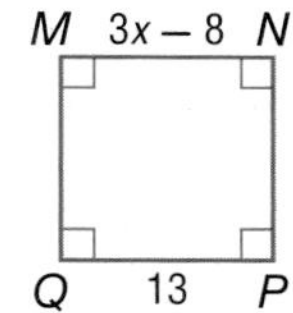


**Refer to the figure at the right.** (Lesson 6-5)

**44.** Name two congruent angles.

**45.** Name two congruent segments.

**46.** Name a pair of congruent triangles.

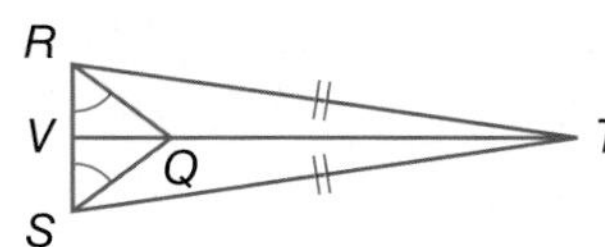


## Skills Review

**Find the distance between each pair of points. Round to the nearest tenth.**

**47.** $X(5, 4)$ and $Y(2, 1)$

**48.** $A(1, 5)$ and $B(-2, -3)$

**49.** $J(-2, 6)$ and $K(1, 4)$

EXTEND 6-6

# Geometry Lab: Two-Dimensional Representations of Three-Dimensional Objects

If you see a three-dimensional object from only one viewpoint, you may not know its true shape. Here are four views of a square pyramid.

The two-dimensional views of the top, left, front, and right sides of an object are called an **orthographic drawing**.

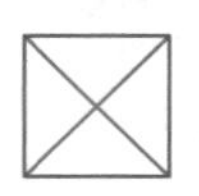

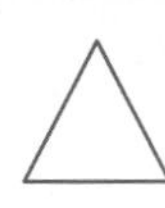


**CCSS Common Core State Standards**
**Content Standards**
**G.MG.1** Use geometric shapes, their measures, and their properties to describe objects (e.g., modeling a tree trunk or a human torso as a cylinder). ★
**Mathematical Practices** 5

## Activity 1

**Make a model of a figure for the orthographic drawing shown.**

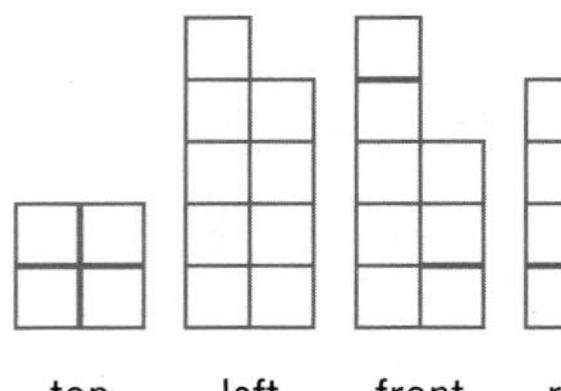

**Step 1** Start with a base that matches the top view.

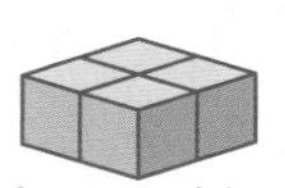


**Step 2** The front view indicates that the front left side is 5 blocks high and that the right side is 3 blocks high. However, the dark segments indicate breaks in the surface.

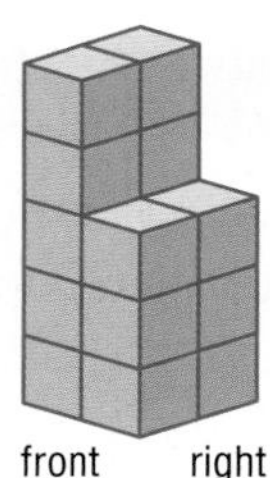

**Step 3** The break on the left side of the front view indicates that the back left column is 5 blocks high, but that the front left column is only 4 blocks high, so remove 1 block from the front left column.

**Step 4** The break on the right side of the front view indicates that the back right column is 3 blocks high, but that the front right column is only 1 block high, so remove 2 blocks from the front right column.

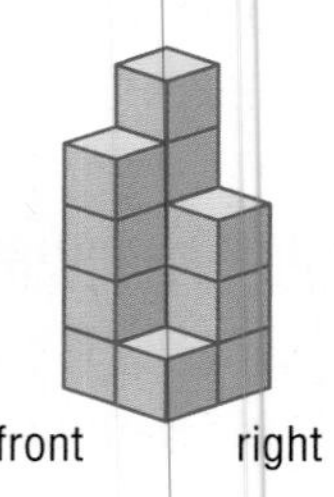


**Step 5** Use the left and right views and the breaks in those views to confirm that you have made the correct figure.

## Model and Analyze

**1.** Make a model of a figure for the orthographic drawing shown.

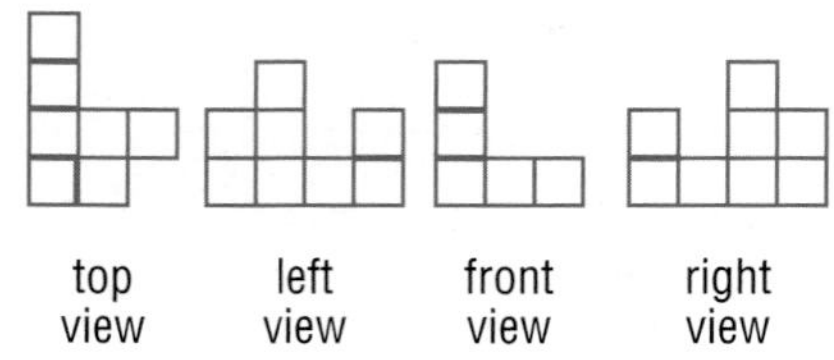


**2.** Make an orthographic drawing of the figure shown.

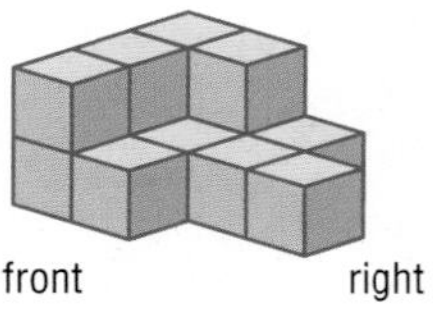

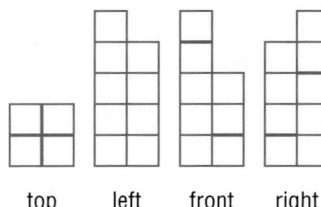

Geometry Lab

# Two-Dimensional Representations of Three-Dimensional Objects *Continued*

If you cut a cardboard box at the edges and lay it flat, you will have a two-dimensional diagram called a **net** that you can fold to form a three-dimensional solid.

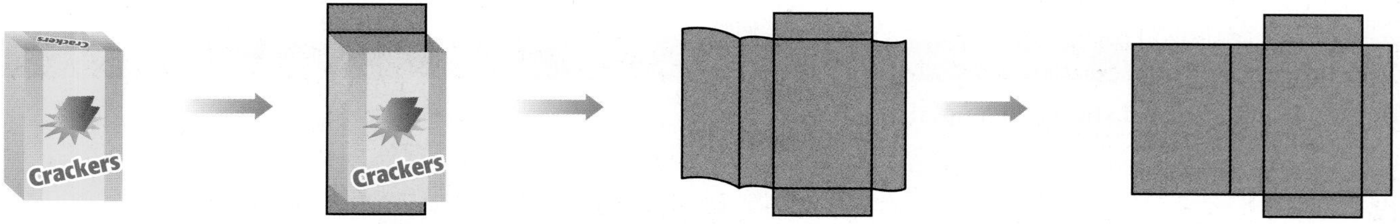


## Activity 2

**Make a model of a figure for the given net. Then identify the solid formed, and find its surface area.**

Use a large sheet of paper, a ruler, scissors, and tape. Draw the net on the paper. Cut along the solid lines. Fold the pattern on the dashed lines and secure the edges with tape. This is the net of a triangular prism.

Use the net to find the surface area $S$.

$S = 2\left[\frac{1}{2}(4)(3)\right] + 4(10) + 3(10) + 5(10)$ Area of two congruent triangles plus area of three rectangles

$= 12 + 40 + 30 + 50$ or $132 \text{ in}^2$ Simplify.

## Model and Analyze

**Make a model of a figure for each net. Then identify the solid formed and find its surface area. If the solid has more than one name, list both.**

**3.**

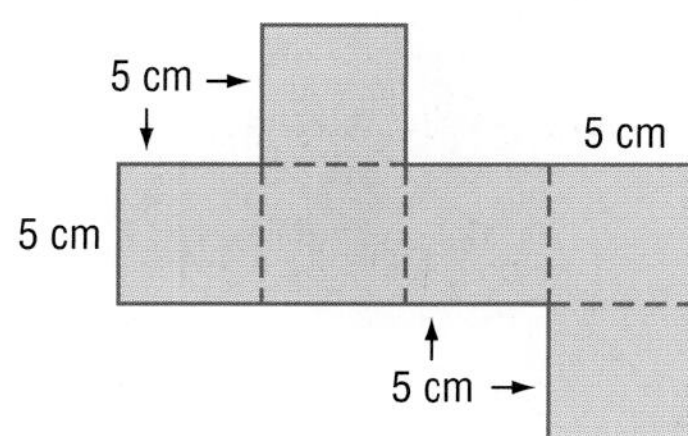


**4.**

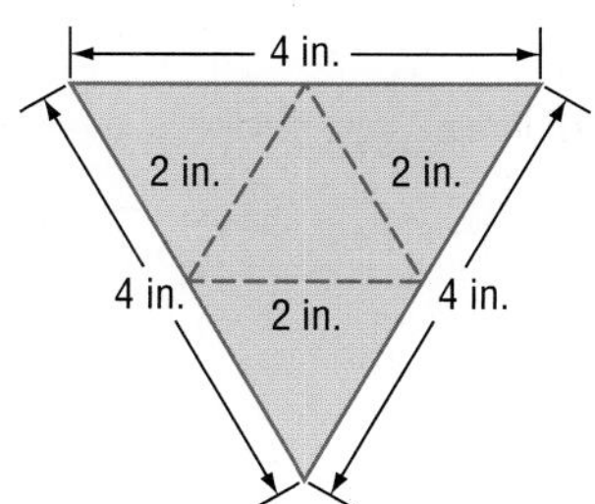


**5.**

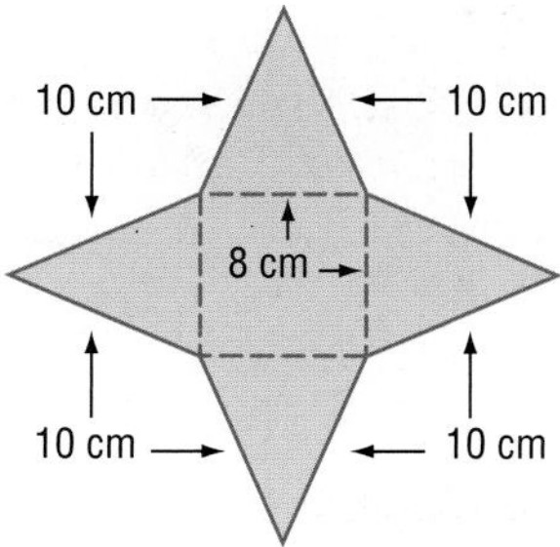


**Identify the Platonic Solid that can be formed by the given net.**

**6.**

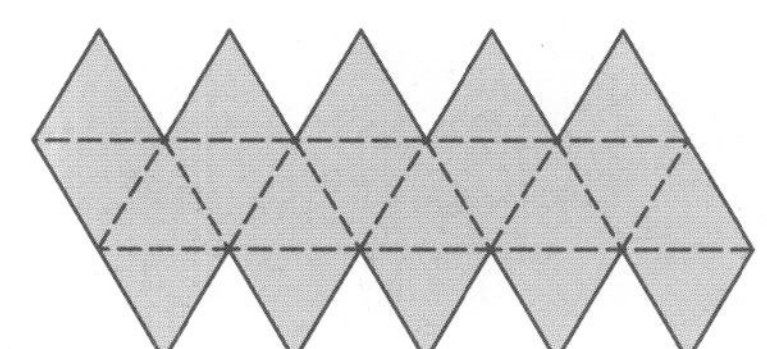

**7.**

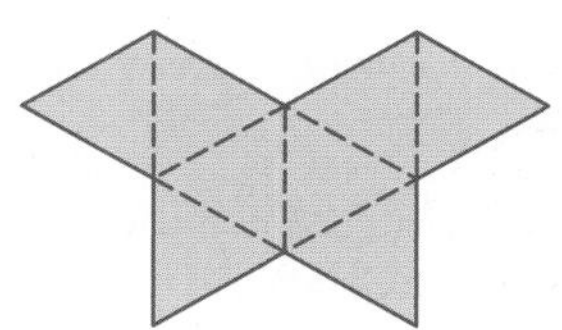

**8.**

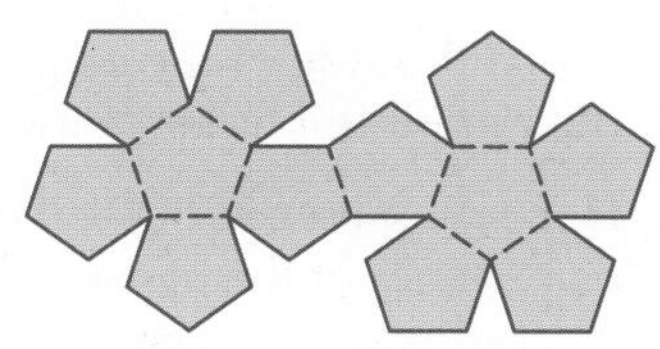

To draw the net of a three-dimensional solid, visualize cutting the solid along one or more of its edges, opening up the solid, and flattening it completely.

## Activity 3

**Draw a net for the solid shown. Then label its dimensions.**

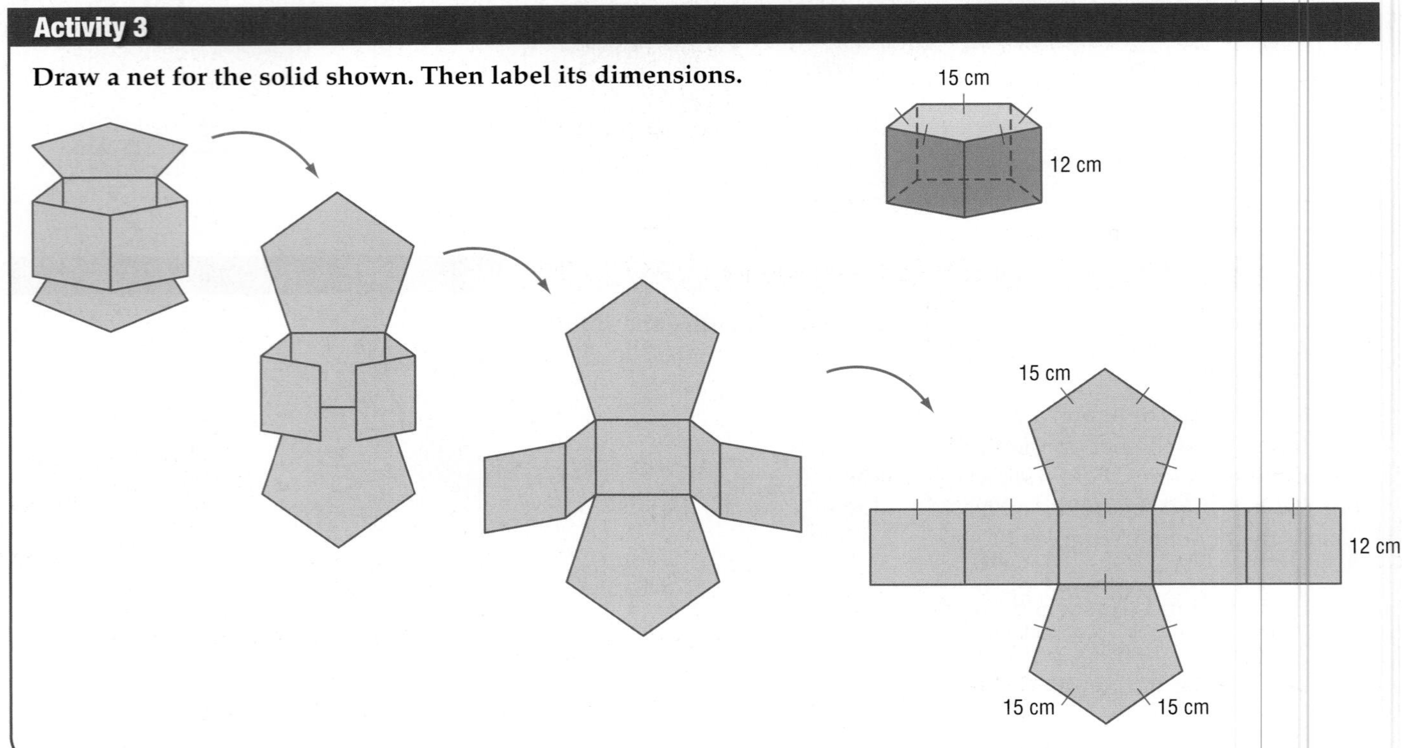


## Model and Analyze

**Draw a net for each solid. Then label its dimensions.**

**9.**

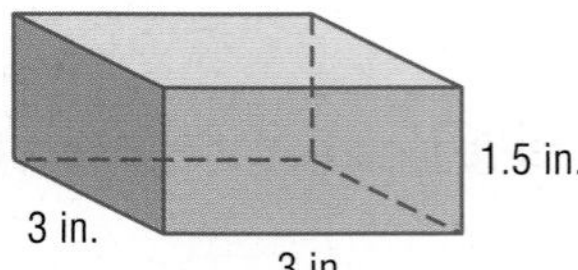


**10.**

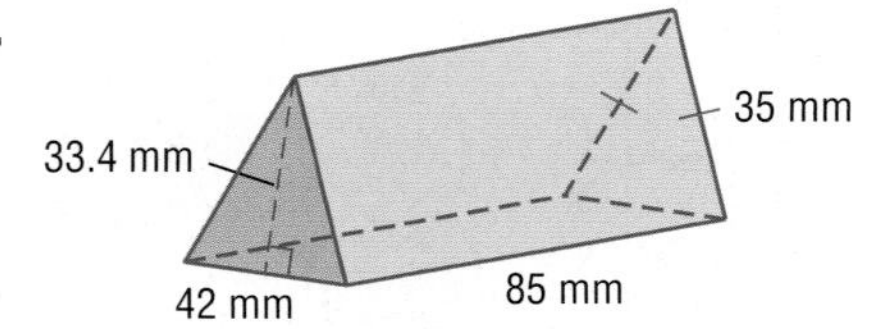


**11.**

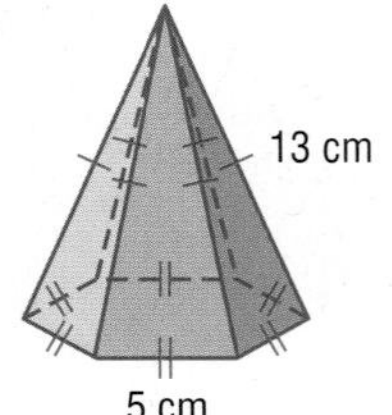


**12. PACKAGING** A can of pineapple is shown.

**a.** What shape are the top and bottom of the can?

**b.** If you remove the top and bottom and then make a vertical cut down the side of the can, what shape will you get when you uncurl the remaining body of the can and flatten it?

**c.** If the diameter of the can is 3 inches and its height is 2 inches, draw a net of the can and label its dimensions. Explain your reasoning.

# Study Guide and Review

## Study Guide

### KeyConcepts

**Angles of Triangles** (Lesson 6-1)

- The measure of an exterior angle is equal to the sum of its two remote interior angles.

**Congruent Triangles** (Lesson 6-2 through 6-4)

- SSS: If all of the corresponding sides of two triangles are congruent, then the triangles are congruent.
- SAS: If two pairs of corresponding sides of two triangles and the included angles are congruent, then the triangles are congruent.
- ASA: If two pairs of corresponding angles of two triangles and the included sides are congruent, then the triangles are congruent.
- AAS: If two pairs of corresponding angles of two triangles are congruent, and a corresponding pair of nonincluded sides is congruent, then the triangles are congruent.

**Isosceles and Equilateral Triangles** (Lesson 6-5)

- The base angles of an isosceles triangle are congruent and a triangle is equilateral if it is equiangular.

**Transformations and Coordinate Proofs** (Lesson 6-6)

- In a congruence transformation, the position of the image may differ from the preimage, but the two figures remain congruent.
- Coordinate proofs use algebra to prove geometric concepts.

### FOLDABLES® StudyOrganizer

Be sure the Key Concepts are noted in your Foldable.

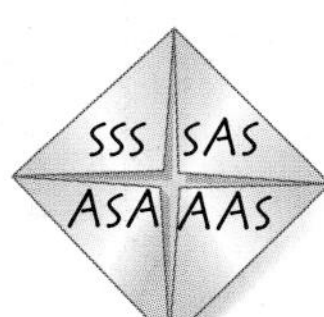


### KeyVocabulary

auxiliary line (p. 335)
base angles (p. 374)
congruent polygons (p. 344)
coordinate proof (p. 383)
corollary (p. 338)
corresponding parts (p. 344)
exterior angle (p. 337)
flow proof (p. 337)
included angle (p. 355)
included side (p. 364)
remote interior angles (p. 337)
vertex angle (p. 374)

### VocabularyCheck

**State whether each sentence is *true* or *false*. If *false*, replace the underlined word or phrase to make a true sentence.**

1. An equiangular triangle is also an example of an <u>acute</u> triangle.
2. A triangle with an angle that measures greater than 90° is a <u>right</u> triangle.
3. An <u>equilateral</u> triangle is always equiangular.
4. A <u>scalene</u> triangle has at least two congruent sides.
5. The <u>vertex</u> angles of an isosceles triangle are congruent.
6. An <u>included</u> side is the side located between two consecutive angles of a polygon.
7. The three types of <u>congruence transformations</u> are rotation, reflection, and translation.
8. A <u>rotation</u> moves all points of a figure the same distance and in the same direction.
9. A <u>flow proof</u> uses figures in the coordinate plane and algebra to prove geometric concepts.
10. The measure of an <u>exterior angle</u> of a triangle is equal to the sum of the measures of its two remote interior angles.

# Study Guide and Review *Continued*

## Lesson-by-Lesson Review

### 6-1 Angles of Triangles

Find the measure of each numbered angle.

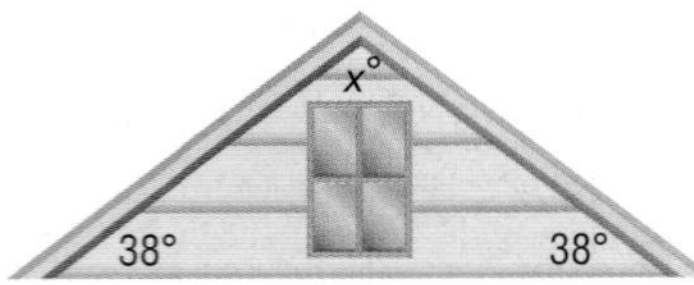


11. $\angle 1$
12. $\angle 2$
13. $\angle 3$
14. **HOUSES** The roof support on Lamar's house is in the shape of an isosceles triangle with base angles of 38°. Find *x*.



**Example 1**

Find the measure of each numbered angle.

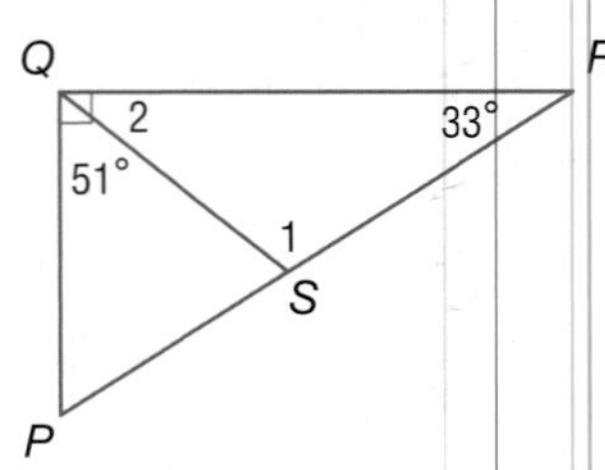


$m\angle 2 + m\angle PQS = 90$

$m\angle 2 + 51 = 90$ Substitution

$m\angle 2 = 39$ Subtract 51 from each side.

$m\angle 1 + m\angle 2 + 33 = 180$ Triangle Sum Theorem

$m\angle 1 + 39 + 33 = 180$ Substitution

$m\angle 1 + 72 = 180$ Simplify.

$m\angle 1 = 108$ Subtract.

### 6-2 Congruent Triangles

Show that the polygons are congruent by identifying all congruent corresponding parts. Then write a congruence statement.

15.

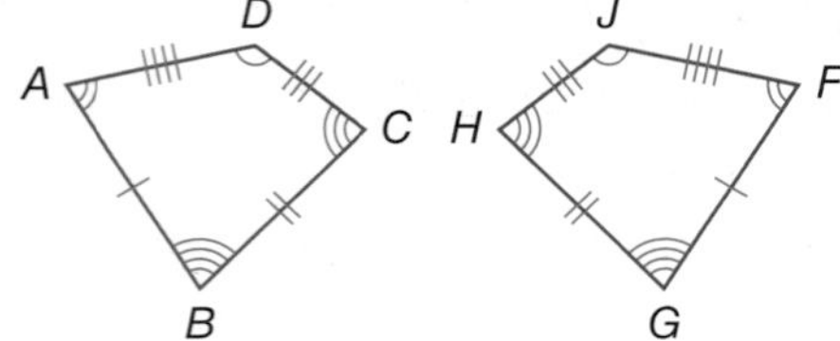


16.

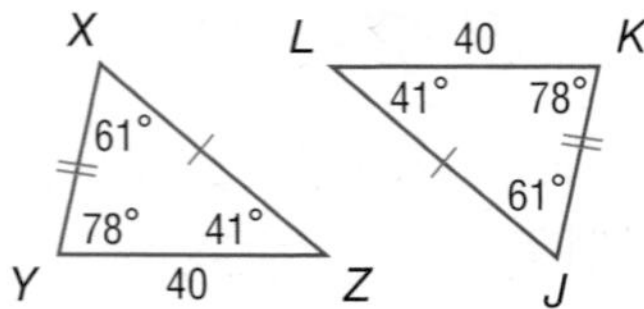


17. **MOSAIC TILING** A section of a mosaic tiling is shown. Name the triangles that appear to be congruent.

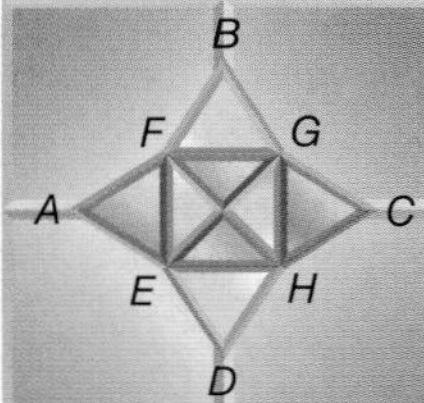


**Example 2**

Show that the polygons are congruent by identifying all the congruent corresponding parts. Then write a congruence statement.

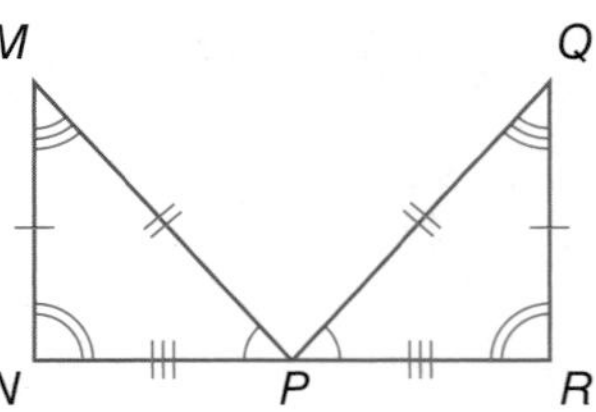


**Angles:** $\angle N \cong \angle R$, $\angle M \cong \angle Q$, $\angle MPN \cong \angle QPR$

**Sides:** $\overline{MN} \cong \overline{QR}$, $\overline{MP} \cong \overline{QP}$, $\overline{NP} \cong \overline{RP}$

All corresponding parts of the two triangles are congruent. Therefore, $\triangle MNP \cong \triangle QRP$.

## 6-3 Proving Triangles Congruent—SSS, SAS

Determine whether $\triangle ABC \cong \triangle XYZ$. Explain.

**18.** $A(5, 2)$, $B(1, 5)$, $C(0, 0)$, $X(-3, 3)$, $Y(-7, 6)$, $Z(-8, 1)$

**19.** $A(3, -1)$, $B(3, 7)$, $C(7, 7)$, $X(-7, 0)$, $Y(-7, 4)$, $Z(1, 4)$

Determine which postulate can be used to prove that the triangles are congruent. If it is not possible to prove that they are congruent, write *not possible*.

**20.**

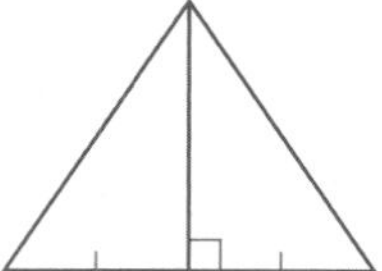

**21.**

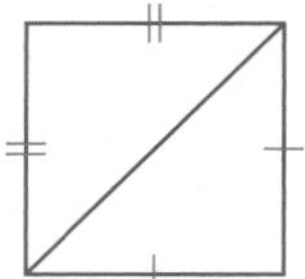

**22.** **PARKS** The diagram shows a park in the shape of a pentagon with five sidewalks of equal length leading to a central point. If all the angles at the central point have the same measure, how could you prove that $\triangle ABX \cong \triangle DCX$?

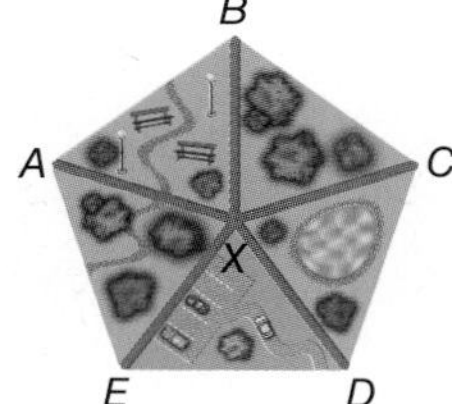


### Example 3

Write a two-column proof.

Given: $\triangle KPL$ is equilateral.
$\overline{JP} \cong \overline{MP}$,
$\angle JPK \cong \angle MPL$

Prove: $\triangle JPK \cong \triangle MPL$

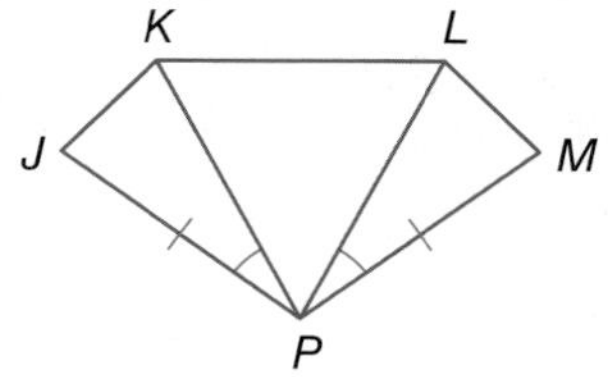


| Statements | Reasons |
|---|---|
| 1. $\triangle KPL$ is equilateral. | 1. Given |
| 2. $\overline{PK} \cong \overline{PL}$ | 2. Def. of Equilateral $\triangle$ |
| 3. $\overline{JP} \cong \overline{MP}$ | 3. Given |
| 4. $\angle JPK \cong \angle MPL$ | 4. Given |
| 5. $\triangle JPK \cong \triangle MPL$ | 5. SAS |

## 6-4 Proving Triangles Congruent—ASA, AAS

Write a two-column proof.

**23.** Given: $\overline{AB} \parallel \overline{DC}$, $\overline{AB} \cong \overline{DC}$

Prove: $\triangle ABE \cong \triangle CDE$

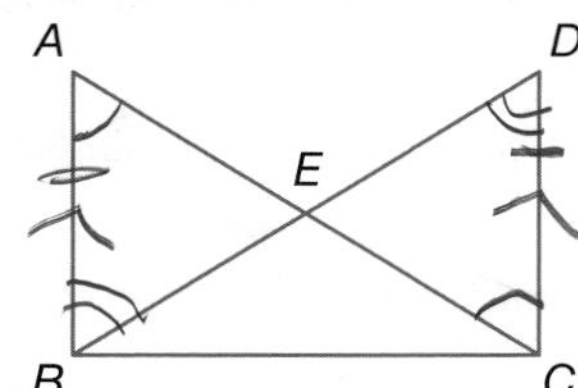


**24.** **KITES** Denise's kite is shown in the figure at the right. Given that $\overline{WY}$ bisects both $\angle XWZ$ and $\angle XYZ$, prove that $\triangle WXY \cong \triangle WZY$.

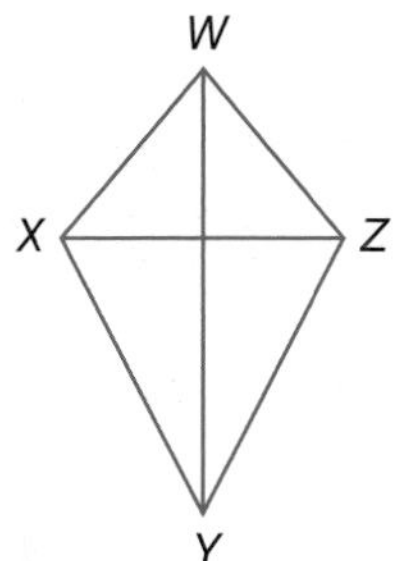


### Example 4

Write a flow proof.

Given: $\overline{PQ}$ *bisects* $\angle RPS$.
$\angle R \cong \angle S$

Prove: $\triangle RPQ \cong \triangle SPQ$

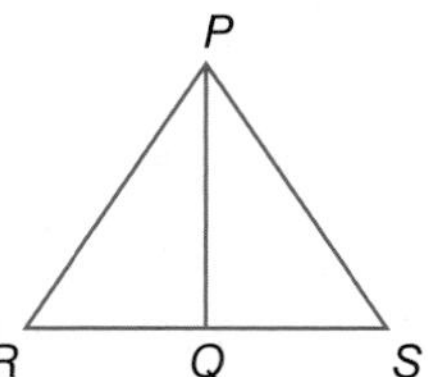


Flow Proof:

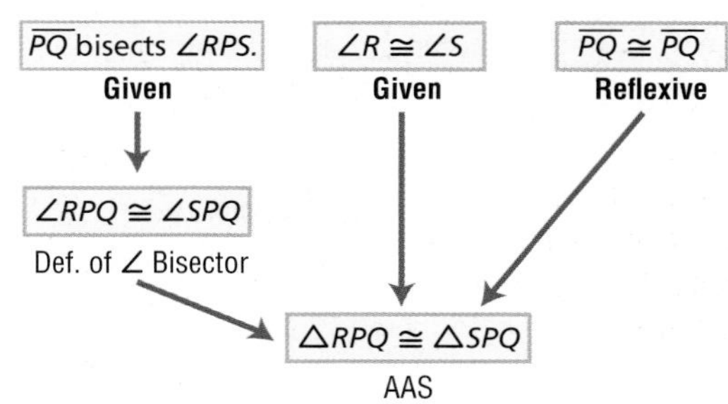

# CHAPTER 6 Study Guide and Review *Continued*

## 6-5 Isosceles and Equilateral Triangles

Find the value of each variable.

**25.**

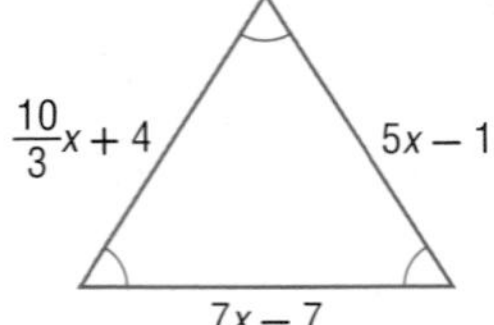


**26.**

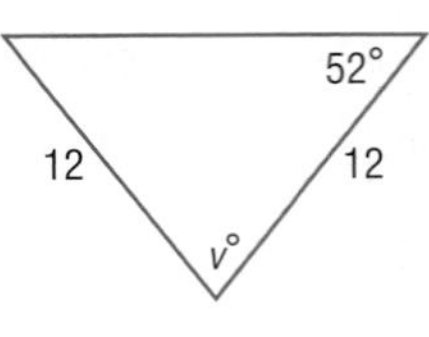


**27. PAINTING** Pam is painting using a wooden easel. The support bar on the easel forms an isosceles triangle with the two front supports. According to the figure below, what are the measures of the base angles of the triangle?

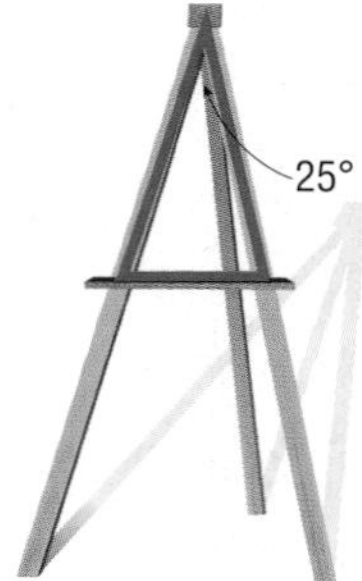


### Example 5

Find each measure.

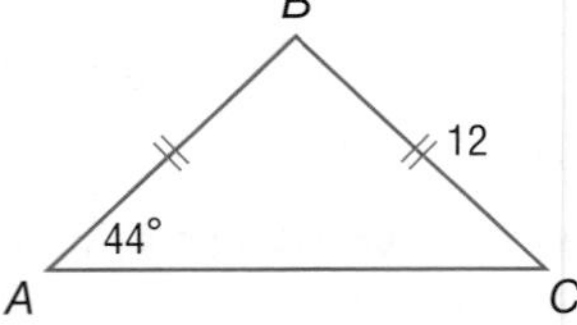


**a.** $m\angle B$

Since $AB = BC$, $\overline{AB} \cong \overline{BC}$. By the Isosceles Triangle Theorem, base angles $A$ and $C$ are congruent, so $m\angle A = m\angle C$. Use the Triangle Sum Theorem to write and solve an equation to find $m\angle B$.

| | |
|---|---|
| $m\angle A + m\angle B + m\angle C = 180$ | △ Sum Theorem |
| $44 + m\angle B + 44 = 180$ | $m\angle A = m\angle C = 44$ |
| $88 + m\angle B = 180$ | Simplify. |
| $m\angle B = 92$ | Subtract. |

**b.** $AB$

$AB = BC$, so $\triangle ABC$ is isosceles. Since $BC = 12$, $AB = 12$ by substitution.

## 6-6 Triangles and Coordinate Proof

Position and label each triangle on the coordinate plane.

**28.** right $\triangle MNO$ with right angle at point $M$ and legs of lengths $a$ and $2a$.

**29.** isosceles $\triangle WXY$ with height $h$ and base $\overline{WY}$ with length $2a$.

**30. GEOGRAPHY** Jorge plotted the cities of Dallas, San Antonio, and Houston as shown. Write a coordinate proof to show that the triangle formed by these cities is scalene.

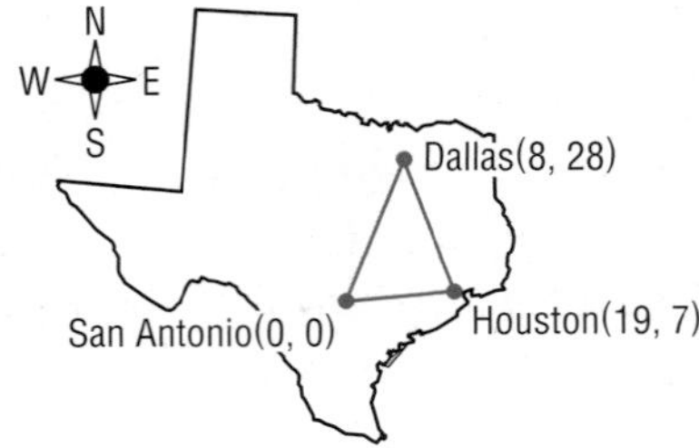


### Example 6

Position and label an equilateral triangle $\triangle XYZ$ with side lengths of $2a$.

- Use the origin for one of the three vertices of the triangle.
- Place one side of the triangle along the positive side of the $x$-axis.
- The third point should be located above the midpoint of the base of the triangle.

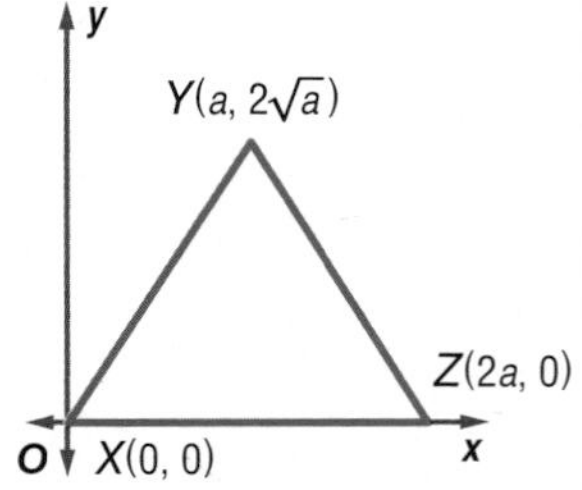

# CHAPTER 6 Practice Test

**Find the measure of each numbered angle.**

1. $\angle 1$
2. $\angle 2$
3. $\angle 3$
4. $\angle 4$

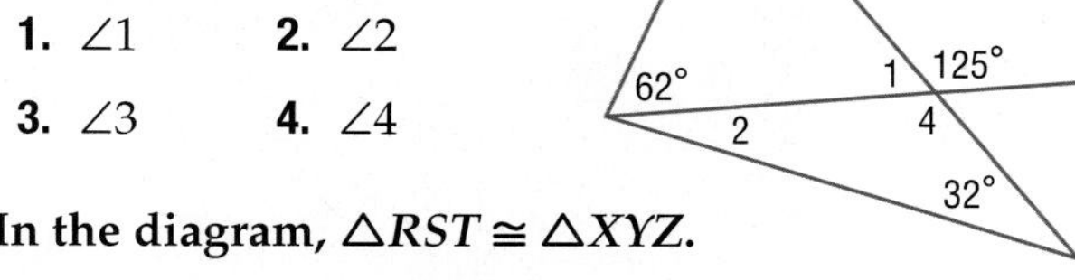


**In the diagram, $\triangle RST \cong \triangle XYZ$.**

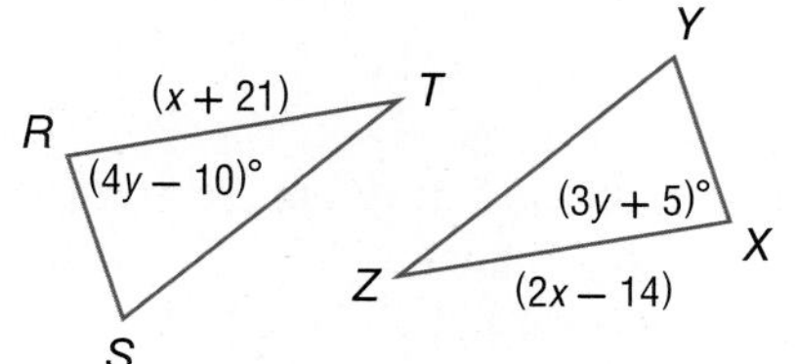


5. Find $x$.

6. Find $y$.

7. **PROOF** Write a flow proof.

   **Given:** $\overline{XY} \parallel \overline{WZ}$ and $\overline{XW} \parallel \overline{YZ}$
   **Prove:** $\triangle XWZ \cong \triangle ZYX$

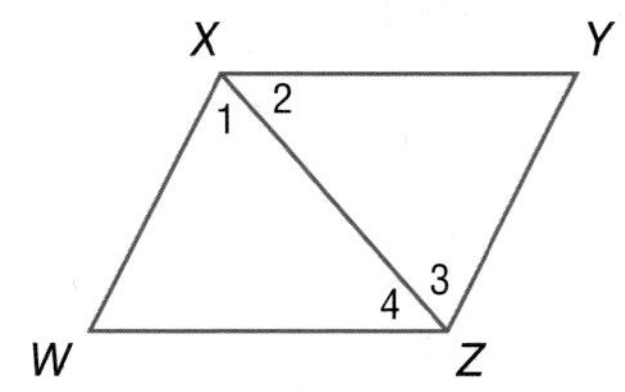


8. **MULTIPLE CHOICE** Find $x$.

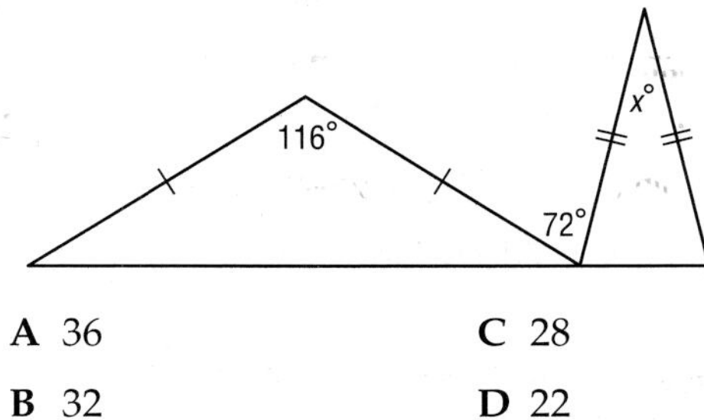


   **A** 36 **C** 28

   **B** 32 **D** 22

9. Determine whether $\triangle TJD \cong \triangle SEK$ given $T(-4, -2)$, $J(0, 5)$, $D(1, -1)$, $S(-1, 3)$, $E(3, 10)$, and $K(4, 4)$. Explain.

**Determine which postulate or theorem can be used to prove each pair of triangles congruent. If it is not possible to prove them congruent, write *not possible*.**

10. 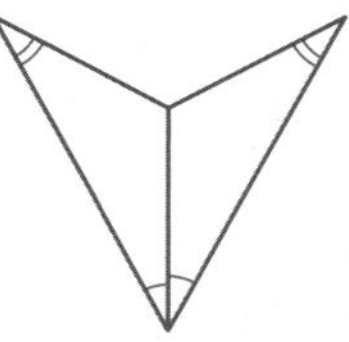

11. 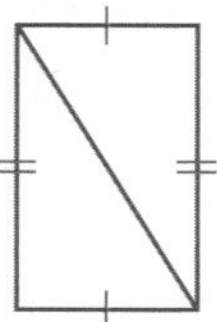

12. 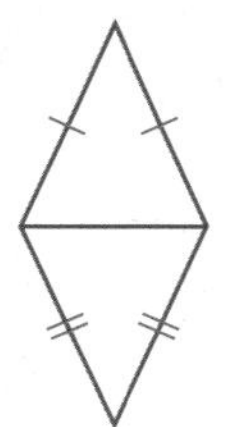

13. 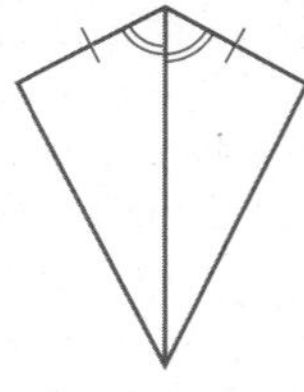

14. **LANDSCAPING** Angie has laid out a design for a garden consisting of two triangular areas as shown below. The points are $A(0, 0)$, $B(0, 5)$, $C(3, 5)$, $D(6, 5)$, and $E(6, 0)$. Name the type of congruence transformation for the preimage $\triangle ABC$ to $\triangle EDC$.

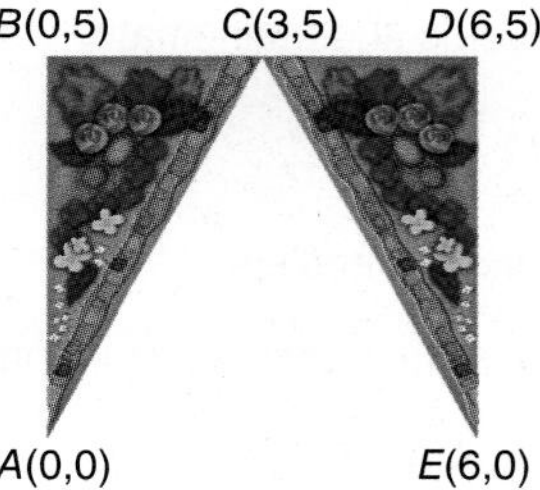


**Find the measure of each numbered angle.**

15. $\angle 1$

16. $\angle 2$

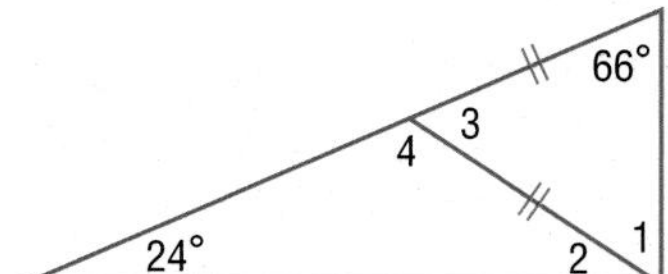


17. **PROOF** $\triangle ABC$ is a right isosceles triangle with hypotenuse $\overline{AB}$. $M$ is the midpoint of $\overline{AB}$. Write a coordinate proof to show that $\overline{CM}$ is perpendicular to $\overline{AB}$.

CHAPTER 6

# Preparing for Standardized Tests

## Short-Answer Questions

Short-answer questions require you to provide a solution to the problem, along with a method, explanation, and/or justification used to arrive at the solution.

Short-answer questions are typically graded using a **rubric**, or a scoring guide.

The following is an example of a short-answer question scoring rubric.

| Scoring Rubric | | |
|---|---|---|
| Criteria | | Score |
| Full Credit | The answer is correct and a full explanation is provided that shows each step. | 2 |
| Partial Credit | • The answer is correct, but the explanation is incomplete.<br>• The answer is incorrect, but the explanation is correct. | 1<br>1 |
| No Credit | Either an answer is not provided or the answer does not make sense. | 0 |

### Strategies for Solving Short-Answer Questions

**Step 1**

Read the problem to gain an understanding of what you are trying to solve.

- Identify relevant facts.
- Look for key words and mathematical terms.

**Step 2**

Make a plan and solve the problem.

- Explain your reasoning or state your approach to solving the problem.
- Show all of your work or steps.
- Check your answer if time permits.

**Standardized Test Example**

**Read the problem. Identify what you need to know. Then use the information in the problem to solve. Show your work.**

Triangle $ABC$ is an isosceles triangle with base $\overline{BC}$. What is the perimeter of the triangle?

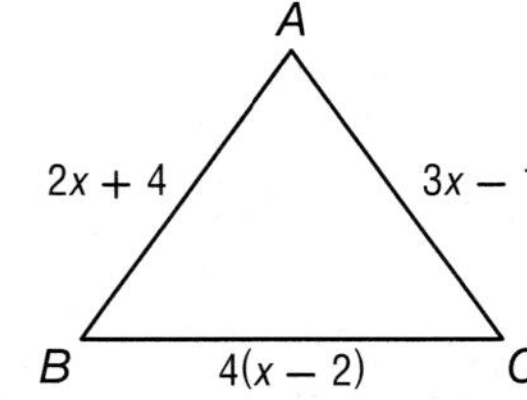


Rob Gage/Taxi/Getty Images

Read the problem carefully. You are told that $\triangle ABC$ is isosceles with base $\overline{BC}$. You are asked to find the perimeter of the triangle.

Make a plan and solve the problem.

The legs of an isosceles triangle are congruent.
So, $\overline{AB} \cong \overline{AC}$ or $AB = AC$. Solve for $x$.

$$AB = AC$$
$$2x + 4 = 3x - 1$$
$$2x - 3x = -1 - 4$$
$$-x = -5$$
$$x = 5$$

Next, find the length of each side.

$AB = 2(5) + 4 = 14$ units
$AC = 3(5) - 1 = 14$ units
$BC = 4(5 - 2) = 12$ units

The perimeter of $\triangle ABC$ is $14 + 14 + 12 = 40$ units.

The steps, calculations, and reasoning are clearly stated. The student also arrives at the correct answer. So, this response is worth the full 2 points.

## Exercises

**Read each problem. Identify what you need to know. Then use the information in the problem to solve. Show your work.**

1. Classify $\triangle DEF$ according to its angle measures.

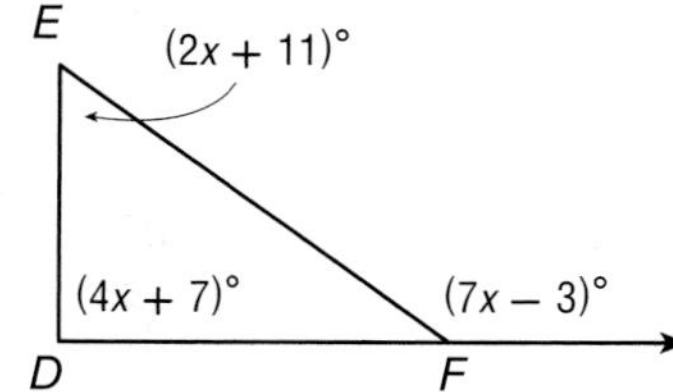


2. In the figure below, $\triangle RST \cong \triangle VUT$. What is the area of $\triangle RST$?

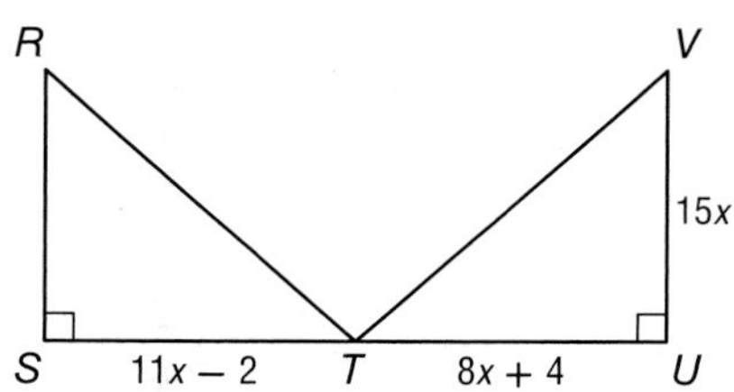


3. A farmer needs to make a 48-square-foot rectangular enclosure for chickens. He wants to save money by purchasing the least amount of fencing possible to enclose the area. What whole-number dimensions will require the least amount of fencing?

4. What is $m\angle 1$ in degrees?

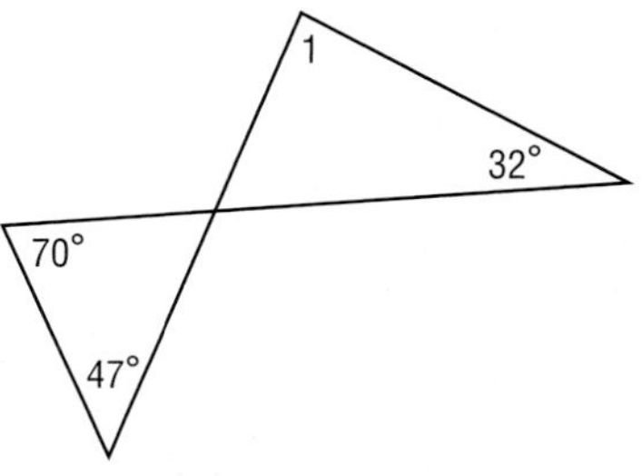


5. Write an equation of the line containing the points $(2, 4)$ and $(0, -2)$.

CHAPTER 6

# Standardized Test Practice

## Cumulative, Chapters 1 through 6

### Multiple Choice

**Read each question. Then fill in the correct answer on the answer document provided by your teacher or on a sheet of paper.**

**1.** If $m\angle 1 = 110°$, what must $m\angle 2$ equal for lines $x$ and $z$ to be parallel?

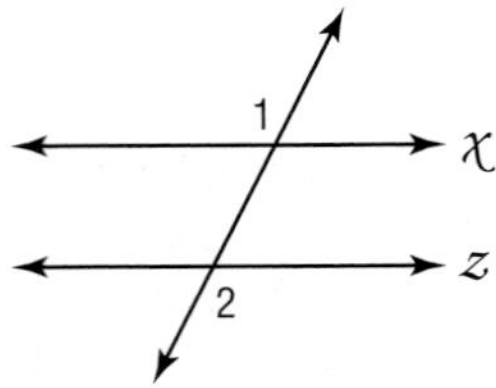


A 30° B 60° C 70° D 110°

**2. Given:** $\overline{WX} \cong \overline{JK}$, $\overline{YX} \cong \overline{IK}$, $\angle X \cong \angle K$

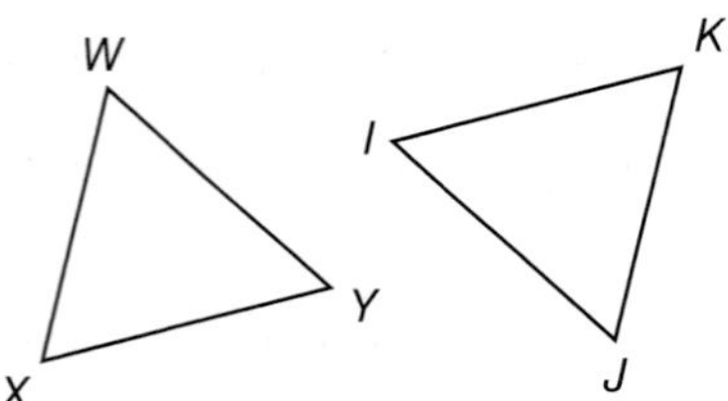


Which of the following lists the correct triangle congruence?

F $\triangle WXY \cong \triangle KIJ$

G $\triangle WXY \cong \triangle IKJ$

H $\triangle WXY \cong \triangle JKI$

J $\triangle WXY \cong \triangle IJK$

**Test-Taking Tip**

Question 3 Read the problem statement carefully to make sure you select the correct answer.

**3.** What is the measure of angle $R$ below?

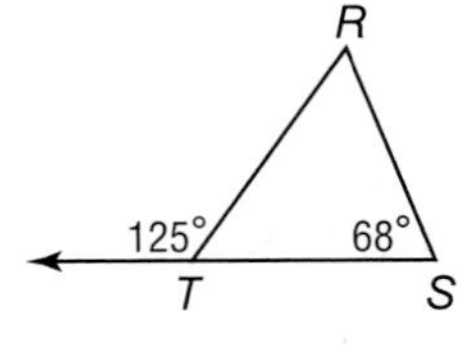


F 57° G 59° H 65° J 68°

**4.** Suppose one base angle of an isosceles triangle has a measure of 44°. What is the measure of the vertex angle?

A 108° C 56°

B 92° D 44°

**5.** If $a \parallel b$, which of the following is *not* true?

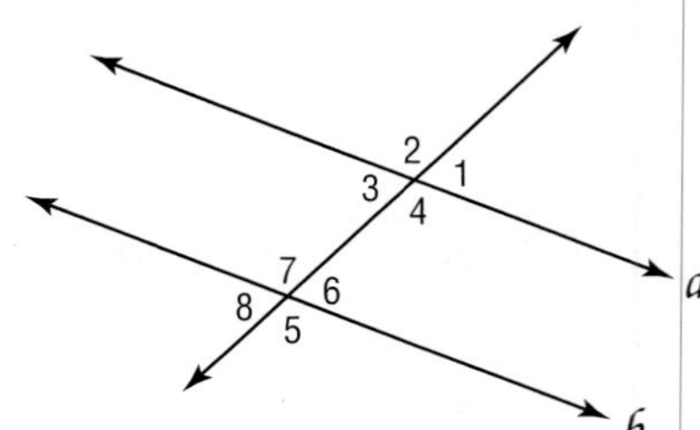


A $\angle 1 \cong \angle 3$ C $\angle 2 \cong \angle 5$

B $\angle 4 \cong \angle 7$ D $\angle 8 \cong \angle 2$

## Short Response/Gridded Response

**Record your answers on the answer sheet provided by your teacher or on a sheet of paper.**

6. **GRIDDED RESPONSE** In the figure below, $\triangle NDG \cong \triangle LGD$. What is the value of $x$?

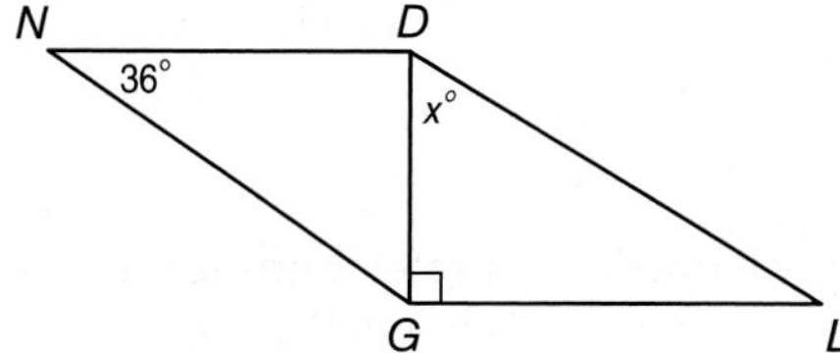


7. Use the figure and the given information below.

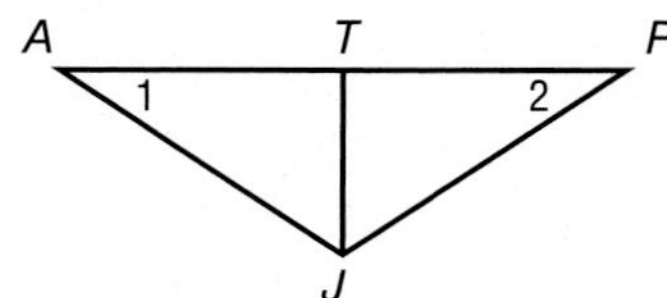


**Given:** $\overline{JT} \perp \overline{AP}$
$\angle 1 \cong \angle 2$

Which congruence theorem could you use to prove $\triangle PTJ \cong \triangle ATJ$ with only the information given? Explain.

8. **PROOF** Prove the following.

**Given:** $\overline{AC} \cong \overline{BD}$
$\overline{EC} \cong \overline{ED}$
**Prove:** $\overline{AE} \cong \overline{BE}$

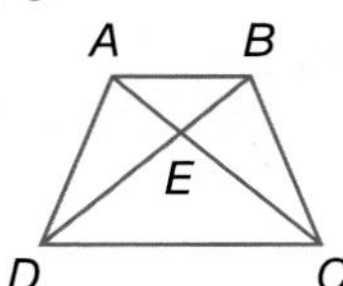


9. **GRIDDED RESPONSE** Find $m\angle TUV$ in the figure.

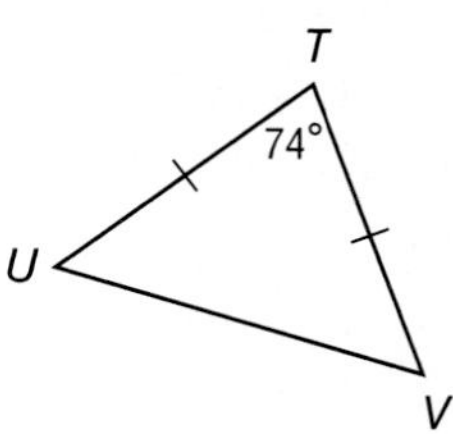


10. Suppose two sides of triangle $ABC$ are congruent to two sides of triangle $MNO$. Also, suppose one of the nonincluded angles of $\triangle ABC$ is congruent to one of the nonincluded angles of $\triangle MNO$. Are the triangles congruent? If so, write a paragraph proof showing the congruence. If not, sketch a counterexample.

## Extended Response

**Record your answers on a sheet of paper. Show your work.**

11. Use a coordinate grid to write a coordinate proof of the following statement.

*If the vertices of a triangle are A(0, 0), B(2a, b), and C(4a, 0), then the triangle* is isosceles.

**a.** Plot the vertices on a coordinate grid to model the problem.

**b.** Use the Distance Formula to write an expression for $AB$.

**c.** Use the Distance Formula to write an expression for $BC$.

**d.** Use your results from parts **b** and **c** to draw a conclusion about $\triangle ABC$.

**Need ExtraHelp?**

| If you missed Question... | 1 | 2 | 3 | 4 | 5 | 6 | 7 | 8 | 9 | 10 | 11 |
|---|---|---|---|---|---|---|---|---|---|---|---|
| Go to Lesson... | 5-5 | 6-2 | 6-1 | 6-5 | 5-5 | 6-2 | 6-4 | 5-3 | 6-5 | 6-3 | 6-6 |

CHAPTER 7

# Relationships in Triangles

## Then

You learned how to classify triangles.

## Now

In this chapter, you will:

- Learn about special segments and points related to triangles.
- Learn about relationships between the sides and angles of triangles.
- Learn to write indirect proofs.

## Why? ▲

**INTERIOR DESIGN** Triangle relationships are used to find and compare angle measures and distances. Interior designers use the relationships in triangles to maximize efficiency and create balance in their designs.

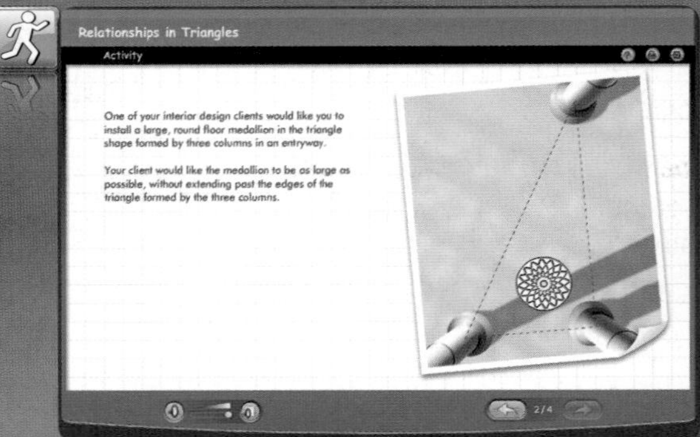


connectED.mcgraw-hill.com **Your Digital Math Portal**

Animation  Vocabulary  eGlossary  Personal Tutor  Virtual Manipulatives  Graphing Calculator  Audio  Foldables  Self-Check Practice  Worksheets

# Get Ready for the Chapter

**Diagnose** Readiness | You have two options for checking prerequisite skills.

**1 Textbook Option** Take the Quick Check below. Refer to the Quick Review for help.

## QuickCheck

Find each measure.

**1.** $BC$

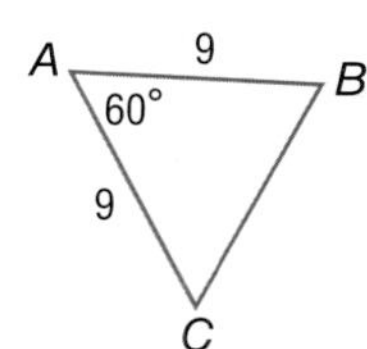


**2.** $m\angle RST$

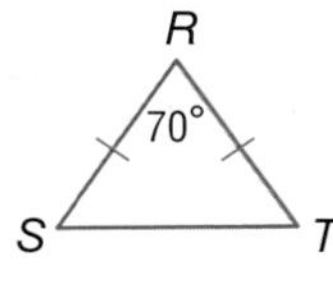


**3. GARDENS** Bronson is creating a right triangular flower bed. If two of the sides of the flower bed are 7 feet long each, what is the length of the third side to the nearest foot?

Make a conjecture based on the given information.

**4.** $\angle 3$ and $\angle 4$ are a linear pair.

**5.** $JKLM$ is a square.

**6.** $\overrightarrow{BD}$ is an angle bisector of $\angle ABC$.

**7. REASONING** Determine whether the following conjecture is *always*, *sometimes*, or *never* true based on the given information. Justify your reasoning.

**Given:** collinear points $D$, $E$, and $F$

**Conjecture:** $DE + EF = DF$

Solve each inequality.

**8.** $x + 13 < 41$

**9.** $x - 6 > 2x$

**10.** $6x + 9 < 7x$

**11.** $8x + 15 > 9x - 26$

**12. MUSIC** Nina added 15 more songs to her MP3 player, resulting in a total of more than 120 songs. How many songs were originally on the player?

## QuickReview

**Example 1**

Find each measure.

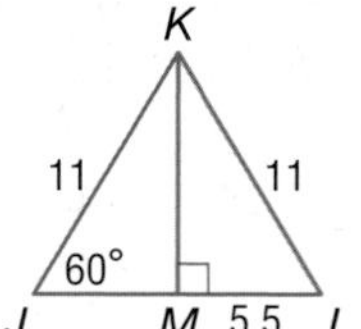


**a.** $JM$

$m\angle J = m\angle L$, so $m\angle L = 60$ and $\overline{JM} \cong \overline{LM}$ by the Converse of the Isosceles Triangle Theorem. Since $LM = 5.5$, $JM = 5.5$ by substitution.

**b.** $m\angle JKL$

| | |
|---|---|
| $m\angle J + m\angle JKL + m\angle L = 180$ | $\triangle$ Sum Theorem |
| $60 + m\angle JKL + 60 = 180$ | $m\angle J = m\angle L = 60$ |
| $120 + m\angle JKL = 180$ | Simplify. |
| $m\angle JKL = 60$ | Subtract. |

**Example 2**

**$K$ is the midpoint of $\overline{JL}$. Make a conjecture based on the given information and draw a figure to illustrate your conjecture.**

**Given:** $K$ is the midpoint of $\overline{JL}$. $J$, $K$, and $L$ are collinear points, and $K$ lies an equal distance between $J$ and $L$.

**Conjecture:** $\overline{JK} \cong \overline{KL}$

**Check:** Draw $\overline{JL}$. This illustrates the conjecture.

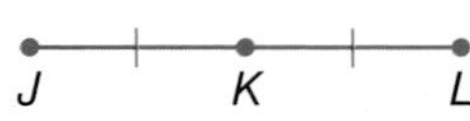


**Example 3**

**Solve $3x + 5 > 2x$.**

| | |
|---|---|
| $3x + 5 > 2x$ | Given |
| $3x - 3x + 5 > 2x - 3x$ | Subtract. |
| $5 > -x$ | Simplify. |
| $-5 < x$ | Divide. |

**2 Online Option** Take an online self-check Chapter Readiness Quiz at connectED.mcgraw-hill.com.

# Get Started on the Chapter

**You will learn several new concepts, skills, and vocabulary terms as you study Chapter 7. To get ready, identify important terms and organize your resources. You may wish to refer to Chapter 0 to review prerequisite skills.**

## FOLDABLES® StudyOrganizer

**Relationships in Triangles** Make this Foldable to help you organize your Chapter 7 notes about relationships in triangles. Begin with seven sheets of grid paper.

1. **Stack** the sheets. Fold the top right corner to the bottom edge to form an isosceles right triangle.

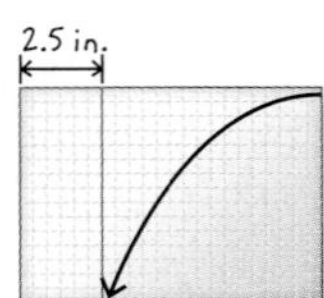


2. **Fold** the rectangular part in half.

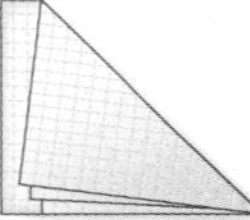

3. **Staple** the sheets along the rectangular fold in four places.

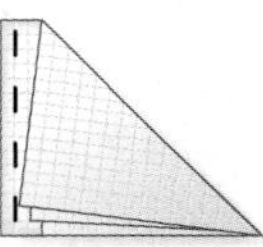

4. **Label** each sheet with a lesson number and the rectangular tab with the chapter title.

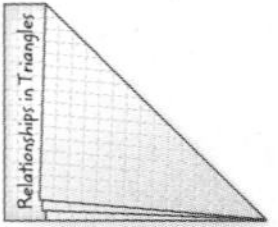


## NewVocabulary

| English | | Español |
|---|---|---|
| perpendicular bisector | p. 406 | mediatriz |
| concurrent lines | p. 407 | rectas concurrentes |
| point of concurrency | p. 407 | punto de concurrencia |
| circumcenter | p. 407 | circuncentro |
| incenter | p. 410 | incentro |
| median | p. 417 | mediana |
| centroid | p. 417 | baricentro |
| altitude | p. 419 | altura |
| orthocenter | p. 419 | ortocentro |
| indirect reasoning | p. 437 | razonamiento indirecto |
| indirect proof | p. 437 | demostración indirecta |
| proof by contradiction | p. 437 | demostración por contradicción |

## ReviewVocabulary

angle bisector bisectriz de un ángulo a ray that divides an angle into two congruent angles

midpoint punto medio the point on a segment exactly halfway between the endpoints of the segment

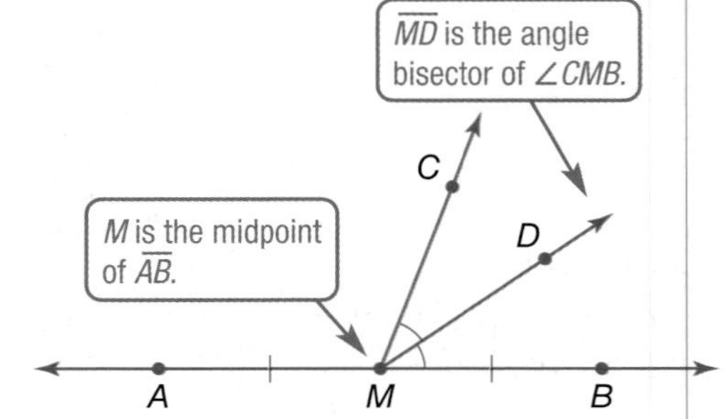

# Geometry Lab
# Constructing Bisectors

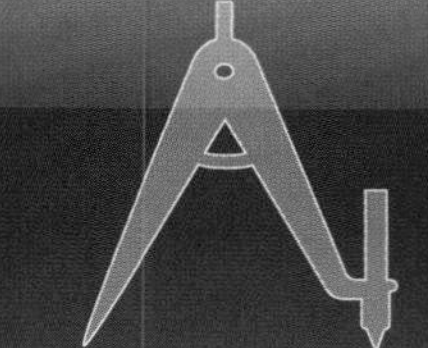

Paper folding can be used to construct special segments in triangles.

**Common Core State Standards**
**Content Standards**
**G.CO.12** Make formal geometric constructions with a variety of tools and methods (compass and straightedge, string, reflective devices, paper folding, dynamic geometric software, etc.).
**Mathematical Practices** 5

## Construction Perpendicular Bisector

**Construct a perpendicular bisector of the side of a triangle.**

**Step 1**

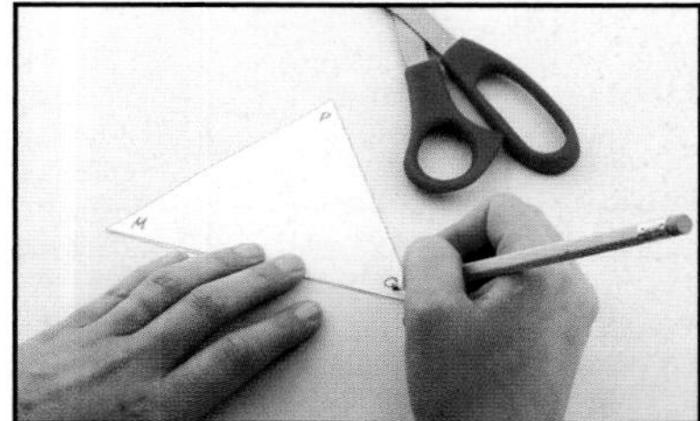

Draw, label, and cut out $\triangle MPQ$.

**Step 2**

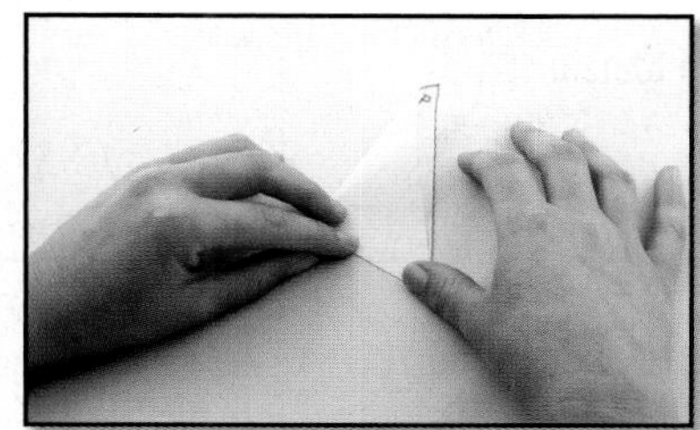

Fold the triangle in half along $\overline{MQ}$ so that vertex $M$ touches vertex $Q$.

**Step 3**

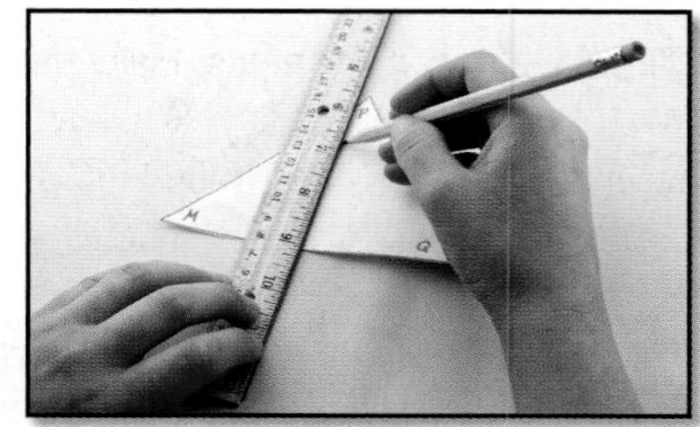

Use a straightedge to draw $\overline{AB}$ along the fold. $\overline{AB}$ is the perpendicular bisector of $\overline{MQ}$.

An angle bisector in a triangle is a line containing a vertex of the triangle and bisecting that angle.

## Construction Angle Bisector

**Construct an angle bisector of a triangle.**

**Step 1**

Draw, label, and cut out $\triangle ABC$.

**Step 2**

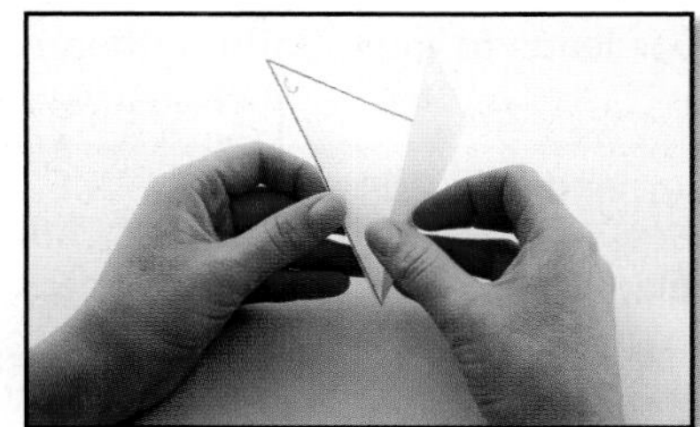

Fold the triangle in half through vertex A, such that sides $\overline{AC}$ and $\overline{AB}$ are aligned.

**Step 3**

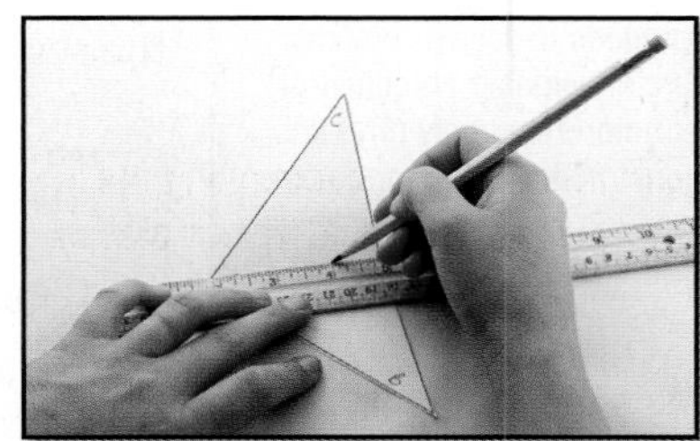

Label point $L$ at the crease along edge $\overline{BC}$. Use a straightedge to draw $\overline{AL}$ along the fold. $\overline{AL}$ is an angle bisector of $\triangle ABC$.

## Model and Analyze

**1.** Construct the perpendicular bisectors of the other two sides of $\triangle MPQ$. Construct the angle bisectors of the other two angles of $\triangle ABC$. What do you notice about their intersections?

**Repeat the two constructions for each type of triangle.**

**2.** acute

**3.** obtuse

**4.** right

LESSON 7-1

# Bisectors of Triangles

**Then**

- You used segment and angle bisectors.

**Now**

1. Identify and use perpendicular bisectors in triangles.
2. Identify and use angle bisectors in triangles.

**Why?**

- Creating a work triangle in a kitchen can make food preparation more efficient by cutting down on the number of steps you have to take. To locate the point that is equidistant from the sink, stove, and refrigerator, you can use the perpendicular bisectors of the triangle.

**NewVocabulary**
perpendicular bisector
concurrent lines
point of concurrency
circumcenter
incenter

**Common Core State Standards**

**Content Standards**
G.CO.10 Prove theorems about triangles.
G.MG.3 Apply geometric methods to solve problems (e.g., designing an object or structure to satisfy physical constraints or minimize cost; working with typographic grid systems based on ratios). ★

**Mathematical Practices**
1 Make sense of problems and persevere in solving them.
3 Construct viable arguments and critique the reasoning of others.

## 1 Perpendicular Bisectors

A segment bisector is any segment, line, or plane that intersects a segment at its midpoint. If a bisector is also perpendicular to the segment, it is called a **perpendicular bisector**.

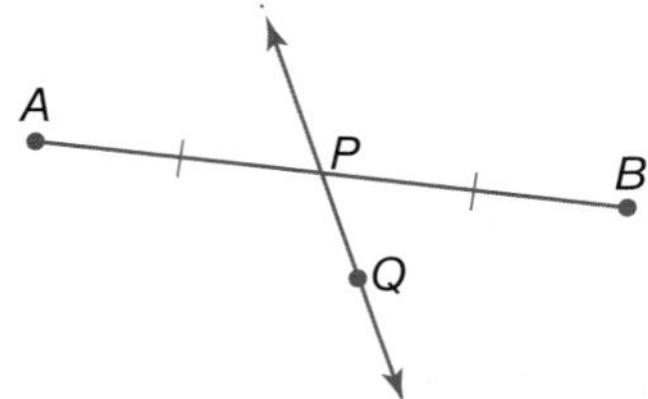


$\overleftrightarrow{PQ}$ is a bisector of $\overline{AB}$.

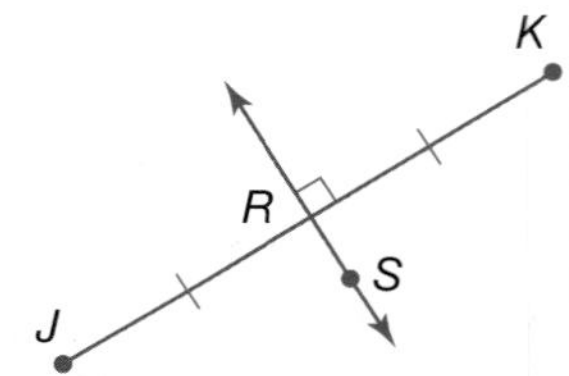


$\overleftrightarrow{RS}$ is a perpendicular bisector of $\overline{JK}$.

Recall that a *locus* is a set of points that satisfies a particular condition. The perpendicular bisector of a segment is the locus of points in a plane equidistant from the endpoints of the segment. This leads to the following theorems.

### Theorems Perpendicular Bisectors

**7.1 Perpendicular Bisector Theorem**

If a point is on the perpendicular bisector of a segment, then it is equidistant from the endpoints of the segment.

**Example:** If $\overline{CD}$ is a $\perp$ bisector of $\overline{AB}$, then $AC = BC$.

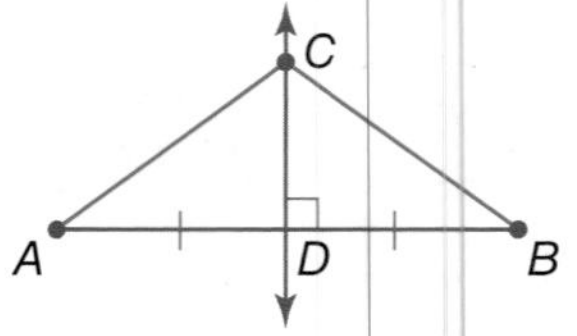


**7.2 Converse of the Perpendicular Bisector Theorem**

If a point is equidistant from the endpoints of a segment, then it is on the perpendicular bisector of the segment.

**Example:** If $AE = BE$, then $E$ lies on $\overline{CD}$, the $\perp$ bisector of $\overline{AB}$.

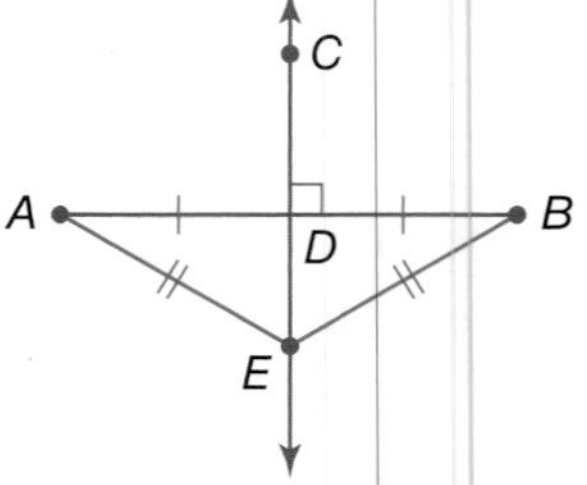


You will prove Theorems 7.1 and 7.2 in Exercises 39 and 37, respectively.

**Study Tip**

**Perpendicular Bisectors** The perpendicular bisector of a side of a triangle does not necessarily pass through a vertex of the triangle. For example, in $\triangle XYZ$ below, the perpendicular bisector of $\overline{XY}$ does not pass through point $Z$.

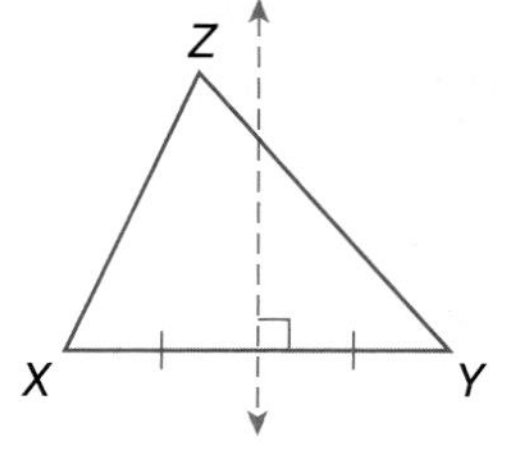

## Example 1 Use the Perpendicular Bisector Theorems

**Find each measure.**

**a.** $AB$

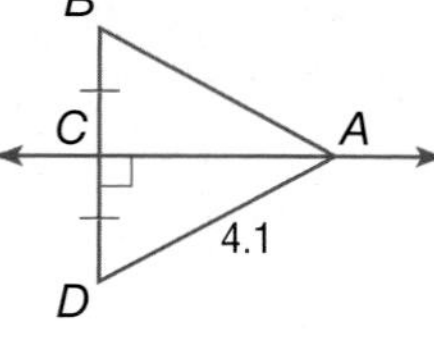


From the information in the diagram, we know that $\overleftrightarrow{CA}$ is the perpendicular bisector of $\overline{BD}$.

$AB = AD$ — Perpendicular Bisector Theorem

$AB = 4.1$ — Substitution

**b.** $WY$

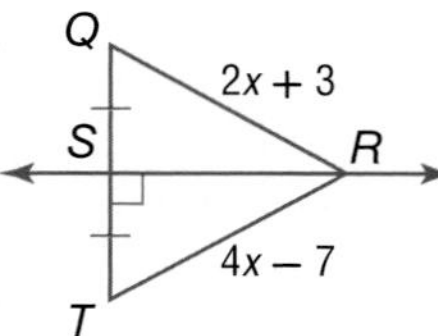


Since $WX = ZX$ and $\overleftrightarrow{XY} \perp \overline{WZ}$, $\overleftrightarrow{XY}$ is the perpendicular bisector of $\overline{WZ}$ by the Converse of the Perpendicular Bisector Theorem. By the definition of segment bisector, $WY = YZ$. Since $YZ = 3$, $WY = 3$.

**c.** $RT$

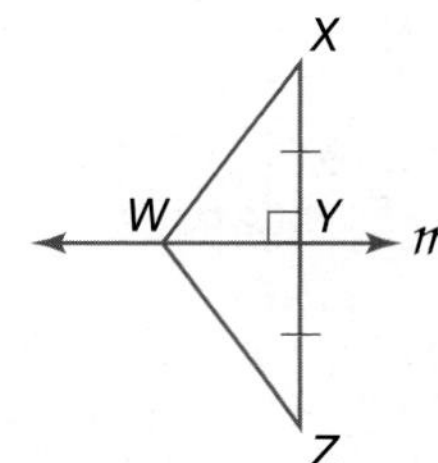

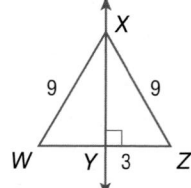


$\overleftrightarrow{SR}$ is the perpendicular bisector of $\overline{QT}$.

| | |
|---|---|
| $RT = RQ$ | Perpendicular Bisector Theorem |
| $4x - 7 = 2x + 3$ | Substitution |
| $2x - 7 = 3$ | Subtract $2x$ from each side. |
| $2x = 10$ | Add 7 to each side. |
| $x = 5$ | Divide each side by 2. |

So $RT = 4(5) - 7$ or 13.

### Guided Practice

**1A.** If $WX = 25.3$, $YZ = 22.4$, and $WZ = 25.3$, find $XY$.

**1B.** If $m$ is the perpendicular bisector of $XZ$ and $WZ = 14.9$, find $WX$.

**1C.** If $m$ is the perpendicular bisector of $XZ$, $WX = 4a - 15$, and $WZ = a + 12$, find $WX$.

When three or more lines intersect at a common point, the lines are called **concurrent lines**. The point where concurrent lines intersect is called the **point of concurrency**.

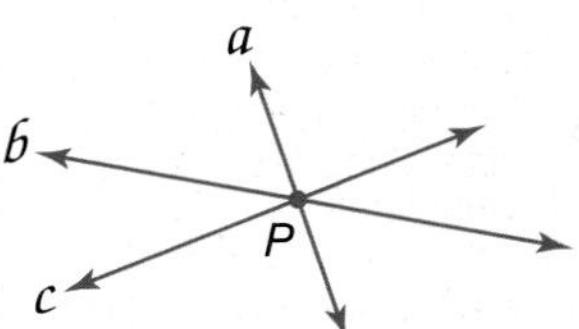


Lines $a$, $b$, and $c$ are concurrent at $P$.

A triangle has three sides, so it also has three perpendicular bisectors. These bisectors are concurrent lines. The point of concurrency of the perpendicular bisectors is called the **circumcenter** of the triangle.

### Theorem 7.3 Circumcenter Theorem

**Words** The perpendicular bisectors of a triangle intersect at a point called the *circumcenter* that is equidistant from the vertices of the triangle.

**Example** If $P$ is the circumcenter of $\triangle ABC$, then $PB = PA = PC$.

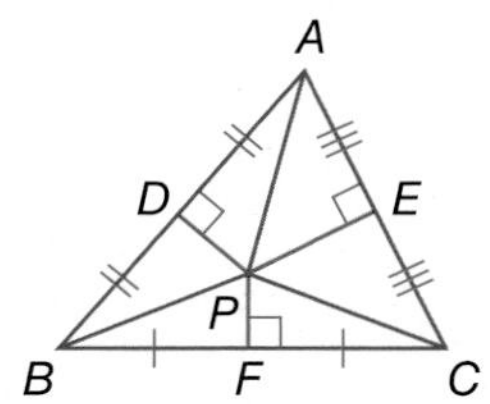

**ReadingMath**

*Circum–* The prefix *circum–* means *about* or *around.* The circumcenter is the center of a circle around a triangle that contains the vertices of the triangle.

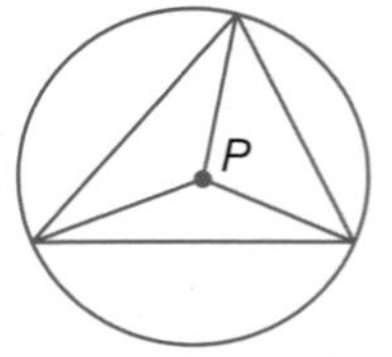


The circumcenter can be on the interior, exterior, or side of a triangle.

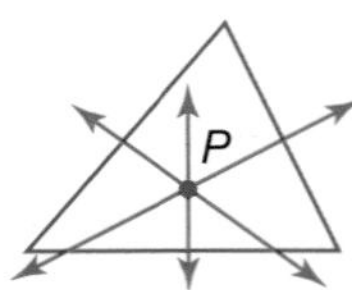


acute triangle

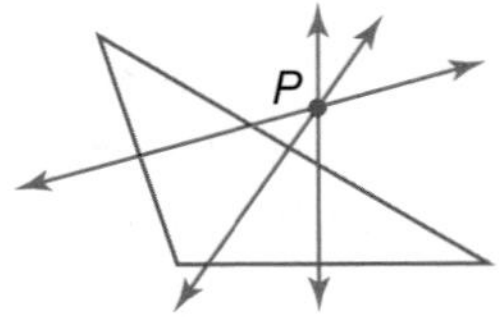


obtuse triangle

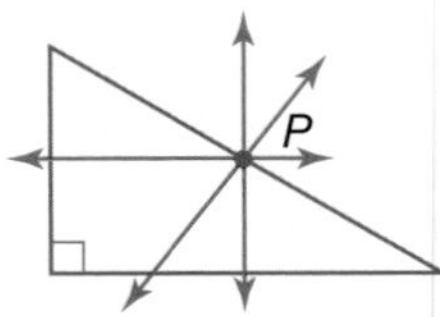


right triangle

**Proof** Circumcenter Theorem

**Given:** $\overline{PD}$, $\overline{PF}$, and $\overline{PE}$ are perpendicular bisectors of $\overline{AB}$, $\overline{AC}$, and $\overline{BC}$, respectively.

**Prove:** $AP = CP = BP$

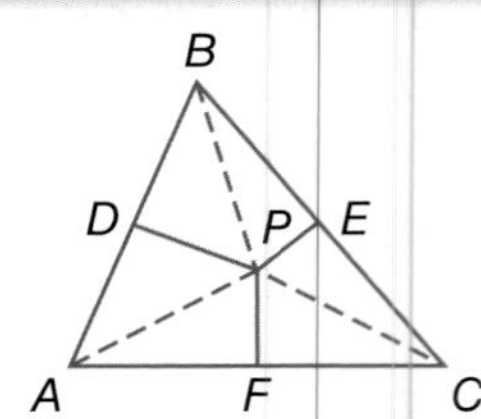


**Paragraph Proof:**

Since $P$ lies on the perpendicular bisector of $\overline{AC}$, it is equidistant from $A$ and $C$. By the definition of equidistant, $AP = CP$. The perpendicular bisector of $\overline{BC}$ also contains $P$. Thus, $CP = BP$. By the Transitive Property of Equality, $AP = BP$. Thus, $AP = CP = BP$.

**Real-WorldLink**

Some basic rules of thumb for a kitchen work triangle are that the sides of the triangle should be no greater than 9 feet and no less than 4 feet. Also, the perimeter of the triangle should be no more than 26 feet and no less than 12 feet.

**Source:** Merillat

**Real-World Example 2** Use the Circumcenter Theorem

PT

**INTERIOR DESIGN A stove *S*, sink *K*, and refrigerator *R* are positioned in a kitchen as shown. Find the location for the center of an island work station so that it is the same distance from these three points.**

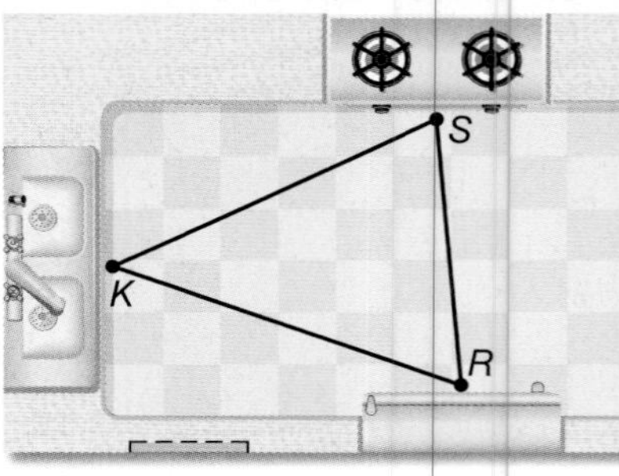


By the Circumcenter Theorem, a point equidistant from three points is found by using the perpendicular bisectors of the triangle formed by those points.

Copy $\triangle SKR$, and use a ruler and protractor to draw the perpendicular bisectors. The location for the center of the island is *C*, the circumcenter of $\triangle SKR$.

S

C

K

R

**GuidedPractice**

**2.** To water his triangular garden, Alex needs to place a sprinkler equidistant from each vertex. Where should Alex place the sprinkler?

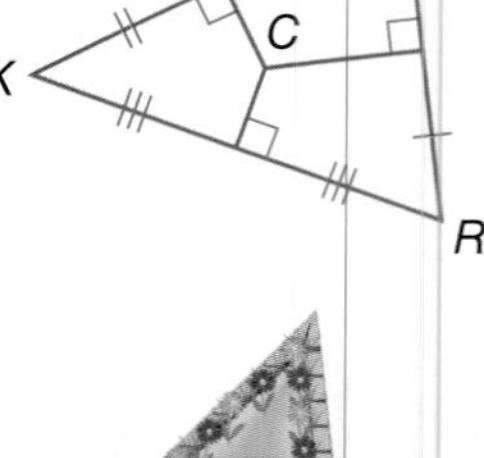

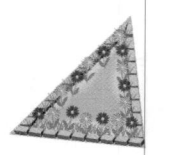

## 2 Angle Bisectors

Recall from Lesson 1-4 that an angle bisector divides an angle into two congruent angles. The angle bisector can be a line, segment, or ray.

The bisector of an angle can be described as the locus of points in the interior of the angle equidistant from the sides of the angle. This description leads to the following theorems.

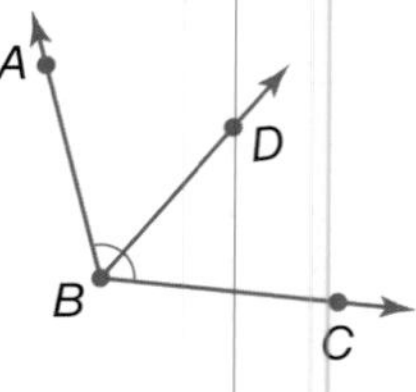


$\overrightarrow{BD}$ is the angle bisector of $\angle ABC$.

Masterfile

## Theorems Angle Bisectors

**7.4 Angle Bisector Theorem**

If a point is on the bisector of an angle, then it is equidistant from the sides of the angle.

**Example:** If $\overrightarrow{BF}$ bisects $\angle DBE$, $\overline{FD} \perp \overrightarrow{BD}$, and $\overline{FE} \perp \overrightarrow{BE}$, then $DF = FE$.

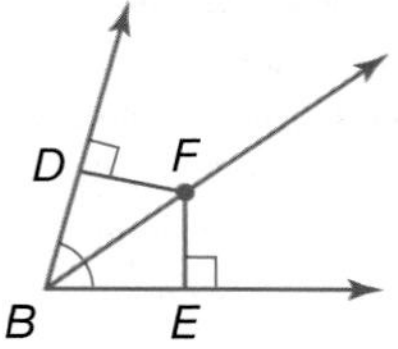


**7.5 Converse of the Angle Bisector Theorem**

If a point in the interior of an angle is equidistant from the sides of the angle, then it is on the bisector of the angle.

**Example:** If $\overline{FD} \perp \overrightarrow{BD}$, $\overline{FE} \perp \overrightarrow{BE}$, and $DF = FE$, then $\overrightarrow{BF}$ bisects $\angle DBE$.

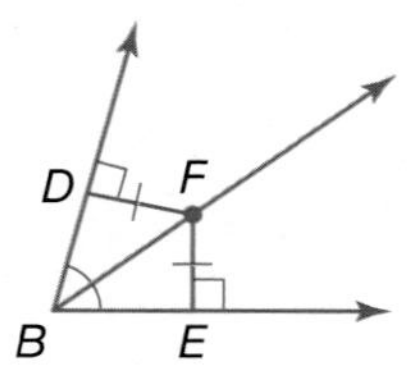


You will prove Theorems 7.4 and 7.5 in Exercises 43 and 40.

## Example 3 Use the Angle Bisector Theorems

**Find each measure.**

**a.** $XY$

| | |
|---|---|
| $XY = XW$ | Angle Bisector Theorem |
| $XY = 7$ | Substitution |

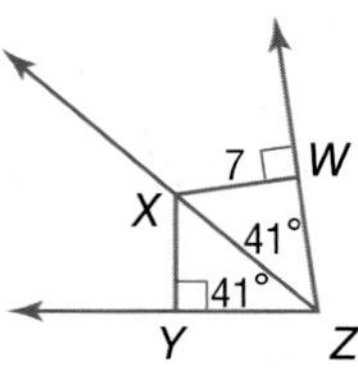


**StudyTip**

**Angle Bisector** For part **b**, only that $JL = LM$ would not be enough information to conclude that $\overrightarrow{KL}$ bisects $\angle JKM$.

**b.** $m\angle JKL$

Since $\overline{LJ} \perp \overrightarrow{KJ}$, $\overline{LM} \perp \overrightarrow{KM}$, $\overline{LJ} \cong \overline{LM}$, $L$ is equidistant from the sides of $\angle JKM$. By the Converse of the Angle Bisector Theorem, $\overrightarrow{KL}$ bisects $\angle JKM$.

| | |
|---|---|
| $\angle JKL \cong \angle LKM$ | Definition of angle bisector |
| $m\angle JKL = m\angle LKM$ | Definition of congruent angles |
| $m\angle JKL = 37$ | Substitution |

**c.** $SP$

| | |
|---|---|
| $SP = SM$ | Angle Bisector Theorem |
| $6x - 7 = 3x + 5$ | Substitution |
| $3x - 7 = 5$ | Subtract $3x$ from each side. |
| $3x = 12$ | Add 7 to each side. |
| $x = 4$ | Divide each side by 3. |

So, $SP = 6(4) - 7$ or 17

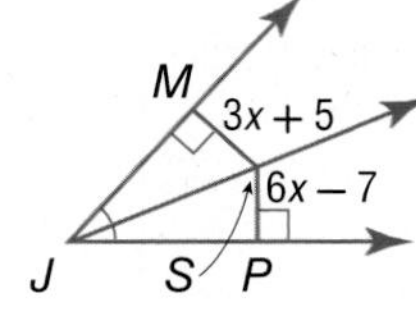


**GuidedPractice**

**3A.** If $m\angle BAC = 38$, $BC = 5$, and $DC = 5$, find $m\angle DAC$.

**3B.** If $m\angle BAC = 40$, $m\angle DAC = 40$, and $DC = 10$, find $BC$.

**3C.** If $\overrightarrow{AC}$ bisects $\angle DAB$, $BC = 4x + 8$, and $DC = 9x - 7$, find $BC$.

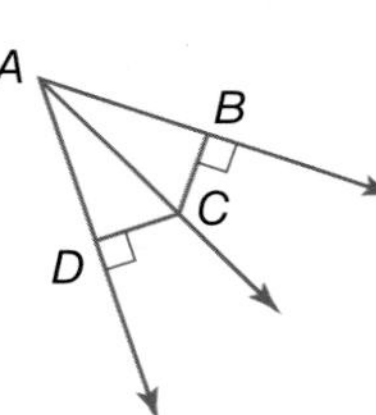

**ReadingMath**

**Incenter** The incenter is the center of a circle that intersects each side of the triangle at one point. For this reason, the incenter always lies in the interior of a triangle.

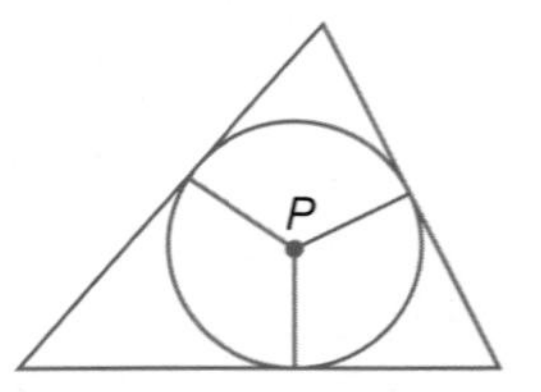


Similar to perpendicular bisectors, since a triangle has three angles, it also has three angle bisectors. The angle bisectors of a triangle are concurrent, and their point of concurrency is called the **incenter** of a triangle.

### Theorem 7.6 Incenter Theorem

**Words** The angle bisectors of a triangle intersect at a point called the *incenter* that is equidistant from the sides of the triangle.

**Example** If $P$ is the incenter of $\triangle ABC$, then $PD = PE = PF$.

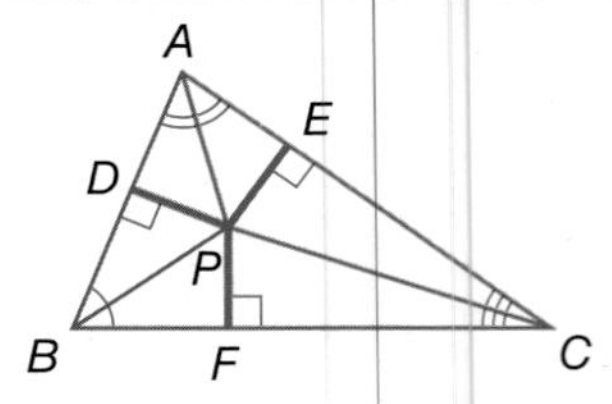


You will prove Theorem 7.6 in Exercise 38.

### Example 4 Use the Incenter Theorem

**Find each measure if $J$ is the incenter of $\triangle ABC$.**

**a.** $JF$

By the Incenter Theorem, since $J$ is equidistant from the sides of $\triangle ABC$, $JF = JE$. Find $JF$ by using the Pythagorean Theorem.

| | |
|---|---|
| $a^2 + b^2 = c^2$ | Pythagorean Theorem |
| $JE^2 + 12^2 = 15^2$ | Substitution |
| $JE^2 + 144 = 225$ | $12^2 = 144$ and $15^2 = 225$. |
| $JE^2 = 81$ | Subtract 144 from each side. |
| $JE = \pm 9$ | Take the square root of each side. |

Since length cannot be negative, use only the positive square root, 9. Since $JE = JF$, $JF = 9$.

**b.** $m\angle JAC$

Since $\overrightarrow{BJ}$ bisects $\angle CBE$, $m\angle CBE = 2m\angle JBE$. So $m\angle CBE = 2(34)$ or 68. Likewise, $m\angle DCF = 2m\angle DCJ$, so $m\angle DCF = 2(32)$ or 64.

| | |
|---|---|
| $m\angle CBE + m\angle DCF + m\angle FAE = 180$ | Triangle Angle Sum Theorem |
| $68 + 64 + m\angle FAE = 180$ | $m\angle CBE = 68$, $m\angle DCF = 64$ |
| $132 + m\angle FAE = 180$ | Simplify. |
| $m\angle FAE = 48$ | Subtract 132 from each side. |

Since $\overrightarrow{AJ}$ bisects $\angle FAE$, $2m\angle JAC = m\angle FAE$. This means that $m\angle JAC = \frac{1}{2}m\angle FAE$, so $m\angle JAC = \frac{1}{2}(48)$ or 24.

**GuidedPractice**

**If $P$ is the incenter of $\triangle XYZ$, find each measure.**

**4A.** $PK$

**4B.** $m\angle LZP$

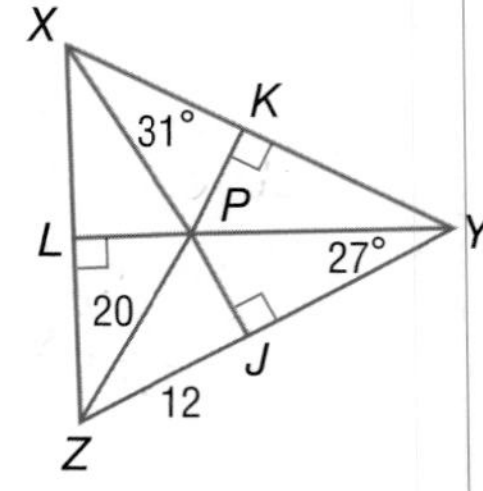

## Check Your Understanding

**Example 1** **Find each measure.**

1. $XW$

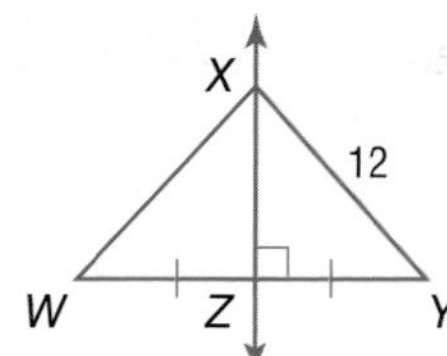


2. $AC$

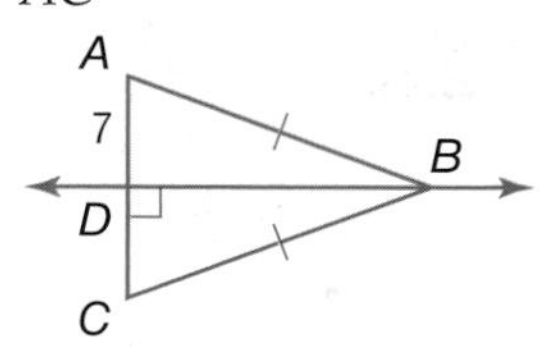


3. $LP$

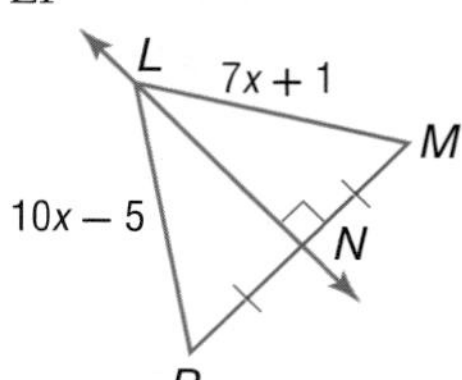


**Example 2**

4. **ADVERTISING** Four friends are passing out flyers at a mall food court. Three of them take as many flyers as they can and position themselves as shown. The fourth one keeps the supply of additional flyers. Copy the positions of points $A$, $B$, and $C$. Then position the fourth friend at $D$ so that she is the same distance from each of the other three friends.

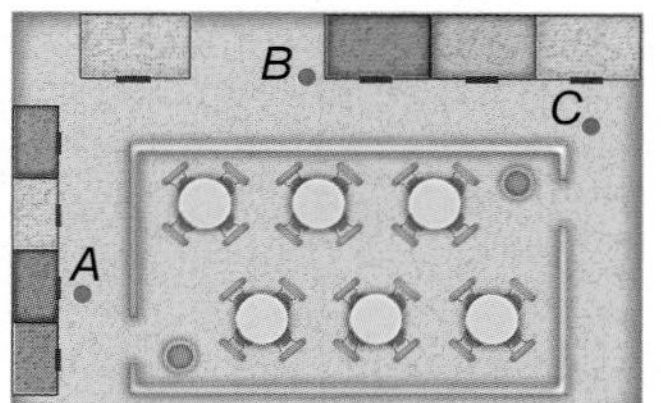


**Example 3** **Find each measure.**

5. $CP$

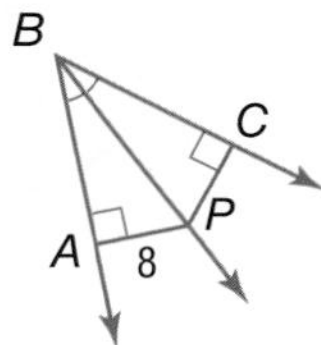


6. $m\angle WYZ$

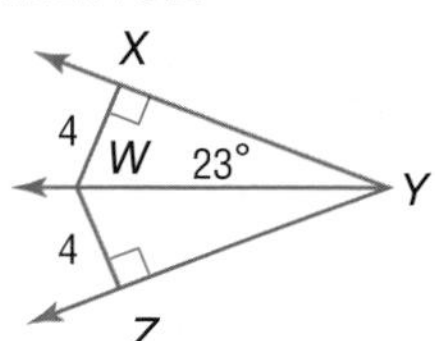


7. $QM$

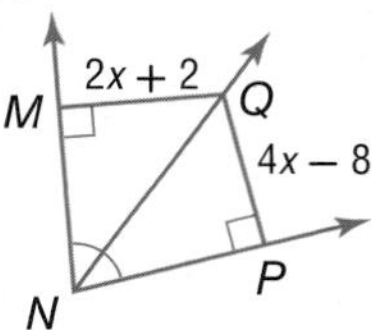


**Example 4**

8. **CCSS SENSE-MAKING** Find $JQ$ if $Q$ is the incenter of $\triangle JLN$.

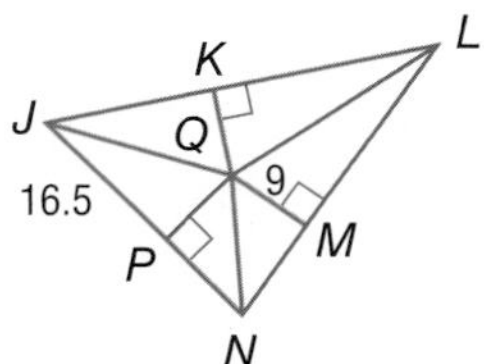


## Practice and Problem Solving

**Example 1** **Find each measure.**

9. $NP$

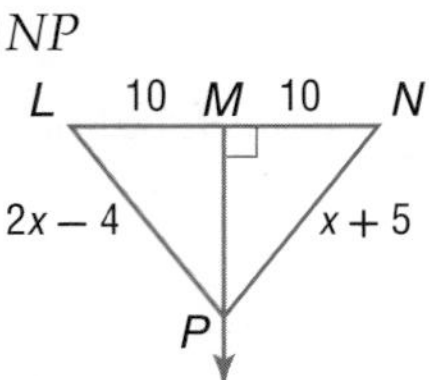


10. $PS$

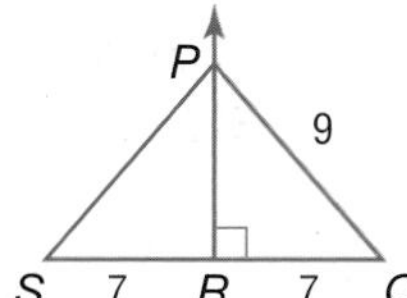


11. $KL$

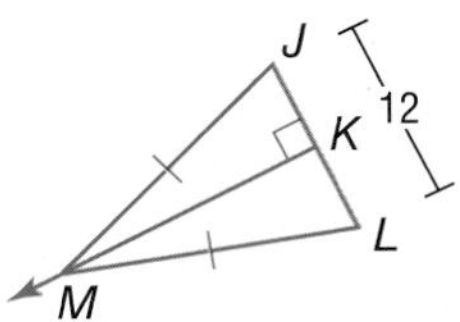


12. $EG$

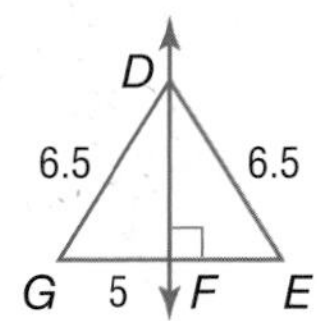


13. $CD$

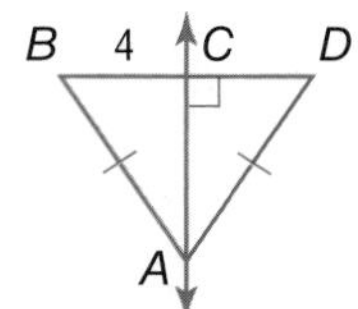


14. $SW$

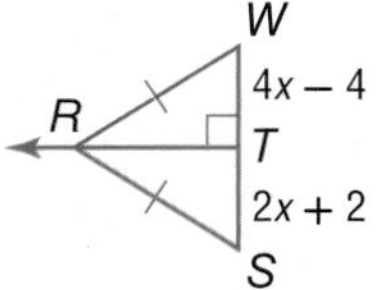

Example 2

15. **STATE FAIR** The state fair has set up the location of the midway, livestock competition, and food vendors. The fair planners decide that they want to locate the portable restrooms the same distance from each location. Copy the positions of points $M$, $L$, and $F$. Then find the location for the restrooms and label it $R$.

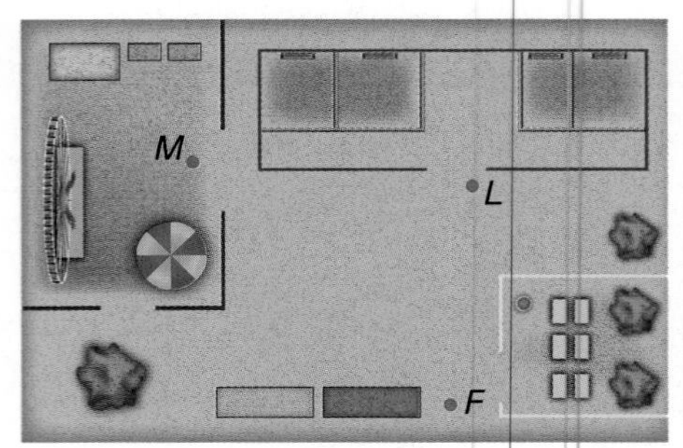


16. **SCHOOL** A school system has built an elementary, middle, and high school at the locations shown in the diagram. Copy the positions of points $E$, $M$, and $H$. Then find the location for the bus yard $B$ that will service these schools so that it is the same distance from each school.

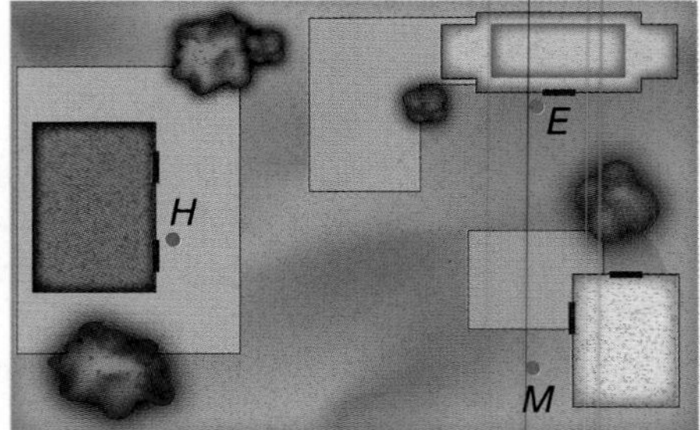


**Point $D$ is the circumcenter of $\triangle ABC$. List any segment(s) congruent to each segment.**

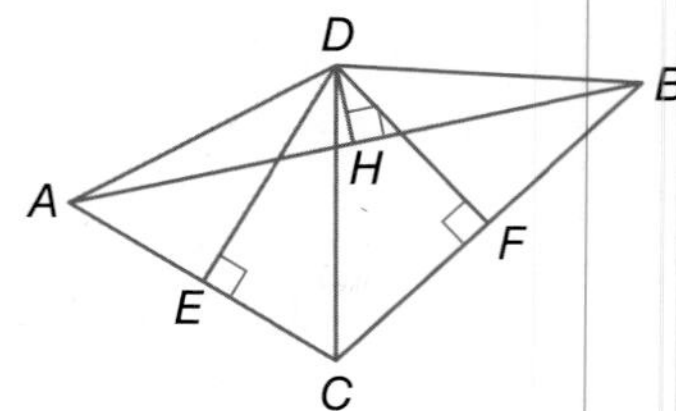


17. $\overline{AD}$

18. $\overline{BF}$

19. $\overline{AH}$

20. $\overline{DC}$

Example 3

**Find each measure.**

21. $AF$

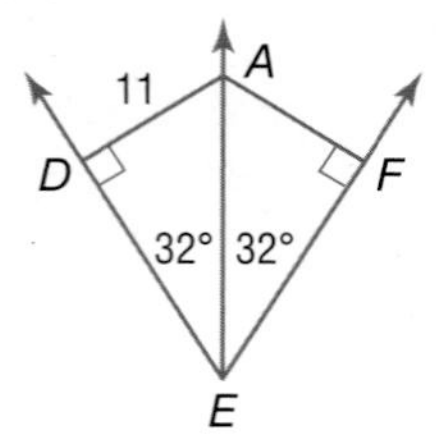


22. $m\angle DBA$

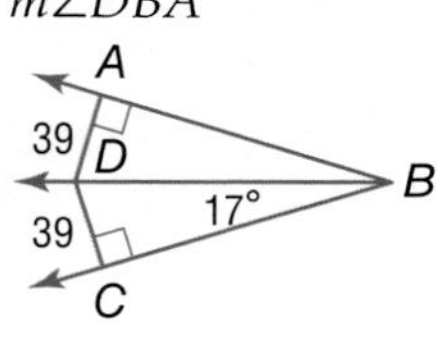


23. $m\angle PNM$

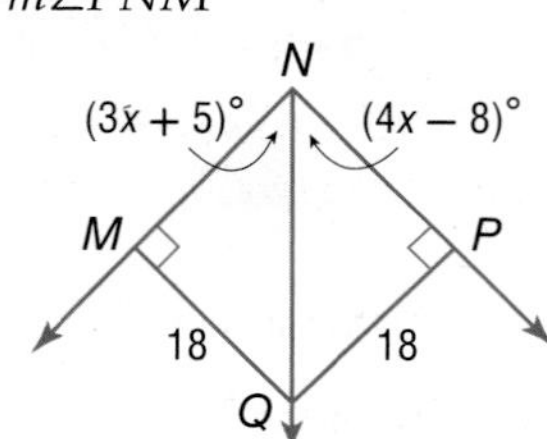


24. $XA$

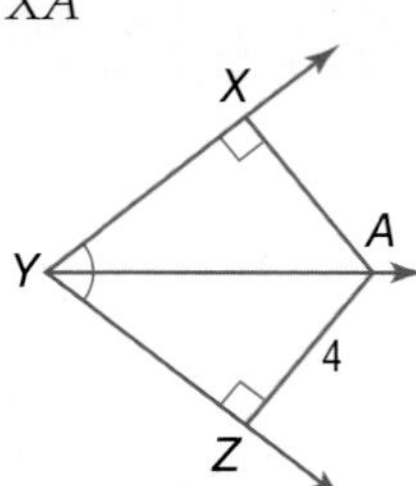


25. $m\angle PQS$

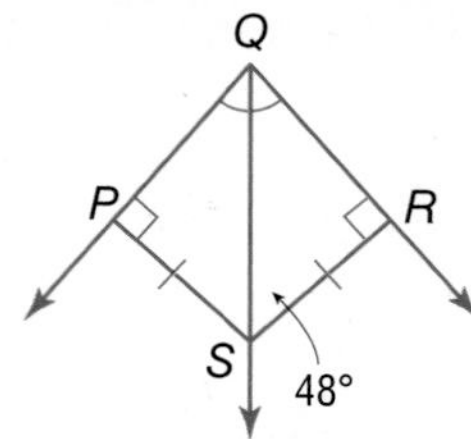


26. $PN$

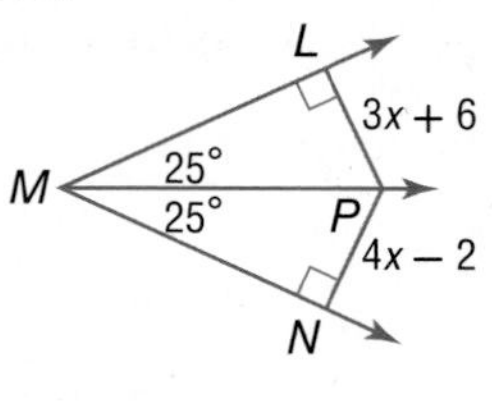


Example 4

**CCSS SENSE-MAKING Point $P$ is the incenter of $\triangle AEC$. Find each measure below.**

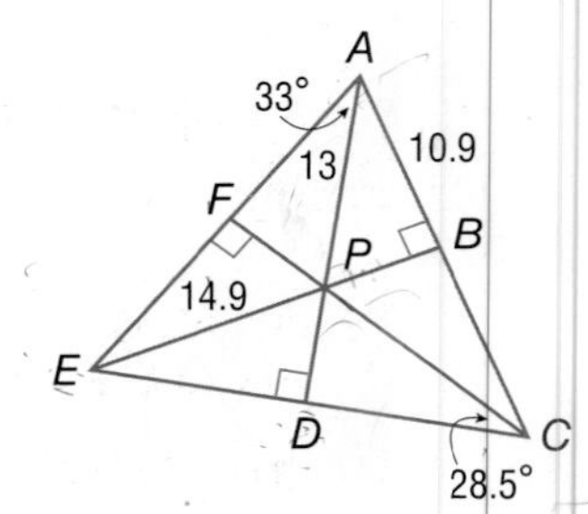


27. $PB$

28. $DE$

29. $m\angle DAC$

30. $m\angle DEP$

**31** **INTERIOR DESIGN** You want to place a centerpiece on a corner table so that it is located the same distance from each edge of the table. Make a sketch to show where you should place the centerpiece. Explain your reasoning.

**Determine whether there is enough information given in each diagram to find the value of $x$. Explain your reasoning.**

**32.**

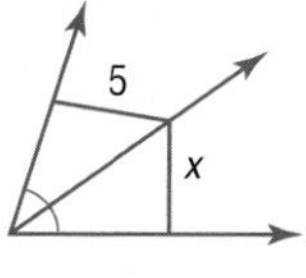


**33.**

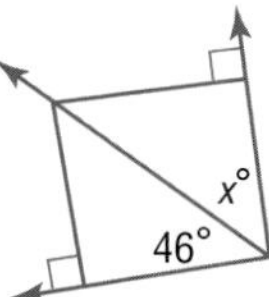


**34.**

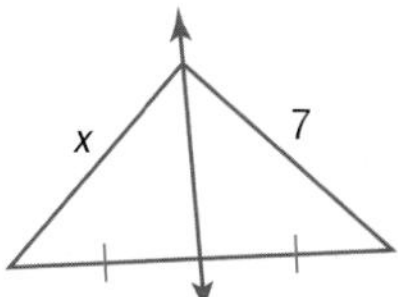


**35.**

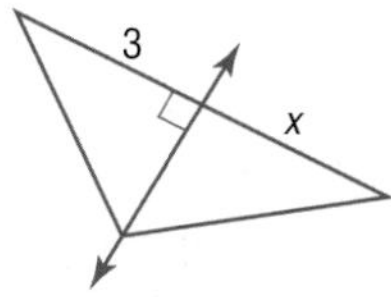


**36. SOCCER** A soccer player $P$ is approaching the opposing team's goal as shown in the diagram. To make the goal, the player must kick the ball between the goal posts at $L$ and $R$. The goalkeeper faces the kicker. He then tries to stand so that if he needs to dive to stop a shot, he is as far from the left-hand side of the shot angle as the right-hand side.

**a.** Describe where the goalkeeper should stand. Explain your reasoning.

**b.** Copy $\triangle PRL$. Use a compass and a straightedge to locate a point $G$ where the goalkeeper should stand.

**c.** If the ball is kicked so it follows the path from $P$ to $R$, construct the shortest path the goalkeeper should take to block the shot. Explain your reasoning.

**PROOF Write a two-column proof.**

**37.** Theorem 7.2

**Given:** $\overline{CA} \cong \overline{CB}$, $\overline{AD} \cong \overline{BD}$

**Prove:** $C$ and $D$ are on the perpendicular bisector of $\overline{AB}$.

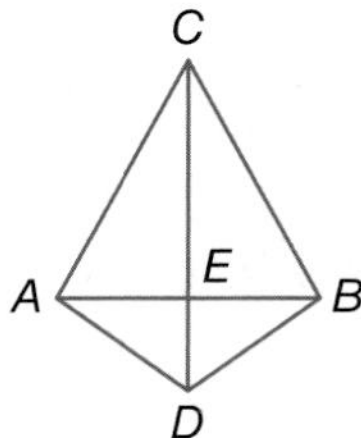


**38.** Theorem 7.6

**Given:** $\triangle ABC$, angle bisectors $\overline{AD}$, $\overline{BE}$, and $\overline{CF}$
$\overline{KP} \perp \overline{AB}$, $\overline{KQ} \perp \overline{BC}$,
$\overline{KR} \perp \overline{AC}$

**Prove:** $KP = KQ = KR$

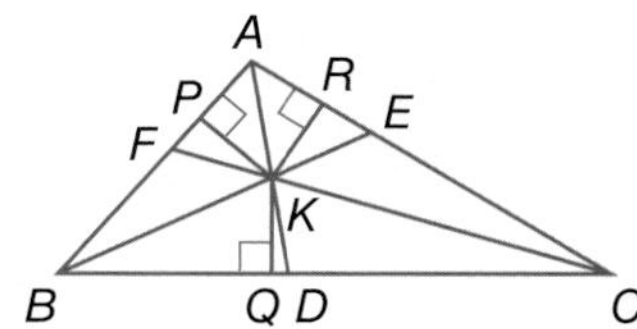


**CCSS ARGUMENTS Write a paragraph proof of each theorem.**

**39.** Theorem 7.1

**40.** Theorem 7.5

**COORDINATE GEOMETRY Write an equation in slope-intercept form for the perpendicular bisector of the segment with the given endpoints. Justify your answer.**

**41.** $A(-3, 1)$ and $B(4, 3)$

**42.** $C(-4, 5)$ and $D(2, -2)$

**43. PROOF** Write a two-column proof of Theorem 7.4.

**44. GRAPHIC DESIGN** Mykia is designing a pennant for her school. She wants to put a picture of the school mascot inside a circle on the pennant. Copy the outline of the pennant and locate the point where the center of the circle should be to create the largest circle possible. Justify your drawing.

**COORDINATE GEOMETRY Find the coordinates of the circumcenter of the triangle with the given vertices. Explain.**

**45.** $A(0, 0)$, $B(0, 6)$, $C(10, 0)$

**46.** $J(5, 0)$, $K(5, -8)$, $L(0, 0)$

**47. LOCUS** Consider $\overline{CD}$. Describe the set of all points in space that are equidistant from $C$ and $D$.

## H.O.T. Problems Use Higher-Order Thinking Skills

**48. ERROR ANALYSIS** Claudio says that from the information supplied in the diagram, he can conclude that $K$ is on the perpendicular bisector of $\overline{LM}$. Caitlyn disagrees. Is either of them correct? Explain your reasoning.

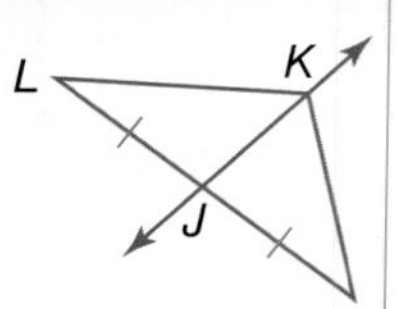


**49. OPEN ENDED** Draw a triangle with an incenter located inside the triangle but a circumcenter located outside. Justify your drawing by using a straightedge and a compass to find both points of concurrency.

**CCSS ARGUMENTS Determine whether each statement is *sometimes, always,* or *never* true. Justify your reasoning using a counterexample or proof.**

**50.** The angle bisectors of a triangle intersect at a point that is equidistant from the vertices of the triangle.

**51.** In an isosceles triangle, the perpendicular bisector of the base is also the angle bisector of the opposite vertex.

**CHALLENGE Write a two-column proof for each of the following.**

**52. Given:** Plane $\mathcal{Y}$ is a perpendicular bisector of $\overline{DC}$.
**Prove:** $\angle ADB \cong \angle ACB$

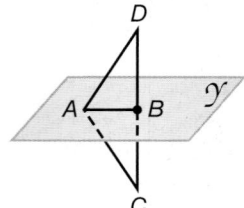

**53. Given:** Plane $\mathcal{Z}$ is an angle bisector of $\angle KJH$, $\overline{KJ} \cong \overline{HJ}$
**Prove:** $\overline{MH} \cong \overline{MK}$

**54. WRITING IN MATH** Compare and contrast the perpendicular bisectors and angle bisectors of a triangle. How are they alike? How are they different? Be sure to compare their points of concurrency.

## Standardized Test Practice

**55. ALGEBRA** An object is projected straight upward with initial velocity $v$ meters per second from an initial height of $s$ meters. The height $h$ in meters of the object after $t$ seconds is given by $h = -10t^2 + vt + s$. Sherise is standing at the edge of a balcony 54 meters above the ground and throws a ball straight up with an initial velocity of 12 meters per second. After how many seconds will it hit the ground?

**A** 3 seconds
**B** 4 seconds
**C** 6 seconds
**D** 9 seconds

**56. SAT/ACT** For $x \neq -3$, $\frac{3x + 9}{x + 3} =$

**F** $x + 12$
**G** $x + 9$
**H** $x + 3$
**J** $x$
**K** 3

**57.** A line drawn through which of the following points would be a perpendicular bisector of $\triangle JKL$?

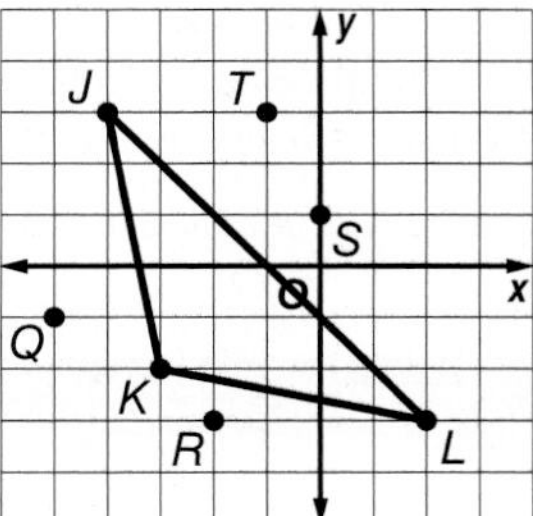


**A** $T$ and $K$
**B** $L$ and $Q$
**C** $J$ and $R$
**D** $S$ and $K$

**58. SHORT RESPONSE** Write an equation in slope-intercept form that describes the line containing the points $(-1, 0)$ and $(2, 4)$.

## Spiral Review

**Name the missing coordinate(s) of each triangle.** (Lesson 6-6)

**59.**

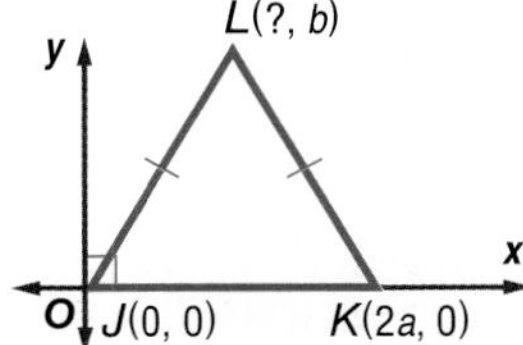


**60.**

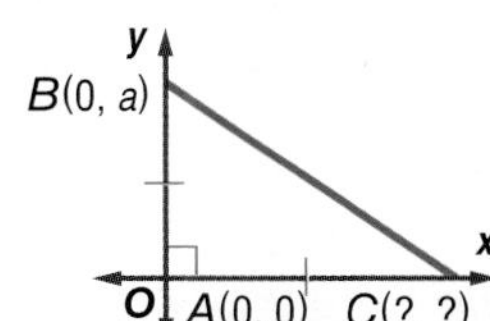


**61.**

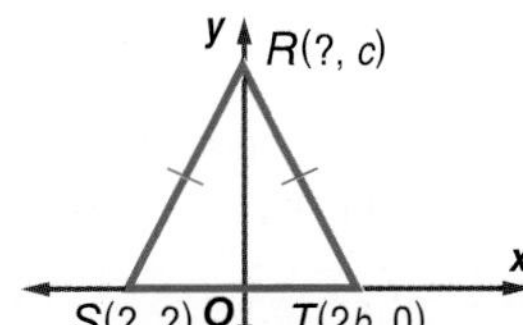


**Solve each equation.** (Lesson 1-2)

**62.** $-4(b + 3) + b(b - 3) = -b(6 - b) + 2(b - 3)$

**63.** $3(a - 3) + a(a - 1) + 12 = a(a - 2) + 3(a - 2) + 4$

**Solve each equation by using the Quadratic Formula. Round to the nearest tenth if necessary.** (Lesson 2-5)

**64.** $3x^2 + 10x = 15$

**65.** $\frac{1}{2}x^2 - 8x + 6 = 0$

## Skills Review

**PROOF Write a two-column proof for each of the following.**

**66. Given:** $\triangle XKF$ is equilateral.
$\overline{XJ}$ bisects $\angle X$.
**Prove:** $J$ is the midpoint of $\overline{KF}$.

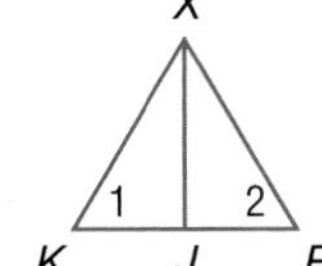


**67. Given:** $\triangle MLP$ is isosceles.
$N$ is the midpoint of $\overline{MP}$.
**Prove:** $\overline{LN} \perp \overline{MP}$

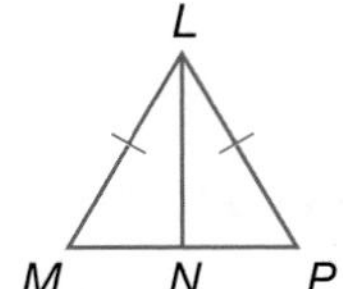

EXPLORE 7-2

# Geometry Lab
# Constructing Medians and Altitudes

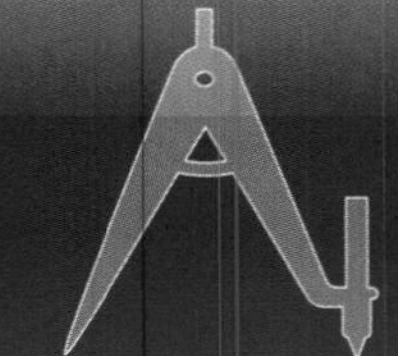

A *median* of a triangle is a segment with endpoints that are a vertex and the midpoint of the side opposite that vertex. You can use the construction for the midpoint of a segment to construct a median.

Wrap the end of string around a pencil. Use a thumbtack to fix the string to a vertex.

**CCSS Common Core State Standards**
**Content Standards**
**G.CO.12** Make formal geometric constructions with a variety of tools and methods (compass and straightedge, string, reflective devices, paper folding, dynamic geometric software, etc.).
**Mathematical Practices** 5

**Construction 1** Median of a Triangle

**Step 1**

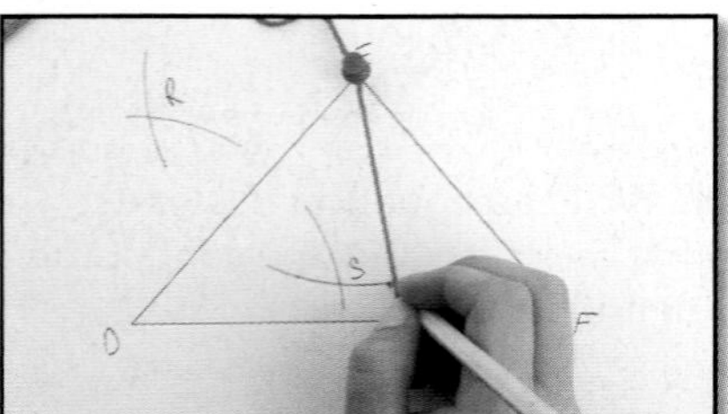

Place the thumbtack on vertex $D$ and then on vertex $E$ to draw intersecting arcs above and below $\overline{DE}$. Label the points of intersection $R$ and $S$.

**Step 2**

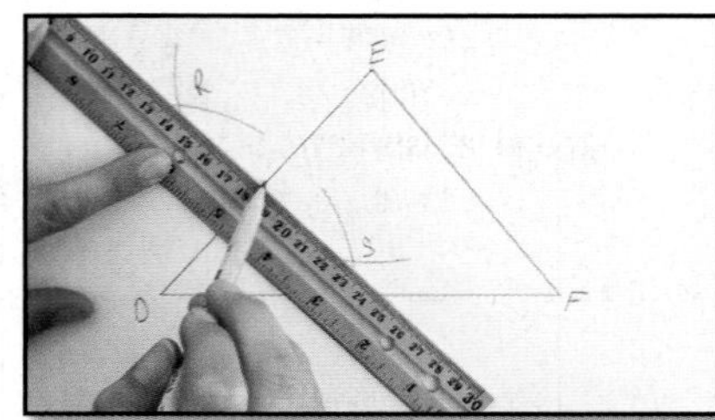

Use a straightedge to find the point where $\overline{RS}$ intersects $\overline{DE}$. Label the point $M$. This is the midpoint of $\overline{DE}$.

**Step 3**

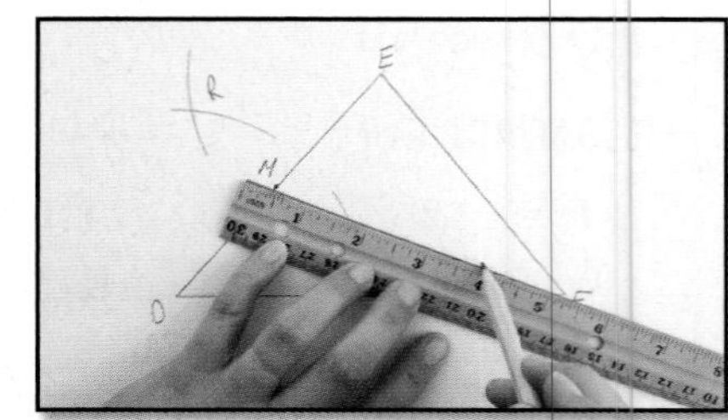

Draw a line through $F$ and $M$. $\overline{FM}$ is a median of $\triangle DEF$.

An *altitude* of a triangle is a segment from a vertex of the triangle to the opposite side and is perpendicular to the opposite side.

**Construction 2** Altitude of a Triangle

**Step 1**

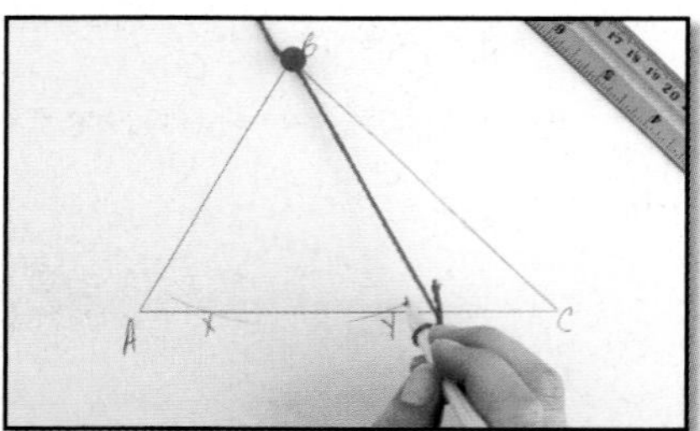

Place the thumbtack on vertex $B$ and draw two arcs intersecting $\overline{AC}$. Label the points where the arcs intersect the sides as $X$ and $Y$.

**Step 2**

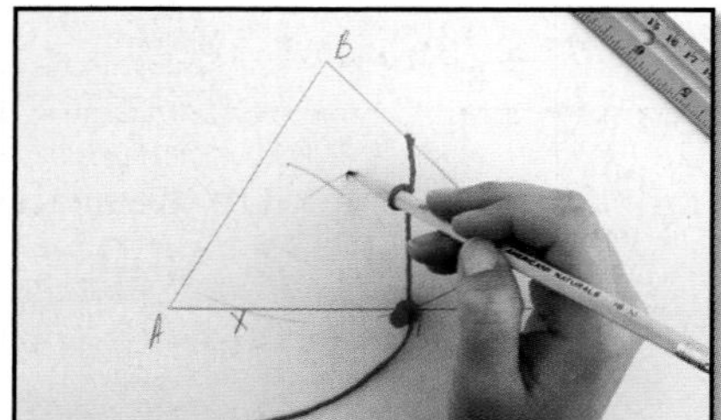

Adjust the length of the string so that it is greater than $\frac{1}{2}XY$. Place the tack on $X$ and draw an arc above $\overline{AC}$. Use the same length of string to draw an arc from $Y$. Label the points of intersection of the arcs $H$.

**Step 3**

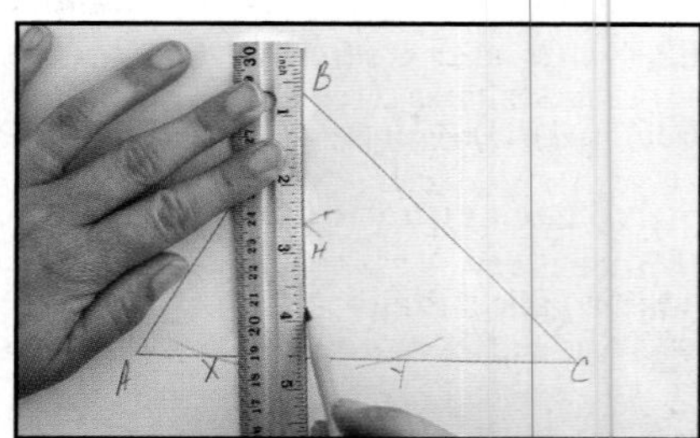

Use a straightedge to draw $\overleftrightarrow{BH}$. Label the point where $\overleftrightarrow{BH}$ intersects $\overline{AC}$ as $D$. $\overline{BD}$ is an altitude of $\triangle ABC$ and is perpendicular to $\overline{AC}$.

## Model and Analyze

1. Construct the medians of the other two sides of $\triangle DEF$. What do you notice about the medians of a triangle?
2. Construct the altitudes to the other two sides of $\triangle ABC$. What do you observe?

LESSON 7-2

# Medians and Altitudes of Triangles

**Then**

- You identified and used perpendicular and angle bisectors in triangles.

**Now**

1. Identify and use medians in triangles.
2. Identify and use altitudes in triangles.

**Why?**

- A mobile is a *kinetic* or moving sculpture that uses the principles of balance and equilibrium. Simple mobiles consist of several rods attached by strings from which objects of varying weights hang. The hanging objects balance each other and can rotate freely. To ensure that a triangle in a mobile hangs parallel to the ground, artists have to find the triangle's balancing point.

**NewVocabulary**
median
centroid
altitude
orthocenter

**Common Core State Standards**

**Content Standards**

G.CO.10 Prove theorems about triangles.

G.MG.3 Apply geometric methods to solve problems (e.g., designing an object or structure to satisfy physical constraints or minimize cost; working with typographic grid systems based on ratios). ★

**Mathematical Practices**

6 Attend to precision.

3 Construct viable arguments and critique the reasoning of others.

## 1 Medians

A **median** of a triangle is a segment with endpoints being a vertex of a triangle and the midpoint of the opposite side.

Every triangle has three medians that are concurrent. The point of concurrency of the medians of a triangle is called the **centroid** and is always inside the triangle.

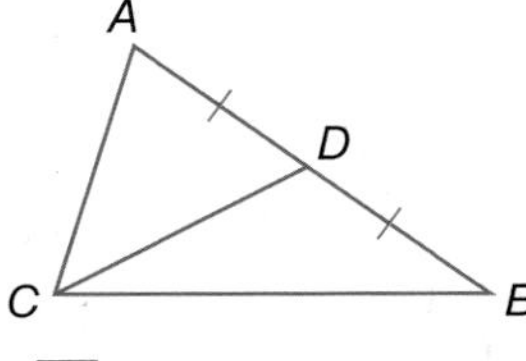


$\overline{CD}$ is a median of $\triangle ABC$.

**Theorem 7.7 Centroid Theorem**

The medians of a triangle intersect at a point called the centroid that is two thirds of the distance from each vertex to the midpoint of the opposite side.

**Example** If $P$ is the centroid of $\triangle ABC$, then

$AP = \frac{2}{3}AK$, $BP = \frac{2}{3}BL$, and $CP = \frac{2}{3}CJ$.

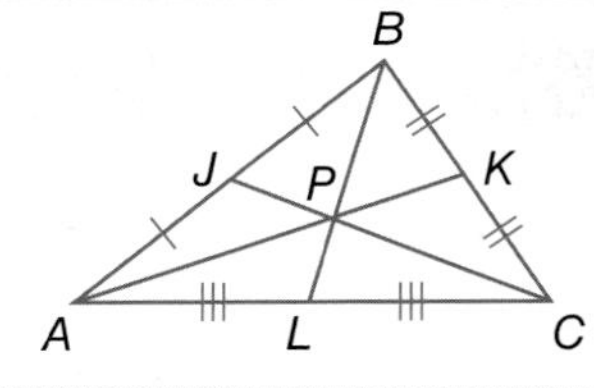


You will prove Theorem 7.7 in Exercise 36.

**Example 1 Use the Centroid Theorem**

**In $\triangle ABC$, $Q$ is the centroid and $BE = 9$. Find $BQ$ and $QE$.**

| | |
|---|---|
| $BQ = \frac{2}{3}BE$ | Centroid Theorem |
| $= \frac{2}{3}(9)$ or 6 | $BE = 9$ |
| $BQ + QE = 9$ | Segment Addition |
| $6 + QE = 9$ | $BQ = 6$ |
| $QE = 3$ | Subtract 6 from each side. |

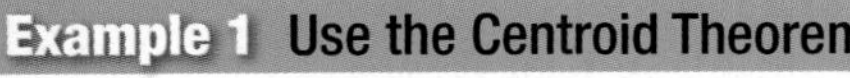

B F Q D A E C

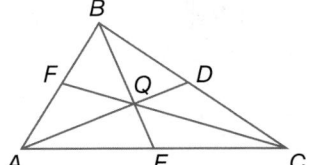

**GuidedPractice** **In $\triangle ABC$ above, $FC = 15$. Find each length.**

**1A.** $FQ$ **1B.** $QC$

James Randklev/Stone/Getty Images

**StudyTip**

**CCSS Reasoning** In Example 2, you can also use number sense to find *KP*. Since $KP = \frac{2}{3}KT$, $PT = \frac{1}{3}KT$ and $KP = 2PT$. Therefore, if $PT = 2$, then $KP = 2(2)$ or 4.

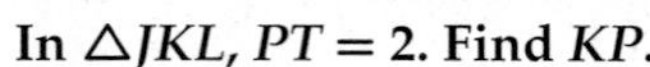

## Example 2 Use the Centroid Theorem

**In $\triangle JKL$, $PT = 2$. Find $KP$.**

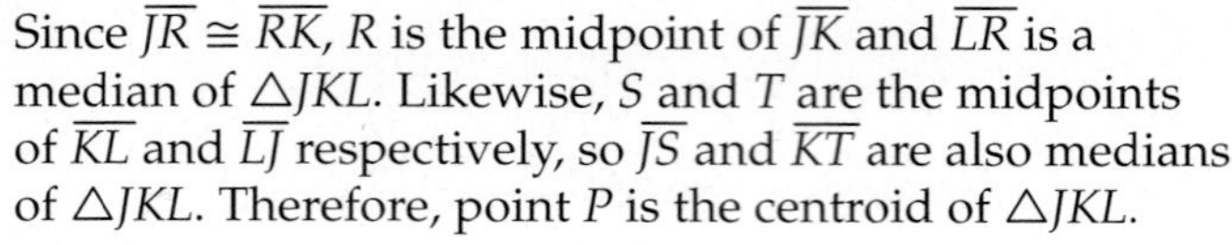

Since $\overline{JR} \cong \overline{RK}$, $R$ is the midpoint of $\overline{JK}$ and $\overline{LR}$ is a median of $\triangle JKL$. Likewise, $S$ and $T$ are the midpoints of $\overline{KL}$ and $\overline{LJ}$ respectively, so $\overline{JS}$ and $\overline{KT}$ are also medians of $\triangle JKL$. Therefore, point $P$ is the centroid of $\triangle JKL$.

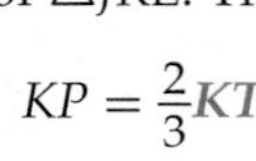

| | |
|---|---|
| $KP = \frac{2}{3}KT$ | Centroid Theorem |
| $KP = \frac{2}{3}(KP + PT)$ | Segment Addition and Substitution |
| $KP = \frac{2}{3}(KP + 2)$ | $PT = 2$ |
| $KP = \frac{2}{3}KP + \frac{4}{3}$ | Distributive Property |
| $\frac{1}{3}KP = \frac{4}{3}$ | Subtract $\frac{2}{3}KP$ from each side. |
| $KP = 4$ | Multiply each side by 3. |

### GuidedPractice

**In $\triangle JKL$ above, $RP = 3.5$ and $JP = 9$. Find each measure.**

**2A.** $PL$ **2B.** $PS$

All polygons have a balance point or centroid. The centroid is also the balancing point or *center of gravity* for a triangular region. The center of gravity is the point at which the region is stable under the influence of gravity.

## Real-World Example 3 Find the Centroid on Coordinate Plane

**PERFORMANCE ART A performance artist plans to balance triangular pieces of metal during her next act. When one such triangle is placed on the coordinate plane, its vertices are located at (1, 10), (5, 0), and (9, 5). What are the coordinates of the point where the artist should support the triangle so that it will balance?**

**Understand** You need to find the centroid of the triangle with the given coordinates. This is the point at which the triangle will balance.

**Plan** Graph and label the triangle with vertices $A(1, 10)$, $B(5, 0)$, and $C(9, 5)$. Since the centroid is the point of concurrency of the medians of a triangle, use the Midpoint Theorem to find the midpoint of one of the sides of the triangle. The centroid is two-thirds the distance from the opposite vertex to that midpoint.

**Math HistoryLink**

Pierre de Fermat **(1601–1665)** Another triangle center is the Fermat point, which minimizes the sum of the distances from the three vertices. Fermat is one of the best-known mathematicians for writing proofs.



**Solve** Graph $\triangle ABC$.

Find the midpoint $D$ of side $\overline{AB}$ with endpoints $A(1, 10)$ and $B(5, 0)$.

$D\left(\frac{1+5}{2}, \frac{10+0}{2}\right) = D(3, 5)$

Graph point $D$. Notice that $\overline{DC}$ is a horizontal line. The distance from $D(3, 5)$ to $C(9, 5)$ is $9 - 3$ or 6 units.

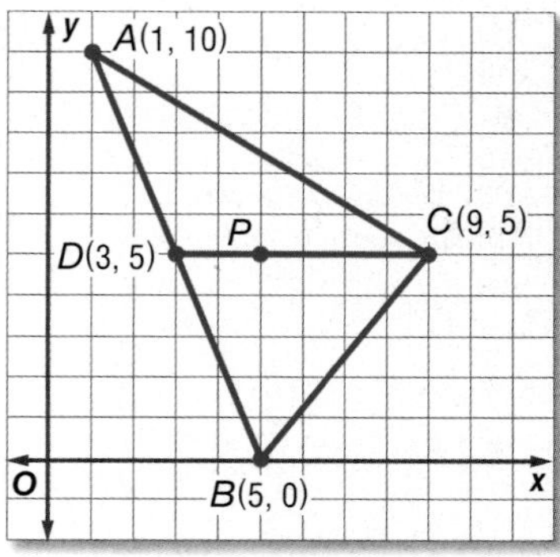


If $P$ is the centroid of $\triangle ABC$, then $PC = \frac{2}{3}DC$. So the centroid is $\frac{2}{3}(6)$ or 4 units to the left of $C$. The coordinates of $P$ are $(9 - 4, 5)$ or $(5, 5)$.

The performer should balance the triangle at the point (5, 5).

**Check** Use a different median to check your answer. The midpoint $F$ of side $\overline{AC}$ is $F\left(\frac{1+9}{2}, \frac{10+5}{2}\right)$ or $F(5, 7.5)$. $\overline{BF}$ is a vertical line, so the distance from $B$ to $F$ is $7.5 - 0$ or 7.5. $\overline{PB} = \frac{2}{3}(7.5)$ or 5, so $P$ is 5 units up from $B$. The coordinates of $P$ are $(5, 0 + 5)$ or $(5, 5)$. ✓

**GuidedPractice**

**3.** A second triangle has vertices at (0, 4), (6, 11.5), and (12, 1). What are the coordinates of the point where the artist should support the triangle so that it will balance? Explain your reasoning.

## 2 Altitudes

An **altitude** of a triangle is a segment from a vertex to the line containing the opposite side and perpendicular to the line containing that side. An altitude can lie in the interior, exterior, or on the side of a triangle.

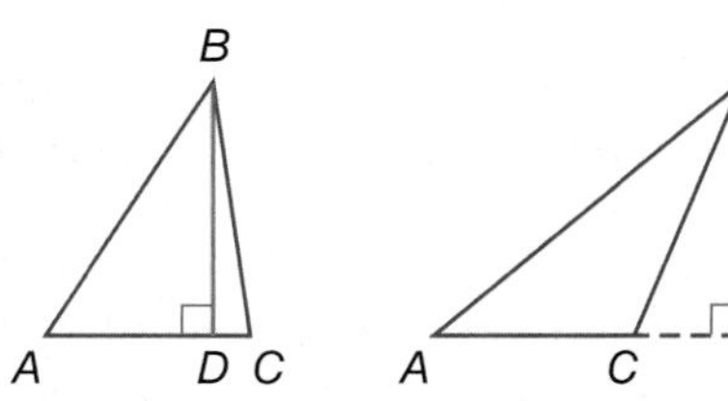


$\overline{BD}$ is an altitude from $B$ to $\overline{AC}$.

**ReadingMath**

Height of a Triangle The length of an altitude is known as the *height* of the triangle. The height of a triangle is used to calculate the triangle's area.

Every triangle has three altitudes. If extended, the altitudes of a triangle intersect in a common point.

**KeyConcept Orthocenter**

The lines containing the altitudes of a triangle are concurrent, intersecting at a point called the **orthocenter**.

**Example** The lines containing altitudes $\overline{AF}$, $\overline{CD}$, and $\overline{BG}$ intersect at $P$, the orthocenter of $\triangle ABC$.

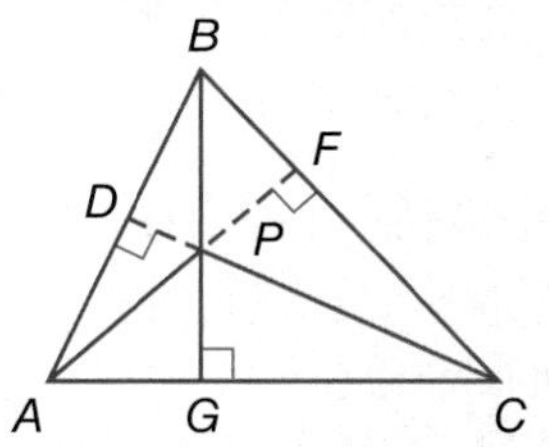

## Example 4 Find the Orthocenter on a Coordinate Plane

**COORDINATE GEOMETRY The vertices of $\triangle FGH$ are $F(-2, 4)$, $G(4, 4)$, and $H(1, -2)$. Find the coordinates of the orthocenter of $\triangle FGH$.**

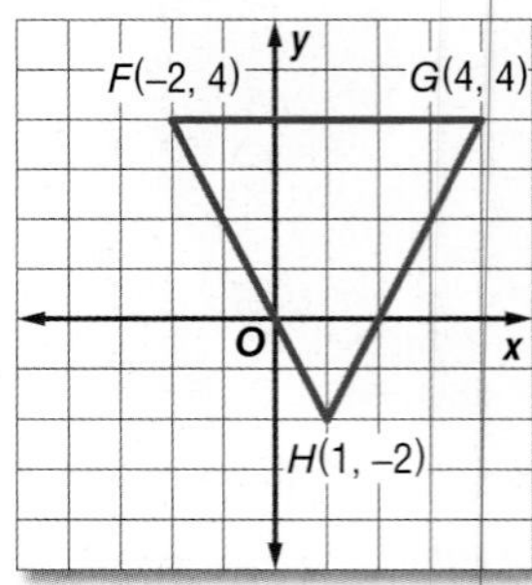


**Step 1** Graph $\triangle FGH$. To find the orthocenter, find the point where two of the three altitudes intersect.

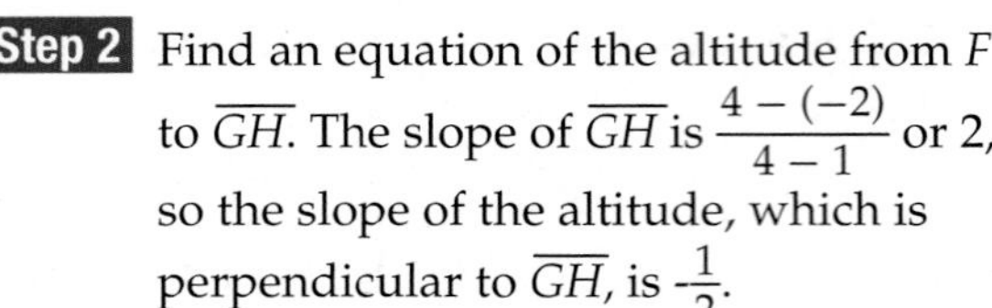

**Step 2** Find an equation of the altitude from $F$ to $\overline{GH}$. The slope of $\overline{GH}$ is $\frac{4-(-2)}{4-1}$ or 2, so the slope of the altitude, which is perpendicular to $\overline{GH}$, is $-\frac{1}{2}$.

| | |
|---|---|
| $y - y_1 = m(x - x_1)$ | Point-slope form |
| $y - 4 = -\frac{1}{2}[x - (-2)]$ | $m = -\frac{1}{2}$ and $(x_1, y_1) = F(-2, 4)$. |
| $y - 4 = -\frac{1}{2}(x + 2)$ | Simplify. |
| $y - 4 = -\frac{1}{2}x - 1$ | Distributive Property |
| $y = -\frac{1}{2}x + 3$ | Add 4 to each side. |

Find an equation of the altitude from $G$ to $\overline{FH}$. The slope of $\overline{FH}$ is $\frac{-2-4}{1-(-2)}$ or $-2$, so the slope of the altitude is $\frac{1}{2}$.

| | |
|---|---|
| $y - y_1 = m(x - x_1)$ | Point-slope form |
| $y - 4 = \frac{1}{2}(x - 4)$ | $m = \frac{1}{2}$ and $(x_1, y_1) = G(4, 4)$ |
| $y - 4 = \frac{1}{2}x - 2$ | Distributive Property |
| $y = \frac{1}{2}x + 2$ | Add 4 to each side. |

**Step 3** Solve the resulting system of equations $\begin{cases} y = -\frac{1}{2}x + 3 \\ y = \frac{1}{2}x + 2 \end{cases}$ to find the point of intersection of the altitudes.

Adding the two equations to eliminate $x$ results in $2y = 5$ or $y = \frac{5}{2}$.

| | |
|---|---|
| $y = \frac{1}{2}x + 2$ | Equation of altitude from $G$ |
| $\frac{5}{2} = \frac{1}{2}x + 2$ | $y = \frac{5}{2}$ |
| $\frac{1}{2} = \frac{1}{2}x$ | Subtract $\frac{4}{2}$ or 2 from each side. |
| $1 = x$ | Multiply each side by 2. |

The coordinates of the orthocenter of $\triangle JKL$ are $\left(1, \frac{5}{2}\right)$ or $\left(1, 2\frac{1}{2}\right)$.

**Study Tip**

**Check for Reasonableness** Use the corner of a sheet of paper to draw the altitudes of each side of the triangle.

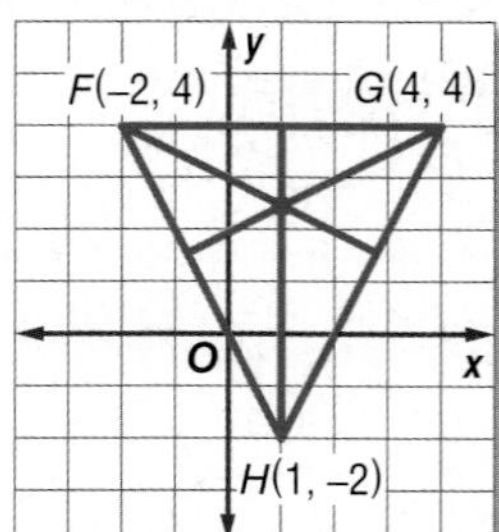


The intersection is located at approximately $\left(1, 2\frac{1}{2}\right)$, so the answer is reasonable.

### Guided Practice

**4.** Find the coordinates of the orthocenter of $\triangle ABC$ graphed at the right.

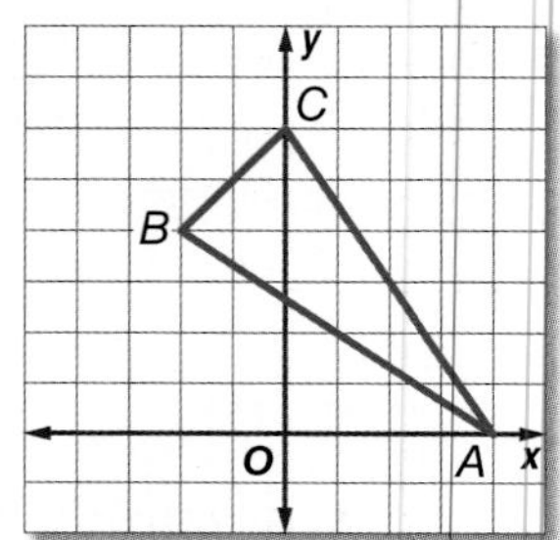

## ConceptSummary Special Segments and Points in Triangles

| Name | Example | Point of Concurrency | Special Property | Example |
|---|---|---|---|---|
| perpendicular bisector | | circumcenter | The circumcenter *P* of $\triangle ABC$ is equidistant from each vertex. | |
| angle bisector | | incenter | The incenter *Q* of $\triangle ABC$ is equidistant from each side of the triangle. | |
| median | | centroid | The centroid *R* of $\triangle ABC$ is two thirds of the distance from each vertex to the midpoint of the opposite side. | |
| altitude | | orthocenter | The lines containing the altitudes of $\triangle ABC$ are concurrent at the orthocenter *S*. | |

## Check Your Understanding

Examples 1–2 **In $\triangle ACE$, $P$ is the centroid, $PF = 6$, and $AD = 15$. Find each measure.**

**1.** $PC$

**2.** $AP$

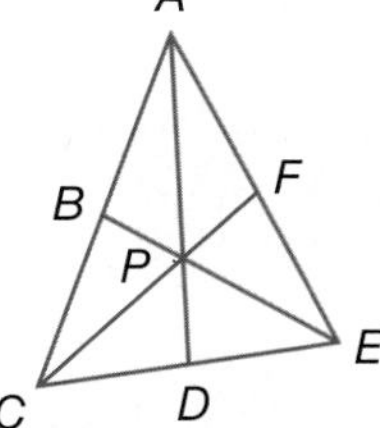


Example 3

**3. INTERIOR DESIGN** An interior designer is creating a custom coffee table for a client. The top of the table is a glass triangle that needs to balance on a single support. If the coordinates of the vertices of the triangle are at (3, 6), (5, 2), and (7, 10), at what point should the support be placed?

Example 4

**4. COORDINATE GEOMETRY** Find the coordinates of the orthocenter of $\triangle ABC$ with vertices $A(-3, 3)$, $B(-1, 7)$, and $C(3, 3)$.

## Practice and Problem Solving

**Examples 1–2** In $\triangle SZU$, $UJ = 9$, $VJ = 3$, and $ZT = 18$. Find each length.

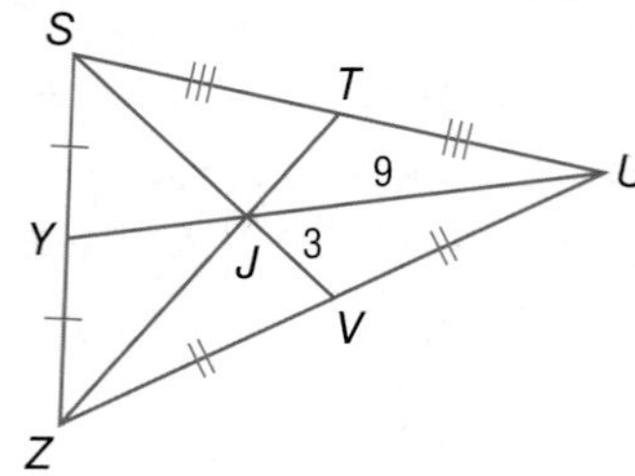


5. $YJ$
6. $SJ$
7. $YU$
8. $SV$
9. $JT$
10. $ZJ$

**Example 3** COORDINATE GEOMETRY **Find the coordinates of the centroid of each triangle with the given vertices.**

11. $A(-1, 11)$, $B(3, 1)$, $C(7, 6)$
12. $X(5, 7)$, $Y(9, -3)$, $Z(13, 2)$

13. INTERIOR DESIGN Emilia made a collage with pictures of her friends. She wants to hang the collage from the ceiling in her room so that it is parallel to the ceiling. A diagram of the collage is shown in the graph at the right. At what point should she place the string?

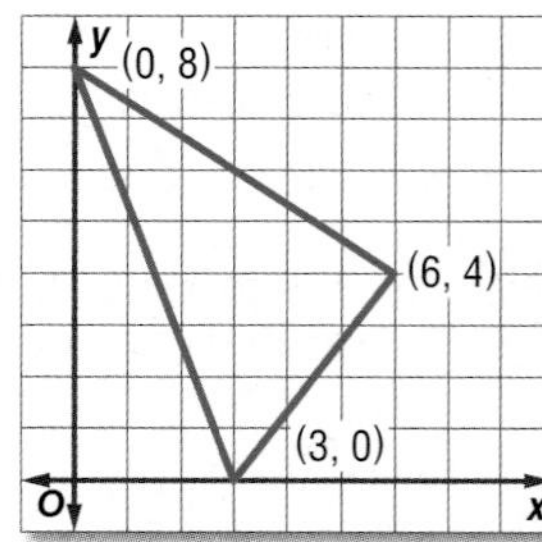


**Example 4** COORDINATE GEOMETRY **Find the coordinates of the orthocenter of each triangle with the given vertices.**

14. $J(3, -2)$, $K(5, 6)$, $L(9, -2)$
15. $R(-4, 8)$, $S(-1, 5)$, $T(5, 5)$

**Identify each segment $\overline{BD}$ as a(n) altitude, median, or perpendicular bisector.**

16.

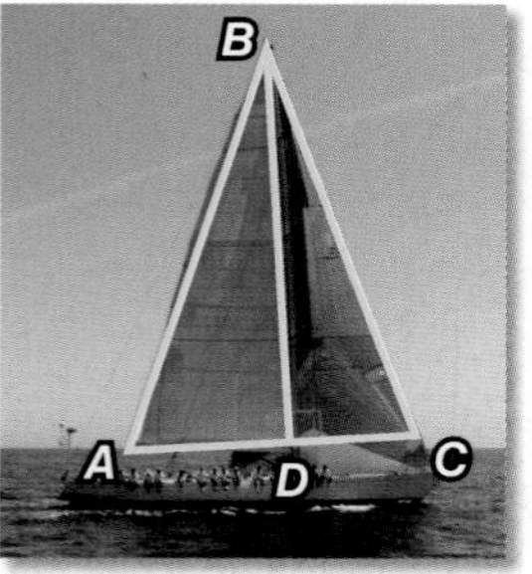


17.

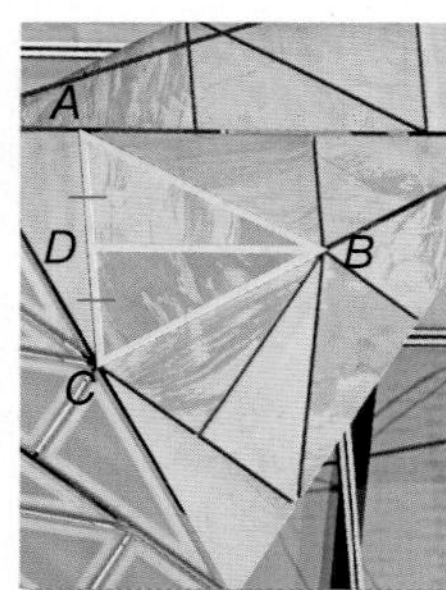


18.

19.

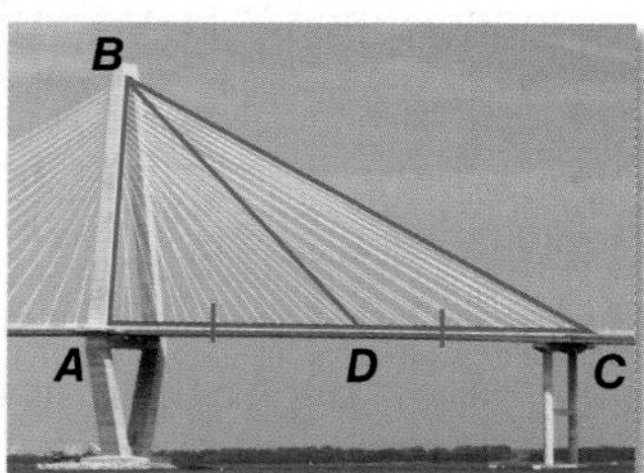


20. CCSS SENSE-MAKING In the figure at the right, if $J$, $P$, and $L$ are the midpoints of $\overline{KH}$, $\overline{HM}$, and $\overline{MK}$, respectively, find $x$, $y$, and $z$.

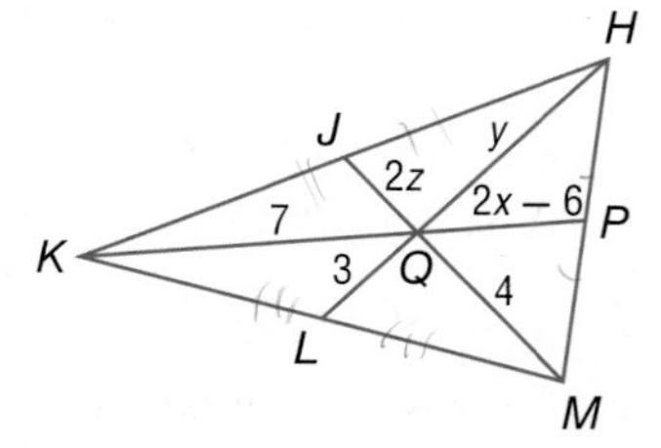

**Copy and complete each statement for $\triangle RST$ for medians $\overline{RM}$, $\overline{SL}$ and $\overline{TK}$, and centroid $J$.**

**21.** $SL = x(JL)$ **22.** $JT = x(TK)$ **23.** $JM = x(RJ)$

**ALGEBRA** **Use the figure at the right.**

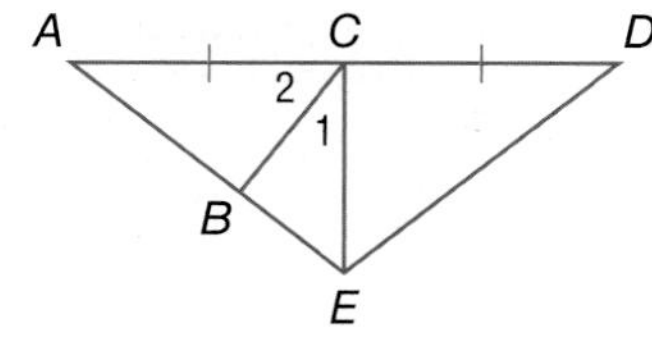


**24.** If $\overline{EC}$ is an altitude of $\triangle AED$, $m\angle 1 = 2x + 7$, and $m\angle 2 = 3x + 13$, find $m\angle 1$ and $m\angle 2$.

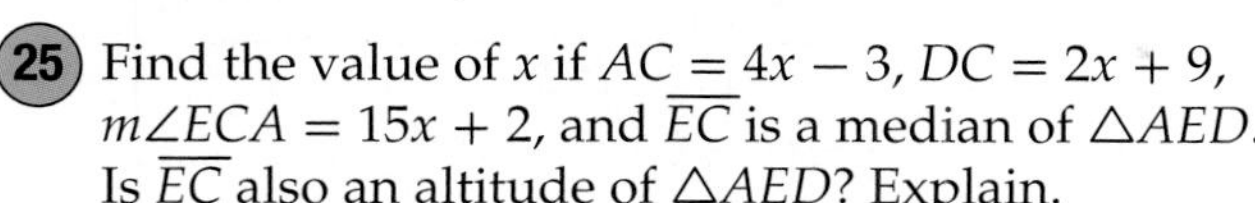

**25** Find the value of $x$ if $AC = 4x - 3$, $DC = 2x + 9$, $m\angle ECA = 15x + 2$, and $\overline{EC}$ is a median of $\triangle AED$. Is $\overline{EC}$ also an altitude of $\triangle AED$? Explain.

**26. GAMES** The game board shown is shaped like an equilateral triangle and has indentations for game pieces. The game's objective is to remove pegs by jumping over them until there is only one peg left. Copy the game board's outline and determine which of the points of concurrency the blue peg represents: *circumcenter, incenter, centroid,* or *orthocenter.* Explain your reasoning.

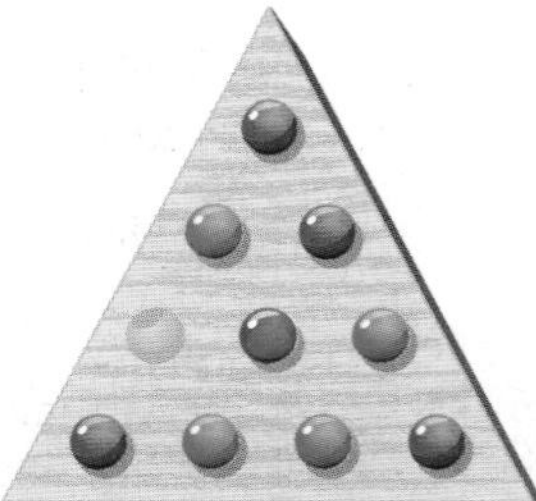

**CCSS ARGUMENTS** **Use the given information to determine whether $\overline{LM}$ is a *perpendicular bisector, median,* and/or an *altitude* of $\triangle JKL$.**

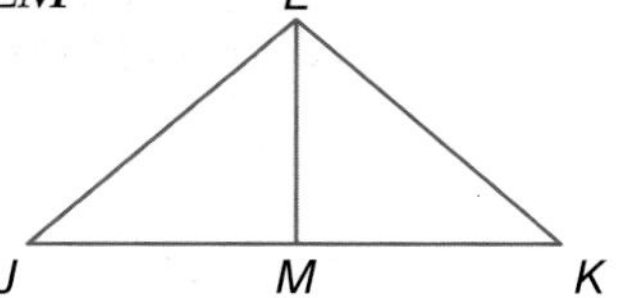


**27.** $\overline{LM} \perp \overline{JK}$ **28.** $\triangle JLM \cong \triangle KLM$

**29.** $\overline{JM} \cong \overline{KM}$ **30.** $\overline{LM} \perp \overline{JK}$ and $\overline{JL} \cong \overline{KL}$

**31. PROOF** Write a paragraph proof.

**Given:** $\triangle XYZ$ is isosceles. $\overline{WY}$ bisects $\angle Y$.

**Prove:** $\overline{WY}$ is a median.

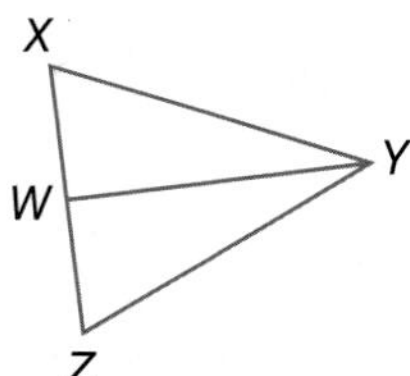


**32. PROOF** Write an algebraic proof.

**Given:** $\triangle XYZ$ with medians $\overline{XR}$, $\overline{YS}$, $\overline{ZQ}$

**Prove:** $\frac{XP}{PR} = 2$

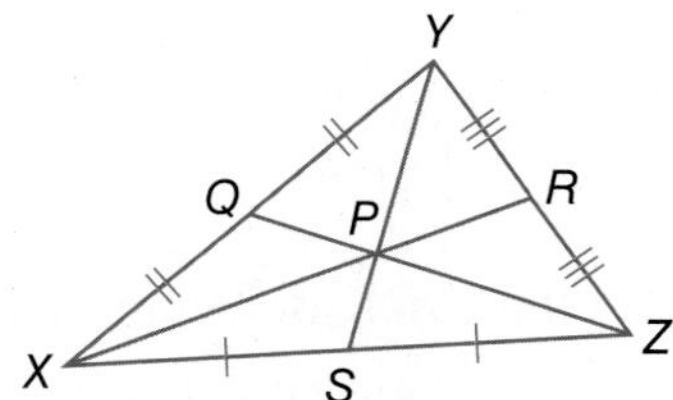


**33. MULTIPLE REPRESENTATIONS** In this problem, you will investigate the location of the points of concurrency for any equilateral triangle.

**a. Concrete** Construct three different equilateral triangles on tracing paper and cut them out. Fold each triangle to locate the circumcenter, incenter, centroid, and orthocenter.

**b. Verbal** Make a conjecture about the relationships among the four points of concurrency of any equilateral triangle.

**c. Graphical** Position an equilateral triangle and its circumcenter, incenter, centroid, and orthocenter on the coordinate plane using variable coordinates. Determine the coordinates of each point of concurrency.

**ALGEBRA** **In $\triangle JLP$, $m\angle JMP = 3x - 6$, $JK = 3y - 2$, and $LK = 5y - 8$.**

**34.** If $\overline{JM}$ is an altitude of $\triangle JLP$, find $x$.

**35.** Find $LK$ if $\overline{PK}$ is a median.

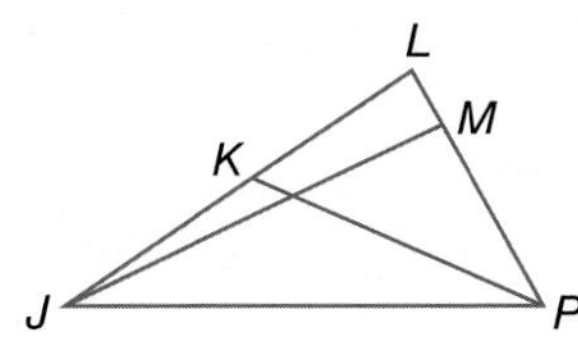

**36. PROOF Write a coordinate proof to prove the Centroid Theorem.**

**Given:** $\triangle ABC$, medians $\overline{AR}$, $\overline{BS}$, and $\overline{CQ}$

**Prove:** The medians intersect at point $P$ and $P$ is two thirds of the distance from each vertex to the midpoint of the opposite side.

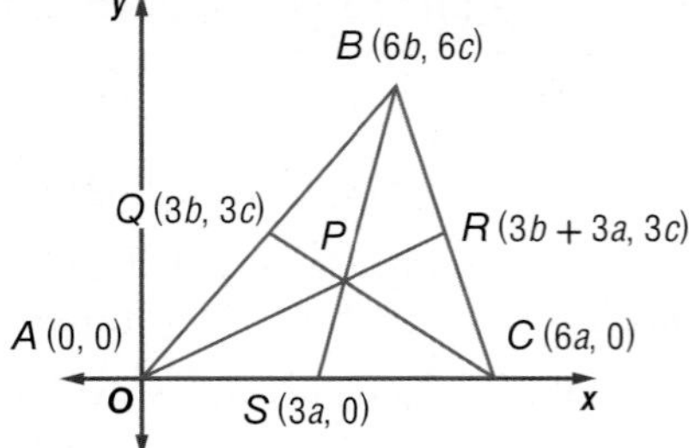


(*Hint*: First, find the equations of the lines containing the medians. Then find the coordinates of point $P$ and show that all three medians intersect at point $P$. Next, use the Distance Formula and multiplication to show $AP = \frac{2}{3}AR$, $BP = \frac{2}{3}BS$, and $CP = \frac{2}{3}CQ$.)

## H.O.T. Problems Use Higher-Order Thinking Skills

**37.** **ERROR ANALYSIS** Based on the figure at the right, Luke says that $\frac{2}{3}AP = AD$. Kareem disagrees. Is either of them correct? Explain your reasoning.

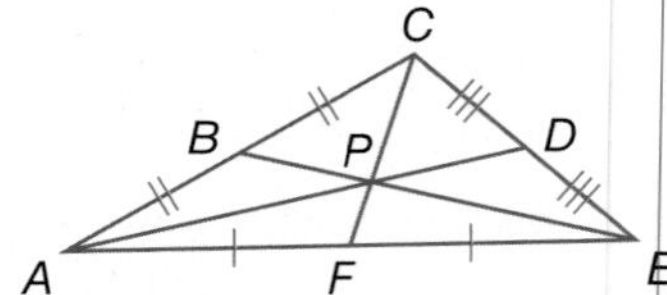


**38.** **CCSS ARGUMENTS** Determine whether the following statement is *true* or *false*. If true, explain your reasoning. If false, provide a counterexample.

*The orthocenter of a right triangle is always located at the vertex of the right angle.*

**39.** **CHALLENGE** $\triangle ABC$ has vertices $A(-3, 3)$, $B(2, 5)$, and $C(4, -3)$. What are the coordinates of the centroid of $\triangle ABC$? Explain the process you used to reach your conclusion.

**40.** **WRITING IN MATH** Compare and contrast the perpendicular bisectors, medians, and altitudes of a triangle.

**41.** **CHALLENGE** In the figure at the right, segments $\overline{AD}$ and $\overline{CE}$ are medians of $\triangle ACB$, $\overline{AD} \perp \overline{CE}$, $AB = 10$, and $CE = 9$. Find $CA$.

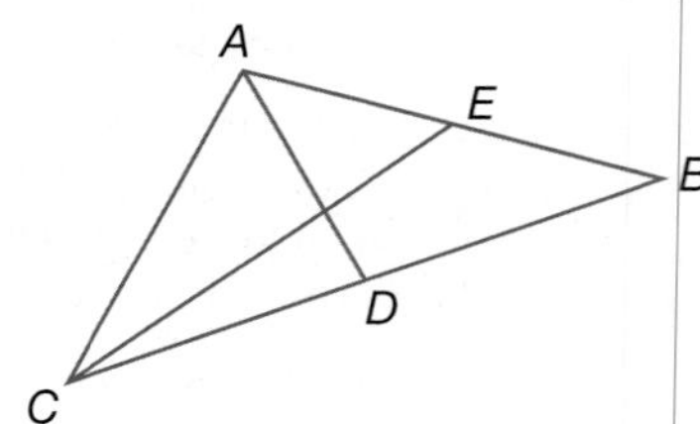


**42.** **OPEN ENDED** In this problem, you will investigate the relationships among three points of concurrency in a triangle.

**a.** Draw an acute triangle and find the circumcenter, centroid, and orthocenter.

**b.** Draw an obtuse triangle and find the circumcenter, centroid, and orthocenter.

**c.** Draw a right triangle and find the circumcenter, centroid, and orthocenter.

**d.** Make a conjecture about the relationships among the circumcenter, centroid, and orthocenter.

**43.** **WRITING IN MATH** Use area to explain why the centroid of a triangle is its center of gravity. Then use this explanation to describe the location for the balancing point for a rectangle.

## Standardized Test Practice

**44.** In the figure below, $\overline{GJ} \cong \overline{HJ}$. Which must be true?

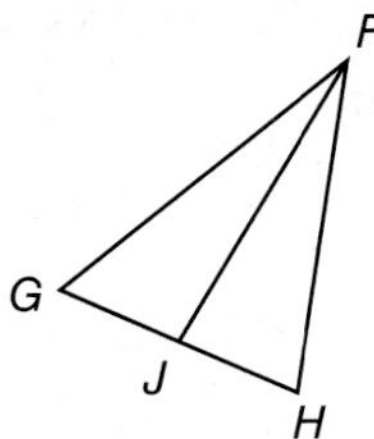


**A** $\overline{FJ}$ is an altitude of $\triangle FGH$.

**B** $\overline{FJ}$ is an angle bisector of $\triangle FGH$.

**C** $\overline{FJ}$ is a median of $\triangle FGH$.

**D** $\overline{FJ}$ is a perpendicular bisector of $\triangle FGH$.

**45. GRIDDED RESPONSE** What is the $x$-intercept of the graph of $4x - 6y = 12$?

**46. ALGEBRA** Four students have volunteered to fold pamphlets for a local community action group. Which student is the fastest?

| Student | Folding Speed |
|---|---|
| Neiva | 1 page every 3 seconds |
| Sarah | 2 pages every 10 seconds |
| Quinn | 30 pages per minute |
| Deron | 45 pages in 2 minutes |

**F** Deron
**G** Neiva
**H** Quinn
**J** Sarah

**47. SAT/ACT** 80 percent of 42 is what percent of 16?

**A** 240
**B** 210
**C** 150
**D** 50
**E** 30

## Spiral Review

**Find each measure.** (Lesson 7-1)

**48.** $LM$

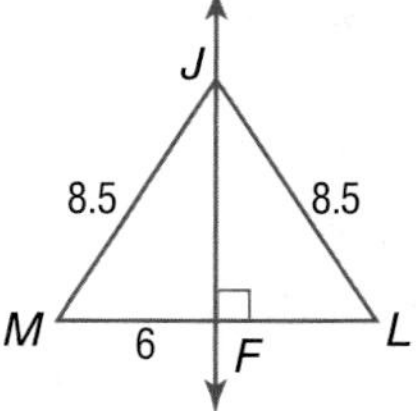


**49.** $DF$

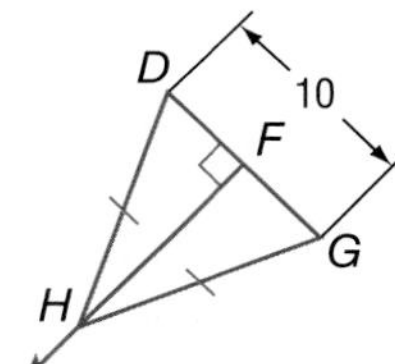


**50.** $TQ$

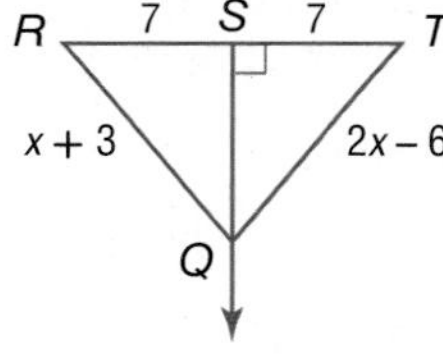


**Position and label each triangle on the coordinate plane.** (Lesson 6-6)

**51.** right $\triangle XYZ$ with hypotenuse $\overline{XZ}$, $ZY$ is twice $XY$, and $\overline{XY}$ is $b$ units long

**52.** isosceles $\triangle QRT$ with base $\overline{QR}$ that is $b$ units long

**53. HIGHWAYS** Near the city of Hopewell, Virginia, Route 10 runs perpendicular to Interstate 95 and Interstate 295. Show that the angles at the intersections of Route 10 with Interstate 95 and Interstate 295 are congruent. (Lesson 5-4)

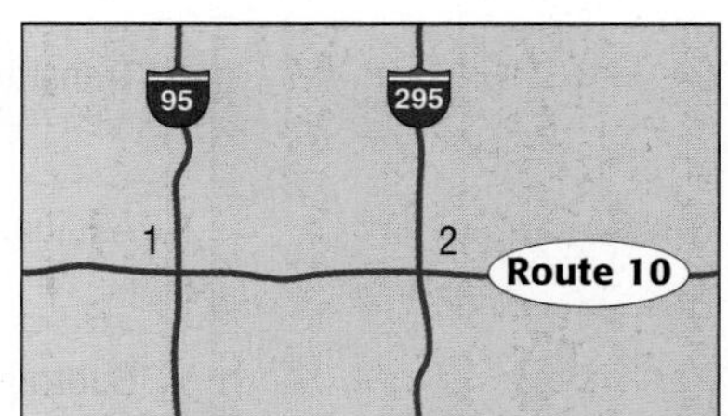


## Skills Review

**PROOF** **Write a flow proof of the Exterior Angle Theorem.**

**54. Given:** $\triangle XYZ$

**Prove:** $m\angle X + m\angle Z = m\angle 1$

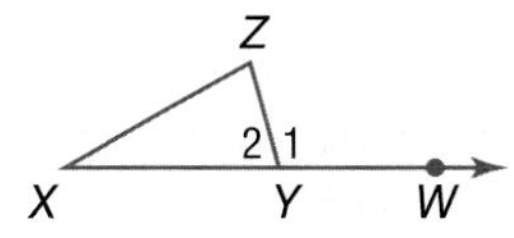

LESSON 7-3

# Inequalities in One Triangle

**Then**

- You found the relationship between the angle measures of a triangle.

**Now**

1. Recognize and apply properties of inequalities to the measures of the angles of a triangle.
2. Recognize and apply properties of inequalities to the relationships between the angles and sides of a triangle.

**Why?**

- To create the appearance of depth in a room, interior designers use a technique called *triangulation.* A basic example of this technique is the placement of an end table on each side of a sofa with a painting over the sofa.

The measures of the base angles of the triangle should be less than the measure of the other angle.

**Common Core State Standards**

**Content Standards**

G.CO.10 Prove theorems about triangles.

**Mathematical Practices**

1 Make sense of problems and persevere in solving them.

3 Construct viable arguments and critique the reasoning of others.

## 1 Angle Inequalities

You learned about the inequality relationship between two real numbers. This relationship is often used in proofs.

### KeyConcept Definition of Inequality

**Words** For any real numbers $a$ and $b$, $a > b$ if and only if there is a positive number $c$ such that $a = b + c$.

**Example** If $5 = 2 + 3$, then $5 > 2$ and $5 > 3$.

The table below lists some of the properties of inequalities you studied in algebra.

### KeyConcept Properties of Inequality for Real Numbers

The following properties are true for any real numbers $a$, $b$, and $c$.

| Property | |
|---|---|
| Comparison Property of Inequality | $a < b$, $a = b$, or $a > b$ |
| Transitive Property of Inequality | **1.** If $a < b$ and $b < c$, then $a < c$.<br>**2.** If $a > b$ and $b > c$, then $a > c$. |
| Addition Property of Inequality | **1.** If $a > b$, then $a + c > b + c$.<br>**2.** If $a < b$, then $a + c < b + c$. |
| Subtraction Property of Inequality | **1.** If $a > b$, then $a - c > b - c$.<br>**2.** If $a < b$, then $a - c < b - c$. |

The definition of inequality and the properties of inequalities can be applied to the measures of angles and segments, since these are real numbers. Consider $\angle 1$, $\angle 2$, and $\angle 3$ in the figure shown.

By the Exterior Angle Theorem, you know that $m\angle 1 = m\angle 2 + m\angle 3$.

Since the angle measures are positive numbers, we can also say that

$$m\angle 1 > m\angle 2 \quad \text{and} \quad m\angle 1 > m\angle 3$$

by the definition of inequality. This result suggests the following theorem.

age fotostock/SuperStock

**ReviewVocabulary**

**Remote Interior Angle** Each exterior angle of a triangle has two *remote interior angles* that are not adjacent to the exterior angle.

## Theorem 7.8 Exterior Angle Inequality

The measure of an exterior angle of a triangle is greater than the measure of either of its corresponding remote interior angles.

**Example:** $m\angle 1 > m\angle A$
$m\angle 1 > m\angle B$

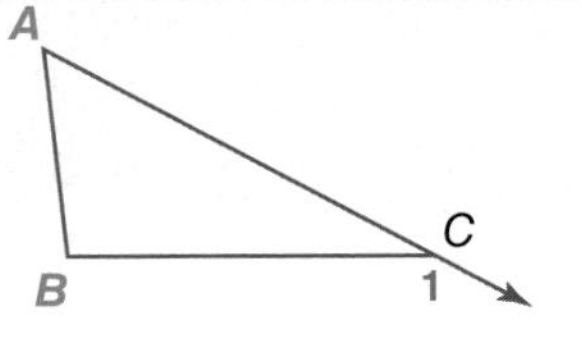


The proof of Theorem 7.8 is in Lesson 7-4.

## Example 1 Use the Exterior Angle Inequality Theorem

PT

**Use the Exterior Angle Inequality Theorem to list all of the angles that satisfy the stated condition.**

**a. measures less than $m\angle 7$**

$\angle 7$ is an exterior angle to $\triangle KML$, with $\angle 4$ and $\angle 5$ as corresponding remote interior angles. By the Exterior Angle Inequality Theorem, $m\angle 7 > m\angle 4$ and $m\angle 7 > m\angle 5$.

$\angle 7$ is also an exterior angle to $\triangle JKL$, with $\angle 1$ and $\angle JKL$ as corresponding remote interior angles. So, $m\angle 7 > m\angle 1$ and $m\angle 7 > m\angle JKL$. Since $m\angle JKL = m\angle 2 + m\angle 4$, by substitution $m\angle 7 > m\angle 2 + m\angle 4$. Therefore, $m\angle 7 > m\angle 2$.

So, the angles with measures less than $m\angle 7$ are $\angle 1$, $\angle 2$, $\angle 4$, $\angle 5$.

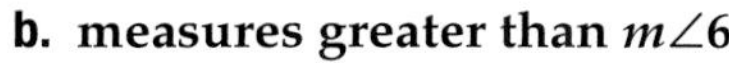

**b. measures greater than $m\angle 6$**

$\angle 3$ is an exterior angle to $\triangle KLM$. So by the Exterior Angle Inequality Theorem, $m\angle 3 > m\angle 6$. Because $\angle 8$ is an exterior angle to $\triangle JKL$, $m\angle 8 > m\angle 6$. Thus, the measures of $\angle 3$ and $\angle 8$ are greater than $m\angle 6$.

### GuidedPractice

**1A.** measures less than $m\angle 1$

**1B.** measures greater than $m\angle 8$

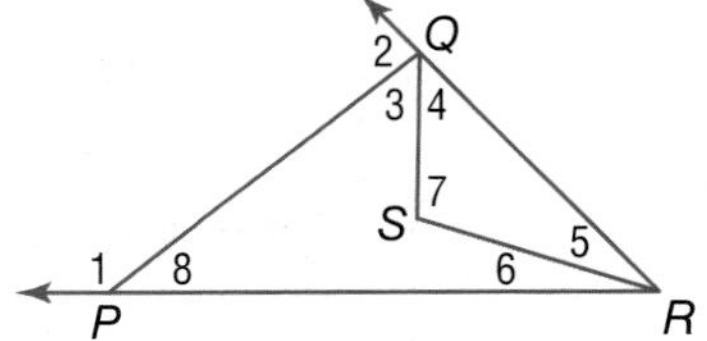


**WatchOut!**

**Identifying Side Opposite** Be careful to correctly identify the side opposite an angle. The sides that form the angle cannot be the sides opposite the angle.

## 2 Angle-Side Inequalities

If two sides of a triangle are congruent, or the triangle is isosceles, then the angles opposite those sides are congruent. What relationship exists if the sides are not congruent? Examine the longest and shortest sides and smallest and largest angles of a scalene obtuse triangle.

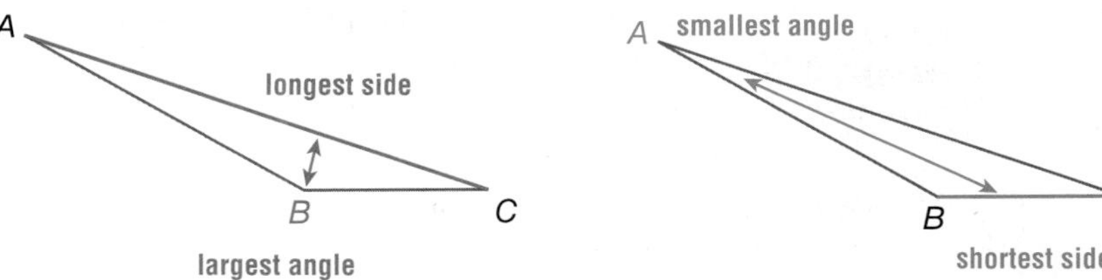


Notice that the longest side and largest angle of $\triangle ABC$ are opposite each other. Likewise, the shortest side and smallest angle are opposite each other.

**WatchOut!**

**Symbols for Angles and Inequalities** The symbol for angle ($\angle$) looks similar to the symbol for less than ($<$), especially when handwritten. Be careful to write the symbols correctly in situations where both are used.

The side-angle relationships in an obtuse scalene triangle are true for all triangles, and are stated using inequalities in the theorems below.

## Theorems Angle-Side Relationships in Triangles

**7.9** If one side of a triangle is longer than another side, then the angle opposite the longer side has a greater measure than the angle opposite the shorter side.

**Example:** $XY > YZ$, so $m\angle Z > m\angle X$.

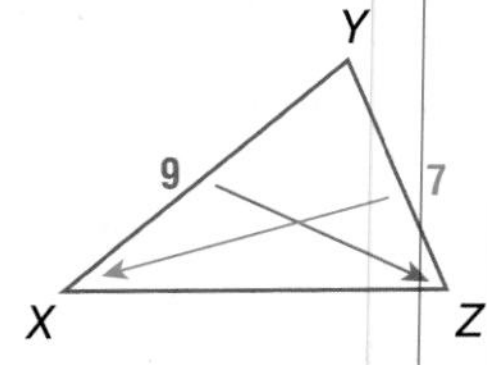


**7.10** If one angle of a triangle has a greater measure than another angle, then the side opposite the greater angle is longer than the side opposite the lesser angle.

**Example:** $m\angle J > m\angle K$, so $KL > JL$.

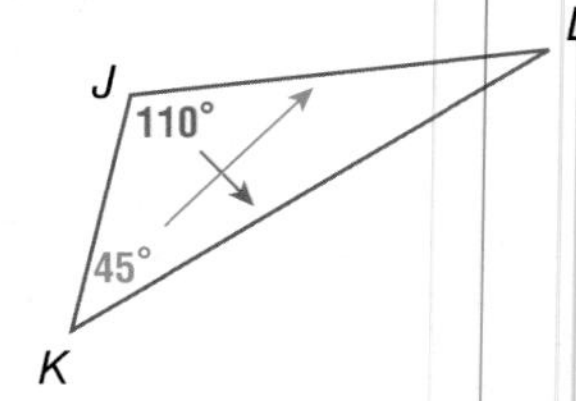


## Proof Theorem 7.9

**Given:** $\triangle ABC$, $AB > BC$

**Prove:** $m\angle BCA > m\angle A$

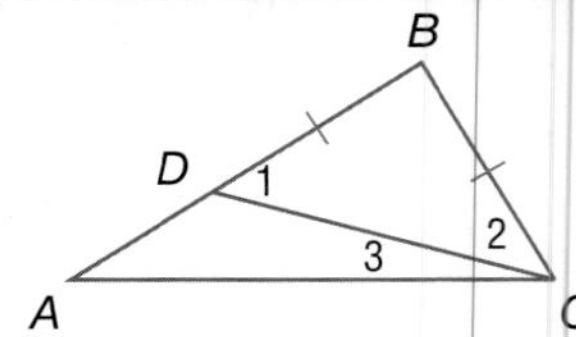


**Proof:**

Since $AB > BC$ in the given $\triangle ABC$, there exists a point $D$ on $\overline{AB}$ such that $BD = BC$. Draw $\overline{CD}$ to form isosceles $\triangle BCD$. By the Isosceles Triangle Theorem, $\angle 1 \cong \angle 2$, so $m\angle 1 = m\angle 2$ by the definition of congruent angles.

By the Angle Addition Postulate, $m\angle BCA = m\angle 2 + m\angle 3$, so $m\angle BCA > m\angle 2$ by the definition of inequality. By substitution, $m\angle BCA > m\angle 1$.

By the Exterior Angle Inequality Theorem, $m\angle 1 > m\angle A$. Therefore, because $m\angle BCA > m\angle 1$ and $m\angle 1 > m\angle A$, by the Transitive Property of Inequality, $m\angle BCA > m\angle A$.

You will prove Theorem 7.10 in Lesson 7-4, Exercise 31.

## Example 2 Order Triangle Angle Measures

**List the angles of $\triangle PQR$ in order from smallest to largest.**

The sides from shortest to longest are $\overline{PR}$, $\overline{PQ}$, $\overline{QR}$. The angles opposite these sides are $\angle Q$, $\angle R$, and $\angle P$, respectively. So the angles from smallest to largest are $\angle Q$, $\angle R$, and $\angle P$.

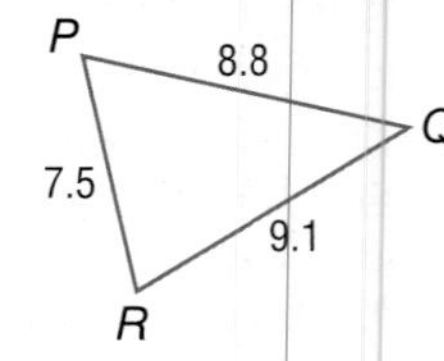


**GuidedPractice**

**2.** List the angles and sides of $\triangle ABC$ in order from smallest to largest.

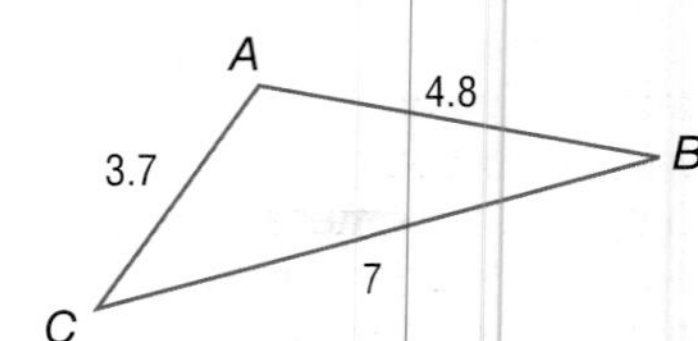

## Example 3 Order Triangle Side Lengths

**List the sides of $\triangle FGH$ in order from shortest to longest.**

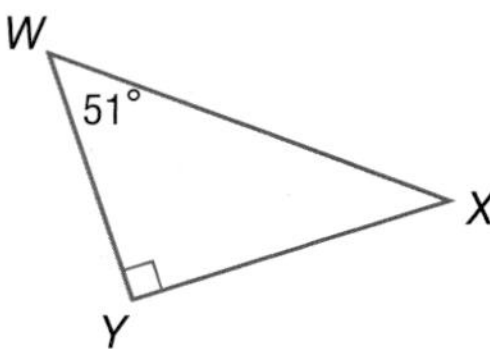


First find the missing angle measure using the Triangle Angle Sum Theorem.

$m\angle F = 180 - (45 + 56)$ or 79

So, the angles from smallest to largest are $\angle G$, $\angle H$, and $\angle F$. The sides opposite these angles are $\overline{FH}$, $\overline{FG}$, and $\overline{GH}$, respectively. So, the sides from shortest to longest are $\overline{FH}$, $\overline{FG}$, $\overline{GH}$.

### Guided Practice

**3.** List the angles and sides of $\triangle WXY$ in order from smallest to largest.

W, X, Y, 51°

You can use angle-side relationships in triangles to solve real-world problems.

## Real-World Example 4 Angle-Side Relationships

**INTERIOR DESIGN An interior designer uses triangulation to create depth in a client's living room. If $m\angle B$ is to be less than $m\angle A$, which distance should be longer—the distance between the two lamps or the distance from the lamp at $B$ to the midpoint of the top of the artwork? Explain.**

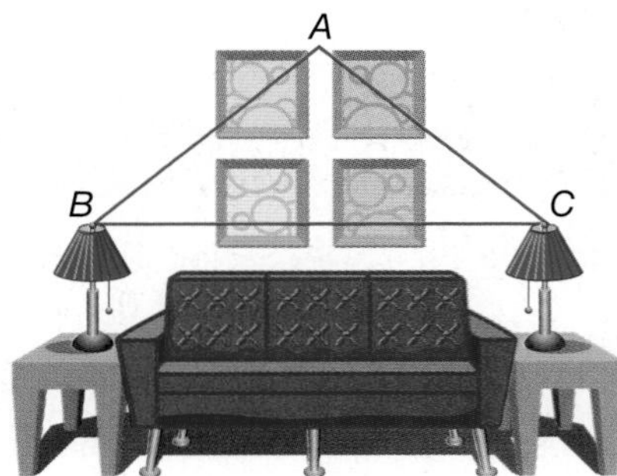


According to Theorem 7.10, in order for $m\angle B < m\angle A$, the length of the side opposite $\angle B$ must be less than the length of the side opposite $\angle A$. Since $\overline{AC}$ is opposite $\angle B$, and $\overline{BC}$ is opposite $\angle A$, then $AC < BC$ and $BC > AC$. So $BC$, the distance between the lamps, must be greater than the distance from the lamp at $B$ to the midpoint of the top of the artwork.

### Guided Practice

**4. LIFEGUARDING** During lifeguard training, an instructor simulates a person in distress so that trainees can practice their rescue skills. If the instructor, Trainee 1, and Trainee 2 are located in the positions shown on the diagram, which of the two trainees is closest to the instructor?

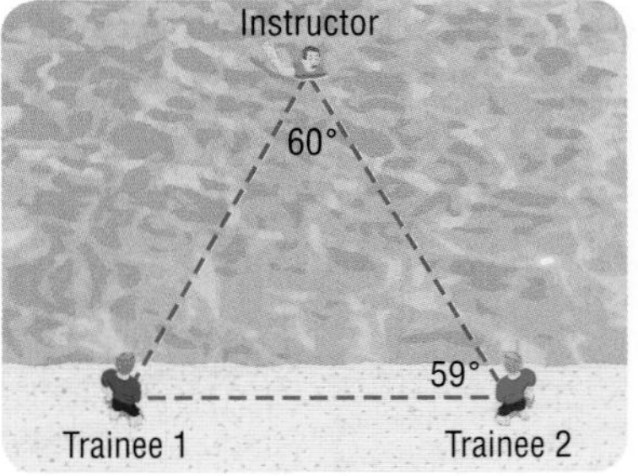

**Real-World Career**

**Interior Designer** An interior designer decorates a space so that it is visually pleasing and comfortable for people to live or work in. Designers must know color and paint theory, lighting design, and space planning. A bachelor's degree is recommended for entry-level positions. Graduates usually enter a 1- to 3-year apprenticeship before taking a licensing exam.

## Check Your Understanding

**Example 1** **Use the Exterior Angle Inequality Theorem to list all of the angles that satisfy the stated condition.**

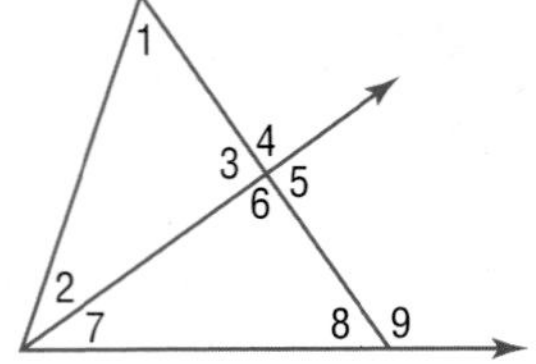


1. measures less than $m\angle 4$
2. measures greater than $m\angle 7$
3. measures greater than $m\angle 2$
4. measures less than $m\angle 9$

**Examples 2–3** **List the angles and sides of each triangle in order from smallest to largest.**

5. 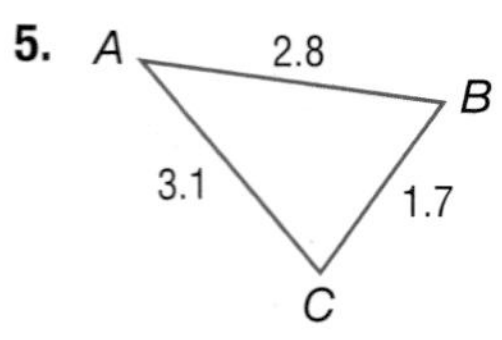


6. 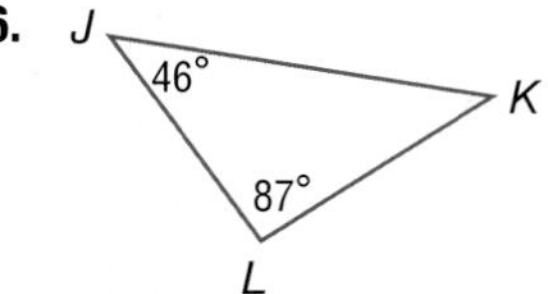


**Example 4**

7. **HANG GLIDING** The supports on a hang glider form triangles like the one shown. Which is longer—the support represented by $\overline{AC}$ or the support represented by $\overline{BC}$? Explain your reasoning.

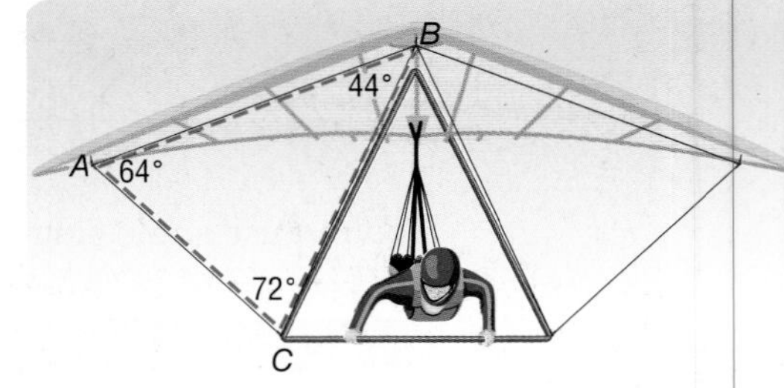


## Practice and Problem Solving

**Example 1** CCSS **SENSE-MAKING** **Use the Exterior Angle Inequality Theorem to list all of the angles that satisfy the stated condition.**

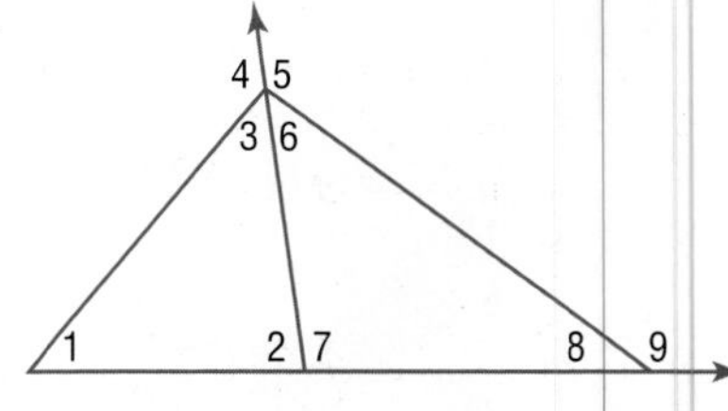


8. measures greater than $m\angle 2$
9. measures less than $m\angle 4$
10. measures less than $m\angle 5$
11. measures less than $m\angle 9$
12. measures greater than $m\angle 8$
13. measures greater than $m\angle 7$

**Examples 2–3** **List the angles and sides of each triangle in order from smallest to largest.**

14. 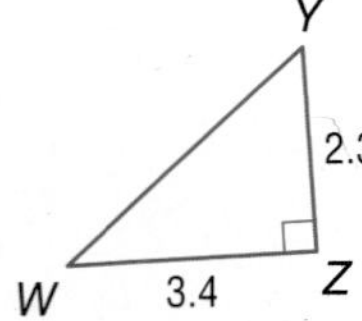


15. 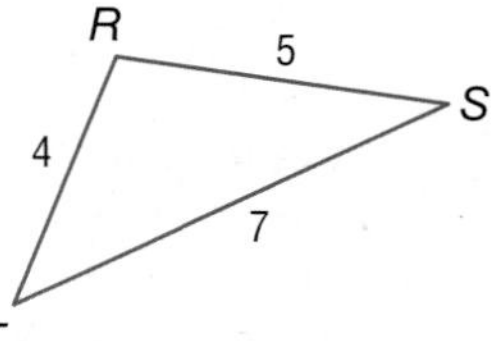


16. 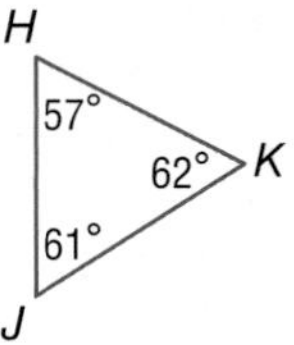


17. 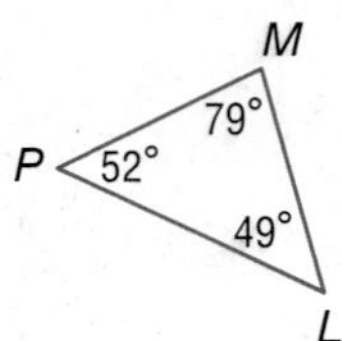


18. 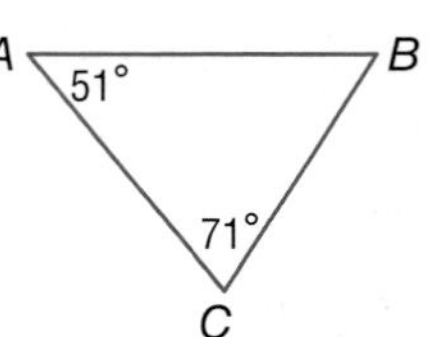


19. 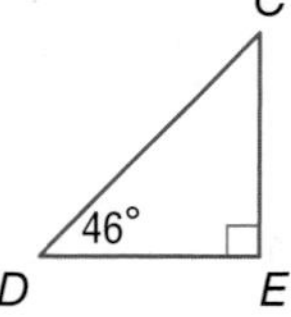

**Example 4**

**20. SPORTS** Ben, Gilberto, and Hannah are playing Ultimate. Hannah is trying to decide if she should pass to Ben or Gilberto. Which player should she choose in order to have the shorter passing distance? Explain your reasoning.

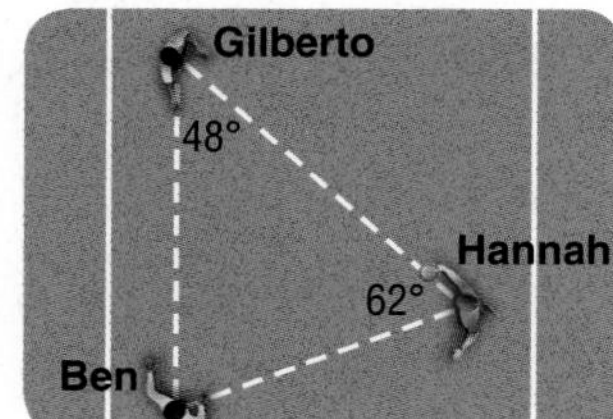

**21 RAMPS** The wedge below represents a bike ramp. Which is longer, the length of the ramp $\overline{XZ}$ or the length of the top surface of the ramp $\overline{YZ}$? Explain your reasoning using Theorem 7.9.

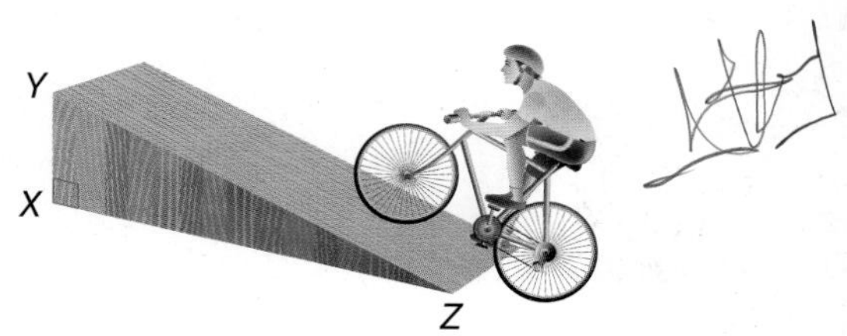


**List the angles and sides of each triangle in order from smallest to largest.**

**22.**

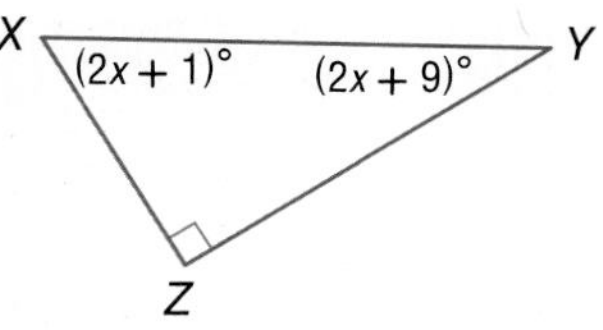

**23.**

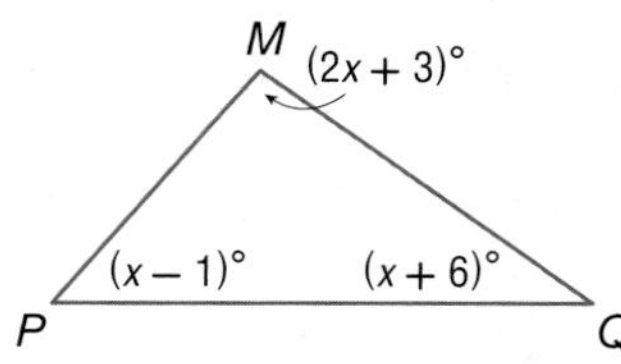


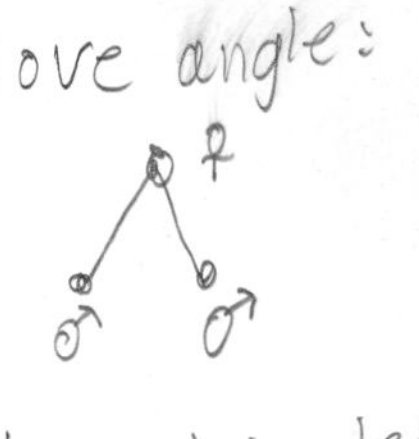

**Use the figure at the right to determine which angle has the greatest measure.**

**24.** $\angle 1, \angle 5, \angle 6$ **25.** $\angle 2, \angle 4, \angle 6$

**26.** $\angle 7, \angle 4, \angle 5$ **27.** $\angle 3, \angle 11, \angle 12$

**28.** $\angle 3, \angle 9, \angle 14$ **29.** $\angle 8, \angle 10, \angle 11$

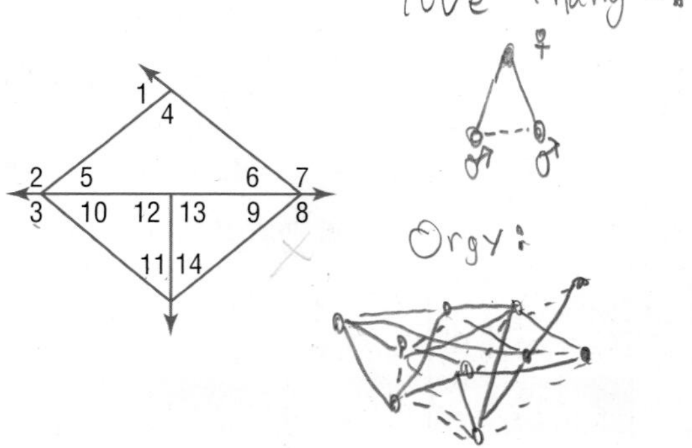

**CCSS SENSE-MAKING** **Use the figure at the right to determine the relationship between the measures of the given angles.**

**30.** $\angle ABD, \angle BDA$ **31.** $\angle BCF, \angle CFB$

**32.** $\angle BFD, \angle BDF$ **33.** $\angle DBF, \angle BFD$

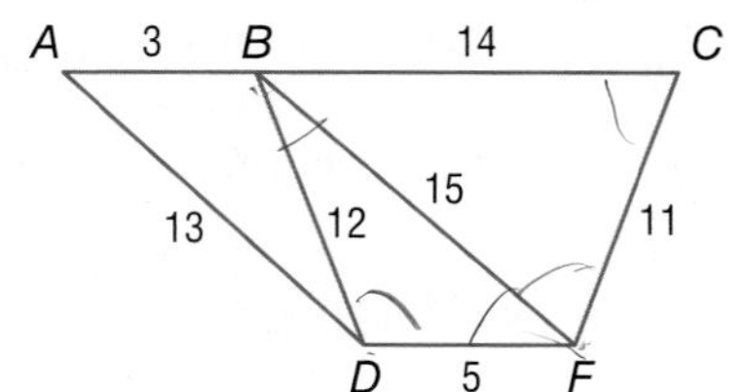


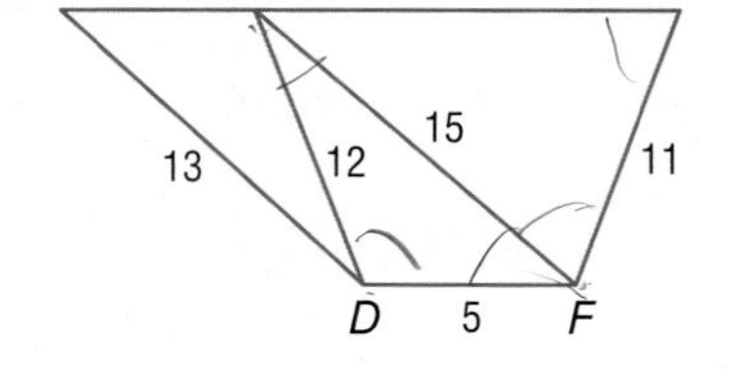

**Use the figure at the right to determine the relationship between the given lengths.**

**34.** $SM, MR$ **35.** $RP, MP$

**36.** $RQ, PQ$ **37.** $RM, RQ$

**38. HIKING** Justin and his family are hiking around a lake as shown in the diagram at the right. Order the angles of the triangle formed by their path from largest to smallest.

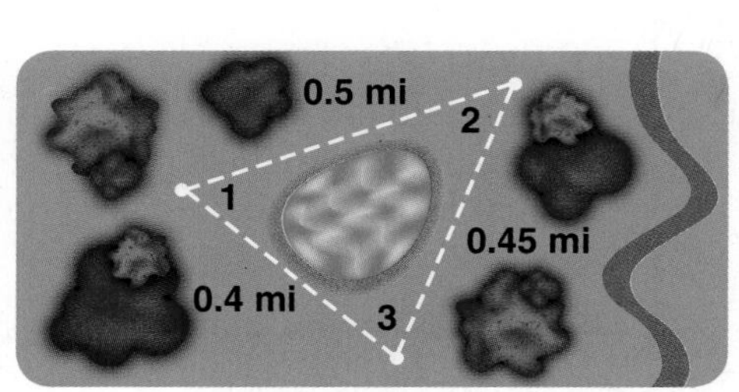

**COORDINATE GEOMETRY List the angles of each triangle with the given vertices in order from smallest to largest. Justify your answer.**

**39.** $A(-4, 6), B(-2, 1), C(5, 6)$

**40.** $X(-3, -2), Y(3, 2), Z(-3, -6)$

**41.** List the side lengths of the triangles in the figure from shortest to longest. Explain your reasoning.

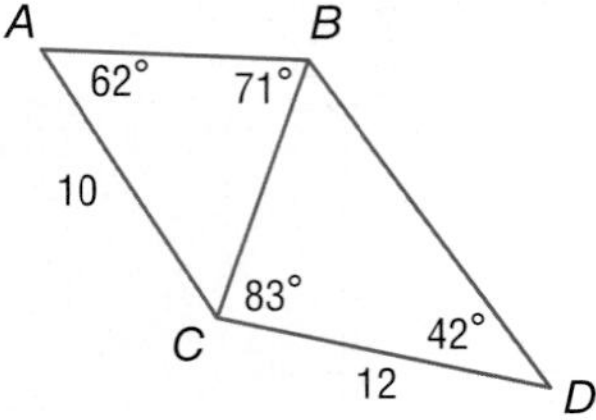


**42.** **MULTIPLE REPRESENTATIONS** In this problem, you will explore the relationship between the sides of a triangle.

**a. Geometric** Draw three triangles, including one acute, one obtuse, and one right angle. Label the vertices of each triangle $A$, $B$, and $C$.

**b. Tabular** Measure the length of each side of the three triangles. Then copy and complete the table.

| Triangle | AB | BC | AB + BC | CA |
|---|---|---|---|---|
| Acute | | | | |
| Obtuse | | | | |
| Right | | | | |

**c. Tabular** Create two additional tables like the one above, finding the sum of $BC$ and $CA$ in one table and the sum of $AB$ and $CA$ in the other.

**d. Algebraic** Write an inequality for each of the tables you created relating the measure of the sum of two of the sides to the measure of the third side of a triangle.

**e. Verbal** Make a conjecture about the relationship between the measure of the sum of two sides of a triangle and the measure of the third side.

## H.O.T. Problems Use Higher-Order Thinking Skills

**43.** **WRITING IN MATH** Analyze the information given in the diagram and explain why the markings must be incorrect.

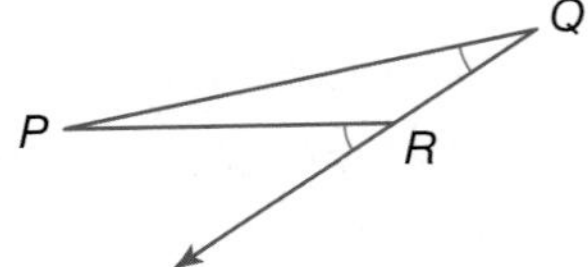


**44.** **CHALLENGE** Using only a ruler, draw $\triangle ABC$ such that $m\angle A > m\angle B > m\angle C$. Justify your drawing.

**45.** **OPEN ENDED** Give a possible measure for $\overline{AB}$ in $\triangle ABC$ shown. Explain your reasoning.

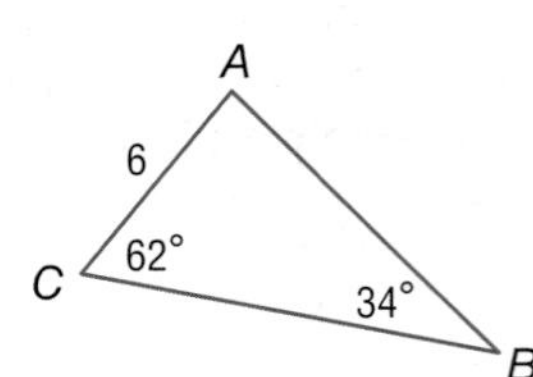


**46.** **CCSS ARGUMENTS** Is the base of an isosceles triangle *always*, *sometimes*, or *never* the longest side of the triangle? Explain.

**47.** **CHALLENGE** Use the side lengths in the figure to list the numbered angles in order from smallest to largest given that $m\angle 2 = m\angle 5$. Explain your reasoning.

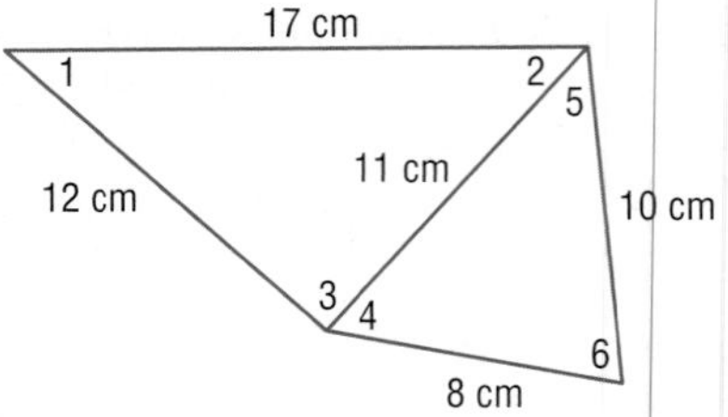


**48.** **WRITING IN MATH** Why is the hypotenuse always the longest side of a triangle?

## Standardized Test Practice

**49. STATISTICS** The chart shows the number and types of DVDs sold at three stores.

| DVD Type | Store 1 | Store 2 | Store 3 |
|---|---|---|---|
| Comedy | 75 | 80 | 92 |
| Action | 54 | 37 | 65 |
| Horror | 30 | 48 | 62 |
| Science Fiction | 21 | 81 | 36 |
| Total | 180 | 246 | 255 |

According to the information in the chart, which of these statements is true?

A The mean number of DVDs sold per store was 56.

B Store 1 sold twice as many action and horror films as store 3 sold of science fiction.

C Store 2 sold fewer comedy and science fiction than store 3 sold.

D The mean number of science fiction DVDs sold per store was 46.

**50.** Two angles of a triangle have measures 45° and 92°. What type of triangle is it?

F obtuse scalene
H acute scalene
G obtuse isosceles
J acute isosceles

**51. EXTENDED RESPONSE** At a five-star restaurant, a waiter earns a total of $t$ dollars for working $h$ hours in which he receives $198 in tips and makes $2.50 per hour.

**a.** Write an equation to represent the total amount of money the waiter earns.

**b.** If the waiter earned a total of $213, how many hours did he work?

**c.** If the waiter earned $150 in tips and worked for 12 hours, what is the total amount of money he earned?

**52. SAT/ACT** Which expression has the *least* value?

A $|-99|$
D $|-28|$
B $|45|$
E $|15|$
C $|-39|$

## Spiral Review

**In $\triangle XYZ$, $P$ is the centroid, $KP = 3$, and $XJ = 8$. Find each length.** (Lesson 7-2)

**53.** $XK$

**54.** $YJ$

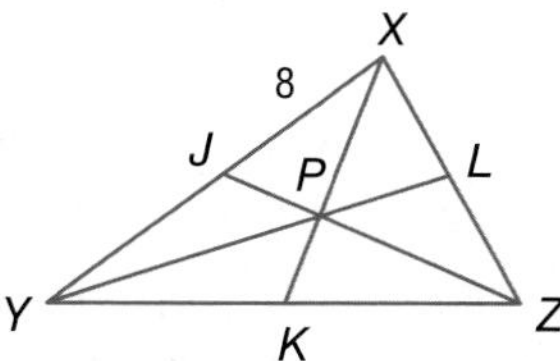


**COORDINATE GEOMETRY Write an equation in slope-intercept form for the perpendicular bisector of the segment with the given endpoints. Justify your answer.** (Lesson 7-1)

**55.** $D(-2, 4)$ and $E(3, 5)$

**56.** $D(-2, -4)$ and $E(2, 1)$

**57. JETS** The United States Navy Flight Demonstration Squadron, the Blue Angels, flies in a formation that can be viewed as two triangles with a common side. Write a two-column proof to prove that $\triangle SRT \cong \triangle QRT$ if $T$ is the midpoint of $\overline{SQ}$ and $\overline{SR} \cong \overline{QR}$. (Lesson 6-3)

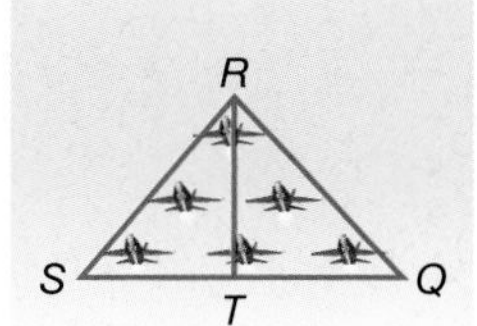


## Skills Review

**Determine whether each statement is true or false if $x = 8$, $y = 2$, and $z = 3$.**

**58.** $z(x - y) = 13$

**59.** $2x = 3yz$

**60.** $x + y > z + y$

CHAPTER 7

# Mid-Chapter Quiz

## Lessons 7-1 through 7-3

Find each measure. (Lesson 7-1)

1. $AB$

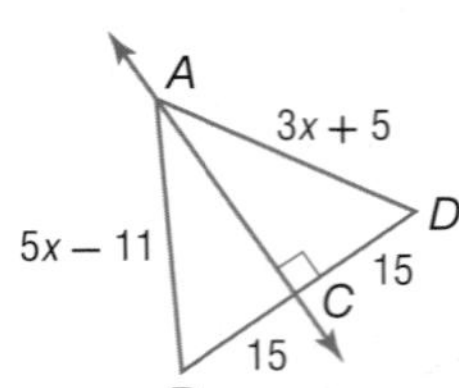


2. $JL$

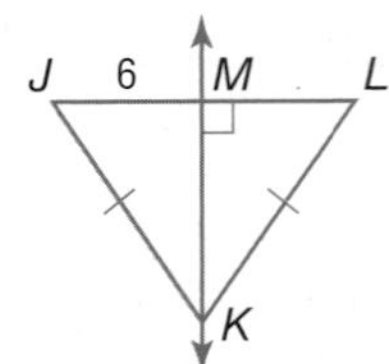


3. **CAMP** Camp Onawatchi ends with a game of capture the flag. If the starting locations of three teams are shown in the diagram below, with the flag at a point equidistant from each team's base, how far from each base is the flag in feet? (Lesson 7-1)

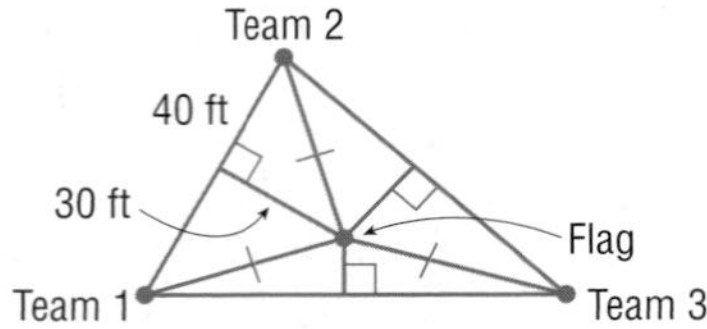


Find each measure. (Lesson 7-1)

4. $\angle MNP$

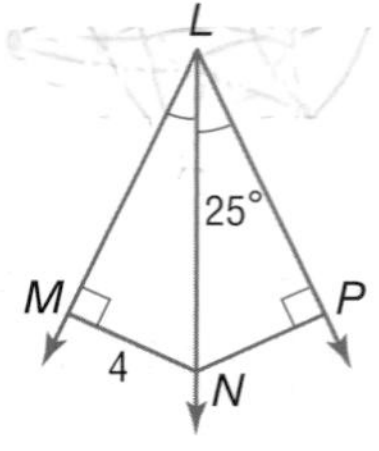


5. $XY$

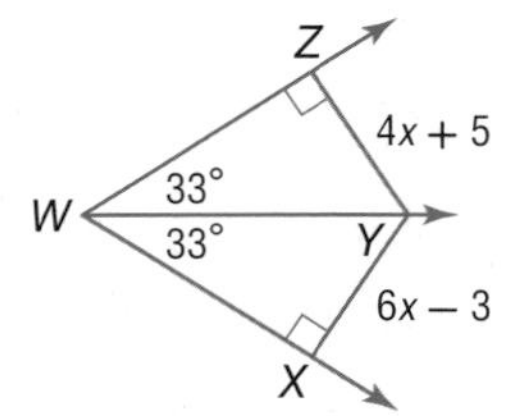


In $\triangle RST$, $Z$ is the centroid and $RZ = 18$. Find each length. (Lesson 7-2)

6. $ZV$
7. $SZ$
8. $SR$

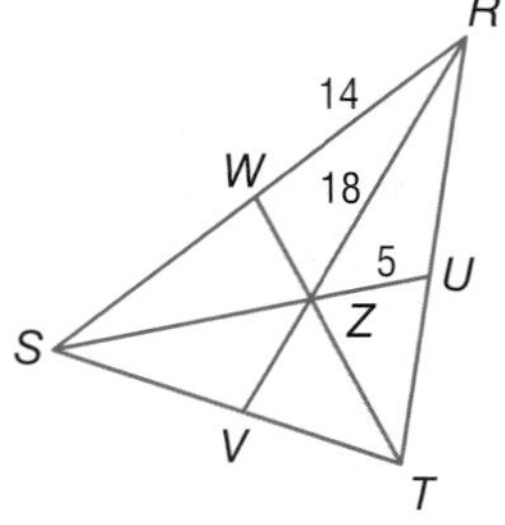


**COORDINATE GEOMETRY** Find the coordinates of the centroid of each triangle with the given vertices. (Lesson 7-2)

9. $A(1, 7)$, $B(4, 2)$, $C(7, 7)$
10. $X(-11, 0)$, $Y(-11, -8)$, $Z(-1, -4)$
11. $R(-6, 4)$, $S(-2, -2)$, $T(2, 4)$
12. $J(-5, 5)$, $K(-5, -1)$, $L(1, 2)$

13. **ARCHITECTURE** An architect is designing a high school building. Describe how to position the central office so that it is at the intersection of each hallway connected to the three entrances to the school. (Lesson 7-2)

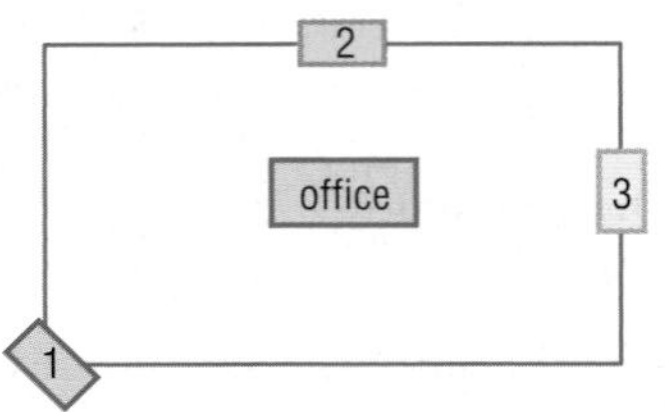


List the angles and sides of each triangle in order from smallest to largest. (Lesson 7-3)

14.

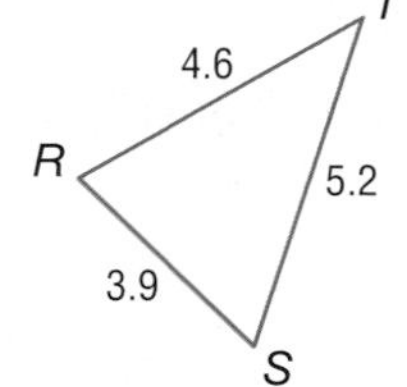


15.

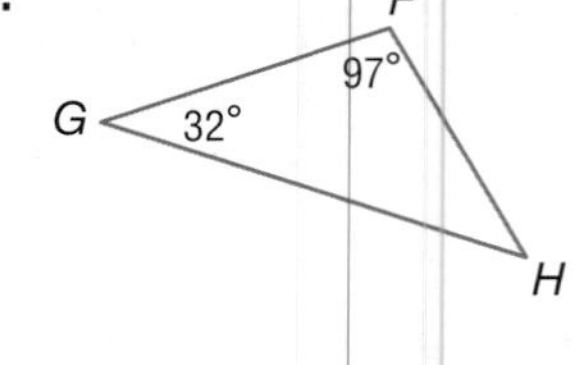


16. **VACATION** Kailey plans to fly over the route marked on the map of Hawaii below. (Lesson 7-3)

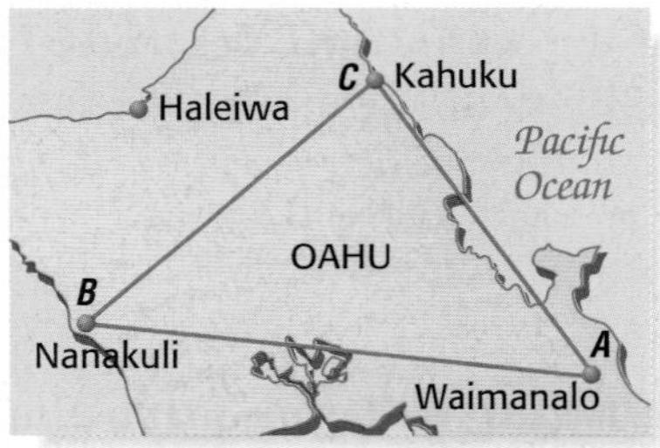


a. If $m\angle A = 2 + m\angle B$ and $m\angle C = 2(m\angle B) - 14$, what are the measures of the three angles?

b. What are the lengths of Kailey's trip in order of least to greatest?

c. The length of the entire trip is about 68 miles. The middle leg is 11 miles greater than one-half the length of the shortest leg. The longest leg is 12 miles greater than three-fourths of the shortest leg. What are the lengths of the legs of the trip?

Use the Exterior Angle Inequality Theorem to list all of the angles that satisfy the stated condition. (Lesson 7-3)

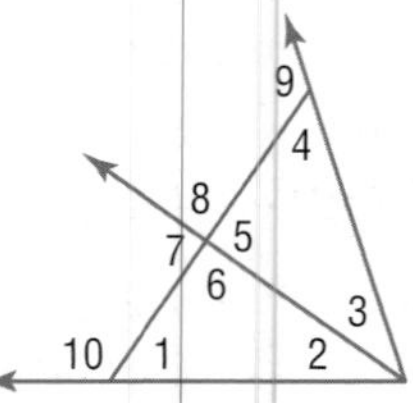


17. measures less than $m\angle 8$
18. measures greater than $m\angle 3$
19. measures less than $m\angle 10$

EXPLORE

# 7-4 Geometry Lab: Matrix Logic

**Matrix logic** uses a rectangular array in which you record what you have learned from clues in order to solve a logic or reasoning problem. Once all the rows and columns are filled, you can deduce the answer.

**Common Core State Standards**
**Content Standards**
**F.IF.4** For a function that models a relationship between two quantities, interpret key features of graphs and tables in terms of the quantities,
**Mathematical Practices** 2, 6

FOOD **Matt, Abby, Javier, Corey, and Keisha go to an Italian restaurant. Each orders their favorite dish: ravioli, pizza, lasagna, manicotti, or spaghetti. Javier loves ravioli, but Matt does not like pasta dishes. Abby does not like lasagna or manicotti. Corey's favorite dish does not end in the letter i. What does each person order?**

**Step 1** Create an appropriate matrix.

Use a 5 × 5 matrix that includes each person's name as the header for each row and their possible favorite foods as the header for each column.

| Name | ravioli | pizza | lasagna | manicotti | spaghetti |
|---|---|---|---|---|---|
| Matt | × | ✓ | × | × | × |
| Abby | × | × | | | |
| Javier | ✓ | × | × | × | × |
| Corey | × | × | | | |
| Keisha | × | × | | | |

**Step 2** Use each clue and logical reasoning to fill in the matrix.

- Since Javier loves ravioli, place a ✓ in Javier's row under ravioli and an × in every other cell in his row. Since only one person likes each dish, you can place an × in every other cell in the ravioli column.
- Since Matt does not like pasta, you know that Matt cannot like manicotti, ravioli, lasagna, or spaghetti, which are all pasta dishes. Therefore, Matt must like pizza. Place a ✓ in Matt's row under pizza. Place an × in every other cell in Matt's row and in every other cell in the pizza column.
- Since Abby does not like lasagna or manicotti, place an × in Abby's row under lasagna and manicotti. This leaves only spaghetti without an × for Abby's row. Therefore, you can conclude that Abby must like spaghetti. Place a ✓ in that cell and an × in every other cell in the spaghetti column.

| Name | ravioli | pizza | lasagna | manicotti | spaghetti |
|---|---|---|---|---|---|
| Matt | × | ✓ | × | × | × |
| Abby | × | × | × | × | ✓ |
| Javier | ✓ | × | × | × | × |
| Corey | × | × | | | × |
| Keisha | × | × | | | × |

- From the matrix, you can see that Corey's favorite dish must be either lasagna or manicotti. However, since Corey's favorite dish does not end in the letter i, you can conclude that Corey must like lasagna. In Corey's row, place a ✓ under lasagna and an × under manicotti.
- This leaves only one empty cell in Keisha's row, so you can conclude that her favorite dish is manicotti.

| Name | ravioli | pizza | lasagna | manicotti | spaghetti |
|---|---|---|---|---|---|
| Matt | × | ✓ | × | × | × |
| Abby | × | × | × | × | ✓ |
| Javier | ✓ | × | × | × | × |
| Corey | × | × | ✓ | × | × |
| Keisha | × | × | × | ✓ | × |

**Step 3** Use your matrix to state the answer to the problem.

From the matrix you can state that Matt orders pizza, Abby orders spaghetti, Javier orders ravioli, Corey orders lasagna, and Keisha orders manicotti.

Geometry Lab

# Matrix Logic *Continued*

## Exercises

**Use a matrix to solve each problem.**

1. **SPORTS** Trey, Nathan, Parker, and Chen attend the same school. Each participates in a different school sport: basketball, football, track, or tennis. Use the following clues to determine in which sport each student participates.
   - Nathan does not like track or basketball.
   - Trey does not participate in football or tennis.
   - Parker prefers an indoor winter sport.
   - Chen scored four touchdowns in the final game of the season.

2. **FAMILY** The Martin family has five children. Use the following clues to determine in what order the children were born.
   - Grace is older than Hannah.
   - Thomas is younger than Sarah.
   - Hannah is older than Thomas and Samuel.
   - Samuel is older than Thomas.
   - Sarah is older than Grace.

3. **PETS** Alejandra, Tamika, and Emily went to a pet store. Each girl chose a different pet to adopt: a dog, a rabbit, or a cat. Each girl named her pet Sweet Pea, Zuzu, or Roscoe. Use the following clues and the matrix shown to determine the animal each girl adopted and what name she gave her pet.
   - The girl who adopted a dog did not name it Sweet Pea.
   - Tamika's pet, who she named Zuzu, is not the type of animal that hops.
   - Roscoe, who is not a cat, was adopted by Emily.
   - The rabbit was not adopted by Alejandra.

| | | Pet | | | Pet Name | | |
|---|---|---|---|---|---|---|---|
| | | dog | rabbit | cat | Sweet Pea | Zuzu | Roscoe |
| Names | Alejandra | | | | | | |
| | Tamika | | | | | | |
| | Emily | | | | | | |
| Pet Name | Sweet Pea | | | | | | |
| | Zuzu | | | | | | |
| | Roscoe | | | | | | |

4. **GEOMETRY** Kasa, Marcus, and Jason each drew a triangle, no two of which share the same side or angle classification. Use the following clues to determine what type of triangle each person has drawn.
   - Kasa did not draw an equilateral triangle.
   - Marcus' triangle has one angle that measures 25 and another that measures 65.
   - Jason drew a triangle with at least one pair of congruent sides.
   - The obtuse triangle has two congruent angles.

# LESSON 7-4 Indirect Proof

**Then**

- You wrote paragraph, two-column, and flow proofs.

**Now**

1. Write indirect algebraic proofs.
2. Write indirect geometric proofs.

**Why?**

- **Matthew:** "I'm almost positive Friday is not a teacher work day, but I can't prove it."
  **Kim:** "Let's assume that Friday *is* a teacher work day. What day is our next Geometry test?"
  **Ana:** "Hmmm . . . according to the syllabus, it's this Friday. But we don't have tests on teacher work days—we're not in school."
  **Jamal:** "Exactly—so that proves it! This Friday can't be a teacher work day."

**NewVocabulary**
indirect reasoning
indirect proof
proof by contradiction

**Common Core State Standards**

**Content Standards**
G.CO.10 Prove theorems about triangles.

**Mathematical Practices**
3 Construct viable arguments and critique the reasoning of others.
2 Reason abstractly and quantitatively.

## 1 Indirect Algebraic Proof

The proofs you have written have been *direct proofs*—you started with a true hypothesis and proved that the conclusion was true. In the example above, the students used **indirect reasoning**, by assuming that a conclusion was false and then showing that this assumption led to a contradiction.

In an **indirect proof** or **proof by contradiction**, you temporarily assume that what you are trying to prove is false. By showing this assumption to be logically impossible, you prove your assumption false and the original conclusion true. Sometimes this is called *proof by negation*.

**KeyConcept How to Write an Indirect Proof**

**Step 1** Identify the conclusion you are asked to prove. Make the assumption that this conclusion is false by assuming that the opposite is true.

**Step 2** Use logical reasoning to show that this assumption leads to a contradiction of the hypothesis, or some other fact, such as a definition, postulate, theorem, or corollary.

**Step 3** Point out that since the assumption leads to a contradiction, the original conclusion, what you were asked to prove, must be true.

### Example 1 State the Assumption for Starting an Indirect Proof

**State the assumption necessary to start an indirect proof of each statement.**

**a. If 6 is a factor of $n$, then 2 is a factor of $n$.**

The conclusion of the conditional statement is *2 is a factor of n*. The negation of the conclusion is *2 is not a factor of n*.

**b. $\angle 3$ is an obtuse angle.**

If *$\angle 3$ is an obtuse angle* is false, then *$\angle 3$ is not an obtuse angle* must be true.

**GuidedPractice**

**1A.** $x > 5$

**1B.** $\triangle XYZ$ is an equilateral triangle.

Digital Vision/Punchstock

Indirect proofs can be used to prove algebraic concepts.

PT

## Example 2 Write an Indirect Algebraic Proof

**Write an indirect proof to show that if $-3x + 4 > 16$, then $x < -4$.**

**Given:** $-3x + 4 > 16$

**Prove:** $x < -4$

**Step 1** **Indirect Proof:**
The negation of $x < -4$ is $x \geq -4$. So, assume that $x > -4$ or $x = -4$ is true.

**Step 2** Make a table with several possibilities for $x$ assuming $x > -4$ or $x = -4$.

| $x$ | $-4$ | $-3$ | $-2$ | $-1$ | $0$ |
|---|---|---|---|---|---|
| $-3x + 4$ | 16 | 13 | 10 | 7 | 4 |

When $x > -4$, $-3x + 4 < 16$ and when $x = -4$, $-3x + 4 = 16$.

**Step 3** In both cases, the assumption leads to the contradiction of the given information that $-3x + 4 > 16$. Therefore, the assumption that $x \geq -4$ must be false, so the original conclusion that $x < -4$ must be true.

**ReadingMath**

**Contradiction** A contradiction is a principle of logic stating that an assumption cannot be both *A* and the opposite of *A* at the same time.

### GuidedPractice

**Write an indirect proof of each statement.**

**2A.** If $7x > 56$, then $x > 8$.

**2B.** If $-c$ is positive, then $c$ is negative.

Indirect reasoning and proof can be used in everyday situations.

PT

## Real-World Example 3 Indirect Algebraic Proof

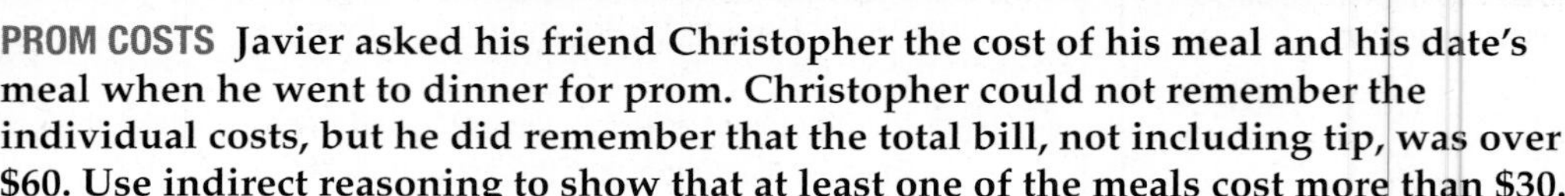

**PROM COSTS Javier asked his friend Christopher the cost of his meal and his date's meal when he went to dinner for prom. Christopher could not remember the individual costs, but he did remember that the total bill, not including tip, was over \$60. Use indirect reasoning to show that at least one of the meals cost more than \$30.**

Let the cost of one meal be $x$ and the cost of the other meal be $y$.

**Step 1** **Given:** $x + y > 60$

**Prove:** $x > 30$ or $y > 30$

**Indirect Proof:**
Assume that $x \leq 30$ and $y \leq 30$.

**Step 2** If $x \leq 30$ and $y \leq 30$, then $x + y \leq 30 + 30$ or $x + y \leq 60$. This is a contradiction because we know that $x + y > 60$.

**Step 3** Since the assumption that $x \leq 30$ and $y \leq 30$ leads to a contradiction of a known fact, the assumption must be false. Therefore, the conclusion that $x > 30$ or $y > 30$ must be true. Thus, at least one of the meals had to cost more than \$30.

**Real-WorldLink**

\$100–\$300 the range in price of a girl's prom dress

\$75–\$125 the range in cost for a tuxedo rental

around \$150 the cost of a fancy dinner for two

\$100–\$200 the range in cost of prom tickets per couple

**Source:** PromSpot

### GuidedPractice

**3. TRAVEL** Cleavon traveled over 360 miles on his trip, making just two stops. Use indirect reasoning to prove that he traveled more than 120 miles on one leg of his trip.

Indirect proofs are often used to prove concepts in number theory. In such proofs, it is helpful to remember that you can represent an even number with the expression $2k$ and an odd number with the expression $2k + 1$ for any integer $k$.

## Example 4 Indirect Proofs in Number Theory

**Write an indirect proof to show that if $x + 2$ is an even integer, then $x$ is an even integer.**

**Step 1** **Given:** $x + 2$ is an even integer.

**Prove:** $x$ is an even integer.

**Indirect Proof:**
Assume that $x$ is an odd integer. This means that $x = 2k + 1$ for some integer $k$.

**Step 2**

| | |
|---|---|
| $x + 2 = (2k + 1) + 2$ | Substitution of assumption |
| $= (2k + 2) + 1$ | Commutative Property |
| $= 2(k + 1) + 1$ | Distributive Property |

Now determine whether $2(k + 1) + 1$ is an even or odd integer. Since $k$ is an integer, $k + 1$ is also an integer. Let $m$ represent the integer $k + 1$.

$2(k + 1) + 1 = 2m + 1$ Substitution

So, $x + 2$ can be represented by $2m + 1$, where $m$ is an integer. But this representation means that $x + 2$ is an odd integer, which contradicts the given statement that $x + 2$ is an even integer.

**Step 3** Since the assumption that $x$ is an odd integer leads to a contradiction of the given statement, the original conclusion that $x$ is an even integer must be true.

**WatchOut!**

**CCSS Arguments** Proof by contradiction and using a counterexample are not the same. A counterexample helps you disprove a conjecture. It cannot be used to prove a conjecture.

**GuidedPractice**

4. Write an indirect proof to show that if the square of an integer is odd, then the integer is odd.

## 2 Indirect Proof with Geometry

Indirect reasoning can be used to prove statements in geometry, such as the Exterior Angle Inequality Theorem.

## Example 5 Geometry Proof

**If an angle is an exterior angle of a triangle, prove that its measure is greater than the measure of either of its corresponding remote interior angles.**

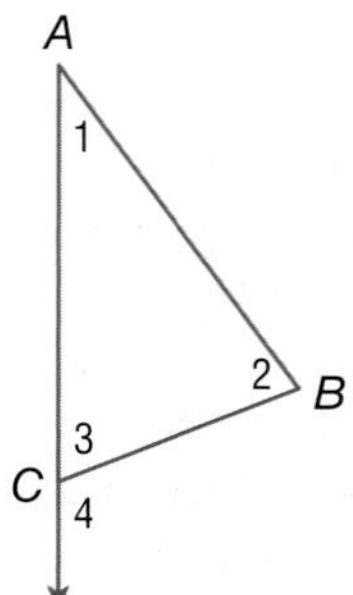


**Step 1** Draw a diagram of this situation. Then identify what you are given and what you are asked to prove.

**Given:** $\angle 4$ is an exterior angle of $\triangle ABC$.

**Prove:** $m\angle 4 > m\angle 1$ and $m\angle 4 > m\angle 2$.

**Indirect Proof:**
Assume that $m\angle 4 \not> m\angle 1$ or $m\angle 4 \not> m\angle 2$.
In other words, $m\angle 4 \leq m\angle 1$ or $m\angle 4 \leq m\angle 2$.

*(continued on the next page)*

**Step 2** You need only show that the assumption $m\angle 4 \leq m\angle 1$ leads to a contradiction. The argument for $m\angle 4 \leq m\angle 2$ follows the same reasoning.

$m\angle 4 \leq m\angle 1$ means that either $m\angle 4 = m\angle 1$ or $m\angle 4 < m\angle 1$.

**Case 1** $m\angle 4 = m\angle 1$

$m\angle 4 = m\angle 1 + m\angle 2$ Exterior Angle Theorem

$m\angle 4 = m\angle 4 + m\angle 2$ Substitution

$0 = m\angle 2$ Subtract $m\angle 4$ from each side.

This contradicts the fact that the measure of an angle is greater than 0, so $m\angle 4 \neq m\angle 1$.

**Case 2** $m\angle 4 < m\angle 1$

By the Exterior Angle Theorem, $m\angle 4 = m\angle 1 + m\angle 2$. Since angle measures are positive, the definition of inequality implies that $m\angle 4 > m\angle 1$. This contradicts the assumption that $m\angle 4 < m\angle 1$.

**Step 3** In both cases, the assumption leads to the contradiction of a theorem or definition. Therefore, the original conclusion that $m\angle 4 > m\angle 1$ and $m\angle 4 > m\angle 2$ must be true.

**StudyTip**

**Recognizing Contradictions** Remember that the contradiction in an indirect proof is not always of the given information or the assumption. It can be of a known fact or definition, such as in Case 1 of Example 5—the measure of an angle must be greater than 0.

### GuidedPractice

**5. Write an indirect proof.**

**Given:** $\overline{MO} \cong \overline{ON}$, $\overline{MP} \not\cong \overline{NP}$

**Prove:** $\angle MOP \not\cong \angle NOP$

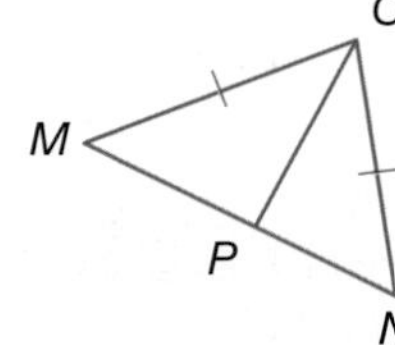


## Check Your Understanding

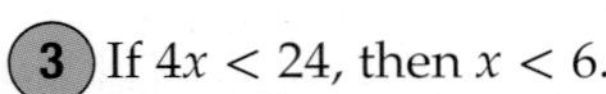

**Example 1** **State the assumption you would make to start an indirect proof of each statement.**

1. $\overline{AB} \cong \overline{CD}$
2. $\triangle XYZ$ is a scalene triangle.
3. If $4x < 24$, then $x < 6$.
4. $\angle A$ is not a right angle.

**Example 2** **Write an indirect proof of each statement.**

5. If $2x + 3 < 7$, then $x < 2$.
6. If $3x - 4 > 8$, then $x > 4$.

**Example 3**

7. **LACROSSE** Christina scored 13 points for her high school lacrosse team during the last six games. Prove that her average points per game was less than 3.

**Example 4**

8. Write an indirect proof to show that if $5x - 2$ is an odd integer, then $x$ is an odd integer.

**Example 5** **Write an indirect proof of each statement.**

9. The hypotenuse of a right triangle is the longest side.
10. If two angles are supplementary, then they both cannot be obtuse angles.

## Practice and Problem Solving

**Example 1** **State the assumption you would make to start an indirect proof of each statement.**

11. If $2x > 16$, then $x > 8$.

12. $\angle 1$ and $\angle 2$ are not supplementary angles.

13. If two lines have the same slope, the lines are parallel.

14. If the consecutive interior angles formed by two lines and a transversal are supplementary, the lines are parallel.

15. If a triangle is not equilateral, the triangle is not equiangular.

16. An odd number is not divisible by 2.

**Example 2** **Write an indirect proof of each statement.**

17. If $2x - 7 > -11$, then $x > -2$.

18. If $5x + 12 < -33$, then $x < -9$.

19. If $-3x + 4 < 7$, then $x > -1$.

20. If $-2x - 6 > 12$, then $x < -9$.

**Example 3**

21. **COMPUTER GAMES** Kwan-Yong bought two computer games for just over \$80 before tax. A few weeks later, his friend asked how much each game cost. Kwan-Yong could not remember the individual prices. Use indirect reasoning to show that at least one of the games cost more than \$40.

22. **FUNDRAISING** Jamila's school is having a Fall Carnival to raise money for a local charity. The cost of an adult ticket to the carnival is \$6 and the cost of a child's ticket is \$2.50. If 375 total tickets were sold and the profit was more than \$1460, prove that at least 150 adult tickets were sold.

**Examples 4–5** **CCSS ARGUMENTS** **Write an indirect proof of each statement.**

23. **Given:** $xy$ is an odd integer.
**Prove:** $x$ and $y$ are both odd integers.

24. **Given:** $n^2$ is even.
**Prove:** $n^2$ is divisible by 4.

25. **Given:** $x$ is an odd number.
**Prove:** $x$ is not divisible by 4.

26. **Given:** $xy$ is an even integer.
**Prove:** $x$ or $y$ is an even integer.

27. **Given:** $XZ > YZ$
**Prove:** $\angle X \not\cong \angle Y$

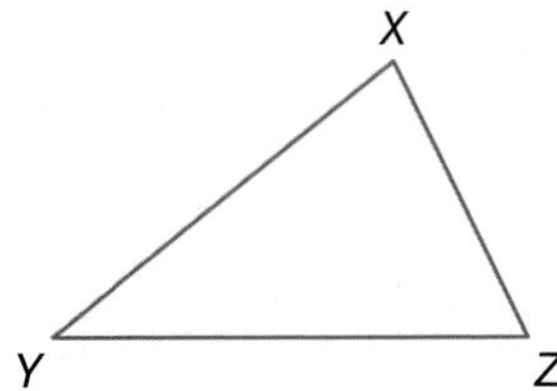


28. **Given:** $\triangle ABC$ is equilateral.
**Prove:** $\triangle ABC$ is equiangular.

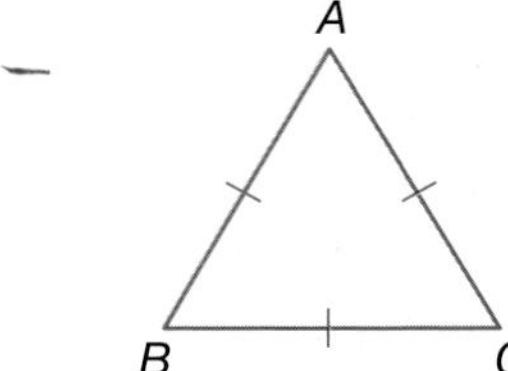


29. In an isosceles triangle neither of the base angles can be a right angle.

30. A triangle can have only one right angle.

31. Write an indirect proof for Theorem 7.10.

32. Write an indirect proof to show that if $\frac{1}{b} < 0$, then $b$ is negative.

**33. BASKETBALL** In basketball, there are three possible ways to score three points in a single possession. A player can make a basket from behind the three-point line, a player may be fouled while scoring a two-point shot and be allowed to shoot one free throw, or a player may be fouled behind the three-point line and be allowed to shoot three free throws. When Katsu left to get in the concession line, the score was 28 home team to 26 visiting team. When she returned, the score was 28 home team to 29 visiting team. Katsu concluded that a player on the visiting team had made a three-point basket. Prove or disprove her assumption using an indirect proof.

**34. GAMES** A computer game involves a knight on a quest for treasure. At the end of the journey, the knight approaches the two doors shown below.

A servant tells the knight that one of the signs is true and the other is false. Use indirect reasoning to determine which door the knight should choose. Explain your reasoning.

**35. SURVEYS** Luisa's local library conducted an online poll of teens to find out what activities teens participate in to preserve the environment. The results of the poll are shown in the graph.

**a.** Prove: *More than half of teens polled said that they recycle to preserve the environment.*

**b.** If 400 teens were polled, verify that 92 said that they participate in Earth Day.

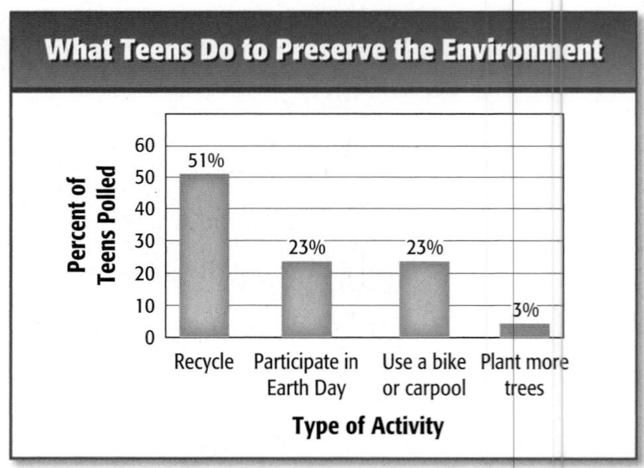


**36. CCSS REASONING** James, Hector, and Mandy all have different color cars. Only one of the statements below is true. Use indirect reasoning to determine which statement is true. Explain.

(1) James has a red car.

(2) Hector does not have a red car.

(3) Mandy does not have a blue car.

**Determine whether each statement about the shortest distance between a point and a line or plane can be proved using a direct or indirect proof. Then write a proof of each statement.**

**37.** **Given:** $\overline{AB} \perp$ line $p$

**Prove:** $\overline{AB}$ is the shortest segment from $A$ to line $p$.

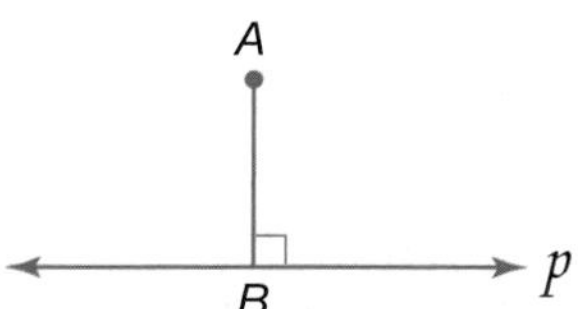


**38. Given:** $\overline{PQ} \perp$ plane $M$

**Prove:** $\overline{PQ}$ is the shortest segment from $P$ to plane $M$.

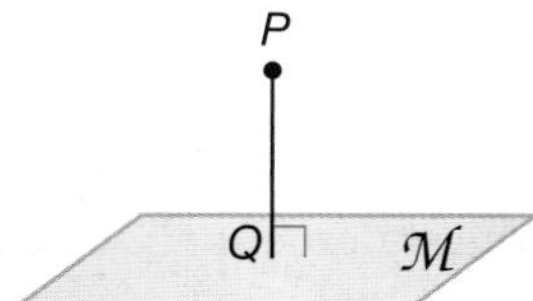


**39. NUMBER THEORY** In this problem, you will make and prove a conjecture about a number theory relationship.

**a.** Write an expression for *the sum of the cube of a number and three.*

**b.** Create a table that includes the value of the expression for 10 different values of $n$. Include both odd and even values of $n$.

**c.** Write a conjecture about $n$ when the value of the expression is even.

**d.** Write an indirect proof of your conjecture.

## H.O.T. Problems Use Higher-Order Thinking Skills

**40. WRITING IN MATH** Explain the procedure for writing an indirect proof.

**41. OPEN ENDED** Write a statement that can be proven using indirect proof. Include the indirect proof of your statement.

**42. CHALLENGE** If $x$ is a rational number, then it can be represented by the quotient $\frac{a}{b}$ for some integers $a$ and $b$, if $b \neq 0$. An irrational number cannot be represented by the quotient of two integers. Write an indirect proof to show that the product of a nonzero rational number and an irrational number is an irrational number.

**43. CCSS CRITIQUE** Amber and Raquel are trying to verify the following statement using indirect proof. Is either of them correct? Explain your reasoning.

*If the sum of two numbers is even, then the numbers are even.*

| Amber | Raquel |
|---|---|
| The statement is true. If one of the numbers is even and the other number is zero, then the sum is even. Since the hypothesis is true even when the conclusion is false, the statement is true. | The statement is true. If the two numbers are odd, then the sum is even. Since the hypothesis is true when the conclusion is false, the statement is true. |

**44. WRITING IN MATH** Refer to Exercise 8. Write the contrapositive of the statement and write a direct proof of the contrapositive. How are the direct proof of the contrapositive of the statement and the indirect proof of the statement related?

## Standardized Test Practice

**45. SHORT RESPONSE** Write an equation in slope-intercept form to describe the line that passes through the point (5, 3) and is parallel to the line represented by the equation $-2x + y = -4$.

**46. Statement:** If $\angle A \cong \angle B$ and $\angle A$ is supplementary to $\angle C$, then $\angle B$ is supplementary to $\angle C$.

Dia is proving the statement above by contradiction. She began by assuming that $\angle B$ is not supplementary to $\angle C$. Which of the following definitions will Dia use to reach a contradiction?

**A** definition of congruence
**B** definition of a linear pair
**C** definition of a right angle
**D** definition of supplementary angles

**47.** List the angles of $\triangle MNO$ in order from smallest to largest if $MN = 9$, $NO = 7.5$, and $OM = 12$.

**F** $\angle N, \angle O, \angle M$
**G** $\angle O, \angle M, \angle N$
**H** $\angle O, \angle N, \angle M$
**J** $\angle M, \angle O, \angle N$

**48. SAT/ACT** If $b > a$, which of the following must be true?

**A** $-a > -b$
**B** $3a > b$
**C** $a^2 < b^2$
**D** $a^2 < ab$
**E** $-b > -a$

## Spiral Review

**49. PROOF** Write a two-column proof. (Lesson 7-3)

**Given:** $\overline{RQ}$ bisects $\angle SRT$.

**Prove:** $m\angle SQR > m\angle SRQ$

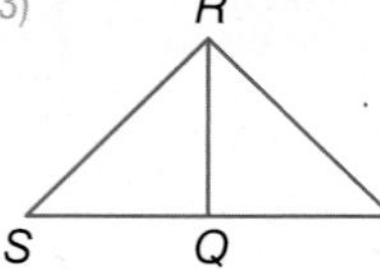


**COORDINATE GEOMETRY** **Find the coordinates of the circumcenter of each triangle with the given vertices.** (Lesson 7-1)

**50.** $D(-3, 3)$, $E(3, 2)$, $F(1, -4)$

**51.** $A(4, 0)$, $B(-2, 4)$, $C(0, 6)$

**Find each measure.** (Lesson 6-1)

**52.** $m\angle 1$

**53.** $m\angle 4$

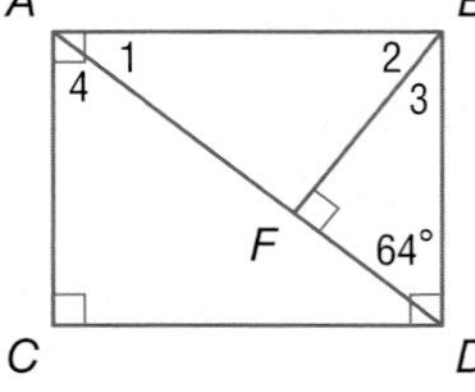


**Find each product.** (Lesson 1-4)

**54.** $\left(\frac{1}{2}m + 3\right)^2$

**55.** $(2n - 6)(2n + 6)$

**56.** $(5a - 4)^2$

**57.** $(x - 2y)(x + 2y)$

## Skills Review

**Determine whether each inequality is *true* or *false*.**

**58.** $23 - 11 > 9$

**59.** $41 - 19 < 21$

**60.** $57 + 68 < 115$

EXPLORE 7-5

# Graphing Technology Lab
# The Triangle Inequality

You can use the Cabri™ Jr. application on a TI-83/84 Plus graphing calculator to discover properties of triangles.

**CCSS Common Core State Standards**
**Content Standards**
**G.CO.12** Make formal geometric constructions with a variety of tools and methods (compass and straightedge, string, reflective devices, paper folding, dynamic geometric software, etc.).
**Mathematical Practices** 5

## Activity 1

**Construct a triangle. Observe the relationship between the sum of the lengths of two sides and the length of the other side.**

**Step 1** Construct a triangle using the triangle tool on the **F2** menu. Then use the **Alph-Num** tool on the **F5** menu to label the vertices as $A$, $B$, and $C$.

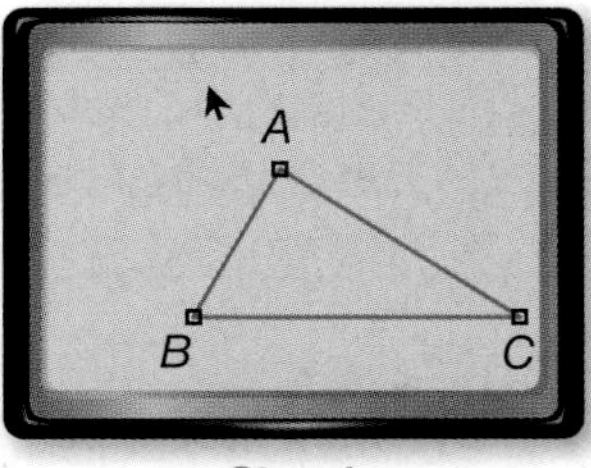

Step 1

**Step 2** Access the **distance & length** tool, shown as **D. & Length**, under **Measure** on the **F5** menu. Use the tool to measure each side of the triangle.

**Step 3** Display $AB + BC$, $AB + CA$, and $BC + CA$ by using the **Calculate** tool on the **F5** menu. Label the measures.

AB+BC 7.5
AB+CA 6.9
BC+CA 9.2
A, B, C; 2.6, 4.3, 4.9

Steps 2 and 3

**Step 4** Click and drag the vertices to change the shape of the triangle.

## Analyze the Results

1. Replace each ● with <, >, or = to make a true statement.
   $AB + BC$ ● $CA$ $\quad$ $AB + CA$ ● $BC$ $\quad$ $BC + CA$ ● $AB$

2. Click and drag the vertices to change the shape of the triangle. Then review your answers to Exercise 1. What do you observe?

3. Click on point $A$ and drag it to lie on line $BC$. What do you observe about $AB$, $BC$, and $CA$? Are $A$, $B$, and $C$ the vertices of a triangle? Explain.

4. **Make a conjecture** about the sum of the lengths of two sides of a triangle and the length of the third side.

5. Do the measurements and observations you made in the Activity and in Exercises 1–3 constitute a proof of the conjecture you made in Exercise 4? Explain.

6. Replace each ● with <, >, or = to make a true statement.
   $|AB - BC|$ ● $CA$ $\quad$ $|AB - CA|$ ● $BC$ $\quad$ $|BC - CA|$ ● $AB$

   Then click and drag the vertices to change the shape of the triangle and review your answers. What do you observe?

7. How could you use your observations to determine the possible lengths of the third side of a triangle if you are given the lengths of the other two sides?

# LESSON 7-5 The Triangle Inequality

**Then**

- You recognized and applied properties of inequalities to the relationships between the angles and sides of a triangle.

**Now**

1. Use the Triangle Inequality Theorem to identify possible triangles.
2. Prove triangle relationships using the Triangle Inequality Theorem.

**Why?**

- On a home improvement show, a designer wants to use scrap pieces of cording from another sewing project to decorate the triangular throw pillows that she and the homeowner have made. To minimize waste, she wants to use the scraps without cutting them. She selects three scraps at random and tries to form a triangle. Two such attempts are shown.

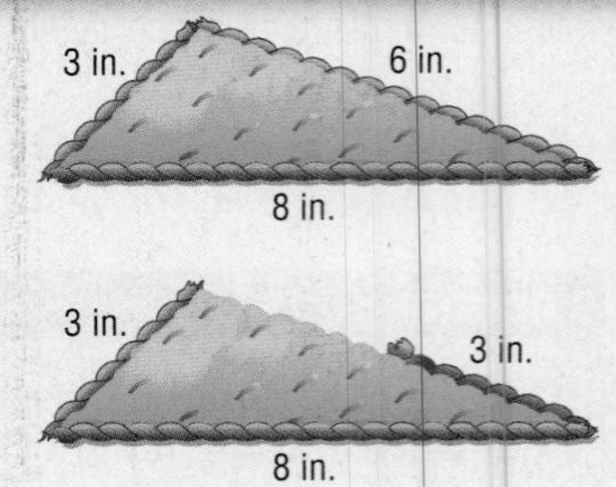

**Common Core State Standards**

**Content Standards**

G.CO.10 Prove theorems about triangles.

G.MG.3 Apply geometric methods to solve problems (e.g., designing an object or structure to satisfy physical constraints or minimize cost; working with typographic grid systems based on ratios). ★

**Mathematical Practices**

1. Make sense of problems and persevere in solving them.
2. Reason abstractly and quantitatively.

## 1 The Triangle Inequality

While a triangle is formed by three segments, a special relationship must exist among the lengths of the segments in order for them to form a triangle.

### Theorem 7.11 Triangle Inequality Theorem

The sum of the lengths of any two sides of a triangle must be greater than the length of the third side.

**Examples** $PQ + QR > PR$
$QR + PR > PQ$
$PR + PQ > QR$

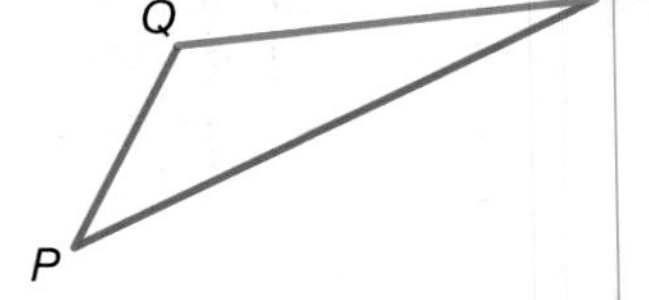


You will prove Theorem 7.11 in Exercise 23.

To show that it is not possible to form a triangle with three side lengths, you need only show that one of the three triangle inequalities is not true.

### Example 1 Identify Possible Triangles Given Side Lengths

**Is it possible to form a triangle with the given side lengths? If not, explain why not.**

**a. 8 in., 15 in., 17 in.**

Check each inequality.

$8 + 15 \overset{?}{>} 17$, $23 > 17$ ✓ $8 + 17 \overset{?}{>} 15$, $25 > 15$ ✓ $15 + 17 \overset{?}{>} 8$, $32 > 8$ ✓

Since the sum of each pair of side lengths is greater than the third side length, sides with lengths 8, 15, and 17 inches will form a triangle.

**b. 6 m, 8 m, 14 m**

$6 + 8 \overset{?}{>} 14$

$14 \not> 14$ ✗

Since the sum of one pair of side lengths is not greater than the third side length, sides with lengths 6, 8, and 14 meters will not form a triangle.

**Guided Practice**

**1A.** 15 yd, 16 yd, 30 yd **1B.** 2 ft, 8 ft, 11 ft

When the lengths of two sides of a triangle are known, the third side can be any length in a range of values. You can use the Triangle Inequality Theorem to determine the range of possible lengths for the third side.

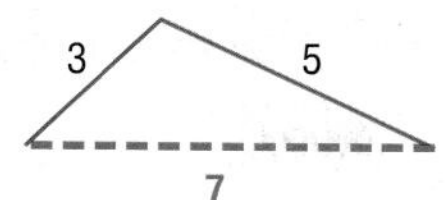


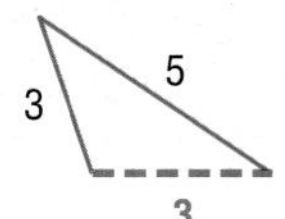

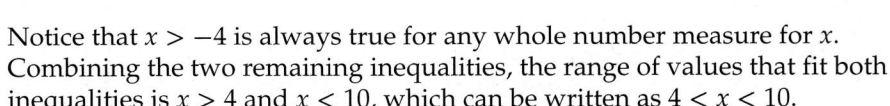

## Standardized Test Example 2 Find Possible Side Lengths

**If the measures of two sides of a triangle are 3 feet and 7 feet, which is the *least* possible whole number measure for the third side?**

**A** 3 ft  **B** 4 ft  **C** 5 ft  **D** 10 ft

> **Test-Taking Tip**
> **Testing Choices** If you are short on time, you can test each choice to find the correct answer and eliminate any remaining choices.

### Read the Test Item

You need to determine which value is the least possible measure for the third side of a triangle with sides that measure 3 feet and 7 feet.

### Solve the Test Item

To determine the least possible measure from the choices given, first determine the range of possible measures for the third side.

Draw a diagram and let $x$ represent the length of the third side.

3 ft
7 ft
$x$ ft

Next, set up and solve each of the three triangle inequalities.

| $3 + 7 > x$ | $3 + x > 7$ | $x + 7 > 3$ |
|---|---|---|
| $10 > x$ or $x < 10$ | $x > 4$ | $x > -4$ |

Notice that $x > -4$ is always true for any whole number measure for $x$. Combining the two remaining inequalities, the range of values that fit both inequalities is $x > 4$ and $x < 10$, which can be written as $4 < x < 10$.

The least whole number value between 4 and 10 is 5. So the correct answer is choice C.

> **Reading Math**
> **Multiple Inequality Symbols** The compound inequality $4 < x < 10$ is read *$x$ is between 4 and 10.*

### Guided Practice

**2.** Which of the following could *not* be the value of $n$?

**F** 7  **H** 13

**G** 10  **J** 22

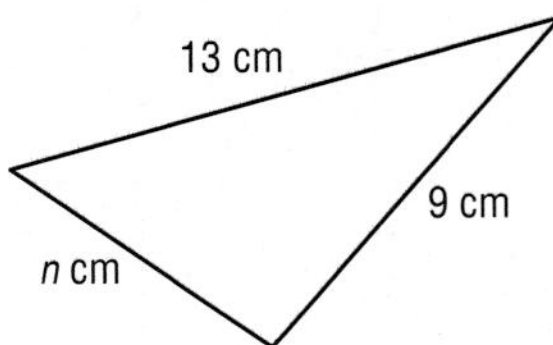


## 2 Proofs Using the Triangle Inequality Theorem

You can use the Triangle Inequality Theorem as a reason in proofs.

## Real-World Example 3 Proof Using Triangle Inequality Theorem

**Real-WorldLink**

A direct flight is not the same as a nonstop flight. For a direct flight, passengers do not change planes, but the plane may make one or more stops before continuing to its final destination.

**TRAVEL The distance from Colorado Springs, Springs, Colorado, to Abilene, Texas, is the same as the distance from Colorado Springs to Tulsa, Oklahoma. Prove that a direct flight from Colorado Springs to Tulsa through Lincoln, Nebraska, is a greater distance than a nonstopflight from Colorado Springs to Abilene.**

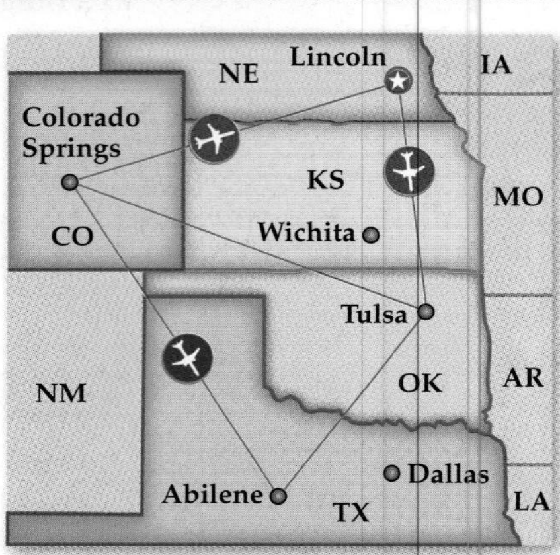


Draw a simpler diagram of the situation and label the diagram. Draw in side $\overline{LT}$ to form $\triangle CTL$.

**Given:** $CA = CT$

**Prove:** $CL + LT > CA$

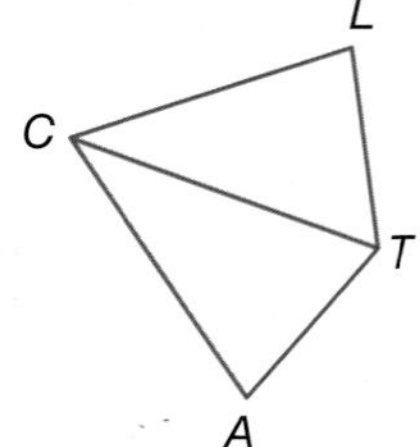


**Proof:**

| Statements | Reasons |
|---|---|
| **1.** $CA = CT$ | **1.** Given |
| **2.** $CL + LT > CT$ | **2.** Triangle Inequality Theorem |
| **3.** $CL + LT > CA$ | **3.** Substitution |

**GuidedPractice**

**3.** Write a two-column proof.

**Given:** $GL = LK$

**Prove:** $JH + GH > JK$

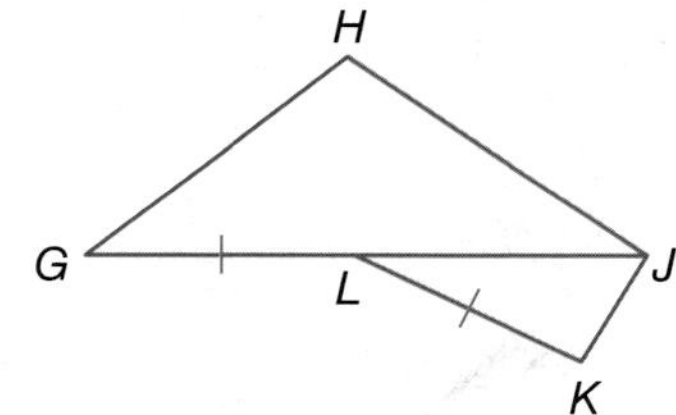


## Check Your Understanding

**Example 1** **Is it possible to form a triangle with the given side lengths? If not, explain why not.**

**1** 5 cm, 7 cm, 10 cm **2.** 3 in., 4 in., 8 in. **3.** 6 m, 14 m, 10 m

**Example 2** **4. MULTIPLE CHOICE** If the measures of two sides of a triangle are 5 yards and 9 yards, what is the least possible measure of the third side if the measure is an integer?

**A** 4 yd **B** 5 yd **C** 6 yd **D** 14 yd

**Example 3** **5. PROOF** Write a two-column proof.

**Given:** $\overline{XW} \cong \overline{YW}$

**Prove:** $YZ + ZW > XW$

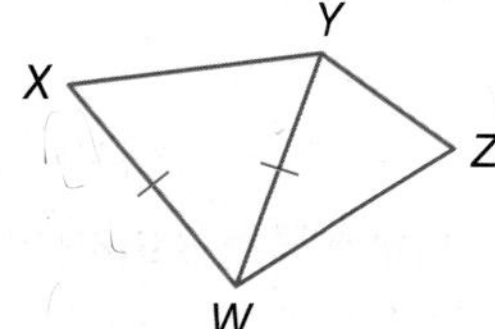

## Practice and Problem Solving

**Example 1** **Is it possible to form a triangle with the given side lengths? If not, explain why not.**

**6.** 4 ft, 9 ft, 15 ft

**7.** 11 mm, 21 mm, 16 mm

**8.** 9.9 cm, 1.1 cm, 8.2 cm

**9.** 2.1 in., 4.2 in., 7.9 in.

**10.** $2\frac{1}{2}$ m, $1\frac{3}{4}$ m, $5\frac{1}{8}$ m

**11.** $1\frac{1}{5}$ km, $4\frac{1}{2}$ km, $3\frac{3}{4}$ km

**Example 2** **Find the range for the measure of the third side of a triangle given the measures of two sides.**

**12.** 4 ft, 8 ft

**13.** 5 m, 11 m

**14.** 2.7 cm, 4.2 cm

**15.** 3.8 in., 9.2 in.

**16.** $\frac{1}{2}$ km, $3\frac{1}{4}$ km

**17.** $2\frac{1}{3}$ yd, $7\frac{2}{3}$ yd

**Example 3** **PROOF** **Write a two-column proof.**

**18. Given:** $\angle BCD \cong \angle CDB$

**Prove:** $AB + AD > BC$

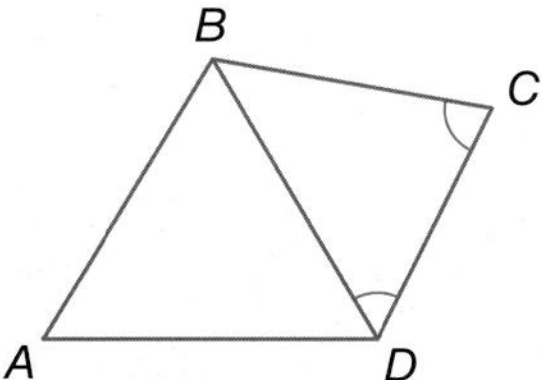


**19. Given:** $\overline{JL} \cong \overline{LM}$

**Prove:** $KJ + KL > LM$

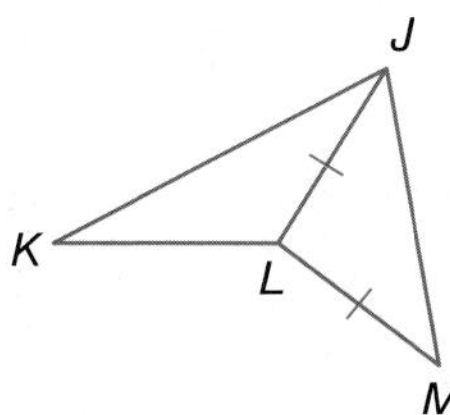


**CCSS SENSE-MAKING** **Determine the possible values of $x$.**

**20.**

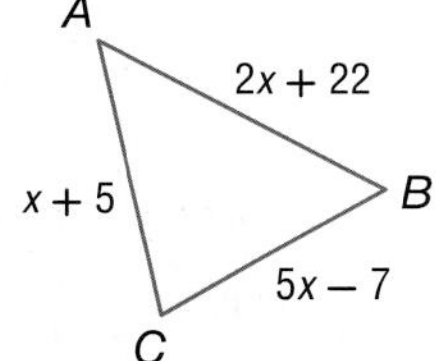


**21.**

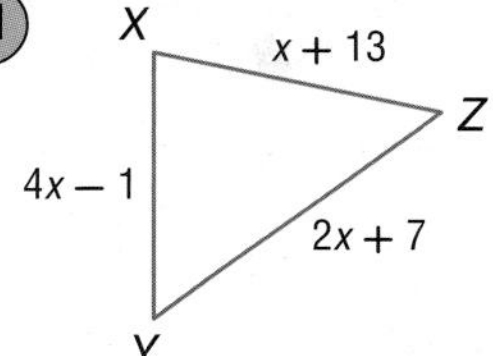


**22. DRIVING** Takoda wants to take the most efficient route from his house to a soccer tournament at The Sportsplex. He can take County Line Road or he can take Highway 4 and then Route 6 to the get to The Sportsplex.

**a.** Which of the two possible routes is the shortest? Explain your reasoning.

**b.** Suppose Takoda always drives below the speed limit. If the speed limit on County Line Road is 30 miles per hour and on both Highway 4 and Route 6 it is 55 miles per hour, which route will be faster? Explain.

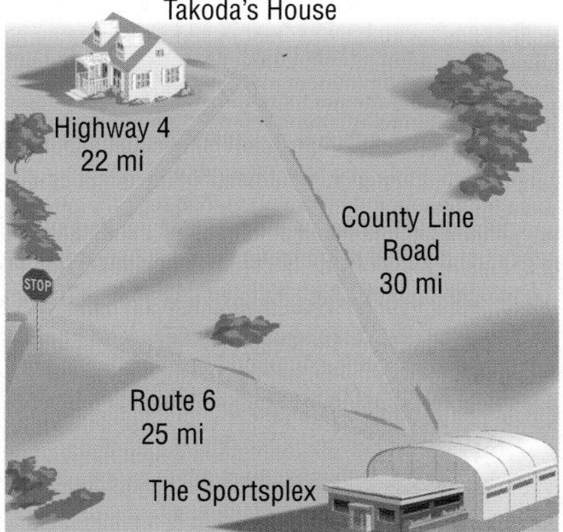


**23. PROOF** **Write a two-column proof.**

**Given:** $\triangle ABC$

**Prove:** $AC + BC > AB$ (Triangle Inequality Theorem)

(*Hint*: Draw auxiliary segment $\overline{CD}$, so that $C$ is between $B$ and $D$ and $\overline{CD} \cong \overline{AC}$.)

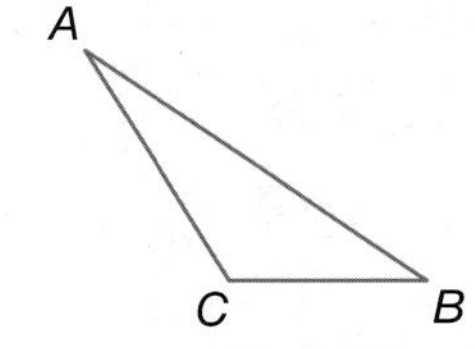

24. **SCHOOL** When Toya goes from science class to math class, she usually stops at her locker. The distance from her science classroom to her locker is 90 feet, and the distance from her locker to her math classroom is 110 feet. What are the possible distances from science class to math class if she takes the hallway that goes directly between the two classrooms?

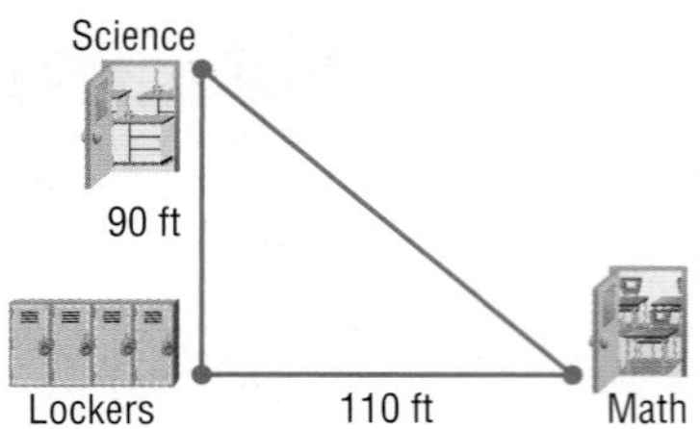


**Find the range of possible measures of $x$ if each set of expressions represents measures of the sides of a triangle.**

25. $x, 4, 6$

26. $8, x, 12$

27. $x + 1, 5, 7$

28. $x - 2, 10, 12$

29. $x + 2, x + 4, x + 6$

30. $x, 2x + 1, x + 4$

31. **DRAMA CLUB** Anthony and Catherine are working on a ramp up to the stage for the drama club's next production. Anthony's sketch of the ramp is shown below. Catherine is concerned about the measurements and thinks they should recheck the measures before they start cutting the wood. Is Catherine's concern valid? Explain your reasoning.

32. **CCSS SENSE-MAKING** Aisha is riding her bike to the park and can take one of two routes. The most direct route from her house is to take Main Street, but it is safer to take Route 3 and then turn right on Clay Road as shown. The additional distance she will travel if she takes Route 3 to Clay Road is between how many miles?

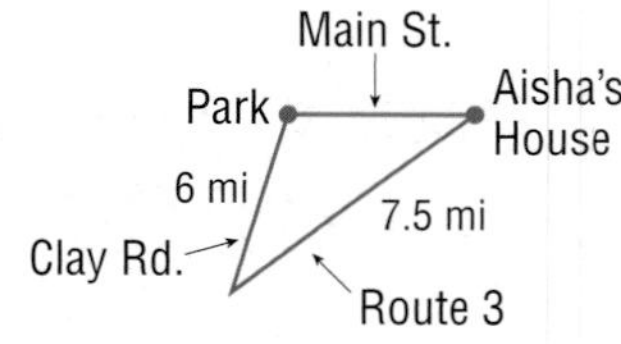


33. **DESIGN** Carlota designed an awning that she and her friends could take to the beach. Carlota decides to cover the top of the awning with material that will drape 6 inches over the front. What length of material should she buy to use with her design so that it covers the top of the awning, including the drape, when the supports are open as far as possible? Assume that the width of the material is sufficient to cover the awning.

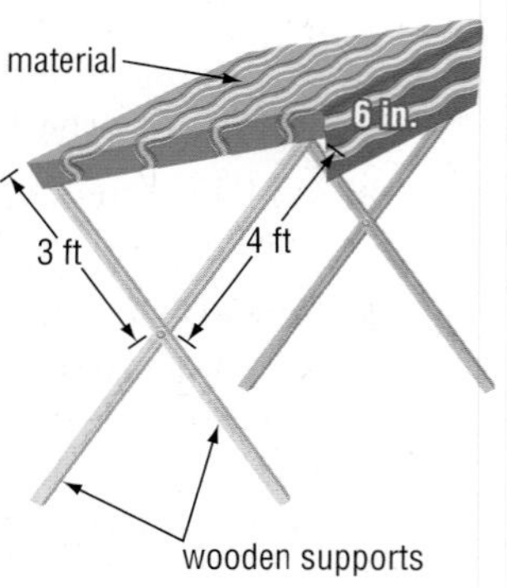


**ESTIMATION** **Without using a calculator, determine if it is possible to form a triangle with the given side lengths. Explain.**

34. $\sqrt{8}$ ft, $\sqrt{2}$ ft, $\sqrt{35}$ ft

35. $\sqrt{99}$ yd, $\sqrt{48}$ yd, $\sqrt{65}$ yd

36. $\sqrt{3}$ m, $\sqrt{15}$ m, $\sqrt{24}$ m

37. $\sqrt{122}$ in., $\sqrt{5}$ in., $\sqrt{26}$ in.

**CCSS REASONING** **Determine whether the given coordinates are the vertices of a triangle. Explain.**

**38.** $X(1, -3)$, $Y(6, 1)$, $Z(2, 2)$

**39.** $F(-4, 3)$, $G(3, -3)$, $H(4, 6)$

**40.** $J(-7, -1)$, $K(9, -5)$, $L(21, -8)$

**41.** $Q(2, 6)$, $R(6, 5)$, $S(1, 2)$

**42.** **MULTIPLE REPRESENTATIONS** In this problem, you will use inequalities to make comparisons between the sides and angles of two triangles.

**a. Geometric** Draw three pairs of triangles that have two pairs of congruent sides and one pair of sides that is not congruent. Mark each pair of congruent sides. Label each triangle pair $ABC$ and $DEF$, where $\overline{AB} \cong \overline{DE}$ and $\overline{AC} \cong \overline{DF}$.

**b. Tabular** Copy the table below. Measure and record the values of $BC$, $m\angle A$, $EF$, and $m\angle D$ for each triangle pair.

| Triangle Pair | $BC$ | $m\angle A$ | $EF$ | $m\angle D$ |
|---|---|---|---|---|
| 1 | | | | |
| 2 | | | | |
| 3 | | | | |

**c. Verbal** Make a conjecture about the relationship between the angles opposite the noncongruent sides of a pair of triangles that have two pairs of congruent legs.

## H.O.T. Problems Use Higher-Order Thinking Skills

**43.** **CHALLENGE** What is the range of possible perimeters for figure $ABCDE$ if $AC = 7$ and $DC = 9$? Explain your reasoning.

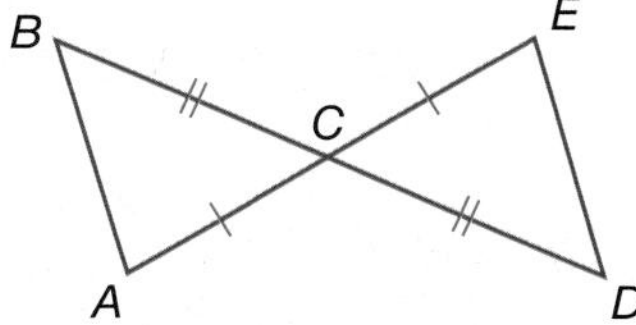


**44.** **REASONING** What is the range of lengths of each leg of an isosceles triangle if the measure of the base is 6 inches? Explain.

**45.** **WRITING IN MATH** What can you tell about a triangle when given three side lengths? Include at least two items.

**46.** **CHALLENGE** The sides of an isosceles triangle are whole numbers, and its perimeter is 30 units. What is the probability that the triangle is equilateral?

**47.** **OPEN ENDED** The length of one side of a triangle is 2 inches. Draw a triangle in which the 2-inch side is the shortest side and one in which the 2-inch side is the longest side. Include side and angle measures on your drawing.

**48.** **WRITING IN MATH** Suppose your house is $\frac{3}{4}$ mile from a park and the park is 1.5 miles from a shopping center.

**a.** If your house, the park, and the shopping center are noncollinear, what do you know about the distance from your house to the shopping center? Explain your reasoning.

**b.** If the three locations are collinear, what do you know about the distance from your house to the shopping center? Explain your reasoning.

## Standardized Test Practice

**49.** If $\overline{DC}$ is a median of $\triangle ABC$ and $m\angle 1 > m\angle 2$, which of the following statements is not true?

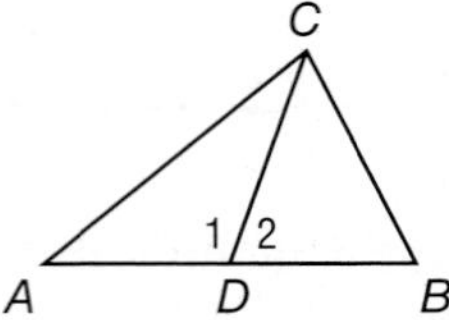


**A** $AD = BD$ **C** $AC > BC$

**B** $m\angle ADC = m\angle BDC$ **D** $m\angle 1 > m\angle B$

**50. SHORT RESPONSE** A high school soccer team has a goal of winning at least 75% of their 15 games this season. In the first three weeks, the team has won 5 games. How many more games must the team win to meet their goal?

**51.** Which of the following is a logical conclusion based on the statement and its converse below?

**Statement:** If a polygon is a rectangle, then it has four sides.

**Converse:** If a polygon has four sides, then it is a rectangle.

**F** The statement and its converse are both true.

**G** The statement is false; the converse is false.

**H** The statement is true; the converse is false.

**J** The statement is false; the converse is true.

**52. SAT/ACT** When 7 is subtracted from $14w$, the result is $z$. Which of the following equations represents this statement?

**A** $7 - 14w = z$

**B** $z = 14w + 7$

**C** $7 - z = 14w$

**D** $z = 14w - 7$

**E** $7 + 14w = 7z$

## Spiral Review

**State the assumption you would make to start an indirect proof of each statement.** (Lesson 7-4)

**53.** If $4y + 17 = 41$, then $y = 6$.

**54.** If two lines are cut by a transversal and a pair of alternate interior angles are congruent, then the two lines are parallel.

**55. GEOGRAPHY** The distance between San Jose, California, and Las Vegas, Nevada, is about 375 miles. The distance from Las Vegas to Carlsbad, California, is about 243 miles. Use the Triangle Inequality Theorem to find the possible distance between San Jose and Carlsbad. (Lesson 7-3)

**Find $x$ so that $m \parallel n$. Identify the postulate or theorem you used.** (Lesson 5-6)

**56.**

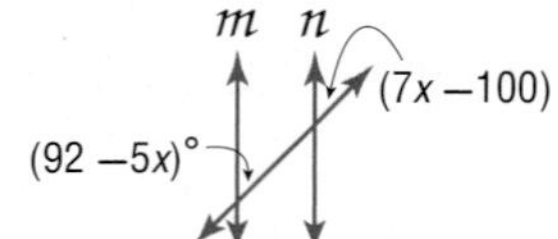


**57.**

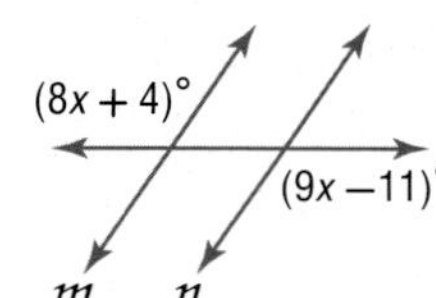


**58.**

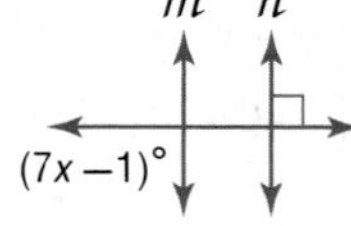


## Skills Review

**Find $x$ and the measures of the unknown sides of each triangle.**

**59.**

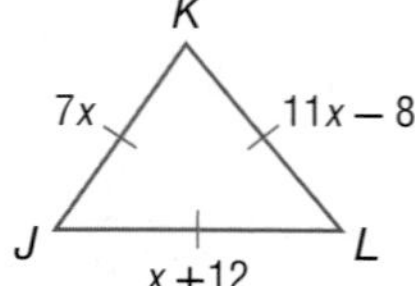


**60.**

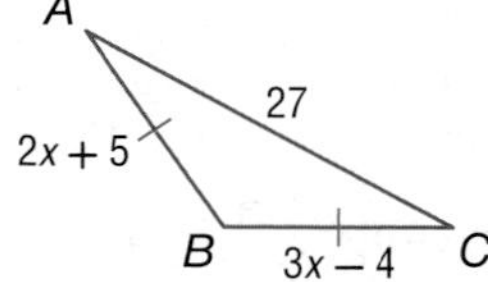


**61.**

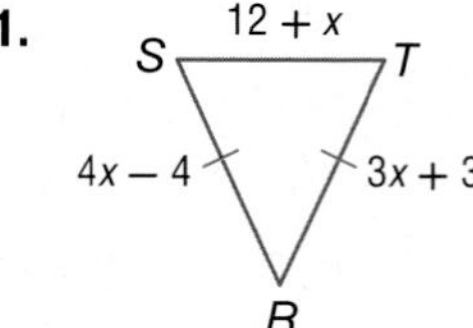

LESSON 7-6

# Inequalities in Two Triangles

**Then**

- You used inequalities to make comparisons in one triangle.

**Now**

1. Apply the Hinge Theorem or its converse to make comparisons in two triangles.
2. Prove triangle relationships using the Hinge Theorem or its converse.

**Why?**

- A car jack is used to lift a car. The jack shown below is one of the simplest still in use today. Notice that as the jack is lowered, the legs of isosceles $\triangle ABC$ remain congruent, but the included angle $A$ widens and $\overline{BC}$, the side opposite $\angle A$, lengthens.

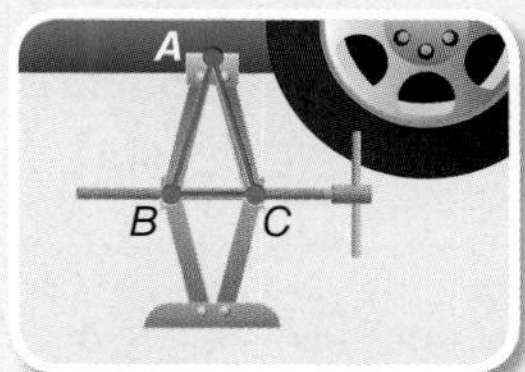

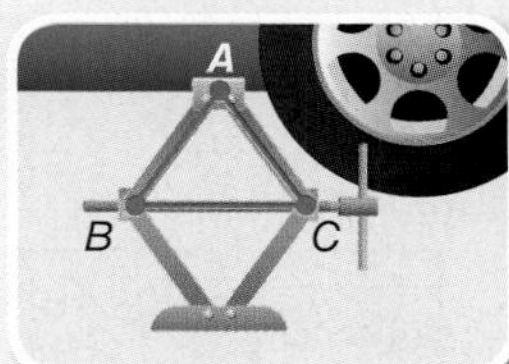

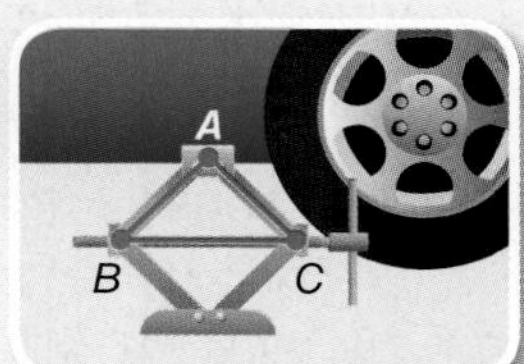

**Common Core State Standards**

**Content Standards**

G.CO.10 Prove theorems about triangles.

**Mathematical Practices**

3 Construct viable arguments and critique the reasoning of others.

1 Make sense of problems and persevere in solving them.

## 1 Hinge Theorem

The observation in the example above is true of any type of triangle and illustrates the following theorems.

### Theorems Inequalities in Two Triangles

**7.13 Hinge Theorem** If two sides of a triangle are congruent to two sides of another triangle, and the included angle of the first is larger than the included angle of the second triangle, then the third side of the first triangle is longer than the third side of the second triangle.

**Example:** If $\overline{AB} \cong \overline{FG}$, $\overline{AC} \cong \overline{FH}$, and $m\angle A > m\angle F$, then $BC > GH$.

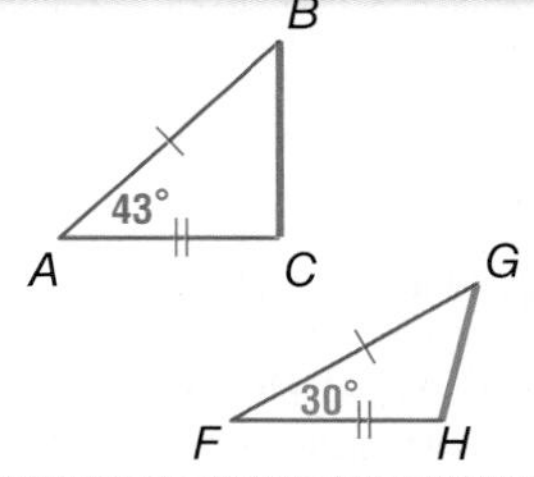


**7.14 Converse of the Hinge Theorem** If two sides of a triangle are congruent to two sides of another triangle, and the third side in the first is longer than the third side in the second triangle, then the included angle measure of the first triangle is greater than the included angle measure in the second triangle.

**Example:** If $\overline{JL} \cong \overline{PR}$, $\overline{KL} \cong \overline{QR}$, and $PQ > JK$, then $m\angle R > m\angle L$.

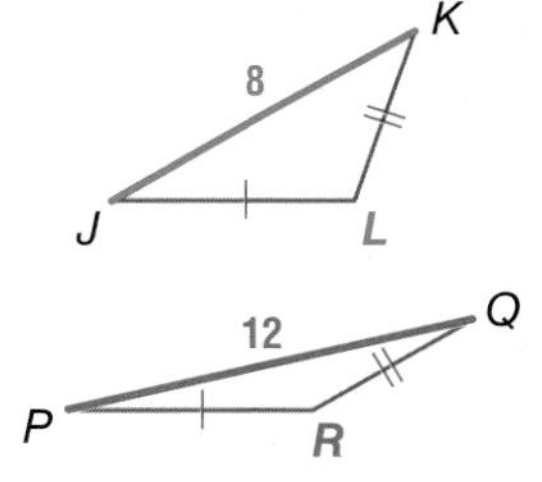


The proof of Theorem 7.13 is on p. 454. You will prove Theorem 7.14 in Exercise 28.

### Example 1 Use the Hinge Theorem and its Converse

**Compare the given measures.**

**a.** $WX$ **and** $XY$

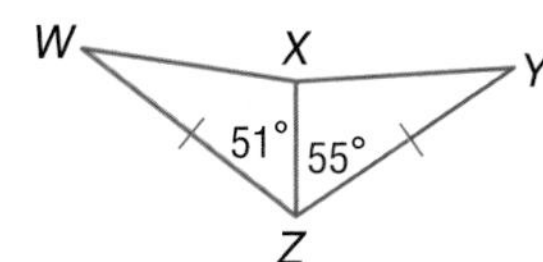


In $\triangle WXZ$ and $\triangle YXZ$, $\overline{WZ} \cong \overline{YZ}$, $\overline{XZ} \cong \overline{XZ}$, and $\angle YZX > \angle WZX$. By the Hinge Theorem, $m\angle WZX < m\angle YZX$, so $WX < XY$.

**b.** $m\angle FCD$ **and** $m\angle BFC$

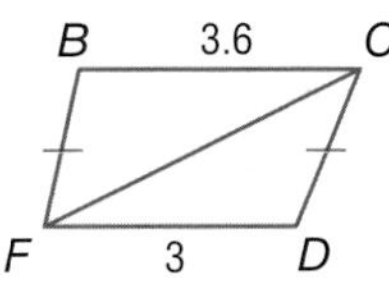


In $\triangle BCF$ and $\triangle DFC$, $\overline{BF} \cong \overline{DC}$, $\overline{FC} \cong \overline{CF}$, and $BC > FD$. By the Converse of the Hinge Theorem, $\angle BFC > \angle DCF$.

## Guided Practice

**Compare the given measures.**

**1A.** $JK$ and $MQ$

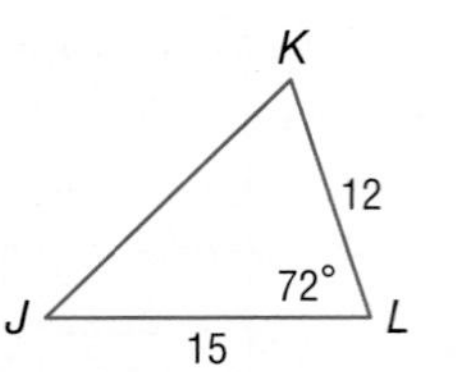


**1B.** $m\angle SRT$ and $m\angle VRT$

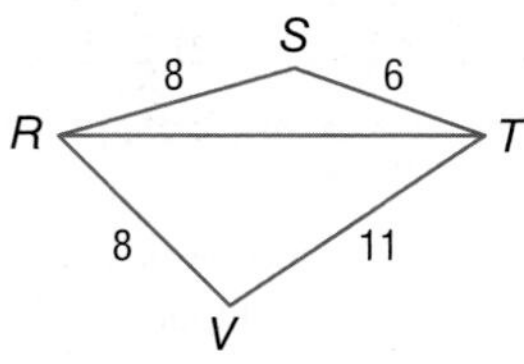


## Proof Hinge Theorem

**Given:** $\triangle ABC$ and $\triangle DEF$,
$\overline{AC} \cong \overline{DF}$, $\overline{BC} \cong \overline{EF}$
$m\angle F > m\angle C$

**Prove:** $DE > AB$

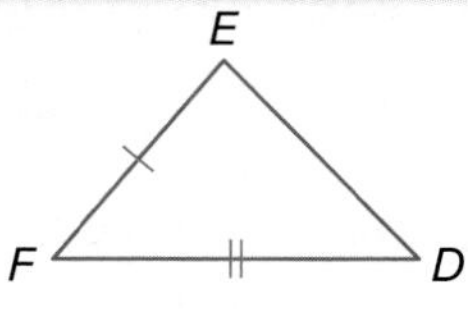


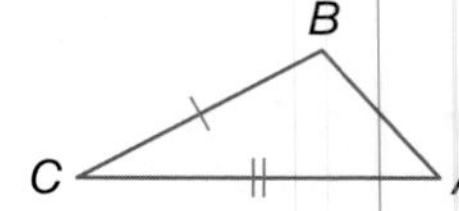


**Proof:**

We are given that $\overline{AC} \cong \overline{DF}$ and $\overline{BC} \cong \overline{EF}$. We also know that $m\angle F > m\angle C$.

Draw auxiliary ray $FP$ such that $m\angle DFP = m\angle C$ and that $\overline{PF} \cong \overline{BC}$. This leads to two cases.

**Case 1** $P$ lies on $\overline{DE}$.

Then $\triangle FPD \cong \triangle CBA$ by SAS. Thus, $PD = BA$ by CPCTC and the definition of congruent segments.

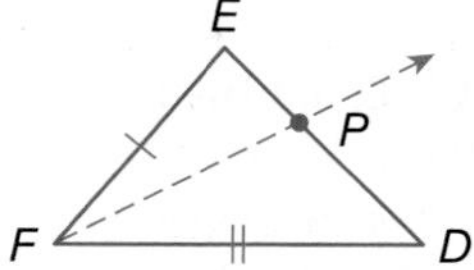


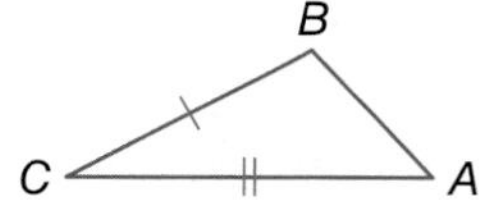


By the Segment Addition Postulate, $DE = EP + PD$. Also, $DE > PD$ by the definition of inequality. Therefore, $DE > AB$ by substitution.

**Case 2** $P$ does not lie on $\overline{DE}$.

Then let the intersection of $\overline{FP}$ and $\overline{ED}$ be point $T$, and draw another auxiliary segment $\overline{FQ}$ such that $Q$ is on $\overline{DE}$ and $\angle EFQ \cong \angle QFP$. Then draw auxiliary segments $\overline{PD}$ and $\overline{PQ}$.

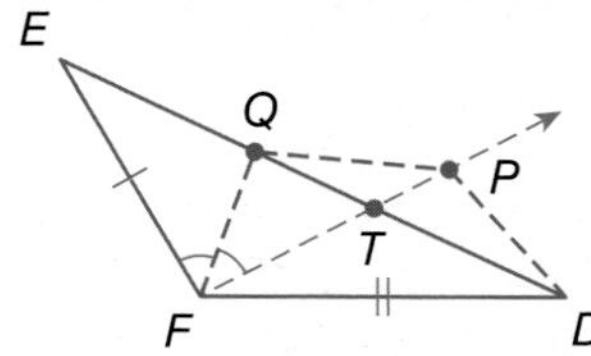


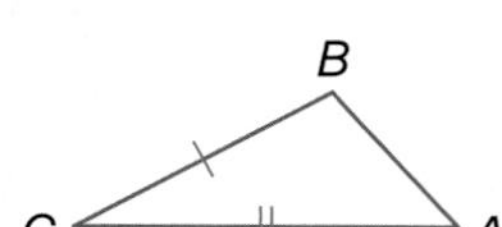


Since $\overline{FP} \cong \overline{BC}$ and $\overline{BC} \cong \overline{EF}$, we have $\overline{FP} \cong \overline{EF}$ by the Transitive Property. Also $\overline{QF}$ is congruent to itself by the Reflexive Property. Thus, $\triangle EFQ \cong \triangle PFQ$ by SAS. By CPCTC, $\overline{EQ} \cong \overline{PQ}$ or $EQ = PQ$. Also, $\triangle FPD \cong \triangle CBA$ by SAS. So, $\overline{PD} \cong \overline{BA}$ by CPCTC and $PD = BA$.

In $\triangle QPD$, $QD + PQ > PD$ by the Triangle Inequality Theorem. By substitution, $QD + EQ >$ PD. Since $ED = QD + EQ$ by the Segment Addition Postulate, $ED > PD$. Using substitution, $ED > BA$ or $DE > AB$.

**StudyTip**

**SAS and SSS Inequality Theorem** The Hinge Theorem is also called the SAS Inequality Theorem. The Converse of the Hinge Theorem is also called the SSS Inequality Theorem.

**Real-WorldLink**

There are over 225,000 miles of groomed and marked snowmobile trails in North America.

**Source:** International Snowmobile Manufacturers Association

You can use the Hinge Theorem to solve real-world problems.

### Real-World Example 2 Use the Hinge Theorem

**SNOWMOBILING Two groups of snowmobilers leave from the same base camp. Group A goes 7.5 miles due west and then turns 35° north of west and goes 5 miles. Group B goes 7.5 miles due east and then turns 40° north of east and goes 5 miles. At this point, which group is farther from the base camp? Explain your reasoning.**

**Understand** Using the sets of directions given in the problem, you need to determine which snowmobile group is farther from the base camp. A turn of 35° north of west is correctly interpreted as shown.

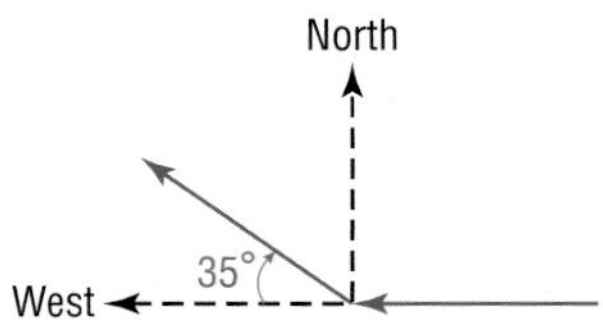


**Plan** Draw a diagram of the situation.

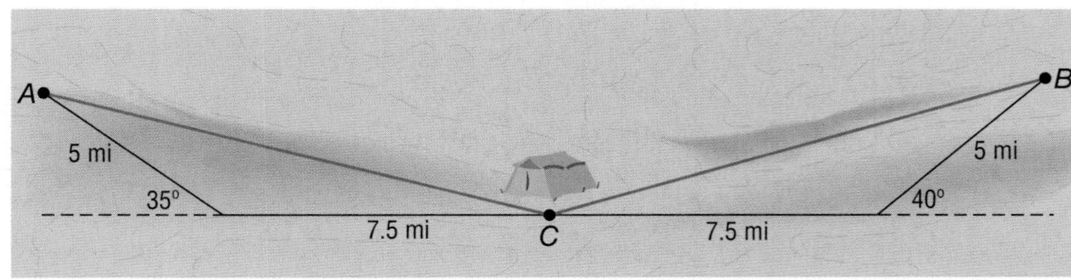


The paths taken by each group and the straight-line distance back to the camp form two triangles. Each group goes 7.5 miles and then turns and goes 5 miles.

Use linear pairs to find the measures of the included angles. Then apply the Hinge Theorem to compare the distance each group is from base camp.

**Solve** The included angle for the path made by Group A measures 180 − 35 or 145. The included angle for the path made by Group B is 180 − 40 or 140.

Since 145 > 140, $AC > BC$ by the Hinge Theorem. So Group A is farther from the base camp.

**Check** Group B turned 5° more than Group A did back toward base camp, so they should be closer to base camp than Group A. Thus, Group A should be farther from the base camp. ✓

**Problem-SolvingTip**

Draw a Diagram Draw a diagram to help you see and correctly interpret a problem that has been described in words.

### GuidedPractice

**2A. SKIING** Two groups of skiers leave from the same lodge. Group A goes 4 miles due east and then turns 70° north of east and goes 3 miles. Group B goes 4 miles due west and then turns 75° north of west and goes 3 miles. At this point, which group is *farther* from the lodge? Explain your reasoning.

**2B. SKIING** In problem 2A, suppose Group A instead went 4 miles west and then turned 45° north of west and traveled 3 miles. Which group would be *closer* to the lodge? Explain your reasoning.

When the included angle of one triangle is greater than the included angle in a second triangle, the Converse of the Hinge Theorem is used.

**StudyTip**

**Using Additional Facts** When finding a range for the possible values for $x$, you may need to use one of the following facts.

- The measure of any angle is always greater than 0 and less than 180.
- The measure of any segment is always greater than 0.

## Example 3 Apply Algebra to the Relationships in Triangles

**ALGEBRA Find the range of possible values for $x$.**

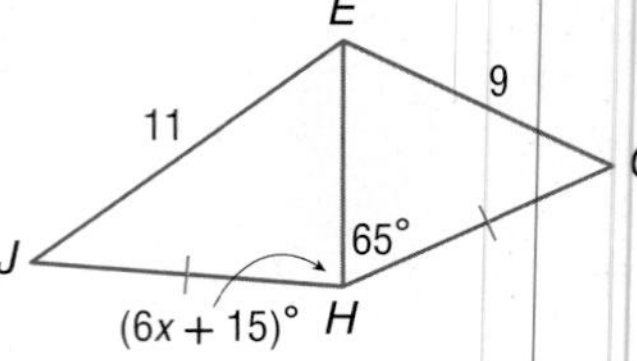


**Step 1** From the diagram, we know that $\overline{JH} \cong \overline{GH}$, $\overline{EH} \cong \overline{EH}$, and $JE > EG$.

| | |
|---|---|
| $m\angle JHE > m\angle EHG$ | Converse of the Hinge Theorem |
| $6x + 15 > 65$ | Substitution |
| $x > 8\frac{1}{3}$ | Solve for $x$. |

**Step 2** Use the fact that the measure of any angle in a triangle is less than 180 to write a second inequality.

| | |
|---|---|
| $m\angle JHE < 180$ | |
| $6x + 15 < 180$ | Substitution |
| $x < 27.5$ | Solve for $x$. |

**Step 3** Write $x > 8\frac{1}{3}$ and $x < 27.5$ as the compound inequality $8\frac{1}{3} < x < 27.5$.

### GuidedPractice

**3.** Find the range of possible values for $x$.

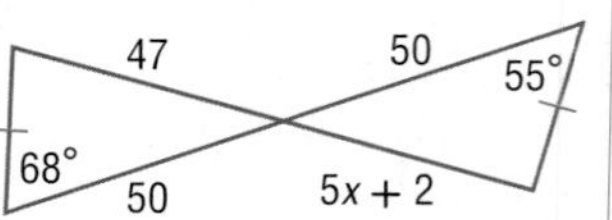


## 2 Prove Relationships In Two Triangles

You can use the Hinge Theorem and its converse to prove relationships in two triangles.

## Example 4 Prove Triangle Relationships Using Hinge Theorem

**Write a two-column proof.**

**Given:** $\overline{AB} \cong \overline{AD}$

**Prove:** $EB > ED$

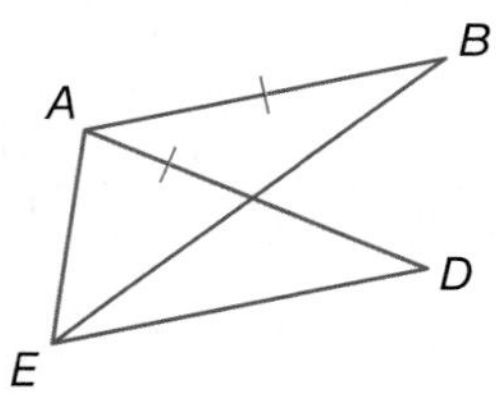


**Proof:**

| Statements | Reasons |
|---|---|
| **1.** $\overline{AB} \cong \overline{AD}$ | **1.** Given |
| **2.** $\overline{AE} \cong \overline{AE}$ | **2.** Reflexive Property |
| **3.** $m\angle EAB = m\angle EAD + m\angle DAB$ | **3.** Angle Addition Postulate |
| **4.** $m\angle EAB > m\angle EAD$ | **4.** Definition of Inequality |
| **5.** $EB > ED$ | **5.** Hinge Theorem |

### GuidedPractice

**4.** Write a two-column proof.

**Given:** $\overline{RQ} \cong \overline{ST}$

**Prove:** $RS > TQ$

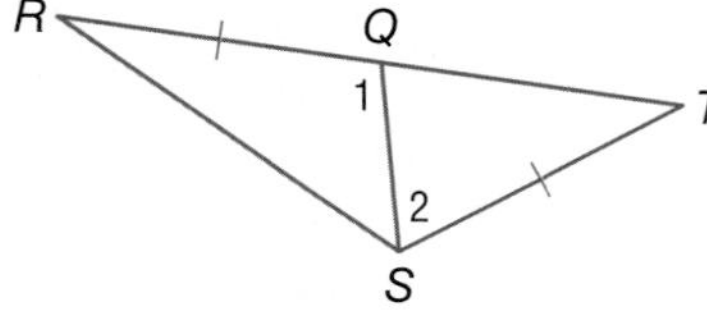

## Example 5 Prove Relationships Using Converse of Hinge Theorem

**Write a flow proof.**

**Given:** $T$ is the midpoint of $\overline{ZX}$.
$\overline{ST} \cong \overline{WT}$
$SZ > WX$

**Prove:** $m\angle XTR > m\angle ZTY$

**Flow Proof:**

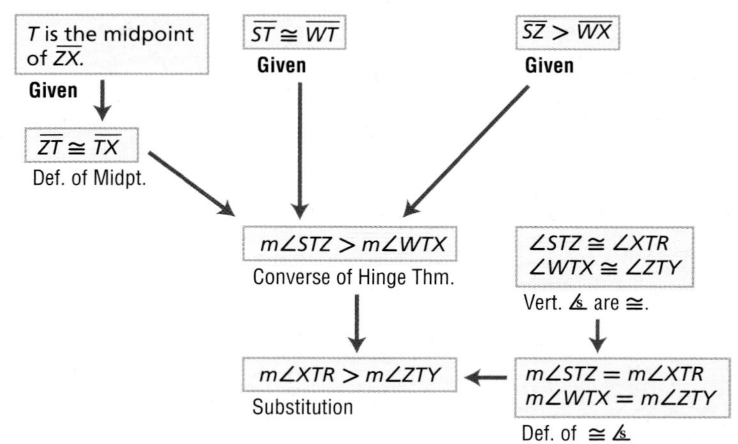


### GuidedPractice

**5.** Write a two-column proof.

**Given:** $\overline{NK}$ is a median of $\triangle JMN$.
$JN > NM$

**Prove:** $m\angle 1 > m\angle 2$

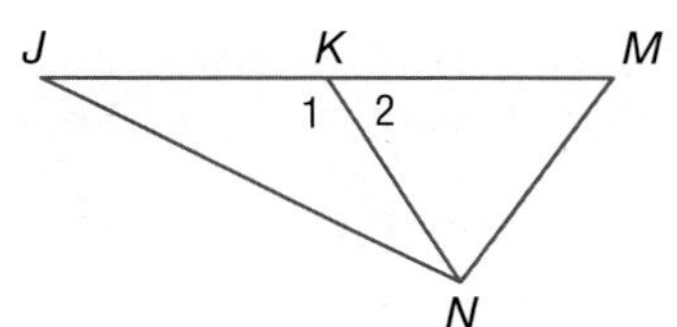


## Check Your Understanding

**Example 1** **Compare the given measures.**

**1.** $m\angle ACB$ and $m\angle GDE$

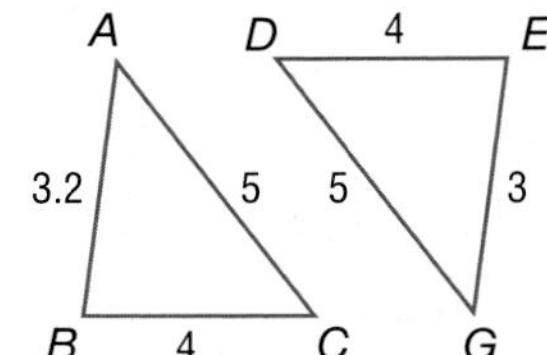


**2.** $JL$ and $KM$

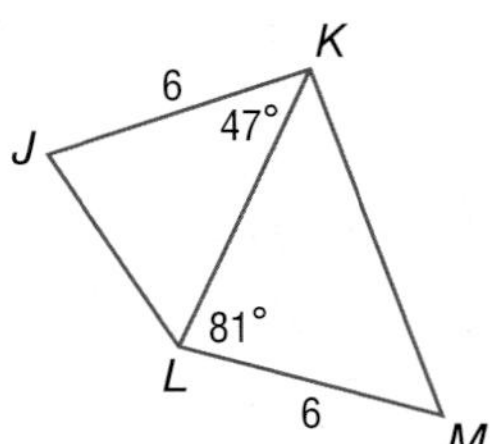


**3** $QT$ and $ST$

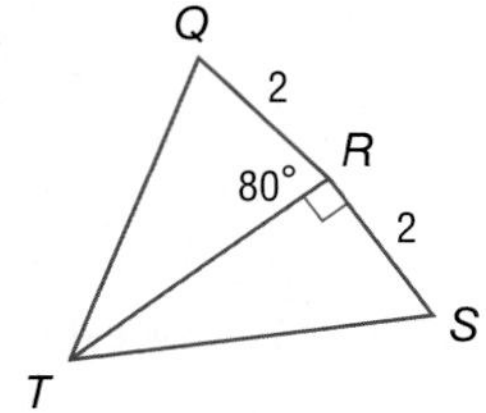


**4.** $m\angle XWZ$ and $m\angle YZW$

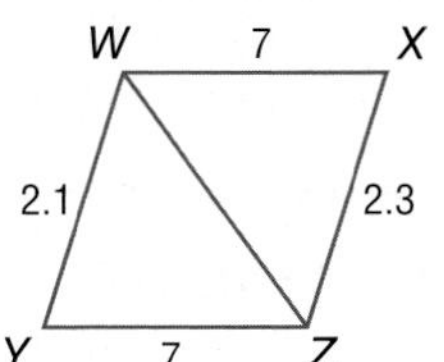

**Example 2**

5. **SWINGS** The position of the swing changes based on how hard the swing is pushed.

   a. Which pairs of segments are congruent?

   b. Is the measure of $\angle A$ or the measure of $\angle D$ greater? Explain.

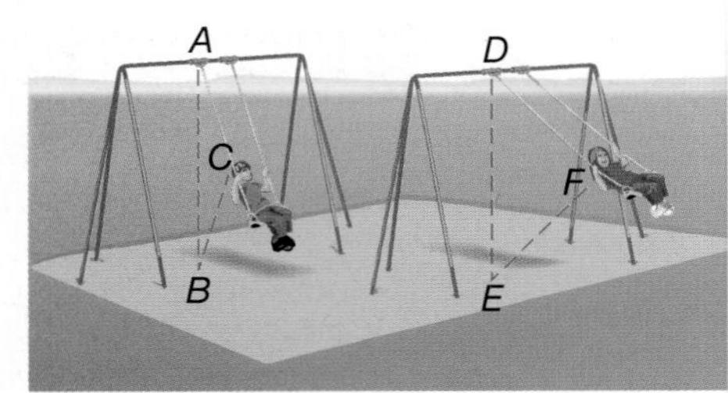


**Example 3** **Find the range of possible values for $x$.**

6.

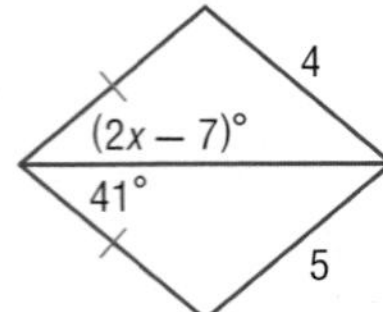


7.

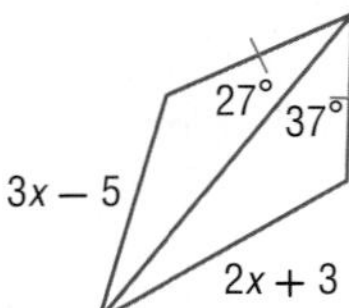


**Examples 4–5** **CCSS ARGUMENTS** **Write a two-column proof.**

8. **Given:** $\triangle YZX$
   $\overline{YZ} \cong \overline{XW}$

   **Prove:** $ZX > YW$

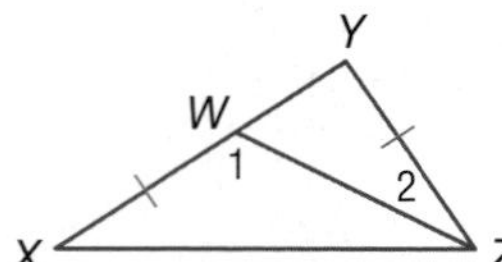


9. **Given:** $\overline{AD} \cong \overline{CB}$
   $DC < AB$

   **Prove:** $m\angle CBD < m\angle ADB$

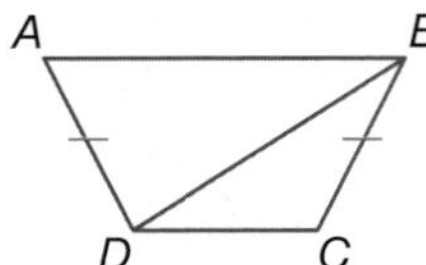


## Practice and Problem Solving

**Example 1** **Compare the given measures.**

10. $m\angle BAC$ and $m\angle DGE$

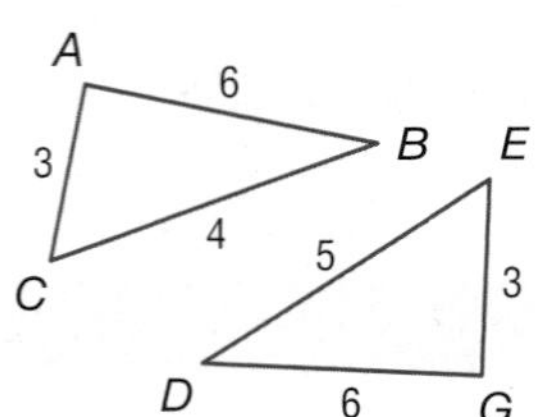


11. $m\angle MLP$ and $m\angle TSR$

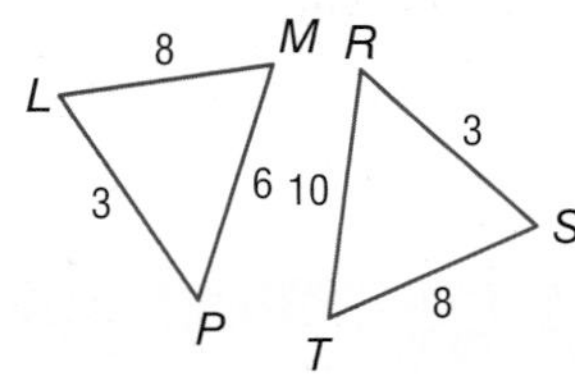


12. $SR$ and $XY$

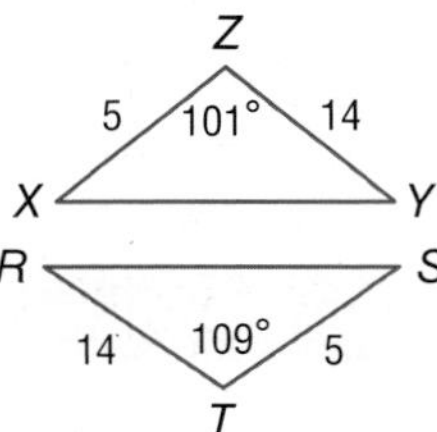


13. $m\angle TUW$ and $m\angle VUW$

14. $PS$ and $SR$

15. $JK$ and $HJ$

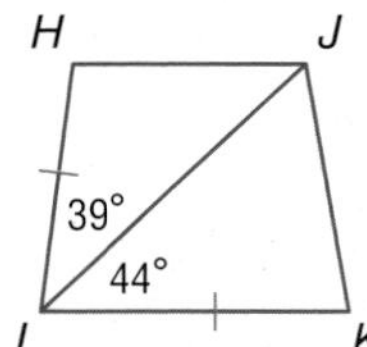

**Example 2**

16. **CAMPING** Pedro and Joel are camping in a national park. One morning, Pedro decides to hike to the waterfall. He leaves camp and goes 5 miles east then turns 15° south of east and goes 2 more miles. Joel leaves the camp and travels 5 miles west, then turns 35° north of west and goes 2 miles to the lake for a swim.

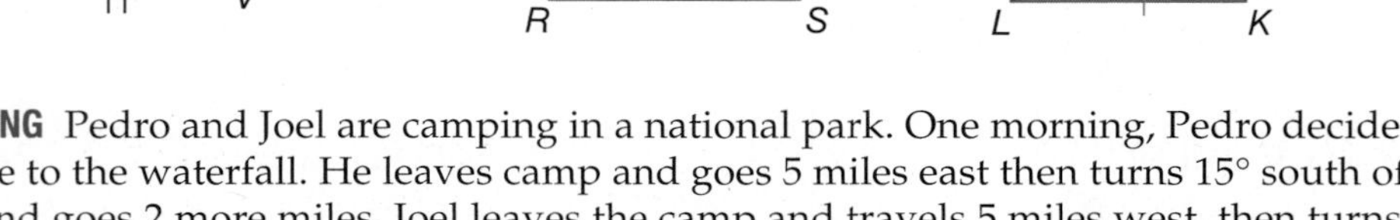

   a. When they reach their destinations, who is closer to the camp? Explain your reasoning. Include a diagram.

   b. Suppose instead of turning 35° north of west, Joel turned 10° south of west. Who would then be farther from the camp? Explain your reasoning. Include a diagram.

**Example 3** **Find the range of possible values for $x$.**

**17.** 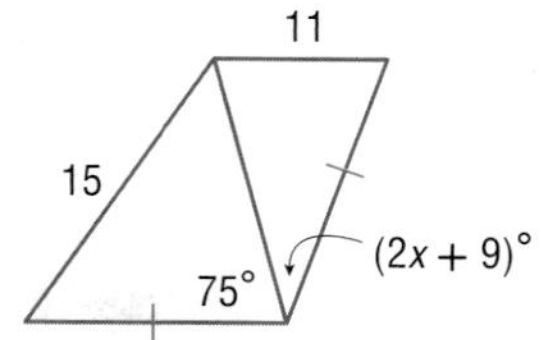


**18.** 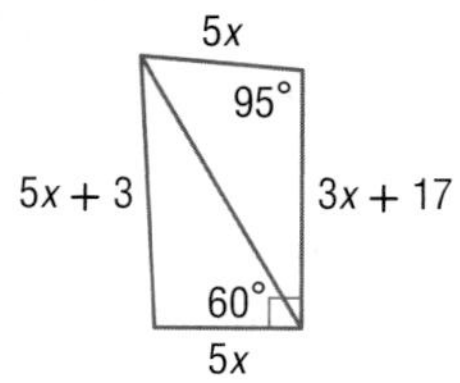


**19**

(x + 20)°
57
28
54
41°
57

**20.**

5x
95°
5x + 3
3x + 17
60°
5x

**21. CRANES** In the diagram, a crane is shown lifting an object to two different heights. The length of the crane's arm is fixed, and $\overline{MP} \cong \overline{RT}$. Is $\overline{MN}$ or $\overline{RS}$ shorter? Explain your reasoning.

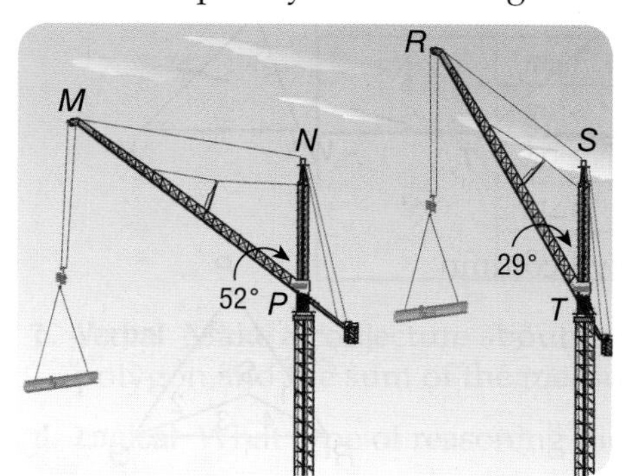


**22. LOCKERS** Neva and Shawn both have their lockers open as shown in the diagram. Whose locker forms a larger angle? Explain your reasoning.

**Examples 4–5** **CCSS ARGUMENTS** **Write a two-column proof.**

**23. Given:** $\overline{LK} \cong \overline{JK}$, $\overline{RL} \cong \overline{RJ}$
$K$ is the midpoint of $\overline{QS}$.
$m\angle SKL > m\angle QKJ$

**Prove:** $RS > QR$

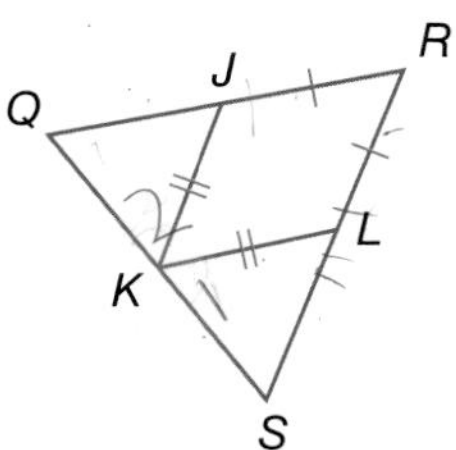

**24. Given:** $\overline{VR} \cong \overline{RT}$, $\overline{WV} \cong \overline{WT}$
$m\angle SRV > m\angle QRT$
$R$ is the midpoint of $\overline{SQ}$.

**Prove:** $WS > WQ$

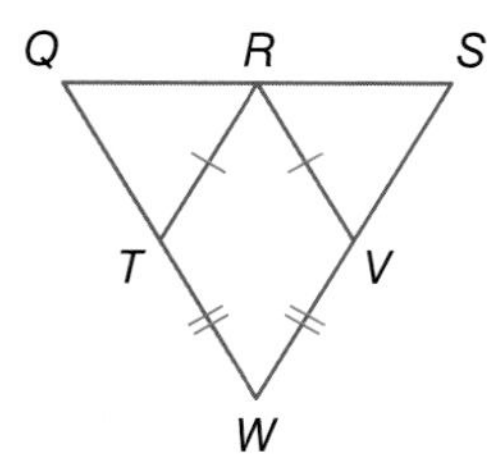

**25. Given:** $\overline{XU} \cong \overline{VW}$, $VW > XW$
$\overline{XU} \parallel \overline{VW}$

**Prove:** $m\angle XZU > m\angle UZV$

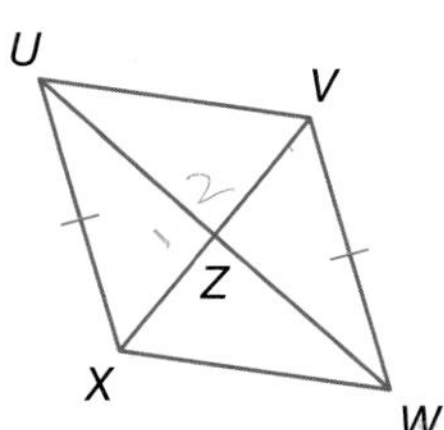


**26. Given:** $\overline{AF} \cong \overline{DJ}$, $\overline{FC} \cong \overline{JB}$
$AB > DC$

**Prove:** $m\angle AFC > m\angle DJB$

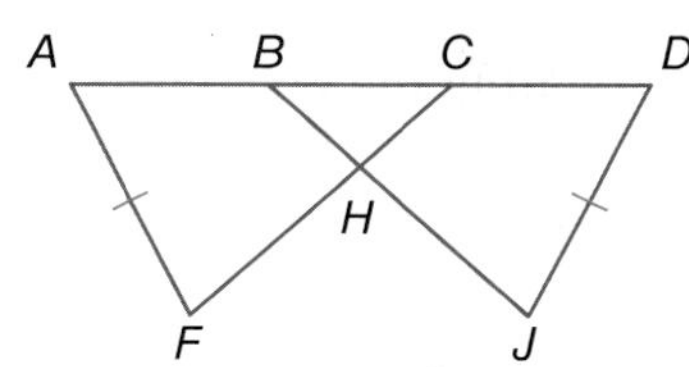

## Standardized Test Practice

**43. SHORT RESPONSE** Write an inequality to describe the possible range of values for $x$.

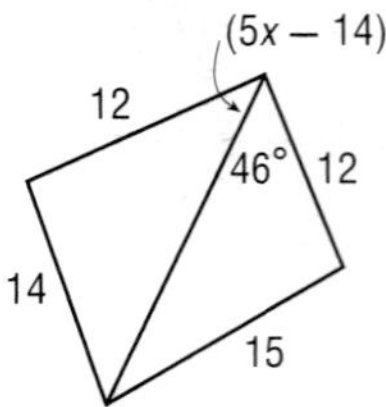


**44.** Which of the following is the inverse of the statement *If it is snowing, then Steve wears his snow boots?*

A If Steve wears his snow boots, then it is snowing.

B If it is not snowing, then Steve does not wear his snow boots.

C If it is not snowing, then Steve wears his snow boots.

D If it never snows, then Steve does not own snow boots.

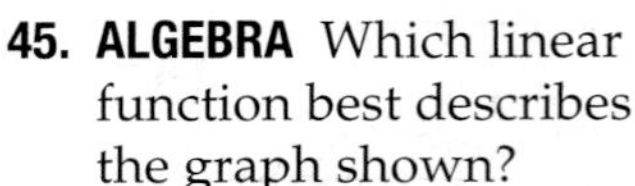

**45. ALGEBRA** Which linear function best describes the graph shown?

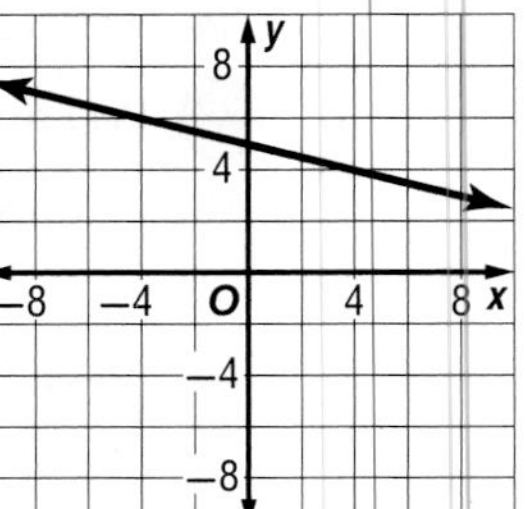


F $y = -\frac{1}{4}x + 5$

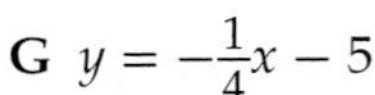

G $y = -\frac{1}{4}x - 5$

H $y = \frac{1}{4}x + 5$

J $y = \frac{1}{4}x - 5$

**46. SAT/ACT** If the side of a square is $x + 3$, then the diagonal of the square is

A $x^2 + 1$

B $x\sqrt{2} + 3\sqrt{2}$

C $2x + 6$

D $x^2\sqrt{2} + 6$

E $x^2 + 9$

## Spiral Review

**Find the range for the measure of the third side of a triangle given the measures of two sides.** (Lesson 7-5)

**47.** 3.2 cm, 4.4 cm

**48.** 5 ft, 10 ft

**49.** 3 m, 9 m

**50. CRUISES** Ally asked Tavia the cost of a cruise she and her best friend went on after graduation. Tavia could not remember how much it cost per person, but she did remember that the total cost was over \$500. Use indirect reasoning to show that the cost for one person was more than \$250. (Lesson 7-4)

**Draw and label a figure to represent the congruent triangles. Then find $x$.** (Lesson 6-2)

**51.** $\triangle QRS \cong \triangle GHJ$, $RS = 12$, $QR = 10$, $QS = 6$, and $HJ = 2x - 4$.

**52.** $\triangle ABC \cong \triangle XYZ$, $AB = 13$, $AC = 19$, $BC = 21$, and $XY = 3x + 7$.

## Skills Review

**Find the value of the variable(s) in each figure. Explain your reasoning.**

**53.**

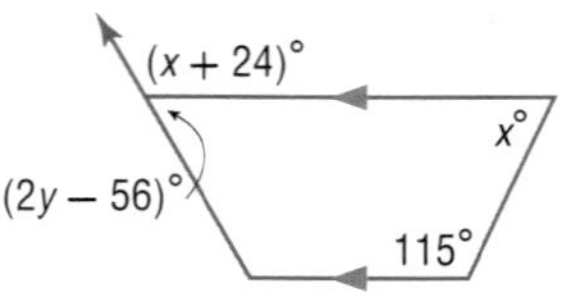


**54.**

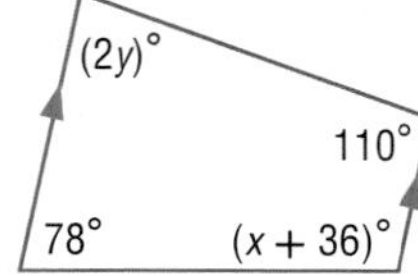


**55.**

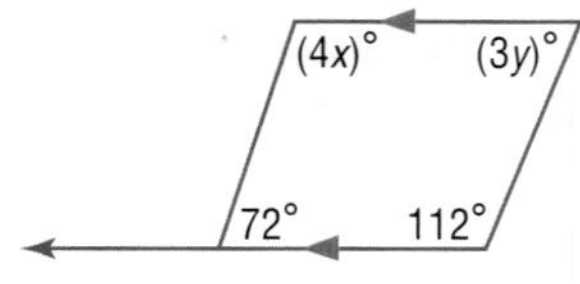

# Study Guide and Review

## Study Guide

### KeyConcepts

**Special Segments in Triangles** (Lessons 7-1 and 7-2)

- The special segments of triangles are perpendicular bisectors, angle bisectors, medians, and altitudes.
- The intersection points of each of the special segments of a triangle are called the points of concurrency.
- The points of concurrency for a triangle are the circumcenter, incenter, centroid, and orthocenter.

**Indirect Proof** (Lesson 7-4)

- Writing an Indirect Proof:
  1. Assume that the conclusion is false.
  2. Show that this assumption leads to a contradiction.
  3. Since the false conclusion leads to an incorrect statement, the original conclusion must be true.

**Triangle Inequalities** (Lessons 7-3, 7-5, and 7-6)

- The largest angle in a triangle is opposite the longest side, and the smallest angle is opposite the shortest side.
- The sum of the lengths of any two sides of a triangle is greater than the length of the third side.
- **SAS Inequality** (Hinge Theorem): In two triangles, if two sides are congruent, then the measure of the included angle determines which triangle has the longer third side.
- **SSS Inequality**: In two triangles, if two corresponding sides of each triangle are congruent, then the length of the third side determines which triangle has the included angle with the greater measure.

### FOLDABLES® StudyOrganizer

Be sure the Key Concepts are noted in your Foldable.

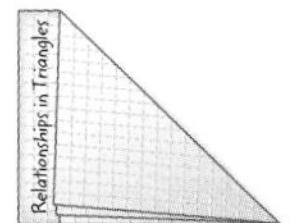


### KeyVocabulary

**altitude** (p. 419)
**centroid** (p. 417)
**circumcenter** (p. 407)
**concurrent lines** (p. 407)
**incenter** (p. 410)
**indirect proof** (p. 437)
**indirect reasoning** (p. 437)
**median** (p. 417)
**orthocenter** (p. 419)
**perpendicular bisector** (p. 406)
**point of concurrency** (p. 407)
**proof by contradiction** (p. 437)

### VocabularyCheck

**State whether each sentence is *true* or *false*. If *false*, replace the underlined term to make a true sentence.**

1. The altitudes of a triangle intersect at the centroid.
2. The point of concurrency of the medians of a triangle is called the incenter.
3. The point of concurrency is the point at which three or more lines intersect.
4. The circumcenter of a triangle is equidistant from the vertices of the triangle.
5. To find the centroid of a triangle, first construct the angle bisectors.
6. The perpendicular bisectors of a triangle are concurrent lines.
7. To start a proof by contradiction, first assume that what you are trying to prove is true.
8. A proof by contradiction uses indirect reasoning.
9. A median of a triangle connects the midpoint of one side of the triangle to the midpoint of another side of the triangle.
10. The incenter is the point at which the angle bisectors of a triangle intersect.

# Study Guide and Review *Continued*

## Lesson-by-Lesson Review

### 7-1 Bisectors of Triangles

**11.** Find $EG$ if $G$ is the incenter of $\triangle ABC$.

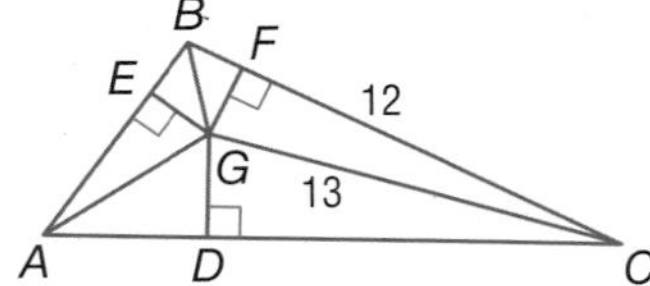


Find each measure.

**12.** $RS$

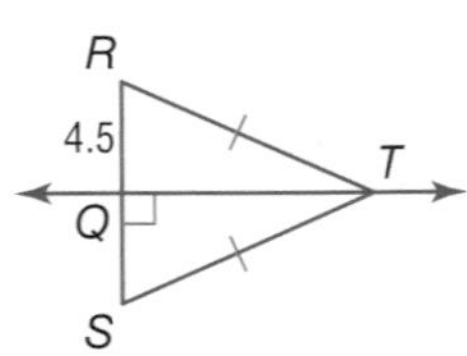


**13.** $XZ$

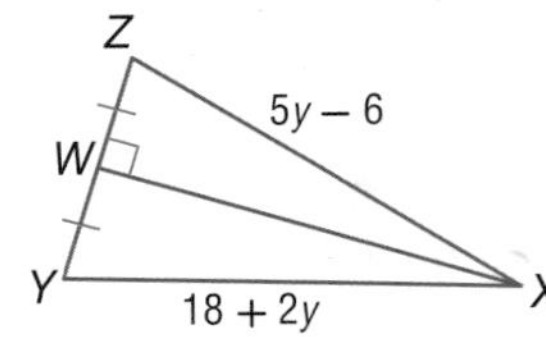


**14. BASEBALL** Jackson, Trevor, and Scott are warming up before a baseball game. One of their warm-up drills requires three players to form a triangle, with one player in the middle. Where should the fourth player stand so that he is the same distance from the other three players?

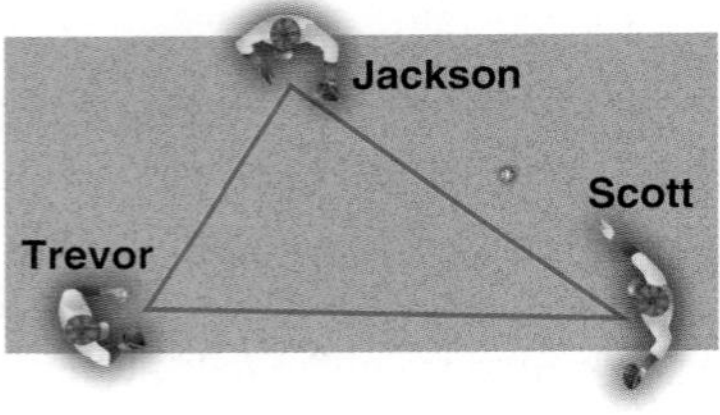


**Example 1**

Find each measure if $Q$ is the incenter of $\triangle JKL$.

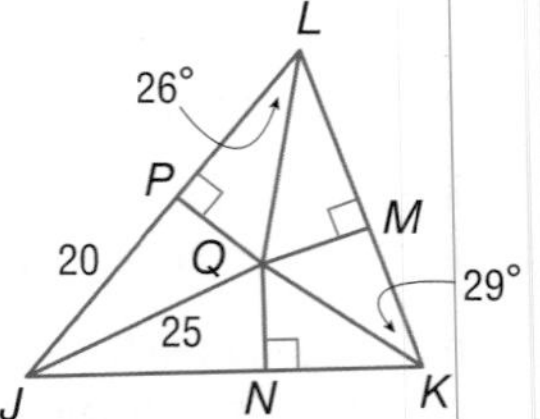


**a.** $\angle QJK$

| | |
|---|---|
| $m\angle KLP + m\angle MKN + m\angle NJP = 180$ | $\triangle$ Sum Theorem |
| $2(26) + 2(29) + m\angle NJP = 180$ | Substitution |
| $110 + m\angle NJP = 180$ | Simplify. |
| $m\angle NJP = 70$ | Subtract. |

Since $\overrightarrow{JQ}$ bisects $\angle NJP$, $2m\angle QJK = m\angle NJP$. So, $m\angle QJK = \frac{1}{2}m\angle NJP$, so $m\angle QJK = \frac{1}{2}(70)$ or 35.

**b.** $QP$

| | |
|---|---|
| $a^2 + b^2 = c^2$ | Pythagorean Theorem |
| $(QP)^2 + 20^2 = 25^2$ | Substitution |
| $(QP)^2 + 400 = 625$ | $20^2 = 400$ and $25^2 = 625$ |
| $(QP)^2 = 225$ | Subtract. |
| $QP = 15$ | Simplify. |

### 7-2 Medians and Altitudes of Triangles

**15.** The vertices of $\triangle DEF$ are $D(0, 0)$, $E(0, 7)$, and $F(6, 3)$. Find the coordinates of the orthocenter of $\triangle DEF$.

**16. PROM** Georgia is on the prom committee. She wants to hang a dozen congruent triangles from the ceiling so that they are parallel to the floor. She sketched out one triangle on a coordinate plane with coordinates (0, 4), (3, 8), and (6, 0). If each triangle is to be hung by one chain, what are the coordinates of the point where the chain should attach to the triangle?

**Example 2**

In $\triangle EDF$, $T$ is the centroid and $FT = 12$. Find $TQ$.

F, 12, T, E, Q, D

| | |
|---|---|
| $FT = \frac{2}{3}FQ$ | |
| $FT = \frac{2}{3}(FT + TQ)$ | |
| $12 = \frac{2}{3}(12 + TQ)$ | $FT = 12$ |
| $12 = 8 + \frac{2}{3}TQ$ | Distributive Property |
| $4 = \frac{2}{3}TQ$ | Subtract. |
| $6 = TQ$ | Multiply. |

## 7-3 Inequalities in One Triangle

List the angles and sides of each triangle in order from smallest to largest.

**17.**

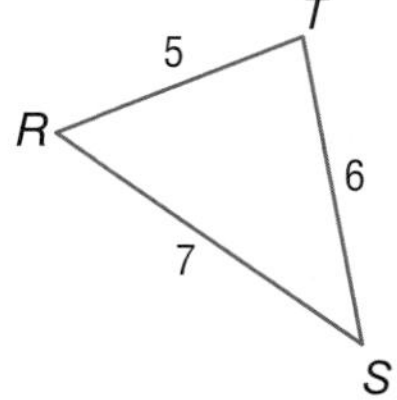


**18.**

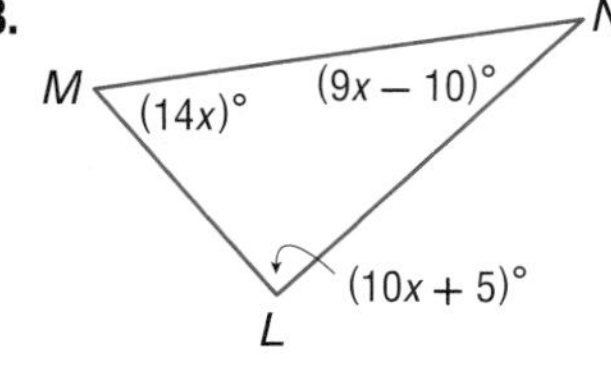


**19.** **NEIGHBORHOODS** Anna, Sarah, and Irene live at the intersections of the three roads that make the triangle shown. If the girls want to spend the afternoon together, is it a shorter path for Anna to stop and get Sarah and go onto Irene's house, or for Sarah to stop and get Irene and then go on to Anna's house?

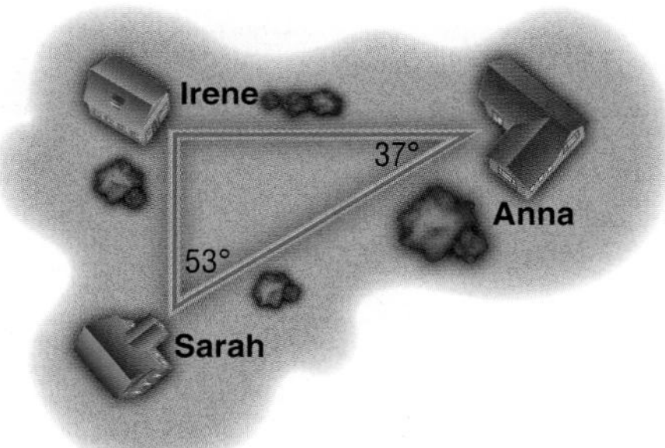


### Example 3

List the angles and sides of $\triangle ABC$ in order from smallest to largest.

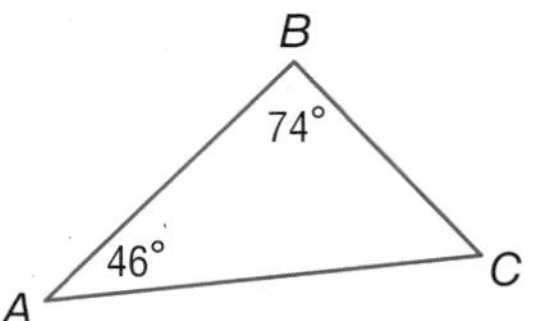


**a.** First, find the missing angle measure using the Triangle Sum Theorem.

$m\angle C = 180 - (46 + 74)$ or 60

So, the angles from smallest to largest are $\angle A$, $\angle C$, and $\angle B$.

**b.** The sides from shortest to longest are $\overline{BC}$, $\overline{AB}$, and $\overline{AC}$.

## 7-4 Indirect Proof

State the assumption you would make to start an indirect proof of each statement.

**20.** $m\angle A \geq m\angle B$

**21.** $\triangle FGH \cong \triangle MNO$

**22.** $\triangle KLM$ is a right triangle.

**23.** If $3y < 12$, then $y < 4$.

**24.** Write an indirect proof to show that if two angles are complementary, neither angle is a right angle.

**25.** **MOVIES** Isaac bought two DVD's and spent over \$50. Use indirect reasoning to show that at least one of the DVD's he purchased was over \$25.

### Example 4

State the assumption necessary to start an indirect proof of each statement.

**a.** $\overline{XY} \ncong \overline{JK}$

$\overline{XY} \cong \overline{JK}$

**b.** If $3x < 18$, then $x < 6$.

The conclusion of the conditional statement is $x < 6$. The negation of the conclusion is $x \geq 6$.

**c.** $\angle 2$ is an acute angle.

If $\angle 2$ *is an acute angle* is false, then $\angle 2$ *is not an acute angle* must be true. This means that $\angle 2$ *is an obtuse or right angle* must be true.

# Study Guide and Review *Continued*

## 7-5 The Triangle Inequality

**Is it possible to form a triangle with the given lengths? If not, explain why not.**

**26.** 5, 6, 9

**27.** 3, 4, 8

**Find the range for the measure of the third side of a triangle given the measure of two sides.**

**28.** 5 ft, 7 ft

**29.** 10.5 cm, 4 cm

**30.** **BIKES** Leonard rides his bike to visit Josh. Since High Street is closed, he has to travel 2 miles down Main Street and turn to travel 3 miles farther on 5th Street. If the three streets form a triangle with Leonard and Josh's house as two of the vertices, find the range of the possible distance between Leonard and Josh's houses when traveling straight down High Street.

### Example 5

**Is it possible to form a triangle with the lengths 7, 10, and 9 feet? If not, explain why not.**

Check each inequality.

$7 + 10 > 9$ — $17 > 9$ ✓

$7 + 9 > 10$ — $16 > 10$ ✓

$10 + 9 > 7$ — $19 > 7$ ✓

Since the sum of each pair of side lengths is greater than the third side length, sides with lengths 7, 10, and 9 feet will form a triangle.

## 7-6 Inequalities in Two Triangles

**Compare the given measures.**

**31.** $m\angle ABC$, $m\angle DEF$

**32.** $QT$ and $RS$

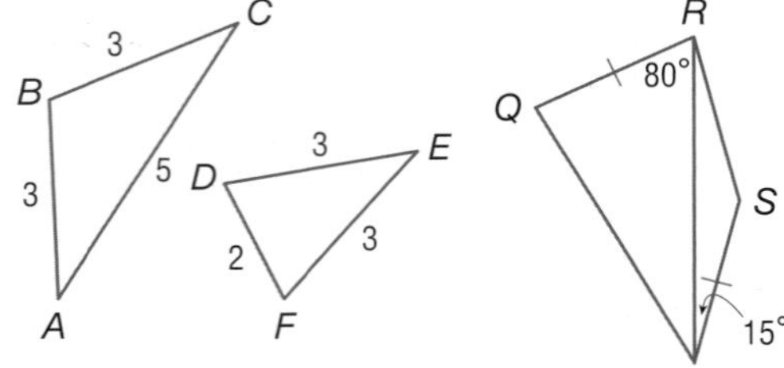


**33.** **BOATING** Rose and Connor each row across a pond heading to the same point. Neither of them has rowed a boat before, so they both go off course as shown in the diagram. After two minutes, they have each traveled 50 yards. Who is closer to their destination?

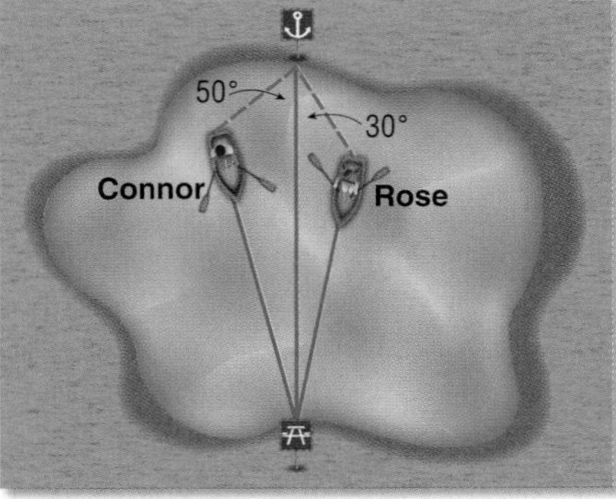


### Example 6

**Compare the given measures.**

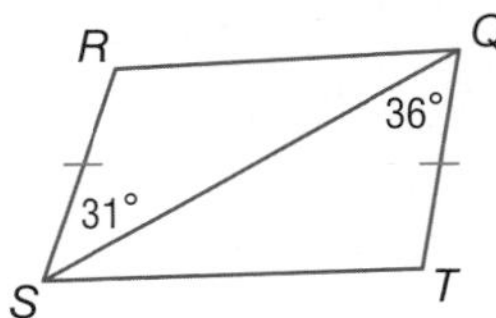


**a.** $RQ$ **and** $ST$

In $\triangle QRS$ and $\triangle STQ$, $\overline{RS} \cong \overline{TQ}$, $\overline{QS} \cong \overline{QS}$, and $\angle SQT > \angle RSQ$. By the Hinge Theorem, $m\angle SQT < m\angle RSQ$, so $RQ < ST$.

**b.** $m\angle JKM$ **and** $m\angle LKM$

In $\triangle JKM$ and $\triangle LKM$, $\overline{JM} \cong \overline{LM}$, $\overline{KM} \cong \overline{KM}$, and $LK > JK$. By the Converse of the Hinge Theorem, $\angle LKM > \angle JKM$.

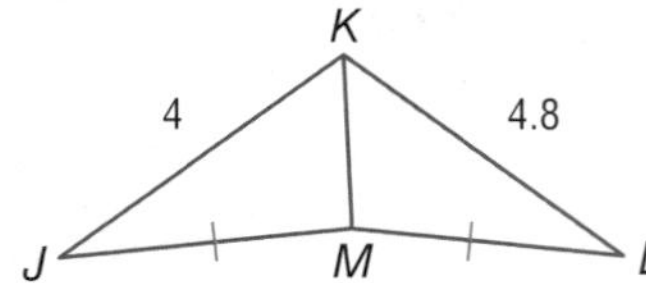

# CHAPTER 7 Practice Test

1. **GARDENS** Maggie wants to plant a circular flower bed within a triangular area set off by three pathways. Which point of concurrency related to triangles would she use for the center of the largest circle that would fit inside the triangle?

**In $\triangle CDF$, $K$ is the centroid and $DK = 16$. Find each length.**

2. $KH$
3. $CD$
4. $FG$

5. **PROOF** Write an indirect proof.

   **Given:** $5x + 7 \geq 52$

   **Prove:** $x \geq 9$

**Find each measure.**

6. $m\angle TQR$

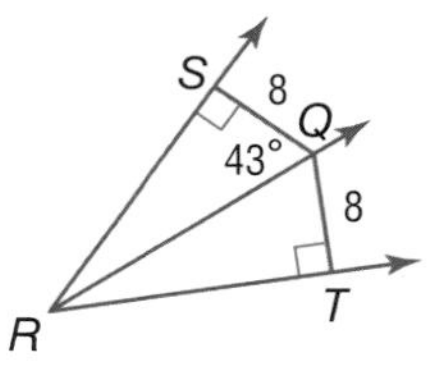


7. $XZ$

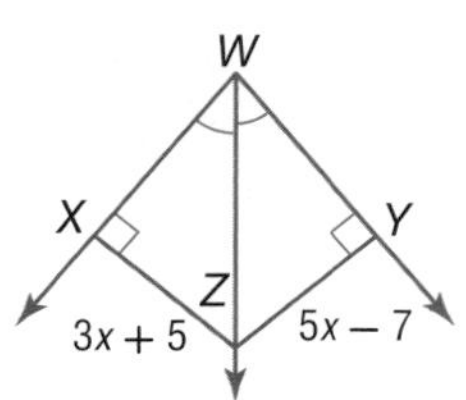


8. **GEOGRAPHY** The distance from Tonopah to Round Mountain is equal to the distance from Tonopah to Warm Springs. The distance from Tonopah to Hawthorne is the same as the distance from Tonopah to Beatty. Determine which distance is greater, Round Mountain to Hawthorne or Warm Springs to Beatty.

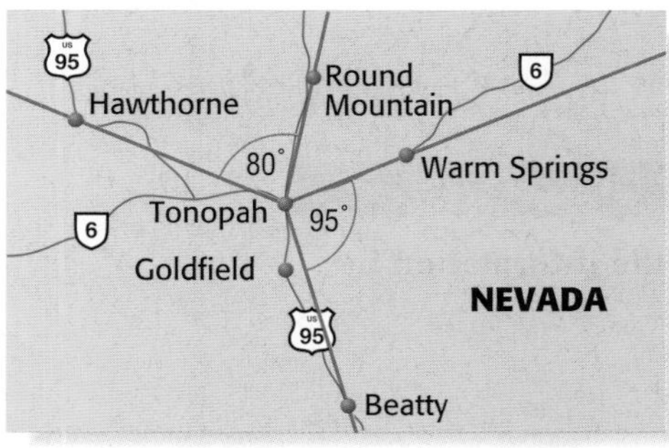


9. **MULTIPLE CHOICE** If the measures of two sides of a triangle are 3.1 feet and 4.6 feet, which is the *least* possible whole number measure for the third side?

   **A** 1.6 feet  **C** 7.5 feet

   **B** 2 feet  **D** 8 feet

**Point $H$ is the incenter of $\triangle ABC$. Find each measure.**

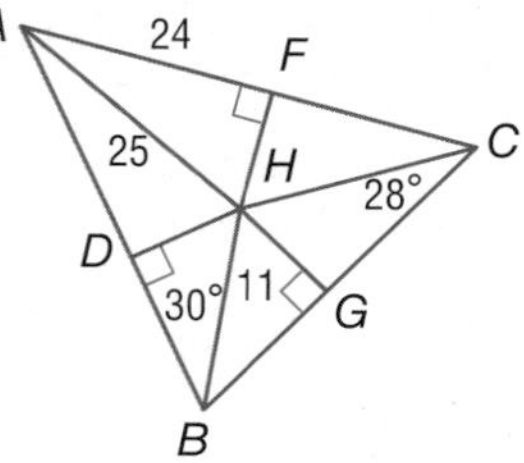


10. $DH$
11. $BD$
12. $m\angle HAC$
13. $m\angle DHG$

14. **MULTIPLE CHOICE** If the lengths of two sides of a triangle are 5 and 11, what is the range of possible lengths for the third side?

    **F** $6 < x < 10$  **H** $6 < x < 16$

    **G** $5 < x < 11$  **J** $x < 5$ or $x > 11$

**Compare the given measures.**

15. $AB$ and $BC$

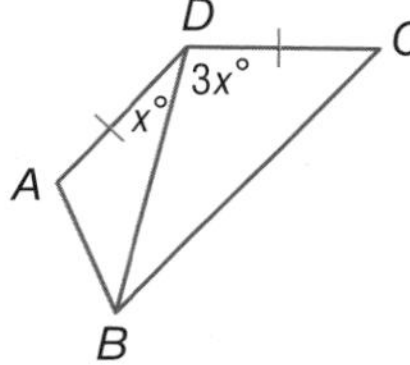


16. $\angle RST$ and $\angle JKL$

**State the assumption necessary to start an indirect proof of each statement.**

17. If 8 is a factor of $n$, then 4 is a factor of $n$.
18. $m\angle M > m\angle N$
19. If $3a + 7 \leq 28$, then $a \leq 7$.

**Use the figure to determine which angle has the greatest measure.**

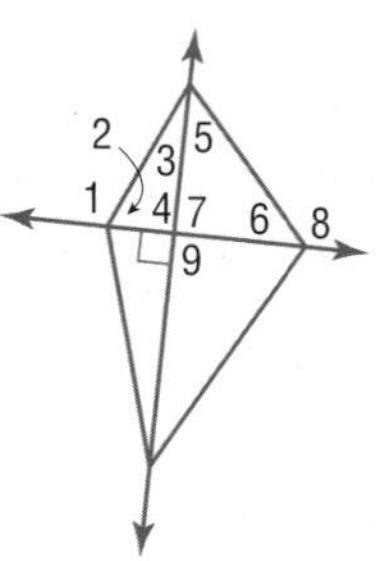

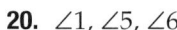

20. $\angle 1, \angle 5, \angle 6$
21. $\angle 9, \angle 8, \angle 3$
22. $\angle 4, \angle 3, \angle 2$

23. **PROOF** Write a two-column proof.

    **Given:** $\overline{RQ}$ bisects $\angle SRT$.

    **Prove:** $m\angle SQR > m\angle SRQ$

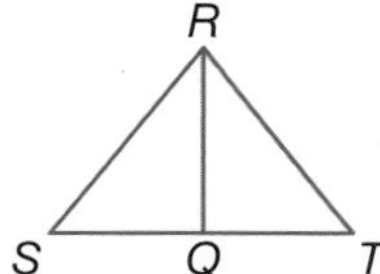


**Find the range for the measure of the third side of a triangle given the measures of the two sides.**

24. 10 ft, 16 ft
25. 23 m, 39 m

CHAPTER 7

# Preparing for Standardized Tests

## Eliminate Unreasonable Answers

You can eliminate unreasonable answers to determine the correct answer when solving multiple choice test items.

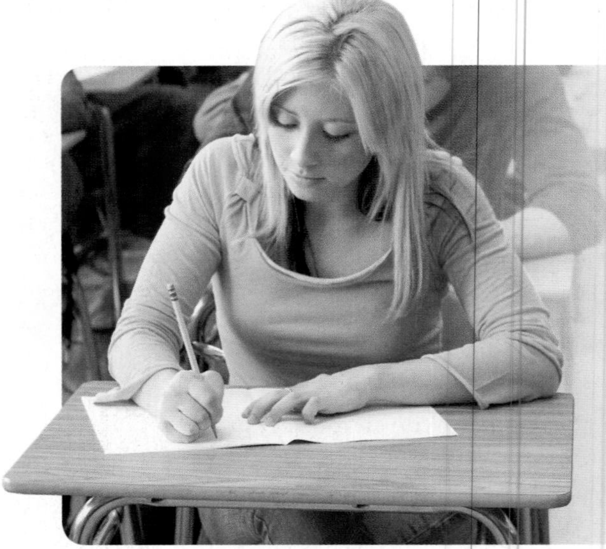

Jack Hollingsworth/Photodisc/Getty Images

### Strategies for Eliminating Unreasonable Answers

**Step 1**

Read the problem statement carefully to determine exactly what you are being asked to find.

- What am I being asked to solve?
- Is the correct answer a whole number, fraction, or decimal?
- Do I need to use a graph or table?
- What units (if any) will the correct answer have?

**Step 2**

Carefully look over each possible answer choice and evaluate for reasonableness. Do not write any digits or symbols outside the answer boxes.

- Identify any answer choices that are clearly incorrect and eliminate them.
- Eliminate any answer choices that are not in the proper format.
- Eliminate any answer choices that do not have the correct units.

**Step 3**

Solve the problem and choose the correct answer from those remaining. Check your answer.

### Standardized Test Example

**Read the problem. Identify what you need to know. Then use the information in the problem to solve.**

What is the measure of $\angle KLM$?

**A** 32

**B** 44

**C** 78

**D** 94

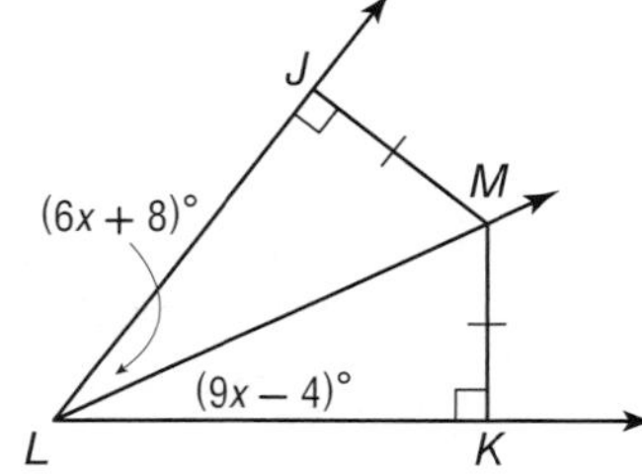

Read the problem and study the figure carefully. Triangle $KLM$ is a right triangle. Since the sum of the interior angles of a triangle is 180°, $m\angle KLM + m\angle LMK$ must be equal to 90°. Otherwise, the sum would exceed 180°. Since answer choice D is an obtuse angle, it can be eliminated as unreasonable. The correct answer must be A, B, or C.

Solve the problem. According to the converse of the Angle Bisector Theorem, if a point in the interior of an angle is equidistant from the sides of the angle, then it is on the bisector of the angle. Point $M$ is equidistant from rays $LJ$ and $LK$, so it lies on the angle bisector of $\angle JLK$. Therefore, $\angle JLM$ must be congruent to $\angle KLM$. Set up and solve an equation for $x$.

$6x + 8 = 9x - 4$
$-3x = -12$
$x = 4$

So, the measure of $\angle KLM$ is $[9(4) - 4]°$, or 32°. The correct answer is A.

## Exercises

**Read each question. Then fill in the correct answer on the answer document provided by your teacher or on a sheet of paper.**

**1.** Point $P$ is the centroid of triangle $QUS$. If $QP =$ 14 centimeters, what is the length of $\overline{QT}$?

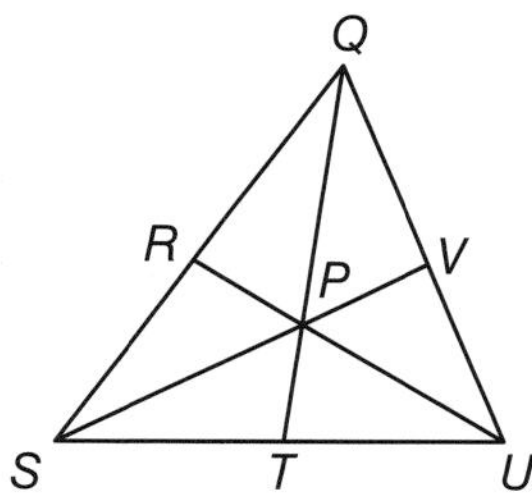


A 7 cm

B 12 cm

C 18 cm

D 21 cm

**2.** What is the area, in square units, of the triangle shown below?

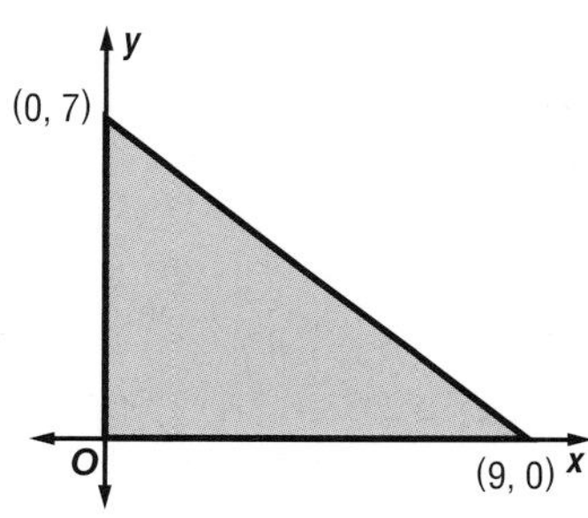


F 8

G 27.4

H 31.5

J 63

**3.** What are the coordinates of the orthocenter of the triangle below?

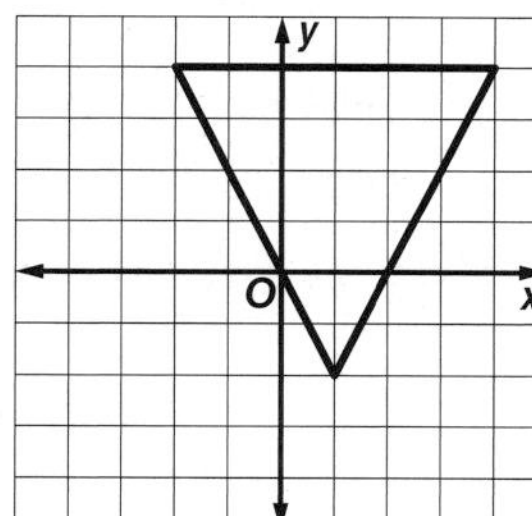


A $\left(-\frac{3}{4}, -1\right)$

B $\left(-\frac{4}{3}, 1\right)$

C $\left(1, \frac{5}{2}\right)$

D $\left(1, \frac{9}{4}\right)$

**4.** If $\triangle ABC$ is isosceles and $m\angle A = 94$, which of the following *must* be true?

F $m\angle B = 94$

G $m\angle B = 47$

H $AB = BC$

J $AB = AC$

**5.** Which of the following could *not* be the dimensions of a triangle?

A 1.9, 3.2, 4

B 1.6, 3, 4.6

C 3, 7.2, 7.5

D 2.6, 4.5, 6

CHAPTER 7

# Standardized Test Practice

## Cumulative, Chapters 1 through 7

### Multiple Choice

**Read each question. Then fill in the correct answer on the answer document provided by your teacher or on a sheet of paper.**

1. Solve for $x$.

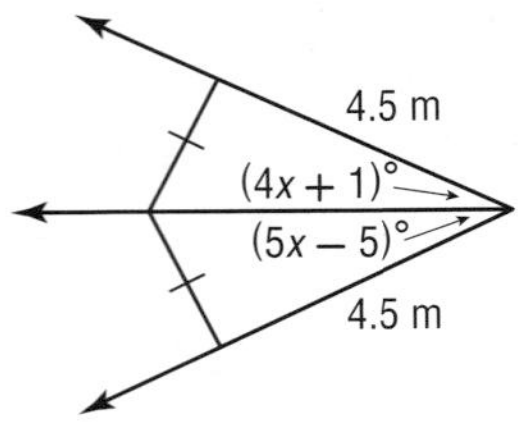


A 3 C 5

B 4 D 6

2. Which of the following could not be the value of $x$?

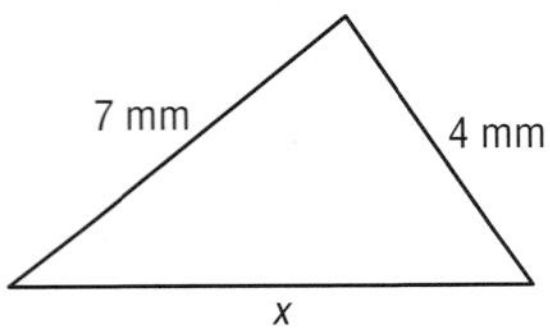


F 8 mm H 10 mm

G 9 mm J 11 mm

3. Jesse claims that if you live in Lexington, then you live in Kentucky. Which assumption would you need to make to form an indirect proof of this claim?

A Suppose someone lives in Kentucky, but not in Lexington.

B Suppose someone lives in Kentucky and in Lexington.

C Suppose someone lives in Lexington and in Kentucky.

D Suppose someone lives in Lexington, but not in Kentucky.

4. Which of the following best describes the shortest distance from a vertex of a triangle to the opposite side?

F altitude H median

G diameter J segment

5. What is the correct relationship between the angle measures of $\triangle PQR$?

F $m\angle R < m\angle Q < m\angle P$

G $m\angle R < m\angle P < m\angle Q$

H $m\angle Q < m\angle P < m\angle R$

J $m\angle P < m\angle Q < m\angle R$

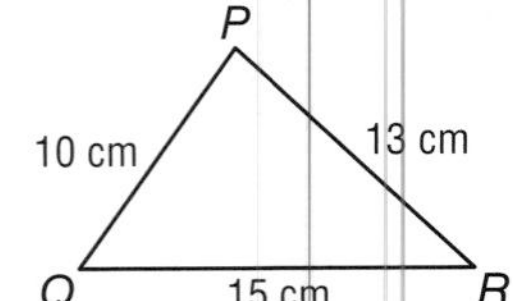


6. Which assumption would you need to make in order to start an indirect proof of the statement?

*Angle S is not an obtuse angle.*

A $\angle S$ is a right angle.

B $\angle S$ is an obtuse angle.

C $\angle S$ is an acute angle.

D $\angle S$ is not an acute angle.

**Test-Taking Tip**

Question 2 The sum of any two sides of a triangle must be greater than the third side.

## Short Response/Gridded Response

**Record your answers on the answer sheet provided by your teacher or on a sheet of paper.**

7. **GRIDDED RESPONSE** If the measures of two sides of a triangle are 9 centimeters and 15 centimeters, what is the least possible measure of the third side in centimeters if the measure is an integer?

8. What are the coordinates of the orthocenter of the triangle below?

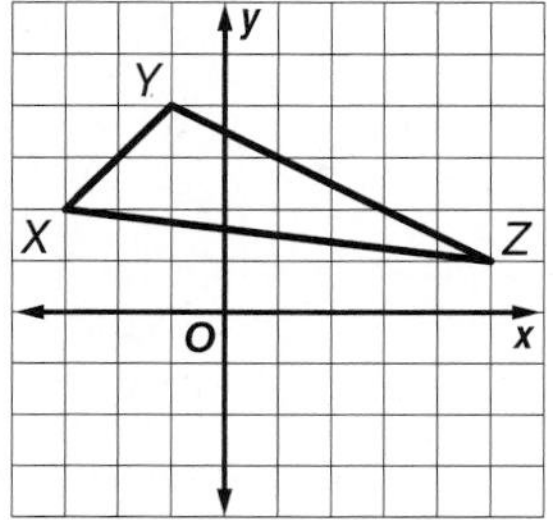


9. List the sides of the triangle below in order from shortest to longest.

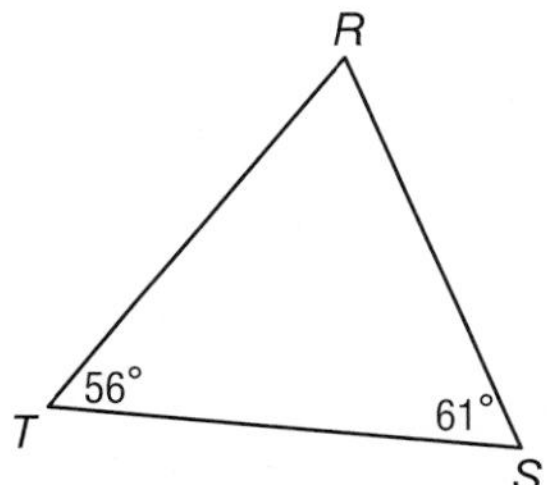


10. **GRIDDED RESPONSE** Suppose you deposit $500 in an account paying 4.5% interest compounded semiannually. Find the dollar value of the account rounded to the nearest penny after 10 years.

11. Eric and Heather are each taking a group of campers hiking in the woods. Eric's group leaves camp and goes 2 miles east, then turns 20° south of east and goes 4 more miles. Heather's group leaves camp and travels 2 miles west, then turns 30° north of west and goes 4 more miles. How many degrees south of east would Eric have needed to turn in order for his group and Heather's group to be the same distance from camp after the two legs of the hike?

12. **GRIDDED RESPONSE** Solve for $x$ in the triangle below.

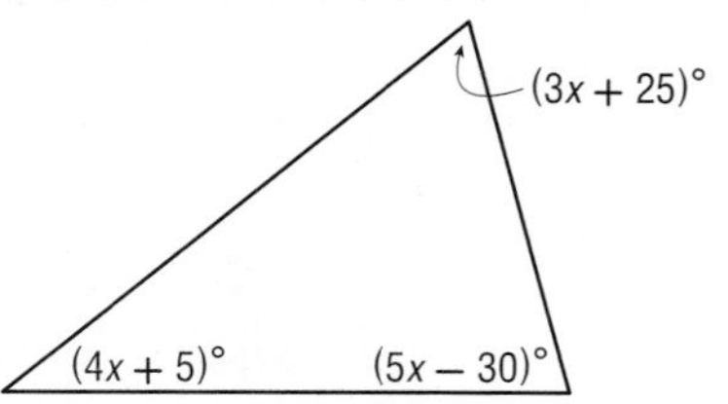


**Need ExtraHelp?**

| If you missed Question... | 1 | 2 | 3 | 4 | 5 | 6 | 7 | 8 | 9 | 10 | 11 | 12 |
|---|---|---|---|---|---|---|---|---|---|---|---|---|
| Go to Lesson... | 7-1 | 7-5 | 7-4 | 7-2 | 7-3 | 7-4 | 7-5 | 7-2 | 7-3 | 4-2 | 7-6 | 6-1 |

CHAPTER 8

# Quadrilaterals

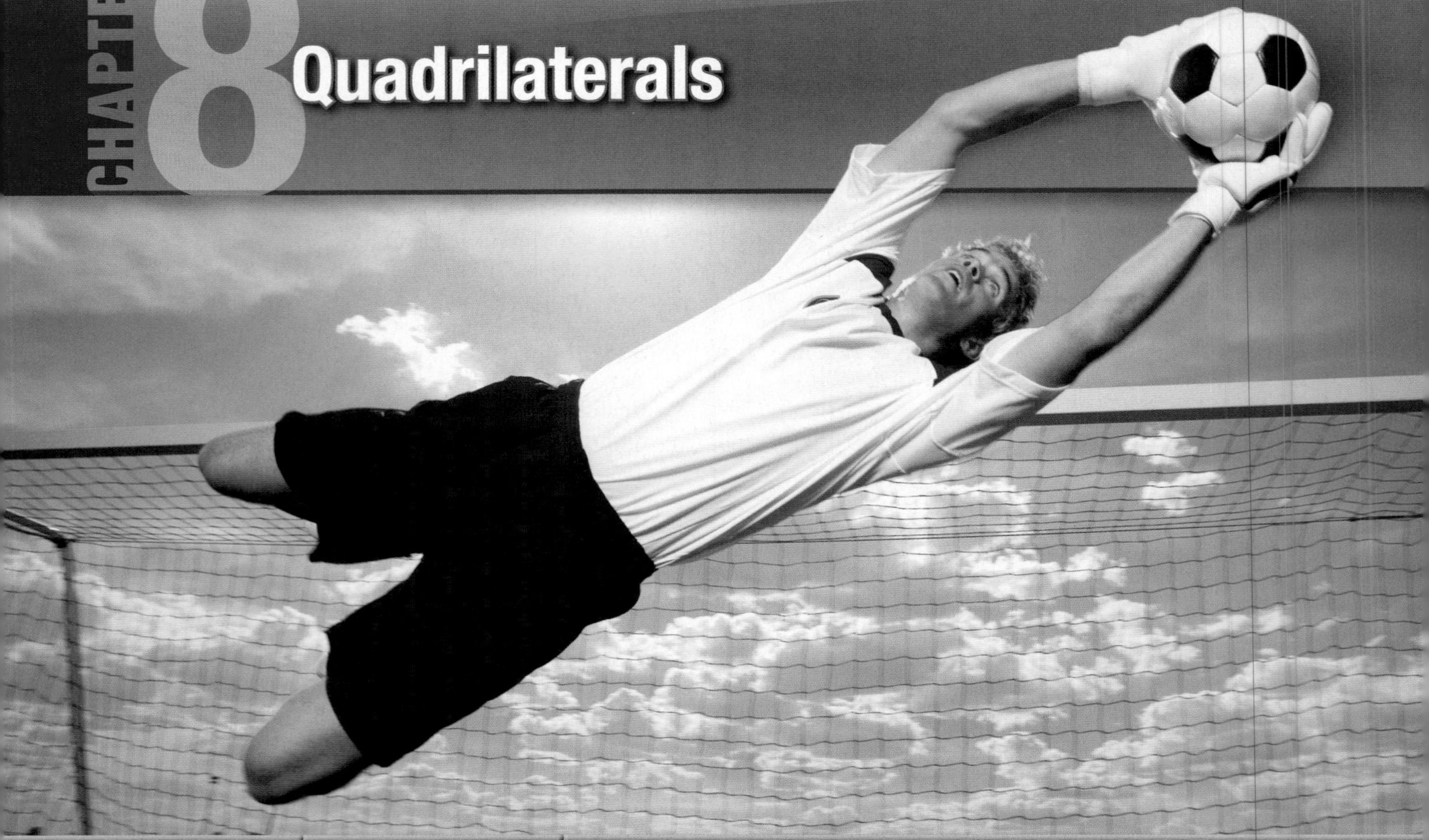

## Then

You classified polygons. You recognized and applied properties of polygons.

## Now

In this chapter, you will:

- Find and use the sum of the measures of the interior and exterior angles of a polygon.
- Recognize and apply properties of quadrilaterals.
- Compare quadrilaterals.

## Why?

**FUN AND GAMES** The properties of quadrilaterals can be used to find various angle measures and side lengths such as the measures of angles in game equipment, playing fields, and game boards.

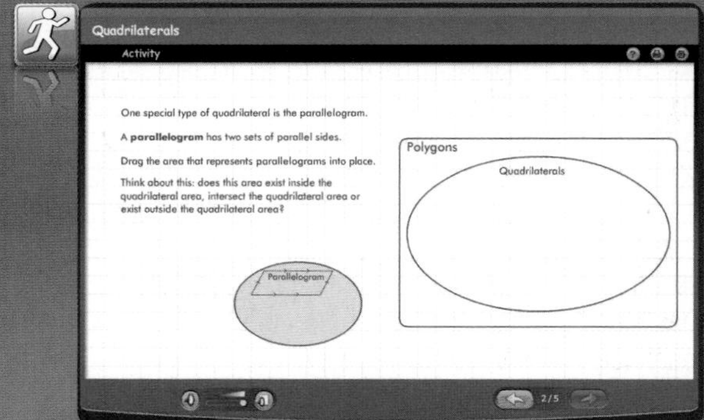


connectED.mcgraw-hill.com **Your Digital Math Portal**

Animation | Vocabulary | eGlossary | Personal Tutor | Virtual Manipulatives | Graphing Calculator | Audio | Foldables | Self-Check Practice | Worksheets

# Get Ready for the Chapter

**Diagnose** Readiness | You have two options for checking Prerequisite Skills.

**1 Textbook Option** Take the Quick Check below. Refer to the Quick Review for help.

## QuickCheck

Find $x$ to the nearest tenth.

1.

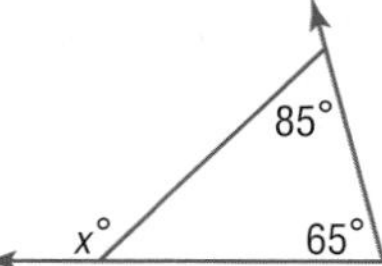


2.

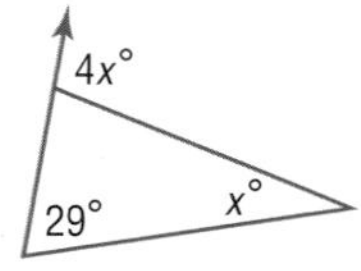


**SPEED SKATING** A speed skater forms at least two sets of triangles and exterior angles as she skates. Find each measure.

3. $m\angle 1$
4. $m\angle 2$
5. $m\angle 3$
6. $m\angle 4$

**ALGEBRA** Find $x$ and the measures of the unknown sides of each triangle.

7.

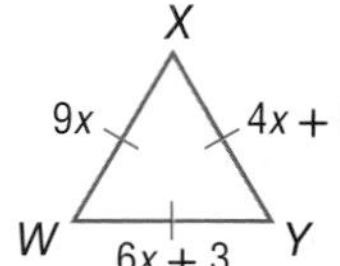


8.

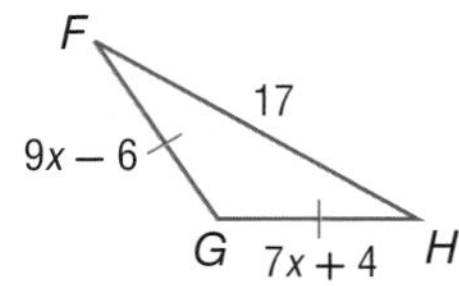


9. **TRAVEL** A plane travels from Des Moines to Phoenix, on to Atlanta, and back to Des Moines, as shown below. Find the distance in miles for each leg of the trip if the total trip was 3482 miles.

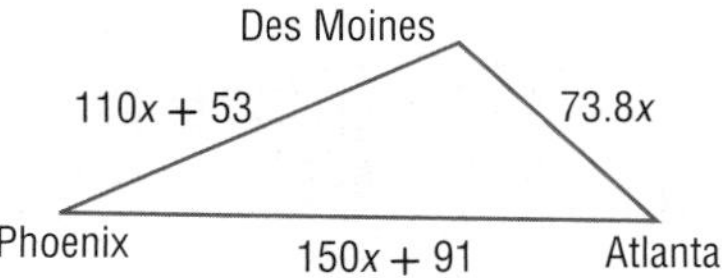


## QuickReview

**Example 1**

Find the measure of each numbered angle.

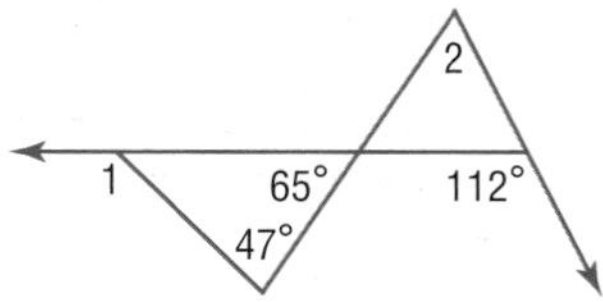


**a.** $m\angle 1$

| | |
|---|---|
| $m\angle 1 = 65 + 47$ | Exterior ∠ Theorem |
| $m\angle 1 = 112$ | Add. |

**b.** $m\angle 2$

| | |
|---|---|
| $180 = m\angle 2 + 68 + 65$ | Triangle Sum Theorem |
| $180 = m\angle 2 + 133$ | Simplify. |
| $m\angle 2 = 47$ | Subtract. |

**Example 2**

**ALGEBRA** Find the measures of the sides of isosceles $\triangle XYZ$.

| | |
|---|---|
| $XY = YZ$ | Given |
| $2x + 3 = 4x - 1$ | Substitution |
| $-2x = -4$ | Subtract. |
| $x = 2$ | Simplify. |
| $XY = 2x + 3$ | Given |
| $= 2(2) + 3$ or $7$ | $x = 2$ |
| $YZ = XY$ | Given |
| $= 7$ | $XY = 7$ |
| $XZ = 8x - 4$ | Given |
| $= 8(2) - 4$ or $12$ | $x = 2$ |

**2 Online Option** Take an online self-check Chapter Readiness Quiz at connectED.mcgraw-hill.com.

# **Get Started** on the Chapter

You will learn several new concepts, skills, and vocabulary terms as you study Chapter 8. To get ready, identify important terms and organize your resources. You may wish to refer to Chapter 0 to review prerequisite skills.

## FOLDABLES StudyOrganizer

**Quadrilaterals** Make this Foldable to help you organize your Chapter 8 notes about quadrilaterals. Begin with one sheet of notebook paper.

**1** **Fold** lengthwise to the holes.

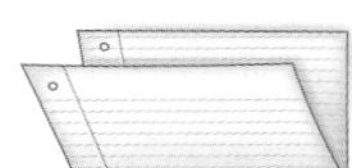

**2** **Fold** along the width of the paper twice and unfold the paper.

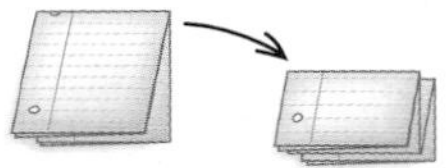

**3** **Cut** along the fold marks on the left side of the paper.

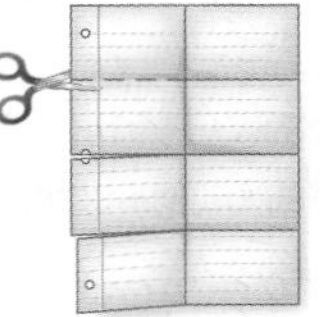

**4** **Label** as shown.

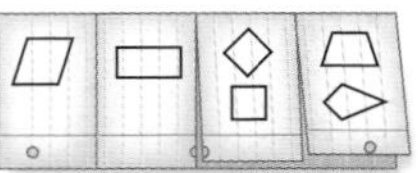

## NewVocabulary

| English | | Español |
|---|---|---|
| diagonal | p. 475 | diagonal |
| parallelogram | p. 485 | paralelogramo |
| rectangle | p. 505 | rectángulo |
| rhombus | p. 512 | rombo |
| square | p. 513 | cuadrado |
| trapezoid | p. 521 | trapecio |
| base | p. 521 | base |
| legs | p. 521 | catetos |
| isosceles trapezoid | p. 521 | trapecio isósceles |
| midsegment of a trapezoid | p. 523 | segmento medio de un trapecio |

## ReviewVocabulary

exterior angle ángulo externo an angle formed by one side of a triangle and the extension of another side

remote interior angle ángulos internos no adyacentes the angles of a triangle that are not adjacent to a given exterior angle

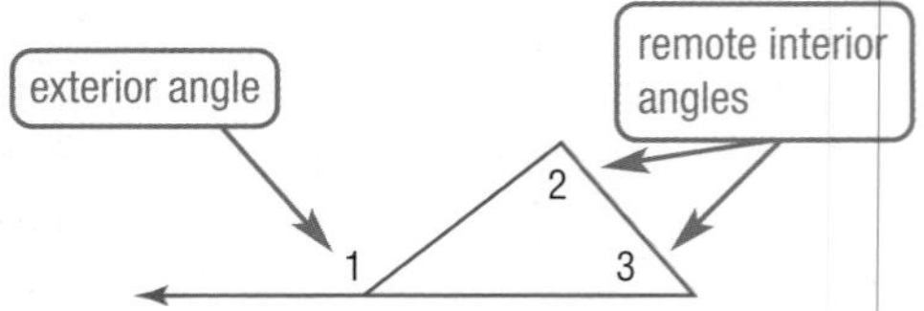


slope pendiente for a (nonvertical) line containing two points $(x_1, y_1)$ and $(x_2, y_2)$, the number $m$ given by the formula $m = \frac{(y_2 - y_1)}{(x_2 - x_1)}$ where $x_2 \neq x_1$

# LESSON 8-1 Angles of Polygons

| Then | Now | Why? |
|---|---|---|
| You named and classified polygons. | 1 Find and use the sum of the measures of the interior angles of a polygon.<br>2 Find and use the sum of the measures of the exterior angles of a polygon. | To create their honeycombs, young worker honeybees excrete flecks of wax that are carefully molded by other bees to form hexagonal cells. The cells are less than 0.1 millimeter thick, but they support almost 25 times their own weight. The cell walls all stand at exactly the same angle to one another. This angle is the measure of the interior angle of a regular hexagon. |

**NewVocabulary**
diagonal

**Common Core State Standards**

**Content Standards**
**G.MG.1** Use geometric shapes, their measures, and their properties to describe objects (e.g., modeling a tree trunk or a human torso as a cylinder). ★

**Mathematical Practices**
4 Model with mathematics.
3 Construct viable arguments and critique the reasoning of others.

## 1 Polygon Interior Angles Sum

A **diagonal** of a polygon is a segment that connects any two nonconsecutive vertices.

The vertices of polygon $PQRST$ that are not consecutive with vertex $P$ are vertices $R$ and $S$. Therefore, polygon $PQRST$ has two diagonals from vertex $P$, $\overline{PR}$ and $\overline{PS}$. Notice that the diagonals from vertex $P$ separate the polygon into three triangles.

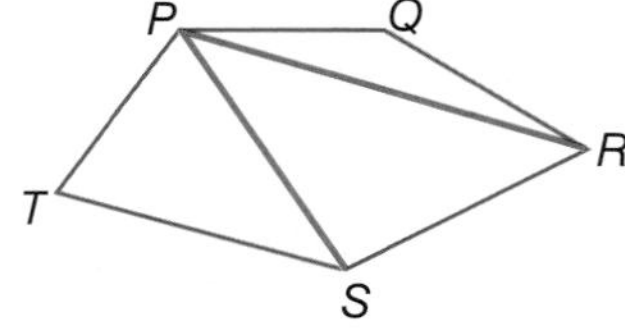


The sum of the angle measures of a polygon is the sum of the angle measures of the triangles formed by drawing all the possible diagonals from one vertex.

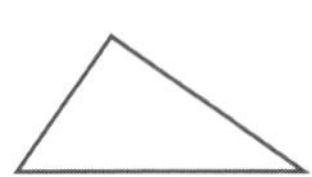

Triangle

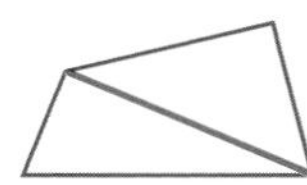

Quadrilateral

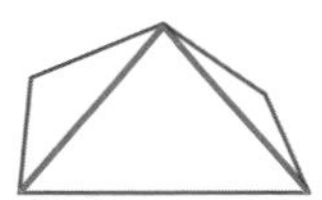

Pentagon

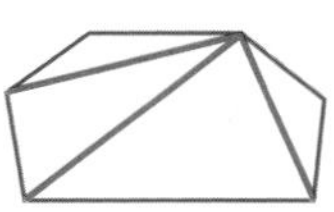

Hexagon

Since the sum of the angle measures of a triangle is 180, we can make a table and look for a pattern to find the sum of the angle measures for any convex polygon.

| Polygon | Number of Sides | Number of Triangles | Sum of Interior Angle Measures |
|---|---|---|---|
| Triangle | 3 | 1 | (1)180 or 180 |
| Quadrilateral | 4 | 2 | (2)180 or 360 |
| Pentagon | 5 | 3 | (3)180 or 540 |
| Hexagon | 6 | 4 | (4)180 or 720 |
| $n$-gon | $n$ | $n-2$ | $(n-2)180$ |

This leads to the following theorem.

**Theorem 8.1 Polygon Interior Angles Sum**

The sum of the interior angle measures of an $n$-sided convex polygon is $(n-2) \cdot 180$.

**Example** $m\angle A + m\angle B + m\angle C + m\angle D + m\angle E = (5-2) \cdot 180$
$= 540$

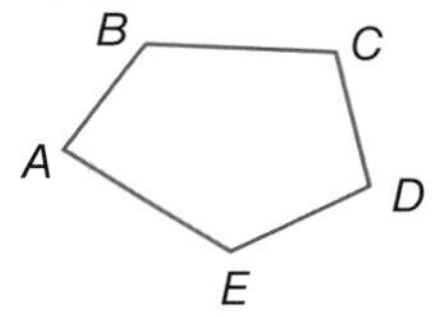


You will prove Theorem 8.1 for octagons in Exercise 42.

Masterfile

You can use the Polygon Interior Angles Sum Theorem to find the sum of the interior angles of a polygon and to find missing measures in polygons.

> **StudyTip**
>
> **Naming Polygons** Remember, a polygon with $n$-sides is an $n$-*gon*, but several polygons have special names.
>
> | Number of Sides | Polygon |
> |---|---|
> | 3 | triangle |
> | 4 | quadrilateral |
> | 5 | pentagon |
> | 6 | hexagon |
> | 7 | heptagon |
> | 8 | octagon |
> | 9 | nonagon |
> | 10 | decagon |
> | 11 | hendecagon |
> | 12 | dodecagon |
> | $n$ | $n$-gon |

## Example 1 Find the Interior Angles Sum of a Polygon

**a. Find the sum of the measures of the interior angles of a convex heptagon.**

A heptagon has seven sides. Use the Polygon Interior Angles Sum Theorem to find the sum of its interior angle measures.

$(n - 2) \cdot 180 = (7 - 2) \cdot 180$ $\quad n = 7$

$= 5 \cdot 180$ or $900$ $\quad$ Simplify.

The sum of the measures is 900.

**CHECK** Draw a convex polygon with seven sides. Use a protractor to measure each angle to the nearest degree. Then find the sum of these measures.

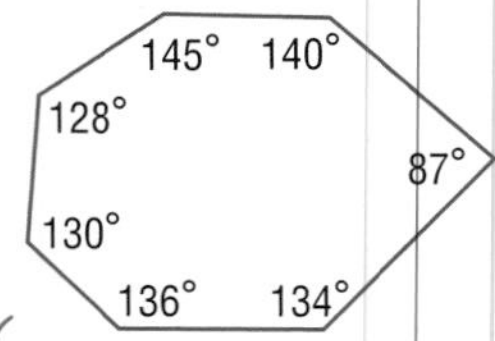


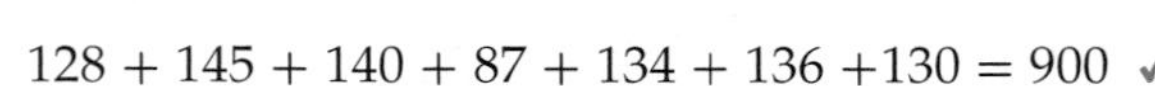

$128 + 145 + 140 + 87 + 134 + 136 + 130 = 900$ ✓

**b. ALGEBRA Find the measure of each interior angle of quadrilateral** ***ABCD*****.**

**Step 1** Find $x$.

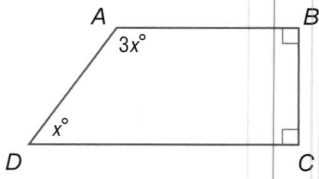

Since there are 4 angles, the sum of the interior angle measures is $(4 - 2) \cdot 180$ or 360.

| | |
|---|---|
| $360 = m\angle A + m\angle B + m\angle C + m\angle D$ | Sum of interior angle measures |
| $360 = 3x + 90 + 90 + x$ | Substitution |
| $360 = 4x + 180$ | Combine like terms. |
| $180 = 4x$ | Subtract 180 from each side. |
| $45 = x$ | Divide each side by 4. |

**Step 2** Use the value of $x$ to find the measure of each angle.

$m\angle A = 3x$ $\quad m\angle B = 90$ $\quad m\angle D = x$

$= 3(45)$ or $135$ $\quad m\angle C = 90$ $\quad = 45$

### GuidedPractice

**1A.** Find the sum of the measures of the interior angles of a convex octagon.

**1B.** Find the measure of each interior angle of pentagon $HJKLM$ shown

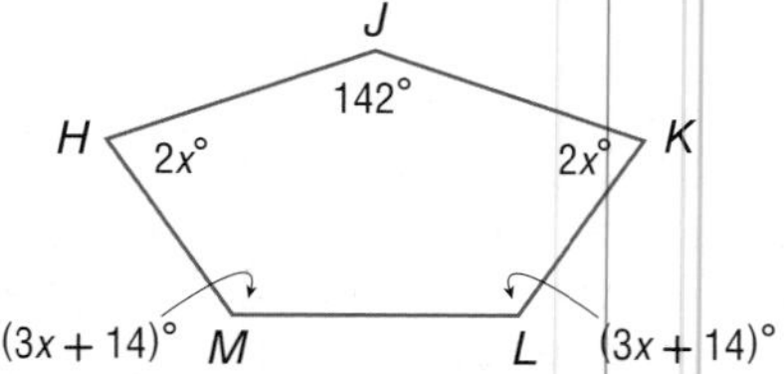


In a regular polygon, all of the interior angles are congruent. You can use this fact and the Polygon Interior Angle Sum Theorem to find the interior angle measure of any regular polygon.

**Review Vocabulary**

**regular polygon**
a convex polygon in which all of the sides are congruent and all of the angles are congruent

## Real-World Example 2 Interior Angle Measure of Regular Polygon

**TENTS The poles for a tent form the vertices of a regular hexagon. When the poles are properly positioned, what is the measure of the angle formed at a corner of the tent?**

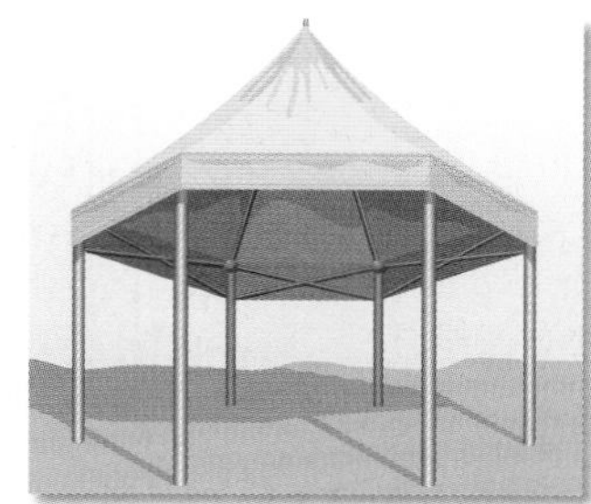

**Understand** Draw a diagram of the situation.

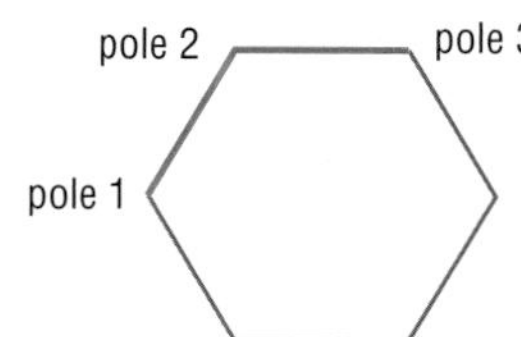


The measure of the angle formed at a corner of the tent is an interior angle of a regular hexagon.

**Plan** Use the Polygon Interior Angles Sum Theorem to find the sum of the measures of the angles. Since the angles of a regular polygon are congruent, divide this sum by the number of angles to find the measure of each interior angle.

**Solve** **Step 1** Find the sum of the interior angle measures.

$(n - 2) \cdot 180 = (6 - 2) \cdot 180$ — $n = 6$

$= 4 \cdot 180 \text{ or } 720$ — Simplify.

**Step 2** Find the measure of one interior angle.

$\frac{\text{sum of interior angle measures}}{\text{number of congruent angles}} = \frac{720}{6}$ — Substitution

$= 120$ — Divide.

The angle at a corner of the tent measures 120.

**Check** To verify that this measure is correct, use a ruler and a protractor to draw a regular hexagon using 120 as the measure of each interior angle. The last side drawn should connect with the beginning point of the first segment drawn. ✓

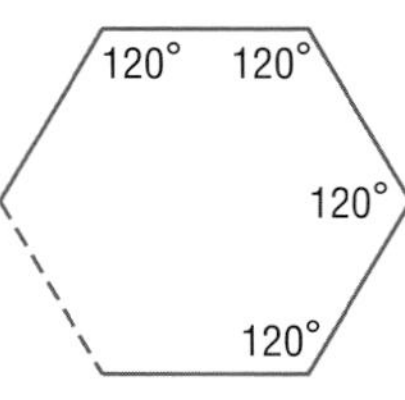


**Real-World Link**

Susan B. Anthony was a leader of the women's suffrage movement in the late 1800s, which eventually led to the Nineteenth Amendment giving women the right to vote. In 1979, the Susan B. Anthony one-dollar coin was first minted, making her the first woman to be depicted on U.S. currency.

**Source:** *Encyclopaedia Britannica*

PhotoSpin, Inc/Alamy

### Guided Practice

**2A. COINS** Find the measure of each interior angle of the regular hendecagon that appears on the face of a Susan B. Anthony one-dollar coin.

**2B. HOT TUBS** A certain company makes hot tubs in a variety of different shapes. Find the measure of each interior angle of the nonagon model.

Given the interior angle measure of a regular polygon, you can also use the Polygon Interior Angles Sum Theorem to find a polygon's number of sides.

## Example 3 Find Number of Sides Given Interior Angle Measure

**The measure of an interior angle of a regular polygon is 135. Find the number of sides in the polygon.**

Let $n$ = the number of sides in the polygon. Since all angles of a regular polygon are congruent, the sum of the interior angle measures is $135n$. By the Polygon Interior Angles Sum Theorem, the sum of the interior angle measures can also be expressed as $(n - 2) \cdot 180$.

| | |
|---|---|
| $135n = (n - 2) \cdot 180$ | Write an equation. |
| $135n = 180n - 360$ | Distributive Property |
| $-45n = -360$ | Subtract $180n$ from each side. |
| $n = 8$ | Divide each side by $-45$. |

The polygon has 8 sides.

### GuidedPractice

3. The measure of an interior angle of a regular polygon is 144. Find the number of sides in the polygon.

**ReviewVocabulary**

exterior angle an angle formed by one side of a polygon and the extension of another side

## 2 Polygon Exterior Angles Sum

Does a relationship exist between the number of sides of a convex polygon and the sum of its exterior angle measures? Examine the polygons below in which an exterior angle has been measured at each vertex.

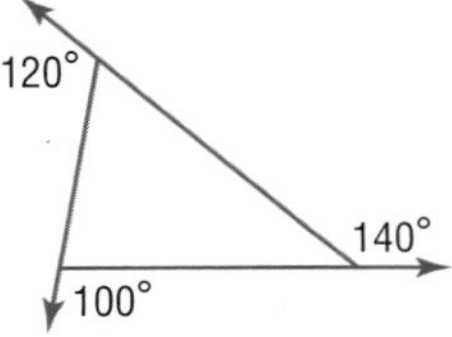


120 + 100 + 140 = **360**

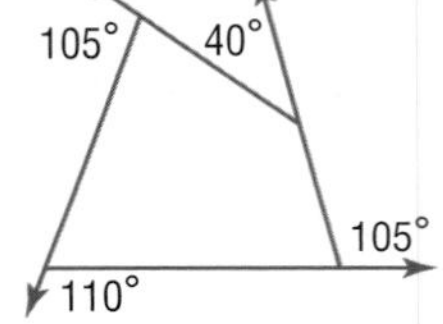


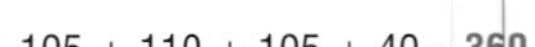


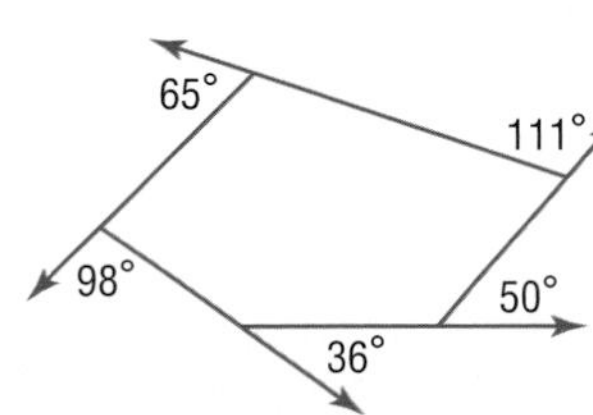


65 + 98 + 36 + 50 + 111= **360**

Notice that the sum of the exterior angle measures in each case is 360. This suggests the following theorem.

### Theorem 8.2 Polygon Exterior Angles Sum

The sum of the exterior angle measures of a convex polygon, one angle at each vertex, is 360.

**Example**

$m\angle 1 + m\angle 2 + m\angle 3 + m\angle 4 + m\angle 5 + m\angle 6 = 360$

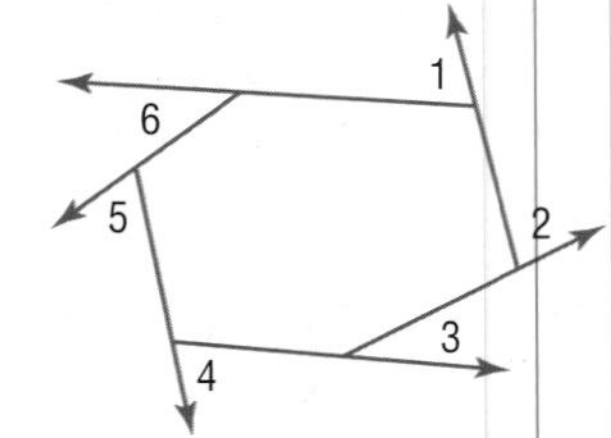


You will prove Theorem 8.2 in Exercise 43.

## Example 4 Find Exterior Angle Measures of a Polygon

**a. ALGEBRA** Find the value of $x$ in the diagram.

Use the Polygon Exterior Angles Sum Theorem to write an equation. Then solve for $x$.

$$(2x - 5) + 5x + 2x + (6x - 5) + (3x + 10) = 360$$

$$(2x + 5x + 2x + 6x + 3x) + [-5 + (-5) + 10] = 360$$

$$18x = 360$$

$$x = \frac{360}{18} \text{ or } 20$$

**StudyTip**

**CCSS Perseverance** To find the measure of each exterior angle of a regular polygon, you can find the measure of each interior angle and subtract this measure from 180, since an exterior angle and its corresponding interior angle are supplementary.

**b.** Find the measure of each exterior angle of a regular nonagon.

A regular nonagon has 9 congruent sides and 9 congruent interior angles. The exterior angles are also congruent, since angles supplementary to congruent angles are congruent. Let $n$ = the measure of each exterior angle and write and solve an equation.

$9n = 360$ Polygon Exterior Angles Sum Theorem

$n = 40$ Divide each side by 9.

The measure of each exterior angle of a regular nonagon is 40.

**4A.** Find the value of $x$ in the diagram.

**4B.** Find the measure of each exterior angle of a regular dodecagon.

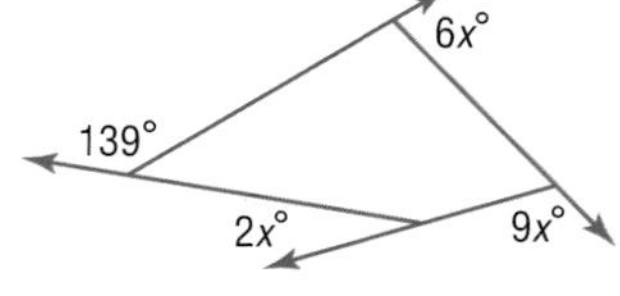


## Check Your Understanding

**Example 1** **Find the sum of the measures of the interior angles of each convex polygon.**

**1.** decagon

**2.** pentagon

**Find the measure of each interior angle.**

**3.**

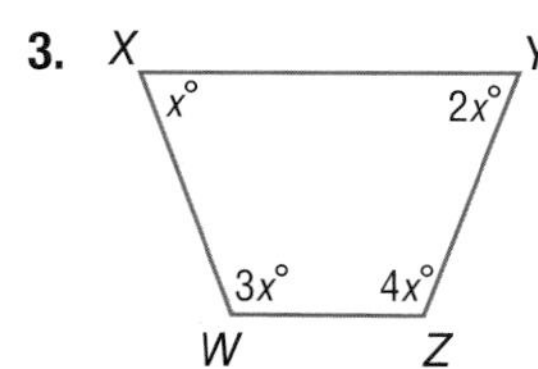


**4.**

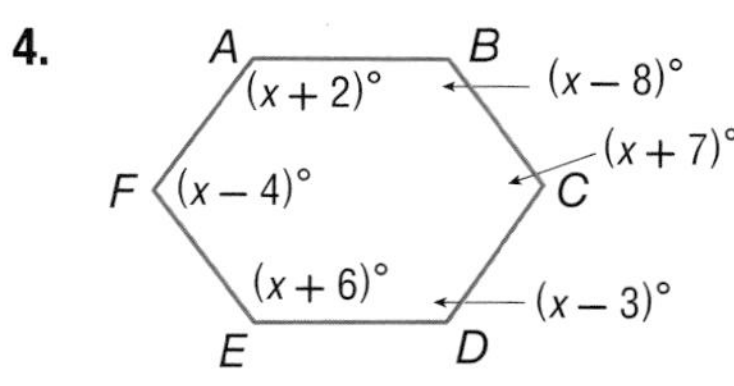


**Example 2** **5** **AMUSEMENT** The Wonder Wheel at Coney Island in Brooklyn, New York, is a regular polygon with 16 sides. What is the measure of each interior angle of the polygon?

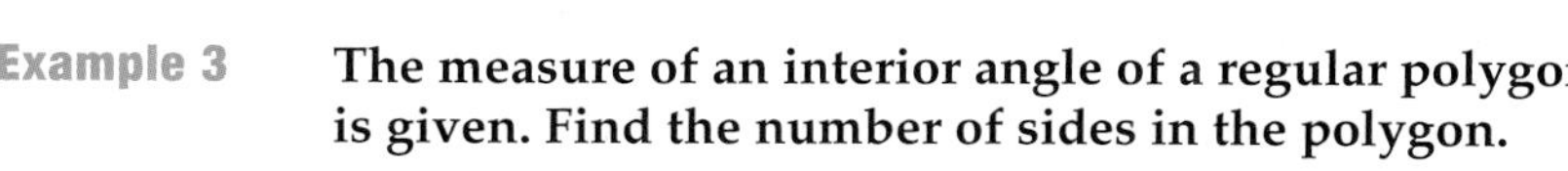

**Example 3** **The measure of an interior angle of a regular polygon is given. Find the number of sides in the polygon.**

**6.** 150

**7.** 170

connectED.mcgraw-hill.com 481

**47** **THEATER** The drama club would like to build a theater in the round, so the audience can be seated on all sides of the stage, for its next production.

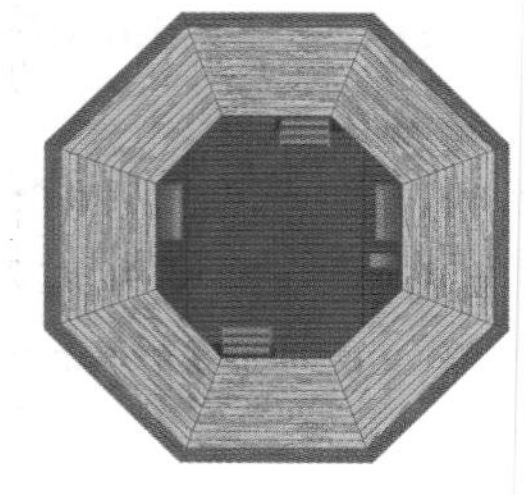

**a.** The stage is to be a regular octagon with a total perimeter of 60 feet. To what length should each board be cut to form the sides of the stage?

**b.** At what angle should each board be cut so that they will fit together as shown? Explain your reasoning.

**48.** **MULTIPLE REPRESENTATIONS** In this problem, you will explore angle and side relationships in special quadrilaterals.

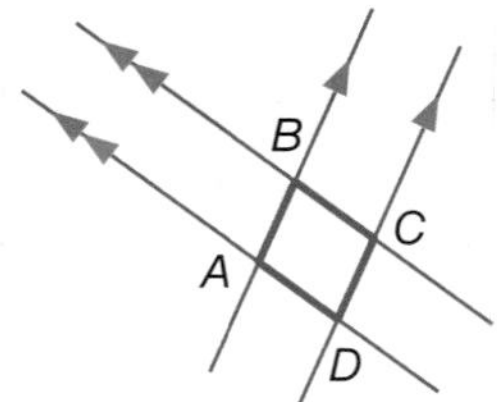


**a. Geometric** Draw two pairs of parallel lines that intersect like the ones shown. Label the quadrilateral formed by *ABCD*. Repeat these steps to form two additional quadrilaterals, *FGHJ* and *QRST*.

**b. Tabular** Copy and complete the table below.

| Quadrilateral | Lengths and Measures | | | | | | | |
|---|---|---|---|---|---|---|---|---|
| *ABCD* | $m\angle A$ | | $m\angle B$ | | $m\angle C$ | | $m\angle D$ | |
| | *AB* | | *BC* | | *CD* | | *DA* | |
| *FGHJ* | $m\angle F$ | | $m\angle G$ | | $m\angle H$ | | $m\angle J$ | |
| | *FG* | | *GH* | | *HJ* | | *JF* | |
| *QRST* | $m\angle Q$ | | $m\angle R$ | | $m\angle S$ | | $m\angle T$ | |
| | *QR* | | *RS* | | *ST* | | *TQ* | |

**c. Verbal** Make a conjecture about the relationship between the angles opposite each other in a quadrilateral formed by two pairs of parallel lines.

**d. Verbal** Make a conjecture about the relationship between two consecutive angles in a quadrilateral formed by two pairs of parallel lines.

**e. Verbal** Make a conjecture about the relationship between the sides opposite each other in a quadrilateral formed by two pairs of parallel lines.

## H.O.T. Problems Use Higher-Order Thinking Skills

**49.** **ERROR ANALYSIS** Marcus says that the sum of the exterior angles of a decagon is greater than that of a heptagon because a decagon has more sides. Liam says that the sum of the exterior angles for both polygons is the same. Is either of them correct? Explain your reasoning.

**50.** **CHALLENGE** Find the values of $a$, $b$, and $c$ if *QRSTVX* is a regular hexagon. Justify your answer.

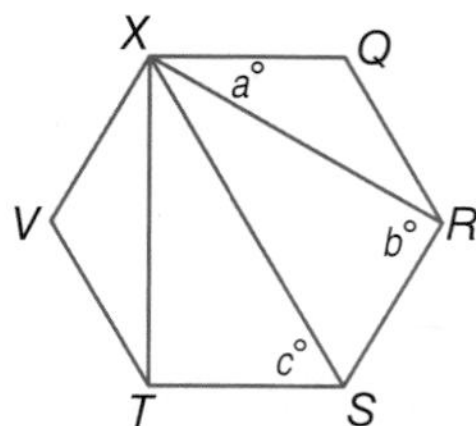


**51.** **CCSS ARGUMENTS** If two sides of a regular hexagon are extended to meet at a point in the exterior of the polygon, will the triangle formed *always, sometimes,* or *never* be equilateral? Justify your answer.

**52.** **OPEN ENDED** Sketch a polygon and find the sum of its interior angles. How many sides does a polygon with twice this interior angles sum have? Justify your answer.

**53.** **WRITING IN MATH** Explain how triangles are related to the Interior Angles Sum Theorem.

## Standardized Test Practice

**54.** If the polygon shown is regular, what is $m\angle ABC$?

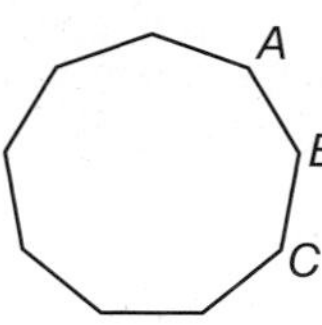


**A** 140

**B** 144

**C** 162

**D** 180

**55. SHORT RESPONSE** Figure $ABCDE$ is a regular pentagon with line $\ell$ passing through side $AE$. What is $m\angle y$?

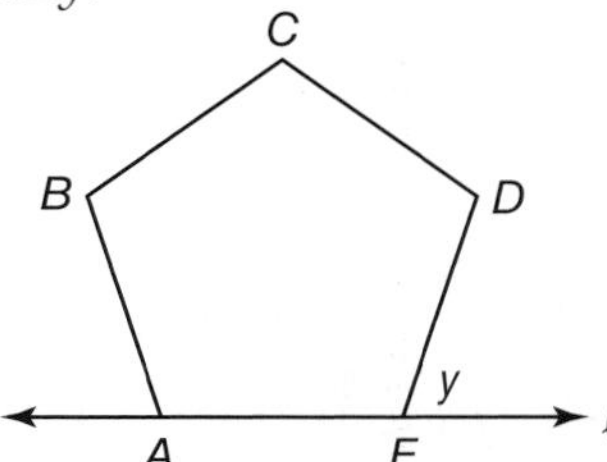


**56. ALGEBRA** $\frac{3^2 \cdot 4^5 \cdot 5^3}{5^3 \cdot 3^3 \cdot 4^6} =$

**F** $\frac{1}{60}$

**G** $\frac{1}{12}$

**H** $\frac{3}{4}$

**J** 12

**57. SAT/ACT** The sum of the measures of the interior angles of a polygon is twice the sum of the measures of its exterior angles. What type of polygon is it?

**A** square

**B** pentagon

**C** hexagon

**D** octagon

**E** nonagon

## Spiral Review

**Compare the given measures.** (Lesson 7-6)

**58.** $m\angle DCE$ and $m\angle SRT$

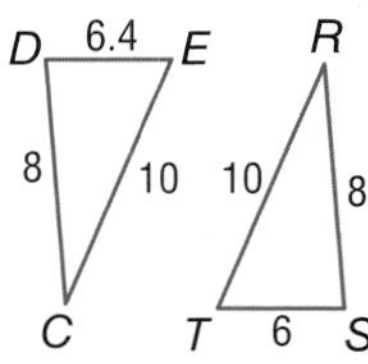


**59.** $JM$ and $ML$

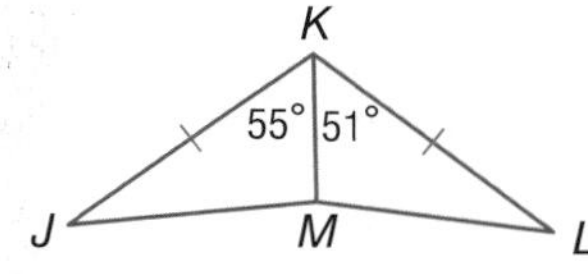


**60.** $WX$ and $ZY$

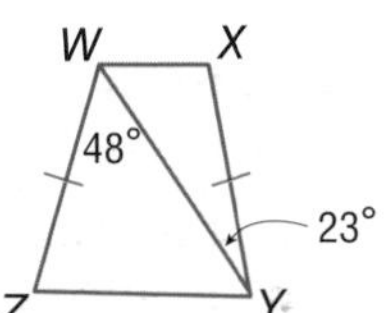


**61. HISTORY** The early Egyptians used to make triangles by using a rope with knots tied at equal intervals. Each vertex of the triangle had to occur at a knot. How many different triangles can be formed using the rope below? (Lesson 7-5)

**Show that the triangles are congruent by identifying all congruent corresponding parts. Then write a congruence statement.** (Lesson 6-2)

**62.**

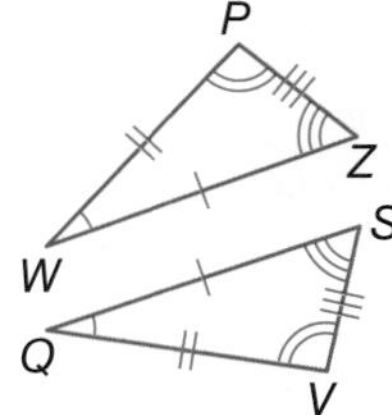


**63.**

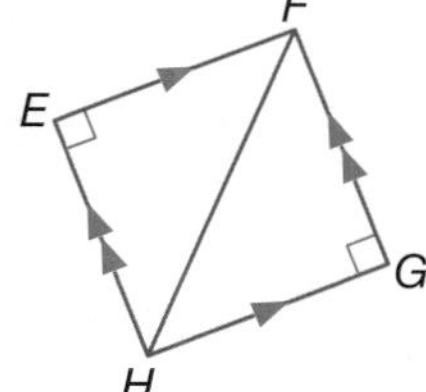


**64.**

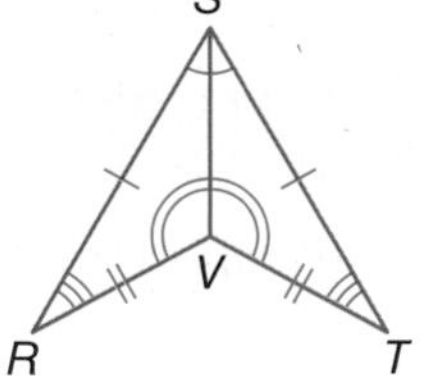


## Skills Review

**In the figure, $\ell \parallel m$ and $\overline{AC} \parallel \overline{BD}$. Name all pairs of angles for each type indicated.**

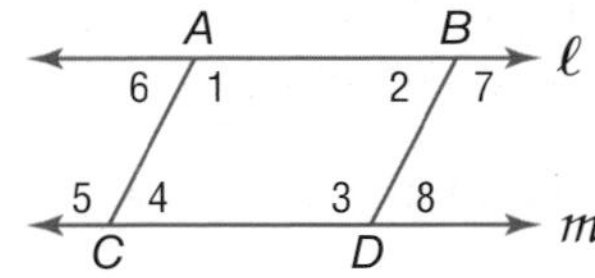


**65.** alternate interior angles

**66.** consecutive interior angles

EXTEND 8-1

# Spreadsheet Lab
# Angles of Polygons

It is possible to find the interior and exterior measurements along with the sum of the interior angles of any regular polygon with *n* number of sides by using a spreadsheet.

PT

## Activity

**Design a spreadsheet using the following steps.**

- Label the columns as shown in the spreadsheet below.
- Enter the digits 3–10 in the first column.
- The number of triangles in a polygon is 2 fewer than the number of sides. Write a formula for Cell B1 to subtract 2 from each number in Cell A1.
- Enter a formula for Cell C1 so the spreadsheet will calculate the sum of the measures of the interior angles. Remember that the formula is $S = (n - 2)180$.
- Continue to enter formulas so that the indicated computation is performed. Then, copy each formula through Row 9. The final spreadsheet will appear as below.

**Polygons and Angles**

| ◇ | A | B | C | D | E | F |
|---|---|---|---|---|---|---|
| 1 | Number of Sides | Number of Triangles | Sum of Measures of Interior Angles | Measure of Each Interior Angle | Measure of Each Exterior Angle | Measures of Exterior Angles |
| 2 | 3 | 1 | 180 | 60 | 120 | 360 |
| 3 | 4 | 2 | 360 | 90 | 90 | 360 |
| 4 | 5 | 3 | 540 | 108 | 72 | 360 |
| 5 | 6 | 4 | 720 | 120 | 60 | 360 |
| 6 | 7 | 5 | 900 | 128.57 | 51.43 | 360 |
| 7 | 8 | 6 | 1080 | 135 | 45 | 360 |
| 8 | 9 | 7 | 1260 | 140 | 40 | 360 |
| 9 | 10 | 8 | 1440 | 144 | 36 | 360 |

Sheet 1 | Sheet 2 | Sheet 3

## Exercises

1. Write the formula to find the measure of each interior angle in the polygon.
2. Write the formula to find the sum of the measures of the exterior angles.
3. What is the measure of each interior angle if the number of sides is 1? 2?
4. Is it possible to have values of 1 and 2 for the number of sides? Explain.

**For Exercises 5–8, use the spreadsheet.**

5. How many triangles are in a polygon with 17 sides?
6. Find the measure of an exterior angle of a regular polygon with 16 sides.
7. Find the measure of an interior angle of a regular polygon with 115 sides.
8. If the measure of the exterior angles is 0, find the measure of the interior angles. Is this possible? Explain.

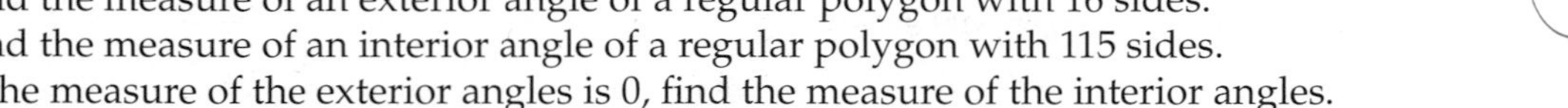

# LESSON 8-2 Parallelograms

**Then**

- You classified polygons with four sides as quadrilaterals.

**Now**

1. Recognize and apply properties of the sides and angles of parallelograms.
2. Recognize and apply properties of the diagonals of parallelograms.

**Why?**

- The arm of the basketball goal shown can be adjusted to a height of 10 feet or 5 feet. Notice that as the height is adjusted, each pair of opposite sides of the quadrilateral formed by the arms remains parallel.

**NewVocabulary**
parallelogram

**Common Core State Standards**

**Content Standards**
G.CO.11 Prove theorems about parallelograms.
G.GPE.4 Use coordinates to prove simple geometric theorems algebraically.

**Mathematical Practices**
4 Model with mathematics.
3 Construct viable arguments and critique the reasoning of others.

## 1 Sides and Angles of Parallelograms

A **parallelogram** is a quadrilateral with both pairs of opposite sides parallel. To name a parallelogram, use the symbol ▱. In ▱*ABCD*, $\overline{BC} \parallel \overline{AD}$ and $\overline{AB} \parallel \overline{DC}$ by definition.

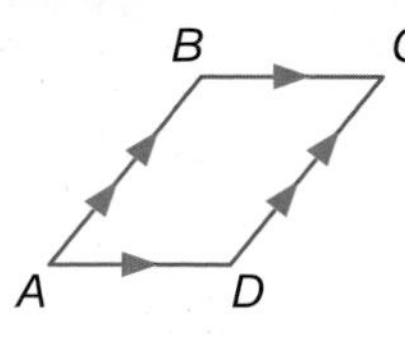

▱*ABCD*

Other properties of parallelograms are given in the theorems below.

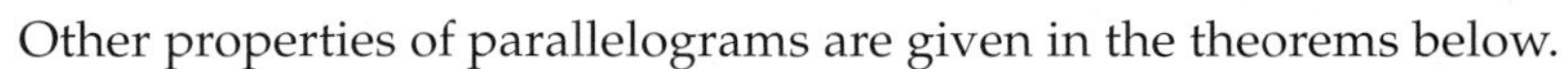

**Theorem** Properties of Parallelograms

**8.3** If a quadrilateral is a parallelogram, then its opposite sides are congruent.

**Abbreviation** *Opp. sides of a ▱ are ≅.*

**Example** If *JKLM* is a parallelogram, then $\overline{JK} \cong \overline{ML}$ and $\overline{JM} \cong \overline{KL}$.

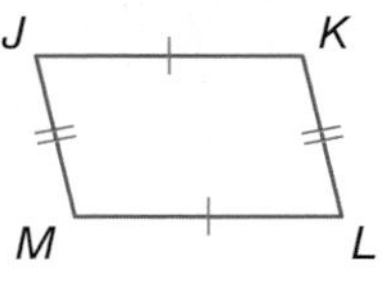


**8.4** If a quadrilateral is a parallelogram, then its opposite angles are congruent.

**Abbreviation** *Opp. ∠s of a ▱ are ≅.*

**Example** If *JKLM* is a parallelogram, then $\angle J \cong \angle L$ and $\angle K \cong \angle M$.

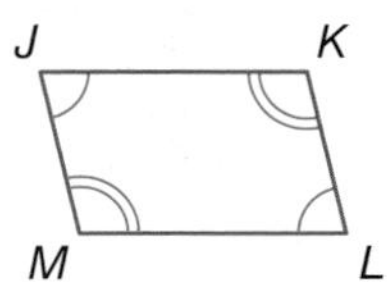


**8.5** If a quadrilateral is a parallelogram, then its consecutive angles are supplementary.

**Abbreviation** *Cons. ∠s in a ▱ are supplementary.*

**Example** If *JKLM* is a parallelogram, then $x + y = 180$.

J K x° y° y° x° M L

**8.6** If a parallelogram has one right angle, then it has four right angles.

**Abbreviation** *If a ▱ has 1 rt. ∠, it has 4 rt. ∠s.*

**Example** In ▱*JKLM*, if $\angle J$ is a right angle, then $\angle K$, $\angle L$, and $\angle M$ are also right angles.

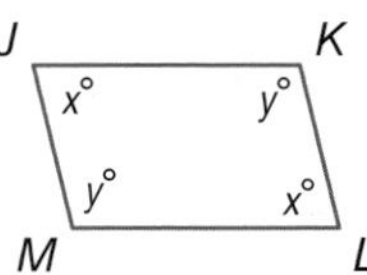


You will prove Theorems 8.3, 8.5, and 8.6 in Exercises 28, 26, and 7, respectively.

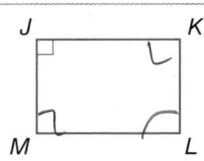

## StudyTip

**Including a Figure** Theorems are presented in general terms. In a proof, you must include a drawing so that you can refer to segments and angles specifically.

## Proof Theorem 8.4

Write a two-column proof of Theorem 8.4.

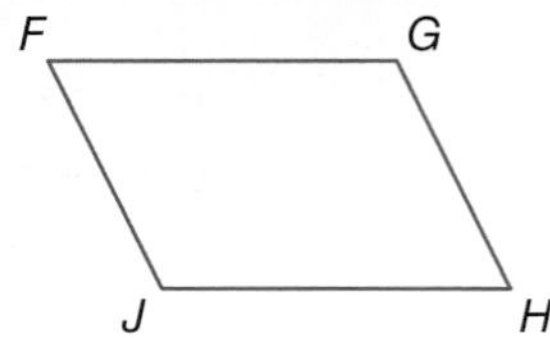


**Given:** ▱$FGHJ$

**Prove:** $\angle F \cong \angle H$, $\angle J \cong \angle G$

**Proof:**

| Statements | Reasons |
|---|---|
| **1.** ▱$FGHJ$ | **1.** Given |
| **2.** $\overline{FG} \parallel \overline{JH}$; $\overline{FJ} \parallel \overline{GH}$ | **2.** Definition of parallelogram |
| **3.** $\angle F$ and $\angle J$ are supplementary. $\angle J$ and $\angle H$ are supplementary. $\angle H$ and $\angle G$ are supplementary. | **3.** If parallel lines are cut by a transversal, consecutive interior angles are supplementary. |
| **4.** $\angle F \cong \angle H$, $\angle J \cong \angle G$ | **4.** Supplements of the same angles are congruent. |

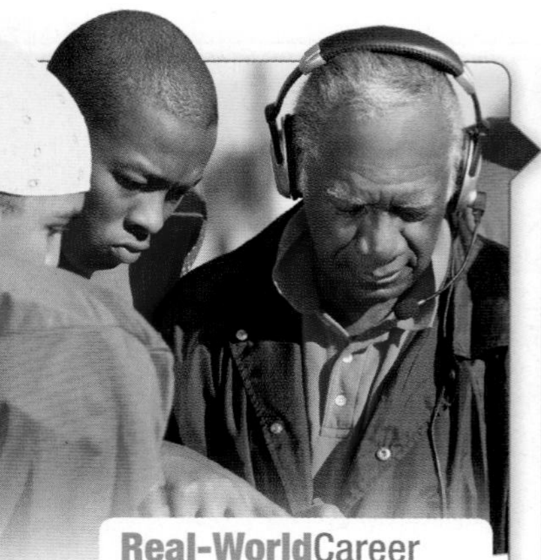

## Real-WorldCareer

**Coach** Coaches organize amateur and professional atheletes, teaching them the fundamentals of a sport. They manage teams during both practice sessions and competitions. Additional tasks may include selecting and issuing sports equiment, materials, and supplies. Head coaches at public secondary schools usually have a bachelor's degree.

## Real-World Example 1 Use Properties of Parallelograms

**BASKETBALL In ▱$ABCD$, suppose $m\angle A = 55$, $AB = 2.5$ feet, and $BC = 1$ foot. Find each measure.**

**a.** $DC$

$DC = AB$ — Opp. sides of a ▱ are ≅.

$= 2.5$ ft — Substitution

**b.** $m\angle B$

$m\angle B + m\angle A = 180$ — Cons. ∠s in a ▱ are supplementary.

$m\angle B + 55 = 180$ — Substitution

$m\angle B = 125$ — Subtract 55 from each side.

**c.** $m\angle C$

$m\angle C = m\angle A$ — Opp. ∠s of a ▱ are ≅.

$= 55$ — Substitution

## GuidedPractice

**1. MIRRORS** The wall-mounted mirror shown uses parallelograms that change shape as the arm is extended. In ▱$JKLM$, suppose $m\angle J = 47$. Find each measure.

**A.** $m\angle L$ **B.** $m\angle M$

**C.** Suppose the arm was extended further so that $m\angle J = 90$. What would be the measure of each of the other angles? Justify your answer.

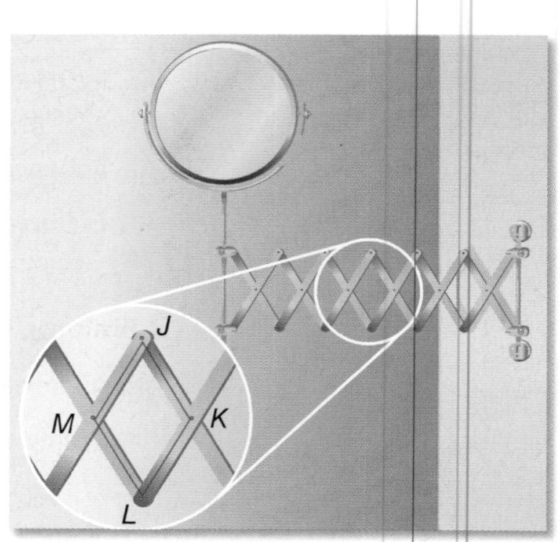


moodboard/SuperStock

## 2 Diagonals of Parallelograms

The diagonals of a parallelogram have special properties as well.

### Theorem Diagonals of Parallelograms

**8.7** If a quadrilateral is a parallelogram, then its diagonals bisect each other.

**Abbreviation** *Diag. of a ▱ bisect each other.*

**Example** If $ABCD$ is a parallelogram, then $\overline{AP} \cong \overline{PC}$ and $\overline{DP} \cong \overline{PB}$.

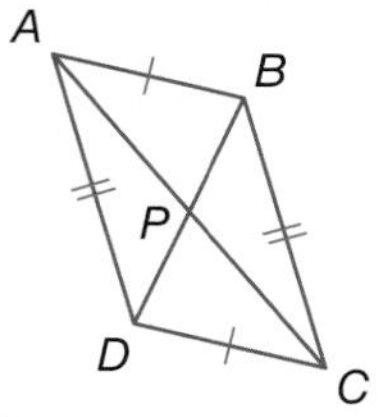


**8.8** If a quadrilateral is a parallelogram, then each diagonal separates the parallelogram into two congruent triangles.

**Abbreviation** *Diag. separates a ▱ into 2 ≅ △s.*

**Example** If $ABCD$ is a parallelogram, then $\triangle ABD \cong \triangle CDB$.

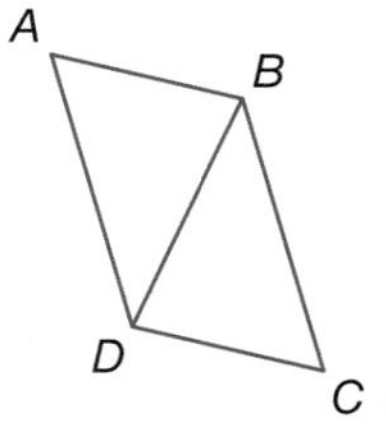


You will prove Theorems 8.7 and 8.8 in Exercises 29 and 27, respectively.

### Example 2 Use Properties of Parallelograms and Algebra

**ALGEBRA If $QRST$ is a parallelogram, find the value of the indicated variable.**

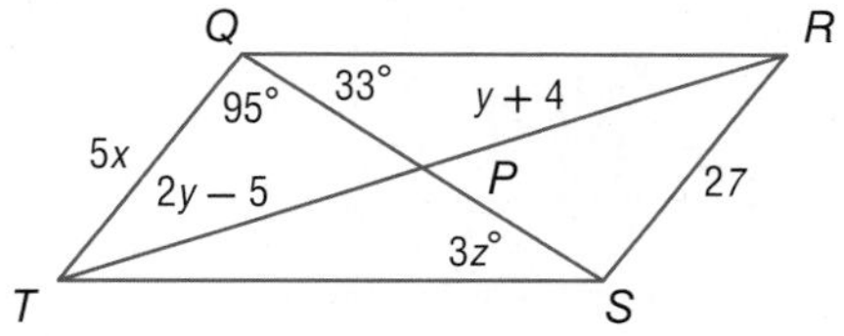


**a.** $x$

| | |
|---|---|
| $\overline{QT} \cong \overline{RS}$ | Opp. sides of a ▱ are ≅. |
| $QT = RS$ | Definition of congruence |
| $5x = 27$ | Substitution |
| $x = 5.4$ | Divide each side by 5. |

**b.** $y$

| | |
|---|---|
| $\overline{TP} \cong \overline{PR}$ | Diag. of a ▱ bisect each other. |
| $TP = PR$ | Definition of congruence |
| $2y - 5 = y + 4$ | Substitution |
| $y = 9$ | Subtract $y$ and add 5 to each side. |

**c.** $z$

| | |
|---|---|
| $\triangle TQS \cong \triangle RSQ$ | Diag. separates a ▱ into 2 ≅ △s. |
| $\angle QST \cong \angle SQR$ | CPCTC |
| $m\angle QST = m\angle SQR$ | Definition of congruence |
| $3z = 33$ | Substitution |
| $z = 11$ | Divide each side by 3. |

**StudyTip**

**Congruent Triangles** A parallelogram with two diagonals divides the figure into two pairs of congruent triangles.

### GuidedPractice

**Find the value of each variable in the given parallelogram.**

**2A.**

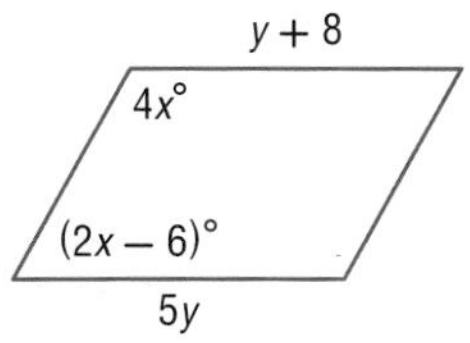


**2B.**

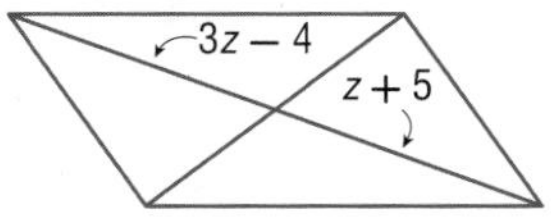

You can use Theorem 8.7 to determine the coordinates of the intersection of the diagonals of a parallelogram on a coordinate plane given the coordinates of the vertices.

## Example 3 Parallelograms and Coordinate Geometry

**COORDINATE GEOMETRY Determine the coordinates of the intersection of the diagonals of ▱*FGHJ* with vertices *F*(−2, 4), *G*(3, 5), *H*(2, −3), and *J*(−3, −4).**

Since the diagonals of a parallelogram bisect each other, their intersection point is the midpoint of $\overline{FH}$ and $\overline{GJ}$. Find the midpoint of $\overline{FH}$ with endpoints (−2, 4) and (2, −3).

$$\left(\frac{x_1 + x_2}{2}, \frac{y_1 + y_2}{2}\right) = \left(\frac{-2 + 2}{2}, \frac{4 + (-3)}{2}\right)$$ Midpoint Formula

$$= (0, 0.5)$$ Simplify.

The coordinates of the intersection of the diagonals of ▱*FGHJ* are (0, 0.5).

**CHECK** Find the midpoint of $\overline{GJ}$ with endpoints (3, 5) and (−3, −4).

$$\left(\frac{3 + (-3)}{2}, \frac{5 + (-4)}{2}\right) = (0, 0.5)\ ✓$$

**StudyTip**

**CCSS Regularity** Graph the parallelogram in Example 3 and the point of intersection of the diagonals you found. Draw the diagonals. The point of intersection appears to be correct.

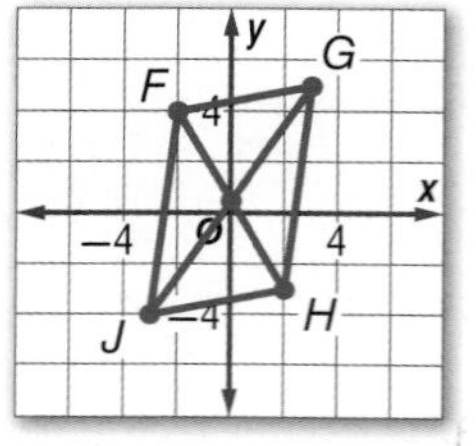


### GuidedPractice

**3. COORDINATE GEOMETRY** Determine the coordinates of the intersection of the diagonals of *RSTU* with vertices *R*(−8, −2), *S*(−6, 7), *T*(6, 7), and *U*(4, −2).

You can use the properties of parallelograms and their diagonals to write proofs.

## Example 4 Proofs Using the Properties of Parallelograms

**Write a paragraph proof.**

**Given:** ▱*ABDG*, $\overline{AF} \cong \overline{CF}$

**Prove:** $\angle BDG \cong \angle C$

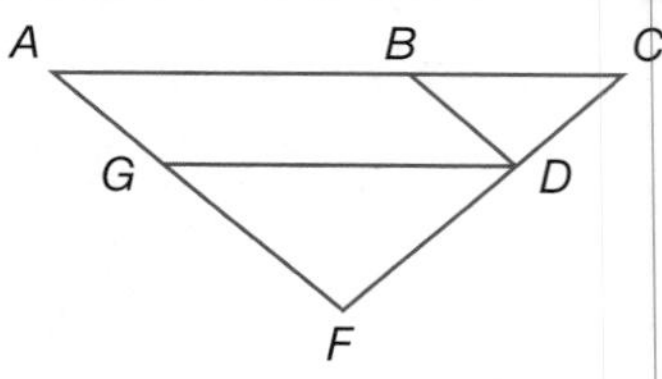


**Proof:**

We are given *ABDG* is a parallelogram. Since opposite angles in a parallelogram are congruent, $\angle BDG \cong \angle A$. We are also given that $\overline{AF} \cong \overline{CF}$. By the Isosceles Triangle Theorem, $\angle A \cong \angle C$. So, by the Transitive Property of Congruence, $\angle BDG \cong \angle C$.

### GuidedPractice

**4.** Write a two-column proof.

**Given:** ▱*HJKP* and ▱*PKLM*

**Prove:** $\overline{HJ} \cong \overline{ML}$

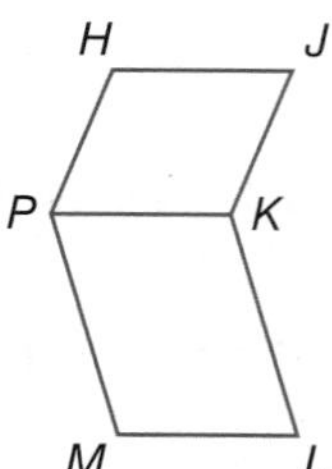

## Check Your Understanding

**Example 1** **1. NAVIGATION** To chart a course, sailors use a *parallel ruler.* One edge of the ruler is placed along the line representing the direction of the course to be taken. Then the other ruler is moved until its edge reaches the compass rose printed on the chart. Reading the compass determines which direction to travel. The rulers and the crossbars of the tool form ▱$MNPQ$.

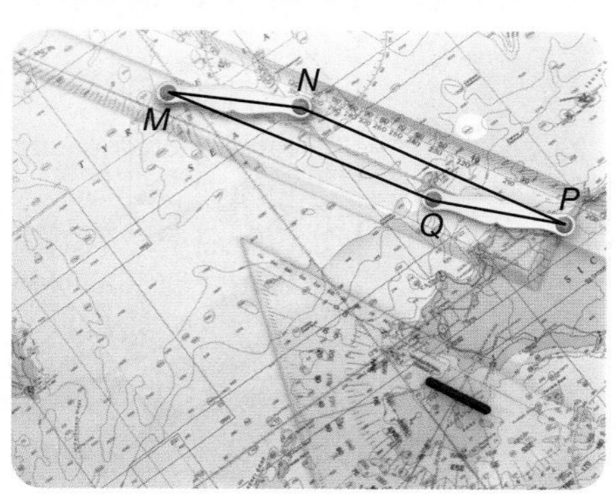


**a.** If $m\angle NMQ = 32$, find $m\angle MNP$.

**b.** If $m\angle MQP = 125$, find $m\angle MNP$.

**c.** If $MQ = 4$, what is $NP$?

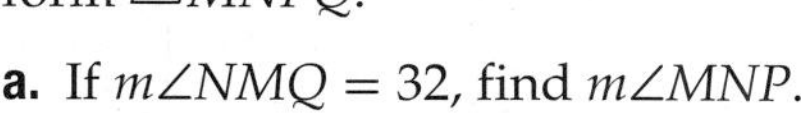

**Example 2** **ALGEBRA** **Find the value of each variable in each parallelogram.**

**2.**

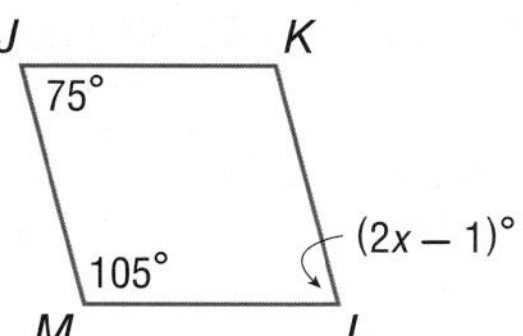


**3.**

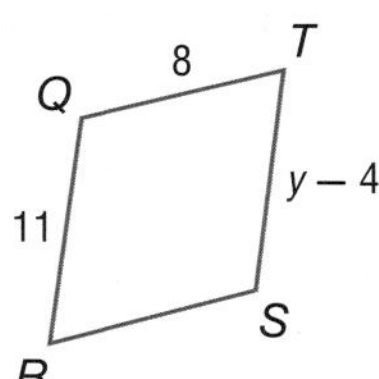


**4.**

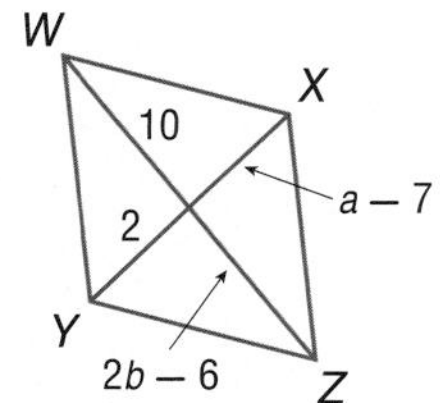


**5.**

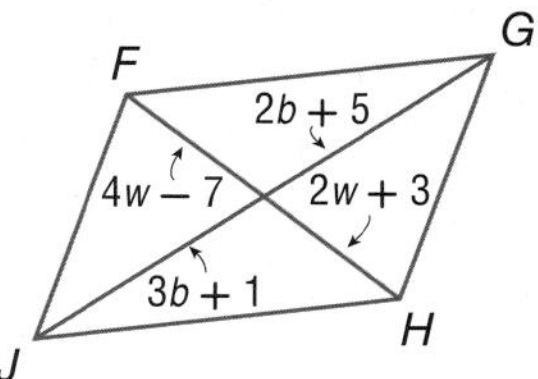


**Example 3** **6. COORDINATE GEOMETRY** Determine the coordinates of the intersection of the diagonals of ▱$ABCD$ with vertices $A(-4, 6)$, $B(5, 6)$, $C(4, -2)$, and $D(-5, -2)$.

**Example 4** **CCSS ARGUMENTS** **Write the indicated type of proof.**

**7.** paragraph

**Given:** ▱$ABCD$, $\angle A$ is a right angle.
**Prove:** $\angle B$, $\angle C$, and $\angle D$ are right angles. (Theorem 6.6)

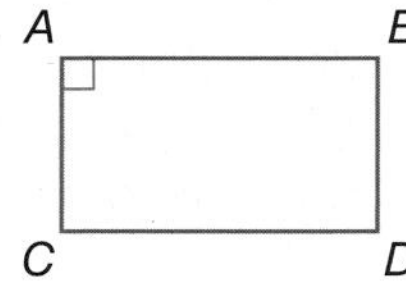


**8.** two-column

**Given:** $ABCH$ and $DCGF$ are parallelograms.
**Prove:** $\angle A \cong \angle F$

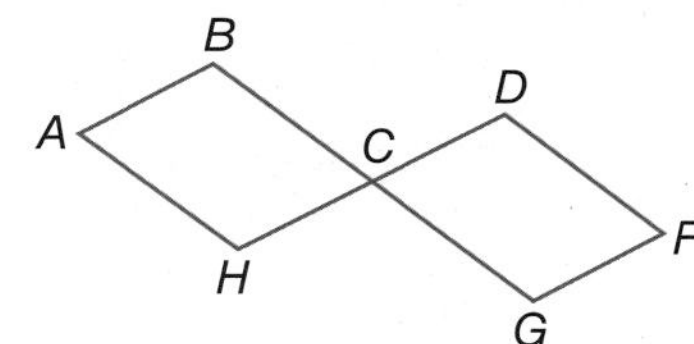


## Practice and Problem Solving

**Example 1** **Use ▱$PQRS$ to find each measure.**

**9** $m\angle R$

**10.** $QR$

**11.** $QP$

**12.** $m\angle S$

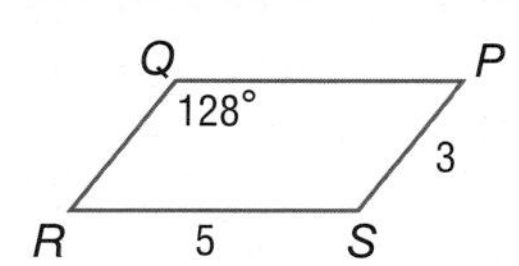


Masterfile

**13** **HOME DECOR** The slats on Venetian blinds are designed to remain parallel in order to direct the path of light coming in a window. In ▱*FGHJ*, $FJ = \frac{3}{4}$ inch, $FG = 1$ inch, and $m\angle JHG = 62$. Find each measure.

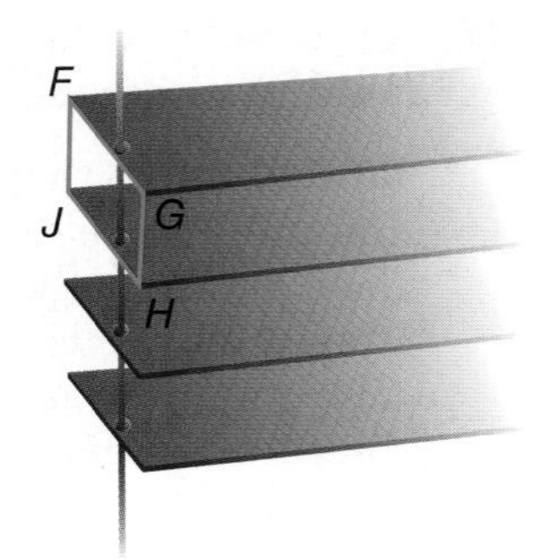


**a.** $JH$

**b.** $GH$

**c.** $m\angle JFG$

**d.** $m\angle FJH$

**14.** **CCSS MODELING** Wesley is a member of the kennel club in his area. His club uses accordion fencing like the section shown at the right to block out areas at dog shows.

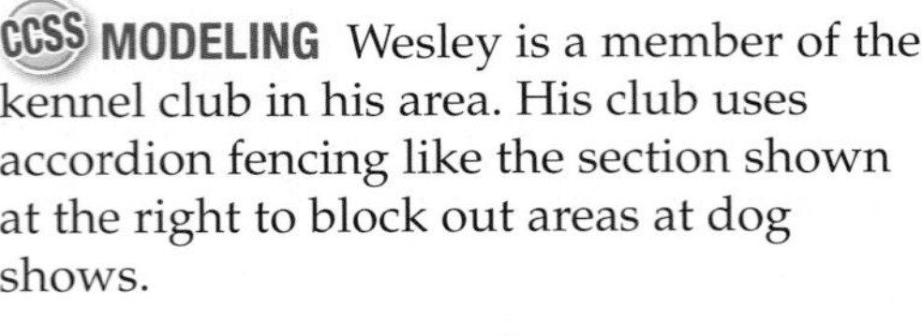

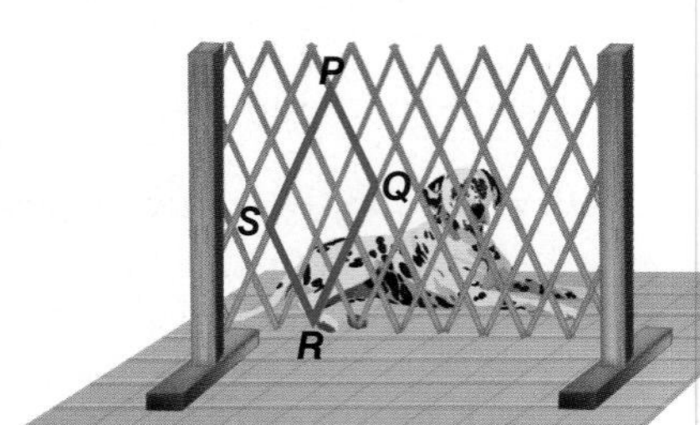


**a.** Identify two pairs of congruent segments.

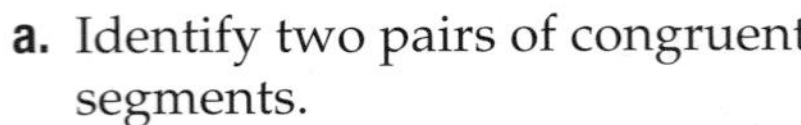

**b.** Identify two pairs of supplementary angles.

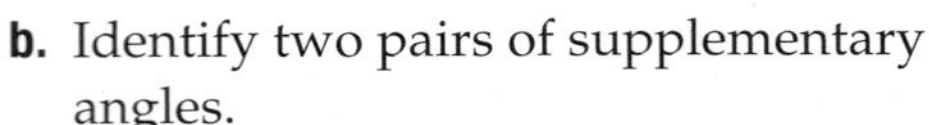

**Example 2** **ALGEBRA** **Find the value of each variable in each parallelogram.**

**15.**

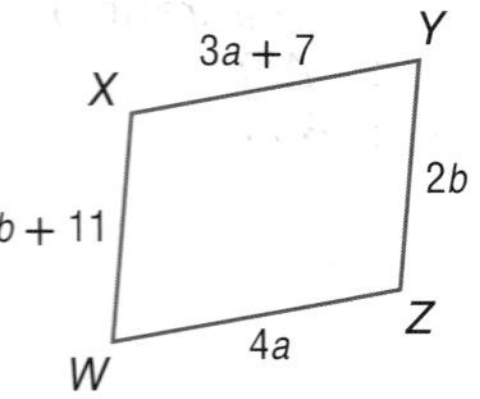


**16.**

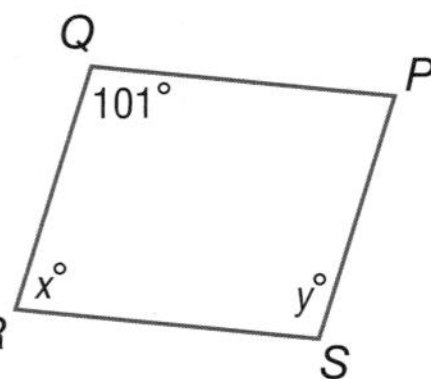


**17.**

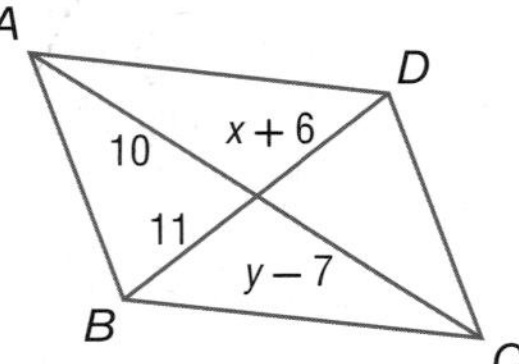


**18.**

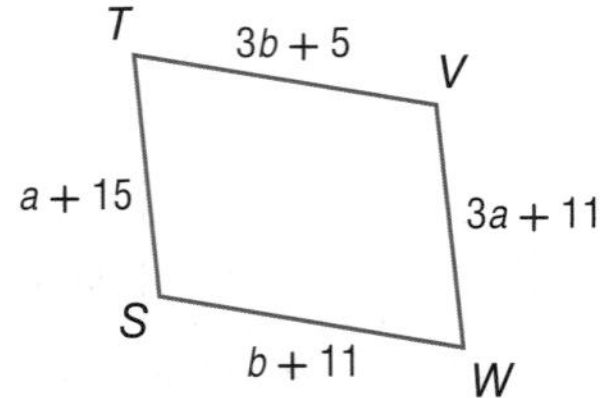


**19.**

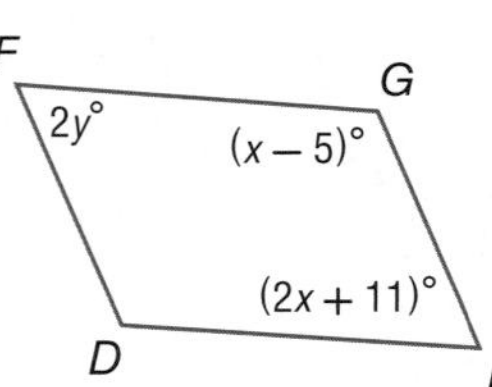


**20.**

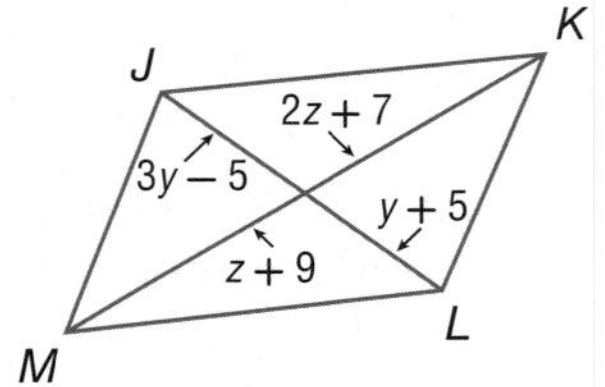


**Example 3** **COORDINATE GEOMETRY** **Find the coordinates of the intersection of the diagonals of ▱*WXYZ* with the given vertices.**

**21.** $W(-1, 7), X(8, 7), Y(6, -2), Z(-3, -2)$

**22.** $W(-4, 5), X(5, 7), Y(4, -2), Z(-5, -4)$

**Example 4** **PROOF** **Write a two-column proof.**

**23. Given:** *WXTV* and *ZYVT* are parallelograms.
**Prove:** $\overline{WX} \cong \overline{ZY}$

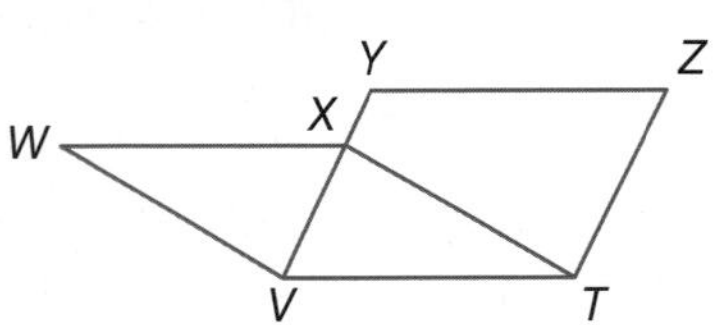


**24. Given:** ▱*BDHA*, $\overline{CA} \cong \overline{CG}$
**Prove:** $\angle BDH \cong \angle G$

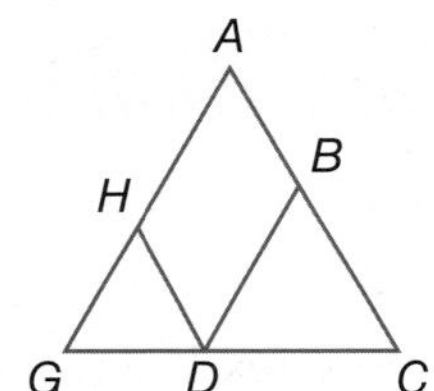

**25. FLAGS** Refer to the Alabama state flag at the right.

**Given:** $\triangle ACD \cong \triangle CAB$

**Prove:** $\overline{DP} \cong \overline{PB}$

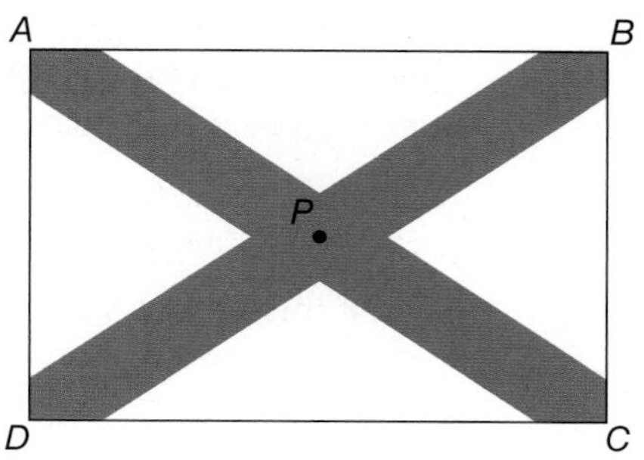


**CCSS ARGUMENTS** **Write the indicated type of proof.**

**26.** two-column
**Given:** ▱$GKLM$
**Prove:** $\angle G$ and $\angle K$, $\angle K$ and $\angle L$, $\angle L$ and $\angle M$, and $\angle M$ and $\angle G$ are supplementary. (Theorem 6.5)

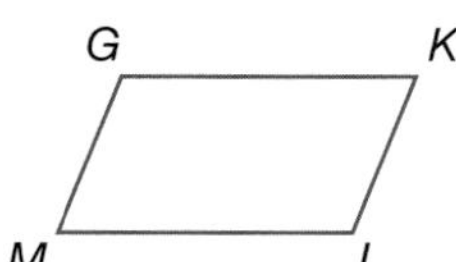


**27.** two-column
**Given:** ▱$WXYZ$
**Prove:** $\triangle WXZ \cong \triangle YZX$ (Theorem 6.8)

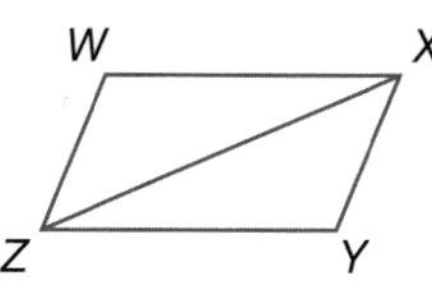


**28.** two-column
**Given:** ▱$PQRS$
**Prove:** $\overline{PQ} \cong \overline{RS}$, $\overline{QR} \cong \overline{SP}$ (Theorem 6.3)

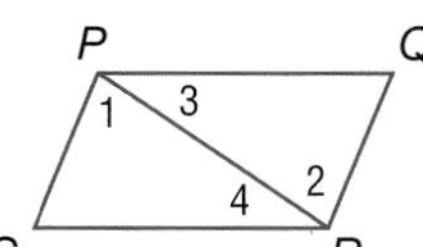


**29.** paragraph
**Given:** ▱$ACDE$ is a parallelogram.
**Prove:** $\overline{EC}$ bisects $\overline{AD}$. (Theorem 6.7)

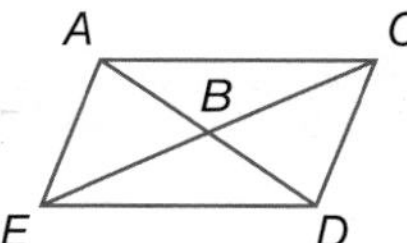


**30. COORDINATE GEOMETRY** Use the graph shown.

**a.** Use the Distance Formula to determine if the diagonals of $JKLM$ bisect each other. Explain.

**b.** Determine whether the diagonals are congruent. Explain.

**c.** Use slopes to determine if the consecutive sides are perpendicular. Explain.

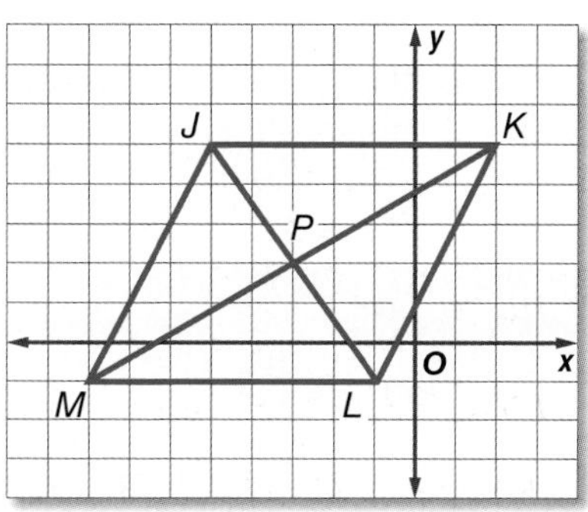


**ALGEBRA** **Use ▱$ABCD$ to find each measure or value.**

**31.** $x$

**32.** $y$

**33** $m\angle AFB$

**34.** $m\angle DAC$

**35.** $m\angle ACD$

**36.** $m\angle DAB$

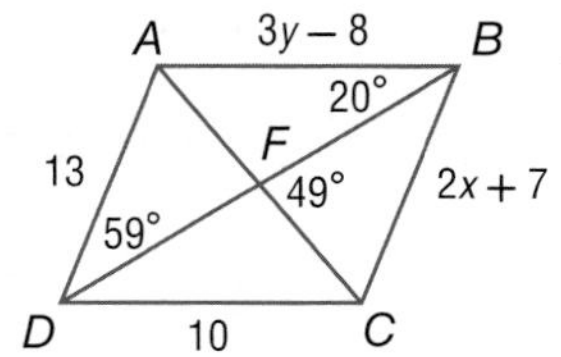


**37. COORDINATE GEOMETRY** ▱$ABCD$ has vertices $A(-3, 5)$, $B(1, 2)$, and $C(3, -4)$. Determine the coordinates of vertex $D$ if it is located in Quadrant III.

38. **MECHANICS** Scissor lifts are variable elevation work platforms. One is shown at the right. In the diagram, *ABCD* and *DEFG* are congruent parallelograms.

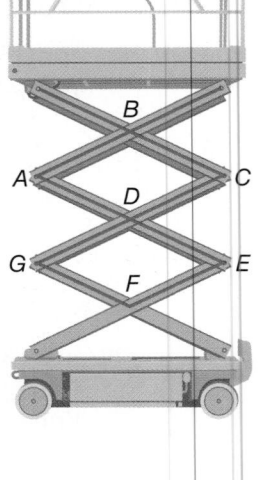

   a. List the angle(s) congruent to $\angle A$. Explain your reasoning.

   b. List the segment(s) congruent to $\overline{BC}$. Explain your reasoning.

   c. List the angle(s) supplementary to $\angle C$. Explain your reasoning.

**PROOF Write a two-column proof.**

39. **Given:** $\square YWVZ$, $\overline{VX} \perp \overline{WY}$, $\overline{YU} \perp \overline{VZ}$
**Prove:** $\triangle YUZ \cong \triangle VXW$

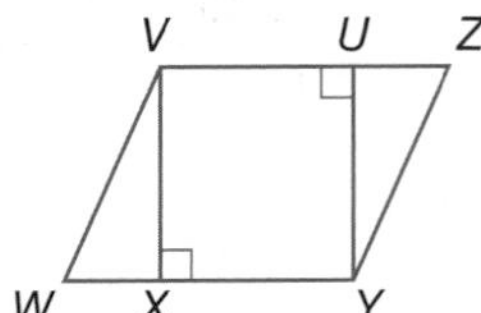


40. **MULTIPLE REPRESENTATIONS** In this problem, you will explore tests for parallelograms.

   a. **Geometric** Draw three pairs of segments that are both congruent and parallel and connect the endpoints to form quadrilaterals. Label one quadrilateral *ABCD*, one *MNOP*, and one *WXYZ*. Measure and label the sides and angles of the quadrilaterals.

   b. **Tabular** Copy and complete the table below.

| Quadrilateral | Opposite Sides Congruent? | Opposite Angles Congruent? | Parallelogram |
|---|---|---|---|
| *ABCD* | | | |
| *MNOP* | | | |
| *WXYZ* | | | |

   c. **Verbal** Make a conjecture about quadrilaterals with one pair of segments that are both congruent and parallel.

## H.O.T. Problems Use Higher-Order Thinking Skills

41. **CHALLENGE** *ABCD* is a parallelogram with side lengths as indicated in the figure at the right. The perimeter of *ABCD* is 22. Find *AB*.

42. **WRITING IN MATH** Explain why parallelograms are *always* quadrilaterals, but quadrilaterals are *sometimes* parallelograms.

43. **OPEN ENDED** Provide a counterexample to show that parallelograms are not always congruent if their corresponding sides are congruent.

44. **CCSS REASONING** Find $m\angle 1$ and $m\angle 10$ in the figure at the right. Explain.

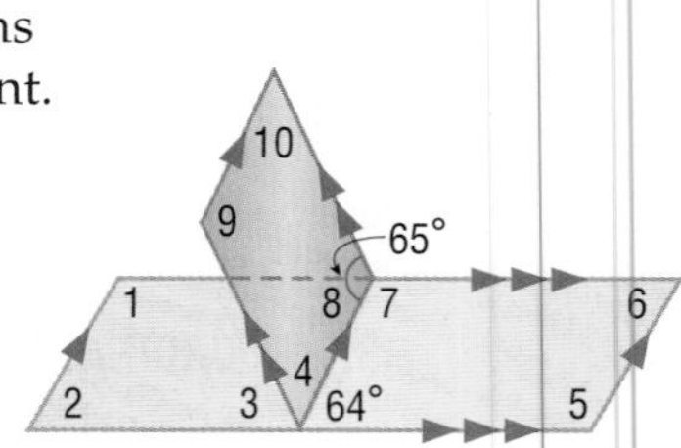


45. **WRITING IN MATH** Summarize the properties of the sides, angles, and diagonals of a parallelogram.

## Standardized Test Practice

**46.** Two consecutive angles of a parallelogram measure $3x + 42$ and $9x - 18$. What are the measures of the angles?

**A** 13, 167 **C** 39, 141

**B** 58.5, 31.5 **D** 81, 99

**47. GRIDDED RESPONSE** Parallelogram *MNPQ* is shown. What is the value of *x*?

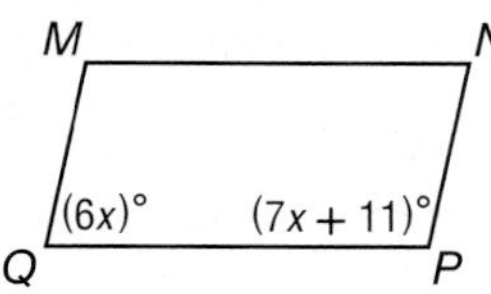


**48. ALGEBRA** In a history class with 32 students, the ratio of girls to boys is 5 to 3. How many more girls are there than boys?

**F** 2 **G** 8 **H** 12 **J** 15

**49. SAT/ACT** The table shows the heights of the tallest buildings in Kansas City, Missouri. To the nearest tenth, what is the positive difference between the median and the mean of the data?

| Name | Height (m) |
|---|---|
| One Kansas City Place | 193 |
| Town Pavillion | 180 |
| Hyatt Regency | 154 |
| Power and Light Building | 147 |
| City Hall | 135 |
| 1201 Walnut | 130 |

**A** 5

**B** 6

**C** 7

**D** 8

**E** 10

## Spiral Review

**The measure of an interior angle of a regular polygon is given. Find the number of sides in the polygon.** (Lesson 8-1)

**50.** 108 **51.** 140 **52.** $\approx 147.3$ **53.** 160 **54.** 135 **55.** 176.4

**56. LANDSCAPING** When landscapers plant new trees, they usually brace the tree using a stake tied to the trunk of the tree. Use the SAS or SSS Inequality to explain why this is an effective method for keeping a newly planted tree perpendicular to the ground. Assume that the tree does not lean forward or backward. (Lesson 7-6)

## Skills Review

**The vertices of a quadrilateral are $W(3, -1)$, $X(4, 2)$, $Y(-2, 3)$ and $Z(-3, 0)$. Determine whether each segment is a side or diagonal of the quadrilateral, and find the slope of each segment.**

**57.** $\overline{YZ}$ **58.** $\overline{YW}$ **59.** $\overline{ZW}$

EXPLORE

# 8-3 Graphing Technology Lab Parallelograms

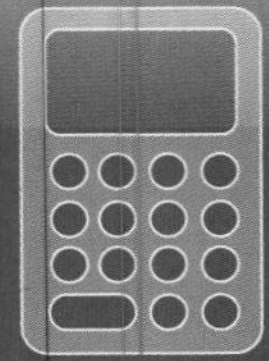

You can use the Cabri™ Jr. application on a TI-83/84 Plus graphing calculator to discover properties of parallelograms.

**Common Core State Standards**
**Content Standards**
**G.CO.12** Make formal geometric constructions with a variety of tools and methods (compass and straightedge, string, reflective devices, paper folding, dynamic geometric software, etc.).
**Mathematical Practices** 5

## Activity

**Construct a quadrilateral with one pair of sides that are both parallel and congruent.**

**Step 1** Construct a segment using the **Segment** tool on the **F2** menu. Label the segment $\overline{AB}$. This is one side of the quadrilateral.

**Step 2** Use the **Parallel** tool on the **F3** menu to construct a line parallel to the segment. Pressing ENTER will draw the line and a point on the line. Label the point $C$.

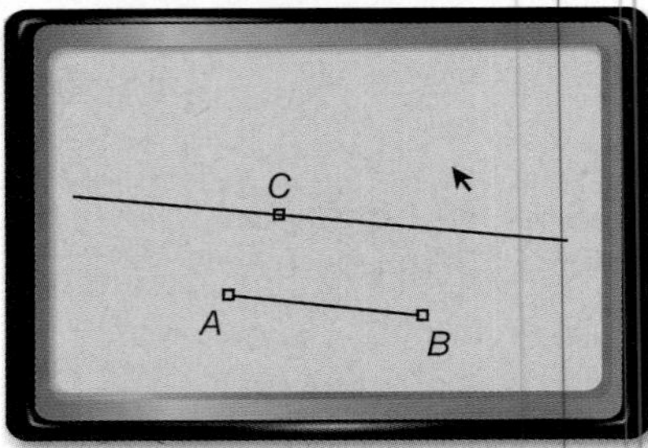


Steps 1 and 2

**Step 3** Access the **Compass** tool on the **F3** menu. Set the compass to the length of $\overline{AB}$ by selecting one endpoint of the segment and then the other. Construct a circle centered at $C$.

**Step 4** Use the **Point Intersection** tool on the **F2** menu to draw a point at the intersection of the line and the circle. Label the point $D$. Then use the **Segment** tool on the **F2** menu to draw $\overline{AC}$ and $\overline{BD}$.

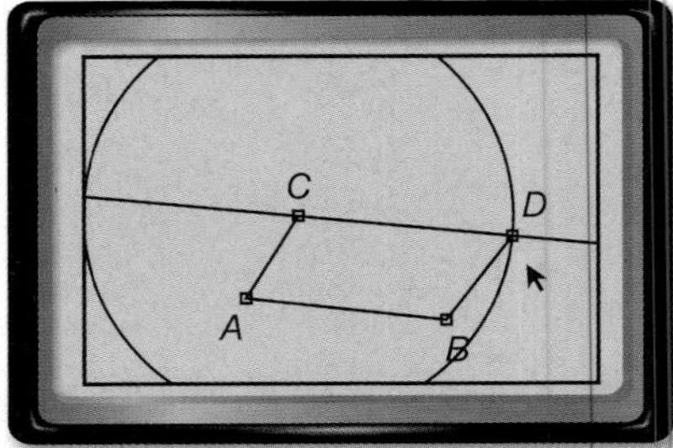


Steps 3 and 4

**Step 5** Use the **Hide/Show** tool on the **F5** menu to hide the circle. Then access **Slope** tool under **Measure** on the **F5** menu. Display the slopes of $\overline{AB}$, $\overline{BD}$, $\overline{CD}$, and $\overline{AC}$.

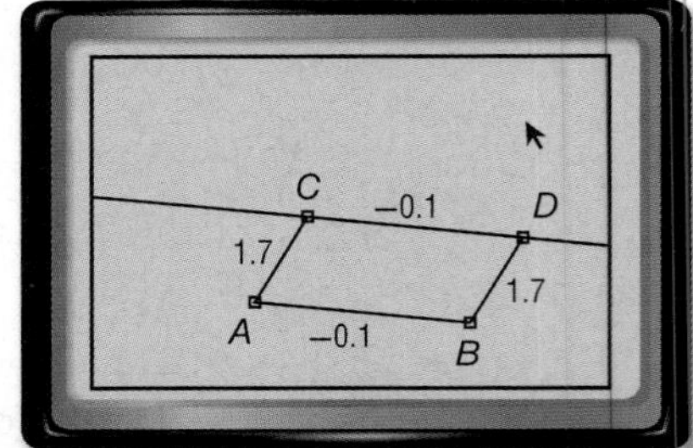


Step 5

## Analyze the Results

1. What is the relationship between sides $\overline{AB}$ and $\overline{CD}$? Explain how you know.
2. What do you observe about the slopes of opposite sides of the quadrilateral? What type of quadrilateral is $ABDC$? Explain.
3. Click on point $A$ and drag it to change the shape of $ABDC$. What do you observe?
4. Make a conjecture about a quadrilateral with a pair of opposite sides that are both congruent and parallel.
5. Use a graphing calculator to construct a quadrilateral with both pairs of opposite sides congruent. Then analyze the slopes of the sides of the quadrilateral. Make a conjecture based on your observations.

# LESSON 8-3 Tests for Parallelograms

**Then**

- You recognized and applied properties of parallelograms.

**Now**

1. Recognize the conditions that ensure a quadrilateral is a parallelogram.
2. Prove that a set of points forms a parallelogram in the coordinate plane.

**Why?**

- Lexi and Rosalinda cut strips of bulletin board paper at an angle to form the hallway display shown. Their friends asked them how they cut the strips so that their sides were parallel without using a protractor.

Rosalinda explained that since the left and right sides of the paper were parallel, she only needed to make sure that the sides were cut to the same length to guarantee that a strip would form a parallelogram.

**Common Core State Standards**

**Content Standards**

G.CO.11 Prove theorems about parallelograms.

G.GPE.4 Use coordinates to prove simple geometric theorems algebraically.

**Mathematical Practices**

3 Construct viable arguments and critique the reasoning of others.

2 Reason abstractly and quantitatively.

## 1 Conditions for Parallelograms

If a quadrilateral has each pair of opposite sides parallel, it is a parallelogram by definition.

This is not the only test, however, that can be used to determine if a quadrilateral is a parallelogram.

### Theorems Conditions for Parallelograms

**8.9** If both pairs of opposite sides of a quadrilateral are congruent, then the quadrilateral is a parallelogram.

**Abbreviation** *If both pairs of opp. sides are ≅, then quad. is a ▱.*

**Example** If $\overline{AB} \cong \overline{DC}$ and $\overline{AD} \cong \overline{BC}$, then *ABCD* is a parallelogram.

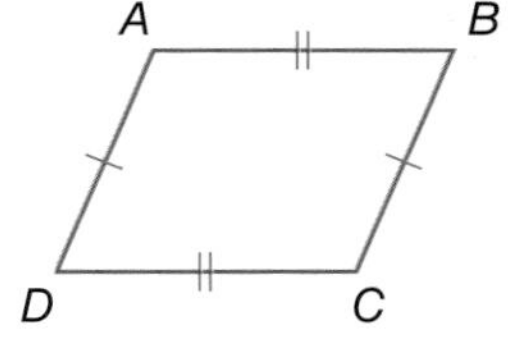


**8.10** If both pairs of opposite angles of a quadrilateral are congruent, then the quadrilateral is a parallelogram.

**Abbreviation** *If both pairs of opp. ∠s are ≅, then quad. is a ▱.*

**Example** If $\angle A \cong \angle C$ and $\angle B \cong \angle D$, then *ABCD* is a parallelogram.

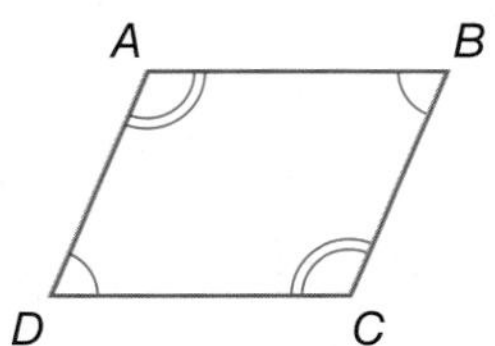


**8.11** If the diagonals of a quadrilateral bisect each other, then the quadrilateral is a parallelogram.

**Abbreviation** *If diag. bisect each other, then quad. is a ▱.*

**Example** If $\overline{AC}$ and $\overline{DB}$ bisect each other, then *ABCD* is a parallelogram.

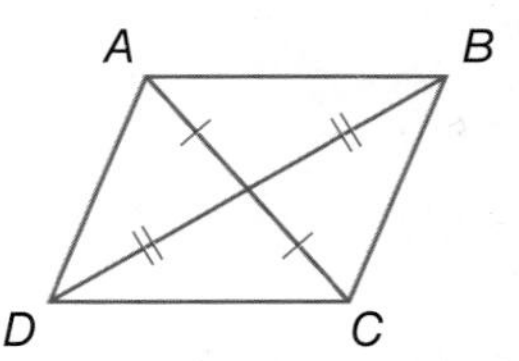


**8.12** If one pair of opposite sides of a quadrilateral is both parallel and congruent, then the quadrilateral is a parallelogram.

**Abbreviation** *If one pair of opp. sides is ≅ and ||, then the quad. is a ▱.*

**Example** If $\overline{AB} \parallel \overline{DC}$ and $\overline{AB} \cong \overline{DC}$, then *ABCD* is a parallelogram.

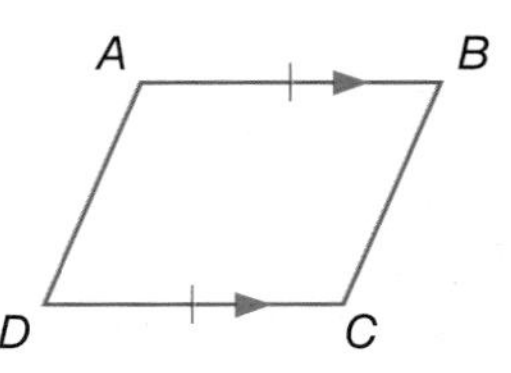


You will prove Theorems 8.10, 8.11, and 8.12 in Exercises 30, 32, and 33, respectively.

## Proof Theorem 8.9

Write a paragraph proof of Theorem 8.9.

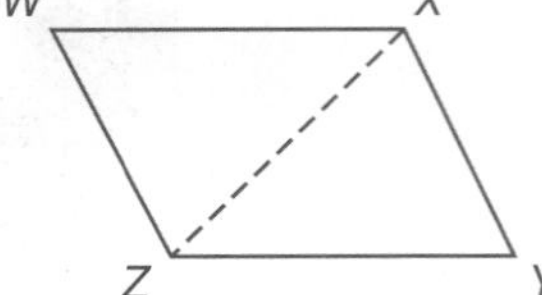


**Given:** $\overline{WX} \cong \overline{ZY}$, $\overline{WZ} \cong \overline{XY}$

**Prove:** *WXYZ* is a parallelogram.

**Paragraph Proof:**

Two points determine a line, so we can draw auxiliary line $\overline{ZX}$ to form $\triangle ZWX$ and $\triangle XYZ$. We are given that $\overline{WX} \cong \overline{ZY}$ and $\overline{WZ} \cong \overline{XY}$. Also, $\overline{ZX} \cong \overline{XZ}$ by the Reflexive Property of Congruence. So $\triangle ZWX \cong \triangle XYZ$ by SSS. By CPCTC, $\angle WXZ \cong \angle YZX$ and $\angle WZX \cong \angle YXZ$. This means that $\overline{WX} \parallel \overline{ZY}$ and $\overline{WZ} \parallel \overline{XY}$ by the Alternate Interior Angles Converse. Opposite sides of *WXYZ* are parallel, so by definition *WXYZ* is a parallelogram.

PT

## Example 1 Identify Parallelograms

**Determine whether the quadrilateral is a parallelogram. Justify your answer.**

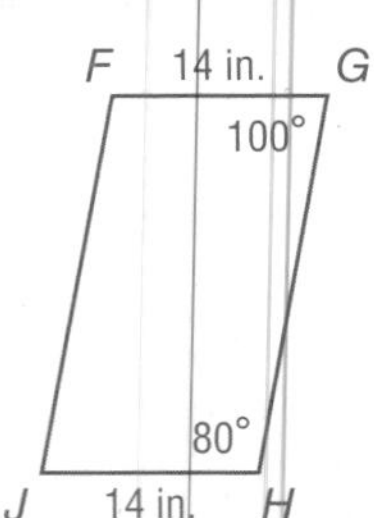


Opposite sides $\overline{FG}$ and $\overline{JH}$ are congruent because they have the same measure. Also, since $\angle FGH$ and $\angle GHJ$ are supplementary consecutive interior angles, $\overline{FG} \parallel \overline{JH}$. Therefore, by Theorem 8.12, *FGHJ* is a parallelogram.

### GuidedPractice

**1A.**

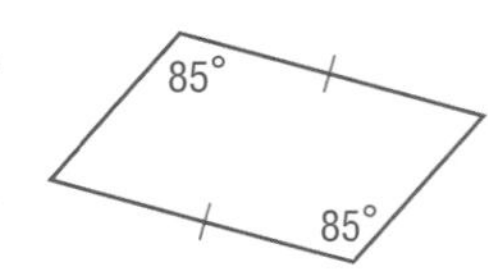


**1B.**

You can use the conditions of parallelograms to prove relationships in real-world situations.

PT

## Real-World Example 2 Use Parallelograms to Prove Relationships

**FISHING The diagram shows a side view of the tackle box at the left. In the diagram, $PQ = RS$ and $PR = QS$. Explain why the upper and middle trays remain parallel no matter to what height the trays are raised or lowered.**

Since both pairs of opposite sides of quadrilateral *PQSR* are congruent, *PQRS* is a parallelogram by Theorem 8.9. By the definition of a parallelogram, opposite sides are parallel, so $\overline{PQ} \parallel \overline{RS}$. Therefore, no matter the vertical position of the trays, they will always remain parallel.

### Real-WorldLink

A 2- or 3-cantilever tackle box is often used to organize lures and other fishing supplies. The trays lift up and away so that all items in the box are easily accessible.

### GuidedPractice

**2. BANNERS** In the example at the beginning of the lesson, explain why the cuts made by Lexi and Rosalinda are parallel.

You can also use the conditions of parallelograms along with algebra to find missing values that make a quadrilateral a parallelogram.

**WatchOut!**

**Parallelograms** In Example 3, if $x$ is 4, then $y$ must be 2.5 in order for $FGHJ$ to be a parallelogram. In other words, if $x$ is 4 and $y$ is 1, then $FGHJ$ is not a parallelogram.

PT

### Example 3 Use Parallelograms and Algebra to Find Values

**If $FK = 3x - 1$, $KG = 4y + 3$, $JK = 6y - 2$, and $KH = 2x + 3$, find $x$ and $y$ so that the quadrilateral is a parallelogram.**

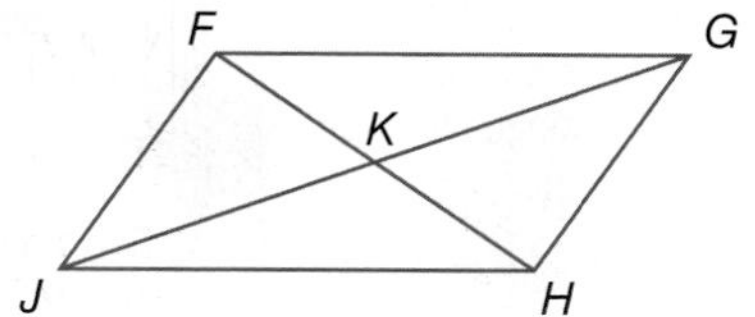


By Theorem 8.11, if the diagonals of a quadrilateral bisect each other, then it is a parallelogram. So find $x$ such that $\overline{FK} \cong \overline{KH}$ and $y$ such that $\overline{JK} \cong \overline{KG}$.

| | |
|---|---|
| $FK = KH$ | Definition of $\cong$ |
| $3x - 1 = 2x + 3$ | Substitution |
| $x - 1 = 3$ | Subtract $2x$ from each side. |
| $x = 4$ | Add 1 to each side. |
| $JK = KG$ | Definition of $\cong$ |
| $6y - 2 = 4y + 3$ | Substitution |
| $2y - 2 = 3$ | Subtract $4y$ from each side. |
| $2y = 5$ | Add 2 to each side. |
| $y = 2.5$ | Divide each side by 2. |

So, when $x$ is 4 and $y$ is 2.5, quadrilateral $FGHJ$ is a parallelogram.

**GuidedPractice**

**Find $x$ and $y$ so that each quadrilateral is a parallelogram.**

**3A.** 56°, $(5y - 26)°$, $(4y + 4)°$, $7x°$

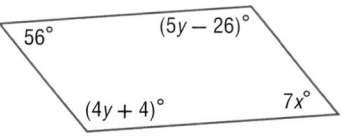

**3B.**

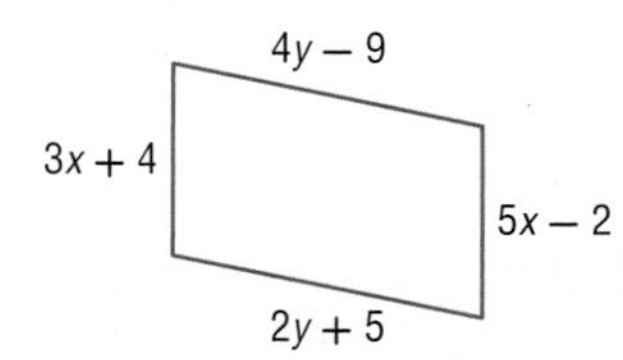


You have learned the conditions of parallelograms. The following list summarizes how to use the conditions to prove a quadrilateral is a parallelogram.

**Concept Summary**

**Prove that a Quadrilateral Is a Parallelogram**

- Show that both pairs of opposite sides are parallel. (Definition)
- Show that both pairs of opposite sides are congruent. (Theorem 8.9)
- Show that both pairs of opposite angles are congruent. (Theorem 8.10)
- Show that the diagonals bisect each other. (Theorem 8.11)
- Show that a pair of opposite sides is both parallel and congruent. (Theorem 8.12)

**StudyTip**

**Midpoint Formula** To show that a quadrilateral is a parallelogram, you can also use the Midpoint Formula. If the midpoint of each diagonal is the same point, then the diagonals bisect each other.

## 2 Parallelograms on the Coordinate Plane

We can use the Distance, Slope, and Midpoint Formulas to determine whether a quadrilateral in the coordinate plane is a parallelogram.

### Example 4 Parallelograms and Coordinate Geometry

**COORDINATE GEOMETRY Graph quadrilateral *KLMN* with vertices *K*(2, 3), *L*(8, 4), *M*(7, −2), and *N*(1, −3). Determine whether the quadrilateral is a parallelogram. Justify your answer using the Slope Formula.**

If the opposite sides of a quadrilateral are parallel, then it is a parallelogram.

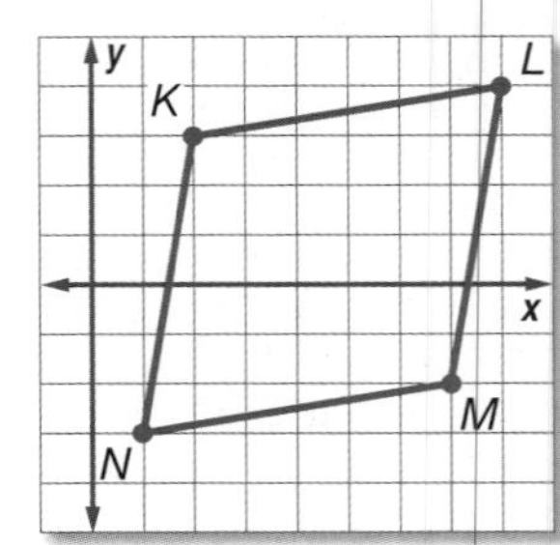


slope of $\overline{KL} = \frac{4-3}{8-2}$ or $\frac{1}{6}$

slope of $\overline{NM} = \frac{-2-(-3)}{7-1}$ or $\frac{1}{6}$

slope of $\overline{KN} = \frac{-3-3}{1-2} = \frac{-6}{-1}$ or 6

slope of $\overline{LM} = \frac{-2-4}{7-8} = \frac{-6}{-1}$ or 6

Since opposite sides have the same slope, $\overline{KL} \parallel \overline{NM}$ and $\overline{KN} \parallel \overline{LM}$. Therefore, *KLMN* is a parallelogram by definition.

**GuidedPractice**

**Determine whether the quadrilateral is a parallelogram. Justify your answer using the given formula.**

**4A.** *A*(3, 3), *B*(8, 2), *C*(6, −1), *D*(1, 0); Distance Formula

**4B.** *F*(−2, 4), *G*(4, 2), *H*(4, −2), *J*(−2, −1); Midpoint Formula

You learned that variable coordinates can be assigned to the vertices of triangles. Then the Distance, Slope, and Midpoint Formulas were used to write coordinate proofs of theorems. The same can be done with quadrilaterals.

**ReviewVocabulary**

**coordinate proof** a proof that uses figures in the coordinate plane and algebra to prove geometric concepts

### Example 5 Parallelograms and Coordinate Proofs

**Write a coordinate proof for the following statement.**

*If one pair of opposite sides of a quadrilateral is both parallel and congruent, then the quadrilateral is a parallelogram.*

**Step 1** Position quadrilateral *ABCD* on the coordinate plane such that $\overline{AB} \parallel \overline{DC}$ and $\overline{AB} \cong \overline{DC}$.

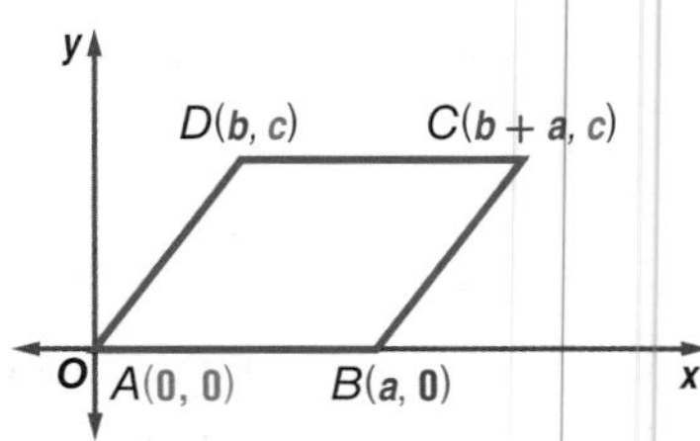


- Begin by placing the vertex *A* at the **origin**.
- Let $\overline{AB}$ have a length of a units. Then *B* has coordinates **(*a*, 0)**.
- Since horizontal segments are parallel, position the endpoints of $\overline{DC}$ so that they have the same *y*-coordinate, *c*.
- So that the distance from *D* to *C* is also *a* units, let the *x*-coordinate of *D* be ***b*** and of *C* be ***b* + *a***.

**Math HistoryLink**

René Descartes
**(1596–1650)**
René Descartes was a French mathematician who was the first to use a coordinate grid. It has been said that he first thought of locating a point on a plane with a pair of numbers when he was watching a fly on the ceiling, but this is a myth.

**Step 2** Use your figure to write a proof.

**Given:** quadrilateral $ABCD$, $\overline{AB} \parallel \overline{DC}$, $\overline{AB} \cong \overline{DC}$

**Prove:** $ABCD$ is a parallelogram.

y, x, O, $D(b, c)$, $C(b + a, c)$, $A(0, 0)$, $B(a, 0)$

**Coordinate Proof:**

By definition, a quadrilateral is a parallelogram if opposite sides are parallel. We are given that $\overline{AB} \parallel \overline{DC}$, so we need only show that $\overline{AD} \parallel \overline{BC}$.

Use the Slope Formula.

slope of $\overline{AD} = \frac{c-0}{b-0} = \frac{c}{b}$ slope of $\overline{BC} = \frac{c-0}{b+a-a} = \frac{c}{b}$

Since $\overline{AD}$ and $\overline{BC}$ have the same slope, $\overline{AD} \parallel \overline{BC}$. So quadrilateral $ABCD$ is a parallelogram because opposite sides are parallel.

**GuidedPractice**

**5.** Write a coordinate proof of this statement: *If a quadrilateral is a parallelogram, then opposite sides are congruent.*

## Check Your Understanding

**Example 1** **Determine whether each quadrilateral is a parallelogram. Justify your answer.**

**1.**

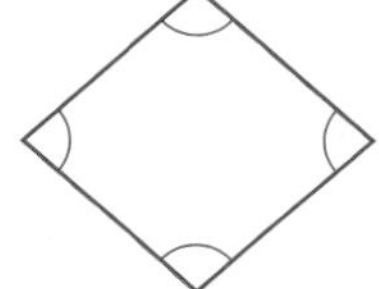

**2.**

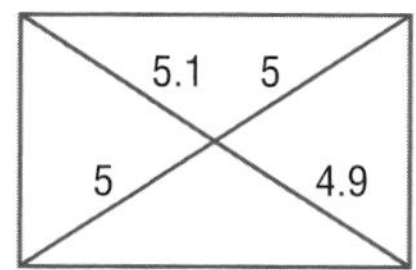


**Example 2** **3. KITES** Charmaine is building the kite shown below. She wants to be sure that the string around her frame forms a parallelogram before she secures the material to it. How can she use the measures of the wooden portion of the frame to prove that the string forms a parallelogram? Explain your reasoning.

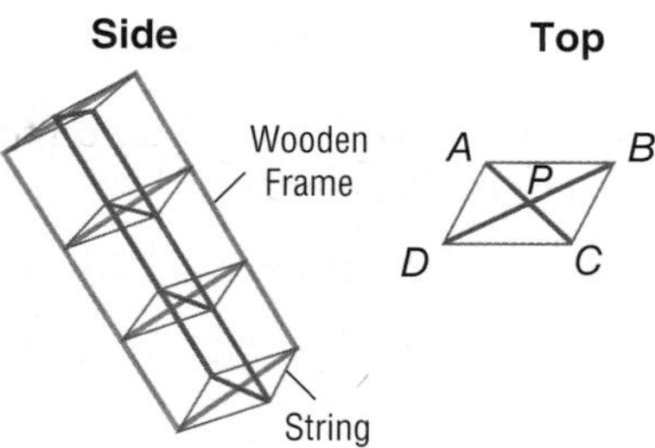


**Example 3** **ALGEBRA** **Find $x$ and $y$ so that the quadrilateral is a parallelogram.**

**4.** $(8x - 8)°$, $(7y + 2)°$, $(6y + 16)°$, $(6x + 14)°$

**5.**

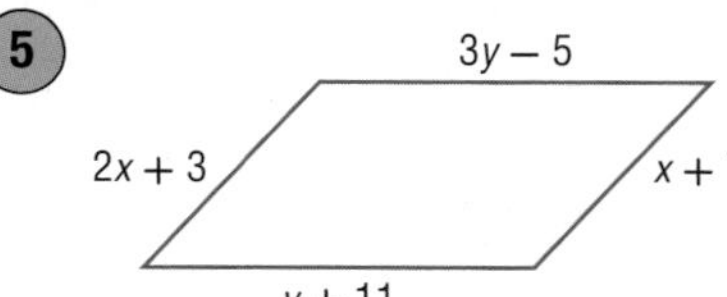

**Example 4** **COORDINATE GEOMETRY** **Graph each quadrilateral with the given vertices. Determine whether the figure is a parallelogram. Justify your answer with the method indicated.**

6. $A(-2, 4)$, $B(5, 4)$, $C(8, -1)$, $D(-1, -1)$; Slope Formula

7. $W(-5, 4)$, $X(3, 4)$, $Y(1, -3)$, $Z(-7, -3)$; Midpoint Formula

**Example 5** 8. Write a coordinate proof for the statement: *If a quadrilateral is a parallelogram, then its diagonals bisect each other.*

## Practice and Problem Solving

**Example 1** **CCSS ARGUMENTS** **Determine whether each quadrilateral is a parallelogram. Justify your answer.**

9. 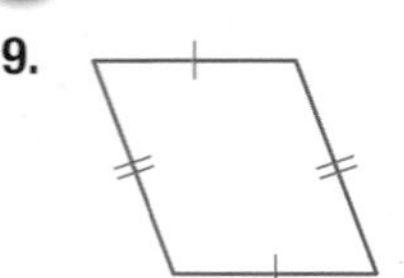

10. 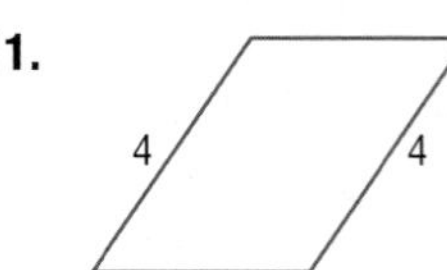


11. 


12. 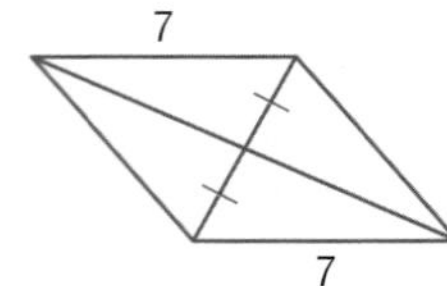


13. 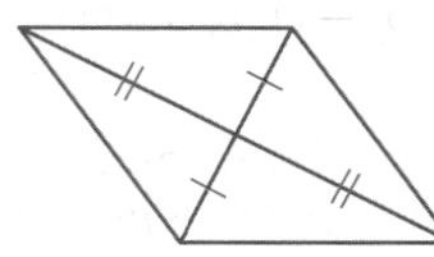

14. 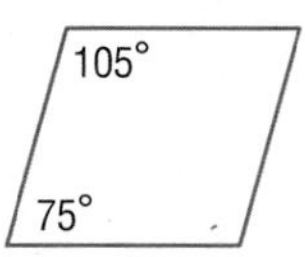


**Example 2** 15. **PROOF** If $ACDH$ is a parallelogram, $B$ is the midpoint of $\overline{AC}$, and $F$ is the midpoint of $\overline{HD}$, write a flow proof to prove that $ABFH$ is a parallelogram.

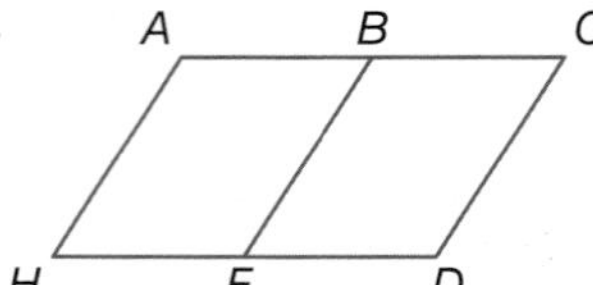


16. **PROOF** If WXYZ is a parallelogram, $\angle W \cong \angle X$, and $M$ is the midpoint of $\overline{WX}$, write a paragraph proof to prove that $ZMY$ is an isosceles triangle.

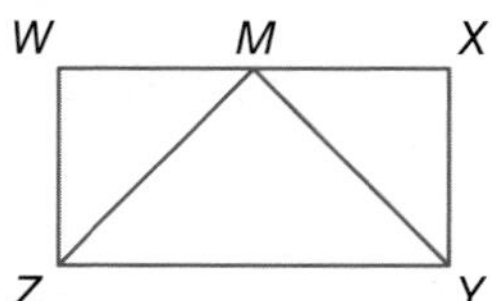


17. **REPAIR** Parallelogram lifts are used to elevate large vehicles for maintenance. In the diagram, $ABEF$ and $BCDE$ are parallelograms. Write a two-column proof to show that $ACDF$ is also a parallelogram.

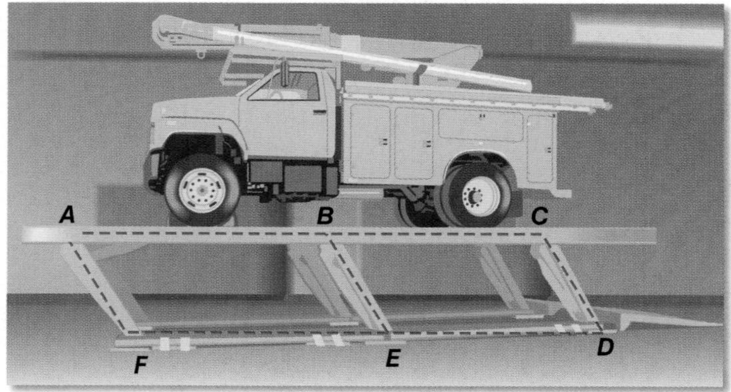


**Example 3** **ALGEBRA** **Find $x$ and $y$ so that the quadrilateral is a parallelogram.**

18. 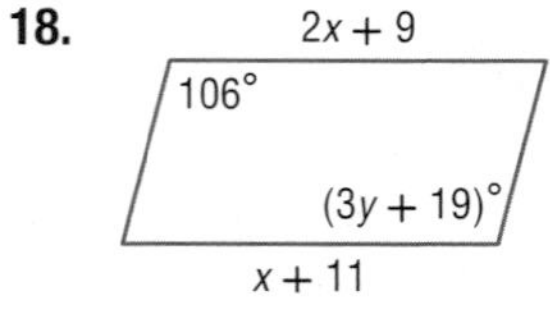


19. 


20. 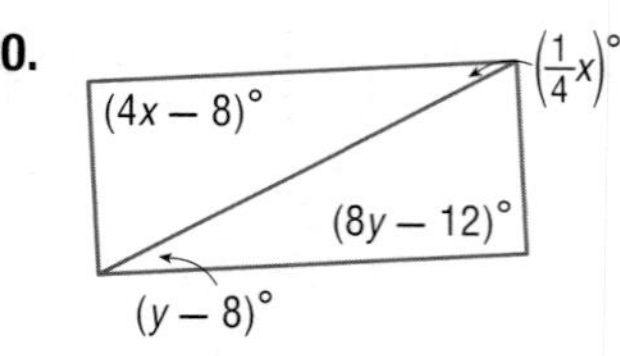

**ALGEBRA Find $x$ and $y$ so that the quadrilateral is a parallelogram.**

**21.**

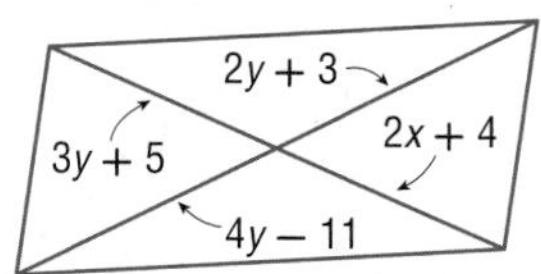


**22.**

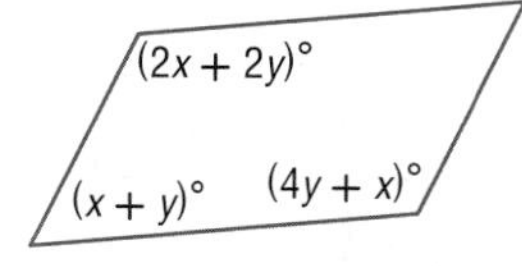


**23.** $2x + 4y$; $21$; $3x + 3y$; $6y + \frac{1}{2}x$

Example 4

**COORDINATE GEOMETRY Graph each quadrilateral with the given vertices. Determine whether the figure is a parallelogram. Justify your answer with the method indicated.**

**24.** $A(-3, 4)$, $B(4, 5)$, $C(5, -1)$, $D(-2, -2)$; Slope Formula

**25.** $J(-4, -4)$, $K(-3, 1)$, $L(4, 3)$, $M(3, -3)$; Distance Formula

**26.** $V(3, 5)$, $W(1, -2)$, $X(-6, 2)$, $Y(-4, 7)$; Slope Formulas

**27.** $Q(2, -4)$, $R(4, 3)$, $S(-3, 6)$, $T(-5, -1)$; Distance and Slope Formulas

Example 5

**28.** Write a coordinate proof for the statement: *If both pairs of opposite sides of a quadrilateral are congruent, then the quadrilateral is a parallelogram.*

**29.** Write a coordinate proof for the statement: *If a parallelogram has one right angle, it has four right angles.*

**30. PROOF** Write a paragraph proof of Theorem 8.10.

**31. PANTOGRAPH** A pantograph is a device that can be used to copy an object and either enlarge or reduce it based on the dimensions of the pantograph.

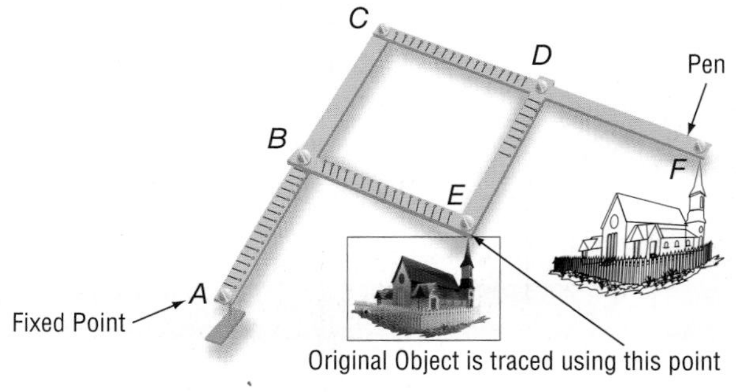


**a.** If $\overline{AC} \cong \overline{CF}$, $\overline{AB} \cong \overline{CD} \cong \overline{BE}$, and $\overline{DF} \cong \overline{DE}$, write a paragraph proof to show that $\overline{BE} \parallel \overline{CD}$.

**b.** The scale of the copied object is the ratio of $CF$ to $BE$. If $AB$ is 12 inches, $DF$ is 8 inches, and the width of the original object is 5.5 inches, what is the width of the copy?

**PROOF Write a two-column proof.**

**32.** Theorem 8.11

**33.** Theorem 8.12

**34. CONSTRUCTION** Explain how you can use Theorem 8.11 to construct a parallelogram. Then construct a parallelogram using your method.

**CCSS REASONING Name the missing coordinates for each parallelogram.**

**35.**

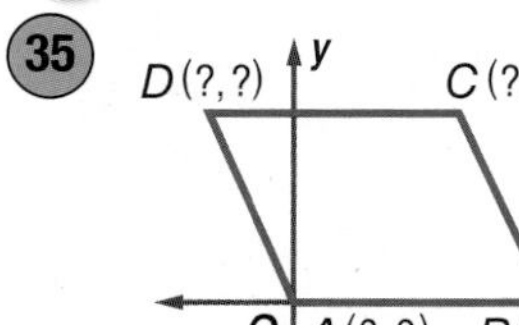


**36.**

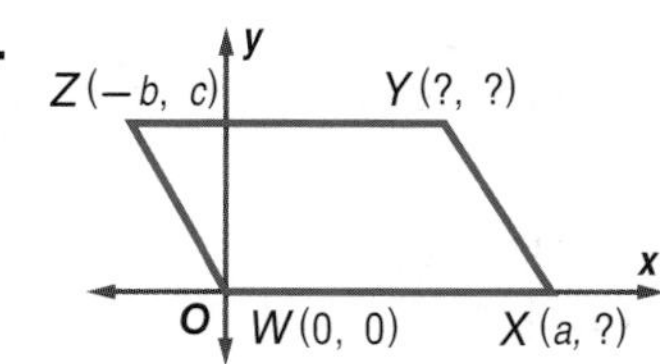

**37** **SERVICE** While replacing a hand rail, a contractor uses a carpenter's square to confirm that the vertical supports are perpendicular to the top step and the ground, respectively. How can the contractor prove that the two hand rails are parallel using the fewest measurements? Assume that the top step and the ground are both level.

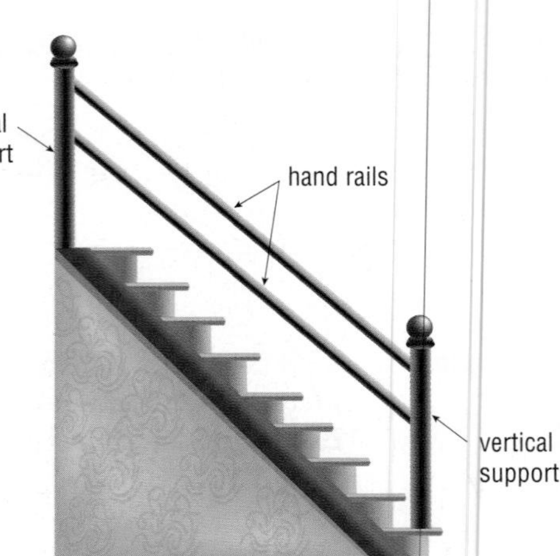


**38.** **PROOF** Write a coordinate proof to prove that the segments joining the midpoints of the sides of any quadrilateral form a parallelogram.

**39.** **MULTIPLE REPRESENTATIONS** In this problem, you will explore the properties of rectangles. A rectangle is a quadrilateral with four right angles.

a. **Geometric** Draw three rectangles with varying lengths and widths. Label one rectangle *ABCD*, one *MNOP*, and one *WXYZ*. Draw the two diagonals for each rectangle.

b. **Tabular** Measure the diagonals of each rectangle, and complete the table at the right.

c. **Verbal** Write a conjecture about the diagonals of a rectangle.

| Rectangle | Side | Length |
|---|---|---|
| *ABCD* | $\overline{AC}$ | |
| | $\overline{BD}$ | |
| *MNOP* | $\overline{MO}$ | |
| | $\overline{NP}$ | |
| *WXYZ* | $\overline{WY}$ | |
| | $\overline{XZ}$ | |

## H.O.T. Problems Use Higher-Order Thinking Skills

**40.** **CHALLENGE** The diagonals of a parallelogram meet at the point (0, 1). One vertex of the parallelogram is located at (2, 4), and a second vertex is located at (3, 1). Find the locations of the remaining vertices.

**41.** **WRITING IN MATH** Compare and contrast Theorem 8.9 and Theorem 8.3.

**42.** CCSS **ARGUMENTS** If two parallelograms have four congruent corresponding angles, are the parallelograms *sometimes*, *always*, or *never* congruent?

**43.** **OPEN ENDED** Position and label a parallelogram on the coordinate plane differently than shown in either Example 5, Exercise 35, or Exercise 36.

**44.** **CHALLENGE** If *ABCD* is a parallelogram and $\overline{AJ} \cong \overline{KC}$, show that quadrilateral *JBKD* is a parallelogram.

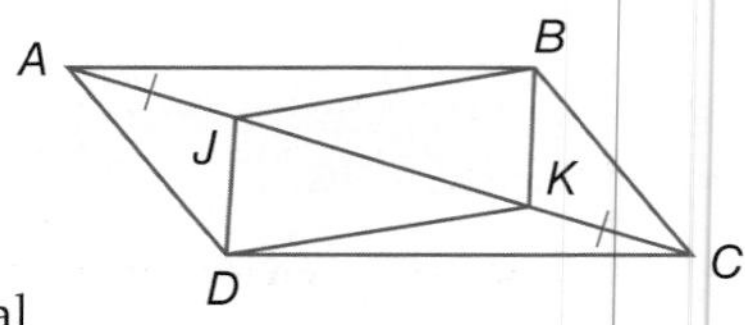


**45.** **WRITING IN MATH** How can you prove that a quadrilateral is a parallelogram?

## Standardized Test Practice

**46.** If sides $\overline{AB}$ and $\overline{DC}$ of quadrilateral $ABCD$ are parallel, which additional information would be sufficient to prove that quadrilateral $ABCD$ is a parallelogram?

**A** $\overline{AB} \cong \overline{AC}$

**B** $\overline{AB} \cong \overline{DC}$

**C** $\overline{AC} \cong \overline{BD}$

**D** $\overline{AD} \cong \overline{BC}$

**47. SHORT RESPONSE** Quadrilateral $ABCD$ is shown. $AC$ is 40 and $BD$ is $\frac{3}{5}AC$. $\overline{BD}$ bisects $\overline{AC}$. For what value of $x$ is $ABCD$ a parallelogram?

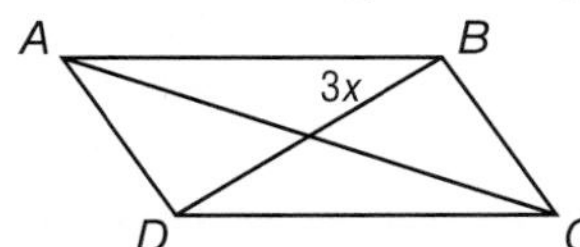


**48. ALGEBRA** Jarod's average driving speed for a 5-hour trip was 58 miles per hour. During the first 3 hours, he drove 50 miles per hour. What was his average speed in miles per hour for the last 2 hours of his trip?

**F** 70

**G** 66

**H** 60

**J** 54

**49. SAT/ACT** A parallelogram has vertices at (0, 0), (3, 5), and (0, 5). What are the coordinates of the fourth vertex?

**A** (0, 3)

**B** (5, 3)

**C** (5, 0)

**D** (0, −3)

**E** (3, 0)

## Spiral Review

**COORDINATE GEOMETRY Find the coordinates of the intersection of the diagonals of ▱$ABCD$ with the given vertices.** (Lesson 8-2)

**50.** $A(-3, 5)$, $B(6, 5)$, $C(5, -4)$, $D(-4, -4)$

**51.** $A(2, 5)$, $B(10, 7)$, $C(7, -2)$, $D(-1, -4)$

**Find the value of $x$.** (Lesson 8-1)

**52.**

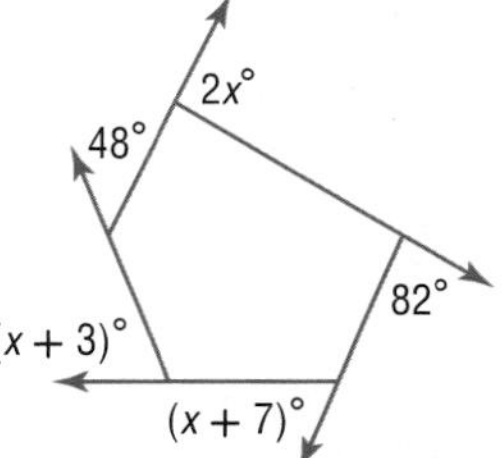


**53.**

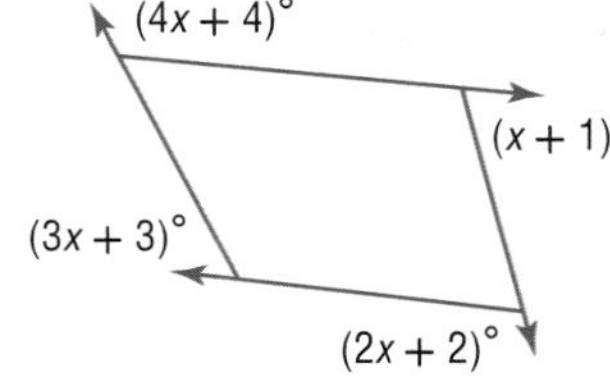


**54.**

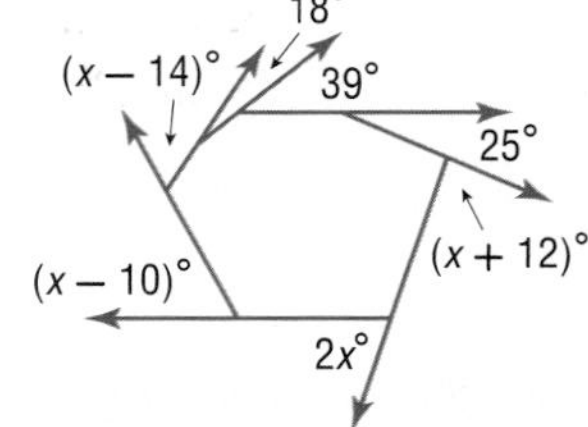


**55. FITNESS** Toshiro was at the gym for just over two hours. He swam laps in the pool and lifted weights. Prove that he did one of these activities for more than an hour. (Lesson 7-4)

**PROOF Write a flow proof.** (Lesson 6-4)

**56. Given:** $\overline{EJ} \parallel \overline{FK}$, $\overline{JG} \parallel \overline{KH}$, $\overline{EF} \cong \overline{GH}$
**Prove:** $\triangle EJG \cong \triangle FKH$

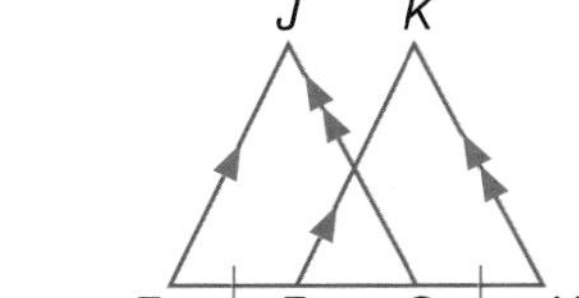


**57. Given:** $\overline{MN} \cong \overline{PQ}$, $\angle M \cong \angle Q$, $\angle 2 \cong \angle 3$
**Prove:** $\triangle MLP \cong \triangle QLN$

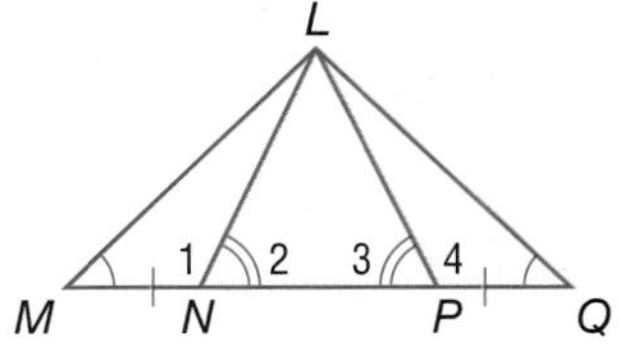


## Skills Review

**Use slope to determine whether $XY$ and $YZ$ are *perpendicular* or *not perpendicular*.**

**58.** $X(-2, 2)$, $Y(0, 1)$, $Z(4, 1)$

**59.** $X(4, 1)$, $Y(5, 3)$, $Z(6, 2)$

# Mid-Chapter Quiz

## Lessons 8-1 through 8-3

**Find the sum of the measures of the interior angles of each convex polygon.** (Lesson 8-1)

**1.** pentagon

**2.** heptagon

**3.** 18-gon

**4.** 23-gon

**Find the measure of each interior angle.** (Lesson 8-1)

**5.**

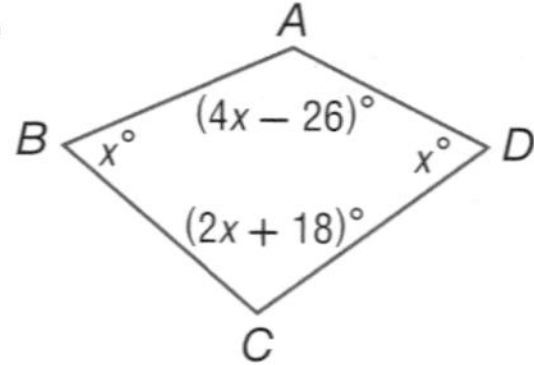


**6.**

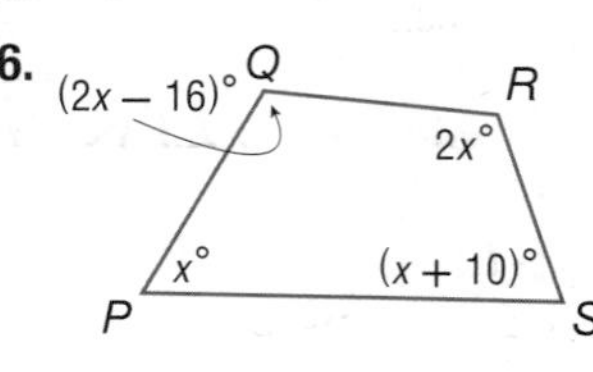


**The sum of the measures of the interior angles of a regular polygon is given. Find the number of sides in the polygon.** (Lesson 8-1)

**7.** 720

**8.** 1260

**9.** 1800

**10.** 4500

**Find the value of $x$ in each diagram.** (Lesson 8-1)

**11.**

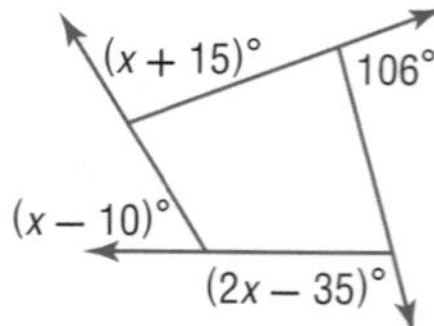


**12.**

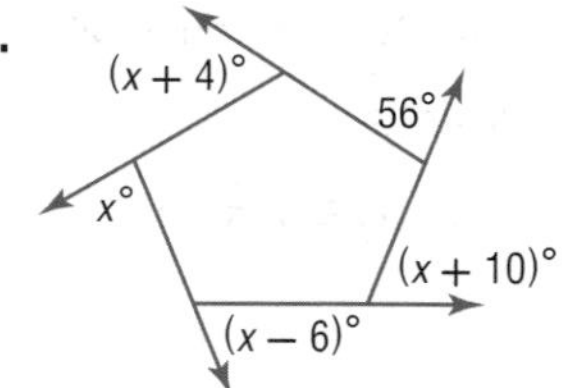


**Use ▱$WXYZ$ to find each measure.** (Lesson 8-2)

**13.** $m\angle WZY$

**14.** $WZ$

**15.** $m\angle XYZ$

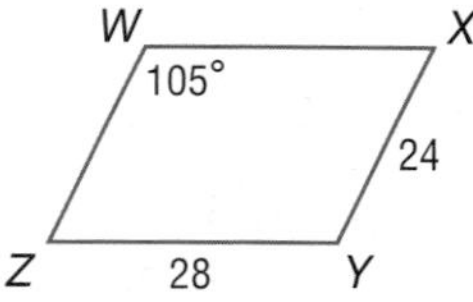


**16. DESIGN** Describe two ways to ensure that the pieces of the design at the right would fit properly together. (Lesson 8-2)

**ALGEBRA Find the value of each variable in each parallelogram.** (Lesson 8-2)

**17.**

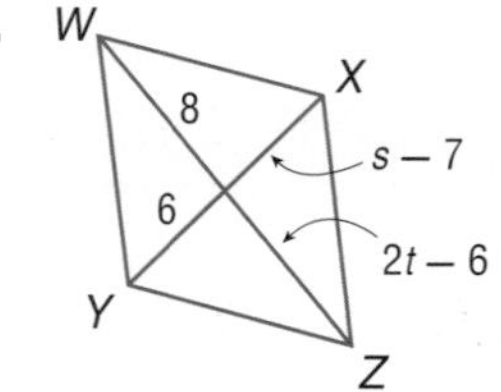


**18.**

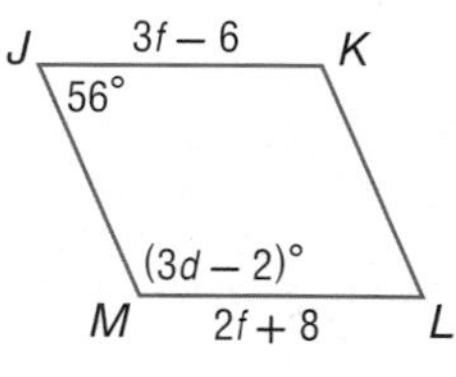


**19. PROOF** Write a two-column proof. (Lesson 8-2)

Given: ▱$GFBA$ and ▱$HACD$

Prove: $\angle F \cong \angle D$

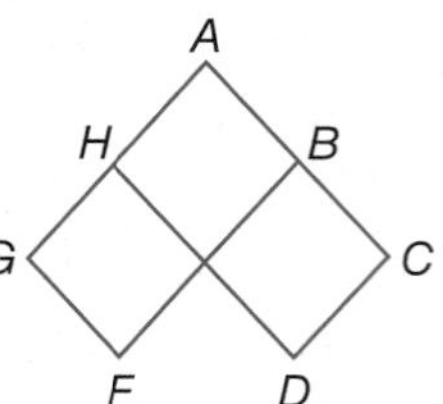


**Find x and y so that each quadrilateral is a parallelogram.** (Lesson 8-3)

**20.**

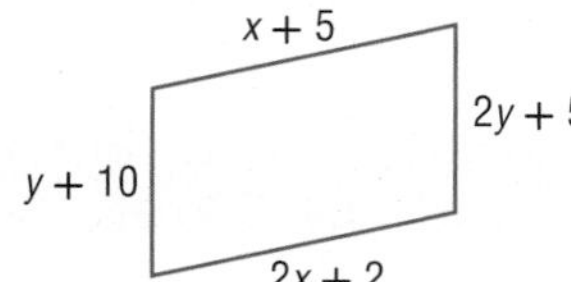


**21.**

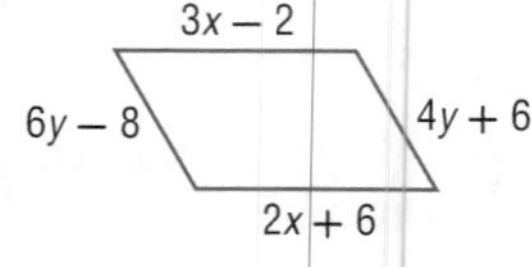


**22. MUSIC** Why will the keyboard stand with legs joined at the midpoints always remain parallel to the floor? (Lesson 8-3)

**23. MULTIPLE CHOICE** Which of the following quadrilaterals is not a parallelogram? (Lesson 8-3)

**A**

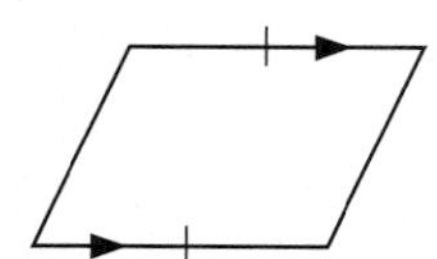

**C**

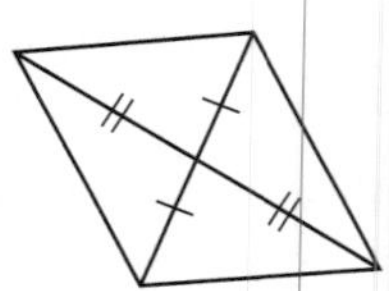

**B**

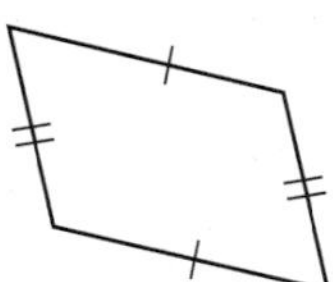

**D**

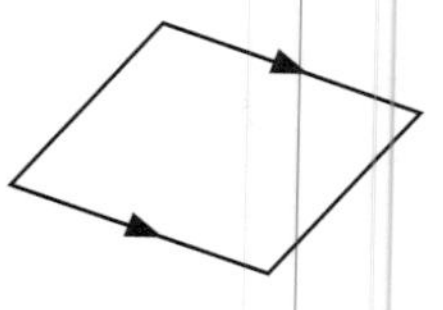

**COORDINATE GEOMETRY Determine whether the figure is a parallelogram. Justify your answer with the method indicated.** (Lesson 8-3)

**24.** $A(-6, -5)$, $B(-1, -4)$, $C(0, -1)$, $D(-5, -2)$; Distance Formula

**25.** $Q(-5, 2)$, $R(-3, -6)$, $S(2, 2)$, $T(-1, 6)$; Slope Formula

# LESSON 8-4 Rectangles

**Then**

- You used properties of parallelograms and determined whether quadrilaterals were parallelograms.

**Now**

1. Recognize and apply properties of rectangles.
2. Determine whether parallelograms are rectangles.

**Why?**

- Leonardo is in charge of set design for a school play. He needs to use paint to create the appearance of a doorway on a lightweight solid wall. The doorway is to be a rectangle 36 inches wide and 80 inches tall. How can Leonardo be sure that he paints a rectangle?

**NewVocabulary**
rectangle

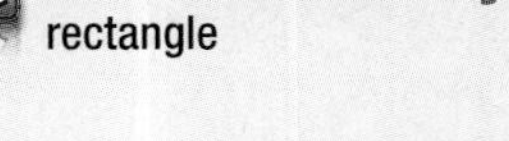

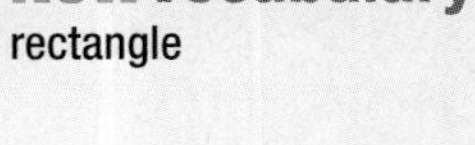

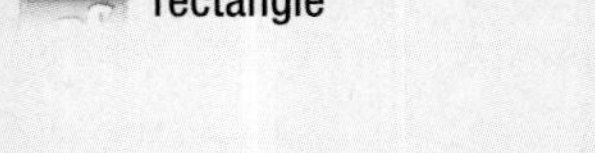

**Common Core State Standards**

**Content Standards**
G.CO.11 Prove theorems about parallelograms.
G.GPE.4 Use coordinates to prove simple geometric theorems algebraically.

**Mathematical Practices**
3 Construct viable arguments and critique the reasoning of others.
5 Use appropriate tools strategically.

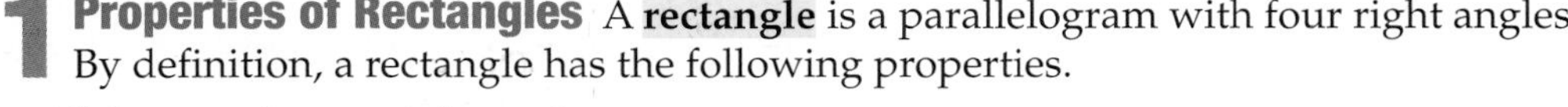

## 1 Properties of Rectangles

A **rectangle** is a parallelogram with four right angles. By definition, a rectangle has the following properties.

- All four angles are right angles.
- Opposite sides are parallel and congruent.
- Opposite angles are congruent.
- Consecutive angles are supplementary.
- Diagonals bisect each other.

In addition, the diagonals of a rectangle are congruent.

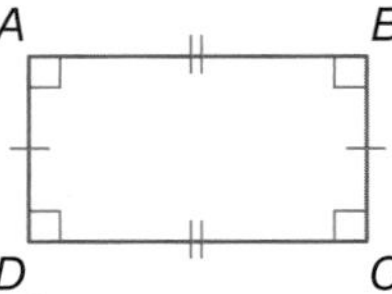

Rectangle *ABCD*

### Theorem 8.13 Diagonals of a Rectangle

If a parallelogram is a rectangle, then its diagonals are congruent.

**Abbreviation** *If a ▱ is a rectangle, diag. are ≅.*

**Example** If ▱*JKLM* is a rectangle, then $\overline{JL} \cong \overline{MK}$.

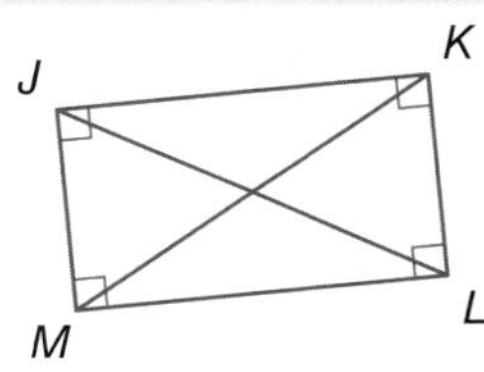


You will prove Theorem 8.13 in Exercise 33.

### Real-World Example 1 Use Properties of Rectangles

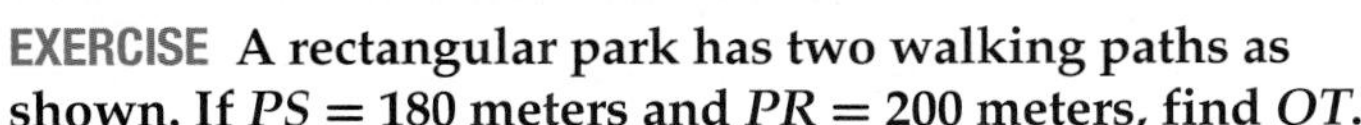

**EXERCISE A rectangular park has two walking paths as shown. If $PS = 180$ meters and $PR = 200$ meters, find $QT$.**

| | |
|---|---|
| $\overline{QS} \cong \overline{PR}$ | If a ▱ is a rectangle, diag. are ≅. |
| $QS = PR$ | Definition of congruence |
| $QS = 200$ | Substitution |

Since *PQRS* is a rectangle, it is a parallelogram. The diagonals of a parallelogram bisect each other, so $QT = ST$.

| | |
|---|---|
| $QT + ST = QS$ | Segment Addition |
| $QT + QT = QS$ | Substitution |
| $2QT = QS$ | Simplify. |
| $QT = \frac{1}{2}QS$ | Divide each side by 2. |
| $QT = \frac{1}{2}(200)$ or 100 | Substitution |

P Q T S R

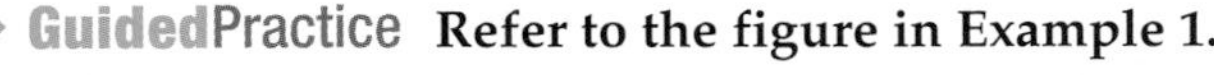

**GuidedPractice** Refer to the figure in Example 1.

**1A.** If $TS = 120$ meters, find $PR$.

**1B.** If $m\angle PRS = 64$, find $m\angle SQR$.

You can use the properties of rectangles along with algebra to find missing values.

## Example 2 Use Properties of Rectangles and Algebra

**ALGEBRA Quadrilateral *JKLM* is a rectangle. If $m\angle KJL = 2x + 4$ and $m\angle JLK = 7x + 5$, find $x$.**

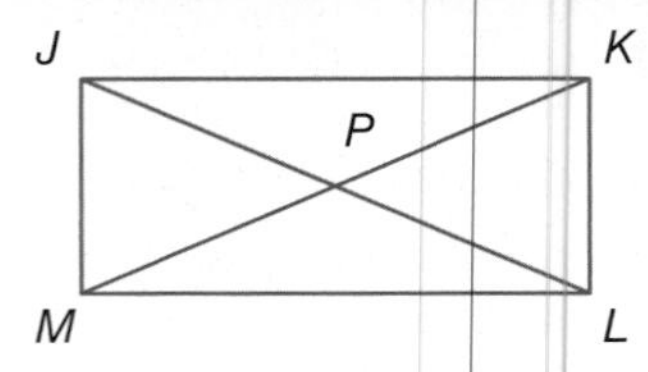


Since *JKLM* is a rectangle, it has four right angles. So, $m\angle MLK = 90$. Since a rectangle is a parallelogram, opposite sides are parallel. Alternate interior angles of parallel lines are congruent, so $\angle JLM \cong \angle KJL$ and $m\angle JLM = m\angle KJL$.

| | |
|---|---|
| $m\angle JLM + m\angle JLK = 90$ | Angle Addition |
| $m\angle KJL + m\angle JLK = 90$ | Substitution |
| $2x + 4 + 7x + 5 = 90$ | Substitution |
| $9x + 9 = 90$ | Add like terms. |
| $9x = 81$ | Subtract 9 from each side. |
| $x = 9$ | Divide each side by 9. |

**Study Tip**

**Right Angles** Recall from Theorem 8-6 that if a parallelogram has one right angle, then it has four right angles.

### Guided Practice

**2.** Refer to the figure in Example 2. If $JP = 3y - 5$ and $MK = 5y + 1$, find $y$.

## 2 Prove that Parallelograms are Rectangles

The converse of Theorem 6.13 is also true.

### Theorem 8.14 Diagonals of a Rectangle

If the diagonals of a parallelogram are congruent, then the parallelogram is a rectangle.

**Abbreviation** *If diag. of a ▱ are ≅, then ▱ is a rectangle.*

**Example** If $\overline{WY} \cong \overline{XZ}$ in ▱*WXYZ*, then ▱*WXYZ* is a rectangle.

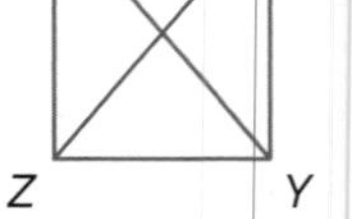


You will prove Theorem 8.14 in Exercise 34.

## Real-World Example 3 Providing Rectangle Relationships

**DODGEBALL A community recreation center has created an outdoor dodgeball playing field. To be sure that it meets the ideal playing field requirements, they measure the sides of the field and its diagonals. If $AB = 60$ feet, $BC = 30$ feet, $CD = 60$ feet, $AD = 30$ feet, $AC = 67$ feet, and $BD = 67$ feet, explain how the recreation center can be sure that the playing field is rectangular.**

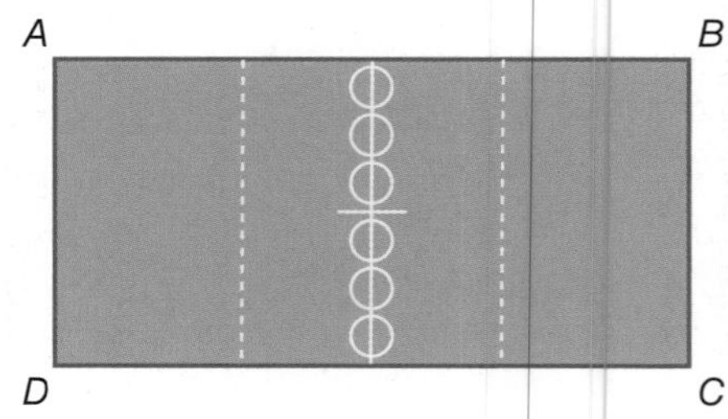


Since $AB = CD$, $BC = AD$, and $AC = BD$, $\overline{AB} \cong \overline{CD}$, $\overline{BC} \cong \overline{AD}$, and $\overline{AC} \cong \overline{BD}$. Because $\overline{AB} \cong \overline{CD}$ and $\overline{BC} \cong \overline{AD}$, *ABCD* is a parallelogram. Since $\overline{AC}$ and $\overline{BD}$ are congruent diagonals in ▱*ABCD*, ▱*ABCD* is a rectangle.

**Real-World Link**

The game of dodgeball is played on a rectangular playing field ideally 60 feet long and 30 feet wide. The field is divided into two equal sections by a center-line and attack-lines that are 3 meters (9.8 feet) from, and parallel to, the centerline.

**Source:** National Amateur Dodgeball Assoc.

Micah Walter/Getty Images

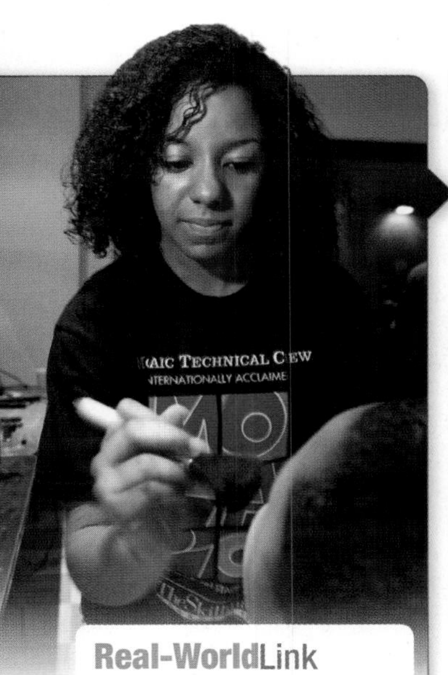

## GuidedPractice

**3. SET DESIGN** Refer to the beginning of the lesson. Leonardo measures the sides of his figure and confirms that they have the desired measures as shown. Using a carpenter's square, he also confirms that the measure of the bottom left corner of the figure is a right angle. Can he conclude that the figure is a rectangle? Explain.

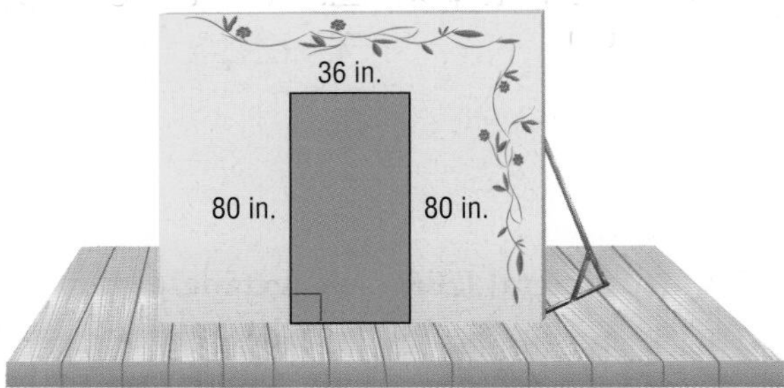


**Real-WorldLink**

The Mosaic Youth Theater in Detroit, Michigan, is a professional performing arts training program for young people ages 12 to 18. Students are involved in all aspects of performances, including set and lighting design, set construction, stage management, sound, and costumes.

You can also use the properties of rectangles to prove that a quadrilateral positioned on a coordinate plane is a rectangle given the coordinates of the vertices.

## Example 4 Rectangles and Coordinate Geometry

**COORDINATE GEOMETRY Quadrilateral *PQRS* has vertices *P*(−5, 3), *Q*(1, −1), *R*(−1, −4), and *S*(−7, 0). Determine whether *PQRS* is a rectangle by using the Distance Formula.**

**Step 1** Use the Distance Formula to determine whether *PQRS* is a parallelogram by determining if opposite sides are congruent.

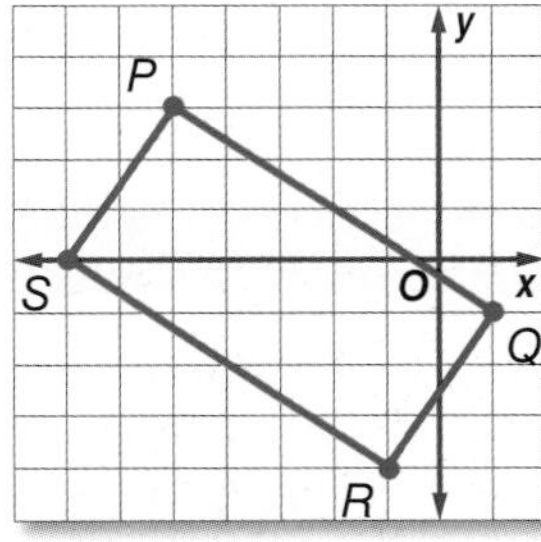


$$PQ = \sqrt{(-5-1)^2 + [3-(-1)]^2} \text{ or } \sqrt{52}$$

$$RS = \sqrt{[-1-(-7)]^2 + (-4-0)^2} \text{ or } \sqrt{52}$$

$$PS = \sqrt{[-5-(-7)]^2 + (3-0)^2} \text{ or } \sqrt{13}$$

$$QR = \sqrt{[1-(-1)^2 + [-1-(-4)]^2} \text{ or } \sqrt{13}$$

Since opposite sides of the quadrilateral have the same measure, they are congruent. So, quadrilateral *PQRS* is a parallelogram.

**StudyTip**

**Rectangles and Parallelograms** A rectangle is a parallelogram, but a parallelogram is not necessarily a rectangle.

**Step 2** Determine whether the diagonals of ▱*PQRS* are congruent.

$$PR = \sqrt{[-5-(-1)]^2 + [3-(-4)]^2} \text{ or } \sqrt{65}$$

$$QS = \sqrt{[1-(-7)]^2 + (-1-0)^2} \text{ or } \sqrt{65}$$

Since the diagonals have the same measure, they are congruent. So, ▱*PQRS* is a rectangle.

## GuidedPractice

**4.** Quadrilateral *JKLM* has vertices *J*(−10, 2), *K*(−8, −6), *L*(5, −3), and *M*(2, 5). Determine whether *JKLM* is a rectangle using the Slope Formula.

## Check Your Understanding

**Example 1** **FARMING** **An X-brace on a rectangular barn door is both decorative and functional. It helps to prevent the door from warping over time. If $ST = 3\frac{13}{16}$ feet, $PS = 7$ feet, and $m\angle PTQ = 67$, find each measure.**

**1.** $QR$  **2.** $SQ$

**3.** $m\angle TQR$  **4.** $m\angle TSR$

**Example 2** **ALGEBRA** **Quadrilateral $DEFG$ is a rectangle.**

**5.** If $FD = 3x - 7$ and $EG = x + 5$, find $EG$.

**6.** If $m\angle EFD = 2x - 3$ and $m\angle DFG = x + 12$, find $m\angle EFD$.

**Example 3** **7. PROOF** If $ABDE$ is a rectangle and $\overline{BC} \cong \overline{DC}$, prove that $\overline{AC} \cong \overline{EC}$.

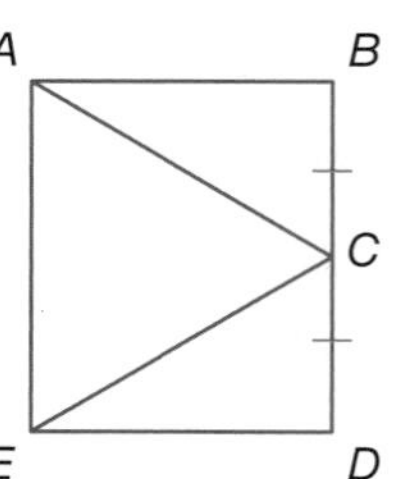


**Example 4** **COORDINATE GEOMETRY** **Graph each quadrilateral with the given vertices. Determine whether the figure is a rectangle. Justify your answer using the indicated formula.**

**8.** $W(-4, 3)$, $X(1, 5)$, $Y(3, 1)$, $Z(-2, -2)$; Slope Formula

**9.** $A(4, 3)$, $B(4, -2)$, $C(-4, -2)$, $D(-4, 3)$; Distance Formula

## Practice and Problem Solving

**Example 1** **FENCING** **X-braces are also used to provide support in rectangular fencing. If $AB = 6$ feet, $AD = 2$ feet, and $m\angle DAE = 65$, find each measure.**

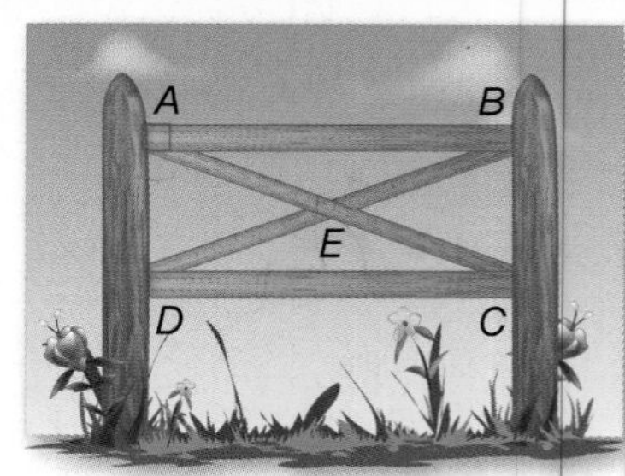


**10.** $BC$  **11.** $DB$

**12.** $m\angle CEB$  **13.** $m\angle EDC$

**Example 2** CCSS **REGULARITY** **Quadrilateral $WXYZ$ is a rectangle.**

Z Y P W X

**14.** If $ZY = 2x + 3$ and $WX = x + 4$, find $WX$.

**15.** If $PY = 3x - 5$ and $WP = 2x + 11$, find $ZP$.

**16.** If $m\angle ZYW = 2x - 7$ and $m\angle WYX = 2x + 5$, find $m\angle ZYW$.

**17.** If $ZP = 4x - 9$ and $PY = 2x + 5$, find $ZX$.

**18.** If $m\angle XZY = 3x + 6$ and $m\angle XZW = 5x - 12$, find $m\angle YXZ$.

**19.** If $m\angle ZXW = x - 11$ and $m\angle WZX = x - 9$, find $m\angle ZXY$.

**Example 3** **PROOF** **Write a two-column proof.**

20. **Given:** $ABCD$ is a rectangle.
**Prove:** $\triangle ADC \cong \triangle BCD$

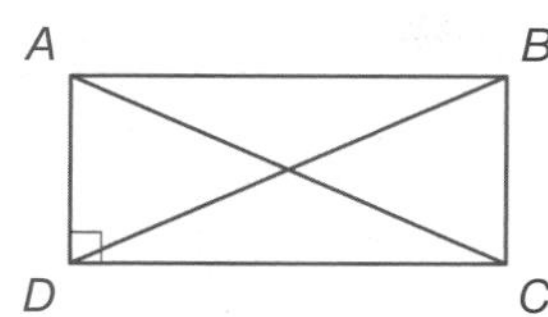


21. **Given:** $QTVW$ is a rectangle.
$\overline{QR} \cong \overline{ST}$
**Prove:** $\triangle SWQ \cong \triangle RVT$

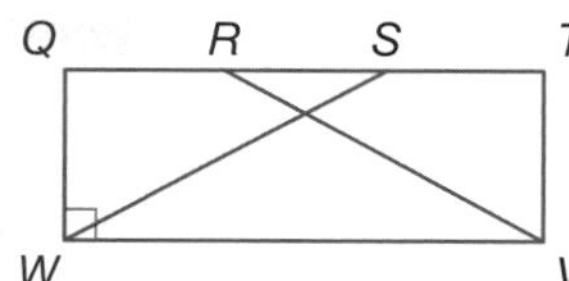


**Example 4** **COORDINATE GEOMETRY** **Graph each quadrilateral with the given vertices. Determine whether the figure is a rectangle. Justify your answer using the indicated formula.**

22. $W(-2, 4)$, $X(5, 5)$, $Y(6, -2)$, $Z(-1, -3)$; Slope Formula

23. $J(3, 3)$, $K(-5, 2)$, $L(-4, -4)$, $M(4, -3)$; Distance Formula

24. $Q(-2, 2)$, $R(0, -2)$, $S(6, 1)$, $T(4, 5)$; Distance Formula

25. $G(1, 8)$, $H(-7, 7)$, $J(-6, 1)$, $K(2, 2)$; Slope Formula

**Quadrilateral $ABCD$ is a rectangle. Find each measure if $m\angle 2 = 40$.**

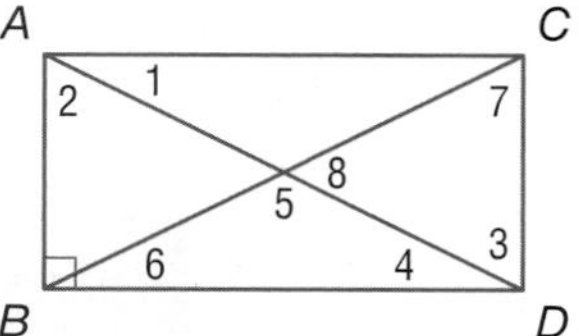


26. $m\angle 1$
27. $m\angle 7$
28. $m\angle 3$
29. $m\angle 5$
30. $m\angle 6$
31. $m\angle 8$

32. **CCSS MODELING** Jody is building a new bookshelf using wood and metal supports like the one shown. To what length should she cut the metal supports in order for the bookshelf to be *square,* which means that the angles formed by the shelves and the vertical supports are all right angles? Explain your reasoning.

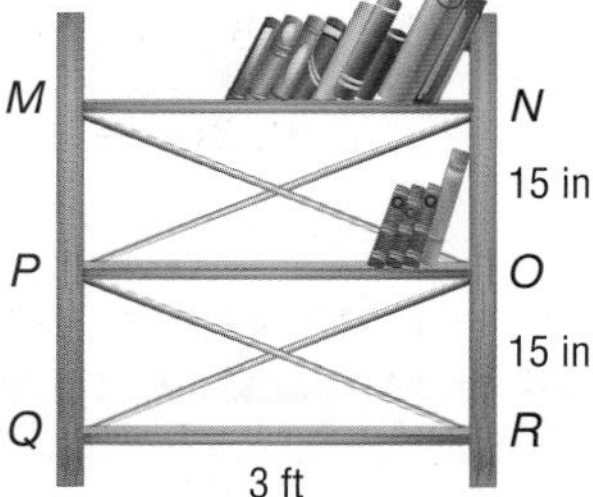


**PROOF** **Write a two-column proof.**

33. Theorem 8.13
34. Theorem 8.14

**PROOF** **Write a paragraph proof of each statement.**

35. If a parallelogram has one right angle, then it is a rectangle.

36. If a quadrilateral has four right angles, then it is a rectangle.

37. **CONSTRUCTION** Construct a rectangle using the construction for congruent segments and the construction for a line perpendicular to another line through a point on the line. Justify each step of the construction.

38. **SPORTS** The end zone of a football field is 160 feet wide and 30 feet long. Kyle is responsible for painting the field. He has finished the end zone. Explain how Kyle can confirm that the end zone is the regulation size and be sure that it is also a rectangle using only a tape measure.

**ALGEBRA** **Quadrilateral $WXYZ$ is a rectangle.**

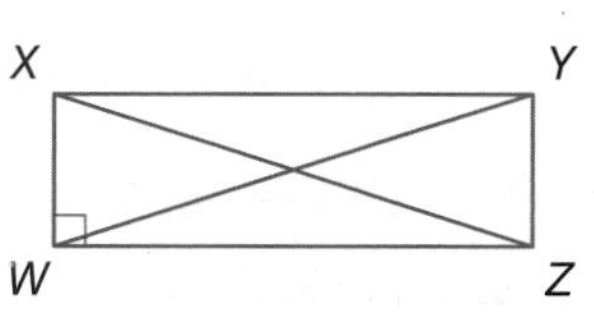


39. If $XW = 3$, $WZ = 4$, and $XZ = b$, find $YW$.

40. If $XZ = 2c$ and $ZY = 6$, and $XY = 8$, find $WY$.

**41.** **SIGNS** The sign below is in the foyer of Nyoko's school. Based on the dimensions given, can Nyoko be sure that the sign is a rectangle? Explain your reasoning.

**PROOF** **Write a coordinate proof of each statement.**

**42.** The diagonals of a rectangle are congruent.

**43** If the diagonals of a parallelogram are congruent, then it is a rectangle.

**44.** **MULTIPLE REPRESENTATIONS** In the problem, you will explore properties of other special parallelograms.

**a.** **Geometric** Draw three parallelograms, each with all four sides congruent. Label one parallelogram *ABCD*, one *MNOP*, and one *WXYZ*. Draw the two diagonals of each parallelogram and label the intersections *R*.

**b.** **Tabular** Use a protractor to measure the appropriate angles and complete the table below.

| Parallelogram | *ABCD* | | *MNOP* | | *WXYZ* | |
|---|---|---|---|---|---|---|
| Angle | $\angle ARB$ | $\angle BRC$ | $\angle MRN$ | $\angle NRO$ | $\angle WRX$ | $\angle XRY$ |
| Angle Measure | | | | | | |

**c.** **Verbal** Make a conjecture about the diagonals of a parallelogram with four congruent sides.

## H.O.T. Problems Use Higher-Order Thinking Skills

**45.** **CHALLENGE** In rectangle *ABCD*, $m\angle EAB = 4x + 6$, $m\angle DEC = 10 - 11y$, and $m\angle EBC = 60$. Find the values of $x$ and $y$.

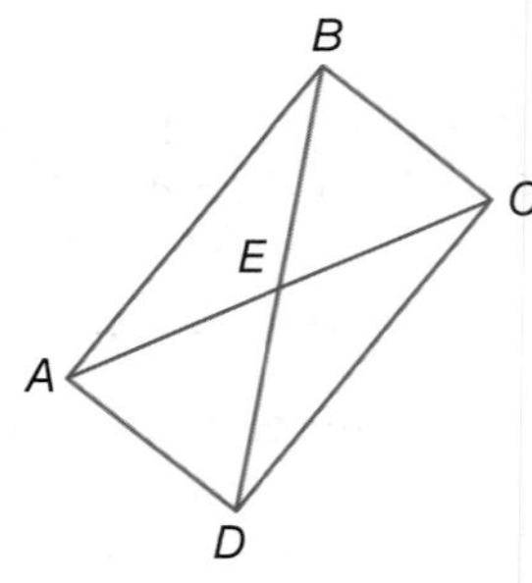


**46.** **CCSS CRITIQUE** Parker says that any two congruent acute triangles can be arranged to make a rectangle. Tamika says that only two congruent right triangles can be arranged to make a rectangle. Is either of them correct? Explain your reasoning.

**47.** **REASONING** In the diagram at the right, lines $n, p, q$, and $r$ are parallel and lines $\ell$ and $m$ are parallel. How many rectangles are formed by the intersecting lines?

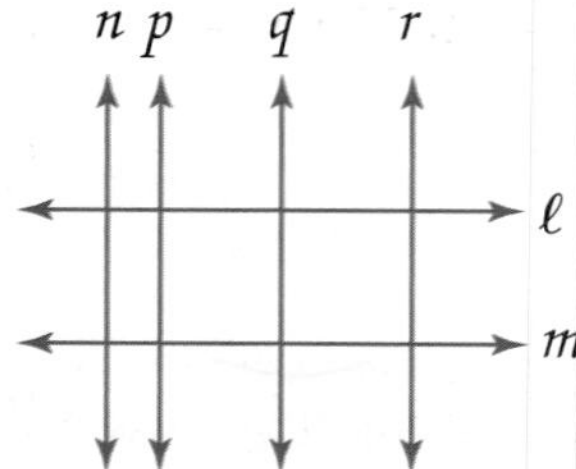


**48.** **OPEN ENDED** Write the equations of four lines having intersections that form the vertices of a rectangle. Verify your answer using coordinate geometry.

**49.** **WRITING IN MATH** Why are all rectangles parallelograms, but all parallelograms are not rectangles? Explain.

## Standardized Test Practice

**50.** If $FJ = -3x + 5y$, $FM = 3x + y$, $GH = 11$, and $GM = 13$, what values of $x$ and $y$ make parallelogram $FGHJ$ a rectangle?

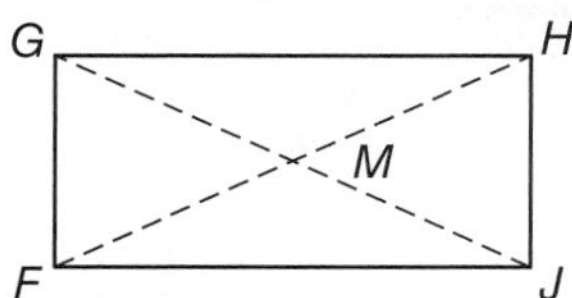


**A** $x = 3, y = 4$
**B** $x = 4, y = 3$
**C** $x = 7, y = 8$
**D** $x = 8, y = 7$

**51. ALGEBRA** A rectangular playground is surrounded by an 80-foot fence. One side of the playground is 10 feet longer than the other. Which of the following equations could be used to find $r$, the shorter side of the playground?

**F** $10r + r = 80$
**G** $4r + 10 = 80$
**H** $r(r + 10) = 80$
**J** $2(r + 10) + 2r = 80$

**52. SHORT RESPONSE** What is the measure of $\angle APB$?

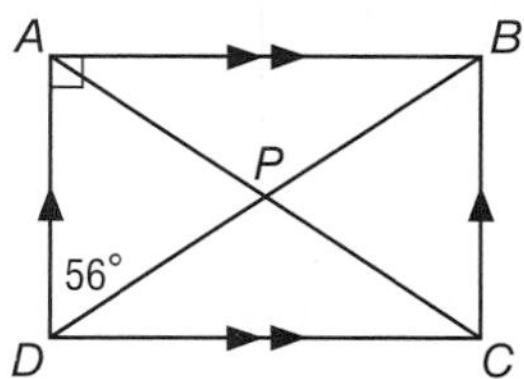


**53. SAT/ACT** If $p$ is odd, which of the following must also be odd?

**A** $2p$
**B** $2p + 2$
**C** $\frac{p}{2}$
**D** $2p - 2$
**E** $p + 2$

## Spiral Review

**ALGEBRA Find $x$ and $y$ so that the quadrilateral is a parallelogram.** (Lesson 8-3)

**54.**

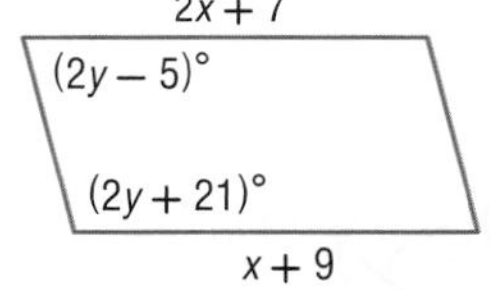


**55.**

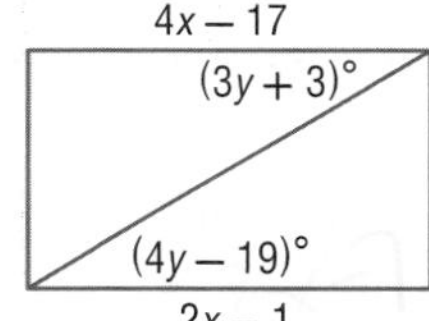


**56.**

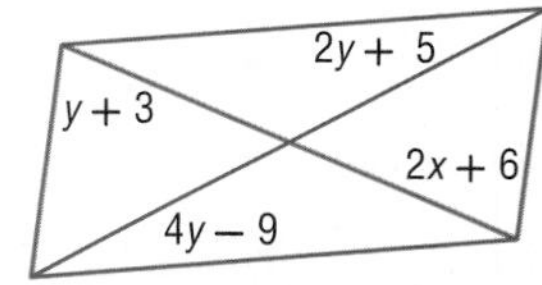


**57. COORDINATE GEOMETRY** Find the coordinates of the intersection of the diagonals of ▱$ABCD$ with vertices $A(1, 3)$, $B(6, 2)$, $C(4, -2)$, and $D(-1, -1)$. (Lesson 8-2)

**Refer to the figure at the right.** (Lesson 6-5)

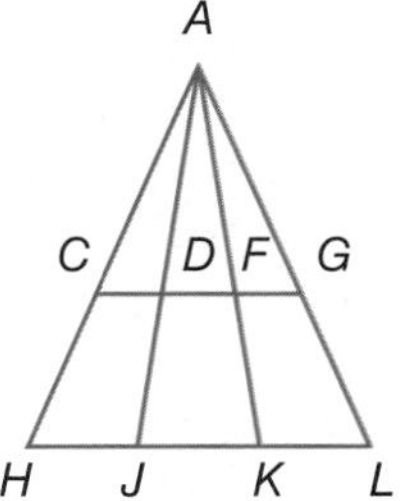


**58.** If $\overline{AC} \cong \overline{AF}$, name two congruent angles.

**59.** If $\angle AHJ \cong \angle AJH$, name two congruent segments.

**60.** If $\angle AJL \cong \angle ALJ$, name two congruent segments.

**61.** If $\overline{JA} \cong \overline{KA}$, name two congruent angles.

## Skills Review

**Find the distance between each pair of points.**

**62.** $(4, 2), (2, -5)$
**63.** $(0, 6), (-1, -4)$
**64.** $(-4, 3), (3, -4)$

LESSON

# 8-5 Rhombi and Squares

**Then**

- You determined whether quadrilaterals were parallelograms and/or rectangles.

**Now**

1. Recognize and apply the properties of rhombi and squares.
2. Determine whether quadrilaterals are rectangles, rhombi, or squares.

**Why?**

- Some fruits, nuts, and vegetables are packaged using bags made out of rhombus-shaped tubular netting. Similar shaped nylon netting is used for goals in such sports as soccer, hockey, and football. A rhombus and a square are both types of equilateral parallelograms.

**NewVocabulary**

rhombus
square

**Common Core State Standards**

**Content Standards**

**G.CO.11** Prove theorems about parallelograms.

**G.GPE.4** Use coordinates to prove simple geometric theorems algebraically.

**Mathematical Practices**

3 Construct viable arguments and critique the reasoning of others.

2 Reason abstractly and quantitatively.

## 1 Properties of Rhombi and Squares

A **rhombus** is a parallelogram with all four sides congruent. A rhombus has all the properties of a parallelogram and the two additional characteristics described in the theorems below.

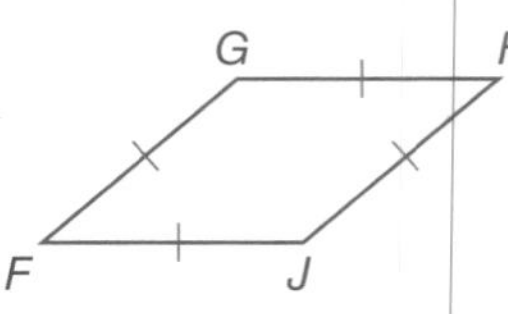


**Theorems** Diagonals of a Rhombus

**8.15** If a parallelogram is a rhombus, then its diagonals are perpendicular.

**Example** If ▱$ABCD$ is a rhombus, then $\overline{AC} \perp \overline{BD}$.

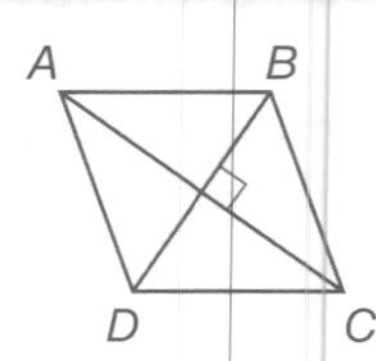


**8.16** If a parallelogram is a rhombus, then each diagonal bisects a pair of opposite angles.

**Example** If ▱$NPQR$ is a rhombus, then $\angle 1 \cong \angle 2$, $\angle 3 \cong \angle 4$, $\angle 5 \cong \angle 6$, and $\angle 7 \cong \angle 8$.

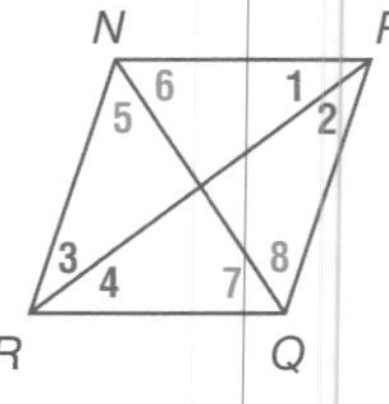


You will prove Theorem 8.16 in Exercise 34.

**Proof** Theorem 8.15

**Given:** $ABCD$ is a rhombus.

**Prove:** $\overline{AC} \perp \overline{BD}$

**Paragraph Proof:**

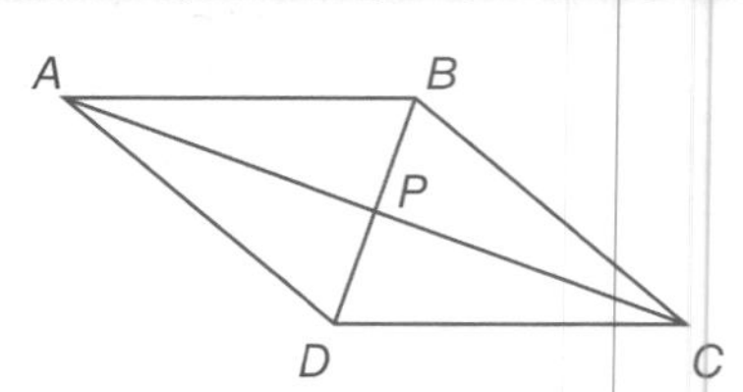


Since $ABCD$ is a rhombus, by definition $\overline{AB} \cong \overline{BC}$. A rhombus is a parallelogram and the diagonals of a parallelogram bisect each other, so $\overline{BD}$ bisects $\overline{AC}$ at $P$. Thus, $\overline{AP} \cong \overline{PC}$. $\overline{BP} \cong \overline{BP}$ by the Reflexive Property. So, $\triangle APB \cong \triangle CPB$ by SSS. $\angle APB \cong \angle CPB$ by CPCTC. $\angle APB$ and $\angle CPB$ also form a linear pair. Two congruent angles that form a linear pair are right angles. $\angle APB$ is a right angle, so $\overline{AC} \perp \overline{BD}$ by the definition of perpendicular lines.

**ReadingMath**

**Rhombi** The plural form of rhombus is *rhombi*, pronounced ROM-bye.

## Example 1 Use Properties of a Rhombus

**The diagonals of rhombus $FGHJ$ intersect at $K$. Use the given information to find each measure or value.**

**a. If $m\angle FJH = 82$, find $m\angle KHJ$.**

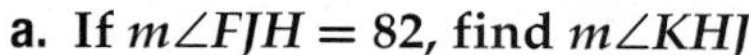

Since $FGHJ$ is a rhombus, diagonal $\overline{JG}$ bisects $\angle FJH$. Therefore, $m\angle KJH = \frac{1}{2}m\angle FJH$. So $m\angle KJH = \frac{1}{2}(82)$ or 41. Since the diagonals of a rhombus are perpendicular, $m\angle JKH = 90$ by the definition of perpendicular lines.

| | |
|---|---|
| $m\angle KJH + m\angle JKH + m\angle KHJ = 180$ | Triangle Sum Theorem |
| $41 + 90 + m\angle KHJ = 180$ | Substitution |
| $131 + m\angle KHJ = 180$ | Simplify. |
| $m\angle KHJ = 49$ | Subtract 131 from each side. |

**b. ALGEBRA If $GH = x + 9$ and $JH = 5x - 2$, find $x$.**

| | |
|---|---|
| $\overline{GH} \cong \overline{JH}$ | By definition, all sides of a rhombus are congruent. |
| $GH = JH$ | Definition of congruence |
| $x + 9 = 5x - 2$ | Substitution |
| $9 = 4x - 2$ | Subtract $x$ from each side. |
| $11 = 4x$ | Add 2 to each side. |
| $2.75 = x$ | Divide each side by 4. |

### GuidedPractice

**Refer to rhombus $FGHJ$ above.**

**1A.** If $FK = 5$ and $FG = 13$, find $KJ$.

**1B. ALGEBRA** If $m\angle JFK = 6y + 7$ and $m\angle KFG = 9y - 5$, find $y$.

A **square** is a parallelogram with four congruent sides and four right angles. Recall that a parallelogram with four right angles is a rectangle, and a parallelogram with four congruent sides is a rhombus. Therefore, a parallelogram that is both a rectangle and a rhombus is also a square.

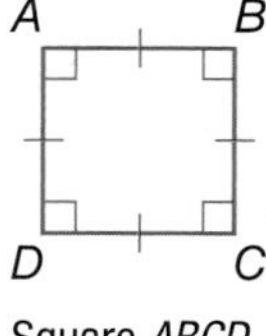

Square $ABCD$

The Venn diagram summarizes the relationships among parallelograms, rhombi, rectangles, and squares.

### ConceptSummary Parallelograms

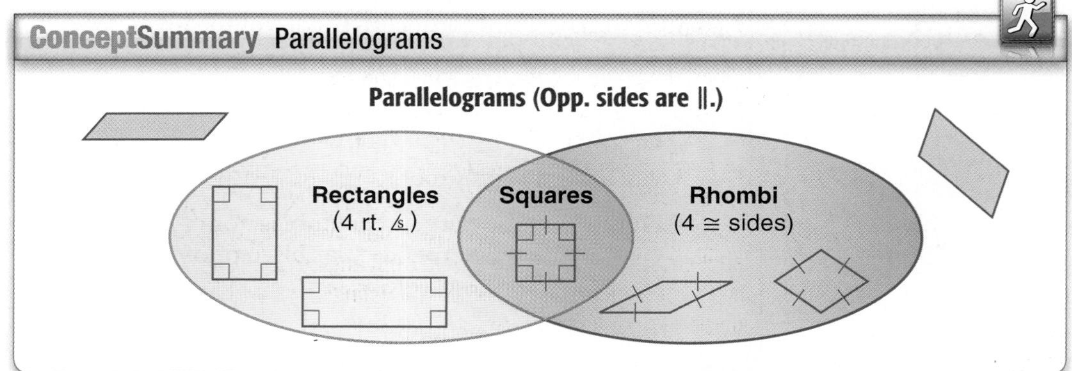

All of the properties of parallelograms, rectangles, and rhombi apply to squares. For example, the diagonals of a square bisect each other (parallelogram), are congruent (rectangle), and are perpendicular (rhombus).

## 2 Prove that Quadrilaterals are Rhombi or Squares

The theorems below provide conditions for rhombi and squares.

> **StudyTip**
> **Common Misconception** Theorems 8.17, 8.18, and 8.19 apply only if you already know that a quadrilateral is a parallelogram.

### Theorems Conditions for Rhombi and Squares

**8.17** If the diagonals of a parallelogram are perpendicular, then the parallelogram is a rhombus. (Converse of Theorem. 8.15)

**Example** If $\overline{JL} \perp \overline{KM}$, then $\square JKLM$ is a rhombus.

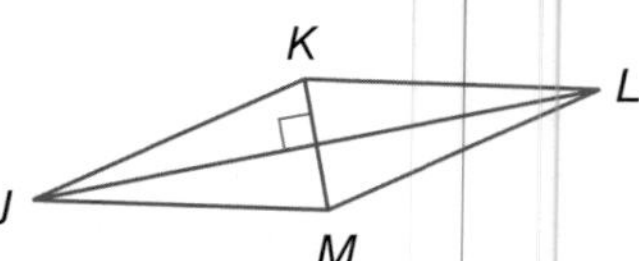


**8.18** If one diagonal of a parallelogram bisects a pair of opposite angles, then the parallelogram is a rhombus. (Converse of Theorem. 8.16)

**Example** If $\angle 1 \cong \angle 2$ and $\angle 3 \cong \angle 4$, or $\angle 5 \cong \angle 6$ and $\angle 7 \cong \angle 8$, then $\square WXYZ$ is a rhombus.

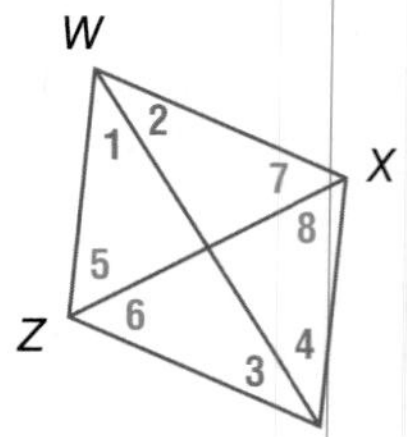


**8.19** If one pair of consecutive sides of a parallelogram are congruent, the parallelogram is a rhombus.

**Example** If $\overline{AB} \cong \overline{BC}$, then $\square ABCD$ is a rhombus.

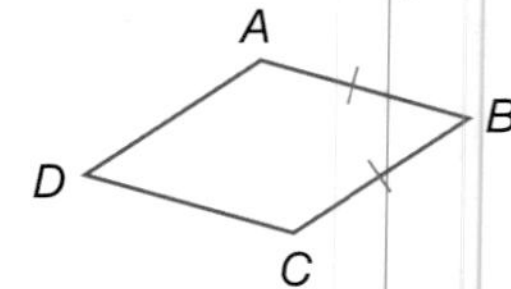


**8.20** If a quadrilateral is both a rectangle and a rhombus, then it is a square.

You will prove Theorems 8.17–8.20 in Exercises 35–38, respectively.

You can use the properties of rhombi and squares to write proofs.

### Example 2 Proofs Using Properties of Rhombi and Squares

> **StudyTip**
> **Congruent Triangles** Since a rhombus has four congruent sides, one diagonal separates the rhombus into two congruent isosceles triangles. Drawing two diagonals separates the rhombus into four congruent right triangles.

**Write a paragraph proof.**

**Given:** $JKLM$ is a parallelogram.

$\triangle JKL$ is isosceles.

**Prove:** $JKLM$ is a rhombus.

**Paragraph Proof:**

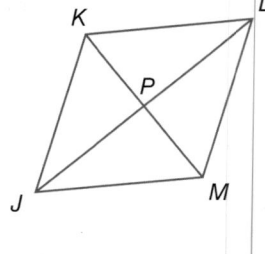

Since it is given that $\triangle JKL$ is isosceles, $\overline{KL} \cong \overline{JK}$ by definition. These are consecutive sides of the given parallelogram $JKLM$. So, by Theorem 8.19, $JKLM$ is a rhombus.

**GuidedPractice**

2. Write a paragraph proof.

**Given:** $\overline{SQ}$ is the perpendicular bisector of $\overline{PR}$.
$\overline{PR}$ is the perpendicular bisector of $\overline{SQ}$.
$\triangle RMS$ is isosceles.

**Prove:** $PQRS$ is a square.

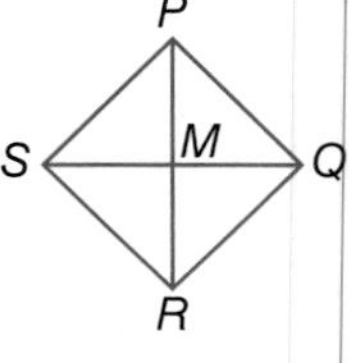

## Real-WorldLink

Archaeology is the study of artifacts that provide information about human life and activities in the past. Since humans only began writing about 5000 years ago, information from periods before that time must be gathered from the objects that archeologists locate.

**Source:** Encyclopaedia Britannica

### Real-World Example 3 Use Conditions for Rhombi and Squares

**ARCHAEOLOGY The key to the successful excavation of an archaeological site is accurate mapping. How can archaeologists be sure that the region they have marked off is a 1-meter by 1-meter square?**

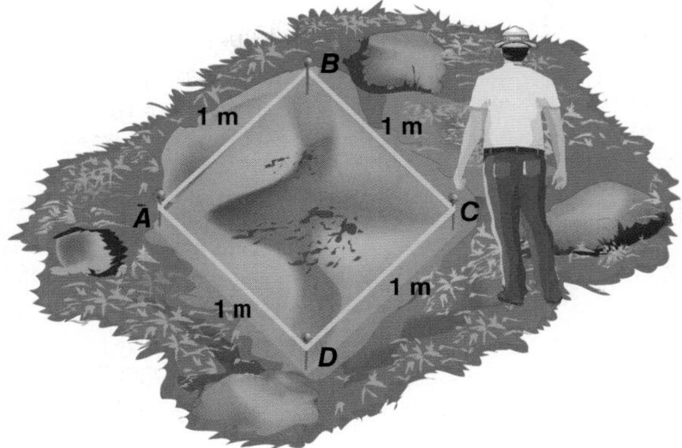


Each side of quadrilateral *ABCD* measures 1 meter. Since opposite sides are congruent, *ABCD* is a parallelogram. Since consecutive sides of ▱*ABCD* are congruent, it is a rhombus. If the archaeologists can show that ▱*ABCD* is also a rectangle, then by Theorem 8.20, ▱*ABCD* is a square.

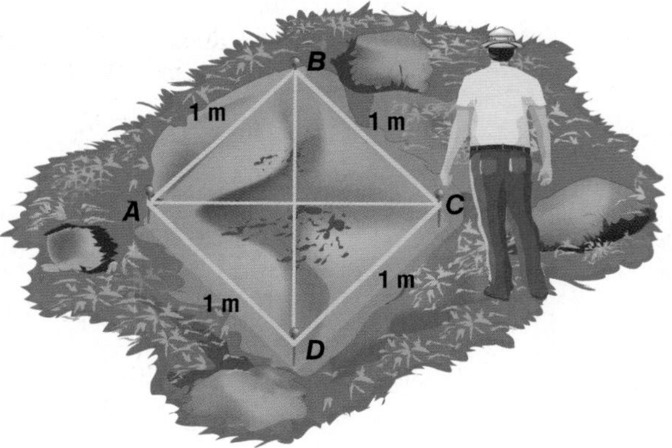


If the diagonals of a parallelogram are congruent, then the parallelogram is a rectangle. So if the archeologists measure the length of string needed to form each diagonal and find that these lengths are equal, then *ABCD* is a square.

### GuidedPractice

**3. QUILTING** Kathy is designing a quilt with blocks like the one shown.

**A.** If she marks the diagonals of each yellow piece and determines that each pair of diagonals is perpendicular, can she conclude that each yellow piece is a rhombus? Explain.

**B.** If all four angles of the green piece have the same measure and the bottom and left sides have the same measure, can she conclude that the green piece is a square? Explain.

In Chapter 6, you used coordinate geometry to classify triangles. Coordinate geometry can also be used to classify quadrilaterals.

## Example 4 Classify Quadrilaterals Using Coordinate Geometry

**COORDINATE GEOMETRY Determine whether ▱*JKLM* with vertices *J*(−7, −2), *K*(0, 4), *L*(9, 2), and *M*(2, −4) is a *rhombus*, a *rectangle*, or a *square*. List all that apply. Explain.**

**Problem-Solving Tip**

Make a Graph When analyzing a figure using coordinate geometry, graph the figure to help formulate a conjecture and also to help check the reasonableness of the answer you obtain algebraically.

**Understand** Plot and connect the vertices on a coordinate plane.

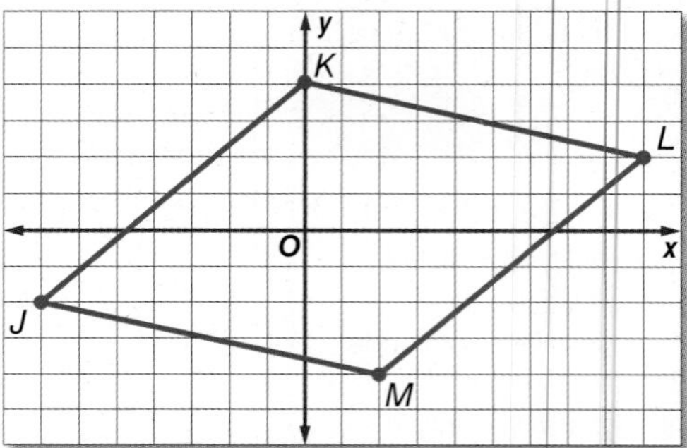


It appears from the graph that the parallelogram has four congruent sides, but no right angles. So, it appears that the figure is a rhombus, but not a square or a rectangle.

**Plan** If the diagonals of the parallelogram are congruent, then it is a rectangle. If they are perpendicular, then it is a rhombus. If they are both congruent and perpendicular, the parallelogram is a rectangle, a rhombus, and a square.

**Solve** **Step 1** Use the Distance Formula to compare the diagonal lengths.

$KM = \sqrt{(2-0)^2 + (-4-4)^2} = \sqrt{68}$ or $2\sqrt{17}$

$JL = \sqrt{[9-(-7)]^2 + [2-(-2)]^2} = \sqrt{272}$ or $4\sqrt{17}$

**Study Tip**

Square and Rhombus
A square is a rhombus, but a rhombus is not necessarily a square.

Since $2\sqrt{17} \neq 4\sqrt{17}$, the diagonals are not congruent. So, ▱*JKLM* is *not* a rectangle. Since the figure is not a rectangle, it also *cannot* be a square.

**Step 2** Use the Slope Formula to determine whether the diagonals are perpendicular.

slope of $\overline{KM} = \frac{-4-4}{2-0} = \frac{-8}{2}$ or $-4$

slope of $\overline{JL} = \frac{2-(-2)}{9-(-7)} = \frac{4}{16}$ or $\frac{1}{4}$

Since the product of the slopes of the diagonals is −1, the diagonals are perpendicular, so ▱*JKLM* is a rhombus.

**Check** $JK = \sqrt{[4-(-2)]^2 + [0-(-7)]^2}$ or $\sqrt{85}$

$KL = \sqrt{(9-0)^2 + (2-4)^2}$ or $\sqrt{85}$

So, ▱*JKLM* is a rhombus by Theorem 8.20.

Since the slope of $\overline{JK} = \frac{4-(-2)}{0-(-7)}$ or $\frac{6}{7}$, the slope of $\overline{KL} = \frac{2-4}{9-0}$ or $-\frac{2}{9}$, and the product of these slopes is not −1, consecutive sides $\overline{JK}$ and $\overline{KL}$ are not perpendicular. Therefore, ∠*JKL* is not a right angle. So ▱*JKLM* is not a rectangle or a square. ✓

### Guided Practice

**4.** Given *J*(5, 0), *K*(8, −11), *L*(−3, −14), *M*(−6, −3), determine whether parallelogram *JKLM* is a *rhombus*, a *rectangle*, or a *square*. List all that apply. Explain.

## Check Your Understanding

**Example 1** **ALGEBRA** **Quadrilateral $ABCD$ is a rhombus. Find each value or measure.**

A B D C

1. If $m\angle BCD = 64$, find $m\angle BAC$.
2. If $AB = 2x + 3$ and $BC = x + 7$, find $CD$.

**Examples 2–3** **3.** **PROOF** Write a two-column proof to prove that if $ABCD$ is a rhombus with diagonal $\overline{DB}$, then $\overline{AP} \cong \overline{CP}$.

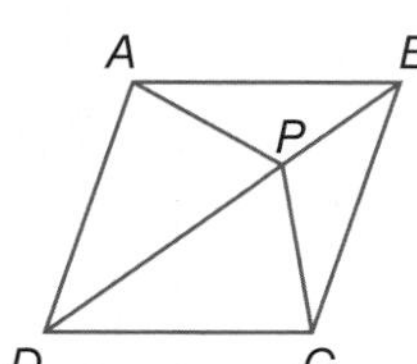


**4.** **GAMES** The checkerboard below is made up of 64 congruent black and red squares. Use this information to prove that the board itself is a square.

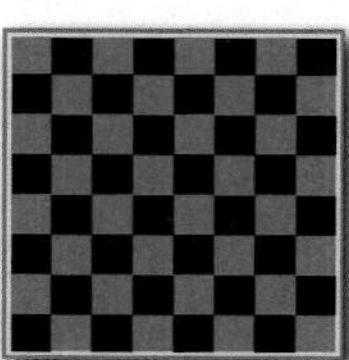

**Example 4** **COORDINATE GEOMETRY** **Given each set of vertices, determine whether ▱$QRST$ is a *rhombus*, a *rectangle*, or a *square*. List all that apply. Explain.**

5. $Q(1, 2)$, $R(-2, -1)$, $S(1, -4)$, $T(4, -1)$
6. $Q(-2, -1)$, $R(-1, 2)$, $S(4, 1)$, $T(3, -2)$

## Practice and Problem Solving

**Example 1** **ALGEBRA** **Quadrilateral $ABCD$ is a rhombus. Find each value or measure.**

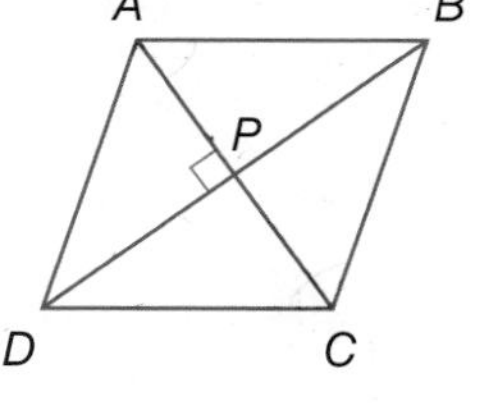


7. If $AB = 14$, find $BC$.
8. If $m\angle BCD = 54$, find $m\angle BAC$.
9. If $AP = 3x - 1$ and $PC = x + 9$, find $AC$.
10. If $DB = 2x - 4$ and $PB = 2x - 9$, find $PD$.
11. If $m\angle ABC = 2x - 7$ and $m\angle BCD = 2x + 3$, find $m\angle DAB$.
12. If $m\angle DPC = 3x - 15$, find $x$.

**Example 2** **CCSS ARGUMENTS** **Write a two-column proof.**

**13. Given:** $\overline{WZ} \parallel \overline{XY}$, $\overline{WX} \parallel \overline{ZY}$
$\overline{WZ} \cong \overline{ZY}$
**Prove:** $WXYZ$ is a rhombus.

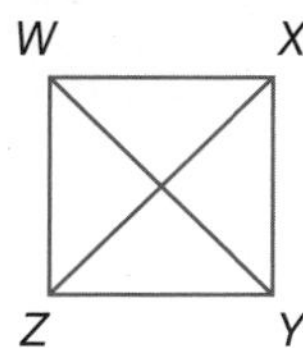


**14. Given:** $QRST$ is a parallelogram.
$\overline{TR} \cong \overline{QS}$, $m\angle QPR = 90$
**Prove:** $QRST$ is a square.

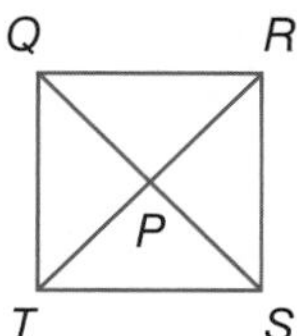


**15. Given:** $JKQP$ is a square.
$\overline{ML}$ bisects $\overline{JP}$ and $\overline{KQ}$.
**Prove:** $JKLM$ is a parallelogram.

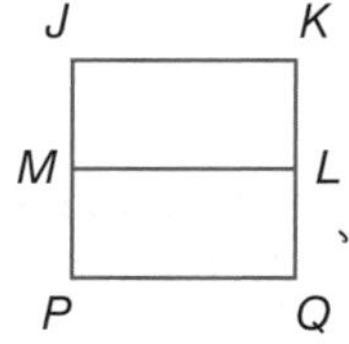


**16. Given:** $ACDH$ and $BCDF$ are parallelograms; $\overline{BF} \cong \overline{AB}$.
**Prove:** $ABFH$ is a rhombus.

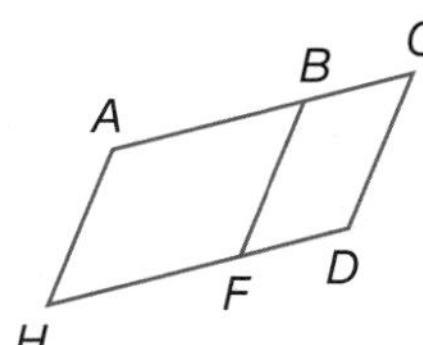

**Example 3**

**17. ROADWAYS** Main Street and High Street intersect as shown in the diagram. Each of the crosswalks is the same length. Classify the quadrilateral formed by the crosswalks. Explain your reasoning.

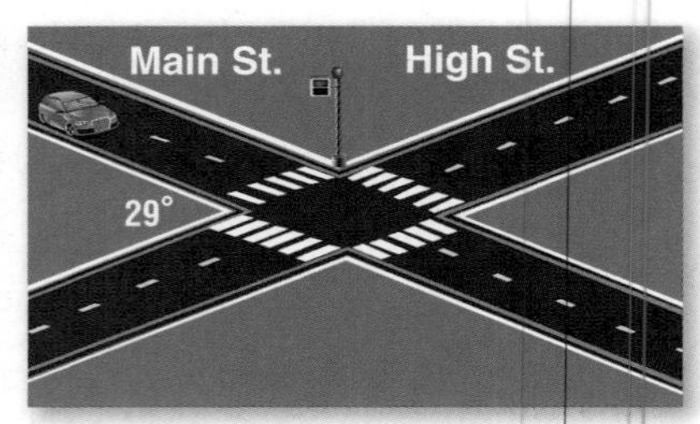


**18. CCSS MODELING** A landscaper has staked out the area for a square garden as shown. She has confirmed that each side of the quadrilateral formed by the stakes is congruent and that the diagonals are perpendicular. Is this information enough for the landscaper to be sure that the garden is a square? Explain your reasoning.

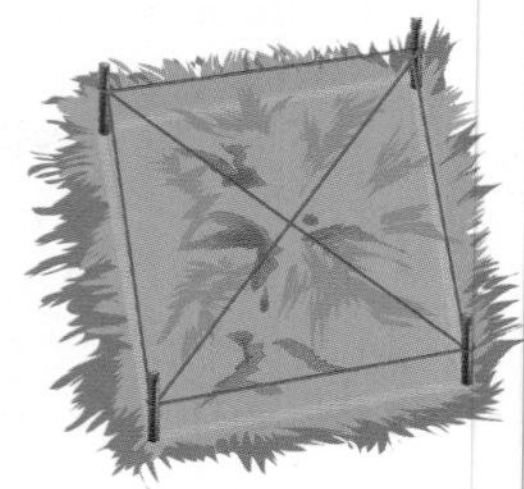

**Example 4**

**COORDINATE GEOMETRY Given each set of vertices, determine whether ▱*JKLM* is a *rhombus*, a *rectangle*, or a *square*. List all that apply. Explain.**

**19.** $J(-4, -1), K(1, -1), L(4, 3), M(-1, 3)$

**20.** $J(-3, -2), K(2, -2), L(5, 2), M(0, 2)$

**21.** $J(-2, -1), K(-4, 3), L(1, 5), M(3, 1)$

**22.** $J(-1, 1), K(4, 1), L(4, 6), M(-1, 6)$

***ABCD* is a rhombus. If $PB = 12$, $AB = 15$, and $m\angle ABD = 24$, find each measure.**

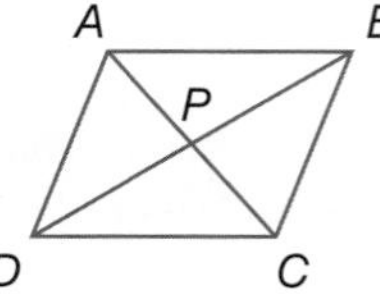


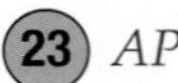

**23.** $AP$

**24.** $CP$

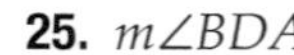

**25.** $m\angle BDA$

**26.** $m\angle ACB$

***WXYZ* is a square. If $WT = 3$, find each measure.**

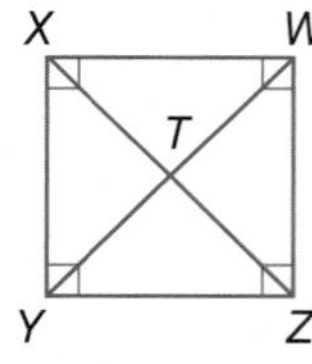


**27.** $ZX$

**28.** $XY$

**29.** $m\angle WTZ$

**30.** $m\angle WYX$

**Classify each quadrilateral.**

**31.**

**32.**

**33.**

**PROOF Write a paragraph proof.**

**34.** Theorem 8.16

**35.** Theorem 8.17

**36.** Theorem 8.18

**37.** Theorem 8.19

**38.** Theorem 8.20

**CONSTRUCTION Use diagonals to construct each figure. Justify each construction.**

**39.** rhombus

**40.** square

**PROOF Write a coordinate proof of each statement.**

**41.** The diagonals of a square are perpendicular.

**42.** The segments joining the midpoints of the sides of a rectangle form a rhombus.

**43.** **DESIGN** The tile pattern below consists of regular octagons and quadrilaterals. Classify the quadrilaterals in the pattern and explain your reasoning.

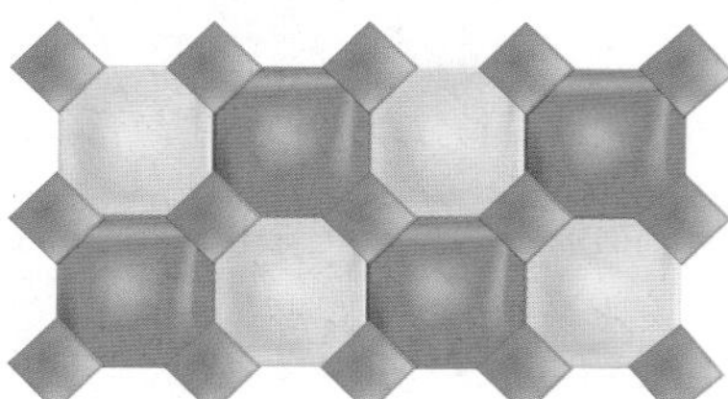

**44.** **REPAIR** The window pane shown needs to be replaced. What are the dimensions of the replacement pane?

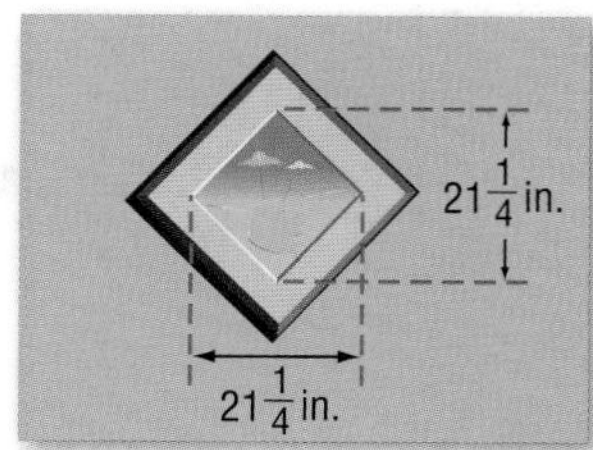


**45.** **MULTIPLE REPRESENTATIONS** In this problem, you will explore the properties of kites, which are quadrilaterals with exactly two distinct pairs of adjacent congruent sides.

**a. Geometric** Draw three kites with varying side lengths. Label one kite *ABCD*, one *PQRS*, and one *WXYZ*. Then draw the diagonals of each kite, labeling the point of intersection *N* for each kite.

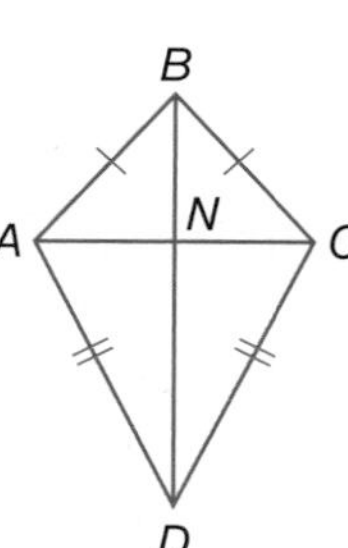

Kite *ABCD*

**b. Tabular** Measure the distance from *N* to each vertex. Record your results in a table like the one shown.

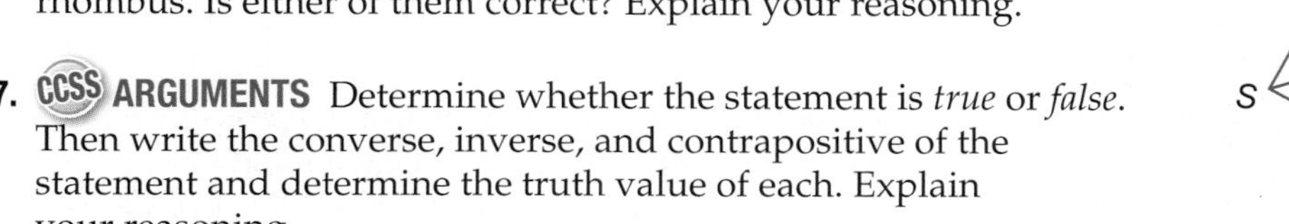

| Figure | Distance from *N* to Each Vertex Along Shorter Diagonal | | Distance from *N* to Each Vertex Along Longer Diagonal | |
|---|---|---|---|---|
| *ABCD* | | | | |
| *PQRS* | | | | |
| *WXYZ* | | | | |

**c. Verbal** Make a conjecture about the diagonals of a kite.

## H.O.T. Problems Use Higher-Order Thinking Skills

**46.** **ERROR ANALYSIS** In parallelogram *PQRS*, $\overline{PR} \cong \overline{QS}$. Lola thinks that the parallelogram is a square, and Xavier thinks that it is a rhombus. Is either of them correct? Explain your reasoning.

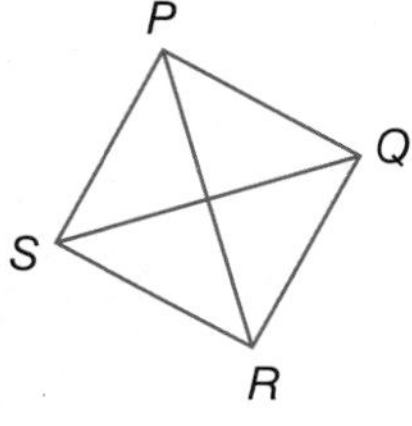


**47.** **CCSS ARGUMENTS** Determine whether the statement is *true* or *false*. Then write the converse, inverse, and contrapositive of the statement and determine the truth value of each. Explain your reasoning.

*If a quadrilateral is a square, then it is a rectangle.*

**48.** **CHALLENGE** The area of square *ABCD* is 36 square units and the area of $\triangle EBF$ is 20 square units. If $\overline{EB} \perp \overline{BF}$ and $\overline{AE} = 2$, find the length of $\overline{CF}$.

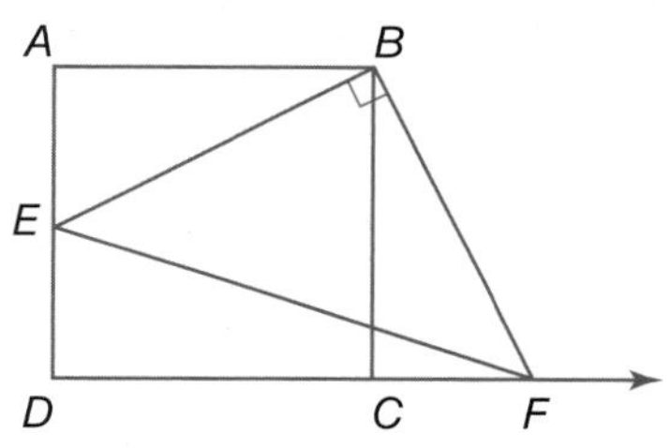


**49.** **OPEN ENDED** Find the vertices of a square with diagonals that are contained in the lines $y = x$ and $y = -x + 6$. Justify your reasoning.

**50.** **WRITING IN MATH** Compare all of the properties of the following quadrilaterals: parallelograms, rectangles, rhombi, and squares.

## Standardized Test Practice

**51.** *JKLM* is a rhombus. If $CK = 8$ and $JK = 10$, find $JC$.

A 4 C 8
B 6 D 10

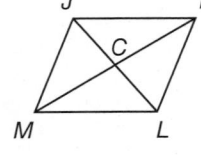

**52. EXTENDED RESPONSE** The sides of square *ABCD* are extended by sides of equal length to form square *WXYZ*.

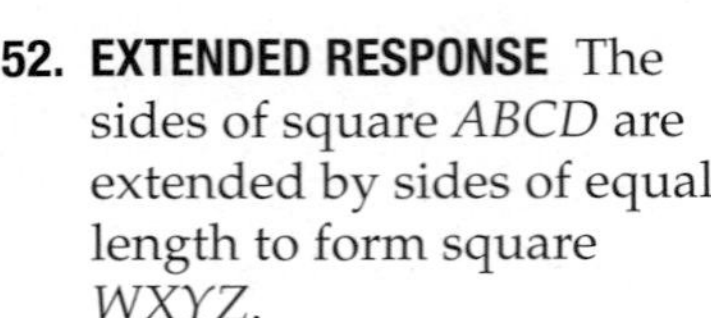

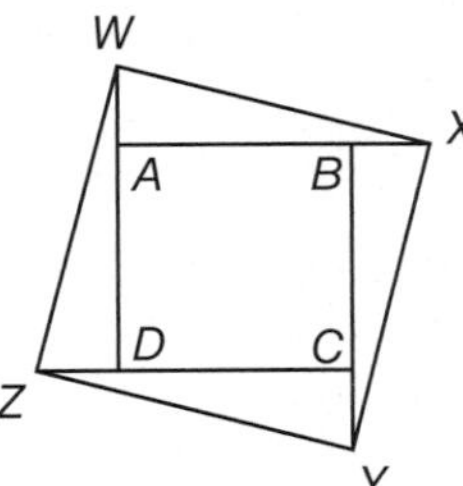


a. If $CY = 3$ cm and the area of *ABCD* is 81 cm$^2$, find the area of *WXYZ*.

b. If the areas of *ABCD* and *WXYZ* are 49 cm$^2$ and 169 cm$^2$ respectively, find *DZ*.

c. If $AB = 2CY$ and the area of $ABCD =$ $g$ square meters, find the area of *WXYZ* in square meters.

**53. ALGEBRA** What values of $x$ and $y$ make quadrilateral *ABCD* a parallelogram?

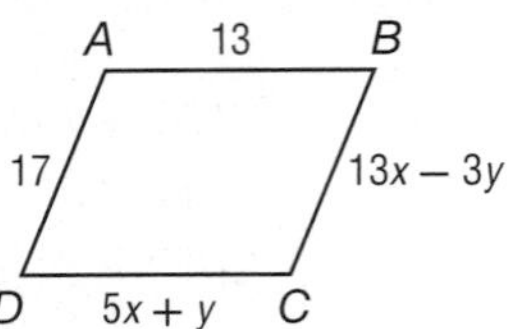


F $x = 3, y = 2$
G $x = \frac{3}{2}, y = -1$
H $x = 2, y = 3$
J $x = 3, y = -1$

**54. SAT/ACT** What is 6 more than the product of $-3$ and a certain number $x$?

A $-3x - 6$ D $-3x + 6$
B $-3x$ E $6 + 3x$
C $-x$

## Spiral Review

**Quadrilateral *ABDC* is a rectangle. Find each measure if $m\angle 1 = 38$.** (Lesson 8-4)

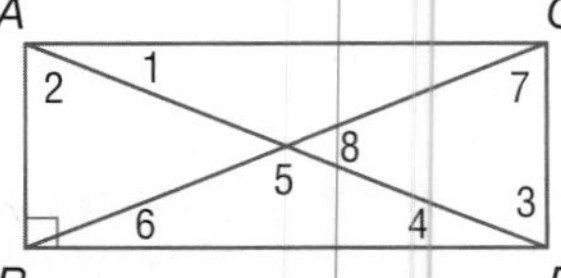


**55.** $m\angle 2$ **56.** $m\angle 5$ **57.** $m\angle 6$

**Determine whether each quadrilateral is a parallelogram. Justify your answer.** (Lesson 8-3)

**58.**

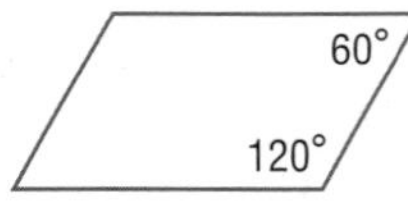


**59.**

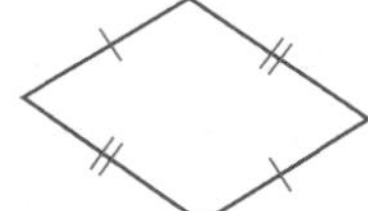

**60.**

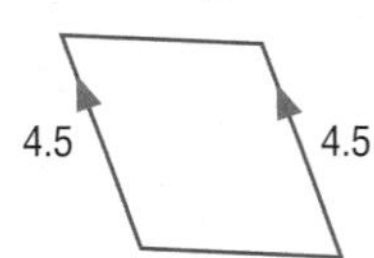


**61. MEASUREMENT** Monifa says that her backyard is shaped like a triangle and that the lengths of its sides are 22 feet, 23 feet, and 45 feet. Do you think these measurements are correct? Explain your reasoning. (Lesson 7-5)

## Skills Review

**Solve each equation.**

**62.** $\frac{1}{2}(5x + 7x - 1) = 11.5$ **63.** $\frac{1}{2}(10x + 6x + 2) = 7$ **64.** $\frac{1}{2}(12x + 6 - 8x + 7) = 9$

# LESSON 8-6 Trapezoids and Kites

| Then | Now | Why? |
|---|---|---|
| You used properties of special parallelograms. | **1** Apply properties of trapezoids.<br>**2** Apply properties of kites. | In gymnastics, vaulting boxes made out of high compression foam are used as spotting platforms, vaulting horses, and steps. The left and right side of each section is a *trapezoid*. |

**NewVocabulary**
- trapezoid
- bases
- legs of a trapezoid
- base angles
- isosceles trapezoid
- midsegment of a trapezoid
- kite

**Common Core State Standards**

**Content Standards**

**G.GPE.4** Use coordinates to prove simple geometric theorems algebraically.

**G.MG.3** Apply geometric methods to solve problems (e.g., designing an object or structure to satisfy physical constraints or minimize cost; working with typographic grid systems based on ratios). ★

**Mathematical Practices**
1. Make sense of problems and persevere in solving them.
2. Reason abstractly and quantitatively.

## 1 Properties of Trapezoids

A **trapezoid** is a quadrilateral with exactly one pair of parallel sides. The parallel sides are called **bases**. The nonparallel sides are called **legs**. The **base angles** are formed by the base and one of the legs. In trapezoid $ABCD$, $\angle A$ and $\angle B$ are one pair of base angles and $\angle C$ and $\angle D$ are the other pair. If the legs of a trapezoid are congruent, then it is an **isosceles trapezoid**.

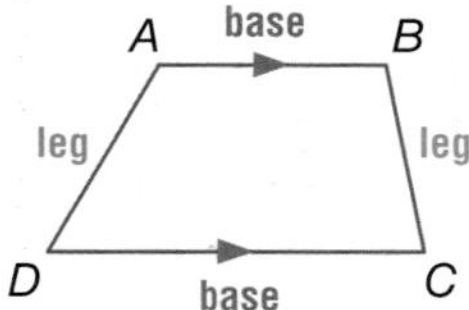


### Theorems Isosceles Trapezoids

**8.21** If a trapezoid is isosceles, then each pair of base angles is congruent.

**Example** If trapezoid $FGHJ$ is isosceles, then $\angle G \cong \angle H$ and $\angle F \cong \angle J$.

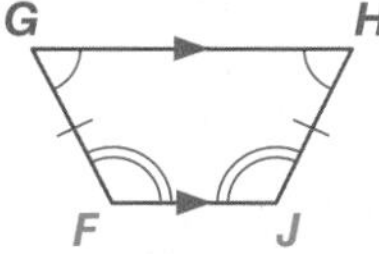


**8.22** If a trapezoid has one pair of congruent base angles, then it is an isosceles trapezoid.

**Example** If $\angle L \cong \angle M$, then trapezoid $KLMP$ is isosceles.

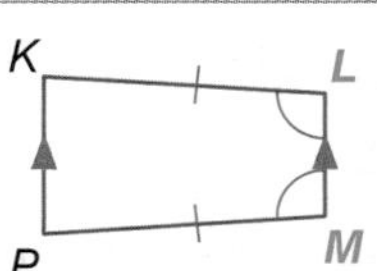


**8.23** A trapezoid is isosceles if and only if its diagonals are congruent.

**Example** If trapezoid $QRST$ is isosceles, then $\overline{QS} \cong \overline{RT}$. Likewise, if $\overline{QS} \cong \overline{RT}$, then trapezoid $QRST$ is isosceles.

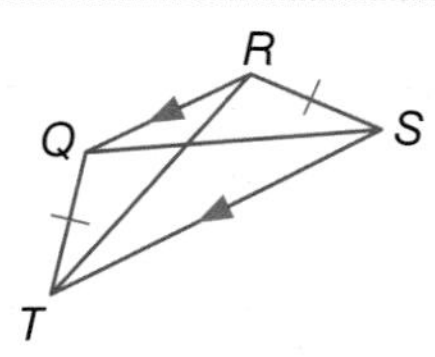


You will prove Theorem 8.21, Theorem 8.22, and the other part of Theorem 8.23 in Exercises 28, 29, and 30.

### Proof Part of Theorem 8.23

**Given:** $ABCD$ is an isosceles trapezoid.

**Prove:** $\overline{AC} \cong \overline{BD}$

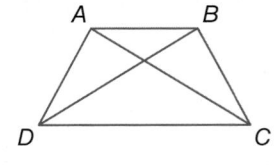

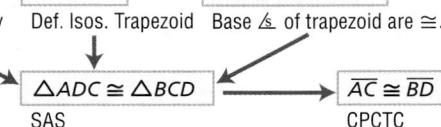

- $ABCD$ is an isosceles trapezoid. — Given
- $\overline{DC} \cong \overline{CD}$ — Reflexive Property
- $\overline{AD} \cong \overline{BC}$ — Def. Isos. Trapezoid
- $\angle ADC \cong \angle BCD$ — Base ∠s of trapezoid are ≅.
- $\triangle ADC \cong \triangle BCD$ — SAS
- $\overline{AC} \cong \overline{BD}$ — CPCTC

### Real-WorldLink

Speakers are amplifiers that intensify sound waves so that they are audible to the unaided ear. Amplifiers exist in devices such as televisions, stereos, and computers.

**Source:** How Stuff Works

## Real-World Example 1 Use Properties of Isosceles Trapezoids

**MUSIC The speaker shown is an isosceles trapezoid. If $m\angle FJH = 85$, $FK = 8$ inches, and $JG = 19$ inches, find each measure.**

F G K J H

**a.** $m\angle FGH$

Since $FGHJ$ is an isosceles trapezoid, $\angle FJH$ and $\angle GHJ$ are congruent base angles. So, $m\angle GHJ = m\angle FJH = 85$.

Since $FGHJ$ is a trapezoid, $\overline{FG} \parallel \overline{JH}$.

| | |
|---|---|
| $m\angle FGH + m\angle GHJ = 180$ | Consecutive Interior Angles Theorem |
| $m\angle FGH + 85 = 180$ | Substitution |
| $m\angle FGH = 95$ | Subtract 85 from each side. |

**b.** $KH$

Since $FGHJ$ is an isosceles trapezoid, diagonals $\overline{FH}$ and $\overline{JG}$ are congruent.

| | |
|---|---|
| $FH = JG$ | Definition of congruent |
| $FK + KH = JG$ | Segment Addition |
| $8 + KH = 19$ | Substitution |
| $KH = 11$ cm | Subtract 8 from each side. |

### GuidedPractice

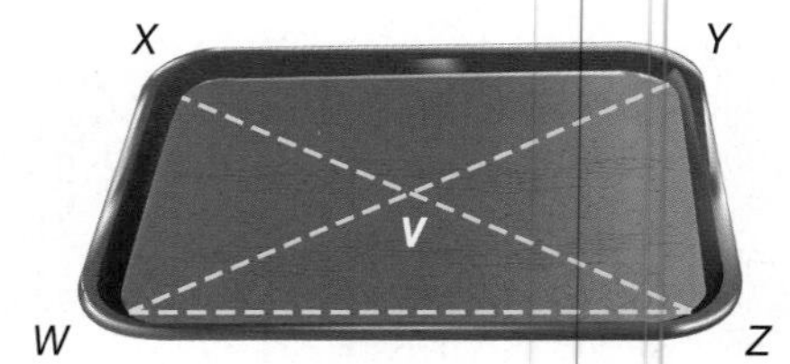


**1. CAFETERIA TRAYS** To save space at a square table, cafeteria trays often incorporate trapezoids into their design. If $WXYZ$ is an isosceles trapezoid and $m\angle YZW = 45$, $WV = 15$ centimeters, and $VY = 10$ centimeters, find each measure.

**A.** $m\angle XWZ$ **B.** $m\angle WXY$

**C.** XZ **D.** XV

### StudyTip

Isosceles Trapezoids
The base angles of a trapezoid are only congruent if the trapezoid is isosceles.

You can use coordinate geometry to determine whether a trapezoid is an isosceles trapezoid.

## Example 2 Isosceles Trapezoids and Coordinate Geomerty

**COORDINATE GEOMETRY Quadrilateral $ABCD$ has vertices $A(-3, 4)$, $B(2, 5)$, $C(3, 3)$, and $D(-1, 0)$. Show that $ABCD$ is a trapezoid and determine whether it is an isosceles trapezoid.**

Graph and connect the vertices of $ABCD$.

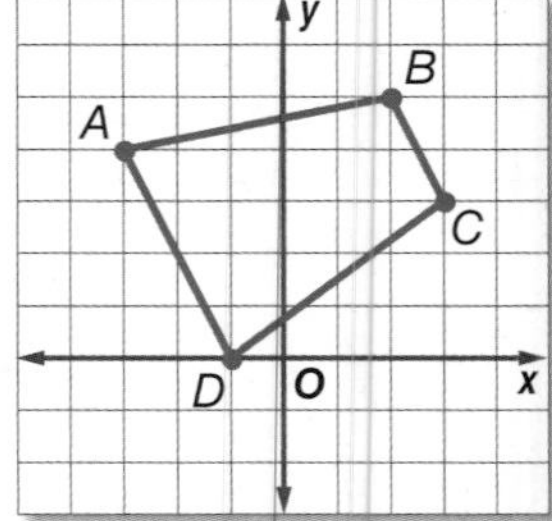


**Step 1** Use the Slope Formula to compare the slopes of opposite sides $\overline{BC}$ and $\overline{AD}$ and of opposite sides $\overline{AB}$ and $\overline{DC}$. A quadrilateral is a trapezoid if exactly one pair of opposite sides are parallel.

Jorg Greuel/Photographer's Choice/Getty Images

Opposite sides $\overline{BC}$ and $\overline{AD}$:

slope of $\overline{BC} = 3 - \frac{5}{3} - 2 = -\frac{2}{1}$ or $-2$

slope of $\overline{AD} = \frac{0 - 4}{-1 - (-3)} = \frac{-4}{2}$ or $-2$

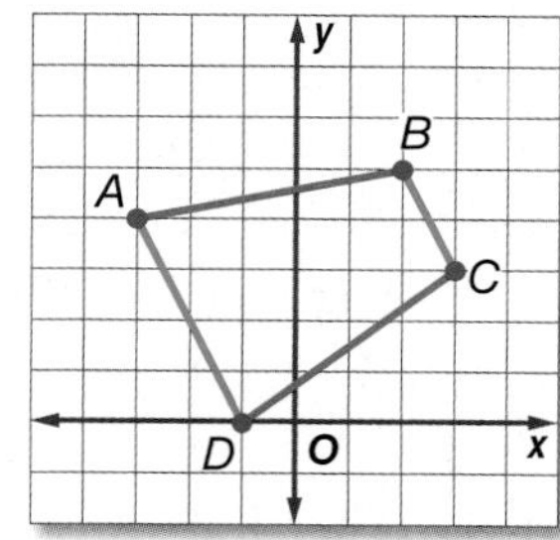


Since the slopes of $\overline{BC}$ and $\overline{AD}$ are equal, $\overline{BC} \parallel \overline{AD}$.

Opposite sides $\overline{AB}$ and $\overline{DC}$:

slope of $\overline{AB} = \frac{5 - 4}{2 - (-3)} = \frac{1}{5}$

slope of $\overline{DC} = \frac{0 - 3}{-1 - 3} = \frac{-3}{-4}$ or $\frac{3}{4}$

**ReadingMath**

**Symbols** Recall that the symbol $\nparallel$ means *is not parallel to.*

Since the slopes of $\overline{AB}$ and $\overline{DC}$ are *not* equal, $\overline{BC} \nparallel \overline{AD}$. Since quadrilateral *ABCD* has only one pair of opposite sides that are parallel, quadrilateral *ABCD* is a trapezoid.

**Step 2** Use the Distance Formula to compare the lengths of legs $\overline{AB}$ and $\overline{DC}$. A trapezoid is isosceles if its legs are congruent.

$AB = \sqrt{(-3 - 2)^2 + (4 - 5)^2}$ or $\sqrt{26}$

$DC = \sqrt{(-1 - 3)^2 + (0 - 3)^2} = \sqrt{25}$ or $5$

Since $AB \neq DC$, legs $\overline{AB}$ abd $\overline{DC}$ are *not* congruent. Therefore, trapezoid *ABCD* is not isosceles.

**GuidedPractice**

**2.** Quadrilateral *QRST* has vertices $Q(-8, -4)$, $R(0, 8)$, $S(6, 8)$, and $T(-6, -10)$. Show that *QRST* is a trapezoid and determine whether *QRST* is an isosceles trapezoid.

**ReadingMath**

**Midsegment** A midsegment of a trapezoid can also be called a *median.*

The **midsegment of a trapezoid** is the segment that connects the midpoints of the legs of the trapezoid.

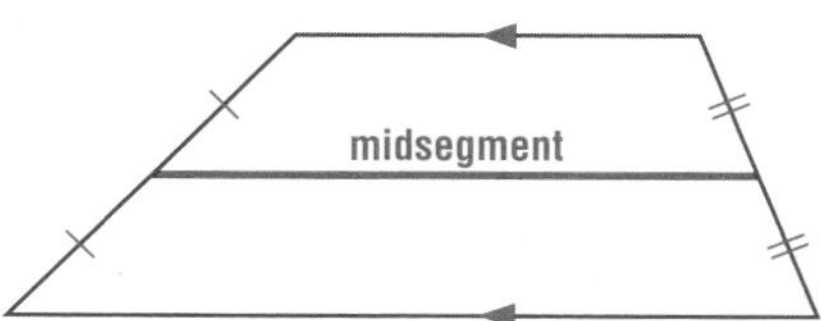


The theorem below relates the midsegment and the bases of a trapezoid.

### Theorem 8.24 Trapezoid Midsegment Theorem

The midsegment of a trapezoid is parallel to each base and its measure is one half the sum of the lengths of the bases.

**Example** If $\overline{BE}$ is the midsegment of trapezoid *ACDF*, then $\overline{AF} \parallel \overline{BE}$, $\overline{CD} \parallel \overline{BE}$, and $BE = \frac{1}{2}(AF + CD)$.

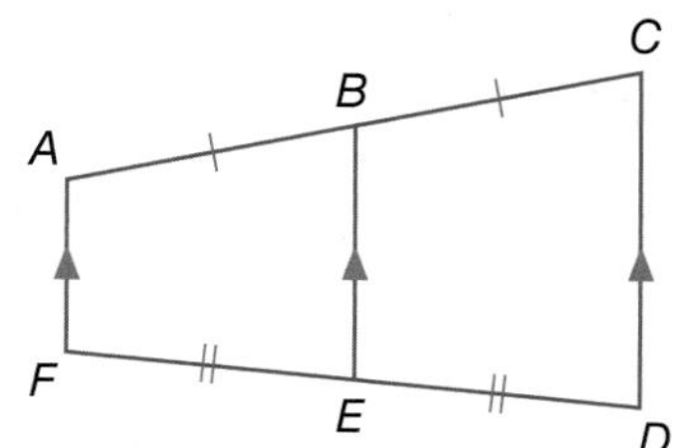


You will prove Theorem 8.24 in Exercise 33.

PT

## Standardized Test Example 3 Midsegment of a Trapezoid

**GRIDDED RESPONSE** **In the figure, $\overline{LH}$ is the midsegment of trapezoid *FGJK*. What is the value of *x*?**

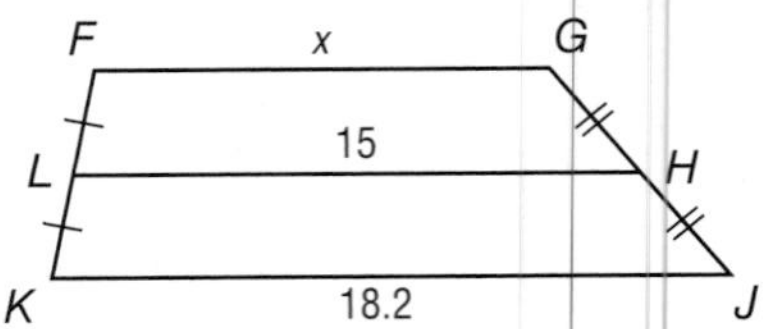

Note: The figure is not drawn to scale.

**Read the Test Item**

You are given the measure of the midsegment of a trapezoid and the measure of one of its bases. You are asked to find the measure of the other base.

**Solve the Test Item**

| | |
|---|---|
| $LH = \frac{1}{2}(FG + KJ)$ | Trapezoid Midsegment Theorem |
| $5 = \frac{1}{2}(x + 18.2)$ | Substitution |
| $30 = x + 18.2$ | Multiply each side by 2. |
| $11.8 = x$ | Subtract 18.2 from each side. |

**Test-TakingTip**

Gridded Responses
Rational answers can often be gridded in more than one way. An answer such as $\frac{8}{5}$ could be gridded as 8/5 or 1.6, but not as 1 3/5.

**Grid In Your Answer**

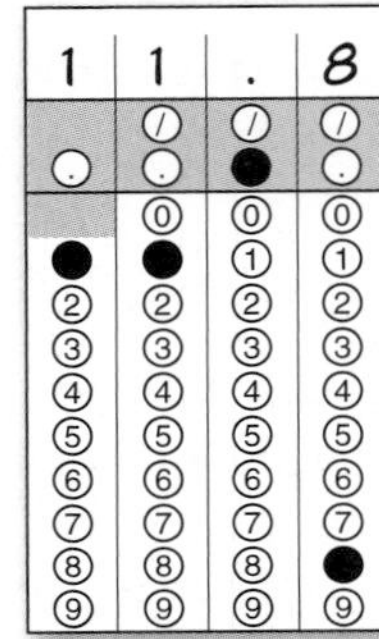


- You can align the numerical answer by placing the first digit in the left answer box or by putting the last digit in the right answer box.
- Do not leave blank boxes in the middle of an answer.
- Fill in **one** bubble for each filled answer box. Do not fill more than one bubble for an answer box. Do not fill in a bubble for blank answer boxes.

### GuidedPractice

3. **GRIDDED RESPONSE** Trapezoid *ABCD* is shown below. If $\overline{FG}$ is parallel to $\overline{AD}$, what is the *x*-coordinate of point *G*?

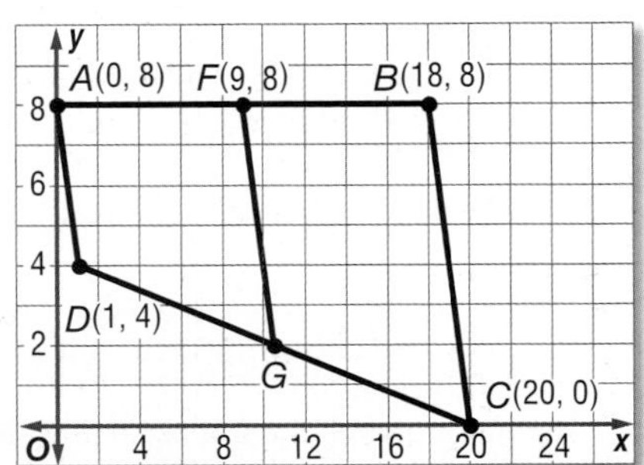


## 2 Properties of Kites

A **kite** is a quadrilateral with exactly two pairs of consecutive congruent sides. Unlike a parallelogram, the opposite sides of a kite are not congruent or parallel.

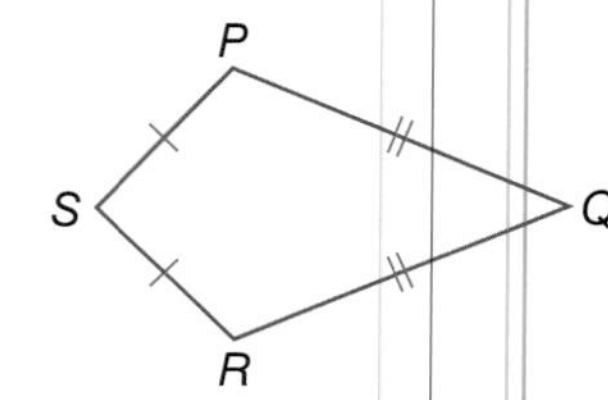

**Study Tip**

Kites The congruent angles of a kite are included by the non-congruent adjacent sides.

## Theorems Kites

**8.25** If a quadrilateral is a kite, then its diagonals are perpendicular.

**Example** If quadrilateral *ABCD* is a kite, then $\overline{AC} \perp \overline{BD}$.

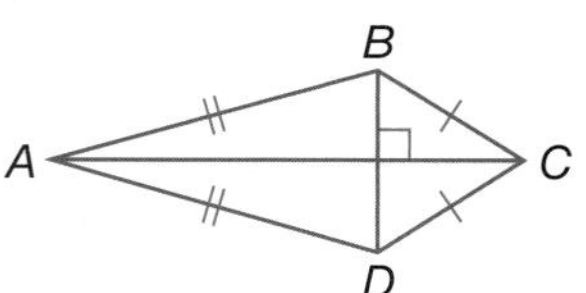


**8.26** If a quadrilateral is a kite, then exactly one pair of opposite angles is congruent.

**Example** If quadrilateral *JKLM* is a kite, $\overline{JK} \cong \overline{KL}$, and $\overline{JM} \cong \overline{LM}$, then $\angle J \cong \angle L$ and $\angle K \ncong \angle M$.

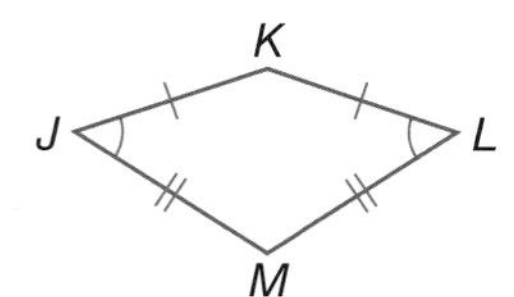


You will prove Theorems 8.25 and 8.26 in Exercises 31 and 32, respectively.

You can use the theorems above, the Pythagorean Theorem, and the Polygon Interior Angles Sum Theorem to find missing measures in kites.

## Example 4 Use Properties of Kites

**a. If *FGHJ* is a kite, find $m\angle GFJ$.**

Since a kite can only have one pair of opposite congruent angles and $\angle G \ncong \angle J$, then $\angle F \cong \angle H$. So, $m\angle F = m\angle H$. Write and solve an equation to find $m\angle F$.

| | |
|---|---|
| $m\angle F + m\angle G + m\angle H + m\angle J = 360$ | Polygon Interior Angles Sum Theorem |
| $m\angle F + 128 + m\angle F + 72 = 360$ | Substitution |
| $2m\angle F + 200 = 360$ | Simplify. |
| $2m\angle F = 160$ | Subtract 200 from each side. |
| $m\angle F = 80$ | Divide each side by 2. |

**b. If *WXYZ* is a kite, find *ZY*.**

Since the diagonals of a kite are perpendicular, they divide *WXYZ* into four right triangles. Use the Pythagorean Theorem to find *ZY*, the length of the hypotenuse of right $\triangle YPZ$.

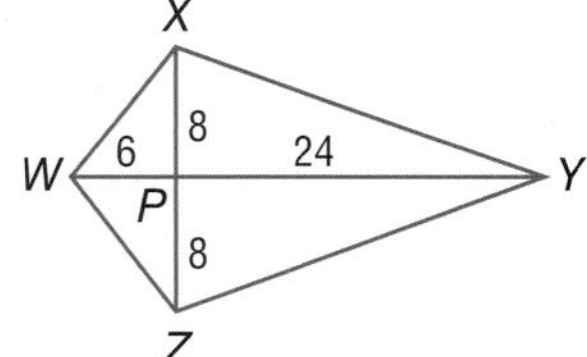


| | |
|---|---|
| $PZ^2 + PY^2 = ZY^2$ | Pythagorean Theorem |
| $8^2 + 24^2 = ZY^2$ | Substitution |
| $640 = ZY^2$ | Simplify. |
| $\sqrt{640} = ZY$ | Take the square root of each side. |
| $8\sqrt{10} = ZY$ | Simplify. |

**Real-World Link**

The fastest recorded speed of a kite is over 120 miles per hour. The record for the highest single kite flown is 12,471 feet.

**Source:** Borealis Kites

### Guided Practice

**4A.** If $m\angle BAD = 38$ and $m\angle BCD = 50$, find $m\angle ADC$.

**4B.** If $BT = 5$ and $TC = 8$, find $CD$.

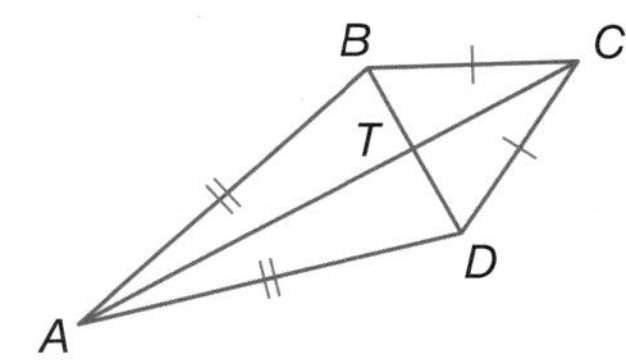

## Check Your Understanding

**Example 1** **Find each measure.**

1. $m\angle D$

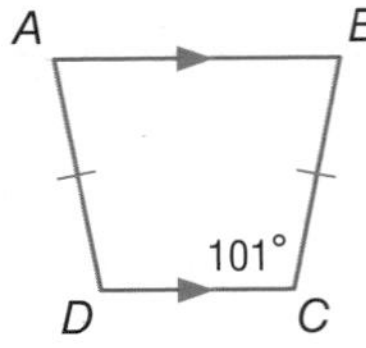


2. $WT$, if $ZX = 20$ and $TY = 15$

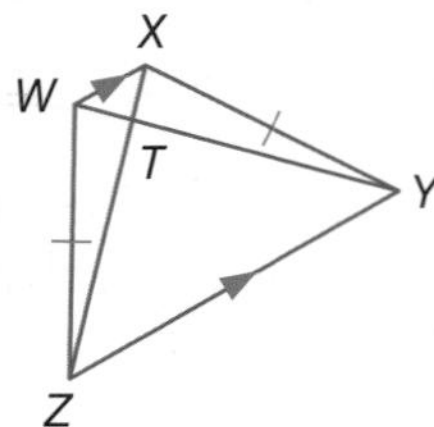


**Example 2** **COORDINATE GEOMETRY** **Quadrilateral $ABCD$ has vertices $A(-4, -1)$, $B(-2, 3)$, $C(3, 3)$, and $D(5, -1)$.**

3. Verify that $ABCD$ is a trapezoid.

4. Determine whether $ABCD$ is an isosceles trapezoid. Explain.

**Example 3**

5. **GRIDDED RESPONSE** In the figure at the right, $\overline{YZ}$ is the midsegment of trapezoid $TWRV$. Determine the value of $x$.

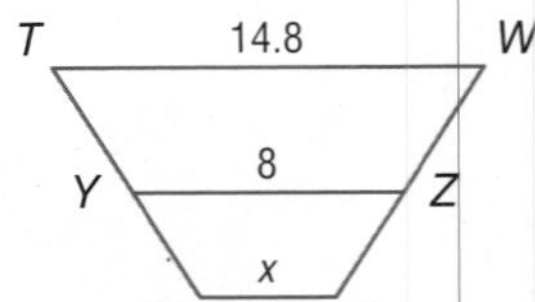


**Example 4** **CCSS SENSE-MAKING** **If $ABCD$ is a kite, find each measure.**

6. $AB$

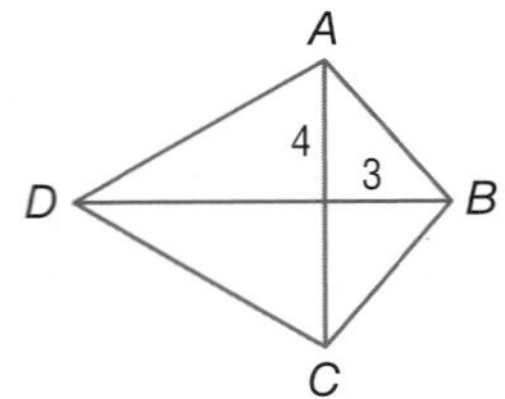


7. $m\angle C$

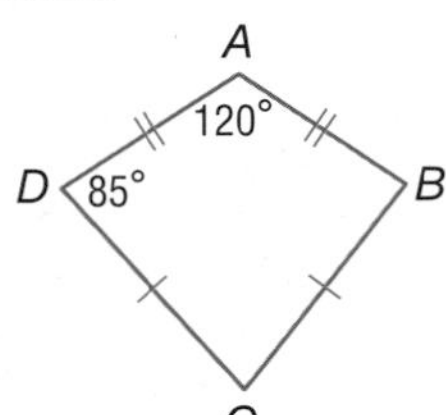


## Practice and Problem Solving

**Example 1** **Find each measure.**

8. $m\angle K$

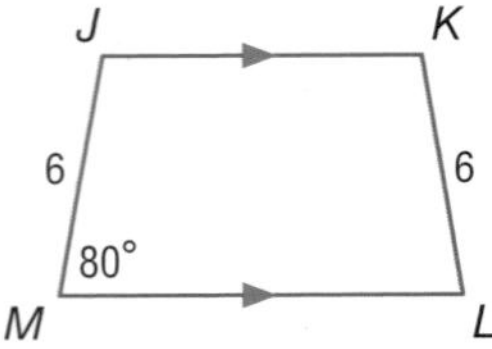


9. $m\angle Q$

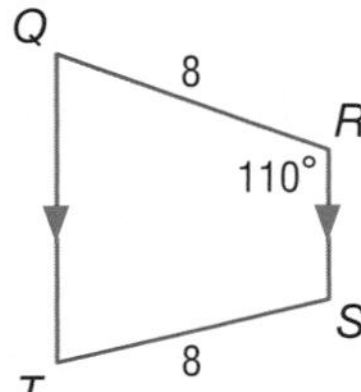


10. $JL$, if $KP = 4$ and $PM = 7$

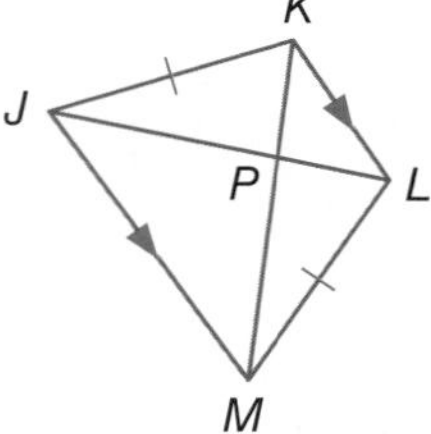


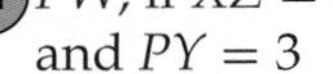

11. $PW$, if $XZ = 18$ and $PY = 3$

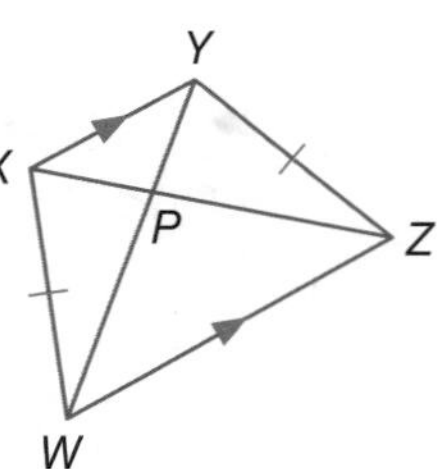


**Example 2** **COORDINATE GEOMETRY** **For each quadrilateral with the given vertices, verify that the quadrilateral is a trapezoid and determine whether the figure is an isosceles trapezoid.**

12. $A(-2, 5)$, $B(-3, 1)$, $C(6, 1)$, $D(3, 5)$

13. $J(-4, -6)$, $K(6, 2)$, $L(1, 3)$, $M(-4, -1)$

14. $Q(2, 5)$, $R(-2, 1)$, $S(-1, -6)$, $T(9, 4)$

15. $W(-5, -1)$, $X(-2, 2)$, $Y(3, 1)$, $Z(5, -3)$

**Example 3** **For trapezoid *QRTU*, *V* and *S* are midpoints of the legs.**

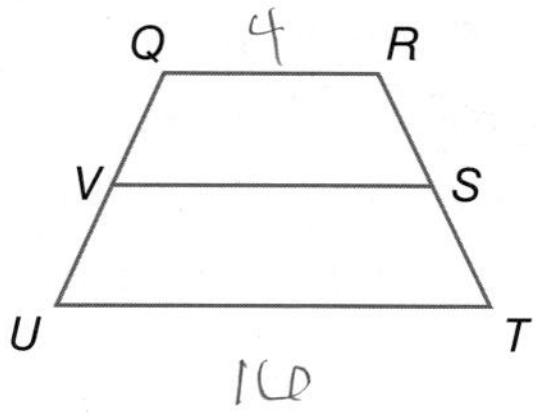

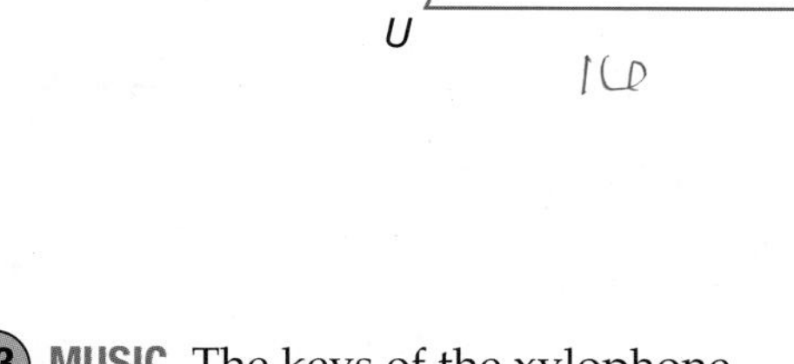


**16.** If $QR = 12$ and $UT = 22$, find $VS$.

**17.** If $QR = 4$ and $UT = 16$, find $VS$.

**18.** If $VS = 9$ and $UT = 12$, find $QR$.

**19.** If $TU = 26$ and $SV = 17$, find $QR$.

**20.** If $QR = 2$ and $VS = 7$, find $UT$.

**21.** If $RQ = 5$ and $VS = 11$, find $UT$.

**22. DESIGN** Juana is designing a window box. She wants the end of the box to be a trapezoid with the dimensions shown. If she wants to put a shelf in the middle for the plants to rest on, about how wide should she make the shelf?

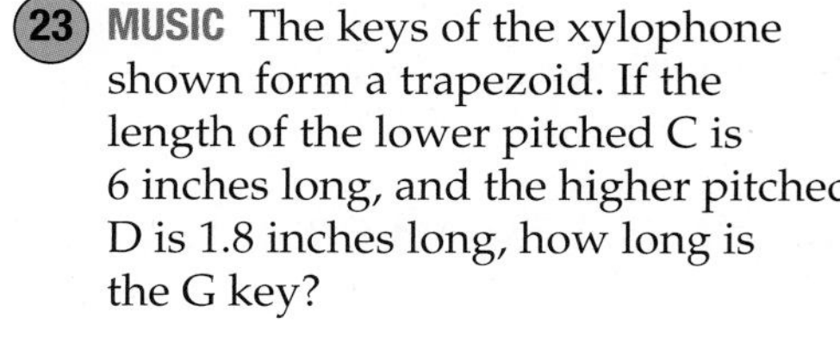

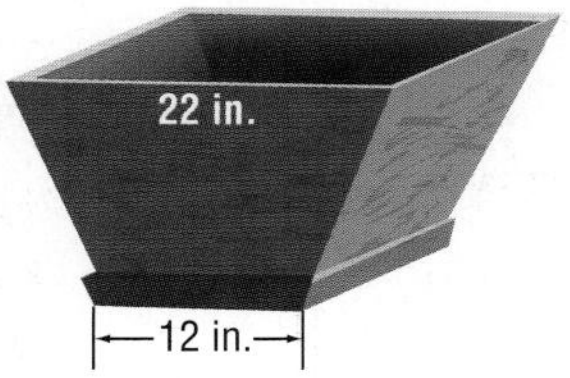

**23** **MUSIC** The keys of the xylophone shown form a trapezoid. If the length of the lower pitched C is 6 inches long, and the higher pitched D is 1.8 inches long, how long is the G key?

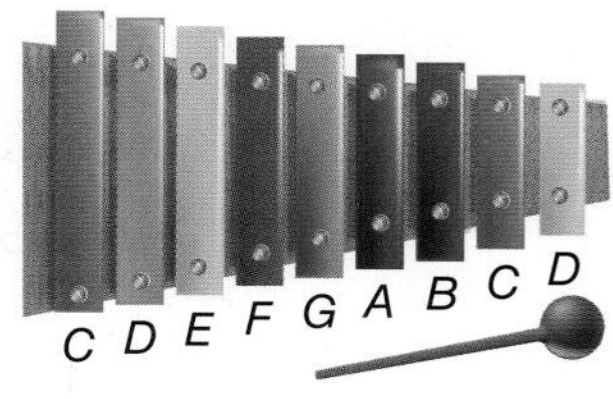


**Example 4** **CCSS SENSE-MAKING** **If *WXYZ* is a kite, find each measure.**

**24.** $YZ$

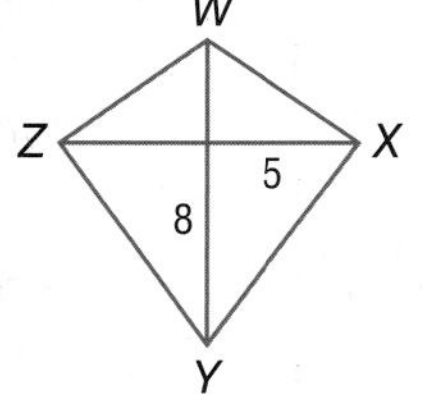


**25.** $WP$

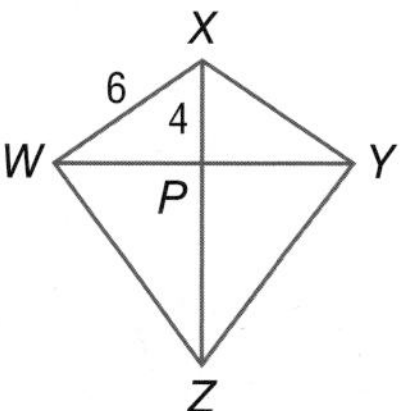


**26.** $m\angle X$

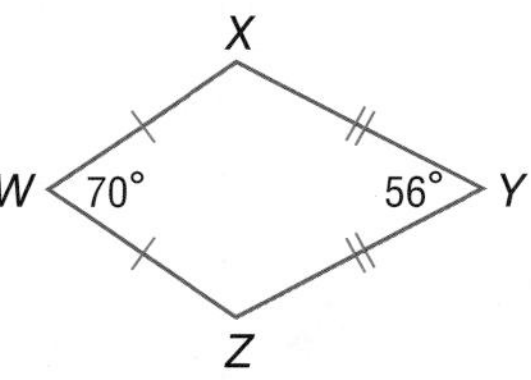


**27.** $m\angle Z$

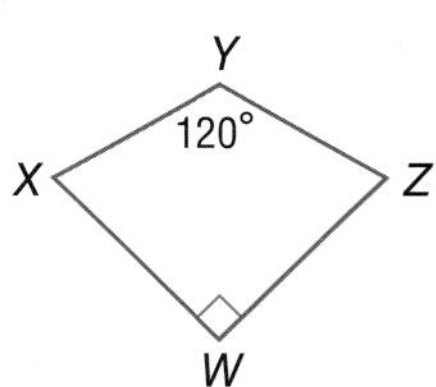


**PROOF** **Write a paragraph proof for each theorem.**

**28.** Theorem 8.21

**29.** Theorem 8.22

**30.** Theorem 8.23

**31.** Theorem 8.25

**32.** Theorem 8.26

**33. PROOF** Write a coordinate proof for Theorem 8.24.

**34. COORDINATE GEOMETRY** Refer to quadrilateral *ABCD*.

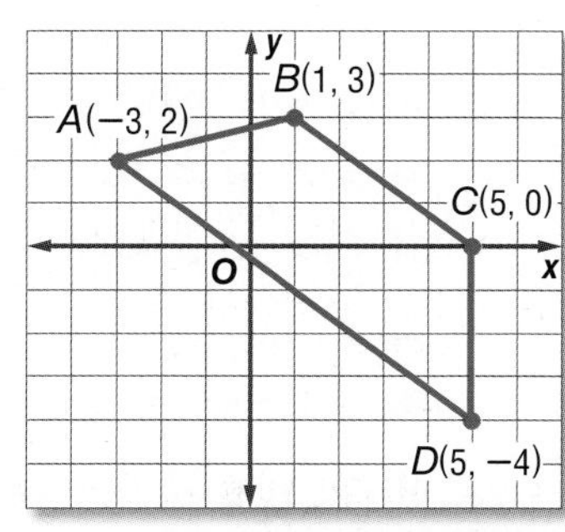


**a.** Determine whether the figure is a trapezoid. If so, is it isosceles? Explain.

**b.** Is the midsegment contained in the line with equation $y = -x + 1$? Justify your answer.

**c.** Find the length of the midsegment.

**ALGEBRA *ABCD* is a trapezoid.**

**35.** If $AC = 3x - 7$ and $BD = 2x + 8$, find the value of $x$ so that $ABCD$ is isosceles.

**36.** If $m\angle ABC = 4x + 11$ and $m\angle DAB = 2x + 33$, find the value of $x$ so that $ABCD$ is isosceles.

**SPORTS The end of the batting cage shown is an isosceles trapezoid. If $PT = 12$ feet, $ST = 28$ feet, and $m\angle PQR = 110$, find each measure.**

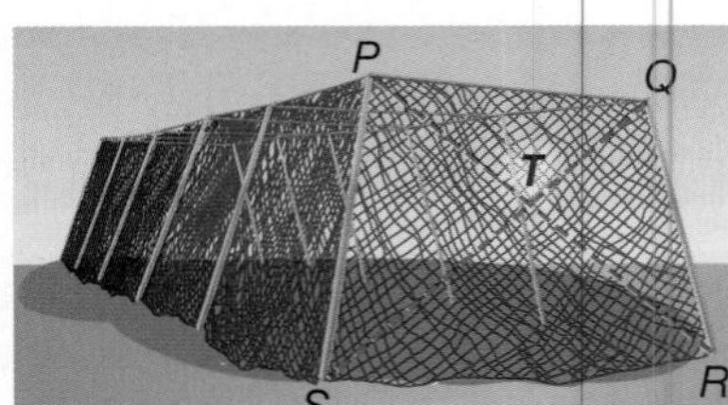

**37.** $TR$

**38.** $SQ$

**39.** $m\angle QRS$

**40.** $m\angle QPS$

**ALGEBRA For trapezoid *QRST*, *M* and *P* are midpoints of the legs.**

**41.** If $QR = 16$, $PM = 12$, and $TS = 4x$, find $x$.

**42.** If $TS = 2x$, $PM = 20$, and $QR = 6x$, find $x$.

**43.** If $PM = 2x$, $QR = 3x$, and $TS = 10$, find $PM$.

**44.** If $TS = 2x + 2$, $QR = 5x + 3$, and $PM = 13$, find $TS$.

**SHOPPING The side of the shopping bag shown is an isosceles trapezoid. If $EC = 9$ inches, $DB = 19$ inches, $m\angle ABE = 40$, and $m\angle EBC = 35$, find each measure.**

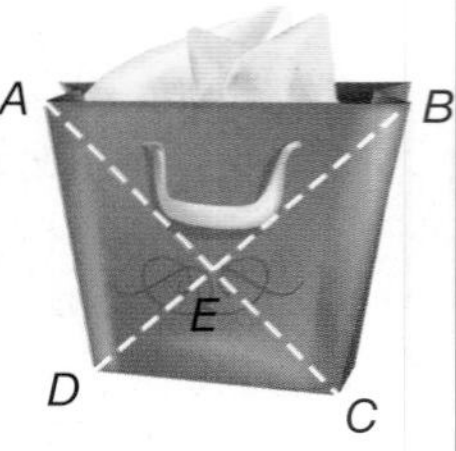


**45.** $AE$

**46.** $AC$

**47.** $m\angle BCD$

**48.** $m\angle EDC$

**ALGEBRA *WXYZ* is a kite.**

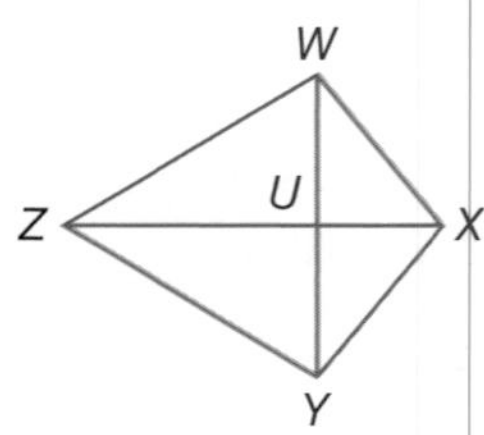


**49.** If $m\angle WXY = 120$, $m\angle WZY = 4x$, and $m\angle ZWX = 10x$, find $m\angle ZYX$.

**50.** If $m\angle WXY = 13x + 24$, $m\angle WZY = 35$, and $m\angle ZWX = 13x + 14$, find $m\angle ZYX$.

**CCSS ARGUMENTS Write a two-column proof.**

**51. Given:** $ABCD$ is an isosceles trapezoid.

**Prove:** $\angle DAC \cong \angle CBD$

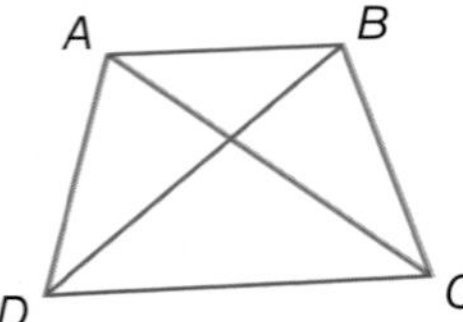


**52. Given:** $\overline{WZ} \cong \overline{ZV}$, $\overline{XY}$ bisects $\overline{WZ}$ and $\overline{ZV}$, and $\angle W \cong \angle ZXY$.

**Prove:** $WXYV$ is an isosceles trapezoid.

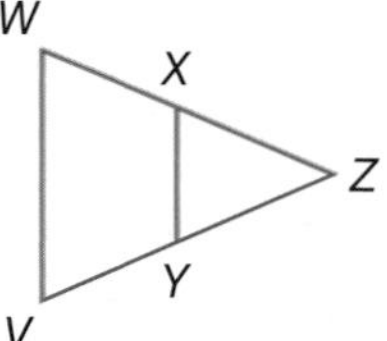


**Determine whether each statement is *always*, *sometimes*, or *never* true. Explain.**

**53.** The opposite angles of a trapezoid are supplementary.

**54.** One pair of opposite sides are parallel in a kite.

**55.** A square is a rhombus.

**56.** A rectangle is a square.

**57.** A parallelogram is a rectangle.

**58. KITES** Refer to the kite at the right. Using the properties of kites, write a two-column proof to show that $\triangle MNR$ is congruent to $\triangle PNR$.

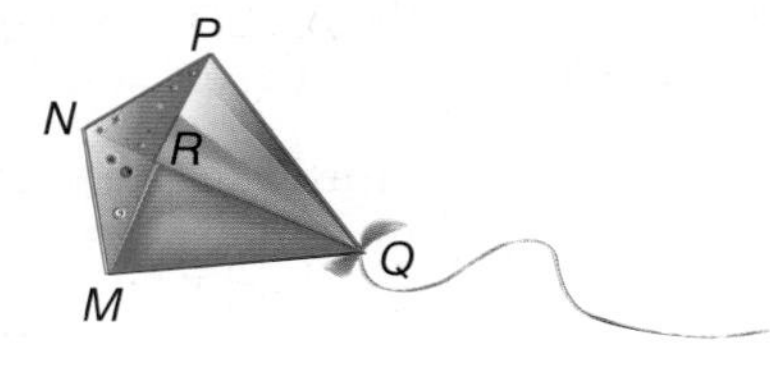


**59. VENN DIAGRAM** Create a Venn diagram that incorporates all quadrilaterals, including trapezoids, isosceles trapezoids, kites, and quadrilaterals that cannot be classified as anything other than quadrilaterals.

**COORDINATE GEOMETRY Determine whether each figure is a *trapezoid*, a *parallelogram*, a *square*, a *rhombus*, or a *quadrilateral* given the coordinates of the vertices. Choose the most specific term. Explain.**

**60.** $A(-1, 4)$, $B(2, 6)$, $C(3, 3)$, $D(0, 1)$

**61** $W(-3, 4)$, $X(3, 4)$, $Y(5, 3)$, $Z(-5, 1)$

**62. MULTIPLE REPRESENTATIONS** In this problem, you will explore proportions in kites.

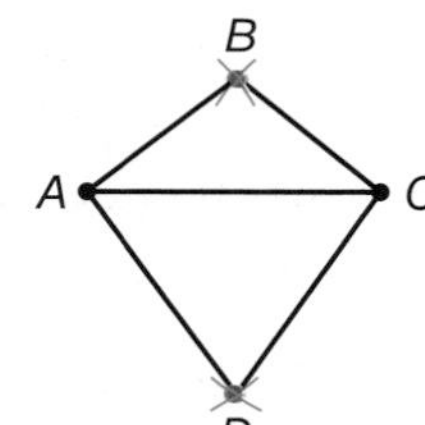


**a. Geometric** Draw a segment. Construct a noncongruent segment that perpendicularly bisects the first segment. Connect the endpoints of the segments to form a quadrilateral *ABCD*. Repeat the process two times. Name the additional quadrilaterals *PQRS* and *WXYZ*.

**b. Tabular** Copy and complete the table below.

| Figure | Side | Length | Side | Length | Side | Length | Side | Length |
|---|---|---|---|---|---|---|---|---|
| *ABCD* | *AB* | | *BC* | | *CD* | | *DA* | |
| *PQRS* | *PQ* | | *QR* | | *RS* | | *SP* | |
| *WXYZ* | *WX* | | *XY* | | *YZ* | | *ZW* | |

**c. Verbal** Make a conjecture about a quadrilateral in which the diagonals are perpendicular, exactly one diagonal is bisected, and the diagonals are not congruent.

**PROOF Write a coordinate proof of each statement.**

**63.** The diagonals of an isosceles trapezoid are congruent.

**64.** The median of an isosceles trapezoid is parallel to the bases.

## H.O.T. Problems Use Higher-Order Thinking Skills

**65. ERROR ANALYSIS** Bedagi and Belinda are trying to determine $m\angle A$ in kite *ABCD* shown. Is either of them correct? Explain.

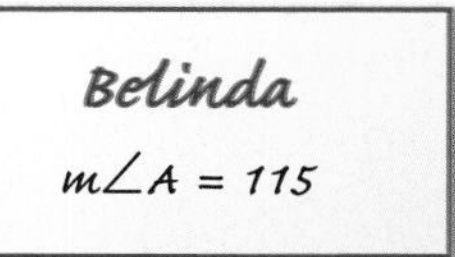


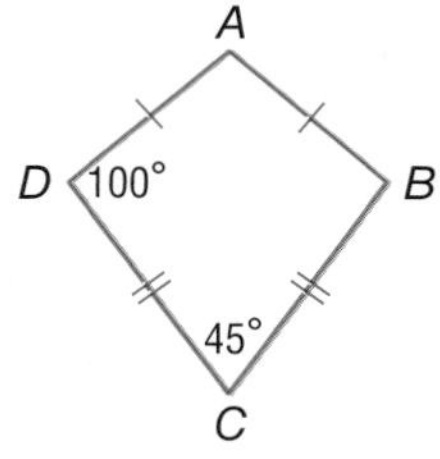


**66. CHALLENGE** If the parallel sides of a trapezoid are contained by the lines $y = x + 4$ and $y = x - 8$, what equation represents the line contained by the midsegment?

**67. CCSS ARGUMENTS** Is it *sometimes*, *always*, or *never* true that a square is also a kite? Explain.

**68. OPEN ENDED** Sketch two noncongruent trapezoids *ABCD* and *FGHJ* in which $\overline{AC} \cong \overline{FH}$ and $\overline{BD} \cong \overline{GJ}$.

**69. WRITING IN MATH** Describe the properties a quadrilateral must possess in order for the quadrilateral to be classified as a trapezoid, an isosceles trapezoid, or a kite. Compare the properties of all three quadrilaterals.

## Standardized Test Practice

**70. ALGEBRA** All of the items on a breakfast menu cost the same whether ordered with something else or alone. Two pancakes and one order of bacon costs \$4.92. If two orders of bacon cost \$3.96, what does one pancake cost?

**A** \$0.96 **C** \$1.98
**B** \$1.47 **D** \$2.94

**71. GRIDDED RESPONSE** If quadrilateral *ABCD* is a kite, what is $m\angle C$?

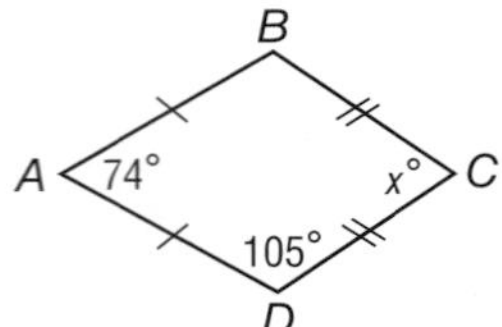


**72.** Which figure can serve as a counterexample to the conjecture below?

*If the diagonals of a quadrilateral are congruent, then the quadrilateral is a rectangle.*

**F** square **H** parallelogram
**G** rhombus **J** isosceles trapezoid

**73. SAT/ACT** In the figure below, what is the value of *x*?

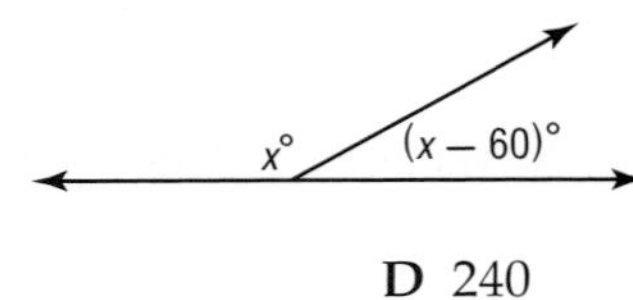


**A** 60 **D** 240
**B** 120 **E** 300
**C** 180

## Spiral Review

**ALGEBRA Quadrilateral *DFGH* is a rhombus. Find each value or measure.** (Lesson 8-5)

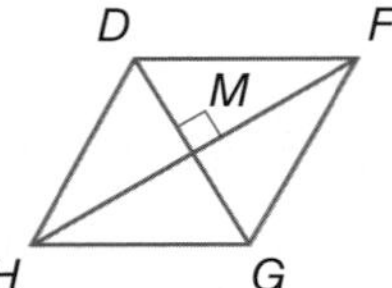


**74.** If $m\angle FGH = 118$, find $m\angle MHG$.

**75.** If $DM = 4x - 3$ and $MG = x + 6$, find *DG*.

**76.** If $DF = 10$, find *FG*.

**77.** If $HM = 12$ and $HD = 15$, find *MG*.

**COORDINATE GEOMETRY Graph each quadrilateral with the given vertices. Determine whether the figure is a rectangle. Justify your answer using the indicated formula.** (Lesson 8-4)

**78.** $A(4, 2)$, $B(-4, 1)$, $C(-3, -5)$, $D(5, -4)$; Distance Formula

**79.** $J(0, 7)$, $K(-8, 6)$, $L(-7, 0)$, $M(1, 1)$; Slope Formula

**80. BASEBALL** A batter hits the ball to the third baseman and begins to run toward first base. At the same time, the runner on first base runs toward second base. If the third baseman wants to throw the ball to the nearest base, to which base should he throw? Explain. (Lesson 7-3)

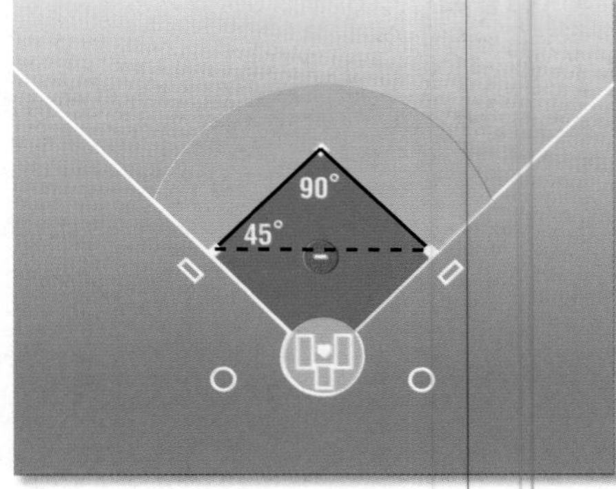


**81. PROOF** Write a two-column proof. (Lesson 6-4)

**Given:** $\angle CMF \cong \angle EMF$,
$\angle CFM \cong \angle EFM$
**Prove:** $\triangle DMC \cong \triangle DME$

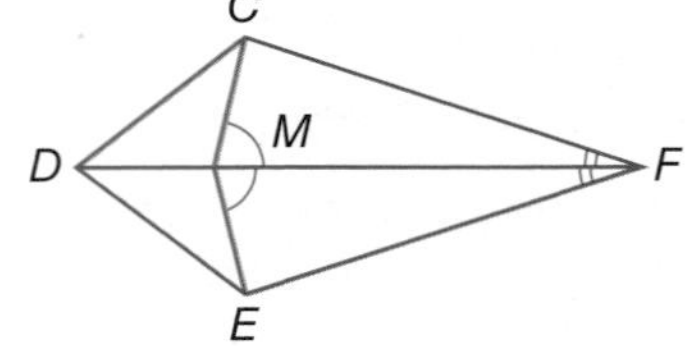


## Skills Review

**Write an expression for the slope of each segment given the coordinates and endpoints.**

**82.** $(x, 4y)$, $(-x, 4y)$ **83.** $(-x, 5x)$, $(0, 6x)$ **84.** $(y, x)$, $(y, y)$

# CHAPTER 8 Study Guide and Review

## Study Guide

### KeyConcepts

**Angles of Polygons** (Lesson 8-1)

- The sum of the measures of the interior angles of a polygon is given by the formula $S = (n - 2)180$.
- The sum of the measures of the exterior angles of a convex polygon is 360.

**Properties of Parallelograms** (Lessons 8-2 and 8-3)

- Opposite sides are congruent and parallel.
- Opposite angles are congruent.
- Consecutive angles are supplementary.
- If a parallelogram has one right angle, it has four right angles.
- Diagonals bisect each other.

**Properties of Rectangles, Rhombi, Squares, and Trapezoids** (Lesson 8-4 through 8-6)

- A rectangle has all the properties of a parallelogram. Diagonals are congruent and bisect each other. All four angles are right angles.
- A rhombus has all the properties of a parallelogram. All sides are congruent. Diagonals are perpendicular. Each diagonal bisects a pair of opposite angles.
- A square has all the properties of a parallelogram, a rectangle, and a rhombus.
- In an isosceles trapezoid, both pairs of base angles are congruent and the diagonals are congruent.

### FOLDABLES® StudyOrganizer

Be sure the Key Concepts are noted in your Foldable.

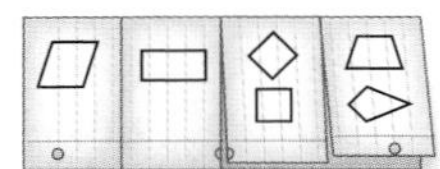

### KeyVocabulary

base (p. 521)

base angle (p. 521)

diagonal (p. 475)

isosceles trapezoid (p. 521)

kite (p. 524)

legs (p. 521)

midsegment of a trapezoid (p. 523)

parallelogram (p. 485)

rectangle (p. 505)

rhombus (p. 512)

square (p. 513)

trapezoid (p. 521)

### VocabularyCheck

**State whether each sentence is *true* or *false*. If *false*, replace the underlined word or phrase to make a true sentence.**

1. No angles in an isosceles trapezoid are congruent.
2. If a parallelogram is a rectangle, then the diagonals are congruent.
3. A midsegment of a trapezoid is a segment that connects any two nonconsecutive vertices.
4. The base of a trapezoid is one of the parallel sides.
5. The diagonals of a rhombus are perpendicular.
6. The diagonal of a trapezoid is the segment that connects the midpoints of the legs.
7. A rectangle is not always a parallelogram.
8. A quadrilateral with only one set of parallel sides is a parallelogram.
9. A rectangle that is also a rhombus is a square.
10. The leg of a trapezoid is one of the parallel sides.

# Lesson-by-Lesson Review

## 8-1 Angles of Polygons

**Find the sum of the measures of the interior angles of each convex polygon.**

**11.** decagon

**12.** 15-gon

**13. SNOWFLAKES** The snowflake decoration at the right suggests a regular hexagon. Find the sum of the measures of the interior angles of the hexagon.

**The measure of an interior angle of a regular polygon is given. Find the number of sides in the polygon.**

**14.** 135

**15.** $\approx 166.15$

### Example 1

**Find the sum of the measures of the interior angles of a convex 22-gon.**

| | |
|---|---|
| $m = (n - 2)180$ | Write an equation. |
| $= (22 - 2)180$ | Substitution |
| $= 20 \cdot 180$ | Subtract. |
| $= 3600$ | Multiply. |

### Example 2

**The measure of an interior angle of a regular polygon is 157.5. Find the number of sides in the polygon.**

| | |
|---|---|
| $157.5n = (n - 2)180$ | Write an equation. |
| $157.5n = 180n - 360$ | Distributive Property |
| $-22.5n = -360$ | Subtract. |
| $n = 16$ | Divide. |

The polygon has 16 sides.

## 8-2 Parallelograms

**Use ▱$ABCD$ to find each measure.**

**16.** $m\angle ADC$

**17.** $AD$

**18.** $AB$

**19.** $m\angle BCD$

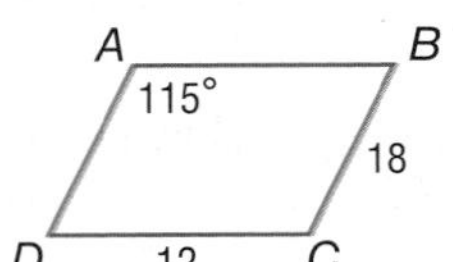


**ALGEBRA Find the value of each variable in each parallelogram.**

**20.**

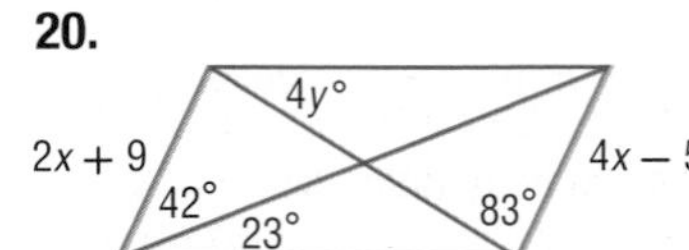


**21.**

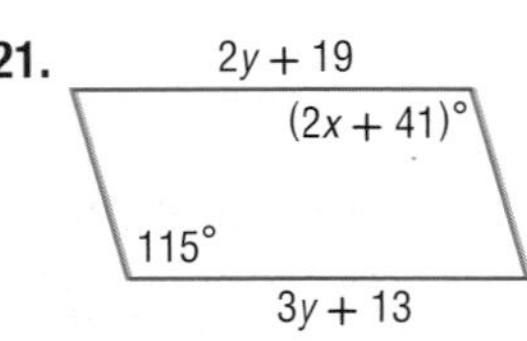


**22. DESIGN** What type of information is needed to determine whether the shapes that make up the stained glass window below are parallelograms?

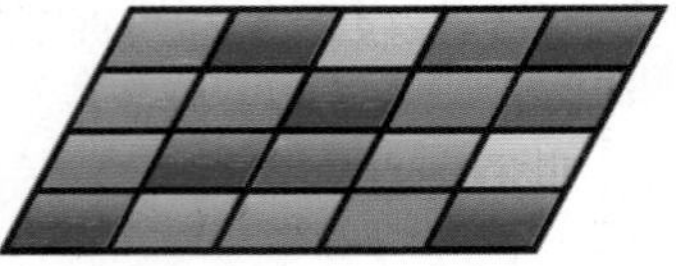

### Example 3

**ALGEBRA If *KLMN* is a parallelogram, find the value of the indicated variable.**

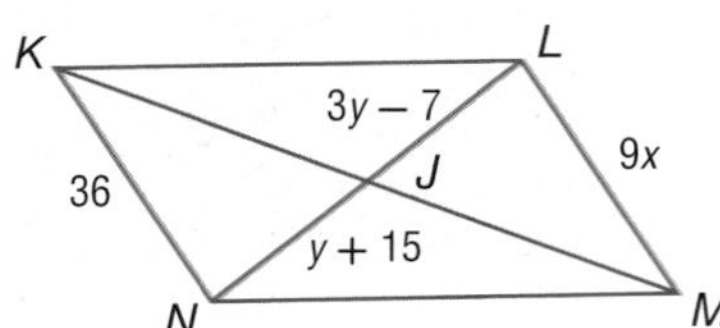


**a.** $x$

| | |
|---|---|
| $\overline{KN} \cong \overline{LM}$ | Opp. sides of a ▱ are ≅. |
| $KN = LM$ | Definition of congruence |
| $36 = 9x$ | Substitution |
| $4 = x$ | Divide. |

**b.** $y$

| | |
|---|---|
| $\overline{NJ} \cong \overline{JL}$ | Diag. of a ▱ bisect each other. |
| $NJ = JL$ | Definition of congruence |
| $y + 15 = 3y - 7$ | Substitution |
| $-2y = -22$ | Subtract. |
| $y = 11$ | Divide. |

## 8-3 Tests for Parallelograms

**Determine whether each quadrilateral is a parallelogram. Justify your answer.**

**23.** 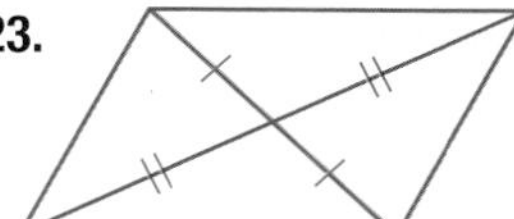

**24.** 

**25. PROOF** Write a two-column proof.

Given: ▱$ABCD$, $\overline{AE} \cong \overline{CF}$

Prove: Quadrilateral $EBFD$ is a parallelogram.

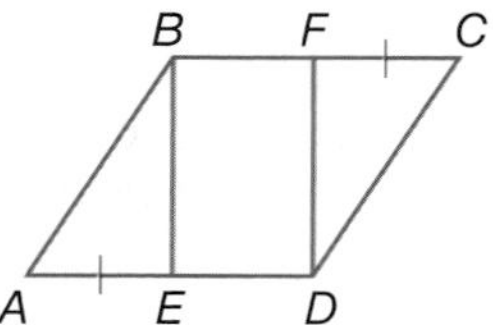


**ALGEBRA Find $x$ and $y$ so that the quadrilateral is a parallelogram.**

**26.** 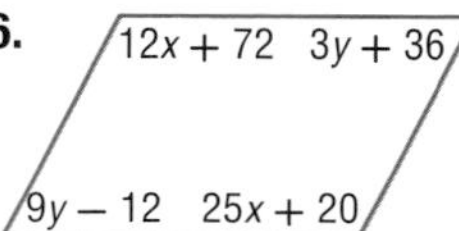


**27.** 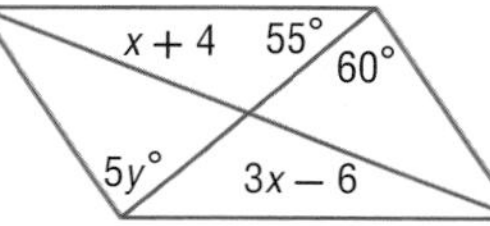


### Example 4

**If $TP = 4x + 2$, $QP = 2y - 6$, $PS = 5y - 12$, and $PR = 6x - 4$, find $x$ and $y$ so that the quadrilateral is a parallelogram.**

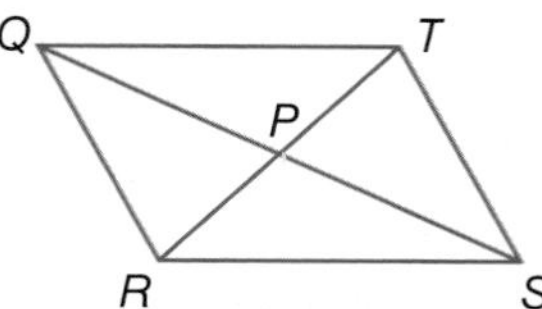


Find $x$ such that $\overline{TP} \cong \overline{PR}$ and $y$ such that $\overline{QP} \cong \overline{PS}$.

| | |
|---|---|
| $TP = PR$ | Definition of ≅ |
| $4x + 2 = 6x - 4$ | Substitution |
| $-2x = -6$ | Subtract. |
| $x = 3$ | Divide. |

| | |
|---|---|
| $QP = PS$ | Definition of ≅ |
| $2y - 6 = 5y - 12$ | Substitution |
| $-3y = -6$ | Subtract. |
| $y = 2$ | Divide. |

## 8-4 Rectangles

**28. PARKING** The lines of the parking space shown below are parallel. How wide is the space (in inches)?

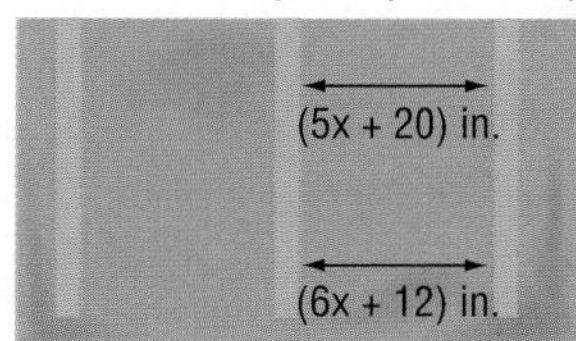


**ALGEBRA** **Quadrilateral $EFGH$ is a rectangle.**

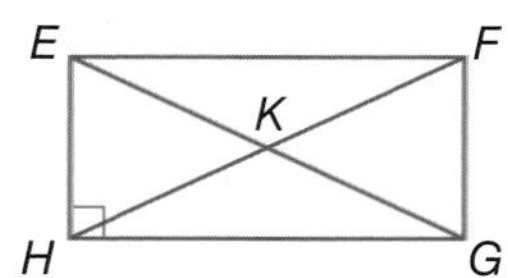


**29.** If $m\angle FEG = 57$, find $m\angle GEH$.

**30.** If $m\angle HGE = 13$, find $m\angle FGE$.

**31.** If $FK = 32$ feet, find $EG$.

**32.** Find $m\angle HEF + m\angle EFG$.

**33.** If $EF = 4x - 6$ and $HG = x + 3$, find $EF$.

### Example 5

**ALGEBRA Quadrilateral $ABCD$ is a rectangle. If $m\angle ADB = 4x + 8$ and $m\angle DBA = 6x + 12$, find $x$.**

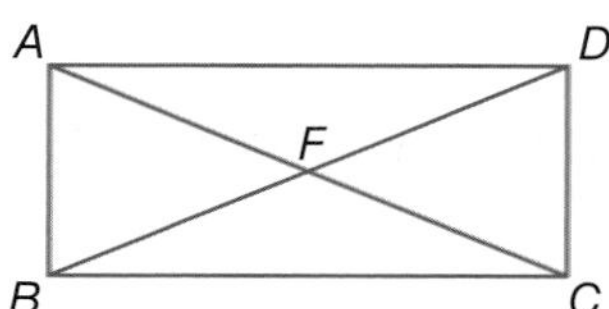


$ABCD$ is a rectangle, so $m\angle ABC = 90$. Since the opposite sides of a rectangle are parallel, and the alternate interior angles of parallel lines are congruent, $\angle DBC \cong \angle ADB$ and $m\angle DBC = m\angle ADB$.

| | |
|---|---|
| $m\angle DBC + m\angle DBA = 90$ | Angle Addition |
| $m\angle ADB + m\angle DBA = 90$ | Substitution |
| $4x + 8 + 6x + 12 = 90$ | Substitution |
| $10x + 20 = 90$ | Add. |
| $10x = 70$ | Subtract. |
| $x = 7$ | Divide. |

# Study Guide and Review *Continued*

## 8-5 Rhombi and Squares

**ALGEBRA** *ABCD* is a rhombus. If $EB = 9$, $AB = 12$ and $m\angle ABD = 55$, find each measure.

**34.** $AE$

**35.** $m\angle BDA$

**36.** $CE$

**37.** $m\angle ACB$

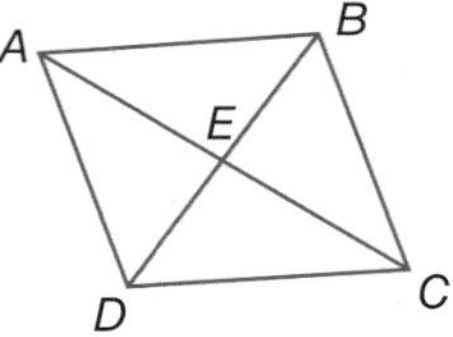


**38. LOGOS** A car company uses the symbol shown at the right for their logo. If the inside space of the logo is a rhombus, what is the length of $FJ$?

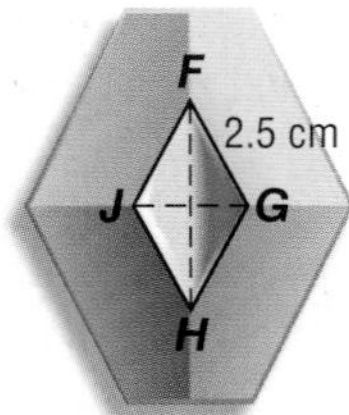


**COORDINATE GEOMETRY** Given each set of vertices, determine whether ▱*QRST* is a *rhombus*, a *rectangle*, or a *square*. List all that apply. Explain.

**39.** $Q(12, 0)$, $R(6, -6)$, $S(0, 0)$, $T(6, 6)$

**40.** $Q(-2, 4)$, $R(5, 6)$, $S(12, 4)$, $T(5, 2)$

### Example 6

The diagonals of rhombus *QRST* intersect at *P*. Use the information to find each measure or value.

**a. ALGEBRA** If $QT = x + 7$ and $TS = 2x - 9$, find x.

| | |
|---|---|
| $\overline{QT} \cong \overline{TS}$ | Def. of rhombus |
| $QT = TS$ | Def. of congruence |
| $x + 7 = 2x - 9$ | Substitution |
| $-x = -16$ | Subtract. |
| $x = 16$ | Divide. |

**b.** If $m\angle QTS = 76$, find $m\angle TSP$.

$\overline{TR}$ bisects $\angle QTS$. Therefore, $m\angle PTS = \frac{1}{2}m\angle QTS$. So $m\angle PTS = \frac{1}{2}(76)$ or 38. Since the diagonals of a rhombus are perpendicular, $m\angle TPS = 90$.

| | |
|---|---|
| $m\angle PTS + m\angle TPS + m\angle TSP = 180$ | △ Sum Thm. |
| $38 + 90 + m\angle TSP = 180$ | Substitution |
| $128 + m\angle TSP = 180$ | Add. |
| $m\angle TSP = 52$ | Subtract. |

## 8-6 Trapezoids and Kites

Find each measure.

**41.** $GH$

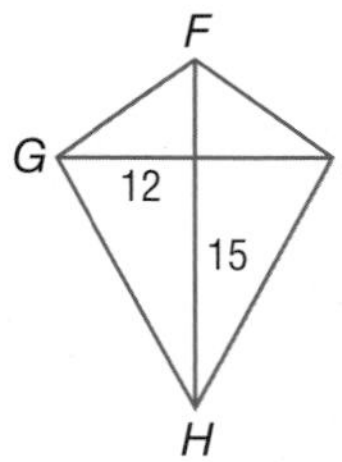


**42.** $m\angle Z$

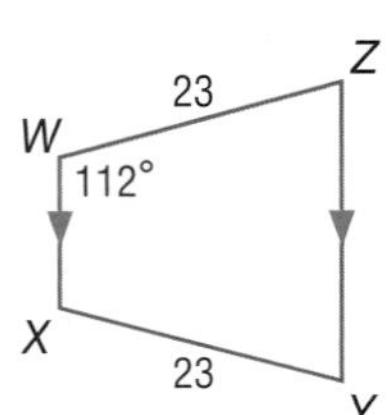


**43. DESIGN** Renee designed the square tile as an art project.

**a.** Describe a way to determine if the trapezoids in the design are isosceles.

**b.** If the perimeter of the tile is 48 inches and the perimeter of the red square is 16 inches, what is the perimeter of one of the trapezoids?

### Example 7

If *QRST* is a kite, find $m\angle RST$.

Since $\angle Q \cong \angle S$, $m\angle Q = m\angle S$. Write and solve an equation to find $m\angle S$.

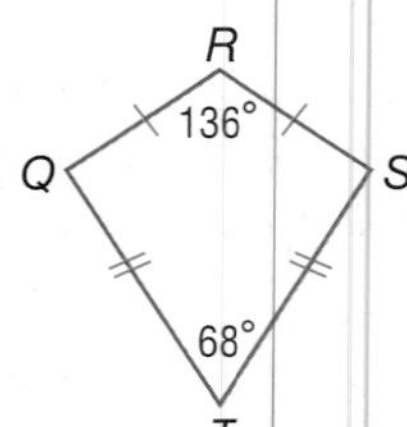


| | |
|---|---|
| $m\angle Q + m\angle R + m\angle S + m\angle T = 360$ | Polygon Int. ⦣s Sum Thm |
| $m\angle Q + 136 + m\angle S + 68 = 360$ | Substitution |
| $2m\angle S + 204 = 360$ | Simplify. |
| $2m\angle S = 156$ | Subtract. |
| $m\angle S = 78$ | Divide. |

# CHAPTER 8 Practice Test

**Find the sum of the measures of the interior angles of each convex polygon.**

1. hexagon
2. 16-gon

3. **ART** Jen is making a frame to stretch a canvas over for a painting. She nailed four pieces of wood together at what she believes will be the four vertices of a square.

   a. How can she be sure that the canvas will be a square?

   b. If the canvas has the dimensions shown below, what are the missing measures?

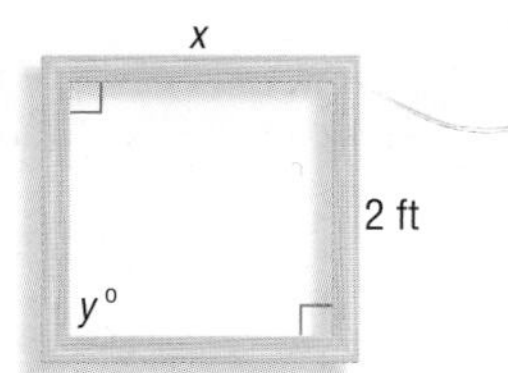


**Quadrilateral *ABCD* is an isosceles trapezoid.**

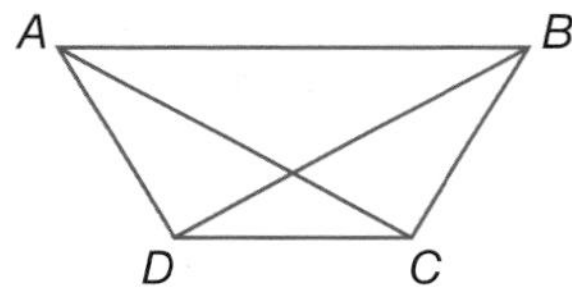


4. Which angle is congruent to $\angle C$?
5. Which side is parallel to $\overline{AB}$?
6. Which segment is congruent to $\overline{AC}$?

**The measure of the interior angles of a regular polygon is given. Find the number of sides in the polygon.**

7. 900
8. 1980
9. 2880
10. 5400

11. **MULTIPLE CHOICE** If *QRST* is a parallelogram, what is the value of $x$?

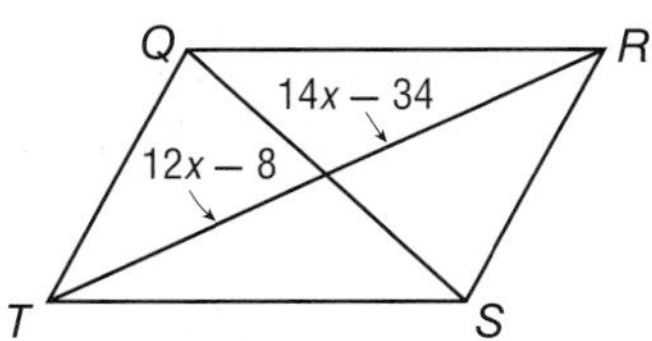


A 11

B 12

C 13

D 14

**If *CDFG* is a kite, find each measure.**

12. $GF$

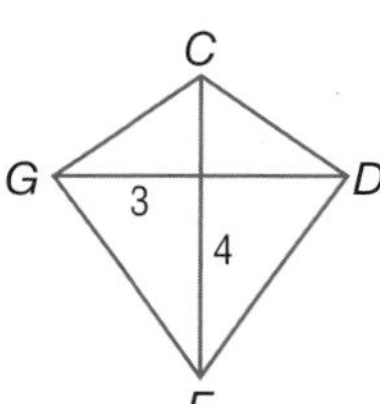


13. $m\angle D$

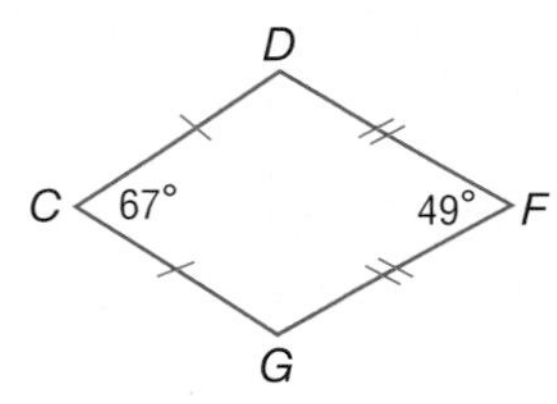


**ALGEBRA Quadrilateral *MNOP* is a rhombus. Find each value or measure.**

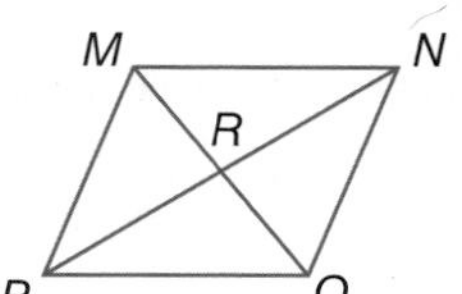


14. $m\angle MRN$
15. If $PR = 12$, find $RN$.
16. If $m\angle PON = 124$, find $m\angle POM$.

17. **CONSTRUCTION** The Smiths are building an addition to their house. Mrs. Smith is cutting an opening for a new window. If she measures to see that the opposite sides are congruent and that the diagonal measures are congruent, can Mrs. Smith be sure that the window opening is rectangular? Explain.

**Use ▱*JKLM* to find each measure.**

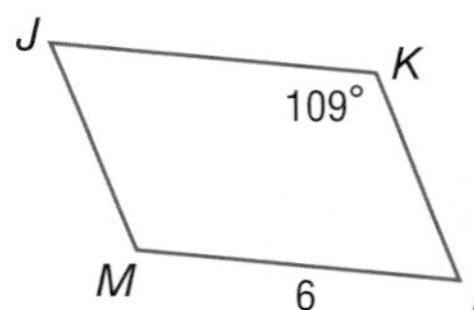


18. $m\angle JML$
19. $JK$
20. $m\angle KLM$

**ALGEBRA Quadrilateral *DEFG* is a rectangle.**

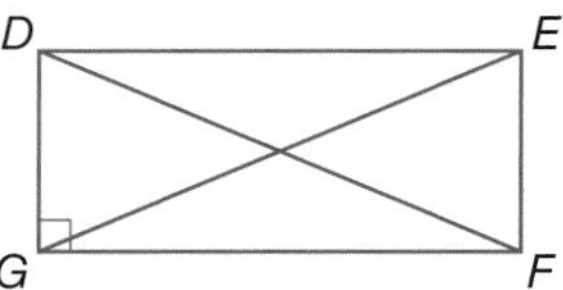


21. If $DF = 2(x + 5) - 7$ and $EG = 3(x - 2)$, find $EG$.
22. If $m\angle EDF = 5x - 3$ and $m\angle DFG = 3x + 7$, find $m\angle EDF$.
23. If $DE = 14 + 2x$ and $GF = 4(x - 3) + 6$, find $GF$.

**Determine whether each quadrilateral is a parallelogram. Justify your answer.**

24.

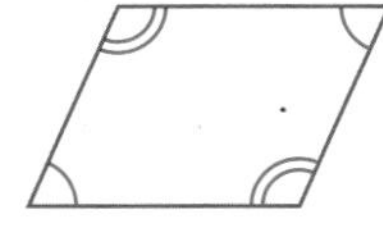

25.

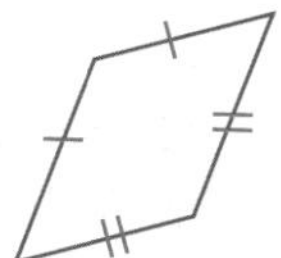

CHAPTER 8

# Preparing for Standardized Tests

## Apply Definitions and Properties

Many geometry problems on standardized tests require the application of definitions and properties in order to solve them. Use this section to practice applying definitions to help you solve extended-response test items.

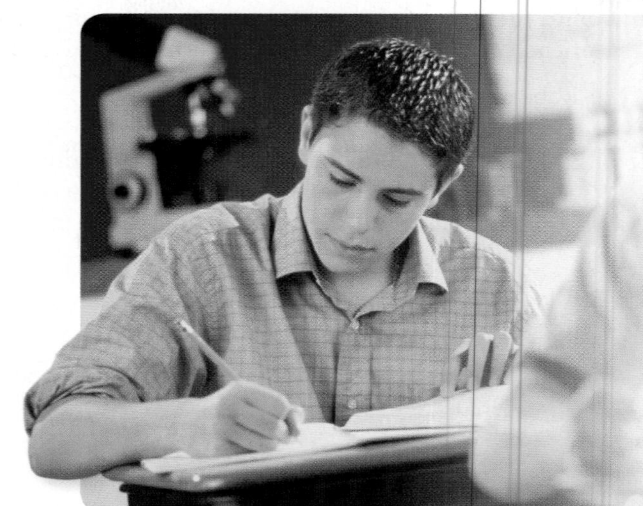

Comstock/Jupiterimages

### Strategies for Applying Definitions and Properties

**Step 1**

Read the problem statement carefully.

- Determine what you are being asked to solve.
- Study any figures given in the problem.
- **Ask yourself:** What principles or properties of this figure can I apply to solve the problem?

**Step 2**

Solve the problem.

- Identify any definitions or geometric concepts you can use to help you find the unknowns in the problem.
- Use definitions and properties of figures to set up and solve an equation.

**Step 3**

- Check your answer.

### Standardized Test Example

**Read the problem. Identify what you need to know. Then use the information in the problem to solve. Show your work.**

A performing arts group is building a theater in the round for upcoming productions. The stage will be a regular octagon with a perimeter of 76 feet.

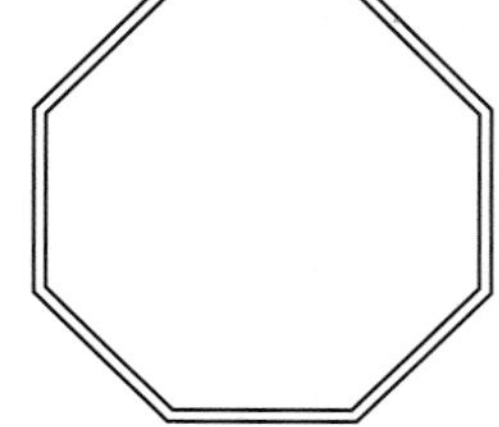

**a.** What length should each board be to form the sides of the stage?

**b.** What angle should the end of each board be cut so that they will fit together properly to form the stage? Explain.

Read the problem carefully. You are told that the boards form a regular octagon with a perimeter of 76 feet. You need to find the length of each board and the angle that they should be cut to fit together properly.

To find the length of each board, divide the perimeter by the number of boards.

$76 \div 8 = 9.5$

So, each board should be 9.5 feet, or 9 feet 6 inches, long.

Use the property of the interior angle sum of convex polygons to find the measure of an interior angle of a regular octagon. First find the sum $S$ of the interior angles.

$$
\begin{aligned}
S &= (n - 2) \cdot 180 \\
&= (8 - 2) \cdot 180 \\
&= 1080
\end{aligned}
$$

So, the measure of an interior angle of a regular octagon is $1080 \div 8$, or $135°$. Since two boards are used to form each vertex of the stage, the end of each board should be cut at an angle of $135 \div 2$, or $67.5°$.

## Exercises

**Read each problem. Identify what you need to know. Then use the information in the problem to solve. Show your work.**

1. $\overline{RS}$ is the midsegment of trapezoid $MNOP$. What is the length of $\overline{RS}$?

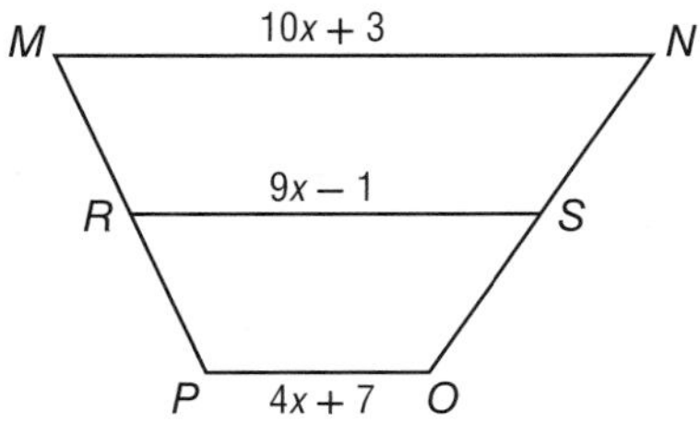


A 14 units
B 19 units
C 23 units
D 26 units

2. If $\overline{AB} \parallel \overline{DC}$, find $x$.

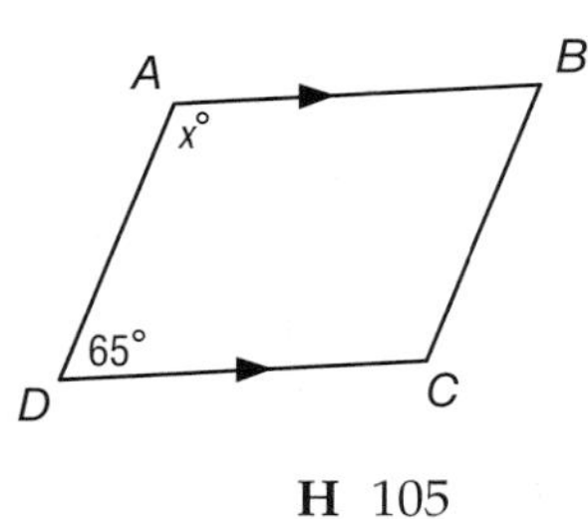


F 32.5
G 65
H 105
J 115

3. Use the graph shown below to answer each question.

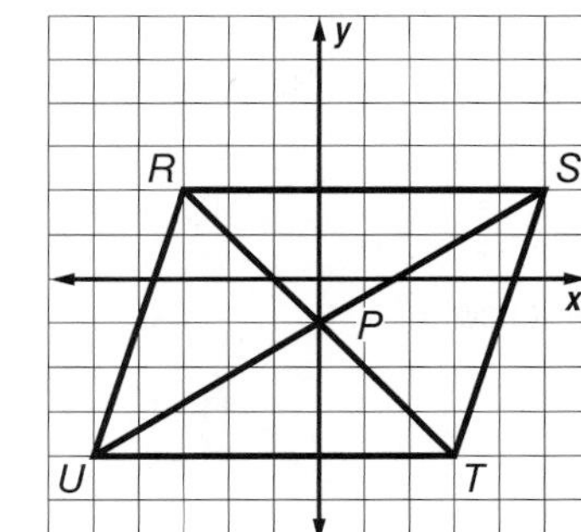


   a. Do the diagonals of quadrilateral $RSTU$ bisect each other? Use the Distance Formula to verify your answer.

   b. What type of quadrilateral is $RSTU$? Explain using the properties and/or definitions of this type of quadrilateral.

4. What is the sum of the measures of the exterior angles of a regular octagon?

A 45
B 135
C 360
D 1080

CHAPTER 8

# Standardized Test Practice

## Cumulative, Chapters 1 through 8

### Multiple Choice

**Read each question. Then fill in the correct answer on the answer document provided by your teacher or on a sheet of paper.**

1. If $a \parallel b$, which of the following might not be true?

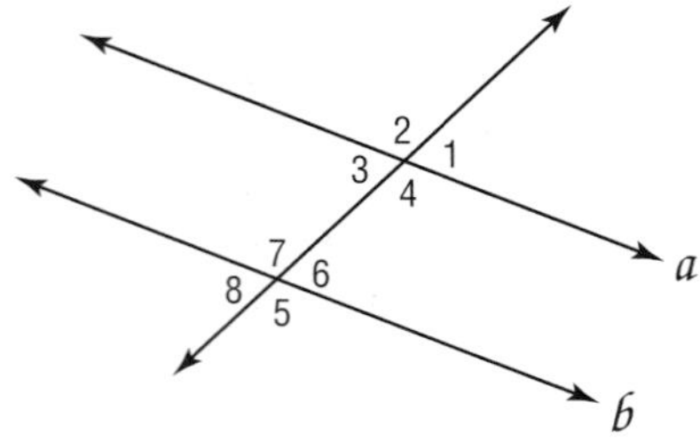


A $\angle 1 \cong \angle 3$

B $\angle 4 \cong \angle 7$

C $\angle 2 \cong \angle 5$

D $\angle 8 \cong \angle 2$

2. Solve for $x$ in parallelogram $RSTU$.

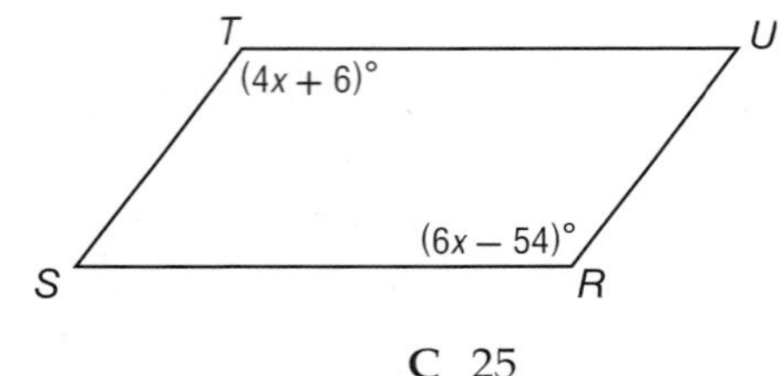


A 12

B 18

C 25

D 30

3. What is the measure of an interior angle of a regular pentagon?

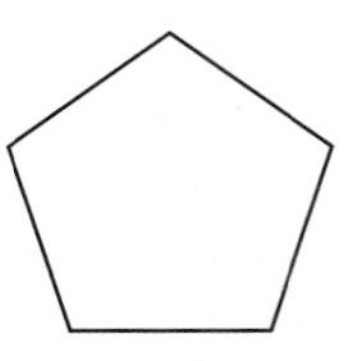

F 96

G 108

H 120

J 135

4. Quadrilateral $ABCD$ is a rhombus. If $m\angle BCD = 120$, find $m\angle DAC$.

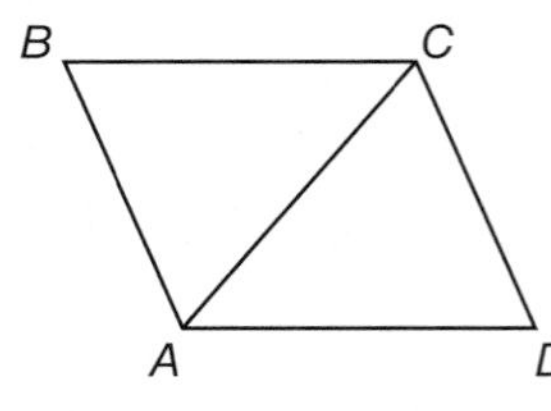


A 30

B 60

C 90

D 120

5. What is the value of $x$ in the figure below?

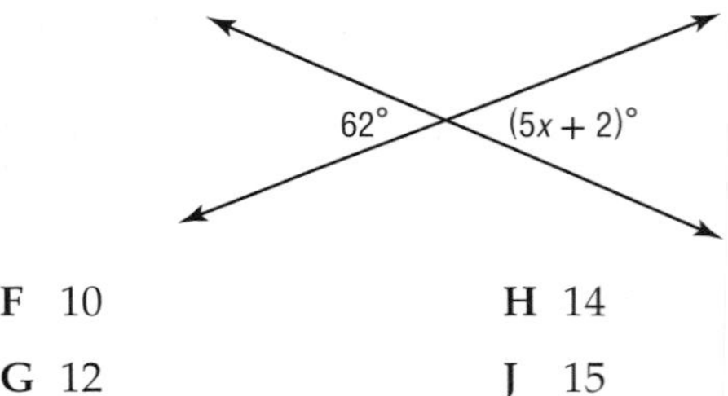


F 10

G 12

H 14

J 15

6. Which of the following statements is true?

A All rectangles are squares.

B All rhombi are squares.

C All rectangles are parallelograms.

D All parallelograms are rectangles.

**Test-TakingTip**

Question 2 Use the properties of parallelograms to solve the problem. Opposite angles are congruent.

## Short Response/Gridded Response

**Record your answers on the answer sheet provided by your teacher or a sheet of paper.**

7. **GRIDDED RESPONSE** The posts for Nancy's gazebo form a regular hexagon. What is the measure of the angle formed at each corner of the gazebo?

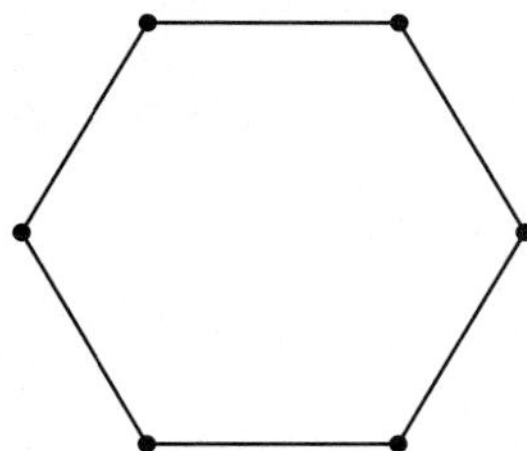

8. What are the coordinates of point $P$, the fourth vertex of an isosceles trapezoid? Show your work.

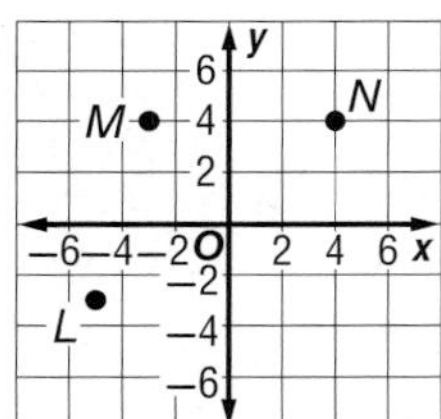


9. What do you know about a parallelogram if its diagonals are perpendicular? Explain.

10. **GRIDDED RESPONSE** Solve for $x$ in the figure below. Round to the nearest tenth if necessary.

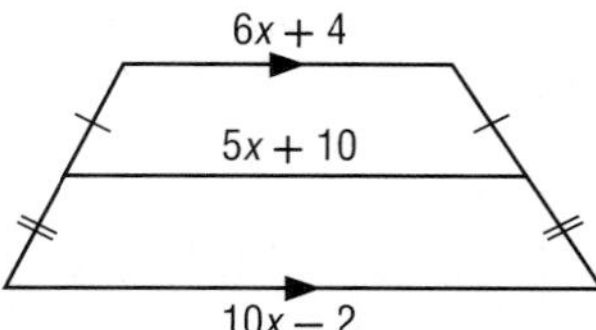


11. What are the coordinates of the circumcenter of the triangle below?

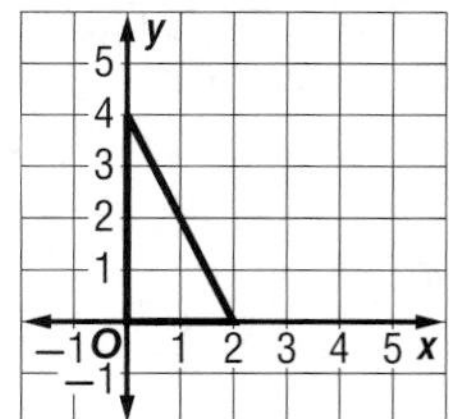


## Extended Response

**Record your answers on a sheet of paper. Show your work.**

12. Determine whether you can prove each figure is a parallelogram. If not, tell what additional information would be needed to prove that it is a parallelogram. Explain your reasoning.

a. 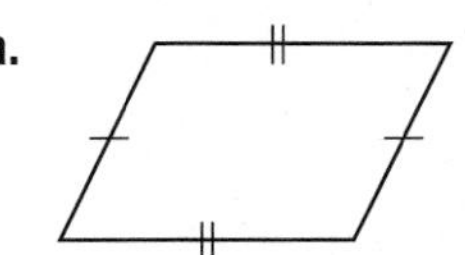

b. 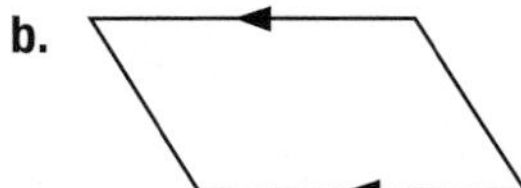

c. 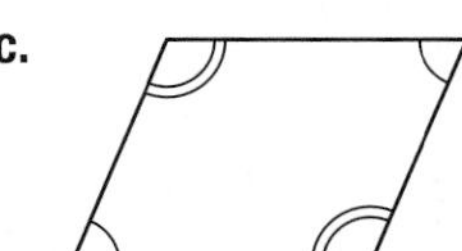

**Need ExtraHelp?**

| If you missed Question... | 1 | 2 | 3 | 4 | 5 | 6 | 7 | 8 | 9 | 10 | 11 | 12 |
|---|---|---|---|---|---|---|---|---|---|---|---|---|
| Go to Lesson... | 5-5 | 8-2 | 8-1 | 8-5 | 5-4 | 8-5 | 8-1 | 8-6 | 8-5 | 8-6 | 7-1 | 8-3 |

CHAPTER 9

# Proportions and Similarity

## Then

You learned about ratios and proportions and applied them to real-world applications.

## Now

In this chapter, you will:

- Identify similar polygons and use ratios and proportions to solve problems.
- Identify and apply similarity transformations.
- Use scale models and drawings to solve problems.

## Why?

**SPORTS** Similar triangles can be used in sports to describe the path of a ball, such as a bounce pass from one person to another.

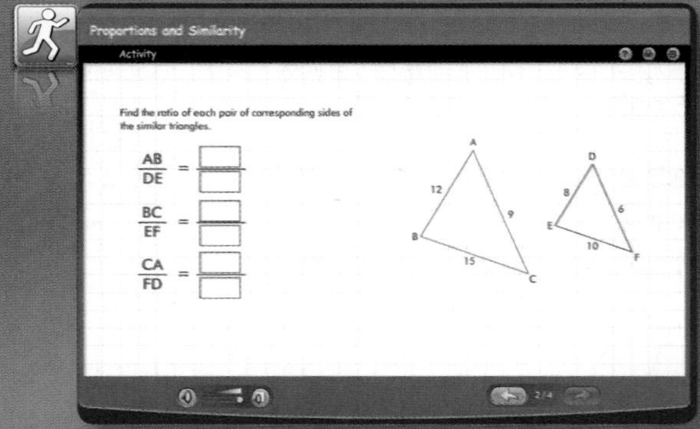


connectED.mcgraw-hill.com Your Digital Math Portal

Animation 
Vocabulary  
eGlossary 
Personal Tutor  
Virtual Manipulatives 
Graphing Calculator 
Audio 
Foldables 
Self-Check Practice 
Worksheets 

Barry Rosenthal/The Image Bank/Getty Images

# Get Ready for the Chapter

**Diagnose** Readiness | You have two options for checking prerequisite skills.

## 1 Textbook Option

Take the Quick Check below. Refer to the Quick Review for help.

### QuickCheck

Solve each equation.

1. $\frac{3x}{8} = \frac{6}{x}$
2. $\frac{7}{3} = \frac{x-4}{6}$
3. $\frac{x+9}{2} = \frac{3x-1}{8}$
4. $\frac{3}{2x} = \frac{3x}{8}$
5. **EDUCATION** The student to teacher ratio at Elder High School is 17 to 1. If there are 1088 students in the school, how many teachers are there?

**ALGEBRA** In the figure, $\overrightarrow{BA}$ and $\overrightarrow{BC}$ are opposite rays and $\overrightarrow{BD}$ bisects $\angle ABF$.

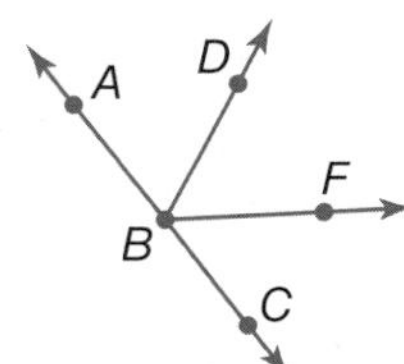


6. If $m\angle ABF = 3x - 8$ and $m\angle ABD = x + 14$, find $m\angle ABD$.
7. If $m\angle FBC = 2x + 25$ and $m\angle ABF = 10x - 1$, find $m\angle DBF$.
8. **LANDSCAPING** A landscape architect is planning to add sidewalks around a fountain as shown below. If $\overrightarrow{BA}$ and $\overrightarrow{BC}$ are opposite rays and $\overrightarrow{BD}$ bisects $\angle ABF$, find $m\angle FBC$.

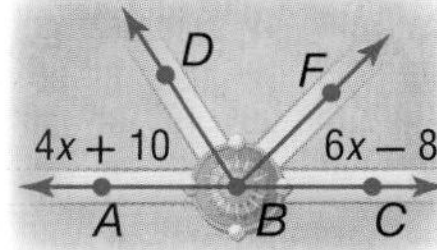


### QuickReview

**Example 1**

Solve $\frac{4x-3}{5} = \frac{2x+11}{3}$.

| | |
|---|---|
| $\frac{4x-3}{5} = \frac{2x+11}{3}$ | Original equation |
| $3(4x-3) = 5(2x+11)$ | Cross multiplication |
| $12x - 9 = 10x + 55$ | Distributive Property |
| $2x = 64$ | Add. |
| $x = 32$ | Simplify. |

**Example 2**

In the figure, $\overrightarrow{QP}$ and $\overrightarrow{QR}$ are opposite rays, and $\overrightarrow{QT}$ bisects $\angle SQR$. If $m\angle SQR = 6x + 8$ and $m\angle TQR = 4x - 14$, find $m\angle SQT$.

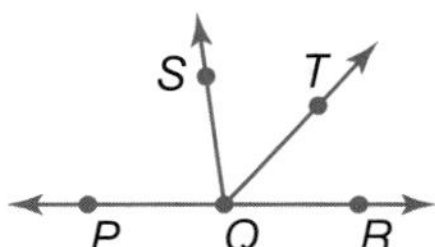


Since $\overrightarrow{TQ}$ bisects $\angle SQR$, $m\angle SQR = 2(m\angle TQR)$.

| | |
|---|---|
| $m\angle SQR = 2(m\angle TQR)$ | Def. of ∠ bisector |
| $6x + 8 = 2(4x - 14)$ | Substitution |
| $6x + 8 = 8x - 28$ | Distributive Property |
| $-2x = -36$ | Subtract. |
| $x = 18$ | Simplify. |

Since $\overrightarrow{TQ}$ bisects $\angle SQR$, $m\angle SQT = m\angle TQR$.

| | |
|---|---|
| $m\angle SQT = m\angle TQR$ | Def. of ∠ bisector |
| $m\angle SQT = 4x - 14$ | Substitution |
| $m\angle SQT = 58$ | $x = 18$ |

## 2 Online Option

Take an online self-check Chapter Readiness Quiz at connectED.mcgraw-hill.com.

# **Get Started** on the Chapter

You will learn several new concepts, skills, and vocabulary terms as you study Chapter 9. To get ready, identify important terms and organize your resources. You may wish to refer to Chapter 0 to review prerequisite skills.

## FOLDABLES StudyOrganizer

**Proportions and Similarity** Make this Foldable to help you organize your Chapter 9 notes about proportions, similar polygons, and similarity transformations. Begin with four sheets of notebook paper.

1. **Fold** the four sheets of paper in half.

2. **Cut** along the top fold of the papers. Staple along the side to form a book.

3. **Cut** the right sides of each paper to create a tab for each lesson.

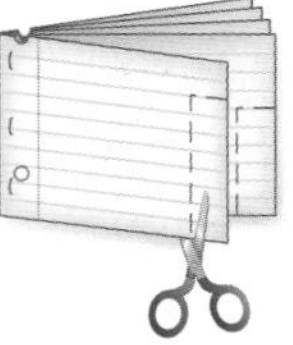

4. **Label** each tab with a lesson number, as shown.

## NewVocabulary

| English | | Español |
|---|---|---|
| ratio | p. 543 | razón |
| proportion | p. 544 | proporción |
| extremes | p. 544 | extremos |
| means | p. 544 | medias |
| cross products | p. 544 | productos cruzados |
| similar polygons | p. 551 | polígonos semejantes |
| scale factor | p. 552 | factor de escala |
| dilation | p. 593 | dilatación |
| similarity transformation | p. 593 | transformación de semejanza |
| enlargement | p. 593 | ampliación |
| reduction | p. 593 | reducción |
| scale model | p. 600 | modelo a escala |
| scale drawing | p. 600 | dibujo a escala |

## ReviewVocabulary

altitude altura a segment drawn from a vertex of a triangle perpendicular to the line containing the other side

angle bisector bisectriz de un ángulo a ray that divides an angle into two congruent angles

median mediana a segment drawn from a vertex of a triangle to the midpoint of the opposite side

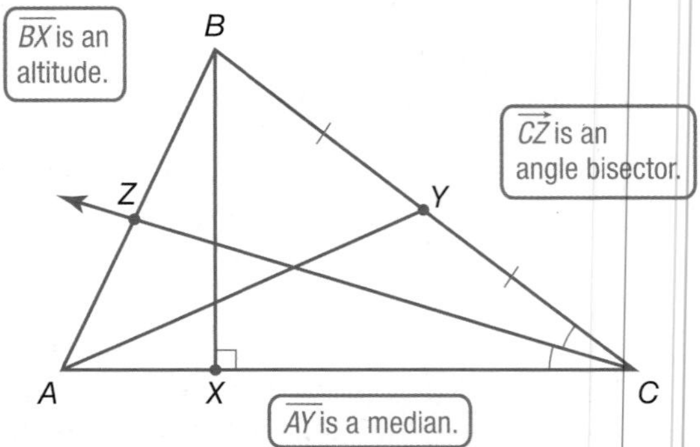

LESSON 9-1

# Ratios and Proportions

| Then | Now | Why? |
|---|---|---|
| You solved problems by writing and solving equations. | 1 Write ratios. 2 Write and solve proportions. | The aspect ratio of a television or computer screen is the screen's width divided by its height. A standard television screen has an aspect ratio of $\frac{4}{3}$ or 4:3, while a high definition television screen (HDTV) has an aspect ratio of 16:9. |

**NewVocabulary**
ratio
extended ratios
proportion
extremes
means
cross products

**Common Core State Standards**

**Content Standards**
G.MG.3 Apply geometric methods to solve problems (e.g., designing an object or structure to satisfy physical constraints or minimize cost; working with typographic grid systems based on ratios). ★

**Mathematical Practices**
7 Look for and make use of structure.
8 Look for and express regularity in repeated reasoning.

## 1 Write and Use Ratios

A **ratio** is a comparison of two quantities using division. The ratio of quantities $a$ and $b$ can be expressed as $a$ to $b$, $a:b$, or $\frac{a}{b}$, where $b \neq 0$. Ratios are usually expressed in simplest form.

The aspect ratios 32:18 and 16:9 are equivalent.

$$\frac{\text{width of screen}}{\text{height of screen}} = \frac{32 \text{ in.}}{18 \text{ in.}}$$ Divide out units.

$$= \frac{32 \div 2}{18 \div 2} \text{ or } \frac{16}{9}$$ Divide out common factors.

### Real-World Example 1 Write and Simplify Ratios

**SPORTS A baseball player's batting average is the ratio of the number of base hits to the number of at-bats, not including walks. Minnesota Twins' Joe Mauer had the highest batting average in Major League Baseball in 2006. If he had 521 official at-bats and 181 hits, find his batting average.**

Divide the number of hits by the number of at-bats.

$$\frac{\text{number of hits}}{\text{number of at-bats}} = \frac{181}{521}$$

$$\approx \frac{0.347}{1}$$ A ratio in which the denominator is 1 is called a *unit ratio.*

Joe Mauer's batting average was 0.347.

**GuidedPractice**

1. **SCHOOL** In Logan's high school, there are 190 teachers and 2650 students. What is the approximate student-teacher ratio at his school?

**Extended ratios** can be used to compare three or more quantities. The expression $a:b:c$ means that the ratio of the first two quantities is $a:b$, the ratio of the last two quantities is $b:c$, and the ratio of the first and last quantities is $a:c$.

(t)Clearviewstock/Alamy, (b)John Capella/Sports Imagery/Getty Images

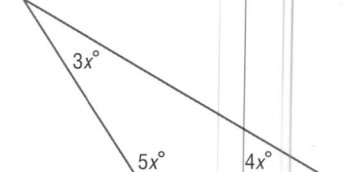

## Example 2 Use Extended Ratios

**The ratio of the measures of the angles in a triangle is 3:4:5. Find the measures of the angles.**

Just as the ratio $\frac{3}{4}$ or 3:4 is equivalent to $\frac{3x}{4x}$ or $3x:4x$, the extended ratio 3:4:5 can be written as $3x:4x:5x$.

Sketch and label the angle measures of the triangle. Then write and solve an equation to find the value of $x$.



$3x + 4x + 5x = 180$ Triangle Sum Theorem

$12x = 180$ Combine like terms.

$x = 15$ Divide each side by 12.

So the measures of the angles are 3(15) or 45, 4(15) or 60, and 5(15) or 75.

**CHECK** The sum of the angle measures should be 180.

$45 + 60 + 75 = 180$ ✓

### GuidedPractice

**2.** In a triangle, the ratio of the measures of the sides is 2:2:3 and the perimeter is 392 inches. Find the length of the longest side of the triangle.

**ReadingMath**

Proportion When a proportion is written using colons, it is read using the word *to* for the colon. For example, 2:3 is read *2 to 3*. The means are the inside numbers, and the extremes are the outside numbers.

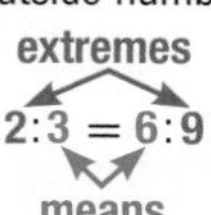


## 2 Use Properties of Proportions

An equation stating that two ratios are equal is called a **proportion**. In the proportion $\frac{a}{b} = \frac{c}{d}$, the numbers $a$ and $d$ are called the **extremes** of the proportion, while the numbers $b$ and $c$ are called the **means** of the proportion.

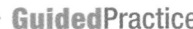

extreme → $\frac{a}{b} = \frac{c}{d}$ ← mean

mean → ← extreme

The product of the extremes $ad$ and the product of the means $bc$ are called **cross products**.

### KeyConcept Cross Products Property

| | |
|---|---|
| Words | In a proportion, the product of the extremes equals the product of the means. |
| Symbols | If $\frac{a}{b} = \frac{c}{d}$ when $b \neq 0$ and $d \neq 0$, then $ad = bc$. |
| Example | If $\frac{4}{10} = \frac{6}{15}$, then $4 \cdot 15 = 10 \cdot 6$. |

You will prove the Cross Products Property in Exercise 41.

The converse of the Cross Products Property is also true. If $ad = bc$ and $b \neq 0$ and $d \neq 0$, then $\frac{a}{b} = \frac{c}{d}$. That is, $\frac{a}{b}$ and $\frac{c}{d}$ form a proportion. You can use the Cross Products Property to solve a proportion.

## Example 3 Use Cross Products to Solve Proportions

**Solve each proportion.**

**a.** $\frac{6}{x} = \frac{21}{31.5}$

| | | |
|---|---|---|
| $\frac{6}{x} = \frac{21}{31.5}$ | Original proportion | $\frac{x+3}{2} = \frac{4x}{5}$ |
| $6(31.5) = x(21)$ | Cross Products Property | $(x + 3)5 = 2(4x)$ |
| $189 = 21x$ | Simplify. | $5x + 15 = 8x$ |
| $9 = x$ | Solve for $x$. | $15 = 3x$ |
| | | $5 = x$ |

**b.** $\frac{x+3}{2} = \frac{4x}{5}$

### StudyTip

Perseverance

Example 3b could also be solved by multiplying each side of the equation by 10, the least common denominator.

$$10\left(\frac{x+3}{2}\right) = \frac{4x}{5}(10)$$
$$5(x + 3) = 2(4x)$$
$$5x + 15 = 8x$$
$$15 = 3x$$
$$5 = x$$

### GuidedPractice

**3A.** $\frac{x}{4} = \frac{11}{-6}$ **3B.** $\frac{-4}{7} = \frac{6}{2y+5}$ **3C.** $\frac{7}{z-1} = \frac{9}{z+4}$

Proportions can be used to make predictions.

## Real-World Example 4 Use Proportions to Make Predictions

**CAR OWNERSHIP Fernando conducted a survey of 50 students driving to school and found that 28 owned cars. If 755 students drive to his school, predict the total number of students who own cars.**

Write and solve a proportion that compares the number of students who own cars to the number who drive to school.

| | |
|---|---|
| $\frac{28}{50} = \frac{x}{755}$ | ← students owning cars<br>← students driving to school |
| $28 \cdot 755 = 50 \cdot x$ | Cross Products Property |
| $21{,}140 = 50x$ | Simplify. |
| $422.8 = x$ | Divide each side by 50. |

Based on Fernando's survey, about 423 students at his school own cars.

### Real-WorldLink

The percent of driving-age teens (ages 15 to 20) with their own vehicles nearly doubled nationwide from 22 percent in 1985 to 42 percent in 2003.

**Source:** CNW Marketing Research

### GuidedPractice

**4. BIOLOGY** In an experiment, students netted butterflies, recorded the number with tags on their wings, and then released them. The students netted 48 butterflies and 3 of those had tagged wings. Predict the number of butterflies that would have tagged wings out of 100 netted.

The proportion shown in Example 4 is not the only correct proportion for that situation. Equivalent forms of a proportion all have identical cross products.

## KeyConcept Equivalent Proportions

**Symbols** The following proportions are equivalent.

$$\frac{a}{b} = \frac{c}{d}, \quad \frac{b}{a} = \frac{d}{c}, \quad \frac{a}{c} = \frac{b}{d}, \quad \frac{c}{a} = \frac{d}{b}$$

**Examples** $\frac{28}{50} = \frac{x}{755}, \frac{50}{28} = \frac{755}{x}, \frac{28}{x} = \frac{50}{755}, \frac{x}{28} = \frac{755}{50}.$

## Check Your Understanding

**Example 1**

1. **PETS** Out of a survey of 1000 households, 460 had at least one dog or cat as a pet. What is the ratio of pet owners to households?

2. **SPORTS** Thirty girls tried out for 15 spots on the basketball team. What is the ratio of open spots to the number of girls competing?

**Example 2**

3. The ratio of the measures of three sides of a triangle is 2:5:4, and its perimeter is 165 units. Find the measure of each side of the triangle.

4. The ratios of the measures of three angles of a triangle are 4:6:8. Find the measure of each angle of the triangle.

**Example 3**

**Solve each proportion.**

5. $\frac{2}{3} = \frac{x}{24}$
6. $\frac{x}{5} = \frac{28}{100}$
7. $\frac{2.2}{x} = \frac{26.4}{96}$
8. $\frac{x-3}{3} = \frac{5}{8}$

**Example 4**

9. **CCSS MODELING** Ella is baking apple muffins for the Student Council bake sale. The recipe that she is using calls for 2 eggs per dozen muffins, and she needs to make 108 muffins. How many eggs will she need?

## Practice and Problem Solving

**Example 1**

**MOVIES For Exercises 10 and 11, refer to the graphic below.**

10. Of the films listed, which had the greatest ratio of Academy Awards to number of nominations?

11. Which film listed had the lowest ratio of awards to nominations?

**Example 2**

12. **GAMES** A video game store has 60 games to choose from, including 40 sports games. What is the ratio of sports games to video games?

13. The ratio of the measures of the three sides of a triangle is 9 : 7 : 5. Its perimeter is 191.1 inches. Find the measure of each side.

14. The ratio of the measures of the three sides of a triangle is 3:7:5, and its perimeter is 156.8 meters. Find the measure of each side.

15. The ratio of the measures of the three sides of a triangle is $\frac{1}{4}:\frac{1}{8}:\frac{1}{6}$. Its perimeter is 4.75 feet. Find the length of the longest side.

16. The ratio of the measures of the three sides of a triangle is $\frac{1}{4}:\frac{1}{3}:\frac{1}{6}$, and its perimeter is 31.5 centimeters. Find the length of the shortest side.

**Find the measures of the angles of each triangle.**

**17.** The ratio of the measures of the three angles is 3:6:1.

**18.** The ratio of the measures of the three angles is 7:5:8.

**19.** The ratio of the measures of the three angles is 10:8:6.

**20.** The ratio of the measures of the three angles is 5:4:7.

Example 3 **Solve each proportion.**

**21.** $\frac{5}{8} = \frac{y}{3}$ **22.** $\frac{w}{6.4} = \frac{1}{2}$ **23.** $\frac{4x}{24} = \frac{56}{112}$ **24.** $\frac{11}{20} = \frac{55}{20x}$

**25.** $\frac{2x+5}{10} = \frac{42}{20}$ **26.** $\frac{a+2}{a-2} = \frac{3}{2}$ **27.** $\frac{3x-1}{4} = \frac{2x+4}{5}$ **28.** $\frac{3x-6}{2} = \frac{4x-2}{4}$

Example 4 **29** **NUTRITION** According to a recent study, 7 out of every 500 Americans aged 13 to 17 years are vegetarian. In a group of 350 13- to 17-year-olds, about how many would you expect to be vegetarian?

**30.** **CURRENCY** Your family is traveling to Mexico on vacation. You have saved \$500 to use for spending money. If 269 Mexican pesos is equivalent to 25 United States dollars, how much money will you get when you exchange your \$500 for pesos?

**ALGEBRA** **Solve each proportion. Round to the nearest tenth.**

**31.** $\frac{2x+3}{3} = \frac{6}{x-1}$ **32.** $\frac{x^2+4x+4}{40} = \frac{x+2}{10}$ **33.** $\frac{9x+6}{18} = \frac{20x+4}{3x}$

**34.** The perimeter of a rectangle is 98 feet. The ratio of its length to its width is 5:2. Find the area of the rectangle.

**35.** The perimeter of a rectangle is 220 inches. The ratio of its length to its width is 7:3. Find the area of the rectangle.

**36.** The ratio of the measures of the side lengths of a quadrilateral is 2:3:5:4. Its perimeter is 154 feet. Find the length of the shortest side.

**37.** The ratio of the measures of the angles of a quadrilateral is 2:4:6:3. Find the measures of the angles of the quadrilateral.

**38.** **SUMMER JOBS** In June of 2000, 60.2% of American teens 16 to 19 years old had summer jobs. By June of 2006, 51.6% of teens in that age group were a part of the summer work force.

**a.** Has the number of 16- to 19-year-olds with summer jobs increased or decreased since 2000? Explain your reasoning.

**b.** In June 2006, how many 16- to 19-year-olds would you expect to have jobs out of 700 in that age group? Explain your reasoning.

**39.** CCSS **MODELING** In a golden rectangle, the ratio of the length to the width is about 1.618. This is known as the *golden ratio*.

**a.** Recall from page 461 that a standard television screen has an aspect ratio of 4:3, while a high-definition television screen has an aspect ratio of 16:9. Is either type of screen a golden rectangle? Explain.

**b.** The golden ratio can also be used to determine column layouts for Web pages. Consider a site with two columns, the left for content and the right as a sidebar. The ratio of the left to right column widths is the golden ratio. Determine the width of each column if the page is 960 pixels wide.

**40.** **SCHOOL ACTIVITIES** A survey of club involvement showed that, of the 36 students surveyed, the ratio of French Club members to Spanish Club members to Drama Club members was 2:3:7. How many of those surveyed participate in Spanish Club? Assume that each student is active in only one club.

EXTEND 9-1

# Graphing Technology Lab
# Fibonacci Sequence and Ratios

Leonardo Pisano (c. 1170–c. 1250), or Fibonacci, was born in Italy but educated in North Africa. As a result, his work is similar to that of other North African authors of that time. His book *Liber abaci*, published in 1202, introduced what is now called the Fibonacci sequence, in which each term after the first two terms is the sum of the two numbers before it.

| Term | 1 | 2 | 3 | 4 | 5 | 6 | 7 |
|---|---|---|---|---|---|---|---|
| Fibonacci Number | 1 | 1 | 2 | 3 | 5 | 8 | 13 |
| | | | ↑ 1 + 1 | ↑ 1 + 2 | ↑ 2 + 3 | ↑ 3 + 5 | ↑ 5 + 8 |

## Activity

**You can use CellSheet on a TI-83/84 Plus graphing calculator to calculate terms of the Fibonacci sequence. Then compare each term with its preceding term.**

**Step 1** Access the **CellSheet** application by pressing the APPS key. Choose the number for **CellSheet** and press ENTER.

**Step 2** Enter the column headings in row 1. Use the **ALPHA** key to enter letters and press [ " ] at the beginning of each label.

**Step 3** Enter **1** into cell **A2.** Then insert the formula **=A2+1** in cell **A3.** Press STO to insert the = in the formula. Then use F3 to copy this formula and use F4 to paste it in each cell in the column. This will automatically calculate the number of the term.

**Step 4** In **column B,** we will record the Fibonacci numbers. Enter 1 in cells **B2** and **B3** since you do not have two previous terms to add. Then insert the formula **=B2+B3** in cell **B4.** Copy this formula down the column.

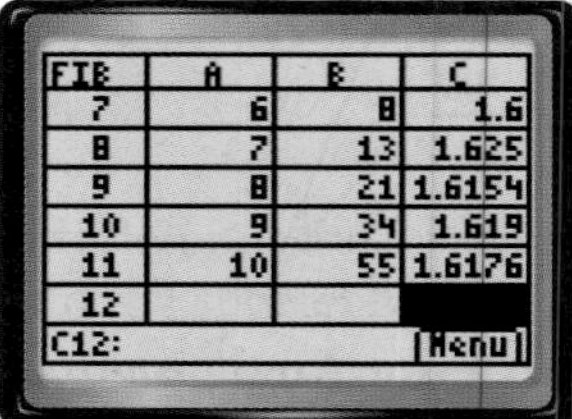

**Step 5** In **column C,** we will find the ratio of each term to its preceding term. Enter 1 in cell **C2** since there is no preceding term. Then enter B3/B2 in cell **C3.** Copy this formula down the column. *The screens show the results for terms 1 through 11.*

## Analyze the Results

1. What happens to the Fibonacci number as the number of the term increases?
2. What pattern of odd and even numbers do you notice in the Fibonacci sequence?
3. As the number of terms gets greater, what pattern do you notice in the ratio column?
4. Extend the spreadsheet to calculate fifty terms of the Fibonacci sequence. Describe any differences in the patterns you described in Exercises 1–3.
5. **MAKE A CONJECTURE** How might the Fibonacci sequence relate to the golden ratio?

# LESSON 9-2 Similar Polygons

**Then**

- You used proportions to solve problems.

**Now**

1. Use proportions to identify similar polygons.
2. Solve problems using the properties of similar polygons.

**Why?**

- People often customize their computer desktops using photos, centering the images at their original size or stretching them to fit the screen. This second method distorts the image, because the original and new images are not geometrically similar.

**NewVocabulary**
similar polygons
scale factor

**Common Core State Standards**

**Content Standards**
G.SRT.2 Given two figures, use the definition of similarity in terms of similarity transformations to decide if they are similar; explain using similarity transformations the meaning of similarity for triangles as the equality of all corresponding pairs of angles and the proportionality of all corresponding pairs of sides.

**Mathematical Practices**
7 Look for and make use of structure.
3 Construct viable arguments and critique the reasoning of others.

## 1 Identify Similar Polygons

**Similar polygons** have the same shape but not necessarily the same size.

### KeyConcept Similar Polygons

Two polygons are similar if and only if their corresponding angles are congruent and corresponding side lengths are proportional.

**Example** In the diagram below, *ABCD* is similar to *WXYZ*.

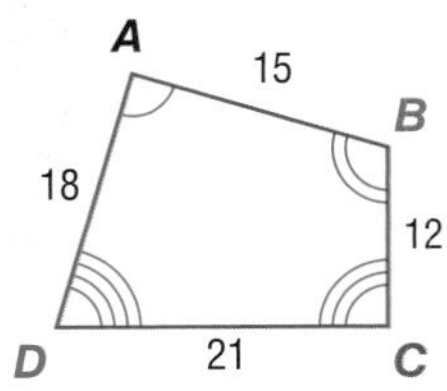


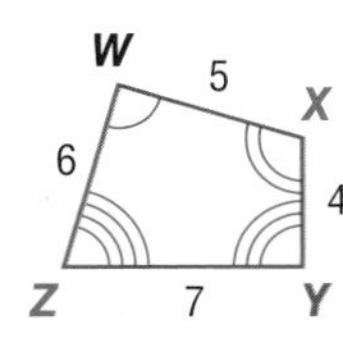


Corresponding angles

$\angle A \cong \angle W$, $\angle B \cong \angle X$, $\angle C \cong \angle Y$, and $\angle D \cong \angle Z$

Corresponding sides

$$\frac{AB}{WX} = \frac{BC}{XY} = \frac{CD}{YZ} = \frac{DA}{ZW} = \frac{3}{1}$$

**Symbols** $ABCD \sim WXYZ$

As with congruence statements, the order of vertices in a similarity statement like $ABCD \sim WXYZ$ is important. It identifies the corresponding angles and sides.

### Example 1 Use a Similarity Statement

**If $\triangle FGH \sim \triangle JKL$, list all pairs of congruent angles, and write a proportion that relates the corresponding sides.**

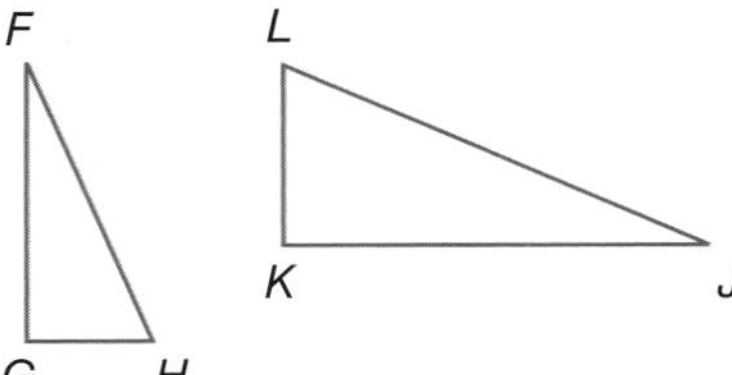


Use the similarity statement.

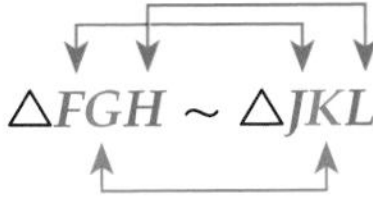


Congruent angles: $\angle F \cong \angle J$, $\angle G \cong \angle K$, $\angle H \cong \angle L$

Proportion: $\frac{FG}{JK} = \frac{GH}{KL} = \frac{HF}{LJ}$

**GuidedPractice**

1. In the diagram, $NPQR \sim UVST$. List all pairs of congruent angles, and write a proportion that relates the corresponding sides.

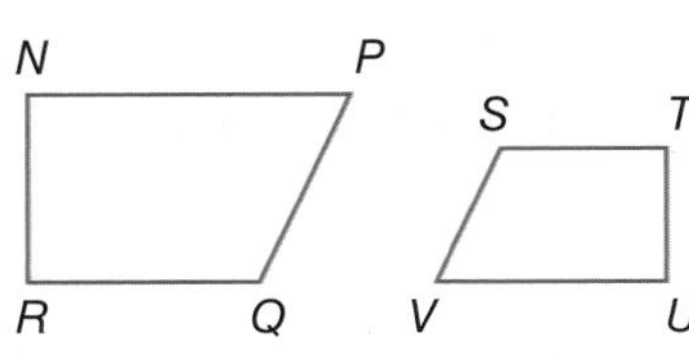

**StudyTip**

**Similarity Ratio** The scale factor between two similar polygons is sometimes called the *similarity ratio.*

The ratio of the lengths of the corresponding sides of two similar polygons is called the **scale factor**. The scale factor depends on the order of comparison.

In the diagram, $\triangle ABC \sim \triangle XYZ$.

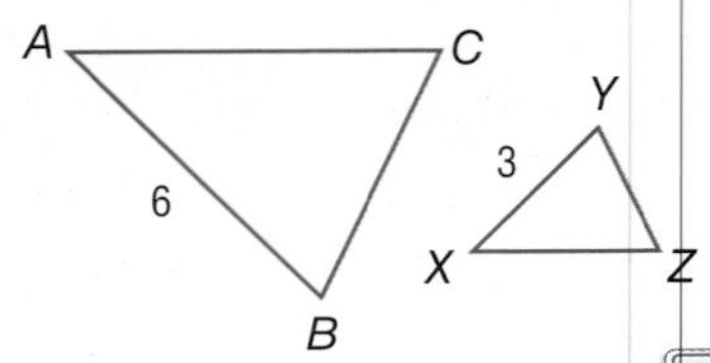


The scale factor of $\triangle ABC$ to $\triangle XYZ$ is $\frac{6}{3}$ or 2.

The scale factor of $\triangle XYZ$ to $\triangle ABC$ is $\frac{3}{6}$ or $\frac{1}{2}$.

## Real-World Example 2 Identify Similar Polygons

**PHOTO EDITING Kuma wants to use the rectangular photo shown as the background for her computer's desktop, but she needs to resize it. Determine whether the following rectangular images are similar. If so, write the similarity statement and scale factor. Explain your reasoning.**

**a.**

**b.**

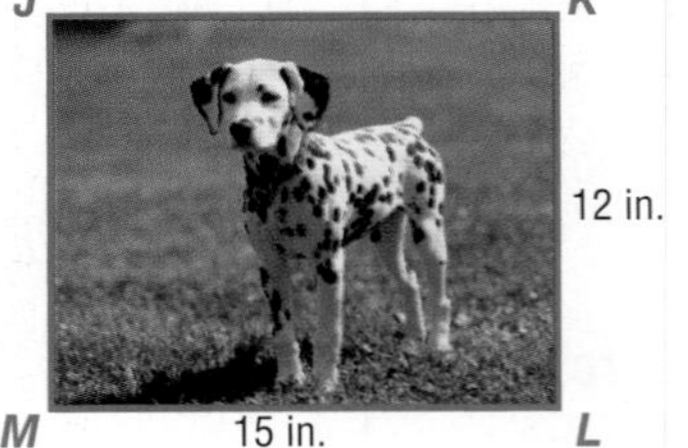


**a. Step 1** Compare corresponding angles.

Since all angles of a rectangle are right angles and right angles are congruent, corresponding angles are congruent.

**Step 2** Compare corresponding sides.

$\frac{DC}{HG} = \frac{10}{14}$ or $\frac{5}{7}$ $\qquad$ $\frac{BC}{FG} = \frac{8}{12}$ or $\frac{2}{3}$ $\qquad$ $\frac{5}{7} \neq \frac{2}{3}$

Since corresponding sides are not proportional, $ABCD \nsim EFGH$. So the photos are not similar.

**ReadingMath**

**Similarity Symbol** The symbol $\nsim$ is read as *is not similar to*.

**b. Step 1** Since *ABCD* and *JKLM* are both rectangles, corresponding angles are congruent.

**Step 2** Compare corresponding sides.

$\frac{DC}{ML} = \frac{10}{15}$ or $\frac{2}{3}$ $\qquad$ $\frac{BC}{KL} = \frac{8}{12}$ or $\frac{2}{3}$ $\qquad$ $\frac{2}{3} = \frac{2}{3}$

Since corresponding sides are proportional, $ABCD \sim JKLM$. So the rectangles are similar with a scale factor of $\frac{2}{3}$.

### GuidedPractice

**2.** Determine whether the triangles shown are similar. If so, write the similarity statement and scale factor. Explain your reasoning.

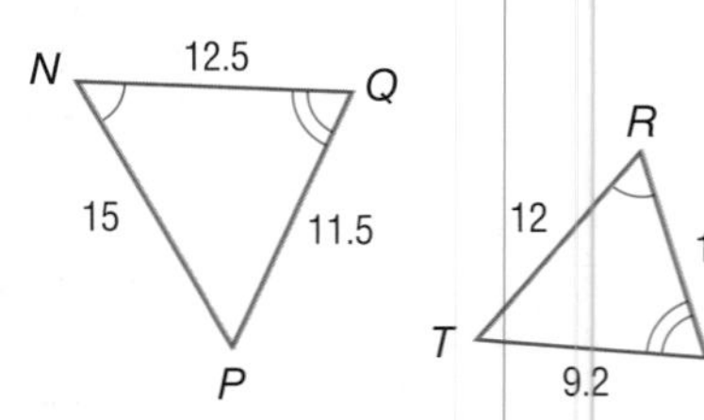


Stockbyte/Getty Images

**StudyTip**

**Similarity and Congruence** If two polygons are congruent, they are also similar. All of the corresponding angles are congruent, and the lengths of the corresponding sides have a ratio of 1:1.

## 2 Use Similar Figures

You can use scale factors and proportions to solve problems involving similar figures.

### Example 3 Use Similar Figures to Find Missing Measures

**In the diagram, $ACDF \sim VWYZ$.**

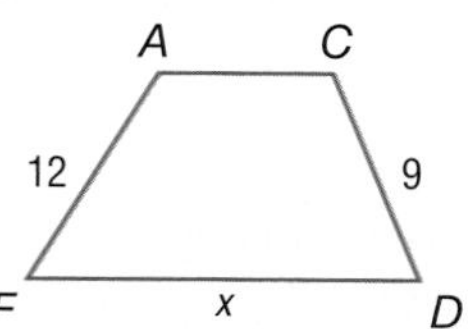


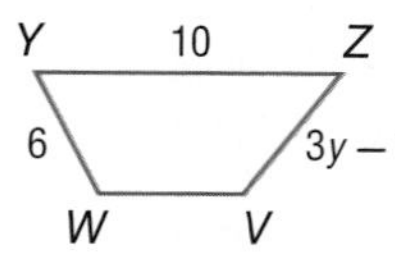


**a. Find $x$.**

Use the corresponding side lengths to write a proportion.

| | |
|---|---|
| $\frac{CD}{WY} = \frac{DF}{YZ}$ | Similarity proportion |
| $\frac{9}{6} = \frac{x}{10}$ | $CD = 9$, $WY = 6$, $DF = x$, $YZ = 10$ |
| $9(10) = 6(x)$ | Cross Products Property |
| $90 = 6x$ | Multiply. |
| $15 = x$ | Divide each side by 6. |

**b. Find $y$.**

| | |
|---|---|
| $\frac{CD}{WY} = \frac{FA}{ZV}$ | Similarity proportion |
| $\frac{9}{6} = \frac{12}{3y - 1}$ | $CD = 9$, $WY = 6$, $FA = 12$, $ZV = 3y - 1$ |
| $9(3y - 1) = 6(12)$ | Cross Products Property |
| $27y - 9 = 72$ | Multiply. |
| $27y = 81$ | Add 9 to each side. |
| $y = 3$ | Divide each side by 27. |

**StudyTip**

**Identifying Similar Triangles** When only two congruent angles of a triangle are given, remember that you can use the Third Angles Theorem to establish that the remaining corresponding angles are also congruent.

**GuidedPractice**

**Find the value of each variable if $\triangle JLM \sim \triangle QST$.**

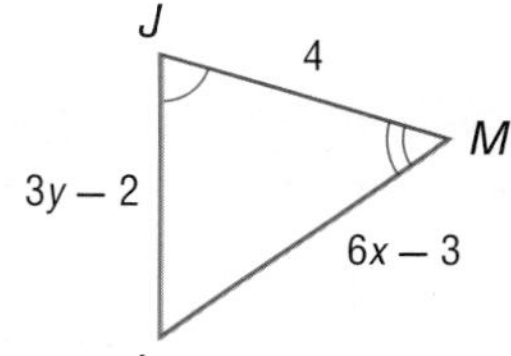


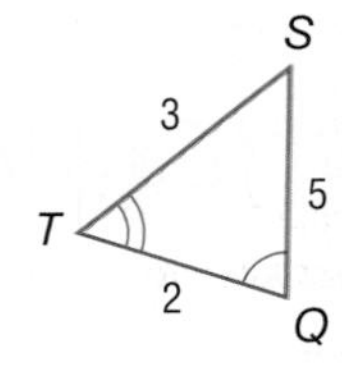


**3A.** $x$

**3B.** $y$

In similar polygons, the ratio of any two corresponding lengths is proportional to the scale factor between them. This leads to the following theorem about the perimeters of two similar polygons.

### Theorem 9.1 Perimeters of Similar Polygons

If two polygons are similar, then their perimeters are proportional to the scale factor between them.

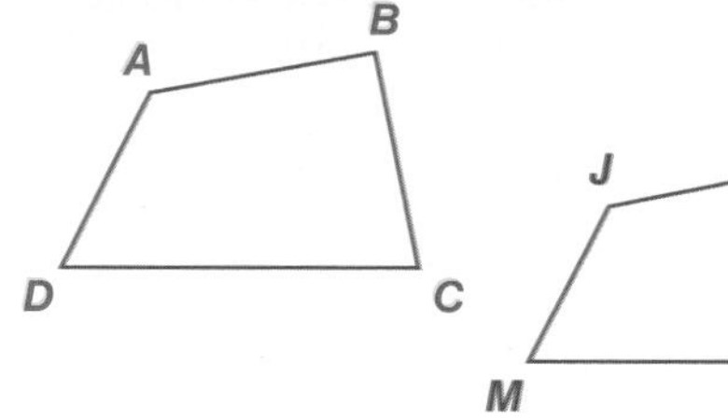


**Example** If $ABCD \sim JKLM$, then

$$\frac{AB + BC + CD + DA}{JK + KL + LM + MJ} = \frac{AB}{JK} = \frac{BC}{KL} = \frac{CD}{LM} = \frac{DA}{MJ}.$$

## Example 4 Use a Scale Factor to Find Perimeter

**If $ABCDE \sim PQRST$, find the scale factor of $ABCDE$ to $PQRST$ and the perimeter of each polygon.**

The scale factor of $ABCDE$ to $PQRST$ is $\frac{CD}{RS}$ or $\frac{4}{3}$.

Since $\overline{BC} \cong \overline{AB}$ and $\overline{AE} \cong \overline{CD}$, the perimeter of $ABCDE$ is $8 + 8 + 4 + 6 + 4$ or 30.

Use the perimeter of $ABCDE$ and the scale factor to write a proportion. Let $x$ represent the perimeter of $PQRST$.

$\frac{4}{3} = \frac{\text{perimeter of } ABCDE}{\text{perimeter of } PQRST}$ Theorem 9.1

$\frac{4}{3} = \frac{30}{x}$ Substitution

$(3)(30) = 4x$ Cross Products Property

$22.5 = x$ Solve.

So, the perimeter of $PQRST$ is 22.5.

**WatchOut!**

**Perimeter** Remember that perimeter is the distance around a figure. Be sure to find the sum of all side lengths when finding the perimeter of a polygon. You may need to use other markings or geometric principles to find the length of unmarked sides.

### GuidedPractice

**4.** If $MNPQ \sim XYZW$, find the scale factor of $MNPQ$ to $XYZW$ and the perimeter of each polygon.

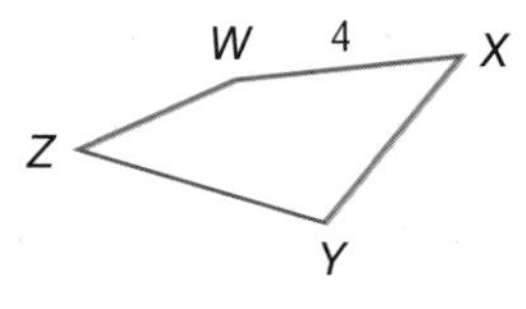


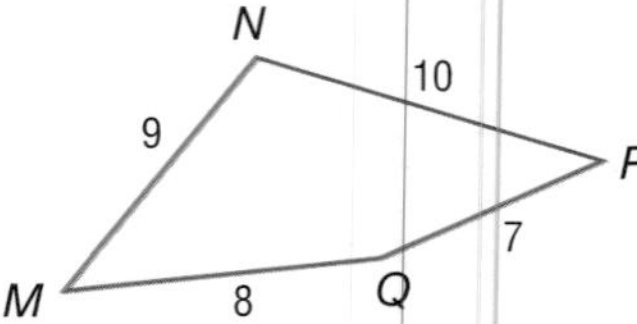

## Check Your Understanding

Example 1 **List all pairs of congruent angles, and write a proportion that relates the corresponding sides for each pair of similar polygons.**

**1.** $\triangle ABC \sim \triangle ZYX$

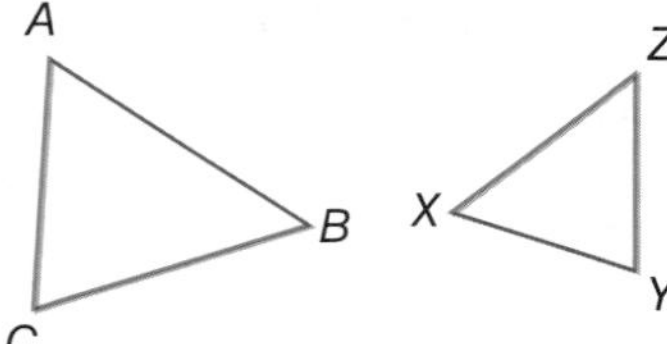


**2.** $JKLM \sim TSRQ$

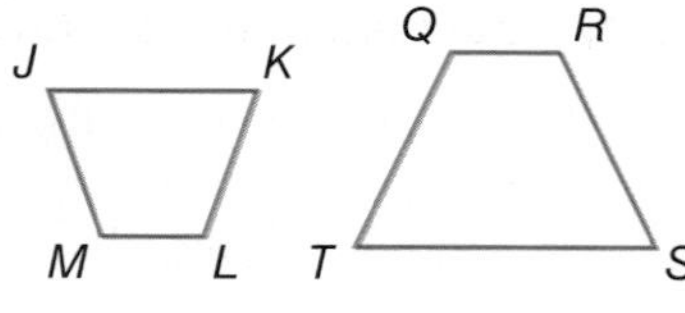


Example 2 **Determine whether each pair of figures is similar. If so, write the similarity statement and scale factor. If not, explain your reasoning.**

**3.**

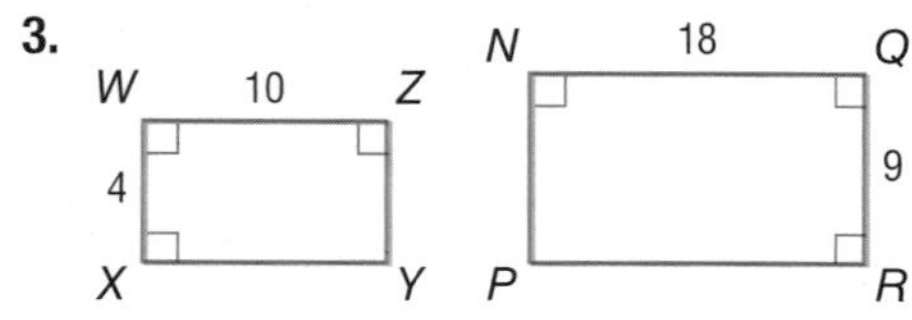


**4.**

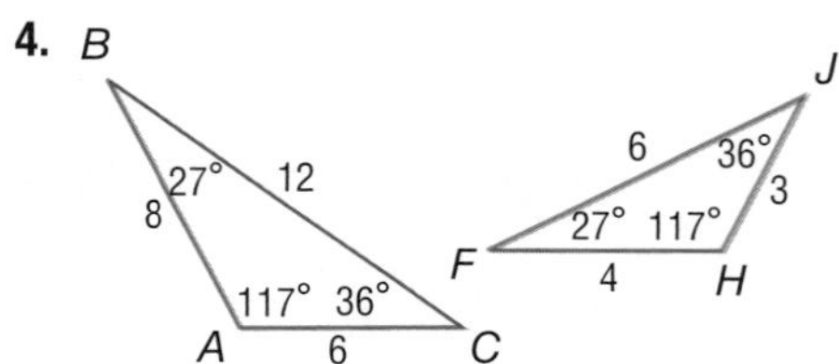

**Example 3** **Each pair of polygons is similar. Find the value of $x$.**

**5.**

**6.**

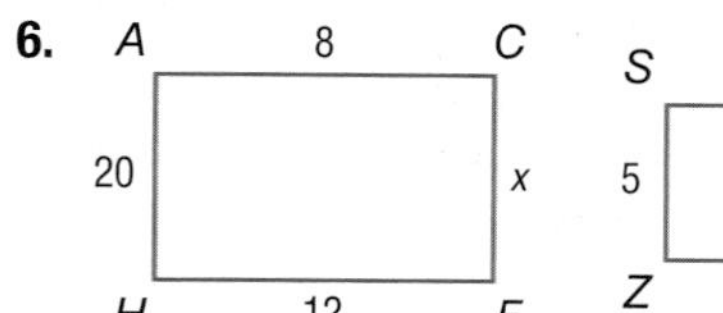


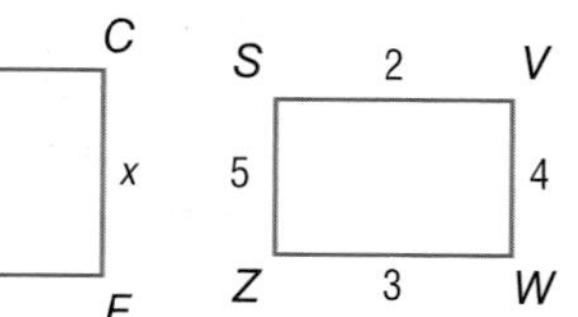


**Example 4**

**7. DESIGN** On the blueprint of the apartment shown, the balcony measures 1 inch wide by 1.75 inches long. If the actual length of the balcony is 7 feet, what is the perimeter of the balcony?

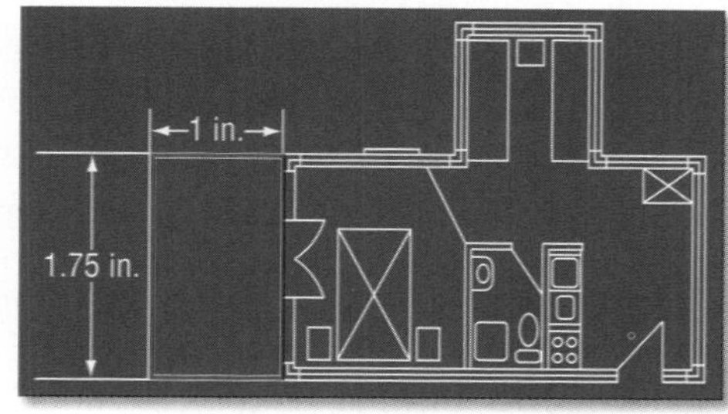


## Practice and Problem Solving

**Example 1** **List all pairs of congruent angles, and write a proportion that relates the corresponding sides for each pair of similar polygons.**

**8.** $\triangle CHF \sim \triangle YWS$

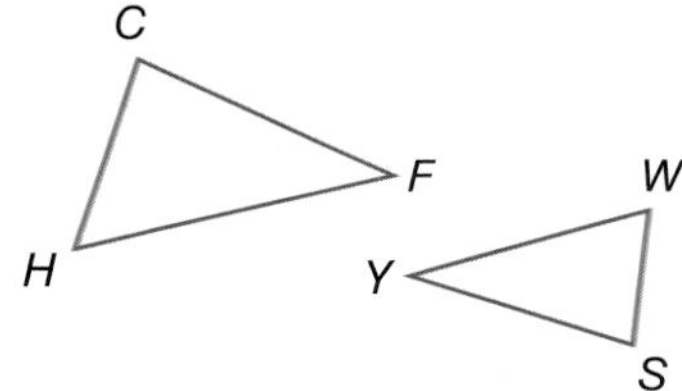


**9.** $JHFM \sim PQST$

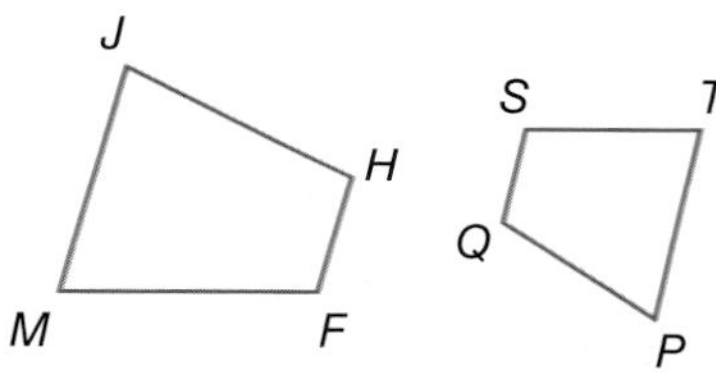


**10.** $ABDF \sim VXZT$

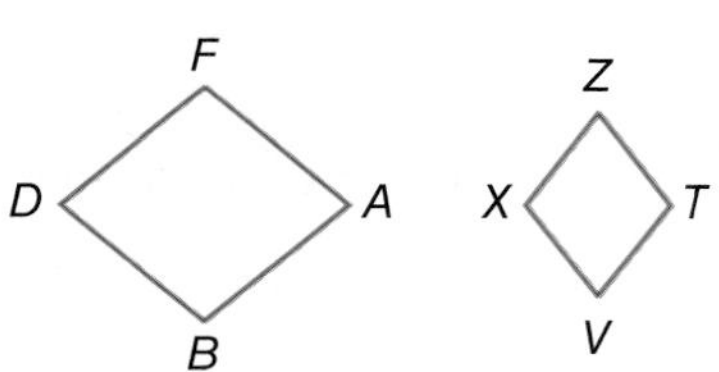


**11.** $\triangle DFG \sim \triangle KMJ$

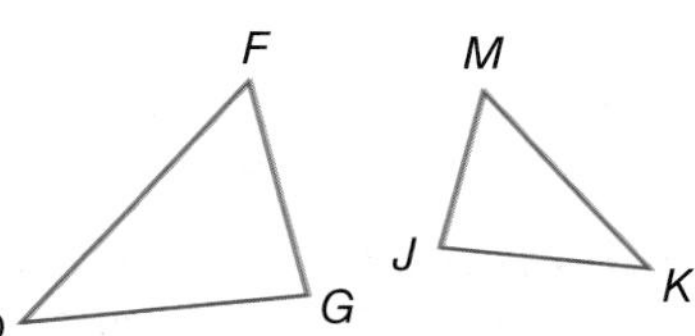


**Example 2** **CCSS ARGUMENTS** **Determine whether each pair of figures is similar. If so, write the similarity statement and scale factor. If not, explain your reasoning.**

**12.**

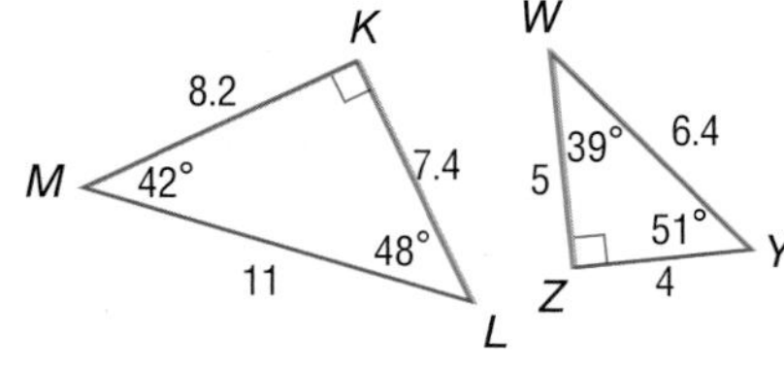


**13.**

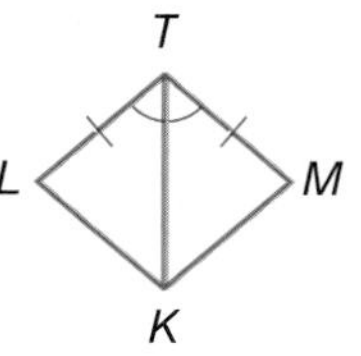


**14.**

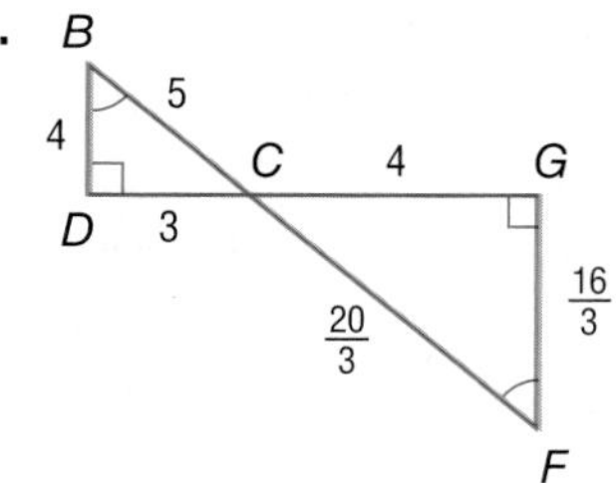


**15.**

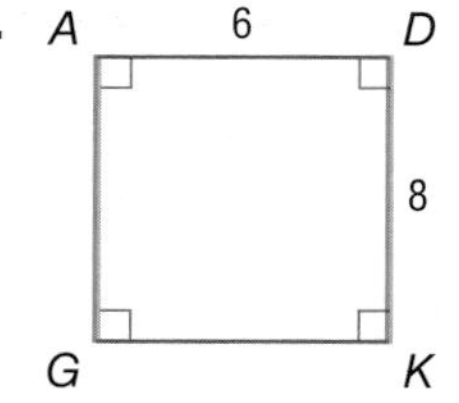


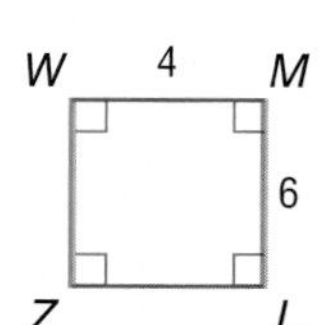

16. **GAMES** The dimensions of a hockey rink are 200 feet by 85 feet. Are the hockey rink and the air hockey table shown similar? Explain your reasoning.

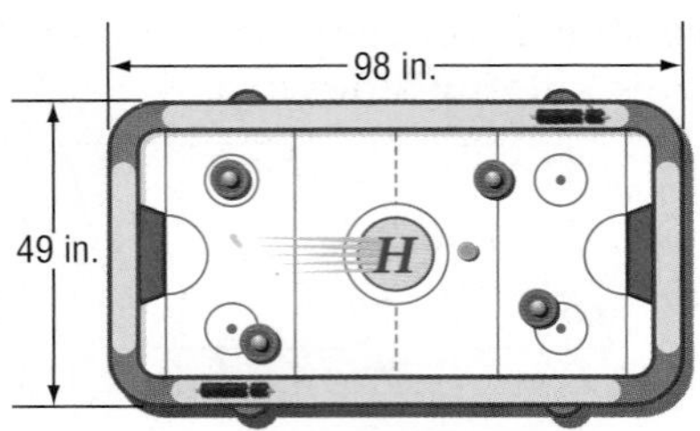


17. **COMPUTERS** The dimensions of a 17-inch flat panel computer screen are approximately $13\frac{1}{4}$ by $10\frac{3}{4}$ inches. The dimensions of a 19-inch flat panel computer screen are approximately $14\frac{1}{2}$ by 12 inches. To the nearest tenth, are the computer screens similar? Explain your reasoning.

**Example 3** CCSS **REGULARITY** **Each pair of polygons is similar. Find the value of $x$.**

18.

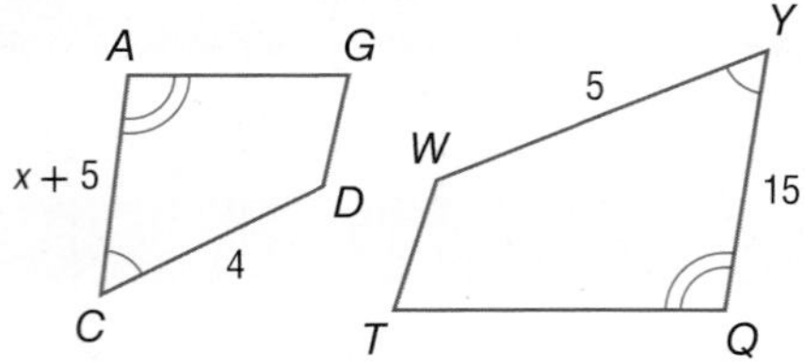


19.

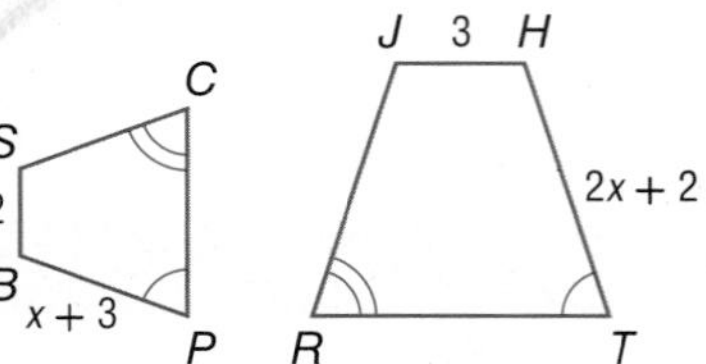


20.

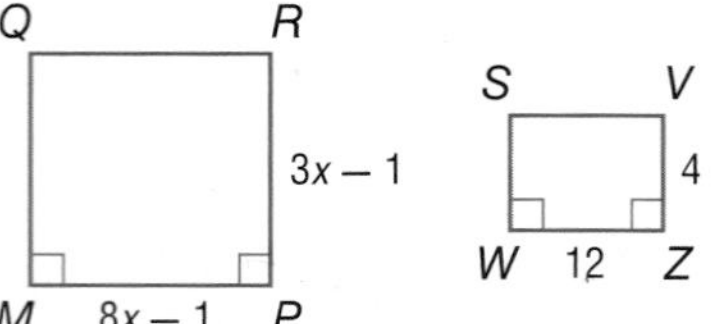


21.

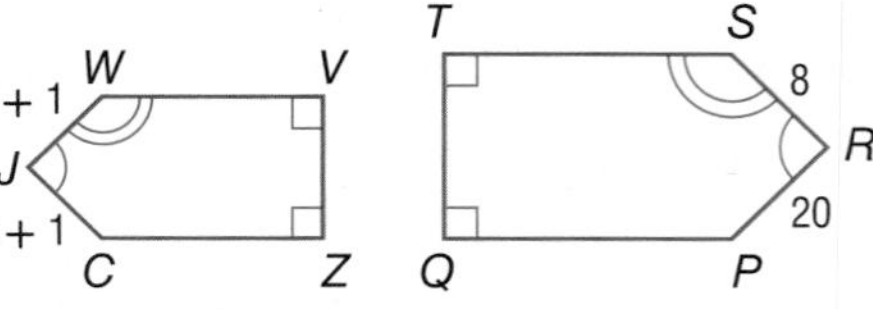


**Example 4** 22. Rectangle $ABCD$ has a width of 8 yards and a length of 20 yards. Rectangle $QRST$, which is similar to rectangle $ABCD$, has a length of 40 yards. Find the scale factor of rectangle $ABCD$ to rectangle $QRST$ and the perimeter of each rectangle.

**Find the perimeter of the given triangle.**

23. $\triangle DEF$, if $\triangle ABC \sim \triangle DEF$, $AB = 5$, $BC = 6$, $AC = 7$, and and $DE = 3$

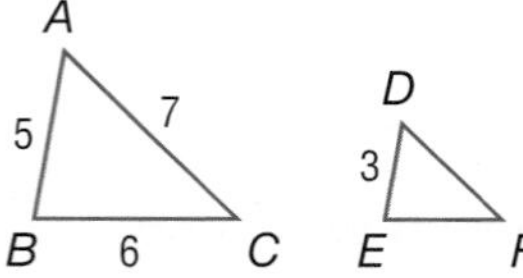


24. $\triangle WZX$, if $\triangle WZX \sim \triangle SRT$, $ST = 6$, $WX = 5$, and the perimeter of $\triangle SRT = 15$

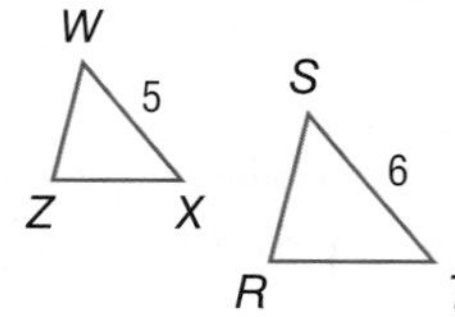


25. $\triangle CBH$, if $\triangle CBH \sim \triangle FEH$, $ADEG$ is a parallelogram, $CH = 7$, $FH = 10$, $FE = 11$, and $EH = 6$

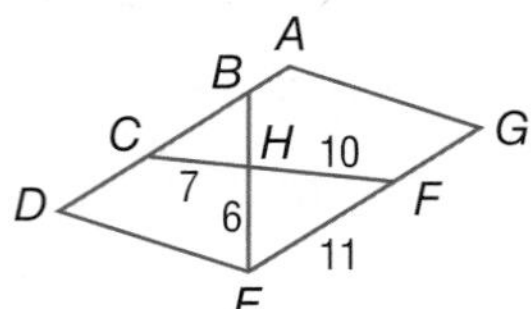


26. $\triangle DEF$, if $\triangle DEF \sim \triangle CBF$, perimeter of $\triangle CBF = 27$, $DF = 6$, $FC = 8$

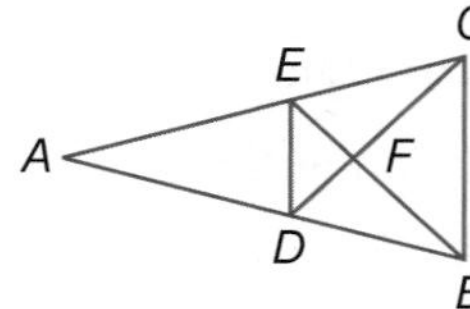


27. Two similar rectangles have a scale factor of 2:4. The perimeter of the large rectangle is 80 meters. Find the perimeter of the small rectangle.

28. Two similar rectangles have a scale factor of 3:2. The perimeter of the small rectangle is 50 feet. Find the perimeter of the large rectangle.

**List all pairs of congruent angles, and write a proportion that relates the corresponding sides.**

**29.**

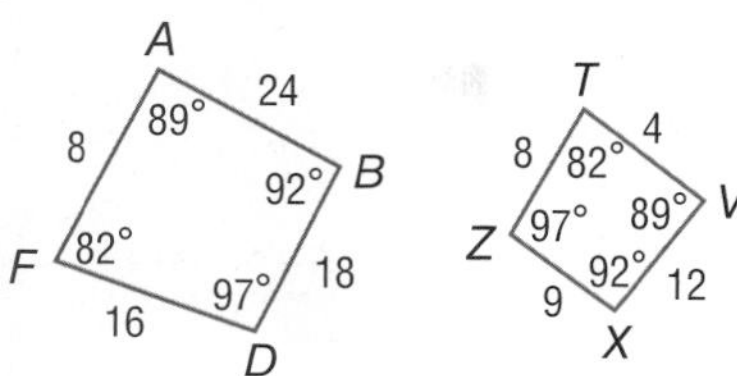


**30.**

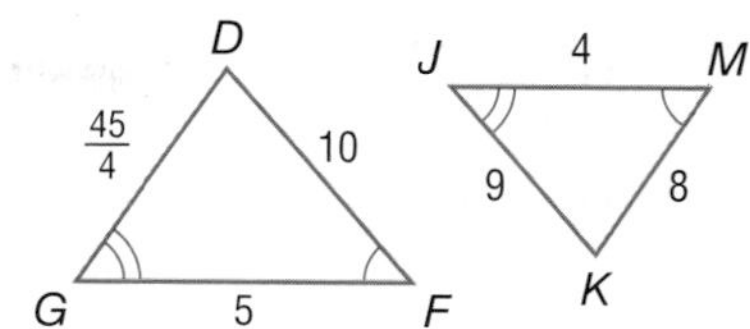


**SHUFFLEBOARD** **A shuffleboard court forms three similar triangles in which $\angle AHB \cong \angle AGC \cong \angle AFD$. For the given sides or angles, find the corresponding side(s) or angle(s) that are congruent.**

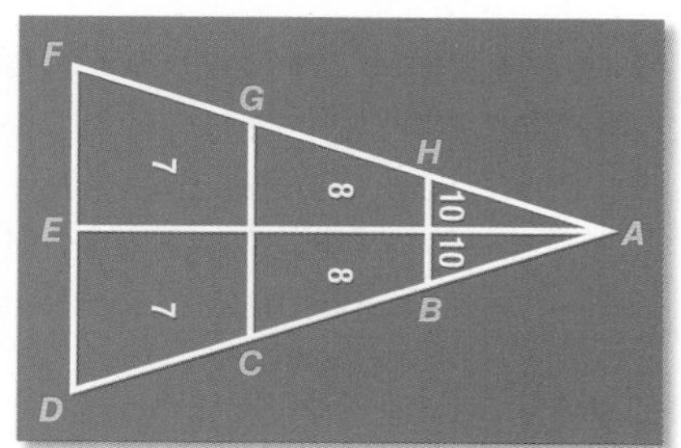


**31.** $\overline{AB}$

**32.** $\overline{FD}$

**33.** $\angle ACG$

**34.** $\angle A$

**Find the value of each variable.**

**35** $ABCD \sim QSRP$

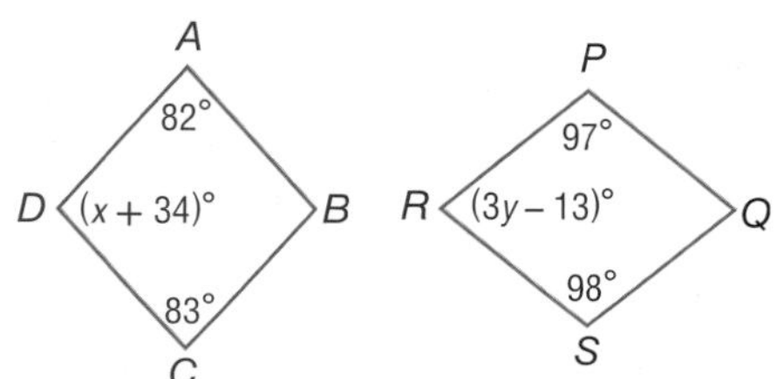


**36.** $\triangle JKL \sim \triangle WYZ$

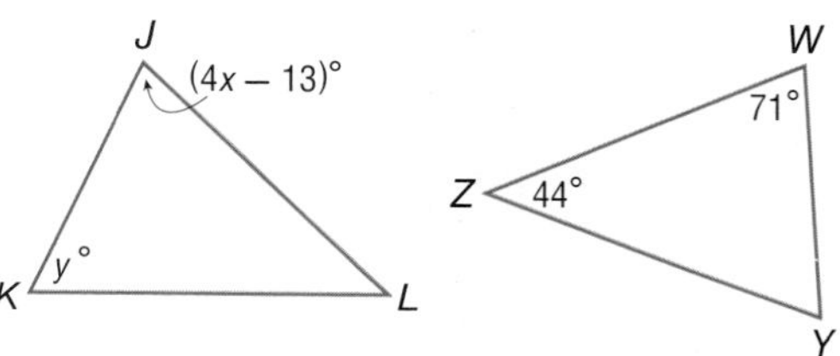


**37.** **SLIDE SHOW** You are using a digital projector for a slide show. The photos are 13 inches by $9\frac{1}{4}$ inches on the computer screen, and the scale factor of the computer image to the projected image is 1:4. What are the dimensions of the projected image?

**COORDINATE GEOMETRY** **For the given vertices, determine whether rectangle *ABCD* is similar to rectangle *WXYZ*. Justify your answer.**

**38.** $A(-1, 5)$, $B(7, 5)$, $C(7, -1)$, $D(-1, -1)$;
$W(-2, 10)$, $X(14, 10)$, $Y(14, -2)$, $Z(-2, -2)$

**39.** $A(5, 5)$, $B(0, 0)$, $C(5, -5)$, $D(10, 0)$;
$W(1, 6)$, $X(-3, 2)$, $Y(2, -3)$, $Z(6, 1)$

**CCSS ARGUMENTS** **Determine whether the polygons are *always*, *sometimes*, or *never* similar. Explain your reasoning.**

**40.** two obtuse triangles

**41.** a trapezoid and a parallelogram

**42.** two right triangles

**43.** two isosceles triangles

**44.** a scalene triangle and an isosceles triangle

**45.** two equilateral triangles

**46.** **PROOF** Write a paragraph proof of Theorem 9.1.

**Given:** $\triangle ABC \sim \triangle DEF$ and $\frac{AB}{DE} = \frac{m}{n}$

**Prove:** $\frac{\text{perimeter of } \triangle ABC}{\text{perimeter of } \triangle DEF} = \frac{m}{n}$

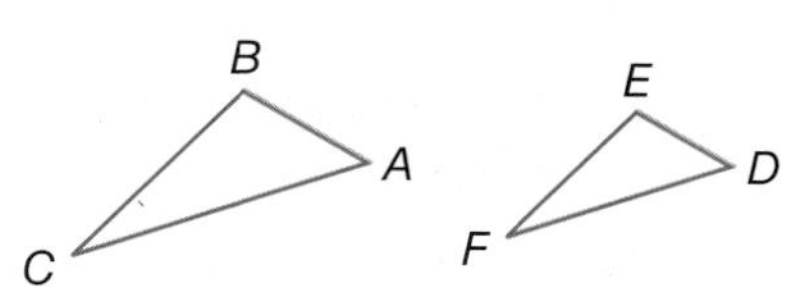

**47.** **PHOTOS** You are enlarging the photo shown at the right for your school yearbook. If the dimensions of the original photo are $2\frac{1}{3}$ inches by $1\frac{2}{3}$ inches and the scale factor of the old photo to the new photo is 2:3, what are the dimensions of the new photo?

**48.** **CHANGING DIMENSIONS** Rectangle *QRST* is similar to rectangle *JKLM* with sides in a ratio of 4:1.

**a.** What is the ratio of the areas of the two rectangles?

**b.** Suppose the dimension of each rectangle is tripled. What is the new ratio of the sides of the rectangles?

**c.** What is the ratio of the areas of these larger rectangles?

**49.** **CHANGING DIMENSIONS** In the figure shown, $\triangle FGH \sim \triangle XYZ$.

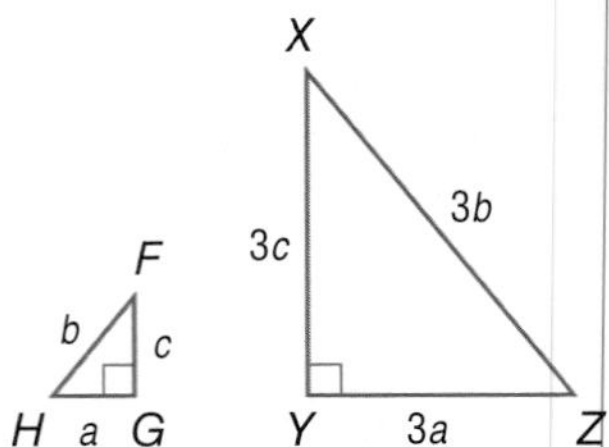


**a.** Show that the perimeters of $\triangle FGH$ and $\triangle XYZ$ have the same ratio as their corresponding sides.

**b.** If 6 units are added to the lengths of each side, are the new triangles similar? Explain.

**50.** **MULTIPLE REPRESENTATIONS** In this problem, you will investigate similarity in squares.

**a.** **Geometric** Draw three different-sized squares. Label them *ABCD*, *PQRS*, and *WXYZ*. Measure and label each square with its side length.

**b.** **Tabular** Calculate and record in a table the ratios of corresponding sides for each pair of squares: *ABCD* and *PQRS*, *PQRS* and *WXYZ*, and *WXYZ* and *ABCD*. Is each pair of squares similar?

**c.** **Verbal** Make a conjecture about the similarity of all squares.

## H.O.T. Problems Use Higher-Order Thinking Skills

**51.** **CHALLENGE** For what value(s) of $x$ is $BEFA \sim EDCB$?

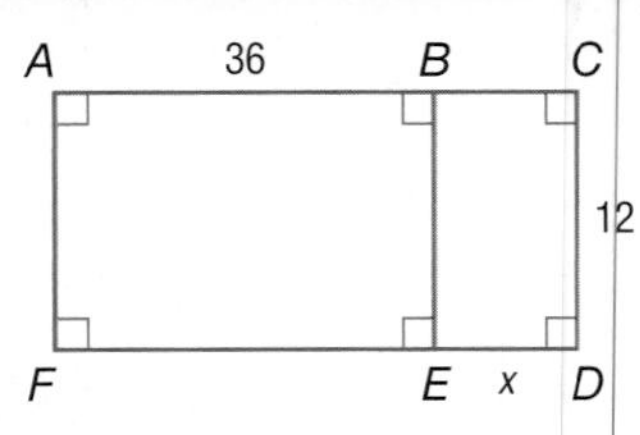


**52.** **REASONING** Recall that an *equivalence relation* is any relationship that satisfies the Reflexive, Symmetric, and Transitive Properties. Is similarity an equivalence relation? Explain.

**53.** **OPEN ENDED** Find a counterexample for the following statement.

*All rectangles are similar.*

**54.** **CCSS** **REASONING** Draw two regular pentagons of different sizes. Are the pentagons similar? Will any two regular polygons with the same number of sides be similar? Explain.

**55.** **WRITING IN MATH** How can you describe the relationship between two figures?

Digital Vision/Masterfile

## Standardized Test Practice

**56. ALGEBRA** If the arithmetic mean of $4x$, $3x$, and 12 is 18, then what is the value of $x$?

A 6   C 4

B 5   D 3

**57.** Two similar rectangles have a scale factor of 3:5. The perimeter of the large rectangle is 65 meters. What is the perimeter of the small rectangle?

F 29 m   H 49 m

G 39 m   J 59 m

**58. SHORT RESPONSE** If a jar contains 25 dimes and 7 quarters, what is the probability that a coin selected from the jar at random will be a dime?

**59. SAT/ACT** If the side of a square is $x + 3$, then what is the diagonal of the square?

A $x^2 + 3$   D $x\sqrt{3} + 3\sqrt{3}$

B $3x + 3$   E $x\sqrt{2} + 3\sqrt{2}$

C $2x + 6$

## Spiral Review

**60. COMPUTERS** In a survey of 5000 households, 4200 had at least one computer. What is the ratio of computers to households? (Lesson 9-1)

**61. PROOF Write a flow proof.** (Lesson 8-6)

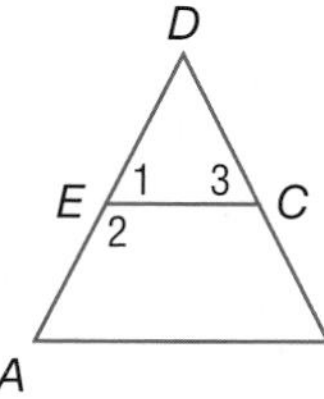


**Given:** $E$ and $C$ are midpoints of $\overline{AD}$ and $\overline{DB}$, $\overline{AD} \cong \overline{DB}$, $\angle A \cong \angle 1$.

**Prove:** $ABCE$ is an isosceles trapezoid.

**62. COORDINATE GEOMETRY** Determine the coordinates of the intersection of the diagonals of $\square JKLM$ with vertices $J(2, 5)$, $K(6, 6)$, $L(4, 0)$, and $M(0, -1)$. (Lesson 8-2)

**State the assumption you would make to start an indirect proof of each statement.** (Lesson 7-4)

**63.** If $3x > 12$, then $x > 4$.

**64.** $\overline{PQ} \cong \overline{ST}$

**65.** The angle bisector of the vertex angle of an isosceles triangle is also an altitude of the triangle.

**66.** If a rational number is any number that can be expressed as $\frac{a}{b}$, where $a$ and $b$ are integers and $b \neq 0$, then 6 is a rational number.

**Find the measures of each numbered angle.** (Lesson 6-1)

**67.** $m\angle 1$

**68.** $m\angle 2$

**69.** $m\angle 3$

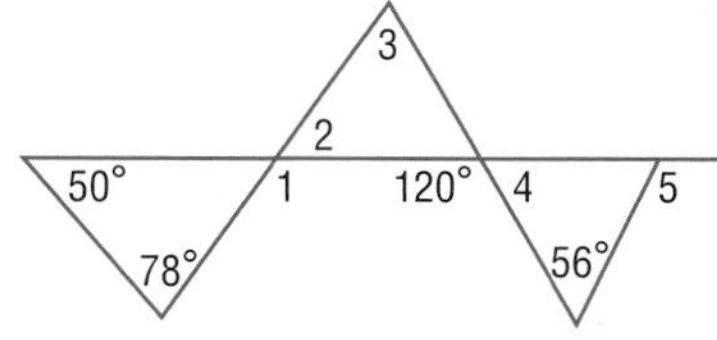


## Skills Review

**ALGEBRA Find $x$ and the unknown side measures of each triangle.**

**70.**

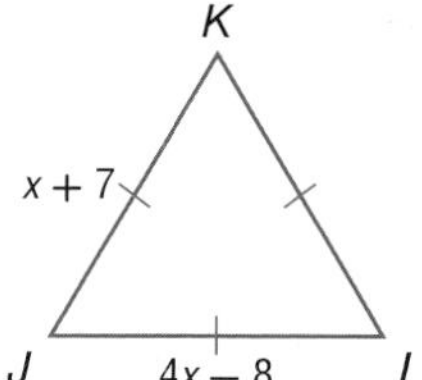


**71.**

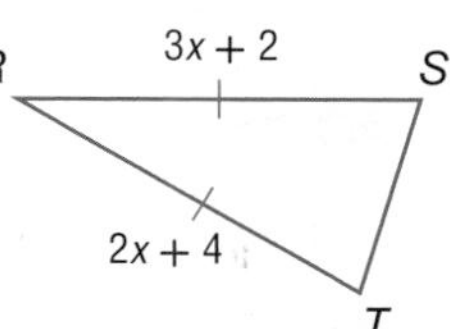


**72.**

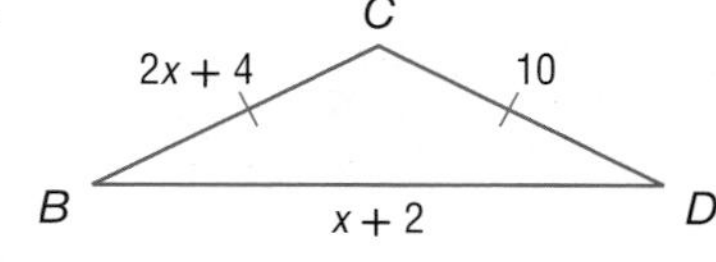

# LESSON 9-3 Similar Triangles

**Then**

- You used the AAS, SSS, and SAS Congruence Theorems to prove triangles congruent.

**Now**

1. Identify similar triangles using the AA Similarity Postulate and the SSS and SAS Similarity Theorems.
2. Use similar triangles to solve problems.

**Why?**

- Julian wants to draw a similar version of his skate club's logo on a poster. He first draws a line at the bottom of the poster. Next, he uses a cutout of the original triangle to copy the two bottom angles. Finally, he extends the noncommon sides of the two angles.

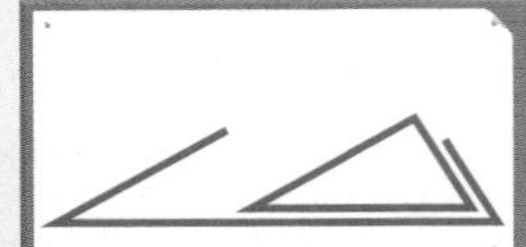

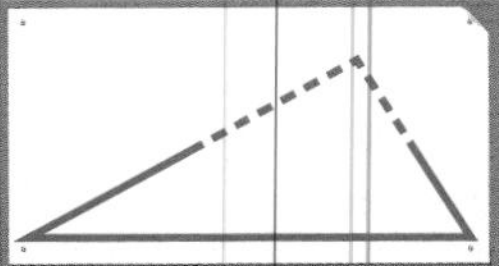

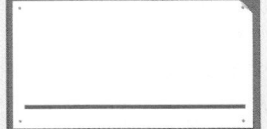

**Common Core State Standards**

**Content Standards**

G.SRT.4 Prove theorems about triangles.

G.SRT.5 Use congruence and similarity criteria for triangles to solve problems and to prove relationships in geometric figures.

**Mathematical Practices**

4 Model with mathematics.

7 Look for and make use of structure.

## 1 Identify Similar Triangles

The example suggests that two triangles are similar if two pairs of corresponding angles are congruent.

### Postulate 9.1 Angle-Angle (AA) Similarity

If two angles of one triangle are congruent to two angles of another triangle, then the triangles are similar.

**Example** If $\angle A \cong \angle F$ and $\angle B \cong \angle G$, then $\triangle ABC \sim \triangle FGH$.

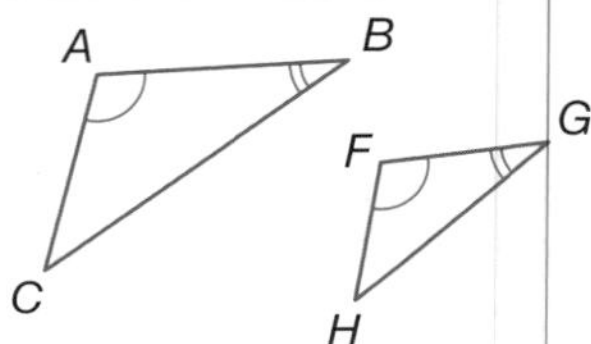

### Example 1 Use the AA Similarity Postulate

**Determine whether the triangles are similar. If so, write a similarity statement. Explain your reasoning.**

**a.**

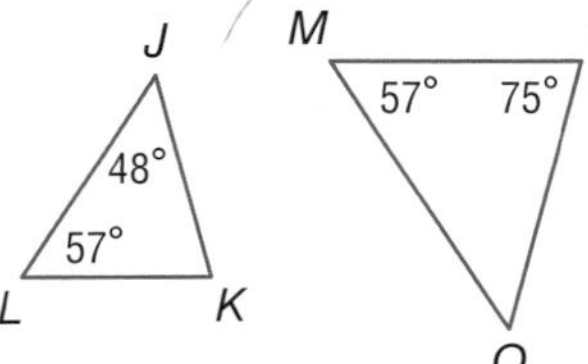


**b.**

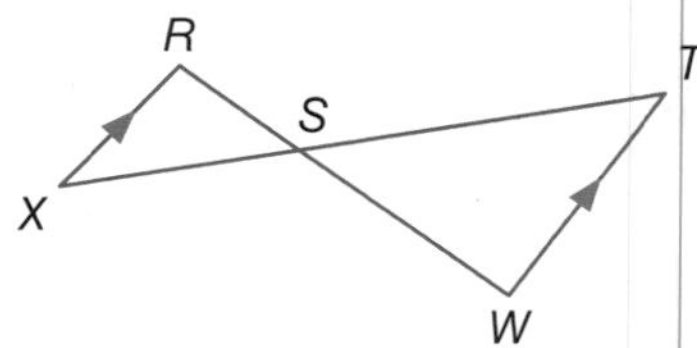


**a.** Since $m\angle L = m\angle M$, $\angle L \cong \angle M$. By the Triangle Sum Theorem, $57 + 48 + m\angle K = 180$, so $m\angle K = 75$. Since $m\angle P = 75$, $\angle K \cong \angle P$. So, $\triangle LJK \sim \triangle MQP$ by AA Similarity.

**b.** $\angle RSX \cong \angle WST$ by the Vertical Angles Theorem. Since $\overline{RX} \parallel \overline{TW}$, $\angle R \cong \angle W$. So, $\triangle RSX \sim \triangle WST$ by AA Similarity.

**GuidedPractice**

**1A.**

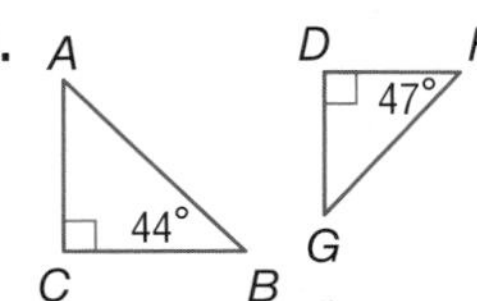


**1B.**

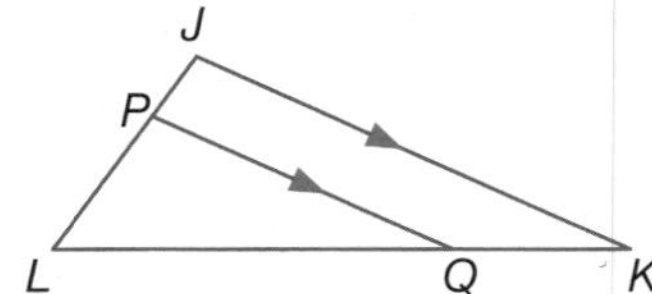

You can use the AA Similarity Postulate to prove the following two theorems.

### Theorems Triangle Similarity

**9.2 Side-Side-Side (SSS) Similarity**
If the corresponding side lengths of two triangles are proportional, then the triangles are similar.

**Example** If $\frac{JK}{MP} = \frac{KL}{PQ} = \frac{LJ}{QM}$, then $\triangle JKL \sim \triangle MPQ$.

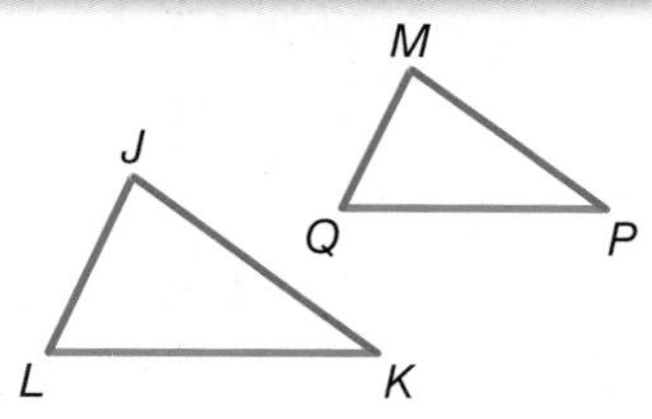


**9.3 Side-Angle-Side (SAS) Similarity**
If the lengths of two sides of one triangle are proportional to the lengths of two corresponding sides of another triangle and the included angles are congruent, then the triangles are similar.

**Example** If $\frac{RS}{XY} = \frac{ST}{YZ}$ and $\angle S \cong \angle Y$, then $\triangle RST \sim \triangle XYZ$.

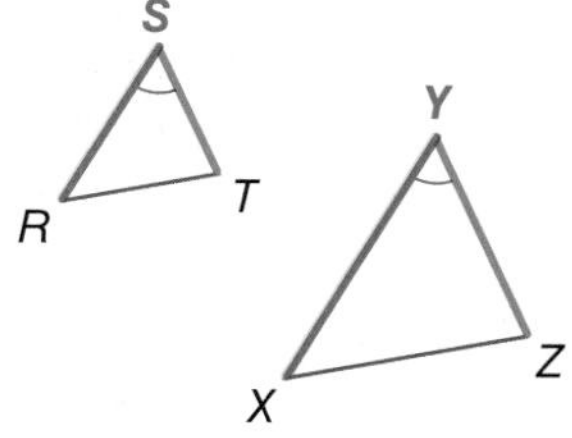


You will prove Theorem 9.3 in Exercise 25.

### Proof Theorem 9.2

**Given:** $\frac{AB}{FG} = \frac{BC}{GH} = \frac{AC}{FH}$

**Prove:** $\triangle ABC \sim \triangle FGH$

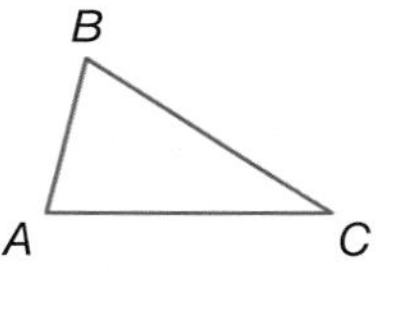


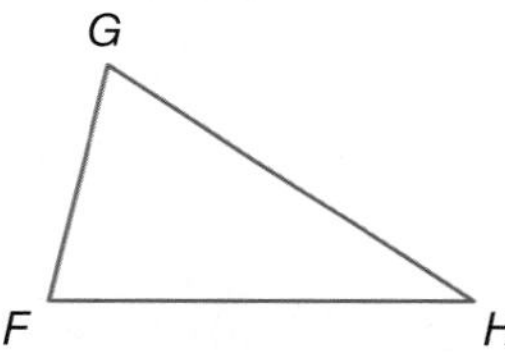


**StudyTip**

Corresponding Sides To determine which sides of two triangles correspond, begin by comparing the longest sides, then the next longest sides, and finish by comparing the shortest sides.

**Paragraph Proof:**

Locate $J$ on $\overline{FG}$ so that $JG = AB$. Draw $\overline{JK}$ so that $\overline{JK} \parallel \overline{FH}$. Label $\angle GJK$ as $\angle 1$.

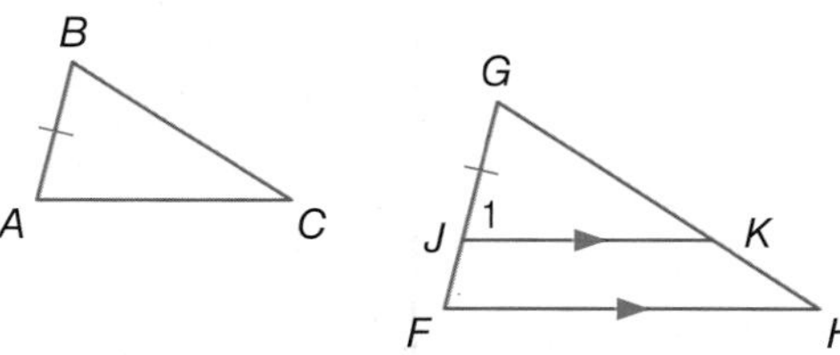


Since $\angle G \cong \angle G$ by the Reflexive Property and $\angle 1 \cong \angle F$ by the Corresponding Angles Postulate, $\triangle GJK \sim \triangle GFH$ by the AA Similarity Postulate.

By the definition of similar polygons, $\frac{JG}{FG} = \frac{GK}{GH} = \frac{JK}{FH}$. By substitution, $\frac{AB}{FG} = \frac{GK}{GH} = \frac{JK}{FH}$.

Since we are also given that $\frac{AB}{FG} = \frac{BC}{GH} = \frac{AC}{FH}$, we can say that $\frac{GK}{GH} = \frac{BC}{GH}$ and $\frac{JK}{FH} = \frac{AC}{FH}$. This means that $GK = BC$ and $JK = AC$, so $\overline{GK} \cong \overline{BC}$ and $\overline{JK} \cong \overline{AC}$.

By SSS, $\triangle ABC \cong \triangle JGK$.

By CPCTC, $\angle B \cong \angle G$ and $\angle A \cong \angle 1$. Since $\angle 1 \cong \angle F$, $\angle A \cong \angle F$ by the Transitive Property. By AA Similarity, $\triangle ABC \sim \triangle FGH$.

## Example 2 Use the SSS and SAS Similarity Theorems

**Determine whether the triangles are similar. If so, write a similarity statement. Explain your reasoning.**

**a.**

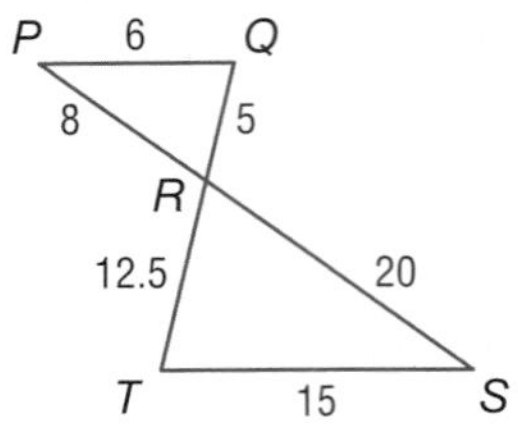


$\frac{PR}{SR} = \frac{8}{20}$ or $\frac{2}{5}$, $\frac{PQ}{ST} = \frac{6}{15}$ or $\frac{2}{5}$, and $\frac{QR}{TR} = \frac{5}{12.5} = \frac{50}{125}$ or $\frac{2}{5}$. So, $\triangle PQR \sim \triangle STR$ by the SSS Similarity Theorem.

**b.**

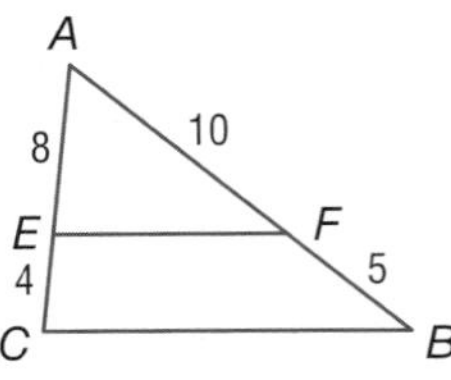


By the Reflexive Property, $\angle A \cong \angle A$.

$\frac{AF}{AB} = \frac{10}{10+5} = \frac{10}{15}$ or $\frac{2}{3}$ and $\frac{AE}{AC} = \frac{8}{8+4} = \frac{8}{12}$ or $\frac{2}{3}$.

Since the lengths of the sides that include $\angle A$ are proportional, $\triangle AEF \sim \triangle ACB$ by the SAS Similarity Theorem.

**Study Tip**

**Draw Diagrams** It is helpful to redraw similar triangles so that the corresponding side lengths have the same orientation.

### Guided Practice

**2A.**

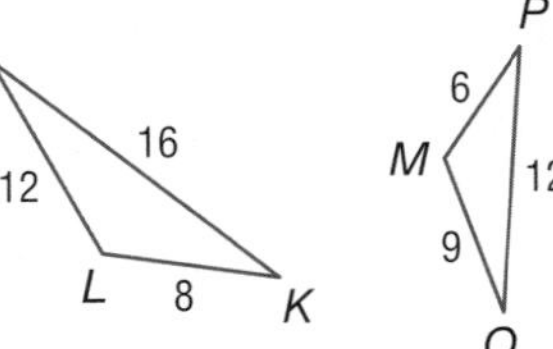


**2B.**

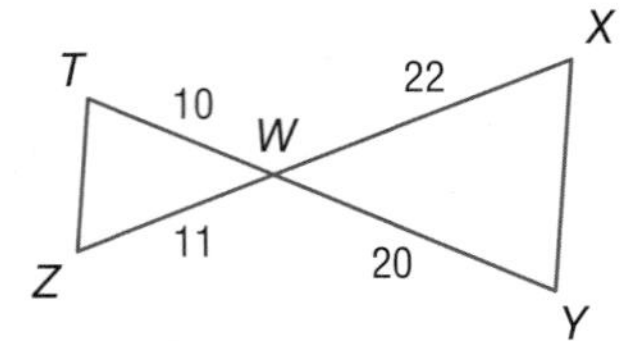


You can decide what is sufficient to prove that two triangles are similar.

## Standardized Test Example 3 Sufficient Conditions

**In the figure, $\angle ADB$ is a right angle. Which of the following would *not* be sufficient to prove that $\triangle ADB \sim \triangle CDB$?**

**A** $\frac{AD}{BD} = \frac{BD}{CD}$

**B** $\frac{AB}{BC} = \frac{BD}{CD}$

**C** $\angle ABD \cong \angle C$

**D** $\frac{AD}{BD} = \frac{BD}{CD} = \frac{AB}{BC}$

### Read the Test Item

You are given that $\angle ADB$ is a right angle and asked to identify which additional information would not be enough to prove that $\triangle ADB \sim \triangle CDB$.

### Solve the Test Item

Since $\angle ADB$ is a right angle, $\angle CDB$ is also a right angle. Since all right angles are congruent, $\angle ADB \cong \angle CDB$. Check each answer choice until you find one that does not supply a sufficient additional condition to prove that $\triangle ADB \sim \triangle CDB$.

**Choice A:** If $\frac{AD}{BD} = \frac{BD}{CD}$ and $\angle ADB \cong \angle CDB$, then $\triangle ADB \sim \triangle CDB$ by SAS Similarity.

**Choice B:** If $\frac{AB}{BC} = \frac{BD}{CD}$ and $\angle ADB \cong \angle CDB$, then we cannot conclude that $\triangle ADB \sim \triangle CDB$ because the included angle of side $\overline{AB}$ and $\overline{BD}$ is not $\angle ADB$. So the answer is B.

**Test-Taking Tip**

**Identifying Nonexamples** Sometimes test questions require you to find a nonexample, as in this case. You must check each option until you find a valid nonexample. If you would like to check your answer, confirm that each additional option is correct.

## GuidedPractice

3. If $\triangle JKL$ and $\triangle FGH$ are two triangles such that $\angle J \cong \angle F$, which of the following would be sufficient to prove that the triangles are similar?

**F** $\frac{KL}{GH} = \frac{JL}{FH}$ **G** $\frac{JL}{JK} = \frac{FH}{FG}$ **H** $\frac{JK}{FG} = \frac{KL}{GH}$ **J** $\frac{JL}{JK} = \frac{GH}{FG}$

## 2 Use Similar Triangles

Like the congruence of triangles, similarity of triangles is reflexive, symmetric, and transitive.

### Theorem 9.4 Properties of Similarity

| | |
|---|---|
| **Reflexive Property of Similarity** | $\triangle ABC \sim \triangle ABC$ |
| **Symmetric Property of Similarity** | If $\triangle ABC \sim \triangle DEF$, then $\triangle DEF \sim \triangle ABC$. |
| **Transitive Property of Similarity** | If $\triangle ABC \sim \triangle DEF$, and $\triangle DEF \sim \triangle XYZ$, then $\triangle ABC \sim \triangle XYZ$. |

You will prove Theorem 9.4 in Exercise 26.

### Example 4 Parts of Similar Triangles

**Find $BE$ and $AD$.**

Since $\overline{BE} \parallel \overline{CD}$, $\angle ABE \cong \angle BCD$, and $\angle AEB \cong \angle EDC$ because they are corresponding angles. By AA Similarity, $\triangle ABE \sim \triangle ACD$.

**StudyTip**

Proportions An additional proportion that is true for Example 4 is $\frac{AC}{CD} = \frac{AB}{BE}$.

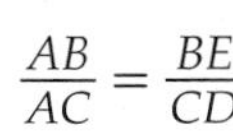

$\frac{AB}{AC} = \frac{BE}{CD}$ Definition of Similar Polygons

$\frac{3}{5} = \frac{x}{3.5}$ $AC = 5$, $CD = 3.5$, $AB = 3$, $BE = x$

$3.5 \cdot 3 = 5 \cdot x$ Cross Products Property

$2.1 = x$ $BE$ is 2.1.

$\frac{AC}{AB} = \frac{AD}{AE}$ Definition of Similar Polygons

$\frac{5}{3} = \frac{y+3}{y}$ $AC = 5$, $AB = 3$, $AD = y + 3$, $AE = y$

$5 \cdot y = 3(y + 3)$ Cross Products Property

$5y = 3y + 9$ Distributive Property

$2y = 9$ Subtract $3y$ from each side.

$y = 4.5$ $AD$ is $y + 3$ or 7.5.

## GuidedPractice

**Find each measure.**

**4A.** $QP$ and $MP$

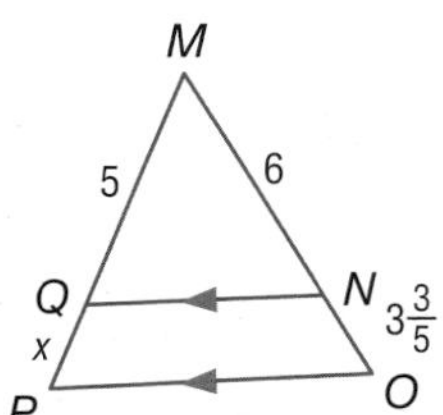


**4B.** $WR$ and $RT$

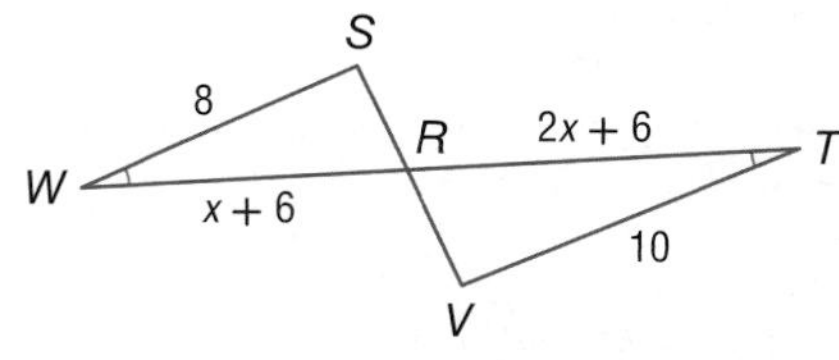

## Real-World Example 5 Indirect Measurement

**ROLLER COASTERS Hallie is estimating the height of the Superman roller coaster in Mitchellville, Maryland. She is 5 feet 3 inches tall and her shadow is 3 feet long. If the length of the shadow of the roller coaster is 40 feet, how tall is the roller coaster?**

**Understand** Make a sketch of the situation. 5 feet 3 inches is equivalent to 5.25 feet.

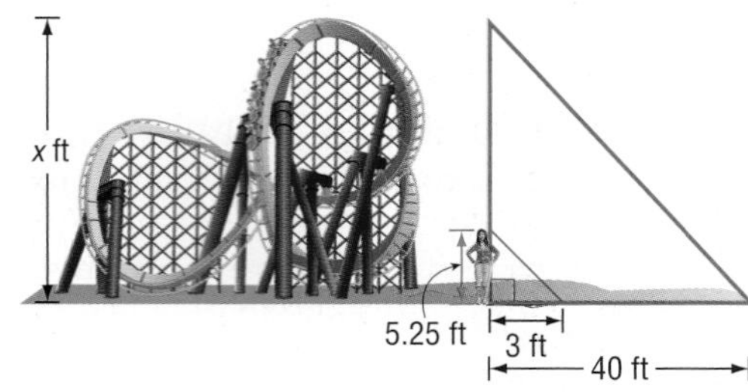


**Plan** In shadow problems, you can assume that the angles formed by the Sun's rays with any two objects are congruent and that the two objects form the sides of two right triangles.

Since two pairs of angles are congruent, the right triangles are similar by the AA Similarity Postulate. So, the following proportion can be written.

$$\frac{\text{Hallie's height}}{\text{coaster's height}} = \frac{\text{Hallie's shadow length}}{\text{coaster's shadow length}}$$

**Solve** Substitute the known values and let $x$ = roller coaster's height.

$\frac{5.25}{x} = \frac{3}{40}$ Substitution

$3 \cdot x = 40(5.25)$ Cross Products Property

$3x = 210$ Simplify.

$x = 70$ Divide each side by 3.

The roller coaster is 70 feet tall.

**Check** The roller coaster's shadow length is $\frac{40 \text{ ft}}{3 \text{ ft}}$ or about 13.3 times Hallie's shadow length. Check to see that the roller coaster's height is about 13.3 times Hallie's height. $\frac{70 \text{ ft}}{5.25 \text{ ft}} \approx 13.3$ ✓

**Problem-Solving Tip**

**Reasonable Answers** When you have solved a problem, check your answer for reasonableness. In this example, Hallie's shadow is a little more than half her height. The coaster's shadow is also a little more than half of the height you calculated. Therefore, the answer is reasonable.

### Guided Practice

5. **BUILDINGS** Adam is standing next to the Palmetto Building in Columbia, South Carolina. He is 6 feet tall and the length of his shadow is 9 feet. If the length of the shadow of the building is 322.5 feet, how tall is the building?

## Concept Summary Triangle Similarity

**AA Similarity Postulate**

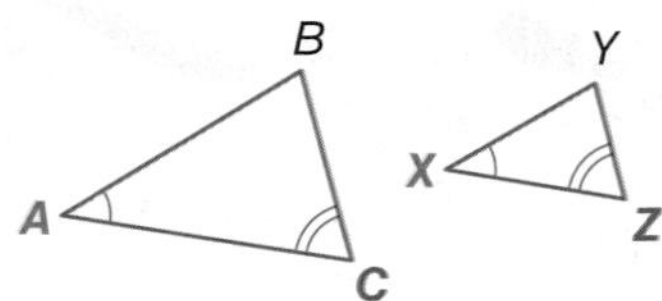


If $\angle A \cong \angle X$ and $\angle C \cong \angle Z$, then $\triangle ABC \sim \triangle XYZ$.

**SSS Similarity Theorem**

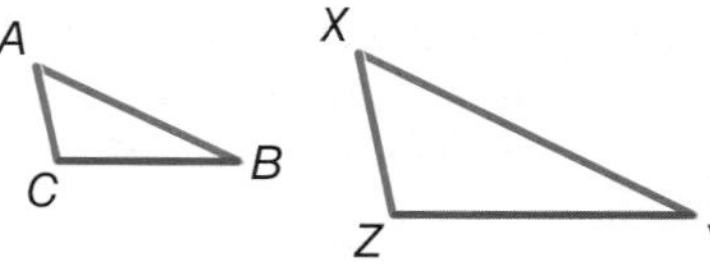


If $\frac{AB}{XY} = \frac{BC}{YZ} = \frac{CA}{ZX}$, then $\triangle ABC \sim \triangle XYZ$.

**SAS Similarity Theorem**

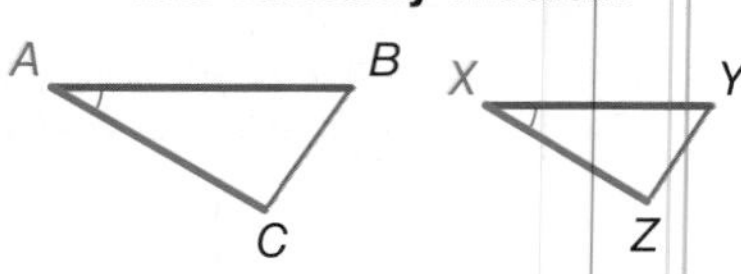


If $\angle A \cong \angle X$ and $\frac{AB}{XY} = \frac{CA}{ZX}$, then $\triangle ABC \sim \triangle XYZ$.

## Check Your Understanding

**Examples 1–2** **Determine whether the triangles are similar. If so, write a similarity statement. Explain your reasoning.**

**1.**

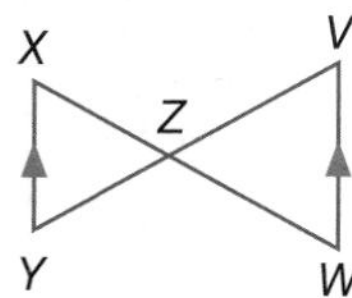


**2.**

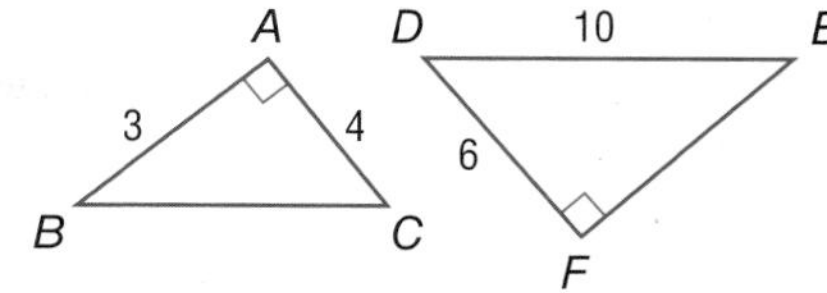


**3.**

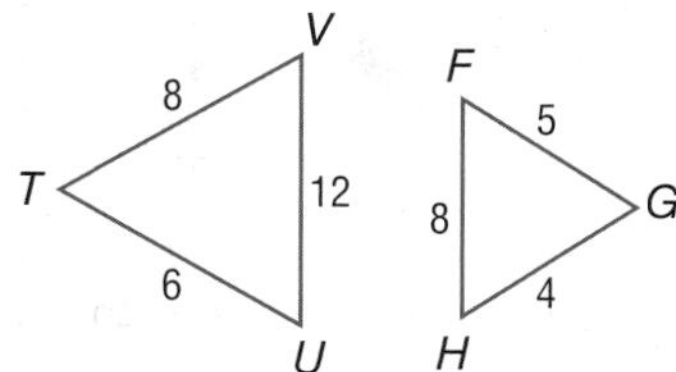


**4.**

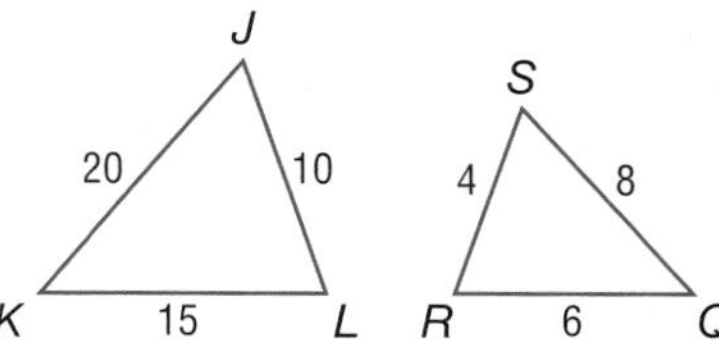


**Example 3**

**5. MULTIPLE CHOICE** In the figure, $\overline{AB}$ intersects $\overline{DE}$ at point $C$. Which additional information would be enough to prove that $\triangle ADC \sim \triangle BEC$?

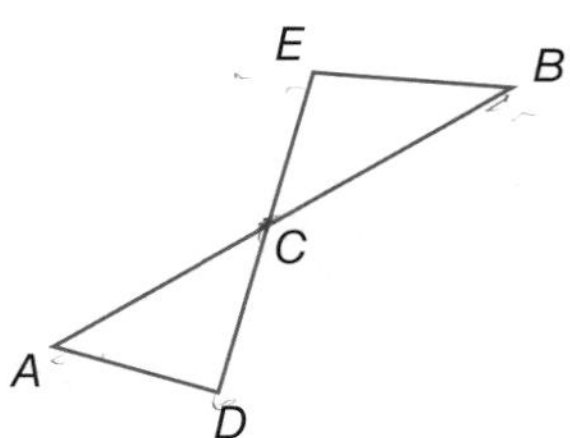


**A** $\angle DAC$ and $\angle ECB$ are congruent.

**B** $\overline{AC}$ and $\overline{BC}$ are congruent.

**C** $\overline{AD}$ and $\overline{EB}$ are parallel.

**D** $\angle CBE$ is a right angle.

**Example 4**

**CCSS STRUCTURE** **Identify the similar triangles. Find each measure.**

**6.** $KL$

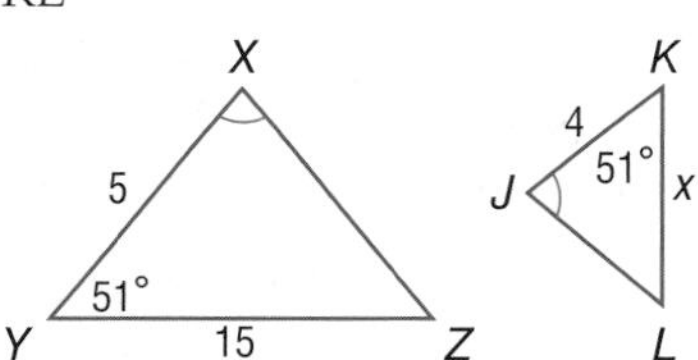


**7.** $VS$

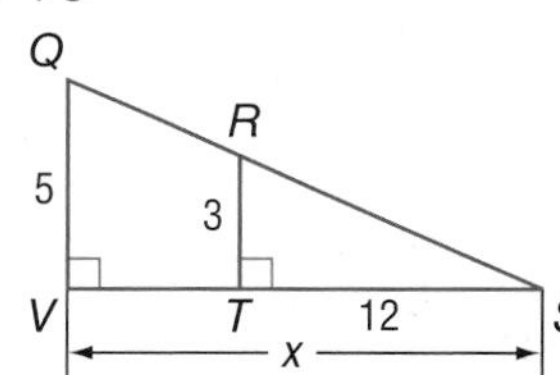


**Example 5**

**8. COMMUNICATION** A cell phone tower casts a 100-foot shadow. At the same time, a 4-foot 6-inch post near the tower casts a shadow of 3 feet 4 inches. Find the height of the tower.

## Practice and Problem Solving

**Examples 1–3** **Determine whether the triangles are similar. If so, write a similarity statement. If not, what would be sufficient to prove the triangles similar? Explain your reasoning.**

**9.**

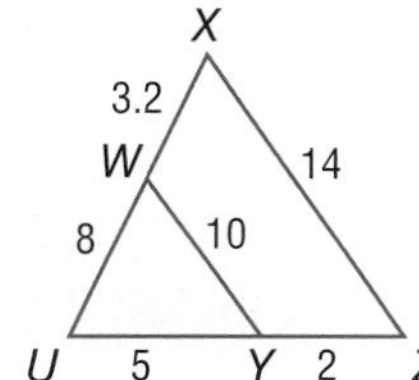


**10.**

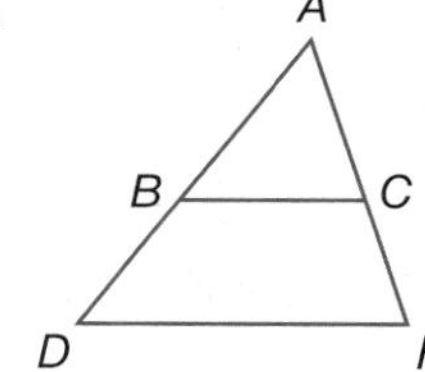


**11.**

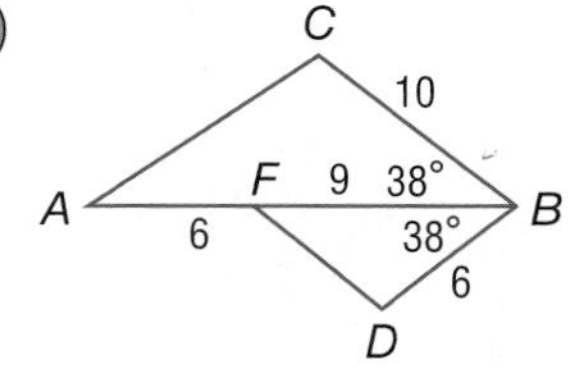

**Examples 1–3** **Determine whether the triangles are similar. If so, write a similarity statement. If not, what would be sufficient to prove the triangles similar? Explain your reasoning.**

**12.**

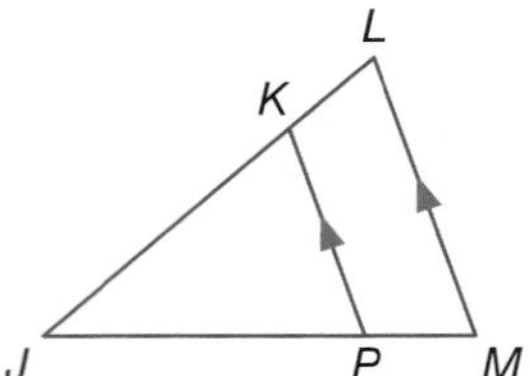


**13.**

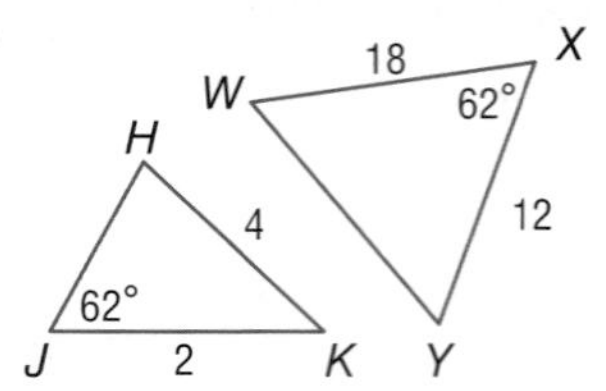


**14.**

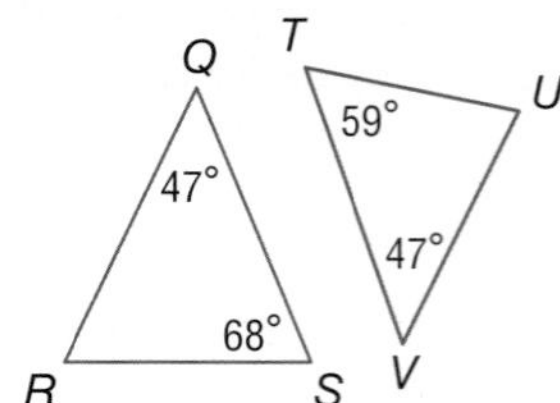


**15.** **CCSS MODELING** When we look at an object, it is projected on the retina through the pupil. The distances from the pupil to the top and bottom of the object are congruent and the distances from the pupil to the top and bottom of the image on the retina are congruent. Are the triangles formed between the object and the pupil and the object and the image similar? Explain your reasoning.

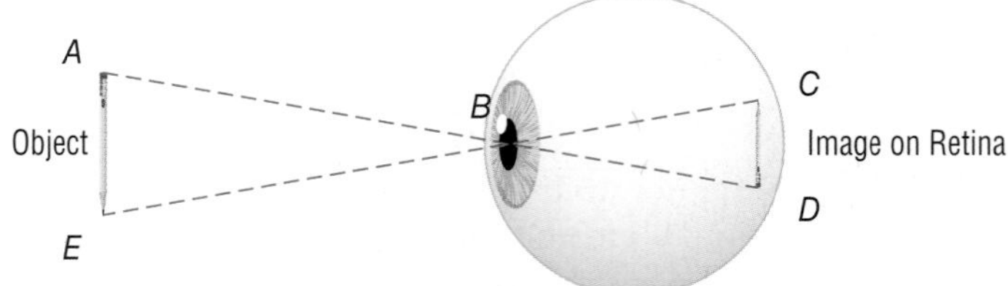


**Example 4** **ALGEBRA** **Identify the similar triangles. Then find each measure.**

**16.** *JK*

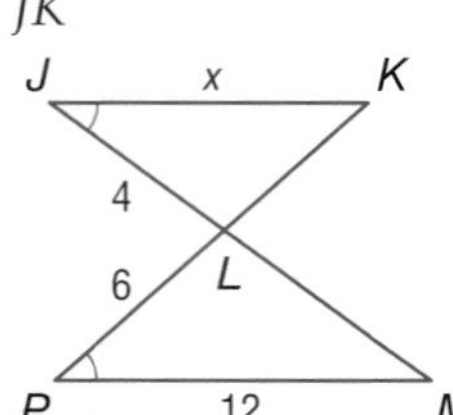


**17** *ST*

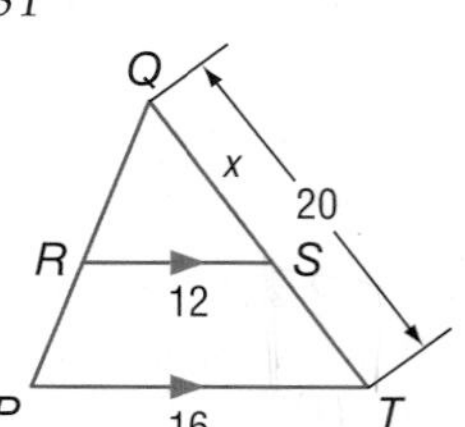


**18.** *WZ, UZ*

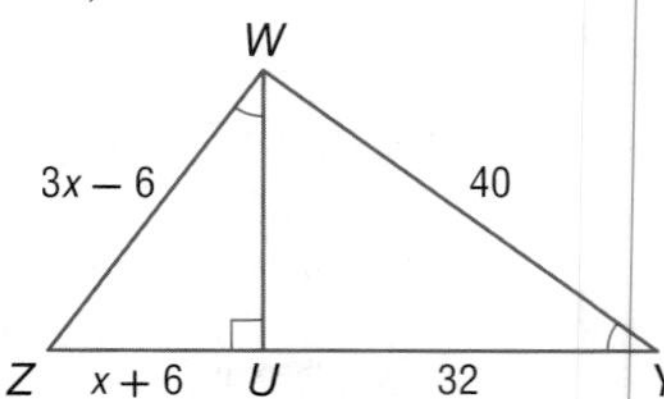


**19.** *HJ, HK*

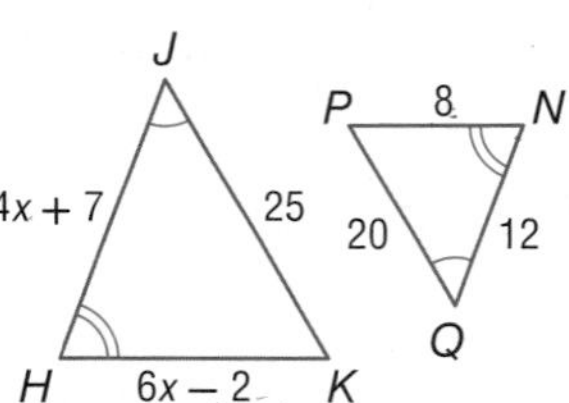


**20.** *DB, CB*

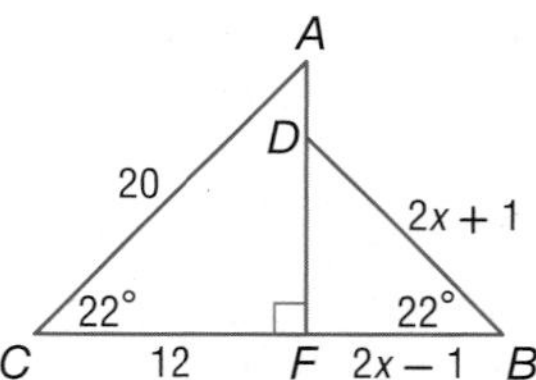


**21.** *GD, DH*

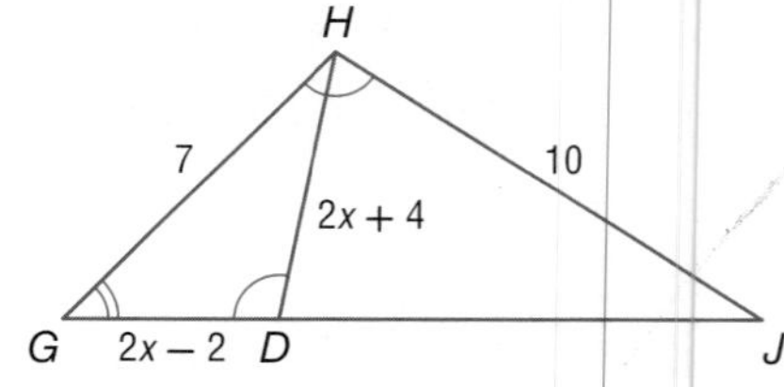


**Example 5**

**22.** **STATUES** Mei is standing next to a statue in the park. If Mei is 5 feet tall, her shadow is 3 feet long, and the statue's shadow is $10\frac{1}{2}$ feet long, how tall is the statue?

**23.** **SPORTS** When Alonzo, who is 5′11″ tall, stands next to a basketball goal, his shadow is 2′ long, and the basketball goal's shadow is 4′4″ long. About how tall is the basketball goal?

**24.** **FORESTRY** A hypsometer, as shown, can be used to estimate the height of a tree. Bartolo looks through the straw to the top of the tree and obtains the readings given. Find the height of the tree.

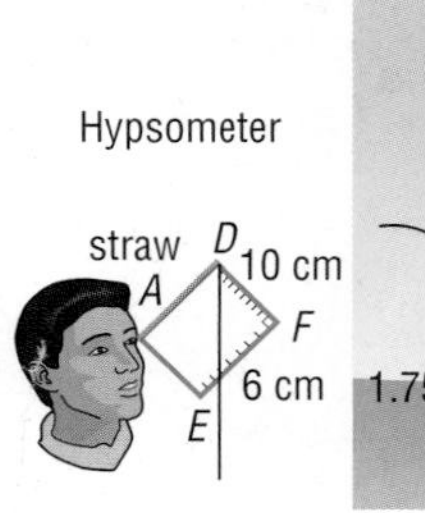


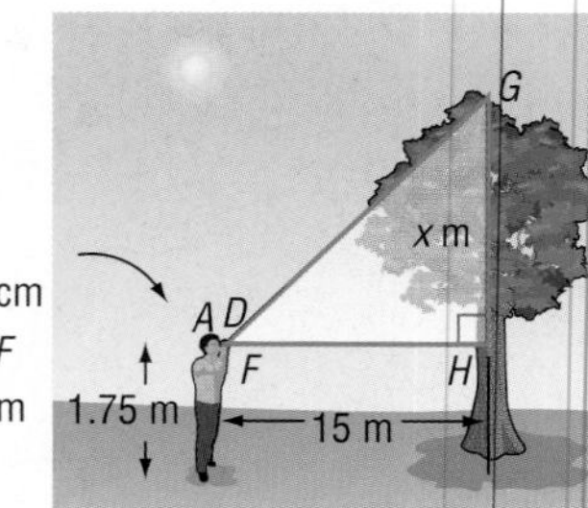


**PROOF** **Write a two-column proof.**

**25.** Theorem 9.3

**26.** Theorem 9.4

**PROOF** Write a two-column proof.

27. **Given:** $\triangle XYZ$ and $\triangle ABC$ are right triangles; $\frac{XY}{AB} = \frac{YZ}{BC}$.

**Prove:** $\triangle YXZ \sim \triangle BAC$

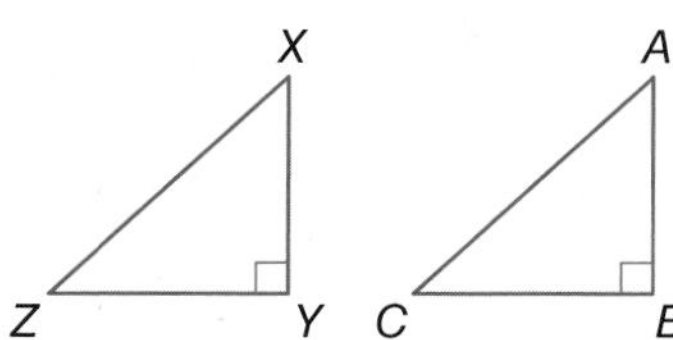


28. **Given:** $ABCD$ is a trapezoid.

**Prove:** $\frac{DP}{PB} = \frac{CP}{PA}$

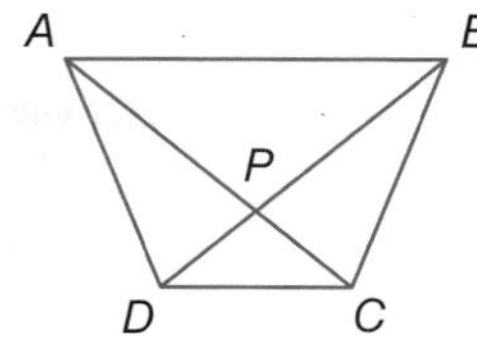


29. **CCSS MODELING** When Luis's dad threw a bounce pass to him, the angles formed by the basketball's path were congruent. The ball landed $\frac{2}{3}$ of the way between them before it bounced back up. If Luis's dad released the ball 40 inches above the floor, at what height did Luis catch the ball?

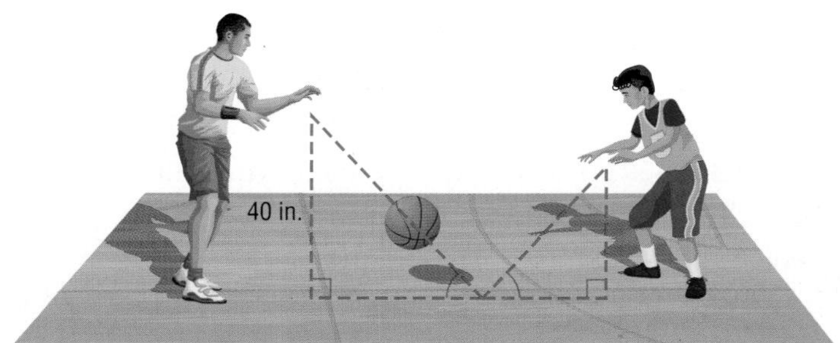


**COORDINATE GEOMETRY** $\triangle XYZ$ **and** $\triangle WYV$ **have vertices** $X(-1, -9)$, $Y(5, 3)$, $Z(-1, 6)$, $W(1, -5)$, **and** $V(1, 5)$.

30. Graph the triangles, and prove that $\triangle XYZ \sim \triangle WYV$.

31. Find the ratio of the perimeters of the two triangles.

32. **BILLIARDS** When a ball is deflected off a smooth surface, the angles formed by the path are congruent. Booker hit the orange ball and it followed the path from $A$ to $B$ to $C$ as shown below. What was the total distance traveled by the ball from the time Booker hit it until it came to rest at the end of the table?

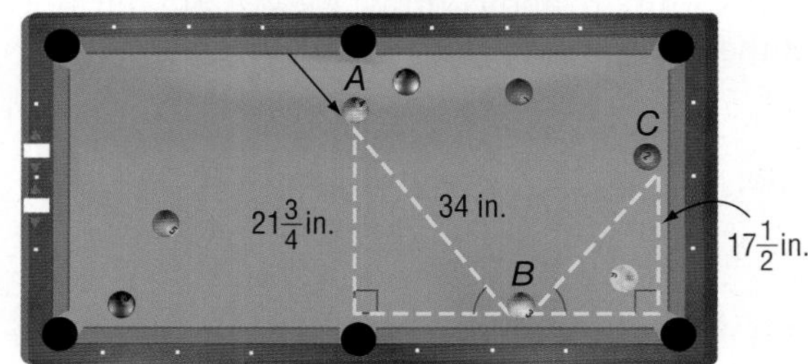


33. **PROOF** Use similar triangles to show that the slope of the line through any two points on that line is constant. That is, if points $A$, $B$, $A'$ and $B'$ are on line $\ell$, use similar triangles to show that the slope of the line from $A$ to $B$ is equal to the slope of the line from $A'$ to $B'$.

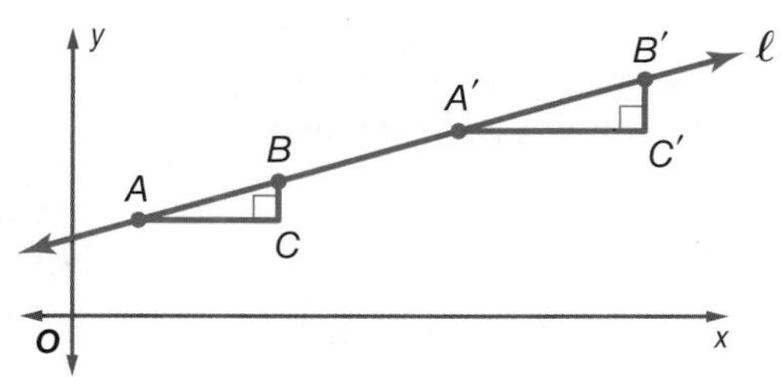

**34. CHANGING DIMENSIONS** Assume that $\triangle ABC \sim \triangle JKL$.

**a.** If the lengths of the sides of $\triangle JKL$ are half the length of the sides of $\triangle ABC$, and the area of $\triangle ABC$ is 40 square inches, what is the area of $\triangle JKL$? How is the area related to the scale factor of $\triangle ABC$ to $\triangle JKL$?

**b.** If the lengths of the sides of $\triangle ABC$ are three times the length of the sides of $\triangle JKL$, and the area of $\triangle ABC$ is 63 square inches, what is the area of $\triangle JKL$? How is the area related to the scale factor of $\triangle ABC$ to $\triangle JKL$?

**35. MEDICINE** Certain medical treatments involve laser beams that contact and penetrate the skin, forming similar triangles. Refer to the diagram at the right. How far apart should the laser sources be placed to ensure that the areas treated by each source do not overlap?

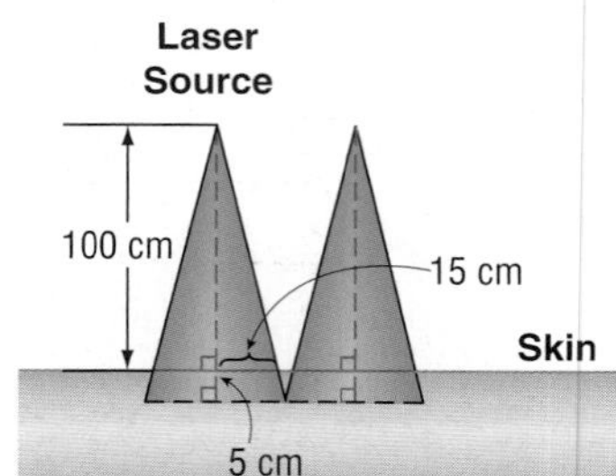


**36. MULTIPLE REPRESENTATIONS** In this problem, you will explore proportional parts of triangles.

**a. Geometric** Draw $\triangle ABC$ with $\overline{DE}$ parallel to $\overline{AC}$ as shown at the right.

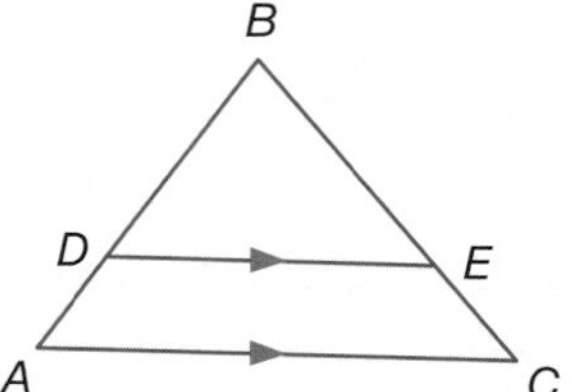


**b. Tabular** Measure and record the lengths $AD$, $DB$, $CD$, and $EB$ and the ratios $\frac{AD}{DB}$ and $\frac{CE}{EB}$ in a table.

**c. Verbal** Make a conjecture about the segments created by a line parallel to one side of a triangle and intersecting the other two sides.

## H.O.T. Problems Use Higher-Order Thinking Skills

**37. WRITING IN MATH** Compare and contrast the AA Similarity Postulate, the SSS Similarity Theorem, and the SAS similarity theorem.

**38. CHALLENGE** $\overline{YW}$ is an altitude of $\triangle XYZ$. Find $YW$.

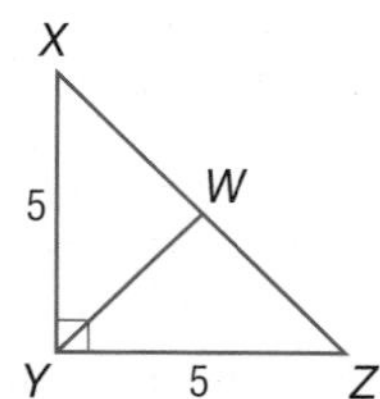


**39. CCSS REASONING** A pair of similar triangles has angle measures of 50°, 85°, and 45°. The sides of one triangle measure 3, 3.25, and 4.23 units, and the sides of the second triangle measure $x - 0.46$, $x$, and $x + 1.81$ units. Find the value of $x$.

**40. OPEN ENDED** Draw a triangle that is similar to $\triangle ABC$ shown. Explain how you know that it is similar.

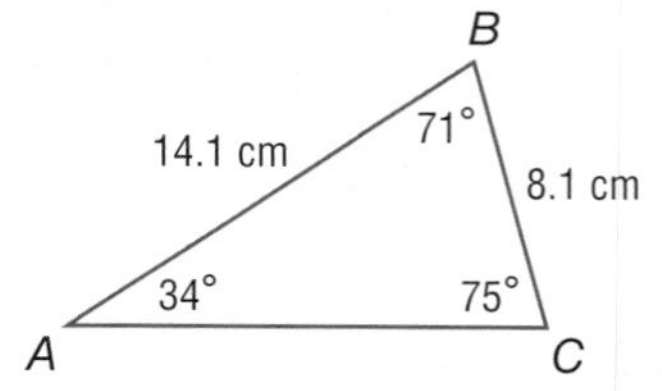


**41. WRITING IN MATH** How can you choose an appropriate scale?

## Standardized Test Practice

**42. PROBABILITY** $\frac{x!}{(x-3)!} =$

A 3.0
B 0.33
C $x^2 - 3x + 2$
D $x^3 - 3x^2 + 2x$

**43. EXTENDED RESPONSE** In the figure below, $\overline{EB} \parallel \overline{DC}$.

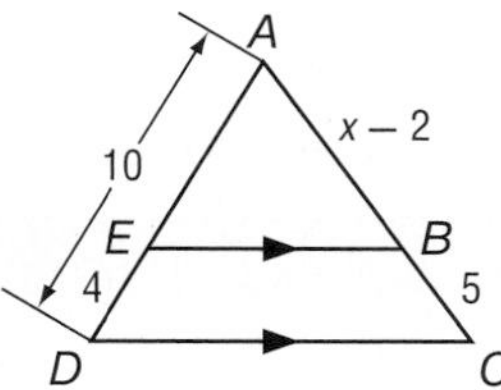


**a.** Write a proportion that could be used to find $x$.

**b.** Find the value of $x$ and the measure of $\overline{AB}$.

**44. ALGEBRA** Which polynomial represents the area of the shaded region?

F $\pi r^2$
G $\pi r^2 + r^2$
H $\pi r^2 + r$
J $\pi r^2 - r^2$

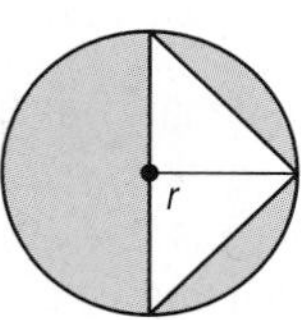


**45. SAT/ACT** The volume of a certain rectangular solid is $16x$ cubic units. If the dimensions of the solid are integers $x$, $y$, and $z$ units, what is the greatest possible value of $z$?

A 32
B 16
C 8
D 4
E 2

## Spiral Review

**List all pairs of congruent angles, and write a proportion that relates the corresponding sides for each pair of similar polygons.** (Lesson 9-2)

**46.** $\triangle JKL \sim \triangle CDE$

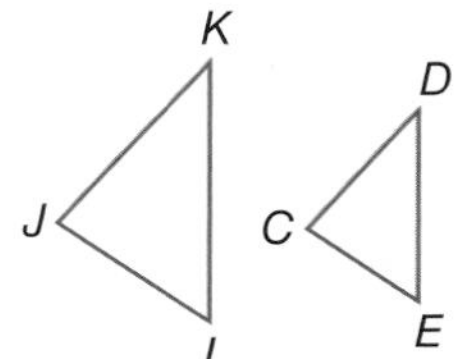


**47.** $WXYZ \sim QRST$

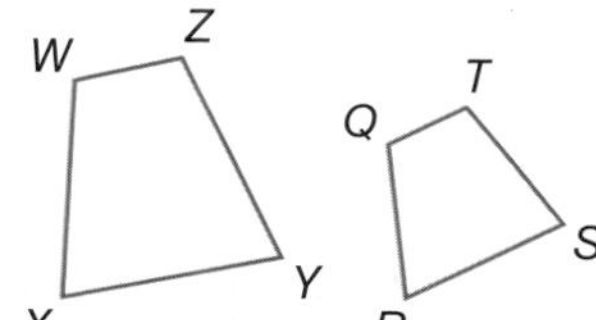


**48.** $FGHJ \sim MPQS$

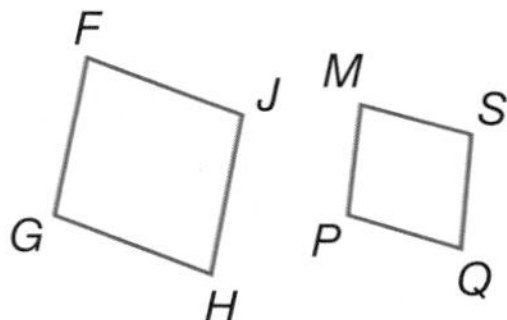


**Solve each proportion.** (Lesson 9-1)

**49.** $\frac{3}{4} = \frac{x}{16}$

**50.** $\frac{x}{10} = \frac{22}{50}$

**51.** $\frac{20.2}{88} = \frac{12}{x}$

**52.** $\frac{x-2}{2} = \frac{3}{8}$

**53. TANGRAMS** A tangram set consists of seven pieces: a small square, two small congruent right triangles, two large congruent right triangles, a medium-sized right triangle, and a quadrilateral. How can you determine the shape of the quadrilateral? Explain. (Lesson 8-3)

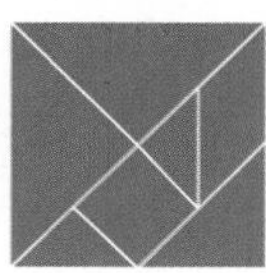

**Determine which postulate can be used to prove that the triangles are congruent. If it is not possible to prove congruence, write *not possible*.** (Lesson 6-3)

**54.**

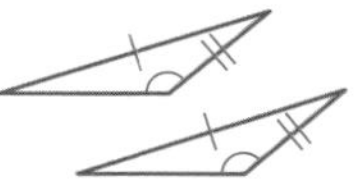

**55.**

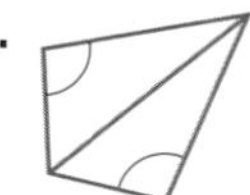

**56.**

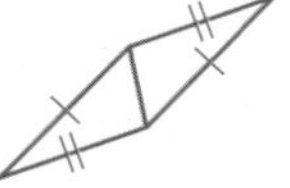

## Skills Review

**Write a two-column proof.**

**57. Given:** $r \parallel t$; $\angle 5 \cong \angle 6$

**Prove:** $\ell \parallel m$

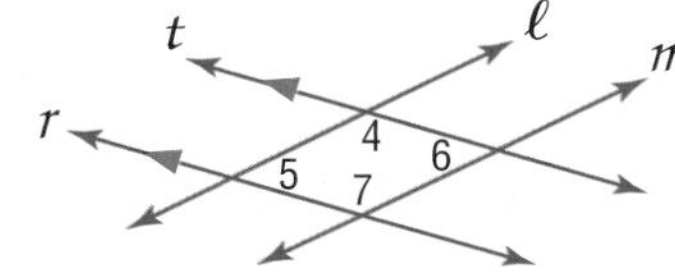

EXTEND 9-3

# Geometry Lab
# Proofs of Perpendicular and Parallel Lines

You have learned that two straight lines that are neither horizontal nor vertical are perpendicular if and only if the product of their slopes is –1. In this activity, you will use similar triangles to prove the first half of this theorem: if two straight lines are perpendicular, then the product of their slopes is –1.

**Common Core State Standards**
**Content Standards**
**G.GPE.5** Prove the slope criteria for parallel and perpendicular lines and use them to solve geometric problems (e.g., find the equation of a line parallel or perpendicular to a given line that passes through a given point).
**Mathematical Practices** 3

## Activity 1 Perpendicular Lines

**Given:** Slope of $\overleftrightarrow{AC} = m_1$, slope of $\overleftrightarrow{CE} = m_2$, and $\overleftrightarrow{AC} \perp \overleftrightarrow{CE}$.
**Prove:** $m_1 m_2 = -1$

**Step 1** On a coordinate plane, construct $\overleftrightarrow{AC} \perp \overleftrightarrow{CE}$ and transversal $\overleftrightarrow{BD}$ parallel to the $x$-axis through $C$. Then construct right $\triangle ABC$ such that $\overline{AC}$ is the hypotenuse and right $\triangle EDC$ such that $\overline{CE}$ is the hypotenuse. The legs of both triangles should be parallel to the $x$-and $y$-axes, as shown.

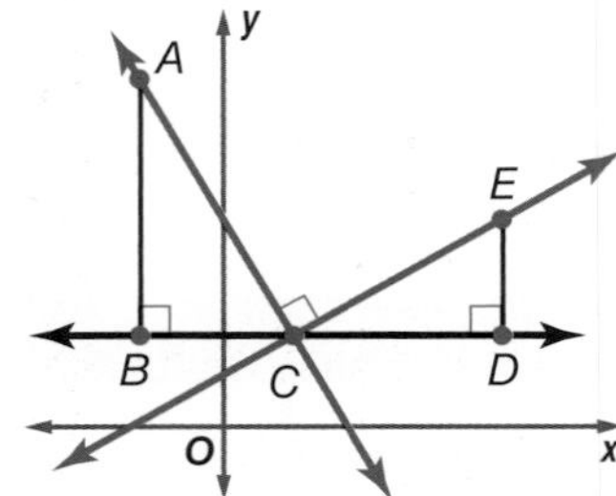


**Step 2** Find the slopes of $\overleftrightarrow{AC}$ and $\overleftrightarrow{CE}$.

| Slope of $\overleftrightarrow{AC}$ | | Slope of $\overleftrightarrow{CE}$ | |
|---|---|---|---|
| $m_1 = \frac{\text{rise}}{\text{run}}$ | Slope Formula | $m_2 = \frac{\text{rise}}{\text{run}}$ | Slope Formula |
| $= \frac{-AB}{BC}$ or $-\frac{AB}{BC}$ | rise = $-AB$, run = $BC$ | $= \frac{DE}{CD}$ | rise = $DE$, run = $CD$ |

**Step 3** Show that $\triangle ABC \sim \triangle CDE$.

Since $\triangle ACB$ is a right triangle with right angle $B$, $\angle BAC$ is complementary to $\angle ACB$. It is given that $\overleftrightarrow{AC} \perp \overleftrightarrow{CE}$, so we know that $\triangle ACE$ is a right angle. By construction, $\angle BCD$ is a straight angle. So, $\angle ECD$ is complementary to $\angle ACB$. Since angles complementary to the same angle are congruent, $\angle BAC \cong \angle ECD$. Since right angles are congruent, $\angle B \cong \angle D$. Therefore, by AA Similarity, $\triangle ABC \sim \triangle CDE$.

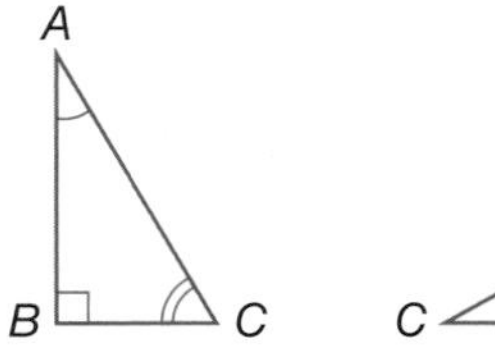

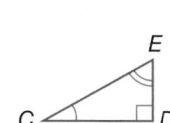

**Step 4** Use the fact that $\triangle ABC \sim \triangle CDE$ to show that $m_1 m_2 = -1$.

Since $m_1 = -\frac{AB}{BC}$ and $m_2 = \frac{DE}{CD}$, $m_1 m_2 = \left(-\frac{AB}{BC}\right)\left(\frac{DE}{CD}\right)$. Since two similar polygons have proportional sides, $\frac{AB}{BC} = \frac{CD}{DE}$. Therefore, by substitution, $m_1 m_2 = \left(-\frac{CD}{DE}\right)\left(\frac{DE}{CD}\right)$ or $-1$.

## Model

1. **PROOF** Use the diagram from Activity 1 to prove the second half of the theorem.

   **Given:** Slope of $\overleftrightarrow{CE} = m_1$, slope of $\overleftrightarrow{AC} = m_2$, and $m_1 m_2 = -1$. $\triangle ABC$ is a right triangle with right angle $B$. $\triangle CDE$ is a right triangle with right angle $D$.

   **Prove:** $\overleftrightarrow{CE} \perp \overleftrightarrow{AC}$

You can also use similar triangles to prove statements about parallel lines.

### Activity 2 Parallel Lines

**Given: Slope of $\overleftrightarrow{FG} = m_1$, slope of $\overleftrightarrow{JK} = m_2$, and $m_1 = m_2$. $\triangle FHG$ is a right triangle with right angle $H$. $\triangle JLK$ is a right triangle with right angle $L$.**

**Prove:** $\overleftrightarrow{FG} \parallel \overleftrightarrow{JK}$

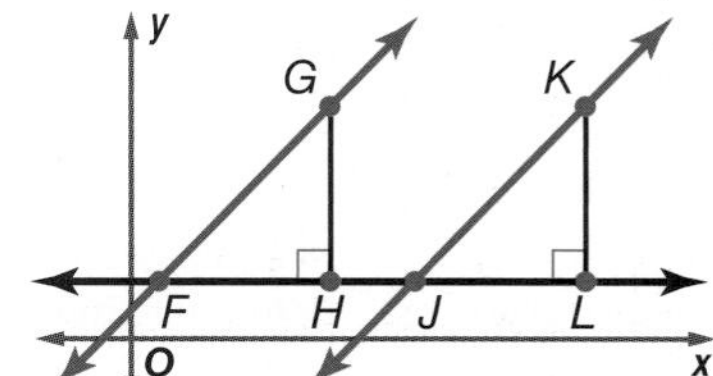


**Step 1** On a coordinate plane, construct $\overleftrightarrow{FG}$ and $\overleftrightarrow{JK}$, right $\triangle FHG$, and right $\triangle JLK$. Then draw horizontal transversal $\overleftrightarrow{FL}$, as shown.

**Step 2** Find the slopes of $\overleftrightarrow{FG}$ and $\overleftrightarrow{JK}$.

| Slope of $\overleftrightarrow{FG}$ | | Slope of $\overleftrightarrow{JK}$ | |
|---|---|---|---|
| $m_1 = \frac{\text{rise}}{\text{run}}$ | Slope Formula | $m_2 = \frac{\text{rise}}{\text{run}}$ | Slope Formula |
| $= \frac{GH}{HF}$ | rise = $GH$, run = $HF$ | $= \frac{KL}{LJ}$ | rise = $KL$, run = $LJ$ |

**Step 3** Show that $\triangle FHG \sim \triangle JLK$.

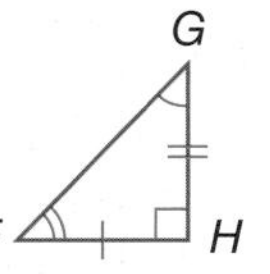

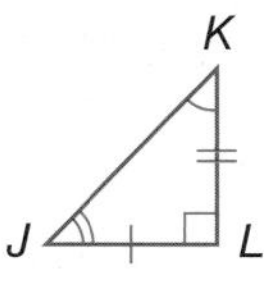


It is given that $m_1 = m_2$. By substitution, $\frac{GH}{HF} = \frac{KL}{LJ}$. This ratio can be rewritten as $\frac{GH}{KL} = \frac{HF}{LJ}$. Since $\angle H$ and $\angle L$ are right angles, $\angle H \cong \angle L$. Therefore, by SAS similarity, $\triangle FHG \sim \triangle JLK$.

**Step 4** Use the fact that $\triangle FHG \sim \triangle JLK$ to prove that $\overleftrightarrow{FG} \parallel \overleftrightarrow{JK}$.

Corresponding angles in similar triangles are congruent, so $\angle GFH \cong \angle KJL$. From the definition of congruent angles, $m\angle GFH = m\angle KJL$ (or $\angle GFH \cong \angle KJL$). By definition, $\angle KJH$ and $\angle KJL$ form a linear pair. Since linear pairs are supplementary, $m\angle KJH + m\angle KJL = 180$. So, by substitution, $m\angle KJH + m\angle GFH = 180$. By definition, $\angle KJH$ and $\angle GFH$ are supplementary. Since $\angle KJH$ and $\angle GFH$ are supplementary and are consecutive interior angles, $\overleftrightarrow{FG} \parallel \overleftrightarrow{JK}$.

## Model

2. **PROOF** Use the diagram from Activity 2 to prove the following statement.

   **Given:** Slope of $\overleftrightarrow{FG} = m_1$, slope of $\overleftrightarrow{JK} = m_2$, and $\overleftrightarrow{FG} \parallel \overleftrightarrow{JK}$.

   **Prove:** $m_1 = m_2$

# LESSON 9-4 Parallel Lines and Proportional Parts

**Then**

- You used proportions to solve problems between similar triangles.

**Now**

1. Use proportional parts within triangles.
2. Use proportional parts with parallel lines.

**Why?**

- Photographers have many techniques at their disposal that can be used to add interest to a photograph. One such technique is the use of a vanishing point perspective, in which an image with parallel lines, such as train tracks, is photographed so that the lines appear to converge at a point on the horizon.

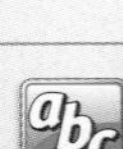

**NewVocabulary**
midsegment of a triangle

**Common Core State Standards**

**Content Standards**

G.SRT.4 Prove theorems about triangles.

G.SRT.5 Use congruence and similarity criteria for triangles to solve problems and to prove relationships in geometric figures.

**Mathematical Practices**

1 Make sense of problems and persevere in solving them.

3 Construct viable arguments and critique the reasoning of others.

## 1 Proportional Parts Within Triangles

When a triangle contains a line that is parallel to one of its sides, the two triangles formed can be proved similar using the Angle-Angle Similarity Postulate. Since the triangles are similar, their sides are proportional.

**Theorem 9.5 Triangle Proportionality Theorem**

If a line is parallel to one side of a triangle and intersects the other two sides, then it divides the sides into segments of proportional lengths.

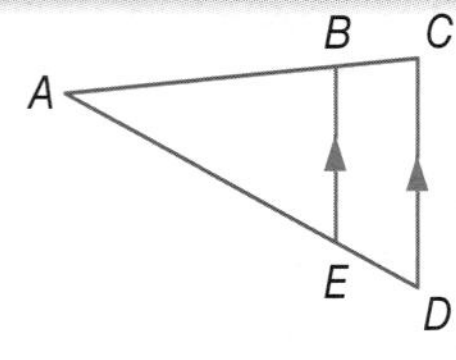

**Example** If $\overline{BE} \parallel \overline{CD}$, then $\frac{AB}{BC} = \frac{AE}{ED}$.

You will prove Theorem 9.5 in Exercise 30.

**Example 1 Find the Length of a Side**

**In $\triangle PQR$, $\overline{ST} \parallel \overline{RQ}$. If $PT = 7.5$, $TQ = 3$, and $SR = 2.5$, find $PS$.**

Use the Triangle Proportionality Theorem.

| | |
|---|---|
| $\frac{PS}{SR} = \frac{PT}{TQ}$ | Triangle Proportionality Theorem |
| $\frac{PS}{2.5} = \frac{7.5}{3}$ | Substitute. |
| $PS \cdot 3 = (2.5)(7.5)$ | Cross Products Property |
| $3PS = 18.75$ | Multiply. |
| $PS = 6.25$ | Divide each side by 3. |

P, T, S, Q, R

**GuidedPractice**

1. If $PS = 12.5$, $SR = 5$, and $PT = 15$, find $TQ$.

### Math HistoryLink

Galileo Galilei **(1564–1642)** Galileo was born in Pisa, Italy. He studied philosophy, astronomy, and mathematics. Galileo made essential contributions to all three disciplines. Refer to Exercise 39.

**Source:** *Encyclopaedia Britannica*

The converse of Theorem 9.5 is also true and can be proved using the proportional parts of a triangle.

## Theorem 9.6 Converse of Triangle Proportionality Theorem

If a line intersects two sides of a triangle and separates the sides into proportional corresponding segments, then the line is parallel to the third side of the triangle.

**Example** If $\frac{AE}{EB} = \frac{CD}{DB}$, then $\overline{AC} \parallel \overline{ED}$.

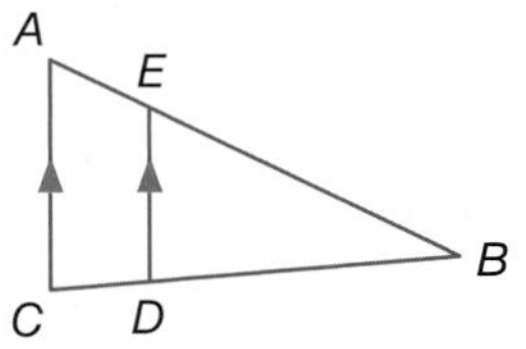


You will prove Theorem 9.6 in Exercise 31.

## Example 2 Determine if Lines are Parallel

**In $\triangle DEF$, $EH = 3$, $HF = 9$, and $DG$ is one-third the length of $\overline{GF}$. Is $\overline{DE} \parallel \overline{GH}$?**

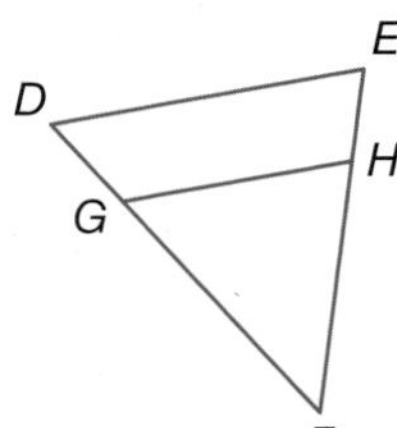


Using the converse of the Triangle Proportionality Theorem, in order to show that $\overline{DE} \parallel \overline{GH}$, we must show that $\frac{DG}{GF} = \frac{EH}{HF}$.

Find and simplify each ratio. Let $DG = x$.
Since $DG$ is one-third of $GF$, $GF = 3x$.

$\frac{DG}{GF} = \frac{x}{3x}$ or $\frac{1}{3}$ $\qquad$ $\frac{EH}{HF} = \frac{3}{9}$ or $\frac{1}{3}$

Since $\frac{1}{3} = \frac{1}{3}$, the sides are proportional, so $\overline{DE} \parallel \overline{GH}$.

### GuidedPractice

**2.** $DG$ is half the length of $\overline{GF}$, $EH = 6$, and $HF = 10$. Is $\overline{DE} \parallel \overline{GH}$?

### StudyTip

Midsegment Triangle The three midsegments of a triangle form the *midsegment triangle.*

A **midsegment of a triangle** is a segment with endpoints that are the midpoints of two sides of the triangle. Every triangle has three midsegments. The midsegments of $\triangle ABC$ are $\overline{RP}$, $\overline{PQ}$, $\overline{RQ}$.

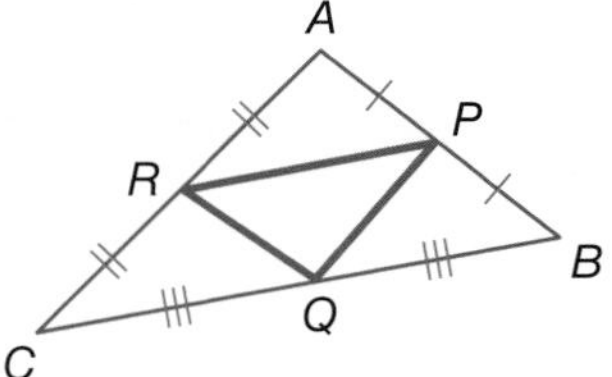


A special case of the Triangle Proportionality Theorem is the Triangle Midsegment Theorem.

## Theorem 9.7 Triangle Midsegment Theorem

A midsegment of a triangle is parallel to one side of the triangle, and its length is one half the length of that side.

**Example** If $J$ and $K$ are midpoints of $\overline{FH}$ and $\overline{HG}$, respectively, then $\overline{JK} \parallel \overline{FG}$ and $JK = \frac{1}{2}FG$.

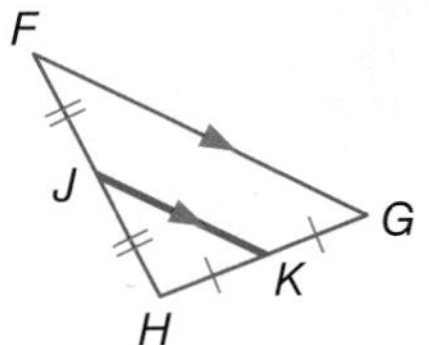


You will prove Theorem 9.7 in Exercise 32.

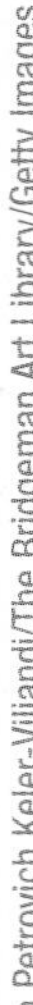

**StudyTip**

**Midsegment** The Triangle Midsegment Theorem is similar to the Trapezoid Midsegment Theorem, which states that the midsegment of a trapezoid is parallel to the bases and its length is one half the sum of the measures of the bases. (Lesson 8-6)

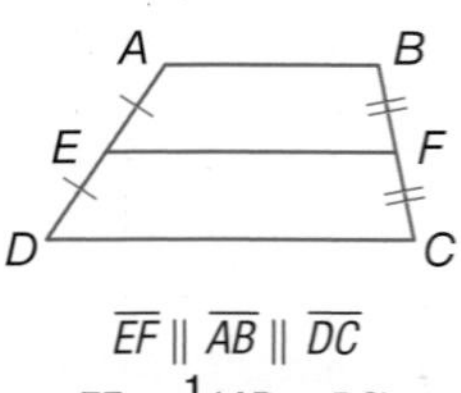


$\overline{EF} \parallel \overline{AB} \parallel \overline{DC}$

$EF = \frac{1}{2}(AB + DC)$

## Example 3 Use the Triangle Midsegment Theorem

**In the figure, $\overline{XY}$ and $\overline{XZ}$ are midsegments of $\triangle RST$. Find each measure.**

**a.** $XZ$

$XZ = \frac{1}{2}RT$ — Triangle Midsegment Theorem

$XZ = \frac{1}{2}(13)$ — Substitution

$XZ = 6.5$ — Simplify.

**b.** $ST$

$XY = \frac{1}{2}ST$ — Triangle Midsegment Theorem

$7 = \frac{1}{2}ST$ — Substitution

$14 = ST$ — Multiply each side by 2.

**c.** $m\angle RYX$

By the Triangle Midsegment Theorem, $\overline{XZ} \parallel \overline{RT}$.

$\angle RYX \cong \angle YXZ$ — Alternate Interior Angles Theorem

$m\angle RYX = m\angle YXZ$ — Definition of congruence

$m\angle RYX = 124$ — Substitution

### GuidedPractice

**Find each measure.**

**3A.** $DE$

**3B.** $DB$

**3C.** $m\angle FED$

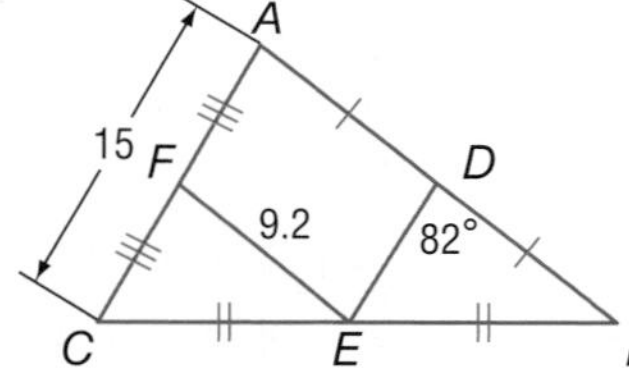


## 2 Proportional Parts with Parallel Lines

Another special case of the Triangle Proportionality Theorem involves three or more parallel lines cut by two transversals. Notice that if transversals $a$ and $b$ are extended, they form triangles with the parallel lines.

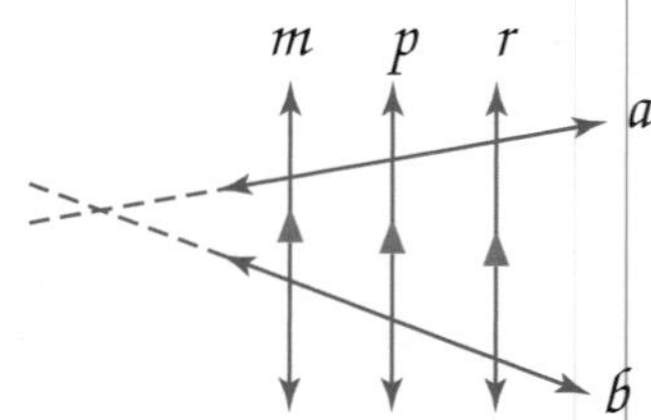


**StudyTip**

**Other Proportions** Two other proportions can be written for the example in Corollary 9.1.

$\frac{AB}{EF} = \frac{BC}{FG}$ and $\frac{AC}{BC} = \frac{EG}{FG}$

### Corollary 9.1 Proportional Parts of Parallel Lines

If three or more parallel lines intersect two transversals, then they cut off the transversals proportionally.

**Example** If $\overline{AE} \parallel \overline{BF} \parallel \overline{CG}$, then $\frac{AB}{BC} = \frac{EF}{FG}$.

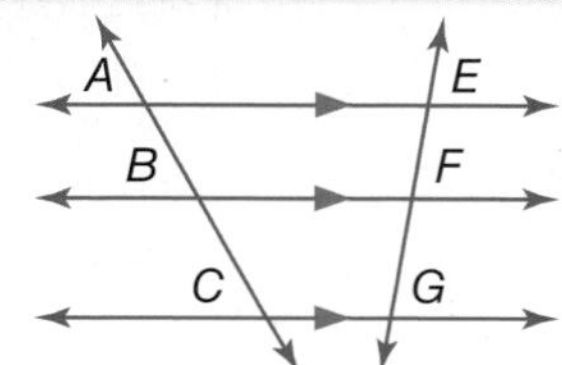


You will prove Corollary 9.1 in Exercise 28.

## Real-WorldLink

To make a two-dimensional drawing appear three-dimensional, an artist provides several perceptual cues.

- *size* - faraway items look smaller
- *clarity* - closer objects appear more in focus
- *detail* - nearby objects have texture, while distant ones are roughly outlined

**Source:** Center for Media Literacy

## Real-World Example 4 Use Proportional Segments of Transversals

**ART Megan is drawing a hallway in one-point perspective. She uses the guidelines shown to draw two windows on the left wall. If segments $\overline{AD}$, $\overline{BC}$, $\overline{WZ}$, and $\overline{XY}$ are all parallel, $AB = 8$ centimeters, $DC = 9$ centimeters, and $ZY = 5$ centimeters, find $WX$.**



By Corollary 9.1, if $\overline{AD} \parallel \overline{BC} \parallel \overline{WZ} \parallel \overline{XY}$,

then $\frac{AB}{WX} = \frac{DC}{ZY}$.

| | |
|---|---|
| $\frac{AB}{WX} = \frac{DC}{ZY}$ | Corollary 9.1 |
| $\frac{8}{WX} = \frac{9}{5}$ | Substitute. |
| $WX \cdot 9 = 8 \cdot 5$ | Cross Products Property |
| $9WX = 40$ | Simplify. |
| $WX = \frac{40}{9}$ | Divide each side by 4. |

The distance between $W$ and $X$ should be $\frac{40}{9}$ or about 4.4 centimeters.

**CHECK** The ratio of $DC$ to $ZY$ is 9 to 5, which is about 10 to 5 or 2 to 1. The ratio of $AB$ to $WX$ is 8 to 4.4 or about 8 to 4 or 2 to 1 as well, so the answer is reasonable. ✓

### GuidedPractice

4. **REAL ESTATE** *Frontage* is the measurement of a property's boundary that runs along the side of a particular feature such as a street, lake, ocean, or river. Find the ocean frontage for Lot A to the nearest tenth of a yard.

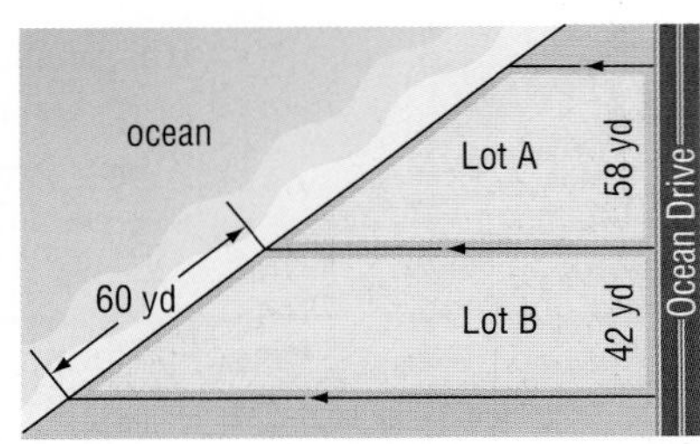


If the scale factor of the proportional segments is 1, they separate the transversals into congruent parts.

### Corollary 9.2 Congruent Parts of Parallel Lines

If three or more parallel lines cut off congruent segments on one transversal, then they cut off congruent segments on every transversal.

**Example** If $\overline{AE} \parallel \overline{BF} \parallel \overline{CG}$, and $\overline{AB} \cong \overline{BC}$, then $\overline{EF} \cong \overline{FG}$.

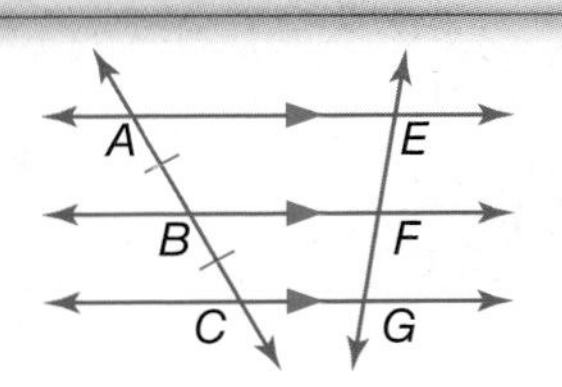


You will prove Corollary 9.2 in Exercise 29.

## Real-World Example 5 Use Congruent Segments of Transversals

**ALGEBRA Find $x$ and $y$.**

Since $\overleftrightarrow{JM} \parallel \overleftrightarrow{KP} \parallel \overleftrightarrow{LQ}$ and $\overline{MP} \cong \overline{PQ}$, then $\overline{JK} \cong \overline{KL}$ by Corollary 7.2.

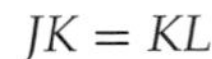

| | |
|---|---|
| $JK = KL$ | Definition of congruence |
| $6x - 5 = 4x + 3$ | Substitution |
| $2x - 5 = 3$ | Subtract $4x$ from each side. |
| $2x = 8$ | Add 5 to each side. |
| $x = 4$ | Divide each side by 2. |
| $MP = PQ$ | Definition of congruence |
| $3y + 8 = 5y - 7$ | Substitution |
| $8 = 2y - 7$ | Subtract $3y$ from each side. |
| $15 = 2y$ | Add 7 to each side. |
| $7.5 = y$ | Divide each side by 2. |

### GuidedPractice

**5A.**

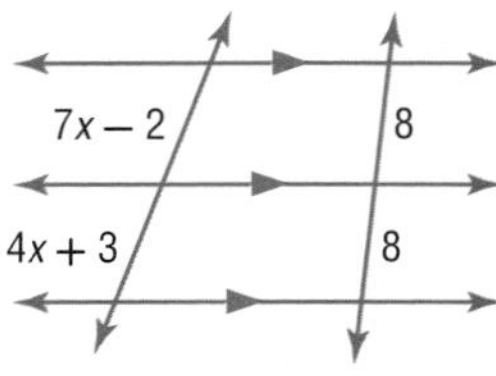


**5B.**

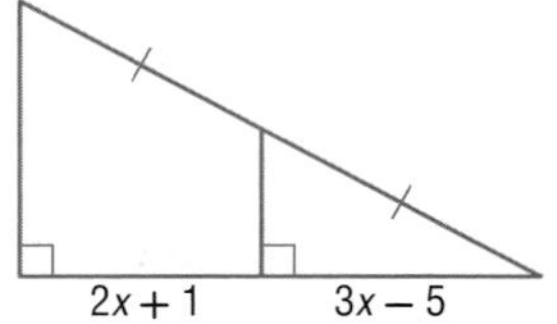


It is possible to separate a segment into two congruent parts by constructing the perpendicular bisector of a segment. However, a segment cannot be separated into three congruent parts by constructing perpendicular bisectors. To do this, you must use parallel lines and Corollary 9.2.

## Construction Trisect a Segment

**Draw a segment $\overline{AB}$. Then use Corollary 9.2 to trisect $\overline{AB}$.**

**Step 1** Draw $\overline{AC}$. Then with the compass at $A$, mark off an arc that intersects $\overline{AC}$ at $X$.

**Step 2** Use the same compass setting to mark off $Y$ and $Z$ such that $\overline{AX} \cong \overline{XY} \cong \overline{YZ}$. Then draw $ZB$.

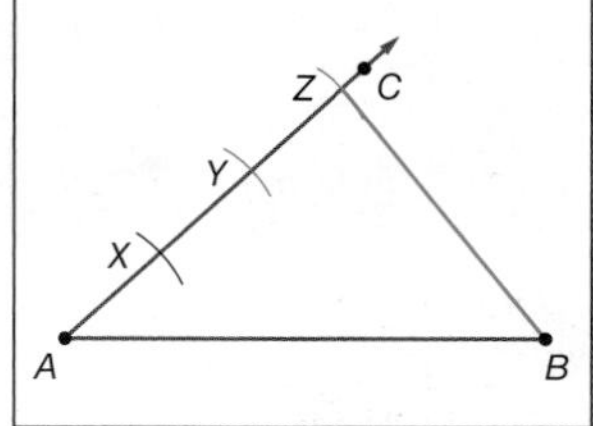


**Step 3** Construct lines through $Y$ and $X$ that are parallel to $\overline{ZB}$. Label the intersection points on $\overline{AB}$ as $J$ and $K$.

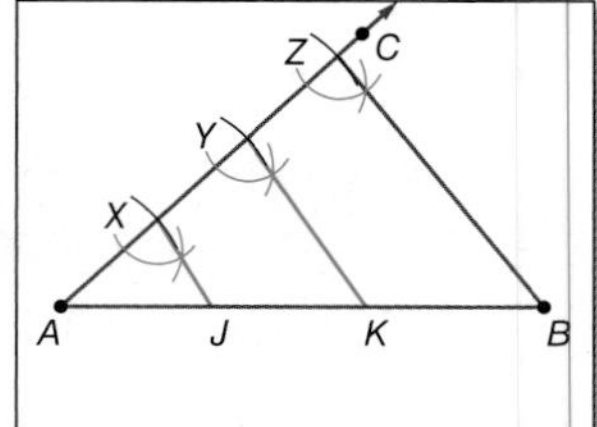


**Conclusion**: Since parallel lines cut off congruent segments on transversals, $\overline{AJ} \cong \overline{JK} \cong \overline{KB}$.

## Check Your Understanding

**Example 1**

1. If $XM = 4$, $XN = 6$, and $NZ = 9$, find $XY$.
2. If $XN = 6$, $XM = 2$, and $XY = 10$, find $NZ$.

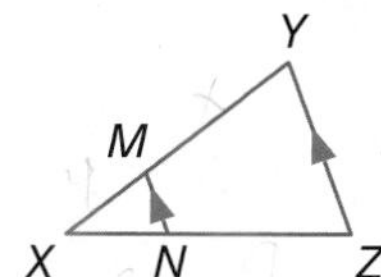

**Example 2**

3. In $\triangle ABC$, $BC = 15$, $BE = 6$, $DC = 12$, and $AD = 8$. Determine whether $\overline{DE} \parallel \overline{AB}$. Justify your answer.

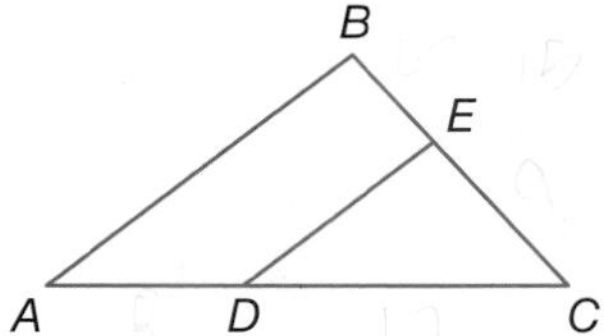

4. In $\triangle JKL$, $JK = 15$, $JM = 5$, $LK = 13$, and $PK = 9$. Determine whether $\overline{JL} \parallel \overline{MP}$. Justify your answer.

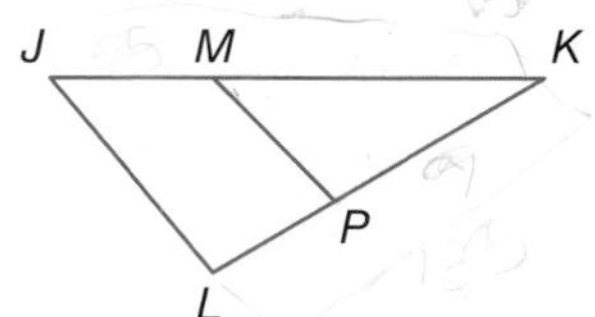

**Example 3** $\overline{JH}$ **is a midsegment of** $\triangle KLM$**. Find the value of** $x$**.**

5.

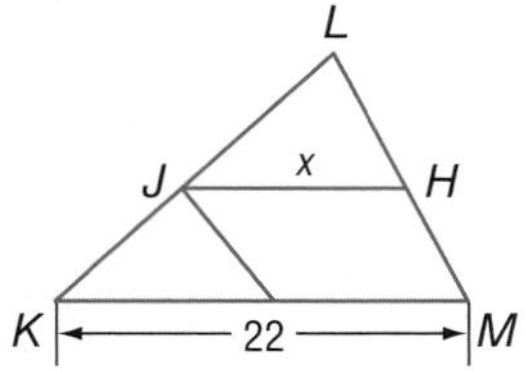

6.

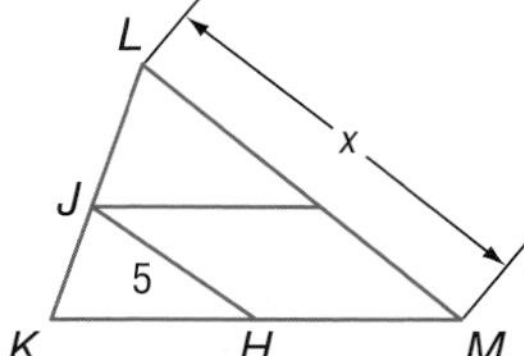

**Example 4**

7. **MAPS** Refer to the map at the right. 3rd Avenue and 5th Avenue are parallel. If the distance from 3rd Avenue to City Mall along State Street is 3201 feet, find the distance between 5th Avenue and City Mall along Union Street. Round to the nearest tenth.

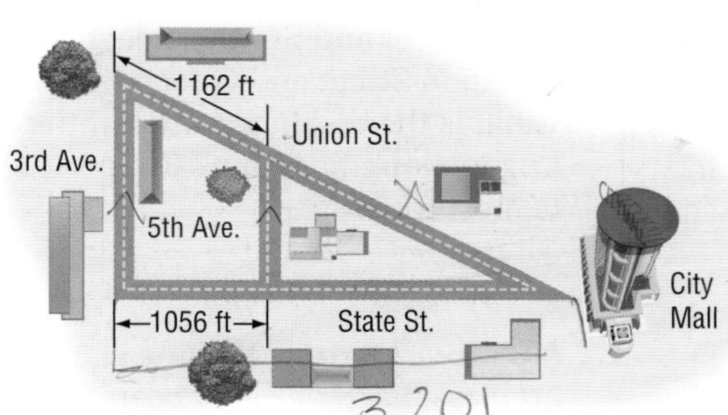


**Example 5** **ALGEBRA** **Find** $x$ **and** $y$**.**

8.

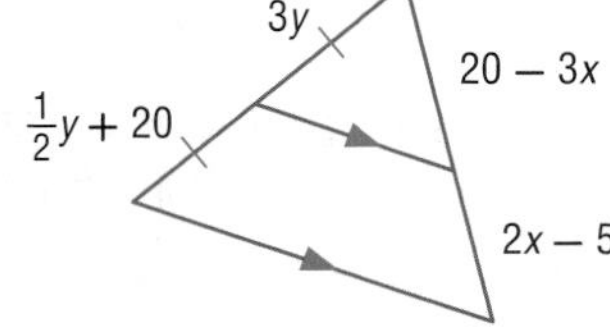

9.

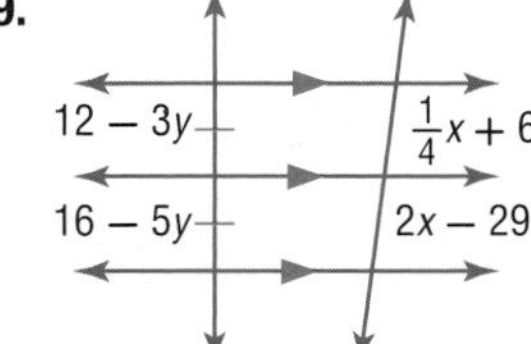

## Practice and Problem Solving

**Example 1**

10. If $AB = 6$, $BC = 4$, and $AE = 9$, find $ED$.
11. If $AB = 12$, $AC = 16$, and $ED = 5$, find $AE$.
12. If $AC = 14$, $BC = 8$, and $AD = 21$, find $ED$.
13. If $AD = 27$, $AB = 8$, and $AE = 12$, find $BC$.

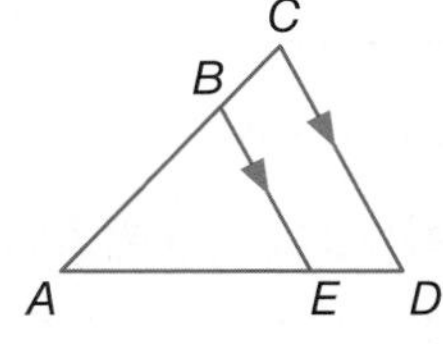

**Example 2** **Determine whether $\overline{VY} \parallel \overline{ZW}$. Justify your answer.**

14. $ZX = 18$, $ZV = 6$, $WX = 24$, and $YX = 16$
15. $VX = 7.5$, $ZX = 24$, $WY = 27.5$, and $WX = 40$
16. $ZV = 8$, $VX = 2$, and $YX = \frac{1}{2}WY$
17. $WX = 31$, $YX = 21$, and $ZX = 4ZV$

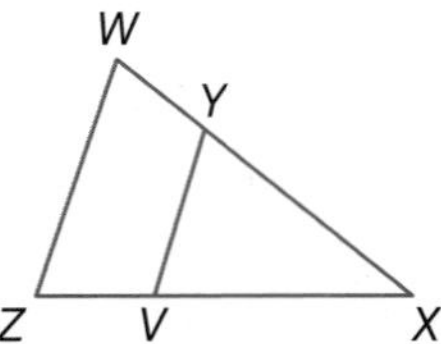


**Example 3** **$\overline{JH}$, $\overline{JP}$, and $\overline{PH}$ are midsegments of $\triangle KLM$. Find the value of $x$.**

18. 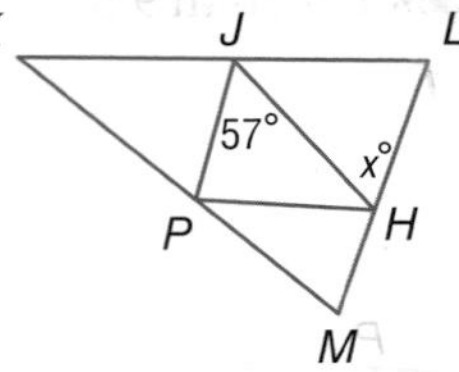


19. 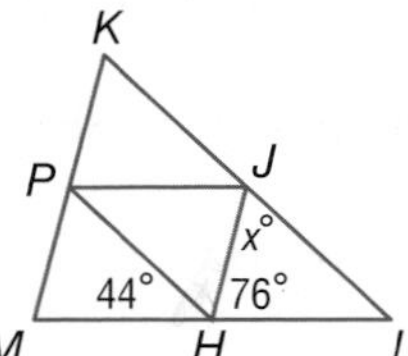


20. 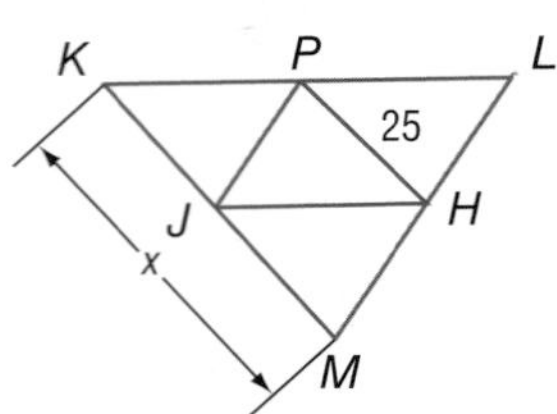


21. 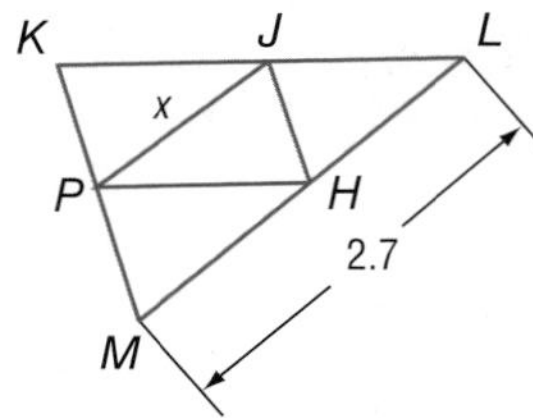


**Example 4**

22. **CCSS MODELING** In Charleston, South Carolina, Logan Street is parallel to both King Street and Smith Street between Beaufain Street and Queen Street. What is the distance from Smith to Logan along Beaufain? Round to the nearest foot.

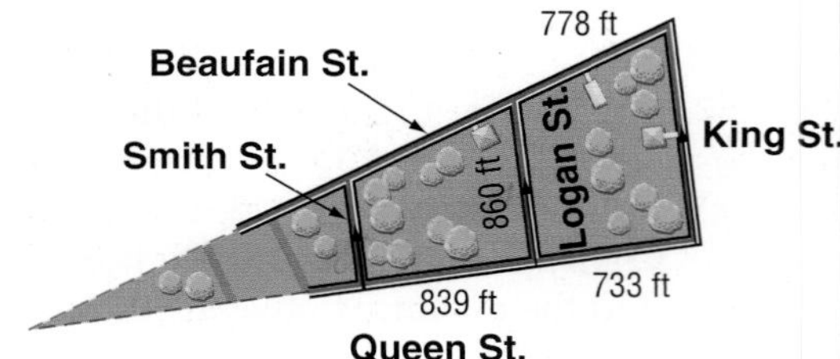


23. **ART** Tonisha drew the line of dancers shown below for her perspective project in art class. Each of the dancers is parallel. Find the lower distance between the first two dancers.

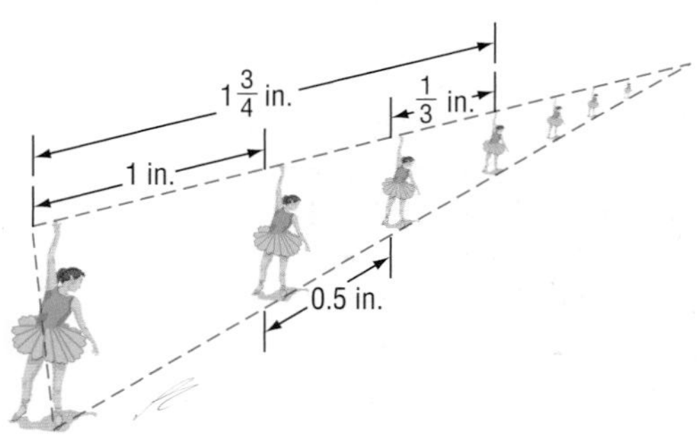


**Example 5** **ALGEBRA Find $x$ and $y$.**

24. 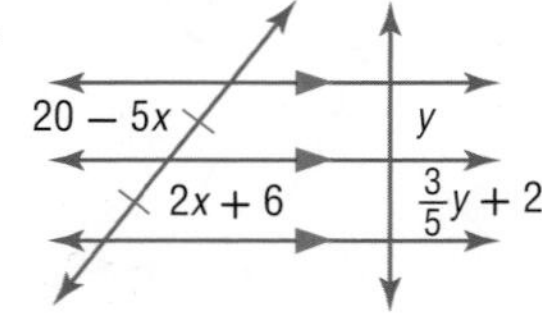


25. 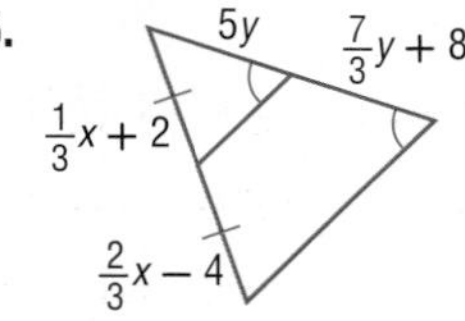

**ALGEBRA** **Find $x$ and $y$.**

**26.**

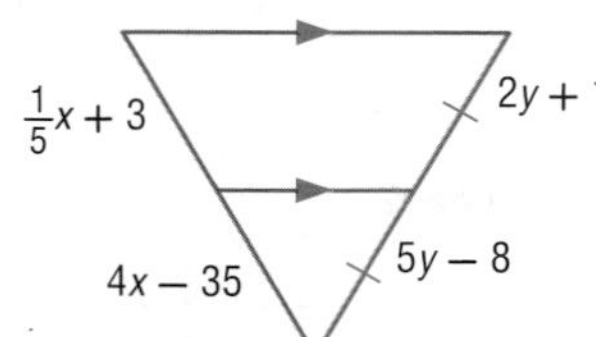


**27.**

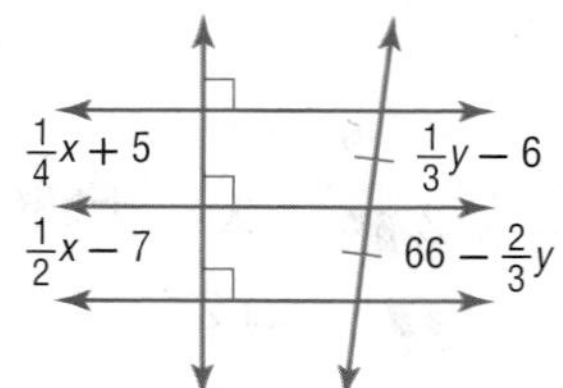


**CCSS ARGUMENTS** **Write a paragraph proof.**

**28.** Corollary 9.1

**29.** Corollary 9.2

**30.** Theorem 9.5

**CCSS ARGUMENTS** **Write a two-column proof.**

**31.** Theorem 9.6

**32.** Theorem 9.7

**Refer to $\triangle QRS$.**

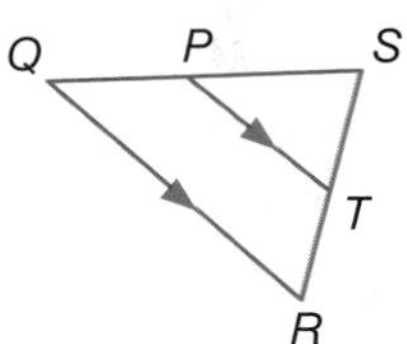


**33.** If $ST = 8$, $TR = 4$, and $PT = 6$, find $QR$.

**34.** If $SP = 4$, $PT = 6$, and $QR = 12$, find $SQ$.

**35** If $CE = t - 2$, $EB = t + 1$, $CD = 2$, and $CA = 10$, find $t$ and $CE$.

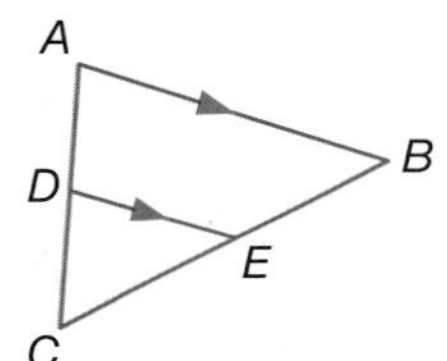


**36.** If $WX = 7$, $WY = a$, $WV = 6$, and $VZ = a - 9$, find $WY$.

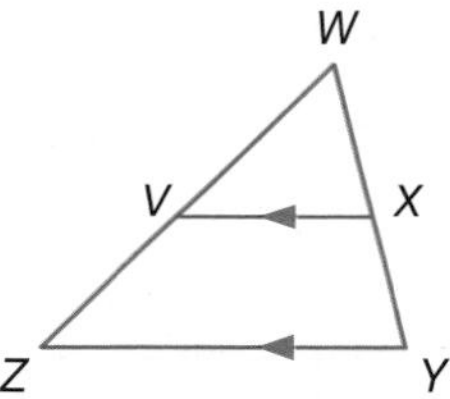


**37.** If $QR = 2$, $XW = 12$, $QW = 15$, and $ST = 5$, find $RS$ and $WV$.

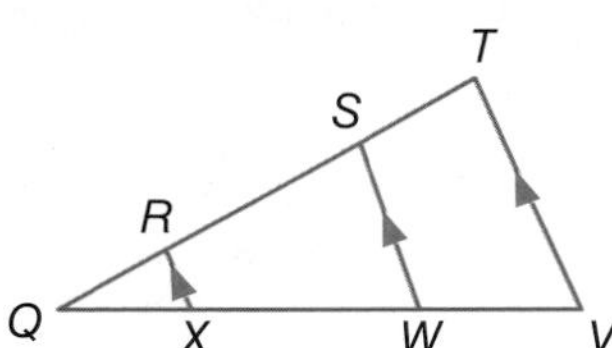


**38.** If $LK = 4$, $MP = 3$, $PQ = 6$, $KJ = 2$, $RS = 6$, and $LP = 2$, find $ML$, $QR$, $QK$, and $JH$.

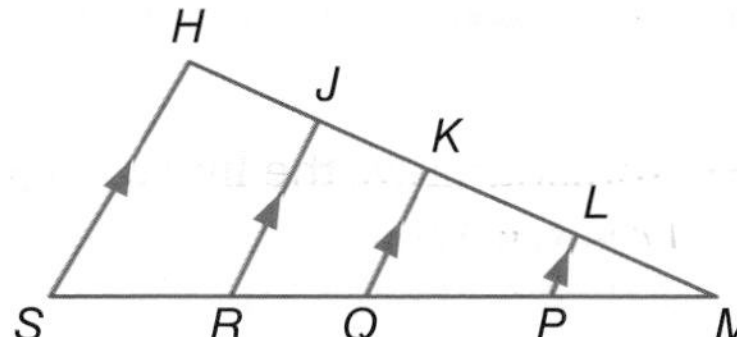


**39.** **MATH HISTORY** The sector compass was a tool perfected by Galileo in the sixteenth century for measurement. To draw a segment two-fifths the length of a given segment, align the ends of the arms with the given segment. Then draw a segment at the 40 mark. Write a justification that explains why the sector compass works for proportional measurement.

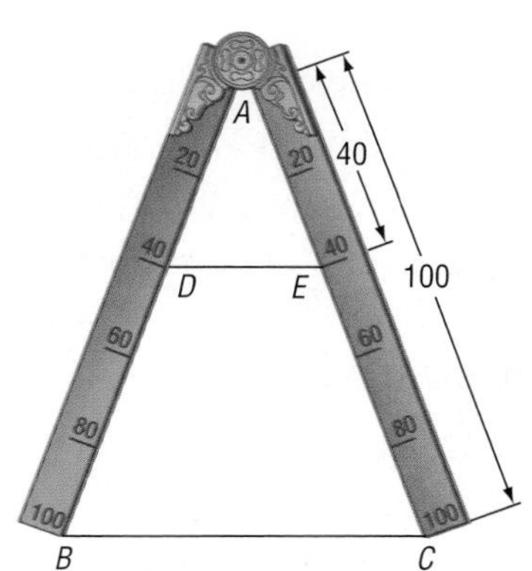


**Determine the value of $x$ so that $\overline{BC} \parallel \overline{DF}$.**

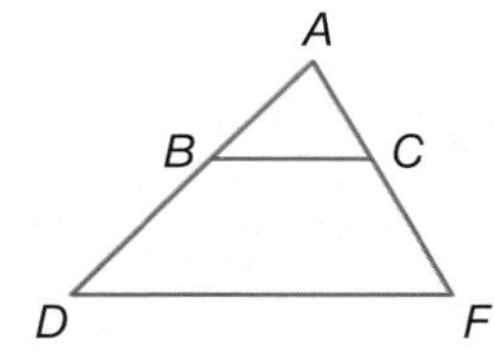


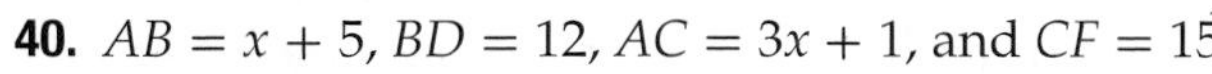

**40.** $AB = x + 5$, $BD = 12$, $AC = 3x + 1$, and $CF = 15$

**41.** $AC = 15$, $BD = 3x - 2$, $CF = 3x + 2$, and $AB = 12$

42. **COORDINATE GEOMETRY** $\triangle ABC$ has vertices $A(-8, 7)$, $B(0, 1)$, and $C(7, 5)$. Draw $\triangle ABC$. Determine the coordinates of the midsegment of $\triangle ABC$ that is parallel to $\overline{BC}$. Justify your answer.

43. **HOUSES** Refer to the diagram of the gable at the right. Each piece of siding is a uniform width. Find the lengths of $\overline{FG}$, $\overline{EH}$, and $\overline{DJ}$.

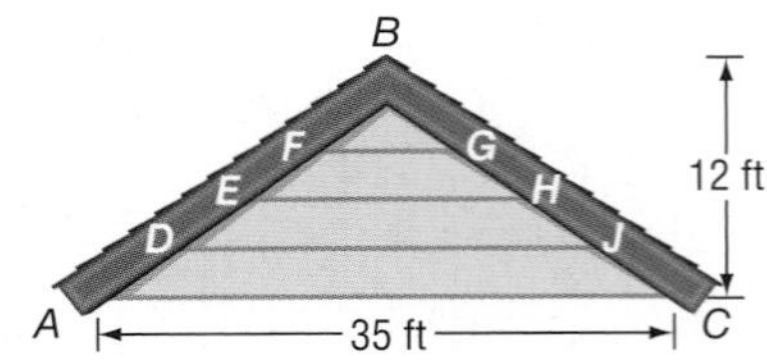


**CONSTRUCTIONS** **Construct each segment as directed.**

44. a segment separated into five congruent segments

45. a segment separated into two segments in which their lengths have a ratio of 1 to 3

46. a segment 3 inches long, separated into four congruent segments

47. **MULTIPLE REPRESENTATIONS** In this problem, you will explore angle bisectors and proportions.

a. **Geometric** Draw three triangles, one acute, one right, and one obtuse. Label one triangle *ABC* and draw angle bisector $\overrightarrow{BD}$. Label the second *MNP* with angle bisector $\overrightarrow{NQ}$ and the third *WXY* with angle bisector $\overrightarrow{XZ}$.

b. **Tabular** Copy and complete the table at the right with the appropriate values.

c. **Verbal** Make a conjecture about the segments of a triangle created by an angle bisector.

| Triangle | Length | | Ratio | |
|---|---|---|---|---|
| ABC | AD | | $\frac{AD}{CD}$ | |
| | CD | | | |
| | AB | | $\frac{AB}{CB}$ | |
| | CB | | | |
| MNP | MQ | | $\frac{MQ}{PQ}$ | |
| | PQ | | | |
| | MN | | $\frac{MN}{PN}$ | |
| | PN | | | |
| WXY | WZ | | $\frac{WZ}{YZ}$ | |
| | YZ | | | |
| | WX | | $\frac{WX}{YX}$ | |
| | YX | | | |

## H.O.T. Problems Use Higher-Order Thinking Skills

48. **CCSS CRITIQUE** Jacob and Sebastian are finding the value of *x* in $\triangle JHL$. Jacob says that *MP* is one half of *JL*, so *x* is 4.5. Sebastian says that *JL* is one half of *MP*, so *x* is 18. Is either of them correct? Explain.

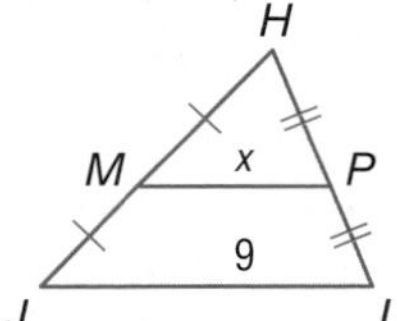


49. **REASONING** In $\triangle ABC$, $AF = FB$ and $AH = HC$. If *D* is $\frac{3}{4}$ of the way from *A* to *B* and *E* is $\frac{3}{4}$ of the way from *A* to *C*, is *DE always, sometimes,* or *never* $\frac{3}{4}$ of *BC*? Explain.

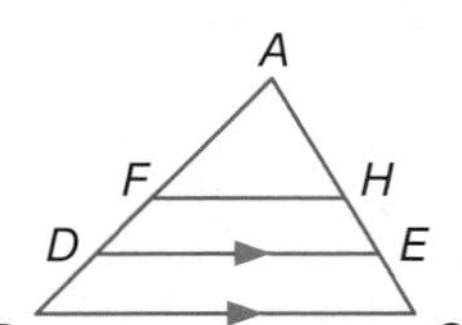


50. **CHALLENGE** Write a two-column proof.

**Given:** $AB = 4$, $BC = 4$, and $CD = DE$

**Prove:** $\overline{BD} \parallel \overline{AE}$

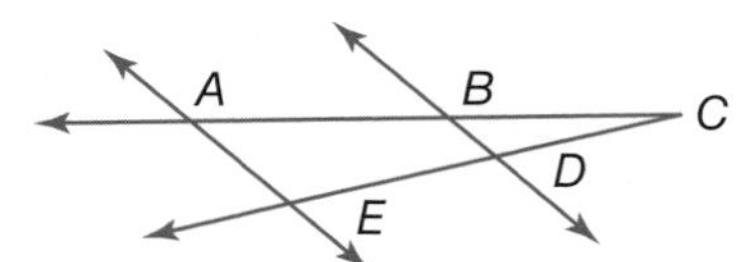


51. **OPEN ENDED** Draw three segments, *a*, *b*, and *c*, of all different lengths. Draw a fourth segment, *d*, such that $\frac{a}{b} = \frac{c}{d}$.

52. **WRITING IN MATH** Compare the Triangle Proportionality Theorem and the Triangle Midsegment Theorem.

## Standardized Test Practice

**53. SHORT RESPONSE** What is the value of $x$?

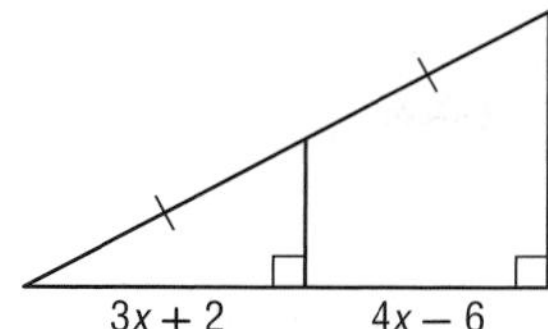


**54.** If the vertices of triangle $JKL$ are (0, 0), (0, 10) and (10, 10), then the area of triangle $JKL$ is

**A** 20 units$^2$  **C** 40 units$^2$
**B** 30 units$^2$  **D** 50 units$^2$

**55. ALGEBRA** A breakfast cereal contains wheat, rice, and oats in the ratio 2:4:1. If the manufacturer makes a mixture using 110 pounds of wheat, how many pounds of rice will be used?

**F** 120 lb  **H** 240 lb
**G** 220 lb  **J** 440 lb

**56. SAT/ACT** If the area of a circle is 16 square meters, what is its radius in meters?

**A** $\frac{4\sqrt{\pi}}{\pi}$  **D** $12\pi$
**B** $\frac{8}{\pi}$  **E** $16\pi$
**C** $\frac{16}{\pi}$

## Spiral Review

**ALGEBRA Identify the similar triangles. Then find the measure(s) of the indicated segment(s).** (Lesson 9-3)

**57.** $\overline{AB}$

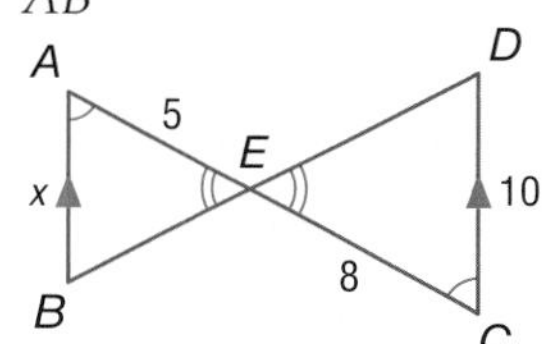


**58.** $\overline{RT}$, $\overline{RS}$

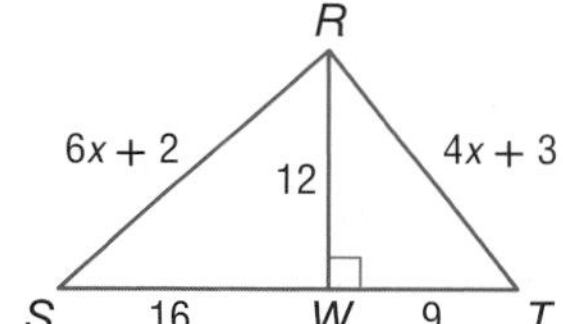


**59.** $\overline{TY}$

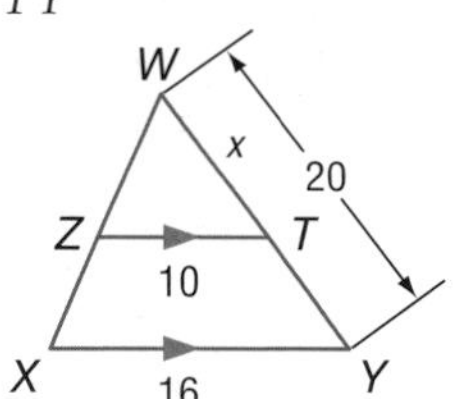


**60. SURVEYING** Mr. Turner uses a carpenter's square to find the distance across a stream. The carpenter's square models right angle $NOL$. He puts the square on top of a pole that is high enough to sight along $\overline{OL}$ to point $P$ across the river. Then he sights along $\overline{ON}$ to point $M$. If $MK$ is 1.5 feet and $OK$ is 4.5 feet, find the distance $KP$ across the stream. (Lesson 9-2)

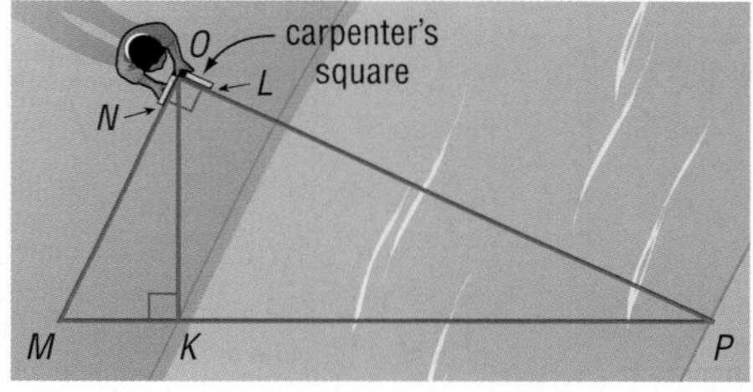


**COORDINATE GEOMETRY For each quadrilateral with the given vertices, verify that the quadrilateral is a trapezoid and determine whether the figure is an isosceles trapezoid.** (Lesson 8-6)

**61.** $Q(-12, 1)$, $R(-9, 4)$, $S(-4, 3)$, $T(-11, -4)$

**62.** $A(-3, 3)$, $B(-4, -1)$, $C(5, -1)$, $D(2, 3)$

**Point $S$ is the incenter of $\triangle JPL$. Find each measure.** (Lesson 7-1)

**63.** $SQ$

**64.** $QJ$

**65.** $m\angle MPQ$

**66.** $m\angle SJP$

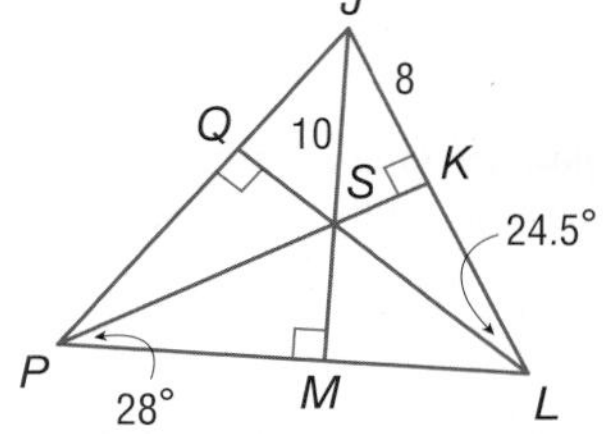


## Skills Review

**Solve each proportion.**

**67.** $\frac{1}{3} = \frac{x}{2}$

**68.** $\frac{3}{4} = \frac{5}{x}$

**69.** $\frac{2.3}{4} = \frac{x}{3.7}$

**70.** $\frac{x-2}{2} = \frac{4}{5}$

**71.** $\frac{x}{12-x} = \frac{8}{3}$

# Mid-Chapter Quiz

## Lessons 9-1 through 9-4

**Solve each proportion.** (Lesson 9-1)

**1.** $\frac{2}{5} = \frac{x}{25}$

**2.** $\frac{10}{3} = \frac{7}{x}$

**3.** $\frac{y+4}{11} = \frac{y-2}{9}$

**4.** $\frac{z-1}{3} = \frac{8}{z+1}$

**5. BASEBALL** A pitcher's earned run average, or ERA, is the product of 9 and the ratio of earned runs the pitcher has allowed to the number of innings pitched. During the 2007 season, Johan Santana of the Minnesota Twins allowed 81 earned runs in 219 innings pitched. Find his ERA to the nearest hundredth. (Lesson 9-1)

**Each pair of polygons is similar. Find the value of *x*.** (Lesson 9-2)

**6.**

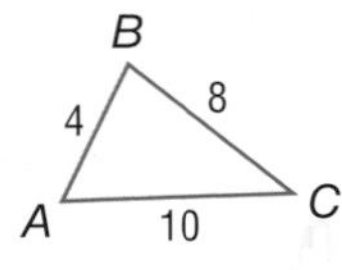


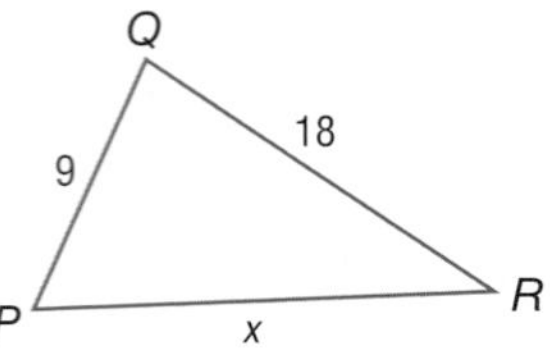


**7.**

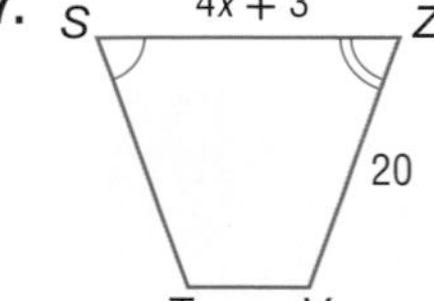


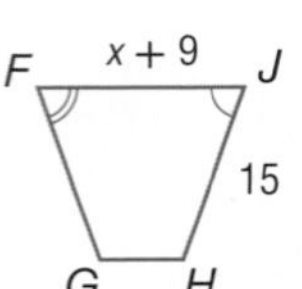


**8. MULTIPLE CHOICE** Two similar polygons have a scale factor of 3:5. The perimeter of the larger polygon is 120 feet. Find the perimeter of the smaller polygon. (Lesson 9-2)

**A** 68 ft

**B** 72 ft

**C** 192 ft

**D** 200 ft

**Determine whether the triangles are similar. If so, write a similarity statement. If not, what would be sufficient to prove the triangles similar? Explain your reasoning.** (Lesson 9-3)

**9.**

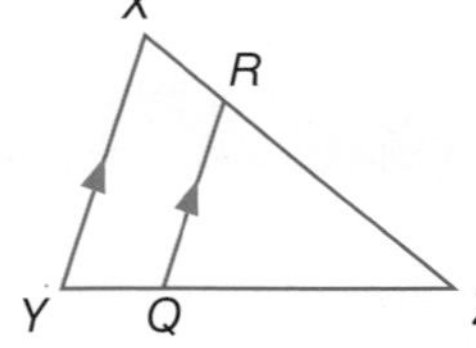


**10.**

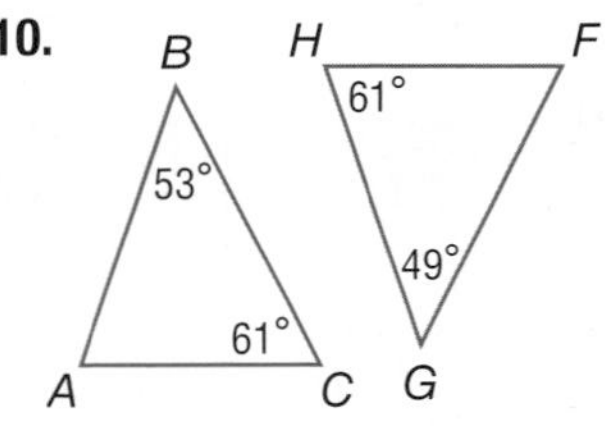


**ALGEBRA Identify the similar triangles. Find each measure.** (Lesson 9-3)

**11.** *SR*

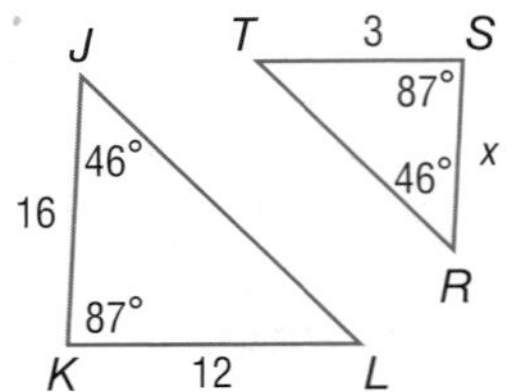


**12.** *AF*

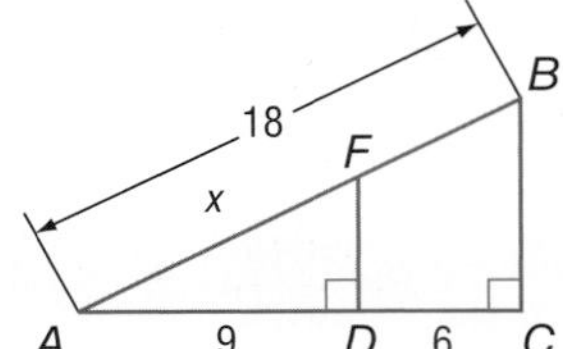


**13. HISTORY** In the fifteenth century, mathematicians and artists tried to construct the perfect letter. A square was used as a frame to design the letter "A," as shown below. The thickness of the major stroke of the letter was $\frac{1}{12}$ the height of the letter. (Lesson 9-4)

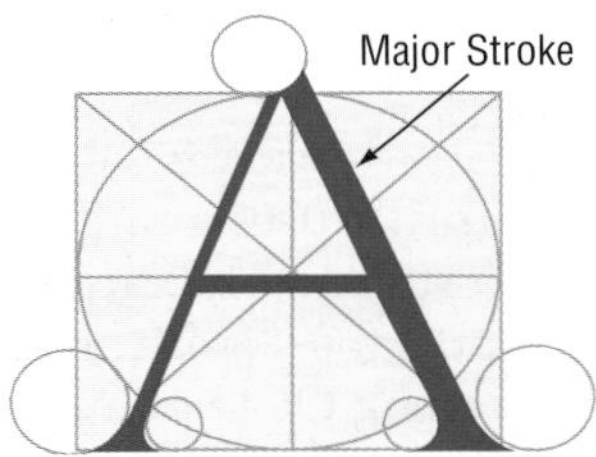


**a.** Explain why the bar through the middle of the A is half the length of the space between the outside bottom corners of the sides of the letter.

**b.** If the letter were 3 centimeters tall, how wide would the major stroke be?

**ALGEBRA Find *x* and *y*.** (Lesson 9-4)

**14.**

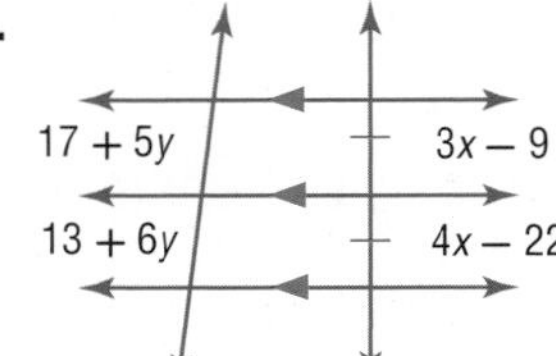


**15.**

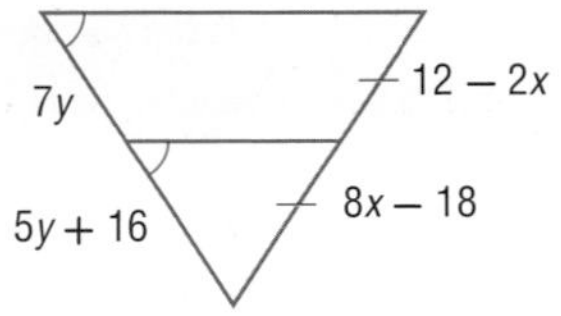

7y
12 − 2x
5y + 16
8x − 18

# LESSON 9-5 Parts of Similar Triangles

**Then**

- You learned that corresponding sides of similar polygons are proportional.

**Now**

1. Recognize and use proportional relationships of corresponding angle bisectors, altitudes, and medians of similar triangles.
2. Use the Triangle Bisector Theorem.

**Why?**

- The "Rule of Thumb" uses the average ratio of a person's arm length to the distance between his or her eyes and the altitudes of similar triangles to estimate the distance between a person and an object of approximately known width.

**CCSS Common Core State Standards**

**Content Standards**

**G.SRT.4** Prove theorems about triangles.

**G.SRT.5** Use congruence and similarity criteria for triangles to solve problems and to prove relationships in geometric figures.

**Mathematical Practices**

1 Make sense of problems and persevere in solving them.

3 Construct viable arguments and critique the reasoning of others.

## 1 Special Segments of Similar Triangles

You learned in Lesson 9-2 that the corresponding side lengths of similar polygons, such as triangles, are proportional. This concept can be extended to other segments in triangles.

### Theorems Special Segments of Similar Triangles

**9.8** If two triangles are similar, the lengths of corresponding altitudes are proportional to the lengths of corresponding sides.

**Abbreviation** ~△s *have corr. altitudes proportional to corr. sides.*

**Example** If $\triangle ABC \sim \triangle FGH$, then $\frac{AD}{FJ} = \frac{AB}{FG}$.

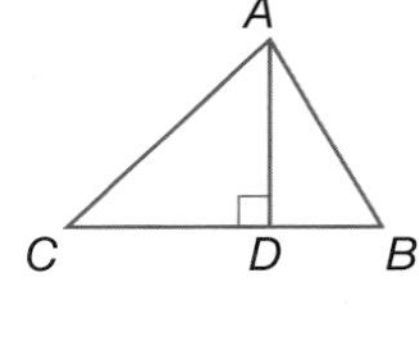


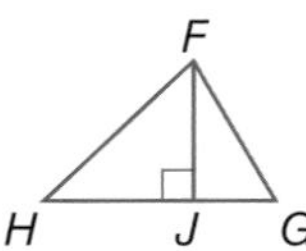


**9.9** If two triangles are similar, the lengths of corresponding angle bisectors are proportional to the lengths of corresponding sides.

**Abbreviation** ~△s *have corr. ∠ bisectors proportional to corr. sides.*

**Example** If $\triangle KLM \sim \triangle QRS$, then $\frac{LP}{RT} = \frac{LM}{RS}$.

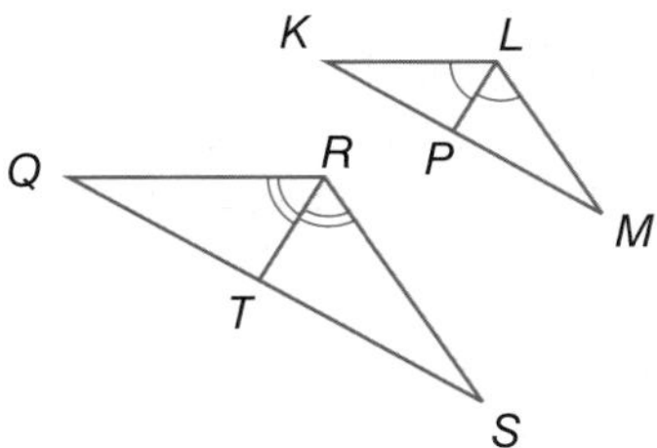


**9.10** If two triangles are similar, the lengths of corresponding medians are proportional to the lengths of corresponding sides.

**Abbreviation** ~△s *have corr. medians proportional to corr. sides.*

**Example** If $\triangle ABC \sim \triangle WXY$, then $\frac{CD}{YZ} = \frac{AB}{WX}$.

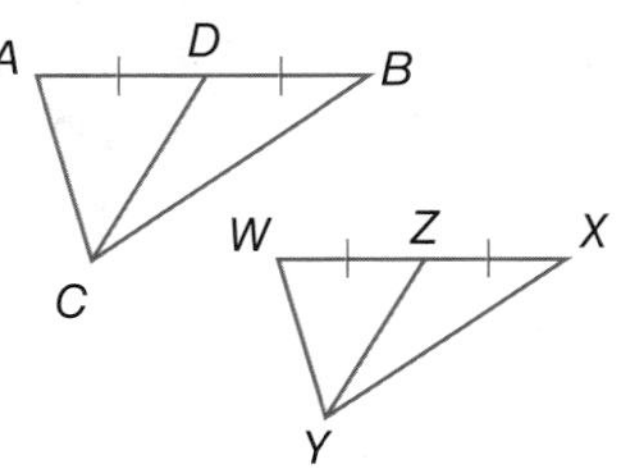


You will prove Theorems 9.9 and 9.10 in Exercises 18 and 19, respectively.

### Real-WorldCareer

**Athletic Trainer** Athletic trainers help prevent and treat sports injuries. They ensure that protective equipment is used properly and that people understand safe practices that prevent injury. An athletic trainer must have a bachelor's degree to be certified. Most also have master's degrees. Refer to Exercise 29.

## Proof Theorem 9.8

**Given:** $\triangle FGH \sim \triangle KLM$
$\overline{FJ}$ and $\overline{KP}$ are altitudes.

**Prove:** $\frac{FJ}{KP} = \frac{HF}{MK}$

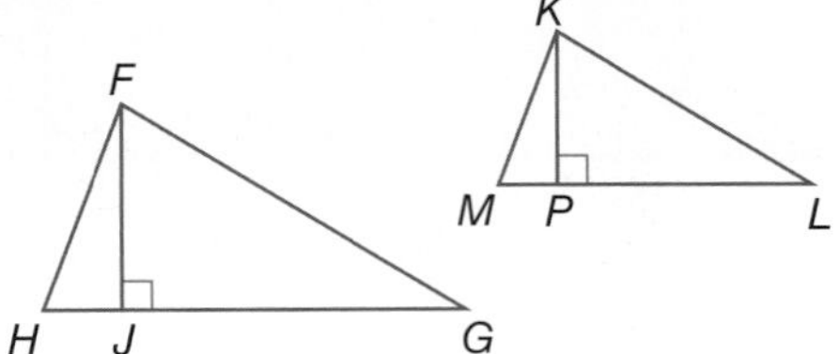


**Paragraph Proof:**
Since $\triangle FGH \sim \triangle KLM$, $\angle H \cong \angle M$. $\angle FJH \cong \angle KPM$ because they are both right angles created by the altitudes drawn to the opposite side and all right angles are congruent.

Thus $\triangle HFJ \sim \triangle MKP$ by AA Similarity. So $\frac{FJ}{KP} = \frac{HF}{MK}$ by the definition of similar polygons.

Since the corresponding altitudes are chosen at random, we need not prove Theorem 7.8 for every pair of altitudes.

You can use special segments in similar triangles to find missing measures.

## Example 1 Use Special Segments in Similar Triangles

**In the figure, $\triangle ABC \sim \triangle FDG$. Find the value of $x$.**

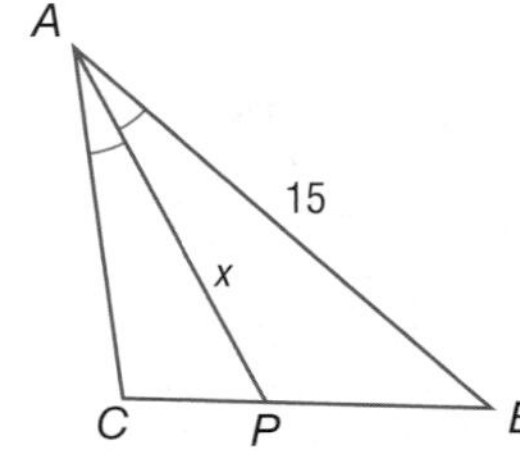


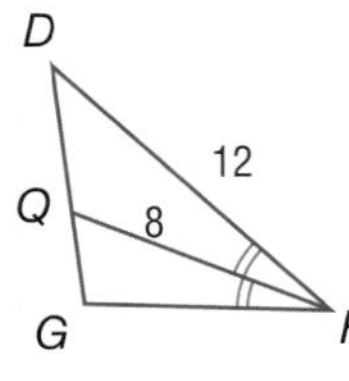


$\overline{AP}$ and $\overline{FQ}$ are corresponding angle bisectors and $\overline{AB}$ and $\overline{FD}$ are corresponding sides of similar triangles $ABC$ and $FDG$.

| | |
|---|---|
| $\frac{AP}{FQ} = \frac{AB}{FD}$ | ~△s have corr. ∠ bisectors proportional to the corr. sides. |
| $\frac{x}{8} = \frac{15}{12}$ | Substitution |
| $8 \cdot 15 = x \cdot 12$ | Cross Products Property |
| $120 = 12x$ | Simplify. |
| $10 = x$ | Divide each side by 12. |

### StudyTip

**Use Scale Factor** Example 1 could also have been solved by first finding the scale factor between $\triangle ABC$ and $\triangle FDG$. The ratio of the angle bisector in $\triangle ABC$ to the angle bisector in $\triangle FDG$ would then be equal to this scale factor.

### GuidedPractice

**Find the value of $x$.**

**1A.**

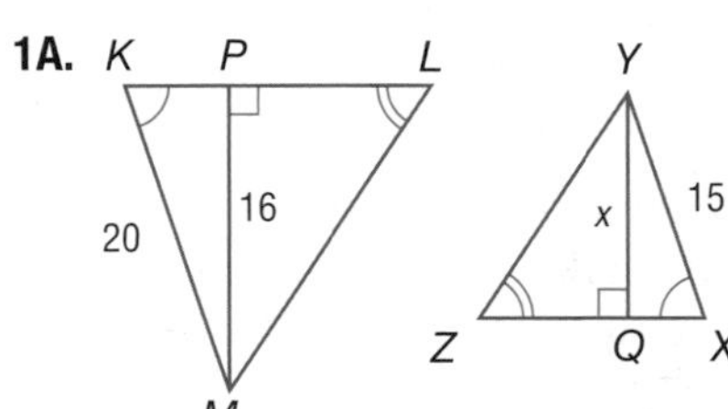


**1B.**

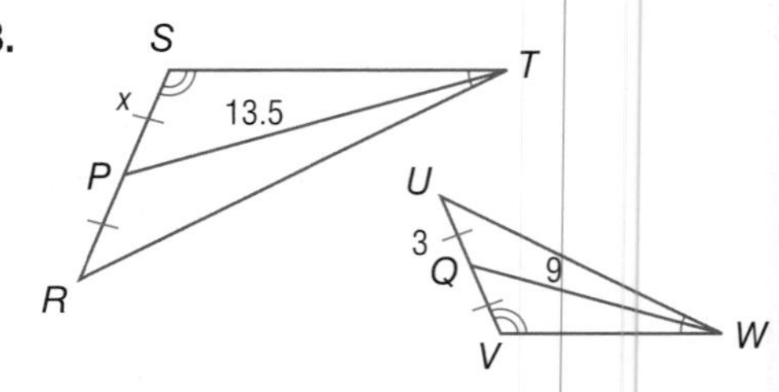


Dennis MacDonald/age fotostock

You can use special segments in similar triangles to solve real-world problems.

**Real-WorldLink**

Hold your outstretched hand horizontal at arm's length with your palm facing you; for each hand width the sun is above the horizon, there is one remaining hour of sunlight.

**Source:** Sail Island Channels

PT

## Real-World Example 2 Use Similar Triangles to Solve Problems

**ESTIMATING DISTANCES Liliana holds her arm straight out in front of her with her elbow straight and her thumb pointing up. Closing one eye, she aligns one edge of her thumb with a car she is sighting. Next she switches eyes without moving her head or her arm. The car appears to jump 4 car widths. If Liliana's arm is about 10 times longer than the distance between her eyes, and the car is about 5.5 feet wide, estimate the distance from Liliana's thumb to the car.**

**Understand** Make a diagram of the situation labeling the given distances and the distance you need to find as $x$. Also, label the vertices of the triangles formed.

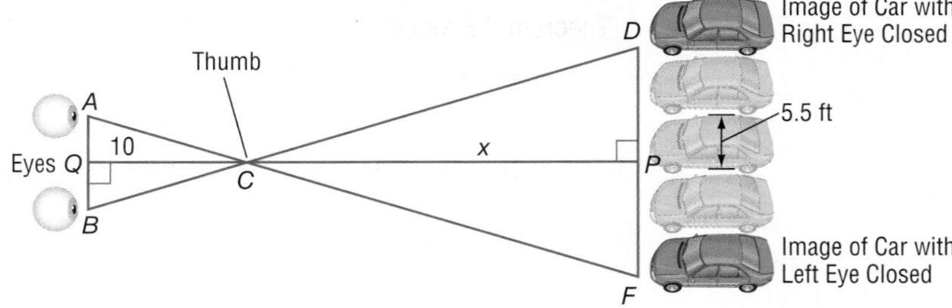


We assume that if Liliana's thumb is straight out in front of her, then $\overline{PC}$ is an altitude of $\triangle ABC$. Likewise, $\overline{QC}$ is the corresponding altitude. We assume that $\overline{AB} \parallel \overline{DF}$.

**Plan** Since $\overline{AB} \parallel \overline{DF}$, $\angle BAC \cong \angle DFC$ and $\angle CBA \cong \angle CDF$ by the Alternate Interior Angles Theorem. Therefore $\triangle ABC \sim \triangle FDC$ by AA Similarity. Write a proportion and solve for $x$.

**Solve**

$$\frac{PC}{QC} = \frac{AB}{DF}$$ Theorem 9.8

$$\frac{10}{x} = \frac{1}{5.5 \cdot 4}$$ Substitution

$$\frac{10}{x} = \frac{1}{22}$$ Simplify.

$$10 \cdot 22 = x \cdot 1$$ Cross Products Property

$$220 = x$$ Simplify.

So the estimated distance to the car is 220 feet.

**Check** The ratio of Liliana's arm length to the width between her eyes is 10 to 1. The ratio of the distance to the car to the distance the image of the car jumped is 22 to 220 or 10 to 1. ✓

### Guided Practice

**2.** Suppose Liliana stands at the back of her classroom and sights a clock on the wall at the front of the room. If the clock is 30 centimeters wide and appears to move 3 clock widths when she switches eyes, estimate the distance from Liliana's thumb to the clock.

## 2 Triangle Angle Bisector Theorem

An angle bisector of a triangle also divides the side opposite the angle proportionally.

### Theorem 9.11 Triangle Angle Bisector

An angle bisector in a triangle separates the opposite side into two segments that are proportional to the lengths of the other two sides.

**Example** If $\overline{JM}$ is an angle bisector of $\triangle JKL$,

then $\frac{KM}{LM} = \frac{KJ}{LJ}$. ← segments with vertex $K$ ← segments with vertex $L$

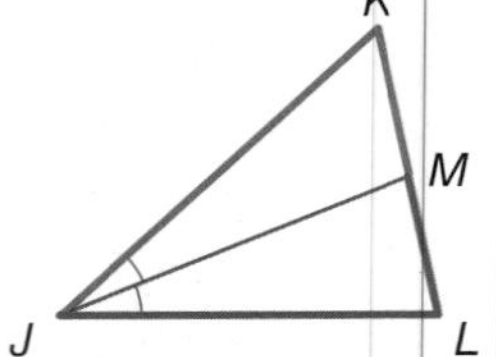


**StudyTip**

**Proportions** Another proportion that could be written using the Triangle Angle Bisector Theorem is $\frac{KM}{KJ} = \frac{LM}{LJ}$.

You will prove Theorem 9.11 in Exercise 25.

### Example 3 Use the Triangle Angle Bisector Theorem

**Find $x$.**

Since $\overline{RT}$ is an angle bisector of $\triangle QRS$, you can use the Triangle Angle Bisector Theorem to write a proportion.

| | |
|---|---|
| $\frac{QT}{ST} = \frac{QR}{SR}$ | Triangle Angle Bisector Theorem |
| $\frac{x}{18-x} = \frac{6}{14}$ | Substitution |
| $(18-x)(6) = x \cdot 14$ | Cross Products Property |
| $108 - 6x = 14x$ | Simplify. |
| $108 = 20x$ | Add $6x$ to each side. |
| $5.4 = x$ | Divide each side by 20. |

**GuidedPractice**

**Find the value of $x$.**

**3A.**

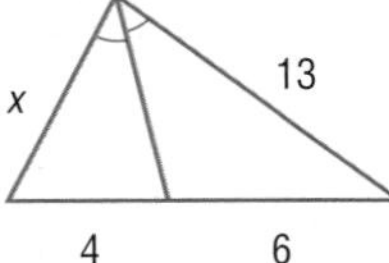


**3B.**

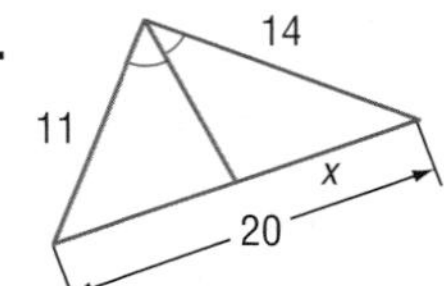


## Check Your Understanding

**Example 1** Find $x$.

**1.**

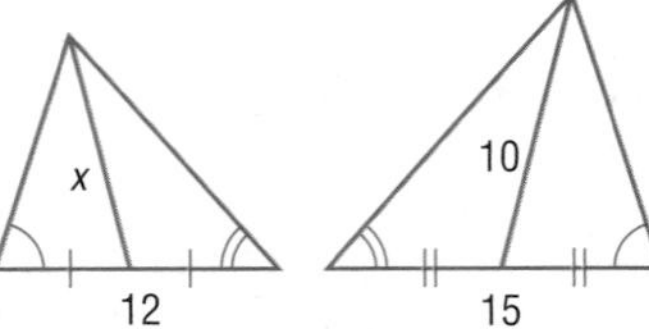


**2.**

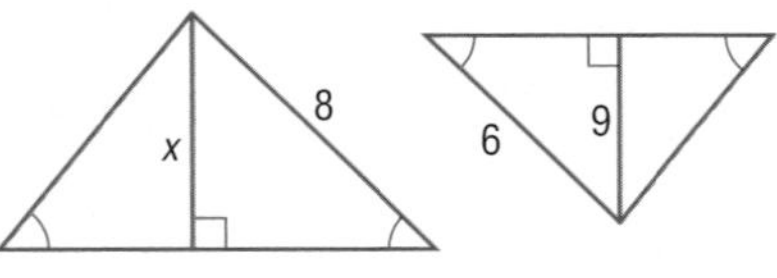


**Example 2**

**3. VISION** A cat that is 10 inches tall forms a retinal image that is 7 millimeters tall. If $\triangle ABE \sim \triangle DBC$ and the distance from the pupil to the retina is 25 millimeters, how far away from your pupil is the cat?

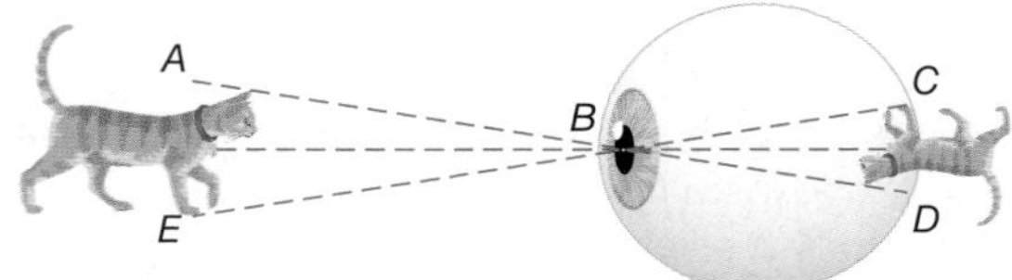

**Example 3** **Find the value of each variable.**

4.

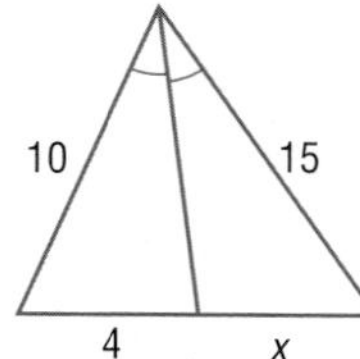


5.

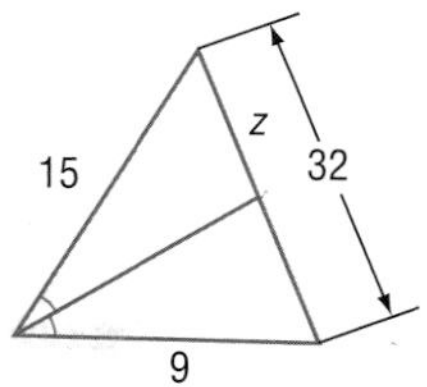


## Practice and Problem Solving

**Example 1** **Find $x$.**

6.

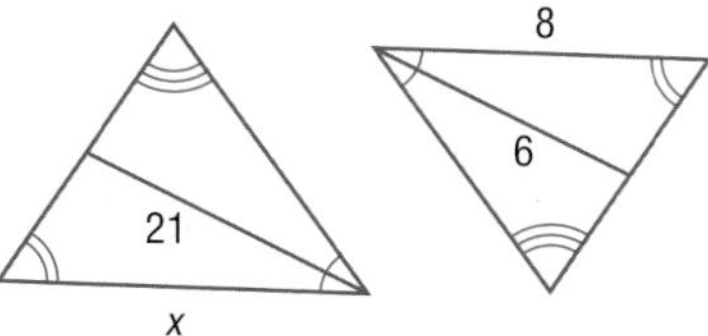


7.

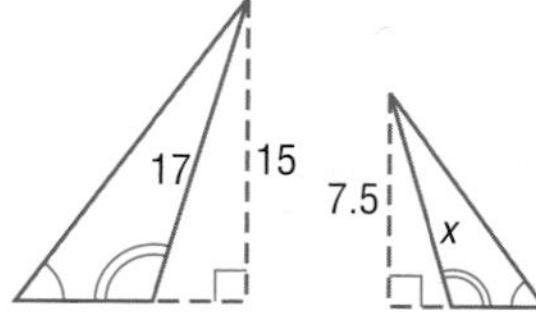


8.

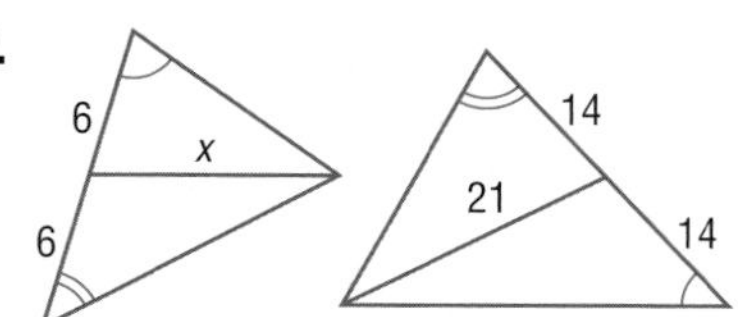


9.

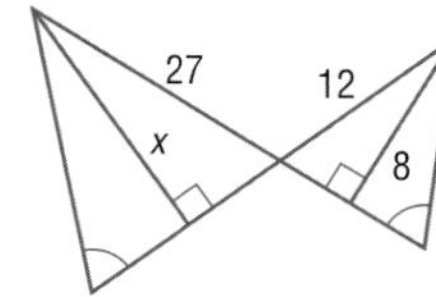


**Example 2**

10. **ROADWAYS** The intersection of the two roads shown forms two similar triangles. If $AC$ is 382 feet, $MP$ is 248 feet, and the gas station is 50 feet from the intersection, how far from the intersection is the bank?

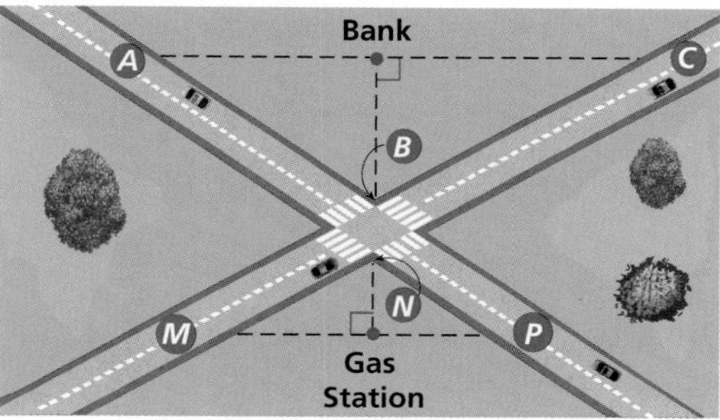


**Example 3** **CCSS SENSE-MAKING** **Find the value of each variable.**

11.

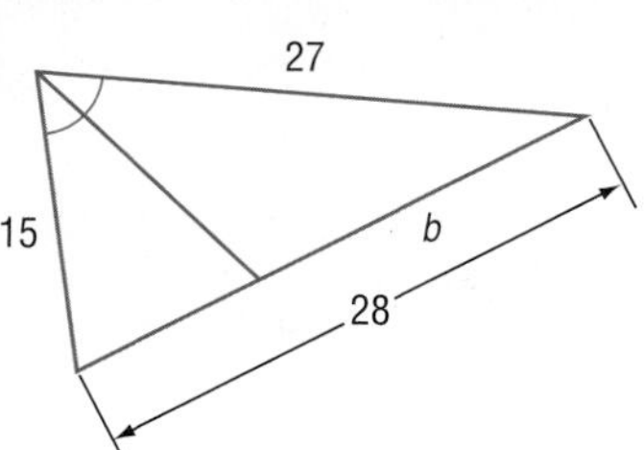


12.

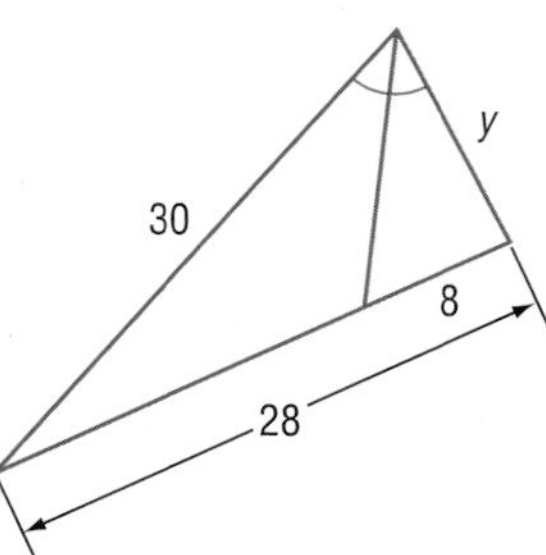


13.

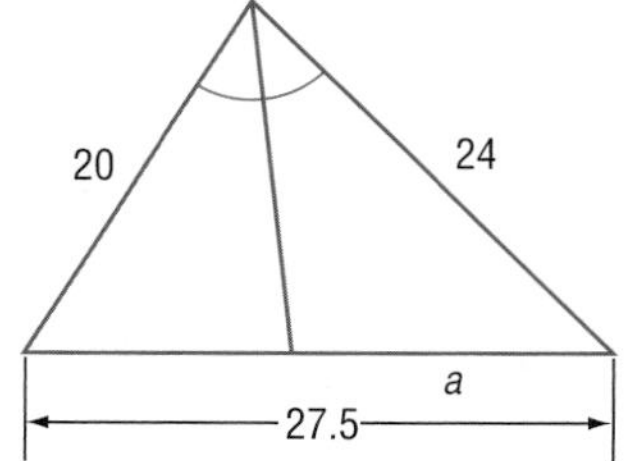


14.

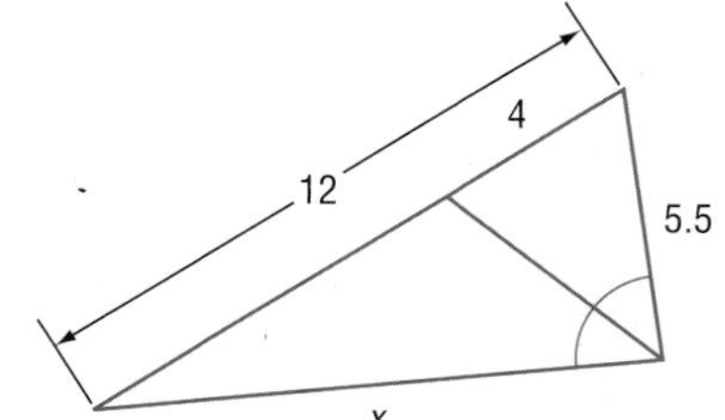

**15.** **ALGEBRA** If $\overline{AB}$ and $\overline{JK}$ are altitudes, $\triangle DAC \sim \triangle MJL$, $AB = 9$, $AD = 4x - 8$, $JK = 21$, and $JM = 5x + 3$, find $x$.

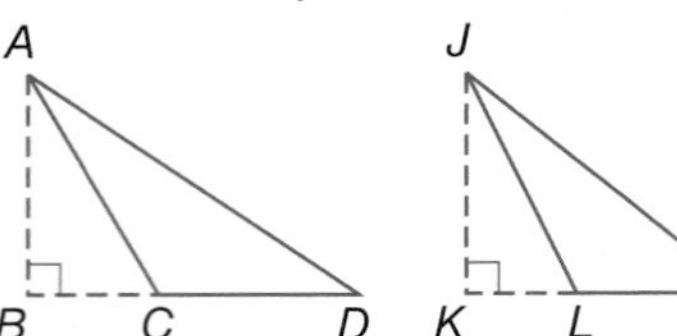


**16.** **ALGEBRA** If $\overline{NQ}$ and $\overline{VX}$ are medians, $\triangle PNR \sim \triangle WVY$, $NQ = 8$, $PR = 12$, $WY = 7x - 1$, and $VX = 4x + 2$, find $x$.

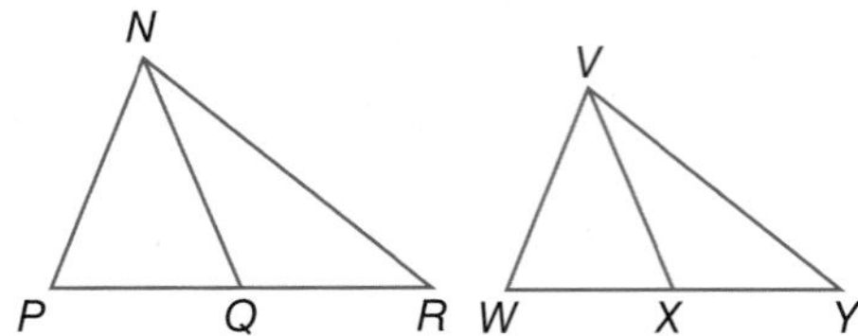


**17.** If $\triangle SRY \sim \triangle WXQ$, $\overline{RT}$ is an altitude of $\triangle SRY$, $\overline{XV}$ is an altitude of $\triangle WXQ$, $RT = 5$, $RQ = 4$, $QY = 6$, and $YX = 2$, find $XV$.

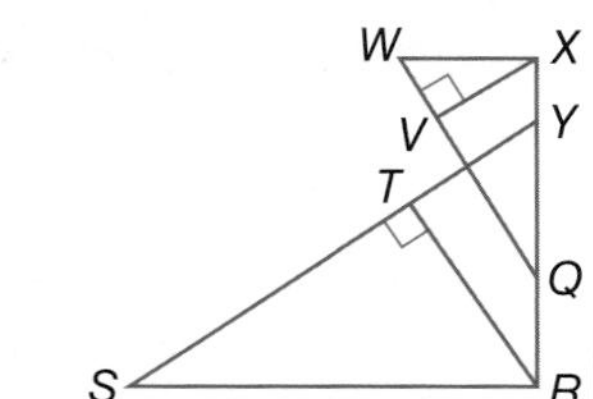


**18.** **PROOF** Write a paragraph proof of Theorem 9.9.

**19.** **PROOF** Write a two-column proof of Theorem 9.10.

**ALGEBRA** **Find $x$.**

**20.**

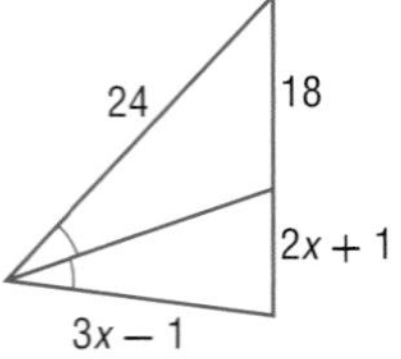


**21.**

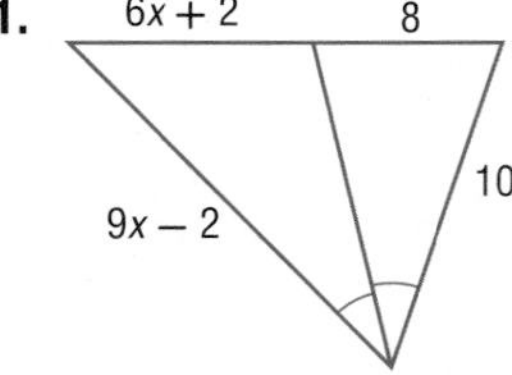


**22.**

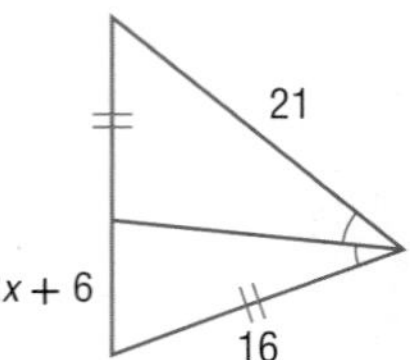


**23.**

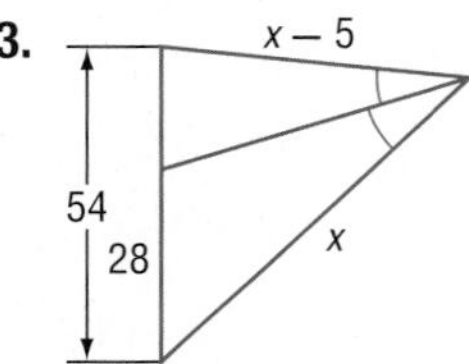


**24.** **SPORTS** Consider the triangle formed by the path between a batter, center fielder, and right fielder as shown. If the batter gets a hit that bisects the triangle at $\angle B$, is the center fielder or the right fielder closer to the ball? Explain your reasoning.

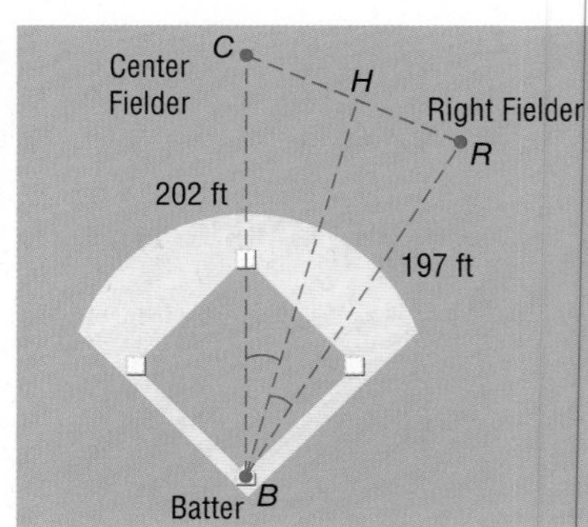


**CCSS ARGUMENTS** **Write a two-column proof.**

**25.** Theorem 9.11

**Given:** $\overline{CD}$ bisects $\angle ACB$.
By construction, $\overline{AE} \parallel \overline{CD}$.

**Prove:** $\frac{AD}{DB} = \frac{AC}{BC}$

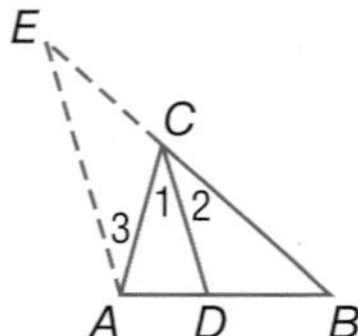


**26.** **Given:** $\angle H$ is a right angle.
$L$, $K$, and $M$ are midpoints.

**Prove:** $\angle LKM$ is a right angle.

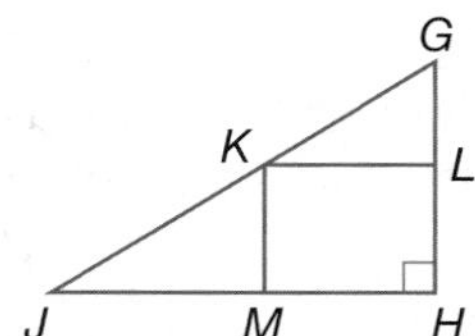

**PROOF** Write a two-column proof.

27. **Given:** $\triangle QTS \sim \triangle XWZ$, $\overline{TR}$ and $\overline{WY}$ are angle bisectors.

**Prove:** $\frac{TR}{WY} = \frac{QT}{XW}$

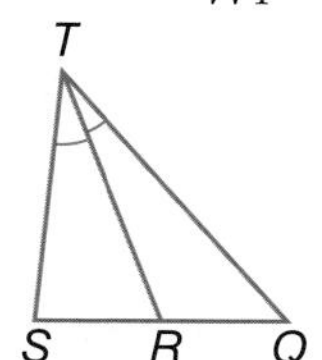


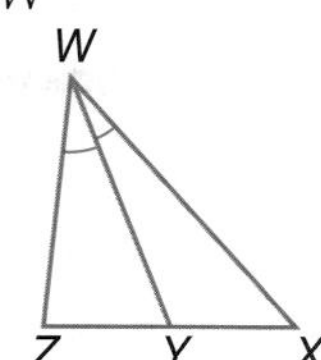


28. **Given:** $\overline{FD} \parallel \overline{BC}$, $\overline{BF} \parallel \overline{CD}$, $\overline{AC}$ bisects $\angle C$.

**Prove:** $\frac{DE}{EC} = \frac{BA}{AC}$

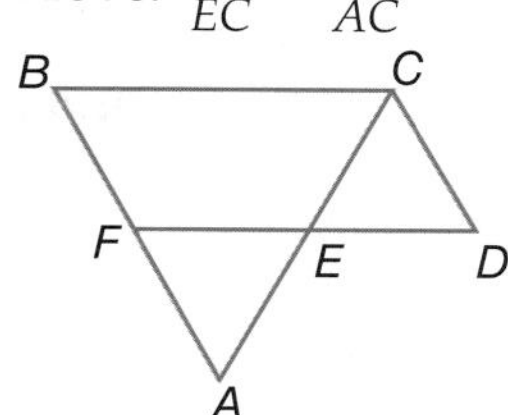


29. **SPORTS** During football practice, Trevor threw a pass to Ricardo as shown below. If Eli is farther from Trevor when he completes the pass to Ricardo and Craig and Eli move at the same speed, who will reach Ricardo to tackle him first?

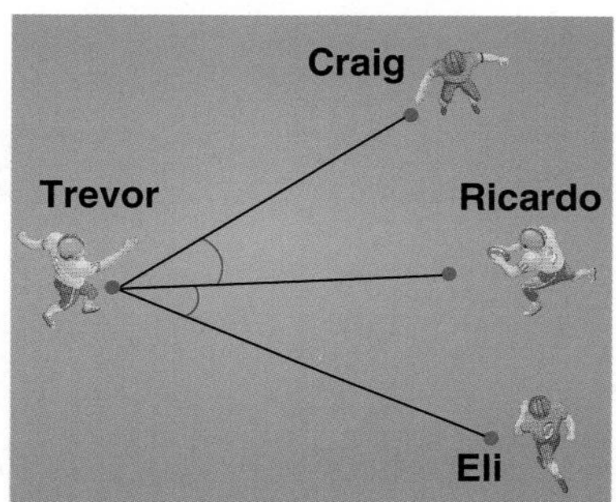


30. **SHELVING** In the bookshelf shown, the distance between each shelf is 13 inches and $\overline{AK}$ is a median of $\triangle ABC$. If $EF$ is $3\frac{1}{3}$ inches, what is $BK$?

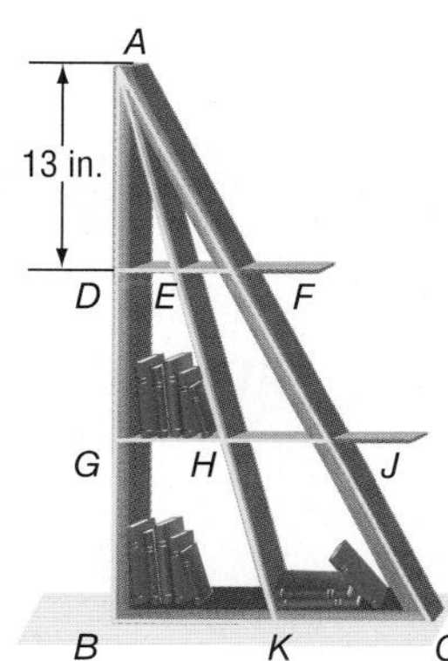


## H.O.T. Problems Use Higher-Order Thinking Skills

31. **ERROR ANALYSIS** Chun and Traci are determining the value of $x$ in the figure. Chun says to find $x$, solve the proportion $\frac{5}{8} = \frac{15}{x}$, but Traci says to find $x$, the proportion $\frac{5}{x} = \frac{8}{15}$ should be solved. Is either of them correct? Explain.

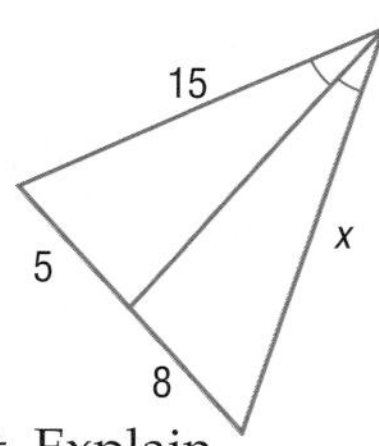


32. **CCSS ARGUMENTS** Find a counterexample to the following statement. Explain.

*If the measure of an altitude and side of a triangle are proportional to the corresponding altitude and corresponding side of another triangle, then the triangles are similar.*

33. **CHALLENGE** The perimeter of $\triangle PQR$ is 94 units. $\overline{QS}$ bisects $\angle PQR$. Find $PS$ and $RS$.

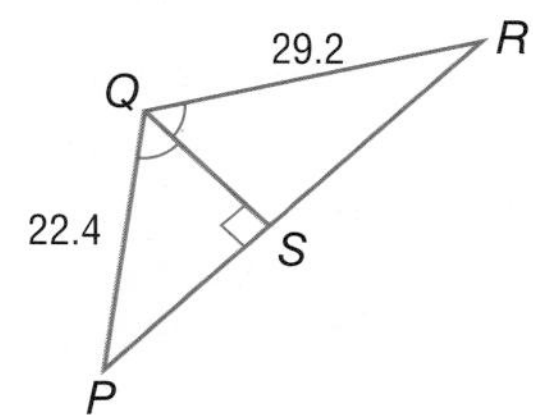


34. **OPEN ENDED** Draw two triangles so that the measures of corresponding medians and a corresponding side are proportional, but the triangles are not similar.

35. **WRITING IN MATH** Compare and contrast Theorem 9.9 and the Triangle Angle Bisector Theorem.

## Standardized Test Practice

**36. ALGEBRA** Which shows 0.00234 written in scientific notation?

**A** $2.34 \times 10^5$ **C** $2.34 \times 10^{-2}$

**B** $2.34 \times 10^3$ **D** $2.34 \times 10^{-3}$

**37. SHORT RESPONSE** In the figures below, $\overline{AB} \perp \overline{DC}$ and $\overline{GH} \perp \overline{FE}$.

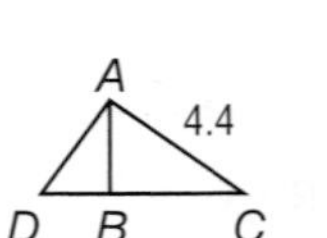


If $\triangle ACD \sim \triangle GEF$, find $AB$.

**38.** Quadrilateral $HJKL$ is a parallelogram. If the diagonals are perpendicular, which statement must be true?

**F** Quadrilateral $HJKL$ is a square.

**G** Quadrilateral $HJKL$ is a rectangle.

**H** Quadrilateral $HJKL$ is a rhombus.

**J** Quadrilateral $HJKL$ is an isosceles trapezoid.

**39. SAT/ACT** The sum of three numbers is 180. Two of the numbers are the same, and each of them is one third of the greatest number. What is the least number?

**A** 15 **D** 45

**B** 30 **E** 60

**C** 36

## Spiral Review

**ALGEBRA** **Find $x$ and $y$.** (Lesson 9-4)

**40.**

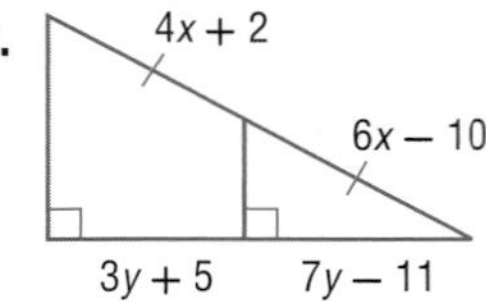


**41.**

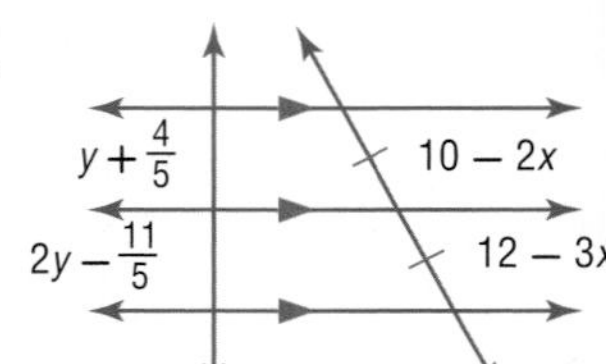


**42.**

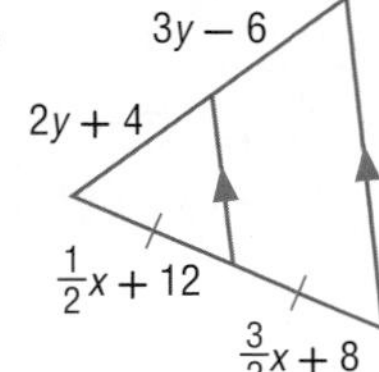


**Find the indicated measure(s).** (Lesson 9-3)

**43.** If $\overline{PR} \parallel \overline{KL}$, $KN = 9$, $LN = 16$, and $PM = 2(KP)$, find $KP$, $KM$, $MR$, $ML$, $MN$, and $PR$.

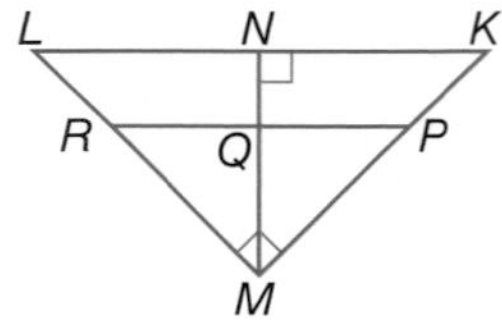


**44.** If $\overline{PR} \parallel \overline{WX}$, $WX = 10$, $XY = 6$, $WY = 8$, $RY = 5$, and $PS = 3$, find $PY$, $SY$, and $PQ$.

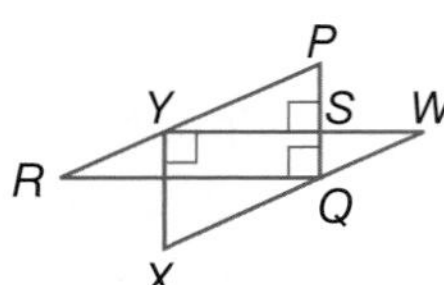


**45. GEESE** A flock of geese flies in formation. Prove that $\triangle EFG \cong \triangle HFG$ if $\overline{EF} \cong \overline{HF}$ and that $G$ is the midpoint of $\overline{EH}$. (Lesson 6-3)

Chase Swift/CORBIS

## Skills Review

**Find the distance between each pair of points.**

**46.** $E(-3, -2)$, $F(5, 8)$

**47.** $A(2, 3)$, $B(5, 7)$

**48.** $C(-2, 0)$, $D(6, 4)$

**49.** $W(7, 3)$, $Z(-4, -1)$

**50.** $J(-4, -5)$, $K(2, 9)$

**51.** $R(-6, 10)$, $S(8, -2)$

# Geometry Lab
# Fractals

A **fractal** is a geometric figure that is created using iteration. **Iteration** is a process of repeating the same operation over and over again. Fractals are **self-similar**, which means that the smaller details of the shape have the same geometric characteristics as the original form.

## Activity 1

**Stage 0** Draw an equilateral triangle on isometric dot paper in which each side is 8 units long.

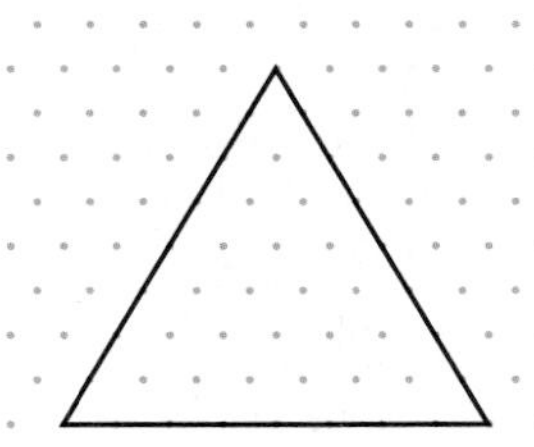

**Stage 1** Connect the midpoints of the sides to form another triangle. Shade the center triangle.

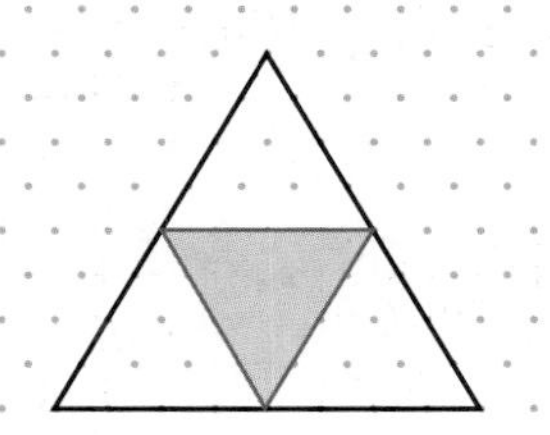

**Stage 2** Repeat the process using the three unshaded triangles. Connect the midpoints of the sides to form three other triangles.

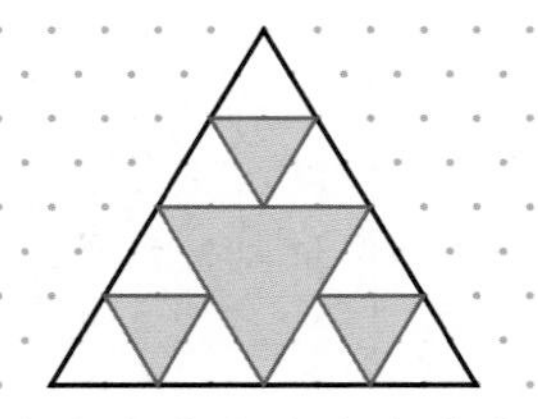

If you repeat this process indefinitely, the figure that results is called the Sierpinski Triangle.

## Analyze the Results

1. If you continue the process, how many unshaded triangles will you have at Stage 3?
2. What is the perimeter of an unshaded triangle in Stage 4?
3. If you continue the process indefinitely, what will happen to the perimeters of the unshaded triangles?
4. **CHALLENGE** Complete the proof below.

   **Given:** $\triangle KAP$ is equilateral. $D, F, M, B, C,$ and $E$ are midpoints of $\overline{KA}, \overline{AP}, \overline{PK}, \overline{DA}, \overline{AF},$ and $\overline{FD}$, respectively.

   **Prove:** $\triangle BAC \sim \triangle KAP$

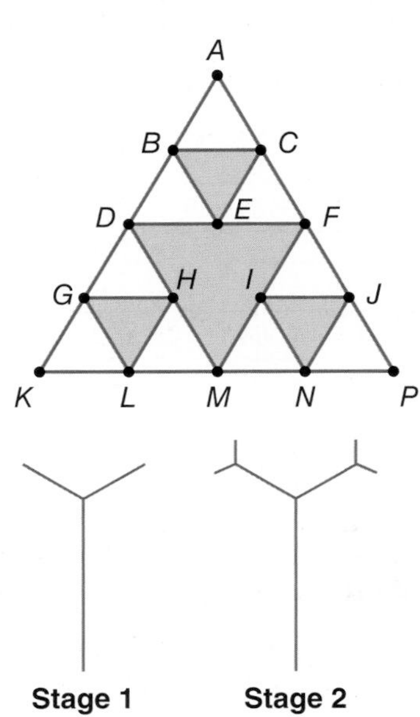


5. A *fractal tree* can be drawn by making two new branches from the endpoint of each original branch, each one-third as long as the previous branch.

   a. Draw Stages 3 and 4 of a fractal tree. How many total branches do you have in Stages 1 through 4? (Do not count the stems.)

   b. Write an expression to predict the number of branches at each stage.

Stage 1 Stage 2

(continued on the next page)

# Fractals *Continued*

Not all iterative processes involve manipulation of geometric shapes. Some iterative processes can be translated into formulas or algebraic equations, similar to the expression you wrote in Exercise 5 on the previous page.

## Activity 2

*Pascal's Triangle* is a numerical pattern in which each row begins and ends with 1 and all other terms in the row are the sum of the two numbers above it. Find a formula in terms of the row number for any row in Pascal's Triangle.

**Step 1** Draw rows 1 through 5 in Pascal's Triangle.

**Step 2** Find the sum of values in each row.

**Step 3** Find a pattern using the **row number** that can be used to determine the sum of any row.

| Row | Pascal's Triangle | Sum | Pattern |
|---|---|---|---|
| 1 | 1 | 1 | $2^0 = 2^{1-1}$ |
| 2 | 1 1 | 2 | $2^1 = 2^{2-1}$ |
| 3 | 1 2 1 | 4 | $2^2 = 2^{3-1}$ |
| 4 | 1 3 3 1 | 8 | $2^3 = 2^{4-1}$ |
| 5 | 1 4 6 4 1 | 16 | $2^4 = 2^{5-1}$ |

### Analyze the Results

**6.** Write a formula for the sum $S$ of any row $n$ in the Pascal Triangle.

**7.** What is the sum of the values in the eighth row of Pascal's Triangle?

## Exercises

**Write a formula for $F(x)$.**

**8.**

| $x$ | 2 | 4 | 6 | 8 | 10 |
|---|---|---|---|---|---|
| $F(x)$ | 3 | 7 | 11 | 15 | 19 |

**9.**

| $x$ | 0 | 5 | 10 | 15 | 20 |
|---|---|---|---|---|---|
| $F(x)$ | 0 | 20 | 90 | 210 | 380 |

**10.**

| $x$ | 1 | 2 | 4 | 8 | 10 |
|---|---|---|---|---|---|
| $F(x)$ | 1 | 0.5 | 0.25 | 0.125 | 0.1 |

**11.**

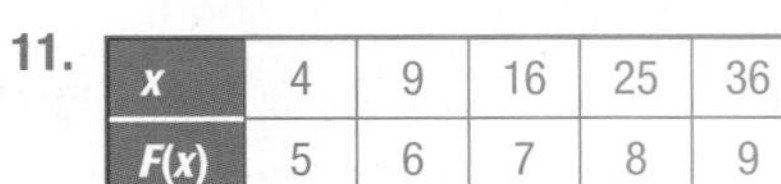

| $x$ | 4 | 9 | 16 | 25 | 36 |
|---|---|---|---|---|---|
| $F(x)$ | 5 | 6 | 7 | 8 | 9 |

**12. CHALLENGE** The figural pattern below represents a sequence of figural numbers called *triangular numbers*. How many dots will be in the 8th term in the sequence? Is it possible to write a formula that can be used to determine the number of dots in the $n$th triangular number in the series? If so, write the formula. If not, explain why not.

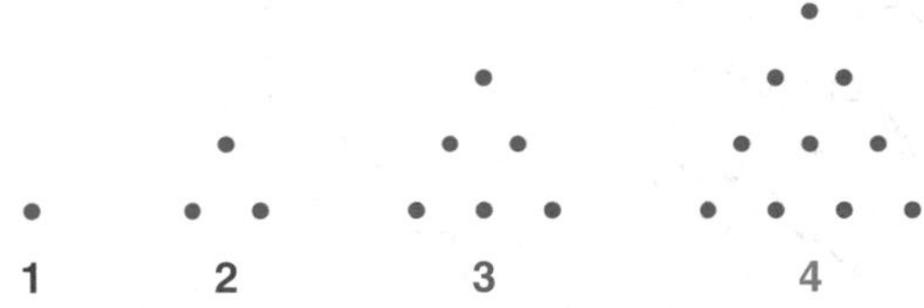

# LESSON 9-6 Similarity Transformations

**Then**

- You identified congruence transformations.

**Now**

1. Identify similarity transformations.
2. Verify similarity after a similarity transformation.

**Why?**

- Adriana uses a copier to enlarge a movie ticket to use as the background for a page in her movie ticket scrapbook. She places the ticket on the glass of the copier. Then she must decide what percentage to input in order to create an image that is three times as big as her original ticket.

Polaris Center 14
Presenting
BEST MOVIE EVER
4:00 PM Sat 1/17/09
MATINEE 11:50
Auditorium 8
00912300050027
01/17/09 2:20 PM

5 cm

6.4 cm

## NewVocabulary

dilation
similarity transformation
center of dilation
scale factor of a dilation
enlargement
reduction

## Common Core State Standards

**Content Standards**

**G.SRT.2** Given two figures, use the definition of similarity in terms of similarity transformations to decide if they are similar; explain using similarity transformations the meaning of similarity for triangles as the equality of all corresponding pairs of angles and the proportionality of all corresponding pairs of sides.

**G.SRT.5** Use congruence and similarity criteria for triangles to solve problems and to prove relationships in geometric figures.

**Mathematical Practices**

6 Attend to precision.
4 Model with mathematics.

## 1 Identify Similarity Transformations

A *transformation* is an operation that maps an original figure, the *preimage*, onto a new figure called the *image*.

A **dilation** is a transformation that enlarges or reduces the original figure proportionally. Since a dilation produces a similar figure, a dilation is a type of **similarity transformation**.

Dilations are performed with respect to a fixed point called the **center of dilation**.

The **scale factor of a dilation** describes the extent of the dilation. The scale factor is the ratio of a length on the image to a corresponding length on the preimage.

The letter $k$ usually represents the scale factor of a dilation. The value of $k$ determines whether the dilation is an enlargement or a reduction.

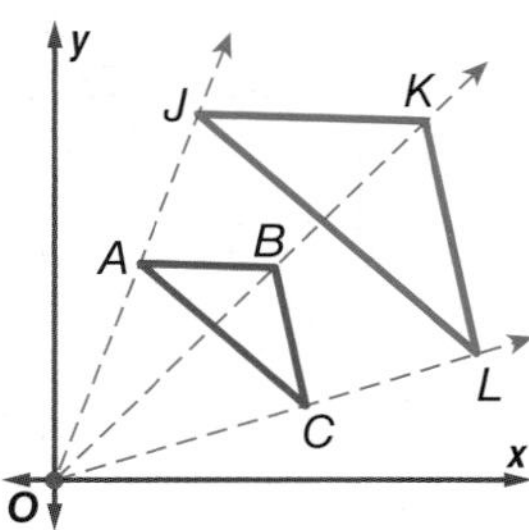


$\triangle JKL$ is a dilation of $\triangle ABC$.
Center of dilation: (0, 0)
Scale factor: $\frac{JK}{AB}$

### ConceptSummary Types of Dilations

A dilation with a scale factor greater than 1 produces an **enlargement**, or an image that is larger than the original figure.

**Symbols** If $k > 1$, the dilation is an enlargement.

**Example** $\triangle FGH$ is dilated by a scale factor of 3 to produce $\triangle RST$. Since $3 > 1$, $\triangle RST$ is an enlargement of $\triangle FGH$.

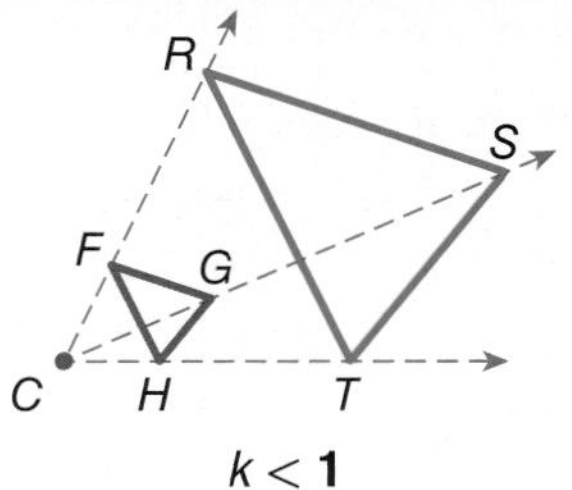


$k < 1$

A dilation with a scale factor between 0 and 1 produces a **reduction**, an image that is smaller than the original figure.

**Symbols** If $0 < k < 1$, the dilation is a reduction.

**Example** *ABCD* is dilated by a scale factor of $\frac{1}{4}$ to produce *WXYZ*. Since $0 < \frac{1}{4} < 1$, *WXYZ* is a reduction of *ABCD*.

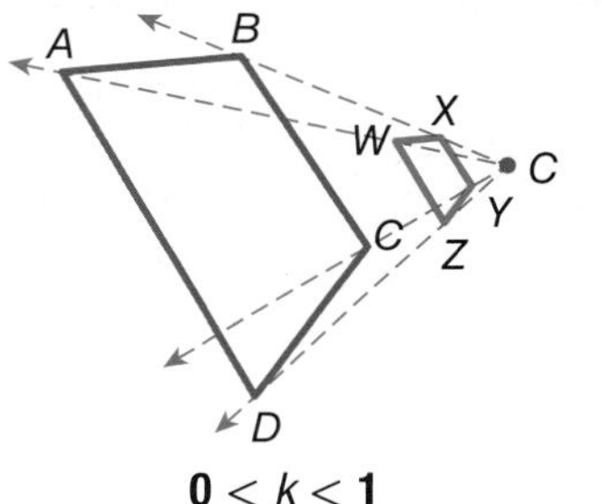


$0 < k < 1$

**StudyTip**

**Multiple Representations** The scale factor of a dilation can be represented as a fraction, a decimal, or as a percent. For example, a scale factor of $\frac{2}{5}$ can also be written as 0.4 or as 40%.

## Example 1 Identify a Dilation and Find Its Scale Factor

**Determine whether the dilation from *A* to *B* is an *enlargement* or a *reduction*. Then find the scale factor of the dilation.**

**a.**

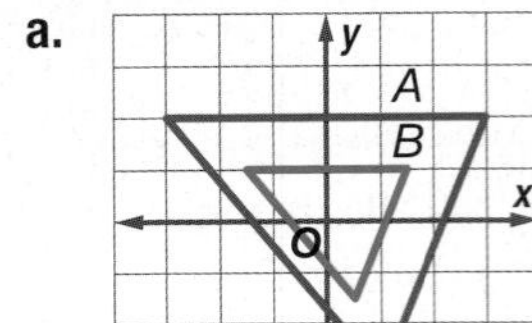


*B* is smaller than *A*, so the dilation is a reduction.

The distance between the vertices at (−3, 2) and (3, 2) for *A* is 6 and from the vertices at (−1.5, 1) and (1.5, 1) for *B* is 3. So the scale factor is $\frac{3}{6}$ or $\frac{1}{2}$.

**b.**

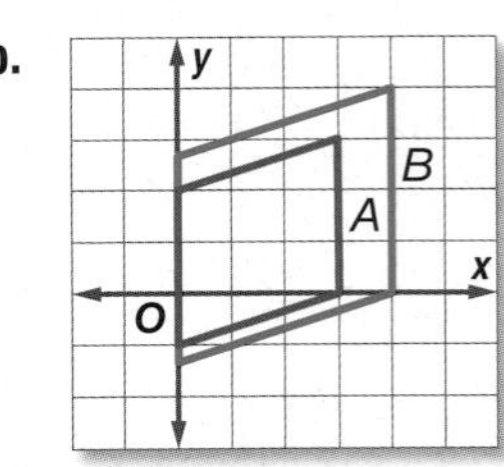


*B* is larger than *A*, so the dilation is an enlargement.

The distance between the vertices at (3, 3) and (3, 0) for *A* is 3 and between the vertices at (4, 4) and (4, 0) for *B* is 4. So the scale factor is $\frac{4}{3}$.

**Guided Practice**

**1A.**

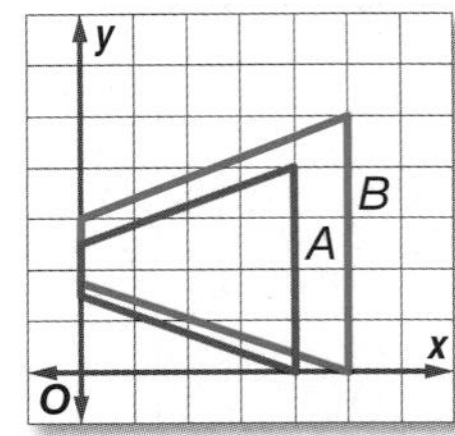


**1B.**

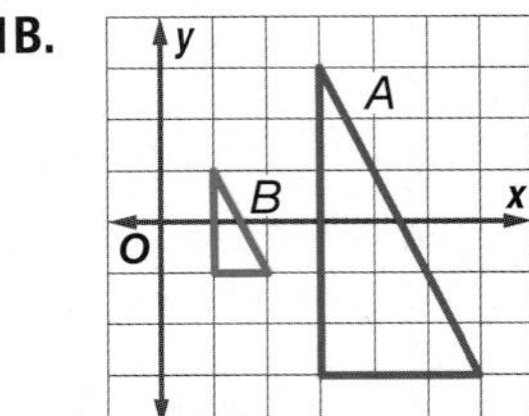


Dilations and their scale factors are used in many real-world situations.

**Real-WorldLink**

Hew Weng Fatt accepted a contest challenge to collect the most movie stubs from a certain popular fantasy movie. He collected 6561 movie stubs in 38 days!

**Source:** *Youth2, Star Publications*

## Real-World Example 2 Find and Use a Scale Factor

**COLLECTING Refer to the beginning of the lesson. By what percent should Adriana enlarge the ticket stub so that the dimensions of its image are 3 times that of her original? What will be the dimensions of the enlarged image?**

Adriana wants to create a dilated image of her ticket stub using the copier. The scale factor of her enlargement is 3. Written as a percent, the scale factor is (3 • 100)% or 300%. Now find the dimension of the enlarged image using the scale factor.

width: 5 cm • 300% = 15 cm     length: 6.4 cm • 300% = 19.2 cm

The enlarged ticket stub image will be 15 centimeters by 19.2 centimeters.

**Guided Practice**

**2.** If the resulting ticket stub image was 1.5 centimeters wide by about 1.9 centimeters long instead, what percent did Adriana mistakenly use to dilate the original image? Explain your reasoning.

CORBIS/SuperStock

## 2 Verify Similarity

**Verify Similarity** You can verify that a dilation produces a similar figure by comparing corresponding sides and angles. For triangles, you can also use SAS Similarity.

### Example 3 Verify Similarity after a Dilation

> **StudyTip**
>
> **Center of Dilation** Unless otherwise stated, all dilations on the coordinate plane use the origin as their center of dilation.

**Graph the original figure and its dilated image. Then verify that the dilation is a similarity transformation.**

**a. original:** $A(-6, -3)$, $B(3, 3)$, $C(3, -3)$; **image:** $X(-4, -2)$, $Y(2, 2)$, $Z(2, -2)$

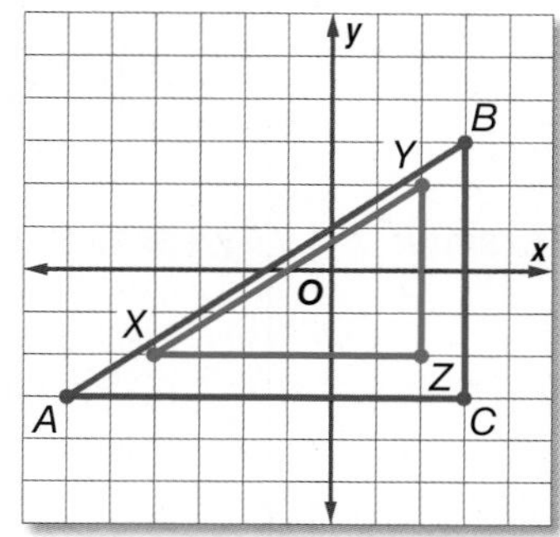


Graph each figure. Since $\angle C$ and $\angle Z$ are both right angles, $\angle C \cong \angle Z$. Show that the lengths of the sides that include $\angle C$ and $\angle Z$ are proportional.

Use the coordinate grid to find the side lengths.

$\frac{XZ}{AC} = \frac{6}{9}$ or $\frac{2}{3}$, and $\frac{YZ}{BC} = \frac{4}{6}$ or $\frac{2}{3}$, so $\frac{XZ}{AC} = \frac{YZ}{BC}$.

Since the lengths of the sides that include $\angle C$ and $\angle Z$ are proportional, $\triangle XYZ \sim \triangle ABC$ by SAS Similarity.

**b. original:** $J(-6, 4)$, $K(6, 8)$, $L(8, 2)$, $M(-4, -2)$;
**image:** $P(-3, 2)$, $Q(3, 4)$, $R(4, 1)$, $S(-2, -1)$

Use the Distance Formula to find the length of each side.

$JK = \sqrt{[6-(-6)]^2 + (8-4)^2} = \sqrt{160}$ or $4\sqrt{10}$

$PQ = \sqrt{[3-(-3)]^2 + (4-2)^2} = \sqrt{40}$ or $2\sqrt{10}$

$KL = \sqrt{(8-6)^2 + (2-8)^2} = \sqrt{40}$ or $2\sqrt{10}$

$QR = \sqrt{(4-3)^2 + (1-4)^2} = \sqrt{10}$

$LM = \sqrt{(-4-8)^2 + (-2-2)^2} = \sqrt{160}$ or $4\sqrt{10}$

$RS = \sqrt{(-2-4)^2 + (-1-1)^2} = \sqrt{40}$ or $2\sqrt{10}$

$MJ = \sqrt{[-6-(-4)]^2 + [4-(-2)]^2} = \sqrt{40}$ or $2\sqrt{10}$

$SP = \sqrt{[-3-(-2)]^2 + [2-(-1)]^2} = \sqrt{10}$

Find and compare the ratios of corresponding sides.

$\frac{PQ}{JK} = \frac{2\sqrt{10}}{4\sqrt{10}}$ or $\frac{1}{2}$ $\quad \frac{QR}{KL} = \frac{\sqrt{10}}{2\sqrt{10}}$ or $\frac{1}{2}$ $\quad \frac{RS}{LM} = \frac{2\sqrt{10}}{4\sqrt{10}}$ or $\frac{1}{2}$ $\quad \frac{SP}{MJ} = \frac{\sqrt{10}}{2\sqrt{10}}$ or $\frac{1}{2}$

$PQRS$ and $JKLM$ are both rectangles. This can be proved by showing that diagonals $\overline{PR} \cong \overline{SQ}$ and $\overline{JL} \cong \overline{KM}$ are congruent using the Distance Formula. Since they are both rectangles, their corresponding angles are congruent.

Since $\frac{PQ}{JK} = \frac{QR}{KL} = \frac{RS}{LM} = \frac{SP}{MJ}$ and corresponding angles are congruent, $PQRS \sim JKLM$.

**GuidedPractice**

**3A.** original: $A(2, 3)$, $B(0, 1)$, $C(3, 0)$
image: $D(4, 6)$, $F(0, 2)$, $G(6, 0)$

**3B.** original: $H(0, 0)$, $J(6, 0)$, $K(6, 4)$, $L(0, 4)$
image: $W(0, 0)$, $X(3, 0)$, $Y(3, 2)$, $Z(0, 2)$

## Check Your Understanding

**Example 1** **Determine whether the dilation from $A$ to $B$ is an *enlargement* or a *reduction*. Then find the scale factor of the dilation.**

1. 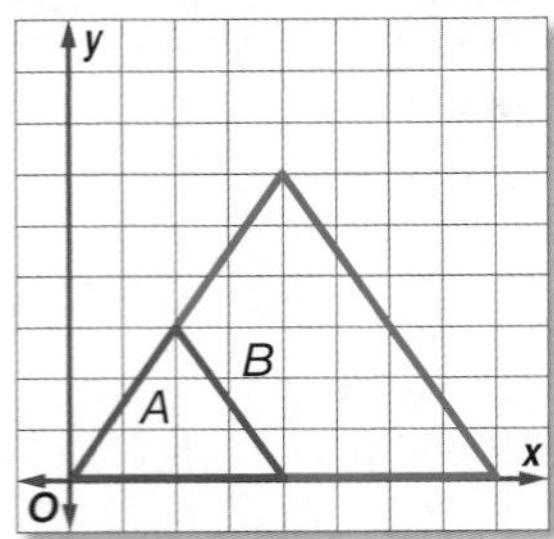

2. 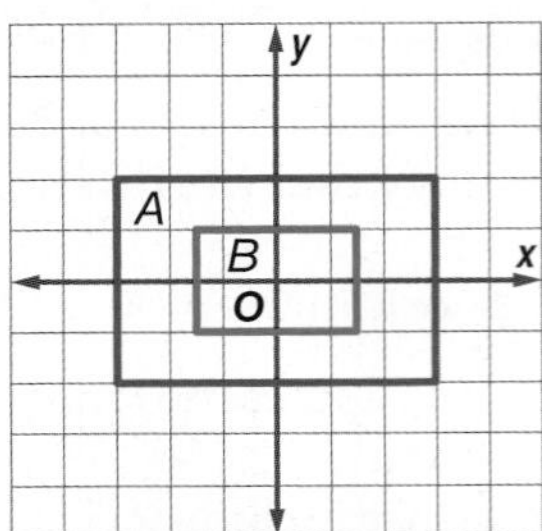

**Example 2** 3. **GAMES** The dimensions of a regulation tennis court are 27 feet by 78 feet. The dimensions of a table tennis table are 152.5 centimeters by 274 centimeters. Is a table tennis table a dilation of a tennis court? If so, what is the scale factor? Explain.

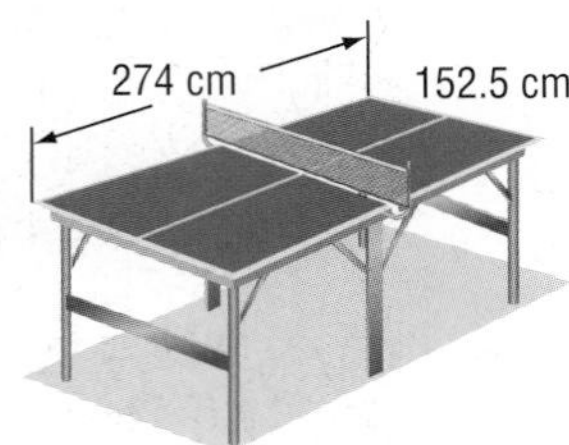


**Example 3** **CCSS ARGUMENTS** **Verify that the dilation is a similarity transformation.**

4. 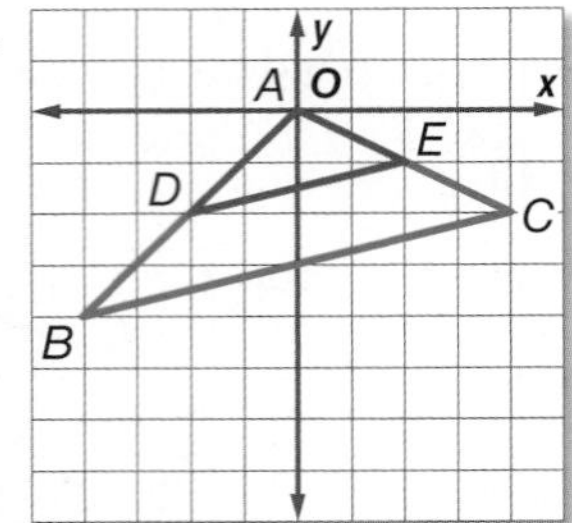

5. 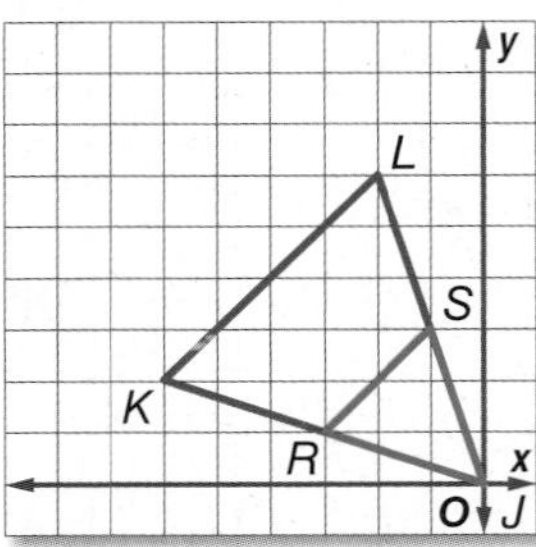

## Practice and Problem Solving

**Example 1** **Determine whether the dilation from $A$ to $B$ is an *enlargement* or a *reduction*. Then find the scale factor of the dilation.**

6. 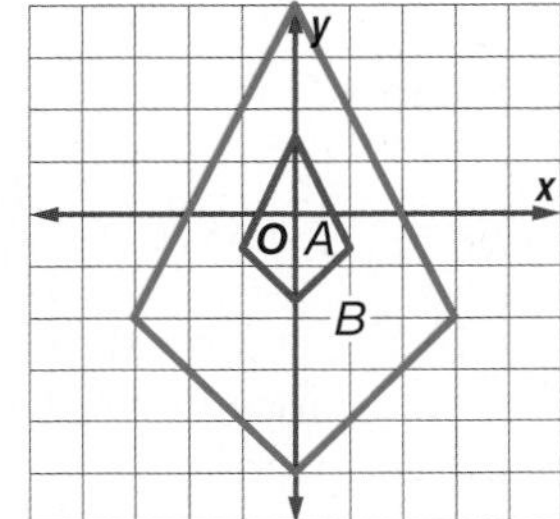

7. 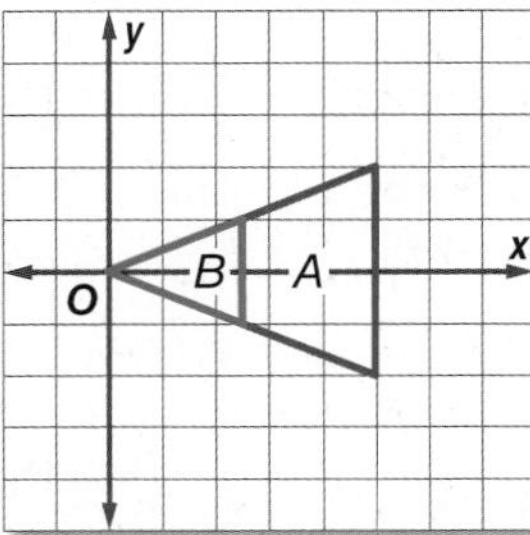

8. 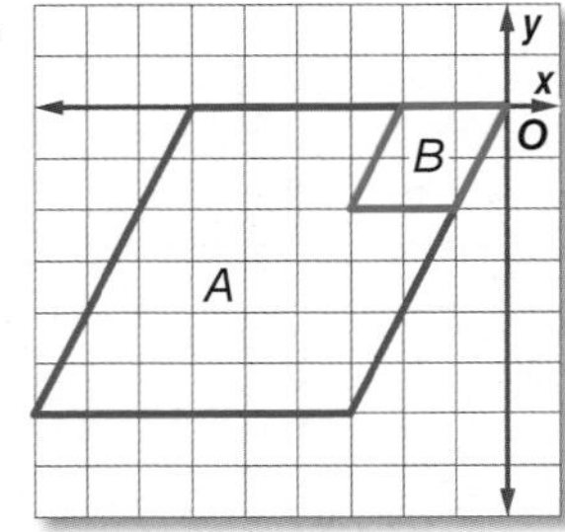

9. 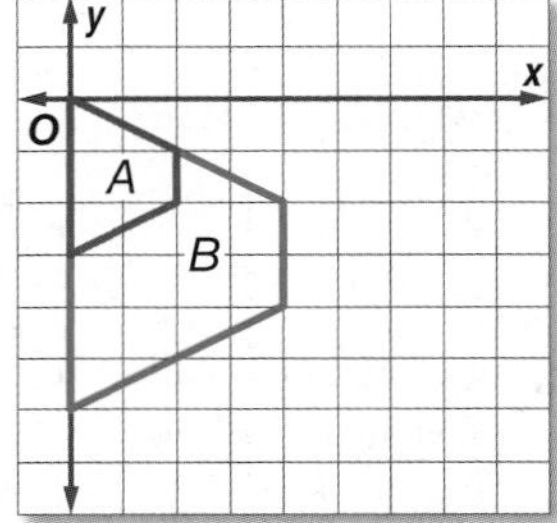

**Determine whether each dilation is an *enlargement* or *reduction*.**

**10.** Before  After

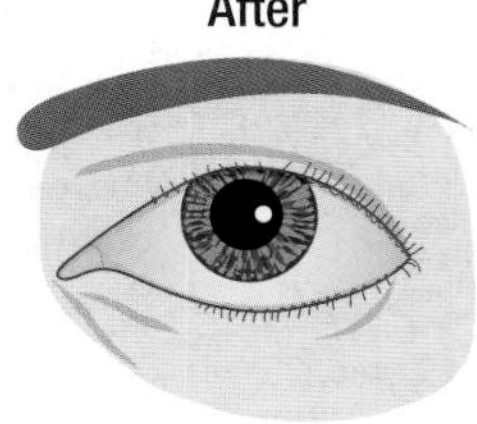

**11.** Painting  Postcard

Example 2

**12. YEARBOOK** Jordan is putting a photo of the lacrosse team in a full-page layout in the yearbook. The original photo is 4 inches by 6 inches. If the photo in the yearbook is $6\frac{2}{3}$ inches by 10 inches, is the yearbook photo a dilation of the original photo? If so, what is the scale factor? Explain.

**13.** CCSS **MODELING** Candace created a design to be made into temporary tattoos for a homecoming game as shown. Is the temporary tattoo a dilation of the original design? If so, what is the scale factor? Explain.

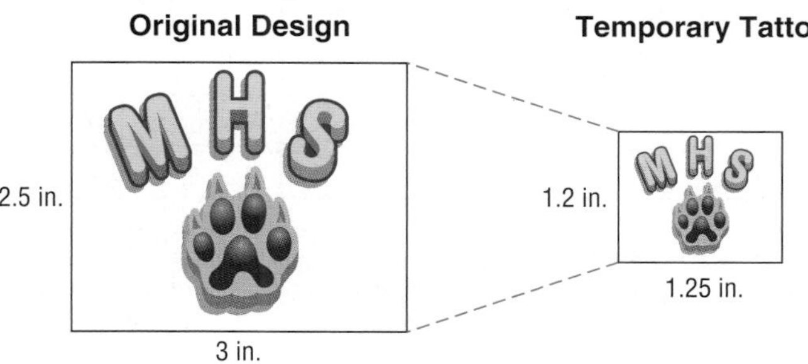


Example 3

**Graph the original figure and its dilated image. Then verify that the dilation is a similarity transformation.**

**14.** $M(1, 4)$, $P(2, 2)$, $Q(5, 5)$; $S(-3, 6)$, $T(0, 0)$, $U(9, 9)$

**15.** $A(1, 3)$, $B(-1, 2)$, $C(1, 1)$; $D(-7, -1)$, $E(1, -5)$

**16.** $V(-3, 4)$, $W(-5, 0)$, $X(1, 2)$; $Y(-6, -2)$, $Z(3, 1)$

**17.** $J(-6, 8)$, $K(6, 6)$, $L(-2, 4)$; $D(-12, 16)$, $G(12, 12)$, $H(-4, 8)$

**If $\triangle ABC \sim \triangle AYZ$, find the missing coordinate.**

**18.**

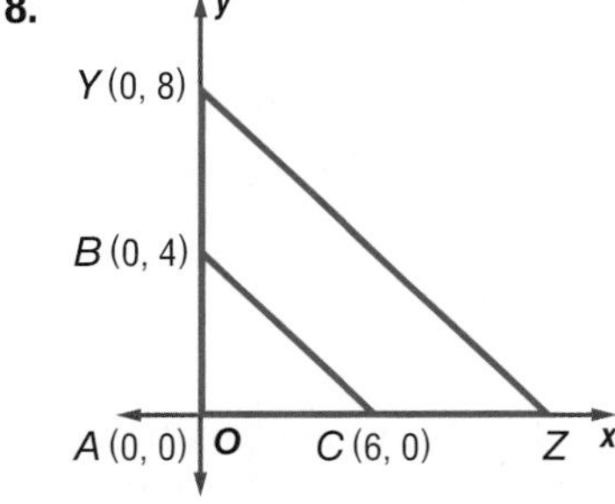


**19**

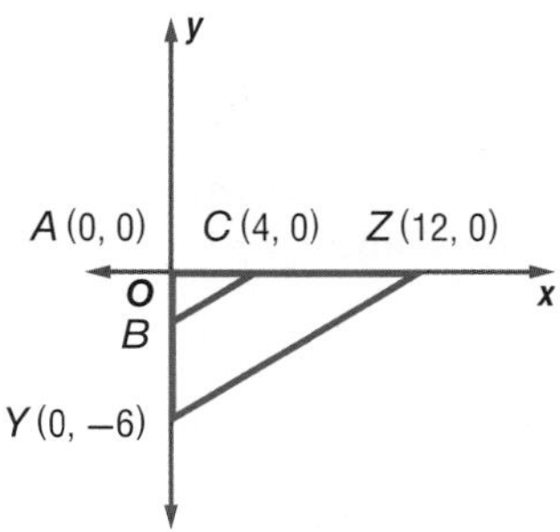

20. **GRAPHIC ART** Aimee painted the sample sign shown using $\frac{1}{2}$ bottle of glass paint. The actual sign she will paint in a shop window is to be 3 feet by $7\frac{1}{2}$ feet.

a. Explain why the actual sign is a dilation of her sample.

b. How many bottles of paint will Aimee need to complete the actual sign?

21. **MULTIPLE REPRESENTATIONS** In this problem, you will investigate similarity of triangles on the coordinate plane.

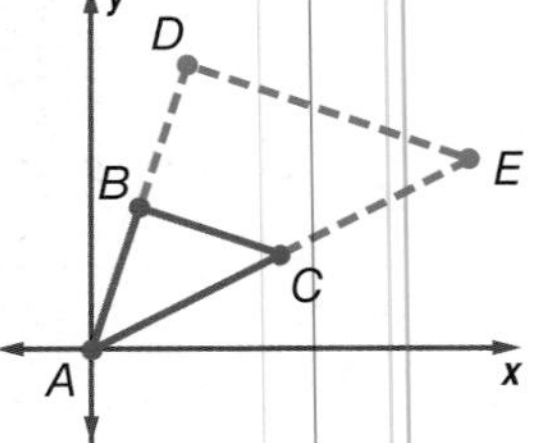


a. **Geometric** Draw a triangle with vertex *A* at the origin. Make sure that the two additional vertices *B* and *C* have whole-number coordinates. Draw a similar triangle that is twice as large as $\triangle ABC$ with its vertex also located at the origin. Label the triangle *ADE*.

b. **Geometric** Repeat the process in part **a** two times. Label the second pair of triangles *MNP* and *MQR* and the third pair *TWX* and *TYZ*. Use different scale factors than part **a**.

c. **Tabular** Copy and complete the table below with the appropriate values.

| Coordinates | | | | | | | | | | | |
|---|---|---|---|---|---|---|---|---|---|---|---|
| $\triangle ABC$ | | $\triangle ADE$ | | $\triangle MNP$ | | $\triangle MQR$ | | $\triangle TWX$ | | $\triangle TYZ$ | |
| A | | A | | M | | M | | T | | T | |
| B | | D | | N | | Q | | W | | Y | |
| C | | E | | P | | R | | X | | Z | |

d. **Verbal** Make a conjecture about how you could predict the coordinates of a dilated triangle with a scale factor of *n* if the two similar triangles share a corresponding vertex at the origin.

## H.O.T. Problems Use Higher-Order Thinking Skills

22. **CHALLENGE** *MNOP* is a dilation of *ABCD*. How is the scale factor of the dilation related to the similarity ratio of *ABCD* to *MNOP*? Explain your reasoning.

23. **CCSS REASONING** The coordinates of two triangles are provided in the table at the right. Is $\triangle XYZ$ a dilation of $\triangle PQR$? Explain.

| $\triangle PQR$ | | $\triangle XYZ$ | |
|---|---|---|---|
| P | $(a, b)$ | X | $(3a, 2b)$ |
| Q | $(c, d)$ | Y | $(3c, 2d)$ |
| R | $(e, f)$ | Z | $(3e, 2f)$ |

**OPEN ENDED Describe a real-world example of each transformation other than those given in this lesson.**

24. enlargement

25. reduction

26. congruence transformation

27. **WRITING IN MATH** Explain how you can use scale factor to determine whether a transformation is an enlargement, a reduction, or a congruence transformation.

## Standardized Test Practice

**28. ALGEBRA** Which equation describes the line that passes through $(-3, 4)$ and is perpendicular to $3x - y = 6$?

**A** $y = -\frac{1}{3}x + 4$ **C** $y = 3x + 4$

**B** $y = -\frac{1}{3}x + 3$ **D** $y = 3x + 3$

**29. SHORT RESPONSE** What is the scale factor of the dilation shown below?

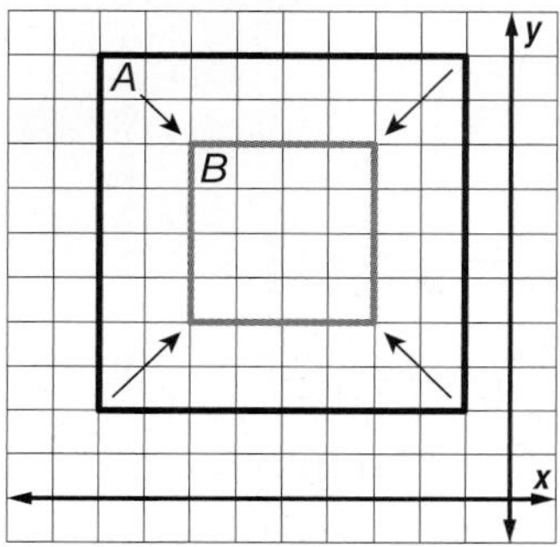


**30.** In the figure below, $\angle A \cong \angle C$.

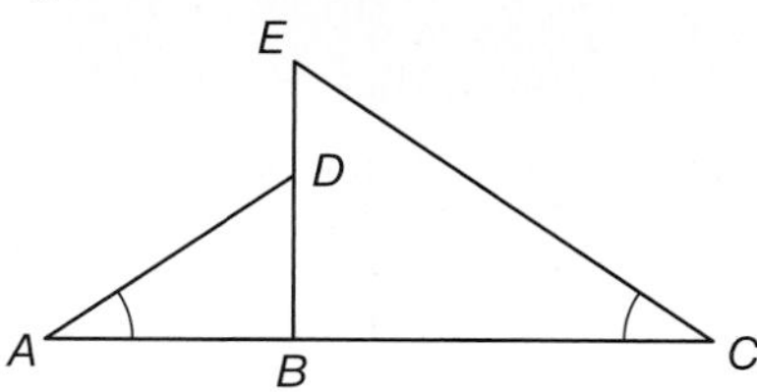


Which additional information would *not* be enough to prove that $\triangle ADB \sim \triangle CEB$?

**F** $\frac{AB}{DB} = \frac{CB}{EB}$ **H** $\overline{ED} \cong \overline{DB}$

**G** $\angle ADB \cong \angle CEB$ **J** $\overline{EB} \perp \overline{AC}$

**31. SAT/ACT** $x = \frac{6}{4p + 3}$ and $xy = \frac{3}{4p + 3}$. $y =$

**A** 4 **C** 1 **E** $\frac{1}{2}$

**B** 2 **D** $\frac{3}{4}$

## Spiral Review

**32. LANDSCAPING** Shea is designing two gardens shaped like similar triangles. One garden has a perimeter of 53.5 feet, and the longest side is 25 feet. She wants the second garden to have a perimeter of 32.1 feet. Find the length of the longest side of this garden. (Lesson 9-5)

**Determine whether $\overline{AB} \parallel \overline{CD}$. Justify your answer.** (Lesson 9-4)

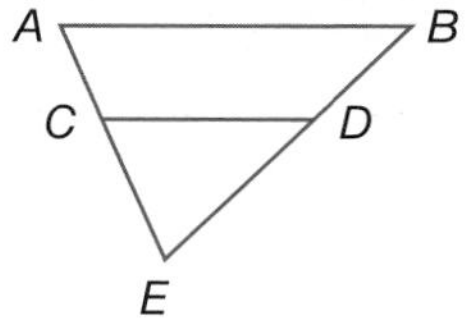

A B C D E

**33.** $AC = 8.4$, $BD = 6.3$, $DE = 4.5$, and $CE = 6$

**34.** $AC = 7$, $BD = 10.5$, $BE = 22.5$, and $AE = 15$

**35.** $AB = 8$, $AE = 9$, $CD = 4$, and $CE = 4$

**If each figure is a kite, find each measure.** (Lesson 8-6)

**36.** $QR$

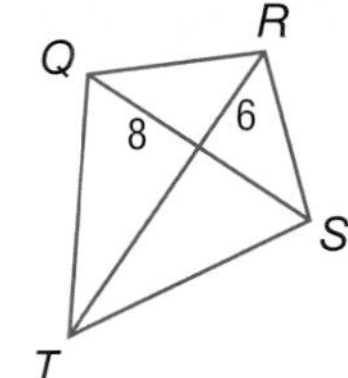


**37.** $m\angle K$

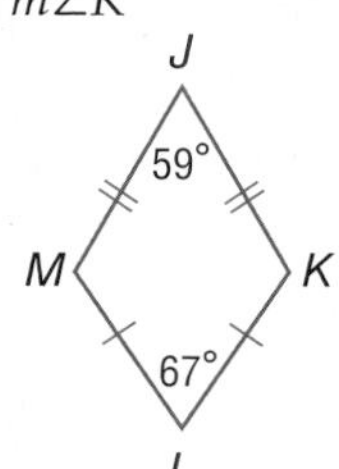


**38.** $BC$

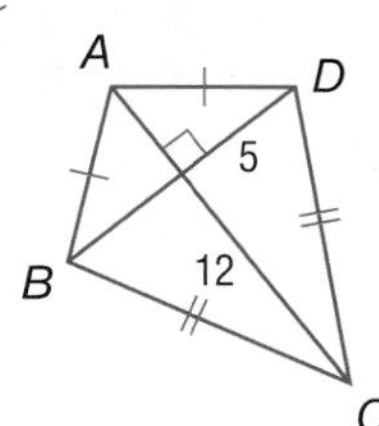


**39. PROOF** Write a coordinate proof for the following statement. (Lesson 6-6)
*If a line segment joins the midpoints of two sides of a triangle, then it is parallel to the third side.*

## Skills Review

**Solve each equation.**

**40.** $145 = 29 \cdot t$

**41.** $216 = d \cdot 27$

**42.** $2r = 67 \cdot 5$

**43.** $100t = \frac{70}{240}$

**44.** $\frac{80}{4} = 14d$

**45.** $\frac{2t + 15}{t} = 92$

LESSON 9-7

# Scale Drawings and Models

**Then**

- You used scale factors to solve problems with similar polygons.

**Now**

1. Interpret scale models.
2. Use scale factors to solve problems.

**Why?**

- In Saint-Luc, Switzerland, Le Chemin des planetes, has constructed a scale model of each planet in the solar system. It is one of the largest complete three-dimensional scale models of the solar system. The diameter of the center of the model of Saturn shown is 121 millimeters; the diameter of the real planet is about 121,000 kilometers.

**NewVocabulary**
scale model
scale drawing
scale

**Common Core State Standards**

**Content Standards**
G.MG.3 Apply geometric methods to solve problems (e.g., designing an object or structure to satisfy physical constraints or minimize cost; working with typographic grid systems based on ratios). ★

**Mathematical Practices**
4 Model with mathematics.
7 Look for and make use of structure.

## 1 Scale Models

A **scale model** or a **scale drawing** is an object or drawing with lengths proportional to the object it represents. The **scale** of a model or drawing is the ratio of a length on the model or drawing to the actual length of the object being modeled or drawn.

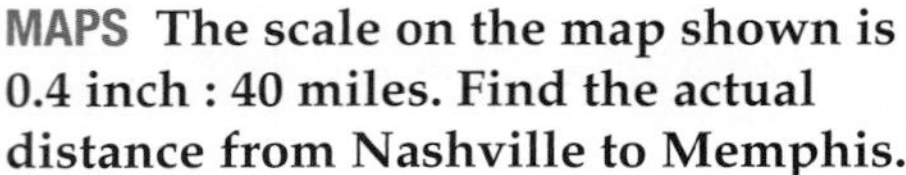

### Example 1 Use a Scale Drawing

**MAPS The scale on the map shown is 0.4 inch : 40 miles. Find the actual distance from Nashville to Memphis.**

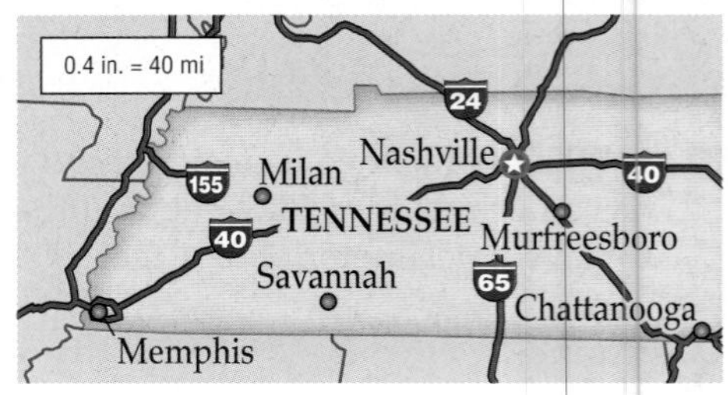


Use a ruler. The distance between Nashville and Memphis is about 1.5 inches.

**Method 1** Write and solve a proportion.

Let $x$ represent the distance between Nashville and Memphis.

| | Scale | | Nashville to Memphis | | |
|---|---|---|---|---|---|
| map → / actual → | $\frac{0.4 \text{ in.}}{40 \text{ mi}}$ | = | $\frac{1.5 \text{ in.}}{x \text{ mi}}$ | ← map / ← actual | |

$0.4 \cdot x = 40 \cdot 1.5$ Cross Products Property

$x = 150$ Simplify.

**Method 2** Write and solve an equation.

Let $a$ = actual distance in miles between Nashville and Memphis and $m$ = map distance in inches. Write the scale as $\frac{40 \text{ mi}}{0.4 \text{ in.}}$, which is $40 \div 0.4$ or 100 miles per inch. So for every inch on the map, the actual distance is 100 miles.

$a = 100 \cdot m$ Write an equation.

$= 100 \cdot 1.5$ $m = 1.5$ in.

$= 150$ Solve.

**CHECK** Use dimensional analysis.

$\text{mi} = \frac{\text{mi}}{\text{in.}} \cdot \text{in.} \Rightarrow \text{mi} = \text{mi}$ ✓

The distance between Nashville and Memphis is 150 miles.

**GuidedPractice**

1. **MAPS** Find the actual distance between Nashville and Chattanooga.

Robert F. Bukaty/AP Images

## 2 Use Scale Factors

**Use Scale Factors** The scale factor of a drawing or scale model is written as a unitless ratio in simplest form. Scale factors are always written so that the model length in the ratio comes first.

### Example 2 Find the Scale

**SCALE MODEL This is a miniature replica of a 1923 Checker Cab. The length of the model is 6.5 inches. The actual length of the car was 13 feet.**

**a. What is the scale of the model?**

To find the scale, write the ratio of a model length to an actual length.

$$\frac{\text{model length}}{\text{actual length}} = \frac{6.5 \text{ in.}}{13 \text{ ft}} \text{ or } \frac{1 \text{ in.}}{2 \text{ ft}}$$

The scale of the model is 1 in.:2 ft.

**b. How many times as long as the actual car is the model?**

To answer this question, find the scale factor of the model. Multiply by a conversion factor that relates inches to feet to obtain a unitless ratio.

$$\frac{1 \text{ in.}}{2 \text{ ft}} = \frac{1 \text{ in.}}{2 \text{ ft}} \cdot \frac{1 \text{ ft}}{12 \text{ in.}} = \frac{1}{24}$$

The scale factor is 1:24. That is, the model is $\frac{1}{24}$ as long as the actual car.

**StudyTip**

**CCSS Regularity** The scale factor of a model that is smaller than the original object is between 0 and 1 and the scale factor for a model that is larger than the original object is greater than 1.

**GuidedPractice**

**2. SCALE MODEL** Mrs. Alejandro's history class made a scale model of the Alamo that is 3 feet tall. The actual height of the building is 33 feet 6 inches.

**A.** What is the scale of the model?

**B.** How many times as tall as the actual building is the model? How many times as tall as the model is the actual building?

### Real-World Example 3 Construct a Scale Model

**SCALE MODEL Suppose you want to build a model of the St. Louis Gateway Arch that is no more than 11 inches tall. Choose an appropriate scale and use it to determine the height of the model. Use the information at the left.**

The actual monument is 630 feet tall. Since 630 feet ÷ 11 inches = 57.3 feet per inch, a scale of 1 inch = 60 feet is an appropriate scale. So, for every inch on the model $m$, let the actual measure $a$ be 60 feet. Write this as an equation.

| | |
|---|---|
| $a = 60 \cdot m$ | Write an equation. |
| $630 = 60 \cdot m$ | $a = 630$ |
| $10.5 = m$ | So the height of the model would be 10.5 inches. |

**Real-WorldLink**

The St. Louis Gateway Arch is the tallest national monument in the United States at 630 feet. The span of the base is also 630 feet. The arch weighs 17,246 tons and can sway a maximum of 9 inches in each direction during high winds.

**Source:** Gateway Arch Facts



**GuidedPractice**

**3. SCALE DRAWING** Sonya is making a scale drawing of her room on an 8.5-by-11-inch sheet of paper. If her room is 14 feet by 12 feet, find an appropriate scale for the drawing and determine the dimensions of the drawing.

## Check Your Understanding

**Example 1** **MAPS Use the map of Maine shown and a customary ruler to find the actual distance between each pair of cities. Measure to the nearest sixteenth of an inch.**

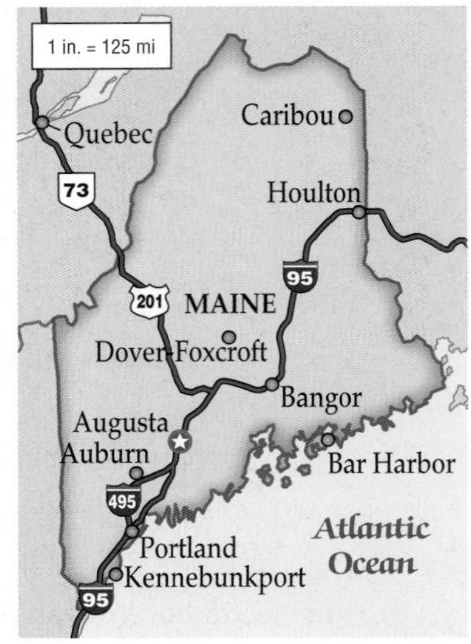


1. Bangor and Portland
2. Augusta and Houlton

**Example 2**

3. **SCALE MODELS** Carlos made a scale model of a local bridge. The model spans 6 inches; the actual bridge spans 50 feet.
   a. What is the scale of the model?
   b. What scale factor did Carlos use to build his model?

**Example 3**

4. **SPORTS** A volleyball court is 9 meters wide and 18 meters long. Choose an appropriate scale and construct a scale drawing of the court to fit on a 3-inch by 5-inch index card.

## Practice and Problem Solving

**Example 1** **CCSS MODELING Use the map of Oklahoma shown and a metric ruler to find the actual distance between each pair of cities. Measure to the nearest centimeter.**

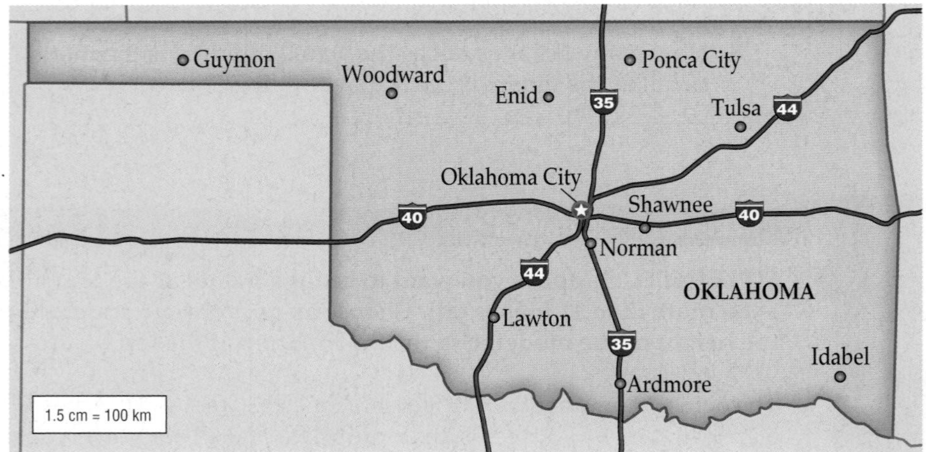


5. Guymon and Oklahoma City
6. Lawton and Tulsa
7. Enid and Tulsa
8. Ponca City and Shawnee

**Example 2**

9. **SCULPTURE** A replica of *The Thinker* is 10 inches tall. A statue of *The Thinker* at the University of Louisville is 10 feet tall.
   a. What is the scale of the replica?
   b. How many times as tall as the actual sculpture is the replica?

Buena Vista Images/Iconica/Getty Images

**10. MAPS** The map below shows a portion of Frankfort, Kentucky.

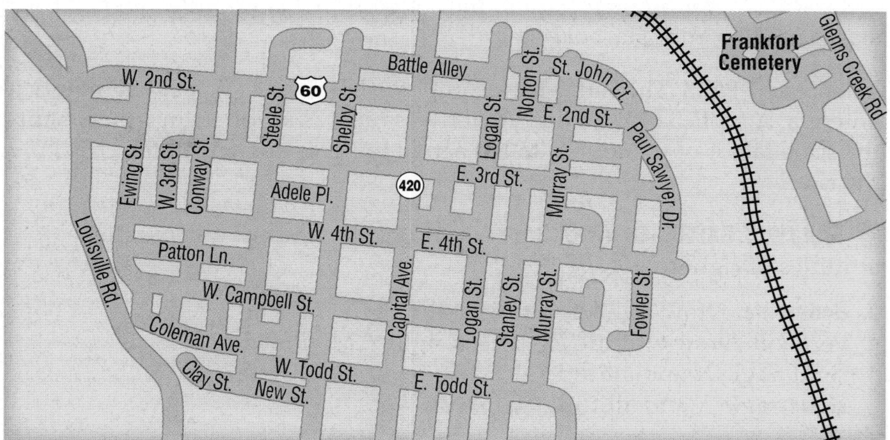


a. If the actual distance from the intersection of Conway Street and 4th Street to the intersection of Murray Street and 4th Street is 0.47 mile, use a customary ruler to estimate the scale of the map.

b. What is the approximate scale factor of the map? Interpret its meaning.

Example 3

**SPORTS Choose an appropriate scale and construct a scale drawing of each playing area so that it would fit on an 8.5-by-11-inch sheet of paper.**

11. A baseball diamond is a square 90 feet on each side with about a 128-foot diagonal.

12. A high school basketball court is a rectangle with length 84 feet and width 50 feet.

**CCSS MODELING Use the map shown and an inch ruler to answer each question. Measure to the nearest sixteenth of an inch and assume that you can travel along any straight line.**

13. About how long would it take to drive from Valdosta, Georgia, to Daytona Beach, Florida, traveling at 65 miles per hour?

14. How long would it take to drive from Gainesville to Miami, Florida, traveling at 70 miles per hour?

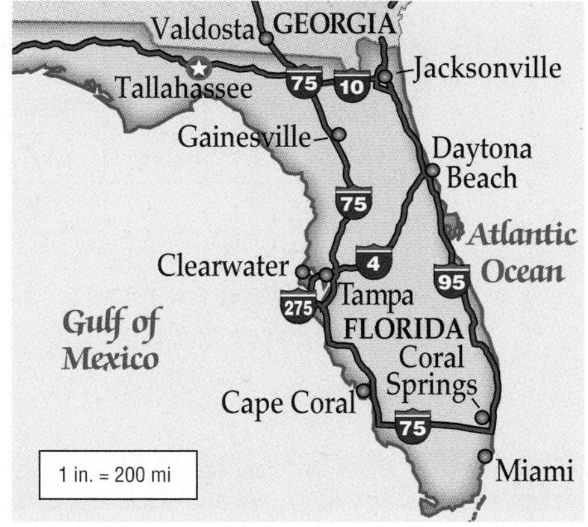


15. **SCALE MODELS** If the distance between Earth and the Sun is actually 150,000,000 kilometers, how far apart are Earth and the Sun when using the 1:93,000,000 scale model?

16. **LITERATURE** In the book, *Alice's Adventures in Wonderland*, Alice's size changes from her normal height of about 50 inches. Suppose Alice came across a door about 15 inches high and her height changed to 10 inches.

a. Find the ratio of the height of the door to Alice's height in Wonderland.

b. How tall would the door have been in Alice's normal world?

17. **ROCKETS** Peter bought a $\frac{1 \text{ in.}}{12 \text{ ft}}$ scale model of the Mercury-Redstone rocket.

a. If the height of the model is 7 inches, what is the approximate height of the rocket?

b. If the diameter of the rocket is 70 inches, what is the diameter of the model? Round to the nearest half inch.

18. **ARCHITECTURE** A replica of the Statue of Liberty in Austin, Texas, is $16\frac{3}{4}$ feet tall. If the scale factor of the replica to the actual statue is 1:9, how tall is the actual statue in New York Harbor?

19. **AMUSEMENT PARK** The Eiffel Tower in Paris, France, is 986 feet tall, not including its antenna. A replica of the Eiffel Tower was built as a ride in an amusement park. If the scale factor of the replica to the actual tower is approximately 1:3, how tall is the ride?

20. **MULTIPLE REPRESENTATIONS** In this problem, you will explore the altitudes of right triangles.



a. **Geometric** Draw right $\triangle ABC$ with the right angle at vertex $B$. Draw altitude $\overline{BD}$. Draw right $\triangle MNP$, with right angle $N$ and altitude $\overline{NQ}$, and right $\triangle WXY$, with right angle $X$ and altitude $\overline{XZ}$.

b. **Tabular** Measure and record indicated angles in the table below.

| | Angle Measure | | | | | |
|---|---|---|---|---|---|---|
| | △*ABC* | | △*BDC* | | △*ADB* | |
| △*ABC* | *ABC* | | *BDC* | | *ADB* | |
| | *A* | | *CBD* | | *BAD* | |
| | *C* | | *DCB* | | *DBA* | |
| | △*MNP* | | △*NQP* | | △*MQN* | |
| △*MNP* | *MNP* | | *NQP* | | *MQN* | |
| | *M* | | *PNQ* | | *NMQ* | |
| | *P* | | *QPN* | | *QNM* | |
| | △*WXY* | | △*WZX* | | △*XZY* | |
| △*WXY* | *WXY* | | *WZX* | | *XZY* | |
| | *W* | | *XWZ* | | *YXZ* | |
| | *Y* | | *ZXW* | | *ZYX* | |

c. **Verbal** Make a conjecture about the altitude of a right triangle originating at the right angle of the triangle.

## H.O.T. Problems Use Higher-Order Thinking Skills

21. **ERROR ANALYSIS** Felix and Tamara are building a replica of their high school. The high school is 75 feet tall and the replica is 1.5 feet tall. Felix says the scale factor of the actual high school to the replica is 50:1, while Tamara says the scale factor is 1:50. Is either of them correct? Explain your reasoning.

22. **CHALLENGE** You can produce a scale model of a certain object by extending each dimension by a constant. What must be true of the shape of the object? Explain your reasoning.

23. **CCSS SENSE-MAKING** Sofia is making two scale drawings of the lunchroom. In the first drawing, Sofia used a scale of 1 inch = 1 foot, and in the second drawing she used a scale of 1 inch = 6 feet. Which scale will produce a larger drawing? What is the scale factor of the first drawing to the second drawing? Explain.

24. **OPEN ENDED** Draw a scale model of your classroom using any scale.

25. **WRITING IN MATH** Compare and contrast scale and scale factor.

## Standardized Test Practice

**26. SHORT RESPONSE** If $3^x = 27^{(x-4)}$, then what is the value of $x$?

**27.** In $\triangle ABC$, $\overline{BD}$ is a median. If $AD = 3x + 5$ and $CD = 5x - 1$, find $AC$.

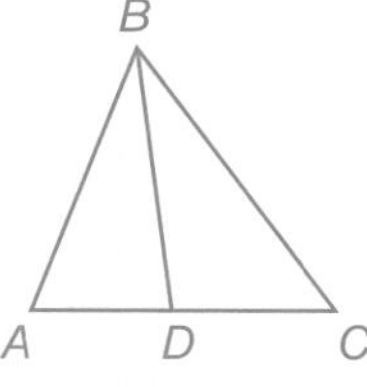


**A** 6 **C** 14

**B** 12 **D** 28

**28.** In a triangle, the ratio of the measures of the sides is 4 : 7 : 10, and its longest side is 40 centimeters. Find the perimeter of the triangle in centimeters.

**F** 37 cm **H** 84 cm

**G** 43 cm **J** 168 cm

**29. SAT/ACT** If Lydia can type 80 words in two minutes, how long will it take Lydia to type 600 words?

**A** 30 min **D** 10 min

**B** 20 min **E** 5 min

**C** 15 min

## Spiral Review

**30. PAINTING** Aaron is painting a portrait of a friend for an art class. Since his friend doesn't have time to model, he uses a photo that is 6 inches by 8 inches. If the canvas is 24 inches by 32 inches, is the painting a dilation of the original photo? If so, what is the scale factor? Explain. (Lesson 9-6)

**Find $x$.** (Lesson 9-5)

**31.**

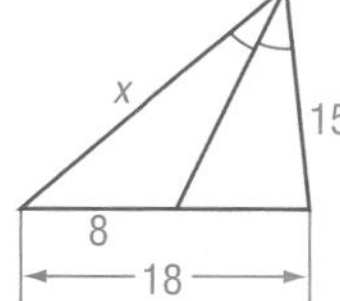


**32.**

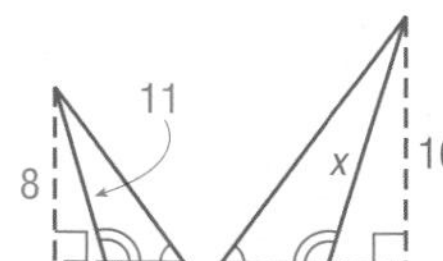


**33.**

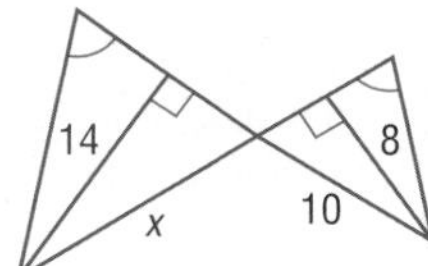


**ALGEBRA Quadrilateral $JKMN$ is a rectangle.** (Lesson 8-4)

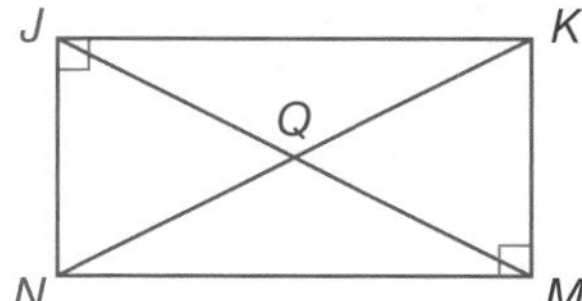


**34.** If $NQ = 2x + 3$ and $QK = 5x - 9$, find $JQ$.

**35.** If $m\angle NJM = 2x - 3$ and $m\angle KJM = x + 5$, find $x$.

**36.** If $NM = 8x - 14$ and $JK = x^2 + 1$, find $JK$.

**In $\triangle ABC$, $MC = 7$, $RM = 4$, and $AT = 16$. Find each measure.** (Lesson 7-2)

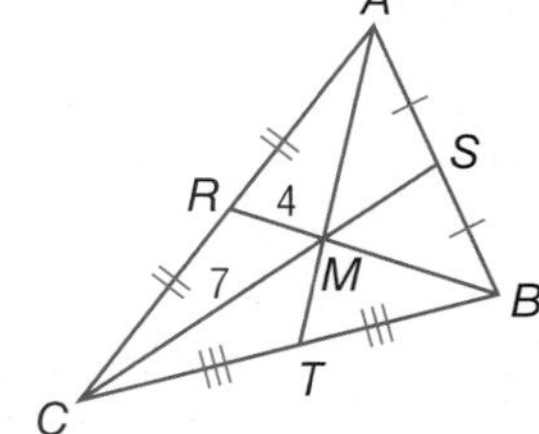


**37.** $MS$ **38.** $AM$ **39.** $SC$

**40.** $RB$ **41.** $MB$ **42.** $TM$

**Determine whether $\triangle JKL \cong \triangle XYZ$. Explain.** (Lesson 6-3)

**43.** $J(3, 9)$, $K(4, 6)$, $L(1, 5)$, $X(1, 7)$, $Y(2, 4)$, $Z(-1, 3)$

**44.** $J(-1, -1)$, $K(0, 6)$, $L(2, 3)$, $X(3, 1)$, $Y(5, 3)$, $Z(8, 1)$

## Skills Review

**Simplify each expression.**

**45.** $\sqrt{4 \cdot 16}$ **46.** $\sqrt{3 \cdot 27}$ **47.** $\sqrt{32 \cdot 72}$ **48.** $\sqrt{15 \cdot 16}$ **49.** $\sqrt{33 \cdot 21}$

# CHAPTER 9 Study Guide and Review

## Study Guide

### KeyConcepts

**Proportions** (Lesson 9-1)

- For any numbers $a$ and $c$ and any nonzero numbers $b$ and $d$, $\frac{a}{b} = \frac{c}{d}$ if and only if $ad = bc$.

**Similar Polygons and Triangles** (Lessons 9-2 and 9-3)

- Two polygons are similar if and only if their corresponding angles are congruent and the measures of their corresponding sides are proportional.
- Two triangles are similar if:

  AA: Two angles of one triangle are congruent to two angles of the other triangle.

  SSS: The measures of the corresponding sides of the two triangles are proportional.

  SAS: The measures of two sides of one triangle are proportional to the measures of two corresponding sides of another triangle and their included angles are congruent.

**Proportional Parts** (Lessons 9-4 and 9-5)

- If a line is parallel to one side of a triangle and intersects the other two sides in two distinct points, then it separates these sides into segments of proportional length.
- A midsegment of a triangle is parallel to one side of the triangle and its length is one-half the length of that side.
- Two triangles are similar when each of the following are proportional in measure: their perimeters, their corresponding altitudes, their corresponding angle bisectors, and their corresponding medians.

**Similarity Transformations and Scale Drawings and Models** (Lessons 9-6 and 9-7)

- A scale model or scale drawing has lengths that are proportional to the corresponding lengths in the object it represents.

### FOLDABLES® StudyOrganizer

Be sure the Key Concepts are noted in your Foldable.

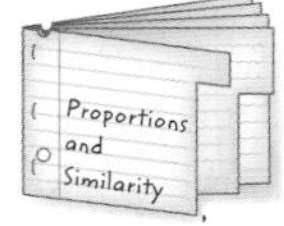


### KeyVocabulary

cross products (p. 544)
dilation (p. 593)
enlargement (p. 593)
extremes (p. 544)
means (p. 544)
midsegment of a triangle (p. 573)
proportion (p. 544)
ratio (p. 543)
reduction (p. 593)
scale (p. 600)
scale drawing (p. 600)
scale factor (p. 552)
scale model (p. 600)
similar polygons (p. 551)
similarity transformation (p. 593)

### VocabularyCheck

**Choose the letter of the word or phrase that best completes each statement.**

**a.** ratio
**b.** proportion
**c.** means
**d.** extremes
**e.** similar
**f.** scale factor
**g.** AA Similarity Post.
**h.** SSS Similarity Theorem
**i.** SAS Similarity Theorem
**j.** midsegment
**k.** dilation
**l.** enlargement
**m.** reduction

1. A(n) __?__ of a triangle has endpoints that are the midpoints of two sides of the triangle.
2. A(n) __?__ is a comparison of two quantities using division.
3. If $\angle A \cong \angle X$ and $\angle C \cong \angle Z$, then $\triangle ABC \sim \triangle XYZ$ by the __?__.
4. A(n) __?__ is an example of a similarity transformation.
5. If $\frac{a}{b} = \frac{c}{d}$, then $a$ and $d$ are the __?__.
6. The ratio of the lengths of two corresponding sides of two similar polygons is the __?__.
7. A(n) __?__ is an equation stating that two ratios are equivalent.
8. A dilation with a scale factor of $\frac{2}{5}$ will result in a(n) __?__.

# Lesson-by-Lesson Review

## 9-1 Ratios and Proportions

**Solve each proportion.**

9. $\frac{x+8}{6} = \frac{2x-3}{10}$
10. $\frac{3x+9}{x} = \frac{12}{5}$
11. $\frac{x}{12} = \frac{50}{6x}$
12. $\frac{7}{x} = \frac{14}{9}$
13. The ratio of the lengths of the three sides of a triangle is 5:8:10. If its perimeter is 276 inches, find the length of the longest side of the triangle.
14. **CARPENTRY** A board that is 12 feet long must be cut into two pieces that have lengths in a ratio of 3 to 2. Find the lengths of the two pieces.

**Example 1**

Solve $\frac{2x-3}{4} = \frac{x+9}{3}$.

| | |
|---|---|
| $\frac{2x-3}{4} = \frac{x+9}{3}$ | Original proportion |
| $3(2x-3) = 4(x+9)$ | Cross Products Property |
| $6x - 9 = 4x + 36$ | Simplify. |
| $2x - 9 = 36$ | Subtract. |
| $2x = 45$ | Add 9 to each side. |
| $x = 22.5$ | Divide each side by 2. |

## 9-2 Similar Polygons

**Determine whether each pair of figures is similar. If so, write the similarity statement and scale factor. If not, explain your reasoning.**

15.

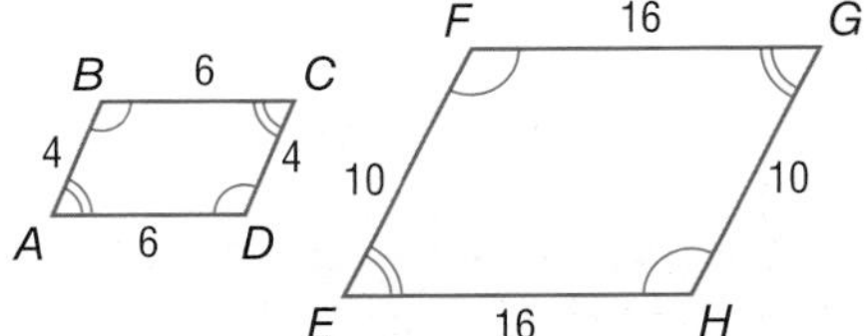


16.

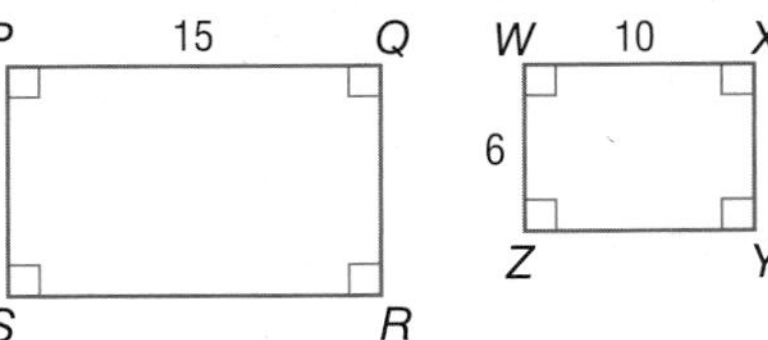


17. The two triangles in the figure below are similar. Find the value of *x*.

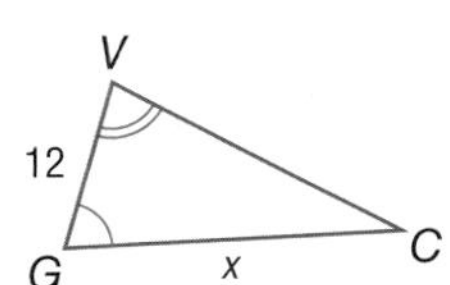


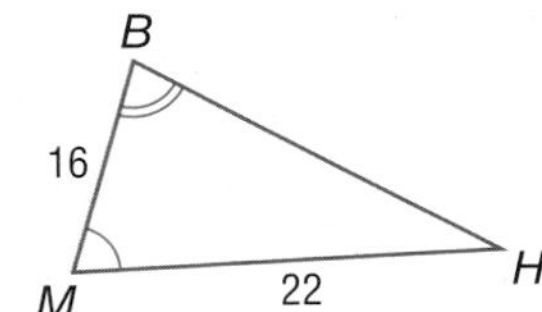


18. **PHOTOS** If the dimensions of a photo are 2 inches by 3 inches and the dimensions of a poster are 8 inches by 12 inches, are the photo and poster similar? Explain.

**Example 2**

**Determine whether the pair of triangles is similar. If so, write the similarity statement and scale factor. If not, explain your reasoning.**

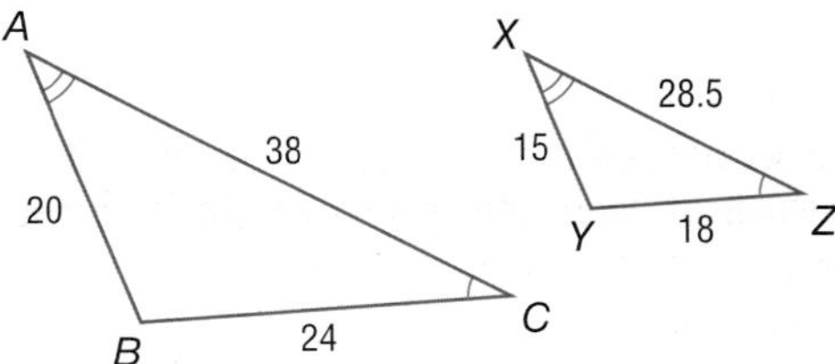


$\angle A \cong \angle X$ and $\angle C \cong \angle Z$, so by the Third Angle Theorem, $\angle B \cong \angle Y$. All of the corresponding angles are therefore congruent.

Similar polygons must also have proportional side lengths. Check the ratios of corresponding side lengths.

$\frac{AB}{XY} = \frac{20}{15}$ or $\frac{4}{3}$ $\quad \frac{BC}{YZ} = \frac{24}{18}$ or $\frac{4}{3}$ $\quad \frac{AC}{XZ} = \frac{38}{28.5}$ or $\frac{4}{3}$

Since corresponding sides are proportional, $\triangle ABC \sim \triangle XYZ$. So, the triangles are similar with a scale factor of $\frac{4}{3}$.

# Study Guide and Review *Continued*

## 9-3 Similar Triangles

Determine whether the triangles are similar. If so, write a similarity statement. Explain your reasoning.

**19.**

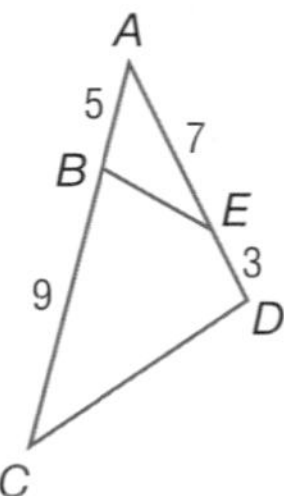


**20.**

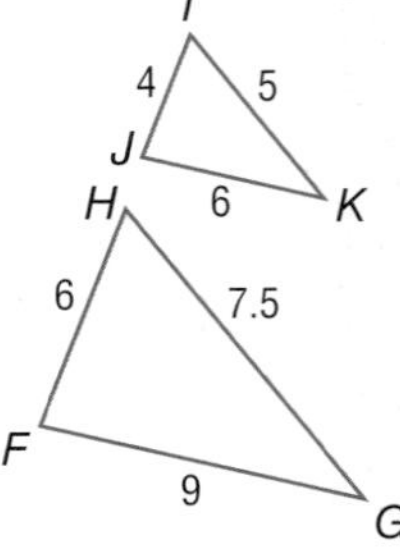


**21.**

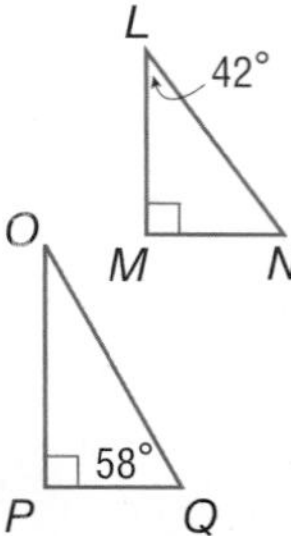


**22.**

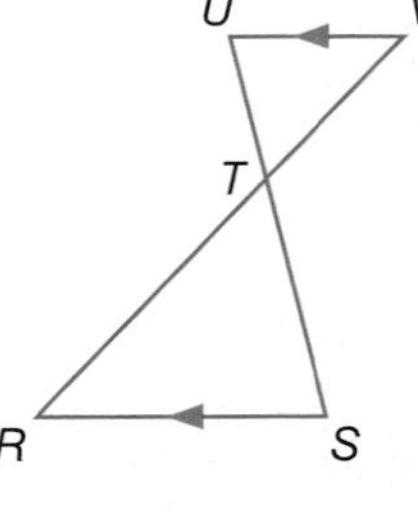


**23. TREES** To estimate the height of a tree, Dave stands in the shadow of the tree so that his shadow and the tree's shadow end at the same point. Dave is 6 feet 4 inches tall and his shadow is 15 feet long. If he is standing 66 feet away from the tree, what is the height of the tree?

### Example 3

Determine whether the triangles are similar. If so, write a similarity statement. Explain your reasoning.

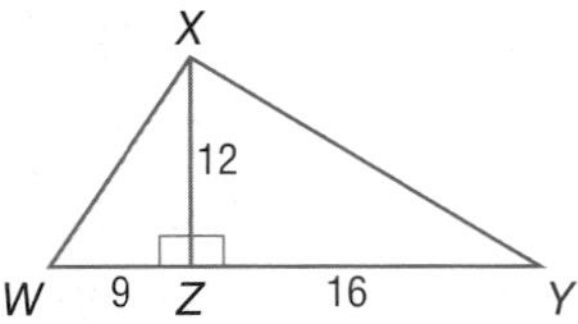


$\angle WZX \cong \angle XZY$ because they are both right angles. Now compare the ratios of the legs of the right triangles.

$$\frac{WZ}{XZ} = \frac{9}{12} = \frac{3}{4} \qquad \frac{XZ}{YZ} = \frac{12}{16} = \frac{3}{4}$$

Since two pairs of sides are proportional with the included angles congruent, $\triangle WZX \sim \triangle XZY$ by SAS Similarity.

## 9-4 Parallel Lines and Proportional Parts

Find $x$.

**24.**

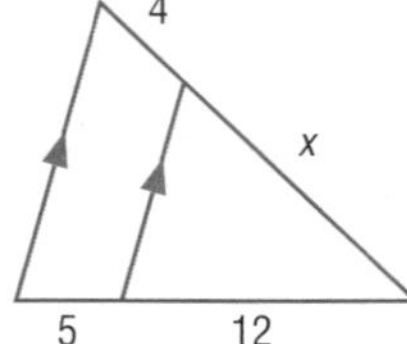


**25.**

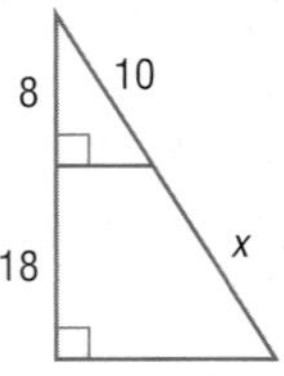


**26. STREETS** Find the distance along Broadway between 37th Street and 36th Street.

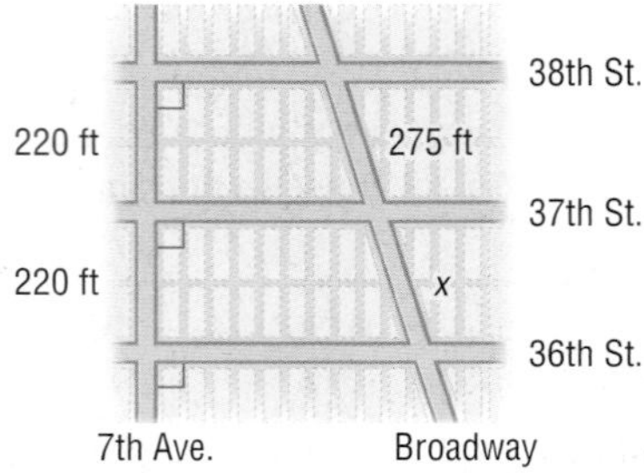


### Example 4

**ALGEBRA** Find $x$ and $y$.

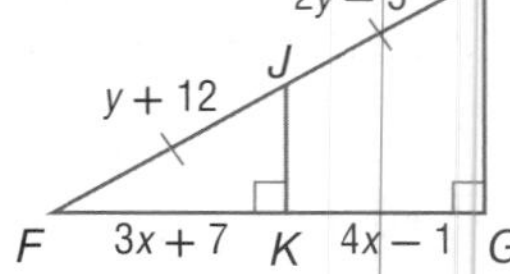


$FK = KG$

$3x + 7 = 4x - 1$

$-x = -8$

$x = 8$

| | |
|---|---|
| $FJ = JH$ | Definition of congruence |
| $y + 12 = 2y - 5$ | Substitution |
| $-y = -17$ | Subtract. |
| $y = 17$ | Simplify. |

## 9-5 Parts of Similar Triangles

**Find the value of each variable.**

**27.**

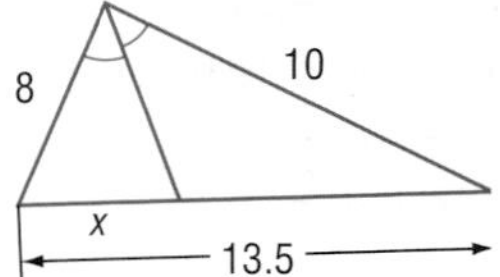


**28.**

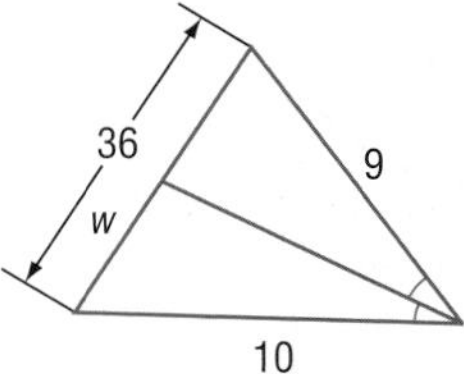


**29.** MAPS The scale given on a map of the state of Missouri indicates that 3 inches represents 50 miles. The cities of St. Louis, Springfield, and Kansas City form a triangle. If the measurements of the lengths of the sides of this triangle on the map are 15 inches, 10 inches, and 13 inches, find the perimeter of the actual triangle formed by these cities to the nearest mile.

### Example 5

**Find $x$.**

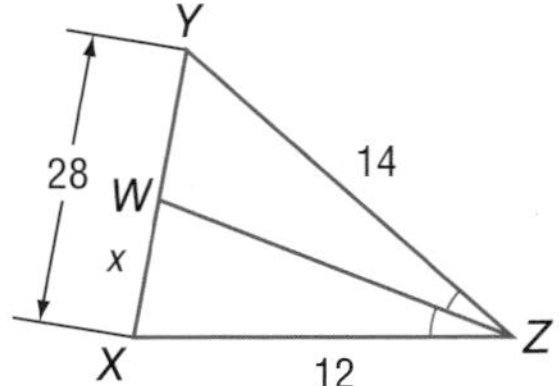


Use the Triangle Angle Bisector Theorem to write a proportion.

| | |
|---|---|
| $\frac{WX}{YW} = \frac{XZ}{YZ}$ | Triangle Angle Bisector Thm. |
| $\frac{x}{28 - x} = \frac{12}{14}$ | Substitution |
| $(28 - x)(12) = x \cdot 14$ | Cross Products Property |
| $336 - 12x = 14x$ | Simplify. |
| $336 = 26x$ | Add. |
| $12.9 = x$ | Simplify. |

## 9-6 Similarity Transformations

**Determine whether the dilation from $A$ to $B$ is an *enlargement* or a *reduction*. Then find the scale factor of the dilation.**

**30.**

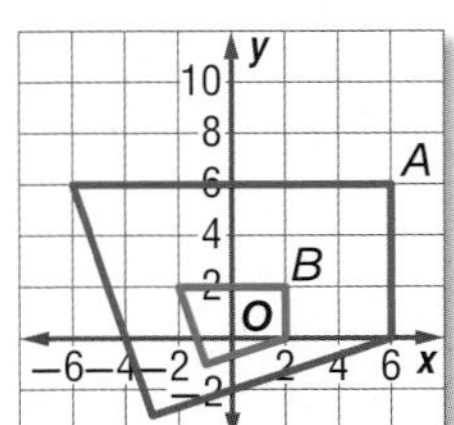


**31.**

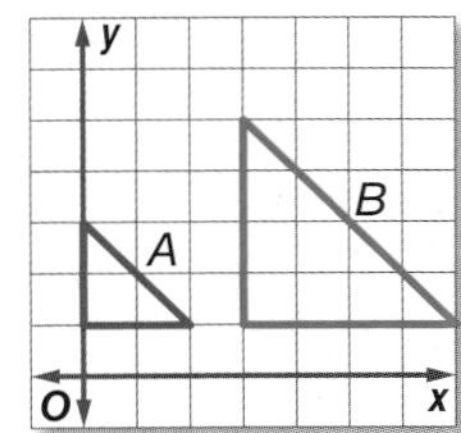


**32.** GRAPHIC DESIGN Jamie wants to use a photocopier to enlarge her design for the Honors Program at her school. She sets the copier to 250%. If the original drawing was 6 inches by 9 inches, find the dimensions of the enlargement.

### Example 6

**Determine whether the dilation from $A$ to $B$ is an *enlargement* or a *reduction*. Then find the scale factor of the dilation.**

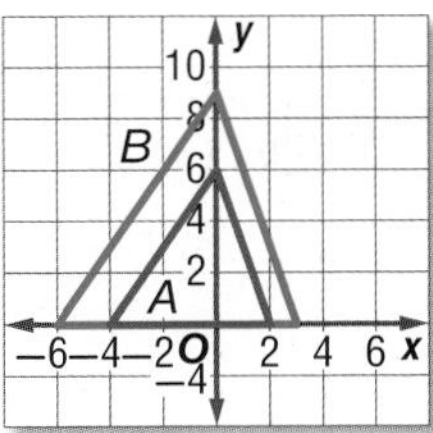


$B$ is larger than $A$, so the dilation is an enlargement. The distance between the vertices at (−4, 0) and (2, 0) for $A$ is 6 and the distance between the vertices at (−6, 0) and (3, 0) for $B$ is 9. So the scale factor is $\frac{9}{6}$ or $\frac{3}{2}$.

# Study Guide and Review *Continued*

## 9-7 Scale Drawings and Models

**33. BUILDING PLANS** In a scale drawing of a school's floor plan, 6 inches represents 100 feet. If the distance from one end of the main hallway to the other is 175 feet, find the corresponding length in the scale drawing.

**34. MODEL TRAINS** A popular scale for model trains is the 1:48 scale. If the actual train car had a length of 72 feet, find the corresponding length of the model in inches.

**35. MAPS** A map of the eastern United States has a scale where 3 inches = 25 miles. If the distance on the map between Columbia, South Carolina, and Charlotte, North Carolina, is 11.5 inches what is the actual distance between the cities?

### Example 7

**In the scale of a map of the Pacific Northwest 1 inch = 20 miles. The distance on the map between Portland, Oregon, and Seattle, Washington, is 8.75 inches. Find the distance between the two cities.**

$\frac{1}{20} = \frac{8.75}{x}$ Write a proportion.

$x = 20(8.75)$ Cross Products Property

$x = 175$ Simplify.

The distance between the two cities is 175 miles.

# CHAPTER 9 Practice Test

**Solve each proportion.**

1. $\frac{3}{7} = \frac{12}{x}$
2. $\frac{2x}{5} = \frac{x+3}{3}$
3. $\frac{4x}{15} = \frac{60}{x}$
4. $\frac{5x-4}{4x+7} = \frac{13}{11}$

**Determine whether each pair of figures is similar. If so, write the similarity statement and scale factor. If not, explain your reasoning.**

5.

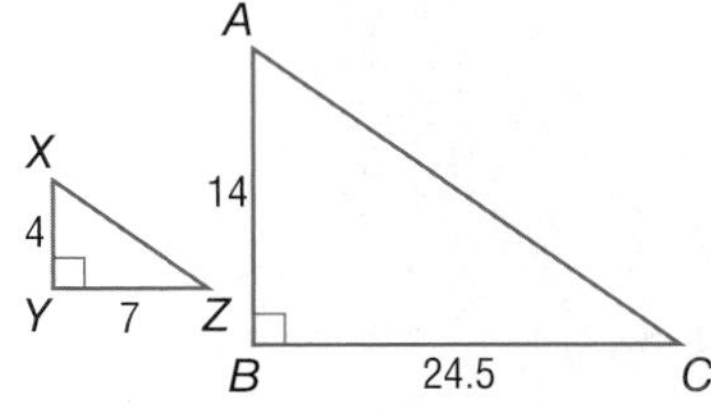


6.

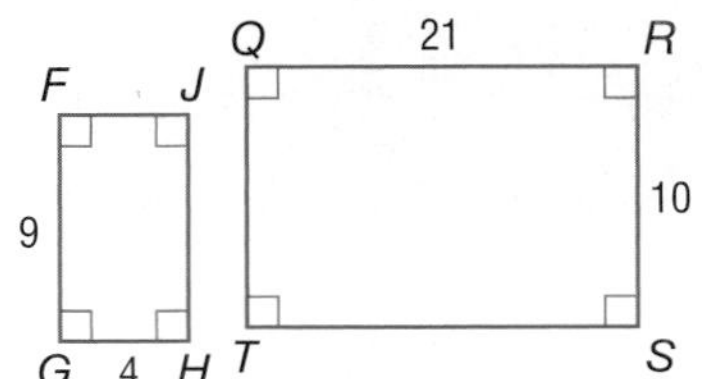


7. **CURRENCY** Jane is traveling to Europe this summer with the French Club. She plans to bring $300 to spend while she is there. If $90 in U.S. currency is equivalent to 63 euros, how many euros will she receive when she exchanges her money?

**ALGEBRA Find $x$ and $y$. Round to the nearest tenth if necessary.**

8.

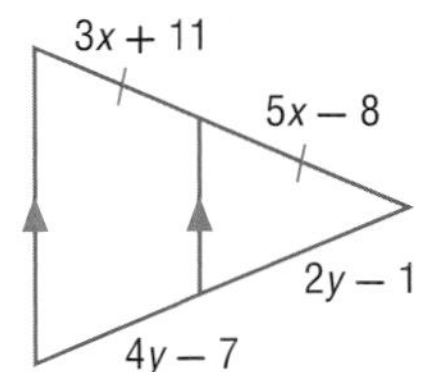


9.

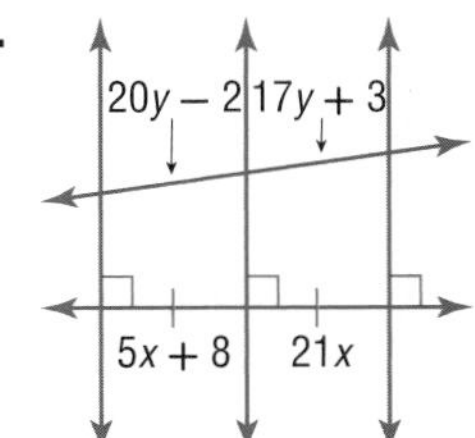


10. **ALGEBRA** Equilateral $\triangle MNP$ has perimeter $12a + 18b$. $\overline{QR}$ is a midsegment. What is $QR$?

11. **ALGEBRA** Right isosceles $\triangle ABC$ has hypotenuse length $h$. $\overline{DE}$ is a midsegment with length $4x$ that is not parallel to the hypotenuse. What is the perimeter of $\triangle ABC$?

12. **SHORT RESPONSE** Jimmy has a diecast metal car that is a scale model of an actual race car. If the actual length of the car is 10 feet and 6 inches and the model has a length of 7 inches, what is the scale factor of model to actual car?

**Find $x$.**

13.

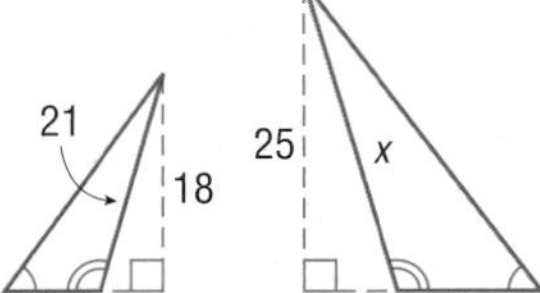


14.

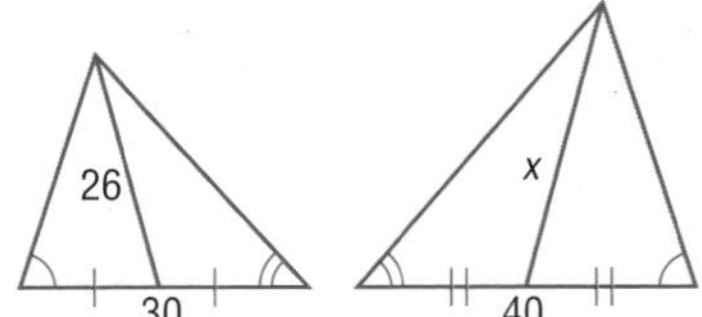


**Determine whether the dilation from $A$ to $B$ is an *enlargement* or a *reduction*. Then find the scale factor of the dilation.**

15.

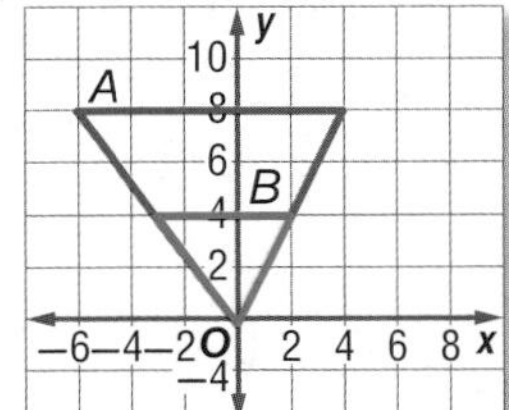


16.

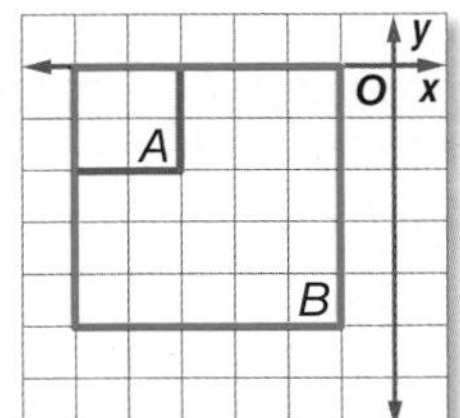


17. **ALGEBRA** Identify the similar triangles. Find $WZ$ and $UZ$.

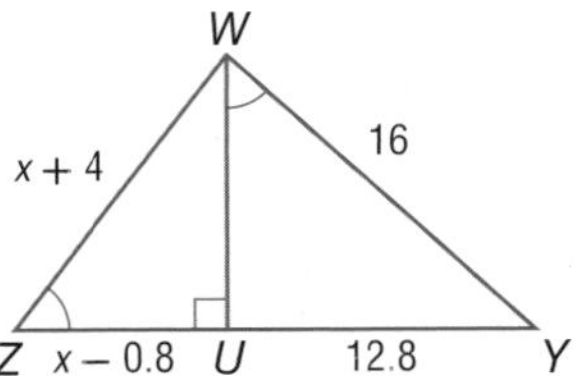

CHAPTER 9

# Preparing for Standardized Tests

## Identifying Nonexamples

Multiple choice items sometimes ask you to determine which of the given answer choices is a nonexample. These types of problems require a different approach when solving them.

Sean Ellis/Photographer's Choice/Getty Images

### Strategies for Identifying Nonexamples

**Step 1**

Read and understand the problem statement.

- **Nonexample:** A nonexample is an answer choice that does not satisfy the conditions of the problem statement.
- **Keywords:** Look for the word *not* (usually bold, all capital letters, or italicized) to indicate that you need to find a nonexample.

**Step 2**

Follow the concepts and steps below to help you identify nonexamples. Identify any answer choices that are clearly incorrect and eliminate them.

- Eliminate any answer choices that are not in the proper format.
- Eliminate any answer choices that do not have the correct units.

### Standardized Test Example

**Read the problem. Identify what you need to know. Then use the information in the problem to solve.**

In the adjacent triangle, you know that $\angle MQN \cong \angle RQS$. Which of the following would *not* be sufficient to prove that $\triangle QMN \sim \triangle QRS$?

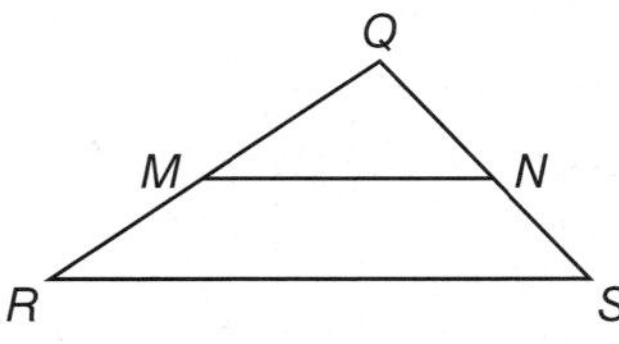


**A** $\angle QMN \cong \angle QRS$

**B** $\overline{MN} \parallel \overline{RS}$

**C** $\overline{QN} \cong \overline{NS}$

**D** $\dfrac{QM}{QR} = \dfrac{QN}{QS}$

The italicized *not* indicates that you need to find a nonexample. Test each answer choice using the principles of triangle similarity to see which one would not prove $\triangle QMN \cong \triangle QRS$.

**Choice A:** $\angle QMN \cong \angle QRS$

If $\angle QMN \cong \angle QRS$, then $\triangle QMN \sim \triangle QRS$ by AA Similarity.

**Choice B:** $\overline{MN} \parallel \overline{RS}$

If $\overline{MN} \parallel \overline{RS}$, then $\angle QMN \cong \angle QRS$, because they are corresponding angles of two parallel lines cut by transversal $\overline{QR}$. Therefore, $\triangle QMN \sim \triangle QRS$ by AA Similarity.

**Choice C:** $\overline{QN} \cong \overline{NS}$

If $\overline{QN} \cong \overline{NS}$, we cannot conclude that $\triangle QMN \sim \triangle QRS$ because we do not know anything about $\overline{QM}$ and $\overline{MR}$. So, answer choice C is a nonexample.

The correct answer is C. You should also check answer choice D to make sure it is a valid example if you have time.

## Exercises

**Read each problem. Identify what you need to know. Then use the information in the problem to solve.**

1. The ratio of the measures of the angles of the quadrilateral below is 6:5:4:3. Which of the following is *not* an angle measure of the figure?

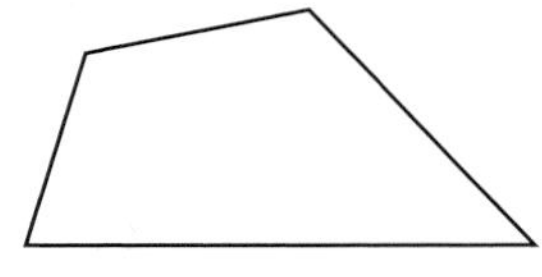

**A** 60° **C** 120°

**B** 80° **D** 140°

2. Which figure can serve as a counterexample to the conjecture below?

> If all angles of a quadrilateral are right angles, then the quadrilateral is a square.

**F** parallelogram

**G** rectangle

**H** rhombus

**J** trapezoid

3. Consider the figure below. Which of the following is *not* sufficient to prove that $\triangle GIK \sim \triangle HIG$?

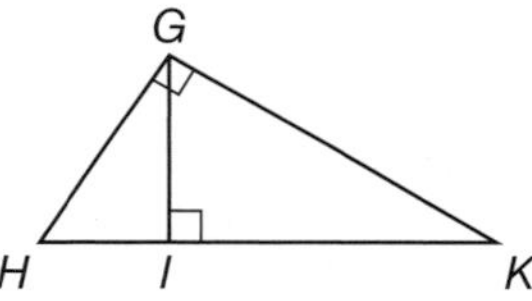


**A** $\angle GKI \cong \angle HGI$

**B** $\frac{HI}{GI} = \frac{GI}{IK}$

**C** $\frac{GH}{GI} = \frac{GK}{IK}$

**D** $\angle IGK \cong \angle IHG$

4. Which triangles are *not* necessarily similar?

**F** two right triangles with one angle measuring 30°

**G** two right triangles with one angle measuring 45°

**H** two isosceles triangles

**J** two equilateral triangles

# CHAPTER 9 Standardized Test Practice

## Cumulative, Chapters 1 through 9

### Multiple Choice

**Read each question. Then fill in the correct answer on the answer document provided by your teacher or on a sheet of paper.**

**1.** Adrian wants to measure the width of a ravine. He marks distances as shown in the diagram.

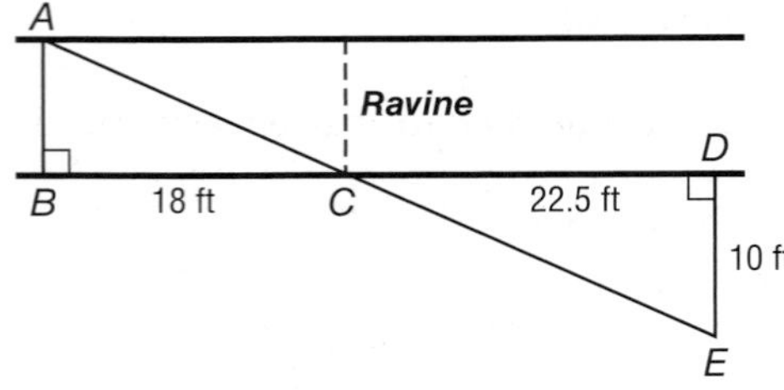


Using this information, what is the *approximate* width of the ravine?

**A** 5 ft **C** 7 ft

**B** 6 ft **D** 8 ft

**2.** Kyle and his family are planning a vacation in Cancun, Mexico. Kyle wants to convert 200 US dollars to Mexican pesos for spending money. If 278 Mexican pesos are equivalent to $25, how many pesos will Kyle get for $200?

**F** 2178 **H** 2396

**G** 2224 **J** 2504

**3.** Refer to the figures below. Which of the following terms *best* describes the transformation?

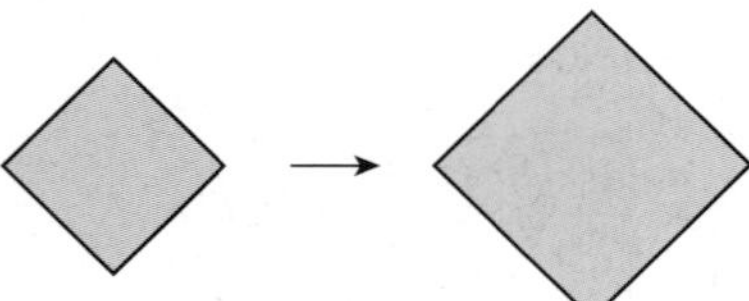

**F** congruent

**G** enlargement

**H** reduction

**J** scale

**4.** The ratio of North Carolina residents to Americans is about 295 to 10,000. If there are approximately 300,000,000 Americans, how many of them are North Carolina residents?

**A** 7,950,000

**B** 8,400,000

**C** 8,850,000

**D** 9,125,000

**5.** Solve for $x$.

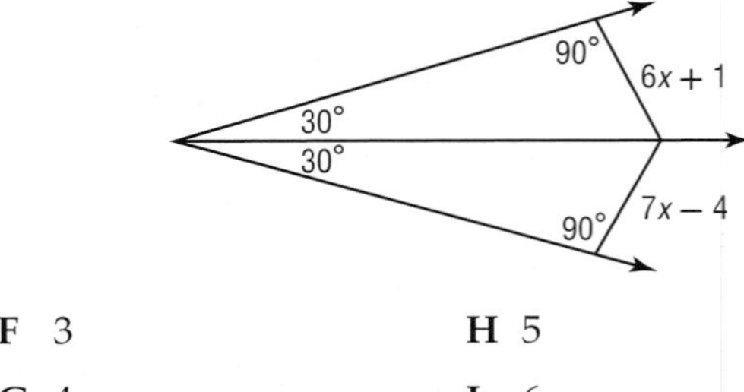


**F** 3 **H** 5

**G** 4 **J** 6

**6.** Two similar trapezoids have a scale factor of 3:2. The perimeter of the larger trapezoid is 21 yards. What is the perimeter of the smaller trapezoid?

**A** 14 yd

**B** 17.5 yd

**C** 28 yd

**D** 31.5 yd

**Test-TakingTip**

**Question 2** Set up and solve the proportion for the number of pesos. Use the ratio pesos : dollars.

## Short Response/Gridded Response

**Record your answers on the answer sheet provided by your teacher or on a sheet of paper.**

**7. GRIDDED RESPONSE** Colleen surveyed 50 students in her school and found that 35 of them have homework at least four nights a week. If there are 290 students in the school altogether, how many of them would you expect to have homework at least four nights a week?

**8. GRIDDED RESPONSE** In the triangle below, $\overline{MN} \parallel \overline{BC}$. Solve for $x$.

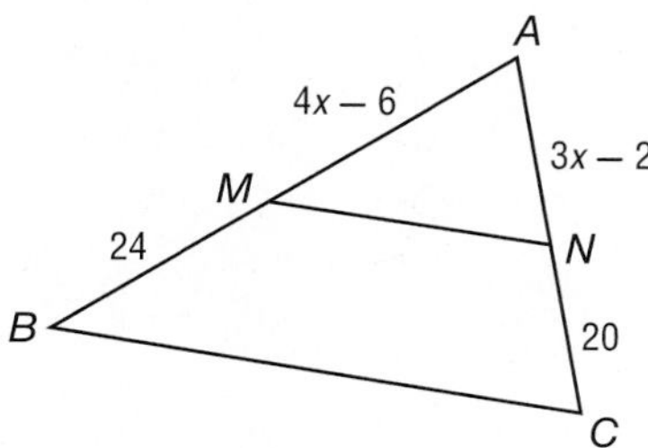


**9.** Quadrilateral $WXYZ$ is a rhombus. If $m\angle XYZ = 110°$, find $m\angle ZWY$.

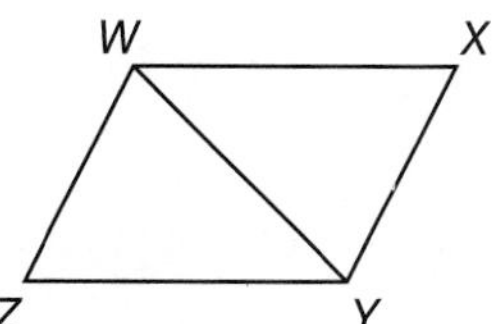


**10. GRIDDED RESPONSE** In the triangle below, $\overline{RS}$ bisects $\angle VRU$. Solve for $x$.

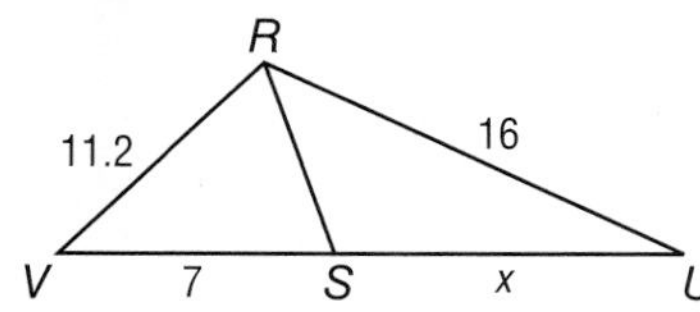


**11. GRIDDED RESPONSE** The scale of a map is 1 inch = 2.5 miles. What is the distance between two cities that are 3.3 inches apart on the map? Round to the nearest tenth, if necessary.

## Extended Response

**Record your answers on a sheet of paper. Show your work.**

**12.** Refer to triangle $XYZ$ to answer each question.

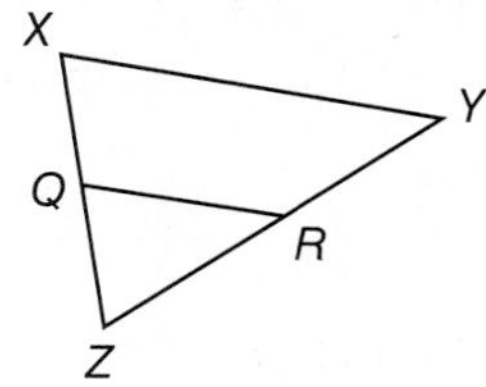


**a.** Suppose $\overline{QR} \parallel \overline{XY}$. What do you know about the relationship between segments $XQ$, $QZ$, $YR$, and $RZ$?

**b.** If $\overline{QR} \parallel \overline{XY}$, $XQ = 15$, $QZ = 12$, and $YR = 20$, what is the length of $\overline{RZ}$?

**c.** Suppose $\overline{QR} \parallel \overline{XY}$, $\overline{XQ} \cong \overline{QZ}$, and $QR = 9.5$ units. What is the length of $\overline{XY}$?

**Need ExtraHelp?**

| If you missed Question... | 1 | 2 | 3 | 4 | 5 | 6 | 7 | 8 | 9 | 10 | 11 | 12 |
|---|---|---|---|---|---|---|---|---|---|---|---|---|
| Go to Lesson... | 9-3 | 9-1 | 9-6 | 9-1 | 7-1 | 9-2 | 9-1 | 9-6 | 8-5 | 9-5 | 9-7 | 9-4 |

CHAPTER 10

# Right Triangles and Trigonometry

## Then

- You solved proportions.

## Now

- In this chapter, you will:
  - Use the Pythagorean Theorem.
  - Use properties of special right triangles.
  - Use trigonometry to find missing measures of triangles.

## Why?

- Properties of triangles can be used in planning and preparation for special events including the height of decorations.

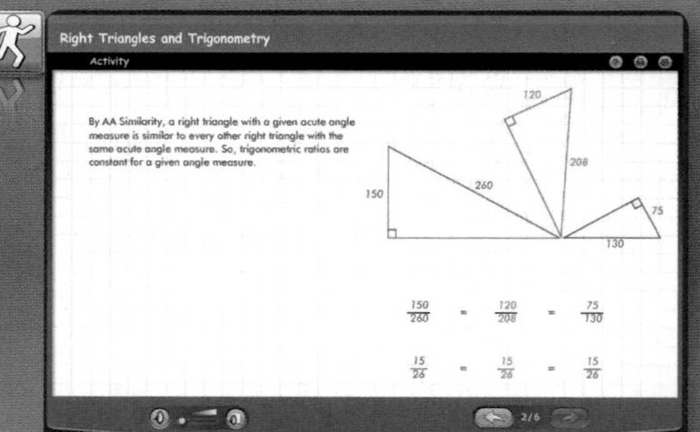


connectED.mcgraw-hill.com **Your Digital Math Portal**

Animation 
Vocabulary 
eGlossary 
Personal Tutor 
Virtual Manipulatives 
Graphing Calculator 
Audio 
Foldables 
Self-Check Practice 
Worksheets

# Get Ready for the Chapter

**Diagnose** Readiness | You have two options for checking prerequisite skills.

**Textbook Option** Take the Quick Check below. Refer to the Quick Review for help.

## QuickCheck

Simplify.

1. $\sqrt{112}$
2. $\dfrac{\sqrt{24}}{2\sqrt{3}}$
3. $\sqrt{15 \cdot 20}$
4. $\dfrac{\sqrt{6}}{\sqrt{3}} \cdot \dfrac{\sqrt{18}}{\sqrt{3}}$
5. $\sqrt{\dfrac{45}{80}}$
6. $\dfrac{8\sqrt{2}}{6 - 3\sqrt{8}}$

Find *x*.

7. 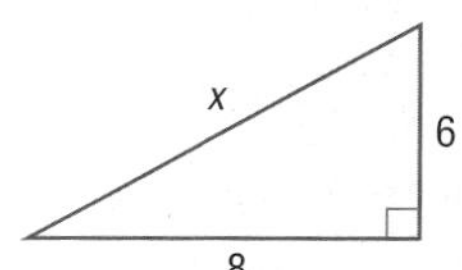


8. 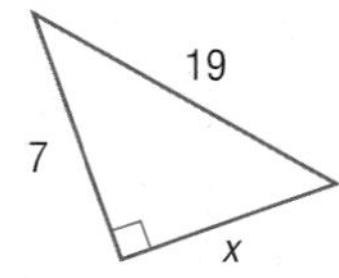


9. **BANNERS** Anna is making a banner out of 4 congruent triangles as shown below. How much blue trim will she need for each side?

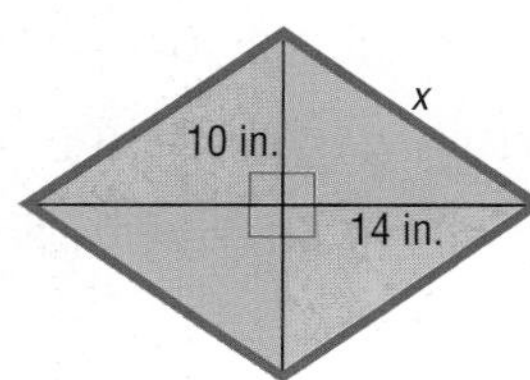


Graph the line segment with the given endpoints.

10. $G(3, -4)$ and $H(3, 4)$
11. $E(-3, 5)$ and $F(4, -3)$
12. **COLLEGES** Quinn is visiting a college campus. He notices from his map that several important buildings are located around a grassy area the students call the Quad. If the library is represented on the map by $L(6, 8)$ and the cafeteria is represented by $C(0, 0)$, graph the line segment that represents the shortest path between the two buildings.

## QuickReview

**Example 1**

Simplify $\dfrac{6}{\sqrt{3}}$.

$\dfrac{6}{\sqrt{3}} = \dfrac{6}{\sqrt{3}} \cdot \dfrac{\sqrt{3}}{\sqrt{3}}$ — Multiply by $\dfrac{\sqrt{3}}{\sqrt{3}}$.

$= \dfrac{6\sqrt{3}}{3}$ or $2\sqrt{3}$ — Simplify.

**Example 2**

Find *x*.

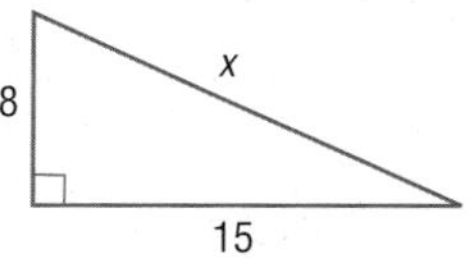


| | |
|---|---|
| $a^2 + b^2 = c^2$ | Pythagorean Theorem |
| $8^2 + 15^2 = x^2$ | $a = 8$ and $b = 15$ |
| $289 = x^2$ | Simplify. |
| $\sqrt{289} = \sqrt{x^2}$ | Take the positive square root of each side. |
| $17 = x$ | Simplify. |

**Example 3**

Graph the line segment with endpoints $A(-4, 2)$ and $B(3, -2)$.

Plot points *A* and *B*.

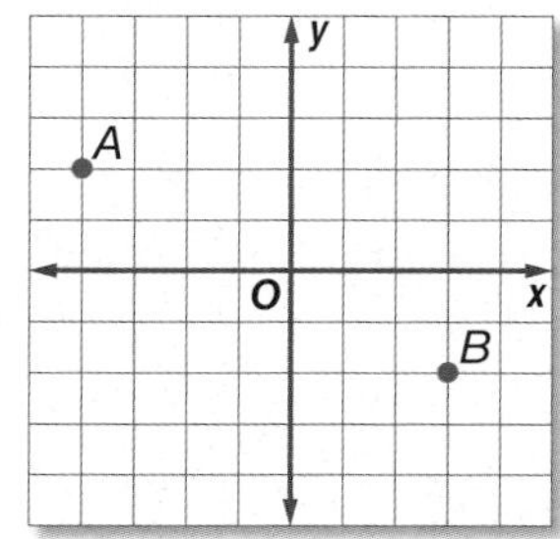


Connect the points.

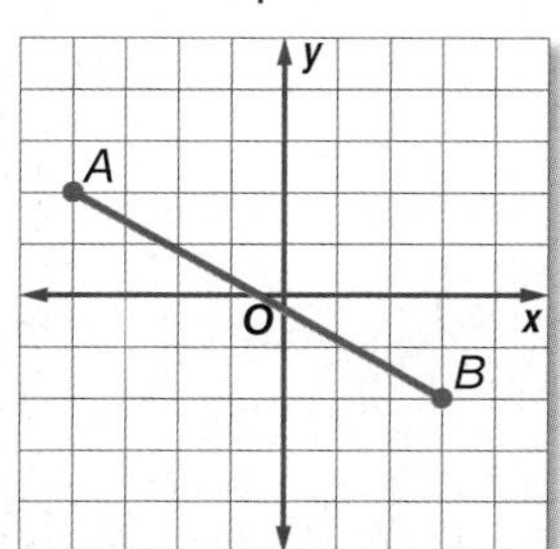


**2 Online Option** Take an online self-check Chapter Readiness Quiz at connectED.mcgraw-hill.com.

# **Get Started** on the Chapter

**You will learn several new concepts, skills, and vocabulary terms as you study Chapter 10. To get ready, identify important terms and organize your resources. You may wish to refer to Chapter 0 to review prerequisite skills.**

## FOLDABLES® StudyOrganizer

**Right Angles and Trigonometry** Make this Foldable to help you organize your Chapter 10 notes about right angles and trigonometry. Begin with three sheets of notebook paper and one sheet of construction paper.

**1** **Stack** the notebook paper on the construction paper.

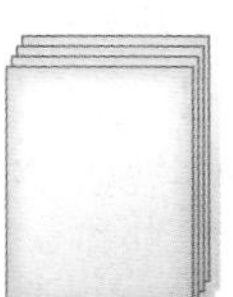

**2** **Fold** the paper diagonally to form a triangle and cut off the excess.

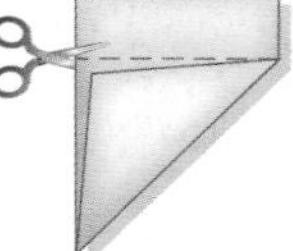

**3** **Open** the paper and staple the inside fold to form a booklet.

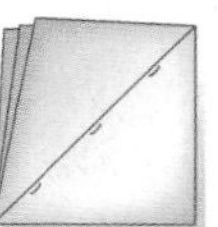

**4** **Label** each page with a lesson number and title.

## NewVocabulary

| English | | Español |
|---|---|---|
| geometric mean | p. 619 | media geométrica |
| Pythagorean triple | p. 630 | triplete pitágorico |
| trigonometry | p. 650 | trigonométria |
| trigonometric ratio | p. 650 | razón trigonométrica |
| sine | p. 650 | seno |
| cosine | p. 650 | coseno |
| tangent | p. 650 | tangente |
| angle of elevation | p. 662 | ángulo de elevación |
| angle of depression | p. 662 | ángulo de depresión |
| Law of Sines | p. 670 | ley de los senos |
| Law of Cosines | p. 671 | ley do los cosenos |
| vector | p. 682 | vector |
| magnitude | p. 682 | magnitud |
| resultant | p. 683 | resultante |
| component form | p. 684 | componente |

## ReviewVocabulary

altitude altura a segment drawn from a vertex of a triangle perpendicular to the line containing the other side

Pythagorean Theorem Teorema de Pitágoras If $a$ and $b$ are the measures of the legs of a right triangle and $c$ is the measure of the hypotenuse, then $a^2 + b^2 = c^2$.

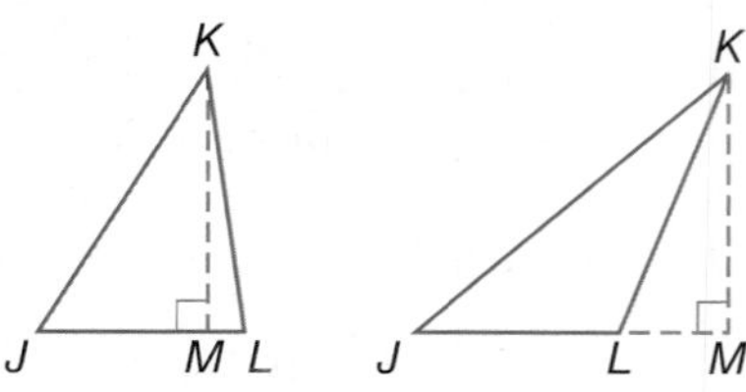


$\overline{KM}$ is an altitude of $\triangle JKL$.

LESSON 10-1

# Geometric Mean

| Then | Now | Why? |
|---|---|---|
| You used proportional relationships of corresponding angle bisectors, altitudes, and medians of similar triangles. | 1 Find the geometric mean between two numbers.<br>2 Solve problems involving relationships between parts of a right triangle and the altitude to its hypotenuse. | Photographing very tall or very wide objects can be challenging. It can be difficult to include the entire object in your shot without distorting the image. If your camera is set for a vertical viewing angle of 90° and you know the height of the object you wish to photograph, you can use the geometric mean of the distance from the top of the object to your camera level and the distance from the bottom of the object to camera level. |

**NewVocabulary**
geometric mean

**Common Core State Standards**

**Content Standards**
**G.SRT.4** Prove theorems about triangles.
**G.SRT.5** Use congruence and similarity criteria for triangles to solve problems and to prove relationships in geometric figures.

**Mathematical Practices**
7 Look for and make use of structure.
3 Construct viable arguments and critique the reasoning of others.

## 1 Geometric Mean

When the means of a proportion are the same number, that number is called the geometric mean of the extremes. The **geometric mean** between two numbers is the positive square root of their product.

$$\text{extreme} \rightarrow \frac{a}{x} \leftarrow \text{mean} = \frac{x}{b} \begin{matrix}\leftarrow \text{mean} \\ \leftarrow \text{extreme}\end{matrix}$$

### KeyConcept Geometric Mean

| | |
|---|---|
| Words | The geometric mean of two positive numbers $a$ and $b$ is the number $x$ such that $\frac{a}{x} = \frac{x}{b}$. So, $x^2 = ab$ and $x = \sqrt{ab}$. |
| Example | The geometric mean of $a = 9$ and $b = 4$ is 6, because $6 = \sqrt{9 \cdot 4}$. |

### Example 1 Geometric Mean

**Find the geometric mean between 8 and 10.**

| | |
|---|---|
| $x = \sqrt{ab}$ | Definition of geometric mean |
| $= \sqrt{8 \cdot 10}$ | $a = 8$ and $b = 10$ |
| $= \sqrt{(4 \cdot 2) \cdot (2 \cdot 5)}$ | Factor. |
| $= \sqrt{16 \cdot 5}$ | Associative Property |
| $= 4\sqrt{5}$ | Simplify. |

The geometric mean between 8 and 10 is $4\sqrt{5}$ or about 8.9.

**GuidedPractice**

**Find the geometric mean between each pair of numbers.**

**1A.** 5 and 45 **1B.** 12 and 15

## 2 Geometric Means in Right Triangles

In a right triangle, an altitude drawn from the vertex of the right angle to the hypotenuse forms two additional right triangles. These three right triangles share a special relationship.

**Review Vocabulary**

altitude (of a triangle)
a segment from a vertex to the line containing the opposite side and perpendicular to the line containing that side

## Theorem 10.1

If the altitude is drawn to the hypotenuse of a right triangle, then the two triangles formed are similar to the original triangle and to each other.

**Example** If $\overline{CD}$ is the altitude to hypotenuse $\overline{AB}$ of right $\triangle ABC$, then $\triangle ACD \sim \triangle ABC$, $\triangle CBD \sim \triangle ABC$, and $\triangle ACD \sim \triangle CBD$.

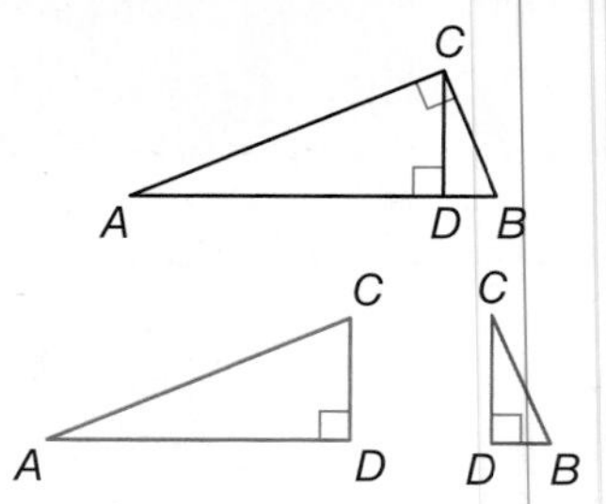


You will prove Theorem 10.1 in Exercise 39.

PT

## Example 2 Identify Similar Right Triangles

**Write a similarity statement identifying the three similar right triangles in the figure.**

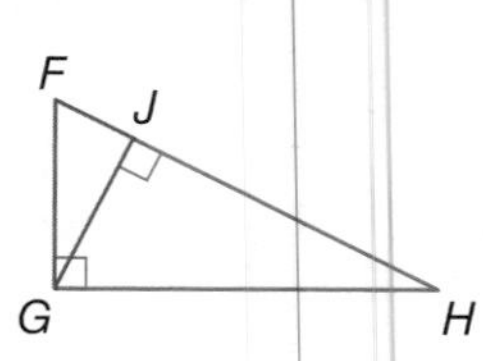


Separate the triangle into two triangles along the altitude. Then sketch the three triangles, reorienting the smaller ones so that their corresponding angles and sides are in the same positions as the original triangle.

**Study Tip**

Reorienting Triangles
To reorient the right triangles in Example 2, first match up the right angles. Then match up the shorter sides.

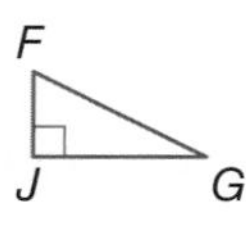


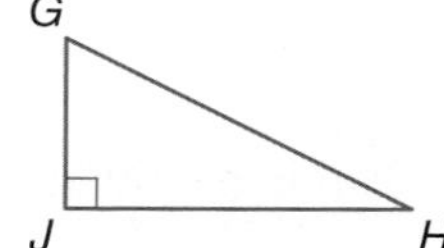


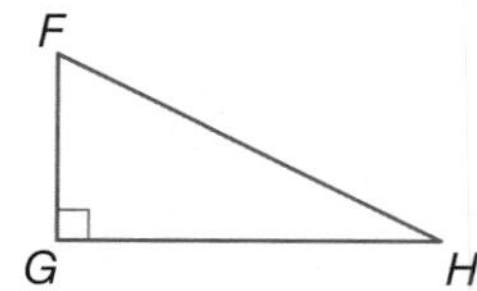


So by Theorem 10.1, $\triangle FJG \sim \triangle GJH \sim \triangle FGH$.

**Guided Practice**

**2A.**

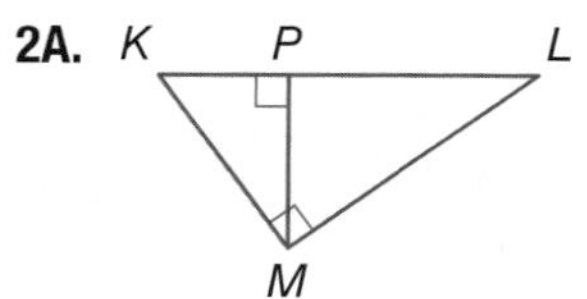


**2B.**

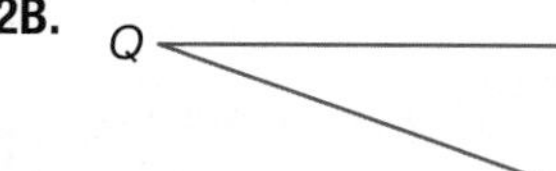

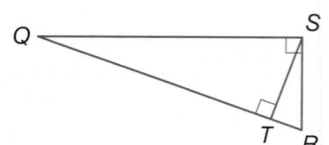


From Theorem 10.1, you know that altitude $\overline{CD}$ drawn to the hypotenuse of right triangle $ABC$ forms three similar triangles: $\triangle ACB \sim \triangle ADC \sim \triangle CDB$. By the definition of similar polygons, you can write the following proportions comparing the side lengths of these triangles.

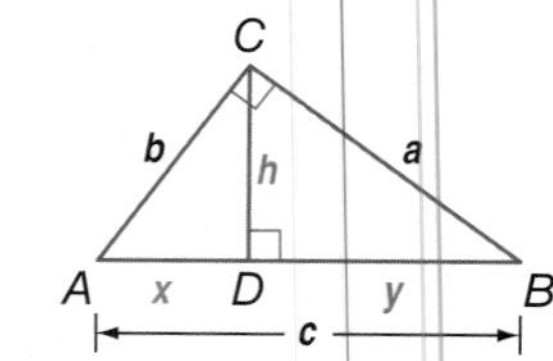


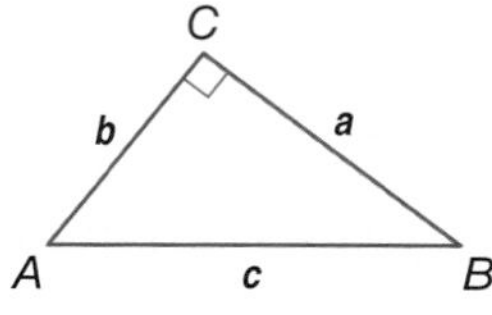


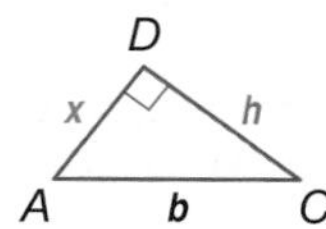


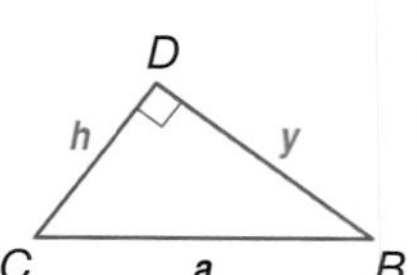


$$\frac{\text{shorter leg}}{\text{longer leg}} = \frac{b}{a} = \frac{x}{h} = \frac{h}{y} \qquad \frac{\text{hypotenuse}}{\text{shorter leg}} = \frac{c}{b} = \frac{b}{x} = \frac{a}{h} \qquad \frac{\text{hypotenuse}}{\text{longer leg}} = \frac{c}{a} = \frac{b}{h} = \frac{a}{y}$$

Notice that the circled relationships involve geometric means. This leads to the theorems at the top of the next page.

## Example 4 Classify Triangles

**Determine whether each set of numbers can be the measures of the sides of a triangle. If so, classify the triangle as *acute, right,* or *obtuse.* Justify your answer.**

**a.** **7, 14, 16**

**Step 1** Determine whether the measures can form a triangle using the Triangle Inequality Theorem.

$7 + 14 > 16$ ✓ $\quad 14 + 16 > 7$ ✓ $\quad 7 + 16 > 14$ ✓

The side lengths 7, 14, and 16 can form a triangle.

**Step 2** Classify the triangle by comparing the square of the longest side to the sum of the squares of the other two sides.

$c^2 \stackrel{?}{=} a^2 + b^2$ Compare $c^2$ and $a^2 + b^2$.

$16^2 \stackrel{?}{=} 7^2 + 14^2$ Substitution

$256 > 245$ Simplify and compare.

Since $c^2 > a^2 + b^2$, the triangle is obtuse.

**b.** **9, 40, 41**

**Step 1** Determine whether the measures can form a triangle.

$9 + 40 > 41$ ✓ $\quad 40 + 41 > 9$ ✓ $\quad 9 + 41 > 40$ ✓

The side lengths 9, 40, and 41 can form a triangle.

**Step 2** Classify the triangle.

$c^2 \stackrel{?}{=} a^2 + b^2$ Compare $c^2$ and $a^2 + b^2$.

$41^2 \stackrel{?}{=} 9^2 + 40^2$ Substitution

$1681 = 1681$ Simplify and compare.

Since $c^2 = a^2 + b^2$, the triangle is a right triangle.

**GuidedPractice**

**4A.** 11, 60, 61 **4B.** $2\sqrt{3}, 4\sqrt{2}, 3\sqrt{5}$ **4C.** 6.2, 13.8, 20

**ReviewVocabulary**

**Triangle Inequality Theorem** The sum of the lengths of any two sides of a triangle must be greater than the length of the third side.

## Check Your Understanding

**Example 1** **Find $x$.**

**1.**

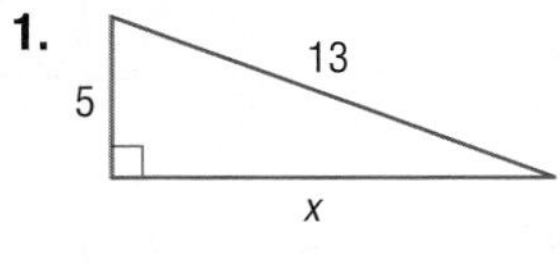


**2.**

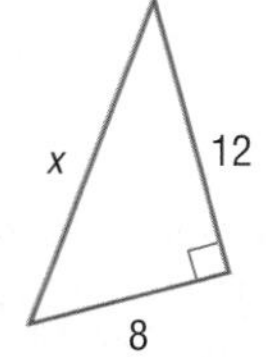


**3**

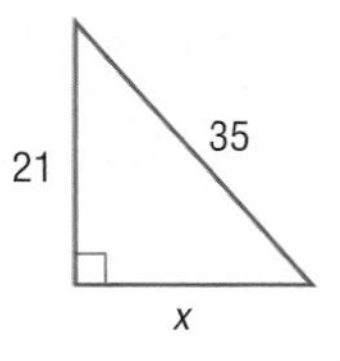

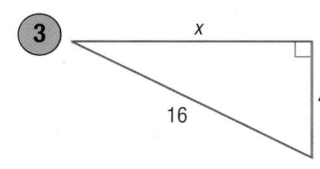


**Example 2** **4.** Use a Pythagorean triple to find $x$. Explain your reasoning.

21
35
x

Example 3

5. **MULTIPLE CHOICE** The mainsail of a boat is shown. What is the length, in feet, of $\overline{LN}$?

A 52.5 C 72.5

B 65 D 75

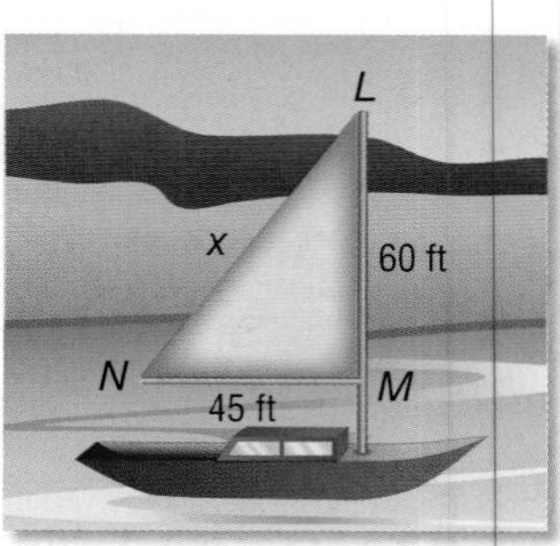


Example 4

**Determine whether each set of numbers can be the measures of the sides of a triangle. If so, classify the triangle as *acute, obtuse,* or *right.* Justify your answer.**

6. 15, 36, 39

7. 16, 18, 26

8. 15, 20, 24

## Practice and Problem Solving

Example 1

**Find $x$.**

9.

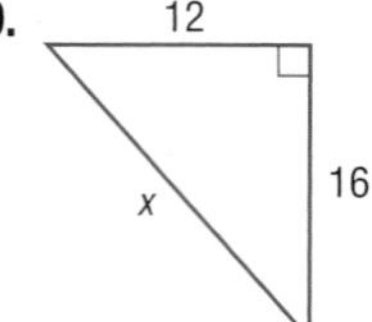


10.

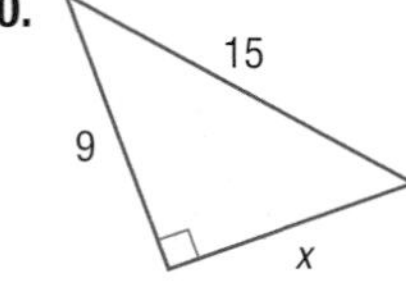


11.

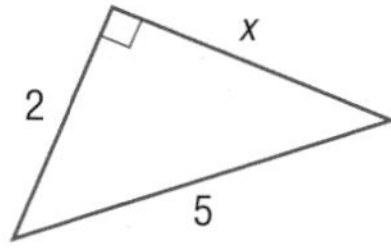


12.

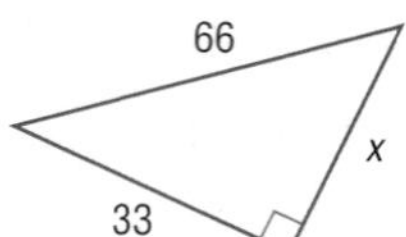


13.

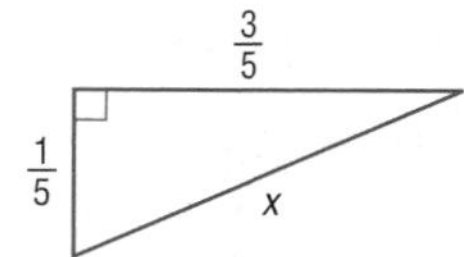


14.

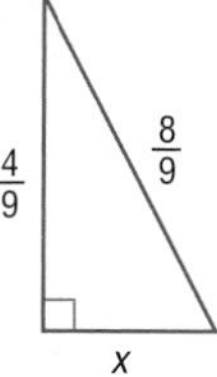


Example 2

**CCSS PERSEVERANCE** **Use a Pythagorean Triple to find $x$.**

15.

16.

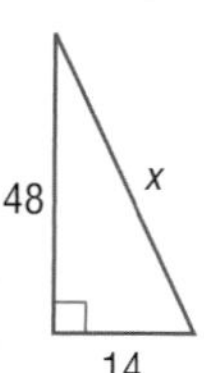


17.

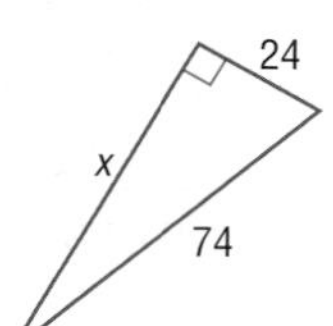


18.

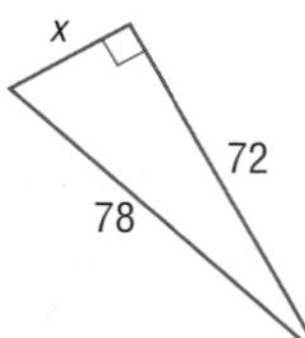


Example 3

19. **BASKETBALL** The support for a basketball goal forms a right triangle as shown. What is the length $x$ of the horizontal portion of the support?

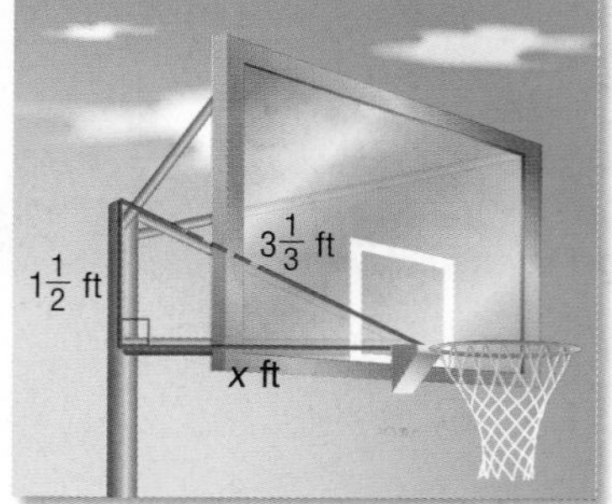


20. **DRIVING** The street that Khaliah usually uses to get to school is under construction. She has been taking the detour shown. If the construction starts at the point where Khaliah leaves her normal route and ends at the point where she re-enters her normal route, about how long is the stretch of road under construction?

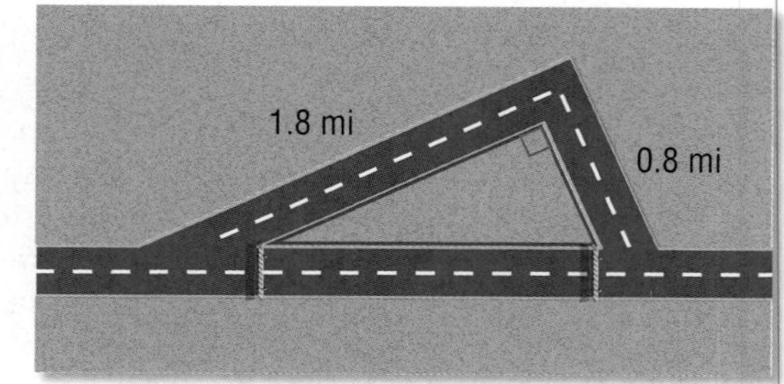

**Example 4** **Determine whether each set of numbers can be the measures of the sides of a triangle. If so, classify the triangle as *acute*, *obtuse*, or *right*. Justify your answer.**

**21.** 7, 15, 21
**22.** 10, 12, 23
**23.** 4.5, 20, 20.5
**24.** 44, 46, 91
**25.** 4.2, 6.4, 7.6
**26.** 4, 12, 14

**Find $x$.**

**27.** 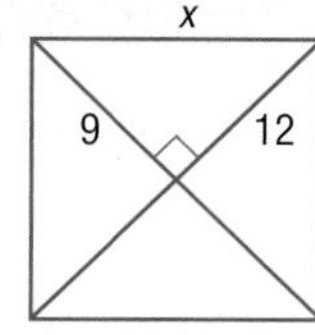


**28.** 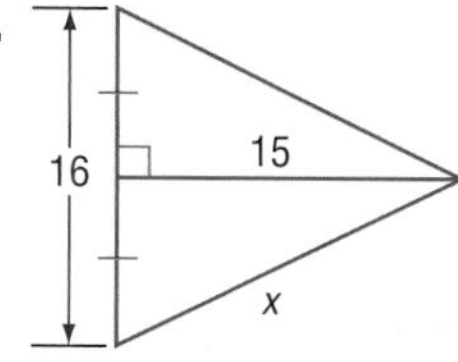


**29.** 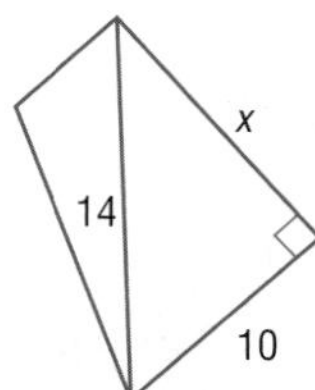


**COORDINATE GEOMETRY** **Determine whether $\triangle XYZ$ is an *acute*, *right*, or *obtuse* triangle for the given vertices. Explain.**

**30.** $X(-3, -2)$, $Y(-1, 0)$, $Z(0, -1)$
**31.** $X(-7, -3)$, $Y(-2, -5)$, $Z(-4, -1)$
**32.** $X(1, 2)$, $Y(4, 6)$, $Z(6, 6)$
**33.** $X(3, 1)$, $Y(3, 7)$, $Z(11, 1)$

**34.** **JOGGING** Brett jogs in the park three times a week. Usually, he takes a $\frac{3}{4}$-mile path that cuts through the park. Today, the path is closed, so he is taking the orange route shown. How much farther will he jog on his alternate route than he would have if he had followed his normal path?

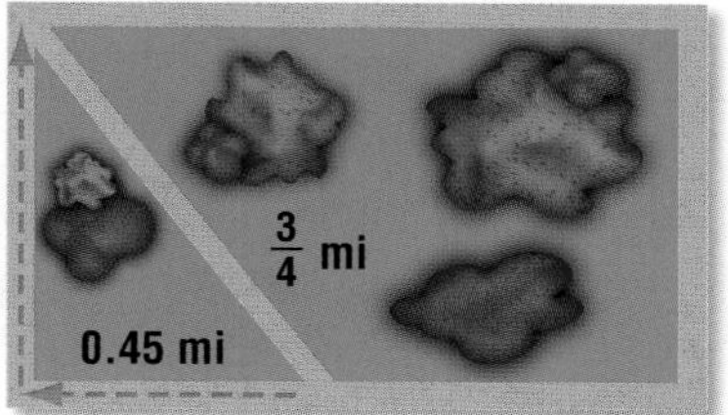


**35.** **PROOF** Write a paragraph proof of Theorem 10.5.

**PROOF** **Write a two-column proof for each theorem.**

**36.** Theorem 10.6
**37.** Theorem 10.7

**CCSS SENSE-MAKING** **Find the perimeter and area of each figure.**

**38.** 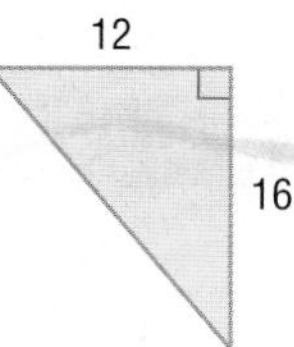


**39.** 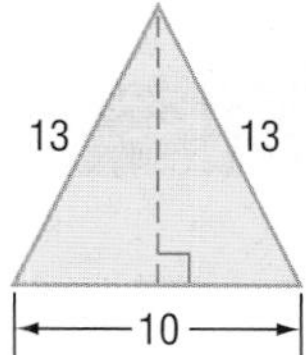


**40.** 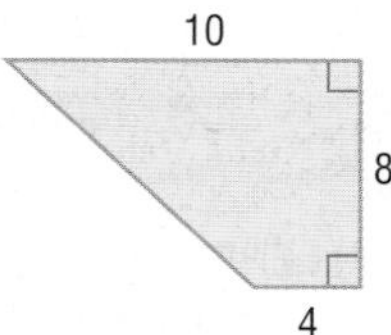


**41.** **ALGEBRA** The sides of a triangle have lengths $x$, $x + 5$, and 25. If the length of the longest side is 25, what value of $x$ makes the triangle a right triangle?

**42.** **ALGEBRA** The sides of a triangle have lengths $2x$, 8, and 12. If the length of the longest side is $2x$, what values of $x$ make the triangle acute?

**43** **TELEVISION** The screen aspect ratio, or the ratio of the width to the height, of a high-definition television is 16:9. The size of a television is given by the diagonal distance across the screen. If an HDTV is 41 inches wide, what is its screen size?

44. **PLAYGROUND** According to the *Handbook for Public Playground Safety,* the ratio of the vertical distance to the horizontal distance covered by a slide should not be more than about 4 to 7. If the horizontal distance allotted in a slide design is 14 feet, approximately how long should the slide be?

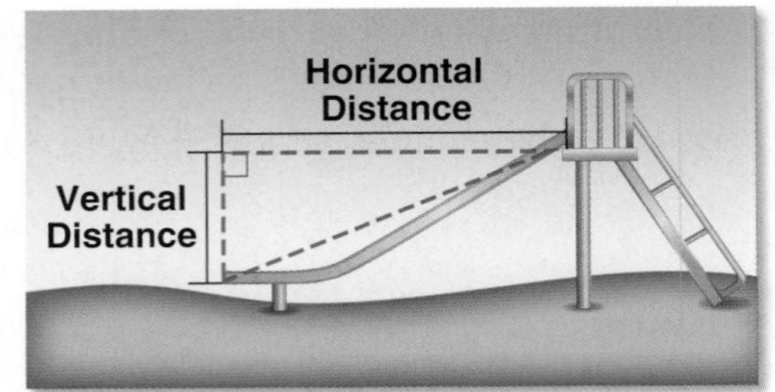


**Find $x$.**

45.

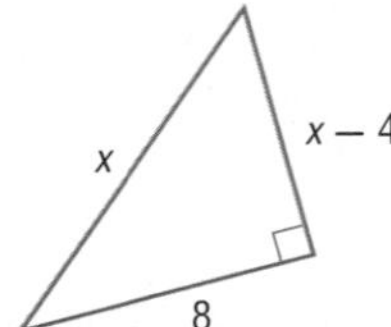


46.

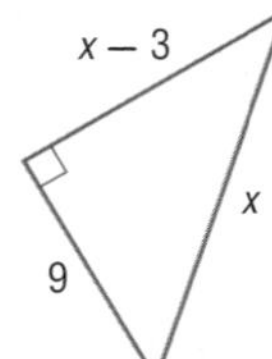


47.

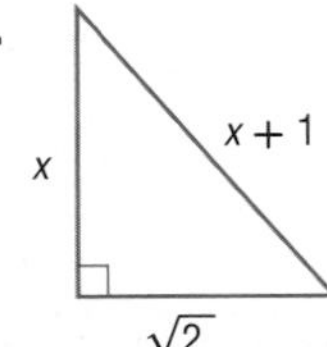


48. **MULTIPLE REPRESENTATIONS** In this problem, you will investigate special right triangles.

a. **Geometric** Draw three different isosceles right triangles that have whole-number side lengths. Label the triangles $ABC$, $MNP$, and $XYZ$ with the right angle located at vertex $A$, $M$, and $X$, respectively. Label the leg lengths of each side, and find the exact length of the hypotenuse.

b. **Tabular** Copy and complete the table below.

| Triangle | Length | | | | Ratio | |
|---|---|---|---|---|---|---|
| $ABC$ | $BC$ | | $AB$ | | $\frac{BC}{AB}$ | |
| $MNP$ | $NP$ | | $MN$ | | $\frac{NP}{MN}$ | |
| $XYZ$ | $YZ$ | | $XY$ | | $\frac{YZ}{XY}$ | |

c. **Verbal** Make a conjecture about the ratio of the hypotenuse to a leg of an isosceles right triangle.

## H.O.T. Problems Use Higher-Order Thinking Skills

49. **CHALLENGE** Find the value of $x$ in the figure at the right.

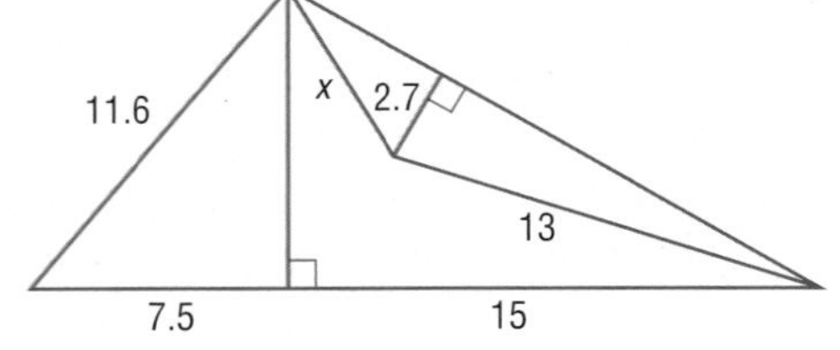


50. **CCSS ARGUMENTS** *True* or *false*? Any two right triangles with the same hypotenuse have the same area. Explain your reasoning.

51. **OPEN ENDED** Draw a right triangle with side lengths that form a Pythagorean triple. If you double the length of each side, is the resulting triangle *acute, right,* or *obtuse*? if you halve the length of each side? Explain.

52. **WRITING IN MATH** Research *incommensurable magnitudes,* and describe how this phrase relates to the use of irrational numbers in geometry. Include one example of an irrational number used in geometry.

## Standardized Test Practice

**53.** Which set of numbers cannot be the measures of the sides of a triangle?

**A** 10, 11, 20 **C** 35, 45, 75
**B** 14, 16, 28 **D** 41, 55, 98

**54.** A square park has a diagonal walkway from one corner to another. If the walkway is 120 meters long, what is the approximate length of each side of the park?

**F** 60 m **H** 170 m
**G** 85 m **J** 240 m

**55. SHORT RESPONSE** If the perimeter of square 2 is 200 units and the perimeter of square 1 is 150 units, what is the perimeter of square 3?

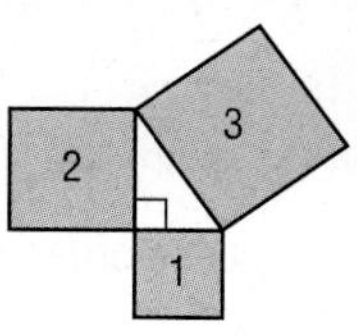


**56. SAT/ACT** In $\triangle ABC$, $\angle B$ is a right angle and $\angle A$ is 20° greater than $\angle C$. What is the measure of $\angle C$?

**A** 30 **C** 40 **E** 70
**B** 35 **D** 45

## Spiral Review

**Find the geometric mean between each pair of numbers.** (Lesson 10-1)

**57.** 9 and 4 **58.** 45 and 5 **59.** 12 and 15 **60.** 36 and 48

**61. SCALE DRAWING** Teodoro is creating a scale model of a skateboarding ramp on a 10-by-8-inch sheet of graph paper. If the real ramp is going to be 12 feet by 8 feet, find an appropriate scale for the drawing and determine the ramp's dimensions. (Lesson 9-7)

**Determine whether the triangles are similar. If so, write a similarity statement. If not, what would be sufficient to prove the triangles similar? Explain your reasoning.** (Lesson 9-3)

**62.**

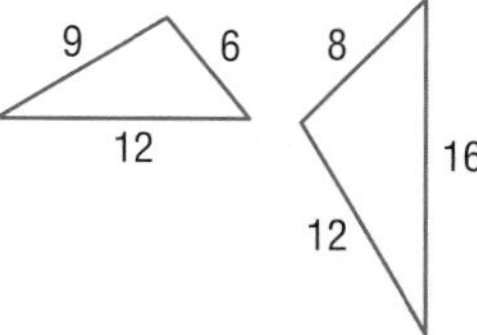


**63.**

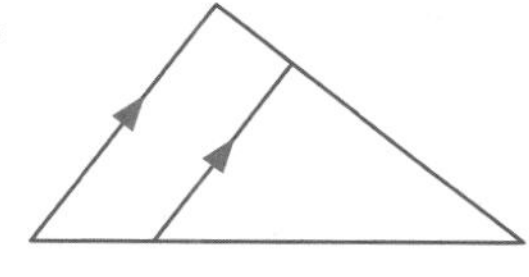

**64.**

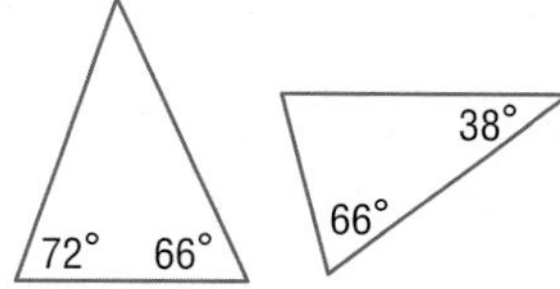


**65. PROOF** Write a two-column proof. (Lesson 7-3)

**Given:** $\overline{FG} \perp \ell$
$\overline{FH}$ is any nonperpendicular segment from $F$ to $\ell$.

**Prove:** $FH > FG$

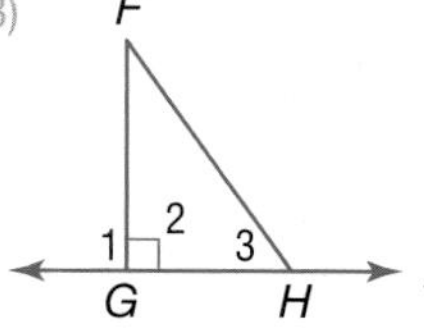


**Find each measure if $m\angle DGF = 53$ and $m\angle AGC = 40$.** (Lesson 6-1)

**66.** $m\angle 1$ **67.** $m\angle 2$

**68.** $m\angle 3$ **69.** $m\angle 4$

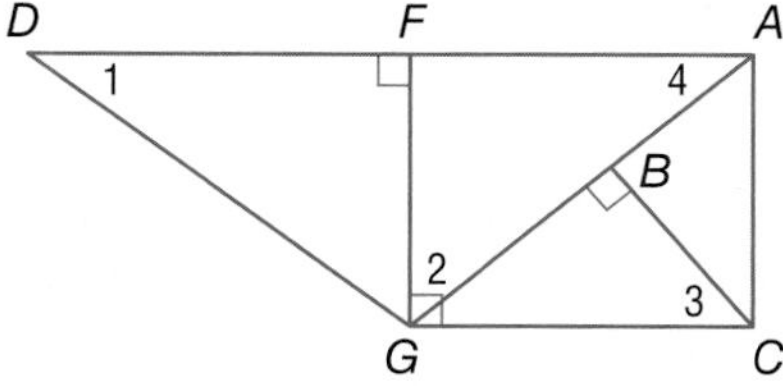


## Skills Review

**Find the value of $x$.**

**70.** $18 = 3x\sqrt{3}$ **71.** $24 = 2x\sqrt{2}$ **72.** $9\sqrt{2} \cdot x = 18\sqrt{2}$ **73.** $2 = x \cdot \frac{4}{\sqrt{3}}$

EXTEND 10-2

# Geometry Lab
# Coordinates in Space

You have used ordered pairs of two coordinates to describe the location of a point on the coordinate plane. Because space has three dimensions, a point requires three numbers, or coordinates, to describe its location in space.

A point in space is represented by an **ordered triple** of real numbers $(x, y, z)$. In the figure at the right, the ordered triple (2, 3, 6) locates point $P$. Notice that a rectangular prism is used to show perspective.

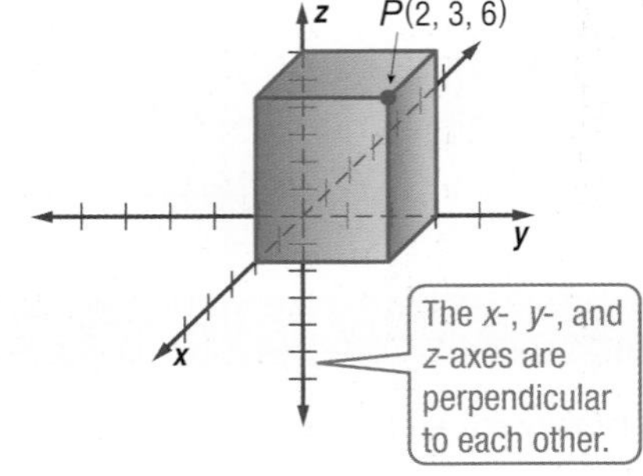


## Activity 1 Graph a Rectangular Solid

**Graph a rectangular solid that has two vertices, $L(4, -5, 2)$ and the origin. Label the coordinates of each vertex.**

**Step 1** Plot the $x$-coordinate first. Draw a segment from the origin 4 units in the positive direction.

**Step 2** To plot the $y$-coordinate, draw a segment five units in the negative direction.

**Step 3** Next, to plot the $z$-coordinate, draw a segment two units long in the positive direction.

**Step 4** Label the coordinate $L$.

**Step 5** Draw the rectangular prism and label each vertex: $L(4, -5, 2)$, $K(0, -5, 2)$, $J(0, 0, 2)$, $M(4, 0, 2)$ $Q(4, -5, 0)$, $P(0, -5, 0)$, $N(0, 0, 0)$, and $R(4, 0, 0)$.

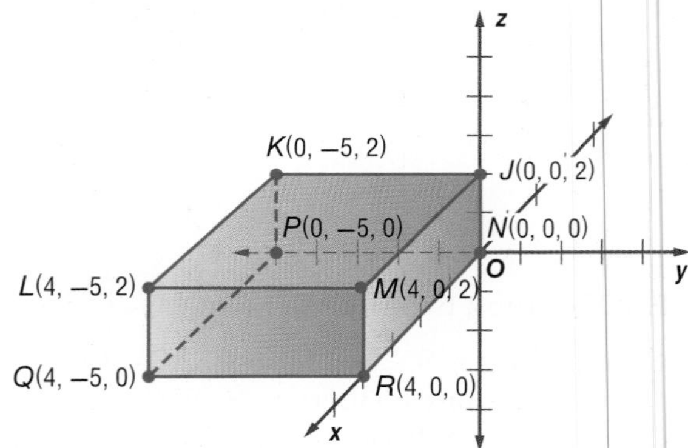


Finding the distance between points and the midpoint of a segment in space is similar to finding distance and a midpoint in the coordinate plane.

### KeyConcept Distance and Midpoint Formulas in Space

If $A$ has coordinates $A(x_1, y_1, z_1)$ and $B$ has coordinates $B(x_2, y_2, z_2)$, then

$$AB = \sqrt{(x_2 - x_1)^2 + (y_2 - y_1)^2 + (z_2 - z_1)^2}.$$

The midpoint $M$ of $\overline{AB}$ has coordinates

$$M\left(\frac{x_1 + x_2}{2}, \frac{y_1 + y_2}{2}, \frac{z_1 + z_2}{2}\right).$$

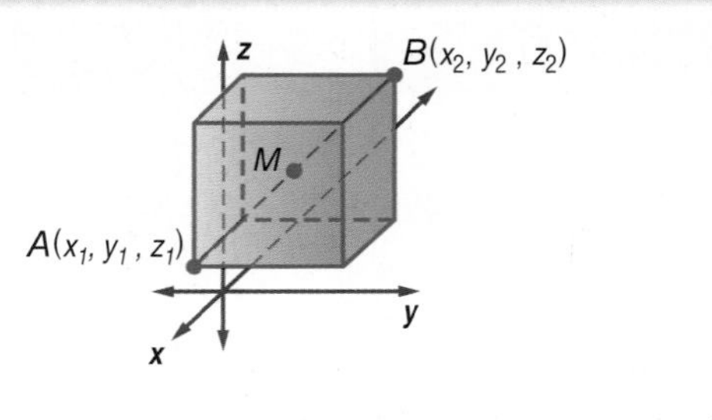

## Activity 2 Distance and Midpoint Formulas in Space

**Consider $J(2, 4, 9)$ and $K(-4, -5, 11)$.**

**a.** Find $JK$.

| | |
|---|---|
| $JK = \sqrt{(x_2 - x_1)^2 + (y_2 - y_1)^2 + (z_2 - z_1)^2}$ | Distance Formula in Space |
| $= \sqrt{(-4 - 2)^2 + (-5 - 4)^2 + (11 - 9)^2}$ | Substitution |
| $= \sqrt{121}$ | Simplify. |
| $= 11$ | Use a calculator. |

**b.** Determine the coordinates of the midpoint $M$ of $\overline{JK}$.

| | |
|---|---|
| $M = \left(\frac{x_1 + x_2}{2}, \frac{y_1 + y_2}{2}, \frac{z_1 + z_2}{2}\right)$ | Midpoint Formula in Space |
| $= \left(\frac{2 + (-4)}{2}, \frac{4 + (-5)}{2}, \frac{9 + 11}{2}\right)$ | Substitution |
| $= \left(-1, -\frac{1}{2}, 10\right)$ | Simplify. |

## Exercises

**Graph a rectangular solid that contains the given point and the origin as vertices. Label the coordinates of each vertex.**

1. $A(2, 1, 5)$
2. $P(-1, 4, 2)$
3. $C(-2, 2, 2)$
4. $R(3, -4, 1)$
5. $P(4, 6, -3)$
6. $G(4, 1, -3)$
7. $K(-2, -4, -4)$
8. $W(-1, -3, -6)$
9. $W(3, 3, 4)$

**Determine the distance between each pair of points. Then determine the coordinates of the midpoint $M$ of the segment joining the pair of points.**

10. $D(0, 0, 0)$ and $E(1, 5, 7)$
11. $G(-3, -4, 6)$ and $H(5, -3, -5)$
12. $K(2, 2, 0)$ and $L(-2, -2, 0)$
13. $P(-2, -5, 8)$ and $Q(3, -2, -1)$
14. $A(4, 7, 9)$ and $B(-3, 8, -8)$
15. $W(-12, 8, 10)$ and $Z(-4, 1, -2)$
16. $F\left(\frac{3}{5}, 0, \frac{4}{5}\right)$ and $G(0, 3, 0)$
17. $G(1, -1, 6)$ and $H\left(\frac{1}{5}, -\frac{2}{5}, 2\right)$
18. $B(\sqrt{3}, 2, 2\sqrt{2})$ and $C(-2\sqrt{3}, 4, 4\sqrt{2})$
19. $S(6\sqrt{3}, 4, 4\sqrt{2})$ and $T(4\sqrt{3}, 5, \sqrt{2})$

20. **PROOF** Write a coordinate proof of the Distance Formula in Space.

    **Given:** $A$ has coordinates $A(x_1, y_1, z_1)$, and $B$ has coordinates $B(x_2, y_2, z_2)$.

    **Prove:** $AB = \sqrt{(x_2 - x_1)^2 + (y_2 - y_1)^2 + (z_2 - z_1)^2}$

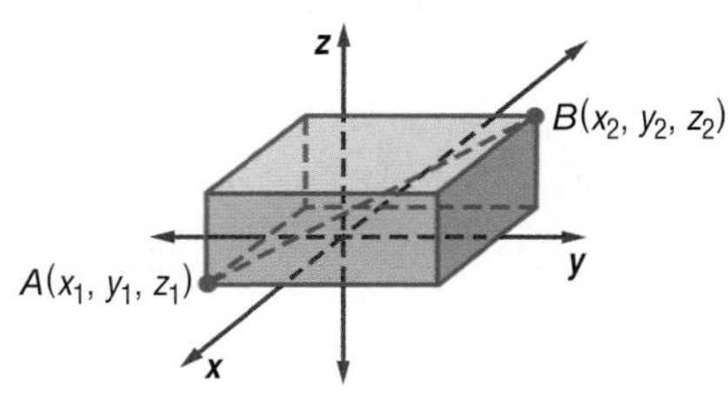


21. **WRITING IN MATH** Compare and contrast the Distance and Midpoint Formulas on the coordinate plane and in three-dimensional coordinate space.

LESSON 10-3

# Special Right Triangles

| Then | Now | Why? |
|---|---|---|
| You used properties of isosceles and equilateral triangles. | 1 Use the properties of 45°-45°-90° triangles.<br>2 Use the properties of 30°-60°-90° triangles. | As part of a packet for students attending a regional student council meeting, Lyndsay orders triangular highlighters. She wants to buy rectangular boxes for the highlighters and other items, but she is concerned that the highlighters will not fit in the box she has chosen. If she knows the length of a side of the highlighter, Lyndsay can use the properties of special right triangles to determine if it will fit in the box. |

**Common Core State Standards**

**Content Standards**

G.SRT.6 Understand that by similarity, side ratios in right triangles are properties of the angles in the triangle, leading to definitions of trigonometric ratios for acute angles.

**Mathematical Practices**

1 Make sense of problems and persevere in solving them.

7 Look for and make use of structure.

## 1 Properties of 45°-45°-90° Triangles

The diagonal of a square forms two congruent isosceles right triangles. Since the base angles of an isosceles triangle are congruent, the measure of each acute angle is 90 ÷ 2 or 45. Such a triangle is also known as a 45°-45°-90° triangle.

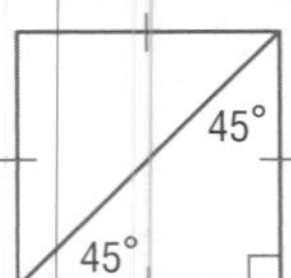


You can use the Pythagorean Theorem to find a relationship among the side lengths of a 45°-45°-90° right triangle.

| | |
|---|---|
| $\ell^2 + \ell^2 = h^2$ | Pythagorean Theorem |
| $2\ell^2 = h^2$ | Simplify. |
| $\sqrt{2\ell^2} = \sqrt{h^2}$ | Take the positive square root of each side. |
| $\ell\sqrt{2} = h$ | Simplify. |

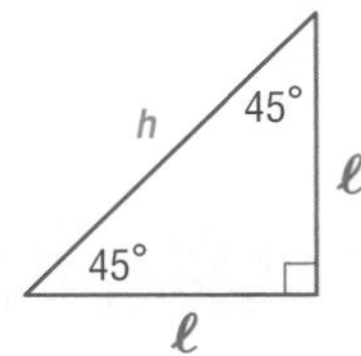


This algebraic proof verifies the following theorem.

### Theorem 10.8 45°-45°-90° Triangle Theorem

In a 45°-45°-90° triangle, the legs $\ell$ are congruent and the length of the hypotenuse $h$ is $\sqrt{2}$ times the length of a leg.

**Symbols** In a 45°-45°-90° triangle, $\ell = \ell$ and $h = \ell\sqrt{2}$.

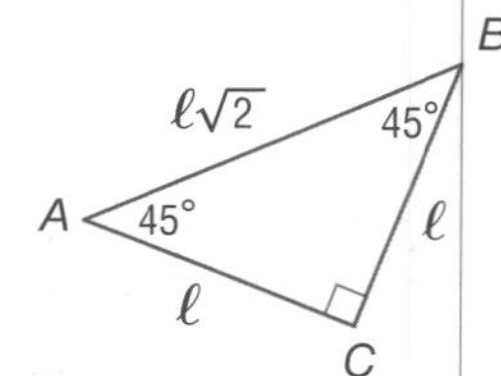

### Example 1 Find the Hypotenuse Length in a 45°-45°-90° Triangle

**Find $x$.**

**a.**

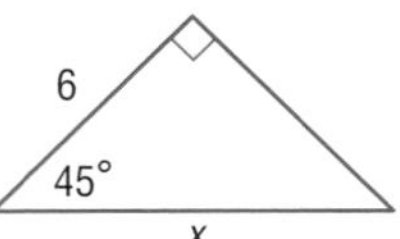


The acute angles of a right triangle are complementary, so the measure of the third angle is 90 − 45 or 45. Since this is a 45°-45°-90° triangle, use Theorem 10.8.

| | |
|---|---|
| $h = \ell\sqrt{2}$ | Theorem 10.8 |
| $x = 6\sqrt{2}$ | Substitution |

**b.**

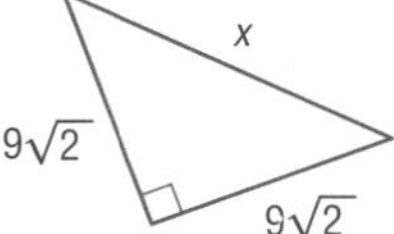


The legs of this right triangle have the same measure, so it is isosceles. Since this is a 45°-45°-90° triangle, use Theorem 10.8.

| | |
|---|---|
| $h = \ell\sqrt{2}$ | Theorem 10.8 |
| $x = 9\sqrt{2} \cdot \sqrt{2}$ | Substitution |
| $x = 9 \cdot 2$ or 18 | $\sqrt{2} \cdot \sqrt{2} = 2$ |

### GuidedPractice

**Find $x$.**

**1A.** **1B.** **1C.**

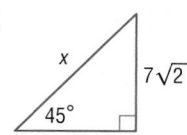

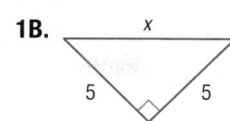

You can also work backward using Theorem 10.8 to find the lengths of the legs of a 45°-45°-90° triangle given the length of its hypotenuse.

### Example 2 Find the Leg Lengths in a 45°-45°-90° Triangle

**Find $x$.**

The legs of this right triangle have the same measure, $x$, so it is a 45°-45°-90° triangle. Use Theorem 10.8 to find $x$.

| | |
|---|---|
| $h = \ell\sqrt{2}$ | 45°-45°-90° Triangle Theorem |
| $12 = x\sqrt{2}$ | Substitution |
| $\frac{12}{\sqrt{2}} = x$ | Divide each side by $\sqrt{2}$. |
| $\frac{12}{\sqrt{2}} \cdot \frac{\sqrt{2}}{\sqrt{2}} = x$ | Rationalize the denominator. |
| $\frac{12\sqrt{2}}{2} = x$ | Multiply. |
| $6\sqrt{2} = x$ | Simplify. |

**ReviewVocabulary**

**rationalizing the denominator** a method used to eliminate radicals from the denominator of a fraction

### GuidedPractice

**2A.**

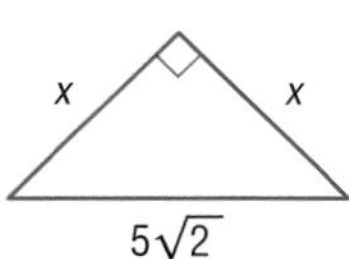


**2B.**

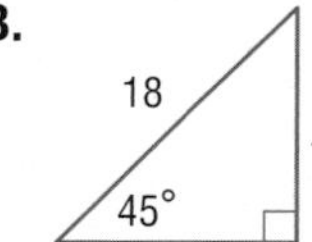


## 2 Properties of 30°-60°-90° Triangles

A 30°-60°-90° triangle is another *special* right triangle or right triangle with side lengths that share a special relationship. You can use an equilateral triangle to find this relationship.

**StudyTip**

**Altitudes of Isosceles Triangles** Notice that an altitude of an isosceles triangle is also a median of the triangle. In the figure at the right, $\overline{BD}$ bisects $\overline{AC}$.

When an altitude is drawn from any vertex of an equilateral triangle, two congruent 30°-60°-90° triangles are formed. In the figure shown, $\triangle ABD \cong \triangle CBD$, so $\overline{AD} \cong \overline{CD}$. If $AD = x$, then $CD = x$ and $AC = 2x$. Since $\triangle ABC$ is equilateral, $AB = 2x$ and $BC = 2x$.

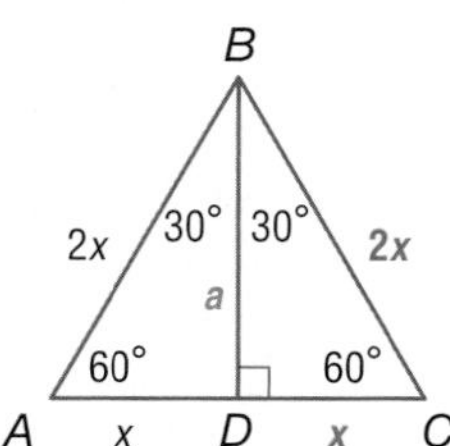


Use the Pythagorean Theorem to find $a$, the length of the altitude $\overline{BD}$, which is also the longer leg of $\triangle BDC$.

| | |
|---|---|
| $a^2 + x^2 = (2x)^2$ | Pythagorean Theorem |
| $a^2 + x^2 = 4x^2$ | Simplify. |
| $a^2 = 3x^2$ | Subtract $x^2$ from each side. |
| $a = \sqrt{3x^2}$ | Take the positive square root of each side. |
| $a = x\sqrt{3}$ | Simplify. |

**StudyTip**

**Use Ratios** The lengths of the sides of a 30°-60°-90° triangle are in a ratio of 1 to $\sqrt{3}$ to 2 or 1 : $\sqrt{3}$ : 2.

This algebraic proof verifies the following theorem.

### Theorem 10.9 30°-60°-90° Triangle Theorem

In a 30°-60°-90° triangle, the length of the hypotenuse $h$ is 2 times the length of the shorter leg $s$, and the length of the longer leg $\ell$ is $\sqrt{3}$ times the length of the shorter leg.

**Symbols** In a 30°-60°-90° triangle, $h = 2s$ and $\ell = s\sqrt{3}$.

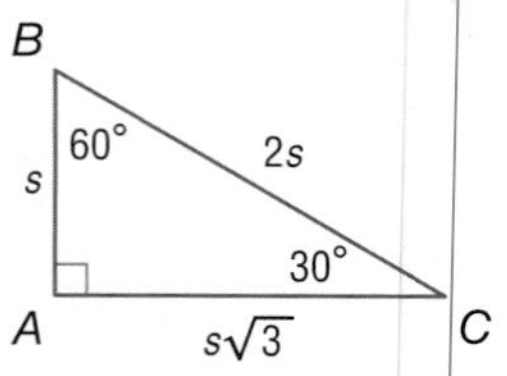


Remember, the shortest side of a triangle is opposite the smallest angle. So the shorter leg in a 30°-60°-90° triangle is opposite the 30° angle, and the longer leg is opposite the 60° angle.

### Example 3 Find Lengths in a 30°-60°-90° Triangle

PT

**Find $x$ and $y$.**

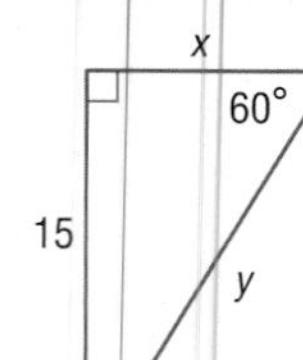


The acute angles of a right triangle are complementary, so the measure of the third angle in this triangle is 90 − 60 or 30. This is a 30°-60°-90° triangle.

Use Theorem 10.9 to find $x$, the length of the shorter side.

| | |
|---|---|
| $\ell = s\sqrt{3}$ | Theorem 10.9 |
| $15 = x\sqrt{3}$ | Substitution |
| $\frac{15}{\sqrt{3}} = x$ | Divide each side by $\sqrt{3}$. |
| $\frac{15}{\sqrt{3}} \cdot \frac{\sqrt{3}}{\sqrt{3}} = x$ | Rationalize the denominator. |
| $\frac{15\sqrt{3}}{\sqrt{3} \cdot \sqrt{3}} = x$ | Multiply. |
| $\frac{15\sqrt{3}}{3} = x$ | $\sqrt{3} \cdot \sqrt{3} = 3$ |
| $5\sqrt{3} = x$ | Simplify. |

Now use Theorem 10.9 to find $y$, the length of the hypotenuse.

| | |
|---|---|
| $h = 2s$ | Theorem 10.9 |
| $y = 2(5\sqrt{3})$ or $10\sqrt{3}$ | Substitution |

**GuidedPractice**

**Find $x$ and $y$.**

**3A.**

12 x 30° 60° y

**3B.**

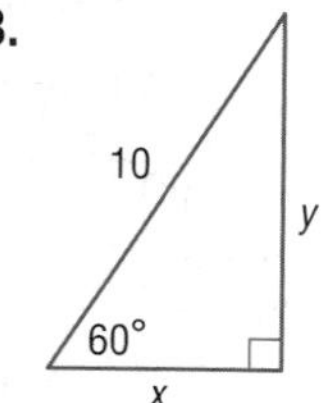


**3C.**

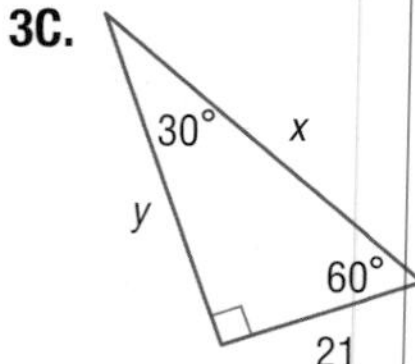

You can use the properties of 30°-60°-90° and 45°-45°-90° triangles to solve real-world problems.

PT

## Real-World Example 4 Use Properties of Special Right Triangles

**INVENTIONS A company makes crayons that "do not roll off tables" by shaping them as triangular prisms with equilateral bases. Sixteen of these crayons fit into a box shaped like a triangular prism that is $1\frac{1}{2}$ inches wide. The crayons stand on end in the box and the base of the box is equilateral. What are the dimensions of each crayon?**

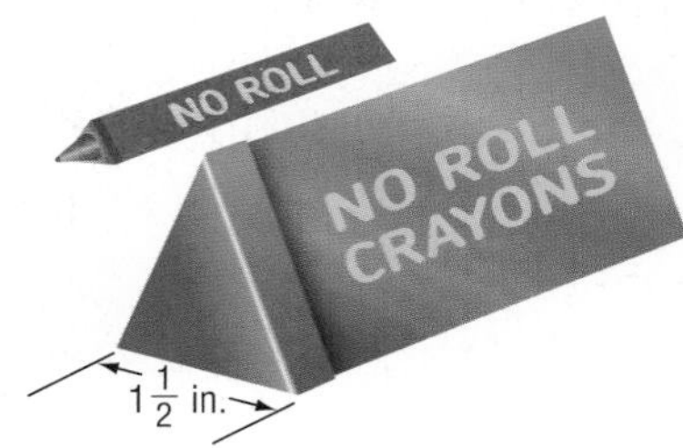


**Understand** You know that 16 crayons with equilateral triangular bases fit into a prism. You need to find the base length and height of each crayon.

**Plan** Guess and check to determine the arrangement of 16 crayons that would stack to fill the box. Find the width of one crayon and use the 30°-60°-90° Triangle Theorem to find its altitude.

**Solve** Make a guess that 4 equilateral crayons will fit across the base of the box. A sketch shows that the total number of crayons it takes to fill the box using 4 crayons across the base is 16. ✓

The width of the box is $1\frac{1}{2}$ inches, so the width of one crayon is $1\frac{1}{2} \div 4$ or $\frac{3}{8}$ inch.

Draw an equilateral triangle representing one crayon. Its altitude forms the longer leg of two 30°-60°-90° triangles. Use Theorem 10.9 to find the approximate length of the altitude $a$.

longer leg length = shorter leg length $\cdot \sqrt{3}$

$$a = \frac{3}{16} \cdot \sqrt{3} \text{ or about } 0.3$$

Each crayon is $\frac{3}{8}$ or about 0.4 inch by about 0.3 inch.

**Check** Find the height of the box using the 30°-60°-90° Triangle Theorem. Then divide by four, since the box is four crayons high. The result is a crayon height of about 0.3 inch. ✓

### Problem-Solving Tip

**Guess and Check** When using the guess and check strategy, it can be helpful to keep a list of those guesses that you have already tried and know do not work.

In Example 4, suppose your first guess had been that the box was 5 crayons wide.

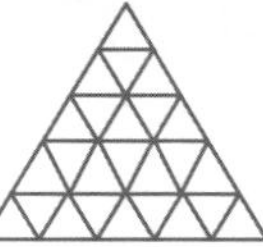

The sketch of this possibility reveals that this leads to a stack of 25, not 16 crayons.

### Guided Practice

**4. FURNITURE** The top of the aquarium coffee table shown is an isosceles right triangle. The table's longest side, $\overline{AC}$, measures 107 centimeters. What is the distance from vertex $B$ to side $\overline{AC}$? What are the lengths of the other two sides?

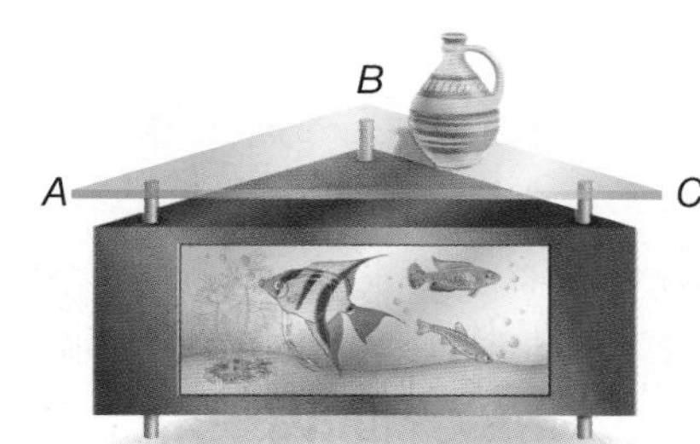

## Check Your Understanding

**Examples 1-2** Find $x$.

1. 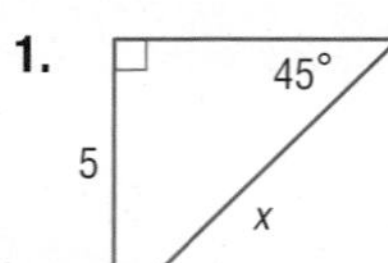


2. 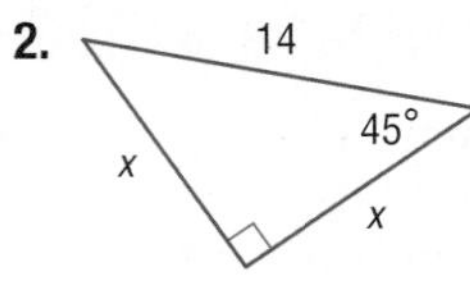


3. 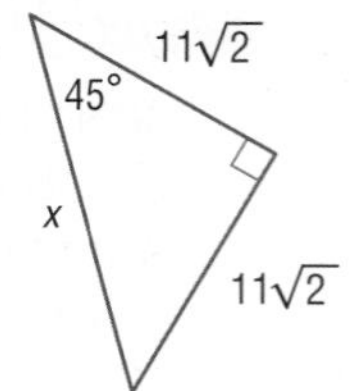


**Example 3** Find $x$ and $y$.

4. 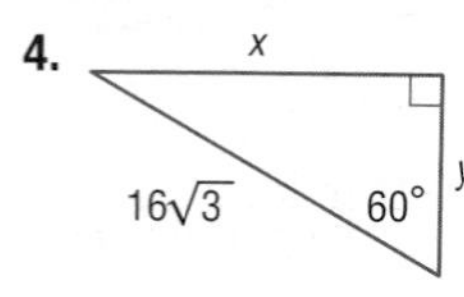


5. 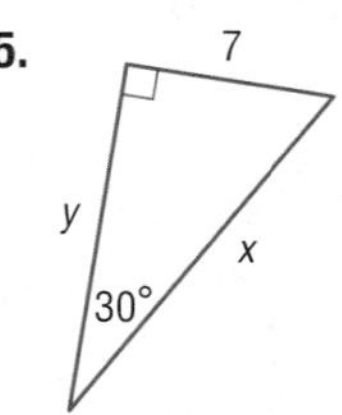


6. 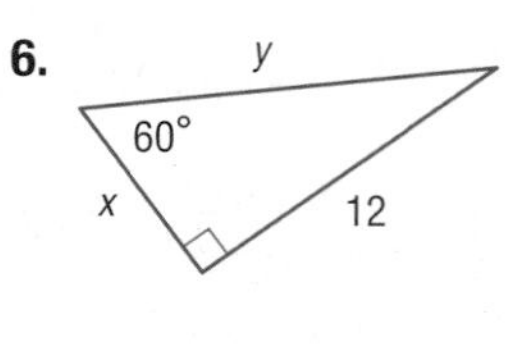


**Example 4**

7. **ART** Paulo is mailing an engraved plaque that is $3\frac{1}{4}$ inches high to the winner of a chess tournament. He has a mailer that is a triangular prism with 4-inch equilateral triangle bases as shown in the diagram. Will the plaque art fit through the opening of the mailer? Explain.

## Practice and Problem Solving

**Examples 1-2** CCSS **SENSE-MAKING** Find $x$.

8. 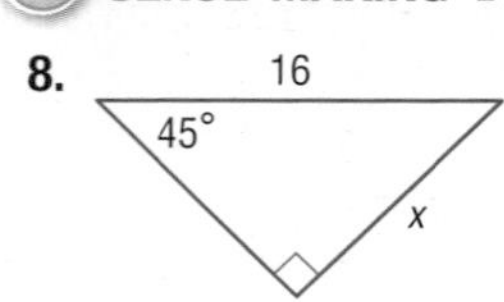


9. 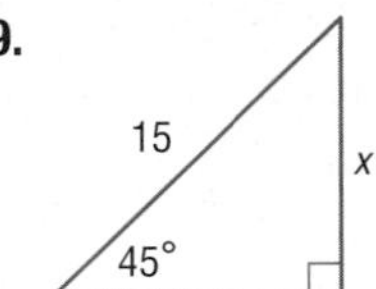


10. 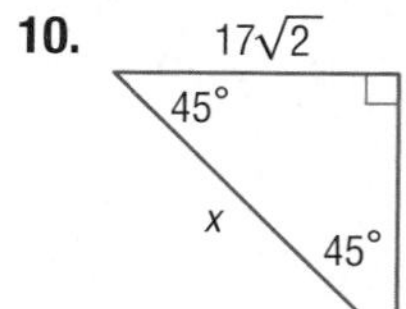


11. 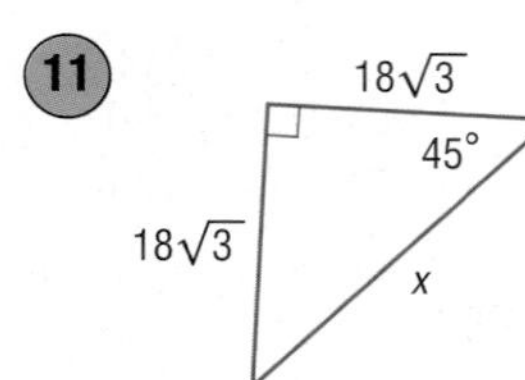


12. 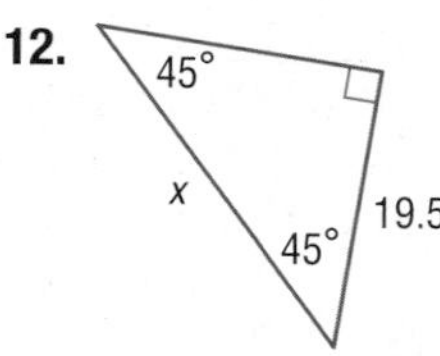


13. 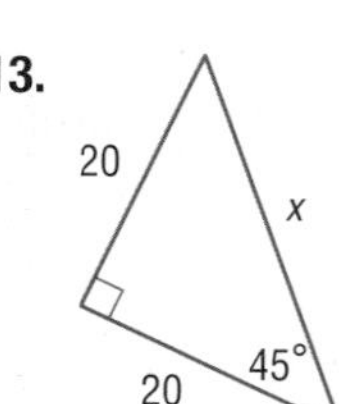


14. If a 45°-45°-90° triangle has a hypotenuse length of 9, find the leg length.

15. Determine the length of the leg of a 45°-45°-90° triangle with a hypotenuse length of 11.

16. What is the length of the hypotenuse of a 45°-45°-90° triangle if the leg length is 6 centimeters?

17. Find the length of the hypotenuse of a 45°-45°-90° triangle with a leg length of 8 centimeters.

**Example 3** **Find $x$ and $y$.**

**18.**
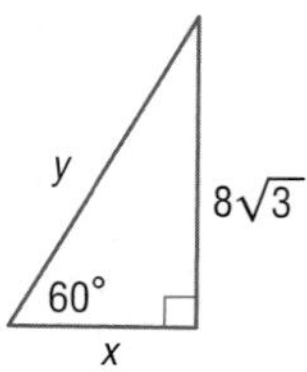


**19.**
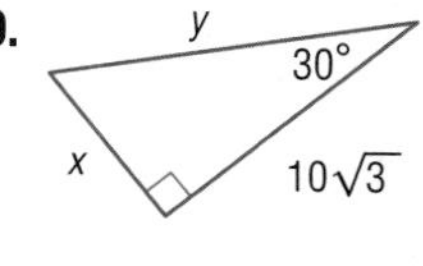


**20.**
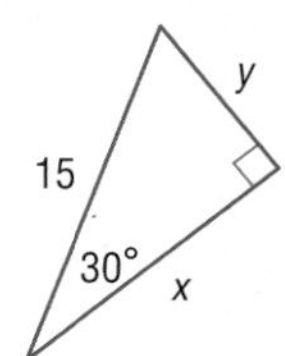


**21.**
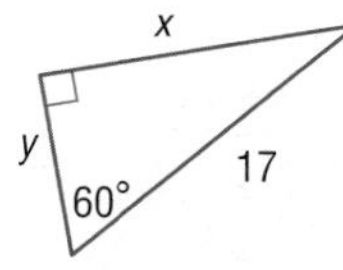


**22.**
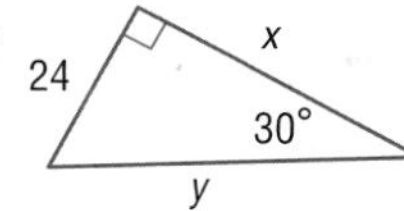


**23.**
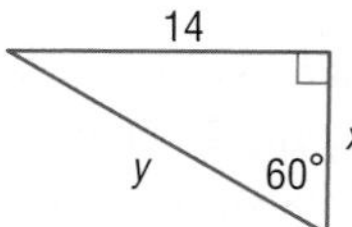


**24.** An equilateral triangle has an altitude length of 18 feet. Determine the length of a side of the triangle.

**25.** Find the length of the side of an equilateral triangle that has an altitude length of 24 feet.

**Example 4** **26.** **CCSS MODELING** Refer to the beginning of the lesson. Each highlighter is an equilateral triangle with 9-centimeter sides. Will the highlighter fit in a 10-centimeter by 7-centimeter rectangular box? Explain.

**27.** **EVENT PLANNING** Grace is having a party, and she wants to decorate the gable of the house as shown. The gable is an isosceles right triangle and she knows that the height of the gable is 8 feet. What length of lights will she need to cover the gable below the roof line?

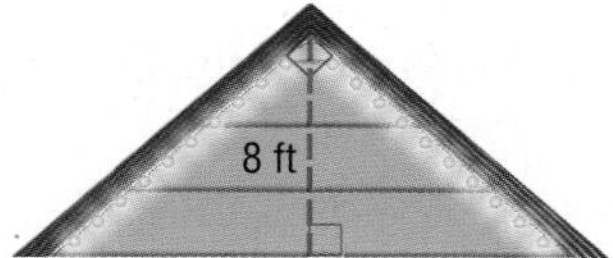


**Find $x$ and $y$.**

**28.**
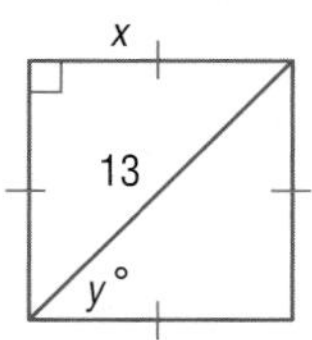


**29.**
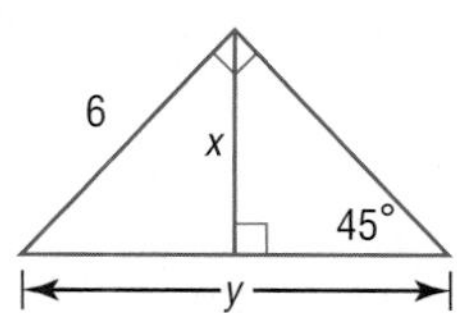


**30.**
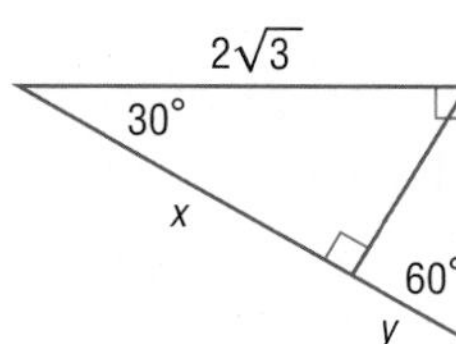


**31.**
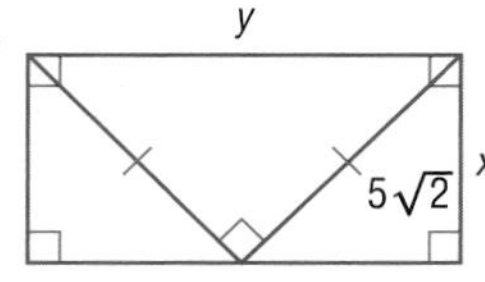


**32.**
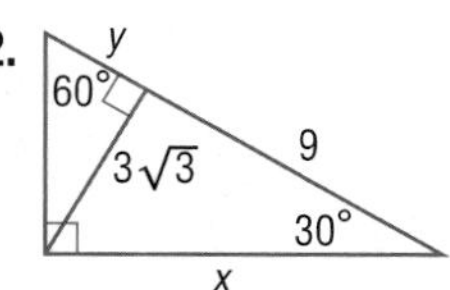


**33.**
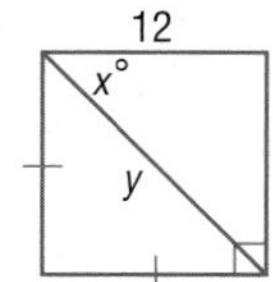


**34.** **QUILTS** The quilt block shown is made up of a square and four isosceles right triangles. What is the value of $x$? What is the side length of the entire quilt block?

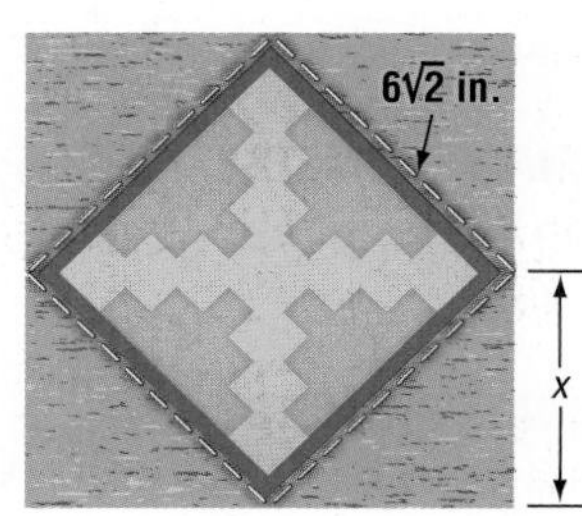

**35.** **ZIP LINE** Suppose a zip line is anchored in one corner of a course shaped like a rectangular prism. The other end is anchored in the opposite corner as shown. If the zip line makes a 60° angle with post $\overline{AF}$, find the zip line's length, $AD$.

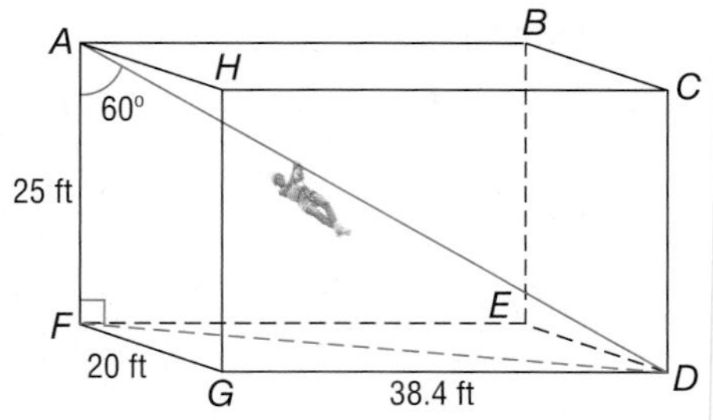


**36.** **GAMES** Kei is building a bean bag toss for the school carnival. He is using a 2-foot back support that is perpendicular to the ground 2 feet from the front of the board. He also wants to use a support that is perpendicular to the board as shown in the diagram. How long should he make the support?

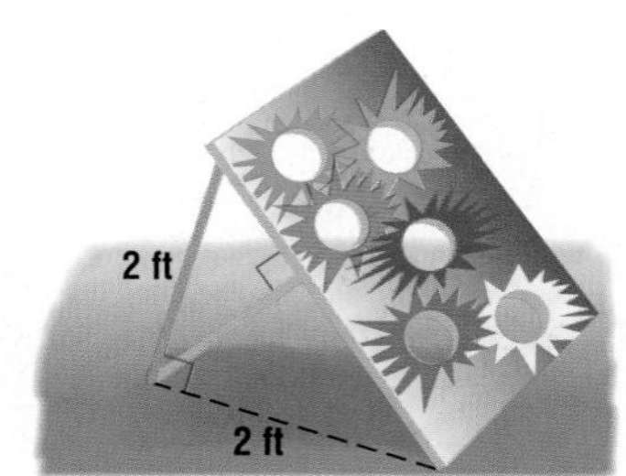


**37.** Find $x$, $y$, and $z$.

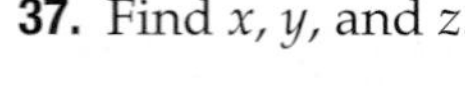

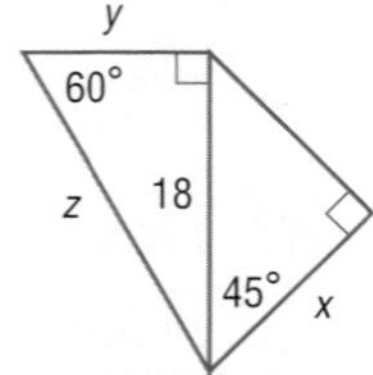


**38.** Each triangle in the figure is a 45°-45°-90° triangle. Find $x$.

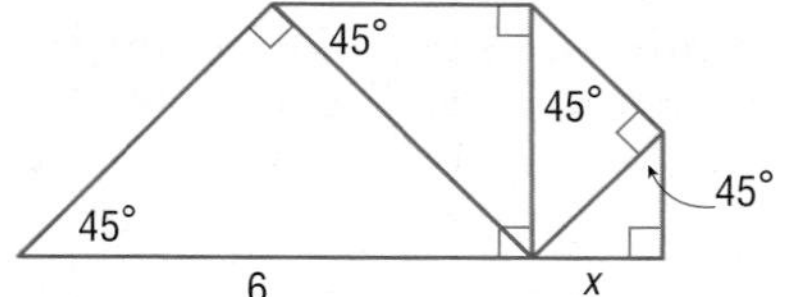


**39.** **CCSS MODELING** The dump truck shown has a 15-foot bed length. What is the height of the bed $h$ when angle $x$ is 30°? 45°? 60°?

**40.** Find $x$, $y$, and $z$, and the perimeter of trapezoid $PQRS$.

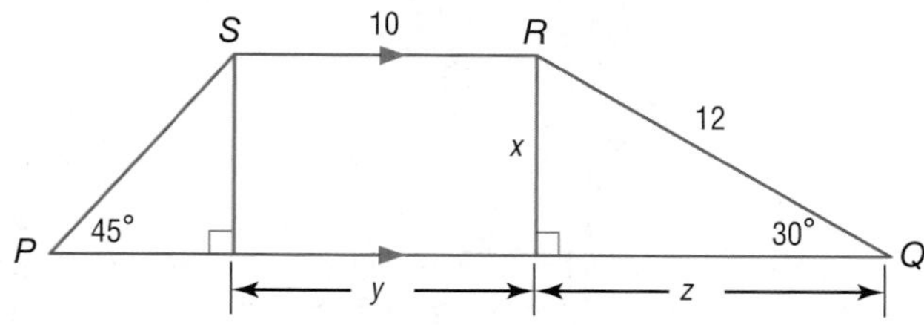


**41.** **COORDINATE GEOMETRY** $\triangle XYZ$ is a 45°-45°-90° triangle with right angle $Z$. Find the coordinates of $X$ in Quadrant I for $Y(-1, 2)$ and $Z(6, 2)$.

**42.** **COORDINATE GEOMETRY** $\triangle EFG$ is a 30°-60°-90° triangle with $m\angle F = 90$. Find the coordinates of $E$ in Quadrant III for $F(-3, -4)$ and $G(-3, 2)$. $\overline{FG}$ is the longer leg.

**43.** **COORDINATE GEOMETRY** $\triangle JKL$ is a 45°-45°-90° triangle with right angle $K$. Find the coordinates of $L$ in Quadrant IV for $J(-3, 5)$ and $K(-3, -2)$.

**44. EVENT PLANNING** Eva has reserved a gazebo at a local park for a party. She wants to be sure that there will be enough space for her 12 guests to be in the gazebo at the same time. She wants to allow 8 square feet of area for each guest. If the floor of the gazebo is a regular hexagon and each side is 7 feet, will there be enough room for Eva and her friends? Explain. (*Hint:* Use the Polygon Interior Angle Sum Theorem and the properties of special right triangles.)

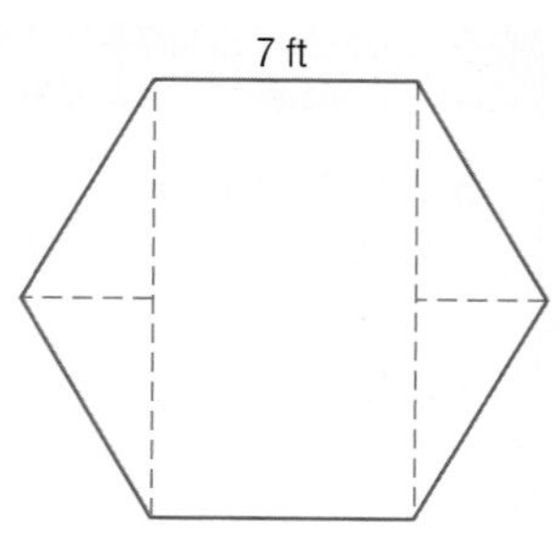


**45. MULTIPLE REPRESENTATIONS** In this problem, you will investigate ratios in right triangles.

**a. Geometric** Draw three similar right triangles with a 50° angle. Label one triangle *ABC* where angle *A* is the right angle and *B* is the 50° angle. Label a second triangle *MNP* where *M* is the right angle and *N* is the 50° angle. Label the third triangle *XYZ* where *X* is the right angle and *Y* is the 50° angle.

**b. Tabular** Copy and complete the table below.

| Triangle | Length | | | | Ratio | |
|---|---|---|---|---|---|---|
| *ABC* | *AC* | | *BC* | | $\frac{AC}{BC}$ | |
| *MNP* | *MP* | | *NP* | | $\frac{MP}{NP}$ | |
| *XYZ* | *XZ* | | *YZ* | | $\frac{XZ}{YZ}$ | |

**c. Verbal** Make a conjecture about the ratio of the leg opposite the 50° angle to the hypotenuse in any right triangle with an angle measuring 50°.

## H.O.T. Problems Use Higher-Order Thinking Skills

**46. CCSS CRITIQUE** Carmen and Audrey want to find *x* in the triangle shown. Is either of them correct? Explain.

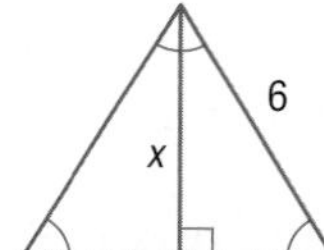


Carmen

$x = \frac{6\sqrt{3}}{2}$

$x = 3\sqrt{3}$

Audrey

$x = \frac{6\sqrt{2}}{2}$

$x = 3\sqrt{2}$

**47. OPEN ENDED** Draw a rectangle that has a diagonal twice as long as its width. Then write an equation to find the length of the rectangle.

**48. CHALLENGE** Find the perimeter of quadrilateral *ABCD*.

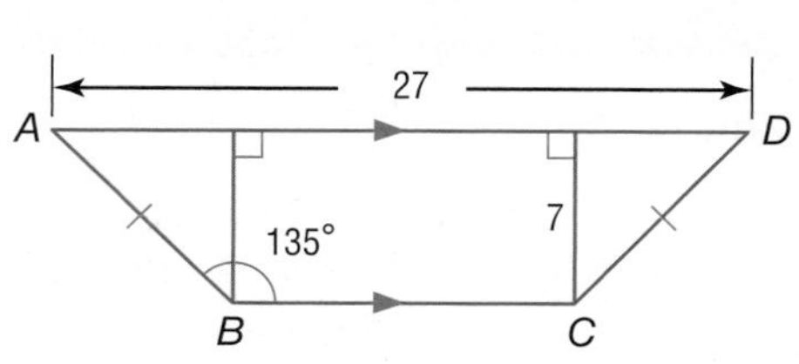


**49. REASONING** The ratio of the measure of the angles of a triangle is 1:2:3. The length of the shortest side is 8. What is the perimeter of the triangle?

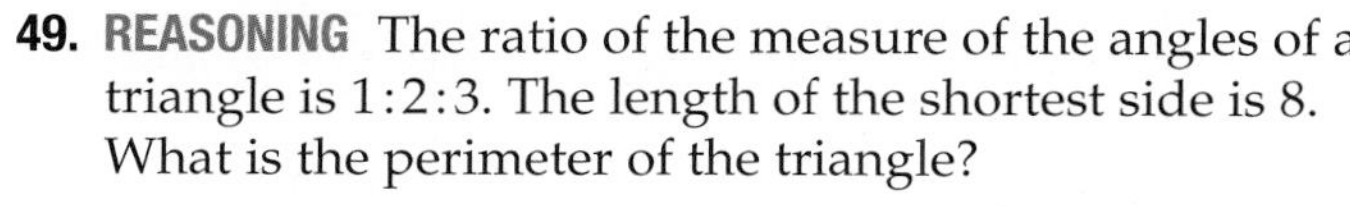

**50. WRITING IN MATH** Why are some right triangles considered *special?*

## Standardized Test Practice

**51.** If the length of the longer leg in a 30°-60°-90° triangle is $5\sqrt{3}$, what is the length of the shorter leg?

A 3
B 5
C $5\sqrt{2}$
D 10

**52. ALGEBRA** Solve $\sqrt{5 - 4x} - 6 = 7$.

F −44
G −41
H 41
J 44

**53. SHORT RESPONSE** $\triangle XYZ$ is a 45°-45°-90° triangle with right angle $Y$. Find the coordinates of $X$ in Quadrant III for $Y(-3, -3)$ and $Z(-3, 7)$.

**54. SAT/ACT** In the figure, below, square $ABCD$ is attached to $\triangle ADE$ as shown. If $m\angle EAD$ is 30° and $AE$ is equal to $4\sqrt{3}$, then what is the area of square $ABCD$?

A $8\sqrt{3}$
B 16
C 64
D 72
E $64\sqrt{2}$

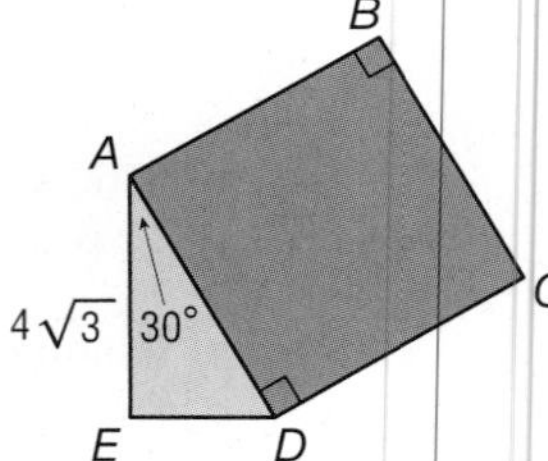


## Spiral Review

**55. SPORTS** Dylan is making a ramp for bike jumps. The ramp support forms a right angle. The base is 12 feet long, and the height is 9 feet. What length of plywood does Dylan need for the ramp? (Lesson 10-2)

**Find $x$, $y$, and $z$.** (Lesson 10-1)

**56.**

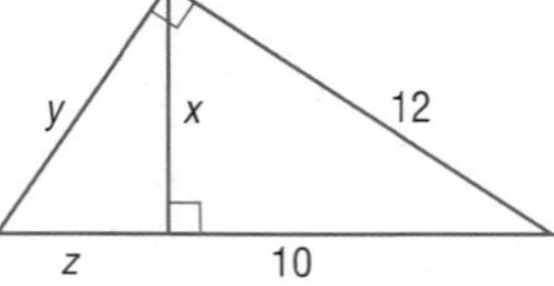


**57.**

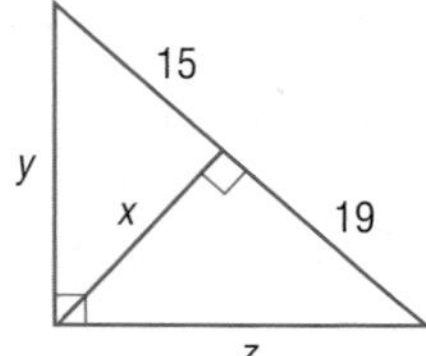


**58.**

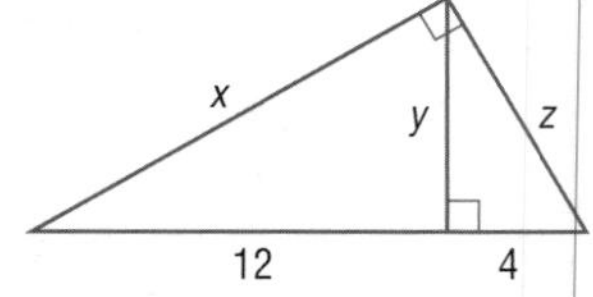


**Find the measures of the angles of each triangle.** (Lesson 9-1)

**59.** The ratio of the measures of the three angles is 2:5:3.

**60.** The ratio of the measures of the three angles is 6:9:10.

**61.** The ratio of the measures of the three angles is 5:7:8.

**Use the Exterior Angle Inequality Theorem to list all of the angles that satisfy the stated condition.** (Lesson 7-3)

**62.** measures less than $m\angle 5$

**63.** measures greater than $m\angle 6$

**64.** measures greater than $m\angle 10$

**65.** measures less than $m\angle 11$

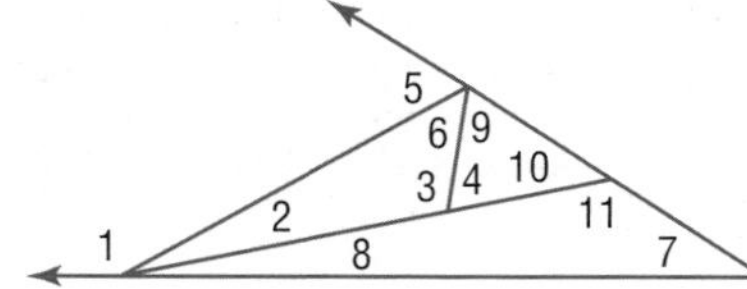


## Skills Review

**Find $x$.**

**66.**

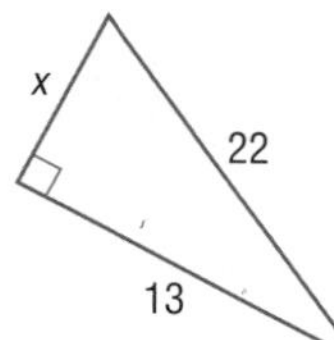


**67.**

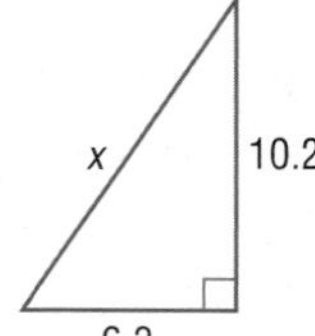


**68.**

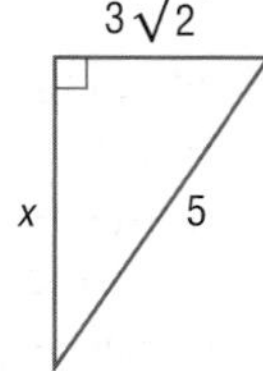

EXPLORE

# 10-4 Graphing Technology Lab: Trigonometry

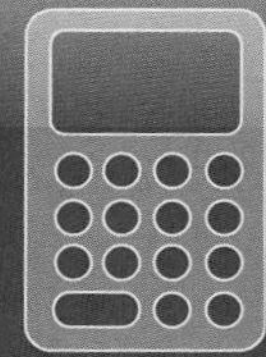

You have investigated patterns in the measures of special right triangles. *Trigonometry* is the study of the patterns in all right triangles. You can use the Cabri™ Jr. application on a graphing calculator to investigate these patterns.

**CCSS Common Core State Standards**
**Content Standards**
**G.SRT.6** Understand that by similarity, side ratios in right triangles are properties of the angles in the triangle, leading to definitions of trigonometric ratios for acute angles.
**Mathematical Practices** 5

## Activity Investigate Trigonometric Ratios

**Step 1** Use the line tool on the **F2** menu to draw a horizontal line. Label the points on the line $A$ and $B$.

**Step 2** Press **F2** and choose the **Perpendicular** tool to create a perpendicular line through point $B$. Draw and label a point $C$ on the perpendicular line.

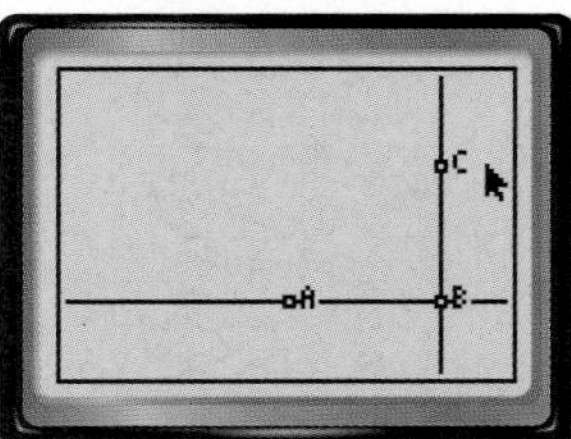


Steps 1 and 2

**Step 3** Use the **Segment** tool on the **F2** menu to draw $\overline{AC}$.

**Step 4** Find and label the measures of $\overline{BC}$ and $\overline{AC}$ using the **Distance** and **Length** tool under **Measure** on the **F5** menu. Use the **Angle** tool to find the measure of $\angle A$.

**Step 5** Calculate and display the ratio $\frac{BC}{AC}$ using the **Calculate** tool on the **F5** menu. Label the ratio as $A/B$.

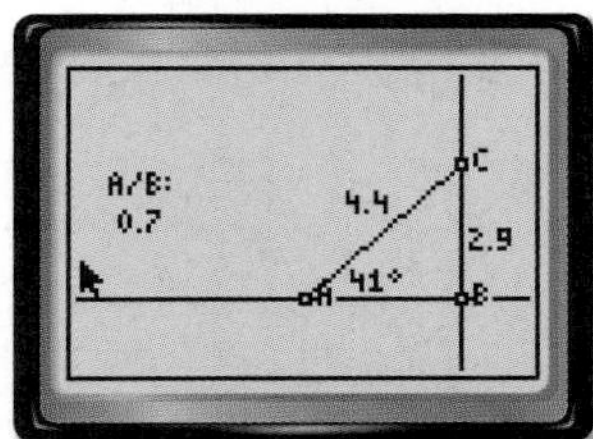


Steps 3 through 5

**Step 6** Press CLEAR. Then use the arrow keys to move the cursor close to point $B$. When the arrow is clear, press and hold the ALPHA key. Drag $B$ and observe the ratio.

## Analyze the Results

1. Discuss the effect on $\frac{BC}{AC}$ by dragging point $B$ on $\overline{BC}$, $\overline{AC}$, and $\angle A$.

2. Use the calculate tool to find the ratios $\frac{AB}{AC}$ and $\frac{BC}{AB}$. Then drag $B$ and observe the ratios.

3. **MAKE A CONJECTURE** The *sine*, *cosine*, and *tangent* functions are trigonometric functions based on angle measures. Make a note of $m\angle A$. Exit Cabri Jr. and use SIN, COS, and TAN on the calculator to find *sine*, *cosine* and *tangent* for $m\angle A$. Compare the results to the ratios you found in the activity. Make a conjecture about the definitions of sine, cosine, and tangent.

# LESSON 10-4 Trigonometry

**Then**

- You used the Pythagorean Theorem to find missing lengths in right triangles.

**Now**

1. Find trigonometric ratios using right triangles.
2. Use trigonometric ratios to find angle measures in right triangles.

**Why?**

- The steepness of a hiking trail is often expressed as a *percent of grade*. The steepest part of Bright Angel Trail in the Grand Canyon National Park has about a 15.7% grade. This means that the trail rises or falls 15.7 feet over a horizontal distance of 100 feet. You can use trigonometric ratios to determine that this steepness is equivalent to an angle of about 9°.

**NewVocabulary**
trigonometry
trigonometric ratio
sine
cosine
tangent
inverse sine
inverse cosine
inverse tangent

**Common Core State Standards**

**Content Standards**

G.SRT.6 Understand that by similarity, side ratios in right triangles are properties of the angles in the triangle, leading to definitions of trigonometric ratios for acute angles.

G.SRT.7 Explain and use the relationship between the sine and cosine of complementary angles.

**Mathematical Practices**

1 Make sense of problems and persevere in solving them.

5 Use appropriate tools strategically.

## 1 Trigonometric Ratios

The word **trigonometry** comes from two Greek terms, *trigon*, meaning triangle, and *metron*, meaning measure. The study of trigonometry involves triangle measurement. A **trigonometric ratio** is a ratio of the lengths of two sides of a right triangle. One trigonometric ratio of $\triangle ABC$ is $\frac{AC}{AB}$.

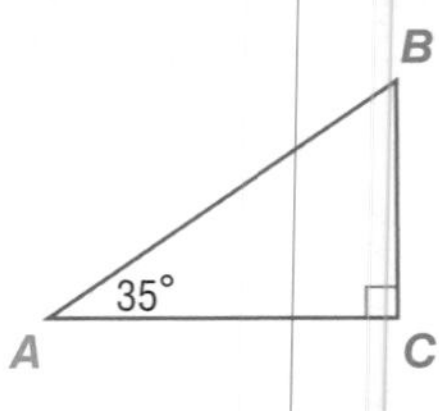


By AA Similarity, a right triangle with a given acute angle measure is similar to every other right triangle with the same acute angle measure. So, trigonometric ratios are constant for a given angle measure.

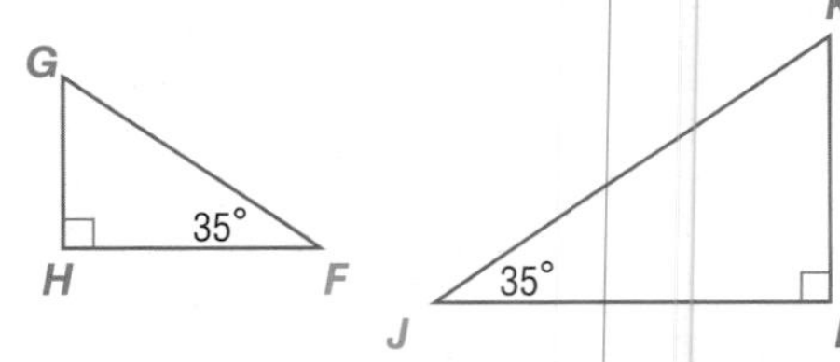


$\triangle ABC \sim \triangle FGH \sim \triangle JKL$, so $\frac{AC}{AB} = \frac{FH}{FG} = \frac{JL}{JK}$

The names of the three most common trigonometric ratios are given below.

**KeyConcept** Trigonometric Ratios

| Words | Symbols |
|---|---|
| If $\triangle ABC$ is a right triangle with acute $\angle A$, then the **sine** of $\angle A$ (written sin $A$) is the ratio of the length of the leg opposite $\angle A$ (opp) to the length of the hypotenuse (hyp). | $\sin A = \frac{\text{opp}}{\text{hyp}}$ or $\frac{a}{c}$<br>$\sin B = \frac{\text{opp}}{\text{hyp}}$ or $\frac{b}{c}$ |
| If $\triangle ABC$ is a right triangle with acute $\angle A$, then the **cosine** of $\angle A$ (written cos $A$) is the ratio of the length of the leg adjacent $\angle A$ (adj) to the length of the hypotenuse (hyp). | $\cos A = \frac{\text{adj}}{\text{hyp}}$ or $\frac{b}{c}$<br>$\cos B = \frac{\text{adj}}{\text{hyp}}$ or $\frac{a}{c}$ |
| If $\triangle ABC$ is a right triangle with acute $\angle A$, then the **tangent** of $\angle A$ (written tan $A$) is the ratio of the length of the leg opposite $\angle A$ (opp) to the length of the leg adjacent $\angle A$ (adj). | $\tan A = \frac{\text{opp}}{\text{adj}}$ or $\frac{a}{b}$<br>$\tan B = \frac{\text{opp}}{\text{adj}}$ or $\frac{b}{a}$ |

## Example 1 Find Sine, Cosine, and Tangent Ratios

**Express each ratio as a fraction and as a decimal to the nearest hundredth.**

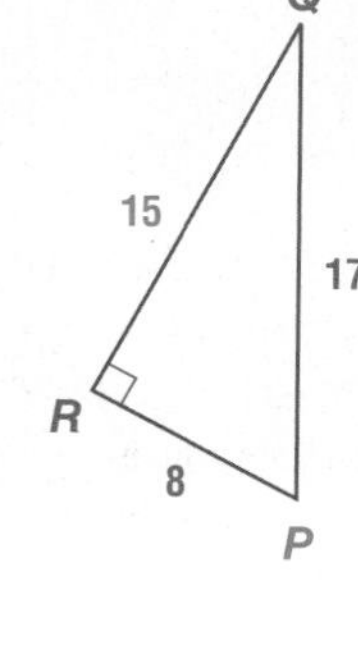


**a. sin $P$**

$\sin P = \frac{\text{opp}}{\text{hyp}}$

$= \frac{15}{17}$ or about 0.88

**b. cos $P$**

$\cos P = \frac{\text{adj}}{\text{hyp}}$

$= \frac{8}{17}$ or about 0.47

**c. tan $P$**

$\tan P = \frac{\text{opp}}{\text{adj}}$

$= \frac{15}{8}$ or about 1.88

**d. sin $Q$**

$\sin Q = \frac{\text{opp}}{\text{hyp}}$

$= \frac{8}{17}$ or about 0.47

**e. cos $Q$**

$\cos Q = \frac{\text{adj}}{\text{hyp}}$

$= \frac{15}{17}$ or about 0.88

**f. tan $Q$**

$\tan Q = \frac{\text{opp}}{\text{adj}}$

$= \frac{8}{15}$ or about 0.53

> **StudyTip**
>
> **Memorizing Trigonometric Ratios** SOH-CAH-TOA is a mnemonic device for learning the ratios for sine, cosine, and tangent using the first letter of each word in the ratios.
>
> $\sin A = \frac{\text{opp}}{\text{hyp}}$
>
> $\cos A = \frac{\text{adj}}{\text{hyp}}$
>
> $\tan A = \frac{\text{opp}}{\text{adj}}$

### GuidedPractice

1. Find sin $J$, cos $J$, tan $J$, sin $K$, cos $K$, and tan $K$. Express each ratio as a fraction and as a decimal to the nearest hundredth.

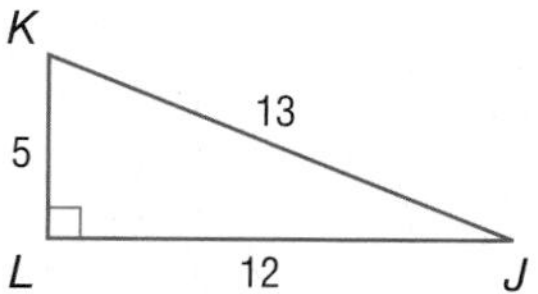


Special right triangles can be used to find the sine, cosine, and tangent of 30°, 60°, and 45° angles.

## Example 2 Use Special Right Triangles to Find Trigonometric Ratios

**Use a special right triangle to express the tangent of 30° as a fraction and as a decimal to the nearest hundredth.**

Draw and label the side lengths of a 30°-60°-90° right triangle, with $x$ as the length of the shorter leg.

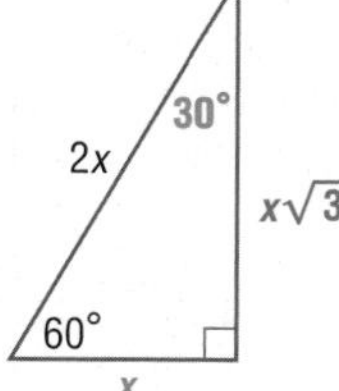


The side opposite the 30° angle has a measure of $x$.

The side adjacent to the 30° angle has a measure of $x\sqrt{3}$.

$\tan 30° = \frac{\text{opp}}{\text{adj}}$ Definition of tangent ratio

$= \frac{x}{x\sqrt{3}}$ Substitution

$= \frac{1}{\sqrt{3}} \cdot \frac{\sqrt{3}}{\sqrt{3}}$ Simplify and rationalize the denominator.

$= \frac{\sqrt{3}}{3}$ or about 0.58 Simplify and use a calculator.

### GuidedPractice

2. Use a special right triangle to express the cosine of 45° as a fraction and as a decimal to the nearest hundredth.

## Real-WorldLink

The grade of a trail often changes many times. Average grade is the average of several consecutive running grades of a trail. Maximum grade is the smaller section of a trail that exceeds the trail's typical running grade. Trails often have maximum grades that are much steeper than a trail's average running grade.

**Source:** Federal Highway Administration

## Real-World Example 3 Estimate Measures Using Trigonometry

**HIKING A certain part of a hiking trail slopes upward at about a 5° angle. After traveling a horizontal distance of 100 feet along this part of the trail, what would be the change in a hiker's vertical position? What distance has the hiker traveled along the path?**

Let $m\angle A = 5$. The vertical change in the hiker's position is $x$, the measure of the leg opposite $\angle A$. The horizontal distance traveled is 100 feet, the measure of the leg adjacent to $\angle A$. Since the length of the leg opposite and the leg adjacent to a given angle are involved, write an equation using a tangent ratio.

$\tan A = \frac{\text{opp}}{\text{adj}}$ Definition of tangent ratio

$\tan 5° = \frac{x}{100}$ Substitution

$100 \cdot \tan 5° = x$ Multiply each side by 100.

Use a calculator to find $x$.

100 [TAN] 5 [ENTER] **8.748866353**

The hiker is about 8.75 feet higher than when he started walking.

The distance $y$ traveled along the path is the length of the hypotenuse, so you can use a cosine ratio to find this distance.

$\cos A = \frac{\text{adj}}{\text{hyp}}$ Definition of cosine ratio

$\cos 5° = \frac{100}{y}$ Substitution

$y \cdot \cos 5° = 100$ Multiply each side by $y$.

$y = \frac{100}{\cos 5°}$ Divide each side by cos 5°.

## StudyTip

Graphing Calculator Be sure your graphing calculator is in degree mode rather than radian mode.

Use a calculator to find $y$.

100 [÷] [COS] 5 [ENTER] **100.3819838**

The hiker has traveled a distance of about 100.38 feet along the path.

### GuidedPractice

**Find $x$ to the nearest hundredth.**

**3A.**

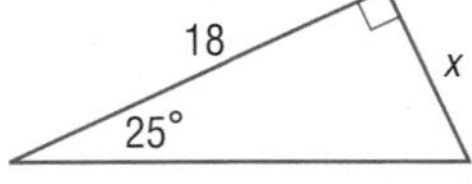


**3B.**

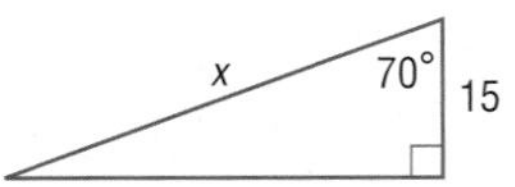


**3C. ARCHITECTURE** The front of the vacation cottage shown is an isosceles triangle. What is the height $x$ of the cottage above its foundation? What is the length $y$ of the roof? Explain your reasoning.

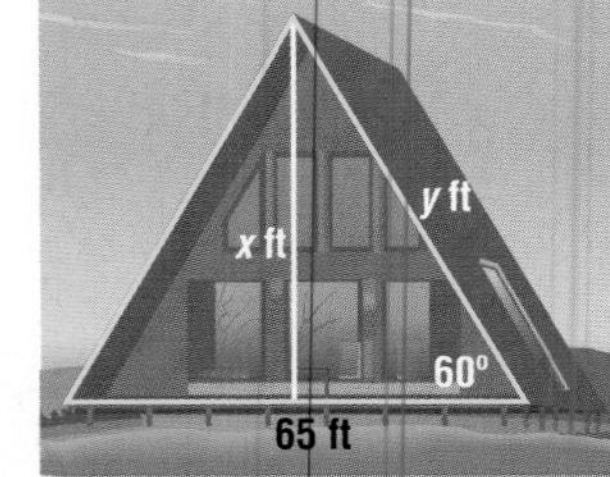


J.A. Kraulis/Masterfile

## 2 Use Inverse Trigonometric Ratios

In Example 2, you found that $\tan 30° \approx 0.58$. It follows that if the tangent of an acute angle is 0.58, then the angle measures approximately 30.

If you know the sine, cosine, or tangent of an acute angle, you can use a calculator to find the measure of the angle, which is the inverse of the trigonometric ratio.

**ReadingMath**

**Inverse Trigonometric Ratios** The expression $\sin^{-1} x$ is read *the inverse sine of x* and is interpreted as the angle with sine $x$. Be careful not to confuse this notation with the notation for negative exponents—$\sin^{-1} x \neq \frac{1}{\sin x}$. Instead, this notation is similar to the notation for an inverse function, $f^{-1}(x)$.

### KeyConcept Inverse Trigonometric Ratios

| | |
|---|---|
| Words | If $\angle A$ is an acute angle and the sine of $A$ is $x$, then the **inverse sine** of $x$ is the measure of $\angle A$. |
| Symbols | If $\sin A = x$, then $\sin^{-1} x = m\angle A$. |
| Words | If $\angle A$ is an acute angle and the cosine of $A$ is $x$, then the **inverse cosine** of $x$ is the measure of $\angle A$. |
| Symbols | If $\cos A = x$, then $\cos^{-1} x = m\angle A$. |
| Words | If $\angle A$ is an acute angle and the tangent of $A$ is $x$, then the **inverse tangent** of $x$ is the measure of $\angle A$. |
| Symbols | If $\tan A = x$, then $\tan^{-1} x = m\angle A$. |

So if $\tan 30° \approx 0.58$, then $\tan^{-1} 0.58 \approx 30°$.

PT

### Example 4 Find Angle Measures Using Inverse Trigonometric Ratios

**Use a calculator to find the measure of $\angle A$ to the nearest tenth.**

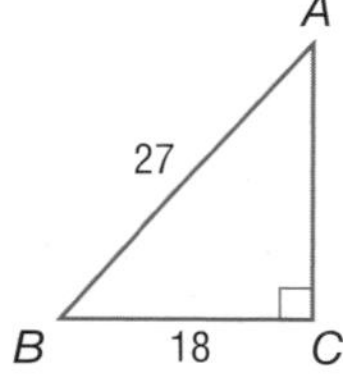


The measures given are those of the leg opposite $\angle A$ and the hypotenuse, so write an equation using the sine ratio.

$\sin A = \frac{18}{27}$ or $\frac{2}{3}$ $\quad \sin A = \frac{\text{opp}}{\text{hyp}}$

If $\sin A = \frac{2}{3}$, then $\sin^{-1} \frac{2}{3} = m\angle A$. Use a calculator.

**KEYSTROKES:** [2nd] [SIN$^{-1}$] [(] 2 [÷] 3 [)] [ENTER] **41.8103149**

So, $m\angle A \approx 41.8°$.

**StudyTip**

**CCSS Tools** Use a graphing calculator. The second functions of the [SIN], [COS], and [TAN] keys are usually the inverses.

**GuidedPractice**

**Use a calculator to find the measure of $\angle A$ to the nearest tenth.**

**4A.**

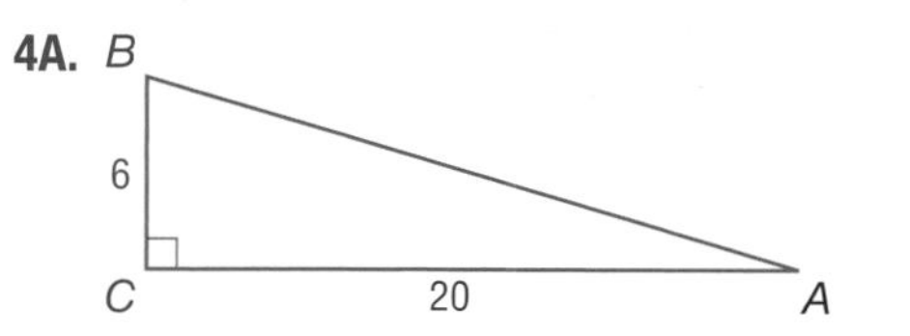


**4B.**

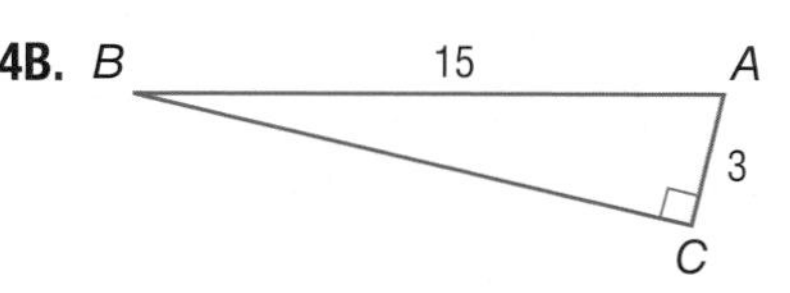

When you use given measures to find the unknown angle and side measures of a right triangle, this is known as *solving a right triangle*. To solve a right triangle, you need to know

- two side lengths or
- one side length and the measure of one acute angle.

PT

## Example 5 Solve a Right Triangle

**Solve the right triangle. Round side measures to the nearest tenth and angle measures to the nearest degree.**

**Step 1** Find $m\angle X$ by using a tangent ratio.

$\tan X = \frac{9}{5}$ — $\tan X = \frac{\text{opp}}{\text{adj}}$

$\tan^{-1}\frac{9}{5} = m\angle X$ — Definition of inverse tangent

$60.9453959 \approx m\angle X$ — Use a calculator.

So, $m\angle X \approx 61$.

**Step 2** Find $m\angle Y$ using Corollary 10.1, which states that the acute angles of a right triangle are complementary.

$m\angle X + m\angle Y = 90$ — Corollary 10.1

$61 + m\angle Y \approx 90$ — $m\angle X \approx 61$

$m\angle Y \approx 29$ — Subtract 61 from each side.

So, $m\angle Y \approx 29$.

**Step 3** Find $XY$ by using the Pythagorean Theorem.

$(XZ)^2 + (ZY)^2 = (XY)^2$ — Pythagorean Theorem

$5^2 + 9^2 = (XY)^2$ — Substitution

$106 = (XY)^2$ — Simplify.

$\sqrt{106} = XY$ — Take the positive square root of each side.

$10.3 \approx XY$ — Use a calculator.

So $XY \approx 10.3$.

**StudyTip**

Alternative Methods Right triangles can often be solved using different methods. In Example 5, $m\angle Y$ could have been found using a tangent ratio, and $m\angle X$ and a sine ratio could have been used to find $XY$.

**WatchOut!**

Approximation If using calculated measures to find other measures in a right triangle, be careful not to round values until the last step. So in the following equation, use $\tan^{-1}\frac{9}{5}$ instead of its approximate value, 61°.

$$XY = \frac{9}{\sin X} = \frac{9}{\sin\left(\tan^{-1}\frac{9}{5}\right)} \approx 10.3$$

### GuidedPractice

**Solve each right triangle. Round side measures to the nearest tenth and angle measures to the nearest degree.**

**5A.**

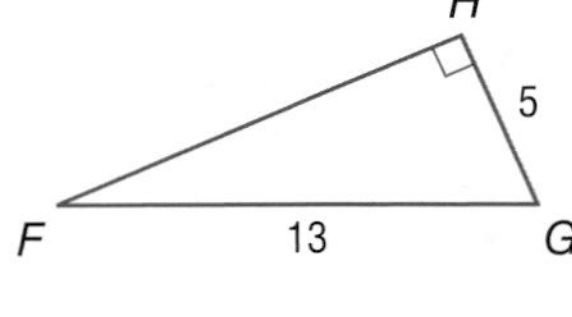


**5B.**

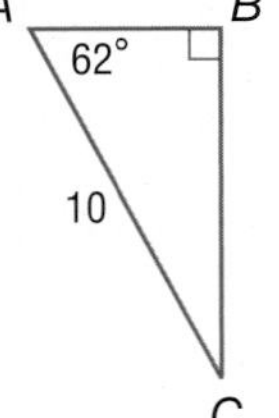


**5C.**

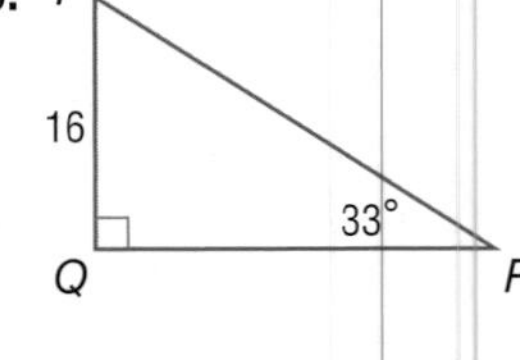

## Check Your Understanding

**Example 1** **Express each ratio as a fraction and as a decimal to the nearest hundredth.**

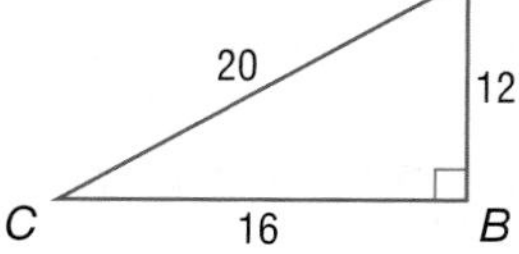


1. $\sin A$
2. $\tan C$
3. $\cos A$
4. $\tan A$
5. $\cos C$
6. $\sin C$

**Example 2**

7. Use a special right triangle to express $\sin 60°$ as a fraction and as a decimal to the nearest hundredth.

**Example 3** **Find $x$. Round to the nearest hundredth.**

8.

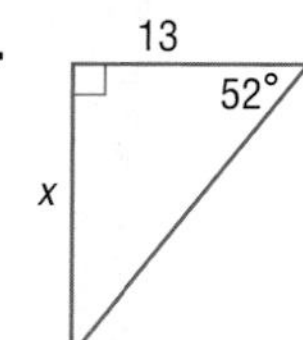


9. x, 24, 61°

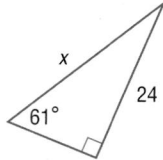

10.

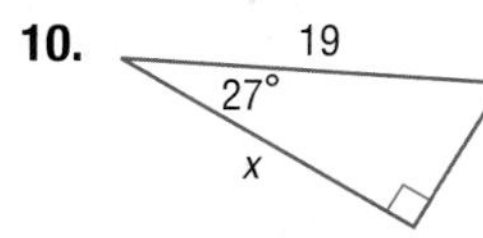


11. **SPORTS** David is building a bike ramp. He wants the angle that the ramp makes with the ground to be 20°. If the board he wants to use for his ramp is $3\frac{1}{2}$ feet long, about how tall will the ramp need to be at the highest point?

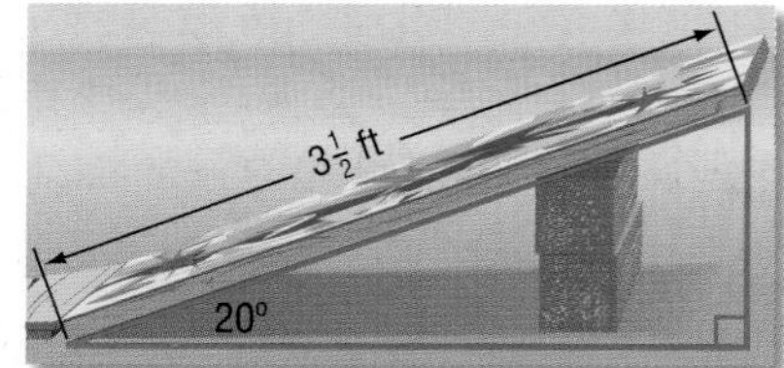


**Example 4** **CCSS TOOLS** **Use a calculator to find the measure of $\angle Z$ to the nearest tenth.**

12.

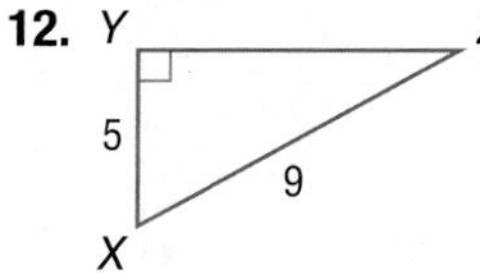


13.

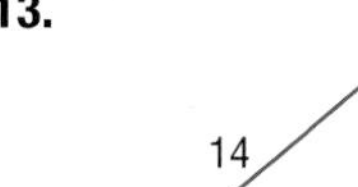
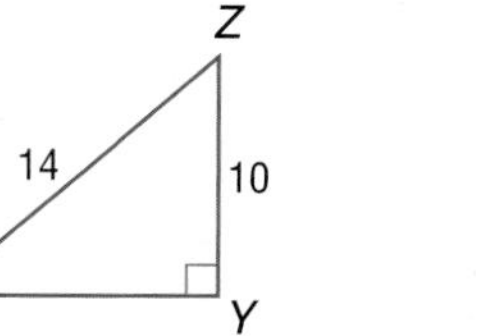


14.

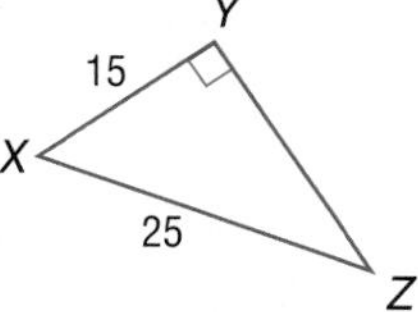


**Example 5**

15. Solve the right triangle. Round side measures to the nearest tenth and angle measures to the nearest degree.

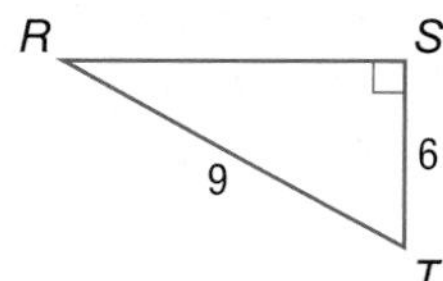


## Practice and Problem Solving

**Example 1** **Find $\sin J$, $\cos J$, $\tan J$, $\sin L$, $\cos L$, and $\tan L$. Express each ratio as a fraction and as a decimal to the nearest hundredth.**

16.

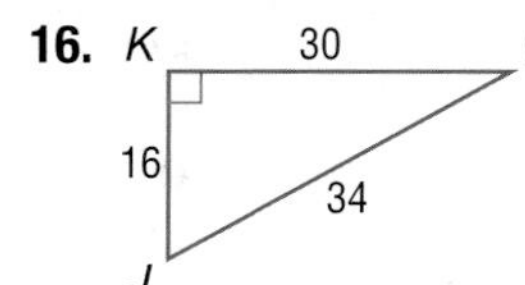


17.

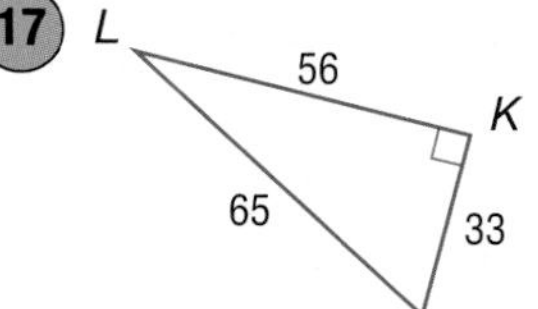


18.

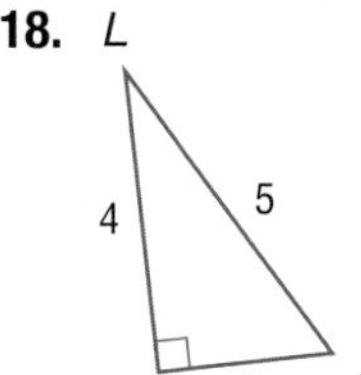


19. L, 85, J, 84, 13, K

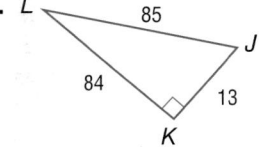

20.

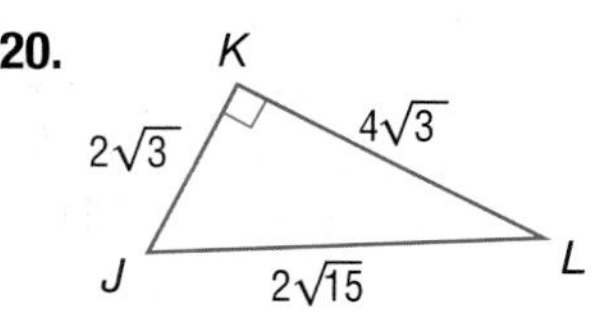


21.

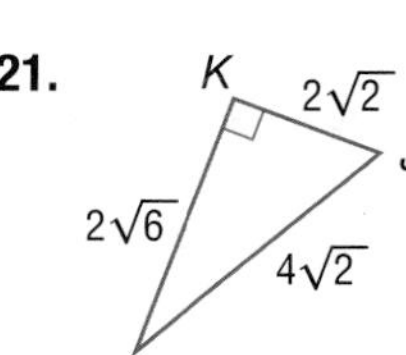

**Example 2** **Use a special right triangle to express each trigonometric ratio as a fraction and as a decimal to the nearest hundredth.**

**22.** tan 60°
**23.** cos 30°
**24.** sin 45°
**25.** sin 30°
**26.** tan 45°
**27.** cos 60°

**Example 3** **Find $x$. Round to the nearest tenth.**

**28.**
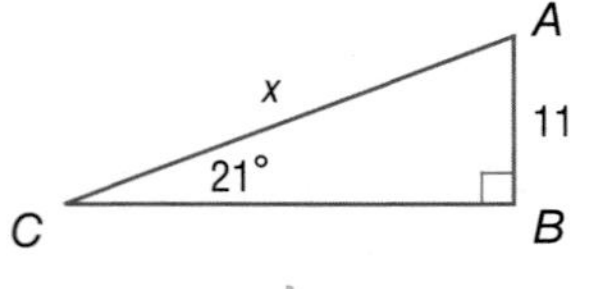


**29.**
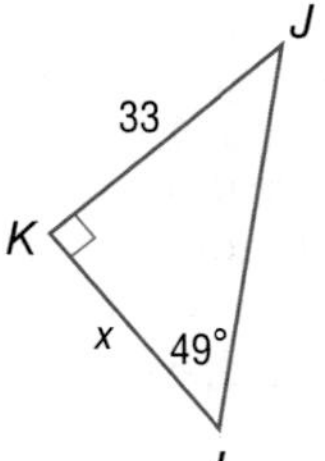


**30.**
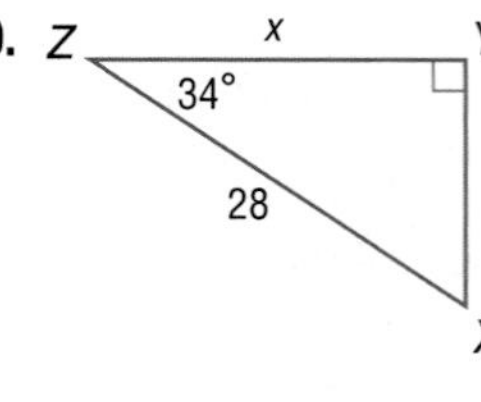


**31.**
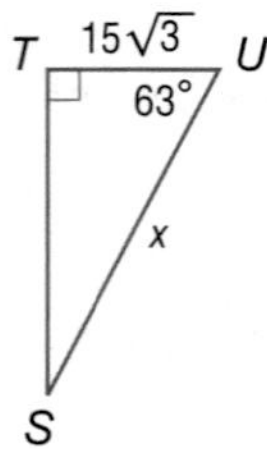


**32.**
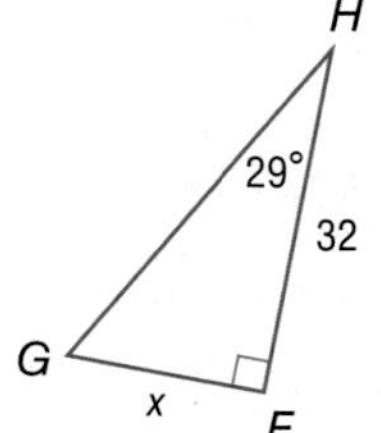


**33.**
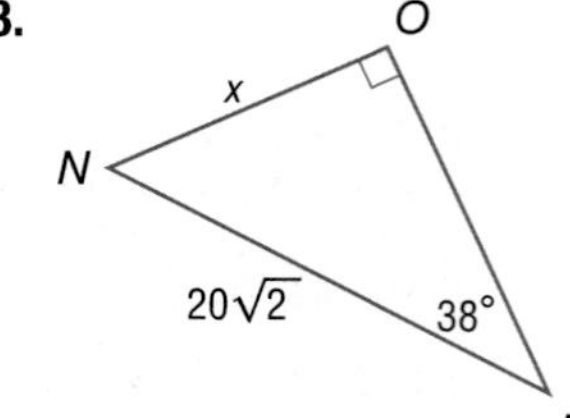


**34. GYMNASTICS** The springboard that Eric uses in his gymnastics class has 6-inch coils and forms an angle of 14.5° with the base. About how long is the springboard?

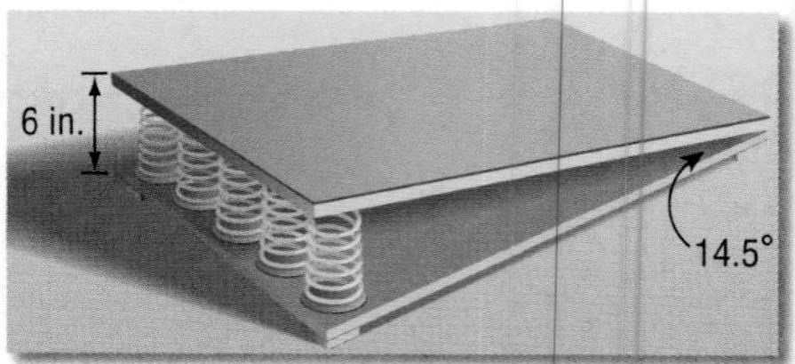


**35** **ROLLER COASTERS** The angle of ascent of the first hill of a roller coaster is 55°. If the length of the track from the beginning of the ascent to the highest point is 98 feet, what is the height of the roller coaster when it reaches the top of the first hill?

**Example 4** **CCSS TOOLS** **Use a calculator to find the measure of $\angle T$ to the nearest tenth.**

**36.**
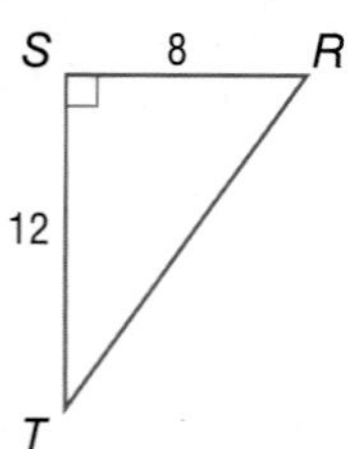


**37.**
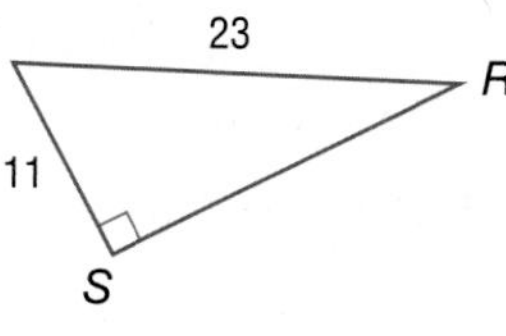


**38.**
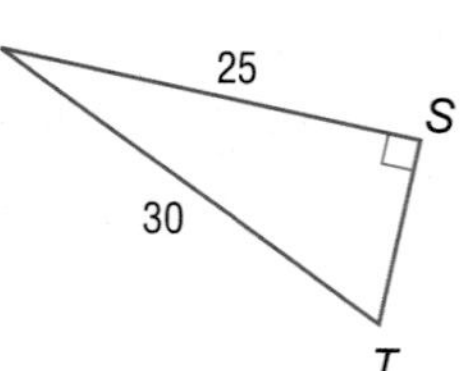


**39.**
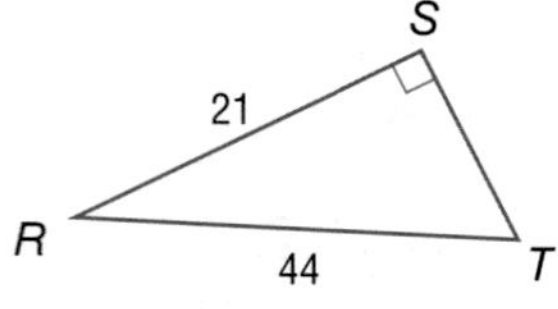


**40.**
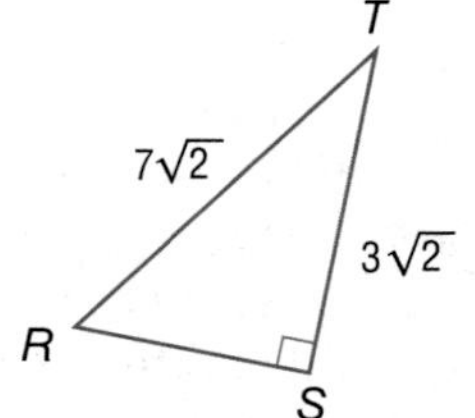


**41.**
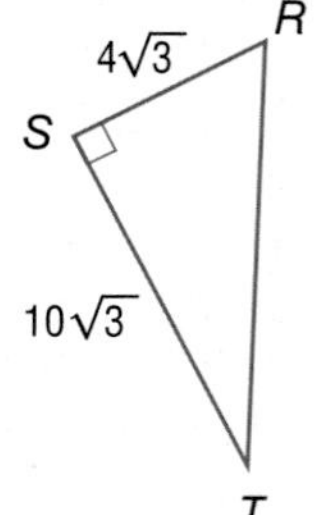

**Example 5** **Solve each right triangle. Round side measures to the nearest tenth and angle measures to the nearest degree.**

**42.**
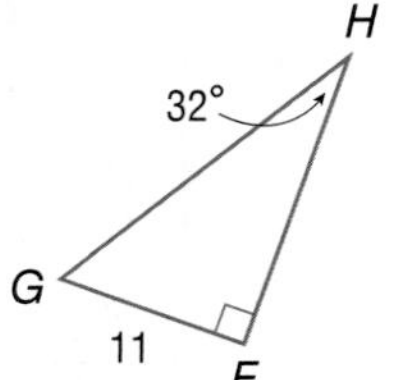


**43.**
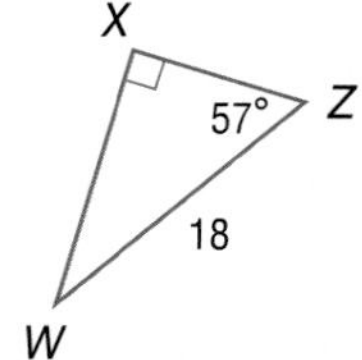


**44.**
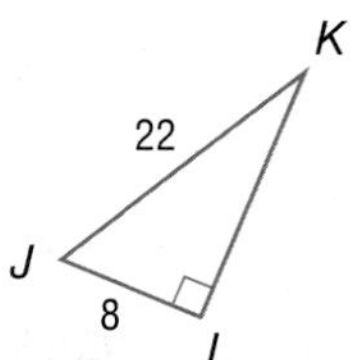


**45.**
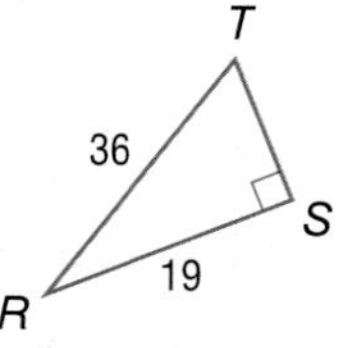


**46. BACKPACKS** Ramón has a rolling backpack that is $3\frac{3}{4}$ feet tall when the handle is extended. When he is pulling the backpack, Ramon's hand is 3 feet from the ground. What angle does his backpack make with the floor? Round to the nearest degree.

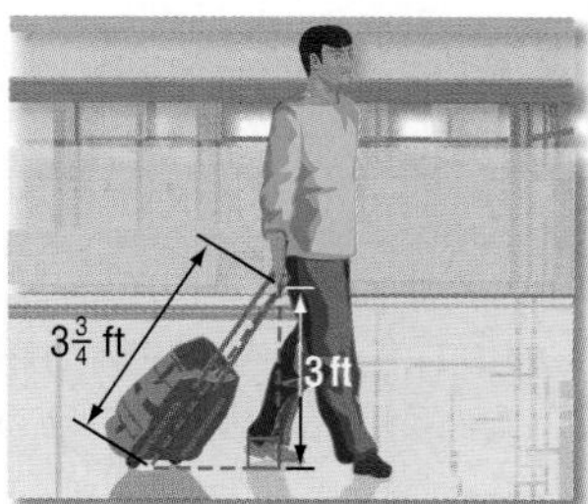


**COORDINATE GEOMETRY** **Find the measure of each angle to the nearest tenth of a degree using the Distance Formula and an inverse trigonometric ratio.**

**47.** $\angle K$ in right triangle $JKL$ with vertices $J(-2, -3)$, $K(-7, -3)$, and $L(-2, 4)$

**48.** $\angle Y$ in right triangle $XYZ$ with vertices $X(4, 1)$, $Y(-6, 3)$, and $Z(-2, 7)$

**49.** $\angle A$ in right triangle $ABC$ with vertices $A(3, 1)$, $B(3, -3)$, and $C(8, -3)$

**50. SCHOOL SPIRIT** Hana is making a pennant for each of the 18 girls on her basketball team. She will use $\frac{1}{2}$-inch seam binding to finish the edges of the pennants.

**a.** What is the total length of seam binding needed to finish all of the pennants?

**b.** If seam binding is sold in 3-yard packages at a cost of \$1.79, how much will it cost?

**CCSS SENSE-MAKING** **Find the perimeter and area of each triangle. Round to the nearest hundredth.**

**51.**
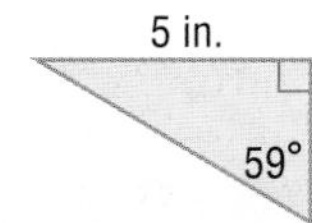


**52.**
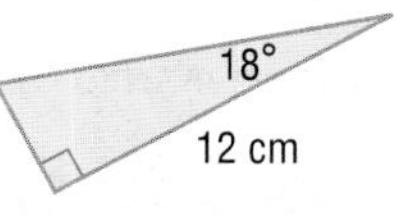


**53.**
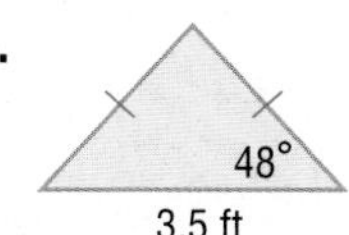


**54.** Find the tangent of the greater acute angle in a triangle with side lengths of 3, 4, and 5 centimeters.

**55.** Find the cosine of the smaller acute angle in a triangle with side lengths of 10, 24, and 26 inches.

**56. ESTIMATION** Ethan and Tariq want to estimate the area of the field that their team will use for soccer practice. They know that the field is rectangular, and they have paced off the width of the field as shown. They used the fence posts at the corners of the field to estimate that the angle between the length of the field and the diagonal is about 40°. If they assume that each of their steps is about 18 inches, what is the area of the practice field in square feet? Round to the nearest square foot.

40°
280 steps

**Find $x$ and $y$. Round to the nearest tenth.**

**57**

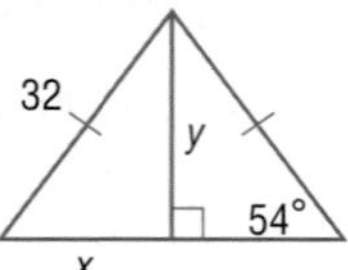


**58.**

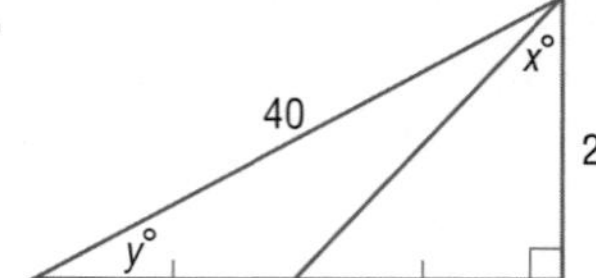


**59.**

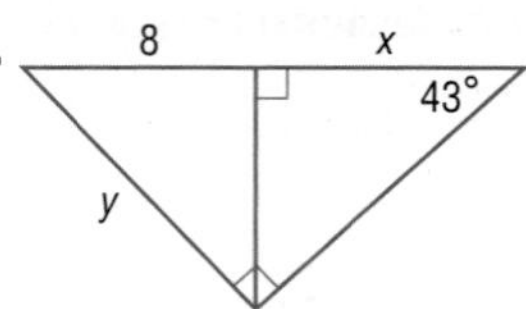


**60. COORDINATE GEOMETRY** Show that the slope of a line at 225° from the $x$-axis is equal to the tangent of 225°.

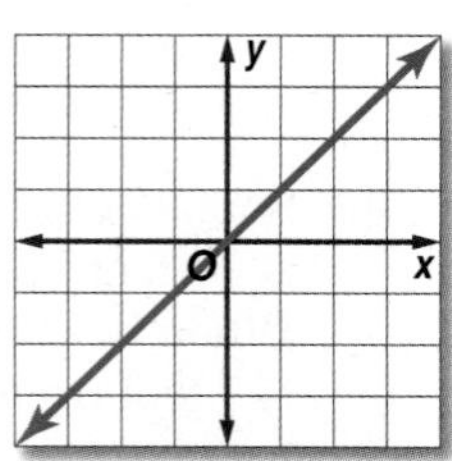


**61. MULTIPLE REPRESENTATIONS** In this problem, you will investigate an algebraic relationship between the sine and cosine ratios.

a. **Geometric** Draw three right triangles that are not similar to each other. Label the triangles $ABC$, $MNP$, and $XYZ$, with the right angles located at vertices $B$, $N$, and $Y$, respectively. Measure and label each side of the three triangles.

b. **Tabular** Copy and complete the table below.

| Triangle | Trigonometric Ratios | | | | Sum of Ratios Squared | |
|---|---|---|---|---|---|---|
| $ABC$ | cos $A$ | | sin $A$ | | $(\cos A)^2 + (\sin A)^2 =$ | |
| | cos $C$ | | sin $C$ | | $(\cos C)^2 + (\sin C)^2 =$ | |
| $MNP$ | cos $M$ | | sin $M$ | | $(\cos M)^2 + (\sin M)^2 =$ | |
| | cos $P$ | | sin $P$ | | $(\cos P)^2 + (\sin P)^2 =$ | |
| $XYZ$ | cos $X$ | | sin $X$ | | $(\cos X)^2 + (\sin X)^2 =$ | |
| | cos $Z$ | | sin $Z$ | | $(\cos Z)^2 + (\sin Z)^2 =$ | |

c. **Verbal** Make a conjecture about the sum of the squares of the cosine and sine of an acute angle of a right triangle.

d. **Algebraic** Express your conjecture algebraically for an angle $X$.

e. **Analytical** Show that your conjecture is valid for angle $A$ in the figure at the right using the trigonometric functions and the Pythagorean Theorem.

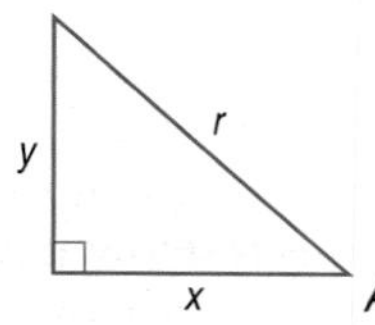


## H.O.T. Problems Use Higher-Order Thinking Skills

**62. CHALLENGE** Solve $\triangle ABC$. Round to the nearest whole number.

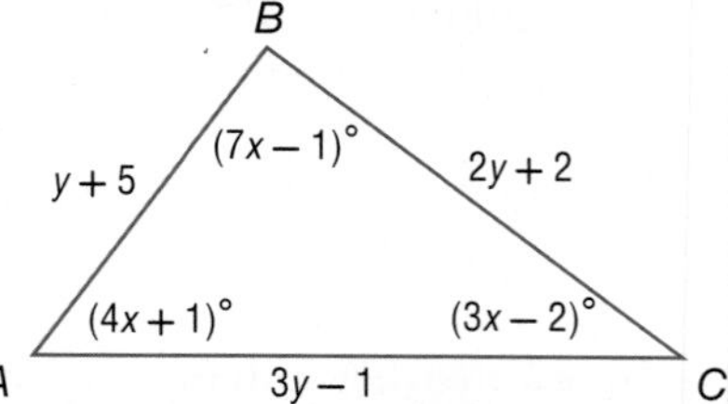


**63. REASONING** Are the values of sine and cosine for an acute angle of a right triangle always less than 1? Explain.

**64. CCSS REASONING** What is the relationship between the sine and cosine of complementary angles? Explain your reasoning and use the relationship to find cos 50 if sin 40 ≈ 0.64.

**65. WRITING IN MATH** Explain how you can use ratios of the side lengths to find the angle measures of the acute angles in a right triangle.

## Standardized Test Practice

**66.** What is the value of $\tan x$?

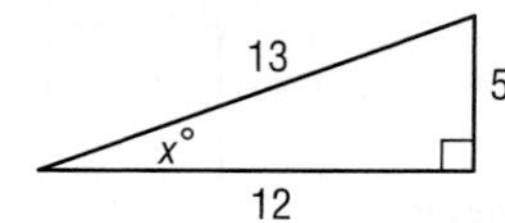


**A** $\tan x = \frac{13}{5}$ **C** $\tan x = \frac{5}{13}$

**B** $\tan x = \frac{12}{5}$ **D** $\tan x = \frac{5}{12}$

**67. ALGEBRA** Which of the following has the same value as $2^{-12} \times 2^3$?

**F** $2^{-36}$ **H** $2^{-9}$

**G** $4^{-9}$ **J** $2^{-4}$

**68. GRIDDED RESPONSE** If $AC = 12$ and $AB = 25$, what is the measure of $\angle B$ to the nearest tenth?

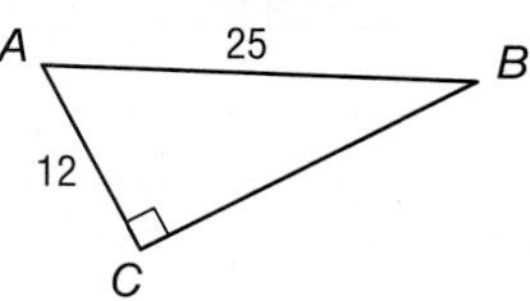


**69. SAT/ACT** The area of a right triangle is 240 square inches. If the base is 30 inches long, how many inches long is the hypotenuse?

**A** 5 **D** $2\sqrt{241}$

**B** 8 **E** 34

**C** 16

## Spiral Review

**Find $x$ and $y$.** (Lesson 10-3)

**70.**

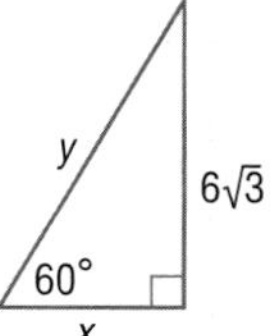


**71.**

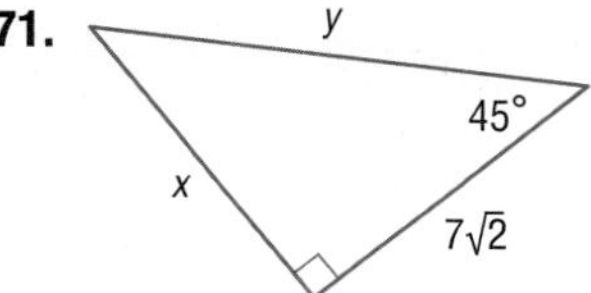


**72.**

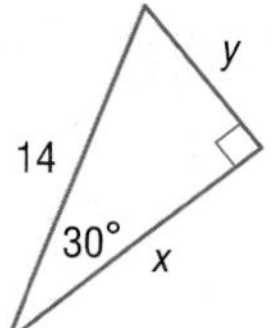


**Determine whether each set of numbers can be the measures of the sides of a triangle. If so, classify the triangle as *acute*, *obtuse*, or *right*. Justify your answer.** (Lesson 10-2)

**73.** 8, 15, 17 **74.** 11, 12, 24 **75.** 13, 30, 35

**76.** 18, 24, 30 **77.** 3.2, 5.3, 8.6 **78.** $6\sqrt{3}$, 14, 17

**79. MAPS** The scale on the map of New Mexico is 2 centimeters = 160 miles. The width of New Mexico through Albuquerque on the map is 4.1 centimeters. How long would it take to drive across New Mexico if you drove at an average of 60 miles per hour? (Lesson 9-7)

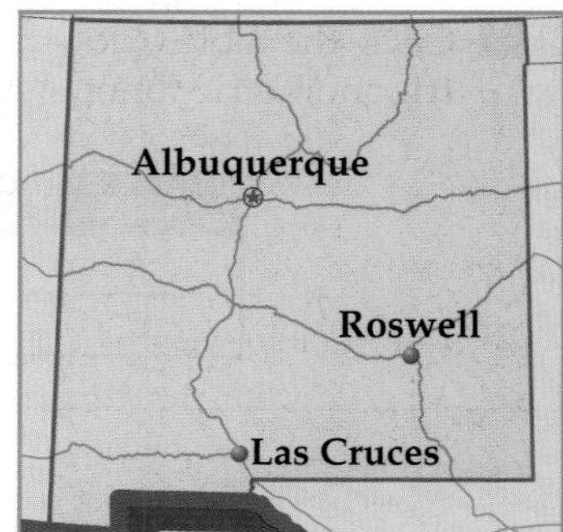


**ALGEBRA Find $x$ and $y$.** (Lesson 9-4)

**80.**

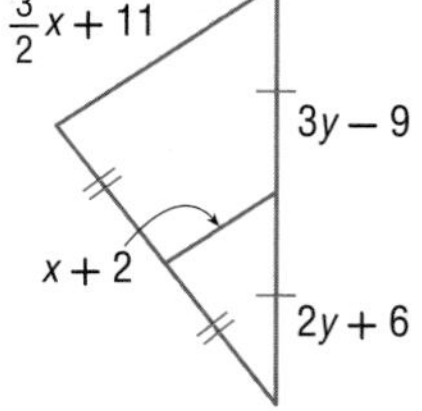


**81.**

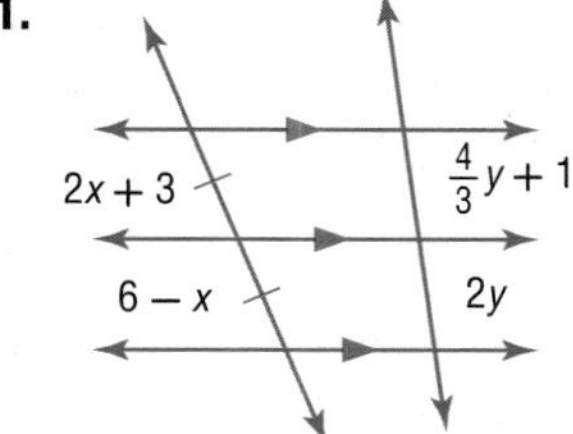


## Skills Review

**Solve each proportion. Round to the nearest tenth if necessary.**

**82.** $2.14 = \frac{x}{12}$ **83.** $0.05x = 13$ **84.** $0.37 = \frac{32}{x}$

**85.** $0.74 = \frac{14}{x}$ **86.** $1.66 = \frac{x}{23}$ **87.** $0.21 = \frac{33}{x}$

# EXTEND 10-4 Graphing Technology Lab: Secant, Cosecant, and Cotangent

In the previous lesson, you used the trigonometric functions sine, cosine, and tangent to find angle relationships in right angles. In this activity, you will use the reciprocals of those functions, cosecant, secant, and cotangent, to explore angle and side relationships in right triangles.

**CCSS Common Core State Standards**
**Content Standards**
**G.SRT.6** Understand that by similarity, side ratios in right triangles are properties of the angles in the triangle, leading to definitions of trigonometric ratios for acute angles.
**Mathematical Practices** 5

### KeyConcept Reciprocal Trigonometric Ratios

| Words | Symbols |
|---|---|
| The **cosecant** of $\angle A$ (written csc $A$) is the reciprocal of sin $A$. | $\csc A = \frac{1}{\sin A}$ or $\frac{c}{a}$ |
| The **secant** of $\angle A$ (written sec $A$) is the reciprocal of cos $A$. | $\sec A = \frac{1}{\cos A}$ or $\frac{c}{b}$ |
| The **cotangent** of $\angle A$ (written cot $A$) is the reciprocal of tan $A$. | $\cot A = \frac{1}{\tan A}$ or $\frac{b}{a}$ |

## Activity Find Trigonometric Values

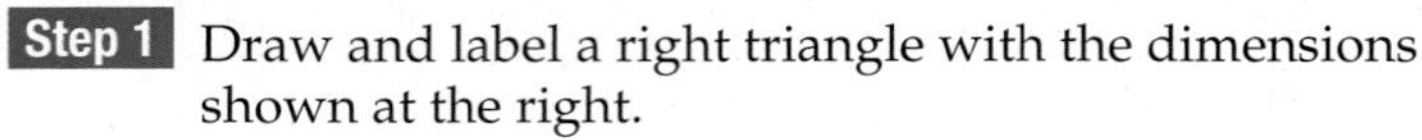

**Step 1** Draw and label a right triangle with the dimensions shown at the right.

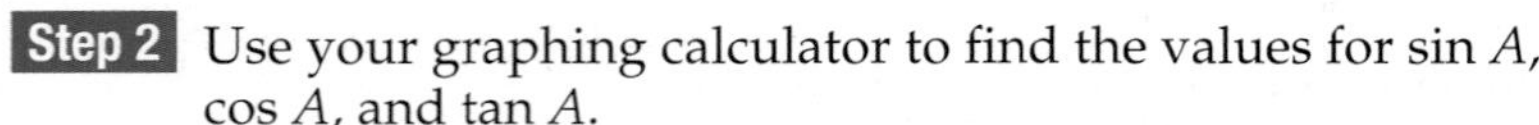

**Step 2** Use your graphing calculator to find the values for sin $A$, cos $A$, and tan $A$.

**Step 3** Next, find the value for csc $A$ by dividing 1 by SIN $A$. Repeat step 3 to find sec $A$ and cot $A$.

**Step 4** Copy the table below and record your results. Next, find the value of each trigonometric function for angle $C$.

| Angle | sin | cos | tan | csc | sec | cot |
|---|---|---|---|---|---|---|
| $A$ | | | | | | |
| $C$ | | | | | | |

## Exercises

1. Find the values of the six trigonometric functions for a 45° angle in a 45°-45°-90° triangle with legs that are 4 cm.

2. In $\triangle FGH$, $\tan F = \frac{5}{12}$. Find cot $F$ and sin $F$ if $\angle G$ is a right angle.

3. Find the values of the six trigonometric functions for angle $T$ in $\triangle RST$ if $m\angle R = 36°$. Round to the nearest hundredth.

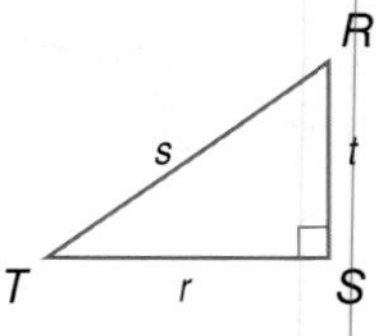

# CHAPTER 10 Mid-Chapter Quiz

## Lessons 10-1 through 10-4

**Find the geometric mean between each pair of numbers.** (Lesson 10-1)

1. 12 and 3
2. 63 and 7
3. 45 and 20
4. 50 and 10

**Write a similarity statement identifying the three similar triangles in each figure.** (Lesson 10-1)

5. 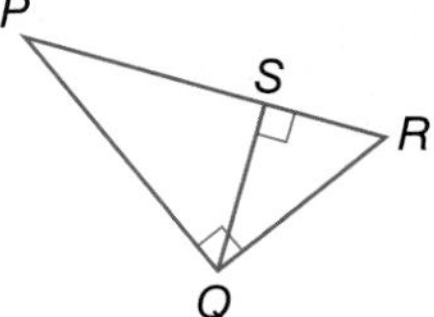


6. 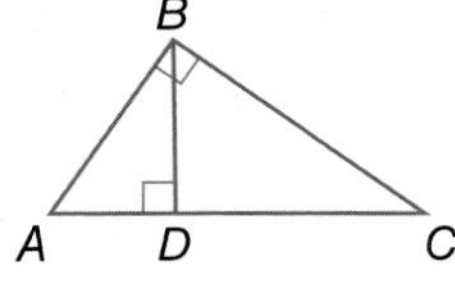


7. Find $x$, $y$, and $z$. (Lesson 10-1)

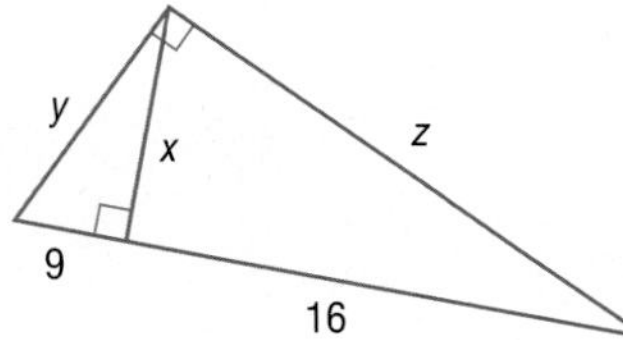


8. **PARKS** There is a small park in a corner made by two perpendicular streets. The park is 100 ft by 150 ft, with a diagonal path, as shown below. What is the length of path $\overline{AC}$? (Lesson 10-2)

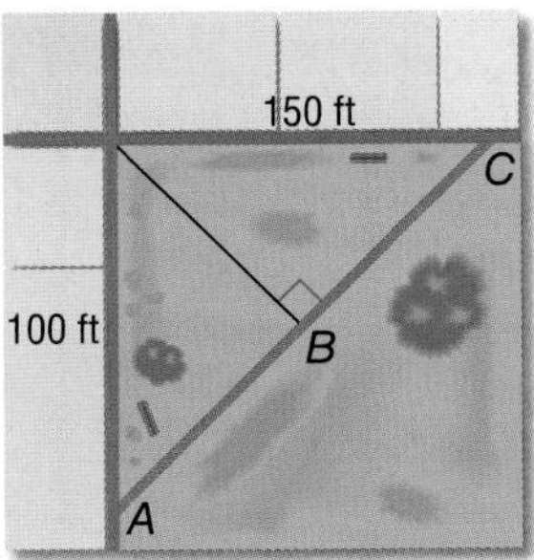


**Find $x$. Round to the nearest hundredth.** (Lesson 10-2)

9. 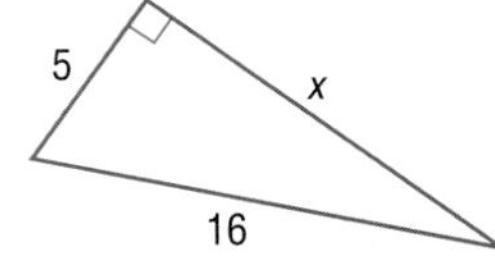


10. 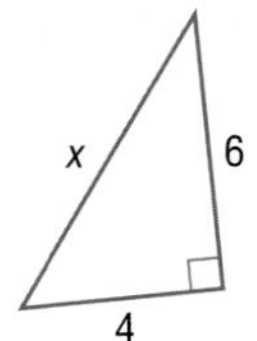


11. **MULTIPLE CHOICE** Which of the following sets of numbers is not a Pythagorean triple? (Lesson 10-2)

**A** 9, 12, 15
**B** 21, 72, 75
**C** 15, 36, 39
**D** 8, 13, 15

**Find $x$.** (Lesson 10-3)

12. 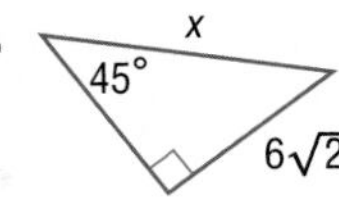


13. 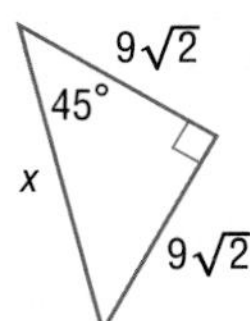


14. **DESIGN** Jamie designed a pinwheel to put in her garden. In the pinwheel, the blue triangles are congruent equilateral triangles, each with an altitude of 4 inches. The red triangles are congruent isosceles right triangles. The hypotenuse of a red triangle is congruent to a side of the blue triangle. (Lesson 10-3)

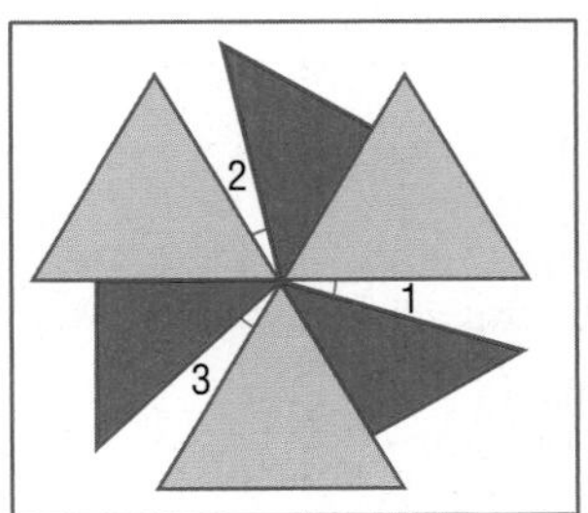


   a. If angles 1, 2, and 3 are congruent, find the measure of each angle.

   b. Find the perimeter of the pinwheel.

**Find $x$ and $y$.** (Lesson 10-3)

15. 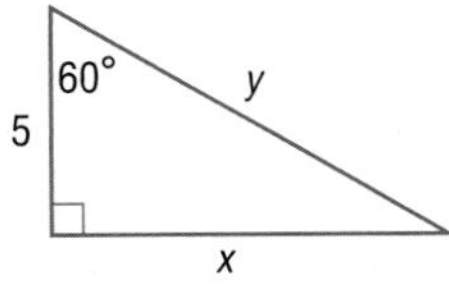


16. 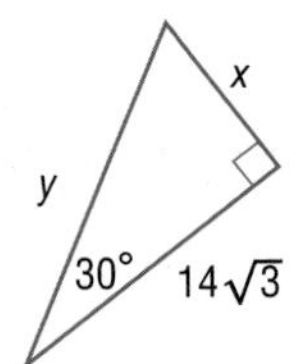


**Express each ratio as a fraction and as a decimal to the nearest hundredth.** (Lesson 10-4)

17. tan $M$
18. cos $M$
19. cos $N$
20. sin $N$

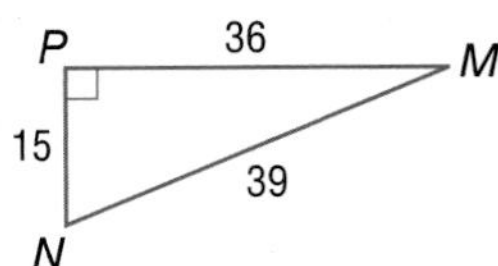


21. Solve the right triangle. Round angle measures to the nearest degree and side measures to the nearest tenth. (Lesson 10-4)

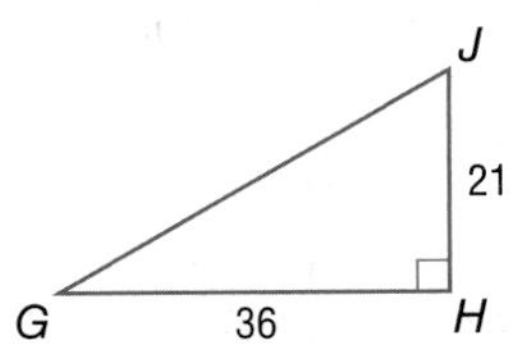

LESSON 10-5

# Angles of Elevation and Depression

**Then**

- You used similar triangles to measure distances indirectly.

**Now**

1. Solve problems involving angles of elevation and depression.
2. Use angles of elevation and depression to find the distance between two objects.

**Why?**

- To make a field goal, a kicker must kick the ball with enough force and at an appropriate angle of elevation to ensure that the ball will reach the goal post at a level high enough to make it over the horizontal bar. This angle must change depending on the initial placement of the ball away from the base of the goalpost.

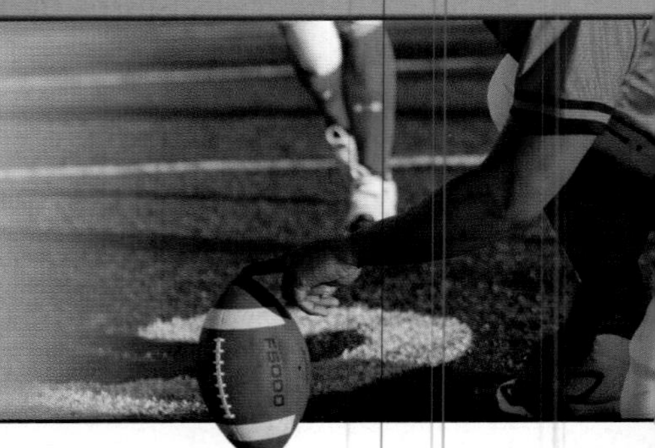

**NewVocabulary**
angle of elevation
angle of depression

**Common Core State Standards**

**Content Standards**
G.SRT.8 Use trigonometric ratios and the Pythagorean Theorem to solve right triangles in applied problems. ★

**Mathematical Practices**
4 Model with mathematics.
1 Make sense of problems and persevere in solving them.

## 1 Angles of Elevation and Depression

An **angle of elevation** is the angle formed by a horizontal line and an observer's line of sight to an object above the horizontal line. An **angle of depression** is the angle formed by a horizontal line and an observer's line of sight to an object below the horizontal line.

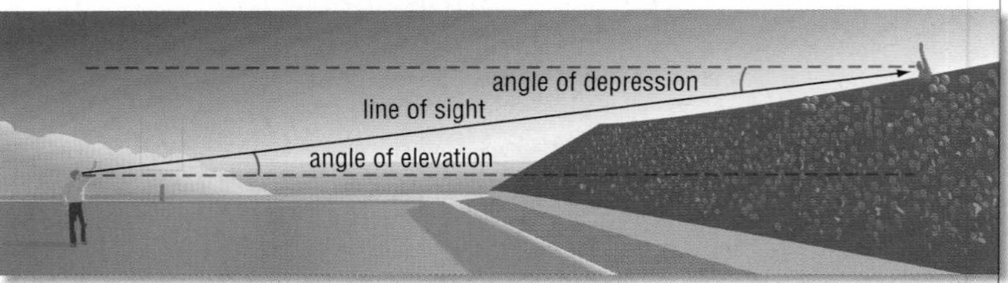


Horizontal lines are parallel, so the angle of elevation and the angle of depression in the diagram are congruent by the Alternate Interior Angles Theorem.

### Example 1 Angle of Elevation

**VACATION Leah wants to see a castle in an amusement park. She sights the top of the castle at an angle of elevation of 38°. She knows that the castle is 190 feet tall. If Leah is 5.5 feet tall, how far is she from the castle to the nearest foot?**

Make a sketch to represent the situation.

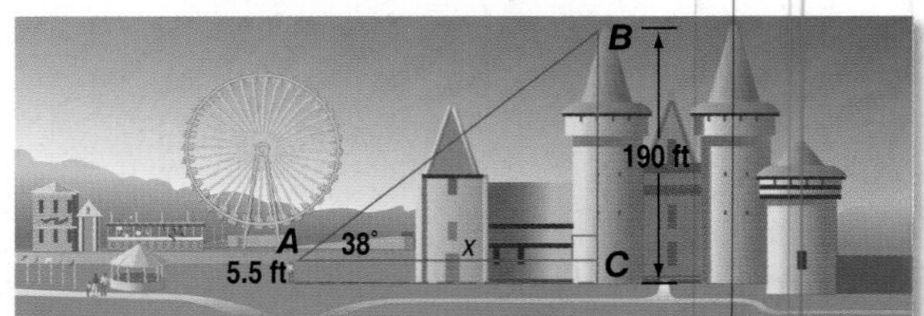


Since Leah is 5.5 feet tall, $BC = 190 - 5.5$ or 184.5 feet. Let $x$ represent the distance from Leah to the castle, $AC$.

$$\tan A = \frac{BC}{AC} \quad \tan = \frac{\text{opposite}}{\text{adjacent}}$$

$$\tan 38° = \frac{184.5}{x} \quad m\angle A = 38,\ BC = 184.5,\ AC = x$$

$$x = \frac{184.5}{\tan 38°} \quad \text{Solve for } x.$$

$$x \approx 236.1 \quad \text{Use a calculator.}$$

Leah is about 236 feet from the castle.

Juice Images/CORBIS

### GuidedPractice

1. **FOOTBALL** The cross bar of a goalpost is 10 feet high. If a field goal attempt is made 25 yards from the base of the goalpost that clears the goal by 1 foot, what is the smallest angle of elevation at which the ball could have been kicked to the nearest degree?

## Example 2 Angle of Depression

**EMERGENCY A search and rescue team is airlifting people from the scene of a boating accident when they observe another person in need of help. If the angle of depression to this other person is 42° and the helicopter is 18 feet above the water, what is the horizontal distance from the rescuers to this person to the nearest foot?**

Make a sketch of the situation.

Since $\overrightarrow{AB}$ and $\overline{DC}$ are parallel, $m\angle BAC = m\angle ACD$ by the Alternate Interior Angles Theorem.

Let $x$ represent the horizontal distance from the rescuers to the person $DC$.

A
B
42° angle of depression
18 ft
42°
C
D
x

Note: Art not drawn to scale.

| | |
|---|---|
| $\tan C = \frac{AD}{DC}$ | $\tan = \frac{\text{opposite}}{\text{adjacent}}$ |
| $\tan 42° = \frac{18}{x}$ | $C = 42$, $AD = 18$, and $DC = x$ |
| $x \tan 42° = 18$ | Multiply each side by $x$. |
| $x = \frac{18}{\tan 42°}$ | Divide each side by tan 42°. |
| $x \approx 20.0$ | Use a calculator. |

The horizontal distance from the rescuers to the person is 20.0 feet.

### GuidedPractice

2. **LIFEGUARDING** A lifeguard is watching a beach from a line of sight 6 feet above the ground. She sees a swimmer at an angle of depression of 8°. How far away from the tower is the swimmer?

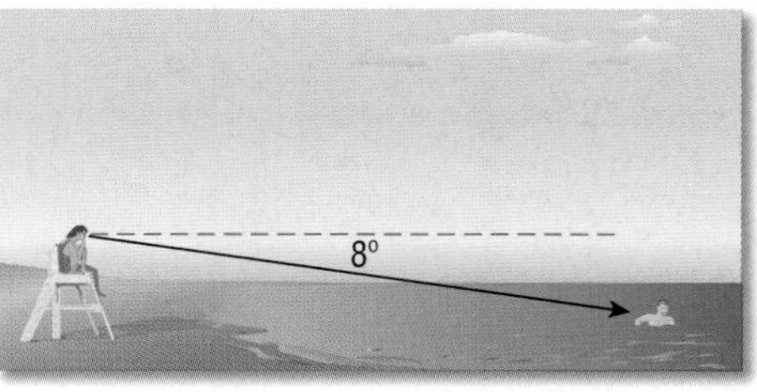


### WatchOut!

**Angles of Elevation and Depression** To avoid mislabeling, remember that angles of elevation and depression are always formed with a horizontal line and never with a vertical line.

### Math HistoryLink

**Eratosthenes (276–194 B.C.)** Eratosthenes was a mathematician and astronomer who was born in Cyrene, which is now Libya. He used the angle of elevation of the Sun at noon in the cities of Alexandria and Syene (now Egypt) to measure the circumference of Earth.

**Source:** *Encyclopaedia Britannica*



## 2 Two Angles of Elevation or Depression

Angles of elevation or depression to two different objects can be used to estimate the distance between those objects. Similarly, the angles from two different positions of observation to the same object can be used to estimate the object's height.

## Example 3 Use Two Angles of Elevation or Depression

**TREE REMOVAL** **To estimate the height of a tree she wants removed, Mrs. Long sights the tree's top at a 70° angle of elevation. She then steps back 10 meters and sights the top at a 26° angle. If Mrs. Long's line of sight is 1.7 meters above the ground, how tall is the tree to the nearest meter?**

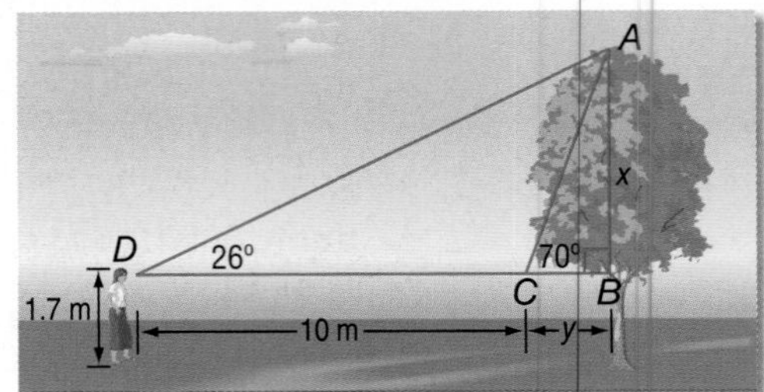


**Understand** $\triangle ABC$ and $\triangle ABD$ are right triangles. The height of the tree is the sum of Mrs. Long's height and $AB$.

**Plan** Since her initial distance from the tree is not given, write and solve a system of equations using both triangles. Let $AB = x$ and $CB = y$. So $DB = y + 10$ and the height of the tree is $x + 1.7$.

**Solve** Use $\triangle ABC$.

$\tan 70° = \frac{x}{y}$ — $\tan = \frac{\text{opposite}}{\text{adjacent}}$; $m\angle ACB = 70$

$y \tan 70° = x$ — Multiply each side by $y$.

Use $\triangle ABD$.

$\tan 26° = \frac{x}{y + 10}$ — $\tan = \frac{\text{opposite}}{\text{adjacent}}$; $m\angle D = 26$

$(y + 10) \tan 26° = x$ — Multiply each side by $y + 10$.

Substitute the value for $x$ from $\triangle ABD$ in the equation for $\triangle ABC$ and solve for $y$.

$$y \tan 70° = x$$

$$y \tan 70° = (y + 10) \tan 26°$$

$$y \tan 70° = y \tan 26° + 10 \tan 26°$$

$$y \tan 70° - y \tan 26° = 10 \tan 26°$$

$$y(\tan 70° - \tan 26°) = 10 \tan 26°$$

$$y = \frac{10 \tan 26°}{\tan 70° - \tan 26°}$$

Use a calculator to find that $y \approx 2.16$. Using the equation from $\triangle ABC$, $x = 2.16 \tan 70°$ or about 5.9.

The height of the tree is 5.9 + 1.7 or 7.6, which is about 8 meters.

**Check** Substitute the value for $y$ in the equation from $\triangle ABD$.

$x = (2.16 + 10) \tan 26°$ or about 5.9. This is the same value found using the equation from $\triangle ABC$. ✓

### Real-WorldLink

In the United States, lumber volume is measured in board-feet, which is defined as a piece of wood containing 144 cubic inches. Woodland owners often estimate the lumber volume of trees they own to determine how many to cut and sell.

**Source:** The Ohio State University School of Natural Resources

### StudyTip

**Indirect Measurement** When using the angles of depression to two different objects to calculate the distance between them, it is important to remember that the two objects must lie in the same horizontal plane. In other words, one object cannot be higher or lower than the other.

### GuidedPractice

**3. SKYSCRAPERS** Two buildings are sited from atop a 200-meter skyscraper. Building A is sited at a 35° angle of depression, while Building B is sighted at a 36° angle of depression. How far apart are the two buildings to the nearest meter?

Compassionate Eye Foundation/Colorblind Images/Getty Images

## Check Your Understanding

**Example 1**

**1. BIKING** Lenora wants to build the bike ramp shown. Find the length of the base of the ramp.

**Example 2**

**2. BASEBALL** A fan is seated in the upper deck of a stadium 200 feet away from home plate. If the angle of depression to the field is 62°, at what height is the fan sitting?



**Example 3**

**3. CCSS MODELING** Annabelle and Rich are setting up decorations for their school dance. Rich is standing 5 feet directly in front of Annabelle under a disco ball. If the angle of elevation from Annabelle to the ball is 40° and Rich to the ball is 50°, how high is the disco ball?

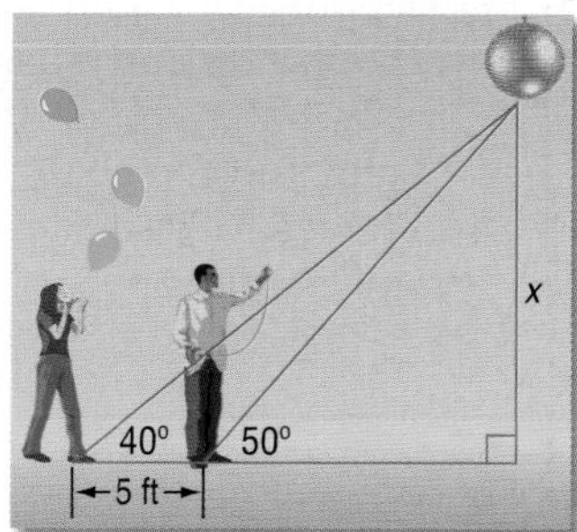


## Practice and Problem Solving

**Example 1**

**4. HOCKEY** A hockey player takes a shot 20 feet away from a 5-foot goal. If the puck travels at a 15° angle of elevation toward the center of the goal, will the player score?

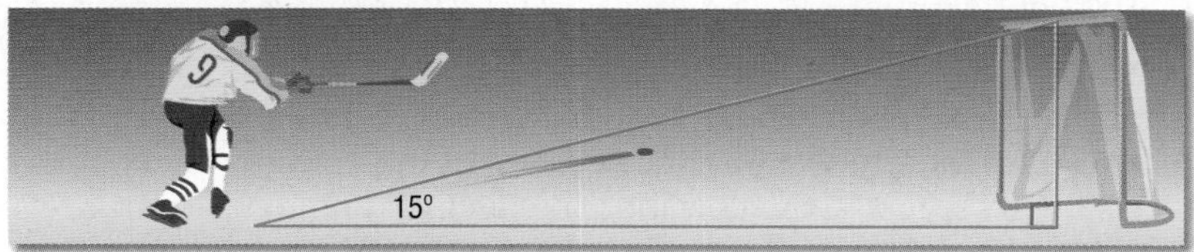


**5. MOUNTAINS** Find the angle of elevation to the peak of a mountain for an observer who is 155 meters from the mountain if the observer's eye is 1.5 meters above the ground and the mountain is 350 meters tall.

**Example 2**

**6. WATERPARK** Two water slides are 50 meters apart on level ground. From the top of the taller slide, you can see the top of the shorter slide at an angle of depression of 15°. If you know that the top of the other slide is approximately 15 meters above the ground, about how far above the ground are you? Round to the nearest tenth of a meter.

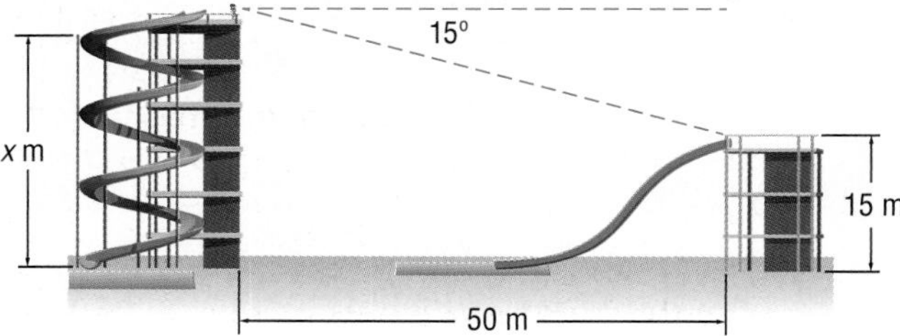


**7. AVIATION** Due to a storm, a pilot flying at an altitude of 528 feet has to land. If he has a horizontal distance of 2000 feet to land, at what angle of depression should he land?

**Example 3**

**8. PYRAMIDS** Miko and Tyler are visiting the Great Pyramid in Egypt. From where Miko is standing, the angle of elevation to the top of the pyramid is 48.6°. From Tyler's position, the angle of elevation is 50°. If they are standing 20 feet apart, and they are each 5 feet 6 inches tall, how tall is the pyramid?

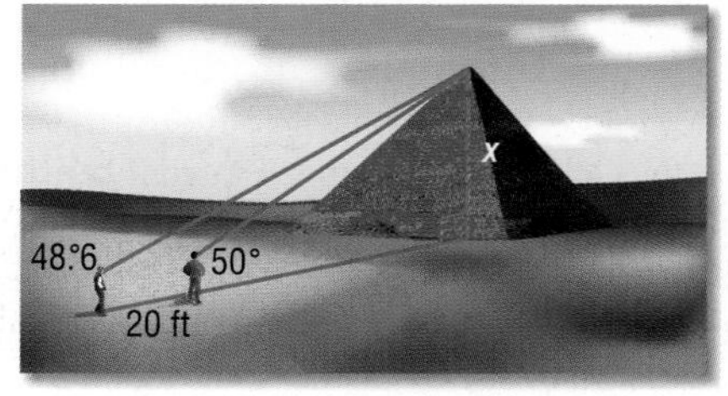

9. **DIVING** Austin is standing on the high dive at the local pool. Two of his friends are in the water as shown. If the angle of depression to one of his friends is 40°, and 30° to his other friend who is 5 feet beyond the first, how tall is the platform?

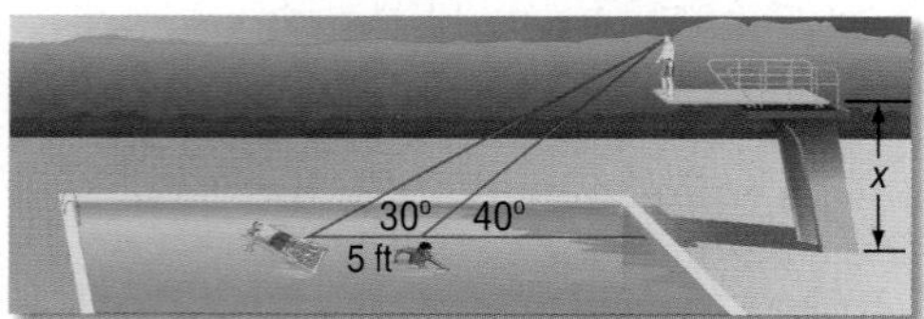


10. **BASKETBALL** Claire and Marisa are both waiting to get a rebound during a basketball game. If the height of the basketball hoop is 10 feet, the angle of elevation between Claire and the goal is 35°, and the angle of elevation between Marisa and the goal is 25°, how far apart are they standing?

11. **RIVERS** Hugo is standing in the top of St. Louis' Gateway Arch, looking down on the Mississippi River. The angle of depression to the closer bank is 45° and the angle of depression to the farther bank is 18°. The arch is 630 feet tall. Estimate the width of the river at that point.

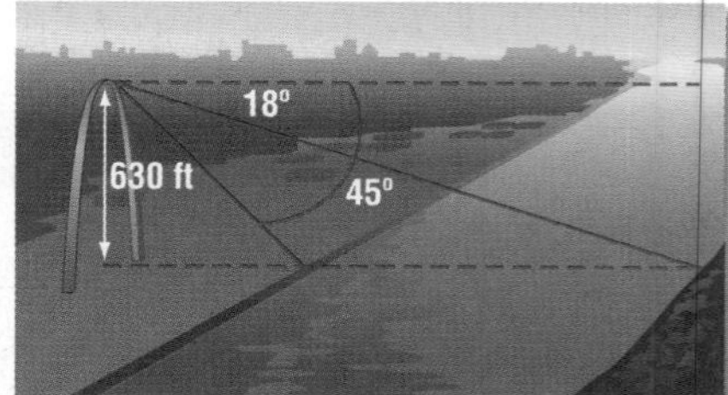


12. **CCSS MODELING** The Unzen Volcano in Japan has a magma reservoir located 15 kilometers beneath the Chijiwa Bay, located east of the volcano. A magma channel, which connects the reservoir to the volcano, rises at a 40° angle of elevation toward the volcano. What length of magma channel is below sea level?

13. **BRIDGES** Suppose you are standing in the middle of the platform of the world's longest suspension bridge, the Akashi Kaikyo Bridge. If the height from the top of the platform holding the suspension cables is 297 meters, and the length from the platform to the center of the bridge is 995 meters, what is the angle of depression from the center of the bridge to the platform?

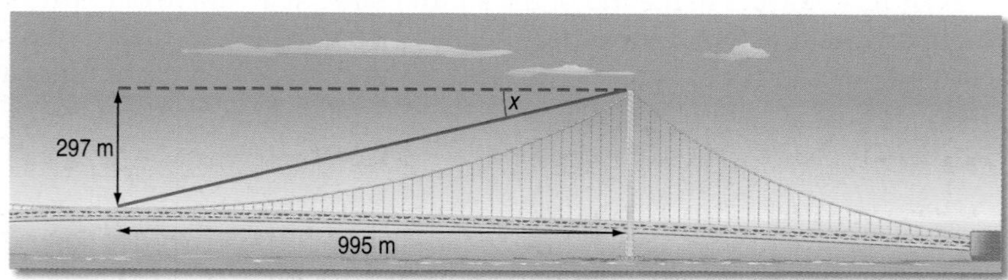


14. **LIGHTHOUSES** Little Gull Island Lighthouse shines a light from a height of 91 feet with a 6° angle of depression. Plum Island Lighthouse, 1800 feet away, shines a light from a height of 34 feet with a 2° angle of depression. Which light will reach a boat that sits exactly between Little Gull Island Lighthouse and Plum Island Lighthouse?

15. **TOURISM** From the position of the bus on the street, the L'arc de Triomphe is at an angle of 34°. If the arc is 162 feet tall, how far away is the bus? Round to the nearest tenth.

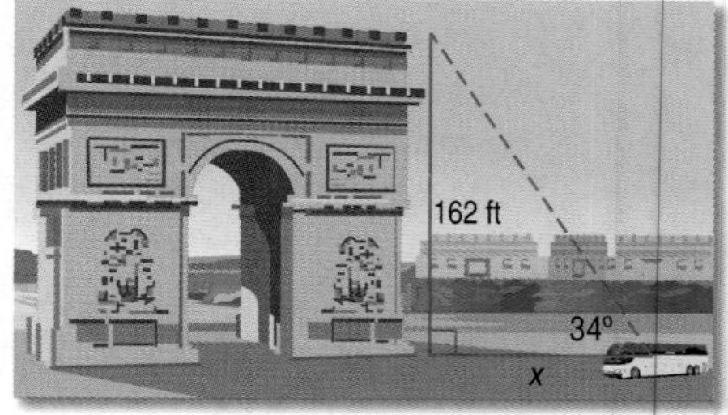

**16. MAINTENANCE** Two telephone repair workers arrive at a location to restore electricity after a power outage. One of the workers climbs up the telephone pole while the other worker stands 10 feet to left of the pole. If the terminal box is located 30 feet above ground on the pole and the angle of elevation from the truck to the repair worker is 70°, how far is the worker on the ground standing from the truck?

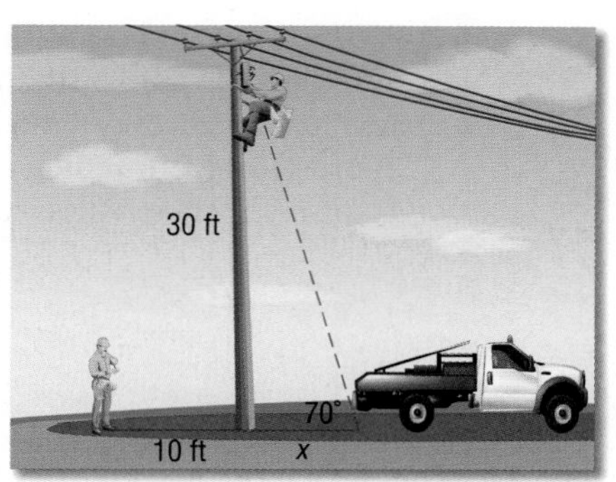


**17. PHOTOGRAPHY** A digital camera with a panoramic lens is described as having a view with an angle of elevation of 38°. If the camera is on a 3-foot tripod aimed directly at a 124-foot-tall monument, how far from the monument should you place the tripod to see the entire monument in your photograph?

**18. CCSS MODELING** As a part of their weather unit, Anoki's science class took a hot air balloon ride. As they passed over a fenced field, the angle of depression of the closer side of the fence was 32°, and the angle of depression of the farther side of the fence was 27°. If the height of the balloon was 800 feet, estimate the width of the field.

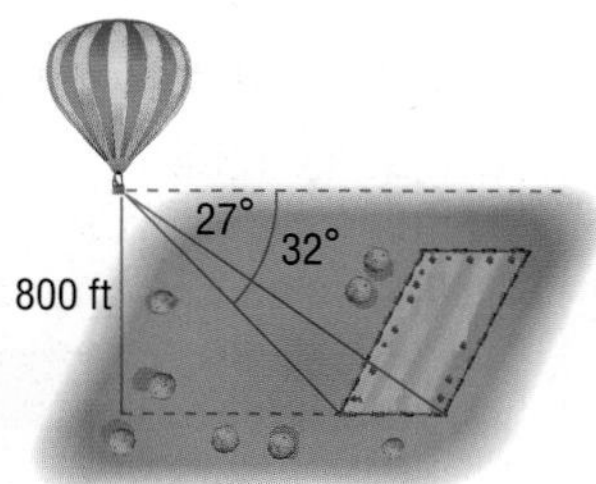


**19. MARATHONS** The Badwater Ultramarathon is a race that begins at the lowest point in California, Death Valley, and ends at the highest point of the state, Mount Whitney. The race starts at a depth of 86 meters above sea level and ends 2530 meters above sea level.

**a.** Determine the angle of elevation to Mount Whitney if the horizontal distance from the base to the peak is 1200 meters.

**b.** If the angle of depression to Death Valley is 38°, what is the horizontal distance from sea level?

**20. AMUSEMENT PARKS** India, Enrique, and Trina went to an amusement park while visiting Japan. They went on a Ferris wheel that was 100 meters in diameter and on an 80-meter cliff-dropping slide.

**a.** When Enrique and Trina are at the topmost point on the Ferris wheel shown below, how far are they from India?

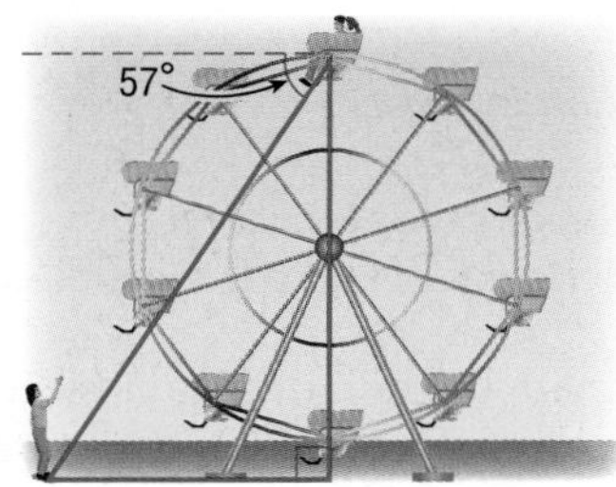


**b.** If the cliff-dropping ride has an angle of depression of 46°, how long is the slide?

**21** **DARTS** Kelsey and José are throwing darts from a distance of 8.5 feet. The center of the bull's-eye on the dartboard is 5.7 feet from the floor. José throws from a height of 6 feet, and Kelsey throws from a height of 5 feet. What are the angles of elevation or depression from which each must throw to get a bull's-eye? Ignore other factors such as air resistance, velocity, and gravity.

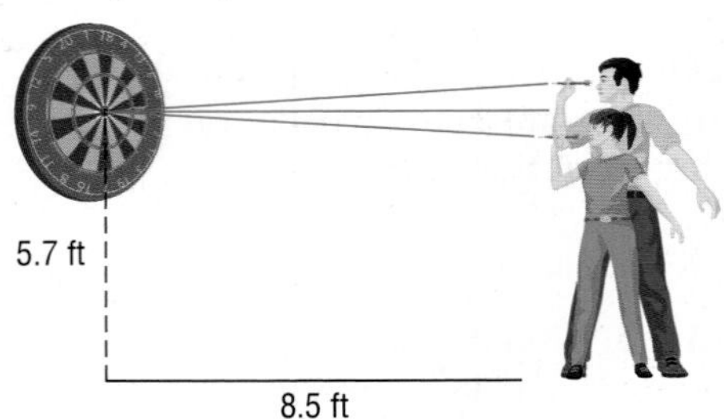


**22.** **MULTIPLE REPRESENTATIONS** In this problem, you will investigate relationships between the sides and angles of triangles.

a. **Geometric** Draw three triangles. Make one acute, one obtuse, and one right. Label one triangle *ABC*, a second *MNP*, and the third *XYZ*. Label the side lengths and angle measures of each triangle.

b. **Tabular** Copy and complete the table below.

| Triangle | Ratios | | |
|---|---|---|---|
| *ABC* | $\frac{\sin A}{BC} =$ | $\frac{\sin B}{CA} =$ | $\frac{\sin C}{AB} =$ |
| *MNP* | $\frac{\sin M}{NP} =$ | $\frac{\sin N}{PM} =$ | $\frac{\sin P}{MN} =$ |
| *XYZ* | $\frac{\sin X}{YZ} =$ | $\frac{\sin Y}{ZX} =$ | $\frac{\sin Z}{XY} =$ |

c. **Verbal** Make a conjecture about the ratio of the sine of an angle to the length of the leg opposite that angle for a given triangle.

## H.O.T. Problems Use Higher-Order Thinking Skills

**23.** **ERROR ANALYSIS** Terrence and Rodrigo are trying to determine the relationship between angles of elevation and depression. Terrence says that if you are looking up at someone with an angle of elevation of 35°, then they are looking down at you with an angle of depression of 55°, which is the complement of 35°. Rodrigo disagrees and says that the other person would be looking down at you with an angle of depression equal to your angle of elevation, or 35°. Is either of them correct? Explain.

**24.** **CHALLENGE** Find the value of $x$. Round to the nearest tenth.

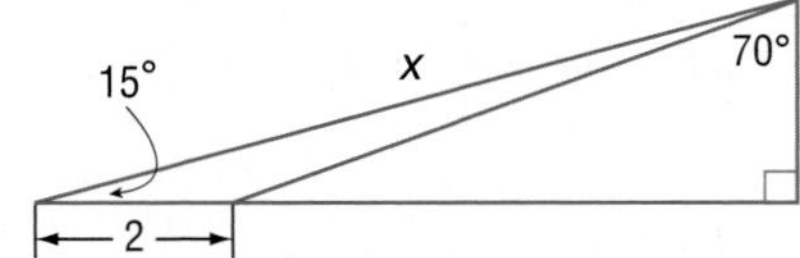


**25.** **CCSS REASONING** Classify the statement below as *true* or *false*. Explain.

*As a person moves closer to an object he or she is sighting, the angle of elevation increases.*

**26.** **WRITE A QUESTION** A classmate finds the angle of elevation of an object, but she is trying to find the angle of depression. Write a question to help her solve the problem.

**27.** **WRITING IN MATH** Describe a way that you can estimate the height of an object without using trigonometry by choosing your angle of elevation. Explain your reasoning.

## Standardized Test Practice

**28.** Ryan wanted to know the height of a cell-phone tower neighboring his property. He walked 80 feet from the base of the tower and measured the angle of elevation to the top of the tower at 54°. If Ryan is 5 feet tall, what is the height of the cell-phone tower?

**A** 52 ft  **C** 110 ft
**B** 63 ft  **D** 115 ft

**29. SHORT RESPONSE** A searchlight is 6500 feet from a weather station. If the angle of elevation to the spot of light on the clouds above the station is 45°, how high is the cloud ceiling?

**30. ALGEBRA** What is the solution of this system of equations?

$$2x - 4y = -12$$
$$-x + 4y = 8$$

**F** (4, 4)  **H** (−4, −4)
**G** (−4, 1)  **J** (1, −4)

**31. SAT/ACT** A triangle has sides in the ratio of 5 : 12 : 13. What is the measure of the triangle's smallest angle in degrees?

**A** 13.34  **D** 42.71
**B** 22.62  **E** 67.83
**C** 34.14

## Spiral Review

**Express each ratio as a fraction and as a decimal to the nearest hundredth.** (Lesson 10-4)

**32.** $\sin C$  **33.** $\tan A$  **34.** $\cos C$

**35.** $\tan C$  **36.** $\cos A$  **37.** $\sin A$

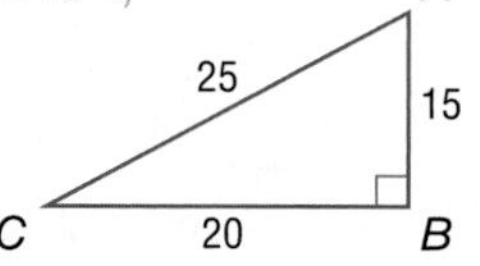


**38. LANDSCAPING** Imani needs to determine the height of a tree. Holding a drafter's 45° triangle so that one leg is horizontal, she sights the top of the tree along the hypotenuse, as shown at the right. If she is 6 yards from the tree and her eyes are 5 feet from the ground, find the height of the tree. (Lesson 10-3)

**PROOF Write a two-column proof.** (Lesson 9-5)

**39. Given:** $\overline{CD}$ bisects $\angle ACB$.

By construction, $\overline{AE} \parallel \overline{CD}$.

**Prove:** $\frac{AD}{DB} = \frac{AC}{BC}$

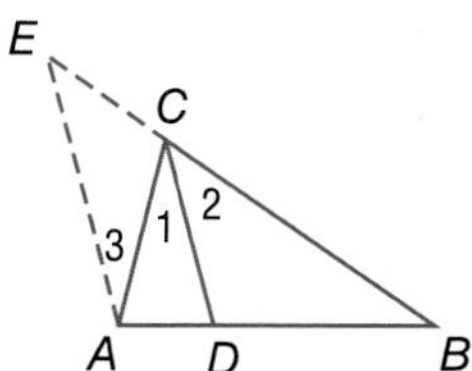


**40. Given:** $\overline{JF}$ bisects $\angle EFG$.

$\overline{EH} \parallel \overline{FG}$, $\overline{EF} \parallel \overline{HG}$

**Prove:** $\frac{EK}{KF} = \frac{GJ}{JF}$

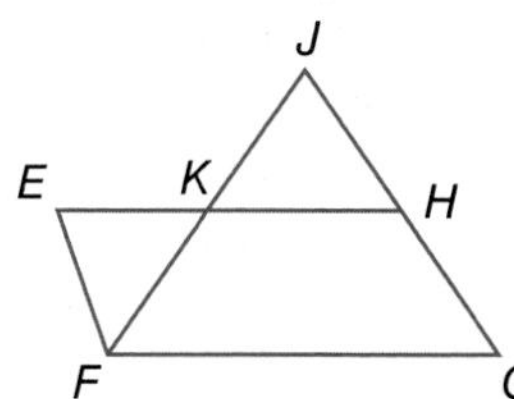

**COORDINATE GEOMETRY Find the coordinates of the centroid of each triangle.** (Lesson 7-2)

**41.** $A(2, 2)$, $B(7, 8)$, $C(12, 2)$

**42.** $X(-3, -2)$, $Y(1, -12)$, $Z(-7, -7)$

**43.** $A(-1, 11)$, $B(-5, 1)$, $C(-9, 6)$

**44.** $X(4, 0)$, $Y(-2, 4)$, $Z(0, 6)$

## Skills Review

**Solve each proportion.**

**45.** $\frac{1}{5} = \frac{x}{10}$  **46.** $\frac{2x}{11} = \frac{3}{8}$  **47.** $\frac{4x}{16} = \frac{62}{118}$  **48.** $\frac{12}{21} = \frac{45}{10x}$

LESSON

# 10-6 The Law of Sines and Law of Cosines

**Then**

- You used trigonometric ratios to solve right triangles.

**Now**

1. Use the Law of Sines to solve triangles.
2. Use the Law of Cosines to solve triangles.

**Why?**

- You have learned that the height or length of a tree can be calculated using *right triangle trigonometry* if you know the angle of elevation to the top of the tree and your distance from the tree. Some trees, however, grow at an angle or lean due to weather damage. To calculate the length of such trees, you must use other forms of trigonometry.

**NewVocabulary**
Law of Sines
Law of Cosines

**Common Core State Standards**

**Content Standards**

G.SRT.9 Derive the formula $A = \frac{1}{2}ab \sin(C)$ for the area of a triangle by drawing an auxiliary line from a vertex perpendicular to the opposite side.

G.SRT.10 Prove the Laws of Sines and Cosines and use them to solve problems.

**Mathematical Practices**

4 Model with mathematics.

1 Make sense of problems and persevere in solving them.

## 1 Law of Sines

In Lesson 10-4, you used trigonometric ratios to find side lengths and acute angle measures in *right* triangles. To find measures for nonright triangles, the definitions of sine and cosine can be extended to obtuse angles.

The **Law of Sines** can be used to find side lengths and angle measures for any triangle.

**Theorem 10.10 Law of Sines**

If $\triangle ABC$ has lengths $a$, $b$, and $c$, representing the lengths of the sides opposite the angles with measures $A$, $B$, and $C$, then

$$\frac{\sin A}{a} = \frac{\sin B}{b} = \frac{\sin C}{c}.$$

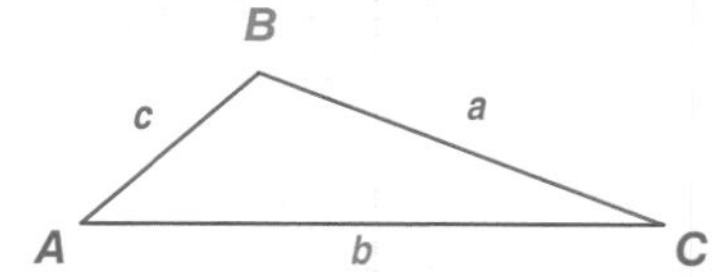


You will prove one of the proportions for Theorem 10.10 in Exercise 45.

You can use the Law of Sines to solve a triangle if you know the measures of two angles and any side (AAS or ASA).

**Example 1 Law of Sines (AAS)**

**Find $x$. Round to the nearest tenth.**

We are given the measures of two angles and a nonincluded side, so use the Law of Sines to write a proportion.

| | |
|---|---|
| $\frac{\sin A}{a} = \frac{\sin C}{c}$ | Law of Sines |
| $\frac{\sin 97°}{16} = \frac{\sin 21°}{x}$ | $m\angle A = 97$, $a = 16$, $m\angle C = 21$, $c = x$ |
| $x \sin 97° = 16 \sin 21°$ | Cross Products Property |
| $x = \frac{16 \sin 21°}{\sin 97°}$ | Divide each side by sin 97°. |
| $x \approx 5.8$ | Use a calculator. |

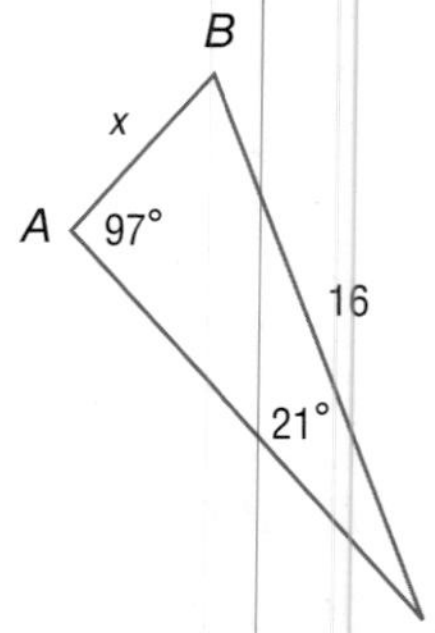


**GuidedPractice**

**1A.**

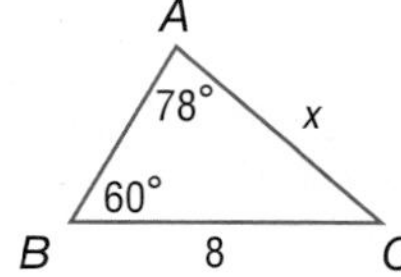


**1B.**

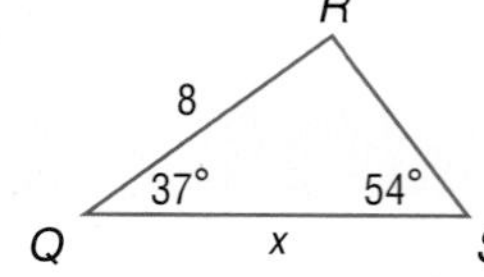


Ben Mangor/SuperStock

If given ASA, use the Triangle Angle Sum Theorem to first find the measure of the third angle.

> **StudyTip**
> Ambiguous Case You can sometimes use the Law of Sines to solve a triangle if you know the measures of two sides and a nonincluded angle (SSA). However, these three measures do not always determine exactly one triangle. You will learn more about this *ambiguous case* in Extend 10-6.

### Example 2 Law of Sines (ASA)

**Find $x$. Round to the nearest tenth.**

By the Triangle Angle Sum Theorem, $m\angle K = 180 - (45 + 73)$ or 62.

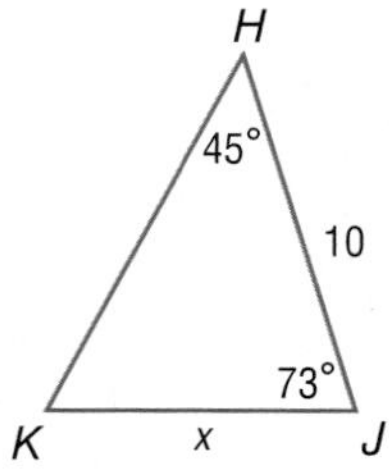


| | |
|---|---|
| $\frac{\sin H}{h} = \frac{\sin K}{k}$ | Law of Sines |
| $\frac{\sin 45°}{x} = \frac{\sin 62°}{10}$ | $m\angle H = 45$, $h = x$, $m\angle K = 62$, $k = 10$ |
| $10 \sin 45° = x \sin 62°$ | Cross Products Property |
| $\frac{10 \sin 45°}{\sin 62°} = x$ | Divide each side by sin 62°. |
| $x \approx 8.0$ | Use a calculator. |

**GuidedPractice**

**2A.**

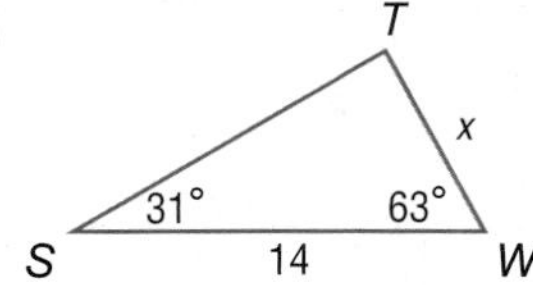


**2B.**

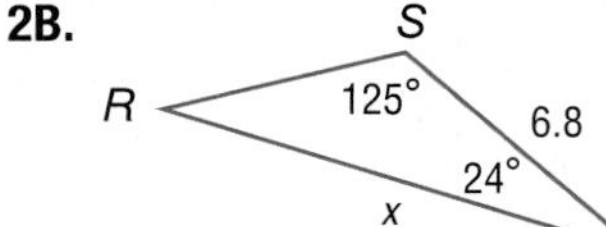


## 2 Law of Cosines

When the Law of Sines cannot be used to solve a triangle, the Law of Cosines may apply.

**Theorem 10.11 Law of Cosines**

If $\triangle ABC$ has lengths $a$, $b$, and $c$, representing the lengths of the sides opposite the angles with measures $A$, $B$, and $C$, then

$a^2 = b^2 + c^2 - 2bc \cos A$,

$b^2 = a^2 + c^2 - 2ac \cos B$, and

$c^2 = a^2 + b^2 - 2ab \cos C$.

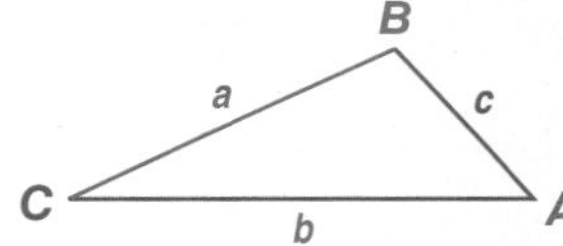


You will prove one of the equations for Theorem 10.11 in Exercise 46.

You can use the **Law of Cosines** to solve a triangle if you know the measures of two sides and the included angle (SAS).

### Example 3 Law of Cosines (SAS)

**Find $x$. Round to the nearest tenth.**

We are given the measures of two sides and their included angle, so use the Law of Cosines.

| | |
|---|---|
| $c^2 = a^2 + b^2 - 2ab \cos C$ | Law of Cosines |
| $x^2 = 9^2 + 11^2 - 2(9)(11) \cos 28°$ | Substitution |
| $x^2 = 202 - 198 \cos 28°$ | Simplify. |
| $x = \sqrt{202 - 198 \cos 28°}$ | Take the square root of each side. |
| $x \approx 5.2$ | Use a calculator. |

> **WatchOut!**
> Order of operations Remember to follow the order of operations when simplifying expressions. Multiplication or division must be performed before addition or subtraction. So, $202 - 198 \cos 28°$ *cannot* be simplified to $4 \cos 28°$.

## Guided Practice

**Find $x$. Round to the nearest tenth.**

**3A.** 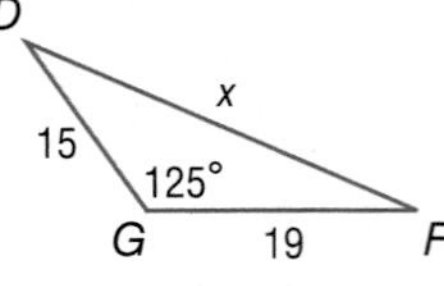


**3B.** 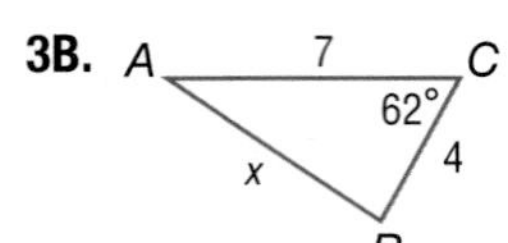


### Study Tip

**Obtuse Angles** There are also values for sin *A*, cos *A*, and tan *A* when $A \geq 90°$. Values of the ratios for these angles can be found using the trigonometric functions on your calculator.

You can also use the Law of Cosines if you know three side measures (SSS).

PT

## Example 4 Law of Cosines (SSS)

**Find $x$. Round to the nearest degree.**

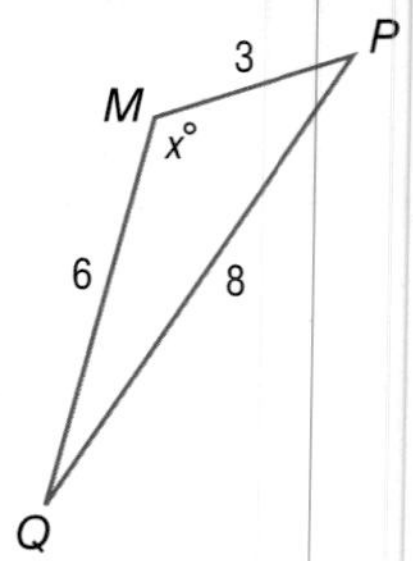

$m^2 = p^2 + q^2 - 2pq \cos M$ — Law of Cosines

$8^2 = 6^2 + 3^2 - 2(6)(3) \cos x°$ — Substitution

$64 = 45 - 36 \cos x°$ — Simplify.

$19 = -36 \cos x°$ — Subtract 45 from each side.

$\frac{19}{-36} = \cos x°$ — Divide each side by −36.

$x = \cos^{-1}\left(-\frac{19}{36}\right)$ — Use the inverse cosine ratio.

$x \approx 122$ — Use a calculator.

M, P, Q triangle: MP = 3, MQ = 6, PQ = 8, angle at M = x°

## Guided Practice

**4A.** 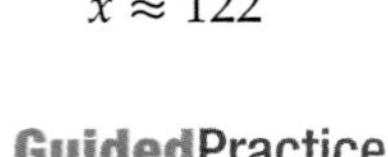 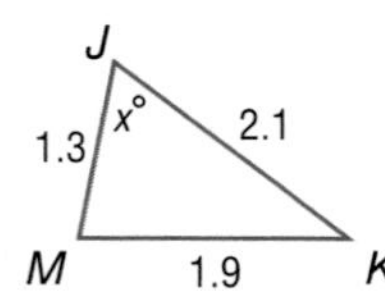


**4B.**  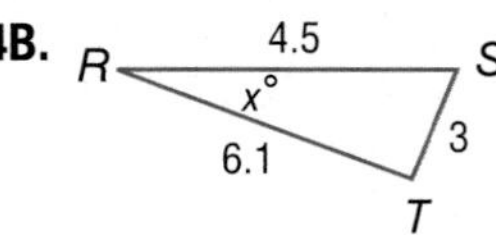


You can use the Law of Sines and Law of Cosines to solve direct and indirect measurement problems.

PT

## Real-World Example 5 Indirect Measurement

**BASKETBALL Drew and Hunter are playing basketball. Drew passes the ball to Hunter when he is 26 feet from the goal and 24 feet from Hunter. How far is Hunter from the goal if the angle from the goal to Drew and then to Hunter is 34°?**

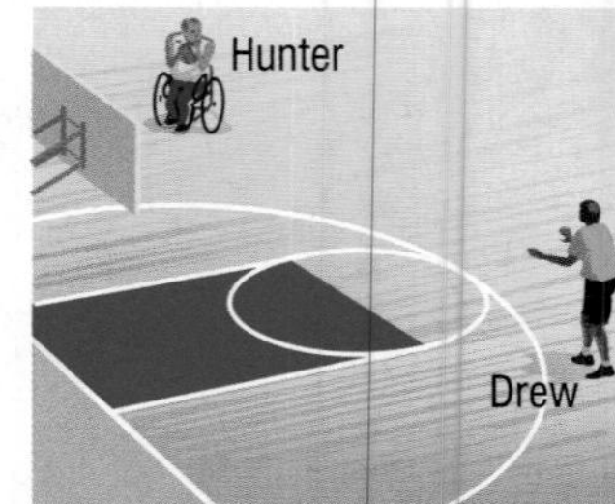


Draw a diagram. Since we know two sides of a triangle and the included angle, use the Law of Cosines.

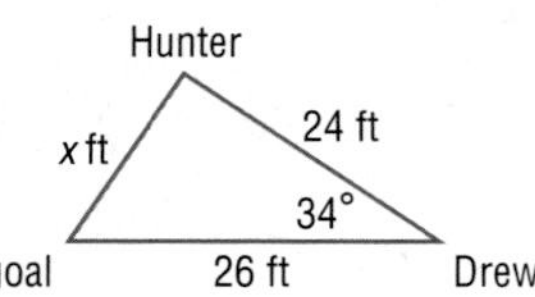


$x^2 = 24^2 + 26^2 - 2(24)(26) \cos 34°$ — Law of Cosines

$x = \sqrt{1252 - 1248 \cos 34°}$ — Simplify and take the positive square root of each side.

$x \approx 15$ — Use a calculator.

Hunter is about 15 feet from the goal when he takes his shot.

### Real-World Link

The first game of basketball was played at a YMCA in Springfield, Massachusetts, on December 1, 1891. James Naismith, a physical education instructor, invented the sport using a soccer ball and two half-bushel peach baskets, which is how the name *basketball* came about.

**Source:** *Encyclopaedia Britannica.*

## GuidedPractice

**5. LANDSCAPING** At 10 feet away from the base of a tree, the angle the top of a tree makes with the ground is 61°. If the tree grows at an angle of 78° with respect to the ground, how tall is the tree to the nearest foot?

When solving right triangles, you can use sine, cosine, or tangent. When solving other triangles, you can use the Law of Sines or the Law of Cosines, depending on what information is given.

**ReadingMath**

**Solve a Triangle** Remember that to *solve* a triangle means to find all of the missing side measures and/or angle measures.

### Example 6 Solve a Triangle

**Solve triangle *ABC*. Round to the nearest degree.**

Since $13^2 + 12^2 \neq 15^2$, this is not a right triangle. Since the measures of all three sides are given (SSS), begin by using the Law of Cosines to find $m\angle A$.

| | |
|---|---|
| $a^2 = b^2 + c^2 - 2bc \cos A$ | Law of Cosines |
| $15^2 = 12^2 + 13^2 - 2(12)(13) \cos A$ | $a = 15$, $b = 12$, and $c = 13$ |
| $225 = 313 - 312 \cos A$ | Simplify. |
| $-88 = -312 \cos A$ | Subtract 313 from each side. |
| $\frac{-88}{-312} = \cos A$ | Divide each side by −312. |
| $m\angle A = \cos^{-1} \frac{88}{312}$ | Use the inverse cosine ratio. |
| $m\angle A \approx 74$ | Use a calculator. |

Use the Law of Sines to find $m\angle B$.

| | |
|---|---|
| $\frac{\sin A}{a} = \frac{\sin B}{b}$ | Law of Sines |
| $\frac{\sin 74°}{15} \approx \frac{\sin B}{12}$ | $m\angle A \approx 74$, $a = 15$, and $b = 12$ |
| $12 \sin 74° = 15 \sin B$ | Cross Products Property |
| $\frac{12 \sin 74°}{15} = \sin B$ | Divide each side by 15. |
| $m\angle B = \sin^{-1} \frac{12 \sin 74°}{15}$ | Use the inverse sine ratio. |
| $m\angle B \approx 50$ | Use a calculator. |

By the Triangle Angle Sum Theorem, $m\angle C \approx 180 - (74 + 50)$ or 56. Therefore $m\angle A \approx 74$, $m\angle B \approx 50$, and $m\angle C \approx 56$.

**WatchOut**

**Rounding** When you round a numerical solution and then use it in later calculations, your answers may be inaccurate. Wait until after you have completed all of your calculations to round.

## GuidedPractice

**Solve triangle *ABC* using the given information. Round angle measures to the nearest degree and side measures to the nearest tenth.**

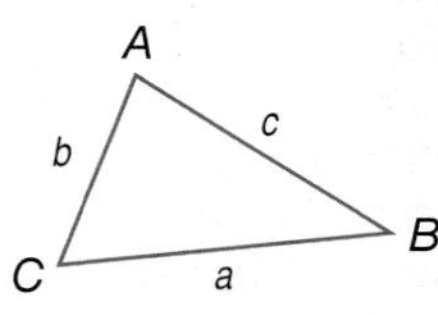


**6A.** $b = 10.2$, $c = 9.3$, $m\angle A = 26$

**6B.** $a = 6.4$, $m\angle B = 81$, $m\angle C = 46$

## ConceptSummary Solving a Triangle

| To solve ... | Given | Begin by using ... |
|---|---|---|
| a right triangle | leg-leg (LL)<br>hypotenuse-leg (HL)<br>acute angle-hypotenuse (AH)<br>acute angle-leg (AL) | tangent ratio<br>sine or cosine ratio<br>sine or cosine ratio<br>sine, cosine, or tangent ratios |
| any triangle | angle-angle-side (AAS)<br>angle-side-angle (ASA)<br>side-angle-side (SAS)<br>side-side-side (SSS) | Law of Sines<br>Law of Sines<br>Law of Cosines<br>Law of Cosines |

## Check Your Understanding

**Examples 1–2** **Find $x$. Round angle measures to the nearest degree and side measures to the nearest tenth.**

1. 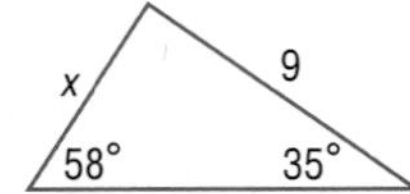


2. 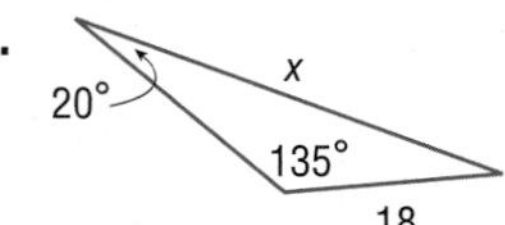


3. 


4. 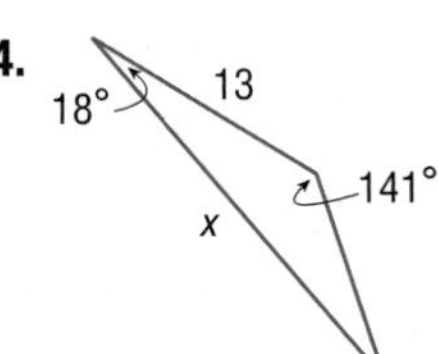


**Examples 3–4**

5. 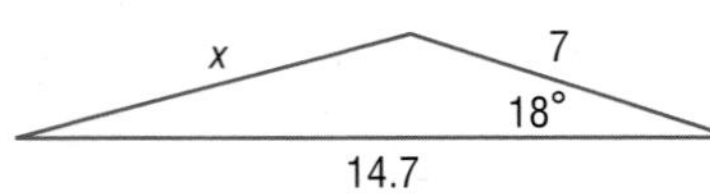


6. 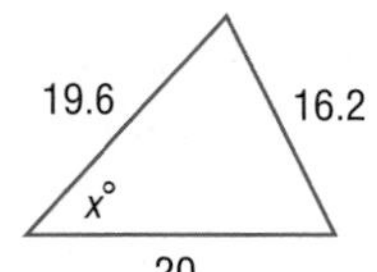


**Example 5**

7. **SAILING** Determine the length of the bottom edge, or foot, of the sail.

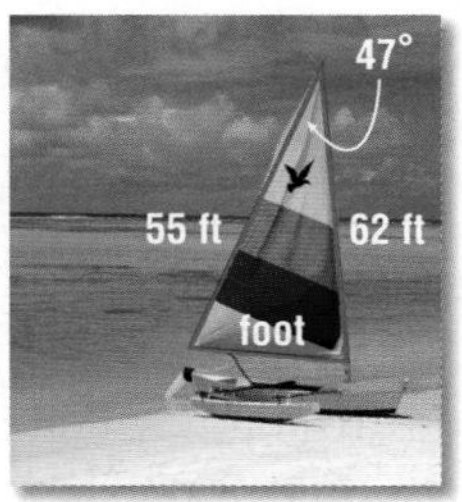


**Example 6**

**CCSS STRUCTURE** **Solve each triangle. Round angle measures to the nearest degree and side measures to the nearest tenth.**

8. 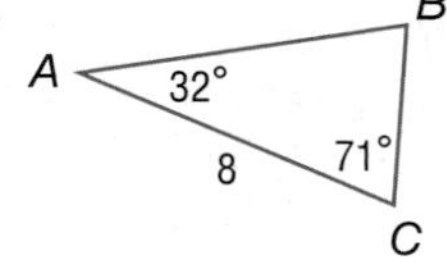


9. 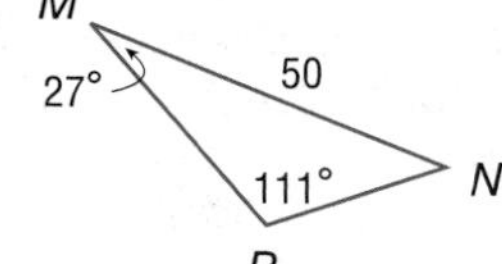


10. 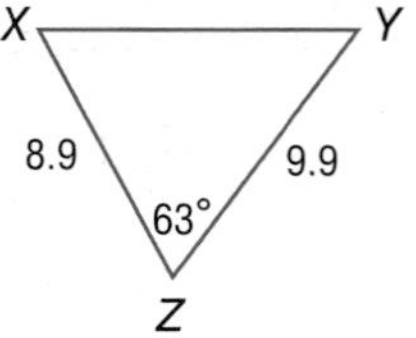


11. Solve $\triangle DEF$ if $DE = 16$, $EF = 21.6$, $FD = 20$.

Paul Miles/Axiom Photographic Agency/Getty Images

## Practice and Problem Solving

**Examples 1–2** **Find $x$. Round side measures to the nearest tenth.**

**12.** 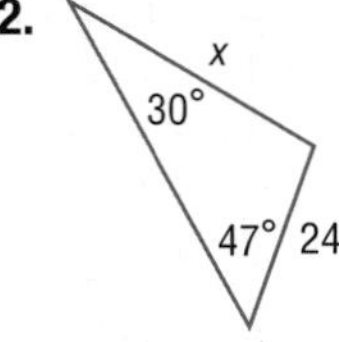


**13.** 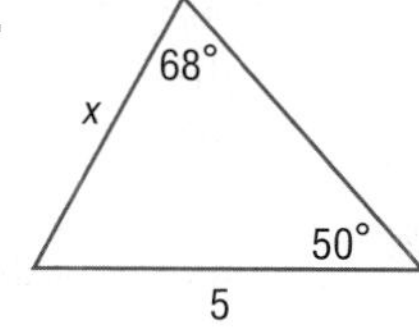


**14.** 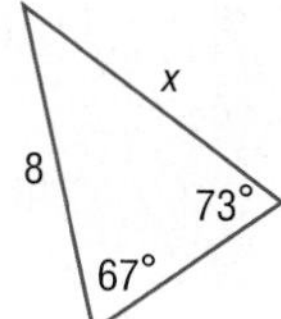


**15.** 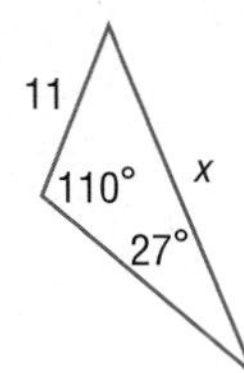


**16.** 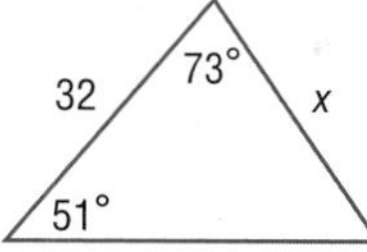


**17.** 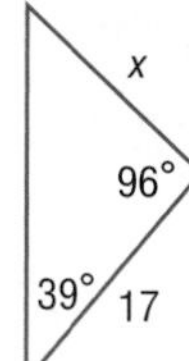


**18.** 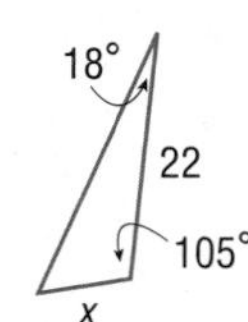


**19.** 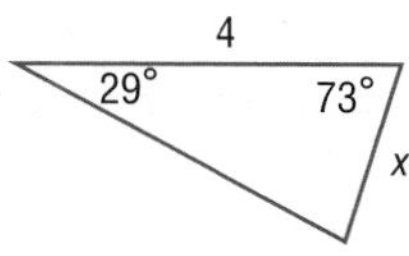


**20.** 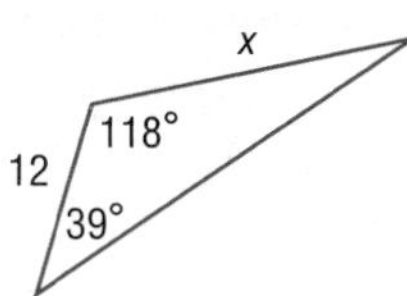


**21.** **CCSS MODELING** Angelina is looking at the Big Dipper through a telescope. From her view, the cup of the constellation forms a triangle that has measurements shown on the diagram at the right. Use the Law of Sines to determine distance between $A$ and $C$.

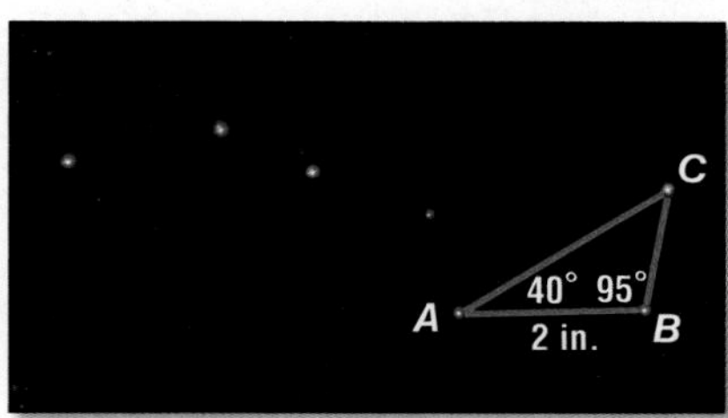


**Examples 3–4** **Find $x$. Round angle measures to the nearest degree and side measures to the nearest tenth.**

**22.** 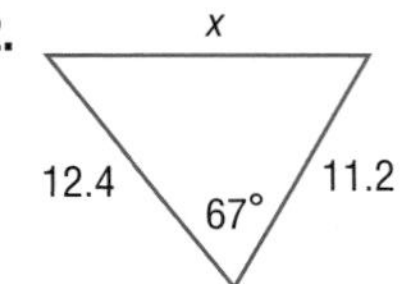


**23.** 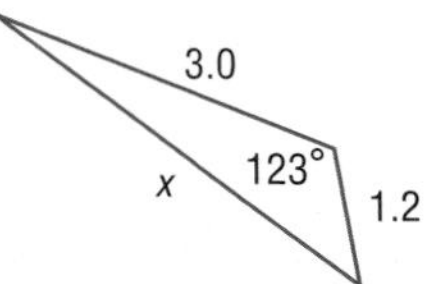


**24.** 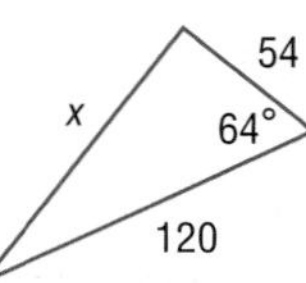


**25.** 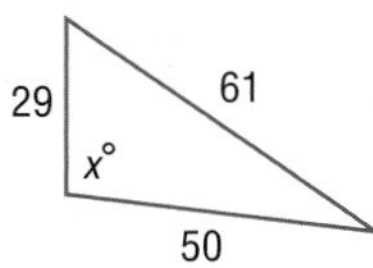


**26.** 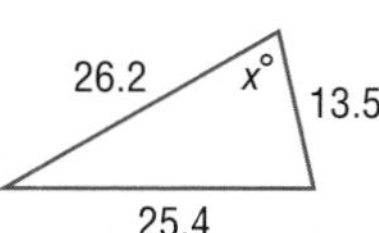


**27.** 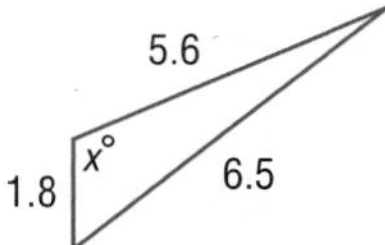


**28.** **HIKING** A group of friends who are camping decide to go on a hike. According to the map shown at the right, what is the measure of the angle between Trail 1 and Trail 2?

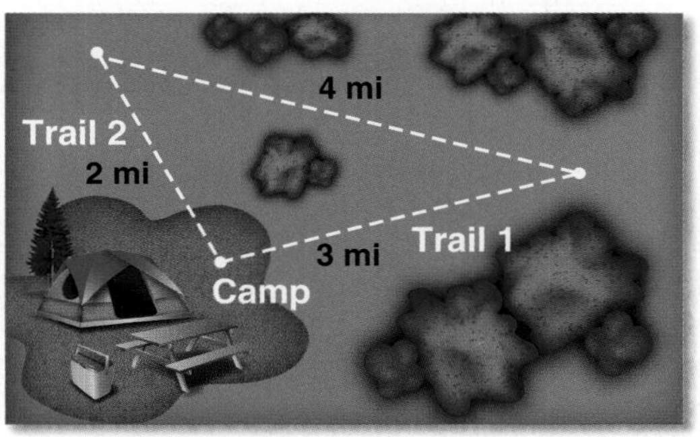

Example 5

**29. TORNADOES** Find the width of the mouth of the tornado shown below.

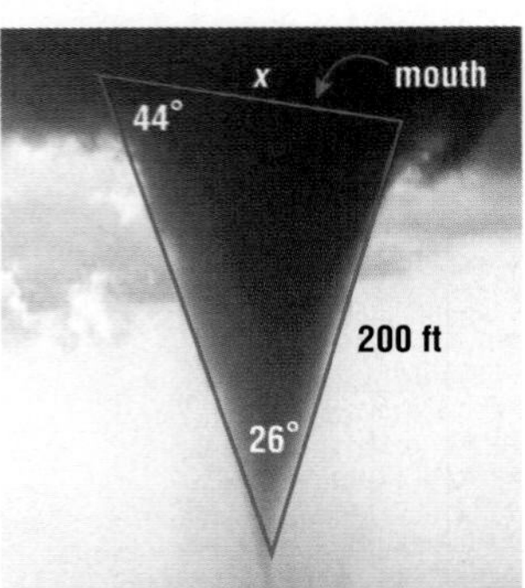


**30. TRAVEL** A pilot flies 90 miles from Memphis, Tennessee, to Tupelo, Mississippi, to Huntsville, Alabama, and finally back to Memphis. How far is Memphis from Huntsville?

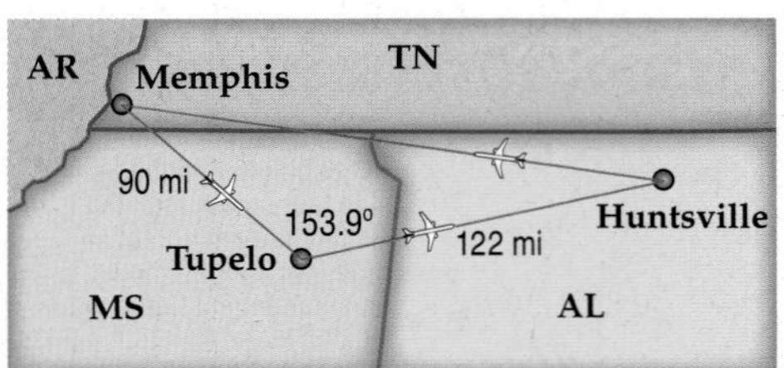


Example 6

**CCSS STRUCTURE Solve each triangle. Round angle measures to the nearest degree and side measures to the nearest tenth.**

**31.**

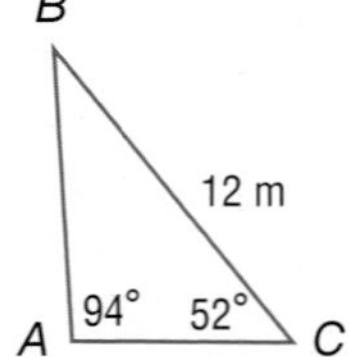


**32.**

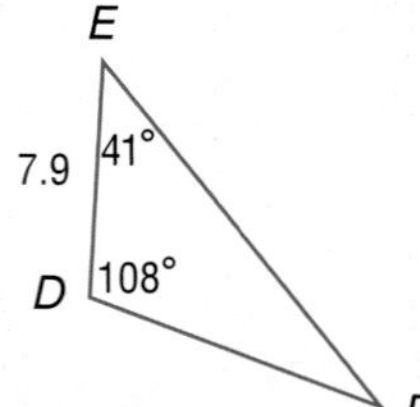


**33**

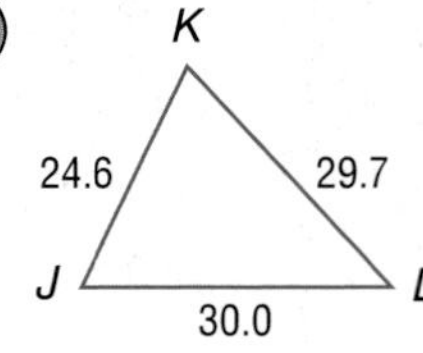


**34.**

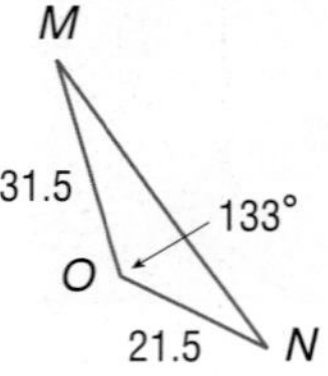


**35.**

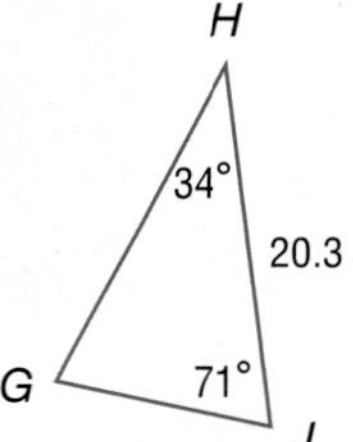


**36.**

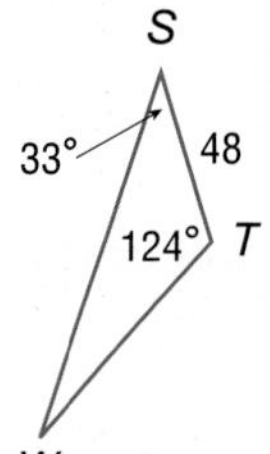


**37.**

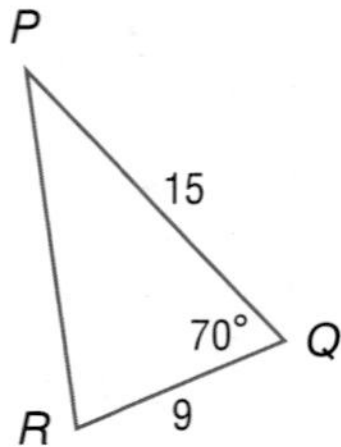


**38.**

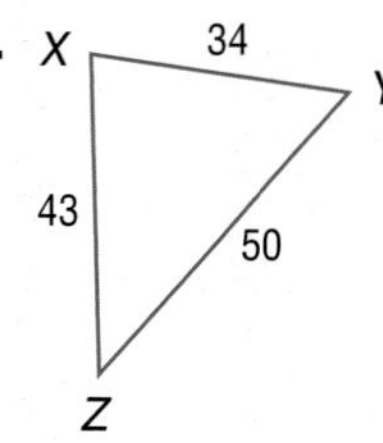


**39.**

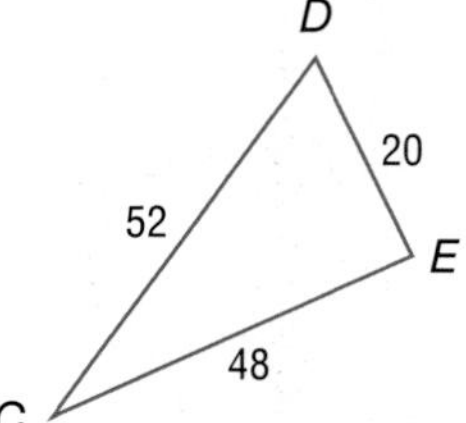


**40.** Solve $\triangle JKL$ if $JK = 33$, $KL = 56$, $LJ = 65$.

**41.** Solve $\triangle ABC$ if $m\angle B = 119$, $m\angle C = 26$, $CA = 15$.

**42.** Solve $\triangle XYZ$ if $XY = 190$, $YZ = 184$, $ZX = 75$.

**43. GARDENING** Crystal has an organic vegetable garden. She wants to add another triangular section so that she can start growing tomatoes. If the garden and neighboring space have the dimensions shown, find the perimeter of the new garden to the nearest foot.

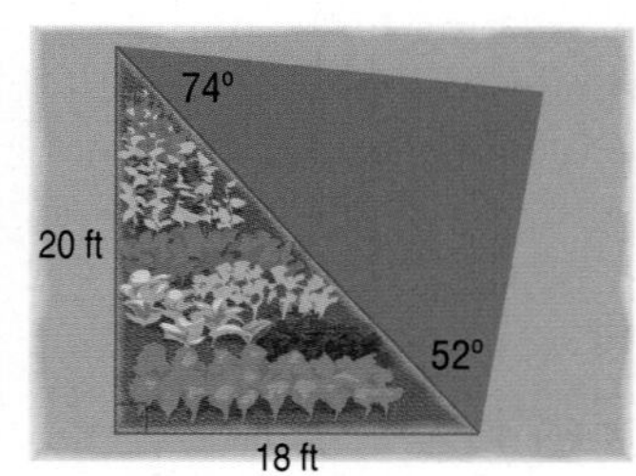


Jason Lafontaine/Painet Inc.

**44. FIELD HOCKEY** Alyssa and Nari are playing field hockey. Alyssa is standing 20 feet from one goal post and 25 feet from the opposite post. Nari is standing 45 feet from one goal post and 38 feet from the other post. If the goal is 12 feet wide, which player has a greater chance to make a shot? What is the measure of the player's angle?

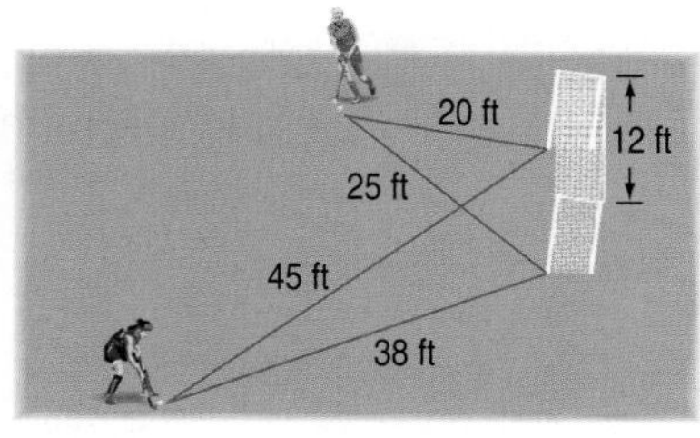


**45. PROOF** Justify each statement for the derivation of the Law of Sines.

**Given:** $\overline{CD}$ is an altitude of $\triangle ABC$.

**Prove:** $\frac{\sin A}{a} = \frac{\sin B}{b}$

**Proof:**

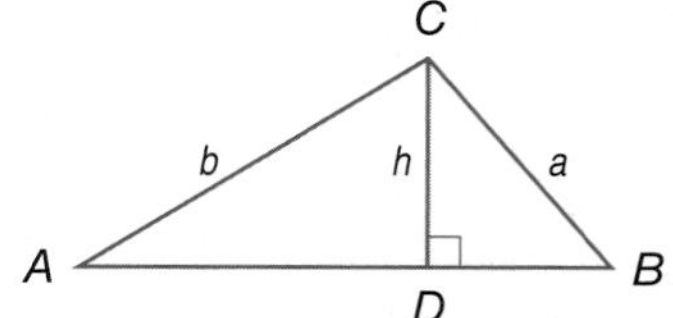


| Statements | Reasons |
|---|---|
| $\overline{CD}$ is an altitude of $\triangle ABC$ | Given |
| $\triangle ACD$ and $\triangle CBD$ are right | Def. of altitude |
| **a.** $\sin A = \frac{h}{b}$, $\sin B = \frac{h}{a}$ | **a.** ___?___ |
| **b.** $b \sin A = h$, $a \sin B = h$ | **b.** ___?___ |
| **c.** $b \sin A = a \sin B$ | **c.** ___?___ |
| **d.** $\frac{\sin A}{a} = \frac{\sin B}{b}$ | **d.** ___?___ |

**46. PROOF** Justify each statement for the derivation of the Law of Cosines.

**Given:** $h$ is an altitude of $\triangle ABC$.

**Prove:** $c^2 = a^2 + b^2 - 2ab \cos C$

**Proof:**

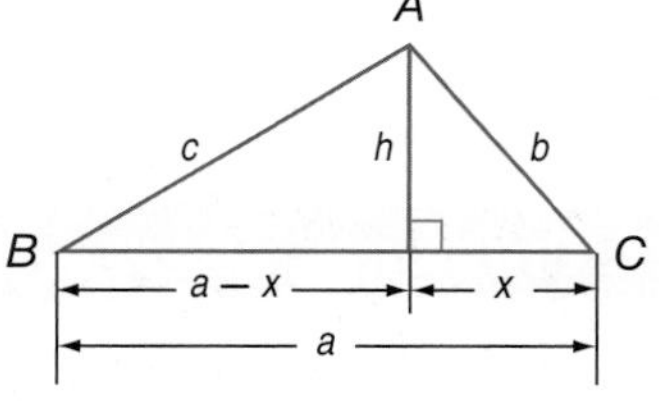


| Statements | Reasons |
|---|---|
| $h$ is an altitude of $\triangle ABC$ | Given |
| Altitude $h$ separates $\triangle ABC$ into two right triangles | Def. of altitude |
| **a.** $c^2 = (a - x)^2 + h^2$ | **a.** ___?___ |
| **b.** $c^2 = a^2 - 2ax + x^2 + h^2$ | **b.** ___?___ |
| **c.** $x^2 + h^2 = b^2$ | **c.** ___?___ |
| **d.** $c^2 = a^2 - 2ax + b^2$ | **d.** ___?___ |
| **e.** $\cos C = \frac{x}{b}$ | **e.** ___?___ |
| **f.** $b \cos C = x$ | **f.** ___?___ |
| **g.** $c^2 = a^2 - 2a(b \cos C) + b^2$ | **g.** ___?___ |
| **h.** $c^2 = a^2 + b^2 - 2ab \cos C$ | **h.** ___?___ |

**CCSS SENSE-MAKING Find the perimeter of each figure. Round to the nearest tenth.**

**47.**

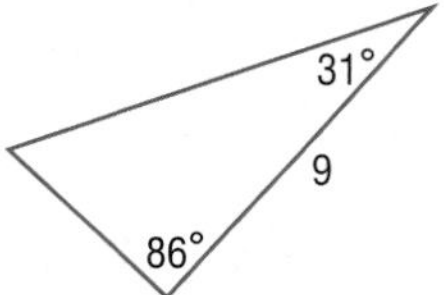


**48.**

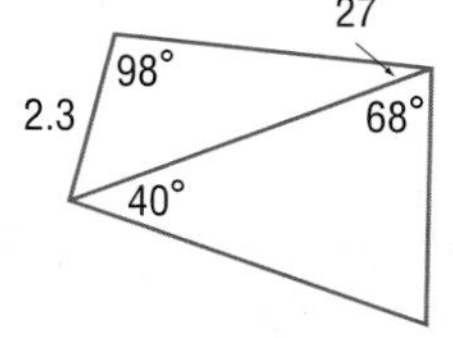


**49**

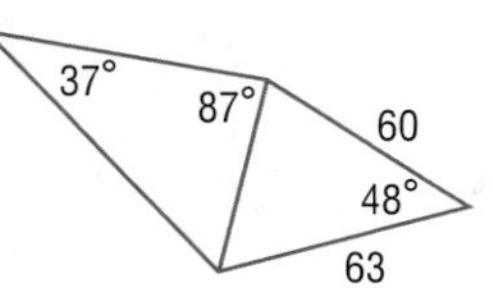


**50.**

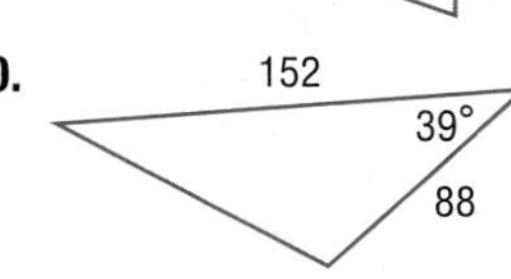

**51** **MODELS** Vito is working on a model castle. Find the length of the missing side (in inches) using the diagram at the right.

**52.** **COORDINATE GEOMETRY** Find the measure of the largest angle in $\triangle ABC$ with coordinates $A(-3, 6)$, $B(4, 2)$, and $C(-5, 1)$. Explain your reasoning.

**53.** **MULTIPLE REPRESENTATIONS** In this problem, you will use trigonometry to find the area of a triangle.

**a.** **Geometric** Draw an acute, scalene $\triangle ABC$ including an altitude of length $h$ originating at vertex $A$.

**b.** **Algebraic** Use trigonometry to represent $h$ in terms of $m\angle B$.

**c.** **Algebraic** Write an equation to find the area of $\triangle ABC$ using trigonometry.

**d.** **Numerical** If $m\angle B$ is 47, $AB = 11.1$, $BC = 14.1$, and $CA = 10.4$, find the area of $\triangle ABC$. Round to the nearest tenth.

**e.** **Analytical** Write an equation to find the area of $\triangle ABC$ using trigonometry in terms of a different angle measure.

## H.O.T. Problems Use Higher-Order Thinking Skills

**54.** **CCSS CRITIQUE** Colleen and Mike are planning a party. Colleen wants to sew triangular decorations and needs to know the perimeter of one of the triangles to buy enough trim. The triangles are isosceles with angle measurements of 64° at the base and side lengths of 5 inches. Colleen thinks the perimeter is 15.7 inches and Mike thinks it is 15 inches. Is either of them correct?

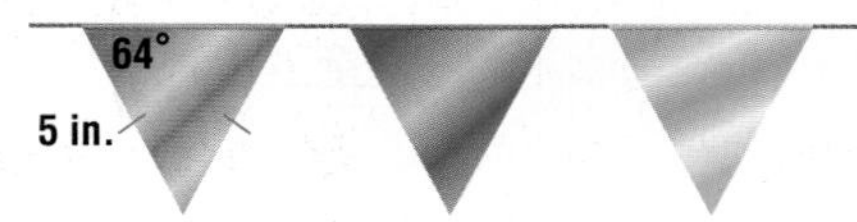


**55.** **CHALLENGE** Find the value of $x$ in the figure at the right.

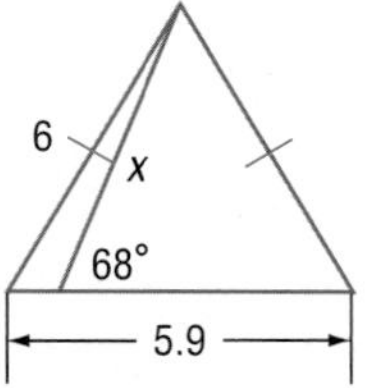


**56.** **REASONING** Explain why the Pythagorean Theorem is a specific case of the Law of Cosines.

**57.** **OPEN ENDED** Draw and label a triangle that can be solved:

**a.** using only the Law of Sines.

**b.** using only the Law of Cosines.

**58.** **WRITING IN MATH** What methods can you use to solve a triangle?

Lenz, G/Arco Images/age fotostock

## Standardized Test Practice

**59.** For $\triangle ABC$, $m\angle A = 42$, $m\angle B = 74$, and $a = 3$, what is the value of $b$?

**A** 4.3 **C** 2.1

**B** 3.8 **D** 1.5

**60. ALGEBRA** Which inequality *best* describes the graph below?

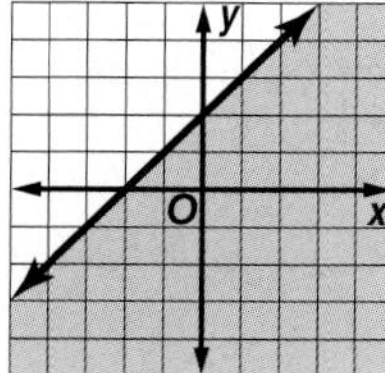


**F** $y \geq -x + 2$ **H** $y \geq -3x + 2$

**G** $y \leq x + 2$ **J** $y \leq 3x + 2$

**61. SHORT RESPONSE** What is the perimeter of the triangle shown below? Round to the nearest tenth.

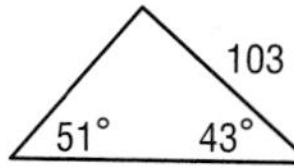


**62. SAT/ACT** If $\sin x = 0.6$ and $AB = 12$, what is the area of $\triangle ABC$?

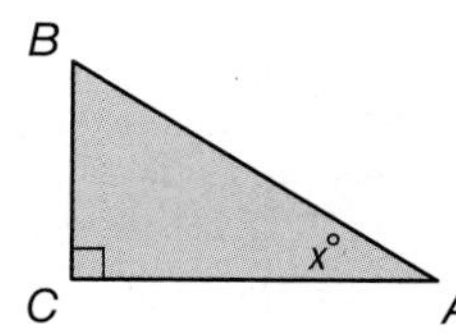


**A** 9.6 units$^2$ **D** 34.6 units$^2$

**B** 28.8 units$^2$ **E** 42.3 units$^2$

**C** 31.2 units$^2$

## Spiral Review

**63. HIKING** A hiker is on top of a mountain 250 feet above sea level with a 68° angle of depression. She can see her camp from where she is standing. How far is her camp from the top of the mountain? (Lesson 10-5)

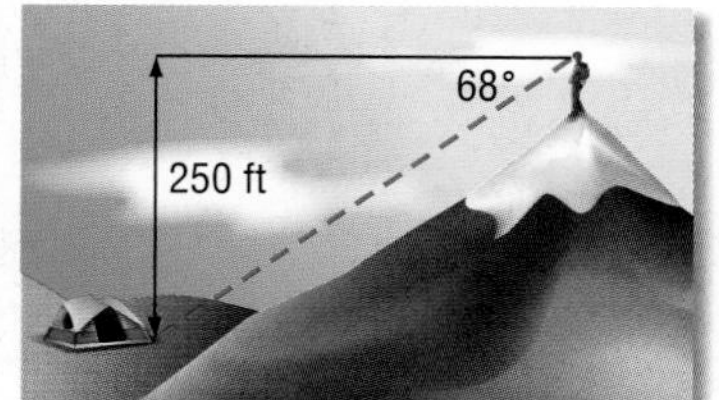


**Use a calculator to find the measure of $\angle J$ to the nearest degree.** (Lesson 10-4)

**64.**

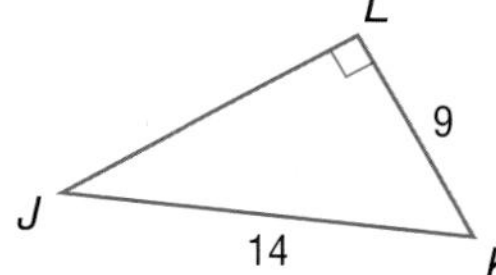


**65.**

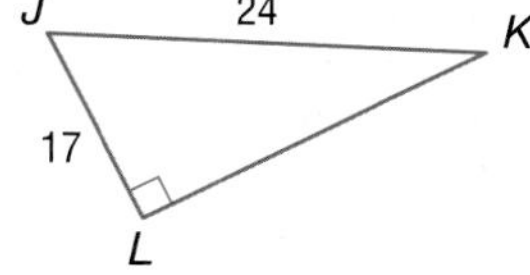


**Determine whether the polygons are *always*, *sometimes*, or *never* similar. Explain your reasoning.** (Lesson 9-2)

**66.** a right triangle and an isosceles triangle

**67.** an equilateral triangle and a scalene triangle

**Name the missing coordinates of each triangle.** (Lesson 6-6)

**68.**

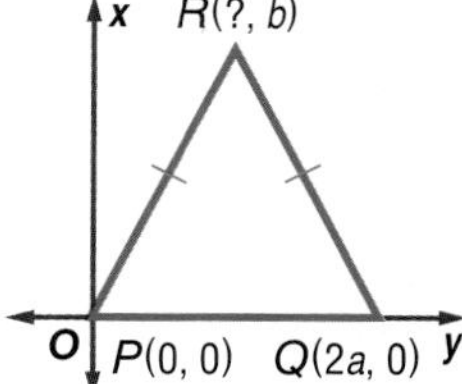


**69.**

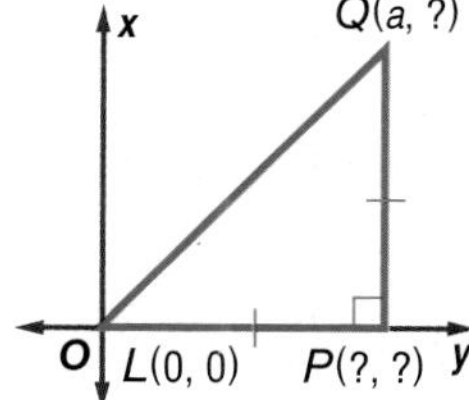


**70.**

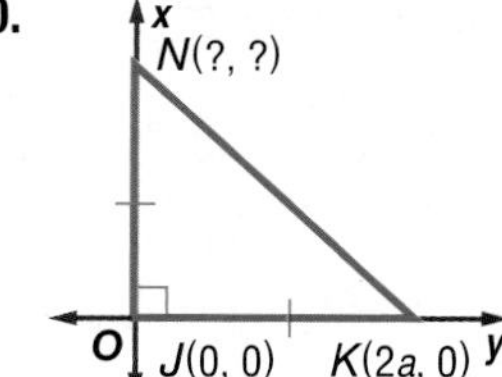


## Skills Review

**Find the distance between each pair of points. Round to the nearest tenth.**

**71.** $A(5, 1)$ and $C(-3, -3)$

**72.** $J(7, 11)$ and $K(-1, 5)$

**73.** $W(2, 0)$ and $X(8, 6)$

EXTEND 10-6

# Geometry Lab: The Ambiguous Case

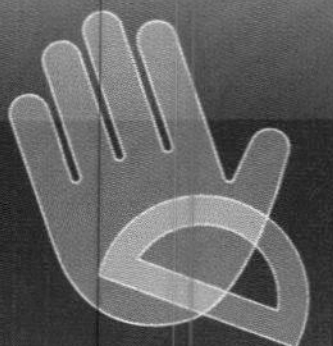

From your work with congruent triangles, you know that three measures determine a unique triangle when the measures are

- three sides (SSS),
- two sides and an included angle (SAS),
- two angles and an included side (ASA), or
- two angles and a nonincluded side (AAS).

A unique triangle is not necessarily determined by three angles (AAA) or by two sides and a nonincluded angle. In this lab, you will investigate how many triangles are determined by this last case (SSA), called the **ambiguous case**.

**CCSS Common Core State Standards**
**Content Standards**
**G.SRT.11** Understand and apply the Law of Sines and the Law of Cosines to find unknown measurements in right and non-right triangles (e.g., surveying problems, resultant forces).
**Mathematical Practices** 2

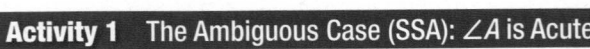

## Activity 1 The Ambiguous Case (SSA): ∠*A* is Acute

**Step 1** On a 5″ × 8″ notecard, draw and label $\overline{AC}$ and a ray extending from *A* to form an acute angle. Label side $\overline{AC}$ as *b*.

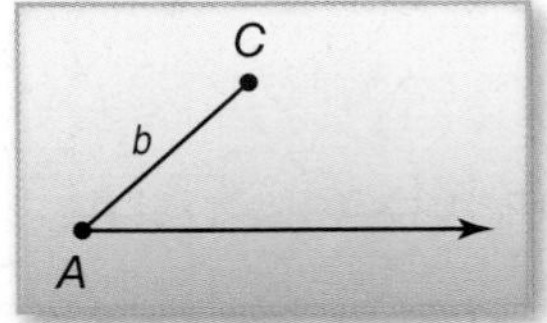


**Step 2** Using a brass fastener, attach one end of a half-inch strip of cardstock to the notecard at *C*. The strip should be longer than *b*. This represents side *a*.

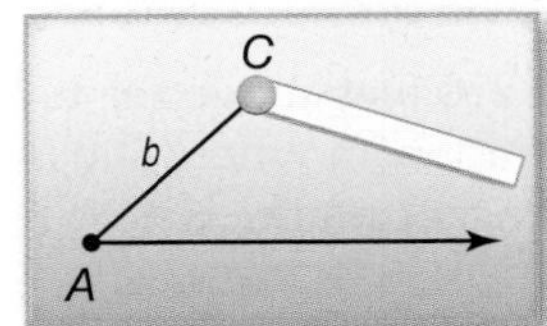


**Step 3** Position side *a* so that it is perpendicular to the ray. Make a black mark on the strip at the point where it touches the ray.

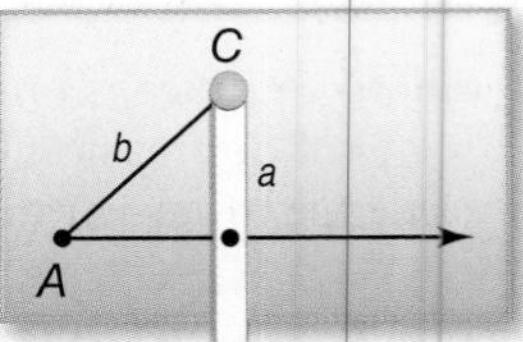


### Model and Analyze

1. If *a* has the given length, how many triangles can be formed? (*Hint*: Rotate the strip to see if the mark can intersect the ray at any other locations to form a different triangle.)
2. Show that if side *a* is perpendicular to the third side of the triangle, then $a = b \sin A$.

**Determine the number of triangles that can be formed given each of the modifications to *a* in Activity 1.**

3. $a < b \sin A$ (*Hint*: Make a green mark above the black mark on the strip, and try to form triangle(s) using this new length for *a*.)
4. $a = b$ (*Hint*: Rotate the strip so that it lies on top of $\overline{AC}$ and mark off this length in red. Then rotate the strip to try to form triangle(s) using this new length for *a*.)
5. $a < b$ and $a > b \sin A$ (*Hint*: Make a blue mark between the black and the red marks. Then rotate the strip to try to form triangle(s) using this new length for *a*.)
6. $a > b$ (*Hint*: Rotate the strip to try to form triangle(s) using the entire length of the strip as the length for *a*.)

**Use your results from Exercises 1–6 to determine whether the given measures define *0, 1, 2,* or *infinitely many* acute triangles. Justify your answers.**

7. $a = 14, b = 16, m\angle A = 55$
8. $a = 7, b = 11, m\angle A = 68$
9. $a = 22, b = 25, m\angle A = 39$
10. $a = 13, b = 12, m\angle A = 81$
11. $a = 10, b = 10, m\angle A = 45$
12. $a = 6, b = 9, m\angle A = 24$

In the next activity, you will investigate how many triangles are determined for the ambiguous case when the angle given is obtuse.

**Activity 2** The Ambiguous Case (SSA): $\angle A$ is Obtuse

**Step 1** On a 5″ × 8″ notecard, draw and label $\overline{AC}$ and a ray extending from $A$ to form an obtuse angle. Label side $\overline{AC}$ as $b$.

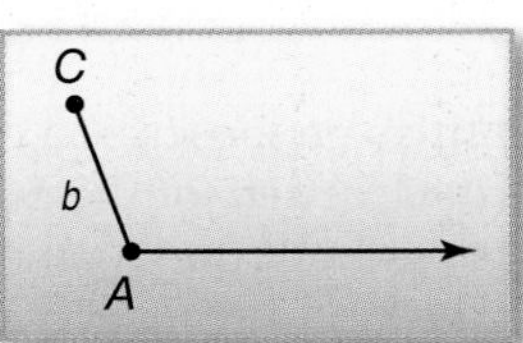


**Step 2** Using a brass fastener, attach one end of a half-inch strip of cardstock to the notecard at C. The strip should be longer than $b$. This represents side $a$.

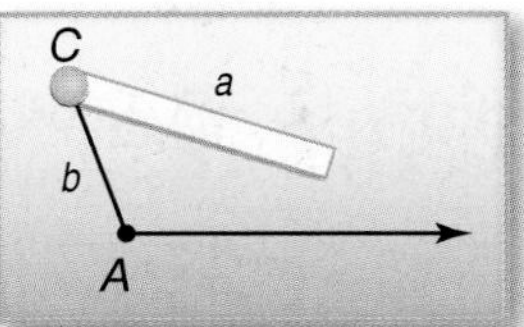


## Model and Analyze

**13.** How many triangles can be formed if $a = b$? if $a < b$? if $a > b$?

**Use your results from Exercise 13 to determine whether the given measures define *0, 1, 2,* or *infinitely many* obtuse triangles. Justify your answers.**

**14.** $a = 10, b = 8, m\angle A = 95$

**15.** $a = 13, b = 17, m\angle A = 100$

**16.** $a = 15, b = 15, m\angle A = 125$

**17.** Explain why three angle measures do not determine a unique triangle. How many triangles are determined by three angles measures?

**Determine whether the given measures define *0, 1, 2,* or *infinitely many* triangles. Justify your answers.**

**18.** $a = 25, b = 21, m\angle A = 39$

**19.** $m\angle A = 41, m\angle B = 68, m\angle C = 71$

**20.** $a = 17, b = 15, m\angle A = 128$

**21.** $a = 13, b = 17, m\angle A = 52$

**22.** $a = 5, b = 9, c = 6$

**23.** $a = 10, b = 15, m\angle A = 33$

**24. OPEN ENDED** Give measures for $a$, $b$, and an acute $\angle A$ that define

**a.** 0 triangles.

**b.** exactly one triangle.

**c.** two triangles.

**25. CHALLENGE** Find both solutions for $\triangle ABC$ if $a = 15, b = 21, m\angle A = 42$. Round angle measures to the nearest degree and side measures to the nearest tenth.

- For Solution 1, assume that $\angle B$ is acute, and use the Law of Sines to find $m\angle B$. Then find $m\angle C$. Finally, use the Law of Sines again to find $c$.
- For Solution 2, assume that $\angle B$ is obtuse. Let this obtuse angle be $\angle B'$. Use $m\angle B$ you found in Solution 1 and the diagram shown to find $m\angle B'$. Then find $m\angle C$. Finally, use the Law of Sines to find $c$.

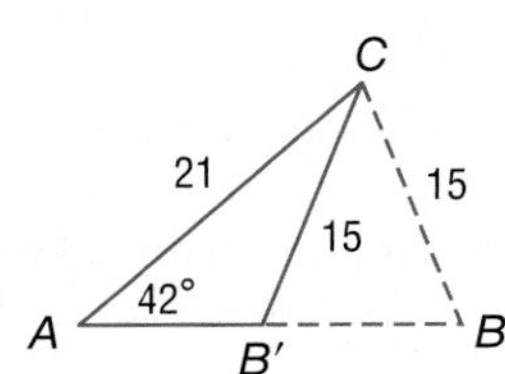

LESSON 10-7

# Vectors

| :: Then | :: Now | :: Why? |
|---|---|---|
| • You used trigonometry to find side lengths and angle measures of right triangles. | • **1** Perform vector operations geometrically. **2** Perform vector operations on the coordinate plane. | • Meteorologists use vectors to represent weather patterns. For example, *wind vectors* are used to indicate wind direction and speed. |

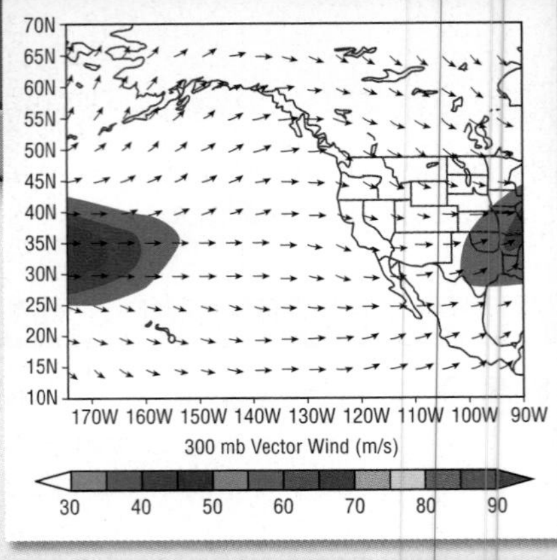

## NewVocabulary

vector
magnitude
direction
resultant
parallelogram method
triangle method
standard position
component form

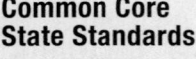

## Common Core State Standards

**Content Standards**

G.GPE.6 Find the point on a directed line segment between two given points that partitions the segment in a given ratio.

**Mathematical Practices**

1 Make sense of problems and persevere in solving them.

4 Model with mathematics.

## 1 Geometric Vector Operations

Some quantities are described by a real number known as a *scalar*, which describes the *magnitude* or size of the quantity. Other quantities are described by a **vector**, which describes both the magnitude and *direction* of the quantity. For example, a speed of 5 miles per hour is a scalar, while a velocity of 5 miles per hour due north is a vector.

A vector can be represented by a directed line segment with an initial point and a terminal point. The vector shown, with initial point $A$ and terminal point $B$, can be called $\overrightarrow{AB}$, $\vec{a}$, or **a**.

terminal point or tip

$B$

$\vec{a}$

$A$

initial point or tail

The **magnitude** of $\overrightarrow{AB}$, denoted $|\overrightarrow{AB}|$, is the length of the vector from its initial point to its terminal point. The **direction** of a vector can be expressed as the angle that it forms with the horizontal or as a measurement between 0° and 90° east or west of the north-south line.

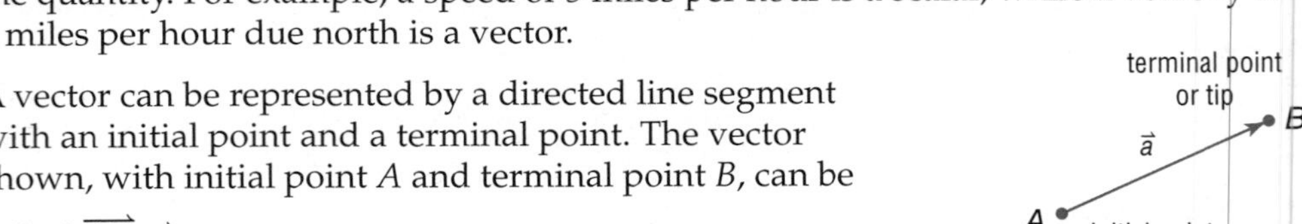


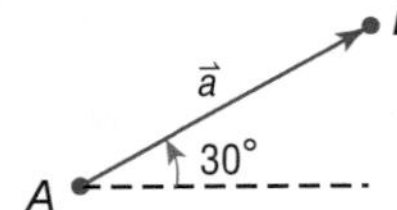


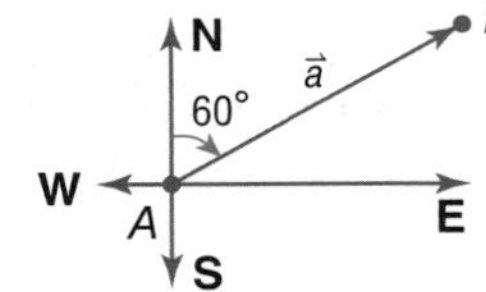

The direction of $\vec{a}$ is 60° east of north.

### Example 1 Represent Vectors Geometrically

**Use a ruler and a protractor to draw each vector. Include a scale on each diagram.**

**a. $\vec{m}$ = 15 miles per hour at 140° to the horizontal**

Using a scale of 1 cm : 5 mi/h, draw and label a 15 ÷ 5 or 3-centimeter arrow at a 140° angle to the horizontal.

$\vec{m}$

140°

1 cm : 5 mi/h

**b. $\vec{c}$ = 55 pounds of force 55° west of south**

Using a scale of 1 in: 25 lbs, draw and label a 55 ÷ 25 or 2.2-inch arrow 55° west of the north-south line on the south side.

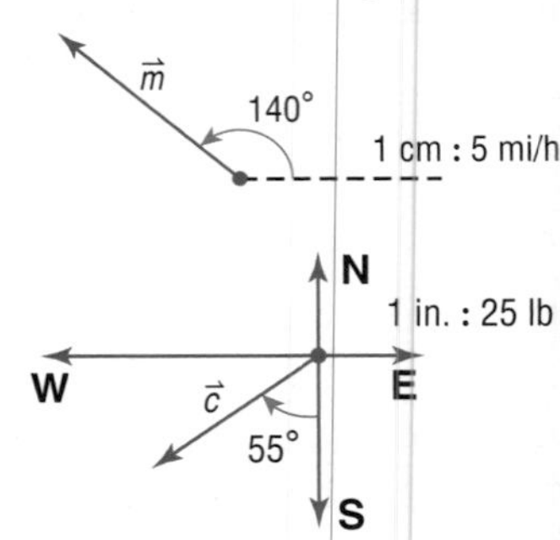


### GuidedPractice

**1A.** $\vec{b}$ = 40 feet per second at 35° to the horizontal

**1B.** $\vec{t}$ = 12 kilometers per hour at 85° east of north

The sum of two or more vectors is a single vector called the **resultant**.

## KeyConcept Vector Addition

To find the resultant of $\vec{a}$ and $\vec{b}$, use one of the following methods.

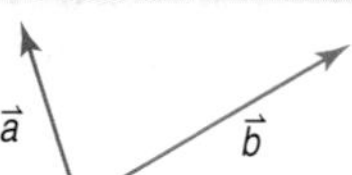


### Parallelogram Method

**Step 1** Translate $\vec{b}$ so that the tail of $\vec{b}$ touches the tail of $\vec{a}$.

**Step 2** Complete the parallelogram. The resultant is the indicated diagonal of the parallelogram.

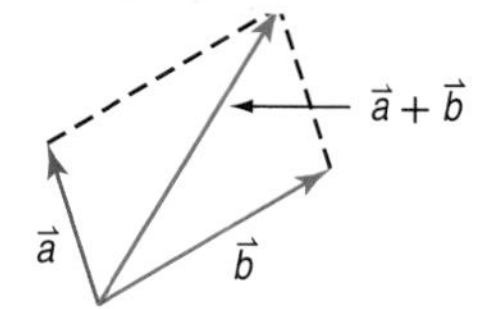


### Triangle Method

**Step 1** Translate $\vec{b}$ so that the tail of $\vec{b}$ touches the tip of $\vec{a}$.

**Step 2** Draw the resultant vector from the tail of $\vec{a}$ to the tip of $\vec{b}$.

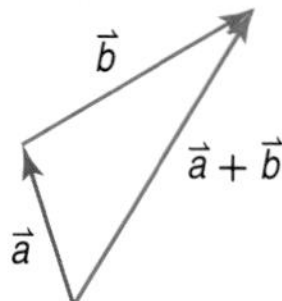


**StudyTip**

Types of Vectors
**Parallel vectors** have the same or opposite direction but not necessarily the same magnitude.

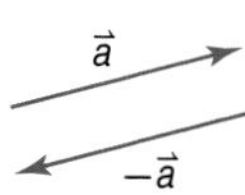


**Opposite vectors** have the same magnitude but *opposite* direction.

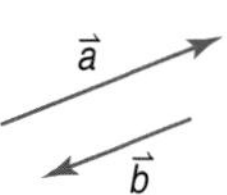


**Equivalent vectors** have the same magnitude and direction.

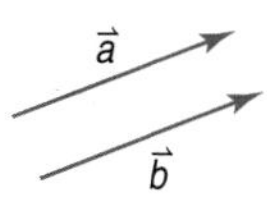


## Example 2 Find the Resultant of Two Vectors

**Copy the vectors. Then find $\vec{c} - \vec{d}$.**

Subtracting a vector is equivalent to adding its opposite vector.

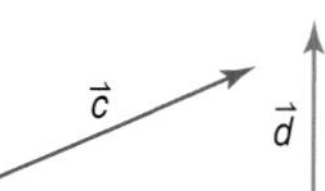


### Parallelogram Method

**Step 1**

Copy $\vec{c}$ and $\vec{d}$. Draw $-\vec{d}$, and translate it so that its tail touches the tail of $\vec{c}$.

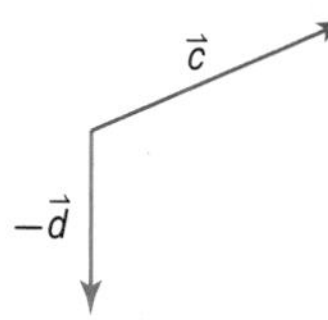


**Step 2**

Complete the parallelogram. Then draw the diagonal.

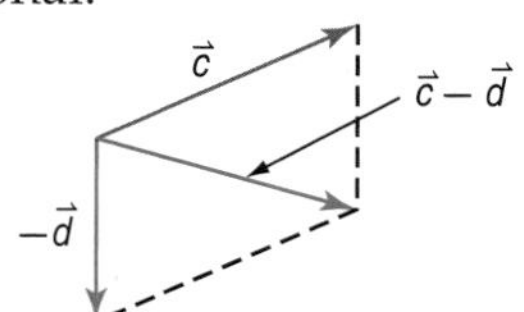


### Triangle Method

**Step 1**

Copy $\vec{c}$ and $\vec{d}$. Draw $-\vec{d}$, and translate it so that its tail touches the tip of $\vec{c}$.

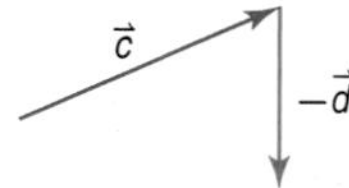


**Step 2**

Draw the resultant vector from the tail of $\vec{c}$ to the tip of $-\vec{d}$.

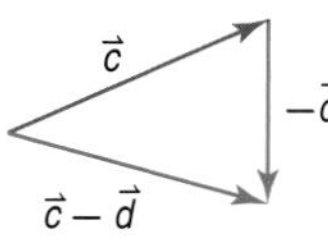


Both methods produce the same resultant vector $\vec{c} - \vec{d}$. You can use a ruler and a protractor to measure the magnitude and direction of each vector to verify your results.

**GuidedPractice**

**2A.** Find $\vec{c} + \vec{d}$.

**2B.** Find $\vec{d} - \vec{c}$.

You can use vectors to solve real-world problems.

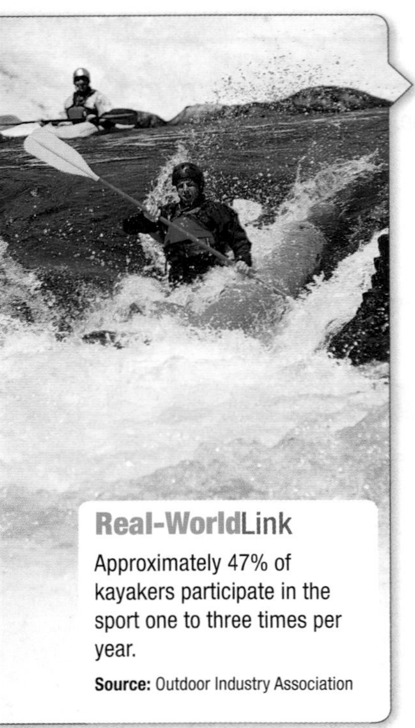

**Real-WorldLink**

Approximately 47% of kayakers participate in the sport one to three times per year.

**Source:** Outdoor Industry Association

PT

## Real-World Example 6 Vector Applications

**KAYAKING Trey is paddling due north in a kayak at 7 feet per second. The river is moving with a velocity of 3 feet per second due west. What is the resultant speed and direction of the kayak to an observer on shore?**

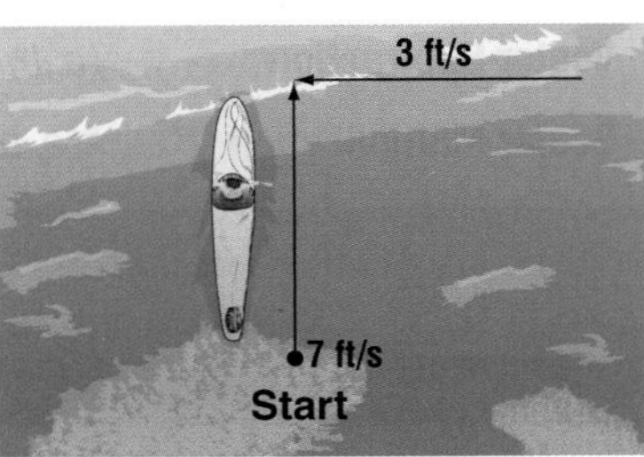


**Step 1** Draw a diagram. Let $\vec{r}$ represent the resultant vector.

The component form of the vector representing the paddling velocity is $\langle 0, 7\rangle$, and the component form of the vector representing the velocity of the river is $\langle -3, 0\rangle$.

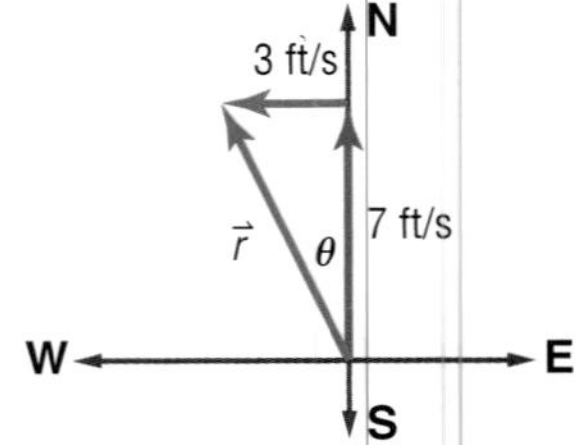


The resultant vector is $\langle 0, 7\rangle + \langle -3, 0\rangle$ or $\langle -3, 7\rangle$. This vector represents the resultant velocity of the kayak, and its magnitude represents the resultant speed.

**Step 2** Use the Distance Formula to find the resultant speed.

$|\vec{r}| = \sqrt{(x_2 - x_1)^2 + (y_2 - y_1)^2}$ Distance Formula

$= \sqrt{(-3 - 0)^2 + (7 - 0)^2}$ $(x_1, y_1) = (0, 0)$ and $(x_2, y_2) = (-3, 7)$

$= \sqrt{58}$ or about 7.6 Simplify.

**Step 3** Use trigonometry to find the resultant direction.

$\tan \theta = \frac{3}{7}$ $\tan \theta = \frac{\text{opp}}{\text{adj}}$

$\theta = \tan^{-1} \frac{3}{7}$ Def. of inverse tangent

$\theta \approx 23.2°$ Use a calculator.

The direction of $\vec{r}$ is about 23.2° west of north.

Therefore, the resultant speed of the kayak is about 7.6 feet per second at an angle of about 23.2° west of north.

### GuidedPractice

**6. KAYAKING** Suppose Trey starts paddling due south at a speed of 8 feet per second. If the river is flowing at a velocity of 2 feet per second due west, what is the resultant speed and direction of the kayak?

## Check Your Understanding

**Example 1** **Use a ruler and a protractor to draw each vector. Include a scale on each diagram.**

1. $\vec{w}$ = 75 miles per hour 40° east of south
2. $\vec{h}$ = 46 feet per second 170° to the horizontal

**Example 2** **Copy the vectors. Then find each sum or difference.**

3. $\vec{c} + \vec{d}$

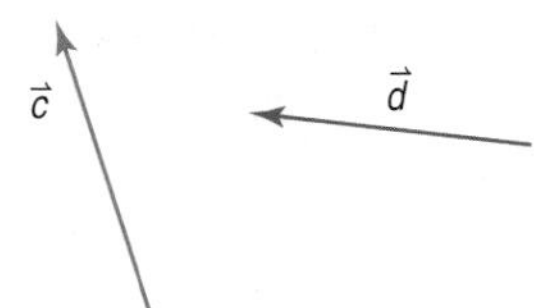


4. $\vec{y} - \vec{z}$

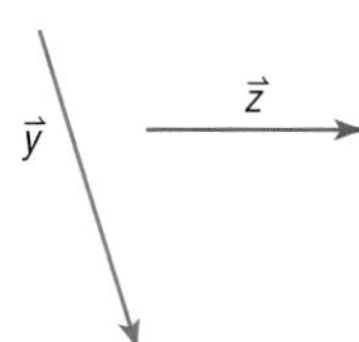


**Example 3** **Write the component form of each vector.**

5.

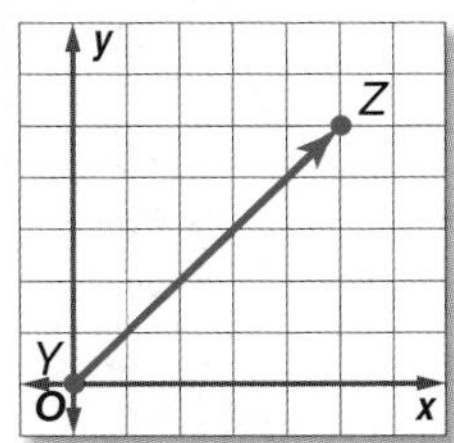


6.

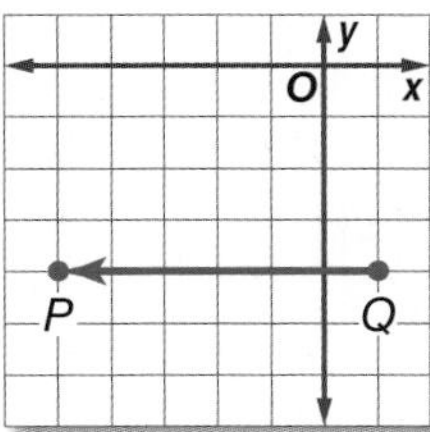


**Example 4** **Find the magnitude and direction of each vector.**

7. $\vec{t} = \langle 2, -4 \rangle$
8. $\vec{f} = \langle -6, -5 \rangle$

**Example 5** **Find each of the following for $\vec{a} = \langle -4, 1 \rangle$, $\vec{b} = \langle -1, -3 \rangle$, and $\vec{c} = \langle 3, 5 \rangle$. Check your answers graphically.**

9. $\vec{c} + \vec{a}$
10. $2\vec{b} - \vec{a}$

**Example 6**

11. **CCSS MODELING** A plane is traveling due north at a speed of 350 miles per hour. If the wind is blowing from the west at a speed of 55 miles per hour, what is the resultant speed and direction that the airplane is traveling?

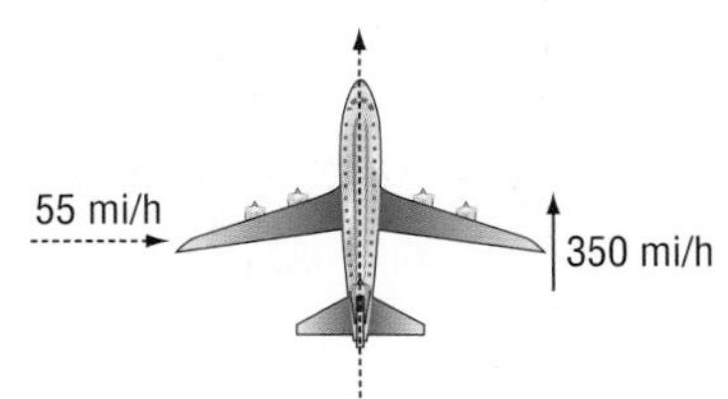


## Practice and Problem Solving

**Example 1** **Use a ruler and a protractor to draw each vector. Include a scale on each diagram.**

12. $\vec{g}$ = 60 inches per second at 145° to the horizontal
13. $\vec{n}$ = 8 meters at an angle of 24° west of south
14. $\vec{a}$ = 32 yards per minute at 78° to the horizontal
15. $\vec{k}$ = 95 kilometers per hour at angle of 65° east of north

**Example 2** **Copy the vectors. Then find each sum or difference.**

**16.** $\vec{t} - \vec{m}$

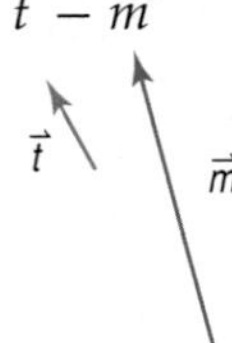

**17.** $\vec{j} - \vec{k}$

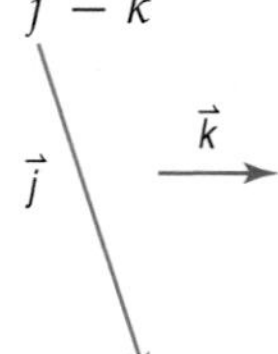

**18.** $\vec{w} + \vec{z}$

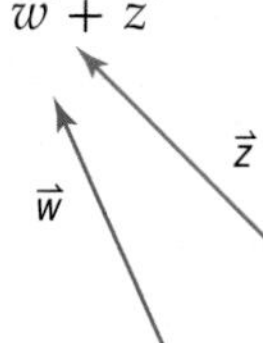

**19.** $\vec{c} + \vec{a}$

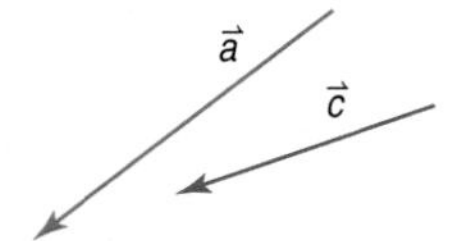

**20.** $\vec{d} - \vec{f}$

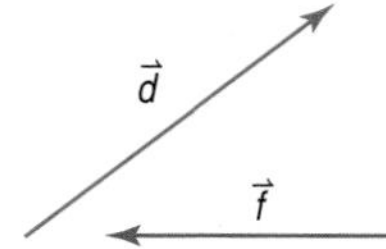

**21.** $\vec{t} - \vec{m}$

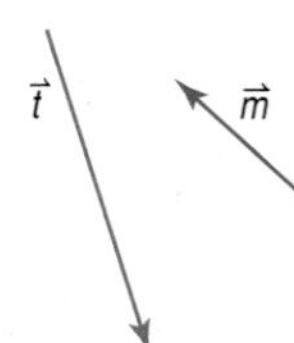

**Example 3** **Write the component form of each vector.**

**22.**

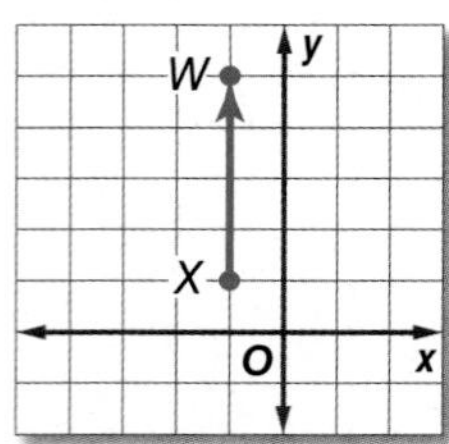

**23.**

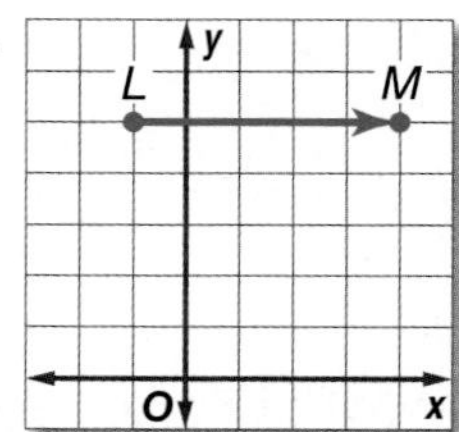

**24.**

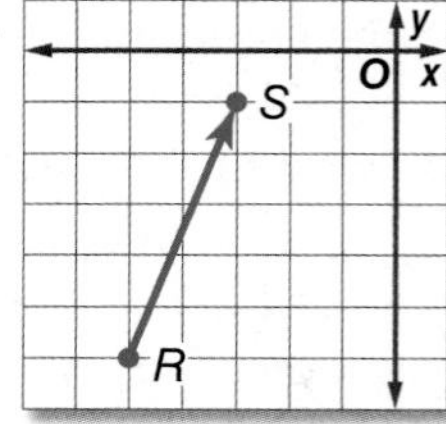

**25.**

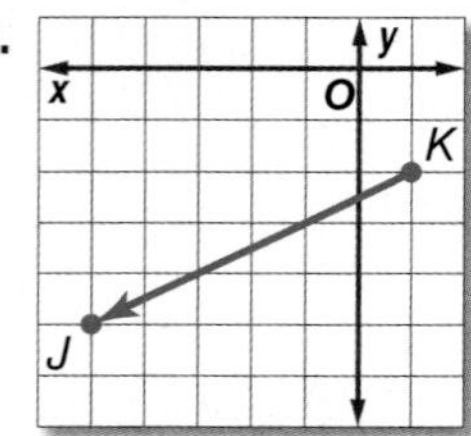

**26.**

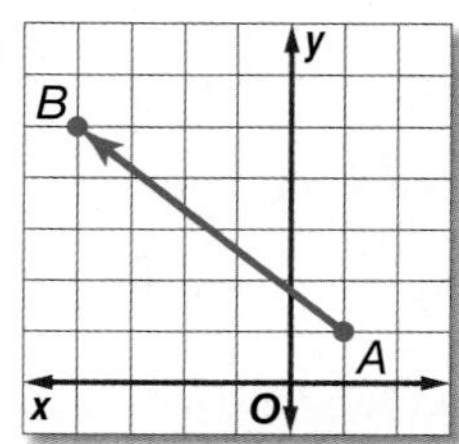

**27.**

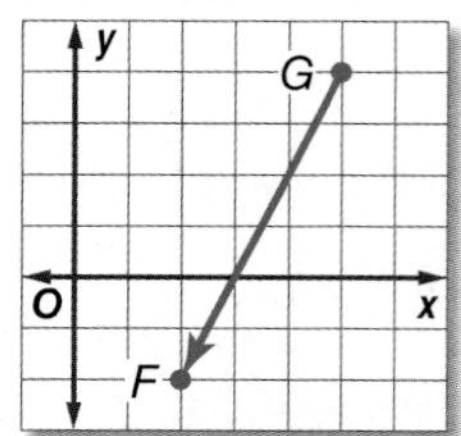

**28. FIREWORKS** The ascent of a firework shell can be modeled using a vector. Write a vector in component form that can be used to describe the path of the firework shown.

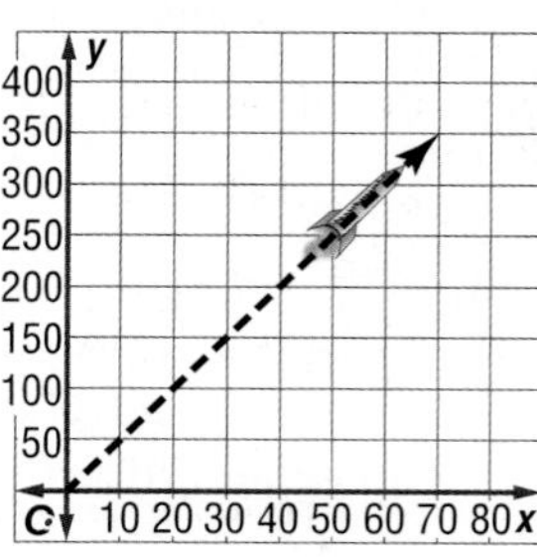

**Example 4** **CCSS SENSE-MAKING** **Find the magnitude and direction of each vector.**

**29.** $\vec{c} = \langle 5, 3 \rangle$

**30.** $\vec{m} = \langle 2, 9 \rangle$

**31.** $\vec{z} = \langle -7, 1 \rangle$

**32.** $\vec{d} = \langle 4, -8 \rangle$

**33.** $\vec{k} = \langle -3, -6 \rangle$

**34.** $\vec{q} = \langle -9, -4 \rangle$

**Example 5** **Find each of the following for $\vec{a} = \langle -3, -5 \rangle$, $\vec{b} = \langle 2, 4 \rangle$, and $\vec{c} = \langle 3, -1 \rangle$. Check your answers graphically.**

**35.** $\vec{b} + \vec{c}$

**36.** $\vec{c} + \vec{a}$

**37.** $\vec{b} - \vec{c}$

**38.** $\vec{a} - \vec{c}$

**39.** $2\vec{c} - \vec{a}$

**40.** $2\vec{b} + \vec{c}$

41. **HIKING** Amy hiked due east for 2 miles and then hiked due south for 3 miles.

  a. Draw a diagram to represent the situation, where $\vec{r}$ is the resultant vector.

  b. How far and in what direction is Amy from her starting position?

42. **EXERCISE** A runner's velocity is 6 miles per hour due east, with the wind blowing 2 miles per hour due north.

  a. Draw a diagram to represent the situation, where $\vec{r}$ is the resultant vector.

  b. What is the resultant velocity of the runner?

**Find each of the following for $\vec{f} = \langle -4, -2 \rangle$, $\vec{g} = \langle 6, 1 \rangle$, and $\vec{h} = \langle 2, -3 \rangle$.**

43. $\vec{f} + \vec{g} + \vec{h}$

44. $\vec{h} - 2\vec{f} + \vec{g}$

45. $2\vec{g} - 3\vec{f} + \vec{h}$

46. **HOMECOMING** Nikki is on a committee to help plan her school's homecoming parade. The parade starts at the high school and continues as shown.

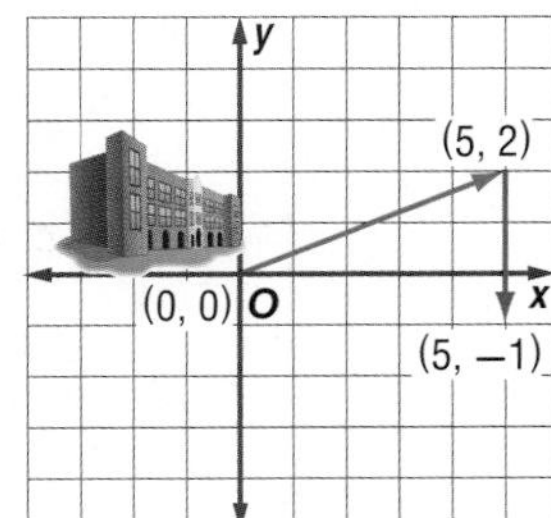


  a. Find the magnitude and direction of the vector formed with an initial point at the school and terminal point at the end of the parade.

  b. Find the length of the parade if 1 unit = 0.25 mile.

47. **SWIMMING** Jonas is swimming from the east bank to the west bank of a stream at a speed of 3.3 feet per second. The stream is 80 feet wide and flows south. If Jonas crosses the stream in 20 seconds, what is the speed of the current?

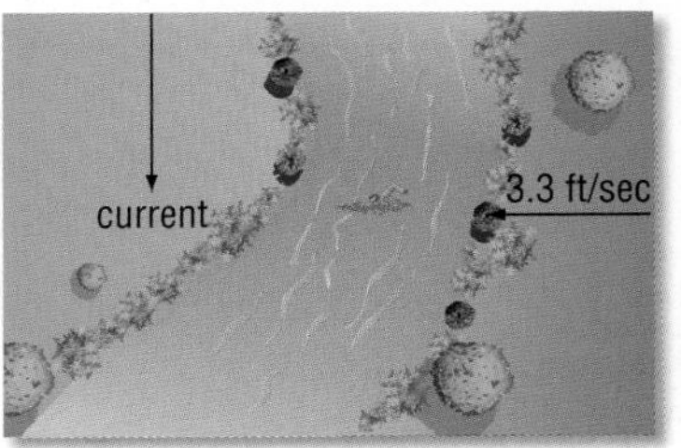


## H.O.T. Problems Use Higher-Order Thinking Skills

48. **CHALLENGE** Find the coordinates of point $P$ on $\overrightarrow{AB}$ that partitions the segment into the given ratio $AP$ to $PB$.

  a. $A(0, 0)$, $B(0, 6)$, 2 to 1

  b. $A(0, 0)$, $B(-15, 0)$, 2 to 3

49. **CCSS PRECISION** Are parallel vectors *sometimes*, *always*, or *never* opposite vectors? Explain.

**PROOF Prove each vector property. Let $\vec{a} = \langle x_1, y_1 \rangle$ and $\vec{b} = \langle x_2, y_2 \rangle$.**

50. commutative: $\vec{a} + \vec{b} = \vec{b} + \vec{a}$

51. scalar multiplication: $k(\vec{a} + \vec{b}) = k\vec{a} + k\vec{b}$, where $k$ is a scalar

52. **OPEN ENDED** Draw a set of parallel vectors.

  a. Find the sum of the two vectors. What is true of the direction of the vector representing the sum?

  b. Find the difference of the two vectors. What is true of the direction of the vector representing the difference?

53. **WRITING IN MATH** Compare and contrast the parallelogram and triangle methods of adding vectors.

## Standardized Test Practice

**54. EXTENDED RESPONSE** Sydney parked her car and hiked along two paths described by the vectors $\langle 2, 3 \rangle$ and $\langle 5, -1 \rangle$.

**a.** What vector represents her hike along both paths?

**b.** When she got to the end of the second path, how far is she from her car if the numbers represent miles?

**55.** In right triangle $ABC$ shown below, what is the measure of $\angle A$ to the nearest tenth of a degree?

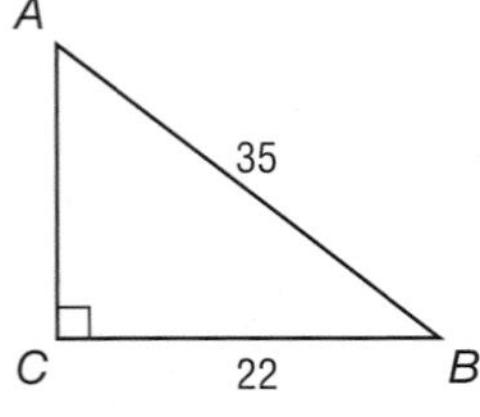


A 32.2    C 51.1
B 38.9    D 57.8

**56. PROBABILITY** A die is rolled. Find the probability of rolling a number greater than 4.

F 0.17    G 0.33    H 0.5    J 0.67

**57. SAT/ACT** Caleb followed the two paths shown below to get to his house $C$ from a store $S$. What is the total distance of the two paths, in meters, from $C$ to $S$?

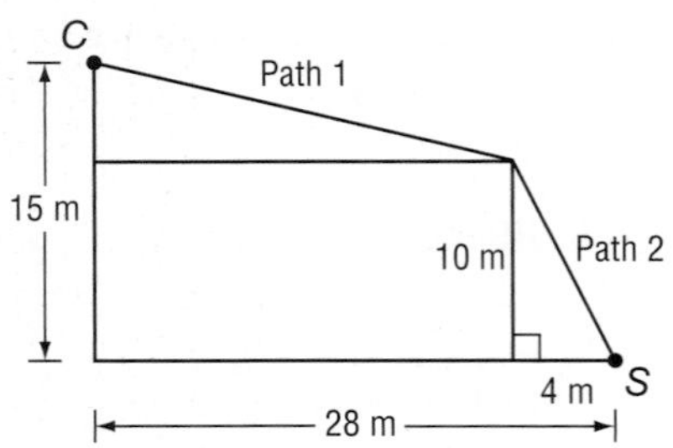


A 10.8 m    D 35.3 m
B 24.5 m    E 38.4 m
C 31.8 m

## Spiral Review

**Find $x$. Round angle measures to the nearest degree and side measures to the nearest tenth.** (Lesson 10-6)

**58.**

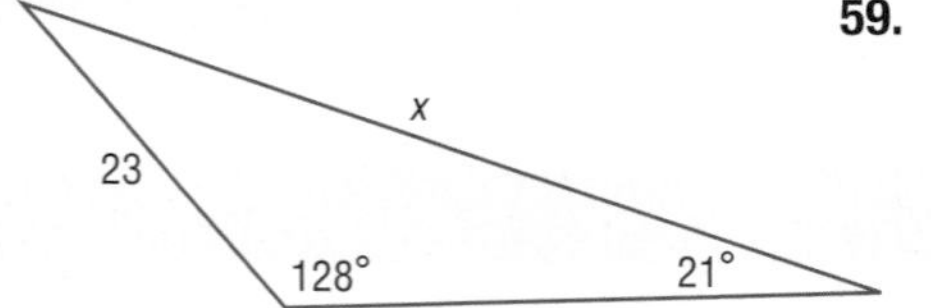


**59.**

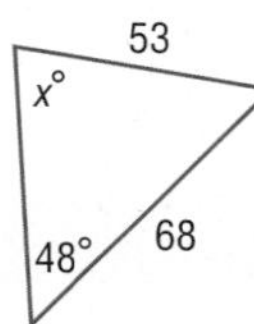


**60.**

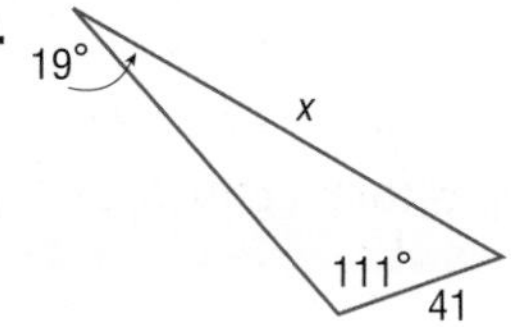


**61. SOCCER** Adelina is in a soccer stadium 80 feet above the field. The angle of depression to the field is 12°. What is the horizontal distance between Adelina and the soccer field? (Lesson 10-5)

**Quadrilateral $WXYZ$ is a rectangle. Find each measure if $m\angle 1 = 30$.** (Lesson 8-4)

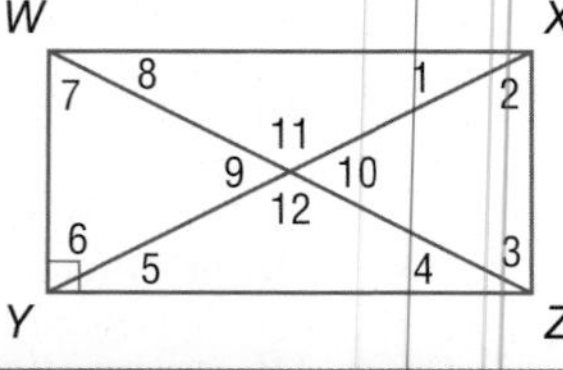


**62.** $m\angle 2$    **63.** $m\angle 8$    **64.** $m\angle 12$

**65.** $m\angle 5$    **66.** $m\angle 6$    **67.** $m\angle 3$

## Skills Review

**Assume that segments and angles that appear to be congruent in each figure are congruent. Indicate which triangles are congruent.**

**68.**

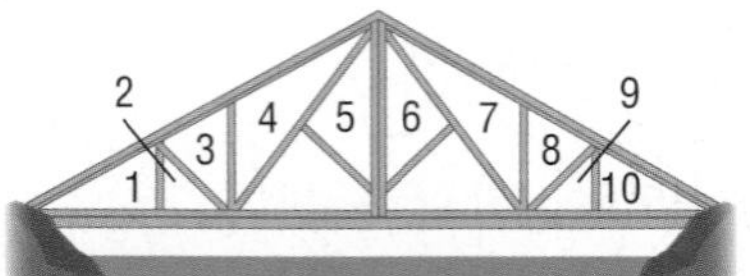

**69.**

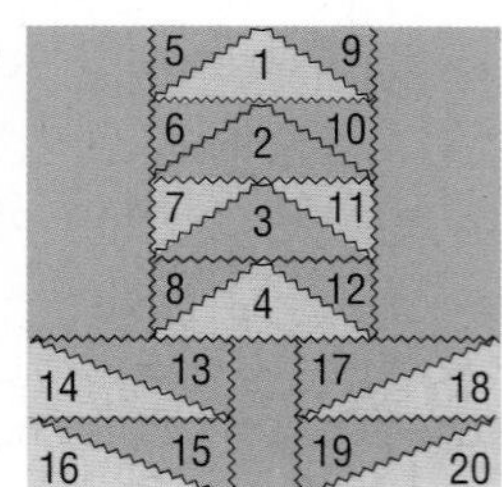

**70.**

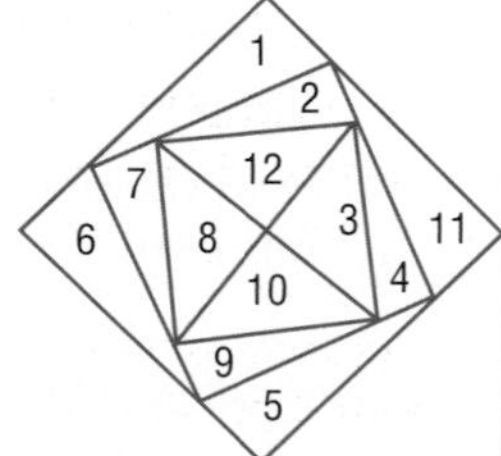

EXTEND 10-7

# Geometry Lab
# Adding Vectors

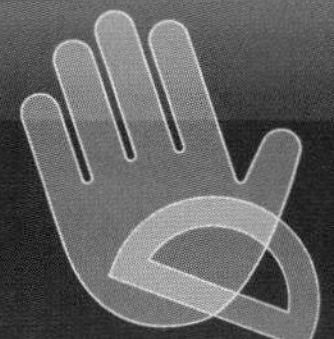

You can use scale drawings to represent vectors and solve problems.

## Activity

**A small aircraft flies due south at an average speed of 175 miles per hour. The wind is blowing 30° south of west at 25 miles per hour. What is the resultant velocity and direction of the plane?**

**Step 1** **Choose a scale.**

Since it is not reasonable to represent the vectors using their actual sizes, you can use a scale drawing. For this activity, let 2 inches represent 100 miles.

**Step 2** **Make a scale drawing.**

Use a ruler and protractor to make a scale drawing of the two vectors.

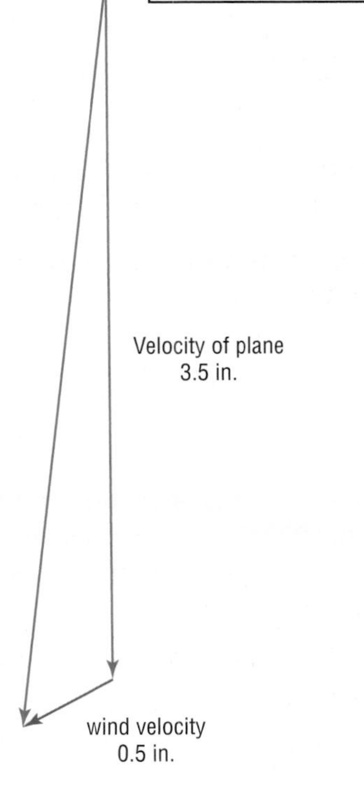


**Step 3** **Find the resultant.**

Find the resultant of the two vectors by using the triangle method or the parallelogram method.

**Step 4** **Measure the resultant.**

Measure the length and angle of the resultant. The resultant length is $3\frac{3}{4}$ inches, and it makes a 7° angle with the vector representing the velocity of the plane.

**Step 5** **Find the magnitude and direction of the resultant.**

Use the scale with the length that you measured in Step 4 to calculate the magnitude of the plane's resultant velocity.

$$3\frac{3}{4}\,\cancel{\text{in.}} \times \frac{100 \text{ mph}}{2\,\cancel{\text{in.}}} = 187.5 \text{ mph}$$

The resultant velocity of the plane is 187.5 miles per hour 7° west of south.

## Exercises

**Make a scale drawing to solve each problem.**

1. **BIKING** Lance is riding his bike west at a velocity of 10 miles per hour. The wind is blowing 5 miles per hour 20° north of east. What is Lance's resultant velocity and direction?

2. **CANOEING** Bianca is traveling due north across a river in a canoe with a current of 3 miles per hour due west. If Bianca can canoe at a rate of 7 miles per hour, what is her resultant velocity and direction?

EXPLORE 10-8

# Graphing Technology Lab Dilations

You can use TI-Nspire Technology to explore properties of dilations.

**CCSS Common Core State Standards**
**Content Standards**
**G.SRT.1** Understand similarity in terms of similarity transformations. Verify experimentally the properties of dilations given by a center and a scale factor:
**a.** A dilation takes a line not passing through the center of the dilation to a parallel line, and leaves a line passing through the center unchanged.
**b.** The dilation of a line segment is longer or shorter in the ratio given by the scale factor.
**Mathematical Practices** 5

## Activity 1 Dilation of a Triangle

**Dilate a triangle by a scale factor of 1.5.**

**Step 1** Add a new **Geometry** page. Then, from the **Points & Lines** menu, use the **Point** tool to add a point and label it *X*.

**Step 2** From the **Shapes** menu, select **Triangle** and specify three points. Label the points *A*, *B*, and *C*.

**Step 3** From the **Actions** menu, use the **Text** tool to separately add the text *Scale Factor* and *1.5* to the page.

**Step 4** From the **Transformation** menu, select **Dilation**. Then select point *X*, $\triangle ABC$, and the text *1.5*.

**Step 5** Label the points on the image *A′*, *B′*, and *C′*.

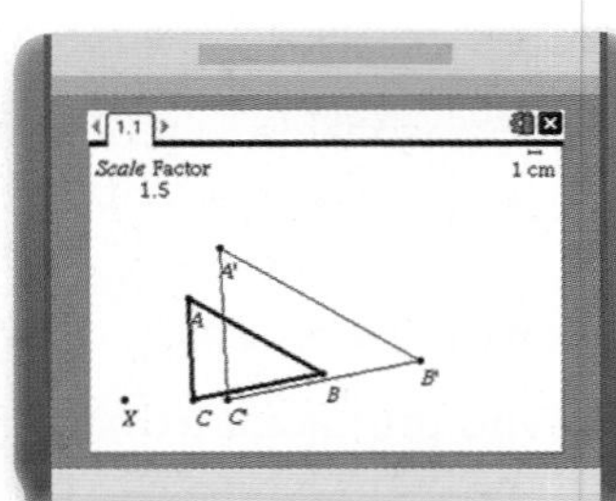


## Analyze the Results

1. Using the **Slope** tool on the **Measurement** menu, describe the effect of the dilation on $\overline{AB}$. That is, how are the lines through $\overline{AB}$ and $\overline{A'B'}$ related?
2. What is the effect of the dilation on the line passing through side $\overline{CA}$?
3. What is the effect of the dilation on the line passing through side $\overline{CB}$?

## Activity 2 Dilation of a Polygon

**Dilate a polygon by a scale factor of −0.5.**

**Step 1** Add a new **Geometry** page and draw polygon *ABCDX* as shown. Add the text *Scale Factor* and *−0.5* to the page.

**Step 2** From the **Transformation** menu, select **Dilation**. Then select point *X*, polygon *ABCDX*, and the text *−0.5*.

**Step 3** Label the points on the image *A′*, *B′*, *C′*, and *D′*.

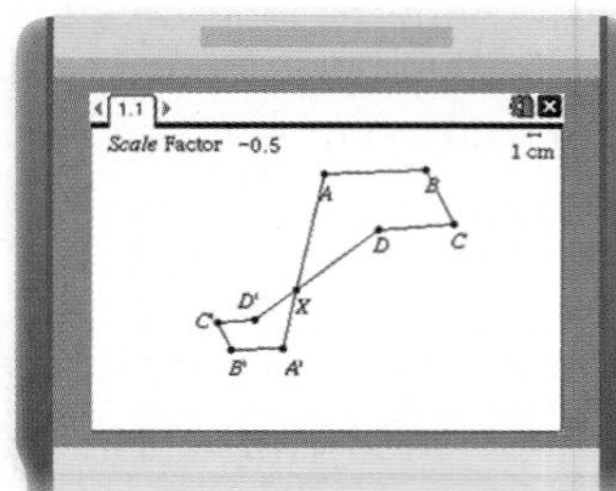


## Model and Analyze

4. Analyze the effect of the dilation in Activity 2 on sides that contain the center of the dilation.
5. Analyze the effect of a dilation of trapezoid *ABCD* shown with a scale factor of 0.75 and the center of the dilation at *A*.
6. **MAKE A CONJECTURE** Describe the effect of a dilation on segments that pass through the center of a dilation and segments that do not pass through the center of a dilation.

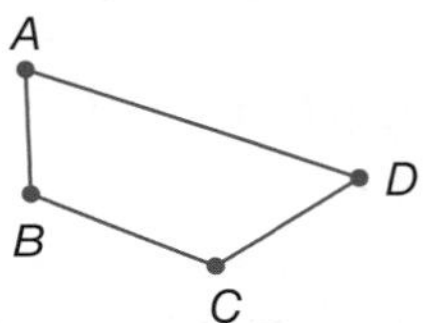

## Activity 3 Dilation of a Segment

**Dilate a segment $\overline{AB}$ by the indicated scale factor.**

**a. scale factor: 0.75**

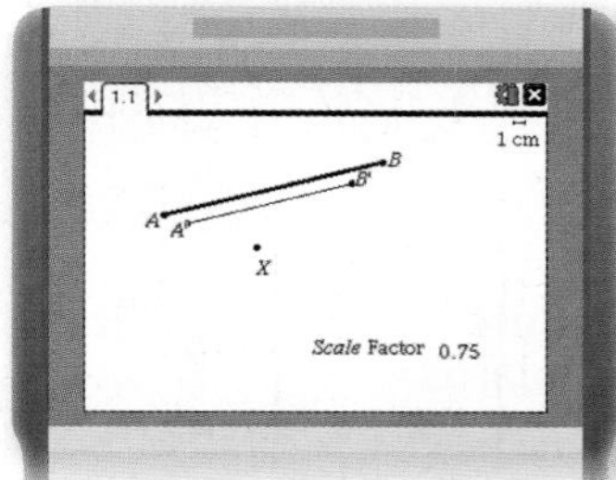


**Step 1** On a new **Geometry** page, draw a line segment using the **Points & Lines** menu. Label the endpoints $A$ and $B$. Then add and label a point $X$.

**Step 2** Add the text *Scale Factor* and *0.75* to the page.

**Step 3** From the **Transformation** menu, select **Dilation**. Then select point $X$, $\overline{AB}$, and the text *0.75*.

**Step 4** Label the dilated segment $\overline{A'B'}$.

**b. scale factor: 1.25**

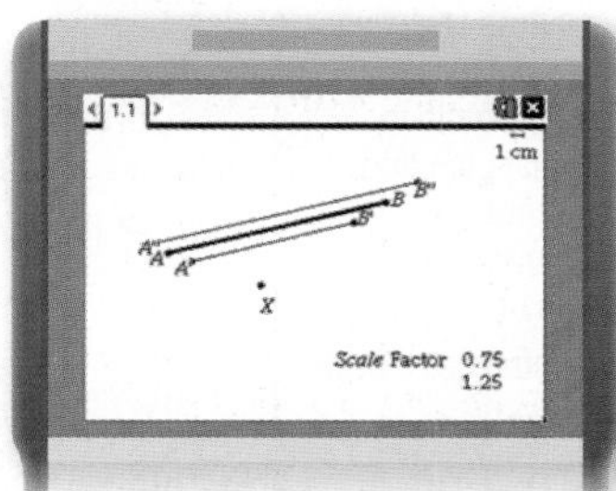


**Step 1** Add the text *1.25* to the page.

**Step 2** From the **Transformation** menu, select **Dilation**. Then select point $X$, $\overline{AB}$, and the text 1.25.

**Step 3** Label the dilated segment $\overline{A''B''}$.

## Model and Analyze

7. Using the **Length** tool on the **Measurement** menu, find the measures of $\overline{AB}$, $\overline{A'B'}$, and $\overline{A''B''}$.

8. What is the ratio of $A'B'$ to $AB$? What is the ratio of $A''B''$ to $AB$?

9. What is the effect of the dilation with scale factor 0.75 on segment $\overline{AB}$? What is the effect of the dilation with scale factor 1.25 on segment $\overline{AB}$?

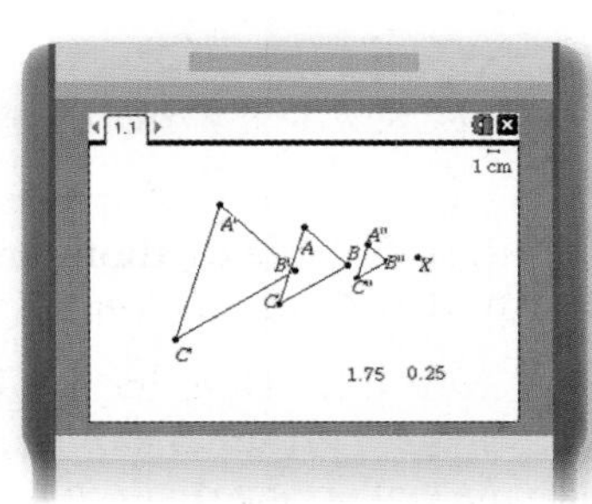


10. Dilate segment $\overline{AB}$ in Activity 3 by scale factors of $-0.75$ and $-1.25$. Describe the effect on the length of each dilated segment.

11. **MAKE A CONJECTURE** Describe the effect of a dilation on the length of a line segment.

12. Describe the dilation from $\overline{AB}$ to $\overline{A'B'}$ and $\overline{A'B'}$ to $\overline{A''B''}$ in the triangles shown.

# LESSON 10-8 Dilations

**Then**

- You identified dilations and verified them as similarity transformations.

**Now**

1. Draw dilations.
2. Draw dilations in the coordinate plane.

**Why?**

- Some photographers still prefer traditional cameras and film to produce negatives. From these negatives, photographers can create scaled reproductions.

**Common Core State Standards**

**Content Standards**

**G.CO.2** Represent transformations in the plane using, e.g., transparencies and geometry software; describe transformations as functions that take points in the plane as inputs and give other points as outputs. Compare transformations that preserve distance and angle to those that do not (e.g., translation versus horizontal stretch).

**G.SRT.1** Understand similarity in terms of similarity transformations. Verify experimentally the properties of dilations given by a center and a scale factor:

a. A dilation takes a line not passing through the center of the dilation to a parallel line, and leaves a line passing through the center unchanged.

b. The dilation of a line segment is longer or shorter in the ratio given by the scale factor.

**Mathematical Practices**

1 Make sense of problems and persevere in solving them.

5 Use appropriate tools strategically.

## 1 Draw Dilations

A dilation or *scaling* is a similarity transformation that enlarges or reduces a figure proportionally with respect to a *center* point and a *scale* factor.

### KeyConcept Dilation

A dilation with center $C$ and positive scale factor $k$, $k \neq 1$, is a function that maps a point $P$ in a figure to its image such that

- if point $P$ and $C$ coincide, then the image and preimage are the same point, or
- if point $P$ is not the center of dilation, then $P'$ lies on $\overrightarrow{CP}$ and $CP' = k(CP)$.

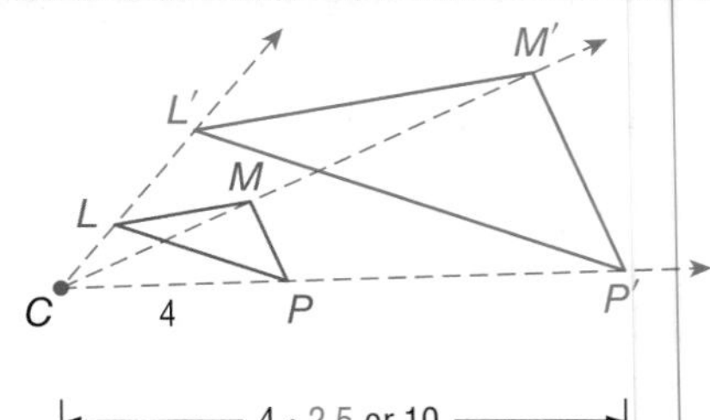


$\triangle L'M'P'$ is the image of $\triangle LMP$ under a dilation with center $C$ and scale factor 2.5.

### Example 1 Draw a Dilation

**Copy $\triangle ABC$ and point $D$. Then use a ruler to draw the image of $\triangle ABC$ under a dilation with center $D$ and scale factor $\frac{1}{2}$.**

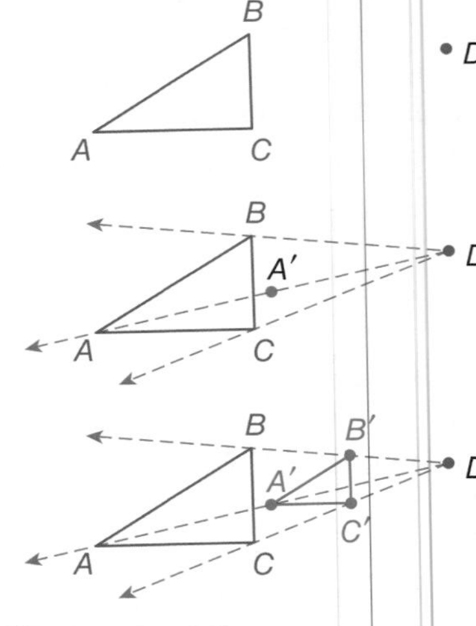


**Step 1** Draw rays from $D$ though each vertex.

**Step 2** Locate $A'$ on $\overrightarrow{DA}$ such that $DA' = \frac{1}{2}DA$.

**Step 3** Locate $B'$ on $\overrightarrow{DB}$ and $C'$ on $\overrightarrow{DC}$ in the same way. Then draw $\triangle A'B'C'$.

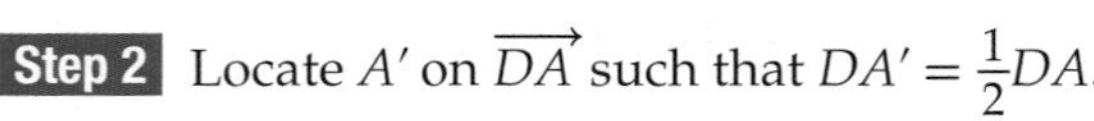

**GuidedPractice**

**Copy the figure and point $J$. Then use a ruler to draw the image of the figure under a dilation with center $J$ and the scale factor $k$ indicated.**

**1A.** $k = \frac{3}{2}$

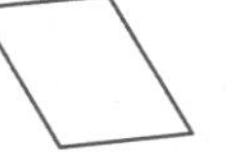


**1B.** $k = 0.75$

John Lund/Stone/Getty Images

In Lesson 9-6, you also learned that if $k > 1$, then the dilation is an *enlargement*. If $0 < k < 1$, then the dilation is a *reduction*. Since $\frac{1}{2}$ is between 0 and 1, the dilation in Example 1 is a reduction.

A dilation with a scale factor of 1 is called an *isometry dilation*. It produces an image that coincides with the preimage. The two figures are congruent.

## Real-World Example 2 Find the Scale Factor of a Dilation

PT

**PHOTOGRAPHY To create different-sized prints, you can adjust the distance between a film negative and the enlarged print by using a photographic enlarger. Suppose the distance between the light source $C$ and the negative is 45 millimeters ($CP$). To what distance $PP'$ should you adjust the enlarger to create a 22.75-centimeter wide print ($X'Y'$) from a 35-millimeter wide negative ($XY$)?**

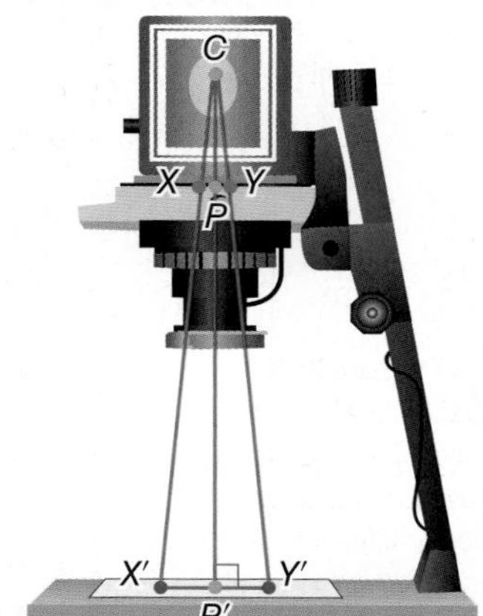


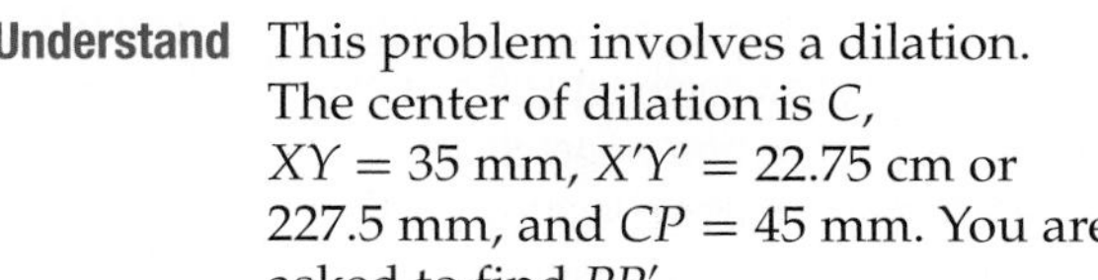

**Understand** This problem involves a dilation. The center of dilation is $C$, $XY = 35$ mm, $X'Y' = 22.75$ cm or 227.5 mm, and $CP = 45$ mm. You are asked to find $PP'$.

**Plan** Find the scale factor of the dilation from the preimage $XY$ to the image $X'Y'$. Use the scale factor to find $CP'$ and then use $CP$ and $CP'$ to find $PP'$.

**Solve** The scale factor $k$ of the enlargement is the ratio of a length on the image to a corresponding length on the preimage.

| | |
|---|---|
| $k = \frac{\text{image length}}{\text{preimage length}}$ | Scale factor of image |
| $= \frac{X'Y'}{XY}$ | image $= X'Y'$, preimage $= XY$ |
| $= \frac{227.5}{35}$ or 6.5 | Divide. |

Use this scale factor of 6.5 to find $CP'$.

| | |
|---|---|
| $CP' = k(CP)$ | Definition of dilation |
| $= 6.5(45)$ | $k = 6.5$ and $CP = 45$ |
| $= 292.5$ | Multiply. |

Use $CP'$ and $CP$ to find $PP'$.

| | |
|---|---|
| $CP + PP' = CP'$ | Segment Addition |
| $45 + PP' = 292.5$ | $CP = 45$ and $CP' = 292.5$ |
| $PP' = 247.5$ | Subtract 45 from each side. |

So the enlarger should be adjusted so that the distance from the negative to the enlarged print ($PP'$) is 247.5 millimeters or 24.75 centimeters.

**Check** Since the dilation is an enlargement, the scale factor should be greater than 1. Since $6.5 > 1$, the scale factor found is reasonable. ✓

**Problem-Solving Tip**

Perseverance

To prevent careless errors in your calculations, estimate the answer to a problem before solving. In Example 2, you can estimate the scale factor of the dilation to be about $\frac{240}{40}$ or 6. Then $CP'$ would be about $6 \cdot 50$ or 300 and $PP'$ about $300 - 50$ or 250 millimeters, which is 25 centimeters. A measure of 24.75 centimeters is close to this estimate, so the answer is reasonable.

## GuidedPractice

**2.** Determine whether the dilation from Figure $Q$ to $Q'$ is an *enlargement* or a *reduction*. Then find the scale factor of the dilation and $x$.

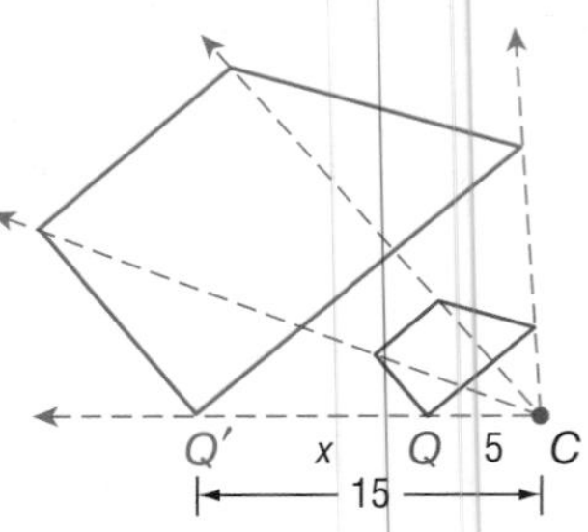


## 2 Dilations in the Coordinate Plane

**Dilations in the Coordinate Plane** You can use the following rules to find the image of a figure after a dilation centered at the origin.

### StudyTip

**Negative Scale Factors** Dilations can also have negative scale factors. You will investigate this type of dilation in Exercise 36.

### KeyConcept Dilations in the Coordinate Plane

**Words** To find the coordinates of an image after a dilation centered at the origin, multiply the *x*- and *y*-coordinates of each point on the preimage by the scale factor of the dilation, *k*.

**Symbols** $(x, y) \rightarrow (kx, ky)$

**Example**

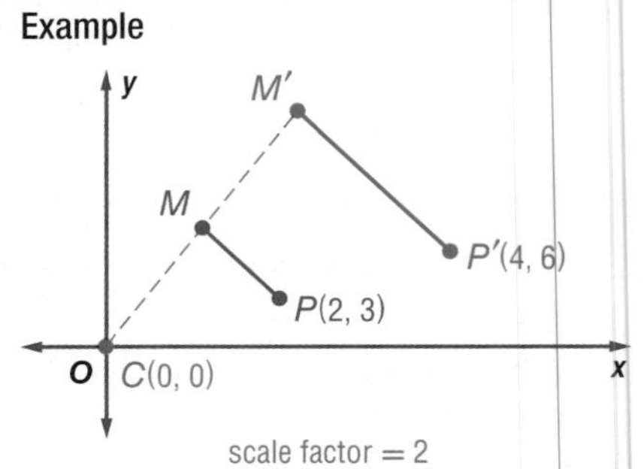

### Example 3 Dilations in the Coordinate Plane

**Quadrilateral *JKLM* has vertices *J*(−2, 4), *K*(−2, −2), *L*(−4, −2), and *M*(−4, 2). Graph the image of *JKLM* after a dilation centered at the origin with a scale factor of 2.5.**

Multiply the $x$- and $y$-coordinates of each vertex by the scale factor, 2.5.

| | | |
|---|---|---|
| $(x, y)$ | $\rightarrow$ | $(2.5x, 2.5y)$ |
| $J(-2, 4)$ | $\rightarrow$ | $J'(-5, 10)$ |
| $K(-2, -2)$ | $\rightarrow$ | $K'(-5, -5)$ |
| $L(-4, -2)$ | $\rightarrow$ | $L'(-10, -5)$ |
| $M(-4, 2)$ | $\rightarrow$ | $M'(-10, 5)$ |

Graph *JKLM* and its image *J'K'L'M'*.

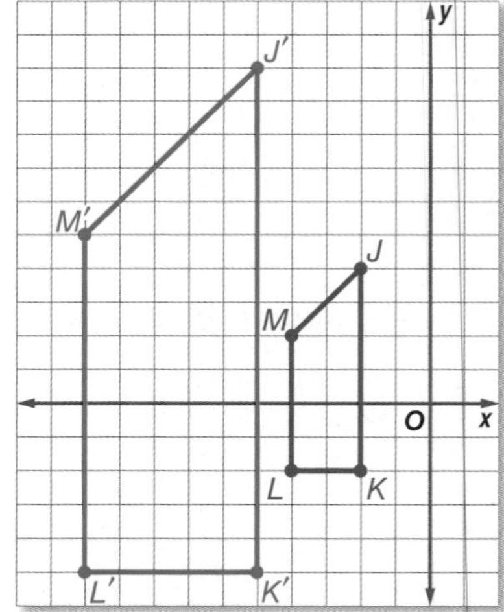


## GuidedPractice

**Find the image of each polygon with the given vertices after a dilation centered at the origin with the given scale factor.**

**3A.** $Q(0, 6), R(-6, -3), S(6, -3); k = \frac{1}{3}$

**3B.** $A(2, 1), B(0, 3), C(-1, 2), D(0, 1); k = 2$

## Check Your Understanding

**Example 1** **Copy the figure and point $M$. Then use a ruler to draw the image of the figure under a dilation with center $M$ and the scale factor $k$ indicated.**

**1.** $k = \frac{1}{4}$

**2.** $k = 2$

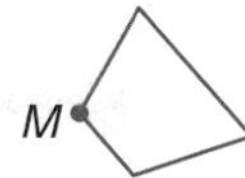


**Example 2** **3** Determine whether the dilation from Figure $B$ to $B'$ is an *enlargement* or a *reduction*. Then find the scale factor of the dilation and $x$.

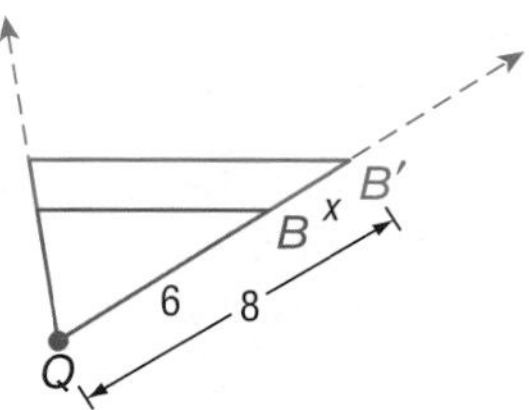


**4.** **BIOLOGY** Under a microscope, a single-celled organism 200 microns in length appears to be 50 millimeters long. If 1 millimeter = 1000 microns, what magnification setting (scale factor) was used? Explain your reasoning.

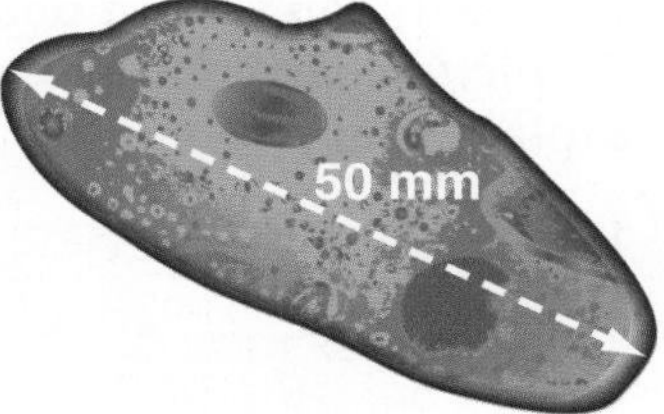


**Example 3** **Graph the image of each polygon with the given vertices after a dilation centered at the origin with the given scale factor.**

**5.** $W(0, 0)$, $X(6, 6)$, $Y(6, 0)$; $k = 1.5$

**6.** $Q(-4, 4)$, $R(-4, -4)$, $S(4, -4)$, $T(4, 4)$; $k = \frac{1}{2}$

**7.** $A(-1, 4)$, $B(2, 4)$, $C(3, 2)$, $D(-2, 2)$; $k = 2$

**8.** $J(-2, 0)$, $K(2, 4)$, $L(8, 0)$, $M(2, -4)$; $k = \frac{3}{4}$

## Practice and Problem Solving

**Example 1** CCSS **TOOLS** **Copy the figure and point $S$. Then use a ruler to draw the image of the figure under a dilation with center $S$ and the scale factor $k$ indicated.**

**9.** $k = \frac{5}{2}$

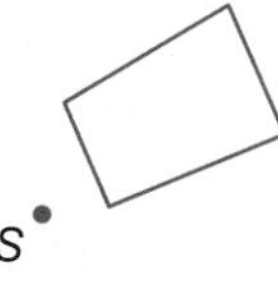


**10.** $k = 3$

**11.** $k = 0.8$

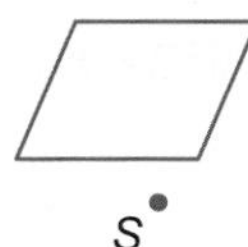


**12.** $k = \frac{1}{3}$

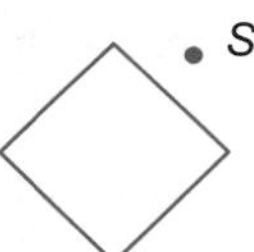


**13.** $k = 2.25$

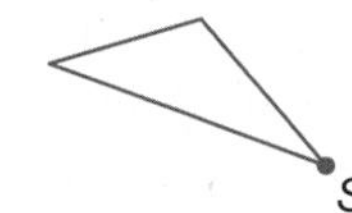


**14.** $k = \frac{7}{4}$

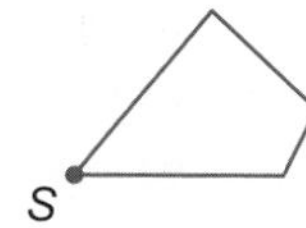

**Example 2** **Determine whether the dilation from figure $W$ to $W'$ is an *enlargement* or a *reduction*. Then find the scale factor of the dilation and $x$.**

**15.**

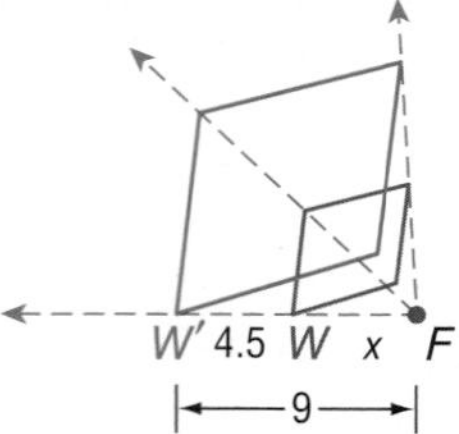


**16.**

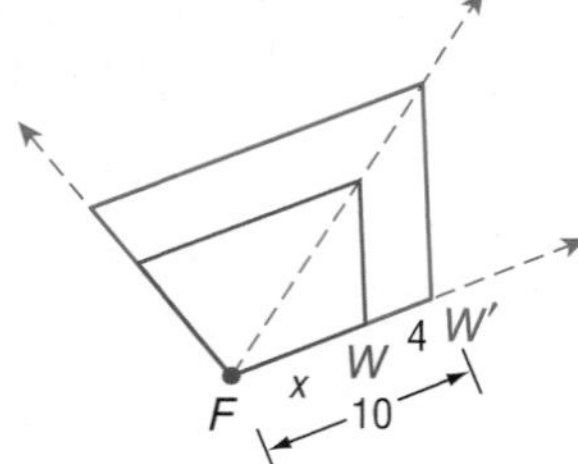


**17.**

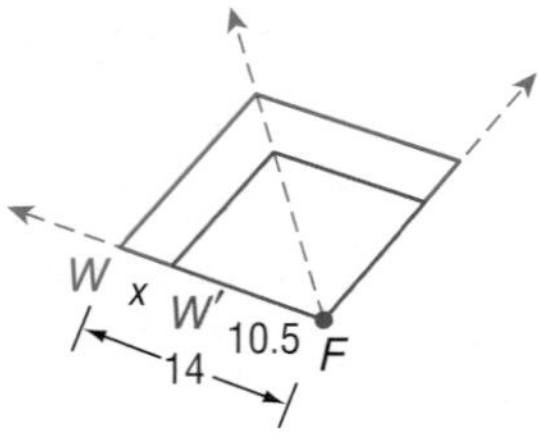


**18.**

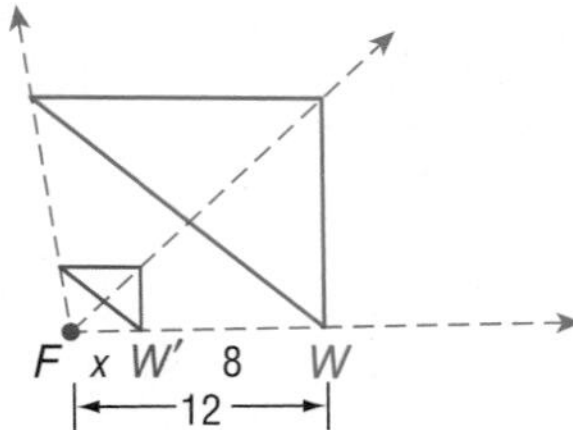


**INSECTS** **When viewed under a microscope, each insect has the measurement given on the picture. Given the actual measure of each insect, what magnification was used? Explain your reasoning.**

**19.**

**Cat Flea**
Actual Length: 2.5 mm

**20.**

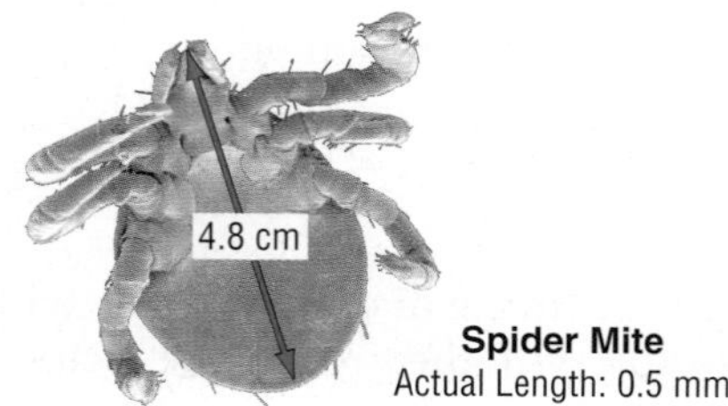


**Spider Mite**
Actual Length: 0.5 mm

**Example 3** **CCSS SENSE-MAKING** **Find the image of each polygon with the given vertices after a dilation centered at the origin with the given scale factor.**

**21.** $J(-8, 0)$, $K(-4, 4)$, $L(-2, 0)$; $k = 0.5$

**22.** $S(0, 0)$, $T(-4, 0)$, $V(-8, -8)$; $k = 1.25$

**23.** $A(9, 9)$, $B(3, 3)$, $C(6, 0)$; $k = \frac{1}{3}$

**24.** $D(4, 4)$, $F(0, 0)$, $G(8, 0)$; $k = 0.75$

**25.** $M(-2, 0)$, $P(0, 2)$, $Q(2, 0)$, $R(0, -2)$; $k = 2.5$

**26.** $W(2, 2)$, $X(2, 0)$, $Y(0, 1)$, $Z(1, 2)$; $k = 3$

**27.** **COORDINATE GEOMETRY** Refer to the graph of $FGHJ$.

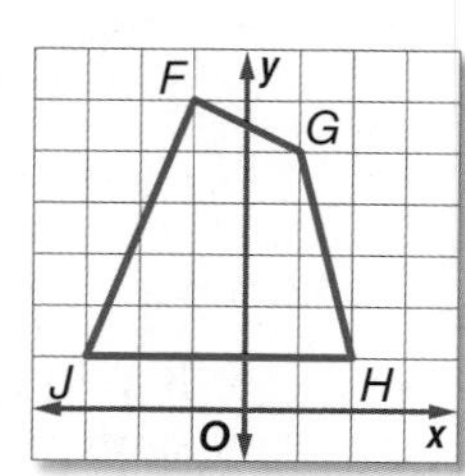


**a.** Dilate $FGHJ$ by a scale factor of $\frac{1}{2}$ centered at the origin, and then reflect the dilated image in the $y$-axis.

**b.** Complete the composition of transformations in part **a** in reverse order.

**c.** Does the order of the transformations affect the final image?

**d.** Will the order of a composition of a dilation and a reflection *always*, *sometimes*, or *never* affect the dilated image? Explain your reasoning.

28. **PHOTOGRAPHY AND ART** To make a scale drawing of a photograph, students overlay a $\frac{1}{4}$-inch grid on a 5-inch by 7-inch high contrast photo, overlay a $\frac{1}{2}$-inch grid on a 10-inch by 14-inch piece of drawing paper, and then sketch the image in each square of the photo to the corresponding square on the drawing paper.

   **a.** What is the scale factor of the dilation?

   **b.** To create an image that is 10 times as large as the original, what size grids are needed?

   **c.** What would be the area of a grid drawing of a 5-inch by 7-inch photo that used 2-inch grids?

29. **MEASUREMENT** Determine whether the image shown is a dilation of *ABCD*. Explain your reasoning.

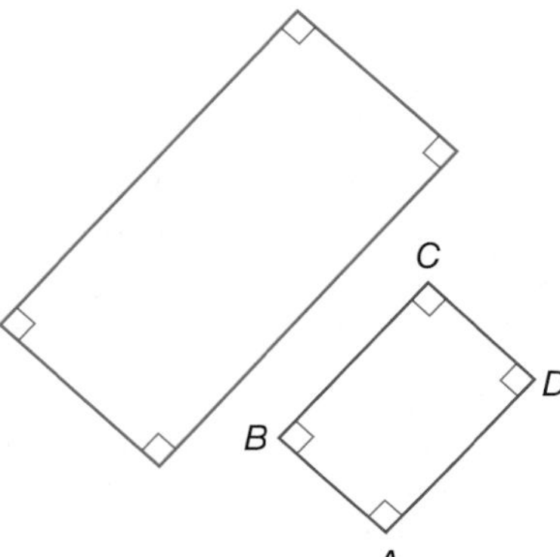


30. **COORDINATE GEOMETRY** *WXYZ* has vertices $W(6, 2)$, $X(3, 7)$, $Y(-1, 4)$, and $Z(4, -2)$.

   **a.** Graph *WXYZ* and find the perimeter of the figure. Round to the nearest tenth.

   **b.** Graph the image of *WXYZ* after a dilation of $\frac{1}{2}$ centered at the origin.

   **c.** Find the perimeter of the dilated image. Round to the nearest tenth. How is the perimeter of the dilated image related to the perimeter of *WXYZ*?

31. **CHANGING DIMENSIONS** A three-dimensional figure can also undergo a dilation. Consider the rectangular prism shown.

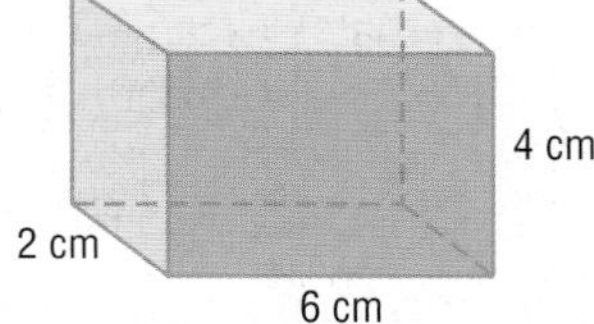


   **a.** Find the surface area and volume of the prism.

   **b.** Find the surface area and volume of the prism after a dilation with a scale factor of 2.

   **c.** Find the surface area and volume of the prism after a dilation with a scale factor of $\frac{1}{2}$.

   **d.** How many times as great is the surface area and volume of the image as the preimage after each dilation?

   **e.** Make a conjecture as to the effect a dilation with a positive scale factor *r* would have on the surface area and volume of a prism.

32. **CCSS PERSEVERANCE** Refer to the graph of $\triangle DEF$.

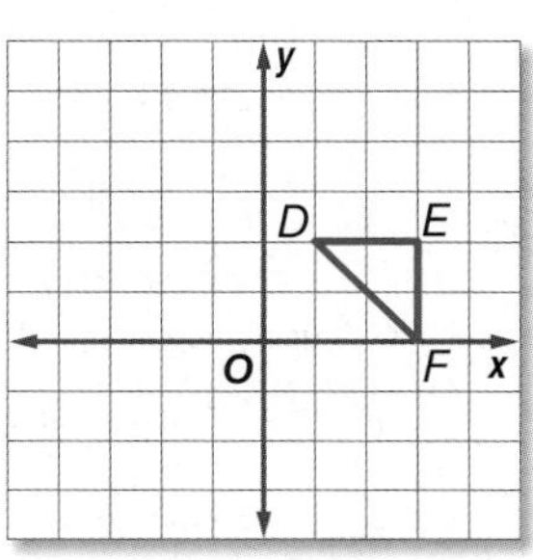


   **a.** Graph the dilation of $\triangle DEF$ centered at point *D* with a scale factor of 3.

   **b.** Describe the dilation as a composition of transformations including a dilation with a scale factor of 3 centered at the origin.

   **c.** If a figure is dilated by a scale factor of 3 with a center of dilation $(x, y)$, what composition of transformations, including a dilation with a scale factor of 3 centered at the origin, will produce the same final image?

**33.** **HEALTH** A coronary artery may be dilated with a balloon catheter as shown. The cross section of the middle of the balloon is a circle.

a. A surgeon inflates a balloon catheter in a patient's coronary artery, dilating the balloon from a diameter of 1.5 millimeters to 2 millimeters. Find the scale factor of this dilation.

b. Find the cross-sectional area of the balloon before and after the dilation.

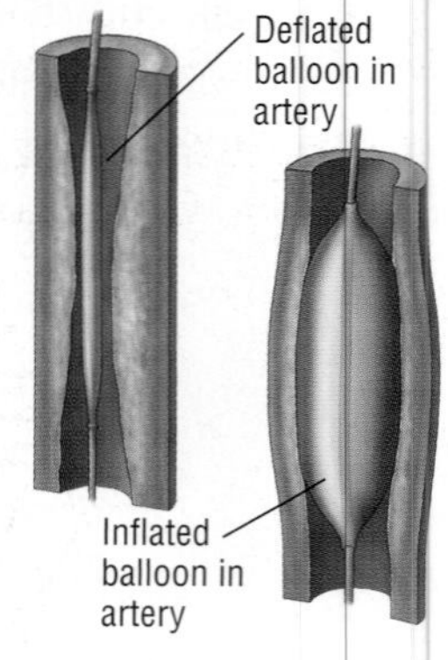


**Each figure shows a preimage and its image after a dilation centered at point *P*. Copy each figure, locate point *P*, and estimate the scale factor.**

**34.**

**35.**

**36.** **MULTIPLE REPRESENTATIONS** In this problem, you will investigate dilations centered at the origin with negative scale factors.

a. **Geometric** Draw $\triangle ABC$ with points $A(-2, 0)$, $B(2, -4)$, and $C(4, 2)$. Then draw the image of $\triangle ABC$ after a dilation centered at the origin with a scale factor of $-2$. Repeat the dilation with scale factors of $-\frac{1}{2}$ and $-3$. Record the coordinates for each dilation.

b. **Verbal** Make a conjecture about the function relationship for a dilation centered at the origin with a negative scale factor.

c. **Analytical** Write the function rule for a dilation centered at the origin with a scale factor of $-k$.

d. **Verbal** Describe a dilation centered at the origin with a negative scale factor as a composition of transformations.

## H.O.T. Problems Use Higher-Order Thinking Skills

**37.** **CHALLENGE** Find the equation for the dilated image of the line $y = 4x - 2$ if the dilation is centered at the origin with a scale factor of 1.5.

**38.** **WRITING IN MATH** Are parallel lines (parallelism) and collinear points (collinearity) preserved under all transformations? Explain.

**39.** **CCSS ARGUMENTS** Determine whether invariant points are *sometimes*, *always*, or *never* maintained for the transformations described below. If so, describe the invariant point(s). If not, explain why invariant points are not possible.

a. dilation of $ABCD$ with scale factor 1

b. rotation of $\overline{AB}$ 74° about $B$

c. reflection of $\triangle MNP$ in the $x$-axis

d. translation of $PQRS$ along $\langle 7, 3 \rangle$

e. dilation of $\triangle XYZ$ centered at the origin with scale factor 2

**40.** **OPEN ENDED** Graph a triangle. Dilate the triangle so that its area is four times the area of the original triangle. State the scale factor and center of your dilation.

**41.** **WRITING IN MATH** Can you use transformations to create congruent figures, similar figures, and equal figures? Explain.

## Standardized Test Practice

**42. EXTENDED RESPONSE** Quadrilateral $PQRS$ was dilated to form quadrilateral $WXYZ$.

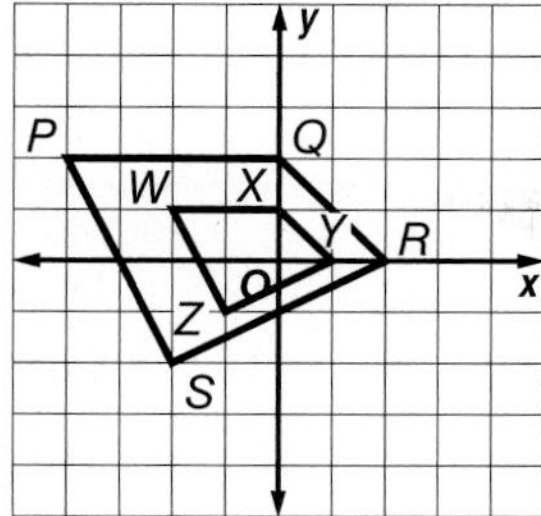


**a.** Is the dilation from $PQRS$ to $WXYZ$ an enlargement or reduction?

**b.** Which number *best* represents the scale factor for this dilation?

**43. ALGEBRA** How many ounces of pure water must a pharmacist add to 50 ounces of a 15% saline solution to make a solution that is 10% saline?

**A** 25 **C** 15

**B** 20 **D** 5

**44.** Tionna wants to replicate a painting in an art museum. The painting is 3 feet wide and 6 feet long. She decides on a dilation reduction factor of 0.25. What size paper should she use?

**F** 4 in. × 8 in. **H** 8 in. × 16 in.

**G** 6 in. × 12 in. **J** 10 in. × 20 in.

**45. SAT/ACT** For all $x$, $(x - 7)^2 = ?$

**A** $x^2 - 49$ **D** $x^2 - 14x + 49$

**B** $x^2 + 49$ **E** $x^2 + 14x - 49$

**C** $x^2 - 14x - 49$

## Spiral Review

**46. REAL ESTATE** A house is built on a triangular plot of land. Two sides of the plot are 160 feet long, and they meet at an angle of 85°. If a fence is to be placed along the perimeter of the property, how much fencing material is needed? (Lesson 10-6)

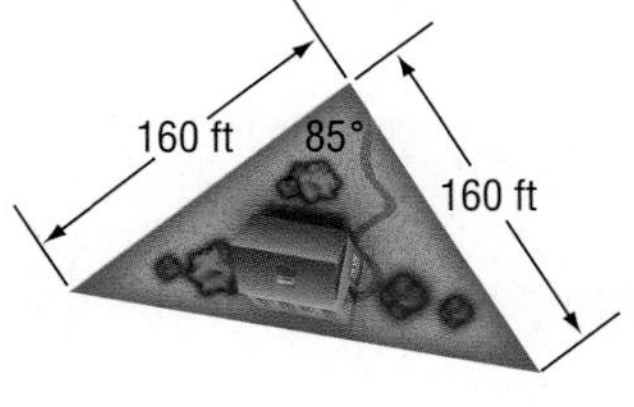


**47. COORDINATE GEOMETRY** In $\triangle LMN$, $\overline{PR}$ divides $\overline{NL}$ and $\overline{MN}$ proportionally. If the vertices are $N(8, 20)$, $P(11, 16)$, and $R(3, 8)$ and $\frac{LP}{PN} = \frac{2}{1}$, find the coordinates of $L$ and $M$. (Lesson 9-4)

**Use the figure at the right to write an inequality relating the given pair of angle or segment measures.** (Lesson 7-6)

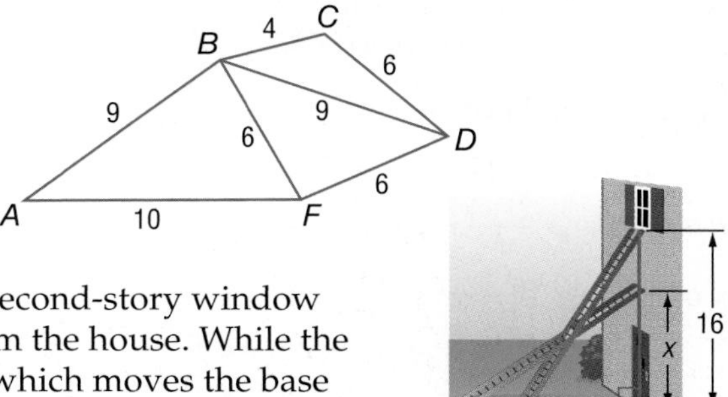


**48.** $AB, FD$

**49.** $m\angle BDC, m\angle FDB$

**50.** $m\angle FBA, m\angle DBF$

**51. PAINTING** A painter sets a ladder up to reach the bottom of a second-story window 16 feet above the ground. The base of the ladder is 12 feet from the house. While the painter mixes the paint, a neighbor's dog bumps the ladder, which moves the base 2 feet farther away from the house. How far up the side of the house does the ladder reach? (Lesson 10-2)

## Skills Review

**Find the value of $x$ to the nearest tenth.**

**52.** $58.9 = 2x$

**53.** $\frac{108.6}{\pi} = x$

**54.** $228.4 = \pi x$

**55.** $\frac{336.4}{x} = \pi$

CHAPTER 10 Study Guide and Review

## Study Guide

### KeyConcepts

**Geometric Mean** (Lesson 10-1)

- For two positive numbers $a$ and $b$, the geometric mean is the positive number $x$ where $a : x = x : b$ is true.

**Pythagorean Theorem** (Lesson 10-2)

- Let $\triangle ABC$ be a right triangle with right angle $C$. Then $a^2 + b^2 = c^2$.

**Special Right Triangles** (Lesson 10-3)

- The measures of the sides of a 45°-45°-90° triangle are $x$, $x$, and $x\sqrt{2}$.
- The measures of the sides of a 30°-60°-90° triangle are $x$, $2x$, and $x\sqrt{3}$.

**Trigonometry** (Lesson 10-4)

- $\sin A = \frac{\text{opposite leg}}{\text{hypotenuse}}$
- $\cos A = \frac{\text{adjacent leg}}{\text{hypotenuse}}$
- $\tan A = \frac{\text{opposite leg}}{\text{adjacent leg}}$

**Angles of Elevation and Depression** (Lesson 10-5)

- An angle of elevation is the angle formed by a horizontal line and the line of sight to an object above.
- An angle of depression is the angle formed by a horizontal line and the line of sight to an object below.

**Laws of Sines and Cosines** (Lesson 10-6)

Let $\triangle ABC$ be any triangle.

- Law of Sines: $\frac{\sin A}{a} = \frac{\sin B}{b} = \frac{\sin C}{c}$
- Law of Cosines: $a^2 = b^2 + c^2 - 2bc\cos A$
  $b^2 = a^2 + c^2 - 2ac\cos B$
  $c^2 = a^2 + b^2 - 2ab\cos C$

**Vectors** (Lesson 10-7)

- A vector is a quantity with both magnitude and direction.

### FOLDABLES StudyOrganizer

Be sure the Key Concepts are noted in your Foldable.

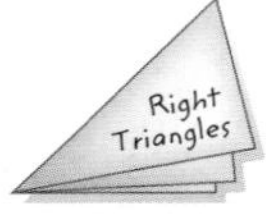


### KeyVocabulary

angle of depression (p. 662)
angle of elevation (p. 662)
component form (p. 684)
cosine (p. 650)
direction (p. 682)
geometric mean (p. 619)
inverse cosine (p. 653)
inverse sine (p. 653)
inverse tangent (p. 653)
Law of Cosines (p. 671)
Law of Sines (p. 670)
magnitude (p. 682)
Pythagorean triple (p. 630)
resultant (p. 683)
sine (p. 650)
standard position (p. 684)
tangent (p. 650)
trigonometric ratio (p. 650)
trigonometry (p. 650)
vector (p. 682)

### VocabularyCheck

**State whether each sentence is *true* or *false*. If *false*, replace the underlined word or phrase to make a true sentence.**

1. The <u>arithmetic</u> mean of two numbers is the positive square root of the product of the numbers.
2. <u>Extended ratios</u> can be used to compare three or more quantities.
3. To find the length of the hypotenuse of a right triangle, take the square root of the <u>difference</u> of the squares of the legs.
4. An angle of <u>elevation</u> is the angle formed by a horizontal line and an observer's line of sight to an object below the horizon.
5. The sum of two vectors is the <u>resultant</u>.
6. Magnitude is the <u>angle a vector makes with the $x$-axis</u>.
7. A vector is in <u>standard position</u> when the initial point is at the origin.
8. The <u>component form</u> of a vector describes the vector in terms of change in $x$ and change in $y$.
9. The <u>Law of Sines</u> can be used to find an angle measure when given three side lengths.
10. A <u>trigonometric ratio</u> is a ratio of the lengths of two sides of a right triangle.

# Lesson-by-Lesson Review

## 10-1 Geometric Mean

Find the geometric mean between each pair of numbers.

**11.** 9 and 4

**12.** $\sqrt{20}$ and $\sqrt{80}$

**13.** $\frac{8\sqrt{2}}{3}$ and $\frac{4\sqrt{2}}{3}$

**14.** Find *x*, *y*, and *z*.

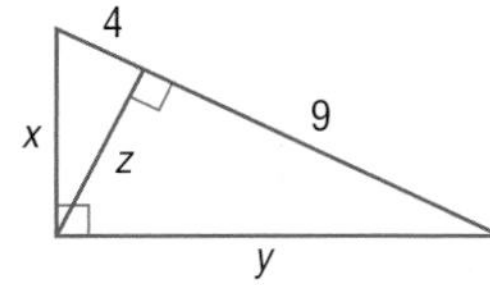


**15.** **DANCES** Mike is hanging a string of lights on his barn for a square dance. Using a book to sight the top and bottom of the barn, he can see he is 15 feet from the barn. If his eye level is 5 feet from the ground, how tall is the barn?

**Example 1**

**Find the geometric mean between 10 and 15.**

| | |
|---|---|
| $x = \sqrt{ab}$ | Definition of geometric mean |
| $= \sqrt{10 \cdot 15}$ | $a = 10$ and $b = 15$ |
| $= \sqrt{(5 \cdot 2) \cdot (3 \cdot 5)}$ | Factor. |
| $= \sqrt{25 \cdot 6}$ | Associative Property |
| $= 5\sqrt{6}$ | Simplify. |

## 10-2 The Pythagorean Theorem and Its Converse

Find *x*.

**16.**

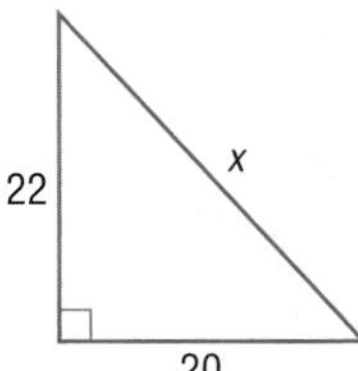


**17.**

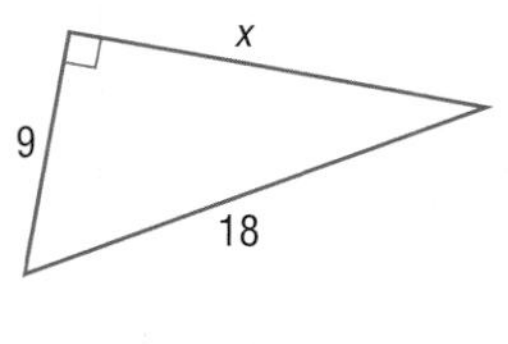


Determine whether each set of numbers can be the measures of the sides of a triangle. If so, classify the triangle as *acute*, *obtuse*, or *right*. Justify your answer.

**18.** 7, 24, 25

**19.** 13, 15, 16

**20.** 65, 72, 88

**21.** **SWIMMING** Alexi walks 27 meters south and 38 meters east to get around a lake. Her sister swims directly across the lake. How many meters to the nearest tenth did Alexi's sister save by swimming?

**Example 2**

**Find *x*.**

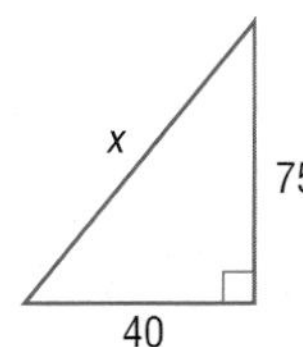


The side opposite the right angle is the hypotenuse, so $c = x$.

| | |
|---|---|
| $a^2 + b^2 = c^2$ | Pythagorean Theorem |
| $40^2 + 75^2 = x^2$ | $a = 40$ and $b = 75$ |
| $7225 = x^2$ | Simplify. |
| $\sqrt{7225} = x$ | Take the positive square root of each side. |
| $85 = x$ | Simplify. |

# Study Guide and Review *Continued*

## 10-3 Special Right Triangles

Find $x$ and $y$.

**22.**

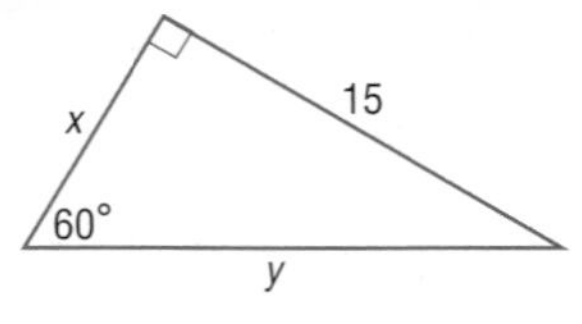


**23.**

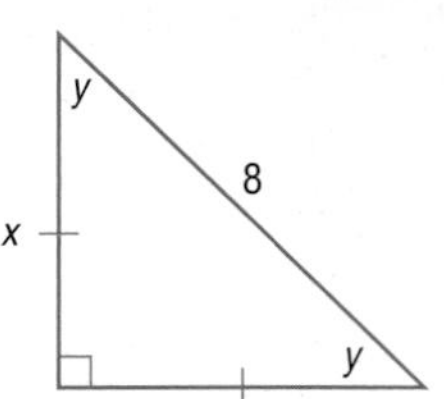


**24.** **CLIMBING** Jason is adding a climbing wall to his little brother's swing-set. If he starts building 5 feet out from the existing structure, and wants it to have a 60° angle, how long should the wall be?

### Example 3

Find $x$ and $y$.

The measure of the third angle in this triangle is 90 – 60 or 30. This is a 30°-60°-90° triangle.

$h = 2s$    30°-60°-90° Triangle Theorem

$20 = 2x$    Substitute.

$10 = x$    Divide.

Now find $y$, the length of the longer leg.

$\ell = s\sqrt{3}$    30°-60°-90° Triangle Theorem

$y = 10\sqrt{3}$    Substitute.

## 10-4 Trigonometry

Express each ratio as a fraction and as a decimal to the nearest hundredth.

**25.** sin $A$    **26.** tan $B$

**27.** sin $B$    **28.** cos $A$

**29.** tan $A$    **30.** cos $B$

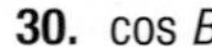

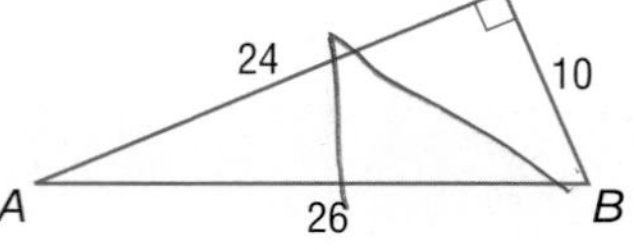


Find $x$.

**31.**

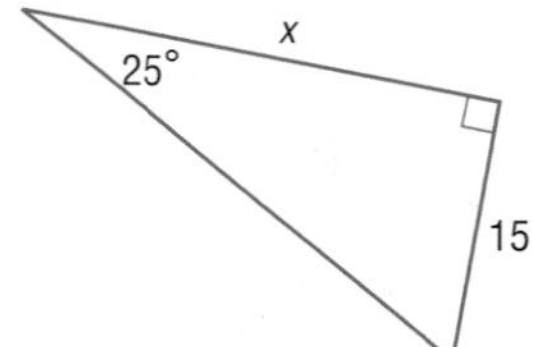


**32.**

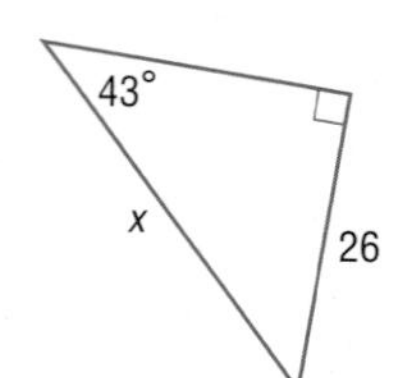


**33.** **GARDENING** Sofia wants to put a flower bed in the corner of her yard by laying a stone border that starts 3 feet from the corner of one fence and ends 6 feet from the corner of the other fence. Find the angles, $x$ and $y$, the fence make with the border.

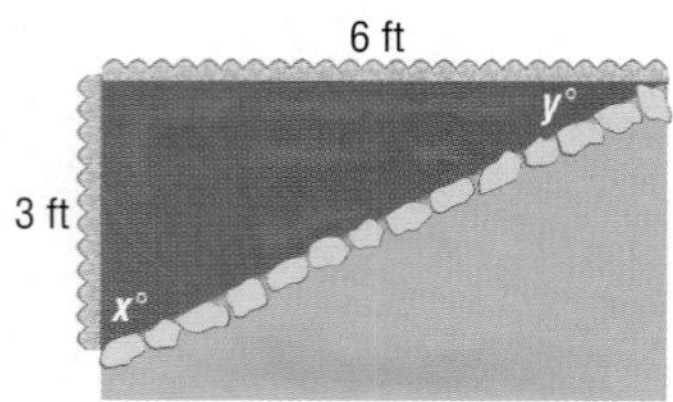


### Example 4

Express each ratio as a fraction and as a decimal to the nearest hundredth.

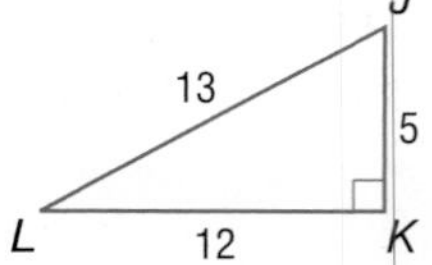


**a.** sin $L$

$\sin L = \frac{5}{13}$ or about 0.38    $\sin L = \frac{\text{opp}}{\text{hyp}}$

**b.** cos $L$

$\cos L = \frac{12}{13}$ or about 0.92    $\cos L = \frac{\text{adj}}{\text{hyp}}$

**c.** tan $L$

$\tan L = \frac{5}{12}$ or 0.42    $\tan L = \frac{\text{opp}}{\text{adj}}$

## 10-5 Angles of Elevation and Depression

**34. JOBS** Tom delivers papers on a rural route from his car. If he throws a paper from a height of 4 feet, and it lands 15 feet from the car, at what angle of depression did he throw the paper to the nearest degree?

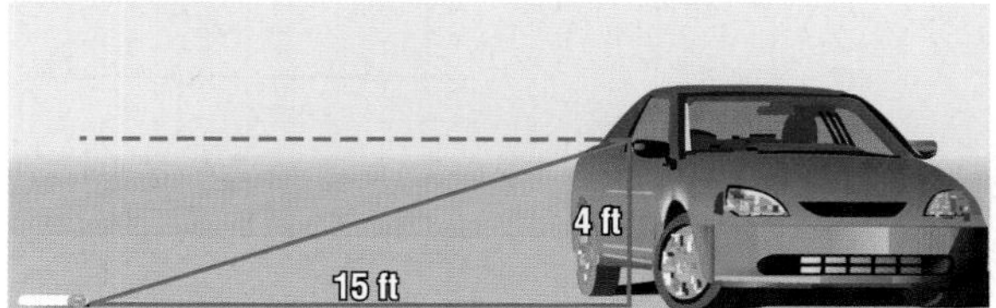


**35. TOWER** There is a cell phone tower in the field across from Jen's house. If Jen walks 50 feet from the tower, and finds the angle of elevation from her position to the top of the tower to be 60°, how tall is the tower?

**Example 5**

**Sarah's cat climbed up a tree. If she sights her cat at an angle of elevation of 40°, and her eyes are 5 feet off the ground, how high up from the ground is her cat?**

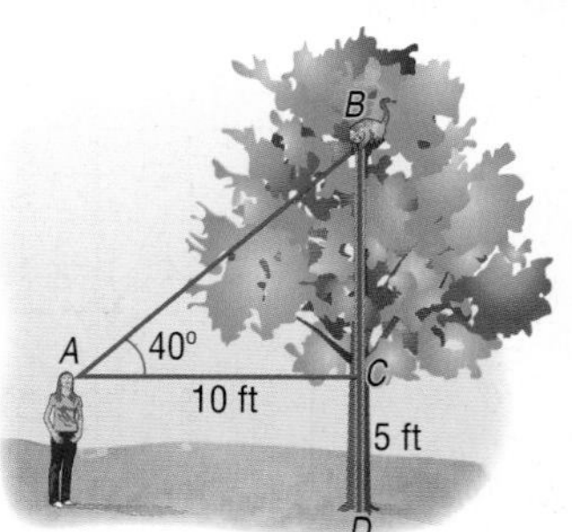


To find the how high the cat is up the tree, find $CB$.

$\tan 40 = \frac{CB}{10}$ — $\tan = \frac{\text{opposite}}{\text{adjacent}}$

$10(\tan 40) = CB$ — Multiply each side by 10.

$8.4 = CB$ — Simplify.

Since Sarah's eyes are 5 feet from the ground, add 5 to 8.4. Sarah's cat is 13.4 feet up.

## 10-6 The Law of Sines and Law of Cosines

**Find $x$. Round angle measures to the nearest degree and side measures to the nearest tenth.**

**36.**

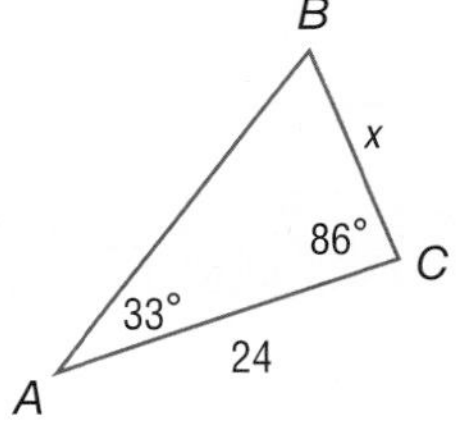


**37.**

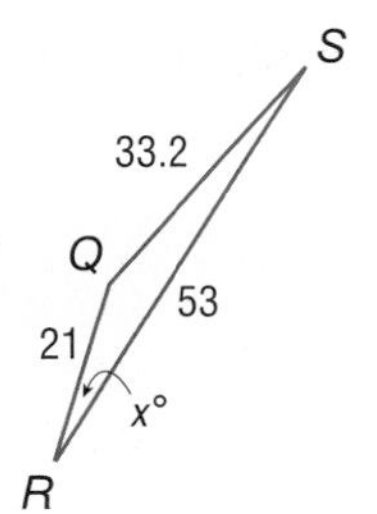


**38. SKIING** At Crazy Ed's Ski resort, Ed wants to put in another ski lift for the skiers to ride from the base to the summit of the mountain. The run over which the ski lift will go is represented by the figure below. The length of the lift is represented by $SB$. If Ed needs twice as much cable as the length of $\overline{SB}$, how much cable does he need?

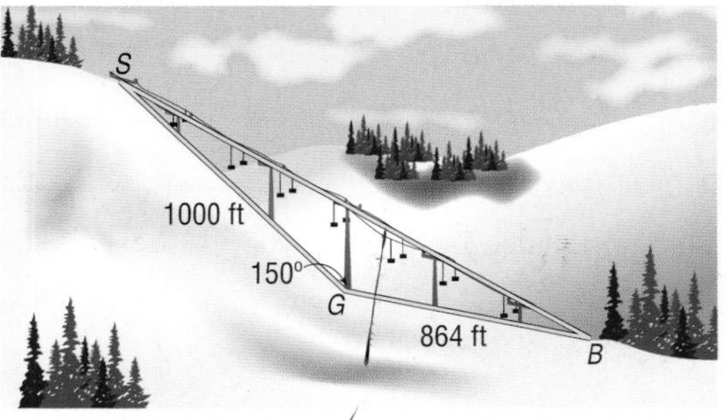


**Example 6**

**Find $x$. Round to the nearest tenth.**

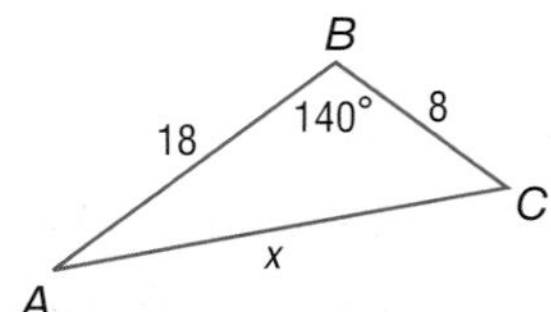


We are given the measures of two sides and their included angle, so use the Law of Cosines.

$b^2 = a^2 + c^2 - 2ac \cos B$ — Law of Cosines

$x^2 = 8^2 + 18^2 - 2(8)(18) \cos 140°$ — Substitution

$x^2 = 388 - 288 \cos 140°$ — Simplify.

$x = \sqrt{388 - 288 \cos 140°} \approx 24.7$ — Take the square root of each side.

**Example 7**

**Find $x$. Round to the nearest tenth.**

60°, 11, x°, 12

$\frac{\sin A}{a} = \frac{\sin C}{c}$ — Law of Sines

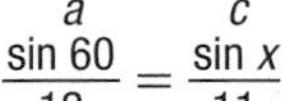

$\frac{\sin 60}{12} = \frac{\sin x}{11}$ — Substitution

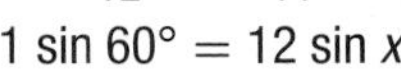

$11 \sin 60° = 12 \sin x$ — Cross Products Property

$\frac{11 \sin 60}{12} = \sin x$ — Divide each side by 12.

$x = \sin^{-1} \frac{11 \sin 60}{12}$ or about 52.5°

CHAPTER 10

# Study Guide and Review *Continued*

## 10-7 Vectors

**39.** Write the component form of the vector shown.

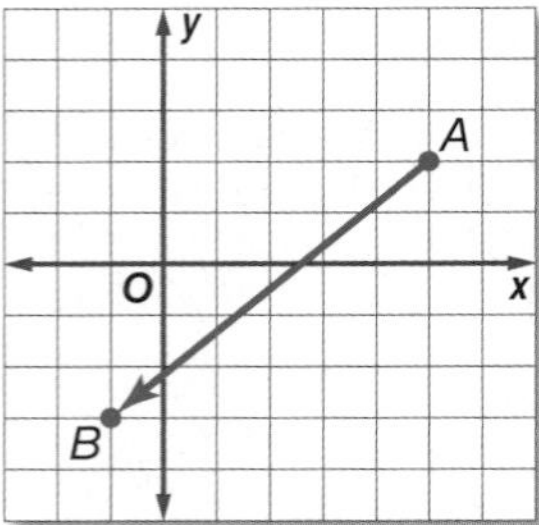


**40.** Copy the vectors to find $\vec{a} + \vec{b}$.

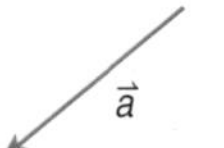


**41.** Given that $\vec{s}$ is $\langle 2, -6 \rangle$ and $\vec{t}$ is $\langle -10, 7 \rangle$, find the component form of $\vec{s} + \vec{t}$.

### Example 8

**Find the magnitude and direction of $\overrightarrow{AB}$ for $A(1, 2)$ and $B(-1, 5)$.**

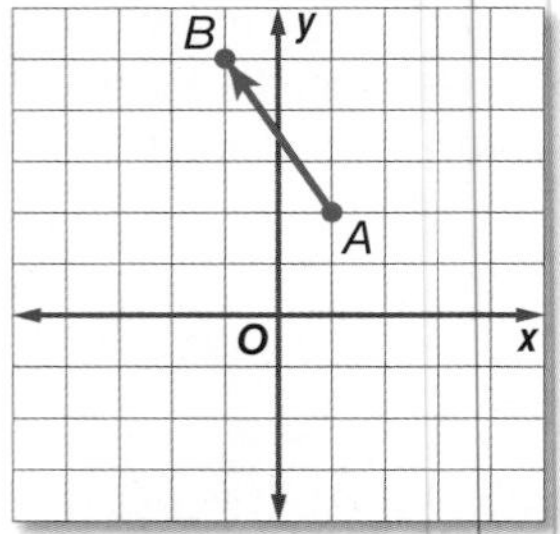


Use the Distance Formula to find the magnitude.

$$\overrightarrow{AB} = \sqrt{(x_2 - x_1)^2 + (y_2 - y_1)^2} \quad \text{Distance Formula}$$

$$= \sqrt{(-1 - 1)^2 + (5 - 2)^2} \quad \text{Substitute.}$$

$$= \sqrt{13} \text{ or about } 3.6 \quad \text{Simplify.}$$

Draw a right triangle with hypotenuse $\overrightarrow{AB}$ and acute angle $A$.

$$\tan A = \left|\frac{5 - 2}{-1 - 1}\right| \text{ or } \frac{3}{2} \quad \tan = \frac{\text{opp}}{\text{adj}}\text{; length cannot be negative.}$$

$$m\angle A = \tan^{-1}\left(-\frac{3}{2}\right) \quad \text{Def. of inverse tangent}$$

$$\approx -56.3 \quad \text{Use a calculator.}$$

The direction of $\overrightarrow{AB}$ is 180 − 56.3 or 123.7°.

## 10-8 Dilations

**42.** Copy the figure and point $S$. Then use a ruler to draw the image of the figure under a dilation with center $S$ and scale factor $r = 1.25$.

**43.** Determine whether the dilation from figure $W$ to $W'$ is an *enlargement* or a *reduction*. Then find the scale factor of the dilation and $x$.

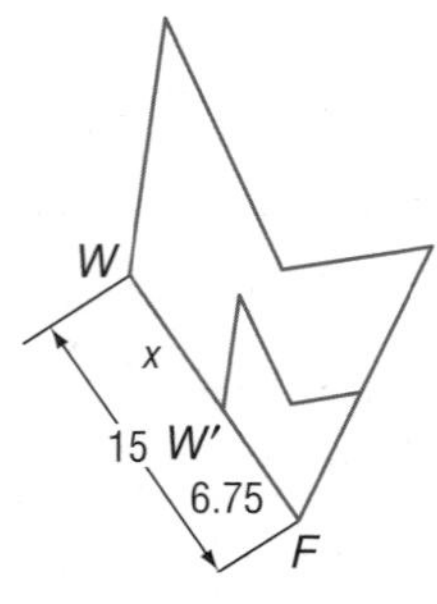


### Example 6

**Square $ABCD$ has vertices $A(0, 0)$, $B(0, 8)$, $C(8, 8)$, and $D(8, 0)$. Find the image of $ABCD$ after a dilation centered at the origin with a scale factor of 0.5.**

Multiply the $x$- and $y$-coordinates of each vertex by the scale factor, 0.5.

| | | |
|---|---|---|
| $(x, y)$ | → | $(0.5x, 0.5y)$ |
| $A(0, 0)$ | → | $A'(0, 0)$ |
| $B(0, 8)$ | → | $B'(0, 4)$ |
| $C(8, 8)$ | → | $C'(4, 4)$ |
| $D(8, 0)$ | → | $D'(4, 0)$ |

B y C B' C' A O A' D' D x

Graph $ABCD$ and its image $A'B'C'D'$.

CHAPTER 10

# Practice Test

**Find the geometric mean between each pair of numbers.**

1. 7 and 11
2. 12 and 9
3. 14 and 21
4. $4\sqrt{3}$ and $10\sqrt{3}$
5. Find $x$, $y$, and $z$.

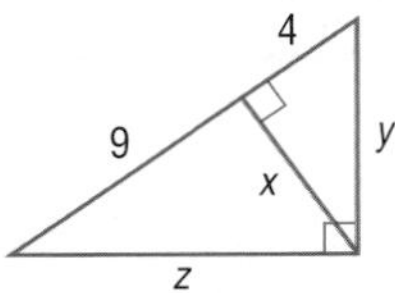


6. **FAIRS** Blake is setting up his tent at a renaissance fair. If the tent is 8 feet tall, and the tether can be staked no more than two feet from the tent, how long should the tether be?

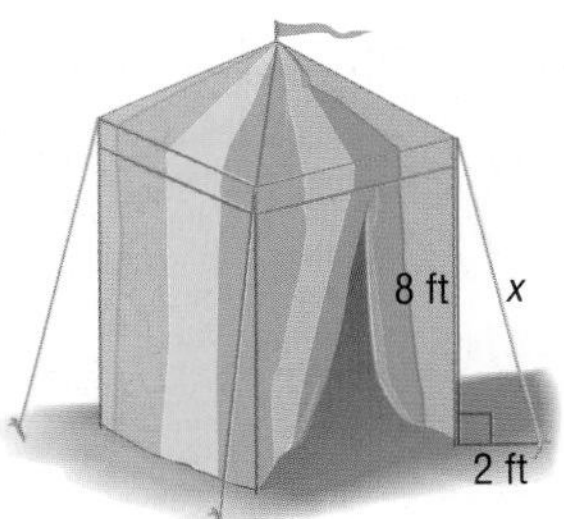


**Use a calculator to find the measure of $\angle R$ to the nearest tenth.**

7. 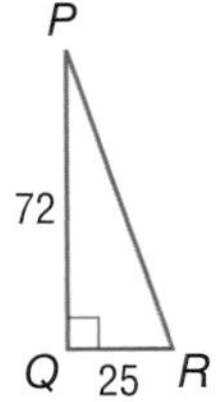


8. 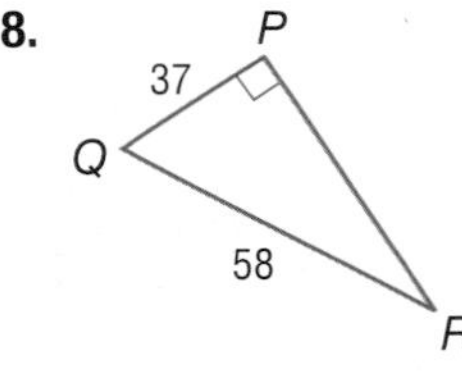


9. Find $x$ and $y$.

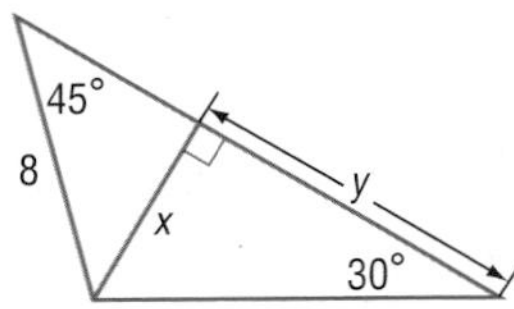


**Express each ratio as a fraction and as a decimal to the nearest hundredth.**

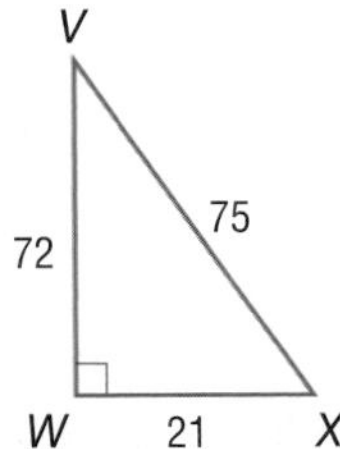


10. $\cos X$
11. $\tan X$
12. $\tan V$
13. $\sin V$

**Find the magnitude and direction of each vector.**

14. $\overrightarrow{JK}$: $J(-6, -4)$ and $K(-10, -4)$
15. $\overrightarrow{RS}$: $R(1, 0)$ and $S(-2, 3)$

16. **SPACE** Anna is watching a space shuttle launch 6 miles from Cape Canaveral in Florida. When the angle of elevation from her viewing point to the shuttle is 80°, how high is the shuttle, if it is going straight up?

**Find $x$. Round angle measures to the nearest degree and side measures to the nearest tenth.**

17. 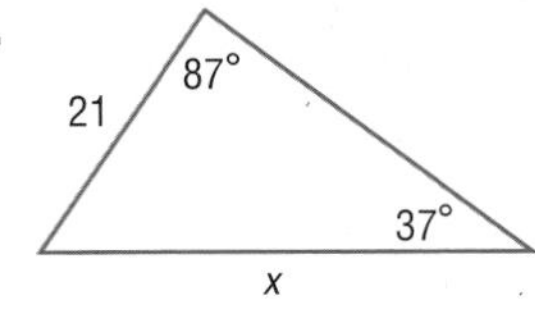


18. 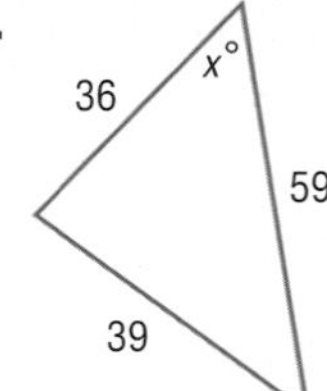


19. **MULTIPLE CHOICE** Which of the following is the length of the leg of a 45°-45°-90° triangle with a hypotenuse of 20?

   **A** 10
   **B** $10\sqrt{2}$
   **C** 20
   **D** $20\sqrt{2}$

**Find $x$.**

20. 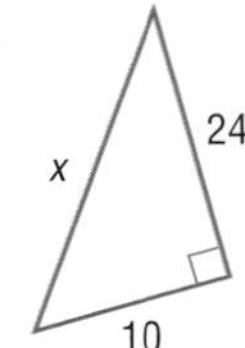


21. 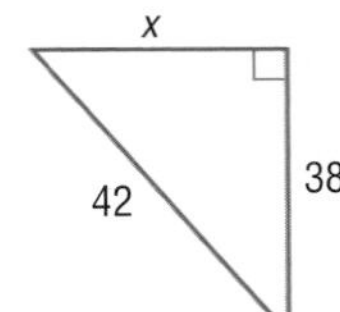


22. **WHALE WATCHING** Isaac is looking through binoculars on a whale watching trip when he notices a sea otter in the distance. If he is 20 feet above sea level in the boat, and the angle of depression is 30°, how far away from the boat is the otter to the nearest foot?

**Write the component form of each vector.**

23. 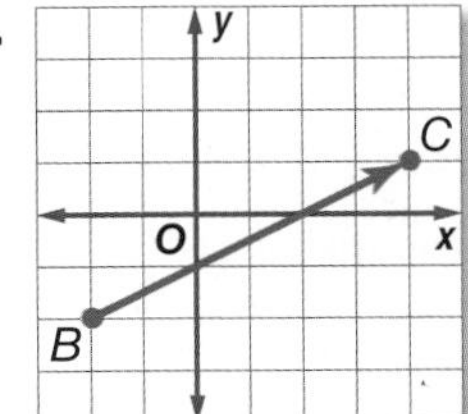


24. 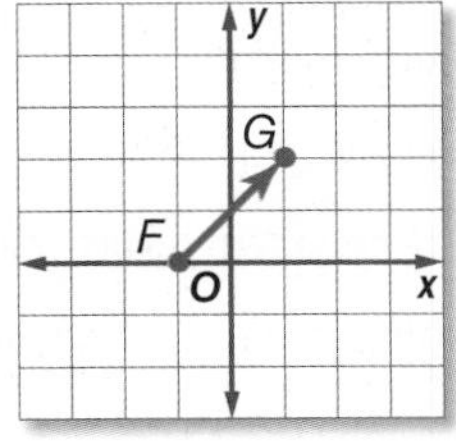


25. Solve $\triangle FGH$. Round to the nearest degree.

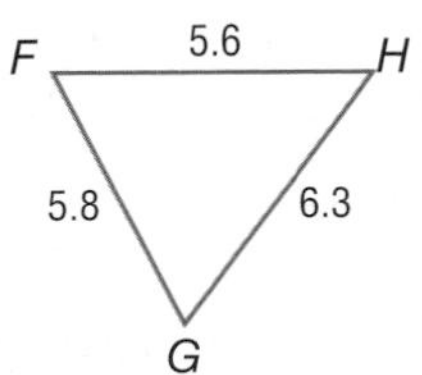

CHAPTER 10

# Preparing for Standardized Tests

## Use a Formula

Sometimes it is necessary to use a formula to solve problems on standardized tests. In some cases you may even be given a sheet of formulas that you are permitted to reference while taking the test.

**Formulas**

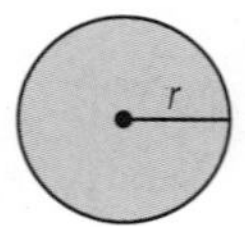


$C = 2\pi r$
$A = \pi r^2$

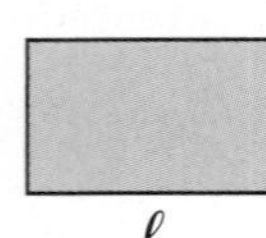


$A = \ell w$

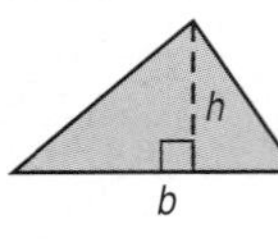


$A = \frac{1}{2}bh$

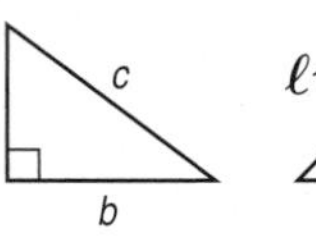


$a^2 + b^2 = c^2$

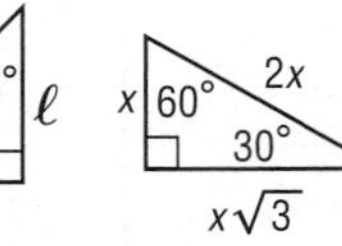


Special Right Triangles

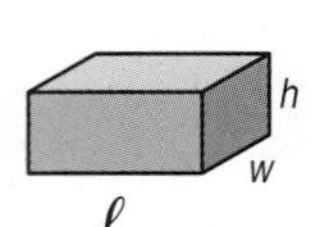


$V = \ell wh$

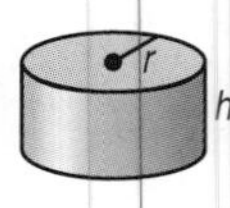


$V = \pi r^2 h$

The are 360 degrees in a circle.
The sum of the measures of the angles of a triangle is 180.

### Strategies for Using a Formula

**Step 1**

Read the problem statement carefully.

**Ask yourself:**

- What am I being asked to solve?
- What information is given in the problem?
- Are there any formulas that I can use to help me solve the problem?

**Step 2**

Solve the problem.

- Substitute the known quantities that are given in the problem statement into the formula.
- Simplify to solve for the unknown values in the formula.

**Step 3**

Check your solution.

- Determine a reasonable range of values for the answer.
- Check to make sure that your answer makes sense.
- If time permits, check your answer.

## Standardized Test Example

**Read the problem. Identify what you need to know. Then use the information in the problem to solve.**

The ratio of the width to the height of a high-definition television is 16:9. This is also called the *aspect ratio* of the television. The size of a television is given in terms of the diagonal distance across the screen. If an HD television is 25.5 inches tall, what is its screen size?

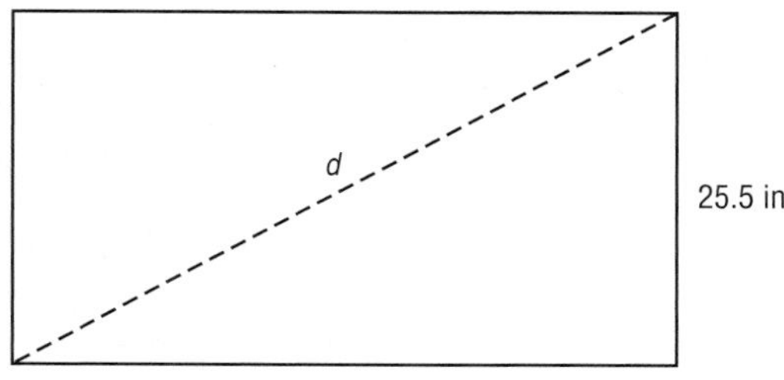


**A** 48 inches

**B** 50 inches

**C** 51 inches

**D** 52 inches

Read the problem statement carefully. You are given the height of the screen and the ratio of the width to the height. You are asked to find the diagonal distance of the screen. You can use the **Pythagorean Theorem** to solve the problem.

Find the width of the screen. Set up and solve a proportion using the aspect ratio 16:9.

$\frac{16}{9} = \frac{w}{25.5}$ ← width of the screen / ← height of the screen

$9w = 408$ Cross Products Property

$w = 45\frac{1}{3}$ Divide each side by 9.

So, the width of the screen is $45\frac{1}{3}$ inches. Use the Pythagorean Theorem to solve for the diagonal distance.

$c^2 = a^2 + b^2$ Pythagorean Theorem

$c^2 = (25.5)^2 + \left(45\frac{1}{3}\right)^2$ Substitute for *a* and *b*.

$c \approx 52.01$ Simplify. Take the square root of both sides to solve for *c*.

The diagonal distance of the screen is about 52 inches. So, the answer is D.

## Exercises

**Read each problem. Identify what you need to know. Then use the information in the problem to solve.**

1. Christine is flying a kite on the end of a taut string. The kite is 175 feet above the ground and is a horizontal distance of 130 feet from where Christine is standing. How much kite string has Christine let out? Round to the nearest foot.

   **A** 204 ft **C** 225 ft

   **B** 218 ft **D** 236

2. What is the value of $x$ below to the nearest tenth?

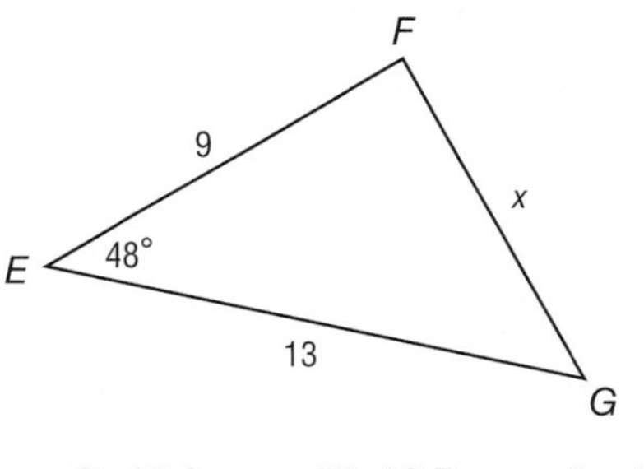


   **F** 9.7 **G** 10.2 **H** 10.5 **J** 11.1

# CHAPTER 10 Standardized Test Practice

## Cumulative, Chapters 1 through 10

### Multiple Choice

**Read each question. Then fill in the correct answer on the answer document provided by your teacher or on a sheet of paper.**

**1.** What is the value of $x$ in the figure below?

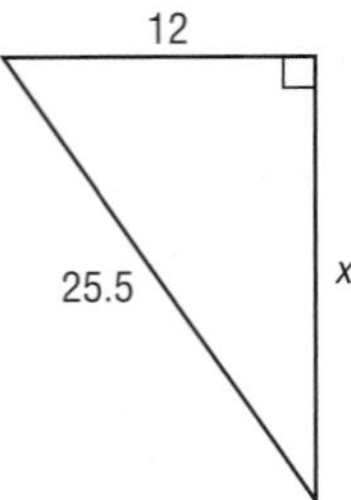


A 22.5

B 23

C 23.5

D 24

**2.** A baseball diamond is a square with 90-ft sides. What is the length from 3rd base to 1st base? Round to the nearest tenth.

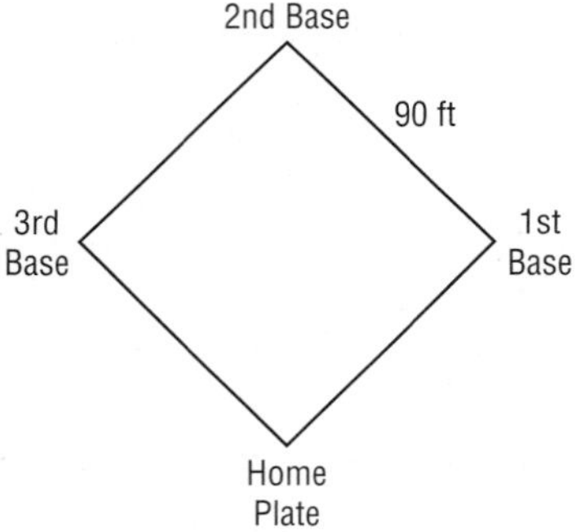


F 155.9 ft

G 141.6 ft

H 127.3 ft

J 118.2 ft

**Test-TakingTip**

Question 1 Some test items require the use of a formula to solve them. Use the Pythagorean Theorem to find $x$.

**3.** The scale of a map is 1 inch = 4.5 kilometers. What is the distance between two cities that are 2.4 inches apart on the map?

A 10.8 kilometers

B 11.1 kilometers

C 11.4 kilometers

D 11.5 kilometers

**4.** What is the value of $x$ in the figure below? Round to the nearest tenth.

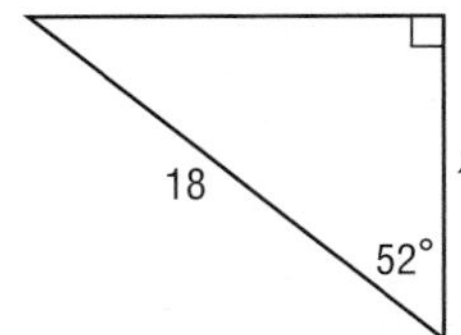


F 10.5

G 11.1

H 13.6

J 14.2

**5.** Which of the following is a side length in isosceles triangle $DEF$?

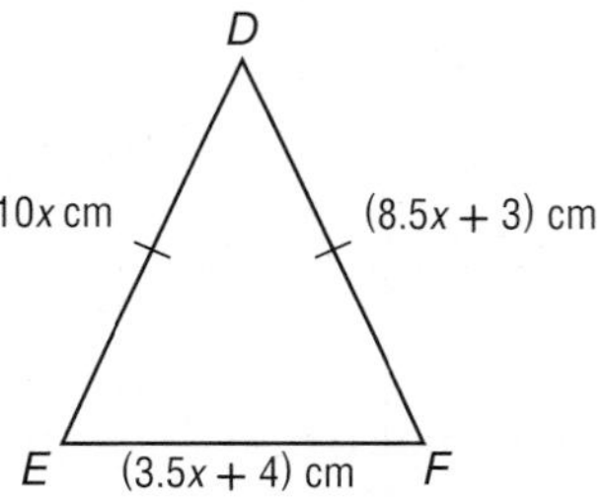


F 2 cm

G 8 cm

H 9 cm

J 11 cm

**6.** Grant is flying a kite on the end of a string that is 350 feet long. The angle elevation from Grant to the kite is 74°. How high above the ground is the kite? Round your answer to the nearest tenth if necessary.

F 336.4 ft

G 295.6 ft

H 141.2 ft

J 96.5 ft

## Short Response/Gridded Response

**Record your answers on the answer sheet provided by your teacher or on a sheet of paper.**

7. **GRIDDED RESPONSE** Find $x$ in the figure below. Round your answer to the nearest tenth if necessary.

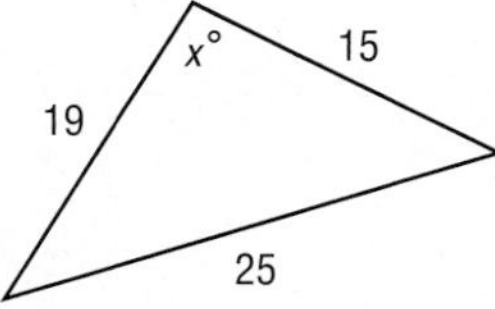


8. Amy is paddling her canoe across a lake at a speed of 10 feet per second headed due north. The wind is blowing 40° east of north with a velocity of 2.8 feet per second. What is Amy's resultant velocity? Express your answer as a vector. Show your work.

9. Janice used a 16-inch dowel and a 21-inch dowel to build a kite as shown below. What is the perimeter of her kite?

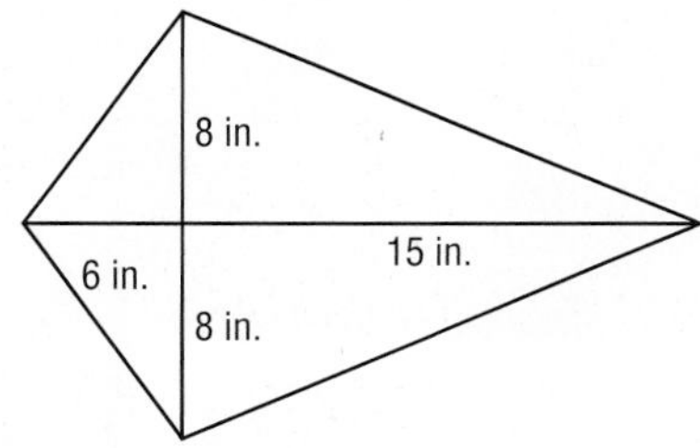


10. **GRIDDED RESPONSE** A model airplane takes off at an angle of elevation of 30°. How high will the plane be after traveling 100 feet horizontally? Round to the nearest tenth. Show your work.

11. According to the Perpendicular Bisector Theorem, what is the length of segment $AB$ below?

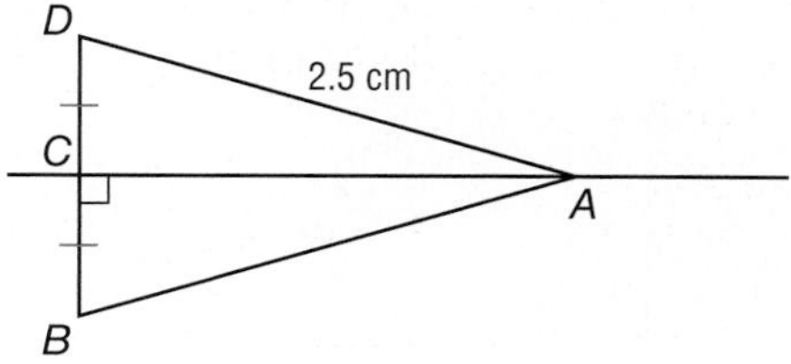


12. Dilate the figure shown on the coordinate grid by a scale factor of 1.5 centered at the origin.

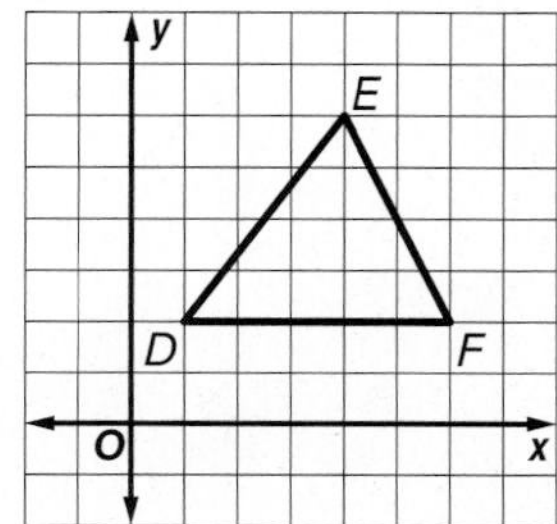


## Extended Response

**Record your answers on a sheet of paper. Show your work.**

13. Refer to the triangle shown below.

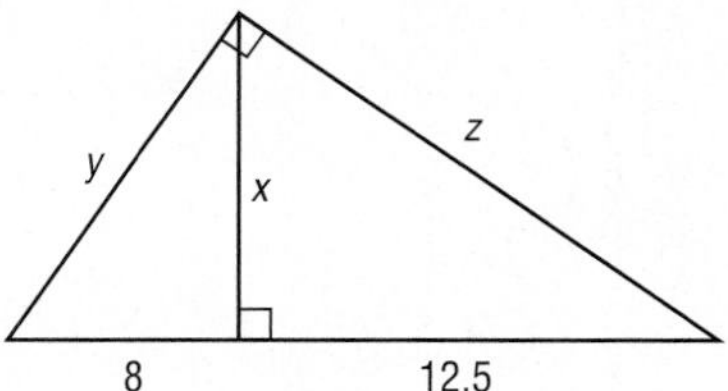


**a.** Find $x$ to the nearest tenth.

**b.** Find $y$ to the nearest tenth.

**c.** Find $z$ to the nearest tenth.

**Need ExtraHelp?**

| If you missed Question... | 1 | 2 | 3 | 4 | 5 | 6 | 7 | 8 | 9 | 10 | 11 | 12 | 13 |
|---|---|---|---|---|---|---|---|---|---|---|---|---|---|
| Go to Lesson... | 10-2 | 10-3 | 9-7 | 10-4 | 6-5 | 10-5 | 10-6 | 10-7 | 8-6 | 10-5 | 7-1 | 10-8 | 10-1 |

CHAPTER 11

# Circles

## Then

You learned about special segments and angle relationships in triangles.

## Now

In this chapter, you will:

- Learn the relationships between central angles, arcs, and inscribed angles in a circle.
- Define and use secants and tangents.
- Use an equation to identify or describe a circle.

## Why? ▲

**SCIENCE** The actual shape of a rainbow is a complete circle. The portion of the circle that can be seen above the horizon is a special segment of a circle called an arc.

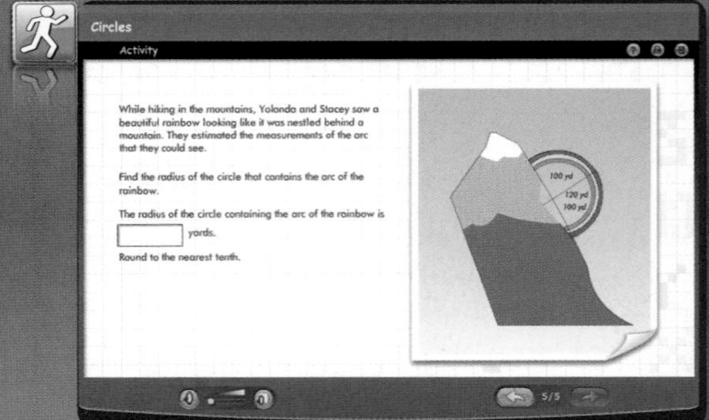


connectED.mcgraw-hill.com Your Digital Math Portal

Animation | Vocabulary | eGlossary | Personal Tutor | Virtual Manipulatives | Graphing Calculator | Audio | Foldables | Self-Check Practice | Worksheets

# Get Ready for the Chapter

**Diagnose** Readiness | You have two options for checking prerequisite skills.

**Textbook Option** Take the Quick Check below. Refer to the Quick Review for help.

## QuickCheck

**Find the percent of the given number.**

1. 26% of 500
2. 79% of 623
3. 19% of 82
4. 10% of 180
5. 92% of 90
6. 65% of 360
7. **TIPPING** A couple ate dinner at an Italian restaurant where their bill was \$32.50. If they want to leave an 18% tip, how much tip money should they leave?

8. Find $x$. Round to the nearest tenth.

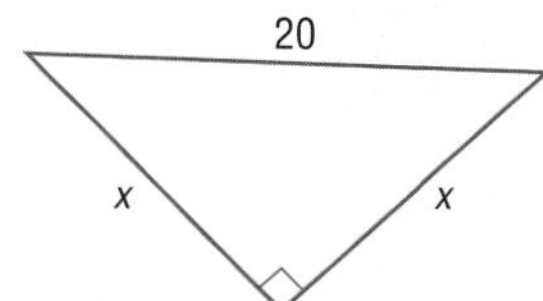


9. **CONSTRUCTION** Jennifer is putting a brace in a board, as shown at the right. Find the length of the board used for a brace.

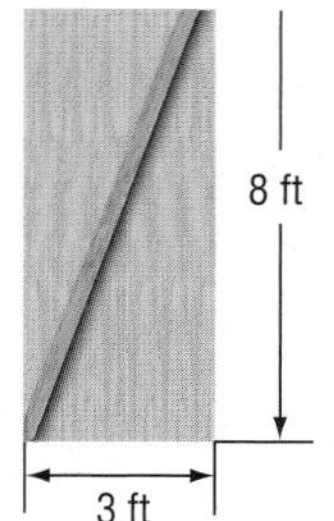


**Solve each equation by using the Quadratic Formula. Round to the nearest tenth if necessary.**

10. $5x^2 + 4x - 20 = 0$
11. $x^2 = x + 12$
12. **FIREWORKS** The Patriot Squad, a professional fireworks company, performed a show during a July 4th celebration. One of the rockets in the show followed the path modeled by $d = 80t - 16t^2$ where $t$ is the time in seconds, but it failed to explode.

## QuickReview

### Example 1

**Find the percent of the given number.**

$15\%$ of $35 = (0.15)(35)$ Change the percent to a decimal.

$= 5.25$ Multiply.

So, 15% of 35 is 5.25.

### Example 2

**Find $x$. Round to the nearest tenth.**

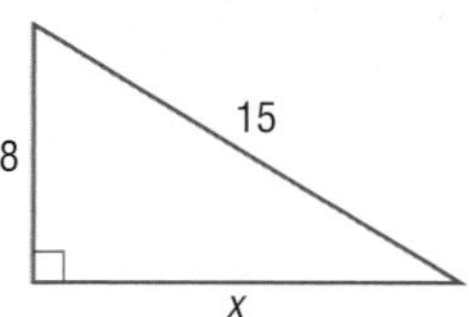


$a^2 + b^2 = c^2$ Pythagorean Theorem

$x^2 + 8^2 = 15^2$ Substitution

$x^2 + 64 = 225$ Simplify.

$x^2 = 161$ Subtract.

$x = \sqrt{161}$ or about 12.7

### Example 3

**Solve $x^2 + 3x - 40 = 0$ by using the Quadratic Formula. Round to the nearest tenth.**

$x = \frac{-b \pm \sqrt{b^2 - 4ac}}{2a}$ Quadratic Formula

$= \frac{-3 \pm \sqrt{3^2 - 4(1)(-40)}}{2(1)}$ Substitution

$= \frac{-3 \pm \sqrt{169}}{2}$ Simplify.

$= 5$ or $-8$ Simplify.

**2 Online Option** Take an online self-check Chapter Readiness Quiz at connectED.mcgraw-hill.com.

# **Get Started** on the Chapter

**You will learn several new concepts, skills, and vocabulary terms as you study Chapter 11. To get ready, identify important terms and organize your resources. You may wish to refer to Chapter 0 to review prerequisite skills.**

## FOLDABLES® StudyOrganizer

**Circles** Make this Foldable to help you organize your Chapter 11 notes on circles. Begin with nine sheets of paper.

**1** **Trace** an 8-inch circle on each paper using a compass.

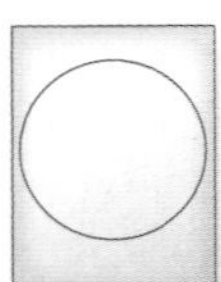

**2** **Cut** out each of the circles.

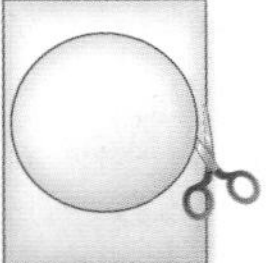

**3** **Staple** an inch from the left side of the papers.

**4** **Label** as shown.

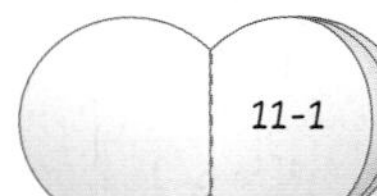


## NewVocabulary

| English | | Español |
|---|---|---|
| circle | p. 715 | círculo |
| center | p. 715 | centro |
| radius | p. 715 | radio |
| chord | p. 715 | cuerda |
| diameter | p. 715 | diámetro |
| circumference | p. 717 | circunferencia |
| pi ($\pi$) | p. 717 | pi ($\pi$) |
| inscribed | p. 718 | inscrito |
| circumscribed | p. 718 | circunscrito |
| central angle | p. 724 | ángulo central |
| arc | p. 724 | arco |
| tangent | p. 750 | tangente |
| secant | p. 759 | secante |
| chord segment | p. 768 | segmento de cuerda |

## ReviewVocabulary

coplanar coplanar points that lie in the same plane

degree grado $\frac{1}{360}$ of the circular rotation about a point

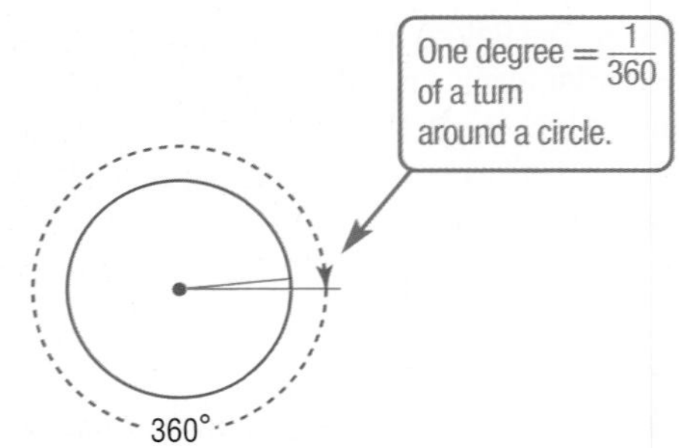

LESSON 11-1

# Circles and Circumference

| Then | Now | Why? |
|---|---|---|
| You identified and used parts of parallelograms. | 1 Identify and use parts of circles. 2 Solve problems involving the circumference of a circle. | The maxAir ride shown speeds back and forth and rotates counterclockwise. At times, the riders are upside down 140 feet above the ground experiencing "airtime"—a feeling of weightlessness. The ride's width, or *diameter,* is 44 feet. You can find the distance that a rider travels in one rotation by using this measure. |

## NewVocabulary

circle
center
radius
chord
diameter
concentric circles
circumference
pi ($\pi$)
inscribed
circumscribed

## Common Core State Standards

**Content Standards**

G.CO.1 Know precise definitions of angle, circle, perpendicular line, parallel line, and line segment, based on the undefined notions of point, line, distance along a line, and distance around a circular arc.

G.C.1 Prove that all circles are similar.

**Mathematical Practices**

4 Model with mathematics.

1 Make sense of problems and persevere in solving them.

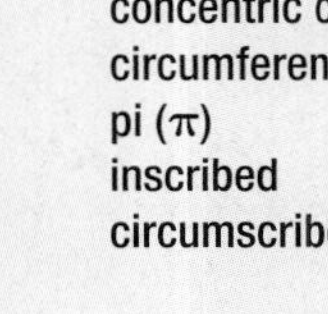

## 1 Segments in Circles

A **circle** is the locus or set of all points in a plane equidistant from a given point called the **center** of the circle.

Segments that intersect a circle have special names.

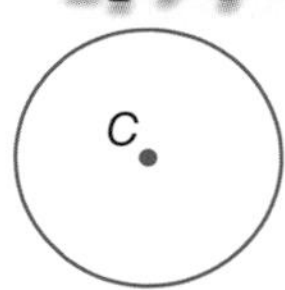

Circle $C$ or $\odot C$

### KeyConcept Special Segments in a Circle

A **radius** (plural radii) is a segment with endpoints at the center and on the circle.

**Examples** $\overline{CD}$, $\overline{CE}$, and $\overline{CF}$ are radii of $\odot C$.

A **chord** is a segment with endpoints on the circle.

**Examples** $\overline{AB}$ and $\overline{DE}$ are chords of $\odot C$.

A **diameter** of a circle is a chord that passes through the center and is made up of collinear radii.

**Example** $\overline{DE}$ is a diameter of $\odot C$. Diameter $\overline{DE}$ is made up of collinear radii $\overline{CD}$ and $\overline{CE}$.

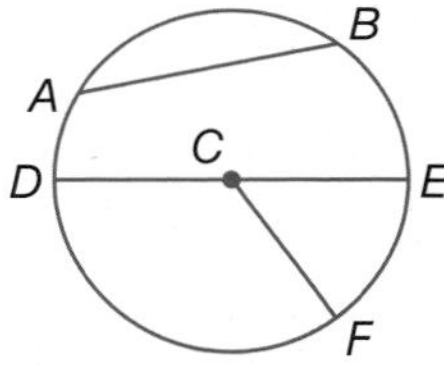


### Example 1 Identify Segments in a Circle

PT

**a. Name the circle and identify a radius.**

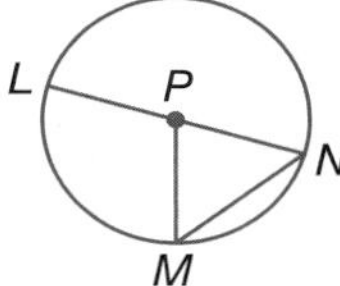


The circle has a center at $P$, so it is named circle $P$, or $\odot P$. Three radii are shown: $\overline{PL}$, $\overline{PN}$, and $\overline{PM}$.

**b. Identify a chord and a diameter of the circle.**

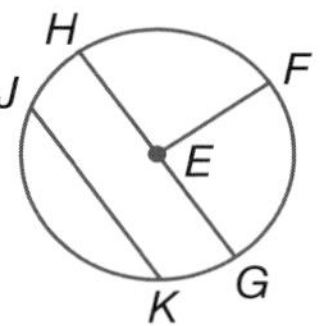


Two chords are shown: $\overline{JK}$ and $\overline{HG}$. $\overline{HG}$ goes through the center, so $\overline{HG}$ is a diameter.

### GuidedPractice

1. Name the circle, a radius, a chord, and a diameter of the circle.

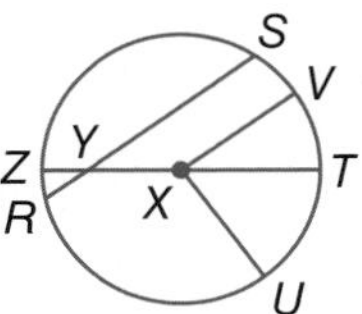


Ilene MacDonald/Alamy

**ReadingMath**

**CCSS Precision** The words *radius* and *diameter* are used to describe lengths as well as segments. Since a circle has many different radii and diameters, the phrases *the radius* and *the diameter* refer to lengths rather than segments.

By definition, the distance from the center of a circle to any point on the circle is always the same. Therefore, all radii $r$ of a circle are congruent. Since a diameter $d$ is composed of two radii, all diameters of a circle are also congruent.

### KeyConcept Radius and Diameter Relationships

If a circle has radius $r$ and diameter $d$, the following relationships are true.

**Radius Formula** $r = \frac{d}{2}$ or $r = \frac{1}{2}d$ **Diameter Formula** $d = 2r$

PT

### Example 2 Find Radius and Diameter

**If $QV = 8$ inches, what is the diameter of $\odot Q$?**

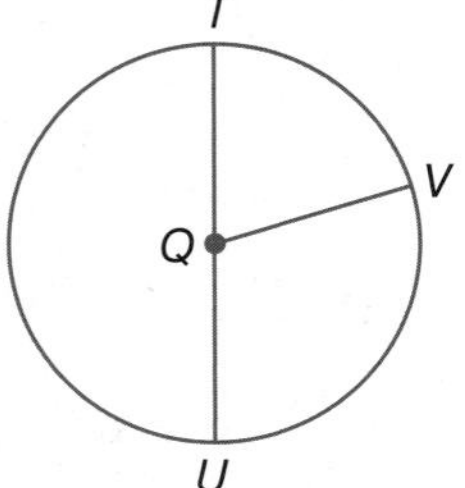


$d = 2r$ Diameter Formula

$= 2(8)$ or $16$ Substitute and simplify.

The diameter of $\odot Q$ is 16 inches.

**GuidedPractice**

**2A.** If $TU = 14$ feet, what is the radius of $\odot Q$?

**2B.** If $QT = 11$ meters, what is $QU$?

As with other figures, pairs of circles can be congruent, similar, or share other special relationships.

**ReviewVocabulary**

**coplanar** points that lie in the same plane

### KeyConcept Circle Pairs

Two circles are congruent if and only if they have congruent radii.

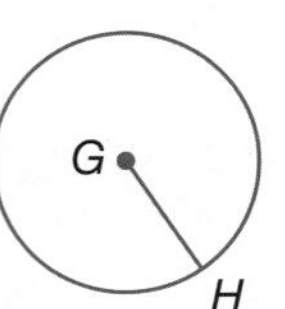


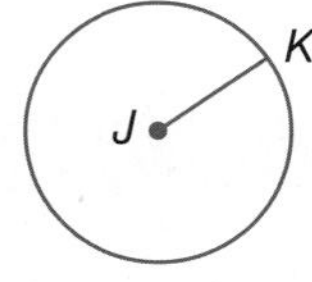


**Example** $\overline{GH} \cong \overline{JK}$, so $\odot G \cong \odot J$.

All circles are similar.

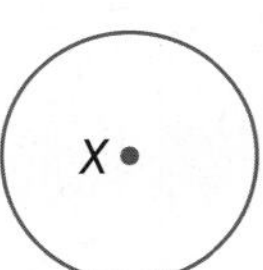


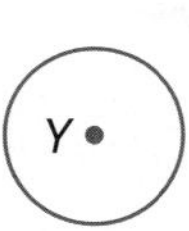


**Example** $\odot X \sim \odot Y$

**Concentric circles** are coplanar circles that have the same center.

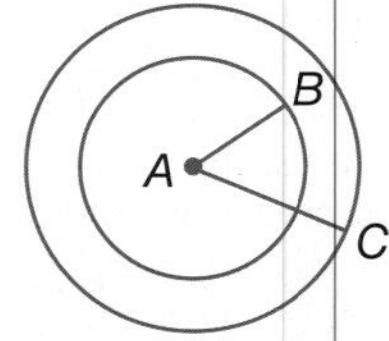


**Example** $\odot A$ with radius $\overline{AB}$ and $\odot A$ with radius $\overline{AC}$ are concentric.

You will prove that all circles are similar in Exercise 52.

Two circles can intersect in two different ways.

| 2 Points of Intersection | 1 Point of Intersection | No Points of Intersection |
| --- | --- | --- |
| | | |

The segment connecting the centers of the two intersecting circles contains the radii of the two circles.

## Example 3 Find Measures in Intersecting Circles

**The diameter of $\odot S$ is 30 units, the diameter of $\odot R$ is 20 units, and $DS = 9$ units. Find $CD$.**

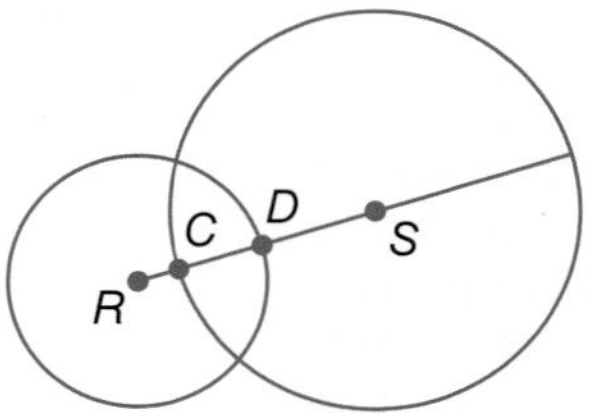


Since the diameter of $\odot S$ is 30, $CS = 15$. $\overline{CD}$ is part of radius $\overline{CS}$.

| | |
|---|---|
| $CD + DS = CS$ | Segment Addition Postulate |
| $CD + 9 = 15$ | Substitution |
| $CD = 6$ | Subtract 9 from each side. |

### Guided Practice

**3.** Use the diagram above to find $RC$.

## 2 Circumference

The **circumference** of a circle is the distance around the circle. By definition, the ratio $\frac{C}{d}$ is an irrational number called **pi ($\pi$)**. Two formulas for circumference can be derived by using this definition.

| | |
|---|---|
| $\frac{C}{d} = \pi$ | Definition of pi |
| $C = \pi d$ | Multiply each side by $d$. |
| $C = \pi(2r)$ | $d = 2r$ |
| $C = 2\pi r$ | Simplify. |

### Key Concept Circumference

**Words** If a circle has diameter $d$ or radius $r$, the circumference $C$ equals the diameter times pi or twice the radius times pi.

**Symbols** $C = \pi d$ or $C = 2\pi r$

David Cannon/Getty Images Sport/Getty Images

**Real-World Link**

In 2005, Roger Federer and Andre Agassi played tennis on the helipad of the Burj Al Arab hotel in the United Arab Emirates. The helipad has a diameter of 79 feet and is nearly 700 feet high.

**Source:** Burj Al Arab, Emporis Buildings

## Real-World Example 4 Find Circumference

**TENNIS Find the circumference of the helipad described at the left.**

| | |
|---|---|
| $C = \pi d$ | Circumference formula |
| $= \pi(79)$ | Substitution |
| $= 79\pi$ | Simplify. |
| $\approx 248.19$ | Use a calculator. |

The circumference of the helipad is $79\pi$ feet or about 248.19 feet.

### Guided Practice

**Find the circumference of each circle described. Round to the nearest hundredth.**

**4A.** radius = 2.5 centimeters

**4B.** diameter = 16 feet

These circumference formulas can also be used to determine the diameter and radius of a circle when the circumference is given.

PT

## Example 5 Find Diameter and Radius

**Find the diameter and radius of a circle to the nearest hundredth if the circumference of the circle is 106.4 millimeters.**

| | | | |
|---|---|---|---|
| $C = \pi d$ | Circumference Formula | $r = \frac{1}{2}d$ | Radius Formula |
| $106.4 = \pi d$ | Substitution | $\approx \frac{1}{2}(33.87)$ | $d \approx 33.87$ |
| $\frac{106.4}{\pi} = d$ | Divide each side by $\pi$. | $\approx 16.94$ mm | Use a calculator. |
| $33.87 \text{ mm} \approx d$ | Use a calculator. | | |

### GuidedPractice

**5.** Find the diameter and radius of a circle to the nearest hundredth if the circumference of the circle is 77.8 centimeters.

**StudyTip**

**Levels of Accuracy** Since $\pi$ is irrational, its value cannot be given as a terminating decimal. Using a value of 3 for $\pi$ provides a quick estimate in calculations. Using a value of 3.14 or $\frac{22}{7}$ provides a closer approximation. For the most accurate approximation, use the $\pi$ key on a calculator. Unless stated otherwise, assume that in this text, a calculator with a $\pi$ key was used to generate answers.

A polygon is **inscribed** in a circle if all of its vertices lie on the circle. A circle is **circumscribed** about a polygon if it contains all the vertices of the polygon.

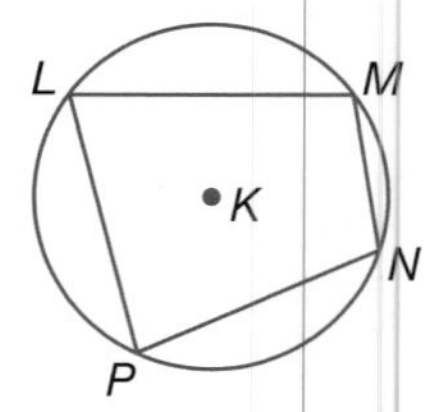


- Quadrilateral *LMNP* is *inscribed in* $\odot K$.
- Circle *K* is *circumscribed about* quadrilateral *LMNP*.

PT

## Standardized Test Example 6 Circumference of Circumscribed Polygon

**SHORT RESPONSE A square with side length of 9 inches is inscribed in $\odot J$. Find the exact circumference of $\odot J$.**

**Read the Test Item**

You need to find the diameter of the circle and use it to calculate the circumference.

**Solve the Test Item**

First, draw a diagram. The diagonal of the square is the diameter of the circle and the hypotenuse of a right triangle.

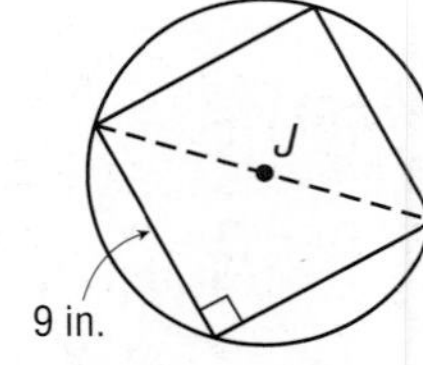


| | |
|---|---|
| $a^2 + b^2 = c^2$ | Pythagorean Theorem |
| $9^2 + 9^2 = c^2$ | Substitution |
| $162 = c^2$ | Simplify. |
| $9\sqrt{2} = c$ | Take the positive square root of each side. |

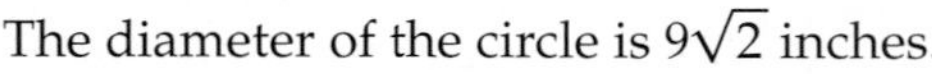

The diameter of the circle is $9\sqrt{2}$ inches.

Find the circumference in terms of $\pi$ by substituting $9\sqrt{2}$ for $d$ in $C = \pi d$. The exact circumference is $9\pi\sqrt{2}$ inches.

**StudyTip**

**Circumcircle** A *circumcircle* is a circle that passes through all of the vertices of a polygon.

### GuidedPractice

**Find the exact circumference of each circle by using the given polygon.**

**6A.** inscribed right triangle with legs 7 meters and 3 meters long

**6B.** circumscribed square with side 10 feet long

## Check Your Understanding

**Examples 1–2** **For Exercises 1–4, refer to $\odot N$.**

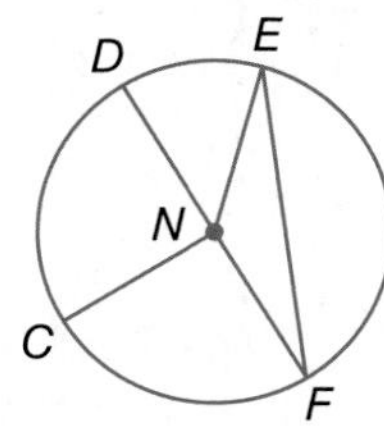


1. Name the circle.
2. Identify each.
   a. a chord  b. a diameter  c. a radius
3. If $CN = 8$ centimeters, find $DN$.
4. If $EN = 13$ feet, what is the diameter of the circle?

**Example 3** **The diameters of $\odot A$, $\odot B$, and $\odot C$ are 8 inches, 18 inches, and 11 inches, respectively. Find each measure.**

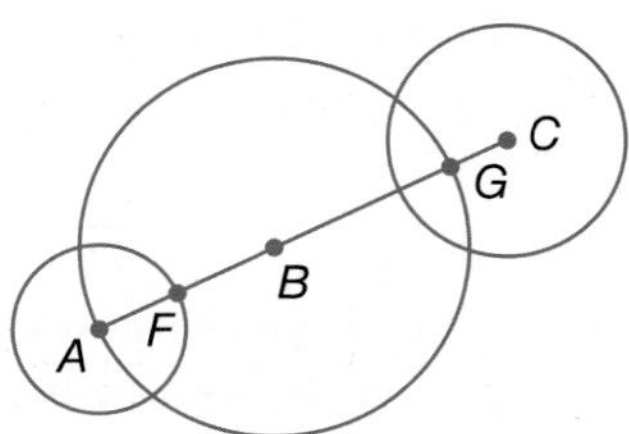


5. $FG$
6. $FB$

**Example 4**

7. **RIDES** The circular ride described at the beginning of the lesson has a diameter of 44 feet. What are the radius and circumference of the ride? Round to the nearest hundredth, if necessary.

**Example 5**

8. **CCSS MODELING** The circumference of the circular swimming pool shown is about 56.5 feet. What are the diameter and radius of the pool? Round to the nearest hundredth.

**Example 6**

9. **SHORT RESPONSE** The right triangle shown is inscribed in $\odot D$. Find the exact circumference of $\odot D$.

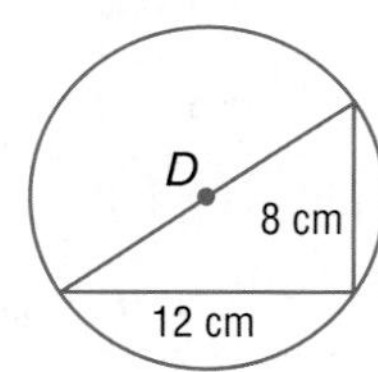


## Practice and Problem Solving

**Examples 1–2** **For Exercises 10–13, refer to $\odot R$.**

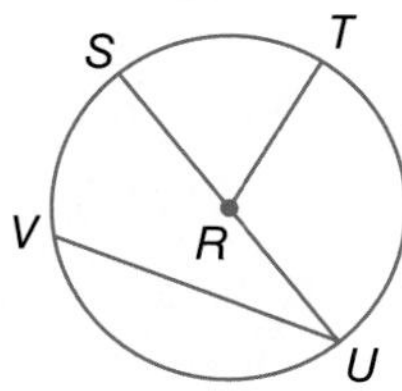


10. Name the center of the circle.
11. Identify a chord that is also a diameter.
12. Is $\overline{VU}$ a radius? Explain.
13. If $SU = 16.2$ centimeters, what is $RT$?

**For Exercises 14–17, refer to $\odot F$.**

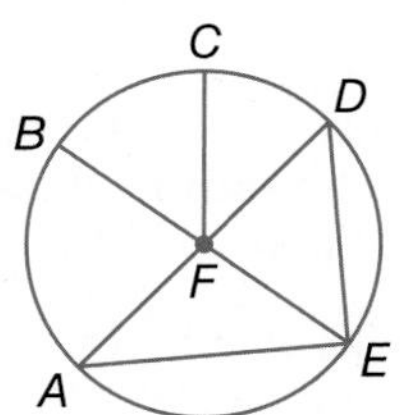


14. Identify a chord that is not a diameter.
15. If $CF = 14$ inches, what is the diameter of the circle?
16. Is $\overline{AF} \cong \overline{EF}$? Explain.
17. If $DA = 7.4$ centimeters, what is $EF$?

**Example 3** **Circle $J$ has a radius of 10 units, $\odot K$ has a radius of 8 units, and $BC = 5.4$ units. Find each measure.**

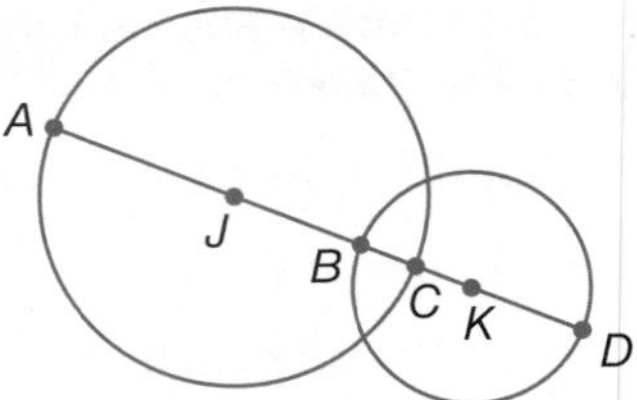


18. $CK$
19. $AB$
20. $JK$
21. $AD$

**Example 4**

22. **PIZZA** Find the radius and circumference of the pizza shown. Round to the nearest hundredth, if necessary.

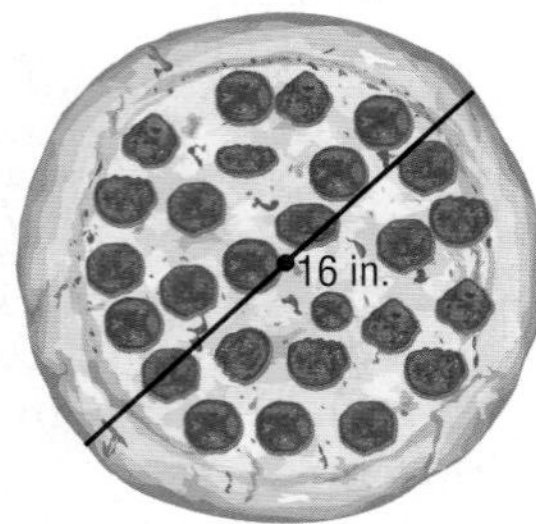


23. **BICYCLES** A bicycle has tires with a diameter of 26 inches. Find the radius and circumference of a tire. Round to the nearest hundredth, if necessary.

**Example 5** **Find the diameter and radius of a circle with the given circumference. Round to the nearest hundredth.**

24. $C = 18$ in.
25. $C = 124$ ft
26. $C = 375.3$ cm
27. $C = 2608.25$ m

**Example 6**  **SENSE-MAKING** **Find the exact circumference of each circle by using the given inscribed or circumscribed polygon.**

28.

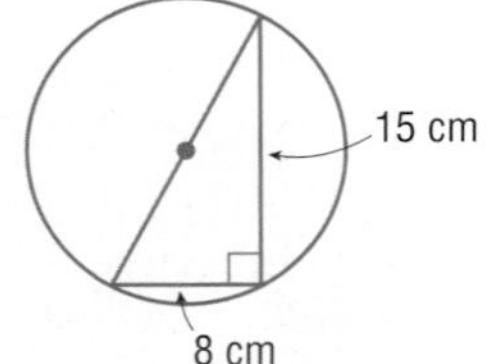


29.

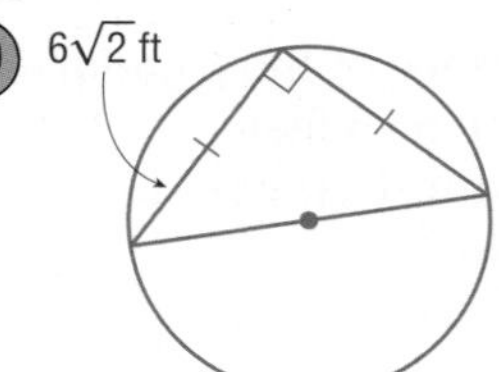


30.

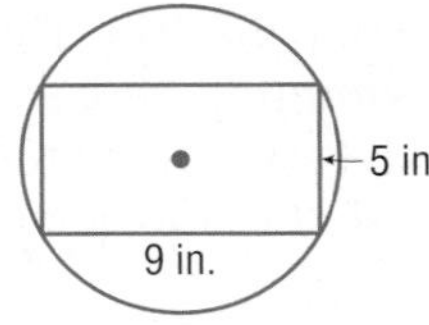


31.

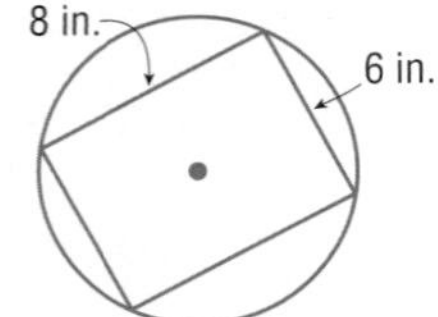


32.

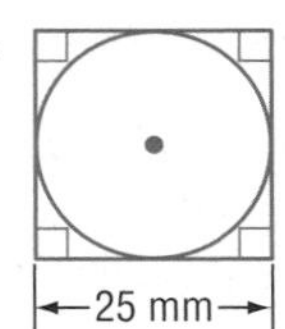


33.

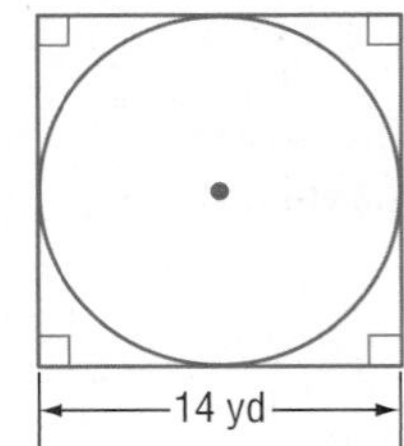


34. **DISC GOLF** Disc golf is similar to regular golf, except that a flying disc is used instead of a ball and clubs. For professional competitions, the maximum weight of a disc in grams is 8.3 times the diameter in centimeters. What is the maximum allowable weight for a disc with circumference 66.92 centimeters? Round to the nearest tenth.

35. **PATIOS** Mr. Martinez is going to build the patio shown.
    a. What is the patio's approximate circumference?
    b. If Mr. Martinez changes the plans so that the inner circle has a circumference of approximately 25 feet, what should the radius of the circle be to the nearest foot?

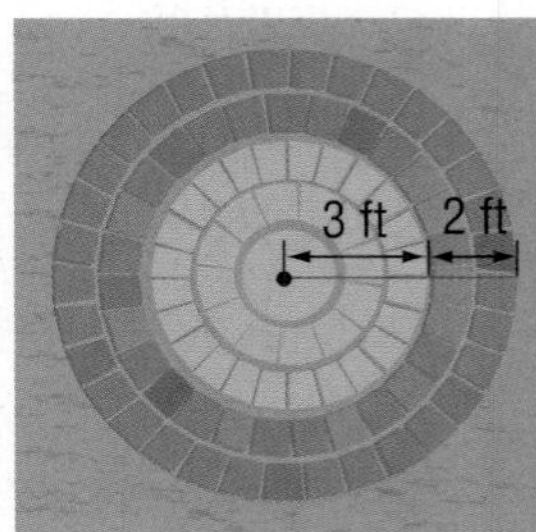


**The radius, diameter, or circumference of a circle is given. Find each missing measure to the nearest hundredth.**

36. $d = 8\frac{1}{2}$ in., $r =$ __?__, $C =$ __?__
37. $r = 11\frac{2}{5}$ ft, $d =$ __?__, $C =$ __?__
38. $C = 35x$ cm, $d =$ __?__, $r =$ __?__
39. $r = \frac{x}{8}$, $d =$ ____, $C =$ __?__

**Determine whether the circles in the figures below appear to be *congruent, concentric,* or *neither.***

**40.** 

**41.** 

**42.** 

**43** **HISTORY** The *Indian Shell Ring* on Hilton Head Island approximates a circle. If each unit on the coordinate grid represents 25 feet, how far would someone have to walk to go completely around the ring? Round to the nearest tenth.

**44.** **CCSS MODELING** A brick path is being installed around a circular pond. The pond has a circumference of 68 feet. The outer edge of the path is going to be 4 feet from the pond all the way around. What is the approximate circumference of the path? Round to the nearest hundredth.

**45.** **MULTIPLE REPRESENTATIONS** In this problem, you will explore changing dimensions in circles.

**a. Geometric** Use a compass to draw three circles in which the scale factor from each circle to the next larger circle is 1:2.

**b. Tabular** Calculate the radius (to the nearest tenth) and circumference (to the nearest hundredth) of each circle. Record your results in a table.

**c. Verbal** Explain why these three circles are geometrically similar.

**d. Verbal** Make a conjecture about the ratio between the circumferences of two circles when the ratio between their radii is 2.

**e. Analytical** The scale factor from $\odot A$ to $\odot B$ is $\frac{b}{a}$. Write an equation relating the circumference ($C_A$) of $\odot A$ to the circumference ($C_B$) of $\odot B$.

**f. Numerical** If the scale factor from $\odot A$ to $\odot B$ is $\frac{1}{3}$, and the circumference of $\odot A$ is 12 inches, what is the circumference of $\odot B$?

**46.** **BUFFON'S NEEDLE** Measure the length $\ell$ of a needle (or toothpick) in centimeters. Next, draw a set of horizontal lines that are $\ell$ centimeters apart on a sheet of plain white paper.

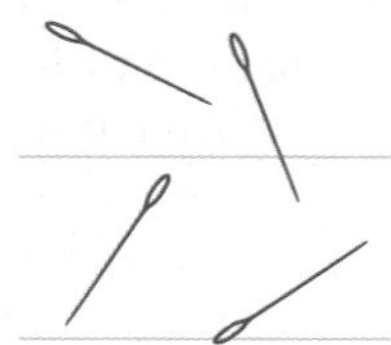

**a.** Drop the needle onto the paper. When the needle lands, record whether it touches one of the lines as a hit. Record the number of hits after 25, 50, and 100 drops.

**b.** Calculate the ratio of two times the total number of drops to the number of hits after 25, 50, and 100 drops.

**c.** How are the values you found in part **b** related to $\pi$?

**47** **MAPS** The concentric circles on the map below show the areas that are 5, 10, 15, 20, 25, and 30 miles from downtown Phoenix.

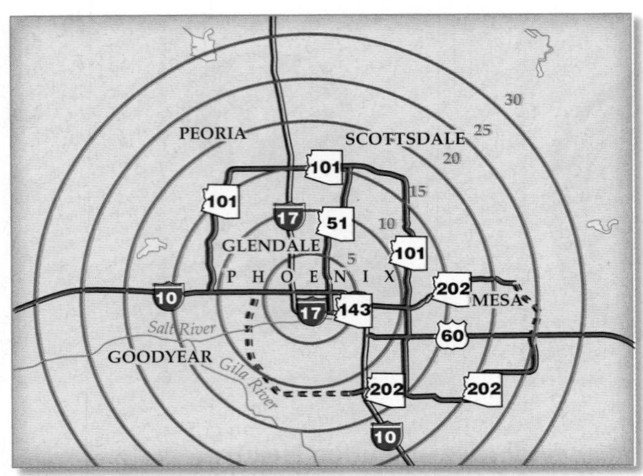


a. How much greater is the circumference of the outermost circle than the circumference of the center circle?

b. As the radii of the circles increase by 5 miles, by how much does the circumference increase?

## H.O.T. Problems Use Higher-Order Thinking Skills

**48.** **WRITING IN MATH** How can we describe the relationships that exist between circles and lines?

**49.** **REASONING** In the figure, a circle with radius $r$ is inscribed in a regular polygon and circumscribed about another.

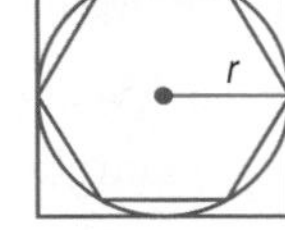


a. What are the perimeters of the circumscribed and inscribed polygons in terms of $r$? Explain.

b. Is the circumference $C$ of the circle greater or less than the perimeter of the circumscribed polygon? the inscribed polygon? Write a compound inequality comparing $C$ to these perimeters.

c. Rewrite the inequality from part **b** in terms of the diameter $d$ of the circle and interpret its meaning.

d. As the number of sides of both the circumscribed and inscribed polygons increase, what will happen to the upper and lower limits of the inequality from part **c**, and what does this imply?

**50.** **CHALLENGE** The sum of the circumferences of circles $H$, $J$, and $K$ shown at the right is $56\pi$ units. Find $KJ$.

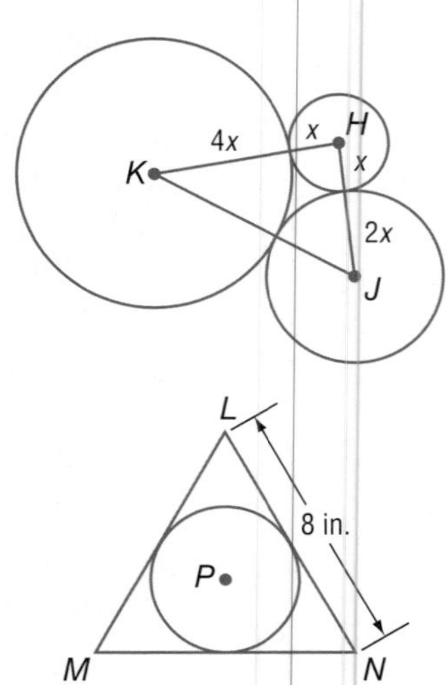


**51.** **REASONING** Is the distance from the center of a circle to a point in the interior of a circle *sometimes, always,* or *never* less than the radius of the circle? Explain.

**52.** **CCSS** **ARGUMENTS** Use the locus definition of a circle and dilations to prove that all circles are similar.

**53.** **CHALLENGE** In the figure, $\odot P$ is inscribed in equilateral triangle $LMN$. What is the circumference of $\odot P$?

**54.** **WRITING IN MATH** Research and write about the history of pi and its importance to the study of geometry.

## Standardized Test Practice

**55. GRIDDED RESPONSE** What is the circumference of $\odot T$? Round to the nearest tenth.

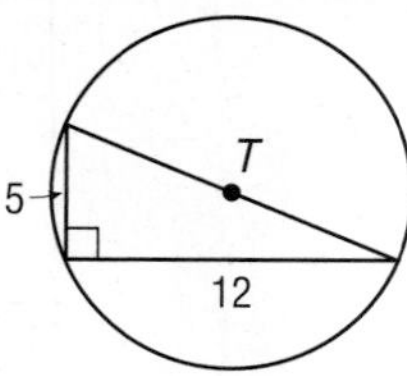


**56.** What is the radius of a table with a circumference of 10 feet?

**A** 1.6 ft
**B** 2.5 ft
**C** 3.2 ft
**D** 5 ft

**57. ALGEBRA** Bill is planning a circular vegetable garden with a fence around the border. If he can use up to 50 feet of fence, what radius can he use for the garden?

**F** 10
**G** 9
**H** 8
**J** 7

**58. SAT/ACT** What is the radius of a circle with an area of $\frac{\pi}{4}$ square units?

**A** 0.4 units
**B** 0.5 units
**C** 2 units
**D** 4 units
**E** 16 units

## Spiral Review

**Copy each figure and point *B*. Then use a ruler to draw the image of the figure under a dilation with center *B* and the scale factor *r* indicated.** (Lesson 10-8)

**59.** $r = \frac{1}{5}$

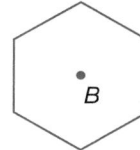

**60.** $r = \frac{2}{5}$

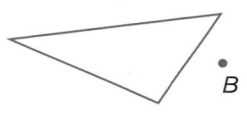

**61.** $r = 2$

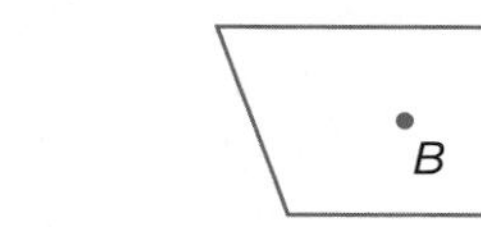


**62.** $r = 3$

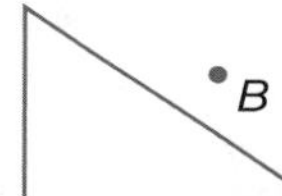


**63. ARCHITECTURE** The Louvre Pyramid is the main entrance to the Louvre Museum in Paris, France. The structure consists mainly of quadrilateral-shaped glass segments, as shown in the photo at the right. Describe one method that could be used to prove that the shapes of the segments are parallelograms. (Lesson 8-3)

## Skills Review

**Find *x*.**

**64.**

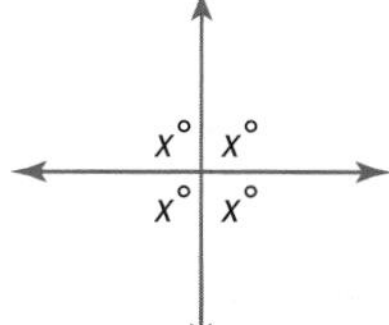


**65.**

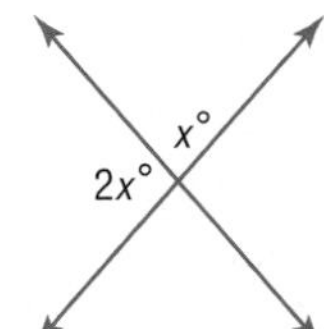


**66.**

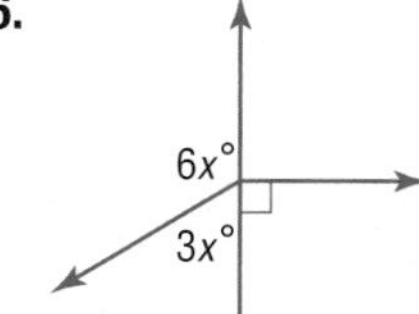


**67.**

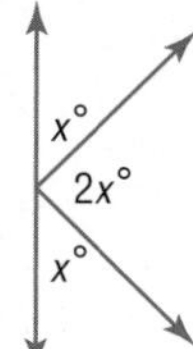


(l)Siede Pries/PhotoDisc/Getty Images, (c)Spike Mafford/PhotoDisc/Getty Images, (r)Dinodia Images/Alamy

# LESSON 11-2 Measuring Angles and Arcs

**Then**

- You measured angles and identified congruent angles.

**Now**

1. Identify central angles, major arcs, minor arcs, and semicircles, and find their measures.
2. Find arc lengths.

**Why?**

- The thirteen stars of the Betsy Ross flag are arranged equidistant from each other and from a fixed point. The distance between consecutive stars varies depending on the size of the flag, but the measure of the central angle formed by the center of the circle and any two consecutive stars is always the same.

**NewVocabulary**

central angle
arc
minor arc
major arc
semicircle
congruent arcs
adjacent arcs
arc length

**Common Core State Standards**

**Content Standards**

G.C.2 Identify and describe relationships among inscribed angles, radii, and chords.

G.C.5 Derive using similarity the fact that the length of the arc intercepted by an angle is proportional to the radius, and define the radian measure of the angle as the constant of proportionality; derive the formula for the area of a sector.

**Mathematical Practices**

6 Attend to precision.

4 Model with mathematics.

## 1 Angles and Arcs

A **central angle** of a circle is an angle with a vertex in the center of the circle. Its sides contain two radii of the circle. $\angle ABC$ is a central angle of $\odot B$.

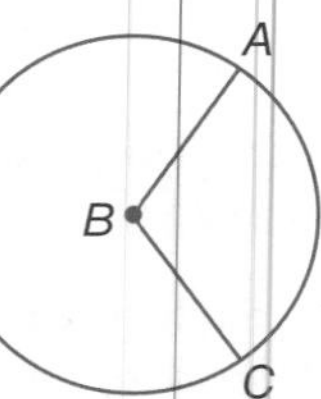


Recall from Lesson 1-4 that a *degree* is $\frac{1}{360}$ of the circular rotation about a point. This leads to the following relationship.

**KeyConcept Sum of Central Angles**

**Words** The sum of the measures of the central angles of a circle with no interior points in common is 360.

**Example** $m\angle 1 + m\angle 2 + m\angle 3 = 360$

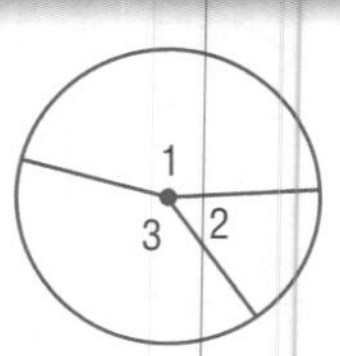

**Example 1 Find Measures of Central Angles**

**Find the value of $x$.**

| | |
|---|---|
| $m\angle GFH + m\angle HFJ + m\angle GFJ = 360$ | Sum of Central Angles |
| $130 + 90 + m\angle GFJ = 360$ | Substitution |
| $220 + m\angle GFJ = 360$ | Simplify. |
| $m\angle GFJ = 140$ | Subtract 220 from each side. |

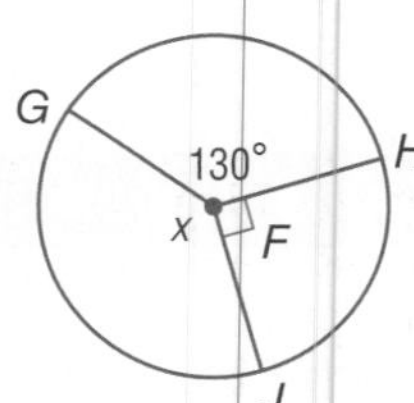


**GuidedPractice**

**1A.**

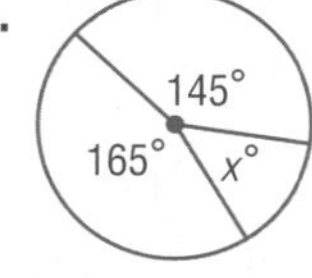


**1B.**

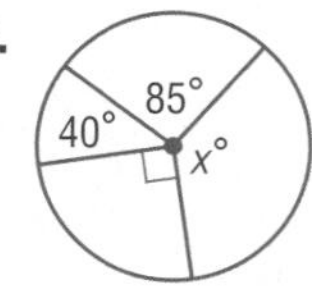


An **arc** is a portion of a circle defined by two endpoints. A central angle separates the circle into two arcs with measures related to the measure of the central angle.

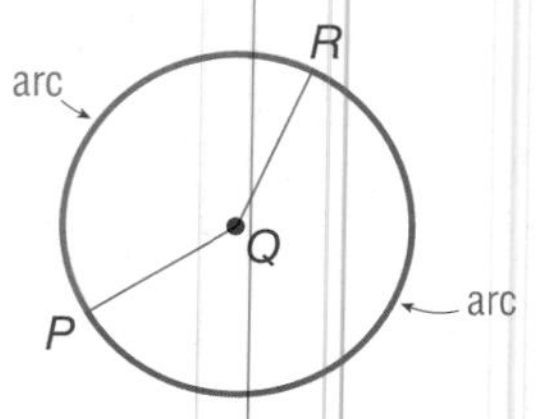


Photodisc/Getty Images

## StudyTip

**Naming Arcs** Minor arcs are named by their endpoints. Major arcs and semicircles are named by their endpoints and another point on the arc that lies between these endpoints.

## KeyConcept Arcs and Arc Measure

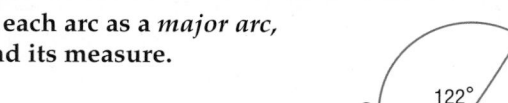

| Arc | Measure |
|---|---|
| A **minor arc** is the shortest arc connecting two endpoints on a circle. | The measure of a minor arc is less than 180 and equal to the measure of its related central angle. $m\widehat{AB} = m\angle ACB = x$ |
| A **major arc** is the longest arc connecting two endpoints on a circle. | The measure of a major arc is greater than 180, and equal to 360 minus the measure of the minor arc with the same endpoints. $m\widehat{ADB} = 360 - m\widehat{AB} = 360 - x$ |
| A **semicircle** is an arc with endpoints that lie on a diameter. | The measure of a semicircle is 180. $m\widehat{ADB} = 180$ |

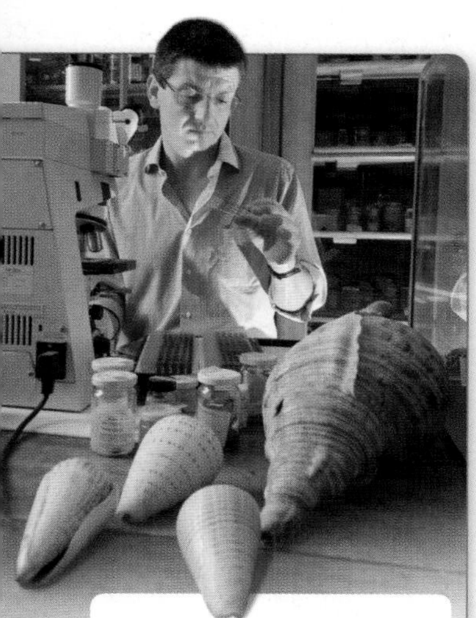

## Real-WorldCareer

**Historical Researcher** Research in museums includes authentication, verification, and description of artifacts. Employment as a historical researcher requires a minimum of a bachelor's degree in history. Refer to Exercises 42–43.

imagebroker/Alamy

## Example 2 Classify Arcs and Find Arc Measures

**$\overline{GJ}$ is a diameter of $\odot K$. Identify each arc as a *major arc, minor arc,* or *semicircle*. Then find its measure.**

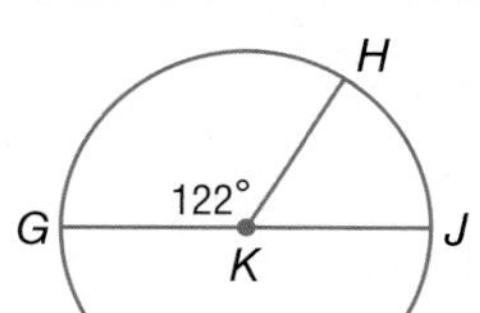


**a.** $m\widehat{GH}$

$\widehat{GH}$ is a minor arc, so $m\widehat{GH} = m\angle GKH$ or 122.

**b.** $m\widehat{GLH}$

$\widehat{GLH}$ is a major arc that shares the same endpoints as minor arc $\widehat{GH}$.

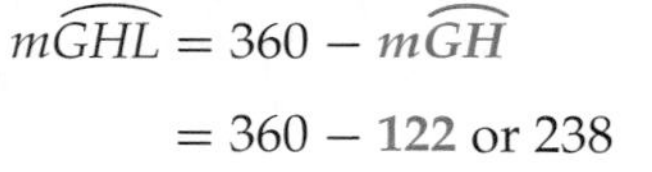

$$m\widehat{GHL} = 360 - m\widehat{GH}$$
$$= 360 - 122 \text{ or } 238$$

**c.** $m\widehat{GLJ}$

$\widehat{GLJ}$ is a semicircle, so $m\widehat{GLJ} = 180$.

### GuidedPractice

**$\overline{PM}$ is a diameter of $\odot R$. Identify each arc as a *major arc, minor arc,* or *semicircle*. Then find its measure.**

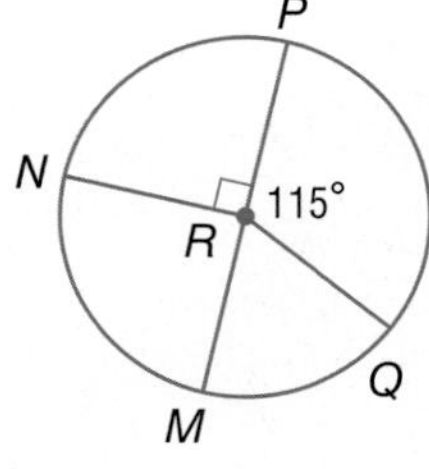


**2A.** $\widehat{MQ}$ **2B.** $\widehat{MNP}$ **2C.** $\widehat{MNQ}$

**Congruent arcs** are arcs in the same or congruent circles that have the same measure.

## Theorem 11.1

**Words** In the same circle or in congruent circles, two minor arcs are congruent if and only if their central angles are congruent.

**Example** If $\angle 1 \cong \angle 2$, then $\widehat{FG} \cong \widehat{HJ}$.
If $\widehat{FG} \cong \widehat{HJ}$, then $\angle 1 \cong \angle 2$.

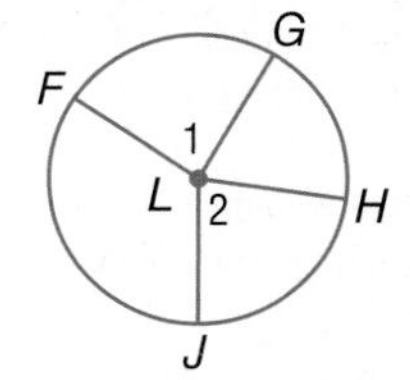


You will prove Theorem 11.1 in Exercise 52.

## Real-World Example 3 Find Arc Measures in Circle Graphs

**SPORTS** Refer to the circle graph. Find each measure.

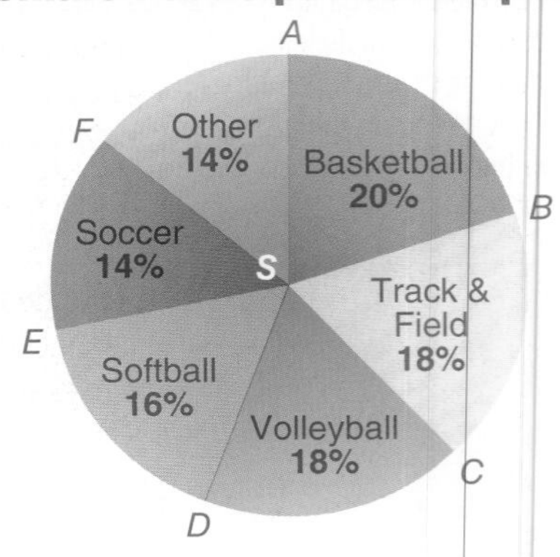


**a.** $m\widehat{CD}$

$\widehat{CD}$ is a minor arc. $m\widehat{CD} = m\angle CSD$

$\angle CSD$ represents 18% of the whole, or 18% of the circle.

$m\angle CSD = 0.18(360)$ Find 18% of 360.

$= 64.8$ Simplify.

**b.** $m\widehat{BC}$

The percents for volleyball and track and field are equal, so the central angles are congruent and the corresponding arcs are congruent.

$m\widehat{BC} = m\widehat{CD} = 64.8$

**GuidedPractice**

**3A.** $m\widehat{EF}$ **3B.** $m\widehat{FA}$

**Adjacent arcs** are arcs in a circle that have exactly one point in common. In $\odot M$, $\widehat{HJ}$ and $\widehat{JK}$ are adjacent arcs. As with adjacent angles, you can add the measures of adjacent arcs.

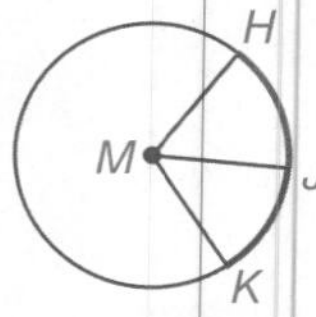


### Postulate 11.1 Arc Addition Postulate

**Words** The measure of an arc formed by two adjacent arcs is the sum of the measures of the two arcs.

**Example** $m\widehat{XYZ} = m\widehat{XY} + m\widehat{YZ}$

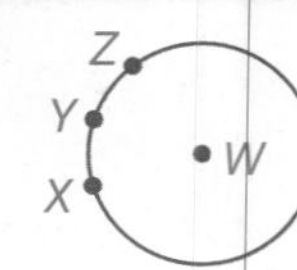

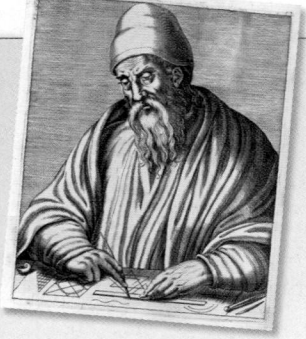

**Math HistoryLink**

Euclid (c. 325–265 B.C.) The 13 books of Euclid's *Elements* are influential works of science. In them, geometry and other branches of mathematics are logically developed. Book 3 of *Elements* is devoted to circles, arcs, and angles.

Hulton Archive/Getty Images

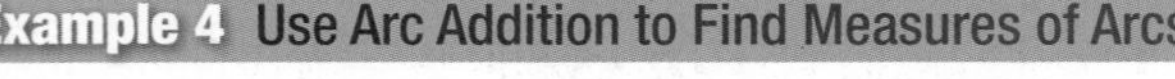

## Example 4 Use Arc Addition to Find Measures of Arcs

**Find each measure in $\odot F$.**

**a.** $m\widehat{AED}$

$m\widehat{AED} = m\widehat{AE} + m\widehat{ED}$ Arc Addition Postulate

$= m\angle AFE + m\angle EFD$ $m\widehat{AE} = m\angle AFE$, $m\widehat{ED} = m\angle EFD$

$= 63 + 90$ or $153$ Substitution

**b.** $m\widehat{ADB}$

$m\widehat{ADB} = m\widehat{AE} + m\widehat{EDB}$ Arc Addition Postulate

$= 63 + 180$ or $243$ $\widehat{EDB}$ is a semicircle, so $m\widehat{EDB} = 180$.

**GuidedPractice**

**4A.** $m\widehat{CE}$ **4B.** $m\widehat{ABD}$

**WatchOut!**

**Arc Length** The length of an arc is given in linear units, such as centimeters. The measure of an arc is given in degrees.

## 2 Arc Length

**Arc length** is the distance between the endpoints along an arc measured in linear units. Since an arc is a portion of a circle, its length is a fraction of the circumference.

### KeyConcept Arc Length

**Words** The ratio of the **length of an arc** $\ell$ to the circumference of the circle is equal to the ratio of the **degree measure of the arc** to 360.

**Proportion** $\frac{\ell}{2\pi r} = \frac{x}{360}$ or

**Equation** $\ell = \frac{x}{360} \cdot 2\pi r$

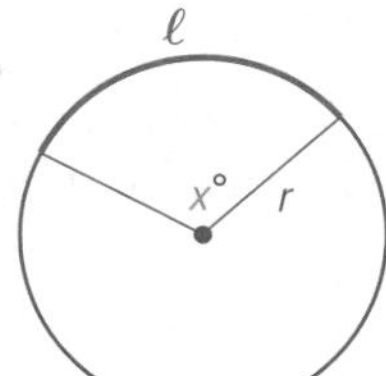


### Example 5 Find Arc Length

**Find the length of $\widehat{ZY}$. Round to the nearest hundredth.**

**a.**

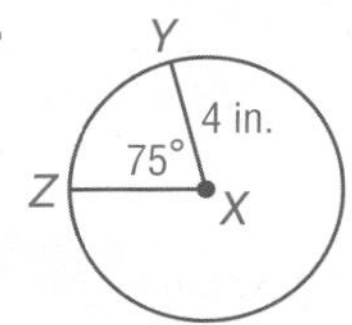


| | |
|---|---|
| $\ell = \frac{x}{360} \cdot 2\pi r$ | Arc Length Equation |
| $= \frac{75}{360} \cdot 2\pi(4)$ | Substitution |
| $\approx 5.24$ in. | Use a calculator. |

**b.**

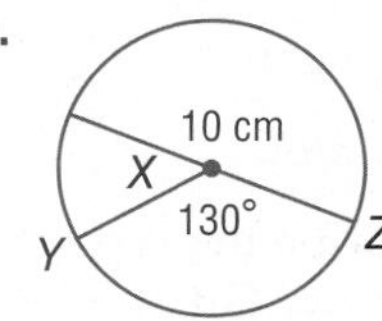


| | |
|---|---|
| $\ell = \frac{x}{360} \cdot 2\pi r$ | Arc Length Equation |
| $= \frac{130}{360} \cdot 2\pi(5)$ | Substitution |
| $\approx 11.34$ cm | Use a calculator. |

**c.**

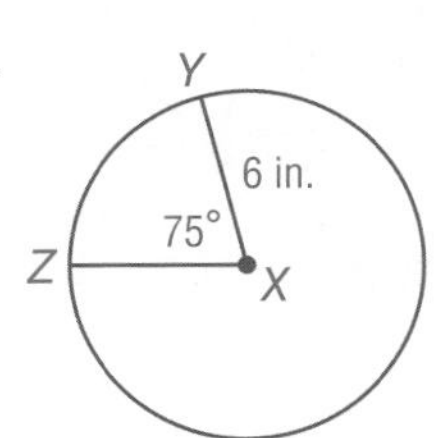


| | |
|---|---|
| $\ell = \frac{x}{360} \cdot 2\pi r$ | Arc Length Equation |
| $= \frac{75}{360} \cdot 2\pi(6)$ | Substitution |
| $\approx 7.85$ in. | Use a calculator. |

Notice that $\widehat{ZY}$ has the same measure, 75, in both Examples 5a and 5c. The arc lengths, however, are different. This is because they are in circles that have different radii.

**StudyTip**

**Alternate Method** The arc lengths in Examples 5a, 5b, and 5c could also have been calculated using the arc length proportion $\frac{\ell}{2\pi r} = \frac{x}{360}$.

### GuidedPractice

**Find the length of $\widehat{AB}$. Round to the nearest hundredth.**

**5A.**

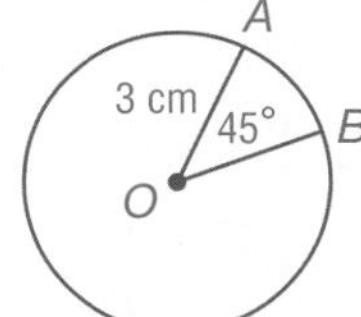


**5B.**

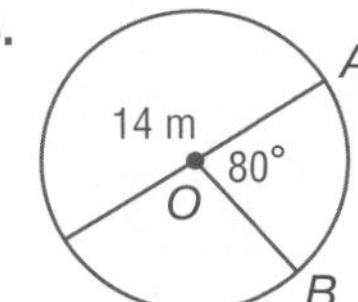


**5C.**

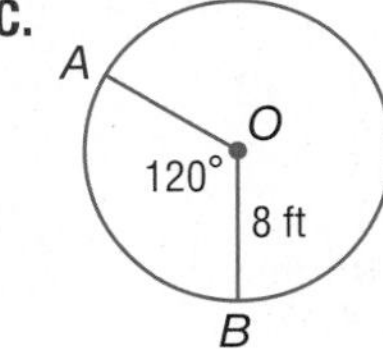

## Check Your Understanding

**Example 1** **Find the value of $x$.**

1. 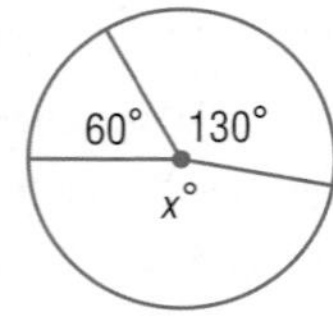


2. 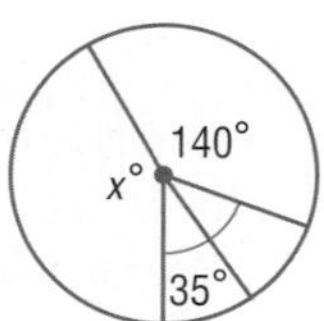


**Example 2** **PRECISION** $\overline{HK}$ **and** $\overline{IG}$ **are diameters of** $\odot L$. **Identify each arc as a *major arc, minor arc,* or *semicircle*. Then find its measure.**

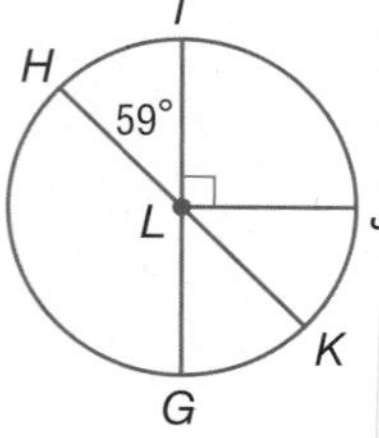


3. $m\widehat{IHJ}$
4. $m\widehat{HI}$
5. $m\widehat{HGK}$

**Example 3**

6. **RESTAURANTS** The graph shows the results of a survey taken by diners relating what is most important about the restaurants where they eat.

   a. Find $m\widehat{AB}$.

   b. Find $m\widehat{BC}$.

   c. Describe the type of arc that the category Great Food represents.

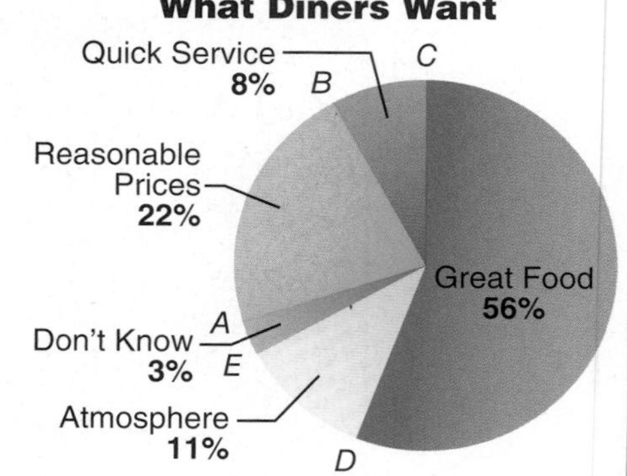


**Example 4** $\overline{QS}$ **is a diameter of** $\odot V$. **Find each measure.**

7. $m\widehat{STP}$
8. $m\widehat{QRT}$
9. $m\widehat{PQR}$

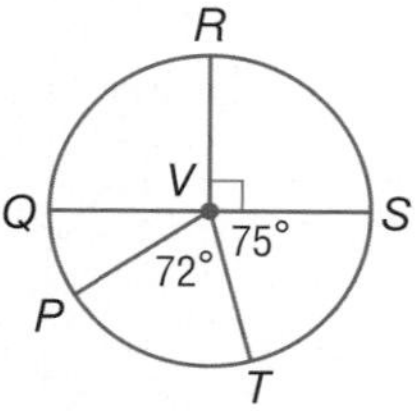


**Example 5** **Find the length of** $\widehat{JK}$. **Round to the nearest hundredth.**

10. 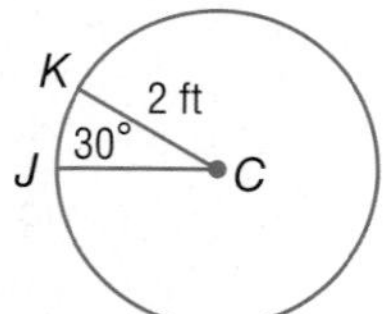


11. 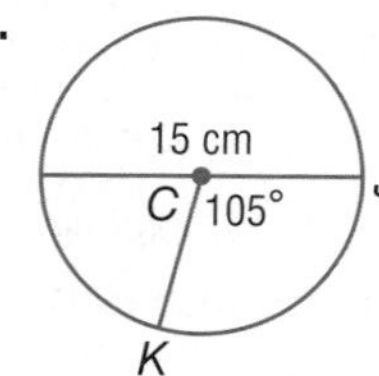


## Practice and Problem Solving

**Example 1** **Find the value of $x$.**

12. 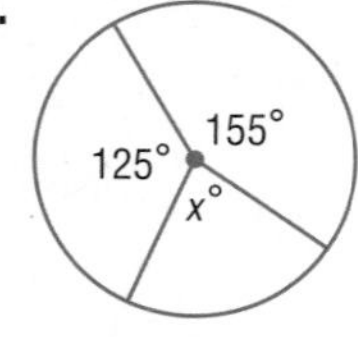


13. 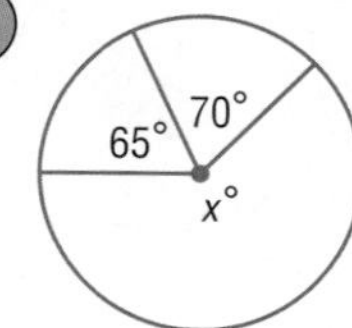


14. 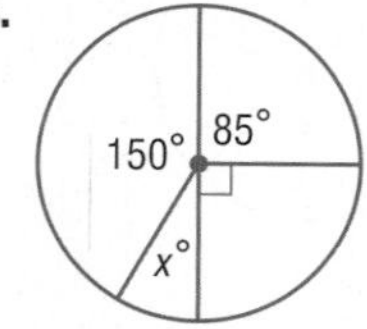


15. 

**Example 2** $\overline{AD}$ **and** $\overline{CG}$ **are diameters of** $\odot B$**. Identify each arc as a** ***major arc, minor arc,*** **or** ***semicircle*****. Then find its measure.**

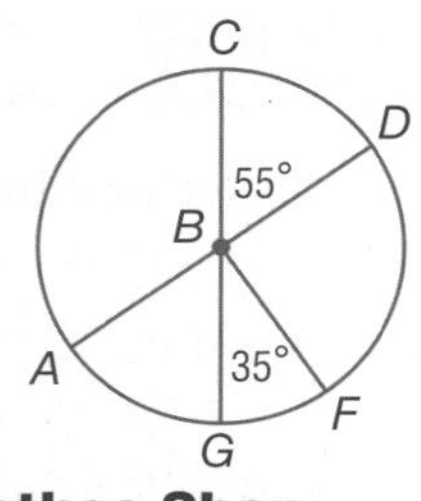


**16.** $m\widehat{CD}$

**17.** $m\widehat{AC}$

**18.** $m\widehat{CFG}$

**19.** $m\widehat{CGD}$

**20.** $m\widehat{GCF}$

**21.** $m\widehat{ACD}$

**22.** $m\widehat{AG}$

**23.** $m\widehat{ACF}$

**Example 3**

**24. SHOPPING** The graph shows the results of a survey in which teens were asked where the best place was to shop for clothes.

**Best Places to Clothes Shop**

None of these 9%
Online 9%
Vintage stores 4%
Flea markets 2%
Mall 76%

**a.** What would be the arc measures associated with the mall and vintage stores categories?

**b.** Describe the kinds of arcs associated with the category "Mall" and the category "None of these."

**c.** Are there any congruent arcs in this graph? Explain.

**25.** **CCSS MODELING** The table shows the results of a survey in which Americans were asked how long food could be on the floor and still be safe to eat.

**a.** If you were to construct a circle graph of this information, what would be the arc measures associated with the first two categories?

**b.** Describe the kind of arcs associated with the first category and the last category.

**c.** Are there any congruent arcs in this graph? Explain.

| Dropped Food | |
|---|---|
| **Do you eat food dropped on the floor?** | |
| Not safe to eat | 78% |
| Three-second rule* | 10% |
| Five-second rule* | 8% |
| Ten-second rule* | 4% |

**Source:** American Diabetic Association
* The length of time the food is on the floor.

**Examples 2, 4** **ENTERTAINMENT** **Use the Ferris wheel shown to find each measure.**

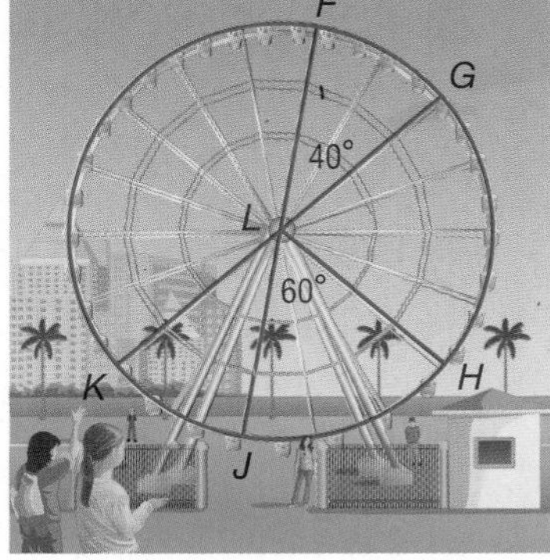


**26.** $m\widehat{FG}$

**27.** $m\widehat{JH}$

**28.** $m\widehat{JKF}$

**29.** $m\widehat{JFH}$

**30.** $m\widehat{GHF}$

**31.** $m\widehat{GHK}$

**32.** $m\widehat{HK}$

**33.** $m\widehat{JKG}$

**34.** $m\widehat{KFH}$

**35.** $m\widehat{HGF}$

**Example 5** **Use** $\odot P$ **to find the length of each arc. Round to the nearest hundredth.**

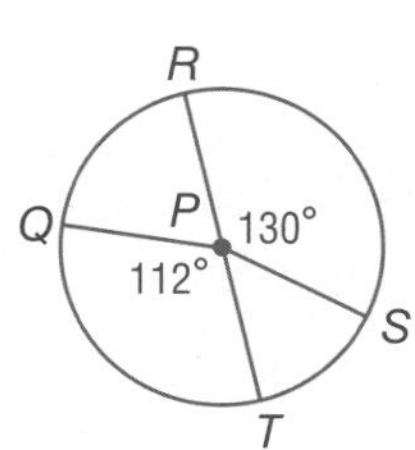


**36.** $\widehat{RS}$, if the radius is 2 inches

**37** $\widehat{QT}$, if the diameter is 9 centimeters

**38.** $\widehat{QR}$, if $PS = 4$ millimeters

**39.** $\widehat{RS}$, if $RT = 15$ inches

**40.** $\widehat{QRS}$, if $RT = 11$ feet

**41.** $\widehat{RTS}$, if $PQ = 3$ meters

**HISTORY** The figure shows the stars in the Betsy Ross flag referenced at the beginning of the lesson.

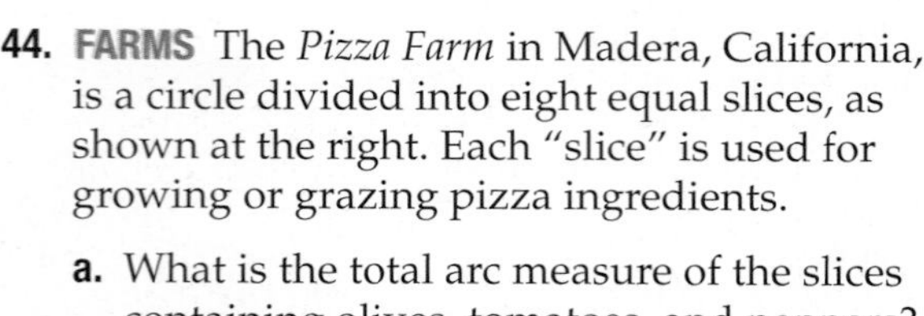

**42.** What is the measure of central angle *A*? Explain how you determined your answer.

**43.** If the diameter of the circle were doubled, what would be the effect on the arc length from the center of one star *B* to the next star *C*?

**44. FARMS** The *Pizza Farm* in Madera, California, is a circle divided into eight equal slices, as shown at the right. Each "slice" is used for growing or grazing pizza ingredients.

**a.** What is the total arc measure of the slices containing olives, tomatoes, and peppers?

**b.** The circle is 125 feet in diameter. What is the arc length of one slice? Round to the nearest hundredth.

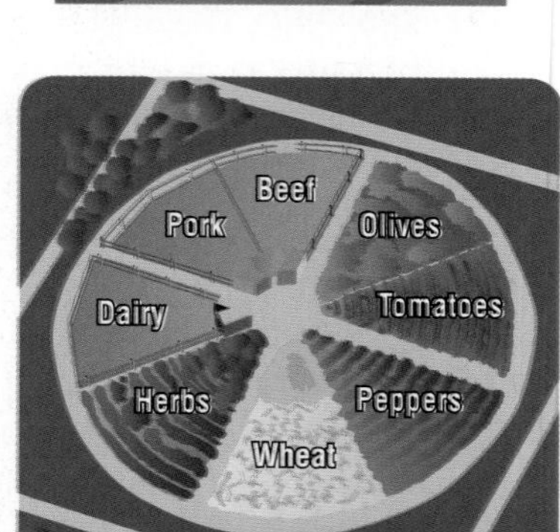


**CCSS REASONING** **Find each measure. Round each linear measure to the nearest hundredth and each arc measure to the nearest degree.**

**45.** circumference of $\odot S$

**46.** $m\widehat{CD}$

**47.** radius of $\odot K$

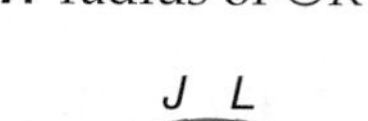

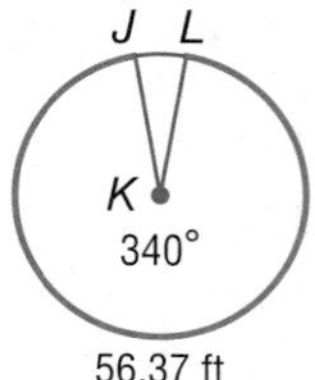


**ALGEBRA** **In $\odot C$, $m\angle HCG = 2x$ and $m\angle HCD = 6x + 28$. Find each measure.**

**48.** $m\widehat{EF}$

**49.** $m\widehat{HD}$

**50.** $m\widehat{HGF}$

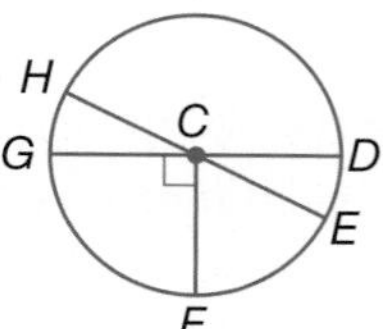


**51) RIDES** A pirate ship ride follows a semicircular path, as shown in the diagram.

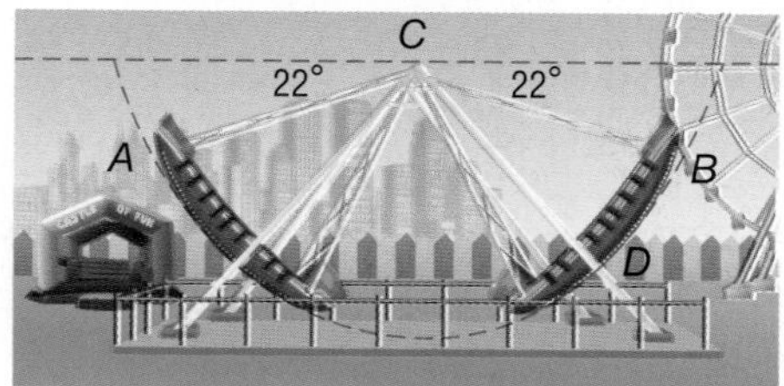


**a.** What is $m\widehat{AB}$?

**b.** If $CD = 62$ feet, what is the length of $\widehat{AB}$? Round to the nearest hundredth.

**52. PROOF** Write a two-column proof of Theorem 11.1.

**Given:** $\angle BAC \cong \angle DAE$

**Prove:** $\widehat{BC} \cong \widehat{DE}$

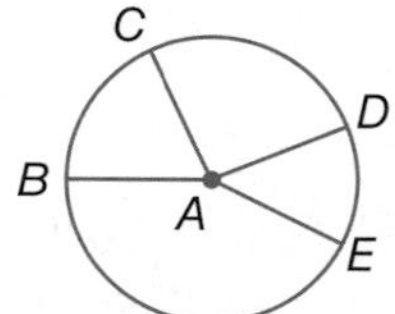

**53.** **COORDINATE GEOMETRY** In the graph, point $M$ is located at the origin. Find each measure in $\odot M$. Round each linear measure to the nearest hundredth and each arc measure to the nearest tenth degree.

**a.** $m\widehat{JL}$ 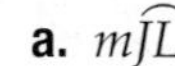 **b.** $m\widehat{KL}$ 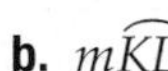 **c.** $m\widehat{JK}$

**d.** length of $\widehat{JL}$ **e.** length of $\widehat{JK}$

y
J (5, 12)
K (12, 5)
M
L (13, 0) x

**54.** **ARC LENGTH AND RADIAN MEASURE** In this problem, you will use concentric circles to show that the length of the arc intercepted by a central angle of a circle is dependent on the circle's radius.

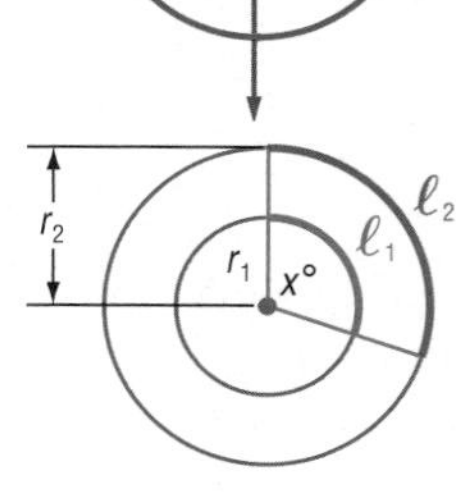


**a.** Compare the measures of arc $\ell_1$ and arc $\ell_2$. Then compare the lengths of arc $\ell_1$ and arc $\ell_2$. What do these two comparisons suggest?

**b.** Use similarity transformations (dilations) to explain why the length of an arc $\ell$ intercepted by a central angle of a circle is proportional to the circle's radius $r$. That is, explain why we can say that for this diagram, $\frac{\ell_1}{r_1} = \frac{\ell_2}{r_2}$.

**c.** Write expressions for the lengths of arcs $\ell_1$ and $\ell_2$. Use these expressions to identify the constant of proportionality $k$ in $\ell = kr$.

**d.** The expression that you wrote for $k$ in part **c** gives the *radian measure* of an angle. Use it to find the radian measure of an angle measuring 90°.

## H.O.T. Problems Use Higher-Order Thinking Skills

**55.** **ERROR ANALYSIS** Brody says that $\widehat{WX}$ and $\widehat{YZ}$ are congruent since their central angles have the same measure. Selena says they are not congruent. Is either of them correct? Explain your reasoning.

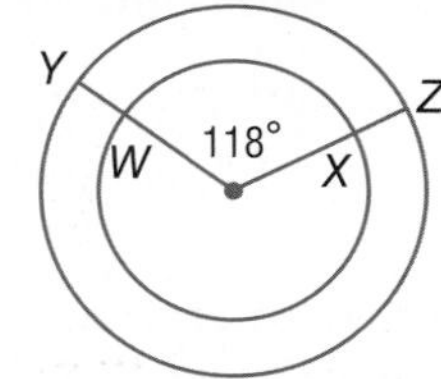


**CCSS ARGUMENTS Determine whether each statement is *sometimes, always,* or *never* true. Explain your reasoning.**

**56.** The measure of a minor arc is less than 180.

**57.** If a central angle is obtuse, its corresponding arc is a major arc.

**58.** The sum of the measures of adjacent arcs of a circle depends on the measure of the radius.

**59.** **CHALLENGE** The measures of $\widehat{LM}$, $\widehat{MN}$, and $\widehat{NL}$ are in the ratio 5:3:4. Find the measure of each arc.

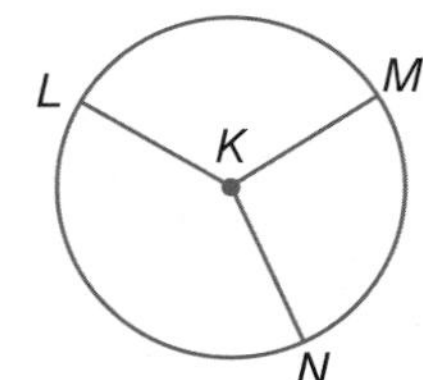


**60.** **OPEN ENDED** Draw a circle and locate three points on the circle. Estimate the measures of the three nonoverlapping arcs that are formed. Then use a protractor to find the measure of each arc. Label your circle with the arc measures.

**61.** **CHALLENGE** The time shown on an analog clock is 8:10. What is the measure of the angle formed by the hands of the clock?

**62.** **WRITING IN MATH** Describe the three different types of arcs in a circle and the method for finding the measure of each one.

## Standardized Test Practice

**63.** What is the value of $x$?

A 120
B 135
C 145
D 160

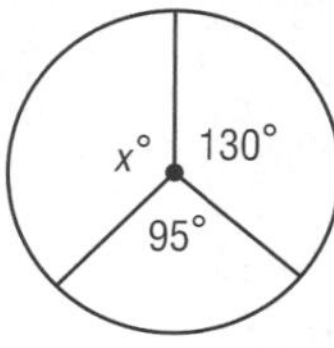


**64. GRIDDED RESPONSE** In $\odot B$, $m\angle LBM = 3x$ and $m\angle LBQ = 4x + 61$. What is the measure of $\angle PBQ$?

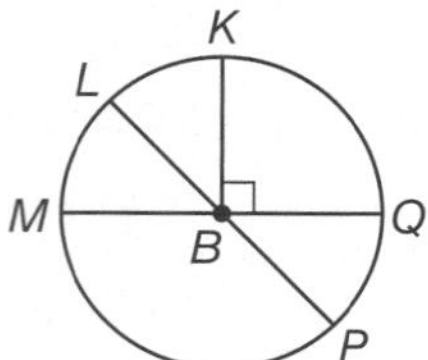


**65. ALGEBRA** A rectangle's width is represented by $x$ and its length by $y$. Which expression best represents the area of the rectangle if the length and width are tripled?

F $3xy$
G $3(xy)^2$
H $9xy$
J $(xy)^3$

**66. SAT/ACT** What is the area of the shaded region if $r = 4$?

A $64 - 16\pi$
B $16 - 16\pi$
C $16 - 8\pi$
D $64 - 8\pi$
E $64\pi - 16$

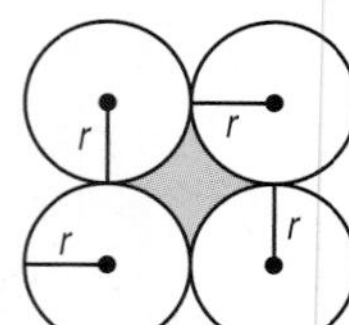


## Spiral Review

**Refer to $\odot J$.** (Lesson 11-1)

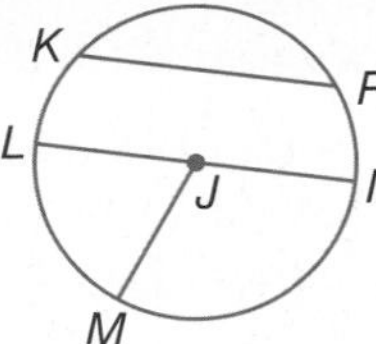


**67.** Name the center of the circle.

**68.** Identify a chord that is also a diameter.

**69.** If $LN = 12.4$, what is $JM$?

**Graph the image of each polygon with the given vertices after a dilation centered at the origin with the given scale factor.** (Lesson 10-8)

**70.** $X(-1, 2)$, $Y(2, 1)$, $Z(-1, -2)$; $r = 3$

**71.** $A(-4, 4)$, $B(4, 4)$, $C(4, -4)$, $D(-4, -4)$; $r = 0.25$

**72. BASEBALL** The diagram shows some dimensions of Comiskey Park in Chicago, Illinois. $\overline{BD}$ is a segment from home plate to dead center field, and $\overline{AE}$ is a segment from the left field foul pole to the right field foul pole. If the center fielder is standing at $C$, how far is he from home plate? (Lesson 10-3)

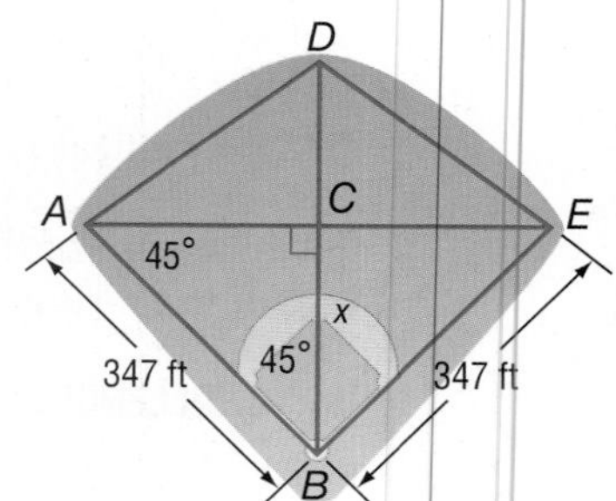


**Find $x$, $y$, and $z$.** (Lesson 10-1)

**73.**

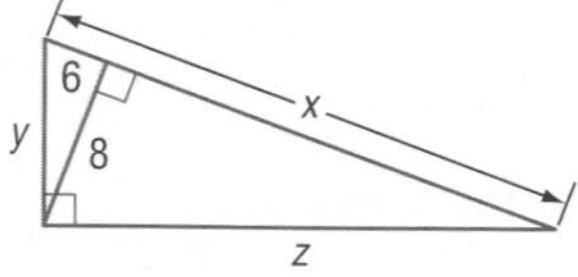


**74.**

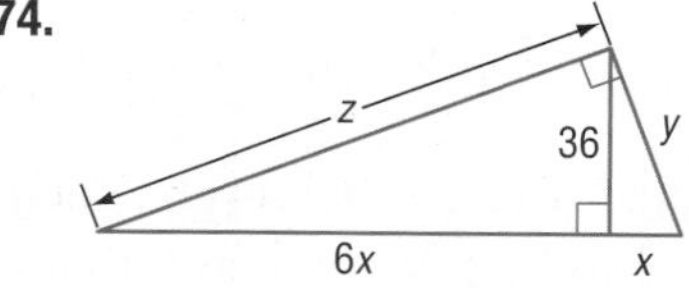


## Skills Review

**Find $x$.**

**75.** $24^2 + x^2 = 26^2$

**76.** $x^2 + 5^2 = 13^2$

**77.** $30^2 + 35^2 = x^2$

LESSON

# 11-3 Arcs and Chords

**Then**

- You used the relationships between arcs and angles to find measures.

**Now**

1. Recognize and use relationships between arcs and chords.
2. Recognize and use relationships between arcs, chords, and diameters.

**Why?**

- Embroidery hoops are used in sewing, quilting, and cross-stitching, as well as for embroidering. The endpoints of the snowflake shown are both the endpoints of a chord and the endpoints of an arc.

**Common Core State Standards**

**Content Standards**

**G.C.2** Identify and describe relationships among inscribed angles, radii, and chords.

**G.MG.3** Apply geometric methods to solve problems (e.g., designing an object or structure to satisfy physical constraints or minimize cost; working with typographic grid systems based on ratios). ★

**Mathematical Practices**

4 Model with mathematics.

3 Construct viable arguments and critique the reasoning of others.

## 1 Arcs and Chords

A *chord* is a segment with endpoints on a circle. If a chord is not a diameter, then its endpoints divide the circle into a major and a minor arc.

### Theorem 11.2

**Words** In the same circle or in congruent circles, two minor arcs are congruent if and only if their corresponding chords are congruent.

**Example** $\widehat{FG} \cong \widehat{HJ}$ if and only if $\overline{FG} \cong \overline{HJ}$.

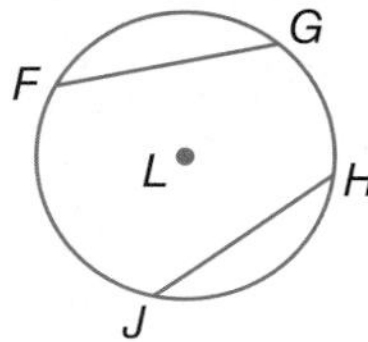


### Proof Theorem 11.2 (part 1)

**Given:** $\odot P$; $\widehat{QR} \cong \widehat{ST}$

**Prove:** $\overline{QR} \cong \overline{ST}$

**Proof:**

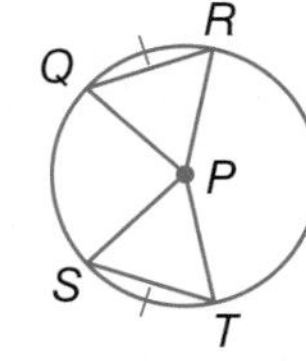


| Statements | Reasons |
|---|---|
| 1. $\odot P$, $\widehat{QR} \cong \widehat{ST}$ | 1. Given |
| 2. $\angle QPR \cong \angle SPT$ | 2. If arcs are $\cong$, their corresponding central $\angle$s are $\cong$. |
| 3. $\overline{QP} \cong \overline{PR} \cong \overline{SP} \cong \overline{PT}$ | 3. All radii of a circle are $\cong$. |
| 4. $\triangle PQR \cong \triangle PST$ | 4. SAS |
| 5. $\overline{QR} \cong \overline{ST}$ | 5. CPCTC |

You will prove part 2 of Theorem 11.2 in Exercise 25.

### Real-World Example 1 Use Congruent Chords to Find Arc Measure

**CRAFTS In the embroidery hoop, $\overline{AB} \cong \overline{CD}$ and $m\widehat{AB} = 60$. Find $m\widehat{CD}$.**

$\overline{AB}$ and $\overline{CD}$ are congruent chords, so the corresponding arcs $\widehat{AB}$ and $\widehat{CD}$ are congruent. $m\widehat{AB} = m\widehat{CD} = 60$

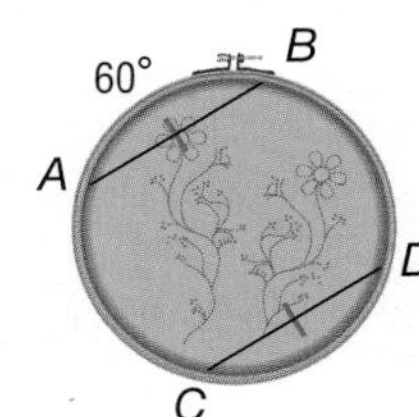


**Guided Practice**

1. If $m\widehat{AB} = 78$ in the embroidery hoop, find $m\widehat{CD}$.

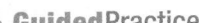

## Example 2 Use Congruent Arcs to Find Chord Lengths

**ALGEBRA In the figures, $\odot J \cong \odot K$ and $\widehat{MN} \cong \widehat{PQ}$. Find $PQ$.**

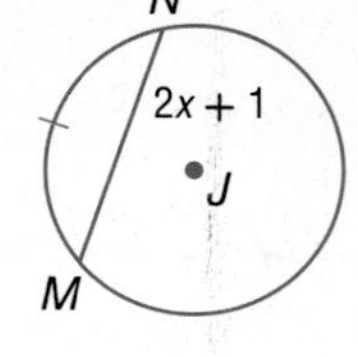

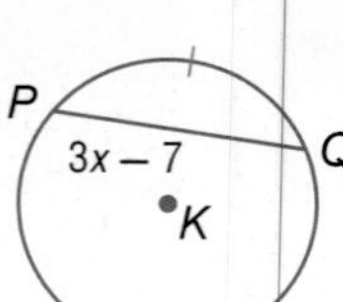


$\widehat{MN}$ and $\widehat{PQ}$ are congruent arcs in congruent circles, so the corresponding chords $\overline{MN}$ and $\overline{PQ}$ are congruent.

$MN = PQ$ — Definition of congruent segments

$2x + 1 = 3x - 7$ — Substitution

$8 = x$ — Simplify.

So, $PQ = 3(8) - 7$ or 17.

### GuidedPractice

**2.** In $\odot W$, $\widehat{RS} \cong \widehat{TV}$. Find $RS$.

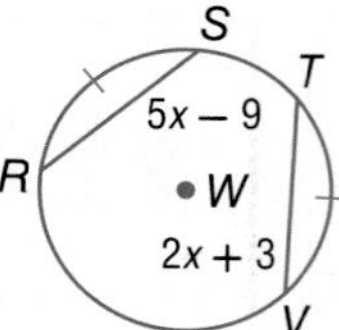


**StudyTip**

**Arc Bisectors** In the figure below, $\overline{FH}$ is an arc bisector of $\widehat{JG}$.

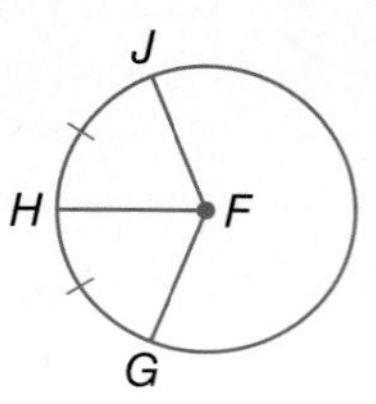


## 2 Bisecting Arcs and Chords

If a line, segment, or ray divides an arc into two congruent arcs, then it *bisects* the arc.

### Theorems

**11.3** If a diameter (or radius) of a circle is perpendicular to a chord, then it bisects the chord and its arc.

**Example** If diameter $\overline{AB}$ is perpendicular to chord $\overline{XY}$, then $\overline{XZ} \cong \overline{ZY}$ and $\widehat{XB} \cong \widehat{BY}$.

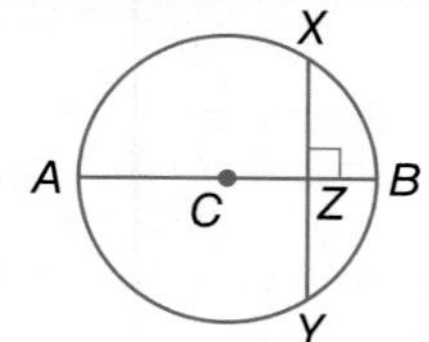


**11.4** The perpendicular bisector of a chord is a diameter (or radius) of the circle.

**Example** If $\overline{AB}$ is a perpendicular bisector of chord $\overline{XY}$, then $\overline{AB}$ is a diameter of $\odot C$.

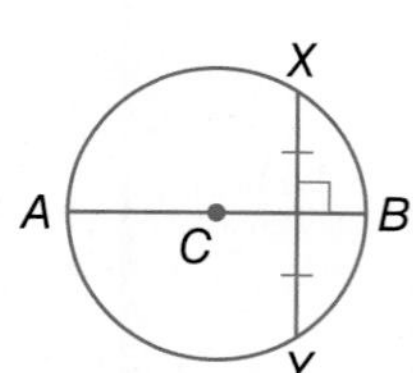


You will prove Theorems 11.3 and 11.4 in Exercises 26 and 28, respectively.

## Example 3 Use a Radius Perpendicular to a Chord

**In $\odot S$, $m\widehat{PQR} = 98$. Find $m\widehat{PQ}$.**

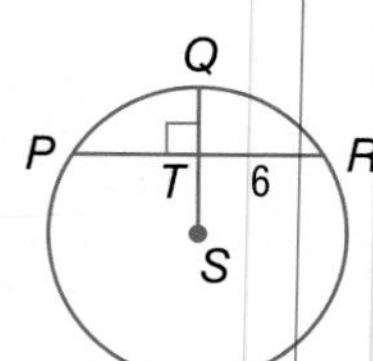


Radius $\overline{SQ}$ is perpendicular to chord $\overline{PR}$. So by Theorem 10.3, $\overline{SQ}$ bisects $\widehat{PQR}$. Therefore, $m\widehat{PQ} = m\widehat{QR}$. By substitution, $m\widehat{PQ} = \frac{98}{2}$ or 49.

### GuidedPractice

**3.** In $\odot S$, find $PR$.

## Real-World Example 4 Use a Diameter Perpendicular to a Chord

**STAINED GLASS In the stained glass window, diameter $\overline{GH}$ is 30 inches long and chord $\overline{KM}$ is 22 inches long. Find $JL$.**

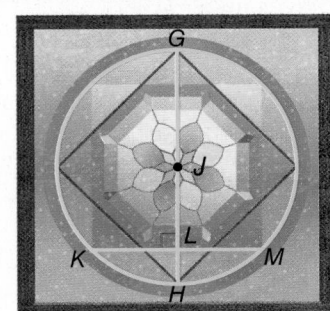

**Step 1** Draw radius $\overline{JK}$.

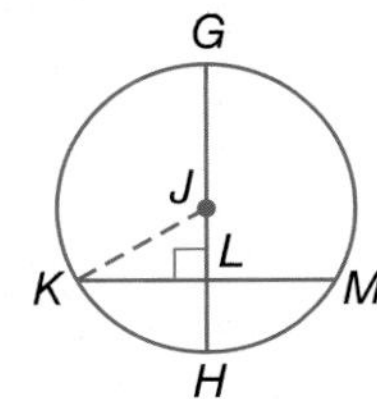


This forms right $\triangle JKL$.

**Step 2** Find $JK$ and $KL$.

Since $GH = 30$ inches, $JH = 15$ inches. All radii of a circle are congruent, so $JK = 15$ inches.

Since diameter $\overline{GH}$ is perpendicular to $\overline{KM}$, $\overline{GH}$ bisects chord $\overline{KM}$ by Theorem 10.3. So, $KL = \frac{1}{2}(22)$ or 11 inches.

**Step 3** Use the Pythagorean Theorem to find $JL$.

| | |
|---|---|
| $KL^2 + JL^2 = JK^2$ | Pythagorean Theorem |
| $11^2 + JL^2 = 15^2$ | $KL = 11$ and $JK = 15$ |
| $121 + JL^2 = 225$ | Simplify. |
| $JL^2 = 104$ | Subtract 121 from each side. |
| $JL = \sqrt{104}$ | Take the positive square root of each side. |

So, $JL$ is $\sqrt{104}$ or about 10.20 inches long.

### Real-WorldLink

To make stained glass windows, glass is heated to a temperature of 2000 degrees, until it is the consistency of taffy. The colors are caused by the addition of metallic oxides.

**Source:** *Artistic Stained Glass by Regg*

### GuidedPractice

**4.** In $\odot R$, find $TV$. Round to the nearest hundredth.

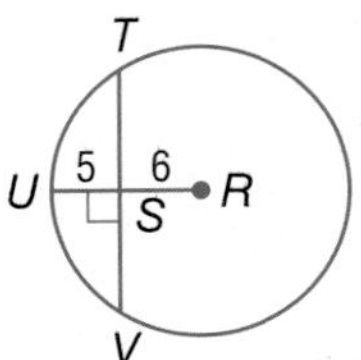


### StudyTip

**Drawing Segments** You can add any known information to a figure to help you solve the problem. In Example 4, radius $\overline{JK}$ was drawn.

In addition to Theorem 11.2, you can use the following theorem to determine whether two chords in a circle are congruent.

### Theorem 11.5

**Words** In the same circle or in congruent circles, two chords are congruent if and only if they are equidistant from the center.

**Example** $\overline{FG} \cong \overline{JH}$ if and only if $LX = LY$.

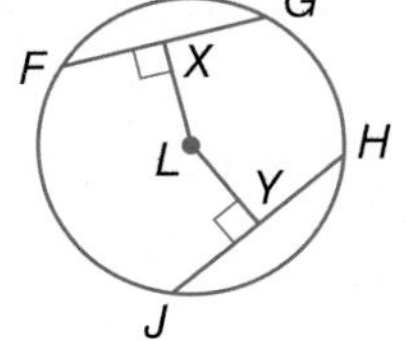


You will prove Theorem 11.5 in Exercises 29 and 30.

## Example 5 Chords Equidistant from Center

**ALGEBRA In ⊙$A$, $WX = XY = 22$. Find $AB$.**

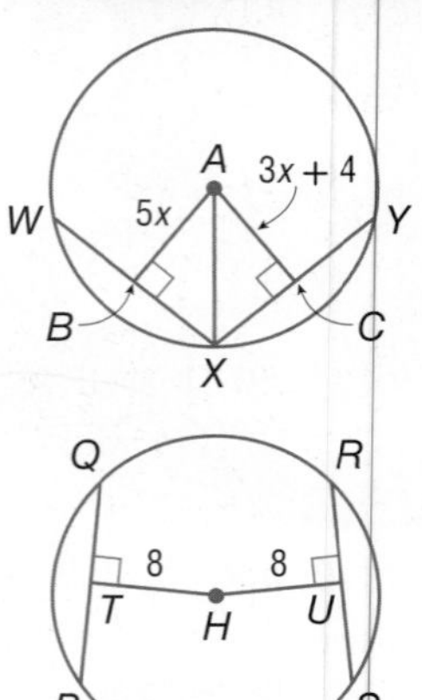


Since chords $\overline{WX}$ and $\overline{XY}$ are congruent, they are equidistant from $A$. So, $AB = AC$.

$AB = AC$

$5x = 3x + 4$ Substitution

$x = 2$ Simplify.

So, $AB = 5(2)$ or 10.

**Guided Practice**

**5.** In ⊙$H$, $PQ = 3x - 4$ and $RS = 14$. Find $x$.

You can use Theorem 11.5 to find the point equidistant from three noncollinear points.

## Construction Circle Through Three Noncollinear Points

**Step 1**

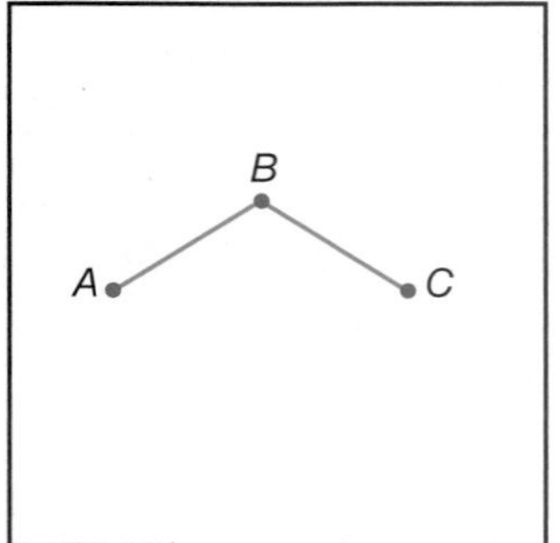


Draw three noncollinear points $A$, $B$, and $C$. Then draw segments $\overline{AB}$ and $\overline{BC}$.

**Step 2**

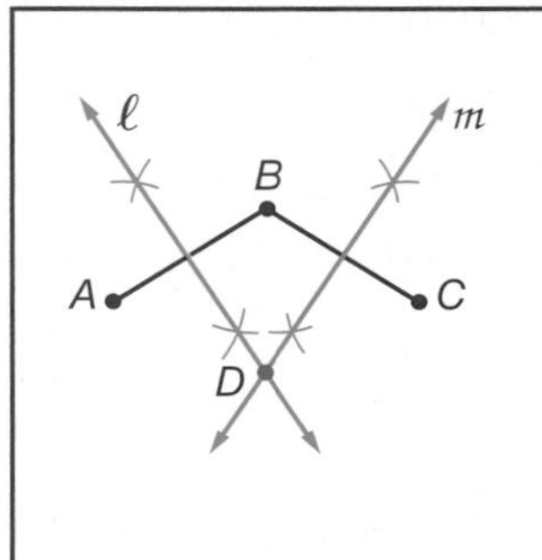


Construct the perpendicular bisectors $\ell$ and $m$ of $\overline{AB}$ and $\overline{BC}$. Label the point of intersection $D$.

**Step 3**

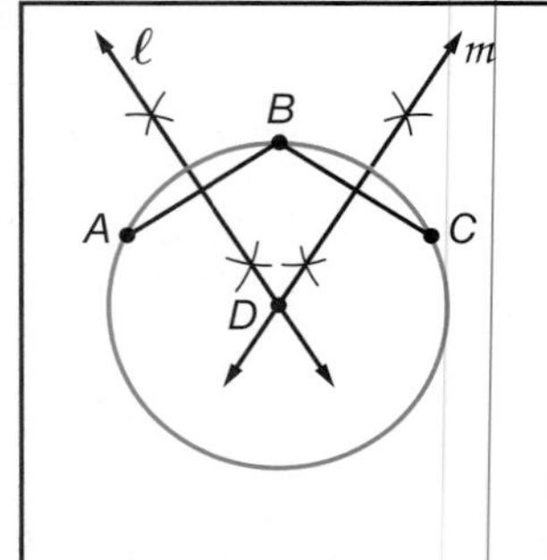


By Theorem 11.4, lines $\ell$ and $m$ contain diameters of ⊙$D$. With the compass at point $D$, draw a circle through points $A$, $B$, and $C$.

## Check Your Understanding

**Examples 1–2 ALGEBRA Find the value of $x$.**

**1**

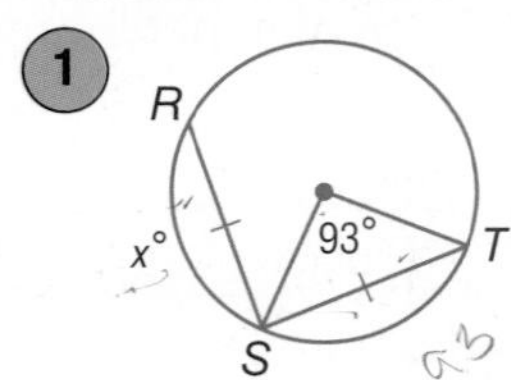


**2.**

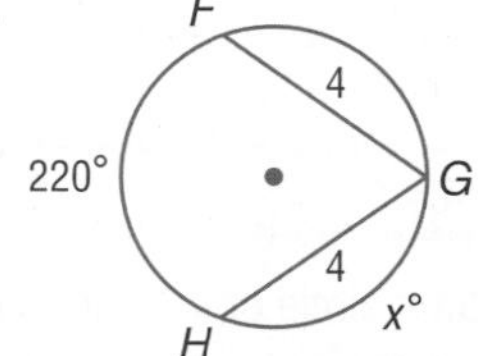


**3.**

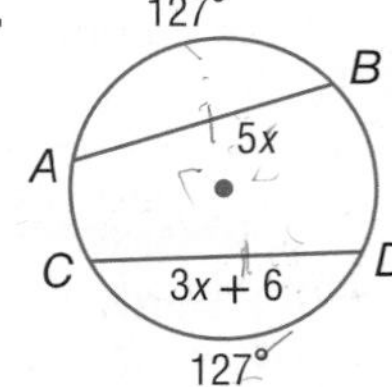


**Examples 3–4** **In ⊙$P$, $JK = 10$ and $m\widehat{JLK} = 134$. Find each measure. Round to the nearest hundredth.**

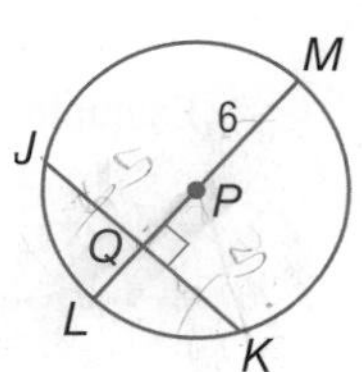


**4.** $m\widehat{JL}$

**5.** $PQ$

Example 5 **6.** In $\odot J$, $GH = 9$, $KL = 4x + 1$. Find $x$.

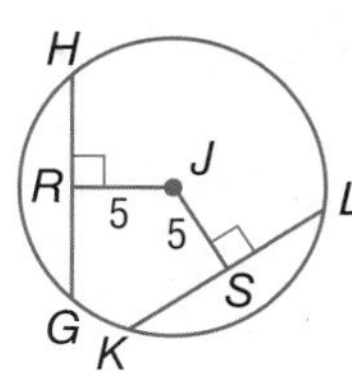


## Practice and Problem Solving

Examples 1–2 **ALGEBRA Find the value of $x$.**

**7.**

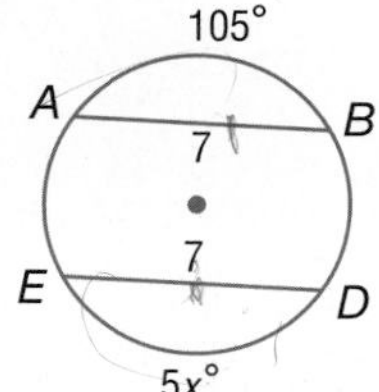


**8.**

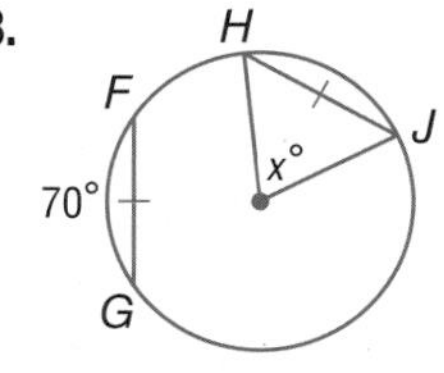


**9.**

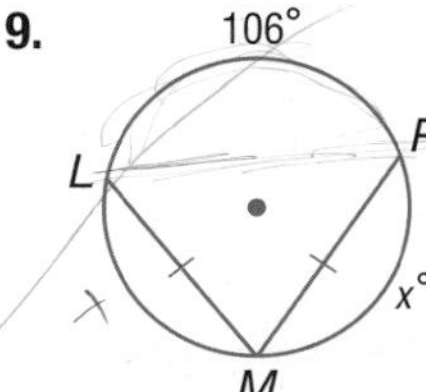


**10.**

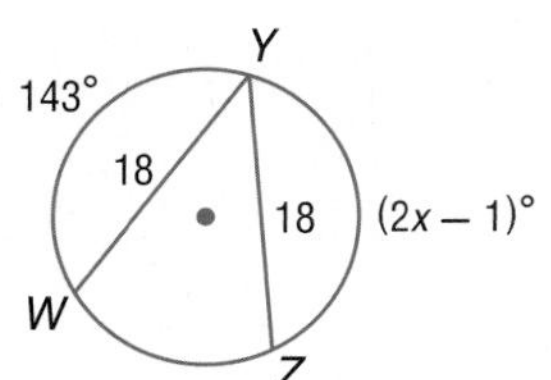


**11.**

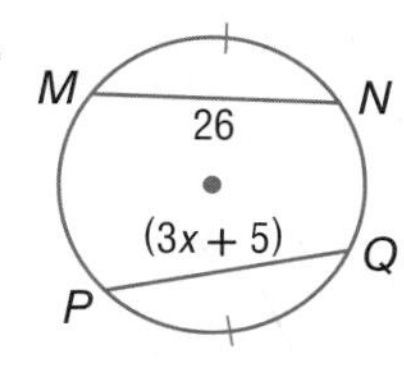


**12.**

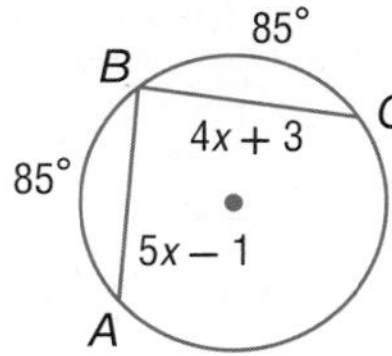


**13** $\odot C \cong \odot D$

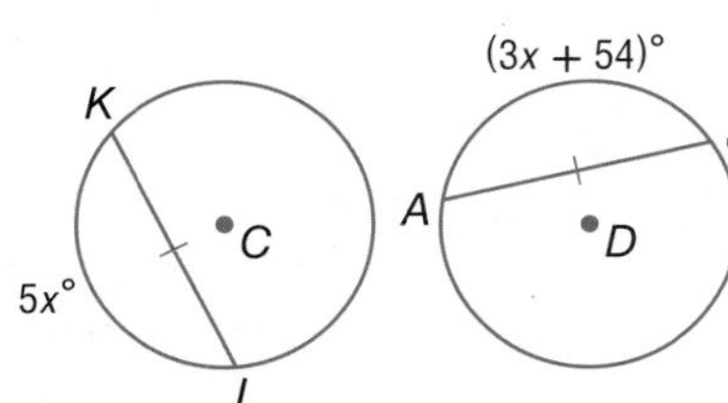


**14.** $\odot P \cong \odot Q$

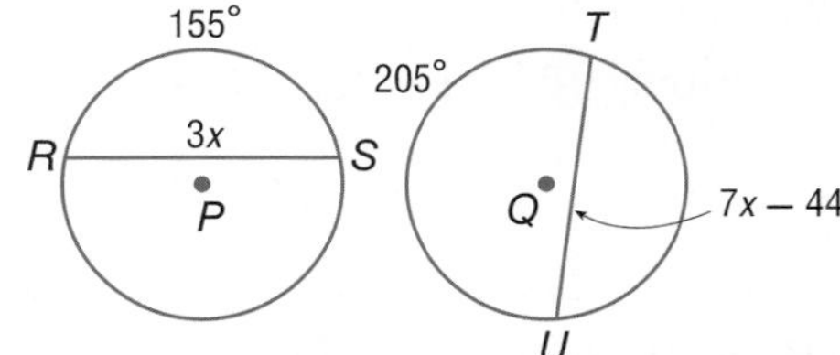


**15.** CCSS **MODELING** Angie is in a jewelry making class at her local arts center. She wants to make a pair of triangular earrings from a metal circle. She knows that $\widehat{AC}$ is 115°. If she wants to cut two equal parts off so that $\widehat{AB} = \widehat{BC}$, what is $x$?

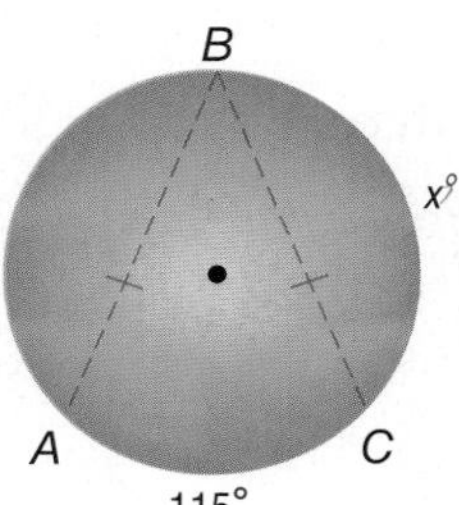


Examples 3–4 **In $\odot A$, the radius is 14 and $CD = 22$. Find each measure. Round to the nearest hundredth, if necessary.**

**16.** $CE$

**17.** $EB$

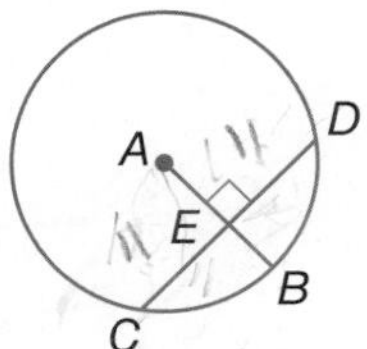


**In $\odot H$, the diameter is 18, $LM = 12$, and $m\widehat{LM} = 84$. Find each measure. Round to the nearest hundredth, if necessary.**

**18.** $m\widehat{LK}$

**19.** $HP$

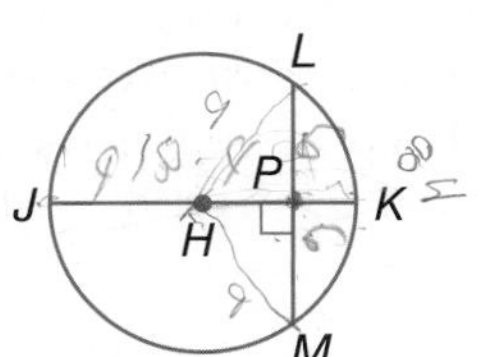

**20. SNOWBOARDING** The snowboarding rail shown is an arc of a circle in which $\overline{BD}$ is part of the diameter. If $\widehat{ABC}$ is about 32% of a complete circle, what is $m\widehat{AB}$?

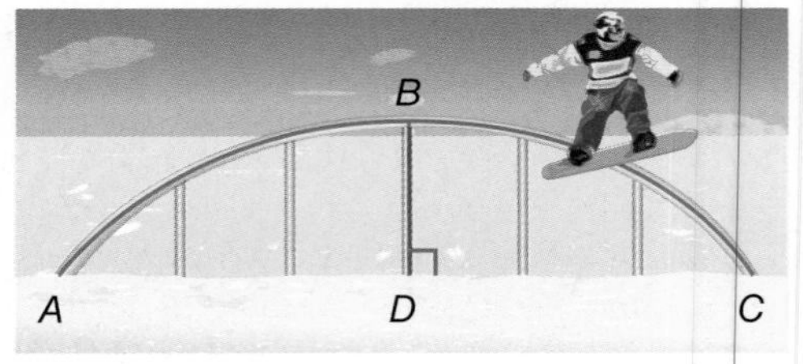


**21) ROADS** The curved road at the right is part of $\odot C$, which has a radius of 88 feet. What is $AB$? Round to the nearest tenth.

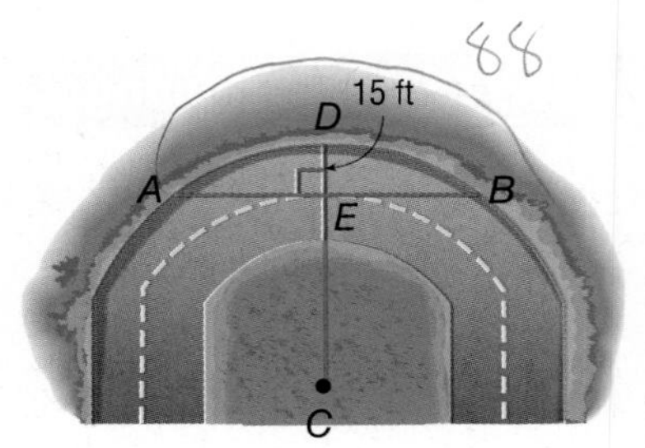


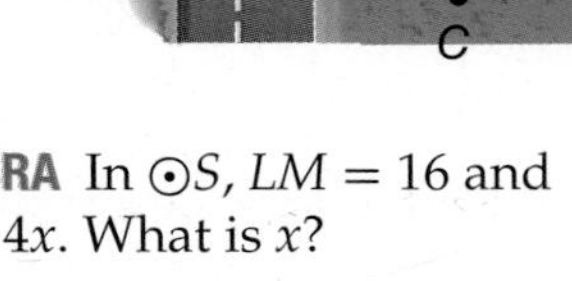

**Example 5**

**22. ALGEBRA** In $\odot F$, $\overline{AB} \cong \overline{BC}$, $DF = 3x - 7$, and $FE = x + 9$. What is $x$?

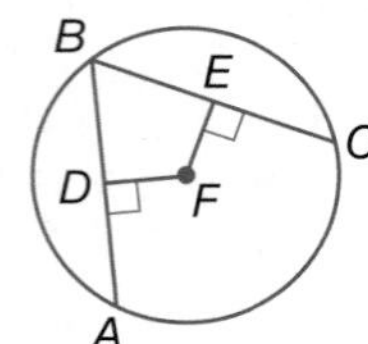


**23. ALGEBRA** In $\odot S$, $LM = 16$ and $PN = 4x$. What is $x$?

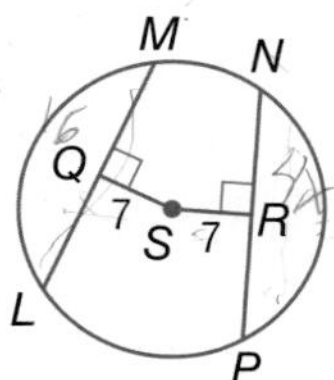


**PROOF** **Write a two-column proof.**

**24. Given:** $\odot P$, $\overline{KM} \perp \overline{JP}$

**Prove:** $\overline{JP}$ bisects $\overline{KM}$ and $\widehat{KM}$.

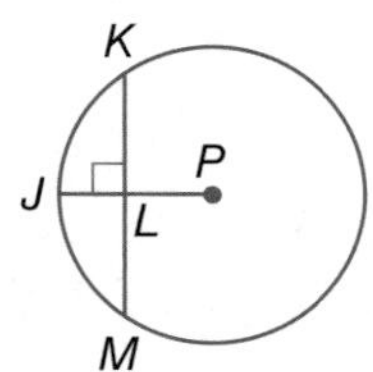


**PROOF** **Write the specified type of proof.**

**25.** paragraph proof of Theorem 11.2, part 2

**Given:** $\odot P$, $\overline{QR} \cong \overline{ST}$

**Prove:** $\widehat{QR} \cong \widehat{ST}$

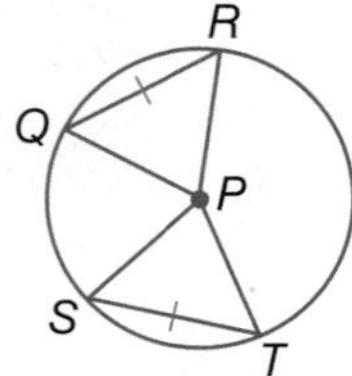


**26.** two-column proof of Theorem 11.3

**Given:** $\odot C$, $\overline{AB} \perp \overline{XY}$

**Prove:** $\overline{XZ} \cong \overline{YZ}$, $\widehat{XB} \cong \widehat{YB}$

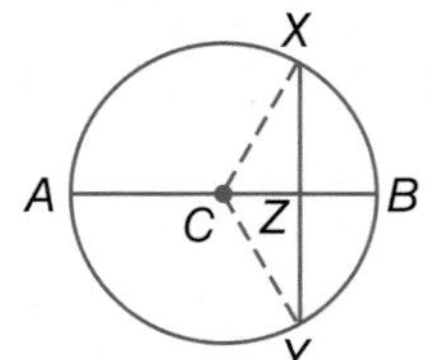


**27. DESIGN** Roberto is designing a logo for a friend's coffee shop according to the design at the right, where each chord is equal in length. What is the measure of each arc and the length of each chord?

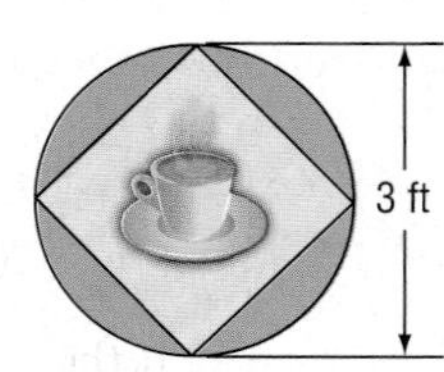


**28. CCSS ARGUMENTS** Write a two-column proof of Theorem 11.4.

**CCSS ARGUMENTS Write a two-column proof of the indicated part of Theorem 11.5.**

29. In a circle, if two chords are equidistant from the center, then they are congruent.

30. In a circle, if two chords are congruent, then they are equidistant from the center.

**ALGEBRA Find the value of $x$.**

31. $\overline{AB} \cong \overline{DF}$

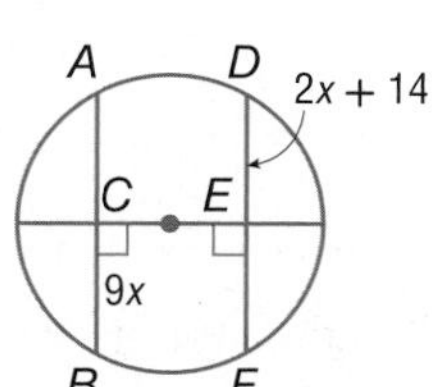


32. $\overline{GH} \cong \overline{KJ}$

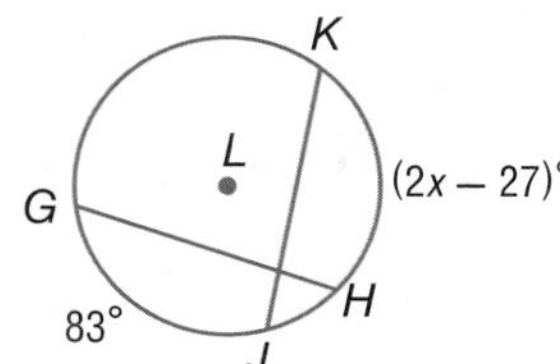


33. $\widehat{WTY} \cong \widehat{TWY}$

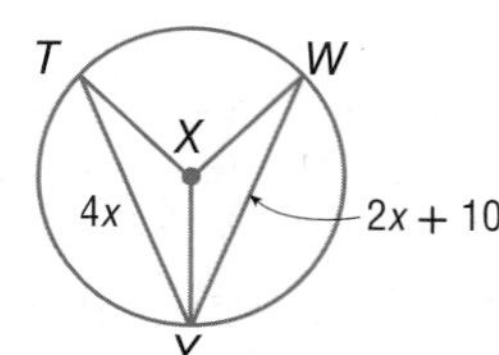


34. **ADVERTISING** A bookstore clerk wants to set up a display of new books. If there are three entrances into the store as shown in the figure at the right, where should the display be to get maximum exposure?

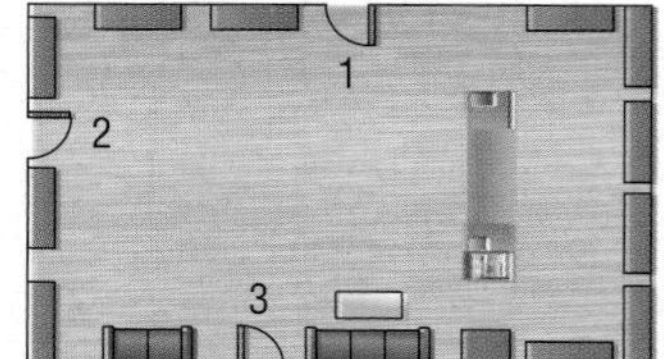


## H.O.T. Problems Use Higher-Order Thinking Skills

35. **CHALLENGE** The common chord $\overline{AB}$ between $\odot P$ and $\odot Q$ is perpendicular to the segment connecting the centers of the circles. If $AB = 10$, what is the length of $\overline{PQ}$? Explain your reasoning.

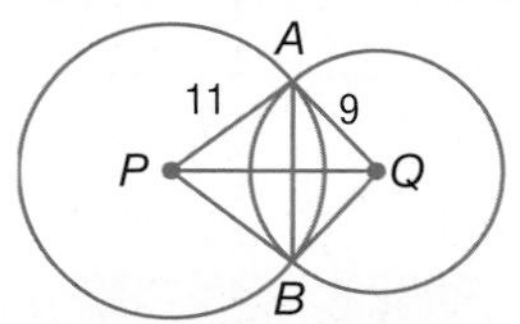


36. **REASONING** In a circle, $\overline{AB}$ is a diameter and $\overline{HG}$ is a chord that intersects $\overline{AB}$ at point $X$. Is it *sometimes*, *always*, or *never* true that $HX = GX$? Explain.

37. **CHALLENGE** Use a compass to draw a circle with chord $\overline{AB}$. Refer to this construction for the following problem.

**Step 1** Construct $\overline{CD}$, the perpendicular bisector of $\overline{AB}$.

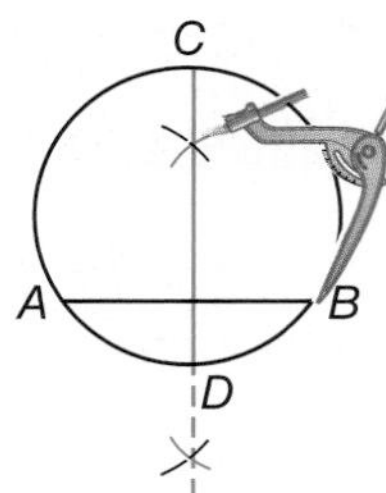


**Step 2** Construct $\overline{FG}$, the perpendicular bisector of $\overline{CD}$. Label the point of intersection $O$.

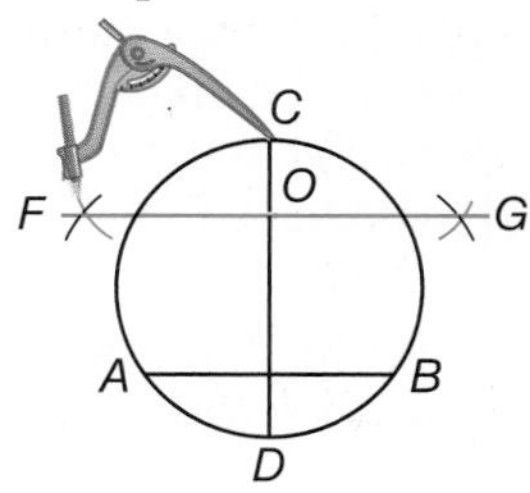


a. Use an indirect proof to show that $\overline{CD}$ passes through the center of the circle by assuming that the center of the circle is *not* on $\overline{CD}$.

b. Prove that $O$ is the center of the circle.

38. **OPEN ENDED** Construct a circle and draw a chord. Measure the chord and the distance that the chord is from the center. Find the length of the radius.

39. **WRITING IN MATH** If the measure of an arc in a circle is tripled, will the chord of the new arc be three times as long as the chord of the original arc? Explain your reasoning.

## Standardized Test Practice

**40.** If $CW = WF$ and $ED = 30$, what is $DF$?

**A** 60
**B** 45
**C** 30
**D** 15

**41. ALGEBRA** Write the ratio of the area of the circle to the area of the square in simplest form.

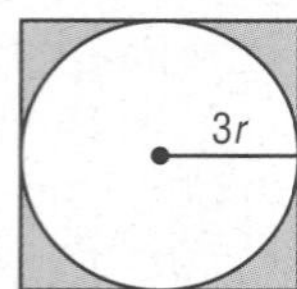


**F** $\frac{\pi}{4}$ **H** $\frac{3\pi}{4}$
**G** $\frac{\pi}{2}$ **J** $\pi$

**42. SHORT RESPONSE** The pipe shown is divided into five equal sections. How long is the pipe in feet (ft) and inches (in.)?

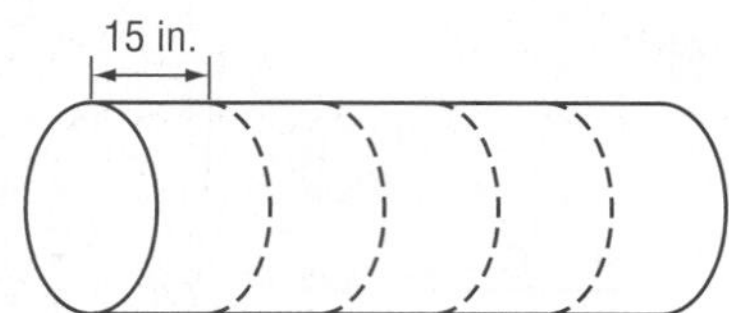


**43. SAT/ACT** Point $B$ is the center of a circle, tangent to the $y$-axis, and the coordinates of Point $B$ are (3, 1). What is the area of the circle?

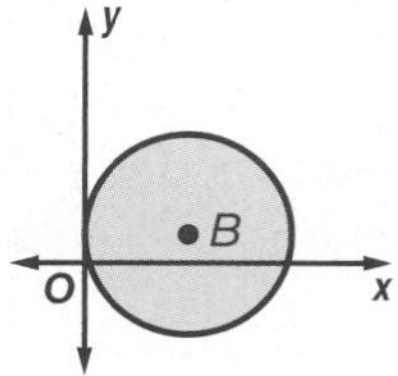


**A** $\pi$ units$^2$
**B** $3\pi$ units$^2$
**C** $4\pi$ units$^2$
**D** $6\pi$ units$^2$
**E** $9\pi$ units$^2$

## Spiral Review

**Find $x$.** (Lesson 11-2)

**44.**

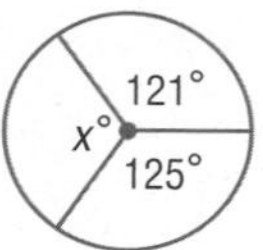


**45.**

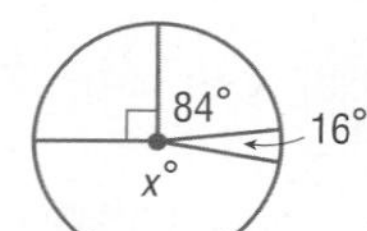


**46.**

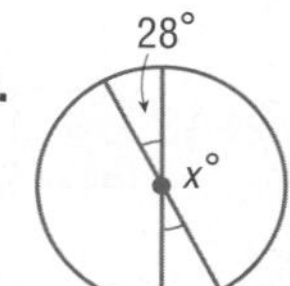


**47. CRAFTS** Ruby created a pattern to sew flowers onto a quilt by first drawing a regular pentagon that was 3.5 inches long on each side. Then she added a semicircle onto each side of the pentagon to create the appearance of five petals. How many inches of gold trim does she need to edge 10 flowers? Round to the nearest inch. (Lesson 11-1)

**Determine whether each set of numbers can be the measures of the sides of a triangle. If so, classify the triangle as *acute*, *obtuse*, or *right*. Justify your answer.** (Lesson 10-2)

**48.** 8, 15, 17 **49.** 20, 21, 31 **50.** 10, 16, 18

## Skills Review

**ALGEBRA Quadrilateral $WXZY$ is a rhombus. Find each value or measure.**

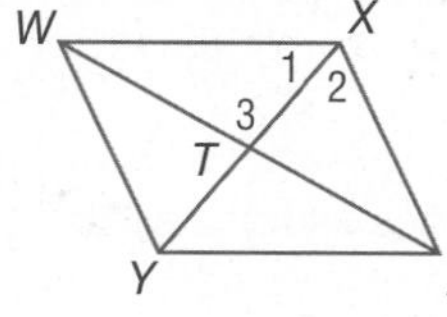


**51.** If $m\angle 3 = y^2 - 31$, find $y$.

**52.** If $m\angle XZY = 56$, find $m\angle YWZ$.

# LESSON 11-4 Inscribed Angles

| Then | Now | Why? |
|---|---|---|
| You found measures of interior angles of polygons. | **1** Find measures of inscribed angles.<br>**2** Find measures of angles of inscribed polygons. | The entrance to a school prom has a semicircular arch. Streamers are attached with one end at point *A* and the other end at point *B*. The middle of each streamer can then be attached to a different point *P* along the arch. |

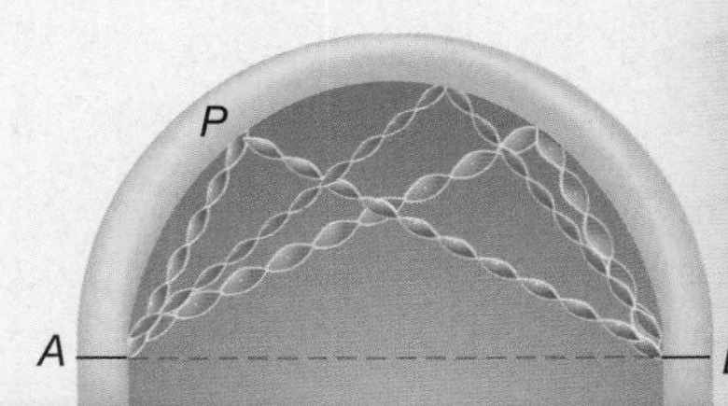

**NewVocabulary**
inscribed angle
intercepted arc

**Common Core State Standards**

**Content Standards**

G.C.2 Identify and describe relationships among inscribed angles, radii, and chords.

G.C.3 Construct the inscribed and circumscribed circles of a triangle, and prove properties of angles for a quadrilateral inscribed in a circle.

**Mathematical Practices**

7 Look for and make use of structure.

3 Construct viable arguments and critique the reasoning of others.

## 1 Inscribed Angles

Notice that the angle formed by each streamer appears to be congruent, no matter where point *P* is placed along the arch. An **inscribed angle** has a vertex on a circle and sides that contain chords of the circle. In $\odot C$, $\angle QRS$ is an inscribed angle.

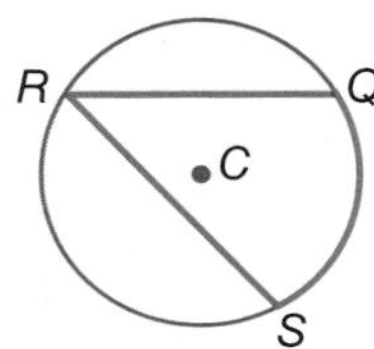


An **intercepted arc** has endpoints on the sides of an inscribed angle and lies in the interior of the inscribed angle. In $\odot C$, minor arc $\widehat{QS}$ is intercepted by $\angle QRS$.

There are three ways that an angle can be inscribed in a circle.

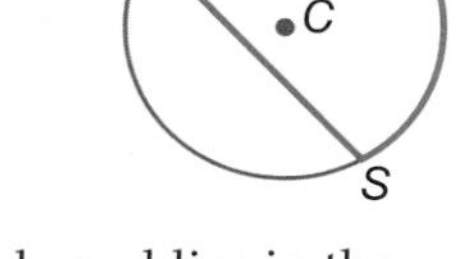

| Case 1 | Case 2 | Case 3 |
|---|---|---|
| Center *P* is on a side of the inscribed angle. | Center *P* is inside the inscribed angle. | The center *P* is in the exterior of the inscribed angle. |

In Case 1, the side of the angle is a diameter of the circle.

For each of these cases, the following theorem holds true.

### Theorem 11.6 Inscribed Angle Theorem

**Words** If an angle is inscribed in a circle, then the measure of the angle equals one half the measure of its intercepted arc.

**Example** $m\angle 1 = \frac{1}{2}m\widehat{AB}$ and $m\widehat{AB} = 2m\angle 1$

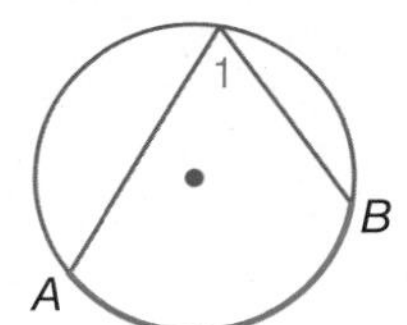


You will prove Cases 2 and 3 of the Inscribed Angle Theorem in Exercises 37 and 38.

moodboard/Alamy

**VocabularyLink**

**Inscribed**

**Everyday Use:** written on or in a surface, such as inscribing the inside of a ring with an inscription

**Math Use:** touching only the sides (or interior) of another figure

## Proof Inscribed Angle Theorem (Case 1)

**Given:** $\angle B$ is inscribed in $\odot P$.

**Prove:** $m\angle B = \frac{1}{2}m\widehat{AC}$

**Proof:**

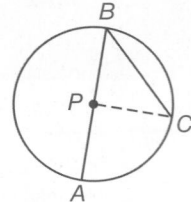

| Statements | Reasons |
|---|---|
| 1. Draw an auxiliary radius $\overline{PC}$. | 1. Two points determine a line. |
| 2. $\overline{PB} \cong \overline{PC}$ | 2. All radii of a circle are $\cong$. |
| 3. $\triangle PBC$ is isosceles. | 3. Definition of isosceles triangle |
| 4. $m\angle B = m\angle C$ | 4. Isosceles Triangle Theorem |
| 5. $m\angle APC = m\angle B + m\angle C$ | 5. Exterior Angle Theorem |
| 6. $m\angle APC = 2m\angle B$ | 6. Substitution (Steps 4, 5) |
| 7. $m\widehat{AC} = m\angle APC$ | 7. Definition of arc measure |
| 8. $m\widehat{AC} = 2m\angle B$ | 8. Substitution (Steps 6, 7) |
| 9. $2m\angle B = m\widehat{AC}$ | 9. Symmetric Property of Equality |
| 10. $m\angle B = \frac{1}{2}m\widehat{AC}$ | 10. Division Property of Equality |

## Example 1 Use Inscribed Angles to Find Measures

**Find each measure.**

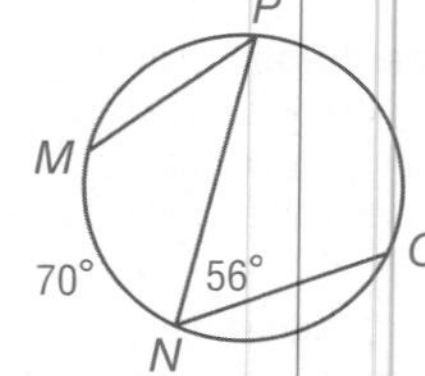


**a.** $m\angle P$

$$m\angle P = \frac{1}{2}m\widehat{MN}$$
$$= \frac{1}{2}(70) \text{ or } 35$$

**b.** $m\widehat{PO}$

$$m\widehat{PO} = 2m\angle N$$
$$= 2(56) \text{ or } 112$$

### GuidedPractice

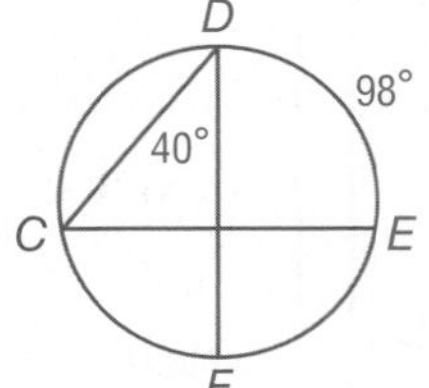


**1A.** $m\widehat{CF}$

**1B.** $m\angle C$

Two inscribed angles that intercept the same arc of a circle are related.

## Theorem 11.7

**Words** If two inscribed angles of a circle intercept the same arc or congruent arcs, then the angles are congruent.

**Example** $\angle B$ and $\angle C$ both intercept $\widehat{AD}$. So, $\angle B \cong \angle C$.

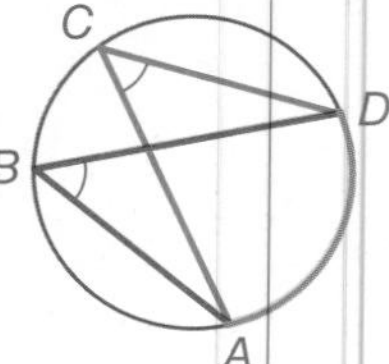


You will prove Theorem 11.7 in Exercise 39.

**Study Tip**

**Inscribed Polygons** Remember that for a polygon to be an inscribed polygon, *all* of its vertices must lie on the circle.

## Example 2 Use Inscribed Angles to Find Measures

**ALGEBRA Find $m\angle T$.**

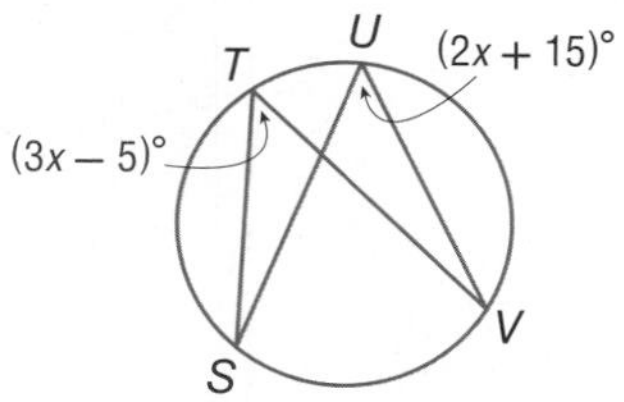


| | |
|---|---|
| $\angle T \cong \angle U$ | $\angle T$ and $\angle U$ both intercept $\widehat{SV}$. |
| $m\angle T = m\angle U$ | Definition of congruent angles |
| $3x - 5 = 2x + 15$ | Substitution |
| $x = 20$ | Simplify. |

So, $m\angle T = 3(20) - 5$ or 55.

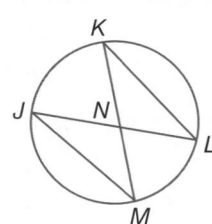

**Guided Practice**

**2.** If $m\angle S = 3x$ and $m\angle V = (x + 16)$, find $m\angle S$.

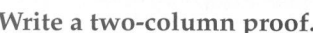

## Example 3 Use Inscribed Angles in Proofs

**Write a two-column proof.**

**Given:** $\widehat{JM} \cong \widehat{KL}$

**Prove:** $\triangle JMN \cong \triangle KLN$

**Proof:**

| Statements | Reasons |
|---|---|
| 1. $\widehat{JM} \cong \widehat{KL}$ | 1. Given |
| 2. $\overline{JM} \cong \overline{KL}$ | 2. If minor arcs are ≅, their corresponding chords are ≅. |
| 3. $\angle M$ intercepts $\widehat{JK}$. $\angle L$ intercepts $\widehat{JK}$. | 3. Definition of intercepted arc |
| 4. $\angle M \cong \angle L$ | 4. Inscribed ∠s of same arc are ≅. |
| 5. $\angle JNM \cong \angle KNL$ | 5. Vertical ∠s are ≅. |
| 6. $\triangle JMN \cong \triangle KLN$ | 6. AAS |

**Guided Practice**

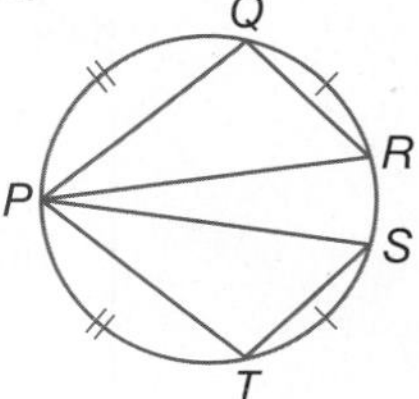


**3. Given:** $\widehat{QR} \cong \widehat{ST}$, $\widehat{PQ} \cong \widehat{PT}$

**Prove:** $\triangle PQR \cong \triangle PTS$

## 2 Angles of Inscribed Polygons

Triangles and quadrilaterals that are inscribed in circles have special properties.

### Theorem 11.8

**Words** An inscribed angle of a triangle intercepts a diameter or semicircle if and only if the angle is a right angle.

**Example** If $\widehat{FJH}$ is a semicircle, then $m\angle G = 90$. If $m\angle G = 90$, then $\widehat{FJH}$ is a semicircle and $\overline{FH}$ is a diameter.

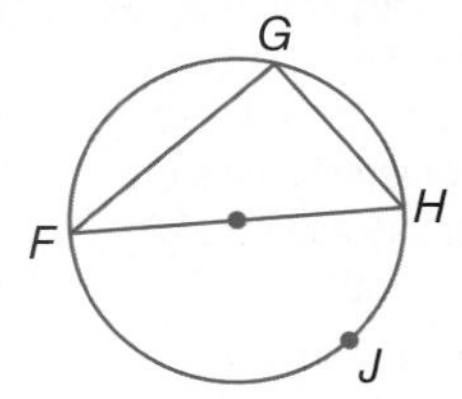


You will prove Theorem 11.8 in Exercise 40.

## Example 4 Find Angle Measures in Inscribed Triangles

**ALGEBRA Find $m\angle F$.**

$\triangle FGH$ is a right triangle because $\angle G$ inscribes a semicircle.

| | |
|---|---|
| $m\angle F + m\angle G + m\angle H = 180$ | Angle Sum Theorem |
| $(4x + 2) + 90 + (9x - 3) = 180$ | Substitution |
| $13x + 89 = 180$ | Simplify. |
| $13x = 91$ | Subtract 89 from each side. |
| $x = 7$ | Divide each side by 13. |

So, $m\angle F = 4(7) + 2$ or 30.

### Guided Practice

**4.** If $m\angle F = 7x + 2$ and $m\angle H = 17x - 8$, find $x$.

While many different types of triangles, including right triangles, can be inscribed in a circle, only certain quadrilaterals can be inscribed in a circle.

**Study Tip**

**CCSS Arguments** Theorem 11.9 can be verified by considering that the arcs intercepted by opposite angles of an inscribed quadrilateral form a circle.

### Theorem 11.9

**Words** If a quadrilateral is inscribed in a circle, then its opposite angles are supplementary.

**Example** If quadrilateral $KLMN$ is inscribed in $\odot A$, then $\angle L$ and $\angle N$ are supplementary and $\angle K$ and $\angle M$ are supplementary.

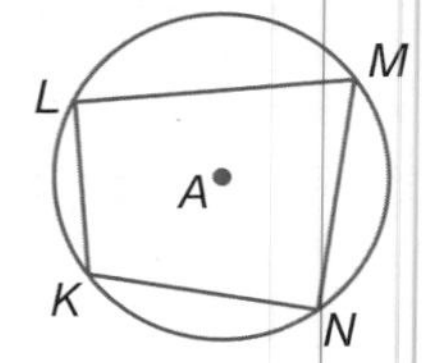


You will prove Theorem 11.9 in Exercise 31.

**Real-World Link**

Charms for jewelry first became popular during the age of the Egyptian Pharaohs. They were repopularized by Queen Victoria in the early 20th century and by Louis Vuitton in 2001.

**Source:** *My Mother's Charms*

## Real-World Example 5 Find Angle Measures

**JEWELRY The necklace charm shown uses a quadrilateral inscribed in a circle. Find $m\angle A$ and $m\angle B$.**

Since $ABCD$ is inscribed in a circle, opposite angles are supplementary.

| | |
|---|---|
| $m\angle A + m\angle C = 180$ | $m\angle B + m\angle D = 180$ |
| $m\angle A + 90 = 180$ | $(2x - 30) + x = 180$ |
| $m\angle A = 90$ | $3x - 30 = 180$ |
| | $3x = 210$ |
| | $x = 70$ |

So, $m\angle A = 90$ and $m\angle B = 2(70) - 30$ or 110.

### Guided Practice

**5.** Quadrilateral $WXYZ$ is inscribed in $\odot V$. Find $m\angle X$ and $m\angle Y$.

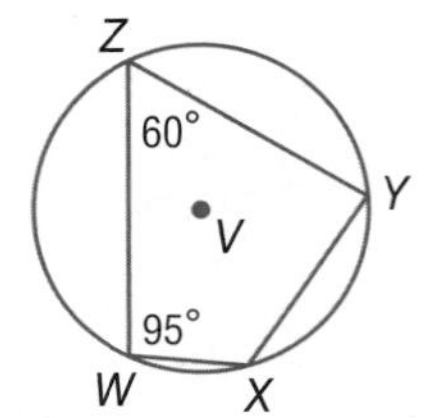

## Check Your Understanding

**Example 1** **Find each measure.**

1. $m\angle B$

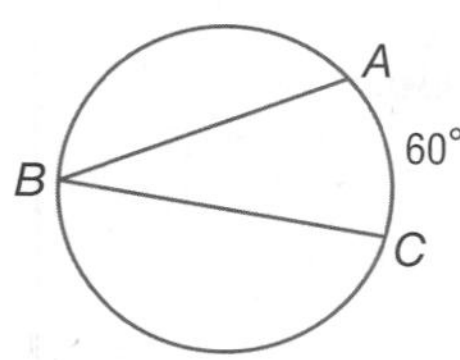


2. $m\widehat{RT}$

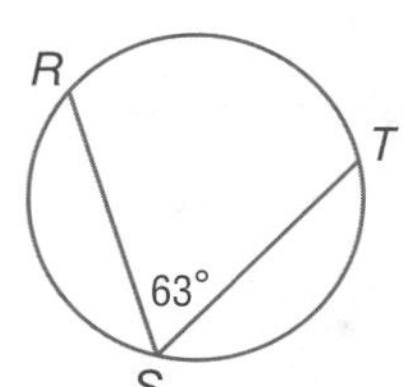


3. $m\widehat{WX}$

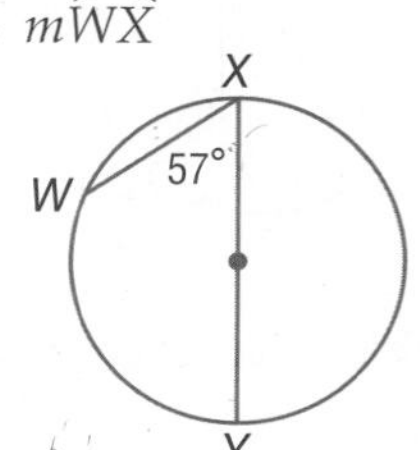


4. **SCIENCE** The diagram shows how light bends in a raindrop to make the colors of the rainbow. If $m\widehat{ST} = 144$, what is $m\angle R$?

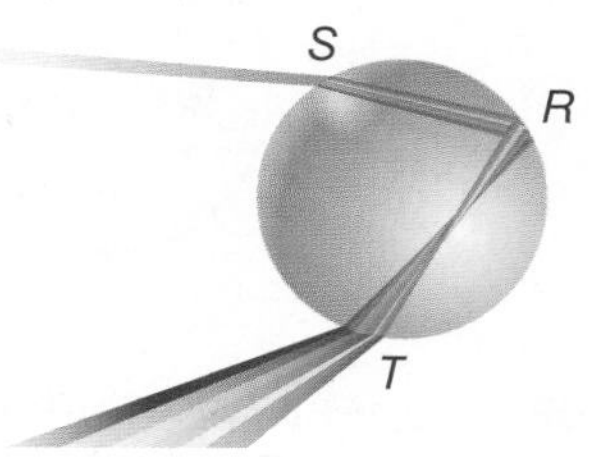


**Example 2** **ALGEBRA** **Find each measure.**

5. $m\angle H$

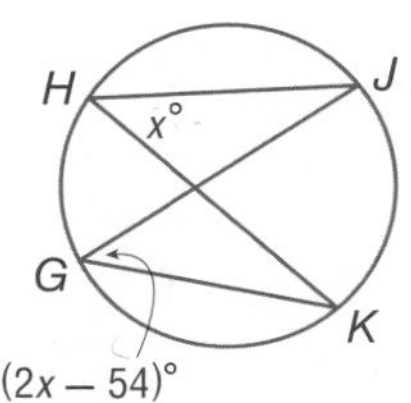


6. $m\angle B$

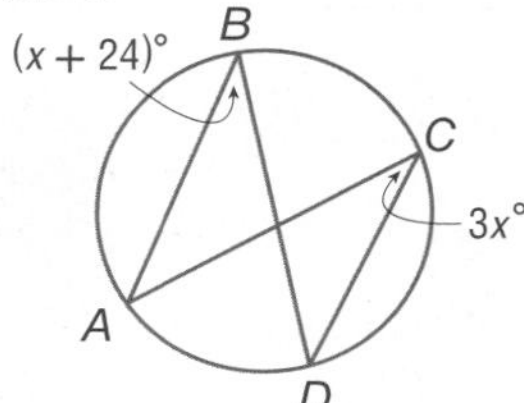


**Example 3** 7. **PROOF** Write a two-column proof.

**Given:** $\overline{RT}$ bisects $\overline{SU}$.

**Prove:** $\triangle RVS \cong \triangle UVT$

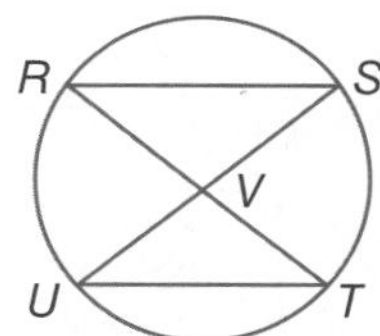


**Examples 4–5** **CCSS STRUCTURE** **Find each value.**

8. $m\angle R$

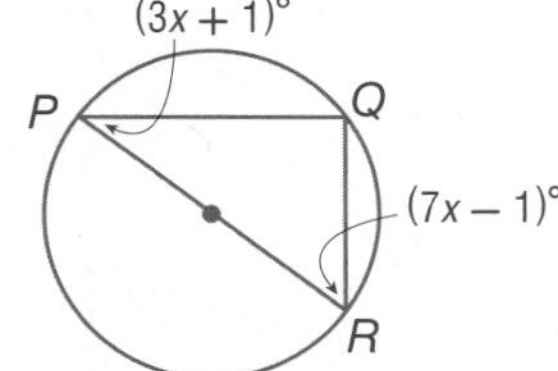


9. $x$

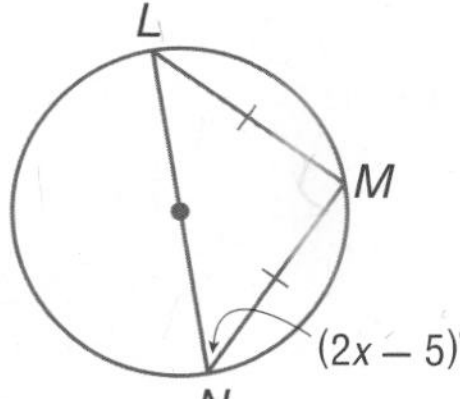


10. $m\angle C$ and $m\angle D$

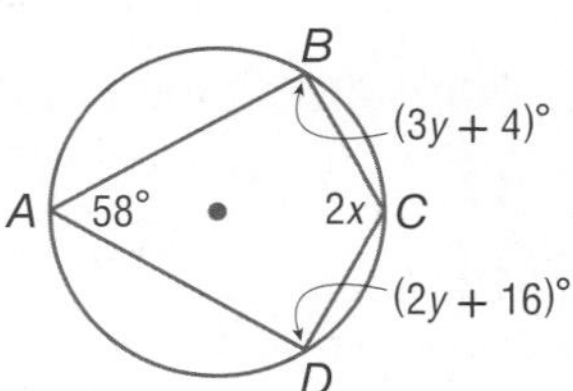


## Practice and Problem Solving

**Example 1** **Find each measure.**

11. $m\widehat{DH}$

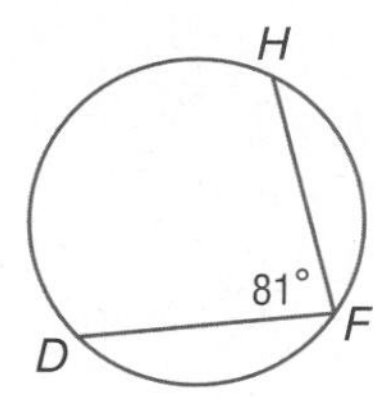


12. $m\angle K$

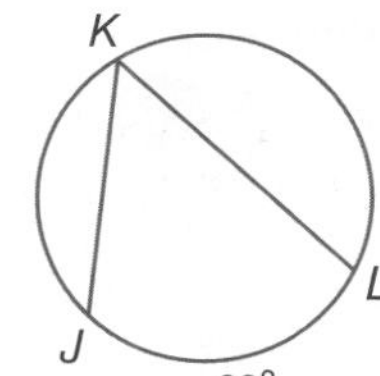


13. $m\angle P$

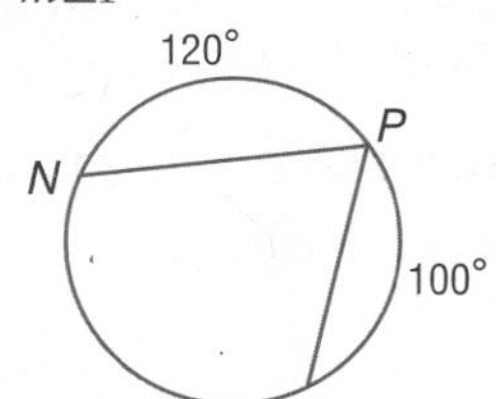

**14.** $m\widehat{AC}$

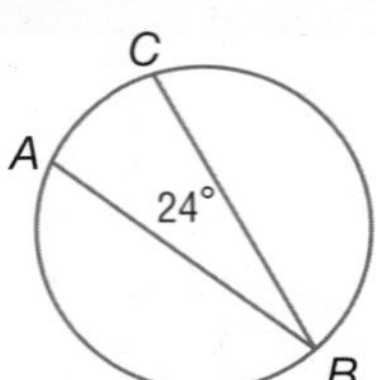


**15.** $m\widehat{GH}$

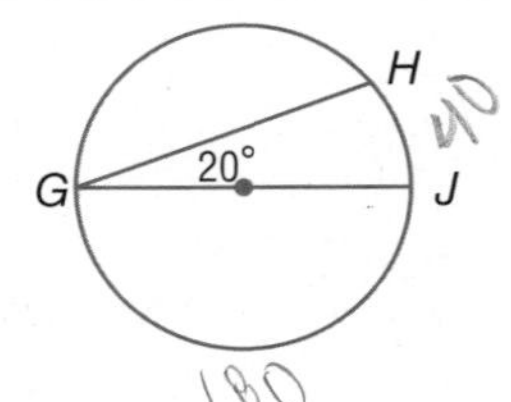


**16.** $m\angle S$

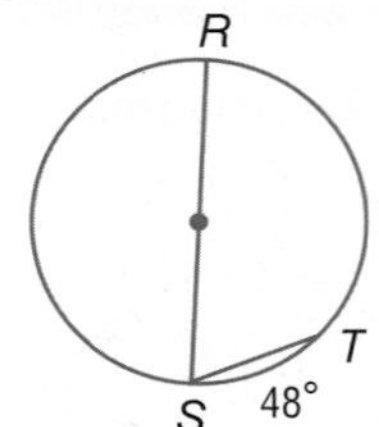


**Example 2** **ALGEBRA** **Find each measure.**

**17.** $m\angle R$

**18.** $m\angle S$

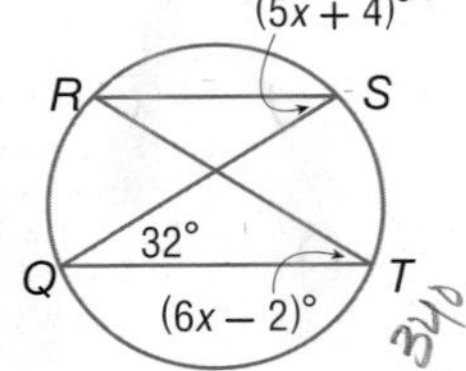


**19.** $m\angle A$

**20.** $m\angle C$

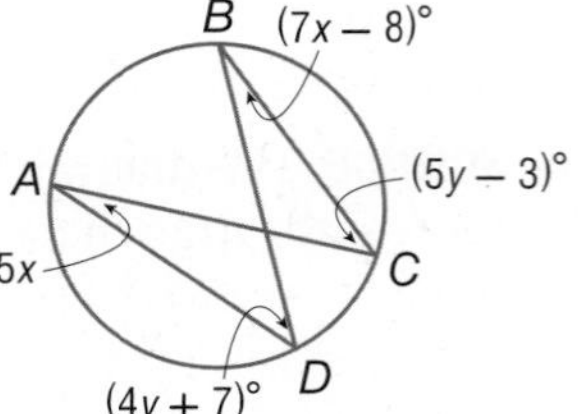


**Example 3** **PROOF** **Write the specified type of proof.**

**21.** paragraph proof

**Given:** $m\angle T = \frac{1}{2}m\angle S$

**Prove:** $m\widehat{TUR} = 2m\widehat{URS}$

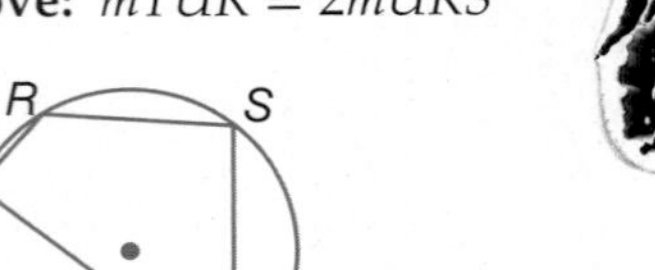


**22.** two-column proof

**Given:** $\odot C$

**Prove:** $\triangle KML \sim \triangle JMH$

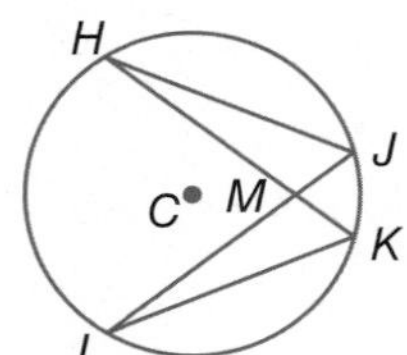


**Example 4** **ALGEBRA** **Find each value.**

**23.** $x$

**24.** $m\angle T$

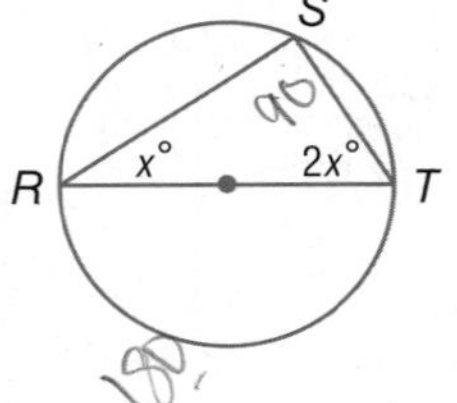


**25.** $x$

**26.** $m\angle C$

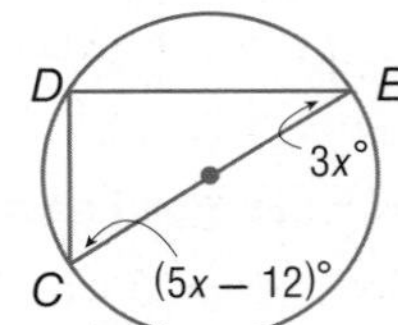


**Example 5** **CCSS STRUCTURE** **Find each measure.**

**27.** $m\angle T$

**28.** $m\angle Z$

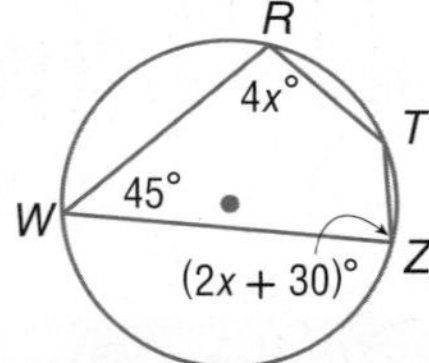


**29.** $m\angle H$

**30.** $m\angle G$

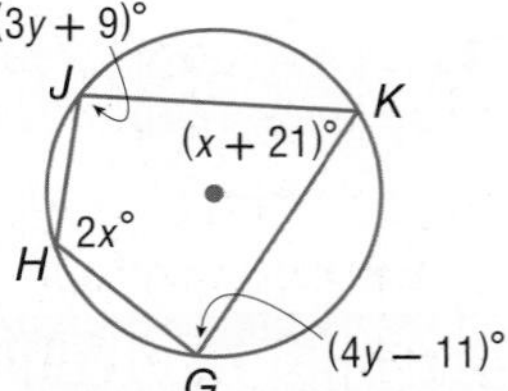


**31.** **PROOF** Write a paragraph proof for Theorem 11.9.

**SIGNS** **A stop sign in the shape of a regular octagon is inscribed in a circle. Find each measure.**

**32.** $m\widehat{NQ}$

**33** $m\angle RLQ$

**34.** $m\angle LRQ$

**35.** $m\angle LSR$

**36. ART** Four different string art star patterns are shown. If all of the inscribed angles of each star shown are congruent, find the measure of each inscribed angle.

**a.** 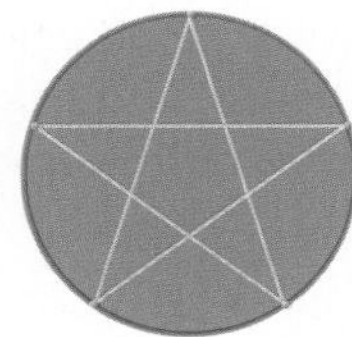 **b.** 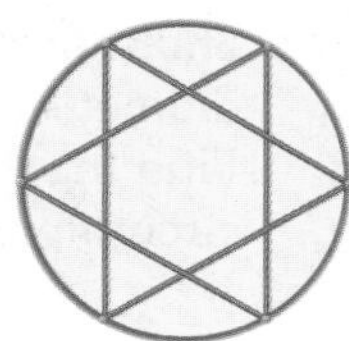 **c.**  **d.** 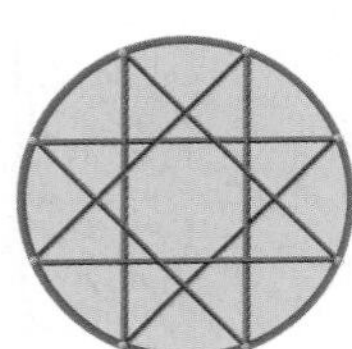

**PROOF Write a two-column proof for each case of Theorem 11.6.**

**37. Case 2**

**Given:** $P$ lies inside $\angle ABC$.
$\overline{BD}$ is a diameter.

**Prove:** $m\angle ABC = \frac{1}{2}m\widehat{AC}$

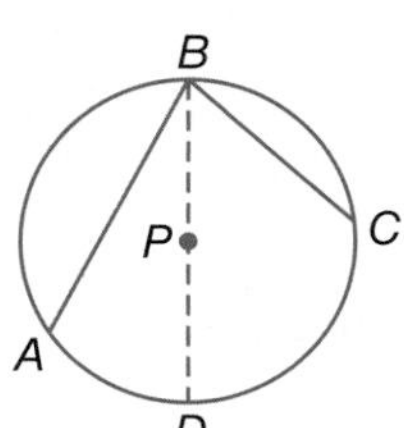


**38. Case 3**

**Given:** $P$ lies outside $\angle ABC$.
$\overline{BD}$ is a diameter.

**Prove:** $m\angle ABC = \frac{1}{2}m\widehat{AC}$

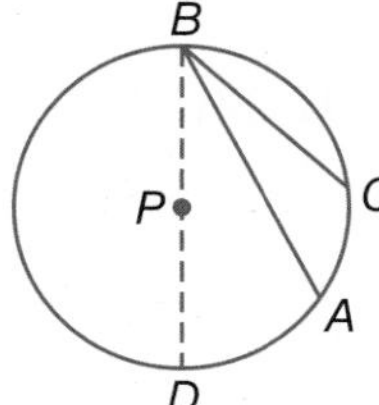


**PROOF Write the specified proof for each theorem.**

**39.** Theorem 11.7, two-column proof

**40.** Theorem 11.8, paragraph proof

**41. MULTIPLE REPRESENTATIONS** In this problem, you will investigate the relationship between the arcs of a circle that are cut by two parallel chords.

**a. Geometric** Use a compass to draw a circle with parallel chords $\overline{AB}$ and $\overline{CD}$. Connect points $A$ and $D$ by drawing segment $\overline{AD}$.

**b. Numerical** Use a protractor to find $m\angle A$ and $m\angle D$. Then determine $m\widehat{AC}$ and $m\widehat{BD}$. What is true about these arcs? Explain.

**c. Verbal** Draw another circle and repeat parts **a** and **b**. Make a conjecture about arcs of a circle that are cut by two parallel chords.

**d. Analytical** Use your conjecture to find $m\widehat{PR}$ and $m\widehat{QS}$ in the figure at the right. Verify by using inscribed angles to find the measures of the arcs.

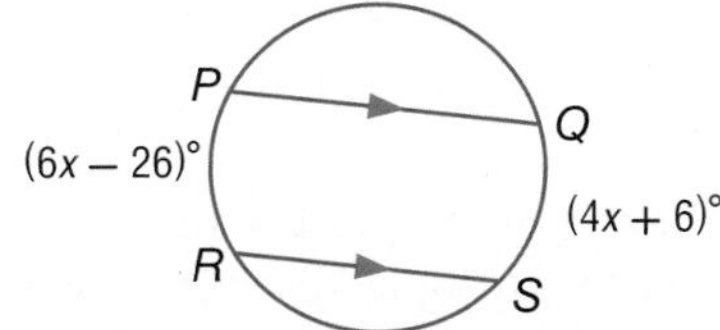


## H.O.T. Problems Use Higher-Order Thinking Skills

**CCSS ARGUMENTS Determine whether the quadrilateral can *always*, *sometimes*, or *never* be inscribed in a circle. Explain your reasoning.**

**42.** square **43.** rectangle **44.** parallelogram **45.** rhombus **46.** kite

**47. CHALLENGE** A square is inscribed in a circle. What is the ratio of the area of the circle to the area of the square?

**48. WRITING IN MATH** A 45°-45°-90° right triangle is inscribed in a circle. If the radius of the circle is given, explain how to find the lengths of the right triangle's legs.

**49. OPEN ENDED** Find and sketch a real-world logo with an inscribed polygon.

**50. WRITING IN MATH** Compare and contrast inscribed angles and central angles of a circle. If they intercept the same arc, how are they related?

## Standardized Test Practice

**51.** In the circle below, $m\widehat{AC} = 160$ and $m\angle BEC = 38$. What is $m\angle AEB$?

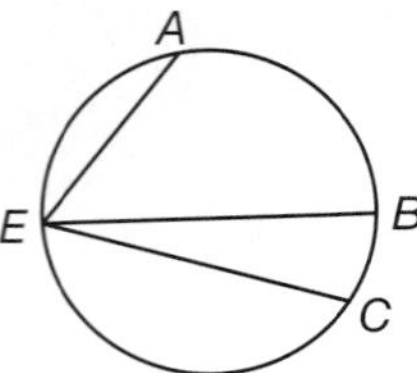


A 42
B 61
C 80
D 84

**52. ALGEBRA** Simplify $4(3x - 2)(2x + 4) + 3x^2 + 5x - 6$.

F $9x^2 + 3x - 14$
G $9x^2 + 13x - 14$
H $27x^2 + 37x - 38$
J $27x^2 + 27x - 26$

**53. SHORT RESPONSE** In the circle below, $\overline{AB}$ is a diameter, $AC = 8$ inches, and $BC = 15$ inches. Find the diameter, the radius, and the circumference of the circle.

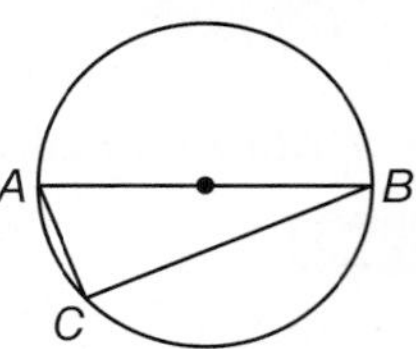


**54. SAT/ACT** The sum of three consecutive integers is −48. What is the least of the three integers?

A −15
B −16
C −17
D −18
E −19

## Spiral Review

**In $\odot M$, $FL = 24$, $HJ = 48$, and $m\widehat{HP} = 65$. Find each measure.** (Lesson 11-3)

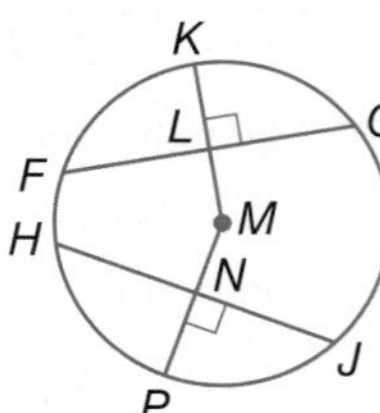
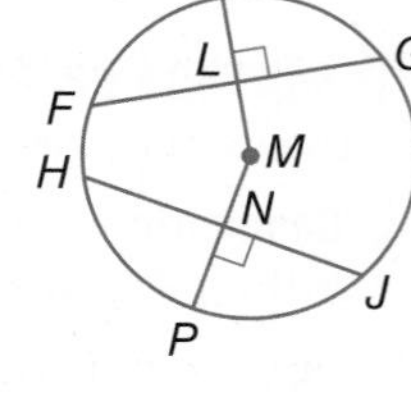


**55.** $FG$

**56.** $m\widehat{PJ}$

**57.** $NJ$

**58.** $m\widehat{HJ}$

**Find $x$.** (Lesson 11-2)

**59.**

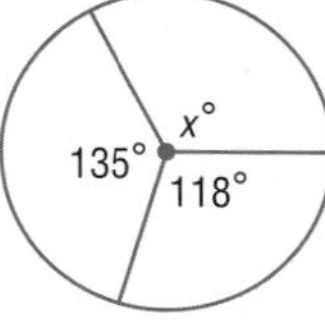


**60.**

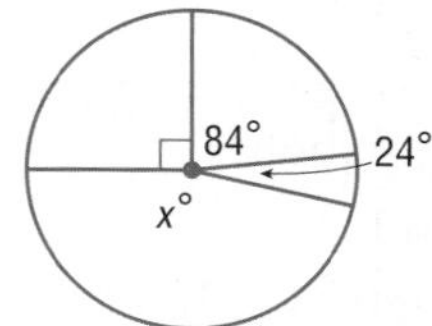


**61.**

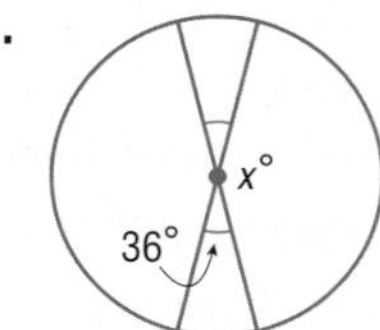


**62. PHOTOGRAPHY** In one of the first cameras invented, light entered an opening in the front. An image was reflected in the back of the camera, upside down, forming similar triangles. Suppose the image of the person on the back of the camera is 12 inches, the distance from the opening to the person is 7 feet, and the camera itself is 15 inches long. How tall is the person being photographed? (Lesson 9-3)

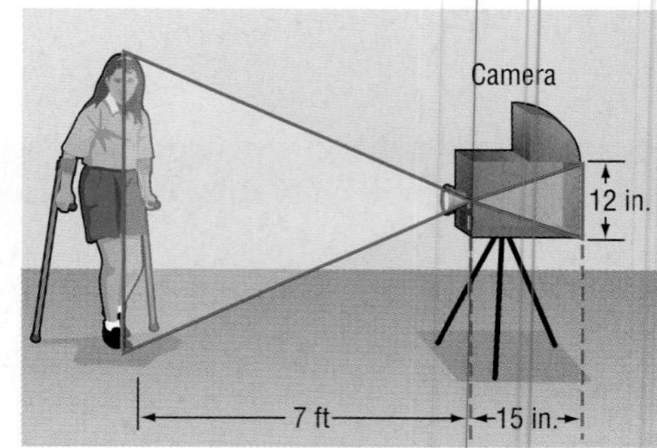


## Skills Review

**ALGEBRA Suppose $B$ is the midpoint of $\overline{AC}$. Use the given information to find the missing measure.**

**63.** $AB = 4x - 5$, $BC = 11 + 2x$, $AC = ?$

**64.** $AB = 6y - 14$, $BC = 10 - 2y$, $AC = ?$

**65.** $BC = 6 - 4m$, $AC = 8$, $m = ?$

**66.** $AB = 10s + 2$, $AC = 40$, $s = ?$

CHAPTER 11

# Mid-Chapter Quiz

## Lessons 11-1 through 11-4

**For Exercises 1–3, refer to ⊙*A*.** (Lesson 11-1)

1. Name the circle.
2. Name a diameter.
3. Name a chord that is not a diameter.

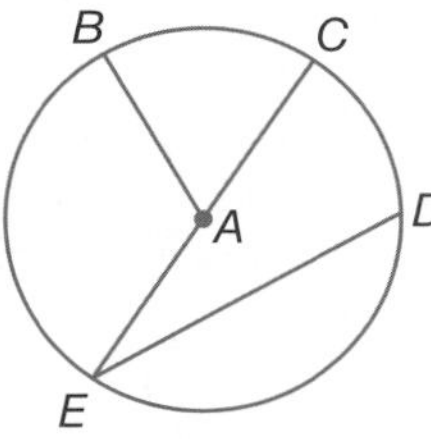


4. **BICYCLES** A bicycle has tires that are 24 inches in diameter. (Lesson 11-1)
   a. Find the circumference of one tire.
   b. How many inches does the tire travel after 100 rotations?

**Find the diameter and radius of a circle with the given circumference. Round to the nearest hundredth.** (Lesson 11-1)

5. $C = 23$ cm
6. $C = 78$ ft

7. **MULTIPLE CHOICE** Find the length of $\widehat{BC}$. (Lesson 11-2)

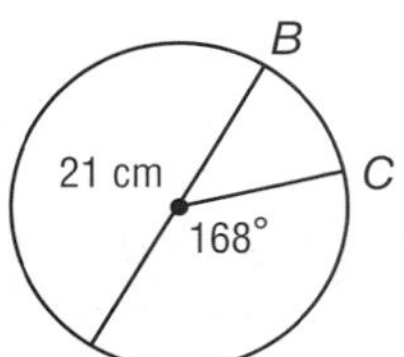


A 18°
B 2.20 cm
C 168°
D 30.79 cm

8. **MOVIES** The movie reel shown below has a diameter of 14.5 inches. (Lesson 11-2)

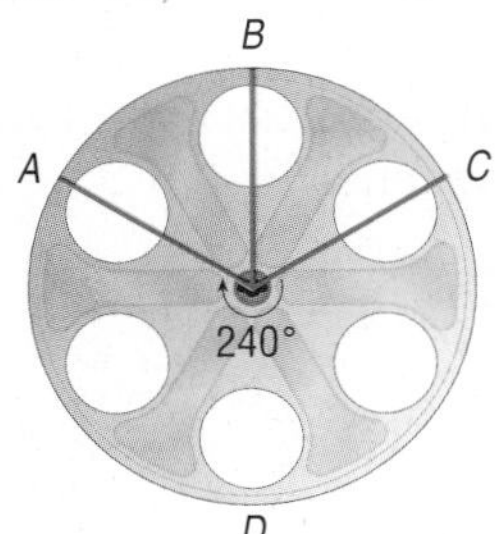


   a. Find $m\widehat{ADC}$.
   b. Find the length of $\widehat{ADC}$.

9. Find the value of *x*. (Lesson 11-3)

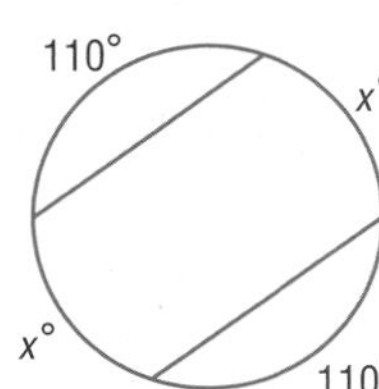


10. In ⊙*B*, $CE = 13.5$. Find *BD*. Round to the nearest hundredth. (Lesson 11-3)

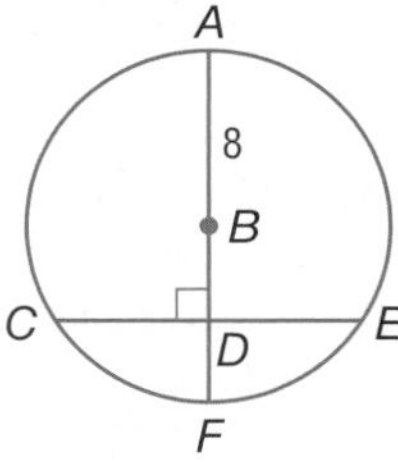


11. The two circles shown are congruent. Find *x* and the length of the chord. (Lesson 11-3)

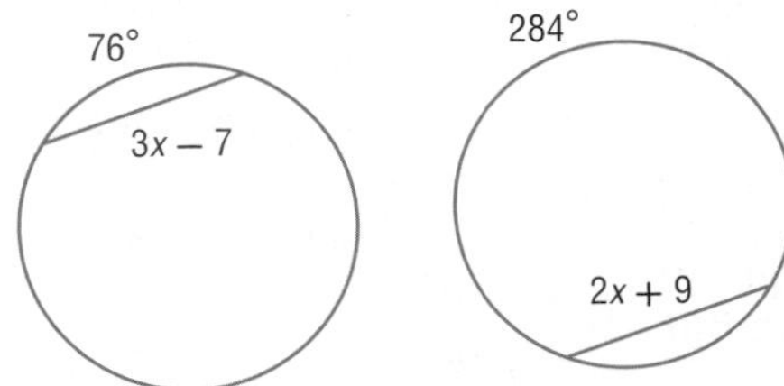


**Find each measure.** (Lesson 11-4)

12. $m\widehat{TU}$

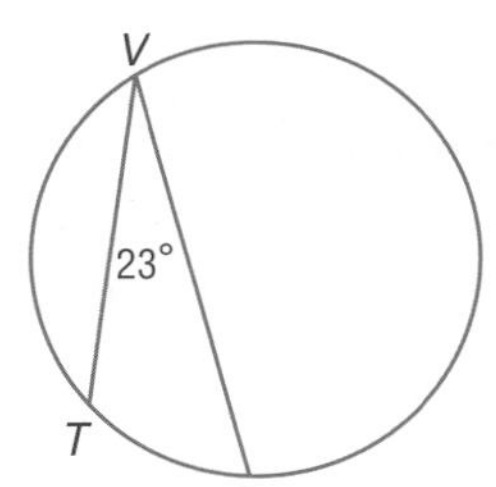


13. $m\angle A$

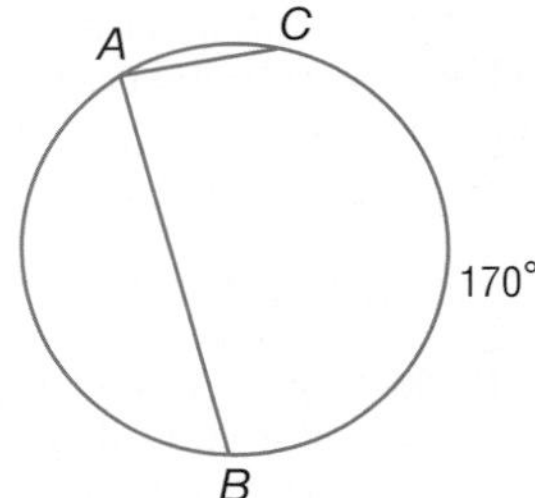


14. **MULTIPLE CHOICE** Find *x*. (Lesson 11-4)

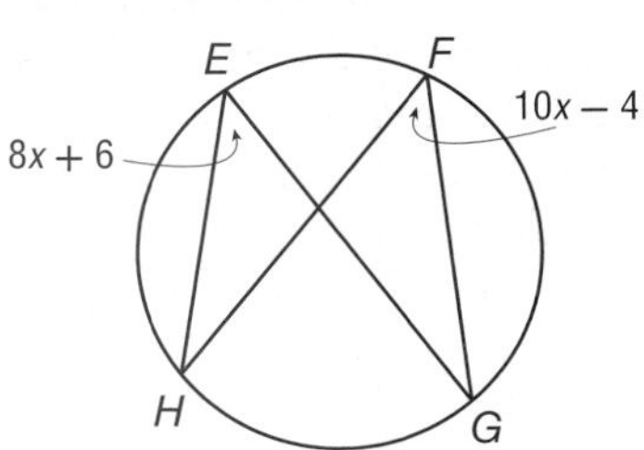


F 1.8
G 5
H 46
J 90

15. If a square with sides of 14 inches is inscribed in a circle, what is the diameter of the circle? (Lesson 11-4)

# LESSON 11-5 Tangents

| Then | Now | Why? |
|---|---|---|
| ● You used the Pythagorean Theorem to find side lengths of right triangles. | ● **1** Use properties of tangents.<br>**2** Solve problems involving circumscribed polygons. | ● The first bicycles were moved by pushing your feet on the ground. Modern bicycles use pedals, a chain, and gears. The chain loops around circular gears. The length of the chain between these gears is measured from the points of tangency. |

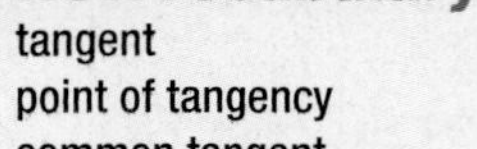

**NewVocabulary**
tangent
point of tangency
common tangent

**Common Core State Standards**

**Content Standards**

G.CO.12 Make formal geometric constructions with a variety of tools and methods (compass and straightedge, string, reflective devices, paper folding, dynamic geometric software, etc.).

G.C.4 Construct a tangent line from a point outside a given circle to the circle.

**Mathematical Practices**

1 Make sense of problems and persevere in solving them.
2 Reason abstractly and quantitatively.

**1 Tangents** A **tangent** is a line in the same plane as a circle that intersects the circle in exactly one point, called the **point of tangency**. $\overleftrightarrow{AB}$ is tangent to $\odot C$ at point $A$. $\overline{AB}$ and $\overrightarrow{AB}$ are also called tangents.

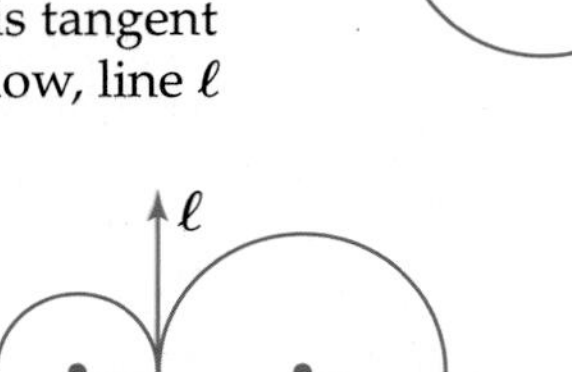


A **common tangent** is a line, ray, or segment that is tangent to two circles in the same plane. In each figure below, line $\ell$ is a common tangent of circles $F$ and $G$.

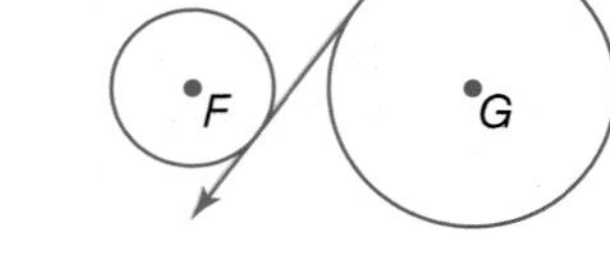


PT

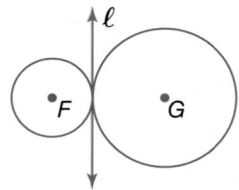

## Example 1 Identify Common Tangents

**Copy each figure and draw the common tangents. If no common tangent exists, state *no common tangent*.**

**a.**

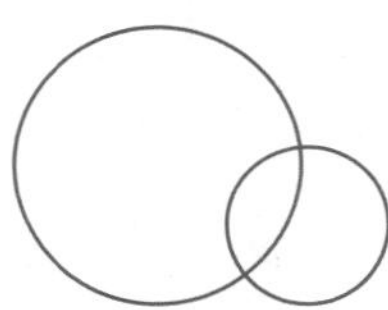

These circles have two common tangents.

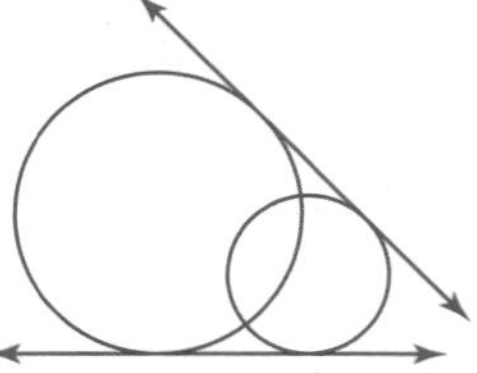

**b.**

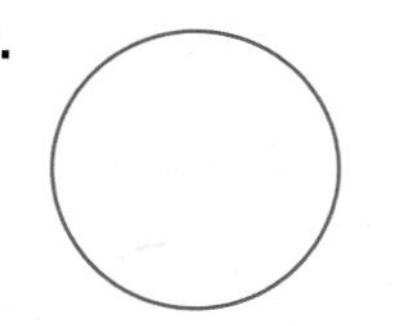

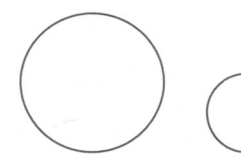

These circles have 4 common tangents.

### GuidedPractice

**1A.**

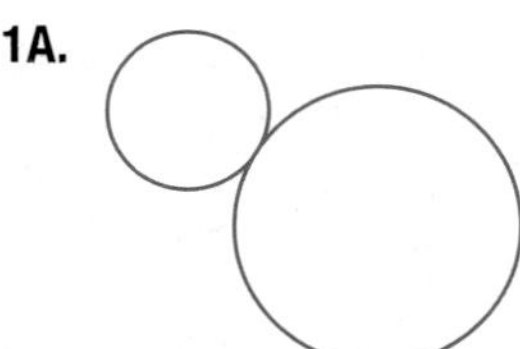

**1B.**

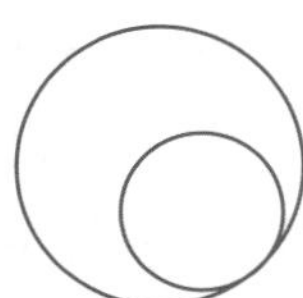

Javier Larrea/age footstock

The shortest distance from a tangent to the center of a circle is the radius drawn to the point of tangency.

**Theorem 11.10**

**Words** In a plane, a line is tangent to a circle if and only if it is perpendicular to a radius drawn to the point of tangency.

**Example** Line $\ell$ is tangent to $\odot S$ if and only if $\ell \perp \overline{ST}$.

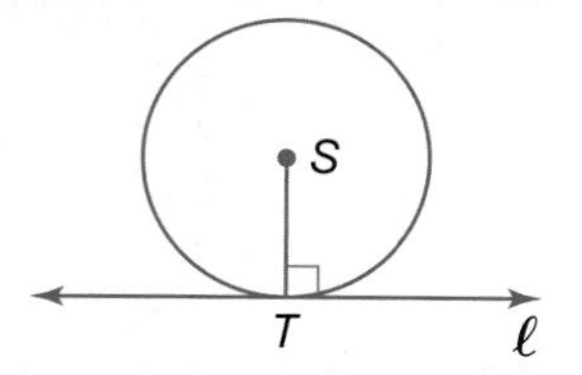


You will prove both parts of Theorem 11.10 in Exercises 32 and 33.

## Example 2 Identify a Tangent

**$\overline{JL}$ is a radius of $\odot J$. Determine whether $\overline{KL}$ is tangent to $\odot J$. Justify your answer.**

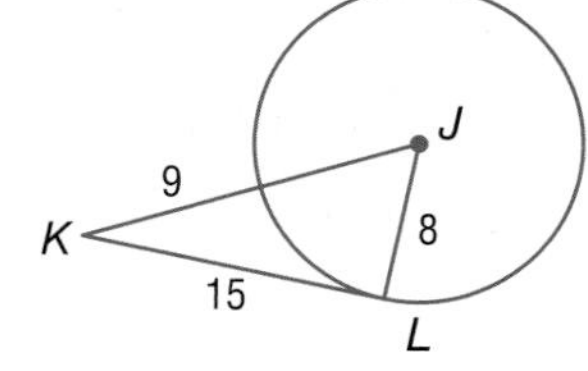


Test to see if $\triangle JKL$ is a right triangle.

$8^2 + 15^2 \stackrel{?}{=} (8 + 9)^2$ Pythagorean Theorem

$289 = 289$ ✓ Simplify.

$\triangle JKL$ is a right triangle with right angle $JLK$. So $\overline{KL}$ is perpendicular to radius $\overline{JL}$ at point $L$. Therefore, by Theorem 11.10, $\overline{KL}$ is tangent to $\odot J$.

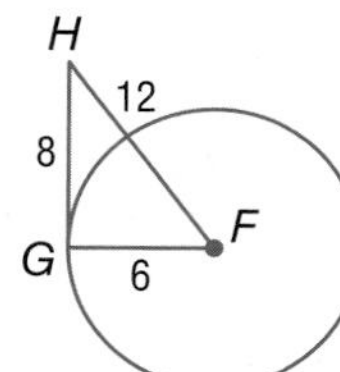


### GuidedPractice

**2.** Determine whether $\overline{GH}$ is tangent to $\odot F$. Justify your answer.

You can also use Theorem 11.10 to identify missing values.

## Example 3 Use a Tangent to Find Missing Values

**$\overline{JH}$ is tangent to $\odot G$ at $J$. Find the value of $x$.**

By Theorem 11.10, $\overline{JH} \perp \overline{GJ}$. So, $\triangle GHJ$ is a right triangle.

$GJ^2 + JH^2 = GH^2$ Pythagorean Theorem

$x^2 + 12^2 = (x + 8)^2$ $GJ = x$, $JH = 12$, and $GH = x + 8$

$x^2 + 144 = x^2 + 16x + 64$ Multiply.

$80 = 16x$ Simplify.

$5 = x$ Divide each side by 16.

**Problem-SolvingTip**

**CCSS Sense-Making** You can use the *solve a simpler problem* strategy by sketching and labeling the right triangles without the circles. A drawing of the triangle in Example 3 is shown below.

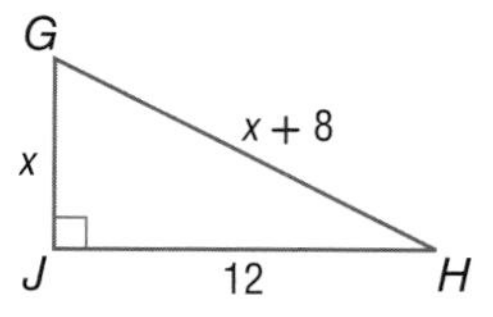


### GuidedPractice

**Find the value of $x$. Assume that segments that appear to be tangent are tangent.**

**3A.**

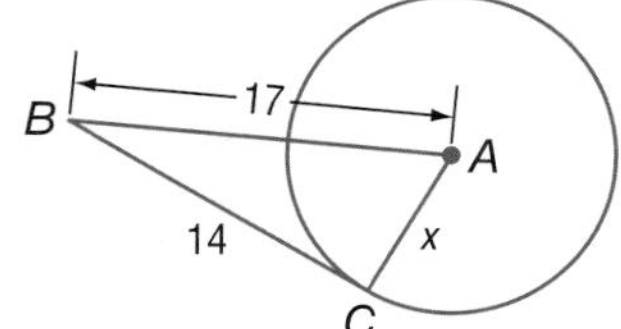


**3B.**

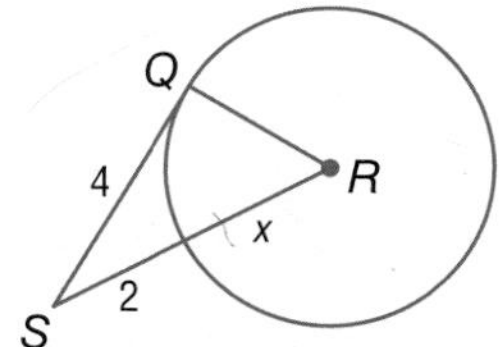

You can use Theorems 11.8 and 11.10 to construct a line tangent to a circle.

## Construction Line Tangent to a Circle Through an External Point

**Step 1** Use a compass to draw circle $C$ and a point $A$ outside of circle $C$. Then draw $\overline{CA}$.

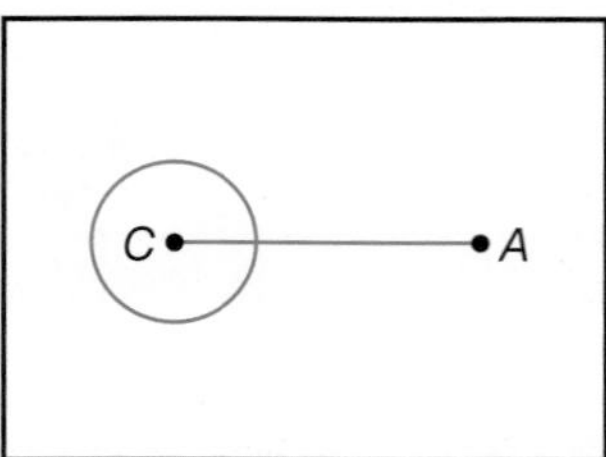


**Step 2** Construct line $\ell$, the perpendicular bisector of $\overline{CA}$. Label the point of intersection $X$.

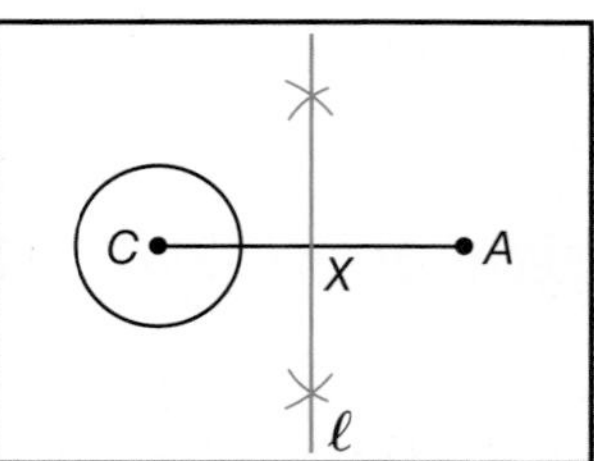


**Step 3** Construct circle $X$ with radius $\overline{XC}$. Label the points of intersection of the two circles $D$ and $E$.

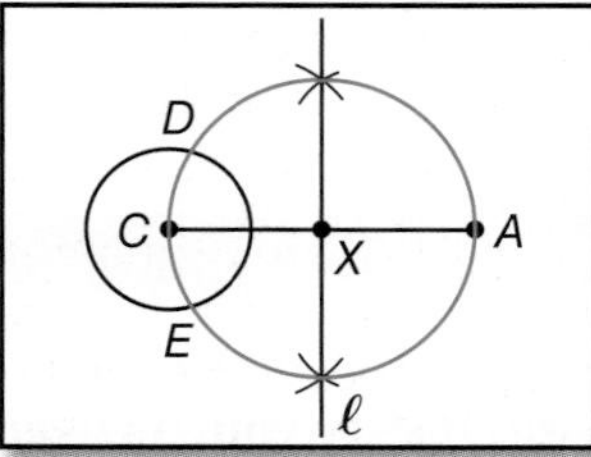


**Step 4** Draw $\overleftrightarrow{AD}$ and $\overline{DC}$. $\triangle ADC$ is inscribed in a semicircle. So, $\angle ADC$ is a right angle and $\overleftrightarrow{AD}$ is tangent to $\odot C$.

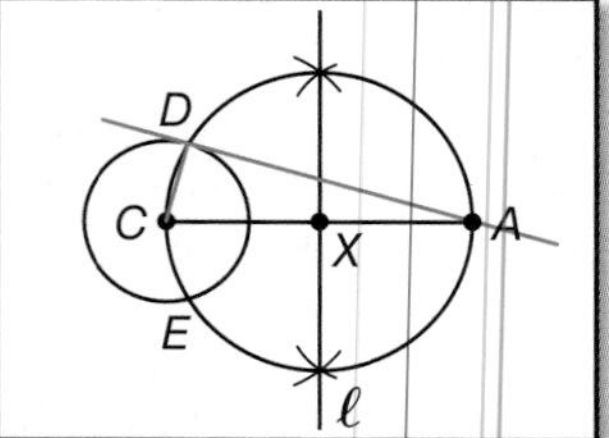


You will justify this construction in Exercise 36 and construct a line tangent to a circle through a point on the circle in Exercise 34.

More than one line can be tangent to the same circle.

## Theorem 11.11

**Words** If two segments from the same exterior point are tangent to a circle, then they are congruent.

**Example** If $\overline{AB}$ and $\overline{CB}$ are tangent to $\odot D$, then $\overline{AB} \cong \overline{CB}$.

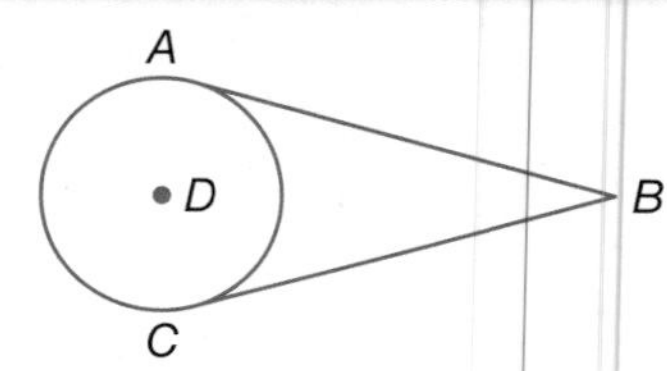


You will prove Theorem 11.11 in Exercise 28.

## Example 4 Use Congruent Tangents to Find Measures

**ALGEBRA $\overline{AB}$ and $\overline{CB}$ are tangent to $\odot D$. Find the value of $x$.**

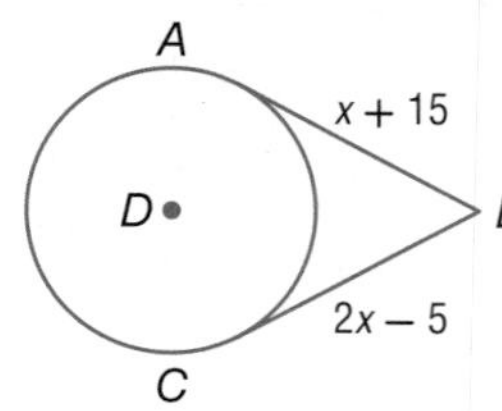


| | |
|---|---|
| $AB = CB$ | Tangents from the same exterior point are congruent. |
| $x + 15 = 2x - 5$ | Substitution |
| $15 = x - 5$ | Subtract $x$ from each side. |
| $20 = x$ | Add 5 to each side. |

### Guided Practice

**ALGEBRA Find the value of $x$. Assume that segments that appear to be tangent are tangent.**

**4A.**

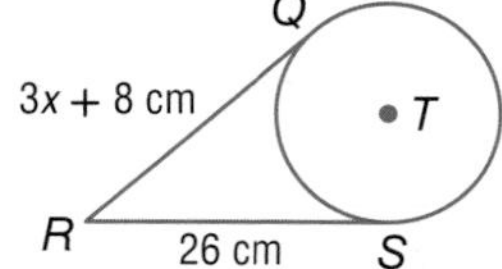


**4B.**

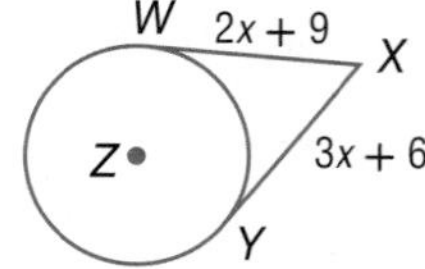

## 2 Circumscribed Polygons

**Circumscribed Polygons** A polygon is circumscribed about a circle if every side of the polygon is tangent to the circle.

**WatchOut!**

**Identifying Circumscribed Polygons** Just because the circle is tangent to one or more of the sides of a polygon does not mean that the polygon is circumscribed about the circle, as shown in the second set of figures.

| Circumscribed Polygons | Polygons Not Circumscribed |
|---|---|
| | |

You can use Theorem 11.11 to find missing measures in circumscribed polygons.

### Real-World Example 5 Find Measures in Circumscribed Polygons

**GRAPHIC DESIGN A graphic designer is giving directions to create a larger version of the triangular logo shown. If $\triangle ABC$ is circumscribed about $\odot G$, find the perimeter of $\triangle ABC$.**

**Step 1** Find the missing measures.

Since $\triangle ABC$ is circumscribed about $\odot G$, $\overline{AE}$ and $\overline{AD}$ are tangent to $\odot G$, as are $\overline{BE}$, $\overline{BF}$, $\overline{CF}$, and $\overline{CD}$. Therefore, $\overline{AE} \cong \overline{AD}$, $\overline{BF} \cong \overline{BE}$, and $\overline{CF} \cong \overline{CD}$.

So, $AE = AD = 8$ feet, $BF = BE = 7$ feet.

By Segment Addition, $CF + FB = CB$, so $CF = CB - FB = 10 - 7$ or 3 feet. So, $CD = CF = 3$ feet.

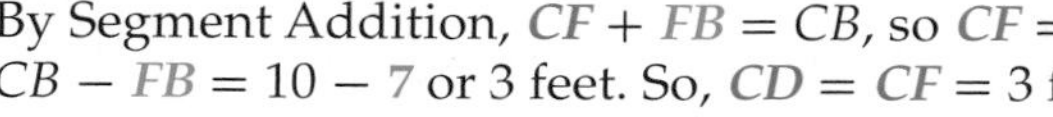

**Step 2** Find the perimeter of $\triangle ABC$.

$$\begin{aligned} \text{perimeter} &= AE + EB + BC + CD + DA \\ &= 8 + 7 + 10 + 3 + 8 \text{ or } 36 \end{aligned}$$

So, the perimeter of $\triangle ABC$ is 36 feet.

**GuidedPractice**

5. Quadrilateral $RSTU$ is circumscribed about $\odot J$. If the perimeter is 18 units, find $x$.

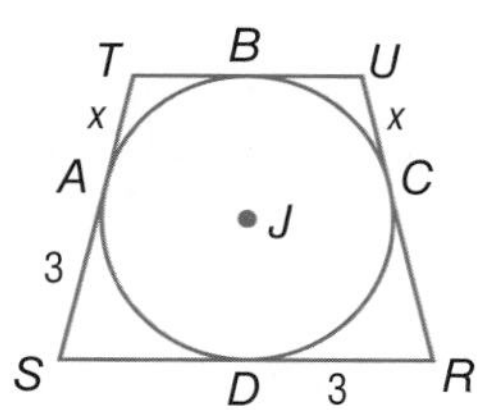


## Check Your Understanding

**Example 1**

1. Copy the figure shown, and draw the common tangents. If no common tangent exists, state *no common tangent*.

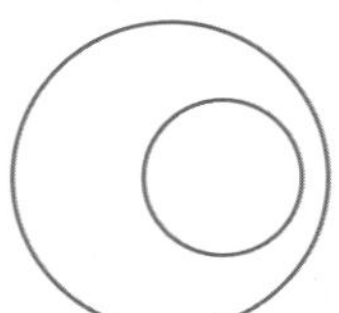

**Example 2**

**Determine whether $\overline{FG}$ is tangent to $\odot E$. Justify your answer.**

2.

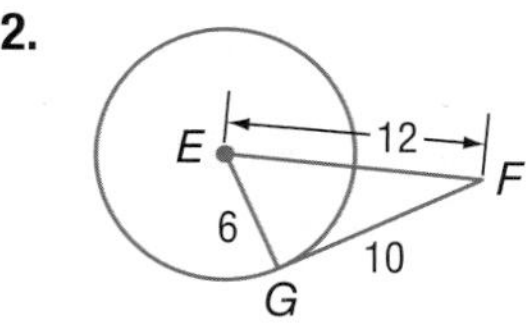


3.

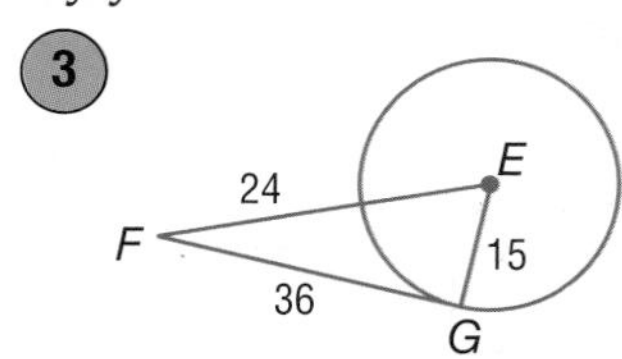

**Examples 3–4** **Find $x$. Assume that segments that appear to be tangent are tangent.**

**4.** 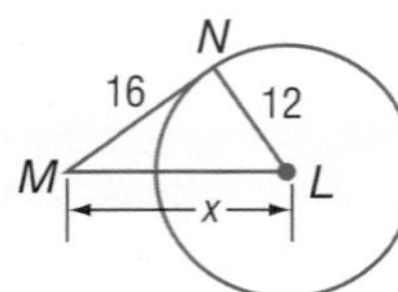


**5.** 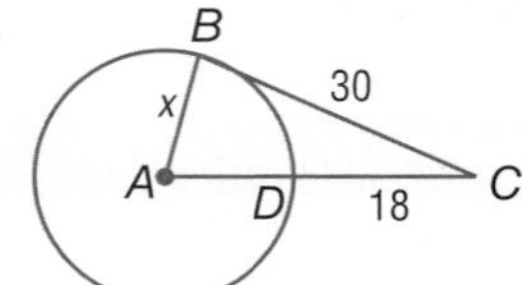


**6.** 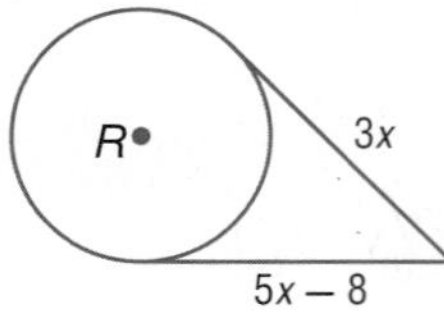


**7. LANDSCAPE ARCHITECT** A landscape architect is paving the two walking paths that are tangent to two approximately circular ponds as shown. The lengths given are in feet. Find the values of $x$ and $y$.

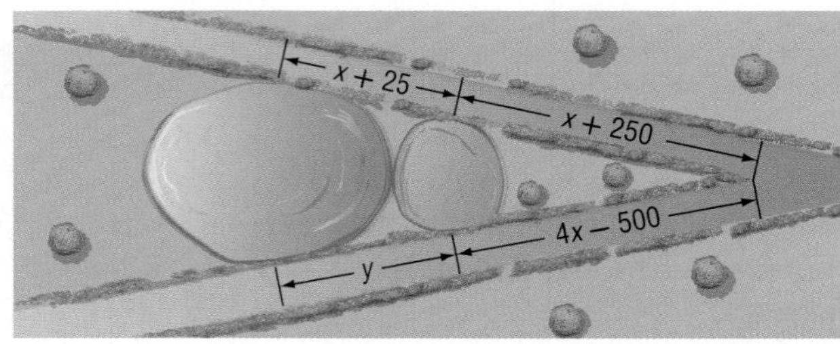


**Example 5** **8. CCSS SENSE-MAKING** Triangle $JKL$ is circumscribed about $\odot R$.

**a.** Find $x$.

**b.** Find the perimeter of $\triangle JKL$.

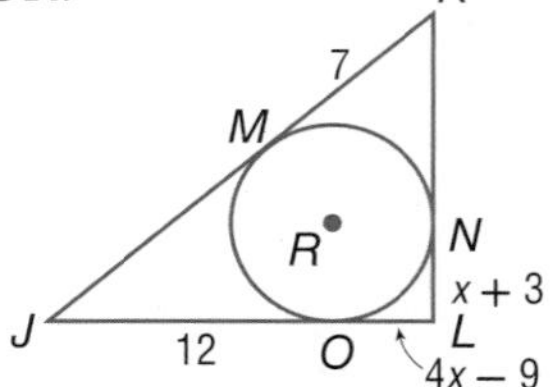


## Practice and Problem Solving

**Example 1** **Copy each figure and draw the common tangents. If no common tangent exists, state *no common tangent*.**

**9.** 

**10.** 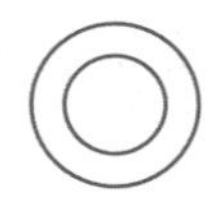

**11.** 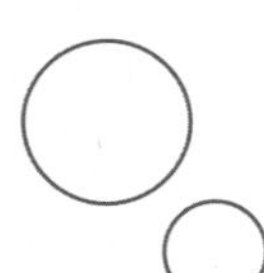

**12.** 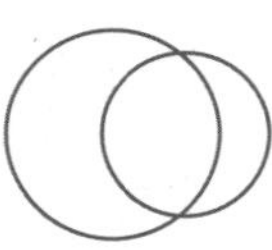

**Example 2** **Determine whether each $\overline{XY}$ is tangent to the given circle. Justify your answer.**

**13.** 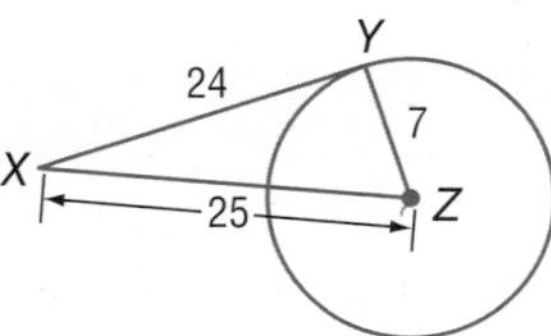


**14.** 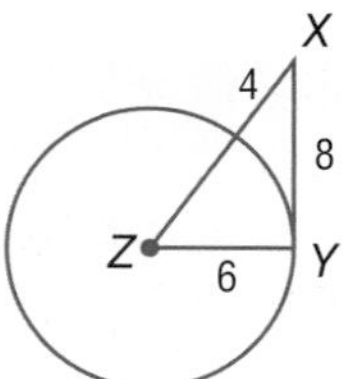


**15.** 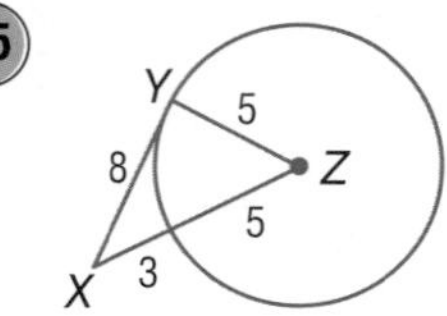


**16.** 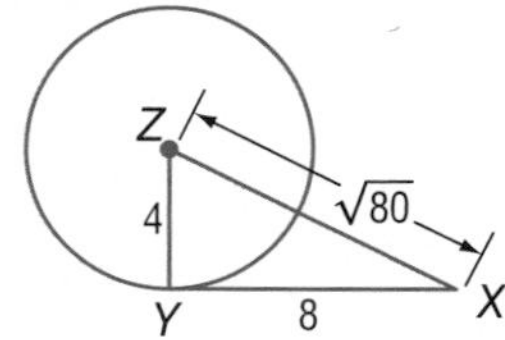

**Examples 3–4** **Find $x$. Assume that segments that appear to be tangent are tangent. Round to the nearest tenth if necessary.**

**17.**

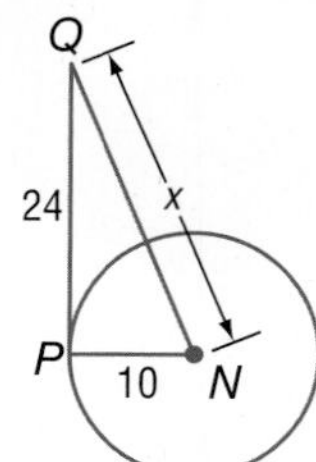


**18.**

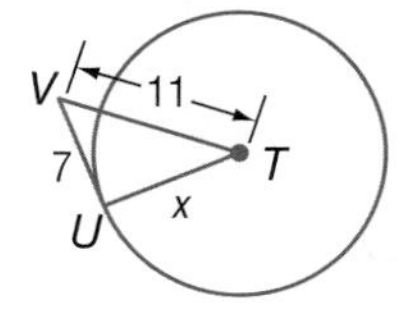


**19.**

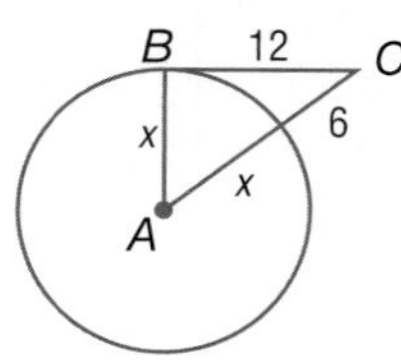


**20.**

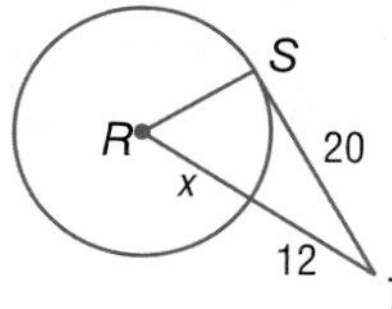


**21.**

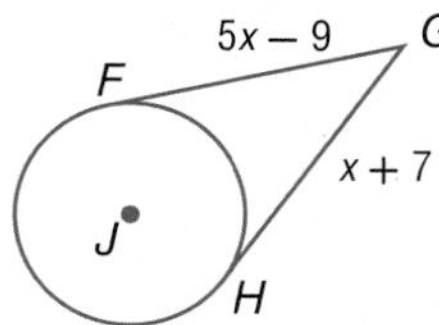


**22.**

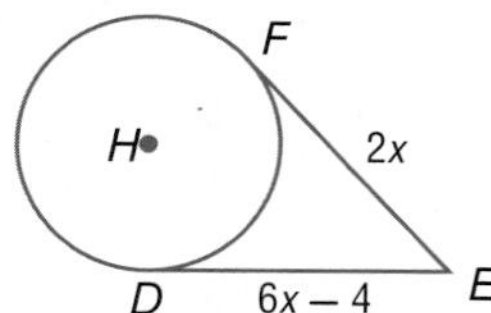


**23. ARBORS** In the arbor shown, $\overline{AC}$ and $\overline{BC}$ are tangents to $\odot D$. The radius of the circle is 26 inches and $EC = 20$ inches. Find each measure to the nearest hundredth.

**a.** $AC$ **b.** $BC$

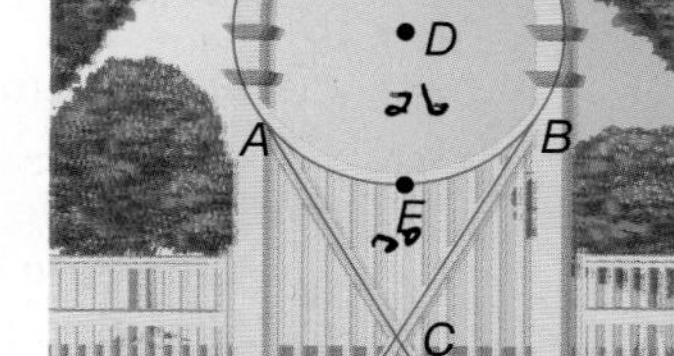

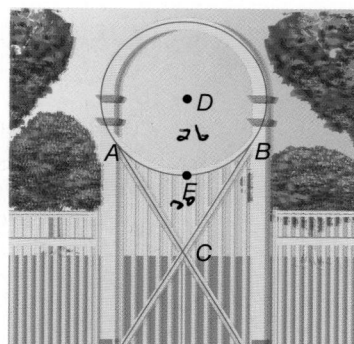

**Example 5** **CCSS SENSE-MAKING** **Find the value of $x$. Then find the perimeter.**

**24.**

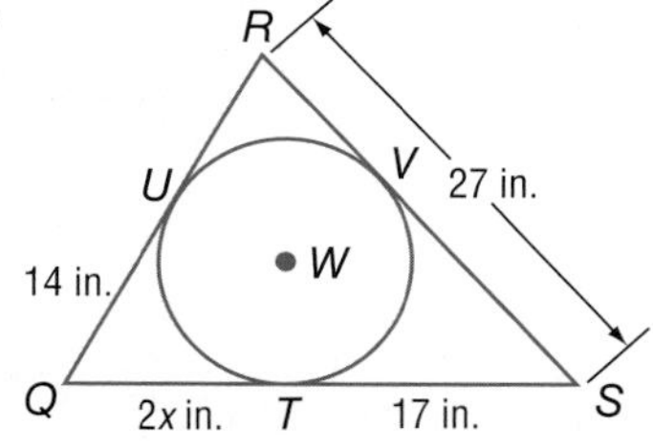


**25.**

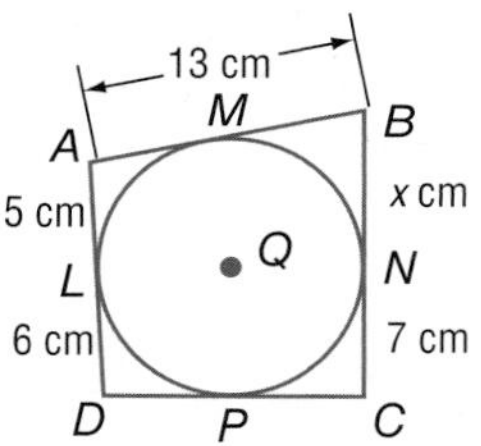


**Find $x$ to the nearest hundredth. Assume that segments that appear to be tangent are tangent.**

**26.**

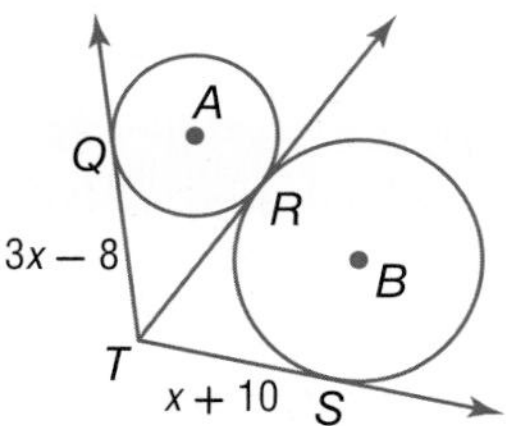


**27.**

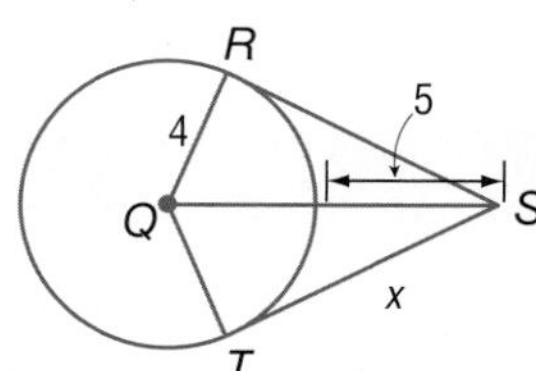


**Write the specified type of proof.**

**28.** two-column proof of Theorem 11.11

**Given:** $\overline{AC}$ is tangent to $\odot H$ at $C$.
$\overline{AB}$ is tangent to $\odot H$ at $B$.

**Prove:** $\overline{AC} \cong \overline{AB}$

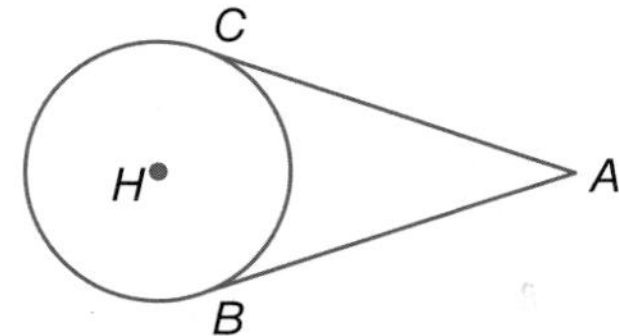


**29.** two-column proof

**Given:** Quadrilateral $ABCD$ is circumscribed about $\odot P$.

**Prove:** $AB + CD = AD + BC$

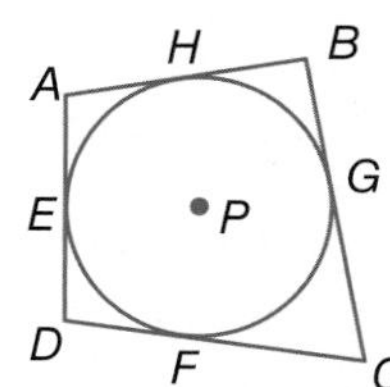

**30. SATELLITES** A satellite is 720 kilometers above Earth, which has a radius of 6360 kilometers. The region of Earth that is visible from the satellite is between the tangent lines $\overline{BA}$ and $\overline{BC}$. What is $BA$? Round to the nearest hundredth.

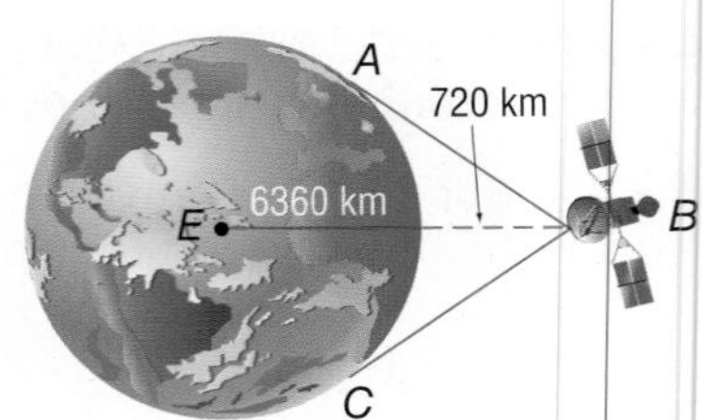


**31 SPACE TRASH** *Orbital debris* refers to materials from space missions that still orbit Earth. In 2007, a 1400-pound ammonia tank was discarded from a space mission. Suppose the tank has an altitude of 435 miles. What is the distance from the tank to the farthest point on Earth's surface from which the tank is visible? Assume that the radius of Earth is 4000 miles. Round to the nearest mile, and include a diagram of this situation with your answer.

**32. PROOF** Write an indirect proof to show that if a line is tangent to a circle, then it is perpendicular to a radius of the circle. (Part 1 of Theorem 11.10)

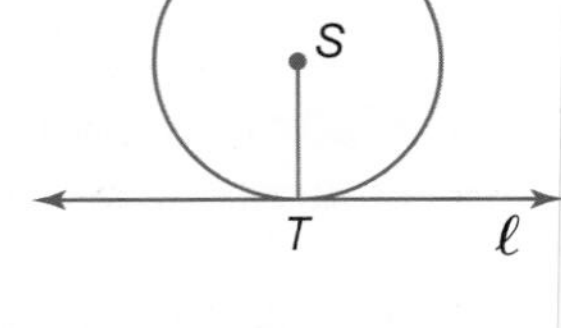


**Given:** $\ell$ is tangent to $\odot S$ at $T$; $\overline{ST}$ is a radius of $\odot S$.

**Prove:** $\ell \perp \overline{ST}$

(*Hint:* Assume $\ell$ is *not* $\perp$ to $\overline{ST}$.)

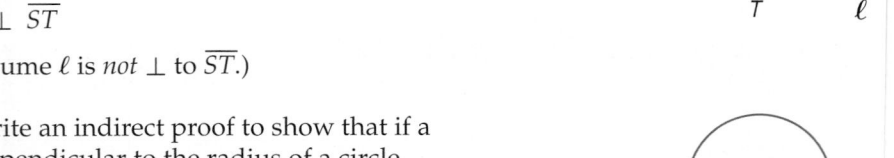

**33. PROOF** Write an indirect proof to show that if a line is perpendicular to the radius of a circle at its endpoint, then the line is a tangent of the circle. (Part 2 of Theorem 11.10)

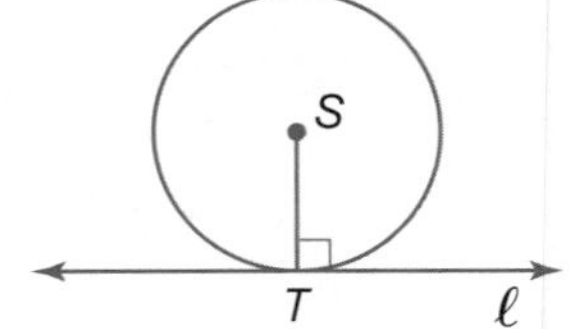


**Given:** $\ell \perp \overline{ST}$; $\overline{ST}$ is a radius of $\odot S$.

**Prove:** $\ell$ is tangent to $\odot S$.

(*Hint:* Assume $\ell$ is *not* tangent to $\odot S$.)

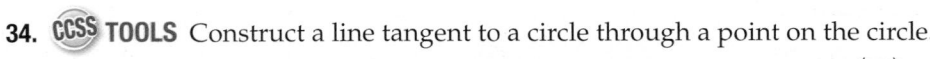

**34. CCSS TOOLS** Construct a line tangent to a circle through a point on the circle.

Use a compass to draw $\odot A$. Choose a point $P$ on the circle and draw $\overleftrightarrow{AP}$. Then construct a segment through point $P$ perpendicular to $\overleftrightarrow{AP}$. Label the tangent line $t$. Explain and justify each step.

## H.O.T. Problems Use Higher-Order Thinking Skills

**35. CHALLENGE** $\overline{PQ}$ is tangent to circles $R$ and $S$. Find $PQ$. Explain your reasoning.

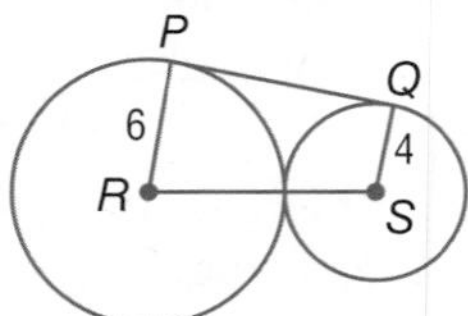


**36. WRITING IN MATH** Explain and justify each step in the construction on page 734.

**37. OPEN ENDED** Draw a circumscribed triangle and an inscribed triangle.

**38. REASONING** In the figure, $\overline{XY}$ and $\overline{XZ}$ are tangent to $\odot A$. $\overline{XZ}$ and $\overline{XW}$ are tangent to $\odot B$. Explain how segments $\overline{XY}$, $\overline{XZ}$, and $\overline{XW}$ can all be congruent if the circles have different radii.

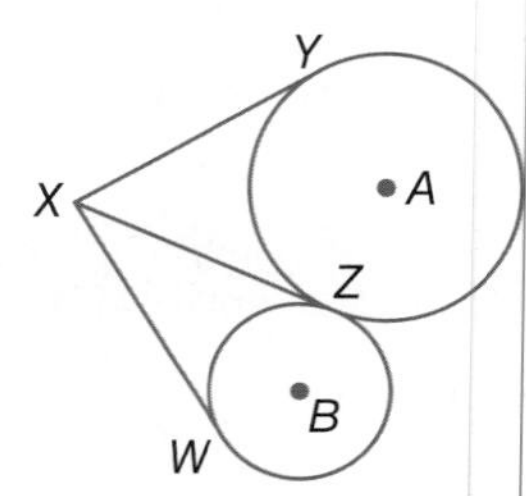


**39. WRITING IN MATH** Is it possible to draw a tangent from a point that is located anywhere outside, on, or inside a circle? Explain.

## Standardized Test Practice

**40.** $\odot P$ has a radius of 10 centimeters, and $\overline{ED}$ is tangent to the circle at point $D$. $F$ lies both on $\odot P$ and on segment $\overline{EP}$. If $ED = 24$ centimeters, what is the length of $\overline{EF}$?

**A** 10 cm
**B** 16 cm
**C** 21.8 cm
**D** 26 cm

**41. SHORT RESPONSE** A square is inscribed in a circle having a radius of 6 inches. Find the length of each side of the square.

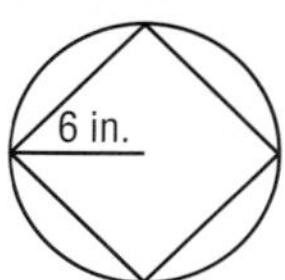


**42. ALGEBRA** Which of the following shows $25x^2 - 5x$ factored completely?

**F** $5x(x)$
**G** $5x(5x - 1)$
**H** $x(x - 5)$
**J** $x(5x - 1)$

**43. SAT/ACT** What is the perimeter of the triangle shown below?

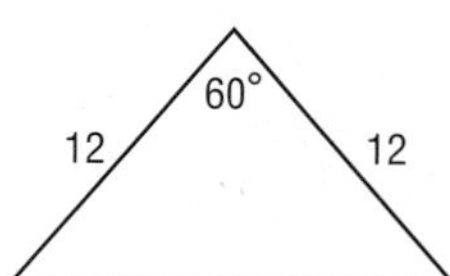


**A** 12 units
**B** 24 units
**C** 34.4 units
**D** 36 units
**E** 104 units

## Spiral Review

**Find each measure.** (Lesson 11-4)

**44.** $m\widehat{JK}$

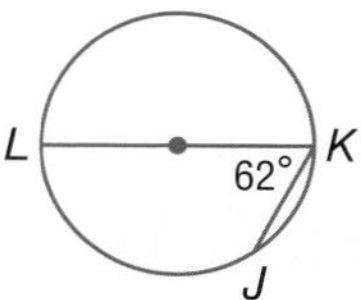


**45.** $m\angle B$

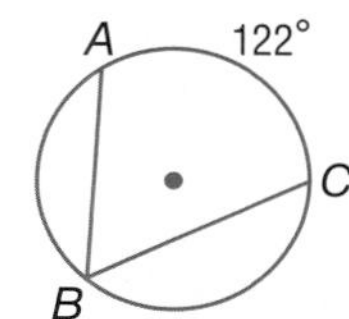


**46.** $m\widehat{VX}$

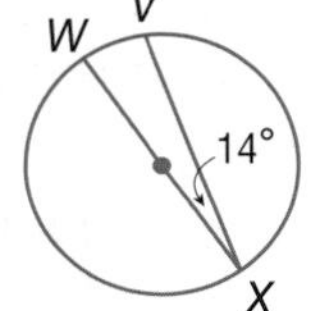


**In $\odot F$, $GK = 14$ and $m\widehat{GHK} = 142$. Find each measure. Round to the nearest hundredth.** (Lesson 11-3)

**47.** $m\widehat{GH}$

**48.** $JK$

**49.** $m\widehat{KM}$

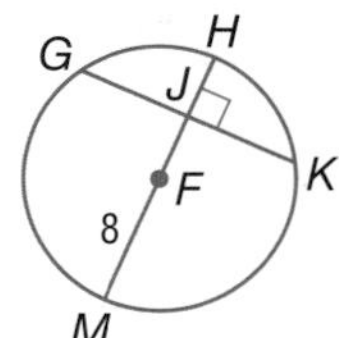


**50. METEOROLOGY** The altitude of the base of a cloud formation is called the *ceiling*. To find the ceiling one night, a meteorologist directed a spotlight vertically at the clouds. Using a theodolite, an optical instrument with a rotatable telescope, placed 83 meters from the spotlight and 1.5 meters above the ground, he found the angle of elevation to be 62.7°. How high was the ceiling? (Lesson 9-5)

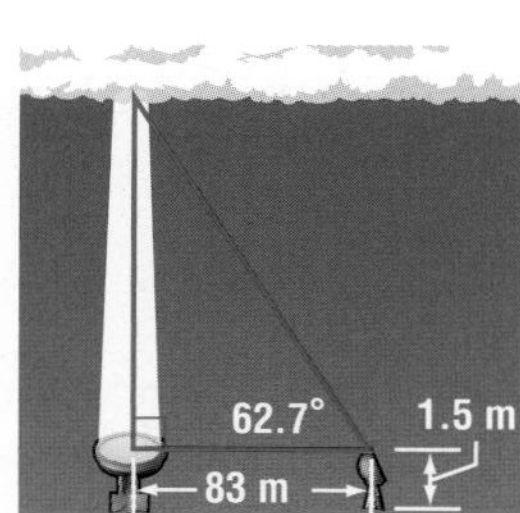


**Determine whether the triangles are similar. If so, write a similarity statement. Explain your reasoning.** (Lesson 8-3)

**51.**

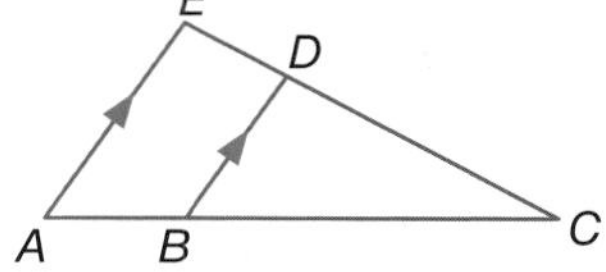


**52.**

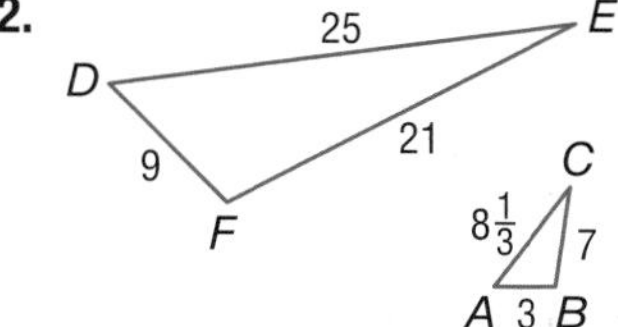


## Skills Review

**Solve each equation.**

**53.** $15 = \frac{1}{2}[(360 - x) - 2x]$

**54.** $x + 12 = \frac{1}{2}[(180 - 120)]$

**55.** $x = \frac{1}{2}[(180 - 64)]$

# EXTEND 11-5 Geometry Lab Inscribed and Circumscribed Circles

In this lab, you will perform constructions that involve inscribing or circumscribing a circle.

**CCSS Common Core State Standards**
**Content Standards**
**G.CO.13** Construct an equilateral triangle, a square, and a regular hexagon inscribed in a circle.
**G.C.3** Construct the inscribed and circumscribed circles of a triangle, and prove properties of angles for a quadrilateral inscribed in a circle.
**Mathematical Practices** 5

## Activity 1 Construct a Circle Inscribed in a Triangle

**Step 1**

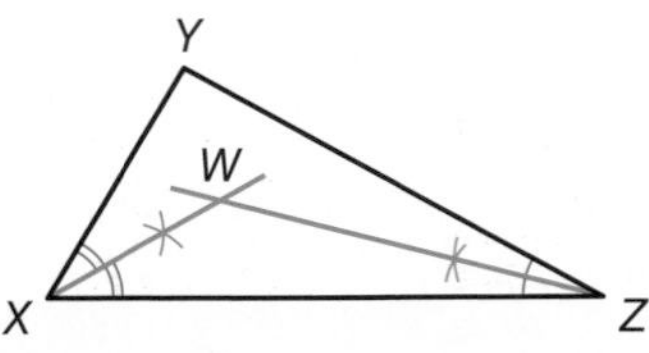


Draw a triangle $XYZ$ and construct two angle bisectors of the triangle to locate the incenter $W$.

**Step 2**

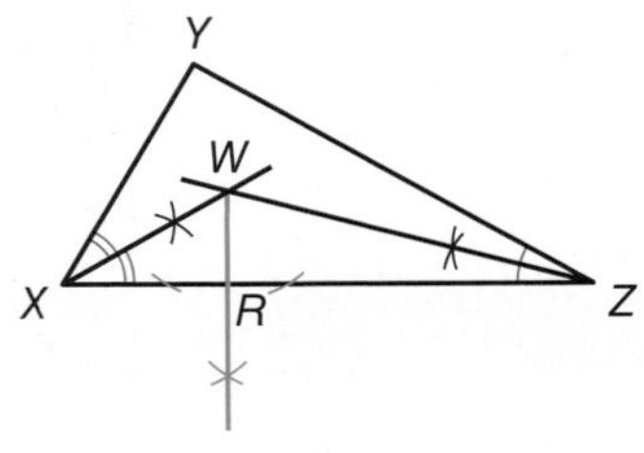


Construct a segment perpendicular to a side through the incenter. Label the intersection $R$.

**Step 3**

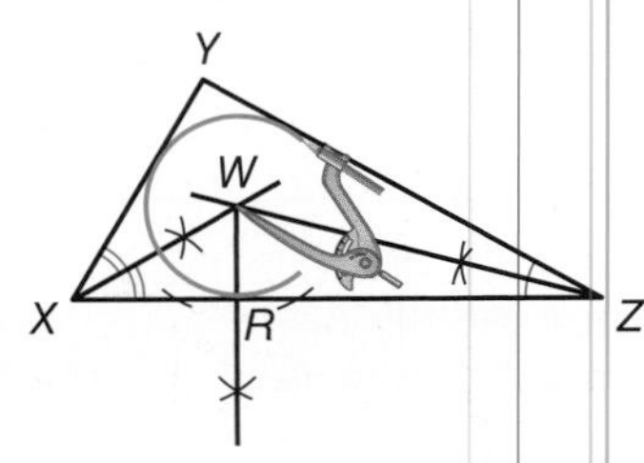


Set a compass of the length of $\overline{WR}$. Put the point of the compass on $W$ and draw a circle with that radius.

## Activity 2 Construct a Triangle Circumscribed About a Circle

**Step 1**

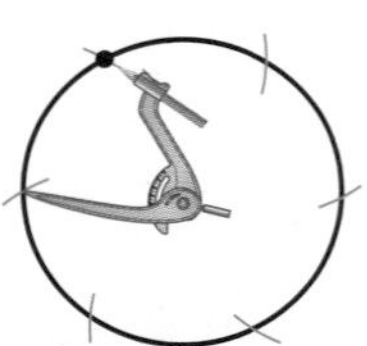

Construct a circle and draw a point. Use the same compass setting you used to construct the circle to construct an arc on the circle from the point. Continue as shown.

**Step 2**

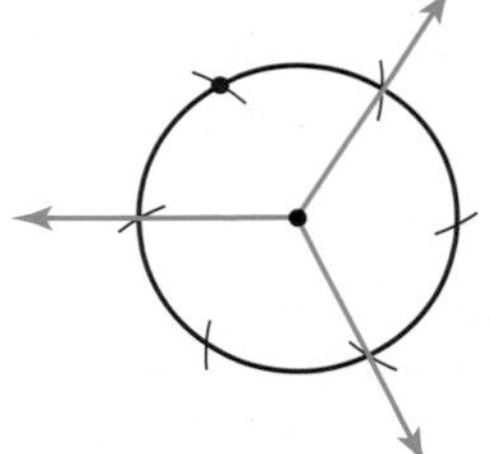

Draw rays from the center through every other arc.

**Step 3**

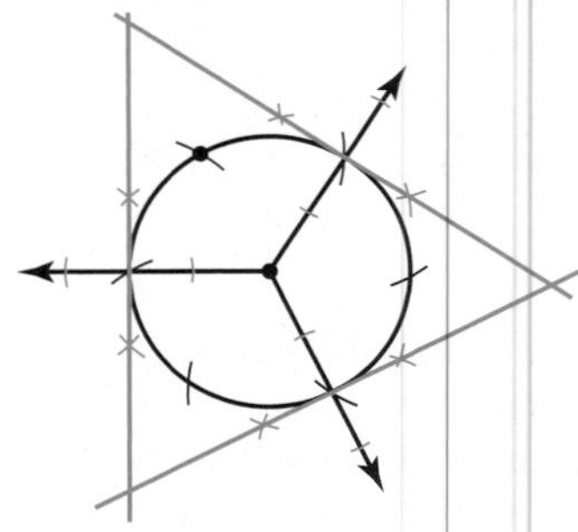

Construct a line perpendicular to each of the rays.

## Model

1. Draw a right triangle and inscribe a circle in it.
2. Inscribe a regular hexagon in a circle. Then inscribe an equilateral triangle in a circle. (*Hint:* The first step of each construction is identical to Step 1 in Activity 2.)
3. Inscribe a square in a circle. Then circumscribe a square about a circle.
4. **CHALLENGE** Circumscribe a regular hexagon about a circle.

# LESSON 11-6 Secants, Tangents, and Angle Measures

| :: Then | :: Now | :: Why? |
|---|---|---|
| You found measures of segments formed by tangents to a circle. | **1** Find measures of angles formed by lines intersecting on or inside a circle.<br>**2** Find measures of angles formed by lines intersecting outside the circle. | An average person's field of vision is about 180°. Most cameras have a much narrower viewing angle of between 20° and 50°. This viewing angle determines how much of a curved object a camera can capture on film. |

**NewVocabulary**
secant

**Common Core State Standards**

**Content Standards**
Reinforcement of G.C.4
Construct a tangent line from a point outside a given circle to the circle.

**Mathematical Practices**
3 Construct viable arguments and critique the reasoning of others.
1 Make sense of problems and persevere in solving them.

## 1 Intersections On or Inside a Circle

A **secant** is a line that intersects a circle in exactly two points. Lines $j$ and $k$ are secants of $\odot C$.

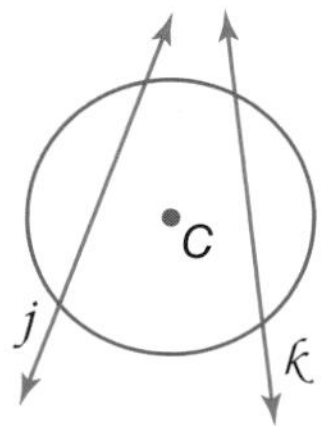


When two secants intersect inside a circle, the angles formed are related to the arcs they intercept.

### Theorem 11.12

**Words** If two secants or chords intersect in the interior of a circle, then the measure of an angle formed is one half the *sum* of the measure of the arcs intercepted by the angle and its vertical angle.

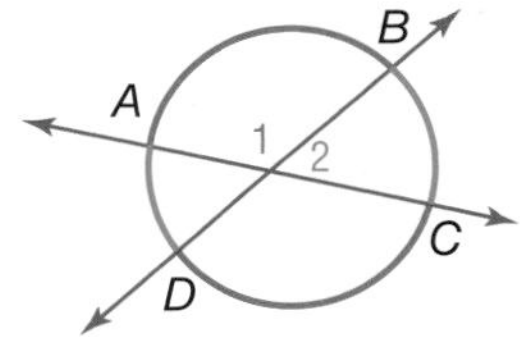


**Example** $m\angle 1 = \frac{1}{2}(m\widehat{AB} + m\widehat{CD})$ and $m\angle 2 = \frac{1}{2}(m\widehat{DA} + m\widehat{BC})$

### Proof

**Given:** $\overleftrightarrow{HK}$ and $\overleftrightarrow{JL}$ intersect at $M$.

**Prove:** $m\angle 1 = \frac{1}{2}(m\widehat{JH} + m\widehat{LK})$

**Proof:**

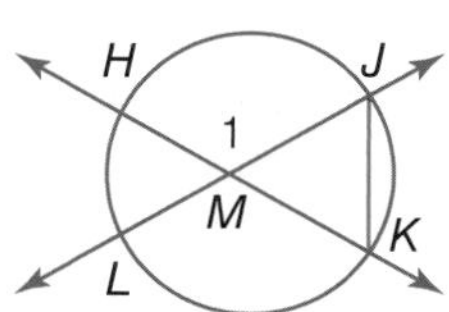


| Statements | Reasons |
|---|---|
| 1. $\overleftrightarrow{HK}$ and $\overleftrightarrow{JL}$ intersect at $M$. | 1. Given |
| 2. $m\angle 1 = m\angle MJK + m\angle MKJ$ | 2. Exterior Angle Theorem |
| 3. $m\angle MJK = \frac{1}{2}m\angle\widehat{LK}$, $m\angle MKJ = \frac{1}{2}m\angle\widehat{JH}$ | 3. The measure of an inscribed ∠ equals half the measure of the intercepted arc. |
| 4. $m\angle 1 = \frac{1}{2}m\angle\widehat{LK} + \frac{1}{2}m\angle\widehat{JH}$ | 4. Substitution |
| 5. $m\angle 1 = \frac{1}{2}(m\widehat{JH} + m\widehat{LK})$ | 5. Distributive Property |

Image Source/Getty Images

## Example 1 Use Intersecting Chords or Secants

**Find $x$.**

**a.**

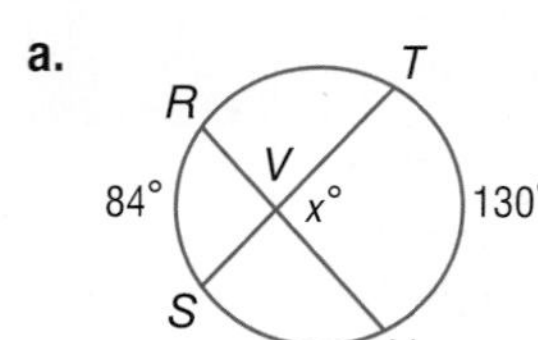


$m\angle TVU = \frac{1}{2}(m\widehat{RS} + m\widehat{TU})$ — Theorem 11.12

$x = \frac{1}{2}(84 + 130)$ — Substitution

$= \frac{1}{2}(214)$ or 107 — Simplify.

**b.**

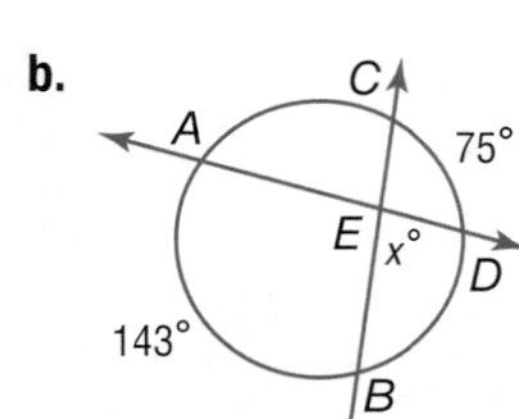


**Step 1** Find $m\angle AEB$.

$m\angle AEB = \frac{1}{2}(m\widehat{AB} + m\widehat{CD})$ — Theorem 11.12

$= \frac{1}{2}(143 + 75)$ — Substitution

$= \frac{1}{2}(218)$ or 109 — Simplify.

**Step 2** Find $x$, the measure of $\angle DEB$.

$\angle AEB$ and $\angle DEB$ are supplementary angles.

So, $x = 180 - 109$ or 71.

**StudyTip**

**Alternative Method** In Example 1b, $m\angle DEB$ can also be found by first finding the sum of the measures of $\widehat{AC}$ and $\widehat{BD}$.

$m\widehat{AC} + m\widehat{BD}$
$= 360 - (m\widehat{AC} + m\widehat{CD})$
$= 360 - (143 + 75)$
$= 142$

$m\angle DEB$
$= \frac{1}{2}(m\widehat{AC} + m\widehat{BD})$
$= \frac{1}{2}(142)$ or 71

**c.**

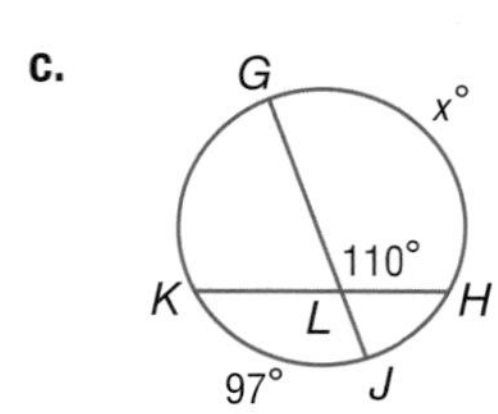


$m\angle GLH = \frac{1}{2}(m\widehat{GH} + m\widehat{KJ})$ — Theorem 11.12

$110 = \frac{1}{2}(x + 97)$ — Substitution

$220 = (x + 97)$ — Multiply each side by 2.

$123 = x$ — Subtract 97 from each side.

### GuidedPractice

**1A.**

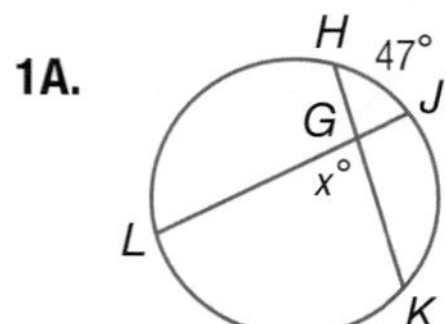


**1B.**



**1C.**

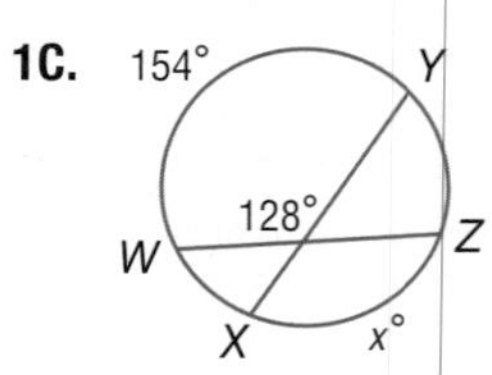


Recall that Theorem 11.6 states that the measure of an inscribed angle is half the measure of its intercepted arc. If one of the sides of this angle is tangent to the circle, this relationship still holds true.

### Theorem 11.13

**Words** If a secant and a tangent intersect at the point of tangency, then the measure of each angle formed is one half the measure of its intercepted arc.

**Example** 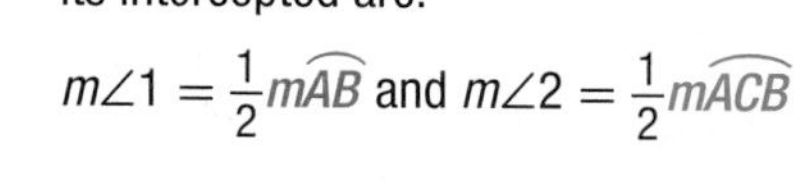 $m\angle 1 = \frac{1}{2}m\widehat{AB}$ and $m\angle 2 = \frac{1}{2}m\widehat{ACB}$

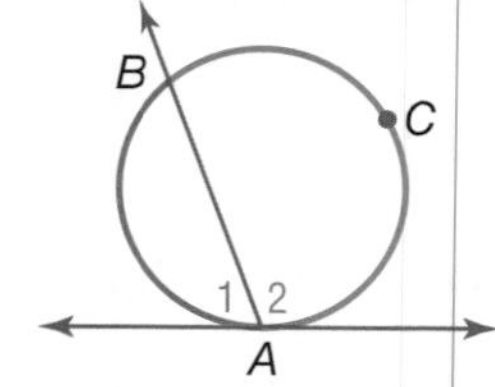


You will prove Theorem 11.13 in Exercise 33.

## Example 2 Use Intersecting Secants and Tangents

**Find each measure.**

**a.** $m\angle QPR$

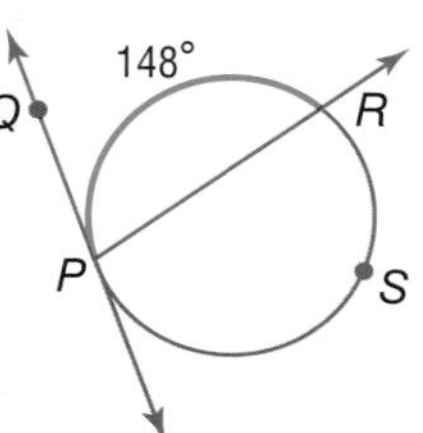


$m\angle QPR = \frac{1}{2}m\widehat{PR}$ — Theorem 11.13

$= \frac{1}{2}(148)$ or 74 — Substitute and simplify.

**b.** $m\widehat{DEF}$

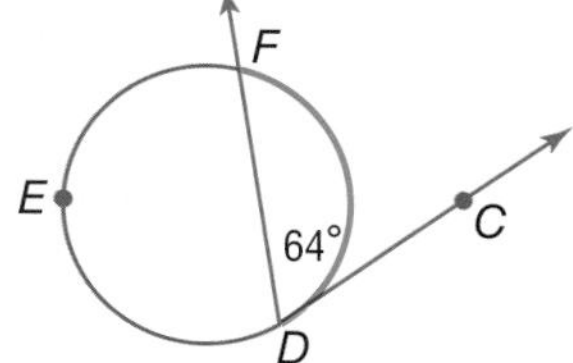


$m\angle CDF = \frac{1}{2}m\widehat{FD}$ — Theorem 11.13

$64 = \frac{1}{2}m\widehat{FD}$ — Substitution

$128 = m\widehat{FD}$ — Multiply each side by 2.

$m\widehat{DEF} = 360 - m\widehat{FD} = 360 - 128$ or 232

### GuidedPractice

**2A.** Find $m\widehat{JLK}$.

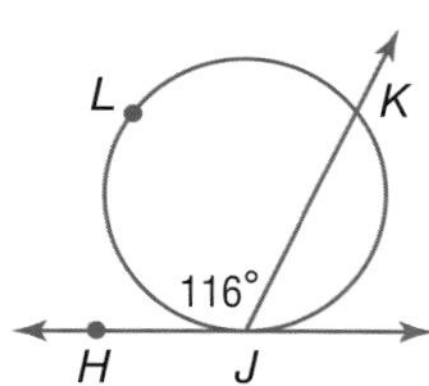


**2B.** Find $m\angle RQS$ if $m\widehat{QTS} = 238$.

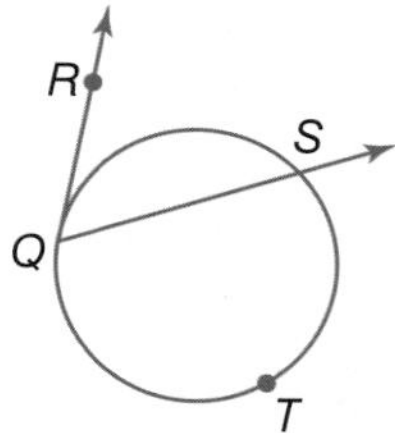


## 2 Intersections Outside a Circle

Secants and tangents can also meet outside a circle. The measure of the angle formed also involves half of the measures of the arcs they intercept.

### Theorem 11.14

**Words** If two secants, a secant and a tangent, or two tangents intersect in the exterior of a circle, then the measure of the angle formed is one half the *difference* of the measures of the intercepted arcs.

**Examples**

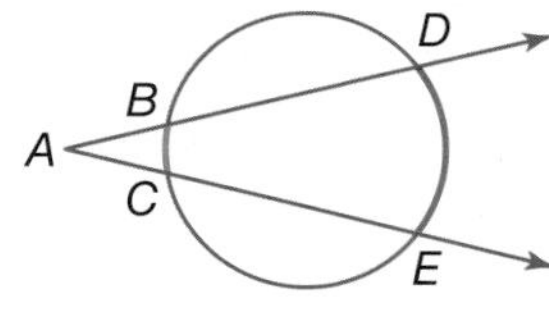


**Two Secants**

$m\angle A = \frac{1}{2}(m\widehat{DE} - m\widehat{BC})$

**Secant-Tangent**

$m\angle A = \frac{1}{2}(m\widehat{DC} - m\widehat{BC})$

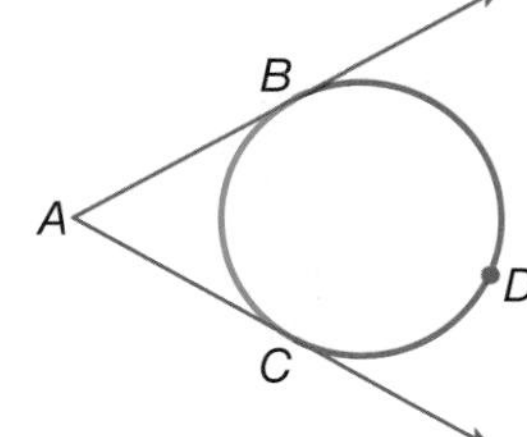


**Two Tangents**

$m\angle A = \frac{1}{2}(m\widehat{BDC} - m\widehat{BC})$

You will prove Theorem 11.14 in Exercises 30–32.

**StudyTip**

**Absolute Value** The measure of each $\angle A$ can also be expressed as half the absolute value of the difference of the arc measure. In this way, the order of the arc measures does not affect the outcome of the calculation.

## Example 3 Use Tangents and Secants that Intersect Outside a Circle

**Find each measure.**

**a.** $m\angle L$

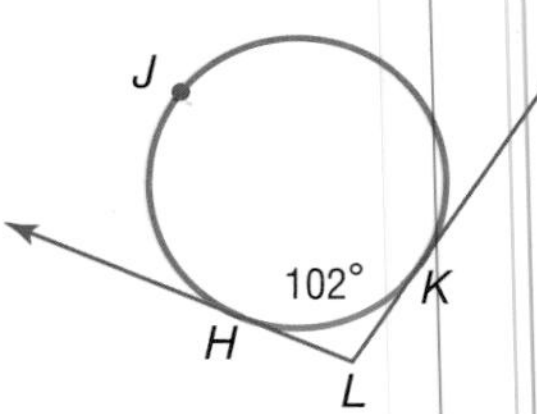


$m\angle L = \frac{1}{2}(m\widehat{HJK} - m\widehat{HK})$ Theorem 11.14

$= \frac{1}{2}(360 - 102) - 102$ Substitution

$= \frac{1}{2}(258 - 102)$ or 78 Simplify.

**b.** $m\widehat{CD}$

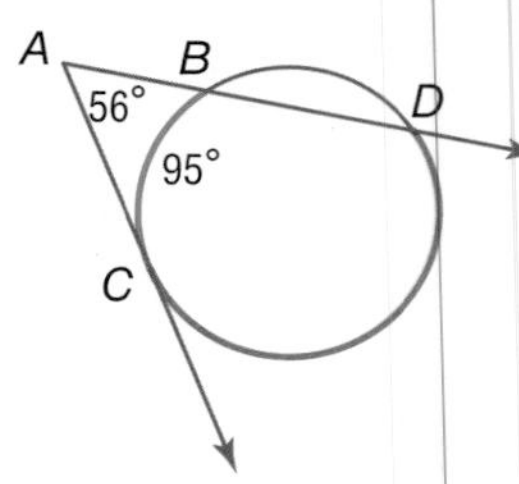


$m\angle A = \frac{1}{2}(m\widehat{CD} - m\widehat{BC})$ Theorem 11.14

$56 = \frac{1}{2}(m\widehat{CD} - 95)$ Substitution

$112 = m\widehat{CD} - 95$ Multiply each side by 2.

$207 = m\widehat{CD}$ Add 95 to each side.

### GuidedPractice

**3A.** $m\angle S$

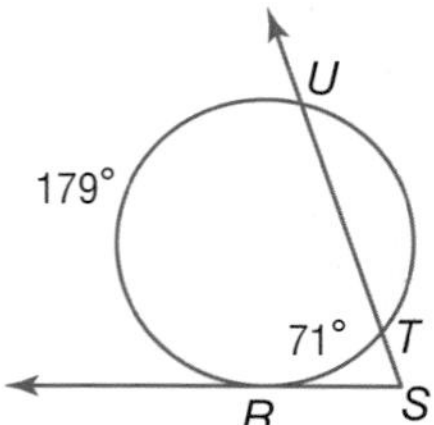


**3B.** $m\widehat{XZ}$

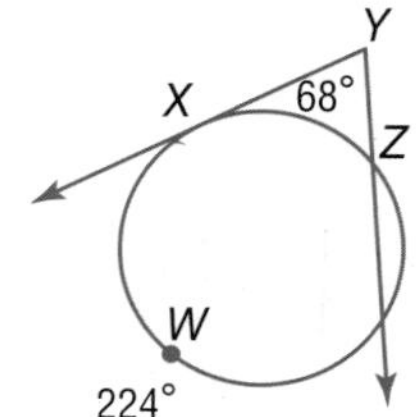


You can apply the properties of intersecting secants to solve real-world problems.

## Real-World Example 4 Apply Properties of Intersecting Secants

**SCIENCE The diagram shows the path of a light ray as it hits a drop of water. The ray is bent, or *refracted*, at points *A*, *B*, and *C*. If $m\widehat{AC} = 128$ and $m\widehat{XBY} = 84$, what is $m\angle D$?**

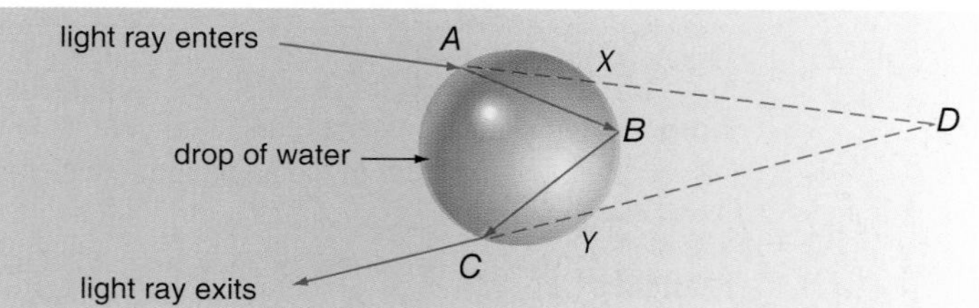


$m\angle D = \frac{1}{2}(m\widehat{AC} - m\widehat{XBY})$ Theorem 11.14

$= \frac{1}{2}(128 - 84)$ Substitution

$= \frac{1}{2}(44)$ or 22 Simplify.

### GuidedPractice

**4.** Find the value of $x$.

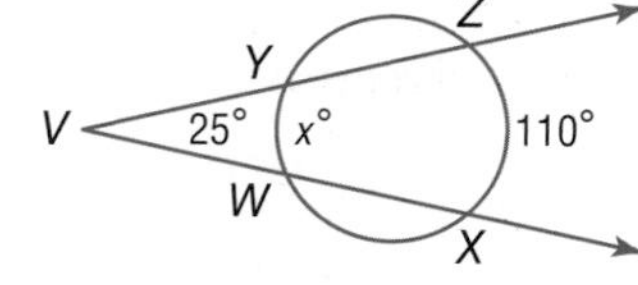

### Real-WorldLink

There is a difference in the *index of refraction* between the two mediums such as air and glass. The index of refraction $N$ is given by the equation $N = \frac{c}{V}$, where $c$ is the speed of light and $V$ is the velocity of light in that material.

**Source:** Microscopy Resource Center

Ruddy Gold/age fotostock

## KeyConcept Circle and Angle Relationships

| Vertex of Angle | Model(s) | Angle Measure |
|---|---|---|
| on the circle | | one half the measure of the intercepted arc<br>$m\angle 1 = \frac{1}{2}x$ |
| inside the circle | | one half the measure of the sum of the intercepted arc<br>$m\angle 1 = \frac{1}{2}(x + y)$ |
| outside the circle | | one half the measure of the difference of the intercepted arcs<br>$m\angle 1 = \frac{1}{2}(x - y)$ |

## Check Your Understanding

**Examples 1–2** **Find each measure. Assume that segments that appear to be tangent are tangent.**

**1.** $m\angle 1$

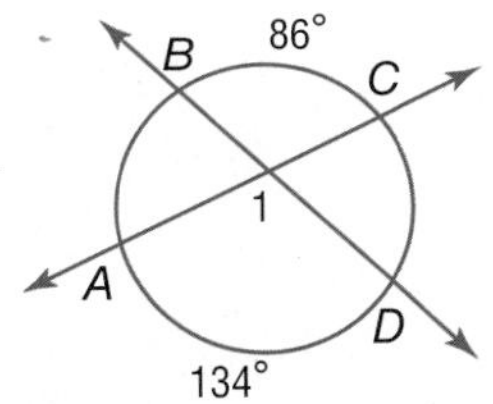


**2.** $m\widehat{TS}$

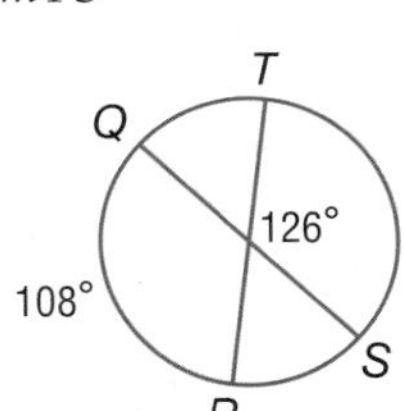


**3.** $m\angle 2$

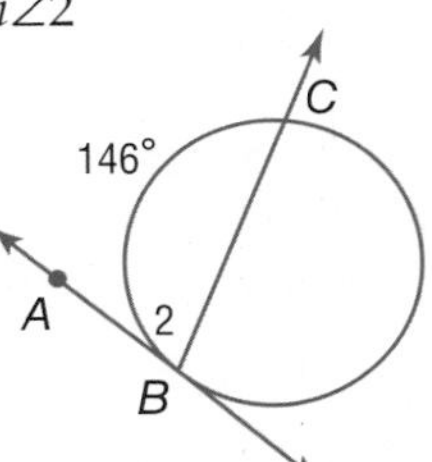


**Examples 3–4** **4.** $m\angle H$

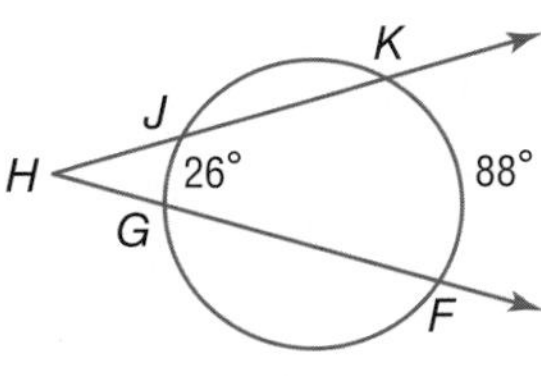


**5.** $m\widehat{QTS}$

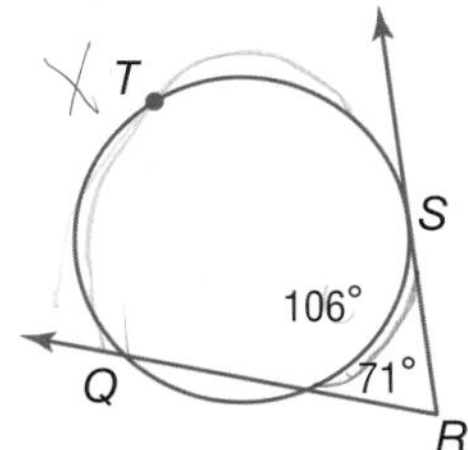


**6.** $m\widehat{LP}$

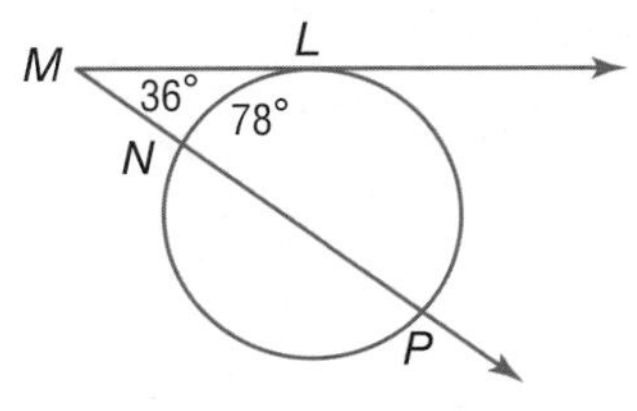


**7** **STUNTS** A ramp is attached to the first of several barrels that have been strapped together for a circus motorcycle stunt as shown. What is the measure of the angle the ramp makes with the ground?

## Practice and Problem Solving

**Examples 1–2** **Find each measure. Assume that segments that appear to be tangent are tangent.**

**8.** $m\angle 3$

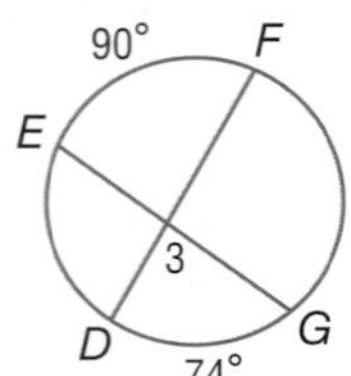


**9.** $m\angle 4$

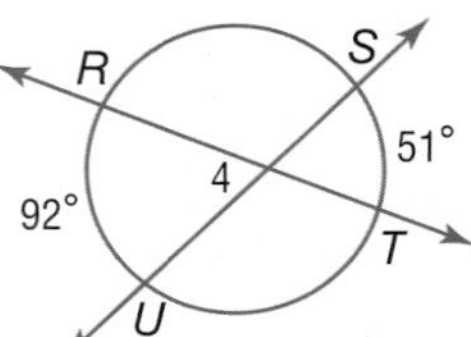


**10.** $m\angle JMK$

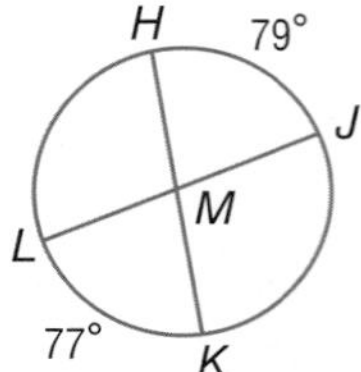


**11** $m\widehat{RQ}$

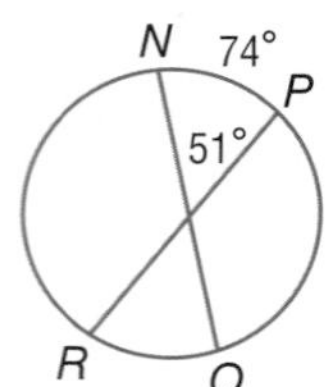


**12.** $m\angle K$

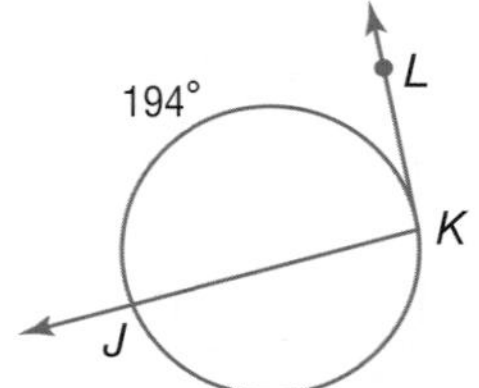


**13.** $m\widehat{PM}$

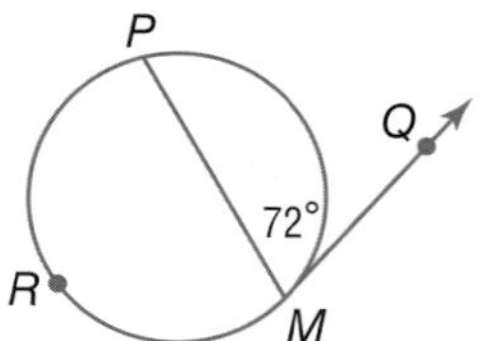


**14.** $m\angle ABD$

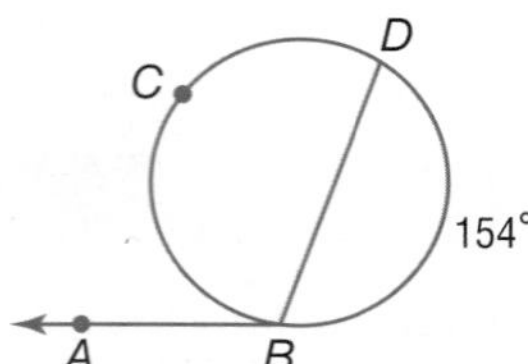


**15.** $m\angle DAB$

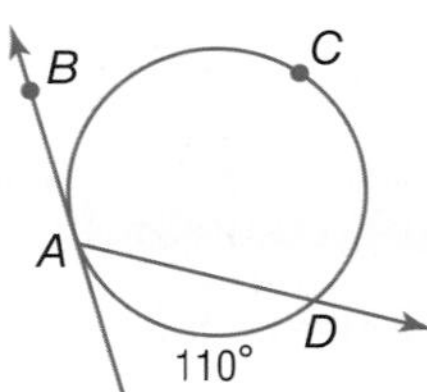


**16.** $m\widehat{GJF}$

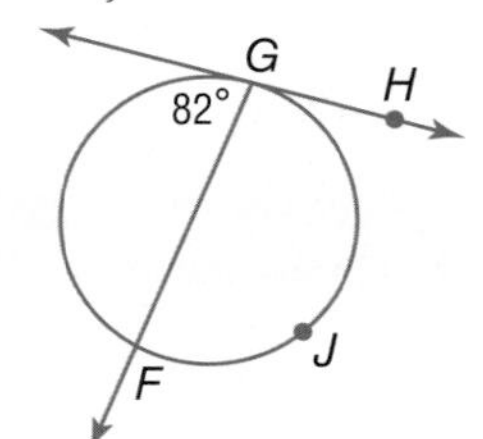


**17. SPORTS** The multi-sport field shown includes a softball field and a soccer field. If $m\widehat{ABC} = 200$, find each measure.

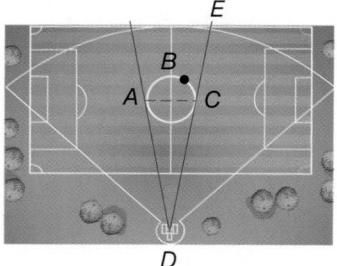

**a.** $m\angle ACE$

**b.** $m\angle ADC$

**Examples 3–4** **CCSS STRUCTURE** **Find each measure.**

**18.** $m\angle A$

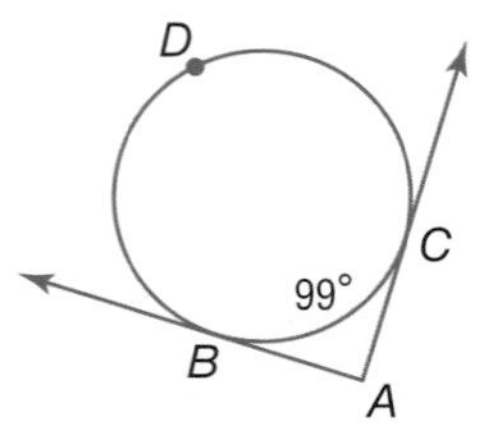


**19.** $m\angle W$

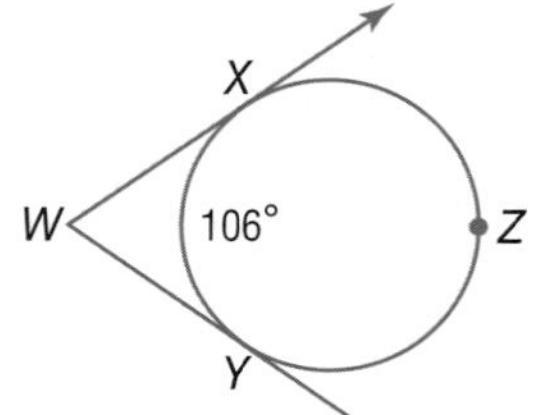


**20.** $m\widehat{JM}$

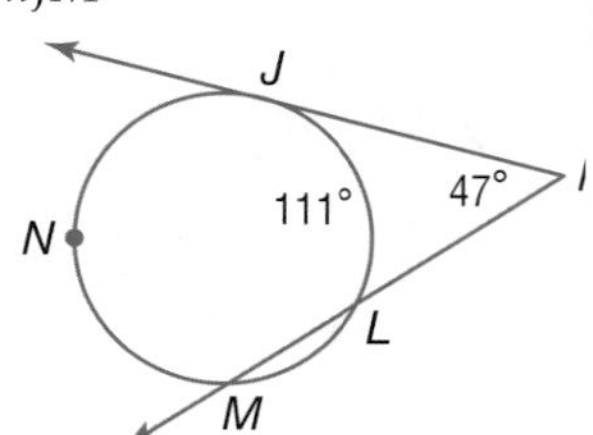


**21.** $m\widehat{XY}$

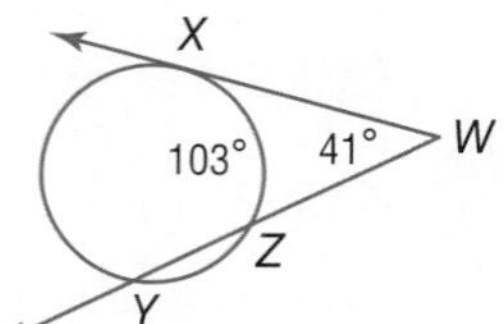


**22.** $m\angle R$

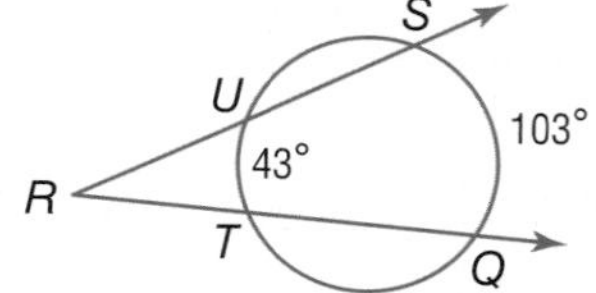


**23.** $m\widehat{SU}$

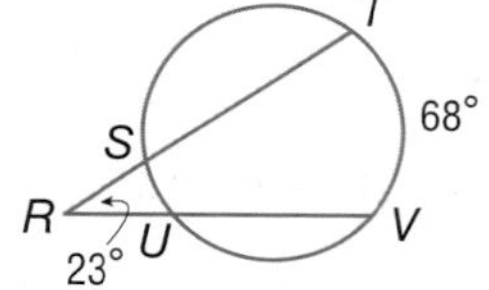

**24. JEWELRY** In the circular necklace shown, $A$ and $B$ are tangent points. If $x = 260$, what is $y$?

**25. SPACE** A satellite orbits above Earth's equator. Find $x$, the measure of the planet's arc, that is visible to the satellite.

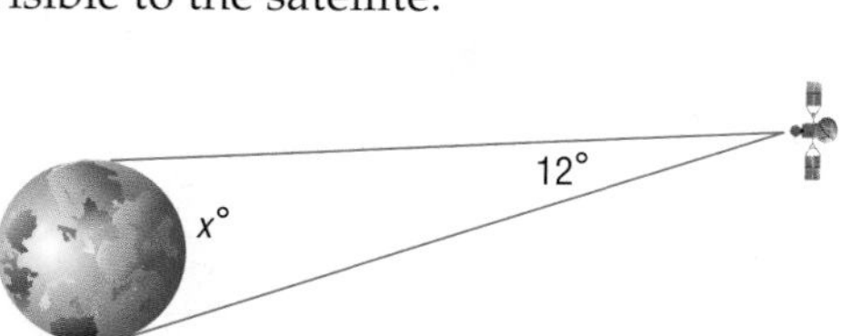


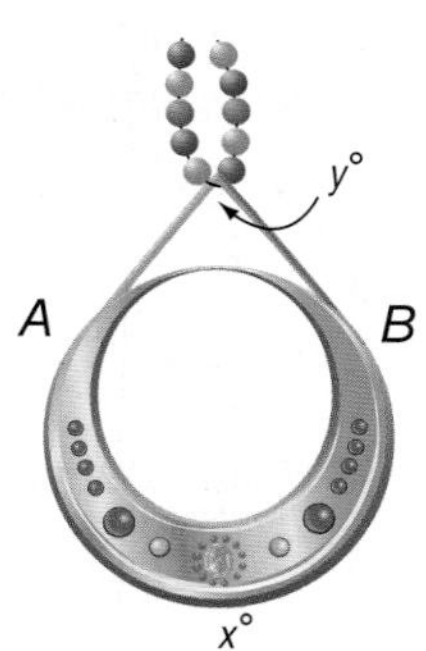


**ALGEBRA Find the value of $x$.**

**26.**

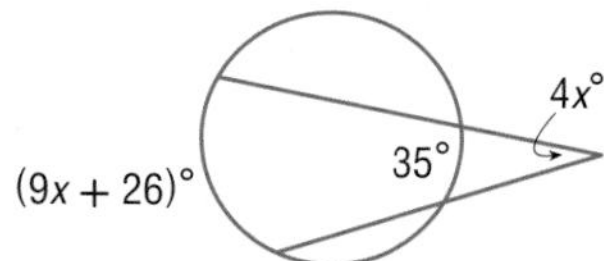


**27**

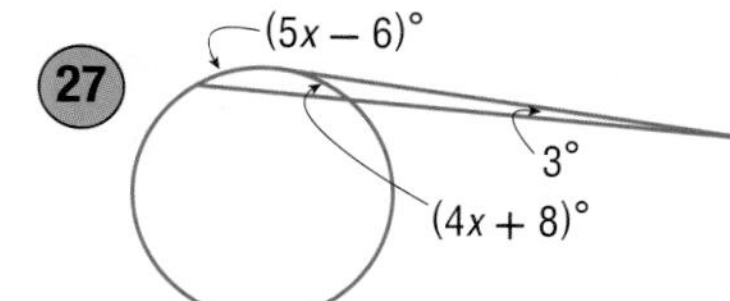


**28.**

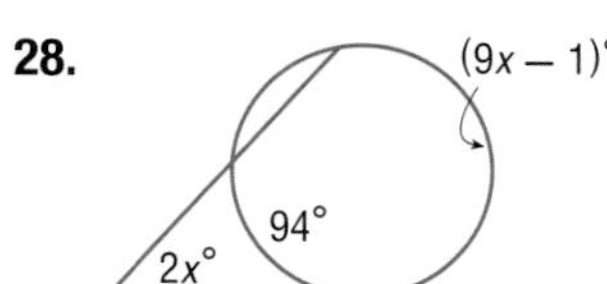


**29. PHOTOGRAPHY** A photographer frames a carousel in his camera shot as shown so that the lines of sight form tangents to the carousel.

**a.** If the camera's viewing angle is 35°, what is the arc measure of the carousel that appears in the shot?

**b.** If you want to capture an arc measure of 150° in the photograph, what viewing angle should be used?

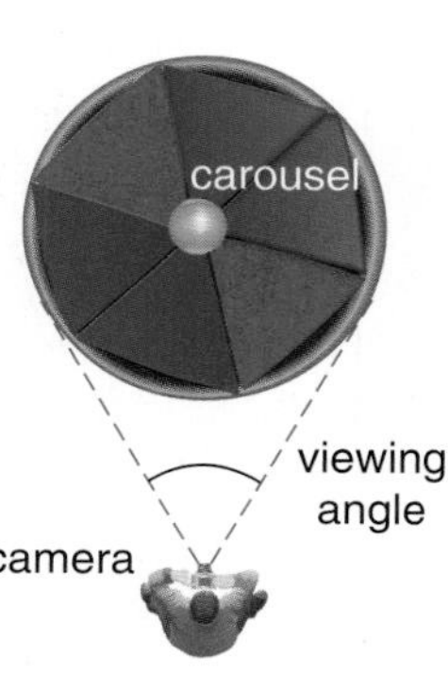


**CCSS ARGUMENTS For each case of Theorem 11.14, write a two-column proof.**

**30. Case 1**

**Given:** secants $\overrightarrow{AD}$ and $\overrightarrow{AE}$

**Prove:** $m\angle A = \frac{1}{2}(m\widehat{DE} - m\widehat{BC})$

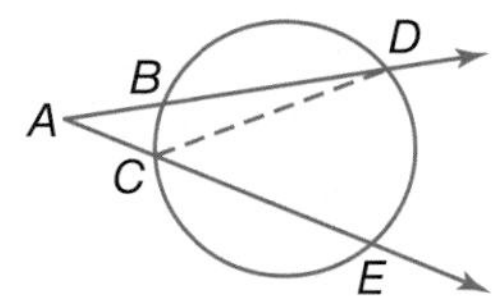


**31. Case 2**

**Given:** tangent $\overrightarrow{FM}$ and secant $\overrightarrow{FL}$

**Prove:** $m\angle F = \frac{1}{2}(m\widehat{LH} - m\widehat{GH})$

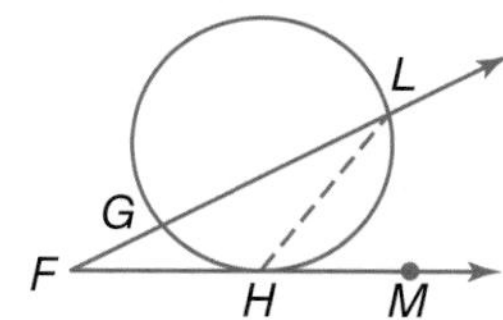


**32. Case 3**

**Given:** tangents $\overrightarrow{RS}$ and $\overrightarrow{RV}$

**Prove:** $m\angle R = \frac{1}{2}(m\widehat{SWT} - m\widehat{ST})$

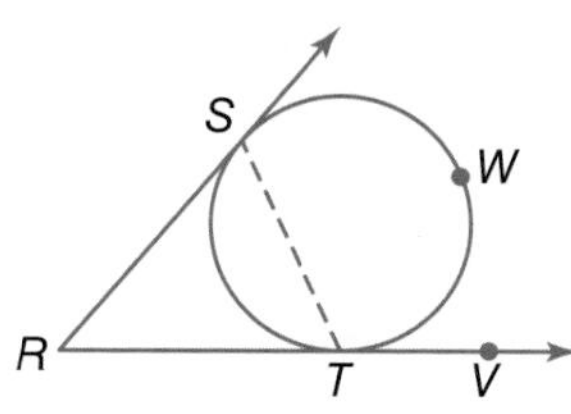


**33. PROOF** Write a paragraph proof of Theorem 11.13.

**a. Given:** $\overleftrightarrow{AB}$ is a tangent of $\odot O$.
$\overleftrightarrow{AC}$ is a secant of $\odot O$.
$\angle CAE$ is acute.

**Prove:** $m\angle CAE = \frac{1}{2}m\widehat{CA}$

**b.** Prove that is $\angle CAB$ is obtuse, $m\angle CAB = \frac{1}{2}m\widehat{CDA}$.

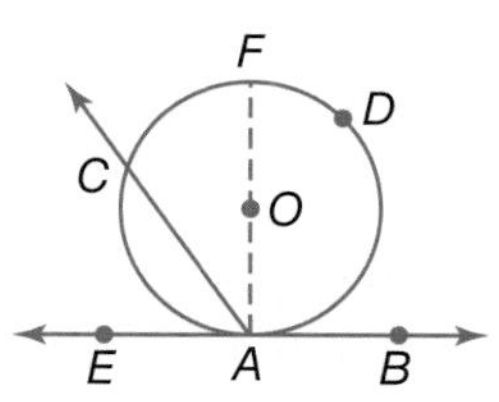

**34. WALLPAPER** In the wallpaper design shown, $\overline{BC}$ is a diameter of $\odot Q$. If $m\angle A = 26$ and $m\widehat{CE} = 67$, what is $m\widehat{DE}$?

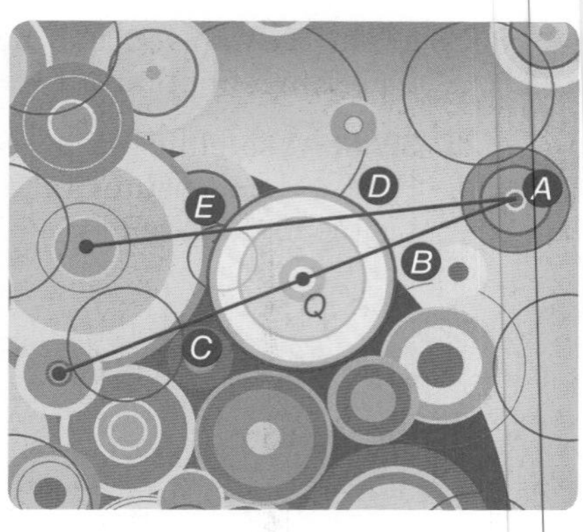


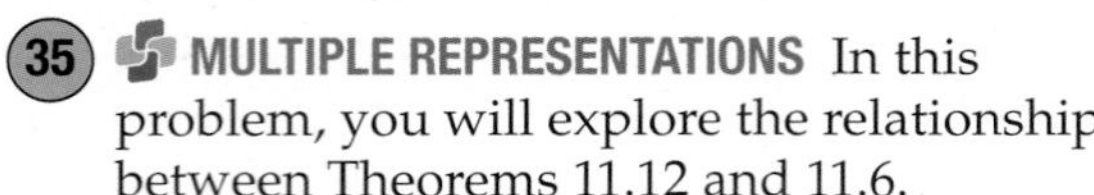

**35. MULTIPLE REPRESENTATIONS** In this problem, you will explore the relationship between Theorems 11.12 and 11.6.

**a. Geometric** Copy the figure shown. Then draw three successive figures in which the position of point $D$ moves closer to point $C$, but points $A$, $B$, and $C$ remain fixed.

**b. Tabular** Estimate the measure of $\widehat{CD}$ for each successive circle, recording the measures of $\widehat{AB}$ and $\widehat{CD}$ in a table. Then calculate and record the value of $x$ for each circle.

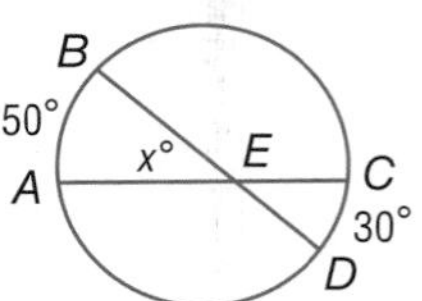


**c. Verbal** Describe the relationship between $m\widehat{AB}$ and the value of $x$ as $m\widehat{CD}$ approaches zero. What type of angle does $\angle AEB$ become when $m\widehat{CD} = 0$?

**d. Analytical** Write an algebraic proof to show the relationship between Theorems 11.12 and 11.6 described in part **c**.

## H.O.T. Problems Use Higher-Order Thinking Skills

**36. WRITING IN MATH** Explain how to find the measure of an angle formed by a secant and a tangent that intersect outside a circle.

**37. CHALLENGE** The circles below are concentric. What is $x$?

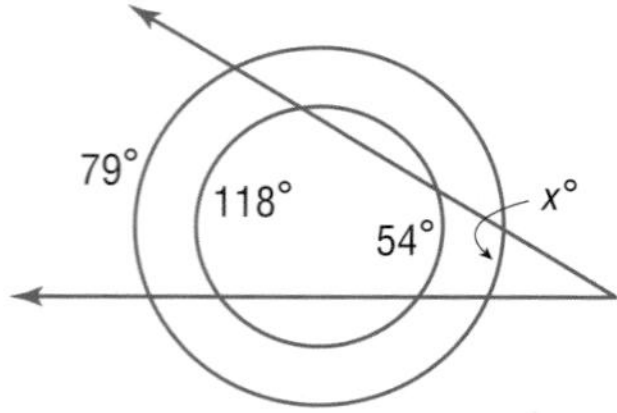


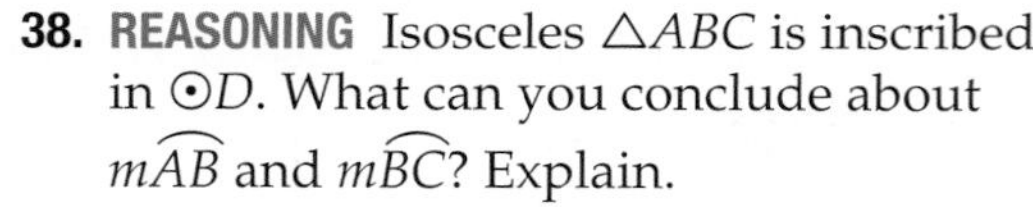

**38. REASONING** Isosceles $\triangle ABC$ is inscribed in $\odot D$. What can you conclude about $m\widehat{AB}$ and $m\widehat{BC}$? Explain.

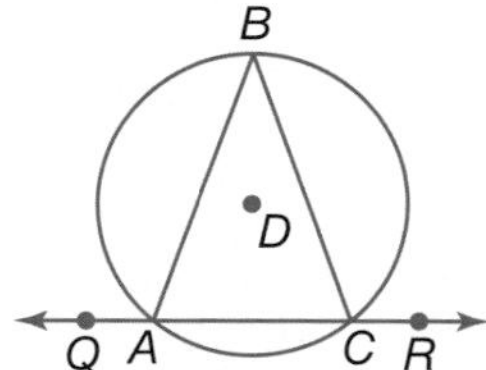


**39. CCSS ARGUMENTS** In the figure, $\overline{JK}$ is a diameter and $\overrightarrow{GH}$ is a tangent.

**a.** Describe the range of possible values for $m\angle G$. Explain.

**b.** If $m\angle G = 34$, find the measures of minor arcs $HJ$ and $KH$. Explain.

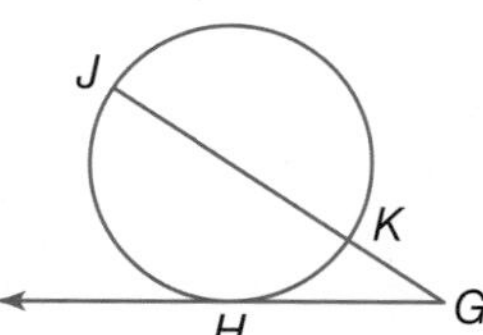


**40. OPEN ENDED** Draw a circle and two tangents that intersect outside the circle. Use a protractor to measure the angle that is formed. Find the measures of the minor and major arcs formed. Explain your reasoning.

**41. WRITING IN MATH** A circle is inscribed within $\triangle PQR$. If $m\angle P = 50$ and $m\angle Q = 60$, describe how to find the measures of the three minor arcs formed by the points of tangency.

## Standardized Test Practice

**42.** What is the value of $x$ if $m\widehat{NR} = 62$ and $m\widehat{NP} = 108$?

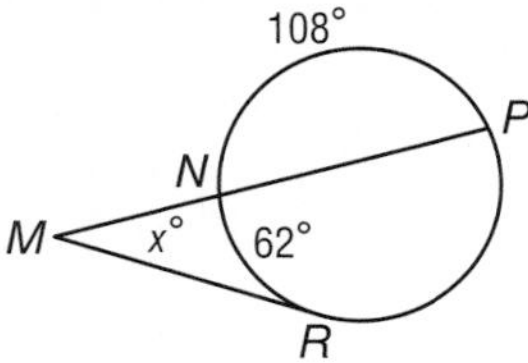


**A** 23° **C** 64°
**B** 31° **D** 128°

**43. ALGEBRA** Points $A(-4, 8)$ and $B(6, 2)$ are both on circle $C$, and $\overline{AB}$ is a diameter. What are the coordinates of $C$?

**F** (2, 10) **H** (5, −3)
**G** (10, −6) **J** (1, 5)

**44. GRIDDED RESPONSE** If $m\angle AED = 95$ and $m\widehat{AD} = 120$, what is $m\angle BAC$?

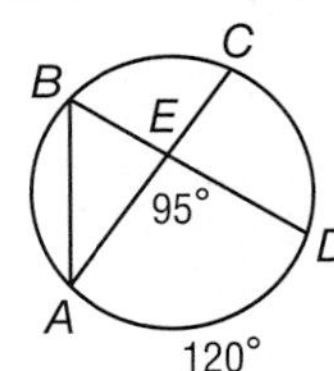


**45. SAT/ACT** If the circumference of the circle below is $16\pi$ units, what is the total area of the shaded regions?

**A** $64\pi$ units$^2$ **D** $8\pi$ units$^2$
**B** $32\pi$ units$^2$ **E** $2\pi$ units$^2$
**C** $12\pi$ units$^2$

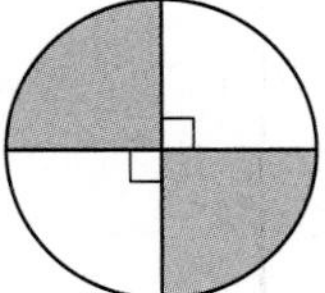

## Spiral Review

**Find $x$. Assume that segments that appear to be tangent are tangent.** (Lesson 11-5)

**46.**

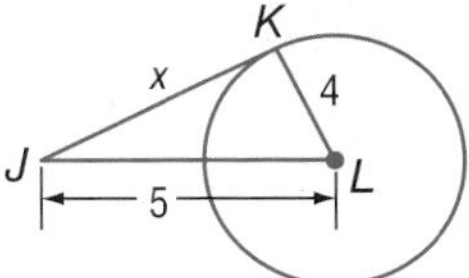


**47.**

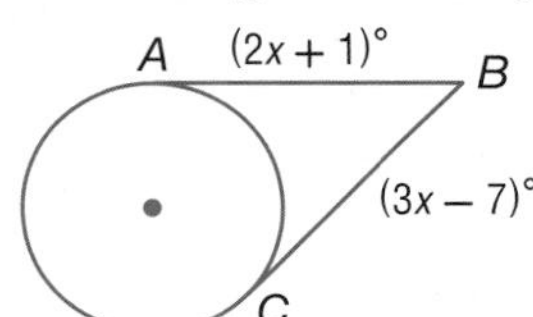


**48.**

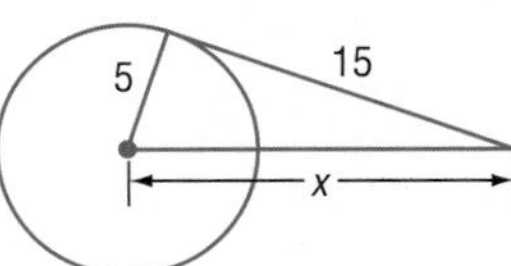


**49. PROOF** Write a two-column proof. (Lesson 11-4)

**Given:** $\widehat{MHT}$ is a semicircle; $\overline{RH} \perp \overline{TM}$.

**Prove:** $\dfrac{TR}{RH} = \dfrac{TH}{HM}$

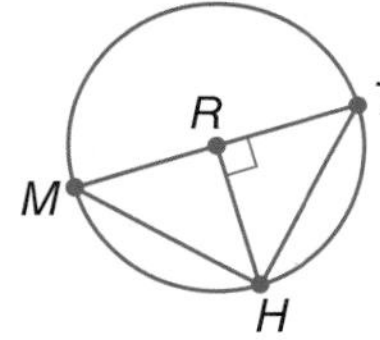


**COORDINATE GEOMETRY Find the measure of each angle to the nearest tenth of a degree by using the Distance Formula and an inverse trigonometric ratio.** (Lesson 10-4)

**50.** $\angle C$ in triangle $BCD$ with vertices $B(-1, -5)$, $C(-6, -5)$, and $D(-1, 2)$

**51.** $\angle X$ in right triangle $XYZ$ with vertices $X(2, 2)$, $Y(2, -2)$, and $Z(7, -2)$

## Skills Review

**Solve each equation.**

**52.** $x^2 + 13x = -36$

**53.** $x^2 - 6x = -9$

**54.** $3x^2 + 15x = 0$

**55.** $28 = x^2 + 3x$

**56.** $x^2 + 12x + 36 = 0$

**57.** $x^2 + 5x = -\dfrac{25}{4}$

LESSON 11-7

# Special Segments in a Circle

| Then | Now | Why? |
|---|---|---|
| You found measures of diagonals that intersect in the interior of a parallelogram. | **1** Find measures of segments that intersect in the interior of a circle.<br>**2** Find measures of segments that intersect in the exterior of a circle. | A large circular cake is cut lengthwise instead of into wedges to serve more people for a party. Only a small portion of the original cake remains. Using the geometry of circles, you can determine the diameter of the original cake. |

**NewVocabulary**
chord segment
secant segment
external secant segment
tangent segment

**Common Core State Standards**

**Content Standards**
Reinforcement of G.C.4 Construct a tangent line from a point outside a given circle to the circle.

**Mathematical Practices**
1. Make sense of problems and persevere in solving them.
7. Look for and make use of structure.

## 1 Segments Intersecting Inside a Circle

When two chords intersect inside a circle, each chord is divided into two segments, called **chord segments**.

### Theorem 11.15 Segments of Chords Theorem

**Words** If two chords intersect in a circle, then the products of the lengths of the chord segments are equal.

**Example** $AB \cdot BC = DB \cdot BE$

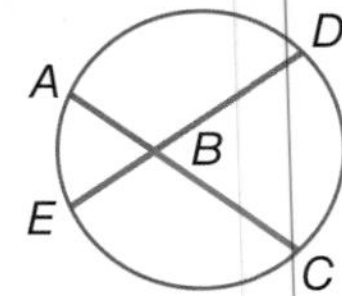


You will prove Theorem 11.15 in Exercise 23.

### Example 1 Use the Intersection of Two Chords

Find $x$.

**a.**

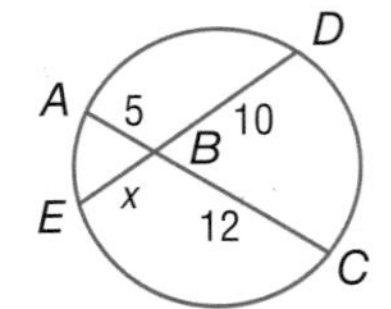


$AB \cdot BC = EB \cdot BD$ — Theorem 11.15

$5 \cdot 12 = x \cdot 10$ — Substitution

$60 = 10x$ — Multiply.

$6 = x$ — Divide each side by 10.

**b.**

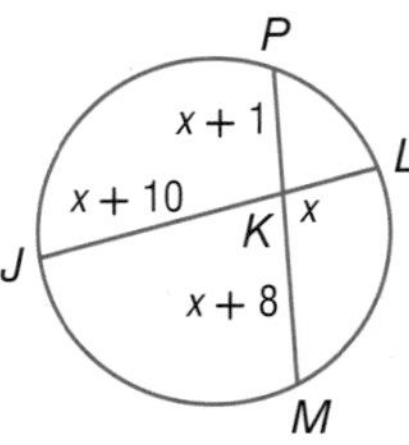


$JK \cdot KL = PK \cdot KM$ — Theorem 11.15

$(x + 10) \cdot x = (x + 1)(x + 8)$ — Substitution

$x^2 + 10x = x^2 + 9x + 8$ — Multiply.

$10x = 9x + 8$ — Subtract $x^2$ from each side.

$x = 8$ — Subtract $9x$ from each side.

**GuidedPractice**

**1A.**

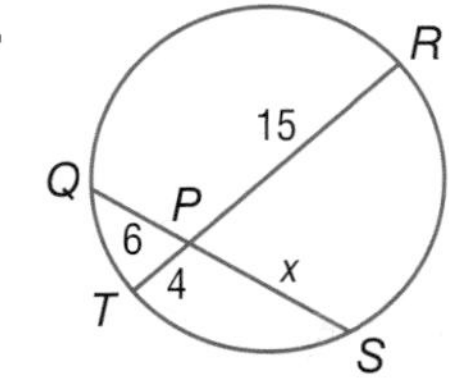


**1B.**

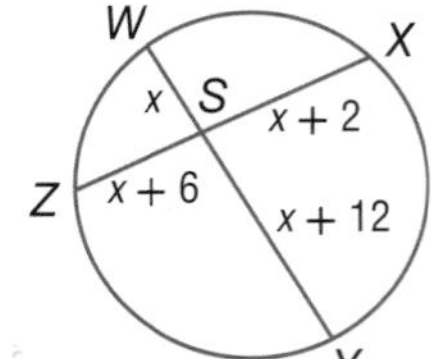

## Real-WorldLink

The lower the Sun is to the horizon, the more of a rainbow you can see. At sunset, you could see a full semicircle of a rainbow with the top of the arch 42 degrees above the horizon.

**Source:** The National Center for Atmospheric Research

## Real-World Example 2 Find Measures of Segments in Circles

PT

**SCIENCE The true shape of a rainbow is a complete circle. However, we see only the arc of the circle that appears above Earth's horizon. What is the radius of the circle containing the arc of the rainbow shown?**

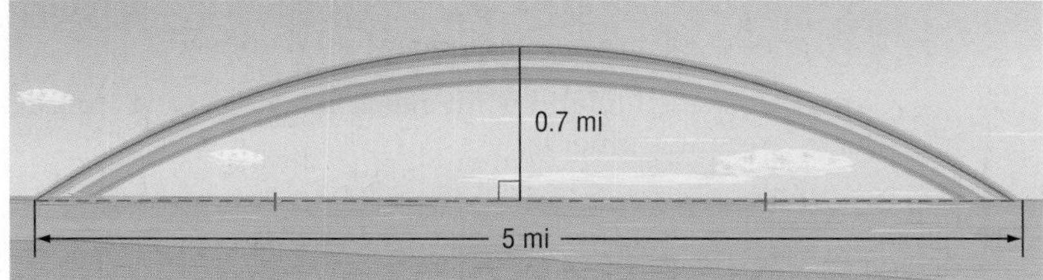


**Understand** You know that the rainbow's arc is part of a whole circle. $\overline{AC}$ is a chord of this circle, and $\overline{DB}$ is a perpendicular bisector of $\overline{AC}$.

**Plan** Draw a model. Since it bisects chord $\overline{AC}$, $\overline{DE}$ is a diameter of the circle. Use the products of the lengths of the intersecting chords to find the length of the diameter.

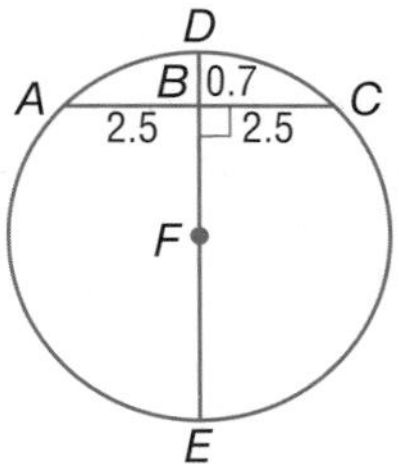

**Solve**

$AB \cdot BC = DB \cdot BE$ — Theorem 11.15

$2.5 \cdot 2.5 = 0.7 \cdot BE$ — Substitution

$6.25 = 0.7BE$ — Multiply.

$8.9 \approx BE$ — Divide each side by 0.7.

$DE = DB + BE$ — Segment Addition Postulate

$= 0.7 + 8.9$ — Substitution

$= 9.6$ — Add.

Since the diameter of the circle is about 9.6 miles, the radius is about $9.6 \div 2$ or 4.8 miles.

**Check** Use the Pythagorean Theorem to check the triangle in the circle formed by the radius, the chord, and part of the diameter.

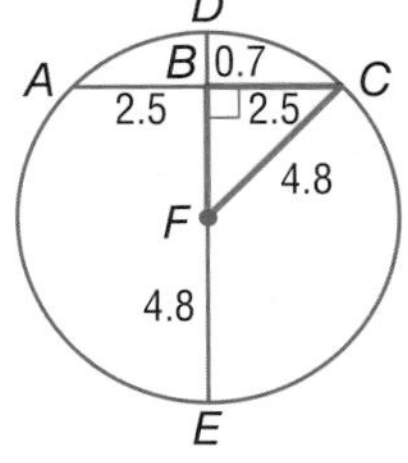


$DB + BF = DF$ — Segment Addition Postulate

$0.7 + BF = 4.8$ — Substitution

$BF = 4.1$ — Subtract 0.7 from each side.

$BF^2 + BC^2 = CF^2$ — Pythagorean Theorem

$4.1^2 + 2.5^2 \stackrel{?}{=} 4.8^2$ — Substitution

$23.06 \approx 23.04$ ✓ — Simplify.

## Problem-SolvingTip

**Make a Drawing** When solving word problems involving circles, it is helpful to make a drawing and label all parts of the circle that are known. Use a variable to label the unknown measure.

## GuidedPractice

**2. ASTRODOME** The highest point, or apex, of the Astrodome is 208 feet high, and the diameter of the circle containing the arc is 710 feet. How long is the stadium from one side to the other?

## 2 Segments Intersecting Outside a Circle

A **secant segment** is a segment of a secant line that has exactly one endpoint on the circle. In the figure, $\overline{AC}$, $\overline{AB}$, $\overline{AE}$ and $\overline{AD}$ are secant segments.

A secant segment that lies in the exterior of the circle is called an **external secant segment**. In the figure, $\overline{AB}$ and $\overline{AD}$ are external secant segments.

A special relationship exists among secants and external secant segments.

> **StudyTip**
>
> **Simplify the Theorem** Each side of the equation in Theorem 11.16 is the product of the lengths of the exterior part and the whole segment.

### Theorem 11.16 Secant Segments Theorem

**Words** If two secants intersect in the exterior of a circle, then the product of the measures of one secant segment and its external secant segment is equal to the product of the measures of the other secant and its external secant segment.

**Example** $AC \cdot AB = AE \cdot AD$

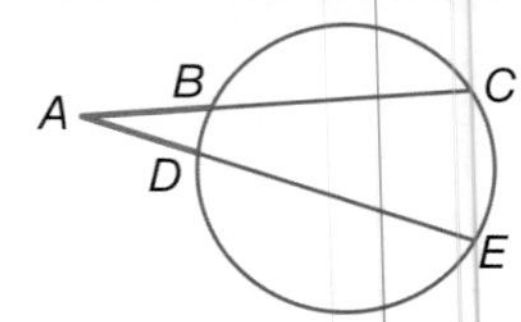


You will prove Theorem 11.16 in Exercise 24.

> **WatchOut!**
>
> **Use the Correct Equation** Be sure to multiply the length of the secant segment by the length of the external secant segment. Do not multiply the length of the internal secant segment, or chord, by the length of the external secant segment.

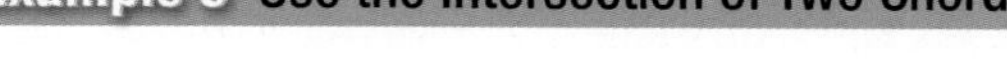

### Example 3 Use the Intersection of Two Chords

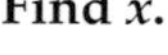

**Find $x$.**

| | |
|---|---|
| $JG \cdot JH = JL \cdot JK$ | Theorem 11.16 |
| $(x + 8)8 = (10 + 6)6$ | Substitution |
| $8x + 64 = 96$ | Multiply. |
| $8x = 32$ | Subtract 64 from each side. |
| $x = 4$ | Divide each side by 8. |

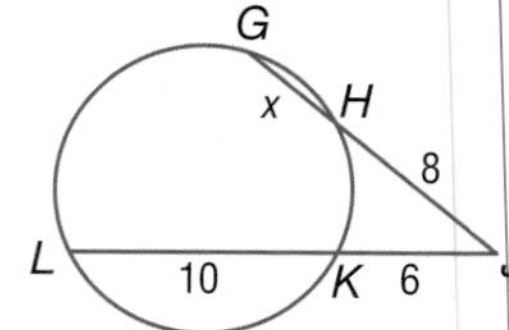


### GuidedPractice

**3A.**

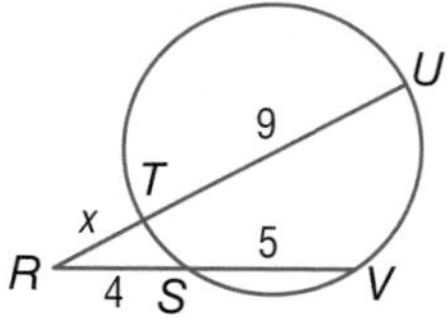


**3B.**

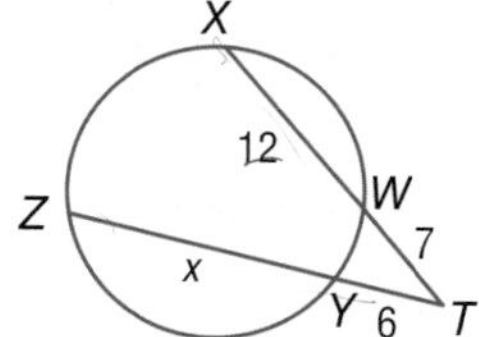


An equation similar to the one in Theorem 11.16 can be used when a secant and a tangent intersect outside a circle. In this case, the **tangent segment**, or segment of a tangent with one endpoint on the circle, is both the exterior and whole segment.

### Theorem 11.17

**Words** If a tangent and a secant intersect in the exterior of a circle, then the square of the measure of the tangent is equal to the product of the measures of the secant and its external secant segment.

**Example** $JK^2 = JL \cdot JM$

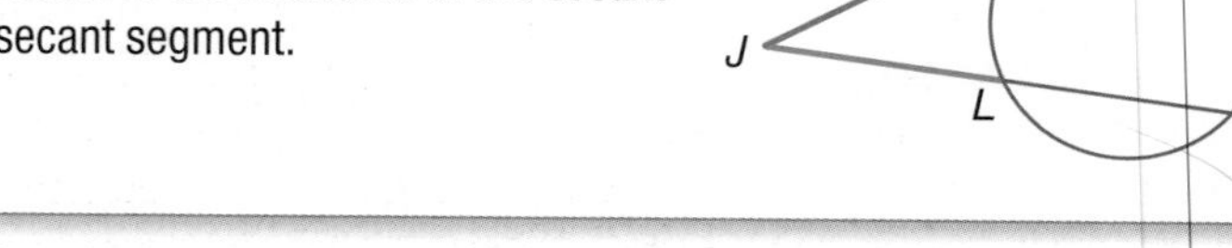


You will prove Theorem 11.17 in Exercise 25.

## Example 4 Use the Intersection of a Secant and a Tangent

**$\overline{PQ}$ is tangent to the circle. Find $x$. Round to the nearest tenth.**

| | |
|---|---|
| $PQ^2 = QR \cdot QS$ | Theorem 11.17 |
| $8^2 = x(x + 7)$ | Substitution |
| $64 = x^2 + 7x$ | Multiply. |
| $0 = x^2 + 7x - 64$ | Subtract 64 from each side. |

Since the expression is not factorable, use the Quadratic Formula.

| | |
|---|---|
| $x = \dfrac{-b \pm \sqrt{b^2 - 4ac}}{2a}$ | Quadratic Formula |
| $= \dfrac{-7 \pm \sqrt{7^2 - 4(1)(-64)}}{2(1)}$ | $a = 1$, $b = 7$, and $c = -64$ |
| $= \dfrac{-7 \pm \sqrt{305}}{2}$ | Simplify. |
| $\approx 5.2$ or $-12.2$ | Use a calculator. |

Since lengths cannot be negative, the value of $x$ is about 5.2.

### GuidedPractice

**4.** $\overline{AB}$ is tangent to the circle. Find $x$. Round to the nearest tenth.

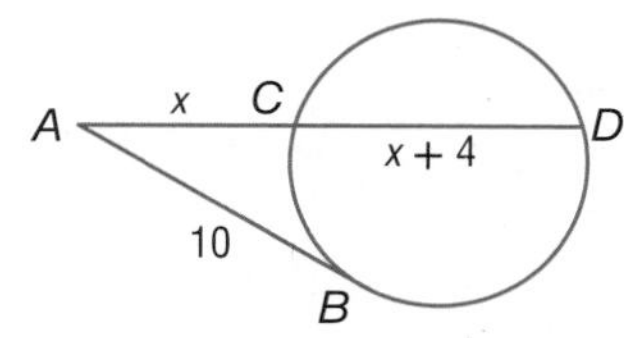

## Check Your Understanding

**Examples 1, 3 and 4** **Find $x$. Assume that segments that appear to be tangent are tangent.**

**1.**

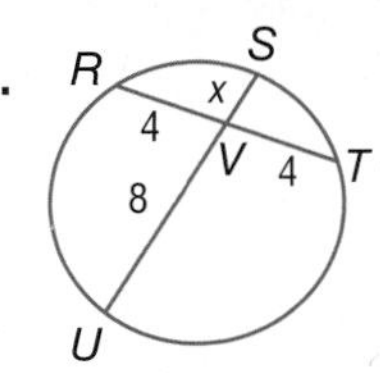


**2.**

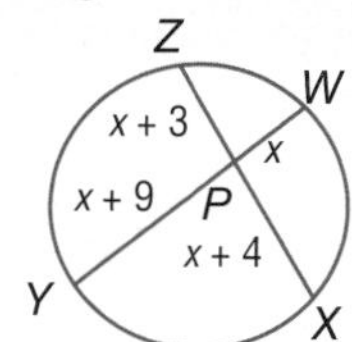


**3.**

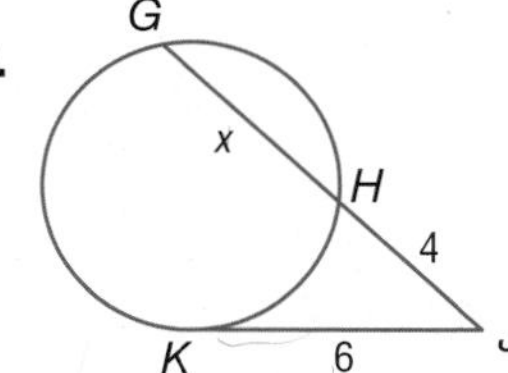


**4.**

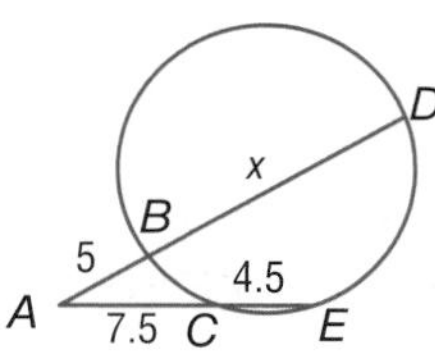


**Example 2**

**5** **SCIENCE** A piece of broken pottery found at an archaeological site is shown. $\overline{QS}$ lies on a diameter of the circle. What was the circumference of the original pottery? Round to the nearest hundredth.

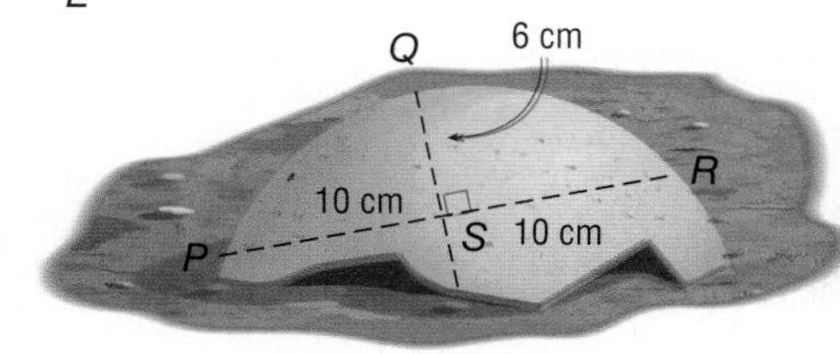

## Practice and Problem Solving

Examples 1, 3 and 4

**Find $x$ to the nearest tenth. Assume that segments that appear to be tangent are tangent.**

**6.**

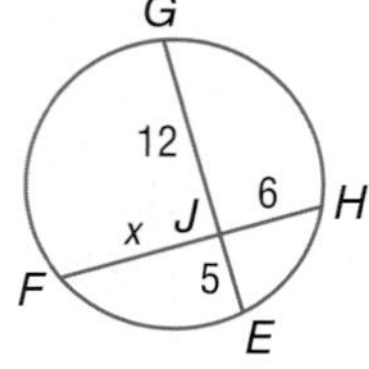


**7.**

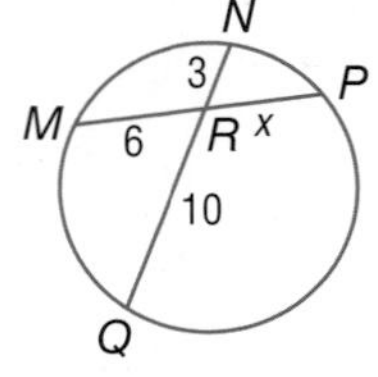


**8.**

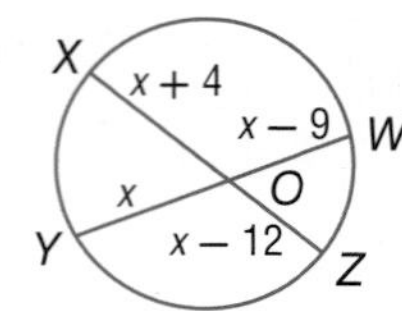


**9.**

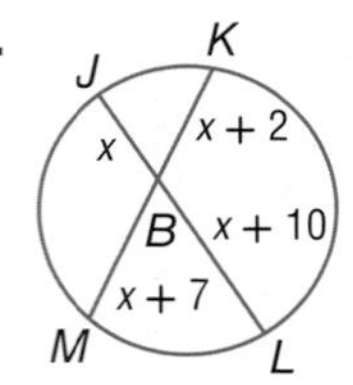


**10.**

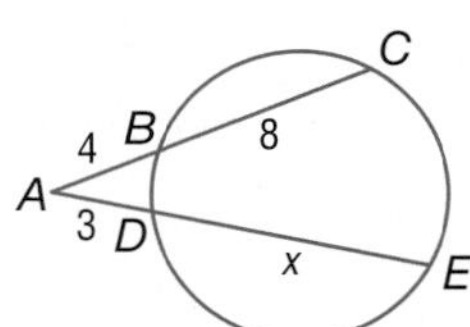


**11.**

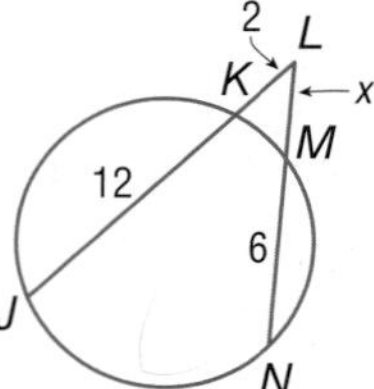


**12.**

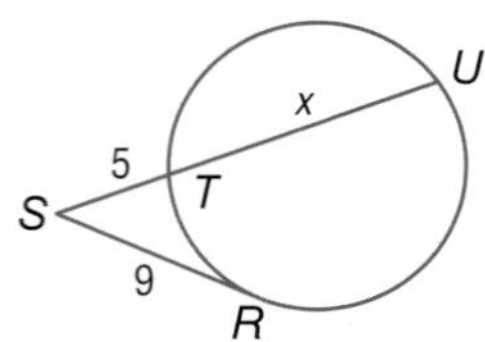


**13**

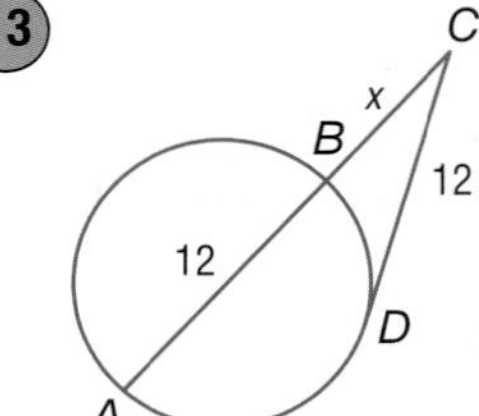


Example 2

**14. BRIDGES** What is the diameter of the circle containing the arc of the Sydney Harbour Bridge shown? Round to the nearest tenth.

**15. CAKES** Sierra is serving cake at a party. If the dimensions of the remaining cake are shown below, what was the original diameter of the cake?

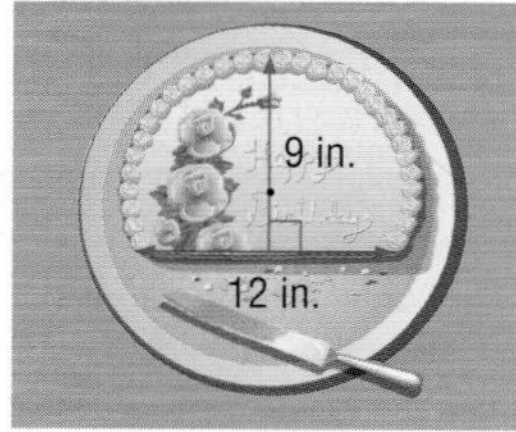


**CCSS STRUCTURE** **Find each variable to the nearest tenth. Assume that segments that appear to be tangent are tangent.**

**16.**

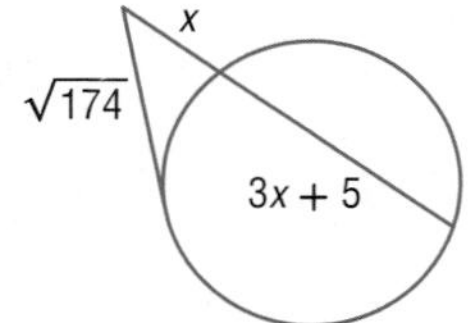


**17.**

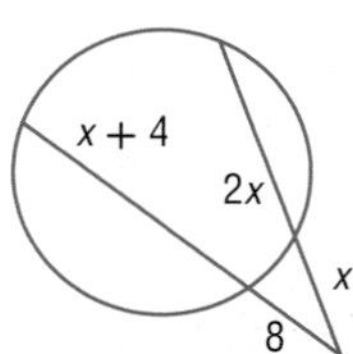


**18.**

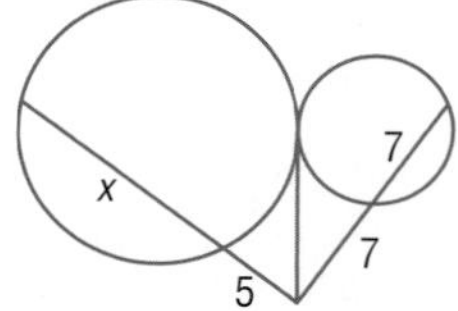


**19.**

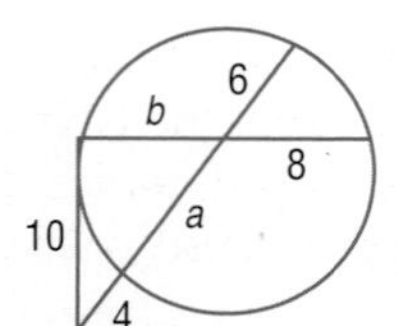


**20.**

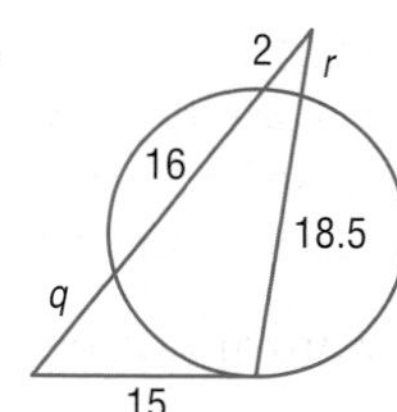


**21.**

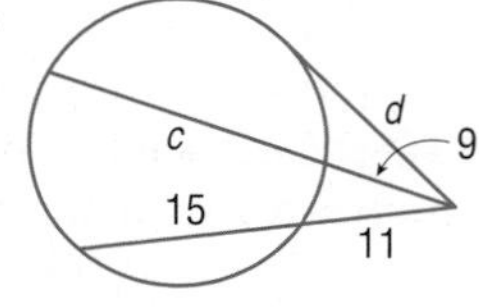


Ilya Genki/Alamy

**22. INDIRECT MEASUREMENT** Gwendolyn is standing 16 feet from a giant sequoia tree and Chet is standing next to the tree, as shown. The distance between Gwendolyn and Chet is 27 feet. Draw a diagram of this situation, and then find the diameter of the tree.

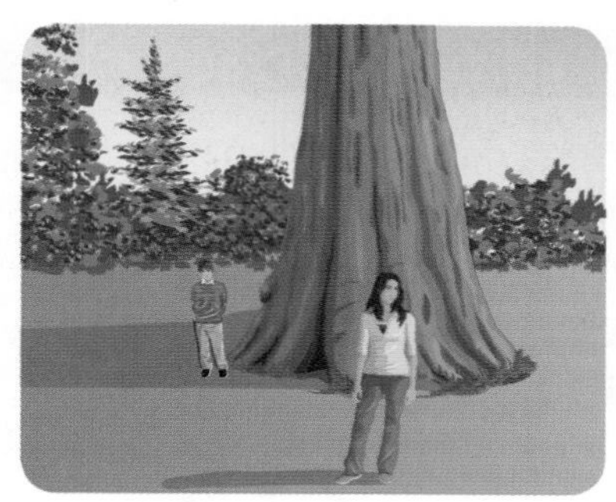

**PROOF** **Prove each theorem.**

**23.** two-column proof of Theorem 11.15

**Given:** $\overline{AC}$ and $\overline{DE}$ intersect at $B$.

**Prove:** $AB \cdot BC = EB \cdot BD$

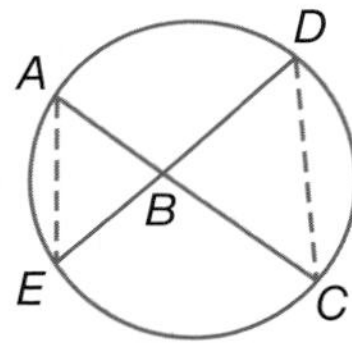


**24.** paragraph proof of Theorem 11.16

**Given:** Secants $\overline{AC}$ and $\overline{AE}$

**Prove:** $AB \cdot AC = AD \cdot AE$

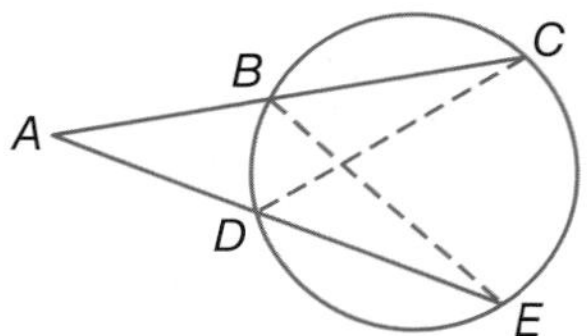


**25.** two-column proof of Theorem 11.17

**Given:** tangent $\overline{JK}$, secant $\overline{JM}$

**Prove:** $JK^2 = JL \cdot JM$

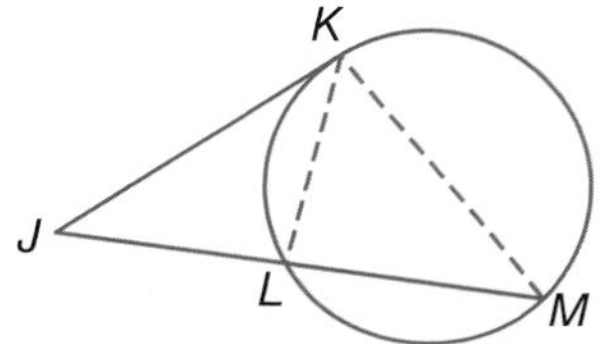


## H.O.T. Problems Use Higher-Order Thinking Skills

**26. CCSS CRITIQUE** Tiffany and Jun are finding the value of $x$ in the figure at the right. Tiffany wrote $3(5) = 2x$, and Jun wrote $3(8) = 2(2 + x)$. Is either of them correct? Explain your reasoning.

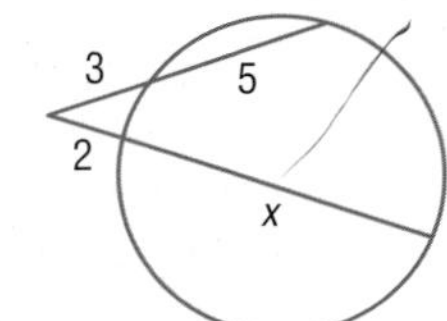


**27. WRITING IN MATH** Compare and contrast the methods for finding measures of segments when two secants intersect in the exterior of a circle and when a secant and a tangent intersect in the exterior of a circle.

**28. CHALLENGE** In the figure, a line tangent to circle $M$ and a secant line intersect at $R$. Find $a$. Show the steps that you used.

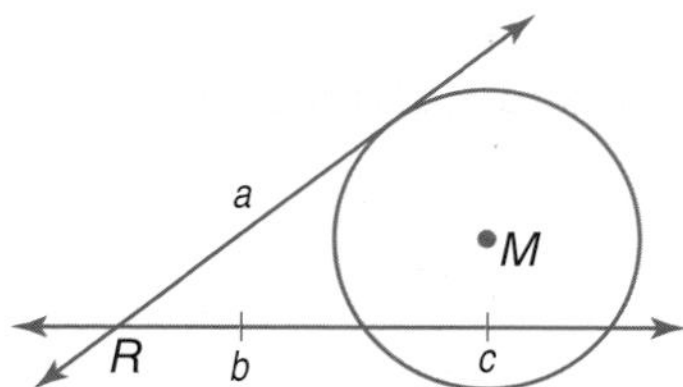


**29. REASONING** When two chords intersect at the center of a circle, are the measures of the intercepting arcs *sometimes*, *always*, or *never* equal to each other?

**30. OPEN ENDED** Investigate Theorem 11.17 by drawing and labeling a circle that has a secant and a tangent intersecting outside the circle. Measure and label the two parts of the secant segment to the nearest tenth of a centimeter. Use an equation to find the measure of the tangent segment. Verify your answer by measuring the segment.

**31. WRITING IN MATH** Describe the relationship among segments in a circle when two secants intersect inside a circle.

## Standardized Test Practice

**32.** $\overline{TV}$ is tangent to the circle, and $R$ and $S$ are points on the circle. What is the value of $x$ to the nearest tenth?

A 7.6  
B 6.4  
C 5.7  
D 4.8

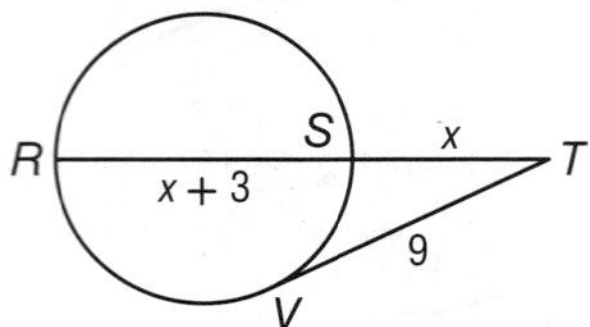


**33. ALGEBRA** A department store has all of its jewelry discounted 40%. It is having a sale that says you receive an additional 20% off the already discounted price. How much will you pay for a ring with an original price of $200?

F $80  
G $96  
H $120  
J $140

**34. EXTENDED RESPONSE** The degree measures of minor arc $\widehat{AC}$ and major arc $\widehat{ADC}$ are $x$ and $y$, respectively.

**a.** If $m\angle ABC = 70°$, write two equations relating $x$ and $y$.

**b.** Find $x$ and $y$.

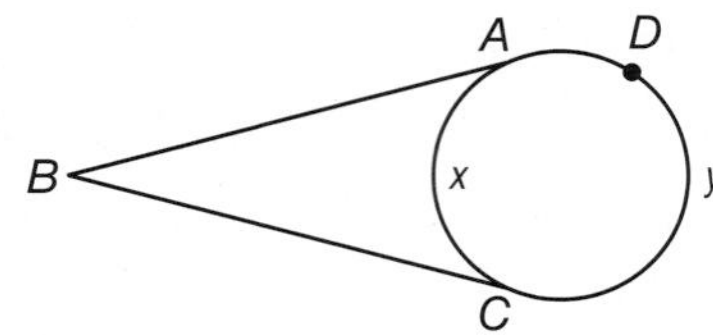


**35. SAT/ACT** During the first two weeks of summer vacation, Antonia earned $100 per week. During the next six weeks, she earned $150 per week. What was her average weekly pay?

A $50  
B $112.50  
C $125  
D $135  
E $137.50

## Spiral Review

**36. WEAVING** Once yarn is woven from wool fibers, it is often dyed and then threaded along a path of pulleys to dry. One set of pulleys is shown. Note that the yarn appears to intersect itself at $C$, but in reality it does not. Use the information from the diagram to find $m\widehat{BH}$. (Lesson 11-6)

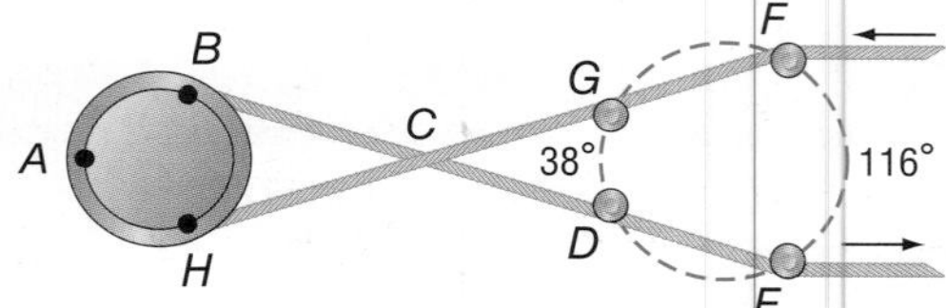


**Copy the figure shown and draw the common tangents. If no common tangent exists, state *no common tangent*.** (Lesson 11-5)

**37.**

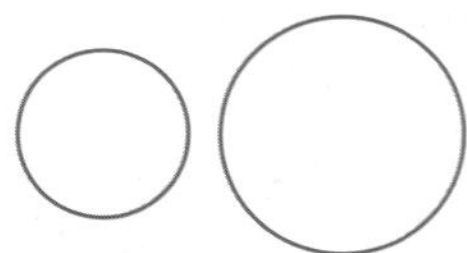

**38.**

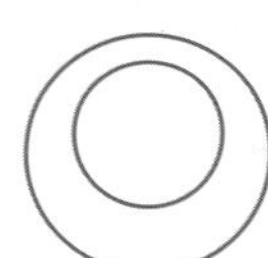

**39.**

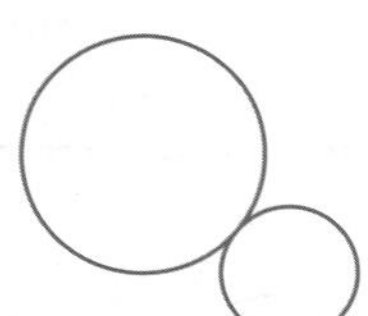

**40.**

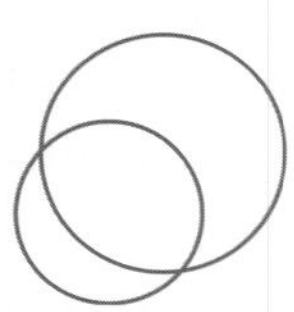

**Copy the vectors to find each sum or difference.** (Lesson 10-7)

**41.** $\vec{c} - \vec{d}$

$\vec{d}$ $\vec{c}$

**42.** $\vec{w} + \vec{x}$

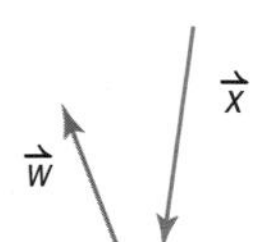


**43.** $\vec{n} - \vec{p}$

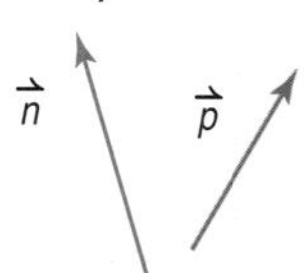


## Skills Review

**Write an equation in slope-intercept form of the line having the given slope and *y*-intercept.**

**44.** $m$: 3, $y$-intercept: $-4$

**45.** $m$: 2, (0, 8)

**46.** $m$: $\frac{5}{8}$, (0, $-6$)

**47.** $m$: $\frac{2}{9}$, $y$-intercept: $\frac{1}{3}$

**48.** $m$: $-1$, $b$: $-3$

**49.** $m$: $-\frac{1}{12}$, $b$: 1

# LESSON 11-8 Equations of Circles

| Then | Now | Why? |
|---|---|---|
| You wrote equations of lines using information about their graphs. | **1** Write the equation of a circle.<br>**2** Graph a circle on the coordinate plane. | Telecommunications towers emit radio signals that are used to transmit cellular calls. Each tower covers a circular area, and towers are arranged so that a signal is available at any location in the coverage area. |

**NewVocabulary**
compound locus

**Common Core State Standards**

**Content Standards**

**G.GPE.1** Derive the equation of a circle of given center and radius using the Pythagorean Theorem; complete the square to find the center and radius of a circle given by an equation.

**G.GPE.6** Find the point on a directed line segment between two given points that partitions the segment in a given ratio.

**Mathematical Practices**

2 Reason abstractly and quantitatively.

7 Look for and make use of structure.

## 1 Equation of a Circle

Since all points on a circle are equidistant from the center, you can find an equation of a circle by using the Distance Formula.

Let $(x, y)$ represent a point on a circle centered at the origin. Using the Pythagorean Theorem, $x^2 + y^2 = r^2$.

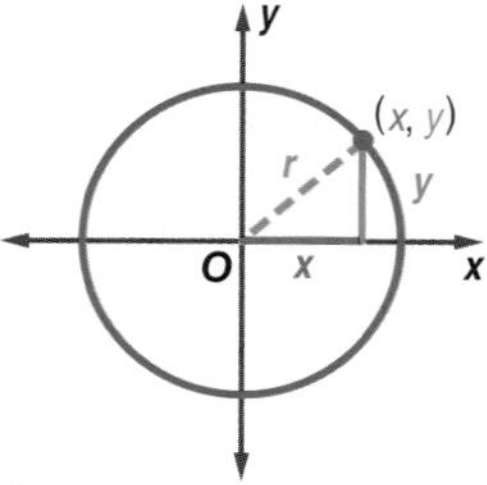


Now suppose that the center is not at the origin, but at the point $(h, k)$. You can use the Distance Formula to develop an equation for the circle.

$d = \sqrt{(x_2 - x_1)^2 + (y_2 - y_1)^2}$ — Distance Formula

$r = \sqrt{(x - h)^2 + (y - k)^2}$ — $d = r$, $(x_1, y_1) = (h, k)$, $(x_2, y_2) = (x, y)$

$r^2 = (x - h)^2 + (y - k)^2$ — Square each side.

### KeyConcept Equation of a Circle in Standard Form

The standard form of the equation of a circle with center at $(h, k)$ and radius $r$ is $(x - h)^2 + (y - k)^2 = r^2$.

The standard form of the equation of a circle is also called the *center-radius* form.

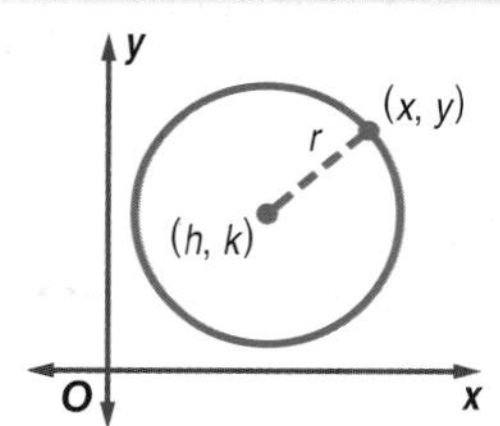

### Example 1 Write an Equation Using the Center and Radius

**Write the equation of each circle.**

**a. center at (1, −8), radius 7**

$(x - h)^2 + (y - k)^2 = r^2$ — Equation of a circle

$(x - 1)^2 + [y - (-8)]^2 = 7^2$ — $(h, k) = (1, -8)$, $r = 7$

$(x - 1)^2 + (y + 8)^2 = 49$ — Simplify.

**b. the circle graphed at the right**

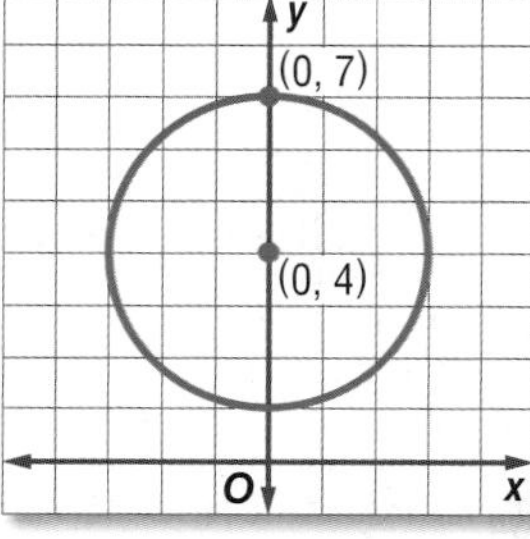


The center is at (0, 4) and the radius is 3.

$(x - h)^2 + (y - k)^2 = r^2$ — Equation of a circle

$(x - 0)^2 + (y - 4)^2 = 3^2$ — $(h, k) = (0, 4)$, $r = 3$

$x^2 + (y - 4)^2 = 9$ — Simplify.

**GuidedPractice**

**1A.** center at origin, radius $\sqrt{10}$

**1B.** center at (4, −1), diameter 8

## Example 2 Write an Equation Using the Center and a Point

**Write the equation of the circle with center at (−2, 4), that passes through (−6, 7).**

**Step 1** Find the distance between the points to determine the radius.

$$r = \sqrt{(x_2 - x_1)^2 + (y_2 - y_1)^2} \quad \text{Distance Formula}$$
$$= \sqrt{[-6 - (-2)]^2 + (7 - 4)^2} \quad (x_1, y_1) = (-2, 4) \text{ and } (x_2, y_2) = (-6, 7)$$
$$= \sqrt{25} \text{ or } 5 \quad \text{Simplify.}$$

**Step 2** Write the equation using $h = -2$, $k = 4$, and $r = 5$.

$$(x - h)^2 + (y - k)^2 = r^2 \quad \text{Equation of a circle}$$
$$[x - (-2)]^2 + (y - 4)^2 = 5^2 \quad h = -2, k = 4, \text{ and } r = 5$$
$$(x + 2)^2 + (y - 4)^2 = 25 \quad \text{Simplify.}$$

### Guided Practice

**2.** Write the equation of the circle with center at (−3, −5) that passes through (0, 0).

## 2 Graph Circles

You can use the equation of a circle to graph it on a coordinate plane. To do so, you may need to write the equation in standard form first.

## Example 3 Graph a Circle

**The equation of a circle is $x^2 + y^2 - 8x + 2y = -8$. State the coordinates of the center and the measure of the radius. Then graph the equation.**

Write the equation in standard form by completing the square.

$$x^2 + y^2 - 8x + 2y = -8 \quad \text{Original equation}$$
$$x^2 - 8x + y^2 + 2y = -8 \quad \text{Isolate and group like terms.}$$
$$x^2 - 8x + 16 + y^2 + 2y + 1 = -8 + 16 + 1 \quad \text{Complete the squares.}$$
$$(x - 4)^2 + (y + 1)^2 = 9 \quad \text{Factor and simplify.}$$
$$(x - 4)^2 + [y - (-1)]^2 = 3^2 \quad \text{Write } +1 \text{ as } -(-1) \text{ and 9 as } 3^2.$$

**StudyTip**

**Completing the Square** To complete the square for any quadratic expression of the form $x^2 + bx$, follow these steps.

**Step 1** Find one half of $b$.

**Step 2** Square the result in Step 1.

**Step 3** Add the result of Step 2 to $x^2 + bx$.

With the equation now in standard form, you can identify $h$, $k$, and $r$.

$$(x - 4)^2 + [y - (-1)]^2 = 3^2$$
$$\uparrow \qquad \uparrow \qquad \uparrow$$
$$(x - h)^2 + (y - k)^2 = r^2$$

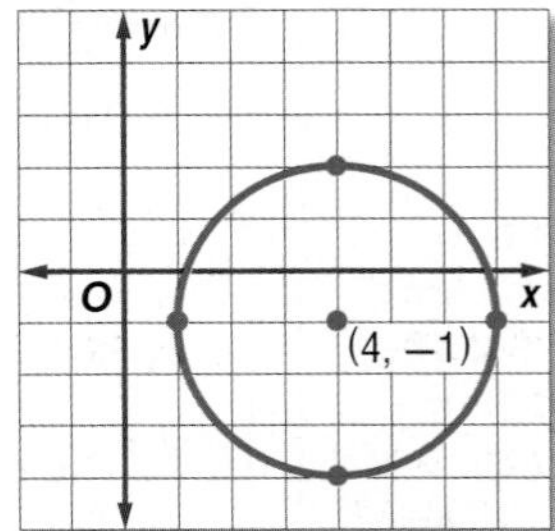


So, $h = 4$, $k = -1$, and $r = 3$. The center is at (4, −1), and the radius is 3. Plot the center and four points that are 3 units from this point. Sketch the circle through these four points.

### Guided Practice

**For each circle with the given equation, state the coordinates of the center and the measure of the radius. Then graph the equation.**

**3A.** $x^2 + y^2 - 4 = 0$

**3B.** $x^2 + y^2 + 8x - 14y + 40 = 0$

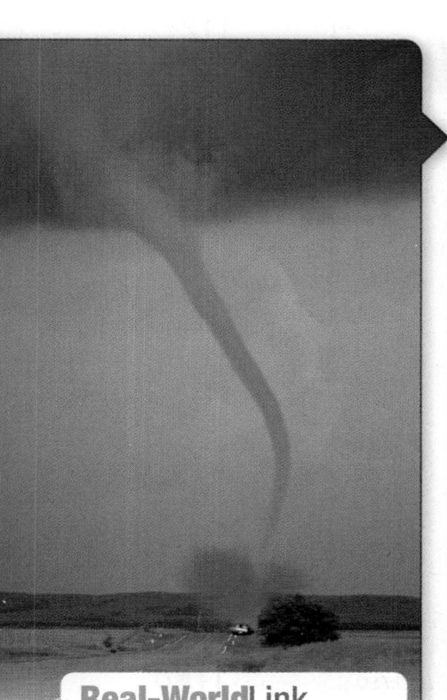

## Real-World Example 4 Use Three Points to Write an Equation

**TORNADOES Three tornado sirens are placed strategically on a circle around a town so they can be heard by all. Write the equation of the circle on which they are placed if the coordinates of the sirens are $A(-8, 3)$, $B(-4, 7)$, and $C(-4, -1)$.**

**Understand** You are given three points that lie on a circle.

**Plan** Graph $\triangle ABC$. Construct the perpendicular bisectors of two sides to locate the center of the circle. Then find the radius.

Use the center and radius to write an equation.

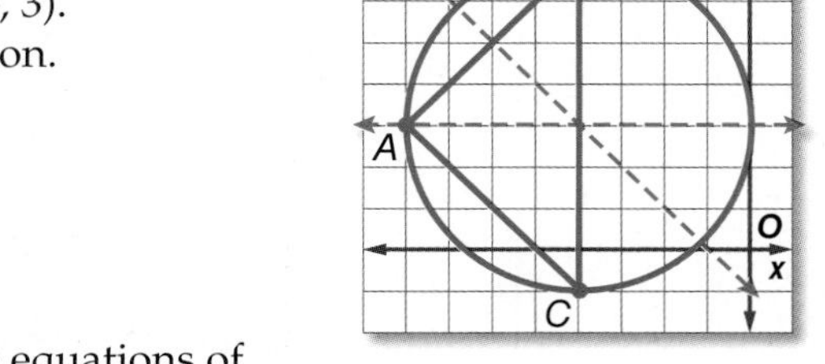


**Solve** The center appears to be at $(-4, 3)$.
The radius is 4. Write an equation.

$$(x - h)^2 + (y - k)^2 = r^2$$

$$[x - (-4)]^2 + (y - 3)^2 = 4^2$$

$$(x + 4)^2 + (y - 3)^2 = 16$$

**Check** Verify the center by finding the equations of the two bisectors and solving the system of equations. Verify the radius by finding the distance between the center and another point on the circle. ✓

### GuidedPractice

**4.** Write an equation of a circle that contains $R(1, 2)$, $S(-3, 4)$, and $T(-5, 0)$.

### Real-WorldLink

About 1000 tornadoes are reported across the United States each year. The most violent tornadoes have wind speeds of 250 mph or more. Damage paths can be a mile wide and 50 miles long.

**Source:** National Oceanic & Atmospheric Administration

A line can intersect a circle in at most two points. You can find the point(s) of intersection between a circle and a line by applying techniques used to find the intersection between two lines and techniques used to solve quadratic equations.

## Example 5 Intersections with Circles

**Find the point(s) of intersection between $x^2 + y^2 = 4$ and $y = x$.**

Graph these equations on the same coordinate plane. The points of intersection are solutions of both equations. You can estimate these points on the graph to be at about $(-1.4, -1.4)$ and $(1.4, 1.4)$. Use substitution to find the coordinates of these points algebraically.

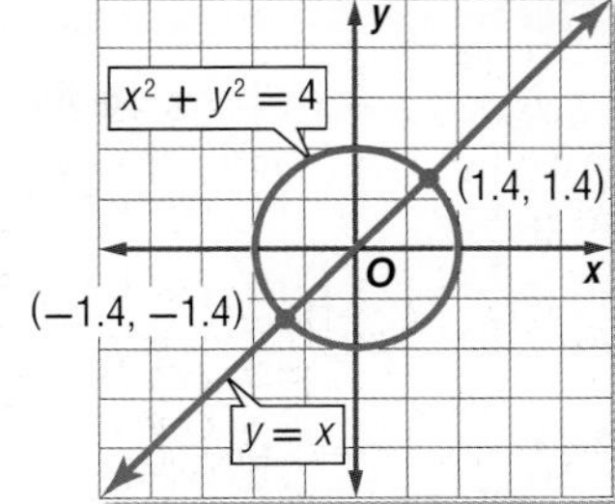


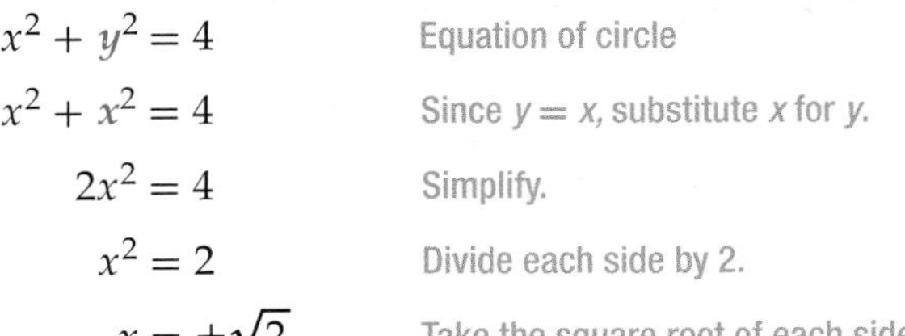

| | |
|---|---|
| $x^2 + y^2 = 4$ | Equation of circle |
| $x^2 + x^2 = 4$ | Since $y = x$, substitute $x$ for $y$. |
| $2x^2 = 4$ | Simplify. |
| $x^2 = 2$ | Divide each side by 2. |
| $x = \pm\sqrt{2}$ | Take the square root of each side. |

So $x = \sqrt{2}$ or $x = -\sqrt{2}$. Use the equation $y = x$ to find the corresponding $y$-values.

| | | |
|---|---|---|
| $y = x$ | Equation of line | $y = x$ |
| $y = \sqrt{2}$ | $x = \sqrt{2}$ or $x = -\sqrt{2}$ | $y = -\sqrt{2}$ |

The points of intersection are located at $(\sqrt{2}, \sqrt{2})$ and $(-\sqrt{2}, -\sqrt{2})$ or at about $(-1.4, -1.4)$ and $(1.4, 1.4)$. Check these solutions in both of the original equations.

### StudyTip

**Quadratic Techniques** In addition to taking square roots, other quadratic techniques that you may need to apply in order to solve equations of the form $ax^2 + bx + c = 0$ include completing the square, factoring, and the Quadratic Formula, $x = \frac{-b \pm \sqrt{b^2 - 4ac}}{2a}$.

### GuidedPractice

**5.** Find the point(s) of intersection between $x^2 + y^2 = 8$ and $y = -x$.

## Check Your Understanding

**Examples 1–2** **Write the equation of each circle.**

1. center at (9, 0), radius 5
2. center at (3, 1), diameter 14
3. center at origin, passes through (2, 2)
4. center at (−5, 3), passes through (1, −4)
5. 

(4, 1)
(2, 1)
O
x
y

6. 

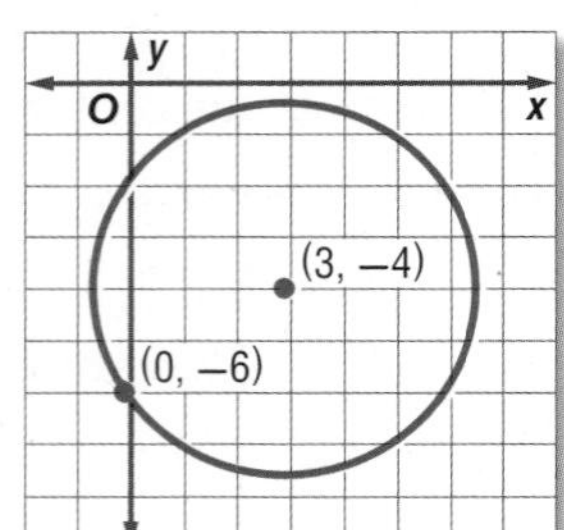


**Example 3** **For each circle with the given equation, state the coordinates of the center and the measure of the radius. Then graph the equation.**

7. $x^2 - 6x + y^2 + 4y = 3$
8. $x^2 + (y + 1)^2 = 4$

**Example 4**

9. **RADIOS** Three radio towers are modeled by the points $R(4, 5)$, $S(8, 1)$, and $T(-4, 1)$. Determine the location of another tower equidistant from all three towers, and write an equation for the circle.

10. **COMMUNICATION** Three cell phone towers can be modeled by the points $X(6, 0)$, $Y(8, 4)$, and $Z(3, 9)$. Determine the location of another cell phone tower equidistant from the other three, and write an equation for the circle.

**Example 5** **Find the point(s) of intersection, if any, between each circle and line with the equations given.**

11. $(x - 1)^2 + y^2 = 4$
    $y = x + 1$
12. $(x - 2)^2 + (y + 3)^2 = 18$
    $y = -2x - 2$

## Practice and Problem Solving

**Examples 1–2** **CCSS STRUCTURE** **Write the equation of each circle.**

13. center at origin, radius 4
14. center at (6, 1), radius 7
15. center at (−2, 0), diameter 16
16. center at (8, −9), radius $\sqrt{11}$
17. center at (−3, 6), passes through (0, 6)
18. center at (1, −2), passes through (3, −4)
19. 

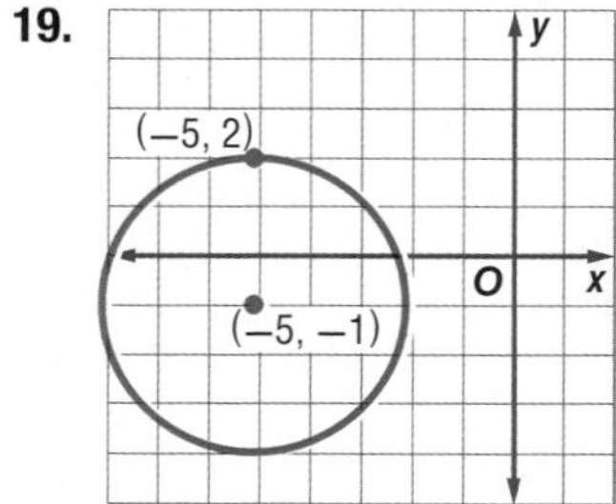


20. 

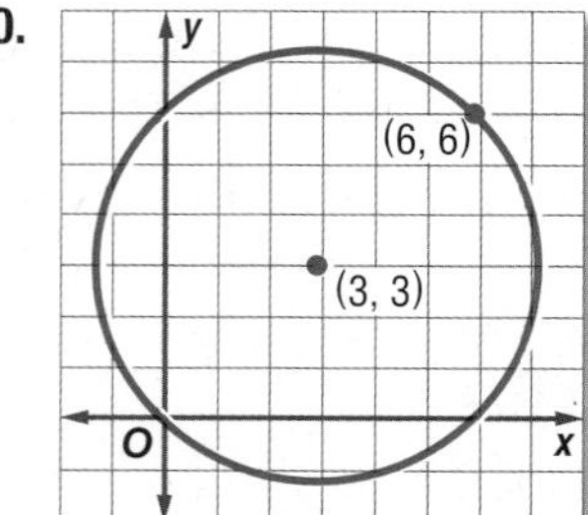


21. **WEATHER** A Doppler radar screen shows concentric rings around a storm. If the center of the radar screen is the origin and each ring is 15 miles farther from the center, what is the equation of the third ring?

22. **GARDENING** A sprinkler waters a circular area that has a diameter of 10 feet. The sprinkler is located 20 feet north of the house. If the house is located at the origin, what is the equation for the circle of area that is watered?

**Example 3** **For each circle with the given equation, state the coordinates of the center and the measure of the radius. Then graph the equation.**

**23.** $x^2 + y^2 = 36$

**24.** $x^2 + y^2 - 4x - 2y = -1$

**25.** $x^2 + y^2 + 8x - 4y = -4$

**26.** $x^2 + y^2 - 16x = 0$

**Example 4** **Write an equation of a circle that contains each set of points. Then graph the circle.**

**27.** $A(1, 6), B(5, 6), C(5, 0)$

**28.** $F(3, -3), G(3, 1), H(7, 1)$

**Example 5** **Find the point(s) of intersection, if any, between each circle and line with the equations given.**

**29.** $x^2 + y^2 = 5$
$y = \frac{1}{2}x$

**30.** $x^2 + y^2 = 2$
$y = -x + 2$

**31.** $x^2 + (y + 2)^2 = 8$
$y = x - 2$

**32.** $(x + 3)^2 + y^2 = 25$
$y = -3x$

**33.** $x^2 + y^2 = 5$
$y = 3x$

**34.** $(x - 1)^2 + (y - 3)^2 = 4$
$y = -x$

**Write the equation of each circle.**

**35.** a circle with a diameter having endpoints at $(0, 4)$ and $(6, -4)$

**36.** a circle with $d = 22$ and a center translated 13 units left and 6 units up from the origin

**37.** **CCSS MODELING** Different-sized engines will launch model rockets to different altitudes. The higher a rocket goes, the larger the circle of possible landing sites becomes. Under normal wind conditions, the landing radius is three times the altitude of the rocket.

**a.** Write the equation of the landing circle for a rocket that travels 300 feet in the air.

**b.** What would be the radius of the landing circle for a rocket that travels 1000 feet in the air? Assume the center of the circle is at the origin.

**38.** **SKYDIVING** Three of the skydivers in the circular formation shown have approximate coordinates of $G(13, -2)$, $H(-1, -2)$, and $J(6, -9)$.

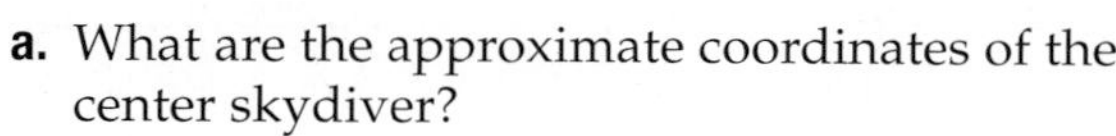

**a.** What are the approximate coordinates of the center skydiver?

**b.** If each unit represents 1 foot, what is the diameter of the skydiving formation?

**39.** **DELIVERY** Pizza and Subs offers free delivery within 6 miles of the restaurant. The restaurant is located 4 miles west and 5 miles north of Consuela's house.

**a.** Write and graph an equation to represent this situation if Consuela's house is at the origin of the coordinate system.

**b.** Can Consuela get free delivery if she orders pizza from Pizza and Subs? Explain.

**40.** **INTERSECTIONS OF CIRCLES** Graph $x^2 + y^2 = 4$ and $(x - 2)^2 + y^2 = 4$ on the same coordinate plane.

**a.** Estimate the point(s) of intersection between the two circles.

**b.** Solve $x^2 + y^2 = 4$ for $y$.

**c.** Substitute the value you found in part **b** into $(x - 2)^2 + y^2 = 4$ and solve for $x$.

**d.** Substitute the value you found in part **c** into $x^2 + y^2 = 4$ and solve for $y$.

**e.** Use your answers to parts **c** and **d** to write the coordinates of the points of intersection. Compare these coordinates to your estimate from part **a**.

**f.** Verify that the point(s) you found in part **d** lie on both circles.

41 Prove or disprove that the point $(1, 2\sqrt{2})$ lies on a circle centered at the origin and containing the point $(0, -3)$.

42. **MULTIPLE REPRESENTATIONS** In this problem, you will investigate a compound locus for a pair of points. A **compound locus** satisfies more than one distinct set of conditions.

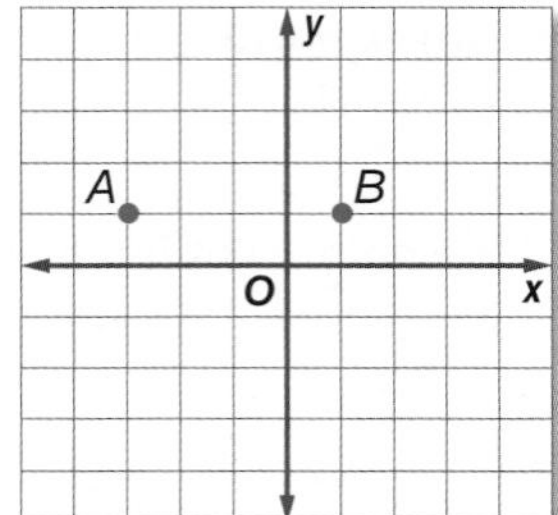


a. **Tabular** Choose two points $A$ and $B$ in the coordinate plane. Locate 5 coordinates from the locus of points equidistant from $A$ and $B$.

b. **Graphical** Represent this same locus of points by using a graph.

c. **Verbal** Describe the locus of all points equidistant from a pair of points.

d. **Graphical** Using your graph from part **b**, determine and graph the locus of all points in a plane that are a distance of $AB$ from $B$.

e. **Verbal** Describe the locus of all points in a plane equidistant from a single point. Then describe the locus of all points that are both equidistant from $A$ and $B$ and are a distance of $AB$ from $B$. Describe the graph of the compound locus.

43. A circle with a diameter of 12 has its center in the second quadrant. The lines $y = -4$ and $x = 1$ are tangent to the circle. Write an equation of the circle.

## H.O.T. Problems Use Higher-Order Thinking Skills

44. **CHALLENGE** Write a coordinate proof to show that if an inscribed angle intercepts the diameter of a circle, as shown, the angle is a right angle.

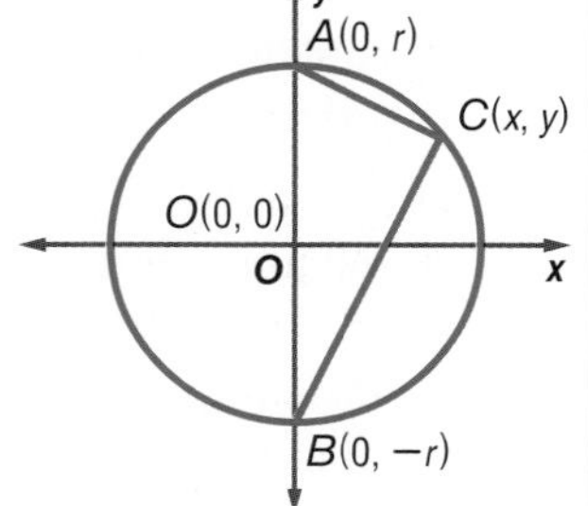


45. **CCSS REASONING** A circle has the equation $(x - 5)^2 + (y + 7)^2 = 16$. If the center of the circle is shifted 3 units right and 9 units up, what would be the equation of the new circle? Explain your reasoning.

46. **OPEN ENDED** Graph three noncollinear points and connect them to form a triangle. Then construct the circle that circumscribes it.

47. **WRITING IN MATH** Seven new radio stations must be assigned broadcast frequencies. The stations are located at $A(9, 2)$, $B(8, 4)$, $C(8, 1)$, $D(6, 3)$, $E(4, 0)$, $F(3, 6)$, and $G(4, 5)$, where 1 unit = 50 miles.

a. If stations that are more than 200 miles apart can share the same frequency, what is the least number of frequencies that can be assigned to these stations?

b. Describe two different beginning approaches to solving this problem.

c. Choose an approach, solve the problem, and explain your reasoning.

**CHALLENGE Find the coordinates of point $P$ on $\overrightarrow{AB}$ that partitions the segment into the given ratio $AP$ to $PB$.**

48. $A(0, 0)$, $B(3, 4)$, 2 to 3

49. $A(0, 0)$, $B(-8, 6)$, 4 to 1

50. **WRITING IN MATH** Describe how the equation for a circle changes if the circle is translated $a$ units to the right and $b$ units down.

## Standardized Test Practice

**51.** Which of the following is the equation of a circle with center (6, 5) that passes through (2, 8)?

**A** $(x - 6)^2 + (y - 5)^2 = 5^2$
**B** $(x - 5)^2 + (y - 6)^2 = 7^2$
**C** $(x + 6)^2 + (y + 5)^2 = 5^2$
**D** $(x - 2)^2 + (y - 8)^2 = 7^2$

**52. ALGEBRA** What are the solutions of $n^2 - 4n = 21$?

**F** 3, 7
**G** 3, −7
**H** −3, 7
**J** −3, −7

**53. SHORT RESPONSE** Solve: $5(x - 4) = 16$.

Step 1: $5x - 4 = 16$
Step 2: $5x = 20$
Step 3: $x = 4$

Which is the first incorrect step in the solution shown above?

**54. SAT/ACT** The center of $\odot F$ is at (−4, 0) and has a radius of 4. Which point lies on $\odot F$?

**A** (4, 0)
**B** (0, 4)
**C** (4, 3)
**D** (−4, 4)
**E** (0, 8)

## Spiral Review

**Find $x$.** (Lesson 11-7)

**55.**
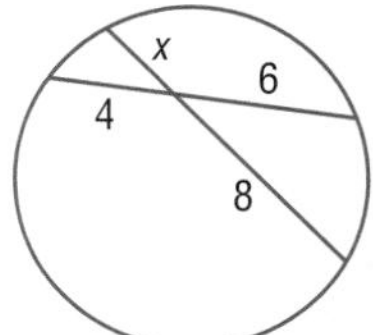


**56.**
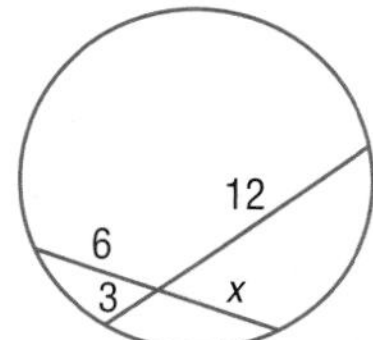


**57.**
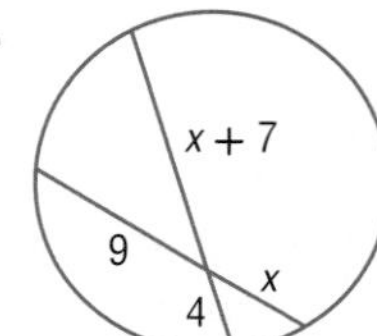


**Find each measure.** (Lesson 11-6)

**58.** $m\angle C$
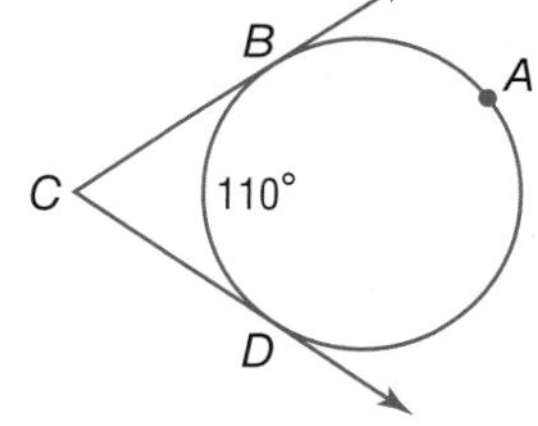


**59.** $m\angle K$
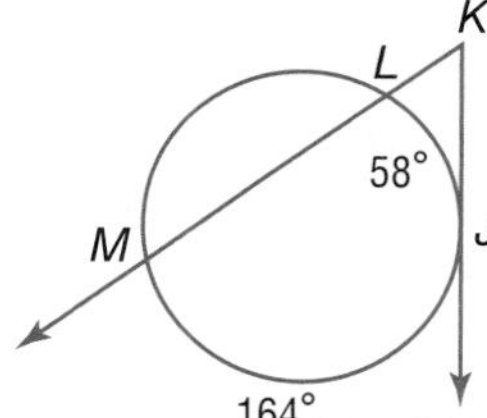


**60.** $m\widehat{YXZ}$
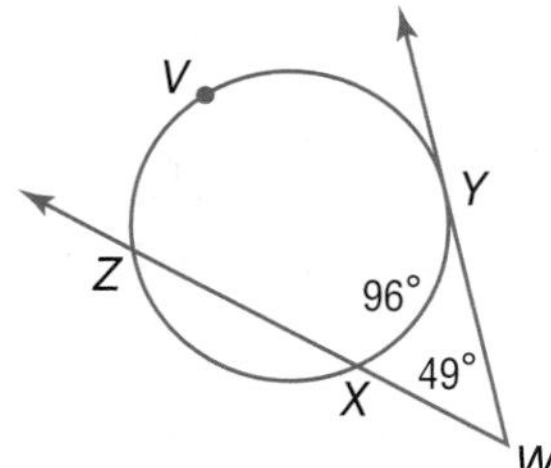


**61. STREETS** The neighborhood where Vincent lives has roundabouts where certain streets meet. If Vincent rides his bike once around the very edge of the grassy circle, how many feet will he have ridden? (Lesson 11-1)

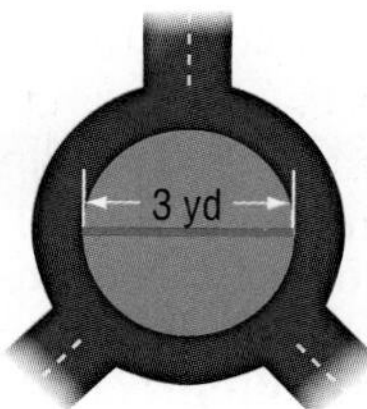


## Skills Review

**Find the perimeter and area of each figure.**

**62.**

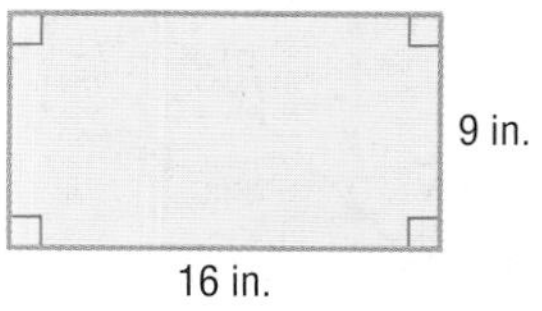

**63.**
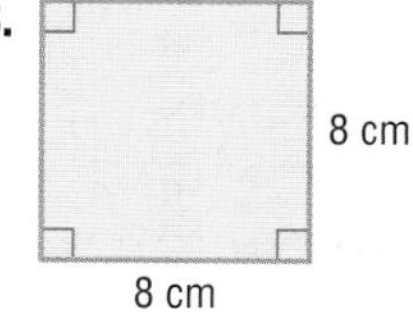


**64.**
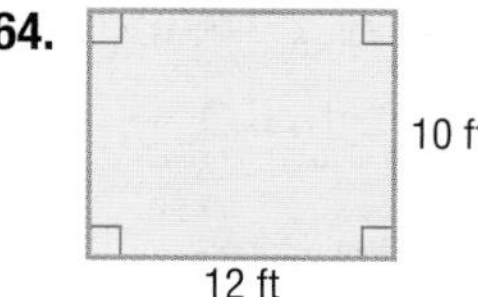

EXTEND 11-8

# Geometry Lab Parabolas

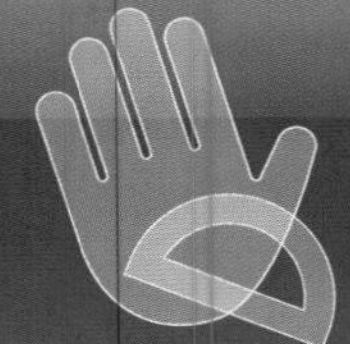

A circle is one type of cross-section of a right circular cone. Such cross-sections are called **conic sections** or **conics**. A circular cross-section is formed by the intersection of a cone with a plane that is perpendicular to the axis of the cone. You can find other conic sections using concrete models of cones.

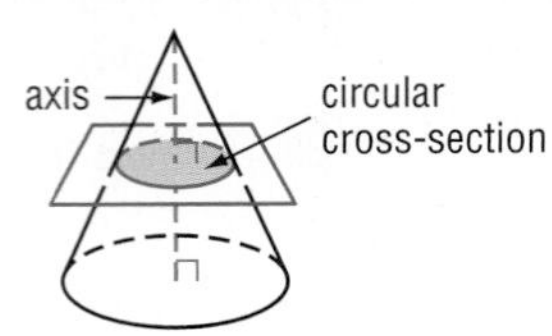


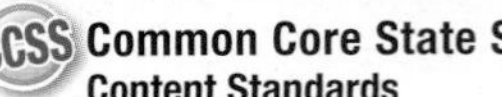

**Common Core State Standards**
**Content Standards**
**G.GPE.2** Derive the equation of a parabola given a focus and directrix.
**Mathematical Practices** 5

## Activity 1 Intersection of Cone and Plane

**Sketch the intersection of a cone and a plane that lies at an angle to the axis of the cone but does not pass through its base.**

**Step 1** Fill a conical paper cup with modeling compound. Then peel away the cup.

**Step 2** Draw dental floss through the cone model at an angle to the axis that does not pass through the base.

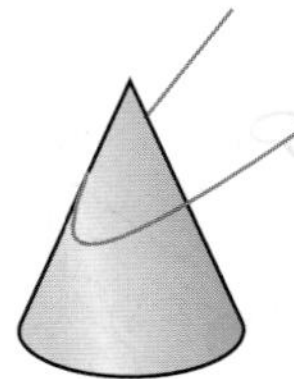

**Step 3** Pull the pieces of the cone apart and trace the cross-section onto your paper.

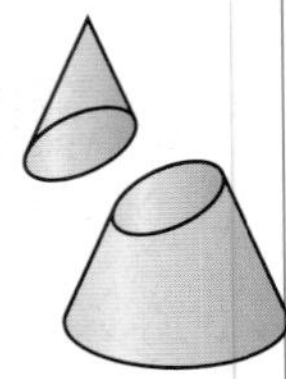

### Model and Analyze

1. The conic section in Activity 1 is called an ellipse. What shape is an ellipse?
2. Repeat Activity 1, drawing the dental floss through the model at an angle parallel to an imaginary line on the side of the cone through the cone's base. Describe the resulting shape.

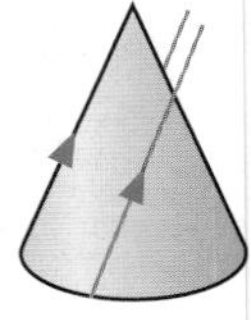

The conic section you found in Exercise 2 is called a **parabola**. In Algebra 1, a parabola was defined as the shape of the graph of a quadratic function, such as $y = x^2$. Like a circle and all conics, a parabola can also be defined as a locus of points. You can explore the loci definition of a parabola using paper folding.

## Activity 2 Shape of Parabola

**Use paper folding to approximate the shape of a parabola.**

**Step 1** Mark and label the bottom edge of a rectangular piece of wax paper $d$. Label a point $F$ at the center.

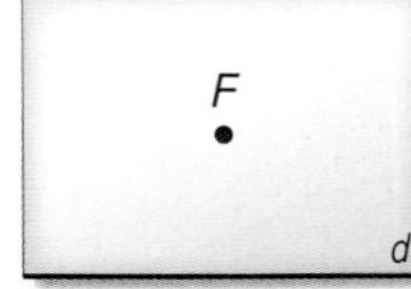


**Step 2** Fold $d$ up so that it touches $F$. Make a sharp crease. Then open the paper and smooth it flat.

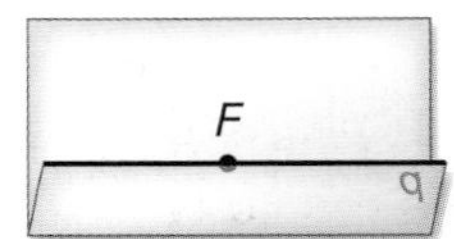


**Step 3** Repeat Step 2 at least 20 times, folding the paper to a different point on $d$ each time. Trace the curve formed.

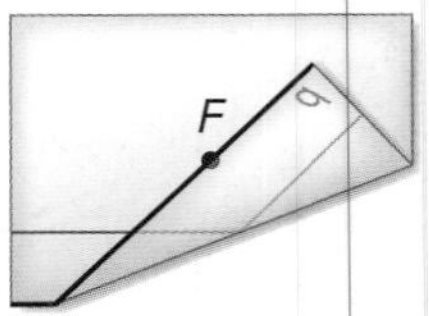

## Model and Analyze

3. Label a point $P$ on the parabola and draw $\overline{PF}$. Then use a protractor to find a point $D$ on line $d$ such that $\overline{PD} \perp d$. Describe the relationship between $\overline{PF}$ and $\overline{PD}$.

**Repeat Activity 2, making the indicated change on a new piece of wax paper. Describe the effect on the parabola formed.**

4. Place line $d$ along the edge above point $F$.
5. Place line $d$ along the edge to the right of point $F$.
6. Place line $d$ along the edge to left of point $F$.
7. Place point $F$ closer to line $d$.
8. Place point $F$ farther away from line $d$.

Geometrically, a parabola is the locus of all points in a plane equidistant from a fixed point, called the **focus**, and a fixed line, called the **directrix**. Recall that the distance between a fixed point and a line is the length of the segment perpendicular to the line through that point. You can find an equation of a parabola on the coordinate plane using its locus definition and the Distance Formula.

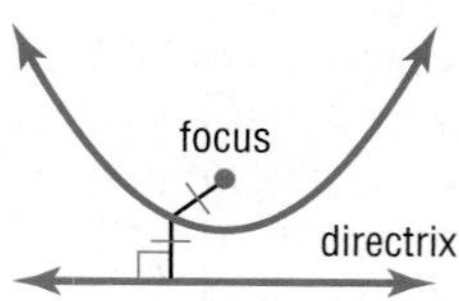


### Activity 3 Equation of Parabola

**Find an equation of the parabola with focus at (0, 1) and directrix $y = -1$.**

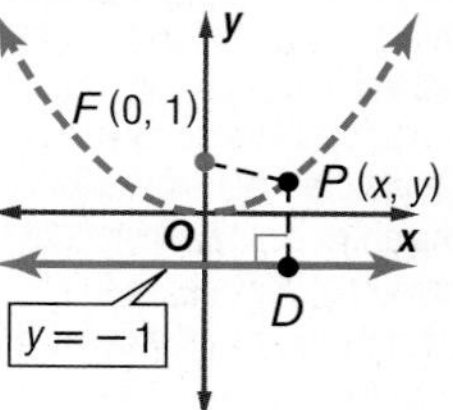


**Step 1** Graph $F(0, 1)$ and $y = -1$. Sketch a U-shaped curve for the parabola between the point and line as shown. Label a point $P(x, y)$ on the curve.

**Step 2** Label a point $D$ on $y = -1$ such that $\overline{PD}$ is perpendicular to the line $y = -1$. The coordinates of this point must therefore be $D(x, -1)$.

**Step 3** Use the Distance Formula to find $PD$ and $PF$.

$$PD = \sqrt{(x-x)^2 + [y-(-1)]^2} \quad D(x, -1), P(x, y), F(0, 1)$$
$$= \sqrt{(y+1)^2} \quad \text{Simplify.}$$

$$PF = \sqrt{(x-0)^2 + (y-1)^2}$$
$$= \sqrt{x^2 + (y-1)^2}$$

**Step 4** Since $PD = PF$, set these expressions equal to each other.

$$\sqrt{(y+1)^2} = \sqrt{x^2 + (y-1)^2} \quad PD = PF$$
$$(y+1)^2 = x^2 + (y-1)^2 \quad \text{Square each side.}$$
$$y^2 + 2y + 1 = x^2 + y^2 - 2y + 1 \quad \text{Square each binomial.}$$
$$4y = x^2 \text{ or } y = \frac{1}{4}x^2 \quad \text{Subtract } y^2 - 2y + 1 \text{ from each side.}$$

An equation of the parabola with focus at (0, 1) and directrix $y = -1$ is $y = \frac{1}{4}x^2$.

## Model and Analyze

**Find an equation of the parabola with the focus and directrix given.**

9. $(0, -2), y = 2$
10. $\left(0, \frac{1}{2}\right), y = -\frac{1}{2}$
11. $(1, 0), x = -1$
12. $(-3, 0), x = 3$

**A line can intersect a parabola in 0, 1, or 2 points. Find the point(s) of intersection, if any, between each parabola and line with the given equations.**

13. $y = x^2, y = x + 2$
14. $y = 2x^2, y = 4x - 2$
15. $y = -3x^2, y = 6x$
16. $y = -(x+1)^2, y = -x$

# LESSON 11-9 Areas of Circles and Sectors

| Then | Now | Why? |
|---|---|---|
| You found the circumference of a circle. | **1** Find areas of circles.<br>**2** Find areas of sectors of circles. | To determine whether a medium or large pizza is a better value, you can compare the cost per square inch. Divide the cost of each pizza by its area. |

**NewVocabulary**
sector of a circle
segment of a circle

**Common Core State Standards**

**Content Standards**

**G.C.5** Derive using similarity the fact that the length of the arc intercepted by an angle is proportional to the radius, and define the radian measure of the angle as the constant of proportionality; derive the formula for the area of a sector.

**G.GMD.1** Give an informal argument for the formulas for the circumference of a circle, area of a circle, volume of a cylinder, pyramid, and cone.

**Mathematical Practices**
1 Make sense of problems and persevere in solving them.
6 Attend to precision.

## 1 Areas of Circles

**Areas of Circles** In Lesson 10-1, you learned that the formula for the circumference $C$ of a circle with radius $r$ is given by $C = 2\pi r$. You can use this formula to develop the formula for the area of a circle.

Below, a circle with radius $r$ and circumference $C$ has been divided into congruent pieces and then rearranged to form a figure that resembles a parallelogram.

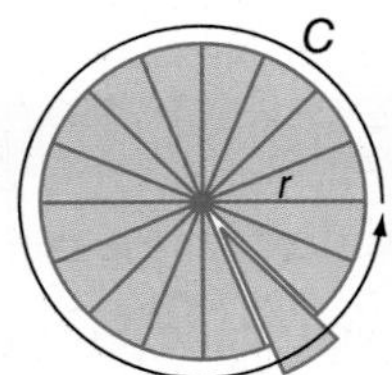


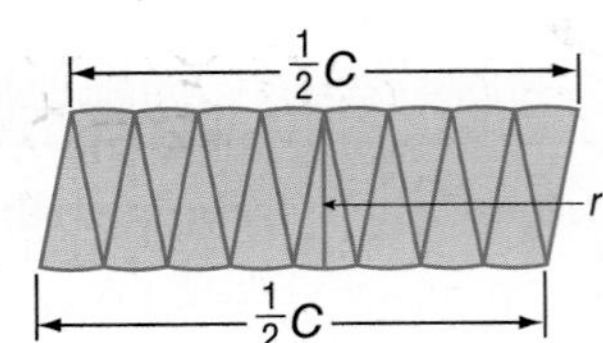


As the number of congruent pieces increases, the rearranged figure more closely approaches a parallelogram. The base of the parallelogram is $\frac{1}{2}C$ and the height is $r$, so its area is $\frac{1}{2}C \cdot r$. Since $C = 2\pi r$, the area of the parallelogram is also $\frac{1}{2}(2\pi r)r$ or $\pi r^2$.

### KeyConcept Area of a Circle

**Words** The area $A$ of a circle is equal to $\pi$ times the square of the radius $r$.

**Symbols** $A = \pi r^2$

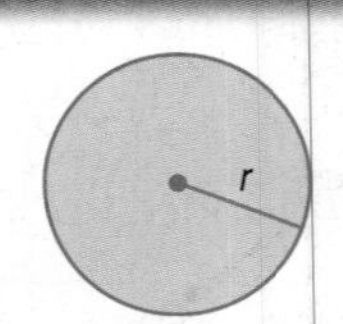


### Real-World Example 1 Area of a Circle

**SPORTS What is the area of the circular putting green shown to the nearest square foot?**

The diameter is 20 feet, so the radius is 10 feet.

| | |
|---|---|
| $A = \pi r^2$ | Area of a circle |
| $= \pi(10)^2$ | $r = 10$ |
| $\approx 314$ | Use a calculator. |

So, the area is about 314 square feet.

**GuidedPractice**

1. **SPORTS** An archery target has a radius of 12 inches. What is the area of the target to the nearest square inch?

Corbis Premium RF/Alamy

## Example 2 Use the Area of a Circle to Find a Missing Measure

**ALGEBRA Find the radius of a circle with an area of 95 square centimeters.**

| | |
|---|---|
| $A = \pi r^2$ | Area of a circle |
| $95 = \pi r^2$ | $A = 95$ |
| $\frac{95}{\pi} = r^2$ | Divide each side by $\pi$. |
| $5.5 \approx r$ | Use a calculator. Take the positive square root of each side. |

The radius of the circle is about 5.5 centimeters.

### GuidedPractice

**2. ALGEBRA** The area of a circle is $196\pi$ square yards. Find the diameter.

### ReviewVocabulary

**central angle** an angle with a vertex in the center of a circle and with sides that contain two radii of the circle

**arc** a portion of a circle defined by two endpoints

## 2 Areas of Sectors

A slice of a circular pizza is an example of a sector of a circle. A **sector of a circle** is a region of a circle bounded by a central angle and its intercepted major or minor arc. The formula for the area of a sector is similar to the formula for arc length.

### KeyConcept Area of a Sector

The ratio of the **area $A$ of a sector** to the **area of the whole circle, $\pi r^2$**, is equal to the ratio of the **degree measure of the intercepted arc $x$** to 360.

Proportion: $\frac{A}{\pi r^2} = \frac{x}{360}$

Equation: $A = \frac{x}{360} \cdot \pi r^2$

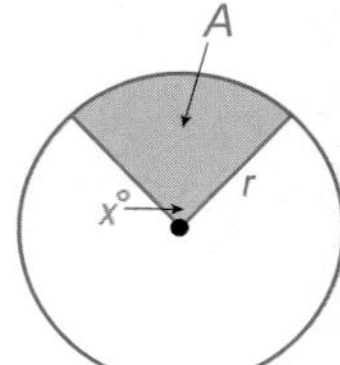


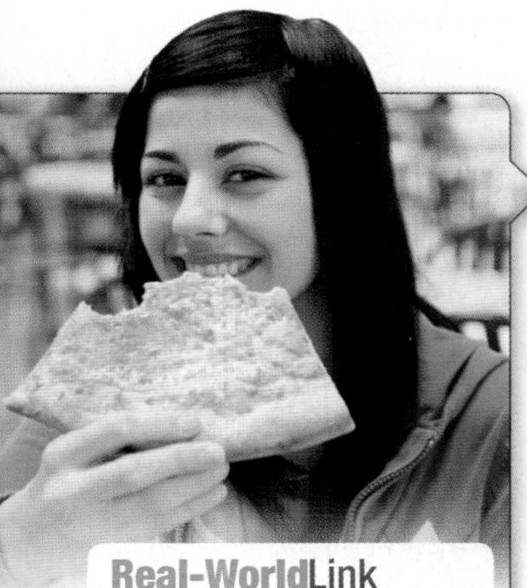

### Real-WorldLink

About 3 billion pizzas are sold each year in the United States. That is equivalent to about 46 slices per person annually.

**Source:** ThinkQuest Library

Picturenet/Blend Images/Getty Images

## Real-World Example 3 Area of a Sector

**PIZZA A circular pizza has a diameter of 12 inches and is cut into 8 congruent slices. What is the area of one slice to the nearest hundredth?**

**Step 1** Find the arc measure of a pizza slice.

Since the pizza is equally divided into 8 slices, each slice will have an arc measure of $360 \div 8$ or 45.

**Step 2** Find the radius of the pizza. Use this measure to find the area of the sector, or slice.

The diameter is 12 inches, so the radius is 6 inches.

| | |
|---|---|
| $A = \frac{x}{360} \cdot \pi r^2$ | Area of a sector |
| $= \frac{45}{360} \cdot \pi(6)^2$ | $x = 45$ and $r = 6$ |
| $\approx 14.14$ | Use a calculator. |

So, the area of one slice of this pizza is about 14.14 square inches.

## GuidedPractice

**Find the area of the shaded sector. Round to the nearest tenth.**

**3A.**

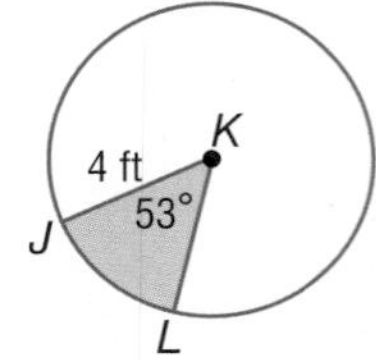


**3B.**

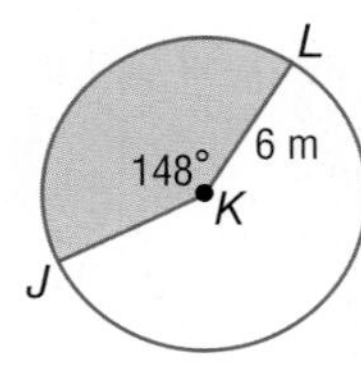


**3C.**

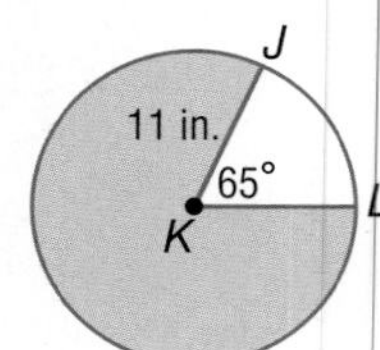


**3D. CRAFTS** The color wheel at the right is a tool that artists use to organize color schemes. If the diameter of the wheel is 10 inches and each of the 12 sections is congruent, find the approximate area covered by green hues.

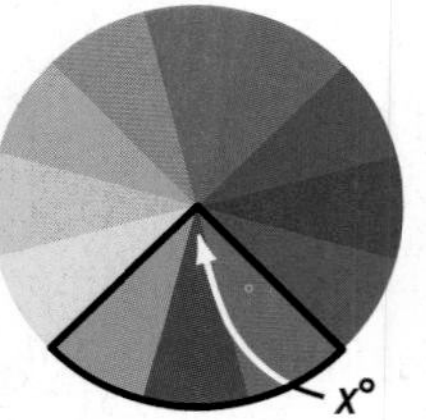


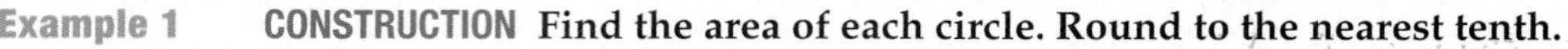

## Check Your Understanding

**Example 1** **CONSTRUCTION** **Find the area of each circle. Round to the nearest tenth.**

**1.**

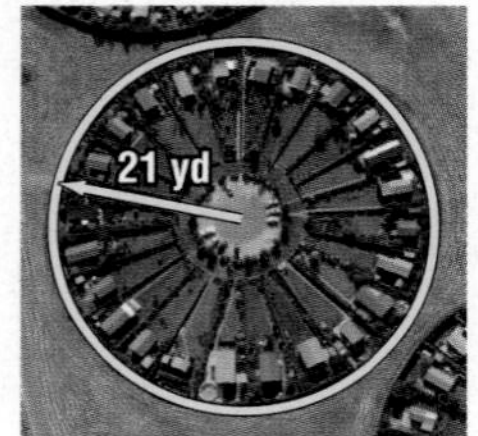


**2.**

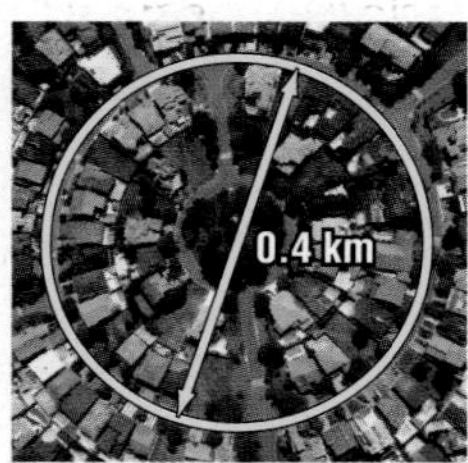


**Example 2** **Find the indicated measure. Round to the nearest tenth.**

**3.** Find the diameter of a circle with an area of 74 square millimeters.

**4.** The area of a circle is 88 square inches. Find the radius.

**Example 3** **Find the area of each shaded sector. Round to the nearest tenth.**

**5.**

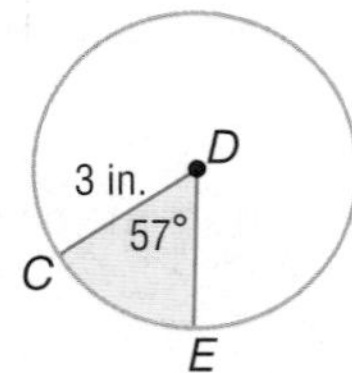


**6.**

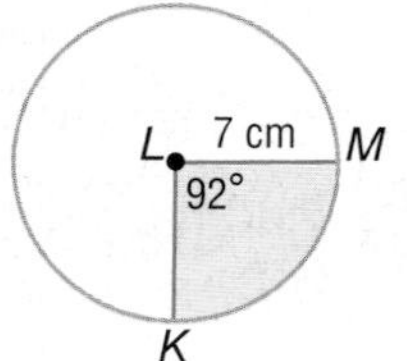


**7. BAKING** Chelsea is baking pies for a fundraiser at her school. She divides each 9-inch pie into 6 equal slices.

**a.** What is the area, in square inches, for each slice of pie?

**b.** If each slice costs $0.25 to make and she sells 8 pies at $1.25 for each slice, how much money will she raise?

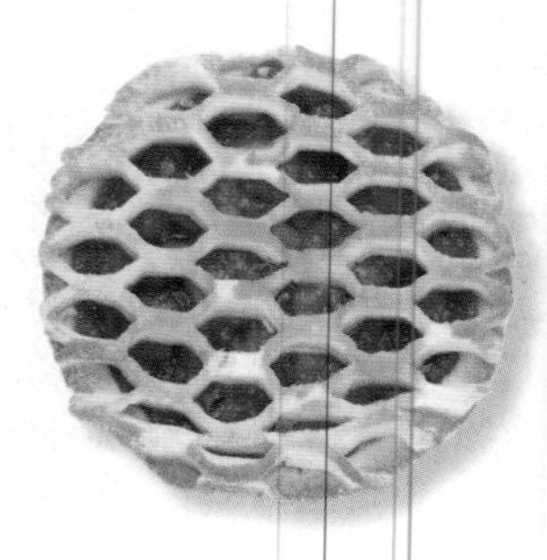

## Practice and Problem Solving

Example 1 — CCSS **MODELING** **Find the area of each circle. Round to the nearest tenth.**

8. 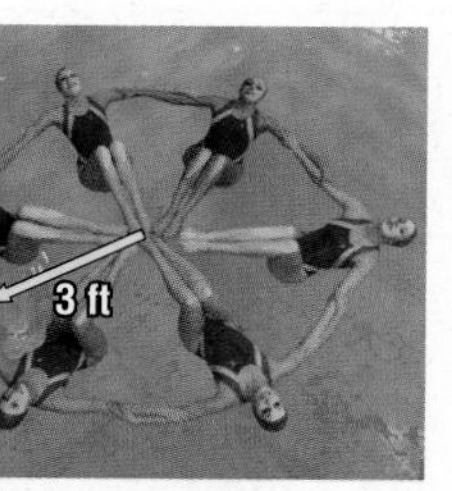


9. 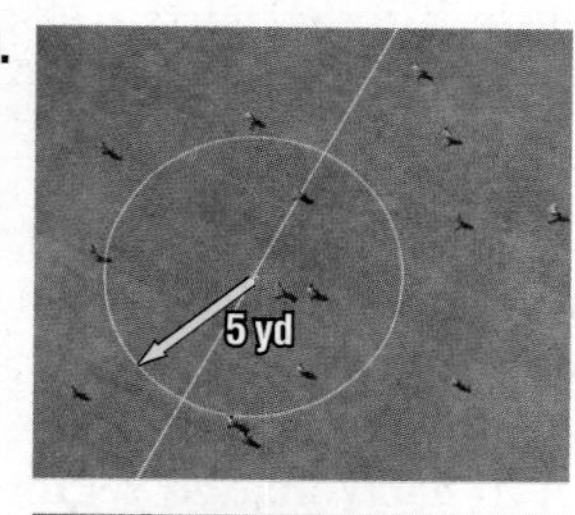


10. 

11. 

12. 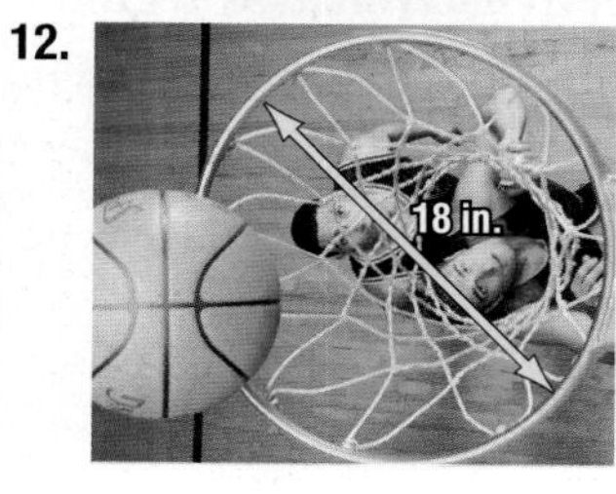


13. 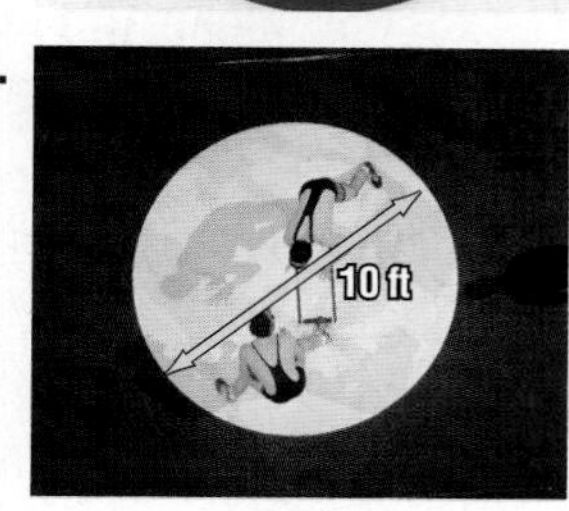


Example 2 — **Find the indicated measure. Round to the nearest tenth, if necessary.**

14. The area of a circle is 68 square centimeters. Find the diameter.
15. Find the diameter of a circle with an area of 94 square millimeters.
16. The area of a circle is 112 square inches. Find the radius.
17. Find the radius of a circle with an area of 206 square feet.

Example 3 — **Find the area of each shaded sector. Round to the nearest tenth, if necessary.**

18. 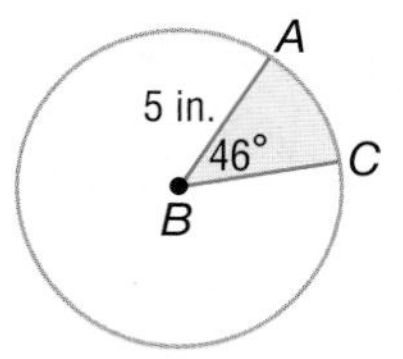


19. 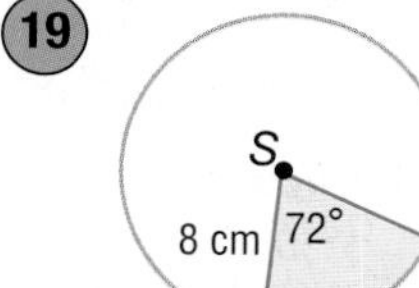


20. 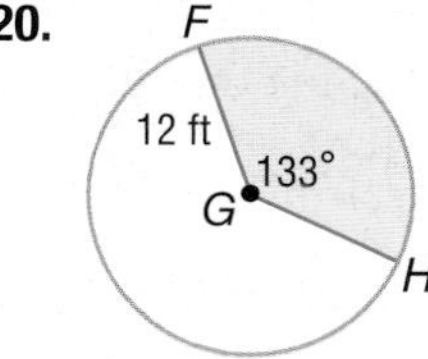


21. 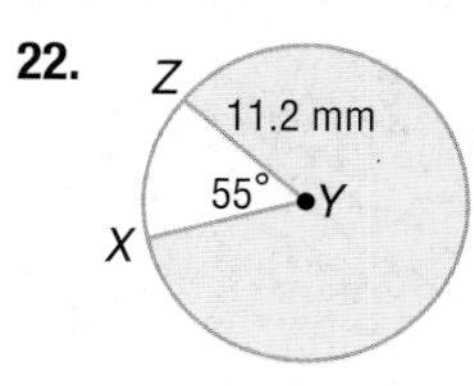


22. 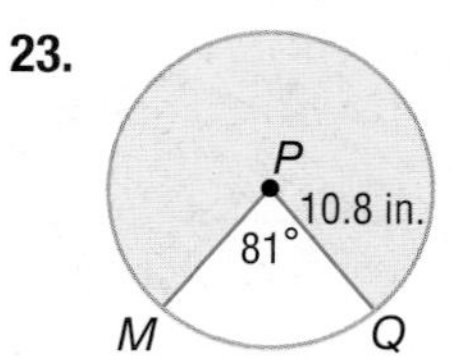


23.


24. **MUSIC** The music preferences of students at Thomas Jefferson High are shown in the circle graph. Find the area of each sector and the degree measure of each intercepted arc if the radius of the circle is 1 unit.

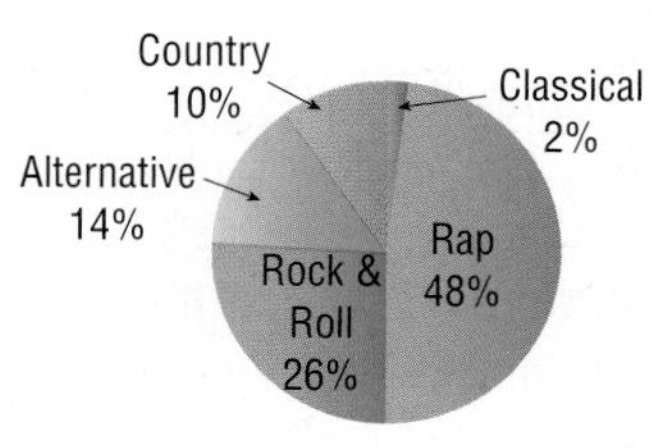


25. **JEWELRY** A jeweler makes a pair of earrings by cutting two 50° sectors from a silver disk.
    a. Find the area of each sector.
    b. If the weight of the silver disk is 2.3 grams, how many milligrams does the silver wedge for each earring weigh?

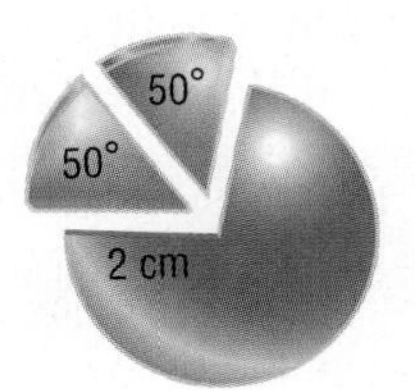


(tl)Pete Saloutos/Corbis, (tc)Stephan Zirwes/Getty Images, (tr)Doug Pensinger/Getty Images Sport/Getty Images, (bl)Reed Kaestner/Corbis, (bc)CORBIS, (br)PhotoLink/Photodisc/Getty Images

26. **PROM** The table shows the results of a survey of students to determine their preference for a prom theme.

| Theme | Percent |
|---|---|
| An Evening of Stars | 11 |
| Mardi Gras | 32 |
| Springtime in Paris | 8 |
| Night in Times Square | 47 |
| Undecided | 2 |

a. Create a circle graph with a diameter of 2 inches to represent these data.

b. Find the area of each theme's sector in your graph. Round to the nearest hundredth of an inch.

**CCSS SENSE-MAKING The area $A$ of each shaded region is given. Find $x$.**

27. $A = 66 \text{ cm}^2$

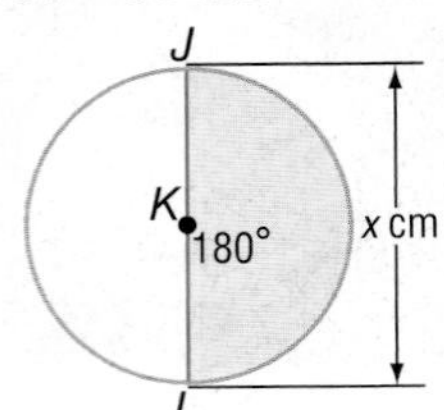


28. $A = 94 \text{ in}^2$

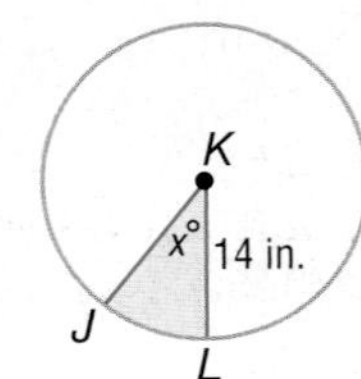


29. $A = 128 \text{ ft}^2$

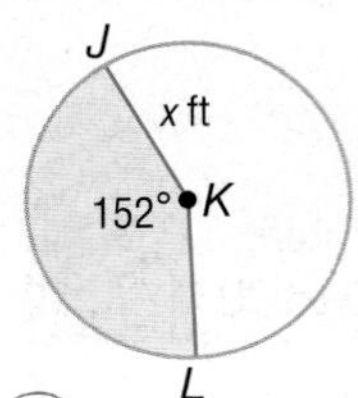


30. **CRAFTS** Luna is making tablecloths with the dimensions shown for a club banquet. Find the area of each tablecloth in square feet if each one is to just reach the floor.

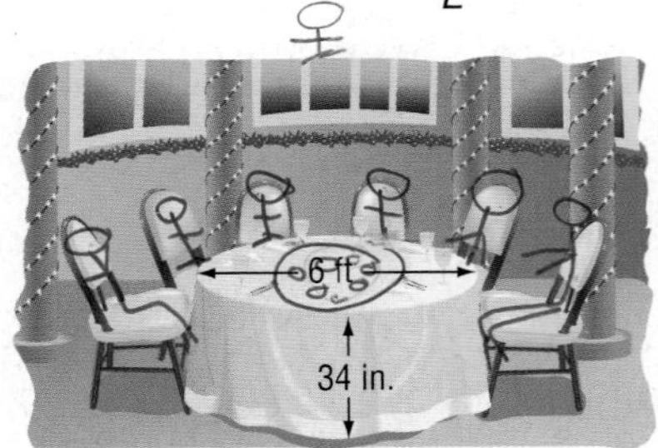


31. **TREES** The age of a living tree can be determined by multiplying the diameter of the tree by its growth factor, or rate of growth.

a. What is the diameter of a tree with a circumference of 2.5 feet?

b. If the growth factor of the tree is 4.5, what is the age of the tree?

**Find the area of the shaded region. Round to the nearest tenth.**

32.

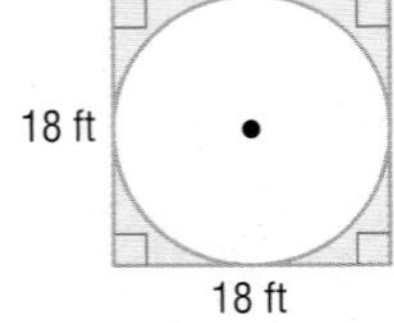


33.

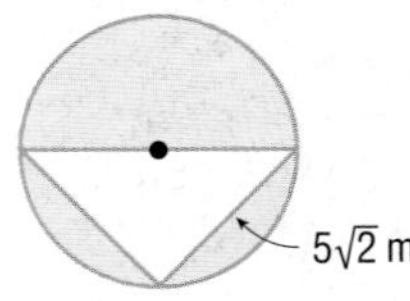


34.

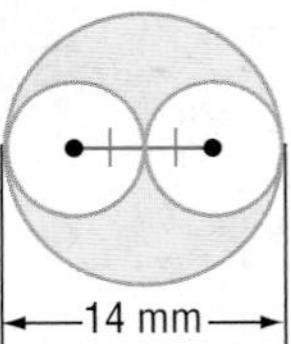


35.

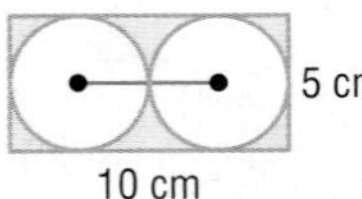


36.

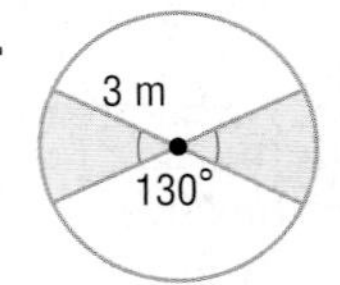


37.

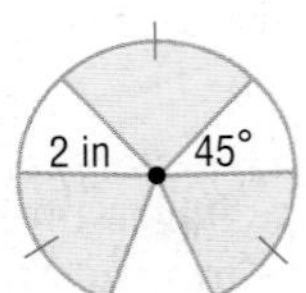


38. **COORDINATE GEOMETRY** What is the area of sector $ABC$ shown on the graph?

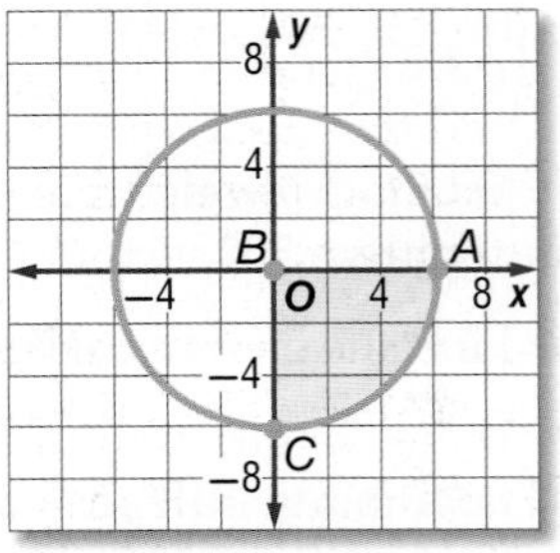

39. **ALGEBRA** The figure shown below is a sector of a circle. If the perimeter of the figure is 22 millimeters, find its area in square millimeters.

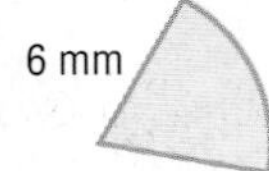

**Find the area of each shaded region.**

**40.**

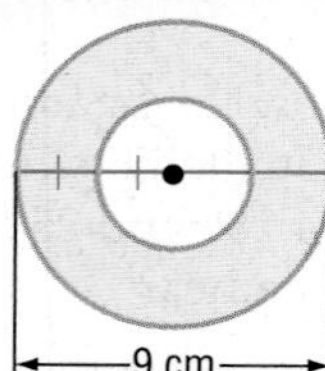


**41**

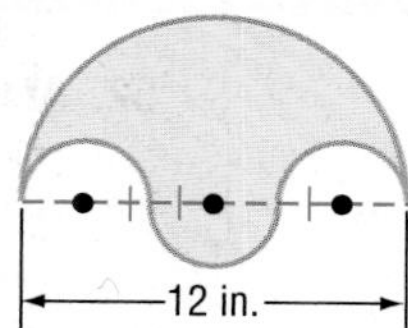


**42.**

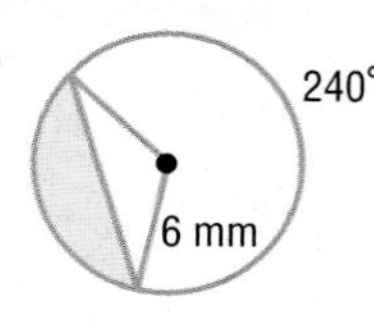


**43.** **MULTIPLE REPRESENTATIONS** In this problem, you will investigate segments of circles. A **segment of a circle** is the region bounded by an arc and a chord.

**a. Algebraic** Write an equation for the area $A$ of a segment of a circle with a radius $r$ and a central angle of $x°$. (*Hint*: Use trigonometry to find the base and height of the triangle.)

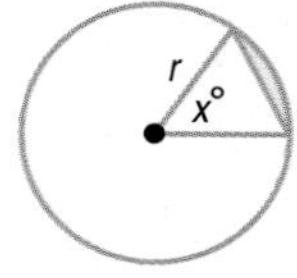


**b. Tabular** Calculate and record in a table ten values of $A$ for $x$-values ranging from 10 to 90 if $r$ is 12 inches. Round to the nearest tenth.

**c. Graphical** Graph the data from your table with the $x$-values on the horizontal axis and the $A$-values on the vertical axis.

**d. Analytical** Use your graph to predict the value of $A$ when $x$ is 63. Then use the formula you generated in part **a** to calculate the value of $A$ when $x$ is 63. How do the values compare?

## H.O.T. Problems Use Higher-Order Thinking Skills

**44.** **ERROR ANALYSIS** Kristen and Chase want to find the area of the shaded region in the circle shown. Is either of them correct? Explain your reasoning.

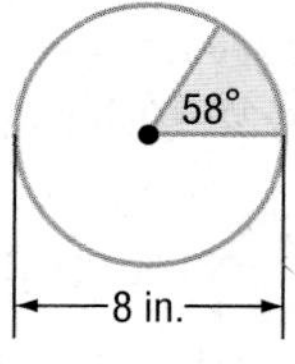


Kristen

$$A = \frac{x}{360} \cdot \pi r^2$$
$$= \frac{58}{360} \cdot \pi(8)^2$$
$$= 32.4 \text{ in}^2$$

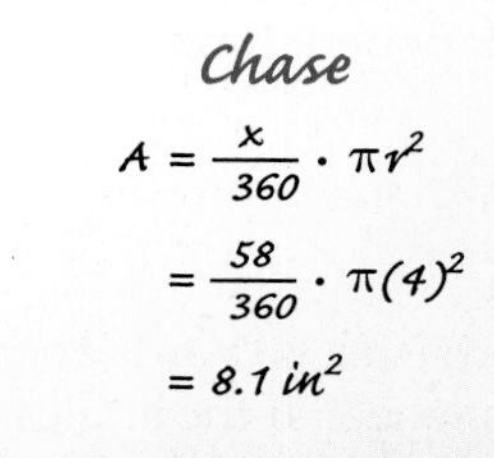

Chase

$$A = \frac{x}{360} \cdot \pi r^2$$
$$= \frac{58}{360} \cdot \pi(4)^2$$
$$= 8.1 \text{ in}^2$$

**45.** **CHALLENGE** Find the area of the shaded region. Round to the nearest tenth.

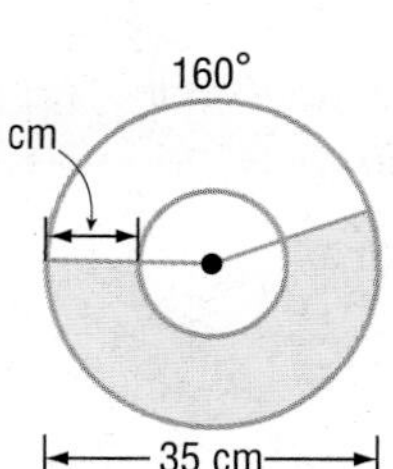


**46.** **CCSS ARGUMENTS** Refer to Exercise 43. Is the area of a sector of a circle *sometimes, always,* or *never* greater than the area of its corresponding segment?

**47.** **WRITING IN MATH** Describe two methods you could use to find the area of the shaded region of the circle. Which method do you think is more efficient? Explain your reasoning.

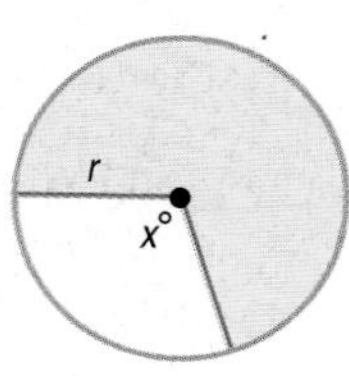


**48.** **CHALLENGE** Derive the formula for the area of a sector of a circle using the formula for arc length.

**49.** **WRITING IN MATH** If the radius of a circle doubles, will the measure of a sector of that circle double? Will it double if the arc measure of that sector doubles?

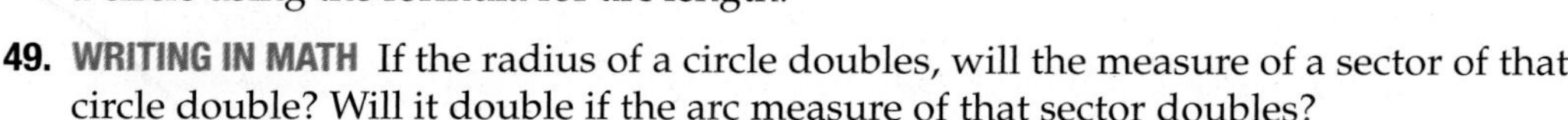

## Standardized Test Practice

**50.** What is the area of the sector?

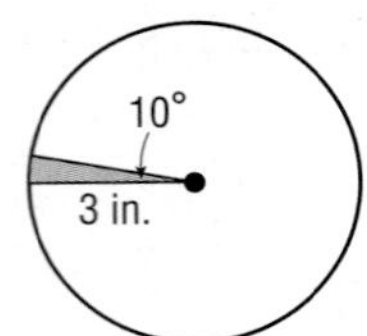


**A** $\frac{9\pi}{10}$ in$^2$

**B** $\frac{3\pi}{5}$ in$^2$

**C** $\frac{\pi}{4}$ in$^2$

**D** $\frac{\pi}{6}$ in$^2$

**51. SHORT RESPONSE** $\overleftrightarrow{MN}$ and $\overleftrightarrow{PQ}$ intersect at $T$. Find the value of $x$ for which $m\angle MTQ = 2x + 5$ and $m\angle PTM = x + 7$. What are the degree measures of $\angle MTQ$ and $\angle PTM$?

**52. ALGEBRA** Raphael bowled 4 games and had a mean score of 130. He then bowled two more games with scores of 180 and 230. What was his mean score for all 6 games?

**F** 90

**G** 155

**H** 180

**J** 185

**53. SAT/ACT** The diagonals of rectangle $ABCD$ each have a length of 56 feet. If $m\angle BAC = 42°$, what is the length of $\overline{AB}$ to the nearest tenth of a foot?

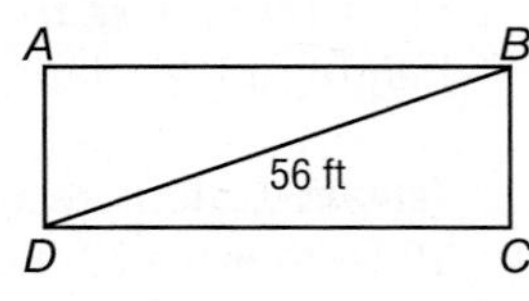


**A** 80.5

**B** 75.4

**C** 56.3

**D** 50.4

**E** 41.6

## Spiral Review

**54. AVIATION** A jet is flying northwest, and its velocity is represented by $\langle -450, 450\rangle$ miles per hour. The wind is from the west, and its velocity is represented by $\langle 100, 0\rangle$ miles per hour. (Lesson 10-7)

**a.** Find the resultant vector for the jet in component form.

**b.** Find the magnitude of the resultant.

**c.** Find the direction of the resultant.

**55. AMUSEMENT PARKS** From the top of a roller coaster, 60 yards above the ground, a rider looks down and sees the merry-go-round and the Ferris wheel. If the angles of depression are 11° and 8° respectively, how far apart are the merry-go-round and the Ferris wheel? (Lesson 10-5)

## Skills Review

**Find each measure.**

**56.** $XT$

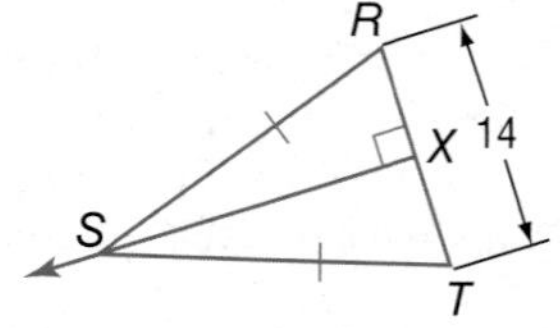


**57.** $AC$

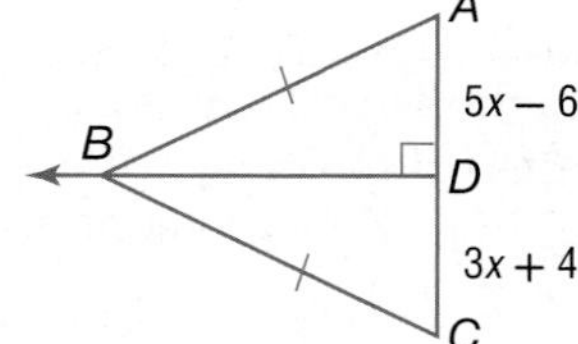


**58.** $JK$

J
9x − 5
L
6x + 7
N
K

# Study Guide and Review

## Study Guide

### KeyConcepts

**Circles and Circumference** (Lesson 11-1)

- The circumference of a circle is equal to $\pi d$ or $2\pi r$.

**Angles, Arcs, Chords, and Inscribed Angles**
(Lessons 11-2 to 11-4)

- The sum of the measures of the central angles of a circle is 360°.
- The length of an arc is proportional to the length of the circumference.
- Diameters perpendicular to chords bisect chords and intercepted arcs.
- The measure of an inscribed angle is half the measure of its intercepted arc.

**Tangents, Secants, and Angle Measures**
(Lessons 11-5 and 11-6)

- A line that is tangent to a circle intersects the circle in exactly one point and is perpendicular to a radius.
- Two segments tangent to a circle from the same exterior point are congruent.
- The measure of an angle formed by two secant lines is half the positive difference of its intercepted arcs.
- The measure of an angle formed by a secant and tangent line is half its intercepted arc.

**Special Segments and Equation of a Circle**
(Lessons 11-7 and 11-8)

- The lengths of intersecting chords in a circle can be found by using the products of the measures of the segments.
- The equation of a circle with center $(h, k)$ and radius $r$ is $(x - h)^2 - (y - k)^2 = r^2$.

### FOLDABLES® StudyOrganizer

Be sure the Key Concepts are noted in your Foldable.

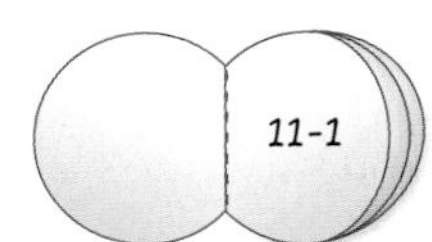


### KeyVocabulary

adjacent arcs (p. 726)
arc (p. 724)
arc length (p. 727)
center (p. 715)
central angle (p. 724)
chord (p. 715)
chord segment (p. 768)
circle (p. 715)
circumference (p. 717)
circumscribed (p. 718)
common tangent (p. 750)
compound locus (p. 780)
concentric circles (p. 716)
congruent arcs (p. 725)
diameter (p. 715)
external secant segment (p. 770)
inscribed (p. 718)
inscribed angle (p. 741)
intercepted arc (p. 741)
major arc (p.725)
minor arc (p. 725)
pi ($\pi$) (p. 717)
point of tangency (p. 750)
radius (p. 715)
secant (p. 759)
secant segment (p. 770)
semicircle (p. 725)
tangent (p. 750)

### VocabularyCheck

**State whether each sentence is *true* or *false.* If *false,* replace the underlined word or phrase to make a true sentence.**

1. Any segment with both endpoints on the circle is a <u>radius</u> of the circle.
2. A chord passing through the center of a circle is a <u>diameter</u>.
3. A <u>central angle</u> has the center as its vertex and its sides contain two radii of the circle.
4. An arc with a measure of less than 180° is a <u>major arc</u>.
5. An <u>intercepted arc</u> is an arc that has its endpoints on the sides of an inscribed angle and lies in the interior of the inscribed angle.
6. A <u>common tangent</u> is the point at which a line in the same plane as a circle intersects the circle.
7. A secant is a line that intersects a circle in exactly <u>one</u> point.
8. A secant segment is a segment of a <u>diameter</u> that has exactly one endpoint on the circle.
9. Two circles are <u>concentric</u> circles if and only if they have congruent radii.

# Study Guide and Review

## Lesson-by-Lesson Review

### 11-1 Circles and Circumference

For Exercises 10–12, refer to $\odot D$.

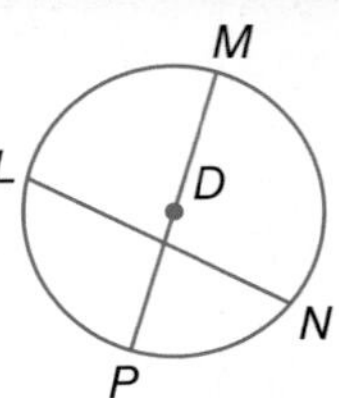


**10.** Name the circle.

**11.** Name a radius.

**12.** Name a chord that is not a diameter.

**Find the diameter and radius of a circle with the given circumference. Round to the nearest hundredth.**

**13.** $C = 43$ cm

**14.** $C = 26.7$ yd

**15.** $C = 108.5$ ft

**16.** $C = 225.9$ mm

**Example 1**

**Find the circumference of $\odot A$.**

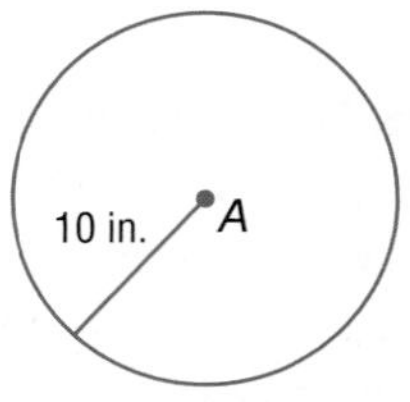


| | |
|---|---|
| $C = 2\pi r$ | Circumference formula |
| $= 2\pi(10)$ | Substitution |
| $\approx 62.83$ | Use a calculator. |

The circumference of $\odot A$ is about 62.83 inches.

### 11-2 Measuring Angles and Arcs

**Find the value of *x*.**

**17.**

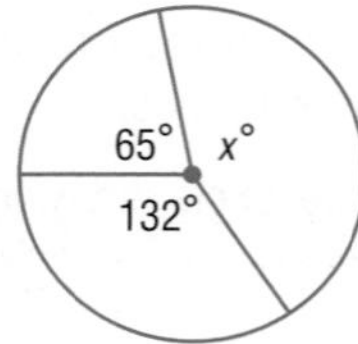


**18.**

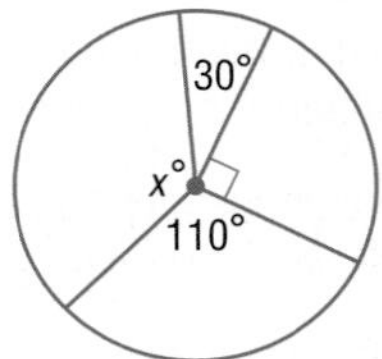


**19. MOVIES** The pie chart below represents the results of a survey taken by Mrs. Jameson regarding her students' favorite types of movies. Find each measure.

**Mrs. Jameson's Students' Favorite Types of Movies**

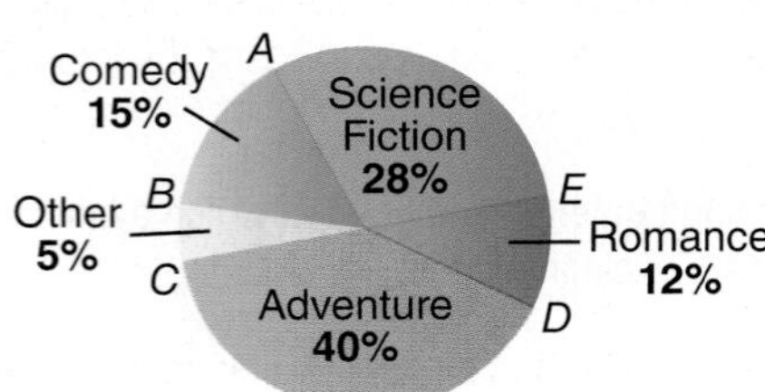


**a.** $m\widehat{AE}$

**b.** $m\widehat{BC}$

**c.** Describe the type of arc that the category Adventure represents.

**Example 2**

**Find the value of *x*.**

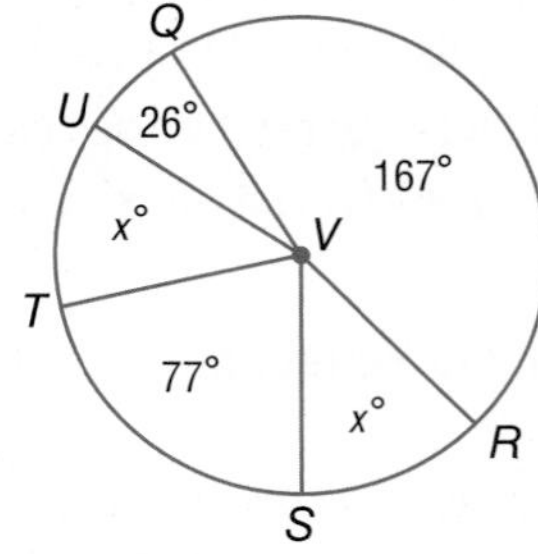


| | |
|---|---|
| $m\angle QVR + m\angle RVS + m\angle SVT + m\angle TVU + m\angle UVQ = 360$ | Sum of Central Angles |
| $167 + x + 77 + x + 26 = 360$ | Substitution |
| $270 + 2x = 360$ | Simplify. |
| $2x = 90$ | Subtract. |
| $x = 45$ | Divide. |

## 11-3 Arcs and Chords

**20.** Find the value of $x$.

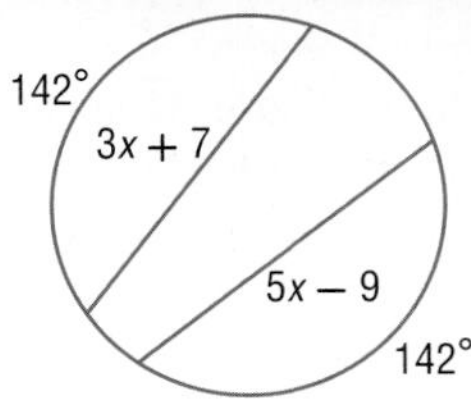


**In $\odot K$, $MN = 16$ and $m\widehat{MN} = 98$. Find each measure. Round to the nearest hundredth.**

**21.** $m\widehat{NJ}$ **22.** $LN$

**23. GARDENING** The top of the trellis shown is an arc of a circle in which $\overline{CD}$ is part of the diameter and $\overline{CD} \perp \overline{AB}$. If $\widehat{ACB}$ is about 28% of a complete circle, what is $m\widehat{CB}$?

### Example 3

**ALGEBRA** In $\odot E$, $EG = EF$. Find $AB$.

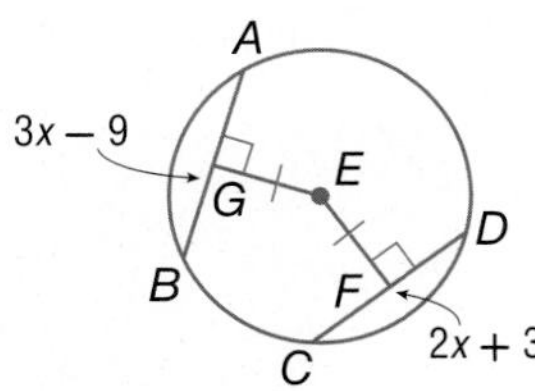


Since chords $\overline{EG}$ and $\overline{EF}$ are congruent, they are equidistant from $E$. So, $AB = CD$.

| | |
|---|---|
| $AB = CD$ | Theorem 10.5 |
| $3x - 9 = 2x + 3$ | Substitution |
| $3x = 2x + 12$ | Add. |
| $x = 12$ | Simplify. |

So, $AB = 3(12) - 9$ or 27.

## 11-4 Inscribed Angles

**Find each measure.**

**24.** $m\angle 1$

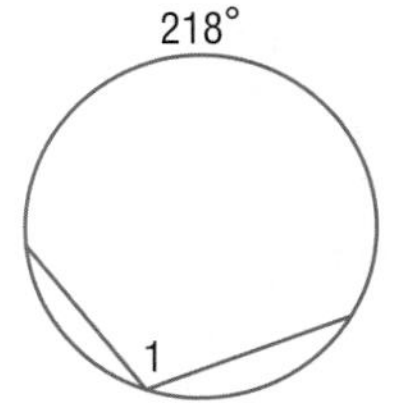


**25.** $m\widehat{GH}$

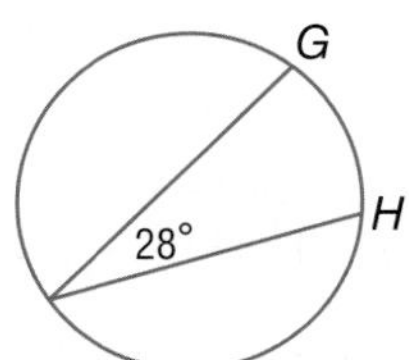


**26. MARKETING** In the logo at the right, $m\angle 1 = 42$. Find $m\angle 5$.

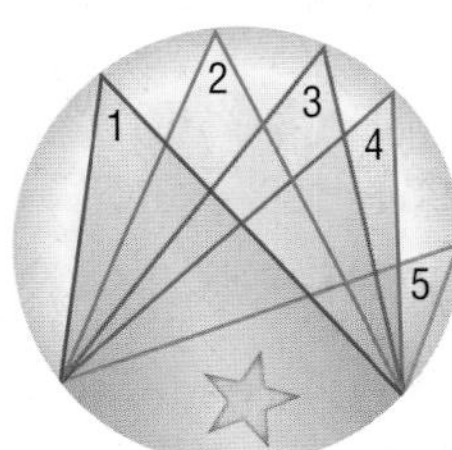


### Example 4

**Find $m\angle D$ and $m\angle B$.**

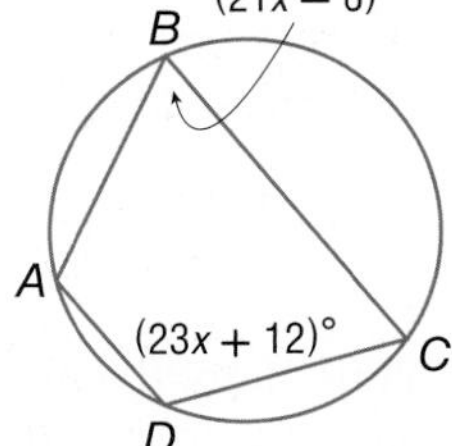


Since $ABCD$ is inscribed in a circle, opposite angles are supplementary.

| | |
|---|---|
| $m\angle D + m\angle B = 180$ | Definition of supplementary |
| $23x + 12 + 21x - 8 = 180$ | Substitution |
| $44x + 4 = 180$ | Simplify. |
| $44x = 176$ | Subtract. |
| $x = 4$ | Divide. |

So, $m\angle D = 23(4) + 12$ or 104 and $m\angle B = 21(4) - 8$ or 76.

# Study Guide and Review *Continued*

## 11-5 Tangents

**27. SCIENCE FICTION** In a story Todd is writing, instantaneous travel between a two-dimensional planet and its moon is possible when the time-traveler follows a tangent. Copy the figures below and draw all possible travel paths.

**28.** Find $x$ and $y$. Assume that segments that appear to be tangent are tangent. Round to the nearest tenth if necessary.

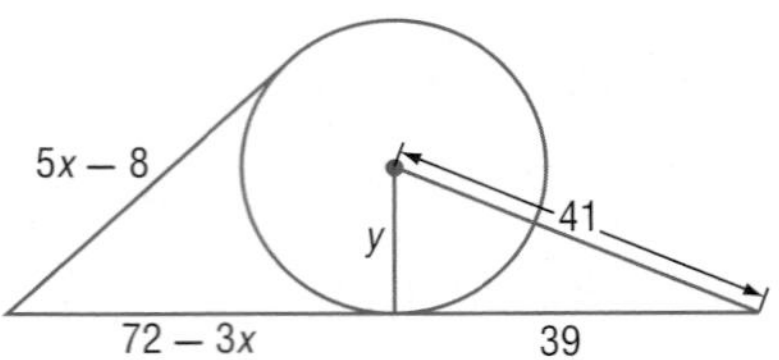


**Example 5**

In the figure, $\overline{KL}$ is tangent to $\odot M$ at $K$. Find the value of $x$.

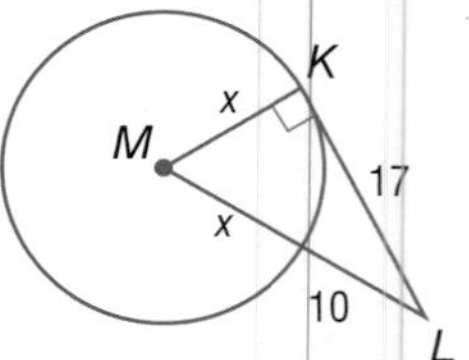


By Theorem 10.9, $\overline{MK} \perp \overline{KL}$. So, $\triangle MKL$ is a right triangle.

| | |
|---|---|
| $KM^2 + KL^2 = ML^2$ | Pythagorean Theorem |
| $x^2 + 17^2 = (x + 10)^2$ | Substitution |
| $x^2 + 289 = x^2 + 20x + 100$ | Multiply. |
| $289 = 20x + 100$ | Simplify. |
| $189 = 20x$ | Subtract. |
| $9.45 = x$ | Divide. |

## 11-6 Secants, Tangents, and Angle Measures

Find each measure. Assume that segments that appear to be tangent are tangent.

**29.** $m\angle 1$

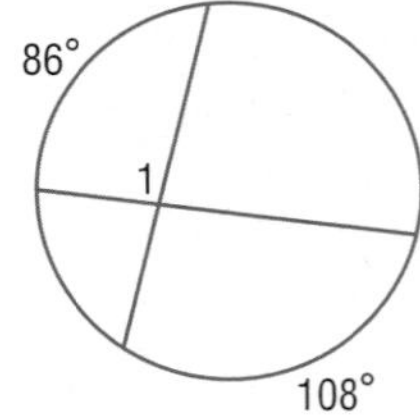


**30.** $m\widehat{AC}$

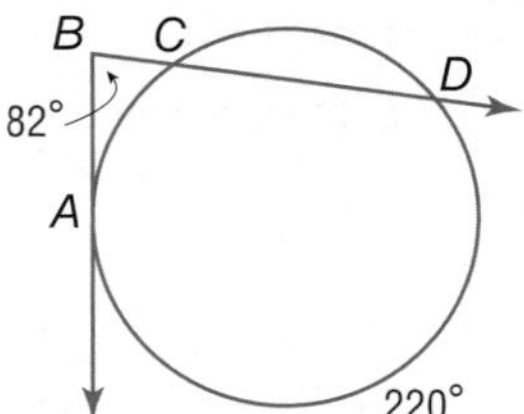


**31. PHOTOGRAPHY** Ahmed needs to take a close-up shot of an orange for his art class. He frames a shot of an orange as shown below, so that the lines of sight form tangents to the orange. If the measure of the camera's viewing angle is 34°, what is $m\widehat{ACB}$?

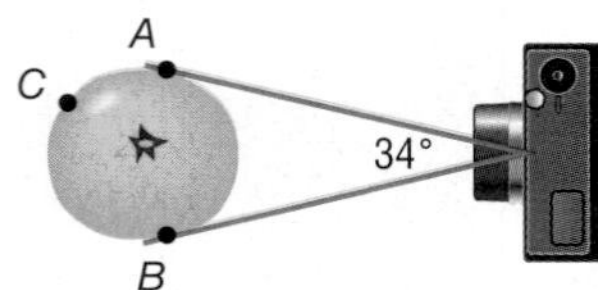


**Example 6**

Find the value of $x$.

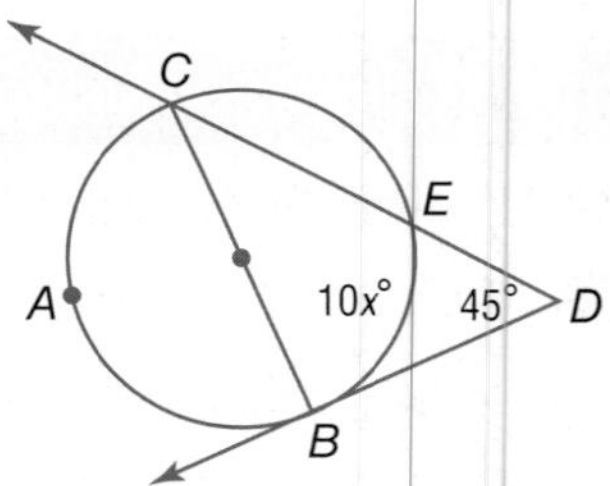


$\widehat{CAB}$ is a semicircle because $\overline{CB}$ is a diameter.

So, $m\widehat{CAB} = 180$.

| | |
|---|---|
| $m\angle D = \frac{1}{2}(m\widehat{CB} - m\widehat{EB})$ | Theorem 10.14 |
| $45 = \frac{1}{2}(180 - 10x)$ | Substitution |
| $90 = 180 - 10x$ | Multiply. |
| $-90 = -10x$ | Subtract. |
| $9 = x$ | Divide. |

# Study Guide and Review *Continued*

## 11-7 Special Segments in a Circle

**Find *x*. Assume that segments that appear to be tangent are tangent.**

**32.**

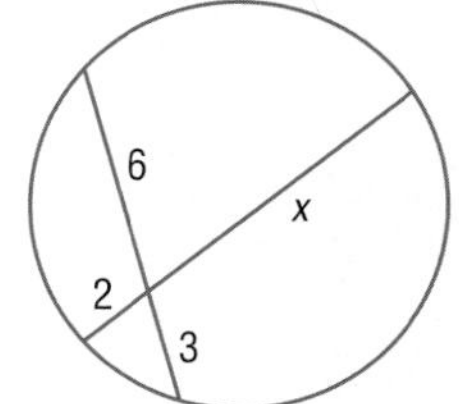


**33.**

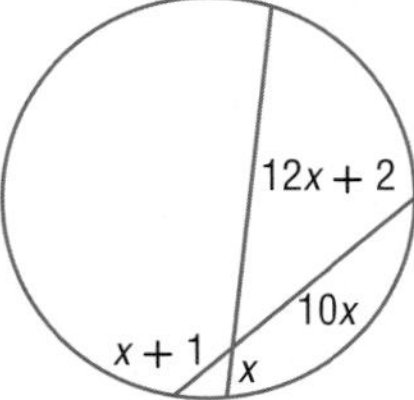


**34. ARCHAEOLOGY** While digging a hole to plant a tree, Henry found a piece of a broken saucer. What was the circumference of the original saucer? Round to the nearest hundredth.

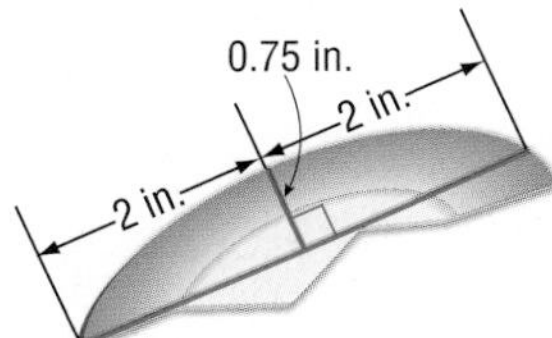


### Example 7

**Find the diameter of circle *M*.**

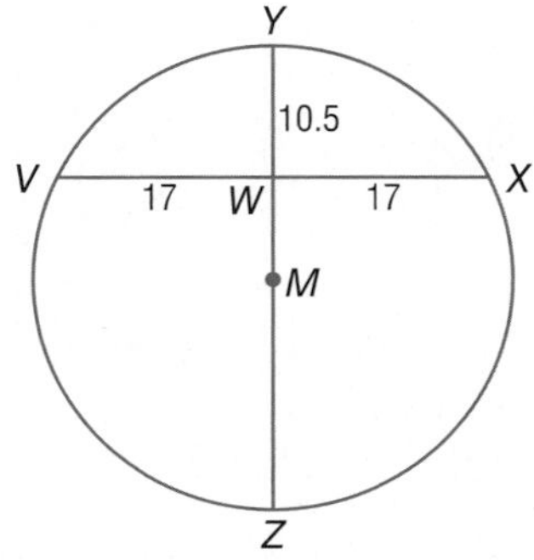


| | |
|---|---|
| $VW \cdot WX = YW \cdot WZ$ | Theorem 11.14 |
| $17 \cdot 17 = 10.5 \cdot WZ$ | Substitution |
| $289 = 10.5 \cdot WZ$ | Simplify. |
| $27.5 \approx WZ$ | Divide each side by 10.5. |
| $YZ = YW + WZ$ | Segment Addition Postulate |
| $YZ = 10.5 + 27.5$ | Substitution |
| $YZ = 38$ | Simplify. |

## 11-8 Equations of Circles

**Write the equation of each circle.**

**35.** center at (−2, 4), radius 5

**36.** center at (1, 2), diameter 14

**37. FIREWOOD** In an outdoor training course, Kat learns a wood-chopping safety check that involves making a circle with her arm extended, to ensure she will not hit anything overhead as she chops. If her reach is 19 inches, the hatchet handle is 15 inches and her shoulder is located at the origin, what is the equation of Kat's safety circle?

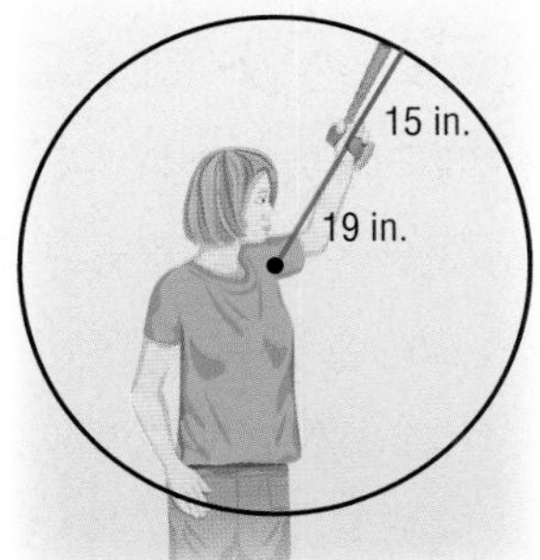


### Example 8

**Write the equation of the circle graphed below.**

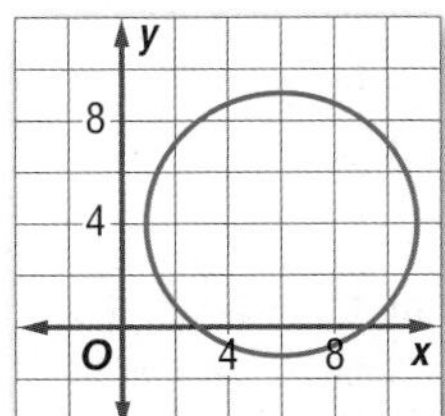


The center is at (6, 4) and the radius is 5.

| | |
|---|---|
| $(x - h)^2 + (y - k)^2 = r^2$ | Equation of a circle |
| $(x - 6)^2 + (y - 4)^2 = 5^2$ | $(h, k) = (6, 4)$ and $r = 5$ |
| $(x - 6)^2 + (y - 4)^2 = 25$ | Simplify. |

## 11-9 Areas of Circles and Sectors

Find the area of each shaded sector. Round to the nearest tenth.

**38.**

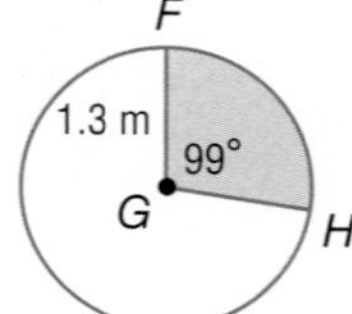


**39.**

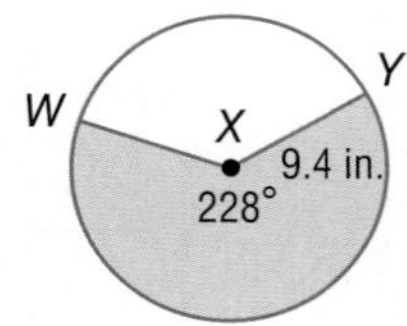


**40.** **BICYCLES** A bicycle tire decoration covers $\frac{1}{9}$ of the circle formed by the tire. If the tire has a diameter of 26 inches, what is the area of the decoration?

**41.** **PIZZA** Charlie and Kris ordered a 16-inch pizza and cut the pizza into 12 slices.

**a.** If Charlie ate 3 pieces, what area of the pizza did he eat?

**b.** If Kris ate 2 pieces, what area of the pizza did she eat?

**c.** What is the area of leftover pizza?

### Example 9

Find the area of the shaded sector. Round to the nearest tenth.

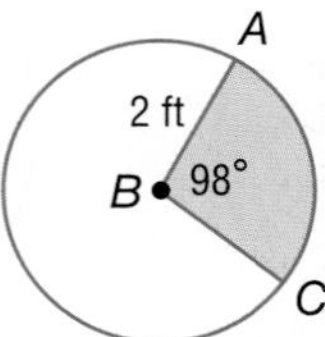


$A = \frac{x}{360} \cdot \pi r^2$ Area of a sector

$= \frac{98}{360} \cdot \pi(2)^2$ Substitution

$\approx 3.4 \text{ ft}^2$ Simplify.

CHAPTER 11

# Practice Test

1. **POOLS** Amanda's family has a swimming pool that is 4 feet deep in their backyard. If the diameter of the pool is 25 feet, what is the circumference of the pool to the nearest foot?

2. Find the exact circumference of the circle below.

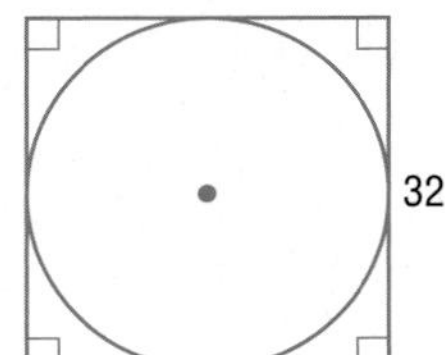


**Find the value of $x$.**

3.

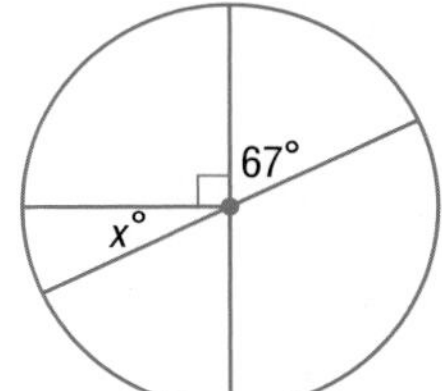


4.

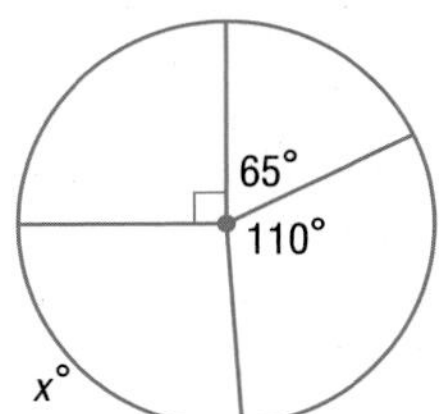


5.

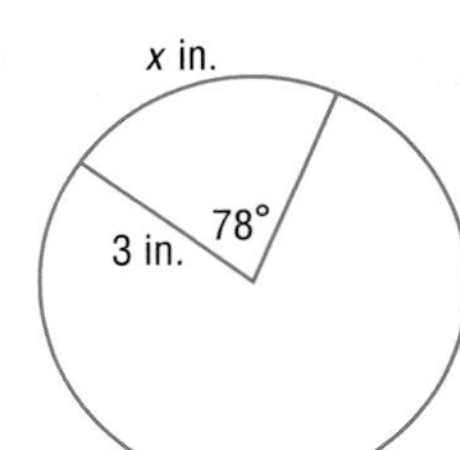


6.

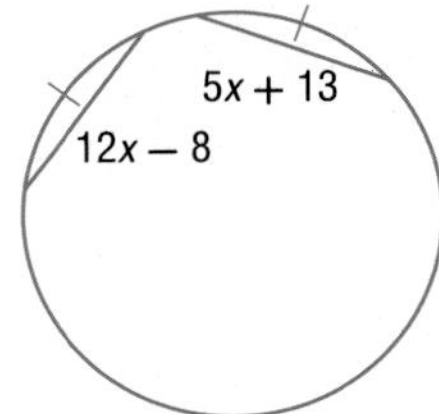


7. **MULTIPLE CHOICE** What is $ED$?

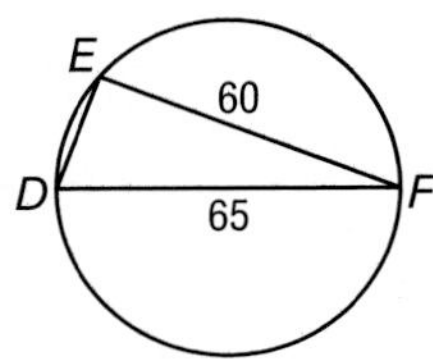


**A** 15 **C** 88.5

**B** 25 **D** not enough information

8. Find $x$ if $\odot M \cong \odot N$.

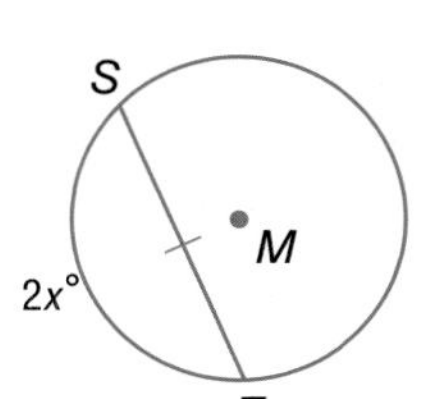


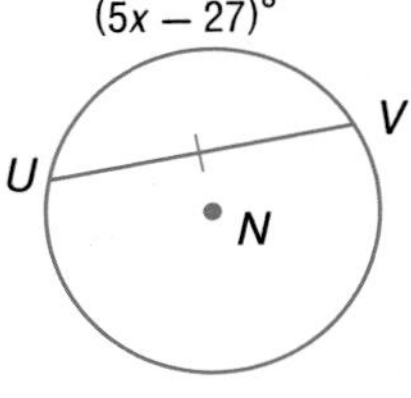


9. **MULTIPLE CHOICE** How many points are shared by concentric circles?

**F** 0 **H** 2

**G** 1 **J** infinite points

10. Determine whether $\overline{FG}$ is tangent to $\odot E$. Justify your answer.

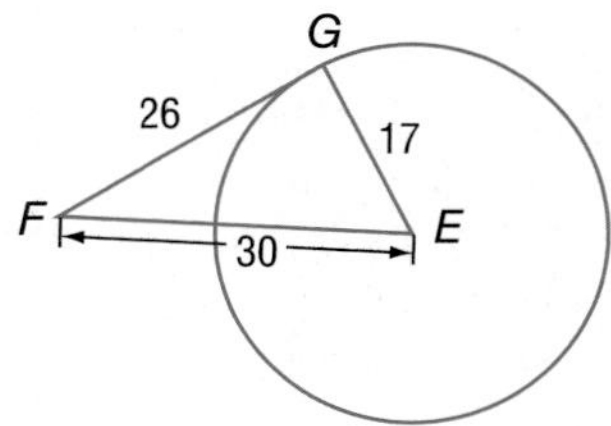


11. **MULTIPLE CHOICE** Which of the figures below shows a polygon circumscribed about a circle?

**A**

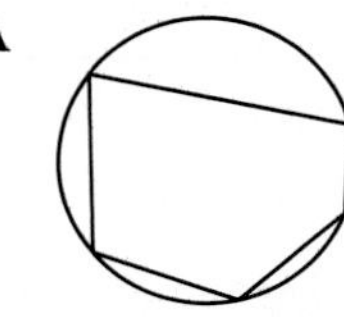

**C**

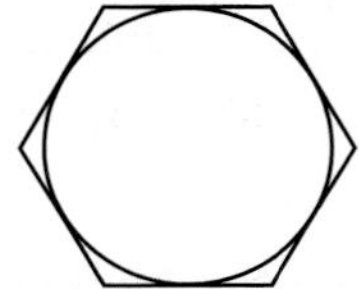

**B**

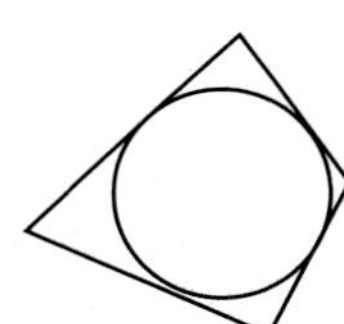

**D**

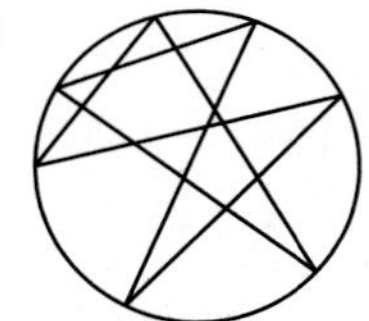

12. Find the perimeter of the triangle at the right. Assume that segments that appear to be tangent are tangent.

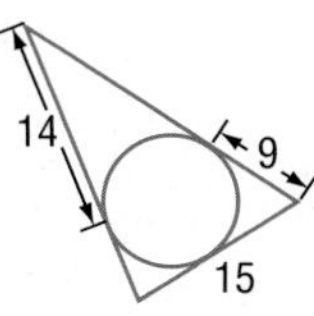


**Find each measure.**

13. $m\angle T$

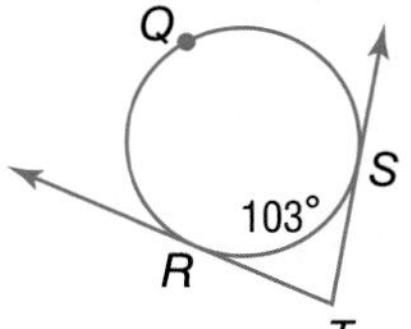


14. $x$

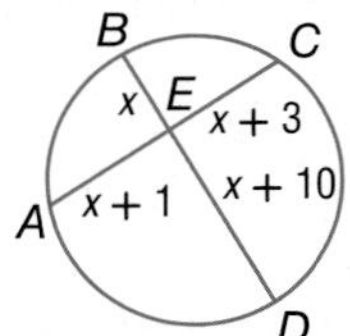


15. **FLOWERS** Hannah wants to encircle a tree trunk with a flower bed. If the center of the tree trunk is the origin and Hannah wants the flower bed to extend to 3 feet from the center of the tree, what is the equation that would represent the flower bed?

# Preparing for Standardized Tests

## Properties of Circles

A circle is a unique shape in which the angles, arcs, and segments intersecting the circle have special properties and relationships. You should be able to identify the parts of a circle, write the equation of a circle, and solve for arc, angle, and segment measures in a circle.

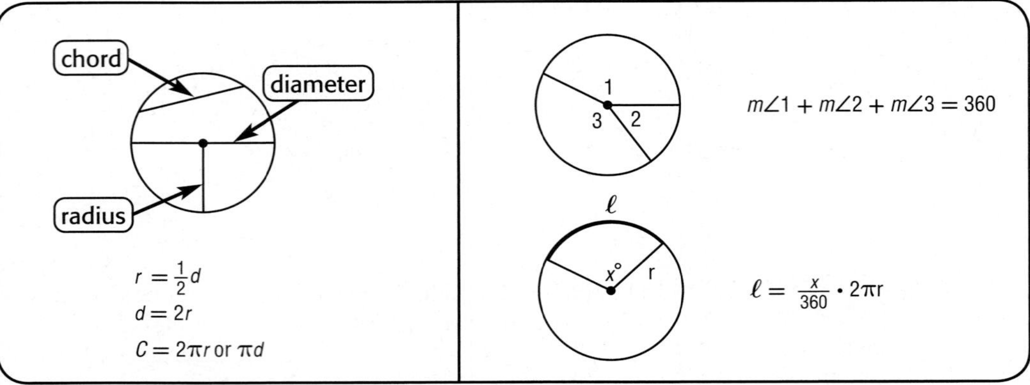


### Strategies for Applying the Properties of Circles

**Step 1**

Review the parts of a circle and their relationships.

- Some key parts include: **radius, diameter, arc, chord, tangent, secant**
- Study the key theorems and the properties of circles as well as the relationships between the parts of a circle.

**Step 2**

Read the problem statement and study any figure you are given carefully.

- Determine what you are being asked to find.
- Fill in any information in the figure that you can.
- Determine which theorems or properties apply to the problem situation.

**Step 3**

Solve the problem and check your answer.

- Apply the theorems or properties to solve the problem.
- Check your answer to be sure it makes sense.

## Standardized Test Example

**Read the problem. Identify what you need to know. Then use the information in the problem to solve.**

Solve for $x$ in the figure.

**A** 2 **B** 3 **C** 4 **D** 6

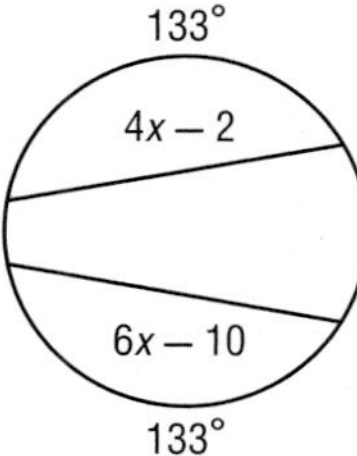


Read the problem statement and study the figure carefully. You are given a circle with two chords that correspond to congruent minor arcs. One important property of circles is that two chords are congruent if and only if their corresponding minor arcs are congruent. You can use this property to set up and solve an equation for $x$.

| | |
|---|---|
| $4x - 2 = 6x - 10$ | Definition of Congruent Segments |
| $4x - 6x = -10 + 2$ | Subtract. |
| $-2x = -8$ | Simplify. |
| $\frac{-2x}{-2} = \frac{-8}{-2}$ | Divide each side by $-2$. |
| $x = 4$ | Simplify. |

So, the value of $x$ is 4. The answer is C. You can check your answer by substituting 4 into each expression and making sure both chords have the same length.

## Exercises

**Read each problem. Identify what you need to know. Then use the information in the problem to solve.**

1. Solve for $x$ in the figure below.

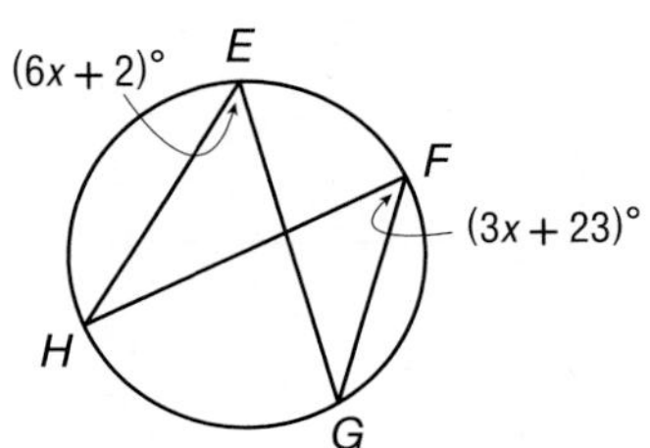


**A** 4 **B** 5 **C** 6 **D** 7

2. Triangle $RST$ is circumscribed about the circle below. What is the perimeter of the triangle?

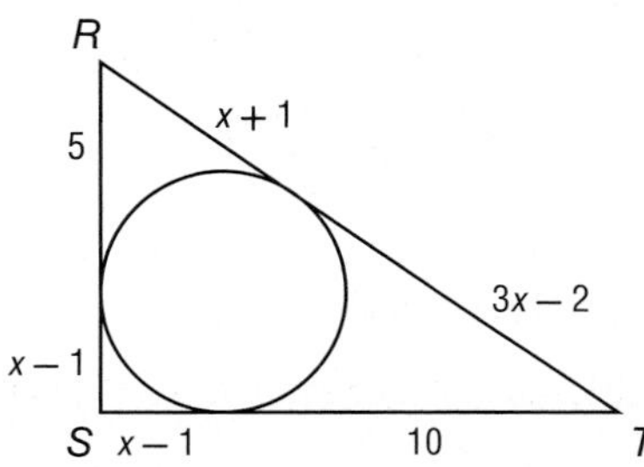


**F** 33 units **G** 36 units **H** 37 units **J** 40 units

CHAPTER 11

# Standardized Test Practice

## Cumulative, Chapters 1 through 11

### Multiple Choice

**Read each question. Then fill in the correct answer on the answer document provided by your teacher or on a sheet of paper.**

1. If $ABCD$ is a rhombus, and $m\angle ABC = 70°$, what is $m\angle 1$?

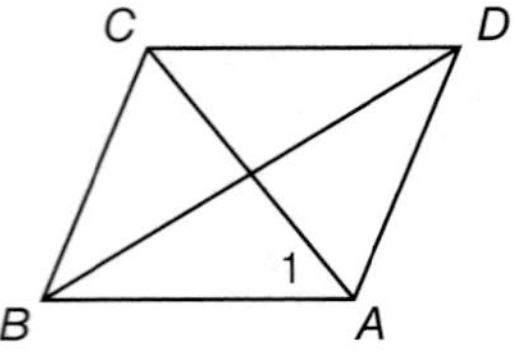


A 45°  C 70°

B 55°  D 125°

2. Karen argues that if you live in Greensboro, North Carolina, then you live in Guilford County. Which assumption would you need to make to form an indirect proof of this claim?

F Suppose someone lives in Guilford County, but not in Greensboro.

G Suppose someone lives in Greensboro, but not in Guilford County.

H Suppose someone lives in Greensboro and in Guilford County.

J Suppose someone lives in Guilford County and in Greensboro.

3. What is the value of $x$ in the figure?

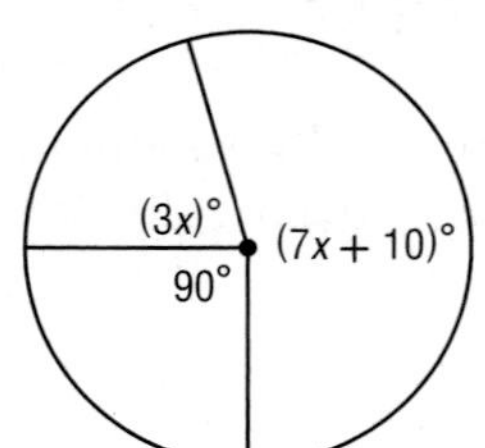


A 19  C 26

B 23  D 28

**Test-Taking Tip**

Question 3 Use the properties of circles to set up and solve an equation to find $x$.

4. Given $a \parallel b$, find $m\angle 1$.

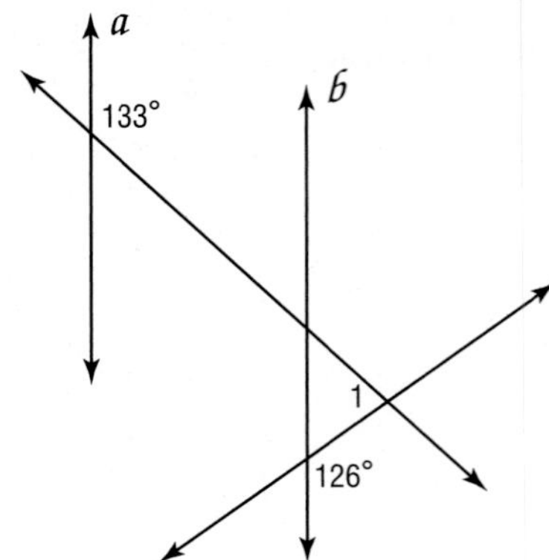


F 47°

G 54°

H 79°

J 101°

5. Which of the following conditions would *not* guarantee that a quadrilateral is a parallelogram?

A both pairs of opposite sides congruent

B both pairs of opposite angles congruent

C diagonals bisect each other

D one pair of opposite sides parallel

6. The ratio of the measures of the angles of the triangle below is 3:2:1. Which of the following is *not* an angle measure of the triangle?

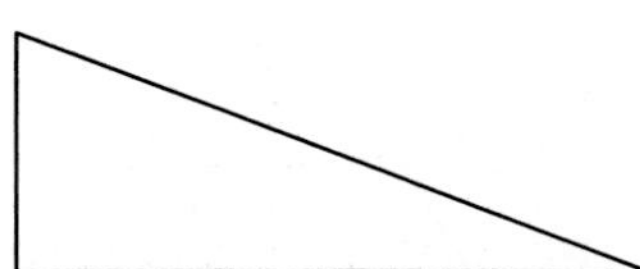

F 30°

G 45°

H 60°

J 90°

## Short Response/Gridded Response

**Record your answers on the answer sheet provided by your teacher or on a sheet of paper.**

**7.** Copy the circles below on a sheet of paper and draw the common tangents, if any exist.

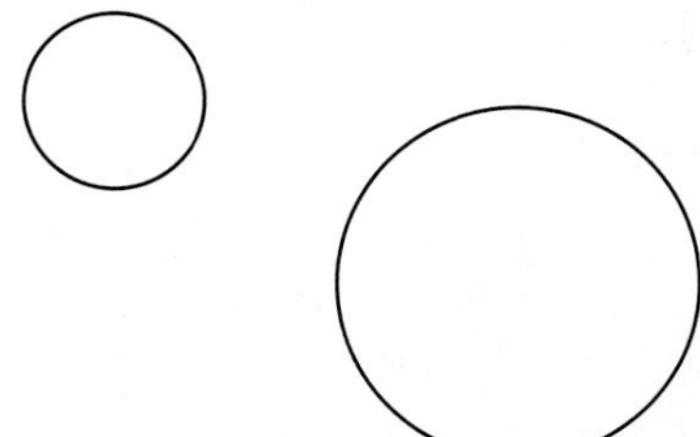

**8. GRIDDED RESPONSE** A square with 5-centimeter sides is inscribed in a circle. What is the circumference of the circle? Round your answer to the nearest tenth of a centimeter.

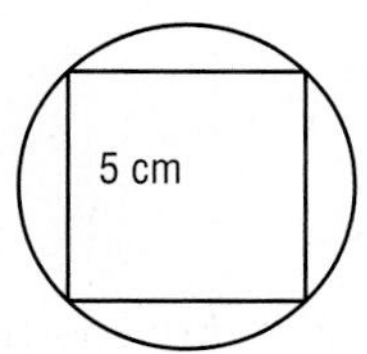


**9.** Solve for $x$ in the figure. Show your work.

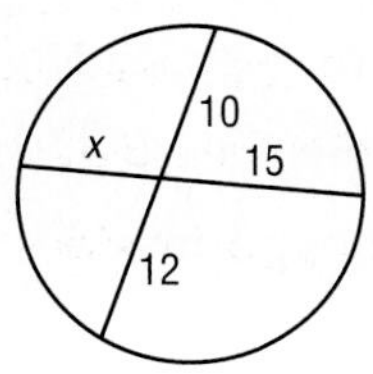


**10. GRIDDED RESPONSE** What is the perimeter of the right triangle below? Round your answer to the nearest tenth if necessary.

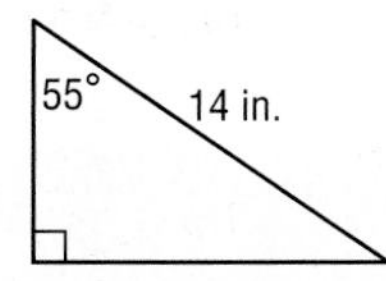


**11. GRIDDED RESPONSE** Solve for $x$ in the figure below.

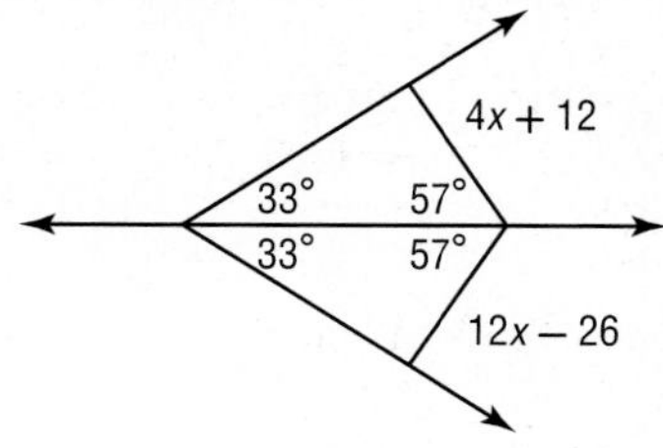


**12.** What is the length of $\overline{EF}$?

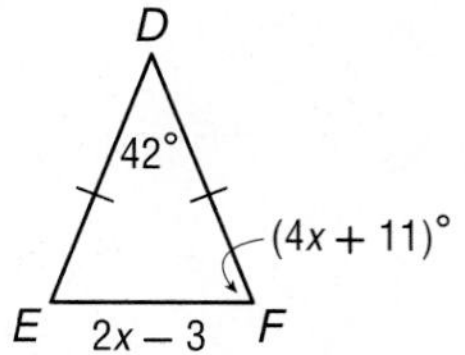


## Extended Response

**Record your answers on a sheet of paper. Show your work.**

**13.** Use the circle shown to answer each question.

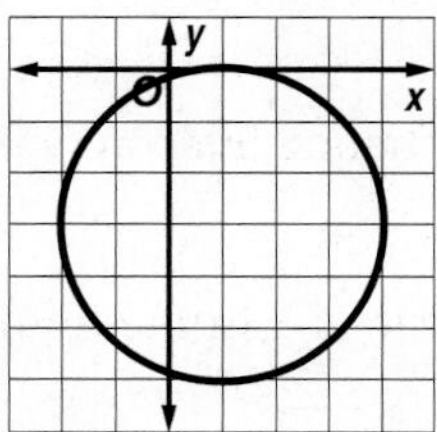


**a.** What is the center of the circle?

**b.** What is the radius of the circle?

**c.** Write an equation for the circle.

### Need ExtraHelp?

| If you missed Question... | 1 | 2 | 3 | 4 | 5 | 6 | 7 | 8 | 9 | 10 | 11 | 12 | 13 |
|---|---|---|---|---|---|---|---|---|---|---|---|---|---|
| Go to Lesson... | 8-5 | 7-4 | 11-2 | 5-5 | 8-3 | 10-3 | 11-5 | 11-1 | 11-7 | 10-4 | 7-1 | 6-5 | 11-8 |

CHAPTER 12

# Extending Surface Area and Volume

## Then

You identified and named three-dimensional figures and calculated surface area and volume for some common solids.

## Now

In this chapter, you will:

- Find lateral areas, surface areas, and volumes of various solid figures.
- Investigate Euclidean and spherical geometries.
- Use properties of similar solids.

## Why?

**ARCHITECTURE** Architects use different types of solids to create designs that are both interesting and functional.

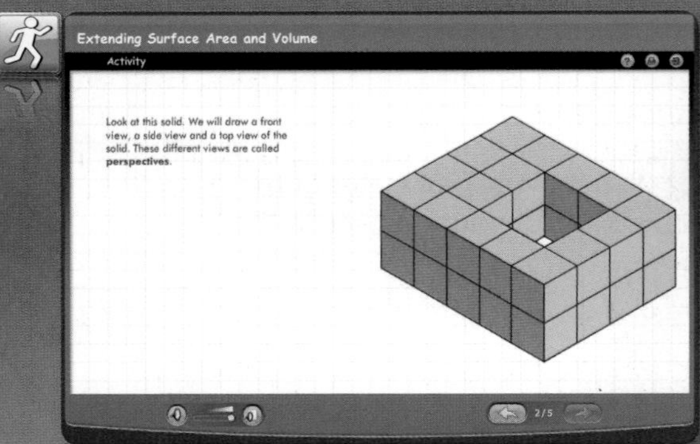


connectED.mcgraw-hill.com Your Digital Math Portal

Animation  Vocabulary  eGlossary  Personal Tutor  Virtual Manipulatives  Graphing Calculator  Audio  Foldables  Self-Check Practice  Worksheets 

# Get Ready for the Chapter

**Diagnose** Readiness | You have two options for checking prerequisite skills.

**1 Textbook Option** Take the Quick Check below. Refer to the Quick Review for help.

## QuickCheck

Determine whether each statement about the figure in Example 1 is *true*, *false*, or *cannot be determined.*

1. ▱$ABCD$ lies in plane $\mathcal{M}$.
2. ▱$CDHG$ lies in plane $\mathcal{N}$.
3. $\overline{AB}$ lies in plane $\mathcal{M}$.
4. $\overline{HG}$ lies in plane $\mathcal{N}$.
5. $\overline{AE} \perp$ to plane $\mathcal{M}$.
6. $\overline{DC} \parallel$ line $\ell$.

Find the area of each figure. Round to the nearest tenth if necessary.

7.

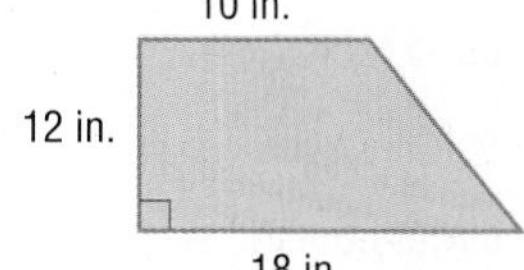


8.

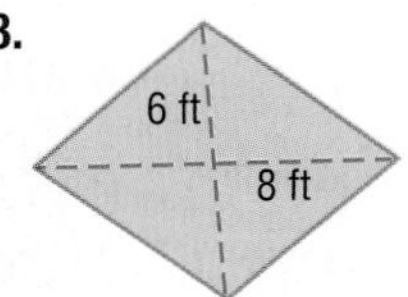


9. **CRAFTS** A seamstress wants to cover a kite frame with cloth. If the length of one diagonal is 16 inches and the other diagonal is 22 inches, find the area of the surface of the kite.

Find the value of the variable in each equation.

10. $a^2 + 40^2 = 41^2$
11. $8^2 + b^2 = 17^2$
12. $a^2 + 6^2 = (7\sqrt{3})^2$

## QuickReview

**Example 1**

In the figure, $\overline{AD} \perp \ell$ and $ABCDEFGH$ is a cube. Determine whether plane $\mathcal{M} \perp$ plane $\mathcal{N}$.

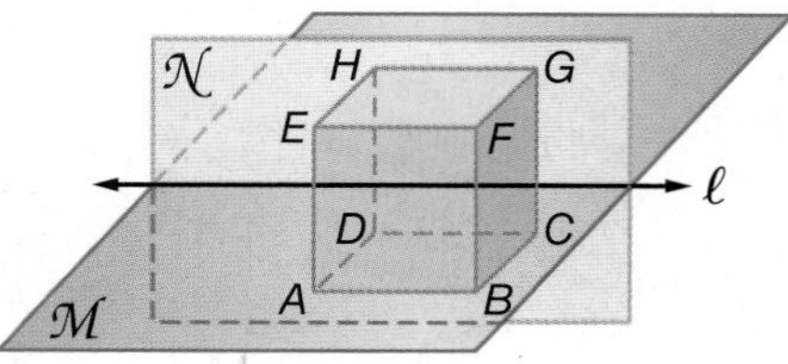


Plane $\mathcal{M} \perp$ plane $\mathcal{N}$ cannot be determined from the given information.

**Example 2**

Find the area of the figure. Round to the nearest tenth if necessary.

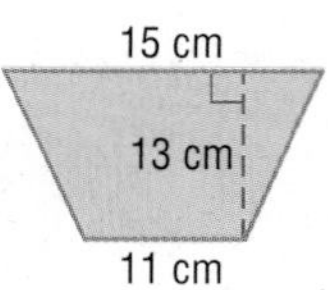


| | |
|---|---|
| $A = \frac{1}{2}h(b_1 + b_2)$ | Area of a trapezoid |
| $= \frac{1}{2}(13)(15 + 11)$ | Substitution |
| $= \frac{1}{2}(13)(26)$ | Simplify. |
| $= 169$ | Multiply. |

The area of the trapezoid is 169 $\text{cm}^2$.

**Example 3**

Find the value of the variable in $8^2 + 7^2 = c^2$.

| | |
|---|---|
| $c^2 = 8^2 + 7^2$ | Original equation |
| $c^2 = 64 + 49$ | Evaluate the exponents. |
| $c^2 = 113$ | Simplify. |
| $c = \pm\sqrt{113}$ | Take the square root of each side. |

**2 Online Option** Take an online self-check Chapter Readiness Quiz at connectED.mcgraw-hill.com.

# Get Started on the Chapter

You will learn several new concepts, skills, and vocabulary terms as you study Chapter 12. To get ready, identify important terms and organize your resources. You may refer to Chapter 0 to review prerequisite skills.

## FOLDABLES® StudyOrganizer

**Surface Area and Volume** Make this Foldable to help you organize your Chapter 12 notes about surface area and volume. Begin with one sheet of notebook paper.

**1** **Fold** the paper in half.

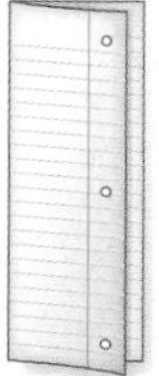

**2** **Fold** the paper again, two inches from the top.

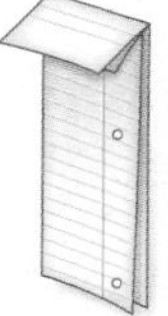

**3** **Unfold** the paper.

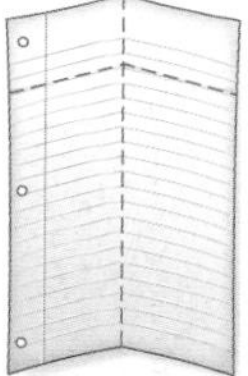

**4** **Label** as shown.

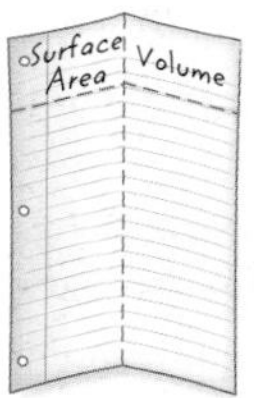


## NewVocabulary

| English | | Español |
|---|---|---|
| right solid | p. 805 | sólido recto |
| oblique solid | p. 806 | sólido oblicuo |
| isometric view | p. 807 | vista isométrica |
| cross section | p. 808 | sección transversal |
| lateral face | p. 814 | cara lateral |
| lateral edge | p. 814 | arista lateral |
| altitude | p. 814 | altura |
| lateral area | p. 814 | área lateral |
| axis | p. 816 | eje |
| regular pyramid | p. 822 | pirámide regular |
| slant height | p. 822 | altura oblicua |
| right cone | p. 824 | cono recto |
| oblique cone | p. 824 | cono oblicuo |
| great circle | p. 849 | círculo mayor |
| Euclidean geometry | p. 857 | geometría euclidiana |
| spherical geometry | p. 857 | geometría esférica |
| similar solids | p. 864 | sólidos semejantes |
| congruent solids | p. 864 | sólidos congruentes |

## ReviewVocabulary

regular polyhedron poliedro regular a polyhedron in which all of the faces are regular congruent polygons

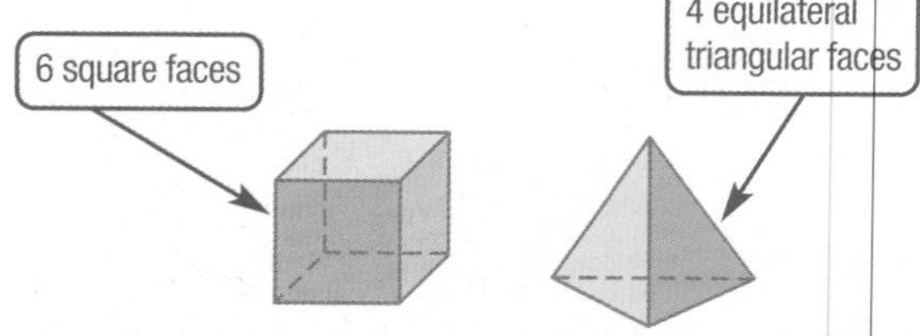

EXPLORE

# 12-1 Geometry Lab: Solids Formed by Translation

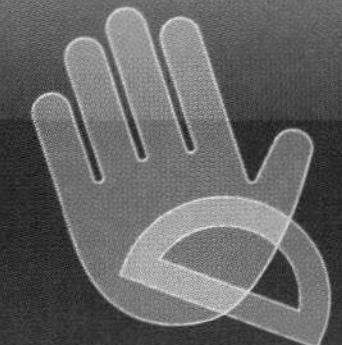

We can relate some three-dimensional solids to two-dimensional figures with which we are already familiar. Some three-dimensional solids can be formed by translating a two-dimensional figure along a vector.

A **right solid** has base(s) that are perpendicular to the edges connecting them or connecting the base and the vertex of the solid. Some right solids are formed by translating a two-dimensional figure along a vector that is perpendicular to the plane in which the figure lies.

## Activity 1

**Identify and sketch the solid formed by translating a horizontal rectangle vertically.**

To help visualize the solid formed, let a playing card represent the rectangle, and lay it flat on a table so that it is horizontal. To show the translation of the rectangle vertically, stack other cards neatly, one by one, on top of the first.

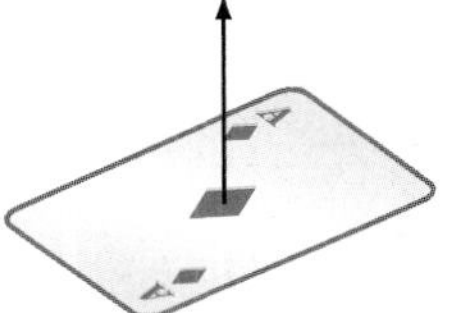

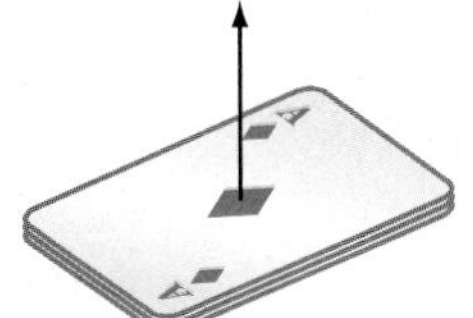

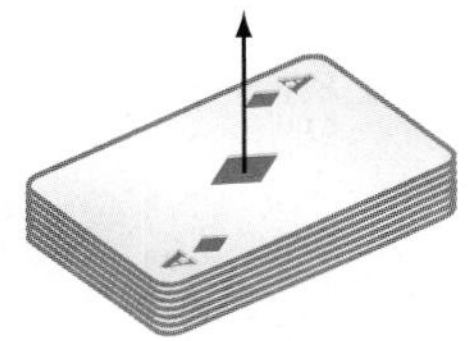

Notice that the solid formed is a right rectangular prism, which has a rectangular base, a translated copy of this base on the opposite side parallel to the base, and four congruent edges connecting the two congruent rectangles. These edges are parallel to each other but perpendicular to the bases. A sketch of the figure is shown.

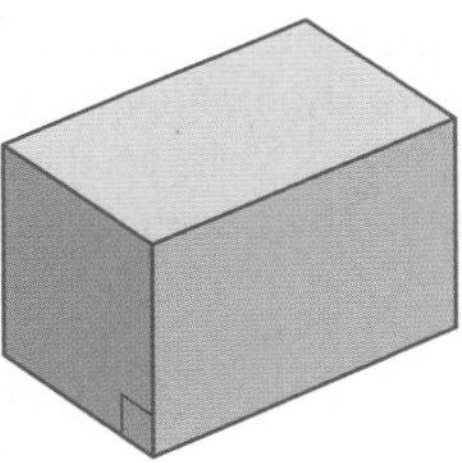

## Model and Analyze

1. Use congruent triangular tangram pieces to identify and sketch the solid formed by translating a horizontal triangle vertically.

2. Use the coins from a roll of quarters to identify and sketch the solid formed by translating a horizontal circle vertically.

**Identify and sketch the solid formed by translating a vertical two-dimensional figure horizontally.**

3. rectangle

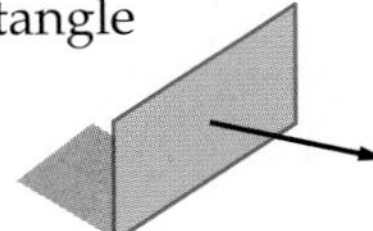

4. triangle

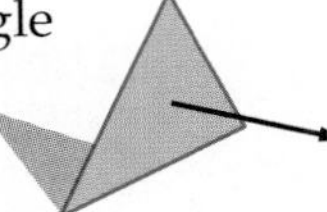

5. circle

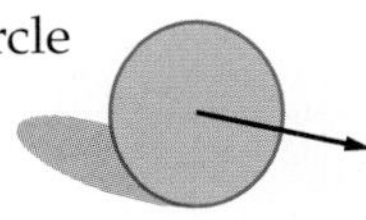

6. **REASONING** Are the solids formed in Exercises 3, 4, and 5 right solids? Explain your reasoning.

(continued on the next page)

## Geometry Lab

# Solids Formed by Translation *Continued*

An **oblique solid** has base(s) that are not perpendicular to the edges connecting the two bases or vertex. An oblique solid can be formed by translating a two-dimensional figure along an oblique vector that is neither parallel nor perpendicular to the plane in which the two-dimensional figure lies.

### Activity 2

**Identify and sketch the solid formed by translating a horizontal rectangle along an oblique vector.**

Let a playing card represent the rectangle. Lay it flat on a table so that it is horizontal. To show the translation of the rectangle along an oblique line, stack other cards one by one on top of the first so that the cards are shifted from the center of the previous card the same amount each time.

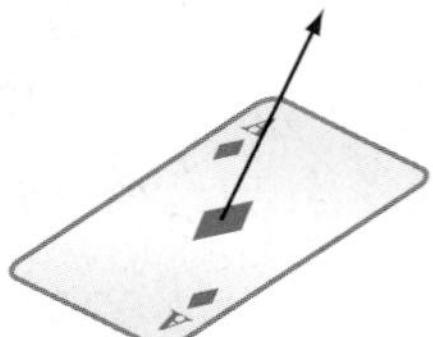

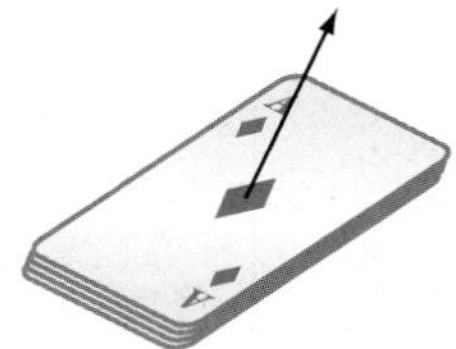

The solid formed is an oblique rectangular prism, which has a rectangular base, a translated copy of this base on the opposite side parallel to the base, and four congruent edges connecting the two congruent rectangles. These edges are parallel to each other but oblique to the bases. A sketch of the figure is shown.

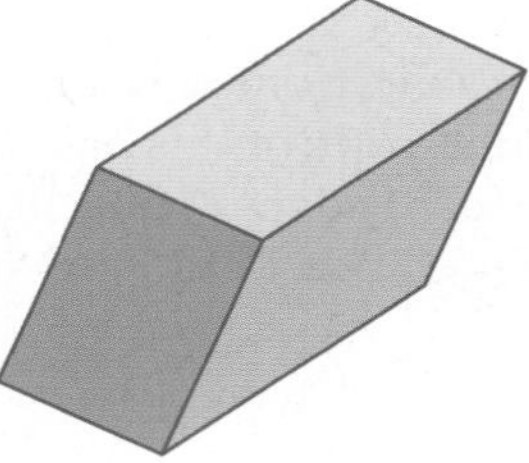

### Model and Analyze

**Identify and sketch the solid formed by translating each vertical two-dimensional figure along an oblique vector. Use concrete models if needed.**

**7.** triangle

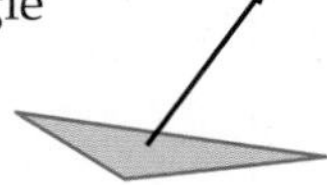

**8.** circle

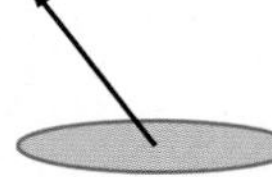

**Identify each solid as *right*, *oblique*, or *neither*.**

**9.**

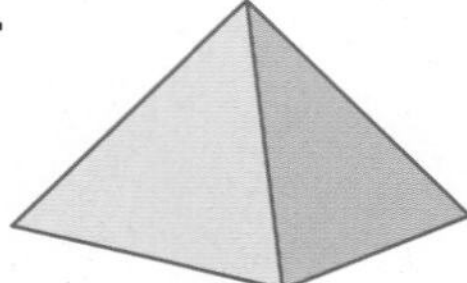

**10.**

**11.**

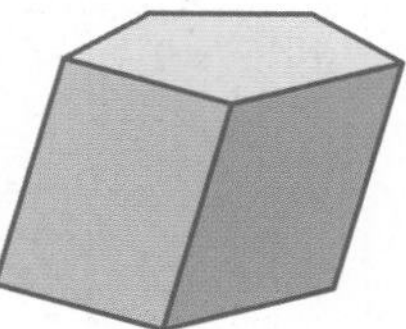

**12. REASONING** Can a pyramid with a square base be formed by translating the base vertically? Explain your reasoning.

# LESSON 12-1 Representations of Three-Dimensional Figures

**Then**

- You identified parallel planes and intersecting planes in three-dimensional figures.

**Now**

1. Draw isometric views of three-dimensional figures.
2. Investigate cross sections of three-dimensional figures.

**Why?**

- Video game programmers use technology to make the gaming environments appear three-dimensional. As players move in the various video game worlds, objects are realistically shown from different perspectives.

**NewVocabulary**
isometric view
cross section

**Common Core State Standards**

**Content Standards**
G.GMD.4 Identify the shapes of two-dimensional cross-sections of three-dimensional objects, and identify three-dimensional objects generated by rotations of two-dimensional objects.

**Mathematical Practices**
5 Use appropriate tools strategically.
1 Make sense of problems and persevere in solving them.

## 1 Draw Isometric Views

In video games, three-dimensional figures are represented on a two-dimensional screen. You can use isometric dot paper to draw **isometric views**, or corner views, of three-dimensional geometric solids on two-dimensional paper.

front view

isometric view

### Example 1 Use Dimensions of a Solid to Sketch a Solid

**Use isometric dot paper to sketch a triangular prism 3 units high with two sides of the base that are 2 units long and 4 units long.**

**Step 1**

Mark the corner of the solid. Draw 3 units down, 2 units to the left, and 4 units to the right. Then draw a triangle for the top of the solid.

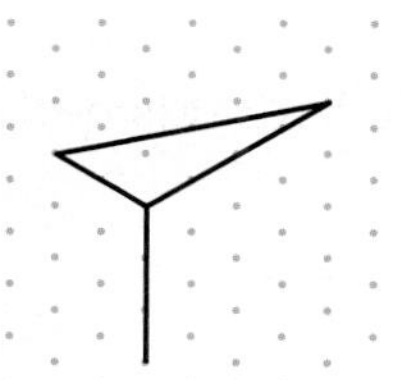

**Step 2**

Draw segments 3 units down from each vertex for the vertical edges. Connect the appropriate vertices using a dashed line for the hidden edge.

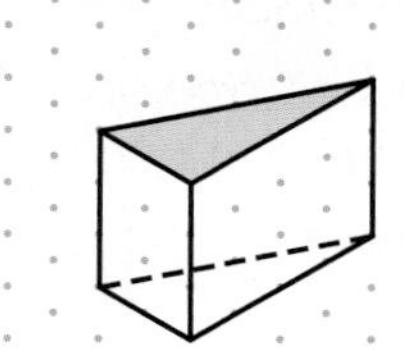

**GuidedPractice**

1. Use isometric dot paper to sketch a rectangular prism 1 unit high, 5 units long, and 4 units wide.

Recall that an *orthographic drawing* shows the top, left, front, and right views of a solid. You can use an orthographic drawing to draw an isometric view of a three-dimensional figure. The top, front, and right views of a cube are shown at the right.

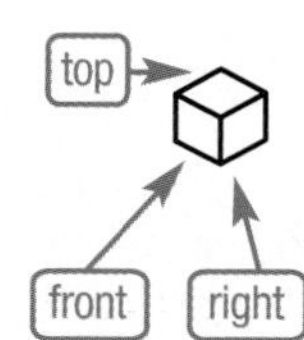

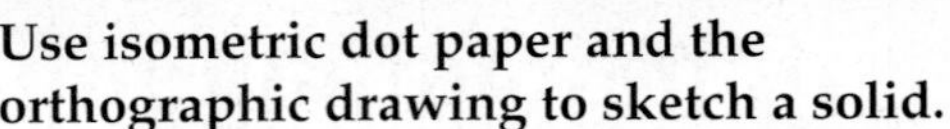

## Example 2 Use an Orthographic Drawing to Sketch a Solid

**Use isometric dot paper and the orthographic drawing to sketch a solid.**

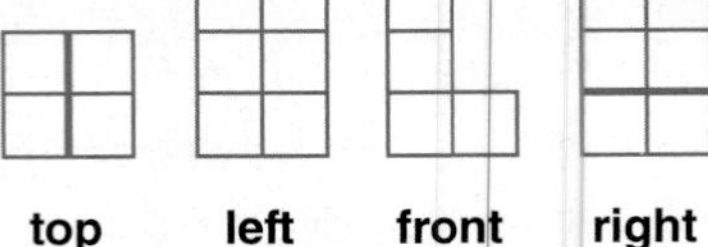

top view | left view | front view | right view

- top view: There are two rows and two columns. The dark segments indicate that there are different heights.
- left view: The figure is 3 units high on the left.
- front view: The first column is 3 units high and the second column is 1 unit high.
- right view: The figure is 3 units high on the right. The dark segments indicate that there are breaks in the surface.

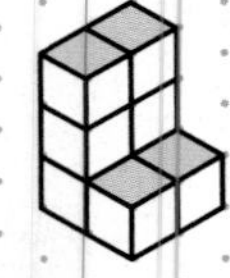

Connect the dots on the isometric dot paper to represent the edges of the solid. Shade the tops of each column.

### GuidedPractice

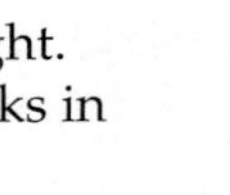

2.

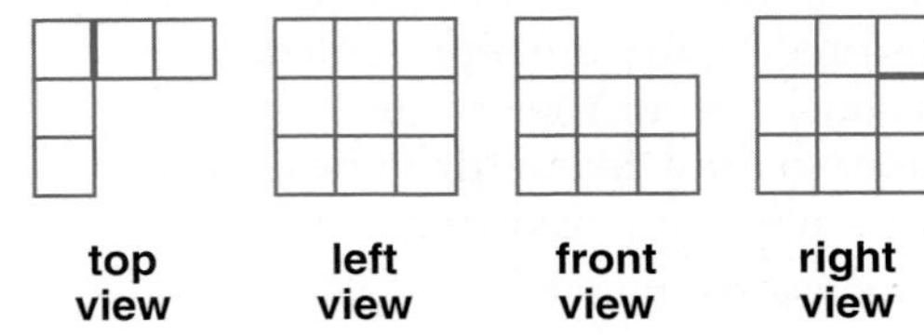

top view | left view | front view | right view

## 2 Investigate Cross Sections

A **cross section** is the intersection of a solid and a plane. The shape of the cross section formed by the intersection of a plane and a three-dimensional figure depends on the angle of the plane.

### Real-World Example 3 Identify Cross Sections of Solids

**PYRAMIDS Scientists are able to use computers to study cross sections of ancient artifacts and structures. Determine the shape of each cross section of the pyramid below.**

horizontal cut | angled cut | vertical cut

The horizontal cross section is a square. The angled cross section is a trapezoid. The vertical cross section is a triangle.

**Real-WorldLink**

The largest pyramid ever constructed is about 63 miles from Mexico City. It is 177 feet tall and its base covers an area of nearly 45 acres.

**Source:** *Guinness World Records*

### GuidedPractice

3. **CAKES** Ramona has a cake pan shaped like half of a sphere, as shown at the right. Describe the shape of the cross sections of cakes baked in this pan if they are cut horizontally and vertically.

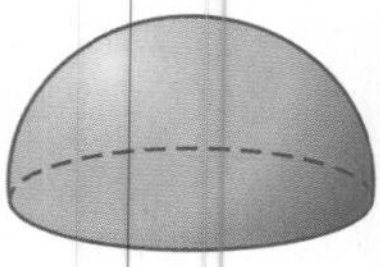

Toño Labra/age fotostock

## Check Your Understanding

**Example 1** **Use isometric dot paper to sketch each prism.**

1. triangular prism 2 units high, with two sides of the base that are 5 units long and 4 units long
2. rectangular prism 2 units high, 3 units wide, and 5 units long

**Example 2** **Use isometric dot paper and each orthographic drawing to sketch a solid.**

3.

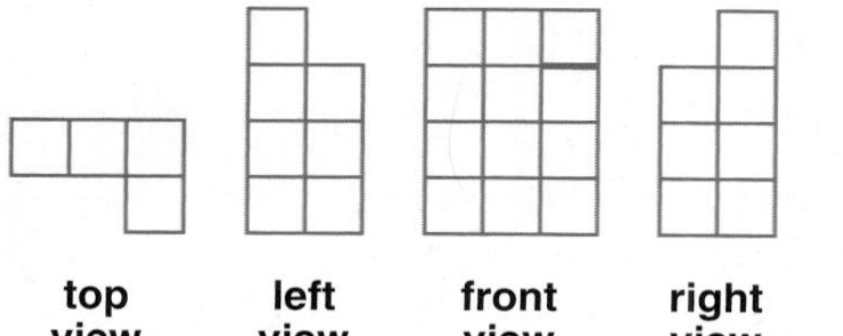

top view | left view | front view | right view

4.

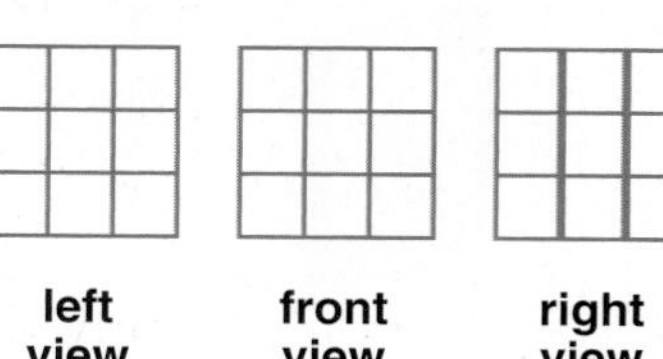

top view | left view | front view | right view

**Example 3**

5. **FOOD** Describe how the cheese at the right can be sliced so that the slices form each shape.
   a. rectangle
   b. triangle
   c. trapezoid

**Describe each cross section.**

6.

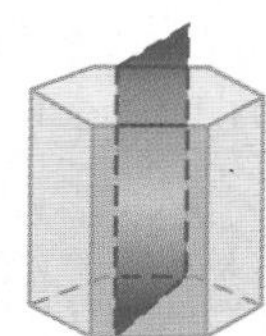

7.

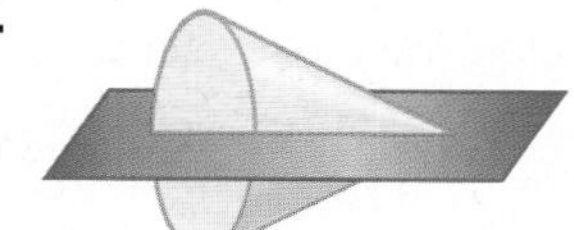

## Practice and Problem Solving

**Example 1** **Use isometric dot paper to sketch each prism.**

8. cube 3 units on each edge
9. triangular prism 4 units high, with two sides of the base that are 1 unit long and 3 units long
10. triangular prism 4 units high, with two sides of the base that are 2 units long and 6 units long

**Example 2** CCSS **TOOLS** **Use isometric dot paper and each orthographic drawing to sketch a solid.**

11.

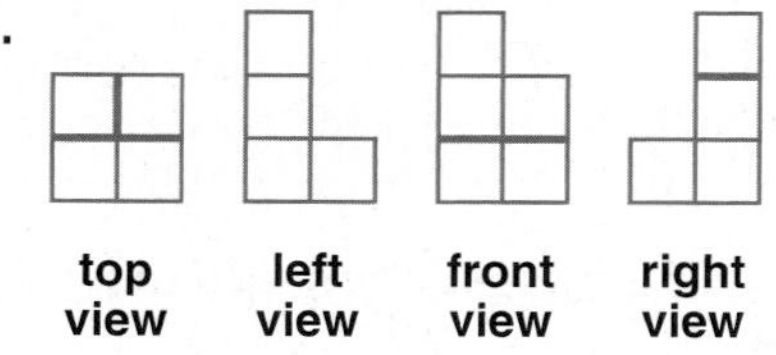

top view | left view | front view | right view

12.

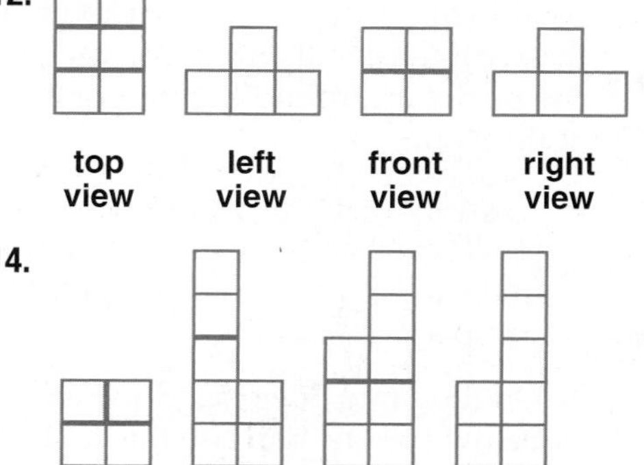

top view | left view | front view | right view

13.

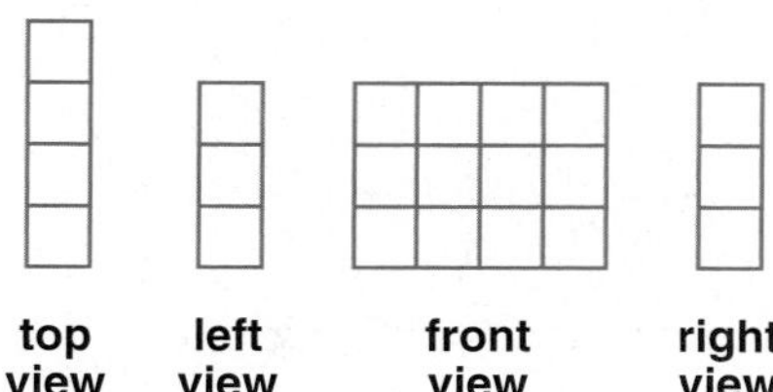

top view | left view | front view | right view

14.

top view | left view | front view | right view

Example 3

**15. ART** A piece of clay in the shape of a rectangular prism is cut in half as shown at the right.

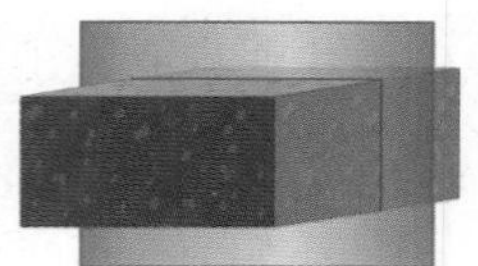

**a.** Describe the shape of the cross section.

**b.** Describe how the clay could be cut to make the cross section a triangle.

**Describe each cross section.**

**16.** 

**17.** 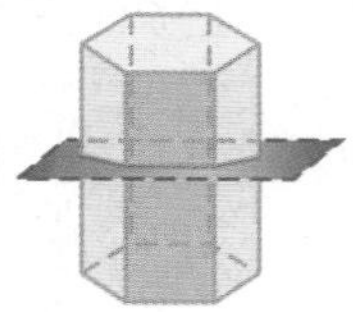

**18.** 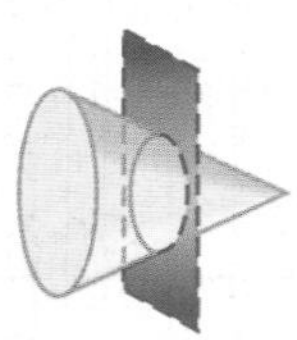

**19.** 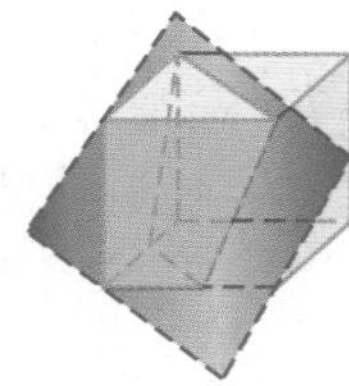

**20. ARCHITECTURE** Draw a top view, front view, and side view of the house at the right.

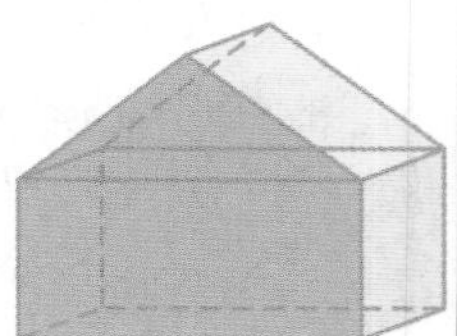

**COOKIES Describe how to make a cut through a roll of cookie dough in the shape of a cylinder to make each shape.**

**21.** circle

**22.** longest rectangle

**23.** oval

**24.** shorter rectangle

**CCSS TOOLS Sketch the cross section from a vertical slice of each figure.**

**25.** 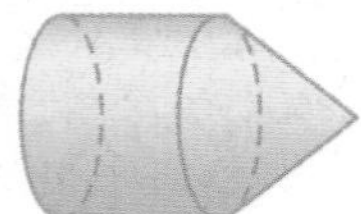

**26.** 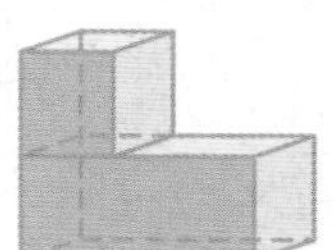

**27.** 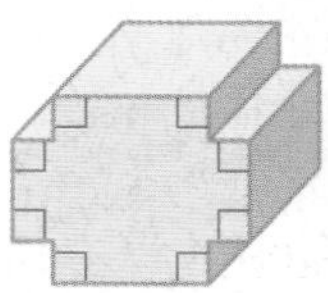

**28. EARTH SCIENCE** Crystals are solids in which the atoms are arranged in regular geometrical patterns. Sketch a cross section from a horizontal slice of each crystal. Then describe the rotational symmetry about the vertical axis.

**a.** tetragonal

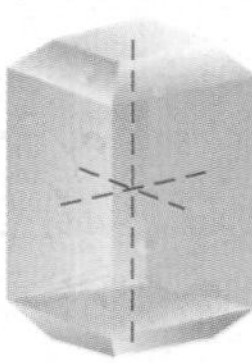

**b.** hexagonal

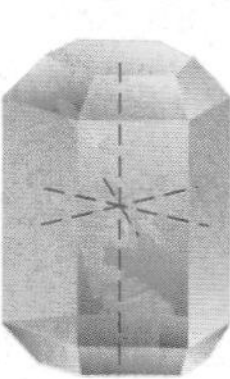

**c.** monoclinic

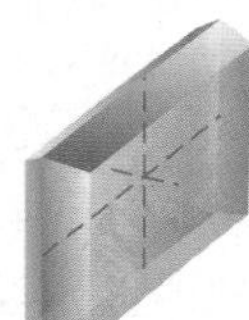

**29. ART** In a *perspective drawing*, a *vanishing point* is used to make the two-dimensional drawing appear three-dimensional. From one vanishing point, objects can be drawn from different points of view, as shown at the right.

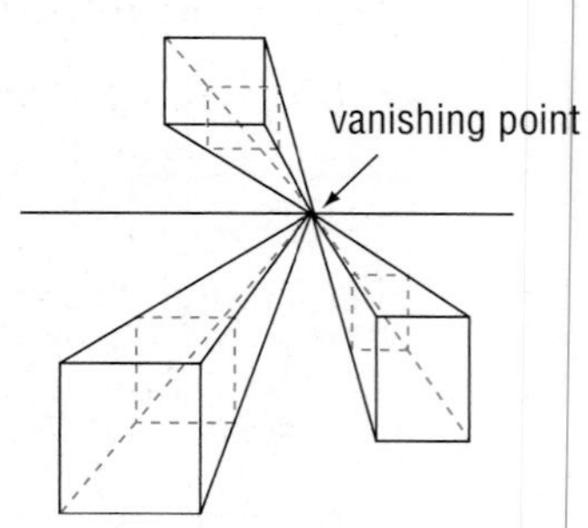


**a.** Draw a horizontal line and a vanishing point on the line. Draw a rectangle somewhere above the line and use the vanishing point to make a perspective drawing.

**b.** On the same drawing, draw a rectangle somewhere below the line and use the vanishing point to make a perspective drawing.

**c.** Describe the different views of the two drawings.

**Draw the top, left, front, and right view of each solid.**

30. 

31. 

32. 

33. The top, front, and right views of a three-dimensional figure are shown at the right.

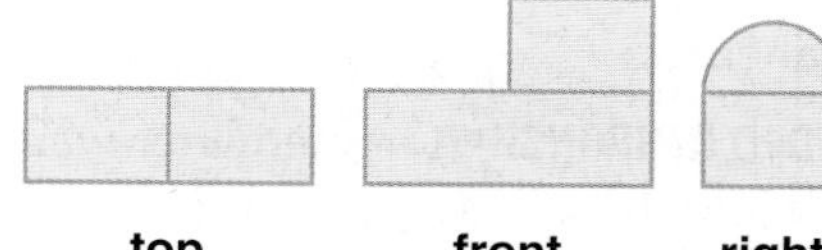


a. Make a sketch of the solid.

b. Describe two different ways that a rectangular cross section can be made.

c. Make a connection between the front and right views of the solid and cross sections of the solid.

34. **MULTIPLE REPRESENTATIONS** In this problem, you will investigate isometric drawings.

a. **Geometric** Create isometric drawings of three different solids.

b. **Tabular** Create a table that includes the number of cubes needed to construct the solid and the number of squares visible in the isometric drawing.

c. **Verbal** Is there a correlation between the number of cubes needed to construct a solid and the number of squares visible in the isometric drawing? Explain.

## H.O.T. Problems Use Higher-Order Thinking Skills

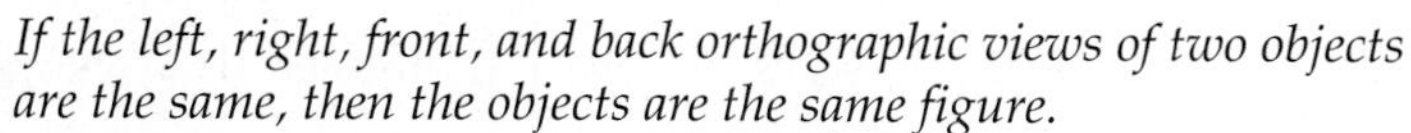

35. **CHALLENGE** The figure at the right is a cross section of a geometric solid. Describe a solid and how the cross section was made.

36. **CCSS ARGUMENTS** Determine whether the following statement is *true* or *false*. Explain your reasoning.

*If the left, right, front, and back orthographic views of two objects are the same, then the objects are the same figure.*

37. **OPEN ENDED** Use isometric dot paper to draw a solid consisting of 12 cubic units. Then sketch the orthographic drawing for your solid.

38. **CHALLENGE** Draw the top view, front view, and left view of the solid figure at the right.

39. **WRITING IN MATH** A hexagonal pyramid is sliced through the vertex and the base so that the prism is separated into two congruent parts. Describe the cross section. Is there more than one way to separate the figure into two congruent parts? Will the shape of the cross section change? Explain.

## Standardized Test Practice

**40.** Which polyhedron is represented by the net shown below?

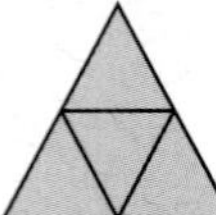

**A** cube
**B** octahedron
**C** triangular prism
**D** triangular pyramid

**41. EXTENDED RESPONSE** A homeowner wants to build a 3-foot-wide deck around his circular pool as shown below.

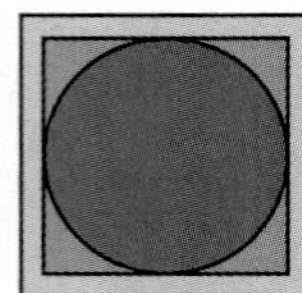

**a.** Find the outer perimeter of the deck to the nearest foot, if the circumference of the pool is about 81.64 feet.

**b.** What is the area of the top of the deck?

**42. ALGEBRA** Which inequality *best* describes the graph shown below?

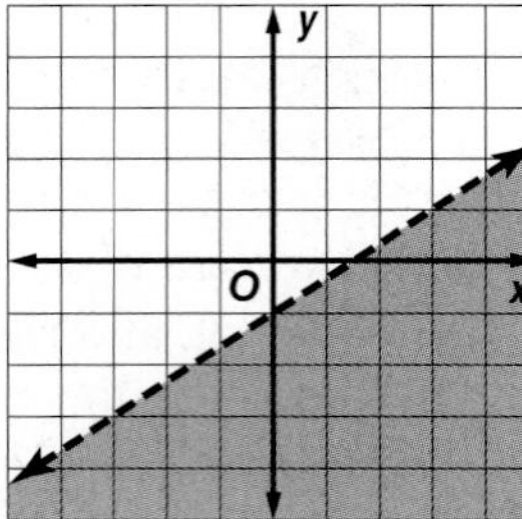


**F** $y < \frac{2}{3}x - 1$
**G** $y \leq \frac{2}{3}x - 1$
**H** $y > \frac{2}{3}x - 1$
**J** $y \geq \frac{2}{3}x - 1$

**43. SAT/ACT** Expand $(4\sqrt{5})^2$.

**A** 20
**B** $8\sqrt{5}$
**C** $16\sqrt{5}$
**D** 40
**E** 80

## Spiral Review

**44. WEATHER** Meteorologists track severe storms using Doppler radar. A polar grid is used to measure distances as the storms progress. If the center of the radar screen is the origin and each ring is 10 miles farther from the center, what is the equation of the fourth ring? (Lesson 11-8)

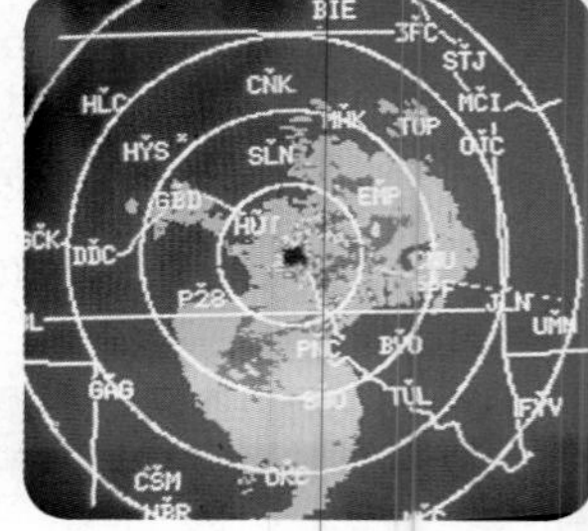

**Find *x* and *y*.** (Lesson 10-3)

**45.**

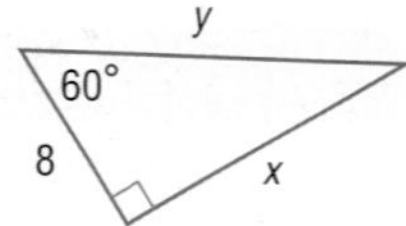


**46.**

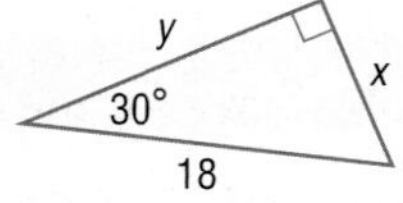


**Use the Venn diagram to determine whether each statement is *always, sometimes,* or *never* true.** (Lesson 8-5)

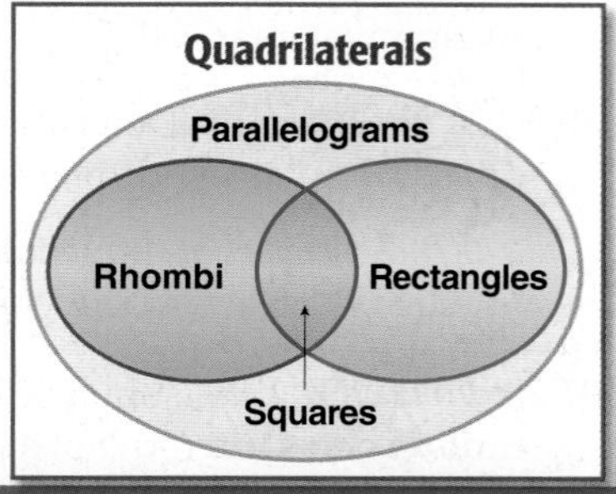


**47.** A parallelogram is a square.

**48.** A square is a rhombus.

**49.** A rectangle is a parallelogram.

**50.** A rhombus is a rectangle but not a square.

**51.** A rhombus is a square.

## Skills Review

**Find the perimeter or circumference and area of each figure. Round to the nearest tenth.**

**52.**

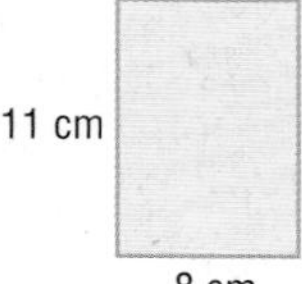


**53.**

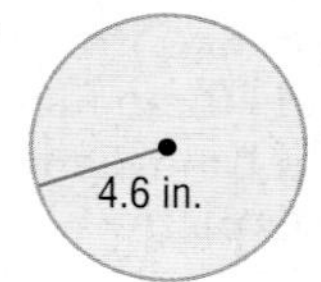


**54.**

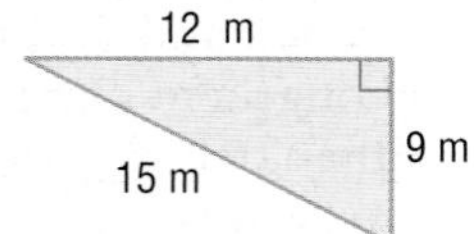

EXTEND 12-1

# Geometry Lab
# Topographic Maps

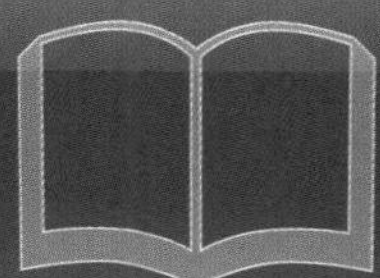

Maps are representations of Earth or some part of Earth. **Topographic maps** are representations of the three-dimensional surface of Earth on a two-dimensional piece of paper. In a topographic map, the *topography*, or shape of Earth's surface, is illustrated through the use of *contours*, which are imaginary lines that join locations with the same elevation.

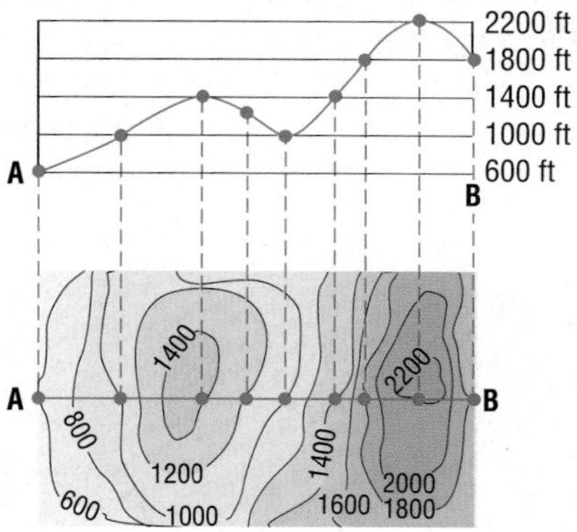


Some topographic maps show more than contours. These maps may include symbols that represent vegetation, rivers, and other landforms, as well as streets and buildings.

Follow these steps to read a topographic map.

- Thin lines represent contours. Since each contour is a line of equal elevation, they never cross. The closer together the contour lines, the steeper the slope.
- Contour lines form V shapes in valleys or riverbeds. The Vs point uphill.
- Most often, closed loops indicate that the surface slopes uphill on the inside and downhill on the outside. The innermost loop is the highest area.
- Pay attention to the colors. Blue represents water; green represents vegetation; red represents urban areas; black represents roads, trails, and railroads.
- The scale on a 1:24,000 map indicates that 1 inch equals 2000 feet.

## Explore the Model

**Use the topographic map above to answer these questions.**

1. According to the scale, what is the vertical distance between each contour line?
2. What is the difference in height between the lowest and highest points?
3. What do you notice about the contour lines for the peaks of the hills?
4. Describe a steep slope on the topographic map. How do you know it is steep?
5. Explain how you would draw a topographic map given a side view of some hills.

## Model and Analyze

6. Draw a topographic map similar to the map below for the side view of the hills from points $A$ to $B$.

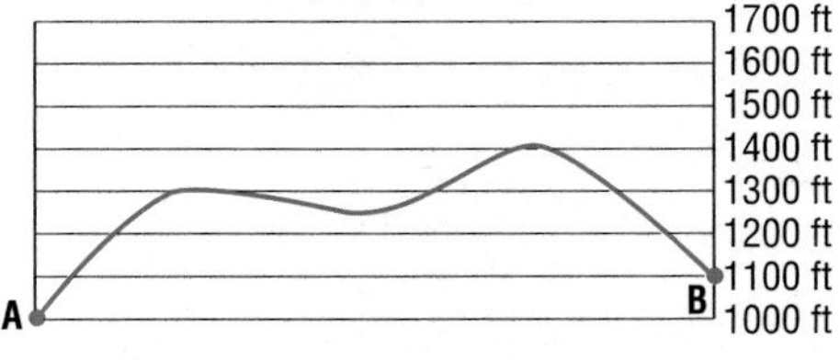


7. Draw a possible side view similar to the map below from points $A$ to $B$ of the hills from the topographic map. Measures are given in feet.

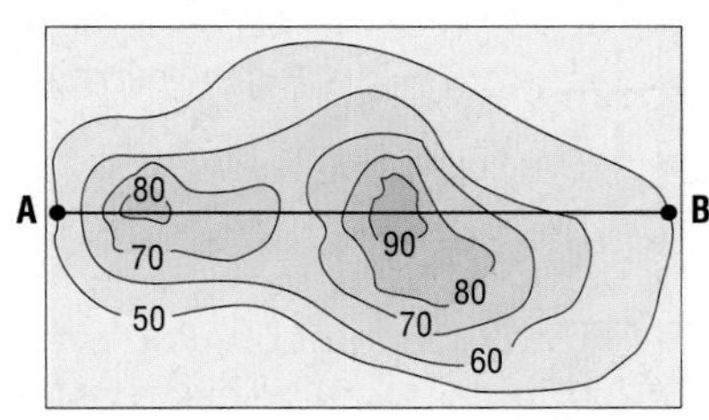

LESSON 12-2

# Surface Areas of Prisms and Cylinders

**Then**

- You found areas of polygons.

**Now**

1. Find lateral areas and surface areas of prisms.
2. Find lateral areas and surface areas of cylinders.

**Why?**

- Atlanta's Georgia Aquarium is the largest aquarium in the world, with more than 8 million gallons of water and more than 500 species from around the world. The aquarium has an underwater tunnel that is 100 feet long with 4574 square feet of viewing windows.

**NewVocabulary**

lateral face
lateral edge
base edge
altitude
height
lateral area
axis
composite solid

**Common Core State Standards**

**Content Standards**

G.MG.3 Apply geometric methods to solve problems (e.g., designing an object or structure to satisfy physical constraints or minimize cost; working with typographic grid systems based on ratios). ★

**Mathematical Practices**

1. Make sense of problems and persevere in solving them.
6. Attend to precision.

## 1 Lateral Areas and Surface Areas of Prisms

In a solid figure, faces that are not bases are called **lateral faces**. Lateral faces intersect each other at the **lateral edges**, which are all parallel and congruent. The lateral faces intersect the base at the **base edges**. The **altitude** is a perpendicular segment that joins the planes of the bases. The **height** is the length of the altitude.

Recall that a prism is a polyhedron with two parallel congruent bases. In a right prism, the lateral edges are altitudes and the lateral faces are rectangles. In an oblique prism, the lateral edges are not perpendicular to the bases. At least one lateral face is not a rectangle.

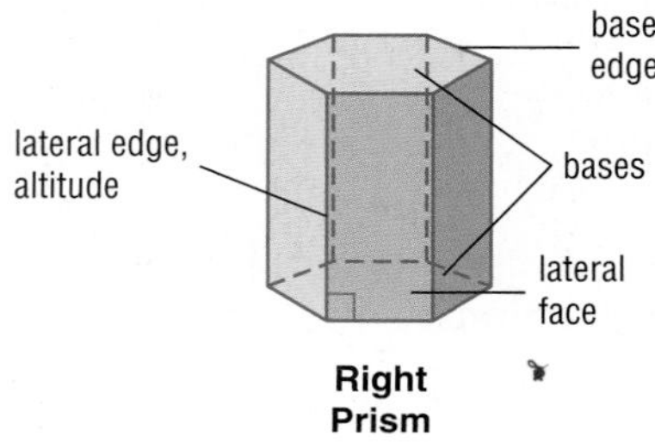


**Right Prism**

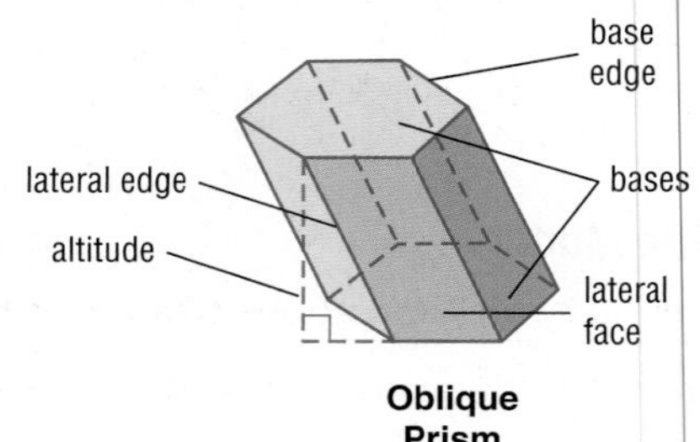


**Oblique Prism**

The **lateral area** $L$ of a prism is the sum of the areas of the lateral faces. The net at the right shows how to find the lateral area of a prism.

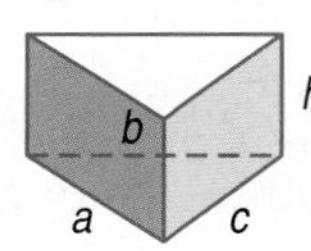


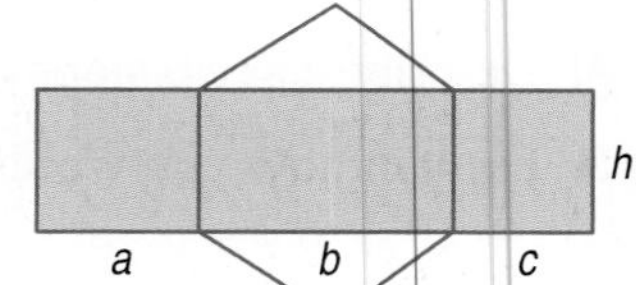


| | |
|---|---|
| $L = a(h) + b(h) + c(h)$ | Sum of areas of lateral faces |
| $= (a + b + c)h$ | Distributive Property |
| $= Ph$ | $P = a + b + c$ |

**KeyConcept** Lateral Area of a Prism

**Words** The lateral area $L$ of a right prism is $L = Ph$, where $h$ is the height of the prism and $P$ is the perimeter of a base.

**Symbols** $L = Ph$

**Model**

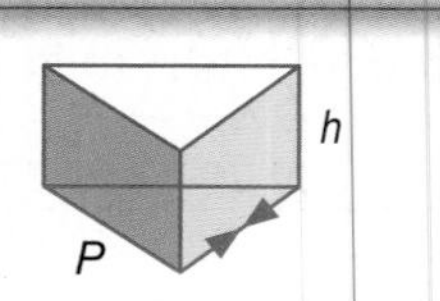


From this point on, you can assume that solids in the text are right solids. If a solid is oblique, it will be clearly stated.

## Example 1 Lateral Area of a Prism

**WatchOut!**

**Right Prisms** The bases of a right prism are congruent, but the faces are not always congruent.

**Find the lateral area of the prism. Round to the nearest tenth.**

7 cm, 5 cm, 6 cm

**Step 1** Find the missing side length of the base.

$c^2 = 6^2 + 5^2$ — Pythagorean Theorem

$c^2 = 61$ — Simplify.

$c \approx 7.8$ — Take the positive square root of each side.

**Step 2** Find the lateral area.

$L = Ph$ — Lateral area of a prism

$\approx (5 + 6 + 7.8)7$ — Substitution

$\approx 131.6$ — Simplify.

The lateral area is about 131.6 square centimeters.

### GuidedPractice

1. The length of each side of the base of a regular octagonal prism is 6 inches, and the height is 11 inches. Find the lateral area.

The surface area of a prism is the sum of the lateral area and the areas of the bases.

### KeyConcept Surface Area of a Prism

| | | |
|---|---|---|
| **Words** | The surface area $S$ of a right prism is $S = L + 2B$, where $L$ is its lateral area and $B$ is the area of a base. | **Model** |
| **Symbols** | $S = L + 2B$ or $S = Ph + 2B$ | |

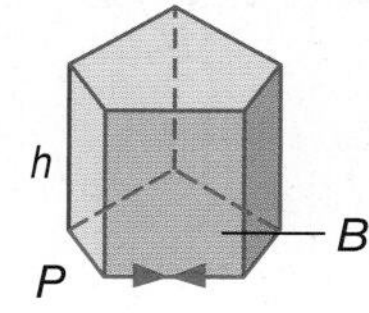


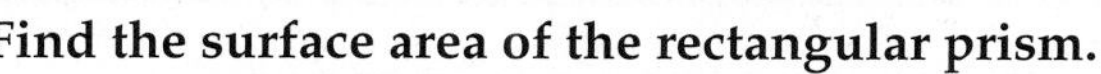

## Example 2 Surface Area of a Prism

**StudyTip**

**CCSS Perseverance** In Example 2, you can also use a 6-foot by 4-foot rectangle as the base. The height would be 9 feet. While choosing a different base does not affect the surface area, it will change the lateral area.

**Find the surface area of the rectangular prism.**

6 ft, 4 ft, 9 ft

Use the 9-foot by 4-foot rectangle as the base.

$S = Ph + 2B$ — Surface area of a prism

$= (2 \cdot 9 + 2 \cdot 4)(6) + 2(9 \cdot 4)$ — Substitution

$= 228$ — Simplify.

The surface area of the prism is 228 square feet.

### GuidedPractice

2. Find the surface area of the triangular prism. Round to the nearest tenth.

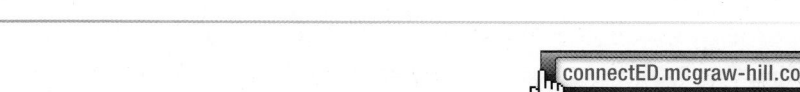

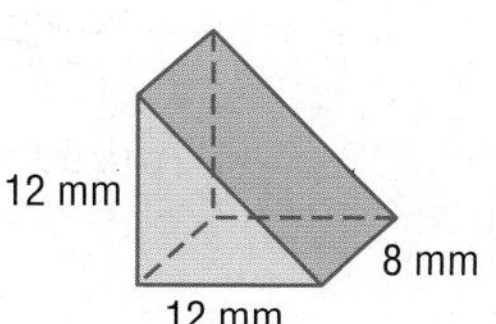

## 2 Lateral Areas and Surface Areas of Cylinders

The **axis** of a cylinder is the segment with endpoints that are centers of the circular bases. If the axis is also an altitude, then the cylinder is a right cylinder. If the axis is not an altitude, then the cylinder is an oblique cylinder.

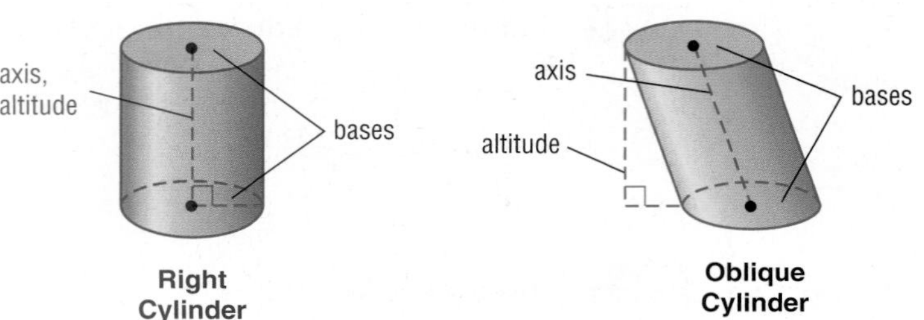


The lateral area of a right cylinder is the area of the curved surface. Like a right prism, the lateral area $L$ equals $Ph$. Since the base is a circle, the perimeter is the circumference of the circle $C$. So, the lateral area is $Ch$ or $2\pi rh$.

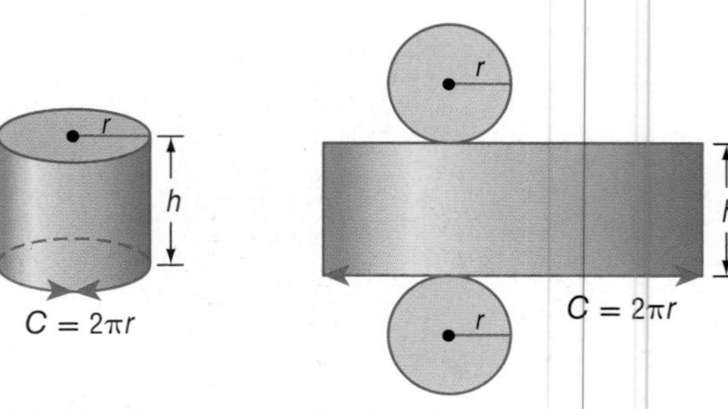


> **StudyTip**
> **Formulas** An alternate formula for the lateral area of a cylinder is $L = \pi dh$, with $\pi d$ as the circumference of a circle.

The surface area of a cylinder is the lateral area plus the areas of the bases.

### KeyConcept Surface Area of a Cylinder

**Words** The lateral area $L$ of a right cylinder is $L = 2\pi rh$, where $r$ is the radius of a base and $h$ is the height.

The surface area $S$ of a right cylinder is $S = 2\pi rh + 2\pi r^2$, where $r$ is the radius of a base and $h$ is the height.

**Model**

**Symbols**
$L = 2\pi rh$
$S = L + 2B$ or
$2\pi rh + 2\pi r^2$

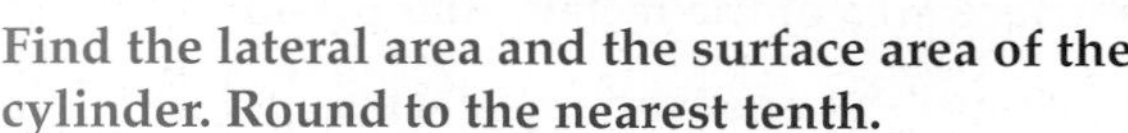

### Example 3 Lateral Area and Surface Area of a Cylinder

**Find the lateral area and the surface area of the cylinder. Round to the nearest tenth.**

15 mm

18 mm

| | |
|---|---|
| $L = 2\pi rh$ | Lateral area of a cylinder |
| $= 2\pi(7.5)(18)$ | Replace $r$ with 7.5 and $h$ with 18. |
| $\approx 848.2$ | Use a calculator. |
| $S = 2\pi rh + 2\pi r^2$ | Surface area of a cylinder |
| $\approx 848.2 + 2\pi(7.5)^2$ | Replace $2\pi rh$ with 848.2 and $r$ with 7.5. |
| $\approx 1201.6$ | Use a calculator. |

The lateral area is about 848.2 square millimeters, and the surface area is about 1201.6 square millimeters.

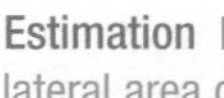

> **StudyTip**
> **Estimation** Before finding the lateral area of a cylinder, use mental math to estimate. To estimate, multiply the diameter by 3 (to approximate $\pi$) and then by the height of the cylinder.

**GuidedPractice**

**3A.** $r = 5$ in., $h = 9$ in.

**3B.** $d = 6$ cm, $h = 4.8$ cm

## Real-World Example 4 Find Missing Dimensions

**CRAFTS Sheree used the rectangular piece of felt shown at the right to cover the curved surface of her cylindrical pencil holder. What is the radius of the pencil holder?**

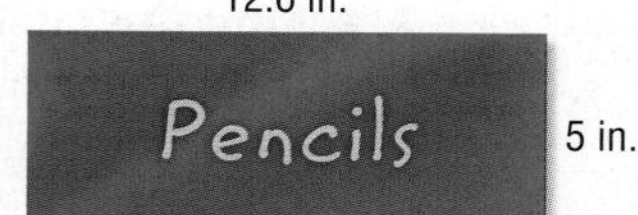


| | |
|---|---|
| $L = 2\pi rh$ | Lateral area of a cylinder |
| $63 = 2\pi r(5)$ | Replace $L$ with $12.6 \cdot 5$ or 63 and $h$ with 5. |
| $63 = 10\pi r$ | Simplify. |
| $2.0 \approx r$ | Divide each side by $10\pi$. |

The radius of the pencil holder is about 2 inches.

### GuidedPractice

**4.** Find the diameter of a base of a cylinder if the surface area is $464\pi$ square centimeters and the height is 21 centimeters.

## Check Your Understanding

Example 1

**1.** Find the lateral area of the prism.

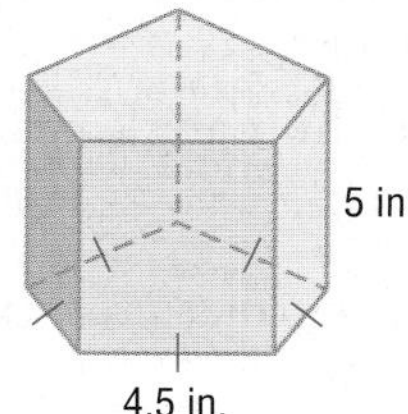


Examples 1–2 **Find the lateral area and surface area of each prism.**

**2.**

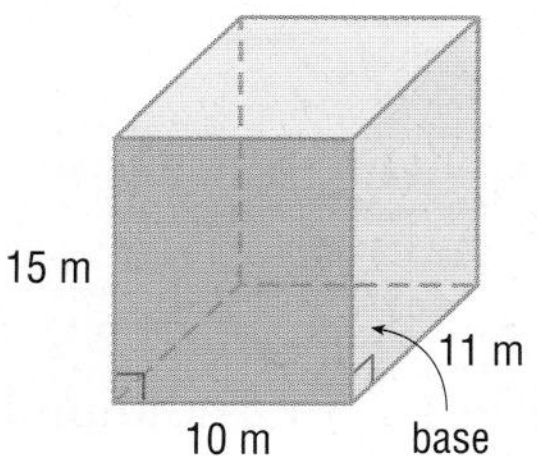


**3.**

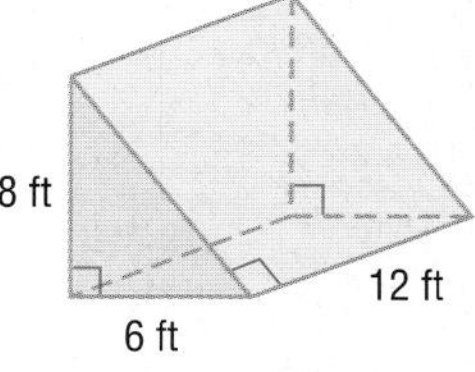


Example 3

**4. CARS** Evan is buying new tire rims that are 14 inches in diameter and 6 inches wide. Determine the lateral area of each rim. Round to the nearest tenth.

**Find the lateral area and surface area of each cylinder. Round to the nearest tenth.**

**5.**

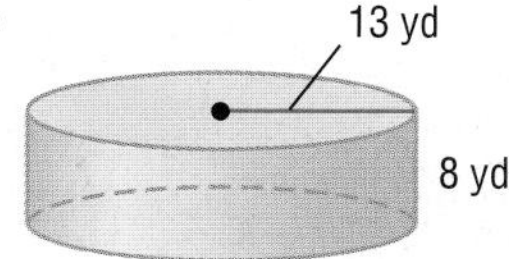


**6.**

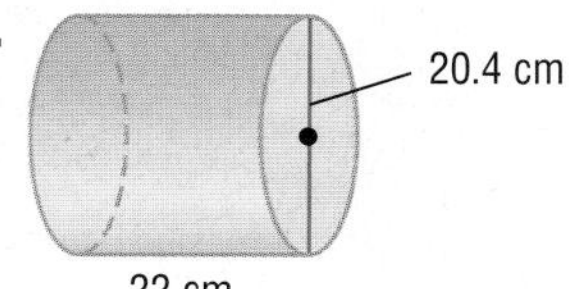


Example 4

**7. FOOD** The can of soup at the right has a surface area of 286.3 square centimeters. What is the height of the can? Round to the nearest tenth.

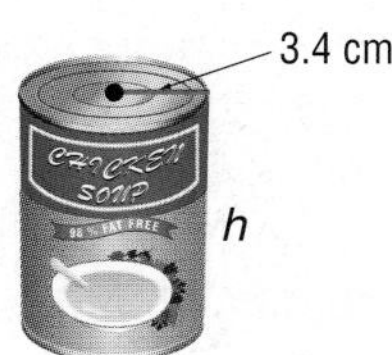


**8.** The surface area of a cube is 294 square inches. Find the length of a lateral edge.

## Practice and Problem Solving

**Examples 1–2** **Find the lateral area and surface area of each prism. Round to the nearest tenth if necessary.**

**9.**

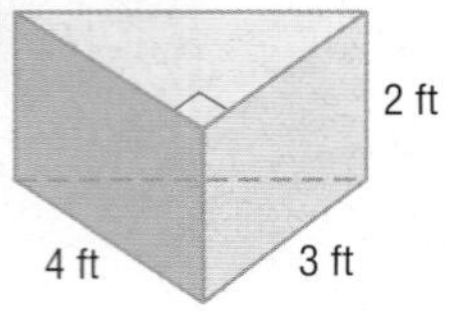

**10.**

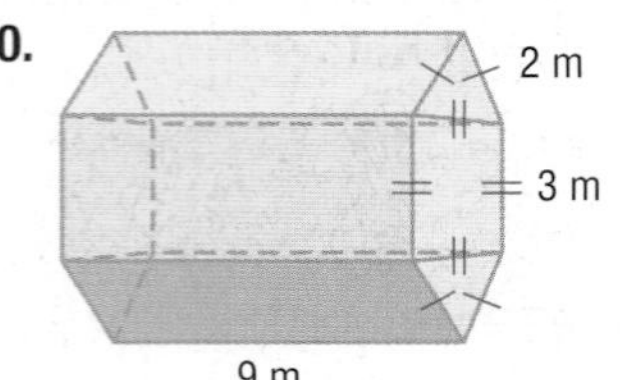


**11.**

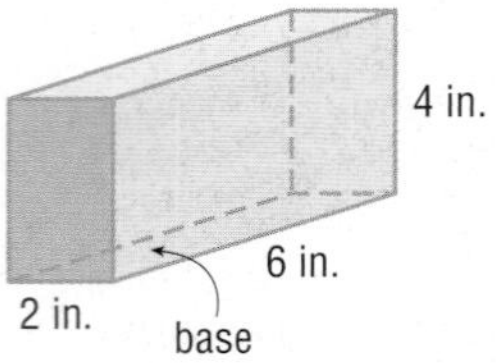

**12.**

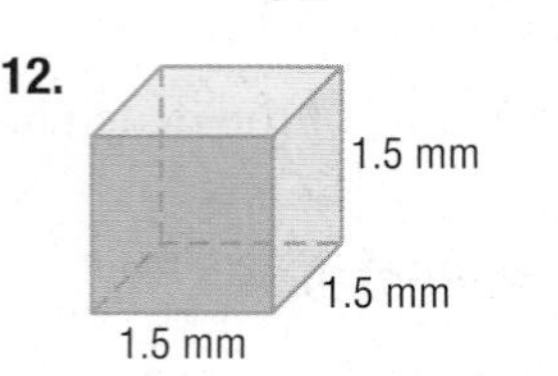


**13.**

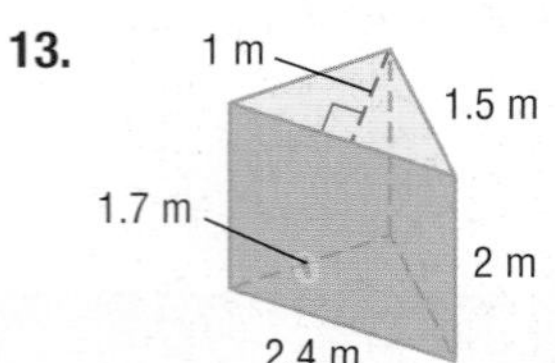


**14.**

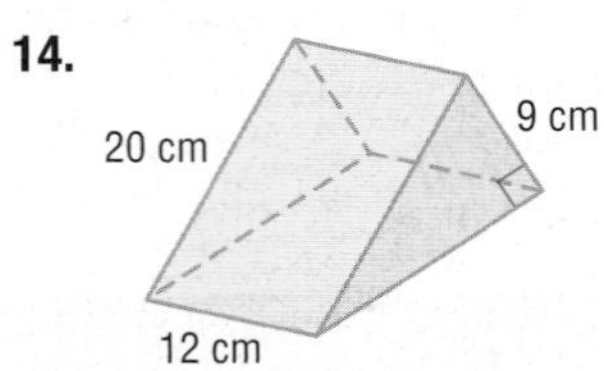


**15.** rectangular prism: $\ell = 25$ centimeters, $w = 18$ centimeters, $h = 12$ centimeters

**16.** triangular prism: $h = 6$ inches, right triangle base with legs 9 inches and 12 inches

**Examples 1–3** **CEREAL** **Find the lateral area and the surface area of each cereal container. Round to the nearest tenth if necessary.**

**17.**

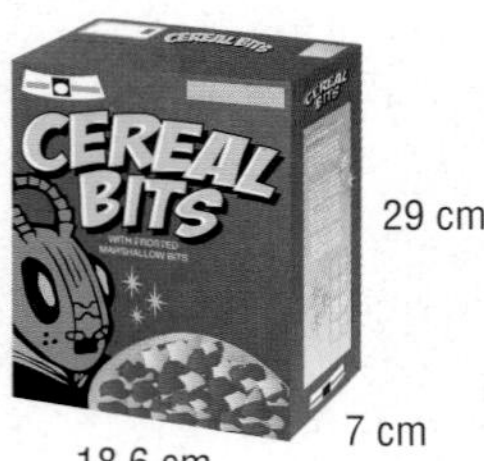


**18.**

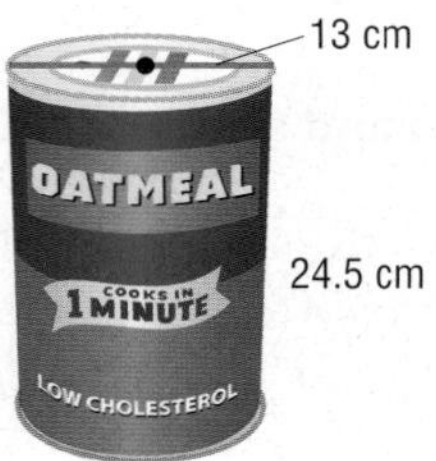


**Example 3** **CCSS SENSE-MAKING** **Find the lateral area and surface area of each cylinder. Round to the nearest tenth.**

**19.**

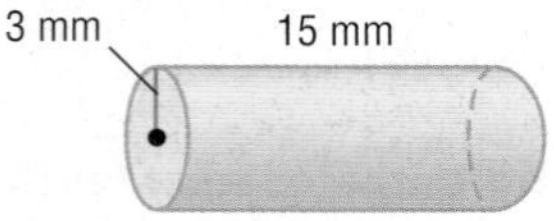

**20.**

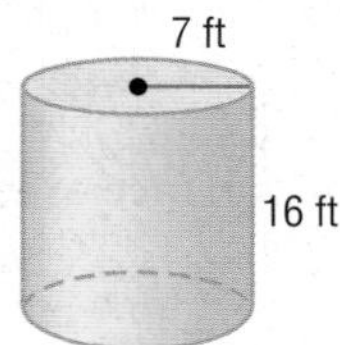


**21.**

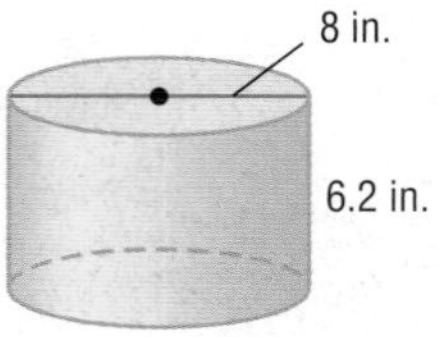

**22.**

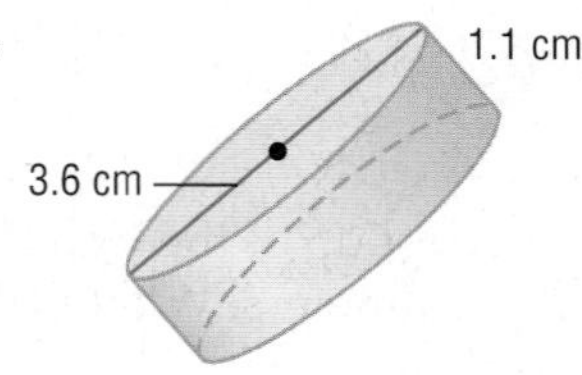


**23.** **WORLD RECORDS** The largest beverage can was a cylinder with height 4.67 meters and diameter 2.32 meters. What was the surface area of the can to the nearest tenth?

**Example 4** **Use the given lateral area and the diagram to find the missing measure of each solid. Round to the nearest tenth if necessary.**

**24.** $L = 48 \text{ in}^2$

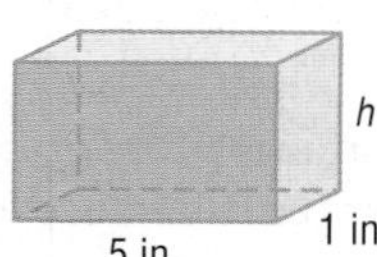


**25.** $L \approx 635.9 \text{ cm}^2$

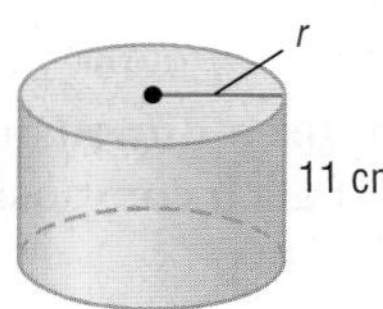


**26.** A right rectangular prism has a surface area of 1020 square inches, a length of 6 inches, and a width of 9 inches. Find the height.

**27** A cylinder has a surface area of $256\pi$ square millimeters and a height of 8 millimeters. Find the diameter.

**28.** **MONUMENTS** A *monolith* mysteriously appeared overnight at Seattle, Washington's Manguson Park. A hollow rectangular prism, the monolith was 9 feet tall, 4 feet wide, and 1 foot deep.

**a.** Find the area in square feet of the structure's surfaces that lie above the ground.

**b.** Use dimensional analysis to find the area in square yards.

**29.** **ENTERTAINMENT** The graphic shows the results of a survey in which people were asked where they like to watch movies.

**a.** Suppose the film can is a cylinder 12 inches in diameter. Explain how to find the surface area of the portion that represents people who prefer to watch movies at home.

**b.** If the film can is 3 inches tall, find the surface area of the portion in part **a**.

**Preferred Places to Watch Movies**

**CCSS SENSE-MAKING** **Find the lateral area and surface area of each oblique solid. Round to the nearest tenth.**

**30.**

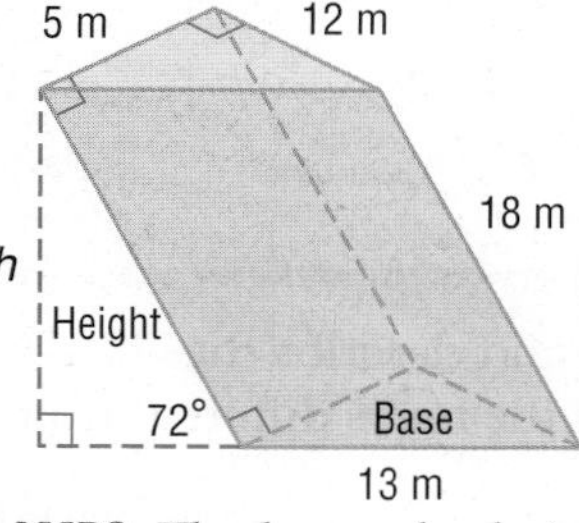


**31.**

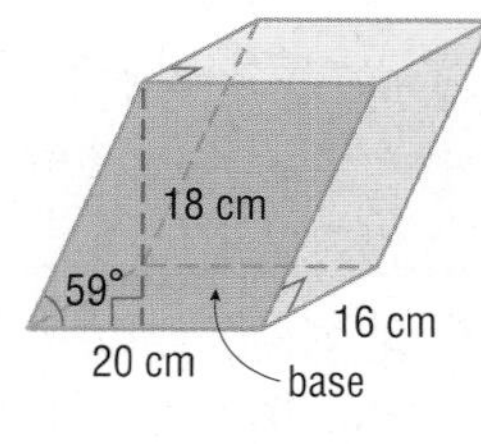


**32.** **LAMPS** The lamp shade is a cylinder of height 18 inches with a diameter of $6\frac{3}{4}$ inches.

**a.** What is the lateral area of the shade to the nearest tenth?

**b.** How does the lateral area change if the height is divided by 2?

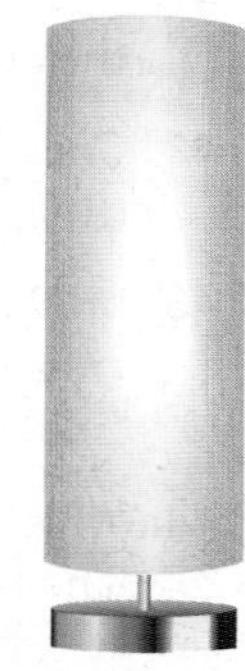

**33.** Find the approximate surface area of a right hexagonal prism if the height is 9 centimeters and each base edge is 4 centimeters. (*Hint*: First, find the length of the apothem of the base.)

Cheryl Hatch/AP Images

34. **DESIGN** A mailer needs to hold a poster that is almost 38 inches long and has a maximum rolled diameter of 6 inches.

 a. Design a mailer that is a triangular prism. Sketch the mailer and its net.

 b. Suppose you want to minimize the surface area of the mailer. What would be the dimensions of the mailer and its surface area?

**A composite solid is a three-dimensional figure that is composed of simpler figures. Find the surface area of each composite solid. Round to the nearest tenth if necessary.**

35.

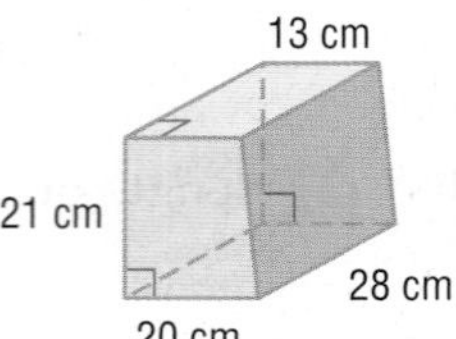


36.

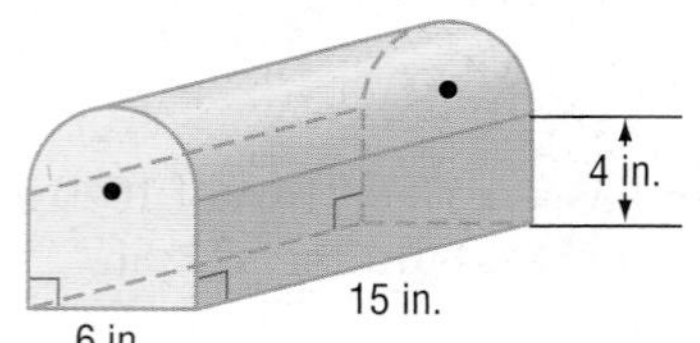


37.

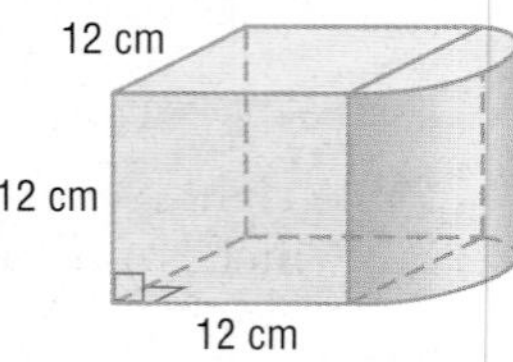


38. **MULTIPLE REPRESENTATIONS** In this problem, you will investigate the lateral area and surface area of a cylinder.

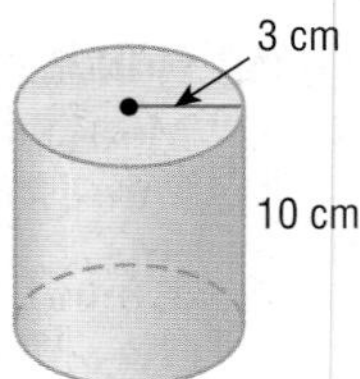


 a. **Geometric** Sketch cylinder $A$ with a radius of 3 centimeters and a height of 5 centimeters, cylinder $B$ with a radius of 6 centimeters and a height of 5 centimeters, and cylinder $C$ with a radius of 3 centimeters and a height of 10 centimeters.

 b. **Tabular** Create a table of the radius, height, lateral area, and surface area of cylinders A, B, and C. Write the areas in terms of $\pi$.

 c. **Verbal** If the radius is doubled, what effect does it have on the lateral area and the surface area of a cylinder? If the height is doubled, what effect does it have on the lateral area and the surface area of a cylinder?

## H.O.T. Problems Use Higher-Order Thinking Skills

39. **ERROR ANALYSIS** Montell and Derek are finding the surface area of a cylinder with height 5 centimeters and radius 6 centimeters. Is either of them correct? Explain.

| Montell | Derek |
|---|---|
| $S = \pi(6)^2 + \pi(6)(5)$ | $S = 2\pi(6)^2 + 2\pi(6)(5)$ |
| $= 36\pi + 30\pi$ | $= 72\pi + 60\pi$ |
| $= 66\pi \text{ cm}^2$ | $= 132\pi \text{ cm}^2$ |

40. **WRITING IN MATH** Sketch an oblique rectangular prism, and describe the shapes that would be included in a net for the prism. Explain how the net is different from that of a right rectangular prism.

41. **CCSS PRECISION** Compare and contrast finding the surface area of a prism and finding the surface area of a cylinder.

42. **OPEN ENDED** Give an example of two cylinders that have the same lateral area and different surface areas. Describe the lateral area and surface areas of each.

43. **CHALLENGE** A right prism has a height of $h$ units and a base that is an equilateral triangle of side $\ell$ units. Find the general formula for the total surface area of the prism. Explain your reasoning.

44. **WRITING IN MATH** A square-based prism and a triangular prism are the same height. The base of the triangular prism is an equilateral triangle, with an altitude equal in length to the side of the square. Compare the lateral areas of the prisms.

## Standardized Test Practice

**45.** If the surface area of the right rectangular prism is 310 square centimeters, what is the measure of the height $h$ of the prism?

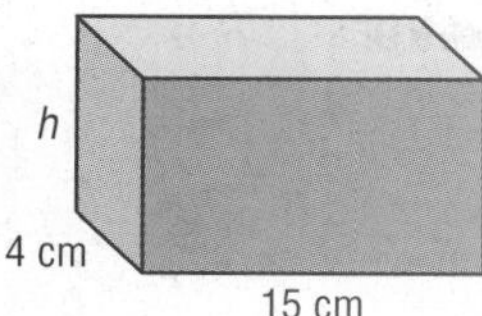


A 5 cm
B $5\frac{1}{6}$ cm
C 10
D $13\frac{3}{9}$ cm

**46. SHORT RESPONSE** A cylinder has a circumference of $16\pi$ inches and a height of 20 inches. What is the surface area of the cylinder in terms of $\pi$?

**47.** Parker Flooring charges the following to install a hardwood floor in a new home.
Subflooring: $2.25 per square foot
Wood flooring : $4.59 per square foot
Baseboards: $1.95 per linear foot around room
Nail & other materials: $25.95 per job
Labor: $99 plus $0.99 square foot
What is the cost to install hardwood flooring in a room that is 18 by 15 feet?

F $2169.75
G $2268.75
H $2367.75
J $2765.55

**48. SAT/ACT** What is the value of $f(-2)$ if $f(x) = x^3 + 4x^2 - 2x - 3$?

A $-31$
B $-\frac{9}{2}$
C 9
D 25
E 28

## Spiral Review

**Use isometric dot paper to sketch each prism.** (Lesson 12-1)

**49.** rectangular prism 2 units high, 3 units long, and 2 units wide

**50.** triangular prism 2 units high with bases that are right triangles with legs 3 units and 4 units long

**The diameters of $\odot R$, $\odot S$, and $\odot T$ are 10 inches, 14 inches, and 9 inches, respectively. Find each measure.** (Lesson 11-1)

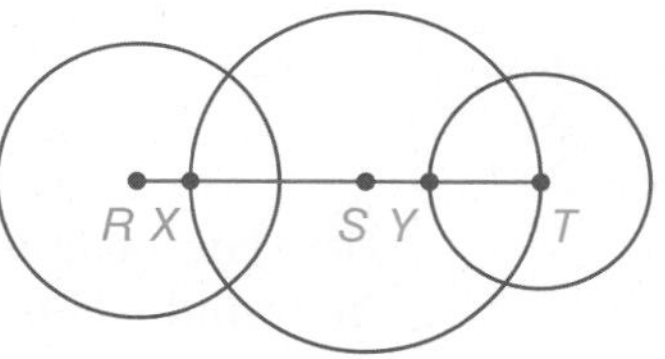


**51.** $YX$

**52.** $SY$

## Skills Review

**Find $x$. Round to the nearest tenth.**

**53.**

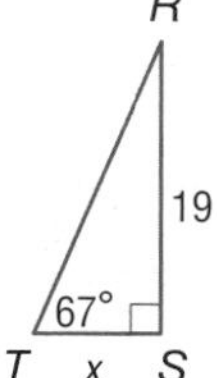


**54.**

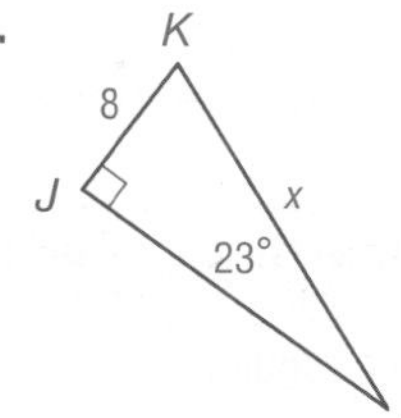


**55.**

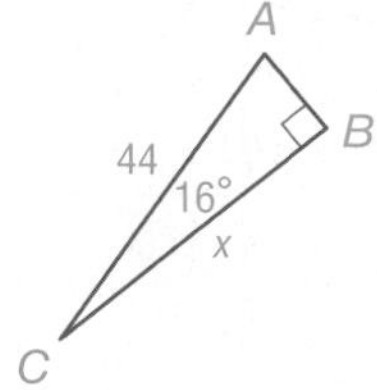

LESSON 12-3

# Surface Areas of Pyramids and Cones

**Then**

- You found areas of regular polygons.

**Now**

1. Find lateral areas and surface areas of pyramids.
2. Find lateral areas and surface areas of cones.

**Why?**

- The Transamerica Pyramid in San Francisco, California, covers nearly one city block. Its unconventional design allows light and air to filter down to the streets around the building, unlike the more traditional rectangular prism skyscrapers.

**NewVocabulary**
regular pyramid
slant height
right cone
oblique cone

**Common Core State Standards**

**Content Standards**
G.MG.1 Use geometric shapes, their measures, and their properties to describe objects (e.g., modeling a tree trunk or a human torso as a cylinder). ★

**Mathematical Practices**
1 Make sense of problems and persevere in solving them.
6 Attend to precision.

## 1 Lateral Area and Surface Area of Pyramids

The *lateral faces* of a pyramid intersect at a common point called the *vertex*. Two lateral faces intersect at a *lateral edge*. A lateral face and the base intersect at a *base edge*. The *altitude* is the segment from the vertex perpendicular to the base.

A **regular pyramid** has a base that is a regular polygon and the altitude has an endpoint at the center of the base. All the lateral edges are congruent and all the lateral faces are congruent isosceles triangles. The height of each lateral face is called the **slant height** $\ell$ of the pyramid.

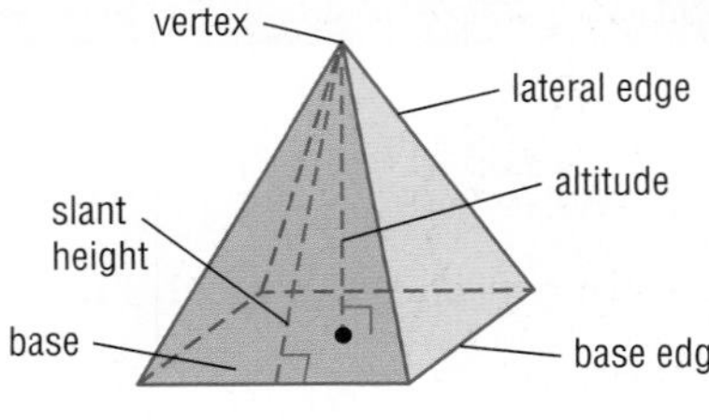


**Regular Pyramid**

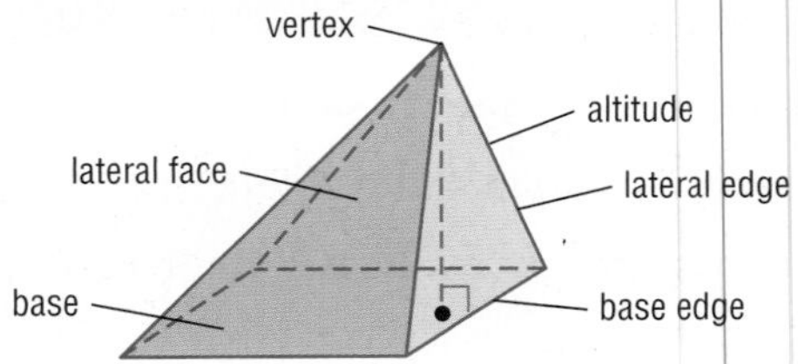


**Nonregular Pyramid**

The lateral area $L$ of a regular pentagonal pyramid is the sum of the areas of all its congruent triangular faces as shown in the net at the right.

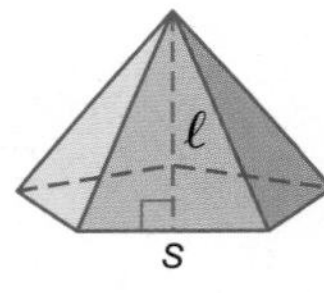


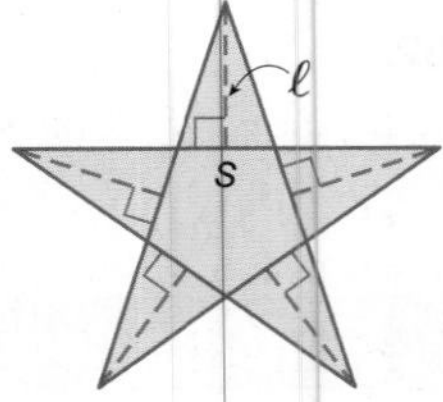


$$L = \frac{1}{2}s\ell + \frac{1}{2}s\ell + \frac{1}{2}s\ell + \frac{1}{2}s\ell + \frac{1}{2}s\ell \quad \text{Sum of the areas of the lateral faces}$$
$$= \frac{1}{2}\ell(s + s + s + s + s) \quad \text{Distributive Property}$$
$$= \frac{1}{2}P\ell \quad P = s + s + s + s + s$$

**KeyConcept Lateral Area of a Regular Pyramid**

**Words** The lateral area $L$ of a regular pyramid is $L = \frac{1}{2}P\ell$, where $\ell$ is the slant height and $P$ is the perimeter of the base.

**Symbols** $L = \frac{1}{2}P\ell$

**Model**

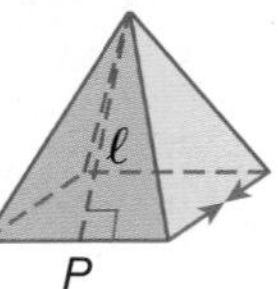


age fotostock/SuperStock

## Example 1 Lateral Area of a Regular Pyramid

**Find the lateral area of the square pyramid.**

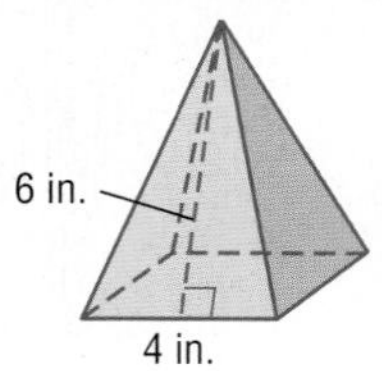


$L = \frac{1}{2}P\ell$ Lateral area of a regular pyramid

$= \frac{1}{2}(16)(6)$ or 48 $P = 4 \cdot 4$ or 16, $\ell = 6$

The lateral area is 48 square inches.

> **StudyTip**
> **Alternative Method** You can also find the lateral area of a pyramid by adding the areas of the congruent lateral faces.
> area of one face:
> $\frac{1}{2}(4)(6) = 12$ in$^2$
> lateral area:
> $4 \cdot 12 = 48$ in$^2$

### GuidedPractice

**1.** Find the lateral area of a regular hexagonal pyramid with a base edge of 9 centimeters and a lateral height of 7 centimeters.

The surface area of a pyramid is the sum of the lateral area and the area of the base.

## KeyConcept Surface Area of a Regular Pyramid

**Words** The surface area $S$ of a regular pyramid is $S = \frac{1}{2}P\ell + B$, where $P$ is the perimeter of the base, $\ell$ is the slant height, and $B$ is the area of the base.

**Model**

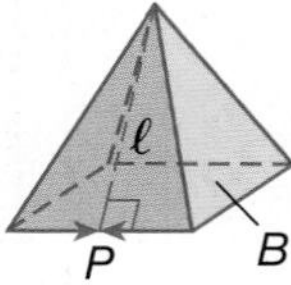


**Symbols** $S = \frac{1}{2}P\ell + B$

## Example 2 Surface Area of a Square Pyramid

**Find the surface area of the square pyramid to the nearest tenth.**

16 cm, $\ell$ cm, 12 cm

**Step 1** Find the slant height.

$c^2 = a^2 + b^2$ Pythagorean Theorem

$\ell^2 = 16^2 + 6^2$ $a = 16$, $b = 6$, and $c = \ell$

$\ell = \sqrt{292}$ Simplify.

**Step 2** Find the perimeter and area of the base.

$P = 4 \cdot 12$ or 48 cm $A = 12^2$ or 144 cm$^2$

**Step 3** Find the surface area of the pyramid.

$S = \frac{1}{2}P\ell + B$ Surface area of a regular pyramid

$= \frac{1}{2}(48)\sqrt{292} + 144$ $P = 48$, $\ell = \sqrt{292}$, and $B = 144$

$\approx 554.1$ Use a calculator.

The surface area of the pyramid is about 554.1 square centimeters.

> **StudyTip**
> **Making Connections** The surface area of a pyramid equals $L + B$, not $L + 2B$, because a pyramid has only one base.

### GuidedPractice

**2A.**

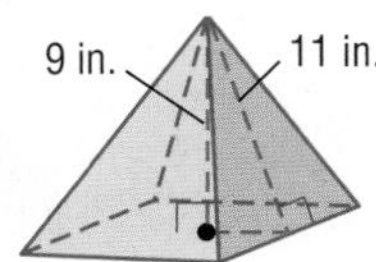


**2B.**

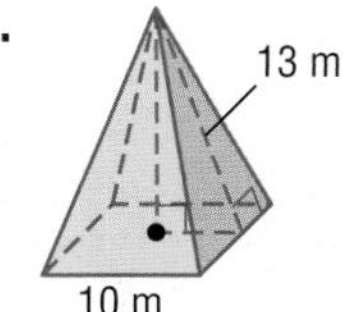

## Example 3 Surface Area of a Regular Pyramid

**Find the surface area of the regular pyramid. Round to the nearest tenth.**

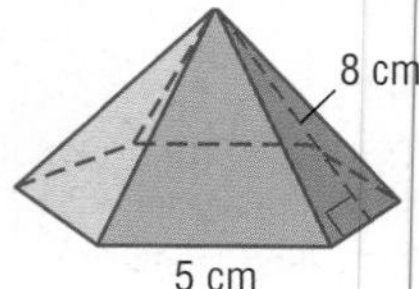


**Step 1** Find the perimeter of the base.
$P = 6 \cdot 5$ or 30 cm

**Step 2** Find the length of the apothem and the area of the base.

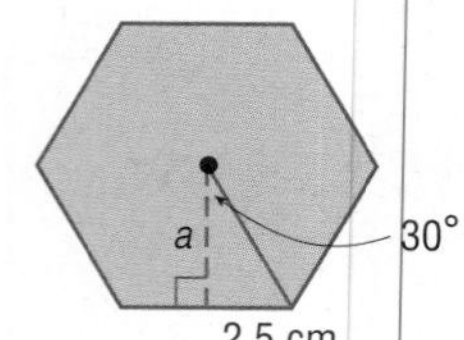


A central angle of the hexagon is $\frac{360°}{6}$ or 60°, so the angle formed in the triangle at the right is 30°.

| | |
|---|---|
| $\tan 30° = \frac{2.5}{a}$ | Write a trigonometric ratio to find the apothem *a*. |
| $a = \frac{2.5}{\tan 30°}$ | Solve for *a*. |
| $\approx 4.3$ | Use a calculator. |
| $A = \frac{1}{2}Pa$ | Area of *a* regular polygon |
| $\approx \frac{1}{2}(30)(4.3)$ | Replace *P* with 30 and *a* with 4.3. |
| $\approx 64.5$ | Multiply. |

So, the area of the base *B* is approximately 64.5 square centimeters.

**Step 3** Find the surface area of the pyramid.

| | |
|---|---|
| $S = \frac{1}{2}P\ell + 1$ | Surface area of a regular pyramid |
| $= \frac{1}{2}(30)(8) + 64.5$ | $P = 30$, $\ell = 8$, and $B \approx 64.5$ |
| $\approx 184.5$ | Simplify. |

The surface area of the pyramid is about 184.5 square centimeters.

**Review Vocabulary**

**Trigonometric Ratios**

$\sin A = \frac{\text{opp}}{\text{hyp}}$

$\cos A = \frac{\text{adj}}{\text{hyp}}$

$\tan A = \frac{\text{opp}}{\text{adj}}$

### Guided Practice

**3A.**

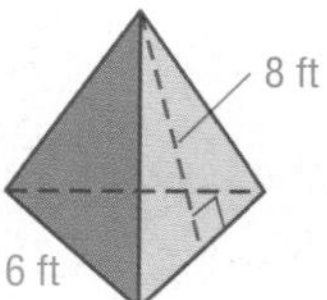


**3B.**

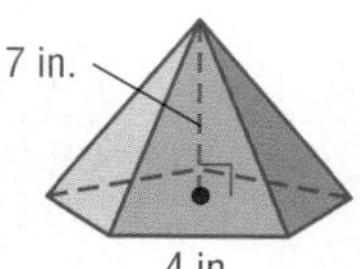


## 2 Lateral Area and Surface Area of Cones

Recall that a cone has a circular base and a vertex. The axis of a cone is the segment with endpoints at the vertex and the center of the base. If the axis is also the altitude, then the cone is a **right cone**. If the axis is not the altitude, then the cone is an **oblique cone**.

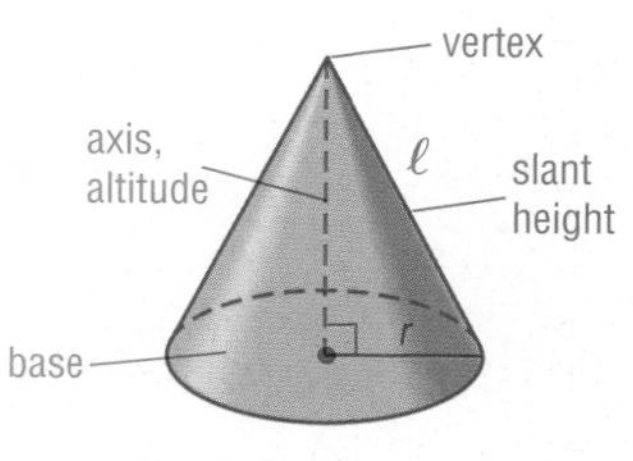


**Right Cone**

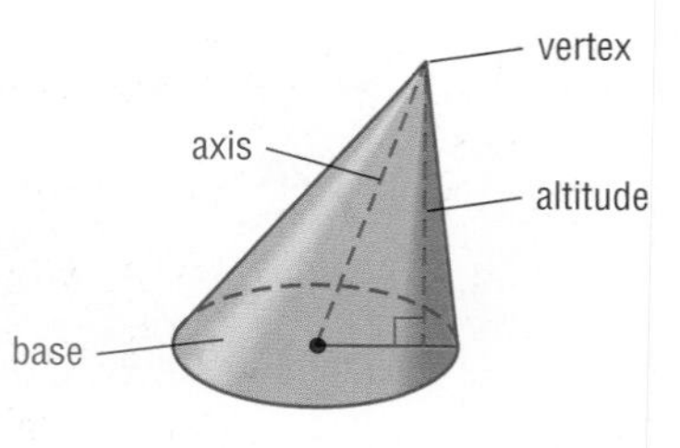


**Oblique Cone**

The net for a cone is shown at the right. The circle with radius $r$ is the base of the cone. It has a circumference of $2\pi r$ and an area of $\pi r^2$. The sector with radius $\ell$ is the lateral surface of the cone. Its arc measure is $2\pi r$. You can use a proportion to find its area.

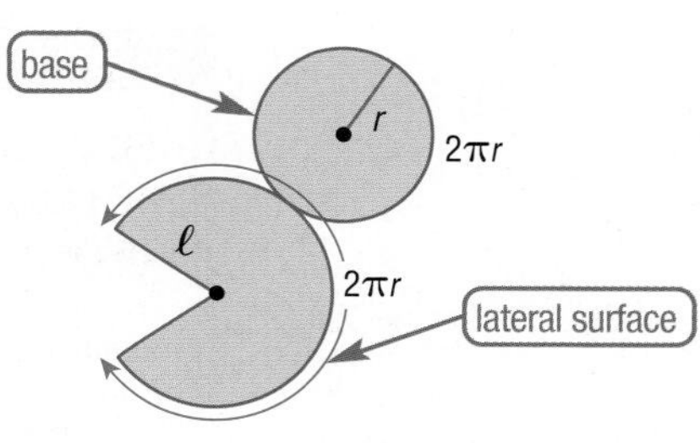


$$\frac{\text{area of sector}}{\text{area of circle}} = \frac{\text{measure of arc}}{\text{circumference of circle}}$$

$$\frac{\text{area of sector}}{\pi \ell^2} = \frac{2\pi r}{2\pi \ell}$$

$$\text{area of sector} = \pi \ell^2 \cdot \frac{2\pi r}{2\pi \ell} \text{ or } \pi r \ell$$

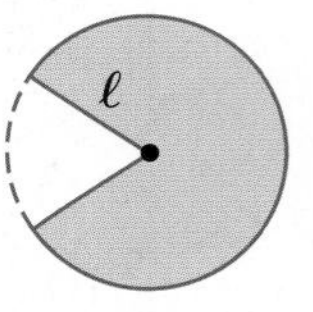


**sector**

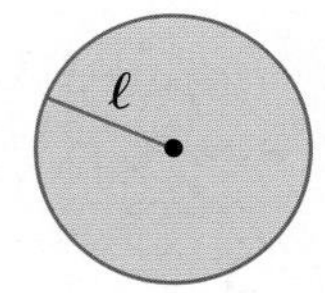


**circle that contains the sector**

**StudyTip**

**CCSS Sense-Making** Like a pyramid, the lateral area of a right circular cone $L$ equals $\frac{1}{2}P\ell$. Since the base is a circle, the perimeter is the circumference of the base $C$. So, the lateral area is $\frac{1}{2}C\ell$.

$$L = \frac{1}{2}C\ell$$
$$= \frac{1}{2}(2\pi r)$$
$$= \pi r \ell$$

## KeyConcept Lateral and Surface Area of a Cone

**Words** The lateral area $L$ of a right circular cone is $L = \pi r \ell$, where $r$ is the radius of the base and $\ell$ is the slant height.

The surface area $S$ of a right circular cone is $S = \pi r \ell + \pi r^2$, where $r$ is the radius of the base and $\ell$ is the slant height.

**Model**

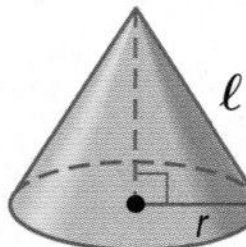


**Symbols** $L = \pi r \ell$ $\quad S = \pi r \ell + \pi r^2$

## Real-World Example 4 Lateral Area of a Cone

**ARCHITECTURE The conical slate roof at the right has a height of 16 feet and a radius of 12 feet. Find the lateral area.**

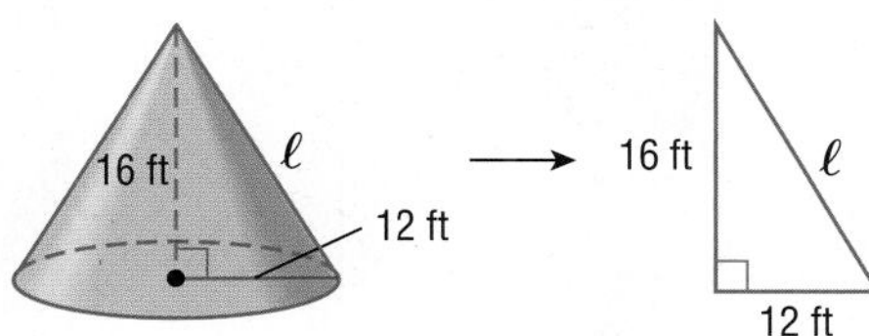

**Step 1** Find the slant height $\ell$.

| | |
|---|---|
| $\ell^2 = 16^2 + 12^2$ | Pythagorean Theorem |
| $\ell^2 = 400$ | Simplify. |
| $\ell = 20$ | Take the positive square root of each side. |

**Step 2** Find the lateral area $L$.

Estimate $L \approx 3 \cdot 12 \cdot 20$ or 720 ft$^2$

| | |
|---|---|
| $L = \pi r \ell$ | Lateral area of a cone |
| $= \pi(12)(20)$ | $r = 12$ and $\ell = 20$ |
| $\approx 754$ | Use a calculator. |

The lateral area of the conical roof is about 754 square feet. The answer is reasonable compared to the estimate.

**StudyTip**

**Draw a Diagram** When solving word problems involving solids, it is helpful to draw a figure and label the known parts. Use a variable to label the measure or measures that you need to find.

### GuidedPractice

**4. ICE CREAM** A waffle cone is $5\frac{1}{2}$ inches tall and the diameter of the base is $2\frac{1}{2}$ inches. Find the lateral area of the cone. Round to the nearest tenth.

## Example 5 Surface Area of a Cone

**Find the surface area of a cone with a diameter of 14.8 centimeters and a slant height of 15 centimeters.**

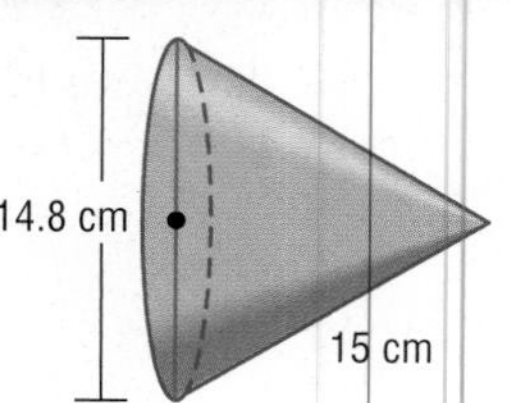


Estimate: $S \approx 3 \cdot 7 \cdot 20 + 3 \cdot 50$ or 570 cm$^2$

$S = \pi r\ell + \pi r^2$ Surface area of a cone

$= \pi(7.4)(15) + \pi(7.4)^2$ $r = 7.4$ and $\ell = 15$

$\approx 520.8$ Use a calculator.

The surface area of the cone is about 520.8 square centimeters. This is close to the estimate, so the answer is reasonable.

### GuidedPractice

**Find the surface area of each cone. Round to the nearest tenth.**

**5A.**

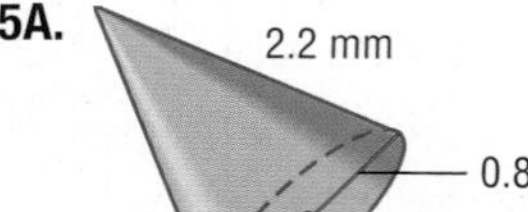


**5B.**

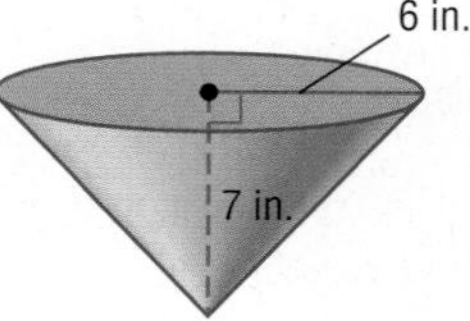


The formulas for lateral and surface area are summarized below.

### ConceptSummary Lateral and Surface Areas of Solids

| Solid | Model | Lateral Area | Surface Area |
|---|---|---|---|
| prism | | $L = Ph$ | $S = L + 2B$ or $S = Ph + 2B$ |
| cylinder | | $L = 2\pi rh$ | $S = L + 2B$ or $S = 2\pi rh + 2\pi r^2$ |
| pyramid | | $L = \frac{1}{2}P\ell$ | $S = \frac{1}{2}P\ell + B$ |
| cone | | $L = \pi r\ell$ | $S = \pi r\ell + \pi r^2$ |

**WatchOut!**

**Bases** The bases of right prisms and right pyramids are not always regular polygons.

## Check Your Understanding

**Examples 1–3** **Find the lateral area and surface area of each regular pyramid. Round to the nearest tenth if necessary.**

1. 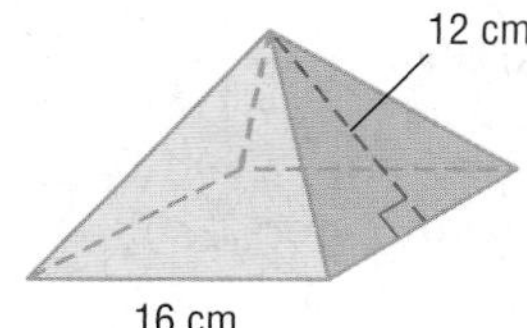


2. 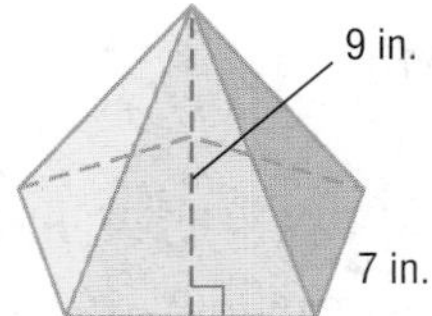


3. 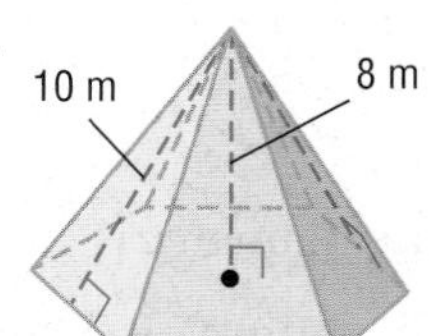


**Examples 4–5** 4. **TENTS** A conical tent is shown at the right. Round answers to the nearest tenth.

a. Find the lateral area of the tent and describe what it represents.

b. Find the surface area of the tent and describe what it represents.

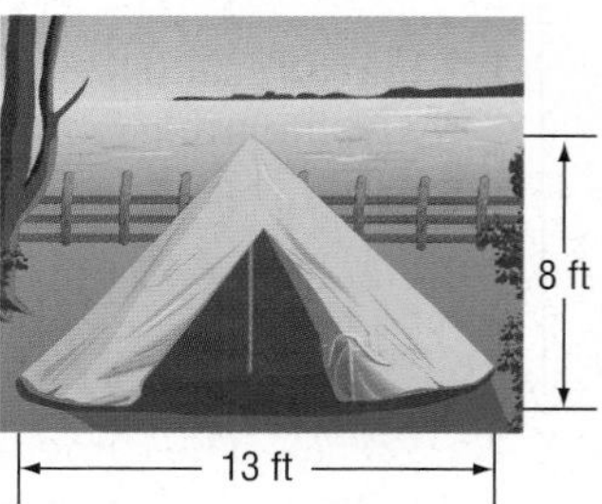


**CCSS SENSE-MAKING** **Find the lateral area and surface area of each cone. Round to the nearest tenth.**

5. 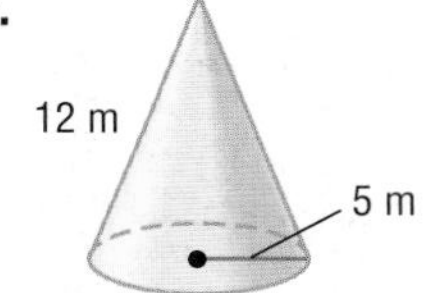


6. 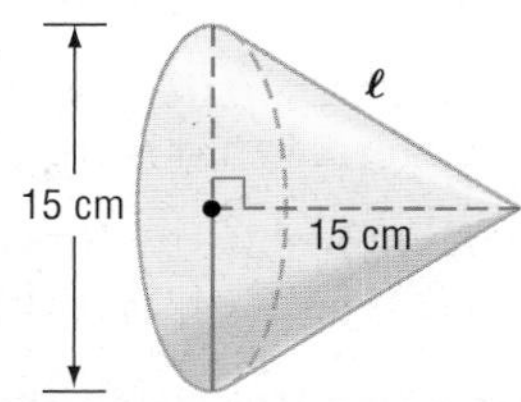


## Practice and Problem Solving

**Examples 1–3** **Find the lateral area and surface area of each regular pyramid. Round to the nearest tenth if necessary.**

7. 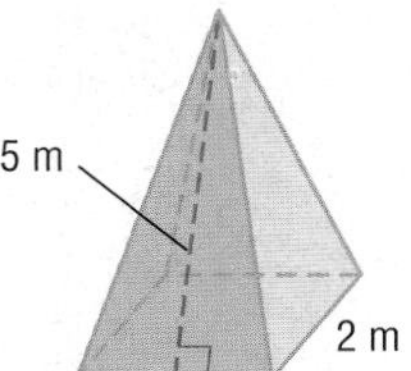


8. 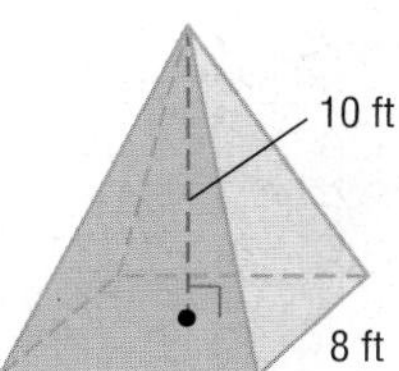


9. 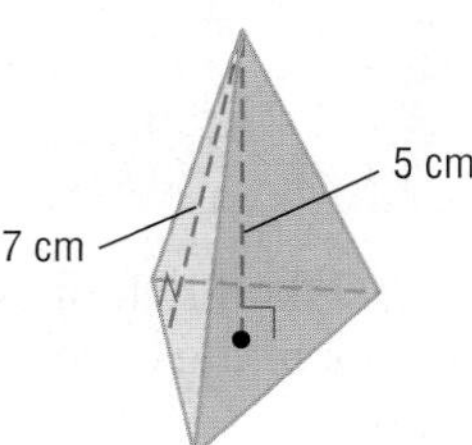


10. 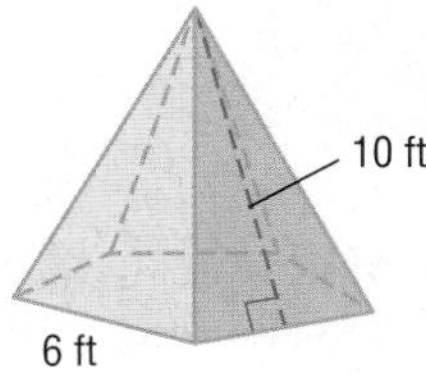


11. square pyramid with an altitude of 12 inches and a slant height of 18 inches

12. hexagonal pyramid with a base edge of 6 millimeters and a slant height of 9 millimeters

13. **ARCHITECTURE** Find the lateral area of a pyramid-shaped building that has a slant height of 210 feet and a square base 332 feet by 332 feet.

**Examples 4–5 Find the lateral area and surface area of each cone. Round to the nearest tenth.**

**14.**

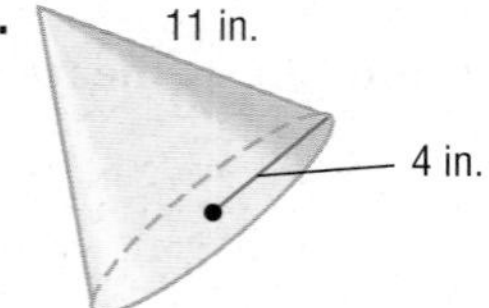


**15.**

12 cm

18 cm

**16.** The diameter is 3.4 centimeters, and the slant height is 6.5 centimeters.

**17.** The altitude is 5 feet, and the slant height is $9\frac{1}{2}$ feet.

**18. MOUNTAINS** A conical mountain has a radius of 1.6 kilometers and a height of 0.5 kilometer. What is the lateral area of the mountain?

**19. HISTORY** Archaeologists recently discovered a 1500-year-old pyramid in Mexico City. The square pyramid measures 165 yards on each side and once stood 20 yards tall. What was the original lateral area of the pyramid?

**20.** Describe two polyhedrons that have 7 faces.

**21.** What is the sum of the number of faces, vertices, and edges of an octagonal pyramid?

**22. TEPEES** The dimensions of two canvas tepees are shown in the table at the right. Not including the floors, approximately how much more canvas is used to make Tepee B than Tepee A?

| Tepee | Diameter (ft) | Height (ft) |
|---|---|---|
| A | 14 | 6 |
| B | 20 | 9 |

**23.** The surface area of a square pyramid is 24 square millimeters and the base area is 4 square millimeters. What is the slant height of the pyramid?

**24.** The surface area of a cone is $18\pi$ square inches and the radius of the base is 3 inches. What is the slant height of the cone?

**25.** The surface area of a triangular pyramid is 532 square centimeters, and the base is 24 centimeters wide with a hypotenuse of 25 centimeters. What is the slant height of the pyramid?

**26.** Find the lateral area of the tent to the nearest tenth.

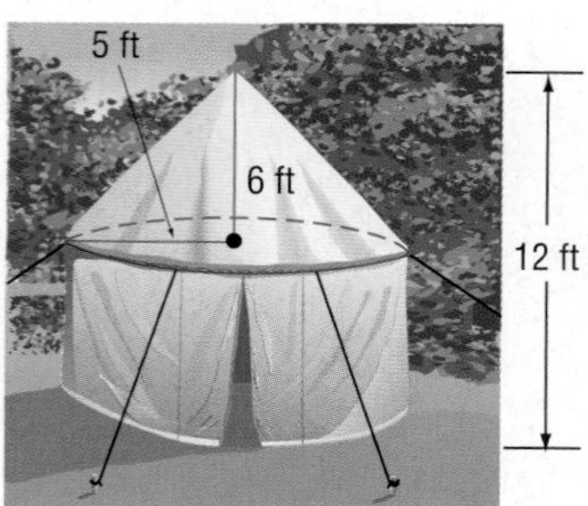


**27.** Find the surface area of the tank. Write in terms of $\pi$.

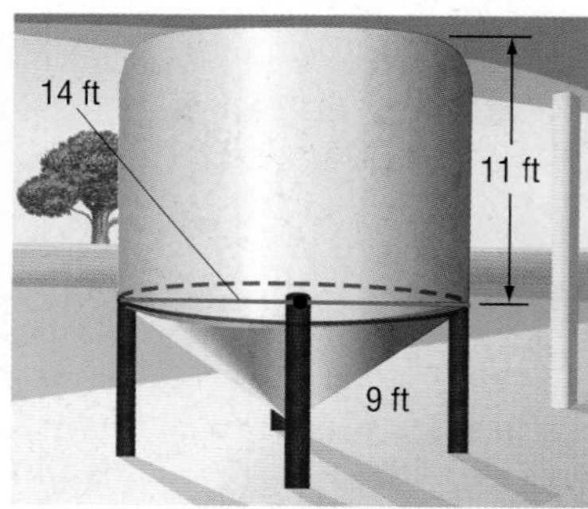


**28. CHANGING DIMENSIONS** A cone has a radius of 6 centimeters and a slant height of 12 centimeters. Describe how each change affects the surface area of the cone.

**a.** The radius and the slant height are doubled.

**b.** The radius and the slant height are divided by 3.

**29. CCSS TOOLS** A solid has the net shown at the right.

**a.** Describe the solid.

**b.** Make a sketch of the solid.

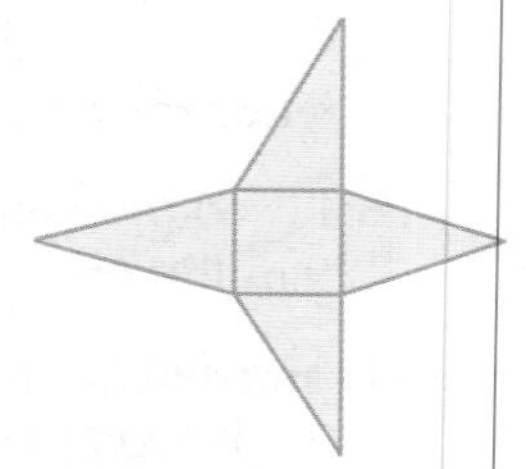

**Sketch each solid and a net that represents the solid.**

**30.** hexagonal pyramid

**31.** rectangular pyramid

**32.** **PETS** A *frustum* is the part of a solid that remains after the top portion has been cut by a plane parallel to the base. The ferret tent shown at the right is a frustum of a regular pyramid.

**a.** Describe the faces of the solid.

**b.** Find the lateral area and surface area of the frustum formed by the tent.

**c.** Another pet tent is made by cutting the top half off of a pyramid with a height of 12 centimeters, slant height of 20 centimeters and square base with side lengths of 32 centimeters. Find the surface area of the frustum.

**Find the lateral area and surface area of each solid. Round to the nearest tenth.**

**33.**

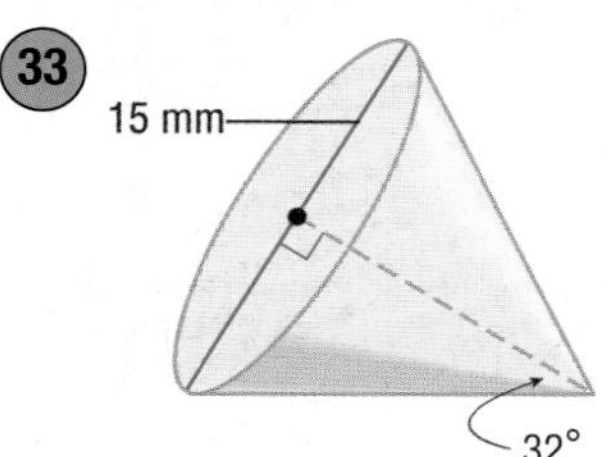


**34.**

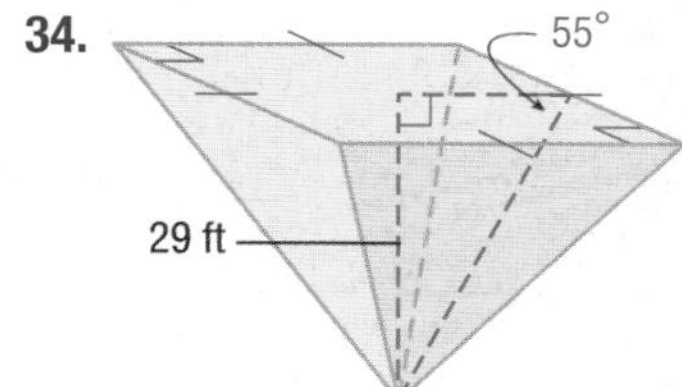


**35.** **MULTIPLE REPRESENTATIONS** In this problem, you will investigate the lateral and surface area of a square pyramid with a base edge of 3 units.

**a. Geometric** Sketch the pyramid on isometric dot paper.

**b. Tabular** Make a table showing the lateral areas of the pyramid for slant heights of 1, 3, and 9 units.

**c. Verbal** Describe what happens to the lateral area of the pyramid if the slant height is tripled.

**d. Analytical** Make a conjecture about how the lateral area of a square pyramid is affected if both the slant height and the base edge are tripled. Then test your conjecture.

## H.O.T. Problems Use Higher-Order Thinking Skills

**36.** **WRITING IN MATH** Why does an oblique solid not have a slant height?

**37.** **REASONING** Classify the following statement as *sometimes*, *always*, or *never* true. Justify your reasoning.

*The surface area of a cone of radius r and height h is less than the surface area of a cylinder of radius r and height h.*

**38.** **REASONING** A cone and a square pyramid have the same surface area. If the areas of their bases are also equal, do they have the same slant height as well? Explain.

**39.** **OPEN ENDED** Describe a pyramid that has a total surface area of 100 square units.

**40.** **CCSS ARGUMENTS** Determine whether the following statement is *true* or *false*. Explain your reasoning.

*A regular polygonal pyramid and a cone both have height h units and base perimeter P units. Therefore, they have the same total surface area.*

**41.** **WRITING IN MATH** Describe how to find the surface area of a regular polygonal pyramid with an $n$-gon base, height $h$ units, and an apothem of $a$ units.

## Standardized Test Practice

**42.** The top of a gazebo in a park is in the shape of a regular pentagonal pyramid. Each side of the pentagon is 10 feet long. If the slant height of the roof is about 6.9 feet, what is the lateral roof area?

**A** 34.5 ft$^2$ **C** 172.5 ft$^2$

**B** 50 ft$^2$ **D** 250 ft$^2$

**43. SHORT RESPONSE** To the nearest square millimeter, what is the surface area of a cone with the dimensions shown?

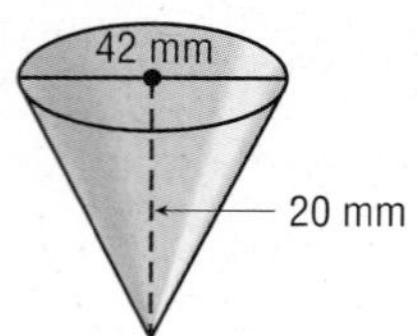


**44. ALGEBRA** Yu-Jun's craft store sells 3 handmade barrettes for $9.99. Which expression can be used to find the total cost $C$ of $x$ barrettes?

**F** $C = \frac{9.99}{x}$ **H** $C = 3.33x$

**G** $C = 9.99x$ **J** $C = \frac{x}{3.33}$

**45. SAT/ACT** What is the slope of a line perpendicular to the line with equation $2x + 3y = 9$?

**A** $-\frac{3}{2}$ **D** $\frac{3}{2}$

**B** $-\frac{2}{3}$ **E** $\frac{9}{2}$

**C** $\frac{2}{3}$

## Spiral Review

**46.** Find the surface area of a cylinder with a diameter of 18 cm and a height of 12 cm. (Lesson 12-2)

**Use isometric dot paper and each orthographic drawing to sketch a solid.** (Lesson 12-1)

**47.**

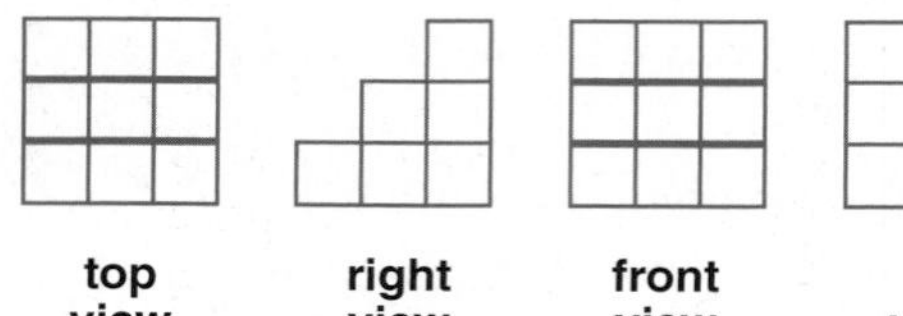


**48.**

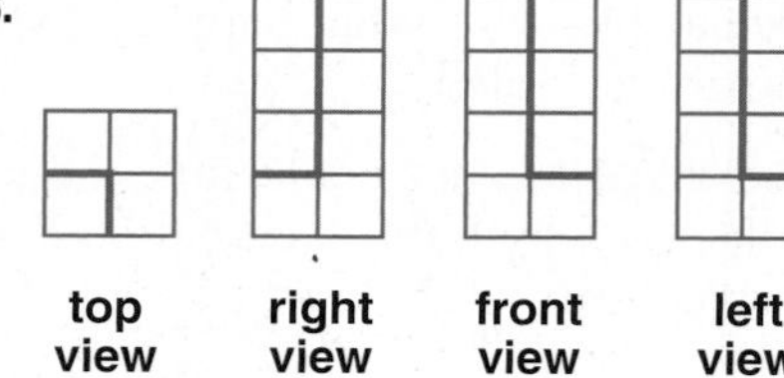


**$\overline{EC}$ and $\overline{AB}$ are diameters of $\odot O$. Identify each arc as a *major arc, minor arc,* or *semicircle* of the circle. Then find its measure.** (Lesson 11-2)

**49.** $m\widehat{ACB}$ **50.** $m\widehat{EB}$ **51.** $m\widehat{ACE}$

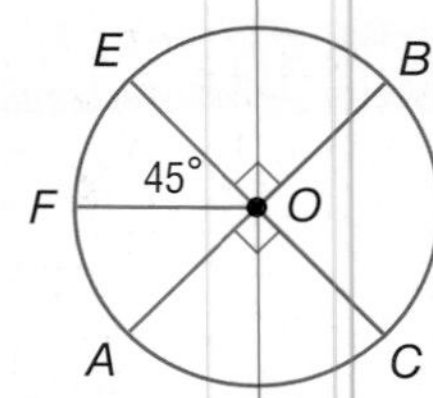


## Skills Review

**Find the perimeter and area of each parallelogram, triangle, or composite figure. Round to the nearest tenth.**

**52.**

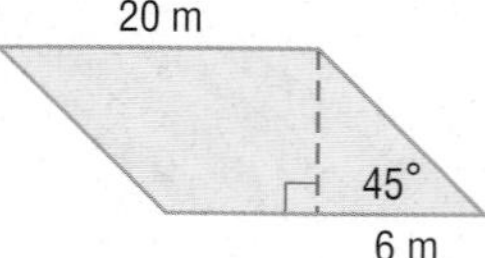


**53.**

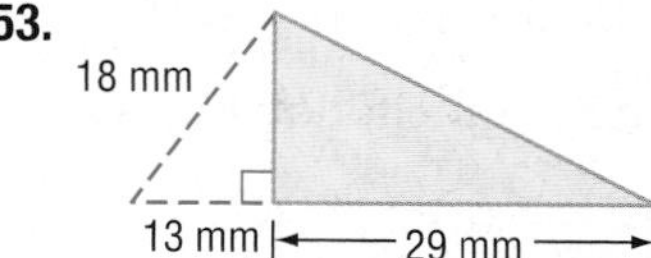


**54.**

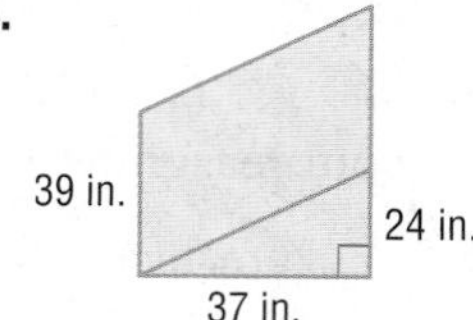

LESSON 12-4

# Volumes of Prisms and Cylinders

**Then**

- You found surface areas of prisms and cylinders.

**Now**

1. Find volumes of prisms.
2. Find volumes of cylinders.

**Why?**

- Planters come in a variety of shapes and sizes. You can approximate the amount of soil needed to fill a planter by finding the volume of the three-dimensional figure that it most resembles.

**Common Core State Standards**

**Content Standards**

**G.GMD.1** Give an informal argument for the formulas for the circumference of a circle, area of a circle, volume of a cylinder, pyramid, and cone.

**G.GMD.3** Use volume formulas for cylinders, pyramids, cones, and spheres to solve problems. ★

**Mathematical Practices**

1 Make sense of problems and persevere in solving them.

7 Look for and make use of structure.

## 1 Volume of Prisms

Recall that the volume of a solid is the measure of the amount of space the solid encloses. Volume is measured in cubic units.

The rectangular prism at the right has 6 • 4 or 24 cubic units in the bottom layer. Since there are two layers, the total volume is 24 • 2 or 48 cubic units.

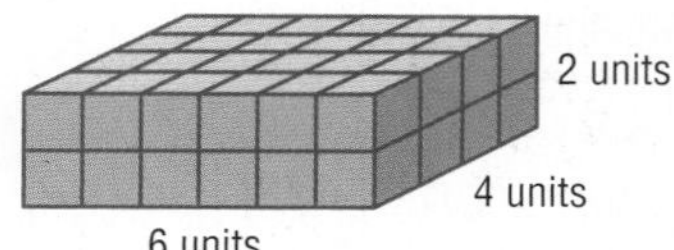


### KeyConcept Volume of a Prism

**Words** The volume $V$ of a prism is $V = Bh$, where $B$ is the area of a base and $h$ is the height of the prism.

**Symbols** $V = Bh$

**Model**

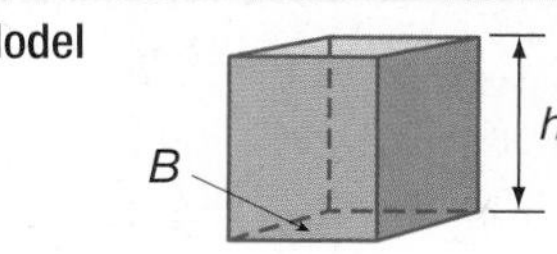

### Example 1 Volume of a Prism

**Find the volume of the prism.**

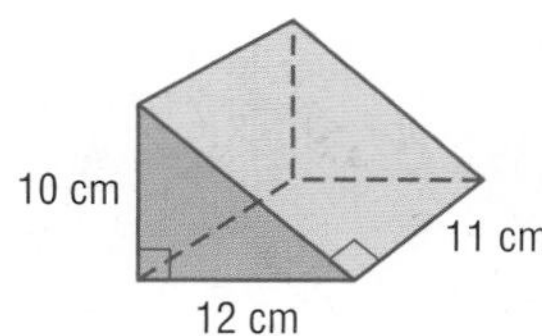


**Step 1** Find the area of the base $B$.

$B = \frac{1}{2}bh$ — Area of a triangle

$= \frac{1}{2}(12)(10)$ or 60 — $b = 12$ and $h = 10$

**Step 2** Find the volume of the prism.

$V = Bh$ — Volume of a prism

$= 60(11)$ or 660 — $B = 60$ and $h = 11$

The volume of the prism is 660 cubic centimeters.

**GuidedPractice**

**1A.**

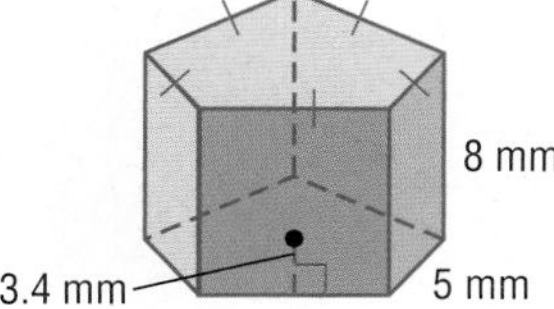


**1B.**

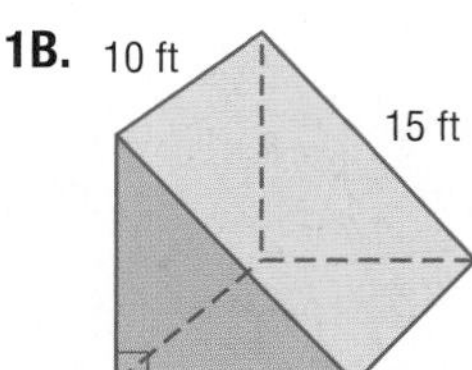

**Real-WorldCareer**

**Architectural Engineer** An architectural engineer applies the technical skills of engineering to the design, construction, operation, maintenance, and renovation of buildings.

Architectural engineers are required to have a bachelor's degree in engineering along with specialized coursework. Refer to Exercise 35.

## 2 Volume of Cylinders

Like a prism, the volume of a cylinder can be thought of as consisting of layers. For a cylinder, these layers are congruent circular discs, similar to the coins in the roll shown. If we interpret the area of the base as the volume of a one-unit-high layer and the height of the cylinder as the number of layers, then the volume of the cylinder is equal to the volume of a layer times the number of layers or the area of the base times the height.

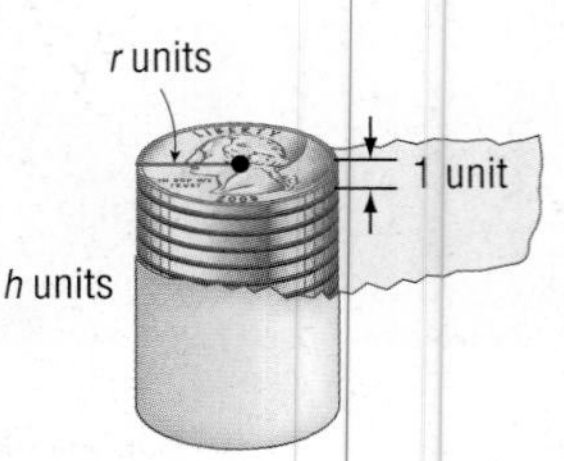


### KeyConcept Volume of a Cylinder

**Words** The volume $V$ of a cylinder is $V = Bh$ or $V = \pi r^2 h$, where $B$ is the area of the base, $h$ is the height of the cylinder, and $r$ is the radius of the base.

**Symbols** $V = Bh$ or $V = \pi r^2 h$

**Model**

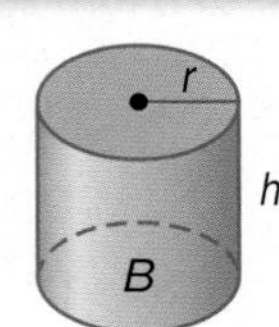


### Example 2 Volume of a Cylinder

**Find the volume of the cylinder at the right.**

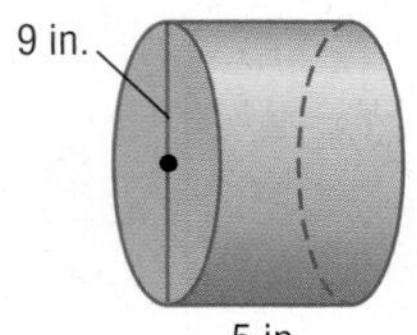


Estimate: $V \approx 3 \cdot 5^2 \cdot 5$ or 375 in$^3$

| | |
|---|---|
| $V = \pi r^2 h$ | Volume of a cylinder |
| $= \pi(4.5)^2(5)$ | $r = 4.5$ and $h = 5$ |
| $\approx 318.1$ | Use a calculator. |

The volume of the cylinder is about 318.1 cubic inches. This is fairly close to the estimate, so the answer is reasonable.

**GuidedPractice**

**2.** Find the volume of a cylinder with a radius of 3 centimeters and a height of 8 centimeters. Round to the nearest tenth.

The first group of books at the right represents a right prism. The second group represents an oblique prism. Both groups have the same number of books. If all the books are the same size, then the volume of both groups is the same.

This demonstrates the following principle, which applies to all solids.

**WatchOut!**

**Cross-Sectional Area** For solids with the same height to have the same volume, their cross-sections must have the same area. The cross sections of the different solids do not have to be congruent polygons.

### KeyConcept Cavalieri's Principle

**Words** If two solids have the same height $h$ and the same cross-sectional area $B$ at every level, then they have the same volume.

**Models**

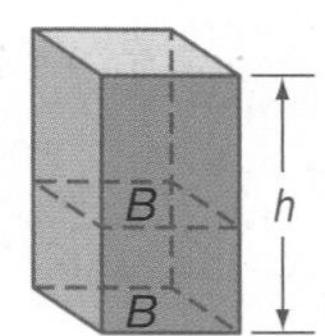


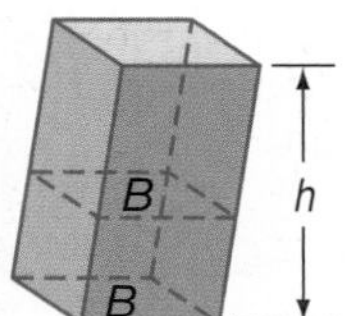


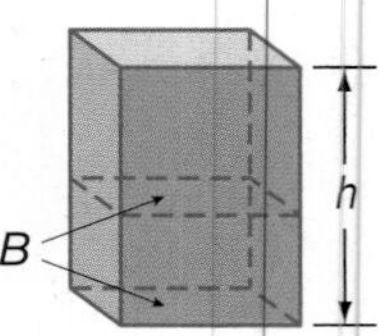


These prisms all have a volume of $Bh$.

Javier Larrea/age fotostock

**Problem-Solving Tip**

**Make a Model** When solving problems involving volume of solids, one way to help you visualize the problem is to make a model of the solid.

## Example 3 Volume of an Oblique Solid

**Find the volume of an oblique hexagonal prism if the height is 6.4 centimeters and the base area is 17.3 square centimeters.**

$V = Bh$ — Volume of a prism

$= 17.3(6.4)$ — $B = 17.3$ and $h = 6.4$

$= 110.72$ — Simplify.

The volume is 110.72 cubic centimeters.

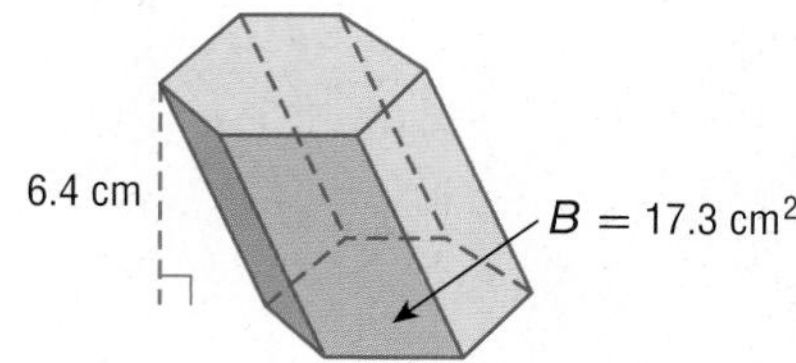


### Guided Practice

**3.** Find the volume of an oblique cylinder that has a radius of 5 feet and a height of 3 feet. Round to the nearest tenth.

## Standardized Test Example 4 Comparing Volumes of Solids

**Prisms A and B have the same length and width, but different heights. If the volume of Prism B is 150 cubic inches greater than the volume of Prism A, what is the length of each prism?**

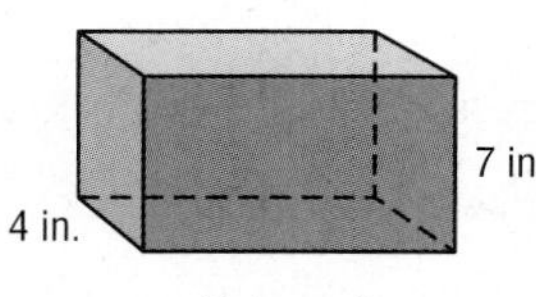


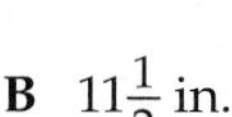

**Prism A**

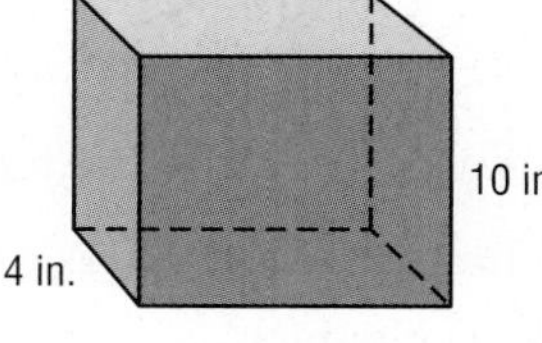


**Prism B**

**A** 10 in. **B** $11\frac{1}{2}$ in. **C** 12 in. **D** $12\frac{1}{2}$ in.

**Read the Test Item**

You know two dimensions of each solid and that the difference between their volumes is 150 cubic inches.

**Test-Taking Tip**

**Writing Equations** By choosing a variable to represent the unknown and setting up an equation, you make it easier to check for mistakes as you work out a problem.

**Solve the Test Item**

Volume of Prism B − Volume of Prism A = 150 — Write an equation.

$4\ell \cdot 10 - 4\ell \cdot 7 = 150$ — Use $V = Bh$.

$12\ell = 150$ — Simplify.

$\ell = 12\frac{1}{2}$ — Divide each side by 12.

The length of each prism is $12\frac{1}{2}$ inches. The correct answer is D.

### Guided Practice

**4.** The containers at the right are filled with popcorn. About how many times as much popcorn does the larger container hold?

**F** 1.6 times as much

**G** 2.5 times as much

**H** 3.3 times as much

**J** 5.0 times as much

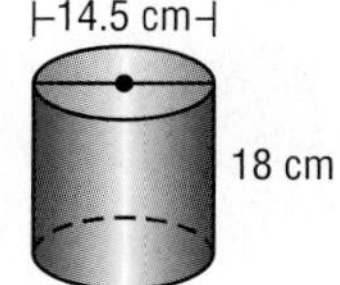


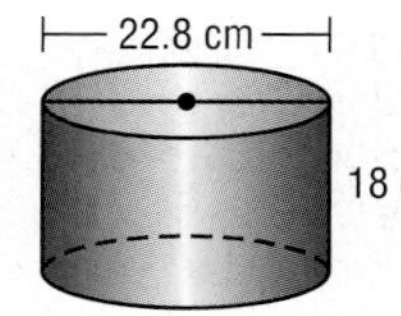

## Check Your Understanding

**Examples 1 and 3** **Find the volume of each prism.**

**1.**

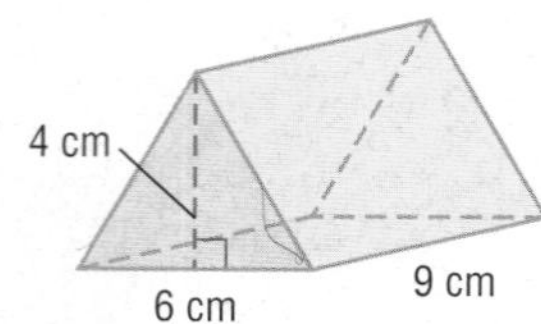


**2.**

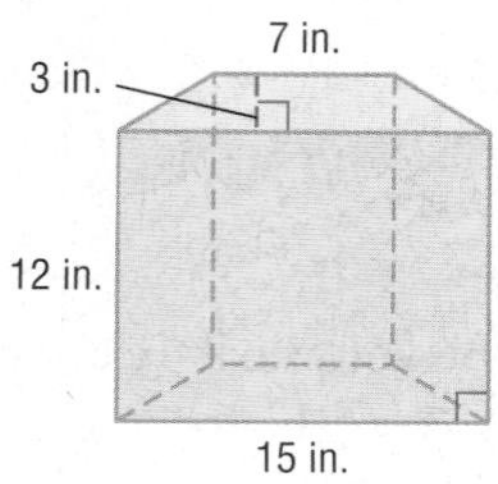


**3.** the oblique rectangular prism shown at the right

**4.** an oblique pentagonal prism with a base area of 42 square centimeters and a height of 5.2 centimeters

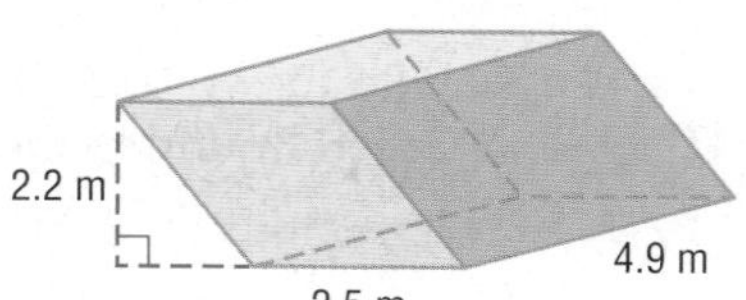


**Examples 2–3** **Find the volume of each cylinder. Round to the nearest tenth.**

**5.**

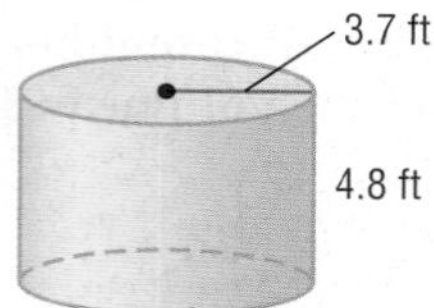


**6.**

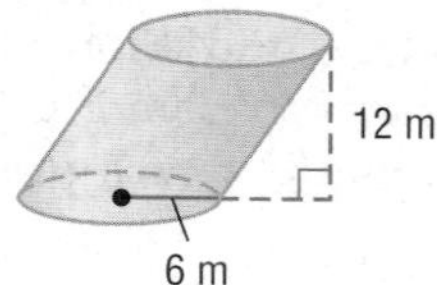


**7.** a cylinder with a diameter of 16 centimeters and a height of 5.1 centimeters

**8.** a cylinder with a radius of 4.2 inches and a height of 7.4 inches

**Example 4** **9. MULTIPLE CHOICE** A rectangular lap pool measures 80 feet long by 20 feet wide. If it needs to be filled to four feet deep and each cubic foot holds 7.5 gallons, how many gallons will it take to fill the lap pool?

**A** 4000 **B** 6400 **C** 30,000 **D** 48,000

## Practice and Problem Solving

**Examples 1 and 3** **CCSS SENSE-MAKING** **Find the volume of each prism.**

**10.**

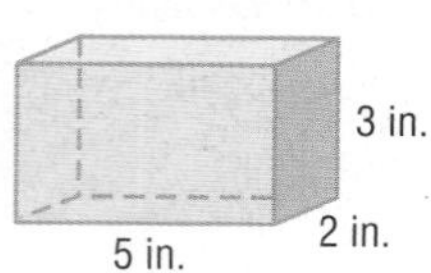


**11**

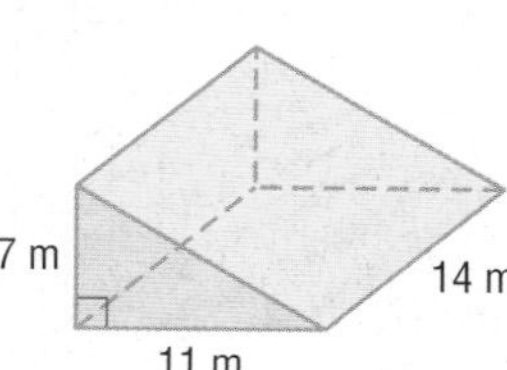


**12.**

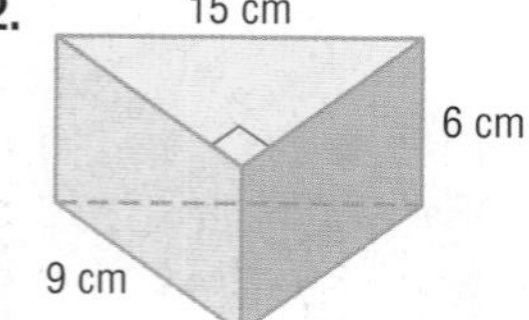


**13.**

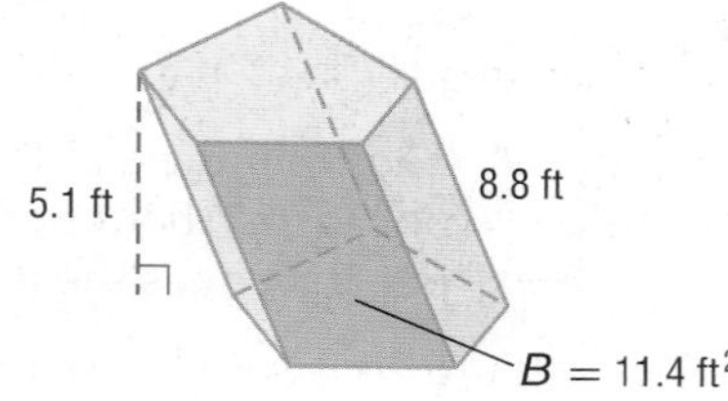


**14.** an oblique hexagonal prism with a height of 15 centimeters and with a base area of 136 square centimeters

**15.** a square prism with a base edge of 9.5 inches and a height of 17 inches

**Examples 2–3** **CCSS SENSE-MAKING** **Find the volume of each cylinder. Round to the nearest tenth.**

**16.**

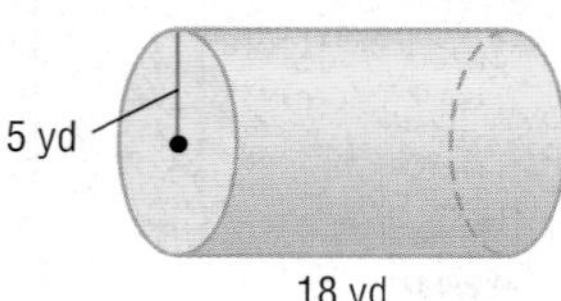


**17**

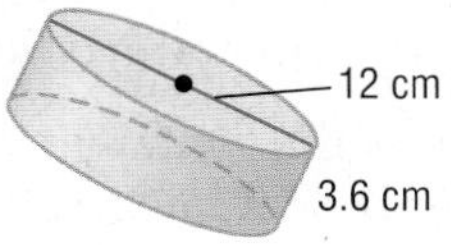

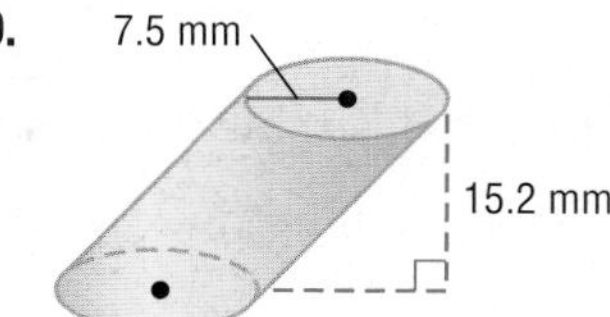


**18.**

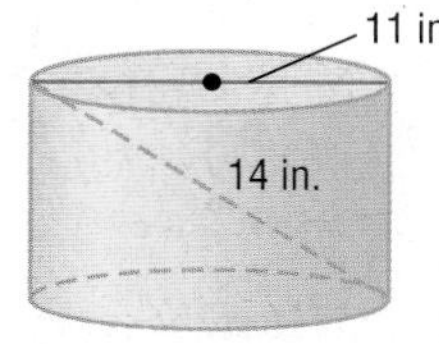


**19.**

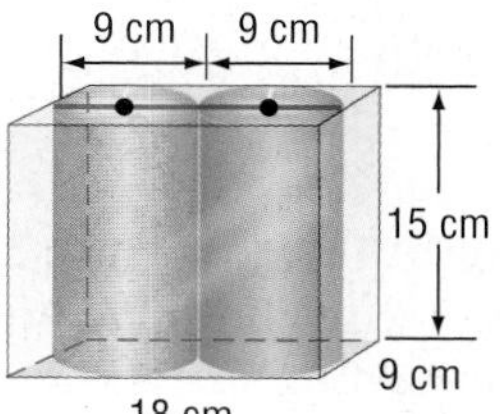


**Example 4**

**20. PLANTER** A planter is in the shape of a rectangular prism 18 inches long, $14\frac{1}{2}$ inches deep, and 12 inches high. What is the volume of potting soil in the planter if the planter is filled to $1\frac{1}{2}$ inches below the top?

**21. SHIPPING** A box 18 centimeters by 9 centimeters by 15 centimeters is being used to ship two cylindrical candles. Each candle has a diameter of 9 centimeters and a height of 15 centimeters, as shown at the right. What is the volume of the empty space in the box?

9 cm 9 cm

15 cm

9 cm

18 cm

**22. SANDCASTLES** In a sandcastle competition, contestants are allowed to use only water, shovels, and 10 cubic feet of sand. To transport the correct amount of sand, they want to create cylinders that are 2 feet tall to hold enough sand for one contestant. What should the diameter of the cylinders be?

**Find the volume of the solid formed by each net.**

**23.**

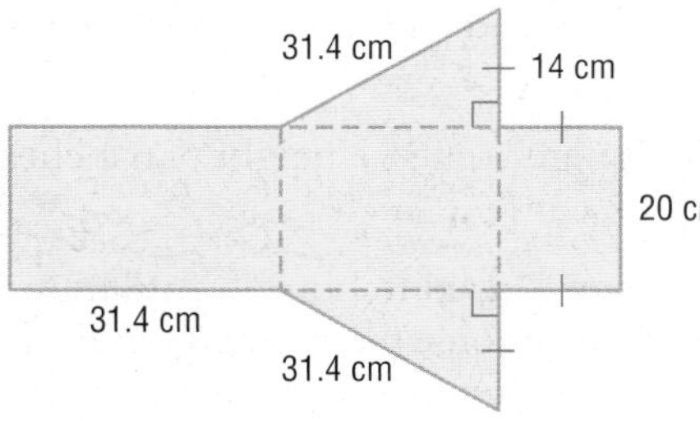


**24.**

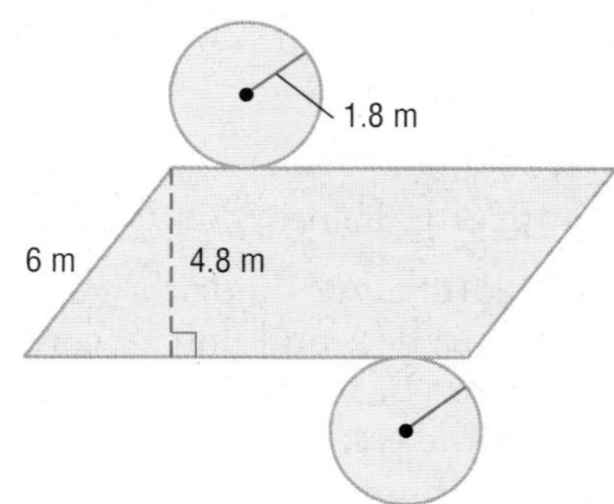


**25. FOOD** A cylindrical can of baked potato chips has a height of 27 centimeters and a radius of 4 centimeters. A new can is advertised as being 30% larger than the regular can. If both cans have the same radius, what is the height of the larger can?

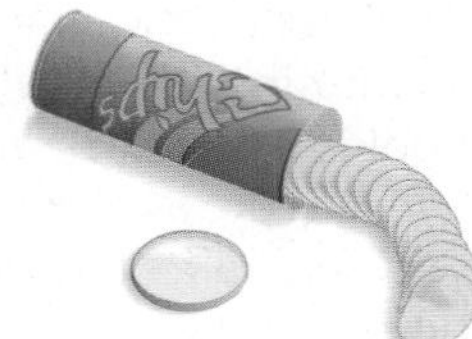

**26. CHANGING DIMENSIONS** A cylinder has a radius of 5 centimeters and a height of 8 centimeters. Describe how each change affects the volume of the cylinder.

**a.** The height is tripled.

**b.** The radius is tripled.

**c.** Both the radius and the height are tripled.

**d.** The dimensions are exchanged.

27. **SOIL** A soil scientist wants to determine the bulk density of a potting soil to assess how well a specific plant will grow in it. The density of the soil sample is the ratio of its weight to its volume.

20 in.

25 in.

a. If the weight of the container with the soil is 20 pounds and the weight of the container alone is 5 pounds, what is the soil's bulk density?

b. Assuming that all other factors are favorable, how well should a plant grow in this soil if a bulk density of 0.0018 pound per square inch is desirable for root growth? Explain.

c. If a bag of this soil holds 2.5 cubic feet, what is its weight in pounds?

**Find the volume of each composite solid. Round to the nearest tenth if necessary.**

28.

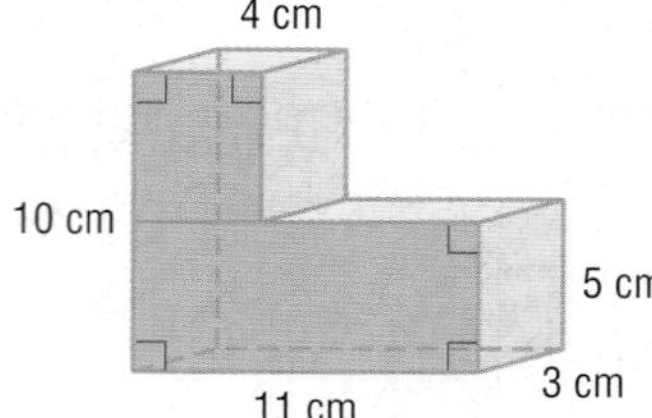


29.

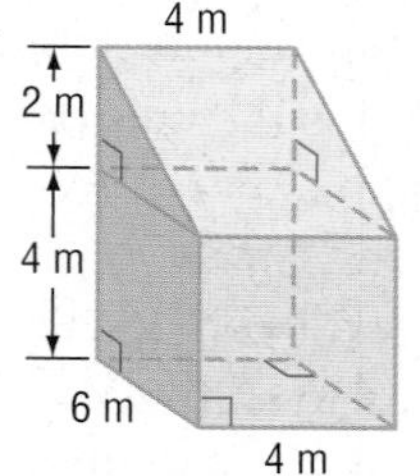


30.

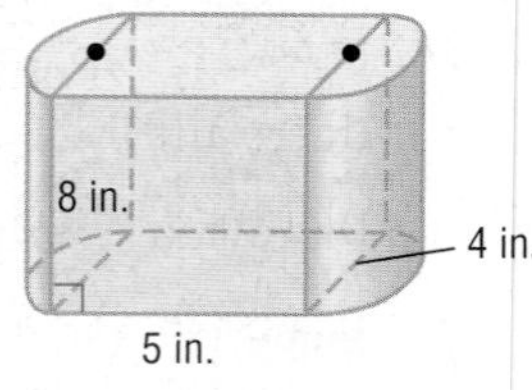


31. **MANUFACTURING** A can 12 centimeters tall fits into a rubberized cylindrical holder that is 11.5 centimeters tall, including 1 centimeter for the thickness of the base of the holder. The thickness of the rim of the holder is 1 centimeter. What is the volume of the rubberized material that makes up the holder?

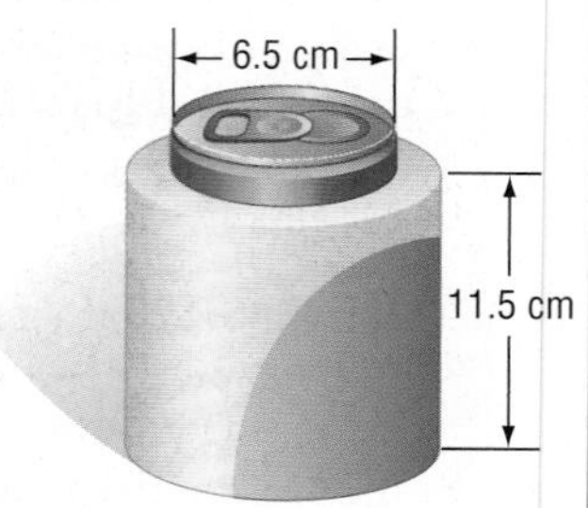


**Find each measure to the nearest tenth.**

32. A cylindrical can has a volume of 363 cubic centimeters. The diameter of the can is 9 centimeters. What is the height?

33. A cylinder has a surface area of $144\pi$ square inches and a height of 6 inches. What is the volume?

34. A rectangular prism has a surface area of 432 square inches, a height of 6 inches, and a width of 12 inches. What is the volume?

35. **ARCHITECTURE** A cylindrical stainless steel column is used to hide a ventilation system in a new building. According to the specifications, the diameter of the column can be between 30 centimeters and 95 centimeters. The height is to be 500 centimeters. What is the difference in volume between the largest and smallest possible column? Round to the nearest tenth cubic centimeter.

36. CCSS **MODELING** The base of a rectangular swimming pool is sloped so one end of the pool is 6 feet deep and the other end is 3 feet deep, as shown in the figure. If the width is 15 feet, find the volume of water it takes to fill the pool.

20 ft

6 ft

10 ft

37. **CHANGING DIMENSIONS** A soy milk company is planning a promotion in which the volume of soy milk in each container will be increased by 25%. The company wants the base of the container to stay the same. What will be the height of the new containers?

38. **DESIGN** Sketch and label (in inches) three different designs for a dry ingredient measuring cup that holds 1 cup. Be sure to include the dimensions in each drawing. (1 cup $\approx$ 14.4375 $in^3$)

39. Find the volume of the regular pentagonal prism at the right by dividing it into five equal triangular prisms. Describe the base area and height of each triangular prism.

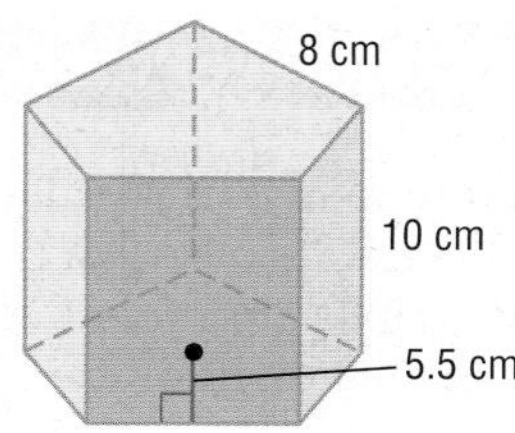


40. **PATIOS** Mr. Thomas is planning to remove an old patio and install a new rectangular concrete patio 20 feet long, 12 feet wide, and 4 inches thick. One contractor bid $2225 for the project. A second contractor bid $500 per cubic yard for the new patio and $700 for removal of the old patio. Which is the less expensive option? Explain.

41. **MULTIPLE REPRESENTATIONS** In this problem, you will investigate cylinders.

  a. **Geometric** Draw a right cylinder and an oblique cylinder with a height of 10 meters and a diameter of 6 meters.

  b. **Verbal** A square prism has a height of 10 meters and a base edge of 6 meters. Is its volume greater than, less than, or equal to the volume of the cylinder? Explain.

  c. **Analytical** Describe which change affects the volume of the cylinder more: multiplying the height by $x$ or multiplying the radius by $x$. Explain.

## H.O.T. Problems Use Higher-Order Thinking Skills

42. **CCSS CRITIQUE** Francisco and Valerie each calculated the volume of an equilateral triangular prism with an apothem of 4 units and height of 5 units. Is either of them correct? Explain your reasoning.

| Francisco | Valerie |
|---|---|
| $V = Bh$<br>$= \frac{1}{2}aP \cdot h$<br>$= \frac{1}{2}(4)(24\sqrt{3}) \cdot 5$<br>$= 240\sqrt{3}$ cubic units | $V = Bh$<br>$= \frac{\sqrt{3}}{2}s^2 \cdot h$<br>$= \frac{\sqrt{3}}{2}(4\sqrt{3})^2 \cdot 5$<br>$= 120\sqrt{3}$ cubic units |

43. **CHALLENGE** The cylindrical can below is used to fill a container with liquid. It takes three full cans to fill the container. Describe possible dimensions of the container if it is each of the following shapes.

  a. rectangular prism

  b. square prism

  c. triangular prism with a right triangle as the base

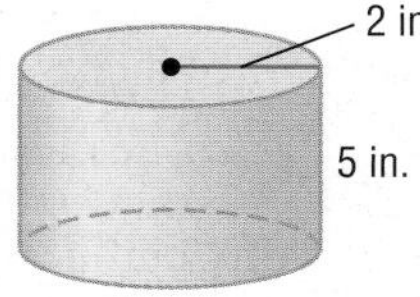


44. **WRITING IN MATH** Write a helpful response to the following question posted on an Internet gardening forum.

*I am new to gardening. The nursery will deliver a truckload of soil, which they say is 4 yards. I know that a yard is 3 feet, but what is a yard of soil? How do I know what to order?*

45. **OPEN ENDED** Draw and label a prism that has a volume of 50 cubic centimeters.

46. **REASONING** Determine whether the following statement is true or false. Explain.

*Two cylinders with the same height and the same lateral area must have the same volume.*

47. **WRITING IN MATH** How are the volume formulas for prisms and cylinders similar? How are they different?

## Standardized Test Practice

**48.** The volume of a triangular prism is 1380 cubic centimeters. Its base is a right triangle with legs measuring 8 centimeters and 15 centimeters. What is the height of the prism?

A 34.5 cm  
B 23 cm  
C 17 cm  
D 11.5 cm

**49.** A cylindrical tank used for oil storage has a height that is half the length of its radius. If the volume of the tank is 1,122,360 $ft^3$, what is the tank's radius?

F 89.4 ft  
G 178.8 ft  
H 280.9 ft  
J 561.8 ft

**50. SHORT RESPONSE** What is the ratio of the area of the circle to the area of the square?

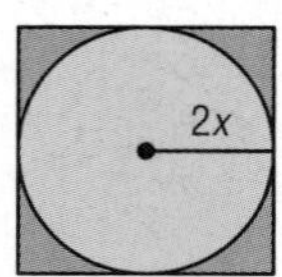


**51. SAT/ACT** A county proposes to enact a new 0.5% property tax. What would be the additional tax amount for a landowner whose property has a taxable value of $85,000?

A $4.25  
B $170  
C $425  
D $4250  
E $42,500

## Spiral Review

**Find the lateral area and surface area of each regular pyramid. Round to the nearest tenth if necessary.** (Lesson 12-3)

**52.**

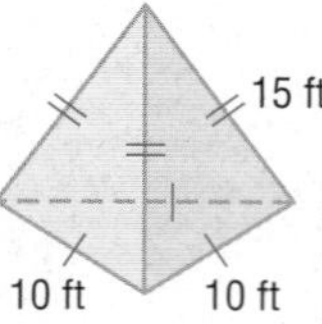


**53.**

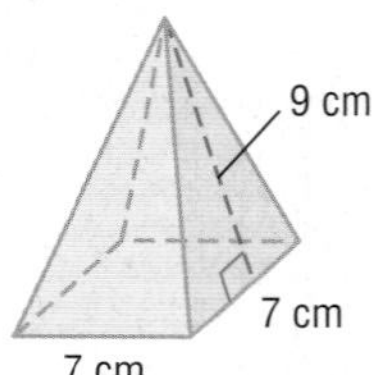


**54.**

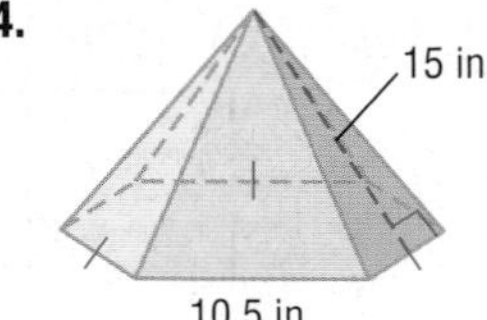


**55. BAKING** Many baking pans are given a special nonstick coating. A rectangular cake pan is 9 inches by 13 inches by 2 inches deep. What is the area of the inside of the pan that needs to be coated? (Lesson 12-2)

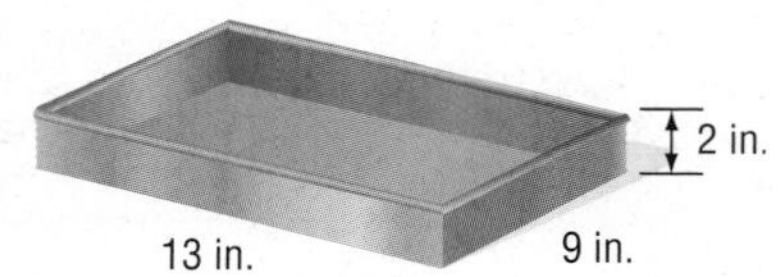


**Find the indicated measure. Round to the nearest tenth.** (Lesson 11-9)

**56.** The area of a circle is 54 square meters. Find the diameter.

**57.** Find the diameter of a circle with an area of 102 square centimeters.

**58.** The area of a circle is 191 square feet. Find the radius.

**59.** Find the radius of a circle with an area of 271 square inches.

## Skills Review

**Find the area of each trapezoid, rhombus, or kite.**

**60.**

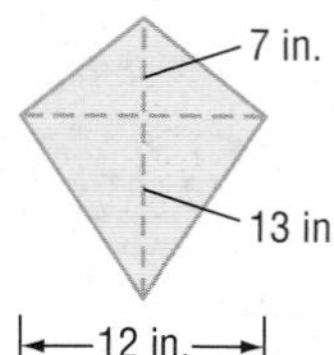


**61.**

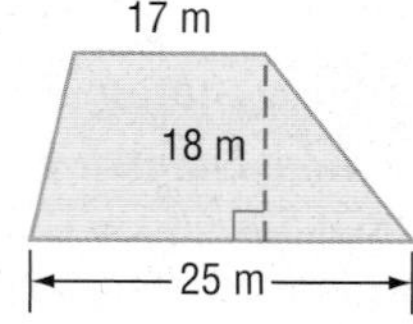


**62.**

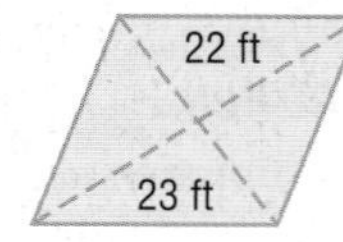

EXTEND 12-4

# Graphing Technology Lab
# Changing Dimensions

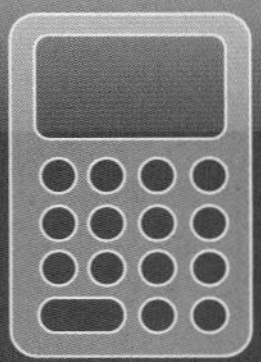

You can use TI-Nspire Technology to investigate how changes in dimension affect the surface area and volume of a rectangular prism.

## Activity

**Step 1** Open a new **Lists & Spreadsheet** page.

**Step 2** Move the cursor to the space beside the letter in each column and label the columns $\ell$ for length, $w$ for width, $h$ for height, $sa$ for surface area, and $v$ for volume.

**Step 3** Insert the values for length, width, and height shown into the table.

1.1 RAD AUTO REAL

| | A l | B w | C h | D sa | E v | F | G |
|---|---|---|---|---|---|---|---|
| 1 | 1 | 2 | 3 | | | | |
| 2 | 1 | 2 | 6 | | | | |
| 3 | 1 | 4 | 6 | | | | |
| 4 | 2 | 4 | 6 | | | | |
| 5 | 3 | 6 | 9 | | | | |

D1

**Step 4** Enter the formula for the surface area in terms of cells A1, B1, and C1 in cell D1.

**Step 5** Enter the formula for the volume in terms of cells A1, B1, and C1 in cell E1.

**Step 6** Highlight cell D1 and select **Fill Down** from the **Data** menu. Scroll down to fill in the surface areas for the other prisms. Repeat the process for volume.

1.1 RAD AUTO REAL

| | A l | B w | C h | D sa | E v | F | G |
|---|---|---|---|---|---|---|---|
| 1 | 1 | 2 | 3 | 22 | 6 | | |
| 2 | 1 | 2 | 6 | | | | |
| 3 | 1 | 4 | 6 | | | | |
| 4 | 2 | 4 | 6 | | | | |
| 5 | 3 | 6 | 9 | | | | |

D2

**Step 7** Add additional values and observe the effect on surface area and volume as one or more of the dimensions changes.

## Analyze the Results

1. How does the surface area change when one of the dimensions is doubled? two of the dimensions? all three of the dimensions?
2. How does the volume change when one of the dimensions is doubled? two of the dimensions? all three of the dimensions?
3. How does the surface area change when all three of the dimensions are tripled?
4. How does the volume change when all three of the dimensions are tripled?
5. **MAKE A CONJECTURE** If the dimensions of a prism are all multiplied by a factor of 5, what do you think the ratio of the new surface area to the original surface area will be? the ratio of the new volume to the original volume? Explain.
6. **CHALLENGE** Write an expression for the ratio of the surface areas and the ratio of the volumes if all three of the dimensions of a prism are increased by a scale factor of $k$. Explain.

CHAPTER 12

# Mid-Chapter Quiz

## Lessons 12-1 through 12-4

1. Describe how to use isometric dot paper to sketch the following figure. (Lesson 12-1)

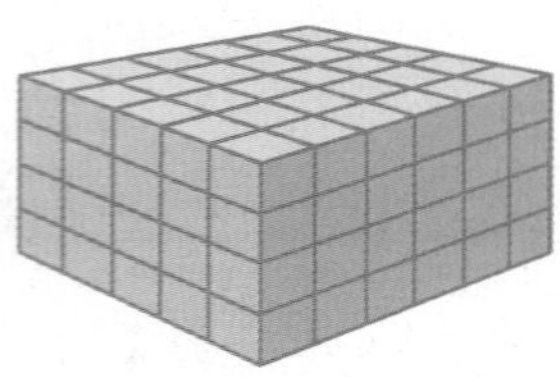

2. Use isometric dot paper to sketch a rectangular prism 2 units high, 3 units long, and 6 units wide. (Lesson 12-1)

3. Use isometric dot paper to sketch a triangular prism 5 units high, with two sides of the base that are 4 units long and 3 units long. (Lesson 12-1)

**Find the lateral area of each prism. Round to the nearest tenth if necessary.** (Lesson 12-2)

4. 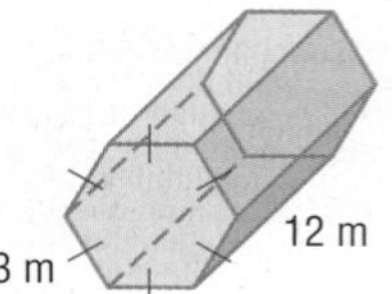


5. 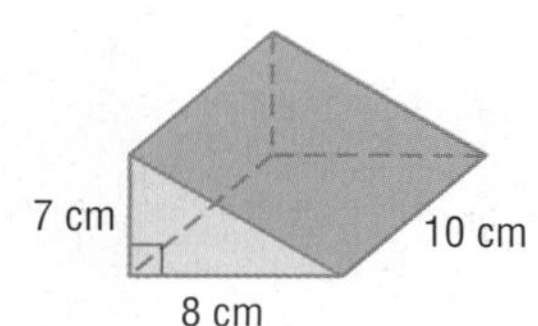


6. **MULTIPLE CHOICE** Coaxial cable is used to transmit long-distance telephone calls, cable television programming, and other communications. A typical coaxial cable contains 22 copper tubes and has a diameter of 3 inches. What is the approximate lateral area of a coaxial cable that is 500 feet long? (Lesson 12-2)

   **A** 16.4 $ft^2$  **C** 294.5 $ft^2$

   **B** 196.3 $ft^2$  **D** 392.7 $ft^2$

**Find the lateral area and surface area of each cylinder. Round to the nearest tenth if necessary.** (Lesson 12-2)

7. 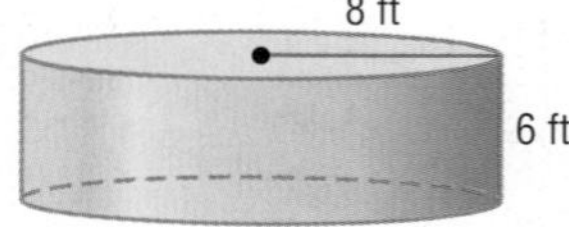


8. 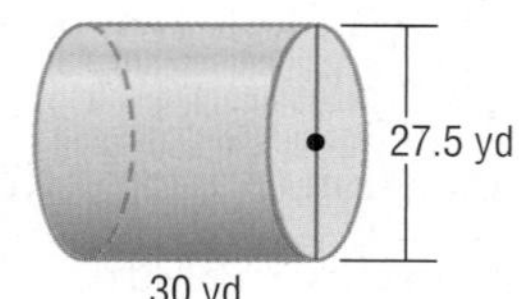


9. 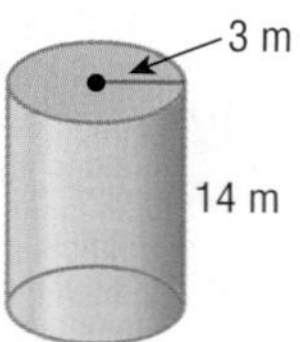


10. 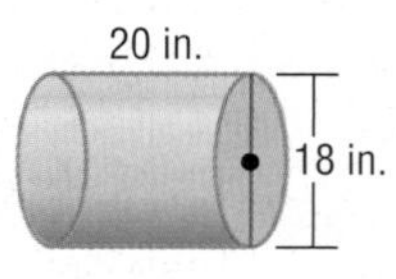


11. **COLLECTIONS** Soledad collects unique salt and pepper shakers. She inherited a pair of tetrahedral shakers from her mother. (Lesson 12-3)

    **a.** Each edge of a shaker measures 3 centimeters. Make a sketch of one shaker.

    **b.** Find the total surface area of one shaker.

**Find the surface area of each regular pyramid or cone. Round to the nearest tenth if necessary.** (Lesson 12-3)

12. 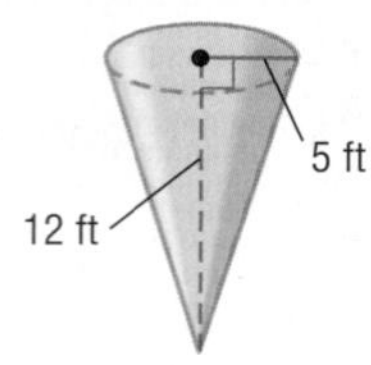


13. 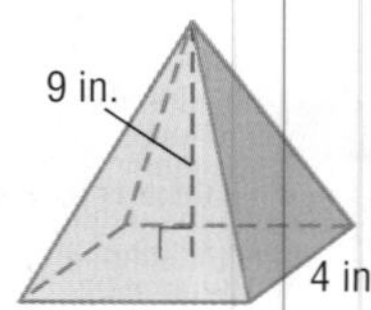


**Find the volume of each prism or cylinder. Round to the nearest tenth if necessary.** (Lesson 12-4)

14. 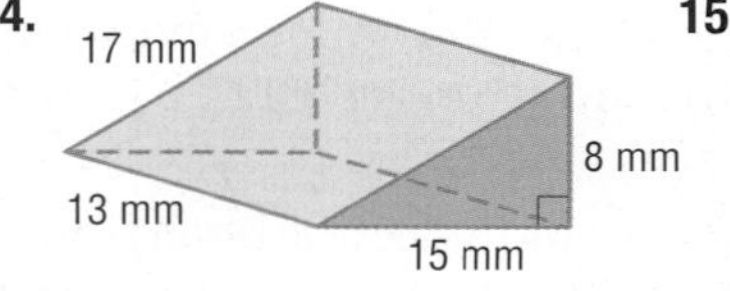


15. 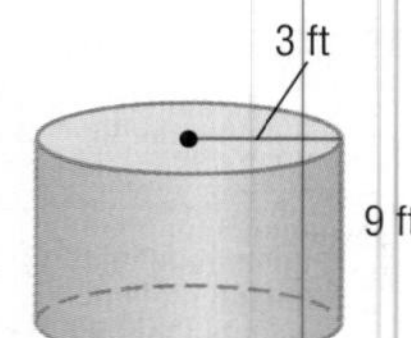


16. 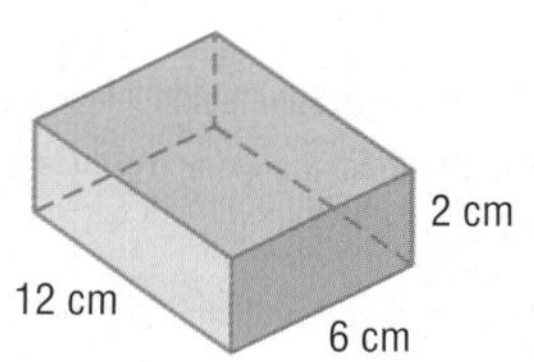


17. 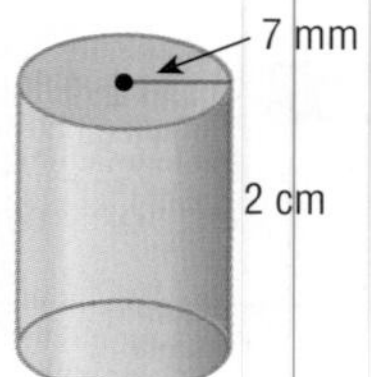


18. 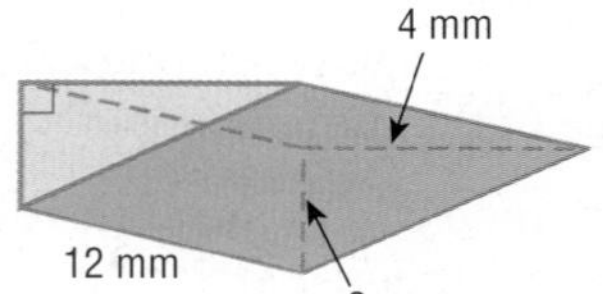


19. 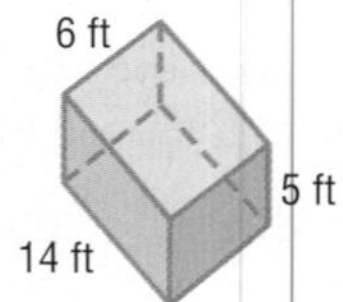


20. **METEOROLOGY** The TIROS weather satellites were a series of weather satellites that carried television and infrared cameras and were covered by solar cells. If the cylinder-shaped body of a TIROS had a diameter of 42 inches and a height of 19 inches, what was the volume available for carrying instruments and cameras? Round to the nearest tenth. (Lesson 12-4)

LESSON 12-5

# Volumes of Pyramids and Cones

| Then | Now | Why? |
|---|---|---|
| You found surface areas of pyramids and cones. | 1 Find volumes of pyramids.<br>2 Find volumes of cones. | Marta is studying crystals that grow on rock formations. For a project, she is making a clay model of a crystal with a shape that is a composite of two congruent rectangular pyramids. The base of each pyramid will be 1 by 1.5 inches, and the total height will be 4 inches. Why is determining the volume of the model helpful in this situation? |

**Common Core State Standards**

**Content Standards**

**G.GMD.1** Give an informal argument for the formulas for the circumference of a circle, area of a circle, volume of a cylinder, pyramid, and cone.

**G.GMD.3** Use volume formulas for cylinders, pyramids, cones, and spheres to solve problems. ★

**Mathematical Practices**

1 Make sense of problems and persevere in solving them.

7 Look for and make use of structure.

## 1 Volume of Pyramids

A triangular prism can be separated into three triangular pyramids as shown. Since all faces of a triangular pyramid are triangles, any face can be considered a base of the pyramid.

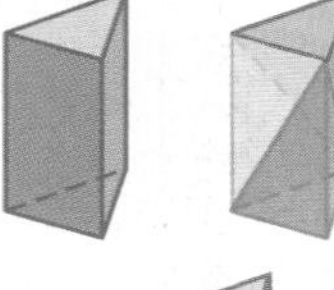

The yellow and orange pyramids have base area $B_1$ and height $h_1$. Therefore, by Cavalieri's Principle, they have the same volume. Likewise, the yellow and green pyramids have base area $B_2$ and height $h_2$, so they have the same volume.

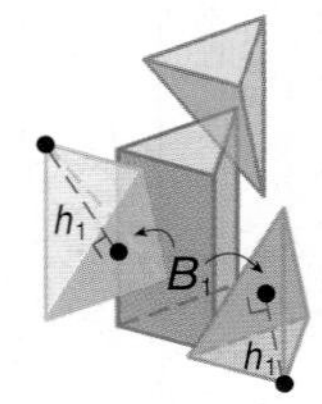


Since the orange and green pyramids have the same volume as the yellow pyramid, it follows that the volumes of all three pyramids are the same. Therefore, each pyramid has one third the volume of the prism with the same base area and height. This is true for a pyramid with any shape base.

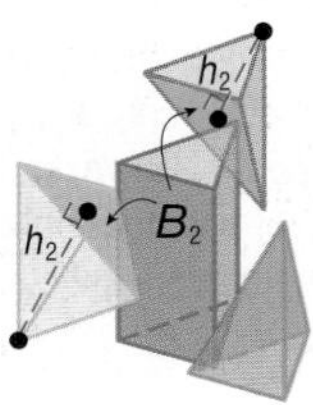


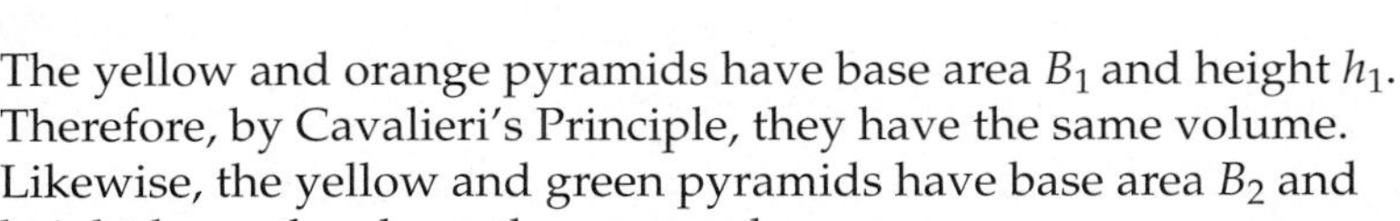

### KeyConcept Volume of a Pyramid

**Words** The volume of a pyramid is $V = \frac{1}{3}Bh$, where $B$ is the area of the base and $h$ is the height of the pyramid.

**Symbols** $V = \frac{1}{3}Bh$

**Models**

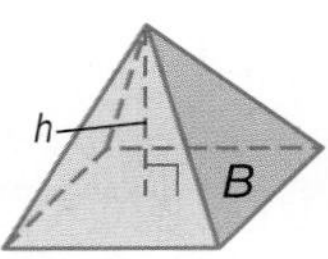


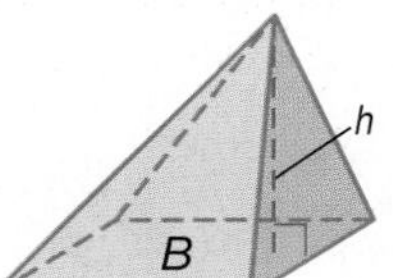


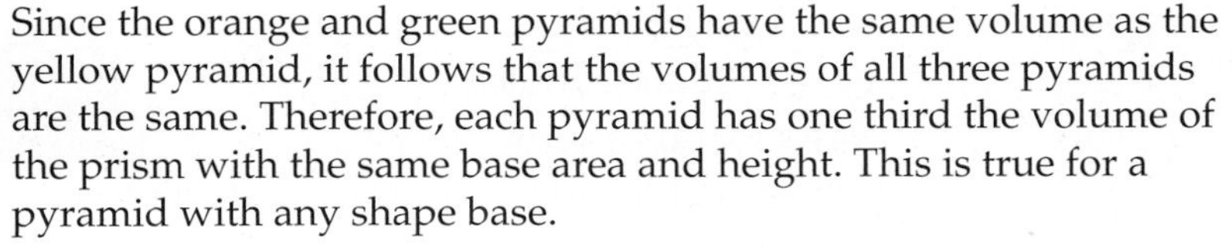

### Example 1 Volume of a Pyramid

**Find the volume of the pyramid.**

$V = \frac{1}{3}Bh$ Volume of a pyramid

$= \frac{1}{3}(9.5 \cdot 8)(9)$ $B = 9.5 \cdot 8$ and $h = 9$

$= 228$ Simplify.

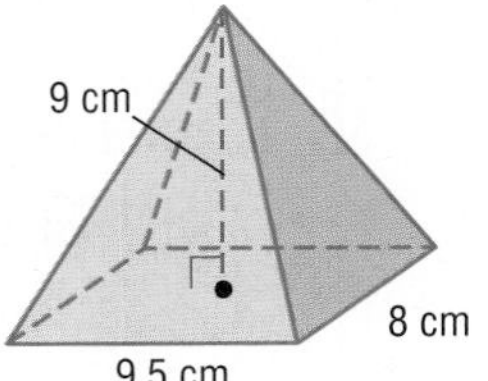


The volume of the pyramid is 228 cubic centimeters.

## GuidedPractice

**1A.**

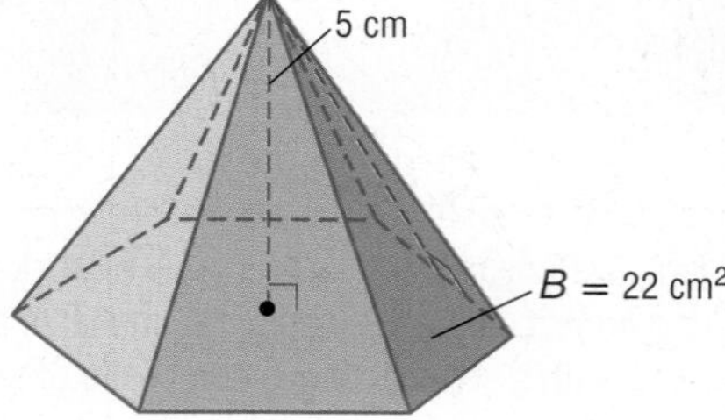


**1B.**

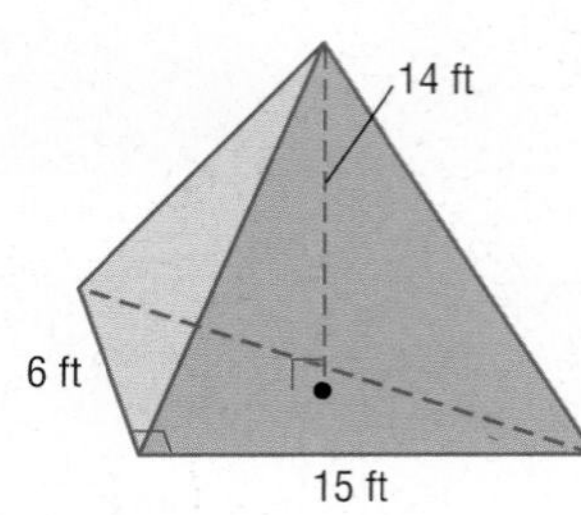


## 2 Volume of Cones

The pyramid and prism shown have the same base area $B$ and height $h$ as the cylinder and cone. Since the volume of the pyramid is one third the volume of the prism, then by Cavalieri's Principle, the volume of the cone must be one third the volume of the cylinder.

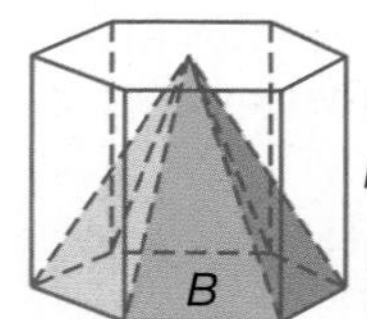


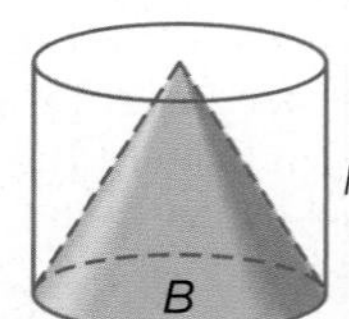


### WatchOut!

**Volumes of Cones** The formula for the surface area of a cone only applies to right cones. However, the formula for volume applies to oblique cones as well as right cones.

### KeyConcept Volume of a Cone

**Words** The volume of a circular cone is $V = \frac{1}{3}Bh$, or $V = \frac{1}{3}\pi r^2 h$, where $B$ is the area of the base, $h$ is the height of the cone, and $r$ is the radius of the base.

**Models**

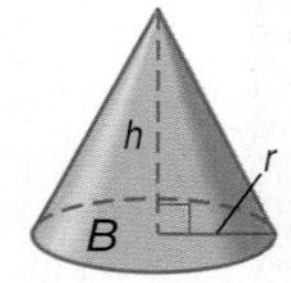


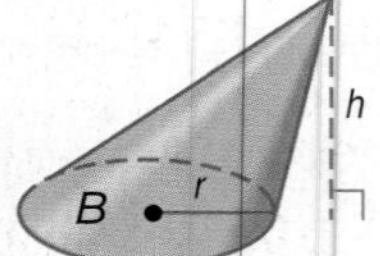


**Symbols** $V = \frac{1}{3}Bh$ or $V = \frac{1}{3}\pi r^2 h$

### Example 2 Volume of a Cone

**a. Find the volume of the cone. Round to the nearest tenth.**

| | |
|---|---|
| $V = \frac{1}{3}\pi r^2 h$ | Volume of a cone |
| $\approx \frac{1}{3}\pi(3.2)^2(5.8)$ | $r = 3.2$ and $h = 5.8$ |
| $\approx 62.2$ | Use a calculator. |

The volume of the cone is approximately 62.2 cubic meters.

**b. Find the volume of the cone. Round to the nearest tenth.**

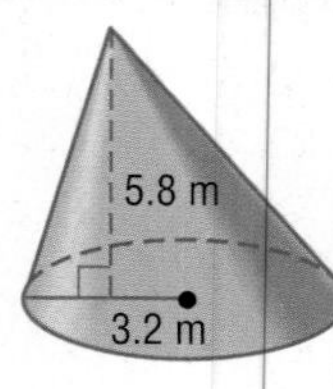


**Step 1** Use trigonometry to find the radius.

| | |
|---|---|
| $\tan 58° = \frac{11}{r}$ | $\tan \theta = \frac{\text{opp}}{\text{adj}}$ |
| $r = \frac{11}{\tan 58°}$ | Solve for $r$. |
| $r \approx 6.9$ | Use a calculator. |

**Step 2** Find the volume.

$V = \frac{1}{3}\pi r^2 h$ — Volume of a cone

$\approx \frac{1}{3}\pi(6.9)^2(11)$ — $r \approx 6.9$ and $h = 11$

$\approx 548.4$ — Use a calculator.

The volume of the cone is approximately 548.4 cubic inches.

### GuidedPractice

**2A.**

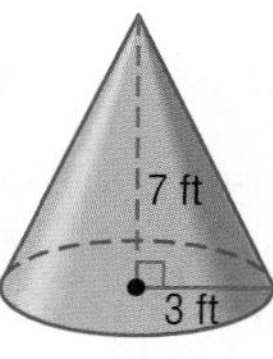


**2B.**

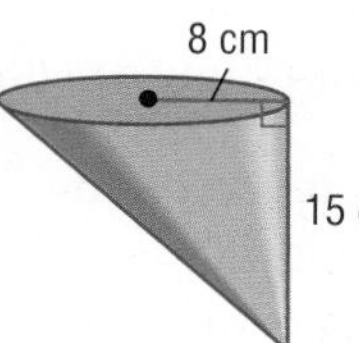


**2C.**

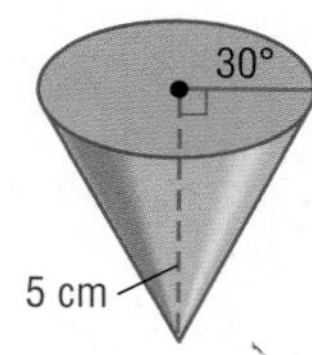


## Real-World Example 3 Find Real-World Volumes

**ARCHITECTURE At the top of the Washington Monument is a small square pyramid, called a *pyramidion*. This pyramid has a height of 55.5 feet with base edges of approximately 34.5 feet. What is the volume of the pyramidion? Round to the nearest tenth.**

Sketch and label the pyramid.

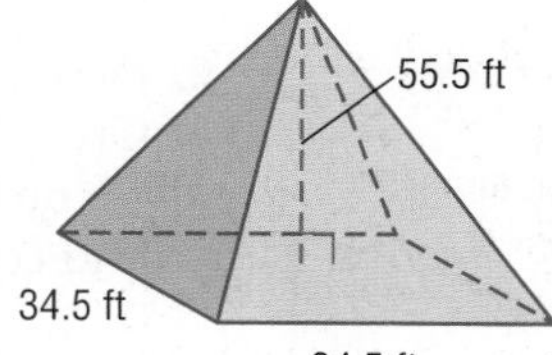


$V = \frac{1}{3}Bh$ — Volume of a pyramid

$= \frac{1}{3}(34.5 \cdot 34.5)(55.5)$ — $B = 34.5 \cdot 34.5$, $h = 55.5$

$\approx 22{,}019.6$ — Simplify.

The volume of the pyramidion atop the Washington Monument is about 22,019.6 cubic feet.

### GuidedPractice

**3. ARCHAEOLOGY** A pyramidion that was discovered in Saqqara, Egypt, in 1992 has a rectangular base 53 centimeters by 37 centimeters. It is 46 centimeters high. What is the volume of this pyramidion? Round to the nearest tenth.

**Real-WorldLink**

The Washington Monument is the largest masonry structure in the world. By law, no other building in D.C. is allowed to be taller than the 555-foot-tall structure.

**Source:** Enchanted Learning

The formulas for the volumes of solids are summarized below.

**ConceptSummary Volumes of Solids**

| Solid | prism | cylinder | pyramid | cone |
|---|---|---|---|---|
| Model | h, B | r, h, B | h, B | h, r, B |
| Volume | $V = Bh$ | $V = Bh$ or $V = \pi r^2 h$ | $V = \frac{1}{3}Bh$ | $V = \frac{1}{3}Bh$ or $V = \frac{1}{3}\pi r^2 h$ |

## Check Your Understanding

**Example 1** **Find the volume of each pyramid.**

1. 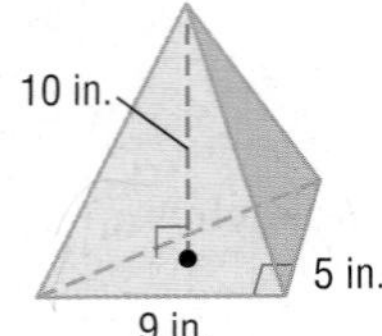


2. 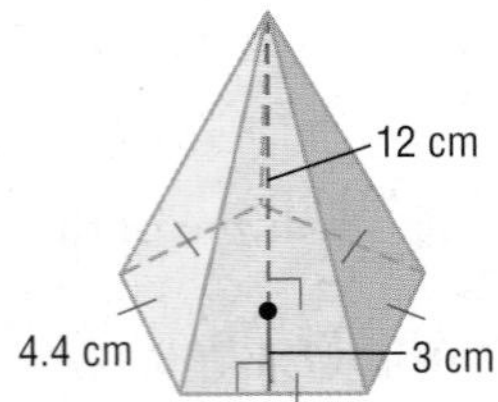


3. a rectangular pyramid with a height of 5.2 meters and a base 8 meters by 4.5 meters

4. a square pyramid with a height of 14 meters and a base with 8-meter side lengths

**Example 2** **Find the volume of each cone. Round to the nearest tenth.**

5. 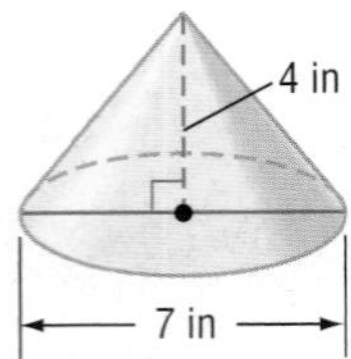


6. 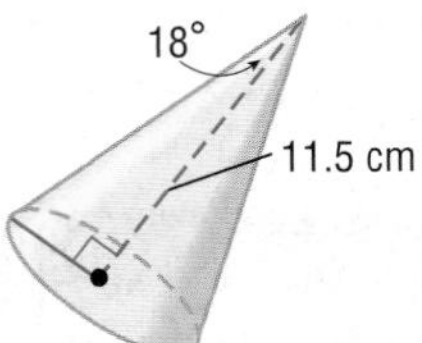


7. an oblique cone with a height of 10.5 millimeters and a radius of 1.6 millimeters

8. a cone with a slant height of 25 meters and a radius of 15 meters

**Example 3**

9. **MUSEUMS** The sky dome of the National Corvette Museum in Bowling Green, Kentucky, is a conical building. If the height is 100 feet and the area of the base is about 15,400 square feet, find the volume of air that the heating and cooling systems would have to accommodate. Round to the nearest tenth.

## Practice and Problem Solving

**Example 1** CCSS **SENSE-MAKING** **Find the volume of each pyramid. Round to the nearest tenth if necessary.**

10. 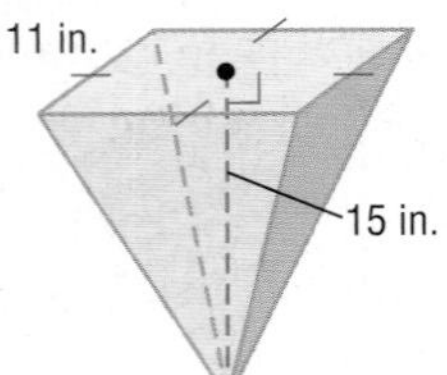


11. 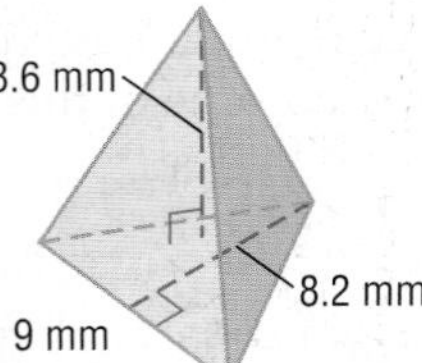


12. 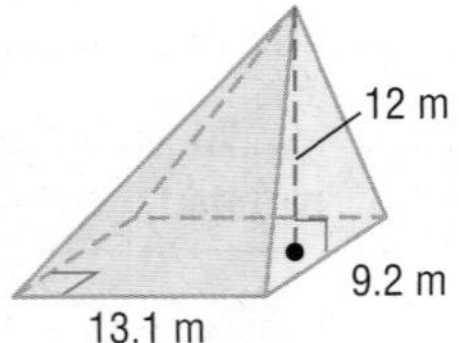


13. 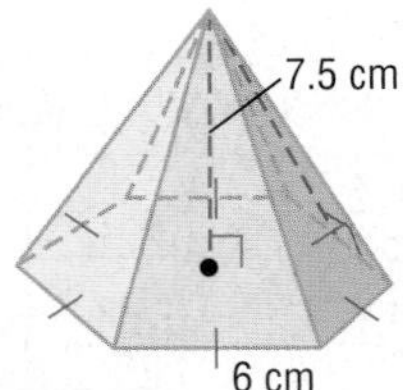


14. a pentagonal pyramid with a base area of 590 square feet and an altitude of 7 feet

15. a triangular pyramid with a height of 4.8 centimeters and a right triangle base with a leg 5 centimeters and hypotenuse 10.2 centimeters

16. A triangular pyramid with a right triangle base with a leg 8 centimeters and hypotenuse 10 centimeters has a volume of 144 cubic centimeters. Find the height.

**Example 2** **Find the volume of each cone. Round to the nearest tenth.**

**17.**

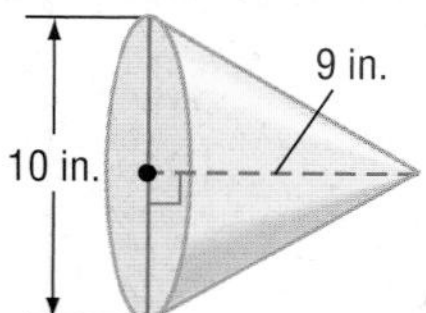


**18.**

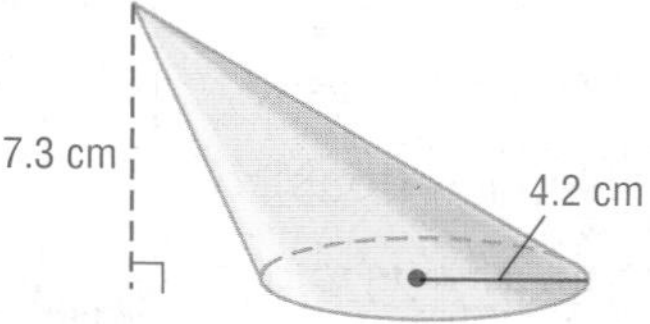


**19.**

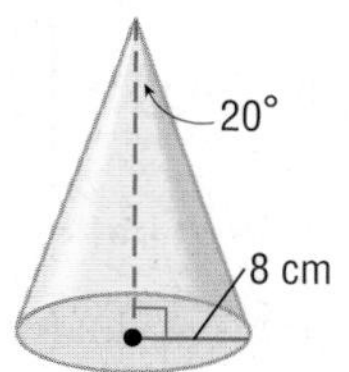


**20.**

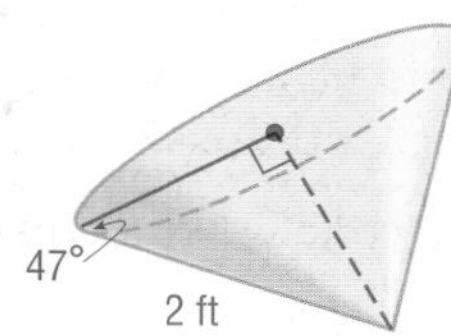


**21.** an oblique cone with a diameter of 16 inches and an altitude of 16 inches

**22.** a right cone with a slant height of 5.6 centimeters and a radius of 1 centimeter

**Example 3** **23** **SNACKS** Approximately how many cubic centimeters of roasted peanuts will completely fill a paper cone that is 14 centimeters high and has a base diameter of 8 centimeters? Round to the nearest tenth.

**24.** **CCSS MODELING** The Pyramid Arena in Memphis, Tennessee, is the third largest pyramid in the world. It is approximately 350 feet tall, and its square base is 600 feet wide. Find the volume of this pyramid.

**25.** **GARDENING** The greenhouse at the right is a regular octagonal pyramid with a height of 5 feet. The base has side lengths of 2 feet. What is the volume of the greenhouse?

**Find the volume of each solid. Round to the nearest tenth.**

**26.**

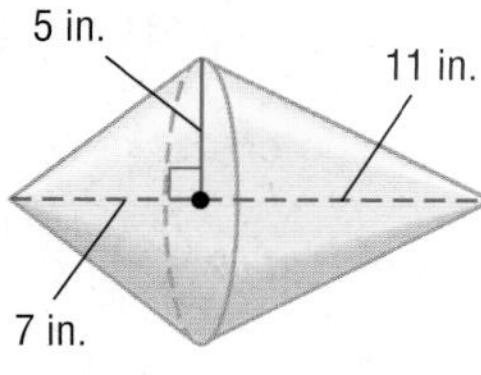


**27.**

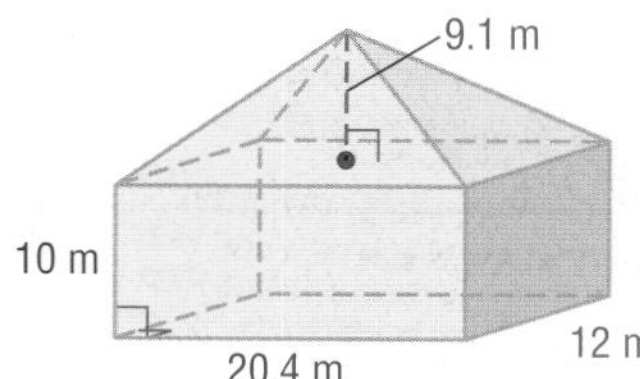


**28.**

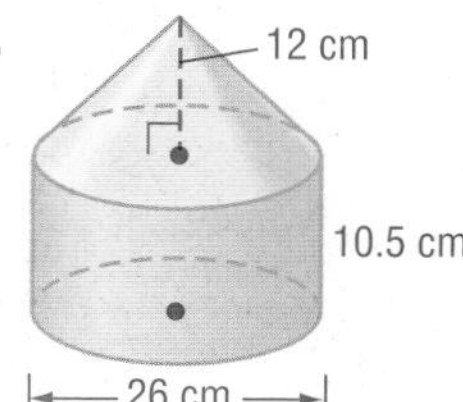


**29.** **HEATING** Sam is building an art studio in her backyard. To buy a heating unit for the space, she needs to determine the BTUs (British Thermal Units) required to heat the building. For new construction with good insulation, there should be 2 BTUs per cubic foot. What size unit does Sam need to purchase?

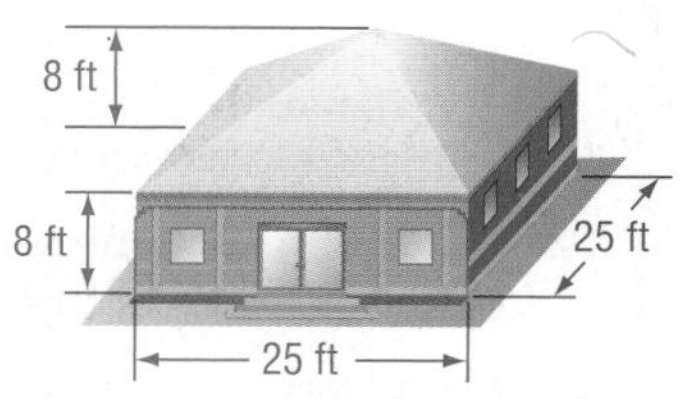


**30.** **SCIENCE** Refer to page 825. Determine the volume of the model. Explain why knowing the volume is helpful in this situation.

31. **CHANGING DIMENSIONS** A cone has a radius of 4 centimeters and a height of 9 centimeters. Describe how each change affects the volume of the cone.

  a. The height is doubled.

  b. The radius is doubled.

  c. Both the radius and the height are doubled.

**Find each measure. Round to the nearest tenth if necessary.**

32. A square pyramid has a volume of 862.5 cubic centimeters and a height of 11.5 centimeters. Find the side length of the base.

33. The volume of a cone is $196\pi$ cubic inches and the height is 12 inches. What is the diameter?

34. The lateral area of a cone is 71.6 square millimeters and the slant height is 6 millimeters. What is the volume of the cone?

35. **MULTIPLE REPRESENTATIONS** In this problem, you will investigate rectangular pyramids.

  a. **Geometric** Draw two pyramids with different bases that have a height of 10 centimeters and a base area of 24 square centimeters.

  b. **Verbal** What is true about the volumes of the two pyramids that you drew? Explain.

  c. **Analytical** Explain how multiplying the base area and/or the height of the pyramid by 5 affects the volume of the pyramid.

## H.O.T. Problems Use Higher-Order Thinking Skills

36. **CCSS ARGUMENTS** Determine whether the following statement is *always*, *sometimes*, or *never* true. Justify your reasoning.

  *The volume of a cone with radius r and height h equals the volume of a prism with height h.*

37. **ERROR ANALYSIS** Alexandra and Cornelio are calculating the volume of the cone at the right. Is either of them correct? Explain your answer.

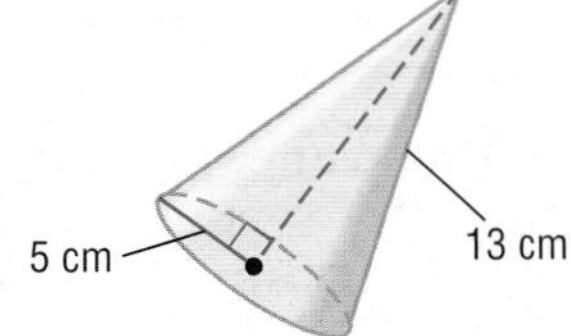


**Alexandra**

$V = \frac{1}{3}Bh$

$= \frac{1}{3}\pi(5^2)(13)$

$\approx 340.3 \text{ cm}^3$

**Cornelio**

$5^2 + 12^2 = 13^2$

$V = \frac{1}{3}Bh$

$= \frac{1}{3}\pi(5^2)(12)$

$\approx 314.2 \text{ cm}^3$

38. **REASONING** A cone has a volume of 568 cubic centimeters. What is the volume of a cylinder that has the same radius and height as the cone? Explain your reasoning.

39. **OPEN ENDED** Give an example of a pyramid and a prism that have the same base and the same volume. Explain your reasoning.

40. **WRITING IN MATH** Compare and contrast finding volumes of pyramids and cones with finding volumes of prisms and cylinders.

## Standardized Test Practice

41. A conical sand toy has the dimensions as shown below. How many cubic centimeters of sand will it hold when it is filled to the top?

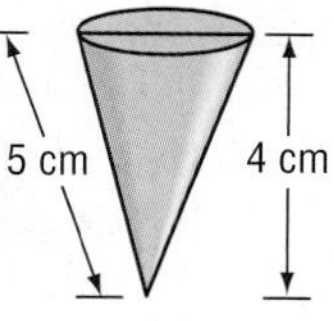


A $12\pi$
B $15\pi$
C $\frac{80}{3}\pi$
D $\frac{100}{3}\pi$

42. **SHORT RESPONSE** Brooke is buying a tent that is in the shape of a rectangular pyramid. The base is 6 feet by 8 feet. If the tent holds 88 cubic feet of air, how tall is the tent's center pole?

43. **PROBABILITY** A spinner has sections colored red, blue, orange, and green. The table below shows the results of several spins. What is the experimental probability of the spinner landing on orange?

| Color | Frequency |
|---|---|
| red | 6 |
| blue | 4 |
| orange | 5 |
| green | 10 |

F $\frac{1}{5}$
G $\frac{1}{4}$
H $\frac{9}{25}$
J $\frac{1}{2}$

44. **SAT/ACT** For all $x \neq -2$ or 0, $\frac{x^2 - 2x - 8}{x^2 + 2x} = ?$

A $-8$
B $x - 4$
C $\frac{-x - 4}{x}$
D $\frac{-8}{x + 2}$
E $\frac{x - 4}{x}$

## Spiral Review

**Find the volume of each prism.** (Lesson 12-4)

45.

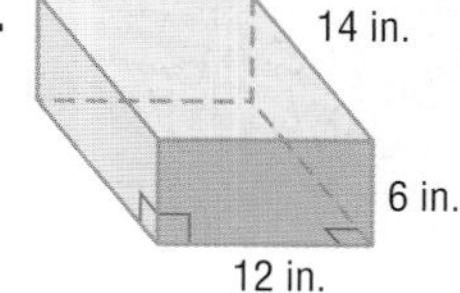


46.

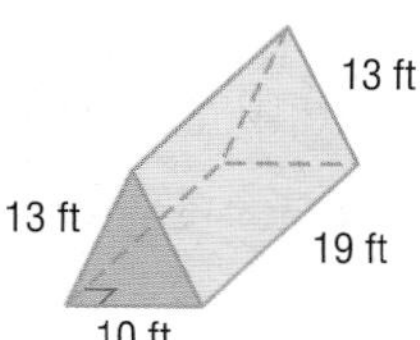


47.

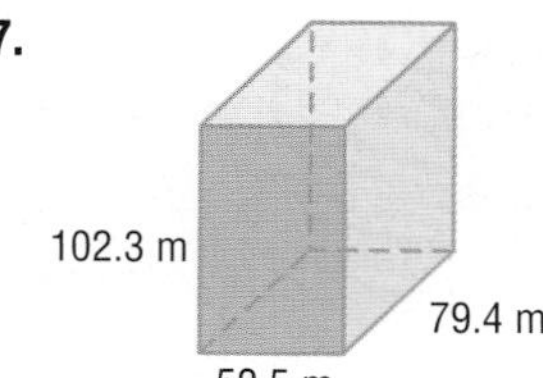


48. **FARMING** The picture shows a combination hopper cone and bin used by farmers to store grain after harvest. The cone at the bottom of the bin allows the grain to be emptied more easily. Use the dimensions in the diagram to find the entire surface area of the bin with a conical top and bottom. Write the exact answer and the answer rounded to the nearest square foot. (Lesson 12-3)

## Skills Review

**Find the area of each shaded region. The polygons in Exercises 50-52 are regular.**

49.

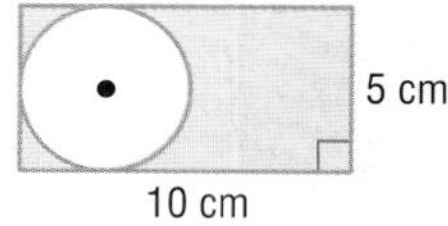


50.

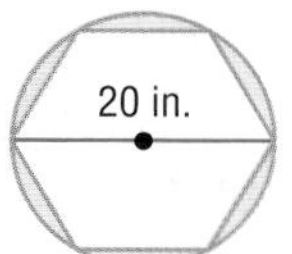


51.

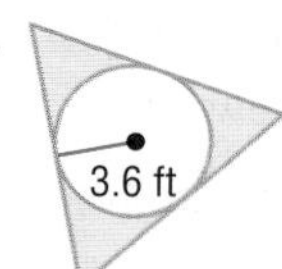


52.

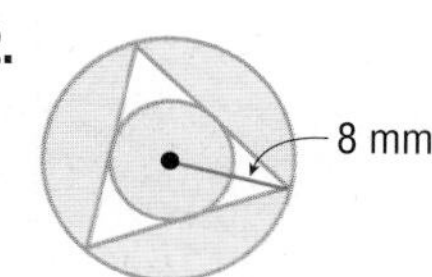

LESSON 12-6

# Surface Areas and Volumes of Spheres

**Then**

- You found surface areas of prisms and cylinders.

**Now**

1. Find surface areas of spheres.
2. Find volumes of spheres.

**Why?**

- When you blow bubbles, soapy liquid surrounds a volume of air. Because of surface tension, the liquid maintains a shape that minimizes the surface area surrounding the air. The shape that minimizes surface area per unit of volume is a sphere.

**NewVocabulary**
great circle
pole
hemisphere

**Common Core State Standards**

**Content Standards**

**G.GMD.1** Give an informal argument for the formulas for the circumference of a circle, area of a circle, volume of a cylinder, pyramid, and cone.

**G.GMD.3** Use volume formulas for cylinders, pyramids, cones, and spheres to solve problems. ★

**Mathematical Practices**

1 Make sense of problems and persevere in solving them.

6 Attend to precision.

## 1 Surface Area of Spheres

Recall that a *sphere* is the locus of all points in space that are a given distance from a given point called the *center* of the sphere.

- A *radius* of a sphere is a segment from the center to a point on the sphere.
- A *chord* of a sphere is a segment that connects any two points on the sphere.
- A *diameter* of a sphere is a chord that contains the center.
- A *tangent* to a sphere is a line that intersects the sphere in exactly one point.

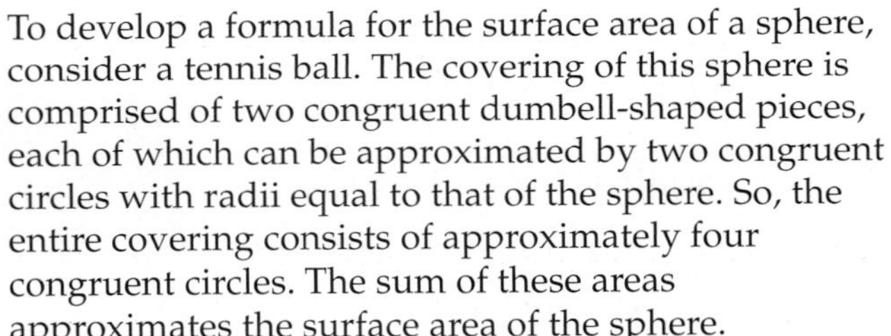


To develop a formula for the surface area of a sphere, consider a tennis ball. The covering of this sphere is comprised of two congruent dumbell-shaped pieces, each of which can be approximated by two congruent circles with radii equal to that of the sphere. So, the entire covering consists of approximately four congruent circles. The sum of these areas approximates the surface area of the sphere.

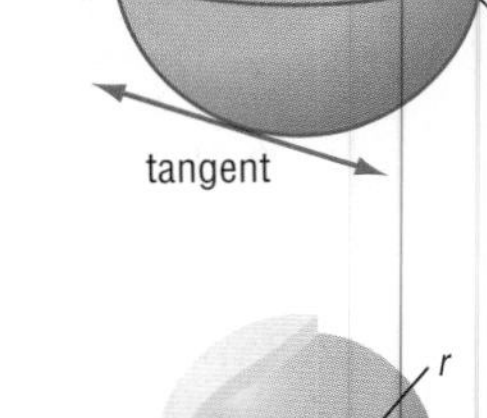


$S \approx 4A$ — Sum of circles with area $A$

$\approx 4(\pi r^2)$ or $4\pi r^2$ — $A = \pi r^2$

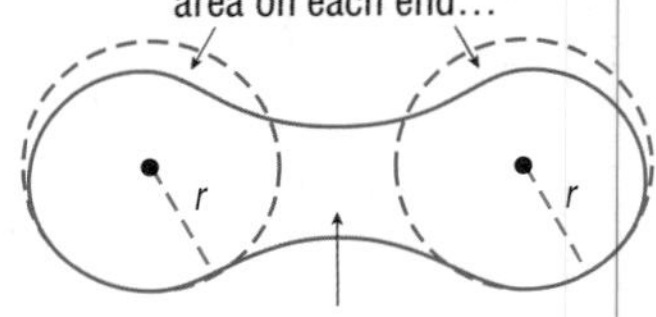


While its derivation is beyond the scope of this course, the exact formula is in fact $S = 4\pi r^2$.

**KeyConcept Surface Area of a Sphere**

| | | |
|---|---|---|
| **Words** | The surface area $S$ of a sphere is $S = 4\pi r^2$, where $r$ is the radius. | **Model** |
| **Symbols** | $S = 4\pi r^2$ | |

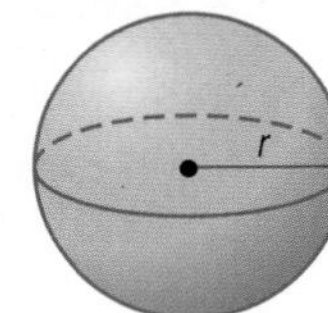


blue jean images/Getty Images

## Example 1 Surface Area of a Sphere

**Find the surface area of the sphere. Round to the nearest tenth.**

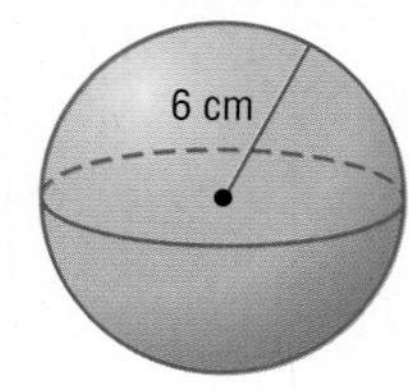


$S = 4\pi r^2$ Surface area of a sphere

$= 4\pi(6)^2$ Replace $r$ with 6.

$\approx 452.4$ Use a calculator.

The surface area is about 452.4 square centimeters.

### Guided Practice

**1A.**

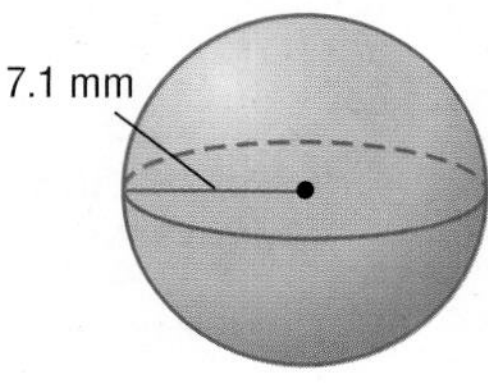


**1B.**

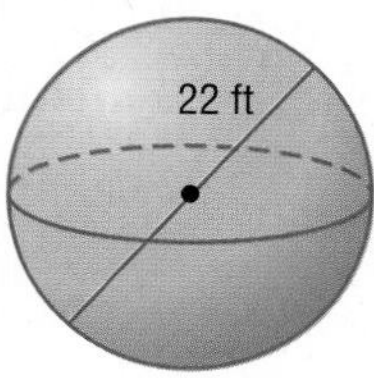


> **StudyTip**
> **Great Circles** A sphere has an infinite number of great circles.

A plane can intersect a sphere in a point or in a circle. If the circle contains the center of the sphere, the intersection is called a **great circle**. The endpoints of a diameter of a great circle are called **poles**.

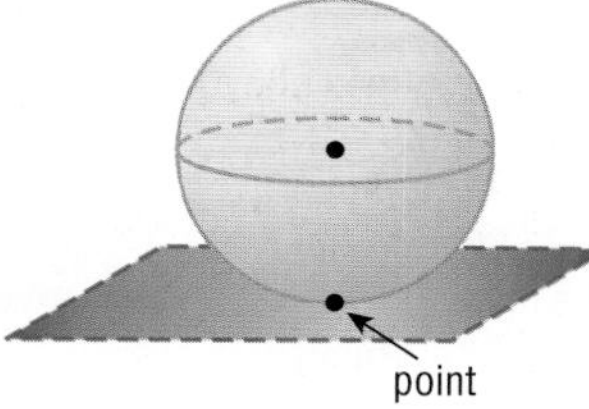


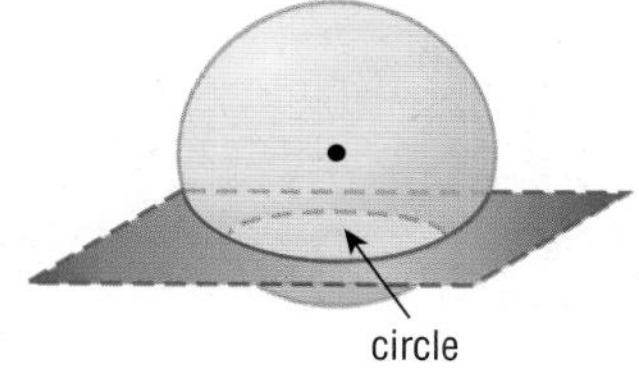


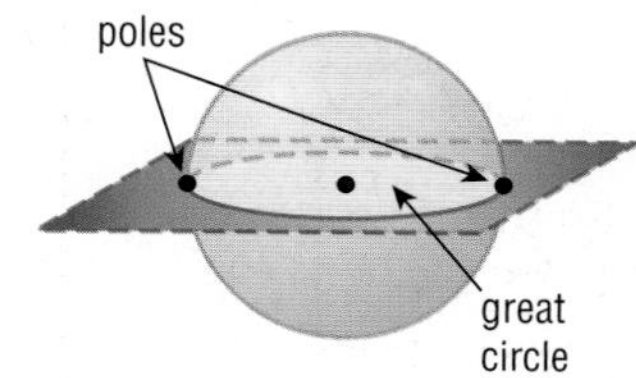


Since a great circle has the same center as the sphere and its radii are also radii of the sphere, it is the largest circle that can be drawn on a sphere. A great circle separates a sphere into two congruent halves, called **hemispheres**.

## Example 2 Use Great Circles to Find Surface Area

> **WatchOut!**
> **Area of Hemisphere** When finding the surface area of a hemisphere, do not forget to include the area of the great circle.

**a. Find the surface area of the hemisphere.**

Find half the area of a sphere with a radius of 2.8 centimeters. Then add the area of the great circle.

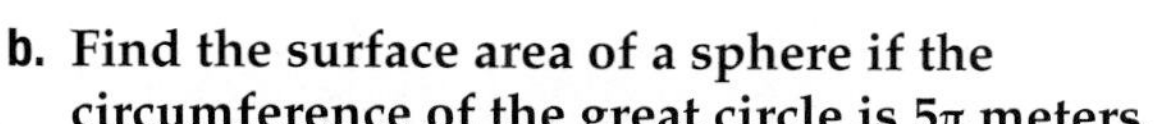

$S = \frac{1}{2}(4\pi r^2) + \pi r^2$ Surface area of a hemisphere

$= \frac{1}{2}[4\pi(2.8)^2] + \pi(2.8)^2$ Replace $r$ with 2.8.

$\approx 73.9 \text{ cm}^2$ Use a calculator.

**b. Find the surface area of a sphere if the circumference of the great circle is 5π meters.**

First, find the radius. The circumference of a great circle is $2\pi r$. So, $2\pi r = 5\pi$ or $r = 2.5$.

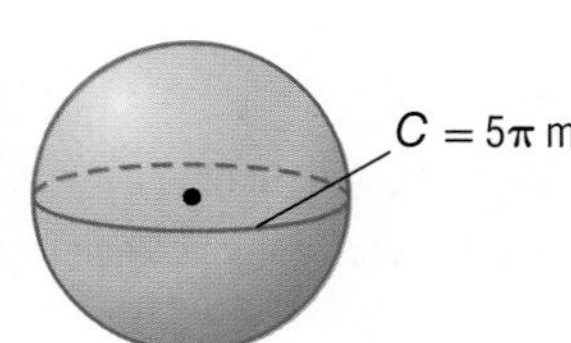


$S = 4\pi r^2$ Surface area of a sphere

$= 4\pi(2.5)^2$ Replace $r$ with 2.5.

$\approx 78.5 \text{ m}^2$ Use a calculator.

**c. Find the surface area of a sphere if the area of the great circle is approximately 130 square inches.**

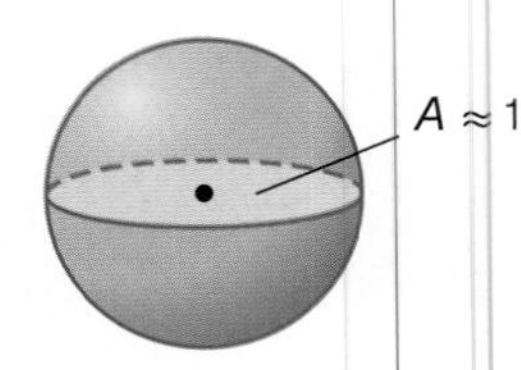


First, find the radius. The area of a great circle is $\pi r^2$. So, $\pi r^2 = 130$ or $r \approx 6.4$.

$S = 4\pi r^2$ — Surface area of a sphere

$\approx 4\pi(6.4)^2$ or about 514.7 in² — Replace $r$ with 6.4. Use a calculator.

**GuidedPractice**

**Find the surface area of each figure. Round to the nearest tenth if necessary.**

**2A.** sphere: circumference of great circle $= 16.2\pi$ ft

**2B.** hemisphere: area of great circle $\approx 94$ mm²

**2C.** hemisphere: circumference of great circle $= 36\pi$ cm

## 2 Volume of Spheres

Suppose a sphere with radius $r$ contains infinitely many pyramids with vertices at the center of the sphere. Each pyramid has height $r$ and base area $B$. The sum of the volumes of all the pyramids equals the volume of the sphere.

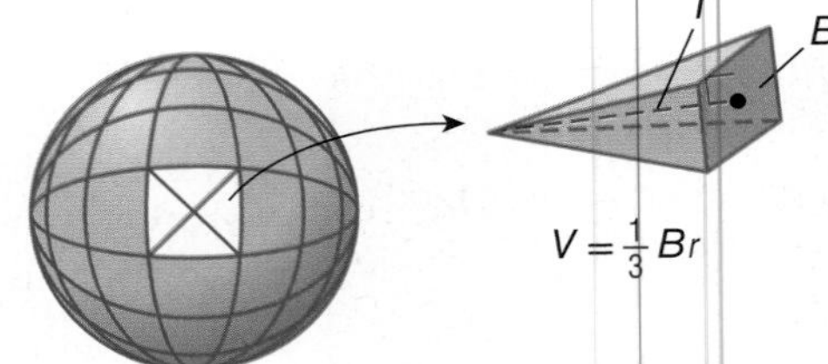


$V = \frac{1}{3}B_1r_1 + \frac{1}{3}B_2r_2 + \ldots + \frac{1}{3}B_nr_n$ — Sum of volumes of pyramids

$= \frac{1}{3}r(B_1 + B_2 + \ldots + B_n)$ — Distributive Property

$= \frac{1}{3}r(4\pi r^2)$ — The sum of the pyramid base areas equals the surface area of the sphere.

$= \frac{4}{3}\pi r^3$ — Simplify.

**StudyTip**

**Draw a Diagram** When solving problems involving volumes of solids, it is helpful to draw and label a diagram when no diagram is provided.

**KeyConcept Volume of a Sphere**

**Words** The volume $V$ of a sphere is $V = \frac{4}{3}\pi r^3$, where $r$ is the radius of the sphere.

**Symbols** $V = \frac{4}{3}\pi r^3$

**Model**

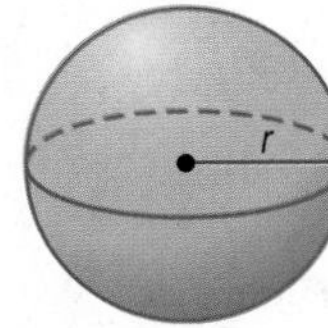

### Example 3 Volumes of Spheres and Hemispheres

**Find the volume of each sphere or hemisphere. Round to the nearest tenth.**

**a. a hemisphere with a radius of 6 meters**

Estimate: $V \approx \frac{1}{2} \cdot \frac{4}{3} \cdot 3 \cdot 6^3$ or 432 m³

$V = \frac{1}{2}\left(\frac{4}{3}\pi r^3\right)$ — Volume of a hemisphere

$= \frac{2}{3}\pi(6)^3$ or about 452.4 m³ — Replace $r$ with 6. Use a calculator.

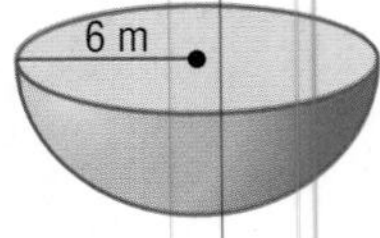


The volume of the hemisphere is about 452.4 cubic meters. This is close to the estimate, so the answer is reasonable.

**StudyTip**

**CCSS Precision** Remember to use the correct units when giving your answers. As with other solids, the surface area of a sphere is measured in square units, and volume is measured in cubic units.

**b. a sphere with a great circle circumference of 18π centimeters**

**Step 1** Find the radius of the sphere.

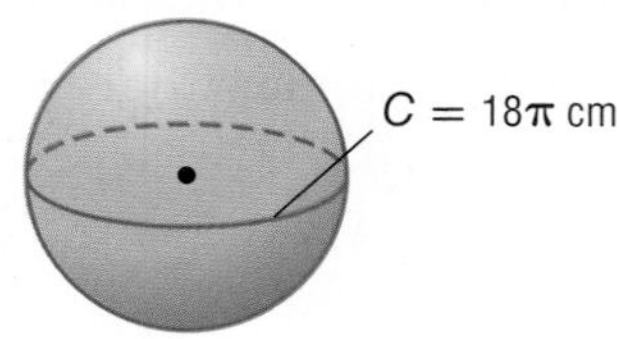


| | |
|---|---|
| $C = 2\pi r$ | Circumference of a circle |
| $18\pi = 2\pi r$ | Replace $C$ with $18\pi$. |
| $r = 9$ | Solve for $r$. |

**Step 2** Find the volume.

| | |
|---|---|
| $V = \frac{4}{3}\pi r^3$ | Volume of a sphere |
| $= \frac{4}{3}\pi(9)^3$ or about 3053.6 $cm^3$ | Replace $r$ with 9. Use a calculator. |

**GuidedPractice**

**3A.** sphere: diameter = 7.4 in.

**3B.** hemisphere: area of great circle ≈ 249 $mm^2$

**Real-WorldLink**

The University of North Carolina has won the greatest number of national championships in women's soccer since the first tournament in 1982. As of 2009, they have won 18 times.

**Source:** Fact Monster

Raleigh News & Observer/Contributor/McClatchy-Tribune/Getty Images

## Real-World Example 4 Solve Problems Involving Solids

**SOCCER The soccer ball globe at the right was constructed for the 2006 World Cup soccer tournament. It takes up 47,916π cubic feet of space. Assume that the globe is a sphere. What is the circumference of the globe?**

**Understand** You know that the volume of the globe is 47,916π cubic feet. The circumference of the globe is the circumference of the great circle.

**Plan** First use the volume formula to find the radius. Then find the circumference of the great circle.

**Solve**

| | |
|---|---|
| $V = \frac{4}{3}\pi r^3$ | Volume of a sphere |
| $47{,}916\pi = \frac{4}{3}\pi r^3$ | Replace $V$ with $47{,}916\pi$. |
| $35{,}937 = r^3$ | Divide each side by $\frac{4}{3}\pi$. |

Use a calculator to find $\sqrt[3]{35{,}937}$.

35937  1 [÷] 3  33

The radius of the globe is 33 feet. So, the circumference is $2\pi r = 2\pi(33)$ or approximately 207.3 feet.

**Check** You can work backward to check the solution.

If $C \approx 207.3$, then $r \approx 33$. If $r \approx 33$, then $V \approx \frac{4}{3}\pi \cdot 33^3$ or about 47,917π cubic feet. The solution is correct. ✓

**GuidedPractice**

**4. BALLOONS** Ren inflates a spherical balloon to a circumference of about 14 inches. He then adds more air to the balloon until the circumference is about 18 inches. What volume of air was added to the balloon?

## Check Your Understanding

**Examples 1–2** **Find the surface area of each sphere or hemisphere. Round to the nearest tenth.**

1.
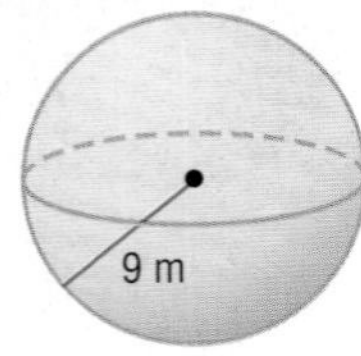


2.
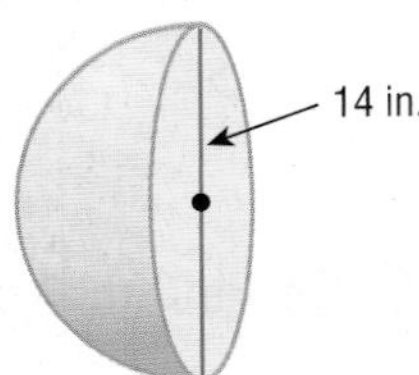


3. sphere: area of great circle = $36\pi$ yd$^2$
4. hemisphere: circumference of great circle ≈ 26 cm

**Example 3** **Find the volume of each sphere or hemisphere. Round to the nearest tenth.**

5. sphere: radius = 10 ft
6. hemisphere: diameter = 16 cm
7. hemisphere: circumference of great circle = $24\pi$ m
8. sphere: area of great circle = $55\pi$ in$^2$

**Example 4**

9. **BASKETBALL** Basketballs used in professional games must have a circumference of $29\frac{1}{2}$ inches. What is the surface area of a basketball used in a professional game?

## Practice and Problem Solving

**Examples 1–2** **Find the surface area of each sphere or hemisphere. Round to the nearest tenth.**

10.
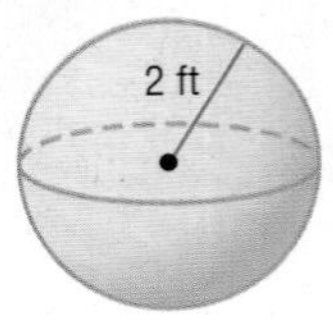


11.
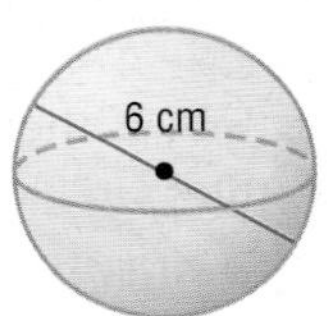


12.
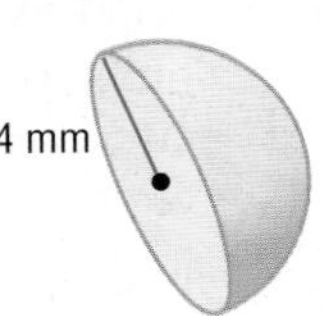


13.
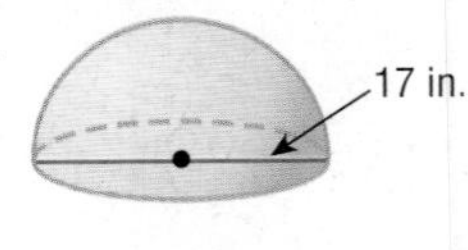


14. sphere: circumference of great circle = $2\pi$ cm
15. sphere: area of great circle ≈ 32 ft$^2$
16. hemisphere: area of great circle ≈ 40 in$^2$
17. hemisphere: circumference of great circle = $15\pi$ mm

**Example 3** **CCSS PRECISION** **Find the volume of each sphere or hemisphere. Round to the nearest tenth.**

18.
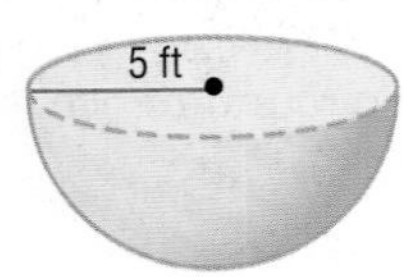


19.
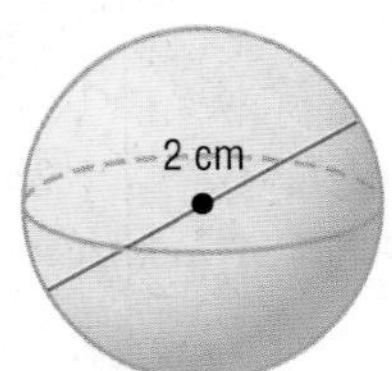


20. sphere: radius = 1.4 yd
21. hemisphere: diameter = 21.8 cm
22. sphere: area of great circle = $49\pi$ m$^2$
23. sphere: circumference of great circle ≈ 22 in.
24. hemisphere: circumference of great circle ≈ 18 ft
25. hemisphere: area of great circle ≈ 35 m$^2$

**Example 4**

**26. FISH** A *puffer fish* is able to "puff up" when threatened by gulping water and inflating its body. The puffer fish at the right is approximately a sphere with a diameter of 5 inches. Its surface area when inflated is about 1.5 times its normal surface area. What is the surface area of the fish when it is *not* puffed up?

**27. ARCHITECTURE** The Reunion Tower in Dallas, Texas, is topped by a spherical dome that has a surface area of approximately $13{,}924\pi$ square feet. What is the volume of the dome? Round to the nearest tenth.

**28. TREE HOUSE** The spherical tree house, or *tree sphere,* shown at the right has a diameter of 10.5 feet. Its volume is 1.8 times the volume of the first tree sphere that was built. What was the diameter of the first tree sphere? Round to the nearest foot.

**CCSS SENSE-MAKING Find the surface area and the volume of each solid. Round to the nearest tenth.**

**29.**

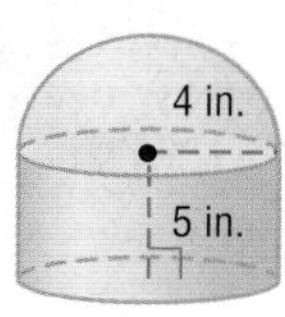


**30.**

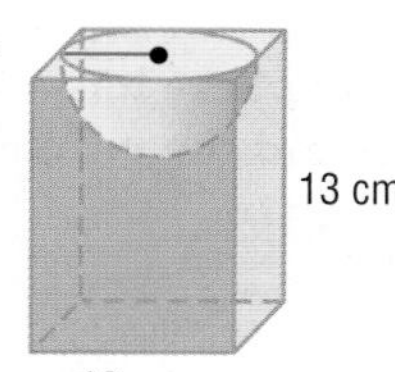


**31. TOYS** The spinning top at the right is a composite of a cone and a hemisphere.

**a.** Find the surface area and the volume of the top. Round to the nearest tenth.

**b.** If the manufacturer of the top makes another model with dimensions that are one-half of the dimensions of this top, what are its surface area and volume?

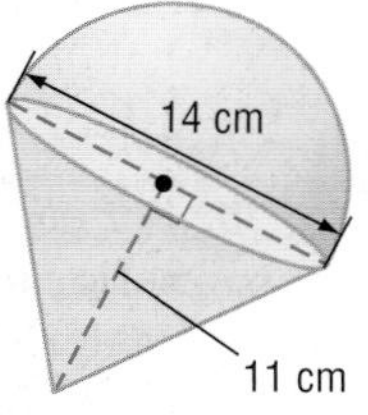


**32. BALLOONS** A spherical helium-filled balloon with a diameter of 30 centimeters can lift a 14-gram object. Find the size of a balloon that could lift a person who weighs 65 kilograms. Round to the nearest tenth.

**Use sphere $S$ to name each of the following.**

**33.** a chord

**34.** a radius

**35.** a diameter

**36.** a tangent

**37.** a great circle

**38. DIMENSIONAL ANALYSIS** Which has greater volume: a sphere with a radius of 2.3 yards or a cylinder with a radius of 4 feet and height of 8 feet?

39. **INFORMAL PROOF** A sphere with radius $r$ can be thought of as being made up of a large number of discs or thin cylinders. Consider the disc shown that is $x$ units above or below the center of the sphere. Also consider a cylinder with radius $r$ and height $2r$ that is hollowed out by two cones of height and radius $r$.

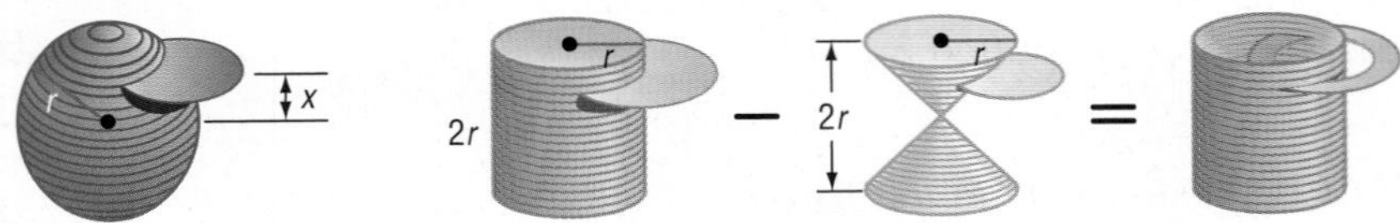


a. Find the radius of the disc from the sphere in terms of its distance $x$ above the sphere's center. (*Hint:* Use the Pythagorean Theorem.)

b. If the disc from the sphere has a thickness of $y$ units, find its volume in terms of $x$ and $y$.

c. Show that this volume is the same as that of the hollowed-out disc with thickness of $y$ units that is $x$ units above the center of the cylinder and cone.

d. Since the expressions for the discs at the same height are the same, what guarantees that the hollowed-out cylinder and sphere have the same volume?

e. Use the formulas for the volumes of a cylinder and a cone to derive the formula for the volume of the hollowed-out cylinder and thus, the sphere.

**CCSS TOOLS Describe the number and types of planes that produce reflection symmetry in each solid. Then describe the angles of rotation that produce rotation symmetry in each solid.**

40. sphere

41. hemisphere

**CHANGING DIMENSIONS A sphere has a radius of 12 centimeters. Describe how each change affects the surface area and the volume of the sphere.**

42. The radius is multiplied by 4.

43. The radius is divided by 3.

44. **DESIGN** A standard juice box holds 8 fluid ounces.

a. Sketch designs for three different juice containers that will each hold 8 fluid ounces. Label dimensions in centimeters. At least one container should be cylindrical. (*Hint:* 1 fl oz $\approx$ 29.57353 $cm^3$)

b. For each container in part **a**, calculate the surface area to volume ($cm^2$ per fl oz) ratio. Use these ratios to decide which of your containers can be made for the lowest materials cost. What shape container would minimize this ratio, and would this container be the cheapest to produce? Explain your reasoning.

## H.O.T. Problems Use Higher-Order Thinking Skills

45. **CHALLENGE** A cube has a volume of 216 cubic inches. Find the volume of a sphere that is circumscribed about the cube. Round to the nearest tenth.

46. **REASONING** Determine whether the following statement is *true* or *false*. If true, explain your reasoning. If false, provide a counterexample.

*If a sphere has radius r, there exists a cone with radius r having the same volume.*

47. **OPEN ENDED** Sketch a sphere showing two examples of great circles. Sketch another sphere showing two examples of circles formed by planes intersecting the sphere that are *not* great circles.

48. **WRITING IN MATH** Write a ratio comparing the volume of a sphere with radius $r$ to the volume of a cylinder with radius $r$ and height $2r$. Then describe what the ratio means.

## Standardized Test Practice

**49. GRIDDED RESPONSE** What is the volume of the hemisphere shown below in cubic meters?

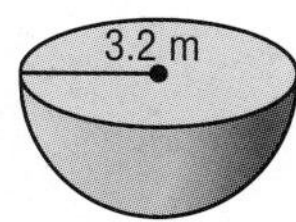


**50. ALGEBRA** What is the solution set of $3z + 4 < 6 + 7z$?

**A** $\{z|z > -0.5\}$ **C** $\{z|z < -0.5\}$
**B** $\{z|z > -2\}$ **D** $\{z|z < -2\}$

**51.** If the area of the great circle of a sphere is 33 $ft^2$, what is the surface area of the sphere?

**F** 42 $ft^2$ **H** 132 $ft^2$
**G** 117 $ft^2$ **J** 264 $ft^2$

**52. SAT/ACT** If a line $\ell$ is a perpendicular bisector of segment $AB$ at $E$, how many points on line $\ell$ are the same distance from point $A$ as from point $B$?

**A** none **D** three
**B** one **E** all points
**C** two

## Spiral Review

**Find the volume of each pyramid. Round to the nearest tenth if necessary.** (Lesson 12-5)

**53.**

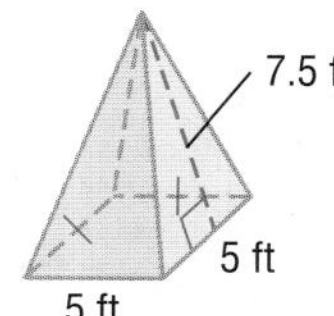


**54.**

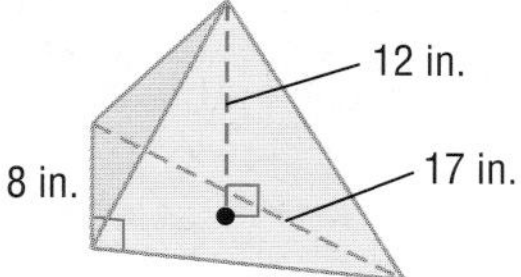


**55.**

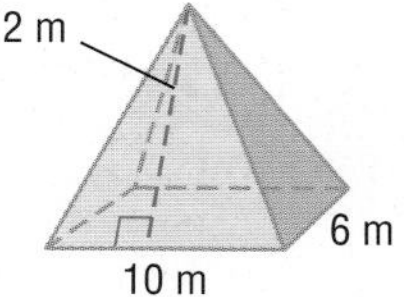


**56. ENGINEERING** The base of an oil drilling platform is made up of 24 concrete cylindrical cells. Twenty of the cells are used for oil storage. The pillars that support the platform deck rest on the four other cells. Find the total volume of the storage cells. (Lesson 12-4)

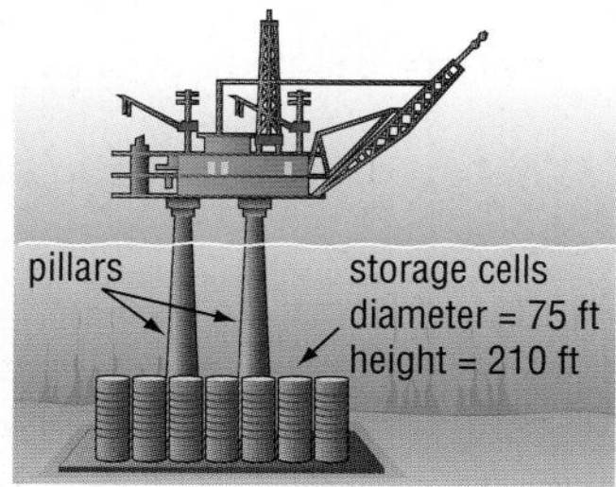


**Find each measure.** (Lesson 11-6)

**57.** $m\angle 5$

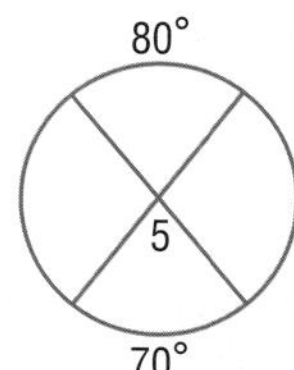


**58.** $m\angle 6$

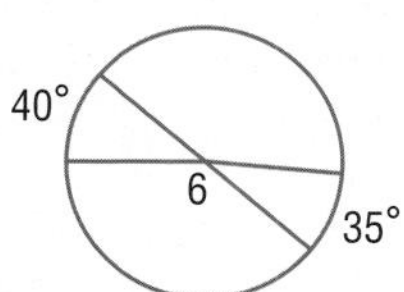


**59.** $m\angle 7$

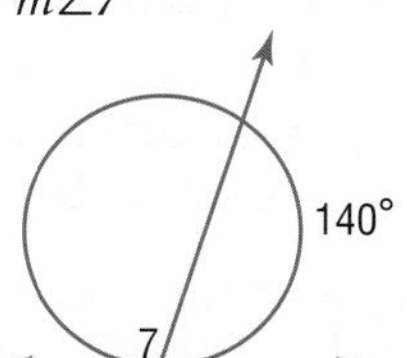


## Skills Review

**Refer to the figure.**

**60.** How many planes appear in this figure?

**61.** Name three points that are collinear.

**62.** Are points $G$, $A$, $B$, and $E$ coplanar? Explain.

**63.** At what point do $\overleftrightarrow{EF}$ and $\overleftrightarrow{AB}$ intersect?

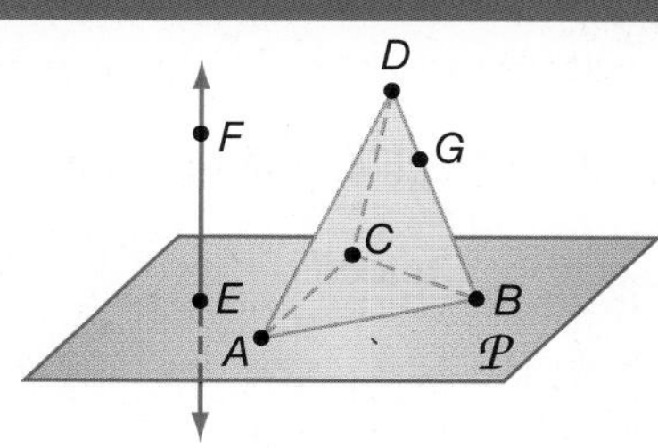

EXTEND 12-6

# Geometry Lab
# Locus and Spheres

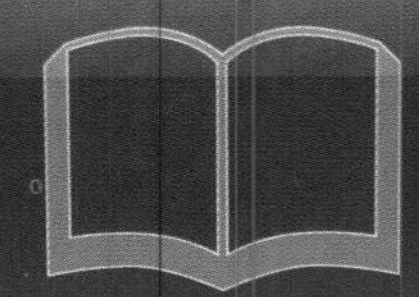

Spheres are defined in terms of a locus of points in space. The definition of a sphere is the set of all points that are a given distance from a given point.

## Activity 1 Locus of Points a Given Distance from Endpoints

**Find the locus of all points that are equidistant from a segment.**

### Collect the Data

- Draw a given line segment with endpoints $J$ and $K$
- Create a set of points that are equidistant from the segment.

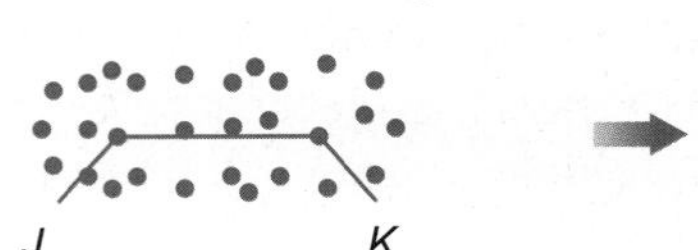


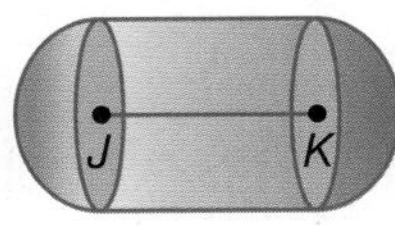


### Analyze

1. Draw a figure and describe the locus of points in space that are 8 units from a segment that is 30 units long.
2. What three-dimensional shapes form the figure?
3. What are the radii and diameters of each hemisphere?
4. What are the diameter and the height of the cylinder?

## Activity 2 Spheres That Intersect

**Find the locus of all points that are equidistant from the centers of two intersecting spheres with the same radius.**

### Collect the Data

- Draw a line segment.
- Draw congruent overlapping spheres, with the centers at the endpoints of the given line segment.

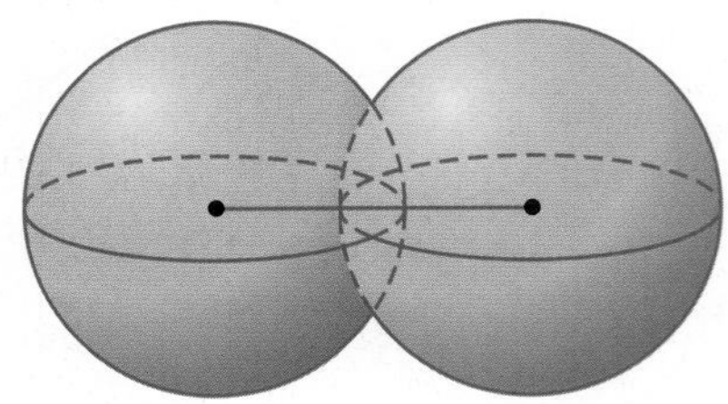

### Analyze

5. What is the shape of the intersection of the upper hemispheres?
6. Can this be described as a locus of points in space or on a plane? Explain.
7. Describe this intersection as a locus.
8. **FIREWORKS** What is the locus of points that describes how particles from a fireworks explosion will disperse in an explosion at 400 feet above ground level if the expected distance a particle could travel is 200 feet?

LESSON 12-7

# Spherical Geometry

**Then**

- You identified basic properties of spheres.

**Now**

1. Describe sets of points on a sphere.
2. Compare and contrast Euclidean and spherical geometries.

**Why?**

- Since Earth has a curved instead of a flat surface,the shortest path between two points on Earth is described by an arc of a great circle instead of a straight line.

**NewVocabulary**

Euclidean geometry
spherical geometry
non-Euclidean geometry

## 1 Geometry on a Sphere

In this text, we have studied **Euclidean geometry**, either in the plane or in space. In plane Euclidean geometry, a *plane* is a flat surface made up of points that extend infinitely in all directions. In **spherical geometry**, or geometry on a sphere, a plane is the surface of a sphere.

Lines are also defined differently in spherical geometry.

### KeyConcept Lines in Plane and Spherical Geometry

| Plane Euclidean Geometry | Spherical Geometry |
|---|---|
| 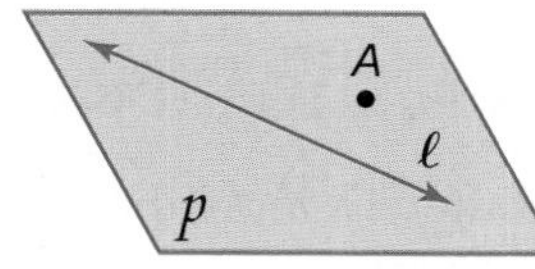  | 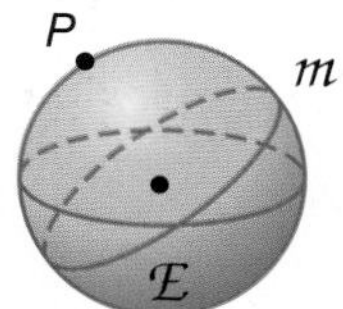  |
| Plane $\mathcal{P}$ contains line $\ell$ and point $A$ not on line $\ell$. | Sphere $\mathcal{E}$ contains great circle $m$ and point $P$ not on $m$. Great circle $m$ is a line on sphere $\mathcal{E}$. |

### Example 1 Describe Sets of Points on a Sphere

**Name each of the following on sphere $\mathcal{F}$.**

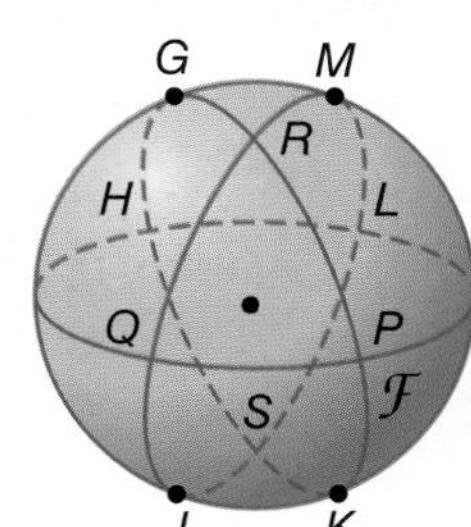


**a. two lines containing point $R$**

$\overleftrightarrow{GP}$ and $\overleftrightarrow{MQ}$ are lines on sphere $\mathcal{F}$ that contain point $R$.

**b. a segment containing point $K$**

$\overline{PS}$ is a segment on sphere $\mathcal{F}$ that contains point $K$.

**c. a triangle**

$\triangle RQP$ is a triangle on sphere $\mathcal{F}$.

**GuidedPractice**

**Name each of the following on sphere $\mathcal{F}$ above.**

**1A.** two lines containing point $P$

**1B.** a segment containing point $Q$

**1C.** a triangle

## Real-World Example 2 Identify Lines in Spherical Geometry

**ENTERTAINMENT Determine whether figure $m$ on the mirror ball shown is a line in spherical geometry.**

Notice that figure $m$ does not go through the poles of the sphere. Therefore figure $m$ is not a great circle and so not a line in spherical geometry.

### GuidedPractice

**2.** Determine whether figure $p$ on the mirror ball shown is a line in spherical geometry.

**StudyTip**

**Elliptical Geometry** *Spherical geometry* is a subcategory of *elliptical geometry.*

## 2 Comparing Euclidean and Spherical Geometries

While some postulates and properties of Euclidean geometry are true in spherical geometry, others are not, or are true only under certain circumstances.

## Example 3 Compare Plane Euclidean and Spherical Geometries

**Tell whether the following postulate or property of plane Euclidean geometry has a corresponding statement in spherical geometry. If so, write the corresponding statement. If not, explain your reasoning.**

**a. Through any two points, there is exactly one line.**

In the figure, notice that there is more than one great circle (line) through polar points $A$ and $B$. However, there is only one great circle through nonpolar points $C$ and $D$.

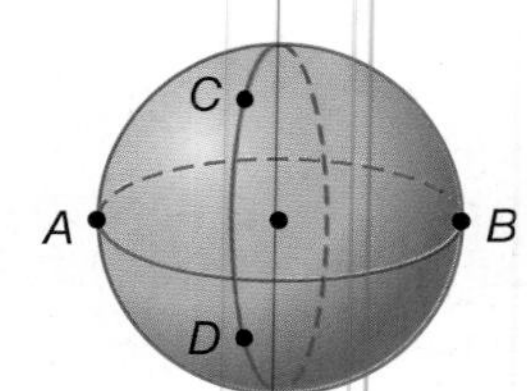


Therefore, a corresponding statement is that through any two nonpolar points, there is exactly one great circle (line).

**b. If given a line and a point not on the line, then there exists exactly one line through the point that is parallel to the given line.**

In the figure, notice that every great circle (line) containing point $A$ will intersect line $\ell$. Thus there exists no great circle through point $A$ that is parallel to line $\ell$.

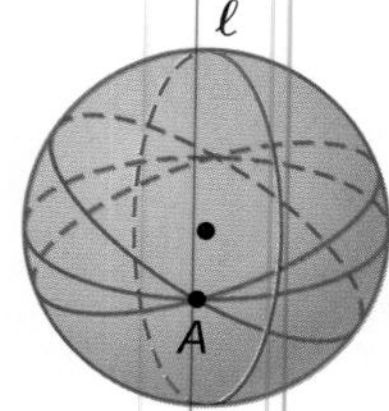


### GuidedPractice

**3A.** A line segment is the shortest path between two points.

**3B.** Through any two points, there is exactly one segment.

**StudyTip**

**Finite Geometries** *Planar networks* are another type of non-Euclidean geometry. You will learn about planar networks in Extend Lesson 13-6.

A **non-Euclidean geometry** is a geometry in which at least one of the postulates from Euclidean geometry fails. Notice in Example 3b that the Parallel Postulate does not hold true on a sphere. Lines, or great circles, cannot be parallel in spherical geometry. Therefore, spherical geometry is non-Euclidean.

## Check Your Understanding

**Example 1** **Name each of the following on sphere $\mathcal{B}$.**

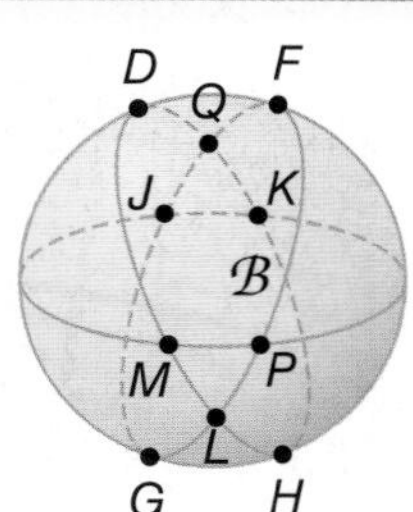


1. two lines containing point $Q$
2. a segment containing point $L$
3. a triangle
4. two segments on the same great circle

**Example 2** **SPORTS Determine whether figure $X$ on each of the spheres shown is a line in spherical geometry.**

5\.

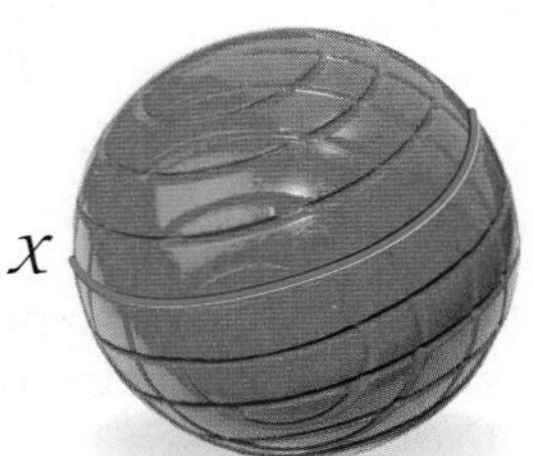

$X$

6\. $X$

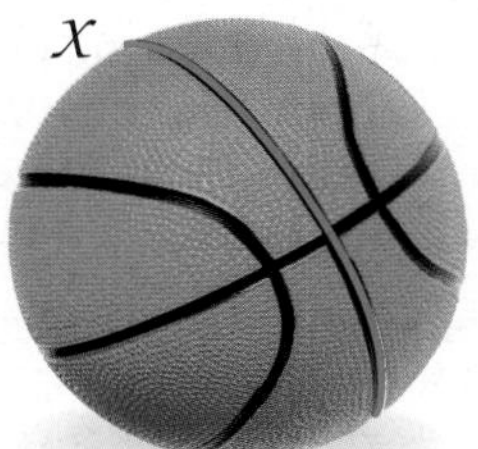

**Example 3** **CCSS REASONING Tell whether the following postulate or property of plane Euclidean geometry has a corresponding statement in spherical geometry. If so, write the corresponding statement. If not, explain your reasoning.**

7. The points on any line or line segment can be put into one-to-one correspondence with real numbers.
8. Perpendicular lines intersect at one point.

## Practice and Problem Solving

**Example 1** **Name two lines containing point $M$, a segment containing point $S$, and a triangle in each of the following spheres.**

9\.

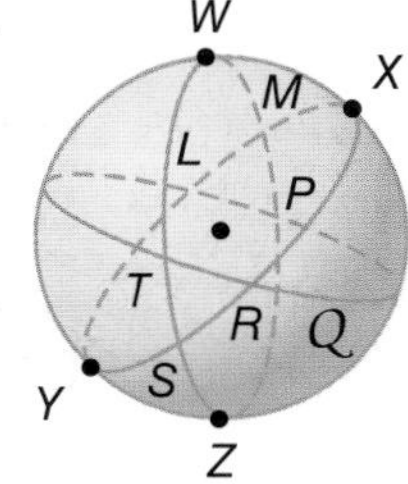


10\.

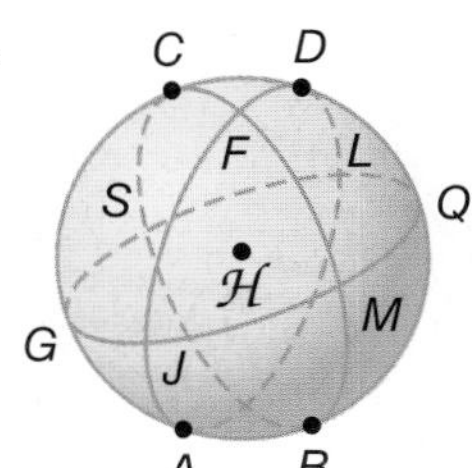


11. **SOCCER** Name each of the following on the soccer ball shown.

   a. two lines containing point $B$
   b. a segment containing point $F$
   c. a triangle
   d. a segment containing point $C$
   e. a line
   f. two lines containing point $A$

**Example 2** **ARCHITECTURE** **Determine whether figure $w$ on each of the spheres shown is a line in spherical geometry.**

**12.**

**13.**

**14.** **CCSS MODELING** Lines of latitude and longitude are used to describe positions on the Earth's surface. By convention, lines of longitude divide Earth vertically, while lines of latitude divide it horizontally.

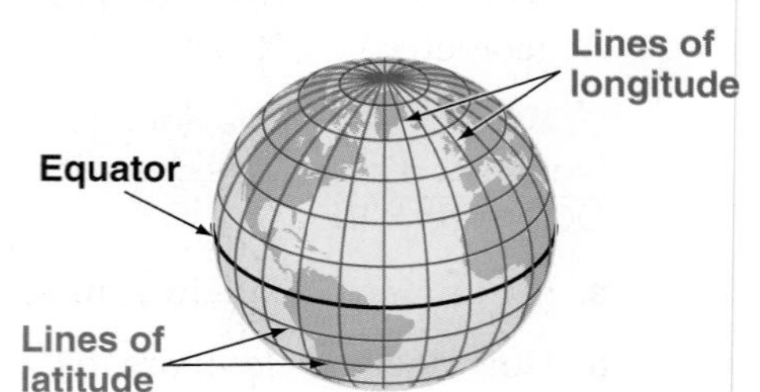


**a.** Are lines of longitude great circles? Explain.

**b.** Are lines of latitude great circles? Explain.

**Example 3** **Tell whether the following postulate or property of plane Euclidean geometry has a corresponding statement in spherical geometry. If so, write the corresponding statement. If not, explain your reasoning.**

**15** A line goes on infinitely in two directions.

**16.** Perpendicular lines form four 90° angles.

**17.** If three points are collinear, exactly one is between the other two.

**18.** If $M$ is the midpoint of $\overline{AB}$, then $\overline{AM} \cong \overline{MB}$.

**On a sphere, there are two distances that can be measured between two points. Use each figure and the information given to determine the distance between points $J$ and $K$ on each sphere. Round to the nearest tenth. Justify your answer.**

**19.**

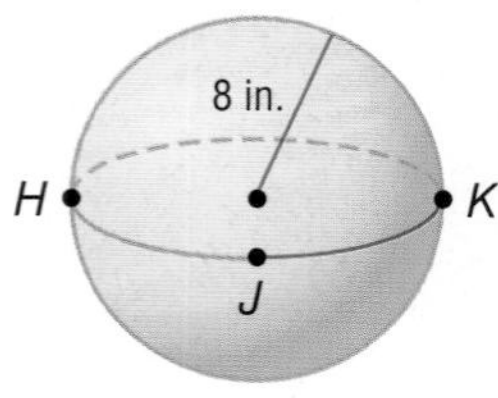


$m\widehat{JK} = 100$

**20.**

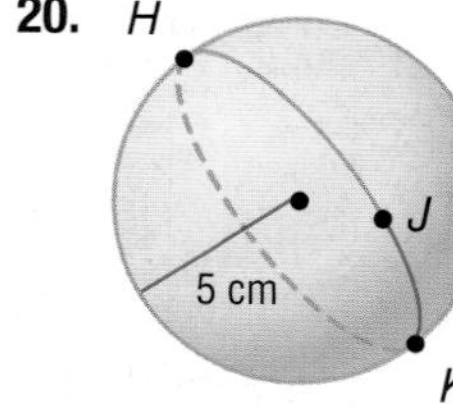


$m\widehat{JK} = 60$

**21.** **GEOGRAPHY** The location of Phoenix, Arizona, is 112° W longitude, 33.4° N latitude, and the location of Helena, Montana, is 112° W longitude, 46.6° N latitude. West indicates the location in terms of the prime meridian, and north indicates the location in terms of the equator. The mean radius of Earth is about 3960 miles.

**a.** Estimate the distance between Phoenix and Helena. Explain your reasoning.

**b.** Is there another way to express the distance between these two cities? Explain.

**c.** Can the distance between Washington, D.C., and London, England, which lie on approximately the same lines of latitude, be calculated in the same way? Explain your reasoning.

**d.** How many other locations are there that are the same distance from Phoenix, Arizona as Helena, Montana is? Explain.

22. **MULTIPLE REPRESENTATIONS** In this problem, you will investigate triangles in spherical geometry.

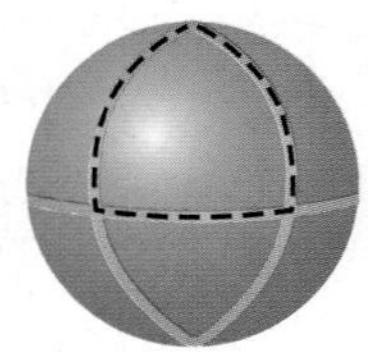

a. **Concrete** Use masking tape on a ball to mark three great circles. At least one of the three great circles should go through different poles than the other two. The great circles will form a triangle. Use a protractor to estimate the measure of each angle of the triangle.

b. **Tabular** Tabulate the measure of each angle of the triangle formed. Remove the tape and repeat the process two times so that you have tabulated the measure of three different triangles. Record the sum of the measures of each triangle.

c. **Verbal** Make a conjecture about the sum of the measures of a triangle in spherical geometry.

23. **QUADRILATERALS** Consider quadrilateral $ABCD$ on sphere $\mathcal{P}$. Note that it has four sides with $\overline{DC} \perp \overline{CB}$, $\overline{AB} \perp \overline{CB}$, and $\overline{DC} \cong \overline{AB}$.

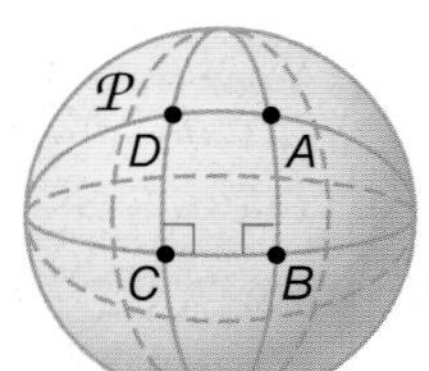


a. Is $\overline{CD} \perp \overline{DA}$? Explain your reasoning.

b. How does $DA$ compare to $CB$?

c. Can a rectangle, as defined in Euclidean geometry, exist in non-Euclidean geometry? Explain your reasoning.

## H.O.T. Problems Use Higher-Order Thinking Skills

24. **WRITING IN MATH** Compare and contrast Euclidean and spherical geometries. Be sure to include a discussion of planes and lines in both geometries.

25. **CHALLENGE** Geometries can be defined on curved surfaces other than spheres. Another type of non-Euclidean geometry is *hyperbolic geometry*. This geometry is defined on a curved saddle-like surface. Compare the sum of the angle measures of a triangle in hyperbolic, spherical, and Euclidean geometries.

**Triangle in plane geometry**

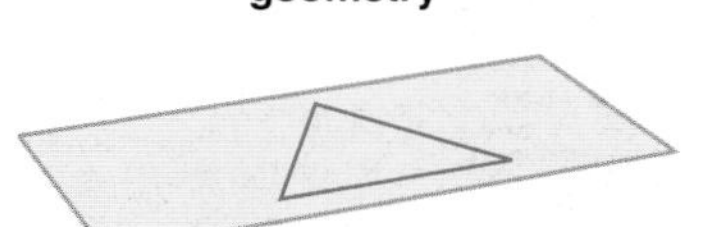

**Triangle in spherical geometry**

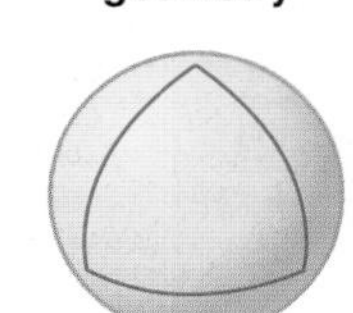

**Triangle in hyperbolic geometry**

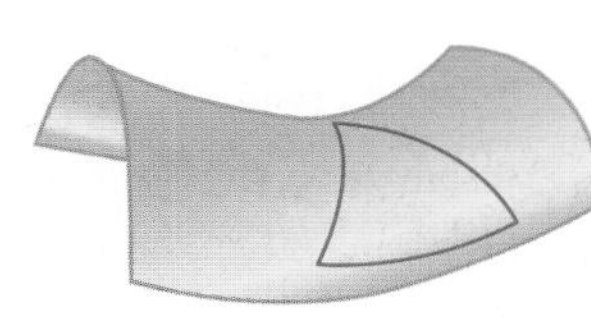

26. **OPEN ENDED** Sketch a sphere with three points so that two of the points lie on a great circle and two of the points do not lie on a great circle.

27. **CCSS ARGUMENTS** A *small circle* of a sphere intersects at least two points, but does not go through opposite poles. Points $A$ and $B$ lie on a small circle of sphere $Q$. Will two small circles *sometimes*, *always*, or *never* be parallel? Draw a sketch and explain your reasoning.

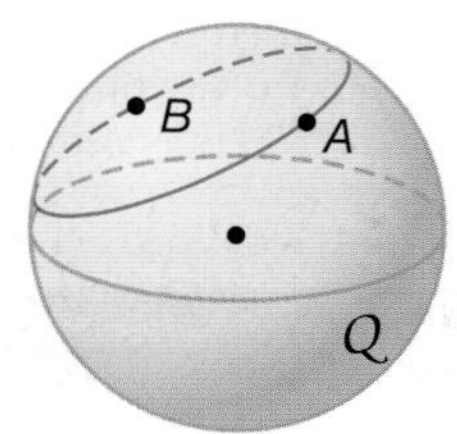


28. **WRITING IN MATH** Do similar or congruent triangles exist in spherical geometry? Explain your reasoning.

29. **REASONING** Is the statement *Spherical geometry is a subset of Euclidean geometry* true or false? Explain your reasoning.

30. **REASONING** Two planes are equidistant from the center of a sphere and intersect the sphere. What is true of the circles? Are they lines in spherical geometry? Explain.

## Standardized Test Practice

**31.** Which of the following postulates or properties of spherical geometry is false?

**A** The shortest path between two points on a circle is an arc.

**B** If three points are collinear, any of the three points lies between the other two.

**C** A great circle is infinite and never returns to its original starting point.

**D** Perpendicular great circles intersect at two points.

**32. SAT/ACT** A car travels 50 miles due north in 1 hour and 120 miles due west in 2 hours. What is the average speed of the car?

**F** 50 mph  **H** 60 mph

**G** 55 mph  **J** none of the above

**33. SHORT RESPONSE** Name a line in sphere $\mathcal{P}$ that contains point $D$.

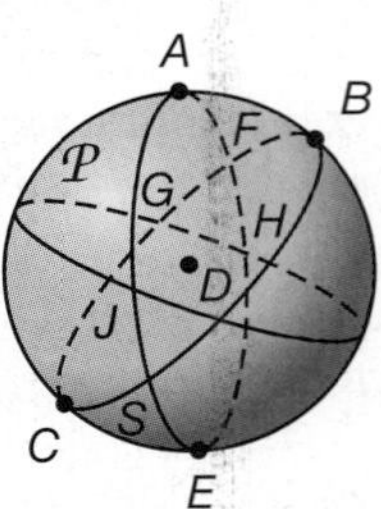


**34. ALGEBRA** The ratio of males to females in a classroom is 3:5. How many females are in the room if the total number of students is 32?

**A** 12  **D** 51

**B** 20  **E** 53

**C** 29

## Spiral Review

**Find the volume of each sphere or hemisphere. Round to the nearest tenth.** (Lesson 12-6)

**35.** sphere: area of great circle $= 98.5\text{ m}^2$

**36.** sphere: circumference of great circle $\approx 23.1$ in.

**37.** hemisphere: circumference of great circle $\approx 50.3$ cm

**38.** hemisphere: area of great circle $\approx 3416\text{ ft}^2$

**Find the volume of each cone. Round to the nearest tenth.** (Lesson 12-5)

**39.**

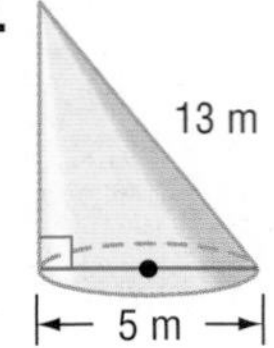


**40.**

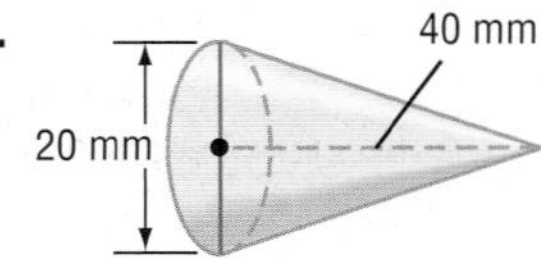


**41.**

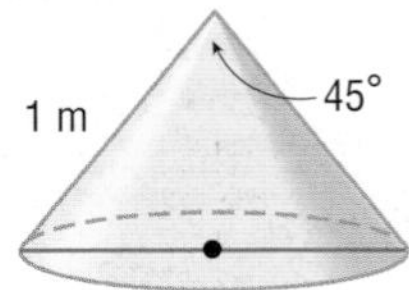


**42. RADIOS** Three radio towers are modeled by the points $A(-3, 4)$, $B(9, 4)$, and $C(-3, -12)$. Determine the location of another tower equidistant from all three towers, and write an equation for the circle which all three points lie on. (Lesson 11-8)

## Skills Review

**For each pair of similar figures, find the area of the green figure.**

**43.**

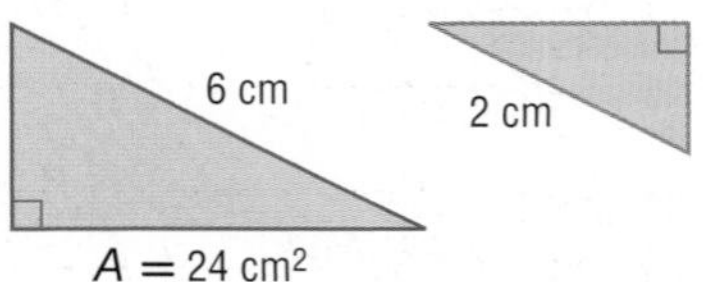


**44.**

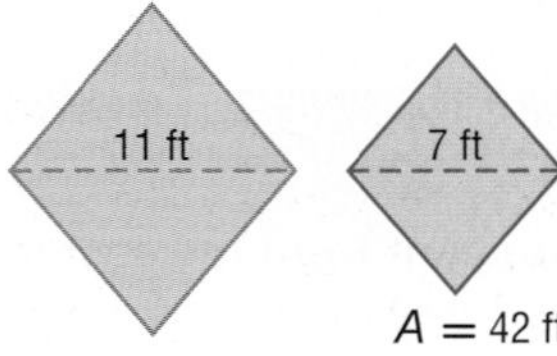


**45.**

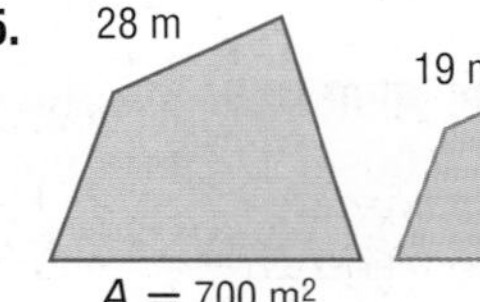

EXTEND 12-7

## Geometry Lab
# Navigational Coordinates

**OBJECTIVE** Use a latitude and longitude measure to identify the hemispheres on which the location lies and estimate the location of a city using a globe or map.

A grid system of imaginary lines on Earth is used for locating places and navigation. Imaginary vertical lines drawn around the Earth through the North and South Poles are called **meridians** and determine the measure of **longitude**. Imaginary horizontal lines parallel to the equator are called **parallels** and determine the measure of **latitude**.

The basic units for measurements are degrees, minutes, and seconds. 1 degree (°) = 60 minutes (‘), and 60 minutes = 60 seconds (″).

| | Location of 0° | Direction | Maximum Degrees |
|---|---|---|---|
| **Latitude (parallels)** | equator | In northern hemisphere, all are degrees north.<br>In southern hemisphere, all are degrees south. | 180° at international dateline |
| **Longitude (meridians)** | Prime Meridian through Greenwich, England | In eastern hemisphere, all are degrees east.<br>In western hemisphere, all are degrees west. | 90° at each pole |

## Activity Investigate Latitude and Longitude

**The table shows the latitude and longitude of three cities.**

| City | Latitude | Longitude |
|---|---|---|
| A | 37°59′N | 84°28′W |
| B | 34°55′S | 138°36′E |
| C | 64°4′N | 21°58′W |

**1.** In which hemisphere is each city located?

**2.** Use a globe or map to name each city.

**3.** Earth is approximately a sphere with a radius of 3960 miles. The equator and all meridians are great circles. The circumference of a great circle is equal to the length of the equator or any meridian. Find the length of a great circle on Earth in miles.

**4.** Notice that the distance between each line of latitude is about the same. The distance from the equator to the North Pole is $\frac{1}{4}$ of the circumference of Earth, and each degree of latitude is $\frac{1}{90}$ of that distance. Estimate the distance between one pair of latitude lines in miles.

## Analyze

**The table shows the latitude and longitude of three cities.**

| City | Latitude | Longitude |
|---|---|---|
| F | 1°28′S | 48°29′W |
| G | 13°45′N | 100°30′E |
| H | 41°17′S | 174°47′E |

**5.** Name the hemisphere in which each city is located.

**6.** Use a globe or map to name each city.

**7.** Find the approximate distance between meridians at latitude of about 22° N. The direct distance between the two cities at the right is about 1646 miles.

| | | |
|---|---|---|
| **Calcutta, India** | 22°34′N | 88°24′E |
| **Hong Kong, China** | 22°20′N | 114°11′E |

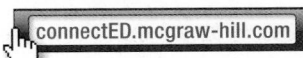

LESSON 12-8

# Congruent and Similar Solids

**Then**

- You compared surface areas and volumes of spheres.

**Now**

1 Identify congruent or similar solids.

2 Use properties of similar solids.

**Why?**

- The gemstones at the right are cut in exactly the same shape, but their sizes are different. Their shapes are *similar.*

**NewVocabulary**
similar solids
congruent solids

## 1 Identify Congruent or Similar Solids

**Similar solids** have exactly the same shape but not necessarily the same size. All spheres are similar and all cubes are similar.

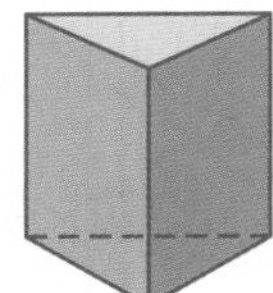
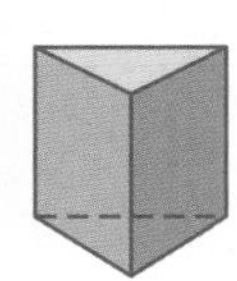
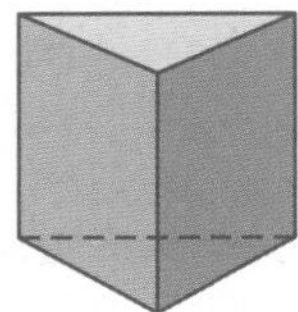
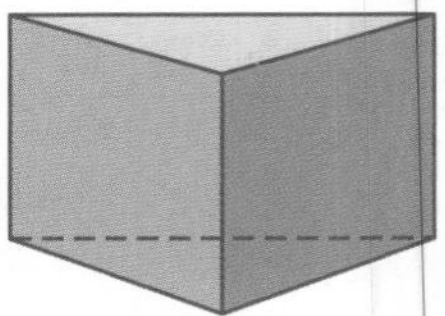

In similar solids, the corresponding linear measures, such as height and radius, have equal ratios. The common ratio is called the *scale factor.* If two similar solids are polyhedrons, their corresponding faces are similar.

### KeyConcept Similar Solids

**Words** Two solids are similar if they have the same shape and the ratios of their corresponding linear measures are equal.

**Models**

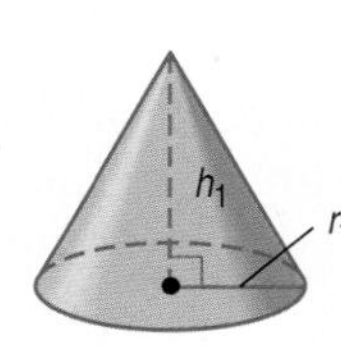

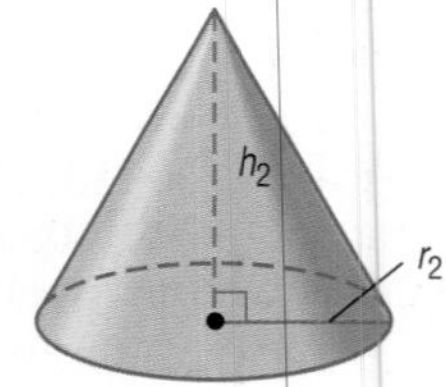


$$\frac{h_1}{h_2} = \frac{r_1}{r_2}$$

**Congruent solids** have exactly the same shape and the same size. Congruent solids are similar solids that have a scale factor of 1:1.

### KeyConcept Congruent Solids

**Words** Two solids are congruent if they have the following characteristics.

- Corresponding angles are congruent.
- Corresponding edges are congruent.
- Corresponding faces are congruent.
- Volumes are equal.

**Models**

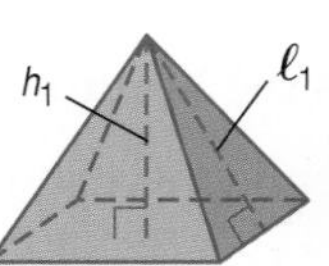

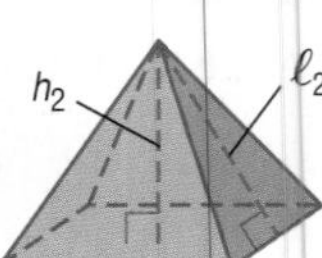


$$\frac{h_1}{h_2} = \frac{\ell_1}{\ell_2} = 1$$

> **StudyTip**
> Similar and Congruent Solids If two solids are similar, then their corresponding linear measures are proportional. If two solids are congruent, then their corresponding linear measures are equal.

### Example 1 Identify Similar and Congruent Solids

**Determine whether each pair of solids is *similar*, *congruent*, or *neither*. If the solids are similar, state the scale factor.**

**a. the square pyramids**

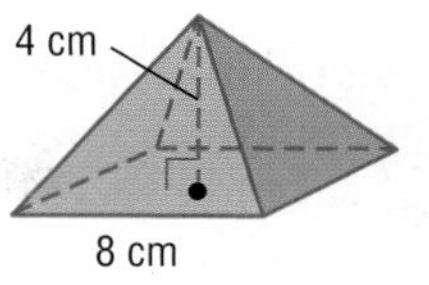


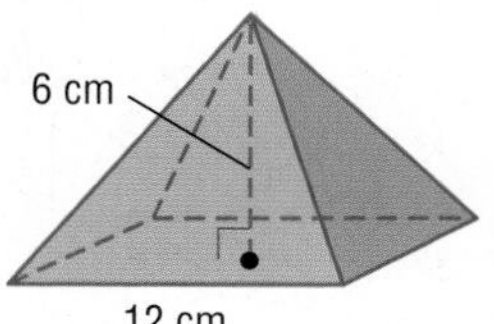


ratio of heights: $\frac{4}{6} = \frac{2}{3}$

ratio of base edges: $\frac{8}{12} = \frac{2}{3}$

The ratios of the corresponding measures are equal, so the pyramids are similar. The scale factor is 2:3. Since the scale factor is not 1:1, the solids are not congruent.

**b. the rectangular prisms**

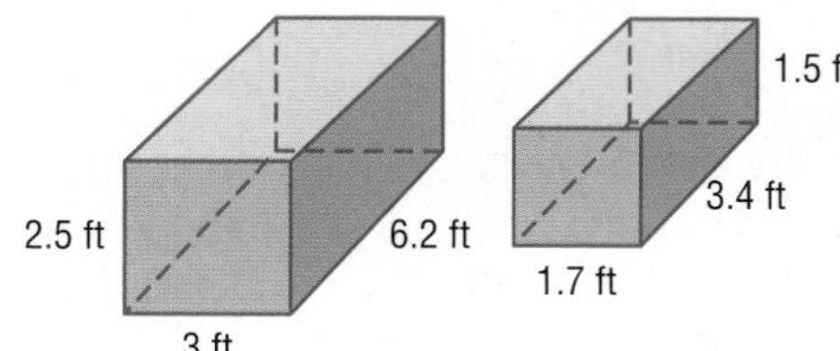


ratio of widths: $\frac{3}{1.7} \approx 1.76$

ratio of lengths: $\frac{6.2}{3.4} \approx 1.82$

ratio of heights: $\frac{2.5}{1.5} \approx 1.67$

Since the ratios of corresponding measures are not equal, the prisms are neither congruent nor similar.

**GuidedPractice**

**1A.**

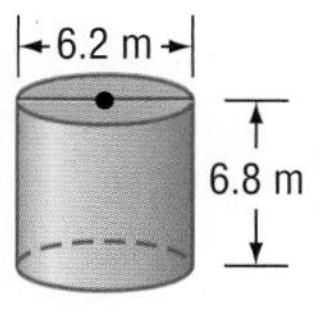

**1B.**

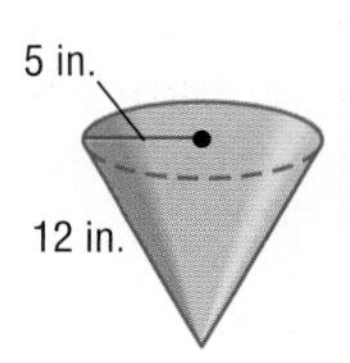


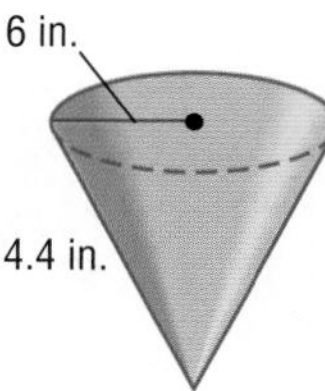


> **StudyTip**
> Check Solutions After finding missing measures of similar solids, work backward to check your solutions.

## 2 Properties of Congruent and Similar Solids

The cubes at the right are similar solids with a scale factor of 3:2.

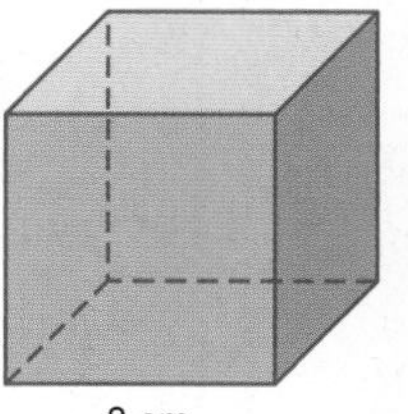


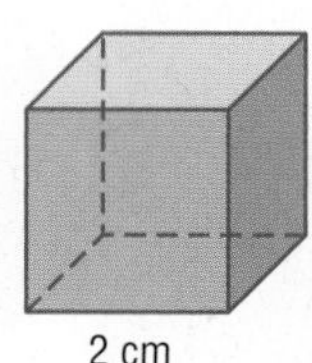


ratio of surface areas: 54:24 or 9:4

ratio of volumes: 27:8

Notice that the ratio of surface areas, 9:4, can be written as $3^2:2^2$. The ratio of volumes, 27:8, can be written as $3^3:2^3$. This suggests the following theorem.

### Theorem 12.1

| | | | Models |
|---|---|---|---|
| **Words** | If two similar solids have a scale factor of $a:b$, then the surface areas have a ratio of $a^2:b^2$, and the volumes have a ratio of $a^3:b^3$. | | 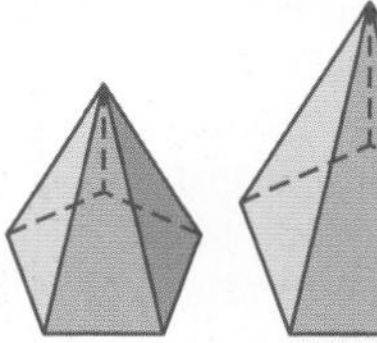 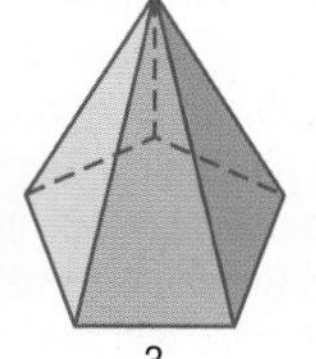  |
| **Examples** | scale factor<br>ratio of surface area<br>ratio of volumes | 2:3<br>4:9<br>8:27 | |

Figures must be similar in order for Theorem 12.1 to apply.

## Example 2 Use Similar Solids to Write Ratios

**Two similar cones have radii of 10 millimeters and 15 millimeters. What is the ratio of the surface area of the small cone to the surface area of the large cone?**

First, find the scale factor.

$\frac{\text{radius of small cone}}{\text{radius of large cone}} = \frac{10}{15}$ or $\frac{2}{3}$ Write a ratio comparing the radii.

The scale factor is $\frac{2}{3}$.

$\frac{a^2}{b^2} = \frac{2^2}{3^2}$ or $\frac{4}{9}$ If the scale factor is $\frac{a}{b}$, then the ratio of surface areas is $\frac{a^2}{b^2}$.

So, the ratio of the surface areas is 4:9.

**StudyTip**

**Similar Solids and Area** If two solids are similar, then the ratio of any corresponding areas is $a^2 : b^2$. In Example 2, the ratio of the lateral areas of the cones is 4:25, and the ratio of the base areas of the cones is 4:25.

### GuidedPractice

**2.** Two similar prisms have surface areas of 98 square centimeters and 18 square centimeters. What is the ratio of the height of the large prism to the height of the small prism?

Many real-world objects can be modeled by similar solids.

## Real-World Example 3 Use Similar Solids to Find Unknown Values

**CONTAINERS The containers at the right are similar cylinders. Find the height *h* of the smaller container.**

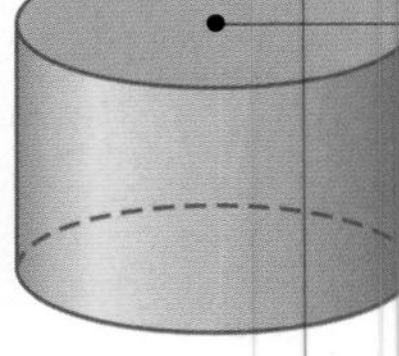


**Understand** You know the height of the larger container and the volumes of both containers.

**Plan** Use Theorem 12.1 to write a ratio comparing the volumes. Then find the scale factor and use it to find $h$.

**Solve**

$\frac{\text{volume of small container}}{\text{volume of large container}} = \frac{270\pi}{640\pi}$ Write a ratio comparing volumes.

$= \frac{27}{64}$ Simplify.

$= \frac{3^3}{4^3}$ Write as $\frac{a^3}{b^3}$.

The scale factor is 3:4.

Ratio of heights → $\frac{h}{10} = \frac{3}{4}$ ← Scale factor

$h \cdot 4 = 10 \cdot 3$ Find the cross products.

$h = 7.5$ Solve for $h$.

So, the height of the smaller container is 7.5 inches.

**Check** Since $\frac{7.5}{10} = 0.75 = \frac{3}{4}$, the solution is correct. ✓

**Math HistoryLink**

**Georg F.B. Riemann (1826–1866)** Spherical geometry is sometimes called *Riemann geometry*, after Georg Reimann, a German mathematician responsible for the Riemannian Postulate, which states that through a point not on a line, there are no lines parallel to the given line.

### GuidedPractice

**3. VOLLEYBALL** A regulation volleyball has a circumference of about 66 centimeters. The ratio of the surface area of that ball to the surface area of a children's ball is approximately 1.6:1. What is the circumference of the children's ball? Round to the nearest centimeter.

## Check Your Understanding

**Example 1** **Determine whether each pair of solids is *similar, congruent,* or *neither*. If the solids are similar, state the scale factor.**

**1.**

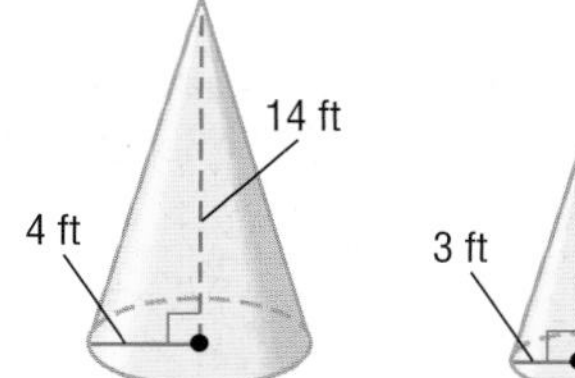


**2.**

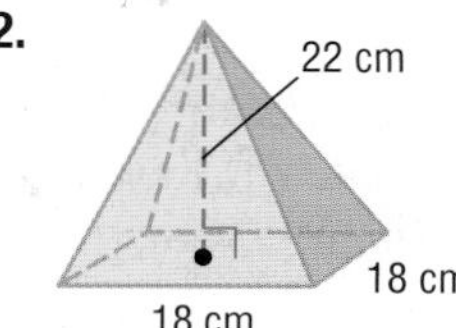


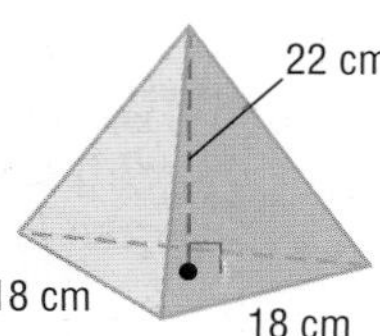


**Example 2**

**3.** Two similar cylinders have radii of 15 inches and 6 inches. What is the ratio of the surface area of the small cylinder to the surface area of the large cylinder?

**4.** Two spheres have volumes of $36\pi$ cubic centimeters and $288\pi$ cubic centimeters. What is the ratio of the radius of the small sphere to the radius of the large sphere?

**Example 3**

**5.** **EXERCISE BALLS** A company sells two different sizes of exercise balls. The ratio of the diameters is 15 : 11. If the diameter of the smaller ball is 55 centimeters, what is the volume of the larger ball? Round to the nearest tenth.

## Practice and Problem Solving

**Example 1** **CCSS REGULARITY** **Determine whether each pair of solids is *similar, congruent,* or *neither*. If the solids are similar, state the scale factor.**

**6.**

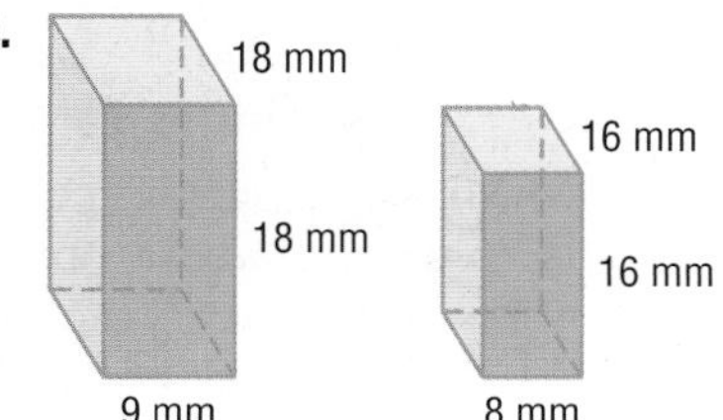


**7.**

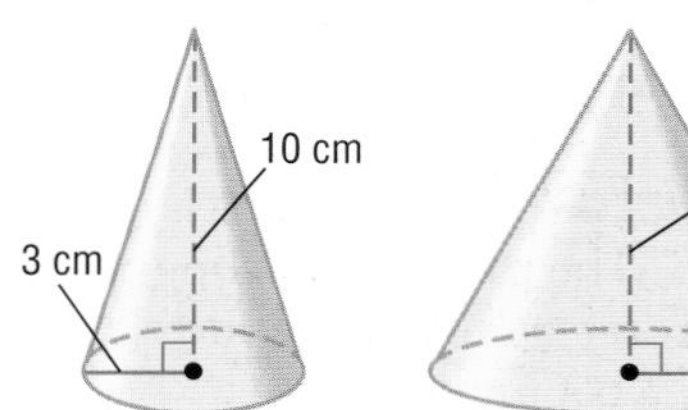


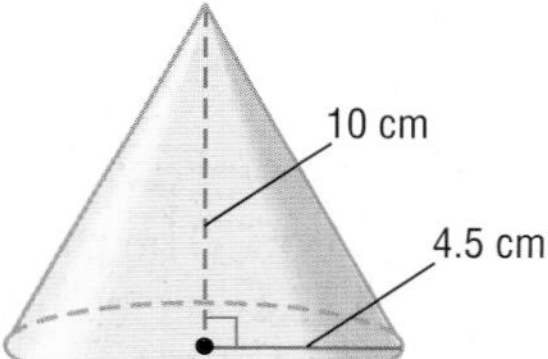


**8.**

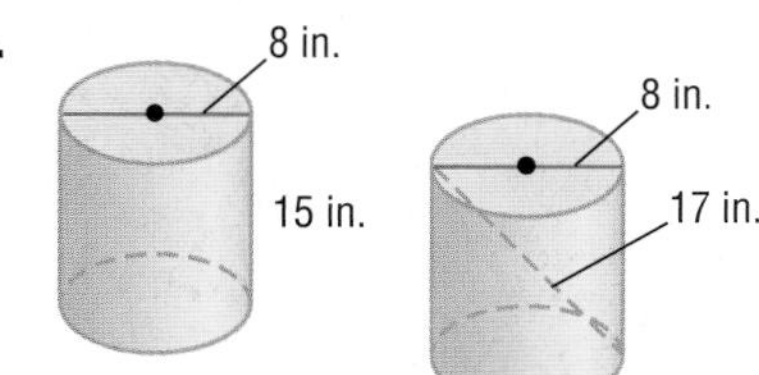


**9.**

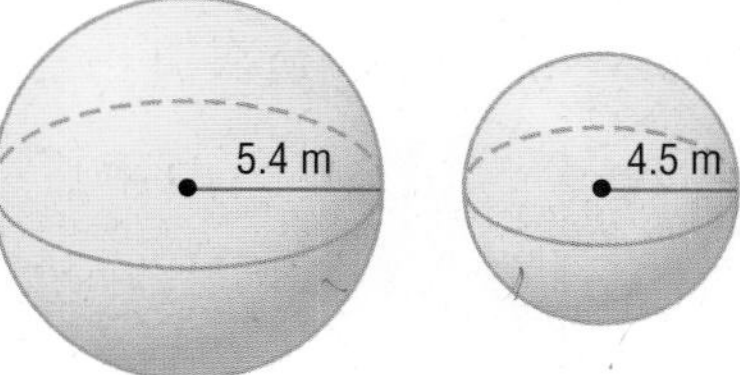


**Example 2**

**10.** Two similar pyramids have slant heights of 6 inches and 12 inches. What is the ratio of the surface area of the small pyramid to the surface area of the large pyramid?

**11.** Two similar cylinders have heights of 35 meters and 25 meters. What is the ratio of the volume of the large cylinder to the volume of the small cylinder?

**12.** Two spheres have surface areas of $100\pi$ square centimeters and $16\pi$ square centimeters. What is the ratio of the volume of the large sphere to the volume of the small sphere?

**13.** Two similar hexagonal prisms have volumes of 250 cubic feet and 2 cubic feet. What is the ratio of the height of the large cylinder to the height of the small cylinder?

**14.** **DIMENSIONAL ANALYSIS** Two rectangular prisms are similar. The height of the first prism is 6 yards and the height of the other prism is 9 feet. If the volume of the first prism is 810 cubic yards, what is the volume of the other prism?

**Example 3**

**15. FOOD** A small cylindrical can of tuna has a radius of 4 centimeters and a height of 3.8 centimeters. A larger and similar can of tuna has a radius of 5.2 centimeters.

**a.** What is the scale factor of the cylinders?

**b.** What is the volume of the larger can? Round to the nearest tenth.

**16. SUITCASES** Two suitcases are similar rectangular prisms. The smaller suitcase is 68 centimeters long, 47 centimeters wide, and 27 centimeters deep. The larger suitcase is 85 centimeters long.

**a.** What is the scale factor of the prisms?

**b.** What is the volume of the larger suitcase? Round to the nearest tenth.

**17. SCULPTURE** The sculpture shown at the right is a scale model of a cornet. If the sculpture is 26 feet long and a standard cornet is 14 inches long, what is the scale factor of the sculpture to a standard cornet?

**18.** The pyramids shown are congruent.

**a.** What is the perimeter of the base of Pyramid A?

**b.** What is the area of the base of Pyramid B?

**c.** What is the volume of Pyramid B?

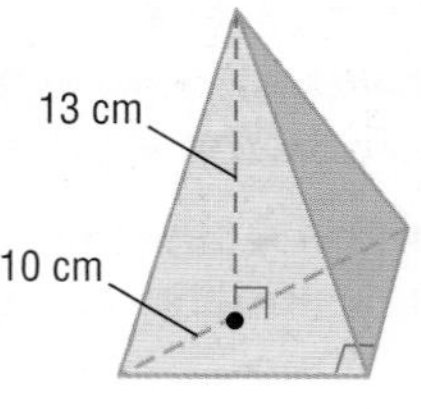


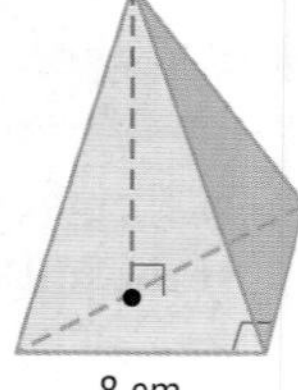


**19. TECHNOLOGY** Jalissa and Mateo each have the same type of MP3 player, but in different colors. The players are congruent rectangular prisms. The volume of Jalissa's player is 4.92 cubic inches, the width is 2.4 inches, and the depth is 0.5 inch. What is the height of Mateo's player?

**CCSS SENSE-MAKING Each pair of solids below is similar.**

**20.** What is the surface area of the smaller solid shown below?

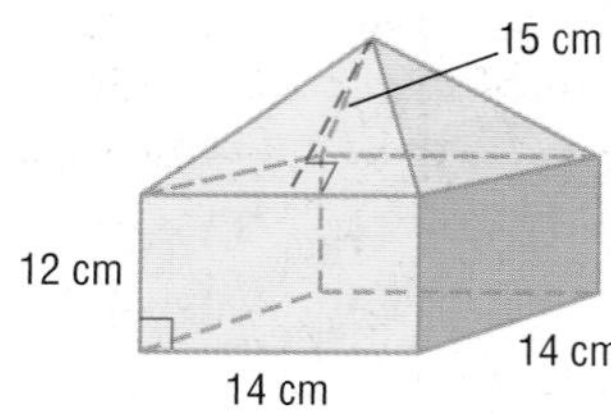


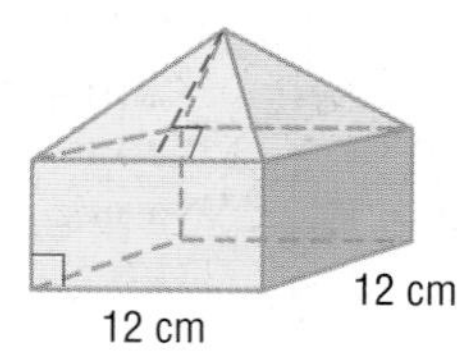


**21.** What is the volume of the larger solid shown below?

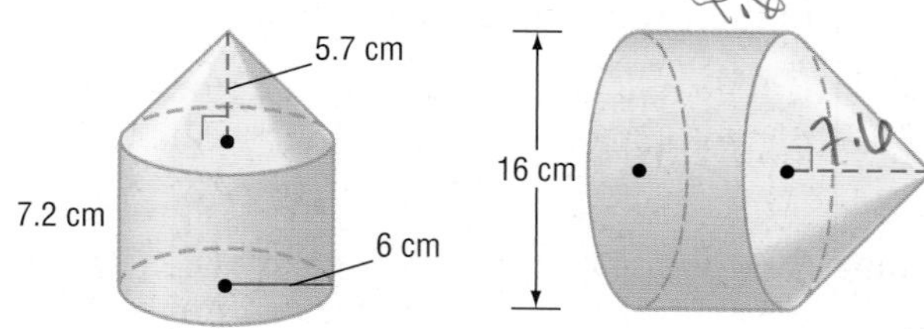


**22. DIMENSIONAL ANALYSIS** Two cylinders are similar. The height of the first cylinder is 23 cm and the height of the other cylinder is 8 in. If the volume of the first cylinder is $552\pi$ cm$^3$, what is the volume of the other prism? Use 2.54 cm = 1 in.

Jim Baron

23. **DIMENSIONAL ANALYSIS** Two spheres are similar. The radius of the first sphere is 10 feet. The volume of the other sphere is 0.9 cubic meters. Use 2.54 cm = 1 in. to determine the scale factor from the first sphere to the second.

24. **ALGEBRA** Two similar cones have volumes of $343\pi$ cubic centimeters and $512\pi$ cubic centimeters. The height of each cone is equal to 3 times its radius. Find the radius and height of both cones.

25. **TENTS** Two tents are in the shape of hemispheres, with circular floors. The ratio of their floor areas is 9 : 12.25. If the diameter of the smaller tent is 6 feet, what is the volume of the larger tent? Round to the nearest tenth.

26. **MULTIPLE REPRESENTATIONS** In this problem, you will investigate similarity. The heights of two similar cylinders are in the ratio 2 to 3. The lateral area of the larger cylinder is $162\pi$ square centimeters, and the diameter of the smaller cylinder is 8 centimeters.

    a. **Verbal** What is the height of the larger cylinder? Explain your method.

    b. **Geometric** Sketch and label the two cylinders.

    c. **Analytical** How many times as great is the volume of the larger cylinder as the volume of the smaller cylinder?

## H.O.T. Problems Use Higher-Order Thinking Skills

27. **ERROR ANALYSIS** Cylinder X has a diameter of 20 centimeters and a height of 11 centimeters. Cylinder Y has a radius of 30 centimeters and is similar to Cylinder X. Did Laura or Paloma correctly find the height of Cylinder Y? Explain your reasoning.

| Laura | Paloma |
|---|---|
| Cylinder X: radius 10, height 11<br>Cylinder Y: radius 30, height a<br>$\frac{10}{30} = \frac{11}{a}$, so $a = 33$. | Cylinder X: diameter 20, height 11<br>Cylinder Y: diameter 20, height a<br>$\frac{20}{20} = \frac{11}{a}$, so $a = 11$. |

28. **CHALLENGE** The ratio of the volume of Cylinder A to the volume of Cylinder B is 1:5. Cylinder A is similar to Cylinder C with a scale factor of 1:2 and Cylinder B is similar to Cylinder D with a scale factor of 1:3. What is the ratio of the volume of Cylinder C to the volume of Cylinder D? Explain your reasoning.

29. **WRITING IN MATH** Explain how the surface areas and volumes of the similar prisms shown at the right are related.

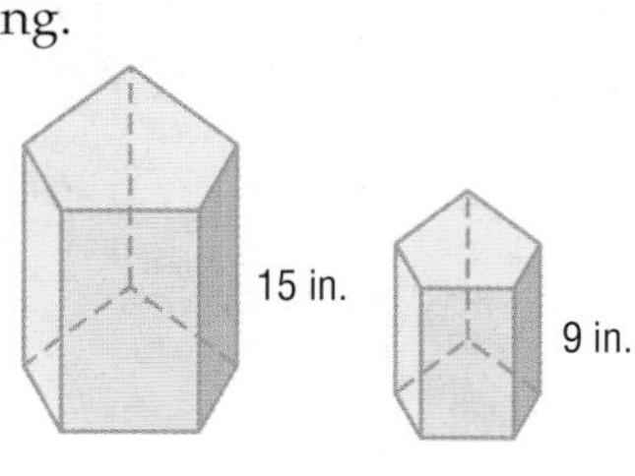


30. **OPEN ENDED** Describe two nonsimilar triangular pyramids with similar bases.

31. **CCSS SENSE-MAKING** Plane $\mathcal{P}$ is parallel to the base of cone C, and the volume of the cone above the plane is $\frac{1}{8}$ of the volume of cone C. Find the height of cone C.

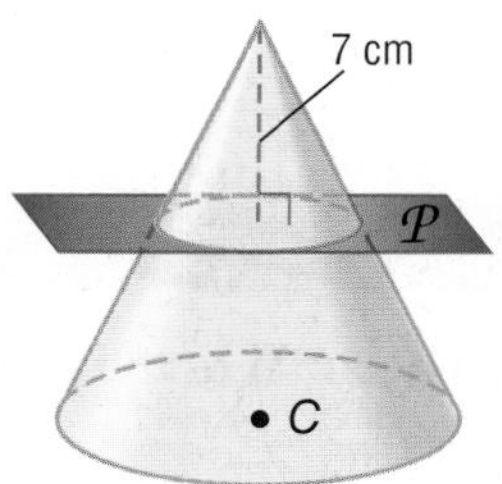


32. **WRITING IN MATH** Explain why all spheres are similar.

## Standardized Test Practice

**33.** Two similar spheres have radii of $20\pi$ meters and $6\pi$ meters. What is the ratio of the surface area of the large sphere to the surface area of the small sphere?

A $\frac{100}{3}$  B $\frac{100}{9}$  C $\frac{10}{3}$  D $\frac{10}{9}$

**34.** What is the scale factor of the similar figures?

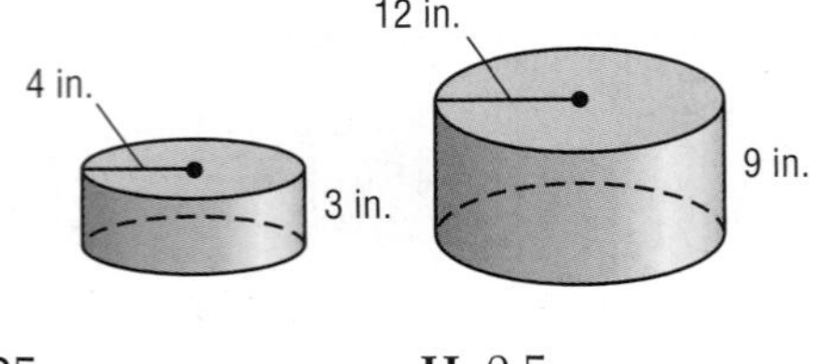


F 0.25
G 0.33
H 0.5
J 0.75

**35. SHORT RESPONSE** Point $A$ and point $B$ represent the locations of Timothy's and Quincy's houses. If each unit on the map represents one kilometer, how far apart are the two houses?

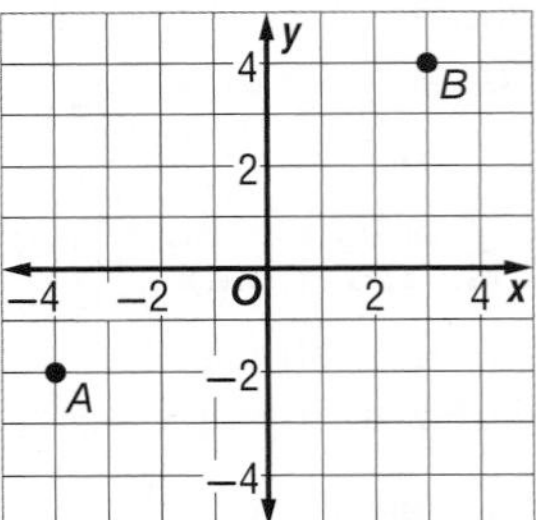


**36. SAT/ACT** If $\frac{x+2}{3} = \frac{(x+2)^2}{15}$, what is one possible value of $x$?

A 0  B 1  C 2  D 3  E 4

## Spiral Review

**Determine whether figure $\chi$ on each of the spheres shown is a line in spherical geometry.** (Lesson 12-7)

**37.**

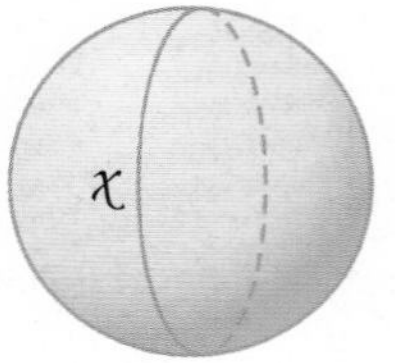


**38.**

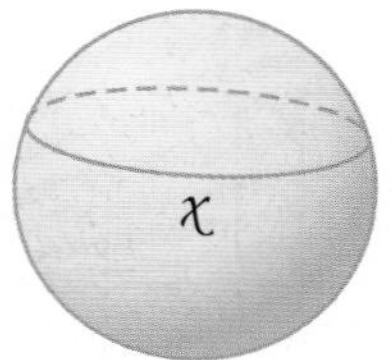


**39.**

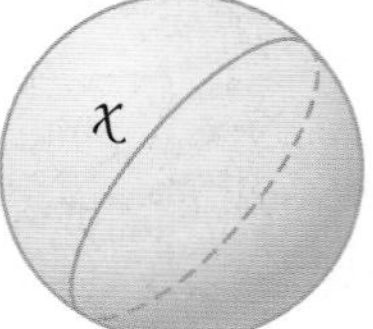


**40. ENTERTAINMENT** Some people think that the Spaceship Earth geosphere at Epcot in Disney World in Orlando, Florida, resembles a golf ball. The building is a sphere measuring 165 feet in diameter. A typical golf ball has a diameter of approximately 1.5 inches. (Lesson 12-6)

**a.** Find the volume of Spaceship Earth to the nearest cubic foot.

**b.** Find the volume of a golf ball to the nearest tenth.

**c.** What is the scale factor that compares Spaceship Earth to a golf ball?

**d.** What is the ratio of the volumes of Spaceship Earth to a golf ball?

**Find $x$. Assume that segments that appear to be tangent are tangent.** (Lesson 11-7)

**41.**

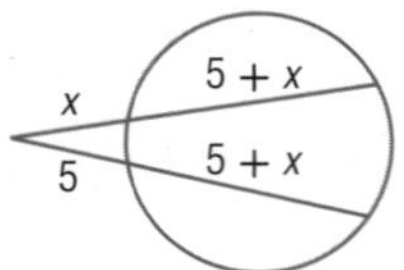


**42.**

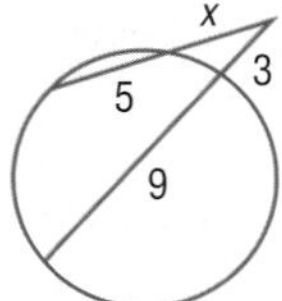


**43.**

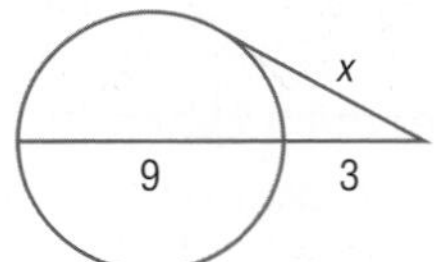


Danielle Gali/age fotostock

## Skills Review

**Express each fraction as a decimal to the nearest hundredth.**

**44.** $\frac{8}{13}$  **45.** $\frac{17}{54}$  **46.** $\frac{11}{78}$  **47.** $\frac{43}{46}$

# Study Guide and Review

## Study Guide

### KeyConcepts

**Representations of Three-Dimensional Figures** (Lesson 12-1)

- Solids can be classified by bases, faces, edges, and vertices.

**Surface Areas of Prisms and Cylinders** (Lesson 12-2)

- Lateral surface area of a right prism: $L = Ph$
- Lateral surface area of a right cylinder: $L = 2\pi rh$

**Surface Areas of Pyramids and Cones** (Lesson 12-3)

- Lateral surface area of a pyramid: $L = \frac{1}{2}P\ell$
- Lateral surface area of a right cone: $L = \pi r\ell$

**Volumes of Prisms and Cylinders** (Lesson 12-4)

- Volume of prism or cylinder: $V = Bh$

**Volumes of Pyramids and Cones** (Lesson 12-5)

- Volume of a pyramid: $V = \frac{1}{3}Bh$
- Volume of a cone: $V = \frac{1}{3}\pi r^2h$

**Surface Areas and Volumes of Spheres** (Lesson 12-6)

- Surface area of a sphere: $S = 4\pi r^2$
- Volume of a sphere: $V = \frac{4}{3}\pi r^3$

**Congruent and Similar Solids** (Lesson 12-8)

- Similar solids have the same shape, but not necessarily the same size.
- Congruent solids are similar solids with a scale factor of 1.

### FOLDABLES® StudyOrganizer

Be sure the Key Concepts are noted in your Foldable.

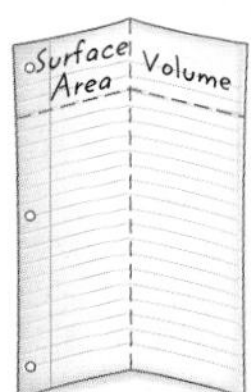


### KeyVocabulary

altitude (p. 814)
axis (p. 816)
base edges (p. 814)
composite solid (p. 820)
congruent solid (p. 864)
cross section (p. 820)
Euclidean geometry (p. 857)
great circle (p. 849)
isometric view (p. 807)
lateral area (p. 814)
lateral edge (p. 814)
lateral face (p. 814)
non-Euclidean geometry (p. 858)
oblique cone (p. 824)
oblique solid (p. 806)
regular pyramid (p. 822)
right cone (p. 824)
right solid (p. 805)
similar solids (p. 864)
slant height (p. 822)
spherical geometry (p. 857)
topographic map (p. 813)

### VocabularyCheck

**State whether each sentence is *true* or *false*. If *false*, replace the underlined term to make a true sentence.**

1. Euclidean geometry deals with a system of points, great circles (lines), and spheres (planes).
2. Similar solids have exactly the same shape, but not necessarily the same size.
3. A right solid has an axis that is also an altitude.
4. The isometric view is when an object is viewed from a corner.
5. The perpendicular distance from the base of a geometric figure to the opposite vertex, parallel side, or parallel surface is the altitude.
6. Rotation symmetry is also called mirror symmetry.
7. The intersection of two adjacent lateral faces is the lateral edge.
8. Euclidean geometry refers to geometrical systems that are not in accordance with the Parallel Postulate.
9. A composite solid is a three-dimensional figure that is composed of simpler figures.
10. The slant height is the height of each lateral face of a pyramid or cone.

# Study Guide and Review *Continued*

## Lesson-by-Lesson Review

### 12-1 Representations of Three-Dimensional Figures

Describe each cross section.

**11.**

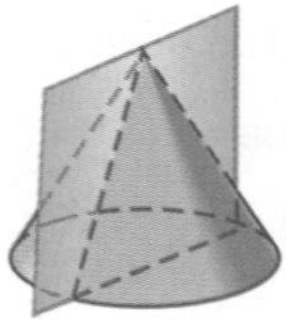

**12.**

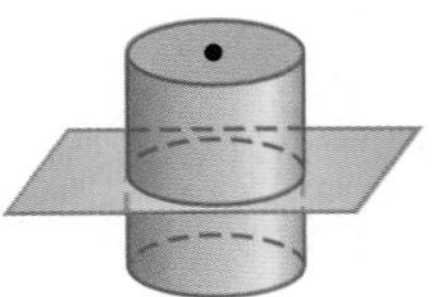

**13.** **CAKE** The cake shown is cut in half vertically. Describe the cross section of the cake.

**Example 1**

**Describe the vertical and horizontal cross sections of the figure shown below.**

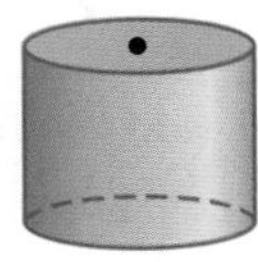

The vertical cross section is a rectangle.
The horizontal cross section is a circle.

### 12-2 Surface Areas of Prisms and Cylinders

Find the lateral area and surface area of each prism. Round to the nearest tenth if necessary.

**14.**

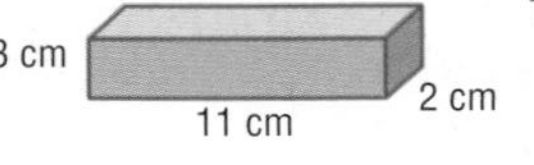


**15.**

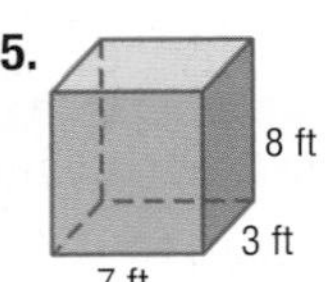


Find the lateral area and surface area of each cylinder. Round to the nearest tenth.

**16.**

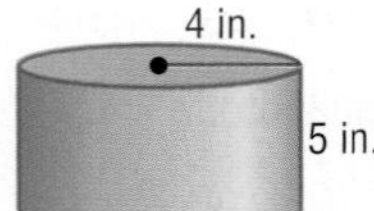


**17.**

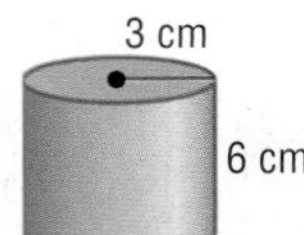


**Example 2**

**Find the surface area of the rectangular prism.**

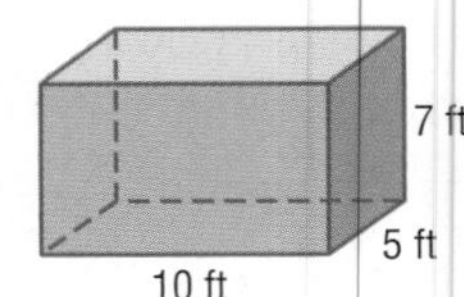


Use the 10-foot by 5-foot rectangle as the base.

| | |
|---|---|
| $S = Ph + 2B$ | Surface area of a prism |
| $= (2 \cdot 10 + 2 \cdot 5)(7) + 2(10 \cdot 5)$ | Substitution |
| $= 310$ | Simplify. |

The surface area is 310 square feet.

### 12-3 Surface Areas of Pyramids and Cones

Find the lateral area and the surface area of each regular pyramid. Round to the nearest tenth.

**18.**

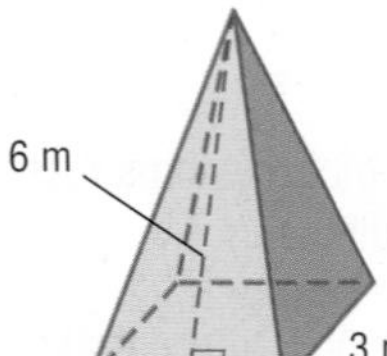


**19.**

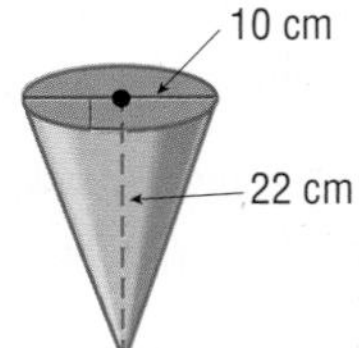


**Example 3**

**Find the surface area of the square pyramid. Round to the nearest tenth.**

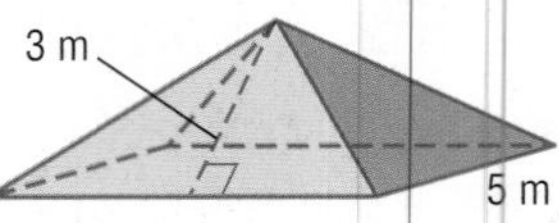


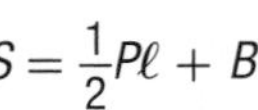

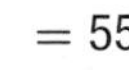

| | |
|---|---|
| $S = \frac{1}{2}P\ell + B$ | Surface area of a regular pyramid |
| $= \frac{1}{2}(4 \cdot 5)3 + 5 \cdot 5$ | $P = 4 \cdot 5$ or 20, $\ell = 3$, $B = 4 \cdot 5$ |
| $= 55$ | Simplify. |

The surface area is 55 square feet.

## 12-4 Volumes of Prisms and Cylinders

**20.** The volume of a cylinder is 770 $cm^3$. It has a height of 5 cm. Find its radius.

**21.** Find the volume of the triangular prism.

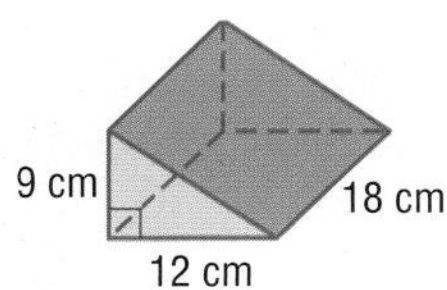


**22.** **TRAILERS** A semi-truck trailer is basically a rectangular prism. A typical height for the inside of these trailers is 108 inches. If the trailer is 8 feet wide and 20 feet long, what is the volume of the trailer?

### Example 4

**Find the volume of the cylinder.**

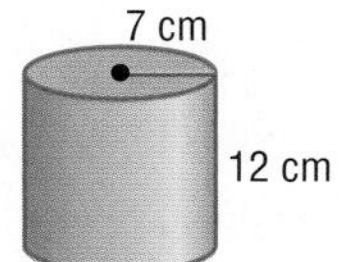


| | |
|---|---|
| $V = \pi r^2 h$ | Volume of a cylinder |
| $= \pi(7)^2(12)$ | $r = 7$ and $h = 12$ |
| $\approx 1847.5$ | Use a calculator. |

The volume is approximately 1847.5 cubic centimeters.

## 12-5 Volumes of Pyramids and Cones

**23.** Find the volume of a cone that has a radius of 1 cm and a height of 3.4 cm.

**24.** Find the volume of the regular pyramid.

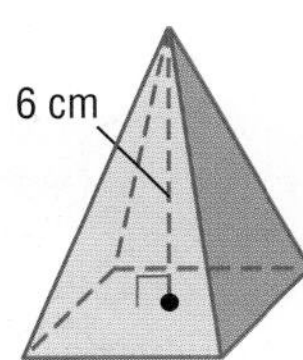


**25.** **ARCHITECTURE** The Great Pyramid measures 756 feet on each side of the base and the height is 481 feet. Find the volume of the pyramid.

### Example 5

**Find the volume of the pyramid.**

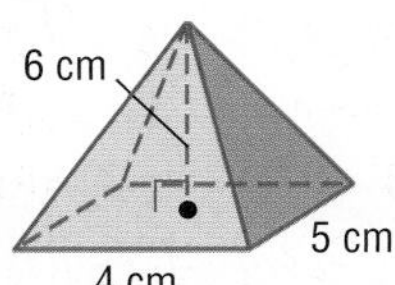


| | |
|---|---|
| $V = \frac{1}{3}Bh$ | Volume of a pyramid |
| $= \frac{1}{3}(4 \cdot 5)(6)$ | $B = 4 \cdot 8$ and $h = 6$ |
| $= 40$ | Simplify. |

The volume is 40 cubic centimeters.

## 12-6 Surface Areas and Volumes of Spheres

**Find the surface area of each figure.**

**26.**

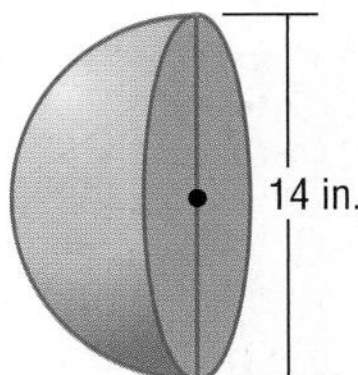


**27.**

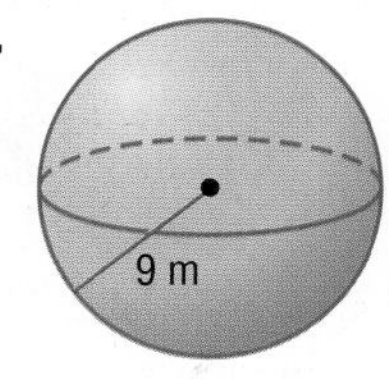


**Find the volume of each sphere or hemisphere. Round to the nearest tenth.**

**28.** hemisphere: circumference of great circle $= 24\pi$ m

**29.** sphere: area of great circle $= 55\pi$ $in^2$

**30.** **CONSTRUCTION** Cement is poured into a hemisphere that is 6 cm across. What is the volume of cement used?

### Example 6

**Find the surface area and volume of the sphere. Round to the nearest tenth.**

14 cm

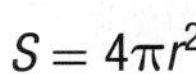
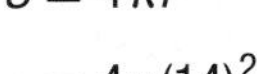

| | |
|---|---|
| $S = 4\pi r^2$ | Surface area of a sphere |
| $= 4\pi(14)^2$ | Substitute. |
| $\approx 2463$ | Use a calculator. |

The surface area is about 2463 square centimeters.

| | |
|---|---|
| $V = \frac{4}{3}\pi r^3$ | Volume of a sphere |
| $= \frac{4}{3}\pi(14)^3$ | Replace $r$ with 9. |
| $\approx 11{,}494$ $cm^3$ | Use a calculator. |

The volume is about 11,494 cubic centimeters.

# CHAPTER 12 Study Guide and Review *Continued*

## 12-7 Spherical Geometry

Name each of the following on sphere $A$.

31. two lines containing point $C$
32. a segment containing point $H$
33. a triangle containing point $B$
34. two lines containing point $L$
35. a segment containing point $J$
36. a triangle containing point $K$

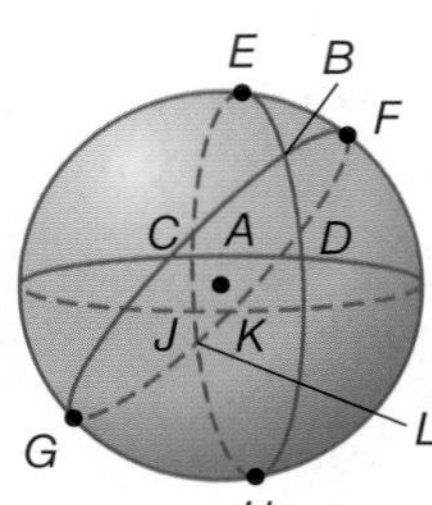


37. **MARBLES** Determine whether figure $y$ on the sphere shown is a line in spherical geometry.

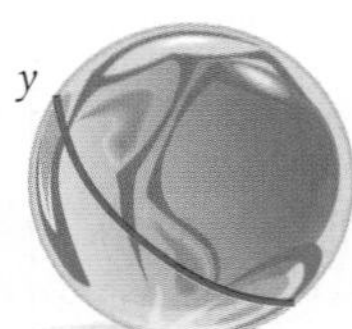


### Example 7

Name each of the following on sphere $A$.

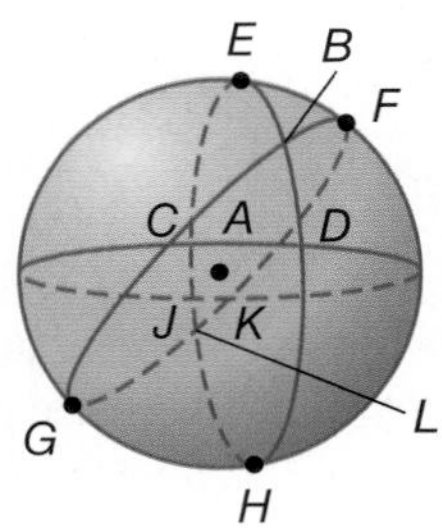


**a.** two lines containing point $D$
$\overleftrightarrow{EH}$, $\overleftrightarrow{CK}$

**b.** a segment containing point $E$
$\overline{DJ}$

## 12-8 Congruent and Similar Solids

Determine whether each pair of solids is *similar*, *congruent*, or *neither*. If the solids are similar, state the scale factor.

38.

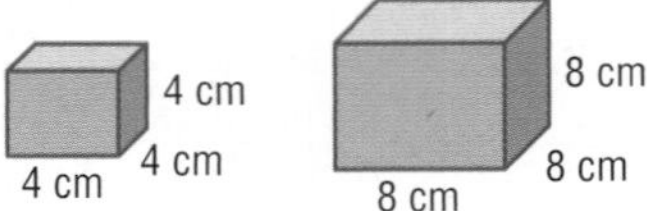


39.

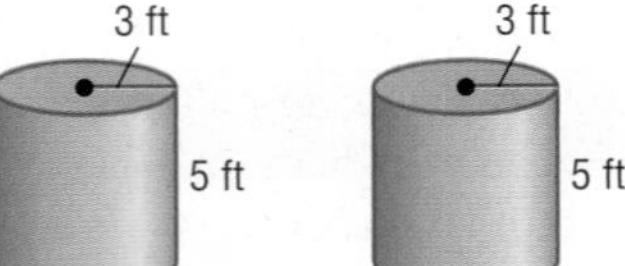


40.

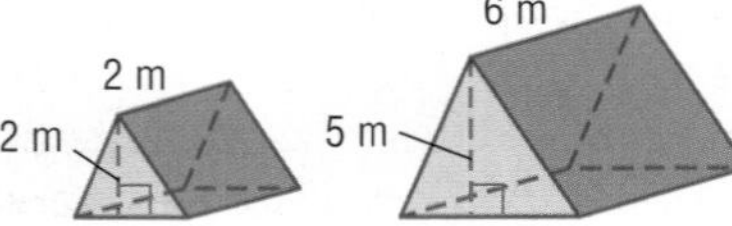


41.

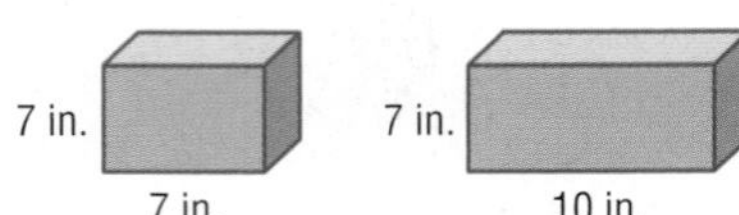


42. **MODELS** A collector's model car is scaled so that 1 inch on the model equals $5\frac{3}{4}$ feet on the actual car. If the model is $\frac{4}{5}$ inches high, how high is the actual car?

### Example 8

Determine whether each pair of solids is similar, congruent, or neither. If the solids are similar, state the scale factor.

**a.**

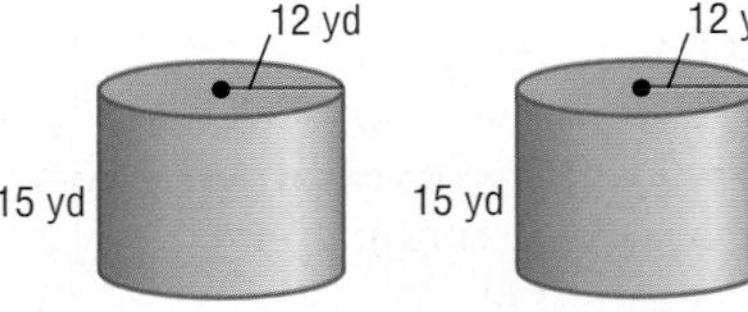


The ratios of the corresponding measures are equal and the scale factor is 1 : 1, so the solids are congruent.

**b.**

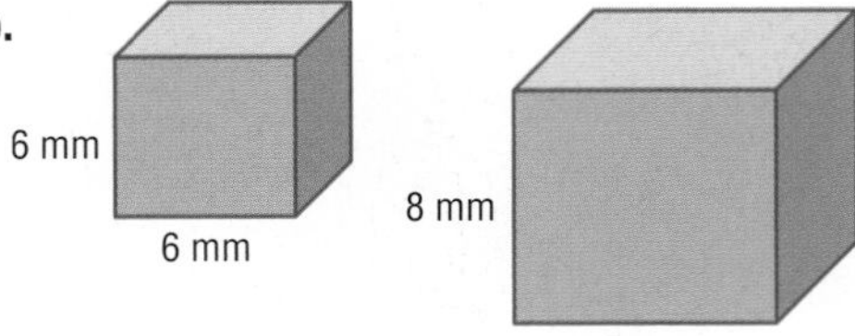


ratio of widths: $\frac{6}{8} = 0.75$

ratio of heights: $\frac{6}{8} = 0.75$

The ratios of the corresponding measures are equal, so the cubes are similar. The scale factor is 3 : 4. Since the scale factor is not 1 : 1, the solids are not congruent.

CHAPTER 12

# Practice Test

1. Use isometric dot paper and the orthographic drawings to sketch the solid.

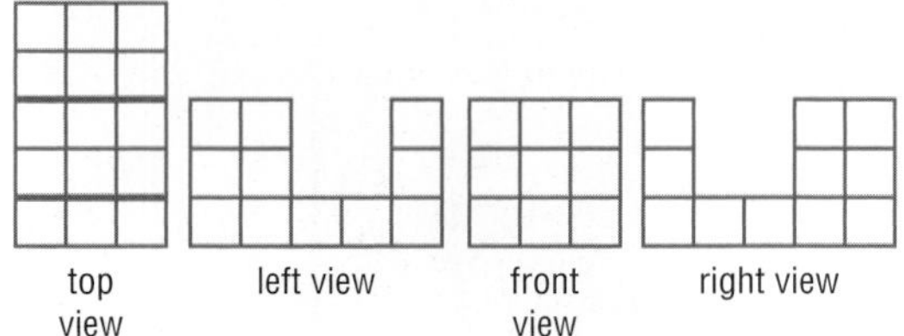


2. Describe the cross section.

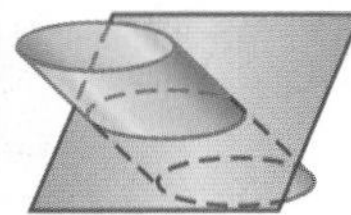

3. **SHORT RESPONSE** Find the surface area of the tent model. Round to the nearest tenth if necessary.

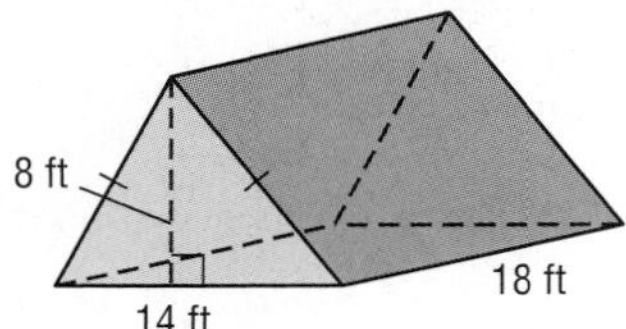


4. **CANDLES** A circular pillar candle is 2.8 inches wide and 6 inches tall. What are the lateral area and surface area of the candle? Round to the nearest tenth if necessary.

5. **TEA** A tea bag is shaped like a regular square pyramid. Each edge of the base is 4 centimeters, and the slant height is 5 centimeters. What is the surface area of the tea bag in square centimeters? Round to the nearest tenth if necessary.

6. **BEEHIVE** Estimate the lateral area and surface area of the Turkish beehive room. Round to the nearest tenth if necessary.

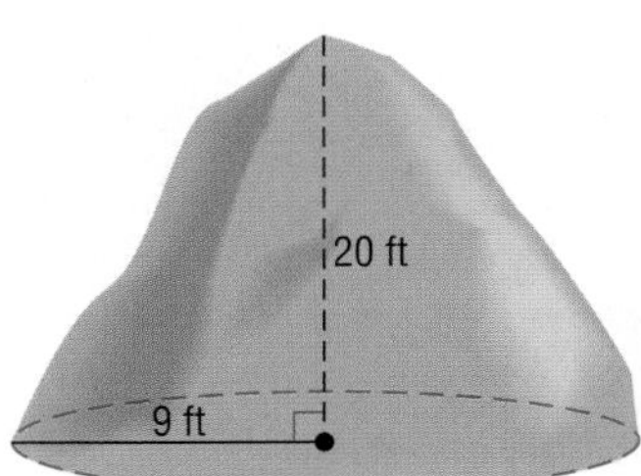


7. Find the volume of the candle in Exercise 4. Round to the nearest tenth if necessary.

8. Find the volume of the tea bag in Exercise 5. Round to the nearest tenth if necessary.

9. **EARTH** Earth's radius is approximately 6400 kilometers. What are the surface area and volume of the Earth? Round to the nearest tenth if necessary.

10. **SOFTBALL** A regulation softball has a circumference of 12 inches. What is the volume of the softball?

**Name each of the following on sphere $\mathcal{A}$.**

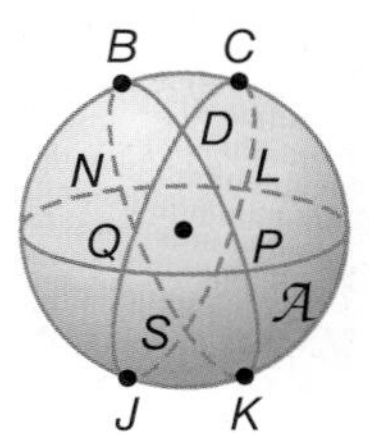


11. two lines containing point $S$
12. a segment containing point $L$
13. a triangle
14. two lines containing point $D$
15. a segment containing point $P$

16. Are these two cubes *similar*, *congruent*, or *neither*? Explain your reasoning.

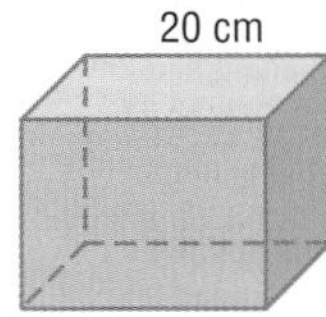


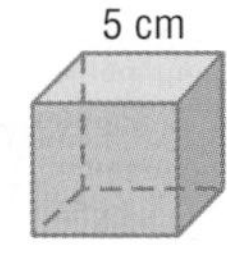


17. Two similar cylinders have heights of 75 centimeters and 25 centimeters. What is the ratio of the volume of the large cylinder to the volume of the small cylinder?

18. **BAKING** Two spherical pieces of cookie dough have radii of 3 centimeters and 5 centimeters, respectively. The pieces are combined to form one large spherical piece of dough. What is the approximate radius of the new sphere of dough? Round to the nearest tenth.

19. **ALGEBRA** A rectangular prism has a base with side lengths $x$ and $x + 3$ and height $2x$. Find the surface area and volume of the prism.

20. **TRANSPORTATION** The traffic cone is 19 inches tall and has a radius of 5 inches.

a. Find the lateral area.

b. Find the surface area.

# Preparing for Standardized Tests

## Make a Drawing

Making a drawing can be a very helpful way for you to visualize how to solve a problem. Sketch your drawings on scrap paper or in your test booklet (if allowed). Do not make any marks on your answer sheet other than your answers.

Ocean/CORBIS

### Strategies for Making a Drawing

**Step 1**

Read the problem statement carefully.

**Ask yourself:**

- What am I being asked to solve? What information is given?
- Would making a drawing help me visualize how to solve the problem?

**Step 2**

Sketch and label your drawing.

- Make your drawing as clear and accurate as possible.
- Label the drawing carefully. Be sure to include all of the information given in the problem statement.
- Fill in your drawing with information that can be gained from intermediate calculations.

### Standardized Test Example

**Solve the problem below. Responses will be graded using the short-response scoring rubric shown.**

A regular pyramid has a square base with 10-centimeter sides and a height of 12 centimeters. What is the total surface area of the pyramid? Round to the nearest tenth if necessary.

| Scoring Rubric | |
|---|---|
| **Criteria** | **Score** |
| **Full Credit:** The answer is correct and a full explanation is provided that shows each step. | 2 |
| **Partial Credit:**<br>• The answer is correct, but the explanation is incomplete.<br>• The answer is incorrect, but the explanation is correct. | 1 |

Read the problem statement carefully. You are given the dimensions of a square pyramid and asked to find the surface area. Sketching a drawing may help you visualize the problem and how to solve it.

Example of a 2-point response:

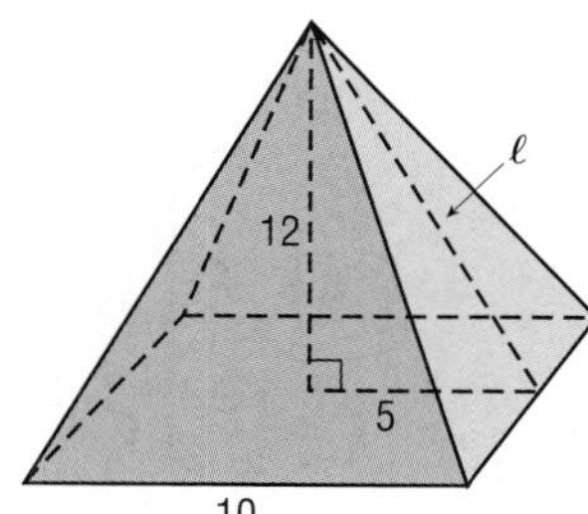


Use the Pythagorean Theorem to find the slant height, $\ell$.

$\ell^2 = 5^2 + 12^2$

$\ell^2 = 169$

$\ell = 13$

Find the lateral area.

$L = \frac{1}{2}P\ell$

$= \frac{1}{2}(40)(13)$

$= 260$

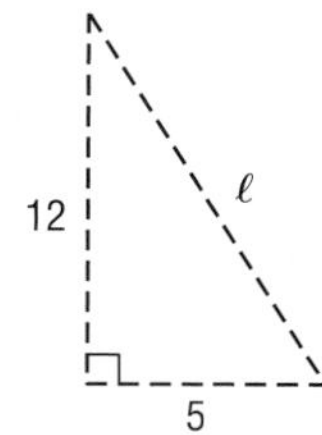


Add the area of the square base.

$S = 260 + 100$ or $360$

The total surface area is 360 square centimeters.

The steps, calculations, and reasoning are clearly stated. The student also arrives at the correct answer. So, this response is worth the full 2 points.

## Exercises

**Solve each problem. Show your work. Responses will be graded using the short-response scoring rubric given at the beginning of the lesson.**

1. A right circular cone has a slant height that is twice its radius. The lateral area of the cone is about 569 square millimeters. What is the radius of the cone? Round to the nearest whole millimeter.

2. From a single point in her yard, Marti measures and marks distances of 18 feet and 30 feet with stakes for two sides of her garden. How far apart should the two stakes be if the garden is to be rectangular shaped?

3. A passing boat is 310 feet from the base of a lighthouse. The angle of depression from the top of the lighthouse is 24°. What is the height of the lighthouse to the nearest tenth of a foot?

4. A regular hexagon is inscribed in a circle with a diameter of 12 centimeters. What is the exact area of the hexagon?

5. Luther is building a model rocket for a science fair project. He attaches a nosecone to a cylindrical body to form the rocket's fuselage. The rocket has a diameter of 4 inches and a total height (including the nosecone) of 2 feet 5 inches. The nosecone is 7 inches tall. What is the volume of the rocket? Give your answer rounded to the nearest tenth cubic inch.

6. Terry wants to measure the height of the top of the backboard of his basketball hoop. At 4:00, the shadow of a 4-foot fence post is 20 inches long, and the shadow of the backboard is 65 inches long. What is the height of the top of the backboard?

CHAPTER 12

# Standardized Test Practice

## Cumulative, Chapters 1 through 12

### Multiple Choice

**Read each question. Then fill in the correct answer on the answer document provided by your teacher or on a sheet of paper.**

**1.** The Great Pyramid of Giza in Egypt originally had a height of about 148 meters. The base of the pyramid was a square with 230-meter sides. What was the original volume of the pyramid? Round to the nearest whole number.

A 1,786,503 m$^3$

B 2,609,733 m$^3$

C 104,128,752 m$^3$

D 122,716,907 m$^3$

**2.** If $\overline{HK}$ is tangent to circle $O$, what is the radius of the circle?

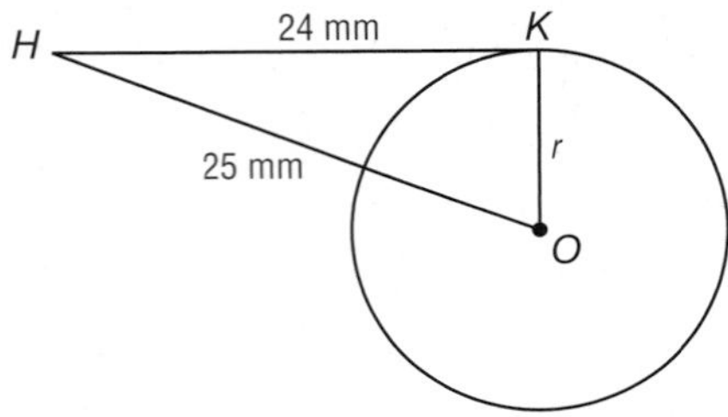


F 7 mm

G 8 mm

H 9 mm

J 10 mm

**3.** What is the sum of the interior angles of the figure?

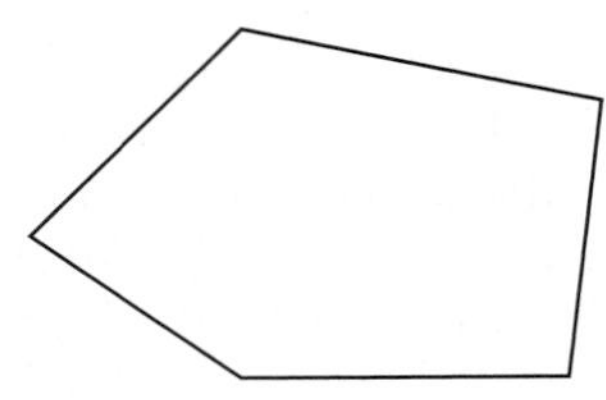

A 450°

B 540°

C 630°

D 720°

**Test-Taking Tip**

Question 1 You can eliminate some unreasonable answers by estimating first. Choices C and D are too large.

**4.** Eddie conducted a random survey of 50 students and found that 14 of them spend more than 2 hours each night doing homework. If there are 421 students at Eddie's school, predict how many of them spend more than 2 hours each night doing homework.

F 118

G 124

H 125

J 131

**5.** $\overline{RS}$ represents the height of Mount Mitchell, the highest point in the state of North Carolina. If $TU = 5013$ feet, $UV = 6684$ feet, and $TV = 8355$ feet, use the ASA Theorem to find the height of Mount Mitchell.

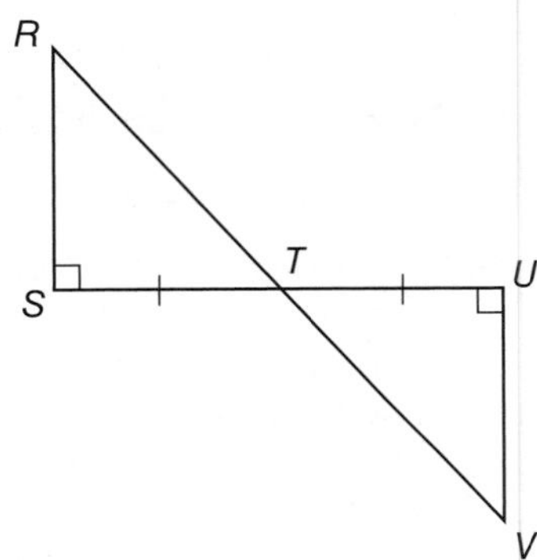


A 5013 ft

B 6684 ft

C 7154 ft

D 8355 ft

**6.** Triangle $DEF$ is shown below.

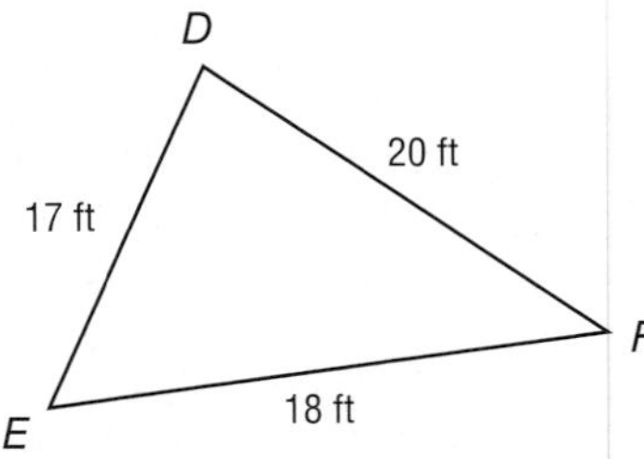


Which statement about this triangle is true?

F $m\angle F > m\angle D$

G $m\angle E > m\angle F$

H $m\angle D < m\angle F$

J $m\angle E < m\angle D$

## Short Response/Gridded Response

**Record your answers on the answer sheet provided by your teacher or on a sheet of paper.**

**7.** What is the value of $x$ in the figure below?

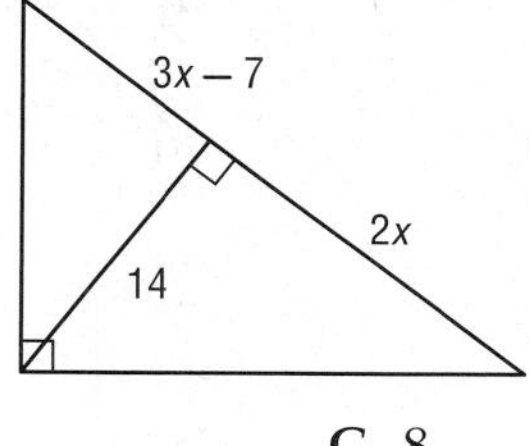


A 5
B 7
C 8
D 10

**8. GRIDDED RESPONSE** What is the perimeter of the isosceles triangle to the nearest tenth of a centimeter?

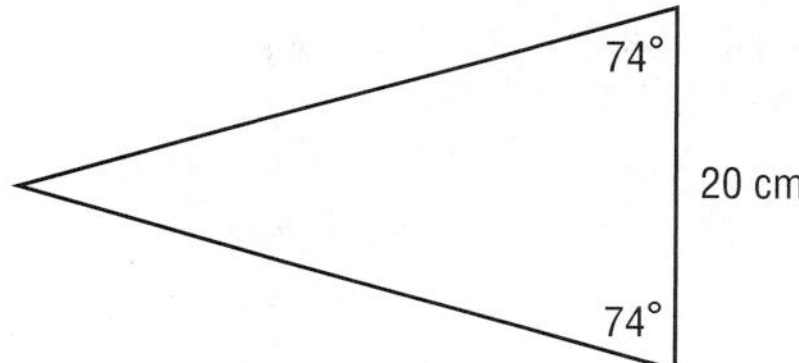


**9.** Determine whether the following statement is *sometimes, always,* or *never* true. Explain.

The orthocenter of a right triangle is located at the vertex of the right angle.

**10. GRIDDED RESPONSE** Given: $c \parallel d$

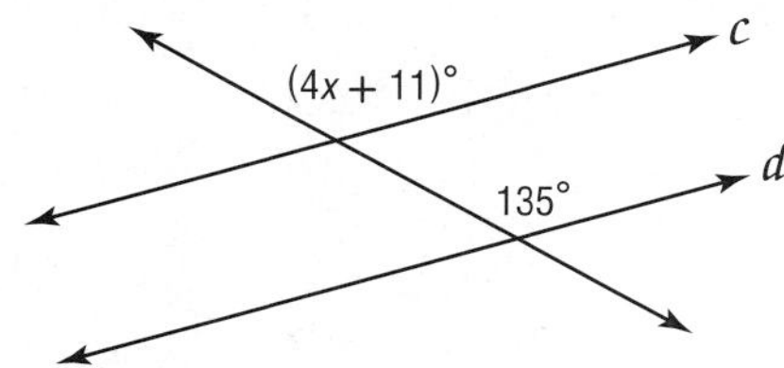


What is the value of $x$ in the figure?

**11.** What is the lateral area of the square pyramid below? Round to the nearest tenth if necessary. Show your work.

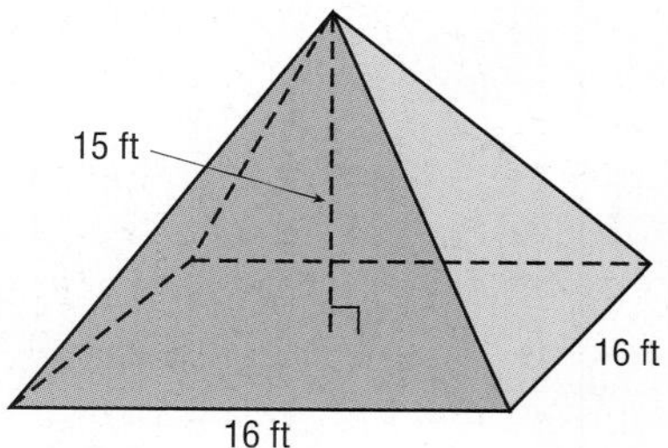


## Extended Response

**Record your answers on a sheet of paper. Show your work.**

**12.** The two prisms below are similar figures.

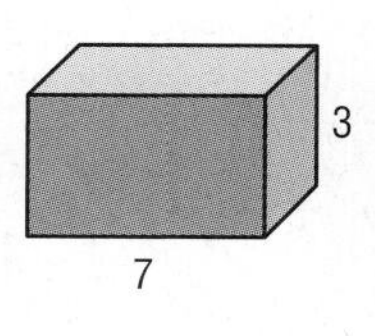


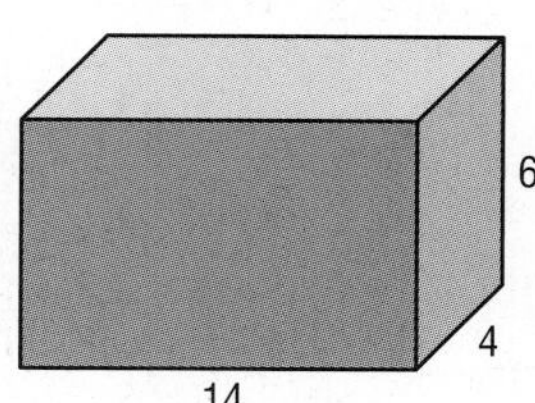


a. What is the scale factor from the smaller prism to the larger one?

b. What are the volumes of the prisms?

c. How many times as great is the volume of the larger prism as the smaller prism?

d. Suppose a solid figure has a volume of 40 cubic units. If its dimensions are scaled by a factor of 1.5, what will the volume of the new figure be?

### Need ExtraHelp?

| If you missed Question... | 1 | 2 | 3 | 4 | 5 | 6 | 7 | 8 | 9 | 10 | 11 | 12 |
|---|---|---|---|---|---|---|---|---|---|---|---|---|
| Go to Lesson... | 12-5 | 11-5 | 8-1 | 9-1 | 6-4 | 7-3 | 10-1 | 10-4 | 7-2 | 5-5 | 12-3 | 12-8 |

CHAPTER 13

# Probability and Measurement

## Then

You learned about experiments, outcomes, and events. You also found probabilities of simple events.

## Now

In this chapter, you will:

- Represent sample spaces.
- Use permutations and combinations with probability.
- Find probabilities by using length and area.
- Find probabilities of compound events.

## Why? ▲

**GAMES** Probability can be used to predict the likelihood of different outcomes of the games that we play.

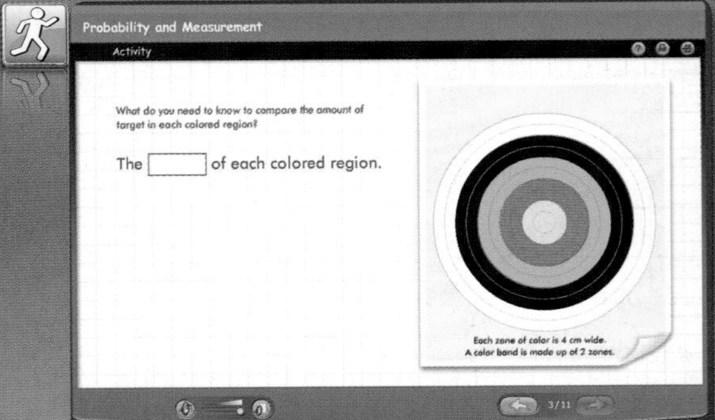


connectED.mcgraw-hill.com Your Digital Math Portal

Animation  Vocabulary   eGlossary  Personal Tutor   Virtual Manipulatives  Graphing Calculator  Audio   Foldables  Self-Check Practice  Worksheets 

# Get Ready for the Chapter

**Diagnose** Readiness | You have two options for checking prerequisite skills.

**1 Textbook Option** Take the Quick Check below. Refer to the Quick Review for help.

## QuickCheck

Simplify.

1. $\frac{1}{2} + \frac{3}{8}$
2. $\frac{7}{9} + \frac{2}{6}$
3. $\frac{2}{5} + \frac{7}{8}$
4. $\frac{2}{9} \cdot \frac{4}{8}$
5. $\frac{3}{7} \cdot \frac{21}{24}$
6. $\frac{3}{10} \cdot \frac{2}{9}$
7. **SOCCER** A soccer team brings a 4.5-gallon cooler of water to their games. How many 4-ounce cups can the team drink per game?

A die is rolled. Find the probability of each outcome.

8. $P$(greater than 1)
9. $P$(odd)
10. $P$(less than 2)
11. $P$(1 or 6)
12. **GAMES** Two friends are playing a game with a 20-sided die that has all of the letters of the alphabet except for Q, U, V, X, Y, and Z. What is the probability that the die will land on a vowel?

The table shows the results of an experiment in which a spinner numbered 1–4 was spun.

| Outcome | Tally | Frequency |
|---|---|---|
| 1 | ||| | 3 |
| 2 | 𝍸 || | 7 |
| 3 | 𝍸 | | 6 |
| 4 | |||| | 4 |

13. What is the experimental probability that the spinner will land on a 4?
14. What is the experimental probability that the spinner will land on an odd number?
15. What is the experimental probability that the spinner will land on an even number?

## QuickReview

**Example 1**

Simplify $\frac{6}{9} \cdot \frac{1}{2}$.

$\frac{6}{9} \cdot \frac{1}{2} = \frac{6 \cdot 1}{9 \cdot 2}$ Multiply the numerators and denominators.

$= \frac{6}{18}$ or $\frac{1}{3}$ Simplify.

**Example 2**

Suppose a die is rolled. What is the probability of rolling less than a five?

$P(\text{less than } 5) = \frac{\text{number of favorable outcomes}}{\text{number of possible outcomes}}$

$= \frac{4}{6}$ or $\frac{2}{3}$

The probability of rolling less than a five is $\frac{2}{3}$ or 67%.

**Example 3**

A spinner numbered 1–6 was spun. Find the experimental probability of landing on a 5.

| Outcome | Tally | Frequency |
|---|---|---|
| 1 | |||| | 4 |
| 2 | 𝍸 || | 7 |
| 3 | 𝍸 ||| | 8 |
| 4 | |||| | 4 |
| 5 | || | 2 |
| 6 | 𝍸 | 5 |

$P(5) = \frac{\text{number of times a 5 is spun}}{\text{total number of outcomes}}$ or $\frac{2}{30}$

The experimental probability of landing on a 5 is $\frac{2}{30}$ or 7%.

**2 Online Option** Take an online self-check Chapter Readiness Quiz at connectED.mcgraw-hill.com.

# Get Started on the Chapter

**You will learn several new concepts, skills, and vocabulary terms as you study Chapter 13. To get ready, identify important terms and organize your resources. You may wish to refer to Chapter 0 to review prerequisite skills.**

## FOLDABLES® StudyOrganizer

**Probability and Measurement** Make this Foldable to help you organize your Chapter 13 notes about probability. Begin with one sheet of paper.

1 **Fold** a sheet of paper lengthwise.

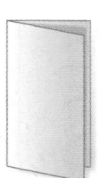

2 **Fold** in half two more times.

3 **Cut** along each fold on the left column.

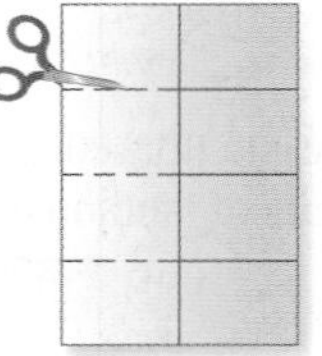

4 **Label** as shown.

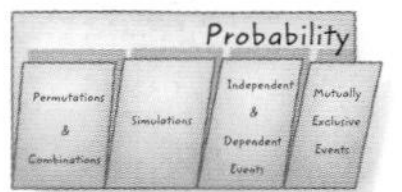


## NewVocabulary

| English | | Español |
|---|---|---|
| sample space | p. 883 | espacio muestral |
| tree diagram | p. 883 | diagrama de árbol |
| permutation | p. 890 | permutación |
| factorial | p. 890 | factorial |
| circular permutation | p. 893 | permutación circular |
| combination | p. 894 | combinacion |
| geometric probability | p. 899 | probabilidad geométrica |
| probability model | p. 907 | modelo de la probabilidad |
| simulation | p. 907 | simulacro |
| random variable | p. 909 | variable aleatoria |
| expected value | p. 909 | valor espenado |
| compound events | p. 915 | eventos compuestos |
| independent events | p. 915 | eventos independientes |
| dependent events | p. 915 | eventos dependientes |
| conditional probability | p. 917 | probabilidad condicional |
| probability tree | p. 917 | árbol de la probabilidad |
| mutually exclusive | p. 924 | mutuamente exclusivos |
| complement | p. 927 | complemento |

## ReviewVocabulary

**event** evento one or more outcomes of an experiment

**experiment** experimento a situation involving chance such as flipping a coin or rolling a die

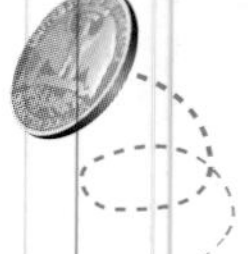

LESSON 13-1

# Representing Sample Spaces

| :: Then | :: Now | :: Why? |
|---|---|---|
| ● You calculated experimental probability. | ● **1** Use lists, tables, and tree diagrams to represent sample spaces.<br>**2** Use the Fundamental Counting Principle to count outcomes. | ● In a football game, a referee tosses a fair coin to determine which team will take possession of the football first. The coin can land on heads or tails. |

**NewVocabulary**
sample space
tree diagram
two-stage experiment
multi-stage experiment
Fundamental Counting Principle

**Common Core State Standards**

**Content Standards**
Preparation for S.CP.9 (+) Use permutations and combinations to compute probabilities of compound events and solve problems.

**Mathematical Practices**
1 Make sense of problems and persevere in solving them.
2 Reason abstractly and quantitatively.

## 1 Represent a Sample Space

**Represent a Sample Space** You have learned the following about experiments, outcomes, and events.

| Definition | Example |
|---|---|
| An *experiment* is a situation involving chance that leads to results called *outcomes*. | In the situation above, the experiment is tossing the coin. |
| An *outcome* is the result of a single performance or *trial* of an experiment. | The possible outcomes are landing on heads or tails. |
| An *event* is one or more outcomes of an experiment. | One event of this experiment is the coin landing on tails. |

The **sample space** of an experiment is the set of all possible outcomes. You can represent a sample space by using an organized list, a table, or a **tree diagram**.

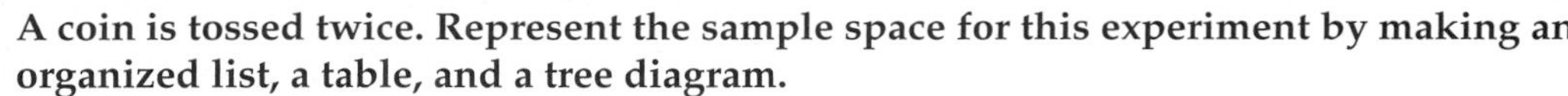

### Example 1 Represent a Sample Space

**A coin is tossed twice. Represent the sample space for this experiment by making an organized list, a table, and a tree diagram.**

For each coin toss, there are two possible outcomes, heads H or tails T.

**Organized List**

Pair each possible outcome from the first toss with the possible outcomes from the second toss.

H, H T, T
H, T T, H

**Table**

List the outcomes of the first toss in the left column and those of the second toss in the top row.

| Outcomes | Heads | Tails |
|---|---|---|
| Heads | H, H | H, T |
| Tails | T, H | T, T |

**Tree Diagram**

Outcomes
First Toss: H, T
Second Toss: H T H T
Sample Space: H, H H, T T, H T, T

**GuidedPractice**

1. A coin is tossed and then a number cube is rolled. Represent the sample space for this experiment by making an organized list, a table, and a tree diagram.

The experiment in Example 1 is an example of a **two-stage experiment**, which is an experiment with two stages or events. Experiments with more than two stages are called **multi-stage experiments**.

PT

## Real-World Example 2 Multi-Stage Tree Diagrams

**WatchOut!**

**CCSS** Sense-Making The words *and/or* in the third question for Example 2 suggest an additional stage in the ordering process. By making separate stages for choosing with or without tomato and with or without pickles, you allow for the possibility of choosing *both* tomato and pickles.

**HAMBURGERS To take a hamburger order, Keandra asks each customer the questions from the script shown. Draw a tree diagram to represent the sample space for hamburger orders.**

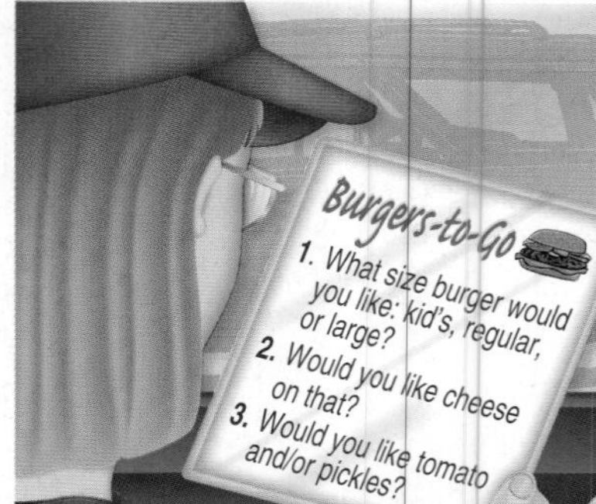


The sample space is the result of four stages.

- Burger size (K, R, or L)
- Cheese (C or NC)
- Tomato (T or NT)
- Pickles (P or NP)

Draw a tree diagram with four stages.

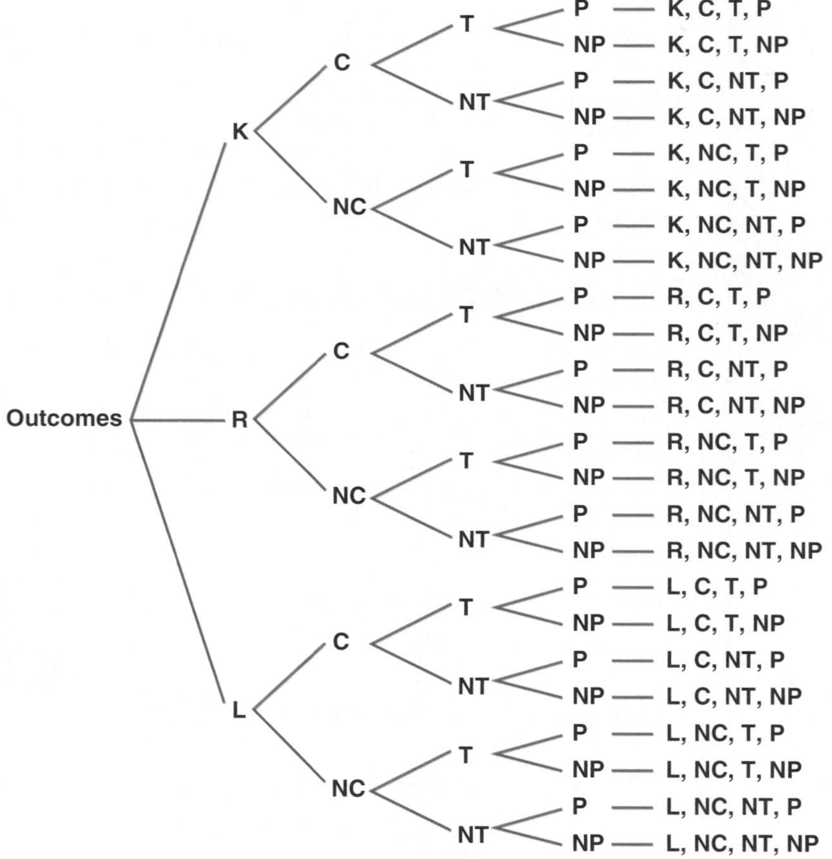


**ReadingMath**

Tree Diagram Notation Choose notation for outcomes in your tree diagrams that will eliminate confusion. In Example 2, *C* stands for *cheese*, while *NC* stands for *no cheese.* Likewise, *NT* and *NP* stand for *no tomato* and *no pickles*, respectively.

### GuidedPractice

2. **MUSIC** Yoki can choose a small MP3 player with a 4- or 8-gigabyte hard drive in black, teal, sage, or red. She can also get a clip and/or a dock to go with it. Make a tree diagram to represent the sample space for this situation.

## 2 Fundamental Counting Principle

For some two-stage or multi-stage experiments, listing the entire sample space may not be practical or necessary. To find the *number* of possible outcomes, you can use the **Fundamental Counting Principle**.

**StudyTip**

**Multiplication Rule** The Fundamental Counting Principle is sometimes called the *Multiplication Rule for Counting* or the *Counting Principle.*

### KeyConcept Fundamental Counting Principle

**Words** The number of possible outcomes in a sample space can be found by multiplying the number of possible outcomes from each stage or event.

**Symbols** In a $k$-stage experiment, let

$n_1$ = the number of possible outcomes for the first stage.

$n_2$ = the number of possible outcomes for the second stage after the first stage has occurred.

⋮

$n_k$ = the number of possible outcomes for the $k$th stage after the first $k - 1$ stages have occurred.

Then the total possible outcomes of this $k$-stage experiment is

$$n_1 \cdot n_2 \cdot n_3 \cdot \ldots \cdot n_k.$$

**Real-WorldLink**

More than 95 percent of high school students order a traditional ring style, which includes the name of the school, a stone, and the graduation year.

**Source:** *Fort Worth Star-Telegram*

### Real-World Example 3 Use the Fundamental Counting System

**CLASS RINGS Haley has selected a size and overall style for her class ring. Now she must choose from the ring options shown. How many different rings could Haley create in her chosen style and size?**

| Ring Options | Number of Choices |
|---|---|
| metals | 10 |
| finishes | 2 |
| stone colors | 12 |
| stone cuts | 5 |
| side 1 activity logos | 20 |
| side 2 activity logos | 20 |
| band styles | 2 |

Use the Fundamental Counting Principle.

| metals | | finishes | | stone colors | | stone cuts | | side 1 logos | | side 2 logos | | band styles | | possible outcomes |
|---|---|---|---|---|---|---|---|---|---|---|---|---|---|---|
| 10 | × | 2 | × | 12 | × | 5 | × | 20 | × | 20 | × | 2 | = | 960,000 |

So, Haley could create 960,000 different rings.

**GuidedPractice**

**3.** Find the number of possible outcomes for each situation.

**A.** The answer sheet shown is completed.

**B.** A die is rolled four times.

**C. SHOES** A pair of women's shoes comes in whole sizes 5 through 11 in red, navy, brown, or black. They can be leather or suede and are available in three different widths.

**Answer Sheet**

1. Ⓐ Ⓑ Ⓒ Ⓓ
2. Ⓐ Ⓑ Ⓒ Ⓓ
3. Ⓐ Ⓑ Ⓒ Ⓓ
4. Ⓐ Ⓑ Ⓒ Ⓓ
5. Ⓐ Ⓑ Ⓒ Ⓓ
6. Ⓐ Ⓑ Ⓒ Ⓓ
7. Ⓣ Ⓕ
8. Ⓣ Ⓕ
9. Ⓣ Ⓕ
10. Ⓣ Ⓕ

## Check Your Understanding

**Example 1** **Represent the sample space for each experiment by making an organized list, a table, and a tree diagram.**

1. For each at bat, a player can either get on base or make an out. Suppose a player bats twice.

2. Quinton sold the most tickets in his school for the annual Autumn Festival. As a reward, he gets to choose twice from a grab bag with tickets that say "free juice" or "free notebook."

**Example 2**

3. **TUXEDOS** Patrick is renting a prom tuxedo from the catalog shown. Draw a tree diagram to represent the sample space for this situation.

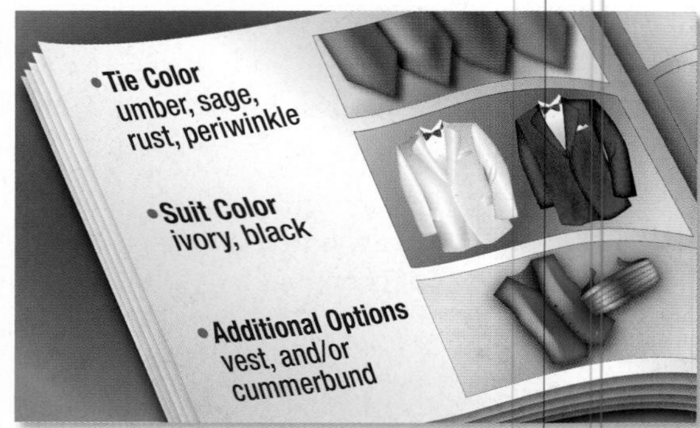


**Example 3** **Find the number of possible outcomes for each situation.**

4. Marcos is buying a cell phone and must choose a plan. Assume one of each is chosen.

| Cell Phone Options | Number of Choices |
|---|---|
| phone style | 15 |
| minutes package | 5 |
| Internet access | 3 |
| text messaging | 4 |
| insurance | 2 |

5. Desirée is creating a new menu for her restaurant. Assume one of each item is ordered.

| Menu Titles | Number of Choices |
|---|---|
| appetizer | 8 |
| soup | 4 |
| salad | 6 |
| entree | 12 |
| dessert | 9 |

## Practice and Problem Solving

**Example 1** **CCSS REASONING** **Represent the sample space for each experiment by making an organized list, a table, and a tree diagram.**

6. Gina is a junior and has a choice for the next two years of either playing volleyball or basketball during the winter quarter.

7. Two different history classes in New York City are taking a trip to either the Smithsonian or the Museum of Natural History.

8. Simeon has an opportunity to travel abroad as a foreign exchange student during each of his last two years of college. He can choose between Ecuador or Italy.

9. A new club is formed, and a meeting time must be chosen. The possible meeting times are Monday or Thursday at 5:00 or 6:00 P.M.

10. An exam with multiple versions has exercises with triangles. In the first exercise, there is an obtuse triangle or an acute triangle. In the second exercise, there is an isosceles triangle or a scalene triangle.

**11.** **PAINTING** In an art class, students are working on two projects where they can use one of two different types of paints for each project. Represent the sample space for this experiment by making an organized list, a table, and a tree diagram.

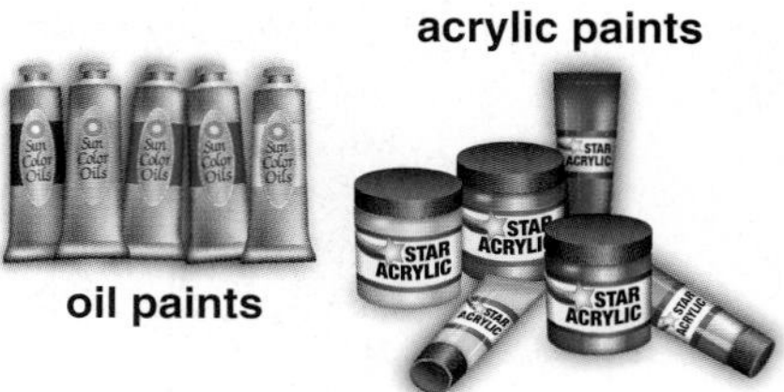


**Example 2** **Draw a tree diagram to represent the sample space for each situation.**

**12.** **BURRITOS** At a burrito stand, customers have the choice of beans, pork, or chicken with rice or no rice, and cheese and/or salsa.

**13.** **TRANSPORTATION** Blake is buying a vehicle and has a choice of sedan, truck, or van with leather or fabric interior, and a CD player and/or sunroof.

**14.** **TREATS** Ping and her friends go to a frozen yogurt parlor which has a sign like the one at the right. Draw a tree diagram for all possible combinations of cones with peanuts and/or sprinkles.

**Example 3** **CCSS PERSEVERANCE In Exercises 15–18, find the number of possible outcomes for each situation.**

**15.** In the Junior Student Council elections, there are 3 people running for secretary, 4 people running for treasurer, 5 people running for vice president, and 2 people running for class president.

**16.** When signing up for classes during his first semester of college, Frederico has 4 class spots to fill with a choice of 4 literature classes, 2 math classes, 6 history classes, and 3 film classes.

**17.** Niecy is choosing one each of 6 colleges, 5 majors, 2 minors, and 4 clubs.

**18.** Evita works at a restaurant where she has to wear a white blouse, black pants or skirt, and black shoes. She has 5 blouses, 4 pants, 3 skirts, and 6 pairs of black shoes.

**19.** **ART** For an art class assignment, Mr. Green gives students their choice of two quadrilaterals to use as a base. One must have sides of equal length, and the other must have at least one set of parallel sides. Represent the sample space by making an organized list, a table, and a tree diagram.

**20.** **BREAKFAST** A hotel restaurant serves omelets with a choice of vegetables, ham, or sausage that come with a side of hash browns, grits, or toast.

**a.** How many different outcomes of omelet and one side are there if a vegetable omelet comes with just one vegetable?

**b.** Find the number of possible outcomes for a vegetable omelet if you can get any or all vegetables on any omelet.

Omelets

All omelets served with your choice of hash browns, grits, or toast.

Vegetable Omelet

Ham Omelet

Sausage Omelet

Vegetable choices: green peppers, tomatoes, onions, mushrooms

**21.** **COMPOSITE FIGURES** Carlito is calculating the area of the composite figure at the right. In how many different ways he can do this?

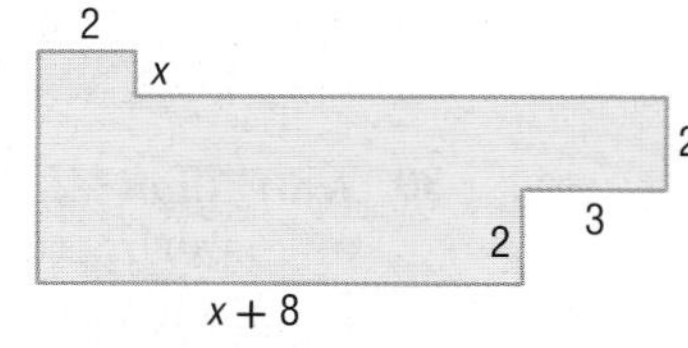

22. **TRANSPORTATION** Miranda got a new bicycle lock that has a four-number combination. Each number in the combination is from 0 to 9.
   a. How many combinations are possible if there are no restrictions on the number of times Miranda can use each number?
   b. How many combinations are possible if Miranda can use each number only once? Explain.

23. **GAMES** Cody and Monette are playing a board game in which you roll two dice per turn.
   a. In one turn, how many outcomes result in a sum of 8?
   b. How many outcomes in one turn result in an odd sum?

24. **MULTIPLE REPRESENTATIONS** In this problem, you will investigate a sequence of events. In the first stage of a two-stage experiment, you spin Spinner 1 below. If the result is red, you flip a coin. If the result is yellow, you roll a die. If the result is green, you roll a number cube. If the result is blue, you spin Spinner 2.

**Spinner 1**

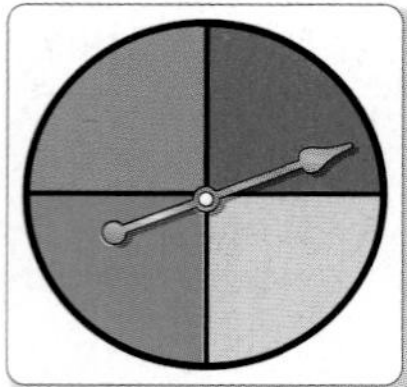

**Spinner 2**

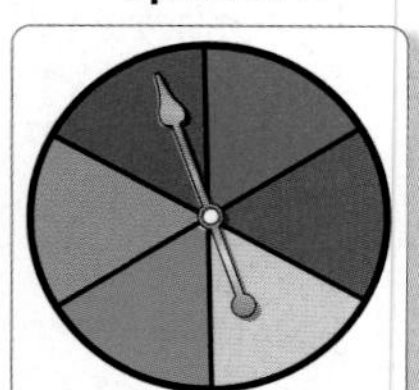

   a. **Geometric** Draw a tree diagram to represent the sample space for the experiment.
   b. **Logical** Draw a Venn diagram to represent the possible outcomes of the experiment.
   c. **Analytical** How many possible outcomes are there?
   d. **Verbal** Could you use the Fundamental Counting Principle to determine the number of outcomes? Explain.

## H.O.T. Problems Use Higher-Order Thinking Skills

25. **CHALLENGE** A box contains $n$ different objects. If you remove three objects from the box, one at a time, without putting the previous object back, how many possible outcomes exist? Explain your reasoning.

26. **OPEN ENDED** Sometimes a tree diagram for an experiment is not symmetrical. Describe a two-stage experiment where the tree diagram is asymmetrical. Include a sketch of the tree diagram. Explain.

27. **WRITING IN MATH** Explain why it is not possible to represent the sample space for a multi-stage experiment by using a table.

Example 3

28. **CCSS ARGUMENTS** Determine if the following statement is *sometimes*, *always*, or *never* true. Explain your reasoning.

*When an outcome falls outside the sample space, it is a failure.*

29. **REASONING** A multistage experiment has $n$ possible outcomes at each stage. If the experiment is performed with $k$ stages, write an equation for the total number of possible outcomes $P$. Explain.

30. **WRITING IN MATH** Explain when it is necessary to show all of the possible outcomes of an experiment by using a tree diagram and when using the Fundamental Counting Principle is sufficient.

## Standardized Test Practice

**31. PROBABILITY** Alejandra can invite two friends to go out to dinner with her for her birthday. If she is choosing among four of her friends, how many possible outcomes are there?

A 4    C 8

B 6    D 9

**32. SHORT RESPONSE** What is the volume of the triangular prism shown below?

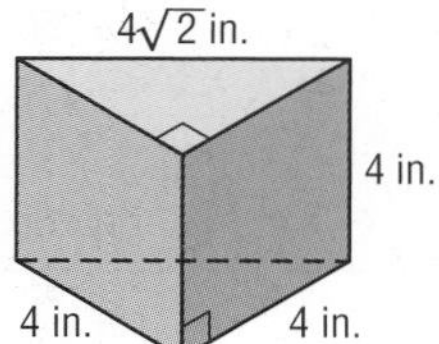


**33.** Brad's password must be five digits long, use the numbers 0–9, and the digits must not repeat. What is the maximum number of different passwords that Brad can have?

F 15,120    H 59,049

G 30,240    J 100,000

**34. SAT/ACT** A pizza shop offers 3 types of crust, 5 vegetable toppings, and 4 meat toppings. How many different pizzas could be ordered by choosing 1 crust, 1 vegetable topping, and 1 meat topping?

A 12    D 60

B 23    E infinite

C 35

## Spiral Review

**35. ARCHITECTURE** To encourage recycling, the people of Rome, Italy, built a model of Basilica di San Pietro from empty beverage cans. The model was built to a 1:5 scale and was a rectangular prism that measured 26 meters high, 49 meters wide, and 93 meters long. Find the dimensions of the actual Basilica di San Pietro. (Lesson 12-8)

**Using spherical geometry, name each of the following on sphere $\mathcal{W}$.** (Lesson 12-7)

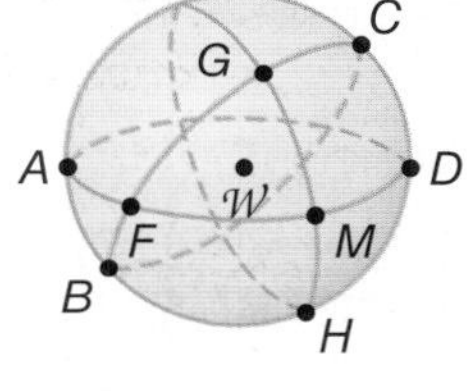


**36.** two lines containing point $F$

**37.** a segment containing point $G$

**38.** a triangle

**Find the lateral area and surface area of each cylinder. Round to the nearest tenth.** (Lesson 12-2)

**39.**

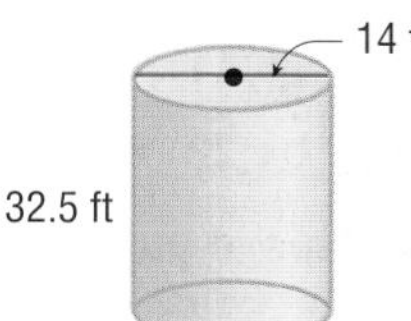


**40.**

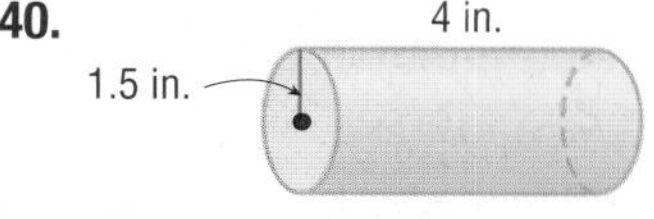


**41.**

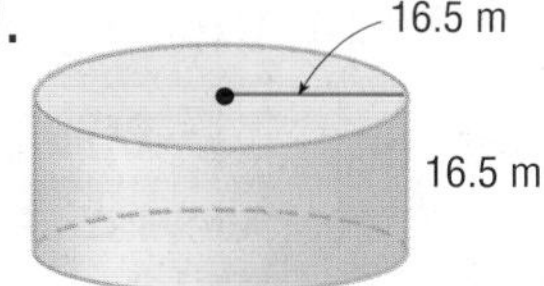


**42. TELECOMMUNICATIONS** The signal from a tower follows a ray that has its endpoint on the tower and is tangent to Earth. Suppose a tower is located at sea level as shown. Determine the measure of the arc intercepted by the two tangents. (Lesson 11-6)

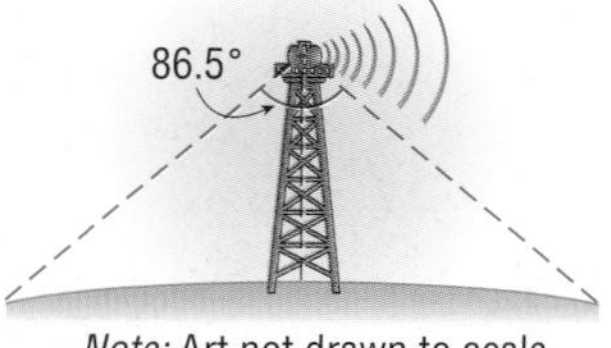


*Note:* Art not drawn to scale

## Skills Review

**Find each quotient.**

**43.** $\frac{5^2}{2}$    **44.** $\frac{3^3}{3 \cdot 2}$    **45.** $\frac{2^4 \cdot 6}{8}$    **46.** $\frac{2^3 \cdot 12}{6}$    **47.** $\frac{4^4 \cdot 3}{24}$

# LESSON 13-2 Probability with Permutations and Combinations

## :: Then

- You used the Fundamental Counting Principle.

## :: Now

1. Use permutations with probability.
2. Use combinations with probability.

## :: Why?

- Lina, Troy, Davian, and Mary are being positioned for a photograph. There are 4 choices for who can stand on the far left, leaving 3 choices for who can stand in the second position. For the third position, just 2 choices remain, and for the last position just 1 is possible.

### NewVocabulary

permutation
factorial
circular permutation
combination

### Common Core State Standards

**Content Standards**

S.CP.9 (+) Use permutations and combinations to compute probabilities of compound events and solve problems.

**Mathematical Practices**

1 Make sense of problems and persevere in solving them.
4 Model with mathematics.

## 1 Probability Using Permutations

A **permutation** is an arrangement of objects in which order is important. One permutation of the four friends above is Troy, Davian, Mary, and then Lina. Using the Fundamental Counting Principle, there are $4 \cdot 3 \cdot 2 \cdot 1$ or 24 possible ordered arrangements of the friends.

The expression $4 \cdot 3 \cdot 2 \cdot 1$ used to calculate the number of permutations of these four friends can be written as 4!, which is read *4 factorial.*

### KeyConcept Factorial

| | |
|---|---|
| Words | The **factorial** of a positive integer $n$, written $n!$, is the product of the positive integers less than or equal to $n$. |
| Symbols | $n! = n \cdot (n-1) \cdot (n-2) \cdot \ldots \cdot 2 \cdot 1$, where $0! = 1$ |

### Example 1 Probability and Permutations of *n* Objects

**SPORTS Chanise and Renee are members of the lacrosse team. If the 20 girls on the team are each assigned a jersey number from 1 to 20 at random, what is the probability that Chanise's jersey number will be 1 and Renee's will be 2?**

**Step 1** Find the number of possible outcomes in the sample space. This is the number of permutations of the 20 girls' names, or 20!.

**Step 2** Find the number of favorable outcomes. This is the number of permutations of the other girls' names given that Chanise's jersey number is 1 and Renee's is 2: $(20-2)!$ or 18!.

**Step 3** Calculate the probability.

$$P(\text{Chanise 1, Renee 2}) = \frac{18!}{20!}$$ ← number of favorable outcomes / ← number of possible outcomes

$$= \frac{\overset{1}{\cancel{18!}}}{20 \cdot 19 \cdot \underset{1}{\cancel{18!}}}$$ Expand 20! and divide out common factors.

$$= \frac{1}{380}$$ Simplify.

**GuidedPractice**

1. **PHOTOGRAPHY** In the opening paragraph, what is the probability that Troy is chosen to stand on the far left and Davian on the far right for the photograph?

Digital Vision/Alamy

In the opening paragraph, suppose 6 friends were available, but the photographer wanted only 4 people in the picture. Using the Fundamental Counting Principle, the number of permutations of 4 friends taken from a group of 6 friends is $6 \cdot 5 \cdot 4 \cdot 3$ or 360.

Another way of describing this situation is the number of permutations of 6 friends taken 4 at a time, denoted ${}_6P_4$. This number can also be computed using factorials.

$$ {}_6P_4 = 6 \cdot 5 \cdot 4 \cdot 3 = \frac{6 \cdot 5 \cdot 4 \cdot 3 \cdot \mathbf{2} \cdot \mathbf{1}}{\mathbf{2} \cdot \mathbf{1}} = \frac{6!}{2!} = \frac{6!}{(6-4)!} $$

This suggests the following formula.

**ReadingMath**

**CCSS Precision** The phrase *distinct objects* means that the objects are distinguishable as being different in some way.

## KeyConcept Permutations

**Symbols** The number of permutations of $n$ distinct objects taken $r$ at a time is denoted by ${}_nP_r$ and given by ${}_nP_r = \frac{n!}{n - r!}$.

**Example** The number of permutations of 5 objects taken 2 at a time is ${}_5P_2 = \frac{5!}{(5-2)!} = \frac{5 \cdot 4 \cdot \cancel{3!}}{\cancel{3!}}$ or 20.

**StudyTip**

**Randomness** When outcomes are decided at random, they are equally likely to occur and their probabilities can be calculated using permutations and combinations.

## Example 2 Probability and ${}_nP_r$

PT

**A class is divided into teams each made up of 15 students. Each team is directed to select team members to be officers. If Sam, Valencia, and Deshane are on a team, and the positions are decided at random, what is the probability that they are selected as president, vice president, and secretary, respectively?**

**Step 1** Since choosing officers is a way of ranking team members, order in this situation is important. The number of possible outcomes in the sample space is the number of permutations of 15 people taken 3 at a time, ${}_{15}P_3$.

$$ {}_{15}P_3 = \frac{15!}{(15-3)!} = \frac{15 \cdot 14 \cdot 13 \cdot \cancel{12!}}{\cancel{12!}} \text{ or } 2730 $$

**Step 2** The number of favorable outcomes is the number of permutations of the 3 students in their specific positions. This is 1!, or 1.

**Step 3** So the probability of Sam, Valencia, and Deshane being selected as the three officers is $\frac{1}{2730}$.

### GuidedPractice

**2.** A student identification card consists of 4 digits selected from 10 possible digits from 0 to 9. Digits cannot be repeated.

**A.** How many possible identification numbers are there?

**B.** Find the probability that a randomly generated card has the exact number 4213.

In a game, you must try to create a word using randomly selected letter tiles. Suppose you select the tiles shown. If you consider the letters O and O to be distinct, then there are 5! or 120 permutations of these letters.

$O_1$ $O_1$ $P_3$ $S_1$ $L_1$

Four of these possible arrangements are listed below.

POOLS POOLS SPOOL SPOOL

Notice that unless the Os are colored, several of these arrangements would look the same. Since there are 2 Os that can be arranged in 2! or 2 ways, the number of permutations of the letters O, P, O, L, and S can be written as $\frac{5!}{2!}$.

## KeyConcept Permutations with Repetition

The number of distinguishable permutations of $n$ objects in which one object is repeated $r_1$ times, another is repeated $r_2$ times, and so on, is

$$\frac{n!}{r_1! \cdot r_2! \cdot \ldots \cdot r_k!}.$$

**Real-WorldLink**

Created in 1956, *The Price is Right* is the longest-running game show in the United States.

Source: IMDB

## Example 3 Probability and Permutations with Repetition

**GAME SHOW On a game show, you are given the following letters and asked to unscramble them to name a U.S. river. If you selected a permutation of these letters at random, what is the probability that they would spell the correct answer of MISSISSIPPI?**

**Step 1** There is a total of 11 letters. Of these letters, I occurs 4 times, S occurs 4 times, and P occurs 2 times. So, the number of distinguishable permutations of these letters is

$$\frac{11!}{4! \cdot 4! \cdot 2!} = \frac{39{,}916{,}800}{1152} \text{ or } 34{,}650. \qquad \text{Use a calculator.}$$

**Step 2** There is only 1 favorable arrangement—MISSISSIPPI.

**Step 3** The probability that a permutation of these letters selected at random spells Mississippi is $\frac{1}{34{,}650}$.

**GuidedPractice**

3. **TELEPHONE NUMBERS** What is the probability that a 7-digit telephone number with the digits 5, 1, 6, 5, 2, 1, and 5 is the number 550-5211?

So far, you have been studying objects that are arranged in *linear* order. Notice that when the spices below are arranged in a line, shifting each spice one position to the right produces a different permutation—curry is now first instead of salt. There are 5! distinct permutations of these spices.

Tuukka Jantti/Stringer/Getty Images Entertainment/Getty Images

**StudyTip**

**Turning the Circle Over** If the circular object looks the same when it is turned over, such as a plain key ring, then the number of permutations must be divided by 2.

In a **circular permutation**, objects are arranged in a circle or loop. Consider the arrangements of these spices when placed on a turntable. Notice that rotating the turntable clockwise one position does *not* produce a different permutation—the order of the spices relative to each other remains unchanged.

Since 5 rotations of the turntable will produce the same permutation, the number of distinct permutations on the turntable is $\frac{1}{5}$ of the total number of arrangements when the spices are placed in a line.

$$\frac{1}{5} \cdot 5! = \frac{5 \cdot 4!}{5} \text{ or } 4!, \text{ which is } (5 - 1)!$$

 **KeyConcept** Circular Permutations

The number of distinguishable permutations of $n$ objects arranged in a circle with no fixed reference point is

$$\frac{n!}{n} \text{ or } (n - 1)!.$$

If the $n$ objects are arranged relative to a fixed reference point, then the arrangements are treated as linear, making the number of permutations $n!$.

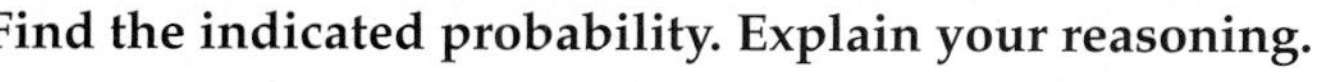

**Example 4** Probability and Circular Permutations

PT

**Find the indicated probability. Explain your reasoning.**

**a. JEWELRY If the 6 charms on the bracelet shown are arranged at random, what is the probability that the arrangement shown is produced?**

Since there is no fixed reference point, this is a circular permutation. So, there are $(6 - 1)!$ or $5!$ distinguishable permutations of the charms. Thus, the probability that the exact arrangement shown is produced is $\frac{1}{5!}$ or $\frac{1}{120}$.

**b. DINING You are seating a party of 4 people at a round table. One of the chairs around this table is next to a window. If the diners are seated at random, what is the probability that the person paying the bill is seated next to the window?**

Since the people are seated around a table with a fixed reference point, this is a linear permutation. So there are $4!$ or 24 ways in which the people can be seated around the table. The number of favorable outcomes is the number of permutations of the other 3 diners given that the person paying the bill sits next to the window, $3!$ or 6.

So, the probability that the person paying the bill is seated next to the window is $\frac{6}{24}$ or $\frac{1}{4}$.

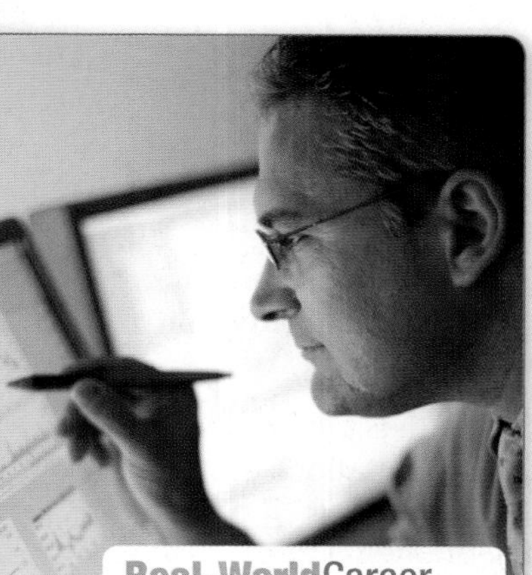

**Real-WorldCareer**

**Statisticians** Statisticians collect statistical data for various subject areas, including sports and games. They use computer software to analyze, interpret, and summarize the data. Most statisticians have a master's degree.

**GuidedPractice**

**4. FOOTBALL** A team's 11 football players huddle together before a play.

**A.** What is the probability that the fullback stands to the right of the quarterback if the team huddles together at random? Explain your reasoning.

**B.** If a referee stands directly behind the huddle, what is the probability that the referee stands directly behind the halfback? Explain your reasoning.

**StudyTip**

**Permutations and Combinations** Use permutations when the order of an arrangement of objects is important and combinations when order is not important.

## 2 Probability Using Combinations

A **combination** is an arrangement of objects in which order is *not* important. Suppose you need to pack 3 of your 8 different pairs of socks for a trip. The order in which the socks are chosen does not matter, so the 3! or 6 groups of socks shown below would *not* be considered different. So, you would use combinations to determine the number of possible different sock choices.

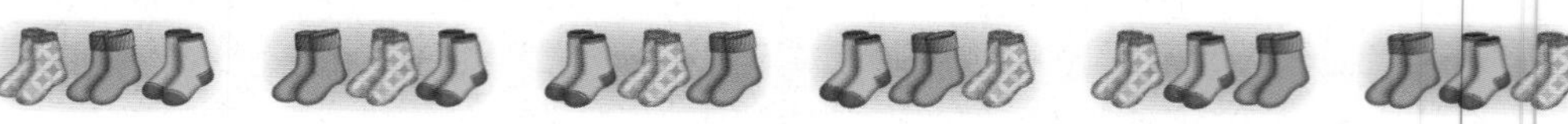

A combination of $n$ objects taken $r$ at a time, or ${}_nC_r$, is calculated by dividing the number of permutations ${}_nP_r$ by the number of arrangements containing the same elements, $r!$.

**KeyConcept** Combinations

**Symbols** The number of combinations of $n$ distinct objects taken $r$ at a time is denoted by ${}_nC_r$ and is given by ${}_nC_r = \frac{n!}{(n-r)!\,r!}$.

**Example** The number of combinations of 8 objects taken 3 at a time is

$${}_8C_3 = \frac{8!}{(8-3)!\,3!} = \frac{8!}{5!3!} = \frac{8 \cdot 7 \cdot \cancel{6} \cdot \cancel{5!}}{\cancel{5!} \cdot \cancel{6}} \text{ or } 56.$$

### Example 5 Probability and ${}_nC_r$

**INVITATIONS For her birthday, Monica can invite 6 of her 20 friends to join her at a theme park. If she chooses to invite friends at random, what is the probability that friends Tessa, Guido, Brendan, Faith, Charlotte, and Rhianna are chosen?**

**Step 1** Since the order in which the friends are chosen does not matter, the number of possible outcomes in the sample space is the number of combinations of 20 people taken 6 at a time, ${}_{20}C_6$.

$${}_{20}C_6 = \frac{20!}{(20-6)!\,6!} = \frac{\cancel{20} \cdot 19 \cdot \cancel{18} \cdot 17 \cdot \cancelto{8}{16} \cdot 15 \cdot \cancel{14!}}{\cancel{14!} \cdot \cancel{6} \cdot \cancel{5} \cdot \cancel{4} \cdot \cancel{3} \cdot \cancel{2}} \text{ or } 38{,}760$$

**Step 2** There is only 1 favorable outcome—that the six students listed above are chosen. The order in which they are chosen is not important.

**Step 3** So the probability of these six friends being chosen is $\frac{1}{38{,}760}$.

**GuidedPractice**

**5. GEOMETRY** If three points are randomly chosen from those named on the rectangle shown, what is the probability that they all lie on the same line segment?

A B C
H D
G F E

## Check Your Understanding

**Example 1**

1. **GEOMETRY** Five students are asked to randomly select and name a polygon from the group shown below. What is the probability that the first two students choose the triangle and quadrilateral, in that order?

 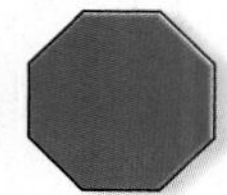  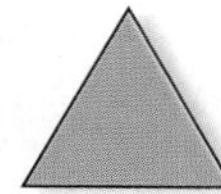 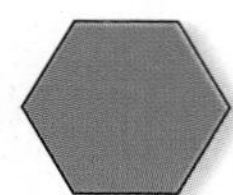

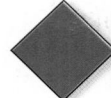

**Example 2**

2. **PLAYS** A high school performs a production of *A Raisin in the Sun* with each freshman English class of 18 students. If the three members of the crew are decided at random, what is the probability that Chase is selected for lighting, Jaden is selected for props, and Emelina for spotlighting?

**Example 3**

3. **DRIVING** What is the probability that a license plate using the letters C, F, and F and numbers 3, 3, 3, and 1 will be CFF3133?

**Example 4**

4. **CHEMISTRY** In chemistry lab, you need to test six samples that are randomly arranged on a circular tray.

   a. What is the probability that the arrangement shown at the right is produced?

   b. What is the probability that test tube 2 will be in the top middle position?

**Example 5**

5. Five hundred boys, including Josh and Sokka, entered a drawing for two football game tickets. What is the probability that the tickets were won by Josh and Sokka?

## Practice and Problem Solving

**Example 1**

6. **CONCERTS** Nia and Chad are going to a concert with their high school's key club. If they choose a seat on the row below at random, what is the probability that Chad will be in seat C11 and Nia will be in C12?

| C6 | C7 | C8 | C9 | C10 | C11 | C12 | C13 | C14 | C15 | C16 | C17 |
|---|---|---|---|---|---|---|---|---|---|---|---|

7. **FAIRS** Alfonso and Colin each bought one raffle ticket at the state fair. If 50 tickets were randomly sold, what is the probability that Alfonso got ticket 14 and Colin got ticket 23?

**Example 2**

8. **CCSS MODELING** The table shows the finalists for a floor exercises competition. The order in which they will perform will be chosen randomly.

   a. What is the probability that Cecilia, Annie, and Kimi are the first 3 gymnasts to perform, in any order?

   b. What is the probability that Cecilia is first, Annie is second, and Kimi is third?

| Floor Exercises Finalists |
|---|
| Eliza Hernandez |
| Kimi Kanazawa |
| Cecilia Long |
| Annie Montgomery |
| Shenice Malone |
| Caroline Smith |
| Jessica Watson |

9. **JOBS** A store randomly assigns their employees work identification numbers to track productivity. Each number consists of 5 digits ranging from 1–9. If the digits cannot repeat, find the probability that a randomly generated number is 25938.

10. **GROUPS** Two people are chosen randomly from a group of ten. What is the probability that Jimmy was selected first and George second?

**Example 3** **11.** **MAGNETS** Santiago bought some letter magnets that he can arrange to form words on his fridge. If he randomly selected a permutation of the letters shown below, what is the probability that they would form the word BASKETBALL?

**12.** **ZIP CODES** What is the probability that a zip code randomly generated from among the digits 3, 7, 3, 9, 5, 7, 2, and 3 is the number 39372?

**Example 4** **13.** **GROUPS** Keith is randomly arranging desks into circles for group activities. If there are 7 desks in his circle, what is the probability that Keith will be in the desk closest to the door?

**14.** **AMUSEMENT PARKS** Sylvie is at an amusement park with her friends. They go on a ride that has bucket seats in a circle. If there are 8 seats, what is the probability that Sylvie will be in the seat farthest from the entrance to the ride?

**Example 5** **15.** **PHOTOGRAPHY** If you are randomly placing 24 photos in a photo album and you can place four photos on the first page, what is the probability that you choose the photos at the right?

**16.** **ROAD TRIPS** Rita is going on a road trip across the U.S. She needs to choose from 15 cities where she will stay for one night. If she randomly pulls 3 city brochures from a pile of 15, what is the probability that she chooses Austin, Cheyenne, and Savannah?

**17.** **CCSS SENSE-MAKING** Use the figure below. Assume that the balls are aligned at random.

**a.** What is the probability that in a row of 8 pool balls, the solid 2 and striped 11 would be first and second from the left?

**b.** What is the probability that if the 8 pool balls were mixed up at random, they would end up in the order shown?

**c.** What is the probability that in a row of seven balls, with three 8 balls, three 9 balls, and one 6 ball, the three 8 balls would be to the left of the 6 ball and the three 9 balls would be on the right?

**d.** If the balls were randomly rearranged and formed a circle, what is the probability that the 6 ball is next to the 7 ball?

**18.** How many lines are determined by 10 randomly selected points, no 3 of which are collinear? Explain your calculation.

**19.** Suppose 7 points on a circle are chosen at random, as shown at the right.

**a.** Using the letters A through E, how many ways can the points on the circle be named?

**b.** If one point on the circle is fixed, how many arrangements are possible?

**20. RIDES** A carousel has 7 horses and one bench seat that will hold two people. One of the horses does not move up or down.

**a.** How many ways can the seats on the carousel be randomly filled by 9 people?

**b.** If the carousel is filled randomly, what is the probability that you and your friend will end up in the bench seat?

**c.** If 6 of the 9 people randomly filling the carousel are under the age of 8, what is the probability that a person under the age of 8 will end up on the horse that does not move up or down?

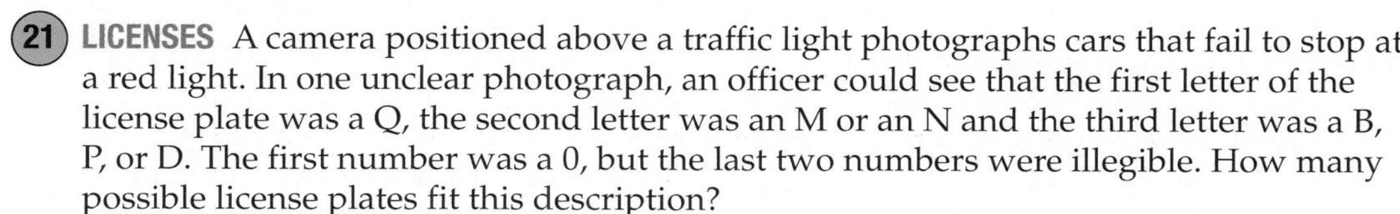

**21. LICENSES** A camera positioned above a traffic light photographs cars that fail to stop at a red light. In one unclear photograph, an officer could see that the first letter of the license plate was a Q, the second letter was an M or an N and the third letter was a B, P, or D. The first number was a 0, but the last two numbers were illegible. How many possible license plates fit this description?

**22. MULTIPLE REPRESENTATIONS** In this problem, you will investigate permutations.

**a. Numerical** Randomly select three digits from 0 to 9. Find the possible permutations of the three integers.

**b. Tabular** Repeat part **a** for four additional sets of three integers. You will use some digits more than once. Copy and complete the table below.

| Integers | Permutations | Average of Permutations | $\frac{\text{Average of Permutations}}{37}$ |
|---|---|---|---|
| 1, 4, 7 | 147, 174, 417, 471, 714, 741 | 444 | 12 |
| | | | |
| | | | |
| | | | |
| | | | |

**c. Verbal** Make a conjecture about the value of the average of the permutations of three digits between 0 and 9.

**d. Symbolic** If the three digits are $x$, $y$, and $z$, is it possible to write an equation for the average $A$ of the permutations of the digits? If so, write the equation. If not, explain why not.

## H.O.T. Problems Use Higher-Order Thinking Skills

**23. CHALLENGE** Fifteen boys and fifteen girls entered a drawing for four free movie tickets. What is the probability that all four tickets were won by girls?

**24. CHALLENGE** A student claimed that permutations and combinations were related by $r! \cdot {}_nC_r = {}_nP_r$. Use algebra to show that this is true. Then explain why ${}_nC_r$ and ${}_nP_r$ differ by the factor $r!$.

**25. OPEN ENDED** Describe a situation in which the probability is given by $\frac{1}{{}_7C_3}$.

**26. CCSS ARGUMENTS** Is the following statement *sometimes*, *always*, or *never* true? Explain.

$${}_nP_r = {}_nC_r$$

**27. PROOF** Prove that ${}_nC_{n-r} = {}_nC_r$.

**28. WRITING IN MATH** Compare and contrast permutations and combinations.

## Standardized Test Practice

**29. PROBABILITY** Four members of the pep band, two girls and two boys, always stand in a row when they play. What is the probability that a girl will be at each end of the row if they line up in random order?

A $\frac{1}{24}$

B $\frac{1}{12}$

C $\frac{1}{6}$

D $\frac{1}{2}$

**30. SHORT RESPONSE** If you randomly select a permutation of the letters shown below, what is the probability that they would spell GEOMETRY?

**31. ALGEBRA** Student Council sells soft drinks at basketball games and makes $1.50 from each. If they pay $75 to rent the concession stand, how many soft drinks would they have to sell to make $250 profit?

F 116

G 117

H 167

J 217

**32. SAT/ACT** The ratio of 12:9 is equal to the ratio of $\frac{1}{3}$ to

A $\frac{1}{4}$

B 1

C $\frac{5}{4}$

D 2

E 4

## Spiral Review

**33. SHOPPING** A women's coat comes in sizes 4, 6, 8, or 10 in black, brown, ivory, and cinnamon. How many different coats could be selected? (Lesson 13-1)

**34.** Two similar prisms have surface areas of 256 square inches and 324 square inches. What is the ratio of the height of the small prism to the height of the large prism? (Lesson 12-8)

**Find $x$. Round to the nearest tenth, if necessary.** (Lesson 11-7)

**35.**

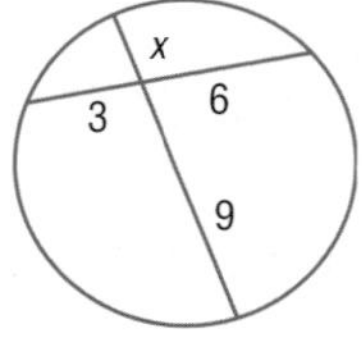


**36.**

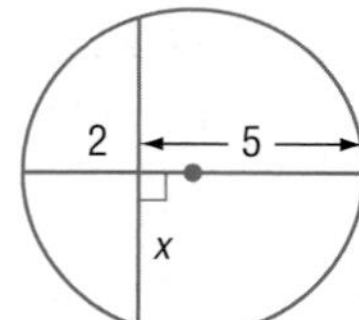


**37.**

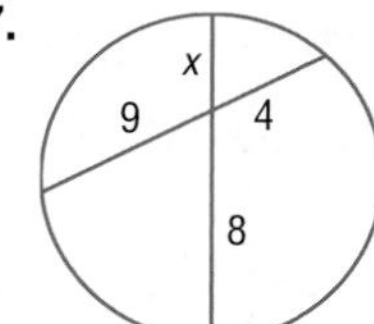


## Skills Review

**Use the number line to find each measure.**

**38.** $DF$

**39.** $AE$

**40.** $EF$

**41.** $BD$

**42.** $AC$

**43.** $CF$

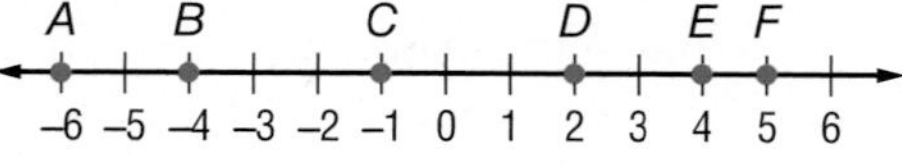

LESSON 13-3

# Geometric Probability

**Then**

- You found probabilities of simple events.

**Now**

1. Find probabilities by using length.
2. Find probabilities by using area.

**Why?**

- The object of the popular carnival game shown is to collect points by rolling a ball up an incline and into one of several circular target areas. The point value of each area is assigned based on the probability of a person landing a ball in that area.

**NewVocabulary**
geometric probability

**Common Core State Standards**

**Content Standards**
S.MD.7 (+) Analyze decisions and strategies using probability concepts (e.g., product testing, medical testing, pulling a hockey goalie at the end of a game).

**Mathematical Practices**
1 Make sense of problems and persevere in solving them.
2 Reason abstractly and quantitatively.

## 1 Probability with Length

The probability of winning the carnival game depends on the area of the target. Probability that involves a geometric measure such as length or area is called **geometric probability**.

### KeyConcept Length Probability Ratio

**Words** If a line segment (1) contains another segment (2) and a point on segment (1) is chosen at random, then the probability that the point is on segment (2) is

$$\frac{\text{length of segment (2)}}{\text{length of segment (1)}}.$$

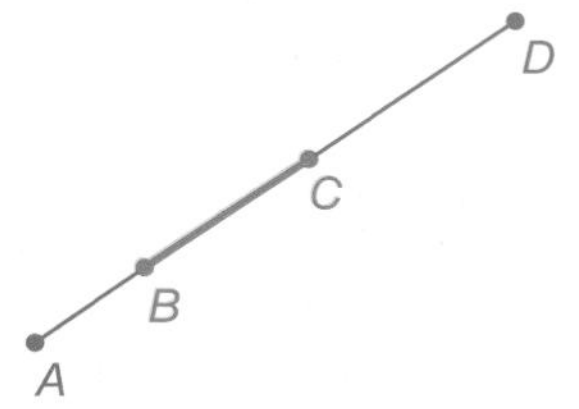


**Example** If a point $E$ on $\overline{AD}$ is chosen at random, then $P(E \text{ is on } \overline{BC}) = \frac{BC}{AD}$.

### Example 1 Use Lengths to Find Geometric Probability

**Point $X$ is chosen at random on $\overline{JM}$. Find the probability that $X$ is on $\overline{KL}$.**

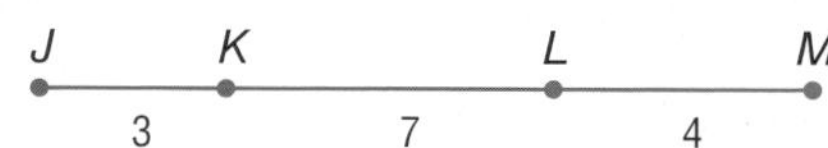


$P(X \text{ is on } \overline{KL}) = \frac{KL}{JM}$ — Length probability ratio

$= \frac{7}{14}$ — $KL = 7$ and $JM = 3 + 7 + 4$ or 14

$= \frac{1}{2}$, 0.5, or 50% — Simplify.

**GuidedPractice**

**Point $X$ is chosen at random on $\overline{JM}$. Find the probability of each event.**

**1A.** $P(X \text{ is on } \overline{LM})$

**1B.** $P(X \text{ is on } \overline{KM})$

Geometric probability can be used in many real-world situations that involve an infinite number of outcomes.

fotog/Getty Images

**Real-WorldLink**

A Chicago Transit Authority train arrives or departs a station like Addison on the Red Line every 15 minutes.

**Source:** Chicago Transit Authority

## Real-World Example 2 Model Real-World Probabilities

**TRANSPORTATION Use the information at the left. Assuming that you arrive at Addison on the Red Line at a random time, what is the probability that you will have to wait 5 or more minutes for a train?**

We can use a number line to model this situation. Since the trains arrive every 15 minutes, the next train will arrive in 15 minutes or less. On the number line below, the event of waiting 5 or more minutes is modeled by $\overline{BD}$.

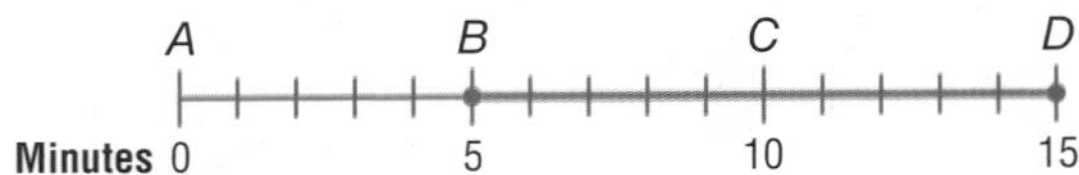


Find the probability of this event.

$P(\text{waiting 5 or more minutes}) = \frac{BD}{AD}$    Length probability ratio

$= \frac{10}{15} \text{ or } \frac{2}{3}$    $BD = 10$ and $AD = 15$

So, the probability of waiting 5 or more minutes for the next train is $\frac{2}{3}$ or about 67%.

### GuidedPractice

**2. TEA** Iced tea at a cafeteria-style restaurant is made in 8-gallon containers. Once the level gets below 2 gallons, the flavor of the tea becomes weak.

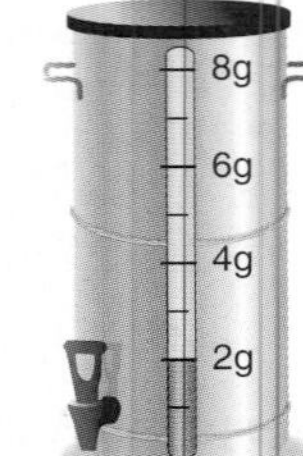


**A.** What is the probability that when someone tries to pour a glass of tea from the container, it is below 2 gallons?

**B.** What is the probability that the amount of tea in the container at any time is between 2 and 3 gallons?

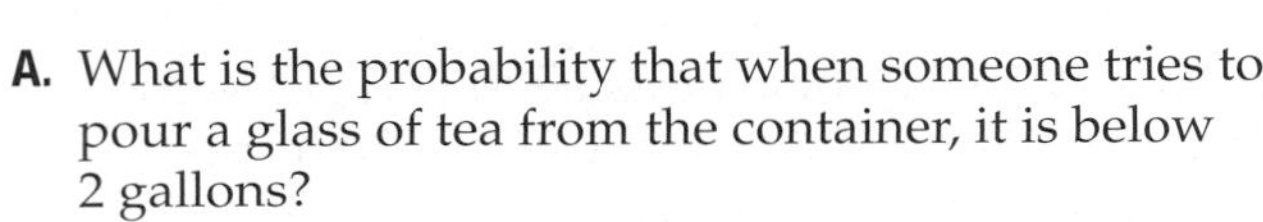

## 2 Probability with Area

**Probability with Area** Geometric probability can also involve area. The ratio for calculating geometric probability involving area is shown below.

### KeyConcept Area Probability Ratio

**Words** If a region $A$ contains a region $B$ and a point $E$ in region $A$ is chosen at random, then the probability that point $E$ is in region $B$ is $\frac{\text{area of region } B}{\text{area of region } A}$.

**Example** If a point $E$ is chosen at random in rectangle $A$, then $P(\text{point } E \text{ is in circle } B) = \frac{\text{area of region } B}{\text{area of region } A}$.

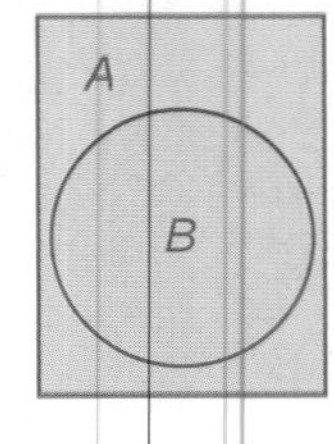


When determining geometric probabilities with targets, we assume

- that the object lands within the target area, and
- it is equally likely that the object will land anywhere in the region.

Ingram Publishing/SuperStock

## Real-World Example 3 Use Area to Find Geometric Probability

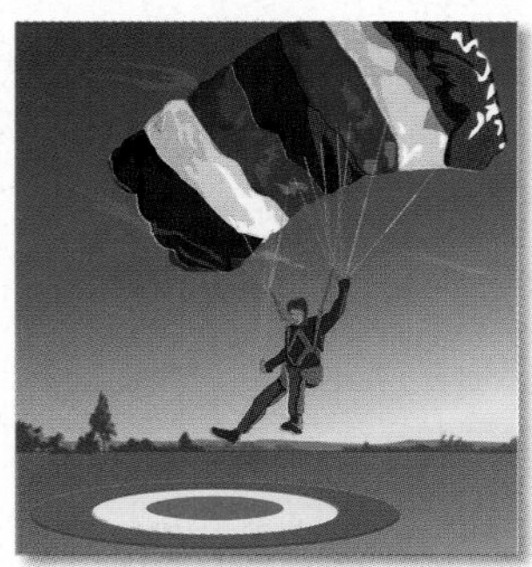

**SKYDIVING Suppose a skydiver must land on a target of three concentric circles. If the diameter of the center circle is 2 yards and the circles are spaced 1 yard apart, what is the probability that the skydiver will land in the red circle?**

You need to find the ratio of the area of the red circle to the area of the entire target. The radius of the red circle is 1 yard, while the radius of the entire target is 1 + 1 + 1 or 3 yards.

$$P(\text{skydiver lands in red circle}) = \frac{\text{area of red circle}}{\text{area of target}}$$ Area probability ratio

$$= \frac{\pi(1)^2}{\pi(3)^2}$$ $A = \pi r^2$

$$= \frac{\pi}{9\pi} \text{ or } \frac{1}{9}$$ Simplify.

The probability that the skydiver will land in the red circle is $\frac{1}{9}$ or about 11%.

**Real-WorldLink**

Champion accuracy skydivers routinely land less than two inches away from the center of a target.

Source: *SkyDiving News*

### Guided Practice

**3. SKYDIVING** Find each probability using the example above.

**A.** $P$(skydiver lands in the blue region)

**B.** $P$(skydiver lands in white region)

You can also use an angle measure to find geometric probability. The ratio of the area of a sector of a circle to the area of the entire circle is the same as the ratio of the sector's central angle to 360. You will prove this in Exercise 27.

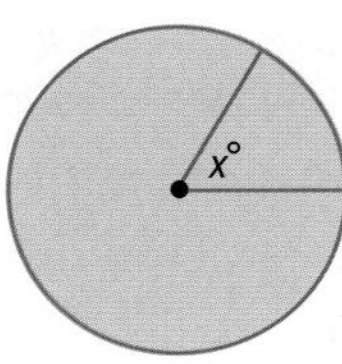

## Example 4 Use Angle Measures to Find Geometric Probability

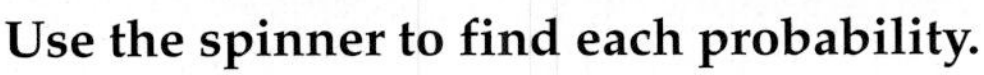

**Use the spinner to find each probability.**

**a. $P$(pointer landing on yellow)**

The angle measure of the yellow region is 45.

$P(\text{pointer landing on yellow}) = \frac{45}{360}$ or 12.5%

**b. $P$(pointer landing on purple)**

The angle measure of the purple region is 105.

$P(\text{pointer landing on purple}) = \frac{105}{360}$ or about 29%

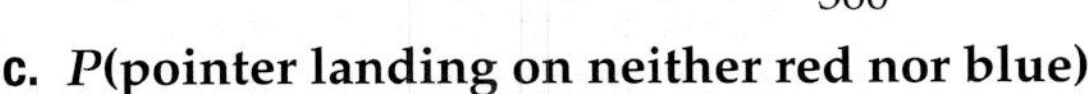

**c. $P$(pointer landing on neither red nor blue)**

The combined angle measures of the red and blue region are 50 + 70 or 120.

$P(\text{pointer landing on neither red nor blue}) = \frac{360 - 120}{360}$ or about 67%

**StudyTip**

**Use Estimation** In Example 4b, the area of the purple sector is a little less than $\frac{1}{3}$ or 33% of the spinner. Therefore, an answer of 29% is reasonable.

### Guided Practice

**4A.** $P$(pointer landing on blue) **4B.** $P$(pointer not landing on green)

## Check Your Understanding

**Example 1**

**Point $X$ is chosen at random on $\overline{AD}$. Find the probability of each event.**

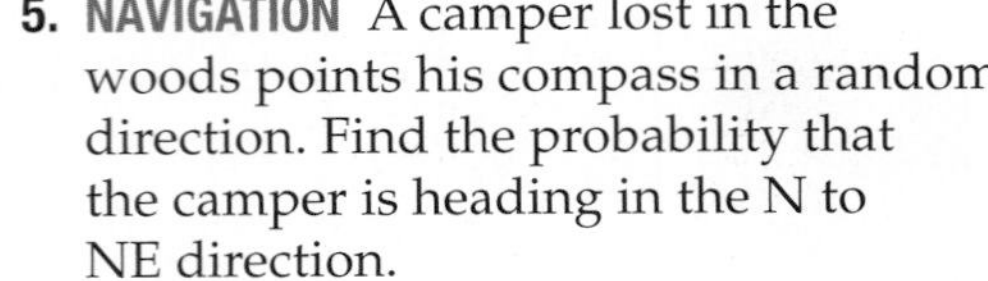


1. $P(X \text{ is on } \overline{BD})$
2. $P(X \text{ is on } \overline{BC})$

**Example 2**

3. **CARDS** In a game of cards, 43 cards are used, including one joker. Four players are each dealt 10 cards and the rest are put in a pile. If Greg doesn't have the joker, what is the probability that either his partner or the pile have the joker?

**Examples 3–4**

4. **ARCHERY** An archer aims at a target that is 122 centimeters in diameter with 10 concentric circles whose diameters decrease by 12.2 centimeters as they get closer to the center. Find the probability that the archer will hit the center.

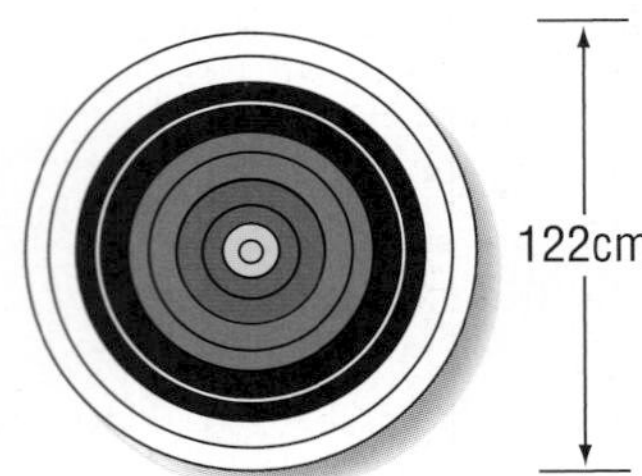


5. **NAVIGATION** A camper lost in the woods points his compass in a random direction. Find the probability that the camper is heading in the N to NE direction.

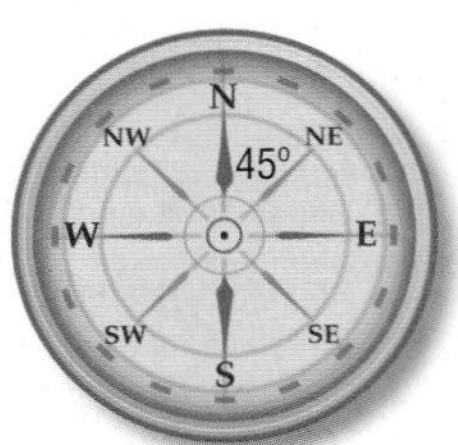


## Practice and Problem Solving

**Example 1**

**CCSS REASONING** **Point $X$ is chosen at random on $\overline{FK}$. Find the probability of each event.**

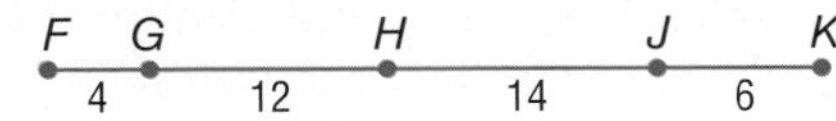


6. $P(X \text{ is on } \overline{FH})$
7. $P(X \text{ is on } \overline{GJ})$
8. $P(X \text{ is on } \overline{HK})$
9. $P(X \text{ is on } \overline{FG})$

10. **BIRDS** Four birds are sitting on a telephone wire. What is the probability that a fifth bird landing at a randomly selected point between birds 1 and 4 will sit at some point between birds 3 and 4?

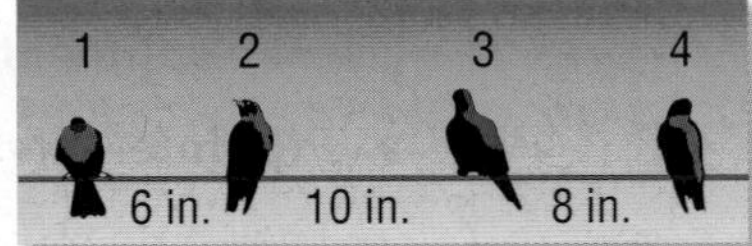


**Example 2**

11. **TELEVISION** Julio is watching television and sees an ad for a CD that he knows his friend wants for her birthday. If the ad replays at a random time in each 3-hour interval, what is the probability that he will see the ad again during his favorite 30-minute sitcom the next day?

**Example 3**

**Find the probability that a point chosen at random lies in the shaded region. Assume that figures that seem to be regular and congruent are regular and congruent.**

12.

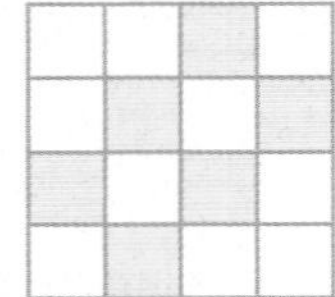

13.

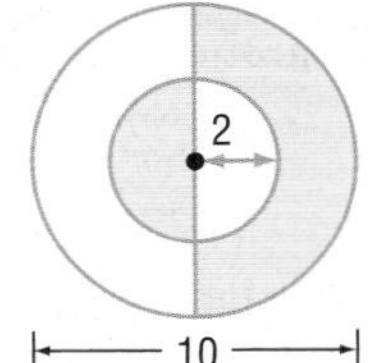


14.

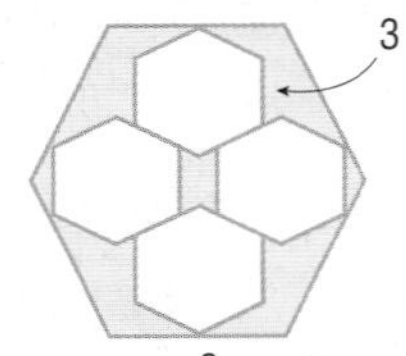

**Example 4** **Use the spinner to find each probability. If the spinner lands on a line it is spun again.**

15. $P$(pointer landing on yellow)
16. $P$(pointer landing on blue)
17. $P$(pointer not landing on green)
18. $P$(pointer landing on red)
19. $P$(pointer landing on neither red nor yellow)

**Describe an event with a 33% probability for each model.**

20. 

21. 10 20 30 40

22. 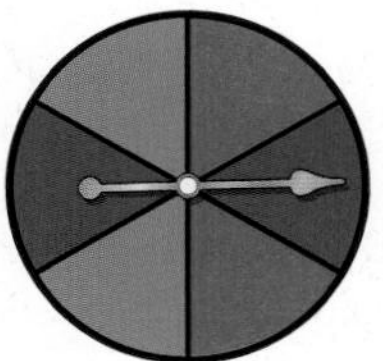

**Find the probability that a point chosen at random lies in the shaded region.**

23. 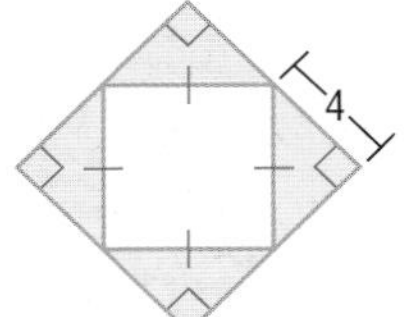


24. 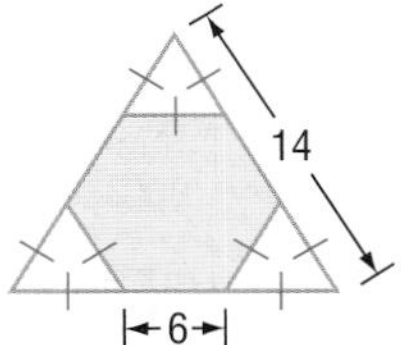


25. 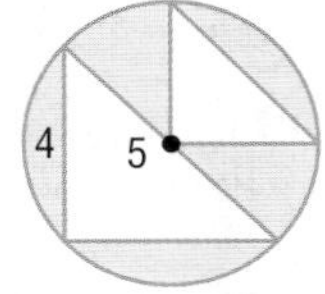


26. **FARMING** The layout for a farm is shown with each square representing a plot. Estimate the area of each field to answer each question.

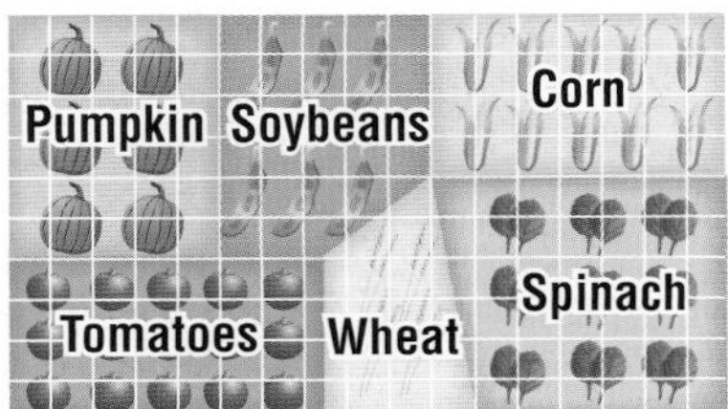


a. What is the approximate combined area of the spinach and corn fields?

b. Find the probability that a randomly chosen plot is used to grow soybeans.

27. **ALGEBRA** Prove that the probability that a randomly chosen point in the circle will lie in the shaded region is equal to $\frac{x}{360}$.

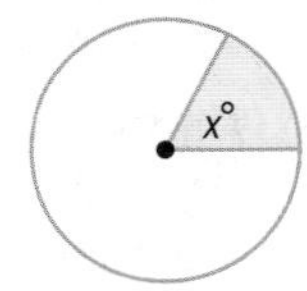


28. **COORDINATE GEOMETRY** If a point is chosen at random in the coordinate grid shown at the right, find each probability. Round to the nearest hundredth.

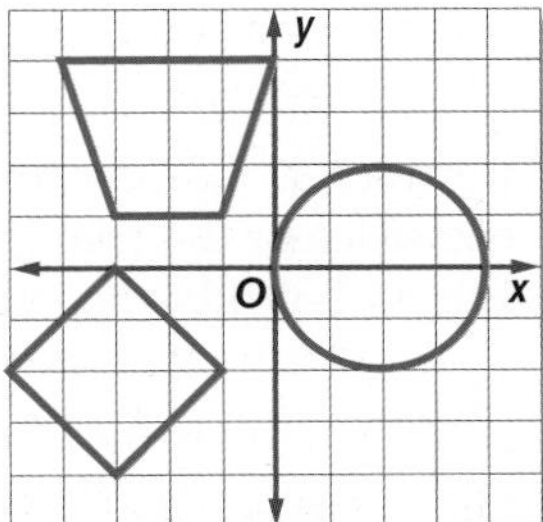


a. $P$(point inside the circle)

b. $P$(point inside the trapezoid)

c. $P$(point inside the trapezoid, square, or circle)

**CCSS SENSE-MAKING** **Find the probability that a point chosen at random lies in a shaded region.**

29. 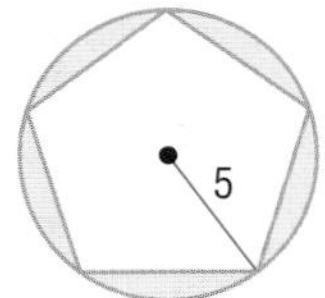


30. 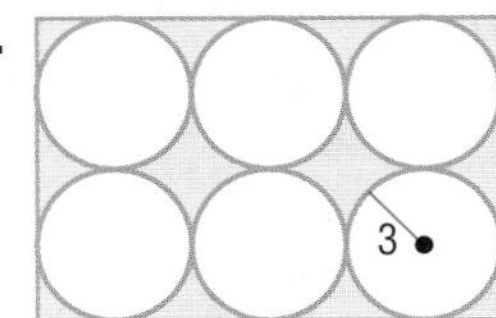


31. 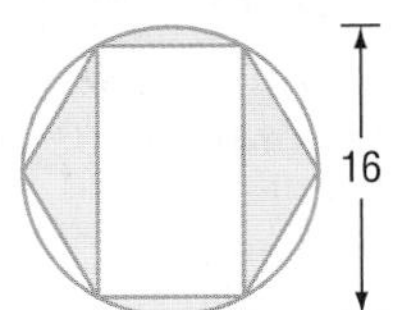

32. **COORDINATE GEOMETRY** Consider a system of inequalities, $1 \leq x \leq 6$, $y \leq x$, and $y \geq 1$. If a point $(x, y)$ in the system is chosen at random, what is the probability that $(x - 1)^2 + (y - 1)^2 \geq 16$?

33. **VOLUME** The polar bear exhibit at a local zoo has a pool with the side profile shown. If the pool is 20 feet wide, what is the probability that a bear that is equally likely to swim anywhere in the pool will be in the incline region?

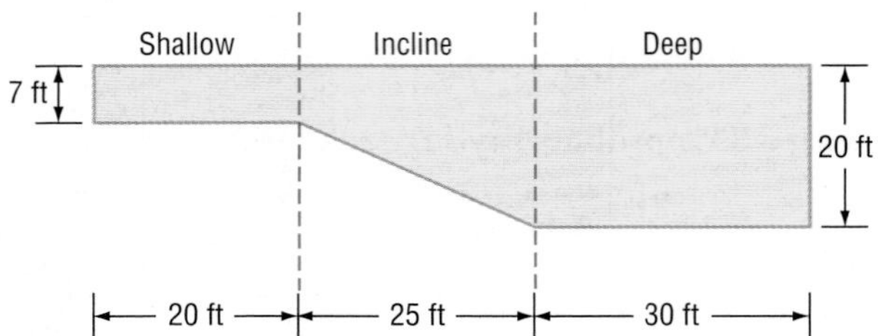


34. **DECISION MAKING** Meleah's flight was delayed and she is running late to make it to a national science competition. She is planning on renting a car at the airport and prefers car rental company A over car rental company B. The courtesy van for car rental company A arrives every 7 minutes, while the courtesy van for car rental company B arrives every 12 minutes.

   a. What is the probability that Meleah will have to wait 5 minutes or less to see each van? Explain your reasoning. (*Hint:* Use an area model.)

   b. What is the probability that Meleah will have to wait 5 minutes or less to see one of the vans? Explain your reasoning.

   c. Meleah can wait no more than 5 minutes without risking being late for the competition. If the van from company B should arrive first, should she wait for the van from company A or take the van from company B? Explain your reasoning.

## H.O.T. Problems Use Higher-Order Thinking Skills

35. **CHALLENGE** Find the probability that a point chosen at random would lie in the shaded area of the figure. Round to the nearest tenth of a percent.

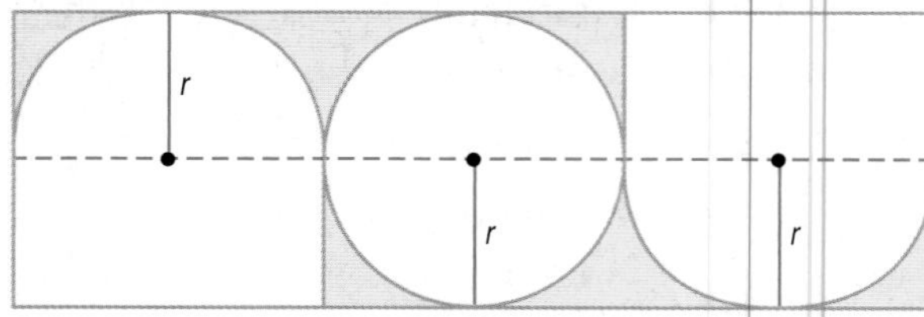


36. **CCSS REASONING** An isosceles triangle has a perimeter of 32 centimeters. If the lengths of the sides of the triangle are integers, what is the probability that the area of the triangle is exactly 48 square centimeters? Explain.

37. **WRITING IN MATH** Can athletic events be considered random events? Explain.

38. **OPEN ENDED** Represent a probability of 20% using three different geometric figures.

39. **WRITING IN MATH** Explain why the probability of a randomly chosen point falling in the shaded region of either of the squares shown is the same.

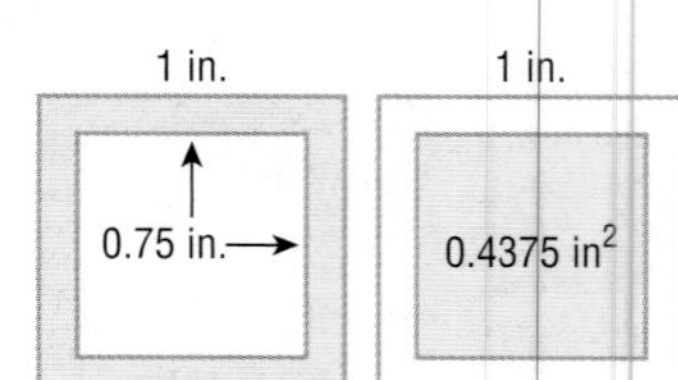

## Standardized Test Practice

**40. PROBABILITY** A circle with radius 3 is contained in a square with side length 9. What is the probability that a randomly chosen point in the interior of the square will also lie in the interior of the circle?

A $\frac{1}{9}$ C $\frac{\pi}{9}$

B $\frac{1}{3}$ D $\frac{9}{\pi}$

**41. ALGEBRA** The area of Miki's room is $x^2 + 8x + 12$ square feet. A gallon of paint will cover an area of $x^2 + 6x + 8$ square feet. Which expression gives the number of gallons of paint that Miki will need to buy to paint her room?

F $\frac{x+6}{x+4}$ H $\frac{x+4}{x+6}$

G $\frac{x-4}{x-6}$ J $\frac{x-4}{x+6}$

**42. EXTENDED RESPONSE** The spinner is divided into 8 equal sections.

**a.** If the arrow lands on a number, what is the probability that it will land on 3?

**b.** If the arrow lands on a number, what is the probability that it will land on an odd number?

**43. SAT/ACT** A box contains 7 blue marbles, 6 red marbles, 2 white marbles, and 3 black marbles. If one marble is chosen at random, what is the probability that it will be red?

A 0.11 D 0.39

B 0.17 E 0.67

C 0.33

## Spiral Review

**44. PROM** Four friends are sitting at a table together at the prom. What is the probability that a particular one of them will sit in the chair closest to the dance floor? (Lesson 13-2)

**Represent the sample space for each experiment by making an organized list, a table, and a tree diagram.** (Lesson 13-1)

**45.** Tito has a choice of taking music lessons for the next two years and playing drums or guitar.

**46.** Denise can buy a pair of shoes in either flats or heels in black or navy blue.

**STAINED GLASS In the stained glass window design, all of the small arcs around the circle are congruent. Suppose the center of the circle is point *O*.** (Lesson 11-4)

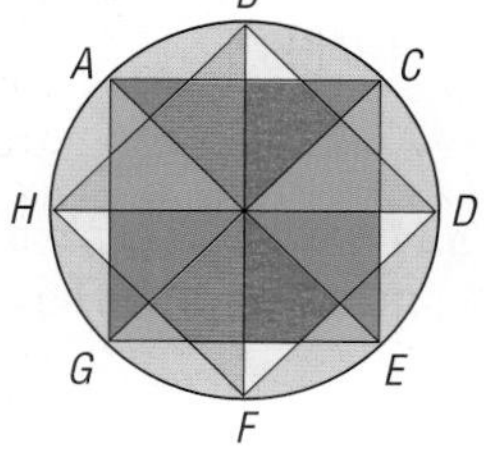


**47.** What is the measure of each of the small arcs?

**48.** What kind of figure is $\triangle AOC$? Explain.

**49.** What kind of figure is quadrilateral *BDFH*? Explain.

**50.** What kind of figure is quadrilateral *ACEG*? Explain.

## Skills Review

**Find the area of the shaded region. Round to the nearest tenth.**

**51.**

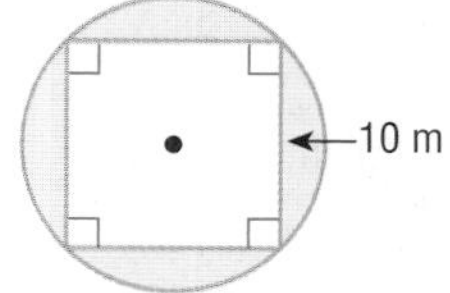


**52.**

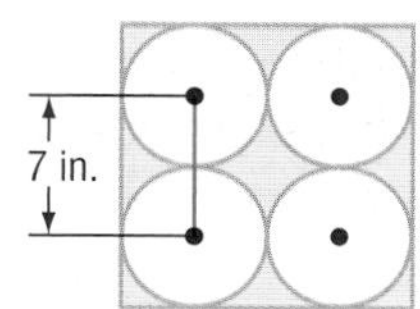


**53.**

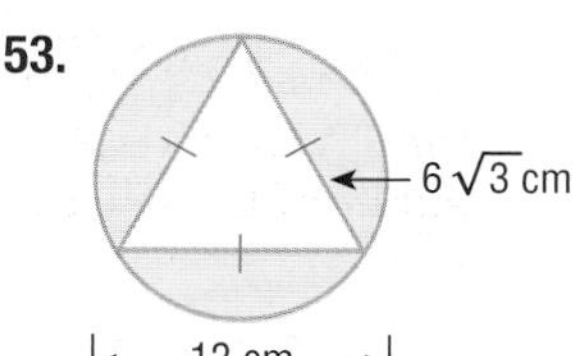

CHAPTER 13

# Mid-Chapter Quiz

## Lessons 13-1 through 13-3

1. **LUNCH** A deli has a lunch special, which consists of a sandwich, soup, dessert, and a drink for $4.99. The choices are in the table below. (Lesson 13-1)

| Sandwich | Soup | Dessert | Drink |
|---|---|---|---|
| chicken salad | tomato | cookie | tea |
| ham | chicken noodle | pie | coffee |
| tuna | vegetable | | cola |
| roast beef | | | diet cola |
| | | | milk |

   a. How many different lunches can be created from the items shown in the table?

   b. If a soup and two desserts were added, how many different lunches could be created?

2. **FLAGS** How many different signals can be made with 5 flags from 8 flags of different colors? (Lesson 13-1)

3. **CLOTHING** Marcy has six colors of shirts: red, blue, yellow, green, pink, and orange. She has each color in short-sleeved and long-sleeved styles. Represent the sample space for Marcy's shirt choices by making an organized list, a table, and a tree diagram. (Lesson 13-1)

4. **SPELLING** A bag contains one tile for each letter of the word TRAINS. If you selected a permutation of these letters at random, what is the probability that they would spell TRAINS? (Lesson 13-2)

5. **CHANGE** Augusto has 3 pockets and 4 different coins. In how many ways can he put one coin in each pocket? (Lesson 13-2)

6. **COINS** Ten coins are tossed simultaneously. In how many of the outcomes will the third coin turn up a head? (Lesson 13-2)

7. Find the probability that a point chosen at random lies in the shaded region. (Lesson 13-3)

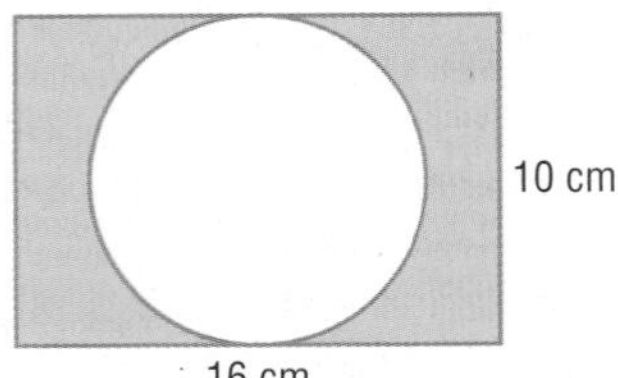


8. **EXTENDED RESPONSE** A 320 meter long tightrope is suspended between two poles. Assume that the line has an equal chance of breaking anywhere along its length. (Lesson 13-3)

   a. Determine the probability that a break will occur in the first 50 meters of the tightrope.

   b. Determine the probability that the break will occur within 20 meters of a pole.

**Point $A$ is chosen at random on $\overline{BE}$. Find the probability of each event.** (Lesson 13-3)

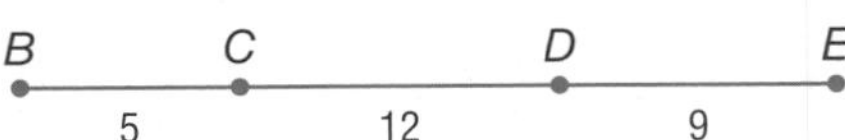


9. $P(A \text{ is on } \overline{CD})$

10. $P(A \text{ is on } \overline{BD})$

11. $P(A \text{ is on } \overline{CE})$

12. $P(A \text{ is on } \overline{DE})$

**Use the spinner to find each probability. If the spinner lands on a line, it is spun again.** (Lesson 13-3)

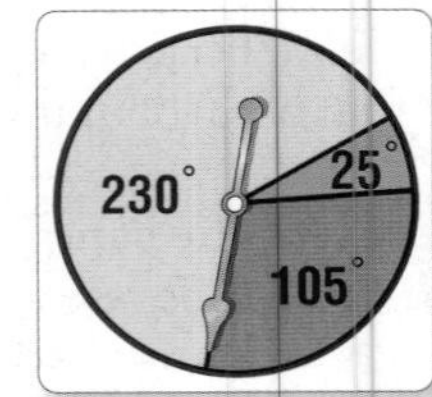


13. $P$(pointer landing on yellow)

14. $P$(pointer landing on blue)

15. $P$(pointer landing on red)

16. **GAMES** At a carnival, the object of a game is to throw a dart at the board and hit region III. (Lesson 13-3)

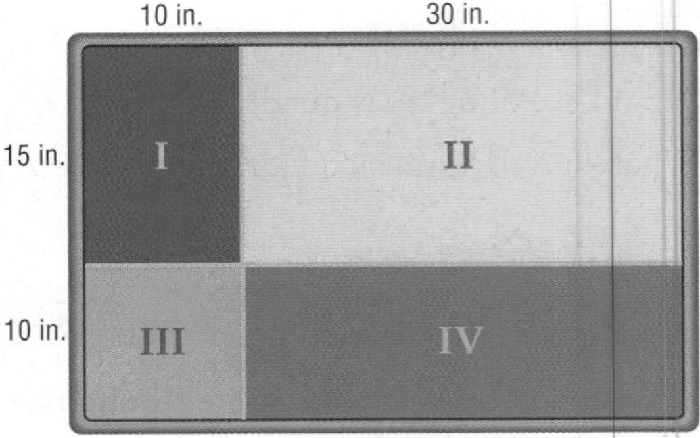


   a. What is the probability that it hits region I?

   b. What is the probability that it hits region II?

   c. What is the probability that it hits region III?

   d. What is the probability that it hits region IV?

LESSON 13-4

# Simulations

| :: Then | :: Now | :: Why? |
|---|---|---|
| ● You found probabilities by using geometric measures. | ● 1 Design simulations to estimate probabilities.<br>2 Summarize data from simulations. | ● Based on practice, Allen knows that he makes 70% of his free throws. He wants to use this information to predict the number of free throws he is likely to make in games. |

**NewVocabulary**
probability model
simulation
random variable
expected value
Law of Large Numbers

**Common Core State Standards**

**Content Standards**
G.MG.3 Apply geometric methods to solve problems (e.g., designing an object or structure to satisfy physical constraints or minimize cost; working with typographic grid systems based on ratios). ★
S.MD.6 (+) Use probabilities to make fair decisions (e.g., drawing by lots, using a random number generator).

**Mathematical Practices**
1 Make sense of problems and persevere in solving them.
4 Model with mathematics.

## 1 Design a Simulation

A **probability model** is a mathematical model used to match a random phenomenon. A **simulation** is the use of a probability model to recreate a situation again and again so that the likelihood of various outcomes can be estimated. To design a simulation, use the following steps.

### KeyConcept Designing a Simulation

**Step 1** Determine each possible outcome and its theoretical probability.

**Step 2** State any assumptions.

**Step 3** Describe an appropriate probability model for the situation.

**Step 4** Define what a trial is for the situation and state the number of trials to be conducted.

An appropriate probability model has the same probabilities as the situation you are trying to predict. Geometric models are common probability models.

### Example 1 Design a Simulation by Using a Geometric Model

BASKETBALL **Allen made 70% of his free throws last season. Design a simulation that can be used to estimate the probability that he will make his next free throw this season.**

**Step 1**

| Possible Outcomes | Theoretical Probability |
|---|---|
| • Allen makes a free throw. → | 70% |
| • Allen misses a free throw. → | (100 − 70)% or 30% |

**Step 2** Our simulation will consist of 40 trials.

**Step 3** One device that could be used is a spinner divided into two sectors, one containing 70% of the spinner's area and the other 30%. To create such a spinner, find the measure of the central angle of each sector.

| Make Free Throw | Miss Free Throw |
|---|---|
| 70% of 360° = 252° | 30% of 360° = 108° |

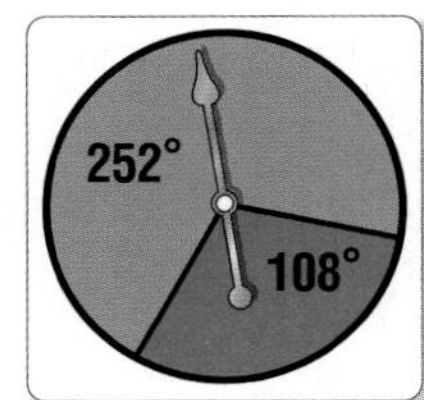

**Step 4** A trial, one spin of the spinner, will represent shooting one free throw. A successful trial will be a made free throw and a failed trial will be a missed free throw. The simulation will consist of 40 trials.

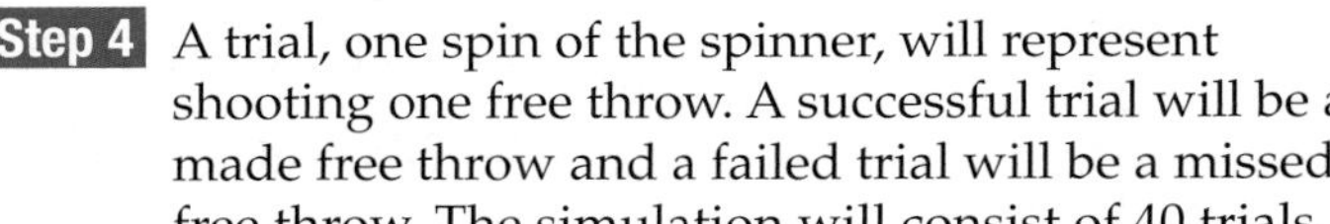

## GuidedPractice

1. **RESTAURANTS** A restaurant attaches game pieces to its large drink cups, awarding a prize to anyone who collects all 6 game pieces. Design a simulation using a geometric model that can be used to estimate how many large drinks a person needs to buy to collect all 6 game pieces.

**Problem-SolvingTip**

Use a Simulation
Simulations often provide a safe and efficient problem-solving strategy in situations that otherwise may be costly, dangerous, or impossible to solve using theoretical techniques. Simulations should involve data that are easier to obtain than the actual data you are modeling.

In addition to geometric models, simulations can also be conducted using dice, coin tosses, random number tables, and random number generators, such as those available on graphing calculators.

## Example 2 Design a Simulation by Using Random Numbers

**EYE COLOR A survey of East High School students found that 40% had brown eyes, 30% had hazel eyes, 20% had blue eyes, and 10% had green eyes. Design a simulation that can be used to estimate the probability that a randomly chosen East High student will have one of these eye colors.**

**Step 1**

| Possible Outcomes | | Theoretical Probability |
|---|---|---|
| Brown eyes | → | 40% |
| Hazel eyes | → | 30% |
| Blue eyes | → | 20% |
| Green eyes | → | 10% |

**Step 2** We assume that a student's eye color will fall into one of these four categories.

**Step 3** Use the random number generator on your calculator. Assign the ten integers 0–9 to accurately represent the probability data. The actual numbers chosen to represent the outcomes do not matter.

| Outcome | Represented by |
|---|---|
| Brown eyes | 0, 1, 2, 3 |
| Hazel eyes | 4, 5, 6 |
| Blue eyes | 7, 8 |
| Green eyes | 9 |

**Step 4** A trial will represent selecting a student at random and recording his or her eye color. The simulation will consist of 20 trials.

**StudyTip**

Random Number Generator
To generate a set of random integers on a graphing calculator, press MATH and select randInt( under the PRB menu. Then enter the beginning and ending integer values for your range and the number of integers you want in each trial.

## GuidedPractice

2. **SOCCER** Last season, Yao made 18% of his free kicks. Design a simulation using a random number generator that can be used to estimate the probability that he will make his next free kick.

## 2 Summarize Data from a Simulation

After designing a simulation, you will need to conduct the simulation and report the results. Include both numerical and graphical summaries of the simulation data, as well as an estimate of the probability of the desired outcome.

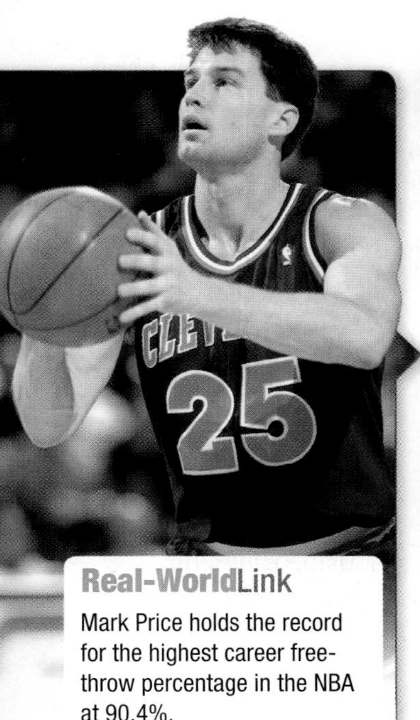

**Real-WorldLink**

Mark Price holds the record for the highest career free-throw percentage in the NBA at 90.4%.

**Source:** National Basketball Association

### Example 3 Conduct and Summarize Data from a Simulation

**BASKETBALL Refer to the simulation in Example 1. Conduct the simulation and report the results using appropriate numerical and graphical summaries.**

Make a frequency table and record the results after spinning the spinner 40 times.

| Outcome | Tally | Frequency |
|---|---|---|
| Make Free Throw | 卌 卌 卌 卌 卌 \| | 26 |
| Miss Free Throw | 卌 卌 \|\|\|\| | 14 |
| Total | | 40 |

Based on the simulation data, calculate the probability that Allen will make his next free throw.

$$\frac{\text{number of made free throws}}{\text{number of free throws attempted}} = \frac{26}{40} \text{ or } 0.65$$

This is an *experimental probability.*

The probability that Allen makes his next free throw is 0.65 or 65%. Notice that this is close to the theoretical probability, 70%. So, the experimental probability of his missing the next free throw is $1 - 0.65$ or 35%.

Make a bar graph of these results.

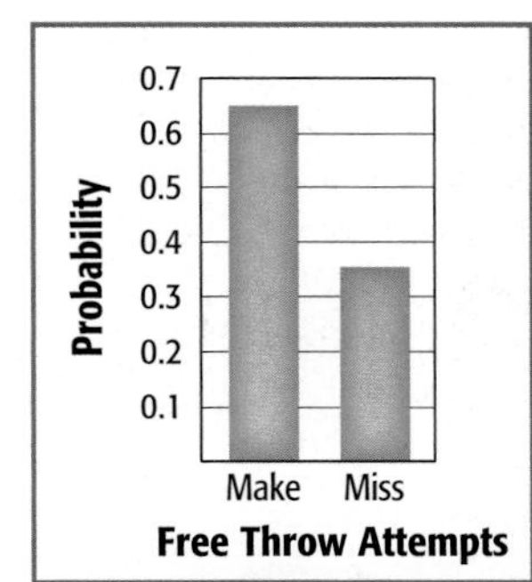


**GuidedPractice**

**3. EYE COLOR** Use a graphing calculator to conduct the simulation in Example 2. Then report the results using appropriate numerical and graphical summaries.

A **random variable** is a variable that can assume a set of values, each with fixed probabilities. For example, in the experiment of rolling two dice, the random variable $X$ can represent the sums of the potential outcomes on the dice. The table shows some of the $X$-values assigned to outcomes from this experiment.

| Sum of Outcomes of Rolling Two Dice | |
|---|---|
| **Outcome** | ***X*-Value** |
| (1, 1) | 2 |
| (1, 2) | 3 |
| (2, 1) | 3 |
| (4, 5) | 9 |
| (6, 6) | 12 |

**Expected value**, also known as mathematical expectation, is the average value of a random variable that one *expects* after repeating an experiment or simulation a theoretically infinite number of times. To find the expected value $E(X)$ of a random variable $X$, follow these steps.

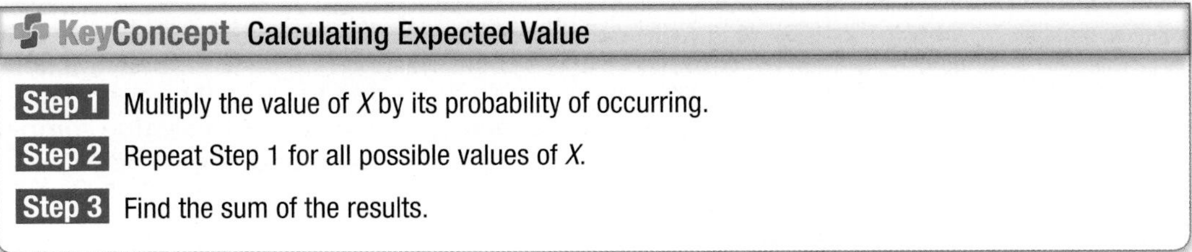

**KeyConcept Calculating Expected Value**

**Step 1** Multiply the value of $X$ by its probability of occurring.

**Step 2** Repeat Step 1 for all possible values of $X$.

**Step 3** Find the sum of the results.

Since it is an average, an expected value does not have to be equal to a possible value of the random variable.

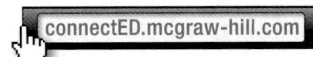

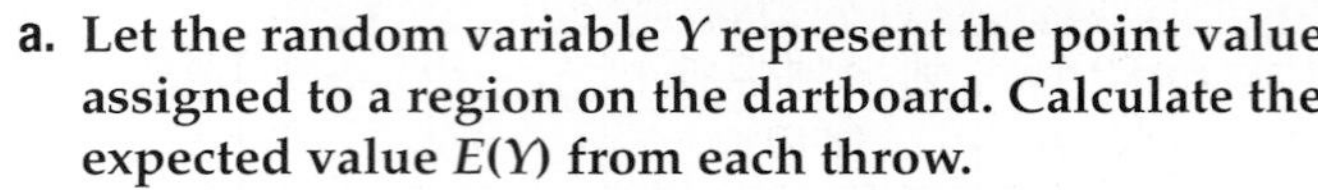

## Example 4 Calculate Expected Value

**DARTS Suppose a dart is thrown at the dartboard. The radius of the center circle is 1 centimeter and each successive circle has a radius 4 centimeters greater than the previous circle. The point value for each region is shown.**

> **StudyTip**
> **Geometric Probability** Remember that when determining geometric probabilities with targets, we assume that the object lands within the target area, and that it is equally likely that the object will land anywhere in the region.

**a. Let the random variable $Y$ represent the point value assigned to a region on the dartboard. Calculate the expected value $E(Y)$ from each throw.**

First calculate the geometric probability of landing in each region.

$$\text{Region 5} = \frac{\pi(1)^2}{\pi(1+4+4+4+4)^2} = \frac{1}{289} \qquad \text{Region 4} = \frac{\pi(4+1)^2 - \pi(1)^2}{\pi(17)^2} = \frac{24}{289}$$

$$\text{Region 3} = \frac{\pi(4+5)^2 - \pi(5)^2}{\pi(17)^2} = \frac{56}{289} \qquad \text{Region 2} = \frac{\pi(4+9)^2 - \pi(9)^2}{\pi(17)^2} = \frac{88}{289}$$

$$\text{Region 1} = \frac{\pi(4+13)^2 - \pi(13)^2}{\pi(17)^2} = \frac{120}{289}$$

$$E(Y) = 1 \cdot \frac{120}{289} + 2 \cdot \frac{88}{289} + 3 \cdot \frac{56}{289} + 4 \cdot \frac{24}{289} + 5 \cdot \frac{1}{289} \text{ or about } 1.96$$

The expected value of each throw is about 1.96.

**b. Design a simulation to estimate the average value, or the average of the results of your simulation, of this game. How does this value compare with the expected value you found in part a?**

Assign the integers 0–289 to accurately represent the probability data.

Region 1 = integers 1–120

Region 2 = integers 121–208

Region 3 = integers 209–264

Region 4 = integers 265–288

Region 5 = integer 289

Use a graphing calculator to generate 50 trials of random integers from 1 to 289. Record the results in a frequency table. Then calculate the average value of the outcomes.

| Outcome | Frequency |
|---|---|
| Region 1 | 16 |
| Region 2 | 13 |
| Region 3 | 13 |
| Region 4 | 8 |
| Region 5 | 0 |

> **Math HistoryLink**
> **Jakob Bernoulli (1654–1705)** Bernoulli was a Swiss mathematician. It seemed obvious to him that the more observations made of a given situation, the better one would be able to predict future outcomes. He provided scientific proof of his Law of Large Numbers in his work *Ars Conjectandi* (Art of Conjecturing), published in 1713.

$$\text{average value} = 1 \cdot \frac{16}{50} + 2 \cdot \frac{13}{50} + 3 \cdot \frac{13}{50} + 4 \cdot \frac{8}{50} + 5 \cdot \frac{0}{50} = 2.26$$

The average value 2.26 is greater than the expected value 1.96.

### GuidedPractice

**4. DICE** If two dice are rolled, let the random variable $X$ represent the sum of the potential outcomes.

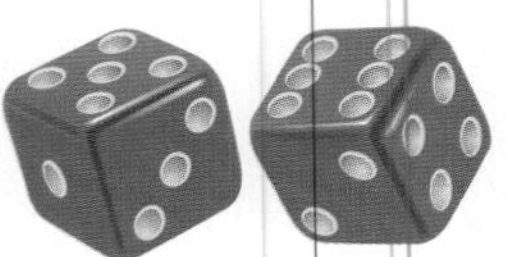

**A.** Find the expected value $E(X)$.

**B.** Design and run a simulation to estimate the average value of this experiment. How does this value compare with the expected value you found in part A?

The difference in the average value from the simulation and the expected value in Example 4 illustrates the **Law of Large Numbers**: as the number of trials of a random process increases, the average value will approach the expected value.

North Wind/North Wind Picture Archives

## Check Your Understanding

**Examples 1, 3**

1. **GRADES** Clara got an A on 80% of her first semester Biology quizzes. Design and conduct a simulation using a geometric model to estimate the probability that she will get an A on a second semester Biology quiz. Report the results using appropriate numerical and graphical summaries.

**Examples 2–3**

2. **FITNESS** The table shows the percent of members participating in four classes offered at a gym. Design and conduct a simulation to estimate the probability that a new gym member will take each class. Report the results using appropriate numerical and graphical summaries.

| Class | Sign-Up % |
|---|---|
| tae kwon do | 45% |
| yoga | 30% |
| swimming | 15% |
| kick-boxing | 10% |

**Example 4**

3. **CARNIVAL GAMES** The object of the game shown is to accumulate points by using a dart to pop the balloons. Assume that each dart will hit a balloon.

| | | | | |
|---|---|---|---|---|
| 25 | 25 | 25 | 25 | 25 |
| 25 | 50 | 50 | 50 | 25 |
| 25 | 50 | 100 | 50 | 25 |
| 25 | 50 | 50 | 50 | 25 |
| 25 | 25 | 25 | 25 | 25 |

   a. Calculate the expected value from each throw.

   b. Design a simulation and estimate the average value of this game.

   c. How do the expected value and average value compare?

## Practice and Problem Solving

**Examples 1, 3** **Design and conduct a simulation using a geometric probability model. Then report the results using appropriate numerical and graphical summaries.**

4. A small community theater sells tickets for a play at four different prices. The prices and the corresponding number of tickets available are shown in the table below. What is the expected value of a randomly selected ticket?

| Ticket Prices | |
|---|---|
| Price | Number Available |
| $8 | 40 |
| $10 | 30 |
| $15 | 20 |
| $20 | 10 |

5. A certain game has a deck of cards labeled 1, 2, 3, or 4. At the beginning of a turn, a player draws a card from the deck and then moves the number of spaces indicated by the number on the card. The table shows the number of each type of card in the deck. What is the expected value of the number on a card drawn from the deck?

| Number on Card | Number of Cards |
|---|---|
| 1 | 20 |
| 2 | 15 |
| 3 | 10 |
| 4 | 5 |

   **A** 1 **C** 3

   **B** 2 **D** 4

6. The winner of a television game show chooses one of ten envelopes containing cash prizes. The number of envelopes containing each cash amount is shown in the table. What is the expected value of the cash prize?

| Cash Amount | Number of Envelopes |
|---|---|
| $1,000 | 4 |
| $5,000 | 3 |
| $10,000 | 2 |
| $25,000 | 1 |

7. The Common Loon is the Minnesota state bird. The table below shows data about loon observations on 700 lakes in 1989. To the nearest tenth, what is the expected number of loons on a randomly chosen lake?

| Number of Loons | Number of Lakes |
|---|---|
| 0 | 365 |
| 1 | 80 |
| 2 | 131 |
| 3 | 41 |
| 4 | 38 |
| 5 | 17 |
| 6 | 14 |
| 7 | 14 |

8. A baseball player at Oregon State has a batting average of .276. If his batting average remains constant, about how many hits can he expect to get in his next 50 times at bat?

A 14 C 28

B 22 D 36

9. Beth has made 3 out of every 7 free throws in the last 10 basketball games she has played. About how many free throws should she expect to make in the next 100 games?

A 30 C 43

B 37 D 70

10. A radio station randomly selected 200 callers to find out what type of music they preferred. Out of the 200 callers, 55 preferred alternative rock, 96 preferred classic rock, and the rest had no preference. How many of the next 50 callers can be predicted to like either classic rock or alternative rock?

A 14 C 24

B 17 D 38

11. Miriam randomly selected 50 plants at a nursery to test for aphids. She discovered 7 plants that had the pests. If the nursery has 780 plants, what could be the predicted number of plants with aphids?

A 109 C 350

B 111 D 430

12. A random sampling of 100 T-shirts in a factory showed that 4 had gaps in the stitching. In a sampling of 1,500 T-shirts, how many may be expected to have similar gaps?

A 25 C 60

B 40 D 375

13. If a random sampling of 6 airlines shows that 18% of 1,200 flights had delays, how many of the next 150 flights on the sampled airlines can be predicted to arrive on time?

A All flights can be predicted to arrive on time.

B None of the flights can be predicted to arrive on time.

C 27 of the flights can be predicted to arrive on time.

D 123 of the flights can be predicted to arrive on time.

14. A random number generator is used to create a series of random numbers from 0 to 9. The results are recorded in the table below.

| Number | Frequency |
|---|---|
| 0 | 2 |
| 1 | 5 |
| 2 | 4 |
| 3 | 2 |
| 4 | 8 |
| 5 | 11 |
| 6 | 0 |
| 7 | 4 |
| 8 | 6 |
| 9 | 2 |

What is the experimental probability of getting an even number on the random number generator?

A $\frac{23}{44}$ C $\frac{21}{44}$

B $\frac{1}{2}$ D $\frac{5}{11}$

15. **CCSS MODELING** Cynthia used her statistics from last season to design a simulation using a random number generator to predict what she would score each time she got possession of the ball.

| Integer Values | Points Scored | Frequency |
|---|---|---|
| 1–14 | 0 | 31 |
| 15 | 1 | 0 |
| 16–28 | 2 | 17 |
| 29–30 | 3 | 2 |

a. Based on the frequency table, what did she assume was the theoretical probability that she would score two points in a possession?

16. **MULTIPLE REPRESENTATIONS** In this problem, you will investigate expected value.

a. **Concrete** Roll two dice 20 times and record the sum of each roll.

b. **Numerical** Use the random number generator on a calculator to generate 20 pairs of integers between 1 and 6. Record the sum of each pair.

c. **Tabular** Copy and complete the table below using your results from parts **a** and **b**.

| Trial | Sum of Die Roll | Sum of Output from Random Number Generator |
|---|---|---|
| 1 | | |
| 2 | | |
| ... | | |
| 20 | | |

d. **Graphical** Use a bar graph to graph the number of times each possible sum occurred in the first 5 rolls. Repeat the process for the first 10 rolls and then all 20 outcomes.

e. **Verbal** How does the shape of the bar graph change with each additional trial?

f. **Graphical** Graph the number of times each possible sum occurred with the random number generator as a bar graph.

g. **Verbal** How do the graphs of the die trial and the random number trial compare?

h. **Analytical** Based on the graphs, what do you think the expected value of each experiment would be? Explain your reasoning.

## H.O.T. Problems Use Higher-Order Thinking Skills

17. **CCSS ARGUMENTS** An experiment has three equally likely outcomes *A*, *B*, and *C*. Is it possible to use the spinner shown in a simulation to predict the probability of outcome *C*? Explain your reasoning.

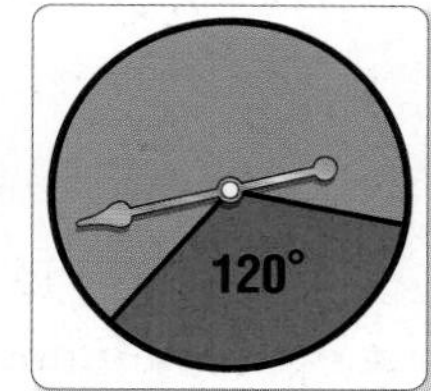


18. **REASONING** Can tossing a coin *sometimes*, *always*, or *never* be used to simulate an experiment with two possible outcomes? Explain.

19. **DECISION MAKING** A lottery consists of choosing 5 winning numbers from 31 possible numbers (0–30). The person who matches all 5 numbers, in any sequence, wins $1 million.

a. If a lottery ticket costs $1, should you play? Explain your reasoning by computing the expected payoff value, which is the expected value minus the ticket cost.

b. Would your decision to play change if the winnings increased to $5 million? if the winnings were only $0.5 million, but you chose from 21 numbers instead of 31 numbers? Explain.

20. **REASONING** When designing a simulation where darts are thrown at targets, what assumptions need to be made and why are they needed?

21. **OPEN ENDED** Describe an experiment in which the expected value is not a possible outcome. Explain.

22. **WRITING IN MATH** How is expected value different from probability?

## Standardized Test Practice

**23. PROBABILITY** Kaya tosses three coins at the same time and repeats the process 9 more times. Her results are shown below where H represents heads and T represents tails. Based on Kaya's data, what is the probability that at least one of the group of 3 coins will land with heads up?

HHH TTT
HHT HTT
TTH TTH
THT HTH
HHH HHT

**A** 0.1 **B** 0.2 **C** 0.3 **D** 0.9

**24. ALGEBRA** Paul collects comic books. He has 20 books in his collection, and he adds 3 per month. In how many months will he have a total of 44 books in his collection?

**F** 5 **G** 6 **H** 8 **J** 15

**25. SHORT RESPONSE** Alberto designed a simulation to determine how many times a player would roll a number higher than 4 on a die in a board game with 5 rolls. The table below shows his results for 50 trials. What is the probability that a player will roll a number higher than 4 two or more times in 5 rolls?

| Number of Rolls Greater Than 4 | Frequency |
|---|---|
| 0 | 8 |
| 1 | 15 |
| 2 | 18 |
| 3 | 9 |
| 4 | 0 |
| 5 | 0 |

**26. SAT/ACT** If a jar contains 150 peanuts and 60 cashews, what is the approximate probability that a nut selected from the jar at random will be a cashew?

**A** 0.25 **C** 0.33 **E** 0.71
**B** 0.29 **D** 0.4

## Spiral Review

**Point $X$ is chosen at random on $\overline{QT}$. Find the probability of each event.** (Lesson 13-3)

Q R S T; QR = 6, RS = 3, ST = 5

**27.** $P(X \text{ is on } \overline{QS})$

**28.** $P(X \text{ is on } \overline{RT})$

**29. BOOKS** Paige is choosing between 10 books at the library. What is the probability that she chooses 3 particular books to check out from the 10 initial books? (Lesson 13-2)

**Find the surface area of each figure. Round to the nearest tenth.** (Lesson 12-6)

**30.**

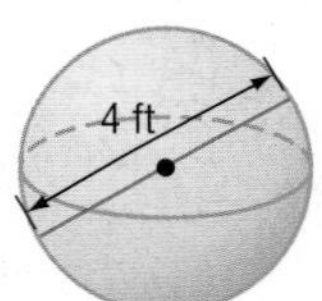


**31.**

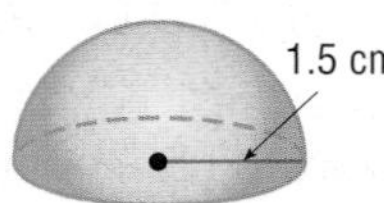


**32.**

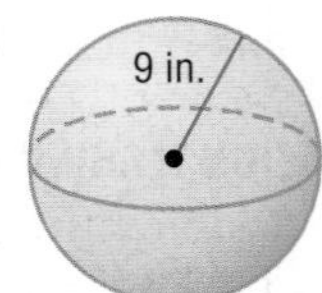


## Skills Review

**33. RECREATION** A group of 150 students was asked what they like to do during their free time.

**a.** How many students like going to the movies or shopping?

**b.** Which activity was mentioned by 37 students?

**c.** How many students did *not* say they like movies?

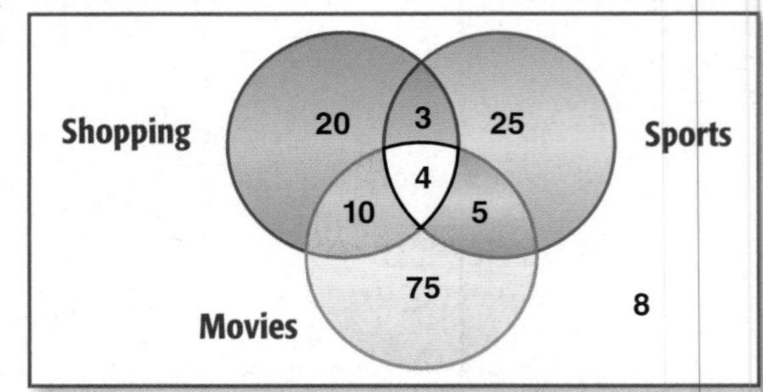

# LESSON 13-5 Probabilities of Independent and Dependent Events

**Then**

- You found simple probabilities

**Now**

1. Find probabilities of independent and dependent events.
2. Find probabilities of events given the occurrence of other events.

**Why?**

- The 18 students in Mrs. Turner's chemistry class are drawing names to determine who will give his or her presentation first. James is hoping to be chosen first and his friend Arturo wants to be second.

**NewVocabulary**
compound event
independent events
dependent events
conditional probability
probability tree

**Common Core State Standards**

**Content Standards**
S.CP.2 Understand that two events $A$ and $B$ are independent if the probability of $A$ and $B$ occurring together is the product of their probabilities, and use this characterization to determine if they are independent.
S.CP.3 Understand the conditional probability of $A$ given $B$ as $\frac{P(A \text{ and } B)}{P(B)}$, and interpret independence of $A$ and $B$ as saying that the conditional probability of $A$ given $B$ is the same as the probability of $A$, and the conditional probability of $B$ given $A$ is the same as the probability of $B$.

**Mathematical Practices**
2 Reason abstractly and quantitatively.
4 Model with mathematics.

CORBIS

## 1 Independent and Dependent Events

A **compound event** or *composite event* consists of two or more simple events. In the example above, James and Arturo being chosen to give their presentations first is a compound event. It consists of the event that James is chosen and the event that Arturo is chosen.

Compound events can be independent or dependent.

- Events $A$ and $B$ are **independent events** if the probability that $A$ occurs does not affect the probability that $B$ occurs.
- Events $A$ and $B$ are **dependent events** if the probability that $A$ occurs in some way changes the probability that $B$ occurs.

Consider choosing objects from a group of objects. If you replace the object each time, choosing additional objects are independent events. If you do not replace the object each time, choosing additional objects are dependent events.

### Example 1 Identify Independent and Dependent Events

**Determine whether the events are *independent* or *dependent*. Explain your reasoning.**

**a. One coin is tossed, and then a second coin is tossed.**

The outcome of the first coin toss in no way changes the probability of the outcome of the second coin toss. Therefore, these two events are *independent.*

**b. In the class presentation example above, one student's name is chosen and not replaced, and then a second name is chosen.**

After the first person is chosen, his or her name is removed and cannot be selected again. This affects the probability of the second person being chosen, since the sample space is reduced by one name. Therefore, these two events are *dependent.*

**c. Wednesday's lottery numbers and Saturday's lottery numbers.**

The numbers for one drawing have no bearing on the next drawing. Therefore, these two events are *independent.*

**GuidedPractice**

**1A.** A card is selected from a deck of cards and put back. Then a second card is selected.

**1B.** Andrea selects a shirt from her closet to wear on Monday and then a different shirt to wear on Tuesday.

Suppose a coin is tossed and the spinner shown is spun. The sample space for this experiment is

{(H, B), (H, R), (H, G), (T, B), (T, R), (T, G)}.

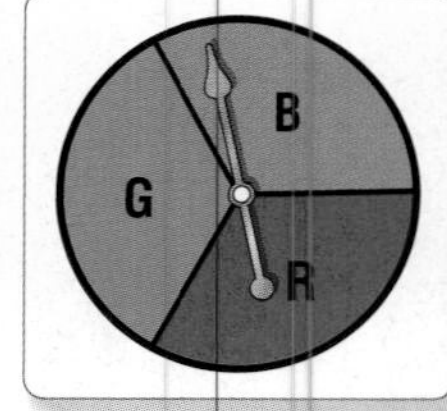


Using the sample space, the probability of the compound event of the coin landing on heads and the spinner on green is $P(\text{H and G}) = \frac{1}{6}$.

Notice that this same probability can be found by multiplying the probabilities of each simple event.

$$P(\text{H}) = \frac{1}{2} \qquad P(\text{G}) = \frac{1}{3} \qquad P(\text{H and G}) = \frac{1}{2} \cdot \frac{1}{3} \text{ or } \frac{1}{6}$$

This example illustrates the first of two Multiplication Rules for Probability.

## KeyConcept Probability of Two Independent Events

**Words** The probability that two independent events both occur is the product of the probabilities of each individual event.

**Symbols** If two events $A$ and $B$ are independent, then

$P(A \text{ and } B) = P(A) \cdot P(B)$.

**ReadingMath**

**and** The word *and* is a key word indicating to multiply probabilities.

This rule can be extended to any number of events.

## Real-World Example 2 Probability of Independent Events

**TRANSPORTATION Marisol and her friends are going to a concert. They put the slips of paper shown into a bag. If a person draws a yellow slip, he or she will ride in the van to the concert. A blue slip means he or she rides in the car.**

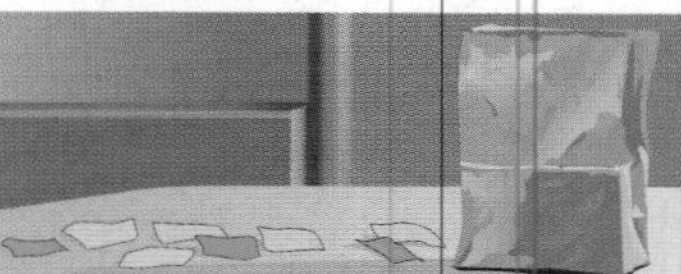

**Suppose Marisol draws a slip. Not liking the outcome, she puts it back and draws a second time. What is the probability that on each draw her slip is blue?**

These events are independent since Marisol replaced the slip that she removed. Let $B$ represent a blue slip and $Y$ a yellow slip.

| | Draw 1 | | Draw 2 | |
|---|---|---|---|---|
| $P(B \text{ and } B) =$ | $P(B)$ | $\cdot$ | $P(B)$ | Probability of independent events |
| $=$ | $\frac{3}{8}$ | $\cdot$ | $\frac{3}{8}$ or $\frac{9}{64}$ | $P(B) = \frac{3}{8}$ |

So, the probability of Marisol drawing two blue slips is $\frac{9}{64}$ or about 14%.

**StudyTip**

**Use an Area Model** You can also use the area model shown below to calculate the probability that both slips are blue. The blue region represents the probability of drawing two successive blue slips. The area of this region is $\frac{9}{64}$ of the entire model.

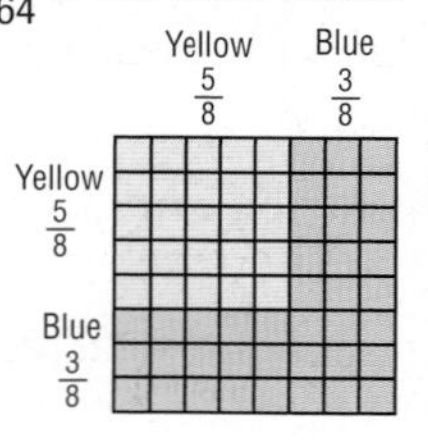


### GuidedPractice

**Find each probability.**

**2A.** A coin is tossed and a die is rolled. What is the probability that the coin lands heads up and the number rolled is a 6?

**2B.** Suppose you toss a coin four times. What is the probability of getting four tails?

The second of the Multiplication Rules of Probability addresses the probability of two dependent events.

## KeyConcept Probability of Two Dependent Events

**Words** The probability that two dependent events both occur is the product of the probability that the first event occurs and the probability that the second event occurs *after* the first event has already occurred.

**Symbols** If two events $A$ and $B$ are dependent, then

$P(A \text{ and } B) = P(A) \cdot P(B|A)$.

This rule can be extended to any number of events.

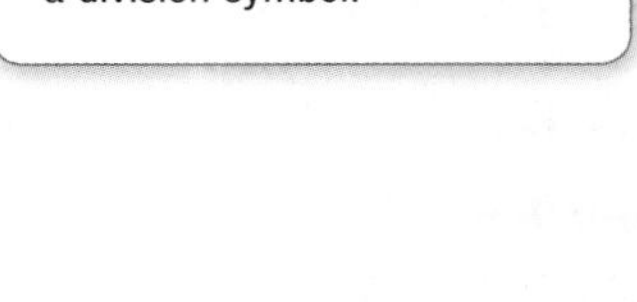

**WatchOut!**

**Conditional Notation** The "|" symbol in the notation $P(B|A)$ should not be interpreted as a division symbol.

The notation $P(B|A)$ is read *the probability that event B occurs given that event A has already occurred.* This is called **conditional probability**.

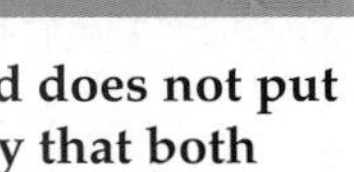

### Example 3 Probability of Dependent Events

**TRANSPORTATION Refer to Example 2. Suppose Marisol draws a slip and does not put it back. Then her friend Christian draws a slip. What is the probability that both friends draw a yellow slip?**

These events are dependent since Marisol does not replace the slip that she removed.

$P(Y \text{ and } Y) = P(Y) \cdot P(Y|Y)$ Probability of dependent events

$= \frac{5}{8} \cdot \frac{4}{7} \text{ or } \frac{5}{14}$ After the first yellow slip is chosen, 7 total slips remain, and 4 of those are yellow.

So, the probability that both friends draw yellow slips is $\frac{5}{14}$ or about 36%.

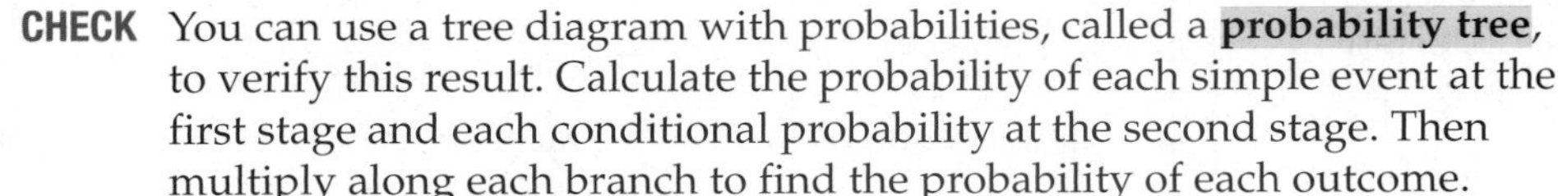

**CHECK** You can use a tree diagram with probabilities, called a **probability tree**, to verify this result. Calculate the probability of each simple event at the first stage and each conditional probability at the second stage. Then multiply along each branch to find the probability of each outcome.

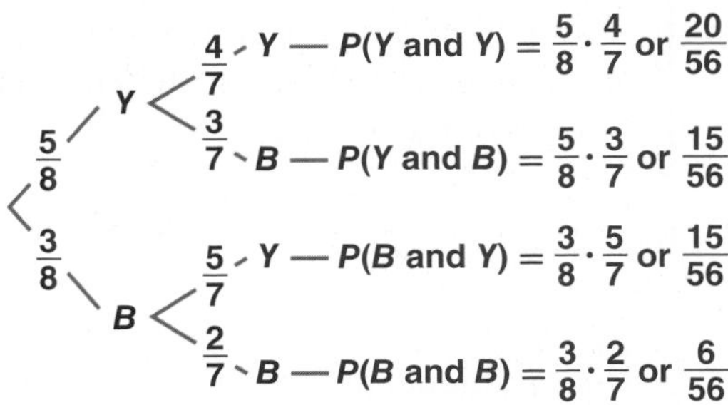


The sum of the probabilities should be 1.

$\frac{20}{56} + \frac{15}{56} + \frac{15}{56} + \frac{6}{56} = \frac{56}{56}$ or 1 ✓

**Real-WorldLink**

A recent study found that with three or more teenage passengers, 85% of fatal crashes of passenger vehicles driven by teens involved driver error, almost 50% involved speeding, and almost 70% involved a single vehicle.

**Source:** National Safety Council

### GuidedPractice

**3.** Three cards are selected from a standard deck of 52 cards. What is the probability that all three cards are diamonds if neither the first nor the second card is replaced?

## 2 Conditional Probabilities

In addition to its use in finding the probability of two or more dependent events, conditional probability can be used when additional information is known about an event.

**ReadingMath**

**Conditional Probability** $P(5|\text{odd})$ is read *the probability that the number rolled is a 5 given that the number rolled is odd.*

Suppose a die is rolled and it is known that the number rolled is odd. What is the probability that the number rolled is a 5?

There are only three odd numbers that can be rolled, so our sample space is reduced from {1, 2, 3, 4, 5, 6} to {1, 3, 5}. So, the probability that the number rolled is a 5 is $P(5|\text{odd}) = \frac{1}{3}$.

PT

### Standardized Test Example 4 Conditional Probability

**Ms. Fuentes' class is holding a debate. The 8 students participating randomly draw cards numbered with consecutive integers from 1 to 8.**

- **Students who draw odd numbers will be on the Proposition Team.**
- **Students who draw even numbers will be on the Opposition Team.**

**If Jonathan is on the Opposition Team, what is the probability that he drew the number 2?**

**A** $\frac{1}{8}$ **B** $\frac{1}{4}$ **C** $\frac{3}{8}$ **D** $\frac{1}{2}$

**Read the Test Item**

Since Jonathan is on the Opposition Team, he must have drawn an even number. So you need to find the probability that the number drawn was 2 given that the number drawn was even. This is a conditional probability problem.

**Solve the Test Item**

**Test-TakingTip**

**Use a Venn Diagram** Use a Venn diagram to help you visualize the relationship between the outcomes of two events.

Let $A$ be the event that an even number is drawn.
Let $B$ be the event that the number 2 is drawn.

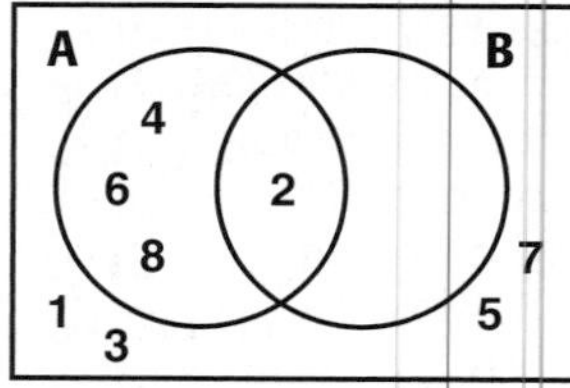


Draw a Venn diagram to represent this situation. There are only four even numbers in the sample space, and only one out of these numbers is a 2. Therefore, the $P(B|A) = \frac{1}{4}$. The answer is B.

**GuidedPractice**

**4.** When two dice are rolled, what is the probability that one die is a 4, given that the sum of the two die is 9?

**F** $\frac{1}{6}$ **G** $\frac{1}{4}$ **H** $\frac{1}{3}$ **J** $\frac{1}{2}$

Since conditional probability reduces the sample space, the Venn diagram in Example 4 can be simplified as shown, with the intersection of the two events representing those outcomes in $A$ and $B$. This suggests the following formula.

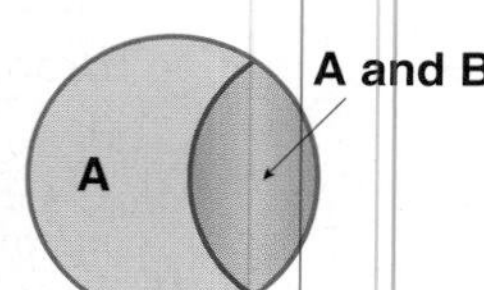


$$P(B|A) = \frac{P(A \text{ and } B)}{P(A)}$$

### KeyConcept Conditional Probability

The conditional probability of $B$ given $A$ is $P(B|A) = \frac{P(A \text{ and } B)}{P(A)}$, where $P(A) \neq 0$.

## Check Your Understanding

**Example 1** **Determine whether the events are *independent* or *dependent*. Explain.**

1. Jeremy took the SAT on Saturday and scored 1350. The following week he took the ACT and scored 23.

2. Alita's basketball team is in the final four. If they win, they will play in the championship game.

**Example 2**

3. **CARDS** A card is randomly chosen from a deck of 52 cards, replaced, and a second card is chosen. What is the probability of choosing both of the cards shown at the right?

**Example 3**

4. **TRANSPORTATION** Isaiah is getting on the bus after work. It costs $0.50 to ride the bus to his house. If he has 3 quarters, 5 dimes, and 2 nickels in his pocket, find the probability that he will randomly pull out two quarters in a row. Assume that the events are equally likely to occur.

**Example 4**

5. **GRIDDED RESPONSE** Every Saturday, 10 friends play dodgeball at a local park. To pick teams, they randomly draw cards with consecutive integers from 1 to 10. Odd numbers are on Team A, and even numbers are Team B. What is the probability that a player on Team B has drawn the number 10?

## Practice and Problem Solving

**Examples 1–3** **CCSS REASONING** **Determine whether the events are *independent* or *dependent*. Then find the probability.**

6. In a game, you roll an even number on a die and then spin a spinner numbered 1 through 5 and get an odd number.

7. An ace is drawn, without replacement, from a deck of 52 cards. Then, a second ace is drawn.

8. In a bag of 3 green and 4 blue marbles, a blue marble is drawn and not replaced. Then, a second blue marble is drawn.

9. You roll two dice and get a 5 each time.

10. **GAMES** In a game, the spinner at the right is spun and a coin is tossed. What is the probability of getting an even number on the spinner and the coin landing on tails?

11. **GIFTS** Tisha's class is having a gift exchange. Tisha will draw first and her friend Brandi second. If there are 18 students participating, what is the probability that Brandi and Tisha draw each other's names?

12. **VACATION** A work survey found that 8 out of every 10 employees went on vacation last summer. If 3 employees' names are randomly chosen, with replacement, what is the probability that all 3 employees went on vacation last summer?

13. **CAMPAIGNS** The table shows the number of each color of Student Council campaign buttons Clemente has to give away. If given away at random, what is the probability that the first and second buttons given away are both red?

| Button Color | Amount |
|---|---|
| blue | 20 |
| white | 15 |
| red | 25 |
| black | 10 |

**Example 4** **14.** A red marble is selected at random from a bag of 2 blue and 9 red marbles and not replaced. What is the probability that a second marble selected will be red?

**15.** A die is rolled. If the number rolled is greater than 2, find the probability that it is a 6.

**16.** A quadrilateral has a perimeter of 12 and all of the side lengths are odd integers. What is the probability that the quadrilateral is a rhombus?

**17.** A spinner numbered 1 through 12 is spun. Find the probability that the number spun is an 11 given that the number spun was an odd number.

**18.** **CLASSES** The probability that a student takes geometry and French at Satomi's school is 0.064. The probability that a student takes French is 0.45. What is the probability that a student takes geometry if the student takes French?

**19** **TECHNOLOGY** At Bell High School, 43% of the students own a CD player and 28% own a CD player and an MP3 player. What is the probability that a student owns an MP3 player if he or she also owns a CD player?

**20.** **PROOF** Use the formula for the probability of two dependent events $P(A \text{ and } B)$ to derive the conditional probability formula for $P(B|A)$.

**21.** **TENNIS** A double fault in tennis is when the serving player fails to land their serve "in" without stepping on or over the service line in two chances. Kelly's first serve percentage is 40%, while her second serve percentage is 70%.

**a.** Draw a probability tree that shows each outcome.

**b.** What is the probability that Kelly will double fault?

**c.** Design a simulation using a random number generator that can be used to estimate the probability that Kelly double faults on her next serve.

**22.** **VACATION** A random survey was conducted to determine where families vacationed. The results indicated that $P(B) = 0.6$, $P(B \cap M) = 0.2$, and the probability that a family did not vacation at either destination is 0.1.

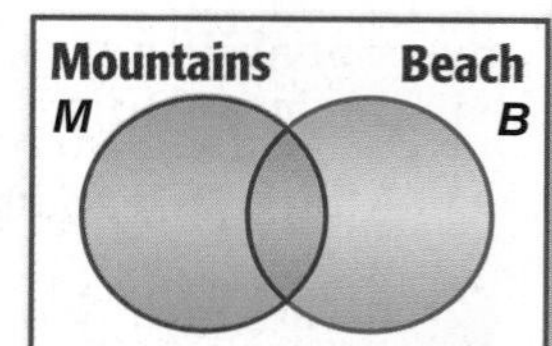


**a.** What is the probability that a family vacations in the mountains?

**b.** What is the probability that a family visiting the beach will also visit the mountains?

**23.** **DECISION MAKING** You are trying to decide whether you should expand a business. If you do not expand and the economy remains good, you expect $2 million in revenue. If the economy is bad, you expect $0.5 million. The cost to expand is $1 million, but the expected revenue after the expansion is $4 million in a good economy and $1 million in a bad economy. You assume that the chances of a good and a bad economy are 30% and 70%, respectively. Use a probability tree to explain what you should do.

## H.O.T. Problems Use Higher-Order Thinking Skills

**24.** **CCSS ARGUMENTS** There are $n$ different objects in a bag. The probability of drawing object $A$ and then object $B$ without replacement is about 2.4%. What is the value of $n$? Explain.

**25.** **REASONING** If $P(A \mid B)$ is the same as $P(A)$, and $P(B \mid A)$ is the same as $P(B)$, what can be said about the relationship between events $A$ and $B$?

**26.** **OPEN ENDED** Describe a pair of independent events and a pair of dependent events. Explain your reasoning.

**27.** **WRITING IN MATH** A medical journal reports the chance that a person smokes given that his or her parent smokes. Explain how you could determine the likelihood that a person's smoking and their parent's smoking are independent events.

## Standardized Test Practice

**28. PROBABILITY** Shannon will be assigned at random to 1 of 6 P.E. classes throughout the day and 1 of 3 lunch times. What is the probability that she will be in the second P.E. class and the first lunch?

A $\frac{1}{18}$  B $\frac{1}{9}$  C $\frac{1}{6}$  D $\frac{1}{2}$

**29. ALGEBRA** Tameron downloaded 2 videos and 7 songs to his digital media player for $10.91. Jake downloaded 3 videos and 4 songs for $9.93. What is the cost of each video?

F $0.99  H $1.42

G $1.21  J $1.99

**30. GRIDDED RESPONSE** A bag of jelly beans contains 7 red, 11 yellow, and 13 green. Victoro picks two jelly beans from the bag without looking. What is the probability as a percent rounded to the nearest tenth that Victoro picks a green one and then a red one?

**31. SAT/ACT** If the probability that it will snow on Tuesday is $\frac{4}{13}$, then what is the probability that it will *not* snow?

A $\frac{4}{9}$  C $\frac{13}{9}$  E $\frac{13}{4}$

B $\frac{9}{13}$  D $\frac{13}{5}$

## Spiral Review

**32. SOFTBALL** Zoe struck out during 10% of her at bats last season. Design and conduct a simulation to estimate the probability that she will strike out at her next at bat this season. (Lesson 13-4)

**Use the spinner to find each probability. The spinner is spun again if it stops on a line.** (Lesson 13-3)

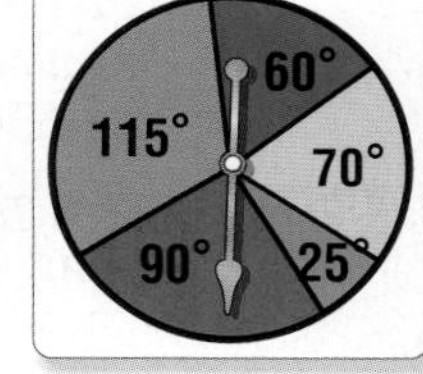


**33.** $P$(pointer landing on red)

**34.** $P$(pointer landing on blue)

**35.** $P$(pointer landing on green)

**36.** $P$(pointer landing on yellow)

**Determine whether each pair of solids is *similar*, *congruent*, or *neither*. If the solids are similar, state the scale factor.** (Lesson 12-8)

**37.**

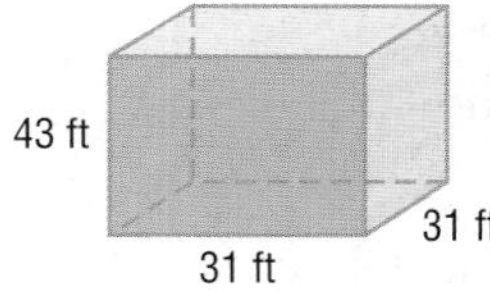


**38.**

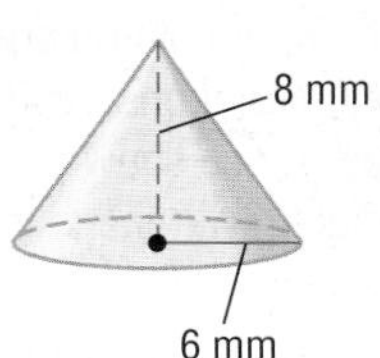


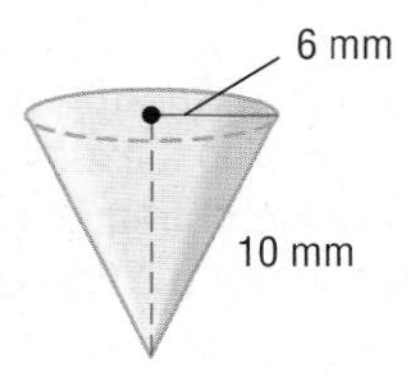


**39. FIREWORKS** Fireworks are shot from a barge on a river. There is an explosion circle inside which all of the fireworks will explode. Spectators sit outside a safety circle 800 feet from the center of the fireworks display. (Lesson 11-1)

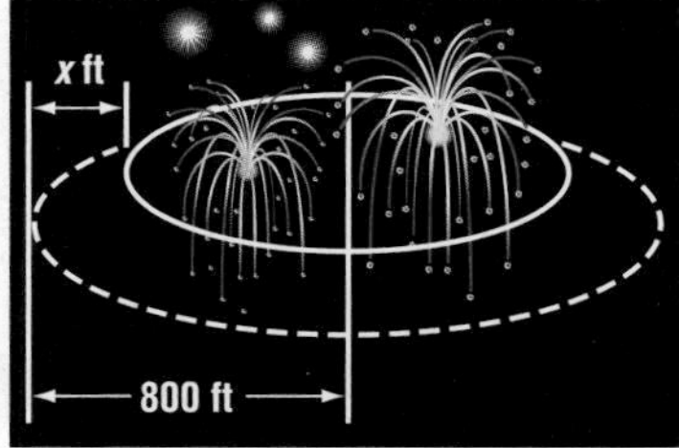


a. Find the approximate circumference of the safety circle.

b. If the safety circle is 200 to 300 feet farther from the center than the explosion circle, find the range of values for the radius of the explosion circle.

c. Find the least and maximum circumferences of the explosion circle to the nearest foot.

## Skills Review

**Find the number of possible outcomes for each situation.**

**40.** Blanca chooses from 5 different flavors of ice cream and 3 different toppings.

**41.** Perry chooses from 6 colors and 2 seat designs for his new mountain bike.

**42.** A rectangle has a perimeter of 12 and integer side lengths.

**43.** Three number cubes are rolled simultaneously.

EXTEND 13-5

# Geometry Lab
# Two-Way Frequency Tables

A **two-way frequency table** or *contingency table* is used to show the frequencies of data from a survey or experiment classified according to two variables, with the rows indicating one variable and the columns indicating the other.

**Common Core State Standards**
**Content Standards**
**S.CP.4** Construct and interpret two-way frequency tables of data when two categories are associated with each object being classified. Use the two-way table as a sample space to decide if events are independent and to approximate conditional probabilities.
**S.CP.6** Find the conditional probability of *A* given *B* as the fraction of *B*'s outcomes that also belong to *A*, and interpret the answer in terms of the model.
**Mathematical Practices** 5

PT

## Activity 1 Two-Way Frequency Table

PROM **Michael asks a random sample of 160 upperclassmen at his high school whether or not they plan to attend the prom. He finds that 44 seniors and 32 juniors plan to attend the prom, while 25 seniors and 59 juniors do not plan to attend. Organize the responses into a two-way frequency table.**

**Step 1** Identify the variables. The students surveyed can be classified according *class* and *attendance.* Since the survey included only upperclassmen, the variable *class* has two categories: senior or junior. The variable *attendance* also has two categories: attending or not attending the prom.

**Step 2** Create a two-way frequency table. Let the rows of the table represent *class* and the columns represent *attendance.* Then fill in the cells of the table with the information given.

**Step 3** Add a *Totals* row and a *Totals* column to your table and fill in these cells with the correct sums.

| Class | Attending the Prom | Not Attending the Prom | Totals |
|---|---|---|---|
| Senior | 44 | 32 | 76 |
| Junior | 25 | 59 | 84 |
| Totals | 69 | 91 | 160 |

The frequencies reported in the *Totals* row and *Totals* column are called **marginal frequencies**, with the bottom rightmost cell reporting the total number of observations. The frequencies reported in the interior of the table are called **joint frequencies**. These show the frequencies of all possible combinations of the categories for the first variable with the categories for the second variable.

## Analyze the Results

1. How many seniors were surveyed?
2. How many of the students that were surveyed plan to attend the prom?

A **relative frequency** is the ratio of the number of observations in a category to the total number of observations.

## Activity 2 Two-Way Relative Frequency Table

PROM **Convert the table from Activity 1 to a table of relative frequencies.**

**Step 1** Divide the frequency reported in each cell by the total number of respondents, 160.

| Class | Attending the Prom | Not Attending the Prom | Totals |
|---|---|---|---|
| Senior | $\frac{44}{160}$ | $\frac{32}{160}$ | $\frac{76}{160}$ |
| Junior | $\frac{25}{160}$ | $\frac{59}{160}$ | $\frac{84}{160}$ |
| Totals | $\frac{69}{160}$ | $\frac{91}{160}$ | $\frac{160}{160}$ |

**Step 2** Write each fraction as a percent rounded to the nearest tenth.

| Class | Attending the Prom | Not Attending the Prom | Totals |
|---|---|---|---|
| Senior | 27.5% | 20% | 47.5% |
| Junior | 15.6% | 36.9% | 52.5% |
| Totals | 43.1% | 56.9% | 100% |

You can use joint and marginal relative frequencies to approximate conditional probabilities.

**Activity 3** Conditional Probabilities

**PROM Using the table from Activity 2, find the probability that a surveyed upperclassman plans to attend the prom given that he or she is a junior.**

The probability that a surveyed upperclassman plans to attend the prom given that he or she is a junior is the conditional probability *P*(attending the prom | junior).

$$P(\text{attending the prom} \mid \text{junior}) = \frac{P(\text{attending the prom and junior})}{P(\text{junior})}$$ Conditional Probability

$$\approx \frac{0.156}{0.525} \text{ or } 29.7\%$$ *P*(attending the prom and junior) = 15.6% or 0.156, *P*(junior) = 52.5% or 0.525

## Analyze and Apply

**Refer to Activities 2 and 3.**

3. If there are 285 upperclassmen, about how many would you predict plan to attend the prom?
4. Find the probability that a surveyed student is a junior and does not plan to attend the prom.
5. Find the probability that a surveyed student is a senior given that he or she plans to attend the prom.
6. What is a possible trend you notice in the data?

When survey results are classified according to variables, you may want to decide whether these variables are independent of each other. Variable *A* is considered independent of variable *B* if $P(A \text{ and } B) = P(A) \cdot P(B)$. In a two-way frequency table, you can test for the independence of two variables by comparing the joint relative frequencies with the products of the corresponding marginal relative frequencies.

**Activity 4** Independence of Events

**PROM Use the relative frequency table from Activity 2 to determine whether prom attendance is independent of class.**

Calculate the expected joint relative frequencies if the two variables were independent. Then compare them to the actual relative frequencies.

For example, if 47.5% of respondents were seniors and 43.1% of respondents plan to attend the prom, then one would expect 47.5% • 43.1% or about 20.5% of respondents are seniors who plan to attend the prom.

Since the expected and actual joint relative frequencies are not the same, prom attendance for these respondents is not independent of class.

| Class | Attending the Prom | Not Attending the Prom | Totals |
|---|---|---|---|
| Senior | 27.5% (20.5%) | 20% (27%) | 47.5% |
| Junior | 15.6% (22.6%) | 36.9% (29.9%) | 52.5% |
| Totals | 43.1% | 56.9% | 100% |

**Note:** The numbers in parentheses are the expected relative frequencies.

**COLLECT DATA Design and conduct a survey of students at your school. Create a two-way relative frequency table for the data. Use your table to decide whether the data you collected indicate an independent relationship between the two variables. Explain your reasoning.**

7. student gender and whether a student's car insurance is paid by the student or the student's parent(s)
8. student gender and whether a student buys or brings his or her lunch

# LESSON 13-6 Probabilities of Mutually Exclusive Events

**Then**

- You found probabilities of independent and dependent events.

**Now**

1. Find probabilities of events that are mutually exclusive and events that are not mutually exclusive.
2. Find probabilities of complements.

**Why?**

- At Wayside High School, freshmen, sophomores, juniors, and seniors can all run for Student Council president. Dominic wants either a junior or a senior candidate to win the election. Trayvon wants either a sophomore or a female to win, but says, "If the winner is sophomore Katina Smith, I'll be thrilled!"

**NewVocabulary**
mutually exclusive events
complement

**Common Core State Standards**

**Content Standards**

S.CP.1 Describe events as subsets of a sample space (the set of outcomes) using characteristics (or categories) of the outcomes, or as unions, intersections, or complements of other events ("or," "and," "not").

S.CP.7 Apply the Addition Rule, $P(A \text{ or } B) = P(A) + P(B) - P(A \text{ and } B)$, and interpret the answer in terms of the model.

**Mathematical Practices**

1 Make sense of problems and persevere in solving them.
4 Model with mathematics.

## 1 Mutually Exclusive Events

In Lesson 13-4, you examined probabilities involving the intersection of two or more events. In this lesson, you will examine probabilities involving the union of two or more events.

$P(A \text{ and } B)$ — Indicates an intersection of two sample spaces.

$P(A \text{ or } B)$ — Indicates a union of two sample spaces.

To find the probability that one event occurs *or* another event occurs, you must know how the two events are related. If the two events cannot happen at the same time, they are said to be **mutually exclusive**. That is, the two events have no outcomes in common.

### Real-World Example 1 Identify Mutually Exclusive Events

**ELECTIONS Refer to the application above. Determine whether the events are *mutually exclusive* or *not mutually exclusive*. Explain your reasoning.**

**a. a junior winning the election or a senior winning the election**

These events are mutually exclusive. There are no common outcomes—a student cannot be both a junior and a senior.

**b. a sophomore winning the election or a female winning the election**

These events are not mutually exclusive. A female student who is a sophomore is an outcome that both events have in common.

**c. drawing an ace or a club from a standard deck of cards.**

Since the ace of clubs represents both events, they are not mutually exclusive.

### GuidedPractice

**Determine whether the events are *mutually exclusive* or *not mutually exclusive*. Explain your reasoning.**

**1A.** selecting a number at random from the integers from 1 to 100 and getting a number divisible by 5 or a number divisible by 10

**1B.** drawing a card from a standard deck and getting a 5 or a heart

**1C.** getting a sum of 6 or 7 when two dice are rolled

Image Source Pink/Alamy

One way of finding the probability of two mutually exclusive events occurring is to examine their sample space.

When a die is rolled, what is the probability of getting a 3 or a 4? From the Venn diagram, you can see that there are two outcomes that satisfy this condition, 3 and 4. So,

$$P(3 \text{ and } 4) = \frac{2}{6} \text{ or } \frac{1}{3}.$$

Notice that this same probability can be found by adding the probabilities of each simple event.

$P(3) = \frac{1}{6}$ $\quad P(4) = \frac{1}{6}$ $\quad P(3 \text{ and } 4) = \frac{1}{6} + \frac{1}{6} = \frac{2}{6} \text{ or } \frac{1}{3}$

This example illustrates the first of two Addition Rules for Probability.

**ReadingMath**

**or** The word *or* is a key word indicating that at least one of the events occurs. $P(A \text{ or } B)$ is read as *the probability that A occurs or that B occurs.*

**KeyConcept Probability of Mutually Exclusive Events**

| | |
|---|---|
| Words | If two events $A$ and $B$ are mutually exclusive, then the probability that $A$ or $B$ occurs is the sum of the probabilities of each individual event. |
| Example | If two events $A$ or $B$ are mutually exclusive, then $P(A \text{ or } B) = P(A) + P(B)$. |

This rule can be extended to any number of events.

**Math HistoryLink**

**Leonhard Euler (1707–1783)** Euler introduced graph theory in 1736 in a paper titled *Seven Bridges of Konigsburg*, a famous solved mathematics problem inspired by an actual place and situation. Also, Euler's formula relating the number of edges, vertices, and faces of a convex polyhedron is the origin of graph theory. Refer to Extend 13-6.

**Real-World Example 2 Mutually Exclusive Events**

**MUSIC Ramiro makes a playlist that consists of songs from three different albums by his favorite artist. If he lets his MP3 player select the songs from this list at random, what is the probability that the first song played is from Album 1 or Album 2?**

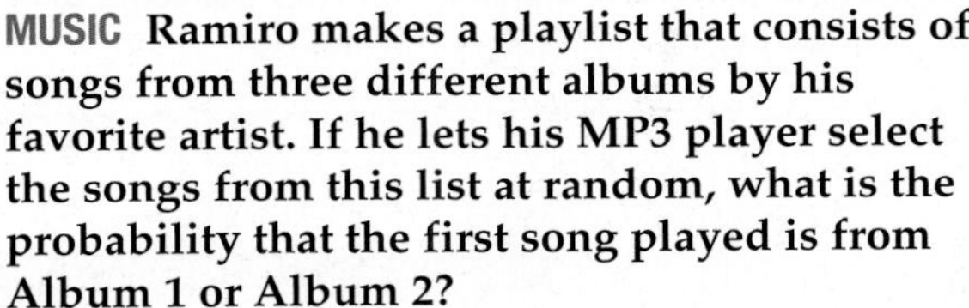

Ramiro's Playlist

| Album | Number of Songs |
|---|---|
| 1 | 10 |
| 2 | 12 |
| 3 | 13 |

These are mutually exclusive events, since the songs selected cannot be from both Album 1 and Album 2.

Let event $A1$ represent selecting a song from Album 1.
Let event $A2$ represent selecting a song from Album 2.
There are a total of $10 + 12 + 13$ or 35 songs.

$P(A1 \text{ or } A2) = P(A1) + P(A2)$ — Probability of mutually exclusive events

$= \frac{10}{35} + \frac{12}{35}$ — $P(A1) = \frac{10}{35}$ and $P(A2) = \frac{12}{35}$

$= \frac{22}{35}$ — Add.

So, the probability that the first song played is from Album 1 or Album 2 is $\frac{22}{35}$ or about 63%.

**GuidedPractice**

**2A.** Two dice are rolled. What is the probability that doubles are rolled or that the sum is 9?

**2B. CARNIVAL GAMES** If you win the ring toss game at a certain carnival, you receive a stuffed animal. If the stuffed animal is selected at random from among 15 puppies, 16 kittens, 14 frogs, 25 snakes, and 10 unicorns, what is the probability that a winner receives a puppy, a kitten, or a unicorn?

When a die is rolled, what is the probability of getting a number greater than 2 or an even number? From the Venn diagram, you can see that there are 5 numbers that are either greater than 2 or are an even number: 2, 3, 4, 5, and 6. So,

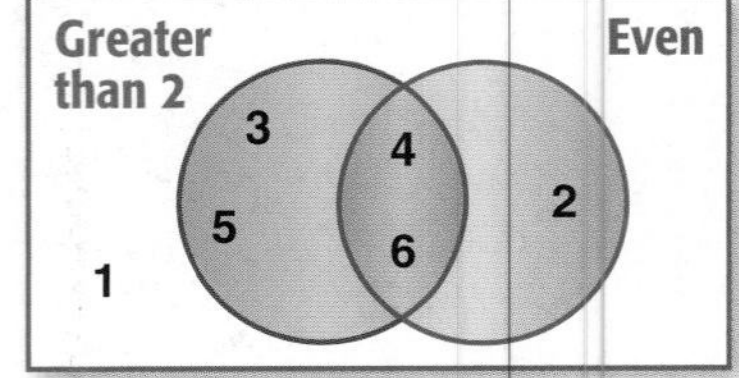


$$P(\text{greater than 2 or even}) = \frac{5}{6}.$$

Since it is possible to roll a number that is greater than 2 *and* an even number, these events are not mutually exclusive. Consider the probabilities of each individual event.

$$P(\text{greater than 2}) = \frac{4}{6} \qquad P(\text{even}) = \frac{3}{6}$$

If these probabilities were added, the probability of two outcomes, 4 and 6, would be counted twice—once for being numbers greater than 2 and once for being even numbers. You must subtract the probability of these common outcomes.

$$P(\text{greater than 2 or even}) = P(\text{greater than 2}) + P(\text{even}) - P(\text{greater than 2 and even})$$

$$= \frac{4}{6} + \frac{3}{6} - \frac{2}{6} \text{ or } \frac{5}{6}$$

This leads to the second of the Addition Rules for Probability.

### KeyConcept Probability of Events That Are Not Mutually Exclusive

**Words** If two events $A$ and $B$ are not mutually exclusive, then the probability that $A$ or $B$ occurs is the sum of their individual probabilities minus the probability that both $A$ and $B$ occur.

**Symbols** If two events $A$ and $B$ are not mutually exclusive, then
$P(A \text{ or } B) = P(A) + P(B) - P(A \text{ and } B)$.

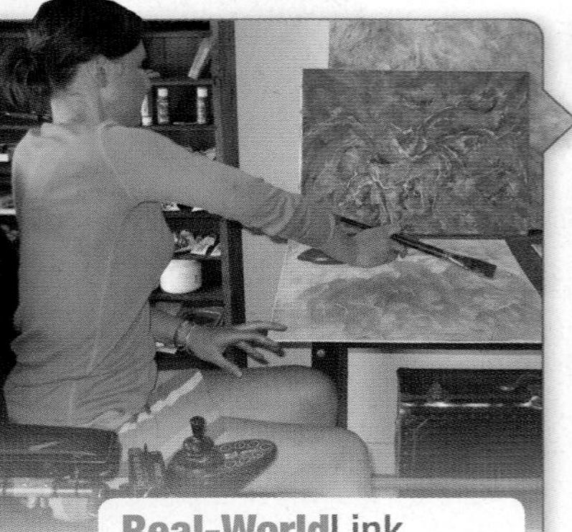

**Real-WorldLink**

Juried art shows are shows in which artists are called to submit pieces and a panel of judges decides which art will be shown. They originated in the early 1800s to exhibit the work of current artists and educate the public.

**Source:** Humanities Web

### Real-World Example 3 Events That Are Not Mutually Exclusive

PT

**ART The table shows the number and type of paintings Namiko has created. If she randomly selects a painting to submit to an art contest, what is the probability that she selects a portrait or an oil painting?**

| Namiko's Paintings | | | |
|---|---|---|---|
| Media | Still Life | Portrait | Landscape |
| watercolor | 4 | 5 | 3 |
| oil | 1 | 3 | 2 |
| acrylic | 3 | 2 | 1 |
| pastel | 1 | 0 | 5 |

Since some of Namiko's paintings are both portraits and oil paintings, these events are not mutually exclusive. Use the rule for two events that are not mutually exclusive. The total number of paintings from which to choose is 30.

$$P(\text{oil or portrait}) = P(\text{oil}) + P(\text{portrait}) - P(\text{oil and portrait})$$

$$= \frac{1+3+2}{30} + \frac{5+3+2+0}{30} - \frac{3}{30} \quad \text{Substitution}$$

$$= \frac{6}{30} + \frac{10}{30} - \frac{3}{30} \text{ or } \frac{13}{30} \quad \text{Simplify.}$$

The probability that Namiko selects a portrait or an oil painting is $\frac{13}{30}$ or about 43%.

**GuidedPractice**

**3.** What is the probability of drawing a king or a diamond from a standard deck of 52 cards?

Realistic Reflections/Getty Images

## 2 Probabilities of Complements

The **complement** of an event $A$ consists of all the outcomes in the sample space that are not included as outcomes of event $A$.

When a die is rolled, the probability of getting a 4 is $\frac{1}{6}$. What is the probability of *not* getting a 4? There are 5 possible outcomes for this event: 1, 2, 3, 5, or 6. So, $P(\text{not } 4) = \frac{5}{6}$. Notice that this probability is also $1 - \frac{1}{6}$ or $1 - P(4)$.

### KeyConcept Probability of the Complement of an Event

**Words** The probability that an event will not occur is equal to 1 minus the probability that the event will occur.

**Symbols** For an event $A$, $P(\text{not } A) = 1 - P(A)$.

### Example 4 Complementary Events

**RAFFLE Francisca bought 20 raffle tickets, hoping to win the $100 gift card to her favorite clothing store. If a total of 300 raffle tickets were sold, what is the probability that Francisca will not win the gift card?**

Let event $A$ represent selecting one of Francisca's tickets. Then find the probability of the complement of $A$.

$P(\text{not } A) = 1 - P(A)$ Probability of a complement

$= 1 - \frac{20}{300}$ Substitution

$= \frac{280}{300}$ or $\frac{14}{15}$ Subtract and simplify.

The probability that one of Francisca's tickets *will not* be selected is $\frac{14}{15}$ or about 93%.

**ReadingMath**

**Complement** The complement of event $A$ can also be noted as $A^C$.

**GuidedPractice**

**4.** If the chance of rain is 70%, what is the probability that it will not rain?

### ConceptSummary Probability Rules

| Types of Events | Words | Probability Rule |
|---|---|---|
| Independent Events | The outcome of a first event *does not affect* the outcome of the second event. | If two events $A$ and $B$ are independent, then $P(A \text{ and } B) = P(A) \cdot P(B)$. |
| Dependent Events | The outcome of a first event *does affect* the outcome of the other event. | If two events $A$ and $B$ are dependent, then $P(A \text{ and } B) = P(A) \cdot P(B \mid A)$. |
| Conditional | Additional information is known about the probability of an event. | The conditional probability of $A$ given $B$ is $P(A \mid B) = \frac{P(A \text{ and } B)}{P(B)}$. |
| Mutually Exclusive Events | Events *do not share* common outcomes. | If two events $A$ or $B$ are mutually exclusive, then $P(A \text{ or } B) = P(A) + P(B)$. |
| Not Mutually Exclusive Events | Events *do share* common outcomes. | If two events $A$ and $B$ are not mutually exclusive, then $P(A \text{ or } B) = P(A) + P(B) - P(A \text{ and } B)$. |
| Complementary Events | The outcomes of one event consist of all the outcomes in the sample space that are not outcomes of the other event. | For an event $A$, $P(\text{not } A) = 1 - P(A)$. |

**Real-WorldLink**

About 81% of American motorists and their right-front passengers use a seat belt.

**Source:** National Highway Traffic Safety Administration

## Real-World Example 5 Identify and Use Probability Rules

**SEAT BELTS Refer to the information at the left. Suppose two people are chosen at random from a group of 100 American motorists and passengers. If this group mirrors the population, what is the probability that at least one of them does not wear a seat belt?**

**Understand** You know that 81% of Americans *do use* a seat belt. The phrase *at least one* means *one or more*. So, you need to find the probability that either

- the first person chosen does not use a seat belt *or*
- the second person chosen does not use a seat belt *or*
- both people chosen do not use a seat belt.

**Plan** The complement of the event described above is the event that both people chosen *do use* a seat belt. Find the probability of this event, and then find the probability of its complement.

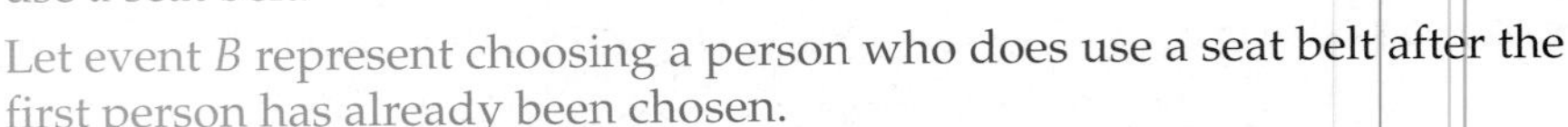

Let event $A$ represent choosing a person who does use a seat belt.

Let event $B$ represent choosing a person who does use a seat belt after the first person has already been chosen.

These are two dependent events, since the outcome of the first event affects the probability of the outcome of the second event.

**Solve**

$$P(A \text{ and } B) = P(A) \cdot P(B|A) \quad \text{Probability of dependent events}$$

$$= \frac{81}{100} \cdot \frac{80}{99} \quad P(A) = \frac{0.81(100)}{100} \text{ or } \frac{81}{100}$$

$$= \frac{6480}{9900} \text{ or } \frac{36}{55} \quad \text{Multiply.}$$

$$P[\text{not } (A \text{ and } B)] = 1 - P(A \text{ and } B) \quad \text{Probability of a complement}$$

$$= 1 - \frac{36}{55} \quad \text{Substitution}$$

$$= \frac{19}{55} \quad \text{Subtract.}$$

So, the probability that at least one of the passengers does not use a seat belt is $\frac{19}{55}$ or about 35%.

**Check** Use logical reasoning to check the reasonableness of your answer. The probability that one person chosen out of 100 *does not* wear his or her seat belt is (100 − 81)% or 19%. The probability that two people chosen out of 100 wear their seat belt should be greater than 19%. Since 35% > 19%, the answer is reasonable.

**StudyTip**

**Key Probability Words** When determining what type of probability you are dealing with in a situation, look for key words and correctly interpret their meaning.

*and* → independent or dependent events

*or* → mutually exclusive or not mutually exclusive

*not* → complementary events

*and then* → conditional

*at least n* → *n* or more

*at most n* → *n* or less

### GuidedPractice

5. **CELL PHONES** According to an online poll, 35% of American motorists routinely use their cell phones while driving. Three people are chosen at random from a group of 100 motorists. What is the probability that

   A. at least two of them use their cell phone while driving?

   B. no more than one use their cell phone while driving?

Digital Vision/Alamy

## Check Your Understanding

**Example 1** **Determine whether the events are *mutually exclusive* or *not mutually exclusive*. Explain your reasoning.**

1. drawing a card from a standard deck and getting a jack or a club

2. adopting a cat or a dog

**Example 2**

3. **JOBS** Adelaide is the employee of the month at her job. Her reward is to select at random from 4 gift cards, 6 coffee mugs, 7 DVDs, 10 CDs, and 3 gift baskets. What is the probability that an employee receives a gift card, coffee mug, or CD?

**Example 3**

4. **CLUBS** According to the table, what is the probability that a student in a club is a junior or on the debate team?

| Club | Soph. | Junior | Senior |
|---|---|---|---|
| Key | 12 | 14 | 8 |
| Debate | 2 | 6 | 3 |
| Math | 7 | 4 | 5 |
| French | 11 | 15 | 13 |

**Example 4** **Determine the probability of each event.**

5. If you have a 2 in 10 chance of bowling a spare, what is the probability of missing the spare?

6. If the chance of living in a particular dorm is 75%, what is the probability of living in another dorm?

**Example 5**

7. **PROM** In Armando's senior class of 100 students, 91% went to the senior prom. If two people are chosen at random from the entire class, what is the probability that at least one of them did not go to prom?

## Practice and Problem Solving

**Examples 1–3** **Determine whether the events are *mutually exclusive* or *not mutually exclusive*. Then find the probability. Round to the nearest tenth of a percent, if necessary.**

8. drawing a card from a standard deck and getting a jack or a six

9. rolling a pair of dice and getting doubles or a sum of 8

10. selecting a number at random from integers 1 to 20 and getting an even number or a number divisible by 3

11. tossing a coin and getting heads or tails

12. drawing an ace or a heart from a standard deck of 52 cards

13. rolling a pair of dice and getting a sum of either 6 or 10

14. **SPORTS** The table includes all of the programs offered at a sports complex and the number of participants aged 14–16. What is the probability that a player is 14 or plays basketball?

| Graceland Sports Complex | | | |
|---|---|---|---|
| Age | Soccer | Baseball | Basketball |
| 14 | 28 | 36 | 42 |
| 15 | 30 | 26 | 33 |
| 16 | 35 | 41 | 29 |

15. CCSS **MODELING** An exchange student is moving back to Italy, and her homeroom class wants to get her a going away present. The teacher takes a survey of the class of 32 students and finds that 10 people chose a card, 12 chose a T-shirt, 6 chose a video, and 4 chose a bracelet. If the teacher randomly selects the present, what is the probability that the exchange student will get a card or a bracelet?

**Example 4** **Determine the probability of each event.**

16. rolling a pair of dice and not getting a 3

17. drawing a card from a standard deck and not getting a diamond

18. flipping a coin and not landing on heads

19. spinning a spinner numbered 1–8 and not landing on 5

20. **RAFFLE** Namid bought 20 raffle tickets. If a total of 500 raffle tickets were sold, what is the probability that Namid will not win the raffle?

21. **JOBS** Of young workers aged 18 to 25, 71% are paid by the hour. If two people are randomly chosen out of a group of 100 young workers, what is the probability that exactly one is paid by the hour?

**Example 5** 22. **RECYCLING** Suppose 31% of Americans recycle. If two Americans are chosen randomly from a group of 50, what is the probability that at most one of them recycles?

**CARDS Suppose you pull a card from a standard 52-card deck. Find the probability of each event.**

23. The card is a 4.

24. The card is red.

25. The card is a face card.

26. The card is not a face card.

27. **MUSIC** A school carried out a survey of 265 students to see which types of music students would want played at a school dance. The results are shown in the Venn Diagram. Find each probability.

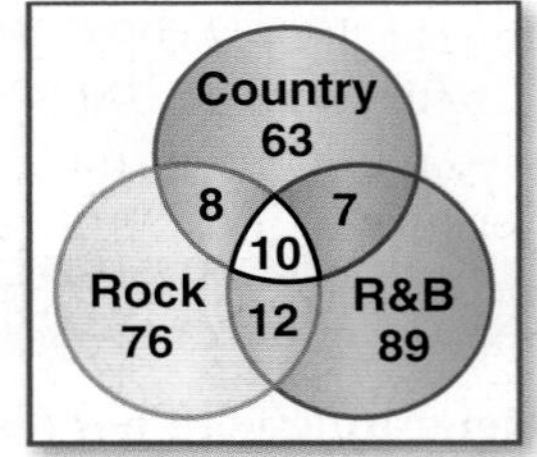


a. $P$(country or R&B)

b. $P$(rock and country or R&B and rock)

c. $P$(R&B but not rock)

d. $P$(all three)

## H.O.T. Problems Use Higher-Order Thinking Skills

28. **CCSS CRITIQUE** Tetsuya and Mason want to determine the probability that a red marble will be chosen out of a bag of 4 red, 7 blue, 5 green, and 2 purple marbles. Is either of them correct? Explain your reasoning.

| Tetsuya | Mason |
|---|---|
| $P(R) = \frac{4}{17}$ | $P(R) = 1 - \frac{4}{18}$ |

29. **CHALLENGE** You roll 3 dice. What is the probability that the outcome of at least two of the dice will be less than or equal to 4? Explain your reasoning.

**REASONING Determine whether the following are mutually exclusive. Explain.**

30. choosing a quadrilateral that is a square and a quadrilateral that is a rectangle

31. choosing a triangle that is equilateral and a triangle that is equiangular

32. choosing a complex number and choosing a natural number

33. **OPEN ENDED** Describe a pair of events that are mutually exclusive and a pair of events that are not mutually exclusive.

34. **WRITING IN MATH** Explain why the sum of the probabilities of two mutually exclusive events is not always 1.

## Standardized Test Practice

**35. PROBABILITY** Customers at a new salon can win prizes during opening day. The table shows the type and number of prizes. What is the probability that the first customer wins a manicure or a massage?

| Prize | Number |
|---|---|
| manicure | 10 |
| pedicure | 6 |
| massage | 3 |
| facial | 1 |

**A** 0.075 **C** 0.5

**B** 0.35 **D** 0.65

**36. SHORT RESPONSE** A cube numbered 1 through 6 is shown.

If the cube is rolled once, what is the probability that a number less than 3 or an odd number shows on the top face of the cube?

**37. ALGEBRA** What will happen to the slope of line $p$ if it is shifted so that the $y$-intercept stays the same and the $x$-intercept approaches the origin?

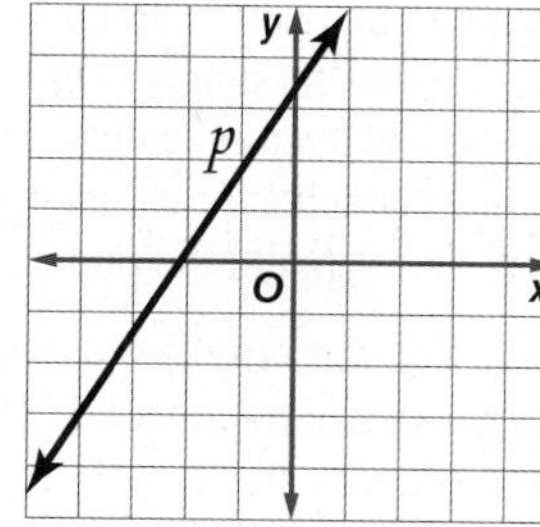


**F** The slope will become negative.

**G** The slope will become zero.

**H** The slope will decrease.

**J** The slope will increase.

**38. SAT/ACT** The probability of choosing a peppermint from a certain bag of candy is 0.25, and the probability of choosing a chocolate is 0.3. The bag contains 60 pieces of candy, and the only types of candy in the bag are peppermint, chocolate, and butterscotch. How many butterscotch candies are in the bag?

**A** 25 **D** 33

**B** 27 **E** 45

**C** 30

## Spiral Review

**Determine whether the events are *independent* or *dependent*. Then find the probability.** (Lesson 13-5)

**39.** A king is drawn, without replacement, from a standard deck of 52 cards. Then, a second king is drawn.

**40.** You roll a die and get a 2. You roll another die and get a 3.

**41. SPORTS** A survey at a high school found that 15% of the athletes at the school play only volleyball, 20% play only soccer, 30% play only basketball, and 35% play only football. Design a simulation that can be used to estimate the probability that an athlete will play each of these sports. (Lesson 13-4)

**Copy the figure and point $P$. Then use a ruler to draw the image of the figure under a dilation with center $P$ and the scale factor $r$ indicated.** (Lesson 10-8)

**42.** $r = \frac{1}{2}$

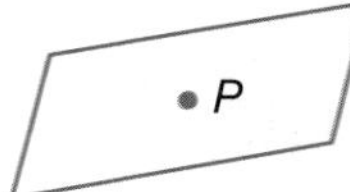


**43.** $r = 3$

**44.** $r = \frac{1}{5}$

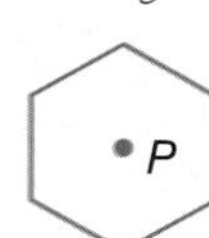

EXTEND 13-6

# Geometry Lab
# Graph Theory

Mathematical structures can be used to model relationships in a set. The study of these graphs is called *graph theory*. These **vertex-edge graphs** are not like graphs that can be seen on a coordinate plane. Each graph, also called a **network**, is a collection of vertices, called **nodes**, and segments, called **edges**, that connect the nodes.

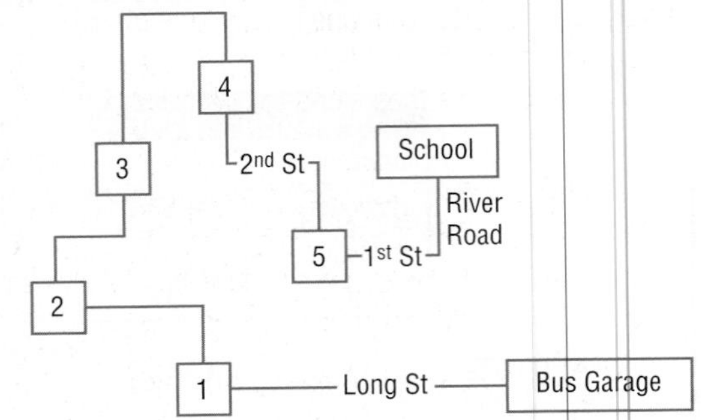


The bus route in the figure is an example of a network. The school, each stop, and the garage are nodes in the network. The connecting streets, such as Long Street, are edges.

This is an example of a **traceable network** because all of the nodes are connected, and each edge is used once in the network.

## Activity 1

**The graph represents the streets on Ava's newspaper route. To complete her route as quickly as possible, how can Ava ride her bike down each street only once?**

**Step 1** Copy the graph onto your paper.

**Step 2** Beginning at Ava's home, trace over her route without lifting your pencil. Remember to trace each edge only once.

**Step 3** Describe Ava's route.

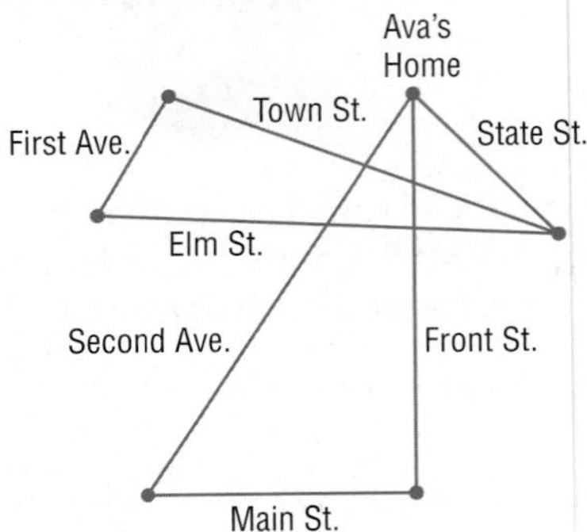


## Analyze

1. Is there more than one traceable route that begins at Ava's house? If so, how many?

2. If it does not matter where Ava starts, how many traceable routes are possible?

**Is each graph traceable? Write *yes* or *no*. Explain your reasoning.**

3. 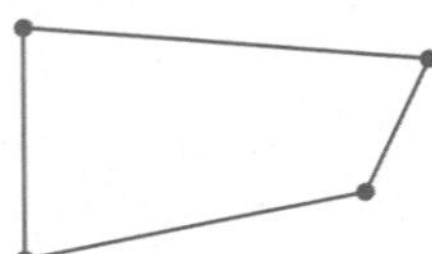

4. 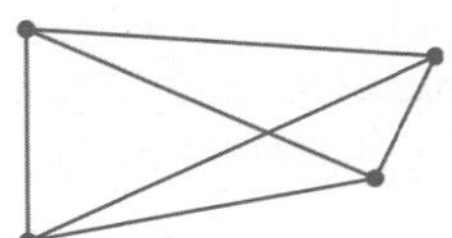

5. 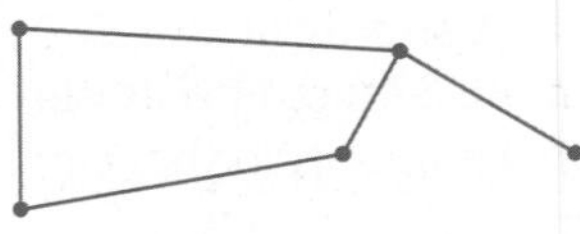

6. The campus for Centerburgh High School has five buildings built around the edge of a circular courtyard. There is a sidewalk between each pair of buildings.
   a. Draw a graph of the campus. Is the graph traceable?
   b. Suppose there are no sidewalks between pairs of adjacent buildings. Is it possible to reach all five buildings without walking down any sidewalk more than once?

7. **REASONING** Write a rule for determining whether a graph is traceable.

In a network, routes from one vertex to another are also called *paths*. **Weighted vertex-edge graphs** are graphs in which a value, or **weight**, is assigned to each edge. The **weight of a path** is the sum of the weights of the edges along the path. The **efficient route** is the path with the minimum weight.

## Activity 2

**The edges of the network have different weights. Find the efficient route from $A$ to $B$.**

**Step 1** Find all of the possible paths from $A$ to $B$. Label each path with the letters of the nodes along the path.

**Step 2** Trace each path and add the weights of each edge. The path with the least weight is the efficient route: $A$-$U$-$X$-$Y$-$Z$-$B$. The weight is 54.

Pay attention to the weights when determining the efficient route. It may not be the path with the fewest edges.

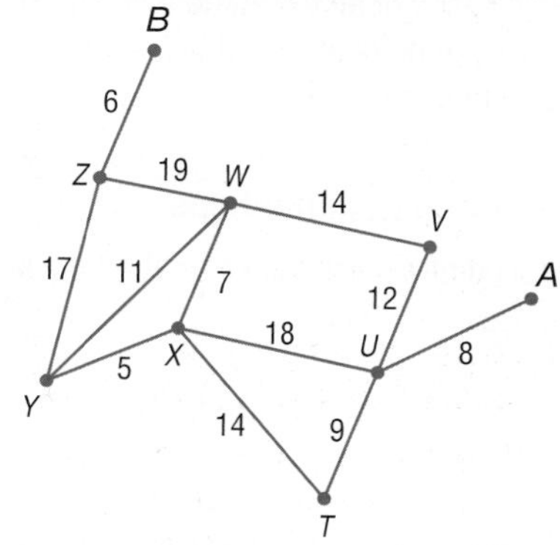


## Model and Analyze

8. What is the longest path from $A$ to $B$ that does not cover any edges more than once?

**Determine the efficient route from $A$ to $B$ for each network.**

9.

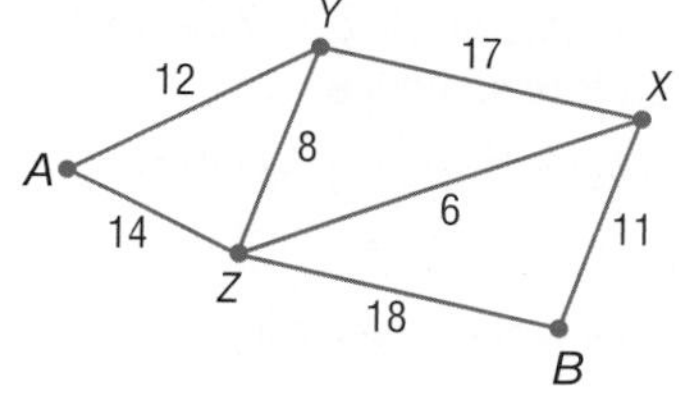


10.

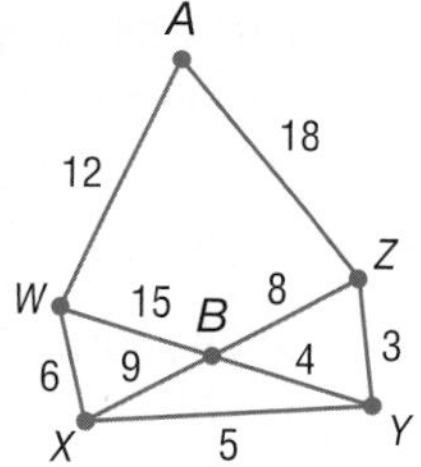


11.

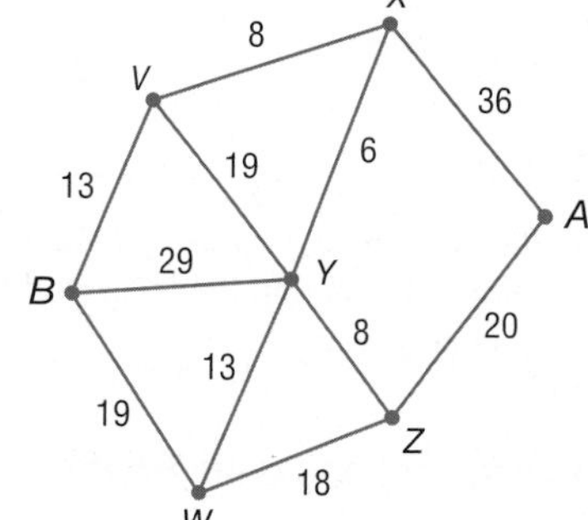


12. **OPEN ENDED** Create a network with 8 nodes and an efficient route with a value of 25.

13. **WRITING IN MATH** Explain your method for determining the efficient route of a network.

14. **TRAVEL** Use the graph at the right to find each efficient route.

    a. from Phoenix to New York

    b. from Seattle to Atlanta

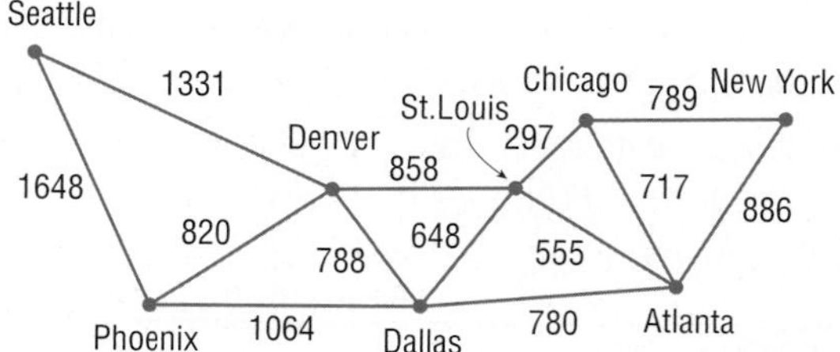


15. *Six Degrees of Separation* is a well-known example of graph theory. In this case, each person is a node and people are linked by an edge when they know each other.

    a. Make a graph of the situation. Directly connect yourself to three other people that you know personally. This represents the first degree of separation.

    b. Expand the graph to show the first three degrees of separation. Name a person who is within 3 degrees of you, and list the path.

# CHAPTER 13 Study Guide and Review

## Study Guide

### KeyConcepts

**Representing Sample Spaces** (Lesson 13-1)

- The sample space of an experiment is the set of all possible outcomes. It can be determined by using an organized list, a table, or a tree diagram.

**Permutations and Combinations** (Lesson 13-2)

- A permutation of $n$ objects taken $r$ at a time is given by $_nP_r = \frac{n!}{(n-r)!}$.
- A combination of $n$ objects taken $r$ at a time is given by $_nC_r = \frac{n!}{(n-r)!r!}$.
- Permutations should be used when order is important, and combinations should be used when order is not important.

**Geometric Probability** (Lesson 13-3)

- If a region $A$ contains a region $B$ and a point $E$ in region $A$ is chosen at random, then the probability that point $E$ is in region $B$ is $\frac{\text{area of region } B}{\text{area of region } A}$.

**Simulations** (Lesson 13-4)

- A simulation uses a probability model to recreate a situation again and again so that the likelihood of various outcomes can be estimated.

**Probabilities of Compound Events** (Lessons 13-5 and 13-6)

- If event $A$ does not affect the outcome of event $B$, then the events are independent and $P(A \text{ and } B) = P(A) \cdot P(B)$.
- If two events $A$ and $B$ are dependent, then $P(A \text{ and } B) = P(A) \cdot P(B|A)$.
- If two events $A$ and $B$ cannot happen at the same time, they are mutually exclusive and $P(A \text{ or } B) = P(A) + P(B)$.
- If two events $A$ and $B$ are not mutually exclusive, then $P(A \text{ or } B) = P(A) + P(B) - P(A \text{ and } B)$.

### FOLDABLES StudyOrganizer

Be sure the Key Concepts are noted in your Foldable.

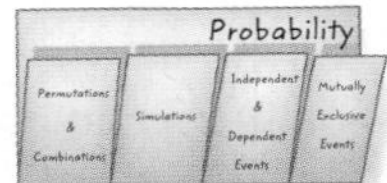


### KeyVocabulary

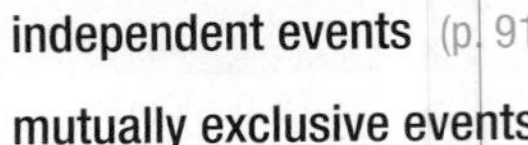

circular permutation (p. 893)
combination (p. 894)
complement (p. 927)
compound events (p. 915)
conditional probability (p. 917)
dependent events (p. 915)
expected value (p. 909)
factorial (p. 890)
Fundamental Counting Principle (p. 885)
geometric probability (p. 899)
independent events (p. 915)
mutually exclusive events (p. 924)
permutation (p. 890)
probability model (p. 907)
probability tree (p. 917)
random variable (p. 909)
sample space (p. 883)
simulation (p. 907)
tree diagram (p. 883)

### VocabularyCheck

**State whether each sentence is *true* or *false*. If *false*, replace the underlined term to make a true sentence.**

1. A tree diagram uses line segments to display possible outcomes.
2. A permutation is an arrangement of objects in which order is NOT important.
3. Determining the arrangement of people around a circular table would require circular permutation.
4. Tossing a coin and then tossing another coin is an example of dependent events.
5. Geometric probability involves a geometric measure such as length or area.
6. $6! = 6 \cdot 5 \cdot 4 \cdot 3 \cdot 2 \cdot 1$, is an example of a factorial.
7. The set of all possible outcomes is the sample space.
8. Combining a coin toss and a roll of a die makes a simple event.
9. Grant flipped a coin 200 times to create a probability tree of the experiment.
10. Drawing two socks out of a drawer without replacing them are examples of mutually exclusive events.

# Lesson-by-Lesson Review

## 13-1 Representing Sample Spaces

11. **POPCORN** A movie theater sells small (S), medium (M), and large (L) size popcorn with the choice of no butter (NB), butter (B), and extra butter (EB). Represent the sample space for popcorn orders by making an organized list, a table, and a tree diagram.

12. **SHOES** A pair of men's shoes comes in whole sizes 5 through 13 in navy, brown, or black. How many different pairs could be selected?

**Example 1**

**Three coins are tossed. Represent the sample space for this experiment by making an organized list.**

Pair each possible outcome from the first toss with the possible outcomes from the second toss and third toss.

HHH, HHT, HTH, HTT, THH, THT, TTH, TTT

## 13-2 Probability with Permutations and Combinations

13. **DINING** Three boys and three girls go out to eat together. The restaurant only has round tables. Fred does not want any girl next to him and Gena does not want any boy next to her. How many arrangements are possible?

14. **DANCE** The dance committee consisted of 10 students. The committee will select three officers at random. What is the probability that Alice, David, and Carlene are selected?

15. **COMPETITION** From 32 students, 4 are to be randomly chosen for an academic challenge team. In how many ways can this be done?

**Example 2**

**For a party, Lucita needs to seat four people at a round table. How many combinations are possible?**

Since there is no fixed reference point, this is a circular permutation.

$P_n = (n - 1)!$ Formula for circular permutation

$P_4 = (4 - 1)!$ $n = 4$

$= 3!$ or 6 Simplify.

So, there are 6 ways for Casey to seat four people at a round table.

## 13-3 Geometric Probability

16. **GAMES** Measurements for a beanbag game are shown. What is the probability of each event?

 a. $P$(hole)

 b. $P$(no hole)

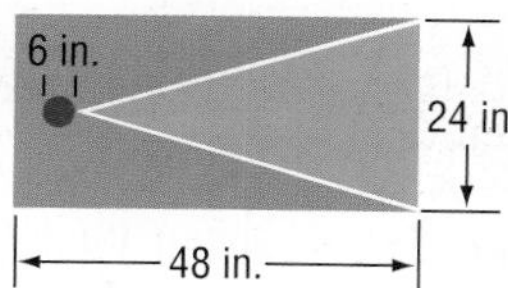


17. **POOL** Morgan, Phil, Callie, and Tyreese are sitting on the side of a pool in that order. Morgan is 2 feet from Phil. Phil is 4 feet from Callie. Callie is 3 feet from Tyreese. Oscar joins them.

 a. Find the probability that Oscar sits between Morgan and Phil.

 b. Find the probability that Oscar sits between Phil and Tyreese.

**Example 3**

**A carnival game is shown.**

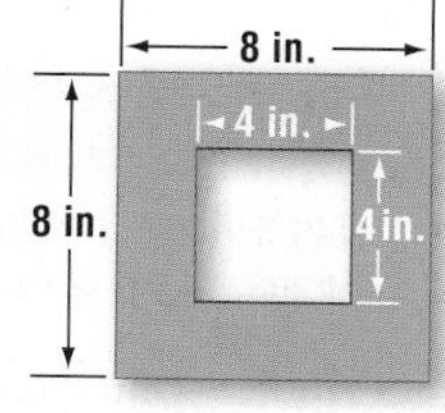


**a. If Khianna threw 10 beanbags at the board, what is the probability that the beanbag went in the hole?**

Area of hole $= 4 \cdot 4 = 16$

Area of board $= (8 \cdot 8) - 16 = 64 - 16$ or 48

$P(\text{hole}) = \frac{16}{64}$ or about 25%

**b. What is the probability that the beanbag did not go in the hole?**

$P(\text{no hole}) = \frac{48}{64}$ or about 75%

# CHAPTER 13 Study Guide and Review *Continued*

## 13-4 Simulations

For each of the following, describe how you would use a geometric probability model to design a simulation.

18. **POLO** Max scores 35% of the goals his team earns in each water polo match.

19. **BOOKS** According to a survey, people buy 30% of their books in October, November, and December, 22% during January, February, and March, 23% during April, May, and June, and 25% during July, August, and September.

20. **OIL** The United States consumes 17.3 million barrels of oil a day. 63% is used for transportation, 4.9% is used to generate electricity, 7.8% is used for heating and cooking, and 24.3% is used for industrial processes.

### Example 4

**Darius made 75% of his field goal kicks last season. Design a simulation that can be used to estimate the probability that he will make his next field goal kick this season.**

Use a spinner that is divided into 2 sectors. Make one sector red containing 75% of the spinner's area and the other blue containing 25% of the spinner's area.

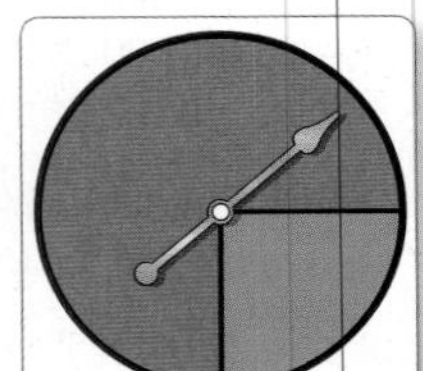

Spin the spinner 50 times. Each spin represents kicking a field goal. A successful trial will be a made field goal, and a failed trial will be a missed field goal.

## 13-5 Probabilities of Independent and Dependent Events

21. **MARBLES** A box contains 3 white marbles and 4 black marbles. What is the probability of drawing 2 black marbles and 1 white marble in a row without replacing any marbles?

22. **CARDS** Two cards are randomly chosen from a standard deck of cards with replacement. What is the probability of successfully drawing, in order, a three and then a queen?

23. **PIZZA** A nationwide survey found that 72% of people in the United States like pizza. If 3 people are randomly selected, what is the probability that all three like pizza?

### Example 5

**A bag contains 3 red, 2 white, and 6 blue marbles. What is the probability of drawing, in order, 2 red and 1 blue marble without replacement?**

Since the marbles are not being replaced, the events are dependent events.

$$P(\text{red, red, blue}) = P(\text{red}) \cdot P(\text{red}) \cdot P(\text{blue})$$

$$= \frac{3}{11} \cdot \frac{2}{10} \cdot \frac{6}{9}$$

$$= \frac{2}{55} \text{ or about } 3.6\%$$

## 13-6 Probabilities of Mutually Exclusive Events

24. **ROLLING DICE** Two dice are rolled. What is the probability that the sum of the numbers is 7 or 11?

25. **CARDS** A card is drawn from a deck of cards. Find the probability of drawing a 10 or a diamond.

26. **RAFFLE** A bag contains 40 raffle tickets numbered 1 through 40.
    a. What is the probability that a ticket chosen is an even number or less than 5?
    b. What is the probability that a ticket chosen is greater than 30 or less than 10?

### Example 6

**Two dice are rolled. What is the probability that the sum is 5 or doubles are rolled?**

These are mutually exclusive events because the sum of doubles can never equal 5.

$$P(\text{sum is 5 or doubles}) = P(\text{sum is 5}) + P(\text{doubles})$$

$$= \frac{4}{36} + \frac{6}{36}$$

$$= \frac{5}{18} \text{ or about } 27.8\%$$

CHAPTER 13

# Practice Test

**Point $X$ is chosen at random on $\overline{AE}$. Find the probability of each event.**

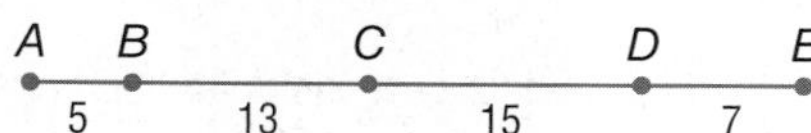


1. $P(X \text{ is on } \overline{AC})$

2. $P(X \text{ is on } \overline{CD})$

3. **BASEBALL** A baseball team fields 9 players. How many possible batting orders are there for the 9 players?

4. **TRAVEL** A traveling salesperson needs to visit four cities in her territory. How many distinct itineraries are there for visiting each city once?

**Represent the sample space for each experiment by making an organized list, a table, and a tree diagram.**

5. A box has 1 red ball, 1 green ball, and 1 blue ball. Two balls are drawn from the box one after the other, without replacement.

6. Shinsuke wants to adopt a pet and goes to his local humane society to find a dog or cat. While he is there, he decides to adopt two pets.

7. **ENGINEERING** An engineer is analyzing three factors that affect the quality of semiconductors: temperature, humidity, and material selection. There are 6 possible temperature settings, 4 possible humidity settings, and 6 choices of materials. How many combinations of settings are there?

8. **SPELLING** How many distinguishable ways are there to arrange the letters in the word "bubble"?

9. **PAINTBALL** Cordell is shooting a paintball gun at the target. What is the probability that he will shoot the shaded region?

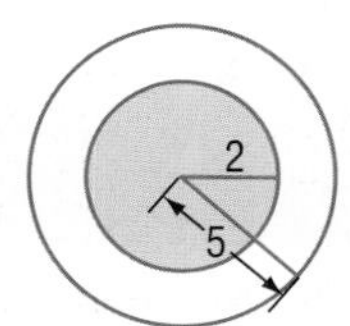


10. **SHORT RESPONSE** What is the probability that a phone number using the numbers 7, 7, 7, 2, 2, 2, and 6 will be 622-2777?

11. **TICKETS** Fifteen people entered the drawing at the right. What is the probability that Jodi, Dan, and Pilar all won the tickets?

Movie Ticket
Giveaway!
Enter for a chance
to win 3 tickets!


**Determine whether the events are *independent* or *dependent*. Then find the probability.**

12. A deck of cards has 5 yellow, 5 pink, and 5 orange cards. Two cards are chosen from the deck with replacement. Find $P$(the first card is pink and the second card is pink).

13. There are 6 green, 2 red, 2 brown, 4 navy, and 2 purple marbles in a hat. Sadie picks 2 marbles from the hat without replacement. What is the probability that the first marble is brown and the second marble is not purple?

**Use the spinner to find each probability. If the spinner lands on a line, it is spun again.**

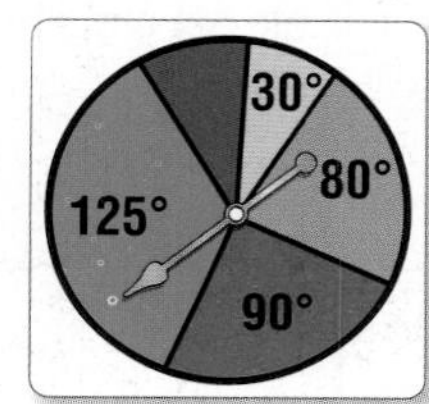


14. $P$(pointer landing on purple)

15. $P$(pointer landing on red)

16. $P$(pointer not landing on yellow)

17. **FOOTBALL** According to a football team's offensive success rate, the team punts 40% of the time, kicks a field goal 30% of the time, loses possession 5% of the time, and scores a touchdown 25% of the time. Design a simulation using a random number generator. Report the results using appropriate numerical and graphical summaries.

**Determine whether the events are *mutually exclusive* or *not mutually exclusive*. Explain your reasoning.**

18. a person owning a car and a truck

19. rolling a pair of dice and getting a sum of 7 and 6 on the face of one die

20. a playing card being both a spade and a club

21. **GRADES** This quarter, Todd earned As in his classes 45% of the time. Design and conduct a simulation using a geometric probability model. Then report the results using appropriate numerical and graphical summaries.

# CHAPTER 13 Preparing for Standardized Tests

## Organize Data

Sometimes you may be given a set of data that you need to analyze in order to solve items on a standardized test. Use this section to practice organizing data and to help you solve problems.

### Strategies for Organizing Data

**Step 1**

When you are given a problem statement containing data, consider:

- **making a list** of the data.
- **using a table** to organize the data.
- **using a data display** (such as a *bar graph, Venn diagram, circle graph, line graph, box-and-whisker plot,* etc.) to organize the data.

**Step 2**

Organize the data.

- Create your table, list, or data display.
- If possible, fill in any missing values that can be found by intermediate computations.

**Step 3**

Analyze the data to solve the problem.

- Reread the problem statement to determine what you are being asked to solve.
- Use the properties of geometry and algebra to work with the organized data and solve the problem.
- If time permits, go back and check your answer.

### Standardized Test Example

**Read the problem. Identify what you need to know. Then use the information in the problem to solve.**

Of the students who speak a foreign language at Marie's school, 18 speak Spanish, 14 speak French, and 16 speak German. There are 8 students who only speak Spanish, 7 who speak only German, 3 who speak Spanish and French, 2 who speak French and German, and 4 who speak all three languages. If a student is selected at random, what is the probability that he or she speaks Spanish or German, but not French?

**A** $\frac{7}{12}$ **B** $\frac{9}{16}$ **C** $\frac{2}{5}$ **D** $\frac{5}{18}$

Mike Kemp/Rubberball Productions/Getty Images

Read the problem carefully. The data is difficult to analyze as it is presented. Use a Venn diagram to organize the data and solve the problem.

**Step 1** Draw three circles, each representing a language.

**Step 2** Fill in the data given in the problem statement.

**Step 3** Fill in the missing values. For example, you know that 18 students speak Spanish and 14 students speak French.

$18 - 8 - 3 - 4 = 3$ (Spanish and German)
$14 - 3 - 4 - 2 = 5$ (only French)

Spanish
French
8
3
5
4
3
2
7
German

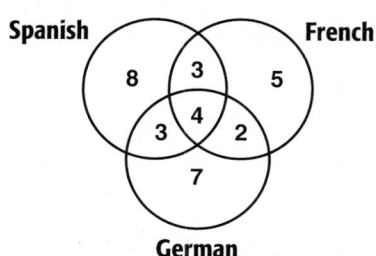

**Step 4** Solve the problem. You are asked to find the probability that a randomly selected student speaks Spanish or German, but not French. From the Venn diagram, you can see that there are 32 total students. Of these, $8 + 3 + 7$, or 18 students speak Spanish or German, but not French. So, the probability is $\frac{18}{32}$ or $\frac{9}{16}$. So, the correct answer is B.

## Exercises

**Read the problem. Identify what you need to know. Then organize the data to solve the problem.**

**1.** Alana has the letter tiles A, H, M, and T in a bag. If she selects a permutation of the tiles at random, what is the probability she will spell the word MATH?

**A** $\frac{1}{4}$

**B** $\frac{1}{12}$

**C** $\frac{3}{50}$

**D** $\frac{1}{24}$

**2.** The table below shows the number of freshmen, sophomores, juniors, and seniors involved in basketball, soccer, and volleyball. What is the probability that a randomly selected student is a junior or plays volleyball?

| Sport | Fr | So | Jr | Sr |
|---|---|---|---|---|
| Basketball | 7 | 6 | 5 | 6 |
| Soccer | 6 | 4 | 8 | 7 |
| Volleyball | 9 | 2 | 4 | 6 |

**F** $\frac{4}{21}$

**G** $\frac{5}{21}$

**H** $\frac{5}{17}$

**J** $\frac{17}{35}$

**3.** Find the probability that a point chosen at random lies in the shaded region.

**A** 0.22

**B** 0.25

**C** 0.28

**D** 0.32

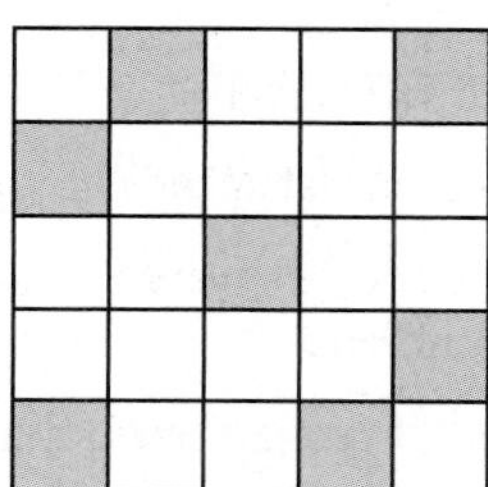

**4.** There are 10 sophomore, 8 junior, and 9 senior members in student council. Each member is assigned to help plan one school activity during the year. There are 4 sophomores working on the field day and 6 working on the pep rally. Of the juniors, 2 are working on the field day and 5 are working on the school dance. There are 2 seniors working on the pep rally. If each activity has a total of 9 students helping to plan it, what is the probability that a randomly selected student council member is a junior or is working on the field day?

**F** $\frac{1}{5}$

**G** $\frac{4}{18}$

**H** $\frac{5}{9}$

**J** $\frac{2}{3}$

# Standardized Test Practice

## Cumulative, Chapters 1 through 13

### Multiple Choice

**Read each question. Then fill in the correct answer on the answer document provided by your teacher or on a sheet of paper.**

1. How much paper is needed to make the drinking cup below? Round to the nearest tenth.

   **F** 73.4 $cm^2$

   **G** 70.7 $cm^2$

   **H** 67.9 $cm^2$

   **J** 58.8 $cm^2$

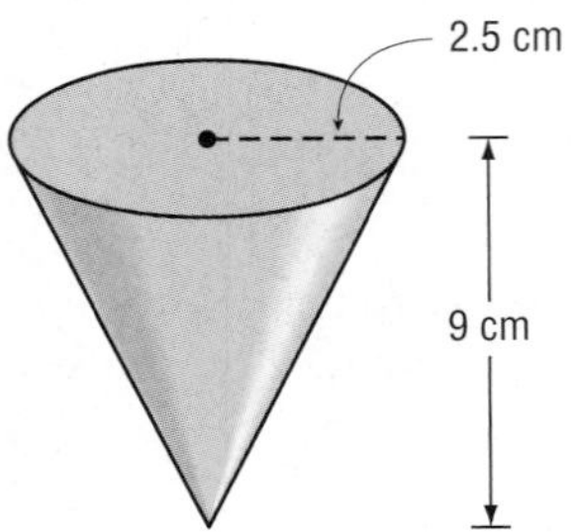


2. Which of the following properties of real numbers justifies the statement below?

   If $3x - 2 = 7x + 12$, then
   $3x - 2 + 2 = 7x + 12 + 2$.

   **A** Addition Property of Equality

   **B** Reflection Property of Equality

   **C** Subtraction Property of Equality

   **D** Symmetric Property of Equality

**Test-Taking Tip**

Question 3 What is the probability of rolling doubles with two number cubes? Multiply this by the number of trials.

3. What is the expected number of times Clarence will roll doubles with two number cubes in 90 trials? (Doubles occur when both number cubes show the same number in a trial.)

   **F** 6

   **G** 9

   **H** 10

   **J** 15

4. Which of the following correctly shows the relationship between the angle measures of triangle $RST$?

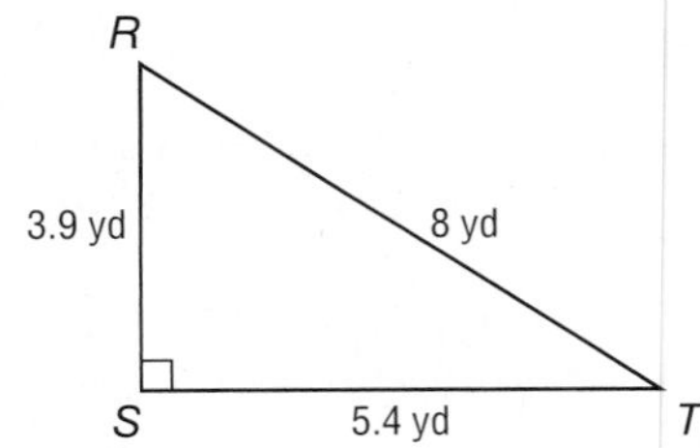


   **F** $m\angle S < m\angle R < m\angle T$

   **G** $m\angle T < m\angle S < m\angle R$

   **H** $m\angle R < m\angle S < m\angle T$

   **J** $m\angle T < m\angle R < m\angle S$

## Short Response/Gridded Response

**Record your answers on the answer sheet provided by your teacher or on a sheet of paper.**

5. **GRIDDED RESPONSE** What is $m\angle S$ in the figure below? Express your answer in degrees.

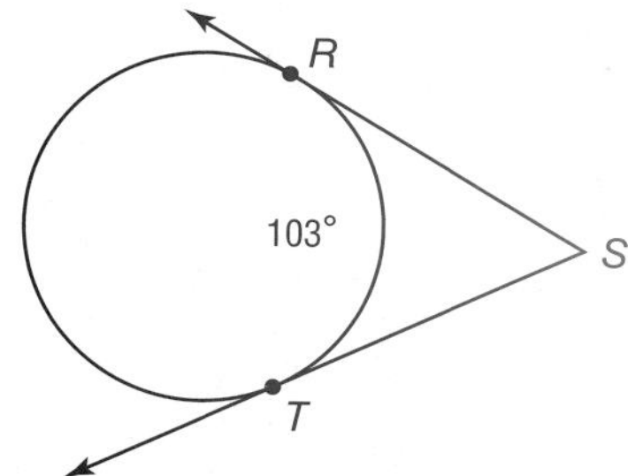


6. **GRIDDED RESPONSE** Segment $AD$ bisects $\angle CAB$ in the triangle below. What is the value of $x$?

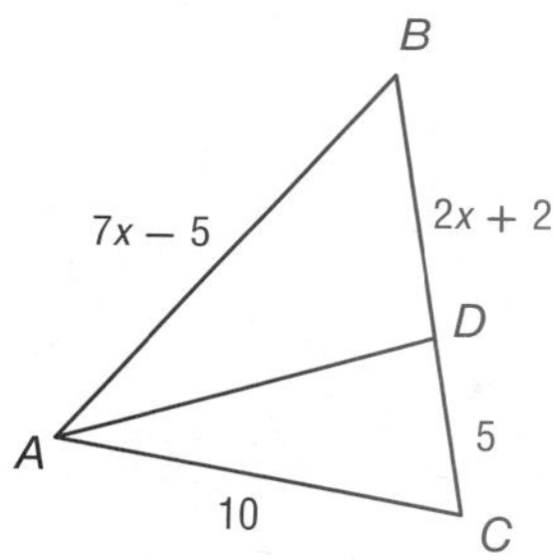


7. **GRIDDED RESPONSE** Armando leans an 18-foot ladder against the side of his house to clean out the gutters. The base of the ladder is 5 feet from the wall. How high up the side of the house does the ladder reach? Express your answer in feet, rounded to the nearest tenth.

8. Solve for $x$ in the triangle below.

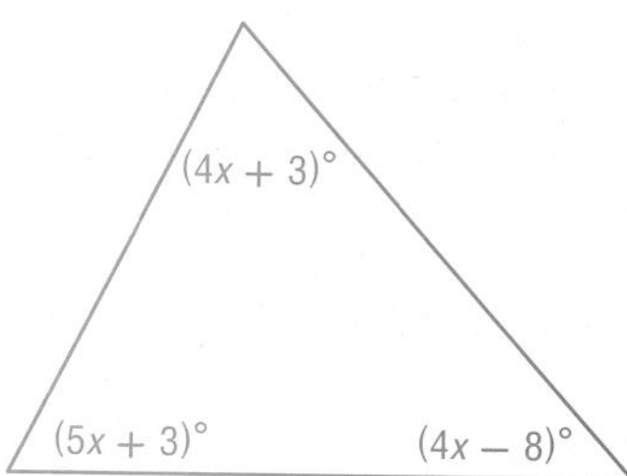


## Extended Response

**Record your answers on a sheet of paper. Show your work.**

9. A bag contains 3 red chips, 5 green chips, 2 yellow chips, 4 brown chips, and 6 purple chips. One chip is chosen at random, the color noted, and the chip returned to the bag.

   a. Suppose two trials of this experiment are conducted. Are the events independent or dependent? Explain.

   b. What is the probability that both chips are purple?

   c. What is the probability that the first chip is green and the second is brown?

**Need ExtraHelp?**

| If you missed Question... | 1 | 2 | 3 | 4 | 5 | 6 | 7 | 8 | 9 |
|---|---|---|---|---|---|---|---|---|---|
| Go to Lesson... | 13-3 | 5-2 | 12-4 | 7-5 | 11-6 | 7-1 | 10-2 | 6-1 | 12-5 |

# Student Handbook

This **Student Handbook** can help you answer these questions.

## What if I Forget a Vocabulary Word?

**Glossary/Glosario** **R1**

The **English-Spanish Glossary** provides a list of important or difficult words used throughout the textbook.

## What if I Need to Find Something Quickly?

**Index** **R24**

The **Index** alphabetically lists the subjects covered throughout the entire textbook and the pages on which each subject can be found.

## What if I Forget a Formula?

**Trigonometric Functions and Identities, Formulas and Symbols** **R47**

Inside the back cover of your math book are several lists of **Formulas, Identities, and Symbols** that are used in the book.

Rubberball/Getty Images

# Glossary/Glosario

**MultilingualeGlossary**

Go to connectED.mcgraw-hill.com for a glossary of terms in these additional languages:

| | | | | |
|---|---|---|---|---|
| Arabic | Chinese | Hmong | Spanish | Vietnamese |
| Bengali | English | Korean | Tagalog | |
| Brazilian Portugese | Haitian Creole | Russian | Urdu | |

## English

## Español

### A

**absolute value function** (p. 149) A function written as $f(x) = |x|$, in which $f(x) \geq 0$ for all values of $x$.

**función del valor absoluto** Una función que se escribe $f(x) = |x|$, donde $f(x) \geq 0$, para todos los valores de $x$.

**adjacent arcs** (p. 726) Arcs in a circle that have exactly one point in common.

**arcos adyacentes** Arcos en un círculo que tienen un solo punto en común.

**algebraic proof** (p. 284) A proof that is made up of a series of algebraic statements. The properties of equality provide justification for many statements in algebraic proofs.

**demostración algebraica** Demostración que se realiza con una serie de enunciados algebraicos. Las propiedades de la igualdad proveen justificación para muchas enunciados en demostraciones algebraicas.

**altitude** **1.** (p. 419) In a triangle, a segment from a vertex of the triangle to the line containing the opposite side and perpendicular to that side. **2.** (p. 814) In a prism or cylinder, a segment perpendicular to the bases with an endpoint in each plane. **3.** (p. 822) In a pyramid or cone, the segment that has the vertex as one endpoint and is perpendicular to the base.

**altura** **1.** En un triángulo, segmento trazado desde uno de los vértices del triángulo hasta el lado opuesto y que es perpendicular a dicho lado. **2.** En un prisma o un cilindro, segmento perpendicular a las bases con un extremo en cada plano. **3.** En una pirámide o un cono, segmento que tiene un extremo en el vértice y que es perpendicular a la base.

**ambiguous case of the Law of Sines** (p. 680) Given the measures of two sides and a nonincluded angle, there exist two possible triangles.

**caso ambiguo de la ley de los senos** Dadas las medidas de dos lados y de un ángulo no incluido, existen dos triángulos posibles.

**angle of depression** (p. 662) The angle between the line of sight and the horizontal when an observer looks downward.

**ángulo de depresión** Ángulo formado por la horizontal y la línea de visión de un observador que mira hacia abajo.

**angle of elevation** (p. 662) The angle between the line of sight and the horizontal when an observer looks upward.

**ángulo de elevación** Ángulo formado por la horizontal y la línea de visión de un observador que mira hacia arriba.

**arc** (p. 724) A part of a circle that is defined by two endpoints.

**arco** Parte d e un círculo definida por dos extremos.

**asymptote** (p. 533) A line that a graph approaches.

**asíntota** Recta a la que se aproxima una gráfica.

**auxiliary line** (p. 335) An extra line or segment drawn in a figure to help complete a proof.

**línea auxiliar** Recta o segmento de recta adicional que es traza en una figura para ayudar a completar una demostración.

**axiom** (p. 276) A statement that is accepted as true.

**axioma** Enunciado que se acepta como verdadero.

**axis** (p. 816) In a cylinder, the segment with endpoints that are the centers of the bases.

**eje** En un cilindro, el segmento cuyos extremos son el centro de las bases.

**axis of symmetry** (p. 93) The vertical line containing the vertex of a parabola.

**eje de simetría** La recta vertical que pasa por el vértice de una parábola.

## B

**base angle of an isosceles triangle** (p. 374) See *isosceles triangle* and isosceles trapezoid.

**ángulo de la base de un triángulo isósceles** Ver *triángulo isósceles* y trapecio isósceles.

**base edges** (p. 814) The intersection of the lateral faces and bases in a solid figure.

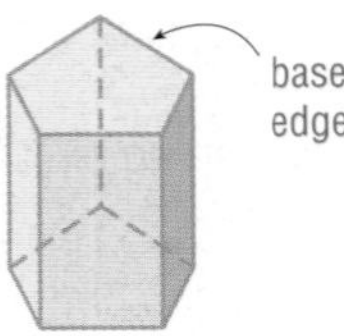


**aristas de las bases** Intersección de las base con las caras laterales en una figura sólida.

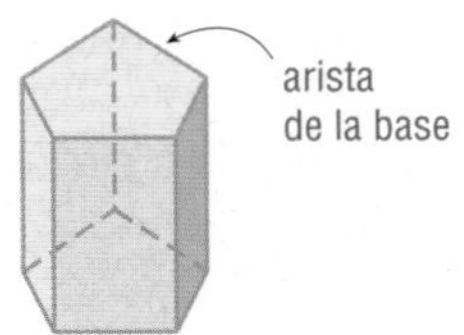


**binomial** (p. 7) The sum of two monomials.

**binomio** La suma de dos monomios.

## C

**center of circle** (p. 715) The central point where radii form a locus of points called a circle.

**centro de un círculo** Punto central desde el cual los radios forman un lugar geométrico de puntos llamado círculo.

**center of dilation** (p. 593) The center point from which dilations are performed.

**centro de la homotecia** Punto fijo en torno al cual se realizan las homotecias.

**central angle** (p. 724) An angle that intersects a circle in two points and has its vertex at the center of the circle.

**ángulo central** Ángulo que interseca un círculo en dos puntos y cuyo vértice está en el centro del círculo.

**centroid** (p. 417) The point of concurrency of the medians of a triangle.

**baricentro** Punto de intersección de las medianas de un triángulo.

**chord** **1.** (p. 715) For a given circle, a segment with endpoints that are on the circle. **2.** (p. 816) For a given sphere, a segment with endpoints that are on the sphere.

**cuerda** **1.** Para cualquier círculo, segmento cuyos extremos están en el círculo. **2.** Para cualquier esfera, segmento cuyos extremos están en la esfera.

**chord segments** (p. 768) Segments that form when two chords intersect inside a circle.

**segmentos de cuerda** Segmentos que se forman cuando dos cuerdas se intersecan dentro de un círculo.

Glossary/Glosario

**circle** (p. 715) The locus of all points in a plane equidistant from a given point called the *center* of the circle.

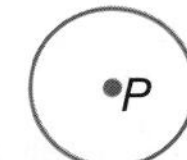


*P* is the center of the circle.

**circular permutation** (p. 893) A permutation of objects that are arranged in a circle or loop.

**circumcenter** (p. 407) The point of concurrency of the perpendicular bisectors of a triangle.

**circumference** (p. 717) The distance around a circle.

**circumscribed** (p. 718) A circle is circumscribed about a polygon if the circle contains all the vertices of the polygon.

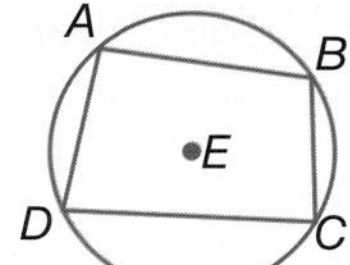


⊙*E* is circumscribed about quadrilateral *ABCD*.

**closed** (p. 251) A set is closed under an operation if for any numbers in the set, the result of the operation is also in the set.

**combination** (p. 894) An arrangement or listing in which order is not important.

**common tangent** (p. 750) A line or segment that is tangent to two circles in the same plane.

**complement** (p. 927) The complement of an event *A* consists of all the outcomes in the sample space that are not included as outcomes of event *A*.

**completing the square** (p. 124) To add a constant term to a binomial of the form $x^2 + bx$ so that the resulting trinomial is a perfect square.

**complex conjugates** (p. 181) Two complex numbers of the form $a + bi$ and $a - bi$.

**complex number** (p. 179) Any number that can be written in the form $a + bi$, where $a$ and $b$ are real numbers and $i$ is the imaginary unit.

**component form** (p. 684) A vector expressed as an ordered pair, ⟨change in $x$, change in $y$⟩.

**círculo** Lugar geométrico formado por todos los puntos en un plano, equidistantes de un punto dado llamado *centro* del círculo.

*P* es el centro del círculo.

**permutación circular** Permutación de objetos que se arreglan en un círculo o un bucle.

**circuncentro** Punto de intersección de las mediatrices de un triángulo.

**circunferencia** Distancia alrededor de un círculo.

**circunscrito** Un polígono está circunscrito a un círculo si todos sus vértices están contenidos en el círculo.

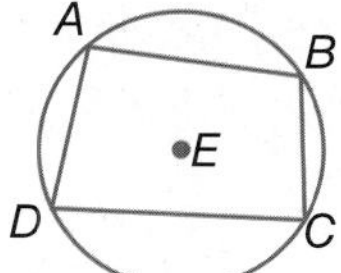


⊙*E* está circunscrito al cuadrilátero *ABCD*.

**cerrado** Un conjunto es cerrado bajo una operación si para cualquier número en el conjunto, el resultado de la operación es también en el conjunto.

**combinación** Arreglo o lista en que el orden no es importante.

**tangente común** Recta o segmento de recta tangente a dos círculos en el mismo plano.

**complemento** El complemento de un evento *A* consiste en todos los resultados en el espacio muestral que no se incluyen como resultados del evento *A*.

**completar el cuadrado** Adición de un término constante a un binomio de la forma $x^2 + bx$, para que el trinomio resultante sea un cuadrado perfecto.

**conjugados complejos** Dos números complejos de la forma $a + bi$ y $a - bi$.

**número complejo** Cualquier número que puede escribirse de la forma $a + bi$, donde $a$ y $b$ son números reales e $i$ es la unidad imaginaria.

**componente** Vector expresado en forma de par ordenado, ⟨cambio en $x$, cambio en $y$⟩.

composite solid (p. 820) A three-dimensional figure that is composed of simpler figures.

solido compuesto Figura tridimensional formada por figuras más simples.

compound event (p. 915) An event that consists of two or more simple events.

evento compuesto Evento que consiste de dos o más eventos simples.

compound interest (p. 238) Interest paid on the principal of an investment and any previously earned interest.

interés compuesto Interés obtenido tanto sobre la inversion inicial como sobre el interes conseguido.

concentric circles (p. 716) Coplanar circles with the same center.

círculos concéntricos Círculos coplanarios con el mismo centro.

concurrent lines (p. 407) Three or more lines that intersect at a common point.

rectas concurrentes Tres o más rectas que se intersecan en un punto común.

conditional probability (p. 917) The probability of an event under the condition that some preceding event has occurred.

probabilidad condicional La probabilidad de un acontecimiento bajo condición que ha ocurrido un cierto acontecimiento precedente.

congruent (p. 344) Having the same measure.

congruente Que tienen la misma medida.

congruent arcs (p. 725) Arcs in the same circle or in congruent circles that have the same measure.

arcos congruentes Arcos que tienen la misma medida y que pertenecen al mismo círculo o a círculos congruentes.

congruent polygons (p. 344) Polygons in which all matching parts are congruent.

polígonos congruentes Polígonos cuyas partes correspondientes son todas congruentes.

congruent solids (p. 864) Two solids with the same shape, size and scale factor of 1:1.

sólidos congruentes Dos sólidos con la misma forma, tamaño y factor de escala de 1:1.

conic section (p. 782) Any figure that can be obtained by slicing a cone.

sección cónica Cualquier figura obtenida mediante el corte de un cono doble.

conjugates (p. 247) Binomials of the form $a\sqrt{b} + c\sqrt{d}$ and $a\sqrt{b} - c\sqrt{d}$.

conjugados Binomios de la forma $a\sqrt{b} + c\sqrt{d}$ and $a\sqrt{b} - c\sqrt{d}$.

coordinate proofs (p. 383) Proofs that use figures in the coordinate plane and algebra to prove geometric concepts.

demostraciones en coordinadas Demostraciones que usan figuras en el plano de coordinados y álgebra para demostrar conceptos geométricos.

corner view (p. 807) The view from a corner of a three-dimensional figure, also called the *isometric view*.

vista de esquina Vista desde una de las esquinas de una figura tridimensional. También se conoce como *vista en perspectiva*.

corollary (p. 338) A statement that can be easily proved using a theorem is called a corollary of that theorem.

corolario Un enunciado que se puede demostrar fácilmente usando un teorema se conoce como corolario de dicho teorema.

corresponding parts (p. 344) Matching parts of congruent polygons.

partes correspondientes Partes que coinciden de polígonos congruentes.

cosecant (p. 496) The reciprocal of the sine of an angle in a right triangle.

cosecante Recíproco del seno de un ángulo en un triángulo rectangulo.

**cosine** (p. 650) For an acute angle of a right triangle, the ratio of the measure of the leg adjacent to the acute angle to the measure of the hypotenuse.

**cotangent** (p. 660) The ratio of the adjacent to the opposite side of a right triangle.

**cross products** (p. 4) In the proportion $\frac{a}{b} = \frac{c}{d}$, where $b \neq 0$ and $d \neq 0$, the cross products are $ad$ and $bc$. The proportion is true if and only if the cross products are equal.

**cross section** (p. 808) The intersection of a solid and a plane.

**coseno** Para cualquier ángulo agudo de un triángulo rectángulo, razón de la medida del cateto adyacente al ángulo agudo a la medida de la hipotenusa.

**cotangente** Razón de la medida del cateto adyacente a la medida de cateto opuesto de un triángulo rectángulo.

**productos cruzados** En la proporción $\frac{a}{b} = \frac{c}{d}$, donde $b \neq 0$ y $d \neq 0$, los productos cruzados son $ad$ y $bc$. La proporción es verdadera si y sólo si los productos cruzados son iguales.

**sección transversal** Intersección de un sólido con un plano.

## D

**decay factor** (p. 230) In exponential decay, the base of the exponential expression, $1 - r$.

**deductive argument** (p. 278) A proof formed by a group of algebraic steps used to solve a problem.

**degree of a monomial** (p. 7) The sum of the exponents of all its variables.

**degree of a polynomial** (p. 7) The greatest degree of any term in the polynomial.

**dependent events** (p. 915) Two or more events in which the outcome of one event affects the outcome of the other events.

**diagonal** (p. 475) In a polygon, a segment that connects nonconsecutive vertices of the polygon.

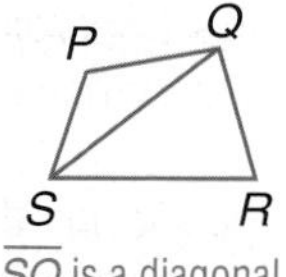

$\overline{SQ}$ is a diagonal.

**diameter** **1.** (p. 715) In a circle, a chord that passes through the center of the circle. **2.** (p. 816) In a sphere, a segment that contains the center of the sphere, and has endpoints that are on the sphere.

**difference of two squares** (p. 58) Two perfect squares separated by a subtraction sign.
$a^2 - b^2 = (a + b)(a - b)$ or
$a^2 - b^2 = (a - b)(a + b)$.

**factor de decaimiento** En decaimiento exponencial, la base de la expresión exponencial, $1 - r$.

**argumento deductivo** Demostración que consta de un conjunto de pasos algebraicos que se usan para resolver un problema.

**grado de un monomio** Suma de los exponentes de todas sus variables.

**grado de un polinomio** El grado mayor de cualquier término del polinomio.

**eventos dependientes** Dos o más eventos en que el resultado de un evento afecta el resultado de los otros eventos.

**diagonal** Recta que conecta vértices no consecutivos de un polígono.

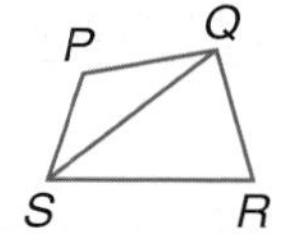

$\overline{SQ}$ es una diagonal.

**diámetro** **1.** En un círculo cuerda que pasa por el centro. **2.** En una estera segmento que incluye el centro de la esfera y cuyos extremos están ubicados en la esfera.

**diferencia de cuadrados** Dos cuadrados perfectos separados por el signo de sustracción.
$a^2 - b^2 = (a + b)(a - b)$ or
$a^2 - b^2 = (a - b)(a + b)$.

**dilation** **1.** (p. 116) A transformation that alters the size of a figure but not its shape. **2.** (pp. 593, 694) A transformation that enlarges or reduces the original figure proportionally. A dilation with center $C$ and positive scale factor $k$, $k \neq 1$, is a function that maps a point $P$ in a figure to its image such that

- if point $P$ and $C$ coincide, then the image and preimage are the same point, or
- if point $P$ is not the center of dilation, then $P'$ lies on $\overrightarrow{CP}$ and $CP' = k(CP)$.

If $k < 0$, $P'$ is the point on the ray opposite $\overrightarrow{CP}$ such that
$CP' = |k|(CP)$.

**homotecia** **1.** Transformacióñ que altera el tamaño de una figure, pero no su forma. **2.** Transformación que amplía o disminuye proporcionalmente el tamaño de una figura. Una homotecia con centro $C$ y factor de escala positivo $k$, $k \neq 1$, es una función que aplica un punto $P$ a su imagen, de modo que si el punto $P$ coincide con el punto $C$, entonces la imagen y la preimagen son el mismo punto, o si el punto $P$ no es el centro de la homotecia, entonces $P'$ yace sobre $\overrightarrow{CP}$ y $CP' = k(CP)$. Si $k < 0$, $P'$ es el punto sobre el rayo opuesto a $\overrightarrow{CP}$, tal que $CP' = |k|(CP)$.

**direction** (p. 150) The measure of the angle that a vector forms with the positive $x$-axis or any other horizontal line.

**dirección** Medida del ángulo que forma un vector con el eje $x$ positivo o con cualquier otra recta horizontal.

**directrix** (p. 783) The fixed line in a parabola that is equidistant from the locus of all points in a plane.

**directriz** Línea fija en una parábola que está equidistante del lugar geométrico de todos los puntos en un plano.

**discriminant** (pp. 192, 356) In the Quadratic Formula, the expression $b^2 - 4ac$.

**discriminante** En la fórmula cuadrática, la expresión $b^2 - 4ac$.

**double root** (p. 638) The roots of a quadratic function that are the same number.

**raíces dobles** Las raíces de una función cuadrática que son el mismo número.

## E

**edge** (p. 932) A line that connects two nodes in a network.

**arista** Recta que conecta dos nodos en una red.

**efficient route** (p. 933) The path in a network with the least weight.

**ruta eficiente** Ruta en una red con el menor peso.

**enlargement** (p. 593) An image that is larger that the original figure.

**ampliación** Imagen que es más grande que la figura original.

**equivalent vectors** (p. 683) Vectors that have the same magnitude and direction.

**vectores iguales** Vectores con la misma magnitud y dirección.

**euclidean geometry** (p. 857) A geometrical system in which a plane is a flat surface made up of points that extend infinitely in all directions.

**geometría euclidiana** Sistema en el cual un plano es una superficie plana formada por puntos que se extienden infinitamente en todas las direcciones.

**expected value** (p. 909) Also *mathematical expectation*, is the average value of a random variable that one expects after repeating an experiment or simulation an infinite number of times.

**valor esperado** También *expectativa matemática*, el valor promedio de una variable aleatoria que uno *espera* después de repetir un experimento o un simulacro un número infinito de veces.

**exponential decay** (p. 229) Exponential decay occurs when a quantity decreases exponentially over time.

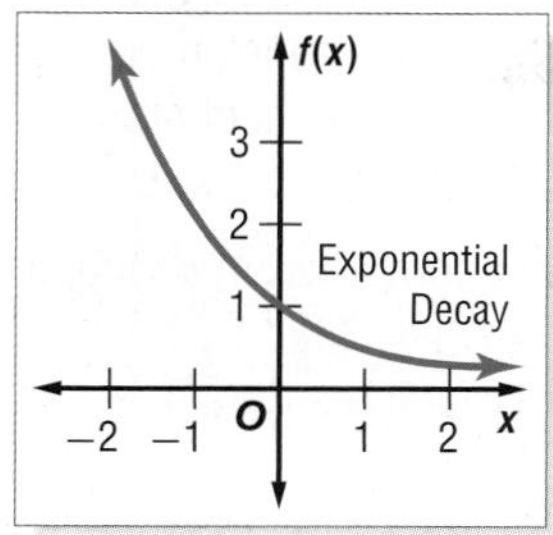


**exponential equation** (p. 237) An equation in which the variables occur as exponents.

**exponential function** (pp. 227, 543) A function of the form $y = ab^x$, where $a \neq 0$, $b > 0$, and $b \neq 1$.

**exponential growth** (p. 227) Exponential growth occurs when a quantity increases exponentially over time.

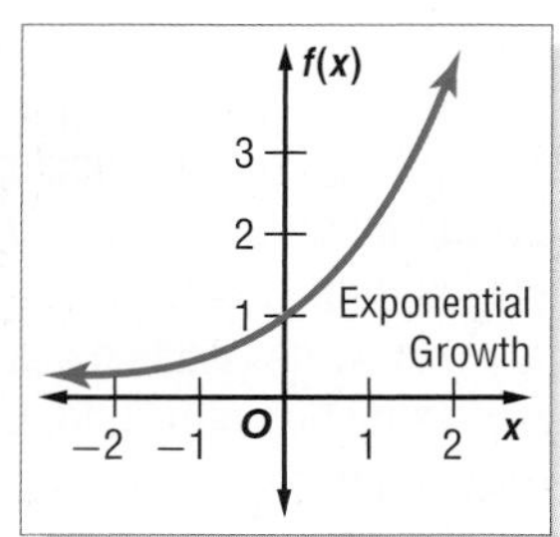


**exponential inequality** (p. 239) An inequality involving exponential functions.

**extended ratios** (pp. 3, 237, 543) Ratios that are used to compare three or more quantities.

**exterior angle** (p. 337) An angle formed by one side of a triangle and the extension of another side.

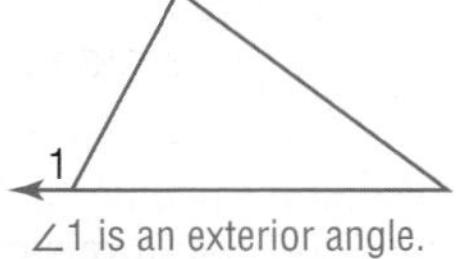

∠1 is an exterior angle.

**external secant segment** (p. 770) A secant segment that lies in the exterior of the circle.

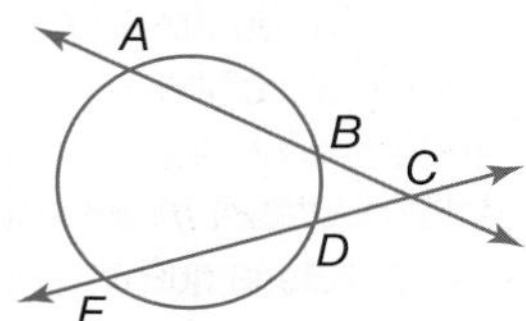

$\overline{BC}$ and $\overline{CD}$ are external secant segments.

**desintegración exponencial** Ocurre cuando una cantidad disminuye exponencialmente con el tiempo.

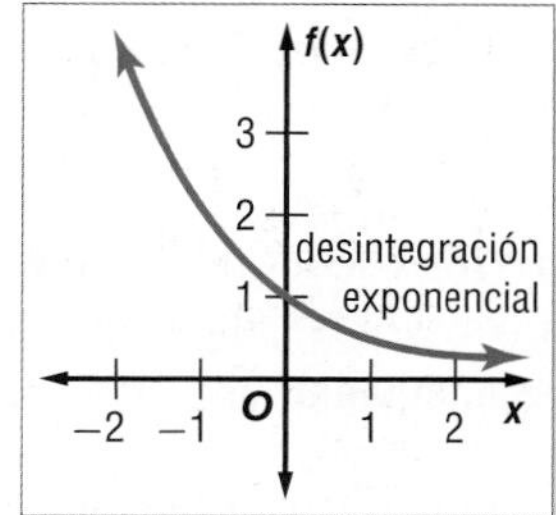


**ecuación exponencial** Ecuación en que las variables aparecen en los exponentes.

**función exponencial** Una función de la forma $y = ab^x$, donde $a \neq 0$, $b > 0$, y $b \neq 1$.

**crecimiento exponencial** El que ocurre cuando una cantidad aumenta exponencialmente con el tiempo.

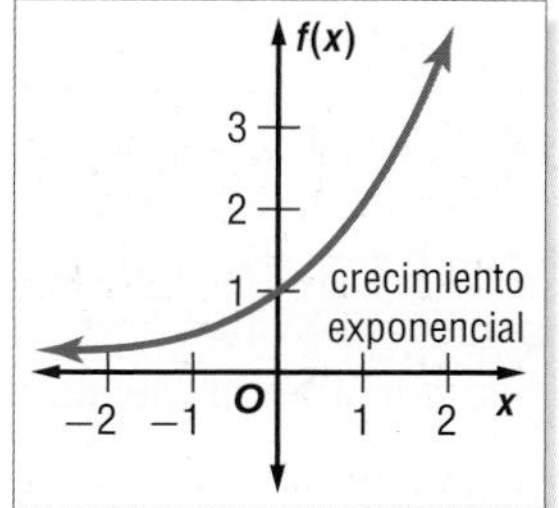


**desigualdad exponencial** Desigualdad que contiene funciones exponenciales.

**razones extendmientes** Razones que se utilizan para comparar tres o más cantidades.

**ángulo externo** Ángulo formado por un lado de un triángulo y la prolongación de otro de sus lados.

∠1 es un ángulo externo.

**segmento secante externo** Segmento secante que yace en el exterior del círculo.

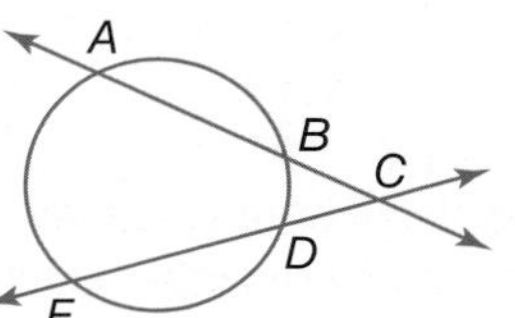

$\overline{BC}$ y $\overline{CD}$ son segmentos secantes externos.

**extraneous solutions** (pp. 260) Results that are not solutions to the original equation.

**soluciones extrañas** Resultados que no son soluciones de la ecuación original.

**extremes** (p. 544) In $\frac{a}{b} = \frac{c}{d}$, the numbers $a$ and $d$.

**extremos** Los números $a$ y $d$ en $\frac{a}{b} = \frac{c}{d}$.

## F

**factored form** (p. 170) The form of a polynomial showing all of its factors. $y = a(x - p)(x - q)$ is the factored form of a quadratic equation.

**forma reducida** La forma de un polinomio que demuestra todos sus factores. $y = a(x - p)(x - q)$ es la forma descompuesta en factores de una ecuación cuadrática.

**factorial** (p. 890) The product of the integers less than or equal to a positive integer $n$, written as $n!$

**factorial** Producto de los enteros menores o iguales a un número positivo $n$, escrito como $n!$

**factoring** (p. 36) To express a polynomial as the product of monomials and polynomials.

**factorización** La escritura de un polinomio como producto de monomios y polinomios.

**factoring by grouping** (p. 37) The use of the Distributive Property to factor some polynomials having four or more terms.

**factorización por agrupamiento** Uso de la Propiedad distributiva para factorizar polinomios que poseen cuatro o más términos.

**flow proof** (p. 337) A proof that organizes statements in logical order, starting with the given statements. Each statement is written in a box with the reason verifying the statement written below the box. Arrows are used to indicate the order of the statements.

**demostración de flujo** Demostración que organiza los enunciados en orden lógico, comenzando con los enunciados dados. Cada enunciado se escribe en una casilla y debajo de cada casilla se escribe el argumento que verifica dicho enunciado. El orden de los enunciados se indica con flechas.

**focus** (p. 783) The fixed point in a parabola that is equidistant from the locus of all points in a plane.

**foco** Punto fijo en una parábola que está equidistante del lugar geométrico de todos los puntos en un plano.

**FOIL method** (p. 23) To multiply two binomials, find the sum of the products of the First terms, the Outer terms, the Inner terms, and the Last terms.

**método FOIL** Para multiplicar dos binomios, busca la suma de los productos de los primeros (First) términos, los términos exteriores (Outer), los términos interiores (Inner) y los últimos términos (Last).

**formal proof** (p. 285) A two-column proof containing statements and reasons.

**demonstración formal** Demonstración en dos columnas que contiene enunciados y razonamientos.

**fractal** (p. 591) A figure generated by repeating a special sequence of steps infinitely often. Fractals often exhibit self-similarity.

**fractal** Figura que se obtiene mediante la repetición infinita de una sucesión particular de pasos. Los fractales a menudo exhiben autosemejanza.

**frustum** (p. 829) The part of a solid that remains after the top portion has been cut by a plane parallel to the base.

**tronco** Parte de un sólido que queda después de que la parte superior ha sido cortada por un plano paralelo a la base.

**fundamental counting principle** (p. 885) A method used to determine the number of possible outcomes in a sample space by multiplying the number of possible outcomes from each stage or event.

**principio fundamental de contar** Método para determinar el número de resultados posibles en un espacio muestral multiplicando el número de resultados posibles de cada etapa o evento.

## G

**geometric mean** (p. 619) For any positive numbers $a$ and $b$, the positive number $x$ such that $\frac{a}{x} = \frac{x}{b}$.

**media geométrica** Para todo número positivo $a$ y $b$, existe un número positivo $x$ tal que $\frac{a}{x} = \frac{x}{b}$.

**geometric probability** (p. 899) Using the principles of length and area to find the probability of an event.

**probabilidad geométrica** Uso de los principios de longitud y área para calcular la probabilidad de un evento.

**great circle** (p. 849) A circle formed when a plane intersects a sphere with its center at the center of the sphere.

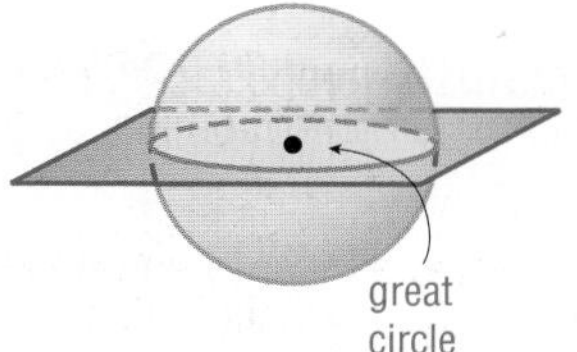


**círculo mayor** Círculo que se forma cuando un plano interseca una esfera y cuyo centro es el mismo que el centro de la esfera.

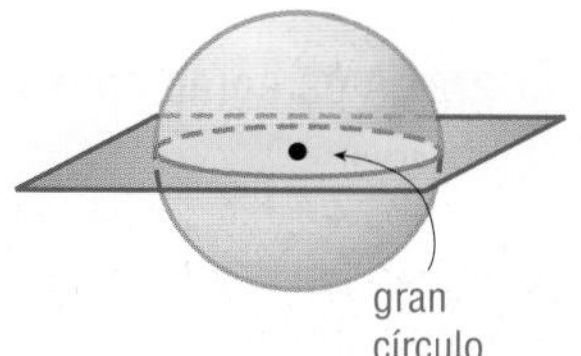


**greatest integer function** (p. 148) A step function, written as $f(x) = [\![x]\!]$, where $f(x)$ is the greatest integer less than or equal to $x$.

**La función más grande del número entero** Una función del paso, escrita como $f(x) = [\![x]\!]$, donde está el número entero $f(x)$ es el número más grande menos que o igual a $x$.

**growth factor** (pp. 229, 535) In exponential growth, the base of the exponential expression, $1 + r$.

**factor del crecimiento** En el crecimiento exponencial, la base de la expresión exponencial, $1 + r$.

## H

**hemisphere** (p. 849) One of the two congruent parts into which a great circle separates a sphere.

**hemisferio** Una de las dos partes congruentes en las cuales un círculo mayor divide una esfera.

## I

**imaginary unit** (pp. 178, 335) $i$, or the principal square root of $-1$.

**unidad imaginaria** $i$, o la raíz cuadrada principal de $-1$.

**incenter** (p. 410) The point of concurrency of the angle bisectors of a triangle.

**incentro** Punto de intersección de las bisectrices interiores de un triángulo.

**included angle** (p. 355) In a triangle, the angle formed by two sides is the included angle for those two sides.

**ángulo incluido** En un triángulo, el ángulo formado por dos lados es el ángulo incluido de esos dos lados.

**included side** (p. 364) The side of a polygon that is a side of each of two angles.

**lado incluido** Lado de un polígono común a dos de sus ángulos.

**independent events** (p. 915) Two or more events in which the outcome of one event does not affect the outcome of the other events.

**eventos independientes** El resultado de un evento no afecta el resultado del otro evento.

**indirect proof** (p. 437) In an indirect proof, one assumes that the statement to be proved is false. One then uses logical reasoning to deduce that a statement contradicts a postulate, theorem, or one of the assumptions. Once a contradiction is obtained, one concludes that the statement assumed false must in fact be true.

**demostración indirecta** En una demostración indirecta, se supone que el enunciado a demostrar es falso. Después, se deduce lógicamente que existe un enunciado que contradice un postulado, un teorema o una de las conjeturas. Una vez hallada una contradicción, se concluye que el enunciado que se suponía falso debe ser, en realidad, verdadero.

**indirect reasoning** (p. 437) Reasoning that assumes that the conclusion is false and then shows that this assumption leads to a contradiction of the hypothesis like a postulate, theorem, or corollary. Then, since the assumption has been proved false, the conclusion must be true.

**razonamiento indirecto** Razonamiento en que primero se supone que la conclusión es falsa y luego se demuestra que esta conjetura lleva a una contradicción de la hipótesis como un postulado, un teorema o un corolario. Finalmente, como se ha demostrado que la conjetura es falsa, la conclusión debe ser verdadera.

**informal proof** (p. 278) A paragraph proof.

**demostración informal** Demostración en forma de párrafo.

**inscribed** (p. 718) A polygon is inscribed in a circle if each of its vertices lie on the circle.

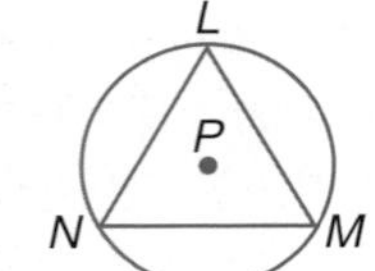

$\triangle LMN$ is inscribed in $\odot P$.

**inscrito** Un polígono está inscrito en un círculo si todos sus vértices yacen en el círculo.

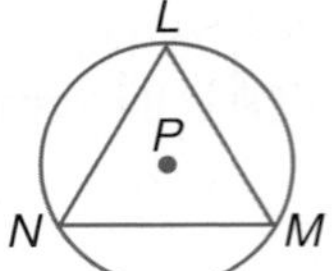

$\triangle LMN$ está inscrito en $\odot P$.

**inscribed angle** (p. 741) An angle that has a vertex on a circle and sides that contain chords of the circle.

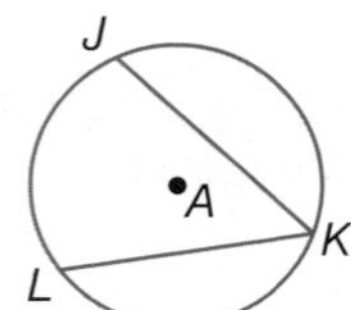

In $\odot A$, $\angle JKL$ is an inscribed angle.

**ángulo inscrito** Ángulo cuyo vértice esté en un círculo y cuyos lados contienen cuerdas del círculo.

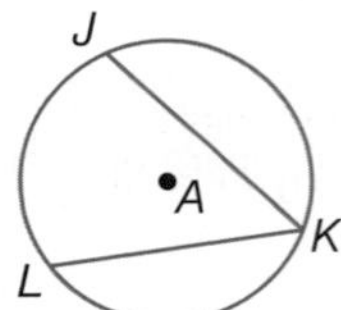

En $\odot A$, $\angle JKL$ es un ángulo inscrito.

**intercepted arc** (p. 741) An angle intercepts an arc if and only if each of the following conditions are met.
**1.** The endpoints of the arc lie on the angle.
**2.** All points of the arc except the endpoints are in the interior of the circle. **3.** Each side of the angle has an endpoint of the arc.

**arco intersecado** Un ángulo interseca un arco si y sólo si se cumple cada una de las siguientes condiciones.
**1.** Los extremos del arco yacen en el ángulo.
**2.** Todos los puntos del arco, excepto los extremos, yacen en el interior del círculo. **3.** Cada lado del ángulo tiene un extremo del arco.

**inverse cosine** (p. 653) The inverse function of cosine, or $\cos^{-1}$. If the cosine of an acute $\angle A$ is equal to $x$, then $\cos^{-1} x$ is equal to the measure of $\angle A$.

**inverso del coseno** Función inversa del coseno, o $\cos^{-1}$. Si el coseno de un $\angle A$ agudo es igual a $x$, entonces co $s^{-1} x$ es igual a la medida del $\angle A$.

**inverse sine** (p. 653) The inverse function of sine, or si $n^{-1}$. If the sine of an acute $\angle A$ is equal to $x$, then $\sin^{-1} x$ is equal to the measure of $\angle A$.

**inverso del seno** Función inversa del seno, o $\sin^{-1}$. Si el seno de un $\angle A$ agudo es igual a $x$, entonces $\sin^{-1} x$ es igual a la medida del $A$.

Glossary/Glosario

**inverse tangent** (p. 653) The inverse function of tangent, or $\tan^{-1}$. If the tangent of an acute $\angle A$ is equal to $x$, then $\tan^{-1} x$ is equal to the measure of $\angle A$.

**inverse del tangente** Función inversa de la tangente, o ta n$^{-1}$. Si la tangente de un $\angle A$ agudo es igual a $x$, entonces $\tan^{-1} x$ es igual a la medida del $\angle A$.

**isometric view** (p. 807) Corner views of three-dimensional objects on two-dimensional paper.

**vista isométrica** Vistas de las esquinas de sólidos geométricos tridimensionales sobre un papel bidimensional.

**isosceles trapezoid** (p. 521) A trapezoid in which the legs are congruent, both pairs of base angles are congruent, and the diagonals are congruent.

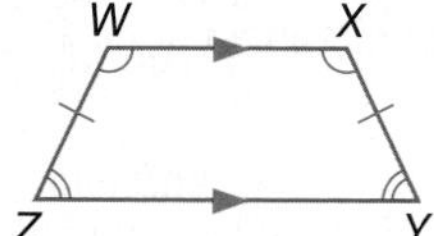


**trapecio isósceles** Trapecio cuyos catetos son congruentes, ambos pares de ángulos de las bases son congruentes y las diagonales son congruentes.

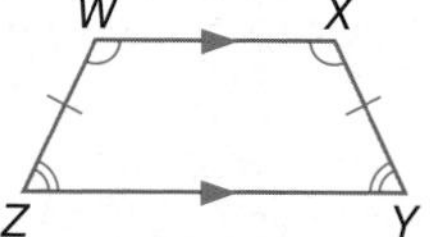


**isosceles triangle** (p. 374) A triangle with at least two sides congruent. The congruent sides are called *legs*. The angles opposite the legs are *base angles*. The angle formed by the two legs is the *vertex angle*. The side opposite the vertex angle is the *base*.

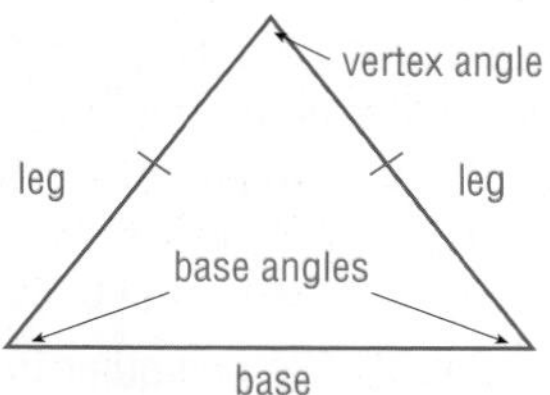


**triángulo isósceles** Triángulo que tiene por lo menos dos lados congruentes. Los lados congruentes se llaman *catetos*. Los ángulos opuestos a los catetos son los *ángulos de la base*. El ángulo formado por los dos catetos es el *ángulo del vértice*. El lado opuesto al ángulo del vértice es la *base*.

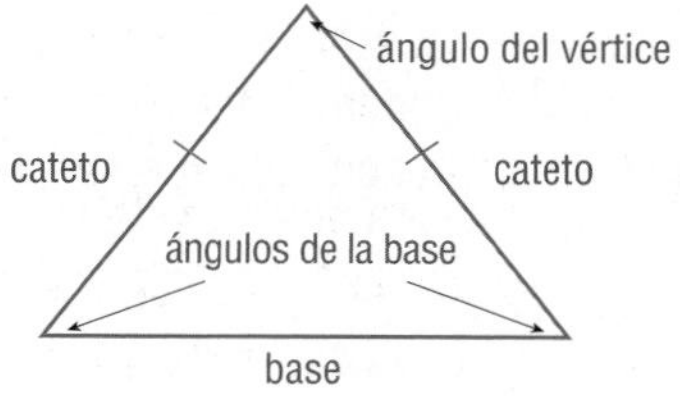


**iteration** (p. 591) A process of repeating the same procedure over and over again.

**iteración** Proceso de repetir el mismo procedimiento una y otra vez.

## J

**joint frequencies** (p. 922) In a two-way frequency table, the frequencies reported in the cells in the interior of the table.

**frecuencias conjuntas** En una tabla de double entrada o de frecuencias, las frecuencias reportadas en las celdas en el interior de la tabla.

## K

**kite** (p. 524) A quadrilateral with exactly two distinct pairs of adjacent congruent sides.

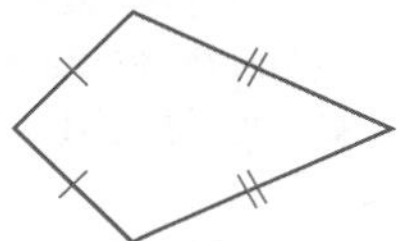

**cometa** Cuadrilátero que tiene exactamente dos pares differentes de lados congruentes y adyacentes.

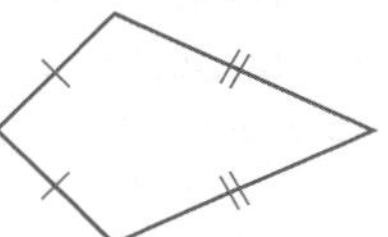

## L

**lateral area** (p. 814) For prisms, pyramids, cylinders, and cones, the area of the faces of the figure not including the bases.

**lateral edges** **1.** (p. 814) In a prism, the intersection of two adjacent lateral face**s.**

**lateral faces** **1.** (p. 814) In a prism, the faces that are not base**s.**

**latitude** (p. 863) A measure of distance north or south of the equator.

**law of cosines** (p. 671) Let $\triangle ABC$ be any triangle with $a$, $b$, and $c$ representing the measures of sides opposite the angles with measures $A$, $B$, and $C$ respectively. Then the following equations are true.

$a^2 = b^2 + c^2 - 2bc \cos A$

$b^2 = a^2 + c^2 - 2ac \cos B$

$c^2 = a^2 + b^2 - 2ab \cos C$

**law of large numbers** (p. 910) Law that states that as the number of trials of a random process increases, the average value will approach the expected value.

**law of sines** (p. 670) Let $\triangle ABC$ be any triangle with $a$, $b$, and $c$ representing the measures of sides opposite the angles with measures $A$, $B$, and $C$ respectively.

Then, $\frac{\sin A}{a} = \frac{\sin B}{b} = \frac{\sin C}{c}$.

**leading coefficient** (p. 8) The coefficient of the term with the highest degree in a polynomial.

**legs of a trapezoid** (p. 521) The nonparallel sides of a trapezoid.

**legs of an isosceles triangle** (p. 374) The two congruent sides of an isosceles triangle.

**longitude** (p. 863) A measure of distance east or west of the Prime Meridian.

**área lateral** En prismas, pirámides, cilindros y conos, es el área de la caras de la figura sin incluir el área de las bases.

**aristas laterales** **1.** En un prisma, la intersección de dos caras laterales adyacentes.

**caras laterales** **1.** En un prisma, las caras que no forman las bases.

**latitud** Medida de la distancia al norte o al sur del ecuador.

**ley de los cosenos** Sea $\triangle ABC$ cualquier triángulo donde $a$, $b$ y $c$ son las medidas de los lados opuestos a los ángulos que miden $A$, $B$ y $C$ respectivamente. Entonces las siguientes ecuaciones son verdaderas.

$a^2 = b^2 + c^2 - 2bc \cos A$

$b^2 = a^2 + c^2 - 2ac \cos B$

$c^2 = a^2 + b^2 - 2ab \cos C$

**ley de los grandes números** Ley que establece que a medida que aumenta el número de ensayos de un proceso aleatorio, el valor promedio se aproximará al valor esperado.

**ley de los senos** Sea $\triangle ABC$ cualquier triángulo donde $a$, $b$ y $c$ representan las medidas de los lados opuestos a los ángulos que miden $A$, $B$ y $C$ respectivamente.

Entonces, $\frac{\text{scn } A}{a} = \frac{\text{scn } B}{b} = \frac{\text{scn } C}{c}$.

**coeficiente inicial** El coeficiente del término con el grado más alto (el primer coeficiente inicial) en un polinomio.

**catetos de un trapecio** Los lados no paralelos de un trapecio.

**catetos de un triángulo isósceles** Las dos lados congruentes de un triángulo isósceles.

**longitud** Medida de la distancia del este o al oeste del Primer Meridiano.

## M

**magnitude** (p. 682) The length of a vector.

**magnitud** Longitud de un vector.

**major arc** (p. 725) An arc with a measure greater than 180. $\widehat{ACB}$ is a major arc.

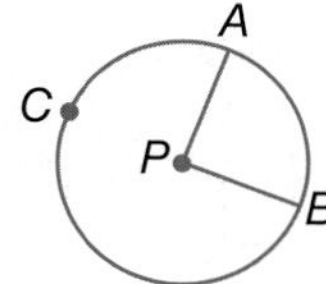


**arco mayor** Arco que mide más de 180. $\widehat{ACB}$ es un arco mayor.

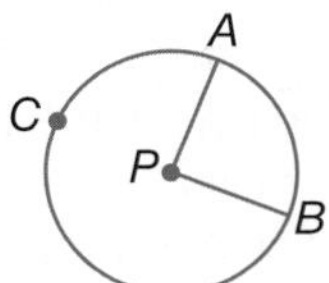


**marginal frequencies** (p. 922) In a two-way frequency table, the accumulated frequencies reported in the Totals row and Totals column.

**frecuencias marginales** En una tabla de double entrada o de frecuencias, las frecuencias acumuladas que se reportan en la hilera de los totales y en la columna de los totales.

**matrix logic** (p. 435) A rectangular array in which learned clues are recorded in order to solve a logic or reasoning problem.

**lógica matricial** Arreglo rectangular en que las claves aprendidas se escriben en orden para resolver un problema de lógica o razonamiento.

**means** (p. 544) In $\frac{a}{b} = \frac{c}{d}$, the numbers $b$ and $c$.

**medias** Los números $b$ y $c$ en la proporción $\frac{a}{b} = \frac{c}{d}$.

**median** (p. 417) In a triangle, a line segment with endpoints that are a vertex of a triangle and the midpoint of the side opposite the vertex.

**mediana** En un triángulo, Segmento de recta de cuyos extremos son un vértice del triángulo y el punto medio del lado opuesto a dicho vértice.

**meridians** (p. 863) Imaginary vertical lines drawn around the Earth through the North and South Poles.

**meridianos** Líneas verticales imaginarias dibujadas alrededor de la Tierra que von del polo norte al polo sur.

**midsegment of trapezoid** (p. 523) A segment that connects the midpoints of the legs of a trapezoid.

**segmento medio de un trapecio** Segmento que conecta los puntos medios de los catetos de un trapecio.

**midsegment of triangle** (p. 573) A segment with endpoints that are the midpoints of two sides of a triangle.

**segmento medio de un triángulo** Segmento cuyas extremos son los puntos medianos de dos lados de un triángulo.

**minor arc** (p. 725) An arc with a measure less than 180. $\widehat{AB}$ is a minor arc.

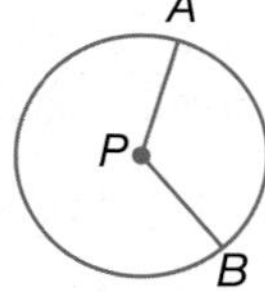


**arco menor** Arco que mide menos de 180. $\widehat{AB}$ es un arco menor.

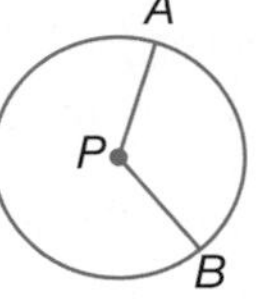


**multi-stage experiments** (p. 884) Experiments with more than two stages.

**experimentos multietápicos** Experimentos con más de dos etapas.

**mutually exclusive** (p. 924) Two events that have no outcomes in common.

**mutuamente exclusivos** Eventos que no tienen resultados en común.

## N

$n$th root (p. 257) If $a^n = b$ for a positive integer $n$, then $a$ is an $n$th root of $b$.

raíz enésima Si $a^n = b$ para cualquier entero positive $n$, entonces $a$ se llama una raíz enésima de $b$.

net (p. 391) A two-dimensional figure that when folded forms the surfaces of a three-dimensional object.

red Figura bidimensional que al ser plegada forma las superficies de un objeto tridimensional.

network (p. 932) A graph of interconnected vertices.

red Gráfico de vértices interconectados.

node (p. 932) A collection of vertices.

nodo Colección de vértices.

non-Euclidean geometry (p. 858) The study of geometrical systems that are not in accordance with the Parallel Postulate of Euclidean geometry.

geometría no euclidiana El estudio de sistemas geométricos que no satisfacen el postulado de las paralelas de la geometría euclidiana.

## O

oblique cone (p. 824) A cone that is not a right cone.

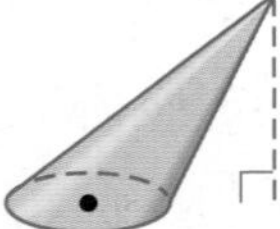

cono oblicuo Cono que no es un cono recto.

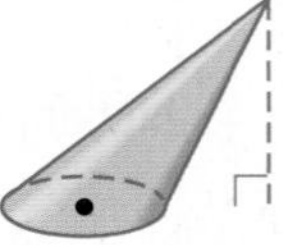

oblique cylinder (p. 816) A cylinder that is not a right cylinder.

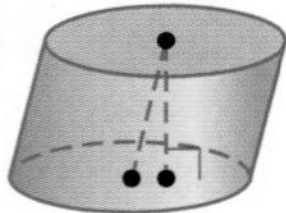

cilindro oblicuo Cilindro que no es un cilindro recto.

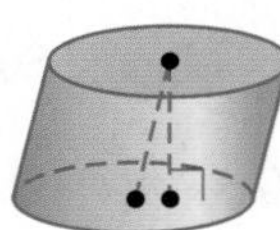

oblique prism (p. 814) A prism in which the lateral edges are not perpendicular to the bases.

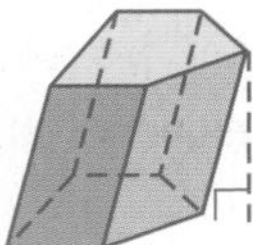

prisma oblicuo Prisma cuyas aristas laterales no son perpendiculares a las bases.

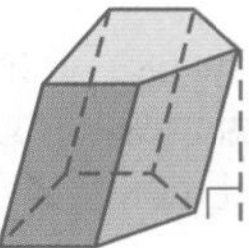

oblique solid (p. 806) A solid with base(s) that are not perpendicular to the edges connecting the two bases or vertex.

sólido oblicuo Sólido con base o bases que no son perpendiculares a las aristas, las cuales conectan las dos bases o vértice.

opposite vectors (p. 683) Vectors that have the same magnitude but opposite direction.

vectores opuestos Vectores que tienen la misma magnitud pero enfrente de la dirección.

ordered triple (p. 638) Three numbers given in a specific order used to locate points in space.

triple ordenado Tres números dados en un orden específico que sirven para ubicar puntos en el espacio.

orthocenter (p. 419) The point of concurrency of the altitudes of a triangle.

ortocentro Punto de intersección de las alturas de un triángulo.

**orthographic drawing** (p. 390) The two-dimensional top view, left view, front view, and right view of a three-dimensional object.

## P

**parabola** **1.** (p. 93) The graph of a quadratic function. parabola **2.** (p. 782) The graph of a quadratic function. The set of all points in a plane that are the same distance from a given point, called the focus, and a given line, called the directrix.

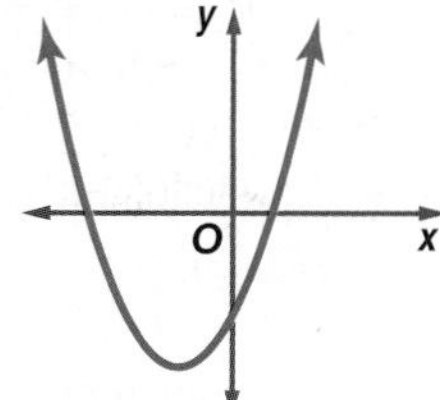


**paragraph proof** (p. 278) An informal proof written in the form of a paragraph that explains why a conjecture for a given situation is true.

**parallel vectors** (p. 683) Vectors that have the same or opposite direction.

**parallelogram** (p. 485) A quadrilateral with parallel opposite sides. Any side of a parallelogram may be called a *base*.

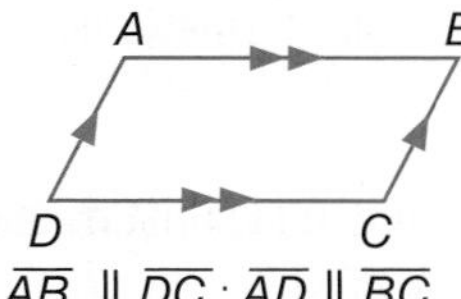


**parallelogram method** (p. 683) A method used to find the resultant of two vectors in which you place the vectors at the same initial point, complete a parallelogram, and draw the diagonal.

**parallels** (p. 863) Imaginary horizontal lines parallel to the equator.

**perfect square trinomial** (p. 64, 604) A trinomial that is the square of a binomial.
$(a + b)^2 = (a + b)(a + b) = a^2 + 2ab + b^2$ or
$(a - b)^2 = (a - b)(a - b) = a^2 - 2ab + b^2$

**permutation** (p. 890) An arrangement of objects in which order is important.

**proyección ortogonal** Vista bidimensional superior, del lado izquierda, frontal y del lado derecho de un objeto tridimensional.

**parábola** **1.** La gráfica de una función cuadrática. **2.** La grafica de una funcion cuadratica. Conjunto de todos los puntos de un plano que estan a la misma distancia de un punto dado, llamado foco, y de una recta dada, llamada directriz.

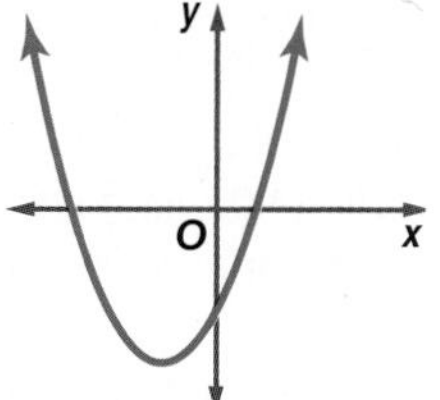


**demostración de párrafo** Demostración informal escrita en párrafo que explica por qué una conjetura para una situación dada es verdadera.

**vectores paralelos** Vectores con la misma dirección o dirección opuesta.

**paralelogramo** Cuadrilátero cuyos lados opuestos son paralelos y cuya *base* puede ser cualquier de sus lados.

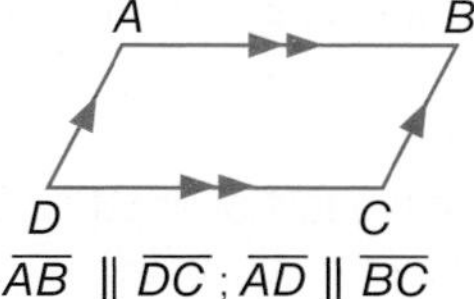


**método del paralelogramo** Método que se usa para hallar la resultante de dos vectores en que se dibujan los vectores con el mismo punto de origen, se completa un paralelogramo y se traza la diagonal.

**paralelos** Rectas horizontales imaginarias paralelas al ecuador.

**trinomio cuadrado perfecto** Un trinomio que es el cuadrado de un binomio.
$(a + b)^2 = (a + b)(a + b) = a^2 + 2ab + b^2$ or
$(a - b)^2 = (a - b)(a - b) = a^2 - 2ab + b^2$

**permutación** Disposicion de objetos en la cual el orden es importante.

**perpendicular bisector** (p. 406) In a triangle, a line, segment, or ray that passes through the midpoint of a side and is perpendicular to that side.

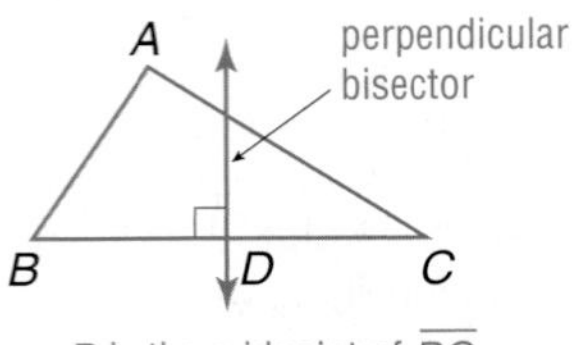


$D$ is the midpoint of $\overline{BC}$.

**mediatriz** Recta, segmento de recta o rayo perpendicular que corta un lado del triángulo en su punto medio.

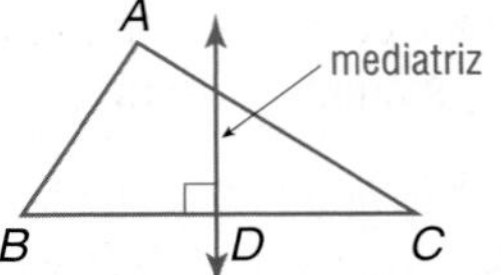


$D$ es el punto medio de $\overline{BC}$.

**pi (π)** (p. 717) An irrational number represented by the ratio of the circumference of a circle to the diameter of the circle.

**pi (π)** Número irracional representado por la razón de la circunferencia de un círculo al diámetro del mismo.

**piecewise-defined function** (pp. 149) A function that is written using two or more expressions.

**función definida por partes** Función que se escribe usando dos o más expresiones.

**piecewise-linear function** (p. 148) A function written using two or more linear expressions.

**función lineal por partes** Función que se escribe usando dos o más expresiones lineal.

**plane Euclidean geometry** (p. 857) Geometry based on Euclid's axioms dealing with a system of points, lines, and planes.

**geometría del plano euclidiano** Geometría basada en los axiomas de Euclides, los cuales abarcan un sistema de puntos, rectas y planos.

**point of concurrency** (p. 407) The point of intersection of concurrent lines.

**punto de concurrencia** Punto de intersección de rectas concurrentes.

**point of tangency** (p. 750) For a line that intersects a circle in only one point, the point at which they intersect.

**punto de tangencia** Punto de intersección de una recta en un círculo en un solo punto.

**poles** (p. 849) The endpoints of the diameter of a great circle.

**postes** Las extremos del diámetro de un círculo mayor.

**polynomial** (p. 7) A monomial or sum of monomials.

**polinomio** Un monomio o la suma de monomios.

**postulate** (p. 276) A statement that describes a fundamental relationship between the basic terms of geometry. Postulates are accepted as true without proof.

**postulado** Enunciado que describe una relación fundamental entre los términos geométricos básicos. Los postulados se aceptan como verdaderos sin necesidad de demostración.

**prime polynomial** (p. 54) A polynomial that cannot be written as a product of two polynomials with integral coefficients.

**polinomio primo** Polinomio que no puede escribirse como producto de dos polinomios con coeficientes enteros.

**probability model** (p. 907) A mathematical model used to match a random phenomenon.

**modelo de probabilidad** Modelo matemático que se usa para relacionar un fenómeno aleatorio.

**probability tree** (p. 917) An organized table of line segments (branches) that shows the probability of each outcome.

**árbol de la probabilidad** Tabla organizada de segmentos de recta (ramas) que muestra la probabilidad de cada resultado.

**proof** (p. 277) A logical argument in which each statement you make is supported by a statement that is accepted as true.

**demostración** Argumento lógico en el cual cada enunciado que se hace está respaldado por un enunciado que se acepta como verdadero.

**proof by contradiction** (p. 437) An indirect proof in which one assumes that the statement to be proved is false. One then uses logical reasoning to deduce a statement that contradicts a postulate, theorem, or one of the assumptions. Once a contradiction is obtained, one concludes that the statement assumed false must in fact be true.

**demostración por contradicción** Demostración indirecta en la cual se supone que el enunciado a demostrarse es falso. Luego, se usa el razonamiento lógico para inferir un enunciado que contradiga el postulado, teorema o una de las conjeturas. Una vez que se obtiene una contradicción, se concluye que el enunciado que se supuso falso es, en realidad, verdadero.

**proportion** (p. 544) An equation of the form $\frac{a}{b} = \frac{c}{d}$ that states that two ratios are equal.

**proporción** Ecuación de la forma $\frac{a}{b} = \frac{c}{d}$ que establece que dos razones son iguales.

**pure imaginary number** (pp. 178, 335) The square roots of negative real numbers. For any positive real number $b$,

$$\sqrt{-b^2} = \sqrt{b^2} \cdot \sqrt{-1}, \text{ or } bi.$$

**número imaginario puro** Raíz cuadrada de un número real negativo. Para cualquier número real positivo $b$,

$$\sqrt{-b^2} = \sqrt{b^2} \cdot \sqrt{-1}, \text{ ó } bi.$$

**pythagorean triple** (p. 630) A group of three whole numbers that satisfies the equation $a^2 + b^2 = c^2$, where $c$ is the greatest number.

**triplete pitágorico** Grupo de tres números enteros que satisfacen la ecuación $a^2 + b^2 = c^2$, donde $c$ es el número mayor.

## Q

**quadratic equation** (pp. 48, 105) An equation of the form $ax^2 + bx + c = 0$, where $a \neq 0$.

**ecuación cuadrática** Ecuación de la forma $ax^2 + bx + c = 0$, donde $a \neq 0$.

**quadratic expression** (p. 23) An expression in one variable with a degree of 2 written in the form $ax^2 + bx + c$.

**expression cuadratica** Una expresión en una variable con un grado de 2, escritos en la forma $ax^2 + bx + c$.

**quadratic Formula** (pp. 133, 264) The solutions of a quadratic equation in the form $ax^2 + bx + c = 0$, where $a \neq 0$, are given by the formula

$$x = \frac{-b \pm \sqrt{b^2 - 4ac}}{2a}.$$

**Fórmula cuadrática** Las soluciones de una ecuación cuadrática de la forma $ax^2 + bx + c = 0$, donde $a \neq 0$, vienen dadas por la fórmula

$$x = \frac{-b \pm \sqrt{b^2 - 4ac}}{2a}.$$

**quadratic function** (p. 93) An equation of the form $y = ax^2 + bx + c$, where $a \neq 0$.

**función cuadrática** Función de la forma $y = ax^2 + bx + c$, donde $a \neq 0$.

**quadratic inequality** (p. 207) An inequality of the form $y > ax^2 + bx + c$, $y \geq ax^2 + bx + c$, $y < ax^2 + bx + c$, or $y \leq ax^2 + bx + c$.

**desigualdad cuadrática** Desigualdad cuadrática de la forma $y > ax^2 + bx + c$, $y \geq ax^2 + bx + c$, $y < ax^2 + bx + c$, o $y \leq ax^2 + bx + c$.

## R

**radical equations** (p. 259) Equations that contain radicals with variables in the radicand.

**ecuaciones radicales** Ecuaciones que contienen radicales con variables en el radicando.

**radical expression** (p. 245) An expression that contains a square root.

**expresión radical** Expresión que contiene una raíz cuadrada.

**radius** **1.** (p. 715) In a circle, any segment with endpoints that are the center of the circle and a point on the circle. **2.** (p. 816) In a sphere, any segment with endpoints that are the center and a point on the sphere.

**radio** **1.** En un círculo, cualquier segmento cuyos extremos son en el centro y un punto del círculo. **2.** En una esfera, cualquier segmento cuyos extremos son el centro y un punto de la esfera.

**random variable** (p. 909) A variable that can assume a set of values, each with fixed probabilities.

**variable aleatoria** Variable que puede tomar un conjunto de valores, cada uno con probabilidades fijas.

**ratio** (p. 543) A comparison of two quantities using division.

**razón** Comparación de dos cantidades mediante división.

**rationalizing the denominator** (p. 247) A method used to eliminate radicals from the denominator of a fraction.

**racionalizar el denominador** Método que se usa para eliminar radicales del denominador de una fracción.

**rectangle** (p. 505) A quadrilateral with four right angles.

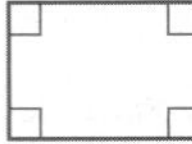

**rectángulo** Cuadrilátero con cuatro ángulos rectos.

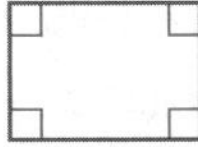

**reduction** (p. 593) An image that is smaller than the original figure.

**reducción** Imagen más pequeña que la figura original.

**reflection** (p. 116) A transformation where a figure, line, or curve, is flipped across a line.

**reflexión** Transformación en que cadapunto de una figura se aplica a través de una recta de simetría a su imagen correspondiente.

**regular pyramid** (p. 822) A pyramid with a base that is a regular polygon.

**pirámide regular** Pirámide cuya base es un polígono regular.

**relative frequency** (p. 922) In a frequency table, the ratio of the number of observations in a category to the total number of observations.

**frecuencia relativa** En una tabla de frecuencias, la razón del número de observaciones en una categoría al número total de observaciones.

**remote interior angles** (p. 337) The angles of a triangle that are not adjacent to a given exterior angle.

**ángulos internos no adyacentes** Ángulos de un triángulo que no son adyacentes a un ángulo exterior dado.

**resultant** (p. 683) The sum of two vectors.

**resultante** Suma de dos vectores.

**rhombus** (p. 512) A quadrilateral with all four sides congruent.

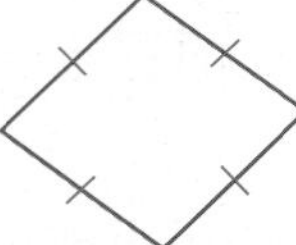

**rombo** Cuadrilátero con cuatro lados congruentes.

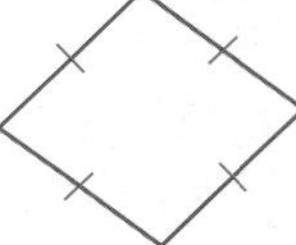

**right cone** (p. 824) A cone with an axis that is also an altitude.

**cono recto** Cono cuyo eje es también su altura.

**right cylinder** (p. 816) A cylinder with an axis that is also an altitude.

**cilindro recto** Cilindro cuyo eje es también su altura.

**right prism** (p. 814) A prism with lateral edges that are also altitudes.

**prisma recto** Prisma cuyas aristas laterales también son su altura.

**right solid** (p. 805) A solid with base(s) that are perpendicular to the edges connecting them or connecting the base and the vertex of the solid.

**sólido recto** Sólido con base o bases perpendiculares a las aristas, conectándolas entre sí o conectando la base y el vértice del sólido.

**sample space** (p. 883) The set of all possible outcomes of an experiment.

**scalar** (p. 688) A constant multiplied by a vector.

**scalar multiplication** (p. 688) Multiplication of a vector by a scalar.

**scale factor** (p. 552) The ratio of the lengths of two corresponding sides of two similar polygons or two similar solids.

**scale factor of dilation** (p. 593) The ratio of a length on an image to a corresponding length on the preimage.

**secant** (p. 759) Any line that intersects a circle in exactly two points.

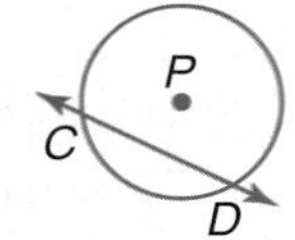

$\overleftrightarrow{CD}$ is a secant of $\odot P$.

**secant segment** (p. 770) A segment of a secant line that has exactly one endpoint on the circle.

**sector of a circle** (p. 785) A region of a circle bounded by a central angle and its intercepted arc.

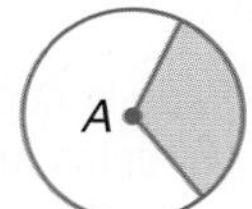

The shaded region is a sector of $\odot A$.

**segment of a circle** (p. 789) The region of a circle bounded by an arc and a chord.

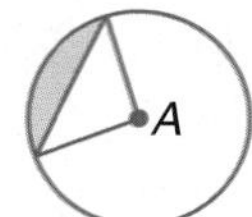

The shaded region is a segment of $\odot A$.

**self-similar** (p. 591) If any parts of a fractal image are replicas of the entire image, the image is self-similar.

**semicircle** (p. 725) An arc that measures 180.

**espacio muestral** El conjunto de todos los resultados posibles de un experimento.

**escalar** Constante multiplicada por un vector.

**multiplicación escalar** Multiplicación de un vector por un escalar.

**factor de escala** Razón entre las longitudes de dos lados correspondientes de dos polígonos o sólidos semejantes.

**factor de escala de homotecia** Razon de una longitud en la imagen a una longitud correspondiente en la preimagen.

**secante** Cualquier recta que interseca un círculo exactamente en dos puntos.

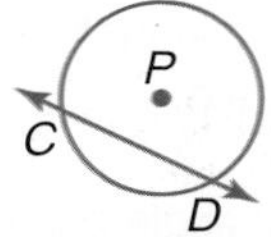

$\overleftrightarrow{CD}$ es una secante de $\odot P$.

**segmento secante** Segmento de una recta secante que tiene exactamente un extremo en el círculo.

**sector circular** Región de un círculo limitada por un ángulo central y su arco de intersección.

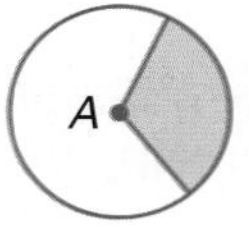

La región sombreada es un sector de $\odot A$.

**segmento de un círculo** Región de un círculo limitada por un arco y una cuerda.

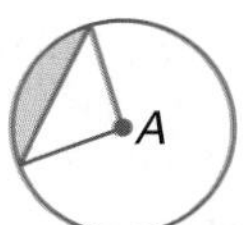

La región sombreada es un segmento de $\odot A$.

**autosemejante** Si cualquier parte de una imagen fractal es una réplica de la imagen completa, entonces la imagen es autosemejante.

**semicírculo** Arco que mide 180.

**sierpinski triangle** (p. 591) A self-similar fractal described by Waclaw Sierpinski. The figure was named for him.

**triángulo de Sierpinski** Fractal autosemejante descrito por el matemático Waclaw Sierpinski. La figura se nombró en su honor.

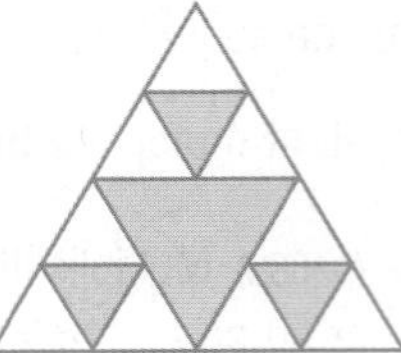

**similar solids** (p. 864) Solids that have exactly the same shape, but not necessarily the same size.

**sólidos semejantes** Sólidos que tienen exactamente la misma forma, pero no necesariamente el mismo tamaño.

**similarity ratio** (p. 552) The scale factor between two similar polygons

**razón de semejanza** Factor de escala entre dos polígonos semejantes.

**similarity transformation** (p. 593) When a figure and its transformation image are similar.

**transformación de semejanza** cuando una figura y su imagen transformada son semejantes.

**simulation** (p. 907) A probability model used to recreate a situation again and again so the likelihood of various outcomes can be estimated.

**simulacro** Modelo de la probabilidad que se usa para reconstruir una situación repetidas veces y así poder estimar la probabilidad de varios resultados.

**sine** (p. 568) For an acute angle of a right triangle, the ratio of the measure of the leg opposite the acute angle to the measure of the hypotenuse.

**seno** Para un ángulo agudo de un triángulo rectángulo, razón entre la medida del cateto opuesto al ángulo agudo a la medida de la hipotenusa.

**slant height** (p. 822) The height of the lateral side of a pyramid or cone.

**altura oblicua** Altura de la cara lateral de una pirámide o un cono.

**solving a triangle** (p. 673) Finding the measures of all of the angles and sides of a triangle.

**resolver un triángulo** Calcular las medidas de todos los ángulos y todos los lados de un triángulo.

**spherical geometry** (p. 857) The branch of geometry that deals with a system of points, great circles (lines), and spheres (planes).

**geometría esférica** Rama de la geometría que estudia los sistemas de puntos, los círculos mayores (rectas) y las esferas (planos).

**square** (p. 513) A quadrilateral with four right angles and four congruent sides.

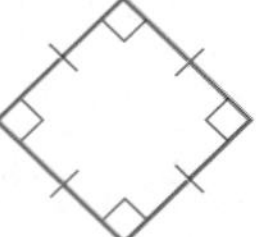

**cuadrado** Cuadrilátero con cuatro ángulos rectos y cuatro lados congruentes.

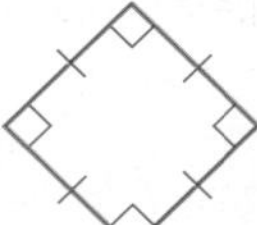

**square root property** (p. 179) For any real number $n$, if $x^2 = n$, then $x = \pm\sqrt{n}$.

**propiedad de la raíz cuadrada** Para cualquier número real $n$, si $x^2 = n$, entonces $x = \pm\sqrt{n}$.

**standard form of a polynomial** (p. 8) A polynomial that is written with the terms in order from greatest degree to least degree.

**forma de estándar de un polinomio** Un polinomio que se escribe con los términos en orden del grado más grande a menos grado.

Glossary/Glosario

**standard position** (p. 684) When the initial point of a vector is at the origin.

**posición estándar** Cuando el punto inicial de un vector es el origen.

**step function** (p. 148) A function with a graph that is a series of horizontal line segments.

**funcion escalonada** Función cuya gráfica es una serie de segmentos de recto.

## T

**tangent** **1.** (p. 568) For an acute angle of a right triangle, the ratio of the measure of the leg opposite the acute angle to the measure of the leg adjacent to the acute angle. **2.** (p. 732) A line in the plane of a circle that intersects the circle in exactly one point. The point of intersection is called the *point of tangency.* **3.** (p. 848) A line that intersects a sphere in exactly one point.

**tangente** **1.** Para un ángulo agudo de un triángulo rectángulo, razón de la medida del cateto opuesto al ángulo agudo a la medida del cateto adyacente al ángulo agudo. **2.** Recta en el plano de un círculo que interseca el círculo en exactamente un punto. El punto de intersección se conoce como *punto de tangencia.* **3.** Recta que interseca una esfera en exatamente un punto.

**tangent segment** (p. 770) A segment of a tangent with one endpoint on a circle that is both the exterior and whole segment.

**segmento tangente** Segmento de la tangente con un extremo en un círculo que es tanto el segmento externo como el segmento completo.

**theorem** (p. 278) A statement or conjecture that can be proven true by undefined terms, definitions, and postulates.

**teorema** Enunciado o conjetura que se puede demostrar como verdadera mediante términos geométricos básicos, definiciones y postulados.

**topographic map** (p. 813) A representation of a three-dimensional surface on a flat piece of paper.

**mapa topográfico** Representación de una superficie tridimensional sobre una hoja del papel.

**traceable network** (p. 932) A network in which all of the nodes are connected and each edge is used once when the network is used.

**red detectable** Red en la cual todos los nodos estén conectados y cada arista se utiliza una vez al usarse la red.

**transformation** **1.** (p. 114) A movement of a geometric figure. **2.** (p. 593) In a plane, a mapping for which each point has exactly one image point and each image point has exactly one preimage point.

**transformación** **1.** Desplazamiento de una figura geométrica. **2.** En un plano, aplicación para la cual cada punto del plano tiene un único punto de la imagen y cada punto de la imagen tiene un único punto de la preimagen.

**translation** (p. 646) A transformation where a figure is slid from one position to another without being turned.

**translación** Transformación en que una figura se desliza sin girar, de una posición a otra.

**trigonometry** (p. 650) The study of the properties of triangles and trigonometric functions and their applications.

**trigonometría** Estudio de las propiedades de los triángulos, de las funciones trigonométricas y sus aplicaciones.

**two-column proof** (p. 285) A formal proof that contains statements and reasons organized in two columns. Each step is called a *statement*, and the properties that justify each step are called *reasons.*

**demostración de dos columnas** Demostración formal que contiene enunciados y razones organizadas en dos columnas. Cada paso se llama *enunciado* y las propiedades que lo justifican son las *razones.*

**two-stage experiment** (p. 884) An experiment with two stages or events.

**experimento de dos pasos** Experimento que consta de dos pasos o eventos.

**trapezoid** (p. 521) A quadrilateral with exactly one pair of parallel sides. The parallel sides of a trapezoid are called *bases*. The nonparallel sides are called *legs*. The pairs of angles with their vertices at the endpoints of the same base are called *base angles*.

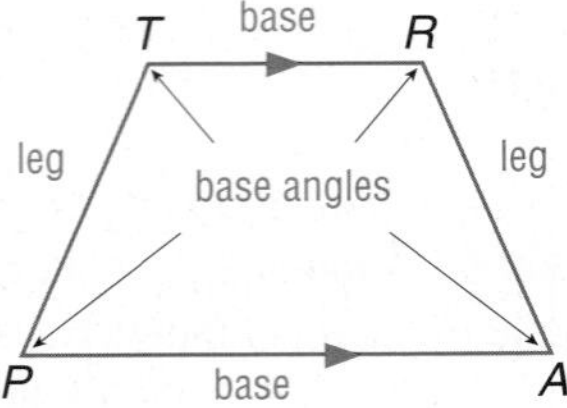


**trapecio** Cuadrilátero con sólo un par de lados paralelos. Los lados paralelos del trapecio se llaman *bases*. Los lados no paralelos se llaman *catetos*. Los pares de ángulos cuyos vértices coinciden en los extremos de la misma base son los *ángulos de la base*.

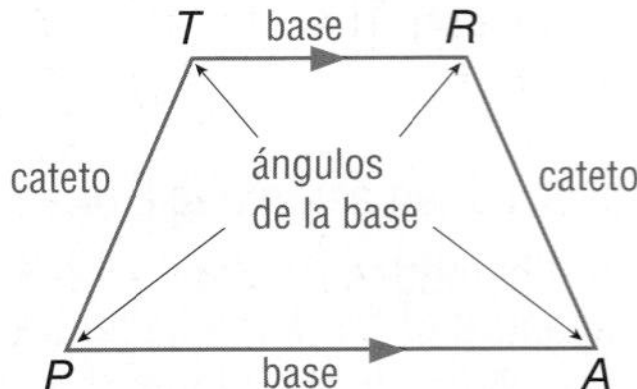


**tree diagram** (p. 883) An organized table of line segments (branches) which shows possible experiment outcomes.

**diagrama del árbol** Tabla organizada de segmentos de recta (ramas) que muestra los resultados posibles de un experimento.

**triangle method** (p. 684) A method used to find the resultant of two vectors in which the second vector is connected to the terminal point of the first and the resultant is drawn from the initial point of the first vector to the terminal point of the second vector.

**método del triangulo** Método para calcular la resultante de dos vectores en cual el segundo vector está conectado al extremo del primer y la resultante se traza desde el punto inicial del primer vector al extremo del segundo vector.

**trigonometric ratio** (p. 650) A ratio of the lengths of sides of a right triangle.

**razón trigonométrica** Razón de las longitudes de los lados de un triángulo rectángulo.

**trinomials** (p. 7) The sum of three monomials.

**trinomios** Suma de tres monomios.

**two-way frequency table** (p. 922) A table used to show the frequencies or relative frequencies of data from a survey or experiment classified according to two variables, with the rows indicating one variable and the columns indicating the other.

**tabla de doble entrada o de frecuencias** Tabla que se usa para mostrar las frecuencias o frecuencias relativas de los datos de una encuesta o experimento clasificado de acuerdo con dos variables, en la cual las hileras indican una variable y las columnas indican la otra variable.

## V

**vector** (p. 600) A directed segment representing a quantity that has both magnitude, or length, and direction.

**vector** Segmento dirigido que representa una cantidad, la cual posee tanto magnitud o longitud como dirección.

**vertex** (p. 93) The maximum or minimum point of a parabola.

**vértice** Punto máximo o mínimo de una parábola.

**vertex angle of an isosceles triangle** (p. 374) See *isosceles triangle*.

**ángulo del vértice un triángulo isosceles** Ver *triángulo isósceles*.

**vertex form** (p. 364) A quadratic function in the form $y = a(x - h)^2 + k$, where $(h, k)$ is the vertex of the parabola and $x = h$ is its axis of symmetry.

**forma de vértice** Función cuadrática de la forma $y = a(x - h)^2 + k$, donde $(h, k)$ es el vértice de la parábola y $x = h$ es su eje de simetría.

**vertex-edge graphs** (p. 932)2 A collection of nodes connected by edges.

**gráficas de vértice-arista** Colección de nodos conectados por aristas.

## W

**weight** (p. 914) The value assigned to an edge in a vertex-edge graph.

**weight of a path** (p. 914) The sum of the weights of the edges along a path.

**weighted vertex-edge graphs** (p. 914) A collection of nodes connected by edges in which each edge has an assigned value.

**peso** Valor asignado a una arista en una gráfica de vértice-arista.

**peso de una ruta** Suma de los pesos de las aristas a lo largo de una ruta.

**gráficas ponderadas de vértice-arista** Colección de nodos conectados por aristas en que cada arista tiene un valor asignado.

# Index


**30°-60°-90° Triangle Theorem,** 642
**45°-45°-90° Triangle Theorem,** 640

**AA (Angle-Angle) Similarity Postulate.** *See* Angle-Angle (AA) Similarity Postulate
**AAS (Angle-Angle-Side) Congruence Theorem.** *See* Angle-Angle-Side (AAS) Congruence Theorem
**Absolute value** 202, 761
complex number, 186
functions, 149–151
**Activities.** *See* Algebra Labs; Graphing Technology Labs; Spreadsheet Labs; TI-Nspire Technology Labs, *See* Geometry Labs; Graphing Technology Labs; Spreadsheet Labs; TI-Nspire[TM] Technology Labs
**Acute angles,** 640
**Addition**
angle, 299–300
arc, 726
Associative Property, 286
Commutative Property, 286
monomials, 7
polynomials, 5–6, 7, 8–13
radical expressions, 252–253
Rules for Probability, 925–926
segment, 292–293
of complex numbers, 180
of vectors, 683, 685, 691
**Addition Property**
of equality, 284
of inequality, 426
**Additive inverse,** 9, 37. *See also* Opposites
**Adjacent arcs,** 726
**Algebra Labs**
Adding and Subtracting Polynomials, 5–6
Drawing Inverses, P22
Factoring Trinomials, 43–44
Factoring Using the Distributive Property, 35
Finding the Maximum or Minimum Value, 130–131
Multiplying Polynomials, 20–21
Polynomials, 5
Rate of Change of a Quadratic Function, 104
Rational and Irrational Numbers, 251
Simplifying *n*th Root Expressions, Solving Equations, 257–258
The Complex Plane, 185
Quadratics and Rate of Change, 206
**Algebra tiles**
adding and subtracting polynomials, 5–6
completing the square, 124
factoring trinomials, 43–44
factoring using distributive property, 35
multiplying polynomials, 20–21
polynomials, 5
**Algebraic expressions,** P10
simplifying, 14–15
**Algebraic equations.** *See* Equations
**Algebraic proofs,** 284–285, 324
**Algebraic vectors,** 685
**Alternate exterior angles,** 308, 309, 316
**Alternate Exterior Angles Converse,** 316
**Alternate Exterior Angles Theorem,** 309
**Alternate interior angles,** 309, 316, 662
**Alternate Interior Angles Converse,** 316
**Alternate Interior Angles Theorem,** 309, 662
**Altitudes**
of cones, 824
of prisms, 814
of pyramids, 822
of right triangles, 619–621
of triangles, 416, 419–421, 620, 641
**Ambiguous case,** 680–681
**Analysis**
functions, 140–142
**Angle Addition Postulate,** 299–300
**Angle-Angle (AA) Similarity Postulate,** 560, 564, 572, 650
**Angle-Angle-Side (AAS) Congruence Theorem,** 365, 670
**Angle bisectors,** 405, 408–409, 421
**Angle Bisector Theorem,** 409
**Angle-Side-Angle (ASA) Congruence Postulate,** 364
**Angle-side inequalities,** 427–428
**Angle-side relationships,** 428–429
**Angle-Side Relationships in Triangles Theorem,** 428
**Angles**
alternate exterior, 309, 316
alternate interior, 309, 316, 662
base, 374, 521–522
bisectors of, 405, 408–409, 421
central, 724, 785
complementary, 299–300, 640
congruent, 301, 308, 316, 374–375, 476, 551
consecutive interior, 309, 316
corresponding, 308, 309, 315, 344, 551–552
of depression, 662–668, 702
of elevation, 662–668, 702
exterior, 308, 309, 316, 337–338, 426, 427–428, 430, 478–480, 484
formed by two parallel lines and a transversal, 308–313
included, 355–356, 671
inscribed, 741–747
interior, 309, 316, 335, 337, 475–477, 484
linear pairs, 300
measures of, 901
pairs, 308, 316
and parallel lines, 308–313
relationships, 299–306
remote interior, 337, 427
right, 303, 372–373, 384, 505, 506
supplementary, 299–300
theorems about, 301, 302, 303, 309, 316, 335–336, 337–338, 346, 347, 365, 374–375, 409, 427–428, 430, 475–477, 478–479, 561, 564, 586, 640, 642, 662, 670, 671
of triangles, 33, 335–342, 393
vertex of, 79
vertical, 302–303

**Applications.** *See also* Real-World Careers; Real-World Examples; Real-World Link, P3, P7, P9, P16, P18–P21, P29, 2, 3, 9–13, 17–19, 24–27, 29–33, 39–42, 48–51, 54–57, 61, 63, 68–71, 78, 79, 81–86, 90, 91, 98, 99, 101, 103, 107–110, 118–121, 126–129, 131, 132, 137–139, 142–145, 147, 158–161, 166, 167, 168, 248–250, 254–256, 260–263, 272, 273, 277, 279, 280, 281, 282, 283, 285, 287, 288, 289, 291, 294, 295, 296, 297, 298, 300, 304, 305, 306, 307, 309, 311, 312, 313, 314, 318, 319, 320, 322, 324, 325, 327, 332, 333, 336, 338, 339, 340, 341, 343, 346, 347, 349, 350, 351, 352, 356, 357, 358, 359, 360, 361, 363, 366, 368, 369, 370, 371, 376, 377, 378, 379, 380, 381, 382, 385, 386, 387, 388, 389, 392, 394, 395, 396, 397, 402, 403, 408, 411, 412, 413, 414, 415, 418, 421, 422, 423, 425, 429, 430, 431, 433, 434, 436, 438, 440, 441, 442, 443, 444, 448, 449, 450, 451, 452, 455, 456, 458, 459, 460, 462, 464, 465, 466, 467, 472, 473, 476, 477, 479, 480, 481, 482, 483, 486, 487, 489, 490, 491, 492, 493, 496, 499, 500, 501, 502, 503, 504, 505, 506, 507, 508, 509, 510, 511, 513, 515, 517, 518, 519, 520, 522, 527, 528, 529, 530, 532, 533, 534, 535, 540, 541, 543, 545, 546, 547, 548, 549, 555, 556, 557, 558, 559, 564, 565, 566, 567, 568, 569, 575, 577, 578, 579, 580, 581, 582, 585, 586, 587, 588, 589, 590, 594, 596, 597, 598, 599, 600, 601, 602, 603, 604, 605, 607, 608, 609, 610, 611, 617, 622, 623, 624, 625, 627, 634, 635, 636, 637, 643, 644, 645, 646, 647, 648, 652, 655, 656, 657, 659, 660, 662, 663, 664, 665, 666, 667, 668, 669, 672, 673, 674, 675, 676, 677, 678, 679, 686, 687, 688, 689, 690, 691, 695, 697, 698, 699, 700, 701, 703, 704, 705, 707, 712, 713, 717, 719, 720, 721, 722, 728, 729, 730, 732, 733, 734, 735, 736, 737, 738, 739, 740, 743, 744, 745, 746, 747, 748, 749, 752, 753, 754, 755, 756, 757, 762, 763, 764, 765, 766, 767, 769, 771, 772, 774, 777, 778, 779, 781, 784, 785, 786, 787, 788, 793, 794, 795, 796, 797, 802, 803, 808, 809, 810, 812, 817, 818, 819, 820, 821, 825, 827, 828, 829, 835, 836, 837, 838, 840, 843, 844, 845, 847, 851, 852, 853, 854, 855, 856, 858, 859, 860, 862, 866, 867, 868, 869, 870, 872, 873, 874, 875, 880, 881, 884, 885, 886, 887, 888, 889, 890, 892, 893, 894, 895, 896, 897, 898, 900, 901, 902, 903, 904, 905, 906, 907, 908, 909, 910, 911, 912, 914, 916, 917, 919, 920, 921, 922, 923, 924, 925, 926, 927, 928, 929, 930, 931, 933, 935, 936, 937

**Arc Addition Postulate,** 726

**Arcs,** 724–731, 733–738
- adding, 726
- adjacent, 726
- bisecting, 734
- congruent, 725, 734
- intercepted, 741
- length of, 727
- major, 725
- measure of, 725
- minor, 725, 733
- naming, 725
- semi circle, 725

**Area** *See also* Surface area
- of circles, 71, 784–789, 851
- cross-sectional, 832
- of cylinders, 816
- geometric probability to find, 900–901
- of a hemisphere, 849
- models, 916
- probability ratio, 900
- of a sector, 785
- of rectangles, P21, 11, 255
- of rhombi, 254
- of similar solids, 866
- of squares, 69
- of triangles, 42, 127, 254

**Argand plane,** 185

**ASA (Angle-Side-Angle) Congruence Postulate.** *See* Angle-Side-Angle (ASA) Congruence Postulate

**Assessment.** *See* Chapter Test; Guided Practice; Mid-Chapter Quizzes; Prerequisite Skills; Quick Checks; Spiral Review; Standardized Test Practice
- Extended Response, 184, 223, 331, 361, 401, 433, 471, 520, 537, 539, 569, 615, 690, 711, 701, 774, 801, 812, 879, 905, 941
- Gridded Response, 213, 223, 234, 243, 283, 307, 331, 352, 389, 401, 425, 471, 493, 530, 539, 549, 615, 659, 711, 723, 732, 767, 801, 855, 879, 921, 941
- Multiple Choice, 222, 283, 291, 298, 307, 314, 329, 330, 322, 343, 352, 361, 371, 382, 389, 400, 415, 425, 433, 444, 452, 462, 467, 468, 470, 483, 493, 503, 511, 520, 530, 537, 538, 549, 559, 569, 581, 590, 599, 605, 613, 614, 627, 637, 648, 659, 669, 679, 690, 701, 709, 710, 723, 732, 740, 748, 757, 767, 774, 781, 790, 799, 800, 812, 821, 830, 838, 847, 855, 862, 870, 878, 889, 898, 905, 914, 921, 931, 939, 940
- Practice Test, 327, 397, 467, 535, 611, 707, 797, 870, 937
- SAT/ACT, 79, 177, 184, 197, 205, 213, 234, 243, 283, 291, 298, 307, 314, 322, 343, 352, 361, 371, 382, 389, 415, 425, 433, 444, 452, 462, 483, 493, 503, 511, 520, 530, 549, 559, 569, 581, 590, 599, 605, 627, 637, 648, 659, 669, 679, 690, 701, 723, 732, 740, 748, 757, 767, 774, 781, 790, 812, 821, 830, 838, 847, 855, 862, 870, 889, 898, 905, 914, 921, 931
- Short Response, 177, 197, 205, 213, 223, 291, 298, 314, 331, 343, 371, 382, 399, 401, 415, 444, 452, 462, 471, 483, 503, 511, 539, 559, 581, 590, 599, 605, 615, 627, 637, 648, 669, 679, 711, 740, 748, 757, 781, 790, 801, 821, 830, 838, 847, 862, 870, 877, 879, 889, 898, 914, 931, 941

**Associative Property**
- of addition, 286
- of multiplication, 178, 180, 286

**Asymptotes**
- of exponential functions, 227

**Auxiliary line,** 335

**Axes** *See also* Axis of symmetry
- of a cone, 824
- of a cylinder, 816

**Axiom,** 276

**Axis of symmetry,** 93–102

## B

**Base angles,** 374, 521–522

**Base edges,** 814, 822

**Bases,** 814, 826
- of trapezoids, 521

**Bernoulli, Jakob,** 910

**Best-fit lines**
regression equations, 146–147

**Binomials,** 7, 35
additive inverse, 9
conjugates, 247
difference of squares, 58–63, 80, 171
factoring, 35
monomials, 80
multiplying, 20–21, 22–23, 170

**Bisecting arcs,** 734

**Bisectors**
of angles, 405, 408–409, 421, 586, 758
of arcs, 734
constructing, 405
perpendicular, 405, 406–407, 421
of triangles, 405, 406–415

**Brahmagupta,** 191

**Buffon's Needle,** 721

## C

**Calculator.** *See* Graphing Technology Labs; TI-Nspire Technology Labs

**Careers.** *See* Real-World Careers

**Cavalieri's Principle,** 832, 841, 842

**Centers**
of circles, 715–716, 736, 775–776
of dilation, 593, 595
of gravity, 418
of spheres, 848–849

**Central angles,** 785
of circles, 724

**Central tendency.** *See* Measures of central tendency

**Centroid Theorem,** 417–418

**Centroid,** 417–418, 419, 421

**Challenge.** *See* Higher-Order-Thinking Problems

**Changing dimensions,** 558, 568, 699, 828, 835, 836, 846, 854

**Chapter 0**
Algebraic Expressions, P10
Changing Units of Measure Between Systems, P6–P7
Changing Units of Measure Within Systems, P4–P5
Linear Equations, P11–P12
Linear Inequalities, P13–P14
Ordered Pairs, P23–P24
Simple Probability, P8–P9
Square Roots and Simplifying Radicals, P27–P28
Systems of Linear Equations, P25–P26

**Check for Reasonableness,** *See* Study Tips, 50, 126, 128

**Chords,** 715, 733–740, 741
congruent, 733
diameter perpendicular to, 735
equidistant from center, 735
intersecting, 759, 760, 768
radius perpendicular to, 734
segments of, 768
of spheres, 848

**Circle Graphs,** 726, 728–730

**Circles,** 715–723
area of, 71, 103, 784–789, 849
centers of, 715–716, 736, 775–776
central angles of, 724
circumference, 71, 717–720, 825, 849, 851
circumscribed, 758
concentric, 716, 766
congruent, 716
degrees of, 724
equations of, 775–781
graphing, 726, 776
great, 849, 857
inscribed, 758
intersections of, 716–717, 779
intersections inside or outside of, 759–762
missing measures, 785
pairs, 716
sectors of, 785–787
segments of, 715
semicircles, 725
similar, 716
small, 861
special segments in, 715, 768–774
standard form of, 775
tangent to, 750–756

**Circular permutations,** 893

**Circumcenter Theorem,** 407–408

**Circumcenter,** 407–408, 421

**Circumference,** 71, 717–720, 825, 849, 851

**Circumscribed circles,** 758

**Circumscribed figures,** 718

**Circumscribed polygons,** 753

**Classifying**
quadrilaterals, 475, 485–530
triangles, 385, 393, 633

**Closed,** 9, 251

**Closure Property,** 251
irrational numbers, 251
rational numbers, 251

**Coefficients**
leading, 8–11, 117, 125

**Combinations,** 894

**Common Core State Standards, Mathematical Content,** P8, P15, P22, 5, 7, 14, 20, 22, 28, 35, 36, 43, 45, 52, 58, 64, 93, 104, 105, 112, 114, 122, 124, 130, 133, 140, 146, 148, 156, 244, 245, 251, 252, 257, 259, 276, 284, 292, 299, 308, 315, 335, 344, 353, 362, 364, 372, 374, 383, 390, 405, 406, 416, 417, 426, 437, 445, 446, 453, 475, 485, 494, 495, 505, 512, 521, 543, 551, 560, 570, 572, 583, 593, 600, 619, 628, 629, 640, 649, 650, 660, 662, 670, 680, 682, 692, 694, 715, 724, 733, 741, 750, 758, 759, 768, 775, 782, 784, 807, 814, 822, 831, 841, 848, 883, 890, 899, 907, 915, 922, 924

**Common Core State Standards, Mathematical Practice,**
Arguments, P21, 50, 120, 144, 255, 279, 280, 281, 285, 287, 288, 289, 295, 296, 304, 305, 306, 312, 341, 349, 350, 351, 359, 360, 368, 386, 387, 413, 414, 423, 424, 432, 439, 441, 458, 459, 482, 489, 491, 500, 502, 517, 519, 528, 529, 555, 557, 579, 588, 589, 596, 625, 636, 700, 722, 731, 738, 739, 744, 747, 765, 766, 789, 811, 829, 846, 861, 888, 897, 913, 920
Critique, 7, 12, 41, 56, 105, 109, 259, 262, 297, 342, 370, 443, 510, 580, 647, 678, 773, 837, 930
Modeling, 17, 52, 55, 108, 124, 126, 137, 148, 151, 259, 311, 369, 480, 481, 490, 509, 518, 546, 547, 566, 567, 578, 597, 602, 603, 623, 624, 645, 646, 665, 666, 667, 675, 687, 719, 721, 729, 737, 779, 787, 836, 845, 860, 895, 911, 912, 929
Perseverance, 18, 58, 62, 128, 479, 545, 634, 695, 699, 815, 887
Precision, P15, P16, 64, 69, 105, 106, 127, 133, 134, 248, 310, 381, 461, 689, 716, 728, 820, 850, 852, 891

Reasoning, P19, 11, 36, 39, 68, 93, 99, 252, 261, 320, 321, 388, 418, 442, 451, 492, 501, 548, 558, 568, 598, 626, 658, 668, 730, 780, 859, 886, 902, 904, 919
Regularity, 26, 28, 29, 45, 46, 70, 114, 119, 245, 339, 340, 341, 348, 379, 380, 488, 508, 556, 601, 867
Sense-Making, 10, 31, 40, 58, 60, 61, 114, 116, 282, 290, 319, 336, 350, 358, 378, 411, 412, 422, 430, 431, 449, 450, 460, 526, 527, 587, 604, 631, 635, 644, 657, 677, 688, 698, 720, 751, 754, 755, 788, 818, 819, 825, 827, 834, 835, 844, 853, 868, 869, 884, 896, 903
Structure, 22, 25, 32, 45, 49, 101, 102, 138, 140, 245, 247, 249, 251
Tools, P20, 14, 15, 146, 565, 674, 676, 745, 746, 764, 772, 778
Tools, 313, 354, 653, 656, 697, 756, 809, 810, 828, 854

**Common tangent,** 750

**Commutative Property**
of addition, 286
of multiplication, 178, 180, 286

**Comparison Property,** 426

**Complement Theorem,** 300

**Complement,** 927–928

**Complementary angles,** 299–300, 640, 658

**Complementary events,** 927

**Completing the square,** 124–129, 219, 777

**Complex Conjugates Theorem,** 76

**Complex conjugates,** 76, 181

**Complex numbers,** 105, 178–184, 192, 216, 218
adding, 180
absolute value, 186
conjugates, 76, 181
dividing, 181
graphing, 183
multiplying, 181
subtracting, 180

**Complex plane,** 185

**Component form,** 684

**Composite events,** 915

**Composite figures,** 820, 887

**Compound events,** 915

**Compound interest,** 238–239

**Concentric circles,** 716, 766

**Concept Summary**
circle and angle relationships, 763
Conjectures, making, 206
factoring methods, 65
lateral and surface areas of solids, 826
linear and nonlinear functions, 140
operations with radical expressions, 254
parallelograms, 513
probability rules, 927
prove that a quadrilateral is a parallelogram, 497
proving triangles congruent, 367
solving a triangle, 674
solving quadratic equations, 136, 194
special functions, 151
special segments and points in triangles, 421
transformations of quadratic functions, 118, 202
triangle similarity, 563
types of dilations, 593
volumes of solids, 843
zeros, factors, roots, and intercepts, 72

**Concurrent lines,** 407
altitudes of triangles, 416, 419–421
angle bisectors of triangles, 421
medians of triangles, 416, 417, 421
perpendicular bisectors of triangles, 421
points of concurrency, 407, 409, 417–418, 419–420, 421, 758

**Conditional probability,** 917–918, 927

**Conditional statements,** 285, 323

**Conditions for Rhombi and Squares Theorems,** 514

**Cones**
altitude of, 824
axis of, 824
lateral area of, 824–825, 826
oblique, 824–826
right, 824–825
surface area of, 824–826
vertex of, 824
volume of, 842–845, 871

**Congruent**
angles, 301, 308, 316, 374–375, 476, 551
arcs, 734
chords, 733
isosceles triangles, 822
lateral edges, 822
polygons, 344–345
segments, 293–294, 374–375
tangents, 752
triangles, 344–351, 377
properties of, 347
proving, 346, 353–361, 364–371

**Congruent arcs,** 725

**Congruent Complements Theorem,** 301–302

**Congruent solids,** 864

**Congruent Supplements Theorem,** 301–302

**Congruent triangles**
Angle-Angle-Side (AAS) Congruence Theorem, 365, 367
Angle-Side-Angle (ASA) Congruence Postulate, 364, 367
Hypotenuse-Angle (HA) Congruence Theorem, 373
Hypotenuse-Leg (HL) Congruence Theorem, 373
Leg-Angle (LA) Congruence Theorem, 373
Leg-Leg (LL) Congruence Theorem, 373
Side-Angle-Side (SAS) Congruence Postulate, 355, 367
Side-Side-Side (SSS) Congruence Postulate, 353, 367

**Conics,** 782

**Conic Sections,** 782

**Conjectures, making,** 372, 445, 550, 649, 839

**Conjugate,** 76, 181, 247

**ConnectED**
Animations, 5, 8, 20, 23, 30, 35, 43, 74, 90, 104, 166, 179, 202, 208, 224, 229, 272, 315, 332, 356, 362, 367, 372, 383, 390, 402, 405, 416, 472, 505, 513, 540, 552, 576, 616, 628, 638, 758, 802, 856, 932
Chapter Readiness Quiz, 3, 91, 167, 273, 333, 403, 473, 541, 617, 713, 803, 881
Graphing Calculator, 111–113, 122–123, 146–147, 156, 169, 187, 198, 215, 235, 445, 494, 550, 649, 660, 839
Personal Tutor, P4–P28, 4, 7, 8, 9, 14, 15, 16, 22, 23, 24, 28, 29, 30, 36, 37, 38, 39, 46, 47, 48, 53, 54, 58, 59, 60, 65, 66, 67, 68, 73, 74, 75, 76, 92, 93, 94, 95, 96, 97, 98, 105, 106, 107, 111, 112, 114, 115, 116, 117, 118, 122, 124, 125, 126, 133, 134, 135,

136, 140, 141, 142, 146, 148, 149, 150, 169, 170, 171, 172, 173, 178, 179, 180, 181, 187, 190, 191, 192, 193, 198, 200, 201, 202, 207, 208, 209, 210, 227, 228, 229, 230, 231, 237, 238, 239, 245, 246, 247, 252, 253, 254, 259, 260, 277, 278, 284, 285, 286, 293, 294, 299, 300, 302, 303, 308, 309, 310, 316, 317, 318, 336, 338, 345, 346, 353, 354, 356, 365, 366, 374, 376, 377, 383, 384, 385, 407, 408, 409, 417, 418, 420, 427, 428, 429, 437, 438, 439, 445, 446, 447, 448, 453, 455, 456, 457, 476, 477, 478, 479, 484, 486, 487, 488, 494, 496, 497, 498, 505, 506, 507, 513, 514, 515, 516, 522, 524, 525, 543, 544, 545, 550, 551, 552, 553, 554, 560, 562, 563, 564, 572, 573, 574, 575, 576, 584, 585, 586, 594, 595, 600, 601, 619, 620, 621, 622, 630, 631, 633, 640, 641, 642, 643, 649, 651, 652, 653, 654, 660, 662, 663, 664, 670, 671, 672, 673, 682, 683, 684, 685, 686, 694, 695, 696, 715, 716, 717, 718, 724, 725, 726, 727, 733, 734, 735, 736, 742, 743, 744, 750, 751, 752, 753, 760, 761, 762, 768, 769, 770, 771, 775, 776, 777, 784, 785, 807, 808, 815, 816, 817, 823, 824, 825, 826, 831, 832, 833, 839, 841, 842, 843, 849, 850, 851, 857, 858, 865, 866, 883, 884, 885, 890, 891, 892, 893, 894, 899, 900, 901, 907, 908, 909, 910, 915, 916, 917, 918, 922, 924, 925, 926, 927, 928

Self-Check Practice, P19, 3, 10, 16, 25, 30, 39, 49, 55, 60, 68, 77, 86, 89, 91, 99, 108, 119, 126, 137, 143, 151, 161, 165, 167, 174, 182, 194, 203, 210, 223, 225, 231, 240, 248, 254, 261, 273, 279, 287, 295, 304, 327, 331, 311, 318, 333, 339, 347, 357, 367, 378, 386, 397, 401, 403, 411, 421, 430, 440, 448, 457, 467, 471, 473, 479, 489, 499, 508, 517, 526, 535, 539, 541, 546, 554, 565, 577, 586, 596, 602, 611, 615, 617, 623, 633, 644, 655, 665, 674, 687, 697, 707, 711, 713, 719, 728, 736, 745, 753, 763, 771, 778, 786, 797, 801, 803, 809, 817, 827, 834, 844, 852, 859, 867, 875, 879, 881, 886, 895, 902, 911, 919, 929, 937, 941

**Connections.** *See* Applications

**Consecutive Interior Angles Converse,** 316

**Consecutive Interior Angles Theorem,** 309

**Consecutive interior angles,** 309, 316

**Constant**
- term, 95

**Constructions**
- altitude of a triangle, 416
- circle through three noncollinear points, 736
- congruent triangles using sides, 355
- congruent triangles using two angles and included side, 364
- congruent triangles using two sides and the included angle, 356
- median of a triangle, 416
- parallel line through a point not on a line, 315
- proving, 362
- trisect a segment, 576
- using paper folding, 405
- using string, 416

**Continuous functions**
- exponential, 227–234, 237, 238, 239
- polynomial, 72–76
- quadratic, 200–204, 206, 217

**Contours,** 813
- Contraction. *See* Dilations

**Contractions.** *See* Dilations

**Converse**
- of Alternate Exterior Angles Theorem, 316
- of Alternate Interior Angles Theorem, 316
- of Angle Bisector Theorem, 409
- of Consecutive Inferior Angles Theorem, 316
- of Hinge Theorem, 453
- of Isosceles Triangle Theorem, 374
- of Perpendicular Bisector Theorem, 406
- of Perpendicular Transversal Theorem, 316
- of Pythagorean Theorem, 632
- of Triangle Proportionality Theorem, 573

**Convex polygon,** 475

**Coordinate geometry,** 343, 352, 363, 413, 414, 415, 420, 421, 422, 432, 433, 434, 444, 488, 489, 490, 491, 498, 500, 501, 503, 504, 507, 508, 509, 516, 517, 518, 520, 522, 526, 527, 529, 530, 534, 557, 559, 567, 580, 581, 635, 646, 657, 658, 669, 678, 698, 699, 731, 767, 774, 788, 855, 889, 903, 904

**Coordinate graph.** *See* Coordinate plane

**Coordinate grid.** *See* Coordinate plane

**Coordinate plane,** 638
- centroid, 418–419
- coordinate proof, 383, 384
- dilations in the, 696
- orthocenter, 420
- parallelograms on, 498

**Coordinate proofs,** 383–389, 393, 498–499

**Coordinates.** *See also* Ordered pairs in space, 638–639

**Coplanar points,** 716

**Corollaries,** 338
- Congruent Parts of Parallel Lines, 575
- Equilateral Triangle, 375
- Proportional Parts of Parallel Lines, 574
- Triangle Angle-Sum, 338

**Corollary to the Fundamental Theorem of Algebra,** 73

**Corresponding Angles Postulate,** 308, 309

**Corresponding parts,** 344–345
- angles, 308, 309, 315, 344, 551–552
- sides, 344, 551–553
- vertices, 344

**Cosecant**
- graphing, 660

**Cosine ratio,** 650–658, 674
- law of, 671–679

**Cosines, Law of,** 671–679

**Cotangent**
- graphing, 660

**Counterexamples,** 70

**CPCTC (corresponding parts of congruent triangles are congruent),** 345

**Critical Thinking.** *See* Higher-Order-Thinking Problems

**Cross Products Property,** 544–545

**Cross products,** 544–545

**Cross section,** 808

**Cross-Curriculum.** *See* Applications

**Cubes**
- surface area of, 29
- Curve fitting, 146–147

**Cylinders**
area of, 816
axis of, 816
lateral area of, 816, 817–818, 826
oblique, 816
right, 816
surface area of, 816, 817–818, 826, 871
volume of, 832, 835, 836, 843, 871

## D

**Data collection device,** 215
**Data.** *See also* Graphs; Statistics; Tables
measures of central tendency. *See* Measures of central tendency, 938
**David, Florence Nightingale,** 150
**Decay**
exponential, 229–231
factor, 229–233
**Decision making,** P9, 904, 912, 913, 920
**Deductive argument,** 278
**Degree** 724
of a monomial, 7
of a polynomial, 7
**Denominators, rationalizing,** P28, 247, 641
**Density**
BTUs, 845
soil, 836
**Dependent events,** 915, 917–918, 927
**Depression, angles of,** 662–668, 702
**Descartes, René,** 499
**Descartes' Rule of Signs,** 74
**Descending order,** 8
**Design,** 820, 836, 854
**Diagnose readiness.** *See* Prerequisite Skills
**Diagonals**
of parallelograms, 487
of polygons, 475
of rectangles, 505–506
of rhombi, 512, 514
of squares, 640
of trapezoids, 521, 522
**Diagonals of a Rectangle Theorem,** 505–506
**Diagonals of a Rhombus Theorem,** 512
**Diagrams,** 360, 455, 827, 846
tree, 883–884
Venn, 180, 529, 918, 925–926, 939
**Diameter,** 715–718
perpendicular to a chord, 735
of spheres, 848
**Difference of squares,** 58–63, 171
**Differences.** *See also* Subtraction
product of a sum and, 30
square of, 28–29
of squares, 58–63, 84
successive, 140–142
**Dilations,** 116, 593–599, 692–693, 694–701
centers of, 593, 595
in the coordinate plane, 696
isometry, 695
scale factor of, 593–594, 695
types of, 593
**Dimensional analysis,** P5, P7, 853, 867, 868, 869
**Direction of vectors,** 682, 684, 686
**Directrix,** 783
**Discrete mathematics.** *See* Combinations; Permutations; Probability; Sequences; Statistics
**Discriminants,** 136–137, 192–195, 219
**Distance Formula,** 523, 638–639, 682, 775
**Distributive Property,** 14–16, 22–24, 180, 284
**Division**
by a negative, P14
of complex numbers, 181
ratios, 543
square roots, 246
**Division Property of Equality,** 284
**Domains**
of exponential functions, 237–239
**Dot plot.** *See* Scatter plots
**Double root,** 106
**Drawings,** 769, 876
justifying, 432
orthographic, 390, 808–809
perspective, 810
scale, 600–605, 606, 637
segments, 735, 769

## E

**Edges**
connecting nodes, 932–933
of polyhedrons, 814, 822
**Elevation, angle of,** 662–668, 702
**Elimination method,** P26
**Ellipse,** 782
**End behavior,** 97, 228
**Endpoints,** 416
**Enlargement,** 593–598, 695–696
**Equations**
of circles, 775–781
exponential, 235–236, 237–239
linear, P11–P12
multi-step, P11
quadratic, 169, 170–177, 179, 217, 218
regression, 169
roots of, 170
solving, 170–177, 178–259, 187–196, 216, 217, 218, 235–236, 237–239, 284–285
standard form of, 174–176, 189
systems of, P25–P26
writing, 833
**Equations**
graphing, 122–123
with polynomial expressions, 16
quadratic, 45–51, 52–57, 58–63, 64–71, 97, 105–110, 122–123, 124–129, 133–139
radical, 259–263
regression, 146–147
roots of, 105, 106, 107
solve
by factoring, 48, 54, 60, 66
systems of, 122–123
write, 142–143
**Equidistant,** 736
**Equilateral triangles,** 375–377, 393
**Equivalence relation,** 290, 558
**Equivalent proportions,** 545
**Eratosthenes,** 663
**Error Analysis.** *See* Higher-Order-Thinking Problems
**Essential Questions,** 12, 56, 138, 144, 212, 282, 290, 321, 370, 381, 432, 451, 502, 510, 558, 647, 678, 700, 722, 756, 829, 837, 904, 913

**Estimation of solutions,** 107, 108, 109, 259
**Estimation,** 450, 585, 657, 695, 816, 901
**Euclid,** 726
**Euclidean geometry,** 857–861
**Euler, Leonhard,** 925
**Events,** P8, 883. *See also* Probability
**Expected value,** 909–910
**Experimental probability,** P9
**Experiments,** P8, 883
**Exponential decay,** 229–231
**Exponential equations,** 235, 236, 237–239
**Exponential functions.** *See also* Functions
end behavior, 97
graphing, 227–234
Property of Equality, 237
Property of Inequality, 239
transformations of, 228
writing, 238
**Expressions.** *See also* Algebraic expressions; Numerical expressions
algebraic, 14–15
polynomial, 16
quadratic, 23
radical, 245–250, 252–253, 254
as radicand, 260
**Exponential growth,** 227–229
compound interest, 238–239
**Exponential inequalities,** 235–236, 239
**Expressions**
algebraic, P10
radical, P27
**Extended ratios,** 543–544
**Extended Response,** *See* Standardized Test Practice, 184, 223, 331, 354, 357, 361, 401, 433, 471, 520, 539, 569, 615, 690, 711, 701, 774, 801, 812, 879, 905, 941
**Exterior Angle Inequality Theorem,** 427–428, 430
**Exterior Angle Theorem,** 337–338, 426
**Exterior angles,** 308, 309, 316, 337–338, 426, 427–428, 430, 478–480, 484
**External secant segments,** 770
**Extraneous solutions,** 260
**Extremes,** 544, 619

## F

**Faces**
lateral, 814, 822
**Factored form,** 36, 170
**Factorials,** 890
**Factoring** 36–42, 58, 494–500, 521, 530, 532, 534
greatest common factor, 171
polynomials, 171–176, 72
quadratic equations, 170–176, 194, 218
trinomials, 171–172
**Factors, greatest common,** 171
**Families of graphs**
of parabolas, 198–199
of quadratic functions, 200
**Fermat, Pierre de,** 419
**Fibonacci Sequence,** 550
**Financial Literacy,** 175, 203, 231, 241, 371, 625
interest, 238–239
money, 240
**Flip.** *See* Reflections
**Flow proofs,** 337, 500, 503
**Focus,** 783
**FOIL method,** 170–181
**Foldables® Study Organizer**
circles, 714, 792
congruent triangles, 334, 393
exponential and logarithmic functions, 226
factoring and quadratic equations, 4, 80
probability and measurement, 882, 934
proportions and similarity, 542, 606
quadratic functions and equations, 92, 157, 168, 216
quadrilaterals, 474, 531
reasoning and proof, 274, 323
relationships in triangles, 404, 463
right angles and trigonometry, 618, 702
surface area and volume, 804, 871
**Formal proofs,** 285
**Formulas** *See also* Interest; Measurement; Rates
area
of circles, 708, 784, 849
of cylinders, 816, 826
of rectangles, 708
of a sector, 785
of triangles, 708
circumference, 708, 784, 851
conditional probability, 917
distance, 523, 638–639, 682, 775
lateral area
of cones, 825, 826
of cylinders, 826
of prisms, 814, 826
of pyramids, 826
of regular pyramids, 823
midpoint, 384, 638–639
quadratic, 133–139, 189–192, 219
rhombi, 254,
standardized tests, 708
surface area
of cones, 825, 826
of cylinders, 826
of prisms, 826
of pyramids, 826
of spheres, 848–850
triangles, 254
using, 162
volume
of cones, 842, 843
of cylinders, 708, 832, 843
of prisms, 708, 831, 843
of pyramids, 841, 843
of spheres, 261, 850, 851
volume of prisms, 70
**Four-step problem solving plan,** 24, 48, 142, 259
**Fractal trees,** 591
**Fractals,** 591–592
**Fractions.** *See also* Rational numbers
in the radicand, 246
**Frustum,** 829
**Functions**
absolute value, 149–151
analyzing using ratios, 141–144
cosine, 650–658, 674
decreasing, 97
end behavior, 97
exponential, 227–234, 237, 238, 239
greatest integer, 148, 151
increasing, 97
inverse, P15–P21
linear, P17
nonlinear, 93, 151
piecewise-defined, 148, 150–151
piecewise-linear, 156
polynomial, 72–76

quadratic, 93–103, 104, 114–121, 157, 158, 200–204, 206, 216–217
sine, 650–658, 674
special, 148–155
step, 148–149, 151
tangent, 650–658, 674
transformations of, 200–204, 216, 228
zeros of, 72–76, 105, 107

**Fundamental Counting Principle,** 885

**Fundamental Theorem of Algebra,** 72–73

## G

**Galilei, Galileo,** 68, 573

**Gauss, Johann Carl Friedrich,** 345

**GCF.** *See* Greatest common factor (GCF)

**General trinomials,** 171

**Geometric Mean (Altitude) Theorem,** 621

**Geometric Mean (Leg) Theorem,** 621

**Geometric means,** 619–627, 702
Pythagorean Theorem, 629
right triangles in, 619–621

**Geometric models,** 907–908

**Geometric probability,** 899–905
angle measures to find, 901
area to find, 900–901
lengths to find, 899

**Geometric proofs,** 286, 439

**Geometry**
angles, 110
area, 69, 71, 254, 255
circles, 71, 103
circumference, 71
cones, 261
parallelogram, 48
perimeter, 50, 255
rectangles, 12, 49, 50, 51, 83, 145, 174, 175, 188, 254, 255
rectangular prisms, 70
rhombus, 254
solids, 256
spheres, 261
squares, 69, 175
surface area, 261
triangles, 19, 26, 42, 49, 127, 254
volume, 70
Get Ready for the Chapter. *See* Prerequisite Skills

**Geometry Labs**
Adding Vectors, 691
The Ambiguous Case, 680–681
Congruence in Right Triangles, 372–373
Constructing Bisectors, 405
Constructing Medians and Altitudes, 406
Coordinates in Space, 638–639
Fractals, 591–592
Graph Theory, 932–933
Inscribed and Circumscribed Circles, 758
Locus and Spheres, 856
Matrix Logic, 435–436
Navigational Coordinates, 863
Necessary and Sufficient Conditions, 275
Orthographic Drawings and Nets, 390–392
Parabolas, 782–783
Proof Without Words, 628
Proofs of Perpendicular and Parallel Lines, 570–571
Proving Constructions, 362
Solids Formed by Translations, 805–806
Topographic Maps, 813
Two-Way Frequency Tables, 922–923

**Geometry Software Labs**
Parallelograms, 494
Triangle Inequality, 445

**Get Ready for the Chapter.** *See* Prerequisite Skills

**Get Ready for the Lesson.** *See* Prerequisite Skills

**Golden rectangle,** 547

**Graph theory,** 932–933
edges, 932–933
efficient route, 933
network, 932–933
nodes, 932–933
traceable network, 932–933
weight, 933
weighted vertex-edge graphs, 933
weight of a path, 933

**Graphing Calculator.** *See* ConnectED, 101, 109

**Graphing Technology Labs**
Angle of Polygons, 484
Changing Dimensions, 839
Curve Fitting, 146–147
Dilations, 692–693
Families of Parabolas, 198–199
Family of Quadratic Functions, 112–113
Fibonacci Sequence and Ratios, 550
Modeling Motion, 215
Modeling Real-World Data, 169
Parallelograms, 494
Piecewise-Linear Functions, 156
Quadratic Inequalities, 111
Secant, Cosecant, and Cotangent, 660
Solving Exponential Equations and Inequalities, 235–236
Solving Quadratic Equations, 187
Systems of Linear and Quadratic Equations, 122–123
Triangle Inequality, 445
Trigonometry, 649
Volume and Changing Dimensions, 839

**Graphs.** *See also* Data; Statistics
circles, 726
of absolute value functions, 149–151
of circles, 776
cosecant, 660
cotangent, 660
estimating solutions, 107
exponential functions, 227–234
families of, 198–199, 200
parabolas, 93
parallelograms, 494
parent, 200
piecewise-linear functions, 156
preparing for standardized tests, 220
quadratic equations, 105–110, 191–193, 217
quadratic functions, 93–103, 112–113, 157, 158, 216, 217
quadratic inequalities, 111, 207, 208–212
secant, 660
vertex form, 202

**Great circle,** 849, 857

**Greatest common factor (GCF)** 53, 59
factoring, 171

**Greatest integer functions,** 148, 151

**Gridded Response,** *See* Standardized Test Practice, 213, 223, 234, 243, 283, 307, 331, 352, 389, 401, 425, 471, 493, 524, 530, 539, 549, 615, 659, 711, 723, 732, 767, 801, 855, 879, 919, 921, 941

**Growth**
exponential, 227–229, 238–239
factor, 229

**Guided Practice,** P15, P16, P17, P18, 7, 8, 9, 14, 15, 16, 23, 24, 28, 29, 30, 36, 37, 38, 39, 46, 47, 48, 53, 54, 58, 59, 60, 65, 66, 67, 68, 73, 74, 75, 76, 93, 94, 95, 96, 97, 98, 105, 106, 107, 114, 115, 116, 117, 118, 125, 126, 133, 134, 135, 136, 142, 148, 149, 150, 170, 171, 172, 173, 178, 179, 180, 181, 190, 191, 192, 193, 200, 201, 202, 207, 208, 209, 210, 245, 246, 247, 252, 253, 254, 260, 227, 228, 229, 230, 231, 237, 238, 239, 277, 278, 285, 286, 293, 294, 299, 300, 302, 303, 308, 309, 310, 317, 318, 336, 338, 345, 346, 347, 353, 354, 356, 357, 365, 366, 367, 375, 376, 377, 383, 384, 385, 407, 408, 409, 410, 417, 418, 419, 420, 427, 428, 429, 437, 438, 439, 440, 446, 447, 448, 454, 455, 456, 457, 476, 477, 478, 479, 486, 487, 488, 496, 497, 498, 499, 505, 506, 507, 513, 514, 515, 516, 522, 523, 524, 525, 543, 544, 545, 551, 552, 553, 554, 560, 562, 563, 564, 572, 573, 574, 575, 576, 584, 585, 586, 594, 595, 600, 601, 619, 620, 622, 630, 631, 632, 633, 641, 642, 643, 651, 652, 653, 654, 663, 664, 670, 671, 672, 673, 682, 683, 684, 685, 686, 694, 696, 715, 716, 717, 718, 724, 725, 726, 727, 733, 734, 735, 736, 742, 743, 744, 750, 751, 752, 753, 760, 761, 762, 768, 769, 770, 771, 775, 776, 777, 784, 785, 786, 807, 808, 815, 816, 817, 823, 824, 825, 826, 831, 832, 833, 842, 843, 848, 849, 850, 851, 857, 858, 865, 866, 883, 884, 885, 890, 891, 892, 894, 899, 900, 901, 908, 909, 910, 915, 916, 917, 918, 924, 925, 926, 927, 928

## H

**H.O.T. Problems.** *See* Higher-Order Thinking Problems

**HA (Hypotenuse-Angle) Congruence Theorem.** *See* Hypotenuse-Angle (HA) Congruence Theorem, 373

**Hands-On.** *See also* Algebra Labs; Algebra Tiles; Manipulatives

**Heights**
- of prisms, 814
- of pyramids, 822, 850
- slant, 822

**Hemispheres**
- surface area of, 849, 852
- volume of, 850, 852

**Higher-Order Thinking Problems**
- Challenge, P21, P22, 12, 18, 26, 41, 50, 56, 62, 70, 78, 102, 104, 109, 120, 138, 144, 176, 183, 196, 204, 212, 233, 242, 249, 255, 262, 282, 290, 297, 306, 313, 321, 342, 351, 360, 370, 381, 388, 414, 424, 432, 443, 451, 461, 482, 492, 502, 510, 519, 529, 548, 558, 568, 580, 589, 598, 604, 626, 628, 636, 647, 658, 668, 678, 681, 689, 700, 722, 731, 739, 747, 756, 766, 773, 780, 789, 811, 820, 829, 837, 846, 854, 861, 869, 888, 897, 904, 920, 930
- Error Analysis, 18, 50, 62, 70, 102, 176, 183, 196, 212, 233, 242, 282, 297, 321, 342, 351, 360, 370, 381, 414, 424, 443, 482, 510, 519, 529, 548, 580, 589, 604, 626, 647, 668, 678, 731, 773, 789, 820, 837, 846, 869, 930
- Open Ended, P21, 18, 26, 32, 41, 50, 62, 70, 78, 102, 109, 120, 128, 138, 144, 176, 183, 196, 204, 212, 233, 242, 249, 255, 262, 282, 290, 297, 306, 313, 321, 342, 360, 370, 381, 388, 414, 424, 432, 443, 451, 461, 482, 492, 502, 510, 519, 529, 548, 558, 568, 580, 589, 598, 604, 626, 636, 647, 678, 681, 700, 731, 739, 747, 756, 766, 773, 780, 811, 820, 829, 837, 846, 854, 861, 869, 888, 897, 904, 913, 920, 930, 933
- Proof, 242
- Reasoning, 12, 18, 26, 32, 41, 50, 56, 62, 70, 78, 102, 109, 120, 128, 138, 154, 176, 183, 196, 204, 212, 233, 242, 249, 262, 282, 290, 297, 306, 313, 321, 342, 351, 360, 370, 381, 388, 414, 424, 432, 451, 461, 482, 492, 502, 510, 519, 529, 548, 549, 558, 568, 580, 589, 598, 604, 626, 636, 647, 658, 668, 678, 689, 700, 722, 731, 739, 747, 756, 766, 773, 780, 789, 805, 806, 811, 820, 829, 837, 846, 854, 861, 869, 888, 897, 904, 913, 920, 930, 932
- Which One Doesn't Belong?, 32, 70, 78, 128, 548
- Writing in Math, P21, 6, 12, 18, 21, 26, 32, 35, 41, 44, 50, 56, 62, 70, 78, 102, 109, 120, 128, 138, 144, 176, 183, 196, 204, 212, 233, 236, 242, 249, 255, 262, 282, 290, 297, 306, 313, 321, 342, 351, 360, 370, 381, 388, 414, 424, 432, 443, 451, 461, 482, 492, 502, 510, 519, 529, 548, 558, 568, 580, 589, 598, 604, 626, 628, 636, 639, 647, 658, 668, 678, 689, 700, 722, 731, 739, 747, 756, 766, 773, 780, 789, 811, 820, 829, 837, 846, 854, 861, 869, 888, 897, 904, 913, 920, 923, 930, 933

**Hinge Theorem,** 453–457

**History, Math,** 68, 150, 191, 345, 419, 499, 573, 663, 726, 866, 910, 925

**HL (Hypotenuse-Leg) Congruence.** *See* Hypotenuse-Leg (HL) Congruence Theorem, 373

**Horizontal lines,** 662

**Horizontal translations,** 228

**Hyperbolic geometry,** 861

**Hypotenuse-Angle (HA) Congruence Theorem,** 373

**Hypotenuse-Leg (HL) Congruence Theorem,** 373

**Hypotenuse,** 619–620, 629–637, 640–641

## I

**If-then statements.** *See* Conditional statements

**Image,** 593, 695

**Imaginary axis,** 185

**Imaginary numbers**
- multiplying, 179

**Imaginary unit,** 178–180

**Incenter,** 410, 421

**Incenter Theorem,** 410

**Included angle,** 355–356, 671

**Included side,** 364–367

**Independent events,** 915–916, 927

**Index,** 257

**Indirect measurement,** 564, 622, 672, 773

**Indirect proofs,** 437–444, 739, 756

**Indirect reasoning,** 437

**Inequalities,** 426
- Addition Property, 426
- Comparison Property, 426
- exponential, 235–236, 239

linear, P13–P14
in one triangle, 426–432
properties of, 239, 426
Pythagorean, 632
quadratic, 111, 207–212
Subtraction Property, 426
solving, 235–236
symbols for, 428, 447
systems of, 904
Transitive Property, 426
and triangles, 426–432, 445, 446–452, 453–462, 463, 633
in two triangles, 453–462

**Informal proofs,** 278, 854

**Inscribed angles,** 741–747

**Inscribed Angle Theorem,** 741–742

**Inscribed circles,** 758

**Inscribed figures,** 718

**Inscribed triangles,** 744

**Intercepted arcs,** 741

**Intercepts.** *See x*-intercepts; *y*-intercepts

**Interdisciplinary connections.** *See* also Applications
archaeology, 205, 515, 796, 843
architecture, 2, 40, 211
art, 57, 83, 535, 575, 578, 644, 747, 810, 887, 926
astronomy, 128, 361, 363, 675
biology, 33
business, 77, 79, 197, 212, 217
earth science, 810
geography, 306, 325, 385, 452, 467, 860
health, 103
meteorology, 757, 840
music, 306, 403, 504, 522, 527, 884, 911, 925, 930
physical science, 29, 54, 68, 69, 188, 219, 263
physics, 159
science, 224, 231, 232, 238, 240, 241, 272, 285, 288, 289, 745, 762, 769, 771, 845
weather, 778

**Interest,**
compound, 238–239

**Interior angles,** 309, 316
remote, 337
sum of,
of a polygon, 475–477, 484
of a triangle, 335

**Internet Connections.** *See* ConnectED

**Interpolation.** *See also* Predictions

**Intersections**
of chords, 759, 760, 768
of circle and line, 777
of circles, 716–717, 779
of lines, 276, 759–762
of planes, 276
of spheres, 856

**Inverse cosine,** 653

**Inverse function,** P16

**Inverse relation,** P15

**Inverse sine,** 653

**Inverse tangent,** 653

**Inverses**
additive, 9, 37
drawing, P22

**Investigations.** *See* Algebra Labs; Graphing Technology Labs; Spreadsheet Labs; TI-Nspire Technology Labs, *See* Geometry Labs; Graphing Technology Labs; TI-Nspire™ Technology Labs

**IQR.** *See* Interquartile range (IQR)

**Irrational numbers,** 134, 251
roots, 107

**Isometric view,** 807–809

**Isometry dilation,** 695

**Isosceles Trapezoids Theorem,** 521

**Isosceles trapezoids,** 521–522

**Isosceles Triangle Theorem,** 374–375

**Isosceles triangles,** 374–376
legs of, 374

**Iteration,** 591–592

## J

**Joint frequencies,** 922

## K

**Key Concepts**
absolute value function, 149
absolute value of a complex number, 186
angles of elevation and depression, 702
angles of polygons, 531
angles of triangles, 393
angles, arcs, chords, and inscribed angles, 792
arc length, 727
arcs and arc measure, 725
area of a circle, 784
area of a sector, 785
area probability ratio, 900
areas of a cylinder, 816
calculating expected value, 909
Cavalieri's Principle, 832
circle and angle relationships, 763
circle pairs, 716
circles and circumference, 792
circular permutations, 893
circumference, 717
classifying triangles, 393
combinations, 894, 934
common Pythagorean triples, 630
completing the square, 124
complex conjugates theorem, 76
complex numbers, 179, 216
compound interest, 238
conditional probability, 918
conditional statement, 323
congruent and similar solids, 871
congruent solids, 864
congruent triangles, 393
corollary to the Fundamental Theorem of Algebra, 73
cross products property, 544
deductive reasoning, 323
definition of congruent polygons, 344
definition of inequality, 426
Descartes' Rule of signs, 74
designing a simulation, 907
difference of squares, 58
dilation, 694
dilations in the coordinate plane, 696
dilations, 116
discriminant, 193
distance and midpoint formulas in space, 638
equation of a circle, 775
equivalent proportions, 545
factorial, 890
factoring $ax^2 + bx + c$, 52
factoring by grouping, 37
factoring perfect square trinomials, 64
factoring trinomials and differences of squares, 80
factoring using the distributive property, 80

Index

factoring $x^2 + bx + c$, 45
finding inverse functions, P17
FOIL method for multiplying binomials, 170
FOIL method, 23
Fundamental Counting Principle, 885
Fundamental Theorem of Algebra, 72
geometric mean, 619, 702
geometric probability, 934
graph quadratic functions, 96
graphing quadratic functions, 157
graphing quadratic functions, 216
greatest integer function, 148
horizontal translations, 115
how to write an indirect proof, 437
indirect proof, 463
inductive reasoning and logic, 323
intersections of lines and planes, 276
inverse relations, P15
inverse trigonometric ratios, 653
isosceles triangles, 393
Labs. *See* Algebra Labs; Graphing Technology Labs; Spreadsheet Labs; TI-Nspire Technology Labs
lateral and surface area of a cone, 825
lateral area of a prism, 814
lateral area of a regular pyramid, 822
laws of sines and cosines, 702
length probability ratio, 899
lines in plane and spherical geometry, 857
maximum and minimum values, 95
models, 606
orthocenter, 419
parent function of exponential decay functions, 229
parent function of exponential growth functions, 227
perfect squares and factoring, 80
permutations with repetition, 892
permutations, 891, 934
placing triangles on a coordinate plane, 383
power property of equality, 259
probabilities of compound events, 934
probability of events that are not mutually exclusive, 926
probability of mutually exclusive events, 925
probability of the complement of an event, 927
probability of two dependent events, 917
probability of two independent events, 916
product of a sum and a difference, 30
product property of square roots, 245
proof process, 278, 323
properties of inequality for real numbers, 426
properties of parallelograms, 531
properties of real numbers, 284
properties of rectangles, rhombi, squares, and trapezoids, 531
property of equality for exponential functions, 237
property of inequality for exponential functions, 239
proportional parts, 606
proportions, 606
Pythagorean Theorem, 702
quadratic formula, 133
quadratic formula, 190
quadratic functions, 93
quotient property of square roots, 246
radius and diameter relationships, 716
reciprocal trigonometric ratios, 660
reflections, 116
representations of three-dimensional figures, 871
representing sample spaces, 934
scale drawing, 606
similar polygons and triangles, 606
similar polygons, 551
similar solids, 864
similarity transformations, 606
simulations, 934
solutions of quadratic equations, 105
solving quadratic equations, 216
solving quadratic functions, 157
special functions, 151
special right triangles, 702
special segments, 792
special segments in a circle, 715
special segments in triangles, 463
square of a difference, 29
square of a sum, 28
square root property, 67
standard form equation of a circle, 775
sum of central angles, 724
surface area of a prism, 815
surface area of a regular pyramid, 823
surface area of a sphere, 848
surface area of prisms and cylinders, 871
surface area of pyramids and cones, 871
surface areas and volumes of spheres, 871
tangents, secants, and angle measures, 792
tests for parallelograms, 531
transformations and coordinate proofs, 393
transformations of exponential functions, 228
transformations of quadratic functions, 157
transformations of quadratic graphs, 216
triangle inequalities, 463
trigonometric ratios, 650
trigonometry, 702
types of dilations, 593
using the discriminant, 136
vector addition, 683
vector operations, 685
vectors, 702
vertical translations, 114
volume of a cone, 842
volume of a cylinder, 832
volume of a prism, 831
volume of a pyramid, 841
volume of a sphere, 850
volumes of prisms and cylinders, 871
volumes of pyramids and cones, 871
zero product property, 171
zero product property, 38

**Kite Theorem,** 525

**Kites,** 524–530

**LA (Leg-Angle) Congruence Theorem.** *See* Leg-Angle (LA) Congruence Theorem, 373

**Labs.** *See* Algebra Labs; Geometry Labs; Graphing Technology Labs; Spreadsheet Labs; TI-Nspire Technology Labs

**Lateral area,** 814
of cones, 824–825, 826
of cylinders, 816, 817–818, 826
of prisms, 814–815, 817–818, 826
of pyramids, 822–824, 826
of regular pyramids, 822–824

**Lateral edges,** 814, 822

**Lateral faces,** 814, 822

**Latitude,** 863

**Law of Cosines,** 671–679

**Law of Cosines Theorem,** 671
**Law of Large Numbers,** 910
**Law of Sines,** 670–679
ambiguous case of, 681
**Law of Sines Theorem,** 670
**LCD.** *See* Least common denominators (LCD)
**LCM.** *See* Least common multiples (LCM)
**Leading coefficient,** 8
**Leg-Angle Congruence Theorem (LA),** 373
**Leg-Leg Congruence Theorem (LL),** 373
**Legs**
of right triangles, 620, 629–637
of trapezoids, 521
**Length**
of arcs, 727
in metric units, P4–P5, P6–P7
probability ratio, 899
in standard units, P4–P5, P6–P7
**Like terms,** 5
**Line segments**
altitudes, 416, 419–421, 619–621, 814, 822, 824
chords, 715, 733–740, 741, 760, 768, 848
diameters, 715–718, 735, 848
edges, 814, 822, 932–933
medians, 416, 417, 421
radii, 715–720, 734, 775, 825, 848–850
slant height, 822–828
**Linear equations,** *See also* Equations, P11–P12
end behavior, 97
writing, 833
**Linear inequalities,** P13–P14
**Linear Pair Theorem,** 300
**Linear regression equation,** 169
**Lines,** 276–277. *See also* Functions; Slopes
auxiliary, 335
concurrent, 407
of symmetry, 93–102
horizontal, 662
parallel, 308–314, 315–322, 573, 574, 575
perpendicular, 310
regression equations, 146–147
secants, 760–761, 770, 771
tangent, 750–757, 761–762
transversals, 575, 576
**LL (Leg-Leg) Congruence Theorem.** *See* Leg-Leg Congruence Theorem (LL), 373
**Locus,** 414
of points a given distance from a point, 856
of points in a plane equidistant from a given point, 715
of points in a plane equidistant from a fixed point and a fixed line, 783
of points in a plane equidistant from the endpoints of a segment, 406
**Logic Puzzles.** *See* Matrix logic
**Logical reasoning,** 328, 330, 435–436
conditional statements, 285, 323
flow proofs, 337, 500, 503
inductive reasoning, 323
logical proofs, 278, 285
paragraph proofs, 278–279, 324, 500, 738, 746
two-column proofs, 285–289, 298, 325, 504, 730, 738, 739, 745, 746, 747, 765, 767
Venn diagrams, 529, 918, 925–926, 939
**Longitude,** 863

## M

**Magnifications.** *See* Dilations
**Magnitudes**
of vectors, 682–689
**Major arcs,** 725
**Manipulatives**
algebra tiles, 5–6, 20–21, 35, 43–44, 124
**Marginal frequencies,** 922
**Materials**
board, 215
books, 215
brass fasteners, 680
data collection device, 215
dental floss, 782
isometric dot paper, 807–809
modeling compound, 782
note card, 680–681
string, 416
toy car, 215
tracing paper, 423, 628
wax paper, 782
**Math History Link,** 68, 150, 345, 419, 499, 573, 663, 726, 866, 910, 925
Bernoulli, Jakob, 910
Brahmagupta, 191
David, Florence Nightingale, 150
Descartes, René, 499
Eratosthenes, 663
Euclid, 726
Euler, Leonhard, 925
Fermat, Pierre de, 419
Galilei, Galileo, 68, 573
Gauss, Johann Carl Friedrich, 345
Riemann, Georg F.B., 866
**Math Online.** *See* ConnectED
**Mathematical expectation,** 909
**Matrices.** *See* also systems of equations
**Matrix logic,** 435–436
**Maximum,** 93–102, 130–131
**Means,** 544, 619
arithmetic, 626
geometric, 619–627, 629, 702
**Measurement.** *See also* Customary system; Metric system
of arc lengths, 727
area, 69, 71, 254, 255, 784–789, 816, 819, 832, 849, 866, 900–901, 916
circumference, 71, 717–720, 825, 849, 851
customary system, P4–P5, P6–P7
indirect, 564, 622, 672, 773
lateral area, 814–815, 816, 817–818, 822–825, 826
length, P4–P5, P6–P7, 727, 734, 894, 899
metric system, P4–P5, P6–P7
Mid-Chapter Quiz, 34, 132
perimeter, 50, 255, 553–554
protractors, 299
rulers, 292
surface area, 261, 815, 816, 817–818, 823–826, 848, 849, 852–853, 871
volume, 70, 831, 832, 833, 834, 835, 836, 839, 841, 842–846, 850, 851, 852, 853, 871, 904
**Medians,** 416, 417, 421
**Mental Math,** 169, 285, 816
**Meridians,** 863
**Metric system,** P4–P5, P6–P7
**Mid-Chapter Quizzes,** 188, 363, 434, 504, 582, 661, 749, 840, 906

**Midpoint Formula,** 384, 638–639
**Midpoint Theorem,** 278
**Midsegment**
of a trapezoid, 523
of a triangle, 573
**Minimum,** 93–102, 130–131
**Minor arcs,** 725, 733
**Minutes,** 863
**Monomials**
adding, 7
degree of, 7
factoring, 36–41
multiplying a polynomial by, 14–19
**Multiple Choice,** *See* Standardized Test Practice, 77, 79, 184, 188, 197, 201, 203, 205, 213, 221, 222, 234, 243, 327, 330, 397, 400, 448, 467, 470, 535, 538, 565, 632, 634, 707, 710, 797, 834, 878, 940
**Multiple representations,** P21, 12, 18, 26, 32, 41, 50, 56, 62, 102, 109, 128, 138, 175, 183, 233, 242, 262, 289, 297, 306, 313, 321, 342, 350, 381, 423, 432, 451, 461, 482, 492, 502, 510, 519, 529, 548, 558, 568, 580, 598, 604, 626, 636, 647, 658, 668, 678, 700, 721, 747, 766, 780, 789, 811, 820, 829, 837, 846, 861, 869, 888, 897, 913
**Multiplication**
Associative Property of, 178, 180
binomials, 20–21, 22–24
Commutative Property of, 178, 180
of binomials, 170
of complex numbers, 181
of imaginary numbers, 179
of a vector by a scalar, 685
FOIL method, 23–24, 28, 170, 181, 254
polynomial by a monomial, 14–19
polynomials, 20–21, 22–27
radical expressions, 253–254
radicands, 253
Rules for Probability, 916–917
square roots, 246
**Multiplication Property of Equality,** 284
**Multi-stage experiment,** 884
**Multi-step equations,** P11–P12
**Mutually exclusive events,** 924–931

## N

**Navigational coordinates,** 863
**Necessary conditions,** 275
**Nets,** 391–392, 814, 825
**Networks,** 932–933
**Nodes,** 932–933
**Non-Euclidean geometry,** 858
hyperbolic geometry, 861
spherical geometry, 857–862
**Non-examples,** 612
**Nonlinear functions,** 93, 151. *See also* Functions
**Not mutually exclusive events,** 926–927
**Note taking.** *See* Foldables™ Study Organizer
***n*th roots,** 257–258
**Number Sense.** *See* Higher-Order Thinking Problems
**Number Theory,** 56, 108, 127, 158, 159, 174, 184, 196, 232, 439
**Numbers**
complex, 105, 178–184, 192, 216, 218
imaginary, 178–179
irrational, 134, 188, 251
figural, 592
Pythagorean triples, 630–631
rational, 251
real, 426, 638
triangular, 592

## O

**Oblique cones,** 824
**Oblique cylinders,** 816
**Oblique prisms,** 814
**Oblique solids,** 806
volume of, 833
**Open Ended.** *See* Higher-Order-Thinking Problems
**Operations**
closed, 251
**Order of operations,** P10, P12
**Ordered pairs,** P23–P24
**Ordered triple,** 638
**Ordinate.** *See y*-coordinate
**Origin,** P23, 74, 696, 775
**Orthocenter,** 419–420, 421
**Orthographic drawings,** 390, 808–809
**Outcome.** *See also* Probability
**Outcomes,** P8, 883–885. *See also* Probability

## P

**Pairs**
angles, 308, 316
ordered, P23–P24
**Paper folding,** 405, 423, 782
**Parabolas ,** 93–100, 782–783
axis of symmetry, 93–100
families of, 198–199
graphing, 93
shading inside and outside, 111
vertices of, 93–100
**Paragraph proofs,** 278–279, 324, 500, 738, 746
**Parallel lines**
angles and, 308–314
congruent parts of, 575
proportional parts of, 574
proving, 315–322, 573
proof, 570–571
and a transversal, 308–313, 316
**Parallel Postulate,** 316
**Parallel vectors,** 683
**Parallelogram method,** 683
**Parallelograms,** 48, 485–493, 513
angles of, 485
conditions for, 495–496
on a coordinate plane, 498
diagonals of, 487, 505–506, 512, 640
graphing, 494
identifying, 496
missing measures, 497
properties of, 485–487, 531
proving relationships, 496
rectangles, 505–511, 531, 547, 708
rhombi, 512–520, 531
sides of, 485
squares, 513–520, 531, 640
tests for, 495–502, 531
**Parallels,** 863
**Parent function**
of exponential decay functions, 229
of exponential growth functions, 227–229

**Parent graphs,** 200
**Part.** *See* Percents
**Pascal's Triangle,** 592
**Paths,** 933
**Patterns**
Fibonacci sequence, 550
figural, 592
**Pentagonal pyramids,** 822
**Perfect square trinomials,** 29, 64–71, 124, 171, 176
**Perimeter** 554
of rectangles, 50, 255
of squares, 13
of triangles, 11
of similar polygons, 553–554
Personal Tutor. *See* ConnectED
**Perimeters of Similar Polygons Theorem,** 553
**Permutations,** 890–893
with repetition, 892
**Perpendicular Bisector Theorem,** 406–407
**Perpendicular bisector,** 206, 405, 406–407, 431
of triangles, 421
**Perpendicular lines**
proof, 570–571
transversal of two parallel lines, 310
**Perpendicular Transversal Converse,** 316
**Perpendicular Transversal Theorem,** 310
**Personal Tutor.** *See* ConnectED
**Perspective drawing,** 810
**Pi (π),** 721, 722, 817
**Piecewise-defined functions,** 148, 150–151
**Piecewise-linear functions,** 150
**Pisano, Leonardo (Fibonacci),** 550
**Planar networks,** 858
**Planes,** 276–277
intersecting, 276
lines in, 857
**Platonic solids,** 391
dodecahedrons, 391
icosahedrons, 391
octahedrons, 391
tetrahedrons, 391
**Plus or minus symbol,** 67
**Point of concurrency**
centroids, 417–418, 421
circumcenter, 407–408, 421
incenter, 410, 421, 758
orthocenter, 419–420, 421
**Points,** 276–277
centroids, 417–418, 421
circumcenter, 407–418, 421
of concurrency, 407, 408, 417–418, 419–420, 421, 758
coplanar, 5, 7, 716
endpoints, 416
incenter, 410, 421, 758
locus of, 406, 715, 783, 856
midpoints, 278, 384, 498, 638–639
ordered pairs, P23–P24
orthocenter, 419–421
of tangency, 750
vanishing, 810
vertices, 384, 822, 824
**Poles,** 849
**Polygon Exterior Angles Sum Theorem,** 478–479
**Polygon Interior Angles Sum Theorem,** 475–477
**Polygons**
angles of, 475–483, 484
circumscribed, 753
congruent, 344–345
convex, 475
diagonals of, 475
exterior angles of, 478–479, 481
inscribed, 743
interior angles of, 475–477, 480–481
missing measures, 476–481, 484
quadrilaterals, 516, 861
regular, 476–477
similar, 551–559, 606
triangles, 335–343, 344–352, 353–361, 364–371, 372–373, 374–382, 383–389, 393, 405, 406–415, 416, 417, 419–421, 427, 428, 429, 445, 446–452, 463, 514, 553, 560–569, 572–577, 583–590, 591, 592, 606, 619–624, 624, 629–637, 640–648, 650–659, 670–679, 683, 702, 744, 758, 822
vertices of, 475–482
**Polynomial functions**
factors of, 72
quadratic, 200–204, 206, 216, 217
roots of, 72–73
zeros of, 72–76
**Polynomials,** 5, 7
binomials, 170, 171
difference of squares, 171
factoring, 171, 172, 174–176
perfect square trinomials, 171, 176
trinomials, 171–172, 176
zeros of, 72–76
**Postulates,** 276–282
angle addition, 299–300
angle-angle (AA) similarity, 560, 564, 572, 650
angle-side-angle (ASA) congruence, 364
arc addition, 726
converse of corresponding angles, 315
corresponding angles, 308
parallel, 316
points, lines, and planes, 276
protractor, 299
ruler, 292
segment addition, 292–293
side-angle-side (SAS) congruence, 355
side-side-side (SSS) Congruence, 353
**Powers**
of equality, 259
**Pre-AP.** *See* Standardized Test Practice
**Predictions,** 545
using polynomials, 10
**Preimage,** 593, 695
**Preparing for Advanced Algebra.** *See* Chapter 0
**Preparing for Algebra.** *See* Chapter 0
**Preparing for Geometry.** *See* Chapter 0
**Preparing for Standardized Tests.** *See* Standardized Test Practice
**Prerequisite Skills**
Get Ready for the Chapter, P2, 3, 91, 167, 225, 273, 333, 403, 473, 541, 617, 713, 803, 881
Skills Review, P15, 7, 14, 22, 28, 36, 45, 52, 58, 64, 79, 93, 105, 114, 124, 133, 140, 148, 177, 184, 197, 205, 213, 234, 243, 245, 252, 259, 283, 291, 298, 307, 314, 322, 343, 352, 361, 371, 382, 389, 415, 425, 433, 444, 452, 462, 483, 493, 503, 511, 520, 530, 549, 559, 569, 581, 590, 599, 605, 627, 637, 648, 659, 669, 679, 690, 701, 723, 732, 740, 748, 757, 767, 774, 781, 790, 812, 821, 830, 838, 847, 855, 862, 870, 889, 898, 905, 914, 921, 931
**Prime polynomial,** 54–56

**Principal square roots,** 246
**Prisms**
altitudes of, 814
heights of, 814
lateral area of, 814–815, 817–818, 826
oblique, 806, 814
rectangular, 818, 831
right, 805, 814, 815
surface area of, 814, 815, 817–818, 826, 871
triangular, 818
volume of, 70, 831, 834, 836, 843, 871
**Probability model,** 907
**Probability tree,** 917
**Probability,** P8–P9
combinations, 894
permutations, 890–893
rules of, 927–928
simple, P8–P9
**Problem-solving strategies.** *See also* Problem-solving
**Problem-Solving Tips** *See also* Problem-solving; Problem-solving strategies
sense-making, 336
determine reasonable answers, 564
draw a diagram, 455
guess and check, 46, 643
make a drawing, 769
make a graph, 516
make a model, 833
**Product Property of Square Roots,** 245
solve a simpler problem, 751
use a simulation, 908
use estimation, 695
work backward, 278
**Problem-solving.** *See also* Conjectures, making; Problem-Solving Tips; Real-World Examples
four-step plan, 76, 239
using similar triangles, 585
**Product Property,** P27, *See also* Multiplication
of a difference, 29–30
special, 28–33
of square roots, 245
of a sum, 30
zero, 38
**Properties**
differences of squares, 58–63
distributive, 15–16, 21–42, 80, 82
double root, 106
end behavior, 97
of equality, 259–260
factoring, 52–53
graphing, 96–98, 112–113
maximum, 93, 95–96, 130
minimum, 93, 95–96, 130
no real root, 106
perfect squares, 64–71
power of equality, 259–260
product of square roots, 245
Quadratic equations, 45–51, 52–57
quotient of square roots, 246
related functions, 112
solve
by completing the square, 124–129, 136
by factoring, 64–67, 136
by graphing, 105–110, 136
using square root property, 67–68, 136
square root, 67
standard form of, 93, 95
systems of, 122–123
two roots, 105
by using the quadratic formula, 133–139
vertex form, 130–131
zero product, 38
**Proof by negation.** *See* Indirect proof
**Proofs**
algebraic, 284–285, 324
by contradiction, 437
coordinate, 383–389, 393, 498–499
direct, 437
flow, 337, 500, 503
formal, 285
geometric, 286, 439
indirect, 437–444, 739, 756
informal, 278, 854
paragraph, 278–279, 324, 500, 738, 746
two-column, 285–289, 298, 325, 504, 730, 738, 739, 745, 746, 747, 765, 767
without words, 628, 784, 841, 854
**Properties**
addition, 284, 426
of angle congruence, 301
associative, 286
of multiplication, 178, 180
commutative, 286
of multiplication, 178–180
comparison, 426
of congruent triangles, 347
cross products, 544–545
distributive, 284
division, 284
multiplication, 284
of parallelograms, 485–487, 531
product, P27
of quadrilaterals, 531
quotient, P27
of real numbers, 284, 426
of rectangles, 505–506, 531
reflexive, 284, 286, 293, 301, 346, 347, 563
of rhombi, 512, 514, 531
of segment congruence, 293
of similarity, 563
of similar solids, 865
of squares, 514, 531
substitution, 284, 293
subtraction, 284, 426
square root, 179
symmetric, 284, 286, 293, 301, 347, 563
transitive, 284, 286, 293–294, 301, 347, 426, 563
of triangle congruence, 347
zero product, 171, 173
**Properties of Angle Congruence Theorem,** 301
**Properties of Segment Congruence Theorem,** 293
**Properties of Similar Solids Theorem,** 865
**Properties of Similarity Theorem,** 563
**Properties of Triangle Congruence Theorem,** 347
**Property of Equality**
for exponential functions, 237
**Property of Inequality**
for exponential functions, 239
**Proportional parts,** 574, 606
**Proportions,** 544–549
similar polygons, 551–559
**Protractor Postulate,** 299
**Pure imaginary numbers,** 178–179
**Pyramids,** 665, 808
altitudes of, 822
height of, 822, 850
lateral area of, 822–824, 826
pentagonal, 822

regular, 822–824, 827–829
surface area of, 822–824, 826, 827–829
vertices of, 822
volume of, 841, 843–844, 871

**Pythagorean Inequality Theorem,** 632

**Pythagorean Theorem,** 290, 628, 629–637, 640–641, 709, 775

**Pythagorean triples,** 630–631

## Q

**Quadrants,** P23

**Quadratic equations**
by factoring, 173–175, 194, 218
by graphing, 194, 217
by inspection, 171
roots of, 174–175
solve by completing the square, 194, 219
standard form of, 174–176
using a calculator, 187
using Quadratic Formula, 187–196
using Square Root Property, 179, 194

**Quadratic expressions,** 23

**Quadratic Formula,** 133–139, 187–192, 194–196, 219
derivation of, 189
discriminant, 136–138

**Quadratic functions,** 157, 158–160. *See also* Functions
families of, 198–199
graphing, 93–103, 216, 217
rate of change for, 206
transformations of, 114–121, 200–204, 216

**Quadratic inequalities,** 111, 207–213
graphing, 207
solve
algebraically, 210–212
by graphing, 208–212
using a table, 212

**Quadratic regression equation,** 169

**Quadratic techniques, apply,** 777, 782

**Quadrilaterals,** 516, 744, 861
parallelograms, 485–493, 513
properties of, 529
rectangles, 505–511
rhombi, 512–520
squares, 513–520, 640
trapezoids, 521–529

**Quantitative reasoning.** *See* Number Sense

**Quick Checks,** 3, 91, 167, 225, 273, 333, 403, 473, 541, 617, 713, 803, 881

**Quotient Property,** P27

**Quotient Property of Square Roots,** 246

**Quotients.** *See also* Division
of square roots, 246

## R

**Radical expressions,** P27, 259–263
adding with like radicands, 252
multiplying, 246, 253–254
simplifying, 245–250, 257–258
subtracting with like radicands, 252
with unlike radicands, 253

**Radicands,** 245
expression as, 260
fractions in, 246
like, 252
multiplying, 253
unlike, 253
variables as, 259
variables on each side, 260

**Radii,** 715–720, 775, 825
perpendicular to a chord, 734
of spheres, 848–850

**Random variables,** 909

**Range,** 94
of exponential function, 227

**Rate of change.** 104. *See also* Rates
for quadratic functions, 206

**Rates**
of change, 104

**Rational approximations,** 134

**Rational numbers.** 251. *See also* Fractions; Percents

**Rationalizing the denominator,** 247

**Ratios,** 543–548
analyzing functions using, 141–144
cosine, 650–658, 674
extended, 543–544
and Fibonacci sequence, 550
golden, 547
length probability, 899
proportions, 544–549, 551–558
scale factors, 552–554, 584, 593–594, 596, 599, 695, 696
sine, 650–659, 674
tangent, 650–658, 674
using similar solids to write, 866

**Reading Math**
3-4-5 right triangle, 631
abbreviations and symbols, 301
and, 916
circum-, 408
complement, 927
conditional probability, 918
contradiction, 438
diameter and radius, 716
flowchart proof, 337
fractions in the radicand, 246
height, 833
height of a triangle, 419
incenter, 410
inverse trigonometric ratios, 653
midsegment, 523
multiple inequality symbols, 447
or, 925
perpendicular, 303, 310
proportion, 544
polynomials as factors, 23
repeated roots, 73
Real axis, 185
rhombi, 513
similarity symbol, 552
solving a triangle, 673
square root solutions, 67
substitution property, 293
symbols, 354, 523
tree diagram notation, 884

**Real numbers**
ordered triple, 638
properties of inequality for, 426

**Real-World Careers**
architectural engineer, 832
athletic trainer, 584
coach, 486
electrical engineer, 179
event planner, 622
historical researcher, 725
interior design, 429
lighting technicians, 356
personal trainer, 338
statistician, 893
urban planner, 53

**Real-World Examples,** P18, 9, 24, 29, 39, 48, 54, 68, 98, 107, 118, 126, 142, 149, 173, 181, 209, 229, 231, 238, 254, 259, 277, 285, 294, 300, 309,

318, 336, 338, 346, 356, 366, 377, 385, 408, 418, 429, 438, 448, 455, 477, 486, 496, 505, 506, 515, 522, 543, 545, 552, 564, 575, 576, 585, 594, 601, 622, 643, 652, 672, 686, 695, 717, 726, 733, 735, 744, 753, 762, 769, 777, 784, 785, 808, 817, 825, 843, 851, 858, 866, 884, 885, 900, 901, 924, 925, 926, 928

**Real-World Links,** P18, 9, 15, 24, 39, 48, 54, 98, 107, 126, 142, 173, 181, 209, 229, 231, 238, 294, 302, 309, 336, 346, 377, 385, 408, 438, 448, 455, 477, 496, 506, 507, 515, 522, 525, 545, 575, 585, 594, 601, 652, 664, 672, 686, 717, 735, 744, 762, 769, 777, 785, 808, 843, 851, 885, 892, 900, 901, 909, 917, 926, 928

**Reasonableness, check for.** *See* Study Tips

**Reasoning.** *See* also Higher-Order-Thinking Problems, *See also* H.O.T. Problems
- indirect, 437
- inductive, 323

**Reciprocal.** *See also* Multiplicative inverse

**Rectangles**
- area of, P21, 254, 255
- diagonals of, 505–506
- golden, 547
- perimeter of, 50, 255
- properties of, 505–506, 531
- proving relationships, 506

**Rectangular prisms,** 818
- surface area of, 815
- volume of, 163, 831

**Rectangular solids,** 638

**Reduction,** 593–598, 695–696

**Reflections**
- of quadratic functions, 116–117

**Reflexive Property**
- of congruence, 293, 301
- of equality, 284, 286
- of similarity, 563
- of triangle congruence, 347

**Regression equation,** 146–147, 169

**Regular pyramids,** 822–824, 827–829
- lateral area of, 822–824
- surface area of, 823–824, 827–829

**Relations**
- inverse, P15–P16

**Relative frequency,** 922

**Remote interior angles,** 337, 427

**Resultant vectors,** 683

**Review Vocabulary**
- absolute value, 47
- altitude of a triangle, 620
- arc, 785
- central angle, 785
- complex conjugates, 76
- complementary angles, 300
- coordinate proof, 498
- coplanar, 716
- domain and range, 96
- equilateral triangle, 375
- exterior angle, 478
- factored form, 254
- FOIL method, 254
- linear pair, 300
- perfect square, 171
- radicand, 190
- rationalizing the denominator, 641
- regular polygon, 477
- remote interior angle, 427
- supplementary angles, 300
- triangle inequality theorem, 633
- trigonometric ratios, 824
- vertical angles, 302

**Review.** *See* Guided Practice; ConnectED; Prerequisite Skills; Spiral Review; Standardized Test Practice; Study Guide and Review; Vocabulary Check

**Rhombi,** 512–520
- conditions for, 514
- diagonals of, 512
- properties of, 512, 514

**Rhombus**
- area of, 254

**Riemann, Georg F.B.,** 866

**Right Angle Theorems,** 303

**Right angles,** 373, 505, 506

**Right cones,** 824

**Right cylinders,** 816

**Right prisms,** 814

**Right solids,** 805

**Right triangles**
- 30°-60°-90°, 641–646
- 45°-45°-90°, 640–646
- altitudes of, 619–621
- congruence, 372–373
- congruent, 372–373
- geometric means in, 619–621
- Hypotenuse-Angle Congruence Theorem (HA), 373
- Hypotenuse-Leg Congruence Theorem (HL), 373
- hypotenuse of, 619–621, 629–637, 640–642
- Leg-Angle Congruence Theorem (LA), 373
- Leg-Leg Congruence Theorem (LL), 373
- legs of, 620, 629–637
- missing measures, 338, 630, 641, 653
- Pythagorean Theorem, 290, 628, 629–637, 640–641, 709, 775
- similar, 620, 650
- solving, 654–655, 657
- special, 640–648, 651, 702

**Rise.** *See* Slopes

**Roots** *See* Square roots
- approximate with a calculator, 107 with a table, 107
- complex, 187, 192
- double, 106
- irrational, 191
- no real, 106
- one rational, 191
- of polynomial functions, 72–73
- of quadratic equations, 174–175
- repeated, 73
- square, 178
- two, 105
- two rational, 190

**Ruler Postulate,** 292

**Run.** *See* Slopes


## S


**Sample spaces,** 883–889

**SAS (Side-Angle-Aide) Similarity Theorem.** *See* Side-Angle-Side (SAS) Similarity Theorem

**SAS (Side-Angle-Side) Congruence Postulate.** *See* Side-Angle-Side (SAS) Congruence Postulate

**SAS Inequality Theorem.** *See* Hinge Theorem

**Scalar multiplication,** 79

**Scalars,** 79

**Scale,** 600–604

**Scale drawings,** 600–604

**Scale factor,** 552–554, 584
of dilation, 593–594, 695
negative, 696

**Scale models,** 600–604

**Scaling,** 694

**Scatter plots**
curve fitting, 146–147
exponential trends, 146
quadratic trends, 146

**Secant segments,** 770

**Secants** 759–762
graphing, 660
intersecting, 760–761, 770
intersecting with a tangent, 762, 771

**Seconds,** 863

**Sectors**
area of, 785–788
of a circle, 785

**Segment Addition Postulate,** 292–293

**Segments**
adding measures, 292–293
altitudes, 416, 419–421, 620–621, 641, 814, 822, 824
bisecting, 576
chords, 715, 733–740, 741, 760, 768, 848
of a circle, 715
congruent, 293–294, 374–375
diameters, 715–718, 735, 848
edges, 814, 822, 932–933
hypotenuse, 619–620, 629–637, 640–641
intersecting inside a circle, 768
intersecting outside a circle, 770
legs, 521, 620, 630–637
medians, 416, 417, 421
radii, 715–720, 734, 775, 825, 848–850
relationships, 292–298
Ruler Postulate, 292
slant height, 822–828

**Segments of Chords Theorem,** 768

**Segments of Secants Theorem,** 770

**Self-similarity,** 591

**Semicircles,** 725

**Sense-Making.** *See* Check for Reasonabless; Reasoning

**Sequences**
Fibonacci, 550

**Sets**
closed, 9, 251

**Short Response,** *See* Standardized Test Practice, 177, 197, 205, 213, 223

**Short Response,** 291, 298, 331, 314, 343, 371, 382, 398, 401, 415, 444, 452, 462, 471, 483, 503, 511, 539, 559, 581, 590, 599, 605, 615, 627, 637, 648, 669, 679, 711, 740, 748, 757, 781, 801, 790, 821, 830, 838, 847, 862, 870, 877, 879, 889, 898, 914, 931, 941

**Side-Angle-Side (SAS) Similarity Theorem,** 561, 564

**Side-Angle-Side (SAS) Similarity,** 561–562

**Side-Side-Side (SSS) Congruence Postulate,** 353

**Side-Side-Side (SSS) Similarity Theorem,** 561, 564

**Sides**
corresponding, 344, 551–553, 562
of parallelograms, 485
sides, 562
of triangles, 429, 446

**Sierpinski Triangle,** 591

**Similar polygons,** 551–559, 606
missing measures, 553
perimeter of, 553–554

**Similar Right Triangles Theorem,** 620

**Similar solids,** 864–870

**Similar triangles,** 560–569, 623
identifying, 553
parts of, 563, 583–590

**Similarity**
ratios, 551–559
statements, 551
transformations, 593–599, 606

**Simplest form**
of ratios, 543

**Simplest form.** *See also* Algebraic expressions

**Simplifying.** *See also* Algebraic expressions
radical expressions, 245–250

**Simulations,** 907–914
designing using a geometric model, 907
using random numbers, 908
summarizing data from, 908

**Sine ratio,** 650–658, 674

**Sines, Law of,** 670–679

**Skills Review,** 13, 19, 27, 33, 42, 51, 57, 63, 71, 103, 110, 121, 129, 139, 145, 155, 177, 184, 197, 205, 213, 79, 234, 243, 250, 256, 263

**Slant height,** 822–828

**Slide.** *See* Translations

**Slopes**
of parallel proof, 570–571
of perpendicular proof, 570–571

**Small circle,** 861

**Solids.** *See also* Three-dimensional figures
cones, 824, 825–826, 842–845, 871
congruent, 864
cross sections of, 808
cylinders, 816, 817–818, 826, 832, 835, 836, 843, 871
formed by translations, 805–806
frustums, 829
lateral area, 814–818, 822–826
nets of, 391–392, 814, 825
orthographic drawings, 390, 808–809
Platonic, 391
prisms, 805, 807, 814–815, 817–818, 826, 831, 834, 836, 843, 871
pyramids, 665, 808, 822–824, 826, 827–829, 841, 843–844, 850
representations of, 807–812
similar, 864–870
spheres, 848–849, 850, 851–853, 856, 857, 871
surface area, 814–821, 822–830, 848, 849, 850, 851–852, 871
volume, 831–838, 839, 841–847, 850–853, 871, 904

**Solutions**
estimating by graphing, 107
pure imaginary, 179

**Solving a right triangle,** 654–655, 657

**Solving a triangle,** 673–674

**Solving equations**
justifying each step, 284–285

**Solving multi-step problems,** 87

**Special functions,** 148–155, 156, 157, 160

**Special products,** 28–33

**Special Segments of Similar Triangles Theorem,** 583

**Spheres,** 261
centers of, 848–849
chords of, 848
diameters of, 848

Index

sets of points on, 857
surface area of, 848–852, 871
that intersect, 856
volume of, 850–853, 871

**Spherical geometry,** 857–862

**Spiral Review,** 13, 19, 27, 33, 42, 51, 57, 63, 71, 79, 103, 110, 121, 129, 139, 145, 155, 177, 184, 197, 205, 213, 234, 243, 250, 256, 263 283, 291, 298, 307, 314, 322, 343, 352, 361, 371, 382, 389, 415, 425, 433, 444, 452, 462, 483, 493, 503, 511, 520, 530, 549, 559, 569, 581, 590, 599, 605, 627, 637, 648, 659, 669, 679, 690, 701, 723, 732, 740, 748, 757, 767, 774, 781, 790, 812, 821, 830, 838, 847, 855, 862, 870, 889, 898, 905, 914, 921, 931

**Spreadsheet Labs,**
Angles of Polygons, 484
Two-Way Frequency Tables, 922–923

**Square Root Property,** 67, 179

**Square roots,** P27–P28, 171
dividing, 246
multiplying, 246
of negative numbers, 178
product property of, 245
quotient property of, 246
simplifying, 245–249

**Squares,** 513–520
area of, 69
conditions for, 512, 514
completing the, 124
diagonals of, 640
of a difference, 28–29
difference of, 58–62, 171
perfect, 29, 64–71, 124, 171
perimeter of, 13
of a sum, 28
sum of, 59
Standard form, 8, 93, 101, 105
of polynomial, 8
of quadratic equations, 93, 95, 105–106, 134
properties of, 514, 531

**SSS (Side-Side-Side) Congruence Postulate.** *See* Side-Side-Side (SSS) Congruence Postulate

**SSS (Side-Side-Side) Inequality Theorem.** *See* Side-Side-Side (SSS) Inequality Theorem

**SSS (Side-Side-Side) Similarity Theorem.** *See* Side-Side-Side (SSS) Similarity Theorem

**Standard form**
of a circle equation, 775
of quadratic equations, 174–176

**Standard position,** 682

**Standardized Test Practice,** 10, 79, 87–89, 162–165, 177, 184, 197, 205, 213, 220–223, 234, 243, 328–331, 398–401, 468–471, 536–539, 612–615, 708–711, 798–801, 876–879, 938–941. *See also* Extended Response; Gridded Response; Short Response; Worked- Out Examples
Extended Response, 60, 63, 89, 110, 165
Gridded Response, 15, 16, 33, 42, 57, 89, 129, 165, 250
Multiple Choice, 13, 19, 27, 33, 42, 51, 57, 63, 71, 89, 103, 110, 119, 121, 129, 132, 139, 145, 147, 161, 164, 248, 250, 256, 263
Short Response, 13, 19, 27, 13, 51, 71, 89, 103, 121, 139, 145, 165, 256, 263
Worked-Out Examples, 15, 60, 117, 247

**Statements**
conditional, 285, 323
congruence, 345
similarity, 551

**Statistics.** *See also* Analysis; Data; Graphs; Measures of central tendency
curve fitting, 146–147

**Step function,** 148–149, 151–154, 156

**String,** 416

**Study Guide and Review,** 80–84, 157–161, 216–294, 323–325, 393–314, 463–466, 531–534, 606–610, 702–706, 792–796, 871–874, 934–936

**Study Organizer.** See Foldables® Study Organizer

**Study Tips**
absolute value, 202, 761
additive inverse, 9
algorithms, 125
alternative method, 654, 727, 760, 823
altitudes of isosceles triangles, 641
ambiguous case, 671
angle bisector, 409
angle relationships, 309
angle-angle-angle, 366
arc bisectors, 734
axiomatic system, 277
center of dilation, 595
check for reasonableness, 338, 420
check your answer, 37, 66
checking solutions, P26, 865
circumcircle, 718
combining like terms, 16
common misconception, 514
commutative and associative properties, 286
completing the square, 776
complex numbers, 180, 192
concentric circles, 766
congruent angles, 487
congruent triangles, 514
coordinate proof, 384
corresponding sides, 561
determining the longest side, 632
diagrams, 360
dimensional analysis, P5, P7
direction angles, 684
discriminant, 136
draw a diagram, 562, 825, 850
drawing segments, 735
elliptical geometry, 858
estimation, 816
Euclid's Postulates, 316
experimental probability, P9
exponential decay, 230
extraneous solutions, 260
finding what is asked for, 317
finite geometries, 858
flow proofs, 337
formulas, 816
function characteristics, 95
geometric probability, 910
graphing calculator, 652
great circles, 849
greatest common factor (GCF), 53
identifying similar triangles, 553
including a figure, 486
indirect measurement, 664
inscribed polygons, 743
interest, 229
isosceles trapezoids, 522
isosceles triangles, 376
key probability words, 928
kites, 525
Law of Cosines, 671
levels of accuracy, 718
linear theorem, 300
lines, P24

locating zeros, 75
location of zeros, 107
making connections, 823
memorizing trigonometric ratios, 651
Mental Math, 285
midpoint formula, 498
midsegment triangle, 573
midsegment, 574
multiple representations, 594
multiplication rule, 885
multiplying polynomials, 24
naming arcs, 725
naming polygons, 476
negative scale factors, 696
nonlinear functions, 151
obtuse angles, 672
other proportions, 574
overlapping figures, 357
patterns, 30
permutations and combinations, 894
perpendicular bisectors, 407
piecewise functions, 150
positive square root, 630
precision, P16, 134
proportions, 563, 586
proposition, 278
proving lines parallel, 318
Pythagorean triples, 630
quadratic formula, 190
quadratic techniques, 777
random number generator, 908
randomness, 891
rational exponents on a calculator, 50
reading math, 180
recognizing contradictions, 440
recognizing perfect square trinomials, 65
rectangles and parallelograms, 507
reflection, 116
reflexive property, 346
regularity, 46
reorienting triangles, 620
right angles, 384, 506
roots, 193
SAS and SSS inequality theorem, 454
scalar multiplication, 685
sense-making, 116
side-side-angle, 355
similar and congruent solids, 865
similar solids and area, 866
similarity and congruence, 553
similarity ratio, 552
simplify first, 253
simplify the theorem, 770
solving an equation by factoring, 48
solving by inspection, 67
solving quadratic inequalities algebraically, 210
solving quadratic inequalities by graphing, 208
square and rhombus, 516
square root, 171
standard form, 776
structure, 35
study notebook, 194
Substitution, 84
symmetry points, 97
synthetic, 75
tangents, 780
technology, 181
testing for zeros, 75
turning the circle over, 893
types of vectors, 683
undefined terms, 277
use a proportion, 621
use an area model, 916
use estimation, 901
use ratios, 642
use scale factor, 584
using a congruence statement, 345
using additional facts, 456
vector subtraction, 685
vertical method, 8
*y*-intercept, 95
zero at the origin, 74

**Substitution method,** P25

**Substitution Property,** 16, 28

**Substitution Property of Equality,** 284

**Subtraction**
monomials, 7
polynomials, 5–6, 9–13
radical expressions, 252, 253

**Subtraction Property**
of equality, 284
of inequality, 426

**Successive differences,** 140–144

**Sufficient conditions,** 275

**Sum.** *See also* Addition
product of, 30

**Supplement Theorem,** 300

**Supplementary angles,** 299–300

**Surface area**
of cones, 824–828, 871
of cylinders, 816, 817–818, 826, 871
of hemispheres, 849, 852
of prisms, 814, 815, 816–818, 826, 871
of pyramids, 822–824, 826, 871
of regular pyramids, 822–824, 827–829
of solids, 826
of spheres, 848–850, 852, 871
using great circles, 849

**Symbols**
angles, 301, 428
combinations, 894
factorial, 890
inequalities, 428, 447
not congruent, 354
not parallel to, 523
not similar to, 552
permutations, 891
plus or minus, 67
probability, 916, 917
scale factor, 593
similar figures, 551
square root, 67

**Symmetric Property**
of congruence, 293, 301
of equality, 284, 286
of similarity, 563
of triangle congruence, 347

**Symmetry,** 93–100

**Synthetic substitution,** 75

**Systems of equations**
with a circle and line, 777
with a circle and circle, 779
linear, P25–P26
with a parabola and line, 783

## T

**Tables.** *See also* Analysis
solving inequalities, 212

**Tangent ratio,** 650–659, 674

**Tangent Theorem,** 751

**Tangent to a Circle Theorem,** 752

**Tangents,** 750–757, 760–766
to circles, 750–756
common, 750, 752–756

intersecting, 761–762
intersecting with a secant, 771
segments, 770
to spheres, 848

**Technology.** *See also* Applications; Graphing Technology Labs; ConnectED; Spreadsheet Labs; TI-Nspire Technology

**Terms**
constant, 95
like, 5

**Test Practice.** *See* Standardized Test Practice

**Test-Taking Tip**
apply definitions and properties, 536, 538
check your answer, 89
coordinate plane, 354
counterexamples, 330
eliminate unreasonable answers, 631, 878
graphing calculator, 164
gridded responses, 524
identifying nonexamples, 562
make a drawing, 876
probability, 940
read carefully, 400
sense-making, 60
set up a proportion, 614
structure, 247
testing choices, 447
the meaning of *a*, 201

**Three-dimensional figures.** *See* Cones; Cubes; Cylinders; Prisms; Spheres
tools, 15
use a Venn diagram, 918
use formulas, 708, 710
use properties, 800
writing equations, 222, 833

**Theorems,** 278
30°-60°-90° Triangle, 642
45°-45°-90° Triangle, 640
Alternate Exterior Angles, 309
Alternate Interior Angles, 309
Angle-Angle-Side (AAS) Congruence, 365
Angle Bisector, 409
Angle-Side Relationships in Triangles, 428
Centroid, 417–418
Circles, 725, 733, 734, 743, 744, 759, 760, 761
Circumcenter, 407–408
Complement, 300
complex conjugates, 76
Conditions for Parallelograms, 495–496
Conditions for Rhombi and Squares, 514
Congruent Complements, 301
Congruent Supplements, 301
Consecutive Interior Angles, 309
Diagonals of a Rectangle, 505–506
Diagonals of a Rhombus, 512
Exterior Angle, 337–338
Exterior Angle Inequality, 427–428
fundamental, 73
Geometric Mean (Altitude), 621
Geometric Mean (Leg), 621
Hinge, 453–457
Hypotenuse-Angle Congruence, 373
Hypotenuse-Leg Congruence, 373
Incenter Theorem, 410
Inscribed Angle, 741–742
Isosceles Trapezoids, 521
Isosceles Triangle, 374
Kite, 525
Law of Cosines, 671
Law of Sines, 670
Leg-Angle Congruence, 373
Leg-Leg Congruence, 373
Linear Pair Theorem, 300
Midpoint Theorem, 278
Perimeters of Similar Polygons, 553
Perpendicular Bisectors, 406
Perpendicular Transversal, 310
Polygon Exterior Angles Sum, 478–479
Polygon Interior Angles Sum, 475–477
Properties of Angle Congruence, 301
Properties of Parallelograms, 485–487
Properties of Segment Congruence, 293
Properties of Similarity, 563
Properties of Similar Solids, 865
Properties of Triangle Congruence, 347
Pythagorean, 629, 632
Pythagorean Inequality, 632
Right Angle, 303
Segments of Chords, 768
Segments of Secants, 770
Side-Angle-Side (SAS) Similarity, 561, 564, 671
Side-Side-Side (SSS) Similarity, 561, 564, 672
Similar Right Triangles, 620
Solving Quadratic Equations, 187
Special Segments of Similar Triangles, 583
Supplement, 300
Tangent to a Circle, 752
Third Angles, 346
Transitive Property of Congruence, 294
Trapezoid Midsegment, 523
Triangle Angle Bisector, 586
Triangle Angle-Sum, 335–336, 346
triangle congruence, 347
Triangle Inequality, 446–448
Triangle Midsegment, 573
Triangle Proportionality, 572, 573
Vertical Angles, 302

**Theoretical probability,** P9

**Third Angles Theorem,** 346

**Three-dimensional figures**
cones, 824, 825–826, 842–845, 871
congruent, 864
cross sections of, 808
cylinders, 816, 817–818, 826, 832, 835, 836, 843, 871
frustums, 829
lateral area, 814–818, 822–826
missing measures, 817
nets of, 391, 814, 825
orthographic drawing of, 390, 392, 808–809
Platonic, 391
prisms, 805, 807, 814–815, 817–818, 826, 831, 834, 836, 843, 871
pyramids, 665, 808, 822–824, 826, 827–829, 841, 843–844, 850
representations of, 807–813
similar, 864–870
spheres, 848–849, 850, 851–853, 856, 857
surface area, 814–821, 822–826, 848, 849, 850, 851–852, 871
volume, 831–838, 839, 841–847, 850–853, 871, 904

**TI-Nspire$^{TM}$ Technology Labs**
Changing Dimensions, 839
Dilations, 692–693

**Topographic maps,** 813

**Topography,** 813

**Traceable network,** 932–933

**Tracing paper,** 423, 628

**Transformations**
dilations, 593–599, 692–693, 694–701
images, 593, 695

isometries, 695
of exponential functions, 228
of quadratic functions, 114–121
preimages, 593, 695
similarity, 593–599, 606
with quadratic functions, 200–204
translations, 228

**Transitive Property**
of congruence, 293–294, 301
of equality, 284, 286
of inequality, 426
of similarity, 563
of triangle congruence, 347

**Translations**
horizontal, 115, 228
of quadratic functions, 114
vertical, 114, 115, 228

**Transversals**
congruent segments of, 576
proportional segments of, 575

**Trapezoid Midsegment Theorem,** 523

**Trapezoids,** 521–529
bases of, 521
diagonals of, 521–522
isosceles, 521–522
legs of, 521
midsegment, 523

**Tree diagram,** 883–884

**Trials,** P8, 883

**Triangle Angle Bisector Theorem,** 586

**Triangle Angle-Sum Theorem,** 335–336, 346

**Triangle Inequality Theorem,** 446–448

**Triangle method,** 683

**Triangle Midsegment Theorem,** 573–574

**Triangle Proportionality Theorem,** 572–573

**Triangles**
30°-60°-90°, 641–646
45°-45°-90°, 640–646
altitudes of, 416, 419–421, 619–621, 641
angle bisectors of, 421
angles of, 335–343, 393
area of, 42, 129, 254
bisectors of, 405, 406–415
circumscribed, 758
classifying, 385, 393, 633
congruent, 344–351, 353–361, 364–371, 377, 393, 514
on a coordinate plane, 383
equilateral, 375–377, 393
exterior angles of, 337
factoring, 43–44, 45–51, 52–54, 80, 83
height of, 419
hypotenuse of, 619–620, 629–637, 640–641
in two, 453–462
in one, 426–432
inequalities, 445, 446–452, 463
inequality theorem, 633
inscribed, 744
isosceles, 374–376, 393, 640, 641, 822
legs of, 620, 629–637
medians of, 416, 417, 421
midsegment of, 573–574
missing measures, 338, 376, 384
number of, 680–681
Pascal's, 592
perfect square, 29, 64–71, 124
perimeter of, 11
perpendicular bisectors of, 421
points in, 421
position on coordinate plane, 383
proportionality, 572–573
reorienting, 620
right, 338, 372–373, 619–621, 630, 631, 640–648, 651, 653, 702
sides of, 429, 446
Sierpinski, 591
similar, 553, 560–569, 583–590, 606, 620, 623, 650
solving, 673–674, 676, 681
special segments in, 463
vertices of, 384

**Triangular numbers,** 592

**Triangular prisms,** 818

**Triangulation,** 426

**Trigonometric ratios,** 650–659
cosine, 650–658, 674
finding direction of a vector, 684, 686
graphing, 649
sine, 650–658, 674
tangent, 650–659, 674

**Trigonometry,** 649, 650–659
calculating height, 663–667
cosine, 650–658, 674
Law of Cosines, 671–679
Law of Sines, 670–679
sine, 650–658, 674
tangent, 650–659, 674

**Trinomials** 7, 97
factoring, 171–172
general, 171
perfect square, 171, 176

**Turn.** *See* Rotations

**Two-column proofs,** 285–289, 298, 325, 504, 730, 738, 739, 745, 746, 747, 765, 767

**Two-dimensional figures**
circles, 715–723
quadrilaterals, 516, 861
triangles, 335–343, 344–352, 353–361, 364–371, 372–373, 374–377, 383–389, 393, 416, 417–425, 426–433, 453–462, 463, 514, 553, 560–569, 572–577, 583–590, 606, 619–627, 629–637, 640–648, 650, 651, 653, 673–674, 676, 683, 702, 744, 758, 822

**Two-stage experiment,** 884

**Two-way frequency table,** 922–923

## U

**Unit analysis.** *See* Dimensional Analysis

**Upper quartile,** P38

## V

**Values**
absolute, 202

**Vanishing point,** 810

**Variables**
on each side of the radicand, 260
isolating, 259

**Variation.** *See also* Measures of variation

**Vectors,** 682–690
adding, 683, 685, 691
algebraic, 685
component form of, 684
commutative property with, 689
direction of, 682–689
equivalent, 683
initial point of, 682
magnitude of, 682–689
multiplying by a scalar, 685
operations with, 685
opposite, 683, 685

parallel, 683
resultant, 683
standard position of, 684
subtracting, 683, 685
terminal point of, 682

**Venn diagram,** 180, 529, 918, 925–926, 939

**Vertex angles,** 374

**Vertex form,** 118, 200–204

**Vertex,** 93–102

**Vertical Angles Theorem,** 302–303

**Vertical angles,** 302–303

**Vertical translations,** 228

**Vertices**
of cones, 824
corresponding, 344
of polygons, 475–482
of pyramids, 822
of triangles, 384

**Vocabulary Check,** 80, 157, 216, 323, 393, 463, 531, 606, 702, 792, 871, 934

**Vocabulary Link**
inscribed, 742
symmetric, 293

**Volume,** 904
of prisms, 70
changing dimensions and, 839
finding a, 142
keep the -1, 54
leading coefficients, 125
minimum and maximum values, 96
multiplying radicands, 253
notation, P17
of cones, 842–845, 871
of cylinders, 832, 835, 836, 843, 871
of hemispheres, 850, 852
of oblique solids, 833
of prisms, 831, 834, 836, 843, 871
of pyramids, 841, 843–844, 871
of rectangular prisms, 831
of solids, 843
of spheres, 850–853, 871
informal proof, 854
precision, 106
regularity, 29
solutions, 135
squaring each side, 260
sum of squares, 59
transformations, 117
unknown value, 38
*x*-values, 141

## W

**Watch Out!**
angles of elevation and depression, 663
approximation, 654
arc length, 727
area of hemisphere, 849
bases, 826
conditional notation, 917
cross-sectional area, 832
dividing by a negative, P14
identifying circumscribed polygons, 753
identifying side opposite, 427
order of operations, P12, 671
parallelograms, 497
percents, 239
perimeter, 554
proof by contradiction vs. counterexample, 439
rationalizing the denominator, P28
right prisms, 815
rounding, 673
symbols for angles and inequalities, 428
use the correct equation, 770
volumes of cones, 842

**Web site.** *See* ConnectED

**Which One Doesn't Belong?.** *See* Higher-Order-Thinking Problems

**Whisker.** *See* Box-and-whisker plot

**Worked-Out Solutions.** *See* Selected Answers and Solutions

**Writing in Math.** *See* Higher-Order-Thinking Problems

***x*-coordinate,** P23

***x*-intercepts,** 72

## Y

***y*-coordinate,** P23

## Z

**Zero pairs,** 5

***Z*ero Product Property,** 38, 173

**Zeros,** 106, 107
of functions, 72–76
locating, 75

Index

## Symbols

| | |
|---|---|
| $\neq$ | is not equal to |
| $\approx$ | is approximately equal to |
| $\sim$ | is similar to |
| $>, \geq$ | is greater than, is greater than or equal to |
| $<, \leq$ | is less than, is less than or equal to |
| $-a$ | opposite or additive inverse of $a$ |
| $\lvert a \rvert$ | absolute value of $a$ |
| $\sqrt{a}$ | principal square root of $a$ |
| $a : b$ | ratio of $a$ to $b$ |
| $(x, y)$ | ordered pair |
| $f(x)$ | $f$ of $x$, the value of $f$ at $x$ |
| $\overline{AB}$ | line segment $AB$ |
| $AB$ | measure of $\overline{AB}$ |
| $\angle$ | angle |
| $\triangle$ | triangle |
| $^\circ$ | degree |
| $\pi$ | pi |
| $\sin x$ | sine of $x$ |
| $\cos x$ | cosine of $x$ |
| $\tan x$ | tangent of $x$ |
| ! | factorial |
| $P(a)$ | probability of $a$ |
| $P(n, r)$ | permutation of $n$ objects taken $r$ at a time |
| $C(n, r)$ | combination of $n$ objects taken $r$ at a time |

## Algebraic Properties and Key Concepts

| | |
|---|---|
| **Identity** | For any number $a$, $a + 0 = 0 + a = a$ and $a \cdot 1 = 1 \cdot a = a$. |
| **Substitution (=)** | If $a = b$, then $a$ may be replaced by $b$. |
| **Reflexive (=)** | $a = a$ |
| **Symmetric (=)** | If $a = b$, then $b = a$. |
| **Transitive (=)** | If $a = b$ and $b = c$, then $a = c$. |
| **Commutative** | For any numbers $a$ and $b$, $a + b = b + a$ and $a \cdot b = b \cdot a$. |
| **Associative** | For any numbers $a$, $b$, and $c$, $(a + b) + c = a + (b + c)$ and $(a \cdot b) \cdot c = a \cdot (b \cdot c)$. |
| **Distributive** | For any numbers $a$, $b$, and $c$, $a(b + c) = ab + ac$ and $a(b - c) = ab - ac$. |
| **Additive Inverse** | For any number $a$, there is exactly one number $-a$ such that $a + (-a) = 0$. |
| **Multiplicative Inverse** | For any number $\frac{a}{b}$, where $a, b \neq 0$, there is exactly one number $\frac{b}{a}$ such that $\frac{a}{b} \cdot \frac{b}{a} = 1$. |
| **Multiplicative (0)** | For any number $a$, $a \cdot 0 = 0 \cdot a = 0$. |
| **Addition (=)** | For any numbers $a$, $b$, and $c$, if $a = b$, then $a + c = b + c$. |
| **Subtraction (=)** | For any numbers $a$, $b$, and $c$, if $a = b$, then $a - c = b - c$. |
| **Multiplication and Division (=)** | For any numbers $a$, $b$, and $c$, with $c \neq 0$, if $a = b$, then $ac = bc$ and $\frac{a}{c} = \frac{b}{c}$. |
| **Addition (>)*** | For any numbers $a$, $b$, and $c$, if $a > b$, then $a + c > b + c$. |
| **Subtraction (>)*** | For any numbers $a$, $b$, and $c$, if $a > b$, then $a - c > b - c$. |
| **Multiplication and Division (>)*** | For any numbers $a$, $b$, and $c$,<br>1. if $a > b$ and $c > 0$, then $ac > bc$ and $\frac{a}{c} > \frac{b}{c}$.<br>2. if $a > b$ and $c < 0$, then $ac < bc$ and $\frac{a}{c} < \frac{b}{c}$. |
| **Zero Product** | For any real numbers $a$ and $b$, if $ab = 0$, then $a = 0$, $b = 0$, or both $a$ and $b$ equal 0. |
| **Square of a Sum** | $(a + b)^2 = (a + b)(a + b) = a^2 + 2ab + b^2$ |
| **Square of a Difference** | $(a - b)^2 = (a - b)(a - b) = a^2 - 2ab + b^2$ |
| **Product of a Sum and a Difference** | $(a + b)(a - b) = (a - b)(a + b) = a^2 - b^2$ |

* *These properties are also true for $<$, $\geq$, and $\leq$.*

## Formulas

| | |
|---|---|
| **Slope** | $m = \frac{y_2 - y_1}{x_2 - x_1}$ |
| **Distance on a coordinate plane** | $d = \sqrt{(x_2 - x_1)^2 + (y_2 - y_1)^2}$ |
| **Midpoint on a coordinate plane** | $M = \left(\frac{x_1 + x_2}{2}, \frac{y_1 + y_2}{2}\right)$ |
| **Pythagorean Theorem** | $a^2 + b^2 = c^2$ |
| **Quadratic Formula** | $x = \frac{-b \pm \sqrt{b^2 - 4ac}}{2a}$ |
| **Perimeter of a rectangle** | $P = 2\ell + 2w$ or $P = 2(\ell + w)$ |
| **Circumference of a circle** | $C = 2\pi r$ or $C = \pi d$ |

### Area

| | | | |
|---|---|---|---|
| **rectangle** | $A = \ell w$ | **trapezoid** | $A = \frac{1}{2}h(b_1 + b_2)$ |
| **parallelogram** | $A = bh$ | **circle** | $A = \pi r^2$ |
| **triangle** | $A = \frac{1}{2}bh$ | | |

### Surface Area

| | | | |
|---|---|---|---|
| **cube** | $S = 6s^2$ | **regular pyramid** | $S = \frac{1}{2}P\ell + B$ |
| **prism** | $S = Ph + 2B$ | **cone** | $S = \pi r\ell + \pi r^2$ |
| **cylinder** | $S = 2\pi rh + 2\pi r^2$ | | |

### Volume

| | | | |
|---|---|---|---|
| **cube** | $V = s^3$ | **regular pyramid** | $V = \frac{1}{3}Bh$ |
| **prism** | $V = Bh$ | **cone** | $V = \frac{1}{3}\pi r^2 h$ |
| **cylinder** | $V = \pi r^2 h$ | | |

## Measures

| Metric | Customary |
|---|---|
| **Length** | |
| 1 kilometer (km) = 1000 meters (m)<br>1 meter = 100 centimeters (cm)<br>1 centimeter = 10 millimeters (mm) | 1 mile (mi) = 1760 yards (yd)<br>1 mile = 5280 feet (ft)<br>1 yard = 3 feet<br>1 foot = 12 inches (in.)<br>1 yard = 36 inches |
| **Volume and Capacity** | |
| 1 liter (L) = 1000 milliliters (mL)<br>1 kiloliter (kL) = 1000 liters | 1 gallon (gal) = 4 quarts (qt)<br>1 gallon = 128 fluid ounces (fl oz)<br>1 quart = 2 pints (pt)<br>1 pint = 2 cups (c)<br>1 cup = 8 fluid ounces |
| **Weight and Mass** | |
| 1 kilogram (kg) = 1000 grams (g)<br>1 gram = 1000 milligrams (mg)<br>1 metric ton (t) = 1000 kilograms | 1 ton (T) = 2000 pounds (lb)<br>1 pound = 16 ounces (oz) |

## Formulas

### Coordinate Geometry

| | |
|---|---|
| **Slope** | $m = \frac{y_2 - y_1}{x_2 - x_1}$ |
| **Distance on a number line:** | $d = \|a - b\|$ |
| **Distance on a coordinate plane:** | $d = \sqrt{(x_2 - x_1)^2 + (y_2 - y_1)^2}$ |
| **Distance in space:** | $d = \sqrt{(x_2 - x_1)^2 + (y_2 - y_1)^2 + (z_2 - z_1)^2}$ |
| **Distance arc length:** | $\ell = \frac{x}{360} \cdot 2\pi r$ |
| **Midpoint on a number line:** | $M = \frac{a + b}{2}$ |
| **Midpoint on a coordinate plane:** | $M = \left(\frac{x_1 + x_2}{2}, \frac{y_1 + y_2}{2}\right)$ |
| **Midpoint in space:** | $M = \left(\frac{x_1 + x_2}{2}, \frac{y_1 + y_2}{2}, \frac{z_1 + z_2}{2}\right)$ |

### Perimeter and Circumference

| | | | | | |
|---|---|---|---|---|---|
| **square** | $P = 4s$ | **rectangle** | $P = 2\ell + 2w$ | **circle** | $C = 2\pi r$ or $C = \pi d$ |

### Area

| | | | |
|---|---|---|---|
| **square** | $A = s^2$ | **triangle** | $A = \frac{1}{2}bh$ |
| **rectangle** | $A = \ell w$ or $A = bh$ | **regular polygon** | $A = \frac{1}{2}Pa$ |
| **parallelogram** | $A = bh$ | **circle** | $A = \pi r^2$ |
| **trapezoid** | $A = \frac{1}{2}h(b_1 + b_2)$ | **sector of a circle** | $A = \frac{x}{360} \cdot \pi r^2$ |
| **rhombus** | $A = \frac{1}{2}d_1 d_2$ or $A = bh$ | | |

### Lateral Surface Area

| | | | |
|---|---|---|---|
| **prism** | $L = Ph$ | **pyramid** | $L = \frac{1}{2}P\ell$ |
| **cylinder** | $L = 2\pi rh$ | **cone** | $L = \pi r\ell$ |

### Total Surface Area

| | | | |
|---|---|---|---|
| **prism** | $S = Ph + 2B$ | **cone** | $S = \pi r\ell + \pi r^2$ |
| **cylinder** | $S = 2\pi rh + 2\pi r^2$ | **sphere** | $S = 4\pi r^2$ |
| **pyramid** | $S = \frac{1}{2}P\ell + B$ | | |

### Volume

| | | | |
|---|---|---|---|
| **cube** | $V = s^3$ | **pyramid** | $V = \frac{1}{3}Bh$ |
| **rectangular prism** | $V = \ell wh$ | **cone** | $V = \frac{1}{3}\pi r^2 h$ |
| **prism** | $V = Bh$ | **sphere** | $V = \frac{4}{3}\pi r^3$ |
| **cylinder** | $V = \pi r^2 h$ | | |

### Equations for Figures on a Coordinate Plane

| | | | |
|---|---|---|---|
| **slope-intercept form of a line** | $y = mx + b$ | **circle** | $(x - h)^2 + (y - k)^2 = r^2$ |
| **point-slope form of a line** | $y - y_1 = m(x - x_1)$ | | |

### Trigonometry

| | | | |
|---|---|---|---|
| **Law of Sines** | $\frac{\sin A}{a} = \frac{\sin B}{b} = \frac{\sin C}{c}$ | **Law of Cosines** | $a^2 = b^2 + c^2 - 2bc \cos A$<br>$b^2 = a^2 + c^2 - 2ac \cos B$<br>$c^2 = a^2 + b^2 - 2ab \cos C$ |
| **Pythagorean Theorem** | $a^2 + b^2 = c^2$ | | |

## Symbols

|  |  |
|---|---|
| $\neq$ | is not equal to |
| $\approx$ | is approximately equal to |
| $\cong$ | is congruent to |
| $\sim$ | is similar to |
| $\angle$, $\measuredangle$s | angle, angles |
| $m\angle A$ | degree measure of $\angle A$ |
| $^\circ$ | degree |
| $\overleftrightarrow{AB}$ | line containing points *A* and *B* |
| $\overline{AB}$ | segment with endpoints *A* and *B* |
| $\overrightarrow{AB}$ | ray with endpoint *A* containing *B* |
| $AB$ | measure of $\overline{AB}$, distance between points *A* and *B* |
| $\sim p$ | negation of *p*, not *p* |
| $p \wedge q$ | conjunction of *p* and *q* |
| $p \vee q$ | disjunction of *p* and *q* |
| $p \longrightarrow q$ | conditional statement, if *p* then *q* |
| $p \longleftrightarrow q$ | biconditional statement, *p* if and only if *q* |
| $\parallel$ | is parallel to |
| $\nparallel$ | is not parallel to |
| $\perp$ | is perpendicular to |
| $\triangle$ | triangle |
| $>, \geq$ | is greater than, is greater than or equal to |
| $<, \leq$ | is less than, is less than or equal to |
| $\square$ | parallelogram |
| *n*-gon | polygon with *n* sides |
| $a : b$ | ratio of *a* to *b* |
| $(x, y)$ | ordered pair |
| $(x, y, z)$ | ordered triple |
| $\sin x$ | sine of *x* |
| $\cos x$ | cosine of *x* |
| $\tan x$ | tangent of *x* |
| $\vec{a}$ | vector *a* |
| $\overrightarrow{AB}$ | vector from *A* to *B* |
| $\lvert\overrightarrow{AB}\rvert$ | magnitude of the vector from *A* to *B* |
| $A'$ | the image of preimage *A* |
| $\rightarrow$ | is mapped onto |
| $\odot A$ | circle with center *A* |
| $\pi$ | pi |
| $\widehat{AB}$ | minor arc with endpoints *A* and *B* |
| $\widehat{ABC}$ | major arc with endpoints *A* and *C* |
| $m\widehat{AB}$ | degree measure of arc *AB* |
| $f(x)$ | *f* of *x*, the value of *f* at *x* |
| ! | factorial |
| ${}_nP_r$ | permutation of *n* objects taken *r* at a time |
| ${}_nC_r$ | combination of *n* objects taken *r* at a time |
| $P(A)$ | probability of *A* |
| $P(A\|B)$ | the probability of *A* given that *B* has already occurred |

## Measures

| Metric | Customary |
|---|---|
| **Length** | |
| 1 kilometer (km) = 1000 meters (m)<br>1 meter = 100 centimeters (cm)<br>1 centimeter = 10 millimeters (mm) | 1 mile (mi) = 1760 yards (yd)<br>1 mile = 5280 feet (ft)<br>1 yard = 3 feet<br>1 yard = 36 inches (in.)<br>1 foot = 12 inches |
| **Volume and Capacity** | |
| 1 liter (L) = 1000 milliliters (mL)<br>1 kiloliter (kL) = 1000 liters | 1 gallon (gal) = 4 quarts (qt)<br>1 gallon = 128 fluid ounces (fl oz)<br>1 quart = 2 pints (pt)<br>1 pint = 2 cups (c)<br>1 cup = 8 fluid ounces |
| **Weight and Mass** | |
| 1 kilogram (kg) = 1000 grams (g)<br>1 gram = 1000 milligrams (mg)<br>1 metric ton (t) = 1000 kilograms | 1 ton (T) = 2000 pounds (lb)<br>1 pound = 16 ounces (oz) |

# Formulas

## Coordinate Geometry

| | | | |
|---|---|---|---|
| **Midpoint** | $M = \left(\frac{x_1 + x_2}{2}, \frac{y_1 + y_2}{2}\right)$ | **Distance** | $d = \sqrt{(x_2 - x_1)^2 + (y_2 - y_1)^2}$ |
| | | **Slope** | $m = \frac{y_2 - y_1}{x_2 - x_1}, x_2 \neq x_1$ |

## Matrices

| | | | |
|---|---|---|---|
| **Adding** | $\begin{bmatrix} a & b \\ c & d \end{bmatrix} + \begin{bmatrix} e & f \\ g & h \end{bmatrix} = \begin{bmatrix} a+e & b+f \\ c+g & d+h \end{bmatrix}$ | **Multiplying by a Scalar** | $k\begin{bmatrix} a & b \\ c & d \end{bmatrix} = \begin{bmatrix} ka & kb \\ kc & kd \end{bmatrix}$ |
| **Subtracting** | $\begin{bmatrix} a & b \\ c & d \end{bmatrix} - \begin{bmatrix} e & f \\ g & h \end{bmatrix} = \begin{bmatrix} a-e & b-f \\ c-g & d-h \end{bmatrix}$ | **Multiplying** | $\begin{bmatrix} a & b \\ c & d \end{bmatrix} \cdot \begin{bmatrix} e & f \\ g & h \end{bmatrix} = \begin{bmatrix} ab+bg & af-bh \\ ce+dg & cf-dh \end{bmatrix}$ |

## Polynomials

| | | | |
|---|---|---|---|
| **Quadratic Formula** | $x = \frac{-b \pm \sqrt{b^2 - 4ac}}{2a}, a \neq 0$ | **Square of a Difference** | $(a - b)^2 = (a - b)(a - b)$ <br> $= a^2 - 2ab + b^2$ |
| **Square of a Sum** | $(a + b)^2 = (a + b)(a + b)$ <br> $= a^2 + 2ab + b^2$ | **Product of Sum and Difference** | $(a + b)(a - b) = (a - b)(a + b)$ <br> $= a^2 - b^2$ |

## Logarithms

| | | | |
|---|---|---|---|
| **Product Property** | $\log_x ab = \log_x a + \log_x b$ | **Power Property** | $\log_b m^p = p \log_b m$ |
| **Quotient Property** | $\log_x \frac{a}{b} = \log_x a - \log_x b, b \neq 0$ | **Change of Base** | $\log_a n = \frac{\log_b n}{\log_b a}$ |

## Conic Sections

| | | | |
|---|---|---|---|
| **Parabola** | $y = a(x - h)^2 + k$ or $x = a(y - k)^2 + h$ | **Ellipse** | $\frac{x^2}{a^2} + \frac{y^2}{b^2} = 1$ or $\frac{y^2}{a^2} + \frac{x^2}{b^2} = 1, a, b \neq 0$ |
| **Circle** | $x^2 + y^2 = r^2$ or $(x - h)^2 + (y - k)^2 = r^2$ | **Hyperbola** | $\frac{x^2}{a^2} - \frac{y^2}{b^2} = 1$ or $\frac{y^2}{a^2} - \frac{x^2}{b^2} = 1, a, b \neq 0$ |

## Sequences and Series

| | | | |
|---|---|---|---|
| ***n*th term, Arithmetic** | $a_n = a_1 + (n - 1)d$ | ***n*th term, Geometric** | $a_n = a_1 r^{n-1}$ |
| **Sum of Arithmetic Series** | $S_n = n\left(\frac{a_1 + a_2}{2}\right)$ or $S_n = \frac{n}{2}[2a_1 + (n - 1)d]$ | **Sum of Geometric Series** | $S_n = \frac{a_1 - a_1 r^n}{1 - r}$ or $S_n = \frac{a_1 - a_n r}{1 - r}, r \neq 1$ |

## Trigonometry

| | | | |
|---|---|---|---|
| **Law of Sines** | $\frac{\sin A}{a} = \frac{\sin B}{b} = \frac{\sin C}{c}, a, b, c \neq 0$ | | |
| **Law of Cosines** | $a^2 = b^2 + c^2 - 2bc \cos A$ | $b^2 = a^2 + c^2 - 2ac \cos B$ | $c^2 = a^2 + b^2 - 2ab \cos C$ |
| **Trigonometric Functions** | $\sin \theta = \frac{\text{opp}}{\text{hyp}}$ | $\cos \theta = \frac{\text{adj}}{\text{hyp}}$ | $\tan \theta = \frac{\text{opp}}{\text{adj}} = \frac{\sin \theta}{\cos \theta}$ |
| | $\csc \theta = \frac{\text{hyp}}{\text{opp}} = \frac{1}{\sin \theta}$ | $\sec \theta = \frac{\text{hyp}}{\text{adj}} = \frac{1}{\cos \theta}$ | $\cot \theta = \frac{\text{adj}}{\text{opp}} = \frac{\cos \theta}{\sin \theta}$ |
| **Pythagorean Identities** | $\cos^2 \theta + \sin^2 \theta = 1$ | $\tan^2 \theta + 1 = \sec^2 \theta$ | $\cot^2 \theta + 1 = \csc^2 \theta$ |

## Symbols

| | | | |
|---|---|---|---|
| $f(x) = \{$ | piecewise-defined function | $\sum$ | sigma, summation |
| $f(x) = \|x\|$ | absolute value function | $\bar{x}$ | mean of a sample |
| $f(x) = [\![x]\!]$ | function of greatest integer not greater than *a* | $\mu$ | mean of a population |
| $f(x, y)$ | *f* of *x* and *y*, a function with two variables, *x* and *y* | $s$ | standard deviation of a sample |
| $\overrightarrow{AB}$ | vector *AB* | $\sigma$ | standard deviation of a population |
| $i$ | the imaginary unit | $P(B \mid A)$ | the probability of *B* given that *A* has already occurred |
| $[f \circ g](x)$ | *f* of *g* of *x*, the composition of functions *f* and *g* | $nPr$ | permutation of *n* objects taken *r* at a time |
| $f^{-1}(x)$ | inverse of $f(x)$ | $nCr$ | combination of *n* objects taken *r* at a time |
| $b^{\frac{1}{n}} = \sqrt[n]{b}$ | *n*th root of *b* | $\text{Sin}^{-1}\, x$ | Arcsin *x* |
| $\log_b x$ | logarithm base *b* of *x* | $\text{Cos}^{-1}\, x$ | Arccos *x* |
| $\log x$ | common logarithm of *x* | $\text{Tan}^{-1}\, x$ | Arctan *x* |
| $\ln x$ | natural logarithm of *x* | | |

## Parent Functions

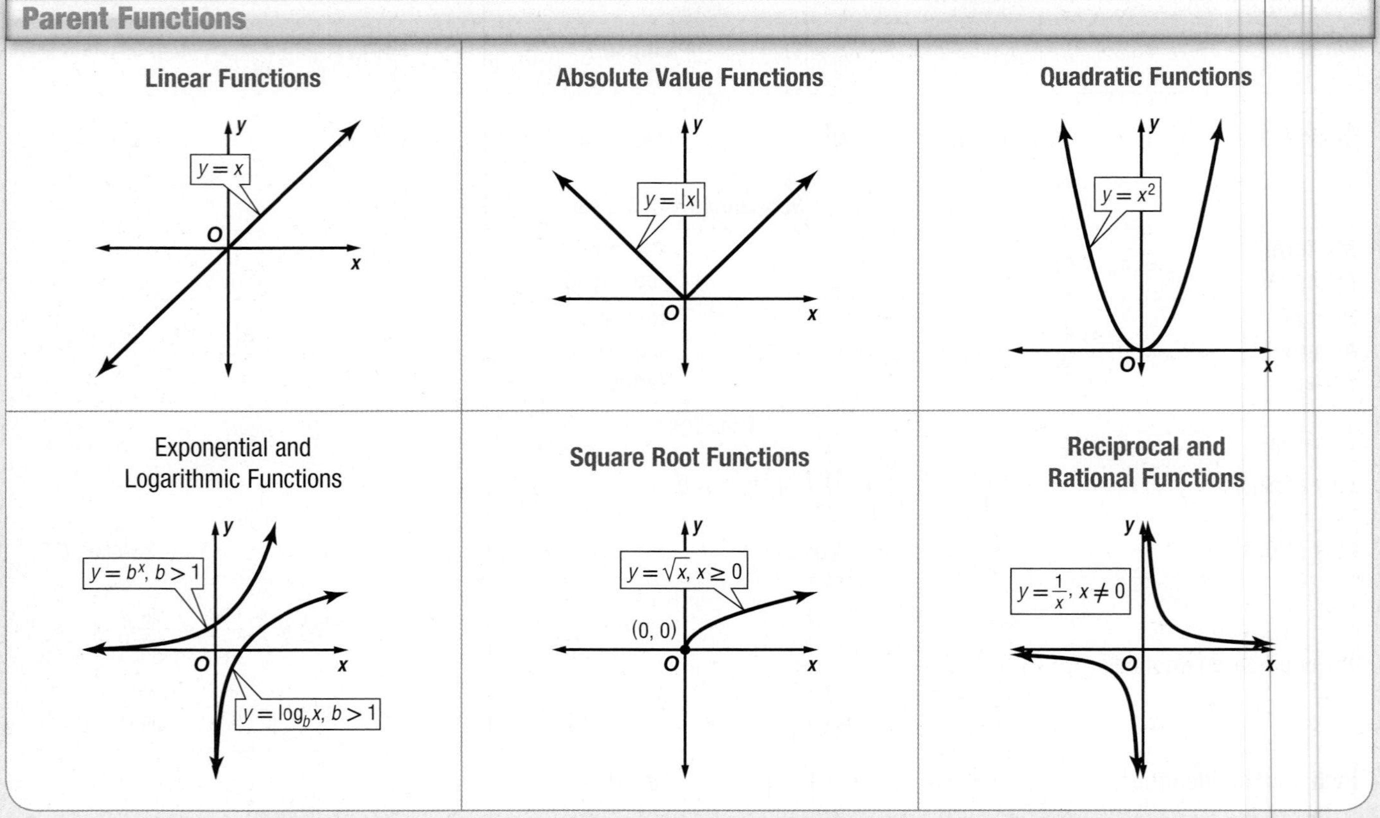